Principles of
Tissue Engineering
Second Edition

PRINCIPLES OF TISSUE ENGINEERING

SECOND EDITION

Edited by

Robert P. Lanza

Tissue Engineering and Transplant Medicine
Advanced Cell Technology
Worcester, Massachusetts

Robert Langer

Department of Chemical and Biomedical Engineering
Harvard–MIT Division of Health, Sciences and Technology
Massachusetts Institute of Technology
Cambridge, Massachusetts

Joseph Vacanti

Harvard Medical School and
Massachusetts General Hospital
Boston, Massachusetts

ACADEMIC PRESS

A Harcourt Science and Technology Company

San Diego   San Francisco   New York   Boston   London   Sydney   Tokyo

On the cover: Deborah Odum Hutchinson is an award winning and nationally published medical artist. After working seven years for the College of Medicine at Texas A&M University, she has been doing freelance art for 12 years. She is also known for her work in watercolors as a fine artist. Her work is a sensitive creation of the delicate anatomical aspects of the body as well as the artistic aspect of the figure in her final illustrations.

This book is printed on acid-free paper.

Copyright ©2000, 1997 Elsevier Science (USA).

All Rights Reserved.
No part of this publication may be reproduced or transmitted in any form or by any means, electronic or mechanical, including photocopy, recording, or any information storage and retrieval system, without permission in writing from the publisher.

Permissions may be sought directly from Elsevier's Science & Technology Rights Department in Oxford, UK: phone: (+44) 1865 843830, fax: (+44) 1865 853333, e-mail: permissions@elsevier.com.uk. You may also complete your request on-line via the Elsevier Science homepage (http://elsevier.com), by selecting "Customer Support" and then "Obtaining Permissions."

Academic Press
An imprint of Elsevier Science
525 B Street, Suite 1900, San Diego, California 92101-4495, USA
http://www.academicpress.com

Academic Press
84 Theobald's Road, London WC1X 8RR, UK
http://www.academicpress.com

Library of Congress Catalog Card Number: 99-68198

International Standard Book Number: 0-12-436630-9

PRINTED IN THE UNITED STATES OF AMERICA
02 03 04 05 06 07   9 8 7 6 5 4 3 2

To Barbara and Eugene

CONTENTS

CONTRIBUTORS

Patrick Aebischer
Division of Surgical Research and Gene Therapy Center
Centre Hospitalier Universitaire Vaudois (CHUV)
Lausanne CH-1011, Switzerland
Chapter 56

Richard A. Altschuler
Auditory Anatomy Laboratory
University of Michigan Medical School
Kresge Hearing Research Institute
Ann Arbor, Michigan 48109
Chapter 53

Pascal Ambrosini
Laboratoire d'Immunologie
Faculté de Médecine de Nancy
Vandoeuvre-les-Nancy 54500, France
Chapter 59

David J. Anderson
Departments of Electrical Engineering and Otolaryngology
University of Michigan
Ann Arbor, Michigan 48109
Chapter 53

Anthony Atala
Laboratory for Tissue Engineering
Children's Hospital
Boston, Massachusetts 02115
Chapter 46

François A. Auger
Department of Surgery
Laval University; and
Laboratoire d'Organogénèse Expérimentale/LOEX
Qúebec, Canada G1S 4L8
Chapter 50

Efstathios S. Avgoustiniatos
Department of Chemical Engineering
Massachusetts Institute of Technology
Cambridge, Massachusetts 02139
Chapter 27

David W. Barnes
American Type Culture Collection
Manassas, Virginia 20110
Chapter 10

Eugene Bell
Department of Biology
Massachusetts Institute of Technology
Cambridge, Massachusetts 02139; and
Tissue Engineering Inc.
Boston, Massachusetts 02210
Chapter 16

Marie C. Béné
Laboratoire d'Immunologie
Faculté de Médecine de Nancy
Vandoeuvre-les-Nancy 54500, France
Chapter 59

John G. Bishop
Center for Biologics Evaluation and Research (CBER)
U. S. Food and Drug Administration (FDA)
Rockville, Maryland 20852
Chapter 65

T. Bohrer
Department of Surgery
Philipps-University of Marburg
Marburg 35033, Germany
Chapter 37

Lawrence J. Bonassar
Department of Anesthesiology
Laboratory for Tissue Engineering
University of Massachusetts Memorial Health Care
Worcester, Massachusetts 01655
Chapter 47

Amy D. Bradshaw
Department of Vascular Biology
Hope Heart Institute
Seattle, Washington 98104
Chapter 11

Kelvin G. M. Brockbank
KGB Associates, Inc.
Charleston, South Carolina 29401
Chapter 33

Leon W. Browder
Department of Biochemistry and Molecular Biology
University of Calgary
Calgary, Alberta, Canada T2N 4N1
Chapter 8

Scott P. Bruder
Orthobiologics
DePuy Orthopaedics
Raynham, Massachusetts 02767
Chapter 48

Mary Bartlett Bunge
The Miami Project to Cure Paralysis
University of Miami School of Medicine
Miami, Florida 33136
Chapter 57

Arnold I. Caplan
Department of Biology
Skeletal Research Center
Case Western Reserve University
Cleveland, Ohio 44106
Chapter 48

Thomas Ming Swi Chang
Artificial Cells and Organs Research Center
Departments of Physiology, Medicine, and
    Biomedical Engineering
McGill University
Montreal, Quebec, Canada H3G 1Y6
Chapter 42

CONTRIBUTORS

Robert G. Chapman
Department of Chemistry
Harvard University
Cambridge, Massachusetts 02138
Chapter 18

Una Chen
Stem Cell Therapy Program
Biochemistry Institute
University of Giessen
Giessen D-35392, Germany
Chapter 43

Richard A. F. Clark
Department of Dermatology
State University of New York
Stony Brook, New York 11794
Chapter 61

Réjean Cloutier
Department of Physical Therapy
Laval University;
Hôpital du Saint-Sacrement; and
Laboratoire d'Organogénèse Expérimentale/LOEX
Qúebec, Canada G1S 4L8
Chapter 50

Clark K. Colton
Department of Chemical Engineering
Massachusetts Institute of Technology
Cambridge, Massachusetts 02139
Chapter 27

Joanne C. Cousins
MRC Clinical Sciences Centre
Imperial College School of Medicine
Hammersmith Hospital
London W12 0NN, England
Chapter 52

Stephen C. Cowin
The Center for Biomedical Engineering
Department of Mechanical Engineering
The School of Engineering of The City College and
     The Graduate School of The City University of New York
New York, New York 10031
Chapter 51

Gislin Dagnelie
Lions Vision Research & Rehabilitation Center
Johns Hopkins University School of Medicine
Baltimore, Maryland 21205
Chapter 54

Thomas F. Deuel
Division of Growth Regulation
Department of Medicine
Beth Israel Deaconess Medical Center and
     Harvard Medical School
Boston, Massachusetts 02215
Chapter 12

Charles N. Durfor
Center for Devices and Radiological Health (CDRH)
U. S. Food and Drug Administration (FDA)
Rockville, Maryland 20852
Chapter 65

Brian E. Edwards
Department of Gynecology and Obstetrics
Division of Developmental Genetics
Johns Hopkins University School of Medicine
Baltimore, Maryland 21287
Chapter 29

Carol A. Erickson
Section of Molecular & Cellular Biology
University of California, Davis
Davis, California 95616
Chapter 3

Gilbert C. Faure
Laboratoire d'Immunologie
Faculté de Médecine de Nancy
Vandoeuvre-les-Nancy 54500, France
Chapter 59

Denise Faustman
Immunobiology Laboratories
Harvard Medical School
Massachusetts General Hospital
Charlestown, Massachusetts 02129
Chapter 25

Dario O. Fauza
Department of Surgery
Children's Hospital
Brookline, Massachusetts 02446; and
Harvard Medical School
Boston, Massachusetts 02115
Chapter 28

Eric G. Fine
Division of Surgical Research and Gene Therapy Center
Centre Hospitalier Universitaire Vaudois (CHUV)
Lausanne CH-1011, Switzerland
Chapter 56

Lee G. Fradkin
Valentis, Inc.
Burlingame, California 94010
Chapter 30

Lisa E. Freed
Division of Health Sciences & Technology
Harvard-Massachusetts Institute of Technology
Cambridge, Massachusetts 02139
Chapter 13

Claudia Gaffey
Center for Devices and Radiological Health (CDRH)
U. S. Food and Drug Administration (FDA)
Rockville, Maryland 20852
Chapter 65

CONTRIBUTORS

John D. Gearhart
Department of Gynecology and Obstetrics
Division of Developmental Genetics
Johns Hopkins University School of Medicine
Baltimore, Maryland 21287
Chapter 29

Lucie Germain
Department of Surgery
Laval University; and
Laboratoire d'Organogénèse Expérimentale/LOEX
Québec, Canada G1S 4L8
Chapter 50

Francine Goulet
Department of Physical Therapy
Laval University; and
Laboratoire d'Organogénèse Expérimentale/LOEX
Québec, Canada G1S 4L8
Chapter 50

Howard P. Greisler
Department of Surgery
Loyola University Medical Center
Maywood, Ilinois 60153
Chapter 32

Alan J. Grodzinsky
Department of Electrical Engineering & Computer Science
Division of Bioengineering & Environmental Health
Center for Biomedical Engineering
Massachusetts Institute of Technology
Cambridge, Massachusetts 02139
Chapter 17

Craig R. Halberstadt
Department of General Surgery Research
Carolinas Medical Center
Charlotte, North Carolina 28232
Chapter 31

Janet Hardin-Young
Organogenesis, Inc.
Canton, Massachusetts 02021
Chapters 23, 62

C. Hasse
Department of Surgery
Philipps-University of Marburg
Marburg 35033, Germany
Chapter 37

Matthias Hebrok*
Department of Molecular and Cellular Biology and
   Howard Hughes Medical Institute
Harvard University
Cambridge, Massachusetts 02138
Chapter 6

Kiki B. Hellman
Center for Devices and Radiological Health (CDRH)
U. S. Food and Drug Administration (FDA)

Rockville, Maryland 20852
Chapter 65

Walter D. Holder
Department of General Surgery Research
Carolinas Medical Center
Charlotte, North Carolina 28232
Chapter 31

Edward Hsu
Department of Biomedical Engineering
Duke University Medical Center
Durham, North Carolina 27710
Chapter 41

Jeffrey A. Hubbell
Institute for Biomedical Engineering
ETH and University of Zürich
Zürich 8044, Switzerland
Chapter 20

Mark S. Humayun
Lions Vision Research & Rehabilitation Center
Johns Hopkins University School of Medicine
Baltimore, Maryland 21205
Chapter 54

H. David Humes
Department of Internal Medicine
University of Michigan Medical School
Ann Arbor, Michigan 48109
Chapter 45

Donald E. Ingber
Department of Pathology
Children's Hospital
Harvard Medical School
Boston, Massachusetts 02115
Chapter 9

Hugo O. Jauregui
Department of Pathology
Division of Pathology Research
Rhode Island Hospital
Providence, Rhode Island 02903
Chapter 39

Roger D. Kamm
Department of Mechanical Engineering
Division of Bioengineering & Environmental Health
Center for Biomedical Engineering
Massachusetts Institute of Technology
Cambridge, Massachusetts 02139
Chapter 17

Ravi S. Kane
Department of Chemistry
Harvard University
Cambridge, Massachusetts 02138
Chapter 18

* *Current address: Diabetes Research Center, Department of Medicine, University of California, San Francisco, California 94143.*

CONTRIBUTORS

Jens O. M. Karlsson
Department of Surgery & Bioengineering
Shriner's Research Center
Cambridge, Massachusetts 02139
Chapter 24

Anne Kessinger
University of Nebraska Medical Center
Omaha, Nebraska 68198
Chapter 44

Byung-Soo Kim
Laboratory for Tissue Engineering
Children's Hospital and Harvard Medical School
Boston, Massachusetts 02115
Chapter 46

Naomi Kleitman
The Miami Project to Cure Paralysis
University of Miami School of Medicine
Miami, Florida 33136
Chapter 57

Joachim Kohn
Department of Chemistry
Rutgers University
Piscataway, New Jersey 08855
Chapter 22

Hiroshi Kubota
Departments of Cell & Molecular Physiology and Medicine
University of North Carolina
Chapel Hill, North Carolina 27514
Chapter 41

Robert P. Lanza
Tissue Engineering & Transplant Medicine
Advanced Cell Technology
Worcester, Massachusetts 01605
Chapter 36

Douglas A. Lauffenburger
Department of Chemical Engineering
Division of Bioengineering & Environmental Health
Center for Biomedical Engineering
Massachusetts Institute of Technology
Cambridge, Massachusetts 02139
Chapter 17

Kuen Yong Lee
Department of Chemical Engineering, Biomedical Engineering, and Biologic & Materials Sciences
University of Michigan
Ann Arbor, Michigan 48109
Chapter 31

Peter I. Lelkes
Department of Medicine
University of Wisconsin Medical School
Milwaukee, Wisconsin 53201
Chapter 14

Robert E. London
Departments of Cell & Molecular Physiology and Medicine
University of North Carolina
Chapel Hill, North Carolina 27514
Chapter 41

Jack W. Love
CardioMend, LLC
Santa Barbara, California 93110
Chapter 34

Thomas L. Luntz
Departments of Cell & Molecular Physiology and Medicine
University of North Carolina
Chapel Hill, North Carolina 27514
Chapter 41

Michael J. Lysaght
Departments of Molecular Pharmacology,
    Physiology & Biotechnology
Division of Biology and Medicine
Brown University
Providence, Rhode Island 02912
Chapter 26

Jeffrey M. Macdonald
Departments of Cell & Molecular Physiology and Medicine
University of North Carolina
Chapel Hill, North Carolina 27514
Chapter 41

Manuela Martins-Green
Department of Cell Biology and Neurosciences
University of California, Riverside
Riverside, California 92521
Chapter 4

Robert W. Massof
Lions Vision Research & Rehabilitation Center
Johns Hopkins University School of Medicine
Baltimore, Maryland 21205
Chapter 54

John M. McPherson
Department of Cell and Protein Therapeutics
Genzyme Corporation
Framingham, Massachusetts 01701
Chapter 49

Douglas A. Melton
Department of Molecular and Cellular Biology and
    Howard Hughes Medical Institute
Harvard University
Cambridge, Massachusetts 02138
Chapter 6

Antonios G. Mikos
Department of Chemical Engineering
Rice University
Houston, Texas 77005
Chapter 21

CONTRIBUTORS

Josef M. Miller
Kresge Hearing Institute
Department of Otolaryngology and Biochemistry
University of Michigan
Ann Arbor, Michigan 48109
Chapter 53

Neal A. Miller
Laboratoire d'Immunologie
Faculté de Médecine de Nancy
Vandoeuvre-les-Nancy 54500, France
Chapter 59

David J. Mooney
Department of Chemical Engineering, Biomedical Engineering,
   and Biologic & Materials Sciences
University of Michigan
Ann Arbor, Michigan 48109
Chapters 31, 46

Jennifer E. Morgan
MRC Clinical Sciences Centre
Imperial College School of Medicine
Hammersmith Hospital
London W12 0NN, England
Chapter 52

Melvin L. Moss
Department of Anatomy and Cell Biology
College of Physicians and Surgeons and
   School of Dental and Oral Surgery
Columbia University
New York, New York 10032
Chapter 51

Claudy Mullon
Research & Development
Circe Biomedical, Inc.
Lexington, Massachusetts 02421
Chapter 40

Christopher S. Muratore
Department of Surgery
Children's Hospital
Boston, Massachusetts 02115
Chapter 64

Gail K. Naughton
Advanced Tissue Sciences
La Jolla, California 92037
Chapter 63

Robert M. Nerem
Department of Bioengineering & Biosciences
Parker H. Petít Institute
Georgia Institute of Technology
Atlanta, Georgia 30332
Chapter 2

Björn Reino Olsen
Department of Cell Biology
Harvard Medical School; and
Harvard–Forsyth Department of Oral Biology

Harvard School of Dental Medicine
Boston, Massachusetts 02115
Chapter 5

Gregory M. Organ
Department of Surgery
Kaiser Permanente Medical Center
University of California, Davis
Oakland, California 94611
Chapter 38

James M. Pachence
Veritas Medical Technologies, Inc.
Princeton, New Jersey 08540
Chapter 22

Nancy L. Parenteau
Organogenesis, Inc.
Canton, Massachusetts 02021
Chapters 23, 62

Terence A. Partridge
MRC Clinical Sciences Centre
Imperial College School of Medicine
Hammersmith Hospital
London W12 0NN, England
Chapter 52

Jacques Penaud
Laboratoire d'Immunologie
Faculté de Médecine de Nancy
Vandoeuvre-les-Nancy 54500, France
Chapter 59

A. Robin Poole
Joint Diseases Laboratory
Shiners Hospital for Crippled Children;
McGill University
Montréal, Canada; and Laboratoire d'Organogénèse
   Expérimentale/LOEX
Qúebec, Canada G1S 4L8
Chapter 50

Denis Rancourt
Department of Mechanical Engineering
Laval University and
Laboratoire d'Organogénèse Expérimentale/LOEX
Québec, Canada G1S 4L8
Chapter 50

Yehoash Raphael
Kresge Hearing Institute
Department of Otolaryngology and Biochemistry
University of Michigan
Ann Arbor, Michigan 48109
Chapter 53

A. H. Reddi
Centre for Tissue Regeneration and Repair
Department of Orthopaedic Surgery
University of California, Davis
Sacramento, California 95817
Chapter 7

CONTRIBUTORS

Lola M. Reid
Departments of Cell & Molecular Physiology and Medicine
University of North Carolina
Chapel Hill, North Carolina 27514
Chapter 41

J. Dezz Ropp
Valentis, Inc.
Burlingame, California 94010
Chapter 30

Robert N. Ross
Medical/Science Analytics
Brookline, Massachusetts 02445
Chapters 23, 62

M. Rothmund
Department of General Surgery
Philipps-University of Marburg
Marburg 35033, Germany
Chapter 37

R. Bruce Rutherford
School of Dentistry
University of Michigan
Ann Arbor, Michigan 48109
Chapter 60

E. Helene Sage
Department of Vascular Biology
Hope Heart Institute
Seattle, Washington 98104
Chapter 11

Jacqueline Sagen
The Miami Project to Cure Paralysis
University of Miami School of Medicine
Miami, Florida 33136
Chapter 57

W. Mark Saltzman
Department of Chemical Engineering
Cornell University
Ithaca, New York 14853
Chapter 19

Gordon H. Sato
Department of Navy
Asmara, Eritrea
Africa
Chapter 10

Jochen Schacht
Kresge Hearing Institute
Department of Otolaryngology and Biochemistry
University of Michigan
Ann Arbor, Michigan 48109
Chapter 53

Michael J. Shamblott
Department of Gynecology and Obstetrics
Division of Developmental Genetics
Johns Hopkins University School of Medicine
Baltimore, Maryland 21287
Chapter 29

Graham Sharp
Department of Internal Medicine
Section of Oncology & Hematology
Omaha, Nebraska 68198
Chapter 44

Albert K. Shung
Department of Chemical Engineering
Rice University
Houston, Texas 77005
Chapter 21

Adam J. Singer
Department of Dermatology
State University of New York
Stony Brook, New York 11794
Chapter 61

Barry A. Solomon
Circe Biomedical, Inc.
Lexington, Massachusetts 02421
Chapter 40

Ruth R. Solomon
Center for Biologics Evaluation and Research (CBER)
U. S. Food and Drug Administration (FDA)
Rockville, Maryland 20852
Chapter 65

Susan J. Sullivan
Organogenesis, Inc.
Canton, Massachusetts 02021
Chapter 33

Shuichi Takayama
Department of Chemistry
Harvard University
Cambridge, Massachusetts 02138
Chapter 18

Jeffrey Teumer
Organogenesis, Inc.
Canton, Massachusetts 02021
Chapter 23

Robert C. Thomson
Medical Products Division
W. L. Gore and Associates
Flagstaff, Arizona 86002
Chapter 21

Mehmet Toner
Department of Surgery & Bioengineering
Shriners Research Center
Cambridge, Massachusetts 02139
Chapter 24

Vickery Trinkaus-Randall
Boston University School of Medicine
Boston, Massachusetts 02118
Chapter 35

Ross Tubo
Department of Cell and Protein Therapeutics
Genzyme Corporation

CONTRIBUTORS

Framingham, Massachusetts 01701
Chapter 49

Brian R. Unsworth
Department of Biology
Marquette University
Milwaukee, Wisconsin 53233
Chapter 14

Charles A. Vacanti
Department of Anesthesiology
Laboratory for Tissue Engineering
University of Massachusetts Memorial Health Care
Worcester, Massachusetts 01655
Chapters 1, 47

Joseph P. Vacanti
Department of Surgery
Harvard Medical School and
Massachusetts General Hospital
Boston, Massachusetts 02114
Chapters 1, 38, 47, Epilogue

Martin P. Vacanti
Department of Anesthesiology
University of Massachusetts Medical School
Worcester, Massachusetts 01655
Chapter 58

Robert F. Valentini
Division of Surgical Research and Gene Therapy Center
Centre Hospitalier Universitaire Vaudois (CHUV)
Lausanne CH-1011, Switzerland
Chapter 56

Gordana Vunjak-Novakovic
Division of Health Sciences & Technology
Harvard-Massachusetts Institute of Technology
Cambridge, Massachusetts 02139
Chapter 13

Lars U. Wahlberg
NeuroSearch A/S
Glostrup DK-2600, Denmark
Chapter 55

Taylor G. Wang
Center for Microgravity Research and Applications
School of Engineering
Vanderbilt University
Nashville, Tennessee 37232
Chapter 36

John F. Warner
Gene-Based Therapeutics
Valentis, Inc.
Burlingame, California 94010
Chapter 30

George M. Whitesides
Department of Chemistry
Harvard University
Cambridge, Massachusetts 02138
Chapter 18

Jay M. Wilson
Department of Surgery
Children's Hospital
Boston, Massachusetts 02115
Chapter 64

Haiyun Wu
Department of Chemical Engineering
Massachusetts Institute of Technology
Cambridge, Massachusetts 02139
Chapter 27

Arron S. L. Xu
Departments of Cell & Molecular Physiology and Medicine
University of North Carolina
Chapel Hilll, North Carolina 27514
Chapter 41

Lian Xue
Department of Surgery
Loyola University Medical Center
Maywood, Ilinois 60153
Chapter 32

Ioannis V. Yannas
Department of Mechanical Engineering
Fibers and Polymers Laboratory
Massachusetts Institute of Technology
Cambridge, Massachusetts 02139
Chapter 15

Michael J. Yaszemski
Departments of Orthopedic Surgery and Bioengineering
Mayo Clinic
Rochester, Minnesota 55901
Chapter 21

Nan Zhang
Division of Growth Regulation
Department of Medicine
Beth Israel Deaconess Medical Center and
Harvard Medical School
Boston, Massachusetts 02215
Chapter 12

Beth A. Zielinski
Departments of Molecular Pharmacology,
  Physiology & Biotechnology
Division of Biology and Medicine
Brown University
Providence, Rhode Island 02912
Chapter 26

A. Zielke
Depatment of Surgery
Phillips University of Marburg
Marburg D-35043, Germany
Chapter 37

U. Zimmerman
Department of Biotechnology
University of Würzburg
Würzburg, Germany
Chapter 37

FOREWORD

Charles A. Vacanti

As we enter the new millennium, it is important to look back and see where we have been. Only with this perspective is it realistic to look toward the future and imagine what is possible.

Tissue engineering will ultimately have a more profound impact than we can now appreciate. It not only will modify the practice of medicine and help elucidate mechanisms of developmental biology, but also has the potential to influence economic development in the industry of biotechnology more than any single advance in science or medicine during the last several decades.

Although we can point to various milestones in history and refer to certain achievements as "tissue engineering," focused efforts to regenerate lost tissue function were made possible only after advances in associated fields of cell biology and material sciences. Consequently, modern tissue engineering was not possible, or even feasible, until advances in cell biology enabled the large-scale production of commercially available enzymes and nutrients to isolate a large number of cells and nourish them in an incubator. It now seems clear that the use of living cells will result in a higher degree of tissue function than the use of chemotactic agents, growth factors, or hormones to stimulate development. Indeed, the ability to isolate cells from many different specialized tissues has existed for only a few decades. Even today, *in vitro* multiplication of certain cell types has been achieved with only limited success.

Early approaches to reintroducing cells into a recipient were quite simplistic and usually unsuccessful. Cells were initially injected as free suspensions in the hope that they would randomly engraft.

A further challenge to the large-scale development and application of tissue engineering is the immunologic barrier. Improvements in the understanding of immunology and the ability to trick the host into thinking foreign cells are "self" may ultimately allow for implantation of allograft or even xenograft cells to generate functional tissue. Unfortunately, at this time in history, the development of a universal donor cell type that can be used to construct the framework of a commercially available cell/polymer construct that will meet the needs of any recipient is still a dream.

After almost 20 years of serious efforts to generate new functional tissue from living cells, we have learned several basic concepts. Most important, "it is not nice to fool Mother Nature." In this respect, many efforts have given the highest degree of success when they have mimicked nature. What we know scientifically is only a minuscule amount of all the processes related to the organization, development, and function of living systems. Although numerous theories have been proposed, explored, and replaced by more modern hypotheses, most of the biologic processes remain a "black box."

At times, tissue engineering as a science and as a medical discipline has been called "unnatural." When one actually considers what physicians have done for centuries, this couldn't be farther from the truth. The basic premise of health care is that the body heals itself. Physicians do nothing more than support a patient's vital functions by optimizing the environment most conducive to healing. Basically, physicians attempt to neutralize hostile factors at the same time that they enhance the supply of oxygen and nutrients that the body needs to heal itself. A basic principle of surgery is to debride the dead tissue that is the source of unfavorable chemical agents and to approximate tissue to protect the tissue from the hostile environment and improve the vascular supply to the injured site. In tissue engineering, we strive to achieve exactly the same goal. Necrotic and scar tissue are first excised. Rather than the remaining tissue being approximated to eliminate the dead space, living cells that belong in the injured area are configured to prevent their dissipation, like a scaffolding that dictates shape and function of the desired tissue by providing structural cues. The scientist then manipulates the environment in precisely the same manner as the physician, that is, by optimizing the delivery of oxygen to the site and the elimination of waste products. Under ideal conditions, this will then enable the body to heal itself. Tissue engineering efforts are focused on the microenvironment as opposed to the macroenvironment being manipulated in traditional health care.

A more recent critical factor in tissue engineering that is only now being more fully explored is the source of the cells to be utilized. Several studies have suggested that more immature cells are able to multiply to a higher degree *in vitro* than fully differentiated cells of specialized tissues. In contrast to the *in vitro* multiplication of fully differentiated cells, these immature or progenitor cells can be induced to differentiate and function after several generations *in vitro*. They also appear to have the ability to differentiate into many of the specialized cells found within specific tissues as a function of the environment in which they are placed.

Finally I believe that it is important to understand the current limitations of the field and the developments in associated fields on which tissue engineering is predicated. Our expectations must be aligned with our abilities as the field emerges. In that capacity, the first human applications may indeed reflect the application of the principles of tissue engineering as a component therapy to replace lost function(s) of specific organs or tissue, rather than replacement of the entire organ. In the instance of liver or pancreas engineering, for example, the patient may be experiencing only a specific enzymatic defect or may lack the ability to manufacture a vital hormone or growth factor. Perhaps only a critical mass of functional tissue need be generated, rather than the entire organ. In this fashion, advances in human application can parallel advances in the science of the field rather than generate unrealistic hopes prematurely.

Although it is believed that developments in this field will result in a tremendous advance in the treatment of many disease processes, the power of tissue engineering as a model for exploring changes associated with developmental biology may equal its special applications to human health care.

In conclusion, a well-thought-out and organized development of the science and application of the field of tissue engineering could generate benefits for humankind far into the new millennium.

PREFACE TO THE SECOND EDITION

The first edition of this textbook, published in 1997, was rapidly recognized as the comprehensive textbook of tissue engineering. This edition is intended to serve as a comprehensive text for the student at the graduate level or the research scientist/physician with a special interest in tissue engineering. It should also function as a reference text for researchers in many disciplines. It is intended to cover the history of tissue engineering and the basic principles involved, as well as to provide a comprehensive summary of the advances in tissue engineering in recent years and the state of the art as it exists today.

Although many reviews had been written on the subject and a few textbooks had been published, none had been as comprehensive in its defining of the field, description of the scientific principles and interrelated disciplines involved, and discussion of its applications and potential influence on industry and the field of medicine in the future as the first edition.

When one learns that a more recent edition of a textbook has been published, one has to wonder if the base of knowledge in that particular discipline has grown sufficiently to justify writing a revised textbook. In the case of tissue engineering, it is particularly conspicuous that developments in the field since the printing of the first edition have been tremendous. Even experts in the field would not have been able to predict the explosion in knowledge associated with this development. The variety of new polymers and materials now employed in the generation of engineered tissue has grown exponentially, as evidenced by data associated with specialized applications. More is learned about cell/biomaterials interactions on an almost daily basis. Since the printing of the last edition, recent work has demonstrated a tremendous potential for the use of stem cells in tissue engineering. While some groups are working with fetal stem cells, others believe that each specialized tissue contains progenitor cells or stem cells that are already somewhat committed to develop into various specialized cells of fully differentiated tissue.

Parallel to these developments, there has been a tremendous "buy in" concerning the concepts of tissue engineering not only by private industry but also by practicing physicians in many disciplines. This growing interest has resulted in expansion of the scope of tissue engineering well beyond what could have been predicted five years ago and has helped specific applications in tissue engineering to advance to human trials.

The chapters presented in this text represent the results of the coordinated research efforts of several hundred scientific investigators internationally. The development of this text in a sense parallels the development of the field as whole and is a true reflection of the scientific cooperation expressed as this field evolves.

Robert P. Lanza
Robert Langer
Joseph Vacanti

PREFACE TO THE FIRST EDITION

Although individual papers on various aspects of tissue engineering abound, no previous work has satisfactorily integrated this new interdisciplinary subject area. *Principles of Tissue Engineering* combines in one volume the prerequisites for a general understanding of tissue growth and development, the tools and theoretical information needed to design tissues and organs, as well as a presentation of applications of tissue engineering to diseases affecting specific organ system. We have striven to create a comprehensive book that, on the one hand, strikes a balance among the diversity of subjects that are related to tissue engineering, including biology, chemistry, materials science, engineering, immunology, and transplantation among others, while, on the other hand, emphasizing those research areas that are likely to be of most value to medicine in the future.

The depth and breadth of opportunity that tissue engineering provides for medicine is extraordinary. In the United States alone, it has been estimated that nearly half-a-trillion dollars are spent each year to care for patients who suffer either tissue loss or end-stage organ failure. Over four million patients suffer from burns, pressure sores, and skin ulcers, over twelve million patients suffer from diabetes, and over two million patients suffer from defective or missing supportive structures such as long bones, cartilage, connective tissue, and intervertebral discs. Other potential applications of tissue engineering include the replacement of worn and poorly functioning tissues as exemplified by aged muscle or cornea; replacement of small caliber arteries, veins, coronary, and peripheral stents; replacement of the bladder, ureter, and fallopian tube; and restoration of cells to produce necessary enzymes, hormones, and other bioactive secretory products.

Principles of Tissue Engineering is intended not only as a text for biomedical engineering students and students in cell biology, biotechnology, and medical courses at advanced undergraduate and graduate levels, but also as a reference tool for research and clinical laboratories. The expertise required to generate this text far exceeded that of its editors. It represents the combined intellect of more than eighty scholars and clinicians whose pioneering work has been instrumental to ushering in this fascinating and important field. We believe that their knowledge and experience have added indispensable depth and authority to the material presented in this book and that in the presentation, they have succeeded in defining and capturing the sense of excitement, understanding, and anticipation that has followed from the emergence of this new field, tissue engineering.

Robert Lanza
Robert Langer
William Chick

TISSUE ENGINEERING IN PERSPECTIVE

Eugene Bell

ORIGINS OF TISSUE ENGINEERING

To write a perspective burdens the writer with the task of relating the many aspects of a subject not only to each other, but also to the subject as a whole. As we begin tracking the origins of tissue engineering, what stands out from the past, and remains with us in the present, is the role played by surgeons in creating one body part for another to meet the needs of individual patients. Improvising anatomic repair using an individual's own organ or tissue is still a technique often used by the reconstructive surgeon. For example, consider the use of a segment of ilium to create an ectopic replacement urinary bladder, to which the ureters are sewn at one end, with the other end everted. The everted end with its exposed mucosa, serving as a stoma, is installed transdermally in the abdominal area, providing an organ through which urine can flow into a sac attached to the surrounding skin. The procedure has been remarkably successful, but in the new age of tissue engineering, our thinking is directed to the fabrication of a substitute bladder that the surgeon can take off the shelf.

It was in the tradition of innovative surgical borrowing to rebuild a body part that new materials were sought to substitute for what the patient's own tissues might provide. The idea of improving on nature by using manmade materials was nurtured by the discovery and availability of the new synthetics during World War II. Since that time of technological expansion, the quest for substitutes for autologous tissues has been a roller-coaster ride. Thermoplastic synthetic polymers such as nylon, Dacron, polyurethane, polypropylene, and many others, not designed for use in the human body as tissue replacement, were introduced by suppliers of surgical paraphernalia and were used by practitioners of the art of surgery, for rebuilding damaged, diseased, aged, or genetically deficient body parts. The synthetic materials were introduced into the human body as components of fabricated replacement parts at a time when the possibility of reconstituting biologic substitutes was still unexplored. Many of the postwar synthetics are still in use today, with major questions regarding their efficacy hanging over us.

What was previously called reconstructive surgery came to be called tissue engineering when the focus of attention became the fabrication of living replacement parts for the body in the laboratory. The technology represented a dramatic shift to the use of the biologic components of actual tissues for reconstructing replacements. Tissue building as it occurs in organisms during development, and for some body parts throughout life, was being imitated. The new approach, coming at a time when biotechnology was achieving widespread acceptance, quickly gained the interest of the financial market and the pharmaceutical companies because of its commercial potential.

THE TISSUE ENGINEERING TRIAD

As we know, biologic tissues consist of the cells, the extracellular matrix (made up of a complex of cell secretions immobilized in spaces continuous with cells), and the signaling systems, which are brought into play through differential activation of genes or cascades of genes whose secreted or transcriptional products are responsible for cueing tissue building and differentiation.

Figure 1 illustrates the triad of tissue engineering, which is based on the three basic components of biologic tissues. The principal components of scaffolds (into which the extracellular matrix is organized in actual tissues)

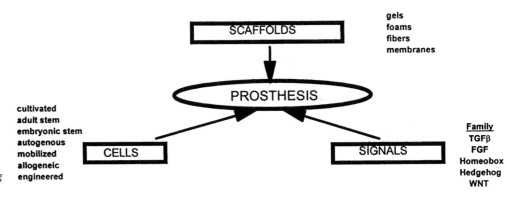

Fig. 1. The tissue engineering triad. See text for discussion.

are collagen biopolymers, mainly in the form of fibers and fibrils. Other forms of polymer organization have also been used (gels, foams, and membranes) for engineering tissue substitutes. The various forms can be combined in the laboratory to create imitations of biopolymer organization in specific tissues. Scaffolds can be enriched with signaling molecules, which may be bound to them or infused into them. Figure 1 shows some of the commonly occurring families of genes whose products play a major role in morphogenesis, pattern formation, and cell differentiation, processes that underlie histogenesis and organogenesis. The focus of the triad is the prosthesis.

The incorporation of cells in reconstituted prosthetic tissue devices often can provide the signals needed for tissue building, but the repertoire of feats of differentiation may be limited (see section on stem cells below). For example, although cultivated allogeneic keratinocytes and dermal fibroblasts, plus a collagen scaffold, can be assembled into a graftable organ that differentiates a fully formed epidermis, having a stratum corneum with barrier properties and a basal lamina, the secondary derivatives such as hair follicles and sebaceous and sweat glands do not develop. Improving the quality and functionality of tissue-engineered skin will mean the introduction of new versions of skin that address the clinical needs in a way better than their precursors have addressed them.

Another deficiency of engineered skin, for a number of significant applications, is its susceptibility to decomposition by metalloproteinases before it can be remodeled into a resistant replacement. Scaffolds of sturdier composition that will slow down enzyme action but still yield to effective rates of remodeling need to be developed. The impetus to improve engineered products will naturally be guided by the ideal, represented by evolution's achievement, the structure and function of the actual body part. It is to be expected that the competitive urges in a growing industry will be an important force in the race to imitate nature faithfully. Surely, the missions of basic scientists will also exert their effect.

Oversimplified materials used for the scaffolding component (the extracellular matrix of the tissue being engineered) may be limiting. If the scaffold cannot provide the developmental signals for tissue building needed by the cells that are seeded into it *in vitro,* or mobilized by it *in vivo,* tissue building might fail, as it does when a Dacron sleeve is used *in vivo* to replace a segment of artery. Man-made biopolymers such as poly(L-lactic acid), poly(glycolic acid), polyglycolide, and poly(L-lactide) have built-in ranges of degradation times that may not be in tune with the required rate of remodeling characteristic of regeneration, because the polymers are not susceptible to breakdown by metalloproteinases and tryptic enzymes, which function normally in the remodeling of collagen-based scaffolds. If they are out of tune with the remodeling activities of cells that occupy the transient scaffolds, including matrix biosynthesis, the process of matrix renewal may be compromised. A potentially valuable attribute of acellular materials installed *in vivo* as precursors of tissue replacements is their ability to mobilize appropriate cells from contiguous tissues, circulating body fluids, or stem cell sources, making it unnecessary to populate the prosthesis with cells before implantation. Because acellular implants of man-made biopolymers are information poor, cell mobilization and vascularization may fail, and so may the organization of the mobilized cells and their secretory output, needed to regenerate a replacement matrix and a functional tissue.

WHAT PARTS OF THE BODY CAN TISSUE ENGINEERING REPLACE?

The breadth of possibilities for creating replacement parts was explored very early in the growth of tissue engineering as a discipline. The reconstitution of living tissues—including skin, arteries consisting of three different cell types, thyroid gland, and adipose and other tissues—was reported between 1979 and 1986 by E. Bell and colleagues. Before that period, tissues for transplantation were available from tissue banks, with allogeneic living tissues coming into use after the discoveries of the immunosuppressants and the significance of matching tissue antigenic composition. The requirement of immunosuppressants in the transfer of allogeneic tissues has turned out not to be absolute. Being able to reconstitute tissues and organs by isolating and serially cultivating the parenchymal cells of which they consist makes it possible to select out cells of the immune system responsible for host initiation of graft rejection. The discovery by Sher *et al.* in 1983 showed that allogeneic skin parenchymal cells, separated from subsets of immune cells that normally populate the respective tissues of the skin, can be allografted without rejection in the absence of immunosuppressants. The work was the principal basis for the success of a non-custom-tailored commercial skin at Organogenesis Inc., a company founded by Bell. Some fresh tissues, such as cartilage (which is not engineered but is considered to be immunologically privileged because it has no blood circulation, and hence no subsets of cells of the immune system found in most tissues), have been grafted successfully across major histocompatibility barriers in work pioneered by A. E. Gross in Toronto and by others. The use of many other cell types for allogeneic grafting needs to be tested, to determine how universal grafting of non-self cell populations, free of immune cells, can become. Since the early work in the 1980s, Bell and colleagues and others have created prototypes of many additional tissues, including cartilage, ligament, bladder, periodontal prostheses, periosteal membrane among the soft tissues, and bone precursor cements for hard tissue replacement. We note that the late Bill Chick was a pioneer in the development of an endocrine pancreas equivalent.

It is not outlandish to consider that one day, replacement parts that are now inanimate devices may be reconstituted as biologic replicas. Some of the tissues needed for making a biologic limb or a part of one already have been fabricated. Others, such as cartilage, tendon, ligament, muscle, and bones, are works in progress. The task ahead is putting some of them together *in vitro* so that the composite becomes integrated for transplantation *in vivo,* in particular, vascularized, and made to function coordinately with the rest of the body, with the help of an artificial neural network.

ACELLULAR PROSTHESES

The use in animal models and in humans of complex allogeneic and xenogeneic tissues, depleted of their living cells by freezing or other methods, has been shown to be immunologically acceptable without the use of immunosuppressants. It is known from studies in experimental animals and humans that acellular allogeneic and xenogeneic implants have been accepted by their hosts. The work of D. W. Metzger with the small intestinal submucosa developed for many applications by S. W. Badylak has shown that immune responses to xenogeneic acellular tissues are of the Th2 class, not involving fixation of complement or graft rejection, but rather inducing tolerance. There has been failure following repeated challenges of grafts by subsequent implants of the antigenic materials, in attempts to drive the immune reaction into a Th1, cell-mediated rejection response. Because of the evolutionary conservation of their structural and compositional likeness to the part being replaced, and the degree to which xenogeneic materials lend themselves to remodeling by host cells, the value of various foreign acellular tissues, many available through tissue banks, is becoming well established. Further, the growing economic success of a number of companies devoted to the use of materials derived from tissue banks is strong testimony to the foregoing. Processing strategies, including control of pathogenic viruses, developed by some of these companies have contributed in a major way to the usefulness of the products derived from allogeneic human tissues.

Acellular collagen matrices, in the form of foams with and without bone precursor minerals, have been used as vehicles for delivering a variety of bone morphogenetic proteins.

THE RELEVANCE OF DEVELOPMENTAL BIOLOGY

The emergence of tissue engineering as a discipline has been stimulated by the necessary reexamination of the origin of tissues in the organism. (The discussions below of stem cells and signaling molecules, which play a role in tissue and organ ontogeny, acknowledge the debt to developmental biology studies.) Analyzing the development of a tissue reveals the significance of the extracellular matrix as a principal source of information for what cells do. In the beginning of embryonic development there are, essentially, just cells, expanding their own numbers in a rapid sequence of cleavages of the fertilized egg. By the blastula stage, or even the earlier morula stage, with as few as 128 cells present, secretion of components of the extracellular matrix has already begun. An early secretory product is collagen. Wessel and McClay (1987) have demonstrated that if collagen biosynthesis is blocked in the morula stage of sea urchin development—or, more pertinent still, if collagen cross-linking, an extracellular function performed by the enzyme lysyl oxidase, is prevented by β-aminopropionitrile—development is brought to a halt. Because development resumes when the blocker is removed, the point is made that the extracellular matrix is a crucial determinant required by cells cooperating in a multicellular complex to carry out their programs for cell division and morphogenesis. As development progresses, biosynthesis and secretion of structural and instructive extracellular matrix molecules continue in parallel. Gradually the extracellular matrix becomes tailored, regionally, to specific tissues and organs as the cells that make them up differentiate, expressing tissue-specific repertoires of matrix secretions. The early output of collagens in embryogenesis is especially significant, because in addition to recognizing integrin receptors and through cell-binding ligands, the collagens bind a large number of different secreted matrix molecules and perform the task of immobilizing many cell-secreted products. If these products were not immobilized, they would escape and fail to contribute to the molecular diversity of the extracellular matrix tapestry. The development of the complex extracellular matrix grows as the molecules that bind to collagen bind other subsets of molecules.

I have dwelt on the origins of the extracellular matrix to bring into focus the design of tissues that are communities of cells in which individual cells, or small groups of cells, as well as large cell masses, are surrounded by networks of collagen fibrils and fibers. Sheets of cells, including epithelial, mesothelial, and endothelial cells, are underlaid by a matrix whose principal component is also fibrillar collagen. In either design the extracellular matrix is the product of the cells that live in it, or on it; it is further enriched by paracrine and endocrine secretions and by products of the wandering cells, such as macrophages, lymphocytes, neutrophils, and mast cells, that move in and out of it. The extracellular matrix is an extremely complex material of high molecular diversity and microarchitectural specificity. It is in developing and regenerating systems that the sequence of genes activated to create the definitive extracellular matrix can be detailed.

CAN NATURE BE IMITATED?

Meeting the challenge of reconstituting tissues is in a way dependent on how the philosophic question "can nature be imitated?" is answered. To begin with, what do we mean by "imitated"? There are three senses that can be discussed: (1) It can mean making an exact or closely approximate biologic replica that exhibits at least some of the basic properties of the original tissue at the time it is implanted. (2) It can mean providing a much less well-developed precursor of a biologic substitute, with the expectation that it will be built into a faithful replica. (3) It can mean using a nonbiologic replacement. Either of the first two approaches might remedy a deficit and restore functionality in the form of a living tissue or organ. Successful examples of nonbiologic replacement parts are artificial limbs, dentures, and hearing aids, which are attached to the body but not grafted to it. Another class of nonbiologic aids includes pacemakers and elements of the skeletal system, including joints with plastic and metallic components, all of which are grafted, but not biologically integrated, remaining inanimate devices. Probably the most risky adventure in engineering an inanimate replacement part was the attempt to construct a total artificial heart (TAH) having connections to the major arterial and venous vessels of the host. Apart from the unwieldy size of the TAH, the incredible difficulty of compliance matching of the artificial to the native arterial vessels may have been given too little attention early on. Mismatches result in great turbulence of flow, as does flow through mechanical chambers of improper design, resulting in blood clot formation. With a heightened awareness of past problems there is a new optimism abroad. New research and development on the TAH is being supported at the Texas Medical Center's Texas Heart Institute, which provides a history of their Mechanical Circulatory Support program at their internet site (http:/www.tmc.edu/thi/mcshist.html). In current development are ventricular assist devices designed to give weak hearts the boost they need to function more adequately and to serve as interim helpers for those awaiting a heart transplant.

We may believe that nature can be imitated biologically, but it is difficult to conceive, however imperfect nature is, that it can be improved. The sense in which this is meant is that it will be difficult if not impossible to improve the genetically and developmentally *normal* endowments that define the character of tissues, such as the tissue scaffolds (the extracellular matrix of tissues) and the signaling systems needed for tissue development. We are reluctant to say absolutely that synthetic materials that precisely imitate body parts will not succeed, but given the number of genes involved in the signaling systems and the tissue scaffold systems, asking for an improvement in the roles of these complex components in the course of tissue development is like asking for a chance to evolve again. What is germaine for tissue engineering is the task of improving the lives of individuals undermined by genetic and developmental aberrations, which are often expressed early, and by the degenerative diseases of aging, such as atherosclerosis, kidney failure, osteoporosis, arthritis, and skeletal wear and tear. In these individuals, degenerative states and the imperfect endowments of nature surely can be improved. Implanting an arterial substitute that resists intimal or medial hyperplasia in an individual with a propensity for atherosclerosis would provide a greatly improved vessel compared to the naturally endowed but disease-ridden vessel for which it is a substitute.

The probability seems low that nonbiologic synthetic materials that precisely imitate body parts and retain their properties without degenerating will be discovered and used successfully as prostheses; it is more likely that alternative strategies are closer at hand. Awakening the capacity of the body to regenerate by giving it assistance in the form of bioabsorbable remodelable scaffolds, instructive molecules, and cells, if they are not available to be mobilized, is the essence of current trends.

THE NEED FOR VASCULARIZATION *IN VITRO* AND *IN VIVO*

It is no small task to design living prosthetic devices so that the cells seeded into them can survive until the implant acquires a circulation. It can be expected that there will be a risk of nonsurvival for cells deep in the prosthesis. Reducing the risk will depend on the use of angiogenic cytokines and tricks of tissue design that can provide access of a medium *in vitro* and can promote development of an adequate circulation sufficiently rapidly *in vivo*. We look forward to the possibility that small arteries and veins can be built into the prosthesis to make contact with a capillary bed *in vitro* under the influence of angiogenic factors. The small arteries and veins would eventually be anastomosed to those of the host. Much experimental work is needed to achieve the foregoing. Obviously the rate of diffusibility of tissue fluids and large molecules allowed by a prosthesis will be one of its important properties.

Although vascular prostheses may be needed to serve other engineered tissues of substance *in vitro* and afterward *in vivo*, a great need for them exists as stand-alone devices for replacing both peripheral and coronary sites affected by arterial disease. There has been a considerable history of preclinical and clinical trials both of natural vessels, mainly veins, and of nonbiologic synthetic prostheses. Those made with nonbiologic synthetic polymers have had limited replacement value and basically have not yet proved themselves. Although the materials were not strongly immunogenic, some elicited persistent inflammatory reactions and others were erodable; none was adequately compliant or able to integrate with host tissues. Remembering that these new polymers have never before been encountered by cells in the course of evolution, it is remarkable that, as foreign bodies, they have been tolerated as well as they have. It is thought by some that self-degrading man-made biopolymers have a future in tissue engineering, even though they are unfamiliar to cells and suffer the shortcomings discussed above. One promising possibility is that they may be replaced *in vivo* in a timely fashion by native scaffolds built by the cells seeded into them. But reconstituting arterial vessels entirely with biologic components—that is, with cells and naturally occurring extracellular matrix polymers such as collagen, which are also bioabsorbable and remodelable—has been in progress for some years, with significant advances being made over the earlier efforts. It has been suggested that genetically engineered smooth muscle cells can be made refractory to mitogenic stimuli that induce medial and intimal hyperplasia. The use of this and other genetic strategies may not be far off.

STEM CELLS

Scaffolds can be populated with adult-derived cells that are capable of undergoing subsequent differentiation after being cultivated *in vitro*. In this category are cells of the skin, cartilage, muscle, tendon, ligament, bone, adipose tissue, endothelium, and many others. Aside from skin, the foregoing cell types are harbored as stem cell populations in the marrow, in addition to those of the hematopoietic and immune systems, but the diversity of mesenchymal and possibly other cell types in the marrow still needs to be probed. Stimulating factors, the cytokines, which move some of the cells into the circulation, will be important for engineering acellular scaffolds. Other stem cells are available to tissue engineering, such as the satellite cells found in striated muscle and to some degree keratinocytes of the skin. Where host cells are available, an acellular scaffold, particularly one enhanced with signals and possessing the binding sites needed for cell attachment, can mobilize host cells that will populate the prosthesis. Already, new sources of stem cells, particularly neuronal stem cells, have been discovered in the adult brain and are opening the door to the reconstitution of nerve tissue for tissue engineering. In addition to the striatum, which harbors extracellular growth factor-responsive stem cells, other central nervous system sources of stem cells in adult vertebrates include the hippocampus and the periventricular subependymal zone. Stem cells giving rise to neuronal and glial phenotypes, from the adult rat hippocampus, are isolated with the help of fibroblast growth factor 2 and are stimulated to differentiate with the help of retinoic acid. Stem cells from these sources are also present in the adult human brain.

The discovery that embryonic stem cells can be recovered from human fetal tissue and propagated for long periods without losing their toti- or pluripotency is of huge importance for tissue engineering. How to direct their differentiation is a subject of high current interest.

Engineered cells, endowed with regulators of cell division, or in which expression of specific gene products may be promoted or on which immunologic neutrality may be conferred, will no doubt become common in the future.

CELL SIGNALING

We currently are witnessing an explosion of information relating to cell signaling, implicated in a broad range of developmental activities, including the establishment of the embryonic axis, the formation and differentiation of somites, the establishment of the central and peripheral nervous systems, and many other histio- and organogenic activities. The National Institute of Health Science Cell Signaling Networks Database (CSNDB) on the internet (http://geo.nihs.go.jp/csndb) lists signaling molecules, their pathways, and original articles. The size of the database conveys the breadth of work in progress in the field.

Responsible for establishing the overall pattern of the body and involved in differentiation of the nervous system and limb development is the homeobox gene family. Its products are transcription factors, which are not secreted but work intracellularly. Mammals, mice, and humans, for example, have four clusters of homeobox genes, each on a different chromosome. The order of gene activation progresses from 3′ to 5′ through the gene sequence, with the expression of each successive gene in the complex being colinear with the anterior–posterior axis of the body plan and the specific organ system undergoing development. What happens to a cell depends on where it is. It is thought that a gradient of secreted Sonic hedgehog (Shh) protein may be responsible for the regionally differential activation of particular genes in the homeobox complex.

Although it appears that cascades of gene activation are implicated in the building of many tissues and organs, there cannot be an infinite regression of gene activation by products of genes that must in turn be expressed. At various points in the causal chain of epigenesis, physical factors and nongenetic chemical factors, distributed asymmetrically in the fertilized egg and partitioned during cleavage, may be the early initiators of gene function. Later in development as well, many events of differentiation may be initiated by nongenetic signals. One principle of signaling has become clear: multiple signals appear to be required for tissue building. For example, at least five signaling molecules appear to direct the events associated with the differentiation of the dorsal compartment of the somite into epaxial muscle, dermis, and hypaxial muscle. Many other multiple signaling pathways also have been identified. For tissue engineering a holistic approach to signaling based on the use of multiple signals identified in developmental studies may yield results superior to those achieved with single growth factors, as exemplified in orthopedic applications. Bell (United States Patent numbers 5,800,537 and 5,893,888) and co-workers have been isolating and testing complexes of tissue-specific signals [animal-derived extracellular matrix (ADMAT)] processed from developing tissues. A dental ADMAT combined with an injectable bone precursor cement has been used successfully to promote bone development around titanium implants in dogs. Along with their co-workers, T. M. Sioussat (of Tissue Engineering Inc.) and D. W. Meraw (of the Mayo Clinic) led development of the company's bone precursor cement and the dental ADMAT in which there were multiple growth factors in picogram quantities, distinguished from individual cytokine products used in microgram quantities, to stimulate new bone formation.

LOOKING FORWARD

The excitement of tissue engineering, conveyed in the chapters of this volume, lies in the challenge of imitating nature, and in the opportunity to witness the outcomes of a discovery or invention, because the interval between conception and clinical realization of a new device or treatment is, in our time, telescoped. The value, too, of viewing cells as they are organized *in vivo,* by developing *in vitro* model tissue and organ systems for studying normal and disease states of cell replication, morphogenesis, and differentiation, represents a new opportunity. To meet the task of reconstituting tissues *in vitro,* understanding how tissues are built *in vivo* is the overwhelming challenge. The general question for those in tissue engineering asks what instructions cells need for organizing into tissues. More specifically, which cells are capable of responding to instructions in the form of signals? If they are hidden stem cells, how can they be harvested? How are responsive cells induced to increase their numbers? And how are these cells primed to engage in the synthesis of tissue-specific products? Which signals are needed to trigger the causal chain of tissue development? Is it indeed an isolated cascade or are tissue interactions required for release of signals? Moving on, what materials are synthesized for extracellular transport as a result of signals received? How are the secreted materials assembled into structures such as collagen fibrils and fibers that persist and serve as the tissue's scaffold, around which the remainder of the extracellular matrix is built? What do the anatomic and chemical fine structures of the extracellular matrix look like? Finally, how do the reconstituted tissues fare in the hosts in which they are implanted? The questions asked are relevant for the task.

The study of these questions is subsumed not by a single discipline, but by many, including microanatomy; cell, molecular, and developmental biology; immunology; materials science; and branches of engineering able to inquire into the modes of action of the physical forces on which many developmental phenomena depend. The present volume is authored by practitioners of a number of the above-mentioned disciplines as well as by others relevant for tissue engineering, reflecting the multidisciplinary character of the new field, which is truly beginning to define itself.

REFERENCES

Acampora, D., D'Esposito, M., Faiella, A., Pannese, M., Migliaccio, E., Morelli, F., Stornaiuolo, A., Nigro, V., Simeone, A., and Boncinelli, E. (1989). The human HOX gene family. *Nucleic Acids Res.* 17, 10385–10402.

Bell, E. (1995). Deterministic models of tissue engineering. *J. Cell Eng.* 1, 28–34.

Bell, E. (1995). Strategy for the selection of scaffolds for tissue engineering. *Tissue Eng.* 1, 163–179.

Bell, E., Ivarsson, B., and Merrill, C. (1979). Production of a tissue-like structure by contraction of collagen lattices by fibroblasts of different proliferative potential *in vitro. Proc. Natl. Acad. Sci. U.S.A.* 76, 1274–1278.

Bell, E., Ehrlich, H. P., and Nakatsuji, T. (1981). Living tissue formed *in vitro* and accepted as skin equivalent tissue *in vitro*. *Science* **211**, 1052–1054.

Bell, E., Moore, H ., Michie, C., Sher, S., Hull, B., and Coon, H. (1984). Reconstitution of a thyroid gland equivalent from cells and matrix materials. *J. Exp. Zool.* **232**, 277–285.

Bell, E., Rosenberg, M., Kemp, P., Gay, R., Green, G., Muthukumaran, N., and Nolte, C. (1991). Recipes for reconstituting skin. *J. Biomech. Eng.* **113**, 113–119.

Bugbee, W. D., and Convery, F. R. (1999). Osteochondral allograft transplantation. *Clin. Sports Med.* **18**, 67–75.

Cell Signaling Networks Database internet site: *http://geo.nihs.go.jp/csndb.*

Gearhart, J. (1998). New potential for human embryonic stem cells. *Science* **282**, 1061–1062.

Ghazavi, M. T., Pritzker, K. P., Davis, A. M., and Gross, A. E. (1997). Fresh osteochondral allografts for post-traumatic osteochondral defects of the knee. *J. Bone Jt. Surg., Br. Vol.* **79**;1008–1013.

Kagan, D., and Bell, E. (1989). Differentiation of a fatty connective tissue equivalent without chemical induction. *J. Cell Biol.* **107**, 603a.

Kalthoff, K. (1996). "Analysis of Biological Development," Chapters 2, 22, and 23. McGraw-Hill, New York.

Kirchenbaum, B., Doetsch, F., Lois, C., and Alvarez-Buylla, A. (1999). Adult subventricular zone neuronal precursors continue to proliferate and migrate in the absence of the olfactory lobe. *J. Neurosci.* **19**, 2171–2180.

Kukekov, V. G,. Laywell, E. D., Suslov, O., Scheffler, B., Thomas, L. B., O'Brien, T. F., Kusakabe, M., and Steindler, D. A. (1999). Multipotent stem/progenitor cells with similar properties arise from two neurogenic regions of adult human brain. *Exp. Neurol.* **156**, 333–344.

Marcelle, C., Stark, M. R., and Bronner-Fraser, M. (1997). Coordinate actions of BMPs, Wnts, Shh and noggin mediate patterning of the dorsal somite. *Development (Cambridge, UK)* **124**, 3955–3963.

Mechanical Circulatory Support internet site: *http://www.tmc.edu/thi/mcshist.html.*

Meraw, S. J., Reeve, C. M., Lohse, B. S., and Sioussat, T. M. (1999). Treatment of peri-implant defects with combination growth factor cement. *J. Periodontol.* (in press).

Metzger, D. W., McPherson, T., Merrill, L. A., Moyad, T. F., and Badylak, S. F. (1998). Immune responses to xenogeneic SIS implants. *In* "2nd SIS Symposium" (S. F. Badylak (ed.), p. 33. Purdue University, West Lafayette, IN.

Reynolds, B. A., and Weiss, S. (1992). Generation of neurons and astrocytes from isolated cells of the adult central nervous system. *Science* **255**, 1707–1710.

Saig, P., Coulomb, B., Lebreton, C., Bell, E., and Dubertret, L. (1985). Psoriatic fibroblasts induce hyperproliferation of normal keratinocytes in a skin equivalent model *in vitro*. *Science* **230**, 669–672.

Scott, J. E. (1992). Supramolecular organization of extracellular matrix glucosaminoglycans, *in vitro* and in the tissues. *FASEB J.* **6**, 2639–2645.

Sher, S., Hull, B., Rosen, S., Church, D., Friedman, L., and Bell, E. (1983). Acceptance of allogeneic fibroblasts in skin equivalent transplants. *Transplantation* **36**, 552–557.

Takahashi, J., Palmer, T. D., and Gage, F. H. (1999). Retinoic acid and the neurotrophins collaborate to regulate neurogenesis in adult derived neural stem cell cultures. *J. Neurobiol.* **38**, 65–81.

Texas Heart Institute email address for information on Mechanical Circulatory Support: *http:/www.tmc.edu/thi/mcshist.html*

Thomson, J. A., Itskovitz-Eldor, J., Shapiro, S. S., Waknitz, M. A., Swiergiel, J. J., Marshall, V. S., and Jones, J. M. (1998). Embryonic stem cell lines derived from human blastocysts. *Science* **282**, 1145–1147.

Weinberg, C., and Bell, E. (1985). A blood vessel model constructed from collagen and vascular cells. *Science* **231**, 397–400.

Wessel, G. M., and McClay, D. R. (1987). Gastrulation in the sea urchin embryo requires the deposition of crosslinked collagen within the extracellular matrix. *Dev. Biol.* **121**(1), 149–165.

Introduction to Tissue Engineering

CHAPTER 1

THE HISTORY AND SCOPE OF TISSUE ENGINEERING

Joseph P. Vacanti and Charles A. Vacanti

INTRODUCTION

During the past decade—in fact, since the publication of the first edition of this book in 1997—the field of tissue engineering has grown at a seemingly logarithmic rate. The expansion has not only been in interest but also in new knowledge. Major efforts in the field are now underway in most of the developed countries of the world. During the past 5 years studies of many scientific investigators have been coordinated in a collaborative effort seldom seen. Research facilities are exchanging both data and scientists to codevelop ideas and projects targeted for human studies. Regulatory agencies are working closely with academic institutions and private industry voluntarily to develop guidelines prospectively in preparation for human applications.

The hope that the concepts and techniques of tissue engineering can be applied to human disease is rapidly becoming reality, driven by the profound need for and realistic potential of this new discipline. To understand the factors that have enabled the emergence and rapid growth of this field, it is perhaps helpful to examine its roots in the context of historical developments. Although the practice of medicine is as old as humankind, its scientific roots have emerged from the development of the scientific method in the Western tradition. When tissues and organs are damaged by infection, disease, or injury, the first priority has always been the development of methods to salvage life. Infection could be controlled by drainage of pus and little else until the revolution of antibiotic therapy. With the nineteenth-century scientific understanding of the germ theory of disease and the introduction of sterile technique, modern surgery had its emergence. The advent of anesthesia by the midnineteenth century enabled the rapid evolution of many surgical techniques. With patients anesthetized, innovative and courageous surgeons could save lives by examining and treating internal areas of the body: the thorax, the abdomen, the brain, and the heart. Initially the surgical techniques were primarily extirpative—for example, removal of tumors, bypass of the bowel in the case of intestinal obstruction, and repair of life-threatening injuries. Maintenance of life without regard to the crippling effects of tissue loss or the psychosocial impact of disfigurement, however was not an acceptable end goal. Techniques that resulted in the restoration of function through structural replacement became integral to the advancement of human therapy.

Artificial or prosthetic materials for replacing limbs, teeth, and other tissues resulted in the partial restoration of lost function. Also, the concept of using one tissue as a replacement for another was developed. In the sixteenth century, Tagliacozzi of Bologna, Italy, reported in his work, "De Custorum Chirurigia per Insitionem," a description of a nose replacement that he constructed from a forearm flap. With the advent of modern concepts of sterility and anesthesia, whole fields of reconstructive surgery have emerged to improve the quality of life by replacing missing function through rebuilding body structures. In our current era, modern techniques of transplanting tissue from one individual into another have been revolutionary and lifesaving.

Principles of Tissue Engineering
Second Edition

The molecular and cellular events of the immune response have been elucidated sufficiently to suppress the response in the clinical setting of transplantation and to produce prolonged graft survival and function in patients. In a sense, transplantation can be viewed as the most extreme form of reconstructive surgery, transferring tissue from one individual into another. As with any successful undertaking, new problems have emerged. Techniques using implantable foreign body materials have produced dislodgment, infection at the foreign body/tissue interface, fracture, and migration over time. Techniques moving tissue from one position to another position have produced biologic changes because of the abnormal interaction of the tissue at its new location. For example, diverting urine into the colon can produce fatal colon cancers 20–30 years later. Making esophageal tubes from the skin can result in skin tumors 30 years later. Using intestine for urinary tract replacement can result in severe scarring and obstruction over time.

Transplantation from one individual into another, although very successful, has severe constraints. The major problem is accessing enough tissue and organs for all of the patients who need them. Currently, 65,000 people are on transplant waiting lists in the United States, and many will die waiting for available organs. Also, problems with the immune system produce chronic rejection and destruction over time. Creating an imbalance of immune surveillance from immunosuppression can cause new tumor formation. These constraints have produced a need for new solutions to provide needed tissue.

It is within this context that the field of tissue engineering has emerged. In essence, new and functional living tissue is fabricated using living cells, which are usually associated in one way or another with a matrix or scaffolding to guide tissue development. Such scaffolds can be natural, man-made, or a composite of both. Living cells can migrate into the implant after implantation or can be associated with the matrix in cell culture before implantation. Such cells can be isolated as fully differentiated cells of the tissue they are hoped to recreate, or they can be manipulated to produce the desired function when isolated from other tissues or stem cell sources. Conceptually, the application of this new discipline to human health care can be thought of as a refinement of previously defined principles of medicine. The physician has historically treated certain disease processes by supporting nutrition, minimizing hostile factors, and optimizing the environment so that the body can heal itself. In the field of tissue engineering, the same thing is accomplished on a cellular level. The harmful tissue is eliminated, and cells necessary for repair are then introduced in a configuration optimizing survival of the cells in an environment that will then permit the body to heal itself. Tissue engineering offers an advantage over cell transplantation alone in that organized three-dimensional functional tissue is designed and developed. This chapter summarizes some of the challenges that must be resolved before tissue engineering can become part of the therapeutic armamentarium of physicians and surgeons. Broadly speaking, the challenges are scientific and social.

SCIENTIFIC CHALLENGES

As a field, tissue engineering has been defined for little more than a decade. Much still needs to be learned and developed to provide a firm scientific basis for therapeutic application. To date, much of the progress in this field has been related to the development of model systems, which have suggested a variety of approaches. Also, certain principles of cell biology and tissue development have been delineated. The field can draw heavily on the explosion of new knowledge from several interrelated well-established disciplines, and, in turn, may promote the coalescence of relatively new, related fields to achieve their potential. The rate of new understanding of complex living systems has been explosive in the past three decades. Tissue engineering can draw on the knowledge gained in the fields of cell biology, biochemistry, and molecular biology and apply it to the engineering of new tissues. Likewise, advances in materials science, chemical engineering, and bioengineering allow the rational application of engineering principles to living systems. Yet another branch of related knowledge is the area of human therapy as applied by surgeons and physicians. In addition, the fields of genetic engineering, cloning, and stem cell biology may ultimately develop hand in hand with the field of tissue engineering in the treatment of human disease, each discipline depending on developments in the others.

We are in the midst of a biologic renaissance. Interactions of the various scientific disciplines can elucidate not only the potential directions of each field of study, but also the right questions to address. The scientific challenge in tissue engineering lies both in understanding cells and their mass transfer requirements, and the fabrication of materials to provide scaffolding and templates.

CELLS

If we postulate that living cells are required to fabricate new tissue substitutes, much needs to be learned in regard to their behavior in two normal circumstances—namely, normal development in morphogenesis and normal wound healing. In both of these circumstances, cells create or recreate functional structures using pre-programmed information and signaling. Some approaches to tissue engineering rely on guided regeneration of tissue using materials that serve as templates for ingrowth of host cells and tissue. Other approaches rely on cells that have been implanted as part of an engineered device. As we understand normal developmental and wound healing gene programs and cell behavior, we can use them to our advantage in the rational design of living tissues.

Acquiring cells for creation of body structures is a major challenge, the solution of which continues to evolve. The ultimate goal in this regard—the large-scale fabrication of structures—may be to create large cell banks composed of universal cells that would be immunologically transparent to any individual. These universal cells could be differentiated cell types that could be accepted by any individual or could be stem cell reservoirs, which could respond to signals to differentiate into differing lineages for specific structural applications. Much is already known about stem cells and cell lineages in the bone marrow and blood. Studies suggest that progenitor cells for many differentiated tissues exist within the marrow and blood, and may very well be ubiquitous. Our knowledge of the existence and behavior of such cells in various mesenchymal tissues (muscle, bone, and cartilage), endodermally derived tissues (intestine and liver), or ectodermally derived tissues (nerves, pancreas, and skin) expands on a daily basis. A new area of stem cell biology involving embryonic stem cells holds promise for tissue engineering. The challenge to the scientific community is to understand the principles of stem and progenitor cell biology and then apply it to tissue engineering. The challenge to the scientific community is to understand the principles of stem and progenitor cell biology and then apply it to tissue engineering. The development of immunologically inert universal cells may come from advances in genetic manipulation.

As intermediate steps, tissue can be harvested as allograft, autograft, or xenograft. The tissues can then be dissociated and placed into cell culture, where proliferation of the cells can be initiated. After expansion to the appropriate cell number, the cells can then be transferred to templates, where further remodeling can occur. Which of these strategies are practical and possibly applicable in humans remains to be explored.

Large masses of cells for tissue engineering need to be kept alive, not only *in vitro* but also *in vivo*. The design of systems to accomplish this, including *in vitro* flow bioreactors and *in vivo* strategies for maintenance of cell mass, presents an enormous challenge in which significant advances have been made. The fundamental biophysical constraint of mass transfer of living tissue needs to be understood and dealt with on an individual basis as we move toward human application.

MATERIALS

There are so many potential applications to tissue engineering that the overall scale of the undertaking is enormous. The field is ripe for expansion, and requires training of a generation of materials scientists and chemical engineers.

The optimal chemical and physical configurations of new biomaterials as they interact with living cells to produce tissue-engineered constructs are under study by many research groups. These biomaterials can be permanent or biodegradable. They can be naturally occurring materials, synthetic materials, or hybrid materials. They need to be developed to be compatible with living systems or with living cells *in vitro* and *in vivo*. Their interface with the cells and the implant site must be clearly understood so that the interface can be optimized. Their design characteristics are major challenges for the field, and should be considered at a molecular chemical level. Systems can be closed, semipermeable, or open. Each design should factor in the specific replacement therapy considered. Design of biomaterials can also incorporate the biologic signaling that the materials may offer. Examples include release of growth and differentiation factors, design of specific receptors and anchorage sites, and three-dimensional site specificity using computer-assisted design and manufacture techniques.

GENERAL SCIENTIFIC ISSUES

As new scientific knowledge is gained, many conceptual issues need to be addressed. Related to mass transfer is the fundamental problem associated with nourishing tissue of large mass as opposed to tissue with a relatively high ratio of surface area to mass. Also, functional tissue equivalents necessitate the creation of composites containing different cell types. For example, all tubes in the body are laminated tubes composed of a vascularized mesodermal element such as smooth muscle, cartilage, or fibrous tissue. The inner lining of the tube, however, is specific to the organ system. Urinary tubes have a stratified transitional epithelium. The trachea has a pseudostratified columnar epithelium. The esophagus has an epithelium that changes along a gradient from mouth to stomach. The intestine has an enormous, convoluted surface area of columnar epithelial cells that migrate from a crypt to the tip of the villus. The colonic epithelium is, again, different for the purposes of water absorption and storage.

Even the well-developed manufacture of tissue-engineered skin used only the cellular elements of the dermis for a long period of time. Attention is now focusing on creating new skin consisting of both the dermis and its associated fibroblasts, as well as an epithelial layer consisting of keratinocytes. Obviously, this is a significant advance, but for truly "normal" skin to be engineered, all of the cellular elements should be contained, so that the specialized appendages can be generated as well. These "simple" composites will indeed prove to be quite complex and require intricate designs. Thicker structures with relatively high ratios of surface area to mass, such as liver, kidney, heart, breast, or central nervous system, will offer other engineering challenges.

Currently, studies for developing and designing materials in three-dimensional space are being developed utilizing both naturally occurring and synthetic molecules. The applications of computer-assisted design and manufacture techniques to the design of these matrices are critically important. Transformation of digital information obtained from magnetic resonance scanning or computerized tomography scanning can then be developed to provide appropriate templates. Some tissues can be designed as universal tissues that will be suitable for any individual, or may be custom-designed tissues specific to one patient. An important area for future study is the entire field of neural regeneration, neural ingrowth, and neural function toward end organ tissues such as skeletal or smooth muscle. Putting aside the complex architectural structure of these tissues, the cells contained in them have a very high metabolic requirement. As such, it is exceedingly difficult to isolate a large number of viable cells. An alternate approach may be the use of less mature progenitor cells, or stem cells, which not only would have a higher rate of survival as a result of their lower metabolic demand, but also would be more able to survive the insult and hypoxic environment of transplantation. As stem cells develop and require more oxygen, their differentiation may stimulate the development of a vascular complex to nourish them. The understanding of and solutions to these problems are fundamentally important to the success of any replacement tissue that needs ongoing neural interaction for maintenance and function.

It has been shown that some tissues can be driven to completion *in vitro* in bioreactors. However, the optimal incubation times will vary from tissue to tissue. Even so, the new tissue will require an intact blood supply at the time of implantation for successful engraftment and function.

Finally, all of these characteristics need to be understood in the fourth dimension of time. If tissues are implanted in a growing individual, will the tissues grow at the same rate? Will cells taken from an older individual perform as young cells in their new "optimal" environment? How will the biochemical characteristics change over time after implantation? Can the strength of structural support tissues such as bone, cartilage, and ligaments be improved in a bioreactor in which force vectors can be applied? When is the optimal timing of this transformation? When does tissue strength take over the biomechanical characteristics as the material degrades?

SOCIAL CHALLENGES

If tissue engineering is to play an important role in human therapy, in addition to scientific issues, fundamental issues that are economic, social, and ethical in nature will arise. Something as simple as a new vocabulary will need to be developed and uniformly applied. A universal problem is funding. Can philanthropic dollars be accessed for the purposes of potential new human therapies? Will industry recognize the potential for commercialization and invest heavily? If this occurs, will the focus be changed from that of a purely academic endeavor? What role will governmental agencies play as the field develops? How will the field be regulated to ensure its safety and efficacy prior to human application? Is the new tissue to be considered transplanted tissue and, therefore, not subject to regulation, or is it a pharmaceutical that must be subjected to the closest scrutiny by regulatory agencies? If lifesaving, should the track be accelerated toward human trials?

There are legal ramifications of this emerging technology as new knowledge is gained. What becomes proprietary through patents? Who owns the cells that will be sourced to provide the living part of tissue fabrication?

In summary, one can see from this brief overview that the challenges in the field of tissue engineering remain significant. All can be encouraged by the progress that has been made in the past few years, but much discovery lies ahead. Ultimate success will rely on the dedication, creativity, and enthusiasm of those who have chosen to work in this exciting but still unproved field.

References

Bronzino, J. D. *et al.* (1995). Section XI: Tissue engineering. *In* "The Biomedical Engineering Handbook" (J. D. Bronzino, ed.), pp. 1580–1824. CRC Press, Hartford, CT.

Langer, R., and Vacanti, J. P. (1993). Tissue engineering. *Science* **260,** 920–926.

Scientific American (1999). Special report: The promise of tissue engineering. *Sci. Am.* **280,** 38–65.

Vacanti, J. P., and Langer, R. (1999). Tissue engineering: The design and fabrication of living replacement devices for surgical reconstruction and transplantation. *Lancet* (in press).

CHAPTER 2

THE CHALLENGE
OF IMITATING NATURE

Robert M. Nerem

INTRODUCTION

Tissue engineering has the potential to address the transplantation crisis caused by the shortage of donor tissues and organs. It is through the imitation of nature that tissue engineering will be able to address this patient need; however, a number of challenges need to be faced. In the area of cell technology, these include cell sourcing, the manipulation of cell function, and the future, effective use of stem cell technology. Next are those issues that are part of construct technology, and these include the design and engineering of tissue like constructs and the manufacturing technology required to provide off-the-shelf availability to the clinician. Finally, there are those issues associated with the integration of a construct into the living system, with the most critical issue being the engineering of immune acceptance. Only if we can meet the challenges posed by these issues and only if we can ultimately address the tissue engineering of the most vital of organs, will we be successful in confronting the crisis in transplantation.

An underlying supposition of tissue engineering is that the employment of the natural biology of the system will allow for greater success in developing therapeutic strategies aimed at the replacement, repair, maintenance, and/or enhancement of tissue function. Another way of saying this is that just maybe the great creator, in whatever form one believes he or she exists, knows something that we mere mortals do not, and if we can only "tap" into a small part of this knowledge base, if we can only imitate nature in some small way, then we will be able to achieve greater success in our efforts to address patient needs in this area.

It is this challenge of imitating nature that has been accepted by those who are providing leadership to this new area of technology called tissue engineering (Langer and Vacanti, 1993; Nerem and Sambanis, 1995). To imitate nature requires that we first understand the basic biology of the tissues of interest; with this we then can develop methods for the control of these biologic processes, and based on the ability to control we finally can develop strategies either for the engineering of living tissue substitutes or for the fostering of tissue repair or remodeling.

It is not only tissues that are of interest, but in many cases entire organs. In fact, even though the initial successes have been substitutes for skin, a relatively simple tissue, in the long term tissue engineering has the potential for creating vital organs such as the kidney, the liver, and the pancreas. Some even believe it will be possible to tissue engineer an entire heart. In addressing the repair and/or replacement of such vital organs, tissue engineering has the potential literally to confront the transplantation crisis, i.e., the shortage in donor tissues and organs available for transplantation.

Although research in what we now call tissue engineering started a quarter of a century ago, the term tissue engineering was not "coined" until 1987 when Professor Y. C. Fung, from the University of California, San Diego, suggested this name at a National Science Foundation meeting.

Since then there has been a large expansion in research efforts in this field, and as the technology has become further developed, an industry has begun to emerge. This industry is still very much a fledgling one, with only a few companies possessing product income streams. A study based on 1997 data documents a total of 35 companies active in the field, with $430 million annually in industrial research and development taking place, and an annual growth rate of 20% (Lysaght *et al.,* 1998). Tissue engineering is literally at the interface of the traditional medical implant industry and the biological revolution (Galletti *et al.,* 1995). By harnessing the products of this revolution and the continuing advances in molecular and cell biology, there will be an entirely new generation of tissue and organ implants. With this there is the potential for a multibillion dollar tissue engineering industry, one that is estimated to be in excess of $20 billion by the year 2020.

Such a revolution in the medical implant industry will only occur, however, if we successfully meet the challenge of imitating nature. Thus, in the remainder of this chapter the critical issues involved in this will be addressed. This is done by first discussing those issues associated with cell technology, i.e., issues important in cell sourcing and in the achievement of the functional characteristics required of the cells to be employed. Next to be discussed are those issues associated with construct technology. These span from the organization of cells into a three-dimensional architecture that functionally mimics tissue to the technologies required to manufacture such products and provide tissue substitutes off the shelf to the clinician. Finally, issues involved in the integration of a living cell construct into, or the fostering of remodeling within, the living system will be discussed. These range from the use of appropriate animal studies to the issues of biocompatibility and immune acceptance. Success in tissue engineering will only be achieved if issues at these three different levels, i.e., cell technology, construct technology, and the technology for integration into the living system, can be addressed.

CELL TECHNOLOGY

CELL SOURCING

The starting point for any attempt to engineer a tissue or organ substitute is a consideration of the cells to be employed. Not only will one need to have a supply of sufficient quantity and one that can be ensured to be free of pathogens and contamination of any type whatsoever, but one will need to decide whether the source to be employed is to provide autologous, allogeneic, or xenogeneic cells. As indicated in Table 2.1, there are both advantages and disadvantages to each of these.

The skin substitutes developed by Organogenesis (Canton, MA) and Advanced Tissue Sciences (La Jolla, CA) represent the first tissue-engineered products, and these in fact use allogeneic cells. The Organogenesis product, Apligraf, is a bilayer model of skin involving fibroblasts and keratinocytes that are obtained from donated human foreskin (Parenteau, 1999). Apligraf already is approved by the Food and Drug Administration (FDA). The first tissue-engineered product approved by the FDA was the Advanced Tissue Sciences product, TransCyte. Approved initially for third-degree burns, TransCyte is made by seeding dermal fibroblasts in a polymeric scaffold; however, once cryopreserved it becomes a nonliving wound covering. The living product of Advanced Tissue Sciences is still undergoing clinical trials, but is a dermis model, also with dermal fibroblasts obtained from donated human foreskin (Naughton, 1999). Even though the cells employed by

Table 2.1. Cell source

Type	Comments
Autologous	Patient's own cells; immune acceptable, does not lend itself to off-the-shelf availability
Allogeneic	Cells from other human sources; lends itself to off-the-shelf availability, but may require engineering immune acceptance
Xenogeneic	From different species; not only requires engineering immune acceptance, but must be concerned with animal virus transmission

both Organogenesis and Advanced Tissue Sciences are allogeneic, immune acceptance did not have to be engineered because both the fibroblast and the keratinocyte do not constitutively express major histocompatibility complex (MHC) II antigens.

For the next generation of tissue-engineered products, which will involve other cell types, however, the immune acceptance of allogeneic cells will be a critical issue in many cases. As an example, consider a blood vessel substitute that employs both endothelial cells and smooth muscle cells. Although allogeneic smooth muscle cells may be immune acceptable, allogeneic endothelial cells certainly would not be. Thus, for the latter, one either uses autologous cells or else engineers the immune acceptance of allogeneic cells, as will be discussed in a later section. Undoubtedly the first human trials will be done using autologous endothelial cells; however, it would appear that the use of such cells would severely limit the availability of a blood vessel substitute. Only by moving away from the use of autologous cells does the opportunity for off-the-shelf availability to the clinician and thus routine use become possible.

CELL FUNCTION AND GENETIC ENGINEERING

Once one has selected the cell type(s) to be employed, then the next issue relates to the manipulation of the functional characteristics of a cell so as to achieve the behavior desired. This can be done either by (1) manipulating a cell's extracellular environment, e.g., its matrix, the mechanical stresses to which it is exposed, or its biochemical environment, or by (2) manipulating a cell's genetic program. In regard to the latter, the manipulation of a cell's genetic program could be used as an ally to tissue engineering in a variety of ways. A partial list of possibilities includes the alteration of matrix synthesis; inhibition of the immune response; enhancement of non-thrombogenicity, e.g., through increased synthesis of antithrombotic agents; engineering the secretion of specific biologically active molecules, e.g., a specific insulin secretion rate in response to a specific glucose concentration; or the alteration of cell proliferation.

Much of the above is in the context of creating a cell-seeded construct that can be implanted as a tissue or organ substitute; however, the fostering of the repair or remodeling of tissue also represents tissue engineering. Here the use of genetic engineering might take a form that is more what we would call gene therapy. An example of this would be the introduction of growth factors to foster the repair of bone defects. In using a gene therapy approach to tissue engineering it should be recognized that in many cases only a transient expression will be required. Because of this, the use of gene therapy as a strategy in tissue engineering may become viable prior to its employment in treating genetically related diseases.

STEM CELL TECHNOLOGY

Returning to the issue of cell selection, there is considerable interest in the use of stem cells, the "mother" cells within the body, as a primary source for therapies based on cell and tissue replacement (Solter and Gearhart, 1999). The excitement about stem cells reached a new height with two articles in the November 6, 1998, issue of *Science*. These reported the isolation of the first lines of human embryonic stem cells.

There are in fact a variety of different stem cells, and Table 2.2 summarizes the types of human stem cells that have been isolated. The data are from the brief review of Vogel (1999) and the question marks in Table 2.2 simply indicate where there is still some uncertainty. Of most interest are the embryonic stem cells. These embryonic stem cells are pluripotent, i.e., capable of differentiating into many cell types, and perhaps are even totipotent, i.e., capable of developing into all cell types.

Although we are quite a long way from being able to use embryonic stem cells, there already is at least one company, Osiris Therapeutics (Baltimore, MD), working with stem cells in the context of tissue engineering. In this case they are using mesenchymal stem cells and the applications on which they are focusing are primarily in the orthopedic area.

To take full advantage of stem cell technology, however, it will be necessary to understand how a stem cell differentiates into a tissue-specific cell. This requires knowledge not just about the molecular pathways of differentiation, but even more importantly the identification of the combination of signals leading to a stem cell becoming a specific type of differentiated tissue cell. Only with this will it be possible to channel a stem cell, as an example, into an endothelial cell for use in a blood vessel substitute as compared to a hepatocyte for a tissue-engineered liver.

Table 2.2. Human stem cells that have been isolated[a]

Type	Source/daughter tissue
Embryonic	Embryo or fetal tissue/all types
Hematopoietic	Adult bone marrow/blood cells, brain (?)
Neuronal	Fetal brain/neurons, glia, blood (?)
Mesenchymal	Adult bone marrow/muscle, bone, cartilage, tendon

[a]Reprinted with permission from Vogel (1999). Copyright 1999 American Association for the Advancement of Science. Question marks indicate uncertainty.

CONSTRUCT TECHNOLOGY

CONSTRUCT DESIGN AND ENGINEERING

With the selection of a source of cells, the next challenge in imitating nature is to develop a model in which these cells are organized in a three-dimensional architecture and with functional characteristics such that a specific tissue is mimicked. This design and engineering of a tissue like substitute is a challenge in its own right.

There are many possible approaches, and these are summarized in Table 2.3. One of these, of course, is a cell-seeded polymeric scaffold, an approach pioneered by Langer and his collaborators (Langer and Vacanti, 1993; Cima *et al.,* 1991). This technology is being used by Advanced Tissue Sciences, and many consider this the classic tissue engineering approach. There are others, however, with one of these being a cell-seeded collagen gel. This approach was pioneered by Bell in the late 1970s and early 1980s (Bell *et al.,* 1979; Weinberg and Bell, 1986), and is being used by Organogenesis in their skin substitute, Apligraf.

A rather intriguing approach is that of Auger and his group in Quebec, Canada (Auger *et al.,* 1995; Heureux *et al.,* 1998). Auger refers to this as cell self-assembly, and it involves a layer of cells secreting their own matrix, which over a period of time becomes a sheet. Originally developed as part of the research on skin substitutes by Auger's group, it has been extended to other applications. For example, the blood vessel substitute developed in Quebec involves rolling up one of these cell self-assembled sheets into a tube. One can in fact make tubes of multiple layers so as to mimic the architecture of a normal blood vessel.

Finally, any discussion of different approaches to the creation of a three-dimensional, functional tissue equivalent would be remiss if acellular approaches were not included. Although in tissue engineering the end result should include functional cells, there are those who are employing a strategy whereby the implant is without cells, i.e., acellular, and the cells are then recruited from the recipient or host. A number of laboratories and companies are developing this approach. One result of this approach is to, in effect, bypass the cell sourcing issue, and replace this with the issue of cell recruitment, i.e., the recruiting of cells from the host in order to populate the construct. Because these are the patient's own cells, there is no need for any engineering of immune acceptance.

Table 2.3. Possible approaches to the engineering of constructs that mimic tissue

Approach	Comments
Acellular matrix	Requires the recruitment of host cells
Cell-seeded collagen gels	Currently used in a skin substitute; has other potential applications
Cell-seeded polymeric scaffolds	Currently used in skin substitutes; has other potential applications
Cell self-assembly	Based on the cells synthesizing their own matrix

Whatever the approach, the engineering of an architecture and of functional characteristics that allow one to mimic a specific tissue is critical to achieving any success and to meeting the challenge of imitating nature. In fact, because of the interrelationship of structure and function in cells and tissues, it would be unlikely to have the appropriate functional characteristics without the appropriate three-dimensional architecture. Thus, many of the chapters in this book will describe in some detail the approach being taken in the design and engineering of constructs for specific tissues and organs, and any further discussion of this will be left to those chapters.

MANUFACTURING TECHNOLOGY

The challenge of imitating nature does not stop with the design and engineering of a specific tissuelike construct or substitute. This is because the patient need that exists cannot be met by making one construct at a time on a bench top in some research laboratory. Accepting the challenge of imitating nature must include the development of cost-effective manufacturing processes. These must allow for a scale-up from making one at a time to a production quantity of 100 or 1000 constructs per week. Anything significantly less would not be cost-effective, and if a product cannot be manufactured in large quantities and cost-effectively, then it will not be widely available for routine use.

Much of the work on manufacturing technology has focused on bioreactor technology. A bioreactor simply represents a controlled environment—both chemically and mechanically—in which a tissuelike construct can be grown. The Massachusetts Institute of Technology MIT and Georgia Tech have large research efforts focused on bioreactor technology (Freed *et al.*, 1993; Neitzel *et al.*, 1998). Although it is generally recognized that a construct, once implanted in the living system, will undergo remodeling, it is equally true that the environment of a bioreactor can be tailored to induce the *in vitro* remodeling of a construct so as to enhance characteristics critical to the success achieved following implantation (Seliktar *et al.*, 1998). Thus, the manufacturing process can be used to influence directly the final product and is part of the overall process leading to the imitation of nature.

Once manufactured, a critical issue will be how the product is delivered and made available to the clinician. The first product of Organogenesis, Apligraf, is delivered fresh and has a 5-day shelf life at room temperature (Parenteau, 1999). On the other hand, Dermagraft, the skin substitute developed by Advanced Tissue Sciences and now in clinical trials, is cryopreserved and shipped and stored at $-70°C$ (Naughton, 1999). This provides for a much more extended shelf life, but introduces other issues that one must address. Ultimately, the clinician will want off-the-shelf availability, and one way or another this will need to be provided if a tissue-engineered product is to have wide use. Although cryobiology is a relatively old field and most cell types can be cryopreserved, there is much that still needs to be learned if we are successfully to cryopreserve three-dimensional tissue-engineered products.

INTEGRATION INTO THE LIVING SYSTEM

THE LIVING SYSTEM

The final challenge is presented by moving a tissue engineering product concept into the living system. Here one starts with animal experiments, and there is a lack of good animal models for use in the evaluation of a tissue-engineered implant. This is despite the fact that a variety of animal models have been developed for the study of different diseases. Unfortunately, these models are still somewhat unproved, at least in many cases, when it comes to their use in evaluating the success of a tissue-engineered implant.

In addition, there is a significant need for the development of methods to evaluate quantitatively the performance of an implant. This is not only the case for animal studies, but is equally true for human clinical trials. In regard to the latter, it may not be enough to show efficacy and long term patency, it may also be necessary to demonstrate the mechanism(s) that lead to the implanted tissue substitute's success. Furthermore, it is not just in clinical trials that there is a need for more quantitative tools for assessment: it also would be desirable to have available technologies to assess periodically the continued viability and functionality of a tissue substitute after implantation.

ENGINEERING IMMUNE ACCEPTANCE

Important to the success of any tissue engineering approach is that it be immune acceptable. This comes naturally with the use of autologous cells; however, if one moves to nonautologous cell systems (as this author believes we must if we are to make the products of tissue engineering widely available for routine use), then the challenge of engineering immune acceptance is critical to our achieving success in the imitation of nature.

It should be recognized that the issues surrounding the immune acceptance of an allogeneic cell-seeded implant are no different than those associated with a transplanted human tissue or organ. Both represent allogeneic cell transplantation, and this means that much of what has been learned in the field of transplant immunology can help us understand implant immunology and the engineering of immune acceptance for tissue-engineered substitutes. For example, it is now known that to have immune rejection there must not only be a recognition by the host of a foreign body, there also must be present what is called the costimulatory signal, or sometimes simply signal 2. It has been demonstrated that, with a donated allogeneic tissue, if one can block the costimulatory signal, one can extend survival of the transplant considerably (Larsen *et al.*, 1996). Thus strategies are under development and may provide greater opportunities in the future for the use of allogeneic cells.

BIOCOMPATIBILITY

Finally, one cannot state that one has successfully met the challenge of imitating nature unless the implanted construct is biocompatible. Even if the implant is immune acceptable, there can still be an inflammatory response. This response can be considered separate from the immune response, although obviously there can be interactions between these two. In addition to any inflammatory response, for some types of tissue-engineered substitutes thrombosis will be an issue. This is certainly an important part of the biocompatibility of a blood vessel substitute.

CONCLUDING DISCUSSION

If we are to meet the challenge of imitating nature, there are a variety of issues. These have been divided here into three different categories. The issue of cell technology includes cell sourcing, the manipulation of cell function, and the future use of stem cell technology. Construct technology includes the engineering of a tissuelike construct and the manufacturing technology required to provide the product and ensure its off-the-shelf availability. Finally, there is the issue of integration into living systems. This has several important facets, with the most critical one being the engineering of immune acceptance.

Much of the discussion here has focused on the challenge of engineering tissuelike constructs for implantation. As noted earlier, however, equally important to tissue engineering are strategies for the fostering of remodeling and ultimately the repair and enhancement of function. As one example, consider a damaged, failing heart. Should the approach be to tissue engineer an entire heart, or should the strategy be to foster the repair of the myocardium? In this latter case, it may be possible to return the heart to relatively normal function through the implantation of a myocardial patch or even through the introduction of growth factors, angiogenic factors, or other biologically active molecules. Which strategy has the highest potential for success? Which approach will have the greatest public acceptance?

The primary issues described here (cell technology, construct technology, and integration into the living system) form the basis of the strategy being implemented in Atlanta, Georgia (Nerem *et al.*, 1998) by the Georgia Tech/Emory Center for the Engineering of Living Tissues, an Engineering Research Center newly funded by the National Science Foundation. These issues are paramount to tissue engineering meeting the challenge of imitating nature. Only if we meet this challenge can the existing patient need be addressed, and will we as a community be able to confront the transplantation crisis.

ACKNOWLEDGMENT

The author acknowledges with thanks the support by the National Science Foundation of the Georgia Tech/Emory Center for the Engineering of Living Tissues.

References

Auger, F. A., Lopez Valle, C. A., Guignard, R., Tremblay, N., Noel, B., Goulet, F., and Germain, L. (1995). Skin equivalent produced with human collagen. *In Vitro Cell Dev. Biol.* **31**, 432–439.

Bell, E., Ivarsson, B., and Merrill, C. (1979). Production of a tissue-like structure by contraction of collagen lattices by human fibroblasts of different proliferative potential *in vitro. Proc. Natl. Acad. Sci. U.S.A.* **76**, 1274–1278.

Cima, L. G., Langer, R., and Vacanti, J. P. (1991). Polymers for tissue and organ culture. *J. Bioact. Compat. Polym.* **6**, 232–239.

Freed, L. E., Vunjak, G., and Langer, R. (1993). Cultivation of cell-polymer cartilage implants in bioreactors. *J. Cell. Biochem.* **51**, 257–264.

Galletti, P. M., Aebischer, P., and Lysaght, M. J. (1995). The dawn of biotechnology in artificial organs. *J. Am. Soc. Artif. Intern. Organs* **41**, 49–57.

Heureux, N. L., Pâquet, S., Labbé, R., Germain, L., and Auger, F. A. (1998). A completely biological tissue-engineered human blood vessel. *FASEB J.* **12**, 47–56.

Langer, R., and Vacanti, J. P. (1993). Tissue engineering. *Science* **260**, 920–926.

Larsen, C. P., Elwood, E. T., Alexander, D. Z., Ritchie, S. C., Hendrix, R., Tucker-Burden, C., Cho, H. R., Aruffo, A., Hollenbaugh, D., Linsley, P. S., Winn, K. J., and Pearson, T. C. (1996). Long-term acceptance of skin and cardiac allografts by blockade of the CD40 and CD28 pathways. *Nature (London)* **381**, 434–438.

Lysaght, M. J., Ngny, N. A. P., and Sullivan, K. (1998). An economic survey of the emerging tissue engineering industry. *Tissue Eng.* **4**(3), 231–238.

Naughton, G. (1999). Skin: The first tissue-engineered products—The Advanced Tissue Sciences story. *Sci. Am.* **280**(4), 84–85.

Neitzel, G. P. *et al.* (1998). Cell function and tissue growth in bioreactors: Fluid mechanical and chemical environments. *J. Jpn. Soc. Microgravity Appl.* **15**(S-11), 602–607.

Nerem, R. M., and Sambanis, A. (1995). Tissue engineering: From biology to biological substitutes. *Tissue Eng.* **1**, 3–13.

Nerem, R. M., Ku, D. N., and Sambanis, A. (1998). Core technologies for the development of tissue engineering. *Tissue Eng. Ther. Use* **2**, 19–29.

Parenteau, N. (1999). Skin: The first tissue-engineered products—The Organogenesis story. *Sci. Am.* **280**(4), 83–84.

Seliktar, D., Black, R. A., and Nerem, R. M. (1998). Use of a cyclic strain bioreactor to precondition a tissue-engineered blood vessel substitute. *Ann. Biomed. Eng.* **26** (Suppl. 1), S-137.

Solter, D., and Gearhart, J. (1999). Putting stem cells to work. *Science* **283**, 1468–1470.

Vogel, G. (1999). Harnessing the power of stem cells. *Science* **283**, 1432–1434.

Weinberg, C. B., and Bell, E. (1986). A blood vessel model constructed from collagen and cultured vascular cells. *Science* **231**, 397–399.

PART I: THE BASIS OF GROWTH AND DIFFERENTIATION

CHAPTER 3

ORGANIZATION OF CELLS INTO HIGHER ORDERED STRUCTURES

Carol A. Erickson

INTRODUCTION

Embryonic tissues are composed of cells that are organized into either epithelia or mesenchyme. Cells in epithelia are held together tenaciously by a series of lateral junctions, they display an apical–basal polarity, and their basal surface adheres to an organized extracellular matrix (ECM), the basal lamina. In contrast, mesenchymal cells are individual cells that do not adhere by way of differentiated junctions, they display a fusiform shape typical of migratory cells, and they are generally enmeshed in a loose ECM. This tissue organization is maintained by cell–cell and cell–matrix adhesive interactions, as well as the ECM in which the cells are embedded (for review, see Armstrong, 1989). The assembly and integrity of tissues thus depend on the timely regulation of the molecular interactions that stabilize the epithelial or mesenchymal phenotype (for review, see Rodriguez-Boulan and Nelson, 1989; Birchmeier and Birchmeier, 1994; Hay and Zuk, 1995; Eaton and Simons, 1995).

In the early embryo, cells are arranged primarily in epithelial sheets. From these sheets, mesenchymal cells are derived by a process known as the epithelial–mesenchymal transformation (EMT), in which individual epithelial cells detach from the epithelium and migrate away. This transformation gives rise to populations such as the cells of the neural crest, cardiac cushion cells, the midline cells of the palate (Fitchett and Hay, 1989), the dermis of the skin, the limb musculature, and the sclerotome. The reverse process, transformation from mesenchyme to epithelium (MET), also occurs in developing organs—for example, in the formation of the kidney tubules from the nephrogenic blastema (Ekblom, 1989), the generation of the endocardium of the heart from cardiac mesenchyme (Eisenberg and Markwald, 1995), or the organization of the somites from the segmental plate. Thus the assembly and composition of tissues that are generated during embryogenesis depend on the regulation of the EMTs. Similarly in the adult, tissue structure is altered by the EMTs that accompany pathological events such as tissue regeneration and repair after wounding, and the invasion and metastasis of cancer cells (Birchmeier *et al.,* 1993; Reichmann, 1994).

In this chapter the focus is on the organization of cells into higher ordered structures, particularly on the regulation of EMTs and METs as fundamental processes that establish tissue architecture. First described are the cellular changes that accompany these transformations in normal development and in tumor cell invasion. These changes include the alteration of cell–cell adhesions, the modulation of the specificity of cell–matrix adhesions, the reorganization of the cytoskeleton, and the expression of proteases. Considered next are the mechanisms that coordinate these cellular changes, thereby triggering transformation. Both growth factors and the ECM have been shown to play regulatory roles in a variety of EMTs and METs, although the signal transduction pathways that mediate responses to growth factors and ECM are still largely unknown.

CELLULAR CHANGES INVOLVED IN THE EMT

ADHESIVE CHANGES

In those EMTs where cellular interactions have been examined in detail, several adhesive changes have been documented to accompany the detachment of single cells from an epithelial sheet. For example, when primary mesenchyme cells detach from the vegetal plate of the sea urchin gastrula, direct measurements indicate that these cells lose adhesion to the vegetal plate epithelium and to the extracellular hyalin layer, and increase adhesion to the ECM of the basal lamina through and on which they migrate (Fink and McClay, 1985). Less direct evidence suggests that multiple adhesive changes accompany the detachment of neural crest cells from the neural epithelium (Newgreen and Gooday, 1985; Delannet and Duband, 1992), ingression of mesoderm through the primitive streak (Burdsal *et al.,* 1993), and invasion of epithelium-derived tumor cells. Considered here is the molecular basis for these adhesive changes.

Modulation of cadherin function

Epithelial cells are held together by a series of differentiated junctions concentrated near their apical surface, as well as other cellular appositions between lateral surfaces. These adhesions are primarily mediated by homophilic interactions between cell adhesion molecules of the cadherin superfamily (Takeichi, 1991, 1995; Kemler, 1992). The most prominent epithelial cadherin is E-cadherin, although N-cadherin is localized in the adherens junctions and in apposing plasma membranes of some embryonic epithelia, such as the neural tube and somite (e.g., Duband *et al.,* 1988; Bronner-Fraser *et al.,* 1992). Numerous experimental studies have demonstrated the importance of cadherins in maintaining the epithelial phenotype. For example, if Madin–Darby canine kidney (MDCK) cells (an epithelial cell line) are treated with antifunctional antibodies against E-cadherin, the cells of the epithelium dissociate into individual mesenchymal cells (Behrens *et al.,* 1985). Similarly, embryonic epithelial structures such as somites (Duband *et al.,* 1987) or epidermis (Levine *et al.,* 1994) come apart when cadherin function is inhibited. Conversely, when some mesenchymal cells are transfected with E-cadherin expression vectors, they assemble into epithelial sheets (Nagafuchi *et al.,* 1987; Ringwald *et al.,* 1987; McNeill *et al,.* 1990). Thus, cadherins stabilize epithelial organization.

Is there evidence that modulation of cadherin function regulates EMTs in the embryo or in pathological conditions? Immunocytochemistry reveals that cadherins are lost or their levels greatly reduced during many developmental EMTs. For example, premigratory neural crest cells express N-cadherin and cad-6, but newly segregated and migrating neural crest cells express very little (Duband *et al.,* 1988; Bronner-Fraser *et al.,* 1992; Akitaya and Bronner-Fraser, 1992; Nakagawa and Takeichi, 1995, 1998). Similarly, cadherins are lost as cells ingress through the primitive streak (Damjanov *et al.,* 1986; Burdsal *et al.,* 1993), as the ventromedial wall of the somite disperses to form mesenchymal sclerotome (Duband *et al.,* 1988), as the midline seam of the palate disperses (Sun *et al.,* 1998), or as the endocardial cushion cells detach from the endocardium (Markwald *et al.,* 1996). Because of low spatial resolution and/or lack of markers for distinguishing the premigratory populations of cells from the rest of the epithelium from which they arise, these studies were unable to determine if cadherins are lost prior to cell detachment, or if the cadherins are down-regulated as a result of the cells becoming mesenchymal. These problems are circumvented in the sea urchin embryo, in which primary mesenchyme cells can be identified prior to ingression using antibody markers and in which the cells are so large that the process can be imaged with excellent clarity using confocal microscopy. In this instance a unique sea urchin cadherin is removed from the surface of the primary mesenchyme cells by endocytosis just prior to ingression (Miller and McClay, 1997b). This is the only instance of which this author is aware that cadherins have been proved to be removed from the surface of the cells that can be identified as those that will undergo the EMT.

In addition to immunocytochemical evidence, experimental studies have further supported a role for loss of cadherins in generating the EMT. Antibodies that interfere with cadherin function have been used to treat murine epiblast, which results in premature primitive streak formation (Burdsal *et al.,* 1993) or somites to generate sclerotome (Duband *et al.,* 1987; Linask *et al.,* 1998). However, such studies do not prove that normally the loss of these cadherins is required for the EMT, nor is it certain that the cellular response is identical to the EMT. In other studies, when the N-cadherin gene was knocked-out, ill-formed somites, abnormal neural tubes, and loosely organized myocardium were observed, but no precocious EMTs were noted (Radice *et al.,* 1997). Interpretation of this mutant phenotype is complicated by the fact that other cadherins are expressed by these tissues and may compensate for the loss of N-cadherin. Conversely, cadherins have also been overexpressed in cells that normally undergo the EMT, resulting in the inhibition of migration. For example, cad-6 has been virally expressed in the dorsal neural epithelium and under these circumstances neural crest cells do not migrate (Nakagawa and Takeichi, 1998). This study shows, at a minimum, that cadherins can stabilize the epithelial phenotype and that their adhesive interactions must be reduced in order for the EMT to proceed, but leaves open the possibility that other cellular changes are required to result in the EMT.

In addition to these normal developmental events, down-regulation of E-cadherin is also observed in pathological circumstances. In many cell lines derived from carcinomas (tumors derived from epithelia), invasion and metastasis are correlated with the loss of E-cadherin (e.g., Frixen *et al.,* 1991; Navarro *et al.,* 1991). Moreover, these invasive carcinoma cell lines lose their ability to invade collagen gels and revert to an epithelial phenotype if they are transfected with E-cadherin (Frixen *et al.,* 1991; Vleminckx *et al.,* 1991).

Cadherins apparently also have an instrumental role in the reverse process of mesenchyme-to-epithelium transition. For example, as kidney mesenchyme coalesces to generate kidney tubules, E-cadherin is reexpressed and probably stabilizes the nephron epithelium (Vestweber *et al.,* 1985; Ekblom, 1989). Similarly, E-cadherin is expressed as mammalian blastomeres undergo compaction, and in null mutations for E-cadherin the trophectoderm fails to form (Larue *et al.,* 1994).

Although the EMT and MET are correlated with cadherin expression in the above instances, cadherin function can also be regulated independently of transcription and translation. For example, the amount of cadherin protein is unchanged during the convergent extension movements of *Xenopus* gastrulation, even though C-cadherin-mediated adhesion is greatly reduced (Brieher and Gumbiner, 1994). Intracellularly, cadherins are bound to a complex of catenins including α-catenin, β-catenin/plakoglobin, and p120^{ctn}, which mediate attachment of cadherins to the cytoskeleton and regulate the adhesive function of cadherin (reviewed by Kemler, 1993; Gumbiner and McCrea, 1993; Ranscht, 1994; Barth *et al.,* 1997; Ben-Ze'ev and Geiger, 1998). Different combinations of the catenins bound to cadherins can effect a change in cadherin function (Bernfield *et al.,* 1992; Nathke *et al.,* 1994), or possibly result in the loss of cadherins from the surface if the catenins dissociate from the complex (Miller and McClay, 1997a). Catenin association with the cadherins, or catenin function, may be regulated, at least in part, by the level of catenin phosphorylation (Volberg *et al.,* 1991; Peifer *et al.,* 1994; Daniel and Reynolds, 1997). For example, when some epithelial cell lines are transfected with a temperature-sensitive *src* gene, which results in a mesenchymal transformation, levels of cadherin expression remain unchanged but β-catenin is phosphorylated (Warren and Nelson, 1987; Matsuyoshi *et al.,* 1992; Behrens *et al.,* 1993; Hamaguchi *et al.,* 1993). It is not known how the state of phosphorylation of the junctional proteins might regulate cadherin function (Daniel and Reynolds, 1997). Indeed, it has not been demonstrated directly that phosphorylation is the essential regulatory event. Nevertheless, we need to keep in mind that cell–cell adhesion can be perturbed without loss of adhesion molecules from the cell surface.

In addition to loss of function of "traditional" cadherins during the EMT, it is now apparent that "mesenchymal" cadherins (or type III cadherins) (Suzuki *et al.,* 1991) are up-regulated simultaneously. For example, cad-7 (Nakagawa and Takeichi, 1995, 1998) and cad-11 (Simonneau *et al.,* 1995) are expressed on neural crest cells as they detach from the neural epithelium. The function of these cadherins is unknown, but their timely expression suggests that they may be essential in stimulating mesenchyme migration.

Change in cell–matrix adhesion

Loss of cadherin function may be essential for producing the EMT, but it is not the only molecular change required—a number of experimental studies show that the EMT will still not occur even after cadherin function has been perturbed (e.g., Nakajima *et al.,* 1998), and the absence of cadherins in knockout mice is not always sufficient to cause the EMT (Radice *et al.,* 1997). As mentioned previously, during ingression of primary mesenchyme cells in the sea urchin embryo, loss of cell–cell adhesion is accompanied by a simultaneous gain in cell–matrix adhesion. Similarly, when premigratory neural crest cells are released precociously from the neural epithelium after cadherin function is disrupted experimentally, they still fail to migrate for at least 18 hr, suggesting that there must be a concomitant change in their adhesion to the ECM or in their motile capability (J.-L. Duband, personal communication; C. A. Erickson, unpublished data). These cell–matrix adhesions are likely to be mediated by the integrin superfamily of adhesion molecules (Hynes, 1992).

Changes in specific integrins have been observed in certain cell types undergoing the EMT. For example, the β1 integrin necessary for attachment of the neural crest cells to fibronectin becomes functional just a few hours before the cells undergo the EMT (Delannet and Duband, 1992). Similarly, mouse primitive streak cells adhere with increased affinity to a variety of ECM molecules as they become mesenchymal, and this adhesion depends on the expression of β1 and β3 integrins (Burdsal *et al.,* 1993). In both of these instances, perturbation of integrin function with antifunctional antibodies inhibits mesenchyme dispersion. The role that the integrins play in the EMT is not clear. They may simply mediate cell motility. Alternatively, they may have a more complex signaling role that controls downstream transcriptional events.

Changes in other cell–matrix adhesion molecules, besides the integrins, also may be important in EMTs. Syndecans are transmembrane heparan sulfate proteoglycans that have a variety of functions, including binding of growth factors, mediating attachment to ECM molecules, and organizing the cytoskeleton (Bernfield *et al.,* 1992). The distribution of the various syndecans suggests that they play a role in maintaining epithelial morphology and perhaps controlling EMTs as well. For example, syndecan-1 is expressed in epithelial sheets in the young mouse embryo but is lost during several EMTs, such as the dispersion of the sclerotome (Sutherland *et al.,* 1991). Similarly, as the palate fuses and the epithelium at the midline undergoes an EMT (Fitchett and Hay, 1989; Griffith and Hay, 1992), the levels of both syndecan-1 and E-cadherin are reduced markedly in the epithelium (McAlmon, 1992). Changes in syndecan expression are also correlated with pathological conditions. For example, at the edge of skin wounds, keratinocytes lose syndecan-1 as they detach from the epithelium and become mesenchymal (Elenius *et al.,* 1991). Experimentally eliminating syndecan-1 expression in NmuMG cells (a normal murine mammary epithelial cell line) with antisense oligonucleotides transforms many clones into mesenchymal cells (Kato *et al.,* 1995). In these same mesenchymal cell lines, cadherins are no longer expressed, the β1 integrins are no longer localized at sites of cell–cell contact, and the actin microfilaments become disorganized. Curiously, mesenchymal cells transfected with syndecan-1 do not assume an epithelial morphology, as might have been predicted if syndecan-1 plays a crucial regulatory role in epithelial stability. Thus the role of syndecan in regulating the epithelial phenotype, as well as the intracellular signals triggered by it, are currently unknown.

PROTEASE EXPRESSION

As cells detach from the epithelium and extend lamellipodia from their basal surface, they must, of necessity, pass through the basal lamina at their basal surface. Similarly, epithelial tumor cells must breach a basal lamina barrier as they invade and metastasize. One way that cells can cross this normally impenetrable barrier (Erickson, 1987) is by secreting matrix-degrading enzymes. There is a rich literature that documents the importance of proteases simulating tumor invasiveness (Werb, 1989, 1997; Alexander and Werb, 1991). Thus it is not surprising that protease production has been observed to accompany some EMTs.

Plasminogen activator, a neutral serine protease, is expressed by neural crest cells after they detach from the neural tube and migrate *in vivo* and in culture (Valinsky and Le Douarin, 1985; Erickson and Isseroff, 1989; Agrawal and Brauer, 1996), by sclerotome cells that form from the ventromedial wall of the somite (McGuire and Alexander, 1992), and by cardiac cushion cells as they detach from the endocardium (McGuire and Orkin, 1992). Metalloproteases are also produced by cells that have undergone an EMT, including the neural crest cells, the dermis, the sclerotome (unpublished studies from our lab), and possibly the mesodermal cells on ingression through the primitive streak (Sanders and Prasad, 1989). Although protease activity is correlated with the EMT, there is only minimal experimental evidence that implicates proteases in the process. Inhibition of urokinase-type plasminogen activator (uPA) activity reduces the number of neural crest cells that migrate from the neural tube, although this response could be due to a toxic effect of the drugs used (Erickson and Isseroff, 1989). Additionally when antisense oligonucleotides are used to inhibit uPA production in endocardial explants, there is a reduction in the number of cushion cells that emigrate (McGuire and Alexander, 1993).

Although proteases are believed to stimulate the EMT by allowing cells to migrate through the basal lamina, some recent studies also show that metalloproteases can cause the shedding of cadherins from the cell surface, thereby reducing cadherin-mediated cell adhesion (Herren *et al.*, 1998). To date, no EMTs that occur during development have been linked with this shedding activity, but it remains an attractive mechanism.

ALTERNATIVE MODELS OF THE EMT

The cellular changes discussed above are those that are commonly associated with the EMT. Nevertheless, there are several other models, as yet unexplored, that could account for this remarkable process (Fig. 3.1).

One alternative model is that cells detach from the epithelium by generating sufficient tractional force to rupture the junctional complexes that keep them adherent to their neighbors, just as fibroblasts rupture their trailing ends and leave bits of themselves as tracks across the substratum as they locomote (Fig. 3.1, model 2). Several reports suggest that as neural crest cells or placode-derived neurons detach from an epithelium, their adherens junctions slide through the lipid bilayer as though they are being pulled along as the cells generate tractional force (Nichols, 1986a,b, 1987). In at least one study (Bilozur and Hay, 1989), transmission electron micrographs reveal vesicles containing microfilaments and adherens junctions near the lumen of the neural tube; these may be remnants of neural crest cells that have detached. Such a model predicts that a simple change in the distribution of integrins, or a reorganization of the cytoskeleton, such as gener-

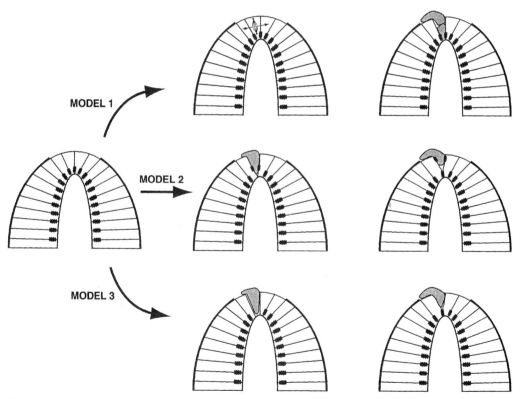

Fig. 3.1. Schematic drawings depicting three models for the EMT that produces neural crest cells. In model 1, an asymmetric mitosis results in an apical cell that remains attached to the lumen of the neural tube by adherens junctions, and a basal daughter that becomes separated from the junctions. Because the dorsal neural tube is not enveloped by a basal lamina, the basal daughter is free to emigrate away. In model 2, the premigratory neural crest cell is able to generate a strong tractional force on the surrounding substratum. As it moves out of the neural tube, it pulls its adherens junctions with it until the junctions reach the basal surface of the epithelium, where the neural crest cell pulls away and severs its connection with the junctions. In model 3, the cadherins maintaining the adherens junction are either down-regulated or endocytosed from the surface, or their function is attenuated, allowing the cell to detach from the epithelium. From Erickson and Reedy (1998), with permission.

ating stress bundles, could activate the motile machinery and would be sufficient to overcome the adhesive interactions in the epithelium. Liu and Jessell (1998) showed that RhoB is expressed in premigratory neural crest cells, and inhibition of RhoB prevents the EMT. Because Rho regulates the assembly and organization of actin filaments (Hall, 1998; Imamura *et al.,* 1998) and induces contractility in fibroblasts (Zhang *et al.,* 1998), these findings suggest that cytoskeleton mobilization is critical in the EMT. [In another model system, the MDCK epithelial cell line, injection of activated Rho inhibits the EMT stimulated by scatter factor (Ridley *et al.,* 1995) and RhoA inactivation is required for cell scattering activity (Imamura *et al.,* 1998), so that the evidence for Rho activity in the EMT is conflicting.] Curiously, Rac and Rho activity is also required for E-cadherin function (Braga *et al.,* 1997, 1999; Zhong *et al.,* 1997) thus stabilizing the epithelial organization, so at present the role of these small GTPases in the EMT is unclear.

A second alternative model for controlling the EMT proposes that asymmetric cell divisions could separate basal daughter cells from apical junctions (Erickson and Reedy, 1998). Specifically, if the mitotic spindle were oriented perpendicular to the surface of an epithelium, resulting in a horizontal cleavage furrow, one of the daughter cells would still be tethered to the apical surface via adherens junctions, whereas the basal daughter cell would be free to migrate away (Fig. 3.1, model 1). This mechanism does not require any change in the apical junction-associated cadherin function. Such an orientation of the mitotic spindle is correlated with the generation of neurons from an epithelium in *Drosophila* (Foe, 1989), and has been proposed to produce the mesenchymal cells that arise from the otic anlage to form the acoustic–vestibular ganglion (Alvarez *et al.,* 1989). Finally, direct time-lapse images of neurons labeled with 3,3′-dioctadecyloxacarbocyanine perchlorate (DiO) in the cerebral cortex of the ferret brain convincingly show that the neurons detach from the apical surface of the ventricular zone only following a horizontal mitosis (Chenn and McConnell, 1995).

STIMULI OF THE TRANSFORMATION

Our current notion of the EMT, as discussed above, is that a variety of cellular changes occur simultaneously. This implies that some regulatory process coordinates these changes (see Hay, 1990, 1991). Growth factors and the ECM have both been proposed to trigger the EMT.

GROWTH FACTORS

Numerous *in vitro* studies have demonstrated that growth factors can activate proteinase cascades, alter cell–cell and cell–substratum adhesiveness, and modulate ECM production. Thus, they are good candidates for coordinating these same events during the EMT. And, in fact, members of the transforming growth factor β (TGF-β) family of growth factors have been shown conclusively to play a role in the detachment of cardiac cushion cells from the endothelium (Potts and Runyan, 1989; Potts *et al.,* 1991; Brown *et al.,* 1999; Boyer *et al.,* 1999), the dispersion of the palate midline seam (Brunet *et al.,* 1993; Brown *et al.,* 1999), the dissociation of the Müllerian duct to form mesenchyme (Trelstad *et al.,* 1982), and the induction and maintenance of carcinoma cell invasiveness and metastasis (Oft *et al.,* 1998). Moreover, Delannet and Duband (1992) have shown that treatment of isolated quail neural tubes with TGF-β accelerates the timing of detachment of neural crest cells from the neural epithelium and may do so by up-regulating integrins. Interestingly, a TGF-β family member, dorsalin-1, is produced by the dorsal neural tube and promotes the appearance of neural crestlike cells from the ventral neural plate (Basler *et al.,* 1993). Bone morphogenetic proteins (BMP4 and BMP7), also members of the TGF-β superfamily, are expressed by the ectoderm contiguous with the neural plate and have been experimentally demonstrated to control neural crest differentiation (Liem *et al.,* 1995). Whether these growth factors directly regulate the EMT or have a more general role in the specification of the neural crest lineage (see Erickson and Reedy, 1998) remains to be determined.

Other families of growth factors have been implicated in the EMT. Neurotrophin-3, a member of the neurotrophin family of growth factors, has been shown to regulate dispersion of the epithelial dermatome to form dermis (Brill *et al.,* 1995).

The "motility-stimulating factors" belong to another class of growth factors that may play a role in regulating the EMT. These can be produced by fibroblasts and act in an autocrine fashion [e.g., migration simulation factor (Grey *et al.,* 1989); autocrine stimulating factor (Liotta *et al.,* 1986)]. Alternatively, some motility factors have been isolated from fibroblast-conditioned medium and can induce epithelial sheets to disperse and are therefore paracrine in function [scatter factor/hepatocyte growth factor (Stoker *et al.,* 1987; Weidner *et al.,* 1991)].

Scatter factor has received a great deal of attention for its potential role in the regulation of the EMT. The first evidence to suggest that it plays a critical role in some EMTs during embryogenesis was the observation that when a source of scatter factor is implanted into chick embryos, ectopic primitive streaks are produced (Stern *et al.,* 1990). Hensen's node has been identified as the probable source of scatter factor in the chick embryo (Streit *et al.,* 1995). When mice are generated that contain a null mutation for c-met, the receptor for scatter factor, the lateral edges of the dermamyotomes from which limb myoblasts emigrate fail to undergo the EMT and therefore myoblasts do not migrate to the limb (Bladt *et al.,* 1995). Interestingly, the limb bud is a source of scatter factor that may induce the EMT in the dermamyotomes located immediately adjacent to it. In some circumstances, scatter factor instead induces the reverse process, the MET. When fibroblasts are transfected with both c-met and scatter factor, the cells assemble into epithelial sheets that form a lumen (Tsarfaty *et al.,* 1994). Moreover, antibodies that inhibit the function of scatter factor also prevent the assembly of kidney mesenchyme into epithelial nephrons (Woolf *et al.,* 1995). These opposing functions of scatter factor are currently confounding. One possibility is that scatter factor works in a paracrine fashion to induce the EMT, but in an autocrine faction in the METs. Despite the confusion about the function of scatter factor in these processes, its involvement seems certain.

Secreted factors other than scatter factor regulate other METs. Members of the Wnt family of secreted glycoproteins play many roles in specifying pattern and cell fate during embryogenesis. One of these, Wnt-4 is expressed in the mouse kidney mesenchyme that subsequently aggregates and assembles to form nephrogenic tubules (Davies, 1996). Mice lacking Wnt-4 activity fail to form pretubular cell aggregates, suggesting that Wnt-4 is an autoinducer of the MET (Stark *et al,.* 1994). Because Wnt signaling increases cadherin-mediated adhesion in established cell lines (Bradley *et al.,* 1993; Hinck *et al.,* 1994), it is reasonable to speculate that Wnt-4 may regulate E-cadherin function in the development of the kidney as well.

EXTRACELLULAR MATRIX

In some circumstances the timely appearance of particular ECM components stimulates the EMT. Cardiac cushion cell migration is the principal example of a matrix-stimulated EMT (Mjaatvedt *et al.,* 1987, 1991). Complexes of ECM produced by the myocardium and known as "adherons" trigger cushion cell detachment from the endocardium (Eisenberg and Markwald, 1995). These matrix complexes may function by up-regulating the production of TGF-β1 in the premigratory cushion cells (Nakajima *et al.,* 1994). There are no data that establish that the ECM is involved in any other developmental EMTs, however. Löfberg and colleagues (1985) implanted filters containing ECM above the dorsal neural tube in axolotl embryos and simulated premature migration of the neural crest. In this situation, however, the neural crest cells have already detached from the neuroepithelium, and the ECM is probably only acting to simulate motility. Stimulation of neural crest detachment by the ECM apparently does not occur in other species, because precocious migration has never been induced with a wide variety of ECM components (e.g., Delannet and Duband, 1992).

Although there are only a few known examples of EMTs regulated by the ECM, the importance of cell–matrix adhesion in this process is suggested by the observation that authentic epithelia that normally never undergo an EMT during embryogenesis will do so if confronted by novel ECM in culture. For example, when lens or thyroid epithelium is embedded in a three-dimensional collagen gel, cellular processes extend from the apical surface and the cells disperse to form mesenchyme (Greenburg and Hay, 1982, 1986, 1988). This dispersion is accompanied by a loss of cadherins, detachment from the basal lamina, a redistribution of integrins from the basal to the apical surface (Zuk and Hay, 1994), as well as a change in integrin subunit expression (loss of α6 and expression of α5 integrin subunits). Importantly, such EMTs are inhibited if antifunctional antibodies to integrins are incorporated into the gel. How the ECM might regulate a host of cellular changes to produce this event is not understood, but this model system may afford us the opportunity to assess the regulatory processes that control the EMT.

The regulated expression of particular ECM components is instrumental in at least one instance of MET as well. When the kidney mesenchyme coalesces to form epithelial nephrons, the condensation is apparently regulated by the expression of E-cadherin, as discussed previously, which establishes the basolateral domain of the epithelium (see Eaton and Simons, 1995). The subsequent apical–basal polarization of that tissue is correlated with the production of laminin A chain, which accumulates around the compacted cells. If adhesion to laminin is inhibited using antifunctional antibodies to either the laminin A E8 domain or to the α6 subunit of integrin, this polarization fails to occur, although the cadherin-mediated condensations of the mesenchyme proceeds normally (Klein *et al.,* 1988; Sorokin *et al.,* 1990). Thus, in this instance, an ECM component has a crucial role in the establishment of epithelial polarization.

INTRACELLULAR SIGNALING MECHANISMS

Although some progress has been made in elucidating what growth factors and ECM molecules are involved in regulating EMTs, the downstream signaling events are not yet known. Because most growth factor receptors are kinases, it seems likely that some EMTs are mediated through the Ras/Raf signaling pathway. Indeed, transfections of MDCK epithelial cell lines with v-Ha-*ras* or v-ki-*ras* result in the dispersion of the cells into mesenchyme. This morphological change is accompanied by the simultaneous loss of E-cadherin, changes in integrin distribution, and increased production of proteases (Behrens *et al.,* 1992, 1993). Conversely, interfering with the TGF-β receptor type III (Brown *et al.,* 1999) or with the downstream signaling of Ras (Lakkis and Epstein, 1998) blocks formation of the cardiac cushions (Epstein *et al.,* 1994). Similarly, inhibition of endogenous Ras proteins prevents the scatter factor-induced EMT in MDCK cells (Ridley *et al.,* 1995).

Transcription factors that are activated by the Ras/Raf pathway include fos/jun (AP-1), and these are likely involved in coordinating the many cellular changes that accompany the EMT. For example, expression of c-fos precedes the EMT in the palate (Smeyne *et al.,* 1993). Induction of c-fos in mammary epithelial cells results in loss of epithelial polarity, loss of cadherins, and increases in the activity of proteases, including stromelysin, collagenase, uPA, and tissue plasminogen activator (tPA) (Reichmann *et al.,* 1992). Interestingly, neural crest cells express high levels of AP-1 as they detach from the neural epithelium (Mitchell *et al.,* 1991). The transcription factor NF-ATc is also regulated by Ras and has been experimentally demonstrated to control cushion cell transformation (Lakkis and Epstein, 1998). Finally, the ETS1 transcription factor is expressed in cells undergoing the EMT, such as the sclerotome and the neural crest; its expression is stimulated by scatter factor, and ETS1 activity is also activated by the Ras signaling cascade (Fafeur *et al.,* 1997).

Members of the Snail family of zinc finger transcription factors are important regulators of the EMT (Leptin and Grunewald, 1990; Kosman *et al.,* 1991; Nieto *et al.,* 1992). *Slug* is expressed in neural crest and primitive streak cells of the chicken embryo just prior to their detachment from the epithelium, and antisense oligo-treated embryos fail to undergo the EMTs (Nieto *et al.,* 1994). This observation provides us with an exciting new handle on how coordinated changes in cells undergoing the EMT are regulated, and may allow us to determine what constellations of genes are the downstream targets of this transcription factor. To date, the only known direct effect of *Slug* is the loss of desmoplakin and desmoglein (Savagner *et al.,* 1997).

Inhibitors of protein kinase C have been reported to stimulate the precocious detachment of neural crest cells from the neural epithelium (Newgreen and Minichiello, 1995). Because this stimulation occurred in the presence of inhibitors of transcription, it suggests that changes in protein phosphorylation can trigger the EMT independent of changes in protein expression.

CONCLUSION

Considerable progress has been made in elucidating the cellular changes that accompany the EMT by combining approaches used to study pathological events, in particular the invasiveness of carcinomas, with those of normal developmental processes. Nevertheless, these data are largely correlative and it is far from clear which of the many cellular changes that are coincident with the EMT are critical. To address these issues, direct observation and perturbation of the events in the intact organism will be essential, because some aspects of tissue culture studies may be irregular owing to culture-induced artifacts or peculiarities of the established cell lines that have been studied.

Several developmental model systems seem especially amenable to experimentation. Invertebrate embryos, such as those of sea urchins, are clear and have large blastomeres, allowing direct observation of the molecular changes as they occur during the EMT. Moreover, the precise cells that will undergo EMTs can be identified prior to onset of the process, thus allowing analysis of the earliest molecular changes in these cells. Perturbation of cadherin and integrin function during the EMT is also crucial. *Xenopus* embryos will prove especially useful in this regard, because injections of RNAs that result in the overexpression of cadherins or in the expression of dominant-negative mutations in cadherin function already have provided insight into the role of these molecules in mediating adhesion (Kintner, 1988; Brieher and Gumbiner, 1994; Levine *et al.,* 1994). The role of phosphorylation of cadherin or the catenins in regulating adhesion is an obvious problem to explore using the *Xenopus* embryo.

The trigger(s) for the EMT remain a mystery, although growth factors will probably figure prominently in regulating the process. One approach to addressing this issue may come from the observation that selected subpopulations of cells are capable of undergoing the EMT. For example, only the dorsal neural tube gives rise to the neural crest, and only the endocardium from the atrioventricular (AV) region is capable of generating cushion mesenchyme (Runyan and Markwald, 1983; Mjaatvedt and Markwald, 1989). It has been established that the TGF-β type III receptor is only found in the AV endothelium and that expression of this receptor in the nontransforming ventricular endocardium allows the cells to undergo the EMT in response to TGF-β2 (Brown *et al.,* 1999). A careful comparison of the signaling molecules in embryonic cells that do and do not undergo the EMT may yield important insights into the control of this process. Moreover, it is becoming apparent that different regulatory pathways may control different aspects of the EMT. For example, loss of cadherin function may be controlled independently of stimulation of motility (Levine *et al.,* 1994; Radice *et al.,* 1997; Zhong *et al.,* 1997; Savagner *et al.,* 1997; Nakajima *et al.,* 1998; Boyer *et al.,* 1999). Only by experimentally dissecting and studying each of the steps in the EMT will the regulation of the whole process be understood.

References

Agrawal, M., and Brauer, P. R. (1996). Urokinase-type plasminogen activator regulates cranial neural crest cell migration *in vitro. Dev. Dyn.* **207**, 281–290.

Akitaya, T., and Bronner-Fraser, M. (1992). Expression of cell adhesion molecules during initiation and cessation of neural crest cell migration. *Dev. Dyn.* **194**, 12–20.

Alexander, C. M., and Werb, Z. (1991). Extracellular matrix degradation. *In* "Cell Biology of Extracellular Matrix" (E. D. Hay, ed.), pp. 255–302. Plenum, New York.

Alvarez, I. S., Martin-Partido, G., Rodriguez-Gallardo, L., Gonzalez-Ramos, C., and Navascues, J. (1989). Cell proliferation during early development of the chick embryo otic anlage: Quantitative comparison of migratory and nonmigratory regions of the otic epithelium. *J. Comp. Neurol.* **290**, 278–288.

Armstrong, P. B. (1989). Cell sorting out: The self-assembly of tissues *in vitro. Crit. Rev. Biochem. Mol. Biol.* **24**, 119–149.

Barth, A. L., Nathke, I. S., and Nelson, W. J. (1997&). Cadherins, catenins and APC protein: Interplay between cytoskeletal complexes and signaling pathways. *Curr. Opin. Cell Biol.* **9**, 683–690.

Basler, K., Edlund, T., Jessell, T. M., and Yamada, T. (1993). Control of cell pattern in the neural tube: Regulation of cell differentiation by dorsalin-1, a novel TGF beta family member. *Cell (Cambridge, Mass.)* **73**, 687–702.

Behrens, J., Birchmeier, W., Goodman, S. L., and Imhof, B. A. (1985). Dissociation of Madin-Darby canine kidney epithelial cells by the monoclonal antibody anti-arc-1: Mechanistic aspects and identification of the antigen as a component related to uvomorulin. *J. Cell Biol.* **101**, 1307–1315.

Behrens, J., Frizen, U., Schipper, J., Weidner, M., and Birchmeier, W. (1992). Cell adhesion in invasion and metastasis. *Semin. Cell Biol.* **3**, 169–178.

Behrens, J., Vakaet, L., Friis, R., Winterhager, E., van Roy, F., Mareel, M. M., and Birchmeier, W. (1993). Loss of epithelial differentiation and gain of invasiveness correlates with tyrosine phosphorylation of the E-cadherin/beta-catenin complex in cells transformed with a temperature-sensitive v-SRC gene. *J. Cell biol.* **120**, 757–766.

Ben-Ze'ev, A., and Geiger, B. (1998). Differential molecular interactions of beta-catenin and plakoglobin in adhesion, signaling and cancer. *Curr. Opin. Cell Biol.* **10**, 629–639.

Bernfield, M., Kokenyesi, R., Kato, M., Hinkes, M. T., Spring, J., Gallo, R. L., and Lose, E. J. (1992). Biology of the syndecans: A family of transmembrane heparan sulfate proteoglycans. *Annu. Rev. Cell Biol.* **8**, 365–393.

Bilozur, M. E., and Hay, E. D. (1989). Cell migration into neural tube lumen provides evidence for the "fixed cortex" theory of cell motility. *Cell Motil. Cytoskel.* **14**, 469–484.

Birchmeier, W., and Birchmeier, C. (1994). Mesenchymal–epithelial transitions. *BioEssays* **16**, 305–307.

Birchmeier, W., Weidner, K. M., Hulsken, J., and Behrens, J. (1993). Molecular mechanisms leading to cell junction (cadherin) deficiency in invasive carcinomas. *Semin. Cancer Biol.* **4**, 231–239.

Bladt, F., Riethmacher, D., Isenmann, S., Aguzzi, A., and Birchmeier, C. (1995). Essential role for the c-met receptor in the migration of myogenic precursor cells into the limb bud. *Nature (London)* **376**, 768–771.

Boyer, A. S., Erickson, C. P., and Runyan, R. B. (1999). Epithelial–mesenchymal transformation in the embryonic heart is mediated through distinct pertussis toxin-sensitive and TGFbeta signal transduction mechanisms. *Dev. Dyn.* **214**, 81–91.

Bradley, R. S., Cowin, P., and Brown, A. M. (1993). Expression of Wnt-1 in PC12 cells results in modulation of plako-globin and E-cadherin and increased cellular adhesion. *J. Cell Biol.* **123**, 1857–1865.

Braga, V. M., Machesky, L. M., Hall, A., and Hotchin, N. A. (1997). The small GTPases Rho and Rac are required for the establishment of cadherin-dependent cell–cell contacts. *J. Cell Biol.* **137**, 1421–1431.

Braga, V. M., Del Maschio, A., Machesky, L., and Dejana, E. (1999). Regulation of cadherin function by Rho and Rac: Modulation by junction maturation and cellular context. *Mol. Biol. Cell* **10**, 9–22.

Brieher, W. M., and Gumbiner, B. M. (1994). Regulation of C-cadherin function during activin induced morphogenesis of *Xenopus* animal caps. *J. Cell Biol.* **126**, 519–527.

Brill, G., Kahane, N., Carmeli, C., von Schack, D., Barde, Y. A., and Kalcheim, C. (1995). Epithelial–mesenchymal conversion of dermatome progenitors requires neural tube-derived signals: Characterization of the role of neurotrophin-3. *Development (Cambridge, UK)* **121**, 2583–2594.

Bronner-Fraser, M., Wolf, J. J., and Murray, B. A. (1992). Effects of antibodies against N-cadherin and N-CAM on the cranial neural crest and neural tube. *Dev. Biol.* **153**, 291–301.

Brown, C. B., Boyer, A. S., Runyan, R. B., and Barnett, J. V. (1999). Requirement of type III TGF-beta receptor for endocardial cell transformation in the heart. *Science* **283**, 2080–2082.

Brunet, C. L., Sharpe, P. M., and Ferguson, M. W. (1993). The distribution of epidermal growth factor binding sites in the developing mouse palate. *Int. J. Dev. Biol.* **37**, 451–458.

Burdsal, C. A., Damsky, C. H., and Pedersen, R. A. (1993). The role of E-cadherin and integrins in mesoderm differentiation and migration at the mammalian primitive streak. *Development (Cambridge, UK)* **118**, 829–844.

Chenn, A., and McConnell, S. K. (1995). Cleavage orientation and the asymmetric inheritance of Notch1 immunoreactivity in mammalian neurogenesis. *Cell (Cambridge, Mass.)* **82**, 631–641.

Damjanov, I., Damjanov, A., and Damsky, C. H. (1986). Developmentally regulated expression of the cell–cell adhesion glycoprotein cell-CAM 120/80 in peri-implantation mouse embryos and extraembryonic membranes. *Dev. Biol.* **116**, 194–202.

Daniel, J. M., and Reynolds, A. B. (1997). Tyrosine phosphorylation and cadherin/catenin function. *BioEssays* **19**, 883–891.

Davies, J. A. (1996). Mesenchyme to epithelium transition during development of the mammalian kidney tubule. *Acta Anat.* **156**, 187–201.

Delannet, M., and Duband, J. L. (1992). Transforming growth factor-beta control of cell-substratum adhesion during avian neural crest cell emigration *in vitro*. *Development (Cambridge, UK)* **116**, 275–287.

Duband, J. L., Dufour, S., Hatta, K., Takeichi, M., Edelman, G. M., and Thiery, J. P. (1987). Adhesion molecules during somitogenesis in the avian embryo. *J. Cell Biol.* **104**, 1361–1374.

Duband, J. L., Volberg, T., Sabanay, I., Thiery, J. P., and Geiger, B. (1988). Spatial and temporal distribution of the adherens-junction-associated adhesion molecule A-CAM during avian embryogenesis. *Development (Cambridge, UK)* **103**, 325–344.

Eaton, S., and Simons, K. (1995). Apical, basal, and lateral cues for epithelial polarization. *Cell (Cambridge, Mass.)* **82**, 5–8.

Eisenberg, L. M., and Markwald, R. R. (1995). Molecular regulation of atrioventricular valvuloseptal morphogenesis. *Circ. Res.* **77**, 1–6.

Ekblom, P. (1989). Developmentally regulated conversion of mesenchyme to epithelium. *FASEB J.* **3**, 2141–2150.

Elenius, K., Vainio, S., Laato, M., Salmivirta, M., Thesleff, I., and Jalkanen, M. (1991). Induced expression of syndecan in healing wounds. *J. Cell Biol.* **114**, 585–595.

Epstein, M. L., Mikawa, T., Brown, A. M., and McFarlin, D. R. (1994). Mapping the origin of the avian enteric nervous system with a retroviral marker. *Dev. Dyn.* **201**, 236–244.

Erickson, C. A. (1987). Behavior of neural crest cells on embryonic basal laminae. *Dev. Biol.* **120**, 38–49.

Erickson, C. A., and Isseroff, R. R. (1989). Plasminogen activator activity is associated with neural crest cell motility in tissue culture. *J. Exp. Zool.* **251**, 123–133.

Erickson, C. A., and Reedy, M. V. (1998). Neural crest development: The interplay between morphogenesis and cell differentiation. *Curr. Top. Dev. Biol.* **40**, 177–209.

Fafeur, V., Tulasne, D., Queva, C., Vercamer, C., Dimster, V., Mattot, V., Stehelin, D., Desbiens, X., and Vandenbunder, B. (1997). The ETS1 transcription factor is expressed during epithelial–mesenchymal transitions in the chick embryo and is activated in scatter factor-stimulated MDCK epithelial cells. *Cell Growth Differ.* **8**, 655–665.

Fink, R. D., and McClay, D. R. (1985). Three cell recognition changes accompany the ingression of sea urchin primary mesenchyme cells. *Dev. Biol.* **107**, 66–74.

Fitchett, J. E., and Hay, E. D. (1989). Medial edge epithelium transforms to mesenchyme after embryonic palatal shelves fuse. *Dev. Biol.* **131**, 455–474.

Foe, V. E. (1989). Mitotic domains reveal early commitment of cells in *Drosophila* embryos. *Development (Cambridge, UK)* **107**, 1–22.

Frixen, U. H., Behrens, J., Sachs, M., Eberle, G., Voss, B., Warda, A., Lochner, D., and Birchmeier, W. (1991). E-Cadherin-mediated cell–cell adhesion prevents invasiveness of human carcinoma cells. *J. Cell Biol.* **113**, 173–185.

Greenburg, G., and Hay, E. D. (1982). Epithelia suspended in collagen gels can lose polarity and express characteristics of migrating mesenchymal cells. *J. Cell Biol.* **95**, 333–339.

Greenburg, G., and Hay, E. D. (1986). Cytodifferentiation and tissue phenotype change during transformation of embryonic lens epithelium to mesenchyme-like cells *in vitro*. *Dev. Biol.* **115**, 363–379.

Greenburg, G., and Hay, E. D. (1988). Cytoskeleton and thyroglobulin expression change during transformation of thyroid epithelium to mesenchyme-like cells. *Development (Cambridge, UK)* **102**, 605–622.

Grey, A. M., Schor, A. M., Rushton, G., Ellis, I., and Schor, S. L. (1989). Purification of the migration stimulating factor produced by fetal and breast cancer patient fibroblasts. *Proc. Natl. Acad. Sci. U.S.A.* **86**, 2438–2442.

Griffith, C. M., and Hay, E. D. (1992). Epithelial–mesenchymal transformation during palatal fusion: Carboxyfluorescein traces cells at light and electron microscopic levels. *Development (Cambridge, UK)* 116, 1087–1099.

Gumbiner, B. M., and McCrea, P. D. (1993). Catenins as mediators of the cytoplasmic functions of cadherins. *J. Cell Sci. Suppl.* 17, 155–158.

Hall, A. (1998). Rho GTPases and the actin cytoskeleton. *Science* 279, 509–514.

Hamaguchi, M., Matsuyoshi, N., Ohnishi, Y., Gotoh, B., Takeichi, M., and Nagai, Y. (1993). p60v-src causes tyrosine phosphorylation and inactivation of the N-cadherin–catenin cell adhesion system. *EMBO J.* 12, 307–314.

Hay, E. D. (1990). Epithelial-mesenchymal transitions. *Semin. Dev. Biol.* 1, 347–356.

Hay, E. D. (1991). Cell Biology: The Extracellular Matrix. Plenum Press, NY.

Hay, E. D., and Zuk, A. (1995). Transformations about epithelium and mesenchyme: Normal, pathological, and experimentally induced. *Am. J. Kidney Dis.* 26, 678–690.

Herren, B., Levkau, B., Raines, E. W., and Ross, R. (1998). Cleavage of beta-catenin and plakoglobin and shedding of VE-cadherin during endothelial apoptosis: Evidence for a role for caspases and metalloproteinases. *Mol. Biol. Cell* 9, 1589–1601.

Hinck, L., Nelson, W. J., and Papkoff, J. (1994). Wnt-1 modulates cell–cell adhesion in mammalian cells by stabilizing beta-catenin binding to the cell adhesion protein cadherin. *J. Cell Biol.* 124, 729–741.

Hynes, R. O. (1992). Integrins: Versatility, modulation, and signaling in cell adhesion. *Cell (Cambridge, Mass.)* 69, 11–25.

Imamura, H., Takaishi, K., Nakano, K., Kodama, A., Oishi, H., Shiozaki, H., Monden, M., Sasaki, T., and Takai, Y. (1998). Rho and Rab small G proteins coordinately reorganize stress fibers and focal adhesions in MDCK cells. *Mol. Biol. Cell* 9, 2561–2575.

Kato, M., Saunders, S., Nguyen, H., and Bernfield, M. (1995). Loss of cell surface syndecan-1 causes epithelia to transform into anchorage-independent mesenchyme-like cells. *Mol. Biol. Cell* 6, 559–576.

Kemler, R. (1992). Classical cadherins. *Semin. Cell Biol.* 3, 149–155.

Kemler, R. (1993). From cadherins to catenins: Cytoplasmic protein interactions and regulation of cell adhesion. *Trends Genet.* 9, 317–321.

Kintner, C. (1988). Effects of altered expression of the neural cell adhesion molecule, N-CAM, on early neural development in *Xenopus* embryos. *Neuron* 1, 545–555.

Klein, G., Langegger, M., Timpl, R., and Ekblom, P. (1988). Role of laminin A chain in the development of epithelial cell polarity. *Cell (Cambridge, Mass.)* 55, 331–341.

Kosman, D., Ip, Y. T., Levine, M., and Arora, K. (1991). Establishment of the mesoderm–neuroectoderm boundary in the *Drosophila* embryo. *Science* 254, 118–122.

Lakkis, M. M., and Epstein, J. A. (1998). Neurofibromin modulation of ras activity is required for normal endocardial–mesenchymal transformation in the developing heart. *Development (Cambridge, UK)* 125, 4359–4367.

Larue, L., Ohsugi, M., Hirchenhain, J., and Kemler, R. (1994). E-Cadherin null mutant embryos fail to form a trophectoderm epithelium. *Proc. Natl. Acad. Sci. U.S.A.* 91, 8263–8267.

Leptin, M., and Grunewald, B. (1990). Cell shape changes during gastrulation in *Drosophila*. *Development (Cambridge, UK)* 110, 73–84.

Levine, E., Lee, C. H., Kintner, C., and Gumbiner, B. M. (1994). Selective disruption of E-cadherin function in early *Xenopus* embryos by a dominant negative mutant. *Development (Cambridge, UK)* 120, 901–909.

Liem, K. F. J., Tremml, G., Roelink, H., and Jessell, T. M. (1995). Dorsal differentiation of neural plate cells induced by BMP-mediated signals from epidermal ectoderm. *Cell (Cambridge, Mass.)* 82, 969–979.

Linask, K. K., Ludwig, C., Han, M. D., Liu, X., Radice, G. L., and Knudsen, K. A. (1998). N-Cadherin/catenin-mediated morphoregulation of somite formation. *Dev. Biol.* 202, 85–102.

Liotta, L. A., Mandler, R., Murano, G., Katz, D. A., Gordon, R. K., Chiang, P. K., and Schiffmann, E. (1986). Tumor cell autocrine motility factor. *Proc. Natl. Acad. Sci. U.S.A.* 83, 3302–3306.

Liu, J. P., and Jessell, T. M. (1998). A role for RhoB in the delamination of neural crest cells from the dorsal neural tube. *Development* 125, 5055–5067.

Löfberg, J., Nynas-McCoy, A., Olsson, C., Jonsson, L., and Perris, R. (1985). Stimulation of initial neural crest cell migration in the axolotl embryo by tissue grafts and extracellular matrix transplanted on microcarriers. *Dev. Biol.* 107, 442–459.

Markwald, R., Eisenberg, C., Eisenberg, L., Trusk, T., and Sugi, Y. (1996). Epithelial–mesenchymal transformations in early avian heart development. *Acta Anat.* 156, 173–186.

Matsuyoshi, N., Hamaguchi, M., Taniguchi, S., Nagafuchi, A., Tsukita, S., and Takeichi, M. (1992). Cadherin-mediated cell–cell adhesion is perturbed by v-src tyrosine phosphorylation in metastatic fibroblasts. *J. Cell Biol.* 118, 703–714.

McAlmon, K. R. (1992). Matrix receptors and development. *Semin. Perinatol.* 16, 90–96.

McGuire, P. G., and Alexander, S. M. (1992). Urokinase expression during the epithelial–mesenchymal transformation of the avian somite. *Dev. Dyn.* 194, 193–197.

McGuire, P. G., and Alexander, S. M. (1993). Inhibition of urokinase synthesis and cell surface binding alters the motile behavior of embryonic endocardial-derived mesenchymal cells *in vitro*. *Development (Cambridge, UK)* 118, 931–939.

McGuire, P. G., and Orkin, R. W. (1992). Urokinase activity in the developing avian heart: A spatial and temporal analysis. *Dev. Dyn.* 193, 24–33.

McNeill, H., Ozawa, M., Kemler, R., and Nelson, W. J. (1990). Novel function of the cell adhesion molecule uvomorulin as an inducer of cell surface polarity. *Cell (Cambridge, Mass.)* 62, 309–316.

Miller, J. R., and McClay, D. R. (1997a). Changes in the pattern of adherens junction-associated beta-catenin accompany morphogenesis in the sea urchin embryo. *Dev. Biol.* 192, 310–322.

Miller, J. R., and McClay, D. R. (1997b). Characterization of the role of cadherin in regulating cell adhesion during sea urchin development. *Dev. Biol.* **192**, 323–339.

Mitchell, P. J., Timmons, P. M., Hebert, J. M., Rigby, P. W., and Tjian, R. (1991). Transcription factor AP-2 is expressed in neural crest cell lineages during mouse embryogenesis. *Genes Dev.* **5**, 105–119.

Mjaatvedt, C. H., and Markwald, R. R. (1989). Induction of an epithelial–mesenchymal transition by an *in vivo* adheron-like complex. *Dev. Biol.* **136**, 118–128, published erratum: **137**(1), 217 (1990).

Mjaatvedt, C. H., Lepera, R. C., and Markwald, R. R. (1987). Myocardial specificity for initiating endothelial–mesenchymal cell transition in embryonic chick heart correlates with a particulate distribution of fibronectin. *Dev. Biol.* **119**, 59–67.

Mjaatvedt, C. H., Krug, E. L., and Markwald, R. R. (1991). An antiserum (ES1) against a particulate form of extracellular matrix blocks the transition of cardiac endothelium into mesenchyme in culture. *Dev. Biol.* **145**, 219–230.

Nagafuchi, A., Shirayoshi, Y., Okazaki, K., Yasuda, K., and Takeichi, M. (1987). Transformation of cell adhesion properties by exogenously introduced E-cadherin cDNA. *Nature (London)* **329**, 341–343.

Nakagawa, S., and Takeichi, M. (1995). Neural crest cell–cell adhesion controlled by sequential and subpopulation-specific expression of novel cadherins. *Development (Cambridge, UK)* **121**, 1321–1332.

Nakagawa, S., and Takeichi, M. (1998). Neural crest emigration from the neural tube depends on regulated cadherin expression. *Development (Cambridge, UK)* **125**, 2963–2971.

Nakajima, Y., Krug, E. L., and Markwald, R. R. (1994). Myocardial regulation of transforming growth factor-beta expression by outflow tract endothelium in the early embryonic chick heart. *Dev. Biol.* **165**, 615–626.

Nakajima, Y., Yamagishi, T., Nakamura, H., Markwald, R. R., and Krug, E. L. (1998). An autocrine function for transforming growth factor (TGF)-beta3 in the transformation of atrioventricular canal endocardium into mesenchyme during chick heart development. *Dev. Biol.* **194**, 99–113.

Nathke, I. S., Hinck, L., Swedlow, J. R., Papkoff, J., and Nelson, W. J. (1994). Defining interactions and distributions of cadherin and catenin complexes in polarized epithelial cells. *J. Cell Biol.* **125**, 1341–1352.

Navarro, P., Gomez, M., Pizarro, A., Gamallo, C., Quintanilla, M., and Cano, A. (1991). A role for the E-cadherin cell–cell adhesion molecule during tumor progression of mouse epidermal carcinogenesis. *J. Cell Biol.* **115**, 517–533.

Newgreen, D. F., and Gooday, D. (1985). Control of the onset of migration of neural crest cells in avian embryos. Role of Ca^{++}-dependent cell adhesions. *Cell Tissue Res.* **239**, 329–336.

Newgreen, D. F., and Minichiello, J. (1995). Control of epitheliomesenchymal transformation. I. Events in the onset of neural crest cell migration are separable and inducible by protein kinase inhibitors. *Dev. Biol.* **170**, 91–101.

Nichols, D. H. (1986a). Mesenchyme formation from the trigeminal placodes of the mouse embryo. *Am. J. Anat.* **176**, 19–31.

Nichols, D. H. (1986b). Formation and distribution of neural crest mesenchyme to the first pharyngeal arch region of the mouse embryo. *Am. J. Anat.* **176**, 221–231.

Nichols, D. H. (1987). Ultrastructure of neural crest formation in the midbrain/rostral hindbrain and preotic hindbrain regions of the mouse embryo. *Am. J. Anat.* **179**, 143–154.

Nieto, M. A., Bennett, M. F., Sargent, M. G., and Wilkinson, D. G. (1992). Cloning and developmental expression of Sna, a murine homologue of the *Drosophila* snail gene. *Development (Cambridge, UK)* **116**, 227–237.

Nieto, M. A., Sargent, M. G., Wilkinson, D. G., and Cooke, J. (1994). Control of cell behavior during vertebrate development by Slug, a zinc finger gene. *Science* **264**, 835–839.

Oft, M., Heider, K. H., and Beug, H. (1998). TGFbeta signaling is necessary for carcinoma cell invasiveness and metastasis. *Curr. Biol.* **8**, 1243–1252.

Peifer, M., Pai, L. M., and Casey, M. (1994). Phosphorylation of the *Drosophila* adherens junction protein Armadillo: Roles for wingless signal and zeste-white 3 kinase. *Dev. Biol.* **166**, 543–556.

Potts, J. D., and Runyan, R. B. (1989). Epithelial–mesenchymal cell transformation in the embryonic heart can be mediated, in part, by transforming growth factor beta. *Dev. Biol.* **134**, 392–401.

Potts, J. D., Dagle, J. M., Walder, J. A., Weeks, D. L., and Runyan, R. B. (1991). Epithelial–mesenchymal transformation of embryonic cardiac endothelial cells is inhibited by a modified antisense oligodeoxynucleotide to transforming growth factor beta 3. *Proc. Natl. Acad. Sci. U.S.A.* **88**, 1516–1520.

Radice, G. L., Rayburn, H., Matsunami, H., Knudsen, K. A., Takeichi, M., and Hynes, R. O. (1997). Developmental defects in mouse embryos lacking N-cadherin. *Dev. Biol.* **181**, 64–78.

Ranscht, B. (1994). Cadherins and catenins: Interactions and functions in embryonic development. *Curr. Opin. Cell Biol.* **6**, 740–746.

Reichmann, E. (1994). Oncogenes and epithelial cell transformation. *Semin. Cancer Biol.* **5**, 157–165.

Reichmann, E., Schwarz, H., Deiner, E. M., Leitner, I., Eilers, M., Berger, J., Busslinger, M., and Beug, H. (1992). Activation of an inducible c-FosER fusion protein causes loss of epithelial polarity and triggers epithelial–fibroblastoid cell conversion. *Cell (Cambridge, Mass.)* **71**, 1103–1116.

Ridley, A. J., Comoglio, P. M., and Hall, A. (1995). Regulation of scatter factor/hepatocyte growth factor responses by Ras, Rac, and Rho in MDCK cells. *Mol. Cell. Biol.* **15**, 1110–1122.

Ringwald, M., Schuh, R., Vestweber, D., Eistetter, H., Lottspeich, F., Engel, J., Dolz, R., Jahnig, F., Epplen, J., and Mayer, S. (1987). The structure of cell adhesion molecule uvomorulin. Insights into the molecular mechanism of Ca^{2+}-dependent cell adhesion. *EMBO J.* **6**, 3647–3653.

Rodriguez-Boulan, E., and Nelson, W. J. (1989). Morphogenesis of the polarized epithelial cell phenotype. *Science* **245**, 718–725.

Runyan, R. B., and Markwald, R. R. (1983). Invasion of mesenchyme into three-dimensional collagen gels: A regional and temporal analysis of interaction in embryonic heart tissue. *Dev. Biol.* **95**, 108–114.

Sanders, E. J., and Prasad, S. (1989). Invasion of a basement membrane matrix by chick embryo primitive streak cells *in vitro*. *J. Cell Sci.* **92**, (Pt. 3), 497–504.

Savagner, P., Yamada, K. M., and Thiery, J. P. (1997). The zinc-finger protein slug causes desmosome dissociation, an initial and necessary step for growth factor-induced epithelial–mesenchymal transition. *J. Cell Biol.* **137**, 1403–1419.

Simonneau, L., Kitagawa, M., Suzuki, S., and Thiery, J. P. (1995). Cadherin 11 expression marks the mesenchymal phenotype: Towards new functions for cadherins? *Cell Adhes. Commun.* **3**, 115–130.

Smeyne, R. J., Vendrell, M., Hayward, M., Baker, S. J., Miao, G. G., Schilling, K., Robertson, L. M., Curran, T., and Morgan, J. I. (1993). Continuous c-fos expression precedes programmed cell death *in vivo. Nature (London)* **363**, 166–169; published erratum: **365**, 279 (1993).

Sorokin, L., Sonnenberg, A., Aumailley, M., Timpl, R., and Ekblom, P. (1990). Recognition of the laminin E8 cell-binding site by an integrin possessing the alpha 6 subunit is essential for epithelial polarization in developing kidney tubules. *J. Cell Biol.* **111**, 1265–1273.

Stark, K., Vainio, S., Vassileva, G., and McMahon, A. P. (1994). Epithelial transformation of metanephric mesenchyme in the developing kidney regulated by Wnt-4. *Nature (London)* **372**, 679–683.

Stern, C. D., Ireland, G. W., Herrick, S. E., Gherardi, E., Gray, J., Perryman, M., and Stoker, M. (1990). Epithelial scatter factor and development of the chick embryonic axis. *Development (Cambridge, UK)* **110**, 1271–1284.

Stoker, M., Gherardi, E., Perryman, M., and Gray, J. (1987). Scatter factor is a fibroblast-derived modulator of epithelial cell mobility. *Nature (London)* **327**, 239–242.

Streit, A., Stern, C. D., Thery, C., Ireland, G. W., Aparicio, S., Sharpe, M. J., and Gherardi, E. (1995). A role for HGF/SF in neural induction and its expression in Hensen's node during gastrulation. *Development (Cambridge, UK)* **121**, 813–824.

Sun, D., Vanderburg, C. R., Odierna, G. S., and Hay, E. D. (1998). TGFβ3 promotes transformation of chicken palate medial edge epithelium to mesenchyme *in vitro. Development* **125**, 95–105.

Sutherland, A. E., Sanderson, R. D., Mayes, M., Seibert, M., Calarco, P. G., Bernfield, M., and Damsky, C. H. (1991). Expression of syndecan, a putative low affinity fibroblast growth factor receptor, in the early mouse embryo. *Development (Cambridge, UK)* **113**, 339–351.

Suzuki, S., Sano, K., and Tanihara, H. (1991). Diversity of the cadherin family: Evidence for eight new cadherins in nervous tissue. *Cell Regul.* **2**, 261–270.

Takeichi, M. (1991). Cadherin cell adhesion receptors as a morphogenetic regulator. *Science* **251**, 1451–1455.

Takeichi, M. (1995). Morphogenetic roles of classic cadherins. *Curr. Opin. Cell Biol.* **7**, 619–627.

Trelstad, R. L., Hayashi, A., Hayashi, K., and Donahoe, P. K. (1982). The epithelial–mesenchymal interface of the male rat Müllerian duct: Loss of basement membrane integrity and ductal regression. *Dev. Biol.* **92**, 27–40.

Tsarfaty, I., Rong, S., Resau, J. H., Rulong, S., da Silva, P. P., and Vande, W. G. (1994). The Met proto-oncogene mesenchymal to epithelial cell conversion. *Science* **263**, 98–101.

Valinsky, J. E., and Le Douarin, N. M. (1985). Production of plasminogen activator by migrating cephalic neural crest cells. *EMBO J.* **4**, 1403–1406.

Vestweber, D., Kemler, R., and Ekblom, P. (1985). Cell-adhesion molecule uvomorulin during kidney development. *Dev. Biol.* **112**, 213–221.

Vleminckx, K., Vakaet, L. J., Mareel, M., Fiers, W., and van Roy, F. (1919). Genetic manipulation of E-cadherin expression by epithelial tumor cells reveals an invasion suppressor role. *Cell (Cambridge, Mass.)* **66**, 107–119.

Volberg, T., Geiger, B., Dror, R., and Zick, Y. (1991). Modulation of intercellular adherens-type junctions and tyrosine phosphorylation of their components in RSV-transformed cultured chick lens cells. *Cell Regul.* **2**, 105–120.

Warren, S. L., and Nelson, W. J. (1987). Nonmitogenic morphoregulatory action of pp60v-src on multicellular epithelial structures. *Mol. Cell. Biol.* **7**, 1326–1337.

Weidner, K. M., Arakaki, N., Hartmann, G., Vandekerckhove, J., Weingart, S., Rieder, H., Fonatsch, C., Tsubouchi, H., Hishida, T., and Daikuhara, Y. (1991). Evidence for the identity of human scatter factor and human hepatocyte growth factor. *Proc. Natl. Acad. Sci. U.S.A.* **88**, 7001–7005.

Werb, Z. (1989). Proteinases and matrix degradation. *In* "Textbook of Rheumatology" (W. N. Kelley, E. D. J. Harris, S. Ruddy, and C. B. Sledge, eds.), pp. 300–321. Saunders, Philadelphia.

Werb, Z. (1997). ECM and cell surface proteolysis: Regulating cellular ecology. *Cell (Cambridge, Mass.)* **91**, 439–442.

Woolf, A. S., Kolatsi-Joannou, M., Hardman, P., Andermarcher, E., Moorby, C., Fine, L. G., Jat, P. S., Noble, M. D., and Gherardi, E. (1995). Roles of hepatocyte growth factor/scatter factor and the met receptor in the early development of the metanephros. *J. Cell Biol.* **128**, 171–184.

Zhang, J., Houston, D. W., King, M. L., Payne, C., Wylie, C., and Heasman, J. (1998). The role of maternal VegT in establishing the primary germ layers in *Xenopus* embryos. *Cell (Cambridge, Mass.)* **94**, 515–524.

Zhong, C., Kinch, M. S., and Burridge, K. (1997). Rho-stimulated contractility contributes to the fibroblastic phenotype of Ras-transformed epithelial cells. *Mol. Biol. Cell* **8**, 2329–2344.

Zuk, A., and Hay, E. D. (1949). Expression of beta 1 integrins changes during transformation of avian lens epithelium to mesenchyme in collagen gels. *Dev. Dyn.* **201**, 378–393.

CHAPTER 4

DYNAMICS OF CELL–ECM INTERACTIONS

Manuela Martins-Green

INTRODUCTION

In recent years the ability to perform tissue and organ replacement has led to an improvement of quality of life and health care. Most of this success can be attributed to the interdisciplinary approaches to tissue engineering. Indeed, today scientists with very diverse backgrounds, including molecular, cellular, and developmental biology, collaborate with biomechanical engineers to develop tissue analogs that allow physicians to improve, maintain, and restore tissue function. Several approaches have been taken toward this goal. One such undertaking involves the use of matrices containing specific cells and growth factors as tissue replacements. Most of the knowledge that has allowed the advancements in preparation of tissue substitutes has come from studies performed during biologic development and development-like processes, such as wound healing, on the basic mechanisms of interaction between cells and the extracellular matrix (ECM). Discussed in this chapter are some of the key findings that led to the understanding of how the dynamics of cell–ECM interactions contribute to cell migration, proliferation, differentiation, and programmed death, all of which are important parameters to consider when preparing and using tissue analogs.

For many years the ECM was thought to serve only as a structural support for tissues. However, as early as 1966, Hauschka and Konigsberg showed that interstitial collagen promoted the conversion of myoblasts to myotubes, and shortly thereafter it was shown that both collagen (Wessells and Cohen, 1968) and glycosaminoglycans (Bernfield *et al.,* 1973) play a crucial role in salivary gland morphogenesis. Based on these and other pieces of indirect evidence, Hay (1977) put forth the idea that the ECM is an important component in embryonic inductions, a concept that implicated the presence of binding sites (receptors) for specific matrix molecules on the surface of cells. By this time, the stage was set to begin to investigate in detail the mechanisms by which extracellular matrix molecules influence cell behavior. Bissell *et al.* (1982) proposed the model of "dynamic reciprocity" between the ECM on the one hand and the cytoskeleton and nuclear matrix on the other. In this model, ECM molecules interact with cell surface receptors, which then transmit signals across the cell membrane to molecules in the cytoplasm; these signals initiate a cascade of events through the cytoskeleton into the nucleus, resulting in the expression of specific genes, whose products, in turn, affect the ECM in various ways. It now has become clear that this concept is essentially correct (Ingber, 1991; Boudreau *et al.,* 1995); cell–ECM interactions participate directly in promoting cell adhesion, migration, growth, differentiation, and programmed cell death (also called apoptosis), as well as in modulation of the activities of cytokines and growth factors, and in directly activating intracellular signaling.

Most of what we know about the molecular basis of cell–ECM interactions in these events comes from studies that have used induced mutations, experimental perturbations *in vivo,* and cell/organ cultures. In this chapter, the composition and diversity of the ECM and its receptors are briefly discussed, then selected examples illustrate the dynamics of cell–ECM interactions during development and wound healing and the potential mechanisms involved in the signal transduction pathways initiated by these interactions. Finally, the implications for tissue engineering are discussed.

COMPOSITION AND DIVERSITY OF THE ECM

The ECM is a molecular complex that has as basic components collagens and other glycoproteins, hyaluronic acid, proteoglycans, glycosaminoglycans, and elastins; the ECM also harbors molecules such as growth factors, cytokines, matrix-degrading enzymes, and their inhibitors. The distribution of these molecules is not static, but rather varies from tissue to tissue and during development from stage to stage (Laurie *et al.,* 1989; Ffrench-Constant and Hynes, 1989; Sanes *et al.,* 1990; Martins-Green and Bissell, 1995; Werb and Chin, 1998). Furthermore, their organization varies with tissue type and has profound implications for the function of the tissues (Scheler *et al.,* 1998). Mesenchymal cells are immersed in an interstitial matrix that confers specific biomechanical and functional properties to connective tissue (Culav *et al.,* 1999). Epithelial and endothelial cells, on the other hand, contact a basement membrane only through their basal surfaces, which confers mechanical strength and specific physiological properties to the epithelia (Fuchs *et al.,* 1997; Dockery *et al.,* 1998).

This diversity of composition, organization, and distribution of the ECM results not only from differential gene expression for the various molecules in specific tissues, but also from the existence of differential splicing and posttranslational modifications of those molecules. For example, differential splicing may change the binding potential of proteins to each other (Ffrench-Constant and Hynes, 1989; Chiquet-Ehrismann *et al.,* 1991; Wallner *et al.,* 1998) or to their receptors (Aota *et al.,* 1994; Akiyama *et al.,* 1995; Cox and Huttenlocher, 1998) and variations in glycosylation can lead to changes in cell adhesion (Dean *et al.,* 1990; Vlodavsky *et al.,* 1996; Schamhart and Kurth, 1997; Cotman *et al.,* 1999). In addition, the presence of divalent cations such as Ca^{2+} (Paulsson, 1988; Ekblom *et al.,* 1994) can affect matrix organization and influence molecular interactions that are important with respect to the way ECM molecules interact with cells (Sjaastad and Nelson, 1997).

Growth factors and cytokines interact with the ECM in a variety of ways, which allows them mutually to affect each other (Nathan and Sporn, 1991; Adams and Watt, 1993). The ECM can serve as a reservoir by binding growth factors and cytokines and protecting them from being degraded, by presenting them more efficiently to their receptors, or by affecting their synthesis (Roberts *et al.,* 1988; Chiquet-Ehrismann, 1991; Flaumenhaft and Rifkin, 1992; Lamszus *et al.,* 1996; Kagami *et al.,* 1998). In this way the ECM can affect the local concentration and biological activity of these factors. For example, when neutrophils adhere to fibronectin they produce higher levels of tumor necrosis factor (Nathan and Sporn, 1991). Conversely, growth factors and cytokines can stimulate cells to alter the production of ECM molecules, their inhibitors, and/or their receptors (Streuli *et al.,* 1993; Schuppan *et al.,* 1998). Transforming growth factor β (TGF-β), for example, up-regulates the expression of matrix molecules and of inhibitors of enzymes that degrade ECM molecules (Bonewald, 1999). In a number of cases, only specific forms of these growth factors and cytokines bind to specific ECM molecules, e.g., platelet-derived growth factor (PDGF) (La Rochelle *et al.,* 1991) and the chicken chemotactic and angiogenic factor (cCAF). The latter protein, chemokine, that is overexpressed during wound repair and in the stroma of tumors (Martins-Green and Bissell, 1990; Martins-Green *et al.,* 1992), is secreted as a 9-kDa protein but can be processed to 7 kDa by plasmin. Both the 9- and 7-kDa forms of the protein are found in association with interstitial collagen, but only the smaller form binds to laminin or tenascin and neither form binds to fibronectin, collagen IV, or heparin (Martins-Green and Bissell, 1995; Martins-Green *et al.,* 1996). Importantly, binding of specific forms of these factors to specific ECM molecules can lead to their localization to particular areas of tissues and can affect their biological activities.

RECEPTORS FOR EXTRACELLULAR MATRIX MOLECULES

In order to establish that ECM molecules directly affect cellular behavior, it was important to identify transmembrane receptors for the specific sequences present on these molecules. As early

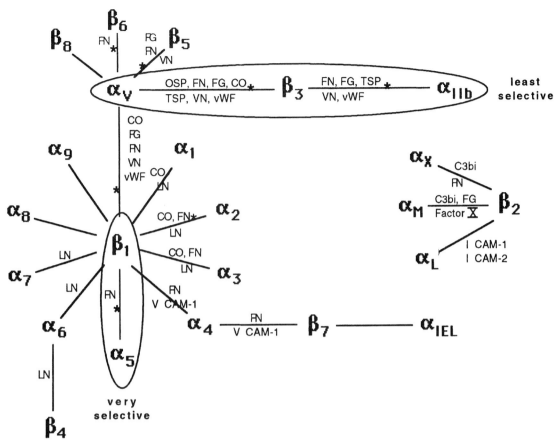

Fig. 4.1. Integrin family of receptors and their extracellular ligands. These heterodimeric receptors are each composed of one α and one β subunit. Extracellular ligands have now been identified for many of the integrin receptors. The β1 and β3 classes mostly mediate cell–matrix adhesion, whereas the β2 class mediates cell–cell adhesions. Moreover, the β1 class is, in general, involved in adhesion to connective tissue macromolecules such as fibronectin, laminin, and collagen, whereas β3 binds vascular ligands such as fibrinogen, von Willebrand factor, thrombospondin and vitronectin. CO, Collagens; C3bi, complement component; FG, fibrinogen; FN, fibronectin; ICAM-1, intercellular adhesion molecule 1; LN, laminin; TSP, thrombospondin; VCAM-1, vascular cell adhesive molecule 1; VN, vitronectin; vWF, von Willebrand factor; OSP, osteopondin; *, RGD-mediated binding.

as 1973, it was observed that during salivary gland morphogenesis near the sites of glycosamino-glycan deposition, the intracellular microfilaments contracted (Bernfield *et al.,* 1973). These investigators proposed that the ECM could "be involved in regulating microfilament function," suggesting that these molecules can specifically interact with cell surface receptors. Using biochemical and molecular biological approaches, it was subsequently shown that various ECM molecules contain specific amino acid motifs that allow them to bind directly to cell surface receptors (Humphries *et al.,* 1991; Hynes, 1992; Gullberg and Ekblom, 1995). The best characterized motif is the tripeptide RGD, first found in fibronectin (Pierschbacher and Ruoslahti, 1984; Yamada and Kennedy, 1984). Peptides containing this amino acid sequence promote adhesion of cells and inhibit the adhesive properties of fibronectin. This and other amino acid adhesive motifs have been found in laminin, entactin, thrombin, tenascin, fibrinogen, vitronectin, type I and VI collagens, bone sialoprotein, and osteopondin (Humphries *et al.,* 1991).

Integrins were the first ECM receptors to be identified (Hynes, 1987). These receptors make up a family of heterodimeric transmembrane proteins composed of α and β subunits. At least 15 α and 8 β subunits have been identified so far; they pair with each other in a variety of combinations, giving rise to a large family that recognizes specific sequences on the ECM molecules (Fig. 4.1). Some integrin receptors are very specific whereas others bind several different epitopes, which may be on the same or different ECM molecules (Fig. 4.1). In this way plasticity and redundancy can be built into specific systems (e.g., Hynes, 1992, 1999; Faull and Ginsberg, 1996; Cotman *et al.,* 1998).

Although the α and β subunits of integrins are unrelated, there is 40–50% homology within each subunit, with the highest divergence in the intracellular domain of the α subunit. All but one of these subunits (β4) have large extracellular domains and very small intracellular domains (Briesewitz *et al.,* 1995; Fornaro and Languino, 1997). The extracellular domain of the α subunits contains four binding sites for divalent cations. These binding sites are homologous to the EF-band of calmodulin and appear to be responsible for the dependence of integrin–ligand binding on the presence of these ions (Gailit and Ruoslahti, 1988). It is now well established that, on ligand binding, integrins can directly induce biochemical signals inside cells (Giancotti, 1997; Kumar, 1998). The cytoplasmic domain of integrins interacts with the cytoskeleton, suggesting that ECM signalling through integrins is transduced via the cytoskeletal elements and can induce cell shape changes, which, in turn, may lead to growth and/or differentiation (Ruoslahti and Yamaguchi, 1992; Hemler, 1998).

Although not as extensively studied as the integrins, it has been found that transmembrane proteoglycans also can be receptors for ECM molecules (Rapraeger *et al.,* 1987; Jalkehen *et al.,* 1991; McFall and Rapraeger, 1998). At least four types of proteoglycan receptors that bind to ECM molecules have been isolated and characterized: syndecan, CD44, RHAMM (receptor for hyaluronate-mediated motility), and thrombomodulin (Rapraeger *et al.,* 1987; Bernfield *et al.,* 1992; Turley, 1992; Spring *et al.,* 1994; Carey, 1997; Naot *et al.,* 1997; McFall and Rapraeger, 1998). Syndecan binds cells to the matrix via chondroitin sulfate and heparan sulfate glycosaminoglycans, sugars that vary depending on the type of tissue in which syndecan is expressed, and which modulate the capability of a particular ligand to interact with it (Kim *et al.,* 1994; Salmivirta and Jalkanen, 1995). In contrast to the binding of matrix to integrins, this binding is independent of divalent cations. Syndecan also associates with the cytoskeleton, showing that these interactions via the cytoskeleton are important in signal transduction through this cell surface receptor (Carey *et al.,* 1996; Couchman *et al.,* 1996). The CD44 receptor also carries chondroitin sulfate and heparan sulfate chains on its extracellular domain. Variation in the glycosaminoglycan composition is tissue specific and may play a role in cell adhesion similar to that shown for syndecan (Brown *et al.,* 1991). The most distant extracellular domain of CD44 contains six cysteine residues that form three disulfide bonds, creating three loops that are structurally similar to the hyaluronan-binding domain of the cartilage link protein and aggrecan. Using a lymphoid cell line and an antibody to CD44 it has been shown that this domain of CD44 can interact directly with hyaluronan (Miyake *et al.,* 1990).

RHAMM has been identified as the major receptor for hyaluronic acid, and in hematopoietic progenitor cells these interactions play an important role in their trafficking (Pilarski *et al.,* 1999). Thrombomodulin is a transmembrane receptor with an extracellular portion containing an N-terminal hydrophobic domain followed by six extracellular growth factor (EGF)-like repeats and a Ser/Thr domain with a glycosaminoglycan attachment site. This protein is expressed in endothelial cells, has anticoagulant properties, and can be found in a proteoglycan or nonproteoglycan form. Its biological activity can be regulated either by its protein or glycosaminoglycan components (Hardingham and Fosang, 1992; Blann and Seigneur, 1997).

Cell surface receptors other than integrins or proteoglycans have also been identified for ECM molecules. A laminin-binding protein of 69 kDa recognizes the YIGSR sequence of laminin that is not recognized by integrins (Mecham, 1991). A second receptor, CD36, binds collagen, thrombospondin, and malaria-infected erythrocytes to endothelial cells and to some types of epithelial cells (Greenwalt *et al.,* 1992). Each of these has a separate binding site, but all are located in the same external loop of CD36 (Asah *et al.,* 1993) and the intracellular signals occurring after ligand binding lead to activation of a variety of signal transduction molecules (Huang *et al.,* 1991; Lipsky *et al.,* 1997).

CELL–ECM INTERACTIONS

Interactions of cells with extracellular matrix molecules play a crucial role during development and wound healing. It is the continuous cross-talk between cells and the surrounding matrix environment that leads to the formation of patterns, the development of form (morphogenesis), and the acquisition and maintenance of differentiated phenotypes during embryogenesis.

Similarly, during wound healing these interactions contribute to the processes of clot formation, inflammation, granulation tissue development, and remodeling. As we will see, many different lines of experimental evidence have shown that the basic cellular mechanisms that result in these events involve cell adhesion/deadhesion, migration, proliferation, differentiation, and programmed cell death.

DEVELOPMENT

Adhesion and migration

Some of the most compelling experiments that demonstrated the direct participation of the ECM in cell adhesion and migration came from studies in gastrulation, migration of neural crest cells (NCCs), angiogenesis, and epithelial organ formation. Cell interactions with fibronectin are important during gastrulation. Microinjection of antibodies to fibronectin into the blastocoel cavity of *Xenopus* embryos causes disruption of normal cell movements and leads to abnormal development (Boucaut *et al.,* 1984a). Furthermore, if RGD-containing peptides are injected during the same stage of development, they cause randomization of the bilateral asymmetry of the heart and gut (Yost, 1992). Similarly, microinjection of RGD-containing peptides and/or antibodies to the β1 subunit of the integrin receptor for fibronectin perturbs gastrulation in embryos of *Pleurodeles* (salamander) (Yost, 1992; Boucaut *et al.,* 1984b; Darribere *et al.,* 1988). These effects are not unique to fibronectin. They can also be introduced by manipulation of other molecules, such as heparan sulfate proteoglycan, which can be competed out by heparin for its target-binding molecule(s) and cause perturbations in gastrulation and neurulation (Erickson and Perris, 1993; Erickson and Reedy, 1998).

The NCCs originate in the dorsal portion of the neural tube just after closure of the tube, migrate extensively throughout the embryo in ECM-filled spaces, and give rise to a variety of phenotypes. The pathways of migration of the NCCs contain collagens, proteoglycans, glycosaminoglycans, and glycoproteins (Erickson and Perris, 1993; Erickson and Reedy, 1998). Microinjection of antibodies to fibronectin (Poole and Thiery, 1986) or to the β1 subunit of the integrin receptor (Bronner-Fraser, 1985, 1986) into the crest pathways in chick embryos reduces the number of NCCs that leave the tube and causes abnormal neural tube development. Other ECM molecules, such as laminin, also affect NCC migration: for example, the YIGSR synthetic peptide, known to inhibit laminin binding to cells, also inhibits NCC migration (Runyan *et al.,* 1986). The importance of cell–ECM interactions in NCC migration is also supported by studies performed in the white mutant of Mexican axolotl embryos. The NCCs that give rise to pigment cells fail to emigrate from the neural tube in these embryos, but when microcarriers containing subepidermal ECM from normal embryos are implanted into the appropriate area in these mutants, the NCC precursors to pigment cells emigrate normally (Perris *et al.,* 1988; Lofberg *et al.,* 1989; Epperlein and Lofberg, 1996). Potentially relevant to these findings is the observation that fibronectin appears between chick NCCs just prior to their emigration from the neural tube (Martins-Green, 1987, 1990; Martins-Green and Bissell, 1995). Perhaps this fibronectin consists predominantly or exclusively of the RGD domain-carrying segment that, when bound to its integrin receptor, promotes secretion of adhesion-degrading enzymes, thereby facilitating emigration (Damsky and Werb, 1992; Grant *et al.,* 1998).

During angiogenesis (the development of blood vessels from preexisting vessels), endothelial cell interactions with ECM molecules and the type and conformation of the matrix play a crucial role in cell migration and the development of blood vessels (Cockerill *et al.,* 1995; Baldwin, 1996; Hanahan, 1997; Kumar *et al.,* 1998). Human umbilical vein endothelial cells migrate and arrange themselves in tubular structures when cultured for 12 hr on a matrix isolated from Engelbreth–Holm–Swarm (EHS) tumors (a basement membranelike matrix that contains mainly laminin but also has collagen type IV, proteoglycans, and entactin/nidogen) (Kleinman *et al.,* 1982). When these cells are cultured on collagen type I, however, tubular structures do not form in this period of time, but if they are grown for a week inside collagen gels, giving the endothelial cells time to deposit their own basement membrane, tubes do develop (Montesano *et al.,* 1983; Madri *et al.,* 1988). The much more rapid tubulogenesis on EHS versus inside collagen I suggests that one or more components of the basement membrane plays an important role in the development of the capillary-like structures, a speculation that has been confirmed both in culture and *in vivo*

(Sakamoto *et al.,* 1991; Grant *et al.,* 1992). Indeed, preincubation of these endothelial cells with antibodies to laminin, the major component of basement membrane, prevents the formation of tubules *in vitro.* Furthermore, synthetic peptides containing the sequence SIKVAV, present in the A chain of laminin, induce endothelial cell adhesion and elongation and promote angiogenesis (Grant *et al.,* 1992). Interestingly, however, the sequence CDPGYIGSR-NH$_2$, a domain present in the laminin B chain, blocks angiogenesis *in vivo* (Sakamoto *et al.,* 1991; Grant *et al.,* 1992) as well as migration of endothelial cells induced by tumor-conditioned medium Sakamoto *et al.,* 1991). The mechanism of action of these sequences is unknown, but it is potentially instructive that sequences in the same molecule have opposite effects. Because endothelial cells have to migrate and penetrate adjacent tissues and matrix during angiogenesis, matrix-degrading enzymes are active, raising the possibility that angiogenic factors are provided by release from the matrix or by appropriate cleavage of ECM molecules such as laminin (Werb *et al.,* 1999). *In vivo,* angiogenic sequences or factors could be provided locally, and when they have served their purpose, inhibition of further action could similarly be initiated by suitable cleavage of CDPGYIGSR–NH$_2$ or some other comparable factor present in the ECM (Sakamoto *et al.,* 1991). Therefore, the way matrix molecules are locally cleaved and/or factors are locally released could have important consequences for the homeostasis of tissues.

Proliferation

Interactions of ECM molecules with cells can also modulate cell proliferation. Some of these effects are well illustrated by laminin. A domain in the A chain of laminin that is rich in EGF-like repeats stimulates proliferation of a variety of different cell lines (Panayotou *et al.,* 1989), and the entire molecule appears to promote proliferation of bone marrow-derived macrophages (Ohki and Kohashi, 1994). Also, thrombospondin exerts its mitogenic activities via its amino-terminal heparin-binding domain (Majack *et al.,* 1986). Contrarily, there also are matrix molecules that are inhibitory of cell proliferation. Heparin and heparin-like molecules, for example, are inhibitors of vascular smooth muscle (VSM) cell proliferation. The conditioned medium of endothelial cells cultured from bovine aortas inhibits the proliferation of VSM cells (Castellot *et al.,* 1987), and this inhibition is obliterated by treatment of the medium with heparinase but not with condroitinases or proteases. This suggests that heparan-type molecules have a direct antiproliferative effect on aortic VSM cells. It has also been proposed (McCaffrey *et al.,* 1989) that this effect is indirect— heparinase treatment could release associated TGF-β, which has been shown to inhibit the proliferation of nonconfluent cultures of VSM cells. However, if the latter were the case, one would expect that treatment of the endothelial cell conditioned medium with proteases should also eliminate the antiproliferative effect. Studies in culture have lent support to this inhibitory effect of the ECM on cell proliferation. For example, normal human breast cells do not growth-arrest when cultured on plastic, but do so if grown in a basement membrane matrix (Petersen *et al.,* 1993; Weaver *et al.,* 1997). Furthermore, growth of a mammary epithelial cell line is stimulated by overexpression of Id-1, a protein that binds to and inhibits the function of basic helix–loop–helix (HLH) transcription factors that are important in cell differentiation. When these overexpressing cells are cultured on EHS they arrest growth and form three-dimensional structures (Desprez *et al.,* 1993). Similarly, in hepatocytes it has been shown that ECM suppresses the expression of immediate-early growth response genes and induces C/EBPα, which is necessary for expression of hepatocyte-specific genes (Rana *et al.,* 1994).

Some of the ECM effects on cell proliferation involve cooperation with growth factors. For example, basic fibroblast growth factor (bFGF), interleukins (IL-1, IL-2, IL-6), hepatocyte growth factor, PDGF-AA, and TGF-β are found in association with the ECM at high concentrations and are released at specific times for interaction with their receptors (Schuppan *et al.,* 1994). An illustrative example of cell–ECM–TGF-β cooperation occurs during the early developmental stages of the mammary gland (Silberstein *et al.,* 1992; Howlett and Bissell, 1993). The development of this gland occurs in two steps: during puberty, the small number of ducts present at birth undergo extensive proliferation and branching, and terminal and lateral buds develop (virgin gland); during pregnancy, further local proliferation of ductal and bud epithelial cells occurs, giving rise to numerous alveoli (pregnant gland). During virgin gland morphogenesis, inductive events taking place between the epithelium and the surrounding mesenchyme are mediated by the basement membrane (basal lamina plus closely associated ECM molecules) and play an important role in epithelial proliferation during branching of the gland. Endogenous TGF-β produced by the ductal epithelium and surrounding mesenchyme forms complexes with mature periductal ECM, stabilizing the epithelium by inhibiting cell proliferation and matrix-degrading enzymes (Silberstein *et al.,* 1990, 1992; Robinson *et al.,* 1991). However, TGF-β is absent from newly synthesized ECM deposited in the branching areas (Silberstein *et al.,* 1992), hence its inhibitory effects on epithelial cell proliferation and on production of matrix-degrading enzymes do not occur, allowing the basement membrane to undergo remodeling. In these regions, proteases that are released locally partially degrade the matrix, thereby promoting cell proliferation and branching morphogenesis. In mice transgenic for the autoactivated isoform of the matrix metalloproteinase stromelysin-1, which is important in basement membrane degradation and tissue remodeling, the virgin glands are morphologically similar to the pregnant glands of normal mice (Sympson *et al.,* 1994, 1995). These results indicate that stromelysin-1 causes matrix degradation leading to the release of factors that stimulate precocious proliferation of the epithelium and development of the alveoli.

Differentiation

Some clear-cut examples of how the ECM can affect cell behavior come from studies of the influences of the ECM on cellular differentiation and maintenance of tissue-specific gene expression. Many of these studies have been performed during keratinocyte, hepatocyte, and mammary gland epithelial differentiation. Keratinocytes form the stratified epidermal layers of the skin. The basal layer is highly proliferative, does not express the markers for terminal differentiation, and is the only cell layer in contact with the basement membrane. As these cells divide, the daughter cells lose contact with the basement membrane, move up to the suprabasal layers, and begin to express differentiation markers such as involucrin. This suggests that close interaction with the basement membrane is responsible for the lack of differentiation of the basal keratinocytes. Indeed, when keratinocytes are grown in suspension, they undergo premature terminal differentiation (Nicholson and Watt, 1991). It has also been shown that human and mouse keratinocytes adhere to fibronectin via its $\alpha_5\beta_1$-integrin receptor and that the expression of both integrin subunits is inversely proportional to the expression of involucrin, a differentiation factor for these cells (Nicholson and Watt, 1991). A major advance in trying to understand the biology of skin development has been the ability to culture keratinocytes on feeder layers of 3T3 cells and form stratified sheets of cells that behave very much like epidermis does *in vivo* (DeVries *et al.,* 1994; Yoshizato and Yoshikawa, 1994; Gross, 1996; Stocum, 1998). This latter development has had profound application in treating patients that have suffered extensive burns.

In culture, expression of hepatocyte-specific genes is observed only when hepatocytes are grown on an EHS matrix (Ben-Ze'ev *et al.,* 1988) or when this matrix or diluted laminin is dripped over the cells (Caron, 1990). The specific cell–ECM interactions have not yet been worked out, but three liver-specific transcription factors eE-TF, e-G-TF/HNF-3, and eH-TF, are activated when cells are cultured on or with matrix molecules, conditions that favor hepatocyte differentiation. In particular, the transcription factor eG-TF/HNF-3 appears to be regulated by the ECM (Liu *et al.,* 1991). In mouse mammary gland, basement membrane and its individual components in conjunction with lactogenic hormones are responsible for the induction of the differentiated phenotype of the epithelial cells (Lin and Bissell, 1993; Blaschke *et al.,* 1994; Werb *et al.,* 1996; M. J. Bissell, 1998). In this system, when midpregnant mammary epithelial cells are cultured on plastic they do not express any of the mammary-specific genes. However, when the same cells are plated and maintained on EHS they form alveolar-like structures and exhibit the fully differentiated phenotype, with expression of the genes for milk proteins (e.g., the expression of β-casein) (Barcellos-Hoff *et al.,* 1989). In the case of β-casein, it has been shown that there are two components to its induction by the ECM: one involves cell rounding (and therefore a change in the cytoskeleton) and the other, a tyrosine kinase signal transduction pathway through the β1 integrin receptor, leading to the activation of elements in the promoter region of the β-casein gene (Roskelly *et al.,* 1994). Using a mammary gland cell line and progressive deletions of the bovine β-casein promoter attached to choline acetyltransferase (CAT), it was found that this promoter contains a 160-bp enhancer for β-casein about 1600 bp upstream from the start site (Schmidhauser *et al.,* 1991; Boudreau *et al.,* 1995; Myers *et al.,* 1998). These ECM-response elements are also present in other tissues. As shown above, they exist in the albumin gene in hepatocytes and they have also been found in lower vertebrates in the LpS1 genes, which are activated in aboral ectoderm cells during early development of the sea urchin *Lytechinus pictus* (Xiang *et al.,* 1991).

Another example that illustrates well how the ECM influences the expression of milk protein genes comes from work performed on the expression of whey acidic protein (WAP) in mouse mammary gland epithelial cells. WAP expression is inhibited when cells are cultured on plastic, but if they are grown on an ECM, expression is up-regulated (Cheng and Bissell, 1989; Lin and Bissell, 1993). It has been found that cells cultured on plastic produce TGF-α, which inhibits the expression of WAP, whereas on EHS the matrix inhibits the production of TGF-α and this leads to up-regulation of the milk protein (Lin *et al.,* 1995).

Programmed cell death (apoptosis)

In the embryogenesis of higher vertebrates, cell death occurs during remodeling, such as during blastocyst formation, during development of the digits, palate, and nervous system, during positive selection of thymocytes in the thymus, during mammary gland involution, and during angiogenesis. In early mouse development a cavity surrounded by a single layer of columnar epithelium forms within the morula. This process is the result of the interplay between two types of signal: one that stimulates cell death of the inner morula cells and the other, mediated by the basal lamina, that is crucial for the survival of the ectodermal cells that line the cavity (Coucouvanis and Martin, 1995).

It has been found that during development of the mammary gland, basement membrane molecules suppress apoptosis of the epithelial cells (Strange *et al.,* 1992; Boudreau *et al.,* 1996; Alexander *et al.,* 1996). Following lactation, however, the numerous alveoli that produce milk regress and are resorbed during involution as a consequence of enzymatic degradation of the alveolar basement membrane and programmed cell death (Talhouk *et al.,* 1992; Strange *et al.,* 1992; Lund *et al.,* 1996; M. J. Bissell, 1998). In this system, the loss of cell–ECM interactions resulting from matrix degradation correlates with an increase in caspase-1 (previously known as interleukin-1β converting enzyme), which promotes apoptosis; inhibitors of this enzyme inhibit cell death (Boudreau *et al.,* 1995). Similarly, an antibody that disrupts the interaction of the β1 integrin with its ECM ligands leads to apoptosis of mammary epithelial cells.

During late developmental stages, development of the circulatory system occurs via angiogenesis. It has been found that αvβ3 integrin interactions with the ECM play a crucial role in angiogenesis during embryogenesis (Brooks *et al.,* 1994; Varner *et al.,* 1995). Disruption of these interactions with an antibody to αvβ3 inhibits the development of new blood vessels in the chorioallantoic membrane (CAM) by causing the endothelial cells to undergo apoptosis (Brooks *et al.,* 1994; Varner *et al.,* 1995).

Wound Healing

Adhesion and migration

Shortly after tissue damage and during the early stages of wound healing, there is a release of blood contents and tissue factors into the area of the wound, leading to platelet activation and adhesion, and the formation of a vascular plug containing mostly platelets, plasma fibronectin, and fibrin (cross-linked by factor XIII), but also including small amounts of tenascin, thrombospondin, and SPARC [secreted protein, acidic and rich in cysteine]. During this process, activated mast cells degranulate, releasing vasodilating and chemotactic factors that will bring to the wound site polymorphonucleocytes. These events constitute the early stages of the inflammatory response. The fibrin–fibronectin meshwork provides a provisional matrix that serves as substrate for the subsequent migration of leukocytes and keratinocytes during the very early stages of healing when inflammation and wound closure are occurring. Leukocyte interactions with ECM molecules via integrin receptors affect many of the functions of these cells, in particular those that lead to cell adhesion and migration or to production of inflammatory mediators (Rosales and Juliano, 1995; Wei *et al.,* 1997). Some of these latter molecules can be damaging to tissues when produced in excess, therefore the course of inflammation can be affected significantly by the types of ECM molecules encountered by these leukocytes (Pakianathan, 1995; Wei *et al.,* 1997).

During closure of cutaneous wounds, keratinocytes migrate over the provisional matrix primarily composed of fibronectin, vitronectin, tenascin, and collagen type III. These cells express α2β1, α3β1, α5β1, α6β1, and αv integrins, which interact with these ECM molecules and in conjunction with matrix metalloproteinases facilitate their migration to close the wound (Juhsz *et al.,* 1993; Gailit *et al.,* 1994). Cell–ECM interactions are equally important in closure of other epithelial wounds. Studies examining the sequential deposition of ECM molecules after wounding of retinal pigment epithelial cells showed that "*de novo*" deposition of fibronectin occurs 24 hr after wounding and is followed by deposition of collagen type IV and laminin. This sequence of matrix deposition is tightly linked to adhesion and migration of cells to close the wound (Kamei *et al.,* 1998). This has also been found to occur during the repair of airway epithelial cells after mechanical injury (White *et al.,* 1999).

As healing progresses, embryonic-type cellular fibronectin produced by macrophages and fibroblasts in the wound bed contributes to formation of the granulation tissue. This matrix molecule serves as substrate for the migration of endothelial cells (which form the vasculature of the wound bed), myofibroblasts, and lymphocytes that are chemoattracted to the wound site by a variety of small cytokines (chemokines), which are secreted by both macrophages and fibroblasts. Several members of the chemokine superfamily have been found in humans, in other mammals, and in avians (Balkwill, 1998; Locati and Murphy, 1999). Chemoattraction by chemokines of cells involved in granulation tissue formation, in conjunction with the interaction of these cells with the ECM via cell surface receptors, results in processes that lead to cell adhesion and migration into the area of the wound to form the granulation tissue (Lukacs and Kunkel, 1998).

One of the most extensively studied chemokines with functions important in wound healing is cCAF (Martins-Green and Bissell, 1990; Martins-Green *et al.,* 1992; Martins-Green, 1999). This chemokine is stimulated to high levels shortly after wounding in the fibroblasts of the wounded tissue (Martins-Green and Bissell, 1990; Martins-Green *et al.,* 1992), and thrombin, an enzyme released on wounding, stimulates these cells to overexpress cCAF (Vaingankar and Martins-Green, 1998; Li *et al.,* 2000). This chemokine, in turn, chemoattracts monocyte/macrophages and lymphocytes (Martins-Green and Feugate, 1998). Expression of cCAF stays elevated during granulation tissue formation, in particular in fibroblasts and endothelial cells of the microvasculature of the wound, and the protein binds to the interstitial collagens, tenascin, and laminin, present in the granulation tissue (Martins-Green and Bissell, 1990; Martins-Green *et al.,* 1992, 1996). Furthermore, cCAF is angiogenic *in vivo* and the angiogenic portion of the molecule is localized in the C terminus of the molecule (Martins-Green and Feugate, 1998; Martins-Green and Kelly, 1998). Based on the pattern of expression and functions of cCAF, it has been proposed that this chemokine participates both in inflammation (via chemotaxis for specific leukocytes) and in formation of the granulation tissue (via stimulation of angiogenesis) (Martins-Green and Hanafusa, 1997; Martins-Green, 1999).

Proliferation

Immediately after wounding, the epithelium undergoes changes that lead to wound closure. During this reepithelialization period, the keratinocytes trailing behind those at the front edge of migration replicate to provide a source of cells to cover the wound. Basement membrane-type ECM, still present in the basal surface of these keratinocytes, may be important in maintaining this proliferative state. In support of this possibility is the finding that during normal skin remodeling, fibronectin associated with the basal lamina of epithelia is crucial for maintaining the basal keratinocyte layer in a proliferative state for constant replenishment of the suprabasal layers (Nicholson and Watt, 1991). It has also been shown, using a dermal wound model, that basement membrane matrices are able to sustain the proliferation of keratinocytes for several days (Dawson *et al.,* 1996).

As reepithelialization is occurring, the granulation tissue begins to form. This latter tissue is composed of fibroblasts, myofibroblasts, monocytes/macrophages, lymphocytes, endothelial cells of the microvasculature, and ECM molecules (embryonic fibronectin, hyaluronic acid, type III collagen, and small amounts of type I collagen) (Clark, 1996). These ECM molecules, in conjunction with growth factors released by the platelets and secreted by the cells present in the granulation tissue, provide signals to the cells, leading to their proliferation (Tuan *et al.,* 1996; D. M. Bissell, 1998). However, the ECM molecules, e.g., fibronectin, specific fragments of laminin and collagen VI, have been shown to stimulate fibroblast and endothelial cell proliferation (Bitterman *et al.,* 1983; Panayotou *et al.,* 1989; Kapila *et al.,* 1998). The cooperation between ECM molecules and growth factors is well illustrated in the proliferation of fibroblasts and the development of new blood vessels (angiogenesis) of the granulation tissue. In the latter case, growth factors such as vascular endothelial growth factors (VEGFs) and FGFs associate with ECM molecules and stimulate proliferation of endothelial cells which then migrate to form the new microvessels (Norrby, 1997).

Differentiation

As healing progresses during the formation of granulation tissue, some of the fibroblasts differentiate into myofibroblasts; they acquire the morphological and biochemical characteristics of smooth muscle cells by expressing smooth muscle α-actin (Desmouliere and Gabbiani, 1994; Grinell, 1994). Matrix molecules are important in this differentiation process. For example, heparin decreases the proliferation of fibroblasts in culture and induces the expression of smooth muscle α-actin in these cells. *In vivo,* the local application of tumor necrosis factor α (TNF-α) leads to the development of granulation tissue, but the presence of cells expressing smooth muscle α-actin was observed only when heparin was also applied (Desmouliere *et al.,* 1992). These results suggest that some of the properties of heparin not related to its anticoagulant effects are important in the induction of smooth muscle α-actin. This function may be related to the ability of heparin to bind cytokines and/or growth factors (such as TNF-α and TGF-β) that regulate cell differentiation (Kirkland *et al.,* 1998; Kim and Mooney, 1998).

Interstitial collagens have also been shown to play a role in the acquisition of the myofibroblastic phenotype. When fibroblasts are cultured on relaxed collagen gels they remain as fibroblasts (Tomasek *et al.,* 1992), but if they are grown on anchored collagen matrices in which the collagen fibers are aligned (much like in the granulation tissue), they show myofibroblast characteristics (Bell *et al.,* 1979; Arora *et al.,* 1999). These and other observations have led to the suggestion that floating collagen matrices mimic dermis whereas anchored collagen matrices mimic granulation tissue (Grinell, 1994; Arora *et al.,* 1999).

Programmed cell death

Apoptosis also plays a role during normal wound healing as the granulation tissue evolves into scar tissue. As the wound heals, the number of fibroblasts, myofibroblasts, endothelial cells, and pericytes decreases dramatically, matrix molecules, especially interstitial collagen, accumulate, and a scar forms (Clark, 1996). In this remodeling phase of healing, cell death by apoptosis leads to elimination of many cells of various types at once without causing tissue damage (Greenhalgh, 1998). For example, studies using transmission electron microscopy and *in situ* end-labeling of DNA fragments have shown that many myofibroblasts and endothelial cells undergo apoptosis during the remodeling process. Morphometric analysis of the granulation tissue shows that the number of cells undergoing apoptosis increases around days 20–25 after injury, resulting in a dramatic reduction in cellularity after day 25 (Desmouliere *et al.,* 1995). Moreover, using model systems that mimic regression of granulation tissue, it has been shown that release of mechanical tension triggers apoptosis of human fibroblasts. Apoptotic cell death was regulated by interstitial-type collagens in combination with growth factors and mechanical tension and did not require differentiation of the fibroblasts into myofibroblasts, strongly suggesting that contractile collagens determine the susceptibility of fibroblasts of the wound tissue to undergo apoptotic cell death (Fluck *et al.,* 1998; Grinell *et al.,* 1999).

In addition to cell death by apoptosis, it has also been shown that during lung remodeling after injury, fibroblasts present in the bronchoalveolar lavage fluid undergo cell death by a process that is distinct from that of necrosis or apoptosis (Polunovsky *et al.,* 1993). Although this process of cell death has not been extensively studied it suggests that there are other processes of programmed cell death that are distinct from apoptosis and occur preferentially in association with wound repair.

SIGNAL TRANSDUCTION EVENTS DURING CELL–ECM INTERACTIONS

From the foregoing examples, it is clear that ECM molecules interact with their receptors and transmit signals directly or indirectly to second messengers that, in turn, unravel a cascade of events, leading to the coordinated expression of a variety of genes involved in cell adhesion, migration, proliferation, differentiation, and death. There is increasing evidence that cell–ECM interactions, especially through integrins, activate a variety of signaling pathways that can be linked to those specific functions. Three potential categories of cell–ECM interactions can lead to these cellular events:

1. Type I interactions involve primarily integrins and proteoglycan receptors, and they are involved in cell adhesion/deadhesion processes during migration (Fig. 4.2A). For example, cell migration is promoted when fibronectin binds simultaneously to integrins through its cell-binding domain and to proteoglycan receptors through its heparin-binding domain (Hynes, 1992; Bernfield *et al.,* 1992; Hardingham and Fosang, 1992; Giancotti, 1997; Schlaepfer and Hunter, 1998; Dedhar, 1999). The proteoglycan receptors interact and colocalize in areas of adhesion, where microfilaments associate with the $\beta 1$ subunit of the integrin receptor via structural proteins such as talin and α-actinin, present in the actin cytoskeleton of the focal adhesions. The cytoplasmic domain of the $\beta 1$ subunit also interacts directly with the focal adhesion tyrosine kinase pp125^{FAK}, which, when activated, undergoes autophosphorylation on tyrosine 397 (Hildebrand *et al.,* 1995); Tyr-397 subsequently serves as the binding site for the SH2 domain of the nonreceptor tyrosine kinase c-Src. In turn, c-Src phosphorylates paxillin, tensin, vinculin, and the protein p130^{cas}, all of which are present in focal adhesion plaques. The specific roles of paxillin and tensin in the process of adhesion/deadhesion during migration are not known but, once phosphorylated, they could mediate signals between the plasma membrane and the cytoskeleton. p130^{cas} is a docking protein that, on activation, binds to the adaptor molecules Crk and Nck, which appear to be involved in migration via the Ras and MAP/JNK kinase pathways. In addition, it has also been shown that c-Src phosphorylates FAK on tyrosine 925, which serves as a site for binding of the Grb2/Sos complex, with subsequent activation of Ras and the MAP kinase cascade (Schlaepfer and Hunter, 1996, 1997; Schlaepfer *et al.,* 1997, 1998). Both pathways are potentially involved in adhesion/deadhesion and migration (Giancotti, 1997; Schlaepfer and Hunter, 1998; Dedhar, 1999).

It is also known that binding of the ECM to integrins can be enhanced by specific cytoplasmic molecules such as the cell adhesion regulator (CAR), a myristoylated protein that can potentially connect the plasma membrane to the cytoskeleton, is activated by phosphorylation, and increases the affinity of $\alpha 2 \beta 1$ to collagen type I (Pullman and Bodmer, 1992). These observations, taken together with the fact that integrins and proteoglycan receptors have no kinase or phosphatase activity in their cytoplasmic domains, suggest that the signals from these receptors leading to adhesion/deadhesion and migration are transduced via regulatory proteins that are part of the actin cytoskeleton (Giancotti, 1997; Schlaepfer and Hunter, 1998; Dedhar, 1999).

2. Type II cell–ECM interactions involve processes that affect proliferation and survival, as well as differentiation and maintenance of the differentiated phenotype (Fig. 4.2B). In these processes, the extracellular matrix interacts with its receptors and cooperates with growth factors or cytokine receptors. These cooperative effects take place, for example, during anchorage dependence of cell growth. Anchorage is required for cells to enter S phase; even in the presence of growth factors, cells will not enter the DNA synthesis phase without being anchored to a substrate (Zhu and Assoian, 1995). Adhesion of cells to ECM molecules plays a very important role in regulating cell survival and proliferation (Mainiero *et al.,* 1997; Giancotti, 1997; Murgia *et al.,* 1998). For example, studies of knockout mice for the $\alpha 1 \beta 1$ integrin, which is the sole receptor for collagen, showed that the fibroblasts of these mice have reduced proliferation even though they attach normally (Pozzi *et al.,* 1998). Collagen binding to $\alpha 1 \beta 1$ in these fibroblasts does not activate the Shc adaptor protein, which in turn is not able to recruit Grb2 and activate the Ras/ERK cascade that leads to the expression of early response genes involved in cell cycle progression (Wary *et al.,* 1998). It has also been shown that cooperation between integrins and growth factors involves the activation of phosphatidylinositol phosphate kinases, which increases the levels of phosphatidylinositol bisphosphate (PIP$_2$). In turn, PIP$_2$ serves as substrate for phospholipase C$_\gamma$ (PLC$_\gamma$), which is activated on growth factor binding to specific receptors on the surface of the cells (Schwartz, 1992). Furthermore, phosphoinositide 3-OH (PI-3) kinase has been shown to rescue cells in suspension from undergoing apoptosis. It appears that in this case PI-3 kinase is activated by Ras, leading to activation of the Akt serine/threonine kinase (Khwaja *et al.,* 1997).

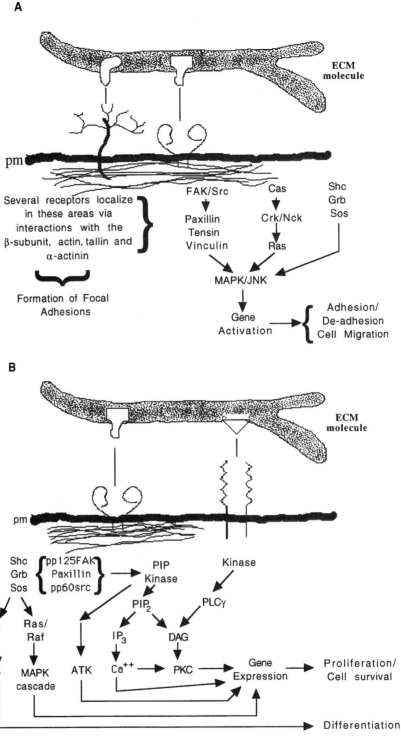

Fig. 4.2. Schematic diagrams (A–C) illustrating the three proposed categories of cell–ECM interactions. (A) At focal adhesions, proteoglycan (treelike) and integrin (heterodimer) receptors on the plasma membrane (pm) bind to different epitopes on the same ECM molecule, leading to cytoskeletal reorganization. A variety of proteins become phosphorylated (e.g., pp125[FAK], Src, PKC), leading through poorly understood pathways to activation of genes important for cell adhesion/deadhesion and for migration. (B) Integrin receptors bind to their ligands, leading to activation of cytoskeletal elements as in A, but growth factors bound to matrix molecules (triangle) also bind to their receptors, which have kinase activity. This kinase activates phospholipase C_γ (PLC$_\gamma$), which, in turn, cleaves phosphatidylinositol bisphosphate (PIP$_2$), leading to inisitol trisphosphate (IP$_3$) and diacylglycerol (DAG); IP$_3$ binds to its receptor on the smooth endoplasmic reticulum, inducing the release of Ca^{2+}, which can lead directly to activation of gene expression or indirectly by co-

C

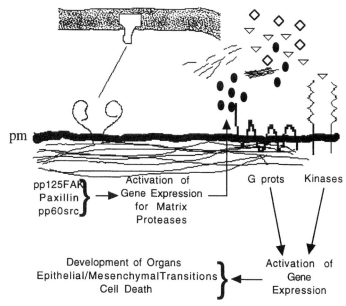

pm

pp125FAK
Paxillin } → Activation of
Gene Expression
pp60src } for Matrix
Proteases

G prots   Kinases

Development of Organs
Epithelial/MesenchymalTransitions } ← Activation of
Cell Death   Gene
                                        Expression

Fig. 4.2. Continued
operation with DAG through protein kinase C (PKC). In this case, the genes activated are
important in cell proliferation, differentiation, and maintenance of the differentiated phenotype.
(C) Integrin receptors bind to fragments of ECM molecules containing specific domains. This leads
to activation of matrix protease genes whose products (represented by black ellipses) degrade the
matrix and release peptides (squiggles) that can further interact with cell surface receptors and/or
release growth factors (triangles and diamonds), which, in turn, bind to their own receptors, acti-
vating G proteins and kinases, leading to expression of genes important in morphogenesis and cell
death.

Similarly, interaction of the cells with ECM molecules, hormones, and growth factors is required to activate genes that are specific for differentiation. Interestingly, in the case of stimulation
of cell differentiation, studies have shown that the cell-ECM interactions that result in the differentiated phenotype are those that fail to activate Shc and the MAP kinase cascade. This has been
shown for endothelial cells in which the interaction of α2β1 with laminin, which does not activate the Shc pathway, leads to formation of capillary-type structures (Kubota *et al.,* 1988), whereas the interaction of α5β1 in the same cells with fibronectin results in proliferation (Wary *et al.,*
1998). Similar observations have been made with primary bronchial epithelial cells when they are
cultured on collagen matrices (Moghal and Neel, 1998).

3. Cell–ECM interactions of Type III involve mostly processes leading to cell death and to
epithelial-to-mesenchymal transitions (Fig. 4.2C). Signal transduction pathways that lead to apoptosis have been delineated for endothelial cells and leukocytes and appear to involve primarily tyrosine kinase activity (Ilan *et al.,* 1998; Fukai *et al.,* 1998; Kettritz *et al.,* 1999). In the case of epithelial-to-mesenchymal transition, remodeling of the matrix is an important component of the
interactions: Enzymatic degradation of the ECM contributes to release of soluble factors and fragments of ECM that contain specific sequences that affect cell behavior. For example, when fibronectin binds only through its cell-binding domain, the cells are stimulated to produce ECM-
remodeling enzymes (Lukashev and Werb, 1998). There are at least three possible ways in which
such a process could be initiated: (1) Changes in expression of fibronectin (FN) receptors would
allow cells to bind fibronectin predominantly through its cell-binding domain and activate α5β1
interactions with the actin cytoskeleton, with subsequent transduction of signals that lead to up-
regulation of ECM-degrading enzymes. The secretion of these enzymes would start a positive feedback loop by degrading additional fibronectin to produce cell-binding fragments, which would
bind to α5β1, activate it, and in this way keep the specific event going. (2) Very localized release
of ECM-degrading enzymes could degrade FN into fragments containing only the cell-binding

domain, which would bind to α5β1 and initiate the positive feedback loop. (3) At a particular time during development, specific cells would produce spliced forms of FN that are capable of interacting only via their cell-binding domain. Binding of these fragments to α5β1 would trigger the feedback loop. This positive feedback loop and consequent runaway process of ECM degradation is advantageous locally during epithelial-to-mesenchymal transitions. However, during other developmental processes and during wound healing, there must be a signal that can break this circle and thereby bring it under control at the appropriate time and place. Without application of such a brake, these processes can lead to abnormal development, to wounds that do not heal, or to pathological situations such as tumor growth and invasion.

Although these three categories may not be exhaustive of the general types of cell–ECM interactions that occur during development and wound healing, they encapsulate the major interactions documented to date. Each category has its place in many developmental and repair events and they may operate in sequence. A compelling example of the latter is the epithelial-to-mesenchymal transition and morphogenesis of the neural crest cell system. These cells originate in the neural epithelium that occupies the crest of the neural folds. After the delamination event that separates the neural epithelium from the epidermal ectoderm, the folds fuse to form the tube (Martins-Green, 1988). At this time the NCCs occupy the dorsalmost portion of the tube, they are not covered by basal lamina, and the subepidermal space above them contains large amounts of fibronectin (Martins-Green and Erickson, 1987). Just before the NCCs emigrate from the neural tube, fibronectin appears between them, they separate from each other, and they migrate away, carrying fibronectin on their surfaces (Martins-Green, 1987, 1990). During the period of emigration at any particular level of the neural tube, the basal lamina is deposited progressively toward the crest from the sides of the tube (Martins-Green and Erickson 1986, 1987). NCC emigration terminates as deposition reaches the crest of the tube. The NCCs then migrate throughout the embryo following specific ECM-paved pathways, arriving at a wide variety of locations where they differentiate into many different phenotypes in response to external cues (Le Douarin, 1982; Erickson and Perris, 1993).

The appearance of fibronectin between the NCCs just before emigration must be the result of secretion by the adjacent cells or introduction from above after loss of cell–cell adhesions. In keeping with the cell–ECM interaction mechanism of Type III, either alternative could initiate a positive feedback loop and release the NCCs, leading to emigration. Enzymatic degradation of the stabilizing domain of fibronectin above the tube could cause enhanced secretion of specific enzymes by the NCCs in response to the effect of the cell-binding domain acting alone, leading to severing of adhesions between cells, and intrusion of fibronectin fragments containing the cell-binding domain. These fragments, in turn, would bind to adjacent cells and stimulate further enzymatic secretion that would be self-perpetuating. NCC emigration occurs in an anterior-to-posterior wave, hence once enzymatic activity was initiated in the head of the embryo, it could propagate posteriorly, triggering NCC emigration in a wave from head to tail.

Clearly some controlling event(s), must terminate NCC emigration at each location along the neural tube, or this scenario could empty the neural tube. Such an event has already been identified. At the time of NCC emigration, the ventral and lateral surfaces of the neural tube are covered by an intact basal lamina, which stabilizes the epithelium and separates it from the fibronectin layer around the tube (Martins-Green and Erickson, 1986). Following the delamination event at any one site (Martins-Green, 1988), the local dorsal portion of the neural tube is temporarily naked of basal lamina. During the few hours of NCC emigration at that site, basal lamina deposition progresses quickly up the side of the tube and terminates local emigration when it becomes complete over the crest of the tube (Martins-Green and Erickson, 1986, 1987). After they have emigrated from the neural tube, the NCCs are in an extracellular space filled with intact fibronectin and other ECM molecules that stimulate the focal adhesions of cell–ECM interactions of Type I, thereby providing the substrate for migration. On arrival at their final destination, further interactions of Type II stimulate differentiation into a wide range of phenotypes (Le Douarin, 1982; Erickson and Perris, 1993; Erickson and Reedy, 1998).

RELEVANCE FOR TISSUE ENGINEERING

The major task when designing tissue and organ replacements is to simulate nature closely. One of the best approaches toward this goal is to study how tissues and organs arise during embryogenesis and during normal processes of repair, and how those functions are maintained. When developing tissue replacements, one must consider several factors (Fig. 4.3):

1. *The need to avoid an immune response that can cause inflammation and/or rejection.* Ideally, one would like to identify and isolate stem cells capable of regenerating each one of the organs and tissues in the body. Selective stimulation and expansion of this cell population would allow regeneration *in vivo* without having to overcome an immune response. Alternatively, one would like to manipulate cells *in vitro* to make them more "universal" and thereby decrease the immune response. These latter cells would then differentiate in the presence of an environment conducive to expression of the appropriate phenotype.

2. *The need to create the proper substrate for cell survival and differentiation.* One of the strategies to fulfill this goal is the use of biocompatible implants composed of extracellular matrix molecules, or the appropriate biodegradable matrix, seeded with autologous cells, "universal" cells, or heterologous cells in conjunction with immunosuppressing drugs. Addition of growth and differentiation factors to these cell-containing matrices as well as agonists or antagonists that favor cell–ECM interactions can potentially increase the rate of successful tissue replacement. One example in which the knowledge obtained in studies of cell–ECM interactions has proved to be useful in tissue engineering was the discovery that most integrins bind to their ECM ligands via the tripeptide RGD. This small sequence of amino acids has been used as an agonist to make synthetic implants more biocompatible and to allow the development of tissue structure, or as an antagonist to prevent or moderate unwanted cell-ECM interactions (Tschopp *et al.,* 1994a; Cheng *et al.,* 1994; Glass *et al.,* 1994). An example of the latter is the use of these peptides to modulate platelet aggregation and formation of thrombi during reconstructive surgery (Tschopp *et al.,* 1993, 1994b; Cheng *et al.,* 1994; Glass *et al.,* 1994).

ECM-derived matrices also have been used for artificial corneas, for skin, for cartilage and bone replacement, and to enhance nerve regeneration (Bellamkonda *et al.,* 1995; Bouhadir and Mooney, 1998; Kim and Mooney, 1998). In the case of skin, patients with extensive burns have been treated with matrices of chondroitin sulfate and collagen (which support the formation of connective tissue and blood vessels) that are covered with an artificial material that prevents dehydration. Three weeks after this transplant, when the connective tissue is well established and vascularized, the superficial layer is removed and replaced with a thin layer of epithelium, which now rests on a substrate that supports the ultimate differentiation of the epidermal phenotype (Langer and Vacanti, 1993). Alternatively, the matrix is seeded with epidermal cells expanded in culture from very small skin grafts taken from the patient.

3. *The need to provide the appropriate environmental conditions for tissue maintenance.* At this stage it is crucial to maintain a balanced environment with the appropriate cues for maintenance of specific cell function(s). Unfortunately, little is known about the cross-talk between cells and

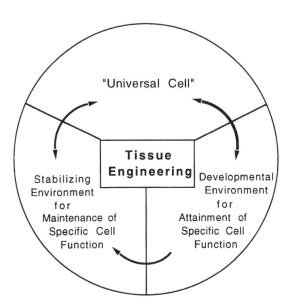

Fig. 4.3. Conceptualization of the interactions of an idealized "universal" cell and environments in which it is conditioned to a particular function (developmental environment) and maintained in that function (stabilizing environment).

ECM under such "normal" conditions. It is important to realize that such stasis on the level of a tissue is achieved via tissue remodeling—the dynamic equilibrium between cells and their environment.

Because organ transplantation is one of the least cost-effective therapies and is not always available, tissue engineering offers hope for more consistent and rapid treatment of those in need of replacement of body parts, and therefore it has the potential of improving the quality of life at a more affordable price. The selected examples presented above illustrate that to achieve this goal there is a need for continuous study of cells and how they interact with their environment. The present challenge is to make the leap in technology that will allow us to sort out complex biological phenomena and to perform simulations of specific biological processes that will guide us in planning subsequent experimental protocols. With the advent of bioinformatics it might be possible for biologists to integrate the vast body of information obtained so far and to make predictions for the future. In this context, a team at the University of Connecticut is developing a promising program called "The Virtual Cell" for modeling cellular processes (Schaff and Loew, 1999).

In summary, the intense research of the past decade on the ECM molecules and on their interactions with cell surface receptors has led to realization that these interactions are many and complex, provide for modulation of fundamental events during development and wound repair, and are crucial for the maintenance of the differentiated phenotype and tissue homeostasis.

REFERENCES

Adams, J. C., and Watt, F. M. (1993). Regulation of development and differentiation by the extracellular matrix. *Development (Cambridge, UK)* **117**, 1183–1198.

Akiyama, S. K., Aota, S., and Yamada, K. M. (1995). Function and receptor specificity of a minimal 20 kilodalton cell adhesive fragment of fibronectin. *Cell Adhes. Commun.* **3**, 13–25.

Alexander, C. M., Howard, E. W., Bissell, M. J., and Werb, Z. (1996). Rescue of mammary epithelial cell apoptosis and entactin degradation by a tissue inhibitor of metalloproteinases-1 transgene. *J. Cell Biol.* **135**, 1669–1677.

Aota, S., Nomizu, M., and Yamada, K. M. (1994). The short amino acid sequence Pro-His-Ser-Arg-Asn in human fibronectin enhances cell adhesive function. *J. Biol. Chem.* **269**, 24756–24761.

Arora, P. D., Narani, N., and McCulloch, C. A. (1999). The compliance of collagen gels regulates transforming growth factor-beta induction of alpha-smooth muscle actin in fibroblasts. *Am. J. Pathol.* **154**, 871–882.

Asah, A. S., Liu, I., Brocetti, F. M., Barnwell, J. W., Kwakye-Berko, F., Dokun, A., Goldberg, J., and Pernambuco, M. (1993). Analysis of CD36 binding domains: Ligand specificity controlled by dephosphorylation of an ectodomain. *Science* **262**, 1436–1440.

Baldwin, H. S. (1996). Early embryonic development. *Cardiovasc. Res.* **31**, E34–E45.

Balkwill, F. (1998). The molecular and cellular biology of the chemokines. *J. Viral Hepatitis* **5**, 1–14.

Barcellos-Hoff, M. H., Aggeler, J., Ram, T. G., and Bissell, M. J. (1989). Functional differentiation and alveolar morphogenesis of primary mammary cultures on reconstituted basement membrane. *Development (Cambridge, UK)* **105**, 223–235.

Bell, E., Ivarsson, B., and Merrill, C. (1979). Production of a tissue-like structure by contraction of collagen lattices by human fibroblasts of different proliferative potential *in vitro. Proc. Natl. Acad. Sci. U.S.A.* **76**, 1274–1278.

Bellamkondra, R., Ranieri, J. P., Bouche, N., and Aebischer, P. (1995). Hydrogel-based three-dimensional matrix for neural cells. *J. Biomed. Mater. Res.* **29**, 633–671.

Ben-Ze'ev, A., Robinson, G. F., Bucher, N.L.R., and Farmer, S. R. (1988). Cell–cell and cell–matrix interactions differentially regulate the expression of hepatic and cytoskeletal genes in primary cultures of rat hepatocytes. *Proc. Natl. Acad. Sci. U.S.A.* **85**, 2161–2165.

Bernfield, M. R., Cohn, R. H., and Banerjee, S. D. (1973). Glycosaminoglycans and epithelial organ formation. *Am. Zool.* **13**, 1067–1083.

Bernfield, M. R., Kokenyesi, R., Kato, M., Hinkes, M. T., Spring, J., Gallo, R. L., and Lose, E. J. (1992). Biology of the syndecans: A family of transmembrane heparan sulfate proteoglycans. *Annu. Rev. Cell Biol.* **8**, 365–393.

Bissell, D. M. (1998). Hepatic fibrosis as wound repair: A progress report. *J. Gastroenterol.* **33**, 295–302.

Bissell, M. J. (1998). Glandular structure and gene expression. Lessons from the mammary gland. *Ann. N. Y. Acad. Sci.* **842**, 1–6.

Bissell, M. J., Hall, H. G., and Parry, G. (1982). How does the extracellular matrix direct gene expression? *J. Theor. Biol.* **99**, 31–68.

Bitterman, P. B., Remard, S. I., Adelberg, S., and Crystal, R. G. (1983). Role of FN as a growth factor for fibroblasts. *J. Cell Biol.* **97**, 1925–1932.

Blann, A., and Seigneur, M. (1997). Soluble markers of endothelial cell function. *Clin. Hemorheol. Microcirc.* **17**, 3–11.

Blaschke, R. J., Howllett, A. R., Desprez, P.-Y., Peterson, O. W., and Bissell, M. J. (1994). Cell differentiation by extracellular matrix components. *In* "Methods in Enzymology" (Ruoslahti, ed.), vol. **245**, pp. 535–556. Academic Press, San Diego, CA.

Bonewald, L. F. (1999). Regulation and regulatory activities of transforming growth factor beta. *Crit. Rev. Eukaryotic Gene Expression* **9**, 33–44.

Boucaut, J. V., Darribere, T., Boulekbache, H., and Thiery, J. P. (1984a). Prevention of gastrulation but not neurulation by antibodies to fibronectin in amphibian embryos. *Nature (London)* **307,** 364–367.

Boucaut, J. C., Darribere, T., Poole, T. J., Aoyama, H., Yamada, K. M., and Thiery, J. P. (1984b). Biologically active synthetic peptides as probes of embryonic development: A competitive peptide inhibitor of fibronectin function inhibits gastrulation in amphibian embryos and neural crest cell migration in avian embryos. *J. Cell Biol.* **99,** 1822–1830.

Boudreau, N.., Myers, C., and Bissell, M. J. (1995). From lamini to lamin: Regulation of tissue-specific gene expression by the ECM. *Trends Cell Biol.* **5,** 1–4.

Boudreau, N., Werb, Z., and Bissell, M. J. (1996). Suppression of apoptosis by basement membrane requires three-dimensional tissue organization and withdrawal from the cell cycle. *Proc. Natl. Acad. Sci. U.S.A.* **93,** 3509–3513.

Bouhadir, K. H., and Mooney, D. J. (1998). *In vitro* and *in vivo* models for the reconstruction of intracellular signaling. *Ann. N. Y. Acad. Sci.* **842,** 188–194.

Briesewitz, R., Kern, A., and Marcantonio, E. E. (1995). Assembly and function of integrin receptors is dependent on opposing alpha and beta cytoplasmic domains. *Mol. Biol. Cell* **6,** 997–1010.

Bronner-Fraser, M. (1985). Alterations in neural crest migration by a monoclonal antibody that affects cell adhesion. *J. Cell Biol.* **101,** 610–617.

Bronner-Fraser, M. (1986). An antibody to a receptor for fibronectin and laminin perturbs cranial neural crest development *in vivo. Dev. Biol.* **117,** 528–536.

Brooks, P. C., Clark, R. A., and Cheresh, D. A. (1994). Requirement of vascular integrin $\alpha_v\beta_3$ for angiogenesis. *Science* **264,** 569–571.

Brown, T. A., Bouchard, T., St. John, T., Wayner, E., and Carter, W. G. (1991). Human keratinocytes express a new CD44 core protein (CD44E) as a heparan-sulfate intrinsic membrane proteoglycan with additional exons. *J. Cell Biol.* **113,** 207–221.

Carey, D. J. (1997). Syndecans: Multifunctional cell-surface co-receptors. *Biochem. J.* **327,** 1–16.

Carey, D. J., Bendt, K. M., and Stahl, R. C. (1996). The cytoplasmic domain of syndecan-1 is required for cytoskeleton association but not detergent insolubility. Identification of essential cytoplasmic domain residues. *J. Biol. Chem.* **271,** 15253–15260.

Caron, J. M. (1990). Induction of albumin gene transcription in hepatocytes by extracellular matrix. *Mol. Cell Biol.* **10,** 1239–1243.

Castellot, J. J., Wright, T. C., and Karnovsky, M. Y. (1987). Regulation of vascular smooth muscle cell growth by heparin and heparan sulfates. *Semin. Thromb. Hemostasis* **13,** 489–503.

Cheng, L.-H., and Bissell, M. J. (1989). A novel regulatory mechanism for whey acidic protein gene expression. *Cell Regul.* **1,** 45–54.

Cheng, S., Craig, W. S., Mullen, D., Tschopp, J. F., Dixon, D., and Pierschbacher, M. D. (1994). Design and synthesis of novel cyclic RGD-containing peptides as highly potent and selective integrin alpha IIb beta 3 antagonists. *J. Med. Chem.* **37,** 1–8.

Chiquet-Ehrismann, R., Matsuoka, Y., Hofer, U., Spring, J., Bernasconi, C., and Chiquet, M. (1991). Tenascin variants: *Differential binding to fibronectin and distinct distribution in cell cultures and tissues. Cell Regul.* **2,** 927–938.

Clark, R.A.F., ed. (1996). "The Molecular and Cellular Biology of Wound Repair." Plenum, New York.

Cockerill, G. W., Gamble, J. R., and Vadas, M. A. (1995). Angiogenesis: Models and modulators. *Int. Rev. Cytol.* **159,** 113–161.

Cotman, C. W., Hailer, N. P., Pfister, K. K., Soltesz, I., and Schachner, M. (1998). Cell adhesion molecules in neural plasticity and pathology: Similar mechanisms, distinct organizations? *Prog. Neurobiol.* **55,** 659–669.

Cotman, S. L., Halfter, W., and Cole, G. J. (1999). Identification of extracellular matrix ligands for the heparan sulfate proteoglycan agrin. *Exp. Cell Res.* **249,** 54–64.

Couchman, J. R., Kapoor, R., Sthanam, M., and Wu, R. R. (1996). Perlecan and basement membrane-chondroitin sulfate proteoglycan (bamacan) are two basement membrane chondroitin/dermatan sulfate proteoglycans in the Engelbreth–Holm–Swarm tumor matrix. *J. Biol. Chem.* **271,** 9595–9602.

Coucouvanis, E., and Martin, G. R. (1995). Signals for death and survival: A two-step mechanism for cavitation in the vertebrate embryo. *Cell (Cambridge, Mass.)* **83,** 279–287.

Cox, E. A., and Huttenlocher, A. (1998). Regulation of integrin-mediated adhesion during cell migration. *Microsc. Res. Tech.* **43,** 412–419.

Culav, E. M., Clark, C. H., and Merrilees, M. J. (1999). Connective tissues: Matrix composition and its relevance to physical therapy. *Phys. Ther.* **79,** 308–319.

Damsky, C. H., and Werb, Z. (1992). Signal transduction by integrin receptors for extracellular matrix: Cooperative processing of extracellular information. *Curr. Opin. Cell Biol.* **4,** 772–781.

Darribere, T., Yamada, K. M., Johnson, K. E., and Boucaut, J. C. (1988). The 140kDa fibronectin receptor complex is required for mesodermal cell adhesion during gastrulation in the amphibian *Pleurodeles waltlii. Dev. Biol.* **126,** 182–194.

Dawson, R. A., Goberdhan, N. J., Freedlander, E., and MacNeil, S. (1996). Influence of extracellular matrix proteins on human keratinocyte attachment, proliferation and transfer to a dermal wound model. *Burns* **22,** 93–100.

Dean, J. W., III, Chanrasekaran, S., and Tanzer, M. L. (1990). A biological role of the carbohydrate moieties of laminin. *J. Biol. Chem.* **265,** 12553–12562.

Dedhar, S. (1999). Integrins and signal transduction. *Curr. Opin. Hematol.* **6,** 37–43.

Desmouliere, A., and Gabbiani, G. (1994). Modulation of fibroblastic cytoskeletal features during pathological situations: The role of extracellular matrix and cytokines. *Cell Motil. Cytoskel.* **29,** 195–203.

Desmouliere, A., Rubbia-Brandt, L., Abdiu, A., Walz, T., Macieira-Coelho, A., and Gabbiani, G. (1992). Heparin induces α-smooth muscle actin expression in cultured fibroblasts and in granulation tissue myofibroblasts. *Lab. Invest.* **67,** 716–726.

Desmouliere, A., Redard, M., Darby, I., and Gabbiani, G. (1995). Apoptosis mediates the decrease in cellularity during the trabssition between granulation tissue and scar. *Am. J. Pathol.* **146**, 56–66.

Desprez, P. Y., Roskelley, C., Campisi, J., and Bissell, M. J. (1993). Isolation of functional cell lines from a mouse mammary epithelial cell strain: The importance of basement membrane and cell–cell interaction. *Mol. Cell Differ.* **1**, 99–110.

De Vries, H.J.C., Middlekoop, E., Mekkes, J. R., Dutrieux, R. P., Wildevuur, C. H., and Westerhof, W. (1994). Dermal regeneration in native non-cross linked collagen sponges with different extracellular molecules. *Wound Repair Regeneration* **2**, 37–47.

Dockery, P., Khalid, J., Sarani, S. A., Bulut, H. E., Warren, M. A., Li, T. C., and Cooke, I. D. (1998). Changes in basement membrane thickness in the human endometrium during the luteal phase of the menstrual cycle. *Hum. Reprod. Update* **4**, 486–495.

Ekblom, P., Ekblom, M., Fecker, L., Klein, G., Zhang, H.-Y., Kadoya, Y., Chu, M.-L., Mayer, U., and Timpl, R. (1994). Role of mesenchymal nidogen for epithelial morphogenesis *in vitro*. *Development (Cambridge, UK)* **120**, 2003–2014.

Epperlein, H. H., and Lofberg, J. (1996). What insights into the phenomena of cell fate determination and cell migration has the study of the urodele neural crest provided? *Int. J. Dev. Biol.* **40**, 695–707.

Erickson, C. A., and Perris, R. (1993). The role of cell–cell and cell–matrix interactions in the morphogenesis of the neural crest. *Dev. Biol.* **159**, 60–74.

Erickson, C. A., and Reedy, M. V. (1998). Neural crest development: The interplay between morphogenesis and cell differentiation. *Curr. Top. Dev. Biol.* **40**, 177–209.

Faull, R. J., and Ginsberg, M. H. (1996). Inside-out signaling through integrins. *J. Am. Soc. Nephrol.* **7**, 1091–1097.

Ffrench-Constant, C., and Hynes, R. O. (1989). Alternative splicing of fibronectin is temporally and spatially regulated in the chicken embryo. *Development (Cambridge, UK)* **106**, 375–388.

Flaumenhaft, R., and Rifkin, D. B. (1992). The extracellular regulation of growth factor action. *Mol. Biol. Cell* **3**, 1057–1065.

Fluck, J., Querfeld, C., Cremer, A., Niland, S., Krieg, T., and Sollberg, S. (1998). Normal human primary fibroblasts undergo apoptosis in three-dimensional contractile collagen gels. *J. Invest. Dermatol.* **110**, 153–157.

Fornaro, M., and Languino, L. R. (197). Alternatively spliced variants: A new view of the integrin cytoplasmic domain. *Matrix Biol.* **16**, 185–93.

Fuchs, E., Dowling, J., Segré, J., Lo, S. H., and Yu, Q. C. (1997). Integrators of epidermal growth and differentiation: Distinct functions for beta 1 and beta 4 integrins. *Curr. Opin. Genet. Dev.* **7**, 672–682.

Fukai, F., Mashimo, M., Akiyama, K., Goto, Y., Tanuma, S., and Katayama, T. (1998). Modulation of apoptotic cell death by extracellular matrix proteins and a fibronectin-derived antiadhesive peptide. *Exp. Cell Res.* **242**, 92–99.,

Gailit, J., and Ruoslahti, E. (1988). Regulation of the fibronectin receptor affinity by divalent cations. *J. Biol. Chem.* **263**, 12927–12933.

Gailit, J., Welch, M. P., and Clark, R.A.F. (1994). TGFβ1 stimulates expression of keratynocytes during reepithelialization of cutaneous wounds. *J. Invest. Dermatol.* **103**, 221–227.

Giancotti, F. G. (1997). Integrin signaling: Specificity and control of cell survival and cell cycle progression. *Curr. Opin. Cell Biol.* **9**, 691–700.

Glass, J., Blevitt, J., Dickerson, K., Pierschbacher, M., and Craig, W. S. (1994). Cell attachment and motility on materials modified by surface-active RGD-containing peptides. *Ann. N. Y. Acad. Sci.* **745**, 177–186.

Grant, D. S., Kinsella, J. L., Fridman, R., Aurbach, R., Piasecki, B. A., Yamada, Y., Zain, M., and Kleinman, H. K. (1992). Interaction of endothelial cells with a laminin A chain peptide (SIKVAV) *in vitro* and induction of angiogenic behavior *in vivo*. *J. Cell. Physiol.* **123**, 117–130.

Grant, M. B., Caballero, S., Bush, S. M., and Spoerri, P. E. (1998). Fibronectin fragments modulate human retinal capillary cell proliferation and migration. *Diabetes* **47**, 1335–1340.

Greenhalgh, D. G. (1998). The role of apoptosis in wound healing. *Int. J. Biochem. Cell Biol.* **30**, 1019–1030.

Greenwalt, D. E., Lipsky, R. H., Ockenhouse, C. F., Ikeda, H., Tandon, N. N., and Jamieson, G. A. (1992). Membrane glycoprotein CD36: A review of its roles in adherence, signal transduction, and transfusion medicine. *Blood* **80**, 1105–1115.

Grinell, F. (1994). Fibroblasts, myofibroblasts and wound contraction. *J. Cell Biol.* **124**, 401–404.

Grinell, F., Zhu, M., Carlson, M. A., and Abrams, J. M. (1999). Release of mechanical tension triggers apoptosis of human fibroblasts in a model of regressing granulation tissue. *Exp. Cell Res.* **248**, 608–619.

Gross, J. (1996). Getting to mammalian wound repair and amphibian limb regeneration: A mechanistic link in the early events. *Wound Repair Regeneration* **4**, 190–202.

Gullberg, D., and Ekblom, P. (1995). Extracellular matrix and its receptors during development. *Int. J. Dev. Biol.* **39**, 845–854.

Hanahan, D. (1997). Signaling vascular morphogenesis and maintenance. *Science* **277**, 48–50.

Hardingham, T. E., and Fosang, A. J. (1992). Proteoglycans: Many forms and many functions *FASEB J.* **6**, 861–870.

Hauschka, S. D., and Konigsberg, I. R. (1966). The influence of collagen on the development of muscle colonies. *Proc. Natl. Acad. Sci. U.S.A.* **55**, 119–126.

Hay, E. D. (1977). Interaction between the cell surface and extracellular matrix in corneal development. *In* "Cell and Tissue Interactions" (J. W. Lash and M. M. Burger, eds.), pp. 115–137. Raven Press, New York.

Hemler, M. E. (1998). Integrin associated proteins. *Curr. Opin. Cell Biol.* **10**, 578–585.

Hildebrand, J. D., Schaller, M. D., and Parsons, T. (1995). Paxillin, a tyrosine phosphorylated focal adhesion-associated protein binds to the carboxyl terminal domain of focal adhesion kinase. *Mol. Biol. Cell* **6**, 637–647.

Howlett, A. R., and Bissell, M. J. (1993). The influence of tissue microenvironment (stroma and extracellular matrix) on the development and function of mammary epithelium. *Epithelial Cell Biol* **2**, 79–89.

Huang, M.-M., Bolen, J. B., Barnwell, J. W., Shattil, S. J., and Brugge, J. S. (1991). Membrane glycoprotein IV (CD36)

is physically associated with the Fyn, Lyn, and Yes protein-tyrosine kinases in human platelets. *Proc. Natl. Acad. Sci. U.S.A.* **88**, 7844–7848.

Humphries, M. J., Mould, A. P., and Yamada, K. M. (1991). Matrix receptors in cell migration. *In* "Receptors for Extracellular Matrix" (J. A. McDonald and R. P. Mecham, eds.), pp. 195–253.

Hynes, R. O. (1987). Integrins: A family of cell surface receptors. *Cell (Cambridge, Mass.)* **48**, 549–554.

Hynes, R. O. (1992). Integrins: Versatility, modulation, and signaling cell adhesion. *Cell (Cambridge, Mass.)* **69**, 11–25.

Hynes, R. O. (1999). The dynamic dialogue between cells and matrices: Implications of fibronectin's elasticity. *Proc. Natl. Acad. Sci. U.S.A.* **96**, 2588–2590.

Ilan, N., Mahooti, S., and Madri, J. A. (1998). Distinct signal transduction pathways are utilized during the tube formation and survival phase of *in vitro* angiogenesis. *J. Cell Sci.* **111**, 3621–3631.

Ingber, D. (1991). Extracellular matrix and cell shape: Potential control points for inhibition of angiogenesis. *J. Cell. Biochem.* **47**, 236–241.

Jalkehen, M., Jalkanen, S., and Bernfield, M. (1991). Binding of extracellular effector molecules by cell surface proteoglycans. *In* "Receptors for Extracellular Matrix" (J. A. McDonald and R. P. Mecham, eds.), pp. 1–38.

Juhsz, I., Murphy, G. F., Yan, H.-C., Herlyn, M., and Albelda, S. M. (1993). Regulation of extracellular matrix proteins and integrin cell-substrate adhesion receptors on epithelium during cutaneous human wound healing *in vivo*. *Am. J. Pathol.* **143**, 1458–1469.

Kagami, S., Kondo, S., Loster, K., Reutter, W., Urushihara, M., Kitamura, A., Kobayashi, S., and Kuroda, Y. (1998). Collagen type I modulates the platelet-derived growth factor (PDGF) regulation of the growth and expression of beta1 integrins by rat mesangial cells. *Biochem. Biophys. Res. Commun.* **252**, 728–732.

Kamei, M., Kawasaki, A., and Tano, Y. (1998). Analysis of extracellular matrix synthesis during wound healing of retinal pigment epithelial cells. *Microsc. Res. Tech.* **42**, 311–316.

Kapila, Y. L., Lancero, H., and Johnson, P. W. (1998). The response of periodontal ligament cells to fibronectin. *J. Periodontol.* **69**, 1008–1019.

Kettritz, R., Xu, Y. X., Kerren, T., Quass, P., Klein, J. B., Luft, F. C., and Haller, H. (1999). Extracellular matrix regulates apoptosis in human neutrophils. *Kidney Int.* **55**, 562–571.

Khwaja, A., Rodriguez-Viciana, P., Wennstrom, S., Warne, P. H., and Downward, J. (1997). Matrix adhesion and Ras transformation both activate a phosphoinositide 3-OH kinase and protein kinase B/Akt cellular survival pathway. *EMBO J.* **16**, 2783–2793.

Kim, B. S., and Mooney, D. J. (1998). Engineering smooth muscle tissue with a predefined structure. *J. Biomed. Mater. Res.* **41**, 322–332.

Kim, C. W., Goldberger, O. A., Gallo, R. L., and Bernfield, M. (1994). Members of the syndecan family of heparan proteoglycans are expressed in distinct cell-, tissue-, and development-specific patterns. *Mol. Biol. Cell* **5**, 797–805.

Kirkland, G., Paizis, K., Wu, L. L., Katerelos, M., and Power, D. A. (1998). Heparin-binding EGF-like growth factor mRNA is upregulated in the peri-infarct region of the remnant kidney model: *In vitro* evidence suggests a regulatory role in myofibroblast transformation. *J. Am. Soc. Nephrol.* **9**, 1464–1473.

Kleinman, H. K., McGarvey, M. L., Liotta, L. A., Robey, P. G., Tryggvason, K., and Martin, G. R. (1982). Isolation and characterization of type IV procollagen, laminin, and heparan sulfate proteoglycan from EHS sarcoma. *Biochemistry* **21**, 6188–6193.

Kubota, Y., Kleinman, H. K., Martin, G. R., and Lawley, T. J. (1988). Role of laminin and basement membrane in the morphological differentiation of human endothelial cells into calpillary-like structures. *J. Cell Biol.* **107**, 1589–1598.

Kumar, C. C. (1998). Signaling by integrin receptors. *Oncogene* **17**, 1365–1363.

Kumar, R., Yoneda, J., Bucana, C. D., and Fidler, I. J. (1998). Regulation of distinct steps of angiogenesis by different angiogenic molecules. *Int. J. Oncol.* **12**, 749–757.

Lamszus, K., Joseph, A., Jin, L., Yao, Y., Chowdhury, S., Fuchs, A., Polverini, P. J., Goldberg, I. D., and Rosen, E. M. (1996). Scatter factor binds to thrombospondin and other extracellular matrix components. *Am. J. Pathol.* **149**, 805–819.

Langer, R., and Vacanti, J. P. (1993). Tissue engineering. *Science* **260**, 920–925.

La Rochelle, W. J., May-Shiroff, M., Robbins, K. C., and Aaronson, S. A. (1991). A novel mechanism regulating growth factor association with the cell surface: Identification of a PDGF retention domain. *Genes Dev.* **5**, 1191–1199.

Laurie, G. W., Horikoshi, S., Killen, P. D., Segui-Real, B., and Yamada, Y. (1989). *In situ* hybridization reveals temporal and spatial changes in cellular expression of mRNA for a laminin receptor, laminin, and basement membrane (type IV) collagen in the developing kidney. *J. Cell Biol.* **109**, 1351–1362.

Le Douarin, N. (1982). "The Neural Crest." Cambridge University Press, Cambridge, UK.

Li, Q. J., Vaingankar, S., Sladek, F., and Martins-Green, M. (2000). Novel nuclear target for thrombin: Activation of the Elk1 transcription factor leads to chemokine gene expression. Submitted for publication.

Lin, C. Q., and Bissell, M. J. (1993). Multi-faceted regulation of cell differentiation by extracellular matrix. *FASEB J.* **7**, 737–743.

Lin, C. Q., Dempsey, P. J., Coffey, R. J., and Bissell, M. J. (1995). Extracellular matrix regulates whey acidic protein gene expression by suppression of TGF-alpha in mouse mammary epithelial cells: Studies in culture and in transgenic mice. *J. Cell Biol.* **129**, 1115–1126.

Lipsky, R. H., Eckert, D. M., Tang, Y., and Ockenhouse, C. F. (1997). The carboxyl-terminal cytoplasmic domain of CD36 is required for oxidized low-density lipoprotein modulation of NF-kappaB activity by tumor necrosis factor-alpha. *Recept. Signal Transduct.* **7**, 1–11.

Liu, J.-K., Di Persio, M. C., and Zaret, K. S. (1991). Extracellular signals that regulate liver transcription factors during hepatic differentiation *in vitro*. *Mol. Cell Biol.* **11**, 773–784.

Locati, M., and Murphy, P. M. (1999). Chemokines and chemokine receptors: Biology and clinical relevance in inflammation and AIDS. *Annu. Rev. Med.* **50**, 425–440.

Lofberg, J., Perris, R., and Epperlein, H. H. (1989). Timing in the regulation of neural crest cell migration: Retarted "maturation" of regional extracellular matrix inhibits pigment cell migration in embryos of the white Axolotl mutant. *Dev. Biol.* **131**, 168–181.

Lukacs, N. W., and Kunkel, S. L. (1998). Chemokines and their role in disease. *Int. J. Clin. Lab. Res.* **28**, 91–95.

Lukashev, M. E., and Werb, Z. (1998). ECM signalling: Orchestrating cell behaviour and misbehaviour. *Trends Cell Biol.* **8**, 437–441.

Lund, L. R., Romer, J., Thomasset, N., Solberg, H., Pyke, C., Bissell, M. J., Dano, K., and Werb, Z. (1996). Two distinct phases of apoptosis in mammary gland involution: Proteinase-independent and -dependent pathways. *Development (Cambridge, UK)* **122**, 181–193.

Madri, J. A., Pratt, B. M., and Tucker, A. M. (1988). Phenotypic modulation of endothelial cells by transforming growth factor-β depends upon the composition and organization of the extracellular matrix. *J. Cell Biol.* **106**, 1375–1384.

Mainiero, F., Murgia, C., Wary, K. K., Curatola, A. M., Pepe, A., Blumemberg, M., Westwick, J. K., Der, C. J., and Giancotti, F. G. (1997). The coupling of alpha6beta4 integrin to Ras-MAP kinase pathways mediated by Shc controls keratinocyte proliferation. *EMBO J.* **16**,2365–2375.

Majack, R. A., Cook, S. C., and Bornstein, P. (1986). Regulation of vascular smooth muscle cell growth by the extracellular matrix: An autocrine role for thrombospondin. *PNAS* **83**, 9050–9054.

Martins-Green, M. (1987). Ultrastructural and immunolabeling studies of the neural crest: Processes leading to neural crest cell emigration. Ph.D. Dissertation, University of California at Davis.

Martins-Green, M. (1988). Origin of the dorsal surface of the neural tube by progressive delamination of epidermal ectoderm and neuroepithelium: Implications for neurulation and neural tube defects. *Development (Cambridge, UK)* **103**, 687–706.

Martins-Green, M. (1990). Transmission electron microscopy and immunolabelling of tissues for light and electron microscopy. *In* "The Postimplantation Mammalian Embryo: A Practical Approach" (A. Copp, ed.), pp. 127–154. IRL Press, Oxford.

Martins-Green, M. (1999). The role of 9E3/cCAF in wound healing and disease. *In* "Cytokine Data Base" (J. Oppenheim and S. Durum, eds.). Academic Press, London (in press).

Martins-Green, M., and Bissell, M. J. (1990). Localization of 9E3/CEF4 in avian tissues: Expression is absent in RSV-induced tumors but is stimulated by injury. *J. Cell Biol.* **110**, 581–595.

Martins-Green, M., and Bissell, M. J. (1995). Cell–extracellular matrix interactions in development. *Semin. Dev. Biol.* **6**, 149–159.

Martins-Green, M., and Erickson, C. A. (1986). Development of neural tube basal lamina during neurulation and neural crest cell emigration in the trunk of the mouse embryo. *J. Embryol. Exp. Morph.* **98**, 219–236.

Martins-Green, M., and Erickson, C. A. (1987). Basal lamina is not a barrier to neural crest cell emigration: Documentation by TEM and by immunofluorescent and immunogold labeling. *Development (Cambridge, UK)* **101**, 517–533.

Martins-Green, M., and Feugate, J. E. (1998). The 9E3/CEF4 gene product is a chemotactic and angiogenic factor that can initiate the wound healing cascade *in vivo*. *Cytokines* **10**, 522–535.

Martins-Green, M., and Hanafusa, H. (1997). The 9E3/CEF4 gene and its product the chicken chemotactic and angiogenic factor (cCAF): Potential roles in wound healing and tumor development. *Cytokines Growth Factor Rev.* **8**, 219–230.

Martins-Green, M., and Kelly, T. (1998). The chicken chemotactic and angiogenic factor (cCAF): Its angiogenic properties reside in the C-terminus of the molecule. *Cytokines* **10**, 819–830.

Martins-Green, M., Aotaki-Keen, A., Hjelmeland, L., and Bissell, M. J. (1992). The 9E3 protein: Immunolocalization *in vivo* and evidence for multiple forms in culture. *J. Cell Sci.* **101**, 701–707.

Martins-Green, M., Stoeckle, M., Wimberly, S., Hampe, A., and Hanafusa, H. (1996). Kinetics of gene activation and protein secretion of the small cytokine 9E3/CEF4 and its post-secretion processing. *Cytokines* **8**, 448–459.

McCaffrey, T. A., Falcone, D. J., Brayton, C. F., Agarwal, L. A., Welt, F. G., and Wekler, B. B. (1989). Transforming growth factor β activity is potentiated by heparin via dissociation of the TGFβ/α-2-macroglobulin inactive complex. *J. Cell Biol.* **109**, 441–448.

McFall, A. J., and Rapraeger, A. C. (1998). Characterization of the high affinity cell-binding domain in the cell surface proteoglycan syndecan-4. *J. Biol. Chem.* **273**, 28270–28276.

Mecham, R. P. (1991). Laminin receptors. *Annu. Rev. Cell Biol.* **7**, 71–91.

Miyake, K., Underhill, C. B., Lesley, J., and Kincade, P. W. (1990). Hyaluronate can function as a cell adhesion molecule and cd44 participates in hyaluronate recognition. *J. Exp. Med.* **172**, 69–75.

Moghal, N., and Neel, B. G. (1998). Integration of growth factor, extracellular matrix, and retinoid signals during bronchial epithelial cell differentiation. *Mol. Cell. Biol.* **18**, 6666–6678.

Montesano, R., Orci, L., and Vassalli, P. (1983). *In vitro* rapid organization of endothelial cells into capillary-like networks is promoted by collagen matrices. *J. Cell Biol.* **97**, 1648–1652.

Murgia, C., Blaikie, P., Kim, N., Dans, M., Petrie, H. T., and Giancotti, F. G. (1998). Cell cycle and adhesion defects in mice carrying a targeted deletion of the integrin beta4 cytoplasmic domain. *EMBO J.* **17**, 3940–3951.

Myers, C. A., Schmidhauser, C., Mellentin-Michelotti, J., Fragoso, G., Roskelley, C. D., Casperson, G., Mossi, R., Pujuguet, P., Hager, G., and Bissell, M. J. (1998). Characterization of BCE-1, a transcriptional enhancer regulated by prolactin and extracellular matrix and modulated by the state of histone acetylation. *Mol. Cell. Biol.* **18**, 2184–2195.

Naot, D., Sionov, R. V., and Ish-Shalom, D. (1997). CD44: Structure, function, and association with the malignant process. *Adv. Cancer Res.* **71**, 241–319.

Nathan, C., and Sporn, M. (1991). Cytokines in context. *J. Cell Biol.* **113**, 981–986.

Nicholson, L. J., and Watt, F. M. (1991). Decreased expression of fibronectin and the $\alpha_5\beta_1$ integrin during terminal differentiation of human keratinocytes. *J. Cell Sci.* **98**, 225–232.

Norrby, K. (1997). Angiogenesis: New aspects relating to its initiation and control. *Apmis* **105**, 417–437.

Ohki, K., and Kohashi, O. (1994). Laminin promotes proliferation of bone marrow-derived macrophages and macrophage cell lines. *Cell Struc. Func.* **19**, 63–71.

Pakianathan, D. R. (1995). Extracellular matrix proteins and leukocyte function. *J. Leukocyte Biol.* **57**, 699–702.

Panayotou, G., End, P., Aumailley, M., Timpl, R., and Engel, J. (1989). Domains of laminin with growth-factor activity. *Cell (Cambridge, Mass.)* **56**, 93–101.

Paulsson, M. (1988). The role of Ca^{2+} binding in the self-aggregation of laminin-nidogen complexes. *J. Biol. Chem.* **263**, 5425–5430.

Perris, R., von Boxberg, Y., and Lofberg, J. (1988). Local embryonic matrices determine region-specific phenotypes in neural crest cells. *Science* **241**, 86–89.

Petersen, O. W., Ronnov-Jessen, L., Howlett, A. R., and Bissell, M. J. (1993). Interaction with basement membrane serves to rapidly distinguish growth and differentiation pattern of normal and malignant human breast epithelial cells *Proc. Natl. Acad. Sci.* **89**, 9064–9088.

Pierschbacher, M. D., and Ruoslahti, E. (1984). Cell attachment activity of fibronectin can be duplicated by small synthetic fragments of the molecule. *Nature (London)* **309**, 30–33.

Pilarski, L. M., Pruski, E., Wizniak, J., Paine, D., Seeberger, K., Mant, M. J., Brown, C. B., and Belch, A. R. (1999). Potential role for hyaluronan and the hyaluronan receptor RHAMM immobilization and trafficking of hematopoietic progenitor cells, *Blood* **93**, 2918–2927.

Polunovsky, V. A., Chen, B., Henke, C., Snover, D., Wendt, C., Ingbar, D. H., and Bitterman, P. B. (1993). Role of mesenchymal cell death in lung remodeling after injury. *J. Clin. Invest.* **92**, 388–397,

Poole, T., and Thiery, J. P. (1986). Antibodies and a synthetic peptide that block cell-fibronectin adhesion arrest neural crest cell migration *in vivo. Prog. Clin. Bio. Res.* 217B, 235–238.

Pozzi, A., Wary, K. K., Giancotti, F. G., and Gardner, H. A. (1998). Integrin alpha1beta1 mediates a unique collagen-dependent proliferation pathway *in vivo. J. Cell Biol.* **142**, 587–594.

Pullman, W. E., and Bodmer, W. F. (1992). Cloning and characterization of a gene that regulates cell adhesion. *Nature (London)* **356**, 529–532.

Rana, B., Mischoulon, D., Xie, Y., Bucher, N. L., and Farmer, S. R. (1994). Cell–extracellular matrix interactions can regulate the switch between growth and differentiation in rat hepatocytes: Reciprocal expression of C/EB Pα and immediate-early growth response transcription factors. *Mol. Cell Biol.* **14**, 5858–5869.

Rapraeger, A., Jalkanen, M., Endo, E., and Bernfield, M. (1987). Integral membrane proteoglycans as matrix receptors: Role in cytoskeleton and matrix assembly at the epithelial cell surface. *In* "Biology of Extracellular Matrix: A Series. Biology of Proteoglycans" (T. N. Wright and R. P. Mecham, eds.), pp. 129–154. Academic Press, Orlando, FL.

Roberts, R., Gallagher, J., Spooncer, E., Allen, T. D., Bloomfield, F., and Dexter, T. M. (1988). Heparan sulphate bound growth factors: A mechanism for stromal cell mediated haemopoiesis. *Nature (London)* **332**, 376–378.

Robinson, S. D., Silberstein, G. B., Roberts, A. B., Flanders, K. C., and Daniel, C. W. (1991). Regulated expression and growth inhibitory effects of transforming growth factor-beta isoforms in mouse mammary gland development. *Development (Cambridge, UK)* **113**, 867–878.

Rosales, C., and Juliano, R. L. (1995). Signal transduction by cell adhesion receptors in leukocytes *J. Leukocyte Biol.* **57**, 189–198.

Roskelly, C. D., Desprez, P. Y., and Bissell, M. J. (1994). Extracellular matrix-dependent tissue-specific gene expression in mammary epithelial cells requires both physical and biochemical signal transduction. *Proc. Natl. Acad. Sci. U.S.A.* **91**, 12378–12382.

Runyan, R. B., Maxwell, G. D., and Shur, B. D. (1986). Evidence for a novel enzymatic mechanism of neural crest cell migration on extracellular glycoconjugate matrices. *J. Cell Biol.* **102**, 431–441.

Ruoslahti, E., and Yamaguchi, Y. (1992). Proteoglycans as modulators of growth factor activities. *Cell (Cambridge, Mass.)* **64**, 867–869.

Sakamoto, N., Iwahana, M., Tanaka, M. G., and Osada, Y. (1991). Inhibition of angiogenesis and tumor growth by a synthetic laminin peptide CDPGYIGSR-NH2. *Cancer Res.* **51**, 903–906.

Salmivirta, M., and Jalkanen, M. (1995). Syndecan family of cell surface proteoglycans: Developmentally regulated receptors for extracellular effector molecules. *Experientia* **51**, 863–873.

Sanes, J. R., Engvall, E., Butkowski, R., and Hunter, D. D. (1990). Molecular heterogeneity of basal laminae: Isoforms of laminin and collagen IV at the neuromuscular junction and elsewhere. *J. Cell Biol.* **111**, 1685–1699.

Schaff, J., and Loew, L. M. (1999). The virtual cell. *Pac. Symp. Biocomput.* **4**, 228–239.

Schamhart, D. H., and Kurth, K. H. (1997). Role of proteoglycans in cell adhesion of prostate cancer cells: From review to experiment. *Urol. Res.* **25**, (Suppl. 2), S89–S96.

Scheler, J. L., Colbett, S. A., Wenk, M. B., and Schwarzbauer, J. E. (1998). Modulation of cell–extracellular matrix interactions. *Ann. N. Y. Acad. Sci.* **857**, 143–154.

Schlaepfer, D. D., and Hunter, T. (1996). Signal transduction from the extracellular matrix—A Role for the focal adhesion protein-tyrosine kinase FAK. *Cell Struct. Func.* **21**, 445–450.

Schlaepfer, D. D., and Hunter, T. (1997). Focal adhesion kinase overexpression enhances ras-dependent integrin signaling to ERK2/mitogen-activated protein kinase through interactions with and activation of c-Src. *J. Biol. Chem.* **272**, 13189–13195.

Schlaepfer, D. D., and Hunter, T. (1998). Integrin signalling and tyrosine phosphorylation: Just the FAKs? *Trends Cell Biol.* **8**, 151–157.

Schlaepfer, D. D., Broome, M. A., and Hunter, T. (1997). Fibronectin-stimulated signaling from a focal adhesion kinase–c-Src complex: Involvement of the Grb2, p130cas, and Nck adaptor proteins. *Mol. Cell. Biol.* **17**, 1702–1713.

Schlaepfer, D. D., Jones, K. C., and Hunter, T. (1998). Multiple Grb2-mediated integrin-stimulated signaling pathways to ERK2/mitogen-activated protein kinase: Summation of both c-Src- and focal adhesion kinase-initiated tyrosine phosphorylation events. *Mol. Cell. Biol.* **18**, 2571–2585.

Schmidhauser, C., Casperson, B. F., and Bissell, M. J. (1991). A bovine β-casein transcriptional enhancer. *J. Cell Biol.* **115**, 292a.

Schuppan, D., Somasundaram, R., Dieterich, W., Ehnis, T., and Bauer, M. (1994). The extracellular matrix in cellular proliferation and differentiation. *Ann. N. Y. Acad. Sci.* **733**, 87–102.

Schuppan, D., Schmid, M., Somasundaram, R., Ackermann, R., Ruehl, M., Nakamura, T., and Riecken, E. O. (1998). Collagens in the liver extracellular matrix bind hepatocyte growth factor. *Gastroenterology* **114**, 139–152.

Schwartz, M. A. (1992). Transmembrane signaling by integrins. *Trends Cell Biol.* **2**, 304–308.

Silberstein, G. B., Daniel, C. W., Coleman, S., and Strickland, P. (1990). Epithelium-dependent induction of mouse mammary gland extracellular matrix by TGF-beta-1. *J. Cell Biol.* **110**, 2209–2219.

Silberstein, G. B., Flanders, K. C., Roberts, A. B., and Daniel, C. W. (1992). Regulation of mammary morphogenesis: Evidence for extracellular matrix-mediated inhibition of ductal budding by transforming growth factor-b1. *Dev. Biol.* **152**, 354–362.

Sjaastad, M. D., and Nelson, W. J. (1997). Integrin-mediated calcium signaling and regulation of cell adhesion by intracellular calcium. *BioEssays* **19**, 47–55.

Spring, J., Paine-Saunders, S. E., Hynes, R. O., and Bernfield, M. (1994). *Drosophila* syndecan: Conservation of a cell surface heparan sulfate proteoglycan. *Proc. Natl. Acad. Sci. U.S.A.* **91**, 3334–3338.

Stocum, D. L. (1998). Regenerative biology and engineering: Strategies for tissue restoration. *Wound Repair Regeneration* **6**, 276–290.

Strange, R., Li, F., Saurer, S., Burkhardt, A., and Friis, R. R. (1992). Apoptotic cell death and tissue remodelling during mouse mammary gland involution. *Development (Cambridge, UK)* **115**, 49–58.

Streuli, C. H., Bailey, N., and Bissell, M. J. (1991). Control of mammary epithelial differentiation: Basement membrane induces tissue specific gene expression in the absence of cell–cell interaction and morphological polarity. *J. Cell Biol.* **115**, 1383–1395.

Streuli, C. H., Schmidhauser, C., Kobrin, M., Bissell, M. J., and Derynck, R. (1993). Extracellular matrix regulates expression of the TGF-β1 gene. *J. Cell Biol.* **120**, 253–260.

Sympson, C. J., Talhouk, R. S., Alexander, C. M., Chin, J. R., Clift, S. M., Bissell, M. J., and Werb, Z. (1994). Targeted expression of stromelysin to the mouse mammary gland provides evidence for a role of proteinases in branching morphogenesis and the requirement for and intact basement membrane for tissue-specific gene expression. *J. Cell Biol.* **125**, 681–721.

Sympson, C. J., Bissell, M. J., and Werb, Z. (1995). Mammary gland tumor formation in transgenic mice overexpressing stromelysin-1. *Semin. Cancer Biol.* **6**, 159–163.

Talhouk, R. S., Bissell, M. J., and Werb, Z. (1992). Coordinated expression of extracellular matrix-degrading proteinases and their inhibitors regulates mammary epithelial function during involution. *J. Cell Biol.* **18**, 1271–1282.

Tomasek, J. J., Haaksma, C. J., Eddy, R. J., and Vaughan, M. B. (1992). Fibroblast contraction occurs on release of tension in attached collagen collagen lattices: Dependency on an organized actin cytoskeleton and serum. *Anat. Rec.* **232**, 358–368.

Tschopp, J. F., Driscoll, E. M., Mu, D. X., Black, S. C., Pierschbacher, M. D., and Lucchesi, B. R. (1993). Inhibition of coronary artery reocclusion after thrombolysis with an RGD-containing peptide with no significant effect on bleeding time. *Coronary Artery Dis.* **4**, 809–817.

Tschopp, J. F., Craig, W. S., Tolley, J., Blevitt, J., Mazur, C., and Pierschbacher, M. D. (1994a). Threrapeutic application of matrix biology. *In* "Methods in Enzymology" **245**, pp. 556–568. Academic Press, San Diego, CA.

Tschopp, J. F., Mazur, C., Gould, K., Connolly, R., and Pierschbacher, M. D. (1994b). Inhibition of thrombosis by a selective fibrinogen receptor antagonist without effect on bleeding time. *Thromb. Haemostasis* **72**, 119–124.

Tuan, T. L., Song, A., Chang, S., Younai, S., and Nimni, M. E. (1996). *In vitro* fibroplasia: Matrix contraction, cell growth, and collagen production of fibroblasts cultured in fibrin gels. *Exp. Cell Res.* **223**, 127–134.

Turley, E. A. (1992). Hyaluronan and cell locolotion. *Cancer Metastasis Rev.* **11**, 21–30.

Vangainkar, S., and Martins-Green, M. (1998). Thrombin activation of the 9E3/CEF4 chemokine involves tyrosine kinases including c-Src and the EGF receptor. *J. Biol. Chem.* **273**, 5226–5234.

Varner, J. A., Brooks, P. C., and Cheresh, D. A. (1995). Review: The integrin alpha V beta 3: Angiogenesis and apoptosis. *Cell Adhes. Commun.* **3**, 367–374.

Vlodavsky, I., Miao, H. Q., Medalion, B., Danagher, P., and Ron, D. (1996). Involvement of heparan sulfate and related molecules in sequestration and growth promoting activity of fibroblast growth factor. *Cancer Metastasis Rev.* **15**, 177–186.

Wallner, E. I., Yang, Q., Peterson, D. R., Wada, J., and Kanwar, Y. S. (1998). Relevance of extracellular matrix, its receptors, and cell adhesion molecules in mammalian nephrogenesis. *Am. J. Physiol.* **275**, F467–F477.

Wary, K. K., Mariotti, A., Zurzolo, C., and Giancotti, F. G. (1998). A requirement for caveolin-1 and associated kinase Fyn in integrin signaling and anchorage-dependent cell growth. *Cell (Cambridge, Mass.)* **94**, 625–634.

Weaver, V. M., Petersen, O. W., Wang, F., Larabell, C. A., Briand, P., Damsky, C., and Bissell, M. J. (1997). Reversion of the malignant phenotype of human breast cells in three-dimensional culture and *in vivo* by integrin blocking antibodies. *J. Cell Biol.* **137**, 231–245.

Wei, J., Shaw, L. M., and Mercurio, A. M. (1997). Integrin signaling in leukocytes: Lessons from the alpha6beta1 integrin. *J. Leukocyte Biol.* **61**, 397–407.

Werb, Z., and Chin, J. R. (1998). Extracellular matrix remodeling during morphogenesis. *Ann. N. Y. Acad. Sci.* **857**, 110–118.

Werb, Z., Sympson, C. J., Alexander, C. M., Thomasset, N., Lund, L. R., MacAuley, A., Ashkenas, J., and Bissell, M. J. (1996). Extracellular matrix remodeling and the regulation of epithelial–stromal interactions during differentiation and involution. *Kidney Int., Suppl.* **54**, S68–S74.

Werb, Z., Vu, T. H., Rinkenberger, J. L., and Coussens, L. M. (1999). Matrix-degrading proteases and angiogenesis during development and tumor formation. *Apmis* **107**, 11–18.

Wessells, N. K., and Cohen, J. H. (1968). Effects of collagenase on developing epithelia *in vitro:* Lung, ureteric bud, and pancreas. *Dev. Biol.* **18**, 294–309.

White, S. R., Dorscheid, D. R., Rabe, K. F., Wojcik, K. R., and Hamann, K. J. (1999). Role of very late adhesion integrins in mediating repair of human airway epithelial cell monolayers after mechanical injury. *Am. J. Respir. Cell Mol. Biol.* **20**, 787–796.

Xiang, M., Tong, G., Tomdinson, C. R., and Klein, W. H. (1991). Structure and promoter activity of the LpS1 genes of *Lytechinus pictus. J. Biol. Chem.* **266**, 10524–10533.

Yamada, K. M., and Kennedy, D. W. (1984). Dualistic nature of adhesive protein function: Fibronectin and its biologically active peptide fragments can autoinhibit fibronectin function. *J. Biol. Chem.* **99**, 29–36.

Yoshizato, K., and Yoshikawa, E. (1993). Development of bilayered gelatin substrate for bioskin: A new structural framework of the skin composed of porous dermal matrix and skin basement membrane. *Mater. Sci. Eng.* **C1**, 95–105.

Yost, H. J. (1992). Regulation of vertebrate left–right asymmetries by extracellular matrix. *Nature (London)* **357**, 158–161.

Zhu, X., and Assoian, R. K. (1995). Integrin-dependent activation of MAP kinase: A link to shape-dependent cell proliferation. *Mol. Biol. Cell* **6**, 273–282.

CHAPTER 5

MATRIX MOLECULES AND THEIR LIGANDS

Björn Reino Olsen

INTRODUCTION

Cellular growth and differentiation, in two-dimensional cell culture as well as in the three-dimensional space of the developing organism, require the presence of a structured environment with which the cells can interact. This extracellular matrix (ECM) is composed of polymeric networks of several types of macromolecules in which smaller molecules, ions, and water are bound. The major types of macromolecules are the fibrous proteins, such as collagens, elastin, fibrillins, fibronectin, and laminins, and the hydrophilic heteropolysaccharides, such as glycosaminoglycan chains in hyaluronan and proteoglycans. It is the combination of fibrous protein polymers and hydrated proteoglycans that gives extracellular matrices their resistance to tensile and compressive mechanical forces.

The macromolecular components of the polymeric assemblies of the ECM are in many cases secreted by cells as precursor molecules that are significantly modified (proteolytically processed, sulfated, oxidized, and cross-linked) before they assemble with other components into functional polymers (Fig. 5.1). The formation of matrix assemblies *in vivo* is therefore in most instances a unidirectional, irreversible process and the disassembly of the matrix is not a simple reversal of assembly, but involves multiple, highly regulated processes. One consequence of this is that polymers reconstituted in the laboratory with components extracted from extracellular matrices do not have all the properties they have when assembled by cells *in vivo*. The ECM *in vivo* is also modified by cells as they proliferate, differentiate, and migrate, and cells in turn continuously interact with the matrix and communicate with each other through it (Hay, 1991).

The ECM is therefore not an inert product of secretory activities, but influences cellular shape, fate, and metabolism in ways that are as important to tissue and organ structure and function as the effects of many cytoplasmic processes. This realization has led to a reassessment of the need for a detailed molecular understanding of the ECM. In the past the ECM was primarily appreciated for its challenge to biochemists interested in protein and complex carbohydrate structure; a detailed characterization of ECM constituents is now considered essential for understanding cell behavior in the context of tissue and organ development and function. Some of these constituents are obviously most important for their structural properties (collagens, elastin, and fibrillin), whereas others (matrix-bound fibroblast growth factors, transforming growth factor β, bone morphogenetic proteins) are signaling molecules. In a third category are multidomain molecules (fibronectin, laminin, thrombospondin, tenascin, syndecans, and other proteoglycans) that are both structural constituents and regulators of cell behavior (Fig. 5.1).

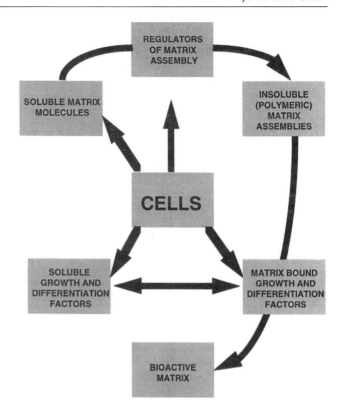

Fig. 5.1. Extracellular matrices are synthesized and assembled through many steps. Soluble matrix molecules are secreted by cells, modified, and assembled into polymeric complexes. These complexes serve as a scaffold for cells and as a reservoir for small molecules such as growth and differentiation factors.

FIBRILLAR COLLAGENS: MAJOR SCAFFOLD PROTEINS IN THE ECM

Collagens constitute a large family of proteins that represent the major proteins (about 25%) in mammalian tissues. A subfamily of these proteins, the fibrillar collagens, contains rigid, rodlike molecules with three subunits, α chains, folded into a right-handed collagen triple helix (Linsenmayer, 1991). Within a fibrillar collagen triple-helical domain, each α chain consists of about 1000 amino acid residues and is coiled into an extended, left-handed polyproline II helix; three α chains are in turn twisted into a right-handed superhelix (Fig. 5.2). The extended conformation of each α chain does not allow the formation of intrachain hydrogen bonds, and the stability of the triple helix is instead due to interchain hydrogen bonds. Such interchain bonds can form only if every third residue of each α chain does not have a side chain and is packed close to the triple-helical axis. Only glycine residues can therefore be accommodated in this position. This explains why the amino acid sequence of each α chain in fibrillar collagens consists of about 300 Gly-X-Y tripeptide repeats, where X and Y can be any residue, but Y is frequently proline or hydroxyproline. It also provides an explanation for why mutations in collagens that lead to a replacement of triple-helical glycine residues with more bulky residues can cause severe abnormalities (Olsen, 1995).

Fibrillar collagen molecules are the major components of collagen fibrils. Their α chains are synthesized as precursors, proα chains, with large propeptide regions flanking the central triple-helical domain. The carboxyl propeptide (C-propeptide) is important for the assembly of trimeric molecules in the rough endoplasmic reticulum. Formation of C-propeptide trimers, stabilized by intra- and interchain disulfide bonds, is the first step in the intracellular assembly and folding of trimeric procollagen molecules (Olsen, 1991). The folding of the triple-helical domain proceeds in a zipperlike fashion from the carboxyl toward the amino end of procollagen molecules, with a rate that is limited by cis–trans isomerization of prolyl peptide bonds. Mutations in fibrillar procollagens that affect the structure and folding of the C-propeptide domain are therefore likely to affect the participation of the mutated chains in triple-helical assemblies. In contrast, mutations upstream of the C-propeptide, such as in-frame deletions or glycine substitutions in the triple-helical domains, exert a dominant negative effect, in that the mutated chains will participate in trimer assembly, but will interfere with subsequent folding of the triple-helical domain.

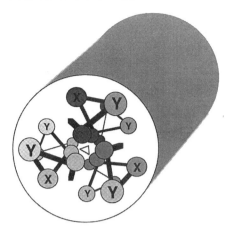

Fig. 5.2. Diagram showing a segment of a triple-helical collagen molecule. The triple helix is composed of three left-handed helices (α chains) that are twisted into a right-handed superhelix. The sequence of each α chain is a repeat of the tripeptide Gly-X-Y. The Gly residues are packed close to the triple-helical axis (indicated by a triangle). Only glycine (without a side chain) can be accommodated in this position. Although any residue can fit into the X and Y positions, Pro is frequently found in the Y position.

Fibrillar procollagen chains are the products of nine genes (Vuorio and de Crombrugghe, 1990). The similarities of these genes suggest that they arose by multiple duplications from a single ancestral gene. Despite their similarities and the high degree of sequence identity of their protein products, they exhibit specificity in the interactions of their C-propeptides during trimeric assemblies. Thus within triple-helical procollagens a relatively small number of chain combinations is found; these combinations represent fibrillar collagen types (Jacenko *et al.,* 1991).

COLLAGENS V/XI—REGULATORS OF FIBRIL ASSEMBLY, SPATIAL ORGANIZATION, AND CELL DIFFERENTIATION

Some collagen types are heterotrimers (types I, V, and XI), whereas others are homotrimers (types II and III). Some chains participate in more than one type: for example, the α1(II) chain forms the homotrimeric collagen II, but is also one of three different chains in collagen XI molecules. Between collagens V and XI there is extensive sharing of polypeptide subunits, and fibrillar collagen molecules previously described as belonging to either collagen V or XI are now referred to as belonging to the V/XI type. Thus, fibrillar procollagen molecules secreted by cells are members of a large group of homologous proteins. They all contain a C-propeptide that is completely removed by an endoproteinase after secretion, and their triple-helical rodlike domains polymerize in a staggered fashion into fibrillar arrays (Fig. 5.3). They differ, however, in the structure of their amino propeptide (N-propeptide) domains and in the extent to which this domain is proteolytically removed. For some collagen types, such as collagen I and II, the N-propeptide processing is complete in molecules within mature fibrils. For other types, such as collagens V/XI, this is not the case, in that a large portion of the N-propeptides in these molecules remains attached to the triple-helical domain (Fig. 5.3). It is believed that the incomplete processing of type V/XI molecules allows them to serve as regulators of fibril assembly (Linsenmayer *et al.,* 1993). Collagen fibrils are heterotypic, i.e., contain more than one collagen type, such that collagen I fibrils contain 5–10% collagen V and collagen II fibrils contain 5–10% collagen XI. The presence of N-propeptide domains on V/XI molecules represents a steric hindrance to addition of molecules at fibril surfaces. This heterotypic/steric hindrance model predicts that collagen fibril diameters in a tissue are determined by the ratio of the "minor" component (V/XI) to the "major" component (I or II). A high ratio results in thin fibrils; a low ratio results in thick fibrils. Direct support for this comes from studies of mutant and transgenic mice. For example, mice that are homozygous for a functional null mutation in α1(XI) collagen and transgenic mice overexpressing collagen II have cartilage collagen fibrils that are abnormally thick (Garofalo *et al.,* 1993; Li *et al.,* 1995).

A characteristic feature of collagen fibrillar scaffolds is their precise three-dimensional patterns. These patterns follow mechanical stress lines and ensure a maximum of tensile strength with a minimum of material. Examples are the criss-crossing lamellae of collagen fibers in lamellar bone or in cornea, the arcades of collagen fibrils under the surface of articular cartilage, and the parallel fiber bundles in tendons and ligaments. Ultimately, cells are responsible for establishing these patterns, but the mechanisms they use probably involve specific molecules and interactions. The

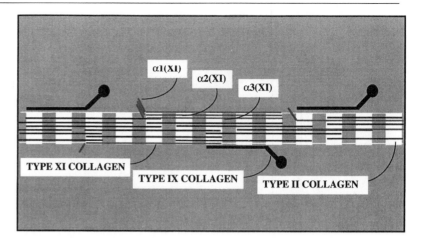

Fig. 5.3. Diagram of a cartilage collagen fibril. Collagen II molecules are the major components. Molecules of collagen XI and IX are located on the surface. Collagen XI molecules, heterotrimers of three different α chains, have amino-terminal domains that are thought to block sterically the addition of collagen II molecules at the fibril surface.

critical pattern-forming interactions are not known; however, the information for spatial organization is not contained within the different fibrillar collagen types, because the same kind of heterotypic fibril can be part of scaffolds with very different spatial organization. Transgenic mice with an alteration in the N-propeptide region of collagen V molecules show a disruption of the lamellar arrangement of fibrils in the cornea, suggesting a role for fibril surface domains in generating the spatial pattern (Andrikopoulos *et al.*, 1995). Finally, members of a unique subfamily of fibril-associated collagens with interrupted triple helices (FACIT) (Olsen *et al.*, 1995) are good candidates for molecules that allow tissue-specific fibril patterns to be generated by modulating the surface properties of fibrils.

The phenotypic consequences of mutations in fibrillar collagen genes indicate that the major function of these proteins is to provide elements of high tensile strength at the tissue level. Mutations in collagen V/XI genes also suggest that fibrillar collagen scaffolds are essential for normal cellular growth and differentiation. For example, a functional null mutation in $\alpha 1(XI)$ collagen in cartilage causes a severe disproportionate dwarfism in mice and perinatal death of homozygotes (Li *et al.*, 1995). Histology of the mutant growth plates reveals a disorganized spatial distribution of cells and a defect in chondrocyte differentiation to hypertrophy. The explanation for this is likely to be related to the fact that in growth plates the proliferation and differentiation of chondrocytes is regulated by locally produced growth factors and cytokines. Cells that produce these factors must therefore be localized close to cells that express the appropriate receptors. The $\alpha 1(XI)$ mutation may disrupt this relationship, because it results in a dramatic decrease in cohesive properties of the matrix and a loss of cellular organization. Transgenic mice with a mutation in $\alpha 2(V)$ collagen have a large number of hair follicles of unusual localization in the hypodermis; this may also be related to a defect in the mechanical properties of the fibrillar collagen scaffold (Andrikopoulos *et al.*, 1995).

FACIT COLLAGENS—MODULATORS OF COLLAGEN FIBRIL SURFACE PROPERTIES

Molecules that are associated with collagen fibrils, contain two or more triple-helical domains, and share characteristic protein domains (modules) are classified as FACIT collagens (Olsen *et al.*, 1995; Shaw and Olsen, 1991). Of the five currently known members in the group (collagens IX, XII, XIV, XVI, and XIX) collagen IX is the best characterized both structurally and functionally. Collagen IX molecules are heterotrimers of three different gene products (van der Rest *et al.*, 1985). Each of the three α chains contains three triple-helical domains separated and flanked by non-triple-helical sequence regions (Fig. 5.4). Between the amino-terminal and central triple-helical domains a flexible hinge gives the molecule a kinked structure with two arms. Type IX molecules are located on the surface of type II/XI-containing fibrils with the long arm parallel to the fibril surface and the short arm projecting into the perifibrillar space (Vaughan *et al.*, 1988) (Fig. 5.3).

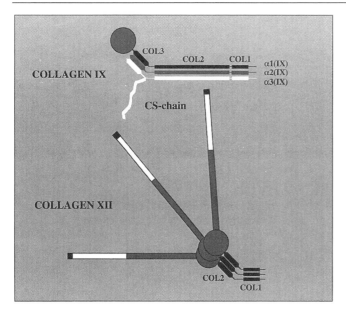

Fig. 5.4. Diagrams of collagen IX and XII (long form) molecules. Collagen IX molecules contain the three chains α1(IX), α2(IX), and α3(IX). Each chain contains three triple-helical domains (COL1, COL2, COL3), interrupted and flanked by non-triple-helical sequences. In cartilage, the α1(IX) chain contains a large globular amino-terminal domain. The α2(IX) chain serves as a proteoglycan core protein in that it contains a chondroitin sulfate (CS) side chain attached to the non-triple-helical region between the COL2 and COL3 domains. Collagen XII molecules are homotrimers of α1(XII) chains. The three chains form two short triple-helical domains separated by a flexible hinge region. A central globule is composed of three globular domains that are homologous to the amino-terminal globular domain of α1(IX) collagen chains. The amino-terminal regions of the three α1(XII) chains contain multiple fibronectin type III repeats and von Willebrand factor A-like domains. These regions form three "fingers" that extend from the central globule. Through alternative splicing a portion of the fingers (white region in diagram) is spliced out in the short form of collagen XII. Hybrid molecules with both long and short fingers can be extracted from tissues.

It is believed that collagen IX may function as a bridging molecule between fibrils or between fibrils and other matrix constituents. Support from this comes from studies of transgenic mice and mutations in humans. Transgenic mice with a dominant negative mutation in the α1(IX) chain (Nakata *et al.,* 1993), as well as mice that are homozygous for α1(IX) null alleles (Faessler *et al.,* 1994), exhibit osteoarthritis in knee joints and mild chondrodysplasia. In humans, a splice site mutation leading to exon skipping and an in-frame deletion in the α2(IX) collagen chain has been shown to cause a form of multiple epiphyseal dysplasia, an autosomal dominant disorder characterized by early-onset osteoarthritis in large joints associated with short stature and stubby fingers (Muragaki *et al.,* 1996).

Molecules of collagens XII and XIV are homotrimers of chains that are made up of several kinds of modules. Some modules are homologous to modules found in collagen IX, whereas others show homology to von Willebrand factor A domains and fibronectin type III repeats. Both types of molecules contain a central globule with three fingerlike extensions and a thin triple-helical tail attached (Fig. 5.4). For collagen XII, two forms that differ greatly in the lengths of the fingerlike extensions are generated by alternative splicing of RNA transcripts. Variations in the carboxyl regions also occur (Olsen *et al.,* 1995). Both collagens are found in connective tissues containing type I collagen fibrils, except mineralized bone matrix, and immunolabeling studies show a fibril-associated distribution. Type XIV collagen has been reported to bind to heparin sulfate and the small fibril-associated proteoglycan decorin (Brown *et al.,* 1993; Font *et al.,* 1993). This would suggest an indirect fibril association. A direct association cannot be ruled out, however, because collagen XII molecules form copolymers with collagen I even in the absence of proteoglycans. A functional interaction between fibrils and collagens XII and XIV is implied by studies showing that addition of the two collagens to type I collagen gels promote gel contraction mediated by fibroblasts (Nishiyama *et al.,* 1994). The effect is dose dependent and can be prevented by denaturation or addition of specific antisera. The association with fibrils of collagens XII and XIV may therefore modulate the frictional properties of fibril surfaces. The synthesis of different isoforms could be important in this context, because they could bind to fibrils with different affinities. Also, because the long form of collagen XII is a proteoglycan, whereas the short form is not, variations in the relative proportion of the two splice variants may serve to modulate the hydrophilic properties of interfibrillar matrix compartments. Finally, the discovery that the collagen I N-propeptide processing enzyme binds to collagen XIV and can be purified as part of a complex with antibodies against collagen XIV raises the possibility that the FACIT collagens provide binding sites for fibril-modifying extracellular matrix enzymes (Colige *et al.,* 1995).

Basement Membrane Collagens and Associated Collagen Molecules

At epithelial (and endothelial)–stromal boundaries, basement membranes serve as specialized areas of ECM for cell attachment. By end-to-end and lateral interactions collagen IV molecules form a networklike scaffold for basement membranes (Yurchenco and O'Rear, 1994). Six different collagen IV genes exist in mammals, and their products interact to form at least three different types of heterotrimeric collagen IV molecules. These different isoforms show characteristic tissue specific expression patterns.

Within basement membranes the collagen IV networks are associated with a large number of noncollagenous molecules, such as various isoforms of laminin, nidogen, and the heparin sulfate proteoglycan perlecan. Additional collagens are also associated with basement membranes. These include the transmembrane collagen XVII in hemidesmosomes and collagen VII in anchoring fibrils. Collagens XVII and VII are important in regions of significant mechanical stress, such as skin, in that they anchor epithelial cells to the basement membrane (collagen XVII) and strap the basement membrane to the underlying stroma (collagen VII). In bullous pemphigoid, autoantibodies against collagen VII cause blisters that separate epidermis from the basement membrane (Liu *et al.,* 1994); dominant and recessive forms of epidermolysis bullosa can be caused by mutations in collagens VII and XVII (McGrath *et al.,* 1995; Uitto *et al.,* 1994). The physiological importance of collagen IV isoforms is highlighted by Alport syndrome (Tryggvason *et al.,* 1993). This disease, characterized by progressive hereditary nephritis associated with sensorineural hearing loss and ocular lesions, can be caused by mutations within $\alpha 3$(IV) and $\alpha 4$(IV) collagen genes (autosomal Alport syndrome) (Mochizuki *et al.,* 1994) or mutations in $\alpha 5$(IV) collagen (X-linked Alport syndrome) (Barker *et al.,* 1990). In cases of large deletions, including both the $\alpha 5$(IV) and the neighboring $\alpha 6$(IV) collagen gene, renal disease is associated with inherited smooth muscle tumors (Zhou *et al.,* 1993).

Two additional basement membrane-associated collagens, collagen VIII and collagen XVIII, are of interest because of their function in vascular physiology and pathology. Collagen VIII is a short-chain, nonfibrillar collagen with significant homology to collagen X, a product of hypertrophic chondrocytes (Muragaki *et al.,* 1991; Yamaguchi *et al.,* 1989, 1991). Its expression appears to be up-regulated during cardiac morphogenesis (Iruela-Arispe and Sage, 1991), in human atherosclerotic lesions (MacBeath *et al.,* 1996), and following experimental lesions to the endothelium in large arteries (Bendeck *et al.,* 1996; Sibinga *et al.,* 1997). It has been suggested that collagen VIII may be important in mediating the migration of smooth muscle cells from the medial layer into the intima during neointimal thickening following endothelial cell injury.

Collagen XVIII, together with collagen XV, belongs to a distinct subfamily of collagens called multiplexins (so named because of their <u>multiple</u> triple-helix domains and <u>interruptions</u>) (Oh *et al.,* 1994; Rehn and Pihlajaniemi, 1994). Because of the alternative utilization of two promoters and alternative splicing, the *COL18A1* gene gives rise to three different transcripts that are translated into three protein variants. These are localized in various basement membranes, including those that separate vascular endothelial cells from the underlying intima in blood vessels (Halfter *et al.,* 1998; Muragaki *et al.,* 1995; Rehn and Pihlajaniemi, 1995; Saarela *et al.,* 1998). When collagen XVIII was initially cloned and sequenced it was noted that the sequence contained multiple consensus sequences for attachment of heparan sulfate side chains (Oh *et al.,* 1994). Subsequent studies have, in fact, confirmed that collagen XVIII forms the core protein of a basement membrane proteoglycan (Halfter *et al.,* 1998). Proteolytic processing of the carboxyl non-triple-helical domain of collagen XVIII in tissues leads to the release of heparin-binding molecules with anti-angiogenic activity (Sasaki *et al.,* 1998).

One of these fragments, named endostatin, represents the carboxyl-terminal 20-kDa portion of collagen XVIII chains (O'Reilly *et al.,* 1997). Recombinant preparations of endostatin have been shown to inhibit the proliferation and migration of vascular endothelial cells, inhibit the growth of tumors in mice and rats, and cause regression of tumors in mice (Boehm *et al.,* 1997; Dhanabal *et al.,* 1999; O'Reilly *et al.,* 1997; Yamaguchi *et al.,* 1999). The limited data available suggest that these effects are mediated by inhibition of tumor-induced angiogenesis. The X-ray crystallographic structure of endostatin has been determined for both the mouse and human proteins (Ding *et al.,* 1998; Hohenester *et al.,* 1998). The compact structure consists of two α helices and a large number of β strands, stabilized by two intramolecular disulfide bonds. A coordinated zinc atom is part of the structure (Ding *et al.*, 1998) and on the surface a patch of arginyl residues forms a potential binding site for heparin (Hohenester *et al.*, 1998). Studies of mutant endostatins have shown that arginines within this patch are required for heparin binding (Yamaguchi *et al.*, 1999). Interestingly, heparin or zinc binding is not required for biological activity (Yamaguchi *et al.*, 1999).

ELASTIC FIBERS AND MICROFIBRILS

Collagen molecules and their fibers evolved as structures of high tensile strength, equivalent to that of steel when compared on the basis of the same cross-sectional area, but three times better on a per unit weight basis. In contrast, elastic fibers, composed of molecules of elastin, provide tissues with elasticity so that they can recoil after transient stretch (Rosenbloom *et al.*, 1993). In organs such as the large arteries, skin, and lungs, elasticity is obviously crucial for normal functioning. Elastin fibers derive their impressive elastic properties, an extensibility that is about five times that of a rubber band with the same cross-sectional area, from the structure of elastin molecules. Each molecule is composed of alternating segments of hydrophobic and α-helical Ala- and Lys-rich sequences. Oxidation of the lysine side chains by the enzyme lysyl oxidase leads to formation of reactive aldehydes and extensive covalent cross-links between neighboring molecules in the fiber. It is thought that the elasticity of the fiber is due to the tendency of the hydrophobic segments to adopt a random-coil configuration following stretch.

On the surface of elastic fibers one finds a cover of microfibrils, beaded filaments with molecules of fibrillin as their major components (Sakai and Keene, 1994). The fibrillins, products of genes on chromosomes 5 (Fib5) and 15 (Fib15) in humans, also form microfibrils that are found in almost all extracellular matrices in the absence of elastin. Fibrillin molecules are composed of multiple repeat domains, the most prominent being calcium-binding extracellular growth factor (EGF)-like repeats; similar repeats in latent transforming growth factor β (TGF-β)-binding proteins suggest that the fibrillins belong to a superfamily of proteins (Kanzaki *et al.*, 1990). The physiological importance of fibrillin is highlighted by mutations causing the Marfan syndrome and congenital contractural arachnodactyly. The Marfan syndrome is caused by mutations in Fib15 and is characterized by dislocation of the eye lens due to weakening of the suspensory ligaments of the zonule, congestive heart failure, aortic aneurysms, and skeletal growth abnormalities resulting in a tall frame, scoliosis, chest deformities, arachnodactyly, and hypermobile joints (Ramirez *et al.*, 1993). In patients with congenital contractural arachnodactyly, mutations in Fib5 lead to similar skeletal abnormalities and severe flexion contractures, but no ophthalmic and cardiovascular manifestations (Viljoen, 1994). The tall stature and arachnodactyly seen in patients with the Marfan syndrome suggest that Fib15 is a negative regulator of longitudinal bone growth. Because fibrillin microfibrils are found in growth plate cartilage it is conceivable that they affect chondrocyte proliferation and/or maturation; the mechanisms, however, are unknown.

FIBRONECTIN: A MULTIDOMAIN, MULTIFUNCTIONAL ADHESIVE ECM GLYCOPROTEIN

Several proteins in the extracellular matrix contain binding sites for structural macromolecules and for cells, thus contributing both to the structural organization of the ECM and its interaction with cells. The prototype of these adhesive glycoproteins is fibronectin, a disulfide-bonded dimer of 220- to 250-kDa subunits (Hynes, 1990). Each subunit is folded into five or six rodlike domains separated by flexible "joints." The domains are composed of three types of multiple repeats or modules, Fn1, Fn2, and Fn3. Fn1 modules are found in the fibrin-binding amino- and carboxyl-terminal regions of fibronectin and in the collagen (gelatin)-binding region (Fig. 5.5). Single copies of Fn1 modules are also found in tissue-type plasminogen activator (t-PA) and coagulation factor XII (Potts and Campbell, 1994).

Nuclear magnetic resonance studies of Fn1 modules demonstrate the presence of two layers of antiparallel β sheets (two strands in one layer and three strands in the other) held together by hydrophobic interactions. The structure is further stabilized by disulfide and salt bridges. Fn2 modules are found together with Fn1 modules in the collagen-binding region of fibronectin and in many other proteins. Their structure, two double-stranded antiparallel β sheets connected by loops, suggests that a ligand such as collagen may bind to this module through interactions of hydrophobic amino acid side chains with its hydrophobic surface. Fn3 modules are the major struc-

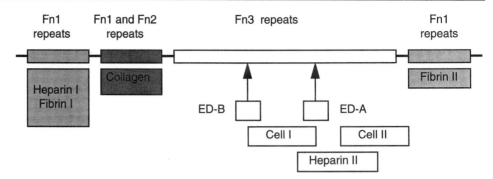

Fig. 5.5. Diagram of a fibronectin polypeptide chain. The polypeptide chain is composed of several repeats (Fn1, Fn2, and Fn3) and contains binding sites for several matrix molecules and cells. Two regions can bind heparin and fibrin and two regions are involved in cell binding as well. By alternative splicing, isoforms are generated that may or may not contain certain Fn3 domains (labeled ED-A and ED-B in the diagram). Additional splice variations in the second cell-binding domain (Cell II) generate other isoforms.

tural units in fibronectin and are found in a large number of other proteins as well. Some of these proteins (for example, the long splice variant of collagen XII) contain more Fn3 modules than does fibronectin. The structure of Fn3 is that of a sandwich of antiparallel β sheets (three strands in one layer and four strands in the other) with a hydrophobic core. The binding of fibronectin to some integrins involves the tripeptide sequence Arg-Gly-Asp in the tenth Fn3 module; these residues lie in an exposed loop between two of the strands in one of the β sheets (Potts and Campbell, 1994).

Fibronectin can assemble into a fibrous network in the ECM through interactions involving cell surface receptors and the amino-terminal region of fibronectin (Mosher *et al.*, 1991). A fibrin binding site is also contained in this region; a second site is in the carboxyl domain. The ability to bind to collagen ensures association between the fibronectin network and the scaffold of collagen fibrils. The collagen-binding domain of fibronectin can be isolated as a discrete proteolytic fragment, consisting of four Fn1 and two Fn2 modules. Binding sites for heparin and chondroitin sulfate further make fibronectin an important bridging molecule between collagens and other matrix molecules (Fig. 5.5).

Transcripts of the fibronectin gene are alternatively spliced in three regions in a cell- and developmental stage-dependent manner. As a result there are many different isoforms of fibronectin (Schwarzbauer, 1990). The main form produced by the liver and circulating in plasma lacks two of the Fn3 repeats found in cell- and matrix-associated fibronectin. One alternatively spliced domain is adjacent to the heparin-binding site, and this region binds to integrins α4β1 and α4β7 (Humphries, 1993). Thus, there is a mechanism for fine tuning of fibronectin structure and interaction properties. Not surprisingly, mice that are homozygous for fibronectin null alleles die early in embryogenesis with multiple defects (George *et al.*, 1993).

The biologically most important activity of fibronectin is its interaction with cells. The ability of fibronectin to serve as a substrate for cell adhesion, spreading, and migration is based on the activities of several modules. The Arg-Gly-Asp sequence in the tenth Fn3 module plays a key role in the interaction with the integrin receptor α5β1, but synergistic interactions with other Fn3 modules are essential for high-affinity binding of cells to fibronectin (Aota *et al.*, 1991). Interactions with integrins containing the β3 subunit, such as αIIbβ3 on platelets, involve a peptide sequence within the ninth Fn3 (Bowditch *et al.*, 1994).

LAMININS: LARGE, ADHESIVE BASEMENT MEMBRANE MOLECULES

Laminins are trimeric molecules of α, β, and γ chains with molecular masses of 140–400 kDa. With a large number of genetically distinct chains (α1, α2, α3, α4, α5, β1, β2, β3, γ1, γ2), several laminin isoforms are now known (Fig. 5.6). Several forms have a cross-shaped structure as visualized by rotary shadowing electron microscopy; some forms contain Y-shaped molecules. The laminins are important components of basement membranes, where they provide interaction sites

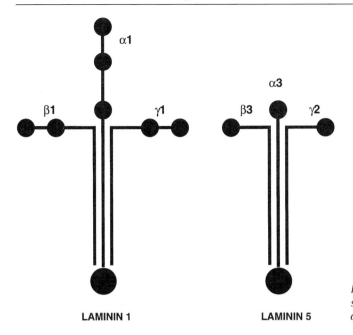

Fig. 5.6. Diagrams of two types of laminins. Laminin 1 has a cross-shaped structure, whereas laminin 5 is Y-shaped, due to a shorter α chain.

for many other constituents, including cell surface receptors (Timpl, 1996; Timpl and Brown, 1994). The functional and structural mapping of these sites and the complete sequencing of many laminin chains have provided detailed insights into the organization of laminin molecules. Within the cross-shaped laminin 1 molecule, three similar short arms are formed by the N-terminal regions of the α1, β1, and γ1 chains, whereas a long arm is composed of the carboxyl regions of all three chains, folded into a coiled-coil structure. The three chains are connected at the center of the cross by three interchain disulfide bridges. The short arms contain multiple EGF-like repeats of about 60 amino acid residues, terminated and interrupted by globular domains. The long arm consists of heptad repeats covering about 600 residues in all three chains. The α1 chain is about 1000 amino acid residues longer than the β1 and γ1 chains and forms five homologous globular repeats at the "base" of the cross; these globular repeats are similar to repeats found in the proteoglycan molecule perlecan (Iozzo, 1994). Calcium-dependent polymerization of laminin is based on interactions between the globular domains at the N termini and is thought to be important for the assembly and organization of basement membranes. Also of significance for the assembly of basement membranes is the high-affinity interaction with nidogen (Timpl and Brown, 1994). The binding site in laminin for nidogen is on the γ1 chain, close to the center of the cross. On nidogen, a rodlike molecule with three globular domains, the binding site for laminin is in the carboxyl globular domains; another globular domain binds to collagen IV. Thus, nidogen is a bridging molecule that connects the laminin and collagen IV networks, and is probably crucial for the assembly of normal basement membranes.

Laminin does not bind directly to collagen IV, but has binding sites for several other molecules besides nidogen. These are heparin, perlecan, and fibulin-1, which bind to the end of the long arm of the laminin cross. The biologically most significant interactions of laminin, however, involve a variety of cell surface receptors, both integrins and nonintegrin receptors. Several integrins are laminin receptors (Kramer 1994). They show distinct preferences for different laminins, and recognize binding sites on either the short or long arms of laminin molecules. In striated muscle cells laminin binds to the dystroglycan–dystrophin complex (Matsumura and Campbell, 1994). A disulfide-linked complex of laminins 5 and 6 is crucial for the firm attachment of keratinocytes to the basement membrane in skin by its interaction with α6β4 integrins in hemidesmosomes (Niessen *et al.,* 1994; Pulkkinen *et al.,* 1994; Sonnenberg *et al.,* 1991). Laminins with β2 chains appear to be important for guiding axons to neuromuscular junctions (Noakes *et al.,* 1995). Through interactions with the integrin αvβ3, laminin is important for angiogenesis (Brooks *et al.,* 1994). Finally, laminin has important roles in induction and maintenance of differentiated cellular phenotypes and in control of cell proliferation (Grant *et al.,* 1989).

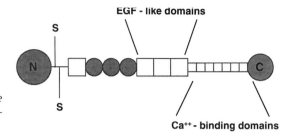

Fig. 5.7. Diagram of thrombospondin-1, showing the multidomain structure and the position of two cysteines that are involved in interchain disulfide (S—S) bonds within the trimeric molecules.

MODULATORS OF CELL–MATRIX INTERACTIONS

Whereas proteins such as fibronectin and laminin are important for adhesion of cells to extracellular matrices, other ECM molecules appear to function as both positive and negative modulators of such adhesive interactions. Examples of such modulators are thrombospondin (Bornstein and Sage, 1994) and tenascin (Erickson, 1993).

Thrombospondins are a group of four homologous trimeric matrix proteins composed of several Ca^{2+}-binding domains, EGF-like repeats, as well as other modules (Bornstein, 1992) (Fig. 5.7). Different members of the group show differences in cellular expression and functional properties. Thrombospondin-1, a component of α granules in platelets, is released when platelets are activated and forms multimers that bind fibrin and fibrinogen in the hemostatic plug. The protein is also produced by many other cell types (fibroblasts, endothelial cells, and smooth muscle cells) and binds to a variety of matrix molecules, including fibrillar collagen, fibronectin, laminin, and heparan sulfate proteoglycan. It has a growth-promoting effect on fibroblasts and may destabilize cell–matrix interactions during wound healing and angiogenesis (Chiquet-Ehrismann, 1995). It therefore promotes cell proliferation and angiogenesis. The cellular receptor for thrombospondin appears to be a chondroitin sulfate proteoglycan; cellular contact sites with thrombospondin have a cytoskeletal organization that is different than the integrin-mediated focal adhesions seen with molecules such as fibronectin (Adams, 1995).

Cartilage oligomeric matrix protein (COMP) is a pentameric protein that is structurally related to thrombospondin and is thought to have evolved from the thrombospondin-3 and -4 branch of the family (Newton, 1994). COMP is a secretory product of chondrocytes and is localized in their territorial matrix in cartilage. Mutations in COMP have been shown to be the cause of pseudoachondroplasia and a form of multiple epiphyseal dysplasia (Briggs *et al.,* 1995).

The three members of the tenascin family (C, R, and X) are large hexameric (tenascin-C and -R) or trimeric (tenascin-X) proteins with subunits composed of multiple protein modules (Erickson, 1993). The modules include fibronectin type III repeats, EGF-like domains, and a carboxyl domain with homology to the carboxyl-terminal domains of β- and γ-fibrinogen chains. These modules form rodlike structures that interact with their amino-terminal domains within the oligomers. Although mice that are homozygous for tenascin-C null alleles have no obvious phenotype (Saga *et al.,* 1992), the multiple interactions and tissue expression patterns defined for tenascins strongly suggest that they have important functions. Tenascin-R is expressed in the central nervous system during development, tenascin-X is expressed particularly in muscle (smooth muscle as well as cardiac and skeletal), and tenascin-C is found in healing wounds, in a wide variety of tumors, and in the brain. In certain experimental conditions tenascin-C can be an adhesive molecule for cells; it can also, however, have antiadhesive effects (Chiquet-Ehrismann, 1995). The adhesive activity can be mediated by either cell surface proteoglycan or integrins, depending on cell type. Tenascin-C can bind heparin and this may be responsible for interactions with cell surface proteoglycans such as glypican (Erickson, 1993; Vaughan *et al.,* 1994). Tenascin-C can also block adhesion by covering up adhesive sites in other matrix molecules, such as fibronectin, sterically blocking their interactions with cells. Interactions with specific "antiadhesive" cell surface receptors are also conceivable. A cell surface molecule on neurons, contactin, binds to both tenascin-C and -R and may mediate their effects on neurons; tenascin can both stimulate and inhibit neurite outgrowth (Vaughan *et al.,* 1994).

PROTEOGLYCANS: MULTIFUNCTIONAL ECM AND CELL SURFACE MOLECULES

A variety of proteoglycans play important roles in cellular growth and differentiation and in matrix structure. They range from the large hydrophilic space-filling complexes of aggrecan and versican with hyaluronan, to the cell surface syndecan receptors. In basement membranes the major heparan sulfate proteoglycan is perlecan (Iozzo, 1994; Timpl, 1994). With three heparan sulfate side chains attached to the amino-terminal region, its core protein is multimodular in structure, having borrowed structural motifs from a variety of other genes. These include a low-density lipoprotein (LDL) receptor-like module, regions with extensive homology to laminin chains, a long stretch of N-CAM-like IgG repeats, and a carboxyl-terminal region with three globular and four EGF-like repeats similar to a region of laminin. Alternative splicing can generate molecules of different lengths. Perlecan is present in a number of basement membranes, but is also found in the pericellular matrix of fibroblasts. In fact, fibroblasts, rather than epithelial cells, appear to be major producers of perlecan (for example, in skin). In liver, perlecan is expressed by sinusoidal endothelial cells and is localized in the perisinusoidal space. The precise functions of perlecan have not been defined, but mutations in the *unc-52* gene in *Caenorhabditis elegans,* encoding a short version of perlecan, cause disruptions of skeletal muscle (Rogalski *et al.,* 1993). This suggests that the molecule, as a component of skeletal muscle basement membranes, is important for assembly of myofilaments and their attachment to cell membranes. Binding of growth factors and cytokines to the heparan sulfate side chains also enables perlecan to serve as a storage vehicle for biologically active molecules such as basic fibroblast growth factor (bFGF).

Several small proteoglycans with homologous core proteins are found in a variety of tissues, where they interact with other matrix macromolecules and regulate their functions. They include decorin, biglycan, lumican, and fibromodulin (Hedbom and Heinegard, 1993; Hildebrand *et al.,* 1994; Kresse *et al.,* 1994). Decorin binds along collagen fibrils and may play a role in fibril assembly. It also has an effect on cell morphology, at least in culture, and may therefore modulate the binding of cells to matrix constituents such as collagen and fibronectin. Through binding of transforming growth factor β (TGF-β) isoforms, the small proteoglycans may help sequester growth factors within the ECM and thus regulate their activities (Hildebrand *et al.,* 1994).

A variety of proteoglycans also have important functions at cell surfaces. These include four members of the syndecan family, transmembrane molecules with highly conserved cytoplasmic domains, and glypican-related molecules that are linked to the cell surface via glycosyl phosphatidylinositol (David, 1993). Through their heparan sulfate side chains these molecules can bind growth factors, protease inhibitors, enzymes, and matrix macromolecules (Kim *et al.,* 1994). They are therefore thought to be important modulators of cell signalling pathways and cell–matrix contacts. Variations in their expression during embryogenesis support this idea (Bernfield *et al.,* 1993;

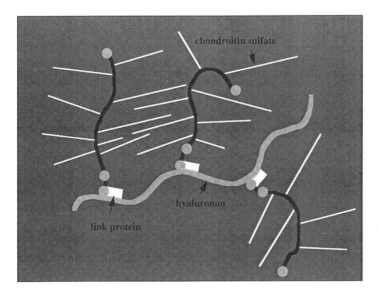

Fig. 5.8. Diagram of a portion of a large proteoglycan complex from cartilage. Monomers of aggrecan, composed of core proteins with glycosaminoglycan side chains (mostly chondroitin sulfate), are bound to hyaluronan. The binding is stabilized by link proteins. For clarity, only some of the glycosaminoglycan side chains are shown in the monomers.

Mali *et al.,* 1994). Syndecan, for example, is transiently lost from many epithelial cells during developmental shape changes, and is expressed in areas of mesenchymal condensations, in support of its function as a bridge between cells and the ECM.

Hyaluronan is an important component of most extracellular matrices (Laurent and Fraser, 1992). It serves as ligand for several proteins, including cartilage link protein and aggrecan and versican core proteins. In cartilage, based on such interactions, it is the backbone for the large proteoglycan complexes responsible for the compressive properties of cartilage (Morgelin *et al.,* 1994) (Fig. 5.8). It also is a ligand for cell surface receptors and regulates cell proliferation and migration (Sherman *et al.,* 1994). One receptor for hyaluronate is the transmembrane molecule CD44. By alternative splicing and variations in posttranslational modifications a family of CD44 proteins is generated (Lesley *et al.,* 1993). These show cell- and tissue-specific expression patterns and are thought to have distinct functional roles. The most widely expressed isoform is important for lymphocyte homing to various destinations. In lymphocyte homing, however, the ligand is not hyaluronate but other matrix molecules. In fact, CD44 is known to bind several molecules, including collagen, fibronectin, and mucosal addressin (Sherman *et al.,* 1994). Hyaluronate-mediated cell motility is based on the interaction of hyaluronan with a cell surface-associated protein called RHAMM (receptor for hyaluronate-mediated motility) (Turley, 1992). As a space-filling molecule and through its interaction with cell surface receptors, hyaluronan is important for several morphogenetic processes during development. It creates cell-free spaces through which cells (for example, neural crest cells) can migrate, and its degradation by hyaluronidase is probably important for processes of cellular condensation.

SUMMARY

Components of the extracellular matrix affect cellular growth and differentiation in a variety of ways (Mooney *et al.,* 1992). The collagen scaffold allows cells to be organized in space and thus provides the basis for spatially defined, short-range interactions between cells. Adhesive glycoproteins bind to cell surface integrins and other receptors regulating cell attachment, cell shape, proliferation, and differentiation (Roskelley *et al.,* 1995).

Large proteoglycans can both create hydrophilic spaces for cellular migration and block cell movements. Cell surface-associated proteoglycans bind growth factors and serve as coreceptors for growth factor–receptor interactions (Mason, 1994).

REFERENCES

Adams, J. C. (1995). Formation of stable microspikes containing actin and the 55 kDa actin bundling protein, fascin, is a consequence of cell adhesion to thrombospondin-1: Implications for the anti-adhesive activities of thrombospondin-1. *J. Cell Sci.* **108**, 1977–1990.

Andrikopoulos, K., Liu, X., Keene, D. R., *et al.* (1995). Targeted mutation in the *col5a2* gene reveals a regulatory role for type V collagen during matrix assembly. *Nat. Genet.* **9**, 31–36.

Aota, S., Nagai, T., and Yamada, K. M. (1991). Characterization of regions of fibronectin besides the arginine-glycine-aspartic acid sequence required for adhesive function of the cell-binding domain using site-directed mutagenesis. *J. Biol. Chem.* **266**, 15938–15943.

Barker, D. F., Hostikka, S. L., Zhou, J., *et al.* (1990). Identification of mutations in the *COL4A5* collagen gene in Alport syndrome. *Science* **248**, 1224–1227.

Bendeck, M. P., Regenass, S., Tom, W. D., *et al.* (1996). Differential expression of α1 type VIII collagen in injured platelet-derived growth factor-BB-stimulated rat carotid arteries. *Circ. Res.* **79**, 524–531.

Bernfield, M., Hinkes, M. T., and Gallo, R. L. (1993). Developmental expression of the syndecans: Possible function and regulation. *Development (Cambridge, UK), Suppl.,* pp. 205–212.

Boehm, T., Folkman, J., Browder, T., *et al.* (1997). Antiangiogenic therapy of experimental cancer does not induce acquired drug resistance. *Nature (London)* **390**, 404–407.

Bornstein, P. (1992). Thrombospondins: Structure and regulation of expression. *FASEB J.* **6**, 3290–3299.

Bornstein, P., and Sage, E. H. (1994). Thrombospondins. *In* "Methods in Enzymology" (E. Rouslahti and E. Engvall, eds.), vol. 245, pp. 62–85. Academic Press, San Diego, CA.

Bowditch, R. D., Hariharan, M., Tominna, E. F., *et al.* (1994). Identification of a novel integrin binding site in fibronectin. *J. Biol. Chem.* **269**, 10856–10863.

Briggs, M. D., Hoffman, S.M.G., King, L. M., *et al.* (1995). Pseudoachondroplasia and multiple epiphyseal dysplasia due to mutations in the cartilage oligomeric matrix protein gene. *Nat. Genet.* **10**, 330–336.

Brooks, P. C., Montgomery, A. M., Rosenfeld, M., *et al.* (1994). Integrin alpha v beta 3 antagonists promote tumor regression by inducing apoptosis of angiogenic blood vessels. *Cell (Cambridge, Mass.)* **79**, 1157–1164.

Brown, J. C., Mann, K., Wiedemann, H., *et al.* (1993). Structure and binding properties of collagen type XIV isolated from human placenta. *J. Cell Biol.* **120**, 557–567.

Chiquet-Ehrismann, R. (1995). Inhibition of cell adhesion by anti-adhesive molecules. *Curr. Opin. Cell Biol.* **7**, 715–719.

Colige, A., Beschin, A., Samyn, B., *et al.* (1995). Characterization and partial amino acid sequencing of a 107-kDa pro-collagen I N-proteinase purified by affinity chromatography on immobilized type XIV collagen. *J. Biol. Chem.* **270**, 16724–16730.

David, G. (1993). Integral membrane heparan sulfate proteoglycans. *FASEB J.* **7**, 1023–1030.

Dhanabal, M., Ramchandran, R., Volk, R., *et al.* (1999). Endostatin: Yeast production, mutants, and antitumor effect in renal carcinoma. *Cancer Res.* **59**, 189–197.

Ding, Y. H., Javaherian, K., Lo, K. M., *et al.* (1998). Zinc-dependent dimers observed in crystals of human endostatin. *Proc. Natl. Acad. Sci. U.S.A.* **95**, 10443–10448.

Erickson, H. P. (1993). Tenascin-C, tenascin-R and tenascin-X: A family of talented proteins in search of functions. *Curr. Opin. Cell Biol.* **5**, 869–876.

Faessler, R., Schnegelsberg, P.N.J., Dausman, J., *et al.* (1994). Mice lacking a1(IX) collagen develop noninflammatory degenerative joint disease. *Proc. Natl. Acad. Sci. U.S.A.* **91**, 5070–5074.

Font, B., Aubert-Foucher, E., Goldschmidt, D., *et al.* (1993). Binding of collagen XIV with the dermatan sulfate side chain of decorin. *J. Biol. Chem.* **268**, 25015–25018.

Garofalo, S., Metsaranta, M., Ellard, J., *et al.* (1993). Assembly of cartilage collagen fibrils is disrupted by overexpression of normal type II collagen in transgenic mice. *Proc. Natl. Acad. Sci. U.S.A.* **90**, 3825–3829.

George, E. L., Georges-Labouesse, E. N., Patel-King, R. S., *et al.* (1993). Defects in mesoderm, neural tube and vascular development in mouse embryos lacking fibronectin. *Development (Cambridge, UK)* **119**, 1079–1091.

Grant, D. S., Tashiro, K., Segui-Real, B., *et al.* (1989). Two different laminin domains mediate the differentiation of human endothelial cells into capillary-like structure *in vitro. Cell (Cambridge, Mass.)* **58**, 933–943.

Halfter, W., Dong, S., Schurer, B., *et al.* (1998). Collagen XVIII is a basement membrane heparan sulfate proteoglycan. *J. Biol. Chem.* **273**, 25404–25412.

Hay, E. D. (1991). "Cell Biology of Extracellular Matrix." Plenum, New York.

Hedbom, E., and Heinegard, D. (1993). Binding of fibromodulin and decorin to separate sites on fibrillar collagens. *J. Biol. Chem.* **268**, 27307–27312.

Hildebrand, A., Romaris, M., Rasmussen, L. M., *et al.* (1994). Interaction of the small interstitial proteoglycans biglycan, decorin and fibromodulin with transforming growth factor beta. *Biochem. J.* **302**, 527–534.

Hohenester, E., Sasaki, T., Olsen, B. R., *et al.* (1998). Crystal structure of the angiogenesis inhibitor endostatin at 1.5 Å resolution. *EMBO J.* **17**, 1656–1664.

Humphries, M. J. (1993). Fibronectin and cancer: Rationales for the use of antiadhesives in cancer treatment. *Semin. Cancer Biol.* **4**, 293–299.

Hynes, R. O. (1990). "Fibronectin." Springer-Verlag, New York.

Iozzo, R. V. (1994). Perlecan: A gem of a proteoglycan. *Matrix Biol.* **14**, 203–208.

Iruela-Arispe, M. L., and Sage, E. H. (1991). Expression of type VIII collagen during morphogenesis of the chicken and mouse heart. *Dev. Biol.* **144**, 107–118.

Jacenko, O., Olsen, B., and Lu Valle, P. (1991). Organization and regulation of collagen genes. *Crit. Rev. Eukaryotic Gene Expression* **1**, 327–353.

Kanzaki, T., Olofsson, A., Moren, A., *et al.* (1990). TGF-beta 1 binding protein: A component of the large latent complex of TGF-beta 1 with multiple repeat sequences. *Cell (Cambridge, Mass.)* **61**, 1051–1061.

Kim, C. W., Goldberger, O. A., Gallo, R. L., *et al.* (1994). Members of the syndecan family of heparan sulfate proteoglycans are expressed in distinct cell-, tissue-, and development-specific patterns. *Mol. Biol. Cell* **5**, 797–805.

Kramer, R. H. (1994). Characterization of laminin-binding integrins. *In* "Methods in Enzymology" (E. Ruoslahti and E. Engvall, eds.), vol. 245, pp. 129–147. Academic Press, San Diego, CA.

Kresse, H., Hausser, H., Schonherr, E., *et al.* (1994). Biosynthesis and interactions of small chondroitin/dermatan sulphate proteoglycans. *Eur. J. Clin. Chem. Clin. Biochem.* **32**, 259–264.

Laurent, T. C., and Fraser, J. R. (1992). Hyaluronan. *FASEB J.* **6**, 2397–2404.

Lesley, J., Hyman, R., and Kincade, P. W. (1993). CD44 and its interaction with extracellular matrix. *Adv. Immunol.* **54**, 271–335.

Li, Y., Lacerda D. A., Warman, M. L., *et al.* (1995). A fibrillar collagen gene. *Coll1a1,* is essential for skeletal morphogenesis. *Cell (Cambridge, Mass.)* **80**, 423–430.

Linsenmayer, T. F. (1991). Collagen. *In* "Cell Biology of Extracellular Matrix" (E. D. Hay, ed.), pp. 7–44. Plenum, New York.

Linsenmayer, T. F., Gibney, E., Igoe, F., *et al.* (1993). Type V collagen: Molecular structure and fibrillar organization of the chicken alpha 1(V) NH$_2$-terminal domain, a putative regulator of corneal fibrillogenesis. *J. Cell Biol.* **121**, 1181–1189.

Liu, Z., Diaz, L. A., and Giudice, G. J. (1994). Autoimmune response against the bullous pemphigoid 180 autoantigen. *Dermatology* **1**, 34–37.

MacBeath, J. R., Kielty, C. M., and Shuttleworth, C. A. (1996). Type VIII collagen is a product of vascular smooth-muscle cells in development and disease. *Biochem. J.* **319**, 993–998.

Mali, M., Andtfolk, H., Miettinen, H. M., *et al.* (1994). Suppression of tumor cell growth by syndecan-1 ectodomain. *J. Biol. Chem.* **269**, 27795–27798.

Mason, I. J. (1994). The ins and outs of fibroblast growth factors. *Cell (Cambridge, Mass.)* **78**, 547–552.

Matsumura, K., and Campbell, K. P. (1994). Dystrophin–glycoprotein complex: Its role in the molecular pathogenesis of muscular dystrophies. *Muscle Nerve* **17**, 2–15.

McGrath, J. A., Gatalica, B., Christiano, A. M., *et al.* (1995). Mutations in the 180-kDa bullous pemphigoid antigen (BPAG2), a hemidesmosomal transmembrane collagen COL17A1), in generalized atrophic benign epidermolysis bullosa. *Nat. Genet.* **11**, 83–86.

Mochizuki, T., Lemmink, H. H., Mariyama, M., *et al.* (1994). Identification of mutations in the alpha 3(IV) and alpha 4(IV) collagen genes in autosomal recessive Alport syndrome. *Nat. Genet.* **8**, 77–81.

Mooney, D., Hansen, L., Vacanti, J., *et al.* (1992). Switching from differentiation to growth in hepatocytes: Control by extracellular matrix. *J. Cell Physiol.* **151**, 497–505.

Morgelin, M., Heinegard, D., Engel, J., *et al.* (1994). The cartilage proteoglycan aggregate: assembly through combined protein–carbohydrate and protein–protein interactions. *Biophys. Chem.* **50**, 113–128.

Mosher, D. F., Fogerty, F. J., Chernousov, M. A., *et al.* (1991). Assembly of fibronectin into extracellular matrix. *Ann. N. Y. Acad. Sci.* **614**, 167–180.

Muragaki, Y., Jacenko, O., Apte, S., et al. (1991). The alpha 2(VIII) collagen gene. A novel member of the short chain collagen family located on the human chromosome 1. *J. Biol. Chem.* **266**, 7721–7727.

Muragaki, Y., Timmons, S., Griffith, C. M., *et al.* (1995). Mouse *Col18a1* is expressed in a tissue-specific manner as three alternative variants and is localized in basement membrane zones. *Proc. Natl. Acad. Sci. U.S.A.* **92**, 8763–8767.

Muragaki, Y., Mariman, E. C., van Beersum, S. E., *et al.* (1996). A mutation in the gene encoding the alpha 2 chain of the fibril-associated collagen IX, *COL9A2,* causes multiple epiphyseal dysplasia (EDM2). *Nat. Genet.* **12**, 103–105.

Nakata, K., Ono, K., Miyazaki, J., *et al.* (1993). Osteoarthritis associated with mild chondrodysplasia in transgenic mice expressing alpha 1(IX) collagen chains with a central deletion. *Proc. Natl. Acad. Sci. U.S.A.* **90**, 2870–2874.

Newton, G., Weremowicz, S., Morton, C. C., *et al.* (1994). Characterization of human and mouse cartilage oligomeric matrix protein. *Genomics* **24**, 435–439.

Niessen, C. M., Hogervorst, F., Jaspars, L. H., *et al.* (1994). The alpha 6 beta 4 integrin is a receptor for both laminin and kalinin. *Exp. Cell Res.* **211**, 360–367.

Nishiyama, T., McDonough, A. M., Bruns, R. R., *et al.* (1994). Type XII and XIV collagens mediate interactions between banded collagen fibers *in vitro* and may modulate extracellular matrix deformability. *J. Biol. Chem.* **269**, 28193–28199.

Noakes, P. G., Gautam, M., Mudd, J., *et al.* (1995). Aberrant differentiation of neuromuscular junctions in mice lacking s-laminin/laminin beta 2. *Nature (London)* **374**, 258–262.

Oh, S. P., Kamagata, Y., Muragaki, Y., *et al.* (1994). Isolation and sequencing of cDNAs for proteins with multiple domains of Gly-X-Y repeats identify a novel family of collagenous proteins. *Proc. Natl. Acad. Sci. U.S.A.* **91**, 4229–4233.

Olsen, B. R. (1991). Collagen biosynthesis. *In* "Cell Biology of Extracellular Matrix" (E. D. Hay, ed.), pp. 177–220. Plenum, New York.

Olsen, B. R. (1995). New insights into the function of collagens from genetic analysis. *Curr. Opin. Cell Biol.* **7**, 720–727.

Olsen, B. R., Winterhalter, K. H., and Gordon, M. K. (1995). FACIT collagens and their biological roles. *Trends Glycosci. Glycotechnol.* **7**, 115–127.

O'Reilly, M. S., Boehm, T., Shing, Y., *et al.* (1997). Endostatin: An endogenous inhibitor of angiogenesis and tumor growth. *Cell (Cambridge, Mass.)* **88**, 277–285.

Potts, J. R., and Campbell, I. D. (1994). Fibronectin structure and assembly. *Curr. Opin. Cell Biol.* **6**, 648–655.

Pulkkinen, L., Christiano, A. M., Gerecke, D., *et al.* (1994). A homozygous nonsense mutation in the beta-3 chain gene of laminin 5 (Lamb3) in Herlitz junctional epidermolysis bullosa. *Genomics* **24**, 357–360.

Ramirez, F., Pereira, L., Zhang, H., *et al.* (1993). The fibrillin-Marfan syndrome connection. *BioEssays* **15**, 589–594.

Rehn, M., and Pihlajaniemi, T. (1994). $\alpha 1$(XVIII), a collagen chain with frequent interruptions in the collagenous sequence, a distinct tissue distribution, and homology with type XV collagen. *Proc. Natl. Acad. Sci. U.S.A.* **91**, 4234–4238.

Rehn, M., and Pihlajaniemi, T. (1995). Identification of three N-terminal ends of type XVIII collagen chains and tissue-specific differences in the expression of the corresponding transcripts. The longest form contains a novel motif homologous to rat and *Drosophila* frizzled proteins. *J. Biol. Chem.* **270**, 4705–4711.

Rogalski, T. M., Williams, B. D., Mullen, G. P., *et al.* (1993). Products of the unc-52 gene in *Caenorhabditis elegans* are homologous to the core protein of the mammalian basement membrane heparan sulfate proteoglycan. *Genes Dev.* **7**, 1471–1484.

Rosenbloom, J., Abrams, W. R., and Mecham, R. (1993). Extracellular matrix 4: The elastic fiber. *FASEB J.* **7**, 1208–1218.

Roskelley, C. D., Srebrow, A., and Bissell, M. J. (1995). A hierarchy of ECM-mediated signalling regulates tissue-specific gene expression. *Curr. Opin. Cell Biol.* **7**, 736–747.

Saarela, J., Rehn, M., Oikarinen, A., *et al.* (1998). The short and long forms of type XVIII collagen show clear tissue specificities in their expression and location in basement membrane zones in humans. *Am. J. Pathol.* **153**, 611–626.

Saga, Y., Yagi, T., Ikawa, Y., *et al.* (1992). Mice develop normally without tenascin. *Genes Dev.* **6**, 1821–1831.

Sakai, L. Y., and Keene, D. R. (1994). Fibrillin: Monomers and microfibrils. *In* "Methods in Enzymology" (E. Ruoslahti and E. Engvall, vol. 245, pp. 29–52. Academic Press, San Diego, CA.

Sasaki, T., Fukai, N., Mann, K., *et al.* (1998). Structure, function and tissue forms of the C-terminal globular domain of collagen XVIII containing the angiogenesis inhibitor endostatin. *EMBO J.* **17**, 4249–4256.

Schwarzbauer, J. (1990). The fibronectin gene. *In* "Extracellular Matrix Genes" (L. J. Sandell and C. D. Boyd, eds.), pp. 195–219. Academic Press, San Diego, CA.

Shaw, L. M., and Olsen, B. R. (1991). Collagens in the FACIT group: Diverse molecular bridges in extracellular matrices. *Trends Biochem Sci.* **16**, 191–194.

Sherman, L., Sleeman, J., Herrlich, P., *et al.* (1994). Hyaluronate receptors: Key players in growth, differentiation, migration and tumor progression. *Curr. Opin. Cell Biol.* **6**, 726–733.

Sibinga, N. E., Foster, L. C., Hsieh, C.-M., *et al.* (1997). Collagen VIII is expressed by vascular smooth muscle cells in response to vascular injury. *Circ. Res.* **80**, 532–541.

Sonnenberg, A., Calafat, J., Janssen, H., *et al.* (1991). Integrin alpha 6/beta 4 complex is located in hemidesmosomes, suggesting a major role in epidermal cell-basement membrane adhesion. *J. Cell Biol.* **113**, 907–917.

Timpl, R. (1994). Proteoglycans of basement membranes. *Exs* **70**, 123–144.

Timpl, R. (1996). Macromolecular organization of basement membranes. *Curr. Opin. Cell Biol.* **8**, 618–624.

Timpl, R., and Brown, J. C. (1994). The laminins. *Matrix Biol.* **14**, 275–281.

Tryggvason, K., Zhou, J., Hostikka, S. L., *et al.* (1993). Molecular genetics of Alport syndrome. *Kidney Int.* **43**, 38–44.

Turley, E. A. (1992). Hyaluronan and cell locomotion. *Cancer Metastasis Rev.* **11**, 21–30.

Uitto, J., Pulkkinen, L., and Christiano, A. M. (1994). Molecular basis of the dystrophic and junctional forms of epidermolysis bullosa: Mutations in the type VII collagen and kalinin (laminin 5) genes. *J. Invest. Dermatol.* **103**, 39S–46S.

van der Rest, M., Mayne, R., Ninomiya, Y., *et al.* (1985). The structure of type IX collagen. *J. Biol. Chem.* **260**, 220–225.

Vaughan, L., Mendler, M., Huber, S., *et al.* (1988). D-Periodic distribution of collagen type IX along cartilage fibrils. *J. Cell Biol.* **106**, 991–997.

Vaughan, L., Weber, P., D'Allessandri, L., *et al.* (1994). Tenascin–contactin/F11 interactions: A clue for a developmental role? *Perspect. Dev. Neurobiol.* **2**, 43–52.

Viljoen, D. (1994). Congenital contractural arachnodactyly (Beals syndrome). *J. Med. Genet.* **31**, 640–643.

Vuorio, E., and de Crombrugghe, B. (1990). The family of collagen genes. *Annu. Rev. Biochem.* **59**, 837–872.

Yamaguchi, N., Benya, P. D., van der Rest, M., *et al.* (1989). The cloning and sequencing of alpha 1(VIII) collagen cDNAs demonstrate that type VIII collagen is a short chain collagen and contains triple-helical and carboxyl-terminal non-triple-helical domains similar to those of type X collagen. *J. Biol. Chem.* **264**, 16022–16029.

Yamaguchi, N., Mayne, R., and Ninomiya, Y. (1991). The alpha 1 (VIII) collagen gene is homologous to the alpha 1 (X) collagen gene and contains a large exon encoding the entire triple helical and carboxyl-terminal non-triple helical domains of the alpha 1 (VIII) polypeptide. *J. Biol. Chem.* **266**, 4508–4513.

Yamaguchi, N., Anand-Apte, B., Lee, M., *et al.* (1999). Endostatin inhibition of endothelial cell migration and tumor growth at picomolar levels does not depend on heparin- or zinc-binding. *EMBO J.* 18, 4414–4423.

Yurchenco, P. D., and O'Rear, J. J. (1994). Basal lamina assembly. *Curr. Opin. Cell Biol.* **6**, 674–681.

Zhou, J., Mochizuki, T., Smeets, H., *et al.* (1993). Deletion of the paired alpha 5(IV) and alpha 6(IV) collagen genes in inherited smooth muscle tumors. *Science* **261**, 1167–1169.

CHAPTER 6

INDUCTIVE PHENOMENA

Matthias Hebrok and Douglas A. Melton

INTRODUCTION

Vertebrate organogenesis is a multistep process involving regulated cell–cell interactions between mesenchymal and epithelial tissues. These interactions, or inductive phenomena, are orchestrated by signaling molecules that govern proliferation, migration, and morphogenesis. In general, two types or inductive phenomena have been distinguished: instructive induction and permissive induction (Slack, 1983). Instructive interactions are defined as those cases in which the responding tissue is naive and its path of differentiation depends on the signal(s) it receives. A permissive interaction takes place when the receiving tissue is already committed to one particular differentiation pathway but requires an additional signal to initiate the process. In this chapter both types of inductive phenomena are described, focusing on tissues and signaling pathways involved during formation of gut-derived organs, including lung and pancreas.

EPITHELIAL TO MESENCHYMAL SIGNALING IN ENDODERM DEVELOPMENT

Organogenesis is initiated after gastrulation when the three germ layers, ectoderm, mesoderm, and endoderm, are formed and the basic body plan is established. During gastrulation endodermal and mesodermal precursors are separated from each other. The endodermal cells form an epithelial sheet that extends along the anteroposterior (AP) axis of the developing embryo, subsequently generating the embryonic gut. The first morphogenetic sign of gut tube formation is the folding of the anterior intestinal portal (AIP) at the rostral end, close to the forming head ectoderm. Soon after, the caudal intestinal portal (CIP) is generated at the posterior end of the embryo. Both structures grow and extend toward each other, eventually meeting and fusing in the middle, where they form the yolk stalk. Initially, the primitive gut is subdivided into fore-, mid-, and hindgut along the AP axis and from each of these regions specific organs will be generated during subsequent development, including thyroid, lungs, stomach, pancreas, liver, and small and large intestine (Gilbert, 1994; Roberts *et al.,* 1998).

Inductive interactions play an important role during morphogenesis and are involved in the formation of visceral organs of the gut. Several studies have described the cross-talk between the mesodermal and endodermal part of visceral organs (Golosow and Grobstein, 1962; Roberts *et al.,* 1998; Shannon, 1994; Shannon *et al.,* 1998; Wessells and Cohen, 1967). Here, the focus is on examples of epithelial–mesenchymal interactions leading to formation of mature organs along the AP axis of the gut. Inductive events during lung development are described and the role of signaling molecules involved in this process is discussed. Furthermore, we suggest that similar inductive mechanisms, mediated by the same signaling pathways, play important roles during several stages of pancreas development.

LUNG DEVELOPMENT AND INSTRUCTIVE SIGNALING

The mammalian respiratory system consists of two separate structures, the trachea and the lung (Hogan and Yingling, 1998; Ten Have-Opbroek, 1981; Wessells, 1970). Both parts are de-

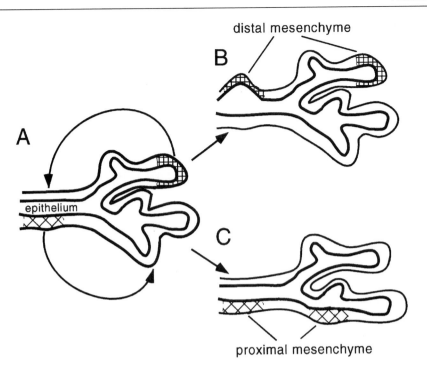

Fig. 6.1. Instructive signaling during lung development. (A) Proximal and distal lung mesenchymes possess different capacities to induce region-specific morphogenesis. Proximal mesenchyme governs tracheal development characterized by growth of a straight tube, whereas distal mesenchyme instructs endodermal budding to assure proper formation of lung morphology. (B) When transplanted to the proximal part of the trachea, distal mesenchyme induces budding in otherwise straight tracheal epithelium, whereas (C) proximal mesenchyme will inhibit bud formation in lung endoderm.

rived from ventral, anterior endoderm and need instructive signals from overlying mesenchyme for proper development (Hogan, 1999). Transplantation of bronchial mesenchyme to tracheal endoderm induces ectopic branching and expression of genes specific of bronchial endoderm, whereas tracheal mesenchyme suppresses epithelial budding when transplanted to distal lung epithelium (Alescio and Cassini, 1962; Shannon, 1994; Shannon *et al.,* 1998; Wessells, 1970) (Fig. 6.1). Specific genes and pathways have been identified that are involved in these processes (Bellusci *et al.,* 1997a,b). Fibroblast growth factors (FGFs) govern cell differentiation and regulation of organ formation (Martin, 1998); inhibition of FGF-signaling, through ectopic expression of a dominant negative FGF receptor, interferes with formation of secondary bronchi and terminal lung buds (Peters *et al.,* 1994). FGF-10, one of the ligands expressed in distal mesenchyme, has emerged as a factor involved in lung budding. In coculture experiments distal epithelium grows toward an FGF-10-soaked bead, reminiscent of the normal budding observed in developing/growing embryos (Bellusci *et al.,* 1997b; Park *et al.,* 1998). Furthermore, reports have described that FGF-10 also induces budding in tracheal endoderm stripped of mesenchyme, suggesting that FGF-10 might mimic the instructive effect of distal mesenchyme when grafted around tracheal endoderm (Hogan and Yingling, 1998; Park *et al.,* 1998). The importance of FGF signaling has been further confirmed in FGF-10 mutant mice, which are deficient in lung bud formation (Min *et al.,* 1998; Sekine *et al.,* 1999).

Another gene involved in lung development encodes Sonic hedgehog (Shh), a secreted glycoprotein and member of the hedgehog family of signaling molecules. Shh is expressed in foregut endoderm and is up-regulated in distal lung epithelium where branching occurs (Bellusci *et al.,* 1997a). Overexpression studies revealed that Shh acts as an endodermal signal that induces cell proliferation in lung mesenchyme (Bellusci *et al.,* 1997a). Loss-of-function experiments have shown that lung development is disturbed in Shh mutant embryos and in mice bearing mutations in downstream targets of hedgehog signaling, including zinc finger transcription factors Gli2 and Gli3 (Litingtung *et al.,* 1998; Motoyama *et al.,* 1998; Pepicelli *et al.,* 1998). Furthermore, Pepicelli *et al.* provided evidence that in lungs of Shh$^{-/-}$ mice, FGF-10 expression is up-regulated, suggesting that high levels of Shh in wild-type embryos might interfere with FGF-10 expression in adjacent mesenchyme. Additional evidence of FGF signaling in respiratory organs comes from studies on tracheal development in *Drosophila,* in which several genes, including FGF (branchless), FGF receptor (breathless), and an inhibitor of FGF signaling (sprouty), have been implicated in this process (reviewed in Hogan and Yingling, 1998).

These studies demonstrate that mesenchymal–epithelial interactions are required for lung development. They indicate that epithelial factors and mesenchymal genes influence cell proliferation and morphogenetic budding, and future work will further clarify how signaling pathways interact to instruct endodermal differentiation in respiratory organs.

PERMISSIVE SIGNALING DURING PANCREAS DEVELOPMENT

For many decades pancreas development has been a preferred object to study general patterns of organ formation. During embryogenesis a dorsal evagination of the duodenum caudal to the stomach is the first morphological sign of pancreas formation. Shortly afterward ventral pancreas buds form adjacent to the liver diverticulum, and subsequently, when the stomach and duodenum move during gut rotation, the ventral bud comes into close contact with the dorsal lobe to fuse and form the mature pancreas (Fig. 6.2) (Slack, 1995).

Before molecular data were available researchers used the pancreas to investigate principles of epithelial–mesenchymal interactions. These experiments showed that mesenchyme is necessary to allow proper formation of mature exocrine structures (Golosow and Grobstein, 1962; Wessells and Cohen, 1967). More recently, numerous transcription factors have been implicated in different aspects of pancreas development (reviewed in Edlund, 1998). At least two of these, Pdx-1 and Isl-1 are involved in epithelial–mesenchymal signaling, and mutations in the genes for these factors interfere with morphogenesis and differentiation of both exocrine and endocrine lineages (Ahlgren *et al.,* 1996, 1997; Jonsson *et al.,* 1994; Offield *et al.,* 1996). Furthermore, mutations in some transcription factors, including Pdx-1, HNF-1β, HNF-1α, and HNF-4α, are linked to formation of maturity-onset diabetes (MODY) (Ahlgren *et al.,* 1998; Horikawa *et al.,* 1997; Stoffers *et al.,* 1997; Yamagata *et al.,* 1996a,b).

Classic experiments suggested that mesenchyme is required for outgrowth and differentiation of pancreas epithelium. At early developmental stages pancreas-specific mesenchyme is required to promote epithelial differentiation. Subsequently, however, heterologous mesenchyme (derived from salivary gland) or embryo extracts are sufficient for epithelial differentiation (Golosow and Grobstein, 1962; Rutter *et al.,* 1964; Wessells and Cohen, 1967), indicating that mesenchyme provides permissive signal(s) that promote induction of pancreas formation. Ronzio and Rutter (1973) did succeed in partially purifying a mesenchyme-specific factor with pancreas-inducing properties, although the identity of this signal(s) is, despite extensive efforts, still unknown. Homologous recombination studies in mice have revealed the influence of specific gene products on mesenchyme-to-epithelial signaling. Isl-1, a homeobox-containing transcription factor expressed in dorsal pancreas mesenchyme and all developing endocrine cells, is involved in two different aspects of pancreas formation. Inactivation studies have shown that Isl-1 promotes endoderm differentiation in a cell-autonomous fashion. Another function requires mesenchymal expression and is crucial for development of dorsal pancreas structures, but is not essential for formation of ventral pancreas tissue (Ahlgren *et al.,* 1997). In addition, a novel gene expressed in ventral but not dorsal pancreas mesenchyme has been identified, although the function of this gene is currently unknown (Hebrok *et al.,* 1999). These studies point to mesenchyme as an important tissue involved in pancreas organogenesis and growing evidence supports the hypothesis that morphogenesis is, at least in part, regulated by separate dorsal–ventral mechanisms.

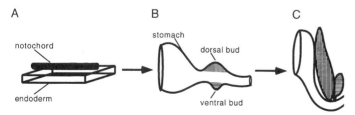

Fig. 6.2. Stages of pancreas development. (A) During early stages of pancreas organogenesis notochord is in immediate contact with dorsal medial endoderm and signals the epithelium to initiate pancreas formation. (B) The first sign of pancreas morphogenesis is an epithelial thickening of the dorsal endoderm caudal to the stomach anlage. Ventral pancreas buds form shortly afterward, close to the liver diverticulum. (C) During the process of gut rotation the ventral lobe turns dorsally and comes into close contact with the dorsal lobe, eventually forming the mature pancreas.

Fig. 6.3. Initiation of pancreas marker genes in isolated endoderm by notochord and notochord factors. Isolated chick endoderm can be induced to express pancreas marker genes, including those for Pdx-1 and insulin, when cocultured either with notochord or with notochord factors such as activin βB and FGF-2 (A, C). Isolated endoderm cultured without any additional factors does not express pancreas markers. β-Tubulin is the positive control for polymerase chain reactions (PCR).

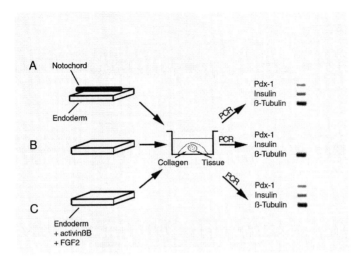

Although mesenchyme–epithelial interactions are indispensable at later stages of development, they cannot be responsible for initial organ induction. At the time when pancreas formation is first induced, prepancreatic dorsal epithelium is not surrounded by mesenchyme but instead contacts the notochord (Kim *et al.*, 1997; Pictet *et al.*, 1972; Wessells and Cohen, 1967), a mesodermally derived structure known to influence patterning of several tissues, including ventral floor plate and somites (Ebensperger *et al.*, 1995; Pourquie *et al.*, 1993). During subsequent development, notochord and endoderm are separated by the dorsal aorta, thereby limiting the time when endoderm can receive notochord signals. To understand if notochord directly influences expression of pancreas-specific genes, an *in vitro* system was exploited, wherein isolated endoderm is cultured alone or in combination with notochord. Results from these studies revealed that expression of pancreas marker genes, including those for insulin or Pdx-1, is initiated in endoderm–notochord cocultures. In contrast, in samples in which endoderm is cultured alone, these markers are not expressed (Hebrok *et al.*, 1998; Kim *et al.*, 1997) (Fig. 6.3). In addition, notochord fails to induce expression of pancreas markers when cultured with posterior endoderm outside the pancreas anlage. This suggests that endoderm is prepatterned at this early stage of development and that the notochord sends permissive rather than instructive signal(s) (Kim *et al.*, 1997).

The notochord is a source of several inducing molecules, including members of the transforming growth factor β (TGF-β) and fibroblast growth factor family. Using the established *in vitro* culture system it was possible to identify specific factors that mimic the notochord's effect on isolated endoderm, namely, inducing expression of insulin and Pdx-1. Two factors, activin βB (a TGF-β molecule) and FGF-2, but not FGF-1 or FGF-7, can induce expression of pancreas marker genes in isolated endoderm (Fig. 6.3). Interestingly, activin increases expression of both markers in a direct concentration-dependent manner, whereas FGF-2 acts positively only at low concentrations, and higher concentrations of FGF-2 even interfere with activin-induced induction (Hebrok *et al.*, 1998). Therefore, at least two separate signaling pathways could be involved during early stages of pancreas development.

In addition to understanding which signals can induce pancreas development in prepatterned endoderm it is also crucial to determine which epithelial genes respond to these cues. Previous work had established that activin inhibits Shh expression in the node region of early chick embryos, thereby initiating asymmetric gene expression involved in left–right asymmetry (Levin *et al.*, 1997). Although Shh is expressed throughout gut epithelium along the AP axis, it is not detected in the pancreas anlage (Ahlgren *et al.*, 1997; Bitgood and McMahon, 1995; Hebrok *et al.*, 1998). During early stages Shh is found in ventral gut epithelium but is excluded from dorsal regions where endoderm contacts notochord, suggesting that notochord signals negatively regulate endodermal Shh expression. Subsequent experiments revealed that ectopic notochord interferes with Shh expression in gut epithelium and that activin βB can substitute for notochord. Furthermore, it was shown that Shh interferes with induction of pancreas markers, arguing that hedge-

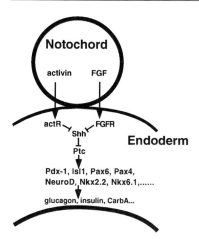

Fig. 6.4. Molecular pathways regulating initiation of pancreas development. Notochord-derived factors, including activin and FGF, presumably act through their respective receptors to inhibit Shh expression in pancreas epithelium. This inhibition allows expression of transcription factors required for activation of endocrine hormones and exocrine enzymes.

hogs have to be excluded from pancreas epithelium to allow proper pancreas differentiation (Hebrok *et al.*, 1998) (Fig. 6.4). This hypothesis is promoted by the fact that ectopic expression of Shh later in development, under the control of the Pdx-1 promoter, leads to aberrant development of adjacent mesenchyme, adopting an intestine-like morphology (Apelqvist *et al.*, 1997).

Another function of firmly controlled hedgehog expression in stomach and duodenum might be to restrict the area of pancreas anlage and thereby regulate organ size. Evidence for this mechanism comes from experiments in which isolated chick stomach rudiments were cultured with cyclopamine, an inhibitor of hedgehog signaling (Cooper *et al.*, 1998). Examination of these rudiments revealed epithelial budding reminiscent of early pancreas buds as well as localized clusters of insulin- and glucagon-positive cells. Cyclopamine treatment of whole embryos results in formation of ectopic pancreas buds in regions expressing pancreatic transcription factor Pdx-1, concordant with the idea that pancreas formation requires epithelial expression of Pdx-1 (Kim and Melton, 1998). Cyclopamine, a steroid alkaloid teratogen similar to cholesterol, interferes with hedgehog signaling, presumably by interacting with its receptor ptc, and Shh mutant mice as well as cyclopamine-treated embryos develop holoprosencephaly (HPE) (Cooper *et al.*, 1998; Marigo *et al.*, 1996). Milder forms of HPE are observed in some patients suffering from Smith–Lemli–Opitz syndrome, a disorder that is linked to a defect in cholesterol synthesis. Interestingly, Smith–Lemli–Opitz syndrome is also characterized by appearance of nesidioblastosis and hypoglycemia (Lachman *et al.*, 1991; McKeever and Young, 1990), further suggesting that blocking of hedgehog signaling positively influences pancreas development and function.

SIGNALING DURING LATE STAGES OF PANCREAS DEVELOPMENT

A variety of factors, including TGF-β, FGF, and hedgehog, are involved during induction and early stages of organogenesis. Experiments from several groups have shown that these pathways also accompany formation of more specialized organ structures. Inactivation studies of FGF ligands proved their involvement in differentiation of several organs, including development of ear, tail, lung, and adult muscle generation (Floss *et al.*, 1997; Mansour *et al.*, 1993; Min *et al.*, 1998; Sekine *et al.*, 1999). FGF ligands act through their receptors, a family of four different transmembrane molecules with cytoplasmic tyrosine kinase activity (FGFR1–4). These receptors are expressed in tissue-specific patterns and inactivation studies using a soluble dominant negative form (dnFGFR) provided evidence that FGF signaling is involved in a variety of multiorgan induction and patterning events. Overexpression of dnFGFR induces an embryonic lethal phenotype and disrupts formation of specific organs, including failure to develop kidneys and proper respiratory system as well as other malformations. In the gastrointestinal region stomach, liver, and pancreas are affected, the latter consisting of a decrease in exocrine acinar cells and an apparent lack of islets of Langerhans (Celli *et al.*, 1998). Furthermore, two of the FGF receptors, FGFR1 and FGFR4, are expressed in pancreas anlage during development, and FGF-2 stimulates proliferation of pancreatic epithelial cells (Le Bras *et al.*, 1998). However, mice bearing mutations in the FGF-2 gene are viable and phenotypically almost indistinguishable from wild-type littermates (Dono *et al.*, 1998; Ortega *et al.*, 1998; Zhou *et al.*, 1998), suggesting that pancreas formation and function are not severely affected in these mutants. Alternatively, it is possible that the remaining members of the FGF family can substitute for FGF-2 function and that pancreas formation is regulated by the combination of these factors.

In addition to FGF signaling, TGF-β molecules have been reported to affect pancreas function. Activin A is expressed during pancreas development (Furukawa *et al.,* 1995a) and is also detected in the mature organ in discrete islet cells including, α and δ cells (Ogawa *et al.,* 1993; Yasuda *et al.,* 1993). Insulin secretion can be stimulated by activin even in the absence of glucose (Furukawa *et al.,* 1995b), and branching morphogenesis of cultured pancreas rudiments is altered after activin treatment (Ritvos *et al.,* 1995). Accumulating evidence points to TGF-β signaling being specifically involved in differentiation of endocrine versus exocrine cells. Follistatin, an inhibitor of TGF-β signaling, is expressed in pancreatic mesenchyme. It has been shown to induce exocrine development and represses endocrine development in isolated pancreas rudiments (Miralles *et al.,* 1998). Furthermore, overexpression of a dominant negative activin receptor under the control of the insulin promoter in transgenic mice leads to smaller islet areas and reduced insulin content when compared to wild-type littermates. Insulin response in glucose-challenged animals is severely affected, suggesting that the remaining β cells are dysfunctional and do not respond properly (Yamaoka *et al.,* 1998). Interestingly, transgenic mice bearing a constitutively active activin receptor also have a reduced number of islets, indicating that the level of activin signaling may require tight regulation for proper pancreas function (Yamaoka *et al.,* 1998). In addition, formation of pancreas structures also depends on related TGF-β factors. Ectopic expression of a dominant negative TGF-β receptor results in increased proliferation of exocrine cells and severely perturbed acinar differentiation, indicating that TGF-β negatively regulates exocrine development and is required for maintenance of the differentiated acinar phenotype (Böttinger *et al.,* 1997).

The identification of distinct signaling pathways has broadened our understanding of general principles of organogenesis. An exciting aspect of developmental biology will be to better understand the puzzle how the same small number of signaling pathways specifies formation of numerous distinct organs. In the case of FGF signaling, we know of at least 17 FGF ligands and four FGF receptors (Martin, 1998), suggesting that generation of individual organs requires signaling from specific combinations of ligands and receptors as well as interactions between different signaling pathways. A future aspect of research will be to investigate how gene products from different pathways, e.g., TGF-β, FGF, and hedgehog molecules, interact during embryonic development and if combination of these factors can govern cell differentiation *in vitro*. These experiments will be useful to explore further the potential of signaling molecules in directing differentiation of undetermined cells, e.g., murine or human stem cells, toward highly specialized tissues and perhaps organs. Studies aimed to investigate the function of FGF, TGF-β, and hedgehog pathways will not only help to understand how organs form during embryogenesis, but also provide knowledge about maintenance of mature organ function. Insights from these experiments might be used to restore functional properties in individuals suffering from diseases, for whom the capacity to perform specific tasks has been lost. In terms of pancreas research a long-term goal will be to help patients suffering from diabetes mellitus and pancreatic cancer.

References

Ahlgren, U., Jonsson, J., and Edlund, H. (1996). The morphogenesis of the pancreatic mesenchyme is uncoupled from that of the pancreatic epithelium in IPF1/PDX1-deficient mice. *Development (Cambridge, UK)* **122,** 1409–1416.

Ahlgren, U., Pfaff, S. L., Jessell, T. M., Edlund, T., and Edlund, H. (1997). Independent requirement for ISL1 in formation of pancreatic mesenchyme and islet cells. *Nature (London)* **385,** 257–260.

Ahlgren, U., Jonsson, J., Jonsson, L., Simu, K., and Edlund, H. (1998). Beta-cell-specific inactivation of the mouse Ipf1/Pdx1 gene results in loss of the beta-cell phenotype and maturity onset diabetes. *Genes Dev.* **12,** 1763–1768.

Alescio, T., and Cassini, A. (1962). Induction *in vitro* of tracheal buds by pulmonary mesenchyme grafted on tracheal epithelium. *J. Exp. Zool.* **150,** 83–94.

Apelqvist, A., Ahlgren, U., and Edlund, H. (1997). Sonic hedgehog directs specialized mesoderm differentiation in the intestine and pancreas. *Curr. Biol.* **7,** 801–804.

Bellusci, S., Furuta, Y., Rush, M. G., Henderson, R., Winnier, G., and Hogan, B. L. M. (1997a). Involvement of Sonic hedgehog (*Shh*) in mouse embryonic lung growth and morphogenesis. *Development (Cambridge, UK)* **124,** 53–63.

Bellusci, S., Grindley, J., Emoto, H., Itoh, N., and Hogan, B. L. (1997b). Fibroblast growth factor 10 (FGF10) and branching morphogenesis in the embryonic mouse lung. *Development (Cambridge, UK)* **124,** 4867–4878.

Bitgood, M. J., and McMahon, A. P. (1995). *Hedgehog* and *BMP* genes are coexpressed at many diverse sites of cell–cell interaction in the mouse embryo. *Dev. Biol.* **172,** 126–138.

Böttinger, E. P., Jakubczak, J. L., Roberts, I. S., Mumy, M., Hemmati, P., Bagnall, K., Merlino, G., and Wakefield, L. M.

(1997). Expression of a dominant-negative mutant TGF-beta type II receptor in transgenic mice reveals essential roles for TGF-beta in regulation of growth and differentiation in the exocrine pancreas. *EMBO J.* **16**, 2621–2633.

Celli, G., LaRochelle, W. J., Mackem, S., Sharp, R., and Merlino, G. (1998). Soluble dominant-negative receptor uncovers essential roles for fibroblast growth factors in multi-organ induction and patterning. *EMBO J.* **17**, 1642–1655.

Cooper, M. K., Porter, J. A., Young, K. E., and Beachy, P. A. (1998). Teratogen-mediated inhibition of target tissue response to Shh signaling. *Science* **280**, 1603–1607.

Dono, R., Texido, G., Dussel, R., Ehmke, H., and Zeller, R. (1998). Impaired cerebral cortex development and blood pressure regulation in FGF-2-deficient mice. *EMBO J.* **17**, 4213–4225.

Ebensperger, C., Wilting, J., Brand-Saberi, B., Mizutani, Y., Christ, B., Balling, R., and Koseki, H. (1995). *Pax-1*, a regulator of sclerotome development is induced by notochord and floor plate signals in avian embryos. *Anat. Embryol.* **191**, 297–310.

Edlund, H. (1998). Transcribing pancreas. *Diabetes* **47**, 1817–1823.

Floss, T., Arnold, H. H., and Braun, T. (1997). A role for FGF-6 in skeletal muscle regeneration. *Genes Dev.* **11**, 2040–2051.

Furukawa, M., Eto, Y., and Kojima, I. (1995a). Expression of immunoreactive activin A in fetal rat pancreas. *Endocr. J.* **42**, 63–68.

Furukawa, M., Nobusawa, R., Shibata, H., Eto, Y., and Kojima, I. (1995b). Initiation of insulin secretion in glucose-free medium by activin A. *Mol. Cell. Endocrinol.* **113**, 83–87.

Gilbert, S. F. (1994). "Developmental Biology." Sinauer Assoc., Sunderland, MA.

Golosow, N., and Grobstein, C. (1962). Epitheliomesenchymal interaction in pancreatic morphogenesis. *Dev. Biol.* **4**, 242–255.

Hebrok, M., Kim, S. K., and Melton, D. A. (1998). Notochord repression of endodermal Sonic hedgehog permits pancreas development. *Genes Dev.* **12**, 1705–1713.

Hebrok, M., Kim, S. K., and Melton, D. M. (1999). Screening for novel pancreatic genes expressed during embryogenesis. *Diabetes* **48**, 1550–1556.

Hogan, B. L. (1999). Morphogenesis. *Cell (Cambridge, Mass.)* **96**, 225–233.

Hogan, B. L., and Yingling, J. M. (1998). Epithelial/mesenchymal interactions and branching morphogenesis of the lung. *Curr. Opin. Genet. Dev.* **8**, 481–486.

Horikawa, Y., Iwasaki, N., Hara, M., Furuta, H., Hinokio, Y., Cockburn, B. N., Lindner, T., Yamagata, K., Ogata, M., Tomonaga, O., Kuroki, H., Kasahara, T., Iwamoto, Y., and Bell, G. I. (1997). Mutation in hepatocyte nuclear factor-1 beta gene (TCF2) associated with MODY. *Nat. Genet.* **17**, 384–385.

Jonsson, J., Carlsson, L., Edlund, T., and Edlund, H. (1994). Insulin-promoter factor 1 is required for pancreas development in mice. *Nature (London)* **371**, 606–609.

Kim, S. K., and Melton, D. A. (1998). Pancreas development is promoted by cyclopamine, a hedgehog signaling inhibitor. *Proc. Natl. Acad. Sci. U.S.A.* **95**, 13036–13041.

Kim, S. K., Hebrok, M., and Melton, D. A. (1997). Notochord to endoderm signaling is required for pancreas development. *Development (Cambridge, UK)* **124**, 4243–4252.

Lachman, M. F., Wright, Y., Whiteman, D. A., Herson, V., and Greenstein, R. M. (1991). Brief clinical report: A 46,XY phenotypic female with Smith–Lemli–Opitz syndrome. *Clin. Genet.* **39**, 136–141.

Le Bras, S., Miralles, F., Basmaciogullari, A., Czernichow, P., and Scharfmann, R. (1998). Fibroblast growth factor 2 promotes pancreatic epithelial cell proliferation via functional fibroblast growth factor receptors during embryonic life. *Diabetes* **47**, 1236–1242.

Levin, M., Pagan, S., Roberts, D. J., Cooke, J., Kuehn, M. R., and Tabin, C. J. (1997). Left/right patterning signals and the independent regulation of different aspects of *situs* in the chick embryo. *Dev. Biol.* **189**, 57–67.

Litingtung, Y., Lei, L., Westphal, H., and Chiang, C. (1998). Sonic hedgehog is essential to foregut development. *Nat. Genet.* **20**, 58–61.

Mansour, S. L., Goddard, J. M., and Capecchi, M. R. (1993). Mice homozygous for a targeted disruption of the proto-oncogene int-2 have developmental defects in the tail and inner ear. *Development (Cambridge, UK)* **117**, 13–28.

Marigo, V., Scott, M. P., Johnson, R. L., Goodrich, L. V., and Tabin, C. J. (1996). Conservation in *hedgehog* signaling: Induction of a chicken *patched* homolog by *Sonic hedgehog* in the developing limb. *Development (Cambridge, UK)* **122**, 1225–1232.

Martin, G. R. (1998). The roles of FGF's in the early development of vertebrate limbs. *Genes Dev.* **12**, 1137–1148.

McKeever, P. A., and Young, I. D. (1990). Smith–Lemli–Opitz syndrome. II: A disorder of the fetal adrenals? *J. Med. Genet.* **27**, 465–466.

Min, H., Danilenko, D. M., Scully, S. A., Bolon, B., Ring, B. D., Tarpley, J. E., DeRose, M., and Simonet, W. S. (1998). Fgf-10 is required for both limb and lung development and exhibits striking functional similarity to *Drosophila* branchless. *Genes Dev.* **12**, 3156–3161.

Miralles, F., Czernichow, P., and Scharfmann, R. (1998). Follistatin regulates the relative proportions of endocrine versus exocrine tissue during pancreatic development. *Development (Cambridge, UK)* **125**, 1017–1024.

Motoyama, J., Liu, J., Mo, R., Ding, Q., Post, M., and Hui, C. C. (1998). Essential function of Gli2 and Gli3 in the formation of lung, trachea and oesophagus. *Nat. Genet.* **20**, 54–7.

Offield, M. F., Jetton, T. L., Labosky, P. A., Ray, M., Stein, R., Magnuson, M. A., Hogan, B. L. M., and Wright, C. V. E. (1996). PDX-1 is required for pancreatic outgrowth and differentiation of the rostral duodenum. *Development (Cambridge, UK)* **122**, 983–995.

Ogawa, K., Abe, K., Kurosawa, N., Kurohmaru, M., Sugino, H., Takahashi, M., and Hayashi, Y. (1993). Expression of alpha, beta A and beta B subunits of inhibin or activin and follistatin in rat pancreatic islets. *FEBS Lett.* **319**, 217–220.

Ortega, S., Ittmann, M., Tsang, S. H., Ehrlich, M., and Basilico, C. (1998). Neuronal defects and delayed wound healing in mice lacking fibroblast growth factor 2. *Proc. Natl. Acad. Sci. U.S.A.* **95**, 5672–5677.

Park, W. Y., Miranda, B., Lebeche, D., Hashimoto, G., and Cardoso, W. V. (1998). FGF-10 is a chemotactic factor for distal epithelial buds during lung development. *Dev. Biol.* **201**, 125–134.

Pepicelli, C. V., Lewis, P. M., and McMahon, A. P. (1998). Sonic hedgehog regulates branching morphogenesis in the mammalian lung. *Curr. Biol.* **8**, 1083–1086.

Peters, K., Werner, S., Liao, X., Wert, S., Whitsett, J., and Williams, L. (1994). Targeted expression of a dominant negative FGF receptor blocks branching morphogenesis and epithelial differentiation of the mouse lung. *EMBO J.* **13**, 3296–301.

Pictet, R. L., Clark, W. R., Williams, R. H., and Rutter, W. J. (1972). An ultrastructural analysis of the developing embryonic pancreas. *Dev. Biol.* **29**, 436–467.

Pourquie, O., Coltey, M., Teillet, M. A., Ordahl, C., and Le Douarin, N. M. (1993). Control of dorso-ventral patterning of somite derivatives by notochord and floorplate. *Proc. Natl. Acad. Sci. U.S.A.* **90**, 5242–5246.

Ritvos, O., Tuuri, T., Eramaa, M., Sainio, K., Hilden, K., Saxén, L., and Gilbert, S. F. (1995). Activin disrupts epithelial branching morphogenesis in developing glandular organs of the mouse. *Mech. Dev.* **50**, 229–245.

Roberts, D. J., Smith, D. M., Goff, D. J., and Tabin, C. J. (1998). Epithelial–mesenchymal signaling during the regionalization of the chick gut. *Development (Cambridge, UK)* **125**, 2791–2801.

Ronzio, R., and Rutter, W. J. (1973). Effects of a partially purified factor from chick embryos on macromolecular synthesis of embryonic pancreatic epithelia. *Dev. Biol.* **30**, 307–320.

Rutter, W. J., Wessells, N. K., and Grobstein, C. (1964). Control of specific synthesis in the developing pancreas. *Nat. Cancer Inst. Monogr.* **13**, 51–61.

Sekine, K., Ohuchi, H., Fujiwara, M., Yamasaki, M., Yoshizawa, T., Sato, T., Yagishita, N., Matsui, D., Koga, Y., Itoh, N., and Kato, S. (1999). Fgf10 is essential for limb and lung formation. *Nat. Genet.* **21**, 138–141.

Shannon, J. M. (1994). Induction of alveolar type II cell differentiation in fetal tracheal epithelium by grafted distal lung mesenchyme. *Dev. Biol.* **166**, 600–614.

Shannon, J. M., Nielsen, L. D., Gebb, S. A., and Randell, S. H. (1998). Mesenchyme specifies epithelial differentiation in reciprocal recombinants of embryonic lung and trachea. *Dev. Dyn.* **212**, 482–494.

Slack, J. M. W. (1983). *In* "From Egg to Embryo. Regional Specification in Early Development" (P. W. Barlow, D. Bray, P. B. Green, and J. M. W. Slack, eds.). Cambridge University Press, Cambridge, UK.

Slack, J. M. W. (1995). Developmental biology of the pancreas. *Development (Cambridge, UK)* **121**, 1569–1580.

Stoffers, D. A., Ferrer, J., Clarke, W. L., and Habener, J. F. (1997). Early-onset type-II diabetes mellitus (MODY4) linked to IPF1. *Nat. Genet.* **17**, 138–139.

Ten Have-Opbroek, A. A. W. (1981). The development of the lung in mammals: An analysis of concepts and findings. *Am. J. Anat.* **162**, 201–219.

Wessells, N. K. (1970). Mammalian lung development: Interactions in formation and morphogenesis of tracheal buds. *J. Exp. Zool.* **175**, 455–466.

Wessells, N. K., and Cohen, J. H. (1967). Early pancreas organogenesis: Morphogenesis, tissue interactions, and mass effects. *Dev. Biol.* **15**, 237–270.

Yamagata, K., Oda, N., Kaisaki, P. J., Menzel, S. Furuta, H., Vaxillaire, M., Southam, L., Cox, R. D., Lathrop, G. M., Boriraj, V. V., Chen, X., Cox, N. J., Oda, Y., Yano, H., Le Beau, M. M., Yamada, S., Nishigori, H., Takeda, J., Fajans, S. S., Hattersley, A. T., Iwasaki, N., Hansen, T., Pedersen, O., Polonsky, K. S., Bell, G. I., *et al.* (1996a). Mutations in the hepatocyte nuclear factor-1alpha gene in maturity-onset diabetes of the young (MODY3). *Nature (London)* **384**, 455–458.

Yamagata, K., Furuta, H., Oda, N., Kaisaki, P. J., Menzel, S., Cox, N. J., Fajans, S. S., Signorini, S., Stoffel, M., and Bell, G. I. (1996b). Mutations in the hepatocyte nuclear factor-4alpha gene in maturity-onset diabetes of the young (MODY1). *Nature (London)* **384**, 458–460.

Yamaoka, T., Idehara, C., Yano, M., Matsushita, T., Yamada, T., Ii, S., Moritani, M., Hata, J., Sugino, H., Noji, S., and Itakura, M. (1998). Hypoplasia of pancreatic islets in transgenic mice expressing activin receptor mutants. *J. Clin. Invest.* **102**, 294–301.

Yasuda, H., Inoue, K., Shibata, H., Takeuchi, T., Eto, Y., Hasegawa, Y., Sekine, N., Totsuka, Y., Mine, T., Ogata, E., and Kojima, I. (1993). Existence of activin-A in A- and D-cells of rat pancreatic islets. *Endocrinology (Baltimore)* **133**, 624–630.

Zhou, M., Sutliff, R. L., Paul, R. J., Lorenz, J. N., Hoying, J. B., Haudenschild, C. C., Yin, M., Coffin, J. D., Kong, L., Kranias, E. G., Luo, W., Boivin, G. P., Duffy, J. J., Pawlowski, S. A., and Doetschman, T. (1998). Fibroblast growth factor 2 control of vascular tone. *Nat. Med.* **4**, 201–207.

CHAPTER 7

MORPHOGENESIS AND TISSUE ENGINEERING

A. H. Reddi

INTRODUCTION

Morphogenesis is the developmental cascade of pattern formation, the establishment of body plan, and the architecture of mirror-image bilateral symmetry of musculoskeletal structures, culminating in the adult form. Tissue engineering is the emerging discipline of fabrication of spare parts for the human body, including the skeleton, to restore function of lost parts due to disease, cancer, and trauma. It is based on rational principles of molecular developmental biology and morphogenesis and is further governed by bioengineering. The three key ingredients for both morphogenesis and tissue engineering are inductive morphogenetic signals, responding stem cells, and the extracellular matrix scaffolding (Reddi, 1998). Advances in the molecular cell biology of morphogenesis will aid in the design principles and architecture for tissue engineering and regeneration.

Regeneration recapitulates embryonic development and morphogenesis. Among the many tissues in the human body, bone has considerable powers for regeneration, and is therefore a prototype model for tissue engineering. Implantation of demineralized bone matrix into subcutaneous sites results in local bone induction. The sequential cascade of bone morphogenesis mimics sequential skeletal morphogenesis in limbs, permitting the isolation of bone morphogens. Although it is traditional to study morphogenetic signals in embryos, bone morphogenetic proteins (BMPs), the primordial inductive signals for bone, can be isolated from demineralized bone matrix from adults. BMPs initiate, promote, and maintain chondrogenesis and osteogenesis and have actions beyond bone. The recently identified cartilage-derived morphogenetic proteins are critical for cartilage and joint morphogenesis. The symbiosis of bone inductive and conductive strategies is critical for tissue engineering, and is in turn governed by the context and biomechanics. The context is the microenvironment, consisting of extracellular matrix scaffolding, and can be duplicated by biomimetic biomaterials such as collagens, hydroxyapatite, proteoglycans, and cell adhesion proteins, including fibronectins and laminins. The rule of architecture for tissue engineering is an imitation and adoption of the laws of developmental biology and morphogenesis, and thus may be universal for all tissues, including bones and joints and associated musculoskeletal tissues.

The traditional approach for identification and isolation of morphogens is to first identify genes in fly and frog embryos by genetic approaches, differential displays, substractive hybridization, and expression cloning. This information is subsequently extended to mice and humans. An alternative approach is to isolate morphogens from bone with known regenerative potential, and this is described in this chapter. The expanding knowledge of bone and cartilage morphogenesis affords a prototype paradigm for all of tissue engineering. The principles gleaned from bone morphogenesis and BMPs can be extended to tissue engineering of a variety of other tissues.

BONE MORPHOGENETIC PROTEINS

Bone grafts have been used by orthopedic surgeons for nearly a century to aid in recalcitrant bone repair. Decalcified bone implants have been used to treat patients with osteomyelitis (Senn, 1889). It was hypothesized that bone might contain a substance (osteogenin) that initiates bone growth (Lacroix, 1945). Urist made the key discovery that demineralized, lyophilized segments of rabbit bone, when implanted intramuscularly, induced new bone formation (Urist, 1965). Bone induction is a sequential multistep cascade (Reddi and Huggins, 1972; Reddi and Anderson, 1976; Reddi, 1981). The key steps in this cascade are chemotaxis, mitosis, and differentiation. Chemotaxis is the directed migration of cells in response to a chemical gradient of signals released from the insoluble demineralized bone matrix. The demineralized bone matrix is predominantly composed of type I insoluble collagen, which binds plasma fibronectin (Weiss and Reddi, 1980). Fibronectin has domains for binding to collagen, fibrin, and heparin. The responding mesenchymal cells attach to the collagenous matrix and proliferate, as indicated by [³H]thymidine autoradiography and incorporation into acid-precipitable DNA on day 3 (Rath and Reddi, 1979). Chondroblast differentiation is evident on day 5; chondrocytes are evident on days 7–8 and cartilage hypertrophy, on day 9 (Fig. 7.1). There is concomitant vascular invasion on day 9, with osteoblast differentiation. On days 10–12 alkaline phosphatase is maximal. Hematopoietic marrow differentiates in the ossicle and is maximal by day 21. This entire sequential bone development cascade is reminiscent of bone and cartilage morphogenesis in the limb bud (Reddi, 1981, 1984). Hence, it has immense implications for isolation of inductive signals initiating cartilage and bone morphogenesis, and a systematic investigation of the chemical components responsible for bone induction has been undertaken.

The foregoing account of the demineralized bone matrix-induced bone morphogenesis in extraskeletal sites demonstrates the potential role of morphogens in the extracellular matrix. Therefore we have embarked on a systematic study of the isolation of putative morphogenetic proteins. A prerequisite for any quest for novel morphogens is the establishment of a battery of bioassays for new bone formation. A panel of *in vitro* assays was established for chemotaxis, mitogenesis, and chondrogenesis, and an *in vivo* bioassay, for bone formation. Although the *in vitro* assays are ex-

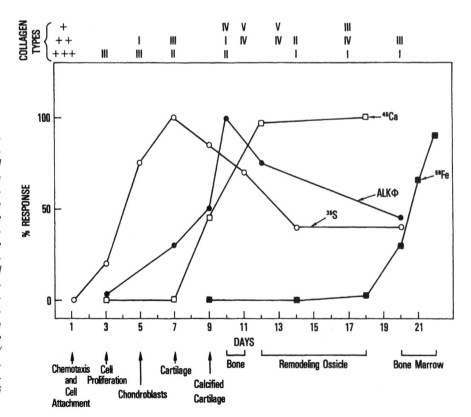

Fig. 7.1. Developmental sequence of the extracellular matrix-induced cartilage, bone, and marrow formation. Changes in ³⁵SO₄ incorporation into proteoglycans and ⁴⁵Ca incorporation into mineral phase indicate peaks of cartilage and bone formation, respectively. The ⁵⁹Fe incorporation into heme is an index of erythropoiesis, as plotted from the data of Reddi and Anderson (1976). The values for alkaline phosphatase indicate early stages of bone formation (Reddi and Huggins, 1972). The transitions in collagen types I–IV (summarized at the top of the figure) are based on immunofluorescent localization. From Reddi (1981), with permission.

Ibm art/JJ
7-2.ai

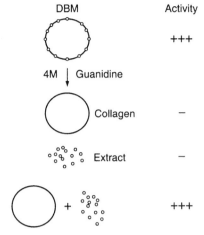

Fig. 7.2. Dissociative extraction by chaotropic reagents such as 4 M guanidine and reconstitution of osteoinductive activity with insoluble collagenous matrix. The results demonstrate a collaboration between a soluble signal and insoluble extracellular matrix. DBM, Demineralized bone matrix.

pedient, we monitored routinely a labor-intensive *in vivo* bioassay, because it was the only bona fide bone induction assay.

A major stumbling block in the approach was that the demineralized bone matrix is insoluble and in the solid state. In view of this, dissociative extractants such as 4 *M* guanidine HCl or 8 *M* urea or 1% sodium dodecyl sulfate (SDS) at pH 7.4 were used (Sampath and Reddi, 1981) to solubilize proteins. Approximately 3% of the proteins were solubilized from demineralized bone matrix, and the remaining residue was mainly insoluble type I bone collagen. The extract alone or the residue alone was incapable of new bone induction. However, addition of the extract to the residue (insoluble collagen) and then implantation in a subcutaneous site resulted in bone induction (Fig. 7.2). Thus, it would appear that for optimal osteogenic activity there was a collaboration between soluble signal in the extract and insoluble substratum (Sampath and Reddi, 1981). This bioassay was a useful advance in the final purification of bone morphogenetic proteins and led to determination of limited tryptic peptide sequences, leading to the eventual cloning of BMPs (Wozney *et al.*, 1988; Luyten *et al.*, 1989; Ozkaynak *et al.*, 1990).

In order to scale up the procedure up, a switch was made to bovine bone. Demineralized bovine bone was not osteoinductive in rats and the results were variable. However, when the guanidine extracts of demineralized bovine bone were fractionated on an S-200 molecular sieve column, fractions less than 50 kDa were consistently osteogenic when bioassayed after reconstitution with allogeneic insoluble collagen (Sampath and Reddi, 1983; Reddi, 1994). Thus, protein fractions inducing bone were not species specific and appear to be homologous in several mammals. It is likely that larger molecular mass fractions and/or the insoluble xenogeneic (bovine and human) collagens were inhibitory or immunogenic. Initial estimates revealed 1 μg of active osteogenic fraction in 1 kg of bone. Hence, over 1 ton of bovine bone was processed to yield optimal amounts for animo acid sequence determination. The amino acid sequences revealed homology to transforming growth factor β_1 (TGF-β_1) (Reddi, 1994). The important work of Wozney and colleagues (1988) resulted in cloned BMP2, BMP2B (now called BMP4), and BMP3 (also called osteogenin). Osteogenic protein-1 and -2 (OP-1 and OP-2) were cloned by Ozkaynak and colleagues (1990). There are nearly 10 members of the BMP family (Table 7.1). The other members of the extended TGF-β/BMP superfamily include inhibins and activins (implicated in follicle-stimulating hormone release from the pituitary), Müllerian duct inhibitory substance (MIS), growth/differentiation factors (GDFs), nodal, and the product of *lefty*, a gene implicated in establishing right/left asymmetry (Reddi, 1997; Cunningham *et al.*, 1995). BMPs are also involved in embryonic induction (Lemaire and Gurdon, 1994; Melton, 1991; Lyons *et al.*, 1995; Reddi, 1997).

BMPs are dimeric molecules and their conformation is critical for biological actions. Reduc-

Table 7.1. Bone morphogenetic proteins

BMP	Other names	Function
BMP2	BMP2A	Bone and cartilage morphogenesis
BMP3	Osteogenin	Bone morphogenesis
BMP3B	GDF-10	Intramembranous bone formation
BMP4	BMP2B	Bone formation
BMP5	—	Bone morphogenesis
BMP6	—	Cartilage hypertrophy
BMP7	Osteogenic protein-1	Bone formation
BMP8	Osteogenic protein-2	Bone formation
BMP8B	—	Spermatogenesis
BMP9	—	Liver differentiation
BMP10	—	?
BMP11	GDF-11	?

tion of the single interchain disulfide bond results in the loss of biological activity. The mature monomer molecule consists of about 120 amino acids, with seven canonical cysteine residues. There are three intrachain disulfides per monomer and one interchain disulfide bond in the dimer. In the critical core of the BMP monomer is the cysteine knot. The crystal structure of BMP7 has been determined (Griffith *et al.*, 1996). It is a good possibility in the near future that the crystal structure of the BMP–receptor complex and receptor contact domains will be determined.

CARTILAGE-DERIVED MORPHOGENETIC PROTEINS

Morphogenesis of the cartilage is the key rate-limiting step in the dynamics of bone development. Cartilage is the initial model for the architecture of bones. Bone can form either directly from mesenchyme, as in intramembranous bone formation, or with an intervening cartilage stage, as in endochondral bone development (Reddi, 1981). All BMPs induce, first, the cascade of chondrogenesis, and therefore in this sense are cartilage morphogenetic proteins. The hypertrophic chondrocytes in the epiphyseal growth plate mineralize and serve as a template for appositional bone morphogenesis. Cartilage morphogenesis is critical for both bone and joint morphogenesis. The two lineages of cartilage are clear-cut. The first, at the ends of bones, forms articulating articular cartilage. The second, the growth plate chondrocytes, hypertrophy and synthesize cartilage matrix destined to calcify prior to replacement by bone; these are the "organizer" centers of longitudinal and circumferential growth of cartilage, setting into motion the orderly program of endochondral bone formation. The phenotypic stability of the articular (permanent) cartilage is at the crux of the osteoarthritis problem. The "maintenance" factors for articular chondrocytes include TGF-β isoforms and the BMP isoforms (Luyten *et al.*, 1992).

An *in vivo* chondrogenic bioassay with soluble purified proteins and insoluble collagen scored for chondrogenesis. A concurrent reverse transcription–polymerase chain reaction (RT–PCR) approach was taken with degenerate oligonucleotide primers. Two novel genes for cartilage-derived morphogenetic proteins (CDMPs) 1 and 2 were identified and cloned (Chang *et al.*, 1994). CDMPs 1 and 2 are also called GDF-5 and GDF-6 (Storm *et al.*, 1994). CDMPs are related to

Table 7.2. Cartilage-derived morphogenetic proteins

CDMP	Other names	Function
CDMP 1	GDF-5	Cartilage condensation
CDMP 2	GDF-6	Cartilage formation, hypertrophy
CDMP 3	GDF-7	Tendon/ligament morphogenesis

bone morphogenetic proteins (Table 7.2). CDMPs are critical for cartilage and joint morphogenesis (Tsumaki *et al.,* 1999)), and stimulate proteoglycan synthesis in cartilage. CDMP 3 (also known as GDF-7) initiates tendon and ligament morphogenesis.

PLEIOTROPY AND THRESHOLDS

Morphogenesis is a sequential multistep cascade. BMPs regulate each of the key steps: chemotaxis, mitosis, and differentiation of cartilage and bone, and initiate chondrogenesis in the limb (Chen *et al.,* 1991; Duboule, 1994). The apical ectodermal ridge is the source of BMPs in the developing limb bud. The intricate dynamic, reciprocal interactions between the ectodermally derived epithelium and mesoderm-derived mesenchyme set into motion the train of events culminating in the pattern of phalanges, radius, ulna, and humerus.

The chemotaxis of human monocytes is optimal at femtomolar concentration (Cunningham *et al.,* 1992). The apparent affinity was $100-200$ pM. The mitogenic response was optimal at the 100-pM range. The initiation of differentiation was in the nanomolar range in solution. However, caution should be exercised, because BMPs may be sequestered by extracellular matrix components and the local concentration may be higher when BMPs are bound on the extracellular matrix. Thus BMPs are pleiotropic regulators that act in concentration-dependent thresholds.

BMPs BIND TO EXTRACELLULAR MATRIX

It is well known that extracellular matrix components play a critical role in morphogenesis. The structural macromolecules and their supramolecular assembly in the matrix do not explain their role in epithelial–mesenchymal interaction and morphogenesis. This riddle can now be explained by the binding of bone morphogenetic proteins to heparan sulfate, heparin, and type IV collagen (Paralkar *et al.,* 1990, 1991, 1992) of the basement membranes. In fact, this might explain in part the necessity for angiogenesis prior to osteogenesis during development. In addition, the actions of activin in the development of the frog, in terms of dorsal mesoderm induction, are modified to neutralization by follistatin (Hemmati-Brivanlou *et al.,* 1994). Similarly, proteins Chordin and Noggin from the Spemann organizer induce neuralization by binding and inactivation of BMP4 (Fig. 7.2). Thus neural induction is likely to be a default pathway when BMP4 is nonfunctional (Piccolo *et al.,* 1996; Zimmerman *et al.,* 1996). An emerging principle in development and morphogenesis is that binding proteins can terminate a dominant morphogen's action and initiate a default pathway. Finally, the binding of a soluble morphogen to the extracellular matrix (ECM) converts it into an insoluble matrix-bound morphogen that acts locally in the solid state (Paralkar *et al.,* 1990).

BMPs: ACTIONS BEYOND BONE

Although BMPs were first isolated and cloned from bone, subsequent work with gene knockouts has revealed a plethora of actions beyond bone. Targeted disruption of BMP2 in mice caused embryonic lethality. Heart development was abnormal, indicating a need for BMP2 (Zhang and Bradley, 1996). BMP4 knockouts exhibit no mesoderm induction, and gastrulation is impaired (Winnier *et al.,* 1996). Transgenic overexpression of BMPs under the control of keratin 10 promoter leads to psoriasis, and targeted deletion of BMP7 revealed the critical role of this molecule in kidney and eye development (Luo *et al.,* 1995; Dudley *et al.,* 1995; Vukicevic *et al.,* 1996). Thus the BMPs are really true morphogens for tissues as disparate as skin, heart, kidney, and eye. In view of this, perhaps they may also be thought of as body morphogenetic proteins.

BMP RECEPTORS

Recombinant human BMP4 and BMP7 bind to BMP receptor IA (BMPR-IA) and BMP receptor IB (BMPR-IB) (ten Dijke *et al.,* 1994). CDMP 1 also binds to both of the type I BMP receptors. There is a collaboration between type I and type II BMP receptors (Nishitoh *et al.,* 1996). The type I receptor serine/threonine kinase phosphorylates a signal-transducing protein substrate called Smad 1 or 5 (Chen *et al.,* 1996). Smad is a term derived from fusion of the *Drosophila* Mad gene and the *Caenorhabditis elegans* (nematode) Sma gene. Smads 1 and 5 signal in partnership with a common co-Smad, Smad 4 (Fig. 7.3). The transcription of BMP-response genes are initiated by Smad 1/Smad 4 heterodimers. Smads are trimeric molecules as determined by X-ray crystallography. The phosphorylation of Smads 1 and 5 by type I BMP receptor kinase is inhibited by

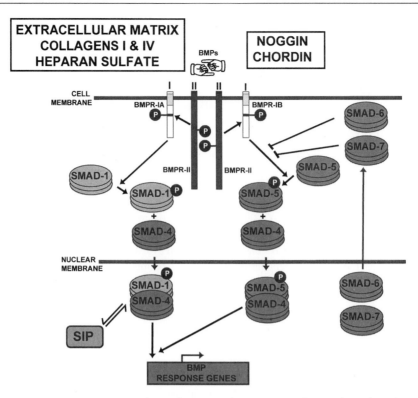

Fig. 7.3. BMP receptors and signaling cascades. BMPs are dimeric ligands with a cysteine knot in each monomer fold. Each monomer has two β sheets (represented as two pointing fingers). In the functional dimer the fingers are oriented in opposite directions. BMPs interact with both type I and type II BMP receptors. The exact stoichiometry of the receptor complex is currently being elucidated. BMPR-II phosphorylates the GS domain of BMPR-I. The collaboration between type I and type II receptors forms the signal-transducing complex. BMP type I receptor kinase complex phosphorylates the trimeric signaling substrate Smad 1 or 5. This phosphorylation is inhibited and modulated by inhibitory Smads 6 and 7. Phosphorylated Smad 1 or 5 interacts with Smad 4 (functional partner) and enters the nucleus to activate the transcriptional machinery for early BMP-response genes. A novel Smad-interacting protein (SIP) may interact and modulate the binding of heteromeric Smad 1/Smad 4 complexes to the DNA.

inhibitory Smads 6 and 7 (Hayashi *et al.,* 1997). Smad-interacting protein (SIP) may interact with Smad 1 and modulate BMP-response gene expression (Heldin *et al.,* 1997; Reddi, 1997). The downstream targets of BMP signaling are likely to be homeobox genes, the cardinal genes for morphogenesis and transcription. BMPs in turn may be regulated by members of the hedgehog family of genes such as Sonic and Indian hedgehog (Johnson and Tabin, 1997), their cognate receptors patched and smoothened, and downstream transcription factors such as Gli1, Gli2, and Gli3. The actions of BMPs can be terminated by specific BMP-binding proteins such as noggin (Zimmerman *et al.,* 1996). Recall that BMPs also bind to extracellular matrix components.

RESPONDING STEM CELLS

It is well known that the embryonic mesoderm-derived mesenchymal calls are progenitors for bone, cartilage, tendons, ligaments, and muscle. However, certain stem cells in adult bone marrow, muscle, and fascia can form bone and cartilage. The identification of stem cells readily sourced from bone marrow may lead to banks of stem cells for cell therapy, and perhaps to gene therapy with appropriate "homing" characteristics to bone marrow and hence to the skeleton. The pioneering work of Friedenstein *et al.* (1968) and Owen and Friedenstein (1988) identified bone marrow stromal stem cells. These stromal cells are distinct from the hematopoietic stem cell lineage. The bone marrow stromal stem cells consist of inducible and determined osteoprogenitors committed to osteogenesis. Determined osteogenic precursor cells have the propensity to form bone calls without any external cues or signals. On the other hand, inducible osteogenic precursors require an inductive signal such as BMP or demineralized bone matrix. It is noteworthy that operational distinctions between stromal stem cells and hematopoietic stem cells are getting increasingly blurry! The stromal stem cells of Friedenstein and Owen are also called mesenchymal stem cells (Caplan, 1991), with potential to form bone, cartilage, adipocytes, and myoblasts in response to cues from environmental and/or intrinsic factors. There is considerable hope and anticipation that these bone marrow stromal cells may be excellent vehicles for cell and gene therapy (Prockop, 1997).

From a practical standpoint, these stromal stem cells can be obtained by bone marrow biopsies and expanded rapidly for use in cell therapy after pretreatment with bone morphogenetic proteins. The potential uses for both cell and gene therapy are very promising. There are continuous improvements in the viral vectors and the efficiency of gene therapy (Bank, 1996; Mulligan, 1993). For example it is possible to use BMP genes transfected in stromal stem cells to target to the bone marrow.

MORPHOGENS AND GENE THERAPY

Advances in the understanding of morphogens are ripe for application to techniques of regional gene therapy for orthopedic tissue engineering. The availability of cloned genes for BMPs and CDMPs and the requisite platform technology of gene therapy may have immediate applications. Whereas protein therapy provides an immediate bolus of morphogen, gene therapy achieves a sustained, prolonged secretion of gene products. Furthermore, improvements in regulated gene expression allow the turning on and off of gene expression. The progress in vectors for delivering genes also bodes well. The use of adenoviruses, adeno-associated viruses, and retroviruses is poised for applications in bone and joint repair (Bank, 1996; Kozarsky and Wilson, 1993; Morsy *et al.,* 1993; Mulligan, 1993). Although gene therapy has some advantages for orthopedic tissue engineering, an optimal delivery system for protein and gene therapy is needed, especially in replacement of large segmented defects and in fibrous nonunions and malunions.

BIOMIMETIC BIOMATERIALS

In addition to inductive signals (BMPs) and responding stem cells (stromal cells), scaffolding (the microenvironment/extracellular matrix) is necessary for optimal tissue engineering. The natural scaffolding biomaterials in the composite tissue of bones and joints are collagens, proteoglycans, cell adhesion glycoproteins such as fibronectin, and the mineral phase. The mineral phase in bone is prediminantly hydroxyapatite. In its native state, the associated citrate, fluoride, carbonate, and trace elements constitute the physiological hydroxyapatite. The high capacity to bind protein makes hydroxyapatite a natural delivery system. Comparison of insoluble collagen, hydroxyapatite, tricalcium phosphate, glass beads, and poly(methyl methacrylate) as carriers reveal collagen also to be an optimal delivery system for BMPs (Ma *et al.,* 1990). It is well known that collagen is an ideal delivery system for growth factors in soft and hard tissue wound repair (McPherson, 1992).

During the course of systematic work on hydroxyapatite of two pore sizes (200 or 500 μm) in two geometric forms (beads or disks), an unexpected observation was made: the geometry of the delivery system is critical for optimal bone induction. The disks were consistently osteoinductive with BMPs in rats, but the beads were inactive (Ripamonti *et al.,* 1992). The chemical composition of the two hydroxyapatite configurations was identical. In certain species the hydroxyapatite alone appears to be osteoinductive (Ripamonti, 1996); in subhuman primates hydroxyapatite alone induces bone, albeit at a much slower rate than with BMPs. One interpretation is that osteoinductive endogenous BMPs in circulation progressively bind to implanted disks of hydroxyapatite. When an optimal threshold concentration of native BMPs is achieved, the hydroxyapatite becomes osteoinductive. Strictly speaking, most hydroxyapatite substrata are ideal osteoconductive materials. This example in certain species also serves to illustrate how an osteoconductive biomimetic biomaterial may progressively function as an osteoinductive substance by binding to endogenous BMPs. Thus, there is a physiological–physicochemical continuum between the hydroxyapatite alone and the progressive composite with endogenous BMPs. Recognition of this experimental nuance will save unnecessary arguments among biomaterials scientists about the osteoinductive action of a conductive substratum such as hydroxyapatite.

Complete regeneration of baboon craniotomy defect was accomplished by recombinant human osteogenic protein (rhOP-1; human BMP7) (Ripamonti *et al.,* 1996). Recombinant BMP2 was delivered by poly(α-hydroxy acid) carrier for calvarial regeneration (Hollinger *et al.,* 1996). A copolymer of poly(lactic acid) and poly(glycolic acid) in a nonunion model in rabbit ulna yielded impressive results (Bostrom *et al.,* 1996).

Sterilization is an important problem in the clinical application of biomimetic biomaterials with BMPs and/or other morphogens. Although gas (ethylene oxide) is used, one always should be concerned about reactive free radicals. Using allogeneic demineralized bone matrix with endogenous native BMPs, as long as low temperature (4°C or less) is maintained, the samples tolerate up to 5–7 MRads of irradiation (Weintroub and Reddi, 1988; Weintroub *et al.,* 1990). The standard dose acceptable to the Food and Drug Administration is 2.5 MRads. This information is important to the biotechnology companies preparing to market recombinant BMP-based osteogenic devices, and perhaps also to the tissue bank industry with an interest in bone grafts (Damien and Parson, 1991). Various forms of freeze-dried and demineralized allogeneic bone may be used in the interim as satisfactory carriers for BMPs. The moral of this experiment is, it is not the irradiation dose but the ambient sample temperature during irradiation that is absolutely critical.

TISSUE ENGINEERING OF BONES AND JOINTS

Unlike bone, with its considerable prowess for repair and even regeneration, cartilage is recalcitrant. But why? In part it may be due to the relative avascularity of hyaline cartilage to the high concentration of protease inhibitors, and perhaps even to growth inhibitors. The wound debridement phase is not optimal to prepare the cartilage wound bed for the optimal *milieu interieur* for repair. Although cartilage has been successfully engineered to predetermined shapes (Kim *et al.,* 1994), true repair of the tissue continues to be a real challenge in part due to hierarchical organization and geometry (Mow *et al.,* 1992). However, considerable excitement in the field has been generated by a group of Swedish workers in Göteborg, using autologous culture-expanded human chondrocytes (Brittberg *et al.,* 1994). A continuous challenge in chondrocyte cell therapy is progressive dedifferentiation and loss of characteristic cartilage phenotype. The redifferentiation and maintenance of the chondrocytes for cell therapy can be aided by BMPs, CDMPs, TGF-β isoforms, and insulin-like growth factors (IGFs). It is also possible to repair cartilage using muscle-derived mesenchymal stem cells (Grande *et al.,* 1995). The problems posed by cartilage proteoglycans in preventing cell immigration for repair were investigated by chondroitinase ABC and trypsin pretreatment in partial-thickness defects (Hunziker and Rosenberg, 1996), with and without TGF-β. Pretreatment with chondroitinase ABC followed by TGF-β revealed a contiguous layer of cells from the synovial membrane, hinting at the potential source of "repair" cells from synovium. Multiple avenues exploring cartilage morphogens, cell therapy with chondrocytes, stem cells from marrow and muscle, and a biomaterial scaffolding may lead to an optimal tissue-engineered articular cartilage.

FUTURE CHALLENGES

It is inevitable that, during aging, humans will confront impaired locomotion due to wear and tear in bones and joints. The need to repair and possibly completely regenerate the musculoskeletal system and other vital organs thus arises. Can we create spare parts for the human body? There is much reason for optimism that tissue engineering can help prolong physical comfort and function. We are living in an extraordinary time in the development of biology, medicine, surgery, and computational and related technology. The confluence of advances in molecular developmental biology and attendant advances in understanding the biology of inductive signals for morphogenesis, stem cells, biomimetic biomaterials, and the extracellular matrix promise future breakthroughs.

The symbiosis of biotechnology and biomaterials has set the stage for systematic advances in tissue engineering (Reddi, 1994; Langer and Vacanti, 1993; Hubbell, 1995). Advances in enabling platform technology include molecular imprinting (Mosbach and Ramstrom, 1996); in principle, specific recognition and catalytic sites are imprinted using templates. The applications include biosensors, catalytic applications to antibodies, and receptor recognition sites. For example, the cell-binding RGD site in fibronectin (Ruoslahti and Pierschbacher, 1987) or the YIGSR domain in laminin can be imprinted in complementary sites (Vukicevic *et al.,* 1990).

The rapidly advancing frontiers in morphogenesis with BMPs, hedgehogs, homeobox genes, and a veritable cornucopia of general and specific transcription factors, coactivators, and repressors will lead to cocrystallization of ligand–receptor complexes, to protein–DNA complexes, and to other macromolecular interactions. This will in turn lead to peptidomimetic agonists for large proteins, as exemplified by erythropoetin (Livnah *et al.*, 1996). To such advances one can add new developments in self-assembly of millimeter-scale structures floating at the interface of perfluorodecalin and water and interacting by capillary forces controlled by the pattern of wettablity (Bowden *et al.*, 1997). The final self-assembly is due to minimization of free energy in the interface. These truly incredible advances will lead to man-made materials that mimic the extracellular matrix in tissues. Superimpose on such chemical progress a biological platform in a bone and joint mold. Let us imagine the head of the femur and a mold that is fabricated with computer-assisted design and manufacture. The mold faithfully reproduces the structural features and may be imprinted with morphogens, inductive signals, and cell adhesion sites. This assembly can be loaded with stem cells and BMPs and other inductive signals, with a nutrient medium optimized for the optimal number of cell cycles, and a predictable exit into differentiation phase, reproducing a totally new bone femoral head. In fact, such a biological approach with vascularized muscle flap and BMPs has yielded new bone with a defined shape and has demonstrated the proof of the principle for further development and validation (Khouri *et al.*, 1991). We indeed are entering a brave new world of prefabricated biological spare parts for the human body based on sound architectural rules of inductive signals for morphogenesis, responding stem cells with lineage control, and growth factors immobilized on a template of biomimetic biomaterial based on the extracellular matrix. Like life, such technologies evolve with continuous refinements to benefit mankind by reducing the agony of human pain and suffering.

ACKNOWLEDGMENTS

This work is supported by the Lawrence Ellison Chair in Musculoskeletal Molecular Biology. I thank Lana Rich for outstanding bibliographic assistance and enthusiastic help. I thank Tim Moseley and Roseline Morgan for assistance with figures.

REFERENCES

Bank, A. (1996). Human somatic cell gene therapy. *BioEssays* **18**, 999–1007.

Bostrom, M., Lane, J. M., Tomin, E., Browne, M., Berbian, W., Turek, T., Smith, J., Wozney, J., and Schildhauer, T. (1996). Use of bone morphogenetic protein-2 in the rabbit ulnar nonunion model. *Clin. Orthop. Relat. Res.* **327**, 272–282.

Bowden, N., Terfort, A., Carbeck, J., and Whitesides, G. M. (1997). Self-assembly of mesoscale objects into ordered two-dimensional assays. *Science* **276**, 233–235.

Brittberg, M., Lindahl, A., Nilsson, A., Ohlsson, C., Isaksson, O., and Peterson, L. (1994). Treatment of deep cartilage defects in the knee with autologous chondrocyte transplantation. *N. Engl. J. Med.* **331**, 889–895.

Caplan, A. I. (1991). Mesenchymal stem cell. *J. Orthop. Res.* **9**, 641–650.

Chang, S. C., Hoang, B., Thomas, J. T., Vukicevic, S., Luyten, F. P., Ryban, N. J. P., Kozak, C. A., Reddi, A. H., and Moos, M., Jr. (1994). Cartilage-derived morphogenetic proteins. *J. Biol. Chem.* **269**, 28227–28234.

Chen, P., Carrington, J. L., Hammonds, R. G., and Reddi, A. H. (1991). Stimulation of chondrogenesis in limb bud mesodermal cells by recombinant human BMP-2B and modulation by TGF-β_1 and TGF-β_2. *Exp. Cell Res.* **195**, 509–515.

Chen, S., Rubbock, M. J., and Whitman, M. (1996). A transcriptional partner for Mad proteins in TGF-β signalling. *Nature (London)* **383**, 691–696.

Cunningham, N. S., Paralkar, V., and Reddi, A. H. (1992). Osteogenin and recombinant bone morphogenetic protein-2B are chemotactic for human monocytes and stimulate transforming growth factor-β_1 mRNA expression. *Proc. Natl. Acad. Sci. U.S.A.* **89**, 11740–11744.

Cunningham, N. S., Jenkins, N. A., Gilbert, D. J., Copeland, N. G., Reddi, A. H., and Lee, S.-J. (1995). Growth/differentiation factor-10: A new member of the transforming growth factor-β superfamily related to bone morphogenetic protein-3. *Growth Factors* **12**, 99–109.

Damien, C. J., and Parson, J. R. (1991). Bone graft and bone graft substitutes: A review of current technology and applications. *J. Appl. Biomater.* **2**, 187–208.

Duboule, D. (1994). How to make a limb? *Science* **266**, 575–576.

Dudley, A. T., Lyons, K. M., and Robertson, E. J. (1995). A requirement for bone morphogenetic protein-7 during development of the mammalian kidney and eye. *Genes Dev.* **9**, 2795–2807.

Friedenstein, A. J., Petrakova, K. V., Kurolesova, A. I., and Frolora, G. P. (1968). Heterotopic transplants of bone marrow: Analysis of precursor cell for osteogenic and hemopoietic tissues. *Transplantation* **6**, 230–247.

Grande, D. A., Southerland, S. S., Manji, R., Pate, D. W., Schwartz, R. E., and Lucas, P. A. (1995). Repair of articular cartilage defects using mesenchymal stem cells. *Tissue Eng.* **1**, 345–353.

Griffith, D. L., Keck, P. C., Sampath, T. K., Rueger, D. C., and Carlson, W. D. (1996). Three-dimensional structure of recombinant human osteogenic protein-1: Structural paradigm for the transforming growth factor-β superfamily. *Proc. Natl. Acad. Sci. U.S.A.* **93**, 878–883.

Hayashi, H., Abdollah, S., Qui, Y., Cai, J., Xu, Y. Y., Grinnell, B. W. *et al.* (1997). The MAD-related protein Smad 7 associates with the TGFβ receptor and functions as an antagonist of TGFβ signalling. *Cell (Cambridge, Mass.)* **89,** 1165–1173.

Heldin, C. H., Miyazono, K., and ten Dijke, P. (1997). TGFβ signaling from cell membrane to nucleus through Smad proteins. *Nature (London)* **390,** 465–471.

Hemmati-Brivanlou, A., Kelly, O. G., and Melton, D. A. (1994). Follistatin an antagonist of activin is expressed in the Spemann organizer and displays direct neuralizing activity. *Cell (Cambridge, Mass.)* **77,** 283–295.

Hollinger, J., Mayer, M., Buck, D., Zegzula, H., Ron, E., Smith, J., Jin, L., and Wozney, J. (1996). Poly (α-hydroxy acid) carrier for delivering recombinant human bone morphogenetic protein-2 for bone regeneration. *J. Controlled Release* **39,** 287–304.

Hubbell, J. A. (1995). Biomaterials in tissue engineering. *Bio Technology* **13,** 565–575.

Hunziker, E. B., and Rosenberg, L. C. (1996). Repair of partial-thickness deficits in articular cartilage: Cell recruitment from the synovial membrane. *J. Bone Jt. Surg., Am. Vol.* **78-A,** 721–733.

Johnson, R. L., and Tabin, C. J. (1997). Molecular models for vertebrate limb development. *Cell (Cambridge, Mass.)* **90,** 979–990.

Khouri, R. K., Koudsi, B., and Reddi, A. H. (1991). Tissue transformation into bone *in vivo. JAMA, J. Am. Med. Assoc.* **266,** 1953–1955.

Kim, W. S., Vacanti, J. P., Cima, L., Mooney, D., Upton, J., Puelacher, W. C., and Vacanti, C. A. (1994). Cartilage engineered in predetermined shapes employing cell transplantation on synthetic biodegradable polymers. *Plast. Reconstr. Surg.* **94,** 233–237.

Kozarsky, K. F., and Wilson, J. M. (1993). Gene therapy: Adenovirus vectors. *Curr. Opin. Genet. Dev.* **3,** 499–503.

Lacroix, P. (1945). Recent investigations on the growth of bone. *Nature (London)* **156,** 576.

Langer, R., and Vacanti, J. P. (1993). Tissue engineering. *Science* **260,** 930–932.

Lemaire, P., and Gurdon, J. B. (1994). Vertebrate embryonic inductions. *BioEssays* **16** (9), 617–620.

Livnah, O., Stura, E. A., Johnson, D. L., Middleton, S. A., Mulcahy, L. S., Wrighton, N. D., Dower, W. J., Jolliffe, L. K., and Wilson, I. A. (1996). Functional mimicry of a protein hormone by a peptide agonist: The EPO receptor complex at 2.8 C. *Science* **273,** 464–471.

Luo, G., Hoffman, M., Bronckers, A. L. J., Sohuki, M., Bradley, A., and Karsenty, G. (1995). BMP-7 is an inducer of morphogens and is also required for eye development, and skeletal patterning. *Genes Dev.* **9,** 2808–2820.

Luyten, F. P., Cunningham, N. S., Ma, S., Muthukumaran, S., Hammonds, R. G., Nevins, W. B., Wood, W. I., and Reddi, A. H. (1989). Purification and partial amino acid sequence of osteogenin, a protein initiating bone differentiation. *J. Biol. Chem.* **265,** 13377–13380.

Luyten, F. P., Yu, Y. M., Yanagishita, M., Vukicevic, S., Hammonds, R. G., and Reddi, A. H. (1992). Natural bovine osteogenin and recombinant BMP-2B are equipotent in the maintenance of proteoglycans in bovine articular cartilage explant cultures. *J. Biol. Chem.* **267,** 3685–3691.

Lyons, K. M., Hogan, B. L. M., and Robertson, E. J. (1995). Colocalization of BMP-2 and BMP-7 RNA suggest that these factors cooperatively act in tissue interactions during murine development. *Mech. Dev.* **50,** 71–83.

Ma, S., Chen, G., and Reddi, A. H. (1990). Collaboration between collagenous matrix and osteogenin is required for bone induction. *Ann. N.Y. Acad. Sci.* **580,** 524–525.

McPherson, J. M. 1992. The utility of collagen-based vehicles in delivery of growth factors for hard and soft tissue wound repair. *Clin. Mater.* **9,** 225–234.

Melton, D. A. (1991). Pattern formation during animal development. *Science* **252,** 234–241.

Morsy, M. A., Mitani, K., Clemens, P., and Caskey, T. (1993). Progress toward human gene therapy. *JAMA, J. Am. Med. Assoc.* **270** (19), 2338–2345.

Mosbach, K., and Ramstrom, O. (1996). The emerging technique of molecular imprinting and its future impact on biotechnology. *Bio Technology* **14,** 163–170.

Mow, V. C., Ratcliffe, A., and Poole, A. R. (1992). Cartilage and diarthrodial joints as paradigms for hierarchical materials and structures. *Biomaterials* **13,** 67–97.

Mulligan, R. C. (1993). The basic science of gene therapy. *Science* **260,** 926–932.

Nishitoh, H., Ichijo, H., Kimura, M., Matsumoto, T., Makishima, F., Yamaguchi, A., Yamashita, H., Enomoto, S., and Miyazono, K. (1996). Identification of type I and type II serine/threonine kinase receptors for growth and differentiation factor-5. *J. Biol. Chem.* **271,** 21345–21352.

Owen, M. E., and Friedenstein, A. J. (1988). Stromal stem cells: Marrow derived osteogenic precursors. *Ciba Found. Symp.* **136,** 42–60.

Ozkaynak, E., Rueger, D. C., Drier, E. A., Corbett, C., Ridge, R. J., Sampath, T. K., and Opperman, H. (1990). OP-1 cDNA encodes an osteogenic protein in the TGF-β family. *EMBO J.* **9,** 2085–2093.

Paralkar, V. M., Nandedkar, A. K. N., Pointers, R. H., Kleinman, H. K., and Reddi, A. H. (1990). Interaction of osteogenin, a heparin binding bone morphogenetic protein, with type IV collagen. *J. Biol. Chem.* **265,** 17281–17284.

Paralkar, V. M., Vukicevic, S., Reddi, A. H. (1991). Transforming growth factor-β type I binds to collagen IV of basement membrane matrix: Implications for development. *Dev. Biol.* **143,** 303–308.

Paralkar, V. M., Weeks, B. S., Yu, Y. M., Kleinman, H. K., and Reddi, A. H. (1992). Recombinant human bone morphogenetic protein 2B stimulates PC12 cell differentiation: Potentiation and binding to type IV collagen. *J. Cell Biol.* **119,** 1721–1728.

Piccolo, S., Sasai, Y., Lu, B., and De Robertis, E. M. (1996). Dorsoventral patterning in *Xenopus:* Inhibition of ventral signals by direct binding of chordin to BMP-4. *Cell (Cambridge, Mass.)* **86,** 589–598.

Prockop, D. J. (1997). Marrow stromal cells and stem cells for nonhematopoietic tissues. *Science* **276,** 71–74.

Rath, N. C., and Reddi, A. H. (1979). Collagenous bone matrix is a local mitogen. *Nature (London)* **278,** 855–857.

Reddi, A. H. (1981). Cell biology and biochemistry of endochondral bone development. *Collagen Relat. Res.* **1,** 209–226.

Reddi, A. H. (1984). Extracellular matrix and development. *In* "Extracellular Matrix Biochemistry" (K. A. Piez and A. H. Reddi, eds.), pp. 375–412. Elsevier, New York.

Reddi, A. H. (1994). Bone and cartilage differentiation. *Curr. Opin. Genet. Dev.* **4**, 937–944.

Reddi, A. H. (1997). Bone morphogenetic proteins: An unconventional approach to isolation of first mammalian morphogens. *Cytokines Growth Factor Rev.* **8**, 11–20.

Reddi, A. H. (1998). Role of morphogenetic proteins in skeletal tissue engineering and regeneration. *Nat. Biotechnol.* **16**, 247–252.

Reddi, A. H., and Anderson, W. A. (1976). Collagenous bone matrix-induced endochondral ossification and hemopoiesis. *J. Cell Biol.* **69**, 557–572.

Reddi, A. H., and Huggins, C. B. (1972). Biochemical sequences in the transformation of normal fibroblasts in adolescent rat. *Proc. Natl. Acad. Sci. U.S.A.* **69**, 1601–1605.

Ripamonti, U. (1996). Osteoinduction in porous hydroxyapatite implanted in heterotopic sites of different animal models. *Biomaterials* **17**, 31–35.

Ripamonti, U., Ma, S., and Reddi, A. H. (1992). The critical role of geometry of porus hydroxyapatite delivery system induction of bone by osteogenin, a bone morphogenetic protein. *Matrix* **12**, 202–212.

Ripamonti, U., Van den Heever, B., Sampath, T. K., Tucker, M. M., Rueger, D. C., and Reddi, A. H. (1996). Complete regeneration of bone in the baboon by recombinant human osteogenic protein-1 (hOP-1, bone morphogenetic protein-7). *Growth Factors* **123**, 273–289.

Ruoslahti, E., and Pierschbacher, M. D. (1987). New perspectives in cell adhesion: RGD and integrins. *Science* **238**, 491–497.

Sampath, T. K., and Reddi, A. H. (1981). Dissociative extraction and reconstitution of bone matrix components involved in local bone differentiation. *Proc. Natl. Acad. Sci. U.S.A.* **78**, 7599–7603.

Sampath, T. K., and Reddi, A. H. (1983). Homology of bone inductive proteins from human, monkey, bovine, and rat extracellular matrix. *Proc. Natl. Acad. Sci. U.S.A.* **80**, 6591–6595.

Senn, N. (1889). On the healing of aseptic bone cavities by implantation of antiseptic decalcified bone. *Am. J. Med. Sci.* **98**, 219–240.

Storm, E. E., Huynh, T. V., Copeland, N. G., Jenkins, N. A., Kingsley, D. M., and Lee, S.-J. (1994). Limb alterations in brachypodism mice due to mutations in a new member of TGF-β superfamily. *Nature (London)* **368**, 639–642.

ten Dijke, P., Yamashita, H., Sampath, T. K., Reddi, A. H., Riddle, D., Helkin, C. H., and Miyazono, K. (1994). Identification of type I receptors for OP-1 and BMP-4. *J. Biol. Chem.* **269**, 16986–16988.

Tsumaki, N., Tanaka, K., Arikawa-Hirasawa, E., Nakase, T., Kimura, T., Thomas, J. T., Ochi, T., Luyten, F. P., and Yamada, Y. (1999). Role of CDMP-1 in skeletal morphogenesis: Promotion of mesenchymal cell recruitment and chondrocyte differentiation. *J. Cell Biol.* **144**, 161–173.

Urist, M. R. (1965). Bone: Formation by autoinduction. *Science* **150**, 893–899.

Vukicevic, S., Luyten, F. P., Kleinman, H. K., and Reddi, A. H. (1990). Differentiation of canalicular cell processes in bone cells by basement membrane matrix component: Regulation by discrete domains of laminin. *Cell (Cambridge, Mass.)* **64**, 437–445.

Vukicevic, S., Kopp, J. B., Luyten, F. P., and Sampath, K. (1996). Induction of nephrogenic mesenchyme by osteogenic protein-1 (bone morphogenetic protein-7). *Proc. Natl. Acad. Sci. U.S.A.* **92**, 9021–9026.

Weintroub, S., and Reddi, A. H. (1988). Influence of irradiation on the osteoinductive potential of demineralized bone matrix. *Calcif. Tissue Int.* **42**, 255–260.

Weintroub, S., Weiss, J. F., Catravas, G. N., and Reddi, A. H. (1990). Influence of whole body irradiation and local shielding on matrix-induced endochondral bone differentiation. *Calcif. Tissue Int.* **46**, 38–45.

Weiss, R. E., and Reddi, A. H. (1980). Synthesis and localization of fibronectin during collagenous matrix mesenchymal cell interaction and differentiation of cartilage and bone *in vivo. Proc. Natl. Acad. Sci. U.S.A.* **77**, 2074–2078.

Winnier, G., Blessing, M., Labosky, P. A., and Hogan, B. L. M. (1996). Bone morphogenetic protein-4 is required for mesoderm formation and patterning in the mouse. *Genes Dev.* **9**, 2105–2116.

Wozney, J. M., Rosen, V., Celeste, A. J., Mitsock, L. M., Whittiers, M., Kriz, W. R., Heweick, R. M., and Wang, E. A. (1988). Novel regulators of bone formation: Molecular clones and activities. *Science* **242**, 1528–1534.

Zhang, H., and Bradley, A. (1996). Mice deficient of BMP-2 are nonviable and have defects in amnion/chorion and cardiac development. *Development (Cambridge, UK)* **122**, 2977–2986.

Zimmerman, L. B., Jesus-Escobar, J. M., and Harland, R. M. (1996). The Spemann organizer signal Noggin binds and inactivates bone morphogenetic protein-4. *Cell (Cambridge, Mass.)* **86**, 599–606.

CHAPTER 8

CELL DETERMINATION AND DIFFERENTIATION

Leon W. Browder

INTRODUCTION

Much of the current research in developmental biology is focused on identifying the genes that are involved in determinative events in development and unraveling the roles of the proteins they encode. Our understanding of the molecular events leading to functional cells and tissues remains quite sketchy. A major advance that has facilitated progress in understanding cell determination and cell differentiation came with the discovery of a family of myogenic regulatory factors (MRFs), which are a group of transcription factors involved in switching on the muscle cell lineage during development. This mechanism serves as a model for the gene control of cell determination and differentiation.

The initial member of the MRF family to be discovered was MyoD, which was identified by its ability to convert cultured fibroblasts into skeletal myoblasts (Davis *et al.,* 1987). The other members of this family in vertebrates are Myf-5, myogenin, and MRF4. Each of these factors has the potential to turn tissue culture cells into myoblasts, which can—in turn—fuse with one another and differentiate into muscle. Myoblast fusion occurs when growth factors become limiting and the myoblasts cease dividing (Olson, 1992).

The genes encoding the MRFs are thought to be master regulatory genes, whose expression initiates a cascade of events that lead to muscle cell differentiation. They are expressed in a hierarchical fashion during myogenesis. Myf-5 and MyoD are expressed in cultured myoblasts (and continue to be expressed after muscle differentiation), whereas myogenin is expressed after myoblast fusion (Braun *et al.,* 1989; Wright *et al.,* 1989; Edmondson and Olson, 1989). Myogenin is an essential intermediate, as shown by the prevention of myoblast differentiation by inhibition of myogenin expression with antisense oligonucleotides (Florini and Ewton, 1990). Myogenin functions as a transcriptional switch to activate all of the genes that are necessary for muscle differentiation. The fourth factor, MRF4, is expressed only after muscle differentiation.

The MRFs share a region of homology with two functionally significant domains: the helix–loop–helix (HLH) domain, which facilitates dimerization, and the basic region, which contains positively charged amino acids that mediate binding to DNA. These characteristics define a large family of proteins that function primarily as transcriptional activators. These are the basic helix–loop–helix (bHLH) proteins.

MRF proteins bind to a sequence in the promoter of target genes called the E box. E boxes contain the sequence CANNTG (where N is any nucleotide). The genes encoding the MRFs contain an E box, which suggests that these proteins may regulate their own and one another's transcription. Each MRF presumably owes its functional distinctiveness to unique sequences outside the bHLH domain. MRFs form functional entities that bind to DNA by dimerizing with a member of the ubiquitously expressed E protein family. This family includes E12, E47, ITF1, and ITF2 (Murre *et al.,* 1989; Davis *et al.,* 1990; Henthorn *et al.,* 1990; Lassar *et al.,* 1991). The most prevalent heterodimers in myotube extracts contain E12, but any of the E proteins can pair with the MRFs to form a functional heterodimer.

Dimerization is essential for bHLH protein function, but their specificity of binding to DNA is due to the basic region. Modifications to this region can either abolish the DNA-binding capability of MyoD or eliminate its ability to activate transcription of muscle-specific genes (Davis *et al.,* 1990). Thus, these mutants act like dominant negative inhibitors of wild-type MyoD by competing with it for binding to its partners and inhibiting its activity.

Nature has produced its own dominant negative inhibitor of the MRFs. The interactions of MRFs with DNA can be prevented by members of a family of HLH factors called "Id," which stands for "inhibitor of DNA binding." Id proteins lack a basic region. When they bind to MRF proteins, they impede their ability to bind to DNA and activate transcription of target genes (Benezra *et al.,* 1990). Likewise, their ability to bind E proteins sequesters these proteins and makes them unavailable for forming functional heterodimers with the MRF factors (Langlands *et al.,* 1997). The inhibitory role of Id proteins is supported by the following observations:

1. Id proteins are expressed in proliferating myoblasts in culture, but disappear when the myoblasts differentiate to form myotubes.
2. Overexpression of Id protein in cultured myoblasts prevents their differentiation into myotubes (Jen *et al.,* 1992).
3. Id transcripts are detected during the gastrula stage of mouse development before MRF transcripts first appear and are down-regulated before MRFs are expressed (Wang *et al.,* 1992).

Although the roles of Id in embryonic development are uncertain, the evidence suggests that Id is initially an inhibitor of myogenesis and its down-regulation then permits myogenesis to proceed by allowing MRFs to bind DNA of target genes.

Another negative regulator of MRF activity is Mist1, which is a novel bHLH protein that accumulates in myogenic precursors and decreases as myoblasts differentiate to form myotubes. Mist1 lacks a transcription activation domain and instead has a repressor region that inhibits transcription. Like Id, Mist1 can act as a dominant negative inhibitor of myogenesis by forming inactive heterodimers with MyoD and E proteins. It can also form homodimers that bind to MRF-specific target sites in DNA (Lemercier *et al.,* 1998).

Lu *et al.* (1999) have identified yet another dominant negative inhibitor of myogenesis. This protein (MyoR, for myogenic repressor) is expressed in undifferentiated myoblasts in the skeletal muscle lineage. MyoR forms heterodimers with E proteins. The MyoR/E protein heterodimers bind MRF-specific target sites in DNA and block transcription of skeletal muscle-specific genes. Myogenesis can proceed only when these proteins are depleted and replaced with MRFs, which bind to E proteins and activate skeletal muscle-specific gene transcription (Lu *et al.,* 1999). This observation begs the question: How is MyoR regulated?

ROLES OF MRFs DURING EMBRYOGENESIS

WHAT DO THE MRFs DO IN THE EMBRYO?

The roles of MRFs in promoting myogenesis in cultured cells suggest that they may also play a role in muscle development during embryogenesis. Most of the sekeltal muscle in vertebrates originates from progenitor cells in the somites. The somites are transient metameric condensations of paraxial mesoderm that later become compartmentalized into the dermamyotome dorsally and the sclerotome ventrally. The dermamyotome subdivides into the dermatome and the myotome. The medial myotomal cells form the axial musculature, and the lateral cells migrate to the limbs to form limb muscle.

A role for MRFs in promoting muscle development during embryonic development is suggested by the location and timing of their expression during development. The MRFs are expressed sequentially in the somites, although details vary somewhat between species. In mice, the first MRF protein detected in trunk somites is Myf-5, which is first seen in medial somite cells (Smith *et al.,* 1994). Myogenin expression follows shortly after the initial detection of Myf-5, and MRF4 is expressed next. MyoD appearance is delayed and is first localized to the lateral portion of the somites.

Initially, Myf-5 and MyoD expression is mutually exclusive and later overlaps (Smith *et al.,* 1994). Myf-5 and MyoD may be involved in establishing the two distinct subdomains of muscle: back musculature and limb musculature (Ordahl and Le Douarin, 1992).

MRF KNOCKOUTS

Data from knockout mice have helped to clarify the roles of the MRF genes in murine development. The initial null mouse experiments produced quite unexpected results. Mice that were null for either *Myf-5* or *MyoD* genes developed normal amounts of skeletal muscle (Rudnicki *et al.,* 1992; Braun *et al.,* 1992). In homozygous *MyoD* null newborn mice, there was a three- to four-fold increase in *Myf-5* expression. This gene is normally down-regulated after day 14 of development. The prolongation and enhancement of *Myf-5* expression suggests that *Myf-5* compensates for the lack of *MyoD.* In the *Myf-5* knockouts, muscle development was delayed until *MyoD* was expressed, and then it proceeded (Braun *et al.,* 1994). These observations suggest that *MyoD* and *Myf-5* may be redundant. If so, does elimination of expression of both genes eliminate muscle development? The *Myf-5* and *MyoD* mutant mice were interbred; the progeny that lacked both of these early-acting MRF genes were unable to initiate myogenesis, produced no myogenin, and were devoid of skeletal muscle (Rudnicki *et al.,* 1993).

If myogenin is an essential intermediate in myogenesis, one would predict that myoblasts would form in myogenin knockout mice, but that skeletal muscle formation would be impaired. This is what has been observed (Hasty *et al.,* 1993; Nabeshima *et al.,* 1993). The myogenin knockout mice had deficient accumulation of transcripts for a number of muscle-specific proteins, including muscle creatine kinase, myosin heavy chain, the α and γ subunits of the acetylcholine receptor, and MRF4. However, normal amounts of MyoD transcripts were present, consistent with the hypothesis that MyoD acts upstream of myogenin (Hasty *et al.,* 1993). During development of myogenin knockouts, somites developed normally and compartmentalized into myotome, dermatome, and sclerotome (Venuti *et al.,* 1995). They even initiated muscle mass differentiation, but myosin heavy chain protein expression was attenuated, and myofibers were diffuse. The disparity between mutant and wild-type embryos widened as development continued (Venuti *et al.,* 1995). Large numbers of myoblasts that failed to differentiate appeared to be present in the mutant muscle masses.

The picture of myogenesis that is emerging is that MyoD and Myf-5 are partially redundant and initiate myogenesis in the myoblasts. They control expression of myogenin, which—in turn—controls muscle cell differentiation via activation of muscle-specific genes and may control expression of MRF4. MRF4 may be responsible for events in fully differentiated myofibers, possibly by maintaining the differentiated state (Rudnicki *et al.,* 1993). According to this scheme, transcription of distinct sets of genes at each stage is regulated by MRFs, which also control the expression of the MRF that initiates the next stage of differentiation (Venuti *et al.,* 1995).

INITIATION OF SKELETAL MUSCLE DEVELOPMENT

WHAT REGULATES THE MRF GENES?

Transplantation experiments with chick embryos have shown that the somites are induced by the neural tube/notochord complex to form muscle (Rong *et al.,* 1992; Buffinger and Stockdale, 1994). Additional evidence has indicated that dermomyotome-specific gene expression is induced by the protein Wnts, produced by the dorsal neural tube, and signals (possibly Wnts) from the surface ectoderm, whereas the protein Sonic hedgehog (Shh) from the notochord and ventral floor-plate of the neural tube induces sclerotome-specific gene expression (Yun and Wold, 1966; Ikeya and Takada, 1998). The activation of the myogenic pathway in the myotome is a consequence of combinatorial effects of the Wnts and Shh. Yet another signal from the lateral mesoderm (possibly bone morphogenetic protein; BMP) inhibits myogenesis (Porquié *et al.,* 1996). The function of BMP is, in turn, antagonized by Noggin, which is expressed within the dorsomedial lip of the dermomyotome (Reshef *et al.,* 1998). These various signals diffuse from their respective origins. The overlapping signals produce discrete concentrations and activities of the secreted factors in the somite and specify cellular identities in the different somite domains (Rawls and Olson, 1997).

How do these signals cause the expression of the MRF genes, which promote myogenesis? The paired-type homeobox gene *Pax-3* has now been shown to be a key trigger of the myogenic program by virtue of its activation of *MyoD* expression (Rawls and Olson, 1997). Pax-3 is expressed in the paraxial mesoderm before the somites form. When the somites become epithelial, Pax-3 is expressed in the dorsal halves of the somites. Later, it becomes restricted to the ventrolateral domain and in the myogenic precursors that migrate from the dermomyotome into the limb buds (Goulding *et al.,* 1991; Jostes *et al.,* 1991).

The *splotch* mutant in the mouse lacks a functional *Pax-3* gene. Consequently, this mutant provides a means to establish whether there is any functional relationship between Pax-3 and myogenesis (Bober *et al.,* 1994; Tremblay *et al.,* 1998). The lack of limb muscles in these mice indicates that Pax-3 expression is necessary to initiate the formation of the migratory limb myogenic precursors that are derived from the ventrolateral edge of the dermomyotome. However, most of the muscles of the back and body wall are unaffected. Thus, this lineage is not dependent on Pax-3.

Evidence on the role of Pax-3 in trunk myogenesis has come from examinations of mice that were deficient for both *Pax-3* and *Myf-5* (Tajbakhsh *et al.,* 1997). These mice developed virtually no muscle and did not express MyoD in the trunk region. It has previously been shown that *Myf-5* mutant mice will express MyoD and form muscle in the trunk. Therefore, this result suggests that there are two pathways for formation of muscle in the trunk: a pathway in which Pax-3 regulates expression of MyoD and a Pax-3-independent pathway in which MyoD is regulated by Myf-5. So, is Pax-3 sufficient to initiate the myogenic program? Maroto *et al.* (1997) have shown that viral-mediated ectopic expression of *Pax-3* in chick embryonic tissue activated *MyoD* expression and myogenesis.

The experiments of Tajbakhsh *et al.* (1997) and Maroto *et al.* (1997) have shown that the "master regulator" MyoD is a component of a continuum of regulatory events. The ultimate aim of contemporary developmental biology is to describe all of the components of that continuum and understand how a fertilized egg can produce muscle (and all other differentiated cell types) in the right place and at the right time.

Myogenic determination and differentiation occur through a complex cascade of events involving a network of factors whose interaction ensures that muscle forms in the right places and at the right times to facilitate orderly embryonic development. Muscle development serves as a valuable paradigm for the understanding of determination and differentiation of other tissue types, whose development likely involves networks of factors similar to those described here.

REFERENCES

Benezra, R., Davis, R. L., Lockshon, D., Turner, D. L., and Weintraub, H. (1990). The protein Id: A negative regulator of helix–loop–helix DNA binding proteins. *Cell (Cambridge, Mass.)* **61,** 49–59.

Bober, E., Franz, T., Arnold, H.-H., Gruss, P., and Tremblay, P. (1994). Pax-3 is required for the development of limb muscles: A possible role for the migration of dermomyotomal muscle progenitor cells. *Development (Cambridge, UK)* **120,** 603–612.

Braun, T., Buschhausen-Denker, G., Bober, E., Tannich, E., and Arnold, H.-H. (1989). A novel muscle factor related to but distinct from MyoD induces myogenic conversion in 10T1/2 fibroblasts. *EMBO J.* **8,** 701–709.

Braun, T., Rudnicki, M. A., Arnold, H.-H., and Jaenisch, R. (1992). Targeted inactivation of the muscle regulatory gene *Myf-5* results in abnormal rib development and perinatal death. *Cell (Cambridge, Mass.)* **71,** 369–382.

Braun, T., Bober, E., Rudnicki, M. A., Jaenisch, R., and Arnold, H.-H. (1994). MyoD expression marks the onset of skeletal myogenesis in *Myf-5* mutant mice. *Development (Cambridge, UK)* **120,** 3083–3092.

Buffinger, N., and Stockdale, F. E. (1994). Myogenic specification in somites: Induction by axial structures. *Development (Cambridge, UK)* **120,** 1443–1452.

Davis, R. L., Weintraub, H., and Lassar, A. B. (1987). Expression of a single transfected cDNA converts fibroblasts to myoblasts. *Cell (Cambridge, Mass.)* **51,** 987–1000.

Davis, R. L., Cheng, P.-F., Lassar, A. B., and Weintraub, H. (1990). The MyoD DNA binding domain contains a recognition code for muscle-specific gene activation. *Cell (Cambridge, Mass.)* **60,** 733–746.

Edmondson, D. G., and Olson, E. N. (1989). A gene with homology to the *myc* similarity region of MyoD is expressed during myogenesis and is sufficient to activate the muscle differentiation program. *Genes Dev.* **3,** 628–640.

Florini, J. R., and Ewton, D. A. (1990). Highly specific inhibition of IGF-I-stimulated differentiation by an antisense oligodeoxyribonucleotide to myogenin mRNA. *J. Biol. Chem.* **265,** 13435–13437.

Goulding, M. D., Chalepakis, G., Deutsch, U., Erselius, J. R., and Gruss, P. (1991). Pax-3, a novel murine DNA binding protein expressed during early myogenesis. *EMBO J.* **10,** 1135–1147.

Hasty, P., Bradley, A., Morris, J. H., Edmondson, D. G., Venuti, J. M., Olson, E. N., and Klein, W. H. (1993). Muscle deficiency and neonatal death in mice with a targeted mutation in the *myogenin* gene. *Nature (London)* **364,** 501–506.

Henthorn, P., Kiledjian, M., and Kadesch, T. (1990). Two distinct transcription factors that bind the immunoglobulin enhancer μE5/κE2 motif. *Science* **247,** 467–470.

Ikeya, M., and Takada, S. (1998). Wnt signaling from the dorsal neural tube is required for the formation of the medial dermomyotome. *Development (Cambridge, UK)* **125**, 4969–4976.

Jen, Y., Weintraub, H., and Benezra, R. (1992). Overexpression of Id protein inhibits the muscle differentiation program: *In vivo* association of Id with E2A proteins. *Genes Dev.* **6**, 1466–1479.

Jostes, B., Walther, C., and Gruss, P. (1991). The murine paired box gene, Pax 7, is expressed specifically during the development of the nervous and muscular system. *Mech. Dev.* **33**, 27–38.

Langlands, K., Yin, X., Anand, G., and Prochownik, E. V. (1997). Differential interactions of Id proteins with basic helix–loop–helix transcription factors. *J. Biol. Chem.* **272**, 19785–19793.

Lassar, A. B., Davis, R. D., Wright, W. E., Kadesch, T., Murre, C., Voronova, A., Baltimore, D., and Weintraub, H. (1991). Functional activity of myogenic HLH proteins requires hetero-oligomerization with E12/E47-like proteins *in vivo*. *Cell (Cambridge, Mass.)* **66**, 305–315.

Lemercier, C., To, R. Q., Carrasco, R. A., and Konieczny, S. F. (1998). The basic helix–loop–helix transcription factor Mist1 functions as a transcriptional repressor of myoD. *EMBO J.* **17**, 1412–1422.

Lu, J., Webb, R., Richardson, J. A., and Olson, E. N. (1999). MyoR: A muscle-restricted basic helix–loop–helix transcription factor that antagonizes the actions of MyoD. *Proc. Natl. Acad. Sci. U.S.A.* **96**, 552–557.

Maroto, M., Reshef, R., Münsterberg, A. E., Koester, S., Goulding, M., and Lassar, A. B. (1997). Ectopic *Pax-3* activates *MyoD* and *Myf-5* expression in embryonic mesoderm and neural tissue. *Cell (Cambridge, Mass.)* **89**, 139–148.

Murre, C., McCaw, P. S., Vaessin, H., Caudy, M., Jan, L. Y., Jan, Y. N., Cabrera, C. V., Buskin, J. N., Hauschka, S. D., Lassar, A. B., Weintraub, H., and Baltimore, D. (1989). Interactions between heterologous helix–loop–helix proteins generate complexes that bind specifically to a common DNA sequence. *Cell (Cambridge, Mass.)* **58**, 537–544.

Nabeshima, Y., Hanaoka, K., Hayasaka, M., Esumi, E., Li, S., Nonaka, I., and Nabeshima, Y. (1993). *Myogenin* gene disruption results in perinatal lethality because of severe muscle defect. *Nature (London)* **364**, 532–535.

Olson, E. N. (1992). Interplay between proliferation and differentiation within the myogenic lineage. *Dev. Biol.* **154**, 261–272.

Ordahl, C. P., and Le Douarin, N. M. (1992). Two myogenic lineages within the developing somite. *Development (Cambridge, UK)* **114**, 339–353.

Pourquié, O., Fan, C.-M., Coltey, M., Hirsinger, E., Watanabe, Y., Bréant, C., Francis-West, P., Brickell, P., Tessier-Lavigne, M., and Le Douarin, N. (1996). Lateral and axial signals involved in avian somite patterning: A role for BMP4. *Cell (Cambridge, Mass.)* **84**, 461–471.

Rawls, A., and Olson, E. N. (1997). MyoD meets its maker. *Cell (Cambridge, Mass.)* **89**, 5–8.

Reshef, R., Maroto, M., and Lassar, A. B. (1998). Regulation of dorsal somitic fates: BMPs and Noggin control the timing and pattern of myogenic regulator expression. *Genes Dev.* **12**, 290–303.

Rong, P. M., Teillet, M.-A., Ziller, C., and Le Douarin, N. M. (1992). The neural tube/notochord complex is necessary for vertebral but not limb and body wall striated muscle differentiation. *Development (Cambridge, UK)* **115**, 657–672.

Rudnicki, M. A., Braun, T., Hinuma, S., and Jaenisch, R. (1992). Inactivation of *MyoD* in mice leads to up-regulation of the myogenic HLH gene *Myf-5* and results in apparently normal muscle development. *Cell (Cambridge, Mass.)* **71**, 383–390.

Rudnicki, M. A., Schnegelsberg, P. N. J., Stead, R. H., Braun, T., Arnold, H. H., and Jaenisch, R. (1993). MyoD or Myf-5 is required for the formation of skeletal muscle. *Cell (Cambridge, Mass.)* **75**, 1351–1359.

Smith, T. H., Kachinsky, A. M., and Miller, J. B. (1994). Somite subdomains, muscle origins, and the four muscle regulatory proteins. *J. Cell Biol.* **127**, 95–105.

Tajbakhsh, S., Rocancourt, D., Cossu, G., and Buckingham, M. (1997). Redefining the genetic hierarchies controlling skeletal myogenesis: *Pax-3* and *Myf-5* act upstream of *MyoD*. *Cell (Cambridge, Mass.)* **89**, 127–138.

Tremblay, P., Dietrich, S., Mericskay, M., Schubert, F. R., Li, Z., and Paulin, D. (1998). A crucial role for *Pax3* in the development of the hypaxial musculature and the long-range migration of muscle precursors. *Dev. Biol.* **203**, 49–61.

Venuti, J. M., Morris, J. H., Vivian, J. L., Olson, E. N., and Klein, W. H. (1995). Myogenin is required for late but not early aspects of myogenesis during mouse development. *J. Cell Biol.* **128**, 563–576.

Wang, Y., Benezra, R., and Sassoon, D. A. (1992). Id expression during mouse development: A role in morphogenesis. *Dev. Dyn.* **194**, 222–230.

Wright, W. E., Sassoon, D. A., and Lin, V. K. (1989). Myogenin, a factor regulating myogenesis, has a domain homologous to MyoD. *Cell (Cambridge, Mass.)* **56**, 607–617.

Yun, K., and Wold, B. (1996). Skeletal muscle determination and differentiation: Story of a core regulatory network and its context. *Curr. Opin. Cell Biol.* **8**, 877–889.

PART II: *In Vitro* CONTROL OF TISSUE DEVELOPMENT

CHAPTER 9

MECHANICAL AND CHEMICAL DETERMINANTS OF TISSUE DEVELOPMENT

Donald E. Ingber

INTRODUCTION

Tissue engineering has as its main goal the fabrication of artificial tissues for use as replacements for damaged body structures. Great advances have been made in terms of the developing prosthetic devices that can repair structural defects (e.g., vascular grafts) and even replace complex mechanical behaviors (e.g., artificial joints). However, the challenge for the future is to develop tissue substitutes that restore the normal biochemical functions of living tissues, in addition to their structural features. To accomplish this feat, we must first establish precise design criteria for tissue fabrication. These design features should be based on a thorough understanding of the molecular and cellular basis of tissue regulation. They also must take into account the important role that insoluble extracellular matrix (ECM) scaffolds and mechanical stresses play in tissue formation and repair. This latter point is critical because the spatial organization of cells and the mechanical constraints imposed on them as they grow appear actively to regulate tissue development (Ingber, 1991; 1997).

The goal of this brief chapter is not to provide an extensive review of literature in the field of ECM biology or tissue development. Rather, the intent is to summarize the known functions of the ECM and describe some recent insights relating to how ECM regulates cell growth and function as well as tissue morphogenesis. The analysis of this regulatory mechanism should be of particular interest to the tissue engineer, because it has led to the identification of critical chemical and mechanical features of the ECM that are responsible for control of the growth and differentiation of many cell types. In addition, the reader will be introduced to some unanswered puzzles in developmental biology, which, if deciphered, could provide powerful new approaches to tissue regeneration and repair.

EXTRACELLULAR MATRIX STRUCTURE AND FUNCTION

COMPOSITION AND ORGANIZATION

One of the most critical elements of tissue engineering is the ability to mimic the ECM scaffolds that normally serve to organize cells into tissues. ECMs are composed of different collagen types, large glycoproteins (e.g., fibronectin, laminin, entactin, osteopontin), and proteoglycans that contain large glycosaminoglycan side chains (e.g., heparan sulfate, chondroitin sulfate, dermatan suflate, keratan sulfate, hyaluronic acid). Although all ECMs share these components, the organization, form, and mechanical properties of ECMs can vary widely in different tissues depending on the chemical composition and three-dimensional organization of the specific ECM

Principles of Tissue Engineering
Second Edition

components that are present. For example, interstitial collagens (e.g., types I and III) self-assemble into a three-dimensional lattice, which, in turn, binds fibronectin and proteoglycans. This type of native ECM hydrogel forms the backbone of loose connective tissues, such as dermis. In contrast, basement membrane collagens (types IV and V) assemble into planar arrays; when these collagenous sheets interact with fibronectin, laminin, and heparan sulfate proteoglycan, a planar ECM results (i.e., the "basement membrane"). The ability of tendons to resist tension and of cartilage and bone to resist compression similarly result from local differences in the organization and composition of the ECM.

In Vivo Foundation for Cell Anchorage

The first and foremost function of the ECM in tissue development is its role as a physiological substratum for cell attachment. This feature is easily visualized by treating whole tissues with ECM-degrading enzymes (collagenase, proteases); cell detachment and loss of cell and tissue morphology rapidly result. Cells that are dissociated in this manner can reattach to an artificial culture substrate (e.g., plastic, glass). However, adhesion is again mediated by cell binding to ECM components that are either experimentally immobilized on the culture surface, deposited *de novo* by the adhering cells, or spontaneously adsorbed from serum (e.g., fibronectin, vitronectin) (Kleinman *et al.,* 1981; Madri and Stenn, 1982; Stenn *et al.,* 1979; Wicha *et al.,* 1979; Salomon *et al.,* 1981; Bissell *et al.,* 1986; Ingber *et al.,* 1987). In fact, standard tissue culture plates are actually bacteriological (nonadhesive) plastic dishes that have been chemically treated using proprietary methods to enhance adsorption of serum- and cell-derived ECM proteins. To summarize, cells do not attach directly to the culture substrate (i.e., plastic or glass), rather they bind to intervening ECM components that are adsorbed (or derivatized) to these substrates. For this reason, cell adhesion can be prevented by coating normally adhesive culture surfaces with polymers that prevent protein adsorption, such as poly(hydroxyethyl methacrylate) (Folkman and Moscona, 1978).

Spatial Organizer of Polarized Epithelium

Living cells exhibit polarized form as well as function (e.g., basal nuclei, supranuclear Golgi complex, apical secretory granules in secretory epithelia). Dissociated cells lose this normal orientation when cultured on standard tissue culture substrata or on interstitial connective tissue. In the case of epithelial cells, normal polarized form is often restored if the cells synthesize and accumulate their own ECM or if they are cultured on exogenous basement membrane (i.e., the specialized epithelial ECM) (Emerman and Pitelka, 1977; Chambard *et al.,* 1981; Ingber *et al.,* 1986b). These types of studies suggest that basement membrane normally serves to integrate and maintain individual cells within a polarized epithelium. Clearly, there are many intracellular and intercellular determinants of polarized cell form and function (e.g., cytoskeletal organization, organelle movement, junctional complex formation). However, anchorage to the ECM appears to provide an initial point of orientation and stability on which additional steps in the epithelial organization cascade can be build. The ECM may regulate the orientiation of other cell types (e.g., chondrocytes, osteoclasts) as well.

Scaffolding for Orderly Tissue Renewal

All tissues are dynamic structures that exhibit continual turnover of all molecular and cellular components. Thus, it is the maintenance of tissue pattern integrity that is most critical to the survival of the organism. Maintenance of specialized tissue form requires that cells that are lost due to injury or aging must be replaced in an organized fashion. Importantly, orderly tissue renewal has been shown to depend on the continued presence of insoluble ECM scaffoldings, which act as templates that maintain the original architectural form and ensure accurate regeneration of preexisting structures (Vracko, 1974). For example, when cells within a tissue are killed by freezing or treatment with toxic chemicals, all of the cellular components die and are removed; however, the basement membranes often remain intact. These residual ECM scaffoldings ensure correct repositioning of cells (e.g., cell polarity) and restore different cell types to their correct locations (e.g., muscle cells within muscle basement membrane, nerve cells in nerve sheaths, endothelium within vessels), in addition to promoting the cell migration and growth that are required for repair of all the component tissues (see below). Conversely, loss of ECM integrity during wound healing results in disorganization of tissue pattern and, thus, scar formation. Uncoupling between basement membrane extension and cell doubling also leads to disorganization of tissue morphology during neoplastic transformation (Ingber *et al.*, 1981).

ESTABLISHMENT OF TISSUE MICROENVIRONMENTS

Specialized ECMs often establish a physical boundary between neighboring tissues. For example, the basement membrane normally restricts mixing between the epithelium and the underlying connective tissue, and compromise of basement membrane integrity is indicative of the onset of malignant invasion when seen in the context of tumor formation (Ingber *et al.*, 1981). The ECM boundary also may regulate macromolecular transport between adjacent tissues given that the basement membrane provides the semipermeable filtration barrier in the kidney glomerulus (Farquhar, 1978). However, little is known about this potential function of the ECM in the local tissue microenvironment.

SEQUESTRATION, STORAGE, AND PRESENTATION OF SOLUBLE REGULATORY MOLECULES

ECMs also may modulate tissue growth and morphogenesis through their ability to bind, store, and eventually release soluble regulators of morphogenesis. For example, basic fibroblast growth factor (bFGF), a mitogen for fibroblasts, smooth muscle cells, and endothelial cells, has been identified within ECMs deposited by cells cultured *in vitro* (Vlodavsky *et al.*, 1987) and within basement membranes in certain normal tissues (e.g., cornea) *in vivo* (Folkman *et al.*, 1988). The low growth rate observed in most normal tissues may result from sequestration of mitogens whereas release of these stored factors ("stormones"), due to injury or hormonally induced changes in ECM turnover, may help to switch growth on locally. Binding of other types of regulatory molecules to the endothelial basement membrane (e.g., plasminogen activator inhibitor) (Pollanen *et al.*, 1987) also may play a role in tissue physiology (e.g., blood coagulation, cell migration).

REGULATOR OF CELL GROWTH, DIFFERENTIATION, AND APOPTOSIS

Most normal (nontransformed) cells grow only when attached and spread on a solid substrate (Folkman and Moscona, 1978). Cells attach and spread *in vitro* either by depositing new ECM components or by binding to exogenous ECM (Kleinman *et al.*, 1981; Madri and Stenn, 1982; Stenn *et al.*, 1979; Wicha *et al.*, 1979; Salomon *et al.*, 1981; Bissell *et al.*, 1986; Ingber *et al.*, 1987). In fact, cell spreading and growth can be suppressed by inhibiting ECM deposition *in vitro* using drugs (Madri and Stenn, 1982; Stenn *et al.*, 1979; Wicha *et al.*, 1979). Cell growth stimulated by soluble mitogens also has been shown to vary depending on the type of ECM component used for cell attachment (e.g., collagen versus fibronectin) (Kleinman *et al.*, 1981; Bissell *et al.*, 1986; Ingber *et al.*, 1987) as well as on the mechanical properties of the ECM (e.g., malleable gel versus rigid ECM-coated dish) (Li *et al.*, 1987; Kubota *et al.*, 1987; Ben Ze'ev *et al.*, 1988; Opas, 1989; Vernon *et al.*, 1992). Furthermore, the substrates that promote growth tend to suppress differentiation and vice versa. For examples, many cells proliferate and lose differentiated features when cultured on attached type I collagen gels that can resist cell tension and promote cell spreading. In contrast, the same cells cease growing and increase expression of tissue-specific functions (e.g., albumin secretion in hepatocytes, milk secretion and acinus formation by mammary cells, capillary tube formation by endothelial cells) if cultured on the same gels that are made flexible by floating them free in medium or on attached ECM gels that exhibit high malleability (e.g., basement membrane gels, such as Matrigel). Under these conditions, the cells exert tension across their adhesions, resulting in contraction of the ECM gel and cell rounding, which, in turn, shut off growth and turn on differentiation-specific gene functions. The differentiation-inducing effects of these malleable ECM substrates also can be suppressed by making the gels rigid through chemical fixation (Li *et al.*, 1987; Opas, 1989), thus confirming the critical role of cell-generated mechanical forces in this response.

Although local changes in ECM turnover may promote tissue remodeling, large-scale breakdown of the ECM may force the same growing tissues to undergo involution. Many cultured cells rapidly lose viability and undergo programmed cell death (i.e., apoptosis) when they are detached from the ECM and maintained in a round form in suspension (Meredith *et al.*, 1993). Loss of basement membrane integrity is also observed in regions of tissues that are actively regressing (Trelstad *et al.*, 1982; Ingber *et al.*, 1986a), and growing tissues (e.g., capillaries, mammary gland) can be induced to involute using pharmacological agents (e.g., proline analogs) that inhibit ECM deposition and lead to basement membrane dissolution *in vivo* (Ingber *et al.*, 1986a; Wicha *et al.*, 1980; Ingber and Folkman, 1988). Transgenic mice studies confirm that growing tissues can be made to involute by shifting the endogenous proteolytic balance such that total ECM breakdown results (Talhouk *et al.*, 1992). These findings suggest that local changes in ECM composition and flexibility may regulate cell sensitivity to soluble mitogens and, thereby, control cell growth, viability, and function in the local tissue microenvironment.

PATTERN FORMATION THROUGH ECM REMODELING

MESENCHYMAL CONTROL OF EPITHELIAL PATTERN

Probably the greatest insight into the role of the ECM in tissue development comes from analysis of embryogenesis. In the embryo, genesis of a tissue's characteristic form (e.g., tubular versus acinar) and deposition of ECM are both controlled by complex interactions between adjacent epithelial and mesenchymal cell societies. The epithelial cell is genetically programmed to express tissue-specific (differentiated) functions and to deposit the insoluble basement membrane, which functions as a common attachment foundation that both separates adjacent tissues and stabilizes tissue form (Banerjee *et al.*, 1977; Dodson and Hay, 1971). However, although production of tissue-specific cell products (cytodifferentiation) is determined by the epithelium, tissue pattern (histodifferentiation) is usually directed by the surrounding mesenchyme. For example, when embryonic mammary epithelium is isolated and combined with salivary mesenchyme, the mammary epithelial cells take on the form of the salivary gland, although they still produce milk proteins (Sakakura *et al.*, 1976). However, the specificity of these epithelial–mesenchymal interactions can vary widely from organ to organ. For instance, embryonic salivary epithelia specifically require salivary mesenchyme for successful development, whereas pancreatic epithelia will undergo normal cytodifferentiation and histodifferentiation in response to mesenchyme isolated from a variety of embryonic tissues (Golosow and Grobstein, 1962).

TISSUE PATTERNING THROUGH LOCALIZED ECM REMODELING

The complex tissue patterns that are generated through epithelial–mesenchymal interactions result from the establishment of local differentials in tissue growth and expansion in a microenvironment that is likely saturated with soluble mitogens. The classic work on salivary gland development by Bernfield and co-workers revealed that the epithelium imposes morphological stability through production of its basement membrane whereas the mesenchyme produces local changes in tissue form, specifically by degrading basement membrane at selective sites (Banerjee *et al.*, 1977; Bernfield and Banerjee, 1978; Smith and Bernfield, 1982; Bernfield *et al.*, 1972; David and Bernfield, 1979). An increased rate of cell division is observed in the tips of growing lobules that also exhibit the highest rate of ECM breakdown and resynthesis (i.e., highest turnover rate). At the same time, the mesenchyme slows basement membrane turnover and suppresses epithelial cell growth in the clefts of the glands. This is accomplished by secretion of fibrillar collagens that slow ECM degradation locally, thereby promoting basement membrane accumulation in these regions. Similar local coupling between ECM turnover, cell growth rates, and tissue expansion is observed in many other developing tissues, including growing capillary blood vessels (Folkman, 1982).

It is important to note that increased ECM turnover involves enhanced rates of matrix synthesis as well as degradation. In fact, net basement membrane accumulation (i.e., increased area available for cell attachment) must result for epithelial tissues to grow and expand laterally, and thus the local rate of ECM synthesis must be greater than that of degradation in these high-turnover regions. If the rate of ECM degradation is significantly greater than synthesis, then net basement membrane dissolution results. As described above, this would lead to cell death and tissue regression rather than expansion.

ROLE OF MECHANICAL STRESSES

Before ending the discussion of the role of ECM in pattern formation, it is critical to emphasize that although chemical regulators mediate tissue morphogenesis, the signals that are actually responsible for dictating tissue pattern are often mechanical in nature. The pattern-generating effects of compression on bone, shear on blood vessels, and tension on muscle are just a few examples. Mechanical stresses are also important for embryological development; however, internal cell-generated forces appear to play a more critical role. For example, mechanical tension that is generated via actomyosin filament sliding within the cytoskeleton of the cells that compose the embryo plays a key role during gastrulation (Beloussov *et al.,* 1975) as well as during tissue morphogenesis (Ash *et al.,* 1973). In fact, the pattern of development can be experimentally altered by applying external stresses to the embryo using micropipettes (Beloussov *et al.,* 1988). Tensile forces that are generated internally within mesenchyme and transmitted across the ECM are also likely required for the "condensation" of mesenchyme that commonly precedes formation of new organ rudiments. In this context, it is interesting to note that the pattern-generating capabilities of mesenchyme isolated from different developing tissues have been shown to correlate with differences in their ability to exert mechanical tension on external substrates (e.g., microbeads) (Nogawa and Nakanishi, 1987). Local changes in ECM turnover may drive morphogenetic patterning of tissues, in part, by altering the mechanical compliance of the ECM and thereby changing cell shape or cytoskeletal tension (Ingber and Jamieson, 1985; Huang and Ingber, 1999). More in-depth discussion of the role of cell-generated mechanical stresses in embryogenesis and would healing can be found elsewhere (Ingber, 1991; Ingber *et al.,* 1994).

MECHANOCHEMICAL SWITCHING BETWEEN GROWTH AND DIFFERENTIATION

Given the pivotal role that the ECM plays in tissue development, many studies have been carried out to analyze how changes in cell–ECM interactions might act locally to regulate cell sensitivity to soluble mitogens and thereby establish the growth differentials that are required for tissue morphogenesis. To accomplish this, simplified *in vitro* model systems have been developed to retain the minimal determinants necessary for maintenance of the physiological functions of interest (i.e., cell growth and differentiation) (Ingber, 1990; Ingber and Folkman, 1989; Mooney *et al.,* 1992b). For example, to determine the effects of varying cell–ECM contacts directly, bacteriological petri dishes that were otherwise nonadhesive were precoated with different densities of purified ECM molecules, such as fibronectin, laminin, or different collagen types. Quiescent, serum-deprived cells were plated on these dishes in a chemically defined medium that contained a constant and saturating amount of soluble growth factor.

When capillary endothelial cells were studied, DNA synthesis and cell doubling rates were found to increase in an exponential fashion as the density of immobilized ECM ligand was raised and cell spreading was promoted (Ingber and Folkman, 1989; Ingber, 1990). When higher cell plating numbers were used to promote cell–cell interactions as well as cell–ECM contact formation, the capillary cells could be switched between growth and differentiation (capillary tube formation) in the presence of saturating amounts of soluble mitogen (FGF) simply by varying the ECM coating density (Ingber and Folkman, 1989). Specifically, when plated on a high ECM density (e.g., >500 ng/cm^2 fibronectin), the cells attached, spread extensively, formed many cell–cell contacts, and organized into a planar cell monolayer. When the same cells were plated on a low ECM density (<100 ng/cm^2), the cells attached but they could not spread, and thus only cell clumps or aggregates were observed. When the same capillary cells were plated on a moderate density, cells attached, spread, and formed cell–cell contacts as they did on the higher ECM densities. However, the tensile forces generated by the cells appeared to overcome the resistance provided by their relatively weak ECM adhesions and thus the cell aggregates began to retract over a period of hours until a mechanical equilibrium was attained. Under these conditions, formation of an extensive network composed of interconnected capillary tubes resulted. Many of these capillary tubes became elevated above the culture surface, although the network remained adherent to the culture dish at discrete points through interconnected multicellular aggregates. The importance of mechanical forces for this switching between growth and differentiation was confirmed by demonstrating that similar capillary tube formation could be induced on the high ECM density, which normally promoted spreading and growth, simply by increasing the cell plating numbers and thereby amplifying the level of cell tension.

The same system was used to demonstrate similar shape (stretch)-dependent switching between growth and differentiation in other cell types. For example, the growth and differentiation of primary rat hepatocytes could be controlled independently of cell–cell contact formation by varying cell–ECM contacts and cell spreading using the method described above (Mooney *et al.,*

1992a). Additional studies confirmed that ECM exerts its regulatory effects at the level of gene expression (Mooney *et al.,* 1992a; Hansen *et al.,* 1994) and that these effects are mediated at least in part through modulation of the cytoskeleton (Mooney *et al.,* 1995). These results are consistent with those from other laboratories demonstrating that malleable ECM gels (e.g., Matrigel, native collagen gels) that promote cell rounding also induce differentiation and suppress growth, whereas the opposite effects are produced when these gels are fixed and made rigid (Li *et al.,* 1987; Kubota *et al.,* 1987; Ben Ze'ev *et al.,* 1988; Opas, 1989).

Altering cell–ECM contacts by varying ECM coating densities appears to influence cell function via two distinct but integrated mechanisms. First, increasing the local density of immobilized ECM ligand promotes clustering of integrins, which are transmembrane ECM receptors on the cell surface (Hynes, 1987). Integrin clustering, in turn, activates a number of different chemical signaling pathways (e.g., tyrosine phosphorylation, inositol lipid turnover, Na^+/H^+ exchange, MAP kinase) that are also utilized by growth factor receptors to alter cellular biochemistry and gene expression (Ingber *et al.,* 1990; Schwartz *et al.,* 1991; McNamee *et al.,* 1993; Plopper *et al.,* 1995). Activation of these signaling pathways likely plays an important role in control of cell differentiation and survival; however, integrin-dependent chemical signaling alone is not sufficient to explain how cells are induced to enter S phase and proliferate (Ingber, 1990; Hansen *et al.,* 1994). A second mechanism that involves tension-dependent changes in cell shape and cytoskeletal organization also comes into play.

The importance of cell tension was demonstrated directly by developing a model system in which cell shape or distortion was varied independently of the local density of immobilized matrix molecule. This was made possible by adapting a technique for forming spontaneously assembled monolayers (SAMs) of alkanethiols (Prime and Whitesides, 1991) to create micropatterned surfaces containing adhesive islands with defined surface chemistry, shape, and position on the cell (micrometer) scale (Singhvi *et al.,* 1994; Chen *et al.,* 1997). The method involves fabrication of a flexible elastomeric stamp that exhibits the particular surface features of interest using photolithographic techniques. The topographic high points on the stamp (e.g., 40×40 μm square plateaus raised above recessed intervening regions) are coated with an alkanethiol ink and the stamp is then apposed to a gold-coated surface. The alkanethiol forms SAMs covering only the regions where the stamp contacts the surface (i.e., 40×40 μm squares). Then the surrounding uncoated regions are filled with a SAM composed of similar alkanethiols that are conjugated to poly(ethylene glycol) (PEG), which prevents protein adsorption. The result is a chemically defined culture surface that is completely covered with a continuous SAM of alkanethiols; however, the local adhesive islands of defined geometry support protein adsorption whereas the surrounding boundary regions coated with PEG do not. Thus, when these substrates are coated with a high density of purified ECM protein, such as laminin or fibronectin, the result is adhesive islands of defined shape and position, coated with a saturating density of matrix molecule. Using this technique, cell position and shape could be precisely controlled because the cells attach only to the ECM-coated adhesive islands. In fact, even square and rectangular cells exhibiting 90° corners could be engineered using this approach (Singhvi *et al.,* 1994).

This micropatterning method was first used to probe the question of cell function: if cells are restricted to a small size, similar to that produced on a low-concentration ECM coating, but the local density of immobilized integrin ligand is increased 1000-fold, which is the critical determinant of cell function, the ECM density or cell shape? The answer was that it was cell shape. Primary hepatocytes remained quiescent on small adhesive islands coated with a high ECM density, even though the cells were stimulated with high concentrations of soluble growth factors (Singhvi *et al.,* 1994). Furthermore, cell growth (DNA synthesis) increased in parallel as the size of the adhesive island was increased, and similar results were obtained with capillary endothelial cells (Chen *et al.,* 1997), thus confirming that cell shape was the critical governor of this response. Inhibition of hepatocyte growth on the small islands also was accompanied by a concomitant increase in albumin secretion. In the case of endothelial cells, the cells were similarly induced to differentiate into capillary tubes when cultured on linear patterns that supported cell–cell contact as well as a moderate degree of cell distension, whereas the cells were induced to undergo apoptosis (cellular suicide) when cultured on the smallest islands that fully prevented cell extension (Chen *et al.,* 1997; Dike *et al.,* 1999). Thus, cell shape and function could be engineered simply by altering the geometry of the adhesive substrate.

Integrin signaling elicited in response to ECM binding has been shown to be critical for control of cell growth and function (Hynes, 1987; Ingber *et al.,* 1990; Schwartz *et al.,* 1991; McNamee *et al.,* 1993; Plopper *et al.,* 1995). Thus, one could argue that cell shape and mechanical distortion of the cytoskeleton are not important. Instead, it might be the increase in total area of cell–ECM contacts and associated integrin binding that dictate whether cells will grow, differentiate, or die on large versus small adhesive islands. To explore this further, substrates were designed in which a single small adhesive island (which would not support spreading or growth) was effectively broken up into many smaller islands (3–5 μm in diameter) that were spread out and separated by nonadhesive barrier regions (Chen *et al.,* 1997). When capillary cells were plated on these substrates, their processes stretched from island to island and the cells exhibited an overall extended form similar to cells on large islands. However, the total area of cell–ECM contact was almost identical to that exhibited by nongrowing cells on the smaller islands. These studies revealed that in the presence of optimal growth factors and high ECM binding, DNA synthesis was high in the cells that spread over multiple small islands whereas apoptosis was completely shut off (Chen *et al.,* 1997). Additional studies have revealed that shape exerts this control over growth by harnessing the cell's molecular machinery (cyclins and cdk inhibitors) that normally controls cell cycle progression late in G_1 phase, and that similar control can be exerted by altering the level of isometric tension within the cytoskeleton, independently of any cell shape change (Huang *et al.,* 1998). Thus, local changes in cell shape and cytoskeletal tension appear to govern how individual cells will respond to chemical signals (soluble mitogens and insoluble ECM molecules) in their microenvironment. This mechanism for establishing local growth differentials may play a critical role in morphogenesis in all developing systems (Ingber and Jamieson, 1985; Huang and Ingber, 1999).

SUMMARY

In summary, our work has shown that the development of functional tissues, such as branching capillary networks, requires both soluble growth factors and insoluble ECM molecules. The ECM appears to be the dominant regulator, however, because it dictates whether individual cells will either proliferate, differentiate, or die locally in response to soluble stimuli. This local control mechanism is likely critical for the establishment of local cell growth differentials that mediate pattern formation in all developing tissues.

Analysis of the molecular basis of these effects revealed that ECM molecules alter cell growth via both biochemical and biomechanical signaling mechanisms. ECM molecules cluster specific integrin receptors on the cell surface and thereby activate intracellular chemical signaling pathways (Ingber *et al.,* 1990; Schwartz *et al.,* 1991; McNamee *et al.,* 1993; Plopper *et al.,* 1995), stimulate expression of early growth response genes (e.g, c-*fos, jun-B*) (Hansen *et al.,* 1994), and induce quiescent cells to pass through the G_0/G_1 transition. However, in addition, the immobilized ECM components must physically resist cell tension and promote changes of cell shape (Ingber, 1990; Ingber and Folkman, 1989; Mooney *et al.,* 1992a,b; Singhvi *et al.,* 1994; Chen *et al.,* 1997) and cytoskeletal tension (Mooney *et al.,* 1995; Huang *et al.,* 1998; Ingber *et al.,* 1995) in order to promote full progression through G_1 and entry into S phase. Studies with living and membrane-permeabilized cells confirm that changes in cell shape result from the action of mechanical tension that is generated within microfilaments and balanced by resistance sites within the underlying ECM (Ingber, 1991; Sims *et al.,* 1992; Ingber *et al.,* 1994). Taken together, this work suggests that the pattern-regulating information the ECM conveys to cells is both chemical and mechanical in nature. Thus, design of future artificial ECMs for tissue engineering applications must take into account both of these features.

The Future

Early tissue engineering efforts by reconstructive surgeons and materials scientists started with a knowledge of the clinical need for and the mechanical behavior of connective tissues on the macroscopic scale, and worked backward. The long-term goal for the field is to design and fabricate tissue substitutes starting from first principles. This includes mechanical and architectural principles as well as an in-depth understanding of the molecular and biophysical bases of tissue regulation. Clearly, given the potent and varied functions of the ECM, fabrication of artificial ECMs will play a central role in all of these future efforts. We and others have already begun to explore the utility of synthetic bioerodible polymers as cell attachment substrates (Cima *et al.,* 1991; Mikos *et al.,* 1993) and the potential usefulness of immobilized synthetic ECM peptides for controlling cell growth and function (Hansen *et al.,* 1994; Mooney *et al.,* 1992a,b; Roberts *et al.,* 1998). These materials offer a major advantage in terms of biocompatibility because the artificial substrates completely disappear over time, and thus the implanted donor cells become fully incorporated into the host. They also provide great chemical versatility as well as the potential for large-scale production at relatively low cost. In addition, use of synthetic chemistry reduces the likelihood of batch-to-batch variation during large-scale production, a problem that can potentially complicate use of purified ECM components. For these reasons, polymer chemistry and novel fabrication techniques will likely lead to development of more effective tissue substitutes.

However, if we understood the fundamental principles that guide ECM remodeling and pattern formation in tissues, perhaps tissue engineering might take a different approach in the future. For example, one could envision entirely new methods of clinical intervention if we understood how tissue-specific mesenchyme generates tissue pattern, how ECM turnover is controlled locally, or how compressing or pulling tissues alters their growth and form. Such knowledge could lead to methods for identifying and isolating relevant "pattern-generating" cells; for developing pharmacological modifiers of ECM remodeling that may be incorporated in local regions of implants to promote or suppress tissue expansion locally; and for fabricating biomimetic scaffolds that mimic the mechanical and architectural features of natural ECMs necessary to switch on or off the function of interest (e.g., growth vs. differentiation or aopotosis) at a particular time or place (Oslakovic *et al.,* 1998). These are just a few of the challenges for the future.

ACKNOWLEDGMENTS

The success of many of the studies summarized would not have been possible without the assistance of my major collaborators, including Judah Folkman, Robert Langer, Jay Vacanti, Martin Schwartz, and George Whitesides, and the multiple students and postdoctoral fellows who executed these experiments in our laboratories. This work was supported by grants from NIH, NASA, and NSF.

REFERENCES

Ash, J. F., Spooner, B. S., and Wessells, N. K. (1973). Effects of papaverine and calcium free medium on salivary gland morphogenesis. *Dev. Biol.* **33**, 463–469.

Banerjee, S. D., Cohn, R. H., and Bernfield, M. R. (1977). Basal lamina of embryonic salivary epithelia. Production by the epithelium and role in maintaining lobular morphology. *J. Cell Biol.* **73**, 445–463.

Beloussov, L. V., Dorfman, J. G., and Cherdantzev, V. G. (1975). Mechanical stresses and morphological patterns in amphibian embryos. *J. Embryol. Exp. Morphol.* **34**, 559–574.

Beloussov, L. V., Lakirev, A. V., and Naumidi, I. I. (1988). The role of external tensions in differentiation of *Xenopus laevis* embryonic tissues. *Cell Differ. Dev.* **25**, 165–176.

Ben Ze'ev, A. G., Robinson, S., Bucher, N. L. *et al.* (1988). Cell–cell and cell–matrix interactions differentially regulate the expression of hepatic and cytoskeletal genes in primary cultures of rat hepatocytes. *Proc. Natl. Acad. Sci. U.S.A.* **85**, 1–6.

Bernfield, M. R., and Banerjee, S. D. (1978). The basal lamina in epithelial–mesenchymal interactions. *In* "Biology and Chemistry of Basement Membranes" (N. Kefalides, ed.), pp. 137–148. Academic Press, New York.

Bernfield, M. R., Banerjee, S. D., and Cohn, R. H. (1972). Dependence of salivary epithelial morphology and branching morphogenesis upon acid mucopolysaccharide-protein proteoglycan at the epithelial cell surface. *J. Cell Biol.* **52**, 674–689.

Bissell, D. M., Stamatoglou, S. C., Nermut, M. V. *et al.* (1986). Interactions of rat hepatocytes with type IV collagen, fibronectin, and laminin matrices. Distinct matrix-controlled modes of attachment and spreading. *Eur. J. Cell Biol.* **40**, 72–78.

Chambard, M., Gabrion, J., and Mauchamp, J. (1981). Influence of collagen gel on the orientation of epithelial cell polarity: Follicle formation from isolated thyroid cells and from preformed monolayers. *J. Cell Biol.* **91**, 157–166.

Chen, C. S., Mrksich, M., Huang, S., Whitesides, G., and Ingber, D. E. (1997). Geometric control of cell life and death. *Science* **276**, 1425–1428.

Cima, L., Vacanti, J., Vacanti, C. *et al.* (1991). Tissue engineering by cell transplantation using degradable polymer substrates. *J. Biomech. Eng.* **113**, 143–151.

David, G., and Bernfield, M. R. (1979). Collagen reduces glycosaminoglycan degradation by cultured mammary epithelial cells: Possible mechanism for basal lamina formation. *Proc. Natl. Acad. Sci. U.S.A.* **76**, 786–790.

Dike, L., Chen, C. S., Mrkisch, M. *et al.* (1999). Geometric control of switching between growth, apoptosis, and differentiation during angiogenesis using micropatternd substrates. *In Vitro* **35**, 441–448.

Dodson, J. W., and Hay, E. D. (1971). Control of corneal differentiation by extracellular materials. Collagen as promoter and stabilizer of epithelial stroma production. *Dev. Biol.* **38**, 249–270.

Emerman, J. T., and Pitelka, D. R. (1977). Maintenance and induction of morphological differentiation in dissociated mammary epithelium on floating collagen membranes. *In Vitro* **13**, 316–328.

Farquhar, M. G. (1978). Structure and function in glomerular capillaries: Role of the basement membrane in glomerular filtration. *In* "Biology and Chemistry of Basement Membranes" (N. Kefalides, ed.), pp. 137–148. Academic Press, New York.

Folkman, J. (1982). Angiogenesis: Initiation and control. *Ann. N.Y. Acad. Sci.* **401**, 212–227.

Folkman, J., and Moscona, A. (1978). Role of cell shape in growth control. *Nature (London)* **273**, 345–349.

Folkman, J., Klagsbrun, M. K., Sasse, J. *et al.* (1988). A heparin-binding angiogenic protein—basic fibroblast growth factor—is stored within basement membrane. *Am. J. Pathol.* **130**, 393–400.

Golosow, N., and Grobstein, C. (1962). Epitheliomesenchymal interaction in pancreatic morphogenesis. *Dev. Biol.* **4**, 242–255.

Hansen, L., Mooney, D., Vacanti, J. P. *et al.* (1994). Integrin binding and cell spreading on extracellular matrix act at different points in the cell cycle to promote hepatocyte growth. *Mol. Biol. Cell* **5**, 967–975.

Huang, S., and Ingber, D. E. (1999). The structural and mechanical complexity of cell-growth control. *Nat. Cell Biol.* **1**, E131–E138.

Huang, S., Chen, C. S., and Ingber, D. E. (1998). Control of cyclin D1, p27^{Kip1} and cell cycle progression in human capillary and endothelial cells by cell shape and cytoskeletal tension. *Mol. Biol. Cell* **9**, 3179–3193.

Hynes, R. O. (1987). Integrins: A family of cell surface receptors. *Cell (Cambridge, Mass.)* **48**, 549–554.

Ingber, D. E. (1990). Fibronectin controls capillary endothelial cell growth by modulating cell shape. *Proc. Natl. Acad. Sci. U.S.A.* **87**, 3579–3583.

Ingber, D. E. (1991). Integrins as mechanochemical transducers. *Curr. Opin. Cell Biol.* **3**, 841–848.

Ingber, D. E. (1997). Tensegrity: The architectural basis of cellular mechanotransduction. *Annu. Rev. Physiol.* **59**, 575–599.

Ingber, D. E., and Folkman, J. (1988). Inhibition of angiogenesis through inhibition of collagen metabolism. *Lab Invest.* **59**, 44–51.

Ingber, D. E., and Folkman, J. (1989). Mechanochemical switching between growth and differentiation during fibroblast growth factor-stimulated angiogenesis *in vitro:* Role of extracellular matrix. *J. Cell Biol.* **109**, 317–330.

Ingber, D. E., and Jamieson, J. D. (1985). Cells as tensegrity structures: Architectural regulation of histodifferentiation by physical forces transduced over basement membrane. *In* "Gene Expression during Normal and Malignant Differentiation" (L. C. Andersson, C. G. Gahmberg, P. Ekblom, eds.), pp. 13–32. Academic Press, Orlando.

Ingber, D. E., Madri, J. A., and Jamieson, J. D. (1982). Role of basal lamina in the neoplastic disorganization of tissue architecture. *Proc. Natl. Acad. Sci. U.S.A.* **78**, 3901–3905.

Ingber, D. E., Madri, J. A., and Folkman, J. (1986a). A possible mechanism for inhibition of angiogenesis by angiostatic steroids: Induction of capillary basement membrane dissolution. *Endocrinology (Baltimore)* **119**, 1768–1775.

Ingber, D. E., Madri, J. A., and Jamieson, J. D. (1986b). Basement membrane as a spatial organizer of polarized epithelia: Exogenous basement membrane reorients pancreatic epithelial tumor cells *in vitro. Am. J. Pathol.* **122**, 129–139.

Ingber, D. E., Madri, J. A., and Folkman, J. (1987). Extracellular matrix regulates endothelial growth factor action through modulation of cell and nuclear expansion. *In Vitro Cell Dev. Biol.* **23**, 387–394.

Ingber, D. E., Prusty, D., Frangione, J. *et al.* (1990). Control of intracellular pH and growth by fibronectin in capillary endothelial cells. *J. Cell Biol.* **110**, 1803–1012.

Ingber, D. E., Dike, L., Hansen, L. *et al.* (1994). Cellular tensegrity: Exploring how mechanical changes in the cytoskeleton regulate cell growth, migration, and tissue pattern during morphogenesis. *Int. Rev. Cytol.* **150**, 173–224.

Ingber, D. E., Prusty, D., Sun, Z. *et al.* (1995). Cell shape, cytoskeletal mechanics and cell cycle control in angiogenesis. *J. Biomech.* **28**, 1471–1484.

Kleinman, H. K., Klebe, R. J., and Martin, G. R. (1981). Role of collagenous matrices in the adhesion and growth of cells. *J. Cell Biol.* **88**, 473–485.

Kubota, Y., Kleinman, H. K., Martin, G. R. *et al.* (1987). Role of laminin and basement membrane in the morphological differentiation of human endothelial cells into capillary-like structures. *J. Cell Biol.* **107**, 1589–1598.

Li, M. L., Aggeler, J., Farson, D. A. *et al.* (1987). Influence of a reconstituted basement membrane and its components on casein gene expression and secretion in mouse mammary epithelial cells. *Proc. Natl. Acad. Sci. U.S.A.* **84**, 136–140.

Madri, J. A., and Stenn, K. S. (1982). Aortic endothelial cell migration: I. Matrix requirements and composition. *Am. J. Pathol.* **106**, 180–186.

McNamee, H., Ingber, D., and Schwartz, M. (1993). Adhesion to fibronectin stimulates inositol lipid synthesis and enhances PDGF-induced inositol lipid breakdown. *J. Cell Biol.* **121**, 673{endash}678.

Meredith, J. E., Jr., Fazeli, B., and Schwartz, M. A. (1993). The extracellular matrix as a cell survival factor. *Mol. Biol. Cell* **4**, 953–961.

Mikos, A. G., Bao, Y., Cima, L. G. *et al.* (1993). Preparation of poly(glycolic acid) bonded fiber structures for cell attachment and transplantation. *J. Biomed. Mater. Res.* **27**, 183–189.

Mooney, D. J., Langer, R., Hansen, L. K. *et al.* (1992a). Induction of hepatocyte differentiation by the extracellular matrix and an RGD-containing synthetic peptide. *Proc. Mater. Res. Soc. Symp.* **252**, 199–204.

Mooney, D. J., Hansen, L. K., Farmer, S. *et al.* (1992b). Switching from differentiation to growth in hepatocytes: Control by extracellular matrix. *J. Cell. Physiol.* **151**, 497–505.

Mooney, D. J., Langer, R., and Ingber, D. E. (1995). Cytoskeletal filament assembly and the control of cell shape and function by extracellular matrix. *J. Cell Sci.* **108**, 2311–2320.

Nogawa, H., and Nakanishi, Y. (1987). Mechanical aspects of the mesenchymal influence on epithelial branching morphogenesis of mouse salivary gland. *Development (Cambridge, UK)* **101**, 491–500.

Opas, M. (1989). Expression of the differentiated phenotype by epithelial cells *in vitro* is regulated by both biochemistry and mechanics of the substratum. *Dev. Biol.* **131**, 281–293.

Oslakovic, K., Matsuura, R., Kumailil, J. *et al.* (1998). Cell-based biomimetic materials. *Int. Conf. Intell. Mater., 4th,* Chiba, Jpn.

Plopper, G., McNamee, H., Dike, L. *et al.* (1995). Convergence of integrin and growth factor receptor signaling pathways within the focal adhesion complex. *Mol. Biol. Cell* **6**, 1349–1365.

Pollanen, J., Saksela, O., Salonen, E. M. *et al.* (1987). Distinct localizations of urokinase-type plasminogen activator and its type 1 inhibitor under cultured human fibroblasts and sarcoma cells. *J. Cell Biol.* **104**, 1086–1096.

Prime, K. L., and Whitesides, G. M. (1991). Self-assembled organic monolayers: Model systems for studying adsorption of proteins at surfaces. *Science* **252**, 1164–1167.

Roberts, C., Chen, C. S., Mrksich, M. *et al.* (1998). Using mixed self-assembled monolayers presenting GRGD and EG3OH groups to characterize long-term attachment of bovine capillary endothelial cells to surfaces. *J. Am. Chem. Soc.* **120**, 6548–6555.

Sakakura, T., Nishizura, Y., and Dawe, C. (1976). Mesenchyme-dependent morphogenesis and epithelium-specific cytodifferentiation in mouse mammary gland. *Science* **194**, 1439–1441.

Salomon, D. S., Liotta, L. S., and Kidwell, W. R. (1981). Differential response to growth factor by rat mammary epithelium plated on different collagen substrates in serum-free medium. *Proc. Natl. Acad. Sci. U.S.A.* **78**, 382–386.

Schwartz, M. A., Lechene, C., and Ingber, D. E. (1991). Insoluble fibronectin activates the Na^+/H^+ antiporter by inducing clustering and immobilization of its receptor, independent of cell shape. *Proc. Natl. Acad. Sci. U.S.A.* **88**, 7849–7853.

Sims, J., Karp, S., and Ingber, D. E. (1992). Altering the cellular mechanical force balance results in integrated changes in cell, cytoskeletal, and nuclear shape. *J. Cell Sci.* **103**, 1215–1222.

Singhvi, R., Kumar, A., Lopez, G. *et al.* (1994). Engineering cell shape and function. *Science* **264**, 696–698.

Smith, R. L., and Bernfield, M. R. (1982). Mesenchyme cells degrade epithelial basal lamina glycosaminoglycan. *Dev. Biol.* **94**, 378–390.

Stenn, K. S., Madri, J. A., and Roll, F. J. (1979). Migrating epidermis produces AB2 collagen and requires continual collagen synthesis for movement. *Nature (London)* **277**, 229–232.

Talhouk, R. S., Bissell, M. J., and Werb, Z. (1992). Coordinate expression of extracellular matrix-degrading proteinases and their inhibitors regulate mammary epithelial function during involution. *J. Cell Biol.* **118**, 1271–1282.

Trelstad, R. L., Hayashi, A., Hayashi, K. *et al.* (1982). The epithelial–mesenchymal interface of the male rat Mullerian duct: Loss of basement membrane integrity and ductal regression. *Dev. Biol.* **92**, 27–40.

Vernon, R. B., Angello, J. C., Iruela-Arispe, L. *et al.* (1992). Reorganization of basement membrane matrices by cellular traction promotes the formation of cellular networks in vitro. *Lab. Invest.* **66**, 536.

Vlodavsky, I., Folkman, J., Sullivan, R. *et al.* (1987). Endothelial cell-derived basic fibroblast growth factor: Synthesis and deposition into subendothelial extracellular matrix. *Proc. Natl. Acad. Sci. U.S.A.* **84**, 2292–2296.

Vracko, R. (1974). Basal lamina scaffold—Anatomy and significance for maintenance of orderly tissue structures. *Am. J. Pathol.* **77**, 314–346.

Wicha, M. S., Liotta, L. A., Garbisa, G. *et al.* (1979). Basement membrane collagen requirements for attachment and growth of mammary epithelium. *Exp. Cell Res.* **124**, 181–190.

Wicha, M. S., Liotta, L. A., Vonderhaar, B. K. *et al.* (1980). Effects of inhibition of basement membrane collagen deposition on rat mammary gland development. *Dev. Biol.* **80**, 253–261.

CHAPTER 10

ANIMAL CELL CULTURE

Gordon H. Sato and David W. Barnes

From the enunciation of the cell theory by Schleiden and Schwann emerged the grand concept that understanding living organisms could come from understanding the fundamental unit— the cell. The concept that organisms are composed of smaller units ("cellulae") derived from the description in 1665 by Robert Hooke of cells in a multicellular organism (Hooke, 1665). What Hooke actually saw was the holes in cork where cells had been, and the relationship of these structures to the "animalcules" observed by van Leeuwenhoek in 1674 using a simple microscope was not immediately clear. More advanced propositions awaited the development of the more sophisticated compound microscope. In 1838, Matthias Schleiden, working with plants, and a year later Theodor Schwann, working with animal cells, concluded that cells were the basic units of life and that complex organisms were composed of "aggregates" of these units (Howland, 1973). For some time this concept competed with Haeckel's idea formulated in 1868 that life consisted of protein aggregates ("monera") that were formed directly from inorganics, a notion that can be traced to Aristotle (Howland, 1973).

The concept that cells contain material vital for life ("protoplasm") was advanced during this era by Purkinje (1825), and the concept that cells were derived from other cells came from Virchow in 1858. Interestingly, both Schleiden and Schwann supported a competing idea to Virchow's, that cellular structure derived directly from the nucleus and the nucleus derived from the nucleolus (Howland, 1973). Virchow also applied his observations to medical theory of the time, giving rise to modern pathology (Florey, 1959).

The cell theory is to biology what the atomic theory is to chemistry or elementary particles are to physics. Our preoccupation has been with the question of how studies of animal cells in culture could lead to an understanding of the whole organism or integrated physiology. Since the first experiments in tissue culture by Ross Harrison (1907), the possibility was raised that knowledge could be gained about these fundamental units in an environment that could be manipulated and controlled. Observations of Harrison concerning the appearance of cellular-proccesses in the cultures helped settle a debate between pioneer cytologist Camillo Golgi and neuroanatomist Ramón y Cajál regarding the genesis of the processes. One proposal was that processes developed independently, then fused with the cells, while the idea supported by Harrison's data called for the cells themselves to produce the processes. Almost immediately outcries were raised about the artificiality of the situation and the artifactual nature of cellular behavior. These objections are still being raised and the purpose of this chapter is to address these questions. Our approach will be general but not comprehensive, as this would be beyond our present ability and inclination (Sato *et al.,* 1960).

The single most important objection to the suitability of studying cells in culture was that for over 50 years after the experiments of Harrison there did not exist cultures that exhibited the specialized properties of the tissue of origin. Cultures from muscle did not contract, from glandular tissue did not secrete hormones, from liver did not secrete bile, etc. How, then, could culture studies contribute to the understanding of integrated physiology? Everything in culture looked like fibroblasts. The predominant view for decades was that the artificial conditions of culture caused cells to "differentiate" into some tissue culture-type cell. This was shown to be due to the selective overgrowth of fibroblasts, which had a selective advantage in the serum-based media available at the time. With the discovery of platelet-derived growth factor (Balk *et al.,* 1973; Ross *et al.,* 1974; Antoniades and Scher, 1977) and hormonally defined media (Sato *et al.,* 1982; Hayashi and Sato, 1976; Barnes and Sato, 1980), the selective advantage of fibroblasts becomes perfectly understandable.

With knowledge of the selective advantage of fibroblasts, methods analogous to the enrichment culture methods of microbiology were devised to culture specialized cells with physiological properties of the tissue of origin (Sato *et al.,* 1970; Sato, 1981). A number of specialized cell lines were developed in this way, mostly from differentiated tumors developed by Jacob Furth (1955). These include adrenal cortical cells that respond to adrenocorticotropic hormone (ACTH) to produce steroids (Y1), pituitary cells that secrete ACTH (ATT20), pituitary cells that secrete growth hormone and prolactin (GH series), neuroblastoma cells that secrete transmitters, teratocarcinoma cells that undergo differentiation in culture, glioma cells with properties of astrocytes (C6), Leydig cells that produce steroids, and pigmented melanoma cells (Sato *et al.,* 1970, 1982; Sato and Ross, 1979). Other labs around this time were establishing gamma globulin-secreting B cell myelomas, muscle cells, mast cells, etc. (Sato, 1981).

Some comments would be appropriate at this time. Certainly cells in culture must exhibit differentiated function if culture studies would be of value in understanding integrated physiology. The establishment of the above-mentioned cell lines is a beginning. The Y1 cells deserve special mention. These adrenal cortical cells secrete steroids in response to ACTH. This means that the cultured cells display at least a part of the repertoire of physiologic responses found in the animal. Since its establishment many physiologic responses have been found in cultured cells. Perhaps someday all the effector substances and responses found in physiology will be displayed and studied in culture. It has been estimated that there are 400 different kinds of cells in the animal body. This estimate is mostly based on histology and surely is an underestimate because of the limits of the methodology. We have a fair way to go before each and every cell type is established in culture and available for study.

In the face of skepticism concerning the relevance of studying cells in culture, it would be useful to draw some generalizations on this matter. The first generalization that might be useful is that the more basic the question the more likely the answer will be relevant for whole animal physiology. Some examples are given. In the 1950s and 1960s Harry Eagle (1955) and associates studied the nutritional requirements of cells in culture and found they required more amino acids and less vitamins than were required by the whole animal. This is because some amino acids, such as arginine, are produced in specialized tissues, such as the liver, and some of the vitamins, such as vitamin E, are required by specialized tissues. The cultured cells gave an accurate picture of the situation in the animal, where there is a division of labor between the tissues. Cultured cells did not, as a result of culturing, assume new biosynthetic properties or suddenly require vitamins useful for specialized function. Vitamins that are involved in coenzymes in basic metabolic pathways are required by all cultured cells. The results of Eagle's experiments are that they point out the stability of differentiated properties throughout experimental manipulation.

In the 1960s, Perry, Penman, and Darnell showed that ribosomal RNA molecules in cultured cells are first formed as large precursors, which are cleaved to the final subunit size. Such a basic process could not be different in cultured cells and cells in the body, and this has proved to be the case. Steiner (1977), using cultured insulinomas, showed that the double-chained insulin molecule was formed as a single chain with a connecting peptide. The connecting peptide is cleaved out and the two chains are joined by disulfide bonds. Such a basic process could not be different in culture and the whole animal and such has been found to be true. The concept of messenger RNA was challenged by Henry Harris when he found that most of the RNA synthesized in cultured cells did not exit the nucleus. This was later shown to be due to the fact that mRNAs are synthesized as large precursors, of which only a portion emerges from the nucleus to direct protein synthesis. Such a phenomenon contributed to the discovery of introns and RNA splicing. The organization of genes into introns and exons and the synthesis of messenger RNA as first a large precursor are too basic to be reinvented by cells when they are put into culture.

In some cases the ease of manipulation of cells in culture makes it possible to obtain information that would be extremely difficult to get from whole animal experiments. Two examples come readily to mind. In the period of intense nuclear bomb testing a controversy raged over the sensitivity of human cells to ionizing irradiation. This was settled by the Puck group using single-cell plating tissue culture methods (Puck and Marcus, 1955). Their results showed that human cells were much more sensitive to irradiation than previously thought. Prior to 1957, the diploid number of human chromosomes was thought to be 48. J. H. Tjio, using cultured cells and the hypotonic method of T. C. Hsu, found the correct number to be 46 (Tjio and Levan, 1956). This led to the Denver classification of chromosomes, the association of chromosomal anomalies with disease, and the detailed characterization of human chromosomes. These tissue culture experiments have had a large impact on medical practice—especially genetic counseling.

In some cases, culture techniques have made possible the discovery of substances that could not be made in whole animal experiments. If a young animal is hypophysectomized it ceases growth and radioactive sulfur is not incorporated into the cartilage of the epiphysis of the long bones. If the animal is injected with growth hormone these processes are restored. If the bones are incubated *in vitro* with hypophysectomized serum, ^{35}S is not incorporated into cartilage. If growth hormone is added, sulfation still does not occur but the serum of hypophysectomized animals injected with growth hormone supports sulfation. This led to the discovery of the insulin-like growth factors that are made in the liver in response to growth hormone (Salmon and Daughaday, 1957). The classic approach of ablating the liver of hypophysectomized animals injected with growth hormone is not feasible.

If an animal is ovariectomized and fragments of its ovary are implanted in its spleen, the implant grows as a large tumorous mass (Biskind and Biskind, 1944). The explanation offered at the time was that ovarian steroids were not produced and delivered to inhibit gonadotropin secretion, because the ovarian tissue in the spleen would deliver its steroids through the hepatic portal system to the liver, where they would be inactivated before reaching the general circulation. As a result the pituitary would oversecrete gonadotrophin, causing the implant to grow. When the implanted ovaries were grown in culture and cloned, the cloned cells would grow in the spleens of animals only if they were ovariectomized. The cells did not exhibit a dependence on gonadotrophin in culture but to a contaminant in preparations of luteinizing hormone. This led to the conclusions that serum was masking hormonal dependencies and that the function of serum in cell culture media might be to provide complexes of hormones and that some of these hormones might be novel. The novel factor was called ovarian growth factor (OGF) and its name was later changed to fibroblast growth factor (FGF) (Clark *et al.,* 1972). Two possible avenues of these studies that were not followed up on were that OGF (FGF) might be a gonadotrophin or a mediator of gonadotrophin. In several instances results from culture were so at odds with conventional wisdom that they were discounted as cell culture artifacts.

In several instances, Levi-Montalcini discovered that animals bearing a certain tumor had enlarged sympathetic ganglia (Levi-Montalcini and Hamburger, 1953). In a culture assay, extracts of the tumor, snake venom, and salivary gland extracts greatly stimulated explanted ganglia. That such an odd assortment and sources of activity in a dubious assay system should evoke suspicion of its relevance to physiology was understandable. The active principle turned out to be nerve growth factor (NGF), and later Levi-Montalcini showed that antibodies to NGF injected into fetal animals prevented the normal development of the sympathetic ganglia. Today there is little doubt that NGF plays a central role in the development and regulation of the sympathetic nervous system. Thyrotropin releasing hormone (TRH) was discovered and isolated by measuring thyrotropin-stimulating hormone (TSH) release from explanted pituitaries. TRH was shown to release prolactin from cultured growth hormone (GH) cells (Tashjian *et al.,* 1968). This was naturally viewed as an artifact of pituitary tumor cultures until it was shown that injection of TRH into animals also released prolactin.

Perhaps the most extreme example of cell culture as the essential means to discovery and characterization of physiological regulators has been the characterization of the heparin-binding fibroblast growth factor family and the transforming growth factor β superfamily. Although members of the two families can be delivered to tissues by blood cells (platelets, leukocytes, etc.) in rare situations, members of the family largely serve as intrinsic mediators of external signals to tissues and cell-to-cell communication within tissues. Both were discovered and characterized by cell culture methods, and to date there have been no tissues in which one or more members of these two families are not present. The FGF family now consists of 9 genetically distinct ligands and upward of 100 receptor variants that arise by combinatorial alternate splicing in four genes (Jaye *et al.*, 1992).

Largely, analyses from cells in culture have revealed that different FGF ligands and splice variants of the receptor are expressed in different cell types within the same tissue at different times and in different combinations. In addition to this diversity, monomeric products of the FGF receptor genes oligomerize to result in the active signal transduction complex. To have impact on this, different monomers must be expressed in the same cell at the same time. On top of all this, results show that the FGF receptor complex is probably a ternary one of FGF ligand, ectodomain of monomeric variants described above, and a heparan sulfate chain that is heterogeneous with respect to length, monosaccharide composition, and sulfation, but may be a third source of diversity in biological function of the FGF family (Kan *et al.*, 1993). Cultured cells in the absence of the confounding complexity of the tissue matrix with respect to carbohydrates are the means to dissect out the latter contribution.

Can the use of culture techniques raise the level of conceptualization of integrated physiology to new heights? Two lines of argument suggest that such is the case. First of all, the classic methods of discovering hormones—glandular ablation and injection of extracts—has probably run its course. This method will not serve for paracrine and autocrine factors, which are becoming recognized as real and important factors both for the regulation of normal cells and for the development of cancer. This point is aptly illustrated by the experiments of Salmon and Daughaday (1957) in the discovery of insulin-like growth factors and the experiment of De Larco, which first discovered the transforming growth factors (De Larco and Todaro, 1976). The second argument comes from the fact that when the serum of a medium is replaced with hormones, it is a complex mixture. Classic experiments tended to reveal a single hormone for a target tissue. What is the significance of this finding? Probably that 10^{13} cells cannot coexist in a coordinated fashion in an organism without the finest subtlety in the regulation of its basic units. The simplicity we seek for intellectual satisfaction is likely to elude us. Probably simple unit processes are assembled in such great combinatorial complexity as to boggle the mind. The complexity of hormone mixtures required by cell cultures is only the tip of a large iceberg. These are the regulatory factors we can find by present methods. As new methods appear, the situation will become more complex.

Complexity and interactions in extracellular influences on cellular functions is not limited to hormones and growth factors, but includes nutrients and vitamins also, as shown by Ham and colleagues (see Bettger and Ham, 1982). These cellular responses and processes are as basic as those by which cells replicate and transcribe nucleic acid or translate proteins, and culture studies have been useful in devising therapeutic approaches to cancer. The human myelogenous leukemia cell line HL60 has played a pivotal role for studying differentiation of human leukemia cells. In particular, differentiation induced by all-*trans* retinoic acid (RA) led directly to the finding that cells from patients with acute promyelocytic leukemia [APL; French–American–British (FAB) classification M3] terminally differentiate in response to RA both *in vitro* and *in vivo* (Breitman *et al.*, 1980). Cytodifferentiation therapy of APL by RA is now an established alternate treatment for this disease. This therapy offers many benefits, including the decreased need for intensive hospital care. Kaoru Miyazaki showed similar effects on lung tumor cells in culture (Barnes *et al.*, 1981).

Takahashi and Breitman discovered retinoylation (retinoic acid acylation) a covalent modification of proteins occurring in a variety of eukaryotic cell lines (Breitman *et al.*, 1980). They also found that proteins in HL60 cells were labeled by 17β-estradiol, progesterone, 1,25-$(OH)_2$ vitamin D, triiodothyronine, thyroxine, and prostaglandin E_2 (PGE_2). All of these hormones, except PGE_2, may be nongenomic and mediated by covalently modifying proteins. Are these reactions found in culture experiments operative in the animal, and what is their physiological significance? How dangerous is it to extrapolate cell culture results to the situation *in vivo*?

We have sought examples of culture experiments that have given erroneous results. Although these are not abundant, some possibilities can be cited. For instance, it is well recognized that cells often become chromosomally unstable and karyotypically abnormal as culture progresses (Todero and Green, 1963). A prediction of the diploid mouse chromosome number based on a 3T3 cell line karyotype, for example, would be grossly inaccurate. As another example, it is often observed that on continual passage in culture, cells are selected that are capable of producing autocrine growth factors. These may confer a selective advantage on the cell in culture and probably result from aberrant expression of genes normally repressed in the cell type from which the cultures are derived. A direct application of observations using such cells to the normal endocrine physiology of the organism thus would be misleading.

It has been suggested that senescence *in vitro,* one of the accepted characteristics of normal cell cultures, may actually be an artifact, resulting from abnormal oxidative damage under routine culture conditions for other nonphysiological stresses, such as exposure to nonphysiological serum components (Ames *et al.,* 1993). Some cell types can be grown without senescence if the endocrine stimulation is appropriate (e.g., thyroid epithelial cells in the presence of thyroid-stimulating hormone *in vitro*), and these serum-free cells are inhibited by serum in the culture medium (Ambesi-Impiombato *et al.,* 1980; Loo *et al.,* 1987). In fact, serum is toxic at 100% for all cells in culture except for vascular endothelial cells. Certainly serum differs from interstitial fluid. What is toxic in serum? Is the factor(s) important in physiology? It is unlikely that we can recreate the *in vivo* environment of a cell sufficiently well as not to produce occasional artifacts. If an ovarian cell is implanted under the kidney capsule, a teratoma results. Surely we are at times placing cells in an unsuitable situation in culture as to evoke unnatural responses such as teratoma formation.

What problems seem important at the moment, related to the question of culture and *in vivo* relevance? We come back to the situation in which, when the serum in the medium of various cell types is replaced by hormones (factors), the cells require a complex mixture. Do these operate *in vivo* and how will we be able to tell? Part of the complexity is likely the result of redundancy of function among extracellular messengers, protecting the organism from catastrophic failure should interference occur with a single physiological signaling pathway. An understanding of the physiological significance of these more complex and interrelated interactions will probably require the most recent and sophisticated techniques: transgenic and gene knockout animals, highly specific synthetic hormone antagonists based on detailed knowledge of ligand–receptor interactions, and dominant negative receptor constructions based on a similarly detailed understanding of receptor interactions and signaling mechanisms. At the present time two very active fields of research using cell culture are the mechanism of hormone action and the nature of cancer. Cancer biology and hormone mechanisms studies converge at several points at surface protein tyrosine kinases, at transcriptional control, and the hormonal stimulation of the transcription of protooncogenes. Hormone mechanism studies and cancer biology studies are most conveniently carried out in culture. The phenomena discovered in culture are so basic as surely to apply to the situation *in vivo.* In the case of cancer the oncogenes discovered *in vitro* are found in the animal and a therapeutic approach suggested by culture experiments is efficacious in humans.

Cohen found that epidermal growth factor (EGF) activates a tyrosine kinase that begins a complex cascade (Ushiro and Cohen, 1980). Hunter and Sefton (1980) found that the *src* oncogene is a tyrosine kinase that begins a cascade that results in overriding growth control. Nishizuka and Berridge have in culture studies elucidated the diacyl glycerol formation by phospholipase, the diacyl glycerol activation of protein kinases, and the mobilization of calcium by the phosphotidyl inositols (Nishizuka, 1986). Rodbell discovered G proteins, etc. Much of the present knowledge in these areas derived from work in animal virology and tumor virus biology, areas that are dependent on cell culture approaches in most respects.

Several neural progenitor cells have been identified first in cultures. Raff *et al.* (1983) showed that proliferating O-2A stem cells were able to differentiate into oligodendrocytes and type 2 astrocytes *in vitro.* Groves *et al.* (1993) demonstrated that week-old cultured O-2A progenitor cells could be successfully transplanted into demyelinated lesions in adult spinal cords and produce extensive remyelination. These data suggest the possibility of expanding O-2A progenitors in culture for transplants to humans afflicted with demyelinating disorders, e.g., multiple sclerosis. In addition, several groups are studying the effects of growth factors of O-2A progenitors, suggesting that endogenous regulation of adult O-2A progenitor proliferation may provide another therapeutic strategy to enhance remyelination.

Neural stem cells with the potential to give rise to differentiated subtypes of central nervous system neurons are of particular importance in future advances in tissue transplantation and engineering. These cultures cannot be established by routine means developed for rodent or human fibroblasts in serum-containing medium. For instance, one such line derived from mouse embryonic brains is dependent on epidermal growth factor for survival, and is growth inhibited by serum (Loo *et al.,* 1987). Cultured under serum-free conditions, these cells maintain a normal karyotype despite extended passage in culture. These cells express astrocyte markers in the presence of transforming growth factor β (Sakai *et al.,* 1990), and express nestin, a marker of neural stem cells, in the neural precursor state in the absence of transforming growth factor β (Loo *et al.,* 1994). A variety of systems of this type exist, each requiring specialized cell culture techniques (see Chapter 58).

Cancer biology studies in culture are instructive to contemplate in regard to the relationship between *in vivo* and *in vitro* behavior. The Rous sarcoma virus assay was based on the formation of foci *in vitro,* which was based on the ability to overcome contact inhibition and grow in soft agar. These properties correlate well with tumorigenicity in the animal (Freedman and Shin, 1974). These aspects of cell culture behavior have been used to identify and isolate oncogenes. Cell culture experiments identifying oncogenes such as *ras* strengthen the argument that basic processes can be identified with assurance in cell culture, because these genes have subsequently been implicated in a variety of human tumors *in vivo* (Shi and Weinberg, 1982). Studies of transforming growth factors were an offshoot of tumor virology studies that depended on cell culture methodology almost entirely from conception to proof (De Larco and Todaro, 1976). Now it is clear that these molecules have important functions in basic processes such as embryonic development. That these molecules had not been discovered in the previous 50 years by scientists using the classic approaches of developmental biology points out the power of the cell culture approach to identify extracellular endocrine-like messengers.

In a related approach, human papilloma virus (HP16) genes E6 and E7 have been used to immortalize cells from fetal cartilage, lung epithelium, skin (keratinocyte), tracheal epithelium, pancreas, kidney, and liver (Thompson *et al.,* 1997; Zabrenetzky *et al.,* 1997). The viral genes are introduced via retrovirus vectors produced with a packaging cell line. Interestingly, human leukocyte antigen (HLA) Class I and Class II molecules are not expressed in these cells, suggesting potential for wide application to transplantation and tissue engineering if the neoplastic potential of the cells can be resolved favorably.

Cell culture approaches have contributed to major modern advances in the development and use of embryonal and lineage-specific stem cells. These provide essential models in studies of normal and abnormal biological processes, as well as potentially unlimited sources of differentiation-competent cells for tissue engineering. Differentiation toward multiple lineages can be induced in culture by treatment with retinoic acid, substratum manipulation, growth factor addition or withdrawal, and other approaches. Stem cell lines now include embryonal stem cells from rhesus monkeys, common marmosets, and humans (Thomson *et al.,* 1995, 1996, 1998; Thomson and Marshall, 1998), as well as human primordial germ cells developed by Gerhardt *et al.* (see Chapter 58). Tissue-specific stem cell precursor culture systems (both tissue and embryonal stem cell derived) exist for a variety of tissues, including cardiomyocyte, endothelial, neural, hematopoietic, liver, pancreas, and other endoderm-derived endocrine cell types.

We are passing from a period when the *in vivo* relevance of culture studies was rejected out of hand to a period of uncritical acceptance. We are counseling critical acceptance. In physics we are told we are nearing the end of knowing all we can about the material universe by knowing all the fundamental particles and their interactions. One day we will approach this situation in biology, when we know all the cells and their interactions. As we progress, we will transition from marveling at the enormity of what we do not know to marveling at the complexity, subtlety, and coordination of what we learn about the biochemistry of life.

Acknowledgment

We are grateful to Emily Amonett for stylistic and scholarly contributions to this chapter and for steadfast friendship over the years.

References

Ambesi-Impiombato, F. S., Parks, L. A., and Coon, H. G. (1980). Culture of hormone-dependent functional epithelial cells from rat thyroids. *Proc. Natl. Acad. Sci. U.S.A.* 77, 3455.

Ames, B. N., Shigenaga, M. K., and Hagan, T. M. (1993). Oxidants, antioxidants and the degenerative diseases of aging. *Proc. Natl. Acad. Sci. U.S.A.* 90, 7915–7922.

Antoniades, H. N., and Scher, C. D. (1977). Radioimmunoassay of a human serum growth factor of Balb/C-3T3 cells: Derivation from platelets. *Proc. Natl. Acad. Sci. U.S.A.* 74, 1973–1978.

Balk, S. D., Whitfield, J. F., Youdale, T., and Braun, A. D. (1973). Roles of calcium, serum, plasma, and folic acid in the control of proliferation of normal and Rous sarcoma virus-infected chicken fibroblasts. *Proc. Natl. Acad. Sci. U.S.A.* 70, 675–680.

Barnes, D. W., and Sato, G. H. (1980). Serum-free cell culture: A unifying approach. *Cell (Cambridge, Mass.)* **22**, 649.

Barnes, D. W., van der Bosch, J., Masui, H. *et al.* (1981). The culture of human tumor cells in serum-free medium. *In* "Methods in Enzymology" (S. Pestka, ed.), Vol. 79, pp. 368–391. Academic Press, New York.

Bettger, W. J., and Ham, R. G. (1982). The nutrient requirements of mammalian cells grown in culture. *Adv. Nutr. Res.* **4**, 249–286.

Biskind, M. S., and Biskind, G. R. (1944). Development of tumors in the rat ovary after transplantation into the spleen. *Proc. Soc. Exp. Biol. Med.* **55**, 176–182.

Breitman, T. R., Selonick, S. E., and Collins, S. J. (1980). Induction of differentiation of the human promyelocytic leukemia cell lin (HL-60) by retinoic acid. *Proc. Natl. Acad. Sci. U.S.A.* **77**, 2936–2941.

Clark, J. L., Jones, K. L., Gospodarowicz, D., and Sato, G. H. (1972). Growth response to hormones by a new rat ovary cell line. *Nature (London), New Biol.* **236**, 180–181.

De Larco, J. E., and Todaro, G. J. (1976). Growth factors from murine sarcoma virus-transformed cells. *Proc. Natl. Acad. Sci. U.S.A.* **75**, 4001–4006.

Eagle, H. (1955). Nutrition needs of mammalian cells in tissue culture. *Science* **122**, 501–504.

Florey, H. (1959). "General Pathology." Saunders, London.

Freedman, V. H., and Shin, S.-I. (1974). Cellular tumorigenicity in nude mice: Correlation with cell growth in semi-solid medium. *Cell (Cambridge, Mass.)* **3**, 355–359.

Furth, J. (1955). Experimental pituitary tumors. *Recent Prog. Horm. Res.* **11**, 221–255.

Groves, A. K., Barnett, S. C., Frnaklin, R. J. M. *et al.* (1993). Repair of demyelinated lesions by transplantation of purified O-2A progenitor cells. *Nature (London)* **362**, 453–455.

Harrison, R. G. (1907). Observations on the living developing nerve fiber. *Proc. Soc. Exp. Biol. Med.* **4**, 140–143.

Hayashi, I., and Sato, G. H. (1976). Replacement of serum by hormones permits the growth of cells in a defined medium. *Nature (London)* **259**, 132–134.

Hooke, R. (1665). "Micrographica." London.

Howland, J. L. (1973). "Cell Physiology." Macmillan, London.

Hunter, T., and Sefton, B. M. (1980). Transforming gene product of Rous sarcoma virus phophorylates tyrosine. *Proc. Natl. Acad. Sci. U.S.A.* **77**, 1311–1315.

Jaye, M., Schlessinger, J., and Dionne, C. A. (1992). Fibroblast growth factor receptor tyrosine kinases: Molecular analysis and signal transduction. *Biochim. Biophys. Acta* **1135**, 185–199.

Kan, M., Wang, F., Xu, J. *et al.* (1993). An essential heparin-binding domain in the fibroblast growth factor receptor kinase. *Science* **259**, 1918–1921.

Levi-Montalcini, R., and Hamburger, V. (1953). A diffusible agent of mouse sarcoma, producing hyperplasia of sympathetic ganglia and hyperneurotization of viscera in the chick embryo. *J. Exp. Zool.* **123**, 233–288.

Loo, D. T., Fuquay, J. I., Rawson, C. L. *et al.* (1987). Extended culture of mouse embryo cells without senescence: Inhibition by serum. *Science* **236**, 200–202.

Loo, D. T., Althoen, M. C., and Cotman, C. W. (1994). Down regulation of nestin by TGF-B or serum in SFME cells accompanies differentiation into astrocytes. *NeuroReport* **5**, 1–5.

Nishizuka, Y. (1986). Studies and perspectives of protein kinase C. *Science* **233**, 305–312.

Puck, T. T., and Marcus, P. I. (1955). A rapid method for viable cell titration and clone production with HeLa cells in tissue culture: The use of X-irradiated cells to supply conditioning factors. *Proc. Natl. Acad. Sci. U.S.A.* **41**, 432–437.

Purkinje, J. E. (1825). "Beobachtungen und Versuche zur Physiologie der Sinne. Neurere Beitrage zur Kenntnis des Sehens in Subjectiver Hinischt." Reimer, Berlin.

Raff, M. C., Miller, R. H., and Nobel, M. (1983). A glial progenitor cell that develops *in vitro* into an astrocyte or an oligodendrocyte depending on culture medium. *Nature (London)* **303**, 390–396.

Ross, R. J., Glomset, B., Kariya, B., and Harker, L. (1974). A platelet-dependent serum factor that stimulates the proliferation of arterial smooth muscle cells in vitro. *Proc. Natl. Acad. Sci. U.S.A.* **71**, 1207–1212.

Sakai, Y., Rawson, C., Lindburg, K., and Barnes, D. (1990). Serum and transforming growth factor beta regulate glial fibrillary acidic protein in serum free-derived mouse embryo cells. *Proc. Natl. Acad. Sci. U.S.A.* **87**, 8378–8382.

Salmon, W. D., and Daughaday, W. H. (1957). A hormonally controlled serum factor which stimulates sulfate incorporation by cartilage *in vitro. J. Lab. Clin. Med.* **49**, 823–834.

Sato, G., ed. (1981). "Functionally Differentiated Cell Lines." Alan R. Liss, New York.

Sato, G., and Ross, R., eds. (1979). "Cold Spring Harbor Conferences on Cell Proliferation," Vol. 6. Cold Spring Harbor Lab., Cold Spring Harbor, NY.

Sato, G., Zaroff, L., and Mills, S. E. (1960). Tissue culture populations and their relation to the tissue of origin. *Proc. Natl. Acad. Sci. U.S.A.* **46**, 963–967.

Sato, G., Augusti-Tocco, G., Kelly, P. *et al.* (1970). Hormone-secreting and hormone responsive cell cultures. *Recent Prog. Horm. Res.* **26**, 539–545.

Sato, G., Pardee, A., and Sirbasku, D., eds. (1982). "Cold Spring Harbor Conferences on Cell Proliferation." Vol. 9. Cold Spring Harbor Lab., Cold Spring Harbor, NY.

Shi, C., and Weinberg, R. A. (1982). Isolation of a transforming sequence from a human bladder carcinoma cell line. *Cell (Cambridge, Mass.)* **29**, 161–169.

Steiner, D. F. (1977). The Banting Memorial Lecture 1976: Insulin today. *Diabetes* **26**, 322–340.

Tashjian, A. H., Yasumura, Y., Levine, L. *et al.* (1968). Establishment of clonal strains of rat pituitary tumor cells that secrete growth hormone. *Endocrinology (Baltimore)* **82**, 342–352.

Thompson, A., Vilner, L., and Hay, R. J. (1997). The immortalization of human fetal cartillage, lung, trachea, pancreas, and liver cell lines by HPV16 E6/E7. *J. Cell Biol.* **H84** (abstr.).

Thomson, J. A., and Marshall, V. S. (1998). Primate embryonic stem cells. *Curr. Top. Dev. Biol.* **38**, 133–165.

Thomson, J. A., Kalishman, J., Golos, T. G., Durning, M., Harris, C. P., Becker, R. A., and Hearn, J. P. (1995). Isolation of a primate embryonic stem cell line. *Proc. Natl. Acad. Sci. U.S.A.* **92**, 7844–7848.

Thomson, J. A., Kalishman, J., Golos, T. G., Durning, M., Harris, C. P., and Hearn, J. P. (1996). Pluripotent cell lines derived from common marmoset (*Callithrix jacchus*) blastocysts. *Biol. Reprod.* **55**, 254–259.

Thomson, J. A., Itskovitz-Eldor, J., Shapiro, S. S., Waknitz, M. A., Swiergiel, J. J., Marshall, V. S., and Jones, J. J. (1998). Embryonic stem cell lines derived from human blastocysts. *Science* **282**, 1145–1147.

Tjio, J. H., and Levan, A. (1956). The chromosome number of man. *Hereditas* **42**, 1–6.

Todaro, G. J., and Green, H. J. (1963). Quantitative studies of the growth of mouse embryo cells in culture and their development into established lines. *J. Cell Biol.* **17**, 299.

Ushiro, H., and Cohen, S. (1980). Identification of phosphotyrosine as a product of epidermal growth factor-activated protein kinase in A431 cell membranes. *J. Biol. Chem.* **255**, 8363–8365.

Virchow, R. (1858). "Die Cellularpathologie in Ihrer Begrundung auf Physiologische und Pathologische Gewebelehre." A. Hirschwald, Berlin.

Zabrenetzky, V., Thompson, A., Vilner, L., and Hay, R. J. (1997). The isolation, characterization and immortalization of human fetal kidney cells. *In Vitro* **33**, 34A.

CHAPTER 11

REGULATION OF CELL BEHAVIOR BY MATRICELLULAR PROTEINS

Amy D. Bradshaw and E. Helene Sage

INTRODUCTION

The extracellular milieu is critical for the control of the behavior of every cell in all tissues. Many factors can contribute to the environment of a cell—for example, cell–cell contact, growth factors, extracellular matrix proteins, and matricellular proteins. All of these components act together to regulate cell surface protein activity, intracellular signal transduction, and subsequent gene expression, which lead to proliferation, migration, differentiation, and ultimately the formation of complex tissues. This chapter focuses on matricellular proteins as modulators of extracellular signals. The matricellular proteins are characterized as secreted, modular proteins that are associated with the extracellular matrix but do not act as structural constituents (Bornstein, 1995). Presumably, the function of matricellular proteins is to provide a link between the extracellular matrix and the cell surface receptors, or the cytokines and proteases localized in the extracellular environment, the activity of which might be affected by this interaction. Thrombospondin-1 and thrombospondin-2, tenascin-C, osteopontin, and SPARC (an acronym for "secreted protein, acidic and rich in cysteine") are representatives of this class of proteins. A growing body of evidence points to these proteins as important mediators of growth factor, extracellular matrix, and cell signaling pathways (Table 11.1). Consequently, *in vitro* systems designed to mimic tissue conditions should consider the influence of matricellular proteins.

MATRICELLULAR PROTEINS

THROMBOSPONDIN-1 AND THROMBOSPONDIN-2

Thrombospondin-1 is a 450,000-Da glycoprotein with seven modular domains (Bornstein and Sage, 1994). To date, five different paralogs of thrombospondin have been identified, thrombospondins-1–5. This chapter reviews the two most characterized forms, thrombospondin-1 and thrombospondin-2. At least five different extracellular matrix-associated proteins are able to bind to thrombospondin-1: collagens I and V, fibronectin, laminin, fibrinogen, and SPARC (Bornstein, 1995; Clezardin *et al.,* 1988). Likewise, cell surface receptors for thrombospondin are multiple and include the integrin family of extracellular matrix receptors (Bornstein, 1995; Hynes, 1992). Given the significant number and variety of thrombospondin-1-binding proteins, it is of little surprise that a wide variety of functions have been attributed to this protein, some of which appear to be contradictory. Many of these disparities may, however, actually reflect the dynamic interaction of thrombospondin-1 with other extracellular factors, which can influence cells in different ways to give rise to distinct cellular outcomes.

Several published studies describe the antiangiogenic properties of thrombospondin-1, whereas other investigators have reported an apparent stimulation of new blood vessel formation by thrombospondin-1. A potential source of variability is the assay used to perform the experiments.

Table 11.1. Matricellular proteins[a]

Protein	Extracellular matrix interaction	Receptor	Adhesive (+) vs. counteradhesive (−)	Growth factor modulation	Extracellular matrix formation
Thrombospondin-1	Col I and V, Fn, Ln, Fg	Integrin, CD-36	−	HGF (−), TGF-β (+)	?
Thrombospondin-2	?	?	−	?	+
Tenascin-C	Fn	Integrin, annexin II	− +	EGF (+), bFGF (+), PDGF (+)	?
Osteopontin	Fn; Col I, II, III, IV, and V	Integrin, CD-44	+	?	+
SPARC	Col I, III, IV, and V	?	−	bFGF (−), VEGF (−), PDGF (−), TGF-β (+)	+

[a]This is not a complete list of all activities and receptors for these proteins. For references, refer to the text. Abbreviations: Col, collagen; Fn, fibronectin; Ln, laminin; Fg, fibrinogen; HGF, hepatocyte growth factor; TGF-β, transforming growth factor β; EGF, epidermal growth factor; PDGF, platelet-derived growth factor; bFGF, basic fibroblast growth factor; VEGF, vascular endothelial growth factor.

A number of the studies that support an antiangiogenic function for thrombospondin-1 were carried out *in vitro* in a defined environment. For example, incubation of both aortic and microvascular endothelial cells with antithrombospondin antibodies significantly inhibited cord formation, a model of angiogenesis *in vitro* (Iruela-Arispe *et al.*, 1991). Likewise, aortic endothelial cells transfected with antisense thrombospondin-1 cDNA generated twice as many capillary-like structures on a gelled basement membrane in comparison to control cells that produce higher levels of thrombospondin-1 (DiPietro *et al.*, 1994). In contrast, the rat aortic explant model, in which thrombospondin-1 has been found to stimulate angiogenesis, is more complex, with many cell types that affect the outcome (Nicosia and Tuszynski, 1994). Rings of rat aorta grown in collagen or fibrin matrices supplemented with increasing amounts of thrombospondin-1 showed a concentration-dependent increase in the number of microvessels. Thrombospondin-1 did not appear to act directly on the endothelial cells in this assay, but rather affected the myofibroblasts present in the aortic rings by stimulation of their proliferation as well as their secretion of a heparin-binding protein(s). The authors speculated that perhaps the secreted protein could orient thrombospondin-1 in the matrix in such a way as to promote microvessel formation. The effect of thrombospondin-1 in angiogenic assays illustrates the versatility of function of this matricellular protein in distinct microenvironments. In all likelihood, the activity of thrombospondin-1 in these experiments reflects the interaction of the matricellular protein with specific cell surface receptors and other extracellular cytokines to elicit defined cellular responses.

One cytokine known to interact with thrombospondin-1 is scatter factor/hepatocyte growth factor (HGF), a known angiogenesis-promoting factor (Lamszuz *et al.*, 1996). Apparently, thrombospondin-1 is able to sequester HGF in the matrix through its association with other matrix molecules and prevent interaction of the cytokine with its cellular receptor. Thus in the rat corneal model of angiogenesis, thrombospondin-1 will inhibit HGF-induced neovascularization in a concentration-dependent manner (Lamszuz *et al.*, 1996).

Thrombospondin-1 has also been shown to interact specifically with another cytokine, transforming growth factor β (TGF-β) (Schultz-Cherry and Murphy-Ullrich, 1993). This interaction leads to activation of the latent form of TGF-β, presumably through a conformational change in the cytokine that allows interaction with cell surface receptors (Schultz-Cherry *et al.*, 1994). Consequently, the presence of thrombospondin-1 in the extracellular milieu can affect the activity of this potent, multifunctional cytokine. In fact, generation of a thrombospondin-1 null mouse has confirmed the importance of thrombospondin-1-mediated activation of TGF-β *in vivo*. The phenotype manifested in the absence of thrombospondin-1 expression mimics to some degree the phe-notype of the TGF-β null mouse (Crawford *et al.*, 1998). Specifically, similar pathologies of the lung and pancreas were observed in both mice; importantly, thrombospondin-1 null mice treated with thrombospondin-1 peptides showed a partial restoration of active TGF-β levels and a rever-sion of the lung and pancreatic abnormalities toward tissues of wild-type mice.

Thrombospondin-1 is also able to influence cell adhesion and cell shape. For example, it will diminish the number of focal adhesions of bovine aortic endothelial cells and thus will promote a migratory phenotype (Murphy-Ullrich and Höök, 1989). Thrombospondin-1, therefore, has been proposed to modulate cell–matrix interaction to allow for cell migration when necessary. Accord-ingly, thrombospondin-1 expression is observed during events such as dermal wound repair, when cell movement is required (DiPietro *et al.*, 1996). Thrombospondin-1 is also expressed by many tumor cells and might also facilitate metastatic migration (Tuszynski and Nicosia, 1996; Weinstat-Saslow *et al.*, 1994). In fact, metastatic human carcinoma cells, transfected with antisense throm-bospondin-1 cDNA to reduce thrombospondin-1 protein expression, lost their capacity to prolif-erate and metastasize in athymic mice (Castle *et al.*, 1991).

Fewer functional studies have been performed with thrombospondin-2, although a throm-bospondin-2 null mouse has been generated that has revealed some interesting insight into this protein. Thrombospondin-2 has been shown to inhibit angiogenesis *in vivo*. Pellets containing re-combinant thrombospondin-2 were implanted in the rat cornea to test whether neovasculariza-tion was enhanced. Thrombospondin-2 was able to inhibit basic fibroblast growth factor (bFGF)-induced vessel invasion to nearly the same degree as thrombospondin-1 (Volpert *et al.*, 1995). Thus, the observation that the thrombospondin-2 null mouse appears to have a greater degree of vascularization in the skin was in accordance with the hypothesis that thrombospondin-2 could serve as an endogenous regulator of angiogenesis in mice (Kyriakides *et al.*, 1998). In fact, the ex-pression pattern of thrombospondin-2 is more consistent with the importance of this protein in vascular formation, because the mRNA for thrombospondin-2 is more closely associated with the vasculature in developing tissues in comparison to that of thrombospondin-1 (Iruela-Arsipe *et al.*, 1993). However, the mechanism by which thrombospondin-2 affects the degree of vascularization is unknown. Interestingly, another observation from the thrombospondin-2 null mouse was that of altered collagen fibrillogenesis in the skin, relative to that seen in wild-type mice. The collagen fibrils in the null mice were larger in diameter, had aberrant contours, and were disordered. Pre-sumably this effect on collagen fibril formation resulted in less tensile strength of the skin in throm-bospondin-2-null mice versus their wild-type counterparts (Kyriakides *et al.*, 1998).

Thrombospondin-1 and -2 can act as negative regulators of cell growth. In particular, en-dothelial cells are susceptible to an inhibition of proliferation by both proteins and, as such, have been classified as inhibitors of angiogenesis. However, the variety of cell surface receptors for thrombospondin allows for diverse signaling events in different cell types; consequently, there might be situations when thrombospondin appears to support angiogenesis as well. Throm-bospondin-1 has been shown to modulate the activity of at least two cytokines, TGF-β and HGF, and thus, either diminish activity, in the case of HGF, or enhance activity, as seen for TGF-β. Fi-nally, thrombospondin-1 can alter cell shape to promote a migratory phenotype. Further charac-terization of the thrombospondins, including closer examination of the remaining family mem-bers (thrombospondins 3–5), will no doubt yield fascinating insight into this multifunctional gene family.

Tenascin-C

Tenascin-C is a matricellular protein with a widespread pattern of developmental expression in comparison to a restricted pattern of expression in adult tissues (Koukoulis *et al.*, 1991). In ad-dition to tenascin-C, two other less characterized forms of tenascin have been identified: tenascin-R and tenascin-X (Erickson, 1993; Jones and Copertino, 1996). This review focuses on tenascin-C, because it is the best characterized of the three known tenascins. Tenascin-C consists of six subunits (or arms) linked by disulfide bonds into a 2000-kDa molecule that can associate with fi-bronectin in the extracellular matrix (Erickson, 1993). Like thrombospondin-1, a number of dif-ferent functions have been attributed to tenascin-C and, accordingly, a number of cell surface re-ceptors appear to mediate distinct properties of this matricellular protein. At least five different integrins are known receptors for tenascin-C, as well as annexin II (Yokosaki *et al.*, 1996; Sriramarao *et al.,* 1993; Chung *et al.,* 1996). Although the integrins appear to support cell adhesion to tenascin-C, annexin II is thought to mediate the counteradhesive function attributed to this protein. Hence, tenascin-C can act as either an adhesive or as a counteradhesive substrate for different cell types, dependent on the profile of receptors expressed on the cell surface (Crossin, 1996).

Tenascin-C has also been shown to modulate the activity of growth factors: specifically, it promotes epidermal growth factor (EGF)-dependent and bFGF-dependent cell growth (End *et al.,* 1992; Jones and Rabinovitch, 1996). In fact, Jones *et al.* (1997) have shown that smooth muscle cells plated in a collagen gel secrete matrix metalloproteinases (MMPs) that degrade the collagen to expose integrin receptor binding sites. Engagement of these receptors induces tenascin-C expression; tenascin-C is subsequently deposited into the extracellular matrix and can itself serve as an integrin ligand. The deposition of tenascin-C leads to cell shape changes initiated by a redistribution of focal adhesion complexes concomitant with a clustering of EGF receptors on the cell surface. Presumably, clustering of the EGF receptors facilitates EGF signaling and thereby enhances the mitogenic effect of EGF. Conversely, when MMP activity is inhibited, tenascin-C expression is decreased and the cells become apoptotic (Jones *et al.,* 1997). Thus, tenascin-C is able to modulate EGF activity such that the presence of this matricellular protein supports cell growth and its absence induces programmed cell death.

Given the widespread expression of tenascin-C in the developing embryo, the lack of an overt phenotype in the tenascin-C null mouse was surprising (Saga *et al.,* 1992). In particular, the high level of tenascin-C expression in the central and peripheral nervous systems had indicated that the absence of this protein might lead to neuronal abnormalities. Although no histologic differences could be detected in the brains of tenascin-C null mice, they do appear to have some behavioral aberrations (Steindler *et al.,* 1995; Fukamauchi *et al.,* 1996). Interestingly, the genetic background of the tenascin-C null mouse is likely to be a major factor in the identification of tissues in which tenascin-C might be functionally important. For example, Nakao *et al.* (1998) used three different congenic mouse lines to study the effect of Habu-snake venom-induced glomerulonephritis in a tenascin-C null background. Although the disease was worse in all tenascin-C null mice in comparison to wild-type controls, each line exhibited a different level of severity. Induction of the disease in one strain, GRS/A, resulted in death from irreversible renal failure. Moreover, mesangial cells cultured from tenascin-C null animals did not respond to cytokines such as platelet-derived growth factor (PDGF) unless exogenous tenascin-C was included in the culture medium (Nakao *et al.,* 1998). Hence, tenascin-C is also able to modulate the activity of this growth factor as observed previously for EGF. Tenascin-C provides another example of a matricellular protein able to affect growth factor efficacy.

Although tenascin-C shows a limited pattern of expression in the adult, an induction of tenascin-C is seen in many tissues undergoing wound repair or neoplasia (Mackie *et al.,* 1988; Higuchi *et al.,* 1993; Giachelli *et al.,* 1998). Thus tenascin-C, like the other matricellular proteins, is ideally suited to act as a modulator of cell shape, migration, and growth. That no developmental abnormalities are manifested in the tenascin-C null mouse points to the greater importance of tenascin-C in remodeling events that take place in response to injury or transformation.

OSTEOPONTIN

As the name implies, osteopontin was originally classified as a bone protein. A more thorough examination, however, reveals a widespread expression pattern for this protein with multiple potential functions (Denhardt and Guo, 1993). Osteopontin associates with the extracellular matrix, in that it binds to fibronectin and to collagens I, II, III, IV, and V (Mukherjee *et al.,* 1995; Chen *et al.,* 1992; Butler, 1995). Osteopontin also affects cellular signaling pathways by virtue of its capacity to act as a ligand for multiple integrin receptors as well as CD44 (Denhardt and Noda, 1998; Weber *et al.,* 1996). Thus osteopontin, like most of the matricellular proteins, is able to act as a bridge between the extracellular matrix and the cell surface. Because matricellular proteins might be synthesized, secreted, and incorporated into the extracellular matrix with greater ease than more complex secreted proteins, which must be incorporated into fibrils and assembled into a network, a bridging function might be useful during remodeling events in the organism when rapid conversion of the cellular substrata is required for cell movement.

In support of this concept, Weintraub *et al.* (1996) showed that transfection of vascular smooth muscle cells with antiosteopontin cDNA reduced adhesion, spreading, and invasion of three-dimensional collagen matrices. Addition of osteopontin to the collagen gel restored the ability of these cells to invade the gel. Interestingly, osteopontin also appears to be susceptible to modification by extracellular proteases, which have revealed cryptic ingetrin binding sites within the sequence. The protease thrombin cleaves osteopontin at the N-terminal domain and exposes an integrin $\alpha9\beta1$ binding site (Smith *et al.,* 1996; Smith and Giachelli, 1998). In addition, Senger *et al.* (1996) reported that endothelial cells treated with vascular endothelial growth factor (VEGF) increased their expression of the integrin $\alpha v\beta3$, another osteopontin cell surface receptor, concomitantly with an increase in osteopontin. These investigators also showed an increase in the amount of thrombin-processed osteopontin in tissues injected with VEGF and radiolabeled osteopontin. The significance of this result lies in the enhanced support of endothelial cell migration *in vitro* by thrombin-cleaved versus full-length protein. Because the migratory activity of the endothelial cells was blocked partially by an anti-$\alpha v\beta3$ antibody, the remainder of the activity was attributed to $\alpha9\beta1$ or to another cell surface receptor (Senger *et al.,* 1996).

The promotion of cell survival is another property ascribed to osteopontin. Denhardt and Noda (1998) have reported that human umbilical vein endothelial cells plated in the absence of growth factors will undergo apoptosis. If these cells are plated on an osteopontin substrate, however, apoptosis is inhibited. Furthermore, rat aortic endothelial cells subjected to serum withdrawal undergo programmed cell death, a response inhibited by an osteopontin substratum. In fact, it is the ligation of integrin $\alpha v\beta3$ by osteopontin at the cell surface that induces NF-κB, a transcription factor that controls a variety of genes through direct binding to their promoter regions. Thus osteopontin and other $\alpha v\beta3$ ligands can protect cells from apoptosis (Scatena *et al.,* 1998).

Finally, osteopontin appears to be involved in inflammatory responses. Expression of osteopontin was found to increase during intradermal macrophage infiltration, and purified osteopontin injected into the rat dermis led to an increase in the number of macrophages at the site of administration. Importantly, antiosteopontin antibodies inhibited macrophage accumulation in a rat intradermal model after a potent macrophage chemotactic peptide was used to induce an inflammatory response (Giachelli *et al.,* 1998).

The phenotype of the osteopontin null mouse supports the hypothesis that osteopontin can affect macrophage activity. Although the number of macrophages did not appear to differ significantly in incisional wounds of wild-type and osteopontin null mice, the amount of cell debris was higher in wounds of the latter animals. Because macrophages are thought to be primary mediators of wound debridement, osteopontin could be important in the regulation of macrophage function (Liaw *et al.,* 1998). Collagen fibril formation in the deeper dermal layers of the wounded osteopontin null mice also appeared to be affected. Osteopontin null mice have smaller collagen fibrils compared to wild-type controls. Similar to thrombospondin-2, osteopontin might affect collagen fibrillogenesis especially at wound sites, because no differences were seen in the size of collagen fibrils in unwounded skin (Liaw *et al.,* 1998).

Osteopontin and its repertoire of cell surface receptors represent components of a pathway used by cells in need of rapid movement or migration. In addition, the ligation of osteopontin by certain cell surface receptors supports cell survival and thus provides a mechanism for a given tissue to protect a subset of cells, expressing the appropriate receptor, from apoptosis. Hence, osteopontin is a useful protein in events that require cell movement and discriminate cell survival, such as wound healing and angiogenesis.

SPARC

SPARC (also known as BM-40 and osteonectin) was first identified as a primary component of bone but has since been known to have a wider distribution (Lane and Sage, 1994). Increased expression of SPARC is observed in many tissues undergoing diverse remodeling events. For example, SPARC is found in the gut epithelium, which normally exhibits rapid turnover, and in healing wounds. An increase in SPARC is observed during glomerulonephritis, liver fibrosis, and in association with many different tumors. Like the other matricellular proteins, SPARC interacts with the extracellular matrix by binding to collagens I, III, IV, and V (Lane and Sage, 1994). SPARC has also been shown to bind to thrombospondin-1 and to a variety of growth factors present in the extracellular space. For example, SPARC binds to PDGF AB and BB and prevents their interaction with PDGF cell surface receptors. PDGF-stimulated mitogenesis is inhibited by the addition of SPARC (Raines *et al.,* 1992). SPARC can also prevent VEGF-induced endothelial cell growth, in that it binds directly to the growth factor and thereby prevents VEGF receptor stimulation of the mitogen-activated protein kinases, Erk-1 and Erk-2 (Kupprion *et al.,* 1998). Interestingly, a variety of mitogenic stimulators are inhibited by SPARC in culture, some of which do not necessarily associate physically with this protein. For example, SPARC is thought not to bind to bFGF, but SPARC inhibits bFGF-stimulated endothelial cell cycle progression (Hasselaar and Sage, 1992). Apparently, the effect of SPARC on cell proliferation is complex and could occur through (1) a direct prevention of receptor activation and/or (2) a pathway mediated by a cell surface receptor recognizing SPARC. Unlike the other matricellular proteins described here, a receptor(s) for SPARC is currently unidentified.

Another significant effect of SPARC on cells in culture is its capacity to elicit changes in cell shape. Many cell types plated on various substrata retract their filopodia and lamellopodia and assume a rounded phenotype after exposure to SPARC. Bovine aortic endothelial cells, for example, are prevented from spreading in the presence of SPARC (Sage *et al.,* 1989). Clearly cell rounding could contribute to the inhibition of cell cycle, mentioned previously; however, these two effects of SPARC appear to be independent. Motamed and Sage (1998) have shown that inhibition of tyrosine kinases reverses the counteradhesive function of SPARC but has no effect on cell cycle inhibition. Thus, at least in endothelial cells, SPARC appears to mediate two aspects of cell behavior through different mechanisms.

The targeted inactivation of the SPARC gene in mice has allowed a more thorough examination of SPARC activity. In support of SPARC as a regulator of cell proliferation, primary mesenchymal cells isolated from SPARC null mice proliferated faster than did their wild-type counterparts. In addition, kidney mesangial cells exhibited a more spread morphology and a higher proportion of intact transcytoplasmic actin cables *in vitro* (Bradshaw *et al.,* 1999). These results indicate that the absence of endogenous SPARC affects the intracellular architecture that gives this cell type its characteristic shape. The identification of a SPARC receptor(s) will elucidate the distinct properties of SPARC in different cell types.

In addition to its effect on cell cycle, evidence shows that SPARC mediates extracellular matrix accumulation through a TGF-β-dependent pathway. Mesangial cells isolated from SPARC null mice showed a decreased collagen I expression accompanied by a decrease in the levels of the cytokine TGF-β1 in comparison to wild-type cells. TGF-β is a known positive regulator of extracellular matrix synthesis in the kidney (MacKay *et al.,* 1989). Addition of recombinant SPARC to SPARC null mesangial cultures restored the levels of collagen I and TGF-β nearly to those of wild-type cells (Francki *et al.,* 1999). Thus it appears that SPARC can act as a regulator of TGF-β and, by extension, collagen I in mesangial cells.

Once again, we see SPARC as an example of a multifunctional protein able to regulate cell shape and modulate growth factor activity. In addition, SPARC appears to govern cell cycle traverse, perhaps in some cases independently of direct growth factor interaction (Funk and Sage, 1991). Surprisingly, a cell surface receptor for SPARC has proved elusive, given that most other matricellular proteins appear to be ligands for a variety of receptors. Characterization of the SPARC receptor(s) will provide valuable information for understanding SPARC activity as well as the action of matricellular proteins in general.

CONCLUSIONS

In addition to the proteins described here, others are also potential candidates for inclusion in the family of matricellular proteins. These include small proteoglycans such as the matrix-associated protein, decorin, which has been shown to be an endogenous regulator of TGF-β activity (Border *et al.,* 1992), as well as the connective tissue growth factor, Cyr 61/Def10, and neuroblastoma overexpressed gene (CCN) gene family (Bork, 1993). The CCN family of proteins are secreted, modular, and show adhesive activity for various cell types. Cyr 61 enhances bFGF-induced mitogenesis in fibroblasts (Kireeva *et al.,* 1996). Furthermore, Cyr 61 protein associates with both the extracellular matrix and cell surfaces (Yang and Lau, 1991). Further analysis of these proteins and their actions will expand our comprehension of matricellular proteins and the functions they serve in regulating cell interaction with components of their immediate environment.

In any cellular environment, numerous extracellular signals are in place to control cell behavior. In adult tissues, injury and disease both lead to a wide-scale release of multiple factors, either secreted from cells or sequestered in the extracellular matrix, all of which are capable of eliciting potent cellular responses. Appropriately, an increase in the expression of many matricellular proteins is associated with pathologic events. Hence, the matricellular proteins appear to be ideally suited to act as modulators of these extracellular signals. They are able to serve as bridges between the extracellular matrix and cell surface receptors, such that cell shape changes or cell movements can be initiated prior to matrix breakdown or synthesis. At least three matricellular proteins, thrombospondin-2, osteopontin, and SPARC, appear to participate in matrix synthesis, either through the promotion of collagen fibrillogenesis or the enhancement of collagen production. Matricellular proteins can either inhibit or potentiate specific growth factor signal transduction pathways. Thus, different growth factor effects may be amplified or subdued by the presence or absence of these proteins. The fact that multiple receptors exist for most of the matricellular proteins allows for diverse functional consequences for different cell types in a complex tissue in response to a matricellular ligand. A given repertoire of cell surface receptors can stimulate (or inhibit) a particular pathway in a cell-type-dependent manner. Apparently, evolution has fine-tuned these proteins to serve as specialized mediators of extracellular signals that provide a coordinated, efficient resolution of tissue injury.

REFERENCES

Border, W. A., Noble, N. A., Yamamoto, T., Harper, J. R., Yamaguchi, Y., Pierschbacher, M. D., and Ruoslahti, E. (1992). Natural inhibitor of transforming growth factor-β protects against scarring in experimental kidney disease. *Nature (London)* **360**, 361–364.

Bork, P. (1993). The modular architecture of a new family of growth regulators related to connective tissue growth factor. *FEBS Lett.* **327**, 125–130.

Bornstein, P. (1995). Diversity of function is inherent in matricellular proteins: An appraisal of thrombospondin 1. *J. Cell Biol.* **130**, 503–506.

Bornstein, P., and Sage, E. H. (1994). Thrombospondins. *Methods Enzymol.* **245**, 62–85.

Bradshaw, A. D., Francki, A., Motamed, K., Howe, C., and Sage, E. H. (1999). Primary mesenchymal cells isolated from SPARC-null mice exhibit altered morphology and rates of proliferation. *Mol. Biol. Cell* **10**, 1569–1579.

Butler, W. T. (1995). Structural and functional domains of osteopontin. *Ann. N.Y. Acad. Sci.* **760**, 6–11.

Castle, V., Varani, J., Fligiel, S., Prochownik, E. V., and Dixit, V. (1991). Anti-sense-mediated reduction in thrombospondin reverses the malignant phenotype of a human squamous carcinoma. *J. Clin. Invest.* **6**, 1883–1888.

Chen, Y., Bal, B. S., and Gorski, J. P. (1992). Calcium and collagen binding properties of osteopontin, bone sialoprotein, and bone acidic glycoprotein-75 from bone. *J. Biol. Chem.* **267**, 24871–24878.

Chung, C. Y., Murphy-Ullrich, J. E., and Erickson, H. P. (1996). Mitogenesis, cell migration, and loss of focal adhesions induced by tenascin-C interacting with its cell surface receptor, annexin II. *Mol. Biol. Cell* **7**, 883–892.

Clezardin, P., Malaval, L., Ehrensperger, A.-S., Delmas, P. D., Dechavanne, M., and McGregor, J. L. (1988). Complex formation of human thrombospondin with osteonectin. *Eur. J. Biochem.* **175**, 275–284.

Crawford, S. E., Stellmach, V., Murphy-Ullrich, J. E., Ribeiro, S. M. F., Lawler, J., Hynes, R. O., Boivin, G. P., and Bouck, N. (1998). Thrombospondin-1 is a major activator of TGF-β *in vivo. Cell* **93**, 1159–1170.

Crossin, K. L. (1996). Tenascin: A multifunctional extracellular matrix protein with a restricted distribution in development and disease. *J. Cell. Biochem.* **61**, 592–598.

Denhardt, D. T., and Noda, M. (1998). Osteopontin expression and function: Role in bone remodeling. *J. Cell. Biochem. Suppl.* **30/31**, 92–102.

Denhardt, D. T., and Guo, X. (1993). Osteopontin: A protein with diverse functions. *FASEB J.* **7**, 1475–148280.

DiPietro, L. A., Negben, D. R., and Polverini, P. J. (1994). Downregulation of endothelial cell thrombospondin 1 enhances *in vitro* angiogenesis. *J. Vasc. Res.* **31**, 178–185.

DiPietro, L. A., Nissen, N. N., Gamelli, R. L., Koch, A. E., Pyle, J. M., and Polverini, P. J. (1996). Thrombospondin 1 synthesis and function in wound repair. *Am. J. Pathol.* **148**, 1851–1860.

End, P., Panayotou, G., Entwhistle, A., Waterfield, M. D., and Chiquet, M. (1992). Tenascin: A modulator of cell growth. *Eur. J. Biochem.* **209**, 1041–1051.

Erickson, H. P. (1993). Tenascin-C, tenascin-R, and tenascin-X: A family of talented proteins in search of functions. *Curr. Opin. Cell Biol.* **5**, 869–876.

Francki, A., Bradshaw, A. D., Bassuk, J. A., Howe, C. C., Couser, W. G., and Sage, E. H. (1999). SPARC regulates the expression of collagen type I and transforming growth factor-beta 1 in mesangial cells. *J. Biol. Chem* **274**, 32145–32152.

Fukamauchi, F., Mataga, N., Wang, Y. J., Sato, S., Youshiki, A., and Kusakabe, M. (1996). Abnormal behavior and neurotransmissions of tenascin gene knockout mouse. *Biochem. Biophys. Res. Commun.* **221**, 151–156.

Funk, S. E., and Sage, E. H. (1991). The Ca^{2+}-binding glycoprotein SPARC modulates cell cycle progression in bovine aortic endothelial cells. *Proc. Natl. Acad. Sci. U.S.A.* **88**, 2648–2652.

Giachelli, C. M., Lombardi, D., Johnson, R. J., Murry, C. E., and Almeida, M. (1998). Evidence for a role of osteopontin in macrophage infiltration in response to pathological stimuli *in vivo. Am. J. Pathol.* **152**, 353–358.

Hasselaar, P., and Sage, E. H. (1992). SPARC antagonizes the effect of basic fibroblast growth factor on the migration of bovine aortic endothelial cells. *J. Cell Biochem.* **49**, 272–283.

Higuchi, M., Ohnishi, T., Arita, N., Hiraga, S., and Hayakawa, T. (1993). Expression of tenascin in human gliomas: Its relation to histological malignancy, tumor dedifferentiation and angiogenesis. *Acta Neuropathol.* **85**, 481–487.

Hynes, R. O. (1992). Integrins: versatility, modulation, and signaling in cell adhesion. *Cell* **69**, 11–2529.

Iruela-Arispe, M. L., Bornstein, P., and Sage, E. H. (1991). Thrombospondin exerts an antiangiogenic effect on cord formation by endothelial cells *in vitro*. *Proc. Natl. Acad. Sci. U.S.A.* **88**, 5026–5030.

Iruela-Arispe, M. L., Liska, D. J., Sage, E. H., and Bornstein, P. (1993). Differential expression of thrombospondin 1, 2, and 3, during murine development. *Dev. Dynam.* **197**, 40–56.

Jones, F. S., and Copertino, D. W. (1996). The molecular biology of tenascin: Structure, splice variants and regulation of gene expression. *In* "Tenascin and Counteradhesive Proteins of the Extracellular Matrix" (K. L. Crossin, ed.), pp. 1–22. Hardwood Acad., Amsterdam, Netherlands.

Jones, P. L., and Rabinovitch, M. (1996). Tenascin-C is induced with progressive pulmonary vascular disease in rats and is functionally related to increased smooth muscle proliferation. *Circ. Res.* **79**, 1131–1142.

Jones, P. L., Crack, J., and Rabinovitch, M. (1997). Regulation of tenascin-C, a vascular smooth muscle cells survival factor that interacts with αvβ3 integrin to promote epidermal growth factor receptor phosphorylation and growth. *J. Cell Biol.* **139**, 279–293.

Kireeva, M. L., Mo, F. E., Yang, G. P., and Lau, L. F. (1996). Cyr61, a product of a growth factor-inducible immediate-early gene, promotes cell proliferation, migration, and adhesion. *Mol. Cell. Biol.* **16**, 1326–1334.

Koukoulis, G. K., Gould, V. E., Bhattacharyya, A., Gould, J. E., Howeedy, A. A., and Virtanen, I. (1991). Tenascin in normal, reactive, hyperplastic, and neoplastic tissues: Biological and pathologic implications. *Human Pathol.* **22**, 636–643.

Kupprion, C., Motamed, K., and Sage, E. H. (1998). SPARC (BM-40, osteonectin) inhibits the mitogenic effect of vascular endothelial growth factor on microvascular endothelial cells. *J. Biol. Chem.* **273**, 29635–29640.

Kyriakides, T. R., Zhu, Y.-H., Smith, L. T., Bain, S. D., Yang, Z., Lin, M. T., Danielson, K. G., Iozzo, R. V., LaMarca, M., McKinney, C. E., Ginns, E. I., and Bornstein, P. (1998). Mice that lack thrombospondin 2 display connective tissue abnormalities that are associated with disordered collagen fibrillogenesis, an increased vascular density, and a bleeding diathesis. *J. Cell Biol.* **140**, 419–430.

Lamszuz, K., Joseph, A., Jin, L., Yao, Y., Chowdhury, S., Fuchs, A., Polverini, P. J., Goldberg, I. D., and Rosen, E. M. (1996). Scatter factor binds to thrombospondin and other extracellular matrix components. *Am. J. Pathol.* **149**, 805–819.

Lane, T. E., and Sage, E. H. (1994). The biology of SPARC, a protein that modulates cell–matrix interactions. *FASEB J.* **8**, 163–17395.

Lawler, J. (1996). "Thrombospondins, Tenascin and Counteradhesive Molecules of the Extracellular Matrix" (K. L. Crossin, ed.), pp. 109–125. Hardwood Acad., Amsterdam, Netherlands.

Liaw, L., Skinner, M. P., Raines, E. W., Ross, R., Cheresh, D. A., and Schwartz, S. M. (1998). The adhesive and migratory effects of osteopontin are mediated via distinct cell surface integrins. *J. Clin. Invest.* **95**, 713–724.

MacKay, K., Striker, L. J., Stauffer, J. W., Doi, T., Agodoa, L. Y., and Striker, G. E. (1989). Transforming growth factor-β: Murine glomerular receptors and responses of isolated glomerular cells. *J. Clin. Invest.* **83**, 1160–1167.

Mackie, E. J., Halfter, W., and Liverani, D. (1988). Induction of tenascin in healing wounds. *J. Cell Biol.* **107**, 2757–2767.

Motamed, K., and Sage, E. H. (1998). SPARC inhibits endothelial cell adhesion but not proliferation through a tyrosine phosphorylation-dependent pathway. *J. Cell. Biochem.* **70**, 543–552.

Murphy-Ullrich, J. E., and Höök, M. (1989). Thrombospondin modulates focal adhesions in endothelial cells. *J. Cell Biol.* **109**, 1309–1319.

Mukherjee, B. B., Nemir, M., Beninati, S., Cordella-Miele, E., Singh, K., Chackalaparampil, I., Shanmugam, V., DeVouge, M. W., and Mukherjee, A. B. (1995). Interaction of osteopontin with fibronectin and other extracellular matrix molecules. *Ann. N.Y. Acad. Sci.* **760**, 201–212.

Nakao, N., Hiraiwa, N., Yoshiki, A., Ike, F., and Kusakabe, M. (1998). Tenascin-C promotes healing of habu-snake venom-induced glomerulonephritis. *Am. J. Pathol.* **152**, 1237–1245.

Nicosia, R. F., and Tuszynski, G. P. (1994). Matrix-bound thrombospondin promotes angiogenesis *in vitro*. *J. Cell Biol.* **124**, 183–193.

Raines, E. W., Lane, T. F., Iruela-Arispe, M. L., Ross, R., and Sage, E. H. (1992). The extracellular glycoprotein SPARC interacts with platelet-derived growth factor (PDGF)-AB and -BB and inhibits the binding of PDGF to its receptor. *Proc. Natl. Acad. Sci. U.S.A.* **89**, 1281–1285.

Saga, Y., Yagi, T., Ikawa, Y., Sakakura, T., and Aizawa, S. (1992). Mice develop normally without tenascin. *Genes Dev.* **6**, 1821–1831.

Sage, H., Vernon, R. B., Funk, S. E., Everitt, E. A., and Angello, J. (1989). SPARC, a secreted protein associated with cellular proliferation, inhibits cell spreading *in vitro* and exhibits Ca^{2+}-dependent binding to the extracellular matrix. *J. Cell Biol.* **109**, 341–356.

Scatena, M., Almeida, M., Chaisson, M. L., Fausto, N., Nicosia, R. F., and Giachelli, C. M. (1998). NF-κB mediates αvβ3 integrin-induced endothelial cell survival. *J. Cell Biol.* **141**, 1083–1093.

Schultz-Cherry, S., Lawler, J., and Murphy-Ullrich, J. E. (1994). The type 1 repeats of thrombospondin 1 activate latent transforming growth factor-beta. *J. Biol. Chem.* **269**, 26783–26788.

Schultz-Cherry, S., and Murphy-Ullrich, J. E. (1993). Thrombospondin causes activation of latent transforming growth factor-beta secreted by endothelial cells by a novel mechanism. *J. Cell Biol.* **122**, 923–932.

Senger, D. R., Ledbetter, S. R., Claffey, K. P., Papdopoulos-Serfgiou, A., Perruzzi, C. A., and Detmar, M. (1996). Stimulation of endothelial cell migration by vascular permeability factor/vascular endothelial growth factor through the cooperative mechanisms involving the αvβ3 integrin, osteopontin and thrombin. *Am. J. Pathol.* **149**, 293–305.

Smith, L. L., Cheung, H.-K., Ling, L. E., Chen, J., Sheppard, D., Pytela, R., and Giachelli, C. M. (1996). Osteopontin N-terminal domain contains a cryptic adhesive sequence recognized by α9β1 integrin. *J. Biol. Chem.* **45**, 28485–28491.

Smith, L. L., and Giachelli, C. M. (1998). Structural requirements for α9β1-mediated adhesion and migration to thrombin-cleaved osteopontin. *Exp. Cell Res.* **242**, 351–360.

Sriramarao, P., Mendler, M., and Bourdon, M. A. (1993). Endothelial cell attachment and spreading on human tenascin is mediated by α2β1 and αvβ3 integrins. *J. Cell Sci.* **105**, 1001–1012.

Steindler, D. A., Settles, D., Erickson, H. P., Laywell, E. D., Yoshiki, A., Faissner, A., and Kusakabe, M. (1995). Tenascin knockout mice: Barrels, boundary molecules, and glial scars. *J. Neurosci.* **15**, 1971–1983.

Tuszynski, G. P., and Nicosia, R. F. (1996). The role of thrombospondin-1 in tumor progression and angiogenesis. *BioEssays* **18**, 71–76.

Volpert, O. V., Tolsma, S. S., Pellerin, S., Feige, J.-J., Chen, H., Mosher, D. F., and Bouck, N. (1995). Inhibition of angiogenesis by thrombospondin-2. *Biochem. Biophys. Res. Commun.* **217**, 326–332.

Weber, G. F., Ashkar, S., Glimcher, M. J., and Cantor, H. (1996). Receptor–ligand interaction between CD44 and osteopontin (Eta-1). *Science* **271**, 509–512.

Weinstat-Saslow, D. L., Zabrenetzky, V. S., VanHoutte, K., Frazier, W. A., Roberts, D. D., and Steeg, P. S. (1994). Transfection of thrombospondin 1 complementary DNA into a human breast carcinoma cell line reduces primary tumor growth, metastatic potential, and angiogenesis. *Cancer Res.* **54**, 6504–6511.

Weintraub, A. S., Giachelli, C. M., Krauss, R. S., Almeida, M., and Taubman, M. B. (1996). Autocrine secretion of osteopontin by vascular smooth muscle cells regulates their adhesion to collagen gels. *Am. J. Pathol.* **149**, 259–272.

Yang, G. P., and Lau, L. (1991). Cyr61, product of a growth factor-inducible immediate early gene, is associated with the extracellular matrix and the cell surface. *Cell Growth Diff.* 7, 351–357.

Yokosaki, Y., Monis, H., Chen, J., and Sheppard, D. (1996). Differential effects of the integrins α9β1, αvβ3, and αvβ6 on cell proliferative responses to tenascin. *J. Biol. Chem.* **271**, 24144–24150.

GROWTH FACTORS

Thomas F. Deuel and Nan Zhang

INTRODUCTION

Tissue remodeling is an essential component in the process of wound healing and thus in the normal maintenance and survival of all organisms. Tissue remodeling occurs throughout the entire temporal span of injury and repair. However, although abundant evidence using histologic methods has accumulated to detail the sequence of appearance of cell types that characterize wound healing, the mechanisms that initiate and sustain the process remain to be established (Ross, 1968; Deuel *et al.,* 1991; Pierce and Mustoe, 1995). What has become increasingly clear from analysis of properties of growth factors and other cytokines is their importance in the initiation and propagation of the processes that begin with injury and end with a healed wound. These factors and their receptors dictate cell type and tissue specificity of response, and the regulation of their genes at the level of transcription has established the time course of their expression. Their properties are those needed in the wound healing process.

Because growth factors applied directly to wounds accelerate the normal process of wound healing *in vivo,* it seems likely that the properties described for the growth factors *in vitro* are recapitulated *in vivo* (Grotendorst *et al.,* 1985; Paulsson *et al.,* 1987; Sprugel *et al.,* 1987; Pierce *et al.,* 1988, 1989a, 1991, 1992; Terracio *et al.,* 1988; Greenhalgh *et al.,* 1990; Hart *et al.,* 1990; Quaglino *et al.,* 1990; Shaw *et al.,* 1990; Antoniades *et al.,* 1991; Deuel *et al.,* 1991; Mustoe *et al.,* 1991; Nagaoka *et al.,* 1991; Soma *et al.,* 1992; Reuterdahl *et al.,* 1993; Sundberg *et al.,* 1993; Pierce and Mustoe, 1995). It is also important to recognize that properties of growth factors may be important in the development of abnormalities in the vascular wall, such as atherosclerosis and in tumor growth, suggesting that deregulation of growth factors and other cytokines may have serious consequences for the host and that tight regulation of their activities is needed for normal function (Broadley *et al.,* 1989; Golden *et al.,* 1991; D'Amore and Smith, 1993; Forsberg *et al.,* 1993). Compelling evidence is therefore available to indicate the importance of growth factors and other cytokines in the wound healing process. However, the specific roles of these endogenous factors remain to be established in normal wound healing, and the signals that regulate the sites and time of their expression are little understood.

This chapter provides an overview of roles of the platelet-derived growth factor (PDGF) and other cytokines as models of factors that initiate and propagate a number of the normal processes required for the orderly progression of inflammation, tissue remodeling, and repair.

PDGF was initially described as a potent mitogen. However, PDGF directs cell migration, activates diverse cell types, and up-regulates expression of genes that otherwise are quiescent and not expressed at significant levels. This latter property of PDGF appears to establish gene activation pathways that are unique to different cell types. It provides a temporal sequence of expression of different pathways that may be of central importance in tissue remodeling and repair, a property required of a signaling molecule in the wound healing process. Experiments in which PDGF has been directly applied to wounds have shown that PDGF accelerates the normal healing process; it does not distort it, indicating that PDGF can up-regulate genes needed for wound healing in a

Principles of Tissue Engineering
Second Edition

precise order. However, there is little if any information to establish the relative contributions of PDGF and different growth factors to the healing of normal wounds, and thus experimental strategies continue to focus on *in vitro* properties of growth factors. Additional models and reagents are needed to directly test these results *in vivo*.

WOUND HEALING

Normal wound healing may be separated operationally into different stages that overlap and proceed in a clear sequence to a healed wound. Initially, platelets release granule products, and products of the coagulation process are deposited locally. The sequential migration of neutrophils, monocytes, and fibroblasts into wounds begins immediately and continues over the first several days. Wound macrophages (derived from circulating monocytes) and fibroblasts become activated, to initiate a cascade of new gene expression that results in *de novo* synthesis of growth factors and other cytokines, synthesis of extracellular matrix proteins (including collagen), and proliferation of fibroblasts. Later, tissue remodeling results in active collagen turnover and cross-linking that lasts from 2 weeks to 1 year postwounding.

Factors that arise locally account for the cell migration and activation of fibroblasts in wounds (Arey, 1936; Dunphy, 1963; Liebovich, 1975). However, circulating platelets are invariably associated with wounds and are now known to release factors originally stored within intracellular compartments, such as the platelet granules, into wounds. These factors attract neutrophils, monocytes, fibroblasts, and other (tissue-specific) cells, such as the smooth muscle cell. Platelets are now considered to be important mediators in the initiation of inflammation and subsequent tissue remodeling and repair. Platelet release is initiated by products of coagulation, such as thrombin, and by platelet exposure to foreign surfaces, such as collagen. Platelet release occurs within seconds of platelet activation. Synthesis of other potent molecules, such as the prostaglandins and leukotrienes, also influences intracellular responses and extracellular activities that contribute significantly to early events within wounds. These factors also initiate the transcriptional activation of genes that encode proteins with other signaling roles, serving to attract inflammatory cells locally. Because inflammatory cells are in close proximity to platelets in a number of models of inflammation (immune complex disease and atherosclerosis), it is likely that platelets are important in the earliest stages of different inflammatory states. Platelets and inflammatory cells appear to "talk" to each other in essential ways, through release of cytokines and growth factors to attract additional inflammatory cells and to activate transcription of factors that regulate or encode additional mediators of inflammation and wound repair. Studies with PDGF in the *in vivo* context support these functional roles for platelet secretory proteins.

PDGF was originally identified as a potent mitogen in serum for mesenchymally derived cells. Purified human PDGF is predominantly a heterodimeric molecule consisting of two separate but highly related polypeptide chains (A and B chains) (Hart *et al.*, 1990). The B chain of PDGF is ~92% identical with the protein product of the v-*sis* gene, the oncogene of simian sarcoma virus, an acute transforming retrovirus, and thus is a protooncogene. Both BB and AA homodimers of PDGF also have been isolated from natural sources and have been shown to be expressed in many cell types of diverse origins, including macrophages, endothelial cells, smooth and skeletal muscle cells, fibroblasts, glial cells, and neurons. These cells secrete molecules with PDGF-like activity. The detection of these molecules has been associated with the expression of the mRNAs for A and B chains that can be specifically associated with chain isoforms of PDGF. However, identification of the specific isoform(s) at the level of the protein has been difficult because reagents to detect each isoform specifically are not readily available. Multiple PDGF-AA transcripts have been detected in cells that appear to be alternative splice variants of a single seven-exon gene (Collins *et al.*, 1987; Tong *et al.*, 1987). Three splice variants result in short and long processed proteins, 110 and 125 amino acids (Collins *et al.*, 1987; Tong *et al.*, 1987; Matoskova *et al.*, 1989) in length, and may be important because of differential regulation of the transcripts and because they appear to bind with different affinities to the extracellular matrix (Ostman *et al.*, 1989; Nagaoka *et al.*, 1991). The transcripts for both proteins are widely expressed in multiple cell types.

Each of the three isoforms (BB, AB, and AA) binds to the PDGF-α receptor, whereas only the BB isoform binds with significant affinity to the type of PDGF receptor-β (Hart *et al.*, 1988; Heldin *et al.*, 1988; Matsui *et al.*, 1989). Both receptor types are highly homologous transmembrane tyrosine kinases and are expressed in a cell type-specific manner, not only governing cell target specificity of the PDGF response (Eriksson *et al.,* 1992), but also adding both complexity and thus diversity to the potential roles of PDGF in inflammation and wound healing. The functions of PDGF *in vivo* thus are regulated by the need for specific ligand–receptor interactions to transduce the signals required for various functions of PDGF and the up-regulation of other genes that propagate the cascade of events initiated by PDGF.

GROWTH FACTORS AND CYTOKINES ACTIVE AS EARLY MEDIATORS OF THE INFLAMMATORY PROCESS

In order to investigate and thus to understand the roles of growth factors and other cytokine mediators of inflammation, tissue remodeling, and wound healing, experimental strategies were designed to first isolate and characterize these factors, to analyze their properties *in vitro,* to determine the sites and levels of expression achievable *in vivo,* and to develop reagents and experimental models to analyze normal functions of growth factors and other cytokines in the intact animal. Early work focused on the roles of PDGF as a mitogen largely because it was the first mitogen to be purified. However, in addition to the striking potency of PDGF as a mitogen, PDGF also is strongly chemotactic for human monocytes and neutrophils, a property of PDGF that may be more important in early phases of inflammation. Optimal concentrations of PDGF that are required for maximum chemotactic and mitogenic activity are 20 and 1 ng/ml, respectively (Deuel *et al.,* 1982), concentrations of PDGF that are well below those measured in human serum (~50 ng/ml) (Deuel *et al.,* 1982). Because platelets aggregate and release at wound sites, the concentrations of PDGF at these sites may be substantially higher than those in serum. Furthermore, PDGF is a chemoattractant for fibroblasts and smooth muscle cells but requires a somewhat higher concentration (Seppa *et al.,* 1981; Senior *et al.,* 1983), suggesting that PDGF may also be involved later in wound healing through its influence on fibroblast and smooth muscle cell movement (Senior *et al.,* 1983).

In contrast to the PDGF AB isoform, the chemotactic potential of PDGF AA has been controversial. This controversy was important to resolve because the A chain homodimers of PDGF (PDGF AA) are widely expressed in normal and transformed cells. However, PDGF AA has been independently established as a strong chemoattractant for human monocytes, granulocytes, and fetal bovine ligament fibroblasts. Relevant to the roles of monocytes/tissue macrophages in the wound healing process, it was shown that highly purified (>98%) monocytes require the addition of lymphocytes or interleukin 1 (IL-1) for chemotactic responsiveness to PDGF AA but not for the chemotactic activity of formyl-methionyl-leucyl-phenylalanine (fMLP) or C5a. Monocytes therefore require activation before becoming responsive to PDGF as a chemoattractant. Activation of lymphocytes or exogenous cytokines is required for monocytes to respond chemotactically to PDGF AA but not to fMLP or C5a, suggesting that regulation of the chemotactic response of the monocyte to PDGF AA by the lymphocyte or cytokines may be another regulatory level of importance in the ultimate pathway to tissue remodeling and wound healing *in vivo* (Shure *et al.,* 1992).

The effect of PDGF on cell migration was extended to other platelet granule proteins when it was shown that β-thromboglobulin (TG-β) is a highly active chemotactic factor for fibroblasts (Senior *et al.,* 1983). Platelet factor 4 also is a potent chemoattractant. Lymphokines, a peptide derived from the fifth component of complement, collagen-derived peptides, fibronectin, tropoelastin, and elastin-derived peptides are wound-associated factors that also are potent chemoattractants with differing cell type specificity (Postlethwaite *et al.,* 1976, 1978, 1979; Senior *et al.,* 1982).

In addition to its roles in cell migration, PDGF also activates the polymorphonuclear leukocyte (Tzeng *et al.,* 1984) and is capable of inducing monocyte activation responses, as evaluated by generation of superoxide anion (O_2) from the membrane-associated oxidase system, release of granule enzymes, and enhanced cell adherence and cell aggregation. Superoxide anion release in monocytes is observed at 10 ng/ml and the levels of PDGF-dependent release are comparable to release induced by 10^{-7} ml/liter fMLP. The potency of PDGF to induce this response in monocytes is of the same magnitude as that observed in polymorphonuclear leukocytes (PMNs). Similarly, lysozyme release and monocyte adherence are increased in a dose-dependent manner in response to PDGF and achieve maximal responses at 40 ng/ml PDGF. The PDGF concentration required to achieve maximal monocyte aggregation was twofold (60 ng/ml) that found for PMNs and, at this concentration of PDGF, it is likely to occur only in the immediate vicinity of platelet aggregates. PDGF thus induces the full sequence of cell activation events in human monocytes, similar to human PMNs, and this property may well be of major significance in the release of enzymes and other factors required to begin the process of tissue breakdown required for remodeling.

Other early cellular effects important to inflammation and repair also may be initiated by the release of PDGF from platelets and cells within the wound. For example, collagenase expression was studied in normal human skin fibroblasts that were cultured for 24 hr in the presence of PDGF. Collagenase release is required for the breakdown of collagen necessary for the later remodeling of collagen and for tissue repair (Gross, 1976; Bauer *et al.,* 1985; Pierce *et al.,* 1989b). A dose-dependent, saturable increase in collagenase activity in culture media was observed with paralleled increases in immunoreactive collagenase protein, suggesting the enhanced synthesis of an immunologically unaltered enzyme. The specificity of this effect was demonstrated by comparing the collagenase stimulatory effect with that on total protein synthesis and DNA synthesis. Under *in vitro* conditions that produced a 2.5-fold increase in collagenase synthesis, there was an ~20% increase in total protein synthesis and no detectable change in DNA synthesis. In addition, as a second control, platelet factor 4, another platelet-derived protein, caused a <20% increase in collagenase expression. In time-course studies, stimulation of collagenase synthesis was first observed 8–10 hr after exposure to the growth factor. Furthermore, when cells were exposed to PDGF for ~24 hr, an increase in the rate of collagenase synthesis was seen for ~6 hr after PDGF was removed, after which the rate reverted to control levels. Growth factor-regulated expression of collagenase may be of unique importance to tissue remodeling, because it occurs somewhat later than the most immediate responses to PDGF at the time collagen degradation may be essential for progression of remodeling and tissue repair (Bauer *et al.,* 1985).

Patients with Werner's syndrome, an autosomal recessive disorder, undergo an accelerated aging process and premature death. Fibroblasts from such patients typically grow poorly in culture but have a markedly attenuated mitogenic response to platelet-derived growth factor and fibroblast growth factor (FGF). In contrast, these cells have a full mitogenic response to fetal bovine serum. The Werner's syndrome cells express high constitutive levels of collagenase *in vitro.* However, no induction of collagenase occurs in the Werner's syndrome fibroblasts in response to PDGF. The coupling of the failure of one or more PDGF-mediated pathways in Werner's syndrome cells with the phenotype of this disorder is perhaps important support for roles of PDGF and other cytokines in tissue remodeling (Bauer *et al.,* 1986).

INFLAMMATORY RESPONSE MEDIATORS ACTIVATE TRANSCRIPTION OF QUIESCENT GENES

Another potentially highly important role of PDGF and other growth factors and cytokines in wound healing is the up-regulation of early-intermediate response genes in cells activated by PDGF. In response to PDGF and other cytokines, fibroblasts and other cells initiate transcription and expression of genes that are dormant in the absence of stimulation. The products of some of these genes are important in the intracellular propagation of the mitogenic signal but others are important for intercellular communication, directed cell migration, and activation of cells of different types. The products of these genes (Cochran *et al.,* 1983; Kohase *et al.,* 1987; Rollins *et al.,* 1988; Pierce *et al.,* 1989a,b) may serve to coordinate a second wave of diverse responses designated to combat wound infection and to initiate and propagate the healing process.

The PDGF early/early-intermediate response genes *JE* and *KC* were first isolated by differential colony hybridization (Cochran *et al.,* 1983) and only later were identified as cytokines belonging to a newly described superfamily of small inducible genes (SIGs) (Kawahara and Deuel, 1989) (now known as the chemokine family). This family surprisingly includes platelet granule proteins, such as the platelet basic protein and platelet factor 4. These PDGF-induced cytokines suggest pathways of gene activation that may account in part for cell type specificity and the normal temporal responses of wound healing. The *JE* gene encodes a basic (pI = 10.4) 148-amino acid polypepetide (Rollins *et al.,* 1988; Kawahara and Deuel, 1989) that shows 82% amino acid identity with the murine *JE* gene (Timmers *et al.,* 1990). Both rat and murine *JE* gene products contain N-terminal hydrophobic leader sequences, virtually identical alternating hydrophobic and hydrophilic domains, a single N-linked glycosylation site at position 126 that is conserved in both the murine and rat *JE* genes, and identical intron–exon splice junctions.

The importance of the *JE* gene product was suggested when the human homolog of the rodent *JE* gene was identified as monocyte chemotactic protein-1 (MCP-1) (Yoshimura *et al.,* 1989b). MCP-1 is identical with monocyte chemotactic and activating factor (MCAF) (Furutani *et al.,* 1989), MCP (Decock *et al.,* 1990), HC11 (Chang *et al.,* 1989), and with the smooth muscle cell chemotactic factor (SMC-CF) (Graves *et al.,* 1989). Both the MCP-1 and rodent *JE* genes are induced in fibroblasts by PDGF [also serum, phorbol myristate acetate (PMA), double-stranded RNA, and interleukin-1] and each cross-hybridizes to the same bands on Southern blots (Rollins *et al.,* 1989). Antibodies raised to the murine *JE* cross-react with MCP-1. MCP-1 is a potent factor chemotactic for monocytes but not for neutrophils (Yoshimura *et al.,* 1989a), a specificity of cell type recognition that is not seen with many of the other chemotactic factors that have been characterized (Senior *et al.,* 1989), suggesting a unique role in the inflammatory process. Other cytokines with sharply defined functional activity also have been identified in the chemokine family. MCP-1/SMC-CF or related molecules appear to be the major chemotactic factor(s) released by a number of tumor cell lines as well (Graves *et al.,* 1989). This relationship calls to attention the surprising parallels that have been found between tumor cells and activated cells such as the fibroblast.

The genes of this family are normally not expressed at detectable levels but can be induced in the presence of the appropriate stimuli. Remarkably, each of the inducers has an important role in cell growth, inflammation, or immune responses. Increasing knowledge of this family of genes and their products has suggested an enlarging role for PDGF and other cytokines in the sequential and cell type-specific development of inflammation and the evolution of the wound healing process (Kawahara *et al.,* 1991).

Biologic Properties of SIG (Chemokine) Family Members

Neutrophil activation protein/IL-8 (NAP-1/IL-8) is a highly cell type-specific activator and chemotactic factor for neutrophils but not monocytes (Yoshimura *et al.,* 1987). NAP-1/IL-8 was identified in conditioned medium of lipopolysaccharide-stimulated monocytes (Walz *et al.,* 1987; Yoshimura *et al.,* 1987) and its gene (*3-10C*) was cloned independently as a staphylococcal enterotoxin A-induced gene from peripheral blood leukocytes (Schmid and Weissmann, 1987). NAP-1 is also chemotactic for T lymphocytes and basophils (Larsen *et al.,* 1989; Leonard *et al.,* 1990) and is the N-terminal processed form of endothelial-derived leukocyte adhesion inhibitor (LAI), an inhibitor of neutrophil adhesion that protects endothelial cells from neutrophil-mediated damage (Gimbrone *et al.,* 1989; Kawahara *et al.,* 1991).

MCP-1 is chemotactic for monocytes. It was purified from the culture media of human glioma cell line U-105MG (Yoshimura *et al.,* 1989a), THP-1 cells (Matsushima *et al.,* 1989), phytohemagglutinin-stimulated mononuclear cells (Robinson *et al.,* 1989), and double-stranded RNA [poly(rI)·poly(rC)]-stimulated fibroblasts (Van Damme *et al.,* 1989; Kawahara *et al.,* 1991).

Macrophage inflammatory protein-1 (MIP-1) was purified in two forms (MIP-1, MIP-2) from lipopolysaccharide-stimulated RAW264.7 cells and is chemotactic for polymorphonuclear cells (Wolpe *et al.,* 1988). MIP-1 induces oxidative burst and degranulation in neutrophils and promotes inflammatory reactions. MIP-1 is also a prostaglandin-independent pyrogen (Davatelis *et al.,* 1989). MIP-2 is a potent chemotactic factor for neutrophils and causes neutrophil degranulation, but does not induce an oxidative burst (Kawahara *et al.,* 1991).

Platelet factor 4 (PF4) is a platelet granule protein that is chemotactic for both neutrophils and monocytes (Deuel *et al.,* 1981). It has immunoregulatory activity (Katz *et al.,* 1986) and can inhibit angiogenesis (Maione *et al.,* 1990). Platelet basic protein, connective tissue-activating peptide III (CTAP-III), β-thromboglobulin, and neutrophil-activating peptide-2 (NAP-2) are N-terminal proteolytic processed forms of a single gene product, CTAP-III is mitogenic for human dermal fibroblasts (Castor *et al.,* 1983), and NAP-2 is induced in monocytes when they are exposed to lipopolysaccharidis and can stimulate the release of elastase from human neutrophils (Walz and Baggiolini, 1986; Kawahara *et al.,* 1991).

Melanoma growth stimulatory activity (MGSA) is a mitogen for cultured melanoma and pigmented (nevus) cells (Richmond *et al.,* 1988). Primary melanoma cells secrete MGSA and for this reason it has been suggested to be an autocrine regulator of growth. The protein product of MGSA is identical to the *gro-α* gene isolated by subtractive hybridization of mRNA from transformed hamster CHEF/16 cells. The rat analog, cytokine-induced neutrophil chemoattractant (CINC), is chemotactic for neutrophils (Watanabe *et al.,* 1989). The murine analog (KC) is a PDGF inducible gene (Cochran *et al.,* 1983; Kawahara *et al.,* 1991).

REGULATION OF *JE* GENE EXPRESSION BY GLUCOCORTICOIDS

The role of PDGF and other cytokines in the induction of otherwise unexpressed genes has resulted in additional insights into the wound healing process. The addition of dexamethasone and other steroid drugs to PDGF- and serum-stimulated BALB/c 3T3 fibroblasts prevents the PDGF- and serum-dependent induction of the *JE* gene in a dose-dependent manner. Furthermore, induction follows the rank order of potency expected for glucocorticoid receptor-mediated antiinflammatory activity. The inhibition appears to be highly specific for inflammatory mediators, suggesting that the role of PDGF in the induction of selective SIG family members may be a highly important step at which glucocorticoids negatively influence both inflammation and repair (Kawahara *et al.,* 1991).

GROWTH FACTORS AND ACCELERATED HEALING

Much new information vis-à-vis the roles of growth factors and other cytokines has resulted from the direct application of PDGF to experimental wounds in animals. Morphometric methods and histologic techniques have been used to compare wounds treated with supraphysiological concentrations of growth factors to untreated wounds. This approach, coupled with the *in vitro* approaches cited above, has greatly clarified roles of PDGF-enhanced tissue repair and remodeling and has stimulated similar experiments with other cytokines as well.

In one set of experiments, exogenous PDGF was applied locally, and once only, to incisional wounds in rats and rabbits. PDGF accelerated tissue repair in a dose-dependent manner (Pierce *et al.,* 1988; Mustoe *et al.,* 1991). The single application of PDGF to rat incisions enhanced by 150 to 170% over a 3-week time course the strength required to "break" the healing wound (Pierce *et al.,* 1988). PDGF accelerated healing by 4–6 days over the first 2 weeks and by 5–10 days between 2 and 7 weeks postwounding (Pierce *et al.,* 1989a). At 89 days postwounding, both PDGF-treated and control wounds had achieved similar wound strength (~90% of unwounded dermis), and doses therefore induced a long-lasting enhancement rate of healing, supporting the view that PDGF and other growth factors initiate events required for normal healing of wounds and enhance the normal rate of wound healing and tissue remodeling.

In these wounds, PDGF significantly increased the rate of increase of cellularity and granulation tissue formation. The early appearance of wounds treated with PDGF is that of a highly exaggerated inflammatory response. An increased neutrophil influx was seen at 12 hr. A marked increase in influx of macrophages and, later, fibroblasts was found from day 2, persisting through 21 days postwounding (Pierce *et al.,* 1988). Furthermore enhanced cellular function of fibroblasts could be established in treated wounds. Increased numbers of procollagen-containing fibroblasts were strikingly apparent as early as 2 days postwounding. At 28 days, it was not possible to distinguish activated from nonactivated fibroblasts in treated and untreated wounds. These results indicate that PDGF accelerates the healing response.

Furthermore, within limits of detection, PDGF also results in a healed wound that cannot be distinguished from its untreated counterpart (Pierce *et al.,* 1989a). PDGF also appears to be multifunctional in diverse ways, perhaps through the recruitment and activation of macrophages and fibroblasts and through up-regulation of expression at different cytokine genes whose functions and cell type specificity provide additional diversity to the normal healing process.

Another polypeptide growth factor that is stored in the granules of platelets, transforming growth factor β (TGF-β), may also be very important in tissue repair (Sporn *et al.,* 1986). Both PDGF and TGF-β increase collagen formation, DNA content, and protein levels in wound chambers implanted in rats (Sporn *et al.,* 1983; Grotendorst *et al.,* 1985). TGF-β enhances the reversible formation of granulation tissue when injected subcutaneously into newborn mice (Roberts *et al.,* 1986). A single application of human TGF-β at the time of wounding advanced by 2–3 days the breaking strength required to rupture the incision margins after when assayed at 1 week. The marked augmentation of wound healing using both human PDGF (hPDGF) and recombinant c-*sis* homodimers (rPDGF-B) revealed a unique pattern of response when compared with the results obtained with TGF-β (Mustoe *et al.,* 1987; Pierce *et al.,* 1988).

Circular excisional wounds also were placed through the dermis of the rabbit ear to the level of underlying cartilage, a model that does not allow contraction from the wound margins (Mustoe *et al.,* 1991). Ingrowth of extracellular matrix and granulation tissue was measured by histomorphometric techniques (Mustoe *et al.,* 1991). A single application of PDGF increased granulation tissue ~200% at 7 days in association with a predominance of fibroblasts and collagen-containing extracellular matrix (Mustoe *et al.,* 1991). The healed wound again was indistinguishable from the control, non-PDGF-treated healed wound. Interestingly, PDGF also nearly doubled the rate at which reepithelialization occurred and neovascularization was prominent in granulation tissue. These unexpected findings in view of the known cell type specificity of PDGF suggested up-regulation either of PDGF receptors or of other cytokine genes with resultant cell type-specific epithelial and/or endothelial cell functions. Whereas again the mechanisms of the PDGF effect are not established, the influence of a single application of a growth factor remarkably accelerated the inflammatory response and the subsequent tissue remodeling to result in a healed wound of normal appearance.

In this model, PDGF accelerated deposition of matrix largely composed of glycosaminoglycans and later of collagen (Pierce *et al.,* 1991). The normal sequence in repair and the acceleration of wound healing, to all appearances, are identical to those in an untreated wound, supporting the view that the role of PDGF is to accelerate normal wound healing in human wounds. It has been demonstrated in human wounds that PDGF AA expression is increased within healing pressure ulcers. The accumulation of PDGF AA is accompanied by activated fibroblasts, extracellular matrix deposition, and active neovessel formation. However, far less PDGF AA is present in chronic nonhealing wounds. Thus, up-regulation of PDGF AA may be important in the normal repair process as well (Robson *et al.,* 1992; Mustoe *et al.,* 1994; Pierce *et al.,* 1994).

PDGF BB, bFGF, and epidermal growth factor (EGF), applied locally at the time of wounding, cause a twofold increase in complete reepithelialization of treated wounds, whereas TGF-β significantly inhibits reepithelialization. Both PDGF BB and TGF-β increased the depth and area of new granulation tissue, the influx of fibroblasts, and the deposition of new matrix into wounds. Explants from treated wounds remained metabolically more active than controls, incorporating 473% more [^{3}H]thymidine into DNA and significantly more [^{3}H]leucine and [^{3}H]proline into collagenase-sensitive protein (Mustoe *et al.,* 1991). The different response to growth factors underscores the importance of tissue and cell type specificity of the growth factors to tissue repair and perhaps also the need for exquisite timing of induction of different factors in the orderly progression of the healing process.

The use of these wound healing models provided the basis to suggest at least one mechanism whereby PDGF may function *in vivo* (Mustoe *et al.,* 1989; Pierce *et al.,* 1989a). Animals that were pretreated with glucocorticoids or total body irradiation to reduce circulating monocytes to nondetectable levels had a sharply reduced influx of wound macrophages. PDGF was tested in these animals. It was shown that the PDGF-induced acceleration of incisional repair was abrogated in glucocorticoid-treated animals (Pierce *et al.,* 1989a). However, PDGF attracted a significant influx of fibroblasts into the compromised wounds of animals. The fibroblasts lacked procollagen type I, suggesting the requirement of the wound macrophage to support stimulation of fibroblast procollagen type I synthesis. PDGF alone does not stimulate procollagen type I synthesis when added directly to fibroblasts *in vitro* (Blatti *et al.,* 1988; Rossi *et al.,* 1988). However, PDGF-activated macrophages synthesize TGF-β (Pierce *et al.,* 1989a), a potent activator of the fibroblast procollagen type I gene (Ignotz and Massague, 1986; Rossi *et al.,* 1988). It is suggested that TGF-β may be a second signal messenger to induce collagen formation in PDGF-treated wounds. This suggestion is supported directly in experiments in which macrophages and fibroblasts in PDGF-treated incisional wounds were shown to contain increased levels of intracellular TGF-β (Pierce *et al.,* 1989a).

The results with wound healing models and PDGF have illustrated much about inflammation, tissue remodeling, and the process of wound healing. Other *in vivo* models of PDGF-treated wounds also indicate that PDGF stimulates normal and reverses deficient dermal repair (Grotendorst *et al.,* 1985; Lynch *et al.,* 1987; Sprugel *et al.,* 1987; Greenhalgh *et al.,* 1990), and have suggested roles for this growth factor in inflammation (Circolo *et al.,* 1990), uterine smooth muscle hypertrophy (Mendoza *et al.,* 1990), lens growth and transparency (Brewitt and Clark, 1988), and central nervous system gliogenesis (Richardson *et al.,* 1988). What may be the most important conclusions are that PDGF has the potential to initiate and accelerate the process and to sustain the process to an end point that is effectively identical to untreated wounds. The most important mechanisms of its activity may be the consequences of its potent chemotactic activity and its ability to induce multiple autocrine loops that lead to an endogenous cascade of new gene activation and cell-specific cytokine synthesis within the healing wound. Clinical trials have now established the efficacy of PDGF in advancing the healing of diabetic ulcers. PDGF is now "in the clinic."

ROLE OF BASIC FIBROBLAST GROWTH FACTOR AND ANGIOGENESIS

Normal wound healing can be divided into three phases: inflammation, fibroplasia, and maturation provoked by liberated angiogenic factors. Vessel-dense granulation tissue is central to the process of tissue repair. The formation of new blood vessels provides a route for oxygen and nutrient delivery, as well as a conduit for components of the inflammatory response. Endogenous bFGF, like PDGF, is also found in the wound site and is presumed to be a necessary part of natural wound healing. Mechanically damaged endothelial cells and burned wound tissue release significant amounts of bFGF into the wound area, suggesting that damaged endothelial cells at the cut edges of blood vessels, as well as other damaged cells at the site of a wound, may provide an early source of bFGF (McNeil *et al.*, 1989; Muthukrishnan *et al.*, 1991; Gibran *et al.*, 1994). A slightly later and more sustained supply of bFGF at the wound site is delivered by the invading activated macrophages (Baird *et al.*, 1985; Rappolee *et al.*, 1988; Fiddes *et al.*, 1991). Moreover, recombinant bFGF has been shown to enhance healing, if added exogenously to a wound site (Brew *et al.*, 1995). It is now clear that bFGF can act as an angiogenic factor *in vitro* (Montesano *et al.*, 1986) and *in vivo* (Folkman and Klagsbrun, 1987), directing endothelial cell migration and proliferation, and that bFGF is a mitogen to endothelial cells. It also up-regulates expression of plasminogen activator and the $\alpha v \beta 3$ integrin that promotes both migration of endothelial cells and luminal formation (Pepper *et al.*, 1991; Friedlander *et al.*, 1995). The effects of stimulating endothelial cell growth, capillary differentiation, and connective tissue cell growth by bFGF contribute to wound healing and tissue regeneration (Gibran *et al.*, 1994; Steenfos, 1994).

Wound healing involves the interactions of many cell types, and is controlled in part by growth factors. Intercellular communication mediated by gap junctions is considered to play an important role in the coordination of cellular metabolism during the growth and development of tissues and organs. Basic fibroblast growth factor, known to be important in wound healing, has been found to increase gap junctional protein annexin 43 expression and intercellular communication in endothelial cells and cardiac fibroblasts (Abdullah *et al.*, 1999). It has been proposed that an increased coupling is necessary for the coordination of these cells in wound healing and angiogenesis, and that one of the actions of bFGF is to modulate intercellular communication. In a variety of animal models, exogenous bFGF has been shown to accelerate wound healing by speeding granulation tissue formation and increasing fibroblast proliferation and collagen accumulation at the site of a subcutaneously implanted sponge in rats (Davidson *et al.*, 1985; Buckley-Sturrock *et al.*, 1989). Fresh wound tensile strength has also been increased (McGee *et al.*, 1988). However, the greatest benefit from the application of exogenous bFGF can probably be obtained in cases in which wound healing is impaired. Research on skin flaps indicates that bFGF has the potential to increase viability by accelerating flap revascularization when administered in a sustained release manner (Hom and Winters, 1998). This may have applications to open or nonunion fractures with impaired wound healing. Despite encouraging experimental results to date coming from studies showing that treatment of genetically diabetic mice with recombinant bFGF not only increased fibroblast and capillary density in the wound, but also accelerated wound closure, clinical application of recombinant human bFGF has fallen short of expectations. In a randomized double-blind placebo-controlled study, direct application of bFGF to nonhealing ulcers of diabetic patients did not accelerate wound healing (Richard *et al.*, 1995). In a subsequent randomized placebo-controlled trial to study the effect of recombinant bovine bFGF on burn healing, all patients treated with bFGF had faster granulation tissue formation and epidermal regeneration, compared with those in the placebo group. Superficial and deep second-degree burns treated with bFGF healed in a mean of 9.9 days (SD 2.5) and 17.0 days (SD 4.6), respectively, compared with 12.4 (2.7) and 21.2 (4.9) days. No adverse effects were seen locally or systematically with bFGF.

It is indicated that bFGF effectively decreased healing time and improved healing quality. Clinical benefits for use of bFGF in burn wounds would be shorter hospital stays and the patients' skin quickly becoming available for harvesting and grafting (Fu *et al.,* 1998).

OTHER ROLES OF GROWTH FACTORS AND CYTOKINES

A growing appreciation for roles of growth factors in many normal and abnormal processes is emerging from numerous other investigations of a diverse nature. For example, TGF-α and TGF-β are expressed with high specificity in developing mouse embryo (Twardzik, 1985; Heine *et al.,* 1987), PDGF may mediate normal gliogenesis (Richardson *et al.,* 1988); maternally encoded FGF, TGF-β, and PDGF have been implicated as important in the developing *Xenopus* embryo (Kimelman and Kirschner, 1987; Weeks and Melton, 1987; Mercola *et al.,* 1988); and bFGF is a potential neurotrophin during development (Anderson *et al.,* 1988). PDGF also has been identified within plaques and is a potent vasoconstrictor and thus has been implicated in the genesis of atherosclerosis (Berk *et al.,* 1986; Libby *et al.,* 1988). Furthermore, it is secreted from endothelial cells and arterial smooth muscle cells (Jaye *et al.,* 1985; Sejersen *et al.,* 1986; Martinet *et al.,* 1987; Tong *et al.,* 1987; Barrett and Benditt, 1988; Majesky *et al.,* 1988; Rubin *et al.,* 1988) and activated monocytes/macrophages (Martinet *et al.,* 1987). Elevated levels of PDGF receptors are found on synovial cells in patients with rheumatoid arthritis (Rubin *et al.,* 1988). In each instance, however, when the *in vitro* studies are considered in the context of roles of the growth factors in inflammation and tissue repair, it seems likely that common roles of growth factors are associated with normal development and the abnormal remodeling is associated with disease states, indicating the importance of attention to mechanisms that regulate cell type and temporal levels of expression of the growth factors and their cognate receptors.

CONCLUSIONS

Much needs to be learned concerning the roles of growth factors and cytokines if we are to establish fully how they function in both normal and dysregulated tissue remodeling. Remarkable progress over the past several years has resulted from the identification, isolation, cloning, and characterization of the properties of these molecules and their use in wound healing models. The most important lesson learned from their use appears to be that the growth factors and cytokines are important to initiate and accelerate the normal processes involved in going from injury to repair. Questions for future avenues of investigation should address mechanisms by which these factors initiate and propagate the processes and the regulatory signals that govern cell type specificity, differentiation responses in cells migrating into and dividing within wounds, and the temporal sequences needed for orderly progression to a healed and ultimately functional tissue.

REFERENCES

Abdullah, K. M., Luthra, G. *et al.* (1999). Cell-to-cell communication and expression of gap junctional proteins in human diabetic and nondiabetic skin fibroblasts: Effects of basic fibroblast growth factor. *Endocrine* 10(1), 35–41.

Anderson, K. J., Dam, D. *et al.,* (1988). Basic fibroblast growth factor prevents death of lesioned cholinergic neurons *in vivo. Nature (London)* 332, 360–361.

Antoniades, H. N., Galanopoulos, T. *et al.* (1991). Injury induces *in vivo* expression of platelet-derived growth factor (PDGF) and PDGF receptor mRNAs in skin epithelial cells and PDGF mRNA in connective tissue fibroblasts. *Proc. Natl. Acad. Sci. U.S.A.* 88(2), 565–569.

Arey, L. (1936). Wound healing. *Physiol. Rev.* 16, 327–406.

Barid, A., Culler, F. *et al.* (1985). Angiogenic factor in human ocular fluid. *Lancet* 2, 563.

Barrett, T. B., and Benditt, E. P. (1988). Platelet-derived growth factor gene expression in human atherosclerotic plaques and normal artery wall. *Proc. Natl. Acad. Sci. U.S.A.* 85(8), 2810–2814.

Bauer, E. A., Cooper, T. W. *et al.* (1985). Stimulation of *in vitro* human skin collagenase expression by platelet-derived growth factor. *Proc. Natl. Acad. Sci. U.S.A.* 82(12), 4132–4136.

Bauer, E. A., Silverman, N. *et al.* (1986). Diminished response of Werner's syndrome fibroblasts to growth factors PDGF and FGF. *Science* 234, 1240–1243.

Berk, B. C., Alexander, R. W. *et al.* (1986). Vasoconstriction: A new activity for platelet-derived growth factor. *Science* 232, 87–90.

Blatti, S. P., Foster, D. N. *et al.* (1988). Induction of fibronectin gene transcription and mRNA is a primary response to growth-factor stimulation of AKR-2B cells. *Proc. Natl. Acad. Sci. U.S.A.* 85(4), 1119–1123.

Brew, E. C., Mitchell, M. B. *et al.* (1995). Fibroblast growth factors in operative wound healing. *J. Am. Coll. Surg.* 180(4), 499–504.

Brewitt, B., and Clark, J. I. (1988). Growth and transparency in the lens, an epithelial tissue, stimulated by pulses of PDGF. *Science* 242, 777–779.

Broadley, K. N., Aquino, A. M. *et al.* (1989). Monospecific antibodies implicate basic fibroblast growth factor in normal wound repair. *Lab. Invest.* **61**(5), 571–575.

Buckley-Sturrock, A., Woodward, S. C. *et al.* (1989). Differential stimulation of collagenase and chemotactic activity in fibroblasts derived from rat wound repair tissue and human skin by growth factors. *J. Cell. Physiol.* **138**(1), 70–78.

Castor, C. W., Miller, J. W. *et al.* (1983). Structural and biological characteristics of connective tissue activating peptide (CTAP-III), a major human platelet-derived growth factor. *Proc. Natl. Acad. Sci. U.S.A.* **80**(3), 765–769.

Chang, H. C., Hsu, F. *et al.* (1989). Cloning and expression of a gamma-interferon-inducible gene in monocytes: A new member of a cytokine gene family. *Int. Immunol.* **1**(4), 388–397.

Circolo, A., Pierce, G. F. *et al.* (1990). Antiinflammatory effects of polypeptide growth factors. Platelet-derived growth factor, epidermal growth factor, and fibroblast growth factor inhibit the cytokine-induced expression of the alternative complement pathway activator factor B in human fibroblasts. *J. Biol. Chem.* **265**(9), 5066–5071.

Cochran, B. H., Reffel, A. C. *et al.* (1983). Molecular cloning of gene sequences regulated by platelet-derived growth factor. *Cell (Cambridge, Mass.)* **33**(3), 939–947.

Collins, T., Bonthron, D. T. *et al.* (1987). Alternative RNA splicing affects function of encoded platelet-derived growth factor A chain. *Nature (London)* **328** 621–624.

D'Amore, P. A., and Smith, S. R. (1993). Growth factor effects on cells of the vascular wall: A survey. *Growth Factors* **8**(1), 61–75.

Davatelis, G., Wolpe, S. D. *et al.* (1989). Macrophage inflammatory protein-1: A prostaglandin-independent endogenous pyrogen. *Science* **243**(Pt. 1), 1066–1068.

Davidson, J. M., Klagsbrun, M. *et al.* (1985). Accelerated wound repair, cell proliferation, and collagen accumulation are produced by a cartilage-derived growth factor. *J. Cell Biol.* **100**(4), 1219–1227.

Decock, B., Conings, R. *et al.* (1990). Identification of the monocyte chemotactic protein from human osteosarcoma cells and monocytes: Detection of a novel N-terminally processed form. *Biochem. Biophys. Res. Commun.* **167**(3), 904–909.

Deuel, T. F., Senior, R. M. *et al.* (1981). Platelet factor 4 is chemotactic for neutrophils and monocytes. *Proc. Natl. Acad. Sci. U.S.A.* **78**(7), 4584–4587.

Deuel, T. F., Senior, R. M. *et al.* (1982). Chemotaxis of monocytes and neutrophils to platelet-derived growth factor. *J. Clin. Invest.* **69**(4), 1046–1049.

Deuel, T. F., Kawahara, R. S. *et al.* (1991). Growth factors and wound healing: Platelet-derived growth factor as a model cytokine. *Annu. Rev. Med.* **42**, 567–584.

Dunphy, J. (1963). The fibroblast—A ubiquitous ally for the surgeon. *N. Engl. J. Med.* **268**, 1367–1377.

Eriksson, A., Siegbahn, A. *et al.* (1992). PDGF alpha- and beta-receptors activate unique and common signal transduction pathways. *EMBO J.* **11**(2), 543–550.

Fiddes, J. C., Hebda, P. A. *et al.* (1991). Preclinical wound-healing studies with recombinant human basic fibroblast growth factor. *Ann. N.Y. Acad. Sci.* **638**, 316–328.

Folkman, J., and Klagsbrun, M. (1987). Angiogenic factors. *Science* **235**, 442–447.

Forsberg, K., Valyi-Nagy, I. *et al.* (1993). Platelet-derived growth factor (PDGF) in oncogenesis: Development of a vascular connective tissue stroma in xenotransplanted human melanoma producing PDGF-BB. *Proc. Natl. Acad. Sci. U.S.A.* **90**(2), 393–397.

Friedlander, M., Brooks, P. C. *et al.* (1995). Definition of two angiogenic pathways by distinct alpha v integrins. *Science* **270**, 1500–1502.

Fu, X., Yang, Y. *et al.* (1998). Ischemia and reperfusion impair the gene expression of endogenous basic fibroblast growth factor (bFGF) in rat skeletal muscles. *J. Surg. Res.* **80**(1), 88–93.

Furutani, Y., Nomura, H. *et al.* (1989). Cloning and sequencing of the cDNA for human monocyte chemotactic and activating factor (MCAF). *Biochem. Biophys. Res. Commun.* **159**(1), 249–255.

Gibran, N. S., Isik, F. F. *et al.* (1994). Basic fibroblast growth factor in the early human burn wound. *J. Surg. Res.* **56**(3), 226–234.

Gimbrone, M. A., Jr., Obin, M. S. *et al.* (1989). Endothelial interleukin-8: A novel inhibitor of leukocyte-endothelial interactions. *Science* **246**, 1601–1603.

Golden, M. A., Au, Y. P. *et al.* (1991). Platelet-derived growth factor activity and mRNA expression in healing vascular grafts in baboons. Association *in vivo* of platelet-derived growth factor mRNA and protein with cellular proliferation. *J. Clin. Invest.* **87**(2), 406–414.

Graves, D. T., Jiang, Y. L. *et al.* (1989). Identification of monocyte chemotactic activity produced by malignant cells. *Science* **245**, 1490–1493.

Greenhalgh, D. G., Sprugel, K. H. *et al.* (1990). PDGF and FGF stimulate wound healing in the genetically diabetic mouse. *Am. J. Pathol.* **136**(6), 1235–1246.

Gross, J. (1976). "Biochemistry of Collagen" (G. N. Ramachandran and A. H. Reddi, eds.). Plenum, NY.

Grotendorst, G. R., Martin, G. R. *et al.* (1985). Stimulation of granulation tissue formation by platelet-derived growth factor in normal and diabetic rats. *J. Clin. Invest.* **76**(6), 2323–2329.

Hart, C. E., Forstrom, J. W. *et al.* (1988). Two classes of PDGF receptor recognize different isoforms of PDGF. *Science* **240**, 1529–1531.

Hart, C. E., Bailey, M. *et al.* (1990). Purification of PDGF-AB and PDGF-BB from human platelet extracts and identification of all three PDGF dimers in human platelets. *Biochemistry* **29**(1), 166–172.

Heine, U., Munoz, E. F. *et al.* (1987). Role of transforming growth factor-beta in the development of the mouse embryo. *J. Cell Biol.* **105**(6, Pt. 2), 2861–2876.

Heldin, C. H., Backstrom, G. *et al.* (1988). Binding of different dimeric forms of PDGF to human fibroblasts: Evidence for two separate receptor types. *EMBO J.* **7**(5), 1387–1393.

Hom, D. B., and Winters, M. (1998). Effects of angiogenic growth factors and a penetrance enhancer on composite grafts. *Ann. Otol., Rhinol., Laryngol.* **107**(9, Pt. 1), 769–774.

Ignotz, R. A., and Massague, J. (1986). Transforming growth factor-beta stimulates the expression of fibronectin and collagen and their incorporation into the extracellular matrix. *J. Biol. Chem.* **261**(9), 4337–4345.

Jaye, M., McConathy, E. *et al.* (1985). Modulation of the *sis* gene transcript during endothelial cell differentiation *in vitro. Science* **228**, 882–885.

Katz, I. R., Thorbecke, G. J. *et al.* (1986). Protease-induced immunoregulatory activity of platelet factor 4. *Proc. Natl. Acad. Sci. U.S.A.* **83**(10), 3491–3495; published erratum: **83**(17), 6664 (1986).

Kawahara, R. S., and Deuel, T. F. (1989). Platelet-derived growth factor-inducible gene JE is a member of a family of small inducible genes related to platelet factor 4. *J. Biol. Chem.* **264**(2), 679–682.

Kawahara, R. S., Deng, Z. W. *et al.* (1991). PDGF and the small inducible gene (SIG) family: Roles in the inflammatory response. *Adv. Exp. Med. Biol.* **305**, 79–87.

Kimelman, D., and Kirschner, M. (1987). Synergistic induction of mesoderm by FGF and TGF-beta and the identification of an mRNA coding for FGF in the early *Xenopus* embryo. *Cell (Cambridge, Mass.)* **51**(5), 869–877.

Kohase, M., May, L. T. *et al.* (1987). A cytokine network in human diploid fibroblasts: Interactions of beta- interferons, tumor necrosis factor, platelet-derived growth factor, and interleukin-1. *Mol. Cell. Biol.* **7**(1), 273–280.

Larsen, C. G., Anderson, A. O. *et al.* (1989). The neutrophil-activating protein (NAP-1) is also chemotactic for T lymphocytes. *Science* **243**, 1464–1466.

Leonard, E. J., Skeel, A. *et al.* (1990). Leukocyte specificity and binding of human neutrophil attractant/activation protein-1. *J. Immunol.* **144**(4), 1323–1330.

Libby, P., Warner, S. J. *et al.* (1988). Production of platelet-derived growth factor-like mitogen by smooth-muscle cells from human atheroma. *N. Engl. J. Med.* **318**(23), 1493–1498.

Liebovich, S. (1975). The role of the macrophage in wound repair. *Am. J. Pathol.* **78**, 71–100.

Lynch, S. E., Nixon, J. C. *et al.* (1987). Role of platelet-derived growth factor in wound healing: Synergistic effects with other growth factors. *Proc. Natl. Acad. Sci. U.S.A.* **84**(21), 7696–7700.

Maione, T. E., Gray, G. S. *et al.* (1990). Inhibition of angiogenesis by recombinant human platelet factor-4 and related peptides. *Science* **247**, 77–79.

Majesky, M. W., Benditt, E. P. *et al.* (1988). Expression and developmental control of platelet-derived growth factor A-chain and B-chain/Sis genes in rat aortic smooth muscle cells. *Proc. Natl. Acad. Sci. U.S.A.* **85**(5), 1524–1528.

Martinet, Y., Rom, W. N. *et al.* (1987). Exaggerated spontaneous release of platelet-derived growth factor by alveolar macrophages from patients with idiopathic pulmonary fibrosis. *N. Engl. J. Med.* **317**(4), 202–209.

Matoskova, B., Rorsman, F. *et al.* (1989). Alternative splicing of the platelet-derived growth factor A-chain transcript occurs in normal as well as tumor cells and is conserved among mammalian species. *Mol. Cell. Biol.* **9**(7), 3148–3150.

Matsui, T., Pierce, J. H. *et al.* (1989). Independent expression of human alpha or beta platelet-derived growth factor receptor cDNAs in a naive hematopoietic cell leads to functional coupling with mitogenic and chemotactic signaling pathways. *Proc. Natl. Acad. Sci. U.S.A.* **86**(21), 8314–8318.

Matsushima, K., Larsen, C. G. *et al.* (1989). Purification and characterization of a novel monocyte chemotactic and activating factor produced by a human myelomonocytic cell line. *J. Exp. Med.* **169**(4), 1485–1490.

McGee, G. S., Davidson, J. M. *et al.* (1988). Recombinant basic fibroblast growth factor accelerates wound healing. *J. Surg. Res.* **45**(1), 145–153.

McNeil, P. L., Muthukrishnan, L. *et al.* (1989). Growth factors are released by mechanically wounded endothelial cells. *J. Cell Biol.* **109**(2), 811–822.

Mendoza, A. E., Young, R. *et al.* (1990). Increased platelet-derived growth factor A-chain expression in human uterine smooth muscle cells during the physiologic hypertrophy of pregnancy. *Proc. Natl. Acad. Sci. U.S.A.* **87**(6), 2177–2181.

Mercola, M., Melton, D. A. *et al.* (1988). Platelet-derived growth factor A chain is maternally encoded in *Xenopus* embryos. *Science* **241**, 1223–1225.

Montesano, R., Vassalli, J. D. *et al.* (1986). Basic fibroblast growth factor induces angiogenesis *in vitro. Proc. Natl. Acad. Sci. U.S.A.* **83**(19), 7297–7301.

Mustoe, T. A., Pierce, G. F. *et al.* (1987). Accelerated healing of incisional wounds in rats induced by transforming growth factor-beta. *Science* **237**, 1333–1336.

Mustoe, T. A., Purdy, J. *et al.* (1989). Reversal of impaired wound healing in irradiated rats by platelet-derived growth factor-BB. *Am. J. Surg.* **158**(4), 345–350.

Mustoe, T. A., Pierce, G. F. *et al.* (1991). Growth factor-induced acceleration of tissue repair through direct and inductive activities in a rabbit dermal ulcer model. *J. Clin. Invest.* **87**(2), 694–703.

Mustoe, T. A., Cutler, N. R. *et al.* (1994). A phase II study to evaluate recombinant platelet-derived growth factor-BB in the treatment of stage 3 and 4 pressure ulcers. *Arch. Surg. (Chicago)* **129**(2), 213–219.

Muthukrishnan, L., Warder, E. *et al.* (1991). Basic fibroblast growth factor is efficiently released from a cytolsolic storage site through plasma membrane disruptions of endothelial cells. *J. Cell. Physiol.* **148**(1), 1–16.

Nagaoka, I., Someya, A. *et al.* (1991). Comparative studies on the platelet-derived growth factor-A and -B gene expression in human monocytes. *Comp. Biochem. Physiol. B* **100B**(2), 313–319.

Ostman, A., Backstrom, G. *et al.* (1989). Expression of three recombinant homodimeric isoforms of PDGF in *Saccharomyces cerevisiae:* Evidence for difference in receptor binding and functional activities. *Growth Factors* **1**(3), 271–281.

Paulsson, Y., Hammacher, A. *et al.* (1987). Possible positive autocrine feedback in the prereplicative phase of human fibroblasts. *Nature (London)* **328**, 715–717.

Pepper, M. S., Ferrara, N. *et al.* (1991). Vascular endothelial growth factor (VEGF) induces plasminogen activators and plasminogen activator inhibitor-1 in microvascular endothelial cells. *Biochem. Biophys. Res. Commun.* **181**(2), 902–906.

Pierce, G. F., and Mustoe, T. A. (1995). Pharmacologic enhancement of wound healing. *Annu. Rev. Med.* **46**, 467–481.

Pierce, G. F., Mustoe, T. A. *et al.* (1988). *In vivo* incisional wound healing augmented by platelet-derived growth factor and recombinant c-*sis* gene homodimeric proteins. *J. Exp. Med.* **167**(3), 974–987.

Pierce, G. F., Mustoe, T. A. *et al.* (1989a). Transforming growth factor beta reverses the glucocorticoid-induced wound-healing deficit in rats: Possible regulation in macrophages by platelet-derived growth factor. *Proc. Natl. Acad. Sci. U.S.A.* **86**(7), 2229–2233.

Pierce, G. F., Mustoe, T. A. *et al.* (1989b). Platelet-derived growth factor and transforming growth factor-beta enhance tissue repair activities by unique mechanisms. *J. Cell Biol.* **109**(1), 429–440.

Pierce, G. F., Vande Berg, J. *et al.* (1991). Platelet-derived growth factor-BB and transforming growth factor beta 1 selectively modulate glycosaminoglycans, collagen, and myofibroblasts in excisional wounds. *Am. J. Pathol.* **138**(3), 629–646.

Pierce, G. F., Tarpley, J. E. *et al.* (1992). Platelet-derived growth factor (BB homodimer), transforming growth factor-beta 1, and basic fibroblast growth factor in dermal wound healing. Neovessel and matrix formation and cessation of repair. *Am. J. Pathol.* **140**(6), 1375–1388.

Pierce, G. F., Tarpley, J. E. *et al.* (1994). Tissue repair processes in healing chronic pressure ulcers treated with recombinant platelet-derived growth factor BB. *Am. J. Pathol.* **145**(6), 1399–1410.

Postlethwaite, A. E., Snyderman, R. *et al.* (1976). The chemotactic attraction of human fibroblasts to a lymphocyte-derived factor. *J. Exp. Med.* **144**(5), 1188–1203.

Postlethwaite, A. E., Seyer, J. M. *et al.* (1978). Chemotactic attraction of human fibroblasts to type I, II, and III collagens and collagen-derived peptides. *Proc. Natl. Acad. Sci. U.S.A.* **75**(2), 871–875.

Postlethwaite, A. E., Snyderman, R. *et al.* (1979). Generation of a fibroblast chemotactic factor in serum by activation of complement. *J. Clin. Invest.* **64**(5), 1379–1385.

Quaglino, D., Jr., Nanney, L. B. *et al.* (1990). Transforming growth factor-beta stimulates wound healing and modulates extracellular matrix gene expression in pig skin. I. Excisional wound model. *Lab. Invest.* **63**(3), 307–319.

Rappolee, D. A., Mark, D. *et al.* (1988). Wound macrophages express TGF-alpha and other growth factors *in vivo:* Analysis by mRNA phenotyping. *Science* **241**, 708–712.

Reuterdahl, C., Sundberg, C. *et al.* (1993). Tissue localization of beta receptors for platelet-derived growth factor and platelet-derived growth factor B chain during wound repair in humans. *J. Clin. Invest.* **91**(5), 2065–2075.

Richard, J. L., Parer-Richard, C. *et al.* (1995). Effect of topical basic fibroblast growth factor on the healing of chronic diabetic neuropathic ulcer of the foot. A pilot, randomized, double-blind, placebo-controlled study. *Diabetes Care* **18**(1), 64–69.

Richardson, W. D., Pringle, N. *et al.* (1988). A role for platelet-derived growth factor in normal gliogenesis in the central nervous system. *Cell (Cambridge, Mass.)* **53**(2), 309–319.

Richmond, A., Balentien, E. *et al.* (1988). Molecular characterization and chromosomal mapping of melanoma growth stimulatory activity, a growth factor structurally related to beta-thromboglobulin. *EMBO J.* **7**(7), 2025–2033.

Roberts, A. B., Sporn, M. B. *et al.* (1986). Transforming growth factor type beta: Rapid induction of fibrosis and angiogenesis *in vivo* and stimulation of collagen formation *in vitro. Proc. Natl. Acad. Sci. U.S.A.* **83**(12), 4167–4171.

Robinson, E. A., Yoshimura, T. *et al.* (1989). Complete amino acid sequence of a human monocyte chemoattractant, a putative mediator of cellular immune reactions. *Proc. Natl. Acad. Sci. U.S.A.* **86**(6), 1850–1854.

Robson, M. C., Phillips, L. G. *et al.* (1992). Platelet-derived growth factor BB for the treatment of chronic pressure ulcers. *Lancet* **339**, 23–25.

Rollins, B. J., Morrison, E. D. *et al.* (1988). Cloning and expression of JE, a gene inducible by platelet-derived growth factor and whose product has cytokine-like properties. *Proc. Natl. Acad. Sci. U.S.A.* **85**(11), 3738–3742.

Rollins, B. J., Stier, P. *et al.* (1989). The human homolog of the JE gene encodes a monocyte secretory protein. *Mol. Cell. Biol.* **9**(11), 4687–4695.

Ross, R. (1968). The fibroblast and wound repair. *Biol. Rev. Cambridge Philos. Soc.* **43**, 51–96.

Rossi, P., Karsenty, G. *et al.* (1988). A nuclear factor 1 binding site mediates the transcriptional activation of a type I collagen promoter by transforming growth factor-beta. *Cell (Cambridge, Mass.)* **52**(3), 405–414.

Rubin, K., Tingstrom, A. *et al.* (1988). Induction of B-type receptors for platelet-derived growth factor in vascular inflammation: Possible implications for development of vascular proliferative lesions. *Lancet* **1**, 1353–1356.

Schmid, J., and Weissmann, C. (1987). Induction of mRNA for a serine protease and a beta-thromboglobulin-like protein in mitogen-stimulated human leukocytes. *J. Immunol.* **139**(1), 250–256.

Sejersen, T., Betsholtz, C. *et al.* (1986). Rat skeletal myoblasts and arterial smooth muscle cells express the gene for the A chain but not the gene for the B chain (c-*sis*) of platelet-derived growth factor (PDGF) and produce a PDGF-like protein. *Proc. Natl. Acad. Sci. U.S.A.* **83**(18), 6844–6848.

Senior, R. M., Griffin, G. L. *et al.* (1982). Chemotactic responses of fibroblasts to tropoelastin and elastin-derived peptides. *J. Clin. Invest.* **70**(3), 614–618.

Senior, R. M., Griffin, G. L. *et al.* (1983). Chemotactic activity of platelet alpha granule proteins for fibroblasts. *J. Cell Biol.* **96**(2), 382–385.

Senior, R. M., Griffin, G. L. *et al.* (1989). Platelet alpha-granule protein-induced chemotaxis of inflammatory cells and fibroblasts. *In* "Methods in Enzymology" (J. Hawlger, ed.), Vol. **169**, pp. 233–244. Academic Press, Orlando, FL.

Seppa, H. E., Yamada, K. M. *et al.* (1981). The cell binding fragment of fibronectin is chemotactic for fibroblasts. *Cell Biol. Int. Rep.* **5**(8), 813–819.

Shaw, R. J., Doherty, D. E. *et al.* (1990). Adherence-dependent increase in human monocyte PDGF(B) mRNA is associated with increases in c-fos, c-jun, and EGR2 mRNA. *J. Cell Biol.* **111**(5, Pt. 1), 2139–2148.

Shure, D., Senior, R. M. *et al.* (1992). PDGF AA homodimers are potent chemoattractants for fibroblasts and neutrophils, and for monocytes activated by lymphocytes or cytokines. *Biochem. Biophys. Res. Commun.* **186**(3), 1510–1514.

Soma, Y., Dvonch, V. *et al.* (1992). Platelet-derived growth factor AA homodimer is the predominant isoform in human platelets and acute human wound fluid. *FASEB J.* **6**(11), 2996–3001.

Sporn, M. B., Roberts, A. B. *et al.* (1983). Polypeptide transforming growth factors isolated from bovine sources and used for wound healing *in vivo. Science* **219**, 1329–1331.

Sporn, M. B., Roberts, A. B. *et al.* (1986). Transforming growth factor-beta: Biological function and chemical structure. *Science* **233**, 532–534.

Sprugel, K. H., McPherson, J. M. *et al.* (1987). Effects of growth factors *in vivo.* I. Cell ingrowth into porous subcutaneous chambers. *Am. J. Pathol.* **129**(3), 601–613.

Steenfos, H. H. (1994). Growth factors and wound healing. *Scand. J. Plast. Reconstr. Surg. Hand Surg.* **28**(2), 95–105.

Sundberg, C., Ljungstrom, M. *et al.* (1993). Microvascular pericytes express platelet-derived growth factor-beta receptors in human healing wounds and colorectal adenocarcinoma. *Am. J. Pathol.* **143**(5), 1377–1388.

Terracio, L., Ronnstrand, L. *et al.* (1988). Induction of platelet-derived growth factor receptor expression in smooth muscle cells and fibroblasts upon tissue culturing. *J. Cell Biol.* **107**(5), 1947–1957.

Timmers, H. T., Pronk, G. J. *et al.* (1990). Analysis of the rat JE gene promoter identifies an AP-1 binding site essential for basal expression but not for TPA induction. *Nucleic Acids Res.* **18**(1), 23–34.

Tong, B. D., Auer, D. E. *et al.* (1987). cDNA clones reveal differences between human glial and endothelial cell platelet-derived growth factor A-chains. *Nature (London)* **328**, 619–621.

Twardzik, D. R. (1985). Differential expression of transforming growth factor-alpha during prenatal development of the mouse. *Cancer Res.* **45**(11, Pt. 1), 5413–5416.

Tzeng, D. Y., Deuel, T. F. *et al.* (1984). Platelet-derived growth factor promotes polymorphonuclear leukocyte activation. *Blood* **64**(5), 1123–1128.

Van Damme, J., Decock, B. *et al.* (1989). Identification by sequence analysis of chemotactic factors for monocytes produced by normal and transformed cells stimulated with virus, double-stranded RNA or cytokine. *Eur. J. Immunol.* **19**(12), 2367–2373.

Walz, A., and Baggiolini, M. (1986). A novel cleavage product of beta-thromboglobulin formed in cultures of stimulated mononuclear cells activates human neutrophils. *Biochem. Biophys. Res. Commun.* **159**, 969–975.

Walz, A., Peveri, P. *et al.* (1987). Purification and amino acid sequencing of NAF, a novel neutrophil-activating factor produced by monocytes. *Biochem. Biophys. Res. Commun.* **149**(2), 755–761.

Watanabe, K., Konishi, K. *et al.* (1989). The neutrophil chemoattractant produced by the rat kidney epithelioid cell line NRK-52E is a protein related to the KC/gro protein. *J. Biol. Chem.* **264**(33), 19559–19563.

Weeks, D. L., and Melton, D. A. (1987). A maternal mRNA localized to the vegetal hemisphere in *Xenopus* eggs codes for a growth factor related to TGF-beta. *Cell (Cambridge, Mass.)* **51**(5), 861–867.

Wolpe, S. D., Davatelis, G. *et al.* (1988). Macrophages secrete a novel heparin-binding protein with inflammatory and neutrophil chemokinetic properties. *J. Exp. Med.* **167**(2), 570–581.

Yoshimura, T., Matsushima, K. *et al.* (1987). Purification of a human monocyte-derived neutrophil chemotactic factor that has peptide sequence similarity to other host defense cytokines. *Proc. Natl. Acad. Sci. U.S.A.* **84**(24), 9233–9237.

Yoshimura, T., Robinson, E. A. *et al.* (1989a). Purification and amino acid analysis of two human monocyte chemoattractants produced by phytohemagglutinin-stimulated human blood mononuclear leukocytes. *J. Immunol.* **142**(6), 1956–1962.

Yoshimura, T., Yuhki, N. *et al.* (1989b). Human monocyte chemoattractant protein-1 (MCP-1). Full-length cDNA cloning, expression in mitogen-stimulated blood mononuclear leukocytes, and sequence similarity to mouse competence gene JE. *FEBS Lett.* **244**(2), 487–493.

CHAPTER 13

TISSUE ENGINEERING BIOREACTORS

Lisa E. Freed and Gordana Vunjak-Novakovic

INTRODUCTION

One approach to tissue engineering is to create an *in vitro* environment that embodies the biochemical and mechanical signals that regulate tissue development and maintenance *in vivo*. Figure 13.1 shows our *in vitro* tissue engineering system, which has three major components (Freed and Vunjak-Novakovic, 1998): (1) metabolically active cells able to express their differentiated phenotype, (2) polymeric scaffolds that provide a three-dimensional (3D) structure for cell attachment and tissue growth, and (3) bioreactor culture vessels that provide an *in vitro* environment in which cell–polymer constructs can develop into functional tissues. These constructs can potentially be used *in vitro,* for controlled studies of tissue growth and function, or *in vivo,* for tissue repair. As compared to *in vivo* implantation of dissociated cells and/or biodegradable materials alone, the implantation of a functional engineered tissue can improve the localization of cell delivery and promote graft fixation and survival. In addition, *in vitro* bioreactor studies can be designed to distinguish the effects of specific biochemical and physical signals from the complex *in vivo* milieu (e.g., host endocrinologic and immunologic responses), thus providing useful, complementary information regarding the process of the formation of 3D tissues starting from isolated cells.

In this chapter, we focus on tissue engineering bioreactors, which will be defined as *in vitro* culture systems designed to perform at least one of the following four functions: (1) establish spatially uniform cell distributions on 3D scaffolds, (2) maintain desired concentrations of gases and nutrients in the culture medium, (3) provide efficient mass transfer to the growing tissue, and (4) expose developing tissues to physical stimuli. We further focus this chapter on engineered tissues based on biodegradable cell culture templates, although bioreactors are also being utilized in conjunction with nondegradable cell culture substrates (e.g., microcarriers, hollow fibers, membranes) to study the assembly of cells into 3D structures (e.g., Jessup *et al.,* 1993; Unsworth and Lelkes, 1998; Qiu *et al.,* 1999) and to develop hybrid bioartificial organs (e.g., Jauregui *et al.,* 1997; Lanza and Chick, 1997; Bader *et al.,* 1995).

The dimensional and functional requirements of the tissue to be engineered determine the specific design requirements for the cell–polymer–bioreactor system. In this chapter, we will use two examples of engineered tissues, cartilage and cardiac muscle, to illustrate general and tissue-specific principles of bioreactor design and operation.

CELL–POLYMER CONSTRUCTS

Cell types studied using our model system (Fig. 13.1) were obtained from cartilage, bone marrow, or cardiac muscle of developmentally immature animals. In particular, we have studied articular chondrocytes from 2- to 4-week-old bovine calves (Freed *et al.,* 1994b) and 2- to 8-month-old rabbits (Freed *et al.,* 1994a), bone marrow stromal cells from 16-day-old embryonic chicks (Martin *et al.,* 1998) and 2- to 4-week-old bovine calves (Martin *et al.,* 1999a), and cardiac cells from 14-day-old embryonic chicks and 1- to 2-day-old neonatal rats (Carrier *et al.,* 1999). Cells

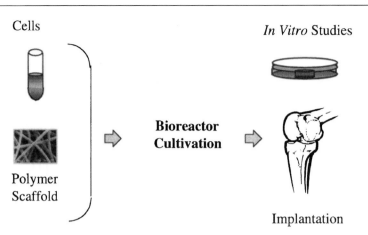

Fig. 13.1. Tissue engineering system. Cells (e.g., isolated from cartilage or cardiac tissue) are seeded onto three-dimensional polymer scaffolds (e.g., fibrous meshes) using bioreactors (e.g., spinner flasks, rotating vessels, or perfused cartridges). The resulting engineered tissue constructs can be used either for in vitro research or in vivo implantation (Freed and Vunjak-Novakovic, 1998).

were isolated by enzymatic digestion, in some cases serially expanded in monolayers, and inoculated into bioreactors containing 3D polymer scaffolds.

The polymer scaffold we have best characterized and most used to date is made of polyglycolic acid (PGA), processed into a 97% porous, nonwoven mesh of fibers 13 μm in diameter (Freed et al., 1994c). The average distance between the fibers was calculated to be 62 μm (Vunjak-Novakovic et al., 1998). This scaffold decreases its mass by 50 or 70% after 4 or 8 weeks of cultivation, respectively (Freed et al., 1994c). In this chapter, we consider our "scaffold" to be a fibrous PGA disk with a diameter of 5 mm and a thickness of 2 mm.

Structural assessments of engineered tissues include (1) presence and spatial arrangement of cells and extracellular matrix (ECM), determined histologically, immunohistochemically, and by electron microscopy (e.g., Riesle et al., 1998; Carrier et al., 1999), (2) quantitative distributions of specific components, determined by computer-based image analysis of histological sections (Martin et al., 1999b; Vunjak-Novakovic et al., 1998), (3) cell number, determined from the amount of DNA (Kim et al., 1988), (4) sulfated glycosaminoglycan (GAG) content, determined by dimethylmethylene blue dye binding (Farndale et al., 1986), (5) total protein content, determined by a Biorad protein assay kit (Carrier et al., 1999), (6) total collagen content, determined from hydroxyproline (Woessner, 1961), (7) type II collagen content, determined by inhibition enzyme-linked immunosorbent assay (ELISA) (Freed et al., 1998), and (8) amounts of type IX collagen, creatine kinase, and sarcomeric myosin, determined by densitometric analysis of Western blots (Riesle et al., 1998; Papadaki et al., 1999).

Functional assessments of engineered tissues include (1) cell metabolism, assessed from glucose consumption and lactate production (Obradovic et al., 1999) and from enzymatic conversion of 3-(4,5-dimethylthiazol-2-yl)-2,5-diphenyl tetrazolium bromide (MTT) (Carrier et al., 1999), (2) cell damage, assessed from medium concentrations of lactate dehydrogenase (LDH) (Carrier et al., 1999), (3) ECM synthesis rates, determined by macromolecular incorporation of radiolabeled tracers (Freed et al., 1998), (4) cartilage mechanical properties (equilibrium modulus, dynamic stiffness, hydraulic permeability, streaming potential), determined in static and dynamic radially confined compression (Vunjak-Novakovic et al., 1999a), and (5) cardiac tissue electrophysiological properties (conduction velocity, signal amplitude, excitation threshold, maximum capture rate), assessed from macroscopic impulse propagation studies using an array of extracellular electrodes (Bursac et al., 1999).

BIOREACTOR TECHNOLOGIES

Cell Seeding

Cell seeding of 3D scaffolds is the first step of bioreactor cultivation of engineered tissues. Seeding requirements include (1) high yield, to maximize cell utilization, (2) high kinetic rate, to minimize the time in suspension for anchorage-dependent and shear-sensitive cells, and (3) high and spatially uniform distribution of attached cells, for rapid and uniform tissue growth.

In the case of engineered cartilage, these requirements are best met using mixed flasks (Fig.

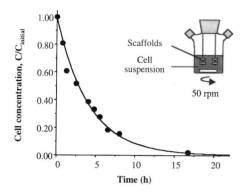

Fig. 13.2. Dynamic cell seeding. Schematic: mixed flask system used to seed cells onto 3D polymer scaffolds. Graph: normalized concentration of cells remaining in the culture medium as a function of time following flask inoculation (Reprinted with permission from Vunjak-Novakovic et al., 1998. Copyright 1998 American Chemical Society).

13.2) containing scaffolds threaded onto needles embedded in the flask stopper (up to 12 scaffolds on four needles per flask). Isolated cells and culture medium are added (120 ml medium, 5×10^6 cells per scaffold) and the flask is magnetically stirred at 50–80 rpm. Under these conditions, essentially all cells attach throughout the scaffold volume in less than 24 hr (Fig. 13.2) and maintain their spherical shape (Vunjak-Novakovic *et al.*, 1998). In this system, magnetic stirring maintains a uniform cell suspension and provides relative velocity between the cells and the scaffolds at an average intensity of turbulence below that causing cell damage or detachment (Cherry and Papoutsakis, 1988). The kinetics and possible mechanisms of cell seeding were related to the formation of cell aggregates using a simple mathematical model, which can be used to optimize seeding conditions for specific scaffold sizes and cell seeding densities (Vunjak-Novakovic *et al.*, 1998).

The above cell seeding requirements can also be met using rotating vessels. These bioreactors are filled with culture medium, to which prewetted scaffolds and cells are added, and the vessel is rotated as a solid body in a horizontal plane at 12–15 rpm (Freed and Vunjak-Novakovic, 1995, 1997a). Cardiac constructs seeded in rotating vessels had higher metabolic activity indexes than those seeded in spinner flasks, implying that optimal hydrodynamic conditions for cell seeding depend on cell type (Carrier *et al.*, 1999).

TISSUE CULTIVATION

Approximately 1–3 days after seeding, cell–polymer constructs are transferred into a variety of cultivation vessels, including static and mixed flasks, and rotating and perfused vessels, as depicted in Fig. 13.3. The flask system contains 120 ml of culture medium and up to 12 tissue tissue constructs that are fixed in place (Fig. 13.3a). Flasks either are operated statically or are mixed at 50–80 rpm using a 4-cm-long magnetic stir bar. Medium is exchanged batchwise (at a rate of 50% every 2–3 days or 3 ml per construct per day), whereas gas exchange is provided by surface aeration of culture medium via loosened side arm caps.

Rotating bioreactors with two different geometries, the slow-turning lateral vessel (STLV) (Fig. 13.3b) (Schwarz *et al.*, 1992) and the high-aspect-ratio vessel (HARV) (Fig. 13.3c) (Prewett *et al.*, 1993), have been used to engineer cartilage and cardiac tissues in ground-based studies (e.g., Freed and Vunjak-Novakovic, 1995, 1997a). The STLV is configured as the annular space between two concentric cylinders, the inner of which is a silicone gas exchange membrane, whereas the HARV is a cylindrical vessel with a gas exchange membrane at its base. These vessels contain 100–110 ml of culture medium and up to 12 tissue constructs, and are operated by solid-body rotation in a horizontal plane at rates of 15–40 rpm. Viscous coupling induces a rotational flow field that can suspend tissue constructs without external fixation (Freed and Vunjak-Novakovic, 1995). Vessel rotational speed is adjusted such that constructs remain suspended close to a stationary point within the vessel, relative to an observer on the ground, due to a dynamic equilibrium between the acting gravitational, centrifugal, and drag forces. Medium is exchanged batchwise (at a rate of 50% every 2–3 days or 3 ml per construct per day), and is equilibrated with gas continuously.

A rotating-wall perfused vessel (RWPV) (Fig. 13.3d) developed at the National Aeronautics and Space Administration (NASA) (Jessup *et al.*, 1996) was used to engineer cartilage in the microgravity environment of space and in a control study on Earth (Freed *et al.*, 1997). The RWPV is configured as an annular space between two concentric cylinders, with the medium inlet at one

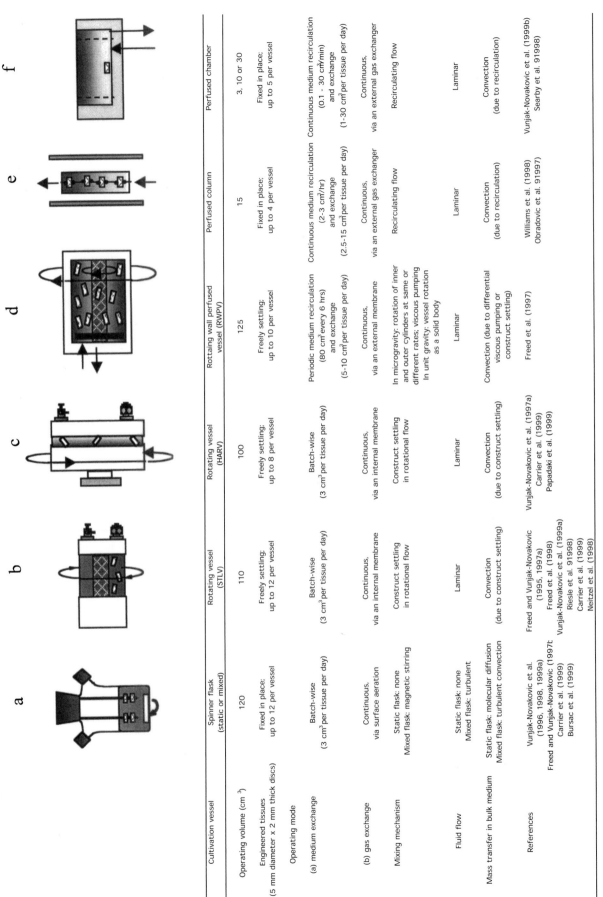

Cultivation vessel	Spinner flask (static or mixed)	Rotating vessel (STLV)	Rotating vessel (HARV)	Rottaing wall perfused vessel (RWPV)	Perfused column	Perfused chamber
	a	b	c	d	e	f
Operating volume (cm³)	120	110	100	125	15	3, 10 or 30
Engineered tissues (5 mm diameter x 2 mm thick discs)	Fixed in place; up to 12 per vessel	Freely settling; up to 12 per vessel	Freely settling; up to 8 per vessel	Freely settling; up to 10 per vessel	Fixed in place; up to 4 per vessel	Fixed in place; up to 5 per vessel
Operating mode						
(a) medium exchange	Batch-wise (3 cm³ per tissue per day)	Batch-wise (3 cm³ per tissue per day)	Batch-wise (3 cm³ per tissue per day)	Periodic medium recirculation (80 cm³ every 6 hrs) and exchange (5-10 cm³ per tissue per day)	Continuous medium recirculation (2-3 cm³/hr) and exchange (2.5-15 cm³ per tissue per day)	Continuous medium recirculation (0.1 - 30 cm³/min) and exchange (1-30 cm³ per tissue per day)
(b) gas exchange	Continuous, via surface aeration	Continuous, via an internal membrane	Continuous, via an internal membrane	Continuous, via an external membrane	Continuous, via an external gas exchanger	Continuous, via an external gas exchanger
Mixing mechanism	Static flask: none; Mixed flask: magnetic stirring	Construct settling in rotational flow	Construct settling in rotational flow	In microgravity: rotation of inner and outer cylinders at same or different rates; viscous pumping. In unit gravity: vessel rotation as a solid body	Recirculating flow	Recirculating flow
Fluid flow	Static flask: none; Mixed flask: turbulent	Laminar	Laminar	Laminar	Laminar	Laminar
Mass transfer in bulk medium	Static flask: molecular diffusion; Mixed flask: turbulent convection	Convection (due to construct settling)	Convection (due to construct settling)	Convection (due to differential viscous pumping or construct settling)	Convection (due to recirculation)	Convection (due to recirculation)
References	Vunjak-Novakovic et al. (1996, 1998, 1999a); Carrier et al. (1999); Bursac et al. (1999)	Freed and Vunjak-Novakovic (1995, 1997a); Freed et al. (1998); Vunjak-Novakovic et al. (1999a); Riesle et al. 91998; Carrier et al. (1999); Neitzel et al. (1998)	Vunjak-Novakovic et al. (1997a); Carrier et al. (1999); Papadaki et al. (1999)	Freed et al. (1997)	Williams et al. (1998); Obradovic et al. 91997	Vunjak-Novakovic et al. (1999b); Searby et al. 91998

Fig. 13.3. Bioreactor culture vessels. (a) Flask (static or mixed), (b) slow-turning lateral vessel (STLV), (c) high-aspect-ratio vessel (HARV), (d) rotating-wall perfused vessel (RWPV), (e) perfused column, and (f) perfused chamber. A summary of operating conditions is given for each vessel type.

end and the medium outlet via a filter on the central cylinder. A flat disk at one end of the central cylinder serves as a viscous pump. The RWPV contains 125 ml of culture medium and holds up to 10 tissue constructs. In microgravity, the inner and outer cylinders were differentially rotated at rates of 10 and 1 rpm, respectively; on Earth, the vessel was rotated as a solid body at 28 rpm. Medium was periodically recirculated between the culture vessel and an external membrane gas exchanger at a rate of 4 ml/min for 20 min four times per day, and was exchanged at a rate of 5–10 ml per construct per day.

Perfused columns were designed to allow nondestructive assessment of tissue development over the course of cultivation using magnetic resonance imaging (Fig. 13.3e) (Obradovic *et al.*, 1997; Williams *et al.*, 1998). The columns have an inlet at the base, an outlet at the top, and contain 15 ml of culture medium and up to four constructs that are fixed to a centrally positioned needle. Medium is continuously recirculated between the column and an external membrane gas exchanger at a rate of 2–3 ml/hr, and is exchanged at rates of 2–15 ml per construct per day.

Perfused chambers (Fig. 13.3f) were designed to allow tissue culture in the microgravity environment of space, and for control studies under conditions of unit gravity on Earth or artificial gravity in space (Vunjak-Novakovic *et al.*, 1999b). The chamber is configured as a central compartment bounded by a porous membrane and an annular space. The medium inlet is in the central tissue culture space and the medium outlet is in the peripheral cell-free space. Chambers contain 3, 10, or 30 ml of medium and up to five tissue constructs. Medium is continuously recirculated between the chamber and an external membrane gas exchanger at rates of 1–30 ml/min, and is exchanged at rates of 1–30 ml per construct per day. Chambers are mounted on a circular tray that allows automated sampling and on-line microscopy.

Hydrodynamic Conditions in Bioreactors

In static flasks, mass transfer in the bulk medium occurs by molecular diffusion, and there is no fluid flow at the surfaces of the tissue constructs. In mixed flasks, mass transfer is by convection, and fluid flow at the surfaces of the tissue constructs is turbulent, such that the smallest turbulent eddies have sizes of 250 μm and velocities of 0.4 cm/sec (Vunjak-Novakovic *et al.*, 1996).

In rotating vessels, laminar flow at the construct surfaces was demonstrated by flow visualization conducted using tracer particles illuminated by a sheet of laser light (Neitzel *et al.*, 1998) (Fig. 13.4a). Particle image velocimetry (PIV) studies, done by either cross-correlation of particle images in window of the flow field or by individual particle tracking, allowed quantitation of the velocity vector at every location in the plane of the light sheet (Fig. 13.4b) (Neitzel *et al.*, 1998). These data were used to calculate maximal shear stresses at the construct surface, which were on the order of 0.8 dyn/cm^2 for a 120-ml rotating vessel containing one model construct and operated such that the inner and outer cylinders were rotated at 13 and 37 rpm, respectively. The observed velocity fields were predicted numerically (Fig. 13.4c) and were used to calculate the rates of mass transport of chemical species through the medium to the construct. Residence time distribution studies demonstrated efficient convective mixing due to gravitational construct settling in rotating vessels (Freed and Vunjak-Novakovic, 1997b). In particular, mixing efficiencies were comparable for an STLV containing 12 model constructs and rotated at 15–28 rpm and a spinner flask containing 12 constructs and stirred at 50–80 rpm.

In perfused columns and chambers, fluid flow is laminar and mass transfer is by convection due to recirculation.

BIOREACTOR MODULATION OF TISSUE FORMATION

In this section, we describe several studies that demonstrate how bioreactors can modulate 3D tissue formation *in vitro*. General design principles for tissue engineering bioreactors include maintaining desired levels of chemical species in the bulk medium, providing efficient mass transfer from the medium to the tissue surfaces, and exposing developing tissues to physical stimuli. These bioreactor functions can result in increased size and improved structure and function of engineered tissues, as follows. The effects of biochemical, hydrodynamic, and mechanical factors were studied for engineered cartilage cultured in rotating bioreactors (Fig. 13.5). In one case, medium oxygen tension, pO_2, and pH were varied using STLVs operated under the same hydrodynamic conditions (Fig. 13.5a); in another case, hydrodynamic and mechanical factors were varied using RWPVs operated with comparable medium compositions (Fig. 13.5b). The specific effects of hy-

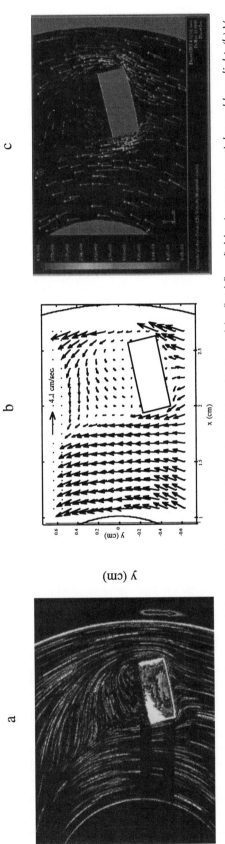

a

b

y (cm)

c

4.1 cm/sec

0.6
0.4
0.2
0
-0.2
-0.4
-0.6

y (cm)

1.5   2   2.5

x (cm)

Fig. 13.4. Flow conditions around a suspended construct in a rotating bioreactor. (a) Visualization of the fluid flow field using tracer particles and laser light. (b) Velocity field determined experimentally by particle image velocimetry. (c) Velocity field determined by numerical simulation (Neitzel et al., 1998).

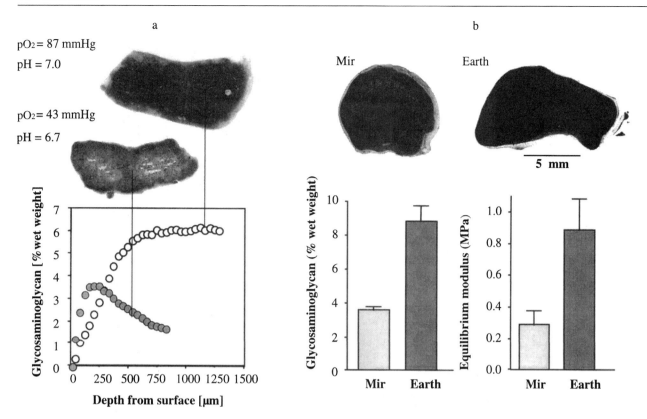

Fig. 13.5. (a) Effects of biochemical factors on engineered cartilage. Comparison of constructs cultivated in medium with relatively high pO₂ and pH, or with lower pO₂ and pH (due to higher or lower rates of bioreactor gas exchange). Histology: safranin-O staining for glycosaminoglycan. Graph: glycosaminoglycan fraction as a function of depth from the construct surface (From Obradovic et al., 1999, Oxygen is essential for bioreactor cultivation of tissue engineered cartilage. Biotechnol. Bioeng. **63,** *197–205. Reprinted by permission of Wiley-Liss, Inc., a subsidiary of John Wiley & Sons, Inc.). (b) Effects of hydrodynamic and mechanical factors on engineered cartilage. Comparison of constructs cultivated in the presence of higher levels of hydrodynamic and mechanical stimuli (due to gravitational settling on Earth) or relatively quiescently (aboard the Mir Space Station). Histology: safranin-O staining for glycosaminoglycan. Graphs: glycosaminoglycan fraction and equilibrium modulus (Freed et al., 1997). Data are the average ± SD of three to four independent measurements.*

drodynamic factors and the accompanying changes in mass transfer were further studied by systematically comparing engineered cartilage and cardiac muscle cultured in different bioreactors (Fig. 13.6).

Biochemical factors were varied by establishing two groups: (1) pO_2 = 87 mm Hg, pH = 7.0 and (2) pO_2 = 43 mm Hg, pH = 6.7 in STLVs operated with or without an internal gas exchange membrane (Obradovic *et al.,* 1999). Cartilaginous constructs were compared with respect to several parameters, including yield of lactate to glucose ($Y_{L/G}$), construct thickness, and the amounts and distributions of glycosaminoglycan. In group 2, low pO_2 and pH were associated with anaerobic cell metabolism ($Y_{L/G}$ of 2.2 mol/mol) as compared to group 1, in which higher values of pO_2 and pH were associated with more aerobic cell metabolism ($Y_{L/G}$ of 1.7–1.8 mol/mol). Constructs grown aerobically for 5 weeks were approximately 2.5 mm thick and had a high fraction [~6% of wet weight (ww)] of uniformly distributed GAG, except for a 460-μm-thick outer capsule (Fig. 13.5a). In contrast, constructs grown anaerobically were only about 1.6 mm thick and had markedly lower GAG fractions (1.5–3.5% ww) that decreased with increasing depth. One key bioreactor function is thus to provide efficient gas exchange for control of medium pO_2 and pH.

Hydrodynamic conditions and mechanical factors were varied by utilizing rotating bioreactor vessels on Earth and in the microgravity environment of space, aboard the Mir Space Station (Freed *et al.,* 1997). In particular, cartilaginous constructs were first cultured in STLVs for 3

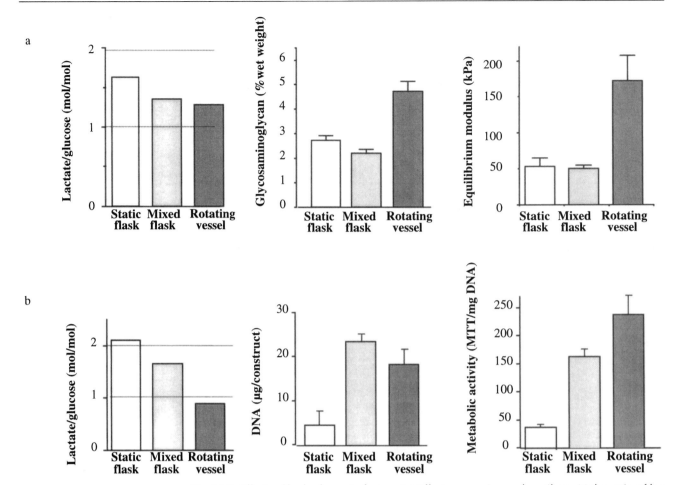

*Fig. 13.6. Effects of hydrodynamic factors. (a) Effects on engineered cartilage. Molar ratio of lactate production and glucose utilization (lactate/glucose), glycosaminoglycan fraction, and equilibrium modulus in constructs cultured in static or mixed flasks, or in rotating vessels (STLVs) (Obradovic et al., 1997; Vunjak-Novakovic et al., 1999a). (b) Effects on engineered cardiac tissue. Glucose/lactate ratio, DNA content, and metabolic activity index (MTT/DNA) of constructs cultured in static or mixed flasks, or in rotating vessels (HARVs) (From Carrier et al., 1999, Cardiac tissue engineering: Cell seeding, cultivation parameters and tissue construct characterization. Biotechnol. Bioeng. **64,** 580–589. Reprinted by permission of Wiley-Liss, Inc., a subsidiary of John Wiley & Sons, Inc.). (a,b) Horizontal lines represent lactate/glucose ratios of 2 and 1, which correspond to theoretical values for anaerobic and aerobic metabolism, respectively. Data are the average ±SD of three to four independent measurements.*

months on Earth and then in RWPVs for an additional 4 months either on Earth or aboard Mir. During cultivation on Earth, constructs were exposed to relatively higher levels of hydrodynamic shear (due to gravitational settling) and mechanical stimuli (due to construct collisions with the rotating vessel walls) as compared to on Mir, where constructs floated freely in the culture medium. Medium values of pH and pO_2 were comparable in the two groups. Cartilaginous constructs were compared with respect to several parameters, including dimensions, wet weight fractions of GAG, and equilibrium moduli in radially confined compression. Constructs grown for 7 months on Earth retained their initial discoid shape, were large (429 mg ww), and had high GAG fractions (8.8% ww) and equilibrium moduli (0.93 MPa) (Fig. 13.5b). In contrast, constructs grown for 3 months on Earth and then 4 months on Mir tended to become more spherical, were smaller (330 mg ww), and had lower GAG fractions (3.6% ww) and equilibrium moduli (0.31 MPa). Other important functions of tissue engineering bioreactors are thus to provide appropriate levels of hydrodynamic and mechanical stimuli.

Cartilaginous constructs seeded in mixed flasks (Fig. 13.2) and then cultured for 6 weeks in static or mixed flasks (Fig. 13.3a) or in rotating vessels (STLVs) (Fig. 13.3b) were compared with respect to several parameters, including $Y_{L/G}$, wet weight fractions of GAG, and equilibrium modulus in radially confined compression (Obradovic *et al.*, 1997; Vunjak-Novakovic *et al.*, 1999a). Values of $Y_{L/G}$ for constructs cultured statically, in mixed flasks, and in rotating vessels were, respectively, 1.7, 1.3, and 1.3 mol/mol, implying that cell metabolism was more aerobic under mixed than static culture conditions (Fig. 13.6a) (Obradovic *et al.*, 1997). GAG fractions and equilibrium moduli were significantly higher for constructs grown in rotating vessels as compared to either static or mixed flasks (Fig. 13.6a) (Vunjak-Novakovic *et al.*, 1999a). In static flasks, diffusional constraints of mass transport resulted in thin, fragile constructs in which GAG accumulated mainly in a 1-mm-thick outer region. In mixed flasks, turbulent conditions apparently induced the formation of capsules at the construct surfaces that were up to 400 μm thick and consisted of multiple layers of elongated cells and little GAG. In rotating vessels, dynamic laminar flow patterns permitted the growth of constructs with high, spatially uniform GAG fractions. These studies demonstrate that hydrodynamic factors, such as mixing pattern and flow regime, play a key role in determining the structure and function of engineered cartilage.

Cardiac-like constructs seeded in mixed flasks (Fig. 13.2) and then cultured for 2 weeks in static or mixed flasks (Fig. 13.3a) or in rotating vessels (HARVs) (Fig. 13.3c) were compared with respect to several parameters, including $Y_{L/G}$, cellularity (DNA per construct), and metabolic activity index (MTT units/mg DNA) (Carrier *et al.*, 1999). Values of $Y_{L/G}$ for constructs cultured statically, in mixed flasks, and in rotating vessels were, respectively, 2.0, 1.6, and 1.0 mol/mol, implying anaerobic, partially aerobic, and aerobic cell metabolism (Fig. 13.6b). Cellularities and metabolic activity indexes of constructs cultured statically were markedly lower than those of constructs cultured in either mixed flasks or rotating vessels (Fig. 13.6b) (Carrier *et al.*, 1999). These studies demonstrate that hydrodynamic factors affect the structure and function of engineered cardiac tissue in a manner consistent with that observed for engineered cartilage (Fig. 13.6a).

BIOREACTOR CULTIVATION OF FUNCTIONAL TISSUES

CARTILAGE

Cartilaginous constructs cultured for 6 weeks in rotating bioreactors (STLVs) were continuously cartilaginous over their entire cross-sectional areas (6.7 mm in diameter $\times$ 5mm thick) (Freed *et al.*, 1998) and contained an interconnected network of collagen fibers that resembled that of articular collagen with respect to overall organization, fiber diameter (Fig. 13.7a) (Riesle *et al.*, 1998), and molecular structure (type II represented more than 90% of the total collagen) (Freed *et al.*, 1998). As compared to native bovine calf articular cartilage, 6-week constructs had comparable cellularities, 68% as much GAG, and 33% as much collagen type II per unit wet weight (Freed *et al.*, 1998). Extracellular matrix synthesis rates decreased to approximately 60% of initial over 6 weeks, and the fraction of newly synthesized macromolecules released into the culture medium decreased from about 25% of total at 4 days to less than 4% at 6 weeks. Over 7 months of *in vitro* cultivation, the GAG fraction and equilibrium modulus reached or exceeded values measured for native cartilage (Fig. 13.7a). However, other properties of 7-month constructs remained subnormal (e.g., collagen fraction, the concentration of pyridinium cross-links, and dynamic stiffness were, respectively, 34, 30, and 46% as high as in native cartilage) (Freed *et al.*, 1997; Riesle *et al.*, 1998). Construct mechanical properties (i.e., equilibrium modulus, hydraulic permeability, dynamic stiffness, and streaming potential) correlated with the wet weight fractions of GAG, collagen, and water (Vunjak-Novakovic *et al.*, 1999a).

CARDIAC TISSUE

Cardiac-like tissues cultivated for 1 week in mixed flasks were 5 mm in diameter and 1.3 mm thick and had a 50- to 70-μm-thick tissuelike region at the construct periphery in which cells expressed the muscle-specific protein sarcomeric tropomyosin and exhibited cardiac-specific ultrastructural features, including intercalated disks and sarcomeres (Bursac *et al.*, 1999) (Fig. 13.7b). As compared to native neonatal rat ventricles, 1-week constructs had 16% as much DNA per unit wet weight (Fig. 13.7b), and similar indexes of cell size (protein/DNA) and metabolic activity (MTT/DNA). The peripheral region of constructs was electrically excitable and could be captured over a wide range of pacing frequencies (80–270 beats/min) (Fig. 13.7b) (Bursac *et al.*, 1999). Impulses propagated at conduction velocities that were half of those in native neonatal rat ventricles (Bursac *et al.*, 1999) (Fig. 13.7b). Contractile cardiac-like tissues were also grown in rotating

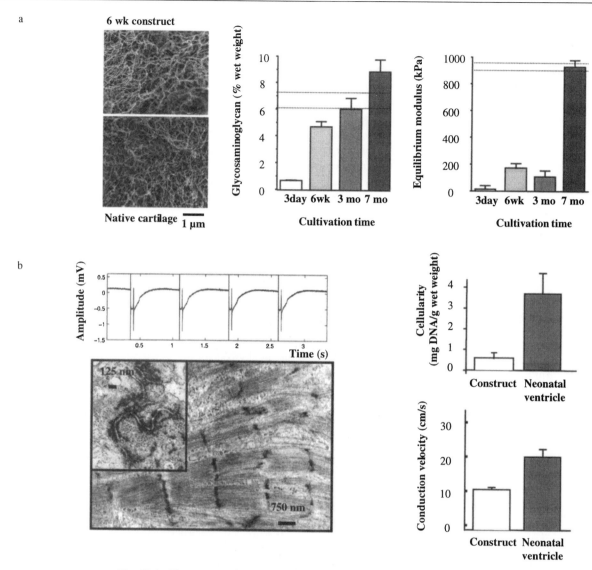

Fig. 13.7. Bioreactor cultivation of functional tissues. (a) Cartilage. Micrographs show the ultra-structural appearance of collagen networks in a 6-week construct and in native calf articular cartilage (Riesle et al., 1998, Collagen in tissue engineered cartilage: Types, structure and cross-links. J. Cell. Biochem. **71***, 313–327. Reprinted by permission of Wiley-Liss, Inc., a subsidiary of John Wiley & Sons, Inc.). Graphs: glycosaminoglycan fraction and equilibrium modulus of constructs cultivated in rotating vessels (STLVs) for 3 days, 6 weeks, and 3 or 7 months (Freed et al., 1997; Vunjak-Novakovic et al., 1999a). Dotted lines represent ranges of parameters measured for native cartilage. (b) Cardiac muscle. Recording: electrophysiological response of construct to electrical stimulation at a rate of 80 bpm. Micrographs: ultrastructural demonstration of sarcomeres (outer photo) and an intercalated disk (inset). Graphs: DNA content and conduction velocity of 1-week constructs and native neonatal rat ventricles (Bursac et al., 1999). Data are the average ±SD of three to six independent measurements.*

vessels (HARVs), and were shown to contain the gap-junctional protein connexin-43, creatine kinase isoform MM, sarcomeric myosin heavy-chain protein (Papadaki *et al.,* 1999), muscle desmin, cardiac troponin T, and sarcomeric tropomyosin (Carrier *et al.,* 1999).

TISSUE ENGINEERING BIOREACTORS: STATE OF THE ART

In contrast to a large number of tissue engineering studies focusing on scaffold design and *in vivo* tissue repair using cells and/or biomaterials, there has been relatively little work in the area of bioreactors. As described in this chapter, specific bioreactor design features can be exploited to improve the structure and function of engineered tissues, including in particular (1) flow and mixing patterns that can result in efficient and spatially uniform seeding of polymer scaffolds, (2) enhanced mass transport of chemical species through the medium and to the tissue surfaces, and (3) physical stimulation of the tissue during its development.

Our early studies demonstrated increased thickness and improved structure of engineered cartilage when scaffolds were seeded in orbitally mixed petri dishes as compared to static dishes (Freed et al., 1994b). The concept of maintaining a uniform suspension of isolated cells and providing relative velocity between cells and the scaffold during seeding was improved by using mixed flasks and rotating vessels (Freed and Vunjak-Novakovic, 1995, 1997a; Vunjak-Novakovic et al., 1996, 1999a). The kinetics and mechanisms of seeding in mixed flasks, which are at this time the preferred system for cell seeding of 3D scaffolds, were characterized in detail (Vunjak-Novakovic et al., 1998).

Engineered cartilage cultured in mixed flasks was structurally superior to that grown in orbitally mixed dishes, which was in turn superior to that grown statically (Vunjak-Novakovic et al., 1996). Efficient convective mixing demonstrated for mixed flasks and rotating vessels (Freed and Vunjak-Novakovic, 1997b) implies that both bioreactors can maintain constant concentrations of chemical species in the bulk medium. The hydrodynamic environment present in rotating vessels (Fig. 13.4) stimulated the development of engineered cartilage that was metabolically and mechanically functional (Fig. 13.5). Engineered tissues (cartilage, cardiac) grown in rotating vessels were structurally and functionally superior to constructs grown in either static or mixed flasks (Fig. 13.6). Although the mechanisms underlying these effects are yet to be determined, others have hypothesized that hydrodynamic forces affect cultured cells via pressure fluctuations, stretching the cell membrane, and/or shear stress (Berthiaume and Frangos, 1993; Smith et al., 1995).

The bioreactor systems we have used to date (Fig. 13.3) promote mass transfer to the construct surface, but do not enhance the relatively slow diffusion of chemical species to the construct interior. This has not been an overwhelming problem in the engineering of cartilage, an avascular tissue with low cellularity (only about 4–5% of the wet weight), which can be cultivated in bioreactors to a thickness exceeding that of native articular cartilage (i.e., 5 mm) (Freed et al., 1998). However, diffusional limitations of mass transport have severely curtailed efforts to engineer tissues that normally have high vascularity and/or cellularity. In particular, only very low thicknesses were reported to date for engineered bone (250–500 μm) (Ishaug et al., 1997; Martin et al., 1998) and cardiac tissue (50–180 μm) (Eschenhagen et al., 1997; Bursac et al., 1999; Carrier et al., 1999).

Other groups have demonstrated advantages of using culture systems that provide continuous perfusion and mechanical stimulation during cultivation. In one system, growing constructs were directly perfused with culture medium within 1.2 ml cartridges (Dunkelman et al., 1995). Engineered cartilage cultivated in these bioreactors for 2–4 weeks repaired osteochondral defects and integrated with underlying bone in a 2-year rabbit study (Schreiber et al., 1999). In other systems, cartilage (Grande et al., 1997) and skeletal muscle (Chromiak et al., 1998) constructs were surrounded by culture medium that was continuously recirculated through a closed loop. In all the above cases, perfusion of medium reportedly improved tissue growth and metabolism by enhancing mass transfer and reducing the variations in medium concentrations of gases, nutrients, metabolites, and regulatory factors that occur in periodically refed cultures. In addition, closed-loop perfused bioreactors reduce the risk of contamination during long-term cell cultivation.

The fact that physical stimuli modulate tissue development has motivated the design of several bioreactor systems in which growing tissues are exposed to mechanical forces. A system that exposed cartilaginous constructs to dynamic compression at physiological frequencies enhanced ECM synthesis rates as compared to statically loaded controls (Buschmann et al., 1995). Likewise, a system that exposed cartilaginous constructs to physiological levels of intermittent hydrostatic pressure enhanced GAG deposition as compared to nonpressurized controls (Carver and Heath, 1999). A system that exposed skeletal muscle constructs to static and dynamic tension was associated with increased size and improved function (Vandenburgh et al., 1991). The concept of tensile loading was extended by exploiting the native contractile properties of collagenous cell culture substrates. Examples include engineered tissue containing genetically modified cells that secreted human growth hormone for up to 3 months in mice (Vandenburgh et al., 1996), engineered cardiac muscle that contracted and in some cases secreted recombinant β-galactosidase for up to 10 days in vitro (Eschenhagen et al., 1997), and engineered tendon that repaired gap defects over 3 months in rabbits (Young et al., 1998). In all of the above studies, the presence of mechanical forces (externally applied or internally generated) during cultivation stimulated the development of engineered tissues, presumably by providing physical stimuli normally present *in vivo.*

A novel bioreactor system for vascular tissue engineering elegantly demonstrates the combined benefits of continuous perfusion and mechanical stimulation (Niklason *et al.,* 1999). Tubular polymer scaffolds seeded with smooth muscle cells were placed in a stirred vessel and cultivated in the presence of pulsatile radial stress, applied via highly distensible silicone tubing in the construct lumen. Under these conditions, small-caliber (<6 mm in diameter) arteries with high rupture strengths and high suture retention strengths could be cultivated over 2 months *in vitro;* these remained patent for up to 24 days following *in vivo* implantation in swine.

SUMMARY AND RESEARCH NEEDS

In this chapter, we have discussed the design and operating principles of tissue engineering bioreactors developed for the cultivation of engineered tissues starting from isolated cells and 3D biodegradable scaffolds. Bioreactor technologies can be utilized to grow functional tissues for *in vivo* implantation, and for controlled *in vitro* studies of how biochemical and physical factors regulate tissue development. In general, the functions of the bioreactor are to seed cells onto 3D scaffolds and to provide appropriate stimuli during *in vitro* tissue development. Specific design requirements depend on the dimensions, complexity, and application of the tissue to be engineered. Further advances in bioreactor technologies that are likely have a major impact on the field of tissue engineering include (1) improved control over culture medium composition, (2) new methods to provide physiological levels of mechanical stimuli to engineered tissues *in vitro,* (3) new techniques to induce and/or mimic vascularization of engineered tissues *in vitro,* (4) determination of appropriate conditions and duration of *in vitro* culture prior to implantation, and (5) extension from animal model systems to human cell sources and scales.

ACKNOWLEDGMENTS

This work was supported by the National Aeronautics and Space Administration (Grants NAG9-836 amd NCC8-174). The authors thank G. P. Neitzel, J. B. Brown, and M. Smith for the bioreactor fluid dynamics data shown in Fig. 13.4, and R. Langer for helpful advice.

REFERENCES

Bader, A., Knop, E., Fruehauf, N., Crome, O., Boeker, K., Christians, U., Oldhafer, K., Ringe, B., Pichlmayr, R., and Sewing, K. F. (1995). Reconstruction of liver tissue *in vitro:* Geometry of characteristic flat bed, hollow fiber, and spouted bed bioreactors with reference to *in vivo* liver. *Artif. Organs* **19,** 941–949.

Berthiaume, F., and Frangos, J. (1993). Effects of flow on anchorage-dependent mammalian cell-secreted products. *In* "Physical Forces and the Mammalian Cell" (J. Frangos, ed.), p. 139. Academic Press, San Diego, CA.

Bursac, N., Papadaki, M., Cohen, R. J., Schoen, F. J., Eisenberg, S. R., Carrier, R., Vunjak-Novakovic, G., and Freed, L. E. (1999). Cardiac muscle tissue engineering: Towards an *in vitro* model for electrophysiological studies. *Am. J. Physiol.* **277,** H433–H444.

Buschmann, M. D., Gluzband, Y. A., Grodzinsky, A. J., and Hunziker, E. B. (1995). Mechanical compression modulates matrix biosynthesis in chondrocyte/agarose culture. *J. Cell Sci.* **108,** 1497–1508.

Carrier, R., Papadaki, M., Rupnick, M., Schoen, F. J., Bursac, N., Langer, R., Freed, L. E., and Vunjak-Novakovic, G. (1999). Cardiac tissue engineering: Cell seeding, cultivation parameters and tissue construct characterization. *Biotechnol. Bioeng.* **64,** 580–589.

Carver, S. E., and Heath, C. A. (1999). Increasing extracellular matrix production in regenerating cartilage with intermittent physiological pressure. *Biotechnol. Bioeng.* **62,** 166–174.

Cherry, R. S., and Papoutsakis, T. (1988). Physical mechanisms of cell damage in microcarrier cell culture bioreactors. *Biotechnol. Bioeng.* **32,** 1001–1014.

Chromiak, J. A., Shansky, J., Perrone, C., and Vandenburgh, H. (1998). Bioreactor perfusion system for the long-term maintenance of tissue-engineered skeletal muscle organoids. *In Vitro Cell. Dev. Biol.: Anim.* **34,** 694–703.

Dunkelman, N. S., Zimber, M. P., LeBaron, R. G., Pavelec, R., Kwan, M., and Purchio, A. F. (1995). Cartilage production by rabbit articular chondrocytes on polyglycolic acid scaffolds in a closed bioreactor system. *Biotechnol. Bioeng.* **46,** 299–305.

Eschenhagen, T., Fink, C., Remmers, U., Scholz, H., Wattchow, J., Weil, J., Zimmermann, W., Dohmen, H. H., Schafer, H., Bishopric, N., Wakatsuku, T., and Elson, E. L. (1997). Three-dimensional reconstitution of embryonic cardiomyocytes in a collagen matrix: A new heart muscle model system. *FASEB J.* **11,** 683–694.

Farndale, R. W., Buttle, D. J., and Barrett, A. J. (1986). Improved quantitation and discrimination of sulphated glycosaminoglycans by the use of dimethylmethylene blue. *Biochim. Biophys. Acta* **883,** 173–177.

Freed, L. E., and Vunjak-Novakovic, G. (1995). Cultivation of cell-polymer constructs in simulated microgravity. *Biotechnol. Bioeng.* **46,** 306–313.

Freed, L. E., and Vunjak-Novakovic, G. (1997a). Microgravity tissue engineering. *In Vitro Cell Dev. Biol.* **33,** 381–385.

Freed, L. E., and Vunjak-Novakovic, G. (1997b). Tissue culture bioreactors: Chondrogenesis a model system. *In* "Principles of Tissue Engineering" (R. P. Lanza, R. Langer, and W. L. Chick, eds.), pp. 151–165. Academic Press, San Diego, CA, and Landes, Austin, TX.

Freed, L. E., and Vunjak-Novakovic, G. (1998). Culture of organized cell communities. *Adv. Drug Delivery Rev.* **33**, 15–30.

Freed, L. E., Grande, D. A., Emmanual, J., Marquis, J. C., Lingbin, Z., and Langer, R. (1994a). Joint resurfacing using allograft chondrocytes and synthetic biodegradable polymer scaffolds. *J. Biomed. Mater. Res.* **28**, 891–899.

Freed, L. E., Marquis, J. C., Vunjak, G., Emmanual, J., and Langer, R. (1994b). Composition of cell–polymer cartilage implants. *Biotechnol. Bioeng.* **43**, 605–614.

Freed, L. E., Vunjak-Novakovic, G., Biron, R., Eagles, D., Lesnoy, D., Barlow, S., and Langer, R. (1994c). Biodegradable polymer scaffolds for tissue engineering. *Bio/Technology* **12**, 689–693.

Freed, L. E., Langer, R., Martin, I., Pellis, N., and Vunjak-Novakovic, G. (1997). Tissue engineering of cartilage in space. *Proc. Natl. Acad. Sci. U.S.A.* **94**, 13885–13890.

Freed, L. E., Hollander, A. P., Martin, I., Barry, J. R., Martin, I., and Vunjak-Novakovic, G. (1998). Chondrogenesis in a cell-polymer-bioreactor system. *Exp. Cell Res.* **240**, 58–65.

Grande, D. A., Halberstadt, C., Naughton, G., Schwartz, R., and Manji, R. (1997). Evaluation of matrix scaffolds for tissue engineering of articular cartilage grafts. *J. Biomed. Mater. Res.* **34**, 211–220.

Ishaug, S. L., Crane, G. M., Miller, M. J., Yasko, A. W., Yaszemski, M. J., and Mikos, A. G. (1997). Bone formation by three-dimensional stromal osteoblast culture in biodegradable polymer scaffolds. *J. Biomed. Mater. Res.* **36**, 17–28.

Jauregui, H. O., Mullon, C. J.-P., and Solomon, B. (1997). Extracorporeal artificial liver support. *In* "Principles of Tissue Engineering" (R. P. Lanza, R. Langer, and W. L. Chick, eds.), pp. 463–479. Academic Press, San Diego, CA, and Landes, Austin, TX.

Jessup, J. M., Goodwin, T. J., and Spaulding, G. (1993). Prospects for use of microgravity-based bioreactors to study three-dimensional host–tumor interactions in human neoplasia. *J. Cell. Biochem.* **51**, 290–300.

Jessup, J. M., Goodwin, T. J., Garcia, R., and Pellis, N. (1996). STS-70: First flight on EDU-1. *In Vitro Cell. Dev. Biol.: Anim.* **32**, 13A.

Kim, Y. J., Sah, R. L., Doong, J. Y. H., and Grodzinsky, A. J. (1988). Fluorometric assay of DNA in cartilage explants using Hoechst 33258. *Anal. Biochem.* **174**, 168–176.

Lanza, R. P., and Chick, W. L. (1997). Endocrinology: Pancreas. *In* "Principles of Tissue Engineering" (R. P. Lanza, R. Langer, and W. L. Chick, eds.), pp. 405–425. Academic Press, San Diego, CA, and Landes, Austin, TX.

Martin, I., Padera, R. F., Vunjak-Novakovic, G., and Freed, L. E. (1998). *In vitro* differentiation of chick embryo bone marrow stromal cells into cartilaginous and bone-like tissues. *J. Orthop. Res.* **16**, 181–189.

Martin, I., Shastri, V. P., Padera, R. F., Langer, R., Vunjak-Novakovic, G., and Freed, L. E. (1999a). Bone marrow stromal cell differentiation on porous polymer scaffolds. *Trans. Orthop. Res. Soc.* **24**, 57.

Martin, I., Obradovic, B., Freed, L. E., and Vunjak-Novakovic, G. (1999b). A method for quantitative analysis of glycosaminoglycan distribution in cultured natural and engineered cartilage. *Ann. Biomed. Eng.* **27**, 1–7.

Neitzel, G. P., Nerem, R. M., Sambanis, A., Smith, M. K., Wick, T. M., Brown, J. B., Hunter, C., Jovanovic, I., Malaviya, P., Saini, S., and Tan, S. (1998). Cell function and tissue growth in bioreactors: Fluid mechanical and chemical environments. *J. Jpn. Soc. Microgr. Appl.* **15** (Suppl. II), 602–607.

Niklason, L. E., Gao, J., Abbott, W. M., Hirschi, K. K., Houser, S., Marini, R., and Langer, R. (1999). Functional arteries grown *in vitro. Science* **284**, 489–493.

Obradovic, B., Freed, L. E., Langer, R., and Vunjak-Novakovic, G. (1997). Bioreactor studies of natural and engineered cartilage metabolism. *Proc. of the Topical Conference on Biomaterials, Carriers for Drug Delivery, and Scaffolds for Tissue Engineering,* 335–337.

Obradovic, B., Carrier, R. L., Vunjak-Novakovic, G., and Freed, L. E. (1999). Oxygen is essential for bioreactor cultivation of tissue engineered cartilage. *Biotechnol. Bioeng.* **63**, 197–205.

Papadaki, M., Bursac, N., Langer, R., Schoen, F. J., Vunjak-Novakovic, G., and Freed, L. E. (1999). Towards a functional tissue engineered cardiac muscle: Effects of cell culture substrate and medium serum concentration. *Soc. Biomater. 25th Annu. Meet. Trans.,* p. 359.

Prewett, T. L., Goodwin, T. J., and Spaulding, G. F. (1993). Three-dimensional modeling of T-24 Human bladder carcinoma cell line: A new simulated microgravity culture system. *J. Tissue Cult. Methods* **15**, 29–36.

Qiu, Q., Ducheyne, P., and Ayyaswamy, P. (1999). Fabrication, characterization and evaluation of bioceramic hollow microspheres used as microcarriers for 3D bone tissue formation in rotating bioreactors. *Biomaterials* **20**, 989–1001.

Riesle, J., Hollander, A. P., Langer, R., Freed, L. E., and Vunjak-Novakovic, G. (1998). Collagen in tissue engineered cartilage: Types, structure and cross-links. *J. Cell. Biochem.* **71**, 313–327.

Schreiber, R. E., Ilten-Kirby, B. M., Dunkelman, N. S., Symons, K. T., Rekettye, L. M., Willoughby, J., and Ratcliffe, A. (1999). Repair of osteochondral defects with allogeneic tissue-engineered cartilage implants. *Clin. Orthop. Relat. Res.* **367S**, S382–S395.

Schwarz, R. P., Goodwin, T. J., and Wolf, D. A. (1992). Cell culture for three-dimensional modeling in rotating-wall vessels: An application of simulated microgravity. *J. Tissue Cult. Methods* **14**, 51–58.

Searby, N. D., de Luis, J., and Vunjak-Novakovic, G. (1998). Design and development testing of a space station cell culture unit. *Proc. Conf. Environ. Microgravity Flight Impl., Danvers, MA, 1998.* 1–4.

Smith, R. L., Donlon, B. S., Gupta, M. K., Mohtai, M., Das, P., Carter, D. R., Cooke, J., Gibbons, G., Hutchinson, N., and Schurman, D. J. (1995). Effect of fluid-induced shear on articular chondrocyte morphology and metabolism *in vitro. J. Orthop. Res.* **13**, 824–831.

Unsworth, B. R., and Lelkes, P. I. (1998). Growing tissues in microgravity. *Nat. Med.* **4**, 901–907.

Vandenburgh, H. H., Swadison, S., and Karlisch, P. (1991). Computer-aided mechanogenesis of skeletal muscle organs from single cells *in vitro. FASEB J.* **5**, 2860–2867.

Vandenburgh, H. H., DelTatto, M., Shansky, J., Lemaire, J., Chang, A., Payumo, F., Lee, P., Goodyear, A., and Raven, L. (1996). Tissue-engineered skeletal muscle organoids for reversible gene therapy. *Hum. Gene Ther.* 7, 2195–2200.

Vunjak-Novakovic, G., Freed, L. E., Biron, R. J., and Langer, R. (1996). Effects of mixing on tissue engineered cartilage. *AIChE J.* **42,** 850–860.

Vunjak-Novakovic, G., Obradovic, B., Bursac, P., Martin, I., Langer, R., and Freed, L. E. (1998). Dynamic seeding of polymer scaffolds for cartilage tissue engineering. *Biotechnol. Prog.* **14,** 193–202.

Vunjak-Novakovic, G., Martin, I., Obradovic, B., Treppo, S., Grodzinsky, A. J., Langer, R., and Freed, L. E. (1999a). Bioreactor cultivation conditions modulate the composition and mechanical properties of tissue engineered cartilage. *J. Orthop. Res.* **17,** 130–138.

Vunjak-Novakovic, G., Preda, C., Bordonaro, J., Pellis, N., de Luis, J., and Freed, L. E. (1999b). Microgravity studies of cells and tissues: From Mir to ISS. *Am. Inst. Phys. Conf. Proc.* **458,** 442–452.

Williams, S. N. O., Burstein, D., Freed, L. E., Gray, M. L., Langer, R., and Vunjak-Novakovic, G. (1998). MRI measurements of fixed charge density as a measure of glycosaminoglycan content and distribution in tissue engineered cartilage. *Trans. Orthop. Res. Soc.* **23,** 203.

Woessner, J. F. (1961). The determination of hydroxyproline in tissue and protein samples containing small proportions of this imino acid. *Arch. Biochem. Biophys.* **93,** 440–447.

Young, R. G., Butler, D. L., Weber, W., Caplan, A. I., Gordon, S. L., and Fink, D. J. (1998). Use of mesenchymal stem cells in a collagen matrix for achilles tendon repair. *J. Orthop. Res.* **16,** 406–413.

CHAPTER 14

TISSUE ASSEMBLY IN MICROGRAVITY

Brian R. Unsworth and Peter I. Lelkes

INTRODUCTION

The increasing disparity between demand and availability of human donor organs for the replacement of dysfunctional organs in patients has raised serious ethical and legal issues and prompted an intensive search for alternatives to human allotransplants or xenotransplantation. Generation of tissue-engineered replacement organs, e.g., by means of three-dimensional (3D) *in vitro* cell culture, may present a suitable substitute for donor organ transplantation. Multicellular spheroids can be generated from one cell type only (homotypic), or comprise different cell types from the same tissue (heterotypic). Such spheroids have provided new insights into the biological responses of *in vivo*-like multicellular assemblies, e.g., on radiation injury (Schwachöfer, 1990), in developing drug resistance (Kerbel *et al.,* 1994), and in understanding phenotypic changes associated with tumor invasiveness (Schuster *et al.,* 1994). Although such multicellular spheroids are a first step toward assembling complex 3D tissue equivalents, there are several inherent limitations, such as their size and the restricted diffusion of nutrients and oxygen; additionally, spheroids >1 mm in diameter generally contain a hypoxic, necrotic center surrounded by a rim of viable cells (Sutherland *et al.,* 1986).

In progressing beyond spheroids, creation of properly assembled and differentiated tissues from single cells in suspension has attracted an interdisciplinary approach, collectively known as tissue engineering. One major aim of this potentially multibillion dollar venture is to generate viable/functional tissue-equivalents that can serve as replacement tissues, in lieu of allotransplantation (Langer and Vacanti, 1993). Resorbable polymeric scaffolds, made, for example, of fibrous polyglycolic acid, serve as provisional, biodegradable matrices for cell seeding (Cima *et al.,* 1991). As detailed elsewhere in this volume, such scaffolds can be molded into complex geometric shapes. Importantly, these scaffolds, in particular when treated with growth factors or adhesive proteins, promote cellular proliferation and 3D assembly (Kim and Mooney, 1998). After the controlled biodegradation of the scaffolds, the 3D cellular constructs form neotissues and neoorgans, such as skin, cartilage, blood vessels, bladders, and heart valves (Langer and Vacanti, 1995; Shinoka *et al.,* 1997, 1998; Atala, 1998; Niklason and Langer, 1997; Mayer *et al.,* 1997). Similarly, RGD peptides conjugated to poly(ethylene glycol) promote aggregation and differentiation of neural cells in suspension, thus eliminating the need for microcarriers and facilitating the use of these aggregates as replacement tissues (Dai *et al.,* 1994). Some of these bioengineered replacement tissues are currently being tested in animals (Krewson and Saltzman, 1996; Atala, 1998; Shinoka *et al.,* 1998; Fauza *et al.,* 1998) and await human trials.

A major focus of tissue engineering is how to best generate functional 3D constructs, large enough to serve as replacement organs or suitable for scientific studies of *in vitro* drug effects/toxicity, oxygen diffusion in tissues, etc. The high shear forces of conventional stirred fermentors/rollers are a disadvantage because they damage the cells and hinder proper tissue-specific differentiation. Decreasing the stir rate while increasing the medium viscosity might partially reduce the hydrodynamic damage (Moreira *et al.,* 1995); however, aggregates formed under these conditions still exhibit necrotic centers. As detailed in Chapter 13, tissue constructs on biodegradable scaffolds have been successfully cultured in stirred bioreactors (Freed and Vunjak-Novakovic, 1997).

MICROGRAVITY AS A NOVEL TISSUE CULTURE VENUE

Efforts at tissue assembly would benefit from a venue that promotes cell–cell association, while avoiding the detrimental effects of high shear stress. Such a venue might be provided by microgravity (Dintenfass, 1986). In an effort to embellish the potential beneficial effects of microgravity and low fluid shear for cell culture here on Earth, scientists at the National Aeronautics and Space Association (NASA) introduced rotating-wall vessel (RWV) cell culture biotechnology. The principles of this system, which combines low fluid shear stress and randomization of the gravitational vectors, are described in detail in Chapter 13.

The use of the different incarnations of the RWV bioreactors is largely dictated by oxygen requirements of the cultured cells. Due to its geometry and the use of a large-surface-area, flat-membrane oxygenator at the rear of the chamber, the oxygenation capacity of the high-aspect-ratio vessel (HARV) is higher than that of the slow-turning lateral vessel (STLV). STLV-type bioreactors, consisting of a cylindrical growth chamber that contains an inner corotating cylinder with a gas-exchange membrane, are suited for cells with low oxygen requirements. HARVs, on the other hand, are mostly used for cell types that require more oxygen per unit volume of culture medium.

In the past, conventional stirred fermentors were used as "controls" for assessing the contributions of impeller-induced shear forces on tissue assembly (Lelkes *et al.,* 1998). Alternatively, Fang *et al.* suggested use of vertically positioned RWVs as dynamic controls, with the axis of the vessels being parallel to gravity. Due to the centrifugal forces of the rotating motion, the vertical position provides a near-normal gravity environment, as opposed to the simulated microgravity in the horizontal position (Fang *et al.,* 1997).

A more sophisticated version of the STLV-type RWV bioreactor, called the rotating-wall perfused vessel, allows the investigator to study the individual contributions of fluid shear stress and (simulated) microgravity. In these vessels, controlled shear stress can be induced by independently rotating the inner and the outer walls. An addition to the family of RWVs, specifically developed by NASA for use in space, is the hydrodynamic focusing bioreactor, which provides new capabilities for long-duration operation in orbit as well as on the ground. Of additional advantage, both of these systems can be fully automated for changing, sampling, and/or modifying the culture medium without stopping the vessel rotation. This capability is very important, in view of our finding that even short cessation of microgravity, or its initiation, can activate cellular signaling pathways (Lelkes and Unsworth, 1998).

The differentiation of cell aggregates grown in the simulated microgravity environment of RWV bioreactors is based on the interplay between two beneficial factors: (1) low shear stress, which promotes close apposition (spatial colocation) of the cells, and (2) randomized gravitational vectors, which either directly affect gene expression or indirectly facilitate paracrine/autocrine intercellular signaling, e.g., through restricted diffusion of differentiative humoral factors (Jessup *et al.,* 1993; Schwarz *et al.,* 1992). To date, numerous groups have been utilizing RWV bioreactor biotechnology to facilitate the assembly of macroscopic, differentiated, tissuelike 3D constructs (also called proto-tissues, tissue equivalents, or organoids) from a variety of normal and transformed cell types. RWV bioreactors provide a suitable *in vitro* venue for culturing otherwise recalcitrant cells, such as viruses (Long *et al.,* 1998) and human tumor cells (Becker *et al.,* 1997; Ingram *et al.,* 1997). The resulting differentiated 3D tissue constructs secrete valuable bioproducts, such as antibodies and antigens (Jessup *et al.,* 1997), and ultimately may serve as possible replacement organs.

RWV bioreactors are also used to further differentiate *in vitro* preassembled bioengineered tissue equivalents, such as skin (D. Dimitrijevich, personal communication) and cartilage (Freed and Vunjak-Novakovic, 1997), as well as to culture intact tissues, such as fragments of liver (Khaoustov *et al.,* 1995), tonsils (Margolis *et al.,* 1997) or kidney (A. Nor and P. I. Lelkes, unpublished). RWV biotechnology in conjunction with an appropriate scaffold/matrix might also provide a suitable venue for generating 3D constructs of neuronal cells, capable of "thinking," i.e., establishing functional neuronal connections (P. I. Lelkes *et al.,* unpublished). For a summary of the diverse cell types (to date more than three dozen) cultured in RWV bioreactors, see Unsworth and Lelkes (1998a). This a very dynamic, rapidly evolving field of research, and a very up-to-date list of publications and other information may be found on the NASA web site at http://idi.usra.edu/peer_review/taskbook/taskbook.html.

In the RWV culture environment, homotypic organoids express tissue-specific differentiation markers beyond the levels seen in conventional tissue culture (Jessup *et al.,* 1997; Unsworth and Lelkes, 1998b). RWV culture also induces significant changes in the repertoire of expressed genes (Hammond *et al.,* 1999). Nonetheless, these prototissues fail to recapitulate faithfully the phenotypic diversity of the parental tissues. Important components, notably mesenchymal/interstitial cells, as well as tissue-specific extracellular matrices, are absent, as is the microvasculature.

Organotypic differentiation of the 3D constructs, both normal and neoplastic, is enhanced in heterotypic cocultures comprising phenotypically diverse cells (Galvan *et al.,* 1995; Goodwin *et al.,* 1997; Zhau *et al.,* 1997). Specifically, coculture of tumor cells and normal stromal cells in RWV bioreactors leads to the formation of "differentiated" tumor masses, which may become valuable tools in cancer research (Zhau *et al.,* 1997). Such 3D tumor cell constructs could also serve as *in vitro* models for studying inter- and intracellular signaling involved in clonal expansion, for studying phenotypic instability, and for testing drug/radiation resistance. On the other hand, differentiated 3D constructs from normal tissue, besides their potential usefulness as replacement organs, could also serve as tissue equivalents for more realistic toxicity studies and thus provide yet another novel alternative to animal testing.

VASCULARIZATION: OVERCOMING SIZE LIMITATIONS OF TISSUE ASSEMBLIES

In conventional fermentors, macroscopic assemblies of cell aggregates that exceed about 1 mm in size invariably develop necrotic cores. In the unique culture conditions in RWV bioreactors, dissociated cells can assemble into macroscopic tissue aggregates up to 1 cm in size, and these are largely devoid of such necrotic cores. Clearly, all *denovo* tissue-engineered constructs will eventually require internal, blood vessellike conduits for the delivery of oxygen and nutrients as well as for the removal of waste products. In spite of sporadic claims to the contrary, inclusion of endothelial cells in the mixture of cells cultured in the RWV bioreactors, with the stated purpose of generating vessellike conduits (Goodwin *et al.,* 1993), has, to the best of our knowledge, so far failed to yield spontaneous *in situ* "vasculogenesis" (Chopra *et al.,* 1998). In these simplistic cocultures, endothelial cells remain viable and organize largely into homotypic clusters without forming tubular structures. In contrast, the expansion of tissue fragments, such as of liver (Khaoustov *et al.,* 1995) and kidney (A. Nor and P. I. Lelkes, unpublished studies), in the RWV environment is accompanied by a commensurate increase in the density of microvessels. This process, suggestive of angiogenesis, indicates that it may be possible to grow blood vessels in RWV bioreactors, provided the cells are seeded onto/into an appropriate permissive environment, such as a branching scaffold or an extracellular matrix construct containing angiogenic growth factors (Eiselt *et al.,* 1998; Griffith *et al.,* 1997). Clearly, organoid vascularization is one of the foremost, yet so far elusive, goals that may determine the medical/clinical usefulness of *in vitro*-generated tissue equivalents.

FROM SINGLE CELLS TO TISSUES IN SPACE

It has been known for more than 20 years that single cells respond to microgravity in space by alterations in cellular morphology and function (Montgomery *et al.,* 1978; Claassen and Spooner, 1994). According to Cogoli and Cogoli-Greuter (1998), more than 25 different cell types have been flown in space. Earlier studies reported that during space flight, cell suspensions preferentially aggregate and yield higher cell densities than they do on Earth (Lorenzi *et al.,* 1995). One reason for enhanced cell–cell interactions may be that microgravity induces a tissue-specific upregulation of cell adhesion molecules, extracellular matrix proteins, and their respective receptors. In line with this notion, studies on board Spacelab indicated a significant increase in collagen synthesis in human dermal fibroblasts in microgravity (Seitzer *et al.,* 1995). However, in whole animals the response might be different and might be determined by hormonal and other physiological factors (Backup *et al.,* 1994; Lelkes *et al.,* 1994; Davidson *et al.,* 1999).

For some cultured cells, for which the expression of certain "glue proteins" such as collagen is up-regulated in simulated microgravity in the RWV environment (Lelkes *et al.,* 1998), one might predict that an excursion into space may be beneficial. Indeed, when PC12 pheochromocytoma

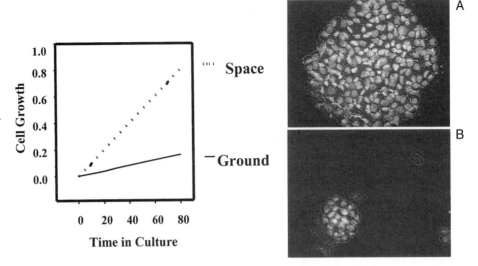

Fig. 14.1. PC12 cell culture maintained for a prolonged period of time (up to 80 days) in space. The left panel shows enhanced "cell growth" in space vs. ground-based controls. Cell growth was inferred from the rate of glucose consumption. The micrographs depict the relative sizes of DNA-stained cell aggregates in space (A) and on Earth (B). Original magnification: ×50.

cells were grown in space and serially passaged over approximately a 10-week period, the degree of aggregation and the rate of proliferation of the samples in space was significantly accelerated in comparison to ground controls (Unsworth and Lelkes, 1998a). The rate of glucose consumption, a measure of cellular metabolic activity and of proliferation, was approximately five times higher in space than on the ground (Fig. 14.1). Similarly, large cellular aggregates were observed in space, but not under static conditions on Earth (Unsworth and Lelkes, 1998a). These results, although preliminary, suggest that the space environment might be advantageous for generating neuronal/neuroendocrine tissue equivalents using PC12 cells. For these cells, the ensuing cultures yielded larger and/or more differentiated organoids, compared with parallel cultures maintained at 1 *g* on Earth.

For other cells, however, which are adversely affected by microgravity, such as bone, cartilage, or muscle, a "space-walk" might yield inferior results. This hypothesis was tested, in part, on Shuttle/Mir missions: in one of these outings, cartilaginous constructs that were flown for 4 months in space proved to be mechanically inferior to the ground controls (Freed *et al.,* 1997). This finding might reflect a direct adverse effect of microgravity on cellular processes involved in the synthesis, secretion, and assembly of glycosaminoglycans and other extracellular matrix proteins, in keeping with the well-known adverse effects of space flight on bone and cartilage (Bonting, 1992). On the other hand, the superiority of cartilaginous constructs on Earth might reflect the effects of mechanical loading on the differentiation of the bioengineered constructs (Freed *et al.,* 1997).

Thus, mass culture in space may not be advantageous for each and every cell type in terms of generating functional replacement tissues. In contrast to the cartilage experiment, other investigators in the cadre of the Office of Life and Microgravity Sciences Applications/Microgravity Research Division (OLMSA/MRD) Cell Culture Biotechnology Group, who have flown cells in the RWV aboard diverse Shuttle missions, reported enhanced cellular responses (e.g., aggregation and secretion of bioactive products) as compared to control cells cultured in RWVs on Earth (Jessup *et al.,* 1996). Taken together, these data suggest a rather cautionary approach toward the potential use of "space factories" for producing medically/scientifically useful tissue equivalents. The outcome might be highly cell type specific.

IN VITRO EMBRYOLOGY

In addition to their potential usefulness as a cell culture venue for mass producing replacement prototissues, simulated microgravity in the RWV bioreactors also offers an innovative, *in vitro* tool for studying instructional cues and mechanisms involved in organogenesis. Importantly, the RWV environment supports both cellular differentiation and tissue expansion. This mimics developmental organogenesis *in vivo,* and is in stark contrast to conventional tissue culture, where cells either differentiate or proliferate (see also Chapter 9).

We are using RWV bioreactors for *in vitro* embryology, in order to understand mechanisms controlling neuroendocrine differentiation of bipotential, sympathoadrenal cells of the adrenal gland (Lelkes *et al.,* 1998). The culture environment of the RWV specifically induces the formation of macroscopic, 3D organoids and causes differentiation of PC12 cells, an adrenal medullary tumor cell line, along the neuroendocrine pathway. This differentiation is associated with rapid activation of intracellular signaling pathways involved in the control of organ-specific differentiation (Lelkes and Unsworth, 1998). The tissue-specific differentiation in the RWV bioreactors seems to mimic closely events occurring during *in vivo* organogenesis. Thus the system offers a promising approach to investigating molecular mechanisms operating during embryogenesis. The repertoire of genes expressed in microgravity-reared 3D tissue constructs, both in RWV cultures on Earth and in space, is significantly altered from that observed in cells cultured at unit gravity and might resemble more closely the panel of genes expressed *in situ* (Kaysen *et al.,* 1999).

GRAVITATIONAL SENSING

The gravity responsiveness of single cells has raised speculations about the existence of gravity-sensing mechanisms and of specialized gravity receptors (Montgomery *et al.,* 1978). A prime candidate for gravitational sensing is the cytoskeleton (Ingber, 1997; Hughes-Fulford and Lewis, 1996). At the single-cell level, alterations in the cytoskeletal architecture might explain changes in the vectorial cascade of intracellular phosphorylation cascades (Schmitt *et al.,* 1996). The functional impairment of space-flown lymphocytes, such as altered capping/patching and locomotion (Cogoli-Greuter *et al.,* 1996; Pellis *et al.,* 1997), which has been attributed to alterations in protein kinase C (PKC)-dependent signaling, may also result from cytoskeletal rearrangement in microgravity. In this context, studies by us and others suggest that the simulated microgravity conditions in RWV bioreactors may mimic some of the alterations in cellular signaling observed in space (Cooper and Pellis, 1998; Lelkes and Unsworth, 1998). For example, when PC12 cells were briefly (15 min to 24 hr) cultured in RWV bioreactors, neurotrophin signaling was impaired at the levels of receptor autophosphorylation and postreceptor signaling, resulting in impaired MAP kinase activation, altered integrin signaling, and changes in the nuclear translocation of transcription factors (Lelkes and Unsworth, 1998).

The question remains: What is fundamentally different between simulated microgravity in RWV bioreactors and the real thing in space? Are the differences due to distinct g levels (10^{-2} g in RWV vs. 10^{-4}–10^{-6} g in space), or due to residual physical forces (sedimentation; low but finite shear stress) on Earth? A definite answer will have to wait until the experiments can be performed on board the International Space Station—in microgravity in the absence of shear stress—as well as under conditions in which gravitational forces can be artificially restored by using an onboard, low-speed centrifuge.

CAVEATS

Microgravity, either simulated or in space, has the potential to become a useful venue for the mass production of replacement tissues, which could help to overcome the shortage of donor organs. For the time being the usefulness of this approach is just a promise, yet to be realized. Similarly, the advantages of RWV-generated 3D tumor cell organoids have to be substantiated scientifically and clinically. And yet, although this field is just barely beyond its infancy, microgravity-based organoid formation should and will find its niche in the wider area of tissue engineering.

It has yet to be proved whether tissue-engineered 3D constructs "from space," once removed from suspension cultures, will remain intact and functional long enough to become fully vascularized by anastomosing with the host tissue. To further enhance their postmicrogravity coherence, the prototissues can be encapsulated into semipermeable poly(lysine)/alginate microcapsules. This encapsulation may also facilitate xenotransplantation by providing an immunoprotective barrier (Sagen *et al.,* 1993).

The anticipated differentiation of certain cell types, such as endothelial cells or neuronal cells, might be thwarted in microgravity by blunted cellular responses to a variety of growth factors, such as basic fibroblast, platelet-derived, or neuronal growth factors (Davidson *et al.,* 1999; Lelkes and Unsworth, 1998). Therefore, the concept of directly applying these growth factors, as intriguing as it sounds, may not work in microgravity. Alternative signaling pathways leading to cellular differentiation may have to be targeted (Cooper and Pellis, 1998).

Using the space environment, rather than the simulated microgravity facilities here on Earth, requires a round-trip ticket. The beginning and the end of this voyage are particularly stressful, both for astronauts and at the level of single cells, for which simulated launch conditions were found to cause alterations in gene expression (Fitzgerald and Hughes-Fulford, 1996; Tjandrawinata *et al.,* 1997). Therefore, when conducting cellular experiments in space, it is mandatory to include both an additional set of ground controls that simulate the conditions of a shuttle launch, and an onboard 1-*g* centrifuge to control for these potentially deleterious forces. At present it is not clear how cellular responses induced by launch conditions will impact subsequent experiments carried out in space.

Another problem may be posed by DNA damage owing to elevated levels of radiation in space. Enhanced DNA damage could translate into cytogenetic alterations of the tissues formed in space (Horneck *et al.,* 1996, 1997).

CONCLUSIONS

Microgravity, either simulated on Earth or experienced in space, provides a venue for the formation of tissue-engineered, highly differentiated 3D constructs. In this venue, recalcitrant cells thrive, even if they otherwise would not grow outside the body. Prototissues generated in this environment will be of importance clinically as potential replacement tissues, thus alleviating the current shortage of donor organs. Moreover, these constructs will serve as differentiated macroscopic tissue equivalents in diverse areas of *in vitro* biomedical research, such as toxicology and tumor biology, thus providing a possible alternative to animal testing. The possibility for recapitulating, *in vitro,* the complex cellular events occurring during organogenesis, for example, opens the door for a comprehensive analysis of the molecular control of developmental processes that depend on cell–cell and cell–extracellular matrix interactions. The advent of the International Space Station early in the next millennium will provide laboratory facilities in which we may be able to expand our current studies from simulated microgravity on the ground to real microgravity in space.

Acknowledgments

This chapter is dedicated to the memory of Gabor J. Lelkes. We would like to thank numerous colleagues who have freely shared unpublished results and/or preprints of their work. On the other hand, we wish to apologize to those colleagues whose work, for space limitations, we were unable to cite. Our own studies cited herein were supported in part by grants from the National Aeronautics and Space Administration (NAG 9-651 and NCC8-173).

References

Atala, A. (1998). Tissue engineering in urologic surgery. *Urol. Clin. North Am.* **25**, 39–50.

Backup, P., Westerlind, K., Harris, S., Spelsberg, T., Kline, B., and Turner, R. (1994). Space flight results in reduced mRNA levels for tissue specific protein in the musculoskeletal system. *Am. J. Physiol.* **266**, E567–E573.

Becker, J. L., Papenhausen, P. R., and Widen, R. H. (1997). Cytogenetic, morphologic and oncogene analysis of a cell line derived from a heterologous mixed müllerian tumor of the ovary. *In Vitro Cell. Dev. Biol.: Anim.* **33**, 325–331.

Bonting, S. L. (1992). Chemical sensors for space applications. *Adv. Space Biol. Med.* **21**, 263–293.

Chopra, V., Dinh, T. V., and Hannigan, E. V. (1998). Three-dimensional endothelial-tumor epithelial cell interactions in human cervical cancers. *In Vitro Cell. Dev. Biol.: Anim.* **33**, 432–442.

Cima, L. G., Vacanti, J. P., Vacanti, C., Ingber, D., Mooney, D., and Langer, R. (1991). Tissue engineering by cell transplantation using degradable polymer substrates. *J. Biomech. Eng.* **113**, 143–151.

Claassen, D. E., and Spooner, B. S. (1994). Impact of altered gravity on aspects of cell biology. *Int. Rev. Cytol.* **156**, 301–373.

Cogoli, A., and Cogoli-Greuter, M. (1998). Activation and proliferation of lymphocytes and other mammalian cells in microgravity. *Adv. Space Biol. Med.* **61**, 33–79.

Cogoli-Greuter, M., Meloni, M. A., Sciola, L., Spano, A., Pippia, P., Monaco, G., and Cogoli, A. (1996). Movements and interactions of leukocytes in microgravity. *J. Biotechnol.* **47**, 279–287.

Cooper, D., and Pellis, N. R. (1998). Suppressed PHA activation of T lymphocytes in simulated microgravity is restored by direct activation of protein kinase C. *J. Leukocyte Biol.* **63**, 550–562.

Dai, W., Belt, J., and Saltzman, W. M. (1994). Cell-binding peptides conjugated to poly(ethylene glycol) promote neural cell aggregation. *Bio/Technology* **12**, 797–801.

Davidson, J. M., Aquino, A. M., Woodward, S. C., and Wilfinger, W. W. (1999). Sustained microgravity reduces intrinsic wound healing and growth factor responses in the rat. *FASEB J.* **13**, 325–329.

Dintenfass, L. (1986). Execution of "ARC" experiment on space shuttle "Discovery" STS 51-C: Some results on aggregation of red blood cells under zero gravity. *Biorheology* **23**, 331–347.

Eiselt, P., Kim, B. S., Chacko, B., Isenberg, B., Peters, M. C., Greene, K. G., Roland, W. D., Loebsack, A. B., Burg, K. J., Culberson, C., Halberstadt, C. R., Holder, W. D., and Mooney, D. J. (1988). Development of technologies aiding large-tissue engineering. *Biotechnol. Prog.* **14**, 134–140.

Fang, A., Pierson, D. L., Mishra, S. K., Koenig, D. W., and Demain, A. L. (1997). Gramicidin S production by *bacillus brevis* in simulated microgravity. *Curr. Microbiol.* **34**, 199–204.

Fauza, D. O., Fishman, S. J., Mehegan, K., and Atala, A. (1998). Videofetoscopically assisted fetal tissue engineering: Skin replacement. *J. Pediatr. Surg.* **33**, 357–361.

Fitzgerald, J., and Hughes-Fulford, M. (1996). Gravitational loading of a simulated launch alters mRNA expression in osteoblasts. *Exp. Cell Res.* **228**, 168–171.

Freed, L. E., and Vunjak-Novakovic, G. (1997). Tissue culture bioreactors: Chondrogenesis as a model system. *In* "Principles of Tissue Engineering" (R. P. Lanza, R. Langer, and W. L. Chick, eds.), pp. 151–165. Academic Press, Austin, TX.

Freed, L. E., Langer, R., Martin, I., Pellis, N. R., and Vunjak-Novakovic, G. (1997). Tissue engineering of cartilage in space. *Proc. Nat. Acad. Sci. U.S.A.* **94**, 13885–13890.

Galvan, D. L., Unsworth, B. R., Goodwin, T. J., Liu, J., and Lelkes, P. I. (1995). Microgravity enhances tissue-specific neuroendocrine differentiation in cocultures of rat adrenal medullary parenchymal and endothelial cells. *In Vitro Cell. Dev. Biol.: Anim.* **31**, 10A.

Goodwin, T. J., Schroeder, W. F., Wolf, D. A., and Moyer, M. P. (1993). Rotating-wall vessel coculture of small intestine as a prelude to tissue modeling: Aspects of simulated microgravity. *Proc. Soc. Exp. Biol. Med.* **202**, 181–192.

Goodwin, T. J., Prewett, T. L., Spaulding, G. F., and Becker, J. L. (1997). Three-dimensional culture of a mixed müllerian tumor of the ovary: Expression of *in vivo* characteristics. *In Vitro Cell. Dev. Biol.: Anim.* **33**, 366–374.

Griffith, L. G., Wu, B., Cima, M. J., Powers, M. J., Chaignaud, B., and Vacanti, J. P. (1997). *In vitro* organogenesis of liver tissue. *Ann. N.Y. Acad. Sci.* **831**, 382–397.

Hammond, T. G., Lewis, F. C., Goodwin, T. J., Linnehan, R. M., Wolf, D. A., Hire, K. P., Campbell, W. C., Benes, E., O'Reilly, K. C., Globus, R. K., and Kaysen, J. H. (1999). Gene expression in space. *Nat. Med.* **5**, 359.

Horneck, G., Rettberg, P., Baumstark-Khan, C., Rink, H., Koszubek, S., Schäfer, M., and Schmitz, C. (1996). DNA repair in microgravity: Studies on bacteria and mammalian cells. *J. Biotechnol.* **47**, 99–112.

Horneck, G., Rettberg, P., Kozubek, S., Baumstark-Khan, C., Rink, H., Schäfer, M., and Schmitz, C. (1997). The influence of microgravity on repair of radiation-induced DNA damage in bacteria and human fibroblasts. *Radiat. Res.* **147**, 376–384.

Hughes-Fulford, M., and Lewis, M. L. (1996). Effects of microgravity on osteoblast growth activation. *Exp. Cell Res.* **224**, 103–109.

Ingber, D. E. (1997). Tensegrity: The architectural basis of cellular mechanotransduction. *Ann. Rev. Physiol.* **59**, 575–599.

Ingram, M., Techy, G. B., Saroufeem, R., Yazan, O., Narayan, K. S., Goodwin, T. J., and Spaulding, G. F. (1997). Three-dimensional growth patterns of various human tumor cell lines in simulated microgravity of a NASA bioreactor. *In Vitro Cell. Dev. Biol.: Anim.* **33**, 459–466.

Jessup, J. M., Goodwin, T. J., and Spaulding, G. (1993). Prospects for use of microgravity-based bioreactors to study three-dimensional host–tumor interactions in human neoplasia. *J. Cell. Biochem.* **51**, 290–300.

Jessup, J. M., Goodwin, T. J., Garcia, R., and Pellis, N. (1996). STS-70: First flight of EDU-1. *In Vitro Cell. Dev. Biol.: Anim.* **32**, 13A.

Jessup, J. M., Brown, D., Fitzgerald, W., Ford, R. D., Nachman, A., Goodwin, T. J., and Spaulding, G. (1997). Induction of cardinoembryonic antigen expression in a three-dimensional culture system. *In Vitro Cell. Dev. Biol.: Anim.* **33**, 352–357.

Kaysen, J. H., Campbell, W. C., Majewski, R. R., Goda, F. O., Navar, G. L., Lewis, F. C., Goodwin, T. J., and Hammond, T. G. (1999). Select *de novo* gene and protein expression during renal epithelial cell culture in rotating wall vessels is shear stress dependent. *J. Membr. Biol.* **168**, 77–89.

Kerbel, R. S., Kobayashi, H., and Graham, C. H. (1994). Intrinsic or acquired drug resistance and metastasis: Are they linked phenotypes? *J. Cell. Biochem.* **56**, 37–47.

Khaoustov, V. I., Darlington, G. J., Soriano, H. E., Krishnan, B., Risin, D., Pellis, N., and Yoffe, B. (1995). Establishment of three dimensional primary hepatocyte cultures in microgravity environment. *Hepatology* **22**, 231A.

Kim, B. S., and Mooney, D. J. (1998). Development of biocompatible synthetic extracellular matrices for tissue engineering. *Trends Biotechnol.* **16**, 224–230.

Krewson, C. E., and Saltzman, W. M. (1996). Nerve growth factor delivery and cell aggregation enhance choline acetyltransferase activity after neural transplantation. *Tissue Eng.* **2**, 183–196.

Langer, R., and Vacanti, J. P. (1993). Tissue engineering. *Science* **260**, 920–926.

Langer, R., and Vacanti, J. P. (1995). Artificial organs. *Sci. Am.* **273**, 130–133.

Lelkes, P. I., and Unsworth, B. R. (1998). Cellular signaling mechanisms involved in the 3-dimensional assembly and differentiation of PC12 pheochromocytoma cells under simulated microgravity in NASA Rotating Wall Vessel Bioreactors. *In* "Advances in Heat and Mass Transfer in Biotechnology" (S. Clegg, ed.), pp. 35–41. Am. Soc. Mech. Eng., New York.

Lelkes, P. I., Ramos, E. M., Chick, D. M., Liu, J., and Unsworth, B. R. (1994). Microgravity decreases tyrosine hydroxylase expression in rat adrenals. *FASEB J.* **8**, 1177–1182.

Lelkes, P. I., Galvan, D. L., Hayman, G. T., Goodwin, T. J., Chatman, D. Y., Cherian, S., Garcia, R. M. G., and Unsworth, B. R. (1998). Simulated microgravity conditions enhance differentiation of cultured PC12 cells towards the neuroendocrine phenotype. *In Vitro Cell. Dev. Biol.: Anim.* **34**, 316–324.

Long, J. P., Pierson, S., and Hughes, J. H. (1998). Rhinovirus replication in HeLa cells cultured under conditions of simulated microgravity. *Aviat. Space Environ. Med.* **69**, 851–856.

Lorenzi, G., Gmunder, F. K., Nordau, C. G., and Cogoli, A. (1995). Cultivation of hamster kidney cells in dynamic cell culture system in space. *In* "BIORACK on Spacelab IML-1" (C. Mattok, ed.), pp. 105–112. ESA Publ. Div., Noordwijk.

Margolis, L. G., Fitzgerald, W., Glushakova, S., Hatfill, S., Amichay, N., Baibakov, B., and Zimmerberg, J. (1997). Lymphocyte trafficking and HIV infection of human lymphoid tissue in a rotating wall vessel bioreactor. *AIDS Res. Hum. Retroviruses* **13**, 1411–1420.

Mayer, J. E., Jr., Shin'oka, T., and Shum-Tim, D. (1997). Tissue engineering of cardiovascular structures. *Curr. Opin. Cardiol.* **12**, 528–532.

Montgomery, P. O., Jr., Cook, J. E., Reynolds, R. C., Paul, J. S., Hayflick, L., Stock, D., Schulz, W. W., Kimsey, S., Thirolf, R. G., Rogers, T., and Campbell, D. (1978). The response of single human cells to zero gravity. *In Vitro Cell. Dev. Biol.* **14**, 165–173.

Moreira, J. L., Santana, P. C., Feliciano, A. S., Cruz, P. E., Racher, A. J., Griffiths, J. B., and Carrondo, M. J. (1995). Effect of viscosity upon hydrodynamically controlled natural aggregates of animal cells grown in stirred vessels. *Biotechnol. Prog.* **11**, 575–583.

Niklason, L. E., and Langer, R. S. (1997). Advances in tissue engineering of blood vessels and other tissues. *Transplant. Immunol.* **5**, 303–306.

Pellis, N. R., Goodwin, T. J., Risin, D., McIntyre, B. W., Pizzini, R. P., Cooper, D., Baker, T. L., and Spaulding, G. F. (1997). Changes in gravity inhibit lymphocyte locomotion through Type I collagen. *In Vitro Cell. Dev. Biol.: Anim.* **33**, 398–405.

Sagen, J., Wang, H., Tresco, P. A., and Aebischer, P. (1993). Transplants of immunologically isolated xenogeneic chromaffin cells provide a long-term source of pain-reducing neuroactive substances. *J. Neurosci.* **13**, 2415–2423.

Schmitt, D. A., Hatton, J. P., Emond, C., Chaput, D., Paris, H., Levade, T., Cazenave, J.-P., and Schaffar, L. (1996). The distribution of protein kinase C in human leukocytes is altered in microgravity. *FASEB J.* **10**, 1627–1634.

Schuster, U., Büttner, R., Hofstädter, F., and Knüchel, R. (1994). A heterologous *in vitro* coculture system to study interaction between human bladder cancer cells and fibroblasts. *J. Urol.* **151**, 1707–1711.

Schwachöfer, J. H. (1990). Multicellular tumor spheroids in radiotherapy research (review). *Anticancer Res.* **10**, 963–969.

Schwarz, R. P., Goodwin, T. J., and Wolf, D. A. (1992). Cell culture for three-dimensional modeling in rotating-wall vessels: An application of simulated microgravity. *J. Tissue Cult. Methods* **14**, 51–58.

Seitzer, U., Bodo, M., Muller, P. K., Acil, Y., and Batge, B. (1995). Microgravity and hypergravity effects on collagen biosynthesis of human dermal fibroblasts. *Cell Tissue Res.* **282**, 513–517.

Shinoka, T., Shum-Tim, D., Ma, P. X., Tanel, R. E., Langer, R., Vacanti, J. P., and Mayer, J. E., Jr. (1997). Tissue-engineered heart valve leaflets—Does cell origin affect outcome? *Circulation* **96**, 102–107.

Shinoka, T., Shum-Tim, D., Ma, P. X., Tanel, R. E., Isogai, N., Langer, R., Vacanti, J. P., and Mayer, J. E., Jr. (1998). Creation of viable pulmonary artery autografts through tissue engineering. *J. Thoracic Cardiov. Surg.* **115**, 536–545.

Sutherland, R. M., Sordat, B., Bamat, J., Gabbert, H., Bourrat, B., and Mueller-Klieser, W. (1986). Oxygenation and differentiation in multicellular spheroids of human colon carcinoma. *Cancer Res.* **46**, 5320–5329.

Tjandrawinata, R. R., Vincent, V. L., and Hughes-Fulford, M. (1997). Vibrational force alters mRNA expression in osteoblasts. *FASEB J.* **11**, 493–497.

Unsworth, B. R., and Lelkes, P. I. (1998a). Growing tissues in microgravity. *Nat. Med.* **4**, 901–907.

Unsworth, B. R., and Lelkes, P. I. (1998b). The use of rotating wall bioreactors for the assembly of differentiated tissue-like organoids. *In* "Advancements in Tissue Engineering" (L. Savage, ed.), pp. 113–132. International Business Communications, Southborough, MA.

Zhau, H. E., Goodwin, T. J., Chang, S.-M., Baker, T. L., and Chung, L. W. K. (1997). Establishment of a three-dimensional human prostate organoid coculture under microgravity-simulated conditions: Evaluation of androgen-induced growth and PSA expression. *In Vitro Cell. Dev. Biol.: Anim.* **33**, 375–380.

PART III: *In Vivo* SYNTHESIS OF TISSUES AND ORGANS

CHAPTER 15

In Vivo Synthesis of Tissues and Organs

Ioannis V. Yannas

SCALE OF FUNCTIONAL DEFICIT: MACROMOLECULE VERSUS ORGAN

Medical treatment in antiquity emphasized the use of herbs, potions, and, eventually, the semipurified chemical substances that were extracted from them. In the twentieth century, advances in the chemistry of natural substances led to the discovery, and later synthesis, of vitamins, hormones, and antibiotics. Inevitably, the ancient predisposition to believe in the healing power of drugs spawned the gigantic pharmaceuticals industry. The successes of chemical therapeutics have been based on the fact that a large number of medical problems are caused by some kind of defect at the macromolecular level—usually an excess or a deficiency of a protein (e.g., insulin), sometimes resulting from an error in a gene—that can be chemically treated.

Failure of an entire organ is a very different medical problem. Rather than originating in a defect at the nanoscale, loss of organ function occurs at the scale of the centimeter. Such failure can occur as a result of a large variety of causes, such as heart failure due to oxygen deficiency of heart muscle (itself resulting frequently from circulatory failure), kidney failure due to cancer, and failure of the central nervous system due to spinal cord injury. Drug treatment is not effective in organ failure because the cause of the problem is loss of an entire mass of functioning tissue, representing loss of a large number of cells, soluble proteins, and insoluble extracellular matrix, rather than being a singular defect in molecules.

Five approaches have been successfully used to treat loss of organ function: transplantation, autografting, implantation of passive prosthesis, *in vitro* synthesis, and *in vivo* synthesis (regeneration). Transplantation from a donor to another person (host) has led to certain spectacular but short-lived successes (Medawar, 1944), unless the host and donor were identical twins, because the host typically rejects the transplanted organ after a period of weeks to a few years. Studies have emphasized development of techniques for increasing host tolerance (Cooper *et al.,* 1997). Autografting circumvents this problem by ensuring that donor and host are the same person, as in patients with serious burns (Burke *et al.,* 1974) or with seriously injured nerves (Millesi, 1967). This procedure has suffered from the trauma that is inflicted on the patient during harvesting of the autograft; also, the procedure falls short when the loss of organ mass exceeds typically 50% of the total. Implantation of passive prostheses provides temporary function, extending to 15 years in some cases; in the long run, the biologic environment destroys, and is itself damaged by, the manmade materials used in these prostheses. Examples are the total hip prosthesis (Kohn and Ducheyne, 1992) and the cardiac pacemaker (Neuman, 1998). *In vitro* synthesis of tissues and organs has been attempted over the past 15 years and the progress that has been made is reviewed in detail in several chapters in this volume. *In vivo* synthesis leads to organ regeneration and is the focus of this chapter. This process was originally developed in the context of treatment of massive skin wounds in animals and humans; however, over the past 15 years it has led to synthesis not only of skin but also of peripheral nerves and the knee meniscus.

The description of *in vivo* organ synthesis will start with an examination of the basic parameters of the living environment (wound bed, animal) in which the desired synthesis takes place. Because it is subject to homeostatic control, the living environment offers unexpectedly controlled conditions for the synthesis.

BASIC PARAMETERS OF THE LIVING ENVIRONMENT DURING *IN VIVO* ORGAN SYNTHESIS

The most successful procedure used to achieve *in vivo* synthesis requires use of an analog of the extracellular matrix (ECM). Examples are grafting of the analog onto a freshly excised skin wound bed or bridging the gap between the tubulated stumps of a freshly transected peripheral nerve with an ECM analog. In each of these two cases the wound bed is typically supplied with a constant flow of exudate that contains cells and cytokines or growth factors. Following a process of induced regeneration, the exudate is converted into physiologic tissue by the process of regeneration; if such conversion does not occur in adult mammals, the exudate will be spontaneously converted into nonphysiologic scar or fibrotic tissue by the process of repair. Unlike in the fetus, the vast majority of tissues and organs in adults are not spontaneously regenerated following removal, e.g., by surgical excision. If the ECM analog has been appropriately selected, i.e., if it is a regeneration template, normal repair processes are blocked and a regenerate (newly synthesized physiologic organ) is synthesized in the wound bed within weeks. The ECM analog may not, by itself, suffice to induce regeneration of physiologic tissue: cells of a given type may have to be seeded into it before grafting. Rigid rules are used to study the structure and function of the new organ and the term "regenerate" is usually reserved for an almost completely faithful replica of the physiologic organ.

The process of *in vivo* synthesis is modeled as if it were taking place inside a bioreactor (wound bed) surrounded by a reservoir (animal) (Fig. 15.1). Conditions inside the animal are subject to homeostatic control; this environment offers approximate constancy in many characteristics of the exudate that flows into the wound bed. Although the flow rate of exudate as well as its composition vary somewhat from day to day during the wound healing process, the temperature and the pH are maintained constant. The exudate and the template (possibly seeded with cells in advance) are the two critically necessary components in the synthetic reaction. The stability of the living environment that supplies this basic reaction component effectively cancels the conventional need of the experimenter to isolate the bioreactor from the reservoir.

The bioreactor feed consists of two components. The flow of exudate from surrounding tissues into the wound bed starts seconds after formation of the wound bed by excision and constitutes the endogenous feed. The ECM analog, occasionally seeded with cells, is supplied immediately after excision as well and constitutes the exogenous feed. The reservoir maintains the bioreactor contents at constant temperature and pH and, in addition, supplies it with a continu-

Fig. 15.1. In vivo synthesis of tissues and organs is modeled as if it were taking place inside a bioreactor (wound bed) that is surrounded by a reservoir (animal). Exudate flows continuously from the reservoir to the bioreactor. The exudate contains cells, cytokines, and several other soluble substances, but not insoluble matrix components. Conditions inside the animal are subject to relatively strict homeostatic control. (Courtesy of M.I.T.)

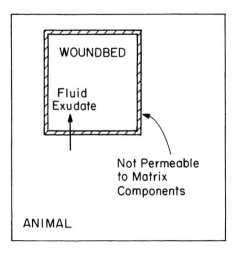

ous flow of exudate consisting of cells, cytokines, and other diffusible components. Exudate typically does not contain ECM components; matrix components that are required for the synthesis are supplied exogenously in the form of an ECM analog. The selection of structural features of the ECM analog supplied to the wound bed is critical and is discussed in the next section. If the appropriate ECM analog (template) is omitted entirely, or if the selection of analog and cell content is inappropriate, the reaction sequence inside the wound bed eventually yields repaired, not regenerated, tissue. The template is therefore required to block the normal repair process in the wound bed. Repaired tissue typically consists of epithelialized, undifferentiated connective tissue, often referred to as scar or fibrotic tissue. An organ consisting mostly of repaired tissue does not function physiologically. The process of *in vivo* synthesis of a physiologic organ is also referred to as an instance of *in situ* organ synthesis, a term used interchangeably with *in vivo* synthesis.

Experimental study of *in vivo* synthesis is routinely complicated by lack of reproducibility from one reaction site to another, nominally identical, anatomic site. The site or tissue substrate on which the reaction is taking place may vary in chemical composition, surface area, or other parameters that affect the kinetics and mechanism of the desired reaction. In the absence of such site-to-site reproducibility it is very difficult to develop results that are statistically significant without performing an inordinately large number of experiments. *In vivo* organ synthesis is potentially beset by the same set of problems, originating either in lack of reproducibility of conditions from one wound bed to another in the same animal or from a wound bed in one animal to that in another. The problem of wound-to-wound variability was dramatically minimized, almost practically eliminated, in the pioneering experiments of Billingham and Medawar (1951, 1955), who developed the concept of the anatomically constant wound bed. In their rodent models, skin was routinely excised down to the layer of muscle that lies under the dermis. With the exception of end effects, the contents of this wound bed, including the exudate that flows into it, are nearly identical from one site to the next. The precise choice of dimensions of an anatomically constant bed varies with the goal of the investigator. In experiments in which the synthesis of peripheral nerve in rats has been studied, an anatomically constant wound bed has consisted of a gap, approximately 10 mm long, along the cylindrical geometry of the nerve fiber; the ECM analog, in the form of a matrix filler inside a cylinder, has been grafted as a bridge for this gap. Studies of regeneration in the knee meniscus have also been based on a careful choice of geometry. Criteria for the selection of an anatomically well-defined wound bed have been published (Yannas, 1997).

The sequence of chemical reactions and biological interactions inside the wound bed is of great complexity. Precise knowledge of such a mechanistic sequence is necessary in order to design a regeneration template without recourse to extensive experimentation. In the absence of such information, the experimenter relies on empirical data relating the structure of the experimental device to the structure or function of the tissue or organ that is under study. An example is measurement of the moisture evaporation rate from a full-thickness skin wound bed in the guinea pig wound model that has been treated with a dermis regeneration template (DRT) seeded with keratinocytes. Following a 14-day period, which is necessary to achieve epidermal confluence over the newly synthesized dermal substrate, the new epidermis becomes keratinized and forms an effective barrier that impedes moisture evaporation from the body. Another example is the continuous measurement of the velocity of an electrical signal that is conducted by a regenerating nerve, previously treated with a nerve regeneration template (NRT). This measurement yields a virtually continuous record of the regeneration process and yields the time required for achievement of nearly physiologic values. Neither of the last two examples provides by itself definitive proof that a new, physiologically functioning organ has been synthesized; additional properties of the new organ need to be determined before its synthesis can be established. However, the noninvasive measurements of function cited above allow the investigator to monitor the progress of the synthesis from beginning to end using the same animal. The less attractive alternative is to use invasive measurements, e.g., histologic data, which provide information on only one time point per animal. Irrespective of the method of measurement, data on the degree of completion of a given property are usefully converted to a physiologic index, a measure of the closeness of a given structural or functional property of the regenerate to the intact adjacent organ.

Mammals typically possess a very small but finite potential for spontaneous regeneration of most of their organs. For example, a cylindrical defect measuring considerably less than about 1 cm in diameter in the long bone of many mammals is spontaneously refilled with apparently physiologic bone (Shapiro, 1988); a gap less than about 5 mm in the transected rat sciatic nerve, previously treated by inserting the two nerve stumps into a silicone tube, is spontaneously synthesized as a fairly successful regenerate (Lundborg, 1987; Lundborg *et al.,* 1982a–d; Dahlin *et al.,* 1995). The minimum mass of dermis that can be spontaneously regenerated in adults appears to be zero (Billingham and Medawar, 1951, 1955; Madden, 1972; Peacock, 1984; Boykin and Molnar, 1992).

Organ mass that has been synthesized using the techniques described in this chapter must be corrected for organ mass that can be synthesized spontaneously at the same anatomically constant wound bed. This correction for "background" regeneration (or spontaneous regeneration) can be determined simply as the amount of organ mass synthesized in the absence of exogenous feed. Tissue or organ synthesis that has resulted from use of an exogenous feed is referred to as induced regeneration. It is clear that the net effect of the exogenous feed on a given physiologic index of the new organ cannot be quantitatively appreciated until the index has been carefully corrected for the contribution of background regeneration.

FUNDAMENTAL DESIGN PRINCIPLES FOR TISSUE AND ORGAN REGENERATION TEMPLATES

By definition, the template essentially interacts with components of the exudate (cells and cytokines) sufficiently to modify drastically the kinetics and mechanism of the spontaneous process that normally converts exudate to scar tissue. Such an interaction requires the template surface to be adequately close and accessible to cells migrating from the exudate; migrating cells that have approached close enough require the template surface to be populated with the appropriate type and density of cell binding sites; such interaction must be allowed to proceed over the necessary time period. Finally, when the interaction has successfully modified the kinetics and mechanism of wound healing away from repair, it is required that the template remove itself even as new tissue is being synthesized adjacent to the surface of the template. Each of these steps in the function of a template must be executed, and each, in turn, must be translated into a design requirement for the template. The structural features that lend function to a regeneration template will be described separately below: features that constitute the porous network structure (scale 1–100 μm) will be discussed first, followed by features that describe the macromolecular network (scale 1–100 nm).

CRITICAL CELL PATH LENGTH: MAXIMUM DIMENSION OF TEMPLATE

Host cells migrating into the template require adequate nutrition during the entire time of residence in it. For modeling purposes, the complexity of nutritional requirements of the cell are simplified by defining a critical nutrient that is required for normal cell function; such a nutrient is assumed to be metabolized by the cell at a rate R in moles/cm^3/sec. The nutrient is pictured being transported from the wound bed, where the concentration of nutrient is assumed to be a constant, C_0, over a distance L through the exudate until it reaches the cell (Fig. 15.2). In the early days following implantation of the template there is no angiogenesis and the nutrient is, therefore, transported exclusively by diffusion, which is characterized by a diffusivity D in cm^2/sec. Dimensional analysis readily yields the cell lifeline number

$$S = RL^2/DC_0, \tag{1}$$

which characterizes the relative magnitude of the rate of nutrient supply to the cell by diffusion and the rate of nutrient consumption by the cell. If the rate of consumption of the critical nutrient exceeds greatly the rate of supply, $S \gg 1$, and the cell must soon die. At steady state the rate of consumption of nutrient by the cell just equals the rate of transport by diffusion over the distance L. Under conditions of steady state, S is of order 1, and the value of L becomes the critical cell path length, L_c, the longest distance away from the wound bed boundary along which the cell can migrate without requiring nutrient in excess of that supplied by diffusion. For many cell nutrients of low molecular weight L_c is of order 100 μm, a distance short enough to suggest the need for very close proximity between wound bed and implant (Yannas and Burke, 1980). The requirement for proximity is critical and suggests that a graft may not "take" unless the template has met this demanding condition of physicochemical contact.

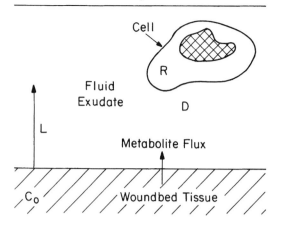

Fig. 15.2. *Model of the interface between host tissue and an implant (or the atmosphere, in the case of an unimplanted wound) adjacent to its surface. A cell from the host interface has migrated into the exudate, which flows out of host tissue. Will the cell receive sufficient nutrient by diffusion from host tissue to migrate all the way to the implant (or the atmosphere)? A critical cell nutrient is transported from the host interface, where the nutrient concentration is assumed to be equal to C_0, over a distance L, through the exudate until it reaches the migrating cell, where it is metabolized at a rate R. Angiogenesis has not yet occurred; therefore, nutrient transport takes place strictly by diffusion, characterized by a diffusivity D. Dimensional analysis allows calculation of the critical cell path length, L_c, an estimate of the maximum permissible distance of separation between the host tissue surface and the implant; it is of order 100 μm. The value of L_c is also an estimate of the minimal thickness of exudate covering an unimplanted wound bed that is required to accommodate migrating cells through its thickness. (Courtesy of M.I.T.)*

Upper and Lower Bounds of Template Pore Diameter

Having successfully migrated onto the template surface a host cell is visualized interacting with an appropriate site on the matrix and binding onto it. Cells commonly make use of cell surface receptors, e.g., the family of integrin receptors, to bind on a matrix surface; typically, each receptor binds on a specific binding site on the matrix surface. A highly detailed model of a particular interaction between a cell type and a matrix surface must specify the nature of the receptor and the binding site, as well as the number of receptors utilized per bound cell. The description of cell–matrix binding given below assumes no such specific knowledge.

We are interested in estimating the structural requirements for a porous solid that binds all or almost all of the cells of a given type in contact with it. We want to estimate the relation between the average pore diameter of a matrix and its capacity for binding cells. The surface density of bound cells can be expressed as Φ_c, equal to the number of bound cells N_c per unit surface A of the porous template. Φ_c is also equal to the number of bound cells per unit volume of porous template, ρ_c, divided by the specific surface of the template, σ. The latter is equal to the surface per unit volume of template and is expressed in units of mm^2/cm^3 template:

$$\Phi_c = N_c/A = \rho_c/\sigma. \tag{2a}$$

If additional information becomes available on the nature of the binding interaction, the model can become more detailed. If χ receptor sites are utilized per bound cell, and if each receptor is bound to one binding site, the number of binding sites per unit surface will be $N_c\chi/A$. Then the volume density of binding sites, ρ_{bs}, will be equal to $\chi\rho_c$ and the surface density of binding sites will be equal to

$$\Phi_{bs} = \chi\Phi_c = \chi N_c/A = \chi\rho_c/\sigma = \rho_{bs}/\sigma. \tag{2b}$$

We will now compare the cell surface density of two matrices that have identical chemical composition but quite different average pore diameter. It is known that when other structural features, such as the pore volume fraction, remain constant, the specific surface increases with a decrease in average pore diameter. The average pore diameter as well as the pore volume fraction of a template can be controlled by careful control of the conditions of freezing prior to drying by sublimation (freeze drying or lyophilization) (Dagalakis *et al.*, 1980). In the example of a matrix that has been experimentally studied for possible template function, the average pore diameter is 10 μm and the volume density, ρ_c, of fibroblasts that are bound on the matrix has been observed by light microscopy to be of order 10^7 cells/cm^3 porous matrix. The number of receptors that are utilized per bound cell is not known. For this matrix, the specific surface σ is calculated, using a simple model that relates the shape and size of pores to their surface, to be about 8×10^4 mm^2/cm^3 matrix volume. According to Eq. (2a), the surface density of bound cells is equal to $\Phi_c = 10^7/8 \times 10^4 = 125$ cells/mm^2 for this matrix. Next we consider a matrix that has an average pore diameter as large as 300 μm. The specific surface in this case is estimated to be only about 3×10^3 mm^2/cm^3. On the other hand, because both matrices have the same chemical composition, they also have the same type and density of binding sites, and it follows that the surface density of bound cells, Φ_c, is the same in both matrices. However, the volume density of bound cells is only $\rho_c = \Phi_c \sigma$, or 125 $(3 \times 10^3) = 3.75 \times 10^5$/cm^3 matrix. The conclusion is that the volume density of bound cells in the matrix with the 300-μm pore diameter is approximately 27 times lower than in the matrix with the 10-μm pore diameter. The calculation suggests the importance of the specific surface of the porous template; if the surface is insufficiently large, an insufficient number of cells in the wound bed will be bound on the matrix. These considerations suggest a maximum pore diameter requirement for the template, solely to meet the binding needs of the cells interacting with it. In the absence of further information, the absolute magnitude of this upper bound for the pore diameter is determined by experiment.

The discussion above makes it obvious that there is also need for a minimum pore diameter, about equal to the characteristic diameter of the cells, of order 10 μm, which must migrate inside the template volume prior to binding on specific sites on its surface. Thus, the specific surface of a regeneration template, and consequently its average pore diameter, are limited both by an upper and a lower bound, as observed experimentally (Yannas, 1998; Yannas *et al.,* 1989).

Template Residence Time

The structural parameters discussed in the preceding section were required to make the cell binding sites on the template accessible to the cells in the exudate. Once such a rendezvous has been accomplished successfully, the next steps in the function of the template require the exudate–matrix interaction to be specific enough to yield new physiologic tissue and eventually the entire organ. While organ synthesis is being pursued at the wound bed the template must leave the site.

A template must be in place long enough for the specific interaction to take place but it must eventually disappear so as not to interfere with the synthetic process. These considerations are consistent with a template residence time, which has both a lower and an upper bound. The appropriate time period is approximately equal to that required for synthesizing relatively mature tissue at the specific organ site by regeneration. The latter is *a priori* unknown; it will be taken to be of the same order as that required to complete the conventional wound healing process at that anatomical site. At the expiration of this period the template is required to remove itself. Because the template is a cross-linked network it cannot diffuse away from the reaction site. The simplest design that can accomplish this disappearing act is one in which the macromolecular network undergoes degradation by enzymes of the wound bed to harmless low-molecular-weight fragments that diffuse away from the site of organ synthesis. These assumptions lead to the hypothesis of isomorphous tissue replacement (Yannas and Burke, 1980; Yannas, 1998):

$$t_b / t_h = O(1). \tag{3}$$

In Eq. (3) t_b denotes a characteristic time constant for degradation of the template at the tissue site where a new organ is synthesized with a time constant of t_h. The ratio of the two constants is on the order of one, as indicated by Eq. (3).

This hypothesis has received some experimental support from two critical observations. When the ratio in Eq. (3) was much smaller than one, indicative of a very rapidly degrading template, the wound healing process resulted in synthesis of scar, as would have been the case if the template was not there. It was also observed that when the ratio in Eq. (3) was much larger than one, the template remained in the wound bed as an intractable implant and wound closure by epithelialization did not occur (Yannas and Burke, 1980).

Adjustment of the degradation rate of the ECM analog to satisfy Eq. (3) requires adjustment of the chemical composition of the ECM analog. Assays have shown that the resistance of the network to degradation by wound bed enzymes, such as the collagenases, increases with its cross-link density (Yannas, 1981; Yannas *et al.,* 1975b). The cross-linking reaction is simply based on a process in which the water in the ECM analog is removed below ~1 wt%; the result of drastic dehydration is formation of interchain amide bonds by condensation (Yannas and Tobolsky, 1967). The advantage of this process is that no cross-linking agent is used. Additional cross-linking can be introduced by reaction with glutaraldehyde (Yannas *et al.,* 1980); however, the cross-linked matrix must subsequently be rinsed exhaustively in order to remove traces of the toxic cross-linking agent (Yannas *et al.,* 1989). Still a third procedure for increasing the resistance to degradation involves incorporation of a glycosaminoglycan (GAG) in the ECM analog (Yannas *et al.,* 1975a).

Use of three independent methods of adjusting the degradation rate provides substantial design flexibility. In this fashion the enzymatic degradation rate can be varied over a wide range, providing implants that, without losing their biological activity during preparation, can be designed effectively to disappear from the new organ site between a few and several days (skin) or between a few and several weeks (peripheral nerve). The time constant t_h varies greatly from one organ site to the next in the same animal species; if an ECM analog of the same composition is to be studied in a different site, the network structure must be adjusted accordingly (Yannas *et al.,* 1975a,b, 1989). There is also species-to-species variation: the half-life of the dermis regeneration template in the full-thickness (full-depth) skin wound in the swine model is 15.5 days but only 7 days in a similar type of wound in the guinea pig model (Orgill *et al.,* 1996).

CHEMICAL COMPOSITION

The chemical composition of the ECM analogs that have induced regeneration is relatively simple. The analogs used successfully so far have been graft copolymers of type I collagen and a sulfated glycosaminoglycan. The GAG component has a dual role: on one hand, it increases the resistance to degradation (Yannas *et al.,* 1975a); on the other, specific GAG components, together with type I collagen, appear to possess specific biological activity (Shafritz *et al.,* 1994). The ratio collagen/GAG is adjusted to meet these two requirements and a covalent reaction is employed to graft the GAG chains onto the chains of the collagen substrate (Yannas *et al.,* 1980).

The mechanistic steps by which the chemical identity of collagen/GAG graft copolymers promotes regeneration are still under study; however, there is sufficient evidence to suggest a nearly complete hypothesis that is based on the observed association between inhibition of wound contraction and regeneration (Yannas, 1998).

It has long been known that the mammalian fetus is capable of regeneration whereas the adult is not (Mast *et al.,* 1992a,b). A rule of thumb suggests that the third trimester of mammalian gestation is the turning point from wound closure primarily by synthesis of physiological tissues and organs (regeneration) to closure by wound contraction and scar synthesis (repair). In amphibians, it has been known that the turning point is metamorphosis from tadpole to young adult (frog) (Wallace, 1981). Basic information on changes in wound healing behavior with development has been, however, less available in mammals, primarily due to experimental difficulties in studying wound healing in mammalian fetal models. The question was tackled in a study of the amphibian frog (*Rana catesbeiana*), in which wound closure was studied quantitatively during development from the larva (tadpole) to the young adult (frog). It was observed that, during larval development, wound contraction gradually became more important and regeneration became increasingly insignificant as a mechanism for wound closure. Following metamorphosis to the young adult, skin wounds closed by a combination of contraction and synthesis of scar tissue whereas regeneration did not contribute to wound closure (Yannas *et al.,* 1996). These observations supported the presence of an antagonistic relation between wound contraction and regeneration.

Other studies have suggested that effective inhibition of wound contraction is necessary for dermal regeneration in full-thickness skin wounds in the guinea pig (Yannas, 1981; Yannas *et al.,* 1982, 1989). In these studies it was shown that contraction was inhibited strongly by grafting the wound with an ECM analog with very specific structure (eventually named the dermal regeneration template because it was observed that it induced regeneration of a nearly physiological dermis); even slight deviations from the structure of the DRT led to strong contraction. Such data cannot be explained by hypothesizing that the DRT acted as a mechanical splint that physically hindered contraction; several other ECM analogs, differing slightly in structure and with mechanical stiffness that was either somewhat higher or lower than that of the DRT, were incapable of inhibiting contraction in the same animal model. Rather, the observations with ECM analogs support further the hypothesis that wound contraction and spontaneous regeneration are mutually exclusive phenomena (Yannas, 1998).

Wound contraction is mediated partly by migratory fibroblasts and partly by contractile fibroblasts (CFBs) (Rudolph *et al.,* 1992). An early hypothesis has suggested that the DRT inhibits contraction by disrupting the large-scale organization of the CFB population, which is involved in formation of a three-dimensional network, capable of supporting contractile mechanical stresses across the macroscopic wound bed (Yannas, 1988; Murphy *et al.,* 1990). But how does the template interfere with formation of this hypothetical network at the scale of the cell? This question has been answered in terms of the specific cell–matrix interactions that are entered into between the CFB and the binding sites on the DRT. It is known that fibroblasts bind to naturally synthesized ECM by formation of highly specialized transmembrane links, involving integrins, which are referred to as fibronexi (Hynes, 1990). Accordingly, it has been suggested that the strong contraction-inhibiting activity of the DRT consists in establishing early connections with contractile cells of the wound bed, thereby preventing eventual formation of cell–ECM connections (fibronexi). By doing so, the DRT inhibits the function of the endogenous ECM, which is synthesized long after the DRT has been grafted, and thereby prevents the formation of an organized CFB network that is capable of contraction (Yannas, 1998). However, the identity of binding interactions between CFBs and the ECM analogs are still not well understood. Although the mechanistic hypothesis presented above has been based primarily on data from skin wounds, there is evidence that similar considerations apply in peripheral nerve wounds as well (Chamberlain *et al.,* 1998a).

EXAMPLES OF *IN VIVO* ORGAN SYNTHESIS

SKIN REGENERATION TEMPLATE

In vivo synthesis of a partly physiological skin organ using the dermis regeneration template has been demonstrated clinically with a population of massively burned patients (Burke *et al,.* 1981; Heimbach *et al.,* 1988) following extensive studies with animal models (Yannas *et al.,* 1975a,b, 1981, 1982). Patients with loss of a substantial fraction of total body surface area (TBSA) face an immediate threat to their survival, originating mainly in the loss of their epidermis. One result of this loss is an increase by an order of magnitude of the moisture evaporation rate from the body, which, if left uncorrected, leads to excessive dehydration and eventually to shock. Another result is a sharp increase in risk for massive bacterial infection, which, if allowed to progress, frequently resists treatment and leads to sepsis. Even when patients manage to survive these immediate threats there is a residual serious problem of quality of life, originating in the occurrence of crippling contractures and disfiguring scars, both of which are sequels of the normal repair process. Conventional treatment is based on use of autografting, which yields excellent results at the treatment site but which is burdened by the trauma caused at the donor site as well as by the unavailability of donor sites when the TBSA exceeds about 40–50%.

Of the two major tissue types that together comprise skin, the epidermis regenerates spontaneously provided there is a dermal substrate underneath. The dermis does not regenerate spontaneously in the adult. In clinical studies of the DRT, the latter has been used as a cell-free device and has induced regeneration of the dermis, thereby providing the essential substrate for spontaneous regeneration of the epidermis. A keratinocyte-seeded DRT that induces simultaneous formation of a dermis and an epidermis in the guinea pig model has been referred to as a skin regeneration template (SRT); however, the evidence has since suggested that the DRT does not require seeding with cells prior to grafting in order to induce regeneration of the dermis in the swine model, which is much more similar to the human than the guinea pig mode. Following dermis regeneration by the unseeded DRT, epidermal cells from the wound edges migrate into the center of the wound and form a mature epidermis (Orgill *et al.,* 1996). In summary, although the seeded DRT induces simultaneous regeneration of skin in the guinea pig model, the unseeded DRT induces sequential regeneration of skin in the swine model. The kinetics of simultaneous regeneration do not depend on the size of the wound whereas those of sequential regeneration do; accordingly, the cell-seeded DRT would appear to be more useful in a clinical setting, where speed of wound closure is of essence. However, the data suggest that *in vivo* synthesis of skin does not require anything more than the DRT, and the latter directly induces synthesis of the dermis only; this explains the use of the term DRT in place of SRT.

The DRT is used as a bilayer device, consisting of an inner layer of the active ECM analog and an outer layer of elastomeric poly(dimethyl siloxane). The layer of silicone is removed about 2 weeks after grafting the device, having served the important temporary role of controlling moisture flux and bacterial invasion at the site of organ synthesis (wound bed) as well as allowing handling, including cutting and suturing, by the surgeon. The structure of the DRT comprises a collagen:chondroitin 6-sulfate ratio of 98:2; lower and upper bounds of pore diameter of 20 ± 4 μm and 125 ± 35 μm, respectively; and lower and upper bounds of residence time in the wound bed of 5 and 15 days, respectively (Yannas *et al.,* 1981, 1982, 1989). Clinical use of the DRT has emphasized treatment of patients with massive burns as well as those who require resurfacing of large or small scars from burns. With wounds of relatively small characteristic dimension, e.g., 1 cm, epithelial cells migrating at speeds of order 0.5 mm/day from each wound edge can provide a confluent epidermis within 10 days. In such cases, the unseeded template fulfills all the design specifications set above for *in vivo* organ synthesis. However, the wounds incurred by a massively burned patient are typically of characteristic dimension of several centimeters, often more than 20–30 cm. These wounds are large enough to preclude formation of a new epidermis by cell migration alone within a clinically acceptable time frame, say 2 weeks. Current clinical protocol favors use of a very thin autoepidermal graft, thin enough not to leave behind a scarred donor site, to cover the neodermis (Burke *et al.,* 1981; Heimbach *et al.,* 1988; Spence, 1998).

NERVE REGENERATION TEMPLATE

Loss of peripheral nerve function typically involves the loss of innervation in the arms, legs, or face, leading to loss of motor and sensory function (paralysis). A common cause of loss of nerve function is accidental mechanical trauma or surgery. In a well-studied experimental model of injury, the nerve trunk is cut along the entire diameter (transection) and the two stumps are inserted in a tube, usually made of silicone elastomer, other synthetic polymers, or, more recently, collagen (Lundborg, 1987; Archibald *et al.,* 1995; Valentini and Aebischer, 1997; Chamberlain *et al.,* 1998b).

The length of the gap that separates the stumps in the tubulated rat model can be adjusted and is an important experimental variable that controls the probability of success in a regeneration process (Lundborg *et al.,* 1982a). It has been observed that spontaneous elongation of the conducting fibers (axons) along the gap is complete, leading to some recovery of function, when the gap length is not much larger than about 5 mm; in contrast, the probability of regeneration is zero when the gap length is as large as 15 mm. This information allows the experimenter to estimate the extent of spontaneous (background) regeneration in the tubulated gap model at about 5 mm; therefore, any recovery of function that occurs across a gap much longer than that can be considered to result from induced regeneration. In practice, a gap length of 15 mm in the rat sciatic nerve is somewhat too large to place in the femur of the rat and, for this reason, a 10-mm gap length has been used extensively and has yielded reliable data provided that the appropriate controls have been used to separate out spontaneous from induced regeneration.

A wound in the sciatic nerve is experimentally attractive because the loss of function is limited primarily to contraction of the plantar muscle in the foot; loss of function or recovery from loss can, accordingly, be monitored by measuring the velocity or amplitude of an electric signal that is experimentally transmitted along the sciatic nerve (Madison *et al.,* 1992). Unlike currently available procedures for monitoring the recovery of skin function following full-thickness loss, this electrophysiological procedure provides for convenient and minimally invasive monitoring of the kinetics of recovery over almost the entire period of healing. Use of electrophysiological procedures has shown that recovery of peripheral nerve function following transection of the rat sciatic nerve extended to beyond 2 years. This lengthy period appears necessary for the slow turnover (remodeling) processes inside the nerve trunk that lead to increase in diameter of conducting fibers (axons) and result in slow but critical increases in conduction velocity (Chamberlain *et al.,* 1998b). On the other hand, the characteristic time constant for the short-term healing events at the site of the transection that lead to wound closure appear to extend to about 6–10 weeks, clearly longer than the time constant for healing in skin, which is about 1.5 weeks (Table 15.1).

The nerve regeneration template (NRT) has typically been studied in a wound bed formed by excising either a 10-mm or a 15-mm gap in the sciatic nerve of the rat. Early versions of the NRT inserted inside a silicone tube led to the first evidence of substantial regeneration across a gap of 15 mm in the rat sciatic nerve (Yannas *et al.,* 1987). The device incorporating the NRT is a collagen tube filled with the porous ECM analog, which has chemical composition similar to that of the DRT. The similarity between the templates ends there. The average pore diameter of the NRT is 5 μm, an order of magnitude smaller than the middle of the range 20–120 μm for pore diameter in the DRT. The residence time is on the order of 6–8 weeks, much shorter than

Table 15.1. Structural properties of two regeneration templates

Design parameter of ECM analog	Regeneration template	
	Dermis	Nerve
Residence time, weeks	1.5	6–8
Average pore diameter, μm	20–120	5
Pore channel orientation	Random	Axial

that for the DRT (about 15 days). Finally, the pore channels of the NRT are highly aligned along the tube axis rather than being randomly oriented, as with the DRT (Chang and Yannas, 1992). The differences between the two templates are highlighted in Table 15.1.

Nerves regenerated in animals with use of the NRT have shown recovery of the velocity of electric signal transmission to an extent of 70% of physiological value. However, recovery of signal amplitude has reached only between 20 and 40% of the intact tissue (Chamberlain *et al.,* 1998b). Neurological studies of performance of regenerated nerves are not complete and it is not yet possible to deduce the extent of recovery of physiological walking behavior (gait).

REGENERATION OF THE KNEE MENISCUS

Joint stabilization and lubrication, as well as shock absorption, are the main functions of the knee meniscus. Its structure is that of fibrocartilage, distinct from that of articular cartilage, consisting primarily of type 1 collagen fibers populated with meniscal fibrochondrocytes. As in articular cartilage, the architecture of collagen fibers is complex. The shape of the meniscus suggests one-half a doughnut, in which the collagen fibers are arranged in a circumferential pattern that is additionally reinforced by radially placed fibers. Meniscus tearing occurs often during use, causing pain and dysfunction. The accepted treatment is partial or complete excision of torn tissue; however, the result is often unsatisfactory, because the treated meniscus has an altered shape that is incompatible with normal joint motion and stability. In the long term such incompatibility leads to joint degeneration, eventually leading to osteoarthritis, a very painful condition.

Regeneration of the surgically removed meniscus was induced following surgical deletion (excision) of the torn segment, amounting to about 80% of total mass, and implantation of a device that was based on a graft copolymer of type I collagen–chondroitin 6-sulfate copolymer (Stone *et al.,* 1989, 1990). The precise structural parameters of this matrix have not been reported in the literature, so it is not possible to discuss the results of this study in terms of possible similarities and differences with the DRT and NRT. The experimental study was based on the canine knee joint, which is particularly sensitive to alterations of structure that lead to biomechanical dysfunction. In this model, joint instabilities rapidly deteriorate to osteoarthritic changes, which can be detected experimentally. The canine meniscus regenerates poorly following excision, resulting in a biomechanically inadequate tissue that does not protect the joint from osteoarthritic changes. In one animal study, knee joints were subjected to resection of the medical meniscus and the lesion was treated either by an autograft or an ECM analog, or was not treated at all. Joint function was evaluated by studying joint stability, gait, and treadmill performance, and extended over 27 months. No evidence was presented to show that the structure of the ECM analog was optimized to deliver a maximal regenerative effect. Two-thirds of the joints implanted with the ECM analog, two-thirds of the joints that were autografted, and only one-fourth of the joints that were resected without further treatment showed regeneration of meniscal tissue. These results were interpreted to suggest that the ECM analog supported significant meniscal regeneration and provided enough biomechanical stability to minimize degenerative osteoarthritis in the canine knee joint (Stone *et al.,* 1989, 1990).

The feasibility of extending this treatment to humans was studied using patients with either an irreparable tear of the meniscal cartilage or a major loss of meniscal cartilage. Patients remained in the study for at least 36 months. Irreparably damaged meniscal tissue was first deleted and the ECM analog was implanted in its place. Regeneration of meniscal cartilage was observed at the site of implantation with the ECM analog while the template was undergoing degradation. The results were compared with those obtained with another group of patients who had been subjected to meniscectomy (removal of damaged meniscal tissue without the treatment), during the same period of observation. Patients who were implanted with the ECM analog showed significant reduction in pain as well as greatly improved resumption of strenuous activity relative to the untreated control group (Stone *et al.,* 1997).

SUMMARY

Studies have examined the process by which tissue and organs can be synthesized *in vivo* through use of ECM analogs; this process is also referred to as induced regeneration. Although several ECM analogs have been studied, only one of these, specified by a narrowly defined structure, has been shown to be capable of partial regeneration of skin in full-thickness skin wounds in animal models and in humans. Another ECM analog with a structure that has been specified equally narrowly has partially regenerated nerves over very large distances. The structural specificity of these templates appears to be related to the requirement for inhibition of wound contraction prior to the incidence of regeneration.

REFERENCES

Archibald, S. J., Sheffner, J., Krarup, C., and Madison, R. D. (1995). Monkey median nerve repaired by nerve graft or collagen nerve guide tube. *J. Neurosci.* **15,** 4109–4123.

Billingham, R. E., and Medawar, P. B. (1951). The technique of free skin grafting in mammals. *J. Exp. Biol.* **28,** 385–402.

Billingham, R. E., and Medawar, P. B. (1955). Contracture and intussusceptive growth in the healing of extensive wounds in mammalian skin. *J. Anat.* **89,** 114–123.

Boykin, J. V., and Molnar, J. A. (1992). Burn scars and skin equivalents. *In* "Wound Healing" (I. K. Cohen, R. F. Diegelmann, and W. J. Lindblad, eds.), pp. 500–509. Saunders, Philadelphia.

Burke, J. F., Bondoc, C. C., and Quinby, W. C. (1974). Primary burn excision and immediate grafting: A method of shortening illness. *J. Trauma* **14,** 389–395.

Burke, J. F., Yannas, I. V., Quimby, W. C., Jr., Bondoc, C. C., and Jung, W. K. (1981). Successful use of a physiologically acceptable artificial skin in the treatment of extensive burn injury. *Ann. Surg.* **194,** 413–428.

Chamberlain, L. J., Yannas, I. V., Arrizabalaga, A., Hsu, H.-P., Norregaard, T. V., and Spector, M. (1998a). Early peripheral nerve healing in collagen and silicone tube implants: Myofibroblasts and the cellular response. *Biomaterials* **19,** 1393–1403.

Chamberlain, L. J., Yannas, I. V., Hsu, H.-P., Strichartz, G., and Spector, M. (1998b). Collagen–GAG substrate enhances the quality of nerve regeneration through collagen tubes up to level of autograft. *Exp. Neurol.* **154,** 315–329.

Chang, A. S., and Yannas, I. V. (1992). Peripheral nerve regeneration. *In* "Neuroscience Year" (Suppl. 2 to the "Encyclopedia of Neuroscience") (B. Smith and G. Adelman, eds.), pp. 125–126. Birkhaeuser, Boston.

Cooper, D. K. C., Kemp, E., Platt, J. L., and White, D. J. G., eds. (1997). "Xenotransplantation," 2nd ed. Springer-Verlag, Berlin.

Dagalakis, N., Flink, J., Stasikelis, P., Burke, J. F., and Yannas, I. V. (1980). Design of an artificial skin. Part III. Control of pore structure. *J. Biomed. Mater. Res.* **14,** 511–528.

Dahlin, L. B., Zhao, Q., and Bjusten, L. M. (1995). Nerve regeneration in silicone tubes: Distribution of macrophages and interleukin-1β in the formed fibrin matrix. *Restorative Neurol. Neurosci.* **8,** 199–203.

Heimbach, D., Luterman, A., Burke, J., Cram, A., Herndon, D., Hunt, J., Jordan, M., McManus, W., Solem, L., Warden, G., and Zawacki, B. (1988). Artificial dermis for major burns. *Ann. Surg.* **208,** 313–320.

Hynes, R. O. (1990). "Fibronectins." Springer-Verlag, New York.

Kohn, D. H., and Ducheyne, P. (1992). Materials for bone and joint replacement. *Mater. Sci. Technol.* **14,** 29–109.

Lundborg, G. (1987). Nerve regeneration and repair. *Acta Orthop. Scand.* **58,** 145–169.

Lundborg, G., Fahlin, L. B., Danielsen, N., Gelberman, R. H., Longo, F. M., Powell, H. C., and Varon, S. (1982a). Nerve regeneration in silicone model chambers: Influence of gap length and of distal stump components. *Exp. Neurol.* **76,** 361–375.

Lundborg, G., Dahlin, L. B., Danielsen, N., Johannesson, A., Hansson, H. A., Longo, F., and Varon, S. (1982b). Nerve regeneration across an extended gap: A neuronobiological view of nerve repair and the possible involvement of neuronotrophic factors. *J. Hand Surg.* **7,** 580–587.

Lundborg, G., Gelberman, R. H., Longo, F. M., Powell, H. C., and Varon, S. (1982c). *In vivo* regeneration of cut nerves encased in silicone tubes: Growth across a six-millimeter gap. *J. Neuropathol. Exp. Neurol.* **41,** 412–422.

Lundborg, G., Longo, F. M., and Varon, S. (1982d). Nerve regeneration model and trophic factors *in vivo.* *Brain Res.* **232,** 157–161.

Madden, J. W. (1972). Wound healing: Biologic and clinical features. *In* "Textbook of Surgery" (D. Sabison, ed.). Saunders, Philadelphia.

Madison, R. M., Archibald, S. J., and Krarup, C. (1992). Peripheral nerve injury. *In* "Wound Healing" (I. K. Cohen, R. F. Diegelmann, and W. J. Lindblad, eds.), pp. 454–455. Saunders, Philadelphia.

Mast, B. A., Diegelmann, R. F., Krummel, T. M., and Cohen, I. K. (1992a). Scarless wound healing in the mammalian fetus. *Surg. Gynecol. Obstet.* **174,** 441–451.

Mast, B. A., Nelson, J. M., and Krummel, T. M. (1992b). Tissue repair in the mammalian fetus. *In* "Wound Healing" (I. K. Cohen, R. F. Diegelmann, and W. J. Lindblad, eds.), pp. 333–334. Saunders, Philadelphia.

Medawar, P. B. (1944). The behavior and fate of skin autografts and skin homografts in rabbits. *J. Anat.* **78**, 176–199.

Millesi, H. (1967). Erfahrungen mit der Mikrochirurgie peripherer Nerven. *Chir. Plast. Reconstr.* **3**, 47–55.

Murphy, G. F., Orgill, D. P., and Yannas, I. V. (1990). Partial dermal regeneration is induced by biodegradable collagen–glycosaminoglycan grafts. *Lab. Invest.* **62**, 305–313.

Neuman, M. R. (1998). Therapeutic and prosthetic devices. *In* "Medical Instrumentation" (J. G. Webster, ed.), pp. 577–622. Wiley, New York.

Orgill, D. P., Butler, C. E., and Regan, J. F. (1996). Behavior of collagen–GAG matrices as dermal replacements in rodent and porcine models. *Wounds* **8**, 151–157.

Peacock, E. E., Jr. (1984). Wound healing and wound care. *In* "Principles of Surgery" (S. I. Schwartz, G. T. Shires, F. C. Spencer, and E. H. Storer, eds.). McGraw-Hill, New York.

Rudolph, R., Vande Berg, J., and Ehrlich, P. (1992). Wound contraction and scar contracture. *In* "Wound Healing" (I. K. Cohen, R. F. Diegelmann, and W. J. Lindblad, eds.), pp. 96–104. Saunders, Philadelphia.

Shafritz, T. A., Rosenberg, L. C., and Yannas, I. V. (1994). Specific effects of glycosaminoglycans in an analog of extracellular matrix that delays wound contraction and induces regeneration. *Wound Repair Regeneration* **2**, 270–276.

Shapiro, F. (1988). Cortical bone repair. *J. Bone J. Surg., Am. Vol.* **70-A**, 1067–1081.

Spence, R. J. (1998). The use of Integra for contracture release. *Proc. Am. Burn Assoc.* **19**, S173.

Stone, K. R., Webber, R. J., Rodkey, W. G., and Steadman, J. R. (1989). Prosthetic meniscal replacement: *In vivo* studies of meniscal regeneration using copolymeric collagen prostheses. *Arthroscopy* **5**, 152.

Stone, K. R., Rodkey, W. G., Webber, R. J., McKineey, L., and Steadman, J. R. (1990). Collagen-based prostheses for meniscal regeneration. *Clin. Orthop.* **252**, 129–135.

Stone, K. R., Steadman, R., Rodkey, W. G., and Li, S.-T. (1997). Regeneration of meniscal cartilage with use of a collagen scaffold. *J. Bone J. Surg., Am. Vol.* **79-A**, 1770–1777.

Valentini, R. F., and Aebischer, P. (1997). Strategies for the engineering of peripheral nervous tissue regeneration. *In* "Principles of Tissue Engineering" (R. Lanza, R. Langer, and W. Chick, eds.), pp. 671–684. R. G. Landes, Austin, TX.

Wallace, H. (1981). "Vertebrate Limb Regeneration." Wiley, New York.

Yannas, I. V. (1981). Use of artificial skin in wound management. *In* "The Surgical Wound" (P. Dineen, ed.), pp. 171–190. Lea & Febiger, Philadelphia.

Yannas, I. V. (1988). Regeneration of skin and nerves by use of collagen templates. *In* "Collagen: Biotechnology" (M. Nimni, ed.), Vol. III, pp. 87–115. CRC Press, Boca Raton, FL.

Yannas, I. V. (1997). Models of organ regeneration processes induced by templates. *Ann. N. Y. Acad. Sci.* **831**, 280–293.

Yannas, I. V. (1998). Studies on the biological activity of the dermal regeneration template. *Wound Repair Regeneration* **6**, 518–524.

Yannas, I. V., and Burke, J. F. (1980). Design of an artificial skin. I. Basic design principles. *J. Biomed. Mater. Res.* **14**, 65–81.

Yannas, I. V., and Tobolsky, A. V. (1967). Crosslinking of gelatine by dehydration. *Nature (London)* **215**, 509–510.

Yannas, I. V., Burke, J. F., Huang, C., and Gordon, P. L. (1975a). Suppression of *in vivo* degradability and of immunogenicity of collagen by reaction with glycosaminoglycans. *Polym. Prepr., Am. Chem. Soc., Div. Polym. Chem.* **16**(2), 209–214.

Yannas, I. V., Burke, J. F., Huang, C., and Gordon, P. L. (1975b). Correlation of *in vivo* collagen degradation rate with *in vitro* measurements. *J. Biomed. Mater. Res.* **9**, 623–628.

Yannas, I. V., Burke, J. F., Gordon, P. L., Huang, C., and Rubinstein, R. H. (1980). Design of an artificial skin. II. Control of chemical composition. *J. Biomed. Mater. Res.* **14**, 107–131.

Yannas, I. V., Burke, J. F., Warpehoski, M., Stasikelis, P., Skrabut, E. M., Orgill, D., and Giard, D. J. (1981). Prompt, long-term functional replacement of skin. *Trans. Am. Soc. Artif. Intern. Organs* **27**, 19–22.

Yannas, I. V., Burke, J. F., Orgill, D. P., and Skrabut, E. M. (1982). Wound tissue can utilize a polymeric template to synthesize a functional extension of skin. *Science* **215**, 174–176.

Yannas, I. V., Orgill, D. P., Silver, J., Norregaard, T., Zervas, N. T., and Schoene, W. C. (1987). Regeneration of sciatic nerve across 15-mm gap by use of a polymeric template. *In* "Advances in Biomedical Materials" (C. G. Gebelein, ed.), pp. 1–9. Am. Chem. Soc., Washington, DC.

Yannas, I. V., Lee, E., Orgill, D. P., Skrabut, E. M., and Murphy, G. F. (1989). Synthesis and characterization of a model extracellular matrix which induces partial regeneration of adult mammalian skin. *Proc. Natl. Acad. Sci. U.S.A.* **86**, 933–937.

Yannas, I. V., Colt, J., and Wai, Y. C. (1996). Wound contraction and scar synthesis during development of the amphibian *Rana catesbeiana*. *Wound Repair Regeneration* **4**, 31–41.

PART IV: MODELS FOR TISSUE ENGINEERING

CHAPTER 16

ORGANOTYPIC AND HISTIOTYPIC MODELS OF ENGINEERED TISSUES

Eugene Bell

INTRODUCTION

Work with intact tissues or tissue fragments in the form of organ cultures has had a long history in experimental embryology in studies devoted to causal relationships between different tissues on which the emergence of differentiated structures depend. It has been recognized that although two-dimensional cell cultures usually lose the characteristic histologic organization and most of the physiologic functions of the original tissue, organ cultures retain both, and can permit the process of development to unfold. But organ culturing in experimental embryology, unlike cell culturing, has consisted of explanting organ anlage, or pieces of intact tissue precursors, under conditions that preserved tissue integrity and discouraged cells from moving out of the explant and from engaging in DNA synthesis in preparation for cell division. Stability of the organ-cultured explant can be attributed, in large measure, to the matrix scaffolds with which the cells in tissues are associated. What cells do in organotypic explants, is for a large part, regulated by instructive scaffolds, already embedded in the extracellular matrix (ECM), and by the flow of information between different contiguous tissues through the ECM. Adequate nutrient medium and physical forces that act on scaffolds and cells may play a role in determining the maintenance and developmental potential of an explant. The disadvantages of organ cultures include their short life, their tendency to undergo central necrosis, the occurrence of vascular occlusion, and the difficulty of preparing standard models, because no two explants are identical.

An alternative approach to the study of tissues *in vitro* consists of reconstituting stable specific tissues with scaffolds, into which cells of the tissue of origin are seeded (Bell *et al.,* 1979). The key requirements for constituting standard models are the appropriate matrix scaffold and a source of cells. Signals may be provided by cells capable of generating them and/or by the addition of signals added to the scaffold or to the culture medium to induce mitogenesis, tissue morphogenesis, and differentiation. Also important may be the application of forces that tissue development requires and, insofar as they are known, the medium or media and other conditions of culture needed to promote survival *in vitro.* The result can be the fabrication of standardized models that closely resemble actual tissues in many respects.

THE COLLAGEN GEL MODEL

A generic collagen gel model of a two-tissue, or organotypic, system is shown in Fig. 16.1. The model uses a collagen gel scaffold prepared by combining, in the cold, a neutralized 0.3–1.0 mg/ml solution of acid extracted collagen with medium, serum, and mesenchymal cells (Bell *et al.,* 1979)—dermal fibroblasts, for example, if the goal is to fabricate a skin equivalent. The bacteriological petri plate or other vessel, to which cells do not attach, into which the mix is poured, is incubated at 37°C in a 5% CO_2 incubator. The collagen polymerizes, when neutralized and warmed, forming a lattice of fine fibrils, 10–20 nm in size, which trap fluid. The result is a gel in

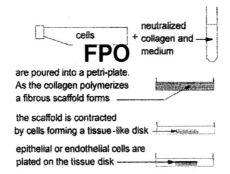

Fig. 16.1. Formation of tissue equivalent. A generic collagen gel model of a two-tissue system.

which the previously added cells are distributed. The mesenchymal cells in the gel extend and attach podial processes to the collagen fibrils and withdraw the processes with the attached fibrils toward the cell body. As the fibrils are bundled by the cells, fluid is squeezed out of the lattice. The process of gel contraction, known as synersis, can reduce the size of the collagen lattice 30- to 40-fold, depending on the cell and collagen concentration used. The condensed gel is tissuelike in its consistency, providing a substrate on which epithelial, endothelial, or mesothelial cells may be plated. What has been described in the foregoing sequence is the formation of an untethered gel, because the tissue equivalent releases from the plate and floats. Tethered gels can be prepared by casting the initial cell–collagen–medium mix in a tissue culture plate or vessel, or, cell-seeded pregel mixes can be poured into chambers in which a peripheral grab-rail is provided, around which gels can form to provide a holdfast, limiting contraction to one dimension (Bell *et al.*, 1989; Bell, 1995a). It was found that restrained tissue equivalents were two to three times stronger than unrestrained tissues and that biosynthesis of collagen by dermal fibroblasts under restrained conditions was greater than that under unrestrained conditions, possibly related to prostaglandin E_2 (PGE_2) output by resident cells, because PGE_2 inhibits collagen biosynthesis. PGE_2 is produced at about half the level, by cells in restrained tissues, of that produced by cells in unrestrained tissue equivalents. Weinberg and Bell (1986) used velcro to restrain the ends of cell-seeded tubular gels that contracted only in the thickness mode.

Porter *et al.* (1998) have studied the mechanisms of gel contraction in restrained and unrestrained gels and attribute contraction in the former to cell migration through the gel. They point out that how a cell behaves in a tissue depends on the mechanical signals it receives from the matrix. The self-organization of tissue equivalents and the movement of cells within collagen gels have been reviewed (Tranquillo, 1999).

MODELS OF CELL INTERACTIONS IN COLLAGEN LATTICES

Coculturing interactive tissue-specific subsets of cells that are combined in an appropriate scaffold will be reviewed in an exemplary way. We will focus first on the skin, because it has been used to model developmental, immunologic, and pathologic events. We will touch on other models by tissue type and discuss other forms of collagen scaffolds and techniques used, in addition to gels, for building tissue and organ models.

THE SKIN EQUIVALENT AS A DEVELOPMENTAL MODEL

As long ago as 1963, Dodson, in Honor Fell's laboratory in Cambridge, isolated dermis from the metatarsus of 12-day chick embryos and subjected it to freeze–thaw cycles before combining it with epidermis that alone was unable to differentiate. Although it lacked living cells, the frozen–thawed dermis, constituting a structurally intact scaffold and probably rich in growth and differentiation factors, was sufficient to support keratinocyte differentiation *in vitro*. A control, consisting of heat-killed dermis, failed to promote keratinocyte differentiation. Similar combinations using adult acellular dermis and keratinocytes, made many years later (Heck *et al.*, 1985; Cuono *et al.*, 1987), were used successfully as transplants to patients requiring skin.

The first skin consisting of "living" dermis and epidermis reconstituted from cultivated cells and collagen scaffolding (Bell *et al.*, 1979, 1981, 1983) was shown to undergo virtually complete differentiation *in vitro*, lacking, however, pigment, sweat glands, neurogenic elements, a microcirculation, and hair follicles. The model can be reproduced faithfully and be kept alive *in vitro* for months, at least. Although collagenolytic activity is high in young dermal equivalents (Nusgens *et al.,* 1984; Rowling *et al.,* 1990), possibly associated with tissue remodeling, it has been observed that the resistance of dermal equivalents to breakdown by collagenase is greatly enhanced by 30 days of cultivation *in vitro,* suggesting that extensive cross-linking (probably by cell-secreted lysyl oxidase) of the collagen fibrils has occurred, as shown by Rowling *et al.* (1990). Continued differentiation of the model *in vitro* and the resemblance of cells in the matrix to their *in vivo* counterparts, rather than to cells grown on plastic in two dimensions, are distinguishing features (Coulomb *et al.,* 1983). Collagen processing by cells grown on two-dimensional (2D) substrates is known to be deficient in that the exported molecule retains the procollagen C- and N-terminal telopeptides normally cleaved by peptidases (Freiberger *et al.,* 1980). In the collagen lattice scaffold, processing and subsequent polymerization, for which it is a requirement, proceed normally (Nusgens *et al.,* 1984).

Populating matrix scaffolds, other than collagen lattices (gels), with dermal fibroblasts to produce dermal equivalents followed the introduction of the original work that reported the fabrication of both dermal and skin equivalents (Bell *et al.,* 1979, 1981). Materials used include nylon, Vicryl (Hansbrough *et al.,* 1993), and Biobrane (Hansbrough *et al.,* 1994). The commercial product Dermagraft (Advanced Tissue Sciences) uses the technique of seeding the Vicryl scaffold with allogeneic fibroblasts.

With the contracted dermal equivalent overplated by keratinocytes, the time required to constitute a skin model is 14 to 21 days, in the course of which at least three different media may be used: (1) An initial medium supports the contractile activity of the mesenchymal cells (dermal fibroblasts) until an equilibrium size is attained and no further contraction occurs. The process of contraction takes about 4 days. (2) At that time a second medium is applied and keratinocytes are plated on the surface of the dermal equivalent. To promote keratinocyte proliferation, low concentrations of Ca^{2+}, epidermal growth factor (EGF), and an up-regulator of cyclic AMP are required. (3) A third medium is used during the second phase of epidermal development, calling for high $[Ca^{2+}]$ and air-lifting to expose the epidermis to the atmosphere to induce keratinization and differentiation of a stratum corneum (Parenteau *et al.,* 1991; Bell *et al.,* 1991). The skin equivalent can be populated with melanocytes (Topol *et al.,* 1986) by plating them on the cell-populated dermal equivalent before the application of keratinocytes. Each melanocyte services about 20–30 surrounding keratinocytes, leading to skin pigmentation *in vitro.* Pigment passes into keratinocytes through a melanocyte podial process that has made contact with the keratinocyte plasma membrane.

THE SKIN EQUIVALENT AS AN IMMUNOLOGICAL MODEL

The skin equivalent can be constituted with cultivated parenchymal cells free of any subsets of immune cells normally found in the dermis and epidermis. Using the X chromosome as a genetic marker, female cells are used to make up skin equivalents, which are then transplanted to male hosts across a major histocompatibility barrier, e.g., from Brown Norway to Fisher rats. Sher *et al.* (1983) demonstrated in the rat model, by karyotyping cells grafted in skin equivalents, that allogeneic fibroblasts were not rejected. It has been reported that clinical trials of skin equivalents made up with human allogeneic keratinocytes as well as allogeneic fibroblasts do not provoke an immune reaction in recipients (Parenteau *et al.,* 1994). The model should be a valuable tool for determining the roles played by cells of the immune system and the microcirculation in allograft rejection of actual skin. It should allow use of cells of any genotype and of human origin to study genetic abnormalities, as well as the contribution of specific genetic loci to skin development by transplanting skin equivalents to immunodeficient rodents.

THE SKIN EQUIVALENT AS A DISEASE MODEL

Because either normal or aberrant skin cells can be incorporated into skin equivalents, models of disease states can be fabricated. A psoriasis model was fabricated to test the contribution of psoriatic dermal fibroblasts to the expression of features of the disease *in vitro* (Saiag *et al.,* 1985). A button of normal keratinocytes suspended in medium was plated in the centers of dermal equivalent disks constituted with normal human or psoriatic dermal fibroblasts, and the rate of spreading of keratinocytes over the dermal substrate was measured. It was observed that the psoriatic fi-

Fig. 16.2. Disease model. Normal
keratinocytes are plated within an
8-mm-diameter circle of a 24-
mm-diameter collagen lattice
seeded with normal or psoriatic
dermal fibroblasts. Keratinocyte
growth and spreading are followed. Psoriatic fibroblasts are responsible for hyperproliferation of
keratinocytes.

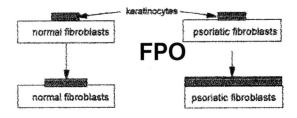

broblasts induced hyperproliferation and greater spreading of keratinocytes compared with the growth and spreading induced by control fibroblasts, suggesting that dermal cells may play a role in the progress of the disease (Fig. 16.2).

In addition to the study of psoriasis, and other epidermal diseases such as epidermolysis bullous, the model should provide an *in vitro* basis for studying dermal connective tissue disorders, including dermatosparaxiis and sclerosis. It is obvious that any pair of populations of mesenchymal cells and epithelial cells, of which one or both is diseased or aberrant, can be used in the three-dimensional coculturing system for studying the expression of features of the disease and testing modalities of treatment.

THE SKIN EQUIVALENT AS A WOUND HEALING MODEL

Two-tissue skin models can be used *in vitro* to analyze the role of dermis in epidermal wound healing (Bell and Scott, 1992). After constituting a differentiated skin equivalent in a 24-mm multiwell well-plate insert, a central disk of the skin is removed with a punch (Fig. 16.3). The acellular layer of collagen in contact with the membrane of the insert is replaced and the remainder of the gap is filled with a collagen scaffold to the level of the interface between the dermis and epidermis. The rate of overgrowth of the neodermis by keratinocytes and the development of the epidermis can be taken as measures of the effectiveness of the design of neodermis as an interacting substrate. The wound healing model can accommodate acellular dermal scaffolds with or without signals to test their effectiveness in attracting dermal fibroblasts from the surrounding matrix.

OTHER TYPES OF EPITHELIAL–MESENCHYMAL MODELS

A generic model for many types of epithelial–mesenchymal combinations is schematized in Fig. 16.1. Tables 16.1 and 16.2 suggest only some of the combinations for organotypic coculturing. Many others are possible. We are not suggesting that combinations need be restricted to ep-

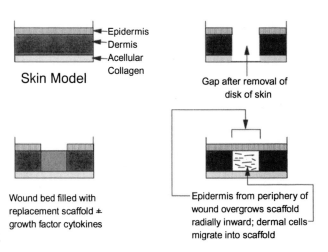

Fig. 16.3. Wound healing model.

Table 16.1. Partner tissues for the reconstitution of organ-equivalent models

Tissue type	Epithelial	Endothelial	Nervous glandular immune	Mesenchymal	Ref.
Skin	Keratinocytes, melanocytes	—	—	Dermal fibroblasts	Bell *et al.* (1979, 1981); Topol *et al.* (1986)
Skin	HaCaT cells, keratinocytes	—	—	Dermal fibroblasts	Schoop *et al.* (1999)
Cornea	Epithelium	Endothelium	—	Corneal fibroblasts (keratocytes)	Minami *et al.* (1993)
Blood–brain barrier	—	Endothelium	Monocytes	Pericytes, fibroblasts	Bell *et al.* (1990), Beigel and Pachter (1994)
Bronchial wall	Bronchial epithelium	—	—	Bronchial wall myofibroblasts	Zhang *et al.* (1999)
Laryngeal mucosa	Laryngeal epithelium	—	—	Lamina propria, fibroblasts	Yamaguchi *et al.* (1996)
Gut	Mucosal cells	—	—	Fibroblasts, smooth muscle cells	Montgomery (1986)
Vascular	—	Endothelium	—	Smooth muscle cells, advential fibroblasts	Weinberg and Bell (1986)
Vascular	—	Somatostatin gene from vascular endothelium	—	Transferred to vascular smooth muscle	Sarkar *et al.* (1999)
Nerve	—	—	Neurons, motor neurons, cells of floorplate, neural precursor cells	Glial cells	Bergsteinsdottir *et al.* (1993), Guthrie and Pini (1995), Tomooka *et al.* (1993)
Nerve	—	—	Microglia and embryonic dopaminergic neurons		Zietlow *et al.* (1999)
Capillary networks	Epithelium from bone marrow	—	—	Adipocytes, fibroblasts from bone marrow	Mori *et al.* (1989)
Capillaries	Aortic endothelium	—	—	Pericytes	Akita *et al.* (1997)
Pancreatic islets	—	—	Islet glandular cells	Islet fibroblasts	Lanza *et al.* (1995), Soon-Shiong *et al.* (1993)
Blood–brain barrier	—	Endothelium	Monocytes	Pericytes, fibroblasts	Bell *et al.* (1990) Beigel and Pachter (1990)
Hair follicles	Hair follicles	—	—	Dermal fibroblasts	Rogers *et al.* (1988)
Thyroid	Glandular epithelium	—	—	Thyroid fibroblasts	Bell *et al.* (1984)

Table 16.2. Models consisting of a single cell type in or on a matrix scaffold

Cell type	Type of scaffold	Ref.
Adipocytes	Collagen gel	Kagan and Bell (1989)
Primary mammary epithelia	Collagen Englebreth–Holm Swarm tumor matrix	Lee et al. (1985) Streuli et al. (1991)
Liver cells	Collagen gel sandwich	Yarmush et al. (1992), Toner et al. (1993), Bader et al. (1995)
Liver cells	Polyvinyl alcohol Poly(lactic acid)– poly(glycolic) acid copolymer	Kneser et al. (1999), Mooney et al. (1994), Shakado et al. (1995)
	Type I collagen gel + laminin or Matrigel	Takeshita et al. (1995) Nishikawa et al. (1996)
	Hollow Fiber + collagen gel, collagen gel	
Bladder urothelium	Non-woven polyglycolic acid	Atala et al. (1992)
Chondrocytes	Alginate beads Bipolymer Collagen gel	Puelacher et al. (1994) Hauselmann et al. (1994) Wakitani et al. (1998)
Mesenchymal stem cells (osteoblasts)	Ceramics	Dennis et al. (1992)
Endothelial cells	Collagen gel + bFGF or vascular endothelial growth factors	Vernon and Sage (1999)
Keratocytes	Collagen gel + SPARC cells	Mishima et al. (1998)
Ocular fibroblasts	Collagen gel	Porter et al. (1998)
Bone marrow stromal cells	Gelatin sponge	Krebsbach et al. (1998)

ithelial–mesenchymal combinations. For example, interactions between populations that are coderivatives embryologically can also be modeled in a three-dimensional system *in vitro*. In the development of the nervous system both attractive and repellent signals guide the course of axons en route to end point targets (Guthrie and Pini, 1995). The authors seeded a collagen gel scaffold with two components, either ventral explants of rat hind brain or spinal cord and floor plate explants from which the former tissues were separated in the scaffold gel. It is shown that outgrowth of motor axons from aspects of hind brain or cord that faced the plate were greatly reduced, suggesting that diffusible factors from the floor plate can exclude motor axons from certain regions of the developing nervous system. The existence of chemorepellants as well as chemoattractants has been discussed (Research News, 1995). They divide into at least two families of proteins, netrins and semaphorins, responsible for growth cone guidance. It is suggested that a single factor can induce attraction or repulsion, depending on the cells involved.

VASCULAR MODELS

VASCULAR MODELS WITH CELLS ADDED

Vascular models that examine the effect of shear and other forces on monolayers of endothelial cells *in vitro* have been developed by Nerem *et al.* (1993), who have shown that the rate of endothelial cell proliferation is decreased by flow and that entry of cells into a cycling state is inhibited. They point out the limitations of the 2D system because the important component of cyclic stretch of basement membrane, which the endothelium normally synthesizes and on which it rests, is absent. They suggest that a coculture system in which the endothelium is supported on smooth muscle tissue would be superior for providing a more physiologic environment. Such a system was developed by Weinberg and Bell (1986), who showed that a basal lamina was laid down between the endothelium and the contiguous smooth muscle tissue, cast in the form of a small-caliber tube *in vitro*. Fabricating the vessel was a three-step procedure. The first tissue layer cast around a small-caliber mandrel was a smooth muscle cell media, whose ends were anchored in a velcro cuff or held fast by ridges and grooves in the mandrel until radial contraction made space for bands that were secured around the ends. Hence the mechanical restraint imposed on the contracting tube allowed contraction to occur radially but not longitudinally, because each end of the vessel equivalent was held fast. The second tissue layer cast was the adventitia surrounding the smooth muscle media. To make room for it, the fluid expressed from the collagen gel scaffold was drawn out of the casting tube, and the mixture of adventitial fibroblasts suspended in medium containing neutralized collagen was introduced into the space between the media and the wall of the cylindrical casting chamber. After the adventitia had contracted radially but not longitudinally, because the media provided a frictional surface that prevented it, the mandrel was extracted, leaving a lumen of the tissue tube that was filled with a suspension of endothelial cells. The cells came to rest on the inner surface of the media as the tube was rotated.

Histologic analysis showed that the smooth muscle cells of the media were oriented in parallel with the long axis of the tissue tube that had formed, rather than circumferentially, as they dispose themselves *in vivo*. By modifying the foregoing procedure to allow the forming vessel to contract freely in its length (L'Heurex *et al.*, 1993; Hirai *et al.*, 1994), simply by eliminating the restraining ties at the ends of the mandrel, the smooth muscle cells orient circumferentially as they do *in vivo*. Similar results are obtained by applying pulsatile pressure, causing the media of the vessel to expand radially (Kanda *et al.*, 1993). Another method (Tranquillo *et al.*, 1996) is based on the observation that polymerizing collagen fibrils align in a magnetic field (Torbert and Ronziere, 1984). By orienting the model devised by Weinberg and Bell (1986), in which the ends of the developing vessel are tethered, with its long axis parallel to a strong magnetic field during fibrillogenesis, it has been shown that collagen fiber orientation is circumferential. The fiber orientation provides contact guidance for smooth muscle cells that encircle the vessel in a direction perpendicular to its length. The use of the new models in experimental animal systems should provide an opportunity to validate their design. Although the more normal organization of the cells of the media should reduce the excessive compliance of the structure that resulted from the longitudinal orientation of the smooth muscle cells, the degree of benefit is still a concern that should be resolvable in animal model trials. The collagen gel lattice may not provide the structural resistance typical of an actual artery, but other collagen scaffolds may (Bell, 1995b). It should also be possible to manipulate *in vitro* conditions by enriching scaffolds with instructive molecules able to induce differentiation of the elastic lamina normally present in arteries.

VASCULAR MODELS WITHOUT CELLS

Vascular prostheses constructed from Dacron and other synthetics have been in use for many years but are known to elicit persistent inflammatory reactions and to become occluded. The ther-

Table 16.3. Acellular models consisting of a matrix scaffold but expected to attract or be populated by host cells

Cell type	Scaffold type	Ref.
Endothelial, smooth muscle, adventiyial fribroblast	Submucosa, including stratum compactum of the jejunum	Sandusky *et al.* (1992)
Same as above	Polyurethane + poly(L-lactic acid)	Leenslag *et al.* (1988)
Same as above	Poly(tetrafluoroethylene)	van der Lei *et al.* (1991), Stronck *et al.* (1992)

mosetting polymers are not biodegradable and do not integrate with host tissues, but some successes have been reported under limited conditions in experimental animals (Table 16.3). For the foregoing reasons, other acellular materials have been proposed and tried as arterial substitutes. Animal tissues that resemble arteries have been used with some success, in particular, the porcine small intestine (Sandusky *et al.,* 1992). The mucosal cells are scraped off the luminal side and the muscular layers are removed from the abluminal side, leaving the stratum compactum, a dense, highly organized fibrillar collagen matrix and the looser connective tissue of the mucosa. The material can be used as a scaffold for cells *in vitro* and has been used in animal experiments. *In vivo,* it is invaded by capillaries that contribute cells that provide an intima, whereas smooth muscle cells migrating from the anastomoses provide a media.

SELECTION OF SCAFFOLDS OTHER THAN GELS FOR MODEL SYSTEMS

A rationale for the selection of collagen scaffolds for tissue engineering has been presented elsewhere (Bell, 1995b). Because *in vitro* models of promise usually lead to trials in experimental animals, which are costly, the argument for the use of a scaffolding material with the evolutionary success credentials of the collagen polymers is strong. Hence we focus on the variety of forms of collagen scaffolds that can be assembled in the laboratory. In addition to gels, foam, fiber, and membrane scaffolds are mentioned.

COLLAGEN FOAM SCAFFOLDS

Collagen foams (often called sponges) were early described by Chvapil (1977). Collagen foams are fabricated by freeze-drying a solution of collagen placed in a mold of some desired configuration. The collagen concentration can be of the order of 1–5 mg/ml. Results depend on the mold design and the choice of freeze–dry cycle. The foams produced should have communicating compartments between 50 and 100 μm in cross-section, and should be free of a surface skin, which will interfere with cell seeding. Foams made from monomeric collagen solutions are fragile, but nonetheless can be seeded with cells without having been cross-linked. Like gels, uncrosslinked or mildly cross-linked foams are readily contracted by mesenchymal cells. Unlike collagen gels, they do not restrict cell division because cells in the foam compartments occupy surfaces, whereas cells in gels are completely surrounded by collagen matrix. After physical or chemical cross-linking [ultraviolet (UV) light, for example] of sufficient intensity and duration, foam scaffolds become resistant to contraction by tissue cells and exhibit decreased or increased resistance to breakdown by collagenase, depending on the cross-linking regimen. Cross-linking with aldehyde produces resilient material but destroys all biologic information intrinsic to collagen polymers.

Foams have been used for the fabrication of numerous tissue types. Sabbagh *et al.* (1998) have shown that the growth and stratification of urothelial cells occur on the collagen foam and conclude that the tissue formed is suitable for grafting. The foam scaffold seeded with bone marrow cells lends itself to medium perfusion, which improves cell survival and differentiation (Glowacki *et al.,* 1998). MacPhee *et al.* (1994) have used foams as a means of harvesting organ-derived endothelial cells by coating foams with fibroblast growth factor-1 (FGF-1) and implanting them in the skin, peritoneum, abdominal mesentery, epimysium, liver, and spleen. Single cells, of which 25% were endothelial, were obtained after collagenase digestion and further concentrated to obtain cultures of high purity. Swope *et al.* (1997) have produced skin equivalents using collagen foams as a dermal scaffold seeded with dermal fibroblasts. Scaffolds were overplated with keratinocytes after mixing with a measured number of melanocytes determined by cell sorting.

COLLAGEN FIBER SCAFFOLDS

Scaffolds consisting of spun collagen fibers have been manufactured at Tissue Engineering Inc. Fibers with diameters of 300 nm, as well as larger fibers (in the micrometer range) have been made on a commercial scale. They can be formed into wools by tangling (Fig. 16.4). Figure 16.5 is a scanning electron micrograph of the wool, into which cells are easily seeded (Fig. 16.6). When cross-linked by methods that do not alter the native 67-nm cross-banding, the fibers are considerably more resistant to collagenase than are foam or gel scaffolds.

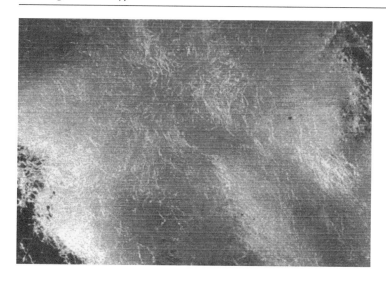

Fig. 16.4. Collagen foam–fiber wool containing fibers 30 μm in diameter (patent pending, 09/282,088).

COLLAGEN MEMBRANE SCAFFOLDS

Collagen membranes can be prepared by allowing collagen in solution to dry on a surface to which it will not bind, like Teflon or polyethylene. To promote formation of fibrils the solution is neutralized and warmed to 37°C, allowing the collagen to polymerize and form the fibrils. Before it begins to gel the solution is spread on a suitable surface and allowed to dry. The thickness of the membrane can be controlled by the concentration of collagen used, and by the thickness of the gel that is allowed to form after neutralized collagen is poured into the mold in which it is dried. Membranes may be cross-linked by a variety of methods (Gorham, 1991), particularly to improve their wet strength. For example, aldehydic cross-linking will prevent cell attachment, and UV cross-linking will reduce resistance to collagenase, so that the ultimate use of the membrane will dictate the mode of cross-linking.

CELL SIGNALING AND THE ENRICHMENT OF SCAFFOLDS

Great progress has been made in developmental and cell biology in the area of cell signaling. Its relevance for the reconstitution of prosthetic devices in tissue engineering cannot be overlooked. Whether scaffolds are seeded with cells before they are implanted, or whether they mobilize cells

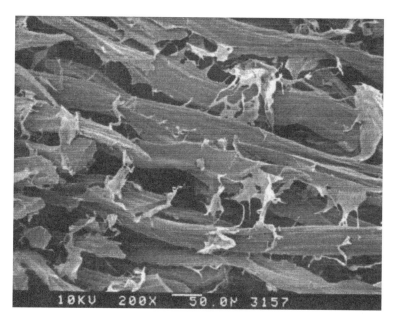

Fig. 16.5. Scanning electron micrograph of fiber–foam wool. The foam appears as irregular membranous elements between fibers that are flattened by the freeze-drying process.

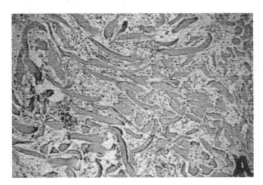
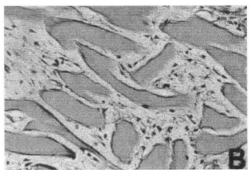

Fig. 16.6. Fiber–foam wool into which dermal fibro-blasts have been seeded. By day 14 the cells are surrounded by newly synthesized extracellular matrix. Magnification: A, 100×; B, 320×.

from adjacent host tissues, the cells, which become associated with a scaffold enriched with signaling molecules, must be able to respond to them in a manner that leads to the differentiation of the tissues designed to be replaced by the prosthesis. Some tissue cells are able to rebuild tissues, even *in vitro,* without the help of signals from a source other than themselves. They must, however, be combined under the right conditions, as described above. The two principal cell types of skin enjoy a paracrine relationship, providing signals for each other that are sufficient for the reconstitution of a well-differentiated two-layered skin. This relationship is expressed during development and throughout postnatal life.

Answering the general question, concerning which adult cells retain the capacity to respond to signals that were responsible for tissue differentiation in the embryo and fetus, is important for *exvivo* tissue engineering—that is, for enriching implantable devices. For example, it has been shown that postnatal bone growth and fracture healing in mammals, in particular, endochondral ossification, is a process driven by the same molecular signals that function in the embryo. One of those signals, the protein Indian hedgehog (Ihh), is involved in regulating prenatal bone development. The Hedgehog gene family, in vertebrates, in addition to Ihh, includes Sonic hedgehog (Shh) and Desert hedgehog (Dhh), coding for proteins that play a role in a broad range of morphogenetic events in the course of development. They are thought to function by binding to and activating their specific receptors, believed to be Patched, and a seven-pass membrane protein, Smoothened. Ihh is secreted by prehypertrophic chondrocytes and in turn activates parathyroid hormone-related protein, which down-regulates the rate of chondrocyte differentiation (Vortkamp *et al.,* 1998). Further evidence that the expression of the Hedgehog family of genes is not limited to prenatal development has been provided by Traiffort *et al.* (1998), who show the presence of Ihh, Shh, and Dhh in various other adult tissues in the rat. Although Shh was found localized in the Purkinje cell layer of the adult brain, it remains to be shown what role the protein plays in that tissue.

Identifying a secreted signal that is responsible for initiating the morphogenetic and functional specialization of a developing or susceptible cell is the first step in the causal sequence leading to gene activation. The second step is the identification of the signal's receptor, and the third step is that of finding the factor or factors responsible for translocating the signal to the nucleus, where the most downstream component will act as a transcription factor. Such a sequence has at least in part been elucidated for a class of secreted signaling proteins called Wnt. The large family of Wnt genes encode proteins (rich in cysteine) that are responsible for the regulation of cell fate. They act in an autocrine or paracrine manner by redirecting the transcriptional programs of target cells. They can also form gradients and activate genes in a concentration-dependent way. The proteins of a class called Frizzled have been identified as the receptors for Wnt proteins, and β-catenine found downstream of Frizzled can translocate to the nucleus and act as a transcription factor (Wodarz and Nusse, 1998). It is well known that the development of tissues *in vitro* is promoted by the use of added signals such as cytokines; keeping up with the stream of newly discovered families of signaling molecules seems a necessity for the future of tissue engineering.

GOALS AND USES OF MODEL TISSUE AND ORGAN BUILDING

One of the major goals of model tissue building is the reconstitution of three-dimensional tissue or organ equivalents that faithfully resemble actual body parts, in order to better understand cell and tissue interactions and functions under controlled and reproducible conditions. From this point of view the model offers an opportunity to study signaling between cell populations of different embryologic origin. It is useful for analysis of disease states and for seeking remedies for them. The model can serve as a test system for toxicity, diagnostic, and cell migration assays; it can be used for the cultivation of viruses such as papilloma virus, which depend on cycling and differentiating cells. It has the advantage, if constituted with human cells, of giving responses to toxins that may be different from those given by animal cells. Cells in an organized tissue exhibit reduced susceptibility to damage as compared with cells grown on a monolayer (Bell *et al.*, 1991).

A second major goal is the development of tissue and organ models that eventually can serve as prostheses. To reach the goal, model making and *in vitro* testing go hand in hand with testing in experimental animals. Model tissues for transplantation can take many forms: (1) they can be seeded with parenchymal cells that normally populate the tissue being replaced; (2) they can be acellular but designed to mobilize cells from contiguous tissues; (3) they can be constituted with degradable biopolymer scaffolds that are permissive, with the expectation that the cells with which they are seeded or which populate them will create a tissue-specific extracellular matrix and associated scaffolding; or (4) they can be instructive, being provided with an enriched scaffold able to influence the activities of cells. If the goal of tissue engineering is the induction of regeneration, the design and selection of models must be based on their success *in vivo*.

References

Akita, M., Murata, E., Merker, H. J., and Kaneko, K. (1997). Morphology of capillary-like structures in a three dimensional aorta/collagen gel culture. *Anat. Anz.* **179,** 127–136.

Atala, A., Vacanti, J. P., Peters, C. A., Mandell, J., Retik, A. B., and Freeman, M. R. (1992). Formation of urothelial structures *in vivo* from dissociated cells attached to biodegradable polymer scaffolds *in vitro*. *J. Urol.* **148,** 658–662.

Bader, A., Knop, E., Boker, K., Fruhauf, N., Schuttler, W., Oldhafer, K., Burkhard, R., Pichlmayr, R., and Sewing, K. F. (1995). A novel bioreactor design for *in vitro* reconstruction of *in vivo* liver characteristics. *Artif. Organs* **19,** 368–374.

Beigel, C., and Pachter, J. S. (1994). Growth of brain microvessel endothelial cells on collagen gels: Applications to the study of blood–brain barrier physiology and CNS inflammation. *In Vitro Cell. Dev. Biol. Anim.* **30A,** 581–588.

Bell, E. (1995a). Deterministic models for tissue engineering. *Mol. Eng.* **1,** 28–34.

Bell, E. (1995b). Strategy for the selection of scaffolds for tissue engineering. *Tissue Eng.* **1,** 163–179.

Bell, E., and Scott, S. (1929). Tissue fabrication: Reconstitution and remodeling *in vitro*. *Mater. Res. Soc. Symp. Proc.* **252,** 141.

Bell, E., Ivarsson, B., and Merrill, C. (1979). Production of a tissue-like structure by contraction of collagen lattices by human fibroblasts of different proliferative potential *in vitro*. *Proc. Natl. Acad. Sci. U.S.A.* **76,** 1274.

Bell, E., Ehrlich, H. P., Buttle, D. J., and Nakatsuji, T. (1981). Living tissue formed *in vitro* and accepted as skin equivalent tissue of full thickness. *Science* **211,** 1052.

Bell, E., Sher, S., Hull, B., Merrill, C., Rosen, S., Chamson, A., Asselineu, D., Dubertret, I., Coulomb, B., Lapiere, C., Nusgens, B., and Neveux, Y. (1983). The reconstitution of living skin. *J. Invest. Dermatol.* **81,** s2–s10.

Bell, E., Moore, H., Mitchie, C., Sher, S., and Coon, H. (1984). Reconstitution of a thyroid gland equivalent from cells and matrix materials. *J. Exp. Zool.* **232,** 277–285.

Bell, E., Rosenberg, M., Kemp, P., Parenteau, N., Haimes, H., Chen, J., Swiderek, M., Kaplan, F., Kagan, D., Mason, V., and Boucher, L. (1989). Reconstitution of living organ equivalents from specialized cells and matrix biomolecules. *Colloq.—Inst. Natl. Sante Rech. Med.* **177,** 13–28.

Bell, E., Kagan, D., Nolte, C., Mason, V., and Hastings, C. (1990). Demonstration of barrier properties exhibited by brain capillary endothelium cultured as a monolayer on a tissue equivalent. *J. Cell Biol.* **5,** 185a.

Bell, E., Rosenberg, M., Kemp, P., Gay, R., Green, G., Muthukumaran, N., and Nolte, C. (1991). Recipes for reconstituting skin. *J. Biomech. Eng.* **113,** 113.

Bergsteinsdottir, K., Hashimoto, Y., Brennan, A., Mirsky, R., and Jessen, K. R. (1993). The effect of three dimensional collagen type I preparation on the structural organization of guinea pig enteric ganglia in culture. *Exp. Cell Res.* **209,** 64–75.

Chvapil, M. (1977). Collagen sponge: Theory and practice of medical applications. *J. Biomed. Mater. Res.* **11,** 721–741.

Coulomb, B., Dubertret, L., Bell, E., Merrill, C., Fosse, M., Breton-Gorius, J., Prost, C., and Touraine, R. (1983). Endogenous peroxidases in normal human dermis: A marker of fibroblast differentiation. *J. Invest. Dermatol.* **8,** 75–78.

Cuono, C. B., Langdon, R., Birchall, N., Barttelbort, S., and McGuire, J. (1987). Composite autologous–allogeneic skin replacement: Development and clinical application. *Plast. Reconstr. Surg.* **80,** 626.

Dennis, J. E., Haynesworth, S. E., Young, R. G., and Caplan, A. I. (1992). Osteogenesis in marrow-derived mesenchymal cell porous ceramic composites transplanted subcutaneously: Effect of fibronectin and larninin on cell retention and rate of osteogenic expression. *Cell Transplant.* **1,** 23–32.

Dodson, J. W. (1963). On the nature of tissue interactions in embryonic skin. *Exp. Cell Res.* **31,** 233.

Freiberger, H., Grove, D., Sivarajah, A., and Pinnell, S. R. (1980). Procollagen I synthesis in human skin fibroblasts: Effect of culture conditions on biosynthesis. *J. Invest. Dermatol.* **75,** 425–430.

Glowacki, J., Mizuno, S., and Greenberger, J. S. (1998). Perfusion enhances functions of bone marrow stromal cells in three-dimensional culture. *Cell Transplant.* 7, 319–326.

Gorham, S. D. (1991). Collagen. *In* "Biomaterials" (D. Byrom, ed.), pp. 57–122. Stockton Press, New York.

Guthrie, S., and Pini, A. (1995). Chemorepulsion of developing motor axons by the floor plate. *Neuron* 14, 1117–1130.

Hansbrough, J. F., Morgan, J. L., Greenleaf, G. E., and Bartel, R. (1993). Composite grafts of human keratinocytes grown on a polyglactin mesh-cultured fibroblast dermal substitute function as a bilayer skin replacement in full-thickness wounds on athymic mice. *J. Burn Care Rehabil.* 14, 485.

Hansbrough, J. F., Morgan, J., Greenleaf, G., and Underwood, J. (1994). Development of a temporary living skin replacement composed of human neonatal fibroblasts cultured in Biobrane, a synthetic dressing material. *Surgery* 115, 633.

Hauselmann, H. J., Fernandes, R. F., Mok, S. S., Schmid, T. M., Block, J. A., Aydelotte, M. B., Kuettner, K. E., and Thonar, E. J. (1994). Phenotypic stability of bovine articular chondrocytes after long-term culture in alginate beads. *J. Cell Sci.* 107, 17.

Heck, E. L., Bergstressgr, P. R., and Baxter, C. R. (1985). Composite skin graft: Frozen dermal allografts support the engraftment and expansion of autologous epidermis. *J. Trauma* 25, 106.

Hirai, J., Kanda, K., Oka, T., and Matsuda, T. (1994). Highly oriented, tubular hybrid vascular tissue for a low pressure circulatory system. *ASAIO J.* 40, M383–M388.

Kagan, D., and Bell, E. (1989). Differentiation of a fatty connective tissue equivalent without chemical induction. *J. Cell Biol.* 107, 603a.

Kanda, K., Matsuda, T., and Oka, T. (1993). Mechanical stress induced cellular orientation and phenotypic modulation of 3-D cultured smooth muscle cells. *ASAIO J.* 39, M686–M690.

Kneser, U., Kaufman, P. M., Fiegel, H. C., Pollok, J. M., Kluth, D., Herbst, H., and Rogiers, X. (1999). Long-term differentiated function of heterotypically transplanted hepatocytes on three-dimensional polymer matrices. *J Biomed. Mater. Res.* 47, 494–503.

Krebsbach, P. H., Mankani, M. H., Satomura, K., Kuznetsov, S. A., and Robey, P. G. (1998). Repair of craniotomy defects using bone marrow stromal cells. *Transplantation* 66, 1272–1278.

Lanza, R. P., Ecker, D., Kuhtreiber, W. M., Staruk, J. E., Marsh, J., and Chick, W. L. (1995). A simple method for transplanting discordant islets into rats using alginate gel spheres. *Transplantation* 59, 1485.

Lee, E. Y.-H., Lee, W. H., Kaetzel, C. S., Parry, G., and Bissell, M. J. (1985). Interaction of mouse mammary epithelial cells with collagen substrata: Regulation of casein gene expression and secretion. *Proc. Natl. Acad. Sci. U.S.A.* 82, 1419–1493.

Leenslag, J. W., Kroes, M. T., Pennings, A. J., and van der Lei, B. (1988). A compliant, biodegradable, vascular graft: Basic aspects of its construction and biological performance. *New Polym. Mater.* 1, 111.

L'Heureux, N., Germain, L., Labbe, R., and Auger, F. A. (1993). *In vitro* construction of a human blood vessel from cultured vascular cells: A morphologic study. *J. Vasc. Surg.* 17, 499–509.

MacPhee, M. J., Ailtrout, R. H., McCormick, K. L., Sayers, T. J., and Pilaro, A. M. (1994). A method for obtaining and culturing large numbers of purified organ-derived murine endothelial cells. *J. Leukocyte Biol.* 55, 467–475.

Minami, Y., Sugihara, H., and Oono, S. (1993). Reconstruction of cornea in three-dimensional collagen gel matrix culture. *Invest. Ophthalmol. Visual Sci.* 34, 2316–2324.

Mishima, H., Hibino, T., Hara, H., Murakami, J., and Otori, T. (1998). SPARC from corneal epithelial cells modulates collagen contraction by keratocytes. *Invest. Ophthalmol. Visual Sci.* 39, 2547–2553.

Montgomery, R. K. (1986). Morphogenesis *in vitro* of dissociated fetal rat small intestine cells upon an open surface and subsequent to collagen gel overlay. *Dev. Biol.* 117, 64–70.

Mooney, D. J., Kaufmann, P. M., Sano, K., McNamara, K. M., Vacanti, J. P., and Langer, R. (1994). Transplantation of hepatocytes using porous, biodegradable sponges. *Transplant. Proc.* 26, 3425–3426.

Mori, M., Sadahira, V., Kawasaki, S. *et al.* (1989). Formation of capillary networks from bone marrow cultured in collagen gel. *Cell. Struct. Funct.* 14, 393–398.

Nerem, R. M., Mitsumata, M., Ziegler, T., Berk, B. C., and Alexander, R. W. (1993). Mechanical stress effects on vascular endothelial cell growth. *In* "Tissue Engineering: Current Perspectives" (E. Bell, ed.), pp. 120–127. Birkhaeuser, Boston.

Nishikawa, Y., Tokusashi, Y., Kadohama, T., Nishimori, H., and Ogawa, K. (1996). Hepatocytic cells form bile duct-like structures within a three dimensional collagen gel matrix. *Exp. Cell Res.* 223, 357–371.

Nusgens, B., Merrill, C., Lapiere, C., and Bell, E. (1984). Collagen biosynthesis by cells in a tissue equivalent matrix *in vitro. Collagen Relat. Res.* 4, 351–364.

Parenteau, N., Nolte, C., Bilbo, P., Rosenberg, M., Wilkins, L., Johnson, E., Watson, S., Mason, V., and Bell, E. (1991). Epidermis generated *in vitro:* Practical considerations and applications. *J. Cell. Biochem.* 45, 245.

Parenteau, N. L., Sabolinski, M. L., Wilkins, L. M., and Rovee, D. T. (1994). Development of a bilayered "skin equivalent": From basic science to clinical use. *J. Cell. Biochem. Suppl.,* pp. 180–273.

Puelacher, W. C., Wisser, J., Vacanti, C. A., Ferraro, N. F., Jaramillo, D., and Vacanti, J. P. (1994). Temporomandibular joint disc replacement made by tissue-engineered growth of cartilage. *J. Oral Maxillofacial Surg.* 52, 1172.

Porter, R. A., Brown, R. A., Eastwood, M., Occleston, N. L., and Khaw, P. T. (1998). Ultrastructural changes during contraction of collagen lattices by ocular fibroblasts. *Wound Repair Regeneration* 6, 157–166.

Research News (1995). Helping neurons find their way. *Science* 268, 971.

Rogers, G. E., Martinent, N., Steinert, P., Wynn, P., Roop, D., Kilkenny, A,. Morgan, D., and Yuspa, S. H. (1988). A procedure for the culture of hair follicles as functionally intact organoids. *Clin. Dermatol.* 6, 36.

Rowling, R. J. E., Raxworthy, M. J., Wood, E. J., and Kearney, J. N. (1990). Fabrication and reorganization of dermal equivalents suitable for skin grafting after cutaneous injury. *Biomaterials* 11, 181–185.

Sabbagh, W., Masters, J. R., Duffy, P. G., Herbage, D., and Brown, R. A. (1998). *In vitro* assessment of a collagen sponge for engineering urothelial grafts. *Br. J. Urol.* **82**, 888–894.

Saiag, P., Coulomb, B., Lebreton, C., Bell, E., and Dubertret, L. (1985). Psoriatic fibroblasts induce hyperproliferation of normal keratinocytes in a skin equivalent model *in vitro*. *Science* **230**, 669.

Sandusky, G. E., Jr., Badylak, S. F., Morff, R. J., Johnson, W. D., and Lantz, G. (1992). Histologic findings after *in vivo* placement of small intestine submucosal vascular grafts and saphenous veins grafts in the carotid artery of dogs. *Am. J. Pathol.* **140**, 317.

Sarkar, R., Dickinson, C. J., and Stanley, J. C. (1999). Effect of somatostatin, somatostatin analogs, and endothelial cell somatostatin gene transfer on smooth muscle cell proliferation *in vitro*. *J. Vasc. Surg.* **29**, 685–693.

Schoop, V. M., Miranecea, N., and Fusenig, N. E. (1999). Epidermal organization and differentiation of HaCaT keratinocytes in organotypic coculture with human dermal fibroblasts. *J. Invest. Dermatol.* **112**, 343–353.

Shakado, S., Sakisaka, S., Noguchi, K., Yoshitake, M., Harada, M., Mimura, Y., Sata, M., and Tanikawa, K. (1995). Effects of extracellular matrices on tube formation of cultured rat hepatic sinusoidal endothelial cells. *Hepatology* **22**, 969–973.

Sher, S., Hull, B., Rosen, S., Church, D., Friedman, L., and Bell, E. (1983). Acceptance of allogeneic fibroblasts in skin equivalent transplants. *Transplantation* **36**, 552.

Soon-Shiong, P., Feldman, E., Nelson, R., Heintz, R., Yao, Q., Yao, Z,. Zheng, T., Merideth, N., Skjak-Braek, G., Espevik, T., Smidsrod, O., and Sandford, P. (1993). Longterm reversal of diabetes by the injection of immunoprotected islets. *Proc. Natl. Acad. Sci. U.S.A.* **90**, 5843–5847.

Streuli, C. H., Bailey, N., and Bissell, M. J. (1991). Control of mammary epithelial differentiation: Basement membrane induces tissue-specific gene expression in the absence of cell–cell interaction and morphological polarity. *J. Cell Biol.* **115**, 1383–1395.

Stronck, J. W. S., van der Lei, B., and Wildevuur, C. R. H. (1992). Improved healing of small-caliber polytetrafluorethylene vascular prostheses by increased hydrophilicity and by enlarged fibril length. *J. Thorac. Cardiovasc. Surg.* **103**, 146–152.

Swope, V. P., Supp, A. P., Cornelius, J. R., Babcock, G. F., and Boyce, S. T. (1997). Regulation of pigmentation in cultured skin substitutes by cytometric sorting of melanocytes and keratinocytes. *J. Invest. Dermatol.* **109**, 289–295.

Takeshita, K., Ishibashi, H., Suzuki, M., Yamamoto, T., Akaike, T., and Kodama, M. (1995). High cell-density culture system of hepatocytes entrapped in a three dimensional hollow fiber module with collagen gel. *Artif. Organs* **19**, 191–193.

Tomooka, Y., Kitani, H., Jing, N., Matsushima, M., and Sakakura, T. (1993). Reconstruction of neural tube-like structures *in vitro* from primary neural precursor cells. *Proc. Acad. Sci. U.S.A.* **90**, 9683–9687.

Toner, M., Tompkins, R. G., and Yarmush, M. L. (1993). Liver support through hepatic tissue engineering. *In* "Tissue Engineering: Current Perspectives" (E. Bell, ed.), pp. 92–110. Birkhaeuser, Boston.

Topol, B. M., Haimes, H. B., Dubertret, L., and Bell, E. (1986). Transfer of melansosomes in a skin equivalent model *in vitro*. *J. Invest. Dermatol.* **87**, 642–647.

Torbert, J., and Ronziere, M. C. (1984). Magnetic alignment of collagen during self-assembly. *biochem. J.* **219**, 1057–1059.

Traiffort, E., Charytoniuk, D. A., Faure, H., and Ruat, M. (1998). Regional distribution of Sonic Hedgehog, patched, and smoothened mRNA in the adult rat brain. *J. Neurochem.* **70**, 1327–1330.

Tranquillo, R. T. (1999). Self-organization of tissue equivalents: The nature and role of contact guidance. *Biochem. Soc. Symp.* **65**, 27–42.

Tranquillo, R. T., Girton, t. S., Bromberek, B. A., Triebes, T. G., and Mooradain, D. L. (1996). Magnetically oriented tissue-equivalent tubes: Application to a circumferentially oriented media-equivalent. *Biomaterials* **17**, 349–357.

van der Lei, B. *et al.* (1991). Experimental microvenous reconstructions with Gor-Tex polytetrafluoroethylene prosthesis implanted by means of the sleeve anastomotic technique. *Microsurgery* **12**, 23–29.

Vernon, R. B., and Sage, E. H. (1999). A novel, quantitative model for study of endothelial cell migration and sprout formation within three dimensional collagen matrices. *Microvasc. Res.* **57**, 118–133.

Vortkamp, A., Pathi, S., Peretti, G. M., Caruso, E. M., Zaleske, D. J., and Tabin, C. J. (1998). Recapitulation of signals regulating embryonic bone formation during postnatal growth and in fracture repair. *Mech. Dev.* **71**, 65–76.

Wakitani, S., Goto, T., Young, R. G., Mansour, J. M., Goldberg, V. M., and Caplan, A. I. (1998). Repair of large full-thickness articular cartilage defects with allograft chondrocytes embedded in a collagen gel. *Tissue Eng.* **4**, 429–444.

Weinberg, C., and Bell, E. (1986). A blood vessel model constructed from collagen and cultured vascular cells. *Science* **231**, 397–400.

Wodarz, A., and Nusse, R. (1998). Mechanisms of Wnt signaling in development. *Annu. Rev. Cell Dev. Biol.* **14**, 59–88.

Yamaguchi, T., Shin, T., and Sugihara, H. (1996). Reconstruction of laryngeal mucosa. A three dimensional collagen gel matrix culture. *Arch. Otolaryngol. Head Neck Surg.* **122**, 649–654.

Yarmush, M. L., Toner, M., Dunn, J. C. Y., Rotern, A., Hubel, A., and Tompkins, R. G. (1992). Hepatic tissue engineering: Development of critical technologies. *Ann. N. Y. Acad. Sci.* **665**, 472.

Zhang, S., Smartt, H., Holgate, S. T., and Roche, W. R. (1999). Growth factors secreted by bronchial epithelial cells control myofibroblast proliferation: An *in vitro* co-culture model of airway remodeling in asthma. *Lab. Invest.* **79**, 395–405.

Zietlow, R., Dunnett, S. B., and Fawcett, J. W. (1999). The effect of microglia on embryonic dopaminergic neuronal survival *in vitro*: Diffusible signals from neurons and glia change microglia from neurotoxic to neuroprotective. *Eur. J. Neurosci.* **11**, 1657–1667.

CHAPTER 17

Quantitative Aspects of Tissue Engineering: Basic Issues in Kinetics, Transport, and Mechanics

Alan J. Grodzinsky, Roger D. Kamm, and Douglas A. Lauffenburger

INTRODUCTION

A chapter on quantitative aspects of tissue engineering remains difficult to write at this point in the development of tissue regeneration and cell therapy processes. Although cell functional behavior and underlying molecular mechanisms are becoming increasingly amenable to quantitative approaches, incorporating these into the complex interactions among multiple cell types in organized tissue is currently beyond the scope of feasibility. A historical parallel may be drawn to the decades-long lag between the advent of the petrochemical industry and the useful introduction of rigorous analysis in terms of fundamental physicochemical theories decades later. Thus, it is clear that today's beginning attempts in tissue engineering must be highly empirical, with design based largely on intuition and experience, much like the petrochemical industry in the 1930s.

Following this parallel, however, it can be recalled that in the 1940s the concept of unit operations appeared, in which each particular chemical production plant could be broken down into some component processes that had some similarities to processes in different types of production plants. In this way, basic principles of chemical reactors, heat exchangers, material separation equipment, and so forth were elucidated in very simple quantitative terms. It was not that mathematical models for these unit operations were immediately combined into a comprehensive quantitative description of the entire plant—which is, of course, common practice today—but instead merely that important design parameters of the components could be identified, and in some cases the direction of their manipulation for process improvement could be indicated.

This, then, appears to be the state of affairs for tissue engineering at the beginning of the twenty-first century: that key parameters governing components of the overall device or procedure might be identifiable, and some design principles for how they could be altered toward an improved device or procedure might be developed. In a manner analogous to the unit operations, simple mathematical descriptions of cell and tissue processes can be constructed for purposes of elucidating what properties matter and how they can be manipulated. Our chapter therefore provides a brief overview of how major cell and tissue properties can be described and quantified in most basic form. These properties include molecular and cell transport through tissue, molecular interactions with cells, and tissue mechanics.

MOLECULAR INTERACTIONS WITH CELLS

There are three main classes of molecules that must be dealt with in the context of cell interactions in tissue: soluble nutrients, soluble signaling molecules, and molecules associated with the extracellular matrix. The latter two classes interact with cells primarily via cell surface receptors, whereas the former class can either bind to cell surface receptors or pass directly across the cell membrane.

NUTRIENTS

For nutrient molecules that pass directly across the cell membrane—either by passive diffusion or via carrier proteins—such as oxygen, glucose, and amino acids, the kinetics of uptake and metabolism generally follow a Michaelis–Menten type dependence. That is, at low concentrations the rate is first order in concentration but as concentration increases the rate asymptotically approaches a constant plateau. In combination with transport rates through tissue, as described below, the consumption of nutrients can be analyzed in a fairly straightforward manner. If transport is modeled as simple diffusion through a medium in which cells are embedded as point sinks, the nutrient concentration distribution, $L(x)$, in the tissue is governed by the combined diffusion/reaction equation:

$$\frac{\partial L}{\partial t} = D\frac{\partial^2 L}{\partial x^2} - \frac{\rho k L}{K_M + L},$$ (1)

where D is the diffusion coefficient, k is the maximal uptake rate constant per cell. K_M is the saturation constant, ρ is the cell density, and x is the spatial distance from the source. For simple nutrients, the source is typically the bloodstream. Significant depletion of nutrient, and hence possible nutrient deprivation, will occur when the ratio of uptake to diffusion becomes small. This ration can be usefully expressed in terms of the Thiele modulus: $\phi = (\rho k X^2 / D)^{1/2}$, where X is the overall distance away from the source—for simple nutrients, this can be considered the mean distance between microcirculatory blood vessels. (this is an approximate expression, assuming that the uptake rate is constant with value k throughout the tissue.) When $\phi < 1$, the steady-state nutrient concentration becomes significantly less than the exogenous level. Transplantation of cells within a polymeric matrix can often give rise to depletion of important nutrients from levels required for sustained cell viability, when ρ is great and X is large because the implant is not adequately vascularized. Nutrient transport limitations have been examined both theoretically and experimentally for the important case of encapsulated cells (Colton, 1995).

GROWTH FACTORS AND OTHER REGULATORY MOLECULES

Many molecules important in tissue engineering, however, cannot be dealt with quite so simply. Regulatory molecules such as growth factors, along with nutrient carriers such as iron-bearing transferrin, bind reversibly to plasma membrane receptors and are then internalized by the cell by means of invaginating membrane structures in a process known as endocytosis. The receptor/ligand complexes are carried to intracellular organelles termed endosomes, from which they are sorted to a variety of fates, including lysosomal degradation and recycling to the cell surface (see Fig. 17.1). This entire process, known as trafficking, is quite complicated, and the distribution of fates for a particular ligand can quantitatively vary as a function of its concentration as well as some of its biochemical and biophysical properties.

In the absence of trafficking, and when there is only a simple one-step reversible binding process, kinetic mass action equations can be written for the number of bound and free surface receptors and their ligands (L). At equilibrium, the number (C) of receptors (R) found in the bound state is given by the expression

$$C = \frac{LR_T}{K_D + L}.$$ (2)

R_T is the total number of surface receptors, constant only when trafficking processes are eliminated. K_D is the dissociation equilibrium constant, equal to the ratio of the receptor/ligand dissociation rate constant to the association rate constant: k_r / k_f. It is essentially the reciprocal of the receptor/ligand binding affinity. Commonly, however, ligands can bind to multiple receptors on the surface of a given cell, or receptors can be found in various states possessing different ligand

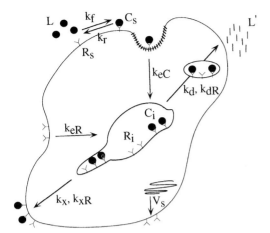

Fig. 17.1. Schematic illustration of receptor/ligand trafficking. L is intact ligand, L' is undegraded ligand; R_s, C_s, R_i, and C_i are cell surface and intracellular free and bound receptors; k_f and k_r are receptor/ligand association and dissociation rate constants; k_{eR} and k_{eC} are internalization rate constants for free and bound receptors; k_d and k_{dR} are degradation rate constants for ligand and receptor; k_x and k_{xR} are recycling rate constants for ligand and receptor; V_s is the receptor synthesis rate.

binding affinities. The interested reader is referred elsewhere for a detailed discussion of how to treat these more complicated situations (Lauffenburger and Linderman, 1996).

However, trafficking is almost universally present under physiological conditions, and the overall dynamics of uptake and metabolism of ligands and their receptors are strongly influenced by internalization, degradation, and recycling. Kinetic mass action equations can be written for the set of events shown in Fig. 17.1, and the dynamics of ligands and receptors in the various cell and tissue compartments can be thereby analyzed for purposes of understanding key design parameters (Lauffenburger and Linderman, 1996).

For instance, consider a molecular ligand delivered to a tissue at rate V_L. For ligands such as growth factors, the source may be other cells in the tissue—perhaps implanted cells expressing a desired growth factor—or a polymeric controlled-release device or the bloodstream. Assuming a spatially homogeneous source, the steady-state ligand concentration will be approximately given by (assuming, for simplicity, that receptor and ligand recycling are negligible) the expression

$$L = K_D \left(\frac{k_{eR}}{k_{eC}} \right) \left[\left(\frac{\rho V_R' + k_{deg} k_{eR}/k_f}{V_L} - 1 \right) \right]^{-1}, \tag{3}$$

where V_R' is the receptor synthesis rate per cell, k_{eR} and k_{eC} are the free receptor and bound receptor internalization rate constants, respectively, and k_{deg} is a rate constant for extracellular proteolytic ligand degradation.

Following Wiley (1985), the corresponding steady-state number of receptor/ligand complexes per cell, which typically governs the functional response, is then approximately given by

$$C = \frac{L R_{max}}{K_{SS} + L}. \tag{4}$$

R_{max} is equal to V_R'/k_{eR}, and K_{ss} is equal to $k_{eC} k_f/[k_{eR}(k_r + k_{eC})]$. Note the similarity in form to Eq. (2), which governs the number of complexes in the absence of trafficking. Thus, trafficking determines the number of available surface receptors and the effective ligand binding affinity.

When ligand concentrations are not spatially homogeneous, then the local value of L at any spatial position within the tissue must be determined by including the effects of diffusion as in Eq. (1). Here x would be the thickness of tissue away from the source. Moreover, in place of the simple metabolic uptake rate constant, k, the net dynamics of trafficking must be incorporated to account for ligand consumption.

Binding and trafficking rate constants can be measured in cell culture assays, using ligands labeled with fluorescent or radioactive tags. Descriptions of basic experimental procedures and corresponding methods of analysis can be found elsewhere (Lauffenburger and Linderman, 1996).

Extracellular matrix components can compete with cells for binding ligand, and in some circumstances may serve as a reservoir due to the reversible nature of protein/protein association. At equilibrium, this binding can be characterized using an expression analogous to Eq. (2). Extracellular matrix components may exhibit binding interactions with cell surface receptors; when they are sufficiently immobilized by their linkage within the matrix, Eq. (2) again can represent these interactions in the simplest case. As with soluble ligands, though, more complicated interactions can arise with multiple receptors for different domains of a particular matrix component.

Expressions such as Eqs. (1)–(4) can be used to estimate the rate of ligand delivery required to maintain a desired level of cell surface complexes and/or free ligand concentration. The number of surface complexes typically is important for determining the cell functional response. We emphasize, though, that it is likely that the rate of ligand binding to receptors to form signaling complexes may govern cell responses as much, or more so, than the static level of complexes (Wiley, 1992). Quantification of the dependence of cell behavior on complex levels and dynamics, however, remains in its infancy at present. A small number of particular examples can be offered as guidelines, including proliferation responses of fibroblasts to epidermal growth factor (Knauer *et al.,* 1984), leading to improvements in cell level principles for growth factor design and delivery (Reddy *et al.,* 1996a,b). A further level of complication in analysis arises from the presence of autocrine ligands in a wide spectrum of tissue physiological processes, though quantitative modeling and experimental studies are now being pursued on this aspect as well (Forsten and Lauffenburger, 1992; Lauffenburger *et al.,* 1998).

MOLECULAR AND CELL TRANSPORT THROUGH TISSUE

Movements of molecular as well as cellular species through tissue is a vital aspect of most tissue engineering processes, because the whole point of the intervention is to alter the composition of a local region from what it had been prior to the therapy. As mentioned above, transport of nutrients and regulatory factors into a cellular implant is generally required. Also, the goal of certain implanted devices is to release regulatory factors into the surrounding tissue. The synthesis of neotissues involves placing cells within a matrix composed of natural materials (e.g., collagen) or synthetic polymers (e.g., biodegradable templates). Within this environment, the cells also secrete and regulate the assembly of their own extracellular matrix composed of collagens, proteoglycans, glycoproteins, and a variety of macromolecules that are important in cell–matrix and cell–cell interactions. Solute transport through this tortuous and often dense matrix plays a critical role in the maintenance of cell viability and in the successful growth and development of the intact tissue. Moreover, colonization of implants or grafts by cultured or host cells is often desirable. And, in some cell therapies, mainly those involving immune white blood cells, the movement of cells from an injection site into certain tissue regions is essential for them to carry out their intended functions.

Molecular Diffusion and Convection

Cell biosynthesis requires transport of low-molecular-weight metabolites and waste products, which occurs primarily by diffusion. Many other solutes play key roles as mediators of cell behavior and in cell production and turnover of extracellular matrix, including growth factors, hormones, cytokines, and endogenous proteinases and their natural inhibitors. Transport of regulatory factors can be crucial in tissue regeneration (Perez *et al.,* 1995) and can be reduced severly by interactions with matrix (Dowd *et al.,* 1999). In addition, the assembly of a physicochemically and mechanically functional matrix requires transport of newly synthesized matrix macromolecules to appropriate locations within the extracellular space. These matrix proteins and proteoglycans (as large as several megadaltons) have such low intratissue diffusivities that even slight fluid convection within the matrix can significantly enhance macromolecular transport (Garcia *et al.,* 1996).

In general, transport depends on the size and charge of the solute, and the density (tortuosity), water content, and charge of the matrix (Garcia *et al.,* 1998) (see Fig. 17.2). Although the principal modes of transport are molecular diffusion and convection (Deen, 1987), electrostatic interactions can also affect the partitioning and transport of charged solutes into and within a charged matrix (Grodzinsky, 1983). In this regard, it is useful to note that matrix charge density at the macroscale is often associated with the content of glycosaminoglycan chains attached to proteoglycans. At the scale of the cell, the dense, negatively charged glycocalyx at the surface of many cell types can also affect transport across the cell membrane.

Each cell and tissue type presents unique challenges for providing adequate transport in tis-

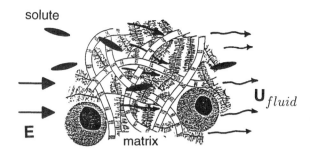

solute

U_{fluid}

E

matrix

Fig. 17.2. The flux of macromolecular and small solutes within a cell-seeded matrix results from diffusion, convection (fluid velocity U), and electrical migration if the solute is charged; an electric field E exists within the tissue (caused by fluid flow or by ion concentration gradients).

sue engineering design. For example, the density of cells in liver and kidney is typically much higher than that in connective tissues such as ligament, tendon, and cartilage. On the other hand, connective tissue cells grow within a very dense, hydrated, negatively charged matrix that may be several millimeters thick but with little or no vascular or lymph supply. Some cells are best grown on surfaces whereas others require three-dimensional matrix encapsulation. The mechanical environment of cells also varies greatly: vascular cells are subjected to fluid shear and to pulsatile stresses and strains within vessel walls; cartilage, bone, and other soft connective tissue cells must sustain peak mechanical stresses as high as 10–20 MPa (100–200 atm). These dynamic stresses produce fluid flows within the matrix that can affect intratissue solute transport and the concentration of solute species at the boundaries of tissues *in vivo* (O'Hare *et al.,* 1990). Such effects must therefore be taken into account in tissue engineering design *in vitro.*

Given these diverse tissue types and geometric constructs, mathematical modeling of transport processes can be critically important in the design of tissue reactors, the basic understanding of the spatial and temporal distribution of small and large solutes in newly developing tissues, and the understanding of failure modes that might result from inadequate transport. We briefly summarize below the governing transport laws for solutes within tissues, and the boundary conditions at the tissue/media interface. These general laws may then be adapted to the particular cell and tissue types of interest.

First, the one-dimensional flux N_i of the ith species within the tissue due to diffusion, convection, and electrical migration (ohmic conduction) is described by the flux equation:

$$N_i = \phi\left(-D_i\frac{\partial c_i}{\partial x} + \frac{z_i}{|z_i|}\mu_i c_i E\right) + W_i c_i U, \tag{5}$$

where c_i is the solute concentration at the position x, ϕ is the matrix (tissue) porosity, z_i is the valence of the solute (if charged), U is the fluid velocity relative to the matrix (averaged over the total matrix area), and E is the electric field within the matrix (e.g., induced by diffusion of ionic species or by electrokinetic effects such as flow-induced streaming potentials through a charged matrix). The intrinsic transport parameters D_i, μ_i, and W_i represent, respectively, the intratissue diffusivity of the solute, the electrical mobility, and a hindrance factor for convection that incorporates hydrodynamic and steric interactions between solute molecules and the matrix molecules forming the "pore" walls. Even in the absence of net fluid convection, frictional hydrodynamic and steric interactions between solute, matrix, and solvent can play a critical role in transport, and may cause significant differences between the intratissue values of D_i, μ_i, and their corresponding values in free solution (Deen, 1987). In general, the flux Eq. (5) is general in that it applies to ionic solutes in the medium, small neutral and charged solutes (e.g., amino acids, glucose), and large neutral and charged macromolecular solutes. For small solutes, $W \rightarrow 1$, and the values of D_i and μ_i may be close to their free solution values; for larger solutes, $W < 1$, and D_i and μ_i as written in Eq. (5) implicitly incorporate associated hindrance factors (Deen, 1987).

It is therefore necessary to measure or estimate values for the intrinsic transport parameters D_i, μ_i, and W_i in order to predict solute distribution profiles within the tissue construct. To predict the spatial and temporal distribution of solutes in the neotissue, the flux equation is incorporated into the continuity law:

$$\frac{\partial c_i}{\partial t} = \frac{\partial N_i}{\partial x} + \sum_i (G_i - R_i), \tag{6}$$

where G_i and R_i represent chemical reaction rates associated with binding of solutes to cells or matrix molecules. In the absence of binding, electric migration, and convection effects, the continuity law, Eq. (6), reduces to Fick's second law for molecular diffusion. For the case of purely diffusive nutrient transport within a matrix containing cells acting as point sinks, the continuity law, Eq. (6), would reduce to the diffusion/reaction Eq. (1) above.

Electroneutrality within the tissue, including charges on all solute (i) and matrix molecules (m), requires that:

$$z_m c_m + \sum_i (z_i c_i) = 0. \tag{7}$$

Although tissues may have a high water content, the concentration of solutes within tissue water may be very different from that in the bathing medium because of size and charge effects. In general, the solute partition coefficient κ is defined as the ratio of the solute concentration within the tissue fluid to that in the bathing medium, $\kappa = c^{tiss}/c^{med}$. This partition coefficient acts as a boundary condition for the flux and continuity equations, which are written specifically in terms of solute concentrations inside the tissue. In addition, for charged solutes in the presence of a charged matrix, this partitioning is further modified due to Donnan equilibrium considerations for positively and negatively charged species, giving the Donnan partition boundary conditions (Maraudas, 1979):

$$\left(c_{i+}^{med}/c_{i+}^{tiss} \right)^{1/|z+|} = \left(c_j^{tiss}/c_{j-}^{med} \right)^{1/|z-|}. \tag{8}$$

Finally, it is useful to recast the continuity law in terms of the dimensionless Peclet number Pe, which is defined as the ratio of solute convection plus migration fluxes to the diffusive flux (Grimshaw *et al.*, 1989):

$$\frac{\partial c_i}{\partial t} = D_i \left[\frac{\partial^2 c_i}{\partial x^2} + \frac{Pe}{X} \left(\frac{\partial c_i}{\partial x} \right) \right], \tag{9}$$

$$Pe = X \left(\frac{W_i U + \phi \mu_i E}{\phi D_i} \right), \tag{10}$$

where X is the characteristic tissue thickness over which transport must occur. Thus, Eq. (9) takes the form of a modified Fick's second law, and puts into perspective the relative importance of the convective and electrical migration terms given values for X, D_i, W_i, and μ_i. It is apparent that for solutes having small enough D_i and/or large enough fluid flow, convection may have a significant effect in determining the solute distribution within the tissue. Said another way, convection and intratissue electrokinetic (e.g., electrophoretic) effects could significantly augment the transport of nutrients and macromolecules compared to diffusion alone, over tissue dimensions X.

Cell Migration

Polymeric scaffolds are often introduced, into which it is hoped that certain cell types will migrate while others do not. For instance, regeneration of bone tissue for enhanced healing of full-thickness injuries requires that osteoprogenitor cells from surrounding tissue colonize an implanted scaffold, and that endothelial cells migrate in, to achieve neovascularization, whereas connective tissue fibroblast influx is undesirable. There are two approaches for controlling migration: (1) surface chemistry of the implanted material for selective cell adhesion interactions with the surface, influencing migratory behavior in a differential manner among various cell types, and (2) release of chemotactic attractants from within the implanted material to induce enhanced migratory responses of particular cell types.

The rate of cell population movement into a tissue can be quantified using a cell transport flux expression analogous to that for molecular diffusion and convection (Tranquillo *et al.*, 1988):

$$J_{cells} = -\mu \frac{\partial N}{\partial x} + \chi \left(\frac{\partial A}{\partial x} \right) N, \tag{11}$$

where μ is the cell random motility coefficient, analogous to a molecular diffusion coefficient and χ is the chemotaxis coefficient; χ is a biphasic function of the attractant concentration, A, because

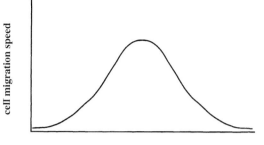

cell-substratum adhesion strength / cell contractile force

Fig. 17.3. Qualitative representation of typical dependence of cell migration speed on the ratio of cell-substratum adhesion strength to cell-generated contractile force.

chemotaxis arises from a spatial difference in the number of receptor/attractant complexes across a cell length. The first term in Eq. (11) represents cell dispersion down the spatial gradient of cell density, whereas the second term yields a directed, "convective" velocity for cell migration in the direction of an attractant concentration gradient. The values μ and χ can be measured in a variety of cell population assays, such as movement through a porous filter or under an agarose gel. They can alternatively be calculated from individual cell tracking assays in which cell speed, directional persistence time, and directional orientation bias are determined, because the cell population parameters are related by theory to the individual cell parameters. Details on these approaches can be found elsewhere (Lauffenburger and Linderman, 1996).

A fair amount of effort has been devoted to quantitative understanding of how substrata properties of materials affect cell migration. A central concept that is beginning to show general verification is that the speed of migration depends in biphasic manner on the strength of cell/substratum adhesiveness (see Fig. 17.3) (DiMilla *et al.*, 1991). This principle can account for a wide range of experimental observations of how migration speed varies with the density of extracellular matrix ligands immobilized on biomaterial surfaces, and for the effects of ligand affinity and competing soluble ligand (Palecek *et al.*, 1997; Wu *et al.*, 1994). This lack of monotonicity is unfortunate, of course, because it makes design of a material quite problematic; unless the full quantitative picture is well characterized one may produce a substratum either too strongly adhesive or too weakly adhesive for effective cell locomotion. Additional modulating approaches may be possible, though, by means of other parameters influencing migration speed. For instance, the dependence illustrated in Fig. 17.3 actually entails the ratio of cell/substratum adhesiveness to effective cell contractile force, so on a given substratum migration can be enhanced or inhibited by factors altering cell force generation or the substratum compliance.

CELL AND TISSUE MECHANICS

The mechanics of a tissue or tissue substrate play a central role in many situations, ranging from the fabrication of cell-based vascular grafts to wound healing. The transmission of stresses between the extracellular matrix and cell, or between the cell membrane and its nucleus, is an important factor in controlling the biological response of the cell to its environment. Even cell motility and adhesion are influenced by the elastic characteristics of the substrate on which they are grown. To appreciate these factors requires an understanding of mechanics, both at the microscale (e.g., a single microtubule and its interaction with the actin filaments) and at the macroscale (e.g., the elastic properties of a cell-based tissue implant). Some of the basic concepts are outlined here, along with some examples of how these concepts are applied in the field of tissue engineering.

ELASTICITY

Before dealing with some of the complexities of real biologic materials, consider first the case of a simple uniform and isotropic elastic medium. The elasticity of a material is typically characterized by its stress–strain relationship, where the stress is the force acting per unit area and the strain is the fractional change in length of the specimen, both of which are tensorial quantities. In the case of uniaxial stress, a material might exhibit a stress–strain relation of the type shown in Fig. 17.4. In this instance, the relationship is linear (or nearly so) for small strains, but becomes nonlinear as strains increase above a certain level. The elastic, or Young's, modulus (E) is defined as

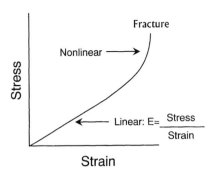

Fig. 17.4. A plot of stress (σ) vs. strain (ε), showing an initial linear region in which the Young's modulus (E) is constant, and a nonlinear region with increasing E, leading ultimately to fracture of the specimen.

the ratio of stress to strain; in this example, *E* would be constant for small strains, but eventually would increase as the material experiences increasing strain.

The elastic modulus of most biologic tissues is highly nonlinear, exhibiting an increasing elastic modulus for higher strains. The fibrous elements that comprise the tissue matrix, however, are more likely to be linearly elastic over an appreciable range of strain. Elastin, for example, which can be stretched up to 300% of its initial length, has a nearly constant modulus. Table 17.1 shows the elastic moduli for several tissues and tissue components. Another important property of the material is the maximum amount of strain it can experience before it fractures, some values of which are also given in the table.

In contrast to elastic deformation in which there exists a unique relation between the applied stress and material strain, in plastic deformation, the material experiences an irreversible deformation, usually at high levels of stress, and fails to return to its original length when the stress is removed.

All the above applies to uniaxial loading. If a material is isotropic, then its elasticity is the same, regardless of the direction in which it is tested, and its mechanical properties can be completely characterized by two parameters, e.g., its elastic modulus and Poisson ratio, *v*. Anisotropic materials, however, exhibit different stress–strain characteristics along the different coordinate directions. The wall of an artery, for example, is extremely stiff in the circumferential and axial directions, but relatively compliant in the radial direction. The description of such materials can lead to considerable complexity as evidenced by the fact that in a truly anisotropic material, 21 constants analogous to the Young's modulus are required to characterize completely a material's elasticity. Intermediate between these two extremes lie materials that are transversely isotropic, that is, materials that exhibit isotropic behavior in two dimensions, but different properties in the third. Many biological materials (blood vessels, ocular sclera) can be well characterized as transversely isotropic because the primary force-supporting structures (collagen and elastin fibers) are oriented primarily in one plane.

The existence of contractile elements, such as smooth muscle cells in vessel walls or the my-

Table 17.1. Elastic constants for a variety of biologic materials[a]

Material	Elastic modulus	Yield stress	Max. strain	Ref.
Cortical bone	6–30 GPa	50–200 MPa	—	Cowin *et al.* (1987)
Collagen fibers	500 MPa	50 MPa	0.1	Kato *et al.* (1989)
Elastin	100 kPa	300 kPa	3.0	Mulcherjee *et al.* (1976)
Cartilage	10 MPa	8–20 MPa	0.7–1.2	Woo *et al.* (1987)
Skin	35 MPa	15 MPa	1.1	Yamada (1970)
Muscle facia	340 MPa	15 MPa	1.17	Yamada (1970)
Tendon	700 MPa	60 MPa	0.10	Yamada (1970)

[a]Note that in many instances, the stress–strain relationship is highly nonlinear; the elastic moduli in those cases represent an approximate, characteristic value.

a

Stress
Removed

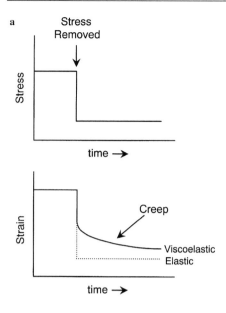

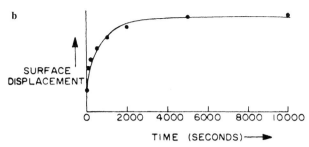

Fig. 17.5. (a) Stress (σ; top) and strain (ε; bottom) plotted against time in a maneuver in which the stress is abruptly reduced. In a purely elastic material (dotted line), the strain immediately adjusts to a new level consistent with the new applied stress. A viscoelastic material (solid line) exhibits a further reduction in strain (creep), indicative of continued deformation following an initial elastic response. (b) The response of a cylindrical sample of cartilage subjected to a sudden compressive stress. The sample continues to compress with time as a result of water being expelled from the tissue through a porous platen. Reproduced with permission from Mow and Mak (1987).

ocardium, or actin filaments in the cytoskeleton, lends further complexity to material properties. As far as the elastic properties of the material are concerned, active constriction can be thought of as influencing both the elastic modulus of the material and the length of the sample under zero stress.

Viscoelasticity and Pseudoelasticity

Although some materials (e.g., elastin, collagen) show an immediate elastic response such that they exhibit an immediate and constant deformation following the application of stress, tissues more often respond in a time-dependent fashion (Fig. 17.5). The initial elastic response of such materials is followed by a period of additional deformation or creep, asymptotically approaching an equilibrium state at long times. Alternatively, when subjected to a periodic load, these materials display a degree of hysteresis when stress is plotted against strain. This is termed a viscoelastic response and is a result of several processes internal to the tissue. One is the transient movement of liquid through the tissue due to the nonuniform distribution of interstitial pressure that results from the application of stress. For example, if a disk or cartilage is subjected to a load as in Fig. 17.5, a gradient in fluid pressure is set up between the central regions and the edge of the specimen, which causes fluid to be expelled. The rate of deformation during this phase is clearly dependent on the permeability of the tissue.

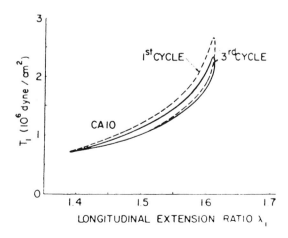

Fig. 17.6. Pseudoelastic behavior of biological tissue. A stress–strain plot for a segment of a canine carotid artery undergoing cyclic loading; first and third cycles shown. T_1 is the axial tension, λ_1 is the axial stretch ratio, and $L(t)/L_0$ with L_0 is the relaxed length. The sample exhibits different curves for loading and unloading, which are highly reproducible. Reproduced with permission, from Lee et al. (1967).

Maxwell body

Voigt body

Kelvin body

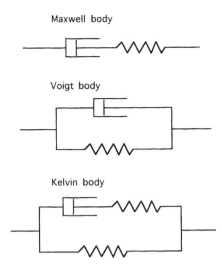

Fig. 17.7. Three commonly used models for viscoelastic behavior composed of viscous elements (dashpots) and elastic elements (springs).

Biological tissues are often referred to as pseudoelastic (Fung, 1981), in that, with periodic loading, they evolve to a stress–strain pattern that is repeatable from one cycle to the next and relatively insensitive to the strain rate. The resulting curve typically differs on extension from relaxation, exhibiting some degree of hysteresis (Fig. 17.6). By treating the loading and unloading maneuvers separately, the viscoelastic characteristics of the material can be incorporated into an otherwise elastic description of the material.

It is often necessary to simulate the true viscoelastic response of a tissue specimen, as, for example, when subjected to an abrupt increase in load (Fig. 17.5). Several simple models have been devised for this purpose. Although these make no attempt to mimic the mechanism responsible for the viscoelasticity, they are useful descriptors that can be used to represent certain viscoelastic characteristics. The most common among these models are the Maxwell, Voight, and Kelvin models. Each is composed of a one or more springs and dashpots as shown in Fig. 17.7.

These simple models are sometimes used as a basis for constructing more complex descriptions of the material. For example, a continuum description has been used to study the mechanical characteristics of leukocytes (Schmid-Schönboin *et al.*, 1981), based on the Kelvin model of Fig. 17.7, by arranging them in various series and parallel networks. Other models for the cytoskeleton take explicit account of the fibrous microstructure (Satcher and Dewey, 1996; MacKintosh and Janmey, 1997; Stamenovic *et al.*, 1996).

MEASUREMENT OF MECHANICAL PROPERTIES

A variety of methods have been employed to determine the mechanical characteristics of biological specimens, the method being determined by the elastic characteristics of the material, its size, and the particular property of interest.

For large-scale samples, the simplest measurement that yields the elastic modulus of a specimen is the uniaxial strain test, in which the sample is grasped at two ends and pulled while axial strain and stress are simultaneously measured. In order to minimize end effects, the sample is often necked down to a lateral dimension in the central section, smaller than the ends, and the strain is measured directly in the necked region. Because stresses in both directions perpendicular to the axial of the specimen are zero, the elastic modulus is determined by the ratio of stress to strain, $E = \sigma_X / \epsilon_X$. Poisson's ratio can be determined from the change of thickness of the specimen in the direction perpendicular to the applied stress. If the sample is anisotropic, additional uniaxial tests in the other two coordinate directions can be used. Alternatively, stresses can be applied in two (biaxial) or three (triaxial) dimensions simultaneously. These tests yield more information about the material, but do so at the cost of greater complexity.

These tests are most often used to measure the properties of a material in tension. The compressive characteristics are also often of interest, especially in materials such as bone and cartilage, which are often subjected to compressive loads *in vivo*. For this purpose, specimens are generally cut into the shape of a short cylinder and compressed between two platens. In the case of cartilage, the platens may be permeable to allow water to escape as the sample is compressed, thereby obtaining information on the permeability of the sample from the time-dependent compression following the application of a load. Both confined and unconfined compression tests are useful. The advantage of confined compression, in which the sample is placed in a rigid cylindrical chamber, typically with nonpermeable side walls, is that the stress, strain, and flow of water are purely axial and the results can be more easily interpreted.

Any of the tests just mentioned are capable of measuring nonlinearities in the stress–strain behavior of a material. As mentioned earlier, biological materials typically become stiffer as the sample is stretched. Therefore, it is important to test the material with loads in the range of those the specimen is likely to experience *in vivo*. It is equally important to know the zero-stress state of a tissue, which, as has been demonstrated for a variety of vessels, is not necessarily achieved simply by reducing the transmural pressure to zero (Liu and Fung, 1988).

All these tests require excised specimens to be cut or machined into a particular shape. Often in the case of biological tissues, it is desirable to obtain measurements from the *in vivo* state so as to avoid the artifacts that are necessarily introduced by specimen removal. Although advantageous in many respects, *in vivo* testing also limits the choice of testing methods available. Fung and Liu (1995) argued for the importance of making measurements *in vivo,* and have demonstrated how, in the case of a blood vessel, a battery of tests may be used to measure the mechanical properties of the arterial wall. Other methods—for example, the simple indentation of the material by a probe (Lai-Fook *et al.,* 1976)—can also be used.

Methods for testing the elastic and viscoelastic properties of cells and even individual molecules are rapidly becoming available. One approach, micropipette aspiration, has been used in several studies (Evans and Yeung, 1989) on various cell types. Introduced over 20 years ago to study red blood cell mechanics and later for leukocytes, the cell is partially drawn into a micropipette by means of a negative pressure. Knowing the pressures acting, and monitoring the extent of cell deformation by microscopy, allow for estimation of cell elasticity. Cell poking, a method analogous to the indentation studies just discussed, can also be applied to individual cells (Zahalak *et al.,* 1990).

Newer methods have been developed that take advantage of our ability to attach ligand-coated microspheres to cell receptors or individual molecules, and then manipulate these microspheres in a specified manner. Microsphere displacement can be controlled by the use of laser tweezers (Block *et al.,* 1989; Kuo and Sheetz, 1993; Kuo, 1995) or a magnetic trap (Glogauer and Ferrier, 1998); rotations can be produced by magnetic fields using a method termed magnetic twisting cytometry (Wang and Ingber, 1994). In these methods, both the applied forces (or torques) and displacements (or rotations) can be monitored, from which elastic and viscoelastic properties can be inferred.

REFERENCES

Block, S. M., Blair, D. F., and Berg, H. C. (1989). Compliance of bacterial flagella measured with optical tweezers. *Nature (London)* **338**, 514–518.
Colton, C. K. (1995). Implantable biohybrid artificial organs. *Cell Transplant.* 4, 415–436.
Cowin, S. C., Van Buskirk, W. C., and Ashman, R. B. (1987). Properties of bone. *In* "Handbook of Bioengineering" (R. Skalak and S. Chien, eds.). McGraw-Hill, New York.
Deen, W. M. (1987). Hindered transport of large macromolecules in liquid-filled pores. *AIChE J. 33,* 1409–1425.
DiMilla, P. A., Barbee, K., and Lauffenburger, D. A. (1991). Mathematical model for the effects of adhesion and mechanics on cell migration speed. *Biophys. J.* **60**, 15–37.
Dowd, C. J., Cooney, C. L., and Nugent M. A. (1999). Heparan sulfate mediates bFGF transport through basement membrane by diffusion with rapid reversible binding. *J. Biol. Chem.* **274**, 5236–5244.
Evans, E., and Yeung, A. (1989). Apparent viscosity and cortical tension of blood granulocytes determined by micropipette aspiration. *Biophys. J.* **56**, 151–160.
Forsten, K. E., and Lauffenburger, D. A. (1992). Autocrine ligand binding to cell receptors. *Biophys. J.* **61**, 518.
Fung, Y. C. (1981). "Biomechanics: Mechanical Properties of Living Tissues." Springer-Verlag, New York.
Fung, Y. C., and Liu, S. Q. (1995). Determination of the mechanical properties of the different layers of blood vessels *in vivo. Proc. Natl. Acad. Sci. U.S.A.* **92**, 2169–2173.
Garcia, A. M., Frank, E. H., Grimshaw, P. E., and Grodzinsky, A. J. (1996). Contribution of fluid convection and electrical migration to molecular transport: Relevance to loading. *Arch. Biochem. Biophys.* **333**, 317–325.
Garcia, A. M., Lark, M. W., Trippel, S. B., and Grodzinsky, A. J. (1998). Transport of TIMP-1 through cartilage. *J. Orthop. Res.* **16**, 734–742.
Glogauer, M., and Ferrier, J. A. (1998). A new method for application of force to cells via ferric oxide beads. *Pfluegers Arch.* **435**(2), 320–327.

Grimshaw, P. E., Grodzinsky, A. J., Yarmush, M. L., and Yarmush, D. M. (1989). Dynamic membranes for protein transport: Chemical and electrical control. *Chem. Eng. Sci.* **44**, 827–840.

Grodzinsky, A. J. (1983). Electromechanical and physicochemical properties of connective tissues. *CRC Crit. Rev. Biomed. Eng.* **9**, 133–199.

Kato, Y. P., Christiansen, D. L., Hahn, R. A., Shieh, J. J., Goldstein, J. D., and Silver, F. H. (1989). Mechanical properties of collagen fibers: A comparison of reconstituted and rat tail tendon fibers. *Biomaterials* **10**, 38–42.

Knauer, D. J., Wiley, H. S., and Cunningham, D. D. (1984). Relationship between epidermal growth factor receptor occupancy and mitogenic response: Quantitative analysis using a steady-state model system. *J. Biol. Chem.* **259**, 5623–5631.

Kuo, S. C. (1995). Optical tweezers: A practical guide. *JMSA* **1**(2), 65–74.

Kuo, S. C., and Sheetz, M. P. (1993). Force of single kinesin molecules measured with optical tweezers. *Science* **260**, 232–234.

Lai-Fook, S. J., Wilson, T. A., Hyatt, R. E., and Rodarte, J. R. (1976). Elastic constants of inflated lobes of dog lungs. *J. Appl. Physiol.* **40**(4), 508–513.

Lauffenburger, D. A., and Linderman, J. J. (1996). "Receptors: Models for Binding, Trafficking, and Signaling." Oxford University Press, New York.

Lauffenburger, D. A., Oehrtman, G. T., Walker, L., and Wiley, H. S. (1998). Real-time quantitative measurement of autocrine ligand binding indicates that autocrine loops are spatially localized. *Proc. Natl. Acad. Sci. U.S.A.* **95**, 15368.

Lee, J. S., Frasher, W. G., and Fung, Y. C. (1967). "Two-dimensional Finite-deformation on Experiments on Dog's Arteries and Veins," Tech. Rep. No. AFOSR 67-1980. University of California at San Diego.

Liu, S. Q., and Fung, Y. C. (1988). Zero-stress states of arteries. *ASME J. Biomech. Eng.* **110**, 82–84.

MacKintosh, F. C., and Janmey, P. A. (1997). Actin gels. *Curr. Opin. Solid State Mater. Sci.* **2**, 350–357.

Maroudas, A. (1979). "Adult Articular Cartilage," pp. 215–290. Pitman Medical, London.

Mow, V. C., and Mak, A. F. (1987). Lubrication of diarthroidal joints. *In* "Handbook of Bioengineering" (R. Skalak and S. Chien, eds.), Chapter 5. McGraw-Hill, New York.

Mukherjee, D. P., Kagan, H. M., Jordan, R. E., and Franzblau, C. (1976). Effect of hydrophobic elastin ligands on the stress–strain properties of elastin fibers. *Connect. Tissue Res.* **4**, 177.

O'Hare, B. P., Urban, J. P. G., and Maroudas, A. (1990). Influence of cyclic loading on the nutrition of articular cartilage. *Ann. Rheum. Dis.* **49**, 536–539.

Palacek, S. P., Loftus, J. C., Ginsberg, M. H., Lauffenburger, D. A., and Horwitz, A. F. (1997). Integrin/ligand binding properties govern cell migration speed through cell/substratum adhesiveness. *Nature (London)* **385**, 537.

Perez, E. P., Merrill, E. W., Miller, D., and Griffith Cima, L. (1995). Corneal epithelial wound healing on bilayer composite hydrogels. *Tissue Eng.* **1**, 263–277.

Reddy, C. C., Niyogi, S. K., Wells, A., Wiley, H. S., and Lauffenburger, D. A. (1996a). Engineering epidermal growth factor for enhanced mitogenic potency. *Nat. Biotechnol.* **14**, 1696.

Reddy, C. C., Wells, A., and Lauffenburger, D. A. (1996b). Receptor-mediated effects on ligand availability influence relative mitogenic potencies of epidermal growth factor and transforming growth factor alpha. *J. Cell. Physiol.* **166**, 512.

Satcher, R. L., Jr., and Dewey, C. F., Jr. (1996). Theoretical estimates of mechanical properties of the endothelial cell cytoskeleton. *Biophys. J.* **71**, 109–118.

Schmid-Schönbein, G. W., Sung, K. L. P., Tözeren, H., Skalak, R., and Chien, S. (1981). Passive mechanical properties of human leukocytes. *Biophys. J.* **36**, 243–256.

Stamenovic, D., Fredberg, J. J., Wang, N., Butler, J. P., and Ingber, D. E. (1996). A microstructural approach to cytoskeletal mechanics based on tensegrity. *J. Theor. Biol.* **181**, 125–136.

Tranquillo, R. T., Zigmond, S. H., and Lauffenburger, D. A. (1988). Measurement of the chemotaxis coefficient for human neutrophils in the under-agarose migration assay. *Cell Motil. Cytoskel.* **11**, 1–15.

Wang, N., and Ingber, D. (1994). Control of cytoskeletal mechanics by extracellular matrix, cell shape, and mechanical tension. *Biophys. J.* **66**, 2181–2189.

Wiley, H. S. (1985). Receptors as models for the mechanisms of membrane protein turnover and dynamics. *Curr. Top. Membr. Transp.* **24**, 369–412.

Wiley, H. S. (1992). Receptors: Topology, dynamics, and regulation. *Fundam. Med. Cell Biol.* **5A**, 113–142.

Woo, S. L.-Y., Mow, V. C., and Lai, W. M. (1987). Biomechanical properties of articular cartilage. *In* "Handbook of Bioengineering" (R. Skalak and S. Chien, eds.). McGraw-Hill, New York.

Wu, P., Hoying, J. B., Williams, S. K., Kozikowski, B. A., and Lauffenburger, D. a. (1994). Integrin-binding peptide in solution inhibits or enhances endothelial cell migration, predictably from cell adhesion. *Ann. Biomed. Eng.* **22**, 144.

Yamada, H. (1970). *In* "Strength of Biological Materials" (F. G. Evans, ed.). Williams & Wilkins, Baltimore, MD.

Zahalak, G. I., McConnaughey, W. B., and Elson, E. L. (1990). Determination of cellular mechanical properties by cell poking, with an application to leukocytes. *ASME J. Biomech. Eng.* **112**, 283–294.

PART V: BIOMATERIALS IN TISSUE ENGINEERING

CHAPTER 18

PATTERNING OF CELLS AND THEIR ENVIRONMENT

Shuichi Takayama, Robert G. Chapman, Ravi S. Kane, and George M. Whitesides

INTRODUCTION

Control of the cell culture environment is crucial for understanding cell behavior and for engineering cell function. This chapter describes the use of a set of microfabrication techniques called "soft lithography" for patterning the substrate to which cells attach, the location and shape of the areas to which cells attach, and the fluid environment surrounding the cells, all with micrometer precision. Some examples wherein these techniques have helped to clarify problems in fundamental cell biology are summarized. The methods described are experimentally simple, inexpensive, and well suited for patterning biological materials.

How do tissues assemble *in vivo?* Once an appropriate community of cells has organized, how does it perform its functions? Answering these fundamental biological questions, and using the information thus obtained for engineering tissues, require the ability to study the behavior of cells in controlled environments. Some of the challenges in trying to control the environment experienced by individual cells lie in the size scales of the stimuli that need to be controlled (from the angstrom scale for molecular detail, through the micrometer scale for an individual cell, to the millimeter and centimeter scale for groups of cells) as well as in the number of the types of stimuli that need to be addressed (the composition of the culture media, the topography and chemical composition of the surface to which the cells attach, the nature of neighboring cells, the temperature, etc.).

Microfabrication and micropatterning using stamps or molds fabricated from elastomeric polymers (soft lithography) provide versatile methods for generating 10- to 100-μm-sized patterns of proteins and ligands on surfaces, 50- to 500-μm-sized culture chambers, and 10- to 100-μm-sized laminar flows of culture media in capillaries (Xia and Whitesides, 1998). Soft lithographic methods are relatively simple and inexpensive; the elastomeric polymer most often used in these procedures—poly(dimethylsiloxane), or PDMS—has several characteristics that make it attractive for biological applications. This chapter gives an overview of the application of soft lithography to the patterning of cells, their substrates, and their fluid environment.

SOFT LITHOGRAPHY

As the need of biologists to control and manipulate materials on the micrometer scale has increased, so has the need for new microfabrication techniques. Our laboratory has developed a set of microfabrication techniques that are useful for patterning on the scale of 0.5 μm and larger. We call these techniques "soft lithography" because they use elastomeric (that is, soft) stamps, molds, membranes, or channels (Xia and Whitesides, 1998).

Many other techniques can and have been used to pattern cells and their environment (Hammarback *et al.,* 1985; López *et al.,* 1993a; Park *et al.,* 1998; Vaidya *et al.,* 1998). The most commonly used method has been photolithography. This technique has, of course, been highly developed for the microelectronics industry; it has also been adapted, with varying degrees of success, for biological studies (Hammarback *et al.,* 1985; Kleinfeld *et al.,* 1988; Ravencroft *et al.,* 1998). As useful and powerful as photolithography is (it is capable of mass production at 200-nm resolution of multilevel, registered structures), it is not always the best or only option for biological studies. It is an expensive technology; it is poorly suited for patterning nonplanar surfaces; it provides almost no control over the chemistry of the surface and hence is not very flexible in generating patterns of specific chemical functionalities or proteins on surfaces; it can generate only two-dimensional microstructures; and it is directly applicable to patterning only a limited set of photosensitive materials (e.g., photoresists).

Soft lithographic techniques are inexpensive, are procedurally simple, are applicable to the complex and delicate molecules often required in biochemistry and biology, can be used to pattern a variety of different materials, are applicable to both planar and nonplanar substrates (Jackman *et al.,* 1995), and do not require stringent control (such as a clean room environment) over the laboratory environment beyond that required for routine cell culture (Xia and Whitesides, 1998). Access to photolithographic technology is required only to create a master for casting the elastomeric stamps or membranes, and even then, the requirement for chrome masks—the preparation of which is one of the slowest and most expensive steps in conventional photolithography—can often be bypassed (Deng *et al.,* 1999; Duffy *et al.,* 1998; Grzybowski *et al.,* 1998; Qin *et al.,* 1996). Soft lithography offers special advantages for biological applications, in that the elastomer most often used (PDMS) is optically transparent and permeable to gases, is flexible and seals conformally to a variety of surfaces (including petri dishes), is biocompatible, and can be implanted if desired. The soft lithographic techniques that we will discuss include microcontact printing, patterning with microfluidic channels, and laminar flow patterning.

SELF-ASSEMBLED MONOLAYERS

Because many of the studies involving the patterning of proteins and cells using soft lithography have been carried out on self-assembled monolayers (SAMs) of alkane thiolates on gold, we give a brief discussion of SAMs (Bain and Whitesides, 1988b; Bishop and Nuzzo, 1996; Delamarche and Michel, 1996; Dubois and Nuzzo, 1992; Merritt *et al.,* 1997; Ostuni *et al.,* 1999; Prime and Whitesides, 1993; Ulman, 1996). SAMs are organized organic monolayer films (Fig. 18.1A) that allow control at the molecular level over the chemical properties of the interface by judicious design and fabrication of derivatized alkane thiol(s) adsorbed to the surface of films of gold or silver. The ease of formation of SAMs, and their ability to present a range of chemical functionality at their interface with aqueous solution, make them particularly useful as model surfaces in studies involving biological components. Furthermore, SAMs can be easily patterned by simple methods such as microcontact printing (μCP) with features down to 500 nm in size and smaller (Xia and Whitesides, 1998). These features of SAMs make them the best structurally defined substrates for use in patterning proteins and cells. SAMs on gold are used for the majority of experiments requiring the patterning of proteins and cells, because they are biocompatible, easily handled, and chemically stable [for example, silver oxidizes relatively rapidly, and Ag(I) ions are cytotoxic].

Methods for Preparing Homogeneous SAMs and Mixed SAMs

Single-component (homogeneous) SAMs of alkanethiolates on gold are formed by exposing a surface of gold to a solution containing, or to the vapors of, an alkane thiol (RSH). The surface properties of these SAMs are determined by the nature of the terminal groups (schematically represented by half-circles at the tips of the SAMs in Figs. 18.1 and 18.2). Gold substrates are prepared on glass cover slips or silicon wafers by evaporating a thin layer of titanium (1–5 nm) to promote the adhesion of gold to the support, followed by a thin layer of gold (10–200 nm) (DiMilla *et al.,* 1994). The SAMs formed on these gold substrates are stable to the conditions used for cell culture, but care should be taken to avoid strong light and temperatures above ~70°C, because both can result in degradation of the SAM (Huang and Hemminger, 1993; Ostuni *et al.,* 1999).

Mixed SAMs, or SAMs composed of two or more types of thiols adsorbed to a surface, can be formed directly by coadsorption of two or more alkane thiols from solution (Fig. 18.2A,B) (Bain

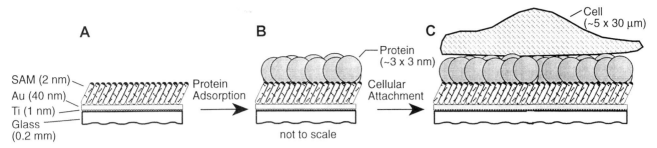

Fig. 18.1. Schematic illustration of the adsorption of proteins and the attachment of a cell to a self-assembled monolayer (SAM). The components of the SAM and the protein molecule are drawn roughly to scale; the cell is much reduced in relative size. See text for details.

and Whitesides, 1988a; Lahiri *et al.*, 1999a; Mrksich *et al.*, 1995), or by the reaction of a nucleophile, such as an amine (H_2NR), with a preformed SAM that presents interchain carboxylic anhydride groups. The anhydride-presenting surface is an exceptionally useful one and is easily prepared by the dehydration of single-component SAMs of $HS(CH_2)_nCO_2H$ ($n = 10$ or 15) (Fig. 18.2C) (Yan *et al.*, 1998). The anhydride method results in mixed SAMs composed of 1:1 mixtures of terminal amide ($-COHNR$) groups and carboxylic acid ($-CO_2H/-CO_2^-$) groups. The surface properties of mixed SAMs are determined by contributions from the various terminal groups present—the amount of each contribution being determined, qualitatively, by the degree of exposure of that terminal group at the surface (Bain *et al.*, 1989; Bain and Whitesides, 1989). In some cases the contribution of one terminal group will predominate, especially if that terminal group is long or large enough to screen the contribution of the shorter terminal group. For example, the characteristics of mixed SAMs prepared by the interchain anhydride methods are often determined predominantly by nature of the ($-COHNR$) groups, because these are longer and shield the ($-CO_2H/-CO_2^-$) groups (Fig. 18.2C) (Yan *et al.*, 1997). The anhydride method makes it possible to generate mixed SAMs with a range of surface characteristics rapidly and conveniently. This method is experimentally simple, and both $HS(CH_2)_{10}CO_2H$ and a variety of amines are commercially available. It is thus unnecessary to synthesize the alkane thiols required in the older methods of making functionalized SAMs.

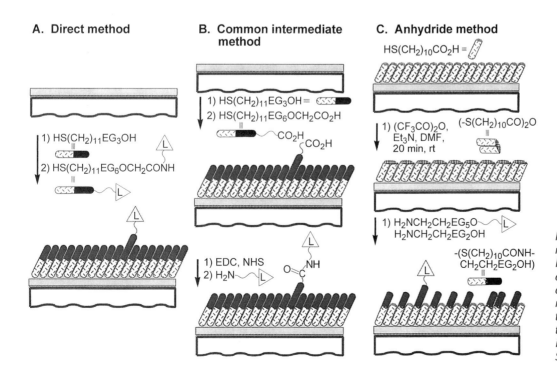

A. Direct method

1) $HS(CH_2)_{11}EG_3OH$

2) $HS(CH_2)_{11}EG_6OCH_2CONH$

B. Common intermediate method

1) $HS(CH_2)_{11}EG_3OH \equiv$

2) $HS(CH_2)_{11}EG_6OCH_2CO_2H$

CO_2H
CO_2H

1) EDC, NHS

2) H_2N

C. Anhydride method

$HS(CH_2)_{10}CO_2H \equiv$

1) $(CF_3CO)_2O$, Et_3N, DMF, 20 min, rt

$(-S(CH_2)_{10}CO)_2O$

1) $H_2NCH_2CH_2EG_5O$
$H_2NCH_2CH_2EG_2OH$

$-(S(CH_2)_{10}CONH-CH_2CH_2EG_2OH)$

Fig. 18.2. Preparation of mixed SAMs that present a ligand by three methods: direct, common intermediate, and anhydride. These mixed SAMs are useful for the specific adsorption of the protein that binds the incorporated ligand (L). See text for details.

Preventing Protein Adsorption: Inert Surfaces

Proteins play an integral part in the adhesion of cells with surfaces: they are the glue that cements the cells to the surface. Thus, if one can control the interaction of proteins with a surface, one can begin to control the interaction of cells with that surface (Fig. 18.1). Most surfaces adsorb proteins. Thus, the main challenge in controlling the interactions of proteins and cells with surfaces lies in finding surfaces that resist nonspecific adsorption of proteins (surfaces that we call "inert," for brevity). Inert surfaces provide the background necessary for spatially restricting protein adsorption or for preparing surfaces that bind only specific proteins, and are used in patterning proteins and cells (Merritt *et al.,* 1997), as biomaterials (Andrade *et al.,* 1987; Helmus and Hubbell, 1993), and in the construction of biosensors (Mrksich and Whitesides, 1995).

Single-component SAMs terminated in oligo(ethylene glycol) (EG_n), oligomers longer than $n = 3$, resist the adsorption from solution of a variety of proteins that range in size (15–340 kDa) and charge ($pI = 6$–12) (Prime and Whitesides, 1991, 1993). Mixed SAMs containing as little as 50% of an $HS(CH_2)_{11}EG_6OH$ [mixed with $HS(CH_2)_{11}CH_3$] show good protein resistance, as do mixed SAMs presenting a 1:1 mixture of terminal $-COHN(CH_2CH_2O)_nR$ ($n = 6$, R = H or CH_3) and $-CO_2H/-CO_2^-$ groups. Other work has established that EG_n groups are not unique in their ability to resist adsorption of proteins, and, as an example, SAMs presenting $-CH_2CH_2CH_2S(=O)-$ groups are also effective in resisting protein adsorption (Deng *et al.,* 1996). R. G. Chapman, L. Yan, S. Takayama, R. E. Holmlin, and G. M. Whitesides (unpublished results) have surveyed a large number of functional groups for their ability to resist the nonspecific adsorption of proteins, and found that polar functional groups that do not contain H-bond donors often make good components of inert surfaces. The combination of inert and adsorptive surfaces with soft lithographic techniques enables the facile patterning of proteins and cells.

Controlled Protein Adsorption

Protein-specific mixed SAMs that present ligands for specific adsorption of a protein of interest, while resisting the nonspecific adsorption of other proteins, have been prepared by several techniques: (1) by coadsorption of a thiol that forms inert surfaces with a thiol that presents a ligand specific for a particular protein (Fig. 18.2A) (Mrksich *et al.,* 1995; Sigal *et al.,* 1996); (2) by the common intermediate method, whereby an amine terminated in a ligand is coupled to a mixed SAM presenting $-EG_3OH$ (the inert surface component) and $-EG_6OCH_2COOH$ (the component to which ligands are coupled) (Fig. 18.2B) (Lahiri *et al.,* 1999a); (3) by the anhydride method, whereby a mixture of two amines—one terminated with a ligand and the other with an "inert" functional group (for example, $H_2NCH_2CH_2EG_2OH$)—is allowed to react with a surface that presents interchain anhydride groups (Fig. 18.2C) (R. G. Chapman, L. Yan, and G. M. Whitesides, unpublished results). The common intermediate method and the anhydride method are more convenient than, and preferable to, the direct method, because they require less organic synthesis.

MICROCONTACT PRINTING

Patterning Ligands, Proteins, and Cells Using Microcontact Printing

Microcontact printing is a technique that uses the relief pattern on the surface of an elastomeric PDMS stamp to form patterns on the surfaces of various substrates (Fig. 18.3A) (Xia and Whitesides, 1998). The stamp is "inked" with a solution containing the patterning component, dried, and brought into conformal contact with a surface for intervals ranging from a few seconds to minutes. The patterning component transfers to the substrate in the regions where the stamp contacts the substrate. The kinds of components used as ink for μCP include thiol derivatives that form SAMs on gold and silver; activated silanes that react with the SiOH groups present on the surface of silicon; various ligands (usually amine containing compounds) that react with activated SAMs (generated using the techniques described in the previous section) (Lahiri *et al.,* 1999b; Yan *et al.,* 1998); and robust proteins that can withstand drying and stamping (Bernard *et al.,* 1998; St. John *et al.,* 1998).

Although the use of proteins as the ink is limited to sturdy proteins, even the more delicate ones can be patterned utilizing microcontact printing. This patterning of proteins is accomplished

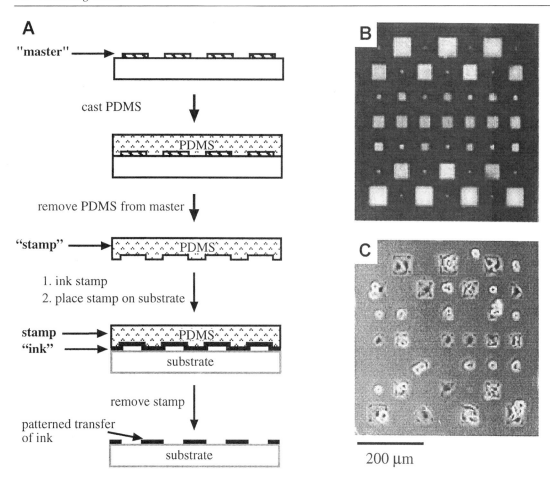

Fig. 18.3. Examples of microcontact printing experiments. (A) Typical procedure. A poly(di-methylsiloxane) (PDMS) stamp is prepared by pouring PDMS prepolymer on a "master," curing the PDMS and removing it from the master. A solution containing the patterning component of in-terest ("ink") is applied to this stamp and the solution allowed to dry. This inked stamp is placed on a substrate to allow the ink to transfer to the substrate. A patterned substrate remains after re-moval of the stamp. (B) Selective adsorption of fibronectin onto a surface patterned into protein-adsorptive and non-protein-adsorptive self-assembled monolayers by microcontact printing as visualized by immunostaining. (C) Patterned attachment and spreading of cells on the protein-patterned substrate in B.

by preparing areas of protein-adsorptive SAMs and allowing proteins to adsorb onto those regions from solutions. For example, López *et al.* (1993b) used microcontact printing to pattern gold sur-faces into regions terminated in methyl groups, when surrounded by inert oligo(ethylene glycol) groups. Immersion of the patterned SAMs in solutions of proteins such as fibronectin, fibrinogen, pyruvate kinase, streptavidin, and immunoglobulins resulted in adsorption of the proteins on the methyl-terminated regions. The pattern of adsorbed proteins could be characterized by scanning electron microscopy; the layers of adsorbed protein appeared to be homogeneous. Alternatively, proteins can be anchored to ligands patterned onto surfaces by μCP. Lahiri *et al.* (1999b), for ex-ample, patterned streptavidin by μCP of its ligand, biotin, onto activated SAMs (using the method illustrated in Fig. 18.2B), and then allowing the protein to bind to the patterned ligand.

All eukaryotic cells, and many prokaryotic cells, are too delicate to be dried or stamped and cannot be patterned directly by μCP. The ability of μCP to create patterns of ligands and proteins, however, allows the patterning of many anchorage-dependent cells (Fig. 18.1), confining them to specific regions of a substrate; these techniques strongly influence the size and shape of the cells (Fig. 18.3C). For example, Mrksich *et al.* (1996) used microcontact printing to pattern gold or sil-ver substrates into regions presenting oligo(ethylene glycol) groups and methyl groups. After coat-ing the substrates with fibronectin, bovine capillary endothelial cells attached only to the methyl-terminated, fibronectin-coated regions of the patterned SAMs. The cells remained attached in the pattern defined by the underlying SAMs for 5–7 days. μCP has also been used to pattern astroglial cells on silicon substrates (Craighead *et al.*, 1998). Astroglial cells attached selectively to 50-μm-wide bars of *N*-1-[3-(trimethoxysilyl)propyl]diethylenetriamine (DETA) SAMs patterned on a silicon surface. Zhang *et al.* (1999) have synthesized oligopeptides containing a cell adhesion motif at the N terminus connected by an oligo(alanine) linker to a cysteine residue at the C terminus. The thiol group of cysteine allowed the oligopeptides to form monolayers on gold-coated surfaces. A combination of microcontact printing and these self-assembling oligopeptide monolayers was used to pattern gold surfaces into regions presenting cell adhesion motifs and oligo(ethylene glycol) groups that resist protein adsorption. Wheeler *et al.* (1999; Branch *et al.*, 1998) created patterns of covalently bound ligands and proteins on glass cover slips and used these patterns to control nerve cell growth.

FUNDAMENTAL STUDIES IN CELL BIOLOGY USING MICROCONTACT-PRINTED SUBSTRATES

The ability to design SAMs to be either protein adsorptive or nonadsorptive, when combined with the ability of μCP to pattern such SAMs routinely on size scales smaller than that of a single cell (2–50 μm), has led to new studies on the effect of patterned surface environments and cell shape on cell behavior (Table 18.1).

Singhvi *et al.* (1994) used μCP to prepare substrates consisting of square and rectangular islands of laminin surrounded by nonadhesive regions, and studied the behavior of rat hepatocytes on them. The cells conformed to the shape of the laminin patterns, allowing one to control cell shape independently of the extracellular matrix (ECM) ligand density. The investigators compared cells grown on various patterns and observed that cell shape, regardless of ECM ligand density, was the major determinant of cell growth and differentiation. Chen *et al.* (1997, 1999) used μCP to prepare substrates that presented circular cell-adhesive islands of various diameters and interisland spacings. Such patterns allowed them to control the extent of cell spreading without varying the total cell–matrix contact area. They found that the extent of spreading (the projected surface area of the cell) and not the area of the adhesive contact controlled whether the cell divided, remained in stationary phase, or entered apoptosis. Dike *et al.* (1999) used μCP to prepare substrates with cell-adhesive lines of varying widths. They found that bovine capillary endothelial (BCE) cells cultured on 10-μm-wide lines underwent differentiation to capillary tubelike structures containing a central lumen. Cells cultured on wider (30 μm) lines formed cell–cell contacts, but these cells continued to proliferate and did not form tubes. These studies indicate that cell growth, function, and differentiation can be controlled, at least in some cases, by patterning the surface of the substrate to which cells adhere.

In another study, Bailly *et al.* (1998) used micropatterned substrates in studies of the regulation of protrusion shape during chemotactic responses of mammalian carcinoma cells. They plated rat mammary carcinoma cells on gold-coated glass cover slips having 10-μm-wide adhesive lanes. On stimulation with epidermal growth factor (EGF), the cells extended lamellipods laterally, over the nonadhesive part of the substrate. These results showed that lamellipod extension could occur independently of any contact with the substrate. Contact formation was, however, necessary for stabilizing the protrusion.

μCP is an excellent method for patterning surfaces with complex and delicate organic groups, and it is the soft lithographic method that has been most utilized for patterning the substrate.

MICROFLUIDIC PATTERNING

Microfluidic channels can be used to pattern surfaces by restricting the flow of fluids to desired regions of a substrate (Fig. 18.4). The patterning components—such as ligands, proteins, and cells—are deposited from the solution to create a pattern on the substrate.

Delamarche *et al.* (1997, 1998) used microfluidic patterning (μFP) to pattern immunoglobulins with submicron resolution on a variety of substrates, including gold, glass, and polystyrene. Only microliters of reagent were required to cover square millimeter-sized areas. Patel *et al.* (1998) developed a method to generate micron-scale patterns of any biotinylated ligand on the surface of a biodegradable polymer. These investigators prepared biotin-presenting polymer films, and patterned the films by allowing solutions of avidin to flow over them through 50-μm channels fab-

Table 18.1. Examples of patterning cells using microcontact printing

Substrate	Features	Initial component patterned	Proteins patterned	Effect observed (cells patterned)	Refs.
Gold	2- to 80-μm rectangles	Hexadecane thiol, hexa(ethylene glycol)–terminated alkane thiol	Laminin	Primary rat hepatocytes adhere and spread only on laminin-coated islands; cell shape controls cell growth and function	Singhvi et al. (1994)
Gold	60-μm lines separated by 120 μm	Hexadecane thiol, tri(propylene sulfoxide)–terminated alkane thiol	Fibronectin	BCE cells were confined to the hexadecane thiol SAM regions for 1–2 days then started to spread into other areas	Deng et al. (1996)
Thin gold film (<12 nm) on polyurethane	25- to 50-μm ridges and grooves	Hexadecane thiol, tri(ethylene glycol)–terminated alkane thiol	Fibronectin	Patterned adsorption of fibronectin; cells are confined to tri(ethylene glycol)-terminated alkane thiol SAM regions for at least 5 days	Mrksich et al. (1996)
Gold	3- to 40-μm squares and circles	Hexadecane thiol, tri(ethylene glycol)-terminated alkane thiol	Fibronectin, vitronectin, collagen, anti-integrin β_1 antibody, or anti-integrin $\alpha_v\beta_3$	Extent of cell spreading (not the area of adhesive contact) controls apoptosis of bovine capillary endothelial cells	Chen et al. (1997, 1999)
Silicon	Features of 1 μm and larger	Octadecyltrichlorosilane, N-1-[3-(trimethoxysilyl)propyl]diethylenetriamine, polylysine	—	LRM 55 cells (astroglial cell line) selectively attached to DETA-patterned surfaces in presence of serum-containing media; rat hippocampal neurons attach to poly(lysine)-patterned regions	Craighead et al. (1998)
Gold and silicon	Circles and squares of 5 to 80 μm	Hexadecane thiol, tri(ethylene glycol)-terminated alkane thiol	Fibronectin	Cell cycle progression of BCE cells were controlled by cell shape and cytoskeletal tension	Huang et al. (1998)
Gold	10-μm lines	Hexadecane thiol, hexa(ethylene glycol)-terminated alkane thiol	Vitronectin	Chemotactic response of MTLn3 metastatic rat mammary adenocarcinoma cells; lamellipod extension is independent of contact with the substratum	Bailly et al. (1998)
Gold	Squares and lines of 20 μm and larger	Peptides containing cell adhesion motifs and cysteine, hexa(ethylene glycol)-terminated alkane thiol	—	Human epidermoid carcinoma A431 cells, primary human embryonic kidney 293 cells, bovine aorta endothelial cells, and NIH 3T3 fibroblasts	Zhang et al. (1999)
Gold	10- or 30-μm-width lines and 5- or 10-μm squares	Hexadecane thiol, tri(ethylene glycol)-terminated alkane thiol	Fibronectin	BCE cells can be switched between growth, apoptosis, and differentiation by altering the geometry of spreading	Dike et al. (1999)
Glass that presents N-hydroxysuccinimide esters or aldehydes	10-μm lines	Poly(lysine), bovine serum albumin laminin	BSA, laminin	On a patterned surface, neural somata and dendrites preferred poly(lysine) surfaces whereas axons preferred surfaces presenting laminin/poly(lysine) mixtures	Branch et al. (1998) Wheeler et al. (1999)

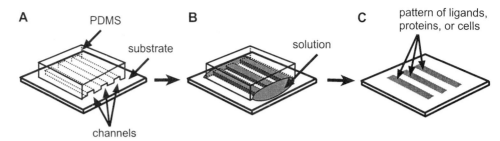

Fig. 18.4. Patterning substrates using microfluidic channels. (A) A poly(dimethylsiloxane) (PDMS) mold is brought into conformal contact with the surface of a substrate to form microfluidic channels. (B) The channels are filled with a solution containing patterning components. (C) The channels are washed clean of the patterning solution and the PDMS mold is removed. This sequence of steps results in generation of patterns of adsorbed proteins on the substrate.

ricated in PDMS. The avidin moieties bound to the biotin groups on the surface, and served as a bridge between the biotinylated polymer and biotinylated ligands. Patterns created with biotinylated ligands containing the RGD or IKVAV oligopeptide sequences determined the adhesion and spreading of bovine aortic endothelial cells and PC12 nerve cells. Folch and Toner (1998) and Folch *et al.* (1999) used µFP to produce patterns of cells on biocompatible substrates. Micropatterns of collagen or fibronectin deposited from fluids in capillaries were used to cause cells to adhere selectively to various biomedical polymers, and to heterogeneous or microtextured substrates. On removing the elastomeric stamp, the bare areas of the substrate could be seeded with more adhesive cell types such as fibroblasts. This procedure produced micropatterned cocultures. By allowing different cell suspensions to flow through different channels, patterns composed of two cell types were also generated.

The mild conditions used with µFP permit the patterning not only of small molecules and proteins, but also of more delicate components such as cells. By filling individual channels with different fluids, multiple components can be patterned at the same time without the need for multiple steps or the accompanying technical concerns of registration.

LAMINAR FLOW PATTERNING

Laminar flow patterning (LFP) is a technique that can pattern surfaces, and the positions of cells on them, in useful ways. It can also pattern fluids (Takayama *et al.*, 1999) themselves. This technique utilizes a phenomenon that occurs in microfluidic systems as a result of their small dimensions—that is, low Reynolds number flow. The Reynolds number (Re) is a nondimensional parameter describing the ratio of inertial to viscous forces in a specific flow configuration. It is a measure of the tendency of a flowing fluid to develop turbulence. The flow of liquids in capillaries often has a low Re and is laminar. Laminar flow allows two or more layers of fluid to flow next to each other without any mixing other than by diffusion of their constituent molecular and particulate components.

Figure 18.5 shows a typical setup for LFP experiments, along with images from some representative work. A network of capillaries is made by bringing a patterned PDMS slab into conformal contact with a petri dish. By flowing different patterning components from the inlets, patterns of parallel stripes are created in the main channel (Fig. 18.5A,B). The positions and environments of cells can be controlled simultaneously in several stripes in the same channel (Fig. 18.5C). Figure 18.5D shows patterning of the substrate with different proteins; Fig. 18.5E shows patterning cell position by deposition from laminar fluid flows; Fig. 18.5F illustrates the use of patterned culture media to deliver chemicals selectively to cells. In this experiment, bovine capillary endothelial cells covered the entire bottom face of the capillary; the fluorescence micrograph visualizes only those cells over which medium containing a fluorescent dye was allowed to flow. Figure 18.5G is another example of patterning the culture media. In this experiment, digestion of fibronectin on the channel surface and sequestering of calcium by ethylenediaminetetraacetic acid (EDTA) cause cells to detach and contract. When a solution of trypsin/EDTA was allowed to flow over only a portion of a cell, the treated part of the cell detached and contracted; the untreated

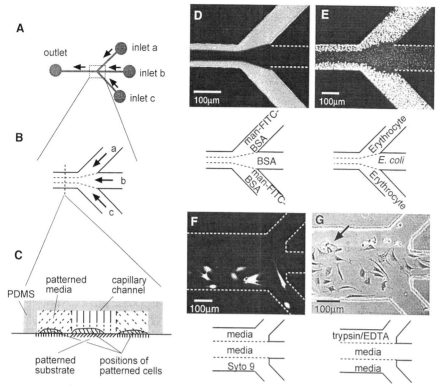

Fig. 18.5 Examples of laminar flow patterning experiments. (A) Top view of the capillary network. A poly(dimethylsiloxane) (PDMS) membrane containing micron-sized channels molded in its surface was placed on the flat surface of a petri dish to form a network of capillaries. (B) A close-up view of the junctions where the inlets converge into a single main channel. Micrographs were obtained for this area of the capillary system. (C) Cross-sectional view of the main channel, looking from the outlet toward the inlet. The patterning of the substrate, the cell positions, and the fluid environment can be controlled simultaneously in the same capillary. (D) Patterning substrate with bovine serum albumin (BSA) and BSA colabeled with mannose and fluorescein (man–FITC–BSA). (E) Patterned cell deposition. Chick erythrocytes and Escherichia coli were deposited selectively in their designated lanes by patterned flow of cell suspensions. Adherent cells were visualized with a fluorescent nucleic acid stain (Syto 9). (F) Using patterned media to selectively stain BCE cells. Syto 9 and media were allowed to flow from the designated inlets. (G) Using patterned media to detach bovine capillary endothelial cells selectively. Trypsin/EDTA and media were allowed to flow from the designated inlets. (D–F) Fluorescence micrographs taken from the top through PDMS. (G) Phase contrast image observed by an inverted microscope looking through the polystyrene petri dish. White dotted lines identify channels not visible with fluorescence microscopy.

part remained spread (for example, see arrow in Fig. 18.5G). Because no physical barriers are required to separate the different liquid streams, different liquids can flow over different portions of a single cell.

LFP has some features that make it complementary to other patterning techniques used for biological applications. It takes advantage of the easily generated multiphase laminar flows to pattern fluids and to deliver components for patterning. The ability to pattern the growth medium is a special feature that cannot be achieved by other processes. This method can pattern over delicate structures, such as a portion of a mammalian cell. This type of patterning is difficult by other techniques. LFP can also give simultaneous control over the surface patterns, cell positioning, and the fluid environment in the same channel.

CONCLUSION AND FUTURE PROSPECTS

Soft lithography brings to microfabrication low-cost, simple procedures, rapid prototyping of custom-designed devices, three-dimensional capability, and biocompatibility. These techniques allow patterning of cells and their environments with great convenience and flexibility at dimensions down to micrometers. We have described three complementary soft lithographic techniques—microcontact printing, patterning using fluids in microfluidic channels, and laminar flow patterning—that are useful in their ability to pattern the cell culture environment.

Microcontact printing is perhaps the simplest method for patterning surfaces and also provides the highest resolution with the greatest flexibility in the shape of the patterns generated. It is most useful when one needs to pattern two types of ligands or proteins, and when the "ink" is able to withstand drying and stamping. Microfluidic channels are well suited for patterning surfaces using delicate objects such as proteins and cells. They are also useful when multiple ligands, proteins, or cells need to be patterned. Laminar flow patterning is similar to patterning with individual microfluidic channels except that the individual flows are kept from mixing with each other by laminar flow, not by physical walls. The ability to pattern the fluid environment is the distinguishing feature of this method, and enables laminar flow to be used to pattern the distribution of different fluids over the surface of a single mammalian cell. This capability allows patterning of portions of a single cell, and remodeling of the cell culture environment, both in the presence of living cells.

The potential applications of soft lithography are just starting to be explored. There are many cell culture environments that we have not discussed in this chapter, but that can potentially be patterned utilizing soft lithographic methods. For example, topographical features created in PDMS affect cell spreading (Flemming *et al.,* 1999). Silicone polymers can sense and affect the mechanical tension within cells (Burton and Taylor, 1997; Sai *et al.,* 1999). With the availability of various three-dimensional fabrication techniques, much more work is expected in three-dimensional patterning of the cell culture substrate (Breen *et al.,* 1999; Jackman *et al.,* 1998; Terfort *et al.,* 1997). The fourth dimension, time, is also an interesting factor. Experiments such as trypsin-mediated remodeling of the exposed surfaces of cells and supports presenting adsorbed proteins using laminar flow demonstrate that real-time temporal changes in the cell culture environment are possible (Takayama *et al.,* 1999). The optical transparency of PDMS makes it possible to pattern the intensity of light in cell cultures (Paul *et al.,* 1999; Xia and Whitesides, 1998). The ability of soft lithography to pattern magnetic materials, and the nonmagnetic character of PDMS, makes it an attractive method for patterning of magnetic fields (Palacin *et al.,* 1996). The gas permeability of PDMS may be useful in patterning the gas of the surrounding cells. PDMS is electrically insulating, and the ability to mold or fabricate electrically conducting wires in it should allow patterning of electric fields (Kenis *et al.,* 1999). Gravitational fields can also be affected: microfluidic culture chambers with adherent cells can be turned upside down without loss of the culture media. Temperature, fluid shear, and other factors may also be accurately patterned.

The potential of a cell is predetermined by its genetics. Realization of that potential depends, inter alia, on whether the cell is exposed to the appropriate environment. Soft lithography provides tools for patterning cells and their environments with a high degree of control. This capability aids efforts to understand fundamental cell biology and advances our ability to engineer cells and tissues. The ease with which electronic components or other nonbiological components can be fabricated with soft lithography also paves the way for the engineering of cells and tissue for use in biosensors and other hybrid systems that combine living and nonliving components.

ACKNOWLEDGMENTS

This work was supported by NIH GM30367, DARPA/Space and Naval Warfare Systems Center San Diego (SPAWAR), DARPA/AFRL. The content of the information does not necessarily reflect the position or the policy of the United States Government, and no official endorsement should be inferred. We thank Emanuele Ostuni for helpful suggestions. S. T. is a Leukemia Society of America Fellow and thanks the society for a fellowship. R. G. C. thanks the Natural Sciences and Engineering Research Council of Canada for a fellowship.

REFERENCES

Andrade, J. D., Nagaoka, S., Cooper, S., Okano, T., and Kim, S. W. (1987). Surfaces and blood compatibility. *Trans. Am. Soc. Artif. Intern. Organs* **33,** 75–84.

Bailly, M., Yan, L., Whitesides, G. M., Condeelis, J. S., and Segall, J. E. (1998). Regulation of protrusion shape and adhesion to the substratum during chemotactic responses of mammalian carcinoma cells. *Exp. Cell Res.* **241,** 285–299.

Bain, C. D., and Whitesides, G. M. (1988a). Formation of two-component surfaces by the spontaneous assembly of monolayers on gold from solutions containing mixtures of organic thiols. *J. Am. Chem. Soc.* **110,** 6560–6561.

Bain, C. D., and Whitesides, G. M. (1988b). Molecular-level control over surface order in self-assembled monolayer films of thiols on gold. *Science* **240,** 62–63.

Bain, C. D., and Whitesides, G. M. (1989). Formation of monolayers by the coadsorption of thiols on gold: Variation in the length of the alkyl chain. *J. Am. Chem. Soc.* **111,** 7164–7175.

Bain, C. D., Evall, J., and Whitesides, G. M. (1989). Formation of monolayers by the coadsorption of thiols on gold: Variation in the head group, tail group, and solvent. *J. Am. Chem. Soc.* **111,** 7155–7164.

Bernard, A., Delamarche, E., Schmid, H., Michel, B., Bosshard, H. R., and Biebuyck, H. (1998). Printing patterns of proteins. *Langmuir* **14,** 2225–2229.

Bishop, A. R., and Nuzzo, R. G. (1996). Self-assembled monolayers: Recent developments and applications. *Curr. Opin. Colloid Interface Sci.* **1,** 127–136.

Branch, D. W., Corey, J. M., Weyhenmeyer, J. A., Brewer, G. J., and Wheeler, B. C. (1998). Micro-stamp patterns of biomolecules for high-resolution neuronal networks. *Med. Biol. Eng. Comput.* **36,** 135–141.

Breen, T. L., Tien, J., Oliver, S. R. J., Hadzic, T., and Whitesides, G. M. (1999). Design and self-assembly of open, 3D lattice mesostructures. *Science* **284,** 948–951.

Burton, K., and Taylor, D. L. (1997). Traction forces of cytokinesis measured with optically modified elastic substrata. *Nature* **385,** 450–454.

Chen, C. S., Mrksich, M., Huang, S., Whitesides, G. M., and Ingber, D. E. (1997). Geometric control of cell life and death. *Science* **276,** 1425–1428.

Chen, C. S., Mrksich, M., Huang, S., Whitesides, G. M., and Ingber, D. E. (1999). Micropatterned surfaces for control of cell shape, position, and function. *Biotechnol. Prog.* **14,** 356–363.

Craighead, H. G., Turner, S. W., Davis, R. C., James, C., Perez, A. M., St. John, P. M., Isaacson, M. S., Kam, L., Shain, W., Turner, J. N., and Banker, G. (1998). Chemical and topographical surface modification for control of central nervous system cell adhesion. *J. Biomed. Microdev.* **1,** 49–64.

Delamarche, E., and Michel, B. (1996). Structure and stability of self-assembled monolayers. *Thin Solid Films* **273,** 54–60.

Delamarche, E., Bernard, A., Schmid, H., Michel, B., and Biebuyck, H. (1997). Patterned delivery of immunoglobulins to surfaces using microfluidic networks. *Science* **276,** 779–781.

Delamarche, E., Bernard, A., Schmid, H., Bietsch, A., Michel, B., and Biebuyck, H. (1998). Microfluidic networks for chemical patterning of substrates: Design and application to bioassays. *J. Am. Chem. Soc.* **120,** 500–508.

Deng, L., Mrksich, M., and Whitesides, G. M. (1996). Self-assembled monolayers of alkanethiolates presenting tri(propylenesulfoxide) groups resist the adsorption of protein. *J. Am. Chem. Soc.* **118,** 5136–5137.

Deng, T., Tien, J., Xu, B., and Whitesides, G. M. (1999). Using patterns in microfiche as photomasks in 10-μm scale microfabrication. *Langmuir* **15,** 6675–6581.

Dike, L. E., Chen, C. S., Mrksich, M., Tien, J., Whitesides, G. M., and Ingber, D. E. (1999). Geometric control of switching between growth, apoptosis, and differentiation during angiogenesis using micropatterned substrates. *In Vitro Cell. Dev. Biol., Anim.* **35,** 441–448.

DiMilla, P., Folkers, J. P., Biebuyck, H. A., Harter, R., Lopez, G., and Whitesides, G. M. (1994). Wetting and protein adsorption of self-assembled monolayers of alkanethiolates supported on transparent films of gold. *J. Am. Chem. Soc.* **116,** 2225–2226.

Dubois, L. H., and Nuzzo, R. G. (1992). Synthesis, structure, and properties of model organic surfaces. *Annu. Rev. Phys. Chem.* **43,** 437–463.

Duffy, D. C., McDonald, J. C., Schueller, O. J. A., and Whitesides, G. M. (1998). Rapid prototyping of microfluidic systems in poly(dimethylsiloxane). *Anal. Chem.* **70,** 4974–4984.

Flemming, R. G., Murphy, C. J., Abrams, G. A., Goodman, S. L., and Nealey, P. F. (1999). Effects of synthetic micro- and nano-structured surfaces on cell behavior. *Biomaterials* **20,** 573–588.

Folch, A., and Toner, M. (1998). Cellular micropatterns on biocompatible materials. *Biotechnol. Prog.* **14,** 388–392.

Folch, A., Ayon, A., Hurtado, O., Schmidt, M. A., and Toner, M. (1999). Molding of deep polydimethylsiloxane microstructures for microfluidics and biological applications. *J. Biomech. Eng.* **121,** 28–34.

Grzybowski, B. A., Haag, R., Bowden, N., and Whitesides, G. M. (1998). Generation of micrometer-sized patterns for microanalytical applications using a laser direct-write method and microcontact printing. *Anal. Chem.* **70,** 4645–4652.

Hammarback, J. A., Palm, S. L., Furcht, L. T., and Letourneau, P. C. (1985). Guidance of neurite outgrowth by pathways of substratum-adsorbed laminin. *J. Neurosci. Res.* **13,** 213–220.

Helmus, M. N., and Hubbell, J. A. (1993). Materials selection. *Cardiovasc. Pathol.* **2,** 53S–71S.

Huang, J. Y., and Hemminger, J. C. (1993). Photooxidation of thiols in self-assembled monolayers on gold. *J. Am. Chem. Soc.* **115,** 3342–3343.

Huang, S., Chen, C. S., and Ingber, D. E. (1998). Control of cyclin D1, p27Kip1, and cell cycle progression in human capillary endothelial cells by cell shape and cytoskeletal tension. *Mol. Biol. Cell* **9,** 3179–3193.

Jackman, R. J., Wilbur, J., and Whitesides, G. M. (1995). Fabrication of submicron features on curved substrates by microcontact printing. *Science* **269,** 664–666.

Jackman, R. J., Brittain, S. T., Adams, A., Prentiss, M. G., and Whitesides, G. M. (1998). Design and fabrication of topologically complex, three-dimensional microstructures. *Science* **280,** 2089–2091.

Kenis, P. J. A., Ismagilov, R. F., and Whitesides, G. M. (1999). Microfabrication inside capillaries using multiphase laminar flow patterning. *Science* **285,** 83–85.

Kleinfeld, D., Kahler, K. H., and Hockberger, P. E. (1988). Controlled outgrowth of dissociated neurons on patterned substrates. *J. Neurosci.* **8,** 4098–4120.

Lahiri, J., Isaacs, L., Tien, J., and Whitesides, G. M. (1999a). A strategy for the generation of surfaces presenting ligands for studies of binding based on an active ester as a common reactive intermediate: A surface plasmon resonance study. *Anal. Chem.* **71,** 777–790.

Lahiri, J., Ostuni, E., and Whitesides, G. M. (1999b). Patterning ligands on reactive SAMs by microcontact printing. *Langmuir* **15**, 2055–2060.

López, G. P., Albers, M. W., Schreiber, S. L., Carroll, R. W., Peralta, E., and Whitesides, G. M. (1993a). Convenient methods for patterning the adhesion of mammalian cells to surfaces using self-assembled monolayers of alkanethiolates on gold. *J. Am. Chem. Soc.* **115**, 5877–5878.

López, G. P., Biebuyck, H. A., Härter, R., Kumar, A., and Whitesides, G. M. (1993b). Fabrication and imaging of two-dimensional patterns of proteins adsorbed on self-assembled monolayers by scanning electron microscopy. *J. Am. Chem. Soc.* **115**, 10774–10781.

Merritt, M. V., Mrksich, M., and Whitesides, G. M. (1997). Using self-assembled monolayers to study the interactions of man-made materials with proteins. *In* "Principles of Tissue Engineering" (R. P. Lanza, W. L. Chick, and R. Langer, eds.), pp. 211–223. R. G. Landes, Austin, TX.

Mrksich, M., and Whitesides, G. M. (1995). Patterning self-assembled monolayers using microcontact printing: A new technology for biosensors? *Trends Biotechnol.* **13**, 228–235.

Mrksich, M., Grunwell, J. R., and Whitesides, G. M. (1995). Bio-specific adsorption of carbonic anhydrase to self-assembled monolayers of alkanethiolates that present benzenesulfonamide groups on gold. *J. Am. Chem. Soc.* **117**, 12009–12010.

Mrksich, M., Chen, C. S., Xia, Y., Dike, L. E., Ingber, D. e., and Whitesides, G. M. (1996). Controlling cell attachment on contoured surfaces with self-assembled monolayers of alkanethiolates on gold. *Proc. Natl. Acad. Sci. U.S.A.* **93**, 10775–10778.

Ostuni, E., Yan, L., and Whitesides, G. M. (1999). The interaction of proteins and cells with self-assembled monolayers of alkanethiolates on gold and silver. *Colloids Surf. B: Biointerfaces* **15**, 3–30.

Palacin, S., Hidber, P. C., Bourgoin, J.-P., Miramond, C., Fermon, C., and Whitesides, G. M. (1996). Patterning with magnetic materials at the micron scale. *Chem. Mater.* **8**, 1316–1325.

Park, A., Wu, B., and Griffith, L. (1998). Integration of surface modification and 3D fabrication techniques to prepare patterned poly(⅝-lactide) substrates allowing regionally selective cell adhesion. *J. Biomater. Sci., Polym. Ed.* **9**, 89–110.

Patel, N., Padera, R., Sanders, G. H. W., Cannizzaro, S. M., Davies, M. C., Langer, R., Roberts, C. J., Tendler, S. J. B., Williams, P. M., and Shakesheff, K. M. (1998). Spatially controlled cell engineering on biodegradable polymer surfaces. *FASEB J.* **12**, 1447–1454.

Paul, K. E., Breen, T. L., Aizenberg, J., and Whitesides, G. M. (1999). Maskless lithography: Embossed photoresist as its own optical element. *Appl. Phys. Lett.* **73**, 2893–2895.

Prime, K. L., and Whitesides, G. M. (1991). Self-assembled organic monolayers: Model systems for studying adsorption of proteins at surfaces. *Science* **252**, 1164–1167.

Prime, K. L., and Whitesides, G. M. (1993). Adsorption of proteins onto surfaces containing end-attached oligo(ethylene oxide): A model system using self-assembled monolayers. *J. Am. Chem. Soc.* **115**, 10714–10721.

Qin, D., Xia, Y., and Whitesides, G. M. (1996). Rapid prototyping of complex structures with feature sizes larger than 20 μm. *Adv. Mater.* **8**, 917–919.

Ravencroft, M. S., Bateman, K. E., Shaffer, K. M., Schessler, H. M., Jung, D. R., Schneider, T. W., Montgomery, C. B., Custer, T. L., Schaffner, A. E., Liu, Q. Y., Li, Y. X., Barker, J. L., and Hickman, J. J. (1998). Developmental neurobiology implications from fabrication and analysis of hippocampal neuronal networks on patterned silane-modified surfaces. *J. Am. Chem. Soc.* **120**, 12169–12177.

Sai, X., Naruse, K., and Sokabe, M. (1999). Activation of pp60src is critical for stretch-induced orienting response in fibroblasts. *J. Cell Sci.* **112**, 1365–1373.

Sigal, G. B., Bamdad, C., Barberis, A., Strominger, J., and Whitesides, G. M. (1996). A self-assembled monolayer for the binding and study of histidine-tagged proteins by surface plasmon resonance. *Anal. Chem.* **68**, 490–497.

Singhvi, R., Kumar, A., Lopez, G. P., Stephanopolous, G. N., Wang, D. I. C., Whitesides, G. M., and Ingber, D. E. (1994). Engineering cell shape and function. *Science* **264**, 696–698.

St. John, P. M., Davis, R., Cady, N., Czajka, J., Batt, C. A., and Craighead, H. G. (1998). Diffraction-based cell detection using a microcontact printed antibody grating. *Anal. Chem.* **70**, 1108–1111.

Takayama, S., McDonald, J. C., Ostuni, E., Liang, M. N., Kenis, P. J. A., Ismagilov, R. F., and Whitesides, G. M. (1999). Patterning cells and their environments using multiple laminar fluid flows in capillary networks. *Proc. Natl. Acad. Sci. U.S.A.* **96**, 5545–5548.

Terfort, A., Bowden, N., and Whitesides, G. M. (1997). Three-dimensional self-assembly of millimetre-scale components. *Nature (London)* **386**, 162–164.

Ulman, A. (1996). Formation and structure of self-assembled monolayers. *Chem. Rev.* **96**, 1533–1554.

Vaidya, R., Tendler, L. M., Bradley, G., O'Brien, M. J., Cone, M., and Lopez, G. P. (1998). Computer-controlled laser ablation: A convenient and versatile tool for micropatterning bifunctional synthetic surfaces for applications in biosensing and tissue engineering. *Biotechnol. Prog.* **14**, 371–377.

Wheeler, B. C., Corey, J. M., Brewer, G. J., and Branch, D. W. (1999). Microcontact printing for precise control of nerve cell growth in culture. *J. Biomech. Eng.* **121**, 73–78.

Xia, Y., and Whitesides, G. M. (1998). Soft lithography. *Angew. Chem., Int. Ed. Engl.* **37**, 550–575.

Yan, L., Marzolin, C., Terfort, A., and Whitesides, G. M. (1997). Formation and reaction of interchain carboxylic anhydride groups on self-assembled monolayers on gold. *Langmuir* **13**, 6704–6712.

Yan, L., Zhao, X.-M., and Whitesides, G. M. (1998). Patterning a preformed, reactive SAM using microcontact printing. *J. Am. Chem. Soc.* **120**, 6179–6180.

Zhang, S., Yan, L., Altman, M., Lässle, M., Nugent, H., Frankel, F., Lauffenburger, D. A., Whitesides, G. M., and Rich, A. (1999). Biological surface engineering: A simple system for cell pattern formation. *Biomaterials* **20**, 1213–1220.

CHAPTER 19

CELL INTERACTIONS WITH POLYMERS

W. Mark Saltzman

INTRODUCTION

Synthetic and naturally occurring polymers are an important element in new strategies for producing engineered tissue (Langer and Vacanti, 1993; Hubbell, 1995). Several classes of polymers have proved to be most useful in biomedical applications, including situations in which the polymer remains in intimate contact with cells and tissues for prolonged periods (Table 19.1). These polymers might be appropriate for tissue engineering applications, as well. But to select appropriate polymers for tissue engineering, it is necessary to understand the influence of the polymer on cell viability, growth, and function.

This chapter briefly reviews previous work on the interactions of tissue-derived cells with polymers, particularly the types of synthetic polymers that have been employed as biomaterials. In flowing blood, the interactions of cells, particularly platelets, with synthetic polymer surfaces are also an important aspect of biomaterials design, but are not considered here.

METHODS FOR CHARACTERIZING CELL INTERACTIONS WITH POLYMERS

Cell interactions with polymers are usually studied using cell culture techniques. Although *in vitro* experiments do not reproduce the entire range of cellular responses observed following implantation of materials, the culture environment provides a level of control and quantification that cannot be easily obtained *in vivo*. To study cell interactions, cells in culture are usually plated over a polymer surface and the extent of cell adhesion and spreading on the surface is measured. By maintaining the culture for longer periods, the influence of the substrate on cell viability, function, and motility can also be determined. Because investigators use different techniques to assess cell interactions with polymers, and because the differences between experimental techniques are potentially important for interpretation of interactions, the most frequently used *in vitro* methods are reviewed in this section.

ADHESION AND SPREADING

Most tissue-derived cells are anchorage dependent and require attachment to a solid surface for viability and growth. For this reason, the initial events that occur when a cell approaches a surface are of fundamental interest. In tissue engineering, cell adhesion to a surface is critical because adhesion precedes other events, such as cell spreading, cell migration, and, often, differentiated cell function.

A number of different techniques for quantifying the extent and strength of cell adhesion have been developed. In fact, so many different techniques are used that it is usually difficult to compare studies performed by different investigators. This situation is further complicated by the fact that cell adhesion depends on a large number of experimental parameters (Lauffenburger and Linderman, 1993), many of which are difficult to control. The simplest methods for quantifying the extent of cell adhesion to a surface involve three steps: (1) suspension of cells over a surface, (2) in-

Table 19.1. Polymers that might be useful in tissue engineering[a]

Polymer	Typical application
Polydimethylsiloxane, silicone elastomers (PDMS)	Breast, penile, and testicular prostheses, catheters, drug delivery devices, heart valves, hydrocephalus shunts, membrane oxygenators
Polyurethanes (PEUs)	Artificial hearts and ventricular assist devices, catheters, pacemaker leads
Poly(tetrafluoroethylene) (PTFE)	Heart valves, vascular grafts, facial prostheses, hydrocephalus shunts, membrane oxygenators, catheters, sutures
Polyethylene (PE)	Hip prostheses, catheters
Polysulfone (PSu)	Heart valves, penile prostheses
Poly(methyl methacrylate) (pMMa)	Fracture fixation, intraocular lenses, dentures
Poly(2-hydroxyethylmethacrylate) (pHEMA)	Contact lenses, catheters
Polyacrylonitrile (PAN)	Dialysis membranes
Polyamides	Dialysis membranes, sutures
Polypropylene (PP)	Plasmapheresis membranes, sutures
Poly(vinyl chloride) (PVC)	Plasmapheresis membranes, blood bags
Poly(ethylene–covinyl acetate)	Drug delivery devices
Poly(L-lactic acid), poly(glycolic acid), and poly(lactide–coglycolide) (PLA, PGA, and PLGA)	Drug delivery devices, sutures
Polystyrene (PS)	Tissue culture
Poly(vinyl pyrrolidone) (PVP)	Blood substitutes

[a]Inclusion in this list is based on past use in biomedical devices, from reviews by Peppas and Langer (1994) and Marchant and Wang (1994).

cubation of the sedimented cells in culture medium for some period of time, and (3) detachment of loosely adherent cells under controlled conditions. The extent of cell adhesion, which is a function of the conditions of the experiment, is determined by quantifying either the number of cells that remain associated with the surface (the "adherent" cells) or the number of cells that were extracted with the washes. Radiolabeled or fluorescently labeled cells can be used to permit measurement of the number of attached cells. Alternatively, the number of attached cells can be determined by direct visualization, by measurement of the concentration of an intracellular enzyme, or by binding of a dye to an intracellular component such as DNA. In many cases, the adherent cells are further categorized based on morphological differences (e.g., extent of spreading, formation of actin filament bundles, presence of focal contacts). This technique is simple, rapid, and because it requires simple equipment, it is commonly performed. Unfortunately, it is often difficult to control the force that is provided to dislodge the nonadherent cells, making it difficult to compare results obtained from different laboratories, even when they are using the same technique.

This disadvantage can be overcome by using centrifuge (McClay *et al.*, 1981) or a flowing fluid (McIntire, 1994) to provide a reproducible detachment force. In centrifugal detachment assays, the technique described above is modified slightly: following the incubation period, the plate is inverted and subjected to a controlled detachment force by centrifugation. In most flow chambers, the fluid is forced between two parallel plates (Lawrence *et al.*, 1987). Prior to applying the flow field, a cell suspension is injected into the chamber, and the cells are permitted to settle onto the surface of interest and to adhere. After some period of incubation, flow is initiated between the plates. These chambers can be used to measure the kinetics of cell attachment, detachment, and rolling on surfaces under conditions of flow. Usually, the overall flow rate is adjusted so that the flow is laminar, and the shear stresses at the wall approximate those found in the circulatory system; however, these chambers can be used to characterize cell detachment under a wide range of conditions.

Radial flow detachment chambers have also been used to measure forces of cell detachment (Cozens-Roberts *et al.,* 1990). Because of the geometry of the radial flow chamber, in which cells are attached uniformly to a circular plate and fluid is circulated from the center to the periphery of the chamber along radial paths, the fluid shear force experienced by the attached cells decreases with radial position from the center to the periphery. Therefore, in a single experiment, the influence of a range of forces on cell adhesion can be determined. A spinning disk apparatus can be used in a similar fashion (Horbett *et al.,* 1988). Finally, micropipette techniques can be used to measure cell membrane deformability or forces of cell–cell or cell–surface adhesion (Evans, 1973; Lamoureux *et al.,* 1989; Tozeren *et al.,* 1989).

MIGRATION

The migration of individual cells within a tissue is a critical element in the formation of the architecture of organs and organisms (Trinkhaus, 1984). Likewise, cell migration is likely to be an important phenomenon in tissue engineering, because the ability of cells to move, either in association with the surface of a material or through an ensemble of other cells, will be an essential part of new tissue formation or regeneration. Cell migration is also difficult to measure, particularly in complex environments. Fortunately, a number of useful techniques for quantifying cell migration in certain situations have been developed. As in cell adhesion, however, no technique has gained general acceptance, so it is often difficult to correlate results obtained by different techniques or different investigators.

Most experimental methods for characterizing cell motility can be divided into two categories. In visual assays, the movements of a small number of cells are observed individually (Gail and Boone, 1970; Zigmond, 1977; Allan and Wilkinson, 1978; Parkhurst and Saltzman, 1992). Population techniques, on the other hand, allow the observation of the collective movements of groups of cells; in filter chamber assays the number of cells migrating through a membrane or filter is measured (Boyden, 1962), whereas in under-agarose assays the leading front of cell movement on a surface under a block of agarose is monitored (Nelson *et al.,* 1975). The experimental details of these assays, as well as their advantages and limitations, have been reviewed (Wilkinson *et al.,* 1982). Both visual and population assays can be quantitatively analyzed, enabling the estimation of intrinsic cell motility parameters, such as the random motility coefficient and the persistence time (Gail and Boone, 1970; Dunn, 1983; Lauffenburger, 1983; Buettner *et al.,* 1989; Farrell *et al.,* 1990).

AGGREGATION

Cell aggregates are important tools in the study of tissue development, permitting correlation of cell–cell interactions with cell differentiation, viability, and migration, as well as subsequent tissue formation. The aggregate morphology permits reestablishment of cell–cell contacts normally present in tissues; therefore, cell function and survival are often enhanced in aggregate culture (Matthieu *et al.,* 1981; Landry *et al.,* 1985; Koide *et al.,* 1989; Cirulli *et al.,* 1993; Parsons-Wingerter and Saltzman, 1993; Peshwa *et al.,* 1994). Because of this, cell aggregates may also be useful in tissue engineering, enhancing the function of cell-based hybrid artificial organs (Nyberg *et al.,* 1993) or reconstituted tissue transplants (Langer and Vacanti, 1993).

Aggregates are usually formed by incubating cells in suspension, using gentle rotational stirring to disperse the cells (Moscona, 1961). Although this method is suitable for aggregation of many cells, serum or serum proteins must be added to promote cell aggregation in many cases (Matsuda, 1988), making it difficult to characterize and control the aggregation process. specialized techniques can be used to produce aggregates in certain cases, principally by controlling cell detachment from a solid substratum (Koide *et al.,* 1989; Parsons-Wingerter and Saltzman, 1993; Takezawa *et al.,* 1993). Synthetic polymers produced by linking cell-binding peptides (such as RGD and YIGSR) to both ends of poly(ethylene glycol) (PEG) have been used to promote aggregation of cells in suspension (Dai *et al.,* 1994).

The kinetics and extent of aggregation can be measured by a variety of techniques. Often, direct visualization of aggregate size is used to determine the extent of aggregation (Moscona, 1961).

The kinetics of aggregation can be monitored by measuring aggregate size distributions over time, which is facilitated by the use of computer image analysis techniques (Munn *et al.,* 1993; Dai *et al.,* 1994) or electronic particle counters (Orr and Roseman, 1969). Specialized aggregometers have been constructed to provide reproducible and rapid measurements of the rate of aggregation; in one such device, small-angle light scattering through rotating sample cuvettes was used to produce continuous records of aggregate growth (Thomas and Steinberg, 1980).

CELL FUNCTION

In tissue engineering, one is usually interested in promotion of some cell-specific function. For example, protein secretion (Cima *et al.,* 1991) and detoxification (Gutsche *et al.,* 1994) are essential functions for hepatocytes. Production of extracellular matrix (ECM) proteins is important in the physiology of many cells, such as chondrocytes (Benya and Shaffer, 1982), osteoblasts (Ishaug *et al.,* 1994), and fibroblasts (Tamada and Ikada, 1994). In some cases, the important cell function involves the coordinated activity of groups of cells, such as the formation of myotubules in embryonic muscle cell cultures (Stol *et al.,* 1985) or the contraction of the matrix surrounding fibroblasts (Bell *et al.,* 1979; Barocas *et al.,* 1995). In these cases, cell function is monitored by watching for changes in the morphology in the culture.

CELL INTERACTION WITH POLYMER SURFACES

EFFECT OF POLYMER CHEMISTRY ON CELL BEHAVIOR

Synthetic polymers

For cells attached to a solid substrate, cell behavior and function depend on the characteristics of the substrate. Consider, for example, the experiments described by Folkman and Moscona (1978), in which cells were allowed to settle onto surfaces formed by coating conventional tissue culture polystyrene (TCPS) with various dilutions of poly(2-hydroxyethyl methacrylate) (pHEMA). As the amount of pHEMA added to the surface was increased, the surface became less adhesive and cell spreading decreased; spreading was quantified by measuring the average cell height on the surface. Average cell height correlated with the rate of cell growth (Fig. 19.1), suggesting that cell shape, which was determined by the adhesiveness of the surface, modulated cell proliferation. In these experiments, two simple polymers (TCPS and pHEMA) were used to produce a series of surfaces with graded adhesivity. These experiments demonstrate that the nature of a polymer surface will have important consequences for cell function, an observation of considerable significance with regard to the use of polymers in tissue engineering.

Following a similar experimental rationale, a number of groups have examined the relationship between chemical or physical characteristics of the substrate and behavior or function of attached cells. Results from a number of similar studies are summarized in Figs. 19.2 and 19.3. Cell adhesion appears to be maximized on surfaces with intermediate wettability (Fig. 19.2) (Horbett

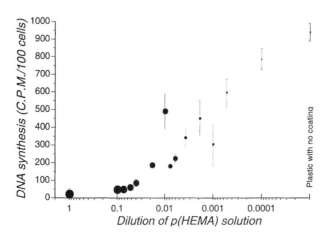

Fig. 19.1. Cell shape and cell growth are modulated by properties of a polymer surface. Cell culture surfaces were produced by evaporating diluted solutions of pHEMA onto TCPS. The uptake of [³H]thymidine was used as a measure of proliferation. The size of the symbol represents the relative cell height; small symbols represent cells with small heights, and therefore significant spreading; large symbols represent cells with large heights, and therefore negligible spreading. Data replotted from Folkman and Moscona (1978).

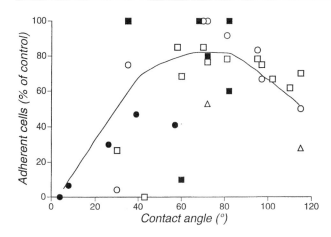

Fig. 19.2. The relationship between cell adhesion and water-in-air contact angle for a variety of polymer surfaces. Data replotted from Tamada and Ikada (1994) for fibroblasts (□), Ikada (1994) for L cells (○), Hasson et al. (1987) for endothelial cells (△) van Wachem et al. (1987) for endothelial cells (●), and Saltzman et al. (1991) for fibroblasts (■).

et al., 1985; van Wachem *et al.,* 1987; Ikada, 1994; Tamada and Ikada, 1994), although there are some obvious exceptions. For most surfaces, adhesion requires the presence of serum and, therefore, this optimum is probably related to the ability of proteins, such as fibronectin (Horbett and Schway, 1988), to absorb to the surface. In the absence of serum, adhesion is enhanced on positively charged surfaces (van Wachem *et al.,* 1987). Fibroblast spreading has been correlated with surface free energy (Fig. 19.3), but the rate of fibroblast growth on polymer surfaces appears to be relatively independent of surface chemistry (van der Valk *et al.,* 1983; Horbett *et al.,* 1985; Saltzman *et al.,* 1991). Cell viability may also be related to interactions with the surface (Ertel *et al.,* 1994). The migration of surface-attached fibroblasts (Saltzman *et al.,*1991), endothelial cells (Hasson *et al.,* 1987), and corneal epithelial cells (Pettit *et al.,* 1994) has been measured as function of polymer surface chemistry; rates of cell migration depend on the nature of the surface, although no general trends have emerged. Collagen synthesis in fibroblasts has been correlated with contact angle, with higher rates of collagen synthesis per cell for the most hydrophobic surfaces (Tamada and Ikada, 1994).

Polymers can frequently be made more suitable for cell attachment and growth by surface modification. In fact, polystyrene (PS) substrates used for tissue culture are usually treated by glow-discharge (Amstein and Hartman, 1975) or exposure to sulfuric acid to increase the number of charged groups at the surface, which improves attachment and growth of many types of cells. Treatment of pHEMA, a nonadhesive polymer, with sulfuric acid also improved adhesion of endothelial cells and permitted cell proliferation on the surface (Hannan and McAuslan, 1987). Modification of PS or PET, poly(ethylene terephthalate), by radiofrequency plasma deposition enhanced attachment and spreading of fibroblasts and myoblasts (Chinn *et al.,* 1989). Again, many of the effects of surface modification appear to be secondary to increased adsorption of cell attachment

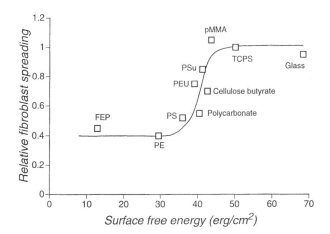

Fig. 19.3. The relationship between cell spreading and surface free energy for fibroblasts cultured on a variety of polymer surfaces. Data replotted from van der Valk et al. (1983) and Schakenraad et al. (1986). FEP, fluoroethylenepropylene copolymer; PE, polyethylene; PS, polystyrene; PEU, polyurethane; PSu, polysulfone; PMMA, poly(methyl methacrylate); TCPS, tissue culture polystyrene.

proteins, such as fibronectin and vitronectin, to the surface. On the other hand, some reports have identified specific chemical groups at the polymer surface [such as hydroxyl ($-$OH) (Curtis et al., 1983; Lydon et al., 1985) or surface C$-$O functionalities (Chinn et al., 1989)] as important factors in modulating the fate of surface-attached cells.

So far, no general principles that would allow prediction of the extent of attachment, spreading, or growth of cultured cells on different polymer surfaces have been identified. For specific cells, however, interesting correlations have been made with parameters such as the density of surface hydroxyl groups (Curtis et al., 1983), density of surface sulfonic groups (Kowalczynska and Kaminski, 1991), surface free energy (van der Valk et al., 1983; Baier et al., 1985; Schakenraad et al., 1986), fibronectin adsorption (Chinn et al., 1989), and equilibrium water content (Lydon et al., 1985), but exceptions to these correlations are always found. Complete characterization of the polymer, including both bulk and surface properties, is critical to understanding the nature of the cell–polymer interactions (Tyler et al., 1992).

Although the interactions of cells with implanted polymers is much more difficult to measure, the surface chemistry of polymers appears to influence cell interactions in vivo. For example, the ability of macrophages to form multinucleated giant cells at a hydrogel surface has been correlated with the presence of certain chemical groups at the surface: macrophage fusion decreases in the order $(CH3)_2N- > -OH = -CO-NH- > -SO_3H > -COOH(-COONa)$ (Smetana et al., 1990; Smetana, 1993). A similar hierarchy has been observed for CHO cell adhesion and growth on surfaces with grafted functional groups: CHO cell attachment and growth decreased in the order $-CH_2NH_2 > -CH_2OH > -CONH_2 > -COOH$ (Lee et al., 1994).

Biodegradable polymers

Biodegradable polymers slowly degrade and then dissolve following implantation. This feature may be important for many tissue engineering applications, because the polymer will disappear as functional tissue regenerates (Vacanti et al., 1988). Biodegradable polymers may provide an additional level of control over cell interactions: during polymer degradation, the surface of the polymer is constantly renewed, providing a dynamic substrate for cell attachment and growth.

Homopolymers and copolymers of poly(L-lactic acid), poly(glycolic acid), and poly(lactide–co-glycolide) (PLA, PGA, PLGA) have been frequently examined as cell culture substrates, because they have been used as implanted sutures for several decades (Schmitt and Polistina, 1967; Schneider, 1967; Kulkarni et al., 1971; Wasserman and Levy, 1975). Chondrocytes proliferate and secrete glycosaminoglycans within porous meshes of PGA and foams of PLA (Freed et al., 1993). Rat hepatocytes attach to blends of biodegradable PLGA polymers and secrete albumin for 5 days in culture (Cima et al., 1991). Neonatal rat osteoblasts also attach to PLA, PGA, and PLGA substrates and synthesize collagen (Ishaug et al., 1994).

Cell adhesion and function have been examined on a variety of other biodegradable polymers. When cells from an osteogenic cell line were seeded onto polyphosphazenes produced with a variety of side groups, the rate of cell growth as well as the rate of polymer degradation depended on side group chemistry (Laurencin et al., 1993). Fibroblasts and hepatocytes attached to poly(phosphoesters) with a variety of side group functionalities (Saltzman et al., 1991). Proliferation of fibroblasts, as well as attachment and neurite outgrowth from PC12 and dorsal root ganglia cells, have been examined on tyrosine-based polycarbonates and polyarylates (Kohn, 1994; Zhou et al., 1994).

Synthetic polymers with adsorbed proteins

Cell attachment, migration, and growth on polymer surfaces appear to be mediated by proteins, either adsorbed from the culture medium or secreted by the cultured cells. Because it is difficult to study these effects in situ during cell culture, often the polymer surfaces are pretreated with purified protein solutions. In this way, the investigators hope that subsequent cell behavior on the surface will represent cell behavior in the presence of a stable layer of surface-bound protein. Cell spreading, but not attachment, has been correlated with fibronectin adsorption to a variety of surfaces (Horbett and Schway, 1988; Chinn et al., 1989; Pettit et al., 1994). Rates of cell migration on a polymer surface have been shown to depend on the concentration of preadsorbed adhesive proteins (Calof and Lander, 1991; DiMilla et al., 1993), as well as the presence of soluble inhibitors to protein-mediated cell adhesion (Wu et al., 1994). It appears that the rate of migration is optimal at intermediate substrate adhesiveness, as one would expect from mathematical models of cell migration (Lauffenburger and Linderman, 1993).

Hybrid polymers with immobilized functional groups

Surface modification techniques have been used to produce polymers for cell attachment (Ikada, 1994). For example, chemical groups can be added to change the wettability of the surface, which often influences cell adhesion (Fig. 19.2), as described above. Alternatively, whole proteins such as collagen can be immobilized to the surface, providing the cell with a substrate that more closely resembles the ECM found in tissues (Tamada and Ikada, 1994). Collagen and other ECM molecules have also been incorporated into hydrogels by either adding the protein to a reaction mixture containing monomers and initiating polymerization (Civerchia-Perez *et al.*, 1980; Carbonetto *et al.*, 1982; Woerly *et al.*, 1993), or mixing the protein with polymerized polymer, such as pHEMA, in appropriate solvents (Stol *et al.*, 1985).

To isolate certain features of ECM molecules, and to produce surfaces that are simpler and easier to characterize, smaller biologically active functional groups have been used to modify surfaces. These biologically active groups can be oligopeptides (Massia and Hubbell, 1989), saccharides (Schnaar *et al.*, 1978), or glycolipids (Blackburn and Schnaar, 1983). Certain short amino acid sequences appear to bind to receptors on cell surfaces and mediate cell adhesion. For example, the cell-binding domain of fibronectin contains the tripeptide RGD (Arg-Gly-Asp).* Cells attach to surfaces containing adsorbed oligopeptides with the RGD sequence and soluble, synthetic peptides containing the RGD sequence reduce the cell-binding activity of fibronectin (Pierschbacher and Ruoslahti, 1984), demonstrating the importance of this sequence in adhesion of cultured cells. A large number of ECM proteins (fibronectin, collagen, vitronectin, thrombospondin, tenascin, laminin, and entactin) contain the RGD sequence. The sequences YIGSR and IKVAV in laminin also have cell-binding activity, and mediate adhesion in certain cells.

Because RGD is critical in cell adhesion to ECM, many investigators have examined the addition of this sequence to synthetic polymer substrates. Synthetic RGD-containing peptides have been immobilized to poly(tetrafluoroethylene) (PTFE) (Massia and Hubbell, 1991a), PET (Massia and Hubbell, 1991b), polyacrylamide (Brandley and Schnaar, 1988), polyurethane (PEU) (H.-B. Lin *et al.*, 1992; H. Lin *et al.*, 1994), poly(carbonate urethane) (Breuers *et al.*, 1991), PEG (Drumheller *et al.*, 1994), poly(vinyl alcohol) (PVA) (Matsuda *et al.*, 1989), PLA (Barrera *et al.*, 1993), and poly(*N*-isopropylacrylamide–co-*N*-*n*-butylacrylamide) (Miura *et al.*, 1991) substrates. The addition of RGD has induced cell adhesion, cell spreading, and focal contact formation on otherwise nonadhesive or weakly adhesive polymers (Massia and Hubbell, 1989; 1991a; Hubbell *et al.*, 1991; Lin *et al.*, 1994). Because cells contain cell adhesion receptors that recognize only certain ECM molecules, use of an appropriate cell-binding sequence has led to cell-selective surfaces as well, whereby the population of the cells that adhered to the polymer was determined by the structure of the peptide (Hubbell *et al.*, 1991).

In previous studies, the presence of serum proteins attenuated the adhesion activity of peptide-grafted PEU surfaces (Lin *et al.*, 1994), highlighting a difficulty in using these materials *in vivo*. This problem may be overcome, however, through the development of base materials that are biocompatible yet resistant to protein adsorption. One of the most successful approaches for reducing protein adsorption or cell adhesion has been production of surfaces rich in PEG (Desai and Hubbell, 1991; Merrill, 1993; Drumheller and Hubbell, 1995). For example, polymer substrates composed of PEG in highly cross-linked matrices of acrylic acid and trimethyloylpropane triacrylate completely resisted protein adsorption and cell adhesion, but readily supported adhesion after derivitization with cell-binding peptides (Drumheller and Hubbell, 1994). Alternatively, matrices formed directly from synthetic polypeptides may also be useful for cell adhesion. Genes coding the β sheet of silkworm silk have been combined with genes coding fragments of fibronectin to produce proteins that form very stable matrices with cell adhesion domains (Pronectin

---

Amino acids are identified by their one letter abbreviations: A, alanine; R, arginine; N, asparagine; D, aspartic acid; B, asparagine or aspartic acid; C, cysteine; Q, glutamine; E, glutamic acid; Z, glutamine or glutamic acid; G, glycine; H, histidine; I, isoleucine; L, leucine; K, lysine; M, methionine; F, phenylalanine; P, proline; S, serine; T, threonine; W, trypotophan; Y, tyrosine; V, valine.

F; Protein Polymer Technologies, Inc.). Synthetic proteins based on peptide sequences from elastin have been used as cell culture substrates: in the presence of serum, fibroblasts and endothelial cells adhered to the surfaces of matrices formed by γ-irradiation cross-linking of polypeptides containing repeated sequences GGAP, GGVP, GGIP, and GVGVP (Nicol *et al.,* 1993).

In addition to cell-binding peptides, a variety of other biologically active molecules have been used to enhance cell adhesion to surfaces. For certain cells, adhesion has been enhanced by adsorption of homopolymers of basic amino acids, such as polylysine and polyornithine. Likewise, covalently bound amine groups have influenced cell attachment and growth (Kikuchi *et al.,* 1992; Massia and Hubbell, 1992). The immobilization of saccharides to polymers has also influenced cell attachment and function. Rat hepatocytes adhered, via asialoglycoprotein receptors, to surfaces derivatized with lactose and remained in a rounded morphology consistent with enhanced function in culture (Kobayashi *et al.,* 1986, 1992; Gutsche *et al.,* 1996). Similar results have been obtained with polymer surfaces derivatized with *N*-acetylglucosamine, which was recognized by a surface lectin on chicken hepatocytes (Gutsche *et al.,* 1994). A variety of carbohydrates have been immobilized on polyacrylamide disks, making them suitable for cell culture (Schnaar *et al.,* 1978).

INFLUENCE OF SURFACE MORPHOLOGY ON CELL BEHAVIOR

The microscale texture of an implanted material can have a significant effect on the behavior of cells in the region of the implant. Fibrosarcomas developed with high frequency, approaching 50% in certain situations, around implanted Millipore filters; the tumor incidence increased with decreasing pore size in the range 450 to 50 μm (Goldhaber, 1961, 1962). In a recent study, porous polymer membranes containing certain structural features (nominal pore size > 0.6 μm and fibers/strands < 5 μm) were associated with enhanced new vessel growth (Brauker *et al.,* 1992).

The behavior of cultured cells on surfaces with edges, grooves, or other textures is different than behavior on smooth surfaces. In many cases, cells oriented and migrated along fibers or ridges in the surface, a phenomenon called contact guidance from early studies on neuronal cell cultures (Weiss, 1934, 1945). Fibroblasts have also been observed to orient on grooved surfaces (Brunette, 1986), particularly when the texture dimensions are 1 to 8 μm (Dunn and Brown, 1986). The degree of cell orientation depended on both the depth (Clark *et al.,* 1991) and pitch (Dunn and Brown, 1986) of the grooves. Not all cells exhibit the same degree of contact guidance when cultured on identical surfaces: baby hamster kidney (BHK) and Madin–Darby canine kidney (MDCK) cells oriented on 100-nm-scale grooves in fused quartz, but cerebral neurons did not (Clark *et al.,* 1991). Fibroblasts, monocytes and macrophages, but not keratinocytes or neutrophils, spread when cultured on silicon oxide with grooves with a 1.2-μm depth and a 0.9-μm pitch (Meyle *et al.,* 1995).

Substrates with peaks and valleys also influenced the function of attached cells. Polydimethylsiloxane (PDMS) surfaces with 2- to 5-μm texture maximized macrophage spreading (Schmidt and von Recum, 1992). Similarly, PDMS surfaces with 4- or 25-μm^2 peaks uniformly distributed on the surface provided better fibroblast growth than did 100-μm^2 peaks or 4-, 25-, or 100-μm^2 valleys (Schmidt and von Recum, 1992).

USE OF PATTERNED SURFACES TO CONTROL CELL BEHAVIOR

A variety of techniques have been used to create chemically patterned surfaces containing cell adhesive and nonadhesive regions. These patterned surfaces have been useful for examining fundamental determinants of cell adhesion, growth, and function. For example, individual fibroblasts were attached to adhesive microislands of palladium that were patterned onto a nonadhesive pHEMA substrate using microlithographic techniques (O'Neill *et al.,* 1986). By varying the size of the microisland, the extent of spreading and hence the cell surface area were controlled. On small islands (~500 μm^2) cells attached but did not spread, whereas on larger islands (4000 μm^2) cells spread to the same extent as in unconfined monolayer culture. Cells on large islands proliferated at the same rate as cells in conventional culture, and most cells attached to small islands proliferated at the same rate as suspended cells. For 3T3 cells, however, contact with the surface enhanced proliferation, suggesting that anchorage can stimulate cell division by simple contact with the substrate as well as by increased spreading.

A number of other studies have employed patterned surfaces in cell culture. Micrometer-scale adhesive islands of self-assembled alkane thiols were created on gold surfaces using a simple stamping procedure (Singhvi *et al.,* 1994), which served to confine cell spreading to islands. When hepatocytes were attached to these surfaces, larger islands (10,000 μm^2) promoted growth, but smaller islands (1600 μm^2) promoted albumin secretion. Stripes of a monoamine-derivatized surface were produced on fluorinated ethylene propylene films by radiofrequency glow discharge (Ranieri *et al.,* 1993). Because proteins adsorbed differently to the monamine-derivatized and the untreated stripes, striped patterns of cell attachment were produced. A similar approach, using photolithography to produce hydrophilic patterns on a hydrophobic surface, produced complex patterns of neuroblastoma attachment and neurite extension (Matsuda *et al.,* 1992). A variety of substrate microgeometries, and cell attachment patterns, were created by photochemical fixation of hydrophilic polymers onto TCPS or hydrophobic polymers onto PVA through patterned photomasks: bovine endothelial cells attached and proliferated preferentially on either the TCPS surface (on TCPS/hydrophilic patterns) or the hydrophobic surface (on PVA/hydrophobic patterns) (Matsuda and Sugawara, 1995). When chemically patterned substrates were produced on self-assembled monolayer films using microlithographic techniques, neuroblastoma cells attached to and remained confined within amine-rich patterns on these substrates (Matsuzawa *et al.,* 1993).

Surfaces containing gradients of biological activity have also been useful tools in cell biology, particularly for examining haptotaxis, the directed migration of cells on surfaces with gradients of immobilized factors (Brandley and Schnaar, 1989; Brandley *et al.,* 1990).

CELL INTERACTIONS WITH POLYMERS IN SUSPENSION

Most of the studies reviewed in the preceding section examined the growth, migration, and function of cells attached to a solid polymer surface. This is a relevant paradigm for a variety of tissue engineering applications, wherein polymers will be used as substrates for the transplantation of cells or as scaffolds to guide tissue regeneration *in situ.* Polymers may be important in other aspects of tissue engineering, as well. For example, polymer microcarriers can serve as substrates for the suspension culture of anchorage-dependent cells, and therefore might be valuable for the *in vitro* expansion of cells or cell transplantation (Demetriou *et al.,* 1986). In addition, immunoprotection of cells suspended within semipermeable polymer membranes is another important approach in tissue engineering, because these encapsulated cells may secrete locally active proteins (Emerich *et al.,* 1994) or function as small endocrine organs (Lacy *et al.,* 1991) within the body.

The idea of using polymer microspheres as particulate carriers for the suspension culture of anchorage-dependent cells was introduced by van Wezel (1967, 1985). As in planar polymer surfaces, the surface characteristics of microcarriers influence cell attachment, growth, and function. In the earliest studies, microspheres composed of diethylaminoethyl (DEAE)–dextran were used; these spheres have a positively charged surface and are routinely used as anion-exchange resins. DEAE–dextran microcarriers support the attachment and growth of both primary cells and cell lines; the best growth occurred when the surface charge was optimized (Levine *et al.,* 1979). In addition to dextran-based microcarriers, microspheres that support cell attachment have been produced from PS (Jacobson and Ryan, 1982; Fairman and Jacobson, 1983; Gutsche *et al.,* 1994), gelatin (Wissemann and Jacobson, 1985), and many of the synthetic and naturally occurring polymers described in the preceding sections. To support cell attachment and growth, the microcarrier surface has been modified chemically, or by immobilization of proteins (Jacobson and Ryan, 1982), peptides (Jacobson and Ryan, 1982) or carbohydrates (Gutsche *et al.,* 1994). Suspension culture techniques can be used to permit cell interactions with complex three-dimensional polymer formulations, as well. For example, cells seeding onto polymer fiber meshes during suspension culture resulted in more uniform cell distribution within the mesh than was obtained by inoculation in static culture (Freed and Vunjak-Novakovic, 1995).

In cell encapsulation techniques, cells are suspended within thin-walled capsules or solid matrices of polymer. Alginate forms a gel with the addition of divalent cations under very gentle conditions and, therefore, has been frequently used for cell encapsulation (Lim and Sun, 1980). Certain synthetic polymers, such as polyphosphazenes, have also been used to encapsulate cells by cation-induced geltion (Bano *et al.,* 1991). Low-melting-temperature agarose has also been studied extensively for cell encapsulation (Nilsson *et al.,* 1983). Methods for the microencapsulation of cells within hydrophilic or hydrophobic polyacrylates by interfacial precipitation have been described (Dawson *et al.,* 1987; Sefton *et al.,* 1987), although the thickness of the capsule can limit the permeation of compounds, including oxygen, through the semipermeable membrane shell. Interfacial polymerization has been used to produce conformal synthetic membranes on cells or cell clusters (Sawhney *et al.,* 1994), thereby providing immunoprotection while dramatically reducing the diffusional distances.

Hollow fibers are frequently used for macroencapsulation; cells and cell aggregates are suspended within thin fibers composed of a porous, semipermeable polymer. Chromaffin cells suspended within hollow fibers formed from a copolymer of vinyl chloride and acrylonitrile, which are commonly used as ultrafiltration membranes, have been studied as potential treatments for cancer patients with pain (Joseph *et al.,* 1994) or for Alzheimer's disease (Emerich *et al.,* 1994). Other materials have been added to the interior of the hollow fibers to provide a matrix that enhances cell function or growth: chitosan (Zienlinski and Aebischer, 1994), alginate (Lacy *et al.,* 1991), and agar (Sullivan *et al.,* 1991) have been used as internal matrices.

CELL INTERACTIONS WITH THREE-DIMENSIONAL POLYMER SCAFFOLDS AND GELS

Cells within tissues encounter a complex chemical and physical environment that is quite different from commonly used cell culture conditions. *In vitro* methods for growing cells in tissuelike environments may have direct application in tissue engineering (Vacanti *et al.,* 1988; Thompson *et al.,* 1989). Previous investigators have used three-dimensional cell culture methods to simulate the chemical and physical environment of tissues (Bell *et al.,* 1979; Yang *et al.,* 1979; Schor, 1980; Schor *et al.,* 1980, 1981; Richards *et al.,* 1983; Chen *et al.,* 1985; Haston and Wilkinson, 1988; Saltzman *et al.,* 1992). White blood cells, fibroblasts, pancreatic cells, epithelial cells, and nervous tissue cells have been cultured in gels containing ECM components, and the dynamics of cell motility (Parkhurst and Saltzman, 1992; Dickinson *et al.,* 1994), aggregation (Krewson *et al.,* 1994), and force generation (Barocas *et al.,* 1995) within gels have been studied. Often, tissue-derived cells cultured in ECM gels have reformed multicellular structures that are reminiscent of tissue architecture.

Gels of agarose have also been used for three-dimensional cell culture. Chondrocytes, which dedifferentiate when cultured as monolayers, reexpressed a differentiated phenotype when cultured in agarose gels (Benya and Shaffer, 1982). When fetal striatal cells were suspended in three-dimensional gels of hydroxylated agarose, ~50% of the cells extended neurites in gels containing between 0.5 and 1.25% agarose, but no cells extended neurites at concentrations above 1.5%, an agarose concentration corresponding to an average pore radius of 150 nm (Bellamkonda *et al.,* 1995). Neurites produced by PC12 cells within agarose gels, even under optimal conditions, are much shorter and fewer in number than neurites produced in gels composed of ECM molecules, however (Krewson *et al.,* 1994).

Macroporous hydrogels have been produced from pHEMA-based materials, using either freeze–thaw or porosigen techniques (Oxley *et al.,* 1993). These materials, when seeded with chondrocytes, have been studied as cartilage replacement (Corkhill *et al.,* 1993). Similar structures have been produced from PVA by freeze–thaw cross-linking (Cascone *et al.,* 1995). Although cells adhere poorly to pHEMA and PVA surfaces, adhesion proteins can be added during the formation to encourage cell attachment and growth. Alternatively, water-soluble polymers containing adhesive peptides, such as RGDS, have been photopolymerized to form a gel matrix around vascular smooth muscle cells: cells remain viable and elongate within the gel during subsequent culture (Moghaddam and Matsuda, 1991).

Fiber meshes (Mikos *et al.,* 1993) and foams (Lo *et al.,* 1995, 1996; Thomson *et al.,* 1995) of PLGA, PLA, and PGA have been used to create three-dimensional environments for cell proliferation and function, and to provide structural scaffolds for tissue regeneration. When cultured on three-dimensional PGA fiber meshes, chondrocytes proliferated, producing glycosaminoglycans and collagen and forming structures that were histologically similar to cartilage (Puelacher *et al.,* 1994). The physical dimensions of the polymer fiber mesh influenced cell growth rate, with slower growth in thicker meshes (Freed *et al.,* 1993). Changing the fluid mechanical forces on the cells during the tissue formation also appeared to influence the development of tissue structure (Freed and Vunjak-Novakovic, 1995). A variety of techniques have been developed for the production of foams of biodegradable polymers, which may also be useful in the production of three-dimensional tissuelike structures.

References

Allan, R. B., and Wilkinson, P. C. (1978). A visual analysis of chemotactic and chemokinetic locomotion of human neutrophil leukocytes: Use of a new assay with *Candida albicans* as gradient source. *Exp. Cell Res.* 111, 191–203.

Amstein, C., and Hartman, P. (1975). Adaptation of plastic surfaces for tissue culture by glow discharge. *J. Clin. Microbiol.* 2, 46–54.

Baier, R., DePalma, V. *et al.* (1985). Human platelet spreading on substrata of known surface chemistry. *J. Biomed. Mater. Res.* 19, 1157–1167.

Bano, M. C., Cohen, S. *et al.* (1991). A novel synthetic method for hybridoma cell encapsulation. *Bio/Technology* 9, 468–471.

Barocas, V. H., Moon, A. G. *et al.* (1995). The fibroblast-populated collagen microsphere assay of cell traction force. Part 2. Measurement of the cell traction parameter. *J. Biomech. Eng.* 117(2), 161–170.

Barrera, D. A., Zylstra, E. *et al.* (1993). Synthesis and RGD peptide modification of a new biodegradable copolymer: Poly(lactic acid–co-lysine). *J. Am. Chem. Soc.* 115, 11010–11011.

Bell, E. B., Ivarsson, B. *et al.* (1979). Production of a tissue-like structure by contraction of collagen lattices by human fibroblasts with different proliferative potential. *Proc. Nat. Acad. Sci. U.S.A.* 75, 1274–1278.

Bellamkonda, R., Ranieri, J. P. *et al.* (1995). Hydrogel-based three-dimensional matrix for neural cells. *J. Biomed. Mater. Res.* 29, 663–671.

Benya, P. D., and Shaffer, J. D. (1982). Dedifferentiated chondrocytes reexpress the differentiated collagen phenotype when cultured in agarose gels. *Cell (Cambridge, Mass.)* 30, 215–224.

Blackburn, C. C., and Schnaar, R. L. (1983). Carbohydrate-specific cell adhesion is mediated by immobilized glycolipids. *J. Biol. Chem.* 258(2), 1180–1188.

Boyden, S. V. (1962). The chemotactic effect of mixtures of antibody and antigen on polymorphonuclear leukocytes. *J. Exp. Med.* 115, 453–466.

Brandley, B., and Schnaar, R. (1988). Covalent attachment of an Arg-Gly-Asp sequence peptide to derivatizable polyacrylamide surfaces: Support of fibroblast adhesion and long-term growth. *Anal. Biochem.* 172, 270–278.

Brandley, B., and Schnaar, R. (1989). Tumor cell haptotaxis on covalently immobilized linear and exponential gradients of a cell adhesion peptide. *Dev. Biol.* 135, 74–86.

Brandley, B. K., Shaper, J. H. *et al.* (1990). Tumor cell haptotaxis on immobilized *N*-acetylglucosamine gradients. *Dev. Biol.* 140, 161–171.

Brauker, J., Martinson, L. A. *et al.* (1992). Neovascularization of immunoisolation membranes: The effect of membrane architecture and encapsulated tissue. *Transplant. Proc.* 24, 2924.

Breuers, W., Klee, D. *et al.* (1991). Immobilization of a fibronectin fragment at the surface of a polyurethane film. *J. Mater. Sci.: Mater. Med.* 2, 106–109.

Brunette, D. (1986). Fibroblasts on micromachined substrata orient hierarchically to grooves of different dimensions. *Exp. Cell Res.* 164, 11–26.

Buettner, H. M., Lauffenburger, D. A. *et al.* (1989). Measurement of leukocyte motility and chemotaxis parameters with the Millipore filter assay. *J. Immunol. Methods* 123, 25–37.

Calof, A. L., and Lander, A. D. (1991). Relationship between neuronal migration and cell-substratum adhesion: Laminin and merosin promote olfactory neuronal migration but are antiadhesive. *J. Cell Biol.* 115(3), 779–794.

Carbonetto, S. T., Gruver, M. M. *et al.* (1982). Nerve fiber growth on defined hydrogel substrates. *Science* 216, 897–899.

Cascone, M. G., Laus, M. *et al.* (1995). Evaluation of poly(vinyl alcohol) hydrogels as a component of hybrid artificial tissues. *J. Mater. Sci.: Mater. Med.* 6, 71–75.

Chen, J., Stuckey, E. *et al.* (1985). Three-dimensional culture of rat exocrine pancreatic cells using collagen gels. *Br. J. Exp. Pathol.* 66, 551–559.

Chinn, J., Horbett, T. *et al.* (1989). Enhancement of serum fibronectin adsorption and the clonal plating efficiencies of Swiss mouse 3T3 fibroblast and MM14 mouse myoblast cells on polymer substrates modified by radiofrequency plasma deposition. *J. Colloid Interface Sci.* 127, 67–87.

Cima, L., Ingber, D. *et al.* (1991). Hepatocyte culture on biodegradable polymeric substrates. *Biotechnol. Bioeng.* 38, 145–158.

Cirulli, V., Halban, P. A. *et al.* (1993). Tumor necrosis factor-α modifies adhesion properties of rat islet B cells. *J. Clin. Invest.* 91, 1868–1876.

Civerchia-Perez, L., Faris, B. *et al.* (1980). Use of collagen–hydroxyethylmethacrylate hydrogels for cell growth. *Proc. Nat. Acad. Sci. U.S.A.* 77, 2064–2068.

Clark, P., Connolly, P. *et al.* (1991). Cell guidance by ultrafine topography *in vitro*. *J. Cell Sci.* 99, 73–77.

Corkhill, P. H., Fitton, J. H. *et al.* (1993). Towards a synthetic articular cartilage. *J. Biomater. Sci., Polym. Ed.* 4, 6150–630.

Cozens-Roberts, C., Quinn, J. *et al.* (1990). Receptor-mediated cell attachment and detachment kinetics: II. Experimental model studies with the radial flow detachment assay. *Biophys. J.* 58, 857–872.

Curtis, A., Forrester, J. *et al.* (1983). Adhesion of cells to polystyrene surfaces. *J. Cell Biol.* 97, 1500–1506.

Dai, W., Belt, J. *et al.* (1994). Cell-binding peptides conjugated to poly(ethylene glycol) promote neural cell aggregation. *Bio/Technology* 12, 797–801.

Dawson, R. M., Broughton, R. L. *et al.* (1987). Microencapsulation of CHO cells in a hydroxyethyl methacrylate–methyl methacrylate copolymer. *Biomaterials* 8, 360–366.

Demetriou, A., Whiting, J. *et al.* (1986). Replacement of liver function in rats by transplantation of microcarrier-attached hepatocytes. *Science* 23, 1190–1192.

Desai, N. P., and Hubbell, J. A. (1991). Biological responses to polyethylene oxide modified polyethylene terephthalate surfaces. *J. Biomed. Mater. Res.* 25, 829–843.

Dickinson, R. B., Guido, S. *et al.* (1994). Biased cell migration of fibroblasts exhibiting contact guidance in oriented collagen gels. *Ann. Biomed. Eng.* **22**, 342–356.

DiMilla, P. A., Stone, J. A. *et al.* (1993). Maximal migration of human smooth muscle cells on fibronectin and type IV collagen occurs at an intermediate attachment strength. *J. Cell Biol.* **122**, 729–737.

Drumheller, P. D., and Hubbell, J. A. (1994). Polymer networks with grafted cell adhesion peptides for highly biospecific cell adhesive surfaces. *Anal. Biochem.* **222**, 380–388.

Drumheller, P. D., and Hubbell, J. A. (1995). Densely crosslinked polymer networks of poly(ethylene glycol) in trimethylolpropane triacrylate for cell-adhesion-resistant surfaces. *J. Biomed. Mater. Res.* **29**, 207–215.

Drumheller, P. D., Ebert, D. L. *et al.* (1994). Multifunctional poly(ethylene glycol) semi-interpenetrating polymer networks as highly selective adhesive substrates for bioadhesive peptide grafting. *Biotechnol. Bioeng.* **43**, 772–780.

Dunn, G. A. (1983). Characterizing a kinesis response: Time averaged measures of cell speed and directional persistence. *Agents Actions, Suppl.* **12**, 14–33.

Dunn, G. A., and Brown, A. F. (1986). Alignment of fibroblasts on grooved surfaces described by a simple geometric transformation. *J. Cell Sci.* **83**, 313–340.

Emerich, D. F., Hammang, J. P. *et al.* (1994). Implantation of polymer-encapsulated human nerve growth factor-secreting fibroblasts attenuates the behavioral and neuropathological consequences of quinolinic acid injections into rodent striatum. *Exp. Neurol.* **130**, 141–150.

Ertel, S. I., Ratner, B. D. *et al.* (1994). *In vitro* study of the intrinsic toxicity of synthetic surfaces to cells. *J. Biomed. Mater. Res.* **28**, 667–675.

Evans, E. (1973). New membrane concept applied to the analysis of fluid shear- and micropipette-deformed red blood cells. *Biophys. J.* **13**, 941–954.

Fairman, K., and Jacobson, B. (1983). Unique morphology of HeLa cell attachment, spreading and detachment from microcarrier beads covalently coated with a specific and non-specific substratum. *Tissue Cell* **15**, 167–180.

Farrell, B., Daniele, R. *et al.* (1990). Quantitative relationships between single-cell and cell-population model parameters for chemosensory migration responses of alveolar macrophages to C5a. *Cell Motil. Cytoskel.* **16**, 279–293.

Folkman, J., and Moscona, A. (1978). Role of cell shape in growth control. *Nature (London)* **273**, 345–349.

Freed, L., Marquis, J. *et al.* (1993). Neocartilage formation *in vitro* and *in vivo* using cells cultured on synthetic biodegradable polymers. *J. Biomed. Mater. Res.* **27**, 11–23.

Freed, L. E., and Vunjak-Novakovic, G. (1995). Cultivation of cell–polymer tissue constructs in simulated microgravity. *Biotechnol. Bioeng.* **46**, 306–313.

Freed, L. E., Vunjak-Novakovic, G. *et al.* (1993). Cultivation of cell–polymer cartilage implants in bioreactors. *J. Cell. Biochem.* **51**, 257–264.

Gail, M., and Boone, C. (1970). The locomotion of mouse fibroblasts in tissue culture. *Biophys. J.* **10**, 980–993.

Goldhaber, P. (1961). The influence of pore size on carcinogenicity of subcutaneously implanted millipore filters. *Proc. Am. Assoc. Cancer Res.* **3**, 228.

Goldhaber, P. (1962). Further observations concerning the carcinogenicity of millipore filters. *Proc. Am. Assoc. Cancer Res.* **4**, 323.

Gutsche, A. T., Parsons-Wingerter, P. *et al.* (1994). *N*-Acetylglucosamine and adenosine derivatized surfaces for cell culture: 3T3 fibroblast and chicken hepatocyte response. *Biotechnol. Bioeng.* **43**, 801–809.

Gutsche, A. T., Zurlo, J. *et al.* (1996). Rat hepatocyte morphology and function on lactose-derivatized polystyrene surfaces. *Biotechnol. Bioeng.* **49**(3), 259–265.

Hannan, G., and McAuslan, B. (1987). Immobilized serotonin: A novel substrate for cell culture. *Exp. Cell Res.* **171**, 153–163.

Hasson, J., Wiebe, D. *et al.* (1987). Adult human vascular endothelial cell attachment and migration on novel bioadsorbable polymers. *Arch. Surg. (Chicago)* **122**, 428–430.

Haston, W., and Wilkinson, P. (1988). Visual methods for measuring leukocyte locomotion. *In* "Methods in Enzymology" (G. Di Sabato, ed.), Vol. 162, pp. 17–38. Academic Press, Orlando, FL.

Horbett, T., and Schway, M. (1988). Correlations between mouse 3T3 cell spreading and serum fibronectin adsorption on glass and hydroxyethylmethacrylate–ethylmethacrylate copolymers. *J. Biomed. Mater. Res.* **22**, 763–793.

Horbett, T., Schway, M. *et al.* (1985). Hydrophilic–hydrophobic copolymers as cell substrates: Effect on 3T3 cell growth rates. *J. Colloid Interface Sci.* **104**, 28–39.

Horbett, T. A., Waldburger, J. J. *et al.* (1988). Cell adhesion to a series of hydrophilic–hydrophobic copolymers studied with a spinning disc apparatus. *J. Biomed. Mater. Res.* **22**, 383–404.

Hubbell, J. A. (1995). Biomaterials in tissue engineering. *Bio/Technology* **13**, 565–576.

Hubbell, J. A., Massia, S. P. *et al.* (1991). Endothelial cell-selective materials for tissue engineering in the vascular graft via a new receptor. *Bio/Technology* **9**, 568–572.

Ikada, Y. (1994). Surface modification of polymers for medical applications. *Biomaterials* **15**(10), 725–736.

Ishaug, S. L., Yaszemski, M. J. *et al.* (1994). Osteoblast function on synthetic biodegradable polymers. *J. Biomed. Mater. Res.* **28**, 1445–1453.

Jacobson, B., and Ryan, U. (1982). Growth of endothelial and HeLa cells on a new multipurpose microcarrier that is positive, negative or collagen coated. *Tissue Cell* **14**, 69–83.

Joseph, J. M., Goddard, M. B. *et al.* (1994). Transplantation of encapsulated bovine chrommafin cells in the sheep subarachnoid space: A preclinical study for the treatment of cancer pain. *Cell Transplant.* **3**, 355–364.

Kikuchi, A., Kataoka, K. *et al.* (1992). Adhesion and proliferation of bovine aortic endothelial cells on monoamine- and diamine-containing polystyrene derivatives. *J. Biomater. Sci. Polym. Ed.* **3**(3), 253–260.

Kobayashi, A., Akaike, T. *et al.* (1986). Enhanced adhesion and survival efficiency of liver cells in culture dishes coated with a lactose-carrying styrene homopolymer. *Makromol. Chem.* **7**, 645–650.

Kobayashi, A., Kobayashi, K. *et al.* (1992). Control of adhesion and detachment of parenchymal liver cells using lactose-carrying polystyrene as substratum. *J. Biomater. Sci. Polym. Ed.* 3(6), 499–508.

Kohn, J. (1994). Tyrosine-based polyarylates: Polymers designed for the systematic study of structure–property correlations. *20th Annu. Meet. Soc. Biomater.*, p. 67.

Koide, N., Shinji, T. *et al.* (1989). Continued high albumin production by multicellular spheroids of adult rat hepatocytes formed in the presence of liver-derived proteoglycans. *Biochem. Biophys. Res. Commun.* 161, 385–391.

Kowalczynska, H. M., and Kaminski, J. (1991). Adhesion of L1210 cells to modified styrene copolymer surfaces in the presence of serum. *J. Cell Sci.* 99, 587–593.

Krewson, C. E., Chung, S. W. *et al.* (1994). Cell aggregation and neurite growth in gels of extracellular matrix molecules. *Biotechnol. Bioeng.* 43, 555–562.

Kulkarni, R. K., Moore, E. G. *et al.* (1971). Biodegradable poly(lactic acid) polymers. *J. Biomed. Mater. Res.* 5, 169–181.

Lacy, P. E., Heger, O. D. *et al.* (1991). Maintenance of normoglycemia in diabetic mice by subcutaneous xenografts of encapsulated islets. *Science* 254, 1782–1784.

Lamoureux, P., Buxbaum, R. *et al.* (1989). Direct evidence that growth cones pull. *Nature (London)* 340, 159–162.

Landry, J., Bernier, D. *et al.* (1985). Spheroidal aggregate culture of rat liver cells: Histotypic reorganization, biomatrix deposition and maintenance of functional activities. *J. Cell Biol.* 101, 914–923.

Langer, R., and Vacanti, J. P. (1993). Tissue engineering. *Science* 260, 920–932.

Lauffenburger, D. (1983). Measurement of phenomenological parameters for leukocyte motility and chemotaxis. *Agents Actions, Suppl.* 12, 34–53.

Lauffenburger, D. A., and Linderman, J. J. (1993). "Receptors: Models for Binding, Trafficking, and Signaling." Oxford University Press, New York.

Laurencin, C. T., Norman, M. E. *et al.* (1993). Use of polyphosphazenes for skeletal tissue regeneration. *J. Biomed. Mater. Res.* 27, 963–973.

Lawrence, M., McIntire, L. *et al.* (1987). Effect of flow on polymorphonuclear leukocyte/endothelial cell adhesion. *Blood* 70, 1284–1290.

Lee, J. H., Jung, H. W. *et al.* (1994). Cell behavior on polymer surfaces with different functional groups. *Biomaterials* 15(9), 705–711.

Levine, D., Wang, D. *et al.* (1979). Optimization of growth surface parameters in microcarrier cell culture. *Biotechnol. Bioeng.* 21, 821–845.

Lim, F., and Sun, A. M. (1980). Microencapulated islets as bioartificial endocrine pancreas. *Science* 210, 908–910.

Lin, H., Sun, W. *et al.* (1994). Synthesis, surface, and cell-adhesion propertiesof polyurethanes containing covalently grafted RGD-peptides. *J. Biomed. Mater. Res.* 28, 329–342.

Lin, H.-B., Zhao, Z.-C. *et al.* (1992). Synthesis of a novel polyurethane co-polymer containing covalently attached RGD peptide. *J. Biomater. Sci., Polym. Ed.* 3, 217–227.

Lo, H., Ponticello, M. *et al.* (1995). Fabrication of controlled release biodegradable foams by phase separation. *Tissue Eng.* 1, 15.

Lo, H., Kadiyala, S. *et al.* (1996). Poly(L-lactic acid) foams with cell seeding and controlled release capacity. *J. Biomed. Mater. Res.* 30(4), 475–484.

Lydon, M., Minett, T. *et al.* (1985). Cellular interactions with synthetic polymer surfaces in culture. *Biomaterials* 6, 396–402.

Marchant, R. E., and Wang, I. (1994). Physical and chemical aspects of biomaterials used in humans. *In* "Implantation Biology" (R. S. Greco, ed.), pp. 13–53. CRC Press, Boca Raton, FL.

Massia, S. P., and Hubbell, J. A. (1989). Covalent surface immobilization of Arg-Gly-Asp- and Tyr-Ile-Gly-Ser-Arg-containing peptides to obtain well-defined cell-adhesive substrates. *Anal. Biochem.* 187, 292–301.

Massia, S. P., and Hubbell, J. A. (1991a). Human endothelial cell interactions with surface-coupled adhesion peptides on a nonadhesive glass substrate and two polymeric biomaterials. *J. Biomed. Mater. Res.* 25, 223–242.

Massia, S. P., and Hubbell, J. A. (1991b). An RGD spacing of 440 nm is sufficient for integrin $\alpha v \beta 3$-mediated fibroblast spreading and 140 nm for focal contact and stress fiber formation. *J. Cell Biol.* 114, 1089–1100.

Massia, S. P., and Hubbell, J. A. (1992). Immobilized amines and basic amino acids as mimetic heparin-binding domains for cell surface proteoglycan-mediated adhesion. *J. Biol. Chem.* 267, 10133–10141.

Matsuda, M. (1988). Serum proteins enhance aggregate formation of dissociated fetal rat brain cells in an aggregating culture. *In Vitro Cell. Dev. Biol.* 24, 1031–1036.

Matsuda, T., and Sugawara, T. (1995). Development of surface photochemical modification method for micropatterning of cultured cells. *J. Biomed. Mater. Res.* 29, 749–756.

Matsuda, T., Kondo, A. *et al.* (1989). Development of a novel artificial matrix with cell adhesion peptide for cell culture and artificial and hybrid organs. *Trans. Am. Soc. Artif. Intern. Organs* 35, 677–679.

Matsuda, T., Sugawara, T. *et al.* (1992). Two-dimensional cell manipulation technology: An artificial neural circuit based on surface microprocessing. *ASAIO J.* 38, M243–M247.

Matsuzawa, M., Potember, R. S. *et al.* (1993). Containment and growth of neuroblastoma cells on chemically patterned substrates. *J. Neurosci. Methods* 50, 253–260.

Matthieu, J., Honegger, P. *et al.* (1981). Aggregating brain cell cultures: A model to study brain development. *In* "Physiological and Biochemical Basis for Perinatal Medicine," pp. 359–366. Karger, Basel.

McClay, D. R., Wessel, G. M. *et al.* (1981). Intercellular recognition: Quantitation of initial binding events. *Proc. Nat. Acad. U.S.A.* 78, 4975–4979.

McIntire, L. V. (1994). 1992 ALZA Distinguished Lecture: Bioengineering and vascular biology. *Ann. Biomed. Eng.* 22, 2–13.

Merrill, E. W. (1993). Poly(ethylene oxide) star molecules: Synthesis, characterization, and applications in medicine and biology. *J. Biomater. Sci. Polym. Ed.* **5**, 1–11.

Meyle, J., Gultig, K. *et al.* (1995). Variation in contact guidance by human cells on a microstructured surface. *J. Biomed. Mater. Res.* **29**, 81–88.

Mikos, A. G., Bao, Y. *et al.* (1993). Preparation of poly(glycolic acid) bonded fiber structures for cell attachment and transplantation. *J. Biomed. Mater. Res.* **27**, 183–189.

Miura, M., Cole, C. A. *et al.* (1991). Application of LCST polymer–cell receptor conjugates for cell culture on hydrophobic surfaces. *17th Annu. Meet. Soc. Biomater.*, Scottsdale, AZ.

Moghaddam, M. J., and Matsuda, T. (1991). Development of a 3D artificial extracellular matrix. *Trans. Am. Soc. Artif. Intern. Organs* **37**, M437–M438.

Moscona, A. A. (1961). Rotation-mediated histogenic aggregation of dissociated cells. A quantifiable approach to cell interactions *in vitro. Exp. Cell Res.* **22**, 455–475.

Munn, L. L., Glacken, M. W. *et al.* (1993). Analysis of lymphocyte aggregation using digital image analysis. *J. Immunol. Methods* **166**, 11–25.

Nelson, R. D., Quie, P. G. *et al.* (1975). Chemotaxis under agarose: A new and simple method for measuring chemotaxis and spontaneous migration of human polymorphonuclear leukocytes and monocytes. *J. Immunol.* **115**, 1650–1656.

Nicol, A., Gowda, D. C. *et al.* (1993). Elastomeric polytetrapeptide matrices: Hydrophobicity dependence of cell attachment from adhesive (GGIP)$_n$ to nonadhesive (GGAP)$_n$ even in serum. *J. Biomed. Mater. Res.* **27**, 801–810.

Nilsson, K., Scheirer, W. *et al.* (1983). Entrapment of animal cells for production of monoclonal antibodies and other biomolecules. *Nature (London)* **302**, 629–630.

Nyberg, S. L., Shatford, R. A. *et al.* (1993). Evaluation of a hepatocyte-entrapment hollow fiber bioreactor: A potential bioartificial liver. *Biotechnol. Bioeng.* **41**, 194–203.

O'Neill, C., Jordan, P. *et al.* (1986). Evidence for two distinct mechanisms of anchorage stimulation in freshly explanted and 3T3 Swiss mouse fibroblasts. *Cell (Cambridge, Mass.)* **44**, 489–496.

Orr, C. W., and Roseman, S. (1969). Intercellular adhesion. I. A quantitative assay for measuring the rate of adhesion. *J. Membr. Biol.* **1**, 109–124.

Oxley, H. R., Corkhill, P. H. *et al.* (1993). Macroporous hydrogels for biomedical application: Methodology and morphology. *Biomaterials* **14**, 1064–1072.

Parkhurst, M. R., and Saltzman, W. M. (1992). Quantification of human neutrophil motility in three-dimensional collagen gels: Effect of collagen concentration. *Biophys. J.* **61**, 306–315.

Parsons-Wingerter, P., and Saltzman, W. M. (1993). Growth versus function in three-dimensional culture of single and aggregated hepatocytes within collagen gels. *Biotechnol. Prog.* **9**, 600–607.

Peppas, N. A., and Langer, R. (1994). New challenges in biomaterials. *Science* **263**, 1715–1720.

Peshwa, M. V., Wu, F. J. *et al.* (1994). Kinetics of hepatocyte spheroid formation. *Biotechnol. Prog.* **10**, 460–466.

Pettit, D. K., Hoffman, A. S. *et al.* (1994). Correlation between corneal epithelial cell outgrowth and monoclonal antibody binding to the cell binding domain of adsorbed fibronectin. *J. Biomed. Mater. Sci.* **28**, 685–691.

Pierschbacher, M. D., and Ruoslahti, E. (1984). Cell attachment activity of fibronectin can be duplicated by small synthetic fragments of the molecule. *Nature (London)* **309**, 30–33.

Puelacher, W. C., Mooney, D. *et al.* (1994). Design of nasoseptal cartilage replacements synthesized from biodegradable polymers and chondrocytes. *Biomaterials* **15**(10), 774–778.

Ranieri, J. P., Bellamkonda, R. *et al.* (1993). Selective neuronal cell adhesion to a covalently patterned monoamine on fluorinated ethylene propylene films. *J. Biomed. Mater. Res.* **27**, 917–925.

Richards, J., Larson, L. *et al.* (1983). Method for culturing mammary epithelial cells in a rat tail collagen gel matrix. *J. Tissue Cult. Methods* **8**, 31–36.

Saltzman, W. M., Parsons-Wingerter, P. *et al.* (1991). Fibroblast and hepatocyte behavior on synthetic polymer surfaces. *J. Biomed. Mater. Res.* **25**, 741–759.

Saltzman, W. M., Parkhurst, M. R. *et al.* (1992). Three-dimensional cell cultures mimic tissues. *Ann. N.Y. Acad. Sci.* **665**, 259–273.

Sawhney, A. S., Pathak, C. P. *et al.* (1994). Modification of islet of Langerhans surfaces with immunoprotective poly(ethylene glycol) coatings via interfacial polymerization. *Biotechnol. Bioeng.* **44**, 383–386.

Schakenraad, J. M., Busscher, H. J. *et al.* (1986). The influence of substratum surface free energy on growth and spreading of human fibroblasts in the presence and absence of serum proteins. *J. Biomed. Mater. Res.* **20**, 773–784.

Schmidt, J. A., and von Recum, A. F. (1992). Macrophage response to microtextured silicone. *Biomaterials* **12**, 385–389.

Schmitt, E. E., and Polistina, R. A. (1967). Surgical sutures. U.S. Pat. 3,297,033.

Schnaar, R. L., Weigel, P. H. *et al.* (1978). Adhesion of hepatocytes to polyacrylamide gels derivitized with *N*-acetylglucosamine. *J. Biol. Chem.* **253**, 7940–7951.

Schneider, M. A. K. (1967). Element de suture adsorbable et son procède de fabrication. Fr. Pat. 1,478,694.

Schor, S. (1980). Cell proliferation and migration on collagen substrata *in vitro. J. Cell Sci.* **41**, 159–175.

Schor, S., Allen, T. *et al.* (1980). Cell migration through three-dimensional gels of native collagen fibers: Collagenolytic activity is not required for the migration of two permanent cell lines. *J. Cell Sci.* **46**, 171–186.

Schor, S., Schor, A. *et al.* (1981). The effects of fibronectin on the migration of human foreskin fibroblasts and Syrian hamster melanoma cells into three-dimensional gels of native collagen fibres. *J. Cell Sci.* **48**, 301–314.

Sefton, M., Dawson, R. *et al.* (1987). Microencapsulation of mammalian cells in a water-insoluble polyacrylate by coextrusion and interfacial precipitation. *Biotechnol. Bioeng.* **29**, 1135–1143.

Singhvi, R., Kumar, A. *et al.* (1994). Engineering cell shape and function. *Science* **264**, 696–698.

Smetana, K. (1993). Cell biology of hydrogels. *Biomaterials* **14**(14), 1046–1050.

Smetana, K., Vacik, J. *et al.* (1990). The influence of hydrogel functional groups on cell behavior. *J. Biomed. Mater. Res.* **24**, 463–470.

Stol, M., Tolar, M. *et al.* (1985). Poly(2-hydroxyethyl methacrylatea)–collagen composites which promote muscle cell differentiation *in vitro. Biomaterials* **6**, 193–197.

Sullivan, S. J., Maki, T. *et al.* (1991). Biohybrid artificial pancreas: Long-term implantation studies in diabetic, pancreatectomized dogs. *Science* **252**, 718–721.

Takezawa, T., Mori, Y. *et al.* (1993). Characterization of morphology and cellular metabolism during the spheroid formation by fibroblasts. *Exp. Cell Res.* **208**, 430–441.

Tamada, Y., and Ikada, Y. (1994). Fibroblast growth on polymer surfaces and biosynthesis of collagen. *J. Biomed. Mater. Res.* **28**, 783–789.

Thomas, W. A., and Steinberg, M. S. (1980). A twelve-channel automatic recording device for continuous recording of cell aggregation by measurement of small-angle light-scattering. *J. Cell Sci.* **41**, 1–18.

Thompson,, J., Haudenschild, C. *et al.* (1989). Heparin-binding growth factor 1 induces the formation of organoid neovascular structures *in vivo. Proc. Nat. Acad. Sci. U.S.A.* **86**, 7928–7932.

Thomson, R. C., Yaszemski, M. J. *et al.* (1995). Fabrication of biodegradable polymer scaffolds to engineer trabecular bone. *J. Biomater. Sci., Polym. Ed.* 7(1), 23–38.

Tozeren, A., Sung, K. *et al.* (1989). Theoretical and experimental studies on cross-bridge migration during cell disaggregation. *Biophys. J.* **55**, 479–487.

Trinkhaus, J. P. (1984). "Cells into Organs. The Forces that Shape the Embryo." Prentice-Hall, Englewood Cliffs, N.J.

Tyler, B. J., Ratner, B. D. *et al.* (1992). Variations between Biomer lots. I. Significant differences in the surface chemistry of two lots of a commercial poly(ether urethane). *J. Biomed. Mater. Res.* **26**, 273–289.

Vacanti, J. P., Morse, M. *et al.* (1988). Selective cell transplantation using bioabsorbable artificial polymers as matrices. *J. Pediatr. Surg.* **23**, 3–9.

van der Valk, P., van Pelt, A. *et al.* (1983). Interaction of fibroblasts and polymer surfaces: Relationship between surface free energy and fibroblast spreading. *J. Biomed. Mater. Res.* **17**, 807–817.

van Wachem, P. B., Hogt, A. H. *et al.* (1987). Adhesion of cultured human endothelial cells onto methacrylate polymers with varying surface wettability and charge. *Biomaterials* **8**, 323–328.

van Wezel, A. L. (1967). Growth of cell strains and primary cells on microcarriers in homogeneous culture. *Nature (London)* **216**, 64–65.

van Wezel, A. L. (1985). Monolayer growth systems: Homogeneous unit processes. *Anim. Cell Biotechnol.* **1**, 265–282.

Wasserman, D., and Levy, A. J. (1975). Nahtmaterials aus weichgemachten Lacttid-Glykolid-Copolymerisaten. Ger. Pat. Offenlegungsschrift 24 06 539.

Weiss, P. (1934). *In vitro* experiments on the factors determining the course of the outgrowing nerve fiber. *J. Exp. Zool.* **68**, 393–448.

Weiss, P. (1945). Experiments on cell and axon orientation *in vitro:* The role of colloidal exudates in tissue organization. *J. Exp. Zool.* **100**, 353–386.

Wilkinson, P., Lackie, J. *et al.* (1982). Methods for measuring leukocyte locomotion. *In* "Cell Analysis" (N. Catsimpoolas, ed.), Vol. 1, pp. 145–194. Plenum, New York.

Wissemann, K., and Jacobson, B. (1985). Pure gelatin microcarriers: Synthesis and use in cell attachment and growth of fibroblast and endothelial cells. *In Vitro Cell. Dev. Biol.* **21**, 391–401.

Woerly, S., Maghami, G. *et al.* (1993). Synthetic polymer derivatives as substrata for neuronal cell adhesion and growth. *Brain Res. Bull.* **30**, 423–432.

Wu, P., Hoying, J. B. *et al.* (1994). Integrin-binding peptide in solution inhibits or enhances endothelial cell migration, predictably from cell adhesion. *Ann. Biomed. Eng.* **22**, 144–152.

Yang, J., Richards, J. *et al.* (1979). Sustained growth and three-dimensional organization of primary mammary tumor epithelial cells embedded in collagen gels. *Proc. Nat. Acad. Sci. U.S.A.* **76**, 3401–3405.

Zhou, J., Ertel, S. I. *et al.* (1994). Evaluation of tyrosine-derived pseudo-poly(amino acids): *In vitro* cell interactions. *20th Annu. Meet. Soc. Biomater.*, p. 371.

Zienlinski, B. A., and Aebischer, P. (1994). Chitosan as a matrix for mammalian cell encapsulation. *Biomaterials* **15**, 1049–1056.

Zigmond, S. H. (1977). Ability of polymorphonuclear leukocytes to orient in gradients of chemotactic factors. *J. Cell Biol.* **75**, 606–616.

CHAPTER 20

MATRIX EFFECTS

Jeffrey A. Hubbell

INTRODUCTION

The extracellular matrix is a complex chemically and physically cross-linked network of proteins and glycosaminoglycans. The matrix serves to organize cells in space, to provide them with environmental signals to direct site-specific cellular regulation, and to separate one tissue space from another. The interaction between cells and the extracellular matrix is bidirectional and dynamic: cells are constantly accepting information on their environment from cues in the extracellular matrix, and cells are frequently remodeling their extracellular matrix. In this chapter, the proteins in the extracellular matrix and their cell surface receptors are introduced, and mechanisms by which cells transduce chemical information in their extracellular matrix are discussed. Methods for spatially displaying matrix recognition factors are described, both in the context of model systems for investigation of cellular behavior and from the perspective of creation of bioactive biomaterials for tissue engineering therapies.

The extracellular matrix serves at least three functions in its role controlling cell behavior: it provides adhesion signals, it provides growth factor binding sites, and it provides degradation sites to give way to the enzymatic activity of cells as they migrate (Roskelley *et al.*, 1995). An understanding of these interactions is important in tissue engineering, because it may be desirable to mimic the biological recognition molecules that control the relationships between cells and their natural biomaterial interface, namely, the extracellular matrix. The components of the extracellular matrix are on one level immobilized, but not necessarily irreversibly. Cell-derived enzymes, such as tissue transglutaminase, serve to cross-link chemically certain components of the extracellular matrix, such as fibronectin chains to other fibronectin chains and to fibrillar collagen chains. Other components are more transiently immobilized, such as growth factors within the extracellular matrix proteoglycan network. This network can be partially degraded, and the growth factors can be proteolytically cleaved, to mobilize the growth factors under cellular control. Not all of the signals of the extracellular matrix are biochemical in nature. A biomechanical interplay exists between cells and their extracellular matrix, and this interaction may play an important role in the functional regulation in many tissues, particular in load-bearing tissues. This chapter considers only the biochemical aspects of biological recognition; the reader is referred elsewhere for treatments of the role of the extracellular matrix as a biomechanical regulator of cell behavior (He and Grinnell, 1994; Dickenson *et al.*, 1994; Grinnell, 1999).

EXTRACELLULAR MATRIX PROTEINS AND THEIR RECEPTORS

Interactions between cells and the extracellular matrix are mediated by cell surface glycoprotein and proteoglycan receptors interacting with proteins bound within the extracellular matrix. This section begins with an introduction of the glycoprotein receptors on cell surfaces involved in cell adhesion. The topic then turns to the extracellular matrix proteins to which those receptors bind, with a discussion of the active domains of those proteins that bind to the cell surface receptors. Finally, the role of cell surface-associated enzymes in the processing and remodeling of the extracellular matrix is addressed.

Principles of Tissue Engineering
Second Edition

There are four major classes of glycoprotein adhesion receptors that are present on cell surfaces, three of which are involved primarily in cell–cell adhesion and one of which is involved in both cell adhesion to other cells and cell adhesion to the extracellular matrix. The first three are introduced briefly, and the fourth more extensively.

The cadherins are a family of cell surface receptors that participate in homophilic binding (i.e., the binding of a cadherin on one cell with an identical cadherin on another cell) (Takeichi, 1991; Adams and Nelson, 1998; Shapiro and Colman, 1998; Alattia *et al.,* 1999). These molecules allow a cell of one type (e.g., endothelial cells) to recognize other cells of the same type and are important in the early stages of organogenesis. These interactions depend on the presence of extracellular Ca^{2+} and may be dissociated by calcium ion chelation. Because all cadherins are present on cell surfaces, cadherins are not involved directly in cell interactions with the extracellular matrix. They may be involved indirectly, in that they may organize cell–cell contacts in concert with another receptor system that is involved in regulation of cell–extracellular matrix binding.

A second class of receptors is the selectin family (Bevilacqua *et al.,* 1991; Lasky, 1995; Vestweber and Blanks, 1999). These membrane-bound proteins are involved in heterophilic binding between blood cells and endothelial cells in a manner that depends, as the cadherins, on extracellular Ca^{2+}. These proteins contain lectinlike features and recognize branched oligosaccharide structures in their ligands, namely the sialyl Lewis X and the sialyl Lewis A structures (Varki, 1994). As with the cadherins, these receptor–ligand interactions are important primarily in cell–cell interactions, and this is particularly true in the context of inflammation.

A third class of receptors is the immunoglobulin family, proteins that are denoted as cell adhesion molecules (CAMs) (Grumet, 1991; Hunkapiller and Hood, 1989). These proteins bind their protein ligands in a manner that is independent of extracellular Ca^{2+} and they participate in both homophilic and heterophilic interactions. As with the cadherins and the selectins, they bind to other cell surface proteins and are thus primarily involved in cell–cell interactions. Selected members of the integrin class of adhesion receptors represent one class of ligand for these receptors, are as discussed below.

The fourth class of adhesion receptors is the integrin family (Ruoslahti, 1991; Hynes, 1992). Although the other three classes of receptors described briefly above are involved primarily in cell–cell recognition, the integrins are involved in both cell–cell and cell–extracellular matrix binding. The integrins are dimeric proteins, consisting of an α and a β subunit assembled noncovalently into an active dimer. There are many known α and β subunits, with at least 15 such α subunits and 8 such β subunits that are capable of assembly into at least 21 αβ combinations. The αβ combinations present in the β1, β2, and β3 subclasses are shown in Table 20.1; the β1, β2, and β3 subclasses are the most commonly expressed integrins and are thus arguably the most generally important. The β2 integrins are involved primarily in cell–cell recognition; for example, the integrin αLβ2 binds to two intercellular adhesion molecules (ICAM-1 and ICAM-2), both members of the immunoglobulin class of adhesion receptors described in the preceding paragraph. By contrast, the β1 and β3 integrins are primarily involved in cell–extracellular matrix interactions. The β1 and β3 integrins bind to numerous proteins present in the extracellular matrix, as illustrated in Table 20.1. These proteins include collagen, fibronectin, vitronectin, von Willebrand factor, and laminin.

Collagen is the primary structural protein of the tissues; the reader is referred elsewhere for a focused review on this extensive topic (Vanderrest and Garrone, 1991; Cremer *et al.,* 1998; Fratzl *et al.,* 1998). Many forms of collagen exist, most of which are multimeric and fibrillar. To these collagens many other adhesion proteins bind, thus collagen has the role of organizing many other proteins that interact with and organize cells. Collagen also interacts directly with integrins, primarily α1β1, α2β1, and α3β1.

Fibronectin is a globular protein that is present in nearly all tissues; fibronectin has been extensively reviewed elsewhere (Schwarzbauer, 1991; Miyamoto *et al.,* 1998; Magnusson and Mosher, 1998). Fibronectin also exists in many forms, depending on the site in the tissues and the regulatory state of the cell that synthesized the fibronectin. Almost all cells interact with fibronectin, primarily through the so-called fibronectin receptor α5β1, and to a lesser extent through the β3 integrin αvβ3 as well as other integrins, as will be described below.

Vitronectin is a multifunctional adhesion protein found in the circulation and in many tissues (Preissner, 1991). The protein is active in promoting the adhesion of numerous cell types and

Table 20.1. Selected members of the integrin receptor class and their ligands

Integrin heterodimer	Ligands
α1β1	Collagen, laminin
α2β1	Collagen, fibronectin, laminin
α3β1	Collagen, fibronectin, laminin
α4β1	Fibronectin, vascular cell adhesion molecule-1
α5β1	Fibronectin
α6β1	Laminin
α7β1	Laminin
αvβ1	Collagen, fibrinogen, fibronectin, vitronectin, von Willebrand factor
αxβ2	Complement protein C3bi, fibrinogen, factor X
αMβ2	Complement protein C3bi
αLβ2	Intercellular adhesion molecule-1, intercellular adhesion molecule-2
αvβ3	Bone sialoprotein, fibrinogen, fibronectin, laminin, thrombospondin, vitronectin, von Willebrand factor
αIIbβ3	Fibrinogen, fibronectin, laminin, thrombospondi, vitronectin, von Willebrand factor

binds primarily to the so-called vitronectin receptor, αvβ3, as well as to αvβ1 and to the platelet receptor αIIbβ3. von Willebrand factor is an adhesion protein that is primarily involved in the adhesion of vascular cells; the reader is referred elsewhere for a detailed review (Ruggeri, 1997; Sadler, 1998). It is synthesized by megakaryocytes, the platelet-generating cells of the bone marrow, and is stored in the α granules of circulating platelets. Activation of the platelet leads to the release of the granule contents, including the von Willebrand factor. von Willebrand factor is also synthesized by and stored within the endothelial cell. A multimeric form of the protein, where tens of copies of the protein may be linked together into insoluble form, is found in the subendothelium and is involved in blood platelet adhesion to the subendothelial tissues on vascular injury.

Laminin is a very complex adhesion protein that is generally present in the basement membrane, the proteins immediately beneath epithelia and endothelia, as well as in many other tissues, as reviewed in detail elsewhere (Timpl, 1996; Timpl and Brown, 1996; Ryan *et al.,* 1996; Aumailley and Smyth, 1998). Laminin is present in a family of forms (Malinda and Kleinman, 1996; Engvall and Wewer, 1996). The classic form was purified from the extracellular matrix of Engelbreth–Holm–Swarm tumor cells and consists of a disulfide cross-linked trimer of one α1 (400,000 Da), one β1 (210,000 Da), and one γ1 (200,000 Da) polypeptide chain. This form binds to the β1 integrins α1β1, α2β1, α3β1, α6β1, and α7β1, and to the β3 integrins αvβ3 and αIIbβ3, as well as to other integrins. A number of other laminin forms exist, composed of αβγ combinations of α1, α2, α3, α4, or α5; β1, β2, or β3; and γ1 or γ2 chains. The details of the differences in function of all of the various laminin forms remain unknown, but it is clear that several of them do stimulate very different behaviors in a variety of cell types. Laminin is a particularly important component of the basal lamina, i.e., the extracellular matrix beneath monolayer structures such as epithelia, mesothelia, and endothelia (Schwarzbauer, 1999). For example, laminin contains numerous domains that bind to endothelial cells (Ponce *et al.,* 1999), and these are undoubtedly important in regulating a variety of cell type-specific functions, as discussed below.

The extracellular matrix proteins described above are very complex. They contain sites responsible for binding to collagen, for binding to glycosaminoglycans (as described below), for cross-linking to other extracellular matrix proteins via transglutaminase activity, for degradation by proteases (as described briefly below), and for binding to integrin and other adhesion receptors (as described in detail immediately below). Because the proteins must be so multifunctional, the sites that serve the singular function of binding to integrins comprise a small faction of the protein mass. In most cases, the receptor-binding domain can be localized to an oligopeptide sequence less than 10 amino acid residues in length, and this site can be mimicked by linear or cyclic oligopeptides of identical or similar sequence as that found in the protein (Humphries, 1990; Yamada, 1991; Ruoslahti, 1991). The first such minimal sequence to be identified was the tripeptide

Table 20.2 Selected cell binding domain sequences of extracellular matrix proteins[a]

Protein	Sequence[b]	Role
Fibronectin	RGDS	Adhesion of most cells, vis α5β1
	LDV	Adhesion
	REDV	Adhesion
Vitronectin	RGDV	Adhesion of most cells, via αvβ3
Laminin A	LRGDN	Adhesion
	SIKVAV	Neurite extension
Laminin B1	YIGSR	Adhesion of many cells, via 67-kDa laminin receptor
	PDSGR	Adhesion
Laminin B2	RNIAEIIKDI	Neurite extension
Collagen I	RGDT	Adhesion of most cells
	DGEA	Adhesion of patelets, other cells
Thrombospondin	RGD	Adhesion of most cells
	VTXG	Adhesion of platelets

[a]From Hubbell (1995), after Yamada and Kleinman (1992).
[b]Single-letter amino acid code: A, alanine, C, cysteine; D, aspartic acid; E, glutamic acid; F, phenylalanine; G, glycine; H, histidine; I, isoleucine; K, lysine; M, methionine; N, asparagine; P, proline; Q, glutamine; R, arginine; S, serine; T, threonine; V, valine; W, trytophan; Y, tyrosine.

RGD (Ruoslahti and Pierschbacher, 1987) (the single-letter amino acid abbreviation nomenclature is shown as a footnote in Table 20.2). Synthetic RGD-containing peptide, when appropriately coupled to a surface or a carrier molecule (see below), is capable of recapitulating most of the adhesive interactions of the RGD site in the protein fibronectin, including integrin binding. At least for the case of integrin binding via αvβ3, the RGD ligand alone is capable of also inducing integrin clustering and, when the signal is presented at sufficient surface concentration, focal contact formation and cytoskeletal organization (Massia and Hubbell, 1991). It is not yet clear if this is the case for RGD binding by all integrins; e.g., in binding by α5β1, it would appear that interactions with cell surface proteoglycans are also necessary (Bloom *et al.*, 1999). Many receptor-binding sequences other than the RGD tripeptide have been identified by a variety of methods, and some of these sequences are shown in Table 20.2. In these receptor-binding sequences, the affinity is highly specific to the particular ordering of the amino acids in the peptide; for example, the peptide RDG, containing the same amino acids but in a different sequence, is completely inactive in binding to integrins.

Of the adhesion peptides, one class contains the central RGD sequence. The sequence is modified by the flanking residues, modifying the receptor specificity of the receptor-binding sequence. For example, the sequence found in fibronectin is RGDS; in vitronectin, it is RGDV; in laminin, it is RGDN; and in collagen, it is RGDT. Other adhesion peptides maintain the central D residue, such as the REDV and LDV sequences of fibronectin. The REDV and LDV sequences are relatively specific in their binding and interact with the integrin α4β1; RGD peptides also bind to α4β1, but the REDV and LDV sequences bind essentially only to α4β1.

In addition to peptides that bind to the integrin adhesion receptors, there are other peptide sequences that bind to other nonintegrin receptors. As an example, laminin bears several such sequences, such as the YIGSR, SIKVAV, and RNIAEIIKDI peptides. The YIGSR (Graf *et al.*, 1987a,b) sequence binds to a 67-kDa monomeric nonintegrin laminin receptor (Meecham, 1991). As do the integrin receptors, this laminin receptor also interacts via its cytoplasmic domain with intracellular proteins involved with linkage to the f-actin cytoskeleton (Massia *et al.*, 1993). The YIGSR sequence is involved in the adhesion and spreading of numerous cell types (see below). The SIKVAV sequence in laminin binds to the neuronal cell surface receptor and stimulates the extension of neurites (Tashiro *et al.*, 1991).

In addition to the highly sequence-specific binding of adhesion peptides to cell surface receptors, most of the adhesion proteins also bind to cell surface components by less specific mechanisms. These proteins contain a heparin-binding domain (so called because of the use of heparin

Table 20.3. Proteoglycan binding domain sequences of extra-cellular matrix proteins[a]

Protein[b]	Sequence
XBBXBX	Consensus sequence
PRRARV	Fibronectin
YEKPGSPPREVVPRPRPGV	Fibronectin
RPSLAKKQRFRHRNRKGYRSQRGHSRGR	Vitronectin
RIQNLLKITNLRIKFVK	Laminin

[a]After Hubbell (1995); references contained in Massia and Hubbell (1992).
[b]An X indicates a hydrophobic amino acid; basic amino acids are underlined.

affinity chromatography in purification of the protein) that binds to cell surface proteoglycans that contain heparin sulfate or chondroitin sulfate glycosaminoglycans (Wight *et al.*, 1992; Jackson *et al.*, 1991; Gallagher, 1997; Lyon and Gallagher, 1998). The peptide sequences that bind to cell surface proteoglycans are rich in cationic residues, such as arginine (R) and lysine (K), relative to their anionic residues, aspartic acid (D) and glutamic acid (E), and they also contain hydrophobic amino acids, such as alanine (A), isoleucine (I), leucine (L), proline (P), and valine (V). Several of these sequences are shown in Table 20.3. For example, the heparin-binding sequence in fibronectin bears a sequence of PRRARV, having a motif of XBBXBX, which is observed in several cell adhesion proteins, X being a hydrophobic residue and B being a basic residue, either K or R. These sites within adhesion proteins such as fibronectin and laminin bind to cell surface proteoglycan in parallel, with interaction by the integrin-binding sites to stabilize the adhesion complex (LeBaron *et al.*, 1988). Cell–cell adhesion molecules also employ cell surface proteoglycan-binding affinity, e.g., neural cell adhesion molecule (NCAM) bears the domain KHKGRDVILKKDVR, which binds to heparan sulfate and chondroitin sulfate proteoglycans (Kallapur and Akeson, 1992). The interactions with cell surface proteoglycans are much less specific than those with integrins; the binding is not as sensitive to the order of the oligopeptide sequence and moreover the effect can be mimicked simply by R or K residues immobilized on a surface, albeit certainly with a great loss in specificity (Massia and Hubbell, 1992).

The extracellular matrix is subject to dynamic remodeling under the influence of cells in contact with it. Cells seeded *in vitro* on an extracellular matrix of one composition may adhere, spread, form focal contacts, remove the initial protein, secrete a new extracellular matrix of different protein composition, and form new focal contacts (Dejana *et al.*, 1988).

Cell surface-bound and cell-derived free enzymes play an important role in this remodeling of the extracellular matrix (Mosher *et al.*, 1992). For example, cell-released protein disulfide isomerases are released from cells to cross-link protein covalently in the extracellular matrix by disulfide bridging (Mayadas and Wagner, 1992). Cell-derived transglutaminases also form an amide linkage between the ε-amino group of lysine and the γ-carboxyl group of glutamic acid to cross-link proteins chemically in the extracellular matrix. These processes are responsible, for example, for the assembly of the globular adhesion protein fibronectin into fibrils within the extracellular matrix beneath cells.

Membrane-bound and cellularly released enzymes are also involved in degradation of the extracellular matrix to permit matrix remodeling and cell migration (Chen, 1992; Birkedall-Hansen, 1995; Blasi, 1993; Shapiro, 1998). Cell-released matrix metalloproteinases such as collagenase and gelatinase, serine proteases such as urokinase plasminogen activator and plasmin, and cathepsins are each involved in both remodeling and degradation during cell migration. Accordingly, the matrix–cell interaction should be understood to be bidirectional: the cell accepting information from the matrix, and the matrix being tailored by the cell.

One of the important roles of cell-associated enzymatic degradation of the extracellular matrix is in mobilization of growth factor activity. Because growth factors are such powerful regulators of biological function, their activity must be highly spatially regulated. One means by which this occurs in nature is by high-affinity binding interactions between growth factors and the three-dimensional extracellular matrix in which they exist. Many such growth factors bind heparin, meaning that they, like adhesion proteins, bear domains that bind extracellular matrix heparan sulfate and chondroitin sulfate proteoglycans. For example, basic fibroblast growth factor binds heparin with high affinity (Faham *et al.*, 1996, 1998). Vascular endothelial cell growth factor is another example of a heparin-binding growth factor (Fairbrother *et al.*, 1998). These growth factors are strongly immobilized by binding to extracellular matrix proteoglycans, and they can be mobilized under local cellular activity, e.g., by degradation of these proteoglycans, or, in the case of vascular endothelial cell growth factor, by cleavage of the main chain of the growth factor away from the heparin-binding domain by plasmin activated at the surface of a nearby cell (Keyt *et al.*, 1996).

MODEL SYSTEMS FOR STUDY OF MATRIX INTERACTIONS

Because the extracellular matrix adhesion proteins may be mimicked, at least to some degree, by small synthetic peptides, it is possible to investigate cell–substrate interactions with well-defined systems. Foundational to them all are the interactions of cells with the surface (other than with adhesion peptides intentionally endowed on the surface) that would produce cell adhesion. These so-called nonspecific interactions are between cell surface receptors and proteins that have adsorbed to the surface. Due to this role played by adsorbing proteins, some introduction to the protein and surface interactions leading to adsorption is warranted. The thermodynamic and kinetic aspects of protein adsorption have been reviewed and the reader is referred elsewhere for a more detailed treatment (Andrade and Hlady, 1986; Wojciechowski and Brash, 1991; norde and Lyklema, 1991). The primary driving force for protein adsorption is the hydrophobic effect: water near a hydrophobic surface fails to hydrogen bond with that surface and thus assumes a more highly ordered structure in which the water is more thoroughly hydrogen bonded to itself than is the case in water far away from the surface. A protein can adsorb to this surface, acting like a surfactant, and thus replace the hydrophobic surface material with a polar surface capable of hydrogen bonding with water. This releases the order in the water, with a net result of a large entropic gain. Electrostatic interactions, e.g., between charges on D, E, K, or R residues on the protein with cationic or anionic functions on the surface material, play a lesser but important role as well; because proteins generally bear a net negative charge, anionic surfaces typically adsorb less protein than do cationic surfaces. These observations, overly generalized here, guide one to examine model surfaces that are hydrophilic and nonionic, as well as being derivatizable, to permit coupling of the adhesion peptide under study.

The tendency for proteins to adsorb to material surfaces has been exploited as a method by which to immobilize peptides onto substrates for study. Ruoslahti and Pierschbacher (1986) have described the peptide Ac-GRGDSPASSKGGGGSRLLLLLLR-NH$_2$ (where the Ac indicates that the N terminus is acetylated and the NH$_2$ indicates that the C terminus is amidated to block the terminal charges) for this purpose. The LLLLLL stretch is hydrophobic and adsorbs avidly to hydrophobic polymer surfaces and effectively immobilizes the cell-binding RGDS sequence from fibronectin. Nonadhesive proteins such as albumin have also been grafted with RGD peptide, e.g., by binding to amine groups on lysine residues on the albumin; adsorption of the albumin conjugate thus immobilizes the attached RGD peptide (Danilov and Juliano, 1989).

Surfaces coated with hydrophilic polymers have also been employed to graft adhesion peptides. One simple system that has been useful is glass modified with a silane, 3-glycidoxypropyl triethoxysilane; once the silane is grafted to the surface the epoxide group is hydrolyzed to produce —CH$_2$CH(OH)CH$_2$OH groups pendant from the surface (glycophase glass). The hydroxyl groups serve as sites for covalent immobilization of adhesion peptide, e.g., via the N-terminal primary amine (Massia and Hubbell, 1990). Titration of the surface density of grafted RGD peptides versus cell response using this system revealed quantitative information on the number density of interactions required to establish morphologically complete cell spreading (Massia and Hubbell, 1991). A surface density of approximately 10 fmol/cm^2 of RGD was required to induce spreading, focal contact formation, integrin αvβ3 clustering, α-actinin and vinculin colocalization with αvβ3, and f-actin cytoskeletal assembly in human fibroblasts cultured on this synthetic extracellular matrix. This surface density corresponds to a spacing of roughly 140 nm between immobilized RGD sites, demonstrating that far less than monolayer coverage is sufficient to promote cell responses. Silane-modified quartz has been employed as a surface to study the role of RGD sites and heparin-binding domains of adhesion proteins in osteoblast adhesion and mineralization, demonstrating a strong benefit for the involvement of both modes of adhesion (Rezania and Healy, 1999). The base material, glycophase glass, is only modestly resistant to protein adsorption and thus nonspecific cell adhesion; accordingly, it is difficult to conduct an investigation of long-term interactions, during which the adherent cells may be synthesizing and secreting their own extracellular matrix to adsorb to the synthetic one experimentally provided. This has motivated exploration with substrates that are more resistant to protein adsorption.

An enormous amount of research has been expended into grafting material surfaces with water-soluble, nonionic polymers such as poly(ethylene glycol), $HO(CH_2CH_2O)_nH$. This vast body of research has been extensively reviewed elsewhere (Harris, 1992; Llanos and Sefton, 1993; Amiji and Park, 1993). Poly(ethylene glycol) has been immobilized on surfaces by numerous means; three particularly effective means will be addressed below.

Thiol compounds chemisorb avidly to gold surfaces (Bain and Whitesides, 1989; Lopez *et al.,* 1993). When those thiols are terminal to an alkane group, $R-(CH_2)_n-SH$, the group adsorbs in perfect monolayers; the thiol–gold interaction contributes about half of the energy of interaction, and the alkane–alkane van der Waals interaction contributes the other half. Accordingly, it is easy to employ alkane ethiols to display, in very regular fashion, some functionality R on a gold-coated substrate (so long as the R group is not so large as to inhibit monolayer packing sterically, in which case it can be diluted with a nonfunctional alkane thiol). Using this approach, Prime and Whitesides (1993) immobilized an oligo(ethylene glycol)-containing alkane thiol, $HS-(CH_2)_{11}(OCH_2CH_2)_nOH$. Protein adsorption was investigated on surfaces formed with this alkane thiol and a hydrophobic coreactant, $HS-(CH_2)_{10}CH_3$. Degrees of polymerization (*n*) as low as 4 were observed to limit dramatically the adsorption of even very large proteins, such as fibronectin. When the oligo(ethylene glycol) monolayer was incomplete, i.e., when the monolayer was mixed with the hydrophobic alkane thiol, longer oligo(ethylene glycol) functions were able to preserve the protein repulsiveness of the surface. Because the background amount of protein adsorption on these materials is so low, one would expect them to be very useful as substrates for peptide attachment for studies with model synthetic extracellular matrices, e.g., with $HS-(CH_2)_{11}(OCH_2CH_2)_n-NH-RGDS$.

Drumheller and Hubbell (1994a,b,c) have developed a polymeric material that was highly resistant to cell adhesion for use in peptide grafting. Materials that contain large amounts of poly(ethylene glycol) generally swell extensively, rendering material properties sometimes unsuitable for cell culture or medical devices. To circumvent this, the poly(ethylene glycol) swelling was constrained by distributing it as a network throughout a densely cross-linked network of a hydrophobic monomer, trimethylolpropane triacrylate. This yielded a material with the surface hydrophilicity of a hydrogel but with mechanical and optical properties of a glass. These materials were highly resistant to protein adsorption, even after an adsorptive challenge to the material with a very large adhesion protein, laminin, and even over multiweek durations. When the polymer network was formed with small amounts of acrylic acid as a comonomer, the polymer still remained cell nonadhesive. The carboxyl groups near the polymer surface were useful, however, as sites for derivatization with adhesion peptides such as the RGD and YIGSR sequences. Because the adsorption of proteins to those surfaces was so low, materials endowed with inactive peptides such as the RDG supported no cell adhesion.

Numerous other approaches are possible. One of particular interest, because of its ease of use, is adsorption. Block copolymers, consisting of adsorbing domains and nonadsorbing domains, can be adsorbed to material surfaces and can be used to regulate biological interactions, e.g., when the nonadsorbing domains are poly(ethylene glycol), surfaces that display very low levels of nonspecific adhesion can be generated (Amiji and Park, 1992). One convenient class of polymers includes ABA block copolymers of poly(ethylene glycol) (the A blocks) and poly(propylene glycol) (B), in which the hydrophilic and cell-repelling poly(ethylene glycol) domains flank the central hydrophobic and adsorbing poly(propylene glycol) block. The central hydrophobic block adsorbs well to hydrophobic surfaces, thus immobilizing the hydrophilic polymer and thereby resisting cell adhesion (Amiji and Park, 1992). Cell adhesion peptides can be displayed at the tips of these hydrophilic chain termini, and a very effective and simple model surface can be obtained (Neff *et al.,* 1998). Similar constructions can be designed for anionic surfaces, e.g., by using a polycationic block as a binding domain, with poly(ethylene glycol) chains attached thereto (Elbert and Hubbell, 1998).

CELL PATTERN FORMATION BY SUBSTRATE PATTERNING

The ability to control material surface properties precisely enables the formation of designed architectures of multiple cells in culture and potentially *in vivo* as well. Large-scale architectures have been formed by patterning adhesive surfaces. Three methods for patterning have been particularly powerful—photolithography, mechanical stamping, and microfluidics.

Photolithographic methods have been employed to impart patterns on cell adhesion surfaces. Alkoxysilanes have been chemisorbed to glass surfaces (using the same grafting chemistry as with the glycophase glass described above), and ultraviolet light was employed to degrade the alkoxy group selectively to yield patterns of surface hydroxyl groups (Healy *et al.*, 1994). These hydroxyl groups were used as sites for reaction with a second layer of an amine-containing alkoxysilane. These aminated regions supported cell adhesion and thus formed the cell-binding regions on the patterned substrate (Kleinfeld *et al.*, 1988). Patterned amines on polymer surfaces have also been employed to induce cell patterning via adhesive domains patterned on a nonadhesive background (Ranieri *et al.*, 1993). These approaches have been combined with the bioactive peptide technology described above. For example, patterned amines have been used as grafting sites for the adhesive peptide YIGSR to pattern neurite extension on material surfaces (Ranieri *et al.*, 1994). One of the goals of this work was to create neuronal networks as a simple system in which to study communication among networks of neurons. A powerful system for such work has been provided by using adhesive aminoalkylsilanes patterned on a nonadhesive perfluoroalkylsilane background (Stenger *et al.*, 1992). Polymers have been synthesized explicitly for the purpose of attaching adhesive peptide sequences such as RGD, and these will be very useful in future studies of cell–cell interactions in neuronal and other cell systems (Herbert *et al.*, 1997). Such patterned surfaces have been formed to control cell shape and size, in order to gain deeper insight into the interplay between cell biomechanics and cell function (Thomas *et al.*, 1999). It is particularly convenient in photopatterning studies to develop photochemistries on materials specifically for the purpose of immobilizing polymers and polymer–peptide adducts (Moghaddam and Matsuda, 1993); this has been addressed, e.g., with phenylazido-derivatized surfaces (Matsuda and Sugawara, 1996; Sugawara and Matsuda, 1996).

Alkane thiols on gold have also been patterned using simple methods. Contact printing has been employed for this purpose (Kumar *et al.*, 1995). Conventional photolithographic etching of silicon was employed to make a master printing stamp, a negative of which was then formed in silicone rubber. Structures as small as 200 nm were preserved in the silicone rubber stamp. The stamp was then wetted with a cell adhesion-promoting alkane thiol, $HS-(CH_2)_{15}-CH_3$. Stamping a gold substrate resulted in creation of a pattern of the hydrophobic alkane group. The stamped gold substrate was then treated with the cell-resistant alkane thiol $HS-(CH_2)_{11}$ $(OCH_2CH_2)_6OH$. Using this system it was possible to create adhesive patches of defined size on a very cell-nonadhesive substrate (Singhvi *et al.*, 1994). Microcontact printing can also be employed with binding approaches other than alkane thiols binding to gold, e.g., adhesion proteins such as laminin have been stamped onto reactive silane-modified surfaces to produce patterns to guide neurite outgrowth in culture (Wheeler *et al.*, 1999). This very flexible and powerful system will be useful in a wide variety of cell biological and tissue engineering applications.

A third powerful method is based on microfluidic systems, in which silicone rubber stamps are formed with silicon masters; the stamps are pressed to a surface, and the thin spaces patterned thereby are employed as capillaries to draw up fluid, containing a treatment compound, onto desired regions of the surface. The fluid can contain a soluble, adsorbing polymer with an attached adhesion peptide (Neff *et al.*, 1998), or it can contain a peptide with some affinity linker for the surface. In the practice of the latter, it is powerful to employ the very high-affinity streptavidin–biotin pair, e.g., by biotinylating the polymer at the surface and exposing, with the aid of the microfluidics channels, to peptide conjugated to streptavidin (Cannizzaro *et al.*, 1998; Patel *et al.*, 1998).

SIGNAL TRANSDUCTION AND FUNCTIONAL REGULATION VIA THE EXTRACELLULAR MATRIX

Cell interaction with adhesive substrates is known to provide information to the cells via numerous means, and this topic has been extensively reviewed (Roskelley *et al.*, 1995; O'Toole, 1997; Katz and Yamada, 1997; Yamada, 1997; Aplin *et al.*, 1998; Sechler *et al.*, 1998; Aota and Yamada, 1997; Miyamoto *et al.,* 1998). The biochemical mechanisms underlying these interactions are likely numerous and have not yet been fully elucidated. One key mechanism involves the focal contact as a site for catalysis. Integrin clustering induces tyrosine phosphorylation of several proteins, many of which still have unknown function (Kornberg *et al.,* 1991; Guan, 1997). One of these proteins is a 125-kDa tyrosine that localizes, after it is tyrosine phosphorylated, at the sites of focal contacts; this protein has been accordingly named pp125 focal adhesion kinase, or pp125[fak] (Schaller *et al.,* 1992; Vuori, 1998; Danen *et al.,* 1998). Thus, although the cytoplasmic domain of integrins bears no direct catalytic activity, clustering of integrins is known to stimulate tyrosine phosphorylation and further specific kinases are known to assemble at the sites of clustered integrins. Interestingly, when cells were permitted to spread via a non-integrin-mediated mechanism, specifically by interaction of cell surface proteoglycans with surface-adsorbed polycations, phosphorylation of intracellular proteins did not occur (Guan *et al.,* 1991). The phosphorylation of these proteins, associated with focal contact formation, is known to be an important signal for survival of a variety of cell types (Ffrisch *et al.,* 1996; Globus *et al.,* 1998).

Cell interactions with extracellular matrix have been demonstrated to exert control over the function of the adherent cells in numerous models; examples in cell spreading and migration, hepatocyte gene expression, mammary cell differentiation, endothelial cell morphological regulation, and neurite extension from neurons will be cited as examples in the following discussions.

Cell spreading, adhesion, and migration are each related to interactions with the extracellular matrix (Huttenlocher *et al.,* 1995). The role of RGD ligand surface density in mediating adhesion morphology, e.g., the formation of focal contacts and the organization of f-actin cytoskeletal structure, was described above. Mooney *et al.* (1995) have demonstrated that the rate of hepatocyte spreading, i.e., the rate at which the hepatocyte projected area increases, was strongly dependent on the surface density of the adhesion ligand, in this case laminin. It has been demonstrated, both mathematically and experimentally, that the surface density of adhesion proteins influences the rate of cell migration (DiMilla *et al.,* 1993). Ligand density too low leads to inadequate traction and adhesion strength for cell migration, and ligand density too high may provide adhesion strength excessive for cell migration (Ward and Hammer, 1993; Palecek *et al.,* 1997). Cell migration can dramatically impact cell proliferation for cells that are contact inhibited, i.e., where cell–cell contact exerts an inhibitory effect on cell proliferation. Zygourakis *et al.* have examined both on theoretical grounds (1991a) and by experiment (1991b) that prolonged cell–cell contact in slowly migrating contact-inhibited cells dramatically limits the rate at which cells proliferate.

Hepatocytes are notoriously difficult to culture over durations extending beyond several days; they begin to lose several of their hepatocyte-specific characteristics, e.g., the synthesis and secretion of albumin (Clayton and Darnell, 1983). The loss or retention of hepatocyte-specific function may relate to interactions with the extracellular matrix. For example, when hepatocytes were permitted to spread on laminin adsorbed to surfaces at differing surface densities, the secretion of albumin was seen to be influenced substantially (Mooney *et al.,* 1995). When hepatocyte spreading was limited by seeding on patterned alkane thiol/alkane thiol–oligo(ethylene oxide) surfaces as described above, cell shape was seen to influence albumin secretion substantially. Cells were seeded on adhesive patches ranging from 1600 to 10,000 μm^2, as well as on unpatterned surfaces. When cell spreading was limited to 1600 μm^2 by the size of the spot, albumin secretion was almost triple that of cells seeded on unpatterned substrates and allowed to spread maximally. When hepatocytes were cultured on laminin substrates, albumin expression was elevated relative to cells cultured on tissue culture plastic; moreover, this elevation of expression was specific to albumin relative to other genes (Bissell *et al.,* 1990). Hepatocytes cultured on fibronectin or laminin at protein surface densities of 1, 50, or 100 ng/cm^2 were seen to be switchable between differentiation (at low protein surface density) and growth (at high surface density), regardless of the identity of the extracellular matrix adhesion protein (Mooney *et al.,* 1992).

Endothelial cells have also been demonstrated to be responsive to their extracellular matrix, particularly to laminin in the basal lamina (Risau, 1995; Bischoff, 1997; Malinda *et al.,* 1999). Most of the relevant experimentation has been performed on Matrigel, reconstituted extracellular matrix from Engelbreth–Holm–Swarm tumor cells; this matrix contains large amounts of laminin of the particular form described above. When endothelial cells were cultured on tissue culture plastic, they formed a polygonal monolayer. When they were cultured on Matrigel, they formed capillary-like tubes in a complicated interconnected network. This tube formation was not inhibited by soluble YIGSR, added to inhibit competitively the interaction of the 67-kDa laminin receptor with the YIGSR site in laminin. When soluble SIKVAV was added, however, the tubes submerged into the protein gel and formed new branches. This observation may relate to regulation of cell-associated protease activity; stimulation with SIKVAV in protein or with soluble SIKVAV increased the activity of matrix metalloproteinases (Kanemoto *et al.,* 1990; Stack *et al.,* 1990). In culture, this peptide stimulated elongation of endothelial cells (Grant *et al.,* 1992). Other receptor-mediated interactions with extracellular matrix proteins have also been shown to be important in endothelial cell functional regulation; for example, angiogenesis *in vivo* was dependent on the integrin αvβ3 (Brooks *et al.,* 1994). Interestingly, purely mechanical means can also be employed to induce capillary-like tube formation; when endothelial cells were seeded on photolithographically formed thin adhesive lines, tube formation was induced (Spargo *et al.,* 1994).

The extracellular matrix has been observed to play an important role in the regulation of neuronal behavior both *in vivo* and *in vitro* (Kapfhammer and Schwab, 1992; McKerracher *et al.,* 1996). For example, neurons in culture extend more neurites on laminin than on fibronectin substrates. This activity has been localized to the active peptide regions of laminin; the sequence YIGSR appears to be involved primarily in neuronal adhesion, whereas the SIKVAV site stimulates neurite extension *in vitro.* Amines that bind to cell surface proteoglycans have also been useful in stimulating neurite outgrowth, although presumably by less specific mechanisms than with the peptides described above. Likewise, patterns on surfaces have been employed to induce and guide neurite outgrowth, e.g., with patterns between hydrophobic and hydrophilic regions (Stenger *et al.,* 1992) or with patterns of the laminin-derived peptides YIGSR and SIKVAV (Ranieri *et al.,* 1993).

TRANSLATION TO WORKING BIOMATERIAL SYSTEMS

Although one goal of biomaterials and tissue engineering research is certainly to develop systems for the quantitative study of biological interactions, another is to develop practical novel therapeutics. Many of the concepts described above, both in terms of development of model surfaces and with regard to manipulating cellular behavior, are directly transferable; however, some cautionary comments should be made. It is not only the chemical identity of an adhesion peptide that determines its biological activity, but also its amount. This was very clearly demonstrated by Palecek *et al.* (1997), who showed that small amounts of an adhesion molecule could enhance cell migration, whereas larger amounts could inhibit it. They further demonstrated that this effect depended on, among other features, the affinity of the receptor–ligand pair, the number of receptors, and the polarization of receptors from the leading to the trailing edge of the cell. Given that many of these features depend on the state of the cell and can be modulated by the cell's biological environment, e.g., by the growth factors to which the cell is exposed (Maheshwari *et al.,* 1999), many confounding features must be considered in translation from model to practical application.

REFERENCES

Adams, C. L., and Nelson, W. J. (1998). Cytomechanics of cadherin-mediated cell–cell adhesion. *Curr. Opin. Cell Biol.* **10,** 572–577.
Alattia, J. R., Kurokawa, H., and Ikura, M. (1999). Structural view of cadherin-mediated cell–cell adhesion. *Cell Mol. Life Sci.* **55,** 359–367.
Amiji, M., and Park, K. (1992). Prevention of protein adsorption and platelet adhesion on surfaces by PEO/PPO/PEO triblock copolymers. *Biomaterials* **13,** 682–692.
Amiji, M., and Park, K. (1993). Surface modification of polymeric biomaterials with poly(ethylene oxide), albumin, and heparin for reduced thrombogenicity. *J. Biomater. Sci., Polym. Ed.* **4,** 217–234.
Andrade, J. D., and Hlady, V. (1986). Protein adsorption and materials biocompatibility: A tutorial review and suggested hypotheses. *Adv. Polym. Sci.* **79,** 1–63.
Aota, S., and Yamada, K. M. (1997). Integrin functions and signal transduction. *Adv. Exp. Med. Biol.* **400B,** 669–682.
Aplin, A. E., Howe, A., Alahari, S. K., and Juliano, R. L. (1998). Signal transduction and signal modulation by cell adhesion receptors: The role of integrins, cadherins, immunoglobulin-cell adhesion molecules, and selectins. *Pharmacol. Rev.* **50,** 197–263.
Aumailley, M., and Smyth, N. (1998). The role of laminins in basement membrane function. *J. Anat.* **193,** 1–21.
Bain, C. D., and Whitesides, G. M. (1989). Modeling organic surfaces with self-assembled monolayers. *Angew. Chem., Int. Ed. Engl.* **28,** 506–512.
Bevilacqua, M., Butcher, E., Furie, B., Furie, B., Gallatin, M., Gimbrone, M., Harlan, J., Kishimoto, K., Lasky, L., McEver, R., Paulson, J., Rosen, S., Seed, B., Siegelman, M., Springer, T., Stoolman, L., Tedder, T., Varki, A., Wagner, D., Weissman, I., and Zimmerman, G. (1991). Selectins: A family of adhesion receptors. *Cell (Cambridge, Mass.)* **67,** 233–233.

Birkedall-Hansen, H. (1995). Proteolytic remodeling of extracellular matrix. *Curr. Opin. Cell Biol.* 7, 728–735.

Bischoff, J. (1997). Cell adhesion and angiogenesis. *J. Clin. Invest.* 100, S37–S39.

Bissell, D. M., Caron, J. M., Babiss, L. E., and Friedman, J. M. (1990). Transcriptional regulation of the albumin gene in cultured rat hepatocytes. *Mol. Biol. Med.* 7, 187–197.

Blasi, F. (1993). Urokinase and urokinase receptor: A paracrine/autocrine system regulating cell migration and invasiveness. *BioEssays* 15, 105–111.

Bloom, L., Ingham, K. C., and Hynes, R. O. (1999). Fibronectin regulates assembly of actin filaments and focal contacts in cultured cells via the heparin-binding site in repeat III13. *Mol. Biol. Cell* 10, 1521–1536.

Brooks, P. C., Clark, R. A., and Cheresh, D. A. (1994). Requirement of vascular integrin alpha v beta 3 for angiogenesis. *Science* 264, 569–571.

Cannizzaro, S. M., Padera, R. F., Langer, R., Rogers, R. A., Black, F. E., Davies, M. C., Tendler, S. J., and Shakesheff, K. M. (1998). A novel biotinylated degradable polymer for cell-interactive applications. *Biotechnol. Bioeng.* 58, 529–535.

Chen, W. T. (1992). Membrane proteases: Roles in tissue remodeling and tumor invasion. *Curr. Opin. Cell Biol.* 4, 802–809.

Clayton, D. F., and Darnell, J. E., Jr. (1983). Changes in liver-specific compared to common gene transcription during primary culture of mouse hepatocytes. *Mol. Cell. Biol.* 3, 1552–1561.

Cremer, M. A., Rosloniec, E. F., and Kang, A. H. (1998). The cartilage collagens: A review of their structure, organization, and role in the pathogenesis of experimental arthritis in animals and in human rheumatic disease. *J. Mol. Med.* 76, 275–288.

Danen, E. H., Lafrenie, R. M., Miyamoto, S., and Yamada, K. M. (1998). Integrin signaling: Cytoskeletal complexes, MAP kinase activation, and regulation of gene expression. *Cell Adhes. Commun.* 6, 217–224.

Danilov, Y. N., and Juliano, R. L. (1989). (Arg-Gly-Asp)$_n$–albumin conjugates as model substratum for integrin-mediated cell adhesion. *Exp. Cell Res.* 182, 186–196.

Dejana, E., Colella, S., Conforti, G., Abbadini, M., Gaboii, M., and Marchisio, P. C. (1988). Fibronectin and vitronectin regulate the organization of their respective Arg-Gly-Asp receptors in cultured human endothelial cells. *J. Cell Biol.* 107, 1215–12223.

Dickenson, R. B., Guido, S., and Tranquillo, R. T. (1994). Biased cell migration of fibroblasts exhibiting contact guidance in oriented collagen gels. *Ann. Biomed. Eng.* 22, 342–356.

DiMilla, P. A., Stone, J. A., Quinn, J. A., Albelda, S. J., and Lauffenberger, D. A. (1993). Maximal migration of human smooth muscle cells on fibronectin and type IV collagen occurs at an intermediate attachment strength. *J. Cell Biol.* 122, 729–737.

Drumheller, P. D., and Hubbell, J. A. (1994a). Phase mixed poly(ethylene glycol)/poly(trimethylolpropane triacrylate) semi-interpenetrating polymer networks obtained by rapid network formation. *J. Polym. Sci. Part A: Polym. Chem.* 32, 2715–2725.

Drumheller, P. D., and Hubbell, J. A. (1994b). Densely cross-linked polymer networks of poly(ethylene glycol) in trimethylolpropane triacrylate for cell adhesion resistant surfaces. *J. Biomed. Mater. Res.* 29, 207–215.

Drumheller, P. D., and Hubbell, J. A. (1994c). Polymer networks with grafted cell adhesive peptides for highly biospecific cell adhesive substrates. *Anal. Biochem.* 222, 380–388.

Elbert, D. L., and Hubbell, J. A. (1998). Self-assembly and steric stabilisation at heterogeneous, biological surfaces via adsorbing block copolymers. *Chem. Biol.* 5, 177–183.

Engvall, E., and Wewer, U. M. (1996). Domains of laminin. *J. Cell Biochem.* 61, 493–501.

Faham, S., Hilemann, R. E., Fromm, J. R., Linhardt, R. J., and Rees, D. C. (1996). Heparin structure and interactions with basic fibroblast growth factor. *Science* 271, 1116–1120.

Faham, S., Linhardt, R. J., and Rees, D. C. (1998). Diversity does make a difference: Fibroblast growth factor–heparin interactions. *Curr. Opin. Struct. Biol.* 8, 578–586.

Fairbrother, W. J., Champe, M. A., Christinger, H. W., Keyt, B. A., and Starovasnik, M. A. (1998). Solution structure of the heparin-binding domain of vascular endothelial growth factor. *Structure* 6, 637–648.

Fath, K. R., Edgell, C. S., and Burridge, K. (1989). The distribution of distinct integrins in focal contacts is determined by the substratum composition. *J. Cell Sci.* 92, 67–75.

Fratzl, P., Misof, K., Zizak, I., Rapp, G., Amenitsch, H., and Bernstorff, S. (1998). Fibrillar structure and mechanical properties of collagen. *Struct. Biol.* 122, 119–122.

Frisch, S. M., Vuori, K., Ruoslahti, E., and Chan-Hui, P. Y. (1996). Control of adhesion-dependent cell survival by focal adhesion kinase. *J. Cell Biol.* 134, 793–799.

Gallagher, J. T. (1997). Structure–activity relationship of heparan sulphate. *Biochem. Soc. Trans.* 25, 1206–1209.

Globus, R. K., Doty, S. B., Lull, J. C., Holmuhamedov, E., Humphries, M. I., and Damsky, C. H. (1998). Fibronectin is a survival factor for differentiated osteoblasts. *J. Cell Sci.* 111, 1385–1393.

Graf, J., Iwamoto, Y., Sasaki, M., Martin, G. R., Kleinman, R. K., Robey, F. A., and Yamada, Y. (1987a). Identification of an amino acid sequence in laminin mediating cell attachment, chemotaxis, and receptor binding. *Cell (Cambridge, Mass.)* 48, 989–996.

Graf, J., Ogle, R. C., Robey, F. A., Sasaki, M., Martin, G. R., Yamada, Y., and Kleinman, H. K. (1987b). A pentapeptide from the laminin B1 chain that mediates cell adhesion and binds the 67000 laminin receptor. *Biochemistry* 26, 6896–6900.

Grant, D. S., Kinsella, J. L., Fridman, R., Aurbach, R., Piasecki, B. A., Yamada, Y., Zain, M., and Kleinman, H. K. (1992). Interaction of endothelial cells with a laminin A chain peptide (SIKVAV) *in vitro* and induction of angiogenic behavior *in vitro*. *J. Cell. Physiol.* 153, 614–625.

Grinnell, F. (1999). Signal transduction pathways activated during fibroblast contraction of collagen matrices. *Curr. Top. Pathol.* 93, 61–73.

Grumet, M. (1991). Cell adhesion molecules and their subgroups in the nervous system. *Curr. Opin. Neurobiol.* **21**, 298–306.

Guan, J. L. (1997). Role of focal adhesion kinase in integrin signaling. *Int. J. Biochem. Cell Biol.* **29**, 1085–1096.

Guan, J. L., Trevethick, J. E., and Hynes, R. L. (1991). Fibronectin/integrin interaction induces tyrosine phosphorylation of a 120 kD protein. *Cell Regul.* **2**, 951–964.

Harris, J. M. (1992). "Poly(Ethylene Glycol) Chemistry." Plenum, New York.

He, Y. J., and Grinnell, F. (1994). Stress relaxation of fibroblasts activates a cyclic AMP signalling pathway. *J. Cell Biol.* **126**, 457–464.

Healy, K. E., Lom, B., and Hockberger, P. E. (1994). Spatial distribution of mammalian cells dictated by material surface chemistry. *Biotechnol. Bioeng.* **43**, 792–800.

Herbert, C. B., McLernon, T. L., Hypolite, C. L., Adams, D. N., Pikus, L., Huang, C. C., Fields, G. B., Letourneau, P. C., Distefano, M. D., and Hu, W. S. (1997). Micropatterning gradients and controlling surface densities of photoactivatable biomolecules on self-assembled monolayers of oligo(ethylene glycol) alkanethiolates. *Chem. Biol.* **4**, 731–737.

Hubbell, J. A. (1995). Biomaterials in tissue engineering. *BioTechnology* **13**, 565–576.

Humphries, M. J. (1990). The molecular basis and specificity of integrin–ligand interactions. *J. Cell Sci.* **97**, 585–592.

Hunkapiller, T., and Hood, L. (1989). Diversity of the immunoglobulin gene superfamily. *Adv. Immunol.* **44**, 1–63.

Huttenlocher, A., Sandborg, R. R., and Horwitz, A. F. (1995). Adhesion in cell migration. *Curr. Opin. Cell Biol.* **7**, 697–706.

Hynes, R. O. (1992). Integrins: Versatility, modulation and signaling in cell adhesion. *Cell (Cambridge, Mass.)* **69**, 11–25.

Izzard, C. S., and Lochner, L. R. (1976). Cell-to-substratum contacts in living fibroblasts: An interference reflection study with an evaluation of the technique. *J. Cell Sci.* **21**, 129–159.

Jackson, R. L., Busch, S. I., and Cardin, A. D. (1991). Glycosaminoglycans: Molecular properties, protein interactions and role in physiological processes. *Physiol. Rev.* **71**, 481–539.

Kallapur, S. G., and Akeson, R. A. (1992). The neural cell adhesion molecule (NCAM) heparin binding domain binds to cell surface heparan sulfate proteoglycans. *J. Neurosci. Res.* **33**, 538–548.

Kanemoto, T., Reich, R., Greatorex, D., Adler, S. H., Yamada, Y., and Kleinman, H. K. (1990). Identification of an amino acid sequence from the laminin A chain which stimulates metastases formation and collagenase IV production. *Proc. Natl. Acad. Sci. U.S.A.* **87**, 2279–2283.

Kapfhammer, J. P., and Schwab, M. E. (1992). Modulators of neuronal migration and neurite growth. *Curr. Opin. Cell Biol.* **4**, 863–868.

Katz, B. Z., and Yamada, K. M. (1997). Integrins in morphogenesis and signaling. *Biochimie* **79**, 467–476.

Keyt, B. A., Berleau, L. T., Nguyen, H. V., Chen, H., Heinsohn, H., Vandlen, R., and Ferrara, N. (1996). The carboxyl-terminal domain (111–165) of vascular endothelial growth factor is critical for its mitogenic potency. *J. Biol. Chem.* **271**, 7788–7795.

Kleinfeld, D., Kahler, K. H., and Hockberger, P. E. (1988). Controlled outgrowth of dissociated neurons on patterned substrates. *J. Neurosci.* **8**, 4098–4120.

Kornberg, L., Earp, H. S., Turner, C., Prokop, C., and Juliano, R. L. (1991). Signal transuction by integrins: Increased protein tyrosine phosphorylation caused by clustering of beta1 integrins. *Proc. Natl. Acad. Sci. U.S.A.* **88**, 8392–8396.

Kumar, A., Abbott, N. L., Kim, E., Biebuyck, H. A., and Whitesides, G. M. (1995). Patterned self assembled monolayers and mesoscale phenomena. *Acc. Chem. Res.* **28**, 219–226.

Lasky, L. A. (1995). Selectin–carbohydrate interactions and the initiation of the inflammatory response. *Annu. Rev. Biochem.* **64**, 113–139.

LeBaron, R. G., Esko, J. D., Woods, A., Johansson, S., and Höök, M. (1988). Adhesion of glycosaminoglycan-deficient Chinese hamster ovary cell mutants to fibronectin substrata. *J. Cell Biol.* **106**, 945–952.

Llanos, G. R., and Sefton, M. V. (1993). Does polyethylene oxide possess a low thrombogenecity? *J. Biomater. Sci., Polym. Ed.* **4**, 381–400.

Lopez, G. P., Albers, M. W., Schreiber, S. L., Carroll, R., Peralta, E., and Whitesides, G. M. (1993). Convenient methods for patterning the adhesion of cells to surfaces using self-assembled monolayers of alkanethiolates on gold. *J. Am. Chem. Soc.* **115**, 5877–5878.

Lyon, M., and Gallagher, J. T. (1998). Bio-specific sequences and domains in heparan sulphate and the regulation of cell growth and adhesion. *Matrix Biol.* **17**, 485–493.

Magnusson, M. K., and Mosher, D. F. (1998). Fibronectin: Structure, assembly, and cardiovascular implications. *Arterioscler. Thromb. Vasc. Biol.* **18**, 1363–1370.

Maheshwari, G., Wells, A., Griffith, L. G., and Lauffenburger, D. A. (1999). Biophysical integration of effects of epidermal growth factor and fibronectin on fibroblast migration. *Biophys. J.* **76**, 2814–2823.

Malinda, K. M., and Kleinman, H. K. (1996). The laminins. *Int. J. Biochem. Cell Biol.* **28**, 957–959.

Malinda, K. M., Nomizu, M., Chung, M., Delgado, M., Kuratomi, Y., Yamada, Y., Kleinman, H. K., and Ponce, M. L. (1999). Identification of laminin α1 and β1 chain peptides active for endothelial cell adhesion, tube formation, and aortic sprouting. *FASEB J.* **13**, 53–62.

Massia, S. P., and Hubbell, J. A. (1990). Covalent surface immobilization of Arg-Gly-Asp- and Tyr-Ile-Gly-Ser-Arg-containing peptides to obtain well-defined cell-adhesive substrates. *Anal. Biochem.* **187**, 292–301.

Massia, S. P., and Hubbell, J. A. (1991). An RGD spacing of 440 nm is sufficient for integrin αvβ3-mediated fibroblast spreading and 140 nm for focal contact and stress fiber formation. *J. Cell Biol.* **114**, 1089–1100.

Massia, S. P., and Hubbell, J. A. (1992). Immobilized amines and basic amino acids as mimetic heparin binding domains for cell surface proteoglycan-mediated adhesion. *J. Biol. Chem.* **267**, 10133–10141.

Massia, S. P., Rao, S. S., and Hubbell, J. A. (1993). Covalently immobilized laminin peptide Tyr-Ile-Gly-Ser-Arg (YIGSR) supports cell spreading and co-localization of the 67-kilodalton laminin receptor with α-actinin and vinculin. *J. Biol. Chem.* **268**, 8053–8059.

Matsuda, T., and Sugawara, T. (1996). Control of cell adhesion, migration, and orientation on photochemically micro-processed surfaces. *J. Biomed. Mater. Res.* **32**, 165–173.

Mayadas, T. N., and Wagner, D. D. (1992). Vicinal cysteines in the prosequence play a role in von Willebrand factor mul-timer assembly. *Proc. Natl. Acad. Sci. U.S.A.* **89**, 3531–3535.

McKerracher, L., Chamoux, M., and Arregui, C. O. (1996). Role of laminin and integrin interactions in growth cone guid-ance. *Mol. Neurobiol.* **12**, 95–116.

Meecham, R. P. (1991). Laminin receptors. *Annu. Rev. Cell Biol.* **7**, 71–91.

Miyamoto, S., Katz, B. Z., Lafrenie, R. M., and Yamada, K. M. (1998). Fibronectin and integrins in cell adhesion, sig-naling, and morphogenesis. *Ann. N.Y. Acad. Sci.* **857**, 119–129.

Moghaddam, M. J., and Matsuda, T. (1993). Molecular design of 3-dimensional artificial extracellular matrix: Photosen-sitive polymers containing cell adhesive peptide. *J. Polym. Sci., Part A: Polym. Chem.* **31**, 1589–1597.

Mooney, D., Hansen, L., Vacanti, J., Langer, R., Farmer, S., and Ingber, D. (1992). Switching from differentiation to growth in hepatocytes: Control by extracellular matrix. *J. Cell. Physiol.* **151**, 497–505.

Mooney, D. J., Langer, R., and Ingber, D. E. (1995). Cytoskeletal filament assembly and the control of cell spreading and function by extracellular matrix. *J. Cell Sci.* **108**, 2311–2320.

Mosher, D. F., Sottile, J., Wu, C., and McDonald, J. A. (1992). Assembly of extracellular matrix. *Curr. Opin. Cell Biol.* **4**, 810–818.

Neff, J. A., Caldwell, K. D., and Tresco, P. A. (1998). A novel method for surface modification to promote cell attachment to hydrophobic substrates. *J. Biomed. Mater. Res.* **40**, 511–519.

Norde, W., and Lyklema, J. (1991). Why proteins prefer interfaces. *J. Biomater. Sci. Polym. Ed.* **2**, 183–202.

O'Toole, T. E. (1997). Integrin signaling: Building connections beyond the focal contact? *Matrix Biol.* **16**, 165–171.

Palecek, S. P., Loftus, J. C., Ginsberg, M. H., Lauffenburger, D. A., and Horwitz, A. F. (1997). Integrin–ligand binding properties govern cell migration speed through cell-substratum adhesiveness. *Nature (London)* **385**, 537–540; erra-tum: **388**, 210 (1997).

Patel, N., Padera, R., Sanders, G. H., Cannizzaro, S. M., Davies, M. C., Langer, R., Roberts, C. J., Tendler, S. J., Williams, P. M., and Shakesheff, K. M. (1998). Spatially controlled cell engineering on biodegradable polymer surfaces. *FASEB J.* **12**, 1447–1454.

Ponce, M. L., Nomizu, M., Delgado, M. C., Kuratomi, Y., Hoffman, M. P., Powell, S., Yamada, Y., Kleinman, H. K., and Malinda, K. M. (1999). Identification of endothelial cell binding sites on the laminin γ1 chain. *Circ. Res.* **84**, 688–694.

Preissner, K. T. (1991). Structure and biological role of vitronectin. *Annu. Rev. Cell Biol.* **7**, 275–310.

Prime, K. L., and Whitesides, G. M. (1993). Adsorption of proteins onto surfaces containing end-attached oligo(ethylene oxide): A model system using self-assembled monolayers. *J. Am. Chem. Soc.* **115**, 10714–10721.

Ranieri, J. P., Bellamkonda, R., Jacob, J., Vargo, T. G., Gardella, J. A., and Aebischer, P. (1993). Selective neuronal cell at-tachment to a covalently patterned monoamine on fluorinated ethylene–propylene films. *J. Biomed. Mater. Res.* **27**, 917–925.

Ranieri, J. P., Bellamkonda, R., Bekos, E. J., Gardella, J. A., Mathieu, H. J., Ruiz, L., and Aebischer, P. (1994). Spatial con-trol of neuronal cell attachment and differentiation on covalently patterned laminin oligopeptide substrates. *Int. J. Dev. Neurosci.* **12**, 725–735.

Rezania, A., and Healy, K. E. (1999). Biomimetic peptide surfaces that regulate adhesion, spreading, cytoskeletal organi-zation, and mineralization of the matrix deposited by osteoblast-like cells. *Biotechnol. Prog.* **15**, 19–32.

Risau, W. (1995). Differentiation of endothelium. *FASEB J.* **9**, 926–933.

Roskelley, C. D., Srebrow, A., and Bissell, M. J. (1995). A hierarchy of ECM-mediated signalling regulates tissue-specific gene expression. *Curr. Opin. Cell Biol.* **7**, 736–747.

Ruggeri, Z. M. (1997). von Willebrand factor. *J. Clin. Invest.* **100**, S41–S46; erratum: **101**, 919 (1998).

Ruoslahti, E. (1991). Integrins. *J. Clin. Invest.* **87**, 1–5.

Ruoslahti, E., and Pierschbacher, M. (1986). Tetrapeptide. U.S. Pat. 4,578,079.

Ruoslahti, M. D., and Pierschbacher, M. D. (1987). New perspectives in cell adhesion: RGD and integrins. *Science* **238**, 491–497.

Ryan, M. C., Christiano, A. M., Engvall, E., Wewer, U. M., Miner, J. H., Sanes, J. R., and Burgeson, R. E. (1996). The functions of laminins: Lessons from *in vivo* studies. *Matrix Biol.* **15**, 369–381.

Sadler, J. E. (1998). Biochemistry and genetics of von Willebrand factor. *Annu. Rev. Biochem.* **67**, 395–424.

Schaller, M. D., Borgman, C. A., Cobb, B. S., Vines, R. R., Reynolds, A. B., and Parsons, J. T. (1992). pp125 FAK, a structurally unique protein tyrosine kinase associated with focal adhesions. *Proc. Natl. Acad. Sci. U.S.A.* **89**, 5192–5196.

Schwarzbauer, J. E. (1991). Fibronectin: From gene to protein. *Curr. Opin. Cell Biol.* **3**, 786–791.

Schwarzbauer, J. E. (1999). Basement membranes: Putting up the barriers. *Curr. Biol.* **9**, R242–R244.

Sechler, J. L., Corbett, S. A., Wenk, M. B., and Schwarzbauer, J. E. (1998). Modulation of cell–extracellular matrix in-teractions. *Ann. N.Y. Acad. Sci.* **857**, 143–1154.

Shapiro, L., and Colman, D. R. (1998). Structural biology of cadherins in the nervous system. *Curr. Opin. Neurobiol.* **8**, 593–599.

Shapiro, S. D. (1998). Matrix metalloproteinase degradation of extracellular matrix: Biological consequences. *Curr. Opin. Cell Biol.* **10**, 602–608.

Singhvi, R., Kumar, A., Lopez, G. P., Stephanopoulos, G. N., Wang, D. I. C., Whitesides, G. M., and Ingber, D. E. (1994). Engineering cell shape and function. *Science* **264**, 696–698.

Spargo, B. J., Testoff, M. A., Nielsen, T. B., Stenger, D. A., Hickman, J. J., and Rudolph, A. S. (1994). Spatially controlled adhesion, spreading, and differentiation of endothelial cells on self-assembled molecular monolayers. *Proc. Natl. Acad. Sci. U.S.A.* **91**, 11070–11074.

Stack, S., Gray, R. D., and Pizzo, S. V. (1990). Modulation of plasminogen activation and type IV collagenase activity by a synthetic peptide derived from the laminin A chain. *Biochemistry* **30**, 2073–2077.

Stenger, D. A., Gerger, J. H., Dulcey, C. S., Hickman, J. J., Rudolph, A. S., Nielsen, T. B., McCort, S. M., and Calvert, J. M. (1992). Coplanar molecular assemblies of aminoalkylsilane and perfluorinated alkylsilane: Characterization and geometric definition of mammalian cell adhesion and growth. *J. Am. Chem. Soc.* **114**, 8435–8442.

Sugawara, T., and Matsuda, T. (1996). Synthesis of phenylazido-derivatized substances and photochemical surface modification to immobilize functional groups. *Biomed. Mater. Res.* **32**, 157–164.

Takeichi, M. (1991). Cadherin cell adhesion receptors as a morphogenetic regulator. *Science* **251**, 1451–1455.

Tashiro, K., Sephel, G. C., Greatorex, D., Sasaki, M., Shirashi, N., Martin, G. R., Kleinman, H. K., and Yamada, Y. (1991). The RGD containing site of the mouse laminin A chain is active for cell attachment, spreading, migration and neurite outgrowth. *J. Cell. Physiol.* **146**, 451–459.

Thomas, C. H., Lhoest, J. B., Castner, D. G., McFarland, C. D., and Healy, K. E. (1999). Surfaces designed to control the projected area and shape of individual cells. *J. Biomech. Eng.* **121**, 40–48.

Timpl, R. (1996). Macromolecular organization of basement membranes. *Curr. Opin. Cell Biol.* **8**, 618–624.

Timpl, R., and Brown, J. (1996). Supramolecular assembly of basement membranes. *BioEssays* **18**, 123–132.

Vanderrest, M., and Garrone, R. (1991). Collagen family of proteins. *FASEB J.* **5**, 2814–2823.

Varki, A. (1994). Selectin ligands. *Proc. Natl. Acad. Sci. U.S.A.* **91**, 7390–7397.

Vestweber, D., and Blanks, J. E. (1999). Mechanisms that regulate the function of the selectins and their ligands. *Physiol. Rev.* **79**, 181–213.

Vuori, K. (1998). Integrin signaling: Tyrosine phosphorylation events in focal adhesions. *Membr. Biol.* **165**, 191–199.

Ward, M. D., and Hammer, D. A. (1992). Morphology of cell-substratum adhesion. Influence of receptor heterogeneity and nonspecific forces. *Cell Biophys.* **20**, 177–222.

Ward, M. D., and Hammer, D. A. (1993). A theoretical analysis for the effect of focal contact formation on cell-substrate attachment strength. *Biophys. J.* **64**, 936–959.

Wheeler, B. C., Corey, J. M., Brewer, G. J., and Branch, D. W. (1999). Microcontact printing for precise control of nerve cell growth in culture. *J. Biomech. Eng.* **121**, 73–78.

Wight, T. N., Kinsella, M. G., and Qwarnström, E. E. (1992). The role of proteoglycans in cell adhesion, migration and proliferation. *Curr. Opin. Cell Biol.* **4**, 793–801.

Wojciechowski, P., and Brash, J. L. (1991). The Vroman effect in tube geometry: The influence of flow on protein adsorption measurements. *J. Biomater. Sci. Polym. Ed.* **2**, 203–216.

Yamada, K. M. (1991). Adhesive recognition sequences. *J. Biol. Chem.* **266**, 12809–12812.

Yamada, K. M. (1997). Integrin signaling. *Matrix Biol.* **16**, 137–411.

Yamada, Y., and Kleinman, H. K. (1992). Functional domains of cell adhesion molecules. *Curr. Opin. Cell Biol.* **4**, 819–823.

Zygourakis, K., Bizioz, R., and Markenscoff, P. (1991a). Proliferation of anchorage-dependent contact-inhibited cells. 1: Development of theoretical models based on cellular automata. *Biotechnol. Bioeng.* **38**, 459–470.

Zygourakis, K., Markenscoff, P., and Bizios, R. (1991b). Proliferation of anchorage-dependent contact-inhibited cells. 2: Experimental results and validation of the theoretical models. *Biotechnol. Bioeng.* **38**, 471–479.

CHAPTER 21

POLYMER SCAFFOLD PROCESSING

Robert C. Thomson, Albert K. Shung, Michael J. Yaszemski,
and Antonios G. Mikos*

INTRODUCTION

Polymer scaffolds must possess many key characteristics, including high porosity and surface area, structural strength, and specific three-dimensional shapes, to be useful as materials for tissue engineering. These characteristics are determined by the scaffold fabrication technique, which must be developed such that it does not adversely affect the biocompatibility of the material of construction. Novel processing techniques have been developed to manufacture scaffolds with desirable and reproducible characteristics, which have been used to demonstrate the feasibility of regenerating human tissue. These polymer scaffold fabrication techniques have been tailored to create scaffolds with particular characteristics that will satisfy the critical tissue engineering requirements of specific organs.

Restoration of organ function by utilizing tissue engineering technologies often requires the use of a temporary porous scaffold. The function of the scaffold is to direct the growth of cells migrating from surrounding tissue (tissue conduction) or the growth of cells seeded within the porous structure of the scaffold. The scaffold must therefore provide a suitable substrate for cell attachment, proliferation, differentiated function, and, in certain cases, cell migration. These critical requirements can be met by the selection of an appropriate material from which to construct the scaffold, although the suitability of the scaffold may also be affected by the processing technique. Many biocompatible materials can be potentially used to construct scaffolds. However, a biodegradable material is normally desired because the role of the scaffold is usually only a temporary one. Many natural and synthetic biodegradable polymers, such as collagen, poly(α-hydroxyesters), and poly(anhydrides), have been widely and successfully used as scaffold materials due to their versatility and ease of processing (Thomson *et al.,* 1995). Many researchers have used poly(α-hydroxyesters) as starting materials from which to fabricate scaffolds using a wide variety of processing techniques. These polymers have proved successful as temporary substrates for a number of cell types, allowing cell attachment, proliferation, and maintenance of differentiated function. Poly(α-hydroxyesters), such as poly(L-lactic acid) (PLLA), poly(lactic–co-glycolic acid) (PLGA) copolymers, and poly(glycolic acid) (PGA), are linear, uncross-linked polymers. These materials are biocompatible, degradable by simple hydrolysis, and are Food and Drug Administration (FDA) approved for certain clinical applications.

The major requirement of any proposed polymer processing technique is not only the utilization of biocompatible materials, but that the process should in no way affect the biocompatibility of the polymer. The processing technique should also allow the manufacture of scaffolds with controlled porosity and pore size, both important factors in organ regeneration. A highly porous scaffold is desirable to allow cell seeding or migration throughout the material. Pore size plays a critical role in both tissue ingrowth and the internal surface area available for cell attachment. A

---

* To whom correspondence should be addressed.

large surface area is required so that a high number of cells, sufficient to replace or restore organ function, can be cultured.

The mechanical properties of the scaffold are often of critical importance especially when regenerating hard tissues such as cartilage and bone. Although the properties of the solid polymer and the porosity of the scaffold have a profound effect on its mechanical properties, polymer processing can also be influential in this respect. The tensile strength may, for example, be enhanced due to the crystallization of polymer chains. Alternatively, the manufacturing process may cause a reduction in the molecular weight of the polymer, resulting in a deleterious effect on mechanical properties. The shape of a hard tissue is often important to its function and in such cases the processing technique must allow the preparation of scaffolds with irregular three-dimensional geometries.

A further design consideration is the incorporation of bioactive molecules into the polymer matrix. Inactivation of bioactive molecules by exposure to high temperatures or a harsh chemical environment is the major obstacle to successful drug incorporation and delivery from a degradable scaffold. However, if localized controlled delivery of bioactive molecules to the site of growing tissue can be achieved, it would provide a powerful tool to aid in organ regeneration.

The technique used to manufacture scaffolds for tissue engineering is dependent on the properties of the polymer and its intended application. Scaffold manufacturing techniques were therefore developed using commercially available starting materials: small solid polymer chunks or in some cases long fibers. The techniques that were developed involved either heating the polymers or dissolving them in a suitable organic solvent. The viscous behavior of the polymers above their glass transition or melting temperatures and their solubility in various organic solvents were two of the more important characteristics to consider during process development.

A plethora of new fabrication methods have been investigated specifically for the purpose of tissue engineering. Several of the more established scaffold fabrication techniques are described in this chapter as well as some of the developments in polymer processing. Each technique has its advantages, but none can be considered as an ideal method of scaffold fabrication to be employed for all tissues. The choice of a fabrication technique, or the development of a new one, is dependent on the critical requirements for the tissue of interest.

FIBER BONDING

Fibers provide a large surface area: volume ratio and are therefore desirable as scaffold materials. One of the first biomedical uses of PGA was as a degradable suture material, which is why it is commercially available in the form of long fibers. PGA fibers in the form of tassels and felts were utilized as scaffolds in organ regeneration feasibility studies (Cima *et al.*, 1991). However, these scaffolds lacked the structural stability necessary for *in vivo* use. A fiber bonding technique was therefore developed to prepare interconnecting fiber networks with different shapes for use as scaffolds in organ regeneration (Mikos *et al.*, 1993a).

PLLA is dissolved in methylene chloride, a nonsolvent for PGA, and the resulting polymer solution is cast over a nonwoven mesh of PGA fibers in a glass container. The solvent is allowed to evaporate and residual amounts are removed by vacuum drying. A composite material is thus produced consisting of nonbonded PGA fibers embedded in a PLLA matrix. The PLLA–PGA composite is then heated to a temperature above the melting point of PGA for a given time period. During heating the PGA fibers join at their cross-points as melting commences, but the two polymers do not join due to their immiscibility in the melt state. The composite is quenched to prevent any further melting of the PGA fibers during cooling. After heat treatment, the PLLA matrix of a PLLA–PGA composite membrane is selectively dissolved in methylene chloride and the resulting bonded PGA fibers are vacuum dried. Using this technique, the fibers are physically joined without any surface or bulk modification and retain their initial diameter. The PLLA matrix is required to prevent collapse of the PGA mesh and to confine the melted PGA to a fiberlike shape (Fig. 21.1). The heating time is also of critical importance because prolonged exposure to the elevated temperature results in the gradual transformation of the PGA fibers into spherical domains.

Bonded PGA fiber structures with high porosities and area:volume ratios are produced using this manufacturing method and utilize only biocompatible materials. In this respect, the structures produced are highly suitable as scaffolds for organ regeneration. In addition, they demonstrate

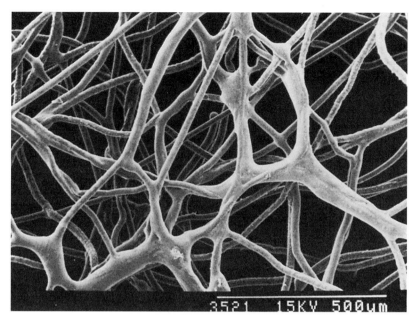

Fig. 21.1. A scanning electron photomicrograph of a PGA-bonded fiber mesh, prepared by a fiber bonding technique, embedded in PLLA, heated at a temperature of 195°C for 90 min and at 235°C for 5 min. From Mikos et al. (1993). Preparation of poly(glycolic acid) bonded fiber structures for cell attachment and transplantation. J. Biomed. Mater. Res. **27,** 183–189. Copyright 1993 John Wiley & Sons, Inc. Reprinted by permission of John Wiley & Sons, Inc.

structural integrity, which is lacking in the PGA tassels and felts used in organ regeneration feasibility studies. However, the stipulations concerning the choice of solvent, immiscibility of the two polymers, and their relative melting temperatures restrict the general application of this technique to other polymers. In addition, the technique does not lend itself to easy and independent control of porosity and pore size.

An alternative method of fiber bonding has been developed to fabricate tubular scaffolds useful in regeneration of tissues such as blood vessels, intestines, and ureters. This method involves coating a nonbonded mesh of PGA fibers with solutions of PLLA or PLGA (Mooney *et al.,* 1996a). The mesh is attached to a rotating Teflon cylinder and is sprayed with an atomized polymer solution. As spraying continues, a coat of PLLA (or PLGA) builds up on the PGA fibers and forms bonds at their cross-points. Longer spraying times result in thickening of the PGA fibers due to PLLA deposition. Although a bonded PGA mesh is formed by this process, the surface of the scaffold presented to transplanted cell populations is a thin coat of PLLA or PLGA. Cell attachment, growth, and function are then determined by the coating rather than by the PGA mesh. This coating may be advantageous if the mechanical properties of PGA fibers are required but the surface properties of PLLA or PLGA are more desirable for the transplanted cell population. The mechanical properties of the bonded mesh are a result of both the PGA fibers and the additional stability supplied by the PLLA or PLGA coating. Thus, if the integrity of the bond created by the coating is compromised, either through mechanical stresses or degradation of the PLLA or PLGA, the stability and mechanical properties of the bonded structure are adversely affected. Unfortunately, this method of fiber bonding does not address the problem of creating scaffolds with complex three-dimensional shapes, but it has proved successful for producing hollow tubes. This method was utilized to create tubular structures for the support of smooth muscle cells and extracellular matrix proteins (Kim *et al.,* 1998). After 7 weeks in culture, these tubes maintained their shape and structure while exhibiting high cellularity with smooth muscle cells and extracellular matrix (ECM) proteins filling the PGA pores.

SOLVENT CASTING AND PARTICULATE LEACHING

In order to overcome some of the drawbacks associated with the fiber bonding preparation, a solvent-casting and particulate-leaching technique has been developed (Mikos *et al.,* 1994). With appropriate thermal treatment, porous constructs of synthetic biodegradable polymers can be prepared with specific porosity, surface:volume ratio, pore size, and crystallinity for different applications. The technique has been validated for PLLA and PLGA scaffolds, but can be applied to any other polymer that is soluble in a solvent such as chloroform or methylene chloride.

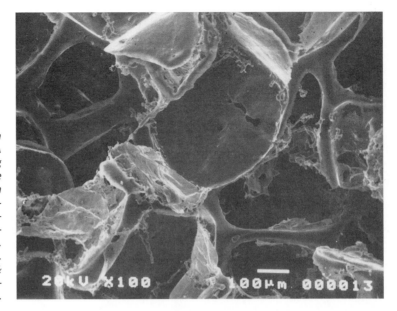

*Fig. 21.2. A scanning electron micrograph of a 75:25 PLGA scaffold with 90% porosity, using sodium chloride of size range 300–500 µm as a porogen. From Ishaug et al. (1997). Bone formation by three-dimensional stromal osteoblast culture in biodegradable poymer scaffolds. J. Biomed. Mater Res. **36,** 17–28. Copyright 1997 John Wiley & Sons, Inc. Reprinted by permission of John Wiley & Sons, Inc.*

Sieved salt particles are dispersed in a PLLA/chloroform solution and cast into a glass container. The salt particles utilized are insoluble in chloroform. The solvent is allowed to evaporate and residual amounts are removed by vacuum drying. The PLLA in the resulting PLLA/salt composite membranes is highly crystalline, but this can be controlled by one of two means depending on the desired crystallinity of the final membrane. In the first processing option, the PLLA/salt composite membranes are immersed in water to leach out the salt. This results in highly crystalline, salt-free PLLA membranes, which are then dried. The alternative option involves a heat treatment stage to produce PLLA membranes with controlled crystallinity. PLLA/salt composite membranes are heated at a temperature above the PLLA melting temperature to ensure complete melting of the polymer crystallites formed during the previous processing step. The melted PLLA membranes with dispersed salt particles are either annealed by cooling at a slow, controlled rate to, produce semicrystalline membranes with specific crystallinity, or quenched, to produce amorphous membranes. Finally, the membranes are immersed in water to leach out the salt and the resulting salt-free PLLA membranes are then dried.

Highly porous membranes with porosities up to 93% and an interconnected pore structure can be produced using this solvent-casting and particulate-leaching technique (Fig. 21.2). The porosity of porous PLLA membranes can be controlled by varying the amount of salt used to construct the composite material. The pore size of the membrane can also be controlled independently of the porosity by altering the size of the salt particles. Scaffolds with pore sizes up to 500 µm have been fabricated using this method. Membranes with high surface area:volume ratios are produced, and the ratio is dependent on both salt weight fraction and particle size. In addition, the crystallinity of PLLA membranes can be tailored to that desired for each application. These characteristics are all desirable properties of a scaffold for organ regeneration. The major disadvantage of this technique is that it can only be used to produce thin wafers or membranes (up to 2 mm in thickness). A three-dimensional scaffold cannot be directly constructed. Membrane lamination or melt molding, however, can circumvent this problem.

One of the problems associated with the aforementioned technique is that the polymer scaffolds produced are brittle and therefore inappropriate for soft tissue applications. To overcome this limitation, PLGA has been blended with poly(ethylene glycol) (PEG) and fabricated into pliable foams using the solvent-casting/particulate-leaching method (Wake *et al.,* 1996). As the amount of PEG in the blend is increased, the pliability of the foams is increased. Thus, thick polymer membranes can be prepared with the ability to be rolled into tubes without damaging the pore structure. Using this method, foam scaffolds can be fabricated in a tubular form. These scaffolds have great potential in soft tissue engineering in a variety of applications, such as regeneration of blood vessels and the esophagus.

A concern with scaffolds made with PLLA or PLGA is the even seeding of cells, which is related to the hydrophobicity of the polymer. To counter this issue, a technique has been developed in which porous PLGA scaffolds are soaked in an aqueous solution of a hydrophilic polymer such as poly(vinyl alcohol) (PVA) (Mooney *et al.*, 1994). Hepatocytes seeded on the treated scaffolds have been shown to be present in much higher densities than in untreated scaffolds.

An alternative method to overcome the difficulty of seeding cells in PLLA or PLGA scaffolds involves the prewetting the scaffolds with ethanol (Mikos *et al.*, 1994). In this approach, porous polymer scaffold disks are soaked in ethanol for 1 hr. The scaffolds are then submersed in water to prewet them. Chondrocytes and hepatocytes seeded on the prewetted scaffolds have been shown to exhibit a uniform distribution throughout the scaffolds. This technique could also be potentially useful for seeding other types of cells as well.

MEMBRANE LAMINATION

The construction of scaffolds with three-dimensional anatomical shapes is necessary for the regeneration of hard tissues such as cartilage and bone, whose function is partially dependent on geometry. Membrane lamination offers a means of constructing highly porous biomaterials with anatomical shapes (Mikos *et al.*, 1993b). However, the lamination procedure is useful only if it preserves the uniform porous structure of the original membranes. Also, the boundary between two layers must be indistinguishable from the bulk of the device. The methodology to process biodegradable polymer scaffolds with anatomical shapes involves the construction of a contour plot of the particular three-dimensional shape. The shapes of the contours are cut from highly porous biodegradable membranes prepared using the solvent-casting and particulate-leaching technique described above. A small quantity of chloroform is then coated onto the contacting surfaces of adjacent membranes and a bond is formed. Thus, the desired three-dimensional shape is constructed layer by layer (Fig. 21.3).

The porous structure of the PLLA and PLGA membranes used to validate the technique was unaffected by the lamination procedure, and no boundary between the layers was observed. This method can be used to prepare three-dimensional anatomically shaped PLLA and PLGA scaffolds with bulk properties identical to those of the individual membranes. This technique can not be applied to the lamination of PGA because it is soluble only in highly toxic solvents. A similar

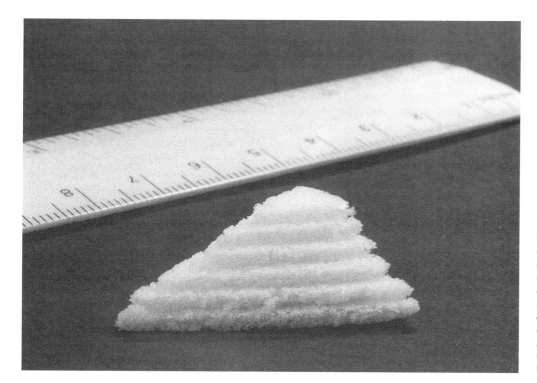

Fig. 21.3. An example of a membrane-laminated PLLA scaffold with a noselike shape. Reprinted from Biomaterials **14,** A. G. Mikos, G. Sarakinos, S. M. Leite et al.; Laminated three-dimensional biodegradable foams for use in tissue engineering, pp. 323–330. Copyright 1993, with permission from Elsevier Science.

method of membrane lamination was used to produce tubular stents (Mooney *et al.*, 1994). Flat porous scaffolds of PLGA are manufactured using the solvent-casting and particulate-leaching method. These are then wrapped around a Teflon cylinder and the overlapping edges are laminated using a small amount of chloroform to fuse the two polymer surfaces together. The Teflon tube is then removed to leave an open-ended tube, which is then capped by laminating end pieces. These constructs have been proposed as scaffolds for the regeneration of tubular tissues such as intestine.

MELT MOLDING

Melt molding, an alternative method of constructing three-dimensional scaffolds, has many advantages over membrane lamination. PLGA scaffolds have been produced by leaching PLGA/gelatin microsphere composites formed using a molding technique (Thomson *et al.*, 1995). A fine PLGA powder is mixed with previously sieved gelatin microspheres and poured into a Teflon mold, which is heated above the glass transition temperature of the polymer. The PLGA/gelatin microsphere composite is subsequently removed from the mold and placed in distilled–deionized water. The gelatin, which is soluble in water, is leached out, leaving a porous PLGA scaffold with a geometry identical to the shape of the mold.

Using this method it is possible to construct PLGA scaffolds of any shape simply by changing the mold geometry. The porosity can be controlled by varying the amount of gelatin used to construct the composite material. The pore size of the scaffold can also be altered, independently of the porosity, by using microspheres of different diameters. In addition, because this technique does not utilize organic solvents and is carried out at relatively low temperatures, it has potential for the incorporation and controlled delivery of bioactive molecules. These molecules may be incorporated either into the polymer or into the gelatin microspheres. Gelatin has been used in the past as a vehicle for controlled delivery (Thomson *et al.*, 1995). This scaffold-manufacturing technique can also be applied to PLLA and PGA. However, higher temperatures are required because these polymers are semicrystalline and must therefore be heated above their melting temperatures. The use of higher temperatures to mold these polymers may exclude the possibility of drug incorporation. It also entails the use of a leachable component other than gelatin, because this material becomes insoluble in water after exposure to the required elevated temperatures. This manufacturing method satisfies many of the scaffold preparation design criteria and offers an extremely versatile means of scaffold preparation. Although gelatin microspheres were utilized to validate the feasibility of this technique, alternative leachable components such as salt or other polymer microspheres may be used.

EXTRUSION

Although extrusion has been widely incorporated in the processing of industrial polymers, it has not been widely used for the fabrication of polymer scaffolds in tissue engineering applications. Widmer *et al.* (1998), however, developed an extrusion process in combination with the aforementioned solvent-casting technique to fabricate porous, biodegradable tubular conduits for the purpose of peripheral nerve regeneration (Fig. 21.4).

Briefly, PLGA or PLLA polymers are first fabricated into composite wafers with sodium chloride as a porogen by the solvent-casting technique. The wafers are cut to the desired length and placed in a customized extrusion tool. The tool is mounted on a hydraulic press and heated to the appropriate processing temperature. Pressure is then applied between the nozzle and the piston of the extrusion tool and the polymer/salt composite is extruded as a tube. After cooling, the salt is leached out by immersion in water and the entire scaffold is then vacuum dried.

The main parameter involved with the extrusion of these polymers is the extrusion temperature. The higher the extrusion temperature, the lower the pressure required to initiate extrusion. The pore morphology of the conduits is not significantly affected by this processing method unless very high extrusion temperatures are used. At high extrusion temperatures, a decrease in molecular weight is most likely caused by thermal degradation of the polymer. Pore diameter is also decreased at very high extrusion temperatures due to the increase in polymer viscosity. However, the most important variables affecting the porosity and pore size of the conduits are the salt weight fraction and salt crystal size, respectively. By varying these two parameters, the porosity and pore size can be easily tailored to incorporate the loading of cells or carriers for growth factors into the conduits.

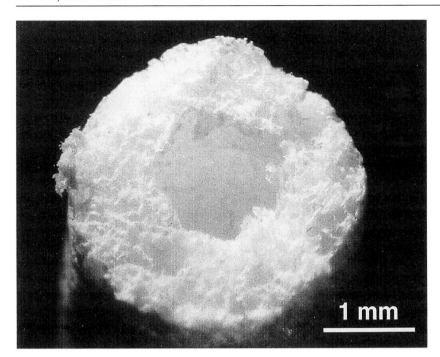

*Fig. 21.4. A photomicrograph of a porous tubular PLLA scaffold processed by an extrusion technique suitable for peripheral nerve regeneration. Reprinted from Biomaterials **19**, M. S. Widmer, P. K. Gupta, L. Lu et al., Manufacture of porous biodegradable polymer conduits by an extrusion process for guided tissue regeneration, pp. 1945–1955. Copyright 1998, with permission from Elsevier Science.*

THREE-DIMENSIONAL PRINTING

One of the major challenges in processing polymer scaffolds for tissue engineering is creating complex three-dimensional shapes to emulate the tissue that the scaffold is designed to repair or replace. The three-dimensional printing (3DP, a trademark name) fabrication technique has been applied in an attempt to fulfill this requirement. The 3DP technique belongs to a family of fabrication techniques collectively known as solid free-form (SFF) methods, including methods such as stereolithography and laser sintering (Park *et al.*, 1998). In 3DP, a thin layer of polymer powder is spread over a piston. An inkjet printer head prints a liquid binder, usually chloroform for polymer systems, onto the powder layer. The position of the printer head is entirely controlled via a computer-assisted design and manufacture (CAD/CAM) program. The piston is dropped and the polymer powder and liquid layering steps are repeated. The previously printed liquid binder then joins this powder layer to the underlying powder layer. This process is repeated until the desired construct is complete. In this manner, intricate three-dimensional shapes can be made layer by layer with a resolution down to 300 μm. The scaffolds can be made porous by incorporating salt in the polymer powder bed. PLLA and PLGA scaffolds have been fabricated using this technique. The strengths of 3DP-constructed polymer scaffolds using PLLA or PLGA are comparable to those fabricated by compression molding using the same base material (Giordano *et al.*, 1996).

Constructs produced with this method have been investigated for several different tissue engineering applications, including organogenesis of liver tissue (Griffith *et al.*, 1997). Surface modification of PLLA substrates for cell adhesion also has been performed using 3DP (Park *et al.*, 1998). Porous scaffolds fabricated using this process have been fashioned into cylinders with interconnected longitudinal channels for the purpose of hepatocyte transplantation (Kim *et al.*, 1998). These scaffolds show great promise in cell transplantation applications due to their branching morphology and porous structure, which may serve to promote angiogenesis and tissue ingrowth.

GAS FOAMING

Although the process of solvent casting/particulate leaching is well established in fabricating porous polymer scaffolds, it does have some drawbacks. One major concern is the use of organic solvents during processing. Organic solvent residues left behind from the processing can have toxic effects *in vitro* and elicit an inflammatory response *in vivo*. To overcome this problem, a fabrication method has been developed to create porous matrices utilizing a gas foaming technique to avoid the use of organic solvents (Mooney *et al.,* 1996b). In this method, PLGA pellets are compression molded into solid disks. After compression, the disks are exposed to high-pressure CO_2 to saturate the polymer. The subsequent reduction in pressure to ambient levels causes the nucleation and formation of pores in the polymer matrix from the CO_2 gas. However, the matrices formed have a closed pore morphology, which may be undesirable for tissue engineering applications.

In order to produce an open pore morphology using the gas foaming method, a technique has been developed to use this method in conjunction with the particulate-leaching technique (Harris *et al.,* 1998). In this process, salt particles and PLGA pellets are mixed together and compressed to form solid disks. The disks are saturated with high-pressure gas and the pressure is subsequently reduced. The salt particles are then leached out. The combination of gas foaming and particulate leaching leads to a porous polymer matrix with an open, interconnected morphology without the use of any organic solvents. Smooth muscle cells have been found to readily adhere and proliferate on scaffolds fabricated with this process. By precluding the use of organic solvents, this technique may see widespread use in cell transplantation applications of many types of cells, including hepatocytes, chondrocytes, and osteoblasts.

FREEZE DRYING

Another method to fabricate polymer scaffolds with variable porosity and pore size utilizes an emulsion/freeze-drying process (Whang *et al.,* 1995). Water is added to a solution of PLGA in methylene chloride to create an emulsion. The mixture is then homogenized, poured into a copper mold, and quenched in liquid nitrogen. After quenching, the polymer scaffold is freeze dried to remove the water and solvent. Scaffolds with porosity of up to 90% and median pore sizes in the range of 15 to 35 μm can be fabricated with an interconnected pore structure. In comparison to solvent casting/particulate leaching, the scaffolds produced with this method offer much higher specific pore surface area as well as the ability to make thick (>1 cm) polymer scaffolds, but the overall size of the pores is smaller.

Freeze drying has also been applied in another fashion to fabricate porous polymer foams for use in drug delivery applications (Hsu *et al.,* 1997). Scaffolds using this technique have been made with several different types of polymers, including PLLA, PLGA, PGA, and a PLGA/poly(propylene fumarate) (PPF) blend. The desired polymer is dissolved in glacial acetic acid or benzene and frozen to $-10°C$. The frozen solution is then freeze dried in a lyophilizer for a week to remove the solvent. The type of polymer used has a large effect on the morphology of the resulting foams. PLGA/PPF foams typically have a leaflet morphology whereas PLLA, PLGA, and PGA foams have a capillary morphology. The type of polymer used also plays a large role in the release of drugs embedded in the polymer matrix. PLGA/PPF foams release drugs at a much faster rate than do foams made with PLGA. The lower molecular weight of the PPF allows more penetration of water into the matrix, which causes greater diffusion of drug out of the foam. Polymer foams made in this fashion, however, may not be well suited for tissue engineering applications, due to their closed pore morphologies and are limited to drug delivery use.

PHASE SEPARATION

Phase separation has been utilized in a novel alternative method of scaffold manufacture, which has been developed primarily to address the problem of drug delivery (Lo *et al.,* 1995). The ability to deliver bioactive molecules from a degrading polymer scaffold to cells within or surrounding the scaffold is an attractive one because it could potentially allow manipulation of tissue growth and cell function. In order to achieve this goal, the scaffold manufacturing process must lend itself to incorporation of bioactive molecules and must not cause any loss of drug activity due to exposure to harsh chemical or thermal environments. Porous PLLA and poly(phosphoester) scaffolds with small hydrophilic and hydrophobic bioactive molecules have been manufactured using the following phase-separation technique. First, the polymer is dissolved in a solvent (molten phenol or naphthalene) at a low temperature (55°C for phenol and 85°C for naphthalene). The bioactive molecule is then dissolved or dispersed in the resulting homogeneous solution, which is then cooled in a controlled fashion until liquid–liquid phase separation is induced. The resulting bicontinuous polymer and solvent phases are then quenched to create a two-phase solid. The solidified solvent is then removed by sublimation, leaving a porous polymer scaffold with bioactive molecules incorporated within the polymer. Drug release experiments have shown that the activity of small bioactive molecules is not adversely affected by the chemical and thermal conditions utilized in this phase-separation technique. This technique has proved useful as a means of incorporating small molecules into polymer scaffolds. However, incorporating and releasing drugs with large protein structures without loss of activity remain a challenge.

POLYMER/CERAMIC COMPOSITE FOAMS

Bone regeneration presents some interesting and unique challenges to engineers and scientists interested in scaffold design. First, because almost all bone defects are irregularly shaped, any proposed scaffold processing technique must be sufficiently versatile to allow the formation of porous polymer-based materials with irregular three-dimensional shapes. Second, the scaffold must have high strength to replace the structural function of bone temporarily until it is regenerated. Although poly(α-hydroxyesters) have been used in a solid form for many orthopedic applications, the compressive strength of foam scaffolds constructed of these materials rapidly declines with increasing porosity (Thomson *et al.,* 1995).

In order to improve the mechanical properties of these foams, short hydroxyapatite fibers were incorporated into the poly(α-hydroxyester) framework in order to reinforce the scaffold (Thomson *et al.,* 1998). A melt-molding technique such as the one previously described would be desirable as a means to produce a three-dimensional scaffold. Unfortunately, melt molding is not suited to the manufacture of a composite foam. The reinforcing effect of fibers within a polymer matrix is effective only if the fibers can be evenly distributed throughout the polymer such that the degree of fiber–polymer contact is maximized and fiber–fiber contact is held to a minimum. Because achieving an even fiber distribution by attempting to mix two solid phases (polymer and fiber) prior to molding is extremely difficult, a solvent-casting technique is employed.

Hydroxyapatite short fibers and a porogen (either gelatin microspheres or salt particles) are first dispersed in a PLGA/dichloromethane solution. The result, after solvent evaporation and drying, is a composite material consisting of hydroxyapatite fibers and a porogen embedded in a PLGA polymer membrane. The required three-dimensional shape is then achieved by compression molding the composite material. Subsequent leaching of the porogen then leaves a porous composite scaffold of PLGA reinforced with short hydroxyapatite fibers (Fig. 21.5). Within a range of fiber contents, these scaffolds have superior compressive strength compared to nonreinforced PLGA scaffolds of the same porosity.

Another process that utilizes hydroxyapatite to reinforce PLGA foams involves the use of particulate hydroxyapatite rather than the use of fibers (Devin *et al.,* 1996). This technique utilizes an emulsion to create the porous scaffold. The NaCl and the particulate hydroxyapatite are first suspended in a PLGA solution in chloroform. An aqueous PVA solution is then added to the suspension to form an emulsion. The emulsion is cast into cylindrical molds and then vacuum dried. The dried matrices are then submersed in water to leach out the salt and form the porous scaffolds. The concentration of the PVA solution and the diameter of the salt crystals both control the diameter of the pores formed, which range from 18 to 150 μm in diameter. Increasing the weight-percent of hydroxyapatite particles in the initial polymer suspension increases the compressive strength of the final matrix. The compressive properties of the composite scaffold are of the same order of magnitude as those of cancellous bone.

An alternative method to formulate polymer/ceramic composites has been proposed using a novel phase transition technique (Zhang and Ma, 1999). In this process, hydroxyapatite powder is added to a PLGA/dioxane solution. The mixture is frozen for several hours to induce phase separation and then freeze dried to sublimate the solvent. The formed composite foams exhibit an interconnected irregular pore morphology with a polymer/hydroxyapatite skeleton. The compressive strength of these foams is significantly higher than that of foams made from pure PLGA. The porosity, pore size, and pore structure can all be controlled by changing the polymer concentration, hydroxyapatite amount, solvent type, and phase-separation temperature. Composite foams with porosity of up to 95% and pore size in the range of 30 to 100 μm can be fabricated with this method.

a

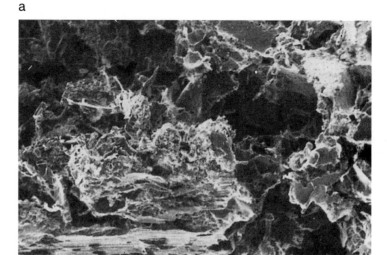

b

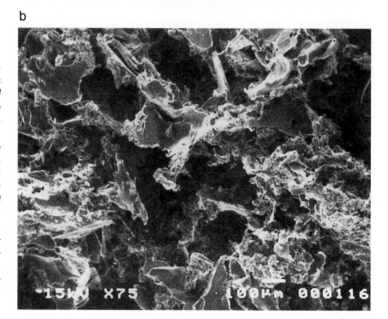

*Fig. 21.5. Scanning electron photomicrographs of highly porous composite scaffolds constructed of PLGA and hydroxyapatite short fibers using a solvent-casting and melt-molding technique. Sodium chloride particles of size ranging from 300 to 500 μm were used as a porogen with an original weight-percent of (a) 70% and (b) 90%. Reprinted from Biomaterials **19,** R. C. Thomson, M. J. Yaszemski, J. M. Powers et al., Hydroxyapatite fiber reinforced poly(alphahydroxy ester) foams for bone regeneration, pp. 1935–1943. Copyright 1998, with permission from Elsevier Science.*

IN SITU POLYMERIZATION

All of the polymer-processing techniques discussed thus far have described methods to manufacture prefabricated scaffolds, which may then be used to regenerate the appropriate tissue. Prefabricated scaffolds are suitable for most tissues engineering applications; however, there are many instances in orthopedic surgery when mechanical stability must be restored immediately. In such cases, a nondegradable bone cement, usually poly(methyl methacrylate) (PMMA), is used to fill the bone defect and provide mechanical stability. A degradable, poly(propylene fumarate)(PPF)-based bone cement has been developed to provide the same function but on a temporary basis (Yaszemski *et al.,* 1996; Peter *et al.,* 1999a). PPF is an unsaturated linear polyester and is a viscous liquid at room temperature. It can be cross-linked at the time of surgery to form a solid degradable bone cement via an addition polymerization with *N*-vinyl pyrrolidone (*N*-VP). As the cross-linking reaction proceeds, the PPF is transformed from a viscous liquid to a puttylike state before finally solidifying. During its liquid and putty states, the cement can be injected or molded into the bone defect. It is therefore well suited for this application because many bone injuries result in defects that are relatively inaccessible without further surgical exposure and are geometrically ill defined. Most addition polymerization reactions are exothermic and generate large quantities of heat. In the case of PMMA, which is polymerized *in situ,* this is sufficient to cause some local tissue necrosis. In contrast, much less heat is generated by the cross-linking reaction between PPF and *N*-VP (Peter *et al.,* 1999b), and no local tissue necrosis has been noted in *in vivo* studies (Yaszemski *et al.,* 1996). Two other components are incorporated into the PPF cement at the time of cross-linking. The first is NaCl, which leaches out to provide pores into which new bone can grow, and the second is β-tricalcium phosphate, which is an osteoconductive material that facilitates new bone growth. Incorporation of β-tricalcium phosphate also enhances the mechanical properties of the cement. By using high-molecular-weight PPF and incorporating β-tricalcium phosphate, the mechanical properties of the scaffold can be closely matched to that of human trabecular bone (Peter *et al.,* 1997, 1998). In this manner, a temporary osteoconductive scaffold may be formed *in situ* to replace the mechanical function of bone until new bone, stimulated to form in the scaffold pores, can assume this structural role.

CONCLUSIONS

The diversity of organ structure and function is such that the design requirements of scaffolds for tissue engineering are specific to the tissue of interest. Consequently, the processing technique employed to manufacture a polymer scaffold with the desired characteristics must be chosen or developed appropriately. Several novel manufacturing techniques have been developed to process synthetic polymers into porous scaffolds with large void volumes for cell seeding and sufficient surface area for cell attachment. Each technique has particular advantages, such as ease of fabrication, superior structural strength, or the ability to incorporate and deliver bioactive molecules, but none can be considered as an ideal method of scaffold fabrication to be employed for all tissues. The choice of a scaffold fabrication technique is therefore a question of setting priorities to determine the critical requirements. Future challenges in polymer scaffold processing include the fabrication of high-strength scaffolds for hard tissue replacement at load-bearing sites and the ability to incorporate and deliver proteins and growth factors without loss of activity. All of the methods discussed in this chapter are viable and promising ways to produce polymer scaffolds for tissue regeneration and repair.

REFERENCES

Cima, L. G., Vacanti, J. P., Vacanti, C., *et al.* (1991). Tissue engineering by cell transplantation using degradable polymer substrates. *J. Biomech. Eng.* **113,** 143–151.

Devin, J. E., Attawia, M. A., and Laurencin, C. T. (1996). Three-dimensional degradable porous polymer–ceramic matrices for use in bone repair. *J. Biomater. Sci. Polym. Ed.* 7, 661–669.

Giordano, R. A., Wu, B. M., Borland, S. W., *et al.* (1996). Mechanical properties of dense polylactic acid structures fabricated by three dimensional printing. *J. Biomater. Sci. Polym.* **8,** 63–75.

Griffith, L. G., Wu, B., Cima, M. J., *et al.* (1997). *In vitro* organogenesis of liver tissue. *Ann. N.Y. Acad. Sci.* **831,** 382–397.

Harris, L. D., Kim, B. S., and Mooney, D. J. (1998). Open pore biodegradable matrices formed with gas foaming. *J. Biomed. Mater. Res.* **42,** 396–402.

Hsu, Y. Y., Gresser, J. D., Trantolo, D. J., *et al.* (1997). Effect of polymer foam morphology and density on kinetics of *in vitro* controlled release of isoniazid from compressed foam matrices. *J. Biomed. Mater. Res.* **35,** 107–116.

Ishaug, S. L., Crane, G. M., Miller, M. J., *et al.* (1997). Bone formation by three-dimensional stromal osteoblast culture in biodegradable polymer scaffolds. *J. Biomed. Mater. Res.* **36,** 17–28.

Kim, B. S., and Mooney, D. J. (1998). Engineering smooth muscle tissue with a predefined structure. *J. Biomed. Mater. Res.* **41,** 322–332.

Kim, S. S., Utsunomiya, H., Koski, J. A., *et al.* (1998). Survival and function of hepatocytes on a novel three-dimensional synthetic biodegradable polymer scaffold with an intrinsic network of channels. *Ann. Surg.* **228,** 8–13.

Lo, H., Ponticiello, M. S., and Leong, K. W. (1995). Fabrication of controlled release biodegradable foams by phase separation. *Tissue Eng.* **1,** 15–27.

Mikos, A. G., Bao, Y., Cima, L. G., *et al.* (1993a). Preparation of poly(glycolic acid) bonded fiber structures for cell attachment and transplantation. *J. Biomed. Mater. Res.* **27,** 183–189.

Mikos, A. G., Sarakinos, G., Leite, S. M., *et al.* (1993b). Laminated three-dimensional biodegradable foams for use in tissue engineering. *Biomaterials* **14,** 323–330.

Mikos, A. G., Lyman, M. D., Freed, L. E., *et al.* (1994). Wetting of poly(L-lactic acid) and poly(DL-lactic–co-glycolic acid) foams for tissue culture. *Biomaterials* **15,** 55–58.

Mooney, D. J., Kaufmann, P. M., Sano, K., *et al.* (1994). Transplantation of hepatocytes using porous, biodegradable sponges. *Transplant. Proc.* **26,** 3425–3426.

Mooney, D. J., Mazzoni, C. L., Breuer, C., *et al.* (1996a). Stabilized polyglycolic acid fibre-based tubes for tissue engineering. *Biomaterials* **17,** 115–124.

Mooney, D. J., Baldwin, D. F., Suh, N. P., *et al.* (1996b). Novel approach to fabricate porous sponges of poly(D,L-lactic–co-glycolic acid) without the use of organic solvents. *Biomaterials* **17**, 1417–1422.

Park, A., Wu, B., and Griffith, L. G. (1998). Integration of surface modification and 3D fabrication techniques to prepare patterned poly(L-lactide) substrates allowing regionally selective cell adhesion. *J. Biomater. Sci., Polym. Ed.* **9**, 89–110.

Peter, S. J., Nolley, J. A., Widmer, M. S., *et al.* (1997). *In vitro* degradation of a poly(propylene fumarate)/β-tricalcium phosphate injectable composite scaffold. *Tissue Eng.* **3**, 207–215.

Peter, S. J., Miller, S. T., Zhu, G., *et al.* (1998). *In vivo* degradation of a poly(propylene fumarate)/β-tricalcium phosphate injectable composite scaffold. *J. Biomed. Mater. Res.* **41**, 1–7.

Peter, S. J., Suggs, L. J., Yaszemski, M. J., *et al.* (1999a). Synthesis of poly(propylene fumarate) by acylation of propylene glycol in the presence of a proton scavenger. *J. Biomater. Sci., Polym. Ed.* **10**, 363–373.

Peter, S. J., Kim, P., Yasko, A. W., *et al.* (1999b). Crosslinking characteristics of an injectable poly(propylene fumarate)/β-tricalcium phosphate paste and mechanical properties of the crosslinked composite for use as a biodegradable bone cement. *J. Biomed. Mater. Res.* **44**, 314–321.

Thomson, R. C., Yaszemski, M. J., Powers, J. M., *et al.* (1995). Fabrication of biodegradable polymer scaffolds to engineer trabecular bone. *J. Biomater. Sci., Polym. Ed.* **7**, 23–28.

Thomson, R. C., Yaszemski, M. J., Powers, J. M., *et al.* (1998). Hydroxyapatite fiber reinforced poly(α-hydroxy ester) foams for bone regeneration. *Biomaterials* **19**, 1935–1943.

Wake, M. C., Gupta, P. K., and Mikos, A. G. (1996). Fabrication of pliable biodegradable polymer foams to engineer soft tissues. *Cell Transplant.* **5**, 465–473.

Whang, K., Thomas, H., and Healy, K. E. (1995). A novel method to fabricate bioabsorbable scaffolds. *Polymer* **36**, 837–841.

Widmer, M. S., Gupta, P. K., Lu, L., *et al.* (1998). Manufacture of porous biodegradable polymer conduits by an extrusion process for guided tissue regeneration. *Biomaterials* **19**, 1945–1955.

Yaszemski, M. J., Payne, R. G., Hayes, W. C., *et al.* (1996). *In vitro* degradation of a poly(propylene fumarate)-based composite material. *Biomaterials* **17**, 2127–2130.

Zhang, R., and Ma, P. X. (1999). Poly(α-hydroxyl acids)/hydroxyapatite porous composites for bone–tissue engineering. I. Preparation and morphology. *J. Biomed. Mater. Res.* **44**, 446–455.

CHAPTER 22

BIODEGRADABLE POLYMERS

James M. Pachence and Joachim Kohn

INTRODUCTION

The design and development of tissue-engineered products has benefited from many years of clinical utilization of a wide range of biodegradable polymers. Newly developed biodegradable polymers and novel modifications of previously developed biodegradable polymers have enhanced the tools available to create clinically important tissue engineering applications The increased demands placed on biomaterials for novel sophisticated medical implants such as tissue engineering constructs continues to fuel the interest in improving the performance of existing medical-grade polymers and in developing new synthetic polymers. This chapter surveys those biologically derived and synthetic biodegradable polymers that have been used or are under consideration for use in tissue engineering applications. The polymers are described in terms of their chemical composition, breakdown products, mechanism of breakdown, mechanical properties, and clinical limitations. Also discussed are product design considerations in processing of biomaterials into a final form (e.g., gel, membrane, matrix) that will effect the desired tissue response.

BIODEGRADABLE POLYMER SELECTION CRITERIA

The selection of biomaterials plays a key role in the design and development of tissue engineering product development. Although the classical selection criterion for a safe, stable implant dictated choosing a passive, "inert" material, it is now understood that any such device will elicit a cellular response (Peppas and Langer, 1994). Therefore, it is now widely accepted that a biomaterial must interact with tissue to repair, rather than act simply as a static replacement. Furthermore, biomaterials used directly in tissue repair or replacement applications (e.g., artificial skin) must be more than biocompatible: they must elicit a desirable cellular response. Consequently, a major focus of biomaterials for tissue engineering applications centers around harnessing control over cellular interactions with biomaterials, often including components to manipulate cellular response within the supporting biomaterial as a key design component. It is important for the tissue engineering product developer to have several biomaterials options available, because each application calls for a unique environment for cell–cell interactions. Such applications include (1) support for new tissue growth (wherein cell–cell communication and cell availability to nutrients, growth factors, and pharmaceutically active agents must be maximized); (2) prevention of cellular activity (when tissue growth, such as in surgically induced adhesions, is undesirable); (3) guided tissue response (enhancing a particular cellular response while inhibiting others); (4) enhancement of cell attachment and subsequent cellular activation (e.g., fibroblast attachment, proliferation, and production of an extracellular matrix of dermis repair); (5) inhibition of cellular attachment and/or activation (e.g., platelet attachment to a vascular graft); and (6) prevention of a biological response (e.g., blocking antibodies against homograft or xenograft cells used in organ replacement therapies).

Biodegradable polymers are applicable to those tissue engineering products for which tissue repair or remodeling is the goal, but not where long-term materials stability is required. Biodegradable polymers must also possess (1) manufacturing feasibility, including sufficient commercial quantities of the bulk polymer; (2) the capability to form the polymer into the final product design; (3) mechanical properties that adequately address short-term function and do not interfere with long-term function; (3) low or negligible toxicity of degradation products, in terms of both local tissue response and systemic response; and (4) drug delivery compatibility in applications that call for release or attachment of active compounds.

BIOLOGICALLY DERIVED BIORESORBABLES

Type I Collagen

Collagen is the major component of mammalian connective tissue, animal protein, accounting for approximately 30% of all protein in the human body. It is found in every major tissue that requires strength and flexibility (e.g., skin, bone). Fourteen types of collagens have been identified, the most abundant being type I (van der Rest *et al.,* 1990). Because of its abundance (it makes up more than 90% of all fibrous proteins) and its unique physical and biological properties, type I collagen has been used extensively in the formulation of biomedical materials (Pachence *et al.,* 1987; Pachence, 1996). Type I collagen is found in high concentrations in tendon, skin, bone, and fascia, which are consequently convenient and abundant sources for isolation of this natural polymer.

The structure, function, and synthesis of type I collagen have been thoroughly investigated (Piez, 1984; Tanzer and Kimura, 1988). Collagen proteins, by definition, are characterized by a unique triple-helix formation extending over a large portion of the molecule. The three peptide subunits that make up the triple helix of collagen have similar amino acid composition, each chain comprising approximately 1050 amino acid residues. The length of each subunit is ~300 nm, and the diameter of the triple helix is ~1.5 nm. The primary structure of collagen (with its high content of proline and hydroxyproline, and with every third amino acid being glycine) shows a strong sequence homology across a genus and the adjacent family line. Because of its phylogenetically well-conserved primary sequence and its helical structure, collagen is only mildly immunoreactive (De Lustro *et al.,* 1987; Anselme *et al.,* 1990). The individual collagen molecules will spontaneously polymerize *in vitro* into strong fibers that can be subsequently formed into larger organized structures (Piez, 1984). The collagen may be further modified to form intro- and intermolecular cross-links, which aid in the formation of collagen fibers, fibrils, and then macroscopic bundles that are used to form tissue (Nimni and Harkness, 1988). For example, tendon and ligaments are composed mainly of oriented type I collagen fibrils, which are extensively cross-linked in the extracellular space. Added strength via *in vivo* cross-linking is imparted onto the collagen fibers by several enzymatic (such as lysyl oxidase) and nonenzymatic reactions; the most extensive cross-linking occurs at the telopeptide portion of the molecule.

Collagen cross-linking can be enhanced after isolation through a number of well-described physical or chemical techniques (Pachence *et al.,* 1987). Increasing the intermolecular crosslinks (1) increases biodegradation time, by making collagen less susceptible to enzymatic degradation; (2) decreases the capacity of collagen to absorb water; (3) decreases its solubility; and (4) increases the tensile strength of collagen fibers. The free ϵ-amines on lysine residues on collagen can be utilized for cross-linking, or can similarly be modified to link or sequester active agents. These simple chemical modifications provide a variety of processing possibilities and consequently the potential for a wide range of tissue engineering applications using type I collagen.

It has long been recognized that substrate attachment sites are necessary for growth, differentiation, replication, and metabolic activity of most cell types in culture. Collagen and its integrin-binding domains (e.g., RGD sequences) assist in the maintenance of such attachment-dependent cell types in culture. For example, fibroblasts grown on collagen matrices appear to differentiate in ways that mimic *in vivo* cellular activity and exhibit nearly identical morphology and metabolism (Silver and Pins, 1992). Chondrocytes can also retain their phenotype and cellular activity when cultured on collagen (Toolan *et al.,* 1996). Such results suggest that type I collagen can serve as tissue regeneration scaffolds for any number of cellular constructs.

The recognition that collagen matrices could support new tissue growth was exploited to develop the original formulations of artificial extracellular matrices for dermal replacements (Yannas and Burke, 1980; Yannas *et al.,* 1980; Burke *et al.,* 1981). Yannas and Burke were the first to show that the rational design and construction of an artificial dermis could lead to the synthesis of a dermislike structure with physical properties that "would resemble dermis more than they resembled scar" (Burke *et al.*, 1981). They created a collagen–chondroitin sulfate composite matrix with a well-described pore structure and cross-linking density that optimizes regrowth while minimizing scar formation (Dagalakis *et al.*, 1980). The reported clinical evidence and its simplicity of concept make this device an important potential tool for the treatment of severely burned patients (Heimbach *et al.*, 1988).

Collagen gels were used by Bell at the Massachusetts Institute of Technology to create a cell-based system for dermal replacement (Bell *et al.*, 1991; Parenteau, 1999). This living skin equivalent (commercially known as Alpligraf) is composed of a mixture of live human fibroblasts and soluble collagen in the form of a contracted gel, which is then seeded with keratinocytes. A number of clinical investigators have tested such cell-based collagen dressings for use as a skin graft substitute for chronic wounds and burn patients.

GLYCOSAMINOGLYCANS

Glycosaminoglycans (GAGs), which consist of repeating disaccharide units in linear arrangement, usually include a uronic acid component (such as glucuronic acid) and a hexosamine component (such as *n*-acetyl-*d*-glucosamine). The predominant types of GAGs attached to naturally occurring core proteins of proteoglycans include chondroitin sulfate, dermatan sulfate, keratan sulfate, and heparan sulfate (Heinegard and Paulsson, 1980; Naeme and Barry, 1993). The GAGs are attached to the core protein by specific carbohydrate sequences containing three or four monosaccharides.

The largest GAG, hyaluronic acid (hyaluronan), is an anionic polysaccharide with repeating disaccharide units of *n*-acetylglucosamine and glucuronic acid, with unbranched units ranging from 500 to several thousand. Hyaluronic acid can be isolated from natural sources (e.g., rooster combs) or via microbial fermentation (Balazs, 1983). Because of its water-binding capacity, dilute solutions of hyaluronic acid are viscous.

Like collagen, hyaluronic acid can be easily chemically modified, as by esterification of the carboxyl moieties, which reduces its water solubility and increases its viscosity (Balazs, 1983; Sung and Topp, 1994). Hyaluronic acid can be cross-linked to form molecular weight complexes in the range 8 to 24×10^6 or to form an infinite molecular network (gels). In one method, hyaluronic acid is cross-linked using aldehydes and small proteins to form bonds between the $C-OH$ groups of the polysaccharide and the amino to imino groups of the protein, thus yielding high-molecular-weight complexes (Balazs and Leshchiner, 1986). Other cross-linking techniques include the use of vinyl sulfone, which reacts to form an infinite network through sulfonyl–bis-ethyl cross-links (Balazs and Leshchiner, 1985). The resultant infinite-network gels can be formed into sheaths, membranes, tubes, sleeves, and particles of various shapes and sizes. No species variations have been found in the chemical and physical structure of hyaluronic acid. The fact that it is not antigenic, eliciting no inflammatory or foreign body reaction, makes it desirable as a biomaterial. Its main drawbacks in this respect are its residence time and the limited range of its mechanical properties.

Because of its relative ease of isolation and modification and its superior ability in forming solid structures, hyaluronic acid has become the preferred GAG in medical device development. It has been used as a viscoelastic during eye surgery since 1976 and has undergone clinical testing as a means of relieving arthritic joints (Weiss and Balazs, 1987). In addition, gels and films made from hyaluronic acid have shown clinical utility to prevent postsurgical adhesion formation (Urmann *et al.*, 1991; Holzman *et al.*, 1994; Medina *et al.*, 1995).

CHITOSAN

Chitosan is a biosynthetic polysaccharide that is the deacylated derivative of chitin. Chitin is a naturally occurring polysaccharide that can be extracted from crustacean exoskeletons or generated via fungal fermentation processes. Chitosan is a β-1,4-linked polymer of 2-amino-2-deoxy-*d*-glucose; it thus carries a positive charge from amine groups (Kaplan *et al.*, 1994). It is hypothesized that the major path for chitin and chitosan breakdown *in vivo* is through lysozyme, which acts slowly to depolymerize the polysaccharide (Taravel and Domard, 1993). The biodegradation rate of the polymer is determined by the amount of residual acetyl content, a parameter that can easily be varied. Chemical modification of chitosan produces materials with a variety of physical and mechanical properties (Muzzarelli *et al.,* 1988; Wang *et al.,* 1988; Laleg and Pikulik, 1991). For example, chitosan films and fibers can be formed utilizing cross-linking chemistries and adapted techniques for altering from other polysaccharides, such as treatment of amylose with epichlorohydrin (Wei *et al.,* 1977). Like hyaluronic acid, chitosan is not antigenic and is a well-tolerated implanted material (Malette *et al.,* 1986).

Chitosan has been formed into membranes and matrices suitable for several tissue engineering applications (Hirano, 1989; Sandford, 1989; Byrom, 1991). Chitosan matrix manipulation can be accomplished using the inherent electrostatic properties of the molecule. At low ionic strength, the chitosan chains are extended via the electrostatic interaction between amine groups, whereupon orientation occurs. As ionic strength is increased, chain–chain spacing is diminished; the consequent increase in the junction zone and stiffness of the matrix result in increased average pore size. Chitosan gels, powders, films, and fibers have been formed and tested for applications such as encapsulation, membrane barriers, contact lens materials, cell culture, and inhibitors of blood coagulations (East *et al.,* 1989).

POLYHYDROXYALKANOATES

Polyhydroxyalkanoate (PHA) polyesters are degradable, biocompatible, thermoplastic materials made by several microorganisms (Miller and Williams, 1987; Gogolewski *et al.,* 1993). They are intracellular storage polymers whose function is to provide a reserve of carbon and energy (Dawes and Senior, 1973). Depending on growth conditions, bacterial strain, and carbon source, the molecular weights of these polyesters can range from tens into the hundreds of thousands. Although the structures of PHAs can contain a variety of *n*-alkyl side chain substituents (see Structure 1), the most extensively studied PHA is the simplest: poly(3-hydroxybutyrate) (PHB).

Imperial Chemical Industries (ICI; London, United Kingdom) developed a biosynthetic process for the manufacture of PHB, based on the fermentation of sugars by the bacterium *Alcaligenes eutrophus.* PHB homopolymers, like all other PHA homopolymers, are highly crystalline, extremely brittle, and relatively hydrophobic. Consequently, the PHA homopolymers have degradation times *in vivo* on the order of years (Holland *et al.,* 1987; Miller and Williams, 1987). The copolymers of PHB with hydroxyvaleric acid are less crystalline, more flexible, and more readily processible, but suffer from the same disadvantage of being too hydrolytically stable to be useful in short-term applications when resorption of the degradable polymer within less than 1 year is desirable.

PHB and it copolymers with up to 30% of 3-hydroxyvaleric acid are now commercially available under the trade name Biopol. It was found previously that a PHA copolymer of 3-hydroxybutyrate and 3-hydroxyvalerate, with a 3-hydroxyvalerate content of about 11%, may have an optimum balance of strength and toughness for a wide range of possible applications. PHB has been found to have low toxicity, in part due to the fact that it degrades *in vivo* to *d*-3-hydroxybutyric acid, a normal constituent of human blood. Applications of these polymers previously tested or now under development include controlled drug release, sutures, and artificial skin as well as industrial applications such as paramedical disposables (Yasin *et al.,* 1989; Doi *et al.,* 1990).

EXPERIMENTAL BIOLOGICALLY DERIVED BIORESORBABLES

Synthetic biomolecules are beginning to find a place in the repertoire of biomaterials for medical applications. Model synthetic proteins structurally similar to elastin have been formulated by Urry and co-workers (Nicol *et al.,* 1992; Urry, 1995). Using a combination of solid-phase peptide chemistry and genetically engineered bacteria, they synthesized several polymers having homologies to the elastin repeat sequences of valine-proline-glycine-valine-glycine repeat (VPGVG). The

Structure 1. Poly(β-hyroxybu-tyrate) and copolymers with hy-droxyvaleric acid. For a ho-mopolymer of HB, Y = 0; commonly used copolymer ra-tios are 7, 11, and 22 mol% of hydroxyvaleric acid.

hydroxybutyric acid (HB)   hydroxyvaleric acid (HV)

constructed amino acid polymers were formed into films and then cross-linked. The resultant films have intriguing mechanical responses, such as a reverse-phase transition: when a film is heated, its internal order increases, translating into substantial contraction with increasing temperature (Urry, 1995). The films can be mechanically cycled many times, and the phase transition of the polymers can be varied by amino acid substitution. Copolymers of VPGVG and VPGXG have been constructed (where X is the substitution) that have been shown to have a wide range of transition temperatures (Urry, 1995). Several medical applications are under consideration for this system, including musculoskeletal repair mechanisms, ophthalmic devices, and mechanical and/or electrically stimulated drug delivery.

Other investigators, notably Tirrell and Cappello, have combined techniques from molecular and fermentation biology to create novel protein-based-biomaterials (Cappello, 1992; Anderson *et al.,* 1994; Tirrell *et al.,* 1994). These protein polymers are based on repeat oligomeric peptide units, which can be controlled via the genetic information inserted into the producing bacteria. It has been shown that the mechanical properties and the biological activities of these protein polymers can be programmed, suggesting a large number of potential biomedical applications (Krejchi *et al.,* 1994).

Another approach to elicit an appropriate cellular response to a biomaterial is to graft active peptides to the surface of a biodegradable polymer. For example, peptides containing the RGD sequence have been grafted to various biodegradable polymers to provide active cell-binding surfaces (Hubbell, 1995).

SYNTHETIC POLYMERS

From the beginnings of the study of materials sciences, the development of highly stable materials has been a major research challenge. Today, many polymers are available that are virtually nondestructible in biological systems, e.g., Teflon, Kevlar, or poly(ether-ether-ketone). On the other hand, the development of degradable biomaterials is a relatively new area of research. The variety of available, degradable biomaterials is still too limited to cover a wide enough range of diverse material properties. Thus, the design and synthesis of new, degradable biomaterials is currently an important research challenge.

Due to the efforts of a wide range of research groups, a large number of different polymeric compositions and structures have been suggested as degradable biomaterials. However, in most cases no attempts have been made to develop these new materials for specific medical applications. Thus, detailed toxicological studies *in vivo,* investigations of degradation rate and mechanism, and careful evaluations of the physicomechanical properties have so far been published for only a very small fraction of those polymers. This leaves the tissue engineer with only a relatively limited number of promising polymeric compositions to choose from. The following section is limited to a review of the most commonly investigated classes of biodegradable, synthetic polymers.

Poly(α-hydroxy acids)

Naturally occurring hydroxy acids, such as glycolic, lactic, and ε-caproic acids, have been utilized to synthesize an array of useful biodegradable polymers for a variety of medical product applications. As an example, bioresorbable surgical sutures made from poly(α-hydroxy acids) have been in clinical use since 1970; other implantable devices made from these versatile polymers (e.g., internal fixation devices for orthopedic repair) are becoming part of standard surgical protocol (Helmus and Hubbell, 1993; Shalaby and Johnson, 1994; Hubbell, 1995).

The ester bonds of the poly(hydroxy acids) are cleaved by hydrolysis, which results in a decrease in the polymer molecular weight (but not mass) of the implant (Vert and Li, 1992). This initial degradation occurs until the molecular weight is less than 5000, at which point cellular degradation takes over. The final degradation and resorption of the poly(hydroxy acid) implants involve inflammatory cells (such as macrophages, lymphocytes, and neutrophils). Although this late-stage inflammatory response can have a deleterious effect on some healing events, these polymers have been successfully employed as matrices for cell transplantation and tissue regeneration (Freed *et al.,* 1994a,b). The degradation rate of these polymers is determined by initial molecular weight, exposed surface area, crystallinity, and (in the case of copolymers) ratio of the hydroxy acid monomers.

The poly(hydroxy acid) polymers have a modest range of mechanical properties and a correspondingly modest range of processing conditions. Nevertheless, these thermoplastics can generally be formed into films, tubes, and matrices using such standard processing techniques as molding, extrusion, solvent casting, and spin casting. Ordered fibers, meshes, and open-cell foams have been formed to fulfill the surface area and cellular requirements of a variety of tissue engineering constructs (Helmus and Hubbell, 1993; Freed *et al.,* 1994b; Hubbell, 1995; Wintermantel *et al.,* 1996). The poly(hydroxy acid) polymers have also been combined with other materials, e.g., poly(ethylene glycol), to modify the cellular response elicited by the implant and its degradation products (Sawhney *et al.,* 1993).

Poly(glycolic acid), poly(lactic acid), and their copolymers

Poly(glycolic acid) (PGA), poly(lactic acid) (PLA), and their copolymers are the most widely used synthetic degradable polymers in medicine. Of this family of linear aliphatic polyesters, PGA has the simplest structure (see Structure 2) and consequently enjoys the largest associated literature base. Because PGA is highly crystalline, it has a high melting point and low solubility in organic solvents. PGA was used in the development of the first totally synthetic absorbable suture (Frazza and Schmitt, 1971). The crystallinity of PGA in surgical sutures is typically in the range 46–52% (Gilding and Reed, 1979). Due to its hydrophilic nature, surgical sutures made of PGA tend to lose their mechanical strength rapidly, typically over a period of 2 to 4 weeks postimplantation (Reed and Gilding, 1981).

In order to adapt the materials properties of PGA to a wider range of possible applications, researchers undertook an intensive investigation of copolymers of PGA with the more hydrophobic poly(lactic acid). Alternative sutures composed of copolymers of glycolic acid and lactic acid are currently marketed under the trade names Vicryl and Polyglactin 910. Due to the presence of an extra methyl group in lactic acid, PLA (Structure 3) is more hydrophobic than PGA. The hydrophobicity of high-molecular-weight PLA limits the water uptake of thin films to about 2% (Gilding and Reed, 1979) and results in a rate of backbone hydrolysis lower than that of PGA (Reed and Gilding, 1981). In addition, PLA is more soluble in organic solvents than is PGA.

It is noteworthy that there is no linear relationship between the ratio of glycolic acid to lactic acid and the physicomechanical properties of their copolymers. Whereas PGA is highly crystalline, crystallinity is rapidly lost in PGA–PLA copolymers. These morphological changes lead to an increase in the rates of hydration and hydrolysis. Thus, copolymers tend to degrade more rapidly compared with either PGA or PLA (Gilding and Reed, 1979; Reed and Gilding, 1981).

Because lactic acid is a chiral molecule, it exists in two stereoisomeric forms that give rise to four morphologically distinct polymers. *d*-PLA and *l*-PLA are the two stereoregular polymers, *d,l*-PLA is the racemic polymer obtained from a mixture of *d*- and *l*-lactic acid, and meso-PLA can be obtained from *d,l*-lactide. The polymers derived from the optically active *d* and *l* monomers are semicrystalline materials, whereas the optically inactive *d,l*-PLA is always amorphous. Generally, *l*-PLA is more frequently employed than *d*-PLA, because the hydrolysis of *l*-PLA yields *l*(+)-lactic acid, which is the naturally occurring stereoisomer of lactic acid.

The differences in the crystallinity of *d,l*-PLA and *l*-PLA have important practical ramifications: because *d,l*-PLA is an amorphous polymer, it is usually considered for applications such as drug delivery, where it is important to have a homogeneous dispersion of the active species within a monophasic matrix. On the other hand, the semicrystalline *l*-PLA is preferred in applications where high mechanical strength and toughness are required—for example, sutures and orthopedic devices (Christel *et al.,* 1982; Leenstag *et al.,* 1987; Vainionpää *et al.,* 1987. PLA, PGA, and their copolymers are also being intensively investigated for a large number of drug delivery applications. This research effort has been comprehensively reviewed by Lewis (1990).

Structure 2. The linear aliphatic polyester with the simplest structure.

Poly(glycolic acid)

(PGA)

$$\left[O - \overset{\overset{\displaystyle CH_3}{|}}{CH} - \overset{\overset{\displaystyle O}{||}}{C} \right]_n$$

Poly(lactic acid)

(PLA)   *Structure 3. The extra methyl group confers hydrophobicity.*

Some controversy surrounds the use of these materials for orthopedic applications. According to one review of short- and long-term response to resorbable pins made from either PGA or PGA:PLA copolymer, in over 500 patients, 1.2% required reoperation due to device failure, 1.7% suffered from bacterial infection of the operative wound, and 7.9% developed a late noninfectious inflammatory response that warranted operative drainage (Böstman, 1991). Subsequently it has become evident that the delayed inflammatory reaction represents the most serious complication of the use of the currently available degradable fixation devices. The mean interval between fixation and the clinical manifestation of this reaction is 12 weeks for PGA and can be as long as 3 years for the more slowly degrading PLA (Böstman, 1991). Whether avoiding reoperation to remove a metal implant outweighs an approximately 8% risk of severe inflammatory reaction is a difficult question; in any event, an increasing number of trauma centers have suspended the use of these degradable fixation devices. It has been suggested that the release of acidic degradation products (glycolic acid for PGA, lactic acid for PLA, and glyoxylic acid for polydioxanone) contributes to the observed inflammatory reaction. Thus, the late inflammatory response appears to be a direct consequence of the chemical composition of the polymer degradation products, for which there is currently no prophylactic measure (Böstman, 1991). A solution to these problems for orthopedic (and perhaps other) applications requires the development of a polymer that is more hydrophobic than PGA or PLA, degrades somewhat more slowly, and does not release acidic degradation products on hydrolysis.

Polydioxanone

The poly(ether-ester) polydioxanone (PDS) is prepared by a ring-opening polymerization of *p*-dioxanone. PDS has gained increasing interest in the medical field and pharmaceutical field due to its degradation to low-toxicity monomers *in vivo*. PDS has a lower modulus than does PLA or PGA, thus it became the first degradable polymer to be used to make a monofilament suture. PDS has also been introduced to the market as a suture clip as well as a bone pin marketed under the name ORTHOSRB in the United States and Ethipin in Europe (Ray *et al.*, 1981; Greisler *et al.*, 1987; Mäkelä *et al.*, 1989).

Poly(ε-caprolactone)

Poly(ε-caprolactone) (PCL) (Structure 4) is an aliphatic polyester that has been intensively investigated as a biomaterial (Pitt, 1990). The discovery that PCL can be degraded by microorganisms led to evaluation of PCL as a biodegradable packaging material (Pitt, 1990). Later, it was discovered that PCL can also be degraded by a hydrolytic mechanism under physiologic conditions (Pitt *et al.*, 1981a). Under certain circumstances, cross-linked PCL, can be degraded enzymatically, leading to "enzymatic surface erosion" (Pitt *et al.*, 1981b). Low-molecular-weight fragments of PCL are reportedly taken up by macrophages and degraded intracellularly, with a tissue reaction similar to that to the other poly(hydroxy acids) (Pitt *et al.*, 1984). Compared with PGA

$$\left[\overset{\overset{\displaystyle O}{||}}{C} - (CH_2)_5 - O \right]_n$$

Poly(ε-caprolactone)   *Structure 4. An aliphatic polyester that can be degraded hydrolytically under physiologic conditions.*

or PLA, the degradation of PCL is significantly slower. PCL is therefore most suitable for the design of long-term, implantable systems such as Capronor, a 1-year implantable contraceptive device (Pitt, 1990).

Poly(ε-caprolactone) exhibits several unusual properties not found among the other aliphatic polyesters. Most noteworthy are its exceptionally low glass transition temperature of $-62°C$ and its low melting temperature of 57°C. Another unusual property of poly(ε-caprolactone) is its high thermal stability. Whereas other tested aliphatic polyesters had decomposition temperatures (T_d) between 235 and 255°C, poly(ε-caprolactone) has a T_d of 350°C, which is more typical of poly(ortho esters) than aliphatic polyesters (Engelberg and Kohn, 1991). PCL is a semicrystalline polymer with a low glass transition temperature of about $-60°C$. Thus, PCL is always in a rubbery state at room temperature. Among the more common aliphatic polyesters, this is an unusual property, which undoubtedly contributes to the very high permeability of PCL for many therapeutic drugs (Pitt *et al.,* 1987).

Another interesting property of PCL is its propensity to form compatible blends with a wide range of other polymers (Koleske, 1978). In addition, ε-caprolactone can be copolymerized with numerous other monomers (e.g., ethylene oxide, chloroprene, tetrahydrofuran, *d*-valerolactone, 4-vinylanisole, styrene, methyl methacrylate, vinylacetate). Particularly noteworthy are copolymers of ε-caprolactone and lactic acid that have been studied extensively (Pitt *et al.,* 1981a; Feng *et al.,* 1983). The toxicology of PCL has been extensively studied as part of the evaluation of Capronor. Based on a large number of tests, the monomer, ε-caprolactone, and the polymer, PCL, are currently regarded as nontoxic and tissue-compatible materials. Consequently, the Capronor system has undergone Food and Drug Administration (FDA)-approved phase I and phase II clinical trials (Pitt, 1990).

It is interesting to note that in spite of its versatility, PCL has so far been predominantly considered for controlled-release drug delivery applications. In Europe, PCL is being used as a biodegradable staple, and it stands to reason that PCL (or blends and copolymers with PCL) will find additional medical applications in the future. The most comprehensive review of the status of PCL has been by Pitt (1990).

Poly(ortho esters)

Poly(ortho esters) are a family of synthetic degradable polymers that have been under development for several years (Heller *et al.,* 1990). Devices made of poly(ortho esters) can be formulated in such a way that the device undergoes "surface erosion"—that is, the polymeric device degrades at its surface only and will thus tend to become thinner over time rather than crumbling into pieces. Because surface-eroding, slablike devices tend to release drugs embedded within the polymer at a constant rate, poly(ortho esters) appear to be particularly useful for controlled-release drug delivery (Heller, 1988); this interest is reflected by the many descriptions of these applications in the literature (Heller and Daniels, 1994).

There are two major types of poly(ortho esters). Originally, poly(ortho esters) were prepared by the condensation of 2,2-diethoxytetrahydrofuran and a dialcohol (Cho and Heller, 1978) and marketed under the trade names Chronomer and Alzamer. On hydrolysis, these polymers release acidic by-products that autocatalyze the degradation process, resulting in degradation rates that increase with time. Later, Heller *et al.* (1980) synthesized a new type of poly(ortho ester) based on the reaction of 3,9-bis(ethylidene-2,4,8,10-tetraoxaspiro{5,5}undecane) (DETOSU) (see Structure 5) with various dialcohols. These poly(ortho esters) do not release acidic by-products on hydrolysis and thus do not exhibit autocatalytically increasing degradation rates.

Structure 5. Poly(ortho esters). The specific composition shown here is a terpolymer of hexadecanol (1,6-HD), trans-cyclohexyldimethanol (t-CDM), and DETOSU.

1,6-HD   DETOSU   t-CDM

sebacic acid (SA)   hexadecandioic acid (HDA)

Structure 6. Poly(SA–HDA anhydride). This composition represents one of many polyanhydrides that have been explored. The clinically relevant polyanhydrides are copolymers of sebacic acid and p-carboxyphenoxy propane.

Polyanhydrides

Polyanhydrides were first investigated in detail by Hill and Carothers (1932) and were considered in the 1950s for possible applications as textile fibers (Conix, 1958). Their low hydrolytic stability, their major limitation for industrial applications, was later recognized as a potential advantage by Langer *et al.* (see Rosen *et al.,* 1983), who suggested the use of polyanhydrides as degradable biomaterials. A study of the synthesis of high-molecular-weight polyanhydrides has been published by Domb *et al.* (1989, 1994).

A comprehensive evaluation of the toxicity of the polyanhydrides showed that, in general, the polyanhydrides possessed excellent *in vivo* biocompatibility (Laurencin *et al.,* 1990). The most immediate applications for polyanhydrides are in the field of drug delivery. Drug-loaded devices are best prepared by compression molding or microencapsulation (Mathiowitz *et al.,* 1988). A wide variety of drugs and proteins, including insulin, bovine growth factors, angiogenesis inhibitors (e.g., heparin and cortisone), enzymes (e.g., alkaline phosphatase and β-galactosidase), and anesthetics, have been incorporated into polyanhydride matrices and their *in vitro* and *in vivo* release characteristics have been evaluated (Chasin *et al.,* 1990). One of the most aggressively investigated uses of the polyanhydrides is for the delivery of chemotherapeutic agents. One particular example of this application is the delivery of bis-chloroethylnitrosourea (BCNU) to the brain for the treatment of glioblastoma multiformae, a universally fatal brain cancer (Langer, 1990). For this application, polyanhydrides derived from bis-(*p*-carboxyphenoxy propane) and sebacic acid (see Structure 6) received FDA regulatory clearance in the fall of 1996 and are currently being marketed under the name Gliadel.

Polyphosphazenes

Polyphosphazenes consist of an inorganic phosphorous–nitrogen backbone (see Structure 7), in contrast to the commonly employed hydrocarbon-based polymers (Scopelianos, 1994). Consequently, the phosphazene backbone undergoes hydrolysis to phosphate and ammonium salts, with the concomitant release of the side group. Of the numerous polyphosphazenes that have been synthesized, those that have some potential use for medical products are substituted with amines of low pK_a, and those with activated alcohol moieties (Allcock, 1990; Crommen *et al.,* 1992; Laurencin *et al.,* 1993). The most extensively studied polyphosphazenes are hydrophobic, having fluoroalkoxy side groups. In part, these materials are of interest because of their expected minimal tissue interaction, which is similar to Teflon.

Aryloxyphosphazenes and closely related derivatives have also been extensively studied. One such polymer can be cross-linked with dissolved cations such as calcium to form a hydrogel matrix because of its polyelectrolytic nature (Allcock and Kwon, 1989). Using methods similar to alginate encapsulation, microspheres of aryloxyphosphazene have been used to encapsulate hybridoma cells without affecting their viability or their capacity to produce antibodies. Interaction

Structure 7. Polyphosphazene. Shown here is a polymer containing an amino acid ester attached to the phosphazene backbone.

HO—⬡—CH₂—CH₂—C(=O)—OH   NH₂—CH—CH₂—⬡—OH
                                                    |
                                                   C=O
                                                    |
                                                    O
                                                    |
                                                    R

desaminotyrosine (Dat)   tyrosine alkyl ester (Tyr-OR)

Structure 8. A poly(amide carbonate) derived from desaminotyrosyl tyrosine alkyl esters. This is an example for a group of new, amino acid-derived polymers.

[—⬡—CH₂—CH₂—C(=O)—NH—CH—CH₂—⬡—O—C(=O)—O—]ₙ
                                          |
                                         C=O
                                          |
                                          O
                                          |
                                          R

with poly(L-lysine) produced a semipermeable membrane. Similar materials have been synthesized that show promise in blood contacting and in novel drug delivery applications.

Poly(amino acids) and pseudo-poly(amino acids)

Because proteins are composed of amino acids, many researchers have tried to develop synthetic polymers derived from amino acids to serve as models for structural, biologic, and immunologic studies (see Structure 8). In addition, many different types of poly(amino acids) have been investigated for use in biomedical applications (Anderson *et al.,* 1985). Poly(amino acids) are usually prepared by the ring-opening polymerization of the corresponding *N*-carboxy anhydrides that are obtained by reaction of the amino acid with phosgene (Bamford *et al.,* 1956).

Poly(amino acids) have several potential advantages as biomaterials. A large number of polymers and copolymers can be prepared from a variety of amino acids. The side chains offer sites for the attachment of small peptides, drugs, cross-linking agents, or pendant groups that can be used to modify the physicomechanical properties of the polymer. Because these polymers release naturally occurring amino acids as the primary products of polymer backbone cleavage, their degradation products may be expected to show a low level of systemic toxicity.

Poly(amino acids) have been investigated as suture materials (Spira *et al.,* 1969), as artificial skin substitutes (Aiba *et al.,* 1985), and as drug delivery systems (Mitra *et al.,* 1979; McCormick-Thomson and Duncan, 1989). Various drugs have been attached to the side chains of poly(amino acids), usually via a spacer unit that distances the drug from the backbone. Poly(amino acid)–drug combinations investigated include poly(L-lysine) with methotrexate and pepstatin (Campbell *et al.,* 1980), and poly(glutamic acid) with adriamycin and norethindrone (van Heeswijk *et al.,* 1985).

Despite their apparent potential as biomaterials, poly(amino acids) have actually found few practical applications. The *N*-carboxy anhydrides, the starting materials, are expensive to make and difficult to handle because of their high reactivity and moisture sensitivity. Most poly(amino acids) are highly insoluble and nonprocessible materials. Because poly(amino acids) degrade via enzymatic hydrolysis of the amide bond, it is difficult to reproduce and control their degradation *in vivo* because the level of enzymatic activity varies from person to person. Furthermore, the antigenicity of polymers containing three or more amino acids excludes their use in biomedical applications (Anderson *et al.,* 1985). Because of these difficulties, only a few poly(amino acids), usually derivatives of poly(glutamic acid) carrying various pendant chains at the γ-carboxylic acid group, are currently being investigated as implant materials (Lescure *et al.,* 1989).

As an alternative approach, James and Kohn (1997) and Kemnitzer and Kohn (1997) have replaced the peptide bonds in the backbone of synthetic poly(amino acids) by a variety of such "nonamide" linkages as ester, iminocarbonate, urethane, and carbonate bonds. The term "pseudo-poly(amino acid)" is used to denote this new family of polymers in which naturally occurring amino acids are linked together by nonamide bonds.

The use of such backbone-modified pseudo-poly(amino acids) as biomaterials was first suggested in 1984 (Kohn and Langer, 1984). The first pseudo-poly(amino acids) investigated were a polyester from *N*-protected *trans*-4-hydroxy-L-proline and a poly(iminocarbonate) from tyrosine dipeptide (Kohn and Langer, 1985, 1987). Several studies indicate that the backbone modification of conventional poly(amino acids) in general improves their physicomechanical properties (Ertel and Kohn, 1994; Fiordeliso *et al.*, 1994; James and Kohn, 1997). This approach is applicable to, among other materials, serine, hydroxyproline, threonine, tyrosine, cysteine, glutamic acid, and lysine; it is only limited by the requirement that the nonamide backbone linkages give rise to polymers with desirable material properties. Additional pseudo-poly(amino acids) can be obtained by considering dipeptides as monomeric starting materials. Hydroxyproline-derived polyesters (Kohn and Langer, 1987; Yu *et al.*, 1987; Yu-Kwon and Langer, 1989), serine-derived polyesters (Zhou and Kohn, 1990), and tyrosine-derived polyiminocarbonates (Pulapura *et al.*, 1990) and polycarbonates (Pulapura and Kohn, 1992) represent specific embodiments of these synthetic concepts.

CREATING MATERIALS FOR TISSUE-ENGINEERED PRODUCTS

As described in this chapter, the bulk polymer properties and the cellular response to biomaterials are important selection criteria for the design of a tissue-engineered product. In addition, the ability to mold the biomaterial into the appropriate cellular-level architecture must be considered, and such architecture must be compatible with the desired tissue response. Hubbell classified approaches to choosing biomaterials for various tissue engineering applications according to type of tissue response sought: (1) conducting tissue responses and architectures, (2) inducing tissue responses and architectures, and (3) blocking tissue responses (Hubbell, 1995; Wintermantel *et al.*, 1996). Implicit in this consideration is that the specific material architecture (e.g., membrane, gel, matrix, tube) is critical to the tissue engineering product design and thus influences the choice of biomaterials.

BARRIERS: MEMBRANES AND TUBES

Design formats requiring cell activity on one surface of a device while precluding transverse movement of surrounding cells onto that surface call for a barrier material. For example, peripheral nerve regeneration must allow for axonal growth and at the same time preclude fibroblast activity that could produce neural-inhibiting connective tissue. Structures such as collagen tubes can be fabricated to yield a structure dense enough to inhibit connective tissue formation along the path of repair while allowing axonal growth through the lumen (Li *et al.*, 1992). Similarly, collagen membranes for periodontal repair provide an environment for periodontal ligament regrowth and attachment while preventing epithelial ingrowth into the healing site (van Swol *et al.*, 1993). Antiadhesion formulations using hyaluronic acid, which prevent ingrowth of connective tissue at a surgically repaired site, also work on this concept (Urmann *et al.*, 1991).

GELS

Gels are used to provide a hydrogel scaffold, to encapsulate, or to provide a specialized environment for isolated cells. For example, collagen gels for tissue engineering were first used to maintain fibroblasts, which were the basis of a living skin equivalent (Parenteau, 1999). Gels have also been used for the maintenance and immunoprotection of xenograft and homograft cells such as hepatocytes, chondrocytes, and islets of Langerhans used for transplantation (Sullivan *et al.*, 1991; Chang, 1992; Lacy, 1995). Semipermeable gels have been created to limit cell–cell communication and interaction with surrounding tissue, and to minimize movement of peptide factors and nutrients through the implant. In general, nondegradable materials are used for cell encapsulation to maximize long-term stability of the implant. In the future, however, it may be possible to formulate novel "smart" gels in which biodegradation is triggered by a specific cellular response instead of simple hydrolysis.

MATRICES

It has been recognized since the mid-1970s that three-dimensional structures are an important component of engineered tissue development (Yannas and Burke, 1980; Yannas *et al.*, 1980). Yannas and co-workers were the first to show that pore size, pore orientation, and fiber structure are important characteristics in the design of cell scaffolds. Several techniques have subsequently been developed to form well-defined matrices from synthetic and biologically derived polymers, and the physical characteristics of these matrices are routinely varied to maximize cellular and tissue responses (Fenkel *et al.,* 1997; Langer and Vacanti, 1993). These engineered matrices have led to several resorbable templates for tissue regeneration.

CONCLUSION

Research in the use of currently available biomaterials and in developing novel bioresorbable polymers has helped to drive the establishment of the field of tissue engineering. Despite a wide range of possible choices, there is a tendency to choose those bioresorbable polymers that have a history of regulatory approval instead of letting the application guide the choice of material. The latter approach, moreover, may require lengthy and costly polymer development work. Nonetheless, in order to gain the sort of precise control over cell response and cell interactions with surrounding tissues that is expected of tissue engineering applications, continued research on new bioresorbable polymers will be necessary.

REFERENCES

Aiba, S., Minoura, N., Fujiwara, Y., Yamada, S., and Nakagawa, T. (1985). Laminates composed of polypeptides and elastomers as a burn wound covering. Physicochemical properties. *Biomaterials* **6,** 290–296.

Allcock, H. R. (1990). Polyphosphazenes as new biomedical and bioactive materials. *In* "Biodegradable Polymers as Drug Delivery Systems" (M. Chasin and R. Langer, eds.), pp. 163–193. Dekker, New York.

Allcock, H. R., and Kwon, S. (1989). An ionically-crosslinkable polyphosphazene: Poly[bis(carboxylatophenoxy)phosphazene] and its hydrogels and membranes. *Macromolecules* **22,** 75–79.

Anderson, J. M., Spilizewski, K. L., and Hiltner, A. (1985). Poly-α-amino acids as biomedical polymers. *In* "Biocompatibility of Tissue Analogs" (D. F. Williams, ed.), Vol. 1, pp. 67–88. CRC Press, Boca Raton, FL.

Anderson, J. P., Cappello, J., and Martin, D. C. (1994). Morphology and primary crystal structure of a silk-like protein polymer synthesized by genetically engineered *E. coli* bacteria. *Biopolymers* **34,** 1049–1058.

Anselme, K., Bacques, C., Charriere, G., Hartmann, D. J., Herbage, D., and Garrone, R. (1990). Tissue reaction to subcutaneous implantation of a collagen sponge. *J. Biomed. Mater. Res.* **24,** 689–703.

Balazs, E. A. (1983). Sodium hyaluronate and viscosurgery. *In* "Healon (Sodium Hyaluronate): A Guide to Its Use in Ophthalmic Surgery" (D. Miller and R. Stegmann, eds.), pp. 5–28. Wiley, New York.

Balazs, E. A., and Leshchiner, A. (1985). Hyaluronate modified polymeric articles. U.S. Pat. 4,500,676.

Balazs, E. A., and Leshchiner, A. (1986). Cross-linked gels of hyaluronic acid and products containing such gels. U.S. Pat. 4,582,865.

Bamford, C. H., Elliot, A., and Hanby, W. E. (1956). Synthetic polypeptides—Preparation, structure and properties. *In* "Physical Chemistry: A Series of Monographs" (E. Hutchinson, ed.), Vol. 5. Academic Press, New York.

Bell, E., Parenteau, R., Gay, R., Rosenberg, M., Kemp, P., Green, G. D., Muthukumaran, N., and Nolte, C. (1991). The living skin equivalent: Its manufacture, its organotypic properties, and its responses to irritants. *Toxicol. In Vitro* **5,** 591–596.

Böstman, O. M. (1991). Absorbable implants for the fixation of fractures. *J. Bone J. Surg.* **73**(1), 148–153.

Burke, J. F., Yannas, I. V., Quinby, W. C., Jr., Bondoc, C. C., and Jung, W. K. (1981). Successful use of a physiologically acceptable artificial skin in the treatment of extensive burn injury. *Ann. Surg.* **194,** 413.

Byrom, D. (1991). Chitosan and chitosan derivatives. *In* "Biomaterials: Novel Materials from Biological Sources" (D. Byron, ed.), pp. 333–359. Stockton Press, New York.

Campbell, P., Glover, G. I., and Gunn, J. M. (1980). Inhibition of intracellular protein degradation by pepstatin, poly(L-lysine) and pepstatinyl-poly(L-lysine). *Arch. Biochem. Biophys.* **203,** 676–680.

Cappello, J. (1992). Genetic production of synthetic protein polymers. *MRS Bull.* **17,** 48–53.

Chang, T.M.S. (1992). Artificial liver support based on artificial cells with emphasis on encapsulated hepatocytes. *Artif. Organs* **16** 71–74.

Chasin, M., Domb, A., Ron, E., Mathiowitz, E., Langer, R., Leone, K., Laurencin, C., Brem, H., and Grossman (1990). Polyanhydrides as drug delivery systems. *In* "Biodegradable Polymers as Drug Delivery Systems" (C. G. Pitt, ed.), pp. 43–69. Dekker, New York.

Cho, N. J., and Heller, J. (1978). Drug delivery devices manufactured from polyorthoesters and polyorthocarbonates. U.S. Pat. 4,078,038.

Christel, P., Chabot, F., Leray, J. L., Morin, C., and Vert, M. (1982). Biodegradable composites for internal fixation. *In* "Biomaterials" (G. O. Winter, D. F. Gibbons, and H. Pienkj, eds.), pp. 271–280. Wiley, New York.

Conix, A. (1958). Aromatic polyanhydrides, a new class of high melting fibre-forming polymers. *J. Polym. Sci.* **29,** 343–353.

Crommen, J.H.L., Schacht, E. H., and Mense, E.H.G. (1992). Biodegradable polymers. II. Degradation characteristics of hydrolysis-sensitive poly[(organo)phosphazenes]. *Biomaterials* **13**(9), 601–611.

Dagalakis, N., Flink, J., Stasikelis, P., Burke, J. F., and Yannas, I. V. (1980). Design of an artificial skin: Control of pore structure. *J. Biomed. Mater. Res.* **14**, 511–528.

Dawes, E. A., and Senior, P. J. (1973). The role and regulation of energy reserve polymers in microorganisms. *Adv. Microb. Physiol.* **10** 135–266.

De Lustro, F., Condell, R. A., Nguyen, M., and McPherson, J. (1987). A comparative study of the biologic and immunologic response to medical devices derived from dermal collagen. *J. Biomed. Mater. Res.* **20**109–120.

Doi, Y., Kanesawa, Y., Kunioka, M., and Saito, T. (1990). Biodegradation of microbial copolyesters: Poly(3-hydroxybutyrate-co-3-hydroxyvalerate) and poly(3-hydroxybutyrate-co-4-hydroxyvalerate). *Macromolecules* **23**, 26–31.

Domb, A. J., Gallardo, C. F., and Langer, R. (1989). Poly(anhydrides). 3. Poly(anhydrides) based on aliphatic–aromatic diacids. *Macromolecules* **22**, 3200–3204.

Domb, A. J., Amselem, S., Langer, R., and Maniar, M. (1994). Polyanhydrides as carriers of drugs. *In* "Biomedical Polymers" (S. Shalaby, ed.), pp. 69–96. Hanser Publishers, Munich and New York.

East, G. C., McIntyre, J. E., and Qin, Y. (1989). Medical use of chitosan. *In* "Chitin and Chitosan" (G. Skjak-Braek, T. Anthonsen, and P. Sandford, eds.), pp. 757–764. Elsevier, London.

Engelberg, I., and Kohn, J. (1991). Physico-mechanical properties of degradable polymers used in medical applications: A comparative study. *Biomaterials* **12**(3), 292–304.

Ertel, S. I., and Kohn, J. (1994). Evaluation of a series of tyrosine-derived polycarbonates as degradable biomaterials. *J. Biomed. Mater. Res.* **28**, 919–930.

Feng, X. D., Song, C. X., and Chen, W. Y. (1983). Synthesis and evaluation of biodegradable block copolymers of ∈-caprolactone and *d,l*-lactide. *J. Polym. Sci., Polym. Lett. Ed.* **21**, 593–600.

Fenkel, S. R., Toolan, B., Menche, D., Pitman, M. I., and Pachence, J. M. (1997). Chondrocyte transplantation using a collagen bilayer matrix for cartilage repair. *J. Bone J. Surg., Br. Vol.* **79**, 831–836.

Fiordeliso, J., Bron, S., and Kohn, J. (1994). Design, synthesis, and preliminary characterization of tyrosine-containing polyarylates: New biomaterials for medical applications. *J. Biomater. Sci., Polym. Ed.* **5**, 497–510.

Frazza, E. J., and Schmitt, E. E. (1971). A new absorbable suture. *J. Biomed. Mater. Res. Symp.* **1**, 43–58.

Freed, L. E., Grande, D. A., Lingbin, Z., Emmanual, J., Marquis, J. C., and Langer, R. (1994a). Joint resurfacing using allograft chondrocytes and synthetic biodegradable polymer scaffolds. *J. Biomed. Mater. Res.* **28**, 891–899.

Freed, L. E., Vunjak-Novakovic, G., Biron, R. J., Eagles, D. B., Lesnoy, D. C., Barlow, S. K., and Langer, R. (1994b). Biodegradable polymer scaffolds for tissue engineering. *Bio/Technology* **12**, 689–693.

Gilding, D. K., and Reed, A. M. (1979). Biodegradable polymers for use in surgery—Poly(glycolic)/poly(lactic acid) homo and copolymers. *Polymer* **20**, 1459–1464.

Gogolewski, S., Jovanovic, M., Perren, S. M., Dillo, J. G., and Hughers, M. K. (1993). Tissue response and *in vivo* degradation of selected polyhydroxyacids: Polylactides (PLA), poly(3-hydroxybutyrate) (PHB), and poly(3-hydroxybutyrate-co-3-hydroxyvalerate) (PHB/VA). *J. Biomed. Mater. Res.* **27**, 1135–1148.

Greisler, H. P., Ellinger, J., Schwarcz, T. H., Golan, J., Raymond, R. M., and Kim, D. U. (1987). Arterial regeneration over polydioxanone prostheses in the rabbit. *Arch. Surg. (Chicago)* **122**, 715–721.

Heimbach, D., Luterman, A., Burke, J., Cram, A., Herndon, D., Hunt, J., and Jordan, M. (1988). Artificial dermis for major burns. *Ann. Surg.* **208**, 313–320.

Heinegard, D., and Paulsson, M. (1980). Proteoglycans and matrix proteins in cartilage. *In* "The Biochemistry of Glycoproteins and Proteoglycans" (W. J. Lennarz, ed.), pp. 297–328. Plenum, New York.

Heller, J. (1988). Synthesis and use of poly(ortho esters) for the controlled delivery of therapeutic agents. *J. Bioact. Compat. Polym.* **3**,(2), 97–105.

Heller, J., and Daniels, A. U. (1994). Poly(ortho esters). *In* "Biomedical Polymers" (S. Shalaby, ed.), pp. 35–67. Hanser Publishers, Munich and New York.

Heller, J., Penhale, D.W.H., and Helwing, R. F. (1980). Preparation of poly(ortho esters) by the reaction of diketen acetals and polyols. *J. Polym. Sci., Polym, Lett, Ed.* **18**, 619–624.

Heller, J., Sparer, R. V., and Zentner, G. M. (1990). Poly(ortho esters) for the controlled delivery of therapeutic agents. *J. Bioact. Compat. Polym.* **3**(2), 97–105.

Helmus, M. N., and Hubbell, J. A. (1993). Materials selection. *Cardiovasc. Pathol.* **2**(3), 53S–71S.

Hill, J. W., and Carothers, W. H. (1932). Studies of polymerizations and ring formations (XIV): A linear superpolyanhydride and a cyclic dimeric anhydride from sebacic acid. *J. Am. Chem. Soc.* **54**, 1569–1579.

Hirano, S. (1989). Chitosan wound dressings. *In* "Chitin and Chitosan" (G. Skjak-Braek, T. Anthonsen, and P. Sandford, eds.), pp. 1–835. Elsevier, London.

Holland, S. J., Jolly, A. M., Yasin, M., and Tighe, B. J. (1987). Polymers for biodegradable medical devices. II. Hydroxybutyrate–hydroxyvalerate copolymers: Hydrolytic degradation studies. *Biomaterials* **8**, 289–295.

Holzman, S., Connolly, R. J., and Schwaitzberg, S. D. (1994). Effect of hyaluronic acid solution on healing of bowel anastomoses. *J. Invest. Surg.* **7**, 431–437.

Hubbell, J. A. (1995). Biomaterials in tissue engineering. *Bio/Technology* **13**, 565–576.

James, K., and Kohn, J. (1997). Pseudo-poly(amino acid)s: Examples for synthetic materials derived from natural metabolites. *In* "Controlled Drug Delivery: Challenges and Strategies" (K. Park, ed.), pp. 389–403. Am. Chem. Soc., Washington, DC.

Kaplan, D. L., Wiley, B. J., Mayer, J. M., Arcidiacono, S., Keith, J., Lombardi, S. J., Ball, D., and Allen, A. L. (1994). Biosynthetic polysaccharides. *In* "Biomedical Polymers" (S. Shalaby, ed.), pp. 189–212. Hanser Publishers, Munich and New York.

Kemnitzer, J., and Kohn, J. (1997). Degradable polymers derived from the amino acid ʟ-tyrosine. *In* "Handbook of Biodegradable Polymers (A. J. Domb, J. Kost, and D. M. Wiseman, eds.), Vol. 7, pp. 251–272. Harwood Academic Publishers, Amsterdam, The Netherlands.

Kohn, J., and Langer, R. (1984). A new approach to the development of bioerodible polymers for controlled release applications employing naturally occurring amino acids. *Polym. Mater., Sci. Eng.* **51**, 119–121.

Kohn, J., and Langer, R. (1985). Non-peptide poly(amino acids) for biodegradable drug delivery systems. *In* "12th International Symposium on Controlled Release of Bioactive Materials Geneva, Switzerland" (N. A. Peppas and R. J. Haluska, eds.), pp. 51–52. Controlled Release Society, Lincolnshire, IL.

Kohn, J., and Langer, R. (1987). Polymerization reactions involving the side chains of α-L-amino acids. *J. Am. Chem. Soc.* **109**, 817–820.

Koleske, J. V. (1978). Blends containing poly-ε-caprolactone and related polymers. *In* "Polymer Blends" (O. R. Paul and S. Newman, eds.), pp. 369–389. Academic Press, New York.

Krejchi, M. T., Atkins, E. D., Waddon, A. J., Fournier, M. J., Mason, T. L., and Tirrell, D. A. (1994). Chemical sequence control of beta-sheet crystals assembly in macromolecular crystals of periodic polypeptides. *Science* **265**, 1427–1432.

Lacy, P. E. (1995). Treating diabetes with transplanted cells. *Sci. Am.,* pp. 51–58.

Laleg, M., and Pikulik, I. (1991). Wet-web strength incease by chitosan. *Nord. Pulp Paper Res. J.* **9**, 99–103.

Langer, R. (1990). Novel drug delivery systems. *Chem. Br.* **26**(3), 232–236.

Langer, R., and Vacanti, J. (1993). Tissue engineering. *Science* **260**, 920–926.

Laurencin, C., Domb, A., Morris, C., Brown, V., Chasin, M., McConnel, R., Lange, N., and Langer, R. (1990). Poly(anhydride) administration in high doses *in vivo*: Studies of biocompatibility and toxicology. *J. Biomed. Mater. Res.* **24**, 1463–1481.

Laurencin, C. T., Norman, M. E., Elgendy, H. M., El-Amin, S. F., Allcock, H. R., Pucher, S. R., and Ambrosio, A. A. (1993). Use of polyphosphazenes for skeletal tissue regeneration. *J. Biomed. Mater. Res.* **27**, 963–973.

Leenstag, J. W., Pennings, A. J., Bos, R.R.M., Roxema, F. R., and Boenng, G. (1987). Resorbable materials of poly-L-lactides. VI. Plates and screws for internal fracture fixation. *Biomaterials* **8**, 70–73.

Lescure, F., Gurny, R., Doelker, E., Pelaprat, M. L., Bichon, D., and Anderson, J. M. (1989). Acute histopathological response to a new biodegradable, polypeptidic polymer for implantable drug delivery system. *J. Biomed. Mater. Res.* **23**, 1299–1313.

Lewis, D. H. (1990). Controlled release of bioactive agents from lactide/glycolide polymers. *In* "Biodegradable Polymers as Drug Delivery Systems (M. Chasin and R. Langer, eds.), pp. 1–41. Dekker, New York.

Li, S. T., Archibald, S. J., Krarup, C., and Madison, R. D. (1992). Peripheral nerve repair with collagen conduits. *Clin. Mater.* **9**, 195–200.

Mäkelä, E. A., Vainionpää, S., Vihtonen, K., Mero, M., Helevirta, P., Törmälä, P., and Rokkanen, P. (1989). The effect of a penetrating biodegradable implant on the growth plate; an experimental study on growing rabbits with special reference to polydioxanone. *Clin. Orthop. Relat. Res.* **241**, 300–308.

Malette, W., Quigley, M., and Adicks, E. (1986). Chitosan effect in vascular surgery, tissue culture and tissue regeneration. *In* "Chitin in Nature and Technology" (R. Muzzarelli, C. Jeuniaux, and G. Gooday, eds.), pp. 435–442. Plenum, New York.

Mathiowitz, E., Saltzman, W. M., Domb, A., Dor, P., and Langer, R. (1988). Polyanhydride microspheres as drug carriers. II. Microencapsulation by solvent removal. *J. Appl. Polym. Sci.* **35**, 755–774.

McCormick-Thomson, L. A., and Duncan, R. (1989). Poly(amino acid) copolymers as a potential soluble drug delivery system. 1. Pinocytic uptake and lysosomal degradation measured *in vitro. J. Bioact. Biocompat. Polym.* **4**, 242–251.

Medina, M., Paddock, H. N., Connolly, R. J., and Schwaitzberg, S. D. (1995). Novel anti-adhesion barrier does not prevent anastomotic healing in a rabbit model. *J. Invest. Surg.* **8**, 179–186.

Miller, N. D., and Williams, D. F. (1987). On the biodegradation of poly-β-hydroxybutyrate (PHB) homopolymer and poly-β-hydroxybutyrate–hydroxyvalerate copolymers. *Biomaterials* **8**, 129–137.

Mitra, S., van Dress, M., Anderson, J. M., Peterson, R. V., Gregonis, D., and Feijen, J. (1979). Pro-drug controlled release from poly(glutamic acid). *Polym. Prepr.* **20**, 32–35.

Muzzarelli, R., Baldassara, V., Conti, F., Ferrara, P., Biagini, G., Gazzarelli, G., and Vasi, V. (1988). Biological activity of chitosan: Ultrastructural study. *Biomaterials* **9**, 247–252.

Naeme, P. J., and Barry, F. P. (1993). The link proteans. *Experimentia* **49**, 393–402.

Nicol, A., Gowda, D. C., and Urry, D. W. (1992). Cell adhesion and growth on synthetic elastomeric matrices containing Arg-Gly-Asp-Ser3. *J. Biomed. Mater. Res.* **26**, 393–413.

Nimni, M. E., and Harkness, R. D. (1988). Molecular structure and functions of collagen. *In* "Collagen" (N. M. E., ed.), Vol. 1, pp. 10–48. CRC Press, Boca Raton, FL.

Pachence, J. M. (1996). Collagen-based devices for soft tissue repair. *J. Appl. Biomater.* **33**, 35–40.

Pachence, J. M., Berg, R. A., and Silver, F. H. (1987). Collagen: Its place in the medical device industry. *MD&DI*, pp. 49–55.

Parenteau, N. (1999). The first tissue-engineered products. *Sci. Am.* **280**, 83–84.

Peppas, N. A., and Langer, R. I. (1994). New challenges in biomaterials. *Science* **263**, 1715–1720.

Piez, K. A. (1984). Molecular and aggregate structures of the collagens. *In* "Extracellular Matrix Biochemistry" (K. A. Piez, and A. H. Reddi, eds.), Chapter 1. Elsevier, New York.

Pitt, C. G. (1990). Poly-ε-caprolactone and its copolymers. *In* "Biodegradable Polymers as Drug Delivery Systems" (M. Chasin and R. Langer, eds.), pp. 71–119. Dekker, New York.

Pitt, C. G., Chasalow, F. I., Hibionada, Y. M., Klimas, D. M., and Schindler, A. (1981a). Aliphatic polyesters. I. The degradation of poly-ε-caprolactone *in vivo. J. Appl. Polym. Sci.* **28**, 3779–3787.

Pitt, C. G., Gratzl, M. M., Kimmel, G. L., Surles, J., and Schindler, A. (1981b). Aliphatic polyesters. II. The degradation of poly-*d,l*-lactide, poly-ε-caprolactone, and their copolymer *in vivo. Biomaterials* **2**, 215–220.

Pitt, C. G., Hendren, R. W., Schindler, A., and Woodward, S. C. (1984). The enzymatic surface erosion of aliphatic polyesters. *J. Controlled Release* **1**, 3–14.

Pitt, C. G., Andrady, A. L., Bao, Y. T., and Sarnuei, N.K.P. (1987). Estimation of the rate of drug diffusion in polymers. *Am. Chem. Soc.* **348**, 49–77.

Pulapura, S., and Kohn, J. (1992). Tyrosine derived polycarbonates: Backbone modified, "pseudo"-poly(amino acids) designed for biomedical applications. *Biopolymers* **32**, 411–417.

Pulapura, S., Li, C., and Kohn, J. (1990). Structure–property relationships for the design of polyiminocarbonates. *Biomaterials* **11**, 666–678.

Ray, J. A., Doddi, M., Regula, D., Williams, J. A., and Melveger, A. (1981). Polydioxanone (PDS), a novel monofilament synthetic absorbable suture. *Surg., Gynecol. Obstet.* **153**, 479–507.

Reed, A. M., and Gilding, D. K. (1981). Biodegradable polymers for use in surgery—Poly(glycolic)/poly(lactic acid) homo and copolymers: 2. *In vitro* degradation. *Polymer* **22**, 494–498.

Rosen, H. B., Chang, J., Wnek, G. E., Linhardt, R. J., and Langer, R. (1983). Bioerodible polyanhydrides for controlled drug delivery. *Biomaterials* **4**, 131–133.

Sandford, P. A. (1989). Chitosan chemistry. *In* "Chitin and Chitosan" (G. Skjak-Braek, T. Anthonsen, and P. Sandford, eds.), pp. 51–69. Elsevier, London.

Sawhney, A. S., Pathak, C. P., and Hubbell, J. A. (1993). Bioerodible hydrogels based on photopolymerized poly(ethylene glycol)-co-poly(α-hydroxy acid) diacrylate macromers. *Macromolecules* **26**, 581–587.

Scopelianos, A. G. (1994). Polyphosphazenes as new biomaterials. *In* "Biomedical Polymers" (S. Shalaby, ed.), pp. 153–171. Hanser Publishers, Munich and New York.

Shalaby, S. W., and Johnson, R. A. (1994). Synthetic absorbable polyesters. *Biomed. Polym.* pp. 2–34.

Silver, F. H., and Pins, G. (1992). Cell growth on collagen: A review of tissue engineering using scaffolds containing extracellular matrix. *J. Long-Term Eff. Med. Implants* **2**,(1), 67–80.

Spira, M., Fissette, J., Hall, C. W., Hardy, S. B., and Gerow, F. J. (1969). Evaluation of synthetic fabrics as artificial skin grafts to experimental burn wounds. *J. Biomed. Mater. Res.* **3**, 213–234.

Sullivan, S. J., Maki, T., Borland, K. M., Mahoney, M. D., Solomon, B. A., Muller, T. E., Monaco, A. P., and Chick, W. L. (1991). Biohybrid artificial pancreas: Long-term implantation studies in diabetic, pancreatectomized dogs. *Science* **252**, 718–721.

Sung, K. C., and Topp, E. M. (1994). Swelling properties of hyaluronic acid ester membranes. *J. Membr. Sci.* **92**, 157–167.

Tanzer, M. L., and Kimura, S. (1988). Phylogenetic aspects of collagen structure and function. *In* "Collagen" (M. E. Nimni, ed.), Vol. 2, pp. 55–98. CRC Press, Boca Raton, FL.

Taravel, M. N., and Domard, A. (1993). Relation between the physiocochemical characteristics of collagen and its interactions with chitosan: I. *Biomaterials* **14**,(12), 930–938.

Tirrell, J. G., Fournier, M. J., Mason, T. L., and Tirrell, D. A. (1994). Biomolecular materials. *Chem. Eng. News,* pp. 40–51.

Toolan, B. C., Frenkel, S. R., Yalowitz, B. S., Pachence, J. M., and Alexander, H. (1996). An analysis of a collagen–chondrocyte composite for cartilage repair. *J. Biomed. Mater. Res.* **31**, 273–280.

Urmann, B., Gomel, V., and Jetha, N. (1991). Effect of hyaluronic acid on post-operative intraperitoneal adhesion prevention in the rat model. *Fertil. Steril.* **56**, 563–567.

Urry, D. (1995). Elastic biomolecular machines. *Sci. Am.,* pp. 64–69.

Vainionpää, S., Kilpukart, J., Latho, J., Heleverta, P., Rokkanen, P., and Törmälä, P. (1987). Strength and strength retention *in vitro,* or absorbable, self-reinforced polyglycolide (PGA) rods for fracture fixation. *Biomaterials* **8**, 45–48.

van der Rest, W. J., Dublet, B., and Champliaud, M. (1990). Fibril-associated collagens. *Biomaterials* **11**, 28.

van Heeswijk, W.A.R., Hoes, C.J.T., Stoffer, T., Eenink, M.J.D., Potman, W., and Feijen, J. (1985). The synthesis and characterization of polypeptide–adriamycin conjugates and its complexes with adriamycin. Part 1. *J. Controlled Release* **1**, 301–315.

van Swol, R. L., Ellinger, R., Pfeifer, J., Barton, N. E., and Blumenthal, N. (1993). Collagen membrane barrier therapy to guide regeneration in class II furcations in humans. *J. Periondontol.* **64**, 622–629.

Vert, M., and Li, S. M. (1992). Bioresorbability and biocompatibility of aliphatic polyesters. *J. Mater. Sci.: Mater. Med.* **3**, 432–446.

Wang, E., Overgaard, S. E., Scharer, J. M., Bols, N. C., and Moo-Young, M. (1988). Occlusion immobilization of hybridoma cells. *Chitosan, Biotechnol. Tech.* **2**, 133–136.

Wei, J. C., Hudson, S. M., Mayer, J. M., and Kaplan, D. L. (1977). A novel method for crosslinking carbohydrates. *J. Polym. Sci.* **30**, 2187–2193.

Weiss, C., and Balazs, E. A. (1987). Arthroscopic viscosurgery. *Arthroscopy* **3**, 138.

Wintermantel, E., Mayer, J., Blum, J., Eckert, K. L., and Luscher, P. (1996). Tissue engineering scaffolds using superstructures. *Biomaterials* **17**(2), 83–91.

Yannas, I. V., and Burke, J. F. (1980). Design of an artificial skin: Basic design principles. *J. Biomed. Mater. Res.* **14**, 65–81.

Yannas, I. V., Burke, J. F., Gordon, P. L., Huang, C., and Rubenstein, R. H. (1980). Design of an artificial skin: Control of chemical composition. *J. Biomed. Mater. Res.* **14**, 107–131.

Yasin, M., Holland, S. J., Jolly, A. M., and Tighe, B. J. (1989). Polymers for biodegradable medical devices. VI. Hydroxybutyrate–hydroxyvalerate copolymers: Accelerated degradation of blends with polysaccharides. *Biomaterials* **10**, 400–412.

Yu, H., Lin, J., and Langer, R. (1987). Preparation of hydroxyproline polyesters. *In* 14th International Symposium on Controlled Release of Bioactive Materials, Toronto, Canada" (P. I. Lee and B. A. Leonhardt, eds.), pp. 109–110. Controlled Release Society, Lincolnshire, IL.

Yu-Kwon, H., and Langer, R. (1989). Pseudopoly(amino acids): A study of the synthesis and characterization of poly(*trans*-4-hydroxy-*N*-acyl-L-proline esters). *Macromolecules* **22**, 3250–3255.

Zhou, Q. X., and Kohn, J. (1990). Preparation of poly(L-serine ester): A structural analogue of conventional poly(L-serine). *Macromolecules* **23**, 3399–3406.

PART VI: TRANSPLANTATION OF ENGINEERED CELLS AND TISSUES

CHAPTER 23

APPROACHES TO TRANSPLANTING ENGINEERED CELLS AND TISSUES

Janet Hardin-Young, Jeffrey Teumer, Robert N. Ross, and Nancy L. Parenteau

INTRODUCTION

Approaches to the use of living cells in tissue engineering is dictated by the primary purpose of the implant, be it metabolic or structural and the availability of a suitable cell source. Whatever the intended mode of action of tissue-engineered grafts, the function is provided either directly or indirectly by the cellular component. Determining where these cells will come from plays a pivotal role. Cellularization of an implant can be achieved through recruitment of host cells *in vivo*, directed *in vivo* tissue formation, *ex vivo* cell propagation, and *in vitro* organotypic tissue culture. Tissue scaffolds and living tissue implants are by no means static devices once inside the body. Therefore the inflammatory response of the host plays a key role in determining the persistence, long-term structure, and functionality of the engineered implant. Another important parameter is the immunologic response when using cells not from the host. Fortunately, certain allogeneic human cells may be transplanted successfully with little or no immune response. However, special measures such as immune therapy, genetic modification, or encapsulation must be taken to introduce xenogeneic cells. Reagents and methods used in the culture of cells should be screened and validated for safety to guard against untoward changes in the cell population or introduction of adventitious agents during processing. This is true regardless of the cell source.

The ultimate goal of tissue engineering is to replace, repair, or enhance biological function in the event of damaged, absent, or dysfunctional elements, often at the scale of a tissue or an organ. Engineered tissues are achieved by using cells that are manipulated through their extracellular environment to develop living biological substitutes for cells and tissues that are lacking. Many different strategies may be used to accomplish this goal, but they all have in common the enlistment of living cells for therapeutic use.

Among the chief factors that dictate the best strategy for developing and utilizing engineered tissues are technical feasibility, required properties of the implant, and the interaction of the host with the graft. This chapter is a general discussion of such factors, and the possible approaches that can be used to transplant cells and reestablish tissues. Specific applications of transplanting engineered cells and tissues will be covered in more detail in subsequent chapters.

DEVELOPING WORKABLE STRATEGIES

Successful development of an engineered biologic substitute depends on contributions from such diverse fields as cell biology, extracellular matrix biochemistry, biomaterials science, biomedical engineering, biochemistry, immunology, cell culture technology, chemical engineering, and physiology. Having identified a clinical need, the decision to go forward with the development of a particular tissue-engineered product is based on a careful evaluation of available scientific knowledge and an assessment of present and near-term technical capabilities. The concept is then tested in the laboratory for its overall scientific merit and for proof of principle. After laboratory evidence suggests proof of principle, the approach is refined and modified by further *in vitro* and animal experimentation, and eventually by clinical trials.

Reproducibility of results is a critical consideration. No tissue engineering process can succeed without the high probability that the product will consistently have the desired characteristics. Reproducibility of the product is achieved through standardization of protocols that are built on a substantial foundation of fundamental technical knowledge, the deepest possible understanding of the basic mechanisms involved, careful experimentation, and direct technical experience. Without these things, the target tissues to be modeled, the laboratory and manufacturing techniques required to develop the product, and the native tissue all remain essentially black boxes. Such unknowns are the antithesis of reproducible results. For *ex vivo* expansion of most cells types, precise regulation of pH, oxygen tension, seeding density, growth factors, and nutrients is required. The development of commercial processes to provide a living cell-based product depends on a thorough understanding of the metabolic and physical requirements of the specific cells and on the technical expertise to provide them in quantities large enough to support a commercially viable enterprise.

The final test of the clinical applicability of a tissue-engineered product is the ability to provide sufficient quantities of reliably reproducible product at a reasonable cost. Even the most promising ideas fail as commercial products if they are proved unfeasible as a result of irreproducibility, cost, unavailability of cell source, or inability to direct the host response to the implant. How an implant is expected to function determines the features that must be designed into it (Table 23.1) and the industrial plan for how it is to be produced. Design approaches range from the relatively simple use of cultured cells (Green *et al.*, 1979; Brittberg *et al.*, 1994) to the use of what could be termed organ equivalents, in which there is an attempt to replicate both structure and function (Parenteau *et al.*, 1992; Bilbo *et al.*, 1993; Wilkins *et al.*, 1994).

The best approach is not always obvious and must often be arrived at through experimentation, including clinical trials. For example, the repair of articular cartilage is being investigated using undifferentiated cultured chondrocytes (Brittberg *et al.*, 1994), chondrocyte precursors in collagen, matrix alone (Reddi, 1994; Speer *et al.*, 1979; von Schroeder *et al.*, 1991; Wakitani *et al.*, 1994), and differentiated cartilage analogs (Vacanti *et al.*, 1991; Freed *et al.*, 1993, 1994a,b). Cartilagenous matrix forms readily *in vitro* and *in vivo*, but its long-term phenotype is difficult to control. Isolated human chondrocytes from a small biopsy seeded on bioresorbable polymer (Cao *et al.*, 1998) have been expanded in an attempt to provide custom-engineered autologous cartilage transplants for the reconstruction of the nose or outer ear (Naumann *et al.*, 1998).

Use of differentiated cartilage analogs for articular cartilage repair has faced problems of inadequate tissue integration with adjacent host cartilage. The difficulties stem from the biology of the host articular cartilage, which is a relatively acellular, avascular, dense matrix that undergoes little interstitial remodeling. Only empirical data will determine which of these is the best approach. The best therapy will be the one that reproducibly forms articular cartilage, effect long-term repair, and is cost effective.

Table 23.1. Types of engineered tissues

Type of tissue	Graft role	Example
Physical	Primarily biomechanical	Bone (Habal and Reddi, 1994; Thomson *et al.*, 1995)
		Blood vessels (Termin *et al.*, 1994)
Physiological	Physiological	Liver (Dunn *et al.*, 1989; Sussman *et al.*, 1994)
		Kidney (Humes, 1995)
Chemical	Produce soluble, diffusible component	Pancreatic islets (Mikos *et al.*, 1994)
		Neuroendocrine substance (Aebischer *et al.*, 1994)
Combination	Multiple	Skin (Sabolinski *et al.*, 1996)

In the case of organotypic analogs of skin, such as the Apligraf human skin equivalent (HSE), the situation of tissue response is quite different from that of cartilage. The development of an organotypic HSE is particularly important because these *in vitro* three-dimensional skin constructs show morphologic and biochemical profiles expressed in living skin. This serves as a benefit rather than a hindrance. Formation of an organotypic analog results in a robust multifunctional tissue that, unlike cartilage, is able to interact and integrate with the wound to affect healing on multiple levels (see Chapter 65).

Tissue structure, whether only minimally achieved prior to implantation or more completely developed in culture, must be achieved quickly *in vivo* for optimum cell function. A vascular implant, for example, must serve immediately as an efficient conduit for blood and not trigger the formation of thrombi. Not only should the implant not be rejected by the host, but the implant should foster the migration of native cells to the site. Implantation must not interfere with normal function of the existing tissue or of the organ. A bone graft must not significantly weaken the site around the implant. Finally, the implant must become, as much as possible, a fully functional analog of the native tissue in the implantation site.

MODE OF ACTION—AN IMPORTANT DESIGN CONSIDERATION

The desired mode of action of a graft dictates the type of implant necessary and determines important functional and mechanical requirements of the graft. Engineered tissue implants presently in use or under investigation can be grouped into four general categories that focus on its primary function (Table 24.1): physical, chemical, physiologic, and a combination of graft roles. These modes of action are discussed below.

PHYSICAL GRAFTS

The primary mode of action of physical grafts (e.g., ligament, cartilage, or vascular replacements) is to perform a biomechanical structural role. The biomechanics depend heavily on the extracellular component of the graft and on the capacity for living cells of the graft (attracted to the site of the graft or implanted with the graft) to maintain that structure. This, however, may be too simplistic a view for tissue engineering. For example, physical properties are of acute concern on implantation of a vascular graft. The surface of the graft must not elicit a thromogenic response and the structure must be able to withstand arterial pressure without threat of aneurysm. In addition, the material must have more than adequate suture retention characteristics. Therefore, it is obvious to focus on the initial physical properties and short-term effects of the grafts. Although this is a necessary first step, and there has been some exciting progress in this area (see Chapter 34), unlike an inert device, a tissue graft, whether cellular or noncellular, will be expected to interact with the body. Because of this, the biology and physiology of the cells become increasingly important postimplant and will ultimately determine the long-term patency of the grafts, just as they can in autologous saphenous vein grafts. This general concept of the physical structure ultimately depending on cell biology and physiology is likely true for most physical tissue-engineered grafts.

There are clinical situations in which a relatively large tissue mass is required physically to fill the space of tissue lost as a result of surgical resection (e.g., tissue replacement after mastectomy) (Eiselt *et al.*, 1998). Living tissues for "bulking" applications are limited by the ability to vascularize the tissue. Problems to solve in designing such a tissue-engineered product with such a mode of action include designing an acellular structural framework to fill the space, providing a suitable matrix for the localization of transplanted cells, and promoting vascularization of the forming tissue. One approach has been to allow a scaffold to prevascularize prior to seeding with functional cells (Wake *et al.*, 1995). Attempts have been made to enhance this process by seeding the matrix with polymer microspheres containing a potent angiogenic molecule [vascular endothelial growth factor (VEGF)] for controlled release to promote vascularization at the site of tissue formation (Eiselt *et al.*, 1998).

PHYSIOLOGIC OR CHEMICAL GRAFTS

In grafts that are intended to have a physiologic or chemical mode of action, the biological or biochemical properties of the graft are more important than the physical structure or mechanical properties of the graft. Products of the constituent cells are of greater concern than properties of the extracellular component of the graft.

A variety of tissue-engineered implants with physiologic or chemical function are under development. For example, biodegradable polymer scaffolds have been developed to support the growth of hepatocytes on implantable microcarriers (Davis and Vacanti, 1996). Biologic function of these hepatocyte transplants has been demonstrated in animal models by production of albumin and other liver products, and by clearing bilirubin and urea metabolites. Pancreatic islet cell transplants are likewise under investigation (Hayashi *et al.,* 1998).

Even in the design of physiologic and chemical grafts, however, the contribution of an extracellular matrix component and mechanical properties must be understood, because they may be important for cell regulation. The function of cultured rat hepatocytes, for example, is improved when cells are cultivated between two collagen layers that provide extracellular matrix contact and permit proper cell polarization (Dunn *et al.,* 1992; Ezzell *et al.,* 1993).

COMBINATION

There are instances in which several aspects of the implant are key elements to its effectiveness. Certain types of grafts are intended to perform two or more equally important functions. For example, cultured human skin has important physical structure and biologic function. The graft skin is a distinct physical structure that covers a physical defect. The stratum corneum serves as an immediate barrier to water loss and infection while the living layers promote wound healing through their biologic interaction with the host wound bed.

ROLE OF THE HOST

It should be clear that the host interaction with the tissue-engineered product is an important consideration. The success of the implant depends on working with the host response in such a way that the metabolic requirements and appropriate regulatory signals are provided by and to the implant. An initial inflammatory response is followed by vascularization and migration of mesenchymal cells into the graft (Kemp *et al.,* 1995). Unlike normal wound healing, however, there seems to be a well-defined period of 5 to 10 weeks during which the host response to an implant can be directed toward integration or toward inflammation and "rejection" of the implant. If the process is not directed toward tissue integration by this time, inflammatory mechanisms of the foreign body reaction are mobilized, and the body attempts to eliminate or isolate the implant. This is an innate, nonspecific immune response. Strategies to avoid this destruction therefore become major design considerations.

It is thought by some that the inflammatory response to an implant is not only beneficial but necessary for the recruitment of host cells. Our experience both in animals and in the clinic suggests that directed tissue integration through signals other than those of acute inflammation might be sufficient and more desirable. Inflammation, when unchecked, can lead to fibrosis, and in some cases, destruction of the implant architecture and/or encapsulation of the implant.

The degree of graft remodeling and cell replacement by the host tissue varies by type of implant. If the graft is to replace a lost or deficient function, the implanted cells should naturally be designed to persist and function for a long time. Premature loss of implanted cells through turnover, lack of ability for self-propagation and/or maintenance, and an inflammatory or immune response by the host would undermine the goals of the therapy. In other cases, however, population by the host cells and remodeling of the structure may be requirements for achieving proper function. In bone grafts, for example, cell population or repopulation and remodeling of the extracellular matrix are important for achieving stable mechanical properties. Another example of this is the host cellularization and remodeling of a small-caliber vascular graft made of collagen (discussed in Chapter 33).

In the case of cultured skin, which will be used in a variety of acute and chronic wounds, flexibility of the implant's response, durability of the engineered tissue, and its ability to direct a positive host response are key to its effectiveness. Remodeling of cultured skin is advantageous and results in improved function and appearance in the dermatologic surgery patient (Eaglstein *et al.,* 1991). The persistence and remodeling of Apligraf (Graftskin) (Apligraf is a registered trademark of Novartis) human skin equivalent, for example, is dependent on the type of application (see Chapter 67). Contribution from the host varies depending on the wound environment and the biology, biochemistry, and physiology of both the cellular elements and the extracellular matrix of the graft.

Whatever the intended mode of action of tissue-engineered grafts, the function is provided either directly or indirectly by the cellular component. Therefore, determining where these cells will come from plays a pivotal role.

SOURCE OF CELLS

Tissue engineering requires a reliable source of viable cells for processing. Which source of cells is best used for a tissue-engineered product depends on may factors, including the functional requirements of the implant as well as the feasibility of collecting, processing, and storing these cells and the tissue-engineered product. Timing is an issue. For example, an off-the-shelf product is more likely to be used in a timely manner compared with a product that requires custom production. Economic considerations are also important. The cost of finding source material, cultivating it, and manipulating some cell types often limits what can be accomplished in tissue engineering.

If the implant relies on *in vivo* cell growth after reimplantation, *in vitro* expansion of cells is less critical. An example of this is bone marrow transplantation, because marrow cells have the potential to reconstitute the recipient's hematopoietic tissue without significant prior *in vitro* culture and expansion. In most cases, however, it is necessary to expand the cell number by some *in vitro* method prior to implantation. Success of the process therefore depends on the ability to scale up the culture system without altering the desirable characteristics of the cultivated cells.

Currently, only a few cell types (e.g., keratinocytes, fibroblasts, chondrocytes, and myoblasts) can be expanded in significant numbers in culture. For other cell types (e.g., normal hepatic, neural, or pancreatic islet cells), it is not currently possible to expand in culture a sufficient number of normal cells for clinical use. For such cell types, alternative strategies (e.g., genetic modification or the use of cultivated cell lines) are being explored. Sources of cells may be autologous (obtained from the same individual for whom the graft is intended), allogeneic (obtained from an individual of the same species who is genetically distinct from the individual for whom the graft is intended), or xenogeneic (obtained from an individual of a species different from the species of individual for whom the graft is intended).

AUTOLOGOUS CELLS

The most direct source of cells for transplant is autologous tissue. The use of autologous tissue obviates the difficulties of overcoming noncompatibility with the host's immune system, which is a challenge in grafting any nonhost material into an immunocompetent host. Cell number can be increased from autologous sources in two ways: (1) autologous cell expansion or (2) autologous cell recruitment. In autologous cell expansion, cells are harvested from a patient biopsy, grown in culture to the desired number, and then reimplanted into the body. In autologous cell recruitment, *in vivo* conditions are manipulated to foster migration of the desired cell population to the site.

For autologous cells to be useful, cultivation of the cell type must permit expansion of the population to a number sufficient to meet the functional need of the implant within a time frame that suits the needs of the patients. The time required to harvest tissue from the patient, expand the cells in culture, and construct the implant can be a critical limitation of autologous material. If the tissue is needed to sustain life, the time needed to manufacture the implant may not be consistent with patient survival. In addition, healthy tissue may be sparse or unavailable. The cost and inherent variability of each autologous, custom-made device may limit clinical feasibility and must be considered in comparison with alternatives that are likely to be available. Cost and variability may be partially alleviated by providing an automated, controlled means of processing patient tissue, which can be placed directly in the hospital laboratories. This strategy is being used for the cultivation of patients' hematopoietic cells (Koller *et al.,* 1993; Van Zant *et al.,* 1994).

Cultured autologous cells were the first clinical application of cellular tissue engineering. Autologous keratinocytes were grown in culture and grafted onto burn victims (Gallico *et al.,* 1984). In this procedure, a small biopsy of skin is taken from an uninjured area of a burn patient. The keratinocytes of the epidermis are then isolated from this biopsy, and their number is increased in tissue culture (Rheinwald and Green, 1975). These cultured keratinocytes are then grafted to the patient as confluent sheets of cells after having been expanded as much as 2000-fold from the initial biopsy (Epicel, Genzyme Tissue Repair, Cambridge, MA) (Green *et al.,* 1979).

Autologous cultured chondrocytes (Carticel, Genzyme Tissue Repair, Cambridge, MA) are being used to treat articular cartilage injuries in humans. Chondrocytes are harvested from an uninvolved area of the injured knee, expanded in culture, and then surgically seeded into the articular defect and held in place with a piece of periosteal sheath, also harvested from the patient. Arthroscopic examination has shown smooth-surfaced articular cartilage regenerated in the site of the original lesions in patients followed for up to 5.5 years after implantation (Brittberg *et al.,* 1994).

Autologous cell recruitment

Recruitment of autologous cells to restore or enhance tissue function can also be achieved *in vivo* by providing an appropriate acellular scaffold designed to foster cell infiltration. Such implants are designed to guide cell migration and thus enable cells recruited from the patient's population of cells to differentiate and replace the implanted regeneration scaffold with a functional, living tissue. The matrix components of the implant are resorbed and eventually replaced by new, host-derived, extracellular matrix. Implanted scaffolds may also contain chemical factors that stimulate chemotaxis and cell differentiation. For example, acellular bone grafts are cellularized by cells recruited from the marrow, blood, or periosteum and are most likely stimulated by bone morphogenic proteins bound to the bone matrix (Reddi and Anderson, 1976).

Because regeneration scaffolds do not contain live cells at the time of implantation, production and storage of the product are simplified. These advantages are balanced by the disadvantage that the clinical success of regeneration scaffolds is totally dependent on appropriate cellular recruitment from the host. In some cases, critical steps in cellular recruitment may be either inadequate or lacking entirely. Recruitment may be aided by the use of bound growth factors or concomitant gene therapy such as the use of VEGF (Takeshita *et al.,* 1996) or other angiogenic or chemotactic factors.

In some cases, the scaffold must provide immediate physical functionality during the early acellular stage and must also induce host regeneration of new tissue. Many such regeneration scaffolds have been produced from components of the natural extracellular matrix. Scaffold grafts in the form of bone powders are used routinely in facial bone repair. Composite scaffolds of controlled porosity and physical characteristics are being developed to direct bone formation (Thomson *et al.,* 1998). Matrix scaffolds have also shown promise in the repair of dermis (Burke *et al.,* 1981; Yannas *et al.,* 1982, 1989; Livesey *et al.,* 1995), tendons (Goldstein *et al.,* 1989; Kato *et al.,* 1991), nerves (Chang and Yannas, 1992), meniscus (Stone *et al.,* 1990), and blood vessels (see Chapter 32).

ALLOGENEIC CELLS

One way of minimizing the limitations of supply inherent in strategies that rely on autologous cells is the development of tissue-engineered products from allogeneic cells. The use of allogeneic cells allows banking of cells. This permits safety and functional performance screening. Moreover, because allogeneic material can be free from the variability associated with patient-dependent autologous sources, it permits the development of a consistently reproducible and cost-effective product. Such standardized allogeneic cell sources also allow the construction of complex tissues (Wilkins *et al.,* 1994) that would otherwise be prohibitively expensive in time, money, and effort if they were to be produced by custom manufacture from autologous sources.

An example of this is the formation of the Apligraf HSE. A method for cultivating human keratinocytes from neonatal foreskin, using a minimally supplemented basal medium (MSBM), was reported in 1992 (Johnson *et al.,* 1992). Three passages in culture provide approximately 10^{10} cells for cryostorage as a cell bank. Subsequent culture of these cells for one passage results in an additional 10-fold increase in cell number, which is then sufficient to produce approximately 100–200 m^2 of HSE constructs from a single source. This allogeneic skin equivalent provides both the advantages of a skin graft without the need for harvesting tissue and the production of large quantities of consistent material.

IMMORTALIZED CELL LINES

Cell lines may be derived from cells that have undergone certain genetic changes that permit them indefinite proliferation as long as their metabolic requirements are satisfied. Such immortalized cell lines can be used to overcome some of the source limitations of current cell culture technology. The genetic changes that immortalize these cell lines are usually accompanied by some loss of differentiated structure and function, however. Although this dedifferentiation may make these cells significantly different from their native counterparts, because these cells remain somewhat plastic, such cell lines are more amenable to gene transfer. Specific properties can thus be engineered into them to produce desired therapeutic proteins, perform selected cell functions, or overcome host immune reaction.

Safety is a concern when using cell lines, however, because immortalized cells exhibit some of the properties of neoplastically transformed cells. Such cell lines are thus considered to be partially transformed cells with a predisposition to become fully neoplastic cells capable of forming tumors in the recipient. Tissue engineering strategies that use cell lines can take this danger into account in a variety of ways. Techniques of encapsulation of the cell line, extracorporeal use, or control of immortalization using conditions that select against gene expression such as temperature or drug sensitivity can limit the cells and risk to the patient. A hepatocyte cell line has been used in an extracorporeal system to treat acute liver failure (Sussman *et al.,* 1994).

XENOGENEIC SOURCES

For some cell types, the tissue demand cannot be met through the use of autologous or allogeneic sources. This has led to strategies that employ xenogeneic tissues. Technology employing cell encapsulation, immune protection, extracorporeal systems, and genetic manipulation is being developed to enable the use of xenogeneic tissues and organs, because one need only breed enough animals to meet demand. The use of xenogeneic tissue requires extra attention to the problem of rejection by the patient's immune system. Xenogeneic cells are feasible only if the tissues can be used effectively. For example, porcine pancreatic islet transplantation will have limited feasibility if the tissue remains difficult to harvest and requires a number of animals for each treatment. Cost and quality-control issues associated with islet preparation could be restrictive.

Microencapsulation techniques have been used to permit transgenic implants (Basic *et al.,* 1996). In general, however, the problem remains unsolved of designing membranes that selectively restrict permeation by some effector molecules of the immune system but permit, or even facilitate, permeation by other biologically active molecules of roughly equivalent size (Edge *et al.,* 1998).

THE IMMUNOLOGY OF NONAUTOLOGOUS CELLS

Although there are significant differences in the mechanisms of allograft and xenograft rejection, there are important common features. Destruction of implanted tissues by the patient's immune system is a hurdle when nonautologous cells are used. These are acquired immune responses, the mechanisms of true graft rejection. For example, the mitigation of complement-mediated rejection is believed to be a major hurdle for the transplantation of xenogeneic cells (Schilling *et al.,* 1976). Once the hyperacute rejection is prevented, transplantation of xenogeneic cells will face many of the same problems associated with allograft rejection.

Allograft rejection is primarily a T cell-mediated immune response. In conventional organ transplantation, cytotoxic T cells (cytotoxic T lymphocytes, CTLs) specific for alloantigens present on the endothelial cells attack the vasculature of the grafted organ, which leads to destruction of the graft (Adams *et al.,* 1994; Pober *et al.,* 1986). Passenger leukocytes contributes to the response by stimulating the expansion of donor-reactive T helper cells, which are necessary for the generation of CTLs.

Not all cell types are equal in their ability to elicit a host immune response. Donor cells can be classified on the basis of their ability to act as target cells for effector CTLs, and their expression of histocompatibility leukocyte antigen (HLA) proteins. Some cell populations (e.g., leukocytes and endothelial cells) are strongly allostimulatory. These professional antigen-presenting cells (APCs) constitutively express major histocompatibility complex (MHC) class I, class II, and costimulatory proteins (e.g., B7 and CD40), are capable of stimulating naive T cells, resulting in the expansion of alloreactive T helper cells, and serve as targets for effector CTLs. Other cell types classified as nonprofessional APCs are less allostimulatory (e.g., keratinocytes, smooth muscle cells, and fibroblasts) (Murray *et al.,* 1995; Laning *et al.,* 1997). Nonprofessional APCs do not constitutively express MHC class II proteins (Gaspari and Katz, 1988) and are unable to induce the activation and expansion of required T helper cells because they lack the functioning costimulatory molecules necessary for proper T cell activation.

One approach to using allogeneic cells in tissue engineering has been to use pure populations of allogeneic nonprofessional APC cells that are weakly or nonallostimulatory and do not contain endothelial cells or passenger leukocytes. The Apligraf HSE, an allogeneic construct containing only human fibroblasts and keratinocytes, is an example of such an engineered graft (Wilkins *et al.*, 1994). This skin construct contains both purified keratinocyte and fibroblast populations and has been shown in both preclinical (Theobald *et al.*, 1993; Laning *et al.*, 1999; Moulton *et al.*, 1999) and clinical (Falanga *et al.*, 1998) studies not to sensitize recipients.

In situations in which the donor cells are strongly allostimulatory, there are several approaches that might be used to control the host immune response, including traditional therapies used in solid organ transplants. Host immune response is not only determined by the allogenicity of the individual cell types that comprise the graft but also by the accessibility to the graft by the host immune system. Developing appropriate animal models to investigate the allogenicity of a graft is critical. Immunocomprised mice such as the severe combined immunodeficient (SCID) mouse lack functioning immune systems, which allows them to accept human tissue transplants. A number of humanized SCID mouse models have been developed to assess the antigenicity of both organ grafts and engineered tissue grafts. For example, SCID mice humanized with human white blood cells reject allogeneic human skin but not an allogeneic HSE (Briscoe *et al.*, 1999).

In most cases, tissue engineered grafts will be avascular when transplanted. Because endothelial cells are a primary target of graft rejection, the absence of allogeneic endothelium in an engineered graft is an important element for the immunologic success of the engineered grafts. In addition, the lack of an established vasculature may beneficially limit the initial interaction between the host immune system and the graft cells.

In summary, the use of allogeneic and xenogeneic cells in tissue engineering provides alternatives to the use of autologous cells in some circumstances. The specific types of cells used in the graft will determine the degree of immunoreactivity of the graft and will therefore determine the immunologic approach used to obtain graft acceptance. In some cases the alloreactivity of the graft will be negligible, so that no special treatment is necessary for graft acceptance. The use of allogeneic or xenogeneic cells does entail particular safety considerations.

SAFETY CONSIDERATIONS

Reagents and methods used in the culture of autologous cells should be screened and validated for safety to guard against untoward changes in the cell population or introduction of adventitious agents during processing. This is true regardless of the cell source. Each autologous sample must be kept isolated throughout the process to avoid possible exposure to cells or tissues from other individuals. This becomes a primary limitation to achieving economies of scale in process, because each tissue must be treated on a "custom" basis (Schaeffer *et al.*, 1999). With allogeneic sources, sufficient safeguards must be in place to minimize the possibility of transmitting an infectious agent from the donor to the host. Although allogeneic sources must be carefully screened to guard against transmission of such infectious diseases as human immune deficiency and hepatitis viruses, there is adequate time and access to the cells permitting thorough screening of cell banks for adventitious agents, abnormal karyotype, tumorigenicity, and phenotypic changes, even well beyond their point of use. This can yield a product with less risk than traditional blood and organ transplants, which already have an excellent safety record. Because the need for batch isolation is eliminated, this permits a scalable process using prescreened material.

The use of xenogeneic cells present a special safety consideration, the possibility of introducing into the human population viruses that are not known or understood, and that might be pathogenic in humans. Therefore animal tissues must be thoroughly screened (Fishman, 1998) and would likely come from tightly controlled, closed animal herds.

CONCLUSION

From this overview, we can see there are many options to consider in the design and production of an engineered tissue grafts. The cultivation and manipulation of cells for the practical purposes of producing functional tissue present a number of formidable but rewarding challenges to both the biologist and the engineer.

References

Adams, P. W., Lee, H.-S., Ferguson, R. M., and Orosz, C. G. (1994). Alloantigenicity of human endothelial cells. II. Analysis of interleukin-2 production and proliferation by T cells after contact with allogeneic epithelia. *Transplantation* **57**(1), 115–122.

Aebischer, P., Goddard, M., Signore, A., and Timpson, R. L. (1994). Functional recovery in hemiparkinsonian primates transplanted with polymer encapsulated PC12 cells. *Exp. Neurol.* **126**(2), 151–158.

Basic, D., Vacek, I., and Sun, A. M. (1996). Microencapsulation and transplantation of genetically engineered cells: A new approach to somatic gene therapy. *Artif. Cells Blood Substitutes Immobil. Biotechnol.* **24**, 219–255.

Bilbo, P. R., Nolte, C.J.M., Oleson, M. A., Mason, V. S., and Parenteau, N. L. (1993). Skin in complex culture: The transition from 'culture' phenotype to organotypic phenotype. *J. Toxicol. Cutaneous Ocul. Toxicol.* **12**, 183–196.

Briscoe, D. M., Dharnidharka, V. R., Isaacs, C., Downing, G., Prosky, S., Parenteau, N. L., and Hardin-Young, J. (1999). The allogeneic response to cultured human skin equivalent in the hu-PBL-SCID mouse model of skin rejection. *Transplantation* **67**(12), 1590–1599.

Brittberg, M., Lindahl, A., Nilsson, A., Ohlsson, C., Isaksson, O., and Peterson, L. (1994). Treatment of deep cartilage defects in the knee with autologous chondrocyte transplantation. *N. Engl. J. Med.* **331**(14), 889–895.

Burke, J. F., Yannas, I. V., Quinby, W. C., Bondoc, C. C., and Jung, W. K. (1981). Successful use of a physiologically acceptable artificial skin in the treatment of extensive burn injury. *Ann. Surg.* **194**, 413–428.

Cao, Y., Rodriguez, A., Ibarra, C., Arevalo, C., and Vacanti, A. (1998). Comparative study of the use of poly(glycolic acid), calcium alginate and pluronics in the engineering of autologous porcine cartilage. *J. Biomater. Sci. Polym. Ed.* **9**(5), 475–487.

Chang, A. S., and Yannas, I. V. (1992). Peripheral nerve regeneration. *In* "Neuroscience Year" (Supplement 2 to the "Encyclopedia of Neuroscience" (B. Smith and G. Adelmand, eds.), pp. 125–126. Birkhaeuser, Boston.

Davis, M. W., and Vacanti, J. P. (1996). Toward development of an implantable tissue engineered liver. *Biomaterials* **17**, 365–372.

Dunn, J. C., Yarmush, M. L., Koebe, H. G., and Tompkins, R. G. (1989). Hepatocyte function and extracellular matrix geometry: Long-term culture in a sandwich configuration. *FASEB J.* **3**, 174–177; erratum: **3**, 1873 (1989).

Dunn, J. C., Tompkins, R. G., and Yarmush, M. L. (1992). Hepatocytes in collagen sandwich: Evidence for transcriptional and translational regulation. *J. Cell Biol.* **116**, 1043–1053.

Eaglstein, W. H., Iriondo, M., and Laszlo, K. (1991). A composite skin substitute (Graftskin) for surgical wounds: A clinical experience. *Dermatol. Surg.* **21**, 839–843.

Edge, A.S.B., Gosse, M. E., and Dinsmore, J. (1998). Xenogeneic cell therapy: Current progress and future developments in porcine cell transplantation. *Cell Transplant.* **7**(6), 525–539.

Eiselt, P., Kim, B. S., Chacko, B., Isenberg, B., Peters, M. C., Greene, K. G., Roland, W. D., Loebsack, A. B., Burg, K. J., Culberson, C., Halberstadt, C. R., Holder, W. D., and Mooney, D. J. (1998). Development of technologies aiding large-tissue engineering. *Biotechnol. Prog.* **14**(1), 134–140.

Ezzell, R. M., Toner, M., Hendricks, K., Dunn, J. C., Tompkins, R. G., and Yarmush, M. L. (1993). Effect of collagen gel configuration on the cytoskeleton in cultured rat hepatocytes. *Exp. Cell Res.* **208**(2), 442–452.

Falanga, V., Margolis, D., Alvarez, O., Auletta, M., Maggiacomo, F., Altman, M., Jensen, J., Sabolinski, M., Hardin-Young, J., and Human Skin Equivalent Investigators Group (1998). Rapid healing of venous ulcers and lack of clinical rejection with an allogeneic cultured human skin equivalent. *Arch. Dermatol.* **134b**, 293–300.

Fishman, J. A. (1998). Infection and xenotransplantation: Developing strategies towards clinical trials. *Graft* **1**(5), 181–186.

Freed, L. E., Vunjak-Novakovic, G., and Langer, R. (1993). Cultivation of cell–polymer cartilage implants in bioreactors. *J. Cell. Biochem.* **51**, 257–264.

Freed, L. E., Grande, D. A., Lingbin, Z., Emmanual, J., Marquis, J. C., and Langer, R. (1994a). Joint resurfacing using allograft chondrocytes and synthetic biodegradable polymer scaffolds. *J. Biomed. Mater. Res.* **28**(8), 891–899.

Freed, L. E., Vunjak-Novakovic, G., Biron, R. J., Eagles, D. B., Lesnoy, D. C., Barlow, S. K., and Langer, R. (1994b). Biodegradable polymer scaffolds for tissue engineering. *Bio/Technology* **12**(7), 689–693.

Gallico, G. G., O'Connor, N. E., Compton, C. C., Kehinde, O., and Green, H. (1984). Permanent coverage of large burn wounds with autologous cultured human epithelium. *N. Engl. J. Med.* **311**(7), 488–451.

Gaspari, A. A., and Katz, S. I. (1988). Induction and functional characterization of MHC class II antigen on murine keratinocytes. *J. Immunol.* **140**, 2956–2963.

Geppert, T. D., and Lipsky, P. E. (1985). Antigen presentation by interferon-γ-treated endothelial cells and fibroblasts: Differential ability to function as antigen-presenting cells despite comparable Ia expression. *J. Immunol.* **135**, 3750–3762.

Goldstein, J. D., Tria, A. J., Zawadsky, J. P., Kato, Y. P., Christiansen, D., and Silver, F. H. (1989). Development of a reconstituted collagen tendon prosthesis. Preliminary implantation study. *J. Bone J. Surg.* **71**(8), 1183–1191.

Green, H., Kehinde, O., and Thomas, J. (1979). Growth of cultured human epidermal cells into multiple epithelia suitable for grafting. *Proc. Natl. Acad. Sci. U.S.A.* **76**, 5665–5668.

Habal, M. B., and Reddi, A. H. (1994). Bone grafts and bone induction substitutes. *Clin. Plast. Surg.* **21**, 525–542.

Hayashi, H., Inoue, K., Shinohara, S., Gu, Y. J., Setoyama, H., Kawakami, Y., Yamasaki, T., Lui, W. X., Kinoshita, N., Imamura, M., Iwata, H., Ikada, Y., and Miyazaki, J. (1998). New approach by tissue engineering for extended selective transplantation with a pancreatic B-cell line (MIN6). *Transplant. Proc.* **30**, 83–85.

Humes, H. D. (1995). Tissue engineering of the kidney. *In* "The Biomedical Engineering Handbook" (J. D. Bronzino, ed.), pp. 1807–1824. CRC Press, Boca Raton, FL.

Johnson, E. W., Meunier, S. F., Roy, C. J., and Parenteau, N. L. (1992). Serial cultivation of normal human keratinocytes: A defined system for studying the regulation of growth and differentiation. *In Vitro Cell. Dev. Biol.* **28**, 429–435.

Kato, Y. P., Dunn, M. G., Zawadsky, J. P., Tria, A. J., and Silver, F. H. (1991). Regeneration of Achilles tendon with a collagen tendon prosthesis. *J. Bone J. Surg.* **73**(4), 561–574.

Kemp, P. D., Cavallaro, J. F., and Hastings, D. N. (1995). Effects of carbodiimide crosslinking and load environment on the remodeling of collagen scaffolds. *Tissue Eng.* **1**, 71–79.

Koller, M. R., Emerson, S. G., and Palsson, B. O. (1993). Large-scale expansion of human stem and progenitor cells from bone marrow mononuclear cells in continuous perfusion cultures. *Blood* **82**, 378–384.

Laning, J. C., Isaacs, C. M., and Hardin-Young, J. (1997). Normal human keratinocytes inhibit the proliferation of unprimed T cells by TGF β and PGE_2, but not IL-10. *Cell. Immunol.* **175**, 16–27.

Laning, J. C., DeLuca, J. E., and Hardin-Young, J. (1999). Effects of immunoregulatory cytokines on the immunogenic potential of the cellular components of a bilayered living skin equivalent. *Tissue Eng.* **5**(2), 171–180.

Livesey, S. A., Herndon, D. N., Hollyoak, M. A., Atkinson, Y. H., and Nag, A. (1995). Transplanted acellular allograft dermal matrix. *Transplantation* **60**, 1–9.

Moulton, K. S., Dharnidharka, V. R., Hardin-Young, J., Melder, R. J., Jain, R. K., and Briscoe, D. M. (1999). Angiogenesis in the hu-PBL-SCID model of human transplant rejection. *Transplantation* **67**(12), 1626–1631.

Murray, A. G., Libby, P., and Pober, J. S. (1995). Human vascular smooth muscle cells poorly costimulate and actively inhibit allogeneic CD4+ T cell proliferation *in vitro*. *J. Immunol.* **154**, 151–161.

Naumann, A., Rotter, N., Bujia, J., and Aigner, J. (1998). Tissue engineering of autologous cartilage transplants for rhinology. *Am. J. Rhinol.* **12**, 59–63.

Parenteau, N. L., Bilbo, P., Nolte, C.J.M., Mason, V. S., and Rosenberg, M. (1992). The organotypic culture of human skin keratinocytes and fibroblasts to achieve form and function. *Cytotechnology* **9**, 163–171.

Pober, J. S., Collins, T., Gimbrone, M. A., Jr., Libby, P., and Reiss, C. S. (1986). Inducible expression of class II major histocompatibility complex antigens and the immunogenicity of vascular endothelium. *Transplantation* **41**(2), 141–146.

Reddi, A. H. (1994). Symbiosis of biotechnology and biomaterials: Applications in tissue engineering of bone and cartilage. *J. Cell. Biochem.* **56**, 192–195.

Reddi, A. H., and Anderson, W. A. (1976). Collagenous bone matrix-induced endochondral ossification and hemopoiesis. *J. Cell Biol.* **69**, 557–572.

Rheinwald, J. G., and Green, H. (1975). Serial cultivation of strains of human epidermal keratinocytes: The formation of keratinizing colonies from single cells. *Cell (Cambridge, Mass.)* **6**, 331–344.

Sabolinski, M. L., Alvarez, O., Auletta, M., Mulder, G., and Parenteau, N. L. (1996). Cultured skin as a "smart material" for healing wounds: Experience in venous ulcers. *Biomaterials* **17**(3), 311–320.

Schaefer, S., Stone, B. B., Wolfrum, J., and du Moulin, G. (1999). *In* "Biopharmaceutical Process Validation" (D. Zabriskie and G. Sofer, eds.). Dekker, New York (in press).

Schilling, A., Land, W., Pratschke, E., Pielsticker, K., and Brendel, W. (1976). Dominant role of complement in the hyperacute xenograft rejection reaction. *Surg. Gynecol. Obstet.* **142**(1), 29–32.

Speer, D. P., Chvapil, M., Volz, R. G., and Holmes, M. D. (1979). Enhancement of healing in osteochondral defects by collagen sponge implants. *Clin. Orthop. Relat. Res.* **144**, 326–335.

Stone, K. R., Rodkey, W. G., Webber, R. J., McKinney, L., and Steadman, J. R. (1990). Future directions. Collagen-based prostheses for meniscal regeneration. *Clin. Orthop.* **252**, 129–135.

Sussman, N. L., Gislason, G. T., Conlin, C. A., and Kelley, J. H. (1994). The Hepatix extracorporeal liver assist device: Initial clinical experience. *Artif. Organs* **18**(5), 390–396.

Takeshita, S., Tsurumi, Y., Couffinahl, T., Asahara, T., Bauters, C., Symes, J., Ferrara, N., and Isner, J. M. (1996). Gene transfer of naked DNA encoding for three isoforms of vascular endothelial growth factor stimulates collateral development *in vivo*. *Lab. Invest.* **75**(4), 487–501.

Termin, P. T., Carr, R. M., and O'Neil, K. D. (1994). A comparison of the healing characteristics of a novel synthetic collagen graft and c-PTFE. *J. Cell. Biochem.* **18**(Suppl. C), 281.

Theobald, V. A., Lauer, J. D., Kaplan, F. A., Baker, K. B., and Rosenberg, M. (1993). "Neutral allografts"—Lack of allogeneic stimulation by cultured human cells expressing MHC class I and class II antigens. *Transplantation* **55**(1), 128–133.

Thomson, R. C., Yaszemski, M. J., Powers, J. M., and Mikos, A. G. (1995). Fabrication of biodegradable polymer scaffolds to engineer trabecular bone. *J. Biomater. Sci., Polym. Ed.* **7**(1), 23–38.

Thomson, R. C., Yaszemski, M. J., Powers, J. M., and Mikos, A. G. (1998). Hydroxyapatite fiber reinforced poly(alpha-hydroxy ester) foams for bone regeneration. *Biomaterials* **19**(21), 1935–1943.

Vacanti, C. A., Langer, R., Schloo, B., and Vacanti, J. P. (1991). Synthetic polymers seeded with chondrocytes provide a template for new cartilage formation. *Plast. Reconstr. Surg.* **88**(5), 753–759.

Van Zant, G., Rummel, S. A., Koller, M. R., Larson, D. B., Drubachevsky, I., Palsson, M., and Emerson, S. G. (1994). Expansion of bioreactors of human progenitor populations from cord blood and mobilized peripheral blood. *Blood Cells* **20**(2–3), 482–490.

von Schroeder, H. P., Kwan, M., Amiel, D., and Coutts, R. D. (1991). The use of polylactic acid matrix and periosteal grafts for the reconstruction of rabbit knee articular defects. *J. Biomed. Mater. Res.* **25**(3), 329–339.

Wake, M. C., Mikos, A. G., Sarakinos, G., Vacanti, J. P., and Langer, R. (1995). Dynamics of fibrovascular tissue ingrowth in hydrogel foams. *Cell Transplant.* **4**(3), 275–279.

Wakitani, S., Goto, T., Pineda, S. J., Young, R. G., Mansour, J. M., Caplan, A. I., and Goldberg, V. M. (1994). Mesenchymal cell-based repair of large, full-thickness defects of articular cartilage. *J. Bone J. Surg. Am. Vol.* **76-A**(4), 579–592.

Wilkins, L. M., Watson, S. R., Prosky, S. J., Meunier, S. F., and Parenteau, N. L. (1994). Development of a bilayered living skin construct for clinical applications. *Biotechnol. Bioeng.* **43**, 747–756.

Yannas, I. V., Burke, J. F., Orgill, D. P., and Skrabut, E. M. (1982). Wound tissue can utilize a polymeric template to synthesize a functional extension of skin. *Science* **215**, 174–176.

Yannas, I. V., Lee, E., Orgill, D. P., Skrabut, E. M., and Murphy, G. F. (1989). Synthesis and characterization of a model extracellular matrix that induces partial regeneration of adult mammalian skin. *Proc. Nat. Acad. Sci. U.S.A.* **86**, 933–937.

CHAPTER 24

CRYOPRESERVATION

Jens O. M. Karlsson and Mehmet Toner

INTRODUCTION

Cold has the power to preserve and the power to destroy. Consider, for instance, the woolly mammoth finds in Siberia—the tissues of these late Pleistocene animals have been sufficiently well preserved to make possible the isolation of intact protein (Prager *et al.*, 1980) and DNA (Johnson *et al.*, 1985) after as much as 50,000 years of frozen storage. On the other hand, the destructive effects of cold temperatures are also well known: frostbite destroys skin tissue, and frost damage to plants causes major crop losses every year. Scientists have exploited the deleterious effects of low temperatures: freeze–thaw cycles are routinely used by biologists to deliberately lyse cells; in cryosurgery, liquid-nitrogen-cooled probes are used to ablate cancerous tissue. In order to also harness the preservative powers of cold, making possible the storage of living biological materials in a state of "suspended animation," it is necessary to prevent cell and tissue damage during the potentially destructive procedures of freezing and thawing. Although the challenges presented by this problem are sometimes daunting, the potential benefits of cryopreservation to medicine and biotechnology are great.

A significant body of knowledge has been accumulated about the mechanisms of freezing-related damage to cells, the key component of bioartificial organs and tissues (Mazur, 1984). Mathematical models have made possible the prediction of the physicochemical response of cells to freezing (Pitt, 1992; Toner, 1993) and the rational design of cell freezing protocols by computer-aided optimization (Karlsson *et al.*, 1996). Although the progress in developing cryopreservation procedures for cells is encouraging, the scale and complexity of tissues and organs present additional problems that must be overcome before cryopreservation can become widely used for these systems (Karlsson and Toner, 1996). Thus, cryopreservation technology is currently evolving in parallel with the field of tissue engineering as a whole: significant advances have been made to date, but many challenges remain. In this chapter, we present an overview of the principles of cryobiology as they apply to both cells and tissues, emphasizing the use of mathematical models to guide the design and optimization of cryopreservation procedures.

APPLICATIONS OF CRYOPRESERVATION TECHNOLOGY IN TISSUE ENGINEERING

Cryobiology is becoming a focus of research in tissue engineering, because preservation is a core technology in bringing cell-based medical devices to market. Effective preservation procedures are required at the following critical steps in the tissue engineering production cycle:

1. Screening of source cells: Nonautologous source cells must be extensively tested for adventitious agents. Such validation can require months of time (Bell and Rosenberg, 1990), during which the source cell pool must be preserved in order to prevent contamination or genotypic changes.

2. Cell banking: To ensure reproducibility in the manufacturing process, the United States

Food and Drug Administration requires the establishment of Master and Working Cell Banks, which must be preserved under conditions that assure genetic stability (Wiebe and May, 1990).

3. Inventory control: Adequate preservation technology is a prerequisite to maintaining a product inventory in order to meet end-user demand. Because cell expansion over several weeks may be necessary to populate a tissue-engineered device (Green *et al.*, 1979), just-in-time manufacturing approaches can be difficult to implement. Thus, the lead time for product delivery may be unacceptably long if long-term storage is not possible.

4. Quality control: Cryopreservation of cell or tissue samples at each step of production can be used for subsequent quality-control testing, and for creating cell and tissue archives documenting the manufacture of each lot of tissue. Because regulation of tissue-engineered devices is still an evolving area, such archives may become necessary for validation of the manufacturing process.

5. Biological manipulations: Freeze–thaw protocols can be used to effect desirable biological changes in tissue. By selectively destroying or impairing immunostimulatory cells, the cryopreservation process can reduce tissue immunogenicity (Taylor *et al.*, 1987; Ingham *et al.*, 1993). In skin grafting applications, freezing can be used to reduce the metabolic activity of fresh tissue to therapeutically optimal levels (Naughton *et al.*, 1997; Naughton, 1999).

6. Product distribution: Shipping of tissue-engineered products from the manufacturing plant to the end-user requires product stability in transit. Adequate technologies for long-term tissue preservation are especially critical for companies with a single manufacturing facility, because constraints on product distribution caused by shelf life limitations will impede access to geographically distant markets.

7. Tissue banking: The need for tissue or organ replacements in hospitals is variable and inherently unpredictable. Thus, to ensure immediate availability of tissue-engineered devices, it will be necessary to establish clinical tissue banks.

Compared with other methods of storing cells and tissue (refrigeration, chemical preservation, *in vitro* culture), freezing to cryogenic temperatures has the benefits of affording long shelf lives with assured genetic stability, virtually no risk of microbial contamination during storage, and improved cost effectiveness (as an illustrative example, the cost of liquid nitrogen used for cryogenic storage is approximately $1/gallon, whereas serum required to maintain a tissue in culture costs well over $1000/gallon). Thus, cryopreservation is currently the most viable approach to meet the requirements of production, distribution, and end-use of tissue-engineered products.

CHALLENGES IN CRYOPRESERVATION PROTOCOL DEVELOPMENT

The cryopreservation process causes significant changes in the thermal, chemical, and physical environment in the tissue, with attendant risks of biological damage. Temperature changes are mainly due to heat-transfer boundary conditions imposed by the freezing and warming methods, but can also be affected by the latent heat of fusion of ice in the tissue. The chemical environment may be altered prior to freezing by addition of cryoprotectant agents, but it is also modulated by the formation of ice in the suspending medium. The growth of this ice phase removes water from the remaining unfrozen solution, thus enriching the medium in solutes, and lowering the chemical potential of the unfrozen water. In addition to effecting these chemical changes, the presence of ice also modifies the physical environment in the tissue, causing alterations in tissue structure. Furthermore, mechanical forces may arise in the tissue due to thermal stress phenomena. Every change imposed on the tissue during cryopreservation disrupts homeostasis, and has the potential to cause irreversible damage.

In order to minimize tissue destruction during freezing and thawing, methods for chemical and thermal processing must be optimized. Chemical manipulations are typically limited to the introduction of one or more cryoprotectant additives into the tissue before freezing, and removal of these additives after thawing. One must choose the types of cryoprotectant to use, and develop a protocol for adding and removing these chemicals with a minimum of deleterious effects to the tissue. Addition and dilution protocols typically involve one or more steps of bathing the tissue in a solution of specified composition and temperature, for a given amount of time (i.e., three parameters must be specified for each step). Thermal processing methods are usually specified in terms of the temperature versus time profile $T(t)$ to be imposed at the boundary of the tissue, because our ability to control the spatial temperature in the tissue interior is limited. The temperature history $T(t)$ may be linear or nonlinear, but for design purposes it is customarily divided into piecewise linear segments, each of which is defined by a constant rate of temperature change, and a limit temperature (or time) at which the next segments begins (thus two parameters must be specified for each segment).

Consider now the development of a cryopreservation procedure. If there are n steps in the chemical-processing protocol, and m steps in the temperature profile, there will be at least $p = 3n + 2m$ protocol parameters to optimize. Because there is interaction between the various processing steps (e.g., the effect of cooling rate depends on the cryoprotectant concentration; the effect of the warming procedure depends on the preceding freezing protocol), all p parameters should be simultaneously optimized. Whereas in the simplest case, $n = 2$ (a single cryoprotectant addition step and a single dilution step) and $m = 2$ (linear cooling and linear warming), optimization using a full factorial design with only two levels for each parameter would require $2^p = 1024$ experiments. Clearly, the number of experiments required for rigorous optimization of cryopreservation procedures is prohibitively large, even for the simplest class of protocols. Thus, it is common practice to optimize sequentially individual parameters in the protocol, using as few as $p + 1$ experiments (for two levels per parameter) to yield a pseudooptimal solution that will be strongly dependent on the initial values used in the search. As a result of the shortcomings of experimental protocol optimization, the empirical development of cryopreservation procedures has been limited to systems that are relatively robust, or those in which damage can be tolerated (McGrath, 1985). The successful design of cryopreservation procedures for complex tissues, however, will probably require the use of theoretical models and computer-aided optimization methods. Thus, the mechanisms of tissue damage during freezing and thawing must be understood, and appropriate mathematical models developed.

CRYOBIOLOGY OF CELLS

Background and Historical Perspective

The study of the behavior of individual cells during freezing and thawing provides a starting point for analysis of tissue damage during cryopreservation. Cells are usually the key component of an engineered tissue, providing the tissue-specific bioactivity required to achieve a therapeutic effect. Even in tissues designed to have a purely biomechanical function (e.g., heart valves), the presence of a viable cell population will reduce the risk of postimplantation graft failure (McNally and McCaa, 1988). Thus, even though there are multiple modes of injury to tissue during cryopreservation, minimizing cell damage must be a major goal in designing protocols for the freezing and thawing of tissue, as well as for cryoprotectant addition and removal.

The first reports of low-temperature biological research are found in the writings of Sir Robert Boyle (1693). However, significant progress toward development of successful cryopreservation protocols was not made until the twentieth century, when two major breakthroughs were achieved. First, the 1949 discovery of the cryoprotective properties of glycerol led to the now ubiquitous and successful use of cryoprotectant additives in freezing media (Polge *et al.,* 1949). Second, the use of mathematical models to predict the behavior of cells during freezing was pioneered by Mazur (1963), and paved the way for subsequent successes in predicting cell damage due to intracellular ice formation (reviewed by Karlsson *et al.,* 1993b). In what follows, an overview of the principles of cell cryobiology and the mathematical models describing the governing phenomena will be given, after which we will turn to the problem of tissue preservation.

Experimental Observations of Cell Freezing

The response of the cell to the freezing process can be observed using a cryomicroscope, a microscope fitted with a low-temperature stage. Modern cryomicroscope designs use digital feedback control systems to achieve precise temperature regulation, and include a video camera for quantitative image analysis (Diller, 1982). As can be seen in Fig. 24.1, the cell response depends on the method of freezing, and, in particular, the cooling rate is a critical parameter in determining the outcome of the freezing protocol. When the extracellular ice forms, a chemical potential difference across cell membranes results, driving water out of the cell by osmosis. However, the plasma

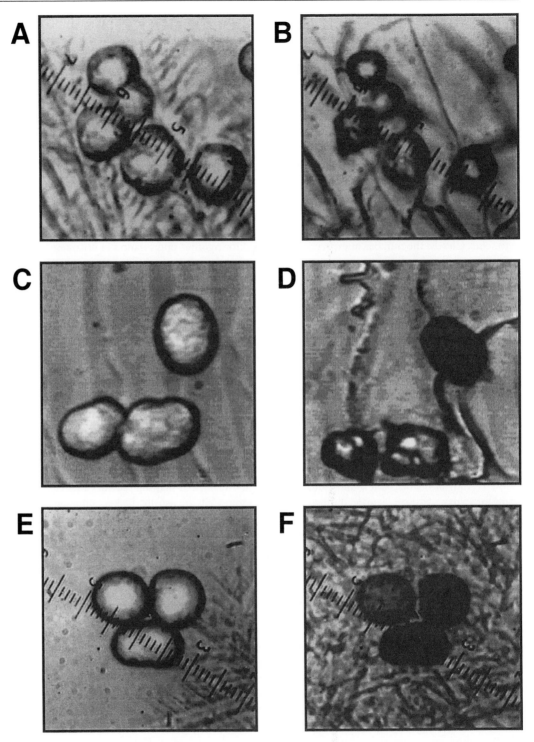

Fig. 24.1. Behavior of cells during freezing, observed using a cryomicroscope. Three separate freezing experiments are shown, in which rat hepatocytes were cooled at different rates: 10°C/min (A, B), 100°C/min (C, D), and 400°C/min (E, F). Cells are shown after seeding of extracellular ice at −1°C (A, C, E) and after cooling to −40°C (B, D, F).

membrane has a finite permeability to water, the magnitude of which determines the rate of water efflux and the corresponding time scale of cell dehydration. Thus, if the rate of cooling is sufficiently slow to allow the intracellular solution to equilibrate with its external environment by expressing water through the cell membrane, the cell will dehydrate extensively with decreasing temperature (see A and B, Fig. 24.1). On the other hand, if the cooling rate is fast compared with the rate of water efflux, low temperatures are reached before significant dehydration can occur. In this latter case, the cell remains largely undeformed, but there is a very high probability of ice formation in the cell, inasmuch as the intracellular solution is in a supercooled nonequilibrium state. Because cryomicroscopes use transmitted light, intracellular ice formation appears as a darkening of the cytoplasm, due to scattering of light by ice crystals in the cell (E and F, Fig. 24.1). At intermediate cooling rates, cells are partially dehydrated, and thus near equilibrium. However, there is still some supercooling of the cytoplasm, and therefore intracellular ice crystals may form in some cells (C and D, Fig. 24.1).

If one measures the survival of cells frozen using various cooling rates, one finds that post-thaw viability decreases with increasing rate of cooling (Mazur *et al.*, 1972). The cooling rate at which the probability of cell survival drops to 50% also corresponds to the cooling rate at which 50% of cells are observed to form intracellular ice (Toner, 1993). However, at very low rates of cooling, a drop-off in survival is seen with *decreasing* rates of cooling (Fig. 24.2). Based on experimental observations of the cooling rate dependence of cell survival, Mazur *et al.* (1972) proposed the two-factor hypothesis of cell freezing injury, which posits that two distinct mechanisms are responsible for cell damage during cryopreservation.

One mode of injury is dominant at high cooling rates, and is associated with intracellular ice formation. The mechanism of damage has not been fully elucidated, but is thought to be due to mechanical disruption of the plasma membrane and/or other structures by intracellular ice crystals (Karlsson *et al.*, 1993b). A dissenting theory holds that the cell membrane is damaged by osmotic stresses arising from intracellular supercooling, and that extracellular ice crystals grow into the cytoplasm following membrane rupture (Muldrew and McGann, 1994). In either case, the extent of damage will be higher with increasing supercooling, and thus the probability of cell survival decreases with increasing cooling rate.

A second mode of injury, unrelated to intracellular ice formation, is hypothesized to be active at low cooling rates. This mechanism of cell damage is usually referred to as "solution effects" injury and is believed to be related to cell dehydration, mechanical deformation of the cell, and/or the prolonged exposure to high intra- and extracellular concentrations of electrolytes and cryoprotectants (Mazur *et al.*, 1972). Because the extent of cell dehydration and the time of exposure to these deleterious conditions both increase with decreasing cooling rate, cell damage due to solution effects should decrease at increasing rates of cooling.

Thus, as shown in Fig. 24.2, there is an optimum cooling rate at which the two mechanisms of damage are balanced, and the probability of cell survival reaches a maximum. Optimal values of other protocol parameters (e.g., in multistep freezing procedures) can similarly be explained in terms of a balance between solution effects and intracellular ice formation (Karlsson *et al.*, 1996). Although the experimental evidence supports the universality of the two-factor hypothesis as a qualitative explanation of cell damage during freezing, the optimum cryopreservation method is known to differ from cell type to cell type (Mazur *et al.*, 1972). This is a result not only of variable sensitivity to freezing injury, but also of variations in cell biophysical properties such as the water permeability of the plasma membrane and the activity of ice nucleating catalytic sites in the

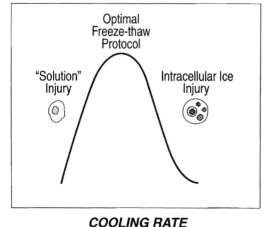

Fig. 24.2. Effect of cooling rate on cell survival. According to the two-factor hypothesis, cell damage at low cooling rates is due to cell dehydration and attendant exposure to concentrated solutions, whereas cell injury at high rates of cooling is due to intracellular ice formation.

cell, which affect the rate of dehydration and the probability of intracellular ice formation, respectively (McGrath, 1988; Toner, 1993). Thus, every cell type requires its own cryopreservation protocol to be developed and optimized, a fact that has spurred efforts to develop rational design methods for the optimization of freezing and warming procedures (Karlsson *et al.*, 1996). Such computer-aided design techniques require theoretical models of the behavior of cells at low temperatures.

MATHEMATICAL MODELING OF THE CELLULAR RESPONSE TO CRYOPRESERVATION

The use of mathematical models to predict the effect of cryopreservation on cells was pioneered by Mazur (1963), who developed the first models of water transport during freezing. Though useful for advancing our understanding of the qualitative behavior of cells during cryopreservation, these early models were unable to predict the probability of intracellular ice formation. This problem was subsequently addressed by Pitt and co-workers, who developed a series of phenomenological models based on fitting probability distributions to the experimental data (Pitt, 1992), and by Cravalho and co-workers, who initiated an effort to develop a mechanistic thermodynamic model of the intracellular water–ice phase transition (Cravalho, 1976; Karlsson *et al.*, 1993b). The latter effort culminated with the development of a mechanistic, fully integrated model of water transport, intracellular ice nucleation, and crystal growth in the presence of cryoprotectants (Karlsson *et al.*, 1994), based in part on the earlier work of Toner and colleagues (1990), in which classical nucleation theory was successfully used to predict the probability of intracellular ice nucleation in the absence of cryoprotectants. The essential model equations are described below.

The description of cell dehydration is based on the original two-compartment, membrane-limited transport model of Mazur (1963), modified here to take into account the presence of cryoprotectants:

$$\frac{dV}{dt} = \left(\frac{L_{pg}A}{v_w}\right) \exp\left[\left(\frac{E_{Lp}}{R}\right)\left(\frac{1}{T_0} - \frac{1}{T}\right)\right]\left(\Delta H_f\left(\frac{T}{T_0} - 1\right) - RT \ln\left[\frac{V - \phi_s V_0}{V + \left(\prod_s v_w - \phi_s\right)V_0}\right]\right), \quad (1)$$

where V is the volume of the cell cytoplasm; V_0, the initial value of V; A, the cell membrane surface area; t, time; ϕ_s and $\prod_s$, the initial values of the volume fraction and osmolarity of intracellular solutes (native electrolytes and cryoprotectant additives), respectively; T, temperature; T_0, the equilibrium melting temperature of water; ΔH_f, the specific heat of fusion of water; v_w, the specific volume of water; L_{pg}, the membrane water permeability at T_0; E_{Lp}, the activation energy for water transport; and R, the gas constant. The main assumptions of this model are that the intracellular solution is ideal and well mixed, and that the cell membrane is impermeable to all species except water. Equation (1) can be solved numerically for a given cooling protocol $T(t)$ to yield predictions of cell volume, water content, solute content, and cytoplasmic supercooling (Fig. 24.3). These parameters are required in order to predict the probability of ice formation in the cell. Figure 24.3 shows predicted values of water content, cryoprotectant concentration, and cell supercooling for rat hepatocytes frozen using various cooling rates. If cooling is fast, the water content and intracellular cryoprotectant concentration remain approximately constant during freezing. As a result, the cytoplasmic supercooling increases almost linearly with decreasing temperature, dramatically increasing the probability of intracellular ice formation. In contrast, for slow cooling rates, the cell dehydrates extensively, allowing the intracellular solution to remain in thermodynamic equilibrium (no supercooling) for temperatures as low as $-40°C$. Thus, the probability of intracellular ice formation will be very small. On the other hand, the loss of cell water caused by slow freezing results in significant increase in the intracellular concentration of cryoprotectant and electrolytes, possibly reaching cytotoxic levels.

The intracellular ice nucleation model is based on classical nucleation theory (Christian, 1975; Turnbull and Fisher, 1949), first applied to intracellular ice formation predictions by Toner (Toner *et al.*, 1990). This model has subsequently been extended to incorporate the effects of cryoprotectant additives on the rate of ice nucleation (Karlsson *et al.*, 1993a, 1994). The total rate of formation of ice nuclei inside one cell is given by

$$J = I_O V + I_S A + I_V V, \quad (2)$$

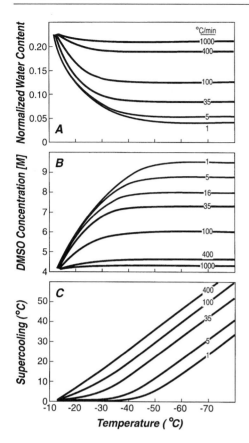

Fig. 24.3. Predicted temperature dependence of intracellular water content (A), DMSO concentration (B), and supercooling (C), during freezing of rat hepatocytes at the indicated cooling rates. Adapted from Karlsson et al. (1993a), with permission from the Biophysical Society.

where J is the nucleation rate per cell; I_O, I_S, and I_V are specific nucleation rates corresponding to three different mechanisms: homogeneous nucleation (I_O), and heterogeneous nucleation mediated by catalytic sites on the cell surface (I_S) or inside the cell volume (I_V), respectively. According to classical nucleation theory, the rate of homogeneous nucleation has a temperature dependence of the form

$$I_O = \Omega_O \exp[-\kappa_O \Delta T^{-2} T^{-3}], \tag{3}$$

where ΔT is the extent of supercooling of the solution, and Ω_O and κ_O are the kinetic and thermodynamic nucleation rate parameters, respectively. The coefficients Ω_O and κ_O are dependent on temperature and on the chemical composition of the solution (Karlsson *et al.*, 1994). The rates of heterogeneous nucleation can be determined as follows (Karlsson *et al.*, 1993a):

$$I_S = w_S \Omega_O \exp[-k_S \kappa_O \Delta T^{-2} T^{-3}], \tag{4}$$

$$I_V = w_V \Omega_O \exp[-k_V \kappa_O \Delta T^{-2} T^{-3}], \tag{5}$$

where w_S, k_S, w_V, and k_V are kinetic and thermodynamic coefficient scaling factors for surface-catalyzed and volume-catalyzed nucleation, respectively, and are constants that can be determined experimentally (Karlsson *et al.*, 1993a).

The crystal growth model assumes that ice nuclei grow spherically, and that growth is rate limited by diffusion of water monomers to the ice–liquid interface. The total volume of ice forming inside a cell during freezing is given by

$$V_{\text{ice}} = \frac{4\pi}{3} \sum_{i=1}^{N} \left[\int_{t_i}^{t} \alpha^2 D \, dt \right]^{3/2}, \tag{6}$$

where N is the number of intracellular ice nuclei; t_i, the time of nucleation event i; α, a nondimensional crystal growth parameter; and D, the effective diffusivity of water in the cytoplasm. N

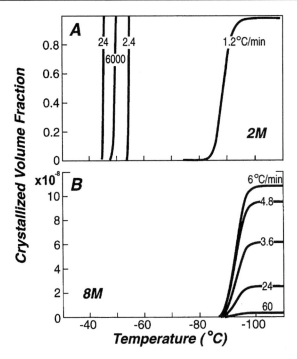

Fig. 24.4. Predicted volume fraction of intracellular ice during freezing of mouse oocytes at the indicated cooling rates. The cells were equilibrated with 2 M (A) or 8 M (B) glycerol prior to freezing. For simplicity, homogeneous nucleation (I_O) was the only ice formation mechanism included in these simulations. Reprinted with permission from Karlsson et al. (1994), J. Appl. Phys. *75(9)*, Fig. 4, p. 4448. Copyright 1994, American Institute of Physics.

and t_i can be determined from Eq. (2); the calculation of α and D has been previously described by Karlsson *et al.* (1994).

Figure 24.4 illustrates the effects of cooling rate and cryoprotectant concentration on ice formation in mouse oocytes. When freezing the cells in 2 *M* glycerol, the cell volume crystallizes completely on appearance of the first ice nucleus. However, the nucleation temperature can be depressed either by using slow cooling to dehydrate the cell, or by using ultrarapid cooling to lower the ice formation temperature kinetically (Karlsson *et al.*, 1994). When 8 *M* glycerol is used, ice crystal growth is arrested before the volume of intracellular ice reaches damaging levels, for all cooling rates. This effect is due to the lowering of the water diffusivity D by highly concentrated cryoprotectant solutions, and is the basis for vitrification technology (Fahy *et al.*, 1984). Innocuous levels of intracellular ice formation can also be achieved with lower, nontoxic concentrations of cryoprotectant, by using a multistep temperature profile with an initial slow cooling step to dehydrate the cell, and a subsequent rapid cooling step to suppress ice nucleation kinetically (Karlsson *et al.*, 1996). Mathematical models are essential in designing and optimizing such protocols.

CRYOPROTECTANT ADDITIVES

Polge and co-workers discovered in 1949 that addition of glycerol to a suspension of spermatozoa prior to freezing resulted in a dramatic increase in the number of viable cells recovered after thawing (Polge *et al.*, 1949). Many other chemicals have since been found to have cryoprotective properties, but glycerol and dimethyl sulfoxide (DMSO) remain the most commonly used to date. Cryoprotectants are typically categorized into two groups according to their ability to enter the cell. Examples of permeating cryoprotectants include glycerol, DMSO, ethylene glycol, and 1,2-propanediol. Nonpermeating cryoprotectants include polyvinyl pyrrolidone, hydroxyethyl starch, and various sugars (e.g., trehalose). The mechanism of cryoprotection by these chemicals is not fully understood. However, both permeating and nonpermeating cryoprotectants reduce the concentration of intracellular water, thus reducing the rate at which the remaining water molecules can form damaging ice crystals. Furthermore, permeating cryoprotectants act as a solvent, diluting the intracellular electrolyte solution when water is removed from the cell during freezing, and thus protecting the cell from damage due to solution effects (Mazur, 1984). There is also evidence that cryoprotectants may directly interact with the plasma membrane and cell proteins to stabilize these structures (Timasheff, 1982; Anchordoguy *et al.*, 1991). Nonetheless, cryoprotectant chemicals may also be cytotoxic, in particular if the time of exposure is long, or if they are used in high concentrations (Fahy, 1986). Cell damage may also occur during the prefreeze addi-

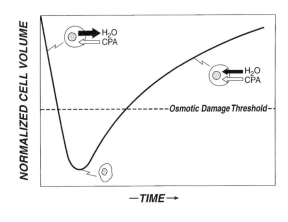

Fig. 24.5. Shrink–swell behavior of cells during exposure to a penetrating cryoprotectant additive (CPA). The cell volume initially decreases due to osmotically driven water loss, then increases as cryoprotectant and water enter the cell. Cell damage can result if the transient volume change is too large.

tion and postthaw removal of cryoprotectant, as a result of osmotic stresses (Levin and Miller, 1981). Thus, care must be taken in choosing the type of additive to use, determining the minimum concentration that will result in effective cryoprotection, and designing protocols for addition and removal of the cryoprotectant.

Because the cell membrane is usually more permeable to water than to cryoprotectants, addition of a cryoprotective chemical to a cell suspension will result in the characteristic "shrink–swell" behavior shown in Fig. 24.5. Assuming the cell is initially suspended in an isotonic solution, the addition of cryoprotectant will create a hypertonic environment. Water will be removed from the cell by osmosis, faster than the rate at which the cryoprotectant can diffuse into the cell, and thus the cell shrinks. The rate of water transport decreases as the cell dehydrates, because the increased intracellular tonicity reduces the osmotic pressure difference across the membrane. When the rate of water efflux equals the rate of cryoprotectant influx, the cell volume is at an extremum, after which cryoprotectant permeation will dominate the transport, and the cell will slowly increase in volume as cryoprotectant diffuses into the cell. During this swelling phase, water also reenters the cell, driven by the osmotic pressure difference generated by the intracellular cryoprotectant.

To effect cryoprotectant removal after thawing, the cell suspension is diluted, resulting in dynamics opposite to those observed during cryoprotectant addition. The cells initially swell as water rapidly enters, then slowly shrink as the cryoprotectant diffuses out. The cell volume reaches its maximum when the water and cryoprotectant fluxes balance each other.

Conditions that cause large excursions of cell volume are deleterious (Gao *et al.,* 1995b). There is thought to be a critical minimum cell volume below which cells are damaged (Meryman, 1970). Similarly, if the cell volume becomes too large, lysis may result. Thus, it is important to predict the volumetric response of cells during cryoprotectant loading and dilution, in order to prevent osmotic damage. The coupled transport of cryoprotectant and water across the cell membrane is often modeled using the Kedem–Katchalsky formalism (Kedem and Katchalsky, 1958), but can be adequately described by a two-parameter model (Kleinhans, 1998): The volume flux of cryoprotectant entering the cell is

$$J_{a} = Pv_{a}(\Pi_{a}^{ex} - \Pi_{a}^{in}), \tag{7}$$

where P is the permeability of the membrane to the cryoprotectant; v_{a}, the specific volume of the cryoprotectant agent; and Π_{a}^{ex} and Π_{a}^{in}, the cryoprotectant osmolarity in the extracellular and intracellular solution, respectively. The water flux into the cell is given by

$$J_{w} = -L_{p}RT[(\Pi_{e}^{ex} + \Pi_{a}^{ex}) - (\Pi_{e}^{in} + \Pi_{a}^{in})], \tag{8}$$

where L_{p} is the membrane water permeability and Π_{e}^{ex} and Π_{e}^{in} are the electrolyte osmolarity in the extracellular and intracellular solution, respectively. The rate of change of the cell volume is then equal to the sum of these two fluxes, multiplied by the instantaneous cell area. The resulting differential equation can be solved numerically (Kleinhans, 1998).

In order to predict the excursions of the cell volume during cryoprotectant loading and removal, the permeability parameters L_{p} and P must be known. They can be determined using standard parameter estimation techniques by fitting the model equations to measurements of cell volume during a shrink–swell experiment. These measurements can be obtained by video microscopy using specially designed microdiffusion or microperfusion stages, or using electronic particle size counters with large numbers of cells (McGrath, 1997).

If the permeability parameters are known, computer simulations can be used to design protocols for adding and removing cryoprotectant (Levin and Miller, 1981; Arnaud and Pegg, 1990; Gao *et al.,* 1995b). Typically, cryoprotectants are added and removed in multiple steps, in which cells are exposed to a series of solutions, the compositions of which have been optimized to minimize cell volume transients. Furthermore, the time of exposure must be chosen to ensure that cells have fully equilibrated with the solution. A third parameter is the temperature, because cryoprotectant toxicity usually decreases with decreasing temperature. However, whereas the membrane permeability also decreases at low temperatures, there is a trade-off between reduced toxicity and the longer equilibration times required.

CRYOBIOLOGY OF TISSUE

TRANSPORT PROCESSES: SIZE EFFECTS

Although the principles of cell cryobiology provide a starting point for the design and optimization of cryopreservation procedures for living tissue and organs, the additional challenges presented by the scale of engineered tissues and organs must be considered (Karlsson and Toner, 1996). On the microscopic length scale of a single biological cell, gradients in temperature, pressure, and solute concentration can typically be neglected; i.e., these quantities can be assumed to be spatially uniform in the cell (Mazur, 1963). However, due to the macroscopic dimensions of tissue, significant spatial variations in the thermal, chemical, and mechanical state of the system are expected. Moreover, because desired changes in flux of mass (i.e., cryoprotectant addition and dilution) and heat (i.e., cooling and warming) can be imposed only at the boundary conditions of the system, our ability to control the state of the tissue during cryopreservation is limited.

The rate of cryoprotectant loading into tissues prior to freezing, and removal after thawing, will be limited by diffusion through the matrix. Cryoprotectant transport kinetics in various types of tissue have been quantified by nuclear magnetic resonance spectroscopy (Fuller *et al.,* 1989), high-performance liquid chromatography (Carpenter and Dawson, 1991), and osmometry (Borel Rinkes *et al.,* 1992a). Typical values for the diffusion coefficient of DMSO in the extracellular matrix are on the order of $D = 10^{-5}$ cm²/sec. For simple tissue geometries, the permeation of cryoprotectant into or out of the tissue can be calculated from solutions of the diffusion equation subject to appropriate boundary conditions (Crank, 1975). In many cases, the penetration depth (λ) of cryoprotectant into the tissue can be approximated by the equation

$$\lambda \approx \sqrt{Dt}. \tag{9}$$

Thus, DMSO penetrates a tissue to a depth of approximately 1 mm after 15 min, and 2 mm after 1 hr. Equilibration of a 1-cm-thick tissue with cryoprotectant should require overnight incubation. Whereas these time scales are often impractical, and may result in cell damage due to excessive exposure to cryoprotectant, incubation times used in practice often result in incomplete equilibration, with low cryoprotectant concentration in the tissue interior (Hu and Wolfinbarger, 1994).

Mass transport in tissue during freezing and thawing also gives rise to nonuniform concentration distributions; their analysis often requires consideration of coupled heat-transfer and phase-transformation processes, in addition to diffusive transport. Cooling and solidification of tissue can be described by a Stefan-like energy equation (Viskanta *et al.,* 1997): during freezing, the preferential exclusion of solutes from the ice lattice lowers the chemical potential of the unfrozen water, creating a driving force for water transport. Models for the ensuing solute redistribution (Körber, 1988) and cell dehydration (Levin *et al.,* 1977; Rubinsky, 1989) have been derived. Because local thermodynamic equilibrium is assumed at the interface between the growing ice crystals and the unfrozen solution, the energy equation and mass transport models are coupled via the phase diagram of the solution (Viskanta *et al.,* 1997); i.e., not only does the rate of cooling affect the distribution of water and solute in the tissue, but the rate of mass diffusion in the unfrozen liquid affects the temperature distribution in the system. Numerical solutions of the equations governing tissue freezing indicate that significant gradients in temperature and solution composition result (Rubinsky, 1989; Viskanta *et al.*, 1997).

We know from cell cryobiology that the cooling rate is a major determinant of the outcome of the cryopreservation procedure (see previous discussion of experimental observations of cell freezing). Due in part to the phase change in the tissue, in part to the low thermal conductivity and high heat capacity of biological materials, the local rate of temperature change experienced by a cell will depend on its location within the tissue (Diller, 1992). Thus, localized cell damage may result in sections of tissue in which the rates of cooling and warming are significantly different from the optimal treatment. Thermal gradients may be significant even in cryopreservation of thin tissues such as skin, due to packaging requirements. There are two possible strategies for minimizing heat-transfer-related problems in tissue cryopreservation. The first is to reduce the sensitivity of cell survival to the rate of change of temperature: it is known that cryoprotectants broaden the "survival signature" (Fig. 24.2), increasing the range of survivable cooling rates (Mazur, 1984). The second approach is to improve the rate of heat transfer within the tissue in order to reduce thermal gradients: one promising method is the use of electromagnetic radiation for warming of cryopreserved tissue (Bai *et al.*, 1992).

DAMAGE MECHANISMS

Cells are the vital element of engineered organs and tissue, and their survival must be a priority in the design of cryopreservation procedures. However, the effects of freezing and thawing on other tissue components, such as extracellular matrix and synthetic elements, must also be considered. Due to the complex structure of organized tissue, there is a larger number of possible damage mechanisms for tissue than for isolated cells (Karlsson and Toner, 1996). The major modes of injury to tissue during cryopreservation are discussed below.

The cooling rate dependence of the survival of cells in cryopreserved tissue is qualitatively similar to the classic survival signature for cells in suspension (Fig. 24.2), indicating that the two-factor hypothesis of freezing injury is also valid for cells in tissue (Borel Rinkes *et al.*, 1992b; Muldrew *et al.*, 1994). However, there are important quantitative differences between cells in tissue culture and isolated cells, and these affect the design of freezing and thawing protocols. For example, corneal keratocytes cultured in monolayers are more sensitive to the cooling rate but less sensitive to cryoprotectant than are disaggregated keratocytes, and the optimal cryopreservation protocols for the cultured and isolated cells are significantly different (Armitage and Juss, 1996). *In vitro* culture has been observed to affect the behavior of other cell types during cryopreservation, typically increasing the probability of intracellular ice formation compared with isolated cells (Karlsson and Toner, 1996). These effects may be due to a number of factors, including facilitation of intracellular ice formation via gap junctions or other cell–cell interactions (Berger and Uhrik, 1996), changes in the biophysical properties of cell membranes by cell isolation procedures or by phenotypic changes in tissue culture (Sandler and Andersson, 1984; Yarmush *et al.*, 1992), and sensitivity to injury dependent on the state of cell adhesion (Hetzel *et al.*, 1973; Hornung *et al.*, 1996).

In addition to these differences between cells in a tissue environment and cells in isolation, there can be differences between cells within the tissue. An obvious but important example is that of tissue in which two or more distinct cell types exist in coculture, such as keratinocytes and fibroblasts in skin (Wilkins *et al.*, 1994) or hepatocytes and fibroblasts in liver (Bhatia *et al.*, 1997). Because the optimal cooling rate (and by extension, the optimal value of other processing parameters) may differ significantly between dissimilar cell types (Karlsson *et al.*, 1993b), development of cryopreservation techniques for tissues with multiple cell species is particularly difficult. Even for a tissue populated with a single cell species, the effective low-temperature response of the cells can depend on their location within the tissue. This effect results from the fact that cell dehydration during freezing and thawing is driven by the low chemical potential of extracellular ice: the larger the distance of a cell from the ice phase, the larger the effective resistance to water transport (Levin *et al.*, 1977; Diller and Raymond, 1990). Thus, in tissues with multiple cell layers, interior cells will retain more water during freezing. Furthermore, physical separation of interior cells from the extracellular ice phase may inhibit surface-catalyzed nucleation of intracellular ice in these cells (Toner, 1993). These factors will give rise to spatially varying requirements for optimal cell survival.

Besides direct injury to cells, cryopreservation procedures may cause damage to molecular structures critical to tissue function. For example, there is evidence that cryoprotectant additives, osmotic stress, and low temperatures adversely affect intercellular tight junctions in epithelial monolayers (Armitage *et al.,* 1995). Furthermore, cell–substrate attachment points may be a target of cryoinjury, resulting in detachment and loss of cells after freezing and thawing (Hornung *et al.,* 1996). The extracellular matrix is deformed by ice crystals during freezing (Persidsky and Luyet, 1975), and is exposed to hypertonic concentrations of electrolytes and cryoprotectant. Nonetheless, the mechanical properties of tissues such as cartilage and skin do not appear to be significantly affected by the cryopreservation procedure (Arnoczky *et al.,* 1988; Foutz *et al.,* 1992).

Last, we must consider damage to the macroscopic tissue structure. One such mode of injury has been identified in tissues with significant luminal spaces, a common example being a vascularized organ. During the freezing process, ice forms first in the vasculature, causing dehydration of the surrounding tissue (Rubinsky, 1989). If cooling is slow, a significant volume of water will then accumulate and solidify in the vascular space, exerting potentially damaging pressure on tissue structures. In liver tissue, overdistention and rupture of blood vessels and sinusoids by this mechanism is an important mode of cryoinjury (Rubinsky *et al.,* 1990). Potential solutions to this problem can be learned from freezing-resistant and freeze-tolerant organisms such as the wood frog *Rana sylvatica* and the winter flounder *Pseudopleuronectes americanus,* which can survive freezing temperatures by secreting glucose (acting as a cryoprotectant) and antifreeze glycoproteins (which bind the surfaces of ice crystals) into their circulation, respectively, in order to control ice growth in the vasculature (Duman and De Vries, 1974; Storey *et al.,* 1992). Such strategies have met with limited success *in vitro* (Rubinsky *et al.,* 1994).

The macroscopic structure of an organ or tissue may also be compromised by fractures formed in the solidified tissue during freezing and thawing. Due in part to thermal gradients in the specimen, in part to volumetric expansion resulting from the water–ice phase transition, significant stress develops during cooling and warming. These mechanical stresses can exceed the strength of the frozen material, causing fractures in the tissue. Such fractures have proved problematic in the cryopreservation of arteries, because grafts must be intact to be useful clinically (Pegg *et al.,* 1997). Likewise, attempts to preserve more complex organs are hampered by structural damage caused by gross fractures (Fahy *et al.,* 1990; Rabin *et al.,* 1997). Current approaches to mitigate mechanical damage include the use of cryoprotectants, which appear to reduce thermal stress in aqueous solutions (Gao *et al.,* 1995a), and the use of slow warming rates to reduce thermal gradients until glassy regions in the tissue undergo softening (Pegg *et al.,* 1997). Advances in mathematical modeling of thermal stresses during freezing and thawing will make possible the design of cryopreservation procedures that minimize the probability of tissue fracture (Rabin and Steif, 1998).

SUMMARY AND FUTURE DIRECTIONS

The challenge in development of cryopreservation technology for living biological systems is twofold. The first challenge is to determine the optimal (minimally damaging) protocol among an essentially infinite number of possibilities. The diversity of potential pathways for damage during freezing, thawing, and the loading and removal of cryoprotectant makes it difficult to find a successful procedure. Thus, theoretical modeling of the dominant damage mechanisms is often required to make the optimization problem tenable. Models of cell damage during freezing are now sufficiently advanced to make possible the rational design of cooling procedures using computer simulations. However, although the mechanisms of damage to tissue during cryopreservation are also becoming better understood, adequate models are not yet available for all relevant processes. The second challenge lies in implementation of the cryopreservation protocol determined to be optimal. Due to the macroscopic dimensions and unfavorable transport properties of biological tissue, it is often impossible to impose the designed chemical and thermal profiles uniformly in the tissue, resulting in suboptimal processing. In addition, differences in the optimal protocols for dissimilar tissue components may lead to conflicting requirements in heterogeneous tissue.

Future research efforts toward the development of cryopreservation methods for tissue engineering must therefore focus on the two problems described above. Fundamental experimental and theoretical research is required to elucidate the mechanisms of damage during cryopreservation, and to develop predictive models of these deleterious processes. Likewise, techniques for accurate measurement of the biophysical properties of tissues must be developed in order to obtain necessary model parameters. Strategies toward enabling implementation of optimal protocols include minimizing temperature and concentration gradients, and improving robustness by decreasing system sensitivity to variations in protocol parameters. Although some of these goals may be achieved by novel technology (e.g., electromagnetic heating to reduce thermal gradients), a complementary approach should be the consideration of cryopreservation requirements in the design of engineered tissues and organs. By suitable choice of device geometry, material properties, and cells, it will be possible to increase system robustness during the chemical and thermal processing required for cryopreservation.

The critical need for effective cryopreservation technology in tissue engineering is an impetus for accelerated efforts toward solving the fundamental problems described in this chapter. Although the remaining scientific and technological challenges are formidable, significant progress has been achieved to date through interdisciplinary research activities, and prospects for the future are bright.

ACKNOWLEDGMENTS

This work was supported in part by a National Science Foundation Faculty Early Career Development (CAREER) Award to J.O.M.K. (Grant BES-9875569) and by a National Institutes of Health Grant to M.T. (Grant DK-46270).

REFERENCES

Anchordoguy, T. J., Cecchini, C. A., Crowe, J. H., and Crowe, L. M. (1991). Insights into the cryoprotective mechanism of dimethyl sulfoxide for phospholipid bilayers. *Cryobiology* 28, 467–473.

Armitage, W. J., and Juss, B. K. (1996). The influence of cooling rate on survival of frozen cells differs in monolayers and in suspensions. *Cryo-Letters* 17, 213–218.

Armitage, W. J., Juss, B. K., and Easty, D. L. (1995). Differing effects of various cryoprotectants on intercellular junctions of epithelial (MDCK) cells. *Cryobiology* 32, 52–59.

Arnaud, F. G., and Pegg, D. E. (1990). Permeation of glycerol and propane-1,2-diol into human platelets. *Cryobiology* 27, 107–118.

Arnoczky, S. P., McDevitt, C. A., Schmidt, M. B., Mow, V. C., and Warren, R. F. (1988). The effect of cryopreservation on canine menisci: A biochemical, morphologic, and biomechanical evaluation. *J. Orthop. Res.* 6, 1–12.

Bai, X., Pegg, D. E., Evans, S., and Penfold, J. D. (1992). Analysis of electromagnetic heating patterns inside a cryopreserved organ. *J. Biomed. Eng.* 14, 459–466.

Bell, E., and Rosenberg, M. (1990). The commercial use of cultivated human cells. *Transplant. Proc.* 22, 971–974.

Berger, W. K., and Uhrik, B. (1996). Freeze-induced shrinkage of individual cells and cell-to-cell propagation of intracellular ice in cell chains from salivary glands. *Experientia* 52, 843–850.

Bhatia, S. N., Yarmush, M. L., and Toner, M. (1997). Controlling cell interactions by micropatterning in co-cultures: hepatocytes and 3T3 fibroblasts. *J. Biomed. Mater. Res.* 34, 189–199.

Borel Rinkes, I.H.M., Toner, M., Ezzell, R. M., Tompkins, R. G., and Yarmush, M. L. (1992a). Effects of dimethyl sulfoxide on cultured rat hepatocytes in sandwich configuration. *Cryobiology* 29, 443–453.

Borel Rinkes, I.H.M., Toner, M., Sheehan, S. J., Tompkins, R. G., and Yarmush, M. L. (1992b). Long-term functional recovery of hepatocytes after cryopreservation in a three-dimensional culture configuration. *Cell Transplant.* 1, 281–292.

Boyle, R. (1683). New experiments and observations touching cold. R. Davis, London.

Carpenter, J. F., and Dawson, P. E. (1991). Quantitation of dimethyl sulfoxide in solutions and tissues by high-performance liquid chromatography. *Cryobiology* 28, 210–215.

Christian, J. W. (1975). "The Theory of Transformations in Metals and Alloys." Pergamon, New York.

Crank, J. (1975). "The Mathematics of Diffusion." Clarendon Press, Oxford.

Cravalho, E. G. (1976). The application of cryogenics to the reversible storage of biomaterials. *Adv. Cryog. Eng.* 21, 399–417.

Diller, K. R. (1982). Quantitative low temperature optical microscopy of biological systems. *J. Microsc. (Oxford)* 126, 9–28.

Diller, K. R. (1992). Modeling of bioheat transfer processes at high and low temperatures. *Bioeng. Heat Transfer* 22, 157–357.

Diller, K. R., and Raymond, J. F. (1990). Water transport through a multicellular tissue during freezing: A network thermodynamic modeling analysis. *Cryo-Letters* 112, 151–162.

Duman, J. G., and De Vries, A. L. (1974). Freezing resistance in winter flounder *Pseudopleuronectes americanus. Nature (London)* 247, 237–238.

Fahy, G. M. (1986). The relevance of cryoprotectant "toxicity" to cryobiology. *Cryobiology* 23, 1–13.

Fahy, G. M., MacFarlane, D. R., Angell, C. A., and Meryman, H. T. (1984). Vitrification as an approach to cryopreservation. *Cryobiology* 21, 407–426.

Fahy, G. M., Saur, J., and Williams, R. J. (1990). Physical problems with the vitrification of large biological systems. *Cryobiology* 27, 492–510.

Foutz, T. L., Stone, E. A., and Abrams, C. F. (1992). Effects of freezing on mechanical properties of rat skin. *Am. J. Vet. Res.* 53, 788–792.

Fuller, B. J., Busza, A. L., and Proctor, E. (1989). Studies on cryoprotectant equilibration in the intact rat liver using nuclear magnetic resonance spectroscopy: A noninvasive method to assess distribution of dimethyl sulfoxide in tissues. *Cryobiology* **26**, 112–118.

Gao, D. Y., Lin, S., Watson, P. F., and Critser, J. K. (1995a). Fracture phenomena in an isotonic salt solution during freezing and their elimination using glycerol. *Cryobiology* **32**, 270–284.

Gao, D. Y., Liu, J., Liu, C., McGann, L. E., Watson, P. F., Kleinhans, F. W., Mazur, P., Critser, E. S., and Critser, J. K. (1995b). Prevention of osmotic injury to human spermatozoa during addition and removal of glycerol. *Hum. Reprod.* **10**, 1109–1122.

Green, H., Kehinde, O., and Thomas, J. (1979). Growth of cultured human epidermal cells into multiple epithelia suitable for grafting. *Proc. Natl. Acad. Sci. U.S.A.* **76**, 5665–5668.

Hetzel, F. W., Kruuv, J., McGann, L. E., and Frey, H. E. (1973). Exposure of mammalian cells to physical damage: Effect of the state of adhesion on colony-forming potential. *Cryobiology* **10**, 206–211.

Hornung, J., Müller, T., and Fuhr, G. (1996). Cryopreservation of anchorage-dependent mammalian cells fixed to structured glass and silicon substrates. *Cryobiology* **33**, 260–270.

Hu, J. F., and Wolfinbarger, L. (1994). Dimethyl sulfoxide concentration in fresh and cryopreserved porcine valved conduit tissues. *Cryobiology* **31**, 461–467.

Ingham, E., Matthews, J. B., Kearney, J. N., and Gowland, G. (1993). The effects of variation of cryopreservation protocols on the immunogenicity of allogeneic skin grafts. *Cryobiology* **30**, 443–458.

Johnson, P. H., Olson, C. B., and Goodman, M. (1985). Isolation and characterization of deoxyribonucleic acid from tissue of the woolly mammoth, *Mammuthus primigenius. Comp. Biochem. Physiol. B* **81B**, 1045–1051.

Karlsson, J.O.M., and Toner, M. (1996). Long-term storage of tissues by cryopreservation: Critical issues. *Biomaterials* **17**, 243–256.

Karlsson, J.O.M., Cravalho, E. G., Borel Rinkes, I.H.M., Tompkins, R. G., Yarmush, M. L., and Toner, M. (1993a). Nucleation and growth of ice crystals inside cultured hepatocytes during freezing in the presence of dimethyl sulfoxide. *Biophys. J.* **65**, 2524–2536.

Karlsson, J.O.M., Cravalho, E. G., and Toner, M. (1993b). Intracellular ice formation: Causes and consequences. *Cryo-Letters* **14**, 323–336.

Karlsson, J.O.M., Cravalho, E. G., and Toner, M. (1994). A model of diffusion-limited ice growth inside biological cells during freezing. *J. Appl. Phys.* **75**, 4442–4455.

Karlsson, J.O.M., Eroglu, A., Toth, T. L., Cravalho, E. G., and Toner, M. (1996). Fertilization and development of mouse oocytes cryopreserved using a theoretically optimized protocol. *Hum. Reprod.* **11**, 1296–1305.

Kedem, O., and Katchalsky, A. (1958). Thermodynamic analysis of the permeability of biological membranes to nonelectrolytes. *Biochim. Biophys. Acta* **27**, 229–246.

Kleinhans, F. W. (1998). Membrane permeability modeling: Kedem-Katchalsky vs a two-parameter formalism. *Cryobiology* **37**, 271–289.

Körber, C. (1988). Phenomena at the advancing ice-liquid interface: Solutes, particles and biological cells. *Q. Rev. Biophys.* **21**, 229–298.

Levin, R. L., and Miller, T. E. (1981). An optimum method for the introduction and removal of permeable cryoprotectants: Isolated cells. *Cryobiology* **18**, 32–48.

Levin, R. L., Cravalho, E. G., and Huggins, C. E. (1977). Water transport in a cluster of closely packed erythrocytes at subzero temperatures. *Cryobiology* **14**, 549–558.

Mazur, P. (1963). Kinetics of water loss from cells at subzero temperatures and the likelihood of intracellular freezing. *J. Gen. Physiol.* **47**, 347–369.

Mazur, P. (1984). Freezing of living cells: Mechanisms and implications. *Am. J. Physiol.* **247**, C125–C142.

Mazur, P., Leibo, S., and Chu, E.H.Y. (1972). A two-factor hypothesis of freezing injury. *Exp. Cell Res.* **71**, 345–355.

McGrath, J. J. (1985). Preservation of biological material by freezing and thawing. *In* "Heat Transfer in Medicine and Biology" (A. Shitzer and R. C. Eberhart, eds.), Vol. 2, pp. 185–238. Plenum, New York.

McGrath, J. J. (1988). Membrane transport properties. *In* "Low Temperature Biotechnology" (J. J. McGrath and K. R. Diller, eds.), Vol. BED-10, pp. 273–330. Am. Soc. Mech. Eng., New York.

McGrath, J. J. (1997). Quantitative measurement of cell membrane transport: Technology and applications. *Cryobiology* **34**, 315–334.

McNally, R. T., and McCaa, C. (1988). Cryopreserved tissues for transplant. *In* "Low Temperature Biotechnology" (J. J. McGrath and K. R. Diller, eds.), Vol. BED-10, pp. 91–106. Am. Soc. Mech. Eng., New York.

Meryman, H. T. (1970). The exceeding of a minimum tolerable cell volume in hypertonic suspension as a cause of freezing injury. *In* "The Frozen Cell" (G.E.W. Wolstenholme and M. O'Connor, eds.), pp. 51–67. Churchill, London.

Muldrew, K., and McGann, L. E. (1994). The osmotic rupture hypothesis of intracellular freezing injury. *Biophys. J.* **66**, 532–541.

Muldrew, K., Hurtig, M., Novak, K., Schachar, N., and McGann, L. E. (1994). Localization of freezing injury in articular cartilage. *Cryobiology* **31**, 31–38.

Naughton, G. (1999). The Advanced Tissue Sciences story. *Sci. Am.* **280**, 84–85.

Naughton, G., Mansbridge, J., and Gentzkow, G. (1997). A metabolically active human dermal replacement for the treatment of diabetic foot ulcers. *Artif. Organs* **21**, 1203–1210.

Pegg, D. E., Wusteman, M. C., and Boylan, S. (1997). Fractures in cryopreserved elastic arteries. *Cryobiology* **34**, 183–192.

Persidsky, M. D., and Luyet, B. J. (1975). Analysis of the freezing process across temperature gradients within gelatin gels. *Cryobiology* **12**, 364–385.

Pitt, R. E. (1992). Thermodynamics and intracellular ice formation. *In* "Advances in Low-Temperature Biology" (P. Steponkus, ed.), Vol. 1, pp. 63–99. JAI Press, London.

Polge, C., Smith, A. U., and Parkes, A. S. (1949). Revival of spermatozoa after vitrification and dehydration at low temperatures. *Nature (London)* **164**, 666.

Prager, E. M., Wilson, A. C., Lowenstein, J. M., and Sarich, V. M. (1980). Mammoth albumin. *Science* **209**, 287–289.

Rabin, Y., and Steif, P. S. (1998). Thermal stresses in a freezing sphere and its application to cryobiology. *J. Appl. Mech.* **65**, 328–333.

Rabin, Y., Olson, P., Taylor, M. J., Steif, P. S., Julian, T. B., and Wolmark, N. (1997). Gross damage accumulation on frozen rabbit liver due to mechanical stress at cryogenic temperatures. *Cryobiology* **34**, 394–405.

Rubinsky, B. (1989). The energy equation for freezing of biological tissue. *J. Heat Transfer* **111**, 988–997.

Rubinsky, B., Lee, C. Y., Bastacky, J., and Onik, G. (1990). The process of freezing and the mechanism of damage during hepatic cryosurgery. *Cryobiology* **27**, 85–97.

Rubinsky, B., Arav, A., Hong, J. S., and Lee, C. Y. (1994). Freezing of mammalian livers with glycerol and antifreeze proteins. *Biochem. Biophys. Res. Commun.* **200**, 732–741.

Sandler, S., and Andersson, A. (1984). The significance of culture for successful cryopreservation of isolated pancreatic islets of Langerhans. *Cryobiology* **21**, 503–510.

Storey, K. B., Bischof, J., and Rubinsky, B. (1992). Cryomicroscopic analysis of freezing in liver of the freeze-tolerant wood frog. *Am. J. Physiol.* **263**, R185–R194.

Taylor, M. J., Bank, H. L., and Benton, M. J. (1987). Selective destruction of leucocytes by freezing as a potential means of modulating tissue immunogenicity: Membrane integrity of lymphocytes and macrophages. *Cryobiology* **24**, 91–102.

Timasheff, S. N. (1982). Preferential interactions in protein-water-cosolvent systems. *In* "Biophysics of Water" (F. Franks and S. Mathias, eds.), pp. 70–72. Wiley, New York.

Toner, M. (1993). Nucleation of ice crystals inside biological cells. *In* "Advances in Low-Temperature Biology" (P. Steponkus, ed.), Vol. 2, pp. 1–51. JAI Press, London.

Toner, M., Cravalho, E. G., and Karel, M. (1990). Thermodynamics and kinetics of intracellular ice formation during freezing of biological cells. *J. Appl. Phys.* **67**, 1582–1592; erratum: **70**, 4653 (1990).

Turnbull, D., and Fisher, J. C. (1949). Rate of nucleation in condensed systems. *J. Chem. Phys.* **17**, 71–73.

Viskanta, R., Bianchi, M.V.A., Critser, J. K., and Gao, D. (1997). Solidification processes of solutions. *Cryobiology* **34**, 348–362.

Wiebe, M. E., and May, L. H. (1990). Cell banking. *Bioprocess. Technol.* **10**, 147–160.

Wilkins, L. M., Watson, S. R., Prosky, S. J., Meunier, S. F., and Parenteau, N. L. (1994). Development of a bilayered living skin construct for clinical applications. *Biotechnol. Bioeng.* **43**, 747–756.

Yarmush, M. L., Toner, M., Dunn, J.C.Y., Rotem, A., Hubel, A., and Tompkins, R. G. (1992). Hepatic tissue engineering. *Ann. N.Y. Acad. Sci.* **665**, 238–252.

CHAPTER 25

IMMUNOMODULATION

Denise Faustman

INTRODUCTION

The application of novel basic science concepts to transplantation may permit a broader and improved therapeutic use of cellular and whole organ transplantation therapies in humans. One embodiment of this concept involves the elimination of immunosuppressive drugs and treatment in the host. For instance, genetically engineered or modified donor cells may be rendered less immunogeneic and the host could potentially receive either no immunosuppressive drugs or decreased dosages of drugs. Advances, especially for cellular transplants have allowed novel therapies to progress to human trials. Furthermore, improvements in the immunomodulations of the donor cells, to render them resistant to graft rejection, have identified previously new immune barriers, such as recurrent disease in immunoprotected islet cells due to type I diabetes. In this chapter, the barriers yet to be overcome will be highlighted, especially as they relate to the specific challenges of each clinical setting. The promise of transplantation therapy will be highlighted as it relates to preliminary human trials, even in the setting of cross-species transplants.

A long-standing goal of the transplant community has been to overcome immunologic rejection of transplanted tissues and organs. Although much of the research has been devoted to identifying new immunosuppressive agents and new combinations of existing agents for allotransplantation, immunosuppression still carries significant short-term side effects, especially enhanced susceptibility to infection. For xenografts, even stronger immunosuppressive regimens are expected to be necessary. One solution to xenograft acceptance may reside in novel therapeutic strategies that strive to prevent the need for immunosuppression.

Our laboratory has attempted to avert transplant rejection by immunologically intervening at a new attack point in the rejection cascade. Tissues and cells were treated before transplantation to conceal or eliminate the antigens that summon immune rejection. This technology is sometimes referred to as "donor antigen modification" or, more stylistically, "designer tissue and organs" (Faustman and Coe, 1991). Our intention was to modify the graft at the molecular level in order to avoid immunosuppression of the host. Earlier efforts at modifying the graft at the cellular level by elimination of donor lymphoid cell populations were successful in mice in the allograft setting, but ineffective for whole organ survival in even lower species or in cellular transplants in higher species or across species barriers. Designer tissues and organs modified at the molecular level offered greater selectivity and latitude in tailoring the graft to escape immune detection. Donor antigen modification contrasts with a protocol of bone marrow chimerism (i.e., a bone marrow transplant and a solid organ transplant from the same donor to ensure acceptance of the organ), whereby host immune manipulations are employed.

Without the risks of systemic immunosuppression, designer tissues and organs could be therapeutic for a broad range of chronic conditions that are not life threatening. They also could be offered to conventional organ transplant patients at earlier stages of their disease when patients are healthier and better able to withstand surgical intervention. The main risk with designer tissues and organs is that the modified donor tissue could be rejected, a risk commonly encountered with any transplant. Because designer tissue can be potentially rendered safer and may ultimately avoid host intervention with toxic drugs, a broader spectrum of diseases could be treated and at earlier time points without induced morbidity.

The purpose of this chapter is to describe the evolution of research on designer tissues and to trace its contribution to a growing body of transplant research. Research in murine hosts demonstrated the successful transplantation of xenogeneic cells and tissues modified at the molecular level without immunosuppression. Transplant success of cells across species by donor antigen modification allowed the extension of these concepts to the genetic level. The ability to modify antigens either at the DNA, RNA, or protein level offers tremendous latitude in the development of new biological therapies. Variations on the designer tissue concept are being probed in animal models of diabetes, solid organ failure, and neurological diseases. A concept launched in the laboratory in the early 1990s has already culminated in landmark human clinical trials for Parkinson's and Huntington's disease using donor-modified pig neurons into humans.

The wide range of applications of designer tissues and organs for allo- and xenotransplantation is readily apparent; modified donor cells, tissues, and organs can be considered potential therapies for an almost limitless number of conditions as long as the dominant antigens are identified and then effectively shielded or eliminated prior to implantation.

ORIGIN OF THE DESIGNER TISSUE CONCEPT

Our idea of targeting donor antigens instead of modifying the host immune system was stimulated by research on the molecular events surrounding the killing of target cells by cytotoxic T lymphocytes (CTLs; hereinafter referred to as T cells). In an elegant study, Spits and co-workers (1986) identified the sequential stages of T cell destruction of the target cell. In contrast to earlier research, which focussed primarily on the T cell, they examined the roles of cell surface markers on both the T cell and the target cell. By using antibodies against different cell surface markers to block distinct stages, they were able to tease apart *in vitro* the interactions between molecules on the T cell and the target cell. They were among the first to identify three stages in T cell cytotoxicity: (1) adhesion between T cell and target cell, (2) T cell receptor activation, and, (3) T cell lysis of the target cell.

A critical finding was that one of the major classes of histocompatibility antigens on the surface of the target cell—the major histocompatibility complex (MHC) class I molecules—was involved in both adhesion to, and activation of, the T cell. Class I antigens and other classes of antigens encoded by the MHC complex serve to distinguish "self" from "non-self" because they differ across species and between members of the same species. Class I antigens have long been eliciting T cell cytotoxicity, but this study offered more detailed insight into their pivotal role. What this study also demonstrated was that antibodies to class I antigens could prevent T cell adhesion and activation, thereby interfering with lysis.

Spits and co-workers crystallized the importance of class I antigens in immune rejection of tumors. With this study's goal centering on enhancing tumor rejection, we departed on the idea of a central role of target class I antigens as pivotal in tumor escape, and perhaps similarly focused on an opposing role for manipulation in transplantation. As a test antigen, we chose to mask class I on donor cells with antibody fragments, and then transplant the donor antigen-modified cells into nonimmunosuppressed hosts.

FIRST DEMONSTRATION OF THE CONCEPT

Using a xenogeneic model, the first successful demonstration of the designer tissue concept was demonstrated (Faustman and Coe, 1991). We found that when we coated the graft with antibody fragments to conceal class I antigens, the graft eluded the host immune system indefinitely and functioned normally. The host developed tolerance to the treated graft because secondary transplants of untreated tissue were later accepted. The mechanism of donor-specific tolerance is still not fully defined, but may involve induction of T cell anergy through altered donor class I density. Altered class I density may be pivotal in T cell shaping in both the periphery as well as the thymus, as defined by Pestano *et al.* (1999).

The graft consisted of purified human cadaveric islets that had been incubated with antibody fragments before being implanted into nonimmunosuppressed mice. We decided to produce very pure antibody fragments because they lacked the portion of the antibody molecule that binds complement, the Fc fragment. When the Fc fragment is enzymatically cleaved from the $F(ab')_2$ fragment, the purified $F(ab')_2$ fragment binds to the graft for several days without fixing complement, which would have destroyed the graft. Grafts survived for 200 days and functioned appropriately, as determined by assays for human C' peptide, a proinsulin processing product. Finally, human liver cells similarly treated with antibody fragments also survived for an extended period.

Treatment with whole antibodies to class I antigens failed to prolong graft survival, as did treatment with whole antibodies and antibody fragments to CD29. CD29 is an antigen uniquely expressed on the passenger lymphocytes that accompany the graft because they cannot be entirely eliminated during the purification process. Finally, treatment of the graft with polyclonal antibody fragments to all antigenic determinants on human islets prolonged graft survival until the point when class I antibody fragments were removed from the mixture. Graft rejection ensued at a slower rate, attesting to the critical role of class I antigens as pivotal for this transplantation barrier, although recognizing the possible importance of other antigens in additional immune responses.

For the sake of rigor, we selected a xenogeneic instead of an allogenic model to test the designer tissue concept. If designer tissues could succeed in the demanding case across a species barrier, then they could be considered for simpler antigeneic models. In other respects, however, our xenogeneic model represented the height of simplicity. First, the tissue contained only one dominant antigen that needed to be masked. We found that human islet tissue highly expressed class I antigens, while displaying only minimal expression of two adhesion molecules, intercellular adhesion molecule-1 (ICAM-1), and lymphocyte function-associated antigen-3 (LFA-3) (which in other tissues are thought to stabilize binding and to contribute to T cell activation). Second, the vasculature could be separated from the tissue to avoid hyperacute rejection, the earliest and most formidable barrier to discordant xenograft acceptance. Third, our model targeted antigens at the protein level rather than at the genetic level. Our goal was to conceal protein antigens that already appeared on the surface of graft cells. Other approaches, discussed later in this chapter, targeted antigens not at the protein level, but at the DNA and RNA level. The use of more sophisticated transgenic and antisense technology, respectively, can similarly prevent antigen expression and validated the donor antigens as a target for the immune response.

In summary, this tough xenogeneic model confirmed the paramount role of class I antigens in islet cell rejection. By providing a challenging test of the designer tissue concept, it also helped to launch a novel therapeutic strategy.

EXPANSION OF RESEARCH ON DESIGNER TISSUES

After our study was published, our laboratory and others began to probe the potential of designer tissues, using a variety of donor antigen modifications techniques as well as masking antigens with protein fragments in varying transplantation settings. The body of early research, in part summarized in Table 25.1, can be classified in four distinct groups based on the technique used to modify the donor tissue. The techniques currently employed for donor antigen modification are presented in Table 25.1 as it relates to the level of interference utilized to prevent normal cell surface protein expression. The methods include antibody masking, gene ablation, gene addition, and RNA ablation.

ANTIBODY MASKING

Antibody masking of xenogeneic tissue is the first of the donor antigen modification techniques to progress to primate and human trials. Pancreatic islet cells and neurons are the most common types of donor cells to be camouflaged with antibody masking, although any type of tissue can theoretically be treated once the dominant antigens have been identified. All of the studies cited below targeted class I antigens using antibody fragments.

Islet cell masking for the treatment of diabetes was early on pursued with mixed results. Osorio and colleagues (1994b) targeted class I antigens in a mouse allograft model. To approximate a diabetic state, the mice first were treated with a drug that chemically induced hyperglycemia. Then they received islet allografts that had been pretreated with antibody fragments. Graft survival was prolonged relative to controls, but within one month the grafts were eventually rejected.

Table 25.1. Designer donor tissues[a]

Technique/tissue	Allo/xeno	Donor/recipient	Target	Ref.
Masking antibody				
Islets	Xeno	Human/mouse	Class I	Faustman and Coe (1991)
Islets	Xeno	Human/mouse	All surface antigens	Faustman and Coe (1991)
Islets	Allo/xeno	Human or monkey/monkey	Class I	Steele *et al.* (1994)
Islets	Allo	Mouse/mouse	Class I	Osorio *et al.* (1994b)
Neurons	Xeno	Pig/rat	Class I	Pakzaban *et al.* (1995)
Neurons	Xeno	Pig/monkey	Class I	Burns *et al.* (1994)
Liver cells	Xeno	Human/mouse	Class I	Faustman and Coe (1991)
Neurons	Xeno	Pig/rat	Class I	Dinsmore *et al.* (1996)
Gene ablation				
Islets	Allo	Mouse/mouse	Class I	Markmann *et al.* (1992)
Islets	Allo	Mouse/mouse	Class I	Osoria *et al.* (1993)
Islets	Allo	Mouse/mouse	Class I	Osoria *et al.* (1994b)
Liver cells	Allo/xeno	Mouse/mouse/ guinea pig/frog	Class I	Li and Faustman (1993)
Kidneys	Allo	Mouse/mouse	Class I	Coffman *et al.* (1993)
Gene addition				
White blood cells	Xeno	Mouse or pig/human[b]	CD59	Platt (1994)
Endothelial cells	Xeno	Pig human[b]	CD46, CD55, CD46/CD55 hybrid	Fodor *et al.* (1994)
Endothelial cells	Xeno	Pig human[b]	CD46/CD55, CD59	Miyagawa *et al.* (1994)
Hearts	Xeno	Pig/baboon	CD55, CD59	Miyagawa *et al.* (1995)
Hearts, kidneys lungs	Xeno	Pig/monkey	CD59	McCurrey *et al.* (1995)
COS cells	Xeno	Pig/human[b]	H transferase	Sandrin *et al.* (1995)

[a]Adapted from Faustman, D. (1997). Designer tissues and organs. In "Xenotransplantation" (D.K.C. Cooper, E. Kemp, J. L. Platt, and D. J. White, eds.), Ch. 28. Springer-Verlag, Germany.
[b]Human serum *in vitro*.

Investigators attributed eventual allograft rejection to a variety of possibilities, including the absence of sufficient quantities of $F(ab')_2$ fragments, antiidiotypic antibodies against the $F(ab')_2$ fragment, and an immune pathway independent of class I activation of T cells.

Steele and colleagues (1994) investigated islet cell transplants in a primate model. Cynomolgus monkeys received either allogeneic or xenogeneic (human) islets. The grafts were pretreated with antibody fragments to class I antigens. Histologic evidence of donor islets were present months after transplantation into nonimmunosuppressed monkeys.

Neuronal xenografts with antibody masking have been investigated for Huntington's disease (Pakzaban *et al.*, 1995) and Parkinson's disease (Burns *et al.*, 1994). In the first study, fetal pig striatal cells were implanted into rats whose striatum had been lesioned 1 week earlier with injections of quinolinic acid. These injections destroy striatal neurons in an attempt to stimulate the dysfunction present in Huntington's disease. Rats received either untreated tissue or tissue pretreated with $F(ab')_2$ fragments against porcine class I antigens. Of the control rats receiving untreated tissue, some were immunosuppressed with cyclosporin A (CsA) whereas others were not. Three to four months later, graft survival was found to be prolonged in animals given $F(ab')_2$-treated grafts and in the CsA-treated animals given unaltered grafts. Grafts did not survive in nonimmunosuppressed controls. Graft volume, determined histologically with the aid of computer image analysis, was significantly larger in the CsA group compared with the $F(ab')_2$ group. Yet in both of these groups, immunohistochemistry revealed graft cytoarchitecture to be well organized and graft axons to have grown correctly in the direction of their target nuclei. It was encouraging that pig neurons appeared to be capable of locating their target.

In a similar study design, Burns and co-workers (1994) applied antibody masking to a primate model of Parkinson's disease. Porcine mesencephalic neuroblasts were implanted into monkeys with 1-methyl-4-phenyl-1,2,3,6-tetrahydropyridine (MPTP)-induced Parkinsonism. Measures of locomotor activity and dopamine fiber density in the host striatum confirmed that pretreatment of donor tissue with $F(ab')_2$ fragments succeeded in restoring motor function and replenishing dopamine fibers at the site of implantation. A control animal, maintained on CsA after receiving untreated cells, showed similar improvement. A nonimmunosuppressed control showed no improvement after transplantation, suggesting graft rejection.

The studies described in this section highlight the range of potential applications of antibody masking in both allo- and xenotransplants. Islet cell masking has encountered unexpected hurdles in the application of type I diabetes. However, neuronal cell masking has been sufficiently effective to usher in human clinical trials.

Gene Ablation

Gene ablation, or gene "knockout" technology, offers another vehicle to modify donor tissue and organs. With gene ablation, the gene or genes encoding protein antigens can be permanently deleted, thereby eliminating antigen expression in all cells. In one common method of gene ablation, the target gene is inactivated in cultured embryonic stem cells via homologous recombination; a target vector containing an inactive version of the gene recombines with, and thereby replaces, the wild-type gene. Through reintroduction of the embryonic cells into a foster mother and through selective breeding, progeny can be produced that are homozygous for the mutation. Rejection by the host immune system is expected to be avoided when the targeted gene encodes a protein antigen slated for expression on the surface of donor cells.

The major advantage of gene ablation over protein modification of donor tissue is that the antigen is permanently eliminated in all cells, tissues, and organs in which it is expressed. The major limitation is that it is available only in mice, which are too phylogenetically distant from humans to be considered as suitable donors. The permanent nature of the modification also can sometimes be a limitation, especially when the protein encoded by the gene has additional functions. Inactivation of the gene might lead to physiological changes that compromise the utility of the donor tissue. Gene ablation can also target genes that encode proteins essential for the target protein expression.

Several transplantation laboratories have exploited the availability of knockout mice deficient in β_2-microglobulin. β_2-Microglobulin is a peptide that performs a chaperone function as part of the class I molecule and is necessary for its assembly and expression. Mice homozygous for the β_2-microglobulin mutation fail to display class I antigens on the cell surface, a feature that makes them highly desirable for transplantation studies. Although much of the research described below has focused on islet cell and liver cell xenotransplants, one project examined whole organ allografts. Coffman and co-workers (1993) found that kidneys from β_2-microglobulin-deficient mice functioned significantly better in allogeneic recipients than did kidneys from normal mice.

The fate of pancreatic islet transplants from β_2-microglobulin-deficient mice has been explored in several studies. The islets showed prolonged survival when implanted into a normal mouse strain (Markmann et al., 1992; Osorio et al., 1993). Most grafts survived indefinitely (>80% beyond 100 days) and were capable of reversing hyperglycemia that had been chemically induced prior to transplantation. Investigators attributed the few instances of graft rejection to three possibilities. First, some surface expression of class I antigens can occur in the absence of β_2-microglobulin. The presence of even a fraction of the total number of class I antigens may be sufficient for T cell recognition and lysis. Second, β_2-microglobulin circulating in the serum of the recipient can be used to reconstitute class I antigens on the donor tissue (because β_2-microglobulin is a highly conserved protein not encoded by genes at the MHC locus). Third, rejection of the tissue from β_2-microglobulin-deficient donors may be mediated by other immune pathways, such as by natural killer cells. Support for the second and third possibilities was presented by Li and Faustman (1993) in their study of liver cell allo- and xenografts.

MECHANISMS OF GRAFT SURVIVAL AFTER CLASS I DONOR ABLATION OR ANTIBODY MASKING

The performance of highly divergent cross-species transplants has allowed the β_2-microglobulin-deficient donor model to be more closely studied. First, as the donor and recipient diverge phylogenetically, so does the homology of the β_2-microglobulin proteins, and donor grafts have slower reconstitution of surface class I with host β_2-microglobulin. These highly divergent cross-species transplants demonstrate delayed β_2-microglobulin reconstitution of surface class I from the serum from host animals. Unfortunately, these grafts demonstrate shortened survival times, suggesting some donor class I antigen expression is beneficial after the transplant is established.

The reconstitution of donor surface class I with host β_2-microglobulin in the gene ablation model or even in the antibody masking model has a marked advantage. The grafts in the majority of cases survive indefinitely. Furthermore, the primary grafts permit secondary nongenetically manipulated or non-antibody-masked grafts to survive similarly if from the same donor (Faustman and Coe, 1991). The data in total suggest donor-specific tolerance occurs in these transplant models.

This important demonstration of tolerance induction was studied by the use of gene ablation models with more permanent class I ablation. With the thought that a little is good so more should be better, other gene ablation models, such as the ablation of other class I assembly genes, were tested, i.e., genes *Tap1, Tap2, Lmp2, Lmp7,* etc. Often these donor gene ablation models were used as F_1 donor mice with two sequential class I processing steps interrupted. Uniformly, these cellular transplants from these donors performed less well and the hosts never developed secondary tolerance sufficient for secondary same-donor transplants (D. Faustman, personal communication).

Insights into the mechanisms of T cell tolerance in hosts receiving class I-ablated grafts may have recently been clarified. In a publication by the Cantor laboratory, peripheral CD8 T cell tolerance was mechanistically characterized from the viewpoint of peripheral tolerance maintenance (Pestano *et al.,* 1999). It has been recognized for years that the persistence of peripheral class I-expressing cells is necessary for peripheral CD8$^+$ T cell tolerance (Vidal-Puig and Faustman, 1994). Using β_2-microglobulin-deficient hosts, transferred and potentially cytotoxic and mature CD8 T cells were transferred into the mutant host. The transferred T cells then failed to engage their T cell receptors. The CD8 cells down-regulated their CD8 gene expression and underwent apoptosis. Thus inhibition or interference of the T cell receptor and CD8 binding to host class I triggered these CD8 cells into a pathway of cell death. This is an extremely important observation and probably underlies the reason why β_2-microglobulin or class I-masked donor cellular transplants escape T cell death. The lack of class I expression immediately after transplantation eliminated potential T cell killing. As the graft gradually reexpresses class I, it is protected from the next layer of the immune response, natural killer cell lysis of totally class I-negative transplants.

PREVENTION OF GRAFT REJECTION DOES NOT PREVENT RECURRENT DISEASE: THE CHALLENGE OF CELLULAR TRANSPLANTS IN TYPE I DIABETES

The fate of pancreatic islet transplants from β_2-microglobulin-deficient mice has been explored in several studies. The islets show prolonged survival when implanted into allogeneic mouse strains (Markmann *et al.,* 1992; Osorio *et al.,* 1993). Graft rejection of donor antigen-modified islets almost invariably occurs when the recipients are already nonobese diabetic (NOD) mice (Markmann *et al.,* 1992). This recurrent disease process is a barrier that is not overcome by xenografts. Soon after implantation, NOD mice, at a slightly delayed rate, reject islet allografts from β_2-microglobulin-deficient mice or class I-masked xenogeneic islets. This is in contrast to the success with recipients whose diabetes is chemically induced. NOD mice, unlike those with chemically induced diabetes, spontaneously develop an autoimmune form of diabetes that destroys the grafts. In this respect, NOD mice more closely resemble humans with type I diabetes.

Previous reports have published many effective therapies that prevent autoimmune diabetes development in animal models. The list is long and experiments are not always consistent, making it difficult to identify therapies that are effective at preventing NOD disease. In contrast to the

Table 25.2. Islet transplantation into diabetic NOD/BB hosts[a]

| Treatments | | Islet survival | | | | Ref. |
Donor islet	Recipient	Isographs %	Days	Allographs %	Days	
0	CFA	66	>100	0	12	Wang et al. (1992)
0	CFA	90	>40	—	—	Ulaeto et al. (1992)
0	BCG	80	>100	0	11	Lakey et al. (1994)
Cultured islets	Nicotin/Desf	—	—	57	38	Nomikos et al. (1986)
0	TNF-α	0	30	—	—	Rabinovitch et al. (1995)
Cultured islets	Mycophenolate	—	—	0	2	Hao et al. (1992)
0	αCD4	—	—	0	22	Gill and Coulombe (1994)
GAD$^{-/-}$	—	100	>40	—	—	Yoon et al. (1999)
Class I$^{-/-}$ (β$_2$microglobulin$^{-/-}$)	—	—	—	0	8	Markmann et al. (1992)
IL-4	0	0	14	—	—	Smith et al. (1997)
IL-10	—	—	—	0	14	
Class I$^{-/-}$ [masking F(ab')$_2$]	—	—	—	0	14	Faustman and Coe (1991)
0	IL-1 Rantag	0	14	—	—	Sandberg et al. (1997)
0	IL-4 + IL-10	0	~20	—	—	Faust et al. (1996)
IL-12	0	0	30	—	—	Yasuda et al. (1998)
IL-10	0	0	14	—	—	Yasuda et al. (1998)
0	αCD4	—	—	0	5[c]	Koulmanda et al. (1998)
0	αCD3	—	—	0	5[c]	Koulmanda et al. (1998)
0	CP, CTLA4 Ig, αCD4 + αCD3	—	—	~50	114[c]	Koulmanda et al. (1998)
0	αCTLA Ig	—	—	0	5[c]	Koulmanda et al. (1998)
0	CP[c], αCD4 + αCD3	—	—	0	5[c]	Koulmanda et al. (1998)
0	Lef	—	—	0	5	Guo et al. (1998)
0	Vitamin D, CsA	80	60	—	—	Casteels et al. (1998)
0	CsA + IL-10	0	34	—	—	Rabinovitch et al. (1997)
0	CsA + IL-4	0	60	—	—	Rabinovitch et al. (1997)

[a]The focus is only on biological approaches, not immunoisolation technologies.
[b]Abbreviations: CFA, complete Freund's adjuvant; BCG, bacillus of Calmette–Guerin; Nicotin, nicotinamides; Desf, desferioxamine; IL, interleukin; GAD, glutamic acid decarboxylase; CP, cyclophosphamide; Lef, leflunomide; CsA, cyclosporin A.
[c]Recipients were prediabetic NOD mice and grafts were xenografts of pig origin.

long list of "cures" for the not-yet diabetic NOD mouse, the list of "cures" for the already diabetic NOD mouse is short (Table 25.2). The transplant barrier of achieving long-term islet survival in diabetic NOD mice is a significant problem that largely stands unconquered.

Since 1992, two research groups have published new therapies for diabetic NOD and BB rodents, combining them with syngeneic islet transplantation. The data are limited and in most cases the work was unsuccessful, except for some approaches utilizing tumor necrosis factor α (TNF-α) or direct inducers or mimics of TNF-α such as the bacillus of Calmette–Guerin (BCG), lipopolysaccharide (LPS), or interleukin-1 (IL-1). As Table 25.2 shows, TNF-α or TNF-α stimulants retard the recurrent and very brisk autoimmune destruction of syngeneic islets. It should be stressed the isogeneic islet survival is not permanent. A histologic examination of the syngeneic grafts in NOD mice reveals that TNF-α-treated mice have transplanted islets surrounded by large clusters of mononuclear lymphocytes and, with extended transplantation times, invading lymphocytes, thus explaining that syngeneic islet survival is prolonged and rarely permanent.

Table 26.2 highlights the next significant problem, i.e., the allograft survival data. If the data are extended to the more clinically relevant allograft islet transplant barrier, islets are uniformly rejected when transplanted into already diabetic NOD mice. In fact, an inspection of the peer-reviewed literature indicates a remarkable paucity of success, other than a single report of a non-clinically applicable approach (lymph mode irradiation with bone marrow transplantation) and encapsulation data published frequently as meeting abstracts and a report using islets with ablated glutamic acid decarboxylase (GAD). The lack of success of islet allografts and xenografts in diabetic NOD or BB rodents still allowed the launching of many of these same new approaches in human clinical trials in type I diabetic patients. Ideally, islet allograft methods should work in relevant autoimmune animal models prior to human clinical trials. As Table 26.2 demonstrates, most if not all of the major islet transplant strategies have been tested in NOD mice or BB rats and failed to demonstrate efficacy. Again, Li and colleagues demonstrated prolonged islet allograft survival in already diabetic NOD mice, but this involved total-body irradiation with bone marrow transplantation (Li *et al.,* 1995).

Graft rejection almost invariably occurs, however, when the recipients are NOD mice (Markmann *et al.,* 1992; Osorio *et al.,* 1994a). Islet allografts from β_2-microglobulin-deficient mice are rejected by NOD mice at a slightly delayed rate, but still soon after implantation. This is in contrast to the success with recipients whose diabetes is chemically induced. NOD mice, unlike those with chemically induced diabetes, spontaneously develop an autoimmune form of diabetes that destroys the graft because it displays a variety of antigens identical to autoantigens. In this respect, NOD mice more closely resemble humans with type I diabetes. These findings suggest that elimination of class I antigens alone will not be sufficient for the treatment of humans with type I diabetes. One of the many directions for future research is to pinpoint on the graft the autoantigens that induce autoimmune rejection.

Elimination of class I antigens also may be insufficient for selected tissues and selected combinations of donor and host species. Liver cells from β_2-microglobulin-deficient donors were implanted into allogeneic recipients and two different of xenogeneic recipients (Li and Faustman, 1993). The allotransplants and the xenotransplants into guinea pigs were not effective. In contrast, xenotransplants of mouse cells into frog recipients survived; when liver cells were transplanted from humans to mice in antibody-masking experiments, they were accepted (Faustman and Coe, 1991). These studies show the results of class I antigen removal to varying extents, according to the species combination of donor and host.

Gene ablation experiments have demonstrated the advantages and limitations of eliminating class I antigens. There are clearly dominant antigens on islet cells and neurons in some allogeneic and xenogeneic combinations. However, class I antigens may play a secondary role depending on the type of donor tissue, the species combination, and/or the disease state of the recipient. In these cases, other dominant antigens need to be identified and targeted.

THE LAUNCHING OF XENOGENEIC HUMAN CLINICAL TRIALS IN THE UNITED STATES USING IMMUNOMODULATION

Few cross-species transplantation technologies have progressed to human clinical trials. In part this has been due to primate models showing minimal efficiency. Also, early whole organ human clinical trials in the 1970s demonstrated minimal success, even with massive dosages of immunosuppression. Finally, some segments of the medical community have been concerned about cross-species infections, therefore necessitating new technologies to avoid the use of donor species closely related to humans, i.e., baboons. The testing of novel transplantation approaches needs to avoid the limitation of a severely compromised host immune system with decreased resistance to fight infections.

Clinically close primate and rat models are available for neurological diseases such as Parkinson's disease. Using the masking class I antibody fragment approach, animal studies show promise for porcine neurons for spinal cord injuries in rat models and procine liver cells for transient liver failure from hypotension or infection. Fetal neuronal xenografts with antibody masking have been investigated for Huntington's disease and Parkinson's disease (Deacon *et al.,* 1997). In the first study, fetal pig striatal cells were implanted into rats whose striatum had been lesioned 1 week earlier with injections of quinolinic acid. These injections destroy striatal neurons, and attempt to stimulate the dysfunction present in Huntington's disease. Rats received either untreated tissue or tissue pretreated with $F(ab')_2$ fragments against porcine class I antigens. Of the control rats receiving untreated tissue, some were immunosuppressed with CSA and others were nonimmunosuppressed. Graft volume, determined histologically with the aid of computer image analysis, was significantly larger and well-organized, and graft axions were correctly grown in the direction of their target nuclei. It was encouraging that pig neurons appeared to be capable of locating their target. Also, because of the use of fetal tissue, the transplanted mass was significantly enlarged at the time of autopsy, demonstrating posttransplantation survival as well as growth of the transplant. Six Parkinson's patients have now been treated with fetal pig neurons masked with class I antibody fragments and have reported long-term survival (>2 years), with patients demonstrating mild to marked functional improvements. Six more patients were similarly treated with fetal pig neurons and CSA. These patients showed less clinical improvement but still had function exceeding base line. At 8 months, one of these patients died of a thromboembolic event and an autopsy was performed. As reported by Deacon *et al.* (1997), similar to the primate studies performed before clinical trials, the fetal pig neurons survived and correctly produced axons over long distances in the brain toward their target nuclei. This confirmed the optimism of the suitability of using pig tissue in this transplant setting, and the masking approach to decrease tissue immunogenicity.

Additional clinical trials of masked neuron transplants have continued in the United States. To date, an additional 11 patients in a Food and Drug Administration (FDA)-scrutinized clinical trial have been enrolled in a phase II/III trial using pig cells for Parkinson's disease. The FDA has approved 36 patients for 18 months in this blinded human study for safety and efficacy. The advantages of the donor antigen modifications methods include the lack of host interventions, thus allowing a broader audience for applications of cellular transplants for disease treatments.

GENE ADDITION

Gene addition is a technique especially promising for transplantation of solid organs, the most challenging type of transplant, and represents another form of donor antigen modification. Solid organs are particularly troublesome from the point of view of transplantation and they are subject to hyperacute rejection. Hyperacute rejection, the first wave of assault against discordant xenografts, results in rapid rejection within minutes to hours. The vascular endothelium, which cannot be removed prior to transplant, expresses the epitope, the target of hyperacute rejection (Platt, 1994). Preformed antibodies that bind to α-Gal activate complement, resulting in loss of vascular integrity, hemorrhage, ischemia, and necrosis.

Gene addition involves the insertion of a foreign gene (or transgene) early in development. The desired gene is microinjected into the nucleus of a fertilized egg, thereby giving rise to a mature animal with the transgene incorporated into the genome of each cell. The desired gene or genes used thus far in xenograft research encode proteins designed to inhibit hyperacute rejection. Gene addition is advantageous because it can be readily undertaken in many kinds of donor animals, particularly swine. Swine are among favored donor animals for solid organ transplants because of their ease of breeding and domestication, low maintenance costs, similarities to humans in organ size and physiology, and fewer ethical objections. The other advantages of gene addition, like those for gene ablation, are permanence and versatility. Many sections of this book highlight these approaches and are discussed in detail in other chapters.

RNA ABLATION

RNA ablation is another promising strategy that strives to prevent antigen expression at the RNA level by blocking transcription or translation. RNA ablation can be achieved through the creation of oligodeoxynucleotides that are complementary to, and hybridize with, DNA or RNA sequences to inhibit transcription or translation, respectively.

RNA ablation in the transplant field has been less frequently pursued to date. For example, Ramanathan and co-workers identified an oligodeoxynucleotide that inhibited induction of class I and ICAM expression by interferon-γ (Ramanathan *et al.,* 1994). The studies were performed *in vitro* in a cell line, K562, that normally has low-level expression of class I antigens. They first postulated that the oligodeoxynucleotide acted in the early stages of interferon-γ induction rather than posttranslationally. In their follow-up study, they showed that the oligodeoxynucleotide acted even earlier via a novel mechanism—it inhibited binding of interferon-γ to the cell surface (Ramanathan *et al.,* 1994; Ramakrishnan and Houston, 1984). This may be an unusual mechanism for an oligodeoxynucleotide, but it only enhances the possibilities for xenotransplant research. What these studies provide is yet another means of blocking expression of class I antigens or other transplantation antigens.

COMMENT

Designer tissues and organs through donor antigen modification hold tremendous promise for xenotransplantation. Research in animal models has already demonstrated that long-term xenograft survival can be achieved without immunosuppression. This achievement has galvanized the transplantation community, for it shows that an overwhelming obstacle to xenograft acceptance can be alleviated in select settings. Immune rejection need not occur if graft antigens can be immunologically masked or genetically eliminated. Researchers now have at their disposal a battery of techniques that operate at the DNA, RNA, or protein level to remove or conceal antigens. Xenotransplanation is within reach, not just for patients with life-threatening conditions, but also for patients with chronic conditions. As testimony to widespread applications for chronic conditions, the first clinical trial of designer tissues is for patients with Parkinson's disease.

The successes with cellular and tissue xenografts still are not the final solution for the far more difficult task of whole organ xenotransplantation or recurrent autoimmune disease. Solid organs have a multiplicity of antigens, particularly those which elicit hyperacute rejection. Once all of the dominant antigens are identified, the therapeutic strategy for whole organs and tissues is conceptually identical—modify the donor, not the host. The barriers for recurrent autoimmunity still stand, but it is hoped that donor antigen modification is beneficial.

References

Burns, L. H., Pakzaban, P., Deacon, T. W., Brownell, A. L., Tatler, S. B., and Isacson, O. (1994). Xenotransplantation of porcine ventral mesencephalic neuroblasts restores function in primates with chronic MPTP-induced Parkinsonism. *Soc. Neurosci.* **19**, 1330–1332.

Casteels, K., Waer, M., Laureys, J., Valckx, D., Depovere, J., Bouillon, R., and Mathieu, C. (1998). Prevention of autoimmune destruction of syngeneic islet grafts in spontaneously diabetic nonobese diabetic mice by a combination of a vitamin D3 analog and cyclosporine. *Transplantation* **65**, 1225–1232.

Coffman, T., Geier, S., Ibrahim, S., Griffiths, R., Spurney, R., Smithies, O., Koller, B., and Sanfilippo, F. (1993). Improved renal function in mouse kidney allografts lacking MHC class I antigens. *J. Immunol.* **151**, 425–435.

Deacon, T., Schumacher, J., Dinsmore, J., Thomas, C., Palmer, P., Koh, S., Edge, A., Penney, D., Kassissieh, S., Dempsey, P., and Isacson, O. (1997). Histological evidence of fetal pig neural cell survival after transplantation into a patient with Parkinson's disease. *Nat. Med.* **3**, 350–353.

Dinsmore, J. H., Pakzaban, P., Deacon, T. W., Burns, L., and Isacson, O. (1996). Survival of transplanted porcine neural cells treated with F(ab')$_2$ antibody fragments directed against donor MHC class-I in a rodent model. *Transplant. Proc.* **28**, 817–818.

Faust, A., Rothe, H., Schade, U., Lampeter, E., and Kolb, H. (1996). Primary nonfunction of islet grafts in autoimmune diabetic nonobese diabetic mice is prevented by treatment with interleukin-4 and interleukin-10. *Transplantation* **62**, 648–652.

Faustman, D., and Coe, C. (1991). Prevention of xenograft rejection by masking donor HLA class I antigens. *Science* **252**, 1700–1702.

Fodor, W. L., Williams, B. L., and Matis, L. A. (1994). Expression of a functional human complement inhibitor in a transgenic pig as a model for the prevention of xenogeneic hyperacute organ rejection. *Proc. Natl. Acad. Sci. U.S.A.* **91**, 11153–11157.

Gill, R. G., and Coulombe, M. (1994). Islet xenografting in autoimmune diabetes. *Transplant. Proc.* **26**, 1140.

Guo, Z., Mital, D., Shen, J., Chang, A. S., Tian, Y., Foster, P., Sankaru, H., McChesney, L., Jensk, S.C., and Williams, J. W. (1998). Immunosuppression preventing concordant xenogeneic islet graft rejection is not sufficient to prevent recurrence of autoimmune diabetes in nonobese diabetic mice. *Transplantation* **65**, 1310–1314.

Hao, L., Wang, Y., Chan, S. M., and Lafferty, K. J. (1992). Effect of mycophenolate mofetil on islet allografting to chemically induced or spontaneously diabetic animals. *Transplant. Proc.* **24**, 2843–2844.

Koulmanda, M., Kovarik, J., and Mandel, T. E. (1998). Cyclophosphamide, but not CTLA4Ig, prolongs survival of fetal pig islet grafts in anti-T cell monoclonal antibody treated NOD mice. *Xenotransplantation* **5**, 215–221.

Lakey, J. R., Singh, B., Warnock, G. L., and Rajotte, R. V. (1994). BCG immunotherapy prevents recurrence of diabetes in islet grafts transplanted into spontaneously diabetic NOD mice. *Transplantation* **57**, 1213–1217.

Li, H., Kaufman, C. L., and Ildstad, S. T. (1995). Allogeneic chimerism induces donor-specific tolerance to simultaneous islet allografts in nonobese diabetic mice. *Surgery* **118**, 192–197; discussion: pp. 197–198.

Li, X., and Faustman, D. (1993). Use of donor β2-microglobulin-deficient transgenic mouse liver cells for isografts, allografts, and xenografts. *Transplantation* **55**, 940–946.

Markmann, J. F., Bassiri, H., Desai, N. M., Odorko, J. S., Kim, J. J., Koller, B. H., Smithies, O., and Barker, C. F. (1992). Indefinite survival of MHC class I-deficient murine pancreatic islet allografts. *Transplantation* **54**, 1085–1089.

McCurrey, K. R., Kooyman, D. L., Alvarado, C. G., Cotterell, A. H., Martin, M. J., Logan, J. S., and Platt, J. L. (1995). Human complement regulatory proteins protect swine-to-primate cardiac xenografts from humoral injury. *Nat. Med.* **1**, 423–427.

Miyagawa, S., Shirakura, R., Iwata, K., Nakata, S., Matsumiya, G., and Izutani, H. (1994). Effects of transfected complement regulatory proteins, MCP, DAF, and MCP/DAE hybrid, on complement-mediated swine endothelial cell lysis. *Transplantation* **58**, 834–840.

Miyagawa, S., Shirakura, R., Nakata, S., Matsuda, H., Terado, A., Matsumoto, M. and Nagasawa, S. (1995). Effect of transfected MACIF (CD59) on complement-mediated swine enothelial cell lysis, compared with those of membrane cofactor protein (CD46) and decay-accelerating factor (CD55). *Transplant. Proc.* **27**, 328–329.

Nomikos, I. N., Prowse, S. J., Carotenuto, P., and Lafferty, K. J. (1986). Combined treatment with nicotinamide and desferrioxamine prevents islet allograft destruction in NOD mice. *Diabetes* **35**, 1302–1304.

Osorio, R. W., Ascher, N. L., Jaenisch, R., Freise, C. E., Roberts, J. P., and Stock, P. G. (1993). Major histocompatibility complex class I deficiency prolongs islet allograft survival. *Diabetes* **42**, 1520–1527.

Osorio, R. W., Asher, N. L., Melzer, J. S., and Stock, P. G. (1994a). Beta-2-microglobulin gene disruption prolongs murine islet allograft survival in NOD mice. *Transplant. Proc.* **26**, 752.

Osorio, R. W., Ascher, N. L. and Stock, P. G. (1994b). Prolongation of *in vivo* mouse islet allograft survival by modulation of MHC class I antigen. *Transplantation* **57**, 783–788.

Pakzaban, P., Deacon, T. W., Burns, L. H., Dinsmore, J., and Isacson, O. (1995). Enhanced survival of neural xenografts after masking of donor major histocompatibility complex class I. *Soc. Neurosci.* **16**, 1708–1709.

Pestano, G. A., Zhou, Y., Trimble, L. A., Daley, J., Weber, G. F., and Canton, H. (1999). Inactivation of misselected CD8 T cells by CD8 gene methylation and cell death. *Science* **284**, 1187–1191.

Platt, J. L. (1994). A perspective on xenograft rejection and accommodation. *Immunol. Rev.* **141**, 127–149.

Rabinovitch, A., Suarez-Pinzon, W. L., Lapchak, P. H., Meager, A., and Power, R. F. (1995). Tumor necrosis factor mediates the protective effect of Freund's adjuvant against autoimmune diabetes in BB rats. *J. Autoimmun.* **8**, 357–366.

Rabinovitch, A., Suarez-Pinzon, W. L., Sorensen, O., Rajotte, R. V., and Power, R. F. (1997). Combination therapy with cyclosporine and interleukin-4 or interleukin-10 prolongs survival of synergeneic pancreatic islet grafts in nonobese diabetic mice: Islet graft survival does not correlate with mRNA levels of Type I or Type II cytokines, or transforming growth factor-beta in the islet grafts. *Transplantation* **64**, 1525–1531.

Ramakrishnan, S., and Houston, L. L. (1984). Inhibition of human acute lymphoblastic leukemia cells by immunotoxins: Potentiation by chloriquine. *Science* **233**, 58–61.

Ramanathan, M., Lantz, M., MacGregor, R. D., Garovoy, M. R., and Hunt, C. A. (1994). Characterization of the oligodeoxynucleotide-mediated inhibition of interferon-gamma-induced major histocompatibility complex class I and intercellular adhesion molecule-1. *J. Biol. Chem.* **269**, 24564–24574.

Sandberg, J. O., Eizirik, D. L., and Sandler, S. (1997). IL-1 receptor antagonist inhibits recurrent of disease after syngeneic pancreatic islet transplantation to spontaneously diabetic non-obese diabetic (NOD) mice. *Clin. Exp. Immunol.* **108**, 314–317.

Sandrin, M. S., Fodor, W. L., Mouhtouris, E., Osman, N., Cohney, S., Rollins, S. A., Guillmette, E. R., Selter, E., Squinto, S. P., and McKenzie, I. F. (1995). Enzymatic remodelling of the carbohydrate surface of xenogenic cell substantially reduces human antibody binding and complement-mediated cytolysis. *Nat. Med.* **1**, 1261–1267.

Smith, D. K., Korbutt, G. S., Suarez-Pinzon, W. L., Kao, D., Kajotte, R. V., and Elliot, J. F. (1997). Interleukin-4 or interleukin-10 expressed from adenovirus-transduced syngeneic islet grafts fails to prevent beta cell destruction in diabetic NOD mice. *Transplantation* **64**, 1040–1049.

Spits, H., van Schooten, W., Keizer, H., van Seventer, G., van de Rijn, M., Terhorst, C., and de Vries, J. E. (1986). Alloantigen recognition is preceded by nonspecific adhesion of cytotoxic T cells and target cells. *Science* **232**, 403–405.

Steele, D. J. R., Hertel-Wulff, B., Chappel, S., and Wallstrom, A. (1994). Long term survival of pancreatic islets in diabetic monkeys. *Cell Transplant.* **3**, 216–218.

Ulaeto, D., Lacy, P. E., Kipnis, D. M., Kanagawa, O., and Unanue, E. R. (1992). A T-cell dormant state in the autoimmune process of nonobese diabetic mice treated with complete Freund's adjuvant. *Proc. Natl. Acad. Sci. U.S.A.* **89**, 3927–3931.

Vidal-Puig, A., and Faustman, D. L. (1994). Tolerance to peripheral tissue is transient and maintained by tissue specific class I expression. *Transplant. Proc.* **26**, 3314–3316.

Wang, T., Singh, B., Warnock, G. L., and Rajotte, R. V. (1992). Prevention of recurrence of IDDM in islet-transplanted diabetic NOD mice by adjuvant immunotherapy. *Diabetes* **41**, 114–117.

Yasuda, H., Nagata, M., Arisawa, K., Yoshida, R., Fujihira, K., Okamoto, N., Moriyama, H., Miki, M., Saito, I., Hamada, H., Yokono, K., and Kasuga, K. (1998). Local expression of immunoregulatory IL-12p40 gene prolonged syngeneic islet graft survival in diabetic NOD mice. *J. Clin. Invest.* **102**, 1807–1814.

Yoon, J. W., Yoon, C. S., Lim, H. W., Huang, Q. Q., Kang, Y., Pyon, K. H., Hirasawa, K., Sherwin, R. S., and Jun, H. S. (1999). Control of autoimmune diabetes in NOD mice by GAD expression or suppression in β cells. *Science* **284**, 1183–1187.

CHAPTER 26

IMMUNOISOLATION

Beth A. Zielinski and Michael J. Lysaght

INTRODUCTION

Serious investigative efforts in the field of immunoisolated cell therapy began in the late 1970s and early 1980s. Since then, cellular encapsulation has evolved into a broad-based approach for the delivery of gene products for treating diseases or conditions that respond to sustained release of biotherapeutic substances. This review chronicles the technological developments underlying the progress and evolution in cell therapy and summarizes results available from the initial clinical evaluations of this approach.

Two decades ago, Chick *et al.* (1977) and later Lim and Sun (1980) successfully maintained glucose homeostasis in chemically induced diabetic rats using encapsulated allogeneic islets. Efforts to develop the artificial pancreas have continued and have fostered the development of an expanding array of device formats relying on a large repository of biomaterials. Cell encapsulation technology has also expanded to the treatment of chronic pain (Sagen *et al.*, 1993; Joseph *et al.*, 1994; Decostard *et al.*, 1998) and to neurodegenerative diseases, including Parkinson's disease (Tresco *et al.*, 1992a,b; Aebischer *et al.*, 1994a; Tseng *et al.*, 1997; Sautter *et al.*, 1998), Alzheimer's (Hoffman *et al.*, 1993) and amyotrophic lateral sclerosis (ALS) (Sagot *et al.*, 1995; Tan *et al.*, 1996). Other disorders that have been addressed with this type of therapy include hypocalcemia (Aebischer *et al.*, 1986), dwarfism (Chang *et al.*, 1993), anemia (Koo and Chang, 1993; Rinsch *et al.*, 1997), hemophilia (Colton, 1996; Brauker *et al.*, 1998), and cancer (Geller *et al.*, 1997a,b). Human trials have been initiated for diabetes (Soon-Shiong *et al.*, 1994; Scharp *et al.*, 1994), chronic pain (Aebischer *et al.*, 1994b; Buscher *et al.*, 1996), and ALS (Ezzell, 1995; Aebischer *et al.*, 1996).

TECHNOLOGY

Encapsulation is based on the premise that cells, once sequestered within a semipermeable membrane, are isolated from the immune system and therefore cannot be recognized and/or destroyed by normal host defenses. The membrane also presents a barrier to any potential outgrowth of cells into the host parenchyma, thus permitting the use of mitotically active cell lines. Bidirectional diffusion through the membrane of small-molecular-weight molecules such as oxygen, glucose, cellular secretogues, and bioactive cellular products allows for continued maintenance of encapsulated cell viability and sustained release of the molecules of interest. Species such as host complement, antibodies, and immune effector cells may be inhibited from passing through the membrane (Fig. 26.1). Available barriers range from tight (approximately 30-kDa molecular-weight cutoff, semipermeable) to open (up to about 0.6-μm pore size, microporous). Immunogenicity of the encapsulated cell population plays a key role in governing the selection of a membrane with the appropriate molecular-weight cutoff (Colton, 1996). Microporous membranes are increasingly being investigated for encapsulated allogeneic cells, i.e., cells isolated from and implanted into members of the same species. Relevant mechanisms of host recognition and immunoreactivity appear cellular in nature and therefore may require only the exclusion of host ef-

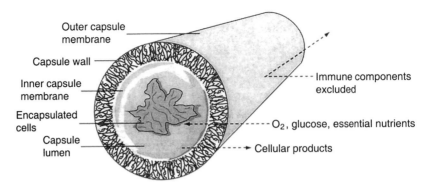

Fig. 26.1. A cross-sectional view of cells encapsulated in a polymer hollow-fiber macrocapsule. Diameter is typically 600 to 1200 μm. Wall thickness is approximately 10% of diameter. Cells grow within the lumen of the hollow fiber and are protected from host immune destruction by the semipermeable polymer membrane. Low-molecular-weight species such as cellular secretogogues, oxygen, glucose, and other essential nutrients pass through the membrane freely, thus maintaining cell viability and allowing for continuous release of bioactive cellular products. Depending on the selectivity of the encapsulating membrane, host antibodies and complement components may be wholly or partially excluded from the capsule lumen. Host immune effector cells are always prevented from passing through immunoisolatory membranes.

fector cells. Tighter membranes, although potentially more restrictive to diffusion of nutrients and secreted product, seem preferable when encapsulating xenogeneic cell types, i.e., cells isolated from tissue derived from a member of a species other than that of the implanted recipient. Membranes of this type include those having lower molecular-weight cutoffs and those alternatively designed to trap host immune complexes and prevent their entry into the capsule lumen (Colton, 1996; Wang, 1998). Xenogeneic cells may induce cellular and humoral immune responses based on immunoreactive secretogues and appear vulnerable to natural and elicited antibodies and to host complement. In addition, transfer of xenogeneic virus from transplant to host is also a concern. Thus, tighter membranes not only provide protection against cellular attack but may also confer protection against immune rejection mediated by macromolecules and retard xenogeneic viral transfer (Colton, 1996; Morris, 1996).

Several capsule formats have been studied. Biohybrid implants directly anastamosed to host vasculature have been examined in both rodents and pancreatectomized diabetic dogs (Sullivan *et al.,* 1991; Maki *et al.,* 1993). Vascular implants have waned in popularity as a result of their intrinsic complexity and the need for long-term anticoagulation when using blood-contacting devices. Nonvascular devices include spherical microcapsules, cylindrical hollow-fiber macrocapsules, flat sheet devices (Geller *et al.,* 1997b), and planar macrocapsules (Ezzell, 1995). Microcapsule implants can be formed by interfacial coacervation of weak polyelectrolytes such as sodium alginate and poly(L-lysine) (Lim and Sun, 1980) or chitosan and carboxymethylcellulose (Yoshioka *et al.,* 1990). In one popular approach, droplets of cells suspended in sodium alginate are dripped into a bath of calcium chloride, resulting in formation around the droplet of an interfacial film of insoluble calcium alginate. The droplets can subsequently be overcoated with several alternating layers of poly(L-lysine) and alginate. In a related technique, water-insoluble polyacrylates can also be formed into microcapsule walls by interfacial precipitation (Boag and Sefton, 1987; Dawson *et al.,* 1987; Sugamori and Sefton, 1989; Mallabone *et al.,* 1989; Crooks *et al.,* 1990; Broughton and Sefton, 1990). Here, cells are suspended in a polymer solution and extruded into a precipitating bath of nonsolvent. Viability of encapsulated cells in this system, however, appears to be marginal possibly because of contact with organic solvent during preparation or because of inadequate diffusive properties of the polymeric membrane (Crooks *et al.,* 1990). A single spherical bead typically contains only a few hundred cells or a single islet, and tens or hundreds of thousands are required to provide a full therapeutic dosage. In contrast, macrocapsules employ much larger depots to contain the full therapeutic dosage in one or a few implants. Macrocapsules are shaped as cylinders or planes and typically one dimension is held below 1 mm. Macrocapsules are fabricated from preformed hollow-fiber or planar membranes prepared by phase inversion of water-insoluble polymers quenched in a nonsolvent bath. Molecular-weight cutoff of the membrane depends on conditions of fabrication and formulation and may be selected to be in a range from 30 kDa to over 10^6 Da. The resulting membranes are then filled with cell suspension and sealed appropriately. In contrast to spherical microcapsules, lumens of macrocapsule implants are usually completely empty prior to cell loading and therefore require the addition of a supportive matrix for optimal cell viability. Both natural and synthetic lumenal matrices have been investigated. Natural matrices include polysaccharides such as alginate and chitosan and cross-linked proteins such as collagen (Lanza et al., 1992; Zielinski and Aebischer, 1994). All are hydrogels with low (2–5%) solid content and hence offer little resistance to diffusive transport. Investigation of synthetic matrices has included the use of poly(vinyl alcohol) (PVA) foams (Li et al., 1998).

Three types of cells may be employed for encapsulation: primary postmitotic cells, mitotically active cell lines, and engineered cell lines (Colton, 1996). Postmitotic cell types, particularly islets of Langerhans, dominate early literature (Chick et al., 1977; Lim and Sun, 1980; Sullivan et al., 1991; Lacy et al., 1991; Lanza et al., 1992, 1993; Gerasimidi-Vazeou et al., 1992; Soon-Shiong et al., 1993; Maki et al., 1993). Primary cells are isolated from excised glands of donor animals. The cells are enzymatically digested, adapted to in vitro tissue culture conditions, and then encapsulated. Primary cells are advantageous when regulated release or the secretion of multiple cellular products is desired. Postmitotic cells also adequately provide initial insight into therapeutic efficacy, although isolated tissue may not be available in sufficient quantity to keep up with demands of most applications in large animal studies or human applications. Both allogeneic and xenogeneic sources have been investigated. Both have been successful in small-animal models; however, xenogeneic cells in large-animal hosts appear to require tighter membranes and more stringent immunoisolatory measures than do their allogeneic counterparts. Inappropriate choice of encapsulating membrane resulting in a lack of immunoprotection and sensitization of the host may explain early problems associated with xenogeneic cell viability.

Mitotically active cell lines provide an alternative to the use of primary cells. Cell lines are easier to maintain in tissue culture and serve as a continuous source of graftable cells. By their very nature, cell lines possess the ability to proliferate within the capsule; however, growth may be constrained by contact inhibition and metabolic factors (Lysaght et al., 1994). Dividing cell lines can also be genetically engineered to produce the desired gene product (Chang et al., 1993; Tan et al., 1996; Rinsch et al., 1997; Sautter et al., 1998). Cell lines of both xenogeneic and allogeneic origin have been successfully encapsulated. Use of mitotically active cell lines for encapsulated cell therapy, although advantageous, requires careful attention to patient safety in the unlikely event of capsule rupture. Allogeneic cell lines, due to similarities in their genetic profile and the profile of the implanted recipient, may possess the ability to proliferate outside the capsule without provoking immunological recognition. Risk of immune evasion may even be greater in immuno-privelaged sites such as the central nervous system (CNS) (Morris, 1996). Allogeneic cell lines would almost certainly need to be transfected with suicide genes prior to encapsulation and implantation, although studies showing the effectiveness of this approach are not yet available. Xenogeneic cell lines are more immunogenic than allogeneic lines and therefore are more susceptible to destruction by a host in the event that they escape the capsule. In studies conducted to date, both peripherally located xenogeneic cells and those transplanted into the CNS have been shown to be recognized and completely destroyed by the host immune system (Fuchs and Bullard, 1988; Somerville and d'Apice, 1993). Cited studies are not exhaustive, nevertheless many investigators believe xenogeneic cell lines may be a wiser choice for encapsulation and implantation.

CLINICAL

The first clinical trials of immunoisolation were initiated in 1993 (Soon-Shiong et al., 1994). By April of 1999, five phase I clinical trials had been reported for diabetes, chronic pain, and ALS (Soon-Shiong et al., 1994; Scharp et al., 1994; Aebischer et al., 1994b; Ezzell, 1995; Buscher et al., 1996), and phase II clinical trials were being initiated for chronic pain in the United States. The trials are summarized in Table 26.1. Trials involving encapsulated allogeneic islets for the treatment of diabetes were initiated to provide proof of principle for encapsulation technology. Those investigations involving chronic pain and ALS have advanced to the next step, which involves the evaluation of human-relevant devices and protocols. If such studies prove successful, this technology may then move directly into large-scale regulatory trials and thence to clinical practice.

Table 26.1. Current clinical trials using cell encapsulation therapy

Application	Investigator/country/year	Device/polymer/cell type	Brief results
Chronic pain	Aebischer *et al.* (Cytotherapeutics)/ Switzerland/1994	Hollow fiber/acrylic, 50 kDa/ calf adrenal chromaffin cells	$n = 10$; histologic and biochemical evidence of cell viability and functionality following explant at 45–155 days; no evidence of complications due to implantation
Chronic pain	Aebischer *et al.* (Cytotherapeutics)/ United States/1995 (Phase I); 1999 (Phase IIA trials being initiated)	Hollow fiber/acrylic, <100 kDa/ calf adrenal chromaffin cells	$n = 15$; histologic and biochemical evidence of cell viability and functionality following explant at 200 days; no evidence of complications due to implantation
ALS	Aebischer *et al.* (Cytotherapeutics)/ Switzerland/1995	Hollow fiber/Polysulfone, 280 kDa/engineered baby hamster kidney cells	$n = 6$; nanogram quantities of CNTF measured in patient CSF up to 17 weeks postimplantation; evidence of cell viability following retrieval
Diabetes	Soon-Shiong *et al.*/United States/ 1994	Microcapsule/alginate, <100 kDa/allogeneic islets	$n = 1$; (immunosuppressed); adequate regulation of patient glucose at 9 months
Diabetes	Scharp *et al.* (Neocrin)/United States/1994	Hollow fiber/acrylic, 65 kDa/ allogeneic islets	$n = 9$; good biocompatibility; no immune complications; cell viability and stimulated insulin release shown following explant at 2 weeks

Based on the previous success of encapsulated islet transplants in rodent and dog models, Soon-Shiong *et al.* (1994) and Scharp *et al.* (1994) have conducted clinical investigations involving type I and type II diabetic patients. The two investigative groups, however, have used very different methods. Soon-Shiong *et al.* implanted modified islet-containing alginate microcapsules into the peritoneum of a 38-year-old diabetic male. The patient, immunosuppressed with 50 mg/ day of cyclosporin for a preexisting kidney transplant, received a total dose of $>10^6$ pooled allogeneic islets administered at two separate time intervals. Soon-Shiong *et al.* utilized purified alginate microcapsules with a high guluronic acid content, which his earlier studies had identified as possibly having improved biocompatibility and increased mechanical strength compared to standard alginate capsules (Otterlei *et al.*, 1991; Soon-Shiong *et al.*, 1991) Insulin secretion from the encapsulated islets was noted 24 hr following implantation of 700,000 microcapsules. Exogenous insulin administration was decreased during this period. Following a supplemental dose of 350,000 encapsulated islets, insulin administration further decreased and at 9 months postimplantation, exogenous insulin administration was discontinued for 1 month then resumed. Peripheral neuropathy, a common complication associated with diabetes and experienced by this patient, was reported to be alleviated posttransplant. Results are difficult to evaluate because other investigators have achieved similar or better results using unencapsulated islets in immunosuppressed recipients. At best this preliminary clinical trial demonstrated that the encapsulation procedure did not adversely affect cell viability or function relative to unencapsulated islets. It offers little insight into the extent of immunoprotection provided by this approach.

Scharp's clinical trial was designed to investigate whether immunoisolation protects against the autoimmune challenge of type I diabetes. This is an important issue because autoimmune rejection of transplanted islets normally destroys unencapsulated islets in a nonimmunosuppressed host. Islets were encapsulated in semipermeable hollow-fiber macrocapsules having a nominal molecular-weight cutoff of approximately 65 kDa. The islets, isolated from a single donor, were suspended in 1% sodium alginate and then loaded into preformed acrylic copolymer capsules. Approximately 50 islets were loaded into each capsule. Four capsules, three containing islets and alginate and one containing just alginate, were implanted subcutaneously into each type I and type

II diabetic patient and nondiabetic control subjects. The net dosage of islets (150) is about 4000× less than would be required for therapeutic implant. During the 2-week implant period, blood glucose levels of all three groups appeared to be unchanged from preimplant values. On retrieval of the capsules, *in vitro* analysis revealed differences in glucose responsiveness between the cell-loaded explants of the diabetic groups and those of the control group. Explanted capsules from the control group responded appropriately to both glucose and theophylline challenges. Those from the experimental groups, on the other hand, responded only to stimulation with theophylline. Differences in responsiveness were attributed to chronic hyperglycemia in the diabetic patients, which likely stressed the islets and impaired insulin secretion. Histologic analysis of retrieved capsules showed substantial cell viability and minimal host tissue reaction in all cases. The investigators conclude that this trial demonstrated the short-term biocompatibility and acute immunoprotective capacity of semipermeable membranes under immunologically challenging conditions. Longer duration studies are planned.

In addition to diabetes, clinical trials involving chronic pain were conducted in Switzerland and the United States by Aebischer and Goddard (Aebischer *et al.*, 1994a; Buscher *et al.*, 1996). These trials flow from earlier collaboration with Sagen *et al.* demonstrating reduction of chronic pain by the implantation of encapsulated bovine adrenal chromaffin cells in rodents. Studies in larger animals confirmed the immunoisolatory properties of the membrane and were used to develop and validate an implant suitable for intrathecal implantation in humans (Joseph *et al.*, 1994). The first clinical trial conducted in Switzerland involved 10 end-stage cancer patients suffering from intractable pain. Results have been reported for the entire series and published for the 7 of the 10 patients (Buscher *et al.*, 1996). Each of the 10 patients received a 5-cm-long tethered device, having a molecular-weight cutoff of approximately 50 kDa, which contained approximately 2 million cells suspended in a supportive alginate matrix. The device was implanted in the lumbar subarachnoid space through a series of expanding catheters during a minimally invasive surgical procedure very analogous to a spinal tap. The silicone tether located at one end of the device allowed for simple withdrawal of the implant if necessary or desirable. Capsules were retrieved for analysis in some patients and recovered postmortem in others. Implant duration ranged from 41 to 172 days. Cell survival in all evaluable cases was confirmed by histology, by catecholamine release postretrieval, and in selected cases by immunohistochemistry. Patients in the open-label study showed a decline in intake of narcotics and reduction in pain ratings. Although preliminary, this protocol represents the first successful clinical trial of a human-relevant immunoisolation device and confirms the immunoprotective function of semipermeable membranes in nonimmunosuppressed xenogeneic hosts.

Based on the results obtained from the Swiss trials, FDA-approved phase I clinical trials were initiated in the United States by Cytotherapeutics (Providence, RI). These trials were designed to test the safety of intrathecal implantation of encapsulating devices into end-stage cancer patients suffering from chronic pain. In addition, survival of the encapsulated xenogeneic bovine chromaffin cells was evaluated following device retrieval. Fifteen patients were enrolled in the trial. Trial protocol paralleled that implemented in the initial clinical trials conducted in Switzerland. No clinically significant safety issues associated with the implantation procedure were observed in any of the patients enrolled. Following retrieval and histologic and biochemical analysis, devices were shown to contain viable and functional encapsulated bovine chromaffin cells for up to 200 days postimplantation. As a result of the success of this phase I trial, phase IIA clinical trials were initiated by Cytotherapeutics and Astra AB of Sweden, and patient enrollment was completed in January, 1999. The trial is being conducted at four clinical sites in the United States and Europe.

A second clinical trial underway in Switzerland involves the delivery of ciliary neurotrophic factor (CNTF) for treatment of amyotrophic lateral sclerosis (Ezzell, 1995; Aebischer *et al.*, 1996). ALS, often called Lou Gehrig's disease in the United States, is characterized by the progressive degeneration of motor neurons, resulting in movement disorder and eventual respiratory failure due to loss of neurons innervating the chest wall and diaphragm. Administration of neurotrophic factors such as CNTF has been shown *in vitro* and in small animal models to protect against death of motor neurons. Systemic delivery of CNTF is unfavorable because of rapid clearance in the serum and toxicity of high doses of purified CNTF. Sagot *et al.* (1995) and Tan *et al.* (1996) have demonstrated efficacy with encapsulated baby hamster kidney (BHK) cells releasing recombinant CNTF in a mouse model of ALS. In studies conducted by Sagot *et al.*, BHK cells, after transfection with the gene for mouse CNTF, produced 100 ng of CNTF per million cells per day in this study. After encapsulation and subcutaneous implantation, the transfected cells prolonged the lives of diseased mice by 40% relative to controls and improved motor function on several behavioral assays. In addition, the phrenic nerves of experimental mice, which innervate the diaphragm, were also shown to contain 50% more myelinated motor neurons than controls. Clinical trials based on these studies were initiated by Aebischer *et al.* (1996) using cells transfected with the gene for human CNTF. Six ALS patients were implanted with polymer capsules containing BHK cells genetically engineered to secrete human CNTF. Encapsulating devices and surgical procedures were modeled after those used for clinical trials involving implantation of encapsulated bovine chromaffin cells for the alleviation of chronic pain (Aebischer *et al.*, 1994a; Buscher *et al.*, 1996). Prior to implantation, devices were shown to release approximately 0.5 μg of human CNTF per day *in vitro*. Following intrathecal implantation, 170–6200 pg/ml of CNTF were measured within the patients' cerebrospinal fluid for at least 17 weeks postimplanation (Aebischer *et al.*, 1996). CNTF levels were undetectable in all patients prior to implantation. On retrieval, all devices showed good viability and CNTF output. This trial represents the first clinical combination of immunoisolation and gene therapy.

FUTURE DIRECTIONS

Future progress of immunoisolated cell therapy will result from both the aggressive clinical application of existing technology and the continued development of new generations of capsules, cells, and clinical application. Techniques may be developed that allow capsule membranes to be modified so as to enhance encapsulated cell viability. Microvascularization of the external portion of the capsule membrane has been suggested by researchers at Baxter Healthcare Corp. as a means of improving the performance of their encapsulation devices for insulin-secreting cells (Brauker *et al.*, 1992, 1995; Danheiser, 1995). Promoting the growth of blood vessels in close proximity to the membrane may increase bidirectional diffusion of essential nutrients and cellular secretogogues, thus enhancing viability and function of the encapsulated cell population. Other modifications in membrane fabrication and capsule design involve the use of biodegradable hydrogel microspheres (Lanza and Chick, 1997). Researchers at Biohybrid Technologies propose the development of biodegradable cell microcapsules that remain intact only for the life cycle of the encapsulated cell population (Danheiser, 1995; Lanza and Chick, 1997). Following degradation of the protective capsule, remains of the grafted cells would be phagocytosed by the host immune system. A new batch of microspheres could be injected or treatment could be ceased. Host sensitization remains a potential problem with this procedure; induction of host tolerance is a potential upside. Other technological advancements include the use of photopolymerized conformal coatings that entrap individual cells or small cell clusters and the use of systems that combine different polyanionic and polycationic moieties to encapsulate cells (Basta *et al.*, 1996; Tun *et al.*, 1996; Wang *et al.*, 1997; Hill *et al.*, 1997).

The future of cell encapsulation is closely entwined with that of human gene therapy. The span of therapeutic applications will increase as cells are tailored to release any of the variety of bioactive molecules whose genes are being identified. In addition, cell lines might be engineered to adapt to an encapsulated environment by incorporation of genes that induce mitotic arrest following exogenous administration of appropriate agents. Other cell types, such as genetically engineered myoblasts, may prove useful as progenitor cells that spontaneously differentiate following tubule formation (Deglon *et al.*, 1996; Rinsch *et al.*, 1997). Such approaches permit cell growth within the capsule but circumvent problems associated with unrestrained cell proliferation. Alternatively, it may be possible to induce quiescent cell lines to proliferate within the capsule when the cell population declines. Modifications of cell lines in these ways, along with optimization of lumenal supportive matrices, have the potential to create capsules possessing an indefinite supply of cells that will not require replacement or reloading.

Future applications of encapsulated cell therapy are likely to include to other CNS disorders such as Huntington's and Alzheimer's disease and a variety of non-CNS related diseases, including hemophilia and malignant cancer (Emerich and Sandberg, 1994; Emerich *et al.*, 1994; Danheiser, 1995; Geller *et al.*, 1997a,b). Emerich *et al.* investigated an animal model of Huntington's disease and its response to encapsulated cell therapy. Kordower *et al.* (1995) reported that poly-

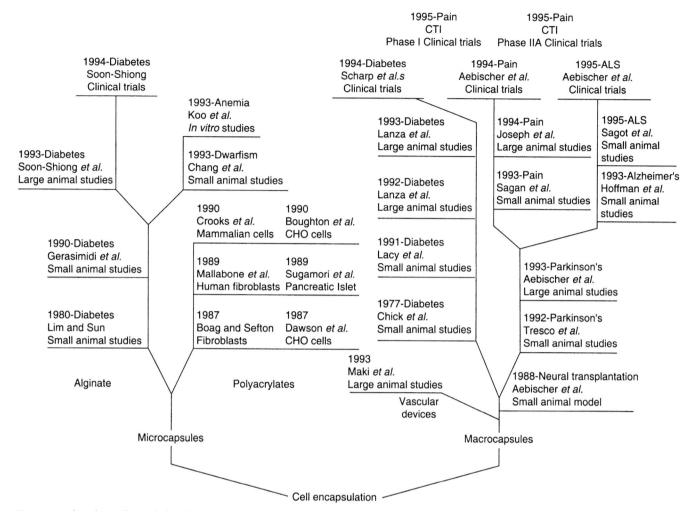

Fig. 26.2. The chronological development of encapsulated cell therapy. The timeline is divided into two main branches—microcapsules and macrocapsules; further bifurcations refer to applications. Side branches represent studies that lead to the current clinical trials in diabetes, chronic pain, and ALS. All cited contributions have been referenced elsewhere in this paper.

mer macrocapsules containing engineered, nerve growth factor-producing BHK cells are capable of rescuing degenerating cholinergic neurons in fornix-lesioned cynomolgus macaques, an animal model of relevance to Alzheimer's disease. Scientists at Baxter Healthcare have initiated preclinical trials using recombinant factor IX produced by encapsulated engineered human fibroblasts for the treatment of hemophilia B (Danheiser, 1995; Brauker *et al.,* 1998) In addition, Baxter's gene therapy group reported studies of encapsulated cell therapy as a delivery system for the slow release of tumor-associated antigen in order to induce specific host immune responses to treat malignant cancer (Geller *et al.,* 1997a,b). Local delivery of cytokines from encapsulated cells for the induction of potent nonspecific immune responses is another potential application that may enhance more specific immune reactions.

In the past two decades, cell encapsulation technology has progressed from *in vitro* investigation, through small-animal and large-animal studies, into full-scale clinical investigation (see Fig. 26.2). Understanding the immunologic and molecular basis of host response to implanted devices, though incomplete, has proved adequate for successful first-generation implants. The underlying portfolio has expanded from initial focus on diabetes to CNS applications and then further into the realm of human gene therapy. The use of genetically engineered cell lines with modified encapsulation devices represents the likely focus of future applications. Immunoisolation is thus poised to represent the first clinical confluence of organ substitution and molecular medicine.

REFERENCES

Aebischer, P., Russell, P. C., Christenson, L. *et al.* (1986). A bioartificial parathyroid. *Trans. Am. Soc. Artif. Intern. Organs* **32**, 134–137.

Aebischer, P., Buchser, E., Joseph, J. M. *et al.* (1994a). Transplantation in humans of encapsulated xenogeneic cells without immunosuppression. *Transplantation* **58**, 1275–1277.

Aebischer, P., Goddard, M., Signore, A. P. *et al.* (1994b). Functional recovery in hemiparkinsonian primates transplanted with polymer-encapsulated PC12 cells. *Exp. Neurol.* **126**, 151–158.

Aebischer, P., Scluep, M., Deglon, N. *et al.* (1996). Intrathecal delivery of CNTF using encapsulated genetically modified xenogeneic cells in amyotrophic lateral sclerosis patients. *Nat. Med.* **2**, 1041.

Basta, G., Rosodivita, M. E., Osticioli, L. *et al.* (1996). Ultrastructural examination of pancreatic islet containing alginate/polyaminoacidic coherent microcapsules. *J. Submicrosc. Cytol. Pathol.* **28**, 209–213.

Boag, A. H., and Sefton, M. V. (1987). Microencapsulation of human fibroblasts in a water-insoluble polyacrylate. *Biotechnol. Bioeng.* **30**, 954–962.

Brauker, J. H., Martinson, L. A., Hill, R. S. *et al.* (1992). Neovascularization of immunoisolation membranes: The effect of membrane architecture and encapsulated tissue. *Transplant. Proc.* **24**, 2924.

Brauker, J. H., Carr-Brendel, V. E., Martinson, L. A. *et al.* (1995). *J. Biomed. Mater. Res.* **29**, 1517–1524.

Brauker, J. H., Frost, G. H., Dwarki, V. *et al.* (1998). Sustained expression of high levels of human factor IX from human cells implanted within an immunoisolation device into athymic rodents. *Hum. Gene Ther.* **9**, 879–888.

Broughton, R. L., and Sefton, M. V. (1990). Effect of capsule permeability on growth of CHO cells in Eudragit RL microcapsules: Use of FITC–dextran as a marker of capsule quality. *Biomaterials* **10**, 462–465.

Buscher, E., Goddard, M., Heyd, B. *et al.* (1996). Immunoisolated xenogeneic chromaffin cell therapy for chronic pain. *Anesthesiology* **85**, 1005–1012.

Chang, P. L., Shen, N., and Westcott, A. J. (1993). Delivery of recombinant gene products with microencapsulated cells *in vivo. Hum. Gene Ther.* **4**, 433–440.

Chick, W. L., Perna, J. J., Lauris, V. *et al.* (1977). Artificial pancreas using living beta cells; effects on glucose homeostasis in diabetic rats. *Science* **197**, 780–782.

Colton, C. K. (1996). Engineering challenges in cell-encapsulation technology. *Trends Biotechnol.* **14**, 158–162.

Crooks, C. A., Douglas, J. A., Broughton, R. L. *et al.* (1990). Microencapsulation of mammalian cells in a HEMA–MMA copolymer: Effects on capsule morphology and permeability. *J. Biomed. Mater. Res.* **24**, 1241–1262.

Danheiser, S. L. (1995). Encapsulated cell therapies target range of application. *Genet. Eng News* **1**, 12–13.

Dawson, R. M., Broughton, R. L., Stevenson, W. T. K. *et al.* (1987). Microencapsulation of CHO cells in a hydroxyethyl methacrylate-methyl methacrylate copolymer. *Biomaterials* **8**, 360–366.

Decostard, I., Buscher, E., Gilliard, N. *et al.* (1998). Intrathecal implants of bovine chromaffin cells alleviate mechanical allodynia in a rat model of neuropathic pain. *Pain* **76**, 159–166.

Deglon, N., Heyd, B., Tan, S. A. *et al.* (1996). Central nervous system delivery of recombinant ciliary neurotrophic factor by polymer encapsulated differentiated C2C12 myoblasts. *Hum. Gene Ther.* **7**, 2135–2146.

Emerich, D. F., and Sandberg, P. R. (1994). Animal models of Huntington's disease. *Neurol. Methods* **21**, 65–133.

Emerich, D. F., Winn, S. R., Harper, J. *et al.* (1994). Implants of polymer-encapsulated human NGF-secreting cells in the nonhuman primate: Rescue and sprouting of degenerating cholinergic basal forebrain neurons. *J. Comp. Neurol.* **349**, 148–164.

Ezzell, C. (1995). Tissue engineering and the human body shop: Encapsulated-cell transplants enter the clinic. *J. NIH Res.* **7**, 47–51.

Fuchs, H. E., and Bullard, D. E. (1988). Immunology of transplantation in the central nervous system. *Appl. Neurophysiol.* **51**, 278–296.

Geller, R. L., Neuenfeldt, S., Levon, S. A. *et al.* (1997a). Immunoisolation of tumor cells: Generation of antitumor immunity through indirect presentation of antigen. *J. Immunother.* **20**, 131–137.

Geller, R. L., Loudovaris, T., Neunfeldt, S. *et al.* (1997b). Use of an immunoisolation device for cell transplantation and tumor immunotherapy. *Ann. N.Y. Acad. Sci.* **831**, 438–451.

Gerasimidi-Vazeou, A., Dionne, K., Gentile, F. *et al.* (1992). Reversal of streptozotocin diabetes in nonimmunosuppressed mice with immunoisolated xenogeneic rat islets. *Transplant. Proc.* **24**, 667–668.

Hill, R. S., Cruise, G. M., Hager, S. R. *et al.* (1997). Immunoisolation of adult porcine islets for the treatment of diabetes mellitus. The use of photopolymerizable polyethylene glycol in the conformal coating of mass-isolated porcine islets. *Ann. N.Y. Acad. Sci.* **831**, 332–343.

Hoffman, D., Breakefield, X. O., Short, M. P. *et al.* (1993). Transplantation of a polymer-encapsulated cell line genetically engineered to release NGF. *Exp. Neurol.* **122**, 100–106.

Joseph, J. M., Goddard, M. B., Mills, J. *et al.* (1994). Transplantation of encapsulated bovine chromaffin cells in the sheep subarachnoid space: A preclinical study for the treatment of cancer pain. *Cell Transplant.* **3**, 355–364.

Koo, J., and Chang, T. M. S. (1993). Secretion of erythropoietin from microencapsulated rat kidney cells: Preliminary results. *Int. J. Artif. Organs* **16**, 557–560.

Lacy, P. E., Hegre, O. D., Gerasimidi-Vazeou, A. *et al.* (1991). Maintenance of normoglycemia in diabetic mice by subcutaneous xenografts of encapsulated islets. *Science* **254**, 1782–1784.

Lanza, R. P., and Chick, W. L. (1997). Immunoisolation: At a turning point. *Immunol. Today* **18**, 135–139.

Lanza, R. P., Butler, D. H., Borland, K. M. *et al.* (1992). Successful xenotransplantation of a diffusion-based biohybrid artificial pancreas: A study using canine, bovine, and porcine islets. *Transplant. Proc.* **24**, 669–671.

Lanza, R. P., Lodge, P., Borland, K. M. *et al.* (1993). Transplantation of islet allografts using a diffusion-based biohybrid artificial pancreas: Long-term studies in diabetic, pancreatectomized dogs. *Transplant. Proc.* **25**, 978–980.

Li, R. H., White, M., Williams, S. *et al.* (1998). Poly(vinyl alcohol) synthetic polymer foams as scaffolds for cell encapsulation. *J. Biomater. Sci., Polym. Ed.* **9**, 239–258.

Lim, F., and Sun, A. M. (1980). Microencapsulated islets as bioartificial endocrine pancreas. *Science* **210**, 908–910.

Lysaght, M. J., Frydel, B., Gentile, F. *et al.* (1994). Recent progress in immunoisolated cell therapy. *J. Cell. Biochem.* **56**, 196–203.

Maki, T., Lodge, J. P. A., Carrot, M. *et al.* (1993). Treatment of severe diabetes mellitus for more than one year using a vascularized hybrid artificial pancreas. *Transplantation* **55**, 713–718.

Mallabone, C. L., Crooks, C. A., and Sefton, M. V. (1989). Microencapsulation of human diploid fibroblasts in cationic polyacrylates. *Biomaterials* **10**, 380–386.

Morris, P. J. (1996). Immunoprotection of therapeutic cell transplants by encapsulation. *Trends Biotechnol.* **14**, 163–167.

Otterlei, M., Ostgaard, K., Skjak-Braek, G. *et al.* (1991). Induction of cytokine production from human monocytes stimulated with alginate. *J. Immunother.* **10**, 286–291.

Rinsch, C., Regulier, E., and Deglon, N. (1997). A gene therapy approach to regulated delivery of erythropoietin as a function of oxygen tension. *Hum. Gene Ther.* **8**, 1881–1889.

Sagen, J., Hama, A. T., Winn, S. R. *et al.* (1993). Pain reduction by spinal implantation of xenogeneic chromaffin cells immunologically-isolated in polymer capsules. *Neurosci. Abstr.* **19**, 234.

Sagot, Y., Tan, S. A., Baetge, E. *et al.* (1995). Polymer encapsulated cell lines genetically engineered to release ciliary neurotrophic factor can slow down progressive motor neuropathy in the mouse. *Eur. J. Neurosci.* **7**, 1313–1322.

Sautter, J., Tseng, J. L., Braguglia, D. *et al.* (1998). Implants of polymer-encapsulated genetically modified cells releasing glial cell line-derived neurotrophic factor improve survival, growth, and function of fetal dopaminergic grafts. *Exp. Neurol.* **149**, 230–236.

Scharp, D. W., Swanson, C. J., Olack, B. J. *et al.* (1994). Protection of encapsulated human islets implanted without immunosuppression in patients with type I or type II diabetes and in nondiabetic control subjects. *Diabetes* **43**, 1167–1170.

Somerville, C. A., and d'Apice, A. J. (1993). Future directions in transplantation: Xenotransplantation. *Kidney Int., Suppl.* **42**, S112–S121.

Soon-Shiong, P., Otterlei, M., Skjak-Braek, G. *et al.* (1991). An immunologic basis for the fibrotic reaction to implanted microcapsules. *Transplant. Proc.* **23**, 758–759.

Soon-Shiong, P., Feldman, E., Nelson, R. *et al.* (1993). Long-term reversal of diabetes by the injection of immunoprotected islets. *Proc. Natl. Acad. Sci. U.S.A.* **90**, 5843–5847.

Soon-Shiong, P., Heintz, R. E., Merideth, N. *et al.* (1994). Insulin independence in a type 1 diabetic patient after encapsulated islet transplantation. *Lancet* **343**, 950–951.

Sugamori, M. E., and Sefton, M. V. (1989). Microencapsulation of pancreatic islets in a water insoluble polyacrylate. *Trans. Am. Soc. Artif. Intern. Organs* **35**, 791–799.

Sullivan, S. J., Maki, T., Borland, K. M. *et al.* (1991). Biohybrid artificial pancreas: Long-term implantation studies in diabetic, pancreatectomized dogs. *Science* **252**, 718–721.

Tan, S. A., Deglon, N., Zurn, A. D. *et al.* (1996). Rescue of motorneurons from axotomy-induced cell death by polymer encapsulated cells genetically engineered to release CNTF. *Cell Transplant.* **5**, 577–587.

Tresco, P. A., Winn, S. R., and Aebischer, P. (1992a). Polymer encapsulated neurotransmitter secreting cells. Potential treatment for Parkinson's disease. *ASAIO J.* **38**, 17–23.

Tresco, P. A., Winn, S. R., Tan, S. *et al.* (1992b). Polymer-encapsulated PC12 cells: Long-term survival and associated reduction in lesion-induced rotational behaviour. *Cell Transplant.* **1**, 255–264.

Tseng, J. L., Baetge, E. E., Zurn, A. D. *et al.* (1997). GDNF reduces drug-induced rotational behavior after medial forebrain bundle transection by a mechanism not involving striatal dopamine. *J. Neurosci.* **17**, 325–333.

Tun, T., Inoue, K., Hayahi, H. *et al.* (1996). A newly developed three-layer agarose microcapsule for a promising biohybrid artificial pancreas: Rat to mouse xenotransplantation. *Cell Transplant.* **5**, S59–S63.

Wang, T., Lacik, I., Brissova, M. *et al.* (1997). An encapsulation system for the immunoisolation of pancreatic islets. *Nat. Biotechnol.* **15**, 358–362.

Wang, T. G. (1998). New technologies for bioartificial organs. *Artif. Organs* **22**, 68–74.

Yoshioka, T., Hirano, R., Shioya, T. *et al.* (1990). Encapsulation of mammalian cell with chitosan–CMC capsule. *Biotechnol. Bioeng.* **35**, 66–72.

Zielinski, B. A., and Aebischer, P. (1994). Chitosan as a matrix for mammalian cell encapsulation. *Biomaterials* **15**, 1049–1056.

CHAPTER 27

ENGINEERING CHALLENGES IN IMMUNOISOLATION DEVICE DEVELOPMENT

Efstathios S. Avgoustiniatos, Haiyan Wu, and Clark K. Colton

INTRODUCTION

One approach to minimizing or eliminating systemic immunosuppression in cell therapies is the use of tissue-engineered implantable immunobarrier (also called immunoisolation) devices, in which the tissue is protected from immune rejection by enclosure within a semipermeable membrane or encapsulant. This chapter deals with those factors that influence the performance and engineering design of immunobarrier devices: (1) tissue supply from primary or cell culture sources, (2) protection from immune rejection, in light of the mechanisms thought to operate in the presence of a semipermeable membrane, the properties of that membrane, and the implications for biology and device design, and (3) maintenance of cell viability and function, its relationship to device design, and the role of, and factors affecting, oxygen-supply limitations. Emphasis is given to the third area dealing with oxygen-supply limitations and its effect on viability and function. After reviewing recent work in this area, a detailed example is provided of how engineering analysis can be applied to the problem of enhancing oxygen supply by *in situ* oxygen generation.

Immunoisolation is the enclosure of transplanted tissue in a semipermeable membrane in order to protect it from immune rejection, thereby creating an implantable biohybrid artificial organ. Immunobarrier devices of this type are under study for the treatment of a variety of diseases. Biohybrid artificial organs have previously been reviewed extensively (Colton and Avgoustiniatos, 1991; Lanza *et al.,* 1992a; Aebischer *et al.,* 1993; Christenson *et al.,* 1993; Bellamkonda and Aebischer, 1994; Lanza and Chick, 1994, 1995a,b; Lysaght *et al.,* 1994; Mikos *et al.,* 1994; Aebischer and Lysaght, 1995; Colton, 1995, 1996). Many of the topics discussed in this chapter have previously been presented in great detail (Colton, 1995). This article focuses on the engineering challenges involved in the attempt to make implantable immunoisolation devices a clinical reality. These challenges are examined largely in the context of a biohybrid artificial pancreas for treatment of diabetes, but the problems and general principles are applicable to other uses of cell therapy as well.

ENGINEERING CHALLENGES

The essential elements of an implanted device that incorporates encapsulated cells are shown in Fig. 27.1, which illustrates the case of a biohybrid artificial pancreas. The implanted tissues are separated from the host by an immunoisolation membrane. Cells can be encapsulated at a high tissuelike density, as illustrated in Fig. 27.1, or dispersed in an extracellular gel matrix, such as agar, alginate, or chitosan. The membrane prevents or retards access of cellular and humoral immune components to the transplanted tissue but permits passage of the secreted product (insulin). At the

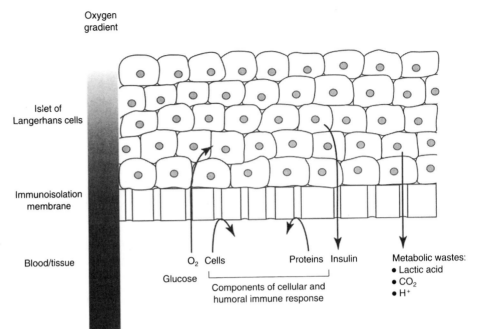

Oxygen
gradient

Islet of
Langerhans cells

Immunoisolation
membrane

Blood/tissue

O₂ Cells

Glucose

Proteins Insulin

Components of cellular and
humoral immune response

Metabolic wastes:
● Lactic acid
● CO₂
● H⁺

Fig. 27.1. Essential elements of an implanted device that incorporates encapsulated cells, illustrating the case of a biohybrid artificial pancreas. Reproduced, with permission, from Colton (1995).

same time there must be sufficient access to nutrients, such as glucose and oxygen, and removal of secreted metabolic waste products, such as lactic acid, carbon dioxide, and hydrogen ions. Transplanted cells must be supplied with nutrients by diffusion from the nearest blood supply, through surrounding tissue, the immunoisolation membrane, and the graft tissue itself.

Engineering challenges exist in three areas: (1) supply of tissue, (2) protection from immune rejection, and (3) maintenance of cell viability and function.

SUPPLY OF TISSUE

The amount of tissue required for transplantation in an immunoisolation device is determined by the secretion rate of the desired agent per cell and the amount of the active agent required by the body (Colton, 1995); it depends on the specific medical application (Avgoustiniatos and Colton, 1997a) and ranges from about 10^6 in central nervous system applications to about 10^9 cells in diabetes. The complexity and difficulty of all aspects of the problem increase with the volume of implanted tissue required. Thus, it is not surprising that some central nervous system applications, which require the least amount of transplanted tissue (Avgoustiniatos and Colton, 1997a) are the first to advance to clinical testing (Aebischer and Lysaght, 1995).

There are various sources of tissue. Most common is primary tissue (terminally differentiated, postmitotic cells) used immediately after isolation or following culture. Cultured, dividing cell lines, either native or genetically modified with recombinant DNA, have been studied and are attractive from the standpoints of harvesting large volumes of cells and of meeting regulatory standards of uniformity, sterility, and safety. In the case of primary tissue, only xenogeneic sources are practical because of the limited supply of human tissue. With cultured cell lines, as with primary tissue, immunoisolation is easier to accomplish with allogeneic than with xenogeneic cells. However, some investigators prefer xenogeneic cell lines because such cells would be rejected by the host if the device broke and released unencapsulated cells (Aebischer and Lysaght, 1995), whereas allogeneic cells might not be so easily rejected, especially if they were tumorigenic.

In applications in which tissue requirements are high, such as for the treatment of diabetes, supplying the required amount of tissue will represent the largest cost component for manufacture of the implant. A cell line that secretes insulin at a sufficiently high rate and that regulates the concentration of glucose in the plasma at the physiological set-point for humans is needed. In the meantime, effort has gone into isolating porcine islets of Langerhans, because porcine insulin has been used in humans for years and the islets regulate the glucose concentration to an appropriate level. Current processes are based on laboratory-scale methodology involving pancreas digestion by collagenase, followed by centrifugation with Ficoll density gradients. Innovative improvements are needed for the development of an economical large-scale process for producing large numbers of viable and functioning islets with high yield.

PREVENTING IMMUNE REJECTION

Possible rejection pathways elicited by immunoisolation are shown in Fig. 27.2. The process may begin with diffusion across the immunobarrier of immunogenic tissue antigens that have been shed from the cell surface, secreted by live cells, or liberated from dead cells. Recognition and display of these antigens by host antigen-presenting cells initiate the cellular and humoral immune responses. The former response leads to activation of cytotoxic cells, macrophages, and other immune cells. Prevention of cells entering the tissue compartment is easily achieved using microporous membranes. This may be the only requirement for immunoisolation of allogeneic tissue, at least for periods up to several months (Korsgren and Jansson, 1994). With xenografts, keeping out components of the humoral immune response is more difficult to accomplish. Such components include cytokines and lymphokines (e.g., interleukin-1), which can have deleterious effects on β cells, as well as newly formed antibodies to immunogenic antigens that have leaked across the barrier. In addition, there may be naturally occurring antibodies, most likely IgM, to cell surface antigens on xenografts. Antibodies produced during preexisting autoimmune disease, such as type I diabetes, might also bind to cell surface antigens. Last, macrophages and certain other immune cells can secrete low-molecular-weight reactive metabolites of oxygen and nitrogen, including free radicals, hydrogen peroxide, and nitric oxide, which are toxic to cells in a nonspecific fashion. The extent to which these agents may play a role in causing rejection of immunoisolated tissue depends on how far they can diffuse before they are inactivated by chemical reactions.

Cytotoxic events occur if antibodies and complement components pass through the membrane. Binding of the first component (C1q) to an IgM, or two or more IgG molecules, initiates a cascade that culminates in the formation of the membrane attack complex, which can lyse a single cell. IgM (910 kDa) and C1q (410 kDa) are both larger than IgG, so if host IgM and C1q can

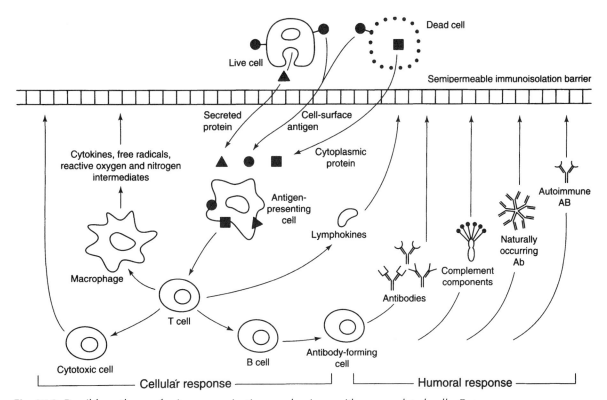

Fig. 27.2. Possible pathways for immune rejection mechanisms with encapsulated cells. Reproduced, with permission, from Colton (1995).

be prevented from crossing the barrier, then a specific, antibody-mediated attack on the islet cell should be averted. If the alternative complement pathway is activated and not inhibited by the implanted tissue, then passage of C3 (200 kDa) across the membrane must also be prevented. Small complement breakdown products, such as C5a (10 kDa), may also pose a problem (Chang, 1998). Both IgM and C1q have a smallest dimension of about 30 nm and should be completely retained by a membrane with a maximum pore diameter of 30 nm. The pores of a hydrophobic membrane in contact with extracellular fluid are likely to be coated with a monolayer of protein ~10 nm thick. Thus, pores with diameters of about 50 nm would be needed to allow C1q and IgM to pass through. Amicon PM30 polysulfone membranes with nominal cutoff of 30 kDa (Hagihara *et al.*, 1997) as well as Nucleopore membranes with 0.1-μm pore size (Ohgawara *et al.*, 1998) have been reported to prevent the permeation of C1q and smaller complement components.

The precise mechanism(s) that may play a role in rejection of encapsulated tissue are, in general, incompletely understood, and probably depend on the specific types of cells present, the species and its phylogenetic distance from humans, and the concentration of humoral immune molecules to which the implanted tissue is exposed. The latter depends on the transport properties of the immunoisolation membrane, as well as on the total mass of cells implanted (which affects the magnitude of the immune response), the tissue density, and the diffusion distances between tissue and immunoisolation membrane (which affect access of the humoral components generated to the graft tissue). If free radicals and other reactive oxygen and nitrogen species pose a significant problem, no passive membrane barrier will be able to provide immunoisolation, and some other approach (e.g., scavenging of free radicals, immunomodulation of transplanted tissue, local suppression of host immune response) will be necessary. However, even if this preceding issue is not a problem, if complete retention of IgG (or even C1q and IgM) coupled with passage of albumin and iron-carrying transferrin (81 kDa) is required, there remains a problem because of the size discrimination properties of membranes in which there is invariably a wide distribution of pore sizes. If cytokine transport must also be prevented, the problem is even more difficult. However, some surprising results suggest it is not hopeless. Alginate poly(lysine) microcapsules, found to be impermeable to IgG (150 kDa) and permeable to the cytokine IL-1b (17.5 kDa), could be made permeable to transferrin but remained impermeable to the lower molecular-weight cytokine tumor necrosis factor (TNF; 51 kDa) by increasing their guluronic acid content (Kulseng *et al.*, 1997). The suggested explanation for this result is that TNF in its active form is a bell-shaped trimeric molecule consisting of three identical subunits (17 kDa), in contrast to transferrin, which consists of one single polypeptide chain, resulting in the trimeric TNF having a more expanded tridimensional structure.

CELL VIABILITY AND FUNCTION

Maintenance of cell viability and function is essential and is limited by the supply of nutrients and oxygen. Diffusion limitations of oxygen in tissue *in vivo* are far more severe than those of glucose because the concentration of glucose in tissue is many fold higher (Tannock, 1972). The requirements of specific tissues for other small molecules and for macromolecules are poorly understood or have not yet been quantified; transport limitations for large molecules are highly dependent on immunoisolation membrane properties, whereas oxygen limitations are always serious. If the local oxygen concentration drops to sufficiently low levels, the cells will die. Hypoxia at levels high enough to keep cells alive can nonetheless have deleterious effects on cell functions that require high cellular ATP concentrations—for example, ATP-dependent insulin secretion (Dionne *et al.*, 1993).

Oxygen supply to encapsulated cells depends in a complicated way on a variety of factors, including (1) the site of implantation and the local oxygen partial pressure (pO_2) in the blood, (2) the spatial distribution of host blood vessels in the vicinity of the implant surface, (3) the oxygen permeability of the membrane or encapsulant, (4) the oxygen consumption rate of the encapsulated cells, (5) the geometric characteristics of the implant device, and (6) the tissue density and spatial arrangement of the encapsulated cells or tissues. The original concept of a biohybrid artificial organ (Colton, 1995), and the design that has received the most extensive study in large animals, is an intravascular arteriovenous shunt in the form of a semipermeable membrane tube through which arterial blood flows and on the outside of which the implanted tissue is contained in a housing. This approach provides the best physiologic oxygen supply (pO_2 = 100 mmHg) but suffers from the need to break the cardiovascular system and may be limited to a small fraction of patients. One alternative is an extravascular device in the form of a planar or cylindrical diffusion chamber implanted, for example, in subcutaneous tissue or intraperitoneally. Such devices are exposed to the mean pO_2 of the microvasculature, around 40 mmHg. Implantation in soft tissue is further disadvantaged if a foreign body response occurs, producing an avascular layer adjacent to the membrane, typically on the order of 100 μm thick. This fibrotic tissue increases the distance between blood vessels and implant, and the fibroblasts in that layer consume oxygen. Discovery of a class of microporous membranes that induce neovascularization at the material–tissue interface by virtue of the membrane microarchitecture (Brauker *et al.,* 1995) may ameliorate the severity of this problem. The angiogenic process takes 2–3 weeks (Padera and Colton, 1996) for completion, and the vascular structures remain indefinitely. By bringing some blood vessels close to the implant, oxygen delivery is improved. Another alternative for an extravascular implant is spherical microcapsules, implanted, for example, in the peritoneal space. The spherical geometry is advantageous from the standpoint of mass transfer, but the large volume of microcapsules employed, and the tendency for most to attach permanently to peritoneal surfaces, may lead to clinical problems.

Despite encouraging results with various tissues and applications (Aebischer and Lysaght, 1995; Colton, 1995), the problem of oxygen transport limitations is one of the major hurdles that remain. The maximum pO_2 (about 40 mmHg for the microvasculature) available for extravascular devices limits the steady-state thickness of viable tissue that can be supported. Potentially even more severe than these steady-state limitations is the hypoxic environment surrounding the device during the first few days after implantation. This situation is ameliorated only as new blood vessels grow toward the device–host tissue interface. During this period, anoxia likely exists within regions of the device, leading to death of a substantial fraction of the initially implanted tissue.

Islets are particularly prone to oxygen supply limitations because they have a relatively high oxygen consumption rate (Dionne, 1990). In the normal physiologic state they are highly vascularized and are supplied with blood at arterial pO_2. When cultured *in vitro* under ambient normoxic conditions, islets develop a necrotic core, the size of which increases with increasing islet size, as is to be expected as a result of oxygen diffusion and consumption within the islet (Dionne, 1990). Despite the extensive literature on immunoisolation of islets (Colton and Avgoustiniatos, 1991; Colton, 1995), only scant attention has been paid to the issue of islet viability within implanted immunobarrier devices. What is clear from these collective studies is that all attempts to support larger volumes of islet tissue in high density within the device (i.e., conditions wherein all or most of the internal device volume is occupied by viable islet tissue) have led to massive islet necrosis, invariably in regions furthest from the oxygen source. With few exceptions, only by suspending islets in an extracellular gel matrix at very low islet volume fractions (e.g., 1–5%), which greatly increases the size of the implanted device, have investigators been able to maintain the viability of initially loaded islets. However, use of such low tissue density puts undesirable constraints on the maximum number of islets that can be supported in a device of a size suitable for surgical implantation. There are some reports about the function and morphologic appearance of islets contained within such devices (Altman *et al.,* 1988; Brunetti *et al.,* 1991; Colton and Avgoustiniatos, 1991; Lacy *et al.,* 1991; Brauker *et al.,* 1992; Hill *et al.,* 1992; Lanza *et al.,* 1992a,b, 1993; Scharp *et al.,* 1994; Colton, 1995). Critical parameters such as the number and volume of viable islets that could be supported, as well as the development of islet necrosis and fibrosis, are being examined (Suzuki *et al.,* 1998a,b, 1999). In one study (Suzuki *et al.,* 1998b) with syngeneic transplantation of islets in a titanium ring chamber into streptozotocin (STZ)-induced diabetic mice, devices containing 250 implanted islets were rarely successful in normalizing blood glucose concentration at 4 weeks after transplantation, whereas the success rates were comparable using devices loaded with 500, 750, and 1000 islets. The highest success rate was obtained by implanting two devices, each loaded with 500 islets. The number of viable islets that remained at explantation between 1 and 12 weeks postimplantation was consistent with the notions that (1) the volume of viable islet tissue that can be supported by a particular device configuration is determined primarily by the rate of supply of oxygen to the tissue contained therein, and (2) most of the loss of viable β cell mass occurs during the first few days after transplantation as a result of the hypoxic environment around the immunobarrier device, which presents the most severe oxygen-supply limitations.

This problem has been observed in naked transplantation (Davalli *et al.,* 1996) in a study in which syngeneic islets were transplanted under the kidney capsule of normal and diabetic mice. In grafts harvested 1 and 3 days after transplantation, islets had undergone substantial damage, as evidenced by the presence of apoptotic nuclei, large fused masses with central necrosis, and reduced graft insulin content. Tissue remodeling was evident after 7 days, and good vascularization and oxygenation of the remaining islets were achieved at day 14. The association of apoptosis with the loss of transplanted islets has also been observed by others (Paraskevas *et al.*, 1997) and may be a general response to severe hypoxia as well as perhaps other factors such as conditions of isolation and culture, glycemic state, and the nonspecific inflammatory reaction associated with the transplantation procedure. In any event, the hypoxic environment for several days following transplantation is a problem common to both naked and immunobarrier transplantation.

OXYGEN-SUPPLY LIMITATIONS

Although supply of high-molecular-weight nutrients and waste elimination may be important factors in immunoisolation, work cited in the previous section suggests that the most critical factors affecting viability and function are the limitations imposed on oxygen supply to tissues in immunobarrier devices. It is therefore not surprising that this area has attracted increased attention over the past few years, and work has been reported in several areas aimed at enhancing oxygen supply or reducing the impact of these limitations.

IMPLANTATION SITE

Gel-encapsulated tissues are often implanted intraperitoneally. The peritoneal cavity provides a large surface area and offers the advantage of insulin delivery in the neighborhood of the portal vein, which is more physiologic than peripheral insulin delivery obtained, for example, by subcutaneous implantation. However, the peritoneal cavity has several disadvantages (Leblond *et al.*, 1999), including delayed and reduced insulin absorption; high peritoneal macrophage population (85% of peritoneal fluid cells), which may induce a fibrotic reaction; high mechanical stress because of continuous movement; possibility of severe pain if a peritoneal reaction occurs; clumping of microcapsules, especially likely in large animals and humans because of the higher number of capsules that have to be implanted; and poor cell oxygenation. The available oxygen partial pressure in the peritoneal cavity is about 40 mmHg, roughly the same as in the microvasculature. This latter factor led investigators to hypothesize that the intrapleural cavity may be a more suitable site for cell implantation (Kaur and Vacanti, 1996) because the oxygen partial pressure of fluids in the pleural cavity is a direct reflection of that in the inspired gas as a result of direct oxygen diffusion across the alveoli to the intrapleural space.

The most successful site for implantation of nonencapsulated islets of Langerhans, particularly in large animals and humans, is the liver, which also has a higher oxygen partial pressure. Large microcapsules (above 700 μm in diameter) can cause large increases in portal vein pressure, which has caused death of animal subjects. However, the use of small microcapsules, about 300 μm in diameter, was found safe and appropriate for intraportal implantation (Leblond *et al.*, 1999). On the other hand, the physiologic suitability of the intrahepatic site for islet transplantation has been questioned (Guan *et al.*, 1998).

IMPLANT–HOST TISSUE INTERFACE

The biologic response to the implant by the host tissue can play an important role in affecting oxygen supply to the transplanted tissue. The details of that response depend in part on the implantation site.

The presence of numerous macrophages adhered to the outer surface of various polymeric hollow-fiber (Hunter *et al.*, 1999) and planar (Jesser *et al.*, 1996) membranes implanted in the peritoneal cavity of rats or mice has been reported. These macrophages may exacerbate oxygen-supply limitations by virtue of their own oxygen consumption, and they have been associated with rat islet graft disintegration or morphologic alteration (Jesser *et al.*, 1996), leading to the suggestion that these membranes are not entirely suitable for immunoisolation in the peritoneal cavity. However, this phenomenon has also been observed with alginate microcapsules. In one study, subcutaneous administration of a macrophage inhibitor eliminated the recurrent hyperglycemia observed in streptozotocin-induced diabetic mice that had been treated with intraperitoneal transplantation of alginate–poly(L-lysine)–alginate-microencapsulated rat islets (Hsu *et al.,* 1999). Those findings are indicative of the detrimental effects associated with macrophage adherence to intraperitoneal implants.

The presence of blood vessels in the vicinity of a device is especially important when implanted subcutaneously because it minimizes the diffusion distance for oxygen. This principle can be illustrated by the results of calculations we presented previously (Avgoustiniatos and Colton, 1997b). We analyzed oxygen diffusion and consumption in a planar diffusion chamber consisting of dense islet tissue encapsulated between composite microporous membranes. We examined two cases: (A) The outer membrane surface is overlaid with a 100-μm-thick avascular fibrotic capsule containing a 10% volume fraction of respiring cells. Blood vessels of the microvasculature are present outside of this layer. (B) Neovascularization is induced to occur at the membrane–host tissue interface, and the outer membrane surface is taken to be uniformly exposed to the 40 mmHg oxygen partial pressure of the microvasculature. In case A, the fibrotic capsule accounts for a 34 mmHg partial pressure drop, 22 mmHg caused by the diffusion resistance of the tissue and 12 mmHg associated with oxygen consumption by the cells in the fibrotic layer. Another 5 mmHg drop occurs across the 30-μm-thick membrane, leaving only about 1 mmHg for support of viable tissue, which amounts to only an 18-μm-thick slab (assuming oxygen is supplied from both sides). A device containing 1 ml of islet tissue for restoring normoglycemia in a human would require an *en face* (surface) area of about 550 cm^2, most likely too large to be practical. In case B, the oxygen partial pressure at the outer surface of the transplanted tissue layer increases from 1 to 15 mmHg, and the thickness of supportable tissue increases from 18 to 70 μm. The required surface area decreases to about 140 cm^2, a more feasible size.

The aforementioned discovery by Brauker *et al.* (1995) of membrane materials that induce neovascularization leads to the presence of blood vessels in closer proximity to the transplanted tissue than is the case with fibrotic tissue. A strategy of implantation involving prevascularization of an immunoisolation device fabricated from such materials has been suggested (Padera and Colton, 1996) in order to prevent the extreme hypoxia that immediately follows implantation. However, it is not known whether the resulting spatial distribution of vessels, especially the density of vascular structures close to the membrane, attained by prevascularization is sufficient to support any particular thickness of tissue that is added to the device. It has been suggested (Vasir *et al.,* 1998) that local hypoxia partially regulates neovascularization of grafted naked islets via up-regulation of expression of vascular endothelial growth factor (VEGF), a potent angiogenic stimulus. VEGF is presumably also secreted by islet tissue in immunoisolation devices, thereby further increasing the extent of neovascularization. However, the critical issue is whether the oxygen supply available to the tissue at the time of implantation is sufficient to maintain all of the tissue in a viable state.

To further enhance the extent of prevascularization, several studies have investigated use of VEGF. In one study (Trivedi *et al.,* 1999) using ported devices fabricated with vascularizing membranes, VEGF in solution was infused at a very low rate through the device into the surrounding tissue for 10 days. Histologic examination revealed a marked increase in vascular structures within 15 μm of the membrane as well as farther away. Two days after infusion ceased, an insulin bolus was infused into the device and its plasma concentration was followed with time. The rate of insulin release into the bloodstream from the device with VEGF infusion was higher than that from control devices (without VEGF) but lower than that resulting from subcutaneous injection. These results are consistent with those of microdialysis studies with a similar device implanted subcutaneously but without angiogenic factor infusion (Rafael *et al.,* 1999). Exchange of glucose with the microcirculation following intravenous glucose injection was poorer when the microdialysis probe was placed within the device than when placed directly in the subcutaneous fat until 3 months after implantation, at which point they became similar.

In another study (Hunter *et al.,* 1999), VEGF (referred to as endothelial cell growth factor; ECGF) was incorporated into calcium alginate that filled the lumen of hollow-fiber membranes (having either a continuous, smooth or a broken, rough outer surface), which were implanted either subcutaneously or intraperitoneally for 65 days. The subcutaneous implant with the discontinuous outer surface displayed marked neovascularization. The others did not. Nonetheless, results from both of these studies suggest that release of angiogenic factors into the surrounding host tissue can increase the neovascularization around the device.

PROPERTIES OF TRANSPLANTED TISSUE

Characterization of metabolic and secretory properties, such as oxygen consumption and insulin secretion rate per unit volume of tissue, by cell lines as well as islet tissue from different donor species is important for improving the performance of biohybrid artificial pancreas devices. One source of insulin-secreting tissue is insulinoma cell lines, which offer the advantage of virtually unlimited uniform tissue. Genetic engineering could be used to modify the insulin secretion and glucose sensing so as to mimic that of native islets while maintaining or imparting new desirable properties, such as a lower oxygen consumption rate per unit volume and less susceptibility to hypoxic conditions, for example, by the action of antiapoptotic genes. Extensive studies with various encapsulated cell lines, such as the βTC3, MIN6, and NOD-RIP-Tag mouse insulinoma cell lines, have been performed for the purpose of characterizing relevant properties (Mukundan *et al.,* 1995; Papas *et al.,* 1996; Benson *et al.,* 1997; Kawakami *et al.,* 1997; Ohgawara *et al.,* 1998; Loudovaris *et al.,* 1999).

Another approach is use of primary tissue that is naturally hypoxia resistant. Tilapia, a tropical teleost fish, have large, anatomically discrete islets that are easily harvested. A study by Wright *et al.* (1998) demonstrated that these tilapia islets are much less sensitive to hypoxia than are mammalian islets. For example, culturing under conditions of bulk medium oxygen partial pressure around 25 mmHg resulted in uniform fragmentation and near-total necrosis of rat islets in less than 24 hr but caused only central necrosis in fish islets at up to 1 week. Fish islet grafts exposed to hypoxic culture for 72 hr prior to transplantation produced normoglycemia in diabetic nude mice for various periods of time. The authors reported starting production of tilapia with "humanized" tilapia insulin genes, which should result in islets that secrete humanized insulin.

In Situ OXYGEN GENERATION

It is now established that limitations of oxygen supply may result in substantial viability loss, especially immediately after implantation, such as observed in isogeneic islets in a planar diffusion immunobarrier device implanted into diabetic mice (Suzuki *et al.,* 1998a,b). One approach to overcome the oxygen limitation problem is to supply implanted tissue with oxygen generated *in situ* adjacent to one side of the immunobarrier device (Wu *et al.,* 1999). On the other side, the exterior of the device is exposed to either culture medium for *in vitro* studies or the host tissue for *in vivo* conditions.

We have previously described *in situ* oxygen generation by electrolytic decomposition of water in an electrolyzer (Wu *et al.,* 1999). The electrolyzer is in the form of a thin, multilayer sheet, within which electrolysis reactions take place on the anode and the cathode to form oxygen and hydrogen, respectively (Erickson and Russell, 1977). The anode side is placed in close contact with the immunobarrier device, which provides a continuous supply of oxygen to the implanted tissue. The oxygen flux N_e (mol/cm^2·sec) generated by the electrolyzer is related to the current I (amperes) applied between the electrodes by

$$I = nFA_eN_e, \tag{1}$$

where $n = 4$ is the number of electrons required to make one molecule of oxygen, F is Faraday's constant (96,500 A·sec/equivalent), and A_e is the cross-sectional area through which oxygen diffuses to the implanted tissue. The oxygen flux can be set at any desired level by appropriate control of the current. In the remainder of this chapter, we show how a theoretical model can be used to analyze oxygen transport in a device of the type we previously employed (Wu *et al.,* 1999). The theoretical model provides estimates of the oxygen profiles in the tissue. In specific numerical examples applicable to *in vivo* conditions, we examine the effects of tissue thickness, oxygen flux, and external partial pressure on oxygen profile characteristics, and on the maximum and minimum oxygen partial pressures to which the tissue is exposed. We define different regimes of operation, depending on the direction of oxygen flux between the exterior and the device and the presence or absence of a nonrespiring region in the middle of the tissue caused by oxygen limitations, and we formulate the oxygen flux requirement to avoid the development of such a region. The results of this analysis can be used to improve the design of experiments for the study of the effectiveness of *in situ* electrochemical oxygen generation by providing estimates for the appropriate value of imposed oxygen flux and the maximum possible viable tissue thickness.

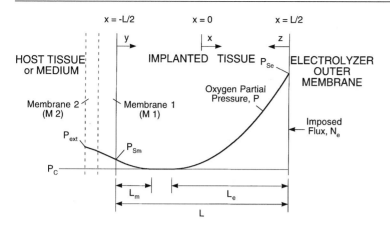

Fig. 27.3. Schematic diagram of a system, corresponding to an immunobarrier device with in situ oxygen electrochemical generation. The heavy curve is the oxygen partial pressure profile. Membranes 1 (M1) and 2 (M2) constitute a laminated composite membrane (Wu et al., 1999). A host tissue layer or a mass-transfer boundary layer (not shown) can be present outside M2. P_{ext} is the external oxygen partial pressure. P_{Sm} and P_{Se} are the oxygen partial pressures at the membrane–tissue interface opposite ($x = -L/2$) and nearest ($x = L/2$) the electrolyzer, respectively. P_C is the critical oxygen partial pressure below which hypoxic damage occurs and respiration ceases, N_e is the oxygen flux generated by the electrolyzer, L is the thickness of the implanted tissue, L_m is the maximum tissue thickness that can be supported by P_{ext}, and L_e is the maximum tissue thickness that can be supported by N_e. Three alternative coordinate systems, x, y, and z, used in our analysis, are also shown.

THEORETICAL ANALYSIS OF *IN SITU* OXYGEN GENERATION

PROBLEM FORMULATION

We consider one-dimensional oxygen diffusion in Cartesian coordinates through a series of layers (Fig. 27.3). Three different coordinates are used in this analysis: x is the distance from the midplane of the implanted tissue space of thickness L (centimeters), considered positive in the direction toward the electrolyzer; y and z are the distances from the interfaces between the implanted tissue and membrane 1 and between the implanted tissue and the electrolyzer outer membrane, respectively, both positive in the direction toward the implanted tissue layer.

The species conservation equation for oxygen in layer i is

$$D_i \frac{d^2 c_i}{dx^2} = V_i,$$ (2)

where D_i (cm^2/sec) is the effective diffusion coefficient of oxygen in layer i, and V_i (mol/cm^3·sec) is the local oxygen consumption rate per unit volume in layer i. The local concentration of oxygen c_i (mol/cm^3) is linearly related to the local oxygen partial pressure P_i (mmHg) in phase i by

$$c_i = \alpha_i P_i,$$ (3)

where α_i is the Bunsen solubility coefficient of oxygen in that phase (mol/cm^3·mmHg). Combination of Eqs. (2) and (3) yields the species conservation equation for oxygen in phase i in terms of the partial pressure P_i:

$$(D\alpha)_i \frac{d^2 P_i}{dx^2} = V_i.$$ (4)

Use of partial pressures instead of concentrations eliminates need for partition coefficients between adjacent phases because the partial pressures at equilibrium are equal across an interface.

Except as noted, we focus on the implanted tissue in this study, and for simplicity we will omit the subscript for this layer. With this simplification, the species conservation equation for oxygen in the live implanted tissue within the chamber becomes

$$D\alpha \frac{d^2 P}{dx^2} = V.$$ (5)

Oxygen consumption rate per unit volume V is assumed to follow Michaelis–Menten kinetics,

$$V = V_{max}(1 - \epsilon) \frac{P}{K_m + P},$$ (6)

for $P > P_C$, where V_{max} is the maximum oxygen consumption rate per unit volume of tissue, $1 - \epsilon$ is the live tissue volume fraction in the tissue layer, K_m is the Michaelis–Menten constant, and P_C is the critical value of P below which oxygen consumption ceases.

Equation (4) must be solved simultaneously in all layers. At each interface between adjacent layers i and j, there is equality in each phase between the oxygen partial pressure

$$P_i = P_j$$ (7)

and oxygen flux

$$(D\alpha)_i \frac{dP_i}{dx} = (D\alpha)_j \frac{dP_j}{dx}.$$ (8)

Equations (7) and (8) supply two boundary conditions necessary to solve Eq. (4) in all internal layers. Two specified external conditions necessary to solve Eq. (4) everywhere are, at $x = L/2$.

$$D\alpha \frac{dP}{dx} = N_e,$$ (9)

and, at $x = -(L/2 + L_{ext})$,

$$P = P_{ext},$$ (10)

where N_e, the oxygen flux imposed by the electrolyzer, is taken to be positive in the negative x direction; L_{ext} is the sum of the layer thicknesses external to the tissue chamber at the side opposite to the electrolyzer; and P_{ext} is the external oxygen partial pressure, i.e., that in the microvasculature for *in vivo* implantation or that of the bulk medium for *in vitro* culture.

We present here the analytical solution to Eq. (5) using zero-order kinetics [K_m set equal to 0, $V = V_{max}(1 - \epsilon) = $ constant]. With $P_C = 0.1$ mmHg (Anundi and De Groot, 1989; Chance *et al.*, 1973; Silver, 1973), the difference between this solution and that obtained with numerical methods using the nonlinear Michaelis–Menten kinetics is negligible here because of the high values of P/K_m attained (Avgoustiniatos and Colton, 1997a; Wu *et al.*, 1999).

For pedagogical reasons, we choose to focus initially on the implanted tissue layer as an isolated problem. In this case, the boundary conditions for the implanted tissue layer are expressed in terms of the oxygen partial pressures P_{Se} and P_{Sm} at the two tissue–membrane interfaces nearest the electrolyzer and nearest the external host tissue or medium side, respectively:
at $x = L/2$,

$$P = P_{Se},$$ (11)

and, at $x = -L/2$,

$$P = P_{Sm},$$ (12)

P_{Se} and P_{Sm}, neither of which is known *a priori*, are shown in Fig. 27.3. Because the problem is not symmetric about $x = 0$, $P_{Se} \neq P_{Sm}$ in general.

Two relations are needed in order to relate P_{Se} and P_{Sm} to the known quantities N_e and P_{ext}, a flux boundary condition at $x = L/2$, Eq. (9), and a flux boundary condition at $x = -L/2$,

$$D\alpha \frac{dP}{dx} = N_m, \tag{13}$$

where N_m is the oxygen flux across the tissue–membrane interface farthest from the electrolyzer, also positive in the negative x direction. An oxygen balance around the tissue layer yields

$$N_m = N_e - VL; \tag{14}$$

N_m can be viewed as the unused portion of the N_e passing through the tissue layer.

If no oxygen consumption occurs external to the implanted tissue, then the oxygen profile between P_{ext} and P_{Sm}, derived using the internal boundary conditions Eqs. (7) and (8), is composed of linear segments, and the flux can be expressed by

$$N_m = \frac{P_{Sm} - P_{ext}}{R_{ext}}, \tag{15}$$

where R_{ext} is the sum of the diffusive mass transfer resistances in series external to the tissue. In an *in vitro* experiment, R_{ext} is associated with each of the two components of the membrane laminate and the mass transfer boundary layer in the medium:

$$R_{ext} = \frac{1}{k_c \alpha_{med}} + \left(\frac{L}{D\alpha} \right)_{M2} + \left(\frac{L}{D\alpha} \right)_{M1}, \tag{16}$$

where M1 and M2 refer to the cell retentive and vascularizing membranes, respectively, $k_c = D_{med}/\delta_c$ is the boundary layer mass transfer coefficient between the stirred medium and membrane M2, and δ_c is the concentration boundary layer thickness (Wu *et al.*, 1999). If the device is implanted, the expression equivalent to Eq. (16) would be

$$R_{ext} = \left(\frac{L}{D\alpha} \right)_{host\ tissue} + \left(\frac{L}{D\alpha} \right)_{M2} + \left(\frac{L}{D\alpha} \right)_{M1}, \tag{17}$$

where the first term on the right-hand side is the diffusional resistance imposed by a host tissue layer between M2 and the closest blood vessels (Avgoustiniatos and Colton, 1997b).

If the external layers contain respiring cells, as is the case with the host tissue layer and with the vascularizing membrane M2 when infiltrated by cells, the effect on oxygen profiles can be consolidated into a correction to P_{ext}:

$$\left(P_{ext} \right)_{corrected} = P_{ext} - \sum_{i=2}^{m} \Delta P_{Vi} \tag{18}$$

where ΔP_{Vi} is the oxygen partial pressure drop through all layers caused by oxygen consumption in layer i, m is the number of layers, and $i = 1$ refers to the live implanted tissue. This problem has been analyzed elsewhere (Avgoustiniatos and Colton, 1997b). Oxygen consumption by cells infiltrating the outer vascularizing membrane layer causes a negligible oxygen partial pressure drop. Hence, the only nonnegligible term in the sum in Eq. (18) is that arising from respiration in the host tissue layer found between the outer membrane and the nearest blood vessels,

$$\left(P_{ext} \right)_{corrected} = P_{ext} - \left(\frac{VL^2}{2D\alpha} \right)_{host\ tissue} \tag{19}$$

for the typical situation where the host tissue is exposed to $P \gg K_m$ so that the local oxygen consumption is independent of partial pressure.

OXYGEN PARTIAL PRESSURE PROFILE

Assuming that all the tissue layer is exposed to $P > P_C$, Eq. (5) can be solved with boundary conditions, Eqs. (11) and (12), to give

$$P(x) = \frac{P_{Se} + P_{Sm}}{2} + (P_{Se} - P_{Sm})\frac{x}{L} - \frac{V}{2D\alpha}\left[\left(\frac{L}{2}\right)^2 - x^2\right]. \tag{20}$$

The first two terms on the right-hand side of Eq. (20) represent the oxygen profile in the absence of reaction but with the same surface partial pressures P_{Se} and P_{Sm}, and the last term represents the symmetrical reduction in local partial pressure resulting from oxygen consumption. These terms are superimposable because V is taken as independent of P.

By differentiating Eq. (20) with respect to x and substituting the result into Eq. (9), we obtain a relationship between P_{Se} and P_{Sm}:

$$P_{Se} - P_{Sm} = N_e \frac{L}{D\alpha} - \frac{VL^2}{2D\alpha}. \tag{21}$$

The first term on the right-hand side of Eq. (21) represents the partial pressure drop in the absence of reaction (i.e., the entire flux N_e passes unchanged through the tissue layer), and the second term represents the reduction in this partial pressure drop arising from oxygen consumption. Alternatively, expressing N_e in Eq. (21) in terms of N_m using Eq. (14) leads to

$$P_{Se} - P_{Sm} = N_m \frac{L}{D\alpha} + \frac{VL^2}{2D\alpha}. \tag{22}$$

The first term on the right-hand side of Eq. (22) represents the partial pressure drop caused by the unused flux N_m passing through the tissue layer, and the second term represents the additional drop resulting from consumption. We can also eliminate V from Eq. (21) by expressing it in terms of N_e and N_m using Eq. (14), which leads to

$$P_{Se} - P_{Sm} = \frac{L}{D\alpha}\frac{N_e + N_m}{2}. \tag{23}$$

Equation (23) shows that the net pressure drop between the two tissue–membrane boundaries is equal to the tissue diffusion resistance multiplied by the average of the flux (a linear function of x) at the two boundaries.

We can also express both P_{Sm} and P_{Se} in terms of the known external oxygen partial pressure P_{ext}. By rewriting Eq. (15) in terms of P_{Sm} and using Eq. (14) we obtain

$$P_{Sm} = P_{ext} - V\left(L - \frac{N_e}{V}\right)R_{ext}. \tag{24}$$

Then, combining Eqs. (21) and (24) we also obtain

$$P_{Se} = P_{ext} - V\left(L - \frac{N_e}{V}\right)\left(\frac{L}{D\alpha} + R_{ext}\right) + \frac{VL^2}{2D\alpha}. \tag{25}$$

By substituting Eqs. (24) and (25) into Eq. (20), we obtain an expression for the oxygen partial pressure profile in terms of N_e and P_{ext}:

$$P(x) = P_{ext} - V\left(L - \frac{N_e}{V}\right)R_{ext} - \frac{V}{2D\alpha}\left[\left(L - \frac{N_e}{V}\right)^2 - \left(\frac{L}{2} - x - \frac{N_e}{V}\right)^2\right]. \tag{26}$$

When oxygen enters the device from both sides, a minimum in the oxygen partial pressure profile should exist. The position of the minimum can be calculated by differentiating Eq. (26) with respect to x and setting the derivative equal to 0, at $x = x_{min}$,

$$\frac{dP}{dx} = 0 \tag{27}$$

to obtain

$$x_{min} = \frac{L}{2} - \frac{N_e}{V}.$$ (28)

A minimum exists when $x_{min} > -L/2$, when $L > N_e/V$, from Eq. (28), or when $N_m < 0$, from Eq. (14). Substituting Eq. (28) into Eq. (26) leads to the minimum oxygen partial pressure P_{min},

$$P_{min} = P_{ext} - V\left(L - \frac{N_e}{V}\right)R_{ext} - \frac{V}{2D\alpha}\left(L - \frac{N_e}{V}\right)^2.$$ (29)

The second term on the right-hand side of Eq. (29) represents the oxygen partial pressure drop through the external layers, and the third term represents the drop through the tissue layer that is supported from the host tissue or culture medium side. By substituting Eqs. (28) and (29) into Eq. (26), a simpler equation for the oxygen profile in terms of P_{min} and x_{min} can be obtained:

$$P(x) = P_{min} + \frac{V}{2D\alpha}(x - x_{min})^2.$$ (30)

Equation (30) implies that the profile is symmetrical around the minimum. However, the profile extends to different lengths, N_e/V and $L - N_e/V$ at the electrolyzer and external sides of the minimum, respectively. Using Eq. (28), we can apply Eq. (30) at the tissue–electrolyzer membrane interface ($x = L/2$) to obtain a relationship between P_{Se} and P_{min},

$$P_{Se} - P_{min} = \frac{N_e^2}{2VD\alpha}.$$ (31)

Maximum Supportable Tissue Thickness

Single layer of live tissue

The maximum thickness of viable tissue L_{max} that can be supported in the device is

$$L_{max} = L_e + L_m,$$ (32)

where L_e and L_m are, respectively, the maximum tissue thickness that can be supported by the imposed oxygen flux, N_e, on one side and by oxygen diffusion from the host tissue or culture medium on the other (Fig. 27.3). If N_m is positive, then, from Eq. (14), N_e is more than large enough to sustain oxygen consumption by all tissue, and there is a positive flux left that exits the device. If N_m is negative, then N_e is not large enough to sustain oxygen consumption by all tissue, and it is complemented by an influx from the opposite side of the device. Setting $N_m = 0$ in Eq. (14) leads to the maximum tissue thickness L_e supportable from N_e:

$$L_e = \frac{N_e}{V}$$ (33)

The maximum tissue thickness supportable by diffusion from the exterior of the device, L_m, may be determined by noting that when

$$L = L_{max} = \frac{N_e}{V} + L_m,$$ (34)

the equations leading to Eq. (30) are still applicable along with

$$P_{min} = P_C.$$ (35)

Substitution of Eqs. (34) and (35) into Eq. (29) results in a quadratic expression in L_m,

$$\left(\frac{V}{2D\alpha}\right)L_{\mathrm{m}}^{2} - \left(VR_{\mathrm{ext}}\right)L_{\mathrm{m}} - (P_{\mathrm{ext}} - P_{\mathrm{C}}) = 0, \tag{36}$$

the positive root of which is

$$L_{\mathrm{m}} = -D\alpha R_{\mathrm{ext}} + \left[\left(D\alpha R_{\mathrm{ext}}\right)^{2} + \frac{2D\alpha}{V}\left(P_{\mathrm{ext}} - P_{\mathrm{C}}\right)\right]^{1/2}; \tag{37}$$

L_{m}, the tissue thickness that can be supported from P_{ext}, is independent of the oxygen flux N_{e}.

In order to have $P > P_{\mathrm{C}}$ everywhere, it is necessary to satisfy

$$L < L_{\mathrm{max}} = \frac{N_{\mathrm{e}}}{V} + L_{\mathrm{m}}, \tag{38}$$

or, in terms of the adjustable oxygen flux N_{e},

$$N_{\mathrm{e}} > V(L - L_{\mathrm{m}}), \tag{39}$$

where L_{m} is given by Eq. (37). If the implanted tissue is proliferating, leading to an increase of L with time, the minimum oxygen flux N_{e} necessary to prevent tissue death also increases.

Presence of nonrespiring tissue

Up to this point, we have considered only $L \leq L_{\mathrm{max}}$, in which case there is a single layer of live tissue. If $L > L_{\mathrm{max}}$, there must exist a nonrespiring region between $x = -L/2 + L_{\mathrm{m}}$ and $x = L/2 - L_{\mathrm{e}}$ in which $P = P_{\mathrm{C}}$ and $V = 0$. Equation (20) cannot be applied in this case because it was derived assuming a constant V throughout the tissue layer. Instead, we break the solution domain into two parts. For convenience, the origin of the coordinate system is placed on one or the other implanted tissue–membrane interface.

At the electrolyzer side, the origin is at $z = 0$, where $z = L/2 - x$. Equation (5) retains the same form in terms of z, and the boundary conditions and associated relationships at $z = 0$ and $z = L_{\mathrm{e}}$ are analogous to their previous form. With $P_{\mathrm{min}} = P_{\mathrm{C}}$, the solution becomes

$$P(z) = P_{\mathrm{C}} + \frac{V}{2D\alpha}(L_{\mathrm{e}} - z)^{2}. \tag{40}$$

P_{Se} can be calculated by setting $z = 0$ in Eq. (40):

$$P_{\mathrm{Se}} = P_{\mathrm{C}} + \frac{VL_{\mathrm{e}}^{2}}{2D\alpha}. \tag{41}$$

Taking the derivative of Eq. (40) with respect to z and applying Eq. (9) leads to Eq. (33), which defines L_{e} in terms of the known parameters N_{e} and V. Equation (40) represents a special case ($P_{\mathrm{min}} = P_{\mathrm{C}}$) of an alternative form of Eq. (30).

At the other side of the device, we solve Eq. (5) in an analogous way in terms of the distance from the tissue–membrane interface, $y = L/2 + x$, in the domain between $y = 0$ and $y = L_{\mathrm{m}}$. The solution for the partial pressure becomes

$$P(y) = P_{\mathrm{C}} + \frac{V}{2D\alpha}(L_{\mathrm{m}} - y)^{2}; \tag{42}$$

P_{Sm} can be calculated by setting $y = 0$ in Eq. (42):

$$P_{\mathrm{Sm}} = P_{\mathrm{C}} + \frac{VL_{\mathrm{m}}^{2}}{2D\alpha}. \tag{43}$$

Taking the derivative of Eq. (42) with respect to y and applying Eqs. (13) and (15) leads to

$$P_{Sm} = P_{ext} - VL_m R_{ext}. \tag{44}$$

When nonrespiring tissue is present, P_{Sm} is always less than P_{ext}, as shown by Eq. (44), because oxygen diffuses from the exterior into the device. In contrast, P_{Sm} calculated from Eq. (24) can be lower, equal to, or higher than P_{ext} in the absence of nonrespiring tissue. Combination of Eqs. (43) and (44) leads to the same quadratic expression, Eq. (36), whose positive root is the physical solution for L_m given by Eq. (37).

CALCULATIONS

Using the equations derived in the previous section, we calculate oxygen profiles in immunoisolation devices supplied with oxygen by *in situ* oxygen generation. We use $V_{max} = 2.76 \times 10^{-8}$ mol/cm^3·sec, applicable to βTC3 cells (Mukundan *et al.,* 1995), tissue density $1 - \epsilon = 0.75$, characteristic of devices seeded with βTC3 cells and examined after *in vitro* culture (Wu *et al.,* 1999), $D\alpha = 1.70 \times 10^{-14}$ mol/cm·sec·mmHg, $(D\alpha)_{M1} = 1.95 \times 10^{-14}$ mol/cm·sec·mmHg, $(D\alpha)_{M2} = 2.79 \times 10^{-14}$ mol/cm·sec·mmHg, $L_{M1} = 30$ μm, $L_{M2} = 15$ μm (Avgoustiniatos and Colton, 1997b; Wu *et al.,* 1999), and $P_C = 0.1$ mmHg (Anundi and De Groot, 1989; Chance *et al.,* 1973; Silver, 1973). We assume $P_{ext} = 40$ mmHg for our base case, corresponding to the oxygen partial pressure of the microvasculature being present on the surface of the outer membrane M2 with $L_{ext} = L_{M1} + L_{M2}$, and we ignore the negligible effect of cells infiltrating M2 (Avgoustiniatos and Colton, 1997b). Using Eq. (6) with $K_m = 0$, we calculate that $V = 2.07 \times 10^{-8}$ mol/cm^3·sec, and using Eq. (17) we calculate that $R_{ext} = 2.07 \times 10^{11}$ mmHg/(mol/cm^2·sec).

EFFECT OF IMPOSED OXYGEN FLUX N_e

Figure 27.4 depicts oxygen partial pressure P for different values of the imposed oxygen flux N_e with a fixed implanted tissue layer thickness $L = 200$ μm. Using Eq. (37), $L_m = 53$ μm in all cases. The five curves correspond to different types of behavior (Table 27.1). For $N_e = 5 \times 10^{-10}$ mol/cm^2·sec, $L < L_e$, P increases monotonically in the y direction, the oxygen generated is sufficient to sustain consumption by all tissue, and there is still some left that exits to the device exterior through the membranes. As N_e decreases, P_{Se} does also. This is consistent with the first derivative of P_{Se} with respect to N_e, as given by Eq. (25) or Eq. (41), always being positive. At $N_e = 4.14 \times 10^{-10}$ mol/cm^2·sec, $L = L_e$, and the oxygen generated is just enough to sustain consumption by all tissue. P increases monotonically with y, and there is no net oxygen transport through the external membrane layers, where $dP/dy = 0$. At $N_e = 3.5 \times 10^{-10}$ mol/cm^2·sec, $L_e < L < L_{max}$, and oxygen supplied by the electrolyzer is not enough to support all tissue; oxygen enters the tissue layer from both sides, resulting in a minimum in the oxygen profile ($P_{min} = 21$

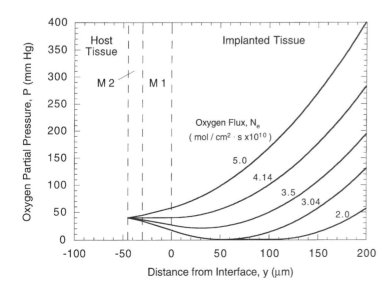

Fig. 27.4. Oxygen partial pressure P as a function of the distance y from the M1–tissue interface for different values of the oxygen flux N_e generated by the electrolyzer. The five curves correspond to five different types of behavior, delineated in Table 27.1. In all cases P_ext = 40 mmHg, L_m = 53 μm, and L = 200 μm.

Table 27.1. Characteristics of curves in Figs. 27.4–27.6

Figure Number	Oxygen flux N_e (mol/cm² · sec × 10¹⁰)	Tissue thickness (μm)			Regime[a]	Oxygen partial pressure (mmHg)[b]			
		L_e	L_m	L		P_{ext}	P_{Sm}	P_{min}	P_{Se}
27.4	5.0	242	53	200	$L_e > L$	40	58	58	403
	4.14	200	53	200	$L_e = L$	40	40	40	284
	3.5	169	53	200	$L_e < L < L_e + L_m$	40	27	21	195
	3.04	147	53	200	$L = L_e + L_m$	40	17	0.1	132
	2.0	97	53	200	$L > L_e + L_m$	40	17	0.1	57
27.5	3.5	169	53	0	$L_e > L$	40	113	113	113
	3.5	169	53	50	$L_e > L$	40	91	91	179
	3.5	169	53	100	$L_e > L$	40	70	70	215
	3.5	169	53	134	$L_e > L = L^*$	40	55	55	222
	3.5	169	53	169	$L_e = L$	40	40	40	214
	3.5	169	53	200	$L_e < L < L_e + L_m$	40	27	21	195
	3.5	169	53	222	$L = L_e + L_m$	40	17	0.1	174
	3.5	169	53	250	$L > L_e + L_m$	40	17	0.1	174
27.6	2.5	121	98	200	$L_e < L < L_e + L_m$	100	66	27	116
	2.5	121	79	200	$L = L_e + L_m$	72.3	38	0.1	89
	2.5	121	53	200	$L > L_e + L_m$	40	17	0.1	89

[a] L_e, L_m, and L^* are given by Eqs. (33), (37), and (45), respectively.
[b] P_{Sm}, P_{min}, and P_{Se} are given by Eqs. (24), (29), and (25), respectively.

mmHg) at $y = L - L_e = 31$ μm. At $N_e = 3.04 \times 10^{-10}$ mol/cm²·sec, the tissue thickness is exactly the maximum that can be supported from both sides, $L = L_{max}$, all the tissue is still viable, and P_{min} drops to P_C at a single point, $y = L_m = L - L_e$. At the lowest flux, 2×10^{-10} mol/cm²·sec, for which $L > L_{max}$, the combination of N_e and N_m is insufficient to sustain consumption throughout the entire tissue thickness, and a nonrespiring region develops between $y = L_m = 53$ μm and $y = L - L_e = 103$ μm, all of which is exposed to $P = P_C$. The latter two profiles are identical at the host side of the device, whereas they differ by a displacement in the y direction at the electrolyzer side. This occurs because, once P_{min} falls to P_C, a nonrespiring region develops, and oxygen transport at the two sides of this region is decoupled.

For a device with fixed $L = 200$ μm having the entire space filled with tissue and alginate ($1 - \epsilon = 0.75$), $N_e = 3.04 \times 10^{-10}$ mol/cm²·sec corresponds to the minimum oxygen flux that prevents development of a nonrespiring region. If the formation of such a region must be avoided, N_e should be adjusted higher. If a cell line is used, its proliferation must be controlled. Assuming that other nutrients or metabolic factors are not restrictive, uncontrollable proliferation should eventually lead to the presence of two layers of live tissue next to the membranes of the device, one supplied with oxygen by the electrolyzer and the other from the exterior of the device, with a layer of nonrespiring tissue in between. Histologic sections of tissue having these characteristics have indeed been described (Colton, 1995) in devices implanted without *in situ* oxygen generation. In general, one should expect proliferation to continue where P is high, and the nonrespiring region to expand where P is lowest, in the middle section of the device, until such time as proliferation is shut off by biological means or the buildup of internal mechanical forces leads to cessation of proliferation or destruction of the device.

EFFECT OF TISSUE THICKNESS L

Variation of the tissue thickness L is examined in Fig. 27.5 for $N_e = 3.5 \times 10^{-10}$ mol/cm²·sec, corresponding to the middle profile in Fig. 27.4. In this case, $L_e = 169$, $L_m = 53$, and $L_{max} = 222$ μm for all curves. For $L < 169$ μm $= L_e$, P increases monotonically in the y direction, the oxygen generated is sufficient to sustain oxygen consumption by all tissue, and there is still some oxygen left that exits to the device exterior through the membrane. This applies to the curves for the four smallest thicknesses shown in Fig. 27.5. The remaining curves correspond to

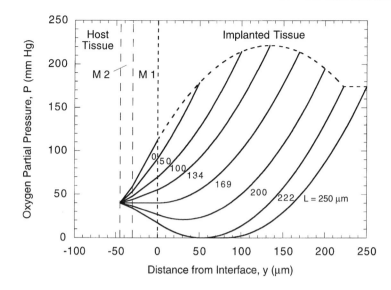

Fig. 27.5. *Oxygen partial pressure P as a function of the distance y from the M1–tissue interface, for different values of the tissue thickness L. The characteristics of the curves are delineated in Table 27.1. The dashed line connecting the highest points of each oxygen profile is a plot of P_{Se} as a function of L. In all cases P_{ext} = 40 mmHg, L_m = 53 μm, N_e = 3.5 × 10^{-10} mol/cm^2·sec, and L_e = 169 μm.*

other types of behavior (see Table 27.1). For $L = 169$ μm $= L_e$, P increases monotonically with y, and there is no net oxygen transport through the external membrane layers, where $dP/dy = 0$. For $L_e < L = 200$ μm $< L_{max}$, oxygen supplied by the electrolyzer is not enough to support all tissue. Oxygen enters the tissue layer from both sides, resulting in a minimum ($P_{min} = 21$ mmHg) in the oxygen profile where the two fronts meet at $y = L - L_e = 31$ μm. This curve is identical to the middle curve in Fig. 27.4. For $L = 222$ μm $= L_{max}$, the tissue thickness is exactly the maximum that can be supported from both sides. All the tissue is still viable, and P_{min} drops to P_C at a single point, $y = L_m = L - L_e$. At 250 μm, for which $L > L_{max}$, the combined oxygen flux from each side is insufficient to sustain consumption throughout the entire tissue thickness, and a nonrespiring region develops between $y = L_m = 53$ μm and $y = L - L_e = 81$ μm. In analogy to curves in Fig. 27.4, the latter two profiles are identical at the host side of the device, while they differ by a displacement in the y direction at the electrolyzer side.

In contrast to Fig. 27.4, where P_{Se} decreased monotonically with N_e, Fig. 27.5 shows that P_{Se} goes through a maximum as L increases. That maximum must occur for $L \leq L_{max}$, because P_{Se} does not vary with L when $L \geq L_{max}$. The value of this maximum partial pressure can be obtained by differentiating Eq. (25) with respect to L and setting the derivative equal to 0, which yields the length L^* at which P_{Se} is maximized:

$$L^* = L_e - D\alpha R_{ext}.\tag{45}$$

If $L^* \leq 0$, the maximum in P_{Se} occurs as $L \to 0$ and, from Eq. (25), is equal to $P_{ext} + N_e R_{ext}$. Substituting L^* for L in Eq. (25) gives, for $L^* > 0$,

$$\left(P_{Se}\right)_{max} = P_{ext} + \frac{VD\alpha R_{ext}^2}{2} + \frac{N_e^2}{2VD\alpha}.\tag{46}$$

For the conditions applying to Fig. 27.5, Eqs. (45) and (46) give $L^* = 134$ μm and $(P_{Se})_{max} = 222$ mmHg, the curve for which is plotted in Fig. 27.5. The dashed line in Fig. 27.5 connects the highest points of all oxygen profiles and is therefore a plot of P_{Se} as a function of L.

The existence of this maximum can be understood by rewriting Eq. (25) as

$$P_{Se} = P_{ext} + N_e \left(\frac{L}{D\alpha} + R_{ext}\right) - VL\left(\frac{L}{2D\alpha} + R_{ext}\right).\tag{47}$$

The second term on the right-hand side of Eq. (47) is the partial pressure drop associated with the imposed oxygen flux N_e in the absence of reaction, and the third term is the reduction in this partial pressure drop associated with oxygen consumption. For $L < L^*$, as L increases, the term associated with N_e increases faster than the term associated with consumption, and P_{Se} increases. For

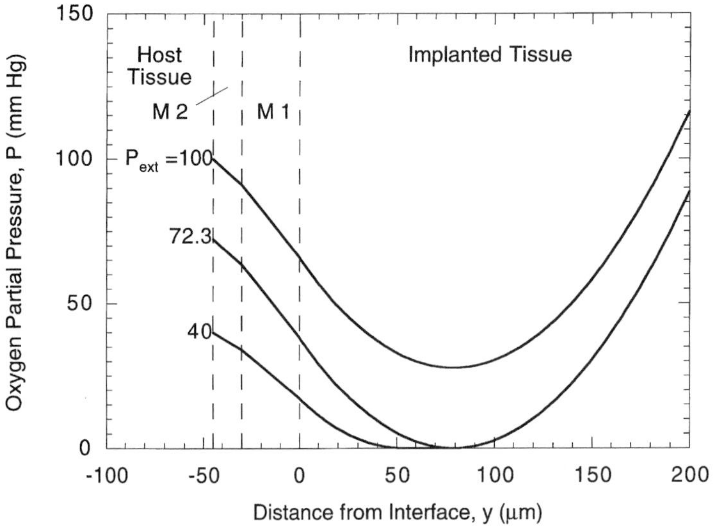

Fig. 27.6. Oxygen partial pressure P as a function of the distance y from the M1–tissue interface for different values of the external oxygen partial pressure P_{ext}. The three curves correspond to three different types of behavior, delineated in Table 27.1. In all cases N_e = 2.5 × 10^{-10} mol/cm²·sec, L_e = 121 μm, and L = 200 μm.

$L > L^*$, the second-order dependence on L in the consumption term prevails and P_{Se} drops with increasing L, until L reaches L_{max}, beyond which point P_{Se} remains constant.

EFFECT OF EXTERNAL OXYGEN PARTIAL PRESSURE P_{ext}

Figure 27.6 shows the effect of P_{ext} on oxygen profiles for a device with L = 200 μm and N_e = 2.5 × 10^{-10} mol/cm²·sec. L_e = 121 μm for all curves, but L_m varies because it depends on P_{ext}, as shown by Eq. (37). The three curves shown correspond to different types of behavior (see Table 27.1). For P_{ext} = 100 mmHg, corresponding to the oxygen partial pressure in arterial blood, L_m = 98 μm, $L_e < L < L_{max}$, and the two fronts of oxygen entering the device from both sides meet, resulting in a minimum (P_{min} = 27 mmHg) at $y = L - L_e$ = 79 μm. For P_{ext} = 72.3 mmHg, L_m = 79 μm, the tissue thickness is the maximum that can be supported from both sides, $L = L_{max}$, and $P_{min} = P_C$ at one point, $y = L_m = L - L_e$. At 40 mmHg, L_m = 53 μm, and $L > L_{max}$. The combined oxygen flux from each side is insufficient to sustain consumption by all tissue, and a nonrespiring region develops between $y = L_m$ = 53 μm and $y = L - L_e$ = 79 μm.

CONCLUDING REMARKS

Our analysis provides estimates of oxygen profiles in implanted tissue within a planar immunobarrier device with *in situ* electrochemical generation for different values of the adjustable oxygen flux N_e, the tissue thickness L, and the external oxygen partial pressure P_{ext}. It can also be used to predict the effects of other parameters, for example, the tissue density 1 − ε and the maximum oxygen consumption rate per unit volume of the implanted tissue, V_{max}. Evaluation of oxygen profile characteristics, including the minimum and maximum values of oxygen partial pressure within the tissue, allows for the appropriate selection of the oxygen flux N_e and the tissue thickness L so as to avoid the loss of tissue viability and function associated with the development of a nonrespiring region. If a proliferating cell line is used, its proliferation must be controlled to avoid the development of such a region. The results of this analysis can be extended to study the possible development of damaging hyperoxic conditions and to assist in the prediction of the secretory function of the implanted tissue.

ACKNOWLEDGMENTS

This work was supported in part by a grant from the Juvenile Diabetes Foundation International (File No. 198300) and through the JDFI Center for Islet Transplantation at Harvard Medical School.

REFERENCES

Aebischer, P., and Lysaght, M. J. (1995). Immunoisolation and cellular xenotransplantation. *Xenotransplantation* 3, 43–48.

Aebischer, P., Goddard, M., and Tresco, P. A. (1993). Cell encapsulation for the nervous system. *In* "Fundamentals of Animal Cell Encapsulation and Immobilization" (M. F. A. Goosen, ed.), pp. 197–224. CRC Press, Boca Raton, FL.

Altman, J. J., Penfornis, A., Boillot, J., and Maletti, M. (1988). Bioartificial pancreas in autoimmune nonobese diabetic mice. *ASAIO Transact.* **34**, 247–249.

Anundi, I., and De Groot, H. (1989). Hypoxic liver cell death: Critical P_{O_2} and dependence of viability on glycolysis. *Am. J. Physiol.* **257**, G58–G64.

Avgoustiniatos, E. S., and Colton, C. K. (1997a). Design considerations in immunoisolation. *In* "Principles of Tissue Engineering" (R. P. Lanza, R. Langer, and W. L. Chick, eds.), pp. 336–346. R. G. Landes, Austin, TX.

Avgoustiniatos, E. S., and Colton, C. K. (1997b). Effect of external oxygen mass transfer resistances on viability of immunoisolated tissue. *Ann. N.Y. Acad. Sci.* **831**, 145–167.

Bellamkonda, R., and Aebischer, P. (1994). Tissue engineering in the nervous system. *Biotechnol. Bioeng.* **43**, 543–554.

Benson, J. P., Papas, K. K., Constantinidis, I., and Sambanis, A. (1997). Towards the development of a bioartificial pancreas: Effects of poly-L-lysine on alginate beads with βTC3 cells. *Cell Transplant.* **6**, 395–402.

Brauker, J. H., Martinson, L. A., Loudovaris, T., Hill, R. S., Carr-Brendel, V., Hodgson, R., Young, S., Mandel, T. E., Charlton, B., and Johnson, R. C. (1992). Immunoisolation with large pore membranes: Allografts are protected under conditions that result in destruction of xenografts. *Cell Transplant.* **1**, 164.

Brauker, J. H., Carr-Brendel, V. E., Martinson, L. A., Crudele, J., Johnston, W. D., and Johnson, R. C. (1995). Neovascularization of synthetic membranes directed by membrane microarchitecture. *J. Biomed. Mater. Res.* **29**, 1517–1524.

Brunetti, P. G., Basta, G., Faloerni, A., Calcinaro, F., Petropaolo, M., and Calafiore, R. (1991). Immunoprotection of pancreatic islet grafts within artificial microcapsules. *Int. J. Artif. Organs* **14**, 789–791.

Chance, B., Oshino, B., Sugano, T., and Mayevsky, A. (1973). Basic principles of tissue oxygen determination from mitochondrial signals. *In* "Oxygen Transport to Tissue. Instrumentation, Methods, and Physiology" (H. I. Bicher and D. F. Bruley, eds.), Vol. 37A, pp. 277–292. Plenum, New York.

Chang, T. M. S. (1998). Artificial cells with emphasis on cell encapsulation of genetically engineered cells. *Artif. Organs* **22**, 958–965.

Christenson, L., Dionne, K. E., and Lysaght, M. J. (1993). Biomedical applications of immobilized cells. *In* "Fundamentals of Animal Cell Encapsulation and Immobilization" (M. F. A. Goosen, ed.), pp. 7–41. CRC Press, Boca Raton, FL.

Colton, C. K. (1995). Implantable bioartificial organs. *Cell Transplant.* **4**, 415–436.

Colton, C. K. (1996). Engineering challenges in cell-encapsulation technology. *Trends Biotechnol.* **14**, 158–162.

Colton, C. K., and Avgoustiniatos, E. S. (1991). Bioengineering in development of the hybrid artificial pancreas. *J. Biomech. Eng.* **113**, 152–170.

Davalli, A., Scaglia, L., Zangen, D., Hollister, J., Bonner-Weir, S., and Weir, G. C. (1996). Vulnerability of islets in the immediate posttransplantation period—Dynamic changes in structure and function. *Diabetes* **45**, 1161–1167.

Dionne, K. E. (1990). Effect of hypoxia on insulin secretion and viability of pancreatic islet tissue. Ph.D. Thesis, Massachusetts Institute of Technology, Cambridge, MA.

Dionne, K. E., Colton, C. K., and Yarmush, M. L. (1993). Effect of hypoxia on insulin secretion by isolated rat and canine islets of Langerhans. *Diabetes* **42**, 12–21.

Erickson, A. C., and Russell, J. H. (1977). Development status of a preprototype water electrolysis system. *Intersoc. Conf. Environ. Syst.,* San Francisco, *1977,* 77-ENAs-34.

Guan, J., Behme, M. T., Zucker, P., Atkinson, P., Hramiak, I., Zhong, R., and Dupré, J. (1998). Glucose turnover and insulin sensitivity in rats with pancreatic islet transplants. *Diabetes* **47**, 1020–1026.

Hagihara, Y., Saitoh, Y., Iwata, H., Taki, T., Hirano, S.-I., Arita, N., and Hayakawa, T. (1997). Transplantation of xenogeneic cells secreting β-endorphin for pain treatment: Analysis of the ability of components of complement to penetrate through polymer capsules. *Cell Transplant.* **6**, 527–530.

Hill, R. S., Young, S. A., Jacobs, S. A., Martinson, L. A., and Johnson, R. C. (1992). Membrane encapsulated islets implanted in epididymal fat pads correct diabetes in rats. *Cell Transplant.* **1**, 168.

Hsu, B. R.-S., Chang, F.-H., Juang, J.-H., Huang, Y.-Y., and Fu, S.-H. (1999). The rescue effect of 15-deoxyspergualin in intraperitoneal microencapsulated xenoislets. *Cell Transplant.* **8**, 307–315.

Hunter, S. K., Kao, J. M., Wang, Y., Benda, J. A., and Rodgers, V. G. J. (1999). Promotion of neovascularization around hollow fiber bioartificial organs using biologically active substances. *ASAIO J.* **45**, 37–40.

Jesser, C., Kessler, L., Lambert, A., Belcourt, A., and Pinget, M. (1996). Pancreatic islet macroencapsulation: A new device for the evaluation of artificial membrane. *Artif. Organs* **20**, 997–1007.

Kaur, S., and Vacanti, C. A. (1996). Evaluation of oxygen delivery in potential cell transplantation sites. *Cell Transplant.* **5**(5S-2), 43.

Kawakami, Y., Inoue, K., Hayashi, H., Wang, W. J., Setoyama, H., Gu, Y. J., Imamura, M., Iwata, H., Ikada, Y., Nozawa, M., and Miyazaki, J.-I. (1997). Subcutaneous xenotransplantation of hybrid artificial pancreas encapsulating pancreatic β cell line (MIN6): Functional and histological study. *Cell Transplant.* **6**, 541–545.

Korsgren, O., and Jansson, J. (1994). Porcine islet-like cell clusters cure diabetic nude rats when transplanted under the kidney capsule, but not when implanted into the liver or spleet. *Cell Transplant.* **3**, 49–54.

Kulseng, B., Thu, B., Espevik, T., and Skjåk-Bræk, G. (1997). Alginate polylysine microcapsules as immune barrier: Permeability of cytokines and immunoglobulins over the capsule membrane. *Cell Transplant.* **6**, 387–394.

Lacy, P. E., Hegre, O. D., Gerasimidi-Vazeou, A., Gentile, F. T., and Dionne, K. E. (1991). Maintenance of normoglycemia in diabetic mice by SC xenografts of encapsulated islets. *Science* **254**, 1782–1784.

Lanza, R. P., and Chick, W. L. (1994). In "Medical Intelligence Unit. Pancreatic Islet Transplantation," Vol. 3. R. G. Landes, Austin, TX.

Lanza, R. P., and Chick, W. L. (1995a). Encapsulated cell therapy. *Sci. Am. Sci. Med.* **2**, 16–25.

Lanza, R. P., and Chick, W. L. (1995b). Encapsulated cell transplantation. *Transplant. Rev.* **9**, 217–230.

Lanza, R. P., Borland, K. M., Lodge, P., Carreta, M., Sullivan, S. J., Muller, T. E., Solomon, B. A., Maki, T., Monaco, A. P., and Chick, W. L. (1992a). Treatment of severely diabetic pancreatectomized dogs using a diffusion-based hybrod pancreas. *Diabetes* **41**, 886–889.

Lanza, R. P., Borland, K. M., Staruk, J. E., Appel, M. C., Solomon, B. A., and Chick, W. L. (1992b). Transplantation of encapsulated canine islets into spontaneously diabetic BB/Wor rats without immunosuppression. *Endocrinology (Baltimore)* **131**, 637–642.

Lanza, R. P., Beyer, A. M., Staruk, J. E., and Chick, W. L. (1993). Biohybrid artificial pancreas. *Transplantation* **56**, 1067–1072.

Leblond, F. A., Simard, G., Henley, N., Rocheleau, B., Huet, P.-M., and Hallé, J.-P. (1999). Studies on smaller (~315 μm) microcapsules: IV. Feasibility and safety of intrahepatic implantations os small alginate poly-L-lysine microcapsules. *Cell Transplant.* **8**, 327–337.

Loudovaris, T., Jacobs, S., Young, S., Maryanov, D., Brauker, J., and Johnson, R. C. (1999). Correction of diabetic NOD mice with insulinomas implanted within Baxter immunoisolation devices. *J. Mol. Med.* **77**, 219–222.

Lysaght, M. J., Frydel, B., Gentile, F., Emerich, D., and Winn, S. (1994). Recent progress in immunoisolated cell therapy. *J. Cell. Biochem.* **56**, 196–203.

Martinson, L., Pauley, R., Brauker, J., Boggs, D., Maryanov, D., Sternberg, S., and Johnson, R. C. (1995). Protection of xenografts with immunoisolation membranes. *Cell Transplant.* **3**, 249.

Mikos, A. G., Papadaki, M. G., Kouvroukoglou, S., Ishaug, S. L., and Thomson, R. C. (1994). Islet transplantation to create a bioartificial pancreas. *Biotechnol. Bioeng.* **43**, 673–677.

Mukundan, N. E., Flanders, P. C., Constantinidis, I., Papas, K. K., and Sambanis, A. (1995). Oxygen consumption rates of free and alginate-entrapped βTC3 mouse insulinoma cells. *Biochem. Biophys. Res. Commun.* **210**, 113–118.

Ohgawara, H., Hirotami, S., Miyazaki, J., and Teraoka, S. (1998). Membrane immunoisolation of a diffusion chamber for a bioartificial pancreas. *Artif. Organs* **22**, 788–794.

Padera, R. F., and Colton, C. K. (1996). Time course of membrane microarchitecture-driven neovascularization. *Biomaterials* **17**, 277–284.

Papas, K. K., Long, R. C., Jr., Constantinidis, I., and Sambanis, A. (1996). Effects of oxygen on metabolic and secretory activities of βTC3 cells. *Biochem. Biophys. Acta* **1291**, 163–166.

Paraskevas, D., Maysinger, D., Feldman, L., Duguid, W., and Rosenberg, L. (1997). Apoptosis occurs in freshly isolated human islets under standard conditions. *Transplant. Proc.* **29**, 750–752.

Rafael, E., Wernerson, A., Arner, P., Wu, G. S., and Tibell, A. (1999). *In vivo* evaluation of glucose permeability of an immunoisolation device implanted for islet transplantation: A novel application of the microdialysis technique. *Cell Transplant.* **8**, 317–326.

Scharp, D. W., Swanson, C. J., Olack, B. J., Latta, P. P., Hegre, O. D., Doherty, E. J., Gentile, F. T., Flavin, K. S., Ansara, M. R., and Lacy, P. E. (1994). Protection of encapsulated human islets implanted without immunosuppression in patients with type I or II diabetes and in nondiabetic control subjects. *Diabetes* **43**, 1167–1170.

Silver, I. A. (1973). Brain oxygen tension and cellular activity. *In* "Oxygen Supply; Theoretical and Practical Aspects of Oxygen Supply and Microcirculation of Tissue" (M. Kessler, D. F. Bruley, L. C. Clark, D. W. Lübbers, I. A. Silver, and J. Strauss, eds.), pp. 186–188. University Park Press, Baltimore, MD.

Suzuki, K., Bonner-Weir, S., Hollister-Lock, J., Colton, C. K., and Weir, G. C. (1998a). Number and volume of islets transplanted in immunobarrier devices. *Cell Transplant.* **7**, 47–52.

Suzuki, K., Bonner-Weir, S., Trivedi, N., Yoon, K.-H., Hollister-Lock, J., Colton, C. K., and Weir, G. C. (1998b). Function and survival of macroencapsulated syngeneic islets transplanted into streptozotocin-diabetic mice. *Transplantation* **66**, 21–28.

Suzuki, K., Colton, C. K., Bonner-Weir, S., Hollister-Lock, J., and Weir, G. C. (1999). Estimating number and volume of islets transplanted within a planar immunobarrier diffusion chamber. *In* "Tissue Engineering Methods and Protocols" (J. R. Morgan and M. L. Yarmush, eds.), Vol. 18, pp. 497–505. Humana Press, Totowa, NJ.

Tannock, I. F. (1972). Oxygen diffusion and the distribution of cellular radiosensitivity in tumors. *Br. J. Radiol.* **45**, 515–524.

Trivedi, N., Steil, G. M., Colton, C. K., Bonner-Weir, S., and Weir, G. C. (1999). Improved vascularization of planar diffusion devices following continuous infusion of vascular endothelial growth factor (VEGF). *Cell Transplant.* **8**, 175.

Vasir, B., Aiello, L. P., Yoon, K. H., Quickel, R. R., Bonner-Weir, S., and Weir, G. C. (1998). Hypoxia induces vascular endothelial growth factor gene and protein expression in cultured rat islet cells. *Diabetes* **47**, 1894–1903.

Wright, J. R., Jr., Yang, H., and Dooley, K. C. (1998). Tilapia—A source of hypoxia-resistant islet cells for encapsulation. *Cell Transplant.* **7**, 299–307.

Wu, H., Avgoustiniatos, E. S., Swette, L., Bonner-Weir, S., Weir, G. C., and Colton, C. K. (1999). *In situ* electrochemical oxygen generation with an immunoisolation device. *Ann. N.Y. Acad. Sci.* **875**, 105–125.

PART VII: FETAL TISSUE ENGINEERING

CHAPTER 28

Fetal Tissue Engineering

Dario O. Fauza

INTRODUCTION

The first reported transplantation of human fetal tissue took place in 1922, when a fetal adrenal graft was transplanted into a patient with Addison's disease (Hurst *et al.,* 1922). A few years later, in 1928, fetal pancreatic cells were transplanted in an attempt to treat diabetes mellitus (Fichera, 1928). In 1957, a fetal bone marrow transplantation program was first undertaken (Thomas *et al.,* 1957). All of those initial experiments involving human fetal tissue transplantation failed, and it is only in the past two decades that fetal tissue transplantation in humans has started to yield favorable outcomes. A number of therapeutic applications of fetal tissue have already been explored, with variable results. Nonetheless, to date, the vast majority of studies have involved simply fetal cell, tissue, or organ transplantation, without any tissue engineering *sensu strictu.*

Fetal tissue has also been utilized as a valuable investigational tool in biomedical science since the 1930s. Embryologists, anatomists, and physiologists have studied fetal metabolism, fetoplacental unit function, premature life support, and brain activity in previable fetuses (Klopper and Diczfalusy, 1969; Diczfalusy, 1974). *In vitro* applications of fetal tissue are well established and somewhat common. Cultures of different fetal cell lines, as well as commercial preparations of human fetal tissue, have been routinely used in the study of normal human development and neoplasias, in genetic diagnosis, in viral isolation and culture, and to produce vaccines (Haase, 1987; Lehrman, 1988; American Medical Association, 1990). Biotechnology, pharmaceutical, and cosmetic companies have employed fetal cells and extraembryonic structures such as placenta, amnion, and the umbilical cord to develop new products and to screen them for toxicity, teratogenicity, and carcinogenicity (Hansen and Sladek, 1989; Vawter *et al.,* 1990). Fetal tissue banks have been operating in the United States and abroad for many years as sources of various fetal cell lines for research (Rojansky and Schenker, 1993).

Considering that a large body of data has emerged from research involving fetal cells or tissues and that attempts at engineering virtually every mammalian tissue have already taken place, it is highly surprising that most fetal cell lines are yet to be explored for actual tissue engineering purposes. The true engineering of fetal tissue, through culture and placement of fetal cells into matrices or membranes, or through other *in vitro* manipulations prior to implantation, has barely begun to be investigated. Human studies are yet to be performed and very few animal experiments have been reported thus far (Vacanti *et al.,* 1988; Cusik *et al.,* 1995; Fauza *et al.,* 1998a,b). Fetal cells were first used experimentally in engineered constructs by Vacanti *et al.* (1988). Interestingly, this investigation was part of the introductory study on selective cell transplantation using bioabsorbable, synthetic polymers as matrices. This same group performed another study involving fetal cells in 1995 (Cusik *et al.,* 1995). Both experiments, in rats, did not include structural replacement or functional studies. The use of fetal constructs as a means of structural and functional replacement in large animal models was first reported experimentally only in 1997 (Fauza *et al.,* 1998a,b). This chapter starts to explore the still exceedingly new field of fetal tissue engineering, along with a general overview of fetal cell and tissue transplantation.

GENERAL CHARACTERISTICS OF FETAL CELLS

Immunologic rejection (in nonautologous applications), growth limitations, differentiation and function restraints, incorporation barriers, and cell/tissue delivery difficulties are well-known complications of tissue engineering. Likely, many of those problems could be better managed, if not totally prevented, if fetal tissue was used. Due to their properties both *in vitro* and *in vivo*, fetal cells are, potentially, an excellent "raw material" for tissue engineering.

In Vitro

Compared with cells harvested postnatally, most fetal cells multiply more rapidly and more often in culture (Vacanti *et al.*, 1988; Fine, 1994; Cusik *et al.*, 1995; Qu *et al.*, 1996; Fauza *et al.*, 1998a,b). Depending on the cell line considered, however, this increased proliferation is more or less pronounced, or, in a few cases, not evident at all (Vacanti *et al.*, 1988; Fauza *et al.*, 1998a,b, 2000).

Because they are very plastic in their differentiation potential, fetal cells respond better than mature cells to environmental cues (Fine, 1994). Limited data from fetal myoblasts and osteoblasts suggest that purposeful manipulations in culture or in a bioreactor can be designed to steer fetal cells to produce improved constructs (Qu *et al.*, 1996; Fauza *et al.*, 2000). Fetal cells can survive at lower oxygen tensions than those tolerated by mature cells and are therefore more resistant to ischemia during *in vitro* manipulations (Wong, 1988). They also commonly lack long extensions and strong intercellular adhesions (Wong, 1988; Dekel *et al.*, 1997). Probably because of those characteristics, fetal cells display better survival after refrigeration and cryopreservation protocols when compared with adult cells (Groscurth *et al.*, 1986; Wong, 1988; Fine, 1994). This enhanced endurance during cryopreservation, however, seems to be tissue specific. For instance, data from primates and humans have shown that fetal hematopoietic stem cells, as well as fetal lung, kidney, intestine, thyroid, and brain tissues, can be well preserved at low temperatures, whereas non-hematopoietic liver and spleen tissues can also be cryopreserved, but not as easily [Gage *et al.*, 1985; Groscurth *et al.*, 1986; National Institutes of Health (NIH), 1988; Crombleholme *et al.*, 1991; Andreoletti *et al.*, 1997; Mychaliska *et al.*, 1998; Surbek *et al.*, 1998].

In Vivo

The expression of major histocompatibility complex (H-2) antigens in the fetus, and hence, fetal allograft survival in immunocompetent recipients, are age and tissue specific (Garvey *et al.*, 1980; Kirkwood and Billington, 1981; Statter *et al.*, 1988; Strong, 1991; Edwards, 1992; Foglia *et al.*, 1986a,b; Bakay *et al.*, 1998). The same applies to fetal allograft growth, maturation, and function (Gonet and Renold, 1965; Mullen *et al.*, 1977; Foglia *et al.*, 1986a; Guvenc *et al.*, 1997). At least in fetal mice, the precise gestational time of detection of H-2 antigen expression and the proportion of cells expressing these determinants depend on inbred strain, specific haplotype, tissue of origin, and antiserum batch employed (Kirkwood and Billington, 1981). Nevertheless, the precise factors governing the timing and tissue-specificity of H-2 antigen expression are yet to be determined in most species, including humans.

In addition to H-2 antigen expression, other mechanisms also seem to govern fetal immunogenicity. A study in mice suggests that, by catabolizing tryptophan, the mammalian conceptus suppresses T cell activity and defends itself against rejection by the mother (Munn *et al.*, 1998). In humans, fetal cells can be found in the maternal circulation in most pregnancies and fetal pro-genitor cells have been found to persist in the circulation of women many years after childbirth (Artlett *et al.*, 1998; Nelson *et al.*, 1998). These preliminary data allow us to suppose that engineered constructs made with fetal cells should be less susceptible to rejection in allologous applications. Xenograft implantations may also become viable, because initial studies suggest that fetal and embryonic cells are also better tolerated in cross-species transplantations, including in humans (Borlongan *et al.*, 1996; Deacon *et al.*, 1997; Dekel *et al.*, 1997; Ourednick *et al.*, 1998; Reinholt *et al.*, 1998). On the other hand, while less immunogenic, some fetal cells may be too immature and functionally limited, if harvested too early (Hullett *et al.*, 1987; Laferty, 1988). Yet, experimental models of fetal islet pancreatic cell transplantation have shown that, with time, the initially immature and functionally limited cells will grow, develop, and eventually function normally (Hullett *et al.*, 1987). Conversely, however, certain cells, such as those from rat striatum, actually function better after implantation if harvested early, as opposed to late, in gestation (Fricker *et al.,* 1997).

Fetal cells may produce high levels of angiogenic and trophic factors that enhance their ability to grow once grafted (Bjorklund *et al.,* 1987). By the same token, those factors may also facilitate regeneration of surrounding host tissues (Bjorklund *et al.,* 1987). Interestingly, significant clinical and hematologic improvements have been described following fetal liver stem cell transplantation in humans, even when there is no evidence of engraftment (Amos and Gordon, 1995). These improvements have been attributed to regeneration of autologous hematopoiesis and inhibition of tumor cell growth promoted by the infused cells, through mechanisms yet to be determined (Thomas, 1993). The underdifferentiated state of fetal cells also optimizes engraftment, by allowing them to grow, elongate, migrate, and establish functional connections with other cells (Fine, 1994).

Applications

Because of all the general benefits derived from the use of fetal cells, along with others specific to each cell line, several types of fetal cellular transplantation have been investigated experimentally or employed in humans for decades now. Clinically, fetal cells have been useful in a number of different conditions, including Parkinson's and Huntington's diseases (Hoffer *et al.,* 1988; Lindvall *et al.,* 1990; Kopyov *et al.,* 1998; Tabbal *et al.,* 1998), diabetes mellitus (Laferty, 1988; Lehrman, 1988), aplastic anemia (Kansal *et al.,* 1979; Harousseau *et al.,* 1980; Lucarelli *et al.,* 1980; Kochupillai *et al.,* 1987a; Lehrman, 1988; Han *et al.,* 1990), Wiskott–Aldrich syndrome (Broxmeyer *et al.,* 1994; Wagner *et al.,* 1995; Lu *et al.,* 1996), thymic aplasia (DiGeorge syndrome) and thymic hypoplasia with abnormal immunoglobulin syndrome (Nezelof syndrome) (Harboe *et al.,* 1966; August *et al.,* 1968; Cleveland *et al.,* 1968; Ammann *et al.,* 1975; NIH, 1988), thalassemia (Fine, 1994), Fanconi anemia (Gluckman *et al.,* 1989); acute myelogenous and lymphoblastic leukemias (Lucarelli *et al.,* 1980; Kochupillai *et al.,* 1987b; NIH, 1988; Vilmer *et al.,* 1992; Wagner *et al.,* 1995), Philadelphia chromosome-positive chronic myeloid leukemia (Bogdanic *et al.,* 1993), X-linked lymphoproliferative syndrome (Vowels *et al.,* 1993; Lu *et al.,* 1996), neuroblastoma (Wagner *et al.,* 1992, 1995; Lu *et al.,* 1996), severe combined immunodeficiency syndrome (Githens *et al.,* 1973; Buckley *et al.,* 1976; O'Reilly *et al.,* 1980; Loudon and Thompson, 1988; NIH, 1988), hemophilia (Fine, 1994), skin reconstruction (Fine, 1994), and neurosensory hypoacusis (Magomedov, 1998). Fetal cells have also been applied to treat inborn errors of metabolism, including Gaucher's disease, Fabry's disease, fucosidosis, Hurler's syndrome, metachromatic leukodystrophy, Hunter's syndrome, glycogenosis, Sanfilippo's syndrome, Morquio syndrome type B, and Niemann–Pick disease (Touraine, 1983, 1989; Touraine *et al.,* 1991; Fine, 1994). Experimentally, fetal cell and organ transplantation has been studied in an ever-expanding array of diseases (Crombleholme and Harrison, 1990; Nozawa *et al.,* 1991; Czech and Sagen, 1995; Fine, 1994; Borlongan *et al.,* 1997; Khan *et al.,* 1997; Scorsin *et al.,* 1997; Winkler *et al.,* 1998). Nonetheless, actual fetal tissue engineering as a therapeutic means has barely started to be explored, with very few studies undertaken thus far; these will be discussed later in more detail (Fauza *et al.,* 1998a,b, 2000; D. O. Fauza, J. Sperling, and J. P. Vacanti, unpublished data, 1999).

THE FETUS AS A TRANSPLANTATION HOST

Although there has been no account yet of a fetus receiving an engineered construct *in utero,* one could envision a number of advantages of such a procedure, not only from a theoretical perspective, but also from clinical and experimental evidence derived from intrauterine cellular transplantation studies already reported. Those potential advantages encompass induction of graft tolerance in the fetus, due to its immunologic immaturity; induction of donor-specific tolerance in the fetus by concurrent or previous intrauterine transplantation of hematopoietic progenitor cells; a completely sterile environment; the presence of hormones, cytokines, and other intercellular signaling factors that may enhance graft survival and development; the unique wound healing properties of the fetus; and early prevention of clinical manifestations of disease, before they can cause irreversible damage. Most of those advantages should be more or less evident, depending on the gestational timing of transplantation.

Fetal Immune Development

Among the potential benefits of *in utero* transplantation, the singularity of the fetal immune system deserves special attention. In that respect, basic research on fetal development, as well as studies involving pre- and postnatal transplantation of lymphohematopoietic fetal cells, have contributed to a better understanding of the fetal immune response.

Fetal tolerance resulting in permanent chimerism has been shown to occur in nature in nonidentical twins with shared placental circulation (Lillie, 1916; Owen, 1945). Little is known, however, about precisely when and by what mechanism this tolerance is lost. During fetal development, the precursors of the hematopoietic stem cells arise in the yolk sac, migrate to the fetal liver, and then migrate to the thymus, spleen, and bone marrow (Metcalf and Moore, 1971). The fetal liver has its highest concentration of hematopoietic stem cells between weeks 4 and the 20 of gestation (Gale, 1987). Because of their cellular immunologic "immaturity," the fetal liver and, to a lesser extent, the fetal thymus have been studied as potential sources of hematopoietic stem cells for major histocompatibility complex-incompatible bone marrow transplantation for almost four decades now (Uphoff, 1958; Bodley *et al.*, 1961; Amos and Gordon, 1995). Umbilical cord blood can also be a source of hematopoietic stem cells, but it has been applied mostly between relatives, although some success with unrelated, mismatched cord blood has also been reported (Gluckman *et al.*, 1989; Vilmer *et al.*, 1992; Wagner *et al.*, 1992; Bogdanic *et al.*, 1993; Vowels *et al.*, 1993; Wengler *et al.*, 1996).

Lymphocytes capable of eliciting graft-versus-host disease (GVHD) are found in the thymus by week 14 of gestation, but are not detectable in the liver until week 18 (O'Reilly *et al.*, 1983). Despite considerable numbers of granulocyte–macrophage colony-forming cells, there is an almost complete absence of mature T cells up to week 14 in human fetal livers (Phan *et al.*, 1989). During gestation, B cell development takes place mostly in the liver and T cell development occurs predominantly in the thymus. This specificity is probably the reason why fetal liver cells are immunoincompetent for cell-mediated and T cell-supported humoral reactions, such as graft rejections and GVHD. Thus, in principle, tissue matching is not necessary in fetal liver transplantation, if the harvest occurs up to a certain point in gestation. In a number of animal models, studies have shown that fetal liver cells induce no or merely moderate GVHD in histoincompatible donor/recipient pairs (Uphoff, 1958; Porter, 1959; Van-Putten *et al.*, 1968; Bortin and Saltztein, 1969; Perryman *et al.*, 198s; Prümmer *et al.*, 1985).

Umbilical cord blood and placental blood, on the other hand, although rich in hematopoietic progenitor cells, contain alloreactive lymphocytes (Clement *et al.*, 1990; Deadcock *et al.*, 1992; Fischer *et al.*, 1992; Harris *et al.*, 1992; Rabian-Herzog *et al.*, 1992; Amos and Gordon, 1995). These lymphocytes are also immature, but it is unclear whether they are more or less reactive than adult ones. Compared with those from adult blood, the proportions of activated T cells and helper–inducer subsets (CD4/29) are significantly reduced, whereas the helper–suppressor (CD4/45A) subset is significantly increased (Amos and Gordon, 1995). Cord blood natural killer cell activity is low or similar to that in adult blood, but lymphokine-activated killer cell activity may be higher (Keever, 1993; Gaddy *et al.*, 1995).

Although fetal liver stem cells should not cause GVHD, they could still be subject to rejection. Because of that, fetal liver stem cell transplantation has been attempted in the clinical setting mainly in patients with depressed immune function, such as in immunodeficiencies, replacement therapy during bone marrow compromise, and during fetal life (*in utero* transplantation). The same principle applies to the use of fetal thymus. No fatal cases of GVHD have been reported in patients who received fetal liver stem cells harvested before week 14 of gestation (Amos and Gordon, 1995). This complication, however, has been reported in a patient who received liver cells from a 16-week-old fetus (Lucarelli *et al.*, 1979). With umbilical cord blood stem cell transplantation, the incidence of GVHD has been minimal (Gluckman *et al.*, 1989; Vilmer *et al.*, 1992; Wagner *et al.*, 1992; Bogdanic *et al.*, 1993; Vowels *et al.*, 1993; Broxmeyer *et al.*, 1994; Lu *et al.*, 1996; Wengler *et al.*, 1996).

Applications of *in utero* transplantation

Cellular intrauterine transplantation has been employed clinically to treat lymphohematopoietic diseases, inborn errors of metabolism, and genetic disorders, with some success (Simpson and Golbus, 1985; Slavin *et al.*, 1989; Touraine *et al.*, 1989, 1991, 1992; Touraine, 1991,

1996; Raudrant *et al.,* 1992; Wengler *et al.,* 1996; Zanjani *et al.,* 1997; Hayward *et al.,* 1998). Determination of the optimal gestational age for transplantation, along with cell selection, route of cell administration, and postinterventional tocolysis, are still evolving (Metcalf and Moore, 1971; Crombleholme *et al.,* 1991; Flake and Harrison, 1995; Zanjani *et al.,* 1997; Fauza *et al.,* 1999).

Through *in utero* transplantation of hematopoietic progenitor cells, both allogeneic and xenogeneic chimerisms have been induced in animal models and allogeneic chimerism has been achieved in humans (Billingham *et al.,* 1953; Binns, 1967; Zanjani *et al.,* 1995; Wengler *et al.,* 1996; Yuh *et al.,* 1997; Hayward *et al.,* 1998; Tanaka *et al.,* 1998). Animal studies have shown induction of donor-specific allogeneic and xenogeneic tolerance in the fetus by intrauterine transplantation of hematopoietic stem cells, improving survival of other grafts later in life (West *et al.,* 1994; Yuh *et al.,* 1997; Tanaka *et al.,* 1998). Tolerance, or lack thereof, of allogeneic or xenogeneic intrauterine implantation of engineered constructs remains to be determined, however. Likewise, potential therapeutical benefits of prenatal autologous implantation of engineered tissue, as well as likely advantages stemming from the commonly scarless fetal wound healing, are yet to be explored.

FETAL TISSUE ENGINEERING

Vacanti *et al.* (1988) were the first to make use of fetal cells in engineered constucts. The experiment, in rats, used fetal cells from the liver, intestine, and pancreas, which were cultured, seeded on bioabsorbable matrices, and later implanted. The fetal constructs were implanted in heterologous fashion and heterotopically, namely, in the interscapular fat, omentum, and mesentery, with no structural replacement. They were removed for histologic analysis no later than 2 weeks after implantation. Successful engraftment was observed in some animals that received hepatic and intestinal constructs, but in none that received pancreatic ones.

Only in 1995 was a second study performed, by the same group, involving fetal liver constructs, also implanted in heterologous and heterotopic fashion in rats (Cusick *et al.,* 1995). Fetal hepatocytes were shown to proliferate to a greater extent than adult ones in culture and to yield higher cross-sectional cell area at the implant. As in the first experiment, neither structural replacement nor functional studies were included. Fetal constructs as a means of structural and functional replacement, in autologous fashion, in large-animal models, were first reported experimentally in 1997 (Fauza *et al.,* 1998a,b). Those studies, which involved treatment of surgically created congenital anomalies with fetal-engineered tissue, are discussed below.

CONGENITAL ANOMALIES

Major congenital anomalies are present in approximately 3% of all newborns (McKusick, 1990). Those diseases are responsible for nearly 20% of deaths occurring in the neonatal period and even higher morbidity rates during childhood (Anonymous, 1989). By definition, birth defects entail loss and/or malformation of tissues or organs. Treatment of many of those congenital anomalies is often hindered by the scarce availability of normal tissues or organs, either in autologous or allologous fashion, mainly at birth. Autologous grafting is frequently not an option in newborns due to donor site size limitations, and the well-known severe donor shortage observed in practically all areas of transplantation is even more critical during the neonatal period.

A novel concept in perinatal surgery has been introduced, involving minimally invasive harvest of fetal tissue, which is then engineered *in vitro* in parallel to the remainder of gestation, so that an infant with a prenatally diagnosed birth defect can benefit from having autologous, expanded tissue readily available for surgical implantation in the neonatal period (Fig. 28.1). (Fauza *et al.,* 1998a,b, 2000). Three studies utilizing this concept have been reported at this time (Fauza *et al.,* 1998a,b, 2000). Each study involved a different model of a congenital anomaly, namely, bladder exstrophy, skin, and diaphragmatic defects. In all experiments, harvest of fetal tissue was by videofetoscopy (Figs. 28.2 and 28.3).

In the bladder reconstruction study, fetal lambs underwent surgical creation of a bladder exstrophy defect. One group of animals had their bladder exstrophy primarily closed and another group underwent bladder augmentation with engineered, autologous fetal bladder tissue at birth. Fetal detrusor cells proliferated in culture significantly faster than adult ones (up to fourfold) (Fig. 28.4). Fetal urothelial cells proliferated at expected postnatal rates. The engineered bladders were radiologically normal at 3 weeks postimplantaiton (Fig. 28.5). At 2 months postopera-

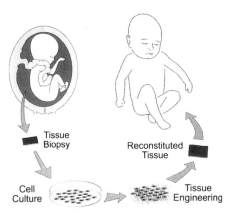

Fig. 28.1. Diagram representing the concept of fetal tissue engineering for surgical treatment of birth defects. A fetus with a prenatally diagnosed congenital anomaly undergoes tissue harvest, preferably though minimally invasive technique. The tissue is then processed and engineered in vitro while pregnancy continues, so that the newborn can benefit from having autologous, expanded tissue promptly available for reconstruction at birth.

tively, after the synthetic matrix had already been reabsorbed, the engineered bladders were more compliant than those primarily closed (Fig. 28.6). Histologic analysis of the engineered tissue showed a multilayered, pseudostratified urothelial lining (transitional epithelium) on its luminal side and overlying layers of smooth muscle cells surrounded by connective tissue. The microscopic architecture of the engineered mucosa was distinct from, but resembled that of, native bladder (Fig. 28.7). The muscular layer was hypertrophied in the exstrophic bladders primarily closed, as expected, but normal in the engineered bladders (Fig. 28.8).

In the skin replacement study, animals were their own controls, each with two experimental skin defects treated either with an engineered, autologous fetal skin tissue, or with an acellular biodegradable scaffold equal to the one used for the engineered skin. Fetal dermal fibroblasts multiplied significantly faster *in vitro* (approximately fivefold) than the same cells harvested postnatally. Fetal keratinocytes multiplied at expected postnatal rates. The engineered skin grafts induced faster epithelization of the wound (partial at 1 week and complete between 2 and 3 weeks postoperatively) compared with the acellular ones (partial at 3 weeks and complete between 3 and 4 weeks postoperatively) (Fig. 28.9). Time-matched histologic analysis of skin architecture showed a higher level of epidermal/dermal organization, richer vascularization, and less dermal scarring in the wounds that received the engineered skin, compared with those that received the acellular matrices (Fig. 28.9). Normal skin annexes were not observed in any grafted area up to 2 months postoperatively, albeit some areas that received engineered skin showed an architectural pattern compatible with ongoing adnexal development at 8 weeks postimplantation (Fig. 28.9).

Given the limitations of autologous grafting and transplantation during the neonatal period, autologous fetal tissue engineering may well become the only alternative for the treatment of a number of life-threatening congenital anomalies in which new tissue is needed to repair a defect

Fig. 28.2. Diagram representing the surgical setup for video-fetoscopic (minimally invasive) harvest of tissue from fetal lambs. Semiflexible, balloon-tipped cannulas are used through the access ports to the uterus in order to prevent dissection of the gestational membranes from the myometrium. Continuous warmed saline amnioinfusion is necessary for proper visualization of the operating field.

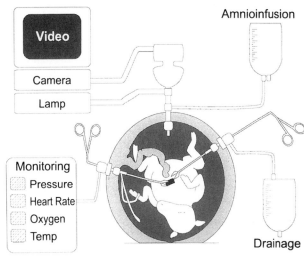

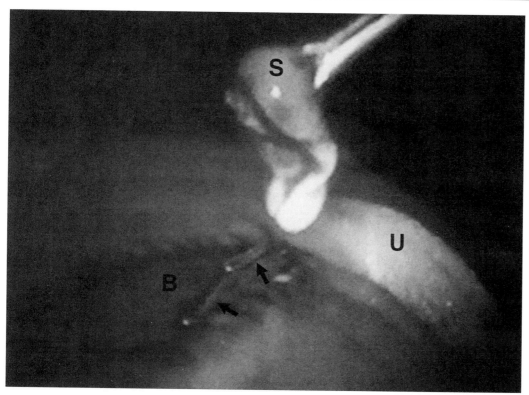

Fig. 28.3. Videofetoscopic view of the harvest of a fetal lamb's bladder specimen for tissue engineering purposes. B, Exstrophic bladder; S, bladder specimen; U, umbilical cord. The arrows point toward titanium clips placed on the harvested bed of the bladder to prevent fetal evisceration. Reproduced with permission from Fauza et al. (1998a).

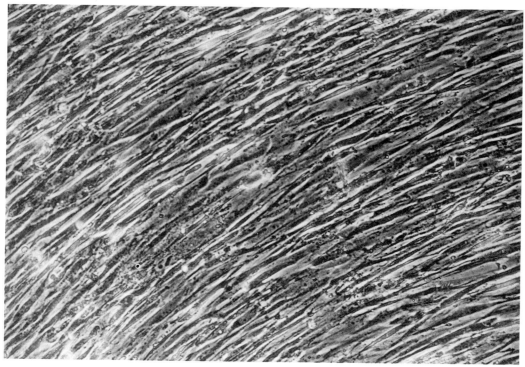

Fig. 28.4. Phase microscopy view of fetal detrusor cells completely filling the culture plate only 3 days after harvest (×100). Reproduced with permission from Fauza et al. (1998a).

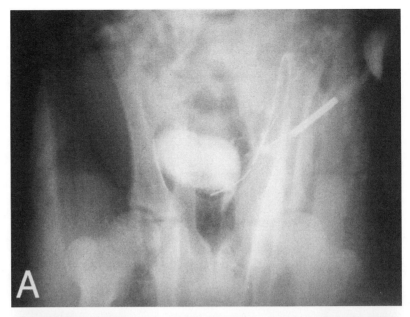

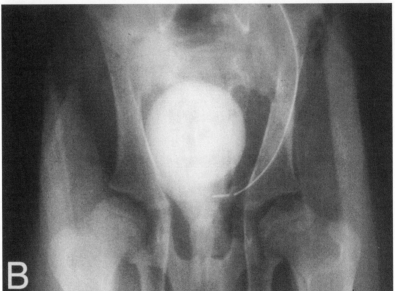

Fig. 28.5. Contrast cystograms performed 3 weeks postoperatively after primary closure of an exstrophic bladder (A) and after augmentation of an exstrophic bladder with autologous engineered fetal tissue (B). Reproduced with permission from Fauza et al. (1998a).

at birth. In many severe but non-life-threatening conditions, this therapeutic concept may also become first choice, as suggested by the functional results of the bladder augmentation study (Fauza *et al.,* 1998a). This concept can be applied to virtually any cell line. It can also include *in utero* implantation of the engineered construct, as well as different forms of cell manipulation *in vitro,* prior to implantation (Fauza *et al.,* 2000; D. O. Fauza, J. Sperling, and J. P. Vacanti, unpublished data, 1999). Fetal cells can also be frozen, stored, and implanted later in life, either in autologous or allologous fashion (D. O. Fauza *et al.,* unpublished data, 2000). All those possibilities are currently being further pursued.

ETHICAL CONSIDERATIONS

The use of fetal tissue has always been the object of intense ethical debate. The ethical controversies primarily arise from the fact that the primary source of fetal tissue is induced abortion. Spontaneous abortion usually does not raise many moral issues. The National Institutes of Health, the American Obstetrical and Gynecological Society, and the American Fertility Society, in accordance with the provisions that control the use of adult human tissue, regulate the use of fetal spec-

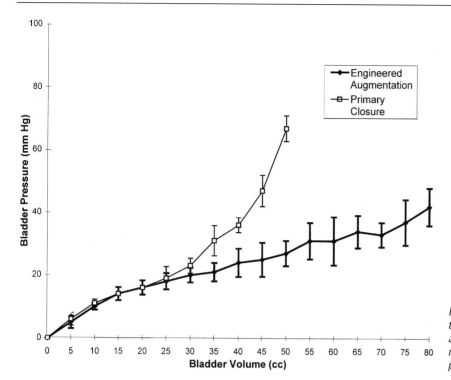

Fig. 28.6. Bladder compliance curves obtained 60 days after either engineered bladder augmentation or primary closure of experimental exstrophic bladders. Reproduced with permission from Fauza et al. (1998a).

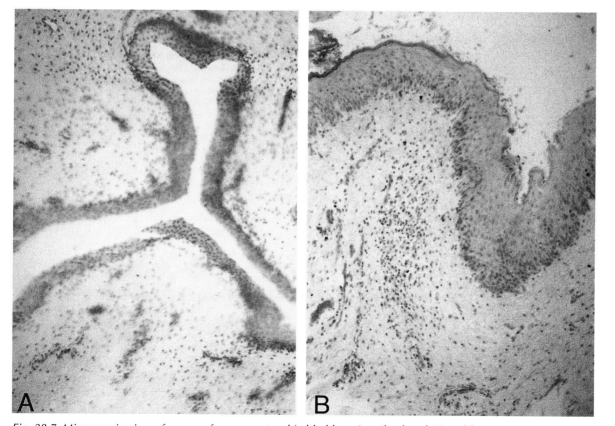

Fig. 28.7. Microscopic view of mucosa from an exstrophic bladder primarily closed (A) and from an engineered fetal bladder tissue (B), 60 days postoperatively (hematoxylin and eosin stain) (×100). Reproduced with permission from Fauza et al. (1998a).

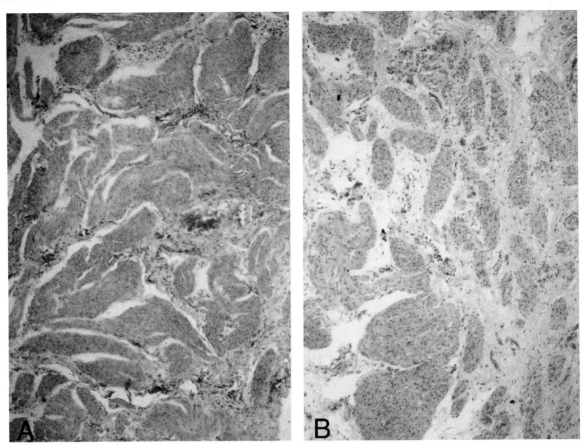

Fig. 28.8. Microscopic view of the muscular layer from an exstrophic bladder primarily closed (A) and from an engineered fetal bladder tissue (B), 60 days postoperatively. Notice the evident hypertrophy of muscular bundles in the exstrophic bladder closed primarily, as expected. Muscular hypertrophy was not observed in the engineered bladders (hematoxylin and eosin stain) (×100). Reproduced with permission from Fauza et al. (1998a).

imens from this latter source (NIH, 1988; Annas and Elias, 1989; Greeley *et al.,* 1989; Gershon, 1991). However, spontaneous abortion generally yields unsuitable fetal tissue, because it is frequently compromised by pathology such as chromosomal abnormalities, infections, and/or anoxia (NIH, 1988; Annas and Elias, 1989). Arguments on the use of fetal tissue from induced abortion are largely based on somewhat limited scientific evidence, along with clashing religious and customary beliefs about the beginning of life. Not surprisingly, despite the efforts of numerous national and international ethical committees and governmental bodies, a consensus has not yet been reached. In the United States, in spite (or perhaps because) of a recent moratorium on federal funding for fetal tissue transplantation research, an agreement on this issue seems to be slowly forming, although a stable solution may still be years away (Gershon, 1991; Strong, 1991).

This polemic notwithstanding, tissue engineering, as a novel development in fetal tissue processing, adds a new dimension to the discussion concerning the use of fetal tissue for therapeutic or research purposes. If specimens from a live, diseased fetus are to be used for the engineering of tissue, which in turn is to be implanted in autologous fashion, no ethical objections should be anticipated, as long as the procedure is a valid therapeutic choice for a given perinatal condition. In that case scenario, ethical considerations are the very same that apply to any fetal intervention. On the other hand, if fetal engineered tissue is to be implanted in heterologous fashion, ethical issues are analogous to the ones involving fetal tissue/organ transplantation, regardless of whether the original specimen comes from a live or deceased fetus. The distinction between autologous and heterologous implantation of engineered fetal tissue is a critical one in that, again, no condemnation of autologous use could be ethically justified.

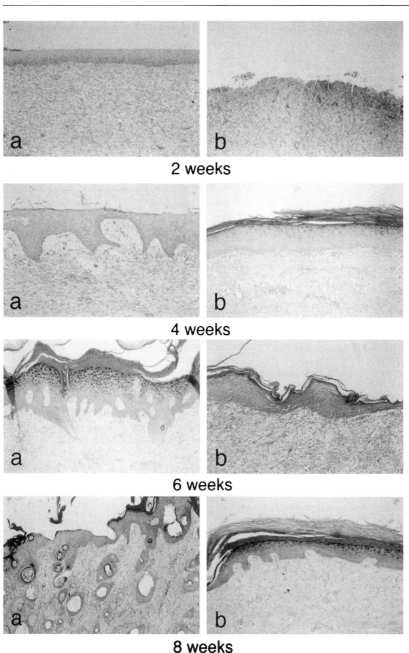

2 weeks

4 weeks

6 weeks

8 weeks

Fig. 28.9. Comparative neoskin histologies from engineered (a) and acellular (b) sites at different times postimplantation, in weeks. Notice the faster epithelization time and higher level of organization of the engineered specimens (hematoxylin and eosin stain) (×100). Reproduced with permission from Fauza et al.(1998b).

FUTURE PERSPECTIVES

Fetal tissue engineering is one of the newest developments of the still immature fields of tissue engineering and fetal surgery. As such, it is but a very promising concept that has hardly started to be explored, with many questions yet to be answered and numerous evolutionary paths, including unsuspected ones, yet to be pursued. Most fetal cell lines are yet to be studied for tissue engineering purposes. The optimal timing for harvest of each kind of fetal cell remains to be determined, along with their growth, differentiation, cryopreservation, and antigenic properties. At the same time, models for a better determination of embryonic and fetal age are only starting to be determined (Evtouchenko *et al.,* 1996). The whole subfield of *in utero* implantation of engineered tissue awaits examination. Banks of diverse fetal cell lines, either for autologous, allologous, or maybe even xenologous applications, are beginning to be optimized (Borel Rinkes *et al.,* 1992; Gluckman *et al.,* 1992, 1993; Rubinstein, 1993; Mychaliska *et al.,* 1998).

Fetal tissue engineering will benefit from the progress expected for tissue engineering in general, which should include several aspects of this multidisciplinary field. For instance, genetic engineering could maximize or modify certain specialized fetal cell functions (Ekhterae *et al.,* 1990). Fetal cells are ideal targets for gene transfer due to their proliferation and differentiation capacities (Johnson *et al.,* 1989; Karson and Anderson, 1990). Cell surface modulation techniques currently under investigation, including, but not restricted to, gene manipulation, could delete immunogenic sites and prevent immunorecognition, maximizing the naturally "permissive" allologous and possibly xenologous implantation of different cryopreserved fetal cells (Pollok and Vacanti, 1996). Methods for studying interfaces between cell and different biomaterials, along with mathematical models and *in vitro* systems that can predict *in vivo* events, should be adapted to fetal cells (Nerem and Girard, 1990; Lauffenberger, 1991; Parkhurst and Saltzman, 1992; Guido and Tranquillo, 1993; Hsieh *et al.,* 1993; Ratner, 1995). Finally, fetal cell and construct manipulation at zero gravity should be examined as tissue engineering in space starts to be explored (Saltzman, 1997).

As a component of the general promise of tissue engineering, fetal tissue engineering may well become the only perinatal alternative for treatment of a number of life-threatening birth defects in the future. As long as progress is made in the ethical debate over the use of fetal cells, its reach, nonetheless, will likely go beyond the perinatal period, offering unique novel perspectives to various aspects of surgery.

REFERENCES

American Medical Association, Council on Scientific Affairs and Council on Ethical and Judicial Affaris (1990). Medical applicatons of fetal tissue transplantation. *JAMA, J. Am. Med. Assoc.* **263,** 565–570.

Ammann, A. J., Wara, W. D., Doyle, N. E. *et al.* (1975). Thymus transplantation in patients with thymic hypoplasia and abnormal immunoglobulin synthesis. *Transplantation* **20,** 457–466.

Amos, T. A. S., and Gordon, M. Y. (1995). Sources of human hematopoietic stem cells for transplantation—A review. *Cell Transplant* **4,** 547–569.

Andreoletti, M., Pages, J. C., Mahieu, D. *et al.* (1997). Preclinical studies for cell transplantation: Isolation of primate fetal hepatocytes, their cryopreservation, and efficient retroviral transduction. *Hum. Gene Ther.* **8,** 267–274.

Annas, G. J., and Elias, S. (1989). The politics of transplantation of human fetal tissue. *N. Engl. J. Med.* **320,** 1079–1082.

Anonymous (1989). Contribution of birth defects to infant mortality—United States, 1986. *MMWR, Morbid. Mortal. Wkly. Rep.* **38,** 633–635.

Artlett, C. M., Smith, J. B., and Jimenez, S. A. (1998). Identification of fetal DNA and cells in skin lesions from women with systemic sclerosis. *N. Engl. J. Med.* **338,** 1186–1191.

August, C. S., Rosen, F. S., Filler, R. M. *et al.* (1968). Implantation of a foetal thymus, restoring immunological competence in a patient with thymic aplasia (DiGeorge's syndrome). *Lancet* **2,** 1210–1211.

Bakay, R. A., Boyer, K. I., Freed, C. R. *et al.* (1998). Immunological responses toi njury and grafting in the central nervous system of nonhuman primates. *Cell Transplant.* **7,** 109–120.

Billingham, R. E., Brent, L., and Medawar, P. B. (1953). "Actively acquired tolerance" of foreign cells. *Nature (London)* **172,** 603–606.

Binns, R. M. (1967). Bone marrow and lymphoid cell injection of the pig foetus resulting in transplantation tolerance or immunity and immunoglobulin production. *Nature (London)* **214,** 179–180.

Bjorklund, A., Lindvall, O., Isacson, O. *et al.* (1987). Mechanisms of action of intracerebral neural implants: Studies on nigral and striatal grafts to the lesioned striatum. *Trends Neurosci.* **10,** 509–516.

Bodley, S. R., Matthias, J. Q., Constandoulakis, M. *et al.* (1961). Hypoplastic anemia treated by transfusion of foetal hemopoietic cells. *Br. Med. J.* 1385–1388.

Bogdanic, V., Nemet, D., Kastelan, A. *et al.* (1993). Umbilical cord blood transplantation in a patient with philadelphia chromosome-positive chronic myeloid leukemia. *Transplantation* **56,** 477–479.

Borel Rinkes, I. H., Toner, M., Sheeha, S. J. *et al.* (1992). Long-term functional recovery of hepatocytes after cryopreservation in a three-dimensional culture configuration. *Cell Transplant.* **1,** 281–292.

Borlongan, C. V., Stahl, C. E., Cameron, D. F. *et al.* (1996). CNS immunological modulation of neural graft rejection and survival. *Neurol. Res.* **18,** 297–304.

Borlongan, C. V., Koutouzis, T. K., Jorden, J. R. *et al.* (1997). Neural transplantation as an experimental treatment modality for cerebral ischemia. *Neurosci. Biobehav. Rev.* **21,** 79–90.

Bortin, M. M., and Saltztein, E. C. (1969). Graft versus host inhibition: Fetal liver and thymus cells to minimize secondary disease. *Science* **164,** 316–318.

Broxmeyer, H. E., Lu, L., Gaddy, J. *et al.* (1994). Human umbilical cord blood transplantation: Expansion and gene therapy of hematopoietic stem and progenitor cells and immunology. *In* "Hematopoietic Stem Cells, Biology and Therapeutic Applications" (D. J. Levitt, ed.), pp. 297–317. Dekker, New York.

Buckley, R. H., Whisnant, J. K., Schiff, R. I. *et al.* (1976). Correction of severe combined immunodeficiency by fetal liver cells. *N. Engl. J. Med.* **294,** 1076–1081.

Clement, L. T., Vink, P. E., and Bradley, G. E. (1990). Novel immunoregulatory funcitons of phenotypically distinct subpopulations of CD4+ cells in the human neonate. *J. Immunol.* **145,** 102–108.

Cleveland, W. W., Fogel, B. J., Brown, W. T. *et al.* (1968). Foetal thymic transplant in a case of DiGeorge's syndrome. *Lancet* **2,** 1211–1214.

Crombleholme, T. M., and Harrison, M. R. (1990). Transplantation of fetal organs. *In* "The Unborn Patient: Prenatal Diagnosis and Treatment" (M. R. Harrison, M. S. Golbus, and R. A. Filly, eds.), pp. 516–525. Saunders, Philadelphia.

Crombleholme, T. M., Langer, J. C., Harrison, M. R. *et al.* (1991). Transplantation of fetal cells. *Am. J. Obstet. Gynecol.* **164**, 218–230.

Cusick, R. A., Sano, K., Lee, H. *et al.* (1995). Heterotopic fetal rat hepatocyte transplantation on biodegradable polymers. *Surg. Forum* **46**, 658–661.

Czech, K. A., and Sagen, J. (1995). Update on cellular transplantation into the CNS as a novel therapy for chronic pain. *Prog. Neurobiol.* **46**, 507–529.

Deacon, T., Schumacher, J., Dinsmore, J. *et al.* (1997). Histological evidence of fetal pig neural cell survival after transplantation into a patient with Parkinson's disease. *Nt. Med.* **3**, 350–353.

Deadcock, S. J., Schwarer, A. P., Bridge, J. *et al.* (1992). Evidence that umbilical cord blood contains a higher frequency of HLA Class II-specific alloreactive T cells than adult peripheral blood. A limiting dilution analysis. *Transplantation* **53**, 1128–1134.

Dekel, B., Burakova, T., Ben-Hur, H. *et al.* (1997). Engraftment of human kidney tissue in rat radiation chimera: II. Human fetal kidneys display reduced immunogenicity to adoptively transferred human peripheral blood mononuclear cells and exhibit rapid growth and development. *Transplantation* **64**, 1550–1558.

Diczfalusy, E. (1974). Reproductive endocrinology in 1974. *In* "Recent Progress in Reproductive Endocrinology" (P. G. Crosignani and V. E. James, eds.), pp. 833–840. Academic Press, London.

Edwards, R. G. (1992). "Fetal Tissue Transplants in Medicine." Cambridge University Press, Cambridge, UK.

Ekhterae, D., Crombleholme, T. M., Karson, E. *et al.* (1990). Retroviral vector-mediated transfer of the bacterial neomycin resistance gene into fetal and adult sheep and human hematopoietic progenitors *in vitro. Blood* **75**, 365–369.

Evtouchenko, L., Studer, L., Spenger, C. *et al.* (1996). A mathematical model for the estimation of human embryonic and fetal age. *Cell Transplant.* **5**, 453–464.

Fauza, D. O., Fishman, S., Mehegan, K. *et al.* (1998a). Videofetoscopically assisted fetal tissue engineering: Bladder augmentation. *J. Pediatr. Surg.* **33**, 7–12.

Fauza, D. O., Fishman, S., Mehegan, K. *et al.* (1998b). Videofetoscopically assisted fetal tissue engineering: Skin replacement. *J. Pediatr. Surg.* **33**, 357–361.

Fauza, D. O., Berde, C., and Fishman, S. (1999). Prolonged local myometrial blockade prevents preterm labor after fetal surgery in a leporine model. *J. Pediatr. Surg.* **34**, 540–542.

Fauza, D. O. Marler, J., Koka, R., *et al.* (2000). Fetal tissue engineering: Diaphragmatic replacement. *J. Pediatr. Surg.* (in press).

Fichera, G. (1928). Implanti omoplastici feto-umani nei cancro e nel diabete. *Tumori* **14**, 434.

Fine, A. (1994). Transplantation of fetal cells and tissue: An overview. *Can. Med. Assoc. J.* **151**, 1261–1268.

Fischer, H. P., Sharrock, C. E. M., and Panayi, G. S. (1992). High frequency of cord blood lymphocytes against mycobacterial 65-kDa heat-shock protein. *Eur. J. Immunol.* **22**, 1667–1669.

Flake, A. W., and Harrison, M. R. (1995). Fetal surgery. *Annu. Rev. Med.* **46**, 67–78.

Foglia, R. P., Dipreta, J., Statter, M. B. *et al.* (1986a). Fetal allograft survival in immunocompetent recipients is age dependent and organ specific. *Ann. Surg.* **204**, 402–410.

Foglia, R. P., LaQuaglia, M., DiPreta, J. *et al.* (1986b). Can fetal and newborn allografts survive in an immunocompetent host? *J. Pediatr. Surg.* **21**, 608–612.

Fricker, R. A., Torres, E. M., Hume, S. P. *et al.* (1997). The effects of donor stage on the survival and function of embryonic striatal grafts in the adult rat brain. II. Correlation between positron emission tomography and reaching behaviour. *Neuroscience* **79**, 711–721.

Gaddy, J., Risdon, G., and Broxmeyer, H. E. (1995). Cord blood natural killer cells are functionally and phenotypically immature but readily respond to IL-2 and IL-12. *J. Interface Cytokine Res.* **15**, 527–536.

Gage, F. H., Brundin, P., Isacson, O. *et al.* (1985). Rat fetal brain tissue grafts survive and innervate host brain following five day pregraft tissue storage. *Neurosci. Lett.* **60**, 133–137.

Gale, R. P. (1987). Development of the immune system in human fetal liver. *Thymus* **10**, 45–56.

Garvey, J. F., McShane, P., Poole, M. D. *et al.* (1980). The effect of cyclosporine A on experimental pancreas allografts in the rat. *Transplant. Proc.* **12**, 266–269.

Gershon, D. (1991). Fetal tissue research. New panel for ethical issues (news). *Nature (London)* **349**, 184.

Githens, J. H., Fulginiti, V. A., Suvatte, V. *et al.* (1973). Grafting of fetal thymus and hematopoietic tissue in infants with immune deficiency syndromes. *Transplantation* **15**, 427–434.

Gluckman, E., Broxmeyer, H. E., Auerbach, A. *et al.* (1989). Hematopoietic reconstitution in a patient with Fanconi's anemia by means of umbilical cord blood from an HLA-identical sibling. *N. Engl. J. Med.* **321**, 1174–1178.

Gluckman, E., Devergie, A., Thierry, D. *et al.* (1992). Clinical applications of stem cell transfusion from cord blood and rationale for cord blood banking. *Bone Marrow Transplant.* **9**, 114–117.

Gluckman, E., Wagner, J., Hows, J. *et al.* (1993). Cord blood banking for hematopoietic stem cell transplantation: An international cord blood transplant registry. *Bone Marrow Transplant.* **11**, 199–200.

Gonet, A. E., and Renold, A. E. (1965). Homografting of fetal rat pancreas. *Diabetologia* **1**, 91–96.

Greeley, H. T., Hamm, T., Johnson, R. *et al.* (1989). The ethical use of human fetal tissue in medicine. *N. Engl. J. Med.* **320**, 1093–1096.

Groscurth, P., Erni, M., Balzer, M. *et al.* (1986). Cryopreservation of human fetal organs. *Anat. Embryol.* **174**, 105–113.

Guido, S., and Tranquillo, R. T. (1993). A methodology for the systematic and quantitative study of cell contact guidance in oriented collagen gels. Correlation of fibroblast orientation and gel birefringence. *J. Cell Sci.* **105**, 317–331.

Guvenc, B. H., Salman, T., Tokar, B. *et al.* (1997). Fetal intestinal transplant as an accessory enteral segment. *Pediatr. Surg. Int.* **12**, 367–369.

Haase, H. (1987). Explanatory memorandum to Council of Europe, Parliamentary Assembly Recommendation 1046

(1986) (1) on the use of human embryos and fetuses for diagnostic, therapeutic, scientific, industrial and commercial purposes. *Hum. Reprod.* **2**, 68–75.

Han, J. R., Yuan, S. W., Ren, Q. F. *et al.* (1990). Clinical and experimental study of treating aplastic anaemia with fetal liver cell suspension and fetal liver cell-free suspension. *J. Chin. Int. Med.* **29**, 347–349.

Hansen, J. T., and Sladek, J. R., Jr. (1989). Fetal research. *Science* **246**, 775–779.

Harboe, M., Pande, H., Brandtzaeg, P. *et al.* (1966). Synthesis of donor type γ-globulin following thymus transplantation in hypo-γ-globulinaemia with severe lymphocytopenia. *Scand. J. Haematol.* **3**, 351–374.

Harousseau, J. L., Devergie, A., Lawler, S. *et al.* (1980). Implant of fetal liver in severe bone marrow aplasia. *Nouv. Rev. Fr. Hematol.* **22**(Suppl.), 572.

Harris, D. T., Schumacher, M. J., Locascio, J. *et al.* (1992). Phenotypic and functional immaturity of human umbilical cord blood T lymphocytes. *Proc. Natl. Acad. Sci. U.S.A.* **89**, 10,006–10,010.

Hayward, A., Ambruso, D., Battaglia, F. *et al.* (1998). Microchimerism and tolerance following intrauterine transplantation and transfusion for alpha-thalassemia-1. *Fetal Diagn. Ther.* **13**, 8–14.

Hoffer, B. J., Granholme, A., Stevens, J. O. *et al.* (1988). Catecholamine containing grafts in Parkinsonism: Past and present. *Clin. Res.* **36**, 189–195.

Hsieh, H. J., Li, N. Q., and Frangos, J. A. (1993). Pulsatile and steady flow induces c-fos expression in human endothelial cells. *J. Cell. Physiol.* **154**, 143–151.

Hullett, D. A., Falany, J. L., Love, R. B. *et al.* (1987). Human fetal pancrease—A potential souce for transplantation. *Transplantation* **43**, 18–22.

Hurst, A. F., Tanner, W. E., and Osman, A. A. (1922). Addison's disease with severe anemia treated by suprarenal grafting. *Proc. R. Soc. Med.* **15**, 19.

Johnson, M. P., Drugan, A., Miller, O. J. *et al.* (1989). Genetic correction of hereditary disease. *Fetal Ther.* **4**(Suppl. 1), 28–39.

Kansal, V., Sood, S. K., Batra, A. K. *et al.* (1979). Fetal liver transplantation in aplastic anemia. *Acta Haematol.* **62**, 128–136.

Karson, E. M., and Anderson, W. F. (1990). Prospects for gene therapy. *In* "The Unborn Patient: Prenatal Diagnosis and Treatment" (M. R. Harrison, M. S. Golbus, and R. A. Filly, eds.), pp. 481–494. Saunders, Philadelphia.

Keever, C. A. (1993). Characterization of cord blood lymphocyte subpopulations. *J. Hematother.* **2**, 203–206.

Khan, A., Sergio, J. J., Zhao, Y. *et al.* (1997). Discordant xenogeneic neonatal thymic transplantation can induce donor-specific tolerance. *Transplantation* **63**, 124–131.

Kirkwood, K. J., and Billington, W. D. (1981). Expression of serologically detectable H-2 antigens on mid-gestation mouse embryonic tissues. *J. Embryol. Exp. Morphol.* **61**, 207–219.

Klopper, A., and Diczfalusy, E. (1969). "Foetus and Placenta." Blackwell, Oxford.

Kochupillai, V., Sharma, S., Francis, S. *et al.* (1987a). Fetal liver infusion in aplastic anaemia. *Thymus* **10**, 95–102.

Kochupillai, V., Sharma, S., Francis, S. *et al.* (1987b). Fetal liver infusion in acute myelogenous leukaemia. *Thymus* **10**, 117–124.

Kopyov, O. V., Jacques, S., Lieberman, A. *et al.* (1998). Safety of intrastriatal neurotransplantation for Huntington's disease patients. *Exp. Neurol.* **149**, 97–108.

Laferty, K. J. (1988). Diabetes islet cell transplant research: Basic science. *In* "Report of the Human Fetal Tissue Transplantation Panel" (Consultants of the Advisory Committee to the Director of the National Institutes of Health), Vol. II (D), p. 142. National Institutes of Health, Bethesda, MD.

Lauffenberger, D. A. (1991). Models for receptor-mediated cell phenomena: Adhesion and migration. *Annu. Rev. Biophys. Chem.* **20**, 387–414.

Lehrman, D. (1988). "A summary: Fetal Research and Fetal Tissue Research." Association of American Medical Colleges, Washington, DC.

Lillie, F. R. (1916). The theory of the free-martin. *Science* **43**, 611–613.

Lindvall, O., Brundin, P., Widner, H. *et al.* (1990). Grafts of fetal dopamine neurons survive and improve motor function in Parkinson' disease. *Science* **247**, 574–577.

Loudon, M. M., and Thompson, E. N. (1988). Severe combined immunodeficiency syndrome, tissue transplant, leukaemia and Q fever. *Arch. Dis. Child.* **63**, 207–209.

Lu, L., Shen, R.-N., and Broxmeyer, H. E. (1996). Stem cells from bone marrow, umbilical cord blood and peripheral blood for clinical application: Current status and future application. *Crit. Rev. Oncol. Hematol.* **22**, 61–78.

Lucarelli, G., Izzi, T., Porcellini, A. *et al.* (1979). Infusion of fetal liver cells in aplastic anemia. *Haematol. Blood Transfus.* **24**, 167–170.

Lucarelli, G., Izzi, T., Porcellini, A. *et al.* (1980). Fetal liver transplantation in aplastic anemia and acute leukaemia. *In* "Fetal Liver Transplantation: Current Concepts and Future Direction" (G. Lucarelli, T. M. Fliedner, and R. P. Gale, eds.), pp. 284–299. Excerpta Medica, Amsterdam.

Magomedov, M. M. (1998). Transplantation of fetal tissues in otorhinolaryngology: Current status and prospects for the future. *Vestn. Otorinol.* **2**, 16–23.

McKusick, V. A. (1990). "Mendelian Inheritance in Man: Catalogs of Autosomal Dominant, Autosomal Recessive, and X-linked Phenotypes." Johns Hopkins University Press, Baltimore, MD.

Metcalf, D., and Moore, M. A. S. (1971). Embryonic aspects of hematopoiesis. *In* "Frontiers of Biology: Haemopoietic Cells" (A. Neuberger, and E. L. Tatum, eds.), pp. 172–271. North-Holland Publ., Amsterdam.

Mullen, Y. S., Clark, W. R., Molnar, I. G. *et al.* (1977). Complete reversal of experimental diabetes mellitus in rats by a single fetal pancreas. *Science* **195**, 68–70.

Munn, D. H., Zhou, M., Attwood, J. T. *et al.* (1998). Prevention of allogeneic fetal rejection by tryptophan catabolism. *Science* **281**, 1191–1193.

Mychaliska, G. B., Muench, M. O., Rice, H. E. *et al.* (1998). The biology and ethics of banking fetal liver hematopoietic stem cells for *in utero* transplantation. *J. Pediatr. Surg.* **33**, 394–399.

National Institutes of Health (NIH) (1988). "Report of the Human Fetal Tissue Transplantation Panel," Vol. I. Consultants of the Advisory Committee to the Director of the NIH, National Institutes of Health, Bethesda, MD.

Nelson, J. L., Furst, D. E., Maloney, S. *et al.* (1998). Microchimerism and HLA-compatible relationships of pregnancy in scleroderma. *Lancet* **351**, 559–562.

Nerem, R. M., and Girard, P. R. (1990). Hemodynamic influences on vascular endothelial biology. *Toxicol. Pathol.* **18**, 572–582.

Nozawa, M., Otsu, I., Ikebukuro, H. *et al.* (1991). Effects of fetal liver transplantation in rats with congenital metabolic disease. *Transplant. Proc.* **23**, 889–891.

O'Reilly, R. J., Kapoor, N., Kirkpatrick, D. *et al.* (1980). Fetal tissue transplants for severe combined immunodeficiency—Their limitations and functional potential. *In* "Primary Immunodeficiencies" (M. Seligmann and W. H. Hitzig, eds.), pp. 419–433. Elsevier/North Holland Biomedical Press, Amsterdam.

O'Reilly, R. J., Pollack, M. S., Kapoor, N. *et al.* (1983). Fetal liver transplantation in man and animals. *In* "Recent Advances in Bone Marrow Transplantation" (R. P. Gale, ed.), pp. 799–830. Alan R. Liss, New York.

Ourednick, J., Ourednick, W., and Mitchell, D. E. (1998). Remodeling of lesioned kitten visual cortex after xenotransplantation of fetal mouse neopallium. *J. Comp. Neurol.* **395**, 91–111.

Owen, R. D. (1945). Immunogenic consequences of vascular anastomosis between bovine twins. *Science* **102**, 400–401.

Parkhurst, M. R., and Saltzman, W. M. (1992). Quantification of human neutrophil motility in three-dimensional collagen gels. Effect of collagen concentration. *Biophys. J.* **61**, 306–315.

Perryman, L. E., McGuire, T. C., Torbeck, R. L. *et al.* (1982). Evaluation of fetal liver cell transplantation for immunoreconstitution of horses with severe combined immunodeficiency. *Clin. Immunol. Immunopathol.* **23**, 1–9.

Phan, D. T., Mihalik, R., Benczur, M. *et al.* (1989). Human fetal liver as a valuable source of haemopoietic stem cells for allogenic bone marrow transplantation. *Haematologia* **22**, 25–35.

Pollok, J.-M., and Vacanti, J. P. (1996). Tissue engineering. *Semin. Pediatr. Surg.* **5**, 191–196.

Porter, K. A. (1959). Use of foetal haemopoietic tissue to prevent late deaths in rabbit radiation chimeras. *Br. J. Exp. Pathol.* **40**, 273–280.

Prümmer, O., Werner, C., Raghavachar, A. *et al.* (1985). Fetal liver transplantation in the dog: II. Repopulation of the granulocyte–macrophage progenitor cell compartment by fetal liver cells from DLA-identical siblings. *Transplantation* **40**, 498–503.

Qu, J., Chehroudi, B., and Brunette, D. M. (1996). The use of micromachined surfaces to investigate the cell behavioural factors essential to osseointegration. *Oral Dis.* **2**, 102–115.

Rabian-Herzog, C., Lesage, S., and Gluckman, E. (1992). Characterization of lymphocyte subpopulations in cord blood. *Bone Marrow Transplant.* **9**(Suppl. 1), 64–67.

Ratner, B. D. (1995). Surface modification of polymers: Chemical, biological and surface analytical challenges. *Biosens. Bioelectron.* **10**, 797–804.

Raudrant, D., Touraine, J. L., and Rebaud, A. (1992). *In utero* transplantation of stem cells in humans: Technical aspects and clinical experience during pregnancy. *Bone Marrow Transplant.* **9**, 98–100.

Reinholt, F. P., Hultenby, K., Tibell, A. *et al.* (1998). Survival of fetal porcine pancreatic islet tissue transplanted to a diabetic patient: Findings by ultrastructural immunocytochemistry. *Xenotransplantation* **5**, 222–225.

Rojansky, N., and Schenker, J. G. (1993). The use of fetal tissue for therapeutic applications. *Int. J. Gynecol. Obstet.* **41**, 233–240.

Rubinstein, P. (1993). Placental blood-derived hematopoietic stem cells for unrelated bone marrow reconstitution. *J. Hematother.* **2**, 207–210.

Saltzman, W. M. (1997). Weaving cartilage at zero g: The reality of tissue engineering in space. *Proc. Natl. Acad. Sci. U.S.A.* **94**, 13380–13382.

Scorsin, M., Hagege, A. A., Marotte, F. *et al.* (1997). Does transplantation of cardiomyocytes improve function of infarcted myocardium? *Circulation* **96**(9, Suppl.), II-188–II-193.

Simpson, T. J., and Golbus, M. S. (1985). *In utero* fetal hematopoietic stem cell transplantation. *Semin. Perinatol.* **9**, 68–74.

Slavin, S., Naparstek, B., Ziegler, M. *et al.* (1989). Intrauterine bone marrow transplantation as a means of correction of genetic disorders through induction of prenatal transplantation tolerance. *In* "Correction of Certain Genetic Diseases by Transplantation" (J. R. Hobbs, ed.), p. 54. Cognet, London.

Statter, M. B., Foglia, R. P., Parks, D. E. *et al.* (1988). Fetal and postnatal testis shows immunoprivilege as donor tissue. *J. Urol.* **139**, 204–210.

Strong, C. (1991). Fetal tissue transplantation: Can it be morally insulated from abortion? *J. Med. Ethics* **17**, 70–76.

Surbek, D. V., Holzgreve, W., Jansen, W. *et al.* (1998). Quantitative immunophenotypic characterization, cryopreservation, and enrichment of second- and third-trimester human fetal cord blood hematopoietic stem cells (progenitor cells). *Am. J. Obstet. Gynecol.* **179**, 1228–1233.

Tabbal, S., Fahn, S., and Frucht, S. (1998). Fetal tissue transplantation in Parkinson's disease. *Curr. Opin. Neurol.* **11**, 341–349.

Tanaka, S. A., Hiramatsu, T., Oshitomi, T. *et al.* (1998). Induction of donor-specific tolerance to cardiac xenografts *in utero*. *J. Heart Lung Transplant.* **17**, 888–891.

Thomas, D. B. (1993). The infusion of human fetal liver cells. *Stem Cells* **11**(Suppl. 1), 66–71.

Thomas, E. D., Lochte, H. L., Lu, W. C. *et al.* (1957). Intravenous infusion of bone marrow in patients receiving radiation and chemotherapy. *N. Engl. J. Med.* **247**, 491.

Touraine, J. L. (1983). Bone-marrow and fetal-liver transplantation in immunodeficiencies and inborn errors of metabolism: Lack of significant restriction of T-cell function in long-term chimeras despite HLA-mismatch. *Immunol. Rev.* **71**, 103–121.

Touraine, J. L. (1989). New strategies in the treatment of immunological and other inheritted diseases: Allogeneic stem cell transplantation. *Bone Marrow Transplant.* **4**(Suppl. 4), 139–141.

Touraine, J. L. (1991). *In utero* transplantation of fetal liver stem cells in humans. *Blood Cells* **17**, 379–387.

Touraine, J. L. (1996). Treatment of human fetuses and induction of immunological tolerance in humans by *in utero* transplantation of stem cells into fetal recipients. *Acta Haematol.* **96**, 115–119.

Touraine, J. L., Raudrant, D., Royo, C. *et al.* (1989). *In-utero* transplantation of stem cells in bare lymphocyte syndrome. *Lancet* **1**, 1382.

Touraine, J. L., Raudrant, D., Vullo, C. *et al.* (1991). New developments in stem cell transplantation with special reference to the first *in utero* transplants in humans. *Bone Marrow Transplant.* **7**(Suppl. 3), 92–97.

Touraine, J. L., Raudrant, D., Rebaud, A. *et al.* (1992). *In utero* transplantation of stem cells in humans: Immunological aspects and clinical follow-up of patients. *Bone Marrow Transplant.* **9**, 121–126.

Uphoff, D. E. (1958). Preclusion of secondary phase of irradiation syndrome by inoculation of fetal hematopoietic tissue following lethal total-body X irradiation. *J. Natl. Cancer Inst. (U.S.)* **20**, 625–632.

Vacanti, J. P., Morse, M. A., Saltzman, W. M. *et al.* (1988). Selective cell transplantation using bioabsorbable artificial polymers as matrices. *J. Pediatr. Surg.* **23**, 3–9.

Van-Putten, L. M., Van Bekkum, D. W., and De Vries, M. J. (1968). Transplantation of foetal haemopoietic cells in irradiated rhesus monkeys. *In* "Radiation and the Control of Immune Response," pp. 41–49. IAEA, Vienna.

Vawter, D. E., Kearney, W., Gervais, K. G. *et al.* (1990). "The Use of Human Fetal Tissue: Scientific, Ethical, and Policy Concerns." Center for Biomedical Ethics, University of Minnesota, Minneapolis.

Vilmer, E., Sterkers, G., and Rahimy, C. (1992). HLA-mismatched cord blood transplantation in a patient with advanced leukemia. *Transplantation* **53**, 1155–1157.

Vowels, M. R., L-P-Tang, R., Berdoukas, V. *et al.* (1993). Brief report: Correction of x-linked lymphoproliferative disease by transplantation of cord-blood stem cells. *N. Engl. J. Med.* **329**, 1623–1625.

Wagner, J. E., Broxmeyer, H. E., Byrd, R. L. *et al.* (1992). Transplantation of umbilical cord blood after myeloablative therapy: Analysis of engraftment. *Blood* **79**, 1874–1881.

Wagner, J. E., Kernan, N. A., Steinbach, M. *et al.* (1995). Allogeneic sibling cord blood transplantation in forty-four children with malignant and non-malignant disease. *Lancet* **346**, 214–219.

Wengler, G. S., Lanfranchi, A., Frusca, T. *et al.* (1996). *In-utero* transplantation of parental CD34 haematopoietic progenitor cells in a patient with X-linked severe combined immunodeficiency (SCIDXI). *Lancet* **348**, 1484–1487.

West, L. J., Morris, P. J., and Wood, K. J. (1994). Fetal liver hematopoietic cells and tolerance to organ allografts. *Lancet* **343**, 148–149.

Winkler, J., Thal, L. J., Gage, F. H. *et al.* (1998). Cholinergic strategies for Alzheimer's disease. *J. Mol. Med.* **76**, 555–567.

Wong, L. (1988). Medical Research Council Tissue Bank (presentations at Sept. 1988 panel meeting). *In* "Report of the Human Fetal Tissue Transplantation Research Panel" (Consultants to the Advisory Committee to the Director), Vol. II, pp. D267–D282. National Institutes of Health, Bethesda, MD.

Yuh, D. D., Gandy, K. L., Hoyt, G. *et al.* (1997). A rodent model of *in utero* chimeric tolerance induction. *J. Heart Lung Transplant.* **16**, 222–230.

Zanjani, E. D., Almeida-Porada, G., and Flake, A. W. (1995). Retention and multilineage expression of human hematopoietic stem cells in human–sheep chimeras. *Stem Cells* **13**, 101–111.

Zanjani, E. D., Almeida-Porada, G., Ascensao, J. L. *et al.* (1997). Transplantation of human hematopoietic stem cells *in utero*. *Stem Cells* **15**(Suppl. 1), 79–92.

CHAPTER 29

Pluripotent Stem Cells

Michael J. Shamblott, Brian E. Edwards, and John D. Gearhart

INTRODUCTION

Pluripotent stem cells, cells that can replicate in culture but also form many or most of the cells comprising the body, have been described. In the mouse and a few other species, embryonic stem and embryonic germ cells can be derived from the inner cell mass of a blastocyst, and from primordial germ cells, respectively. These cells differentiate *in vitro* into a mixture of cell types, including heart muscle cells, blood cells, neural cells, and vascular cells. This repertoire can be greatly expanded by using powerful enrichment and selection strategies, and purified populations of differentiated cells can be generated. Reports of human embryonic stem and germ cells have received significant attention in both the scientific and popular press. The capacity of these cells to differentiate *in vitro* into a wide variety of cell types suggests their eventual use in the generation of cells for tissue engineering and transplantation therapies.

DEFINITION OF A STEM CELL

A stem cell can replicate itself and produce cells that take on more specialized functions. The breadth of function adopted by the more differentiated daughter cells and their progeny is commonly referred to as the developmental potential, or potency, of the stem cell. Stem cells that give rise to only one type of differentiated cell are termed unipotent. In common usage, the relative terms oligopotent, multipotent, and pluripotent represent an increase in the number of differentiated cell types from few to many or most. A totipotent cell is one that can generate the totality of cell types that comprise the organism. In practice, these few terms poorly describe a continuum of possibilities.

The ability to form a wide variety of cell types makes pluripotent stem cells a promising resource for tissue engineering and transplantation. Through the use of enrichment, selection, expression, and sorting technologies, *in vitro* differentiation of stem cells will certainly contribute to future transplantation therapies.

DEVELOPMENTAL POTENTIAL

In order to describe more fully the potential of a stem cell, one must consider the process by which a fertilized egg develops into a complex multicellular organism. Within several days following fertilization, the processes of cell cleavage, compaction, and cavitation generate a mass of cells called the blastocyst. The blastocyst consists of two cell populations, an outer layer of cells surrounding a fluid-filled cavity and a mass of cells within the cavity. The outer cells (trophectoderm) contribute to the placenta but not the embryo proper. The inner cell mass (ICM) cells further differentiate during gastrulation to form the three germ layers of the embryo: ectoderm, mesoderm, and endoderm. Cells of the germ layers generate all the tissues of the embryo. During gastrulation, cells that will go on to form the germ cells (eggs and sperm) are also allocated.

TYPES OF PLURIPOTENT STEM CELLS

There are only a few stem cell types that can be cultured and qualify for true pluripotent/totipotent status. These include embryonal carcinoma (EC), embryonic stem (ES), and embryonic germ (EG) cells. EC cells were first described as the stem cell component of teratocarcinomas that arose following the transfer of a mouse postimplatation embryo or embryonic gonad to an ectopic site (Stevens, 1958). Later, similar stem cells were derived from spontaneous teratocarcinomas formed in both mouse (mouse EC cells) and humans (human EC cell). EC cultures are heteroploid and some are multipotent.

Pluripotent/totipotent stem cells have been derived from two embryonic sources. ES cells are derived from the ICM of preimplantation embryos (Evans and Kaufman, 1981; Martin, 1981), whereas EG cells are derived from primordial germ cells (PGCs) that (during normal development) migrate to and colonize the gonad, eventually forming eggs and sperm (Matsui *et al.,* 1991; Resnick *et al.,* 1992). Both mouse ES and EG cells demonstrate germ-line transmission in experimentally produced chimeras (Labosky *et al.,* 1994; Stewart *et al.,* 1994). Mouse ES and EG cells share several morphological characteristics such as high levels of alkaline phoshatase (AP) activity and the presence of some specific cell surface antigens. Other important shared characteristics include growth as multicellular colonies, normal and stable karyotypes, the ability to be continuously passaged, and the capability to differentiate into cells derived from all three embryonic germ layers. Pluripotent stem cell lines that share most of these characteristics have also been reported for chicken (Pain *et al.,* 1996), mink (Sukoyan *et al.,* 1993), hamster (Doetschman *et al.,* 1988), pig (Shim *et al.,* 1997; Wheeler, 1994), rhesus monkey (Thomson *et al.,* 1995), and common marmoset (Thomson *et al.,* 1996). Human pluripotent stem cells derived from PGCs (Shamblott *et al.,* 1998) and ICM (Thomson *et al.,* 1998) have been described.

Multipotent stem cells can be cultured from a number of fetal and adult sources. Perhaps the best known source is bone marrow, which contains both hematopoietic stem cells (Civin *et al.,* 1984) and mesenchymal stem cells (Caplan, 1991; Pittenger *et al.,* 1999), some of which can apparently also form hepatocytes (Petersen *et al.,* 1999). Neural stem cells have also been cultured from the ependymal cells lining the brain ventricles (Johansson *et al.,* 1999) and cells found in the cerebellum (Snyder *et al.,* 1995). It is likely that of the many lineage-restricted stem cell populations that exist *in vivo,* some will be amenable to *in vitro* growth and analysis.

METHODS OF ASSAY

Histologic, immunocytochemical, and molecular markers have been used to indirectly assess pluripotency. These include high levels of AP and presentation of specific cell surface glycolipids (Kannagi *et al.,* 1983; Solter and Knowles, 1978) and glycoproteins (Andrews *et al.,* 1984). These properties are characteristic of pluripotent stem cells, but are neither necessary nor sufficient as a definition.

The only direct measure of developmental potential is the analysis of cell differentiation. There are both *in vitro* and *in vivo* methods to establish broad developmental potential. For example, some types of stem cells can be observed (and even directed to some extent) to differentiate in culture. When differentiation occurs, structures termed embryoid bodies (EBs) are formed; these were initially described as arising in human (Peyron, 1939) and mouse (Stevens, 1958, 1959, 1970) germ cell tumors called teratomas and teratocarcinomas.* These cell aggregates range from a cluster of pluripotent stem cells enclosed by a layer of endoderm to complex structures in which inductive and stochastic developmental processes lead to a seemingly haphazard array of cell types, including representatives of all three germ layers. The presence of differentiated cell types within EBs has been used as evidence of cell pluripotency, as a means to study some early events in development, and as a source of differentiated cells (Doetschman *et al.,* 1985; Evans and Kaufman, 1981). Although useful in the production of cells for tissue engineering and transplantation, the *in vitro* differentiation of pluripotent stem cells is limited by our understanding of the genes, molecules, and physical events involved and cannot easily rival *in vivo* differentiation.

When some types of stem cells are implanted at ectopic sties within an isogenic or immunologically compromised mouse, teratomas and teratocarcinomas form. The complex physiological milieu of the tumor provides a wide but difficult to control array of stimuli. Analysis of the vari-

---

* *Teratocarcinomas are malignant tumors arising from relatively undifferentiated germ cells. A wide variety of highly differentiated cells and structures can be found within the tumor, including gut and respiratory epithelia, nerves, muscle, cartilage, and bone. The presence of stem cells within the tumor distinguishes teratocarcinomas from teratomas*

ety of cell types that comprise the tumor can be used as a measure of developmental potential. If cells that derive from all three germ layers are identified, the term pluripotent can be used. This property alone may not be a definitive test of stem cell pluripotency, because it has been demonstrated that visceral endoderm (yolk sac) of the rat and mouse is capable of forming highly differentiated teratomas containing cells of all three embryonic germ layers (Sobis *et al.,* 1991, 1993). The ultimate and definitive test of developmental potency is the generation of chimeric animals. Under the least restrictive definition, only a stem cell that can contribute to all the tissues of the embryonic and adult organism can be said to be totipotent. This test can be performed for only a few species, mainly the mouse.

In practice, the assignment of pluripotent and totipotent status is more difficult than the analysis of differentiation. Bone marrow and peripheral blood contain hematopoietic stem cells that generate all of the types of blood cells. These cells are often referred to simply as pluripotent stem cells, in acknowledgment of their capacity to form cells of erythroid, lymphoid, and myeloid lineages. This usage does not allow for stem cells that can be demonstrated to form representatives of all three germ layers, but where totipotency cannot be established.

Totipotency is also a problematic term. ICM cells are capable of generating every cell of the embryonic and adult organism, but have lost the capability to form trophoblasts and therefore cannot entirely reconstitute a conceptus. Therefore, the ICM and stem cells derived from it are only totipotent with respect to the embryo. Another important consideration in the use of the term totipotent is that it is sometimes used to describe the capacity of cells form an embryo, that is, that the cells are capable of self-organization into an organism. No known stem cell possesses this property, so under this definition, stem cells are at best pluripotent.

A resolution to these nomenclature dilemmas is a matter of preference and will certainly be a subject for honest debate. Under the strictest definitions, only fertilized eggs and early blastomeres (2–4 cell stage) qualify as totipotent; however, they are not stem cells because they cannot continuously replicate. Stem cells with the demonstrated ability to form representatives of all three germ layers are pluripotent, with special acknowledgment for cells such as mouse ES and EG cells, which can be demonstrated to form all of the cells of the embryonic and adult organism.

DIFFERENTIATION *IN VITRO*

BASIC TECHNIQUES

In vitro culture and differentiation of stem cells offer the hope of precise control of the process and the products, but require responsibility for its direction. Most ES and EG cells require leukemia inhibitory factor (LIF; ∼1000 U/ml) to proliferate and remain pluripotent.[†] The simple withdrawal of LIF from the growth media enables differentiation, albeit within a relatively narrow range of cell types produced. The formation of EBs has been the most common and successful paradigm for more complex and varied differentiation of pluripotent stem cells. Although EBs are spontaneously formed in mouse and human ES/EG cell cultures, their formation can be enhanced by withdrawing LIF from the media then plating cells in an environment conducive to aggregation, rather than adhesion to a substrate. This is achieved by plating cells in suspension culture in bacteriologic petri dishes or in hanging drops. A viscosity-increasing agent such as methylcellulose is often added to the culture media to promote EB aggregation. After several days of growth and influence by added factors and compounds, EBs are disaggregated and plated onto an adhesive substrate and allowed to grow and further differentiate. With the proper combinations of growth and differentiation factors, mouse ES (Keller *et al.,* 1993; Wiles and Keller, 1991) and EG (unpublished results) cell cultures can generate cells of the hematopoietic lineage and cardiomyocytes (Klug *et al.,* 1995; Rohwedel *et al.,* 1996). In addition, mouse ES cells have been used to generate *in vitro* cultures of neurons (Bain *et al.,* 1995), skeletal muscle (Rohwedel *et al.,* 1994), vascular endothelial cells (Rohwedel *et al.,* 1994), adipocytes (Dani *et al.,* 1997), and visceral endoderm (Abe *et al.,* 1996; Doetschman *et al.,* 1985), which is the precursor to structures such as the pancreas, liver, and lung.

[†] *LIF is a member of a cytokine family that heterodimerizes with the gp130 molecule. Other members of the family, such as oncostatin M, or strategies that directly activate gp130-mediated signal transduction, can replace LIF.*

TARGET CELL ENRICHMENT

A number of exogenous factors have been used to direct EB differentiation. The most commonly used chemical reagents are retinoic acid (RA), dimethyl sulfoxide (DMSO), hexamethyl bisacetamide (HMBA), and reagents that elevate intracellular cAMP levels, such as dibutryl-cAMP (db-cAMP), forskolin, and 3-isobutyl-1-methyl-xanthine (IBMX). RA, cAMP elevators, and HMBA have been used to influence the formation of neurons from EC and ES cells (Andrews *et al.*, 1986; McBurney *et al.*, 1982). DMSO has been used to enhance the production of mesodermal derivatives such as muscle cells (Habara-Ohkubo, 1996; McBurney *et al.*, 1982) and endodermal derivatives (Edwards and Adamson, 1986). The modes of action by which RA and DMSO act are not fully understood, although it is clear that the effects of RA are mediated through RA receptors and this receptor impacts the regulation of genes that contain RA-responsive elements.

Growth factors have been used in the differentiation of pluripotent stem cells. The most sophisticated application of growth factors is in the derivation of hematopoietic cells from mouse ES cells. Included are stem cell factor (known as c-kit ligand, SCF, steel factor, and mast cell factor), vascular endothelial growth factor (VEGF), erythropoitin (EPO), interleukin-1 (IL-1), IL-3, IL-6, IL-11, macrophage colony-stimulating factor (M-CSF), granulocyte colony-stimulating factor (G-CSF), granulocyte—macrophage colony-stimulating factor (GM-CSF), and bone morphogenic protein 4 (BMP4) (Biesecker and Emerson, 1993; Keller *et al.*, 1993; Keller, 1995; Kennedy *et al.*, 1997; Palacios *et al.*, 1995).

The exact component(s) that cause a desired differentiation effect are often unknown, except for their activity in a bioassay. Typically, these factors are secreted by a support or feeder culture. These feeder cultures "condition" the media in which they grow. These media are then harvested (and sometimes concentrated) and used as a growth supplement. One example is the use of bone marrow stromal cell conditioned medium (CM) as a component of hematopoeitic cell differentiation. This CM is required for eventual production and long-term repopulation of the lymphoid, myeloid, and erythroid lineages (Palacios *et al.*, 1995).

SELECTIVE MATRICES

The nature of the surface a cell attaches to can have a profound impact on its characteristics. Most ES and EG cells grow best on a feeder layer of mouse embryonic fibroblasts (primary mouse embryo fibroblasts or STO fibroblasts) that have been irradiated or treated with mitomycin C to make them mitotically inactive. Instead of live cells, the extracellular matrix (ECM) of these fibroblasts or a layer of gelatin will suffice, but often with suboptimal performance. Several adhesion matrices have been employed in the growth of stem cell-derived cells. Collagens I and IV have been used to promote adhesion and cell division in a number of cell types, including vascular endothelial cells (Hirashima *et al.*, 1999), and laminin has been shown to promote neurite outgrowth in many neuronal culture systems, including ES cell-derived neuronal cells (Bain *et al.*, 1995). Cell-derived extracellular matrices such as Matrigel and human ECM (Collaborative Biomedical Products) have been used to support the differentiation of various cell types.

Deposition of growth factors within extracellular matrices and basement membranes is a critical phenomenon of development. Signals deposited by one cell type are subsequently read by naive cells that grow in contact with the conditioned ECM or degrade a conditioned basement membrane. It is likely that complex differentiation and tissue engineering will benefit from the use of these types of inductive processes.

GENETIC SELECTION

Genetic selection is a powerful strategy for generating pure or nearly pure populations of differentiated cell types. In order to develop this strategy, a gene that is expressed specifically in the desired cell type must be identified. This gene can be specific to a terminally differentiated cell type or to a lineage-restricted stem cell. The promoter of the specific gene is then joined to a gene that encodes resistance to an antibiotic drug such as neomycin (G418), hygromycin, bleomycin, or zeocin. This construct is transfected into stem cells that are then caused to differentiate. If the differentiating cells are treated with the antibiotic drug, all cells that cannot express the fusion gene construct will die (Fig. 29.1). An example of genetic selection is the production of a highly pure (>99%) population of cardiomyocytes from differentiated mouse ES cells using the α-cardiac myosin heavy chain promoter joined to a cDNA coding for neomycin resistance (Klug *et al.*, 1996).

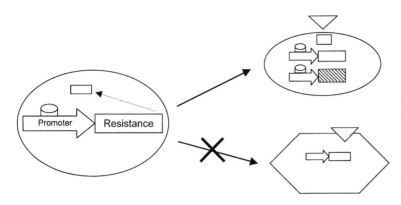

Fig. 29.1. Genetic selection used to obtain a specific population of differentiated cells. A target cell-specific promoter requiring a transcription factor (○) is joined to a gene imparting drug resistance (□) and transferred into pluripotent stem cells. When the stem cells differentiate, only cells that can activate, transcribe, and translate (dashed arrow) the target cell-specific promoter along with their cell-specific genes (▧) are resistant to the antibiotic (▽) and survive.

Transcription Factor Expression

Another genetic strategy is to transfect pluripotent stem cells with a construct that causes the expression of a factor known (or thought) to play an important role in the commitment to a desired lineage. Usually, these are transcription factors. As commitment decisions are usually hierarchical, it is beneficial to choose a transcription factor that is influential in the earliest stages. The promoter driving the expression can be a strong constitutive promoter, a tissue-specific promoter, or an exogenously regulated promoter such as the commercially available tetracycline (Clontech) and ecdysone (Invitrogen) expression systems. Examples of ES cell differentiation influenced by transcription factor expression are the generation of skeletal muscle cells following the constitutive expression of MyoD1 (Dekel *et al.,* 1992; Shani *et al.,* 1992) and the generation of neurons following the constitutive expression of neuroD2 or neuroD3 (O'Shea *et al.,* 1997).

Fluorescence-Activated Cell Sorting

Fluorescence-activated cell sorting (FACS) is a powerful method to physically select a subset of target cells from a mixed population. This strategy is used when the target cell expresses a specific cell surface molecule.[‡] The molecules can be receptors or protein, glycoprotein, or lipid surface antigens. An antibody that is specific to the surface molecule is used to identify the target cell in solution. This antibody is then tagged with a fluorophore. Target cells that fluoresce are physically segregated and saved. Negative cells are washed away or collected separately. One or more antibody/fluorophore combination(s) can be used to impose highly selective criteria, allowing fine-tuning of cell populations or the ability to recover extremely rare cell types. One complication with FACS is that adherent cell types need to be effectively disaggregated prior to sorting. Special precaution must be taken during this process to retain the specific surface molecule(s). Damage caused by enzymatic treatment such as trypsin digestion must be minimized and should be followed by a period of cellular recovery and resynthesis. FACS has been used as both an analytical and a preparative tool to select many cell types. One example of its use is in the study and preparation of ES cell-derived hematopoietic stem cells prior to mouse immune system repopulation experiments (Palacios *et al.,* 1995).

Markers of Differentiation

Markers are used to monitor the process of differentiation. Markers can be specific proteins, glycoproteins, glycolipids, or other metabolites made by the target cell of interest. Methods of detection include histologic analysis using dyes; immunoreactivity of cells or cell homogenates to specific antibodies; and mRNA expression by using Northern blot analysis, reverse transcriptase and polymerase chain reaction (RT-PCR), or RNase protection assay. The loss of markers of pluripotency can also be used to monitor differentiation. The mouse transcription factor Oct3/4[§]

[‡] *Cells can also be loaded with enzyme substrates that fluoresce when acted on by specific endoenous enzymes or reporter enzymes such as β-galactosidase (Metzger* et al., *1996).*

[§] *A commonly used abbreviation for mouse Oct3 (GenBank Accession No. MUSOCT3) and Oct4 (MMOCT4P), although their reported nucleotide and predicted amino acid sequences are not identical.*

is expressed in unfertilized eggs, ICM, PGCs, and pluripotent stem cells. The loss of Oct3 expression or other markers such as AP or SSEA-3 (Andrews, 1984) can be used to gauge the effectiveness and progress of differentiation schemes.

EXAMPLES OF *IN VITRO* DIFFERENTIATION AND MARKERS

HEMATOPOEISIS

Through the use of the selective growth media, exogenously added compounds and growth factors, selective growth matrices, and genetic strategies, a number of different cell types have been derived from mouse pluripotent stem cells. These include hematopoietic cells, vascular endothelium, muscle cells, neurons, and visceral endoderm (Fig. 29.2).

Hematopoiesis within EBs was one of the first differentiation phenomena observed, and remains one of the most thoroughly studied. Mouse ES cells are grown in a standard media containing LIF. The cells are then removed from the plates by incubation in trypsin, then plated into bacteriologic dishes and grown in 1% methylcellulose in the presence of various cytokines and other factors. After 3 days EBs form and are allowed to grow for about 2 weeks. During this time, erythroblasts accumulate in endothelium-lined channels or cysts, forming blood islands that are red due to the presence of hemoglobin. EBs are then disaggregated using trypsin or other enzymes and plated onto adhesive tissue culture plates or are harvested for analysis. Through the use of specific markers such as GATA-1, c-*myb*, SCL, and embryonic and adult globin, it has been shown that most, if not all, of the hematopoietic linages are represented among the differentiated cells produced (Keller *et al.*, 1993). Not unexpectedly, hematopoietic stem cells are produced by this process, because EB-derived cells can be used to repopulate the hematopoietic system of a mouse that has had its own immune system damaged or destroyed by radiation (Palacios *et al.*, 1995).

VASCULAR ENDOTHELIUM

Hematopoiesis and the development of a vascular system are developmentally connected. Both are mesodermal derivatives, and are colocalized in the developing aorta–gonad–mesonephros (AGM) region of embryo (Medvinsky and Dzierzak, 1996). It is not clear however, if there is a common precursor (hemangioblast) or if erythropoiesis requires endothelial cells (Kennedy *et al.*, 1997). Vascular endothelium is effectively produced in mouse EBs using standard culture condi-

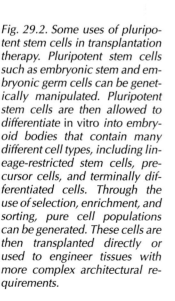

Fig. 29.2. Some uses of pluripotent stem cells in transplantation therapy. Pluripotent stem cells such as embryonic stem and embryonic germ cells can be genetically manipulated. Pluripotent stem cells are then allowed to differentiate in vitro into embryoid bodies that contain many different cell types, including lineage-restricted stem cells, precursor cells, and terminally differentiated cells. Through the use of selection, enrichment, and sorting, pure cell populations can be generated. These cells are then transplanted directly or used to engineer tissues with more complex architectural requirements.

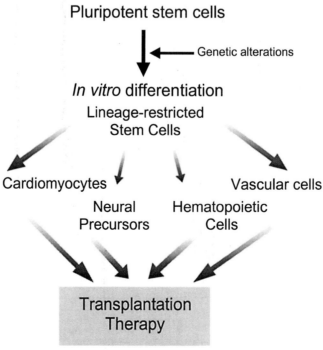

tions, and can organize into vessel-like channels filled with blood cells. This process can apparently be enhanced through the expression of transforming growth factor β1 (TGF-β1) (Zhang *et al.,* 1998). Vascular endothelial cells can be grown out from disaggregated EBs onto collagen-coated surfaces and studied or selected by using antibodies to CD34, platelet–endothelial cell adhesion molecule (PECAM), VEGF receptor (flk-1), VE–cadherin and von Willebrand factor (vWf), as well as by binding of acetylated low-density lipoprotein (Ac-LDL) (Hirashima *et al.,* 1999; Wang *et al.,* 1992).

MUSCLE

The spontaneous contraction of ES cell-derived cardiomyocytes (heart muscle cells) is one of the most visually and conceptually striking examples of ES cell differentiation *in vitro.* The rhythmic contractile movements seen in these cultures suggest a great deal of specialized differentiation and cellular organization. It is surprising, therefore, that this occurs when LIF is withdrawn from the culture medium for ~10 days. The time course of muscle-specific structural gene expression in these cells roughly parallels that seen in the embryo. Sarcomeres with functional gap junctions and Ca^{2+} channels, as well as organized myofibrils, are formed (T. Doetschman *et al.,* 1993; T. C. Doetschman *et al.,* 1985; Rohwedel *et al.,* 1994; Westfall *et al.,* 1997). Cardiomyocytes, which precede skeletal and smooth muscle developmentally, are formed in the greatest abundance in EBs and have been identified through the use of specific markers such as myosin light chain, myosin heavy chain, muscle-specific actin, and myogenin (Robbins *et al.,* 1990). Furthermore, ventricular specification of EB-derived cells occurs, as evidenced by ventricular myosin light chain-2 (Miller-Hance *et al.,* 1993). DMSO, RA, db-cAMP, and expression of the muscle-specific transcription factor MyoD have been used to enhance the production of smooth and skeletal muscle (Dinsmore *et al.,* 1996; Drab *et al.,* 1997). Expression of the M-twist transcription factor has the opposite effect, to inhibit myogenic differentiation (Rohwedel *et al.,* 1995).

NEURAL DIFFERENTIATION

ES cells (Bain *et al.,* 1995; Fraichard *et al.,* 1995) and some EC cells (Andrews, 1984; Lang *et al.,* 1989; Pleasure and Lee, 1993) are readily differentiated into neuroectodermal derivatives such as neurons and glia. Typically, ES cells are used to form EBs in the presence of RA. The EBs are then disaggregated (or plated directly as EBs) and plated onto an adherent plastic, laminin, or poly(D-lysine)-coated surface, where mixed populations can then be further identified and selected for. General markers such as neurofilament heavy and light isoforms, microtubule-associated protein (MAP), Tau, and β-tubulin III have been used to identify neuronal cells, and Pax-6 and nestin have been used to identify early neuroepithelial cells. Distinct morphological characteristics such as neurite elongation and dendrite formation, and more specific markers including tyrosine hydroxylase (TH), 5-hydroxytryptamine (5-HT), synaptophysin, Mash1, Math4a, neurotransmitters, and neurocan have been used to refine neuronal identification. Patch clamping has also been used for functional identification.

The general identification of glial cells can be done by examining cells for the expression of glial fibrillary acidic protein (GFAP). Specific identification of astrocytes can be made by using the marker OP-4, whereas oligodendroglial cells can be identified by the markers GalC and O4.

Several different approaches have been used to influence the outcome of RA-induced EB differentiation. Nerve growth factor (NGF) has been used to accelerate the formation of neurons within an EB (Wobus *et al.,* 1988), and expression of the transcription factor Noggin (O'Shea and Gratsch, 1998) and neuroD2 and neuroD3 (O'Shea *et al.,* 1997) lead to significant enhancement of neuronal production. Genetic selection using an early neuroepithelial-specific transcription factor (Sox2) promoter to drive neomycin resistance resulted in efficient purification of neuroepithelial progenitor cells that can then differentiate into neurons (Li *et al.,* 1998).

VISCERAL ENDODERM

Visceral endodermal cells give rise to a number of cell and tissue types, including the epithelial lining of the digestive tract, respiratory tracts, and bladder, as well as the parenchyma of the liver, pancreas, thyroid gland, and parathyroid gland. Visceral endoderm is generated in RA-treated mouse EC cell aggregates (Hogan *et al.,* 1981) and EBs derived from mouse (Doetschman *et al.,* 1985) and human (Shamblott *et al.,* 1998) pluripotent stem cells. Mouse EB-derived visceral endoderm has been shown to express many genes in common with liver, including transcription factors such as hepatocyte nuclear factor-1 (HNF-1), HNF-3, and HNF-4 forms; enzymes such as aldolase B (ALDO-B) and liver-type pyruvate kinase (L-PK), as well as the serum proteins α-fetoprotein (AFP), transthyretin (TTR), albumin, and several apolipoproteins (Abe *et al.,* 1996; Duncan *et al.,* 1997, 1998). This extensive representation of metabolic and regulatory molecules suggests that visceral endoderm may be a useful cell type for the generation of several target tissues, including liver and pancreas.

IN VIVO APPLICATIONS

CURRENT THERAPEUTIC MODELS

Several reports of stem cell therapy have energized and encouraged the entire field of stem cell biology. Most of the work to date has attempted to use lineage-restricted or multipotent stems cells rather than pluripotent stem cells. Some of this work may have a profound effect on our compartmentalized notions of germ layer origins. One outstanding example of stem cell-derived cell therapy is the engrafting of neurons generated from human EC cells (NTERA-2) to rat cerebrum damaged by induced transient ischemia (Borlongan *et al.,* 1998). The significant cellular integration and functional recovery observed in these rats suggest the potential use of these cells to treat stroke, Parkinson's disease, and other human neurodegenerative disorders. Importantly, grafts of postmitotic neurons generated from RA-treated NTERA-2 into the CNS of rodents do not lead to tumors (>1 year of observation) (Trojanowski *et al.,* 1997).

Multipotent neural stem cell (NSC) cultures derived from neonatal mouse cerebellum (clone C17.2) have a broad capacity to participate in the development of the brain when implanted (Snyder *et al.,* 1995). It has been reported that intracerebroventricular injection of these cells can result in widespread engraftment and correction of dysmyelination caused by a genetic lack of myelin basic protein (MBP) in *shiverer* mice (Yandava *et al.,* 1999). One of the most important aspects of these two transplantation studies is that cells placed at one location migrated widely and correctly to the desired location and engrafted in a way that led to the amelioration of a neuropathologic condition.

Cells of the central nervous system are of ectodermal origin, and, it was thought, are capable of repopulating cells of this dermal type only. However, this developmental exclusivity has been brought into question (Tajbakhsh *et al.,* 1994). The demonstration that NSCs derived from mouse forebrain striata engraft into the hematopoietic system (which is of mesodermal origin) of sublethally irradiated mice suggests a greater developmental plasticity than is commonly afforded multipotent stem cells (Bjornson *et al.,* 1999).

A similar suggestion has been made concerning bone marrow stem cells. These cells develop into hematopoietic and mesenchymal lineages, both of which are of mesodermal origin. Demonstration that these stem cells can also form liver cells, which are of endodermal origin, may further exemplify the ability of some cell types to cross dermal-origin boundaries (Petersen *et al.,* 1999).

The use of pluripotent stem cells as a source of cells for transplantation holds great promise due not only to the breadth of cells that could potentially be generated, but also to the ability to manipulate the cells genetically prior to differentiation. The risks inherent with their use must also be considered, however. As described earlier, grafting of undifferentiated mouse pluripotent stem cells to various sites such as the mouse testis, kidney capsule, or brain leads to the formation of a tumor. Transplantation models and eventual therapies will have to ensure that pluripotent stem cells have been eliminated from the implant.

GENETIC MANIPULATION

One significant advantage of the use of pluripotent stem cells in developmental studies or in transplantation paradigms is the ease with which they can be genetically manipulated. Mouse ES and EG cells are easily and routinely cultured and are capable of continuous passage, so tremendous cell expansion is possible. This is not the case for most nontransformed euploid (chromosomally normal) differentiated mouse cell lines. The human pluripotent stem cell cultures to date are not as easily grown as their mouse counterparts, and there are no examples of the successful introduction of foreign DNA into them. Fortunately, many nontransformed euploid differentiated cell lines have the capacity for extensive cell division, so genetic alterations and selections may be carried out on lineage-restricted and/or differentiated human cells that are derived from pluripotent stem cells.

Mouse pluripotent stem cells, and to a lesser extent human EC cells, are amenable to most routine methods of transfection. Electroporation, lipofection, and chemically medicated and virally mediated gene delivery have all been used successfully on mouse ES cells. These methods and novel viral/retroviral approaches are currently being investigated for use on human pluripotent stem cells.

Mouse ES cells have been the most extensively manipulated pluripotent stem cell type. Gene "knockout" and disruption experiments require gene targeting and homologous recombination, both of which are routine. Adding genes and artificial genetic constructs is also easily achievable. As described above, these strategies have been used successfully to select cell populations genetically or to direct differentiation by expression of transcription factors. In future stem cell differentiation experiments, combinations of gene disruption and the addition of novel gene function may be used to create truly customized cell populations and cells that are particularly amenable to the assembly and function of organ structures.

CELL DELIVERY

Developmental potential of transplanted cells

If cell or organ transplantation is the goal of a pluripotent stem cell differentiation experiment, the issues of which cells to deliver, the method of delivery, and the location of delivery will eventually arise. The type of cell to deliver may not obvious, however. Fully differentiated and post-mitotic cells are the most predictable and probably the safest choice, as they have function(s) that can be defined and characterized *in vitro*. The finite capacity of these cells to repopulate and the need to create many narrowly defined cell classes limits the usefulness of this approach. At the other end of the spectrum are undifferentiated pluripotent stem cells, with an enormous capacity to repopulate and differentiate. The high probability of tumor formation will certainly limit the medical application of this approach. The most obvious middle ground is the use of lineage-restricted stem cells. Current efforts detailed above using multipotent stem cells are an excellent first approximation of this approach. Future uses may include the selection or enrichment for stem cells that can repopulate and form several different terminally differentiated cell types. Future practices may include the design of fully repopulating organs.

Method and location of delivery

The optimal method and location of cell delivery are dependent on several factors, including target location, desired function, ease of access, and patient safety. Cells can be introduced as a suspension or implanted as a pellet or tissue-engineered organoid. These cells and structures can be implanted within a damaged organ or at an ectopic site.

IMMUNORECOGNITION AND REJECTION

Immune rejection is one of the most important issues in allogeneic transplantation therapy, whether it is whole organ, tissue, or cellular transplantation. At its most basic level, immune rejection is simply the result of the recognition of the graft as different from the host. Strategies to counteract rejection involve either ways to prevent the recognition of the graft as non-self or to prevent the physiologic reaction to the alloantigens present on the graft.

Currently, there are several pharmacologic means to prevent immune rejection. These chemicals tend to depress the entire immune response pathway, leading to negative side effects, namely, an increased risk of infection. There are areas of immune privilege such as the CNS, seminiferous tubule, or the eye. Transplantation into these regions may lessen the need for immunosuppressive agents.

Other alternative methods to prevent graft rejection include genetic modification of the graft tissue. In these scenarios, genes involved in antigen presentation can be disrupted, modified, or added to in pluripotent stem cells. This strategy has been attempted with mouse ES cells. Major histocompatibility complex (MHC) class I recognition molecules are found on all nucleated cells and require β_2-microglobulin to function. Genetic disruption of this gene highly reduces class I antigen presentation and cytotoxic T cell response (Zijlstra *et al.*, 1989). MHC class II molecules are constitutively expressed in immune cells but only rarely and weakly in nonimmune cells. Disruption of several α and β genes in mouse ES cells has been used to generate mice with limited MHC class II antigen presentation (Cosgrove *et al.*, 1991; Grusby *et al.*, 1991). By combining these disrupted alleles, class I- and II-deficient mice have been created (Grusby *et al.*, 1993). However, grafts derived from these mice have been subject to rejection following transplantation, though at a slower rate than normal controls. Minor histocompatibility antigens, tissue-specific antigens, and hybrid histocompatibility antigens are also capable of causing cell-mediated graft rejection and may require modification before some pluripotent stem cell-derived cells can be used for transplantation.

A technically feasible method of preventing graft rejection is to use autologous tissue. In some cases, lineage-restricted stem cells collected from the patient, such as those found in bone marrow, may be useful. Autologous pluripotent stem cell lines may be generated through therapeutic cloning (Solter and Gearhart, 1999), which involves somatic cell nuclear transfer technology, similar to the technique used in the production of the famous cloned sheep, Dolly (Wilmut *et al.*, 1997). A somatic nucleus would be taken from the patient and introduced into an enucleated oocyte. Pluripotent stem cells would then be derived from the resulting blastocyst. The resulting stem cell line could then be transplanted without rejection. This may be the most straightforward method to prevent graft rejection, although it would require somatic cell nuclear transplantation for every patient.

SUMMARY AND FUTURE PROSPECTS

STATUS AND CHALLENGES

Since the first reports of human pluripotent stem cells, research emphasis has been placed on the isolation and characterization of lineage-restricted human cells that comprise EBs. Scientific and political challenges lie ahead with respect to this research. If these stem cells are to be of general use in tissue engineering and transplantation, genes and factors involved in developmental and differentiation processes will need to be elucidated and utilized. Differentiation cascades will also have to be defined in order to generate cells that are appropriately restricted in developmental potential. For pluripotent stem cell-derived tissue engineering to be practical, the above challenges must be overcome, but will also be augmented by difficult issues of tissue architecture. In a more practical, but no less important, sense, we must develop improved techniques for pluripotent stem cell colony disaggregation and growth.

Pluripotent stem cell research faces political challenges as well, due to the origin of the cells. Federal funding of the research, which is currently being debated, would be necessary to realize rapidly the theraputic potential of these cells. Our current system of peer review and governmental oversight of scientific research funding has served scientific research well. The open communication and sharing of technologies and reagents that are possible with governmental funding stimulates and encourages vigorous research. Oversight provided by the federal funding agencies can serve to set guidelines for the pursuance of the work. There is also a place for the biotechnology industry in this equation, with their ability to focus considerable financial resources on practical and profitable approaches.

Significant progress has been made in mouse and human pluripotent stem cell growth, differentiation, and transplantation. The challenges ahead are difficult but will eventually provide rewards by uncovering developmental mechanisms and providing cells and tissues for transplantation.

REFERENCES

Abe, K., Niwa, H., Iwase, K., Takiguchi, M., Mori, M., Abe, S.-I., Abe, K., and Yamura, K.-I. (1996). Endoderm-specific gene expression in embryonic stem cells differentiated to embryoid bodies. *Exp. Cell Res.* **229**, 27–34.

Andrews, P. W. (1984). Retinoic acid induces neuronal differentiation of a cloned human embryonal carcinoma cell line *in vitro. Dev. Biol.* **103**, 285–293.

Andrews, P. W., Banting, G., Damjanov, I., Arnaud, D., and Avner, P. (1984). Three monoclonal antibodies defining distinct differentiation antigens associated with different high molecular weight polypeptides on the surface of human embryonal carcinoma cells. *Hybridoma* **3**, 347–361.

Andrews, P. W., Gonczol, E., Plotkin, S. A., Dignazio, M., and Oosterhuis, J. W. (1986). Differentiation of TERA-2 human embryonal carcinoma cells into neurons and HCMV permissive cells. Induction by agents other than retinoic acid. *Differentiation (Berlin)* **31**, 119–126.

Bain, G., Kitchens, D., Yao, M., Huettner, J., and Gottlieb, D. (1995). Embryonic stem cells express neuronal properties *in vitro*. *Dev. Biol.* **168,** 342–357.

Biesecker, L., and Emerson, S. (1993). Interleukin-6 is a component of human umbilical cord serum and stimulates hematopoiesis in embryonic stem cells *in vitro*. *Exp. Hematol.* **21,** 774–778.

Bjornson, C. R., Rietze, R. L., Reynolds, B. A., Magli, M. C., and Vescovi, A. L. (1999). Turning brain into blood: A hematopoietic fate adopted by adult neural stem cells *in vivo*. *Science* **283,** 534–537.

Borlongan, C., Tajima, Y., Trojanowski, J., Lee, V.-Y., and Sanberg, P. (1998). Transplantation of cryopreserved human embryonal carcinoma-derived neurons (NT2N cells) promotes functional recovery in ischemic rats. *Exp. Neurol.* **149,** 310–321.

Caplan, A. I. (1991). Mesenchymal stem cells. *J. Orthop. Res.* **9,** 641–650.

Civin, C. I., Strauss, L. C., Brovall, C., Fackler, M. J., Schwartz, J. F., and Shaper, J. H. (1984). Antigenic analysis of hematopoiesis. III. A hematopoietic progenitor cell surface antigen defined by a monoclonal antibody reaised against KG-1a cells. *J. Immunol.* **133,** 157–165.

Cosgrove, D., Gray, D., Dierich, A., Kaufman, J., Lemeur, M., Benoist, C., and Mathis, D. (1991). Mice lacking MHC class II molecules. *Cell (Cambridge, Mass.)* **66,** 1051–1066.

Dani, C., Smith, A. G., Dessolin, S., Leroy, P., Staccini, L., Villageois, P., Darimont, C., and Ailhaud, G. (1997). Differentiation of embryonic stem cells into adipocytes *in vitro*. *J. Cell Sci.* **110,** 1279–1285.

Dekel, I., Magal, Y., Pearson-White, S., Emerson, C. P., and Shani, M. (1992). Conditional conversion of ES cells to skeletal muscle by an exogenous MyoD1 gene. *New Biol.* **4,** 217–224.

Dinsmore, J., Ratliff, J., Deacon, T., Pakzaban, P., Jacoby, D., Galpern, W., and Isacson, O. (1996). Embryonic stem cells differentiated in vitro as a novel source of cells for transplantation. *Cell Transplant.* **5,** 131–143.

Doetschman, T., Williams, P., and Maeda, N. (1988). Establishment of hamster blastocyst-derived embryonic stem (ES) cells. *Dev. Biol.* **127,** 224–227.

Doetschman, T., Shull, M., Kier, A., and Ciffin, J. (1993). Embryonic stem cell model systems for vascular morphogenesis and cardiac disorders. *Hypertension* **22,** 618–629.

Doetschman, T. C., Eistetter, H., Katz, M., Schmidt, W., and Kemler, R. (1985). The *in vitro* development of blastocyst-derived embryonic stem cell lines: Formation of visceral yolk sac, blood islands and myocardium. *J. Embryol. Exp. Morphol.* **87,** 27–45.

Drab, M., Haller, H., Bychkov, R., Erdmann, B., Lindschau, C., Haase, H., Morano, I., Luft, F. C., and Wobus, A. M. (1997). From totipotent embryonic stem cells to spontaneously contracting smooth muscle cells: A retinoic acid and db-cAMP *in vitro* differentiation model. *FASEB J.* **11,** 905–915.

Duncan, S. A., Nagy, A., and Chan, W. (1997). Murine gastrulation requires HNF-4 regulated gene expression in the visceral endoderm: Tetraploid rescue of Hnf-4($-/-$) embryos. *Development (Cambridge, UK)* **124,** 279–287.

Duncan, S. A., Navas, M. A., Dufort, D., Rossant, J., and Stoffel, M. (1998). Regulation of a transcription factor network required for differentiation and metabolism. *Science* **281,** 692–695.

Edwards, S. A., and Adamson, E. D. (1986). Induction of c-fos and AFP expression in a differentiating teratocarcinoma cell line. *Exp. Cell Res.* **165,** 473–480.

Evans, M. J., and Kaufman, M. H. (1981). Establishment in culture of pluripotential cells from mouse embryos. *Nature (London)* **292,** 154–156.

Fraichard, A., Chassande, O., Bilbaut, G., Dehay, C., Savatier, P., and Samrut, J. (1995). *In vitro* differentiation of embryonic stem cells into glial cells and functional neurons. *J. Cell Sci.* **108,** 3161–3188.

Grusby, M. J., Johnson, R. S., Papaioannou, V. E., and Glimcher, L. H. (1991). Depletion of CD4$^+$ T cells in major histocompability complex class II-deficient mice. *Science* **253,** 1417–1420.

Grusby, M., Auchincloss, j., H., Lee, R., Johnson, R., Spencer, J., Zijlstra, M., Jaenisch, R., Papaioannou, V., and Glimcher, L. (1993). Mice lacking major histocompatibility complex class I and class II molecules. *Proc. Natl. Acad. Sci. U.S.A.* **90,** 3913–3917.

Habara-Ohkubo, A. (1996). Differentiation of beating cardiac muscle cells from a derivative of P19 embryonal carcinoma cells. *Cell Struct. Funct.* **21,** 101–110.

Hirashima, M., Kataoka, H., Nishikawa, S., and Matsuyoshi, N. (1999). Maturation of embryonic stem cells into endothelial cells in an *in vitro* model of vasculogenesis. *Blood* **93,** 1253–1263.

Hogan, B. L., Taylor, A., and Adamson, E. (1981). Cell interactions modulate embryonal carcinoma cell differentiation into parietal or visceral endoderm. *Nature (London)* **291,** 235–237.

Johansson, C. B., Momma, S., Clarke, D. L., Risling, M., Lendahl, U., and Frisen, J. (1999). Identification of a neural stem cell in the adult mammalian central nervous system. *Cell (Cambridge, Mass.)* **96,** 25–34.

Kannagi, R., Cochran, N., Ishigami, F., Hakomori, S., Andrews, P., Knowles, B., and Solter, D. (1983). Stage-specific embryonic antigens (SSEA-3 and -4) are epitopes of a unique globo-series ganglioside isolated from human teratocarcinoma cells. *EMBO J.* **2,** 2355–2361.

Keller, G., Kennedy, M., Papayannopoulou, T., and Wiles, M. V. (1993). Hematopoietic Commitment during embryonic stem cell differentiation in culture. *Mol. Cell. Biol.* **13,** 473–486.

Keller, G. M. (1995). *In vitro* differentiation of embryonic stem cells. *Curr. Opin. Cell. Biol.* **7,** 862–869.

Kennedy, M., Firpo, M., Choi, K., Wall, C., Robertson, S., Kabrun, N., and Keller, G. (1997). A common precursor for primitive erythropoiesis and definitive haematopoiesis. *Nature (London)* **386,** 488–493.

Klug, M. G., Soonpaa, M. H., and Field, L. J. (1995). DNA synthesis and multinucleation in embryonic stem cell-derived cardiomyocytes. *Am. J. Physiol.* **269,** H1913–H1921.

Klug, M. G., Soonpaa, M. H., Koh, G. Y., and Field, L. J. (1996). Genetically selected cardiomyocytes from differentiating embronic stem cells form stable intracardiac grafts. *J. Clin. Invest.* **98,** 216–224.

Labosky, P., Barlow, D., and Hogan, B. (1994). Mouse embryonic germ (EG) cell lines: Transmission through the germline and differences in the methylation imprint of insulin-like growth factor 2 receptor (IGF2r) gene compared with embryonic stem (ES) cell lines. *Development (Cambridge, UK)* **120**, 3197–3204.

Lang, F., Mazaurti-Sticker, M., and Machieke, A. (1989). States of developmental commitment of a mouse embryonal carcinoma cell line differentiated along a neuronal pathway. *J. Cell Biol.* **109**, 2481–2493.

Li, M., Pevny, L., Lovell-Badge, R., and Smith, A. (1998). Generation of purified neural precursors from embryonic stem cells by lineage selection. *Curr. Biol.* **8**, 971–974.

Martin, G. R. (1981). Isolation of a pluripotent cell line from early mouse embryos cultured in media conditioned by teratocarcinoma stem cells. *Proc. Natl. Acad. Sci. U.S.A.* **78**, 7634–7638.

Matsui, Y., Toksoz, D., Nishikawa, S., Nishikawa, S., Williams, D., Zsebo, K., and Hogan, B. L. (1991). Effect of Steel factor and leukaemia inhibitory factor on murine primordial germ cells in culture. *Nature (London)* **353**, 750–752.

McBurney, M. W., Jones-Villeneuve, E. M., Edwards, M. K., and Anderson, P. J. (1982). Control of muscle and neuronal differentiation in a cultured embryonal carcinoma cell line. *Nature (London)* **299**, 165–167.

Medvinsky, A., and Dzierzak, E. (1996). Definitive hematopoiesis in autonomously initated by the AGM region. *Cell (Cambridge, Mass.)* **86**, 897–906.

Metzger, J. M., Lin, W. I., and Samuelson, L. C. (1996). Vital staining of cardiac myocytes during embryonic stem cell cardiogenesis *in vitro*. *Circ. Res.* **78**, 547–552.

Miller-Hance, W., LaCorbiere, M., Fuller, S., Evans, S., Lyons, G., Schmidt, C., Robbins, J., and Chien, K. (1993). *In vitro* chamber specification during embryonic stem cell cardiogenesis. *J. Biol. Chem.* **268**, 25244–25252.

O'Shea, K., and Gratsch, T. (1998). Noggin induces a neural phenotype in ES cells, which is antagonized by BMP-4. *Soc. Neurosci. Abstr.* **24**, 1526.

O'Shea, K., Gratsch, T., Tapscott, S., and McCormick, M. (1997). Neuronal differentiation of embryonic stem (ES) cells constituitively expressing neuroD2 or neuroD3. *Soc. Neurosci. Abstr.* **23**, 1144.

Pain, B., Clark, M. E., Shen, M., Nakazawa, H., Sakurai, M., Samarut, J., and Etches, R. J. (1996). Long-term *in vitro* culture and characterization of avian embryonic stem cells with multiple morphogenetic potentialities. *Development (Cambridge, UK)* **122**, 2339–2348.

Palacios, R., Golunski, E., and Samaridis, J. (1995). *In vitro* generation of hematopoietic stem cells from an embryonic stem cell line. *Proc. Natl. Acad. Sci. U.S.A.* **92**, 7530–7534.

Petersen, B. E., Bowen, W. C., Patrene, K. D., Mars, W. M., Sullivan, A. K., Murase, N., Boggs, S. S. Greenberger, J. S., and Goff, J. P. (1999). Bone marrow as a potential source of hepatic oval cells. *Science* **284**, 1168–1170.

Peyron, A. (1939). *Bull. Assoc. Fr. Etude Cancer* **28**, 658–681.

Pittenger, M. F., Mackay, A. M., Beck, S. C., Jaiswal, R. K., Douglas, R., Mosca, J. D., Moorman, M. A., Simonetti, D. W., Craig, S., and Marshak, D. R. (1999). Multilineage potential of adult human mesenchymal stem cells. *Science* **284**, 143–147.

Pleasure, S. J., and Lee, V. M. (1993). NTera 2 cells: A human cell line which displays characteristics expected of a human committed neuronal progenitor cell. *J. Neurosci. Res.* **35**, 585–602.

Resnick, J. L., Bixler, L. S., Cheng, L., and Donovan, P. J. (1992). Long-term proliferation of mouse primordial germ cells in culture. *Nature (London)* **359**, 550–551.

Robbins, J., Gulick, J., Sanchez, A., Howles, P., and Doetschman, T. (1990). Mouse embryonic stem cells express the cardiac myosin heavy chain genes during development *in vitro*. *J. Biol. Chem.* **265**, 11905–11909.

Rohwedel, J., Maltsev, V., Bober, E., Arnold, H.-H., Hescheler, J., and Wobus, A. (1994). Muscle cell differentiation of embryonic stem cells reflects myogensis *in vivo*: Developmentally regulated expression of myogenic determination genes and functional expression of ionic currents. *Dev. Biol.* **164**, 87–101.

Rohwedel, J., Horak, V., Hebrok, M., Fuchtbauer, E. M., and Wobus, A. M. (1995). M-twist expression inhibits mouse embryonic stem cell-derived myogenic differentiation *in vitro*. *Exp. Cell Res.* **220**, 92–100.

Rohwedel, J., Sehlmeyer, U., Shan, J., Meister, A., and Wobus, A. (1996). Primordial germ cell-derived mouse embryonic germ (EG) cells *in vitro* resemble undifferentiated stem cells with respect to differentiation capacity and cell cycle distribution. *Cell Biol. Int. Rep.* **20**, 579–587.

Shamblott, M. J., Axelman, J., Wang, S., Bugg, E. M., Littlefield, J. W., Donovan, P. J., Blumenthal, P. D., Huggins, G. R., and Gearhart, J. D. (1998). Derivation of pluripotent stem cells from cultured human primordial germ cells. *Proc. Natl. Acad. Sci. U.S.A.* **95**, 13726–13731.

Shani, M., Faerman, A., Emerson, C. P., Pearson-White, S., Dekel, I., and Magal, Y. (1992). The consequences of a constitutive expression of MyoD1 in ES cells and mouse embryos. *Symp. Soc. Exp. Biol.* **46**, 19–36.

Shim, H., Gutierrez-Adan, A., Chen, L., BonDurant, R., Behboodi, E., and Anderson, G. (1997). Isolation of pluripotent stem cells from cultured porcine primordial germ cells. *Biol. Reprod.* **57**, 1089–1095.

Snyder, E. Y., Taylor, R. M., and Wolfe, J. H. (1995). Neural progenitor cell engraftment corrects lysosomal storage throughout the MPS VII mouse brain. *Nature (London)* **374**, 367–370.

Sobis, H., Verstuyf, A., and Vandeputte, M. (1991). Endodermal origin of yolk-sac-derived teratomas. *Development (Cambridge, UK)* **111**, 75–78.

Sobis, H., Verstuyf, A., and Vandeputte, M. (1993). Visceral yolk sac-derived tumors. *Int. J. Dev. Biol.* **37**, 155–168.

Solter, D., and Gearhart, J. (1999). Putting stem cells to work. *Science* **283**, 1468–1470.

Solter, D., and Knowles, B. (1978). Monoclonal antibody defining a stage-specific mouse embryonic antigen (SSEA-1). *Proc. Natl. Acad. Sci. U.S.A.* **75**, 5565–5569.

Stevens, L. C. (1958). Studies on transplantable testicular teratomas of Strain 129 mice. *J. Natl. Cancer Inst. (U.S.)* **20**, 1257–1270.

Stevens, L. C. (1959). Embryology of testicular teratomas in Strain 129 mice. *J. Natl. Cancer Inst. (U.S.)* **23**, 1249–1295.

Stevens, L. C. (1970). The development of transplantable teratocarcinomas from intratesticular grafts of pre- and postimplantation mouse embryos. *Dev. Biol.* **21**, 364–382.

Stewart, C., Gadi, I., and Bhatt, H. (1994). Stem cells from primordial germ cells can reenter the germ line. *Dev. Biol.* **161**, 626–628.

Sukoyan, M. A., Vatolin, S. Y., Golubitsa, A. N., Zhelezova, A. I., Semenova, L. A., and Serov, O. L. (1993). Embryonic stem cells derived from morulae, inner cell mass, and blastocysts of mink: Comparisons of their pluripotencies. *Mol. Reprod. Dev.* **36**, 148–158.

Tajbakhsh, S., Vivarelli, E., Cusella-De Angelis, G., Rocancourt, D., Buckingham, M., and Cossu, G. (1994). A population of myogenic cells derived from the mouse neural tube. *Neuron* **13**, 813–821.

Thomson, J. A., Kalishman, J., Golos, T. G., Durning, M., Harris, C. P., Becker, R. A., and Hearn, J. P. (1995). Isolation of a primate embryonic stem cell line. *Proc. Natl. Acad. Sci. U.S.A.* **92**, 7844–7848.

Thomson, J. A., Kalishman, J., Golos, T. G., Durning, M., Harris, C. P., and Hearn, J. P. (1996). Pluripotent cell lines derived from common marmoset (*Callithirx jacchus*) blastocysts. *Biol. Reprod.* **55**, 254–259.

Thomson, J. A., Itskovitz-Eldor, J., Shapiro, S. S., Waknitz, M. A., Swiergiel, J. J., Marshall, V. S., and Jones, J. M. (1998). Embryonic stem cell lines derived from human blastocysts. *Science* **282**, 1145–1147; published erratum: p. 1827.

Trojanowski, J., Kleppner, S., Hartley, R., Miyazono, M., Fraser, N., Kesari, S., and Lee, V.-Y. (1997). Transfectable and transplantable postmitotic human neurons: A potential "platform" for gene therapy of nervous system diseases. *Exp. Neurol.* **144**, 92–97.

Wang, R., Clark, R., and Bautch, V. (1992). Embryonic stem cell-derived cystic embryoid bodies form vascular channels: An *in vitro* model of blood vessel development. *Development (Cambridge, UK)* **114**, 303–316.

Westfall, M., Paysk, K., Yule, D., Samuelson, L., and Metzger, J. (1997). Ultrastructure and cell-cell coupling of cardiac myocytes differentiating in embryonic stem cell cultures. *Cell Motil. Cytoskel.* **36**, 43–54.

Wheeler, M. B. (1994). Development and validation of swine embryonic stem cells: A review. *Reprod. Fertil. Dev.* **6**, 563–568.

Wiles, M. V., and Keller, G. (1991). Multiple hematopoietic lineages develop from embryonic stem (ES) cells in culture. *Development (Cambridge, UK)* **111**, 259–267.

Wilmut, I., Schnieke, A. E., McWhir, J., Kind, A. J., and Campbell, K. H. (1997). Viable offspring derived from fetal and adult mammalian cells. *Nature (London)* **385**, 810–813; published erratum: **386**, 200.

Wobus, A. M., Grosse, R., and Schoneich, J. (1988). Specific effects of nerve growth factor on the differentiation pattern of mouse embryonic stem cells *in vitro. Biomed. Biochim. Acta* **47**, 965–973.

Yandava, B. D., Billinghurst, L. L., and Snyder, E. Y. (1999). "Global" cell replacement is feasible via neural stem cell transplantation: Evidence from the dysmyelinated shiverer mouse brain. *Proc. Natl. Acad. Sci. U.S.A.* **96**, 7029–7034.

Zhang, X. J., Tsung, H. C., Caen, J. P., Li, X. L., Yao, Z., and Han, Z. C. (1998). Vasculogenesis from embryonic bodies of murine embryonic stem cells transfected by Tgf-beta1 gene. *Endothelium* **6**, 95–106.

Zijlstra, M., Li, E., Sajjadi, F., Subramani, S., and Jaenisch, R. (1989). Germ-line transmission of a disrupted beta 2-microglobulin gene produced by homologous recombination in embryonic stem cells. *Nature (London)* **342**, 435–438.

PART VIII: GENE THERAPY

CHAPTER 30

GENE-BASED THERAPEUTICS

Lee G. Fradkin, J. Dezz Ropp, and John F. Warner

INTRODUCTION

Many inherited and acquired disease processes, characterized at the genetic level, are known to be due to protein deficiencies or defects. Potential gene-based therapeutic strategies can now be designed to address these diseases. The initial premise of gene therapy was to correct heritable genetic disorders through gene replacement. However, the true potential of gene-based therapies can now be viewed in a broader context of biologics delivery for applications involving cancer, vaccines, genetic and metabolic disorders, viral infections, as well as cardiovascular and other organ-based interventions. The basic goal of gene-based therapeutics is the safe and effective delivery of genes to the disease site with subsequent expression, under physiologic conditions, of a therapeutic protein.

Biotechnology, in existence for approximately two decades, has traditionally focused on the production of therapeutic proteins. This industry was formed on the concept that synthetic (i.e., *in vitro*) manufacture of therapeutic proteins offers safety, control, and cost savings over isolation from natural sources. The diverse biologic activities provided by therapeutic proteins present great opportunities for the treatment of disease. Despite the immense advances in this field, the individual qualities of recombinant therapeutic proteins also present obstacles in their development, manufacture, and utilization. The concept of gene-based therapeutics arises from a desire to minimize the complexity presented by utilizing protein as the delivered therapeutic and to capitalize, instead, on the manufacture and delivery of a more generic molecule, namely, DNA encoding the therapeutic protein. Although the protein is still the therapeutic entity per se, delivery of genes requires the development of a single manufacturing process for the gene delivery vehicle that should be generally applicable to similar vehicles bearing different genes. As with any novel technology, however, the theory of gene-based therapeutics is more straightforward than its application. This chapter will identify advances and current shortcomings in the technology, with highlights from recent literature. Throughout the text the reader is referred to other recent specific reviews to avoid their repetition. Rather than a comprehensive review of the literature, the focus of this chapter is on those areas in which advancement is required to bring gene-based therapeutics to fruition.

The large number of genes for which the sequence has been determined through recent advances in genomics research presents a staggering potential for the future of gene-based therapeutics. Further work on determining the functional activities of many of these new gene candidates is necessary; however, in all cases, well-understood, robust gene delivery systems are required. Despite the infancy of gene-based therapeutics, over 300 clinical trials have been initiated in the past decade. The maturation of this therapeutic approach will come about through the results of these clinical trials, preclinical experimentation, and basic research in the fields of virology, cell biology, and molecular biology.

The primary considerations for successful gene transfer technologies are manufacture of the gene delivery vehicle, its delivery to the target tissue and cell surface, cellular internalization, intracellular trafficking, nuclear uptake, and functional gene expression with appropriate, controlled levels of duration. In addition, safety of the delivery system, the expressed gene product, and associated immune responses are potential issues requiring evaluation. As the field of gene-based therapeutics establishes a more solid footing, in part through addressing the issues discussed below, the full potential of gene-based therapeutics in the treatment of human disease will be realized.

GENE DELIVERY

Although the various viral- and plasmid-based gene delivery vehicles differ in composition and function, their ultimate goal is the production of a therapeutic protein in sufficient quantity at the appropriate site to elicit the desired biological responses. Viral delivery systems consist of modified, usually nonreplicating, viral genomes carrying a specific transgene. Plasmid-based gene delivery systems utilize a variety of agents (lipids, polymers, peptides) complexed with DNA encoding a transgene or utilize the "naked" DNA alone (reviewed in Maurer *et al.,* 1999; Rolland, 1998, and references therein). The most commonly used complexing agents are cationic liposomes and condensing agents such as poly(ethyleneimine) (PEI) and poly(L-lysine). Gene delivery vehicles must successfully traverse multiple barriers as they transit from the site of administration to their final destination, the nucleus of target cells (Mahato *et al.,* 1997a,b). Layered over these delivery barriers is the added complexity that each class of vehicle (adenoviral, retroviral, plasmid-based, etc.) has its own advantages and disadvantages that affect these potentially rate-limiting steps. To generate functional gene-based therapeutics, there must be a clear understanding of the requirements pertinent to the route of administration and the capabilities (as well as limitations) of the chosen delivery vehicle.

Although the barriers encountered along the gene delivery development path differ greatly depending on the route of administration, issues of gene delivery can be dissected temporally and spatially into three steps: (1) extracellular trafficking, (2) uptake into target cells, and (3) intracellular trafficking. Although the relative needs will vary by application, all delivery vehicles will require the ability to function at each of these steps. For instance, highly efficient DNA transport through the cytoplasm and into the nucleus is of little consequence if the delivery vehicle never reaches the target cells. Work in all categories of gene-based delivery has focused, therefore, more on a mechanistic understanding of delivery and the requisite improvements necessary to overcome these delivery barriers (Deshmukh and Huang, 1997; Lee and Huang, 1997; Meyer *et al.,* 1997; Onodera *et al.,* 1999; Palsson and Andreadis, 1997). Consequently, a more thorough understanding of current gene delivery system capabilities has begun to emerge. In this section we address some of the considerations regarding current gene delivery systems and outline some strategies employed to overcome these barriers.

EXTRACELLULAR TRAFFICKING

On administration to the patient, gene delivery vehicles will encounter a physiologic milieu quite different from that presented under *in vitro* conditions. For example, systemic intravenous (IV) administration, perhaps the most challenging route of administration, presents several extracellular trafficking barriers that are a major limitation for this delivery route. Delivery systems will require stability within a complex array of serum proteins as well as an ability to avoid clearance by the phagocytic cells of the reticuloendothelial system (RES). Immune clearance presents further considerations more specific for viral systems (Anderson, 1998; Kuriyama *et al.,* 1998; Worgall *et al.,* 1997). Poly(ethylene glycol) (PEG) and lipids conjugated to adenovirus abrogate the effect of neutralizing antibodies *in vitro* and *in vivo* (Chillon *et al.,* 1998; O'Riordan *et al.,* 1999). In the case of plasmid-based systems, serum affects the biophysical and biochemical properties of lipid–DNA complexes by altering both size and charge, leading to an increase in complex disintegration, DNA release, and ultimately degradation (S. Li *et al.,* 1999). It is likely that the delivery vehicles that arrive at the target site could differ substantially from that originally formulated and administered to the patient. The composition of lipids in plasmid-based delivery systems affects the transfection efficiency (S. Li *et al.,* 1998; Zhao *et al.,* 1997). Data suggest the lipid composition is also an important factor in the recruitment of serum proteins (S. Li *et al.,* 1998, 1999). There is likely a direct relationship between the serum stability of lipid–DNA-based systems and their relative transfection efficiency (S. Li *et al.,* 1999). In general, there is a clear need for understanding the subtle balance of aggregation, disassembly, and DNA degradation that are critical for gene delivery.

Although the interaction of plasmid delivery systems with serum components is unavoidable following IV administration, the ultimate effect in practical terms is less obvious. Complement depletion in mice prior to IV injection of cationic lipid–DNA complexes suggests that, although the complexes interact with serum complement proteins, the interactions do not influence the transfection efficiency or systemic distribution of the complexes (Barron *et al.*, 1998). Clearly, we need to develop a better understanding of the array of interactions delivery vehicles encounter in order to optimize their IV administration. Serum interaction and stability studies are obviously an important part of developing new plasmid-based delivery systems and further exploration may lead to predictive models for *in vivo* activity (Goula *et al.*, 1998; Vitiello *et al.*, 1998: Yang and Huang, 1997, 1998).

Retroviruses are also sensitive to opsonization and inactivation by serum components following systemic delivery (Agrawal *et al.*, 1999; Rother *et al.*, 1995; Takeuchi *et al.*, 1994, and references therein). Furthermore, packaging cell line origin can impact immune response activation and profoundly affect viral clearance and stability (reviewed in Anderson, 1998). Strategies designed to minimize immune response activation by utilizing alternate packaging cell lines are currently ongoing (Forestell *et al.*, 1997; Pensiero *et al.*, 1996; Rigg *et al.*, 1996). These issues of *in vivo* stability are partially responsible for limiting the primary usage of retroviruses in humans to *ex vivo* applications. However, direct *in vivo* intratumoral administration of retroviral vectors has been accomplished using the p53 gene in the treatment of non-small-cell lung cancer (Roth *et al.*, 1996) and the interferon-γ (IFN-γ) gene for metastatic melanoma (Nemunaitis *et al.*, 1999). Overall aspects of retroviral delivery have been reviewed (Anderson, 1998; Onodera *et al.*, 1999; Robbins *et al.*, 1998).

Targeting a specific disease requires knowledge of the appropriate tissue and cell types necessary to express the therapeutic protein as well as an understanding of the delivery system and route of delivery that will achieve the clinical goal. Biodistribution studies have provided a foundation for IV delivery of lipid–DNA complexes (Mahato *et al.*, 1998; McLean *et al.*, 1997; Niven *et al.*, 1998; Osaka *et al.*, 1996). A detailed comparison of IV versus intratracheal (IT) plasmid-based delivery of the cystic fibrosis transductance regulator (CFTR) gene (Griesenbach *et al.*, 1998) provides an elegant demonstration of differential uptake and expression based on the delivery route. Following IV administration of a cationic lipid-based delivery system, DNA was delivered to distal lung in the alveolar region, including alveolar type II epithelial cells, whereas, following IT administration, DNA was found in the epithelial lining of the bronchioles (i.e., clara cells) (Griesenbach *et al.*, 1998).

Differences in gene expression profiles based on delivery route highlight another major issue confounding development of gene-based delivery vehicles, namely, the lack of *in vitro* to *in vivo* correlation with available models. The relevance of the *in vitro* model to the *in vivo* situation is essential. As in the case of differences in expression pattern for IT versus IV delivery, the extracellular barriers differ immensely for each route of administration and ideally should be reflected in the model. The extracellular barriers presented by blood components for IV delivery are quite different from those presented by mucus, which is relevant for IT or aerosol lung delivery. The latter is likely the delivery route of choice for the CFTR gene (McDonald *et al.*, 1998), and the CF sputum presents its own unique delivery barriers (Kitson *et al.*, 1999; Stern *et al.*, 1998).

In an attempt to quantitate the barriers for airway delivery, investigators (Kitson *et al.*, 1999) have developed an *ex vivo* sheep tracheal epithelium model to analyze directly lipid- and adenoviral-based extra- and intracellular barriers. Overall, the results suggest that extracellular barriers are a major impediment to pulmonary delivery for both lipid-based and adenoviral systems. Normal mucus is a barrier for plasmid-based gene delivery, with a 25-fold increase in expression following its removal; no effect of normal mucus removal was observed for adenovirus infectivity (Kitson *et al.*, 1999). Expression following both adenovirus and plasmid-based gene delivery was significantly reduced in the presence of fresh CF sputum. These results suggested that the extracellular and plasma membrane barrier issues should be the primary focus for increased delivery efficiency of airway tissue. Once lipid–DNA complexes have traversed the apical surface of the respiratory epithelial membrane, however, it was suggested that the intracellular barriers are likely similar for *in vivo* and *in vitro* systems.

Previous attempts have been made to overcome the extracellular barrier presented by CF sputum by applying surface active agents (Parsons *et al.*, 1998) or DNase (Stern *et al.*, 1998) to the airway passages for enhanced transfection by adenoviral and lipidic delivery systems. Such "assisted delivery" by coadministration of an agent to provide a more hospitable environment for the delivery system is an attractive option for airway delivery due to ease of administration. The combined use of traditional pharmaceutics or biologics and gene-based therapeutics could potentially provide a greater benefit in such applications.

Several delivery strategies have been employed to avoid the potentially deleterious extracellular barriers of IV and IT delivery through direct administration of the gene delivery vehicle to the tissue of interest. Examples include intratumoral injection (e.g., Nomura *et al.,* 1999), intramuscular (IM) injection (e.g., Kessler *et al.,* 1996; Marshall and Leiden, 1998; Monahan *et al.,* 1998), and particle bombardment (i.e., gene gun) (e.g., Mahvi *et al.,* 1997; Rakhmilevich *et al.,* 1999). Delivery by the IM route can be accompanied by electroporation to enhance transfection efficiency (Mir *et al.,* 1999). Moreover, other invasive direct application strategies such as direct instillation of adenovirus within the pericardial sac are in development (Lamping *et al.,* 1997; March *et al.,* 1999). While minimizing the extracellular barriers the delivery vehicles are required to overcome, the downstream barriers beginning with cell uptake still exist.

CELLULAR UPTAKE

The identification of the coxsackie and adenovirus receptor (CAR) (Bergelson *et al.,* 1997) as the receptor for adenovirus serotypes A, C, D, E, and F (Roelvink *et al.,* 1998) has furthered our understanding of the requirements for adenovirus infection. Decisions for adenoviral gene delivery strategies can now be based on a critical assessment of receptor levels in target tissues. The importance of receptor levels was highlighted by analyses demonstrating the variability seen in gene expression levels (Y. Li *et al.,* 1999) and distribution of adenovirus receptors (Hemmi *et al.,* 1998; Nalbantoglu *et al.,* 1999; Walters *et al.,* 1999). Variability in receptor availability could profoundly affect the outcome of adenovirus-based gene delivery and therefore should be a primary consideration in developing gene delivery strategies.

The increasing ability to overcome CAR deficiency by modification of adenovirus fiber proteins to provide alternative cell-binding epitopes (i.e., retargeting viral infectivity) should help to alleviate limitations of viral delivery based solely on receptor availability (Hidaka *et al.,* 1999; Sandig and Strauss, 1996). Retargeting strategies, in addition to potentially overcoming the lack of a receptor, provide an opportunity to generate the desired specificity for target cells. Without specificity, all cells possessing the CAR will be transduced by adenovirus. This leads to potential toxicities due to the increased dosing needed to overcome this lack of specific targeting as well as potentially dangerous side effects due to the expression of the therapeutic gene in an inappropriate cell population. Specific targeting is especially desired for the delivery of therapeutic genes preferentially into tumor cells (reviewed in Bilbao *et al.,* 1998). Several strategies aimed at eliminating natural viral tropism while providing novel interaction, typically through antibodies (or fragments thereof), have shown promise in proof-of-principle studies for adenovirus as well as adeno-associated virus (AAV) (Bartlett *et al.,* 1999; Douglas and Curiel, 1997; Goldman *et al.,* 1997; Miller *et al.,* 1998; Rogers *et al.,* 1997). These strategies also effect an enhancement in transfection in cases in which the CAR is limiting, and consequently reduce toxicity associated with the high levels of adenovirus required to achieve a therapeutic effect (Rancourt *et al.,* 1998). As with any secondary modification, manufacture of these viral–antibody hybrids will present yet another challenge to overcome before the full utility of this approach is recognized. Similarly, genetic manipulation of the critical fiber knob, by adding alternate binding epitopes directly during viral production, presents its own manufacturing concerns related to producer cell line specificity (Dmitriev *et al.,* 1998; Douglas and Curiel, 1997; Krasnykh *et al.,* 1998). In addition, potential immune recognition will be a concern for any retargeting strategy.

Retroviral vectors, based on the prototype murine leukemia virus (MuLV), possess a natural tropism based on receptor levels as well. Cells competent for retroviral transduction present the natural MuLV (amphotropic) receptor and are actively dividing. Although cell division is required for retroviral transduction (Miller *et al.,* 1990), rapid proliferation alone is not sufficient for transduction efficiency. Transduction may also be limited at the receptor level (Bauer *et al.,* 1995; Movassagh *et al.,* 1998; Orlic *et al.,* 1996; Sabatino *et al.,* 1997). As with adenovirus, there is a requirement for high-level expression of the appropriate viral receptor. Increasing the *Pit2* receptor level from 18,000 to 150,000 per cell increased the MuLV transduction efficiency from 10 to 50% in rat 208F embryo fibroblasts (Kurre *et al.,* 1999). Retargeting strategies for retroviral gene delivery have been the subject of reviews (Anderson, 1998; Gunzburg and Salmons, 1996; Sandig and Strauss, 1996).

In addition to the more standard reengineering of genetic and biochemical functionalities of retrovirus, novel analyses have emerged that focus more on the physicochemical forces involved in retroviral–cell interaction (Palsson and Andreadis, 1997). This study explored the physicochemical forces that determine the binding of the retroviral vector to the target cell and the kinetic interplay between cell cycle and retroviral life cycle, events that determine the intracellular fate of the virus and ultimately constrain the efficiency of the gene transfer process (Palsson and Andreadis, 1997).

Plasmid-based systems rely on ionic charge-based interactions for initial cell binding and subsequent endocytosis (Hope *et al.,* 1998, and references therein). Direct membrane fusion is the likely alternate route for cationic lipid–DNA-based systems; however, the relative contribution of these two pathways to nuclear delivery is unknown at this time. The differences in serum interactions and transfection efficiency from one lipid formulation to another make it difficult to generalize regarding uptake mechanisms (S. Li *et al.,* 1998, 1999; Zhao *et al.,* 1997). Furthermore, polymer-based systems display behaviors that differ from lipid systems. Consequently, much of the plasmid-based formulation technology development to date has relied on empirical assessments. Future formulation development will require evaluation in well-defined therapeutic models.

Targeting strategies have been employed for plasmid-based systems in an attempt to increase efficiency and specificity of delivery (e.g., Cristiano, 1998; Erbacher *et al.,* 1999; Harbottle *et al.,* 1998; Hung *et al.,* 1998; Kawakami *et al.,* 1998; Mohr *et al.,* 1999). Although the addition of targeting elements or other protein components to plasmid-based systems is an attempt to mimic the positive qualities of viral gene delivery systems, some of the negative consequences, such as increased immunogenicity, may become more apparent. Overall, a variety of strategies and considerations exist to improve the cellular uptake efficiency of gene-based therapeutics.

INTRACELLULAR TRAFFICKING

The viral systems have naturally evolved highly efficient intracellular trafficking mechanisms and for this reason have been exploited for gene delivery with minimal attempts to modify their inherent functionality. Detailed understanding of the intracellular trafficking pathway for adenovirus has continued to evolve, however, in order to capitalize fully on this gene delivery workhorse (Leopold *et al.,* 1998; E. Li *et al.,* 1998; Miyazawa *et al.,* 1999; Suomalainen *et al.,* 1999; K. Wang *et al.,* 1998). Understanding the multiple functions afforded by the coat proteins of adenovirus in the intracellular phase of its infection cycle (Greber and Kasamatsu, 1996; Greber *et al.,* 1993, 1994) has provided a foundation on which to build plasmid-based delivery systems. Engineering of plasmid-based systems has focused on overcoming intracellular trafficking barriers such as endosomal entrapment and nuclear uptake (reviewed in Meyer *et al.,* 1997; Wagner, 1998).

It is difficult to generalize regarding the intracellular trafficking barriers and approaches for plasmid-based systems due to the multitude of lipids, polymers, and peptides employed. Reviews (Hope *et al.,* 1998; Meyer *et al.,* 1997) have covered this subject, and therefore the present discussion will be limited to common trafficking barriers (Zabner *et al.,* 1995) and strategies to overcome them. Endosomal release and nuclear uptake are the primary foci for improvement in transfection efficiency (i.e., expression) following entry into the cell for plasmid-based delivery. Decomplexation and cytoplasmic transport are perceived as the most likely secondary barriers that need to be addressed. Increased efficiency of intracellular trafficking will result in an improved therapeutic index and, consequently, allow reduced dosing of the plasmid delivery system to obtain the desired biologic response. Such improvements should also minimize potential toxicities related to high dosing.

Although endosomal entrapment may or may not practically limit expression (Brisson *et al.,* 1999; El Ouahabi *et al.,* 1997; Laurent *et al.,* 1999; Meyer *et al.,* 1997; Mizuguchi *et al.,* 1997; Zabner *et al.,* 1995), several endosomolytic agents (mostly modeled after viral elements) have been incorporated into plasmid-based systems (Baru *et al.,* 1998; Wagner, 1998). A body of evidence has begun to emerge suggesting that plasmid-based systems, analogous to retroviral systems (Miller *et al.,* 1990), require cell proliferation for successful transfection (Fasbender *et al.,* 1997; Mortimer *et al.,* 1999; Ropp, 1999; Tseng *et al.,* 1999; Wilke *et al.,* 1996). The prevailing hypothesis is that nuclear membrane breakdown during mitosis is required for uptake of plasmid DNA into the nucleus. In contrast, adenoviral vectors are able to transduce nondividing cells (Greber and Kasamatsu, 1996), suggesting that the viral genome has evolved a means to pass through the nuclear membrane.

A number of strategies to overcome the nuclear membrane are currently in development for plasmid-based systems. Incorporation of peptide nuclear localization signals (NLSs) into plasmid delivery systems to transport plasmid into the nucleus of quiescent cells has produced mixed results to date (Ludtke et al., 1999; Meyer et al., 1997; Sebestyen et al., 1998; Zanta et al., 1999). Further development of such approaches will focus on appropriate presentation of the NLSs as well as feasibility of scale-up and production. Alternatively, a nuclear transport cis-acting DNA element from SV40 may prove useful in plasmid-based systems (Dean, 1997; Graessmann et al., 1989). An approach that circumvents the need for nuclear transport utilizes cytoplasmic expression via the T7 polymerase system (reviewed in Meyer et al., 1997). Increased expression of a marker gene in mouse brain tumors (Mizuguchi et al., 1997) and direct intratumoral injection of the T7/tk gene (Chen et al., 1998) have illustrated the potential for this approach. In vivo utility of such strategies may be limited, however, due to potential immunogenicity issues related to the use of a xenogenic protein (e.g., phage T7 polymerase).

The cytoplasm of the cell presents transport and stability barriers to gene delivery independent of whether expression is nuclear or cytoplasmic. Lechardeur et al. (1999), for example, demonstrated the potential negative impact of DNA instability within the cytoplasm for plasmid-based systems. Clearly, an important role for the gene delivery agent, therefore, is protecting the stability and integrity of the nucleic acid cargo during all stages of delivery. Aside from DNA loss due to nucleases and other degradative enzymes, the cytoplasm is a viscous environment, which limits free diffusion of plasmid-sized molecules (Luby-Phelps, 1994; and reviewed in Meyer et al., 1997). Adenovirus overcomes this environment by utilizing specific endogenous cellular molecular motors to facilitate transport through the cytoplasm (Suomalainen et al., 1999). The understanding of plasmid-based transport through the cytoplasm is, however, relatively unexplored. Only recently has the potential involvement or exploitation of cytoskeletal components for enhancement of plasmid-based gene delivery been addressed (Hamm-Alvarez, 1999; Kitson et al., 1999). Understanding the intracellular trafficking of DNA from cellular uptake through delivery into the nucleus should allow increases in the efficiencies of gene transfer. Increased efficiency of delivery and expression will ultimately affect dosing regimens, therapeutic indices, and safety profiles.

PERSISTENT GENE EXPRESSION

The duration of gene expression and the impact of immunologic responses directed against the vector and/or gene product are important considerations for gene-based therapeutics. The desired level of persistence of therapeutic gene expression varies with each specific therapeutic application. Expression of rapidly acting agents with known systemic toxicities, such as cytokines, would be preferentially expressed in a controlled short-term manner. Therapeutic or prophylactic vaccination, similarly, may not benefit from long-term expression of an antigen due to the potential danger of inducing immune tolerance to an organismal or tumor epitope. A study by Ochsenbein et al. (1999) indicates, however, the need for sustained or repeated immunization in the treatment of established tumors.

Germ-line gene therapies, theoretically capable of stably directing appropriate tissue-specific expression of therapeutic proteins prior to the onset of disease, are the subject of ongoing ethical and legislative debate and are unlikely to be pursued vigorously in the near future. Short-term expression (i.e., several days to months) is within the capability of several gene delivery approaches. The following section will focus on the technical aspects of achieving sustained gene expression and begins by defining those treatments that will benefit from persistent expression.

THERAPEUTIC APPLICATIONS

Therapeutic applications requiring prolonged gene expression include both gene replacement by somatic cell gene therapy and sustained, but nonpermanent, expression of a gene product whose therapeutic window is restricted to its duration of expression. Although the gene replacement types of applications represent the "holy grail" of gene therapy, e.g., the introduction of the CFTR gene into cystic fibrosis patients to effect remission of disease, this type of treatment is unlikely to represent the first generation of significant successes in somatic cell gene therapy. There are several factors that make it unlikely that the initial successes in gene therapy will occur through replacement of a missing or defective chromosomal gene (the exception of genes encoding secreted proteins is discussed below): (1) the current gene transfer technologies may not be efficient enough to transduce a sufficiently high percentage of affected cells in a single tissue to restore function, (2) many diseases affect multiple organs and may require widespread gene transfer throughout the patient's body, (3) it is unclear whether gene replacement will reverse long-established symptoms presented by the patient, and (4) the technologies for transferring large segments of DNA capable of bearing appropriately regulated promoters are only beginning to be developed. Such regulation may be necessary to restrict expression to the desired tissues. The current lack of a safe and reproducible means by which to express stably a newly introduced gene over the human life-span by either single or multiple doses of a vector is also an important consideration for replacement-type applications.

Gene replacement approaches based on the transient expression of a secreted protein, however, are likely to come to fruition in the near future. Secreted proteins act both on transduced and nontransduced cells, therefore the requirement for a high percentage of cells to be transduced is reduced. In some cases, these diseases have been treated successfully with systemically administered human- or animal-derived protein therapeutics as well as recombinant protein therapies (reviewed in Buckel, 1996). Recombinant production of proteins has reduced reliance on naturally derived proteins for a number of therapeutics, thus alleviating the risks of associated adventitious agents. However, the relatively short duration of protein activity presents practical limitations and, hence, provides a potential need for longer duration protein production at the disease site via gene-based approaches. Although significant progress has been made toward increasing the effective circulating time of recombinant proteins through derivatization with agents such as PEG (Francis *et al.,* 1998), the effective half-lives of stabilized proteins may not allow sufficiently infrequent treatment schedules to reduce significantly the cost or patient burden of frequent administration in some cases. The use of slow-release matrix–protein complexes could be promising in this regard (e.g., Putney and Burke, 1998).

Current gene-based therapeutics technologies appear adequate to allow the moderately sustained expression and, in some cases, infrequent repetitive delivery needed to ameliorate certain genetic diseases. Although direct comparisons of the therapeutic efficacies of protein-based versus gene-based approaches have not yet been extensively pursued, there are indications that protein production from a newly introduced gene may be more physiologically relevant, and hence effective in some cases (Morsy *et al.,* 1998b; Rakhmilevich *et al.,* 1999). Among the approaches being pursued are expression of factors VIII and IX for the treatment of hemophilia (reviewed in Herzog and High, 1998), α1-Antitrypsin for the treatment of emphysema and liver cirrhosis (reviewed in Geddes and Beckles, 1995), leptin for the treatment of obesity (Morsy *et al.,* 1998b), and apo A-1 for the treatment of atherosclerosis (Benoit *et al.,* 1999, and references therein).

A second type of therapeutic applications requiring relatively persistent gene expression includes those involving the treatment of diseases not necessarily caused by the absence of a protein, but which can be treated through expression of a therapeutic protein. These therapeutic proteins are often normally expressed in tissue- and/or condition-specific manners. Use of a gene-based approach allows expression of a therapeutic protein in a controlled spatial and temporal manner. The applications of both protein- and gene-based approaches to deliver antiangiogenic factors to established tumors to effect regression by reducing their vascularization provide examples of this type of approach.

The antiangiogenic protein, angiostatin, a proteolytic fragment of plasminogen, was initially laboriously purified from the urine and serum of tumor-bearing mice (O'Reilly *et al.,* 1994), but is now produced as a recombinant protein in quantities adequate for clinical trials. Through mechanisms not yet clear, angiostatin suppresses the angiogenesis necessary for tumor growth beyond ~1 mm (Folkman, 1972), while apparently leaving normal angiogenic processes sufficiently intact to preclude systemic toxicity. The activities of angiostatin and the unrelated antiangiogenic protein, endostatin (O'Reilly *et al.,* 1997), require the continued presence of the therapeutic protein for efficacy; cessation of therapy leads to outgrowth of residual tumor cells.

Several reports have demonstrated the successful use of angiostatin and endostatin gene-based therapeutics in mice (Blezinger *et al.,* 1999; Cao *et al.,* 1998; Griscelli *et al.,* 1998; Liu *et al.,* 1999; Nguyen *et al.,* 1998; Tanaka *et al.,* 1998). Because expression of angiostatin or endostatin causes regression of tumors to micrometastatic states without completely eradicating tumor cells, both protein and gene therapeutic approaches may benefit from the inclusion of adjuvant therapies such as radiation (Gorski *et al.,* 1998; Mauceri *et al.,* 1998) or chemotherapy (Harris, 1998). Of particular interest will be whether vector-mediated antiangiogenic protein expression *in situ* differs from systemic administration of recombinant proteins with respect to efficacy, toxicity, and cost.

CURRENT APPROACHES FOR PERSISTENCE

Persistent therapeutic protein gene expression can be achieved in one of two general ways: (1) a single dose of a persistently expressing vector or (2) repeated delivery of a less persistently expressing vector. Retroviral, adenoviral, adeno-associated viral, and plasmid-based gene delivery systems are the principal vectors being evaluated in a variety of animal and clinical models for their ability to express therapeutic proteins following administration by several different routes. Current considerations regarding these delivery systems involve their persistence of expression, mechanisms limiting that persistence, and the ability to administer repeat doses.

Immediately after infection, innate immune responses play a role in eliminating the vast majority of recombinant adenovirus (Ad) virions (Worgall *et al.,* 1997). However, the onset of antigen-driven immunity directed against the viral open reading frames (ORFs) remaining in the first generation (E1-deleted or E1/E3-deleted) replication-deficient adenoviral vectors is chiefly responsible for limiting their duration of expression (e.g., Tripathy *et al.,* 1996, Y. Yang *et al.,* 1994a, 1996a). Loss of expression generally occurs by 1 month posttransduction and can be ameliorated by codelivery of pharmacologic and vector-encoded immunosuppressive agents (Bouvet *et al.,* 1998; Cassivi *et al.,* 1999; Cichon and Strauss, 1998; Elshami *et al.,* 1995; Ilan *et al.,* 1997; Jani *et al.,* 1999; Jooss *et al.,* 1998; Kaplan and Smith, 1997; Kay *et al.,* 1997; Lochmuller *et al.,* 1996; T. A. Smith *et al.,* 1996; Vilquin *et al.,* 1995; Yap *et al.,* 1997) or by infection of immunodeficient strains of mice (Barr *et al.,* 1995; Dai *et al.,* 1995; Elshami *et al.,* 1995; Schowalter *et al.,* 1999; Yang *et al.,* 1994b; Zsengeller *et al.,* 1995). Although the potential safety issues regarding the use of immunosuppressive agents in human gene-based therapeutics have not yet been addressed, these approaches have been important in determining the mechanisms limiting the duration of gene expression and the ability to deliver vectors repetitively.

Partially, in an attempt to generate more persistent expression by reducing vector immunogenicity, the so-called "gutless" or "helper-dependent" adenovirus vectors have recently been developed (reviewed in Morsy and Caskey, 1999). All viral ORFs are deleted from these viral vectors, leaving only the transgene, the viral terminal repeats required for replication in packaging cells, a packaging sequence, and "stuffer" DNA to increase the genome to sufficient length for packaging into viral capsids. Recombinant virus is generally produced by cocultivation of a plasmid of the recombinant genome and a helper virus expressing all required trans-acting replication and packaging factors. Although contaminating helper virus remains an issue, studies using these vectors demonstrate improved performance with respect to immunogenicity, safety, and duration of expression (Chen *et al.,* 1997; Morsy *et al.,* 1998a; Schiedner *et al.,* 1998).

Adeno-associated viral (AAV) vectors have been "gutless" since their inception, bearing only the transgene flanked by the viral inverted terminal repeats necessary for replication and packaging (reviewed in Carter, 2000; Hallek and Wendtner, 1996; Rabinowitz and Samulski, 1998). Plasmids containing the recombinant viral genome or the replication and packaging proteins are cointroduced into tissue culture cells by transient or stable transfection. These cells are then infected with adenovirus or transfected with appropriate Ad gene expression plasmids to provide the helper functions necessary to initiate and sustain the replication and packaging of AAV virions. The duration of expression of AAV vectors introduced into the muscle of nonimmunosuppressed animals can be in excess of several months (e.g., Snyder *et al.,* 1997).

Similar to the helper-dependent Ad and AAV recombinant vectors, plasmid-based vehicles encode only the desired therapeutic gene. Additionally, most lack protein components, reducing the likelihood of antigen-specific immune responses that potentially shorten the duration of expression. However, in a manner essentially independent of the specific type of plasmid-based gene delivery vehicle employed, expression declines precipitously from a peak level 8–12 hr following IV administration to undetectable levels after 2 to 3 days (e.g., Liu *et al.,* 1997).

The majority of plasmid-based systems utilize the human cytomegalovirus immediate-early (hCMV-IE) promoter–enhancer due to its relative lack of cell type specificity and high activity.

Although extensive evaluation of a variety of cell type-specific and viral promoters has been reported (reviewed in Miller and Whelan, 1997; Walther and Stein, 1996), it has not become apparent that other promoters express more persistently than the hCMV-IE promoter–enhancer. This sequence is also commonly used to express transgenes in recombinant adenoviral vectors when expression becomes lost due to cell-mediated immune responses at approximately 1 month post-IV administration. The discrepancy between hCMV-IE promoter–enhancer persistence in plasmid-based and adenoviral contexts may reflect different cellular responses to these vehicles and suggests that the promoter is capable of persistent expression in some heterologous constructs. Presumably, some as yet undefined aspect of the viral genome structure, not shared by most plasmids, facilitates persistence.

Plasmid-based gene delivery using the IM route of administration is notably different from that of IV administration in its degree of persistent expression. Introduction of uncomplexed, "naked", or complexed DNAs by IM administration can result in prolonged expression (i.e., greater than 1 year) (e.g., Wolff *et al.*, 1992). The mechanism by which muscle fibers persistently express newly introduced transgenes is not yet clear. The route provides a simple means of systemically expressing therapeutic proteins, but may not be suitable for all applications. For example, IM administration of the vascular endothelial growth factor (VEGF) gene resulted in substantial gene-dependent injury at the site of injection (Springer *et al.*, 1998). Comparisons of IM route-dependent toxicity or pathology at the site of introduction should therefore be made for each therapeutic protein-encoding vector.

To achieve persistent gene expression, several groups have focused on the incorporation of cis- and trans-acting viral DNA elements into plasmid vectors. Sequences derived from viral replication origins and trans-acting origin binding factor ORFs derived from the Epstein–Barr Virus (EBV) (Guanghuan *et al.*, 1999; and reviewed in Sclimenti and Calos, 1998) and bovine papilloma virus (BPV) (Piirsoo *et al.*, 1996, and references therein) genomes have been used. These plasmid vectors are prevented from replicating by deletion of part of the replication origin necessary for the initiation of DNA synthesis. The relative persistence of these vectors is thought to involve the interaction of the origin binding protein with both the plasmid-born replication origin sequence and the host cell chromatin, tethering the plasmid and preventing its loss during cell division (Calos, 1998). Although this model predicts that the effect of the viral elements should be manifest predominantly in dividing cells, this has not yet been tested. An unresolved potential safety issue arises from the epidemiological association of EBV with several human cancers (Rickinson and Kieff, 1996) and the demonstration that B cell-directed expression of the EBV origin binding protein, EBNA-1, in transgenic mice results in the development of lymphomas (Wilson *et al.*, 1996). Sufficiently detailed structure–function analyses may allow the generation of mutant EBNA-1 species that facilitate nuclear persistence while lacking oncogenic potential, if these two activities are indeed separable.

Persistent therapeutic gene expression may be achievable through use of replicating vectors. In general, however, their use has been avoided due to the safety issues associated with uncontrolled *in vivo* viral replication. A modified replicating SV40-based plasmid system (Cooper *et al.*, 1997) is under development. The introduction of a specific set of mutations that reduce the oncogenicity of large T antigen, without blocking its ability to bind to the origin, potentially reduce the risk of tumorigenesis. If sufficiently limited distribution can be maintained, for example, by direct intratumoral injection or by engineering cell type-specific replication, certain therapeutic applications may be achievable.

The IM route of administration has proved to result in relatively persistent *in vivo* expression with both plasmid-based and viral gene delivery systems. Additionally, the use of viral elements in plasmid-based vectors shows promise for long-term expression by other routes. With these promising advances comes the concern that expression needs to be controlled to avoid adverse responses associated with prolonged expression. Ideally, vector transcription could be regulated in a positive manner by administration of a short-lived nontoxic drug (Clackson, 1997; Harvey and Caskey, 1998; Rossi and Blau, 1998). In the absence of the inducer, little or no expression of the therapeutic gene would occur and, in this manner, gene expression can be modulated. Several such systems have been described: regulation of transcription through use of the antibiotic tetracycline (reviewed in Blau and Rossi, 1999), the immunosuppressive rapamycin (Magari *et al.*, 1997; Rivera *et al.*, 1996; Ye *et al.*, 1999), the insect hormone ecdysone (No *et al.*, 1996), and the antiprogestins

(Burcin *et al.,* 1999, and references therein). Although promising, current outstanding issues include limited *in vivo* demonstration of efficacy, high basal levels of transcription in the absence of inducer, and the possibility that these prokaryotic or chimeric repressor proteins may prove immunogenic, precluding repeat delivery. However, the perceived need for regulated gene expression is, in part, an acknowledgment of the substantial progress that has been made regarding the levels and persistence of gene expression systems.

REPEAT ADMINISTRATION AND IMMUNE REACTIVITY

Prolonged therapeutic expression of proteins *in vivo* may also be accomplished by repetitive delivery of vectors. Unfortunately, the impact of the innate and adaptive immune responses following administration of either viral or plasmid-based gene delivery particles on the effectiveness of subsequent doses is becoming more apparent. Adenoviral (Benihoud *et al.,* 1998; Dong *et al.,* 1996; Kagami *et al.,* 1998; T. A. Smith *et al.,* 1996; Yang *et al.,* 1996b; Yei *et al.,* 1994) and AAV (Halbert *et al.,* 1998; Manning *et al.,* 1998) delivery systems express efficiently on repeated administration only in mice lacking functional adaptive immune systems [e.g., nude or severe combined immunodeficient (SCID) mice] or immunocompetent mice tolerized or transiently immunosuppressed by pharmacologic or vector-encoded agents. Although the "gutless" Ad vectors have been designed to avoid elicitation of adaptive immune responses by elimination of viral ORFs, the inability to deliver repetitively AAV vectors, which lack all viral ORFs, to nonimmunosuppressed animals bodes poorly for this approach with respect to repeat delivery. Furthermore, a study has demonstrated that prior administration of an Ad vector rendered transcriptionally inactive by UV irradiation precluded expression from a subsequently administered Ad vector, suggesting that the capsid proteins associated with the incoming viral genomes are sufficient to generate a neutralizing immune response (Kafri *et al.,* 1998).

Another means by which to avoid the adaptive immune responses that may block secondary delivery is "serotype switching," which involves the alternating use of recombinant viruses that are genetically divergent so as to not elicit cross-neutralizing immune responses. Such an approach has been reported for both Ad (Kass-Eisler *et al.,* 1996; Mack *et al.,* 1997; Mastrangeli *et al.,* 1996; Roy *et al.,* 1998) and AAV (Rutledge *et al.,* 1998; Xiao *et al.,* 1999). Applications are likely to be restricted, however, due to the limited number of Ad and AAV serotypes available in cloned form.

The ability to deliver repetitively plasmid-based gene expression constructs differs from the viral systems in that the innate (i.e., cytokine release) immune system, rather than the adaptive (i.e., T and B cell activation) immune sytem, appears to mediate interference. Productive IV delivery of a second plasmid-based construct after prior delivery of a first was shown to require a delay of 9–11 days (Song *et al.,* 1997). This "refractory period," the length of time required between doses for productive secondary transfection, is dependent on the amount of the first treatment dose (Fradkin *et al.,* 2000). When the interval between IV doses was arbitrarily set as 3 days, the initial dose had to be reduced from a typical dose of 50 μg to 0.5–5 μg of liposome-complexed DNA per mouse to allow subsequent gene expression. Delivery of liposome-complexed DNA via intraperitoneal injection also showed a refractory period (Vernachio *et al.,* unpublished data). Induction of the refractory period does not require expression of a protein or mRNA by the first vector, e.g., a promoterless plasmid effectively induces a refractory period of length similar to that of an expressing plasmid. Administration of liposome-complexed plasmid by the IV route elicits rapid IFN-γ, type I IFN, TNF-α, IL-6, and IL-12 expression (Dow *et al.,* 1999; Dow *et al.,* unpublished data; Whitmore *et al.,* 1999), similar to that seen following immunostimulation by oligonucleotides containing hypomethylated CpG motifs (reviewed in Krieg, 1996; Pisetsky, 1996). Further studies have implicated the interferon pathway in establishment of the refractory period (Fradkin *et al.,* 2000). Thus, the innate immune responses to lipid–DNA complexes *in vivo* appear to have a profound impact on repeat gene expression.

SAFETY CONSIDERATIONS FOR GENE-BASED THERAPEUTICS

One of the primary issues surrounding any new emerging medical technology, particularly one involving gene transfer, is safety. The development of safe and effective gene-based therapeutics will require the careful choice and design of vectors for particular therapeutic applications, the controlled assessment of toxicology in compliance with good laboratory practices, and validated manufacturing processes following current good manufacturing practices. Regulatory groups will need to address certain environmental release issues due to the technology's use of genetically modified organisms, pathogenicity of the delivery agents (e.g., viral systems), and characterization and quality assurance of the master seed banks and vector stocks (K. T. Smith *et al.,* 1996). Moreover, gene transfer has potential safety and toxicity considerations associated with the delivery system and expression of the delivered gene within cells. These considerations are all part of normal drug development and are a sign of the maturation of gene-based therapeutics from a research technology to clinical utility.

The basic nature of gene-based therapeutic strategies to deliver and express therapeutic proteins at the disease site under physiologically relevant conditions defines a new pharmacology compared to more established small-molecule and recombinant protein drug treatments. Thus, novel evaluation technologies will be employed to measure these new pharmacologic parameters to establish proper dosing regimens. For example, therapeutic proteins produced at the disease site may be undetectable using current expression detection systems, although the biologic effect occurs. Therefore, our sense of appropriate quantitative potency specifications to ensure biologic activity may need to be redefined. Although numerous clinical trials have been completed or are underway, no gene-based therapeutics have yet been approved for commercial clinical use. Hence, with limited published preclinical data available, pharmacologists and toxicologists will be challenged to design relevant treatment regimens using predictive models to identify potential acute and long-term adverse effects. However, guideline documents dealing with the preclinical safety evaluation of biotechnology-derived products (International Conference on Harmonization, 1998) and safety aspects of gene therapy products (CPMP/SWP, 1998) have been released. Furthermore, preclinical safety requirements for human somatic cell therapy and gene therapy products have been defined by the Food and Drug Administration (FDA) (1998). Thus, initial safety guidelines are in place to direct the safe development of gene-based therapeutics.

In spite of the tremendous potential of gene-based therapeutics to treat a variety of disease conditions, there remain a number of safety and scientific concerns regarding the technology (Verdier and Descotes, 1999). Some key safety considerations for gene-based therapeutics include the toxicity associated with the delivery systems employed, the expression of novel genes in new tissue types, interactions of the vectors with host genetic sequences, and the impact of host immune responses (i.e., innate and adaptive) directed against the delivery vehicle and/or the gene product.

The sequelae of host inflammatory and immune responses activated by any gene delivery particle (i.e., viral or plasmid-based) may play an important role in toxicity as well as impact the functional duration of the therapy. The induction of various cytokines could likely contribute to different degrees of acute toxicity depending on the gene delivery system employed and species evaluated, particularly with viral gene delivery (Muruve *et al.,* 1999). Toxicity issues may be more pronounced on systemic administration of vehicles compared to regional (e.g., intratumoral) delivery strategies. Although limited, preliminary studies would suggest that some gene therapeutics can be administered safely with appropriate dosing.

Safety issues concerning activation of protooncogenes through insertional mutagenesis or inactivation of tumor suppressor genes are important considerations, depending on the particular delivery system employed. Concerns regarding generation of replication-competent viruses and gene integration into germ-line tissue must be addressed. Because most viral delivery systems have been designed to be replication defective for safety purposes, replication-competent species potentially arising during manufacture must be detected. The *cis-* or *trans*-complementation of defective functions inherent in the vector by either viral or cellular gene products is an additional consideration for possible vector mobilization within the host. In the case of retroviral gene delivery, no replication-competent retrovirus (RCR) has been detected clinically following single or multiple intratumoral injections (Nemunaitis *et al.,* 1999). Through appropriate vector and packaging cell construction, it is hoped that recombination events leading to the generation of some of these species can be minimized.

The variety of gene delivery systems each provides a unique set of tissue specificities and properties that have advantages and disadvantages, dependent on the therapeutic application. The development of any new therapeutic technology needs to evolve through thorough demonstrations of preclinical safety. Although animal models can provide a certain level of safety evaluation, the critical safety profiles need to be defined through well-designed human studies.

LARGE-SCALE PRODUCTION ISSUES

The successful development of gene-based therapeutic approaches for treating human disease requires progression through a well-defined set of clinical trials. Through each stage of clinical testing (i.e., phases I, II, and II) larger and more diverse patient populations are required. As a result, increasing amounts of gene-based product, processed under stringently controlled current good manufacturing practices, will be required. The safe application of gene-based therapeutics necessitates that production conditions be clearly defined with assurances that the manufacturing process can be validated and the reagent components can be certified for safety and conformity.

There are a number of issues that have arisen in the large-scale manufacture of these novel viral and plasmid-based gene therapeutics. Viral vectors must be produced in tissue culture packaging cells and are therefore subject to several specific considerations. Tissue culture cell lines used for manufacturing vectors need to be tested for infectious agents and any growth medium components (e.g., serum or growth factors) certified to be free of adventitious agents. Purification schemes must be appropriate for the removal of potential contaminants or product interference, e.g., retroviral transduction is adversely affected by proteoglycans secreted by producer cells in the preparation of viral stocks (Le Doux *et al.*, 1996, 1998).

Recombinant viruses are generally produced in tissue culture cells that have either been stably transfected or are transiently transfected with plasmid constructs expressing the viral ORFs required for viral DNA replication and packaging. Stably introduced expression constructs are generally maintained through drug selection of a contemporaneously introduced drug resistance gene. The gene-based therapeutic should be purified in such a manner as to render it demonstrably free of the selective drug. Ideally, the stability of the viral ORF under drug selection should be monitored over time to ensure homogeneity in production. The use of transient transfection into unmodified tissue culture cells reduces the burden of cell line stability characterization. However, although adequate to generate materials necessary for preclinical and phase I/II trials, transient transfection protocols have not yet been reported to be scaled up to yield the amounts of vector required for later stage trials and commercialization.

Another issue in the large-scale manufacture of viral vectors is the variability in the number of useful vector particles generated, i.e., the particle-to-infectivity (P:I) ratio, and their subsequent stability on storage. These variabilities can render production lots differentially efficacious for gene transfer and therapeutic effect. Most aspects of viral packaging are poorly understood and the P:I ratio cannot therefore be controlled or manipulated. Additionally, titers used to calculate infectivity ratios are generally determined using tissue culture cells under conditions optimized for *in vitro* infection; the relevant *in vivo* potencies are not well characterized (e.g., Forestell *et al.*, 1995). Methodologies to reduce P:I ratios should actively be sought because undesired inflammatory responses to both viral and plasmid-based particles are likely a function of total particle numbers delivered.

The generation of replication-competent virus (RCV) arising during vector production from the recombination of the vector genome and the viral replication protein encoding ORFs has been problematic for most viral-based gene delivery systems. Generation of RCV has been reported for adenovirus (Lochmuller *et al.*, 1994; Zhu *et al.*, 1999), AAV (Allen *et al.*, 1997; X. S. Wang *et al.*, 1998), and retrovirus vectors (Chong *et al.*, 1998; Miller and Buttimore, 1986, and references therein). Replication-competent retrovirus arising during manufacture was found to be responsible for the onset of lymphomas in vector-treated primates (Donahue *et al.*, 1992), demonstrating the reality of the dangers associated with RCVs. Although the possibility of recombination can be reduced by elimination of regions of homology between the vector genome and viral ORFs (Fallaux *et al.*, 1998), evidence that AAV replication-competent virus can arise through nonhomologous recombination has been presented (Allen *et al.*, 1997). The large amounts of actively replicating DNAs in packaging cells may preclude complete prevention of the generation of RCV, particularly during large-scale production. Acceptable safe limits, if any, of such contaminants may need to be established through rigorous preclinical testing.

Plasmid-based gene delivery vectors, due to their *in vitro* formulation from small-molecule compounds (e.g., cationic lipids), do not necessarily require animal-derived growth products in their manufacture. However, manufacturing issues exist for this class of delivery system as well. Some aspects of plasmid DNA preparation can involve the utilization of animal-derived reagents, i.e., the bovine pancreatic RNase often used to degrade contaminating *Escherichia coli* RNA in plasmid preparations. These components can be purified from animals free from known adventitious agents; however, other unknown biologic agents may be present. A current problem in the manufacture of plasmid-based gene delivery vehicles is our limited understanding about the *in vivo* transfection-active species of formulated DNA. This has rendered the development of physical assays for ascertaining batch-to-batch variation difficult. The performance of vectors *in vitro* has proved to be a poor predictor of their *in vivo* function; different barriers to their activity are likely to exist in the *in vivo* environment. Issues involving the manufacture of gene-based products will be resolved as the technology and its product development continue to evolve.

CONCLUSION

The use of proteins or biologics for the therapeutic treatment of various diseases is well established. Gene-based therapeutics provides an alternative, and potentially superior, means of delivering therapeutic proteins to diseased sites. Although gene-based therapeutics has tremendous potential for treating a variety of diseases, there are a number of issues arising from preclinical and clinical studies remaining to be addressed before clinical products are approved for routine use in humans.

With the availability of varied types of gene delivery systems, the challenge will be to understand the most appropriate delivery vehicle for each therapeutic application. Suitable manufacturing processes, compliant with regulatory guidelines, and well-designed clinical studies with clear end points need to be implemented. In addition, the safety profiles need be clearly delineated. In particular, toxicities associated with the different delivery systems (i.e., viral and plasmid-based), novel gene expression in tissues, and host immune responses directed against the delivery vehicles and/or gene products need to be fully understood. Preliminary indications of safety and efficacy can be derived from preclinical studies utilizing inbred mice and rabbits. Relatively more outbred animals, e.g., nonhuman primates and companion animals, provide models potentially more closely related to the human with respect to dosing regimens, toxicity, and immune responses. Evaluation of the safety and efficacy of human gene therapeutics, however, can be achieved only through human clinical trials.

A greater understanding of cell biology, virology, immunology, and molecular biology is required to achieve the full potential of gene-based therapeutics. Fortunately, ongoing advances in these fields of basic scientific inquiry are mirroring the challenges arising in the development of gene-based therapeutics. Additionally, through a well-defined clinical development process, a great deal of knowledge will be gained regarding various pragmatic aspects in the utilization of gene-based therapeutics. The research technology of gene-based therapeutics is now positioned to move to the next stage, intervention in human disease.

ACKNOWLEDGMENTS

L.G.F. and J.D.R. made equal contributions in the writing of this chapter. We wish to thank Denny Liggitt, Qing Wang, Meg Snowden, Steve Dow, and Rodney Pearlman for their comments on the manuscript and encouragement. We also thank our colleagues at Valentis, Inc., current and past, in particular, John Vernachio and Jeff Fairman, for their contributions and discussions. We apologize to the many scientists whose work was not cited due to space limitations.

REFERENCES

Agrawal, R. S., Karhu, K., Laukkanen, J., Kirkinen, P., Yla-Herttuala, S., and Agrawal, Y. P. (1999). Complement and anti-alpha-galactosyl natural antibody-mediated inactivation of murine retrovirus occurs in adult serum but not in umbilical cord serum. *Gene Ther.* **6**, 146–148.

Allen, J. M., Debelak, D. J., Reynolds, T. C., and Miller, A. D. (1997). Identification and elimination of replication-competent adeno-associated virus (AAV) that can arise by nonhomologous recombination during AAV vector production. *J. Virol.* **71**, 6816–6822.

Anderson, W. F. (1998). Human gene therapy. *Nature (London)* **392**, 25–30.

Barr, D., Tubb, J., Ferguson, D., Scaria, A., Lieber, A., Wilson, C., Perkins, J., and Kay, M. A. (1995). Strain related variations in adenovirally mediated transgene expression from mouse hepatocytes *in vivo*: Comparisons between immunocompetent and immunodeficient inbred strains. *Gene Ther.* **2**, 151–155.

Barron, L. G., Meyer, K. B., and Szoka, F. C., Jr. (1998). Effects of complement depletion on the pharmacokinetics and gene delivery mediated by cationic lipid–DNA complexes. *Hum. Gene Ther.* **9**, 315–323.

Bartlett, J. S., Kleinschmidt, J., Boucher, R. C., and Samulski, R. J. (1999). Targeted adeno-associated virus vector transduction of nonpermissive cells mediated by a bispecific F(Ab')$_2$ antibody. *Nat. Biotechnol.* **17**, 181–186.

Baru, M., Nahum, O., Jaaro, H., Sha'anani, J., and Nur, I. (1998). Lysosome-disrupting peptide increases the efficiency of *in-vivo* gene transfer by liposome-encapsulated DNA. *J. Drug Target.* **6**, 191–199.

Bauer, T. R., Jr., Miller, A. D., and Hickstein, D. D. (1995). Improved transfer of the leukocyte integrin CD18 subunit into hematopoietic cell lines by using retroviral vectors having a gibbon ape leukemia virus envelope. *Blood* **86**, 2379–2387.

Benihoud, K., Saggio, I., Opolon, P., Salone, B., Amiot, F., Connault, E., Chianale, C., Dautry, F., Yeh, P., and Perricaudet, M. (1998). Efficient, repeated adenovirus-mediated gene transfer in mice lacking both tumor necrosis factor alpha and lymphotoxin alpha. *J. Virol.* **72**, 9514–9525.

Benoit, P., Emmanuel, F., Caillaud, J. M., Bassinet, L., Castro, G., Gallix, P., Fruchart, J. C., Branellec, D., Denefle, P., and Duverger, N. (1999). Somatic gene transfer of human ApoA-I inhibits atherosclerosis progression in mouse models. *Circulation* **99**, 105–110.

Bergelson, J. M., Cunningham, J. A., Droguett, G., Kurt-Jones, E. A., Krithivas, A., Hong, J. S., Horwitz, M. S., Crowell, R. L., and Finberg, R. W. (1997). Isolation of a common receptor for Coxsackie B viruses and adenoviruses 2 and 5. *Science* **275**, 1320–1323.

Bilbao, G., Gomez-Navarro, J., and Curiel, D. T. (1998). Targeted adenoviral vectors for cancer gene therapy. *Adv. Exp. Med. Biol.* **451**, 365–374.

Blau, H. M., and Rossi, F. M. (1999). Tet B or not tet B: Advances in tetracycline-inducible gene expression. *Proc. Natl. Acad. Sci. U.S.A.* **96**, 797–799.

Blezinger, P., Wang, J., Gondo, M., Quezada, A., Mehrens, D., French, M., Singhal, A., Sullivan, S., Rolland, A., Ralston, R., and Min, W. (1999). Systemic inhibition of tumor growth and tumor metastases by intramuscular administration of the endostatin gene. *Nat. Biotechnol.* **17**, 343–348.

Bouvet, M., Fang, B., Ekmekcioglu, S., Ji, L., Bucana, C. D., Hamada, K., Grimm, E. A., and Roth, J. A. (1998). Suppression of the immune response to an adenovirus vector and enhancement of intratumoral transgene expression by low-dose etoposide. *Gene Ther.* **5**, 189–195.

Brisson, M., He, Y., Li, S., Yang, J. P., and Huang, L. (1999). A novel T7 RNA polymerase autogene for efficient cytoplasmic expression of target genes. *Gene Ther.* **6**, 263–270.

Buckel, P. (1996). Recombinant proteins for therapy. *Trends Pharmacol. Sci.* **17**, 450–456.

Burcin, M. M., Schiedner, G., Kochanek, S., Tsai, S. Y., and O'Malley, B. W. (1999). Adenovirus-mediated regulable target gene expression *in vivo. Proc. Natl. Acad. Sci. U.S.A.* **96**, 355–360.

Calos, M. P. (1998). Stability without a centromere. *Proc. Natl. Acad. Sci. U.S.A.* **95**, 4084–4085.

Cao, Y., O'Reilly, M. S., Marshall, B., Flynn, E., Ji, R. W., and Folkman, J. (1998). Expression of angiostatin cDNA in a murine fibrosarcoma suppresses primary tumor growth and produces long-term dormancy of metastases. *J. Clin. Invest.* **101**, 1055–1063.

Carter, B. J. (2000). Adeno-associated virus and AAV vectors for gene delivery. *In* "Gene Therapy: Therapeutic Mechanisms and Strategies" (D. Lasic and N. Templeton-Smith, eds.). Dekker, New York (in press).

Cassivi, S. D., Liu, M., Boehler, A., Tanswell, A. K., Slutsky, A. S., Keshavjee, S., and Todd, S. (1999). Transgene expression after adenovirus-mediated retransfection of rat lungs is increased and prolonged by transplant immunosuppression. *J. Thorac. Cardiovasc. Surg.* **117**, 1–7.

Chen, H. H., Mack, L. M., Kelly, R., Ontell, M., Kochanek, S., and Clemens, P. R. (1997). Persistence in muscle of an adenoviral vector that lacks all viral genes. *Proc. Natl. Acad. Sci. U.S.A.* **94**, 1645–1650.

Chen, X., Li, Y., Xiong, K., Aizicovici, S., Xie, Y., Zhu, Q., Sturtz, F., Shulok, J., Snodgrass, R., Wagner, T. E., and Platika, D. (1998). Cancer gene therapy by direct tumor injections of a nonviral T7 vector encoding a thymidine kinase gene. *Hum. Gene Ther.* **9**, 729–736.

Chillon, M., Lee, J. H., Fasbender, A., and Welsh, M. J. (1998). Adenovirus complexed with polyethylene glycol and cationic lipid is shielded from neutralizing antibodies *in vitro. Gene Ther.* **5**, 995–1002.

Chong, H., Starkey, W., and Vile, R. G. (1998). A replication-competent retrovirus arising from a split-function packaging cell line was generated by recombination events between the vector, one of the packaging constructs, and endogenous retroviral sequences. *J. Virol.* **72**, 2663–2670.

Cichon, G., and Strauss, M. (1998). Transient immunosuppression with 15-deoxyspergualin prolongs reporter gene expression and reduces humoral immune response after adenoviral gene transfer. *Gene Ther.* **5**, 85–90.

Clackson, T. (1997). Controlling mammalian gene expression with small molecules. *Curr. Opin. Chem. Biol.* **1**, 210–218.

Cooper, M. J., Lippa, M., Payne, J. M., Hatzivassiliou, G., Reifenberg, E., Fayazi, B., Perales, J. C., Morrison, L. J., Templeton, D., Piekarz, R. L., and Tan, J. (1997). Safety-modified episomal vectors for human gene therapy. *Proc. Natl. Acad. Sci. U.S.A.* **94**, 6450–6455.

CPMP/SWP (1998). "Safety Studies for Gene Therapy Products—Annex to Note for Guidance on Gene Therapy Product Quality Aspects in the Production of Vectors and Genetically Modified Somatic Cells," Volume CPMP/SWP/112/98 draft, European Agency for the Evaluation of Medicinal Products-Committee for Proprietary Medicinal Products.

Cristiano, R. J. (1998). Targeted, non-viral gene delivery for cancer gene therapy. *Front. Biosci.* **3**, D1161–D1170.

Dai, Y., Schwarz, E. M., Gu, D., Zhang, W. W., Sarvetnick, N., and Verma, I. M. (1995). Cellular and humoral immune responses to adenoviral vectors containing factor IX gene: Tolerization of factor IX and vector antigens allows for long-term expression. *Proc. Natl. Acad. Sci. U.S.A.* **92**, 1401–1405.

Dean, D. A. (1997). Import of plasmic DNA into the nucleus is sequence specific. *Exp. Cell Res.* **230**, 293–302.

Deshmukh, H. M., and Huang, L. (1997). Liposome and polylysine mediated gene transfer. *New J. Chem.* **21**, 113–121.

Dmitriev, I., Krasnykh, V., Miller, C. R., Wang, M., Kashentseva, E., Mikheeva, G., Belousova, N., and Curiel, D. T. (1998). An adenovirus vector with genetically modified fibers demonstrates expanded tropism via utilization of a coxsackievirus and adenovirus receptor-independent cell entry mechanism. *J. Virol.* **72**, 9706–9713.

Donahue, R. E., Kessler, S. W., Bodine, D., McDonagh, K., Dunbar, C., Goodman, S., Agricola, B., Byrne, E., Raffeld,

M., Moen, R., *et al.* (1992). Helper virus induced T cell lymphoma in nonhuman primates after retroviral mediated gene transfer. *J. Exp. Med.* **176**, 1125–1135.

Dong, J. Y., Wang, D., Van Ginkel, F. W., Pascual, D. W., and Frizzell, R. A. (1996). Systematic analysis of repeated gene delivery into animal lungs with a recombinant adenovirus vector. *Hum. Gene Ther.* **7**, 319–331.

Douglas, J. T., and Curiel, D. T. (1997). Strategies to accomplish targeted gene delivery to muscle cells employing tropism-modified adenoviral vectors. *Neuromuscul. Disord.* **7**, 284–298.

Dow, S. W., Fradkin, L. G., Liggitt, H. D., Willson, A. P., Heath, T. D., and Potter, T. A. (1999). Lipid-DNA complexes induce potent activation of innate immune responses and antitumor activity when administered intravenously. *J. Immunol.* **163**, 1552–1561.

El Ouahabi, A., Thiry, M., Pector, V., Fuks, R., Ruysschaert, J. M., and Vandenbranden, M. (1997). The role of endosome destabilizing activity in the gene transfer process mediated by cationic lipids. *FEBS Lett.* **414**, 187–192.

Elshami, A. A., Kucharczuk, J. C., Sterman, D. H., Smythe, W. R., Hwang, H. C., Amin, K. M., Litzky, L. A., Albelda, S. M., and Kaiser, L. R. (1995). The role of immunosuppression in the efficacy of cancer gene therapy using adenovirus transfer of the herpes simplex thymidine kinase gene. *Ann. Surg.* **222**, 298–307, 307–310.

Erbacher, P., Remy, J. S., and Behr, J. P. (1999). Gene transfer with synthetic virus-like particles via the integrin-mediated endocytosis pathway. *Gene Ther.* **6**, 138–145.

Fallaux, F. J., Bout, A., van der Velde, I., van den Wollenberg, D. J., Hehir, K. M., Keegan, J., Auger, C., Cramer, S. J., van Ormondt, H., van der Eb, A. J., Valerio, D., and Hoeben, R. C. (1998). New helper cells and matched early region 1-deleted adenovirus vectors prevent generation of replication-competent adenoviruses. *Hum. Gene Ther.* **9**, 1909–1917.

Fasbender, A., Zabner, J., Zeiher, B. G., and Welsh, M. J. (1997). A low rate of cell proliferation and reduced DNA uptake limit cationic lipid-mediated gene transfer to primary cultures of ciliated human airway epithelia. *Gene Ther.* **4**, 1173–1180.

Folkman, J. (1972). Anti-angiogenesis: new concept for therapy of solid tumors. *Ann. Surg.* **175**, 409–416.

Food and Drug Administration (FDA) (1998). "Guidance for Human Somatic Cell Therapy and Gene Therapy." Center for Biologics Evaluation and Research, FDA, Washington, DC. Available at: *http://www.fda.gov/cber/guidelines.htm*

Forestell, S. P., Bohnlein, E., and Rigg, R. J. (1995). Retroviral end-point titer is not predictive of gene transfer efficiency: Implications for vector production. *Gene Ther.* **2**, 723–730.

Forestell, S. P., Dando, J. S., Chen, J., de Vries, P., Bohnlein, E., and Rigg, R. J. (1997). Novel retroviral packaging cell lines: Complementary tropisms and improved vector production for efficient gene transfer. *Gene Ther.* **4**, 600–610.

Fradkin, L. G., *et al.* (2000). In preparation.

Francis, G. E., Fisher, D., Delgado, C., Malik, F., Gardiner, A., and Neale, D. (1998). PEGylation of cytokines and other therapeutic proteins and peptides: The importance of biological optimisation of coupling techniques. *Int. J. Hematol.* **68**, 1–18.

Geddes, D. M., and Beckles, M. A. (1995). Gene therapy for alpha 1-antitrypsin deficiency. *Monaldi Arch. Chest Dis.* **50**, 388–393.

Goldman, C. K., Rogers, B. E., Douglas, J. T., Sosnowski, B. A., Ying, W., Siegal, G. P., Baird, A., Campain, J. A., and Curiel, D. T. (1997). Targeted gene delivery to Kaposi's sarcoma cells via the fibroblast growth factor receptor. *Cancer Res.* **57**, 1447–1451.

Gorski, D. H., Mauceri, H. J., Salloum, R. M., Gately, S., Hellman, S., Beckett, M. A., Sukhatme, V. P., Soff, G. A., Kufe, D. W., and Weichselbaum, R. R. (1998). Potentiation of the antitumor effect of ionizing radiation by brief concomitant exposures to angiostatin. *Cancer Res.* **58**, 5686–5689.

Goula, D., Benoist, C., Mantero, S., Merlo, G., Levi, G., and Demeneix, B. A. (1998). Polyethylenimine-based intravenous delivery of transgenes to mouse lung. *Gene Ther.* **5**, 1291–1295.

Graessmann, M., Menne, J., Liebler, M., Graeber, I., and Graessmann, A. (1989). Helper activity for gene expression, a novel function of the SV40 enhancer. *Nucleic Acids Res.* **17**, 6603–6612.

Greber, U. F., and Kasamatsu, H. (1996). Nuclear targeting of SV40 and adenovirus. *Trends Cell Biol.* **6**, 189–195.

Greber, U. F., Willetts, M., Webster, P., and Helenius, A. (1993). Stepwise dismantling of adenovirus 2 during entry into cells. *Cell (Cambridge, Mass.)* **75**, 477–486.

Greber, U. F., Singh, I., and Helenius, A. (1994). Mechanisms of virus uncoating. *Trends Microbiol.* **2**, 52–56.

Griesenbach, U., Chonn, A., Cassady, R., Hannam, V., Ackerley, C., Post, M., Tanswell, A. K., Olek, K., O'Brodovich, H., and Tsui, L. C. (1998). Comparison between intratracheal and intravenous administration of liposome-DNA complexes for cystic fibrosis lung gene therapy. *Gene Ther.* **5**, 181–188.

Griscelli, F., Li, H., Bennaceur-Griscelli, A., Soria, J., Opolon, P., Soria, C., Perricaudet, M., Yeh, P., and Lu, H. (1998). Angiostatin gene transfer: Inhibition of tumor growth *in vivo* by blockage of endothelial cell proliferation associated with a mitosis arrest. *Proc. Natl. Acad. Sci. U.S.A.* **95**, 6367–6372.

Guanghuan, T., Korchmaier, A. L., Liggitt, H. D., Yu, W.-H., Heath, T. D., and Debs, R. J. (1999). Non-replicating EBV-based plasmids produce long-term gene expression *in vivo*. Submitted for publication.

Gunzburg, W. H., and Salmons, B. (1996). Development of retroviral vectors as safe, targeted gene delivery systems. *J. Mol. Med.* **74**, 171–182.

Halbert, C. L., Standaert, T. A., Wilson, C. B., and Miller, A. D. (1998). Successful readministration of adeno-associated virus vectors to the mouse lung requires transient immunosuppression during the initial exposure. *J. Virol.* **72**, 9795–9805.

Hallek, M., and Wendtner, C. M. (1996). Recombinant adeno-associated virus (rAAV) vectors for somatic gene therapy: Recent advances and potential clinical applications. *Cytokines Mol. Ther.* **2**, 69–79.

Hamm-Alvarez, S. F. (1999). Targeting endocytosis and motor proteins to enhance DNA persistence. *Pharm. Sci. Technol. Today* **2**, 190–196.

Harbottle, R. P., Cooper, R. G., Hart, S. L., Ladhoff, A., McKay, T., Knight, A. M., Wagner, E., Miller, A. D., and Coutelle,

C. (1998). An RGD-oligolysine peptide: A prototype construct for integrin-mediated gene delivery. *Hum. Gene Ther.* **9,** 1037–1047.

Harris, A. L. (1998). Anti-angiogenesis therapy and strategies for integrating it with adjuvant therapy. *Recent Results Cancer Res.* **152,** 341–352.

Harvey, D. M., and Caskey, C. T. (1998). Inducible control of gene expression: Prospects for gene therapy. *Curr. Opin. Chem. Biol.* **2,** 512–518.

Hemmi, S., Geertsen, R., Mezzacasa, A., Peter, I., and Dummer, R. (1998). The presence of human coxsackievirus and adenovirus receptor is associated with efficient adenovirus-mediated transgene expression in human melanoma cell cultures. *Hum. Gene Ther.* **9,** 2363–2373.

Herzog, R. W., and High, K. A. (1998). Problems and prospects in gene therapy for hemophilia. *Curr. Opin. Hematol.* **5,** 321–326.

Hidaka, C., Milano, E., Leopold, P. L., Bergelson, J. M., Hackett, N. R., Finberg, R. W., Wickham, T. J., Kovesdi, I., Roelvink, P., and Crystal, R. G. (1999). CAR-dependent and CAR-independent pathways of adenovirus vector-mediated gene transfer and expression in human fibroblasts. *J. Clin. Invest.* **103,** 579–587.

Hope, M. J., Mui, B., Ansell, S., and Ahkong, Q. F. (1998). Cationic lipids, phosphatidylethanolamine and the intracellular delivery of polymeric, nucleic acid-based drugs. *Mol. Membr. Biol.* **15,** 1–14.

Hung, M. C., Chang, J. Y. J., and Xing, X. M. (1998). Preclinical and clinical study of HER-2/neu-targeting cancer gene therapy. *Adv. Drug Delivery Rev.* **30,** 219–227.

Ilan, Y., Jona, V. K., Sengupta, K., Davidson, A., Horwitz, M. S., Roy-Chowdhury, N., and Roy-Chowdhury, J. (1997). Transient immunosuppression with FK506 permits long-term expression of therapeutic genes introduced into the liver using recombinant adenoviruses in the rat. *Hepatology* **26,** 949–956.

International Conference on Harmonization (ICH) (1998). ICH harmonized tripartite guideline S6 endorsed by the ICH steering committee at step 4 of the ICH process, 16, July, 1997: Preclinical safety evaluation of biotechnology-derived pharmaceuticals. *In* "Proceedings of the Fourth International Conference on Harmonization" (P. F. D'Arcy and D. W. G. Harron, eds.), pp. 1045–1056. Greystone Books, Antrim, England.

Jani, A., Menichella, D., Jiang, H., Chbihi, T., Acsadi, G., Shy, M. E., and Kamholz, J. (1999). Modulation of cell-mediated immunity prolongs adenovirus-mediated transgene expression in sciatic nerve. *Hum. Gene Ther.* **10,** 787–800.

Jooss, K., Turka, L. A., and Wilson, J. M. (1998). Blunting of immune responses to adenoviral vectors in mouse liver and lung with CTLA4Ig. *Gene Ther.* **5,** 309–319.

Kafri, T., Morgan, D., Krahl, T., Sarvetnick, N., Sherman, L., and Verma, I. (1998). Cellular immune response to adenoviral vector infected cells does not require de novo viral gene expression: Implications for gene therapy. *Proc. Natl. Acad. Sci. U.S.A.* **95,** 11377–11382.

Kagami, H., Atkinson, J. C., Michalek, S. M., Handelman, B., Yu, S., Baum, B. J., and O'Connell, B. (1998). Repetitive adenovirus administration to the parotid gland: Role of immunological barriers and induction of oral tolerance. *Hum. Gene Ther.* **9,** 305–313.

Kaplan, J. M., and Smith, A. E. (1997). Transient immunosuppression with deoxyspergualin improves longevity of transgene expression and ability to readminister adenoviral vector to the mouse lung. *Hum. Gene Ther.* **8,** 1095–1104.

Kass-Eisler, A., Leinwand, L., Gall, J., Bloom, B., and Falck-Pedersen, E. (1996). Circumventing the immune response to adenovirus-mediated gene therapy. *Gene Ther.* **3,** 154–162.

Kawakami, S., Yamashita, F., Nishikawa, M., Takakura, Y., and Hashida, M. (1998). Asialoglycoprotein receptor-mediated gene transfer using novel galactosylated cationic liposomes. *Biochem. Biophys. Res. Commun.* **252,** 78–83.

Kay, M. A., Meuse, L., Gown, A. M., Linsley, P., Hollenbaugh, D., Aruffo, A., Ochs, H. D., and Wilson, C. B. (1997). Transient immunomodulation with anti-CD40 ligand antibody and CTLA4Ig enhances persistence and secondary adenovirus-mediated gene transfer into mouse liver. *Proc. Natl. Acad. Sci. U.S.A.* **94,** 4686–4691.

Kessler, P. D., Podsakoff, G. M., Chen, X., McQuiston, S. A., Colosi, P. C., Matelis, L. A., Kurtzman, G. J., and Byrne, B. J. (1996). Gene delivery to skeletal muscle results in sustained expression and systemic delivery of a therapeutic protein. *Proc. Natl. Acad. Sci. U.S.A.* **93,** 14082–14087.

Kitson, C., Angel, B., Judd, D., Rothery, S., Severs, N. J., Dewar, A., Huang, L., Wadsworth, S. C., Cheng, S. H., Geddes, D. M., and Alton, E. W. F. W. (1999). The extra- and intracellular barriers to lipid and adenovirus-mediated pulmonary gene transfer in native sheep airway epithelium. *Gene Ther.* **6,** 534–546.

Krasnykh, V., Dmitriev, I., Mikheeva, G., Miller, C. R., Belousova, N., and Curiel, D. T. (1998). Characterization of an adenovirus vector containing a heterologous peptide epitope in the HI loop of the fiber knob. *J. Virol.* **72,** 1844–1852.

Krieg, A. M. (1996). An innate immune defense mechanism based on the recognition of CpG motifs in microbial DNA. *J. Lab. Clin. Med.* **128,** 128–133.

Kuriyama, S., Tominaga, K., Kikukawa, M., Nakatani, T., Tsujinoue, H., Yamazaki, M., Nagao, S., Toyokawa, Y., Mitoro, A., and Fukui, H. (1998). Inhibitory effects of human sera on adenovirus-mediated gene transfer into rat liver. *Anticancer Res.* **18,** 2345–2351.

Kurre, P., Kiem, H. P., Morris, J., Heyward, S., Battini, J. L., and Miller, A. D. (1999). Efficient transduction by an amphotropic retrovirus vector is dependent on high-level expression of the cell surface virus receptor. *J. Virol.* **73,** 495–500.

Lamping, K. G., Rios, C. D., Chun, J. A., Ooboshi, H., Davidson, B. L., and Heistad, D. D. (1997). Intrapericardial administration of adenovirus for gene transfer. *Am. J. Physiol.* **272,** H310–H317.

Laurent, N., Coninck, S. W. D., Mihaylova, E., Leontieva, E., WarnierPirotte, M. T., Wattiaux, R., and Jadot, M. (1999). Uptake by rat liver and intracellular fate of plasmid DNA complexed with poly-L-lysine or poly-D-lysine. *FEBS Lett.* **443,** 61–65.

Lechardeur, D., Sohn, K.-J., Haardt, M., Joshi, P. B., Monck, M., Graham, R. W., Beatty, B., Squire, J., O'Brodovich, H., and Lukacs, G. L. (1999). Metabolic instability of plasmid DNA in the cytosol: A potential barrier to gene transfer. *Gene Ther.* **6,** 482–497.

Le Doux, J. M., Morgan, J. R., Snow, R. G., and Yarmush, M. L. (1996). Proteoglycans secreted by packaging cell lines inhibit retrovirus infection. *J. Virol.* **70**, 6468–6473.

Le Doux, J. M., Morgan, J. R., and Yarmush, M. L. (1998). Removal of proteoglycans increases efficiency of retroviral gene transfer. *Biotechnol. Bioeng.* **58**, 23–34.

Lee, R. J., and Huang, L. (1997). Lipidic vector systems for gene transfer. *Crit. Rev. Ther. Drug Carrier Syst.* **14**, 173–206.

Leopold, P. L., Ferris, B., Grinberg, I., Worgall, S., Hackett, N. R., and Crystal, R. G. (1998). Fluorescent virions: Dynamic tracking of the pathway of adenoviral gene transfer vectors in living cells. *Hum. Gene Ther.* **9**, 367–378.

Li, E., Stupack, D., Bokoch, G. M., and Nemerow, G. R. (1998). Adenovirus endocytosis requires actin cytoskeleton reorganization mediated by Rho family GTPases. *J. Virol.* **72**, 8806–8812.

Li, S., Rizzo, M. A., Bhattacharya, S., and Huang, L. (1998). Characterization of cationic lipid–protamine–DNA (LPD) complexes for intravenous gene delivery. *Gene Ther.* **5**, 930–937.

Li, S., Tseng, W.-C., Beer Stolz, D., Wu, S.-P., Watkins, S. C., and Huang, L. (1999). Dynamic changes in the characteristics of cationic lipidic vectors after exposure to mouse serum: Implications for intravenous lipofection. *Gene Ther.* **6**, 585–594.

Li, Y., Pong, R. C., Bergelson, J. M., Hall, M. C., Sagalowsky, A. I., Tseng, C. P., Wang, Z., and Hsieh, J. T. (1999). Loss of adenoviral receptor expression in human bladder cancer cells: A potential impact on the efficacy of gene therapy. *Cancer Res.* **59**, 325–330.

Liu, F., Qi, H., Huang, L., and Liu, D. (1997). Factors controlling the efficiency of cationic lipid-mediated transfection *in vivo* via intravenous administration. *Gene Ther.* **4**, 517–523.

Liu, Y., Thor, A., Shtivelman, E., Cao, Y., Tu, G., Heath, T. D., and Debs, R. J. (1999). Systemic gene delivery expands the repertoire of effective antiangiogenic agents. *J. Biol. Chem.* **274**, 13338–13344.

Lochmuller, H., Jani, A., Huard, J., Prescott, S., Simoneau, M., Massie, B., Karpati, G., and Acsadi, G. (1994). Emergence of early region 1-containing replication-competent adenovirus in stocks of replication-defective adenovirus recombinants (delta E1 + delta E3) during multiple passages in 293 cells. *Hum. Gene Ther.* **5**, 1485–1491.

Lochmuller, H., Petrof, B. J., Pari, G., Larochelle, N., Dodelet, V., Wang, Q., Allen, C., Prescott, S., Massie, B., Nalbantoglu, J., and Karpati, G. (1996). Transient immunosuppression by FK506 permits a sustained high-level dystrophin expression after adenovirus-mediated dystrophin minigene transfer to skeletal muscles of adult dystrophic (mdx) mice. *Gene Ther.* **3**, 706–716.

Luby-Phelps, K. (1994). Physical properties of cytoplasm. *Curr. Opin. Cell Biol.* **6**, 3–9.

Ludtke, J. J., Zhang, G., Sebestyen, M. G., and Wolff, J. A. (1999). A nuclear localization signal can enhance both the nuclear transport and expression of 1 kb DNA. *J. Cell Sci.* **112**, 2033–2041.

Mack, C. A., Song, W. R., Carpenter, H., Wickham, T. J., Kovesdi, I., Harvey, B. G., Magovern, C. J., Isom, O. W., Rosengart, T., Falck-Pedersen, E., Hackett, N. R., Crystal, R. G., and Mastrangeli, A. (1997). Circumvention of anti-adenovirus neutralizing immunity by administration of an adenoviral vector of an alternate serotype. *Hum. Gene Ther.* **8**, 99–109.

Magari, S. R., Rivera, V. M., Iuliucci, J. D., Gilman, M., and Cerasoli, F. (1997). Pharmacologic control of a humanized gene therapy system implanted into nude mice. *J. Clin. Invest.* **100**, 2865–2872.

Mahato, R. I., Takakura, Y., and Hashida, M. (1997a). Development of targeted delivery systems for nucleic acid drugs. *J. Drug Target.* **4**, 337–357.

Mahato, R. I., Takakura, Y., and Hashida, M. (1997b). Nonviral vectors for *in vivo* gene delivery: Physicochemical and pharmacokinetic considerations. *Crit. Rev. Ther. Drug Carrier Syst.* **14**, 133–172.

Mahato, R. I., Anwer, K., Tagliaferri, F., Meaney, C., Leonard, P., Wadhwa, M. S., Logan, M., French, M., and Rolland, A. (1998). Biodistribution and gene expression of lipid/plasmid complexes after systemic administration. *Hum. Gene Ther.* **9**, 2083–2099.

Mahvi, D. M., Sheehy, M. J., and Yang, N. S. (1997). DNA cancer vaccines: A gene gun approach. *Immunol. Cell Biol.* **75**, 456–460.

Manning, W. C., Zhou, S., Bland, M. P., Escobedo, J. A., and Dwarki, V. (1998). Transient immunosuppression allows transgene expression following readministration of adeno-associated viral vectors. *Hum. Gene Ther.* **9**, 477–485.

March, K. L., Woody, M., Mehdi, K., Zipes, D. P., Brantly, M., and Trapnell, B. C. (1999). Efficient *in vivo* catheter-based pericardial gene transfer mediated by adenoviral vectors. *Clin. Cardiol.* **22**, 1-23-9.

Marshall, D. J., and Leiden, J. M. (1998). Recent advances in skeletal-muscle-based gene therapy. *Curr. Opin. Genet. Dev.* **8**, 360–365.

Mastrangeli, A., Harvey, B. G., Yao, J., Wolff, G., Kovesdi, I., Crystal, R. G., and Falck-Pedersen, E. (1996). "Sero-switch" adenovirus-mediated *in vivo* gene transfer: Circumvention of anti-adenovirus humoral immune defenses against repeat adenovirus vector administration by changing the adenovirus serotype. *Hum. Gene Ther.* **7**, 79–87.

Mauceri, H. J., Hanna, N. N., Beckett, M. A., Gorski, D. H., Staba, M. J., Stellato, K. A., Bigelow, K., Heimann, R., Gately, S., Dhanabal, M., Soff, G. A., Sukhatme, V. P., Kufe, D. W., and Weichselbaum, R. R. (1998). Combined effects of angiostatin and ionizing radiation in antitumour therapy. *Nature (London)* **394**, 287–291.

Maurer, N., Mori, A., Palmer, L., Monck, M. A., Mok, K. W., Mui, B., Akhong, Q. F., and Cullis, P. R. (1999). Lipid-based systems for the intracellular delivery of genetic drugs. *Mol. Membr. Biol.* **16**, 129–140.

McDonald, R. J., Liggitt, H. D., Roche, L., Nguyen, H. T., Pearlman, R., Raabe, O. G., Bussey, L. B., and Gorman, C. M. (1998). Aerosol delivery of lipid:DNA complexes to lungs of rhesus monkeys. *Pharm. Res.* **15**, 671–679.

McLean, J. W., Fox, E. A., Baluk, P., Bolton, P. B., Haskell, A., Pearlman, R., Thurston, G., Umemoto, E. Y., and McDonald, D. M. (1997). Organ-specific endothelial cell uptake of cationic liposome–DNA complexes in mice. *Am. J. Physiol.* **42**, (*Heart Circ Phys.*) H387–H404.

Meyer, K. E. B., Uyechi, L. S., and Szoka, F. C. (1997). Manipulating the intracellular trafficking of nucleic acids. *In* "Gene Therapy for Diseases of the Lung" (K. L. Brigham, ed.), pp. 135–180. Dekker, New York.

Miller, A. D., and Buttimore, C. (1986). Redesign of retrovirus packaging cell lines to avoid recombination leading to helper virus production. *Mol. Cell. Biol.* **6**, 2895–2902.

Miller, C. R., Buchsbaum, D. J., Reynolds, P. N., Douglas, J. T., Gillespie, G. Y., Mayo, M. S., Raben, D., and Curiel, D. T. (1998). Differential susceptibility of primary and established human glioma cells to adenovirus infection: Targeting via the epidermal growth factor receptor achieves fiber receptor-independent gene transfer. *Cancer Res.* **58**, 5738–5748.

Miller, D. G., Adam, M. A., and Miller, A. D. (1990). Gene transfer by retrovirus vectors occurs only in cells that are actively replicating at the time of infection. *Mol. Cell. Biol.* **10**, 4239–4242.

Miller, N., and Whelan, J. (1997). Progress in transcriptionally targeted and regulatable vectors for genetic therapy. *Hum. Gene Ther.* **8**, 803–815.

Mir, L. M., Bureau, M. F., Gehl, J., Rangara, R., Rouy, D., Caillaud, J. M., Delaere, P., Branellec, D., Schwartz, B., and Scherman, D. (1999). High-efficiency gene transfer into skeletal muscle mediated by electric pulses. *Proc. Natl. Acad. Sci. U.S.A.* **96**, 4262–4267.

Miyazawa, N., Leopold, P. L., Hackett, N. R., Ferris, B., Worgall, S., Falck-Pedersen, E., and Crystal, R. G. (1999). Fiber swap between adenovirus subgroups B and C alters intracellular trafficking of adenovirus gene transfer vectors. *J. Virol.* **73**, 6056–6065.

Mizuguchi, H., Nakagawa, T., Morioka, Y., Imazu, S., Nakanishi, M., Kondo, T., Hayakawa, T., and Mayumi, T. (1997). Cytoplasmic gene expression system enhances the efficiency of cationic liposome-mediated *in vivo* gene transfer into mouse brain. *Biochem. Biophys. Res. Commun.* **234**, 15–18.

Mohr, L., Schauer, J. I., Boutin, R. H., Moradpour, D., and Wands, J. R. (1999). Targeted gene transfer to hepatocellular carcinoma cells *in vitro* using a novel monoclonal antibody-based gene delivery system. *Hepatology* **29**, 82–89.

Monahan, P. E., Samulski, R. J., Tazelaar, J., Xiao, X., Nichols, T. C., Bellinger, D. A., Read, M. S., and Walsh, C. E. (1998). Direct intramuscular injection with recombinant AAV vectors results in sustained expression in a dog model of hemophilia. *Gene Ther.* **5**, 40–49.

Morsy, M. A., and Caskey, C. T. (1999). Expanded-capacity adenoviral vectors—The helper-dependent vectors. *Mol. Med. Today* **5**, 18–24.

Morsy, M. A., Gu, M., Motzel, S., Zhao, J., Lin, J., Su, Q., Allen, H., Franlin, L., Parks, R. J., Graham, F. L., Kochanek, S., Bett, A. J., and Caskey, C. T. (1998a). An adenoviral vector deleted for all viral coding sequences results in enhanced safety and extended expression of a leptin transgene. *Proc. Natl. Acad. Sci. U.S.A.* **95**, 7866–7871.

Morsy, M. A., Gu, M. C., Zhao, J. Z., Holder, D. J., Rogers, I. T., Pouch, W. J., Motzel, S. L., Klein, H. J., Gupta, S. K., Liang, X., Tota, M. R., Rosenblum, C. I., and Caskey, C. T. (1998b). Leptin gene therapy and daily protein administration: A comparative study in the ob/ob mouse. *Gene Ther.* **5**, 8–18.

Mortimer, I., Tam, P., MacLachlan, I., Graham, R. W., Saravolac, E. G., and Joshi, P. B. (1999). Cationic lipid-mediated transfection of cells in culture requires mitotic activity. *Gene Ther.* **6**, 403–411.

Movassagh, M., Desmyter, C., Baillou, C., Chapel-Fernandes, S., Guigon, M., Klatzmann, D., and Lemoine, F. M. (1998). High-level gene transfer to cord blood progenitors using gibbon ape leukemia virus pseudotype retroviral vectors and an improved clinically applicable protocol. *Hum. Gene Ther.* **9**, 225–234.

Muruve, D. A., Barnes, M. J., Stillman, I. E., and Libermann, T. A. (1999). Adenoviral gene therapy leads to rapid induction of multiple chemokines and acute neutrophil-dependent hepatic injury *in vivo*. *Hum. Gene Ther.* **10**, 965–976.

Nalbantoglu, J., Pari, G., Karpati, G., and Holland, P. C. (1999). Expression of the primary coxsackie and adenovirus receptor is downregulated during skeletal muscle maturation and limits the efficacy of adenovirus-mediated gene delivery to muscle cells. *Hum. Gene Ther.* **10**, 1009–1019.

Nemunaitis, J., Fong, T., Burrows, F., Bruce, J., Peters, G., Ognoskie, N., Meyer, W., Wynne, D., Kerr, R., Pippen, J., Oldham, F., and Ando, D. (1999). Phase I trial of interferon gamma retroviral vector administered intratumorally with multiple courses in patients with metastatic melanoma. *Hum. Gene Ther.* **10**, 1289–1298.

Nguyen, J. T., Wu, P., Clouse, M. E., Hlatky, L., and Terwilliger, E. F. (1998). Adeno-associated virus-mediated delivery of antiangiogenic factors as an antitumor strategy. *Cancer Res.* **58**, 5673–5677.

Niven, R., Pearlman, R., Wedeking, T., Mackeigan, J., Noker, P., Simpson-Herren, L., and Smith, J. G. (1998). Biodistribution of radiolabeled lipid-DNA complexes and DNA in mice. *J. Pharm. Sci.* **87**, 1292–1299.

No, D., Yao, T. P., and Evans, R. M. (1996). Ecdysone-inducible gene expression in mammalian cells and transgenic mice. *Proc. Natl. Acad. Sci. U.S.A.* **93**, 3346–3351.

Nomura, T., Yasuda, K., Yamada, T., Okamoto, S., Mahato, R. I., Watanabe, Y., Takakura, Y., and Hashida, M. (1999). Gene expression and antitumor effects following direct interferon (IFN)-gamma gene transfer with naked plasmid DNA and DC-chol liposome complexes in mice. *Gene Ther.* **6**, 121–129.

Ochsenbein, A. F., Klenerman, P., Karrer, U., Ludewig, B., Pericin, M., Hengartner, H., and Zinkernagel, R. M. (1999). Immune surveillance against a solid tumor fails because of immunological ignorance. *Proc. Natl. Acad. Sci. U.S.A.* **96**, 2233–2238.

Onodera, M., Nelson, D. M., Sakiyama, Y., Candotti, F., and Blaese, R. M. (1999). Gene therapy for severe combined immunodeficiency caused by adenosine deaminase deficiency: Improved retroviral vectors for clinical trials. *Acta Haematol.* **101**, 89–96.

O'Reilly, M. S., Holmgren, L., Shing, Y., Chen, C., Rosenthal, R. A., Moses, M., Lane, W. S., Cao, Y., Sage, E. H., and Folkman, J. (1994). Angiostatin: A novel angiogenesis inhibitor that mediates the suppression of metastases by a Lewis lung carcinoma. *Cell (Cambridge, Mass.)* **79**, 315–328.

O'Reilly, M. S., Boehm, T., Shing, Y., Fukai, N., Vasios, G., Lane, W. S., Flynn, E., Birkhead, J. R., Olsen, B. R., and Folkman, J. (1997). Endostatin: An endogenous inhibitor of angiogenesis and tumor growth. *Cell (Cambridge, Mass.)* **88**, 277–285.

O'Riordan, C. R., Lachapelle, A., Delgado, C., Parkes, V., Wadsworth, S. C., Smith, A. E., and Francis, G. E. (1999). PE-

Gylation of adenovirus with retention of infectivity and protection from neutralizing antibody *in vitro* and *in vivo*. *Hum. Gene Ther.* **10**, 1349–1358.

Orlic, D., Girard, L. J., Jordan, C. T., Anderson, S. M., Cline, A. P., and Bodine, D. M. (1996). The level of mRNA encoding the amphotropic retrovirus receptor in mouse and human hematopoietic stem cells is low and correlates with the efficiency of retrovirus transduction. *Proc. Natl. Acad. Sci. U.S.A.* **93**, 11097–11102.

Osaka, G., Carey, K., Cuthbertson, A., Godowski, P., Patapoff, T., Ryan, A., Gadek, T., and Mordenti, J. (1996). Pharmacokinetics, tissue distribution, and expression efficiency of plasmid [^{33}P]DNA following intravenous administration of DNA/cationic lipid complexes in mice: Use of a novel radionuclide approach. *J. Pharm. Sci.* **85**, 612–618.

Palsson, B., and Andreadis, S. (1997). The physico-chemical factors that govern retrovirus-mediated gene transfer. *Exp. Hematol.* **25**, 94–102.

Parsons, D. W., Grubb, B. R., Johnson, L. G., and Boucher, R. C. (1998). Enhanced *in vivo* airway gene transfer via transient modification of host barrier properties with a surface-active agent. *Hum. Gene Ther.* **9**, 2661–2672.

Pensiero, M. N., Wysocki, C. A., Nader, K., and Kikuchi, G. E. (1996). Development of amphotropic murine retrovirus vectors resistant to inactivation by human serum. *Hum. Gene Ther.* **7**, 1095–1101.

Piirsoo, M., Ustav, E., Mandel, T., Stenlund, A., and Ustav, M. (1996). Cis and trans requirements for stable episomal maintenance of the BPV-1 replicator. *EMBO J.* **15**, 1–11.

Pisetsky, D. S. (1996). Immune activation by bacterial DNA: A new genetic code. *Immunity* **5**, 303–310.

Putney, S. D., and Burke, P. A. (1998). Improving protein therapeutics with sutained-release formulations. *Nat. Biotechnol.* **16**, 153–157.

Rabinowitz, J. E., and Samulski, J. (1998). Adeno-associated virus expression systems for gene transfer. *Curr. Opin. Biotechnol.* **9**, 470–475.

Rakhmilevich, A. L., Timmins, J. G., Janssen, K., Pohlmann, E. L., Sheehy, M. J., and Yang, N. S. (1999). Gene gun-mediated IL-12 gene therapy induces antitumor effects in the absence of toxicity: A direct comparison with systemic IL-12 protein therapy. *J. Immunother.* **22**, 135–144.

Rancourt, C., Rogers, B. E., Sosnowski, B. A., Wang, M., Piche, A., Pierce, G. F., Alvarez, R. D., Siegal, G. P., Douglas, J. T., and Curiel, D. T. (1998). Basic fibroblast growth factor enhancement of adenovirus-mediated delivery of the herpes simplex virus thymidine kinase gene results in augmented therapeutic benefit in a murine model of ovarian cancer. *Clin. Cancer Res.* **4**, 2455–2461.

Rickinson, A. B., and Kieff, E. (1996). Epstein-Barr virus. *In* "Fields Virology" (B. N. Fields, D. M. Knipe, and P. M. Howley, eds.), pp. 2397–2446. Lippincott-Raven, Philadelphia.

Rigg, R. J., Chen, J., Dando, J. S., Forestell, S. P., Plavec, I., and Bohnlein, E. (1996). A novel human amphotropic packaging cell line: High titer, complement resistance, and improved safety. *Virology* **218**, 290–295.

Rivera, V. M., Clackson, T., Natesan, S., Pollock, R., Amara, J. F., Keenan, T., Magari, S. R., Phillips, T., Courage, N. L., Cerasoli, F., Jr., Holt, D. A., and Gilman, M. (1996). A humanized system for pharmacologic control of gene expression. *Nat. Med.* **2**, 1028–1032.

Robbins, P. D., Tahara, H., and Ghivizzani, S. C. (1998). Viral vectors for gene therapy. *Trends Biotechnol.* **16**, 35–40.

Roelvink, P. W., Lizonova, A., Lee, J. G., Li, Y., Bergelson, J. M., Finberg, R. W., Brough, D. E., Kovesdi, I., and Wickham, T. J. (1998). The coxsackievirus-adenovirus receptor protein can function as a cellular attachment protein for adenovirus serotypes from subgroups A, C, D, E, and F. *J. Virol.* **72**, 7909–7915.

Rogers, B. E., Douglas, J. T., Ahlem, C., Buchsbaum, D. J., Frincke, J., and Curiel, D. T. (1997). Use of a novel cross-linking method to modify adenovirus tropism. *Gene Ther.* **4**, 1387–1392.

Rolland, A. P. (1998). From genes to gene medicines: Recent advances in nonviral gene delivery. *Crit. Rev. Ther. Drug Carrier Syst.* **15**, 143–198.

Ropp, J. D. (1999). Cell division is a prerequisite for successful *in vitro* transfection. Submitted for publication.

Rossi, F. M., and Blau, H. M. (1998). Recent advances in inducible gene expression systems. *Curr. Opin. Biotechnol.* **9**, 451–456.

Roth, J. A., Nguyen, D., Lawrence, D. D., Kemp, B. L., Carrasco, C. H., Ferson, D. Z., Hong, W. K., Komaki, R., Lee, J. J., Nesbitt, J. C., Pisters, K. M., Putnam, J. B., Schea, R., Shin, D. M., Walsh, G. L., Dolormente, M. M., Han, C. I., Martin, F. D., Yen, N., Xu, K., Stephens, L. C., McDonnell, T. J., Mukhopadhyay, T., and Cai, D. (1996). Retrovirus-mediated wild-type p53 gene transfer to tumors of patients with lung cancer. *Nat. Med.* **2**, 985–991.

Rother, R. P., Fodor, W. L., Springhorn, J. P., Birks, C. W., Setter, E., Sandrin, M. S., Squinto, S. P., and Rollins, S. A. (1995). A novel mechanism of retrovirus inactivation in human serum mediated by anti-alpha-galactosyl natural antibody. *J. Exp. Med.* **182**, 1345–1355.

Roy, S., Shirley, P. S., McClelland, A., and Kaleko, M. (1998). Circumvention of immunity to the adenovirus major coat protein hexon. *J. Virol.* **72**, 6875–6879.

Rutledge, E. A., Halbert, C. L., and Russell, D. W. (1998). Infectious clones and vectors derived from adeno-associated virus (AAV) serotypes other than AAV type 2. *J. Virol.* **72**, 309–319.

Sabatino, D. E., Do, B. Q., Pyle, L. C., Seidel, N. E., Girard, L. J., Spratt, S. K., Orlic, D., and Bodine, D. M. (1997). Amphotropic or gibbon ape leukemia virus retrovirus binding and transduction correlates with the level of receptor mRNA in human hematopoietic cell lines. *Blood Cells Mol. Dis.* **23**, 422–433.

Sandig, V., and Strauss, M. (1996). Liver-directed gene transfer and application to therapy. *J. Mol. Med.* **74**, 205–212.

Schiedner, G., Morral, N., Parks, R. J., Wu, Y., Koopmans, S. C., Langston, C., Graham, F. L., Beaudet, A. L., and Kochanek, S. (1998). Genomic DNA transfer with a high-capacity adenovirus vector results in improved *in vivo* gene expression and decreased toxicity. *Nat. Genet.* **18**, 180–183.

Schowalter, D. B., Himeda, C. L., Winther, B. L., Wilson, C. B., and Kay, M. A. (1999). Implication of interfering antibody formation and apoptosis as two different mechanisms leading to variable duration of adenovirus-mediated transgene expression in immune-competent mice. *J. Virol.* **73**, 4755–4766.

Sclimenti, C. R., and Calos, M. P. (1998). Epstein-Barr virus vectors for gene expression and transfer. *Curr. Opin. Biotechnol.* **9**, 476–479.

Sebestyen, M. G., Ludtke, J. J., Bassik, M. C., Zhang, G. F., Budker, V., Lukhtanov, E. A., Hagstrom, J. E., and Wolff, J. A. (1998). DNA vector chemistry: The covalent attachment of signal peptides to plasmid DNA. *Nat. Biotechnol.* **16**, 80–85.

Smith, K. T., Shepherd, A. J., Boyd, J. E., and Lees, G. M. (1996). Gene delivery systems for use in gene therapy: An overview of quality assurance and safety issues. *Gene Ther.* **3**, 190–200.

Smith, T. A., White, B. D., Gardner, J. M., Kaleko, M., and McClelland, A. (1996). Transient immunosuppression permits successful repetitive intravenous administration of an adenovirus vector. *Gene Ther.* **3**, 496–502.

Snyder, R. O., Spratt, S. K., Lagarde, C., Bohl, D., Kaspar, B., Sloan, B., Cohen, L. K., and Danos, O. (1997). Efficient and stable adeno-associated virus-mediated transduction in the skeletal muscle of adult immunocompetent mice. *Hum. Gene Ther.* **8**, 1891–1900.

Song, Y. K., Liu, F., Chu, S., and Liu, D. (1997). Characterization of cationic liposome-mediated gene transfer *in vivo* by intravenous administration. *Hum. Gene Ther.* **8**, 1585–1594.

Springer, M. L., Chen, A. S., Kraft, P. E., Bednarski, M., and Blau, H. M. (1998). VEGF gene delivery to muscle: Potential role for vasculogenesis in adults. *Mol. Cell* **2**, 549–558.

Stern, M., Caplen, N. J., Browning, J. E., Griesenbach, U., Sorgi, F., Huang, L., Gruenert, D. C., Marriot, C., Crystal, R. G., Geddes, D. M., and Alton, E. (1998). The effect of mucolytic agents on gene transfer across a CF sputum barrier *in vitro*. *Gene Ther.* **5**, 91–98.

Suomalainen, M., Nakano, M. Y., Keller, S., Boucke, K., Stidwill, R. P., and Greber, U. F. (1999). Microtubule-dependent plus- and minus end-directed motilities are competing processes for nuclear targeting of adenovirus. *J. Cell Biol.* **144**, 657–672.

Takeuchi, Y., Cosset, F. L., Lachmann, P. J., Okada, H., Weiss, R. A., and Collins, M. K. (1994). Type C retrovirus inactivation by human complement is determined by both the viral genome and the producer cell. *J. Virol.* **68**, 8001–8007.

Tanaka, T., Cao, Y., Folkman, J., and Fine, H. A. (1998). Viral vector-targeted antiangiogenic gene therapy utilizing an angiostatin complementary DNA. *Cancer Res.* **58**, 3362–3369.

Tripathy, S. K., Black, H. B., Goldwasser, E., and Leiden, J. M. (1996). Immune responses to transgene-encoded proteins limit the stability of gene expression after injection of replication-defective adenovirus vectors. *Nat. Med.* **2**, 545–550.

Tseng, W. C., Haselton, F. R., and Giorgio, T. D. (1999). Mitosis enhances transgene expression of plasmid delivered by cationic liposomes. *Biochim. Biophys. Acta* **1445**, 53–64.

Verdier, F., and Descotes, J. (1999). Preclinical safety evaluation of human gene therapy products. *Toxicol. Sci.* **47**, 9–15.

Vilquin, J. T., Guerette, B., Kinoshita, I., Roy, B., Goulet, M., Gravel, C., Roy, R., and Tremblay, J. P. (1995). FK506 immunosuppression to control the immune reactions triggered by first-generation adenovirus-mediated gene transfer. *Hum. Gene Ther.* **6**, 1391–1401.

Vitiello, L., Bockhold, K., Joshi, P. B., and Worton, R. G. (1998). Transfection of cultured myoblasts in high serum concentration with DODAC:DOPE liposomes. *Gene Ther.* **5**, 1306–1313.

Wagner, E. (1998). Effects of membrane-active agents in gene delivery. *J. Controlled Release* **53**, 155–158.

Walters, R. W., Grunst, T., Bergelson, J. M., Finberg, R. W., Welsh, M. J., and Zabner, J. (1999). Basolateral localization of fiber receptors limits adenovirus infection from the apical surface of airway epithelia. *J. Biol. Chem.* **274**, 10219–10226.

Walther, W., and Stein, U. (1996). Cell type specific and inducible promoters for vectors in gene therapy as an approach for cell targeting. *J. Mol. Med.* **74**, 379–392.

Wang, K., Huang, S., Kapoor-Munshi, A., and Nemerow, G. (1998). Adenovirus internalization and infection require dynamin. *J. Virol.* **72**, 3455–3458.

Wang, X. S., Khuntirat, B., Qing, K., Ponnazhagan, S., Kube, D. M., Zhou, S., Dwarki, V. J., and Srivastava, A. (1998). Characterization of wild-type adeno-associated virus type 2-like particles generated during recombinant viral vector production and strategies for their elimination. *J. Virol.* **72**, 5472–5480.

Whitmore, M., Li, S., and Huang, L. (1999). LPD lipopolyplex initiates a potent cytokine response and inhibits tumor growth. *Gene Therap.* **6**, 1867–1875.

Wilke, M., Fortunati, E., van den Broek, M., Hoogeveen, A. T., and Scholte, B. J. (1996). Efficacy of a peptide-based gene delivery system depends on mitotic activity. *Gene Ther.* **3**, 1133–1142.

Wilson, J. B., Bell, J. L., and Levine, A. J. (1996). Expression of Epstein–Barr virus nuclear antigen-1 induces B cell neoplasia in transgenic mice. *EMBO J.* **15**, 3117–3126.

Wolff, J. A., Ludtke, J. J., Acsadi, G., Williams, P., and Jani, A. (1992). Long-term persistence of plasmid DNA and foreign gene expression in mouse muscle. *Hum. Mol. Genet.* **1**, 363–369.

Worgall, S., Wolff, G., Falck-Pedersen, E., and Crystal, R. G. (1997). Innate immune mechanisms dominate elimination of adenoviral vectors following *in vivo* administration. *Hum. Gene Ther.* **8**, 37–44.

Xiao, W., Chirmule, N., Berta, S. C., McCullough, B., Gao, G., and Wilson, J M. (1999). Gene therapy vectors based on adeno-associated virus type 1. *J. Virol.* **73**, 3994–4003.

Yang, J. P., and Huang, L. (1997). Overcoming the inhibitory effect of serum on lipofection by increasing the charge ratio of cationic liposome to DNA. *Gene Ther.* **4**, 950–960.

Yang, J. P., and Huang, L. (1998). Time-dependent maturation of cationic liposome-DNA complex for serum resistance. *Gene Ther.* **5**, 380–387.

Yang, Y., Nunes, F. A., Berencsi, K., Furth, E. E., Gonczol, E., and Wilson, J. M. (1994a). Cellular immunity to viral antigens limits E1-deleted adenoviruses for gene therapy. *Proc. Natl. Acad. Sci. U.S.A.* **91**, 4407–4411.

Yang, Y., Nunes, F. A., Berencsi, K., Gonczol, E., Engelhardt, J. F., and Wilson, J. M. (1994b). Inactivation of E2a in re-combinant adenoviruses improves the prospect for gene therapy in cystic fibrosis. *Nat. Genet.* **7**, 362–369.

Yang, Y., Jooss, K. U., Su, Q., Ertl, H. C., and Wilson, J. M. (1996a). Immune responses to viral antigens versus trans-gene product in the elimination of recombinant adenovirus-infected hepatocytes *in vivo*. *Gene Ther.* **3**, 137–144.

Yang, Y., Greenough, K., and Wilson, J. M. (1996b). Transient immune blockade prevents formation of neutralizing an-tibody to recombinant adenovirus and allows repeated gene transfer to mouse liver. *Gene Ther.* **3**, 412–420.

Yap, J., O'Brien, T., Tazelaar, H. D., and McGregor, C. G. (1997). Immunosuppression prolongs adenoviral mediated transgene expression in cardiac allograft transplantation. *Cardiovasc. Res.* **35**, 529–535.

Ye, X., Rivera, V. M., Zoltick, P., Cerasoli, F., Jr., Schnell, M. A., Gao, G., Hughes, J. V., Gilman, M., and Wilson, J. M. (1999). Regulated delivery of therapeutic proteins after *in vivo* somatic cell gene transfer. *Science* **283**, 88–91.

Yei, S., Mittereder, N., Tang, K., O'Sullivan, C., and Trapnell, B. C. (1994). Adenovirus-mediated gene transfer for cystic fibrosis: Quantitative evaluation of repeated *in vivo* vector administration to the lung. *Gene Ther.* **1**, 192–200.

Zabner, J., Fasbender, A. J., Moninger, T., Poellinger, K. A., and Welsh, M. J. (1995). Cellular and molecular barriers to gene transfer by a cationic lipid. *J. Biol. Chem.* **270**, 18997–19007.

Zanta, M. A., Belguise-Valladier, P., and Behr, J. P. (1999). Gene delivery: A single nuclear localization signal peptide is sufficient to carry DNA to the cell nucleus. *Proc. Natl. Acad. Sci. U.S.A.* **96**, 91–96.

Zhao, D. D., Watarai, S., Lee, J. T., Kouchi, S., Ohmori, H., and Yasuda, T. (1997). Gene transfection by cationic lipo-somes: Comparison of the transfection efficiency of liposomes prepared from various positively charged lipids. *Acta Med Okayama* **51**, 149–154.

Zhu, J., Grace, M., Casale, J., Chang, A. T., Musco, M. L., Bordens, R., Greenberg, R., Schaefer, E., and Indelicato, S. R. (1999). Characterization of replication-competent adenovirus isolates from large-scale production of a recombinant adenoviral vector. *Hum. Gene Ther.* **10**, 113–121.

Zsengeller, Z. K., Wert, S. E., Hull, W. M., Hu, X., Yei, S., Trapnell, B. C., and Whitsett, J. A. (1995). Persistence of repli-cation-deficient adenovirus-mediated gene transfer in lungs of immune-deficient (nu/nu) mice. *Hum. Gene Ther.* **6**, 457–467.

PART IX: BREAST

CHAPTER 31

BREAST RECONSTRUCTION

Kuen Yong Lee, Craig R. Halberstadt, Walter D. Holder, and David J. Mooney

INTRODUCTION

There are many patients who need reconstructive surgery after mastectomy or lumpectomy due to breast cancer. Tissue engineering may provide an approach to create new soft tissue masses for these applications. Various cell types can be used for breast reconstruction, including preadipocytes, fibroblasts, smooth muscle cells, muscle myocytes, and chondrocytes. To introduce cell–polymer constructs into the body, proper selection of polymeric materials should provide three-dimensional support for engineered new tissue. There are two strategies for introducing the cell–polymer constructs into the body. The first is to surgically implant and the second approach is to deliver in a minimally invasive manner using injectable forms of materials. A critical challenge to engineer large tissue masses is vascularization, and approaches to optimize this process include delivery of angiogenic molecules or endothelial cells.

There is a significant need for breast reconstruction due to cancer; the American Cancer Society estimates that approximately 182,000 new cases of breast cancer are reported each year. Direct medical costs for treating new and existing cases approach $6 billion each year, according to the National Cancer Institute (American Cancer Society, 1997). Historically, breast cancer was treated with radical mastectomy. This procedure removed the breast, underlying pectoral muscles, and axillary lymph nodes. Radical breast surgery was developed at a time when breast cancers typically were very large before detection and were thus often unresectable by current standards. With the advent of mammography and a rising awareness of the importance of breast self-examination programs, earlier and smaller breast cancers are detected and do not warrant radical surgery. Radical mastectomy was abandoned for most patients when it became increasingly apparent that equivalent survival occurred after modified radical mastectomy, which preserved the pectoral muscles and removed less skin and fewer lymph nodes. During the past 30 years the National Surgical Adjuvant Breast Program (NSABP) has conducted numerous randomized prospective clinical trials that have explored newer and less deforming treatments for breast cancer (Harris *et al.*, 1997). These studies have unequivocally demonstrated that for most women with small breast cancers, simple excision (lumpectomy) of the breast cancer with sampling of the axillary lymph nodes followed by radiation is equivalent in survival to modified radical mastectomy. However, in some women mastectomy remains the best treatment due to extensive tumor involvement within the breast such that removal of all or most of the breast tissue is required.

The current approaches to reconstruct breast tissue include reconstructive surgery utilizing autologous tissue flaps, or implants of synthetic materials. Complex flap reconstructions using transversus rectus abdominal myocutaneous (TRAM) flaps, latissimus dorsi flaps, or muscle-free flaps from the buttock or thigh are used to reconstruct large volumes of lost breast tissue. These are very extensive procedures with long recovery periods, and require the harvest of tissue from another site in the body with its attendant side effects and complications. The advantage of this approach is that a patient's own tissues are utilized, and the use of a tissue results in a more natural tactile tissue quality. A more commonly used procedure involves the use of synthetic materials. In

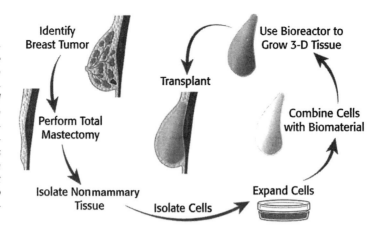

Fig. 31.1. Schematic of the approach to engineer a soft tissue replacement for women with breast cancer. The first steps include identification and removal of the breast tissue, then isolation and multiplication in vitro of appropriate cells derived from another site in the body. These cells are then combined with a biodegradable polymer matrix and implanted into the body to recreate a soft tissue that emulates the lost breast tissue.

this process a tissue expander is placed, usually beneath the pectoral muscles, and is gradually inflated over several weeks to the desired size. It is subsequently removed by a relatively simple procedure and a saline-filled permanent silastic implant is placed in the cavity. Because the implant is a foreign body there may be a substantial inflammatory response producing fibrosis, thickening, and an unnatural shape and tactile quality. Also, implants may leak and require replacement. Silicone-filled implants are now rarely used due to a possible induction of autoimmune disease caused by the leakage of silicone.

Tissue engineering is potentially a potent approach to treat the loss of tissues or organs (Langer and Vacanti, 1993; Mooney and Mikos, 1999), and could find wide utility in breast tissue reconstruction. Tissues can be engineered with a number of different strategies, but a particularly appealing approach for breast reconstruction utilizes a combination of a patient's own cells combined with a polymeric scaffold. In this approach, tissue-specific cells are isolated from a small tissue biopsy and expanded *in vitro*. The cells are subsequently placed onto polymeric scaffolds, which act as synthetic extracellular matrices. These scaffolds deliver the cells to the desired site in body, define a space for tissue formation, and potentially control the structure and function of the engineered tissue (Putnam and Mooney, 1996; Kim and Mooney, 1998a). Utilization of biodegradable polymer scaffolds leads ultimately to a new tissue mass with no permanent synthetic element. New soft tissues, which can be transplanted into a patient and reconstruct lost breast tissue, can potentially be created with this process (Fig. 31.1). This approach may circumvent the main limitations of reconstruction of breast tissue using tissue transplantation, while eliminating the need for the permanent implantation of synthetic prosthetic materials.

In this chapter, we summarize the various cell types that may be useful for breast reconstruction, the polymers that can potentially be utilized in this tissue engineering approach, and the development of suitable animal models that allow one to test this concept for breast reconstruction. In addition, we review approaches being developed to promote vascularization of the engineered tissue, because this will be a critical challenge to engineer a large mass of soft tissue with cell transplantation.

CELL TYPES FOR SOFT TISSUE ENGINEERING

Many considerations must be addressed when choosing which cell type(s) to be used in human breast tissue engineering. The first consideration is that there is substantial variability in the size, shape, and consistency of the breast. All three factors change over time, with a tendency for breast parenchyma (glands and ducts) to involute or regress as a woman ages, after pregnancy, and particularly after menopause, and to be replaced by fat. Also, comparing breast tissue among women of any given age, there is considerable variability in size, shape, tactile, elastic, and tensile characteristics of the tissue. The tensile and elastic characteristics of the breast are produced by three major factors: (1) the amount and quality of fat in the breast, (2) the amount and quality of glandular and duct tissue in the breast, and (3) the mechanical characteristics of the fibrous support structures of the breast (Cooper's ligaments). In breast reconstruction, the creation of a functional breast with lactational ability is not needed and, in fact, may add to a woman's breast cancer risk by introducing mammary epithelial cells that may be predisposed to cancer development. The major goal of breast reconstruction is to produce a breast mound with all of the aesthetic properties of a normal breast. A major concern for a woman undergoing breast reconstruction for cancer or augmentation is that the right side should match the left side and that the reconstruction should be aesthetically correct. Also, it is important to understand that there is considerable variability in the expectations of women who undergo reconstructive or augmentation breast surgery in terms of the desired size, shape, and elastic and tactile qualities of the breast reconstruction. With these observations, a projection can be made that fat cells would need to be a major part of the tissue-engineered construct and other cell types (e.g., muscle) would be needed to contribute appropriate conformational, tensile, elastic, and tactile characteristics of the construct. Also, due to the variability in normal breast, some individual manipulation of these constituents may be required.

A major cell type comprising normal breast tissue, and likely the engineered replacement, is the adipocyte. Lipid-laden fat cells (adipocytes) are terminally differentiated and will not divide further *in vivo* or under cell culture conditions. If these cells are to be used, they would have to be harvested from fat in an equivalent volume required for the construct (Nguyen *et al.*, 1990; Kononas *et al.*, 1993; Hang-Fu *et al.*, 1995). Alternatively, preadipocytes could be used for engineering soft tissue (Patrick *et al.*, 1999), because these cells can potentially be expanded in culture. These cells are fibroblast-like in structure and often contain small lipid droplets. More research is required to best determine how to maintain the preadipocyte phenotype with growth factors or growth medium supplementation to optimally expand and maintain these cells.

Because breast tissue is more than fat alone in most cases, additional cell types must be considered to contribute to the shape, tactile, elastic, and tensile characteristics of a breast reconstructed by tissue engineering methods. Possible other cell types include smooth muscle cells, fibroblasts, skeletal muscle, and elastic cartilage. Fibroblasts contribute greatly to the support structure of the breast by laying down bands of collagen that connect the breast tissue to the skin and to the pectoral muscle, as well as helping maintain the overall shape. Fibroblasts are an integral component of tissue-engineered skin products, and transplantation of fibroblast-containing tissues has been demonstrated successfully to replace lost skin tissue in several situations (Landeen *et al.*, 1992; Eaglstein and Falanga, 1998). The density and firmness of the breast are primary effects of the glandular epithelium and ductal structure. Tissues that have similar tactile and elastic properties are almost exclusively muscle. Smooth muscle cells can be readily isolated from a number of organs and expanded in culture. Techniques have been developed to seed these cells efficiently onto three-dimensional scaffolds fabricated from various polymers (Kim *et al.*, 1998), and to grow new tissues from these cell–polymer constructs with a defined size and shape (Kim and Mooney, 1998b). Implantation of smooth muscle-containing polymers can lead to the reformation of significant tissue masses, with reorganization of the smooth muscle tissue into appropriate three-dimensional structures (Oberpenning *et al.*, 1999). Muscle myocytes can also be greatly expanded *in vitro,* and have been demonstrated to reform functional tissue masses under appropriate conditions (Vandenburgh and Kaufman, 1979; Vandenburgh *et al.*, 1991). However, it remains to be demonstrated that smooth or skeletal muscle myocytes will maintain a tissue mass over long periods of time without neural stimulation. An alternative approach is to utilize chondrocytes as a component of an engineered breast tissue. Elastic cartilage has many of the mechanical properties of glandular breast tissue that are potentially important for tissue engineering of breast (e.g., elasticity). Chondrocytes can be readily expanded in culture and are able to survive in low oxygen tensions. Chondrocytes have been used extensively in tissue engineering to engineer a variety of tissue constructs both *in vitro* and *in vivo* (Atala *et al.*, 1994; Breinan *et al.*, 1998; Wakitani *et al.*, 1998).

A number of issues must be ultimately addressed in relation to cell expansion for breast tissue reconstruction. In considering the number of cell types that may potentially be used, it is very likely that a variety of growth factors will be needed for the *in vitro* expansion of cells. Implanted cell-bearing polymers may also need to be created to provide the short- or long-term stimulation of growth factors and other substances to enhance vascularization, or to provide or sustain a local milieu that will maintain a stable, healthy tissue mass in an area that may be foreign to the tissue (Mooney *et al.*, 1996c; Peters *et al.*, 1999a; Sheridan *et al.*, 2000). Standard isolation and expansion protocols must also be developed, and critical factors include aseptic technique in the harvest, routine quality control testing of all cultures, and long-term cell storage. It may also be possible to isolate all the cell types required for breast tissue engineering from a single tissue source. For example, fat contains, in addition to adipocytes, a large vascular network, composed primarily of capillary endothelial cells and some vascular smooth muscle cells as well as a collagen stromal structure produced by fibroblasts (Williams *et al.,* 1994). Hence, multiple cell types can potentially be obtained from this tissue. In attempting to produce fat tissue, it may be beneficial to purify and expand the cellular components of fat without isolating each component. This may be particularly important if, as suggested by several authors, vascular endothelial cells and others exhibit organ specificity (Bassenge, 1996; Craig *et al.,* 1998; Murphy *et al.,* 1998).

MATERIALS

There are two general strategies for introducing the cell–polymer constructs back into the body in order to engineer tissues. The constructs can be implanted using an open surgical procedure (implantable material), or introduced in a minimally invasive manner utilizing syringes or endoscopic delivery (injectable materials). Implantable forms of materials are typically foams, sponges, films, and other solid devices. The typical injectable forms of materials include hydrogels and microbeads. Most synthetic polymers, such as aliphatic polyesters, polyanhydrides, poly(amino acids), and poly(ortho esters), are suitable for fabrication of implantable devices. On the other hand, natural polymers, including alginate, chitosan, hyaluronic acid, and collagen, are good examples of injectable materials. Hydrogels are highly attractive injectable forms due to the potential to implant into the body in a minimally invasive manner. Their properties can be engineered for biocompatibility, selective permeability, mechanical and chemical stability, and other requirements as specified by the application (Jen *et al.,* 1996). Certain polymers such as the polyanhydrides and collagen can be used in both types of applications (Anseth *et al.,* 1999; Doillon *et al.,* 1994).

IMPLANTABLE MATERIALS

A variety of polymers can be utilized to form solid or macroporous scaffolds for tissue engineering applications (Fig. 31.2). Aliphatic polyesters of poly(glycolic acid) (PGA) and poly(lactic acid) (PLA) are well-characterized synthetic biodegradable polymers that have been widely applied to the biomedical field generally, and to the tissue engineering arena specifically (Wong and Mooney, 1997). PGA has a high crystallinity, a high melting temperature, and low solubility in organic solvents. PLA has a much more hydrophobic character than PGA due to the introduction of the methyl group. PLA has low water uptake and its ester bond is less labile to hydrolysis owing to steric hindrance of the methyl group. Therefore, PLA degrades more slowly and has higher solubility in organic solvents compared with PGA. Copolymers of PLA and PGA (PLGA) can be readily synthesized and their physical properties are regulated by the ratio of glycolic acid to lactic acid. This enables these copolymers to be used in various applications as biodegradable matrices in tissue engineering (Mooney *et al.,* 1995).

These aliphatic polyesters can be readily processed into various physical forms appropriate for tissue engineering applications. A number of techniques have been proposed to generate highly porous scaffolds, including solvent casting/particulate leaching (Mikos *et al.,* 1994), phase separation (Lo *et al.,* 1995), emulsion freeze drying (Whang *et al.,* 1995), fiber extrusion and fabric formation (Freed *et al.,* 1994), and gas foaming (Mooney *et al.,* 1996b; Harris *et al.,* 1998). Nonwoven fabrics of PGA, stabilized by spraying with PLA or PLGA solution, thus providing significant resistance to compressive forces, were successfully used to culture smooth muscle cells (Kim and Mooney, 1998b), even with a tubular structure (Mooney *et al.,* 1996a). Gas foaming and particulate leaching techniques have proved efficient for generating open pore structures (Fig. 31.3), and because the gas foaming process does not require any organic solvents or high temperature, biologically active molecules can be incorporated into these matrices without denaturation (Shea *et al.,* 2000; Sheridan *et al.,* 2000). Implantation of porous polymer scaffolds leads to host tissue ingrowth and formation of granulation tissue throughout the scaffold (Fig. 31.4).

A number of other synthetic polymers can also be utilized to fabricate scaffolds for breast tissue reconstruction. These include polycaprolactone, polyanhydrides, poly(amino acids), and poly(ortho esters). Polycaprolactone (PCL) is also one of the aliphatic polyesters. PCL is a semicrystalline polymer with high solubility in organic solvents and low melting temperature. Because

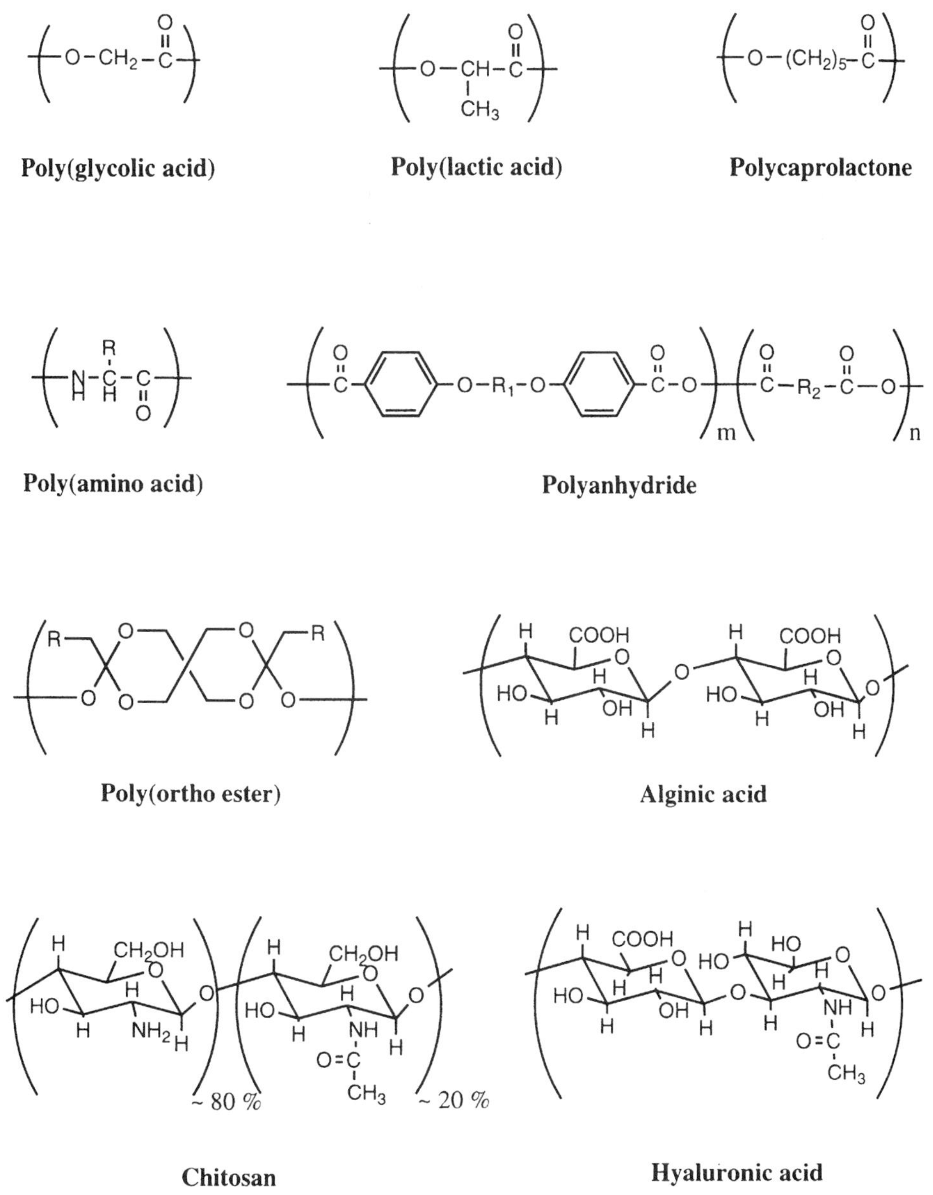

Poly(glycolic acid)   **Poly(lactic acid)**   **Polycaprolactone**

Poly(amino acid)   **Polyanhydride**

Poly(ortho ester)   **Alginic acid**

Chitosan   **Hyaluronic acid**

Fig. 31.2. Chemical structures of various implantable and injectable materials for tissue engineering scaffolds.

the degradation rate of PCL is much slower than that of PGA or PLA, it has been used as a long-term drug delivery carrier (Pitt, 1990). The biocompatibility of PCL has been investigated in rat (Yamada *et al.,* 1997) and rabbit models (Lowry *et al.,* 1997). Polyanhydrides are usually copolymers of aromatic diacids and aliphatic diacids. They usually degrade by surface erosion, and their degradation rate can be controlled, depending on the choice of diacids (Domb *et al.,* 1989). The degradation rates of polyanhydrides are much faster than those of poly(ortho esters) in the absence of any additives. Therefore, polyanhydrides have been widely used as biodegradable implants for local drug delivery of cefazolin, bupivacaine (Park *et al.,* 1998), methotrexate (Dang *et al.,* 1994), and taxol (Fung *et al.,* 1998). Poly[bis(*p*-carboxyphenoxy)propane-sebacic acid] was approved by the FDA for clinical trials (Engelberg and Kohn, 1991). It is biocompatible and nontoxic, even in the rat brain, when compared to standard neurosurgical implants (Tamargo *et al.,* 1989). Poly(amino acids) have been studied due to their similarity to proteins, and have been widely investigated for biomedical applications such as sutures and artificial skin (Anderson *et al.,* 1985). Poly(amino acids) are usually polymerized by ring-opening of *N*-carboxyanhydrides, and versatile copolymers can be prepared from various combinations of amino acids. However, due to their highly insoluble and nonprocessible properties, "pseudo"-poly(amino acids) have been also devel-

a

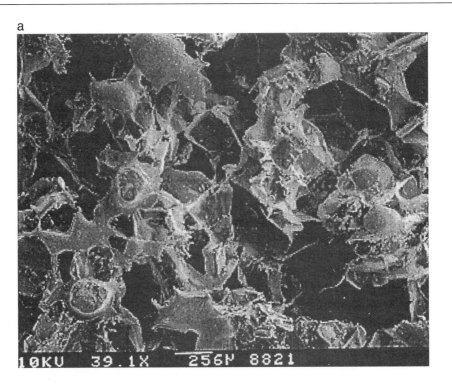

b

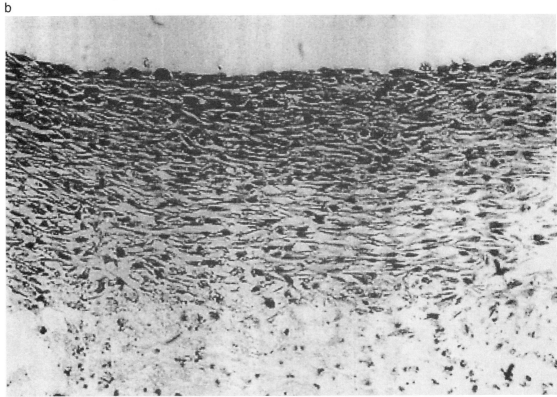

Fig. 31.3. (a) Scanning electron photomicrograph of the surface of a PLGA matrix formed with the gas foaming/particulate leaching methods. (b) Photomicrograph of hematoxylin- and eosin-stained cross-section of new tissue formed by smooth muscle cells on the matrix shown in a after 2 weeks in culture (×400). From "Open pore biodegradable matrices formed with gas foaming;" L. D. Harris, B.-S. Kim, and D. J. Mooney; J. Biomed. Mater. Res. Copyright © 1998 John Wiley & Sons, Inc. Reprinted by permission of John Wiley & Sons, Inc.

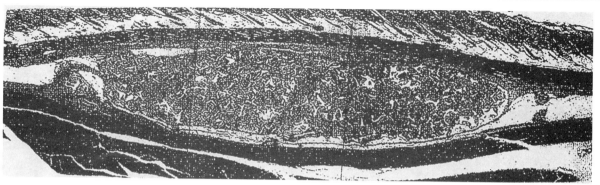

Fig. 31.4. Formation of new three-dimensional tissue following implantation of a highly porous biodegradable polymer matrix. Rat adipocytes were seeded for 24 hr onto a 13-mm-diameter, 3-mm-thick PLA matrix and then implanted into the subcutaneous space of a female Lewis rat. The implants were harvested 4 weeks postimplantation, fixed in formalin, embedded and 4-μm sections were stained with hematoxylin and eosin. A ×2.5 magnification of the graft is shown. A well-defined tissue is present throughout the implant site, with the establishment of a vascular bed, and the matrix size and shape predetermine the size and shape of the engineered tissue mass. Unpublished data from W. D. Holder and C. R. Halberstadt.

oped by Kohn and Langer (1987). It was also reported that poly(amino acids) containing L-arginine, L-lysine, or L-ornithine caused endothelium-dependent relaxation of bovine intrapulmonary artery and vein, and stimulated the formation and/or release of an endothelium-derived relaxing factor identified as nitric oxide (Ignarro *et al.*, 1989). Poly(ortho esters) are biodegradable polymers that degrade by gradual surface erosion and have been known as useful materials for controlled drug delivery. Poly(ortho ester) membranes containing indomethacin (Solheim *et al.*, 1992), gentamycin (Pinholt *et al.*, 1992), or insulin-like growth factor (Busch *et al.*, 1996) have been prepared and implanted into rats without any noticeable inflammation after degradation.

INJECTABLE MATERIALS

A number of polymers, mainly naturally derived, can be utilized in an injectable form (Fig. 32.2). Alginate has found the widest use to date as an injectable scaffold material for tissue engineering (Atala *et al.*, 1994) due to its simple gelation properties when ionically cross-linked with divalent cations such as Ca^{2+}, Mg^{2+}, Ba^{2+}, and Sr^{2+} (Fig. 31.5a). Because of favorable properties, including biocompatibility, nonimmunogenicity, hydrophilicity, and relatively low cost, there have been many attempts to utilize this material as wound dressings, dental impressions, immobilization matrices, and scaffolds for cultured and transplanted cells (Klock *et al.*, 1997; Shapiro and Cohen, 1997; Gombotz and Wee, 1998). Alginate can be used in an injectable form either by being preformed into small beads or by injection before or during the solidifying process. Due to the ease of implantation in a minimally invasive manner, alginate beads have been prepared and used for transplantation of chondrocytes (Lemare *et al.*, 1998; Gregory *et al.*, 1999) or hepatocytes (Joly *et al.*, 1997).

Alginate is not an ideal material for a tissue engineering approach because it loses divalent ions into the surrounding medium, and then dissolves. This process is uncontrollable and unpredictable. In addition, the molecular weight of many alginates is typically above the renal clearance threshold of the kidney (Al-Shamkhani and Duncan, 1995). Alginate is also known to discourage protein adsorption due to its hydrophilic character, and this may decrease the survival of many cell types in alginate hydrogels (Smentana, 1993). Hydrolytically degradable covalently cross-linked hydrogels derived from alginate have been reported (Bouhadir *et al.*, 1999). To avoid the degradation problem of alginate, polyguluronate blocks with molecular masses of 6000 Da were isolated from alginate, oxidized, and covalently cross-linked with adipic dihydrazide. The gelling of these polymers could be readily controlled, and their mechanical properties depended on the cross-linking density. An RGD-containing cell adhesion ligand has also been covalently coupled to enhance the cellular interaction with alginate hydrogels (Rowley *et al.*, 1999). These modified alginate hydrogels provided for adhesion, proliferation, and expression of a differentiated phenotype of mouse skeletal myoblasts (Fig. 31.5b).

a

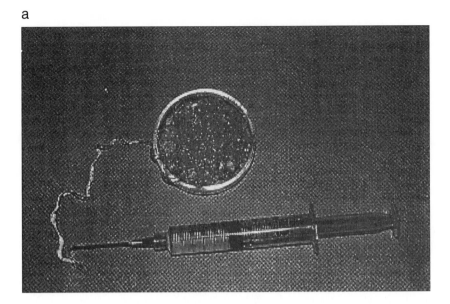

b

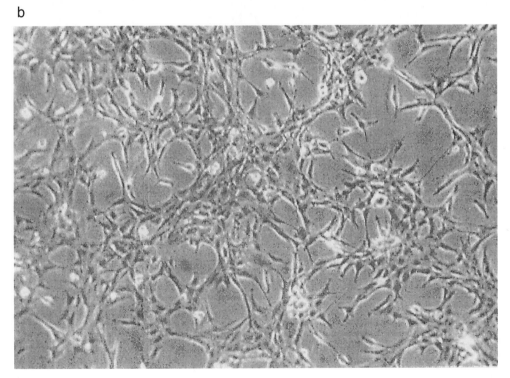

Fig. 31.5. (a) Alginate hydrogel, ionically cross-linked with Ca²⁺, as an injectable material for the tissue engineering approach. From K. H. Bouhadir and D. J. Mooney (unpublished). (b) Photograph of myoblasts cultured on GRGDY-modified alginate hydrogel surfaces for 3 days. Reprinted from Biomaterials 20; J. A. Rowley, G. Madlambayan, and D. J. Mooney; "Alginate hydrogels as synthetic extracellular matrix materials," pp. 45–53. Copyright 1999, with permission from Elsevier Science.

Other naturally occurring polymers can also be utilized as an injectable material to achieve breast tissue reconstruction. These include chitosan, hyaluronic acid, and collagen. Chitosan is the second most plentiful biomass, and various derivatives of chitosan have been formed by coupling various molecules to the reactive amino groups in the backbone (Muzzarelli, 1983; Lee *et al.,* 1995). Chitosan has been known to be biocompatible and biodegradable and has a low toxicity (Chandy and Sharma, 1990; Tomihata and Ikada, 1997). This enables chitosan to be useful for versatile biomedical applications such as plastic surgery (Biagini *et al.,* 1991) and wound healing (Muzzarelli *et al.,* 1993). Chitosan matrices were fabricated and proposed as cell substrates for bovine chromaffin cells and porcine hepatocytes due to their structural similarity to glycosaminoglycans (Eser-Elcin *et al.,* 1998; Elcin *et al.,* 1999). Hyaluronic acid is known to be one of the glycosaminoglycan components of the extracellular matrix and plays a significant role in wound healing and cell proliferation. Hyaluronic acid is degraded by hyaluronidase, which exists in cells and serum (Afify *et al.,* 1993). Hyaluronic acid and its derivatives have been used as a drug delivery system (Larsen and Balazs, 1991). Hydrogels of hyaluronic acid have also been prepared by covalent cross-linking with various kinds of hydrazides (Vercruysse *et al.,* 1997). Human fibroblasts and keratinocytes were cultured on hyaluronic acid-derived biomaterials (Zacchi *et al.,* 1998) and cross-linked nonanimal hyaluronic acid gel was used for facial intradermal implants (Duranti *et al.,* 1998). Collagen is the best known tissue-derived natural polymer. It is the main component of all mammalian tissues, including skin, bone, cartilage, tendon, and ligament. Collagen has been used as a tissue culture scaffold or artificial skin due to the easy attachment to it by many different cell types. However, collagen offers a limited range of physical properties, and can be expensive (Pulapura and Kohn, 1992). The attachment of cells to collagen can be altered by chemical modification and incorporation of either fibronectin, chondroitin sulfate, or low levels of hyaluronic acid into the collagen matrix (Srivastava *et al.,* 1990).

ANIMAL MODELS

A primary problem with the development of tissue-engineered human breast reconstruction or augmentation is that there are no similar animal models. Human beings are the only animals to have developed breasts. Other primates may have small breast mounds, particularly when lactating, but these are inadequate for developing large tissue constructs for partial or complete mastectomy equivalents to what occurs in human patients. Cattle and goats have udders that have no structural or functional similarity to humans. Rats and mice have multiple teats along mammary lines extending from the neck to the inguinal region, with small mammary glands beneath each teat. Pigs, dogs, and sheep similarly have subcutaneous mammary glands that are inconspicuous unless the animal is lactating. Despite the lack of an ideal model, major development of concepts can be achieved and important research questions answered using several animal models.

There has been excellent success using inbred female Lewis rats as a small animal for the development of transplantable tissues on absorbable polymers such as PGA, PLA, PGLA, and alginate (Holder *et al.,* 1998). This model allows for transplantation of cells between individuals without concern for immunologic rejection, which parallels the likely autologous nature of cell transplantation for breast engineering. In addition, the Lewis rat is larger than many other strains and allows for the testing of larger (1–2 cm) or multiple constructs in the same animal. In order to prevent migration of implanted materials and control the local formation of the engineered tissue, a "purse-string" technique has been developed to secure implants at specific sites (Roland *et al.,* 1998). Briefly, a single nonabsorbable suture is passed in and out of the skin to form a circle. The skin is lifted with a forcep, forming a pocket, and the implant is placed into the subcutaneous tissue in the center of the circle through a small incision that is then sutured together. The purse-string suture is then tightened to hold the implant in place. The suture may be removed 2 weeks later and the implant will remain in place.

Another attractive animal model that allows for the transplantation of human-derived cells is Nude (*nu/nu*) mice or severe combined immunodeficient (SCID) mice. These animals have a compromised immune systems to the extent that they will often accept xenogeneically transplanted tissues, and these mice have been particularly useful for transplantation and immunologic studies of human tumors, bone marrow, skin, and other tissues (Mortensen *et al.,* 1976; Lopez-Valle *et al.,* 1996; Ullmann *et al.,* 1998). They are potential models for evaluating various polymer constructs with and without human cells *in vivo* without the adverse effects of a major immunologic reaction. Further, basic questions about human cells and polymers *in vivo* may be answered without going to human trials prematurely. Development of these models may not be straightforward, because there are subtle differences between strains regarding acceptance of various tissues and growth of the implanted tissues in different sites (e.g., a tissue may grow in one mouse strain but not another or grow in a subcutaneous site but not in an internal location). Nude mice basically lack a T lymphocyte response whereas SCID mice lack both T and B lymphocyte responses. However, both types have natural killer cells that may interact with some transplanted materials. These mouse colonies must be monitored closely for changes in the immune status of the mice, which at times occur spontaneously. Human cells must be routinely screened for human immunodeficiency virus (HIV), hepatitis, and mycoplasma before transplanting into immunosuppressed mice to assure the safety of the animals and workers, and the validity of the experiments.

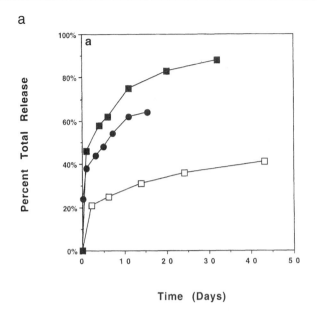

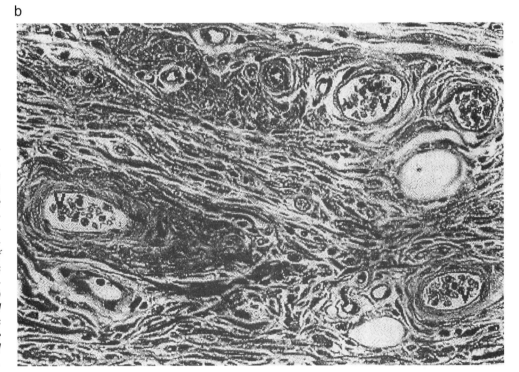

Fig. 31.6. (a) Release of ^{125}I-*labeled growth factors (□, VEGF; ●, FGF-2; ■, EGF) from PLGA matrices formed with the gas foaming/particulate leaching methods. From Peters et al. (1999a). Photomicrographs of tissue cross-section after 4 weeks of implantation into Lewis rats for a matrix-releasing (b) plasmid-encoding PDGF (×100) and (c) control plasmid (×400). Photomicrographs have labels for adipose tissue (A), polymer (P), and vessel (V). From Shea et al. (2000).*

In evaluating larger animals as models for breast tissue engineering, animals with skin and subcutaneous tissues that are similar to those of humans are required in order to evaluate larger constructs in the subcutaneous position. For this reason, the same animal must be used as a tissue donor and recipient. Dogs generally have little fat and very loose skin and present many of the problems of the rat in terms of migration of implanted materials. Pig skin and subcutaneous tissues are very similar to human skin and tissues, but most pigs continue to undergo rapid gain weight throughout their lives, which makes monitoring implants very difficult. We have had success with adult miniature Yucatan pigs that tend to maintain their size without massive weight gain (C. R. Halberstadt and W. D. Holder, unpublished data). Implanted materials can be readily monitored and sites easily marked with a skin tattoo. These animals have a pectoral musculature similar to that in humans so that subpectoral muscle sites may also be used. Primates, although relat-c

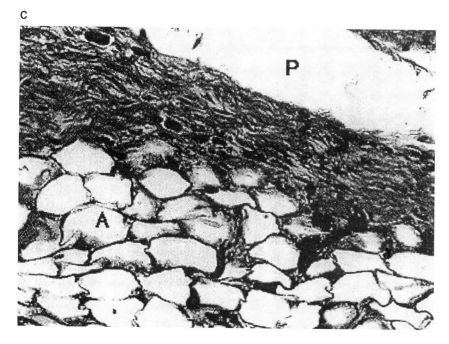

Fig. 31.6.c

ed to humans, anatomically do not have any particular advantage over sheep or pigs as a model. Further research and development will be required to establish a technology for biopsies and cellular construct implantation that is efficient, technically straightforward, and minimally invasive in any of these large animal models.

STRATEGIES TO ENHANCE THE VASCULARIZATION OF ENGINEERED TISSUE

A critical challenge to engineer breast tissue, or any tissue of significant thickness, is to develop a vascular network that can support the metabolic needs of the engineered tissue and integrate it with the rest of the body. The need for a vascular network has been strikingly demonstrated in studies of hepatocyte transplantation, in which over 90% of the cells transplanted even on relatively thin (3mm thick) scaffolds died within days following implantation (Mooney *et al.,* 1997). This finding of significant cell death prior to tissue vascularization is not unique to hepatocytes, but has also been noted in smooth muscle cell transplantation as well (Cohn *et al.,* 1997), and is likely a general finding in all efforts to engineer tissues (Colton, 1995). The presence of vascularized networks in natural metabolic organs results in a short diffusion distance between the nutrient source and the cells (Thomson *et al.,* 1995), and these vascular networks must be created in engineered tissues as well.

There are three general approaches taken to date to promote angiogenesis in engineered tissues. The first relies on vascularization, which accompanies the inflammatory response to any implanted foreign material. Optimization of the porosity and pore size of tissue engineering scaffolds can increase the rate of granulation tissue formation in engineered tissues (Mikos *et al.,* 1993; Mooney *et al.,* 1994). However, the blood vessel ingrowth occurs slowly with this approach, and will likely not be sufficient to engineer large tissue volumes. The second two approaches attempt actively to modulate the vascularization process by either delivering angiogenic molecules or blood vessel-forming cells (endothelial cells) to the site at which the tissue is being engineered.

Site-specific delivery of angiogenic molecules may provide an efficient means of stimulating localized vessel formation. Controlled and sustained release of angiogenic molecules from the tissue engineering scaffolds may allow one to optimize this process (Eiselt *et al.,* 1998; Peters *et al.,* 1999b; Sheridan *et al.,* 2000). Various growth factors, including vascular endothelial growth factor (VEGF), basic fibroblast growth factor (FGF-2), and epidermal growth factor (EGF), could be incorporated into a polymeric matrix and released at a controlled and sustained rate for an extended period of time (Fig. 31.6a). Alternatively, delivery of plasmid DNA encoding the angiogenic proteins may be another approach to generate the vascular networks in engineered tissues (Shea *et al.,* 2000). PLGA scaffolds containing a plasmid encoding for platelet-derived growth factor (PDGF), a potent angiogenesis promoter, greatly increased the number of blood vessels and granulation tissue that formed compared to scaffolds, which did not deliver the plasmid DNA (Fig. 31.6, b and c).

Another potential approach to enhance angiogenesis in engineered tissues is to cotransplant endothelial cells along with the primary cell type of interest. This approach is suggested by the observation that endothelial cells will spontaneously form capillary-like structures *in vitro* if cultured in an appropriate environment (Ingber and Folkman, 1989; Pepper *et al.,* 1992). Endothelial cells seeded into a tissue engineering scaffold may be able to form capillaries, or capillary-like structures, which then can merge with capillaries growing into the scaffold from the host tissue. This possibility is supported by our previous finding that transplantation of syngeneic endothelial cells on polymer scaffolds led to a statistically significant enhancement in the density of blood vessels in the scaffolds following 2 weeks implantation into rats (Holder *et al.,* 1997). In addition, we have demonstrated that the new capillaries that form in these scaffolds are composed of both the transplanted endothelial cells and the ingrowing cells from the host (Nör *et al.,* 1999).

One attractive approach to engineering tissues in these models is to deliver cellular constructs that are approximately 500 μm in thickness or less. This may optimize diffusional transport of nutrients to the cells while each small cell–polymer unit becomes vascularized. We have taken this approach by designing porous alginate–RGD beads (500 μm to 3mm in diameter) for cell transplantation. Because the beads are macroporous, the vascular endothelial cells can migrate into the beads and establish a three-dimensional vascular bed in a short time. This approach can potentially be combined with angiogenic factors and cells to enhance the timing of endothelialization to promote the survival of the transplanted cells.

CONCLUDING REMARKS

Tissue engineering may provide a new therapy to reconstruct breast tissue for woman undergoing lumpectomy or mastectomy. A number of cell types and polymeric materials may be useful to reconstruct breast tissue. In addition, it is clear that optimization of vascularization is critical to engineer large tissues. This approach may ultimately allow one to engineer and reconstruct breast tissue in the future with less pain and less invasive methods than those currently utilized clinically.

REFERENCES

Afify, A. M., Stern, M., Guntenhoner, M., and Stern, R. (1993). Purification and characterization of human serum hyaluronidase. *Arch. Biochem. Biophys.* **305**, 434–441.

Al-Shamkhani, A., and Duncan, R. (1995). Radioiodination of alginate via covalently-bound tyrosinamide allows for monitoring of its fate *in vivo. J. Bioact. Compat. Polym.* **10**, 4–13.

American Cancer Society (1997). "Cancer Facts and Figures: Breast Cancer." ACS, Washington, DC.

Anderson, J. M., Spilizewski, K. L., and Hiltner, A. (1985). Poly-α-amino acids as biomedical polymers. *In* "Biocompatibility of Tissue Analogues" (D. F. Williams, ed.), pp. 67–88. CRC Press, Boca Raton, FL.

Anseth, K. S., Shastri, V. S., and Langer, R. (1999). Photopolymerizable degradable polyanhydrides with osteocompatibility. *Nat. Biotechnol.* **17**, 156–159.

Atala, A., Kim, W., Paige, K. T., Vacanti, C. A., and Retik, A. B. (1994). Endoscopic treatment of vesicoureteral reflux with a chondrocyte-alginate suspension. *J. Urol.* **152**, 641–643.

Bassenge, E. (1996). Endothelial function in different organs. *Prog. Cardiovasc. Dis.* **39**, 209–228.

Biagini, G., Pugnaloni, A., Damadei, A., Bertani, A., Belligolli, A., Bicchiega, V., and Muzzarelli, R. A. A. (1991). Morphological study of the capsular organization around tissue expanders coated with *N*-carboxybutyl chitosan. *Biomaterials* **12**, 287–291.

Bouhadir, K. H., Hausman, D. S., and Mooney, D. J. (1999). Synthesis of cross-linked poly(aldehyde guluronate) hydrogels. *Polymer* **40**, 3575–3584.

Breinan, H. A., Minas, T., Barone, L., Tubo, R., Hsu, H. P., Shortkroff, S., Nehrer, S., Sledge, C. B., and Spector, M. (1998). Histological evaluation of the course of healing of canine articular cartilage defects treated with cultured autologous chondrocytes. *Tissue Eng.* **4**, 101–114.

Busch, O., Solheim, E., Bang, G., and Tornes, K. (1996). Guided tissue regeneration and local delivery of insulin-like growth factor I by bioerodible polyorthoester membranes in rat calvarial defects. *J. Oral Maxillofacial Implants* **11**, 498–505.

Chandy, T., and Sharma, C. P. (1990). Chitosan—As a biomaterial. *Biomater., Artif. Cells, Artif. Organs* **18**, 1–24.

Cohn, N. A., Kim, B. S., Mooney, D. J., Emelianov, S. Y., and O'Donnell, M. (1997). Layer imaging in tissue engineering using an elasticity microscope. *Proc. IEEE Ultrasonics Symp.,* pp. 1431–1434.

Colton, C. K. (1995). Implantable biohybrid artificial organs. *Cell Transplant.* **4**, 415–436.

Craig, L. E., Spelman, J. P., Strandberg, J. D., and Zink, M. C. (1998). Endothelial cells from diverse tissues exhibit differences in growth and morphology. *Microvasc. Res.* **55**, 65–76.

Dang, W., Colvin, O. M., Brem, H., and Saltzman, W. M. (1994). Covalent coupling of methotrexate to dextran enhances the penetration of cytotoxicity into a tissue-like matrix. *Cancer Res.* **54**, 1729–1735.

Doillon, C. J., Deblois, C., Cote, M. F., and Fournier, N. (1994). Bioactive collagen sponges as connective-tissue substitute. *Mater. Sci. Eng. C: Biomim.* **2**, 43–49.

Domb, A. J. Gallardo, C. F., and Langer, R. (1989). Polyanhydrides. 3. Polyanhydrides based on aliphatic-aromatic diacids. *Macromolecules* **22**, 3200–3204.

Duranti, F., Salti, G., Bovani, B., Calandra, M., and Rosati, M. L. (1998). Injectable hyaluronic acid gel for soft tissue augmentation. A clinical and histological study. *Dermatol. Surg.* **24**, 1317–1325.

Eaglstein, W. H., and Falanga, V. (1998). Tissue engineering for skin: An update. *J. Am. Acad. Dermatol.* **39**, 1007–1010.

Eiselt, P., Kim, B.-S., Chacko, B., Isenberg, B., Peters, M. C., Greene, K. G., Roland, W. D., Loebsack, A. B., Burg, K. J. L., Culberson, C., Halberstadt, C. R., Holder, W. D., and Mooney, D. J. (1998). Development of technologies aiding large-tissue engineering. *Biotechnol. Prog.* **14**, 134–140.

Elcin, Y. M., Dixit, V., Lewin, K., and Gitnick, G. (1999). Xenotransplantation of fetal porcine hepatocytes in rats using a tissue engineering approach. *Artif. Organs* **23**, 146–152.

Engelberg, I., and Kohn, J. (1991). Physico-mechanical properties of degradable polymers used in medical applications: A comparative study. *Biomaterials* **12**, 292–304.

Eser-Elcin, A., Elcin, Y. M., and Pappas, G. D. (1998). Neural tissue engineering: Adrenal chromaffin cell attachment and viability on chitosan scaffolds. *Neurol. Res.* **20**, 648–654.

Freed, L. E., Vunjak-Novakovic, G., Biron, R. J., Eagles, D. B., Lesnoy, D. C., Barlow, S. K., and Langer, R. (1994). Biodegradable polymer scaffolds for tissue engineering. *Bio/Technology* **12**, 689–693.

Fung, L. K., Ewend, M. G., Sills, A., Sipos, E. P., Thompson, R., Watts, M., Colvin, O. M., Brem, H., and Saltzman, W. M. (1998). Pharmacokinetics of interstitial delivery of carmustine, 4-hydroperoxycyclophosphamide, and paclitaxel from a biodegradable polymer implant in the monkey brain. *Cancer Res.* **58**, 672–684.

Gombotz, W. R., and Wee, S. F. (1998). Protein release from alginate matrices. *Adv. Drug Delivery Rev.* **31**, 267–285.

Gregory, K. E., Marsden, M. E., Anderson-MacKenzie, J., Bard, J. B., Bruckner, P., Farjanel, J., Robins, S. P., and Hulmes, D. J. (1999). Abnormal collagen assembly, though normal phenotype, in alginate bead cultures of chick embryo chondrocytes. *Exp. Cell Res.* **246**, 98–107.

Hang-Fu, L., Marmolya, G., and Feiglin, D. H. (1995). Liposuction fat-fillant implant for breast augmentation and reconstruction. *Aesthetic Plast. Surg.* **19**, 427–437.

Harris, J., Morrow, M., and Norton, L. (1997). Malignant tumors of the breast. *In* "Cancer Principles and Practice of Oncology" (V. T. DeVita, S. Hellman, and S. A. Rosenberg, eds), pp. 1570–1584. Lippincott-Raven, Philadelphia.

Harris, L. D., Kim, B.-S., and Mooney, D. J. (1998). Open pore biodegradable matrices formed with gas foaming. *J. Biomed. Mater. Res.* **42**, 396–402.

Holder, W. D., Gruber, H. E., Roland, W. D., Moore, A. L., Culberson, C. R., Loebsack, A. B., Burg, K. J. L., and Mooney, D. J. (1997). Increased vascularization and heterogeneity of vascular structures occurring in polyglycolide matrices containing aortic endothelial cells implanted in the rat. *Tissue Eng.* **3**, 149–160.

Holder, W. D., Gruber, H., Moore, A. L., Culberson, C. C., Mooney, D. J., and Anderson, W. (1998). Cellular ingrowth and thickness changes in polylactide and polyglycolide matrices implanted subcutaneously in the rat. *J. Biomed. Mater. Res.* **41**, 412–421.

Ignarro, L. J., Gold, M. E., Buga, G. M., Byrns, R. E., Wood, K. S., Chaudhuri, G., and Frank, G. (1989). Basic polyamino acids rich in arginine, lysine, or ornithine cause both enhancement of and refractoriness to formation of endothelium-derived nitric oxide in pulmonary artery and vein. *Circ. Res.* **64**, 315–329.

Ingber, D. E., and Folkman, J. (1989). Mechanochemical switching between growth and differentiation during fibroblast growth factor-stimulated angiogenesis *in vitro:* Role of extracellular matrix. *J. Cell Biol.* **109**, 317–330.

Jen, A. C., Wake, M. C., and Mikos, A. G. (1996). Hydrogels for cell immobilization. *Biotechnol. Bioeng.* **50**, 357–364.

Joly, A., Desjardins, J. F., Fremond, B., Desille, M., Campion, J. P., Malledant, Y., Lebreton, Y., Semana, G., Edwards-Levy, F., Levy, M. C., and Clement, B. (1997). Survival, proliferation, and functions of porcine hepatocytes encapsulated in coated alginate beads: A step toward a reliable bioartificial liver. *Transplantation* **63**, 795–803.

Kim, B.-S., and Mooney, D. J. (1998a). Development of biocompatible synthetic extracellular matrices for tissue engineering. *Trends Biotechnol.* **16**, 224–230.

Kim, B.-S., and Mooney, D. J. (1998b). Engineering smooth muscle tissue with a predefined structure. *J. Biomed. Mater. Res.* **41**, 322–332.

Kim, B.-S., Putnam, A. J., Kulik, T. J., and Mooney, D. J. (1998). Optimizing seeding and culture methods to engineer smooth muscle tissue on biodegradable polymer matrices. *Biotechnol. Bioeng.* **57**, 46–54.

Klock, G., Pfeffermann, A., Ryser, C., Grohn, P., Kuttler, B., Hahn, H. J., and Zimmermann, U. (1997). Biocompatibility of mannuronic acid-rich alginates. *Biomaterials* **18**, 707–713.

Kohn, J., and Langer, R. (1987). Polymerization reactions involving the side chains of α-L-amino acids. *J. Am. Chem. Soc.* **109**, 817–820.

Kononas, T. C., Bucky, L. P., Hurley, C., and May, J. W., Jr. (1993). The fate of suctioned and surgically removed fat after reimplantation for soft-tissue augmentation: A volumetric and histologic study in the rabbit. *Plast. Reconstr. Surg.* **91**, 763–768.

Landeen, L. K., Ziegler, F. C., Halberstadt, C. R., Cohen, C., and Slivka, S. R. (1992). Characterization of a human dermal replacement. *Wounds* **4**, 167–175.

Langer, R., and Vacanti, J. P. (1993). Tissue engineering. *Science* **260**, 920–926.

Larsen, N. E., and Balazs, E. A. (1991). Drug delivery systems using hyaluronan and its derivatives. *Adv. Drug Delivery Rev.* 7, 279–308.

Lee, K. Y., Ha, W. S., and Park, W. H. (1995). Blood compatibility and biodegradability of partially *N*-acylated chitosan derivatives. *Biomaterials* 16, 1211–1216.

Lemare, F., Steimberg, N., Le Griel, C., Demignot, S., and Adolphe, M. (1998). Dedifferentiated chondrocytes cultured in alginate beads: Restoration of the differentiated phenotype and of the metabolic responses to interleukin-1β. *J. Cell. Physiol.* 176, 303–313.

Lo, H., Ponticiello, M. S., and Leong, K. W. (1995). Fabrication of controlled release biodegradable foams by phase separation. *Tissue Eng.* 1, 15–28.

Lopez-Valle, C. A., Germain, L., Rouabhia, M., Xu, W., Guignard, R., Goulet, F., and Auger, F. A. (1996). Grafting on nude mice of living skin equivalents produced using human collagens. *Transplantation* 62, 317–323.

Lowry, K. J., Hamson, K. R., Bear, L., Peng, Y. B., Calaluce, R., Evans, M. L., Anglen, J. O., and Allen, W. C. (1997). Polycaprolactone/glass bioabsorbable implant in a rabbit humerus fracture model. *J. Biomed. Mater. Res.* 36, 536–541.

Mikos, A. G., Sarakinos, G., Lyman, M. D., Ingber, D. E., Vacanti, J. P., and Langer, R. (1993). Prevascularization of porous biodegradable. *Biotechnol. Bioeng.* 42, 716–723.

Mikos, A. G., Thorsen, A. J., Czerwonka, L. A., Bao, Y., and Langer, R. (1994). Preparation and characterization of poly(L-lactic acid) foams. *Polymer* 35, 1068–1077.

Mooney, D. J., and Mikos, A. G. (1999). Growing new organs. *Sci. Am.* 280, 60–65.

Mooney, D. J., Kaufmann, P. M., Sano, K., McNamara, K. M., Vacanti, J. P., and Langer, R. (1994). Transplantation of hepatocytes using porous, biodegradable sponges. *Transplant. Proc.* 26, 3425–3426.

Mooney, D. J., Breuer, M. D., McNamara, K., Vacanti, J. P., and Langer, R. (1995). Fabricating tubular devices from polymers of lactic and glycolic acid for tissue engineering. *Tissue Eng.* 1, 107–118.

Mooney, D. J., Mazzoni, C. L., Breuer, C., McNamara, K., Hern, D., Vacanti, J. P., and Langer, R. (1996a). Stabilized polyglycolic acid fibre based tubes for tissue engineering. *Biomaterials* 17, 115–124.

Mooney, D. J., Baldwin, D. F., Suh, N. P., Vacanti, J. P., and Langer, R. (1996b). Novel approach to fabricate to porous sponges of poly(D,L-lactic-co-glycolic acid) without the use of organic solvents. *Biomaterials* 17, 1417–1422.

Mooney, D. J., Kaufmann, P. M., Sano, K., Schwendeman, S. P., Majahod, K., Schloo, B., Vacanti, J. P., and Langer, R. (1996c). Localized delivery of epidermal growth factor improves the survival of transplanted hepatocytes. *Biotechnol. Bioeng.* 50, 422–429.

Mooney, D. J., Sano, K., Kaufmann, P. M., Majahod, K., Schloo, B., Vacanti, J. P., and Langer, R. (1997). Long-term engraftment of hepatocytes transplanted on biodegradable polymer sponges. *J. Biomed. Mater. Res.* 37, 413–420.

Mortensen, N. B., Romert, P., and Ballegaard, S. (1976). Transplantation of human adipose tissue to nude mice. *Acta Pathol. Microbiol. Scand.* 84, 283–289.

Murphy, H. S., Bakopoulos, N., Dame, M. K., Varani, J., and Ward, P. A. (1998). Heterogeneity of vascular endothelial cells: Differences in susceptibility to neutrophil-mediated injury. *Microvasc. Res.* 56, 203–211.

Muzzarelli, R. A. A. (1983). Chitin and its derivatives: New trends of applied research. *Carbohydr. Polym.* 3, 53–75.

Muzzarelli, R. A. A., Zucchini, C., Ilari, P., Pugnaloni, A., Mattioli-Belmonte, M., Biagini, G., and Castaldini, C. (1993). Osteoconductive properties of methylpyrrolidinone chitosan in an animal model. *Biomaterials* 14, 925–929.

Nguyen, A., Pasyk, K. A., Bouvier, T. N., Hassett, C. A., and Argenta, L. C. (1990). Comparative study of survival of autologous adipose tissue taken and transplanted by different techniques. *Plast. Reconstr. Surg.* 85, 378–386.

Nör, J. E., Christensen, J., Mooney, D. J., and Polverini, P. J. (1999). Vascular endothelial growth factor (VGEF)-mediated angiogenesis is associated with enhanced endothelial cell survival and induction of Bcl-1 expression. *Am. J. Pathol.* 154, 375–384.

Oberpenning, F., Meng, J., Yoo, J. J., and Atala, A. (1999). *De novo* reconstitution of a functional mammalian urinary bladder by tissue engineering. *Nat. Biotechnol.* 17, 149–155.

Park, E. S., Maniar, M., and Shah, J. C. (1998). Biodegradable polyanhydride devices of cefazolin sodium, bupivacaine, and taxol for local drug delivery: Preparation, and kinetics and mechanism of *in vitro* release. *J. Controlled Release* 52, 179–189.

Patrick, C. W., Chauvin, P. B., Hobley, J., and Reece, G. P. (1999). Preadipocyte seeded PLGA scaffolds for adipose tissue engineering. *Tissue Eng.* 5, 139–151.

Pepper, M. S., Ferrara, N., Orci, L., and Montesano, R. (1992). Potent synergism between vascular endothelial growth factor and basic fibroblast growth factor in the induction of angiogenesis *in vitro*. *Biochem. Biophys. Res. Commun.* 189, 824–831.

Peters, M. C., Shea, L. D., and Mooney, D. J. (1999a). Protein and plasmid DNA delivery from tissue engineering matrices. *Polym. Prep.* 40, 273–274.

Peters, M. C., Isenberg, B. C., Rowley, J. A., and Mooney, D. J. (1999b). Release from alginate enhances the biological activity of vascular endothelial growth factor. *J. Biomater. Sci. Polym. Ed.* 9, 1267–1278.

Pinholt, E. M., Solheim, E., Bang, G., and Sudmann, E. (1992). Bone induction by composites of bioresorbable carriers and demineralized bone in rats: A comparative study of fibrin–collagen paste, fibrin sealant, and polyorthoester with gentamicin. *J. Oral Maxillofacial Surg.* 50, 1300–1304.

Pitt, C. G. (1990). Poly-ε-caprolactone and its copolymers. *In* "Biodegradable Polymers as Drug Delivery Systems" (M. Chasin and R. Langer, eds.), pp. 71–120. Dekker, New York.

Pulapura, S., and Kohn, J. (1992). Trends in the development of bioresorbable polymers for medical applications. *J. Biomater. Appl.* 6, 216–250.

Putnam, A. J., and Mooney, D. J. (1996). Tissue engineering using synthetic extracellular matrices. *Nat. Med.* 2, 824–826.

Roland, W. D., Holder, W. D., Culberson, C. R., Beiler, R. J., Burg, K. J. L., Greene, K. G., Loebsack, A. B., Wyatt, S., and Halberstadt, C. R. (1998). Optimizing cell culture time and seeding density on porous, absorbable constructs. *Tissue Eng.* **4**, 497 (Abstr. P-30).

Rowley, J. A., Madlambayan, G., and Mooney, D. J. (1999). Alginate hydrogels as synthetic extracellular matrix materials. *Biomaterials* **20**, 45–53.

Shapiro, L., and Cohen, S. (1997). Novel alginate sponges for cell culture and transplantation. *Biomaterials* **18**, 583–590.

Shea, L. D., Smiley, E., Bonadio, J., and Mooney, D. J. (2000). DNA delivery from polymer matrices for tissue engineering. *Nat. Biotechnol.* **17**, 551–554.

Sheridan, M. H., Shea, L. D., Peters, M. C., and Mooney, D. J. (2000). Bioabsorbable polymer scaffolds for tissue engineering capable of sustained growth factor delivery. *J. Controlled Release* **64**, 91–102.

Smentana, K. (1993). Cell biology of hydrogels. *Biomaterials* **14**, 1046–1050.

Solheim, E., Pinholt, E. M., Bang, G., and Sudmann, E. (1992). Inhibition of heterotopic osteogenesis in rats by a new bioerodible system for local delivery of indomethacin. *J. Bone Jt. Surg., Am. Vol.* **74-A**, 705–712.

Srivastava, S., Gorham, S. D., and Courtney, J. M. (1990). The attachment and growth of an established cell line on collagen, chemically modified collagen, and collagen composite surfaces. *Biomaterials* **11**, 162–168.

Tamargo, R. J., Epstein, J. I., Reinhard, C. S., Chasin, M., and Brem, H. (1989). Brain biocompatibility of a biodegradable, controlled-release polymer in rats. *J. Biomed. Mater. Res.* **23**, 253–266.

Thomson, R. C., Wake, M. C., Yaszemski, M. J., and Mikos, A. G. (1995). Biodegradable polymer scaffolds to regenerate organs. *Adv. Polym. Sci.* **122**, 245–274.

Tomihata, K., and Ikada, Y. (1997). *In vitro* and *in vivo* degradation of films of chitin and its deacetylated derivatives. *Biomaterials* **18**, 567–575.

Ullmann, Y., Hyams, M., Ramon, Y., Beach, D., Peled, I. J., and Lindenbaum, E. S. (1998). Enhancing the survival of aspirated human fat injected into nude mice. *Plast. Reconstr. Surg* **101**, 1940–1944.

Vandenburgh, H., and Kaufman, S. (1979). *In vitro* model for stretch-induced hypertrophy of skeletal muscle. *Science* **203**, 265–268.

Vandenburgh, H., Swasdison, S., and Karlisch, P. (1991). Computer-aided mechanogenesis of skeletal muscle organs from single cells *in vitro*. *FASEB J.* **5**, 2860–2867.

Vercruysse, K. P., Marecak, D. M., Marecek, J. F., and Prestwich, G. D. (1997). Synthesis and *in vitro* degradation of new polyvalent hydrazide cross-linked hydrogels of hyaluronic acid. *Bioconjugate Chem.* **8**, 686–694.

Wakitani, S., Goto, T., Young, R. G., Mansour, J. M., Goldberg, V. M., and Caplan, A. I. (1998). Repair of large full-thickness articular cartilage defects with allograft articular chondrocytes embedded in a collagen gel. *Tissue Eng.* **4**, 429–444.

Whang, K., Thomas, C. H., Healy, K. E., and Nuber, G. (1995). A novel method to fabricate bioabsorbable scaffolds. *Polymer* **36**, 837–842.

Williams, S. K., Wang, T. F., Castrillo, R., and Jarrell, B. E. (1994). Liposuction-derived human fat used for vascular graft sodding contains endothelial cells and not mesothelial cells as the major cell type. *J. Vasc. Surg.* **19**, 916–923.

Wong, W. H., and Mooney, D. J. (1997). Synthesis and properties of biodegradable polymers used as synthetic matrices for tissue engineering. *In* "Synthetic Biodegradable Polymer Scaffolds" (A. Atala and D. J. Mooney, eds.), pp. 49–80. Birkhaeuser, Boston.

Yamada, K., Miyamoto, S., Nagata, I., Kikuchi, H., Ikada, Y., Iwata, H., and Yamamoto, K. (1997). Development of a dural substitute from synthetic bioabsorbable polymers. *J. Neurosurg.* **86**, 1012–1017.

Zacchi, V., Soranzo, C., Cortivo, R., Radice, M., Brun, P., and Abatangelo, G. (1998). *In vitro* engineering of human skin-like tissue. *J. Biomed. Mater. Res.* **40**, 187–194.

PART X: CARDIOVASCULAR SYSTEM

CHAPTER 32

BLOOD VESSELS

Lian Xue and Howard P. Greisler

INTRODUCTION

Tissue-engineered blood vessels have a long history, beginning with early developments in polymeric vascular graft technology and in cell and tissue transplantation. More recent biologic manipulation is geared toward modulating the tissue reaction to implants and to thereby impacting on the common failure modes of thrombosis and intimal hyperplasia. This chapter reviews the biologic reactions elicited by commercially available and by experimental vascular grafts and discusses various strategies designed to induce endothelialization and/or to inhibit intimal hyperplasia. These strategies include endothelial cell seeding, impregnation with slow-release cytokines and growth factors, and the use of bioresorbable polymers. Novel techniques for the *in vitro* production of totally tissue-engineered blood vessels are addressed. A better understanding of physiologic and pathologic processes occurring after implantation of prosthetic grafts, combined with the development of novel technologies, will lead the road toward the development of ideal vascular substitutes.

The search for artificial vascular grafts dates back to 1952, when Voorhees developed Vinyon N, the first fabric graft (Voorhees *et al.,* 1952). The 550,000 vascular bypass surgeries performed annually in the United States make the development of vascular substitutes a continuing priority.

Initial graft research was directed at developing inert materials that passively transported blood and minimally interacted with blood and tissue. Poly(ethylene terephthalate) (Dacron) and expanded poly(tetrafluoroethylene) (e-PTFE) are the products of this research and are currently the most widely used prosthetic vascular grafts. However, the development of completely nonreactive substances has proved unrealistic. Both Dacron and e-PTFE react with blood components and perigraft tissues in both beneficial and detrimental fashions. An extraordinary expansion of our knowledge of mechanisms of vascular response to physiologic and pathologic conditions and the concomitant advance in the field of biomaterials have led to a more sophisticated strategy, optimizing tissue–biomaterial reactions to elicit desirable results. Synthetic grafts can be coated with proteins, anticoagulants, and antibiotics to potentially improve graft function. Similarly, synthetic polymers or biologically derived structural proteins can be bonded to various bioactive cytokines and growth factors to induce a favorable host response. Resorbable polymers comprise another type of graft in which extensive *in vivo* tissue ingrowth provides the biomechanical and functional integrity of the vascular conduit. Following over two decades of intensive research, a better understanding of the clinical value of endothelial cell (EC) seeding is starting to emerge. The use of tissue engineering techniques to construct a biologically viable vascular substitute is becoming feasible. Genetic manipulation of vascular cells may enhance the potential function of these biologic grafts (Conte, 1998).

The efficacy of vascular grafts depends on the intrinsic properties of the graft and the hemodynamic environment in which the graft is placed, as well as patient variables such as clinical indication, outflow resistance, progress of primary disease, and comorbidities. An understanding of graft pathophysiologic processes and failure modes will lay the basis for enhancement of long-term patency rates and development of more efficacious vascular substitutes.

GRAFT HEALING

All grafts, regardless of their composition and structure, evoke complex but predictable host responses. Blood–biomaterial surface reactions start immediately after restoration of circulation. The cellular and humoral responses to synthetic materials include the deposition of plasma proteins and platelets, the infiltration of neutrophils and monocytes, and the migration and proliferation of endothelial and smooth muscle cells. Even the most inert substances thus far developed are still recognized as "foreign." The tissue/graft/blood interfaces are highly complex microenvironments that are ultimately responsible for graft patency.

PROTEIN ADSORPTION

Serum proteins—such as albumin, fibrinogen, and immunoglobulin G (IgG), the most abundant serum proteins—adsorb to the graft almost instantaneously following exposure to the systemic circulation. Then, according to the Vroman effect, there is a redistribution of proteins regulated by each protein's relative biochemical and electrical affinity for the graft surface and by its relative abundance (Vroman, 1987). Because platelets and blood cells interact predominantly with the bound proteins and not with the prosthetic material, the constitution and concentration of bound protein largely determine the interaction of blood and graft and therefore affect graft survival. For example, fibrinogen, laminin, fibronectin, and vitronectin have an RGD sequence (Arg-Gly-Asp) region that binds to the glycoprotein gIIb/IIIa receptor complex on platelet membranes. Thus, platelet deposition and activation are influenced by prior protein adsorption. The binding of RGD sequence to $\beta 2$ integrins expressed on the cell surface also mediates the recruitment of the circulating leukocyte to the graft. The neutrophils bound to immobilized IgG and fibrinogen express a deficiency in their ability to kill bacteria, which may be partly responsible for the susceptibility of grafts to infections (De La Cruz *et al.*, 1998). Additional plasma proteins, including complement components, can be activated directly by synthetic surfaces to varying degrees. The generation of complement C5a, a monocyte chemoattractant, is greater following implantation of Dacron compared to e-PTFE grafts in an animal model (Shepard *et al.*, 1984). In addition, the rapid accumulation of coagulant proteins such as thrombin and factor Xa on the luminal surface after implantation contributes to the graft-associated procoagulant activity (Toursarkissian *et al.*, 1997).

PLATELET DEPOSITION

Early platelet disposition occurs primarily via receptor-mediated interactions with adsorbed proteins and, to a lesser extent, by direct adherence to the graft. The platelet/protein complex is mediated through von Willebrand factor and platelet membrane glycoproteins. After adherence, the platelets undergo conformational changes and degranulate, thereby releasing a variety of bioactive substances, including serotonin, epinephrine, ADP, and thromboxane A2 (TXA2), which activate additional platelets and increase thrombin production. Activated platelets also release growth factors, including platelet-derived growth factor (PDGF), epidermal growth factor (EGF), and transforming growth factor β (TGF-β), which modulate smooth muscle cell (SMC) migration and proliferation and extracellular matrix synthesis and degradation. In addition, platelets release monocyte chemoattractants such as platelet factor 4 and β-thromboglobulin, which mediate the recruitment of monocytes to the graft. Platelet deposition and activation continue chronically after graft implantation. Animal experiments demonstrate increased levels of thromboxane and decreased systemic platelet counts 1 year after Dacron graft implantation (Ito *et al.*, 1990). Human studies have also revealed an increase in In-[III] labeled platelet adhesion to grafts at late times following implantation (McCollum *et al.*, 1981; Stratton *et al.*, 1982, 1983).

Because the deposition and activation of platelets elicit a cascade of histopathologic events, the thrombogenic nature of the synthetic graft surface is therefore a major determinant of both early and late graft patency. A myriad of interventions have been tried in order to attenuate platelet deposition, aggregation, and activation. Antiplatelet agents directly targeting platelet/graft-binding molecules, such as platelet surface glycoprotein IIb/IIIa (αIIbβ3 integrin) and different functional domains of thrombin, have been shown to at least transiently decrease the platelets on Dacron grafts (Kelly *et al.*, 1992; Mazur *et al.*, 1994). A variety of techniques used to alter surface thrombogenicity have been studied experimentally, including the application of antiplatelet or anticoagulant substances to biomaterial surfaces, the alteration of surface electronegativity, and the radio frequency glow discharge of plasma-polymerized monomers to surfaces, in the hope of enhancing surface thromboresistance without alteration of the material's bulk properties.

NEUTROPHIL AND MONOCYTE INFILTRATION

The acute inflammatory response is mediated through potent chemoattractants such as C5a and leukotriene B4 (LTB4), which draw neutrophils to the synthetic graft surface. Neutrophils are attracted to the fibrin coagulum of the inner and outer capsules of the graft. The neutrophil binds to fibrinogen deposited on the graft surface by β2 integrins, mainly CD11b/CD18, which then mediate adhesion-dependent functions such as chemotaxis and promote transmigration. IgG, another protein deposited on the graft surface, is also important for neutrophil function. IgG can bind to Fcγ receptors expressed on neutrophils. Normally, this binding triggers a series of effector functions. However, the recruitment of phagocytic cells around prosthetic grafts does not sufficiently increase infection resistance. On the contrary, neutrophils adherent to biomaterials express a deficiency in their ability to kill and phagocytose bacteria. It has been proposed that Fc receptor ligation by immobilized IgG on graft surface may account for this deficiency (De La Cruz *et al.,* 1998). Neutrophils can also interact with other deposited proteins, such as C3bi and factor X.

Neutrophils adhere to the endothelial cells in the perianastomotic region through selectin- and integrin-mediated mechanisms. L-Selectin is thought to modulate neutrophil/endothelial cell interactions by presenting neutrophil ligands to the E-selectin and P-selectin on the vascular endothelium. Selectin–carbohydrate bonds are important in initial contact whereas integrin–peptide bonds are responsible for firm adhesion and transmigration of neutrophils (Jones *et al.,* 1996). Both intercellular adhesion molecular-1 (ICAM-1) and vascular cell adhesion molecule-1 (VCAM-1) on the endothelial cell (EC) surface can bind integrins. ECs up-regulate ICAM-1 and are induced to express VCAM-1 when stimulated by agonists such as interleukin 1 (IL-1) tumor necrosis factor (TNF), lipopolysaccharide, and thrombin (Bevilacqua *et al.,* 1985; Gamble *et al.,* 1985). Activated neutrophils release oxygen free radicals and various proteases, resulting in matrix degradation and possible inhibition of complete endothelialization and tissue incorporation of the vascular graft.

Circulating monocytes are also attracted to areas of injured or regenerating endothelium, especially in areas preactivated by IL-1 and TNF. There are many plasma monocyte recruitment and activating factors, including LTB4, platelet factor 4, and PDGF (Ford-Hutchinson *et al.,* 1980; Deuel *et al.,* 1981). In the presence of these plasma activating factors, monocytes differentiate into macrophages and become the major participant in the body's chronic inflammatory response. Macrophages release proteases and oxygen free radicals, which have long-term effects on the tissue/graft/blood interface.

A variety of cytokines are released from inflammatory cells activated by biomaterials. When cultured in the presence of polygalactin, a bioresorbable material phagocytized both *in vivo* and *in vitro* by macrophages, macrophages release into the culture media mitogens capable of stimulating the proliferation of quiescent smooth muscle cells, endothelial cells, and fibroblasts as compared to similar macrophages cultured in the presence of Dacron (Greisler *et al.,* 1991b). This mitogenic activity appears to be related to the secretion of fibroblast growth factor-2 (FGF-2) because pretreatment of the culture media with a neutralizing anti-FGF-2 antibody diminishes the stimulatory effect on smooth muscle cell growth (Greisler *et al.,* 1991b, 1996b). Cultured monocytes and macrophages incubated with Dacron and e-PTFE have been demonstrated to produce different amounts of IL-1β, IL-6, and TNF-α, concentrations being higher with Dacron than with e-PTFE (Swartbol *et al.,* 1997). TNF-α is one of the factors that may contribute to the enhanced proliferation of SMCs caused by a leukocyte–biomaterial interaction, and IL-1 may be partly responsible for the increased SMC proliferation caused by the leukocyte–EC reaction (Mattana *et al.,* 1997). IL-1 induces up-regulation of the expression of IGF-1 by ECs (Glazebrook *et al.,* 1998). Coincubation of polymorphonuclear leukocytes and IL-1β-treated human vascular endothelial cells (HUVECs) dramatically increased PDGF release (Totani *et al.,* 1998). Because the inflammatory reaction elicits a cascade of growth processes, it has been proposed that approaches attenuating the initial inflammatory reaction may improve the long-term patency of graft.

ENDOTHELIAL AND SMOOTH MUSCLE CELL INGROWTH

Native uninjured blood vessels possess an endothelial lining that constantly secretes bioactive substances; these factors inhibit thrombosis, promote fibrinolysis, and inhibit SMC proliferation to help maintain normal blood flow. After graft implantation, tissue ingrowth originates primarily from the cut edge of the adjacent artery. Unlike most animal models, in which there is often complete endothelialization of synthetic grafts, currently available grafts implanted into humans manifest only limited endothelial cell ingrowth, not extending beyond 1–2 cm of both anastomoses. However, endothelial islands have been described in the midportions of grafts at significant distances from the anastomosis, which suggests that other EC sources for graft endothelialization may exist. Interstitial tissue ingrowth accompanied by microvessels from the perigraft tissue is another major source for graft flow surface endothelialization on high-porosity e-PTFE and on bioresorbable prostheses (Clowes *et al.,* 1986; Golden *et al.,* 1990; Greisler *et al.,* 1988; Onuki *et al.,* 1998). There is also a possibility of circulating endothelial cell or stem cell deposition and proliferation based on the observation of scattered islands of endothelial cells isolated from pannus and unaccompanied by transmural microvessel ingrowth, present on impervious Dacron grafts in a canine model and also observed in a 24-month human Dacron graft explant (Shi *et al.,* 1994; Kouchi *et al.,* 1998; Wu *et al.,* 1995).

The ECs growing onto prosthetic graft surfaces undergo phenotypic modulation. These phenotypically altered "activated" cells may secrete bioactive substances that promote thrombogenesis and SMC growth. Activated ECs increase their PDGF synthesis and secretion, and subintimal smooth muscle cell proliferation occurs predominantly in the areas with an endothelial lining, i.e., the perianastomotic region, providing a link between endothelial cell turnover and intimal thickening (Greisler, 1996). This perianastomotic region is highly complex with chronic endothelial cell injury and has complex biomechanical characteristics and chronic inflammation. Although the time course may differ among species, in most animal models SMC proliferation starts as early as 2 days, peaks by 2 weeks after implantation of the graft, and thereafter diminishes. But neointimal volume keeps increasing by extracellular matrix synthesis and ultimately may form a stenotic lesion (Hamdan *et al.,* 1996).

SMCs found within the pseudointima of prosthetic grafts are also functionally altered. These SMCs produce significantly higher amounts of PDGF than do those of the adjacent vessel, which may contribute to the development of intimal hyperplasia (Pitsch *et al.,* 1997). In addition to SMCs, inflammatory mononuclear phagocytes and foreign body giant cells produce a variety of growth-modulating substances, including PDGF, FGF, and TGF-β, which perpetuate SMC proliferation and production of extracellular matrix components.

GRAFT FAILURE AND PATENCY RATES

A variety of mechanisms can lead to vascular graft occlusion. Immediate graft failure is usually the result of technical problems during surgery or may be due to the host's hypercoagulable status. Failure in the first month following graft placement is most likely the result of thrombosis in the face of high distal resistance. Anastomotic pseudointimal hyperplasia is the most common reason for graft failure from 6 months to 3 years after graft insertion. Late graft occlusions are frequently secondary to the progression of distal atherosclerotic disease. Thus, thrombogenicity and intimal hyperplasia represent the most common "intrinsic" causes of graft failure amenable to intervention by manipulation of graft construction, composition, and/or induced alterations in blood and tissue reactions to the implanted graft.

Clinical restenosis and patency rates reflect these "intrinsic" problems of synthetic grafts. Aortic reconstruction involves the placement of large-caliber Dacron or e-PTFE grafts. Because of the relatively high flow rates and low outflow resistance, these grafts have an 85-95% 5-year patency rate (Lind *et al.,* 1982; Moore *et al.,* 1968). When synthetic grafts are used in the infrainguinal, and especially the infrapopliteal, positions, the results are less than optimal. The 1- and 3-year patency rates for e-PTFE grafts used in infrapopliteal bypasses are 43 and 30%, respectively (Veith *et al.,* 1986). These small-caliber grafts are especially susceptible to anastomotic intimal hyperplasia and are more prone to early thrombosis because of lower flow rates and higher resistance outflow vessels.

Graft infection

Vascular graft infection occurs in only 1–5% of implanted grafts (Leikweg and Greenfield, 1977; Szilagyi *et al.,* 1972). However, the catastrophic effects of graft infection resulting in a 50% amputation rate and a 25–75% mortality rate justify experimental attempts aimed at increasing prosthetic material resistance to bacterial infection (Goldstone and Moore, 1974). Penicillins and cephalosporins have been noncovalently bound to Dacron and e-PTFE surfaces with subsequent increased resistance to *Staphylococcus aureus* infection in animal models. Rifampin-bonded gelatin-sealed Dacron grafts have also been shown *in vitro* to provide increased protection against bacterial colonization (Goeau-Brissonniere *et al.,* 1994; Vicaretti *et al.,* 1998).

GRAFT CHARACTERISTICS AFFECTING VASCULAR HEALING

Chemical composition, construction parameters, and biomechanics are the intrinsic characteristics of a vascular graft that predominantly influence its interaction with its host and, in turn, determine its fate.

Porosity

The rate of tissue ingrowth is dependent on graft porosity or permeability (over limited ranges that differ in various graft type). Clowes has shown, in a baboon model, enhanced tissue ingrowth with complete reendothelialization in e-PTFE grafts with a 60- or 90-μm internodal distance (Clowes *et al.,* 1986). Transinterstitial capillary ingrowth was not seen with the more commonly used 30-μm internodal distance in e-PTFE. In a canine study, 90-μm internodal e-PTFE grafts also demonstrated greater endothelial coverage and higher patency rates than similarly treated 30-μm internodal distance e-PTFE grafts (Cameron *et al.,* 1993). However, human trials using this more widely expanded e-PTFE have failed to show any advantage in platelet disposition as compared to standard 30-μm internodal distance e-PTFE grafts (Clowes and Kohler, 1991). In addition, transinterstitial ingrowth may differ when comparing two materials with equal porosity, e.g., PGA and Dacron (Greisler *et al.,* 1985). Greisler (Greisler *et al.,* 1987a, 1988a) has demonstrated significantly greater transinterstitial capillary ingrowth resulting in reendothelialization in lactide/glycolide woven prostheses as compared to Dacron prostheses when implanted into both rabbit and canine models. High-porosity grafts require preclotting.

Compliance

The compliance mismatch between arteries and grafts causes flow disruption, which has been thought to influence anastomotic pseudointimal hyperplasia (Abbott and Cambria, 1982). There have been attempts to design more compliant grafts, including the use of more flexible materials or by changing construction parameters. It is reported that grafts made of polyurethane may possess better compliance than e-PTFE (Stansby *et al.,* 1994). Surgeons have reported improved patency by interposing a piece of vein between the synthetic graft and artery at the distal anastomosis, a type of composite graft that may increase compliance at the distal anastomosis. Animal experiments have demonstrated a decreased neointimal thickness in composite grafts at the perianastomotic area (Gentile *et al.,* 1998). This method is now used by some as an alternative choice for femopopliteal arterial reconstruction, especially below-knee, when sufficient autologous vein is not available, although the benefit remains controversial. The value of more compliant grafts, however, is unclear.

Difficulties arise from the fact that there is great variability in the arterial tree in that the distal small-caliber arteries are often less complaint than the larger more central arteries. With progressive disease, wall thickening and calcification can occur throughout the arteries of atherosclerotic individuals. Another issue is that even experimental, relatively elastic grafts become less compliant over time because of the fibroplastic reaction that occurs following implantation. Even if a compliance match were possible, a final complicating factor may be the development of a hypercompliant zone in the region of the artery near the anastomosis (Hasson *et al.,* 1985).

Flow surface characteristics

The interface reaction between blood and the graft surface is dependent on surface chemical and physical properties such as surface charge, surface energy, and the degree of roughness. For example, a negative surface charge attenuates platelet adhesion and a positive charge promotes it. Heterogeneity of charge density distribution is also thought to be thrombogenic. A myriad of approaches have been made to stabilize and/or passivate the thrombotic reaction, including modification of surface properties, incorporation of antiplatelet or anticoagulant substances onto the graft, and endothelialization of the blood-contacting surface.

ANGIOGENESIS

ECs have only limited capacity for regeneration. After about 70 cell cycles, ECs can no longer divide. In order to achieve complete reendothelialization of the large surface areas required clinically, it is likely necessary to recruit ECs from other than an adjacent artery. In response to local angiogenic stimuli, ECs preexisting in capillaries in perigraft tissue become activated and move toward stimuli in a process that involves secretion of proteases, degradation of basement membrane and extracellular matrix, migration and proliferation of ECs, and reconstruction of tubular structures.

The platelet disposition and the inflammatory reaction around the graft result in release of a number of cytokines and proteolytic enzymes, which in turn incite secretion of a cascade of growth factors. Among these factors, some are angiogenic, such as tissue factor, vascular endothelial growth factor (VEGF), and FGFs, and some are angiogenic inhibitors, such as platelet factor 4, interferons, and thrombospondin; some exhibit biphasic activities, such as TNF-α and TGF-β. All of these bioactive substances create a microenvironment in which angiogenesis is a function of balance between positive and negative regulators. In addition, another balance between proteases and proteolytic inhibitors, such as plasminogen activator/plasminogen activator inhibitor (PA/PAI), is involved in cell migration and angiogenesis (Pepper, 1997). Exogenous intervention may change these balances so as to facilitate the endothelialization of prosthetic graft.

Local delivery of exogenous angiogenic factors may induce transmural capillary ingrowth *in vivo* and stimulate endothelialization of prosthetic grafts. In this context, a matrixlike fibrin providing spatial support is essential and can be used as a reservoir from which exogenous cytokines and growth factors can be released into the local microenvironment. Bioresorbable grafts may also facilitate local angiogenesis by stimulating macrophage release of endogenous angiogenic factors (see below).

Analogous strategies have been used to stimulate the development of collateral blood flow into ischemic tissue. In animal models, systemic or local applications of exogenous angiogenic factors, including VEGF, aFGF, and bFGF, may increase the capillary density of ischemic myocardial or limb tissue, prompt development of collateral circulation, and improve tissue hypoxia. A gleam of hope may be ahead for those severe ischemic extremities with no runoff channels for bypass surgery, a frustrating situation that usually ends with amputation.

STRATEGIES FOR THE DEVELOPMENT OF VASCULAR GRAFTS

To improve the long-term patency rate of arterial substitutes, especially in small-caliber arterial and venous systems, the following approaches have been pursued: (1) modification of the blood contact surface to obviate the thrombogenicity of synthetic grafts, (2) manipulation of graft-induced *in vivo* responses by controlled delivery of bioactive substances, (3) development of bioresorbable vascular prostheses, (4) endothelialization of the graft blood-contacting surface, and (5) development of completely tissue-engineered vascular substitutes.

SURFACE MODIFICATIONS

The simplest modification is to coat the graft with an inert polymer. Carbon coating, first done in the 1960s, has been found to decrease surface thrombosis, conjectured to be due to its negative charge and hydrophobic nature. The alternative carbon-impregnated prosthetic graft was also found experimentally to reduce platelet disposition. However, the advantage of these grafts is not confirmed by clinical comparative studies (Bacourt, 1997; Tsuchida *et al.,* 1992). A prospective multicenter clinical study of 81 carbon-impregnated e-PTFE and 79 standard e-PTFE grafts for below-knee popliteal and distal bypass showed no difference in patency rate between the two groups at up to 12 months after implantation. Silicone polymer coating, which produces a smooth surface that is devoid of the usual e-PTFE graft permeability and texture, followed by plasma glow discharge polymerization, effectively abolished pannus tissue ingrowth and graft surface neointimal hyperplasia in a baboon arterial interposition graft model and in a canine arteriovenous graft model (Lumsden *et al.,* 1996). Nojiri *et al.* (1995) have developed a three-layered graft consisting of Dacron for the outer layer (to promote perigraft tissue incorporation), nonporous polyurethane in the middle layer (to obtain a smooth surface), and a 2-hydroxyethyl methacrylate and styrene (HEMA–st) copolymer coating for the inner layer (to establish a nonthrombogenic blood interface). These grafts, with an inner diameter of 3 mm, were implanted in canine carotid arteries and remained patent for over 1 year. Only a monolayer of adsorbed proteins was found on the luminal surface of the grafts, with no pannus ingrowth from the adjacent artery, no thrombus, and no endothelial lining or neointimal formation.

Another approach is to cover the flow surface with proteins. Protein coating was originally introduced to decrease the initial porosity of Dacron grafts as an alternative to preclotting with blood. When the impregnated protein is degraded, the graft undergoes tissue ingrowth (Quarmby *et al.,* 1998). The compounds most often used have been albumin, gelatin, and collagen. Knitted Dacron prostheses coated with albumin, gelatin, and collagen have been available for clinical use. The albumin coating, created in the 1970s, was found in animal models to diminish platelet and leukocyte adhesion. In the canine thoracoabdominal aortic model, knitted Dacron grafts impregnated with carbodiimide cross-linked human albumin were compared to identical Dacron grafts with the recipients' blood preclotting. Albumin-impregnated grafts displayed less transinterstitial blood loss at implantation, and significantly thinner inner capsules at 20 weeks (Kang *et al.,* 1997).

Native collagen is intrinsically thrombogenic, whereas cross-linking of collagen is thought to diminish its thrombogenicity. In addition, collagen coating establishes a good matrix for cell ingrowth and induces neointimal formation, thereby potentially improving graft long-term patency. Promising results have been demonstrated by several groups (Bos *et al.,* 1998). But another clinical study reported that there was no appreciable difference in graft patency between woven and collagen-impregnated knitted Dacron aortoiliac grafts (Quarmby *et al.,* 1998).

Gelatin is a derivative of collagen but is more easily degraded when applied as a graft coating. Modification of coating techniques could control the degradation time (Bos *et al.,* 1998). Various techniques have been utilized in protein impregnation processes, including alkylation, plasma discharge, nonspecific cross-linking and covalent binding, and the application of thin polymer films (Dempsey *et al.,* 1998; Eberhart *et al.,* 1987; Ishikawa *et al.,* 1984; Kottke-Marchant *et al.,* 1989; Phaneuf *et al.,* 1998; Rumisek *et al.,* 1986; Tsai *et al.,* 1990).

Antithrombogenic substances have been affixed experimentally to synthetic grafts. For example, heparin binding to Dacron prostheses transiently reduces its thrombogenicity in some models (Mohamed *et al.,* 1998). In canine studies, heparin coupled to polyurethane and poly(dimethylsiloxane) surfaces improved graft patency rates (Park *et al.,* 1988). Thrombolytic substances, such as urokinase, have also been affixed to synthetic surfaces with promising results. Relatively little is known considering the long-term efficacy of these modifications because the duration of effective release is limited (Nojiri *et al.,* 1996).

Manipulating *In Vivo* Healing Processes

Tissue incorporation is an unavoidable and desirable process for implanted prostheses. But excessive cell proliferation as well as extracellular matrix deposition result in intimal hyperplasia, leading to vascular graft failure. The ideal healing process of vascular grafts would be rapid endothelialization of blood-contacting surfaces, spatially and temporally limited subendothelial SMC growth, followed by phenotypic and functional differentiation of cell components, as well as controlled remodeling of the extracellular matrix. The recent expansion of knowledge concerning mechanisms responsible for the migration and proliferation of ECs and SMCs, angiogenesis, extracellular matrix deposition and remodeling, and physiologic parameters provides the possibility of manipulating the healing process by optimizing the microenvironments of graft and perigraft tissue.

Fibroblast growth factors, notably FGF-1 (acidic FGF) and FGF-2 (basic FGF), known as strong mitogens for mesenchymal and neuroectodermal cells, also possess potent angiogenic activities. Both FGFs have been experimentally applied to grafts in combination with heparin, which potentates their mitogenic activity and protects them from proteolytic degradation. In order to achieve a controlled healing response, it is important to have a defined delivery system with which to apply bioactive substances to the graft and predictably release them locally, with bioactivities preserved for a desired period. Greisler has evaluated the affixation of FGF-1 to synthetic surfaces. In early attempts, FGF-1 was applied to various synthetic grafts by means of sequential application of fibronectin, heparin, FGF-1, and a second heparin layer, utilizing known binding affinities between successively applied agents. Using [^{125}I]FGF-1, the retention of the growth factor was quantitated. After 1 week in an animal model, retention was 44% in Dacron grafts and 23% in polydioxanone (PDS) grafts (Greisler *et al.,* 1986a, 1987b). After graft explantation, the FGF-1 was eluted and was shown to have retained its mitogenic activity on quiescent murine lung capillary endothelial cells. However, this technique failed to enhance endothelialization *in vivo.* It was proposed based on *in vitro* studies that FGF-1 bound to immobilized heparin did not possess mitogenic activity until the bond between the FGF-1 and heparin or between the heparin and fibronectin was broken (Greisler *et al.*, 1996a).

Further studies developed a method in which FGF-1 and heparin were delivered by fibrin glue (FG), fibrinogen polymerized by thrombin. *In vivo* release kinetics were quantitated using [^{125}I]FGF-1 and heparin impregnated into e-PTFE grafts implanted into rabbit infrarenal aortas. There was 13% retention of FGF-1 on the surface following 1 week of circulation, diminishing to 4% after 1 month. Equal distribution of the [^{125}I]FGF-1 throughout the walls of the prostheses was achieved (Greisler *et al.*, 1992a). Since fibrin glue is potentially a thrombogenic substance, the acute thrombogenicity of this FG/FGF-1/heparin-treated e-PTFE graft was assessed in canine aortoiliac model using autologous platelets radiolabeled with ^{111}In. Compared to untreated grafts, there was no difference in platelet disposition at 5 and 30 min after implantation, but there was significantly less platelet deposition in the FG/FGF-1/heparin-treated grafts at 120 min (Gosselin *et al.*, 1995). There was also no differential deposition of platelets on the anastomotic versus middle graft segments over time. Later *in vitro* studies have shown a decreased platelet adherence on either FG- or FG/FGF-1/heparin-treated grafts when compared to blood-preclotted e-PTFE grafts. The orientation and state of polymerization likely alter the affinity of fibrin to platelet adherence (Zarge *et al.*, 1997).

Utilizing the FG delivery system, FGF-1- and heparin-impregnated e-PTFE grafts, 60 μm internodal distance, have been evaluated in both canine aortoiliac and thoracoabdominal aortic models. When compared to FG-treated (no FGF-1) e-PTFE grafts or uncoated e-PTFE grafts, the FG/FGF-1/heparin graft treatment resulted in a significant increase in endothelial cell proliferation as assayed by *en face* autoradiography, with a more rapid development of a confluent factor VIII-positive endothelial blood-contacting surface, fully confluent in 28 days (Gray *et al.*, 1994; Greisler *et al.*, 1992a). There was also extensive transinterstitial capillary ingrowth observed throughout the graft wall. Cross-sectional autoradiography showed a similar significant increase in subendothelial myofibroblast proliferation within FG/FGF-1/heparin-treated grafts at 1 month, returning to base line at later times. When analyzed at 140 days, the FG/FGF-1/heparin grafts had developed significantly thicker inner capsules consisting of myofibroblasts and collagen compared to both untreated and FG/heparin-treated grafts.

Coimmobilization of FGF-2 and heparin in a microporous polyurethane graft by cross-linked gelatin gel has been demonstrated to accelerate tissue regeneration on synthetic grafts, associated with a greater extent of endothelialization via perianastomotic and transmural capillary ingrowth, in a rat aortic grafting model (Doi and Matsuda, 1997). A consistent "neointima" approximately 40 μm thick, with intermittent endothelialization as well as SMCs and fibroblasts underneath the luminal surface, was observed in the middle portion of treatment grafts, compared to the control grafts covered with only a fibrin layer. In response to exogenous stimuli, ECs, SMCs, and fibroblasts turn over and enter the cell cycle. Transformed cells achieve the ability of dividing, synthesizing ECM, and producing growth factors by autocrine and paracrine mechanisms. A number of growth factors and cell-cycling proteins are involved in this process. Intimal hyperplasia caused by excessive cellular and ECM proliferation could be a sequela of such intervention. The key point is fine control of this proliferation process. Although studies to date indicate that the induced cell proliferation is transient, the incorporation of a later released inhibitor aimed at cessation of cell cycling could be beneficial.

BIORESORBABLE GIFTS

Current clinically available synthetic vascular grafts, made of either e-PTFE or Dacron, are permanent prostheses within the host after implantation. Theoretically, it is possible to stimulate a rapid and controlled ingrowth of tissue to assume the load bearing sufficient to resist dilation, and including cellular and extracellular components with desirable physiologic characteristics, forming a "neoartery." The synthetic material may no longer be necessary following tissue ingrowth. Another strategy in the design of vascular grafts therefore is the use of bioresorbable materials.

Wesolowski *et al.* (1963) and Ruderman *et al.* (1972) were the first to describe the concept that a slowly absorbable vascular graft could induce a host regenerative process, producing a "new" functional artery. These early partially resorbable grafts were composed of Dacron and catgut yarns or Dacron and polylactide yarns. The first published report of a fully bioresorbable grafting was by Bowald in 1979 and described the use of a rolled sheet of Vicryl (a copolymer of polyglycolide and polylactide) (Bowald *et al.,* 1979, 1980). These early grafts were prone to aneurysmal dilation and rupture.

Early studies done by Greisler (1982) and Greisler *et al.* (1985) evaluated woven poly(glycolic acid) (PGA) grafts in a rabbit model. Four weeks after implantation, these 24-mm by 4-mm grafts contained an inner capsule composed of a confluent layer of endothelial cells and smooth musclelike myofibroblasts amid dense collagen fibers. Similarly constructed and implanted Dacron grafts demonstrated an inner capsule containing solely fibrin coagulum, with minimal cellularity. Macrophage infiltration and phagocytosis were in parallel with the resorption of the PGA, which was totally resorbed by 3 months. In this initial experiment, 10% of the PGA grafts showed aneurysmal dilation, with no difference in the incidence of dilation between 1 and 3 months and 3 and 12 months, suggesting that the critical time for the development of aneurysms is during prosthetic resorption before the ingrowth of tissue with adequate strength to resist hemodynamic pressures. Later studies evaluated grafts composed of polydioxanone, a more slowly resorbed compound. These grafts showed similar endothelialization of the regenerated luminal surface and produced prostacyclin and thromboxane in concentrations similar to those in control rabbit aorta. PDS remained present up to 6 months. Only 1/28 PDS grafts exhibited aneurysmal dilation, with explant times as late as 1 year. The explanted specimens of these PDS grafts demonstrated some biomechanical characteristics similar to native arteries and were able to withstand static bursting pressures of 6000 and 2000 mmHg mean pressure without fatigue (Greisler *et al.,* 1987a).

Additional experiments were done with a variety of resorbable materials, some combined with Dacron (Greisler *et al.,* 1986). The results demonstrated that transanastomotic pannus ingrowth is not a critical source of cells replacing bioresorbable vascular prostheses but rather transinterstitial ingrowth of myofibroblasts, capillaries, and endothelial cells in the principal mechanism (Fig. 32.1). In addition, tissue ingrowth was observed to parallel the kinetics of macrophage phagocytosis and prosthetic resorption (Greisler *et al.,* 1987a, 1988a). *In vivo,* the rate of cell proliferation and collagen deposition in the inner capsule parallels the kinetics of macrophage-mediated prosthetic resorption (Greisler *et al.,* 1991a, 1993b). These studies also confirmed the inhibitory effect of Dacron and the stimulatory effect of Polyglactin 910 (PG910) on tissue ingrowth and inner capsule cellularity. As described earlier in this chapter, activated macrophages release a variety of growth factors, including PDGF, interleukin-1, basic FGF, TGF-β, and tumor necrosis factor. The phagocytosis of various biomaterials can lead to macrophage activation. Rabbit peritoneal macrophages were cultured in the presence of PG910, Dacron, or neither, and the mitogenic activity in the resulting conditioned media was compared using growth assays of quiescent BALB/c 3T3 fibroblasts, rabbit aortic SMCs, and murine capillary ECs. The media of the macrophages grown in the presence of the bioresorbable polymer stimulated significantly more proliferation in all cell types than did the media of the macrophages grown in the absence of the material or in the presence of Dacron (Greisler, 1991; Greisler *et al.,* 1993a, 1996b). Preincubation of the conditioned media with neutralizing antibody against FGF-2 blocked 50–80% of that mitogenic activity.

A bioresorbable graft must regenerate a tissue complex of efficient strength prior to loss of prosthetic integrity so as to minimize the possibility of aneurysmal dilation. There are several theoretical ways to obviate this problem. One is to compound the bioresorbable material with a nonresorbable material that remains behind as a mechanical strut. Another solution involves the combination of two or more bioresorbable materials with different resorption rates so that the more rapidly resorbed material evokes a rapid tissue ingrowth while the second material provides temporary structural integrity to the graft. Third, growth factors, chemoattractants, and/or cells can be applied to the graft to enhance tissue ingrowth and organization.

Partially resorbable grafts have also been investigated in our laboratory. Dacron was found to inhibit the macrophage-mediated arterial regeneration stimulated by the resorbable component PG910. Polypropylene was then chosen as the nonresorbable component because of it's high tensile strength, low fatigability, low degradation *in vivo,* and minimal inhibitory effect on cellular regeneration of grafts (Greisler *et al.,* 1992b). The composite grafts constructed from yarns containing 69% PG910 and 31% polypropylene were implanted into rabbit and dog arteries. One year after implantation, these grafts showed 100% patency with no aneurysms (Greisler *et al.,* 1987c). In a different study, totally bioresorbable composite grafts woven from yarns of 74%

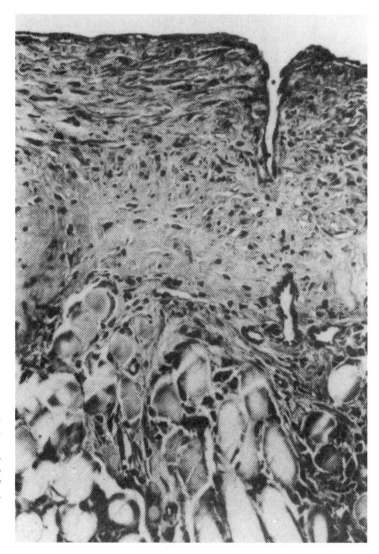

Fig. 32.1. Midportion of a poly-dioxanone prosthesis/tissue complex explanted after 2 months, showing capillaries invading the inner capsule, with one capillary communicating with the luminal surface (hematoxylin and eosin; ×25). From Arch. Surg., 1987, **122,** 715–721. Copyrighted 1987, American Medical Association.

PG910 and 26% PDS also demonstrated a 100% 1-year patency with no aneurysms in the rabbit aorta model (Greisler *et al.,* 1988b). The PG910 was totally resorbed by 2 months and the PDS by 6 months. The regenerated arteries withstood 800 mmHg of pulsatile systolic pressure *ex vivo* without bursting. A confluent, functional, von Willebrand factor (vWF)-positive endothelial cell layer over circumferentially oriented smooth musclelike myofibroblasts was present in the inner capsule of both these grafts.

Galletti *et al.* (1988) at Brown University used Vicryl prostheses coated with retardant polyesters to protect the Vicryl temporarily from hydrolytic and cellular degradation. When implanted into the canine aorta, prosthetic resorption was noticed at 4 weeks and was complete by 24 weeks. No coated grafts developed aneurysmal dilatation, whereas one of the uncoated grafts ruptured. It was theorized that the low pH generated in the microenvironment of the degrading bioresorbable polymers (polyglycolide and polylactide) stimulated macrophages to secrete growth factors that induce fibroblast proliferation. In separate experiments, another group in The Netherlands evaluated grafts prepared from a mixture of 95% polyurethane and 5% polylactide (van der Lei *et al.,* 1987; Yue *et al.,* 1988). They found that only relatively compliant grafts that induced circumferential smooth muscle development contained elastin and remained mechanically stable without dilating. They concluded that modifications of the graft preparation, including smooth muscle cell seeding, may help enhance optimal orientation of the smooth muscle cells and prevent aneurysm formation.

With further research, a small-diameter, totally resorbable vascular graft may be able to improve the current dismal long-term patency rates of small-caliber e-PTFE grafts.

ENDOTHELIAL CELL SEEDING

Normally, endothelial cells perform a variety of physiologic functions, among them being the promotion of thromboresistance. Endothelial cells possess a negative outer charge that repels platelet adherence. They also produce glycosaminoglycans that bind antithrombin III, and produce prostacyclins and tissue plasminogen activator, all of which facilitate the anticoagulant and fibrinolytic activities of endothelial cells. Theoretically, the presence of a confluent monolayer of endothelial cells could improve graft thromboresistance. In addition, a confluent EC monolayer may also prevent the development of pseudointimal hyperplasia by preventing the deposition of platelets, which release bioactive factors responsible for SMC migration, proliferation, and production of ECM, and by assuming a quiescent phenotype in which ECs may release SMC growth inhibitors.

In 1978, Herring (Herring *et al.,* 1978) first reported that endothelial cell seeding onto a graft surface enhanced graft survival. Since then, this subject has been intensively investigated. Considerable progress has been achieved, especially related to technical problems. Initial difficulties in cell harvesting, cell seeding, cell adhesion, and prevention of desquamation have all been largely overcome (Graham *et al.,* 1982; Jarrell and Williams, 1991). The ideal seeded grafts should have a confluent endothelial cell lining at the time of implantation and the cells should be able to resist sheer stress after the restoration of circulation, with the more desirable of their physiologic function intact. One of the difficulties is the relatively low cell density initially applied to the graft. The cell density of the endothelial cell lining on a normal vein is approximately 10^3 ECs/mm^2 (Sipehia *et al.,* 1996). In a completely lined vascular graft, cell density has to approach this value. The initial attachment of at least 5×10^3 ECs/mm^2 is required for immediate confluent human EC coverage of a small-caliber vascular graft, because some of the cells will be washed off after exposure to the flow. To maximize immediate cell inoculation density, a two-stage seeding procedure is often performed in which endothelial cells are harvested, allowed to proliferate *in vitro,* and then seeded and grown to confluence on the vascular graft prior to implantation (Prendiville *et al.,* 1991; Sentissi *et al.,* 1986; Shindo *et al.,* 1987). Disadvantages of this technique are the potential for infection, the alterations of EC phenotype and function, the requirement of a waiting period of 3–4 weeks for expansion of cell population, and the necessity for two operative procedures. An alternative method involves the use of microvascular endothelial cells. Small-diameter Dacron grafts seeded with enzymatically harvested omental microvascular cells in a single-stage technique showed confluent endothelial linings, larger thrombus-free surface area, and improved patency rates at 1 year in a canine model (Pasic *et al.,* 1994).

Another technical issue related to EC seeding is cell adhesion and retention. Endothelial cells adhere very poorly to synthetic graft materials. Many adhesive proteins, such as fibronectin, collagen, fibrin, laminin, RGD-containing peptides, and plasma, have been applied to the graft surface to improve the seeding efficiency. Nitrogen-containing plasma treatment, a novel surface modification technique, generates basic groups on the graft surface and enhances EC resistance to shear stress-induced desquamation (Sipehia *et al.,* 1996; Tseng and Edelman, 1998). The studies on the kinetics of EC seeding showed that at least 20% of initially adherent cells were lost during the first hour and even 60% within the first 24 hr (Rosenman *et al.,* 1985b; Schneider *et al.,* 1988; Vohra *et al.,* 1991). Prolongation of incubation time before exposure to flow has been suggested to allow maturation of the cytoskeleton so as to improve the cell retention on the graft surface (Miyata *et al.,* 1991; Prendiville *et al.,* 1991; Sugawara *et al.,* 1997). Preconditioning the seeded EC monolayer with graded shear stress promotes reorganization of the EC cytoskeleton and production of extracellular matrix, which in turn enhances the EC retention at flow exposure (Ott and Ballermann, 1995). The properties of prostheses affect EC attachment. Dacron expressed higher and easier cell attachment compared to e-PTFE when both grafts were coated with the same matrix (Sugawara *et al.,* 1997; Vohra *et al.,* 1990). Polyurethane also showed better cell attachment than e-PTFE (Giudiceandrea *et al.,* 1998). PTFE has a negative surface charge, which is an important property against platelet deposition. Meanwhile, the negative charges also obstruct the negatively charged ECs from accessing the graft surface because of natural repulsive forces. A technique called electrostatic transplantation has been developed to induce a temporary positive, or less negative, graft luminal surface charge, entailing EC attraction to the graft surface and accelerating their morphologic maturation. Using this technique, Bowlin *et al.* (1998) achieved no cell loss after a 30-min exposure to 15 dynes/cm^2 shear stress *in vitro* compared to 30% cell loss of control grafts and a cell density of $78,420 \pm 6274$ cells/cm^2, which was 2.3 times more than control after 120 min of physiologic shear stress exposure. The seeded cells maintained the ratio of prostacyclin/TXA2 production. A theoretical advantage of this technique is that when the graft is removed from the device, the luminal surface charge immediately reverts to its natural negative state, rendering any nonendothelialized areas less thrombogenic. This overcomes the limitation of adhesive protein coating, which may allow exposure of thrombogenic surfaces to flow when ECs do not reach complete confluence or desquamate from the surface. Another desirable aspect of this technique is its simplicity, which may further bring EC seeding time to a clinically acceptable range.

Zilla demonstrated a considerable cell loss 9 days after the implantation into baboons of EC-seeded e-PTFE grafts using fibrin glue coating for cell attachment, but the endothelium was kept confluent by compensatory overspreading of the remaining cells (Zilla *et al.*, 1994b). Animal studies have suggested that endothelial cell-seeded Dacron and e-PTFE grafts elicit decreased platelet deposition and increased patency rates (Allen *et al.*, 1984; Graham *et al.*, 1985; Stanley *et al.*, 1982; Wakefield *et al.*, 1986; Whitehouse *et al.*, 1983). Endothelial cell-seeded grafts have also been shown to be more resistant to bacterial contamination (Rosenman *et al.*, 1985a). However, the seeded grafts have not been reproducibly shown to reduce anastomotic pseudointimal hyperplasia significantly (Graham *et al.*, 1991).

Although a myriad of encouraging results have been achieved experimentally, the clinical data have been limited and have shown mixed results. Herring *et al.* (1989) reported an increased 1-year patency rate for endothelial-seeded e-PTFE femoral–popliteal grafts as compared to unseeded grafts in the nonsmoking population. But some later studies using the same single-staged endothelial seeding technique have failed to show a long-term advantage of the seeded grafts (Jensen *et al.*, 1994). Magometschnigg *et al.* (1992) showed a twofold increase in early patency and a twofold decrease in late amputation rates using cultured endothelial cells seeded onto fibrin glue-coated e-PTFE grafts placed as infrapopliteal (femoral–tibial) bypasses. Utilizing a similar two-stage endothelial cell-seeding technique, Zilla *et al.* (1994a) demonstrated increased patency and decreased platelet deposition in clinically implanted endothelial cell-seeded e-PTFE femoro-popliteal bypass grafts over 3 years as compared to unseeded grafts. This group reported (Meinhart *et al.*, 1997) very encouraging long-term results in a clinical follow-up study involving 108 *in vitro* endothelialized e-PTFE grafts, 107 of which were implanted in the infrainguinal position. In their initial randomized study, the primary patency rates were 84.7% for endothelialized grafts and 55.4% for control e-PTFE grafts after 3 years, 73.8 versus 31.3% after 5 years, and 73.8 versus 0% after 7 years. Combining these data with their more recent additional patients, who received EC-seeded grafts, the overall 7-year patency was 66% for 107 of endothelialized femoropopliteal e-PTFE grafts.

A major concern with EC-seeded grafts in humans is the potential for intimal thickening, supported by histopathologic observations from two separate case reports. In one case with bilateral above-knee grafts seeded with cephalic vein ECs, one of the grafts developed stenosis and had to be replaced 41 months after implantation. The central part of this graft was explanted and investigated. The graft had a confluent endothelium lining on a collagen IV-positive basement membrane with subintimal tissue of 1.21 ± 0.19 mm thickness (Deutsch *et al.*, 1997). The unusually thick subendothelial layer has also been found in another case in which a microvascular EC-seeded Dacron graft was placed as a mesoatrial bypass and had to be resected because of external mechanical stricture 9 months after implantation (Park *et al.*, 1990).

Besides technical problems, concern exists as to the ultimate function of those endothelial cells on the graft surface, the cells having been injured by the process of manipulation and/or chronic exposure to an unphysiologic environment. Injured endothelial cells produce a variety of procoagulants, such as von Willebrand factor, plasminogen activator inhibitor, thrombospondin, and collagen. Higher levels of PDGF and bFGF have been measured in endothelial-seeded grafts, which is of particularly concern given their potential role in stimulating the migration and proliferation of SMCs, and thereby stimulating the development of pseudointimal hyperplasia (Graham and Fox, 1991; Sapienza *et al.*, 1998).

Advances in molecular biology have provided the possibility of applying genetically modified endothelial cells with desired functions to synthetic grafts. The first application of this technology to vascular grafting was reported by Wilson *et al.* (1989). In this initial study, Dacron grafts seeded with retrovirally transduced ECs containing a *lacZ* marker gene were implanted into canine carotid artery. Gene expression from modified ECs on the graft surface was identified for a period of 5 weeks. Tissue plasminogen activator (tPA) has been successfully transfected into endothelial cells (Dichek *et al.*, 1991). These modified cells effectively express this fibrinolytic agent after being seeded onto synthetic graft surfaces (Shayani *et al.*, 1994). It may also be possible to transfect endothelial cells so that they oversecrete growth factors and growth inhibitors in order to encourage further endothelial cell proliferation and migration and to prevent smooth muscle cell hyperplasia.

Gene transduction is, however, a developing technology. The effects of gene transduction and of the vector employed on EC biological behavior and on the host response introduce new variables to EC seeding, which remain to be elucidated. Controversial results have been reported in the literature related to the proliferation, adhesion, and retention of genetically modified ECs on the surface of synthetic grafts. Brothers *et al.* (1990) observed a lower proliferation rate of *lacZ* and Tn5 gene transduced canine ECs versus wild type. In contrast, Jaklitsh *et al.* (1993) and Huber *et al.* (1995) separately reported that there were no changes in the proliferation rate of tPA retrovirally transduced ECs, but the migration of cells was decreased. A major concern is that modified ECs express poor retention on graft surfaces *in vivo*. A significantly lower percentage of surface endothelialization was detected at 6 weeks on canine thoracoabdominal aortic e-PTFE grafts seeded with *lacZ*-transduced ECs compared with those seeded with nontransduced control ECs (Baer *et al.*, 1996; Podrazik *et al.*, 1993). Dunn *et al.* (1996) reported only 6% retention of ECs that had been tPA transduced on Dacron grafts after 2 hr of exposure to the flow *in vivo*. The similar poor retention of tPA-transduced ECs on e-PTFE grafts has been described by other groups (Falk *et al.*, 1998; Huber *et al.*, 1995). For this reason, little has been documented so far concerning the long-term benefit of genetically modified EC-seeded grafts *in vivo*.

The introduction of genetic engineering offers considerable potential to manipulate the function of seeded cells on graft surface and of perigraft cells, which may in turn further augment the efficiency of EC seeding. It merits consideration for the development of a new generation of vascular grafts in the future.

Totally Tissue-Engineered Vascular Grafts

In addition to the continuing emphasis on the endothelialization of the flow surface, the function of other cell types in the vascular wall has attracted more and more attention. It has been suggested that ECs by themselves cannot produce a stable intima without SMCs or fibroblasts underneath. Tissue fragments containing multiple cell types, including venous tissue, adipose tissue, and bone marrow, have been seeded onto grafts and were found to accelerate the graft healing process (Noishiki *et al.*, 1998). Interestingly, bone marrow cell seeding was reported to induce an abundant capillary ingrowth in the graft wall and a rapid, complete endothelialization of the inner surface without intimal hyperplasia. The primitive stem cells in the bone marrow, having the ability to differentiate in response to their microenvironment and to proliferate as well as to secrete cytokines for supporting their survival, may provide a useful cell source for tissue engineering.

Driven by the desire to develop an ideal vascular substitute, attempts to construct a neoartery have been carried out. Vascular cells are seeded into a three-dimensional ECM or onto polymer scaffolds. After implantation, these cells proliferate and produce ECM while the scaffolds degrade, and eventually are replaced by self-tissue. The newly formed conduits are viable vessels with the ability to remodel to fit the hemodynamic environment and maintain many of the normal functions of the cell components.

SMCs seeded onto biodegradable grafts composed of 95% polyurethane and 5% poly(L-lactic acid) were demonstrated to result in fast and uniform neomedia development (Yue *et al.*, 1988). Shinoka *et al.* (1998) generated a graft from autologous mixed vascular cells expanded *in vitro* by culture of carotid artery segments and then seeded onto polygalactin or PGA scaffolds. After 7 days of *in vitro* culture, the grafts were implanted as pulmonary artery interpositions in lambs. The scaffolds no longer existed at 11 weeks, and the newly generated resembled native artery architecture, with ECs lining the luminal surface and with the development of ECM, including collagen and elastin fibers. However, an increase in the diameter of the grafts was noticed. Niklason *et al.*

(1999) reported encouraging results with a similar graft produced by seeding SMCs onto PGA scaffolds that were sodium hydroxide-modified to endure better cell attachment, and the graft was cultured in an *in vitro* pulsatile radial stress environment for 8 weeks before implantation. Engineered grafts showed contractile responses to serotonin, endothelin-1, and prostaglandin $F_{2\alpha}$ and expressed the SMC differentiation marker myosin heavy chain. The grafts cultured under pulsatile condition produced more collagen than those grown without pulsatile stress and exhibited a mechanical strength comparable to native human saphenous veins (ruptured at 2150 ± 909 mmHg versus 1680 ± 307 mmHg). Autologous ECs were seeded onto the luminal surface and were cultured for another 3 days before implantation. Four of the grafts were implanted into swine saphenous arteries, of which two were generated under pulsatile stress and two under static conditions. Two pulsed grafts remained patent up to 24–28 days without dilation or rupture, and two non-pulsed grafts remained open for 3 weeks and then developed thrombosis. The polymer remnants were no longer visible at 4 weeks.

When using bioresorbable polymers as scaffolds, in parallel to the resorption of polymer is an inflammatory process, which not only may play a major role in inducing tissue ingrowth but may also provide a pathologic microenvironment for all cell components. The complete understanding and fine control of this process will be the key point for the success of this kind of graft.

The first attempt to create totally biologic vascular structures *in vitro* without synthetic polymers was made by Weinberg and Bell (1986). They constructed a three-layered blood vessel model with a collagen matrix as a scaffold for ECs, SMCs, and fibroblasts. Similar efforts have been made by other groups (Langer and Vacanti, 1993; L'Heureux *et al.,* 1993; Ziegler and Nerem, 1994). However, this model failed to yield requisite mechanical strength, and later grafts were reinforced with Dacron meshes. This weakness is presumed to be, at least in part, a result of lack of organization of the ECM. As an alternative, L'Heureux *et al.* (1998), rather than use exogenous ECM components, generated a vessel exclusively from cultured vascular cells in a well-defined environment; ECM was produced, with the organization resembling that in natural vessels. SMCs were cultured *in vitro* to form a cellular sheet, then wrapped around a tubular support to produce the "media," and subsequently a similar sheet of fibroblasts was placed around the media to provide the "adventitia." After maturation, the tubular support was removed and ECs were seeded onto the luminal surface. This constructed "vessel" displayed a burst strength over 2000 mmHg. The SMCs expressed circumferential and longitudinal orientations with a differentiation marker, desmin, and even showed contractile responses when challenged with vasoactive agonists. Abundant ultrastructurally organized collagen and elastin fibers were present in the ECM. However, in short-term studies, after these vessels were implanted in canine femoral arteries, intramural blood infiltration was noticed. Although many more efforts will be required to conquer technical obstacles and optimize the manufacturing system, and as well examine the long-term efficacy, the possibility is open for creating novel viable substitutes for vascular replacement.

ENDOVASCULAR STENTS AND STENT GRAFTS

The progress in vascular surgery and interventional radiology has led to the birth of intravascular interventions for vascular disease. Stents and stent grafts are the products of this developing technique, which minimizes invasive procedures and thus especially benefits high-risk patients.

Endovascular stents were initially designed to provide structural support following angioplasty, originally described by Dotter in 1969 (Dotter, 1969), but have become clinically widespread. There are three basic types of stents: balloon-expandable stents, which need balloon inflation to expand the stent into the arterial wall; self-expanding stents, allowing delivery in a collapsed form with the stent expanding to its predetermined size after release from the delivery device; and thermal-expanding stents, made by a shape-memory metal alloy that is in an easily manipulated form that assumes a memorized shape at a certain transition temperature. All these stents have been successfully used in pelvic arteries with a 2-year patency of approximately 84% (Müller-Hülsbeck *et al.,* 1998).

The stent graft is a combination in which a prosthetic graft is fixed to the arterial wall by an intravascular stent. The graft is usually either Dacron or e-PTFE and is combined with different stents. The primary indications for endovascular grafting are arterial aneurysm, arterial–venous fistulae, and vascular trauma. There have been numerous attempts to extend this technique to the treatment of arterial occlusive disease. Theoretically, the graft creates a barrier to exclude diseased arterial wall and provides a smooth flow conduit, while the stent support affixes the graft and may enhance luminal patency by resisting external compression. Although the short-term results show promise in carefully selected cases (Allen *et al.,* 1998), little is known about the long-term consequences of this new technology.

There are limited data regarding the pathologic and hemodynamic influences of the stent or stent graft anchored to the arterial wall. It has been reported that stent placement may cause a variety of flow disturbances (Müller-Hülsbeck *et al.,* 1998). The stent components may stimulate a nonspecific inflammatory reaction and induce neointimal formation (Müller-Hülsbeck *et al.,* 1997; Schürmann *et al.,* 1996). Stent grafts possess even more complexities in respect to the healing process, which is different from either the stent or the graft. Obvious compliance changes between stent/unsupported graft/artery interfaces yield a remarkable hemodynamic disturbance. Delivery procedures, such as balloon dilation, which may alter graft intrinsic characteristics such as porosity, wall thickness, etc., as well as create mechanical injury to the surrounding artery, should also be taken into account (Salzmann *et al.,* 1997, 1998). In addition, unlike conventional bypass, the endovascular graft is placed into luminal arteries with perigraft exposure to diseased arterial intima or to thrombus. All these factors change the healing characteristics of prostheses. An inflammatory reaction and progressive thickening of neointima have been observed in both e-PTFE- and Dacron-based stent grafts (Hussain *et al.,* 1998; Yee *et al.,* 1998). A controlled study in the canine iliac artery model showed that endovascular stent grafts, composed of e-PTFE grafts and balloon-expandable stents, resulted in both greater endothelialization and an approximately five times thicker neointima in the midportion of the graft and a higher percentage of stenosis at the distal anastomosis, when compared to conventional e-PTFE grafts. Yee *et al.* investigated a similar stent graft in the porcine iliac artery and concluded that intravascular placement dramatically altered the healing of this type of graft (Ohki *et al.,* 1997).

Endovascular intervention alleviates surgical morbidity and mortality, thus having considerable clinical potential. The accumulation of experience, optimization of devices, and a better understanding of resultant pathologic processes will enhance their efficacy.

REFERENCES

Abbott, W. M., and Cambria, R. P. (1982). Control of physical characteristics elasticity and compliances of vascular grafts. *In* "Biological and Synthetic Vascular Prostheses" (J. C. Stanley, ed.), p. 189. Grune & Stratton, New York.

Allen, B. T., Long, J. A., Clark, R. E., Sicard, G. A., Clark, R. D., Hopkins, K. T., and Welch, M. J. (1984). Influence of endothelial cell seeding on platelet deposition and patency in small-diameter Dacron arterial grafts. *J. Vasc. Surg.* **1,** 224–233.

Allen, B. T., Hovsepian, D. M., Reilly, J. M., Rubin, B. G., Malden, E., Keller, C. A., Picus, D. D., and Sicard, G. A. (1998). Endovascular stent grafts for aneurysmal and occlusive vascular disease. *Am. J. Surg.* **176,** 574–580.

Bacourt, F. (1997). Prospective randomized study of carbon-impregnated polytetrafluoroethylene grafts for below-knee popliteal and distal bypass; results at 2 years. *Ann. Vasc. Surg.* **11,** 569–603.

Baer, R. P., Whitehill, T. E., Sarkar, R., Sarkar, M., Messina, L. M., Komorowski, T. A., and Stanley, J. C. (1996). Retroviral-mediated transduction of endothelial cells with the lacZ gene impairs cellular proliferation *in vitro* and graft endothelialization *in vivo. J. Vasc. Surg.* **24,** 892–899.

Bevilacqua, M. P., Pober, J. S., Wheeler, M. E., Cotran, R. S., and Gimbrone, M. A., Jr. (1985). Interleukin-1 acts on cultured human vascular endothelium to increase the adhesion of polymorphonuclear leukocytes, monocytes, and related leukocyte cell lines. *J. Clin. Invest.* **76,** 2003–2011.

Bos, G. W., Poot, A. A., Beugeling, T., van Aken, W. G., and Feijen, J. (1998). Small-diameter vascular graft prostheses; current status. *Arch. Physiol. Biochem.* **106,** 100–115.

Bowald, S., Busch, C., and Eriksson, I. (1979). Arterial regeneration following polyglactin 910 suture mesh grafting. *Surgery* **86,** 722–729.

Bowald, S., Busch, C., and Eriksson, I. (1980). Absorbable material in vascular prostheses: A new device. *Acta Chir. Scand.* **146,** 391–395.

Bowlin, G. L., Rittgers, S. E., Milsted, A., and Schmidt, S. P. (1998). *In vitro* evaluation of electrostatic endothelial cell transplantation onto 4 mm interior diameter expanded polytetrafluoroethylene grafts. *J. Vasc. Surg.* **27,** 504–511.

Brothers, T. E., Judge, L. M., Wilson, J. M., Burkel, W. E., and Stanley, J. C. (1990). Effect of genetic transduction on *in vitro* canine endothelial cell prostanoid production and growth. *Surg. Forum.* **41,** 337–339.

Cameron, B. L., Tsuchida, H., Connall, T. P., Nagae, T., Furukawa, K., and Wilson, S. E. (1993). High porosity PTFE improves endothelialization of arterial grafts without increasing early thrombogenicity. *J. Cardiovasc. Surg.* **34,** 281–285.

Clowes, A. W., and Kohler, T. (1991). Graft endothelialization. The role of angiogenic mechanisms. *J. Vasc. Surg.* **13,** 734–736.

Clowes, A. W., Kirkman, T. R., and Reidy, M. A. (1986). Mechanisms of arterial graft healing: Rapid transmural capillary ingrowth provides a source of intimal endothelium and smooth muscle in porous PTFE prostheses. *Am. J. Pathol.* **123,** 220–230.

Conte, M. S. (1998). The ideal small arterial substitute, a search for the Holy Grail? *FASEB J.* **12,** 43–45.

De La Cruz, C., Haimovich, B., and Greco, R. S. (1998). Immobilized IgG and fibrinogen differentially affect the cytoskeletal organization and bactericidal function of adherent neutrophils. *J. Surg. Res.* **80**, 28–34.

Dempsey, D. J., Phaneuf, M. D., Bide, M. J., Szycher, M., Quist, W. C., and Logerfo, F. W. (1998). Synthesis of a novel small diameter polyurethane vascular graft with reactive binding sites. *ASAIO J.* **44**, M506–M510.

Deuel, T. F., Senior, R. M., Chang, D., Griffin, G. L., Heinrikson, R. L., and Kaiser, E. T. (1981). Platelet factor 4 is chemotactic for neutrophils and monocytes. *Proc. Natl. Acad. Sci. U.S.A.* **78**, 4584–4587.

Deutsch, M., Meinhart, J., Vesely, M., Fischlein, T., Groscurth, P., von Oppell, U., and Zella, P. (1997). *In vitro* endothelialization of expanded polytetrafluoroethylene grafts; a clinical case report after 41 months of implantation. *J. Vasc. Surg.* **25**, 757–763.

Dichek, D. A., Nussbaum, O., Degen, S. J., and Anderson, W. F. (1991). Enhancement of the fibrinolytic activity of sheep endothelial cells by retroviral-mediated gene transfer. *Blood* **77**, 533–541.

Doi, K., and Matsuda, T. (1997). Enhanced vascularization in a microporous polyurethane graft impregnated with basic fibroblast growth factor and heparin. *J. Biomed. Mater. Res.* **34**, 361–370.

Dotter, C. T. (1969). Transluminally-placed coilspring endarterial tube grafts. Long-term patency in canine popliteal artery. *Invest. Radiol.* **4**, 329–332.

Dunn, P. F., Newman, K. D., Jones, M., Yamada, I., Shayani, V., Virmani, R., and Dichek, D. A. (1996). Seeding of vascular grafts with genetically modified endothelial cells; secretion of recombinant tPA results in decreased seeded cell retention *in vitro* and *in vivo*. *Circulation* **93**, 1439–1446.

Eberhart, R. C., Munro, M. S., Williams, G. B., Kulkarni, P. V., Shannon, W. A., Jr., Brink, B. E., and Fry, W. J. (1987). Albumin adsorption and retention on C18-alkyl-derivatized polyurethane vascular grafts. *Artif. Organs* **11**, 375–382.

Falk, J., Townsend, L. E., Vogel, L. M., Boyer, M., Olt, S., Wease, G. L., Trevor, K. T., Seymour, M., Glover, J. L., and Bendick, P. J. (1998). Improved adherence of genetically modified endothelial cells to small-diameter expanded polytetrafluoroethylene grafts in a canine model. *J. Vasc. Surg.* **27**, 902–909.

Ford-Hutchinson, A. W., Bray, M. A., Doig, M. V., Shipley, M. E., and Smith, M. J. (1980). Leukotriene B, a potent chemokinetic and aggregating substance released from polymorphonuclear leukocytes. *Nature (London)* **286**, 264–265.

Galletti, P. M., Aebischer, P., Sasken, H. F., Goddard, M. B., and Chiu, T. H. (1988). Experience with fully bioresorbable aortic grafts in the dog. *Surgery* **103**, 231–241.

Gamble, J. R., Harlan, J. M., Klebanoff, S. J., and Vadas, M. A. (1985). Stimulation of the adherence of neutrophils to umbilical vein endothelium by human recombinant tumor necrosis factor. *Proc. Natl. Acad. Sci. U.S.A.* **82**, 8667–8671.

Gentile, A. T., Mills, J. L., Gooden, M. A., Hagerty, R. D., Berman, S. S., Hughes, J. D., Kleinhart, L. B., and Williams, S. K. (1998). Vein patching reduces neointimal thickening associated with prosthetic graft implantation. *Am. J. Surg.* **176**, 601–607.

Giudiceandrea, A., Seifalian, A. M., Krijgsman, B., and Hamilton, G. (1998). Effect of prolonged pulsatile shear stress *in vitro* on endothelial cell seeded PTFE and compliant polyurethane vascular grafts. *J. Vasc. Endovasc. Surg.* **15**, 147–154.

Glazebrook, H., Hatch, T., and Brindle, N. P. (1998). Regulation of insulin-like growth factor-1 expression in vascular endothelial cells by the inflammatory cytokines interleukin-1. *J. Vasc. Res.* **35**, 143–149.

Golden, M. A., Hanson, S. R., Kirkman, T. R., Schnerder, P. A., and Clowes, A. W. (1990). Healing of polytetrafluoroethylene arterial grafts is influenced by graft porosity. *J. Vasc. Surg.* **11**, 838–844.

Goldstone, J., and Moore, W. S. (1974). Infections in vascular prostheses. *Am. J. Surg.* **128**, 225–233.

Goeau-Brissonniere, O., Mercier, F., Nicolas, M. H., Bacourt, F., Coggia, M., Lebrault, C., and Pechere, J. C. (1994). Treatment of vascular graft infection by *in situ* replacement with a rifampin-bonded gelatin sealed Dacron graft. *J. Vasc. Surg.* **19**, 739–741.

Gosselin, C., Ren, D., Ellinger, J., and Greisler, H. P. (1995). *In vivo* platelet deposition on polytetrafluoroethylene coated with fibrin glue containing fibroblast growth factor 1 and heparin in a canine model. *Am. J. Surg.* **170**, 126–130.

Graham, L. M., and Fox, P. L. (1991). Growth factor production following prosthetic graft implantation. *J. Vasc. Surg.* **13**, 742–746.

Graham, L. M., Burkel, W. E., Ford, J. W., Vinter, D. W., Kahn, R. H., and Stanley, J. C. (1982). Expanded polytetrafluoroethylene vascular prostheses seeded with enzymatically derived and cultured canine endothelial cells. *Surgery* **91**, 550–559.

Graham, L. M., Stanley, J. C., and Burkel, W. E. (1985). Improved patency of endothelial-cell-seeded, long, knitted Dacron and ePTFE vascular prostheses. *ASAIO J.* **8**, 65–73.

Graham, L. M., Brothers, T. E., Vincent, C. K., Burkel, W. E., and Stanley, J. C. (1991). The role of an endothelial cell lining in limiting distal anastomotic intimal hyperplasia of 4-mm-I.D. Dacron grafts in a canine model. *J. Biomed. Mater. Res.* **25**, 525–533.

Gray, J. L., Kang, S. S., Zenni, G. C., Kim, D. U., Kim, P. I., Burgess, W. H., Drohan, W., Winkles, J. A., Hauderschild, C. C., and Greisler, H. P. (1994). FGF-1 affixation stimulates ePTFE endothelialization without intimal hyperplasia. *J. Surg. Res.* **57**, 596–612.

Greisler, H. P. (1982). Arterial regeneration over absorbable prostheses. *Arch. Surg. (Chicago)* **117**, 1425–1431.

Greisler, H. P. (1991). Bioresorbable materials and macrophage interactions. *J. Vasc. Surg.* **13**, 748–750.

Greisler, H. P. (1996). Regulation of vascular graft healing by induction of tissue incorporation. *In* "Human Biomaterials Applications" (D. L. Wise, D. E. Altobelli, M. J. Yaszemski, and H. P. Greisler, eds.), p. 227. Humana Press, Totowa, NJ.

Greisler, H. P., Kim, D. U., Price, J. B., and Voorhee, A. B., Jr. (1985). Arterial regenerative activity after prosthetic implantation. *Arch. Surg. (Chicago)* **120**, 315–323.

Greisler, H. P., Klosak, J. J., Dennis, J. W., Ellinger, J., Kim, D. U., Burgess, W., and Maciag, T. (1986a). Endothelial cell growth factor attachment to biomaterials. *ASAIO Trans.* **32**, 346–349.

Greisler, H. P., Schwarcz, T. H., Ellinger, J., and Kim, D. U. (1986b). Dacron inhibition of arterial regenerative activity. *J. Vasc. Surg.* **3**, 747–756.

Greisler, H. P., Ellinger, J., Schwarcz, T. H., Golan, J., Raymond, R. M., and Kim, D. U. (1987a). Arterial regeneration over a polydioxanone prosthesis in the rabbit. *Arch. Surg. (Chicago)* **122**, 715–721.

Greisler, H. P., Klosak, J. J., Dennis, J. W., Karesh, S. M., Ellinger, J., and Kim, D. U. (1987b). Biomaterial pretreatment with ECGF to augment endothelial cell proliferation. *J. Vasc. Surg.* **5**, 393–399.

Greisler, H. P., Kim, D. U., Dennis, J. W., Klosak, J. J., Widerborg, K. A., Endean, E. D., Taymond, R. M., and Ellinger, J. (1987c). Compound polyglactin 910/polypropylene small vessel prostheses. *J. Vasc. Surg.* **5**, 572–583.

Greisler, H. P., Dennis, J. W., Endean, E. D., Ellinger, J., Burrle, K. F., and Kim, D. U. (1988a). Derivation of neointima in vascular grafts. *Circulation* **78**, 16–112.

Greisler, H. P., Endean, E. D., Klosak, J. J., Ellinger, J., Dennis, J. W., Buttle, K., and Kim, D. U. (1988b). Polyglactin 910/polydioxanone bicomponent totally resorbable vascular prostheses. *J. Vasc. Surg.* **7**, 697–705.

Greisler, H. P., Cabusao, E. B., Lam, T. M., Ellinger, J., and Kim, D. U. (1991a). Kinetics of collagen deposition within bioresorbable and nonresorbable vascular prostheses. *ASAIO Trans.* **37**, M472–M475.

Greisler, H. P., Ellinger, J., Henderson, S. C., Shaheen, A. M., Burgess, W. H., and Lam, T. M. (1991b). The effects of an atherogenic diet on macrophage/biomaterial interactions. *J. Vasc. Surg.* **14**, 10–23.

Greisler, H. P., Cziperle, D. J., Kim, D. U., Garfield, J. D., Petsikas, D., Murchan, P. M., Applegren, E. O., Drohan, W., and Burgess, W. H. (1992a). Enhanced endothelialization of expanded polytetrafluorethylene grafts by fibroblast growth factor type 1 pretreatment. *Surgery* **112**, 244–254.

Greisler, H. P., Tattersall, C. W., Henderson, S. C., Cabusao, E. A., Garfield, J. D., and Kim, D. U. (1992b). Polypropylene small-diameter vascular grafts. *J. Biomed. Mater. Res.* **26**, 1383–1394.

Greisler, H. P., Henderson, S. C., and Lam, T. M. (1993a). Basic fibroblast growth factor production *in vitro* by macrophages exposed to Dacron and polyglactin 910. *J. Biomater. Sci., Polym. Ed.* **4**, 415–430.

Greisler, H. P., Petsikas, D., Lam, T. M., Patel, N., Ellinger, J., Cabusao, E., Tattersall, C. W., and Kim, D. U. (1993b). Kinetics of cell proliferation as a function of vascular graft material. *J. Biomater. Res.* **27**, 955–961.

Greisler, H. P., Gosselin, C., Ren, D., Kang, S. S., and Kim, D. U. (1996a). Biointeractive polymers and tissue engineered blood vessels. *Biomaterials* **17**, 329–336.

Greisler, H. P., Petsikas, D., Cziperle, D. J., Murchan, P., Henderson, S. C., and Lam, T. M. (1996b). Dacron stimulation of macrophage transforming growth factor-beta release. *Cardiovasc. Surg.* **4**, 169–173.

Hamdan, A. D., Misare, B., Contreras, M., Logerfo, F. W., and Quist, W. C. (1996). Evaluation of anastomotic hyperplasia progression using the cyclin specific antibody MIB-1. *Am. J. Surg.* **172**, 168–171.

Hasson, J. E., Megerman, J., and Abbott, W. M. (1985). Increased compliance near vascular anastomoses. *J. Vasc. Surg.* **2**, 419–423.

Herring, M., Gardner, A., and Glover, J. (1978). A single-staged technique for seeding vascular grafts with autogenous endothelium. *Surgery* **84**, 498–504.

Herring, M. B., Gardner, A., and Glover, J. (1989). Seeding human arterial prostheses with mechanically derived endothelium. The detrimental effect of smoking. *J. Vasc. Surg.* **1**, 279–289.

Huber, T. S., Welling, T. H., Sarkar, R., Messina, L. M., and Stanley, J. C. (1995). Effects of retroviral-mediated tissue plasminogen activator gene transfer and expression on adherence and proliferation of canine endothelial cells seeded onto expanded polytetrafluoroethylene. *J. Vasc. Surg.* **22**, 795–803.

Hussain, F. M., Kopchok, G., Heilbron, M., Daskalakis, T., Donayre, C., and White, R. A. (1998). Wallgraft endoprosthesis, initial canine evaluation. *Am. Surg.* **64**, 1002–1006.

Ishikawa, Y., Sasakawa, S., Takase, M., and Osada, Y. (1984). Effect of albumin immobilization by plasma polymerization on platelet reactivity. *Thromb. Res.* **35**, 193–202.

Ito, R. K., Rosenblatt, M. S., Contreras, M. A., Beophy, X. M., and LoGerfo, F. W. (1990). Monitoring platelet interactions with prosthetic graft implants in a canine model. *ASAIO Trans.* **36**, M175–M178.

Jaklitsch, M. T., Biro, S., Casscells, W., and Dichek, D. A. (1993). Transduced endothelial cells expressing high levels of tissue plasminogen activator have an unaltered phenotype *in vitro. J. Cell. Physiol.* **154**, 207–216.

Jarrell, B. E., and Williams, S. K. (1991). Microvessel derived endothelial cell isolation, adherence, and monolayer formation for vascular grafts. *J. Vasc. Surg.* **13**, 733–734.

Jensen, N., Lindblad, B., and Bergovist, D. (1994). Endothelial cell seeded Dacron aortobifurcated grafts: Platelet deposition and long-term follow-up. *J. Cardiovasc. Surg.* **35**, 425–429.

Jones, D. A., Smith, C. W., and McIntire, L. V. (1996). Leukocyte adhesion under flow conditions; principles important in tissue engineering. *Biomaterials* **17**, 337–347.

Kang, S. S., Petsikas, D., Murchan, P., Cziperle, D. J., Ren, D., Kim, D. U., and Greisler, H. P. (1997). Effects of albumin coating of knitted Dacron grafts on transinterstitial blood loss and tissue ingrowth and incorporation. *Cardiovasc. Surg.* **5**, 184–189.

Kelly, A. B., Maragamore, J. M., Bourdon, P., Hanson, S. R., and Harker, L. A. (1992). Antithrombotic effects of synthetic peptides targeting various functional domains of thrombin. *Proc. Natl. Acad. Sci. U.S.A.* **89**, 6040–6044.

Kottke-Marchant, K., Anderson, J. M., Umemura, Y., and Marchant, R. E. (1989). Effect of albumin coating of the *in vitro* blood compatibility of Dacron arterial prostheses. *Biomaterials* **10**, 147–155.

Kouchi, Y., Onuki, Y., Wu, M. H., Shi, O., Ghali, R., Wechezak, A. R., Kaplan, S., Walker, M., and Sauvage, L. R. (1998). Apparent blood stream origin of endothelial and smooth muscle cells in the neointima of long, impervious carotid-femoral grafts in the dog. *Ann. Vasc. Surg.* **12**, 46–54.

Langer, R., and Vacanti, J. P. (1993). Tissue engineering. *Science* **260**, 920–926.

Leikweg, W. G., and Greenfield, L. J. (1977). Vascular prostheses graft infections: Collected experiences and results of treatment. *Surgery* **81**, 335–342.

L'Heureux, N., Germain, L., Labb, J. T., and Auger, F. A. (1993). *In vitro* construction of a human blood vessel from cultured vascular cells. A morphologic study. *J. Vasc. Surg.* **17**, 499–509.

L'Heureux, N., Paquet, S., Labb, J. T., Germain, L., and Auger, F. A. (1998). A completely biological tissue-engineered human blood vessel. *FASEB J.* **12**, 47–56.

Lind, R. E., Wright, C. B., Lynch, T. G., Lamberth, W. C., Jr., Slaymaker, E. E., and Brandt, B., 3rd (1982). Aortofemoral bypass grafting: Microvel. *Am. Surg.* **48**, 89–92.

Lumsden, A. B., Chen, C., Coyle, K. A., Ofenloch, J. C., Wang, J. H., Yasuda, H. K., and Hanson, S. R. (1996). Nonporous silicone polymer coating of expanded polytetrafluoroethylene grafts reduces graft neointimal hyperplasia in dog and baboon models. *J. Vasc. Surg.* **24**, 825–833.

Magometschnigg, H., Kadletz, M., Vodrazka, M., Dock, W., Grimm, M., Grabenwoger, M., Minar, E., Staudacher, M., Fenzi, G., and Wolner, E. (1992). Prospective clinical study with *in vitro* endothelial cell lining of expanded polytetrafluoroethylene grafts in crural repeat reconstruction. *J. Vasc. Surg.* **15**, 527–535.

Mattana, J., Effiong, C., Kapasi, A., and Singhal, P. C. (1997). Leukocyte–polytetrafluoroethylene interaction enhances proliferation of vascular smooth muscle cells via tumor necrosis factor-alpha secretion. *Kidney Int.* **52**, 1478–1485.

Mazur, C., Tschopp, J. F., Faliakou, E. C., Gould, K. E., Diehl, J. T., Pierschbacher, M. D., and Connolly, R. J. (1994). Selective allbb3 receptor blockage with peptide TP9201 prevents platelet uptake on Dacron vascular grafts without significant effect on bleeding time. *J. Lab. Clin. Med.* **124**, 589–599.

McCollum, C. N., Kester, R. C., Rajah, S. M., Learoyd, P., and Pepper, M. (1981). Arterial graft maturation, the duration of thrombotic activity in Dacron aortobifemoral grafts measured by platelet and fibrinogen kinetics. *Br. J. Surg.* **68**, 61–64.

Meinhart, J., Deutsch, M., and Zilla, P. (1997). Eight years of clinical endothelial cell transplantation closing the gap between prosthetic grafts and vein grafts. *ASAIO J.* **43**, M515–M521.

Miyata, T., Conte, M. S., Trudell, L. A., Mason, D., Wittemore, A. D., and Birinyi, L. K. (1991). Delayed exposure to pulsatile shear stress improves retention of human saphenous vein endothelial cells on seeded ePTFE grafts. *J. Surg. Res.* **50**, 485–493.

Mohamed, M. S., Mukherjee, M., and Kakkar, V. V. (1998). Thrombogenicity of heparin and non-heparin bound arterial prostheses; an *in vitro* evaluation. *J. R. Coll. Surg. Edinb.* **43**, 155–157.

Moore, W. S., Cafferata, H. T., Hall, A. D., and Blaisdell, F. W. (1968). In defense of grafts across the inguinal ligaments. An evaluation of early and late results of aorto-femoral bypass grafts. *Ann. Surg.* **168**, 207–214.

Müller-Hülsbeck, S., Link, J., Schwarzenberg, H., Steffens, J. C., Brossmann, J., Hulsbeck, A., and Heller, M. (1997). MR imaging signal intensity abnormality after placement of arterial endoprostheses. *AJR, Am. J. Roentgenol.* **169**, 743–748.

Müller-Hülsbeck, S., Schwarzenberg, H., Wesner, F., Drost, R., Glher, C. C., and Heller, M. (1998). Visualization of flow patterns from stents and stent-grafts in an *in vitro* flow-model. *Invest. Radiol.* **33**, 762–770.

Niklason, L. E., Gao, J., Abbott, W. M., Hirschi, K. K., Houser, S., and Langer, R. (1999). Functional arteries grown *in vitro*. *Science* **284**, 489–493.

Noishiki, Y., Yamane, Y., Okoshi, T., Tomizawa, Y., and Satoh, S. (1998). Choice, isolation, and preparation of cells for bioartificial vascular grafts. *Artif. Organs* **22**, 50–62.

Nojiri, C., Senshu, K., and Okano, T. (1995). Nonthrombogenic polymer vascular prosthesis. *Artif. Organs* **19**, 32–38.

Nojiri, C., Kido, T., Sugiyama, T., Horiuchi, K., Kijima, T., Hagiwara, K., Kuribayashi, E., Nogawa, A., Ogiwara, K., and Akutsu, T. (1996). Can heparin immobilized surfaces maintain nonthrombogenic activity during *in vivo* long-term implantation? *ASAIO J.* **42**, M468–M475.

Ohki, T., Martin, M. L., Veith, F. J., Yuan, J. G., Soundararajan, K., Sanchez, L. A., Parsons, R. E., Lyon, R. T., and Yamazaki, Y. (1997). Anastomotic intimal hyperplasia; a comparison between conventional and endovascular stent graft techniques. *J. Surg. Res.* **69**, 255–267.

Onuki, Y., Kouchi, Y., Yoshida, H., Wu, M. H. D., Shi, O., Wechezak, A. R., Coan, D., and Sauvage, L. R. (1998). Early flow surface endothelialization before microvessel ingrowth in accelerated graft healing, with BrdU identification of cellular proliferation. *Ann. Vasc. Surg.* **12**, 207–215.

Ott, M. L., and Ballermann, B. J. (1995). Shear stress-conditioned, endothelial cell-seeded vascular grafts, improved cell adherence in response to *in vitro* shear stress. *Surgery* **117**, 334–339.

Park, K. D., Okano, T., Nojiri, C., and Kim, S. W. (1988). Heparin immobilization onto segmented polyurethane–urea surfaces—Effect of hydrophilic spacers. *J. Biomed. Mater. Res.* **22**, 977–992.

Park, P. K., Jarrell, B. E., Williams, S. K., Carter, T. L., Rose, D. G., Martinez-Hernandez, A., and Carabasi, R. A., 3rd (1990). Thrombus free, human endothelial surface in midregion of a Dacron vascular graft in the splanchnic venous circuit—Observations after nine months of implantation. *J. Vasc. Surg.* **11**, 468–475.

Pasic, M., Muller-Glauser, W., von Segesser, L. K., Lachat, M., Mihaljevic, T., and Turina, M. (1994). Superior late patency of small-diameter Dacron grafts seeded with omental microvascular cells: An experimental study. *Ann. Thorac. Surg.* **58**, 677–683.

Pepper, M. S. (1997). Manipulation angiogenesis—From basic science to the bedside. *Arterioscler. Thromb. Vasc. Biol.* **17**, 605–619.

Phaneuf, M. D., Dempsey, D. J., Bide, M. J., Szycher, M., Quist, W. C., and LoGerfo, F. W. (1998). Bioengineering of a novel small diameter polyurethane vascular graft with covalently bound recombinant hirudin. *ASAIO J.* **44**, M653–M658.

Pitsch, R. J., Minion, D. J., Goman, M. L., van Aalst, J. A., Fox, P. L., and Graham, L. M. (1997). Platelet-derived growth factor production by cells from Dacron grafts implanted in a canine model. *J. Vasc. Surg.* **26**, 70–78.

Podrazik, R. M., Whitehill, T. A., Komorowski, T. A., Karo, K. H., Messina, L. M., and Stanley, J. C. (1993). *In vivo* fate of lacZ-transduced endothelial cells-seeded ePTFE thoracoabdominal vascular prostheses in dogs. *Surg. Forum.* **44**, 334–337.

Prendiville, E. J., Colemam, J. E., Callow, A. D., Gould, K. E., Laliberte-Verdon, S., Ramberg, K., and Connolly, R. J. (1991). Increased *in vitro* incubation time of endothelial cells on fibronectin-treated ePTFE increases cell retention in blood flow. *Eur. J. Vasc. Surg.* **5**, 311–319.

Quarmby, J. W., Burnand, K. G., Lockhart, S. J. M., Donald, A. E., Sommerville, K. M., Jamieson, C. W., and Browse, N. L. (1998). Prospective randomized trial of woven versus collagen-impregnated knitted prosthetic Dacron grafts in aortoiliac surgery. *Br. J. Surg.* **85**, 775–777.

Rosenman, J. E., Kempczinski, R. F., Berlatzky, Y., Pearce, W. H., Ramalanjaona, G. R., and Bjornson, H. S. (1985). Bacterial adherence to endothelial-seeded polytetrafluoroethylene grafts. *Surgery* **98**, 816–823.

Rosenman, J. E., Kempczinski, R. F., Pearce, W. H., and Silberstein, E. B. (1985). Kinetics of endothelial cell seeding. *J. Vasc. Surg.* **2**, 778–784.

Ruderman, R. J., Hegyeli, A. F., Hattler, B. G., and Leonard, F. (1972). A partially biodegradable vascular prosthesis. *ASAIO Trans.* **18**, 30–37.

Rumisek, J. D., Wade, C. E., Brooks, D. E., Okerberg, C. V., Barry, M. J., and Clarke, J. S. (1986). Heat-denatured albumin-coated Dacron vascular grafts: Physical characteristics and *in vivo* performance. *J. Vasc. Surg.* **4**, 136–143.

Salzmann, D. L., Yee, D. C., Roach, D. J., Berman, S. S., and William, S. K. (1997). Effects of balloon dilatation on ePTFE structural characteristics. *J. Biomed. Mater. Res.* **36**, 498–507.

Salzmann, D. L., Yee, D. C., Roach, D. J., Berman, S. S., and William, S. K. (1998). Healing response associated with balloon-dilated ePTFE. *J. Biomed. Mater. Res.* **41**, 364–370.

Sapienza, P., di Marzo, L., Cucina, A., Corvino, V., Mingoli, A., Giustiniani, O., Ziparo, E., and Cavallaro, A. (1998). Release of PDGF-BB and bFGF by human endothelial cells seeded on expanded polytetrafluoroethylene vascular grafts. *J. Surg. Res.* **75**, 24–29.

Schneider, P. A., Hanson, S. R., Price, T. M., and Harker, L. A. (1988). Preformed confluent endothelial cell monolayers prevent early platelet deposition on vascular prostheses in baboons. *J. Vasc. Surg.* **8**, 229–235.

Schürmann, K., Vorwerk, D., Kulisch, A., Stroehmer-Kulisch, E., Biesterfield, S., Stopinski, T., and Gunther, R. W. (1996). Neointimal hyperplasia in low-profile Nitinol stents, Palmaz stents, and Wallstents; a comparative experimental study. *Cardiovasc. Intervent. Radiol.* **19**, 248–254.

Sentissi, J. M., Ramberg, K., O'Donnell, T. F., Jr., Connolly, R. J., and Callow, A. D. (1986). The effect of flow on vascular endothelial cells grown in tissue culture on polytetrafluoroethylene grafts. *Surgery* **99**, 337–343.

Shayani, V., Newman, K. D., and Dichek, D. A. (1994). Optimization of recombinant t-PA secretion from seeded vascular grafts. *J. Surg. Res.* **57**, 495–504.

Shepard, A. D., Gelfand, J. A., Callow, A. D., and O'Donnell, T. F., Jr. (1984). Complement activation by synthetic vascular prostheses. *J. Vasc. Surg.* **1**, 829–838.

Shi, O., Wu, M. H., Hayashida, N., Wechezak, A. R., Clowes, A. W., and Sauvage, L. R. (1994). Proof of fallout endothelialization of impervious Dacron grafts in the aorta and inferior vena cava of the dog. *J. Vasc. Surg.* **20**, 546–556.

Shindo, S., Takagi, A., and Whittemore, A. D. (1987). Improved patency of collagen-impregnated grafts after *in vitro* autogenous endothelial cell seeding. *J. Vasc. Surg.* **6**, 325–332.

Shinoka, T., Shum-Tim, D., Ma, P. X., Tanel, R. E., Isogai, N., Langer, R., Vacanti, J. P., and Mayer, J. E. (1998). Creation of viable pulmonary artery autografts through tissue engineering. *J. Thorac. Cardiovasc. Surg.* **115**, 536–546.

Sipehia, R., Martucci, G., and Lipscombe, J. (1996). Transplantation of human endothelial cell monolayer on artificial vascular prosthesis: The effect of growth-support surface chemistry, cell seeding density, ECM protein coating, and growth factors. *Artif. Cells, Blood Substitutes* and *Immob Biotechnol.* **24**, 51–63.

Stanley, J. C., Burkel, W. E., Ford, J. W., Vinter, D. W., Kahn, R. H., Whitehouse, W. M., Jr., and Graham, L. M. (1982). Enhanced patency of small-diameter, externally supported Dacron iliofemoral grafts seeded with endothelial cells. *Surgery* **92**, 994–1005.

Stansby, G., Berwanger, C., Shukla, N., Schmitz-Rixen, T., and Hamilton, G. (1994). Endothelial seeding of compliant polyurethane vascular graft material. *Br. J. Surg.* **81**, 1286–1289.

Stratton, J. R., Thiele, B. L., and Ritchie, J. L. (1982). Platelet deposition on Dacron aortic bifurcation grafts in man; quantitation with indium-111 platelet imaging. *Circulation* **66**, 1287–1293.

Stratton, J. R., Thiele, B. L., and Ritchie, J. L. (1983). Natural history of platelet deposition on Dacron aortic bifurcation grafts in the first year after implantation. *Am. J. Cardiol.* **52**, 371–374.

Sugawara, Y., Miyata, T., Sato, O., Kimura, H., Namba, T., and Makuuchi, M. (1997). Rapid postincubation endothelial retention by Dacron grafts. *J. Surg. Res.* **67**, 132–136.

Swartbol, P., Truedsson, L., Parsson, H., and Norgren, L. (1997). Tumor necrosis factor-a and interleukin-6 release from white blood cells induced by different graft materials *in vitro* are affected by pentoxifylline and iloprost. *J. Biomed. Mater. Res.* **36**, 400–406.

Szilagyi, D. E., Smith, R. F., Elliot, J. P., and Vrandecic, M. P. (1972). Infection in arterial reconstruction with synthetic vascular graft. *Ann. Surg.* **176**, 321–333.

Totani, L., Cumashi, A., Piccoli, A., and Lorenzet, R. (1998). Polymorphonuclear leukocytes induce PDGF release from IL-1 beta-treated endothelial cells, role of adhesion molecules and serine proteases. *Arterioscler. Thromb. Vasc. Biol.* **18**, 1534–1540.

Toursarkissian, B., Eisenberg, P. R., Abendschein, D. R., and Rubin, B. G. (1997). Thrombogenicity of small-diameter prosthetic grafts: Relative contributions of graft-associated thrombin and factor Xa. *J. Vasc. Surg.* **25**, 730–735.

Tsai, C. C., Huo, H. H., Kulkarni, P. V., and Eberhart, R. C. (1990). Biocompatible coatings with high albumin affinity. *Trans. Am. Soc. Artif. Intern. Organs* **36**, M307–M310.

Tseng, D. Y., and Edelman, E. R. (1998). Effects of amide and amine plasma-treated ePTFE vascular grafts on endothelial cell lining in an artificial circulatory system. *J. Biomed. Mater. Res.* **42,** 188–198.

Tsuchida, H., Cameron, B. L., Marcus, C. S., and Wilson, S. E. (1992). Modified polytetrafluoroethylene: Indium 111-labeled platelet deposition on carbon-lined and high porosity polytetrafluoroethylene grafts. *J. Vasc. Surg.* **16,** 643–649.

van der Lei, B., Nieuwenhuis, P., Molenaar, I., and Wildevuur, C. R. (1987). Long-term biologic fate of neoarteries regenerated in microporous, compliant, biodegradable, small-caliber vascular grafts in rats. *Surgery* **101,** 459–467.

Veith, F. J., Gupta, S. K., Ascer, E., White-Flores, S., Samson, R. H., Scher, L. A., Towne, J. B., Bernhard, V. M., Bonier, P., and Flinn, W. R. (1986). Six-year prospective multicenter randomized comparison of autologous saphenous vein and expanded polytetrafluoroethylene grafts in infrainguinal arterial reconstructions. *J. Vasc. Surg.* **3,** 104–114.

Vicaretti, M., Hawthorne, W. J., Ao, P. Y., and Fletcher, J. P. (1998). An increased concentration of rifampicin bonded to gelatin-sealed Dacron reduces the incidence of subsequent graft infections following a staphylococcal challenge. *Cardiovasc. Surg.* **6,** 268–273.

Vohra, R. K., Thomson, G. J., Sharma, H., Carr, H. M., and Walker, M. G. (1990). Effects of shear stress on endothelial cell monolayer on expanded polytetrafluoroethylene grafts using preclot and fibronectin matrices. *Eur. J. Vasc. Surg.* **4,** 33–41.

Vohra, R. K., Thomson, G. J., Carr, H. M., Sharma, H., Welch, M., and Walker, M. G. (1991). *In vitro* adherence and kinetics studies of adult human endothelial cell-seeded polytetrafluoroethylene and gelatin-impregnated Dacron grafts. *Eur. J. Vasc. Surg.* **5,** 93–103.

Voorhees, A. B., Jr., Jaretzki, A., and Blakemore, A. H. (1952). The use of tubes constructed of Vinyon N cloth in bridging arterial defects. *Ann. Surg.* **135,** 332–336.

Vroman, L. (1987). Methods of investigating protein interaction on artificial and natural surfaces. *Ann. N. Y. Acad. Sci.* **516,** 300–305.

Wakefield, T. W., Lindblad, B., Graham, L. M., Whitehouse, W. M., Jr., Ripley, S. D., Petry, N. A., Spaulding, S. A., Burkel, W. E., and Stanley, J. C. (1986). Nuclide imaging of vascular graft–platelet interactions; comparison of indium excess and technetium subtraction techniques. *J. Surg. Res.* **40,** 388–394.

Weinberg, C. B., and Bell, E. (1986). A blood vessel model constructed from collagen and cultured vascular cells. *Science* **231,** 397–400.

Wesolowski, S. A., Fries, C. C., Domingo, R. T., Liebig, W. J., and Sawyer, P. N. (1963). The compound prosthetic vascular graft. A pathologic survey. *Surgery* **53,** 19–44.

Whitehouse, W. M., Jr., Wakefield, T. W., Vinter, D. W., Ford, J. W., Swanson, D. P., Thrall, J. H., Froelich, J. W., Brown, L. E., Burkel, W. E., and Graham, L. M. (1983). Indium-111 oxide labeled platelet imaging of endothelial seeded Dacron thoracoabdominal vascular prostheses in a canine experimental model. *ASAIO Trans.* **29,** 183–187.

Wilson, J. M., Birinyi, L. K., Salomon, R. N., Libby, P., Callow, A. D., and Mulligan, R. C. (1989). Implantation of vascular grafts lined with genetically modified endothelial cells. *Science* **244,** 1344–1346.

Wu, M. H., Shi, O., Wechezak, A. R., Clowes, A. W., Gordon, I. L., and Sauvage, L. R. (1995). Definitive proof of endothelialization of a Dacron arterial prosthesis in a human being. *J. Vasc. Surg.* **21,** 862–867.

Yee, D. C., Williams, S. K., Salzmann, D. L., Pond, G. D., Patula, V., Berman, S. S., and Roach, D. J. (1998). Stent versus endovascular graft healing characteristics in the porcine iliac artery. *J. Vasc. Intervent. Radiol.* **9,** 609–617.

Yue, X., van der Lei, B., Schakenraad, J. M., van Oene, G. H., Kuit, J. H., Feijen, J., and Wildevuur, C. R. (1988). Smooth muscle cell seeding in biodegradable grafts in rats: A new method to enhance the process of arterial wall regeneration. *Surgery* **103,** 206–212.

Zarge, J. I., Gosselin, C., Huang, P., and Greisler, H. P. (1997). Platelet deposition on ePTFE grafts coated with fibrin glue with or without FGF-1 and heparin. *J. Surg. Res.* **67,** 4–8.

Ziegler, T., and Nerem, R. M. (1994). Tissue engineering a blood vessel, regulation of vascular biology by mechanical stresses. *J. Cell. Biochem.* **56,** 204–209.

Zilla, P., Deutsch, M., Meinhart, J., Puschmann, R., Eberl, T., Minar, E., Dudczak, R., Lugmaier, H., Schmidt, P., and Noszian, I. (1994a). Clinical *in vitro* endothelialization of femoropopliteal bypass grafts; an actuarial follow-up over three years. *J. Vasc. Surg.* **19,** 540–548.

Zilla, P., Preiss, P., Groscurth, P., Rosemeier, F., Deutsch, M., Odell, J., Heidinger, C., Fasol, R., and von Oppell, U. (1994b). *In vitro*-lined endothelium, initial integrity and ultrastructural events. *Surgery* **116,** 524–534.

CHAPTER 33

SMALL-DIAMETER VASCULAR GRAFTS

Susan J. Sullivan and Kelvin G. M. Brockbank

INTRODUCTION

Surgical treatment of vascular disease is now a common medical procedure. However, to date the use of synthetic materials is limited to grafts larger than 5 mm due to the frequency of occlusion observed with synthetic vessels of smaller diameter. Consequently, significant efforts in the past 15 years have focused on the development of a small-diameter blood vessel equivalent using tissue engineering. The seeding of synthetic grafts with endothelial cells has been investigated as a means to increase patency, but has been limited by the challenges associated with maintaining effective surface coverage. As an alternative to the use of synthetic materials, two approaches have been taken to create a blood vessel using cell and matrix components. Several groups have established techniques for creating a cellular vessel through coculture of smooth muscle cells with collagen or other matrix proteins and lining the lumen with endothelial cells. The second approach is designed to provide a graft constructed of a material, such as collagen, that would provide the required mechanical properties on implant but would also facilitate remodeling and infiltration of host cells into a cellular vessel. Our group has evaluated an acellular construct composed of porcine intestinal collagen matrix that provides the requisite mechanical properties, can be remodeled *in vivo,* and has shown promising patency in the rabbit carotid model.

In the early 1900s, clinicians began investigating the use of vein segments as arterial bypass grafts, but these procedures did not gain popular acceptance until the development of contrast angiography and the introduction of heparin as an anticoagulant (Carrell, 1910; Schloss and Shumacker, 1949; Charles and Scott, 1933; Murray and Jones, 1940). Arterial replacement is now a common treatment for vascular disease, with over 400,000 autologous coronary grafts (either arteries or veins) and over 200,000 vein grafts into peripheral arteries performed each year. A marketing analysis has shown that at least 300,000 coronary artery bypass procedures are performed annually in the United States involving in excess of 1 million vascular grafts (Frost and Sullivan, 1997). Autogenous vessels, particularly internal mammary arteries and saphenous veins, remain the "gold standard" for coronary grafting. However, many patients do not have veins suitable for grafting due to preexisting vascular disease, vein stripping, or vein harvesting in prior vascular procedures. Studies estimates that as many as 30% of the patients who require arterial bypass procedures will have saphenous veins unsuitable for use in vascular reconstruction (Edwards *et al.,* 1996). In addition, it has been shown that 2–5% of saphenous veins considered for bypass procedures were unusable on the basis of gross pathology and that up to 12% were subsequently classified as diseased, a classification that was associated with patency rates less than half that of nondiseased veins (Panetta *et al.,* 1992). Finally, significant morbidity and surgical costs are associated with the harvest of autologous vessels. All of these factors contribute to a widely recognized clinical need for a readily available, small-diameter vascular graft.

Chemically cross-linked xenogeneic and allogeneic blood vessels have been assessed as vascular grafts. Early attempts to use bovine arterial grafts failed due to aneurysmal degeneration and infection (Rosenberg *et al.,* 1970). Cross-linked human umbilical vein grafts are still in use clinically despite the need for routine monitoring for early detection of aneurysms and poor patency rates in more challenging graft positions, such as across and below the knee (Brewster *et al.,* 1983; Edwards and Mulherin, 1980).

SYNTHETIC GRAFTS

The use of synthetic materials to fabricate a blood vessel substitute began in the 1950s and has led to vascular prostheses made from a wide variety of materials, including nylon, Teflon, Orlon, Dacron, polyethylene, and polyurethane (Greisler, 1991). Expanded poly(tetrafluoroethylene) (e-PTFE) was introduced about 20 years ago, and Dacron and PTFE remain the most widely used synthetic materials. Both of these materials have been very successful in applications requiring large-diameter (greater than 5–6 mm) vascular substitutes in areas of high blood flow (Pratt, 1984). However, in low flow or smaller diameter applications such as peripheral vascular repair below the knee or coronary bypass, these grafts are not effective and there are no Food and Drug Administration (FDA)-approved products for such applications. Possis Medical, Inc., has received a humanitarian-use investigational device exemption (IDE) for an e-PTFE graft that acts as an arteriovenous shunt running from the aorta to the vena cava. When this graft is employed, coronary anastomoses are made in a side-to-side manner and pressure is maintained by placement of a Venturi flow restrictor near the vena cava. The Venturi flow restriction maintains blood flow through the graft at 500–600 ml/min and back pressure is maintained between the Venturi restriction and the aorta. Previous e-PTFE grafts often failed as a result of anastomotic hyperplasia, which leads to stenosis, and Dacron grafts were susceptible to thrombosis (Callow, 1983; O'-Donnell *et al.,* 1984; Stephen *et al.,* 1977). Composite devices that incorporate biological substances into the synthetic graft in order to modulate the host response and help prevent graft failure have also been developed. Synthetic grafts have been impregnated with biodegradable components such as albumin, collagen, gelatin, elastin, and fibrin (Gundry and Behrendt, 1987; Kottke-Marchant *et al.,* 1989) designed to decrease the surface reactivity to platelets; however, long-term efficacy has not been achieved using this approach. Agents that are known to inhibit thrombogenesis or promote anticoagulation, such as heparin, prostaglandin E_1 (PGE_1), hirudin, and aspirin, have also been bound to the lumen of the synthetic vessels but have met with limited success (Park *et al.,* 1988; Phaneuf *et al.,* 1998).

IN VITRO BLOOD VESSEL TISSUE ENGINEERING

Initial attempts at tissue engineering a blood vessel substitute involved seeding the lumen of a synthetic graft with endothelial cells. Modifying the prosthetic graft surface with a monolayer of autologous endothelial cells was initially suggested by Herring *et al.* (1987) as a means to provide a more biocompatible surface with the potential to decrease thrombosis and intimal hyperplasia. Some advances have been made as a result of extensive experimentation in endothelialization of grafts (Williams *et al.,* 1994; Pasic *et al.,* 1994; Meinhart *et al.,* 1997). Progress was initially limited by the challenges associated with the harvest and culture of endothelial cells. However, human umbilical vein endothelial cells are now commonly used as a model for vascular endothelial cells in studies evaluating methods for graft seeding. In addition to the issues of cell sourcing, defining techniques for establishing a confluent layer of cells on the luminal surface has also been challenging. Unfortunately, methods that promote endothelial cell adhesion can, in turn, lead to a more thrombogenic surface. If the surface is not completely covered or some cells are lost on exposure to flow, the patency rates would be reduced rather than improved. Results from Bowlin and Rittgers (1997) suggest that electrostatic methods used to endothelialize e-PTFE grafts may be a promising approach. In this process, the surface is induced to become positively charged, which promotes endothelial cell adhesion. Once the seeding process is complete, however, the graft reverts to its original, highly negatively charged state so that any exposed areas will no longer promote cell adhesion. Endothelialization of synthetic grafts has shown some feasibility in animal models, but clinical trials have not shown significant differences in patency as a result of endothelialization, and the source of cells for widespread clinical application remains problematic (Meinhart *et al.,* 1977; Welch *et al.,* 1992).

Another approach to a tissue-engineered blood vessel is based on the coculture of cells with natural biologic and/or synthetic materials in order to "grow" a cellular vessel. Weinberg and Bell (1986) demonstrated *in vitro* development of a model blood vessel with three layers, corresponding to an intima, media, and adventitia. A confluent layer of endothelial cells was grown in culture onto the lumen of a tubular collagen construct consisting of an outer layer of fibroblasts and a middle layer of smooth muscle cells. In this study, the cells and matrix were cast in an annular mold and an external Dacron mesh was used to provide additional mechanical support. The model grossly resembled a muscular artery, except for the Dacron mesh; electron microscopy demonstrated well-differentiated bipolar smooth muscle cells. Endothelial cells formed a monolayer of flattened cells interconnected by cellular junctions with numerous cytoplasmic vesicles, Weibel–Palade bodies, and patches of basement membrane. In addition to mechanical properties that necessitated the use of a support mesh, there were a number of differences between the model and normal arteries. A major difference was that elastin, the principal arterial tissue matrix protein besides collagen, was not present in the model. Matsuda and Miwa (1995) also created a hybrid construct using an artificial scaffold of polyurethane seeded with smooth muscle and endothelial cells. This construct was shown to remodel *in vivo* such that the endothelial cells on the lumen became oriented in parallel to the direction of blood flow, and the smooth muscle cells in the medial layer redistributed and became circumferentially oriented. Implants were successful in the canine model for up to 1 year.

The constructs of Weinberg and Matsuda used a composite approach to reinforce the strength of the cellular layers. A key challenge in the development of a cellular model blood vessel was to create a construct with the required mechanical properties. Auger and colleagues (L'Heureux *et al.,* 1993) expanded on this coculture approach and were able to produce a vessel equivalent using human cells and collagen without using any synthetic supports. L'Heureux *et al.* (1998) have now demonstrated that their construct could mature in culture into a vessel with excellent mechanical properties. This construct had a burst strength exceeding 2000 mmHg. Umbilical cord-derived smooth muscle and skin-derived fibroblast cells were cultured separately in the presence of ascorbic acid and were formed into a vessel by wrapping successive layers around a mandrel. Constructs that were lined with umbilical cord-derived endothelial cells demonstrated feasibility in short-term studies in a xenogeneic model. Others, particularly the laboratory groups of Nerem at Georgia Tech and Tranquillo at the University of Minnesota (Ziegler *et al.,* 1995; Tranquillo *et al.,* 1996), have investigated the effect of culture conditions and *in vitro* mechanical forces on the development of cellularized constructs. Girton *et al.* (1999) have generated blood vessel medial layer equivalents with smooth muscle cells and collagen and have shown that the mechanical strength can be enhanced by nonenzymatic cross-linking of proteins with reducing sugars such as glucose and ribose. This glycation of the media equivalents significantly increased both stiffness and tensile strength of the constructs. The work at Georgia Tech has focused on the development of a cellular construct with a confluent endothelium and has evaluated the physiological responses of the smooth muscle and endothelial cells to shear and mechanical stresses associated with flow conditions.

Bovine vessels containing vascular biopsy-derived smooth muscle and endothelial cells have now been cultured on tubes of partially hydrolyzed poly(glycolic acid) under pulsatile conditions (Niklason *et al.,* 1999). These vessels also had burst strengths greater than 2000 mmHg and exhibited the beginnings of vascular contractile responses. Four engineered arteries were implanted in the saphenous arteries of miniature swine; both of the vessels exposed to pulsatile conditions *in vitro* prior to implantation were patent for up to 24 days postimplantation, whereas both control vessels that were not exposed to pulsatile conditions were occluded within 3 weeks. It is interesting to note that, just as in the studies of Weinberg and Bell more than a decade ago, these constructs were notably lacking in elastin content.

These advances in the development of cellular blood vessel equivalents through *in vitro* tissue engineering are promising. However, generating a blood vessel equivalent composed of autologous cells requires harvesting the patient's cells months in advance of the required graft procedure and extensive culture time, and would be a difficult technology to commercialize.

IN VIVO TISSUE ENGINEERING OF BLOOD VESSELS

An alternative to generating a cellularized vascular construct *in vitro* is to construct a prosthesis that is remodeled *in vivo.* Zarge *et al.* (1997) reviewed progress in the development of bioresorbable synthetic grafts that induce tissue ingrowth and remodeling after implantation to form a "neoartery." The clinical efficacy of such grafts depend on a balance between rapid graft resorption during tissue ingrowth and slower degradation rates for maintenance of structural stability.

Structural stability may also be maintained by incorporation of a nonresorbable material that acts as a mechanical strut or combinations of materials with different resorption rates. Bioresorbable graft performance may also be modified by incorporation of growth factors, chemoattractants, or cells within or on the graft.

There have also been efforts to develop resorbable grafts composed of naturally occurring materials. Collagen and collagen-based materials have been increasingly used in a variety of medical products since the mid-1960s. However, in general, the collagen constructs were designed to be similar to the synthetic polymer prostheses in terms of their ability to persist. Consequently, purification and cross-linking methods were developed to optimize mechanical strength and enhance the stability of the collagen *in vivo* (Sabelman, 1985; Khor, 1997). In these cases, the material is seen as a foreign body and is essentially encapsulated by inflammatory cells and organized scar tissue rather than being integrated into the surrounding tissue.

Wilson *et al.* (1990) investigated the use of a collagen-based biomaterial as a vascular prosthesis. In these studies, canine iliac and carotid arteries were processed using a multistep detergent and enzymatic extraction method. This method generated an acellular matrix of collagen, elastin, and some glycosaminoglycans. These acellular matrix allografts were also used in coronary artery bypass and compared with autogenous saphenous veins (Wilson *et al.,* 1995). The patency rates were 44% (4 of 9) for the matrix allografts and 57% (4 of 7) for the saphenous vein grafts. Implantation of these acellular prostheses into the carotid and femoral arteries in dogs demonstrated over 90% patency without administration of postoperative anticoagulants. The implants were followed for up to 4 years and showed no evidence of aneurysm formation or dystropic calcification. Even after 4 years, however, the collagen remained intact and there was no significant cellular infiltration. Inhibition of cellular infiltration may have been attributable to the presence of residual sodium dodecylsulfate remaining in the graft after processing.

In contrast, investigators at Organogenesis have developed methods for using the intestinal collagen layer derived from the submucosa of the small intestine as a biomaterial with the potential for true integration by remodeling. Over 30 years ago, the submucosal layer of the small intestine was proposed as a vascular substitute (Lawler *et al.,* 1971; Egusa, 1968; Matsumoto *et al.,* 1966). This collagenous material was initially of interest because of its inherent mechanical strength and natural intubation. The studies employed inverted segments of autologous small intestine submucosa (SIS) to replace portions of the inferior vena cava or aorta and demonstrated over 80% patency in a small series of dogs. Badylak and colleagues (1989) extended these studies using autogenous SIS as a large-diameter (10-mm) vascular graft in the canine infrarenal aorta. Grafts were prepared by inverting the intestine, removing the serosal and muscularis layers by manual abrasion, and then returning the tube to its original orientation. The luminal diameter was adjusted to match that of the recipient blood vessel by inserting a glass mandrel of the desired diameter and suturing to gather any excess SIS. Of the 12 grafts, 11 remained patent until explant, at times ranging from 1 week to 1 year postimplant. Using similar methods for constructing the grafts, Badylak's group also investigated an autogenous, small-diameter SIS graft in the canine carotid and femoral positions and reported patency of 75%. In these studies, however, all the dogs received both aspirin and warfarin for the first 8 weeks postsurgery (Lantz *et al.,* 1990). Histological evaluation of the explanted grafts suggested that the SIS was remodeled and Badylak's research group has also pursued SIS applications in soft tissue repair and orthopedics (Badylak, 1993).

In our early studies exploring the use of this material in soft tissue repair and vascular prosthetics, the submucosal layer was obtained after mechanical cleaning using a customized, commercial gut-cleaning machine. Our results with 4-mm vascular grafts indicated that without aggressive anticoagulant therapy, the grafts were thrombogenic when evaluated in a canine *ex vivo* shunt or implanted as an aortic graft in rabbits (Termin *et al.,* 1993). In subsequent grafts, the luminal surface was modified by treatment with a benzalkonium chloride/heparin complex to reduce thrombogenicity, and these grafts showed promise when implanted in the infrarenal rabbit aorta (Termin *et al.,* 1994). In addition, we demonstrated that this material could be minimally cross-linked with 1-ethyl-3-(3-dimethylaminopropyl)carbodiimide (EDC), a water-soluble carbodiimide, and sterilized with peracetic acid. Unlike the more traditional methods that used fixatives such as glutaraldehyde, this process reduced antigenicity without preventing cell ingrowth *in vivo* (Hardin-Young *et al.,* 1996). In addition, using a water-soluble carbodiimide also eliminates the problems associated with the residual toxicity seen with aldehyde fixation of biomaterials.

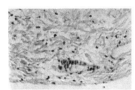

Fig. 33.1. Effect of chemically cleaning the ICL.

Although these studies were promising, it also became clear that the SIS material obtained after mechanical cleaning was quite heterogeneous (Fig. 33.1). The amount of residual cellular material in the submucosal layer varied between preparations from different donor pigs as well as within different sections of the small intestine. This variability complicated graft fabrication and *in vitro* characterization as well as analysis of *in vivo* results. Consequently, extensive efforts at Organogenesis in the past 2 years have focused on material processing and characterization. Mechanical and chemical cleaning of the proximal porcine jejunum using a proprietary process that removes cells and cellular debris while maintaining the native collagen structure generates the intestinal collagen layer (ICL), a sheet of predominantly type I collagen.

ICL is prepared from the intestines of swine (about 450 lb) obtained from a closed herd (Parsons Farm, Hadley, MA). The mesenteric layer is manually removed from the small intestine prior to mechanical stripping of the mucosal and membranous layers. It is mechanically processed through three series of rollers on the Ernest Bitterling Model M34 (Nottingham, England) under a constant flow of water. The remaining submucosal collagen layer is longitudinally slit between the lymphatic tags and cut into 6-inch lengths, which are then processed to remove any residual cells. The chemical cleaning of the intestinal collagen layer is a proprietary process involving several washes in chelators and salts at specific pH ranges. Once the process is complete, the ICL is stored at $-80°C$ until use. All lots of material are tested for sterility and residual endotoxin levels prior to release. In related work, an ICL laminate has been shown to be effective in hernia repair in animal models. This construct, GraftPatch, received premarket approval [known as 510(k) clearance] from the FDA in 1998 for use in soft tissue repair (see the FDA web site, http://www.fda.gov/cdrh/manual/idemanul.html). Unlike the synthetic mesh products traditionally used for hernia repair, GraftPatch became well-integrated into the surrounding abdominal wall tissue and developed a vascularized, mesothelial-like layer on the luminal surface that presumably contributes to the resistance of the patch to adhesion formation (Abraham *et al.*, 1999).

Having developed methods for producing a material that can be made reproducibly and could be used in a commercial product, we are currently evaluating the *in vivo* efficacy of 4-mm vascular grafts constructed from the ICL. To manufacture a vascular graft, the ICL is aseptically entubated using a mandrel of the desired diameter. In contrast to the suturing method used in the Lantz studies (1990), the collagen tube is created by dehydration and cross-linking of the ICL wrapped around the mandrel. A thin layer of bovine collagen is deposited on the graft lumen followed by cross-linking in a 1 mM 1-ethyl-3-(3-dimethylaminopropyl)carbodiimide hydrochloride. After rinsing to remove all residual cross-linker, the grafts are sterilized overnight with 0.1% peracetic acid. The grafts are rinsed again with sterile water and subsequently coated with heparin (benzalkonium chloride complex) in isopropanol.

In using the ICL as a vascular graft, the goal is to process the tissue so that native collagen structure is retained while ensuring that the necessary mechanical properties are provided. For a vascular conduit, it is critical that the material can be sutured and has sufficient burst strength to withstand physiologic blood pressure. In addition, the porosity of the material should allow cellular infiltration from the adventitial side while maintaining hemostasis. The protocol developed for graft fabrication was designed to identify the lowest amount of chemical cross-linking that provided the requisite mechanical characteristics. Thus, in contrast to synthetic polymers or highly cross-linked biomaterials, the ICL graft fulfills the hemodynamic and hemostasis requirements while still having the potential to be remodeled by the host tissue. The graft characteristics are summarized in Table 33.1 ($n = 25$).

Having determined that the grafts had the requisite mechanical properties for a vascular graft, the *in vivo* efficacy was evaluated using a small animal model in collaboration with Dr. Otto Hagen at Duke University Medical Center (Davies and Hagen, 1995). Through a right cervical

Table 33.1. Mechanical analysis of ICL grafts

Mechanical test	Result
Suture retention test	3.7 ± 0.5 Newtons
Burst test	12.4 ± 5.5 psi
Porosity	4.26×10^{-3} ml/cm^2/min
Compliance (change between 80 and 120 mmHg $\times 10^{-2}$)	5.0%

oblique incision, the jugular vein of New Zealand White rabbits (2.0 to 2.5 kg) was mobilized, ligated, and divided to expose the carotid artery. Following complete mobilization of the right carotid artery and prior to arterial clamping, heparin (200 IU/kg) was given intravenously. Atraumatic arterial clamps were placed proximal and distal to the arteriotomy sites to occlude flow prior to the proximal arteriotomy. An end-to-side proximal anastomosis of artery to graft was done with 10-0 Ethilon sutures and the distal anastomosis was done in a similar manner. The segment of carotid artery between anastomoses was excised, leaving an interposition graft 3–4 cm long. No postoperative anticoagulants were used and the animals were monitored for 1 to 3 months prior to explant. All animal care and handling was in compliance with the "Principles of Laboratory Animal Care" as formulated by the National Society for Medical Research (1996) and the "Guide for the Care and Use of Laboratory Animals" published by the National Research Council (National Academy Press, 1996).

The ICL vascular grafts showed excellent patency in this pilot study, with all 20 grafts patent on explant after 14 ($n = 2$), 28 ($n = 9$), 53 ($n = 4$), or 90 ($n = 5$) days. Preliminary histological evaluation has shown that the graft does undergo remodeling by host cells and studies are underway to identify the cells infiltrating the graft as well as the changes in those populations as a function of time of implant. These studies will also investigate how the mechanical properties of the graft are affected by *in vivo* remodeling. Although short-term patency is highly dependent on the luminal surface properties, long-term patency is equally dependent on mechanical properties of the graft. It is increasingly evident that remodeling of blood vessels is strongly affected by the loading and deformation to which they are subjected. For example, the remodeling of veins used in arterial bypass is modulated by distension and shear stress, both of which are a function of the graft mechanical properties (Dobrin, 1995).

Ideally, the ICL grafts will provide a collagen scaffold that is remodeled *in vivo* into a functional neovessel, as illustrated in Fig. 33.2. The use of an acellular vascular prosthesis would provide an "off the shelf" option to address the clinical need for a small-caliber graft. These data are

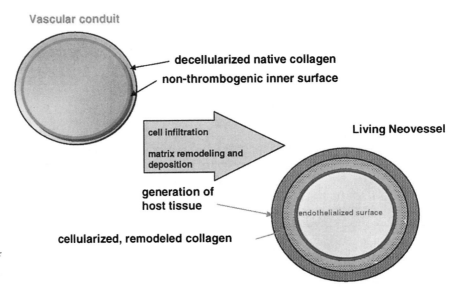

Fig. 33.2. Schematic diagram of in vivo remodeling of an ICL graft.

very encouraging, but much more work will be needed to determine the long-term efficacy of the ICL grafts in large animal models.

SUMMARY

Surgical treatment of vascular disease is now a common medical procedure. However, to date, the use of synthetic materials is limited to grafts larger than 5–6 mm due to the frequency of occlusion observed with synthetic vessels of smaller diameter. Significant progress has been made during the past 10 years in the development of blood vessel equivalents using tissue engineering. The first approach is designed to produce constructs that are cellularized *in vitro*. Cocultures of smooth muscle cells with matrices such as collagen under a variety of conditions have been used to optimize the cellular orientation and mechanical properties of these cellular constructs. An alternative method that would have the advantage of being a more readily available commercial product is to produce a vascular graft that is cellularized *in vivo*. In our studies, the intestinal collagen layer has shown promise as an acellular graft that will remodel over time. ICL was used to form 4-mm vascular grafts with appropriate mechanical properties for blood vessels *in vivo* and initial studies employing heparinized ICL tubes as carotid interposition bypass grafts in rabbits have demonstrated 100% patency after periods ranging from 1 to 3 months.

Although several approaches to the tissue engineering of a small-diameter vascular graft have begun to show feasibility, demonstrating potential clinical efficacy will require much more data from long-term implant studies. Clearly, the grafts will be influenced by the interaction with the host tissue and the remodeling will be dependent on a number of factors, including the hemodynamic environment, the degree of cellular infiltration and graft endothelialization, and the potential for development of cellular hyperplasia. The requirement for a construct with sufficient mechanical strength, a nonthrombogenic surface, long-term resistance to hyperplasia and aneurysm formation, as well as compatibility with the host tissue, remains a major challenge for tissue engineering.

References

Abraham, G. A., Murray, J., Wong, K. H., Connolly, R. J., and Sullivan, S. J. (1999). Evaluation of GraftPatch for soft tissue reinforcement. *Trans. Soc. Biomater* **XXII**, 589.

Badylak, S. F. (1993). Small intestinal submucosa: A biomaterial conducive to smart tissue remodeling. *In* "Tissue Engineering: Current Perspectives," pp. 179–189. Birkhaeuser, Cambridge, MA.

Badylak, S. F., Lantz, G. C., Coffey, A., and Geddes, L. A. (1989). Small intestinal submucosa as a large diameter vascular graft in the dog. *J. Surg. Res.* **47**, 74–80.

Bowlin, G. L., and Rittgers, S. E. (1997). Electrostatic endothelial cell seeding technique for small diameter vascular prostheses: Feasibility testing. *Cell Transplant.* **6**, 623–629.

Brewster, D. C., LaSalle, A. J., Robinson, J. G., and Strayhorn, E. C. (1983). Factors affecting patency of femoropopliteal bypass grafts. *Surg. Gynecol. Obstet.* **157**, 437–442.

Callow, A. D. (1983). Historical overview of experimental and clinical development of vascular grafts. *In* "Biologic and Synthetic Vascular Prosthesis" (J. Stanley, ed.), pp. 11–26. Grune & Stratton, New York.

Carrell, A. (1910). Graft of the vena cava on the abdominal aorta. *Ann. Surg.* **52**, 462–470.

Charles, A. F., and Scott, D. A. (1933). Studies on heparin. *J. Biol. Chem.* **102**, 425–438.

Davies, M. G., and Hagen, P.-O. (1995). Pathophysiology of vein graft failure: A review. *Eur. J. Vasc. Surg.* **9**, 7–18.

Dobrin, P. B. (1995). Mechanical factors associated with the development of intimal and medial thickening in vein grafts subjected to arterial pressure. A model of arteries exposed to hypertension. *Hypertension* **26**, 38–43.

Edwards, W. H., and Mulherin, J. L., Jr. (1980). The role of graft material in femoropopliteal bypass grafts. *Ann. Surg.* **191**, 721–726.

Edwards, W. S., Holdefer, W. F., and Motashemi, M. (1996). The importance of proper caliber of lumen in femoral popliteal artery reconstruction. *Surg. Gynecol. Obstet.* **122**, 37–42.

Egusa, S. (1968). Replacement of inferior vena cava and abdominal aorta with the autogenous segment of small intestine. *Acta Med.* **22**, 153–165.

Frost and Sullivan (1997). "World Cell Therapy Markets," 5413–43 Revision No. 1.

Girton, T. S., Oegema, T. R., and Tranquillo, R. T. (1999). Exploiting glycation to stiffen and strengthen tissue equivalents for tissue engineering. *J. Biomed. Mater. Res.* **46**, 87–92.

Greisler, H. P. (1991). New biologic and synthetic biological prostheses. *In* "Medical Intelligence Unit," Vol. 1, pp. 1–91. R. G. Landes, Austin, TX.

Gundry, S. R., and Behrendt, D. M. (1987). A comparison of fibrin glue, albumin and blood as agents to pretreat porous grafts. *J. Surg. Res.* **43**, 75–77.

Hardin-Young, J., Carr, R. M., Downing, G. J., Condon, K. D., and Termin, P. L. (1996). Modification of native collagen reduces antigenicity but preserves cell compatibility. *Biotechnol. Bioeng.* **49**, 675–682.

Herring, M. B., Gardner, A. L., and Glover, J. (1987). A single-staged technique for seeding vascular grafts with autogenous endothelium. *Surgery* **84**, 498–502.

Khor, E. (1997). Methods for the treatment of collagenous tissues for bioprostheses. *Biomaterials* **18**, 95–105.

Kottke-Marchant, K., Anderson, J. M., Umemura, Y., and Marchant, R. E. (1989). Effect of coating on the *in vitro* blood compatibility of Dacron arterial prostheses. *Biomaterials* **10**, 14–17.

Lantz, G. C., Badylak, S. F., Coffey, A. C., Geddes, L. A., and Blevins, W. E. (1990). Small intestinal submucosa as a small-diameter arterial graft in the dog. *J. Invest. Surg.* **3**, 217–227.

Lawler, M. R., Foster, J. H., and Scott, H. W. (1971). Evaluation of canine intestinal submucosa as a vascular substitute. *Am. J. Surg.* **122**, 517–519.

L'Heureux, N., Germain, L., Labbe, R., and Auger, F. A. (1993). *In vitro* construction of a human blood vessel from cultured vascular cells: A morphologic study. *J. Vasc. Surg.* **17**, 499–509.

L'Heureux, N., Paquet, S., Germain, L., Labbe, R., and Auger, F. A. (1998). A completely biological tissue-engineered human blood vessel. *FASEB J.* **12**, 47–56.

Matsuda, T., and Miwa, H. (1995). A hybrid vascular model biomimicking the hierarchic structure of arterial wall. *J. Thorac. Cardiovasc. Surg.* **110**, 988–997.

Matsumoto, T., Holmes, R. H., and Burdick, C. D. (1966). The fate of the inverted segment of small bowel used for the replacement of major veins. *Surgery* **60**, 739–743.

Meinhart, J., Deutch, M., and Zilla, P. (1997). Eight years of clinical endothelial cell transplantation. *ASAIO J.* **43**, M515–M521.

Murray, G., and Jones, J. M. (1940). Prevention of acute failure of circulation following injuries to large arteries. Experiments with glass cannulae kept patent by administration of heparin. *Br. Med. J.* **2**, 6–7.

National Academy Press (1996). "Guide for the Care and Use of Laboratory Animals." National Research Council, Washington, DC.

National Society for Medical Research (1996). "Principles of Laboratory Animal Care." National Research Council, Washington, DC.

Niklason, L. E., Gao, J., Abbott, W. M., Hirschi, K. K., Houser, S., Marini, R., and Langer, R. (1999). Functional arteries grown *in vitro*. *Science* **284**, 489–493.

O'Donnell, T. F., Mackey, W., McCullough, J. L., Maxwell, S. L., Farber, S. P., Dejerling, R. A., and Callow, R. D. (1984). Correlation of operative findings with angiographic and non-invasive hemodynamic factors associated with failure of PTFE grafts. *J. Vasc. Surg.* **1**, 136–148.

Panetta, T. F., Marin, M. L., Veith, F. J., Goldsmith, J., Gordon, R. E., Jones, A. M., Schwartz, M. L., Gupta, S. K., and Wengerter, K. R. (1992). Unsuspected pre-existing saphenous vein disease: An unrecognized cause of vein bypass failure. *J. Vasc. Surg.* **15**, 102–112.

Park, K. D., Okano, T., Jojitri, C., and Kim, S. W. (1988). Heparin immobilization onto segmented polyurethane urea surfaces. *J. Biomed. Mater. Res.* **22**, 977–980.

Pasic, M., Muller-Glauser, W., and von Segesser, L. (1994). Superior late patency of small-diameter Dacron grafts seeded with omental microvascular cells: An experimental study. *Ann. Thorac. Surg.* **58**, 677–682.

Phaneuf, M. D., Dempsey, D. J., Bide, M. J., Szycher, M., Quist, W. C., and LoGerfo, F. W. (1998). Bioengineering of a novel small diameter polyurethane vascular graft with covalently-bound recombinant hirudin. *ASAIO J.* **44**, M653–M658.

Pratt, G. H. (1984). Use of Teflon in replacement aortic and arterial segments. *Am. J. Surg.* **1**, 136–148.

Rosenberg, N., Lord, G. H., Henderson, J., Bothwell, J. W., and Gaughran, E. R. (1970). Collagen arterial graft of bovine origin: Seven year observations in the dog. *Surgery* **671**, 951–956.

Sabelman, E. E. (1985). Biology, biotechnology and biocompatibility of collagen. *In* "Biocompatibility of Tissue Analogs" (D. F. Williams, ed.), Vol. 1, pp. 27–64. CRC Press, Boca Raton, FL.

Schloss, G., and Shumacker, H. R., Jr. (1949). Studies in vascular repair, the use of free vascular transplants for bridging arterial defects. An historical review with particular reference to histological observations. *Yale J. Biol. Med.* **22**, 273–290.

Stephen, M., Loewenthal, J., Little, J. M., May, J., and Sheil, A. G. R. (1977). Autogenous veins and velour Dacron in femoropopliteal arterial bypass. *Surgery* **81**, 314–318.

Termin, P. L., Carr, R. M., O'Neil, K. D., and Connolly, R. J. (1993). Thrombogenicity of intestinal submucosa: Results of canine *ex vivo* shunt and acute rabbit implant studies. *AAMI Cardiovasc. Sci. Technol. Conf. Proc.*

Termin, P. L., Carr, R. M., and O'Neil, K. D. (1994). Remodeling of an absorbable synthetic collagen graft. *20th Annu. Meet. Soc. Biomater.*

Tranquillo, R. T., Girton, T. S., Bromberek, B. A., Triebes, T. G., and Mooradian, D. L. (1996). Magnetically-oriented tissue-equivalent tubes. *Biomaterials* **17**, 349–353.

Weinberg, C. B., and Bell, E. (1986). A blood vessel model constructed from collagen and cultured vascular cells. *Science* **231**, 397–399.

Welch, M., Durrans, D., Carr, H. M., Vohra, R., Rooney, O. B., and Biol, M. I. (1992). Endothelial cell seeding: A review. *Ann. Vasc. Surg.* **6**, 473–483.

Williams, S. K., Jarrell, B. E., and Kleinert, L. B. (1994). Endothelial cell transplantation onto porcine arteriovenous grafts evaluated using a canine model. *J. Invest. Surg.* **7**, 503–517.

Wilson, G. J., Yeger, H., Lement, P., Lee, M. J., and Courtman, D. W. (1990). Acellular matrix allograft small caliber vascular prostheses. *ASAIO Trans.* **36**, M340–M343.

Wilson, G. J., Courtman, D. W., Klement, P., Lee, J. M., and Yeger, H. (1995). Acellular matrix: A biomaterials approach for coronary artery bypass and heart valve replacement. *Ann. Thorac. Surg.* **60**, S353–S358.

Zarge, J. L., Huang, P., and Greisler, H. P. (1997). Blood vessels. *In* "Principles of Tissue Engineering" (R. Lanza, R. Langer, W. Chick, eds.), pp. 349–364. R. G. Landes, Georgetown, TX.

Ziegler, T., Robinson, K. A., Alexander, R. W., and Nerem, R. M. (1995). Co-culture of endothelial and smooth muscle cells in a flow environment: An improved culture model of the vascular wall? *Cells Mater.* **5**, 115–124.

CHAPTER 34

CARDIAC PROSTHESES

Jack W. Love

INTRODUCTION

At the time of writing this chapter, in 1999, the treatment of symptomatic valvular heart disease is based on two different methodologies, the use of mechanical devices and prostheses or the use of tissue, for repair or replacement of diseased native heart valves. This chapter focuses on the applications of tissue for such treatment, but it is important to mention, in passing, mechanical devices and prostheses, because they are used more widely than tissue prostheses. The major reason for the dominance of mechanical devices and prostheses is their greater durability. Investigators working in the field must be aware of the limitations of tissue methods and devices in order to advance the design and performance of tissue in the treatment of valvular heart disease.

Although there were clever and ingenious indirect attempts to treat valvular heart disease before 1960, success proved to be elusive until it was possible to repair or replace valves in their orthotopic locations. The achievement that made possible a direct approach to the diseased heart valve was the development of the heart–lung machine in 1953. In that year, Professor John Gibbon of Philadelphia performed the world's first open-heart operation on a patient with congenital heart disease. His machine, the culmination of 17 years of research and development, gave the surgeon the ability to divert, or bypass, the blood around the heart and lungs so that the heart could be opened and repaired under direct vision. The original Gibbon heart–lung machine was large and unwieldy, and required large amounts of donor blood to prime the circuit, but it was a start. Great improvements in technology for pumping and oxygenating blood followed rapidly. The first operations for valvular heart disease in which the valves were exposed within the heart were performed in 1960. Those first operations, in Boston, Massachusetts, and Portland, Oregon, were performed by Dwight Harken and Albert Starr, respectively. Harken and Starr replaced diseased valves with mechanical, ball-in-cage prostheses of their own design. Two years later, Åke Senning in Zurich began repairing and replacing diseased aortic valves with free-hand techniques using the patients' own (autologous) fascia lata (Senning, 1967), and Ross (1962) in London replaced a diseased aortic valve with a cadaveric human aortic valve. Figures 34.1 and 34.2 show Senning's methods for replacing and repairing aortic valves with fresh, untreated autologous fascia lata. Out of this pioneering work by men of genius and determination sprang the modern era of successful treatment of valvular heart disease. Throughout the world in 1999, mechanical valvular prostheses are more widely used than are tissue prostheses, in spite of significant drawbacks to their use. The drawbacks include noise, nonphysiologic flow patterns, catastrophic failure modes, and the need for lifelong anticoagulant therapy. It can be argued that the ideal treatment for repair or replacement of diseased human heart valves should be based on tissue technology rather than on the use of mechanical devices. Tissue valves are quiet and are associated with physiologic flow patterns, typically have slowly developing rather than catastrophic failure modes, and usually do not require anticoagulant therapy. The major, unsolved problem with tissue heart valves is their limited durability, generally 5 to 15 years. If the durability problem can be solved, tissue valves will be the clear choice for treatment of valvular heart disease. To clarify the issue, it is necessary to say something about the varieties of tissue valves and techniques that have been used for the past 35 years.

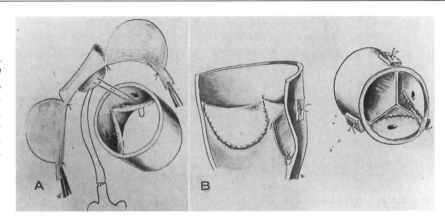

Fig. 34.1. (A) The fascia lata strip cut to form three cusps is sutured to the aortic annulus. (B) The commissures are anchored with sutures passed through the aortic wall and tied over Ivalon sponge pledgets. From Å. Senning (1967). Fascia lata replacement of aortic valves. J. Thorac. Cardiovasc. Surg. 54, 465–470, with permission from Mosby, Inc.

A SHORT HISTORY OF TISSUE HEART VALVES

As noted above, the first tissue heart valves were made of human tissue from autologous and homologous sources. Autologous tissue was an obvious first choice, to avoid an immune response. Unfortunately, fresh, untreated autologous tissues, both fascia lata and pericardium, were found to undergo shrinkage and thickening when used for valve repair or replacement. Figure 34.3 is a photograph of one of Senning's fascia lata valves explanted after 6 years, showing the shrinkage and thickening that occurred. That early finding of the fate of fresh, untreated autologous tissue has been reconfirmed by Kumar *et al.* (1995). These changes occurred within months or years, although Senning had some patients who survived up to 20 years with their autologous tissue fascia lata valves. One unanswered question is why Senning could have long-term success with some patients but not with all patients. There was a different outcome with homologous, cadaveric human aortic valves transplanted to the aortic position of host patients. Those valves did not under-

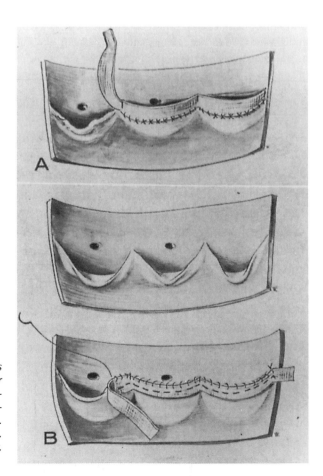

Fig. 34.2. Senning's techniques for leaflet extension and repair of leaflet prolapse. From Å. Senning (1967). Fascia lata replacement of aortic valves. J. Thorac. Cardiovasc. Surg. 54, 465–470, with permission from Mosby, Inc.

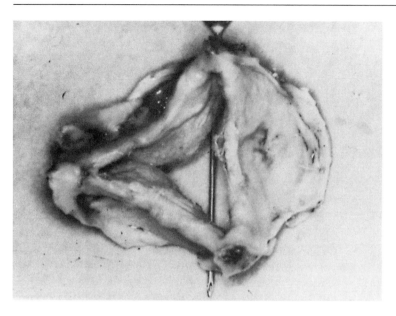

Fig. 34.3. Aortic valve replacement with fascia lata 6 years postimplantation. All three leaflets became thick and stiff, their bases aneurysmal. The valve become incompetent and was replaced. From Å. Senning (1984). Alterations in valvular surgery: Biologic valves. In "Proceedings of the Second International Symposium on Cardiac Bioprostheses" (L. H. Cohn and V. Gallucci, eds.). Yorke Medical Books, New York.

go shrinkage and thickening, but they were also found to have limited durability. The surgical procedure for implanting cadaveric aortic valves can be performed with variations, but all of the variations require a level of surgical expertise that has discouraged many surgeons from using them (Love, 1993).

The problems encountered with both autologous and homologous tissue valves led surgeons in the United States and Europe to consider the use of animal, or heterologous, tissue valves. Looking back on the history of tissue valves, one is tempted to question the logic of a progression from autologous to homologous to heterologous tissue valves; the reverse sequence would seem to have been more reasonable from an immunologic point of view. In simplest terms, the determination of surgeons to use tissue valves, combined with the apparent failure of autologous tissue valves, and the difficulties of both supply and implantation of homologous tissue valves, led to this development. The first recorded use of animal tissue valves was by French surgeons who implanted valves taken from the abattoir and implanted without treatment, other than attempted sterilization with antibiotic solutions (Binet *et al.,* 1965). The results were, predictably, unsatisfactory, and caused most of the first users to abandon the use of animal tissue valves.

One of the pioneers from that early work persisted with the concept of making animal tissue valves suitable for use in human patients. Carpentier in Paris explored different chemical treatments of animal tissue valves, and settled on an agent from the leather-tanning industry, glutaraldehyde (Carpentier *et al.,* 1969). Glutaraldehyde was found to reduce the immunogenicity of the untreated animal tissue by producing collagen cross-links that reduced solubility, and thus the immune stimulus to the host. Tanning with glutaraldehyde renders the tissue stiffer than it is in the untreated state. From 1968 to the present, heterologous tissue valves have occupied an important place in the cardiac surgeon's armamentarium for treating valvular heart disease. Carpentier, with engineers from the Baxter Edwards company, took porcine aortic valves from the slaughterhouse, tanned them with glutaraldehyde, and mounted them on stents made of Elgiloy wire covered with Dacron fabric for implantation in humans, and in so doing launched the bioprosthetic industry. An American engineer/entrepreneur, Warren Hancock, worked in parallel with Carpentier and founded his own company to manufacture porcine bioprostheses.

Another type of heterograft bioprosthesis was introduced in 1974 by Ionescu of Romania, while he was working in Leeds (Bartek *et al.,* 1974). He used bovine pericardium mounted on a stent machined from titanium covered with Dacron fabric. One of his valves is shown in Fig. 34.4. Initially, his valve was well received, primarily because it had low gradients in even the smallest sizes, in contrast to the porcine bioprostheses. The pig aortic valve has a ridge of muscle across one leaflet that becomes an impediment to flow in the smaller sizes, after the tissue has been stiffened by glutaraldehyde. However, the rigid titanium stent of the Ionescu valve was a design flaw. In the early days of bioprosthetic development, it became apparent that a flexible stent, whether made of

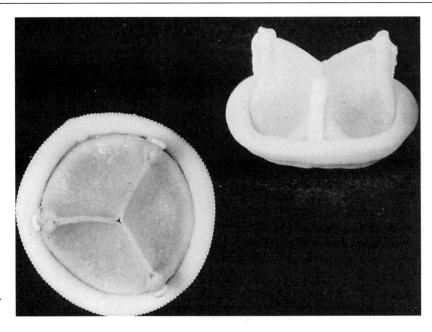

Fig. 34.4. The Ionescu–Shiley bovine pericardial bioprosthesis.

wire as in the Carpentier valve or of thermoplastic as in the Hancock valve, was an essential component of a successful bioprosthesis. The tissue is prone to mechanical failure when mounted on a rigid stent (Reis *et al.,* 1971; Christie and Barratt-Boyes, 1991). The Ionescu valve had another problem, a suture at each of the three leaflet commissures that proved to be another point of tissue failure due to stress. The bad experience with the Ionescu valve caused some investigators to think that the problem was the tissue, not the stent. However, bovine pericardium tanned with glutaraldehyde and mounted on a flexible stent performs as well, or better than, porcine aortic valves (Marchand *et al.,* 1998; Frater *et al.,* 1998).

The use of cadaveric homograft valves, pioneered by Ross in London, has had two devoted proponents from the earliest days to the present. They are Brian Barrett-Boyes of Auckland, New Zealand, and Mark O'Brien of Brisbane, Australia. Their persistence kept the concept alive for three decades until it became more widely accepted. A key factor important for the wider acceptance of the concept was the founding of the CryoLife Company in Marietta, Georgia. CryoLife began its existence as a company devoted to the harvesting and the preservation of cadaveric aortic roots and valves that could be used for the homograft operation. Before CryoLife, surgeons wanting to use homograft valves needed to have their own tissue banks in their medical centers. Those surgeons were required to harvest the valves from the autopsy suite, sterilize the valves, and preserve them until they were used. That burden undoubtedly contributed to the lack of early acceptance of the homograft operation. Another factor that was important for wider acceptance was the evidence provided by O'Brien that cryopreserved homografts might retain viability after implantation (O'Brien *et al.,* 1991). Although the matter has provoked significant controversy and skepticism, the debate has focused attention on the possibility of a viable tissue valve that might have the property of self-repair. That dream comes closer to the ideal replacement valve. Homograft valves appear in some series to have better durability than heterograft valves, but the difference has not been of an order of magnitude, in even the best reported results. Homograft valves have shown superior performance in the subset of patients with infective endocarditis. Stent-mounted bioprosthetic valves do not fare well in the presence of infection, whereas the unstented homograft valve can be expected to survive in the presence of infection. It is the presence or absence of the stent that probably makes the difference; foreign body material has long been known to require removal when surrounded by infection. However, there are downsides to the homograft valve. The problem of supply, once a severe limitation, has been greatly eased by the CryoLife technology. There is a small but real concern for the possibility of transmission of the autoimmune deficiency syndrome (AIDS) virus. Perhaps the biggest problem with the homograft valve is that it can be used only in the aortic or pulmonary position. Work is in progress to mount homograft valves on stents to make them available for use in the atrioventricular position, but this is not routinely possible.

THE STATUS OF TISSUE VALVES IN 1999

Following this brief history of tissue valves, the status of tissue valves in 1999 can be described. Commercially available tissue valves in 1999 include heterografts, homografts, and autografts. The heterograft valves are of two kinds, porcine and bovine, both tanned with glutaraldehyde. The porcine heterografts are aortic valves from slaughterhouse pigs, and they are available both stent mounted and unstented. The bovine heterograft is available as a stent-mounted pericardial bioprosthesis, with glutaraldehyde-tanned bovine pericardium from slaughterhouse material. Homograft valves are available commercially and are also implanted in some centers from tissue banks maintained at those centers. The tissue is treated with antibiotic solution, but is not tanned with glutaraldehyde. One company, Autogenics, has marketed a stent-mounted autograft, a bioprosthesis made in the operating room at the time of the valve replacement operation with the patient's own pericardium prepared, cut, and mounted on a stent system in 10 min. The pericardium is immersed for 10 min in 0.625% glutaraldehyde solutions buffered to pH 7.4 with phosphate. The rationale for using glutaraldehyde with the autograft is quite different than it is with heterograft tissue. The heterograft tissue is clearly immunogenic without tanning, and it is completely tanned for more than a week to produce collagen cross-linking to minimize solubility. The tanning was found by Carpentier to greatly attenuate, but not eliminate, immunogenicity of the tissue. With the autograft, there is no concern for immunogenicity; the light tanning is done to make the tissue stiffer and thus easier to use in the fabrication process. The investigators who developed this valve also reported that the lightly tanned autologous tissue did not display the thickening and shrinkage that had been noted by early investigators who used fresh, untreated tissue (Love *et al.*, 1992).

Limited durability has been the drawback to all of the tissue valves used for the past 25 years. Heterograft and homograft valves are subject to a usually gradual deterioration from a process known as calcific degeneration. As the process continues, there is an associated loss of tensile strength of the tissue. Tissue valves fail over time either because of excessive stiffening of the tissue, leading to stenosis and regurgitation, or from primary tissue failure with tearing. The autograft valve has been in clinical use for over 6 years, and it is not yet clear if it will prove to have superior durability and freedom from calcific degeneration and primary tissue failure. The inventors of the autograft valve believe that the immune response to heterograft and homograft tissue plays a role, perhaps the major role, in determining the fate of the tissue valve (Love, 1993). This point of view is not widely accepted in 1999, and only longer follow-up of the clinical series will show if the autograft is a superior tissue valve. There is one piece of intriguing evidence from long-term follow-up of a clinical series of porcine heterograft bioprostheses that may support the concept of immune response as a determinant of the fate of those valves. Jamieson has reported the durability of porcine valves related to the age of the patient at the time of implantation (Jamieson *et al.*, 1994). His data are displayed as a family of curves that confirm what has long been known, that durability is directly proportional to the patient's age at the time of implantation. Porcine valves, which are known to fail early in young patients, may last 10 to 15 years in older patients. The durability of the heterograft valve in children and adolescents is so unacceptable that such valves are used in that age cohort only when no other kind of valve can be used. Several interpretations can be made of Jamieson's data, but one possible explanation is that a more active immune response can be expected in young patients contrasted with older patients.

There is an accepted animal model for studying the fate of tissue valves, and data on the performance of tissue valves in that animal model are required by the U.S. Food and Drug Administration for approval. If a heterograft tissue valve is implanted in a juvenile weanling sheep, defined as one less than 6 months of age and generally less than 30 kg in weight, the heterograft valve is typically heavily calcified in 3 to 5 months (Barnhart *et al.*, 1982a,b). The calcification seen in the juvenile sheep model is apparently identical to the same process seen in humans over a period of several years. The model has been widely used to investigate the efficacy of anticalcifcation treatments that have been proposed by valve manufacturers. It has been reported that the autologous tissue heart valve shows minimal to no calcification in that model (Love *et al.*, 1994). That evidence is the primary reason for the expectation that the autograft will prove to be a more durable

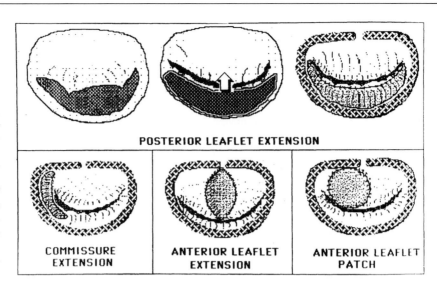

POSTERIOR LEAFLET EXTENSION

COMMISSURE EXTENSION   ANTERIOR LEAFLET EXTENSION   ANTERIOR LEAFLET PATCH

Fig. 34.5. Carpentier's techniques, using pericardium, for patching the mitral valve. From S. Chauvaud, V. Jebara, J.-C. Chachques, B. E. Asmar, S. Mihaileanu, P. Perier, G. Dreyfus, J. Relland, J.-P. Couetil, and A. Carpentier (1991). Valve extension with glutaraldehyde-preserved autologous pericardium. J. Thorac. Cardiovasc. Surg. **102,** *171–178, with permission from Mosby, Inc.*

bioprosthesis, less subject to calcific degeneration and primary tissue failure in humans. A study correlating the immune response in juvenile sheep to heterograft, homograft, and autograft valves is waiting to be done. Such a study could settle the issue of the importance of the immune response in determining the fate of the tissue valve.

The author of this chapter has been one of the pioneer investigators of the autograft valve. His group, along with four other groups in different parts of the world, has provided evidence that the autograft valve may behave differently than heterograft and homograft valves. In 1967, Ross in London transplanted the autologous pulmonary valve of a child to the aortic position to treat aortic valve disease. That operation has become known as the Ross procedure, or the pulmonary switch operation, and is currently enjoying a wave of popularity because of the remarkable durability of the transplanted autologous pulmonary valve. Patients have survived more than 25 years with no evidence of degeneration in their transplanted valves. The concept of this operation was developed in the animal laboratory at Stanford University; Ross applied the concept to the clinical setting. The downside to the Ross procedure is that there is only one pulmonary valve that can be transplanted, it can only be used in the aortic position, and when it has been moved to the aortic position, it must be replaced with an imperfect bioprosthesis. For the sake of this discussion, it is important to remember that this most perfect of all tissue valves to date is an autologous valve. Since 1980, Carpentier in Paris has used autologous pericardium treated with a brief immersion in glutaraldehyde for the repair of mitral valve leaflets damaged mainly by rheumatic endocarditis (Chauvaud *et al.,* 1991); the techniques are shown in Fig. 34.5. He and his colleagues have reported that autologous pericardial patches used for leaflet extension procedures have shown no calcific degeneration for more than 10 years. Using autologous pericardium for patching or leaflet extension is, admittedly, different from using that material to construct an entire valve, but the excellent long-term behavior of lightly tanned autologous pericardium is further evidence for the superiority of autologous tissue.

Duran, a Spanish surgeon working in Saudi Arabia, and most recently in the United States, resurrected Senning's original concepts for repair and replacement of aortic valves, using autologous pericardium treated with a brief immersion in glutaraldehyde, as shown in Fig. 34.6. He initially used tanned bovine pericardium, but switched to autologous pericardium when other groups began reporting better results with the autologous tissue. His method is essentially a free-hand technique that requires a significant element of surgical skill. He has reported his medium-term follow-up of 51 patients, mean age 31.2 years, observed up to 5 years with an impressive lack of degenerative changes (Duran *et al.,* 1995). Duran's work constitutes yet another piece of evidence for the superiority of autologous tissue for valve repair or replacement.

Deac in Romania has an interesting clinical series in progress that adds further support to the use of autologous tissue. He modified a technique described by Mickleborough for intraoperative

 Normal

 Rheumatic A. R.

 Resection

 Pericardial Graft for Cusp Extension

 Cusp Extension

Fig. 34.6. Duran's methods for aortic valve reconstruction with pericardium. Reprinted with permission from the Society of Thoracic Surgeons (The Annals of Thoracic Surgery, 1991, Vol. 52, pp. 447–454).

construction of a stentless atrioventricular valve, mitral or tricuspid (Deac *et al.*, 1995). Mickleborough's method is shown in Fig. 34.7. Whereas Mickleborough used tanned bovine pericardium, Deac has used lightly tanned autologous pericardium. Deac's results with 18 patients, mean age 42 years, from 1989 to 1994, are encouraging. It is especially noteworthy that he, too, has not seen evidence of calcific degeneration in any of his patients.

Finally, the work of this author and his group with a stent-mounted bioprosthesis made in the operating room at the time of valve replacement was encouraging in both the animal laboratory model and early clinical experience since the first human implant in 1992 (Fabiani *et al.*, 1995). An Autogenics valve, made with autologous pericardium treated with a brief immersion in glutaraldehyde, is seen in Fig. 34.8. The valve could be made within 10 min with disposable, single-use instruments, in the operating room at the time of the valve replacement operation. Love (1993) has given a detailed account of early work on the autologous pericardial valve. The Autogenics valve is not being marketed in 1999.

THE ISSUE OF WHETHER TO USE A STENT

Another issue that must be considered by investigators of tissue valves is whether to mount the tissue on a stent. With mechanical valves, this is not an issue because, by definition, the me-

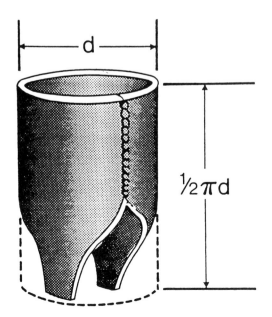

Fig. 34.7. Mickelborough's experimental method for constructing a stentless mitral valve from tanned bovine pericardium. Deac has modified her method and used lightly tanned autologous pericardium in a growing clinical series. Pericardium is wrapped around a cylinder of the appropriate diameter. Total length of the leaflets equals $\frac{1}{2}\pi d$. A single suture line is needed to construct the valve. Triangular portions of the pericardium are removed, as shown, leaving chordal tips of the prosthesis. From Mickleborough et al. (1988).

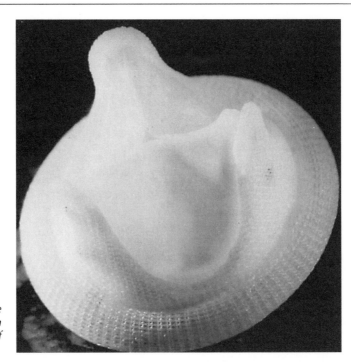

Fig. 34.8. The autologous tissue cardiac valve (ATCV) made in the operating room at the time of valve replacement.

chanical valve is made entirely of metal and plastic and fabric, part of which is a stent. With tissue, however, a decision can be made to use or not to use a stent. There are arguments that can be mode for both sides of the issue. A stent confers precise dimensions to the bioprosthesis, and the implantation of stent-mounted valves is a relatively straightforward surgical procedure, well within the capabilities of most cardiac surgeons. Stentless valves, on the other hand, whether heterograft, homograft, or autograft, generally require a higher level of surgical skill for their successful implantation. Dimensions are less precise, and more subjective judgments must be made in choosing and implanting the valve. It is a perfect example of the old surgical saw that good results come from experience, and experience comes from bad results. Those surgeons in 1999 who are able to report the best results with stentless valves are surgeons of extensive experience. Pediatric cardiac surgeons argue for stentless tissue valves for another reason. A stent, no matter how cleverly designed, occupies part of the valve orifice, and thus reduces the effective orifice area, and eliminates any potential for growth. These are important considerations because reduction of effective orifice area increases the transvalvular pressure gradient or energy loss, and a stent retains its fixed dimensions. For infants and children who require valve replacement, growth must be considered. A stented valve of fixed dimensions cannot grow with the patient, and the valve must be replaced one or more times as a child with a stented valve matures. There is some early evidence from both clinical and laboratory experience that the transplanted autologous pulmonary valve retains viability and can grow with the patient. If this proves to be the case, it will be a powerful argument in favor of the pulmonary switch operation, and the use of any other kind of unstented autologous tissue valve that might be expected to retain viability and the capacity for growth.

It should be noted that what are being marketed as stentless valves are not really stentless. The heterograft and allograft stentless valve both have stents, i.e., the biologic stents of the donor. They are stentless in the sense that neither has a rigid, mechanical stent.

Whether to use a stent is a more debatable issue for aortic or pulmonic valve repair than it is for a mitral or tricuspid repair. That is because the aortic and pulmonic valves have their own native stents from their attachments to the aorta and pulmonary artery, respectively, whereas the atrioventricular valves gain support from the chordal apparatus that attaches the leaflets to the papillary muscles of the interior ventricular walls. No entirely satisfactory bioprosthetic replacement valve has been found to mimic the unique and delicate chordal apparatus. Deac's work with a stentless valve, which substitutes two tongues of leaflet tissue for the chordae tendineae, is an attempt to solve this problem. Following animal implants in the laboratory, he is reporting success in a clinical series underway in Tirgu-Mures, Romania (Deac *et al.,* 1995). Deac uses autologous pericardium treated with a brief immersion in glutaraldehyde for his valve.

THE ISSUE OF REPAIR VERSUS REPLACEMENT

The use of tissue of any kind to treat valvular heart disease involves another consideration and choice, whether to repair or replace a damaged valve. As noted at the beginning of this chapter, the earliest work with tissue to treat valvular heart disease involved techniques for both repair and replacement. Senning's seminal paper of 1967, written after 7 years of work with fascia lata for both repair and replacement of aortic valves, should be consulted for an introduction to, and discussion of, this issue. He was a surgeon of uncommon skill and great imagination who led the way for many who followed, and he should be acknowledged as one of the key figures in the treatment of valvular heart disease. After his pioneering work beginning in 1962, there were many groups around the world that used untreated autologous fascia lata to make valves. All of those groups, including Senning, abandoned the work because of early failures and unpredictable results. It should be remembered, however, that Senning (1984) was able to report a group of his patients with normally functioning valves 12 to 20 years postoperatively.

Carpentier revisited the concept of repair and replacement with his repairs of mitral valves damaged by rheumatic heart disease and degenerative processes. As noted earlier in this chapter, he evolved his methods along several different lines, which include plastic procedures on the valve leaflets and chordae, the use of annuloplasty rings to repair dilated annuli, and the use of gusset patches to increase the area and mobility of scarred leaflets. For this latter work, he used fully tanned bovine pericardium initially, but in 1980 he switched to the use of autologous pericardium treated with a brief immersion in glutaraldehyde (Chauvaud *et al.*, 1991). His vast experience with repair techniques, disseminated to the current generation of cardiac surgeons, has established the principle that it is always preferable to repair a diseased heart valve, whenever possible, rather than to replace that valve. That principle derives from *a priori* thinking that a natural valve is better than any man-made substitute, from the knowledge that there is no perfect prosthetic heart valve, either tissue or mechanical, and now from accumulated clinical experience showing better clinical results in series of valve repairs compared with valve replacements.

Duran has provided additional evidence for the superiority of repair over replacement with his work in Saudi Arabia treating aortic valvular heart disease in a population of primarily young patients afflicted with rheumatic heart disease. Like Carpentier, he initially used fully tanned bovine pericardium for his aortic valve repairs, but switched to autologous pericardium treated with a brief immersion in glutaraldehyde because of evidence for the superiority of the autologous tissue (Duran *et al.*, 1995).

The issue is still being debated in 1999, but there is mounting evidence to support the view that repair should always be performed whenever possible, and that replacement should be reserved for those patients who are not candidates for repair with current techniques.

SPECULATION ABOUT THE FUTURE

With regard to the use of tissue for heart valve repair or replacement, it is necessary to consider which tissue to use. The source of the tissue is one issue, and the specific tissue is another issue. The sources in 1999 are domestic animals (cattle and pigs), human cadavers, and autologous tissue taken from the host. The specific tissues available in 1999 are either natural valves or pericardium used to make valves. The natural valves used commercially are either porcine aortic valves, fully tanned with glutaraldehyde and stent mounted, or untanned human cadaver valves for aortic replacement, with or without accompanying aortic root replacement. Bovine pericardium, fully tanned with glutaraldehyde, is used for replacement when fashioned into stent-mounted or stentless bioprostheses, or for repair purposes. Autologous pericardium treated with a brief immersion in glutaraldehyde is being used for repair of mitral valves by Carpentier, for repair of aortic valves by Duran, and for mitral replacements by Deac. The Autogenics stented autologous tissue bioprosthesis has been used for aortic and mitral replacement. It is not yet clear which tissue, from which source, will prove to be the tissue of choice. Nor is it clear how the tissue will be best used, either for repair techniques or for replacement techniques, with or without a stent.

In discussing the use of tissue for repair or replacement of diseased heart valves, it is germane to consider disease transmission with animal or cadaveric tissue. The risk is extremely small, but definite and proved. With animal tissue, there is concern for the transmission of what have come to be known as the prion diseases. Bovine spongiform encephalopathy, or BSE, is the manifestation of prion disease in cattle. In sheep, it is known as scrapie. There is no known equivalent in domestic pigs, and thus the porcine heterograft is not known to carry the risk of prion transmission. The prion diseases, although fortunately rare, are formidable afflictions. There is no *in vivo* test for the diseases, there is no known treatment, and the diseases are uniformly fatal. Prion disease has been epidemic among domestic sheep and cattle in Europe, but it is not clear if the diseases have appeared in North American animals. Transmission from animals to man has been documented from the use of animal tissue products. The extensive review by Prusiner (1995), who coined the term "prion," should be consulted for details.

With cadaveric tissue, the rare but definite risk is transmission of the human immunodeficiency virus (HIV), leading to AIDS. Donors can be tested for HIV, but the problem is the window of unknown length that can exist between infection and development of a positive test for the virus. One case has been documented of a donor who tested negative for the virus, whose organs transmitted the virus to multiple recipients of his heart, liver, kidneys, and bone (Simonds *et al.,* 1992). Apart from the problem of the window between infection and test positivity, another aspect of the risk of HIV transmission is the imperfect sensitivity and specificity of the test for HIV antibodies. Until a more reliable, early test for HIV is developed, or until there is an effective treatment for AIDS, the use of cadaveric tissue will carry risk of transmission of the disease.

Autologous tissue carries no risk of disease transmission, and that fact must be recognized as a strong argument for its use whenever possible. From the standpoint of disease transmission alone, it may be speculated that autologous tissue will be more widely used for both repair and replacement of diseased heart valves in the future. Patients will always demand that risks of all kinds associated with surgical treatments be minimized. In the very litigious environment that exists in the United States, risk minimization is a major consideration for physicians.

Autologous tissue can be used for repair or replacement, and autologous tissue replacement valves can be either stent mounted or unstented. The use of autologous tissue for both repair and replacement of diseased heart valves is an old idea, unsuccessful when first tried with fresh, untreated tissue, and now successful with tissue that has been treated with a brief immersion in glutaraldehyde. To date, the only autologous tissue that has been used successfully is pericardium. Pericardium, both fully tanned bovine and lightly tanned autologous, now has a proved record of successful use for valve repair and replacement. Patients having second or third cardiac operations may not have suitable pericardium available, and it thus becomes important to consider other candidate tissues. Further, materials scientists are concerned about isotropy and anisotropy in tissues used for valve repair or replacement. Valvular leaflets are anisotropic, whereas pericardium is isotropic (Zioupos *et al.,* 1994). Fascia lata, used by Senning, should be reevaluated as a candidate tissue. Venous tissue, peritoneum, and rectus fascia have all been used or suggested for use in the past as fresh, untreated tissue (Love, 1993). These tissues should also be reexamined for use after light tanning with glutaraldehyde. A tissue other than pericardium could prove to be best for valvular applications.

Another topic of speculation about the future is the potential for development of techniques for repairing or replacing heart valves nonsurgically. The potential is for techniques that would enable the treating physician, surgeon or cardiologist, to introduce a folded valve by way of a catheter into the heart or great vessels, there to unfold and attach the valve in the orthotopic position by fixation means yet to be developed. With constantly improving catheter technology and imaging techniques, it is not beyond the realm of near possibility that nonsurgical treatment of valvular heart disease may become feasible. It is obvious that only a tissue valve could be used in this way; mechanical valves cannot be compressed in size and thus could not be used this way.

RECENT DEVELOPMENTS

In the most literal sense of the phrase "tissue engineering," it is now possible to grow heart valve leaflets *in vitro,* and then implant them in experimental animals. This impressive achievement has been reported by a team from the Harvard Medical School and the Massachusetts Institute of Technology in a series of presentations and articles since 1995. The concept was introduced at a meeting of the European Association for Cardiothoracic Surgery in 1995, and subsequently published in 1997 (Zund *et al.,* 1997). Fibroblast and endothelial cell cultures derived from human, bovine, and ovine sources were seeded on biodegradable poly(glycolic acid) meshes to create valve leaflets. Xenograft leaflets made of human fibroblasts and bovine endothelial cells, and allograft leaflets made of ovine cells, were engineered *in vitro.* The authors found that elastin and collagen production by the fibroblasts was a function of how long the biodegradable mesh survived.

In another study, ovine endothelial cells and fibroblasts were isolated from arteries, cultured, and then seeded onto biodegradable scaffolds of poly(glycolic acid). Both allograft and autograft leaflets were grown by this method and were used to replace one leaflet of the pulmonary valve in lambs (Shinoka *et al.,* 1995). An inflammatory response was seen with the implanted allograft leaflets that was not seen with the autograft leaflets. Evidence based on cell labeling suggested that the cells seeded *in vitro* persisted in the leaflets up to 11 weeks postoperatively (Shinoka *et al.,* 1996). Tensile strength of the leaflets increased with time, and was comparable to that of the native leaflets after 9–10 weeks. Additionally, the authors compared fibroblasts of dermal origin with those of femoral arterial wall origin to determine which would provide tissue more like normal valve leaflets (Shinoka *et al.,* 1997). Based on measurement of collagen and elastin content, and tensile strength, the arterial wall fibroblasts provided tissue sheets more like normal valve leaflets. The seminal work of this group of investigators is bringing us ever closer to the possibility of growing whole valves *in vitro* that could be used to replace diseased heart valves. It is interesting that these investigators have found that autologous cells are better accepted *in vivo* than are allogeneic cells.

The quest for more durable xenograft valves continues. A group in Hanover, Germany, has reported the use of a solvent to remove cells from the porcine aortic valve (Gratzer *et al.,* 1998). An acellular matrix was produced, onto which cultured human endothelial cells were seeded. The authors did not study the tensile strength of their modified porcine valves, nor did they implant any valves in animals. Their untanned, modified valves contain xenograft collagen and allograft endothelium, and it can be questioned if such a valve could be expected to be more durable than a glutaraldehyde-tanned xenograft.

Love (1998) has described methods and instruments that can be used to reconstruct diseased semilunar heart valves with lightly tanned autologous pericardium, without the use of a stent. The key to the method is the use of a precisely sized and cut unitary trefoil tissue pattern, and instruments to hold the tissue in anatomical orientation while the reconstruction is being performed. By eliminating the stent, the patient is left with only a piece of his or her own tissue and the sutures used to accomplish the reconstruction. There is tissue-to-tissue interface at the line of attachment, which allows firm healing, and full effective valve orifice area. Love and colleagues have optimized the geometry of the trefoil with computer-assisted design, and they are validating the geometry with finite-element analysis. Additionally, they are developing an instrument for preuse intraoperative testing of tissue that provides measurements of thickness, strength, and isotropy/anisotropy. The system has been tested in experimental animals and is being readied for clinical use.

THE IDEAL VALVE REPAIR OR REPLACEMENT

By way of summing up the foregoing discussion, it may be useful to describe some of the design goals for the ideal repair or replacement in the light of 39 years of experience with surgery for valvular heart disease. Mechanical prostheses have enjoyed great success in the treatment of valvular heart disease, and it cannot be disputed that hundreds of thousands of lives have been saved and extended by their use. Mechanical devices cannot be used to repair valve leaflets; only tissue will serve that purpose, unless a synthetic material becomes available. No synthetic material has proved to have the properties needed to endure bidirectional flexing some 40 million times a year without failing or producing thrombosis. The limited durability of currently available tissue valves has restricted their use. It seems clear that a tissue valve with better durability would more closely approach the ideal replacement valve. The reasons for this are given in the beginning of this chapter. Better, more durable tissue is clearly needed for valve repair or reconstruction. Tissue-engineered heart valves grown *in vitro* with autologous cells are on the horizon, and may be a near-perfect solution to the need for more durable tissue valves. It may be that lightly tanned autologous pericardium will prove to have better durability and freedom from calcific degeneration. If so, it may qualify as a preferred tissue for use with both repair and replacement techniques. Manufacturers of heterograft bioprostheses have used a variety of chemical treatments, many proprietary and unpublished, to prevent the calcific degeneration of porcine and bovine tissue valves. One is reminded of the old adage that it is not possible to make a silk purse from a sow's ear. A chemical treatment of autologous tissue that prevented the thickening and shrinkage of the fresh, untreated tissue, but allowed the tissue to remain viable and self-repairing, would be a great advance in tissue engineering.

The argument has been made that valve repair is always preferable to valve replacement. An extension of that argument is the use of extensive repair techniques, such as those used by Duran, that blur the distinction between repair and replacement. Such extensive repairs accomplish the goal of giving the patient an essentially new valve, while avoiding the need for a stent, with the disadvantages imposed by the stent. The goal is near achievement with operations on aortic and pulmonic valves, but more elusive with operations on mitral and tricuspid valves because of their complicated anatomy, including the chordal apparatus. Deac's method is a noteworthy approach, but may not lend itself to the kind of reproducibility that is required for widespread acceptance by the surgical community.

It should be clear that many opportunities exist for tissue engineering in the field of cardiac valvular repairs and prostheses. The needs are for ways to treat existing tissue to make it suitable for valvular applications, or for growing valvular tissue *in vitro,* and for designs of improved techniques and methods for repairing and replacing valves that eliminate or minimize the use of stents. Nonsurgical methodology for treating valvular heart disease, such as valve delivery and placement by a catheter, is another opportunity for tissue engineering.

REFERENCES

Barnhart, G. R., Jones, M., Ishihara, T., Chavez, A. M., Rose, D. M., and Ferrans, V. J. (1982a). Bioprosthetic valvular failure: Clinical and pathological observations in an experimental animal model. *J. Thorac. Cardiovasc. Surg.* **83,** 618–631.

Barnhart, G. R., Jones, M., Ishihara, T., Rose, D. M., Chavez, A. M., and Ferrans, V. J. (1982b). Degeneration and calcification of bioprosthetic cardiac valves. Bioprosthetic tricuspid valve implantation in sheep. *Am. J. Pathol.* **106,** 136–139.

Bartek, I. T., Holden, M. P., and Ionescu, M. I. (1974). Frame-mounted tissue heart valves: Technique of construction. *Thorax* **29,** 51–55.

Binet, J. P., Carpentier, A., Langlois, J., Duran, C., and Colvez, P. (1965). Implantation de valves hétérogènes dans le traitement de cardiopathies. *C. R. Hebd. Seances Acad. Sci.* **261,** 5733–5734.

Carpentier, A., Lemaigre, G., Robert, L., Carpentier, S., and Dubost, C. (1969). Biological factors affecting long-term results of valvular heterografts. *J. Thorac. Cardiovasc. Surg.* **58,** 467–483.

Chauvaud, S., Jebara, V., Chachques, J.-C., Asmar, B. E., Mihaileanu, S., Perier, P., Dreyfus, G., Relland, J., Couetil, J.-P., and Carpentier, A. (1991). Valve extension with glutaraldehyde-preserved autologous pericardium. *J. Thorac. Cardiovasc. Surg.* **102,** 171–178.

Christie, G. W., and Barratt-Boyes, B. G. (1991). On stress reduction in bioprosthetic heart valve leaflets by the use of a flexible stent. *J. Cardiac Surg.* **6,** 476–481.

Deac, R. F., Simionescu, D., and Deac, D. (1995). New evolution in mitral physiology and surgery: Mitral stentless pericardial valve. *Ann. Thorac. Surg.* **60,** S433–S438.

Duran, C. M. G., Gometza, B., Kumar, N., Gallo, R., and Martin-Duran, R. (1995). Aortic valve replacement with freehand autologous pericardium. *J. Thorac. Cardiovasc. Surg.* **110,** 511–516.

Fabiani, J.-N., Dreyfus, G. D., Marchand, M., Jourdan, J., Aupart, M., Latrémouille, C., Chardigny, C., and Carpentier, A. F. (1995). The autologous tissue cardiac valve: A new paradigm for heart valve replacement. *Ann. Thorac. Surg.* **60,** 189–194.

Frater, R. W. M., Furlong, P., Cosgrove, D. M., Okies, J. E., Colburn, L. Q., Katz, A. S., Lowe, N. L., and Ryba, E. A. (1998). Long-term durability and patient functional status of the Carpentier–Edwards PERIMOUNT bioprosthesis in the aortic position. *J. Heart Valve Dis.* **7,** 48–53.

Gratzer, P. F., Pereira, C. A., and Lee, J. M. (1998). Solvent environment modulates effects of glutaraldehyde crosslinking on tissue-derived biomaterials. *J. Biomed. Mater. Res.* **31,** 533–543.

Jamieson, W. R. E., Burr, L. H., Tyers, G. F. O., and Munro, A. I. (1994). Carpentier–Edwards standard and supra-annular porcine bioprostheses: 10 year comparison of structural valve deterioration. *J. Heart Valve Dis.* **3,** 59–65.

Kumar, P. K., Prabhakar, G., Kumar, M., Kumar, N., Shahid, M., Liaqat, A., Becker, A., and Duran, C. M. G. (1995). Comparison of fresh and glutaraldehyde-treated autologous stented pericardium as pulmonary valve replacement. *J. Cardiac Surg.* **10,** 545–551.

Love, C. S., Willems, P. W., and Love, J. W. (1994). An autologous tissue bioprosthetic heart valve. *In* "New Horizons and the Future of Heart Valve Bioprostheses" (S. Gabbay and R. W. M. Frater, eds.). Silent Partners, Austin, TX.

Love, J. W. (1993). "Autologous Tissue Heart Valves." R. G. Landes, Austin, TX.

Love, J. W. (1998). Autologous pericardial reconstruction of semilunar heart valves. *J. Heart Valve Dis.* **7,** 40–47.

Love, J. W., Schoen, F. J., Breznock, E. M., Shermer, S. P., and Love, C. S. (1992). Experimental evaluation of an autologous tissue heart valves. *J. Heart Valve Dis.* **1,** 232–241.

Marchand, M., Aupart, M., Norton, R., Goldsmith, I. R. A., Pelletier, C., Pellerin, M., Dubiel, T., Daenen, W., Casselman, F., Holden, M., David, T. E., and Ryba, E. A. (1998). Twelve-year experience with Carpentier–Edwards PERIMOUNT pericardial valve in the mitral position: A multicenter study. *J. Heart Valve Dis.* **7,** 292–298.

Mickleborough, L., *et al.* (1988). A simplified concept for a bileaflet atrioventricular valve that maintains annular-papillary muscle continuity. *J. Card. Surg.* **4,** 58–68.

O'Brien, M., McGiffin, E. C., Stafford, E. G., Gardner, M. A. H., Pohlner, P. F., McLachlan, G. J., Gall, K., Smith, S., and Murphy, E. (1991). Allograft aortic valve replacement: Long-term comparative clinical analysis of the viable cryopreserved and antibiotic 4°C stored valves. *J. Cardiac Surg.* **6**(4, Suppl.), 534–543.

Prusiner, S. B. (1959). The prion diseases. *Sci. Am.* **272**(1), 48–57.

Reis, R. L., Hancock, W. D., Yarbrough, J. W., Glancy, D. L., and Morrow, A. G. (1971). The flexible stent. A new concept in the fabrication of tissue heart valve prostheses. *J. Thorac. Cardiovasc. Surg.* **62**, 683–689.

Ross, D. N. (1962). Homograft replacement of the aortic valve. *Lancet* **2**, 487.

Ross, D. N. (1967). Replacement of the aortic and mitral valve with a pulmonary autograft. *Lancet* **2**, 956–958.

Senning, Å. (1967). Fascia lata replacement of aortic valves. *J. Thorac. Cardiovasc. Surg.* **54**, 465–470.

Senning, Å. (1984). Alterations in valvular surgery: Biologic valves. *In* "Proceedings of the Second International Symposium on Cardiac Bioprostheses" (L. H. Cohn and V. Gallucci, eds.). Yorke Medical Books, New York.

Shinoka, T., Breuer, M. D., Tanel, R. E., Zund, G., Miura, T., Ma, P. X., Langer, R., Vacanti, J. P., and Mayer, J. E. (1995). Tissue engineering heart valves: Valve leaflet replacement in a lamb model. *Ann. Thorac. Surg.* **60**, S513–S516.

Shinoka, T., Ma, P. X., Shum-Tim, D., Breuer, C. K., Cusick, R. A., Zund, G., Langer, R., Vacanti, J. P., and Mayer, J. E. (1996). Tissue-engineering heart valves. Autologous valve leaflet replacement study in a lamb model. *Circulation* **94**(Supp. II), II-164–II-168.

Shinoka, T., Shum-Tim, D., Ma, P. X., Tanel, R. E., Langer, R., Vacanti, J. P., and Mayer, J. E. (1997). Tissue-engineered heart valve leaflets. Does cell origin affect outcome? *Circulation* **96**(Suppl. II), II-102–II-107.

Simonds, R. J., Holmberg, S. D., Hurwitz, R. I., Coleman, T. R., Bottenfield, S., Conley, L. J., Kohlenberg, S. H., Castro, K. G., Dahan, B. A., Schable, M. S., Rayfield, M. A., and Rogers, M. F. (1992). Transmission of immunodeficiency virus type 1 from a seronegative organ and tissue donor. *N. Engl. J. Med.* **326**, 726–732.

Zioupos, P., Barbenel, J. C., and Fisher, J. (1994). Anisotropic elasticity and strength of glutaraldehyde fixed bovine pericardium for use in pericardial bioprosthetic valves. *J. Biomed. Mater. Res.* **28**, 49–57.

Zund, G., Breuer, C. K., Shinoka, T., Ma, P. X., Langer, R., Mayer, J. E., and Vacanti, J. P. (1997). The *in vitro* construction of a tissue engineered bioprosthetic heart valve. *Eur. J. Cardiothorac. Surg.* **11**, 493–497.

PART XI: CORNEA

CHAPTER 35

CORNEA

Vickery Trinkaus-Randall

INTRODUCTION

The cornea is an excellent candidate for tissue engineering: however, the biologic constraints that impose restrictions on alterations of corneal shape or curvature, or on the placement of a prosthesis, make the tissue very challenging to engineer. Devices to alter corneal curvature to improve vision are much in demand, although some procedures have resulted in major structural changes to the eye, actually causing damage. Attention to the biology of the cornea is necessary for such interventions to succeed. On the other hand, the need for synthetic corneas is real and the effort to develop a keratoprosthesis has come a long way in the past 5 years. To date several intact models are being evaluated at various levels of stringency. The replacement of a human cornea requires an interdisciplinary approach, where the goal of the materials scientists is to design materials that are accepted and ultimately integrated by the cornea. There is a great need in underdeveloped countries for synthetic corneas because, for cultural or medical reasons, transplants are not feasible.

The cornea has two major functions: (1) to protect the intraocular contents of the eye and (2) to serve as the principal optical element (i.e., refract light). The elasticity and thickness of the cornea allow it to maintain the intraocular pressure (IOP) within the eye. The cornea comprises one-sixth of the outer wall of the eye and junctures with the sclera, imparting an oval appearance to the anterior surface of the cornea (Fig. 35.1A).

To function as the major optical element the cornea must remain transparent, allowing for the formation of an image on the retina. In addition, light scatter can alter the ability of the retina to obtain a clear image. Therefore any opacity in the cornea occurring from disease or trauma can scatter light and degrade the image. To provide or improve vision, the development of synthetic corneas and of other devices and procedures that alter the refractive properties of the cornea has been evaluated. The cornea appears on the surface to be a simple tissue, but the development of synthetic models has been quite challenging. For example, the cornea is avascular and must remain that way; its thickness is regulated by a series of pumps and cell junctions. If the integrity of the latter is lost, the eye becomes edematous. In addition, the corneal matrix may be altered by components from the tears and/or from the aqueous humor located posterior to the endothelium. All of these limitations are compounded by the observation that the cornea is one of the most highly innervated tissues and that repair is impeded in the absence of innervation.

When diseases or trauma render the cornea opaque, the cornea loses its ability to refract light. This disability can be treated with surgery, called a "penetrating keratoplasty," whereby the cornea is replaced with a donor cornea. When corneal transplants have failed repeatedly, synthetic corneas have been sutured in place in an attempt to restore some vision. The goal of this chapter is to address the progress that investigators have made in creating a synthetic cornea and in reshaping the cornea. We will follow the course of history, from designing prosthetic devices that were intended merely to provide vision, to developing models that are designed to be biocompatible. The first part of the chapter describes the morphology and function of the cornea to help elucidate what is required in engineering the cornea and what are the limitations of the tissue. The latter part of the chapter examines various methods of modifying the cornea.

Principles of Tissue Engineering
Second Edition

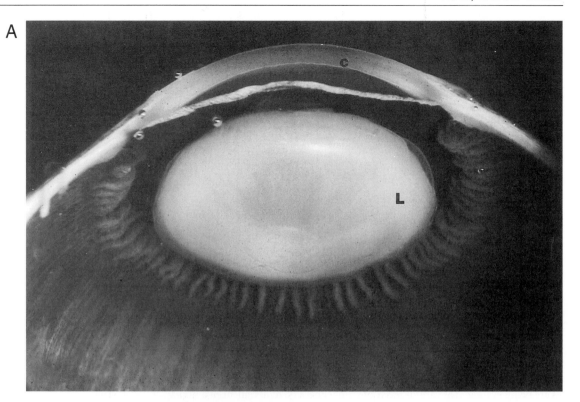

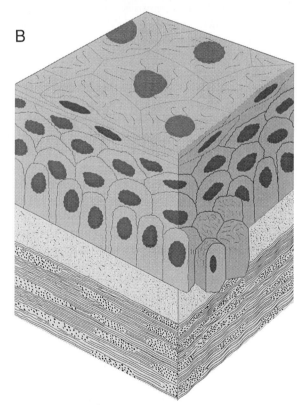

Fig. 35.1. (A) A cross-section of the eye illustrates the curvature of the cornea (C). The endothelium is anterior to the aqueous humor, and the cornea is anterior to the lens (L). (B) A schematic illustrating the three regions of the cornea. The epithelium is most anterior and is composed of five to seven layers. Bowman's membrane is an acellular domain in between the epithelium and the stroma. The stroma is the major structural component and is populated by stromal keratocytes. The endothelium is not represented in this diagram.

The cornea is divided into three regions. It is characterized by five to seven layers of epithelium on the anterior surface, a stroma (which is the major structural component), and comprises an extensive matrix located between the two basal laminae and a single layer of endothelial cells on the posterior side of Descement's membrane (Fig. 35.1B) (Poise and Mandell, 1970). All regions of the cornea obtain O_2 from tears through the anterior corneal surface. Other nutrients, such as glucose, amino acids, carbohydrates, and lipids, are derived from the aqueous humor. Growth factors are both endogenous and supplied by the aqueous humor via the posterior corneal surface and the tears. The limbal vessels also play a minor role in supplying nutrients. CO_2 is eliminated from the anterior surface whereas the waste product, lactate, is removed mainly via diffusion across the stroma and endothelium into the aqueous humor. Regulation of lactate concentration is essential because an excess can cause edema of the stroma, resulting in opacity.

The transparency of the cornea depends in part on maintenance of the structure and integrity of cells and the surrounding proteoglycans and collagen. In addition, the cornea must regulate the diffusion of water and ions into the stroma. The normal cornea is 78% H_2O by weight. The epithelium prevents swelling by inhibiting the diffusion of water and ions from the tears across tight junctions. The endothelium uses both a barrier function consisting of tight junctions and a metabolic pump located in the basolateral membrane to prevent swelling of the stroma. The stroma, composed of highly charged proteoglycans, must in turn remain relatively dehydrated. When excess water enters the stroma, glycosaminoglycans take up water. The consequence is an increase in corneal thickness (edema), a disruption in the alignment of collagen fibrils, and ultimately interference with light transmission through the cornea. The latter are all critical components of transparency because edema alters the corneal refractive power, which is determined by both the index of refraction and the radius of curvature of the cornea. The index of refraction in a normal cornea is 1.376 and the radius of curvature averages 7.7 mm on the central anterior surface. The shorter the radius of curvature, the steeper the cornea and the more refracting power it has. Clinically the term "diopters" (corneal power) is used rather than corneal shape (millimeters radii). About 75% of the diopteric power of the eye depends on the interface of the cornea and the surrounding air.

Another critical component of vision is glare. The healthy cornea has the ability to absorb UV radiation (280–300 nm), which protects the posterior part of the eye. Glare is produced by bright light that is scattered within the eye, causing discomfort and decreased vision. As a consequence, opacities may decrease the visual acuity, not only by disrupting the geometric image but also by producing light scatter within the globe. The light scatter reduces the ability to discriminate images on the retina (Miller and Benedek, 1973). This loss may occur when a person has cataracts. In this case a person can see images without loss of acuity in dim light but is bothered by bright light. Light scatter can also occur when the stroma is injured or when inflammation lies within or near the visual axis. Because the cornea is absent in synthetic corneas, the choice of materials or UV blockers is an important component in the design of a synthetic cornea.

Corneal topography is a critical component of vision. To best understand the design of a cornea, consider topographic maps of a mountain. An aspheric surface is composed of compound curves that differ in various meridians. The cornea flattens across the center and periphery, increasing the radius of curvature about 2 mm paracentrally and about 4 mm peripherally. The change is more dramatic along the nasal portion of the cornea than along the temporal portion. This is an important aspect to clinicians who are interested in reshaping the cornea by performing refractive surgery or by placing optics or rings within the cornea to alter the curvature. Persons from outside the field of ophthalmology should be aware that the most commonly corrected corneal property is a change of curvature. Compensation for subtle alterations in curvature and corneal power (diopters) has in the past been performed using eye glasses. It remains the least invasive methodology for correcting vision.

BIOLOGY OF THE CORNEA

EPITHELIUM AND ADHESION STRUCTURES

The role of the corneal epithelium is threefold: (1) it must act as a mechanical barrier to foreign organisms, (2) it should have a smooth, transparent optical surface caused by adsorption of the tear film, and (3) it must mediate the diffusion of water, solutes, and drugs on the anterior surface. The epithelium of a healthy normal cornea is not keratinized and is composed of five to seven layers. There is a single layer of basal columnar cells, two to three layers of wing-shaped cells, and two to three layers of superficial cells. The predominantly mitotically active cell layer is the basal layer and the resulting daughter cells move apically and are ultimately sloughed into the tear

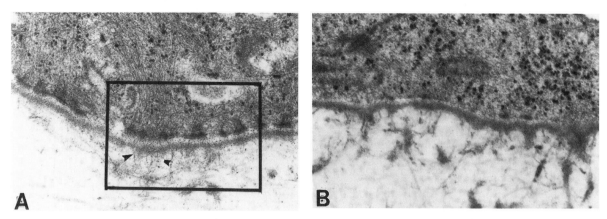

Fig. 35.2. (A) An electron micrograph of hemidesmosomes along the basal lamina of a cornea. Note the electron-dense areas on the cytoplasmic face of the cell and the anchoring fibrils (arrowheads) extending into the anterior stroma. Hemidesmosomes are shown within the rectangle. (B) Hemidesmosomes fail to form in the presence of inhibitors of tyrosine phosphorylation.

film (Lavker *et al.,* 1991). Cell turnover is low and only 5% of the basal cells mitose at any one time. A small number of the basal cells become apoptotic. Because the epithelium is subject to large shear forces caused both by blinking and by the movement of the eye during blinking, epithelial cells must remain tightly adherent to neighboring cells via specialized cell–cell junctions (i.e., desmosomes, zonula adherens, tight junctions, and gap junctions). They must also remain adherent to the basal lamina via cell–substrate junctions (hemidesmosomes) (Fig. 35.2A).

Hemidesmosomes can be thought of as the root system of the basal epithelial cells and in intact corneas the entire complex penetrates the basal lamina and terminates in the anterior stroma in the form of anchoring fibrils. It has been shown that the complete formation of hemidesmosomes requires phosphorylation of tyrosine residues. In the absence of phosphorylation, filaments penetrate the lamina lucida of the basal lamina, but the adhesion complex, including intermediate filaments, is not detected (Fig. 35.2B). The highly organized complex is disrupted in disease (i.e., Bullous pemphigoid), trauma (chemical burn or wound), or after surgery, because the cells become motile, using other junctions to migrate along the basal lamina to repair the wound. During wound repair the genomic expression is altered, with some proteins becoming down-regulated (i.e., Bullous pemphigoid antigen) and others up-regulated (i.e., vinculin) (Zieske *et al.,* 1989). It is also known that these proteins are components of signal transduction pathways. A more indepth discussion of this process can be found in reviews by Gipson and Sugrue (1949) and Trinkaus-Randall *et al.* (1998).

The surface or basal lamina to which the epithelial cells adhere changes with age and/or disease, causing a decrease in adherence of cells and a change in the expression of epithelial adhesion proteins. The distribution of $\alpha 6$ and $\beta 4$ integrin subunits along the basal surface of the cells is altered with age, indicating that the basal lamina no longer adequately supports the tissue. In fact, when these cells are cultured *in vitro* under enriched conditions, the proteins are synthesized normally (Trinkaus-Randall *et al,.* 1993). Another example of change occurs when the basal lamina is thickened in diabetes, leading to a decrease in the ability of anchoring fibrils to penetrate (Kenyon, 1983). These morphologic and biochemical changes may explain why the epithelium is often disrupted following vitrectomy surgery. These findings illustrate the importance of the surface chemistry of the basal lamina. Any design for a synthetic device requires that the surface chemistry of the anterior surface support the adherence and proliferation of epithelial cells and the deposition of extracellular matrix proteins. In contrast, the posterior surface should possess a surface chemistry that discourages cellular attachment and epithelial downgrowth.

A discussion on the epithelium is incomplete without at least a mention of the tear film. The plasma membrane of superficial epithelial cells consists of a hydrophobic lipid bilayer, which if unaltered would prevent the tear film from coating the apical cells evenly. However, mucin secreted by goblet cells of the conjunctiva, and glycoproteins secreted by epithelial cells, coat the epithelium, decrease the surface tension, and render it wettable (Rismondo *et al.,* 1989). As a result the tears spread over the apical epithelium and maintain an intact tear film for 20–30 sec between blinking. Abnormalities of the mucin–glycoprotein layer cause the tear film to break up into dry spots after resurfacing. In addition, the preocular tear film is hypertonic in comparison to the extracellular fluid within the corneal stroma, and when the epithelium is intact the fluid is drawn from the stroma. When the endothelial barrier function is compromised, the tear film may actually play a role in corneal dehydration.

STROMA

The stroma is the major structural component of the cornea and is composed of an extracellular matrix rich in collagen and sulfated proteoglycans. The distribution of proteins is not uniform from anterior to posterior, and the anterior part of the stroma contains less water than the posterior region (3.04 mg H_2O/mg dry weight vs. 3.85% mg H_2O/mg dry weight) (Castoro *et al.,* 1988). The change in hydration plays a role in corneal edema (Kangas *et al.,* 1994). The major cells of the stroma, keratocytes, are fibroblast-like and are dispersed throughout the stroma, contacting each other via gap junctions. These cells become myofibroblast-like after injury and express α-smooth muscle actin (Jester *et al.,* 1995). This cell phenotype can be "turned on" by high concentrations of transforming growth factor β1 (TGF-β1), suggesting that the bioavailability of growth factors in the stroma changes with injury and repair. Other evidence suggests that a fibroblast growth factor (FGF-2) and TGF-β are present transiently at the wound (Trinkaus-Randall and Nugent, 1998). *In vitro* experiments demonstrate that receptor binding and the TGF-β1 mRNA level are enhanced at the migrating cells along the wound margin (Song *et al.,* 1999). FGF-2, TGF-β1, and epidermal growth factor (EGF) mRNAs have been detected in the human stroma and in the aqueous humor (McAvoy and Charlain, 1990).

The major collagens of the stroma are types I, V, and VI, and XII (Cintron *et al.,* 1981; Poschi and Klaus, 1980; Doane *et al.,* 1991). The main collagens (type I collagen) are stacked in orderly sheets that form lamellae and regulate the fibril diameter of type V collagen and the elongation of type I collagen. The orthogonal array in the posterior cornea is caused by fibrils in one lamella running at right angles to an adjacent lamella. The lamellae in the anterior stroma display an oblique orientation, with bundles of fibrils interdigitating from one lamella to the next fibril. Proteoglycans, the other major component, maintain the spacing between collagen fibrils. The glycosaminoglycan (GAG) chains on the protein cores are sulfated and play a role in maintaining the cornea in a hydrated state. The two major proteoglycans are decorin, which contains chondroitin/dermatan sulfate side chains, and lumican, which possesses keratan sulfate side chains. Keratocan is another keratan sulfate proteoglycan that is cornea and sclera specific. Decorin binds to collagen fibrils and is hypothesized to act as a spacer of type I collagen fibrils. Furthermore, deletional analysis and computer modeling have shown that fibril organization is disrupted in the absence of decorin (Midura *et al.,* 1989; Cintron *et al.,* 1990; Funderburgh and Chandler, 1989; Weber *et al,.* 1996). Changes in the relative biochemical composition of the GAGs occur in response to injury. The change is seen as an increase in the relative synthesis of dermatan sulfate compared to that of keratan sulfate (Cintron *et al.,* 1981; Brown *et al.,* 1995). There is an accompanying change in sulfation and hydration, resulting in an alteration in the alignment of fibrils, leading to haze.

ENDOTHELIUM

Endothelial dysfunction can occur when there is an increase in the hydration of GAGs. The most clinically significant corneal edema occurs in the posterior stroma, where the ability to absorb H_2O is greater than that in the anterior stroma due to the higher concentration of keratan sulfate. In the posterior stroma the H_2O is not bound as tightly and consequently is easily released. On the other hand, the H_2O absorbed by the anterior stroma is bound more tightly, potentially explaining why anterior stromal edema is more persistent. The force required to prevent the stroma from swelling is called the swelling pressure (SP) and at a normal hydration the SP = 80 g/cm^2, or 55 mmHg (Hedsby and Dohlman, 1963). A healthy, intact endothelium counteracts the tendency toward hydrophilicity and maintains transparency by (1) possessing a barrier function that decreases the flow of H_2O into the stroma and (2) possessing a metabolic pump that transports ions from the stroma to the aqueous humor, with H_2O following by diffusion. Under normal conditions there is a balance between the pump and endothelial functions that maintains a stromal hydration of 78% and a stromal thickness of 0.52 ± 0.002 mm.

Table 35.1 Chemical composition of human aqueous humor and basic salt solution[a]

Component	Concentration (mmoles/liter) Aqueous humor	Basic salt solution
Sodium	162.9	155.7
Potassium	2.2–3.9	10.1
Calcium	1.8	3.3
Magnesium	1.1	1.5
Chloride	131.6	128.9
Bicarbonate	20.15	—
Phosphate	0.62	—
Lactate	2.5	—
Glucose	2.7–3.7	—
Ascorbate	1.06	—
Glutathione	0.002	—
Citrate	—	5.8
Acetate	—	28.6
pH	7.38	7.6
Osmolality (mOsm)	304	298

[a]As an irrigating solution.

Endothelial cells are present as a single layer of cells adherent to the posterior basal lamina, called Descement's membrane (Waring *et al.,* 1982). Endothelial cell number is monitored clinically using a specular microscope and is thought to decrease with age or trauma (Laing *et al.,* 1976). When cells are lost, the remaining cells reorganize by changing their shape while continuing to maintain a tight apposition with adjacent cells. This barrier function is composed of endothelial cells attached via discontinuous tight junctions located at the apical portion of the lateral plasma membrane. Gap junctions are located at the lateral plasma membranes. When there is a major loss in cell number the barrier becomes leaky and water from the aqueous humor passes into the stroma (Rao *et al.,* 1982; Yee *et al.,* 1985; O'Neal and Poise, 1986; Carlson and Bourne, 1988). The driving force of the metabolic pump is controlled by Na^+,K^+-ATPase, which is located in the lateral membranes and transports Na^+ from the stroma to the aqueous humor (McCartney *et al.,* 1987; Gersoki *et al.,* 1985). The tight junctions prevent bulk flow of fluid by keeping the Na^+,K^+-ATPase site located along the lateral membranes.

The function of the junctions can be evaluated clinically by measuring fluorescein permeability (Carlson and Bourne, 1988). Drugs and irrigating solutions applied to the eye have been shown to compromise the barrier function because the junctions are pH and Ca^{2+} dependent. Preservatives that function as antimicrobials or antioxidants that are found in solutions of epinephrine can harm the junctions when administered at high concentrations (Edelhauser *et al.,* 1976; Green *et al.,* 1977). This explains the importance of selecting irrigating solutions that resemble the aqueous humor (McDermott *et al.,* 1988) (see Table 35.1). This system of junctions and pumps may seem inefficient, but most nutrients necessary for the cornea are derived from the aqueous humor and must diffuse across. Therefore flexibility in the system is necessary.

REFRACTION

Different regions of the eye are composed of different materials (i.e., possess different indices of refraction), thus it is highly surprising that the cornea is transparent, because the light must strike tissue of varied structure and density, increasing the potential for light scatter. However, there is evidence that suggests that refractile elements (<2000 Å) of small dimensions, compared to the wavelength of light (<2000 Å), do not scatter light extensively (Farrell, 1994). In fact, investigators have demonstrated that only 1% of the incident light is scattered in the cornea. Transparency is regulated by the arrangement of collagen fibrils (the diameter of collagen fibrils is 300 Å and the fibrils are uniformly spaced 550 Å apart) and the transparency of keratocytes is due to cytoplasmic crystallin proteins. In addition, the changes in the refractive index that take place within the cornea occur over a distance less than half the wavelength of light. This evidence refutes the older hypothesis that light scatter was cancelled by destructive interference (Benedek, 1971).

SUMMARY

In summary, our goal is to explain why the cornea is an excellent candidate for tissue engineering and why the biologic constraints that impose restrictions on alterations of corneal shape and curvature, or the placement of a prosthesis, make the tissue very challenging to engineer. The need for synthetic corneas is real in all parts of the world.

CRITERIA FOR KERATOPROSTHESES

The transplantation of human donor corneal grafts is performed for patients needing a replacement cornea due to degenerated, scarred, or opacified corneas. As will be discussed later, as early as the 1700s synthetic corneas were designed and implanted in persons who had bilateral loss of eyesight such that the patients regained "finger-counting" vision. These surgeries were conducted prior to the era when corneas were available for performing corneal transplants. Modern techniques now allow donor corneal transplants, but in some countries transplantation of tissue remains culturally unacceptable or tissue is unavailable. Although transplantation or replacement of corneas (penetrating keratoplasty) has a high success rate, there remain problems. Persons affected by specific conditions, such as chemical burns, ocular pemphigoid, Stevens–Johnson syndrome, or recurrent graft rejections, have a poor level of success, with the donor cornea losing its transparency. Transmission of disease has also become a factor, although it is a rare occurrence (Duffy *et al.*, 1974; Hort *et al.*, 1988). These factors have led to a recent increase in research and development of synthetic corneas. The first devices were designed to give a patient vision without consideration to biologic compatibility or aesthetics. Currently, the biologic constraints of the tissue are considered in the design and shape of the device.

Laboratories working on the design of synthetic devices have arrived at a series of criteria (Leibowitz *et al.*, 1994; Chirila, 1994a,b). To date the three major criteria agreed on by both clinicians and scientists are (1) that the device must remain stable in the eye for extended periods of time and must not act as an inflammatory or immunologic stimulus; (2) that the central optic must remain transparent and form a high-quality image at or near the retina, and (3) that the peripheral component must allow for the ingrowth of stromal keratocytes, the subsequent deposition of extracellular matrix proteins, and ultimately be "permanently anchored" in place. There are several additional criteria that should result in the development of a device that would require minimal clinical follow-up. The interface of the cells and material must remain colonized to develop sufficient tensile strength to maintain the device in place. The continuity will prevent aqueous humor leakage and inhibit epithelial downgrowth and eventual extrusion. The peripheral component must not become encapsulated and diffusion of nutrients must occur. A small number of investigators have additional goals in the development of the optimal device. The most important additional goal is to create an anterior surface that will be covered by a continuous confluent sheet of epithelium. The resulting epithelium must remain stable and capable of recovering the surface if it is injured. Ultimately coverage is thought to prevent epithelial downgrowth and enzymatic digestion. Methods to achieve this surface will be discussed. Epithelial surface covering is an additional goal and has been one of the most difficult challenges. As a result, many groups have decided alternatively to clean the anterior surface periodically to prevent protein deposits and epithelial downgrowth. Another goal is the modification of the surface chemistry of the posterior surface so that cells or protein do not attach to the posterior surface. Failure to modify the surface chemistry has resulted in the failure of several synthetic epikeratoplasty models.

One difference in the design of a synthetic cornea from that of prostheses intended for skin or other vascular tissue is the lack of use of biodegradable polymers. Although these polymers have certain advantages, to date they have not been able to fit within the biologic constraints imposed by the cornea. Many of the mechanisms of repair are similar to those of skin and other ectodermally derived tissues, but the cornea must remain avascular to be transparent. Because the center must remain transparent, there are no known techniques to ensure that the central transparent material can be held in place if the peripheral porous skirt material degrades and fails to hold the device in place. The only potential porous material is that used by Legeais's group; the material they use becomes translucent after fibroplasia or when the spaces between fibers are filled with fluid (Legeais *et al.,* 1994).

DEVELOPMENT OF THE KERATOPROSTHESIS

The successful development of a synthetic cornea (keratoprosthesis) has been a difficult goal to accomplish. The major reason for the failure of keratoprostheses was extrusion (the spontaneous removal of a device from the cornea). This was generally preceded by erosion, necrosis of adjacent tissue, inflammation, and epithelialization of the device. Other complications that arose were secondary glaucoma, retroprosthetic membranes, retinal detachment, and ultimately poor visual acuity (Dohlman *et al.,* 1974). Unfortunately, the development of the device has met with many setbacks and its advances have been separated by some 50 years. The recent advance is correlated with an increase in the understanding of the biology of the eye and developments in polymer science.

In the late 1700s one of the first documented efforts at inserting a device was performed by Pellier de Quengsy (1789). A glass plate was inserted into an opaque eye. The insertion of glass plates into corneas was also reported in the 1800s. As one would predict, the glass plate was not stable and quickly extruded. Over time, efforts were made to stabilize the central optic with celluloid plates and egg membranes (Dimmer, 1891). Plexiglass and acrylate (Binder and Binder, 1956; Gyorffy, 1951), or supporting plates of solid or fenestrated poly(methyl methacrylate) (PMMA) (Castroviejo *et al.,* 1969). At that time, although an effort was made to support the central optic, there was no effort made to achieve the correct refractive index or to optimize the corneal curvature. The peripheral constructs were not biocompatible but increased the surface area available to secure the cornea to the device. As a consequence, cellular ingrowth into the device did not occur and the device was not anchored. These initial prostheses led to the development by Cardona (1964) of "through and through" devices.

Cardona was a pioneer in the development of keratoprostheses and demonstrated that highly purified and polymerized materials could be maintained in the cornea. The materials first used by him were rigid and impermeable and were secured with sutures. Neither keratocyte ingrowth nor deposition of matrix proteins was reported. In fact, some of the devices that he designed protruded through the eyelid and used the surrounding tissue for stabilization. Devices of this style are currently used when lubrication of the cornea is minimal. Consequently the device never "healed in place." More importantly, the devices often gave the person "finger-counting" vision. The more current Cardona devices are complex and include a PMMA optic surrounded by a perforated skirt of poly(tetrafluoroethylene) (PTFE) reinforced with a mesh composed of poly(ethylene terephthalate) (Dacron) and autologous tissue.

The complexity of the device is matched by the surgery—the device is inserted through the eyelid, often requiring the removal of the lens, iris, and ocular muscles. Because there is no UV-absorbing tissue remaining, the optic is tinted to absorb UV irradiation. After all of this, the device often gives no more than 30° of visual field. The two-compartment model, consisting of a core and a peripheral skirt, was designed to reduce extrusion, but the devices continued to be extruded. One of the underlying reasons was that the material did not encourage cellular ingrowth and the synthesis of matrix proteins. In contrast, the free edge permitted necrosis and allowed for epithelial downgrowth. The latter occurs because epithelium preferentially follows the stroma (rather than migrate onto the anterior surface), resulting in extrusion. Attempts to achieve the interpenetration or anchoring of the material biologically in place led to the development of both the osteo-odontoprosthesis and other porous polymers for the outer rim by Strampelli and Marchi (1970).

The premise of the Italian group was that a biologic material would be more easily tolerated than a synthetic material. To manufacture their device, an acrylic optical core was embedded within the core of a tooth. The insertion of the device was complicated and was often accompanied by infiltration with vascularization, abscess formation, and extrusion. Blencke *et al.* (1978) altered the osteo-odontoprosthesis by replacing the tooth portion with a glass ceramic. Caizza *et al.* (1990) implanted a modified device using bone. The main concern regarding these devices is the high potential for bone resorption that diminishes not only the strength of the device but also the tensile strength between the device and surrounding tissue. One device examined 12 years after implantation did suffer from extensive bone resorption. Although the concerns are valid, these devices are surgically implanted into humans. Taloni and his collaborators (1995) in Italy have altered the complexity of the operation and followed the central vision function in patients with these devices. To date there is no reported long-term follow-up *in vivo*. Although this is an interesting concept there are major drawbacks as to the long-term durability of the device and the extensiveness of the surgery.

RECENT DEVELOPMENTS OF POROUS MATERIAL

By the 1980s investigators proposed that acceptance of the device and lack of extrusion depended on the ability of stromal fibroblasts to migrate into the device and deposit a matrix. Fibrovascular ingrowth into Proplast was demonstrated by Barber *et al.* (1980), into a melt-blown poly(butylene):poly(probylene) web by Trinkaus-Randall *et al.* (1991), and into expanded poly(tetrafluoroethylene) (Impra) by Neel (1983). Kain (1990) developed a model in which carbon fibers projected from the optic haptic as rays and demonstrated that keratocytes migrated along the fiber. However, the carbon fibers were friable. Barber *et al.* (1980) were the first to evaluate the biologic response of the cornea to Proplast, which ranges in pore size from 100 to 500 μm. Severe inflammation was not detected over a period of 4 months *in vitro*. Although collagen was identified in histologic specimens, extrusion from interlamellar pockets occurred unless conjunctival flaps were used. In fact, when Proplast was used as the skirt material for intact devices there was an 18% extrusion rate from corneas after 6 months. White and Gona (1988) developed a similar keratoprosthesis with a central cylinder of poly(methyl methacrylate) and a periphery of Proplast, with pore sizes ranging from 80 to 400 mm. The two materials were joined by either cyanoacrylate or by "screwing" the PMMA into the trephined Proplast disk. They, too, reported a high extrusion rate, indicating that the device was not anchored within the stroma. The use of Proplast was discontinued because of its inflammatory responses when used in other devices.

The development of the poly(butylene):poly(propylene) web by Trinkaus-Randall *et al.* (1990) was based on surgical mesh fabrics used in skin, muscle, and arterial wall (Beahan and Hull, 1982). The optimal pore size determined in these tissues was used as the base line for determining the optimal pore size in the cornea (Bobyn *et al.*, 1982; Pourdeyhimi, 1989; Cook *et al.*, 1989). Trinkaus-Randall *et al.* (1990, 1991) evaluated fibroblast ingrowth into many webs with different pore and fiber sizes. A fiber diameter that ranged from 2 to 12 μm with a void volume of 88% was found to be optimal in *in vitro* and *in vivo* studies (Fig. 35.3A). Scanning electron micrographs demonstrated that keratocytes migrated into the interstitial spaces and along the fibers (Fig. 35.3B) (Trinkaus-Randall *et al.*, 1990, 1991). To achieve the optimal stability *in vivo* the polyolefinic webs (3M; St. Paul, MN) were modified with antioxidants to prevent the breakdown of fibers and maintain the strength and resiliency of the stroma–device interface so that the implantation of the device would not weaken the strength of the cornea (Trinkaus-Randall *et al.*, 1992). The porous material was placed in intrastromal pockets and evaluated over 2 years and neither degradation nor extrusion was detected. Several years later Legeais *et al.* (1994) evaluated a skirt material fabricated from a porous PTFE (Impra) (pore size of 18–22 μm) that permitted fibroplasia. To date this group has evaluated the retention of a synthetic cornea in human patients (skirt, 200 μm thick; pore size, 80 μm) (Legeais *et al.*, 1995a).

Trinkaus-Randall's group took a different approach and characterized and quantitated the biologic response to each material separately under *in vitro* and *in vivo* conditions. As a consequence the biologic response of the cornea was analyzed and the biochemical composition of the stroma and epithelium was studied in normal and wounded conditions. Using these parameters we have been able to determine whether the tissue responded as predicted under normal trauma or if there was a change in the regulation of proteins. There are now several groups extending this approach and the data will help enhance the overall understanding of the response of the cornea to synthetic materials.

Evaluating the response to the porous skirt material was performed using an intrastromal pocket whereby the cells migrate into the disk at its anterior, posterior, and lateral edges (Fig. 35.4). This requires cells to move into an empty zone (unlike a wound bed) and deposit a matrix. Fibroplasia is assessed using a combination of phase-contrast and fluorescent microscopy and biochemical analysis. To evaluate the deposition of collagen and proteoglycans, both the accumulated matrix and the matrix synthesized at various times up to 3 months was determined. Type I

Fig. 35.3. (A) A scanning electron micrograph of the porous material [a poly(butylene):poly(propylene) melt-blown web, 80:20] used in the construct. (B) A scanning electron micrograph of fibroblasts migrating along and between fibers (arrowheads).

collagen was determined from amino acid analysis of acid hydrolysates of the disks at various times and compared to intact and wounded corneas (Trinkaus-Randall *et al.,* 1991). The collagen, calculated as a ratio of hydroxyproline to proline, was standardized to total protein. Trinkaus-Randall *et al.* (1991) demonstrated that the collagen deposited in the disk did not increase significantly until day 21.

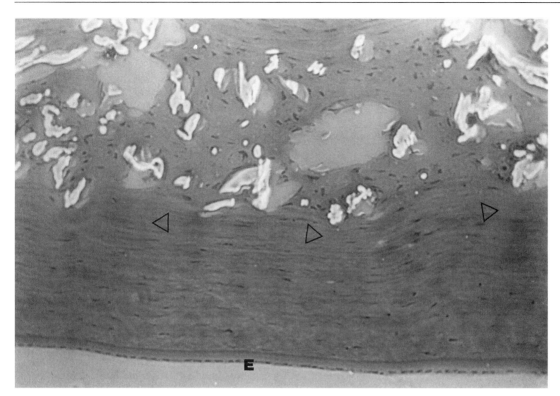

Fig. 35.4. Phase-contrast micrograph of the porous construct inserted between collagen lamellae and evaluated after 42 days. The micrograph is a cross-section of the cornea. Note that the endothelium (E) is intact. The porous material is white. Cells have migrated from the posterior stroma (open arrowheads) and have deposited an extracellular matrix. Open holes indicate regions where deposition of matrix is incomplete.

These investigators also analyzed the optimum pretreatment regimes of the porous materials and found that by day 28 there was less edema and 5.5-fold more collagen deposited when the disks were preseeded with keratocytes, compared to untreated disks. The change could not be attributed to the collagen present at the time of implantation, because the values of collagen in all cases (no pretreatment, 0.05 mg/disk; collagen type I, 0.07 mg/disk; preseeding of cells, 0.04 mg/disk; collagen type I and preseeding of cells, 0.05 mg/disk) were negligible at 8 days (Trinkaus-Randall *et al.*, 1994). When the deposition and synthesis of glycosaminoglycans were evaluated

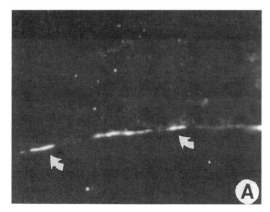
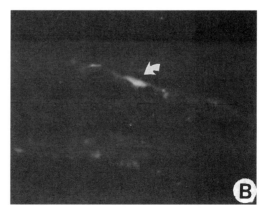

Fig. 35.5. Immunofluorescent micrograph of a cross-section of the stroma where the porous construct has been inserted. Frozen sections of the cornea were stained with antibodies to (A) FGF-2 and (B) TGF-β. Note that the growth factors were detected at the margin between the disk and the stroma (white arrows).

they found that the migration of cells into the disks mimics the scenario of wound repair, but took longer to occur (Brown *et al.*, 1995). The change in dermatan sulfate into an iduronate form and the decrease in the relative amount of keratan sulfate are hallmarks of stromal repair (Funderburgh and Chandler, 1989; Cintron *et al.*, 1990). Interestingly, heparan sulfate, which is absent in control corneas, was detected, indicating that growth factors (such as FGF-2, which binds heparan sulfate) may play a role in stromal wound repair. Specifically, Trinkaus-Randall and Nugent (1998) demonstrated that although FGF-2 was not detected immunohistochemically in unwounded stroma, it was present transiently at the margin between the disks and stroma (Fig. 35.5). Other developers of synthetic corneas have begun to assess the biological response using similar criteria. Legeais's group (Legeais *et al.*, 1994) examined fibroplasia using electron microscopy and Drubaix (Drubaix *et al.*, 1995) reported changes in collagen types after insertion of synthetic material into the stroma. Chirila's group has been examining the production of collagen in hydrogel sponges and reported that the hydrogel sponge did not possess a high enough tensile strength until cellular ingrowth occurred (Chirila *et al.*, 1993).

RECENT DEVELOPMENTS OF OPTIC MATERIAL

The development of optic material has had a long and laborious history. As mentioned previously, the use of glass and other polymers has been attempted. The material used for the central optical material must allow light transmission in the visible range. However, to protect the retina from ultraviolet damage the material needs to be modified to absorb light at wavelengths below 300 nm. This is even more critical in devices in which the lens has been removed. Optimally, the optic should have tensile properties similar to that of a cornea and allow for continuous reshaping of both anterior and posterior curvatures. The material must either be impermeable to keratocytes or the index of refraction must not be altered. The optic must support the diffusivity of nutrients and should either support epithelial cells or be capable of being modified to support and maintain epithelial cells.

The most commonly used material, PMMA, has both advantages and disadvantages. It transmits light, has been made in ultraviolet-absorbing forms for intraocular lenses, can be shaped, retains clarity with age, exhibits low toxicity, and is well tolerated as long as polymerization has been carried to completion. Its disadvantages include its rigidity and hydrophobicity, its lack of diffusivity, and the subsequent potential lack of supporting epithelial cells.

One of the earlier materials that continued to be used by Van Andel *et al.* (1988) is glass held in place with a platinum holder. The material was designed as a low-cost device. Glass has many of the same advantages and disadvantages as PMMA yet requires more frequent clinical follow-up.

Another polymer, silicone, was used in temporary devices by Duncker and Eckardt (1988) and in designs intended for permanent devices by Kain (1990). In the latter design the silicon optic was supported by peripheral carbon fibers. Legeais *et al.* (1994) discussed the use of a silicon optic coated with poly(vinylpyrrolidone) (PVP). Their evaluation of the model did not include any mention of reepithelialization. However, this group has made advances in shaping the optic.

Other potential candidates are the hydrogels, which eliminate many of the disadvantages of PMMA. Hydrogels are elastic and can be manufactured to have a high tensile strength. They allow for the diffusivity of nutrients, maintain transparency, and can be modified to support epithelial cell growth by coating, by altering the composition of the hydrogel, or by chemically modifying the surface (Trinkaus-Randall *et al.*, 1994; Rosen, 1976; Wu *et al.*, 1994; Latkany *et al.*, 1997). Poly(2-hydroxyethyl methacrylate) (pHEMA) has been used extensively in soft contact lenses. The laboratory of Chirila (1994a,b) developed a device in which both components were cross-linked polymers of HEMA. The central optic was a HEMA-2-ethoxyethyl methacrylate acrylamide. Epithelial downgrowth was observed along with stromal necrosis in these devices, so, unfortunately, the device did not support epithelialization. Tensile strength was not increased above the strength of the homopolymer when the HEMA was copolymerized with 4-*t*-butyl-2-hydroxycyclohexyl methacrylate (Chirila *et al.*, 1995). More recently, there have been advances in the development of intact devices, which will be discussed later.

A poly(vinyl alcohol) hydrogel (refractive index, 1.42) was developed for use as an optic by Trinkaus-Randall *et al.* (1988). The hydrogel was prepared from a poly(vinyltrifluoroacetate) precursor (Ofstead and Posner, 1987). The precursor was molded and attached to the porous component using either thermal fusion or a solvent adhesive. After conversion to a poly(vinyl alcohol), the hydrogel retained the shape of the precursor. This hydrogel possessed a higher elasticity (436 vs. 15%) and tensile strength (110 vs. 35 kg/cm^2) than that of the intact cornea. It possessed a water content around 70%, which is similar to the water content of the cornea (79%). The high water content and the structure of the hydrogel allow for the diffusivity of ions and glucose through the material. The elasticity is important, because it is critical that the material reseals after passage of a suture. When ultraviolet absorbers (phenylbenzotriazole and a benzophenome derivative) were copolymerized, transmission of ultraviolet light was not detected below 340 nm. (Tsuk *et al.,* 1997). The surface of the poly(vinyl alcohol) hydrogel also has the ability to be modified to support epithelial cell growth and maintenance. Trinkaus-Randall and co-workers (Wu *et al.,* 1994; Latkany *et al.,* 1997) and Laroche *et al.* (1989) studied the response of epithelial cells to hydrogels. In the laboratory of Trinkaus-Randall, when the surfaces were coated with extracellular matrix proteins and the epithelium was cultured, the cells became confluent, multilayered, displayed typical cytoarchitecture, showed a basal–apical polarity, and synthesized a typical repertoire of adhesion proteins (Fig. 35.6). However, when the disks were implanted *in vivo* the apical surface of the cultured cells was altered, with the shearing force associated with blinking (Fig. 35.7, A and B).

Techniques have been used to modify the surface chemistry of the poly(vinyl alcohol) hydrogels with argon radio frequency (RF) plasma treatments. This treatment repeatedly enhanced cell adhesion and proliferation *in vitro* and *ex vivo,* and limbal epithelium migrated over the treated surfaces of implanted devices (Fig. 35.8) (Latkany *et al.,* 1997). This was an exciting observation but the long-term prognosis *in vivo* is not known at this time.

The hydrogel evaluated by Laroche *et al.* (1989) *in vitro* has a refractive index of 1.368, is oxygen permeable, and also demonstrated diffusivity of nutrients. Their group conducted preliminary

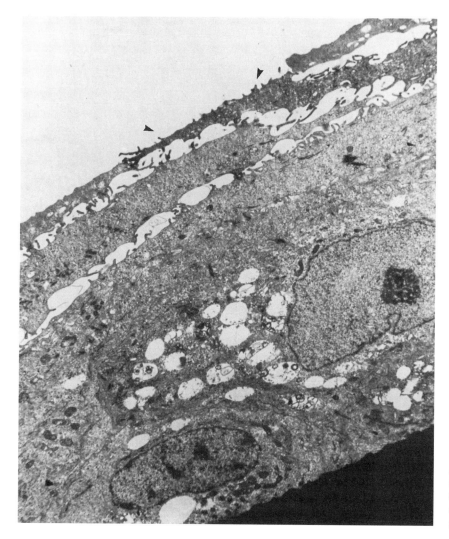

Fig. 35.6. Transmission electron micrograph of epithelial cells cultured on a hemagel, with the surface treated with matrix proteins. The cells are multilayered and the apical cells have characteristic microvilli (arrowheads).

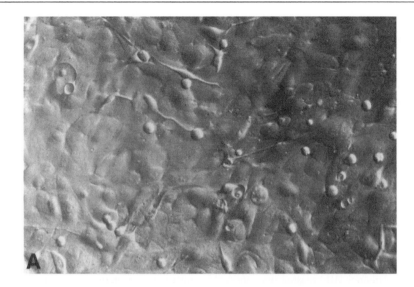

Fig. 35.7. (A) Phase-contrast micrographs of confluent cells cultured on HEMA modified with plasma treatments. Cells maintain typical epithelial cytoarchitecture. (B) HEMA constructs sutured into corneas as a keratoplasty, demonstrating the roughening of the apical surface that occurs in the absence of tear film.

experiments using the adhesion peptide, RGD; however, no results have been described *in vivo.* The use of the adhesion peptide has been shown to be successful *in vitro* by Massia and Hubbell (1991), with cells showing varied morphologic responses to small concentrations of peptide. These elegant studies helped to explain the cell surface interaction *in vitro,* but they do not ensure adhesivity and/or maintain cytoarchitecture *in vivo.* This has continued to be one of the most challenging tasks in the development of keratoprostheses.

EVALUATION OF INTACT DEVICES

Although some groups have evaluated the material components of the devices separately, other groups have fabricated intact devices and sutured them *in vivo.* The devices from many laboratories all have peripheral skirts, but the material, the method of bonding, and the surface treatment differ. The criteria for success in these devices are that they (1) have sufficient strength and resiliency to permit surgical manipulation, (2) remain intact without any degradation at the interface between the porous and transparent haptics, (3) can be anchored in place via cellular ingrowth and protein deposition, (4) are not susceptible to enzymatic degradation, and (5) remain transparent over an extended time period. To date very little work has been conducted on molding the devices to achieve an optimal curvature or on attempting to correct vision by altering the refracting power of the optic. Instead, the advancements have been in the manufacture of a device that is retained in the cornea for extended periods of time and that does not result in epithelial downgrowth and stromal necrosis.

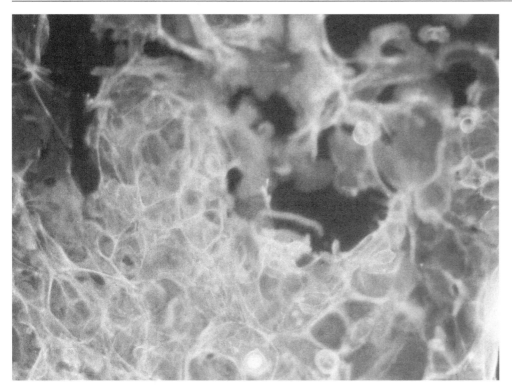

Fig. 35.8. HEMA constructs (radio frequency plasma modified) sutured into corneas ex vivo *and cultured for 3 weeks on a rotating surface. Epithelia are seen migrating onto the HEMA. Epithelia are labeled with fluorescein isothiocyanate—phalloidin, staining F-actin filaments.*

The results from Legeais' group are well worth reporting. The porous material is 200 μm thick with a pore size ranging from 50 to 80 μm (Legeais *et al.,* 1995a,b). Legeais reports that the PTFE porous peripheral skirt became translucent on fibroplasia. However, it has been observed that translucency of Impra is achieved when the pores of the material are filled with fluid and that any solvent can render the PTFE (Impra) translucent. The device was implanted in 24 human patients and five failures were reported in the first 6 months. The haptic was inserted into lamellar pockets and the optic was positioned through a hole made in the central cornea. The devices gave the patients a visual field of 110–130°, far superior to the earlier models. In a second model reported from this group the optic was silicon coated with PVP and the junction between the two polymers was produced chemically. They found that after implantation in rabbits the silicon remained hydrophilic and the IOP could be measured with a Goldman tonometer. Their studies lasted for 3 months. Unfortunately, the design of the device is not available for examination. Several other designs have been developed and implanted, and these aim to enhance fibroplasia and minimize downgrowth. These devices have been implanted in eyes with dystrophy, pseudophakic edema, pemphigus, and burns. Unfortunately, signs of decolonization 9–12 months after implantation have been noted. Dohlman, who has been instrumental in the development of keratoprostheses, has reevaluated which conditions have the best prognosis and what factors can be used as predictors of outcome (Dohlman, 1997). Dohlman has continued to use a plastic device that is not colonized; however, he has demonstrated that with meticulous follow-up the patients can have an excellent outcome (Table 35.2). He found that eyes with graft failure had the best outcome, whereas those with chemical burn and Steven–Johnson syndrome had the worst outcome. Patients with pemphigus ranged in the middle in terms of predictability of outcome. In addition, Dohlman has demonstrated that in some dry-eye patients the device must be implanted through the lid.

Another laboratory in China (Chang *et al.,* 1995) used a silicone membrane altered by plasma-induced graft copolymerization-treated HEMA. Chang's device was sutured in place using the same procedure used in corneal transplantation (penetrating keratoplasty). They found that the graft did not remain in the place of the host cornea and was extruded by 3 months. It is likely that stromal downgrowth occurred, i.e., the fibroblasts grew under and essentially formed a second cornea, gradually extruding the cornea. It is also possible that epithelial downgrowth occurred, aiding in the extrusion of the device. No other reports from this group can be found, so it is difficult to do more than hypothesize about the reasons for failure.

Table 35.2. Factors that improve success of retention

Temporary tissue coverage with conjunctiva
Applications of collagenase inhibitor
Implementation of YAG laser
Glaucoma shunt
New repair technology
Meticulous and frequent follow-up of individuals

Chirila's group has employed an exciting idea whereby both the peripheral rim and the central optic are cross-linked polymers of HEMA, with the advantage that both are chemically similar. In this model the periphery was made from a hydrogel sponge produced by phase-separation polymerization in aqueous solution. The sponge has a high water content (80% w/w), facilitating

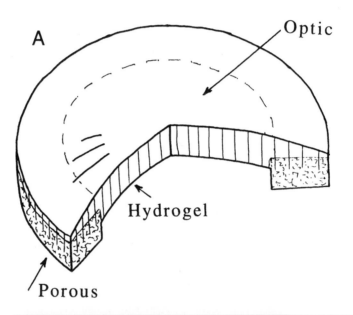

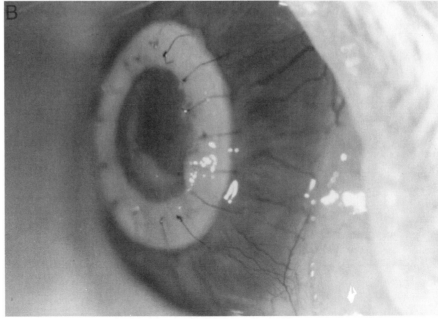

Fig. 35.9. (A) Design of a synthetic corneal construct to be sutured into the cornea as a full-thickness device, showing optical and porous regions. (B) Side view of device placed into the cornea and held in place with single sutures around the circumference of the device until the porous region (white) is anchored, after keratocyte invasion and matrix deposition.

A

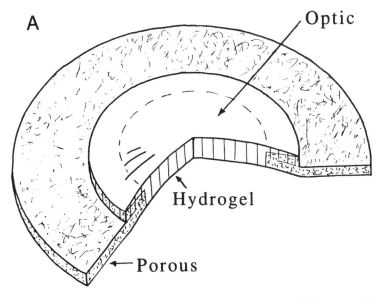

Optic

Hydrogel

← Porous

B

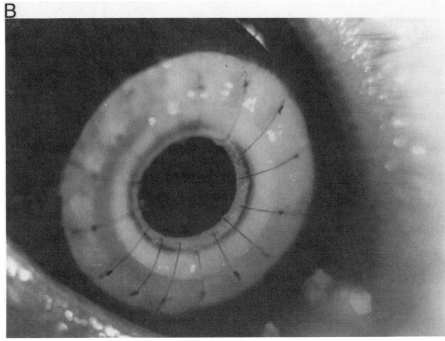

Fig. 35.10. (A) Schematic of a construct to be sutured into the cornea as a partial-thickness device. Note the flange of the porous material that extends into the stroma between lamellae. (B) View of a partial-thickness device placed into a stromal bed and secured in the cornea with a series of individual sutures. Fibroplasia occurs more rapidly in this model and inflammation is less, compared to the full-thickness device.

diffusivity through the material. Fibroplasia was detected when the devices were pretreated with type I collagen (Chirila *et al.,* 1993; Chirila, 1994a,b). More recently they have detected ingrowth of keratocytes in compromised corneal models; however, epithelialization remains a challenge.

We have produced several models, varying in size (4.5–9 mm) and design (Trinkaus-Randall *et al.,* 1997). The devices were evaluated in rabbits using both penetrating keratoplasty and lamellar keratoplasty models, whereby the device was implanted leaving Descemet's membrane intact. The time course of fibroplasia depended on the model and on the size of the porous skirt. The response to the device sutured in as a penetrating keratoplasty was the least successful. In this model the device has no flanges to increase the surface area for fibroplasia (Fig. 35.9). In the models in which Descemet's membrane was left intact, the device was placed as posteriorly as possible in the stromal lamellar bed and the flanges were inserted into the stromal lamellae (Fig. 35.10). Fibroplasia was detected as early as 15 days after surgery and as late as 21 days, the difference being the diameter of the skirt. Although fibroplasia occurred earlier when the area of the skirt was increased,

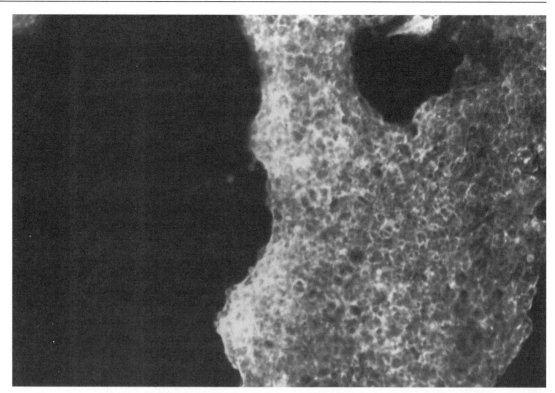

Fig. 35.11. Epithelialization occurs in the partial-thickness models (Fig. 35.10). It is generally seen after 3 weeks. The epithelium that has migrated from the limbus has been stained with fluorescein isothiocyanate—phalloidin, staining F-actin, and delineates epithelial cytoarchitecture.

the cornea could not support the device in the long term, because there was less diffusivity of nutrients. Smaller models (7 mm in diameter) have been retained in rabbit corneas for a minimum of 1 year with epithelialization in some of the cases (Fig. 35.11). Collagen types III and VI were detected after 6 weeks using immunohistochemical techniques.

Interestingly, heparan sulfate absent in the unwounded stroma was detected and coincided with the appearance of growth factors. The appearance of heparan sulfate suggested that proteoglycans might modulate activity of growth factors so that those cells would become responsive to FGF-2 (Trinkaus-Randall and Nugent, 1998). In all of these models the anterior chamber remained intact and no signs of leakage was detected.

These observations led us to develop methods of control release to enhance the rate of fibroplasia and wound repair. Experiments were conducted to demonstrate that small volumes of sodium alginate injected into the stroma without cytokines did not elicit an inflammatory response over a period of 14 days (a typical time for response in a rabbit cornea). There was also no long-term perturbation of collagen lamellae. Injection of FGF-2 elicited vascularization whereas TGF-β did not. Release of ^{125}I-labelled FGF-2 and TGF-β from alginate gels was measured in *ex vivo* and *in vivo* experiments and a first-order release was detected for a period of 3–5 days. However, there was a large vascularization response. Poly(lactic–co-glycolic acid) (PLGA) microspheres were also used and we demonstrated that 1.3% of the encapsulated TGF-β was released/day and the release persisted with nearly linear kinetics for 4 weeks. The released growth factor possessed biologic activity. Our goal is to place PLGA capsules containing TGF-β into porous materials and insert them into the stromas and evaluate the release. This would allow us to understand the degradation in PLGA, release of growth factor, and change in diffusivity that occurs within the stroma (Trinkaus-Randall and Nugent, 1998).

The advances in the past 5 years have accelerated our level of knowledge and resulted in several models that are currently being evaluated *in vivo*. Long-term studies are needed to follow fibroplasia of the porous material and epithelialization of the anterior surface in the models that have been described.

OTHER DEVICES USED TO CHANGE THE CURVATURE OF THE CORNEA

To date many technologies have been developed to change the curvature of the cornea. The scope of the field is too vast to cover in this chapter. Briefly, the recent procedures that are commonly used are laser *in situ* keratomileusis (LASIK) and the insertion of rings into the stroma. These developments stem from the radial keratotomy developed in Russia and subsequent excimer laser radial keratotomy and photoreactive radial keratotomy to correct severe myopic patients (as reviewed by Waring and the PERK) (Brint *et al.*, 1994; Waring, 1992, 1998; Nizam *et al.*, 1992; Sastry *et al.*, 1993). Laser *in situ* keratomileusis combines several techniques, but changes the curvature without removing the epithelium or Bowman's membrane. The goal is to avoid epithelial hyperplasia and haze, which have been typical sequelae of these procedures. This procedure is now considered an acceptable treatment after a corneal transplant when vision cannot be rehabilitated with eyeglasses or contact lenses (Ophthalmology Times, 1999). A more reversible procedure to alter curvature is the insertion of intrastromal corneal ring segments. The rings are fabricated from PMMA and are inserted into the lamellae. The advantages of these devices are that they can be removed and the cornea has not been reshaped in the procedure.

ACKNOWLEDGMENTS

Departmental grants have been received from Research to Prevent Blindness and Lion's Eye Research Foundation and NIH-EY-06000. I would also like to thank Dan Orlow for his schematic drawing of the cornea, and all of the present and past members in my laboratory and Matthew Nugent, for their work, insight, and critical discussions.

REFERENCES

Barber, J. C., Feaster, F. T., and Priour, D. J. (1980). Acceptance of aitreous-carbon alloplastic material, proplast, into the rabbit eye. *Invest. Ophthalmol. Visual Sci.* **19**, 182–190.

Beahan, P., and Hull, D. (1982). A study of the interface between a fibrous polyurethane arterial prosthesis and natural tissue. *J. Biomed. Mater. Res.* **16**, 827–838.

Benedek, G. B. (1971). Theory of transparency of the eye. *Appl. Opt.* **10**, 459.

Binder, H. F., and Binder, R. F. (1956). Experiments on Plexiglas; corneal implants. *Am. J. Ophthalmol.* **41**, 793–797.

Blencke, B. A., Hagan, P., Bromer, H., and Deurscher, K. (1978). Study on the use of glass ceramics in osteo-odonto-keratoplasty. *Ophthalmologica* **176**, 105–112.

Bobyn, J. D., Wilson, G. J., MacGregor, D. C., Piliar, R. M., and Weatherly, G. C. (1982). Effect of pore size in the peel strength of attachment of fibrous tissue to porous-surface implants. *J. Biomed. Mater. Res.* **16**, 571–584.

Brint, S. F., Ostrick, D. M., Fisher, C., Slade, S. G., Maloney, R. K., Epstein, R., Stulting, R. D., and Thompson, K. P. (1994). Six-month results of the multi-center phase I study of excimer laser myopic keratomileusis. *J. Cataract Refract. Surg.* **20**, 610–615.

Brown, C. T., Applebaum, E., Banwatt, R., and Trinkaus-Randall, V. (1995). Synthesis of stromal glycosaminoglycans in response to injury. *J. Cell. Biochem.* **59**, 57–68.

Caizza, S., Falcinelli, G., and Pintucci, S. (1990). Exceptional case of bone resorption in an osteo-odontokeratoprosthesis. A scanning electron microscopy and x-ray microanalysis study. *Cornea* **9**, 23–27.

Cardona, H. (1964). Plastic keratoprosthesis—A description of the plastic material and comparative, histologic study of recipient corneas. *Am. J. Ophthalmol.* **58**, 247–252.

Carlson, K. H., and Bourne, W. M. (1988). The clinical measurement of endothelial permeability. *Cornea* **7**, 183.

Castoro, J. A., Bettelheim, A. A., and Beitelheim, F. A. (1988). Water concentration gradients across bovine cornea. *Invest. Ophthalmol. Visual Sci.* **29**, 963.

Castroviejo, R., Cardona, H., and Devoe, A. G. (1969). The present status of prosthokeratoplasty. *Trans. Am. Ophthalmol. Soc.* **67**, 207–231.

Chang, P. C. T., Lee, S. D., and Huang, J. D. (1995). Biocompatibility of an artificial corneal membrane *in vivo* animal study. *Invest. Ophthalmol. Visual Sci. Suppl.* **36**, 1467.

Chirila, T. V. (1994a). Pat.

Chirila, T. V. (1994b). Modern artificial corneas; the use of porous polymers. *Trends Polym. Sci.* **2**, 296–300.

Chirila, T. V., Constable, I. J., Crawford, G. J., Vlayasekaran, S., Thompson, D. E., Chen, V. C., Fletcher, W. A., and Griffin, B. J. (1993). Poly92-hydroxyethyl methacrylate) sponges as implant materials: *In vivo* and *in vitro* evaluation of cellular invasion. *Biomaterials* **14**, 26–38.

Chirila, T. V., Dao-Yi Yu, Yi-Chi Chen, and Crawford, G. (1995). Enhancement of mechanical strength of poly(2-hydroxyethylmethacrylate) sponges. *J. Biomed. Mater. Res.* **29**, 1029–1032.

Cintron, C., Hong, B., and Kublin, C. (1981). Quantitative analysis of collagen from normal developing corneas and corneal scars. *Curr. Eye Res.* **1**, 1.

Cintron, C., Gregory, J. D., Damie, S. P., and Kublin, C. (1990). Biochemical analyses of proteoglycans in rabbit corneal scars. *Invest. Ophthalmol. Visual Sci.* **31**, 1975–1981.

Cook, S. D., Barrack, R. L., Thomas, K. A., and Haddad, R. J. (1989). Quantitative histological analysis of tissue growth into porous total knee components. *J. Arthroplasty Suppl.* pp. 533–542.

de Quengsy, P. (1789). "Précis au cours d'opérations sur la chirugie des yeux." Didot.

Dimmer, F. (1891). Zwei Falle von Celluloidplatten der Homhear. *Klin. Monatsbl. Augenheilkd.* **29**, 104–105.

Doane, K., Vang, G., and Birk, D. (1991). Corneal type VI collagen may function in cell matrix interactions during stromal development. *Invest. Ophthalmol. Visual Sci.* **32**, 1012.

Dohlman, C. H. (1997). Keratoprostheses. *In* "Cornea" (J. H. Krachmer, M. J. Mannis, and E. J. Holland, eds.), Vol. 3, pp. 1855–1863. Mosby Year Book, St. Louis, MO.

Dohlman, C. H., Schneider, H. A., and Doane, M. G. (1974). Prosthokeratoplasty. *Am. J. Ophthalmol.* **77**, 694–700.

Drubaix, I., Legeais, J. M., Menasche, M., Savoldelli, M., Robert, L., Renard, G., and Pouliquen, Y. (1995). Collagen synthesis in an microporous ePTFE implanted in rabbit corneal stroma. *Invest. Ophthalmol. Visual Sci. Suppl.* **36**, 98.

Duffy, P., Wolf, J., Collins, G., Devoe, A. G., Streeten, B., and Cowen, D. (1974). Possible person to person transmission of Creutzfeld–Jakob Disease. *N. Engl. J. Med.* **290**, 692.

Duncker, G., and Eckardt, C. (1988). Modified temporary keratoprosthesis in the triple procedure: A new surgical technique. *J. Cataract Refract. Surg.* **14**, 434–436.

Edelhauser, H. F., Van Horn, D. L., Miller, P., and Pederson, H. J. (1976). Effect of thiol-oxidation of glutathione with diamide on corneal endothelial function, junctional complexes, and microfilaments. *J. Cell Biol.* **68**, 567.

Farrell, R. A. (1994). Corneal transparency. *In* "Principles and Practice of Ophthalmology" (D. M. Albert and F. A. Jacobiec, eds.), Vol. 6, pp. 64–81. Saunders, Philadelphia.

Funderburgh, J. L., and Chandler, J. W. (1989). Proteoglycans of normal corneas with nonperforating wounds. *Invest. Ophthalmol. Visual Sci.* **30**, 435.

Gersoki, D. H., Matsuda, M., Yee, R. W. (1985). Pump function of the human corneal endothelium: Effects of age and corneal guttata. *Ophthalmologica* **92**, 759.

Gipson, I., and Sugrue, S. P. (1994). Cell biology of the corneal epithelium. *In* "Principles and Practice of Ophthalmology" (D. M. Albert and F. A. Jakobiec, eds.), pp. 3–16. Saunders, Philadelphia.

Green, K., Hull, D. S., Vaughn, E. D., Malizia, A. A., and Bowman, K. (1977). Rabbit endothelial response to ophthalmic preservatives. *Arch. Ophthalmol. (Chicago)* **95**, 2218.

Gyorffy, L. (1951). Acrylic corneal implants in keratoplasty. *Am. J. Ophthalmol.* **34**, 757–760.

Hanna, K. D., Hayward, J. M., Hagen, K. B., Simon, G., Parel, J. M., and Waring, G. O. (1993). Keratometry for astigmatism using an arcuate keratome. *Arch. Ophthalmol. (Chicago)* **111**, 998–1004.

Hedsby, B. U., and Dohlman, C. H. (1963). A new method for the determination of the swelling pressure of the corneal stroma *in vitro*. *Exp. Eye Res.* **2**, 122.

Hort, R. H., Pflugfelder, S. C., Ulman, S., Forster, R., Poack, F. M., and Schifl, E. R. (1988). Clinical evidence for hepatitis B transmission resulting from corneal transplantation. *Ophthalmologica Suppl.* **95**, 162.

Jester, J. V., Peiroll, W. M., Barry, P. A., and Cavanaugh, H. D. (1995). Expression of alpha-smooth muscle actin during corneal stromal wound healing. *Invest. Ophthalmol. Visual Sci.* **36**, 809–819.

Kain, H. L. (1990). A new concept for keratoprosthesis. *Klin. Monatsbl. Augenheilk* **197**, 386–392.

Kangas, T. A., Edelhauser, H. F., Twining, S. S., and O'Brien, W. J. (1994). Loss of stromal glycosaminoglycans during corneal edema. *Invest. Ophthalmol. Visual Sci.* **31**, Suppl.

Kenyon, K. R. (1983). Morphology and pathologic responses of the cornea to disease. *In* "The Cornea" (G. Smolin and R. A. Thoft, eds.), pp. 43–76. Little, Brown, Boston.

Laing, R. A., Sanscrom, M. M., Berrospi, A. R., and Leibowitz, H. M. (1976). Changes in the corneal endothelium as a function of age. *Exp. Eye Res.* **22**, 587.

Laroche, L., Honiger, J., Thenot, J. C., Scarano, M., and Assouline, M. (1989). Keratophakie alloplastique. *Ophthalmologica* **3**, 227–228.

Latkany, R., Tsuk, A., Sheu, M.-S., Loh, I.-H., and Trinkaus-Randall, V. (1997). Plasma surface modification of artificial corneas for optimal epithelialization. *J. Biomed. Mater. Res.* **34**, 29–37.

Lavker, R. M., Dong, G., Cheng, S. Z., Kudoh, K., Cotsarelis, G., and Sun, T. T. (1991). Relative proliferative rates of limbal and corneal epithelia: Implications for corneal epithelial migration, circadian rhythm and suprabasally located DNA-synthesizing keratinocytes. *Invest. Ophthalmol. Visual Sci.* **32**, 1864.

Legeais, J. M., Renard, G., Parel, J. M., Serdarevic, O., Mei-Mui, M., and Pouliquen, Y. (1994). Expanded fluorocarbon polymer for keratoprosthesis. Cellular ingrowth and transparency. *Exp. Eye Res.* **58**, 41–51.

Legeais, J. M., Renard, G., Parel, J. M., Salvoldelli, M., and Pouliquen, Y. (1995a). Keratoprosthesis with biocolonization microporous fluorocarbon haptic. Preliminary results in a 24-patient study. *Arch. Ophthalmol. (Chicago)* **113**, 757–763.

Legeais, J. M., Renard, G., Thevenin, D., and Pouliquen, Y. (1995b). Advances in artificial corneas. *Invest. Ophthalmol. Visual Sci. Suppl.* **36**, 1466.

Leibowitz, H. M., Trinkaus-Randall, V., Tsuk, A., and Franzblau, C. (1994). Progress in the development of a synthetic cornea. *Prog. Retinal Eye Res.* **13**, 605–621.

Massia, S. P., and Hubbell, J. A. (1991). An RGD spacing of 440 nm is sufficient for alpha3beta1-mediated fibroblast spreading and 140 nm for focal contact and stress fiber formation. *J. Cell Biol.* **114**, 1089–1100.

McAvoy, J. W., and Charlain, C. G. (1990). Growth factors in the eye. *Prog. Growth Factor Res.* **2**, 29.

McCartney, M., Wood, T. O., and McLaughlin, B. J. (1987). Immunohistochemical localization of ATPase in human dysfunctional corneal endothelium. *Curr. Eye Res.* **6**, 1479.

McDermott, M. L., Edelhauser, H. F., Hack, H. M., and Langston, R. H. S. (1988). Ophthalmic irrigants: A current review and update. *Ophthalmic. Surg.* **19**, 724.

Midura, R. J., Toledo, O. M. S., Yanagishita, M., and Hascall, V. C. (1989). Analysis of proteoglycan synthesis by corneal explants from embryonic chicken. *J. Biol. Chem.* **264**, 1414.

Miller, D., and Benedek, G. (1973). "Intracellular Light Scattering." Thomas, Springfield, IL.

Neel, H. B., III (1983). Implants of Gore-Tex; comparisons with Teflon coated polytetrafluoroethylene carbon and porous polyethylene implants. *Arch. Otolaryngol.* **109**, 427–433.

Nizam, A., Waring, G. O., Lynn, M. J., Ward, M. A., Asbell, P. A., Balyeac, H. D., Cohen, E., Culbertson, W., Doughman, D. J., Fecko, P. (1992). Stability of refraction and visual acuity during five years in eyes with simple myopia. The PERK Study Group. *Refract. Corneal Surg.* **8**, 439–447.

Ofstead, R. F., and Posner, C. J. (1987). Semi-crystalline polyvinyl alcohol hydrogels: Synthesis and characterization. *Proc. ACS Div. Polym. Mater.* **57**, 361–365.

O'Neal, M. R., and Poise, K. A. (1986). Decreased endothelial pump function with aging. *Invest. Ophthalmol. Visual Sci.* **27**, 457.

Ophthalmology Times (1999) April 1, Vol. 24, p. 1.

Poise, K. A., and Mandell, R. B. (1970). Critical tension at the corneal surface. *Arch. Ophthalmol.* **84**, 505.

Poschi, A., and Klaus, V. M. (1980). Synthesis of type V collagen by chick corneal fibroblasts *in vivo* and *in vitro*. *FEBS Lett.* **115**, 100.

Pourdeyhimi, B. (1989). Porosity of surgical mesh fabrics: New technology. *J. Biomed. Mater. Res.* **23**, 145.

Rao, G. N., Lehman, L. E., and Aquavella, J. V. (1982). Cell size-shape relationships in corneal endothelium. *Invest. Ophthalmol. Visual Sci.* **22**, 271.

Rismondo, V., Osgood, T. B., Learing, P., Haltenhauer, M. A., Ubels, J. L., Edelhauser, H. F. (1989). Electrolyte composition of lacrimal gland flow and tears of normal and vitamin A deficient rabbits. *CLAO* **15**, 222.

Rosen, J. J. (1976). Cellular interactions at hydrogel interfaces. Ph.D. Thesis, Case Western Reserve University, Cleveland, OH.

Sastry, S. M., Sperduto, R. D., Waring, G. O., Remaley, N. A., Lynn, M. J., Blanco, E., and Miller, D. N. (1993). Radial keratometry does not effect intraocular pressure. *Refract. Corneal Surg.* **9**, 459–464.

Song, Q. H., Nugent, M. A., and Trinkaus-Randall, V. (1999). Injury induces TGF-β1 in rabbit corneal fibroblasts. *Invest. Ophthalmol. Visual Sci. Suppl.* **40**, 3281.

Strampelli, B., and Marchi, V. (1970). Osteo-odonto-keratoprosthesis. *Ann. Ophthalmol. Clin. Ocul.* **96**, 1–57.

Taloni, M., Falsini, B., Caselli, M., Wicozzi, I., Falcinelli, G., and Falcinelli, G. C. (1995). Parameters of central visual function in patients with osteo-odonto-keratoprosthesis. *Invest. Ophthalmol. Visual Sci. Suppl.* **36**(4), 1473.

Trinkaus-Randall, V., and Nugent, M. A. (1998). Biological response to a synthetic cornea. *J. Controlled Release* **53**, 205–214.

Trinkaus-Randall, V., Capecchi, J., Newton, A., Vadasz, A., Leibowitz, H. M., and Franzblau, C. (1988). Development of a biopolymeric keratoprosthesis material: Evaluation *in vitro* and *in vivo*. *Invest. Ophthalmol. Visual Sci.* **29**, 293–400.

Trinkaus-Randall, V., Capecchi, J., Sammon, L., Leibowitz, H. M., and Franzblau, C. (1990). *In vitro* evaluation of fibroplasia in a porous polymer. *Invest. Ophthalmol. Visual Sci.* **31**, 1321–1326.

Trinkaus-Randall, V., Banwatt, R., Capecchi, J., Leibowitz, H. M., and Franzblau, C. (1991). *In vivo* fibroplasia of a porous polymer in the cornea. *Invest. Ophthalmol. Visual Sci.* **32**, 3245–3251.

Trinkaus-Randall, V., Johnson-Wini, B., Banwatt, R. S., Gibbons, D., Capecchi, J., and Franzblau, C. (1992). Healing a porous polymer in cornea and matrix metalloproteinase involvement. *Invest. Ophthalmol. Visual Sci.* **33**, 983.

Trinkaus-Randall, V., Tong, M., Thomas, P., and Cornell-Bell, A. (1993). Confocal imaging of the α6 and β4 integrin subunits in the human cornea with aging. *Invest. Ophthalmol. Visual Sci.* **34**, 3103–3109.

Trinkaus-Randall, V., Banwatt, R., Wu, X. Y., Leibowitz, H. M., and Franzblau, C. (1994). Effect of penetrating porous webs on stromal fibroplasia *in vivo*. *J. Biomed. Mater. Res.* **28**, 195–202.

Trinkaus-Randall, V., Wu, X. Y., Tablante, R., and Tsuk, A. (1997). Implantation of a synthetic cornea: Design, development and biological response. *Artif. Organs* **21**, 1185–1191.

Trinkaus-Randall, V., Edelhauser, H. F., Leibowitz, H. M., and Freddo, T. F. (1998). Corneal structure and function. *In* "Corneal Disorders: Clinical Diagnosis and Management" (H. M. Leibowitz, ed.), Chapter 1, pp. 2–32. Saunders, Philadelphia.

Tsuk, A., Trinkaus-Randall, V., and Leibowitz, H. M. (1997). Ultraviolet light absorbing hydrogel materials for keratoprosthesis. *J. Biomed. Mater. Res.* **34**, 299–304.

van Andel, P., Cuperus, P., Kolenbrander, M., Singh, D., and Worst, J. (1988). A glass-platinum keratoprosthesis with peripheral episcleral fixation: Results in the rabbit eye. *Third World Biomater. Cong.* Kyoto, Jpn. Abstr. 5E-45.

Waring, G. O. (1992). The third international congress on international technology in ophthalmology. *Refract. Corneal Surg.* **8**, 101–106.

Waring, G. O. (1998). Refractive surgery. *In* "Corneal Disorders: Clinical Diagnosis and Management" (H. M. Leibowitz, ed.), Chapter 39, pp. 1023–1106. Saunders, Philadelphia.

Waring, G. O., Bourne, W. M., Edelhauser, H. F., and Kenyon, K. R. (1982). The corneal endothelium: Normal and pathologic structure and function. *Ophthalmologica* **89**, 531.

Weber, I. T., Harrison, R. W., and Iozzo, R. V. (1996). Model structure of decorin and implications for collagen fibrillogenesis. *J. Biol. Chem.* **271**, 31767–31770.

White, J. H., and Gona, O. (1988). "Proplast" for keratoprosthesis. *Ophthalmol. Surg.* **19**, 331–333.

Wu, X. Y., Cornell-Bell, A., Davies, T. A., Simons, E. R., and Trinkaus-Randall, V. (1994). Expression of integrin and organization of F-actin in epithelial cells depends on the underlying surface. *Invest. Ophthalmol. Visual Sci.* **35**, 878–890.

Yee, R. W., Geroski, D. H., and Matsuda, M. (1985). Correlation of corneal endothelial pump site density, barrier function and morphology in wound repair. *Invest. Ophthalmol. Visual Sci.* **26**, 1191.

Zieske, J. D., Bukusoglu, G., and Gipson, I. K. (1989). Enhancement of vinculin synthesis by migrating stratified squamous epithelium. *J. Cell Biol.* **109**, 571–576.

PART XII: ENDOCRINOLOGY AND METABOLISM

CHAPTER 36

BIOARTIFICIAL PANCREAS

Taylor G. Wang and Robert P. Lanza

INTRODUCTION

The World Health Organization (WHO) estimates that over 143 million people suffer from diabetes mellitus worldwide, and that number will more than double by the year 2025. In patients with insulin-dependent diabetes mellitus (IDDM), there is a marked decrease in the number of insulin-producing islet cells in the pancreas. In the past, animal or recombinant human sources of insulin have provided the replacement therapy required by these patients. Unfortunately, injected insulin cannot precisely mimic the ability of the normal pancreas to regulate blood glucose concentrations. The concentration of insulin in the blood is normally closely linked to the blood glucose concentration. The moment-to-moment adjustments in insulin secretion by the beta cells of the pancreatic islets, which serve to control blood glucose concentrations, depend on a complex series of biochemical pathways in the living beta cell. These adjustments are extremely difficult or impossible to simulate by insulin injections. The results of the Diabetes Control and Complications Trial (DCCT) suggest that failure to achieve physiologic glucose control with injected insulin is responsible for the serious complications of this disease, including diabetic nephropathy, retinopathy, and neuropathy (Diabetes Control and Complication Trial Research Group, 1993).

There is hope that transplantation of islets of Langerhans will eliminate the need for daily insulin injections, and will prove effective in preventing or retarding the development of complications associated with diabetes. Restoration of normal glucose metabolism has already been achieved in several patients with IDDM by islet transplantation (Bretzel *et al.*, 1996; Hering *et al.*, 1996; Sutherland *et al.*, 1994). Success, unfortunately, has been sporadic, and the requirements for immunosuppressive drugs expose these patients to a wide variety of serious complications (Fung *et al.*, 1992; Kahan, 1989). In addition, many of these immunosuppressive drugs, including glucocorticoids and cyclosporin, have dose-dependent deleterious effects on glucose homeostasis and beta cell function (Alejandro *et al.*, 1989; Gunnarsson *et al.*, 1983; Schlump *et al.*, 1986; Van Schilfgaarde *et al.*, 1986). A number of immunoisolation systems have been developed and refined during the past several years in which the transplanted tissue can be separated from the immune system of the host by a selectively permeable membrane. Low-molecular-weight substances such as nutrients, electrolytes, oxygen, and bioactive secretory products are exchanged across the membrane while immunocytes, antibodies, and other transplant rejection effector mechanisms are excluded (Darquy and Reach, 1985; Soon-Shiong *et al.*, 1990). This approach has the potential not only to allow allogeneic transplantation of cells and tissues without immunosuppression, but also to permit the use of xenografts. The latter, by most accounts, will be essential before any immunoisolation device can be applied on a wide scale.

Like other human organs, donor pancreases are in very short supply. In the United States, it is estimated that only approximately 1000 pancreases are recovered each year. Even with improved procurement of human organs, the supply of donor tissue would remain quite inadequate if pancreatic islet transplantation were to be developed as an effective therapy. For the immediate future, the logical alternative is to use nonhuman donor islets. For example, methods have been developed to isolate pancreatic islets from porcine and bovine glands; insulins from these animals are fully active in humans and have been used to treat diabetics for over 70 years. Pig and cow islets are an attractive option because they are readily available and the amino acid sequences of these insulins are similar to those in human insulin. Further, herds of specific pathogen-free (SPF) animals would be available as a potential safeguard against the transfer of infectious organisms to human islet recipients.

Although fundamental advances in our understanding of the human immune system and the immune rejection process may eventually lead to means for developing technologies to overcome the vigorous humoral and cellular immune responses associated with transplantation of xenogeneic tissues, these solutions are likely to be decades in the future. In the meantime, encapsulation may be the only way to establish prolonged survival of cell and tissue transplants in patients with diseases caused by the loss of specific vital metabolic functions. In addition to diabetes, encapsulation has broad application to treating major diseases such as cancer and autoimmune deficiency syndrome (AIDS) and a wide range of other disorders resulting from functional defects of native cell systems (Lanza and Chick, 1995). Applications include the use of cells such as hepatocytes for the treatment of liver failure and enzymatic defects (Cai *et al.,* 1989; Eguchi *et al.,* 1996; Hori, 1988; Makowka *et al.,* 1980; Mito and Sawa, 1996; Sun *et al.,* 1986, 1987); adrenal cells in Parkinson's disease (Freed, 1996; Sagen, 1996; Widner *et al.,* 1991); cells to produce nerve growth factors in Alzheimer's disease (Hefti and Knusel, 1988; Phelps *et al.,* 1989a,b), epilepsy, Lou Gehrig's disease, Huntington's disease (Frim *et al.,* 1993; Schumacher *et al.,* 1991; Winn *et al.,* 1994), spinal cord injuries (Tessler, 1996), and strokes; cells to produce clotting factors VIII and IX in hemophilia (Hellman *et al.,* 1989; Roberts, 1989); and endocrine cells in other disorders resulting from hormone deficiency (Fu and Sun, 1989; Lanza *et al.,* 1992). Moreover, by using recombinant DNA and cell engineering technologies in conjunction with encapsulation technologies, it should also prove possible to treat patients suffering from chronic pain (Sagen, 1996), Kaposi's sarcoma, and various hematologic disorders.

To date, most of the research in the area of encapsulated cell transplantation has been carried out with pancreatic islets. There are several reasons for this: (1) diabetes is a leading cause of morbidity and mortality in the world, at an annual cost of approximately $134 billion in the United States alone; (2) pancreatic islets, which comprise only 1–2% of the human adult pancreative volume, can be isolated from animal sources; and (3) the quantity of differentiated islet tissue to be transplanted is within a reasonable range (less than 1 g).

Many types of encapsulation systems have been studied. These include devices anastomosed to the vascular system as arteriovenous shunts, diffusion chambers (both hollow fibers and wider-bore tubular membranes), and spherical micro- and macrocapsules (Fig. 36.1). Results in diabet-

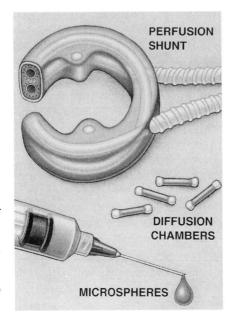

Fig. 36.1. Biohybrid artificial devices. In the form of a vascular implant, islets can be distributed in a chamber surrounding a permselective membrane, and the device implanted as a shunt in the vascular system. Alternatively, islets can be immunoisolated within membrane diffusion chambers or microcapsules and placed intraperitoneally, intramuscularly, subcutaneously or in other sites. The islets within these biohybrid devices are generally immobilized in hydrogels such as alginate or agar. One of the important functions of these gels is to provide more uniform islet distribution by preventing settling and subsequent aggregation of the islets into larger, necrotic masses. From Lanza and Chick (1995), with permission.

PERFUSION SHUNT

DIFFUSION CHAMBERS

MICROSPHERES

ic rodents and dogs indicate that these systems can function for periods of several months to more than 1 year. Currently, spherical micro- and macrocapsules appear to have the highest therapeutic potential.

THEORETICAL BACKGROUND

Over the past decade, several methods for encapsulating islets have been investigated. Micro- and macrocapsules offer a number of distinct advantages over the use of other biohybrid devices, including (1) greater surface-to-volume ratio, (2) ease of implantation (can simply be injected), and (3) retrievability by lavage and needle aspiration (or alternatively, fabrication from biodegradable materials).

IMMUNOISOLATION

The effectiveness of immunoisolation of a polymer capsule is closely tied to its membrane's permeability, or more precisely, the pore size of the capsule membrane. It has been suggested, and argued forcefully, that with proper selection of a cutoff pore size, the polymer membrane can serve as a gatekeeper (Fig. 36.2), keeping the higher molecular weight (MW) components of immune system out, and allowing the smaller MW molecules such as oxygen and hormones to pass through without much impedance.

Although most of the current polymer capsule designs are based on this principle (De Vos *et al.,* 1993; Goosen, 1994; Lim and Sun, 1980; Soon-Shiong *et al.,* 1993), there is a fundamental weakness in this model. Pore sizes of polymer membranes are not homogeneous by nature, but rather consist of a broad spectrum of sizes and have very long tails. This flaw was first suggested by Colton as early as 1991, but was largely ignored by capsule designers (Colton and Avgoustiniatos, 1991). Ignoring this may have contributed to inconsistent conclusions in the literature (Brissova *et al.,* 1998; Cole *et al.,* 1992; Fan *et al.,* 1990; Iwata *et al.,* 1994; Lanza and Chick, 1994; Lacik *et al.,* 1998; Lum *et al.,* 1992). For the capsule to provide total immunoprotection for living cells, the polymer membrane must not have pores larger than the antibody complement component (Colton and Avgoustiniatos, 1991). This suggests that unless a gatekeeper capsule has a step function cutoff at less than 300 Å in pore diameter and is defect free, the effectiveness of the capsule as an immunoisolation device will be compromised.

A new mathematical model incorporating the pore size distribution for immunoisolation of living cells has been proposed (Wang, 1997). In this new model, the pore size distributions of the membrane are taken into consideration, the capsule wall is thicker, and the pores are bigger than in the gatekeeper model. Figure 36.3 is the schematic drawing of the new model. The large pores will allow the nutrients, proteins, and immune system components to enter the membrane. However, the small pores inside the membrane will prevent or delay most of the immune system components from passing all the way through to the inner volume of the capsule, where the cells reside. The capsule membrane operates more as an immune system barrier than as a gatekeeper. In calculations, we will focus on preventing the antibody-mediated attack on living cells. We assume the immune system density is very diluted, and antibodies and complement components pass

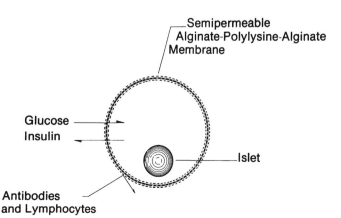

Fig. 36.2. Gatekeeper model. The immune system is excluded from entering the capsule, yet nutrients, oxygen, and proteins pass through without impedance.

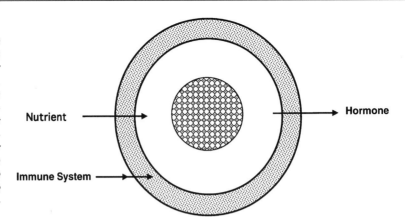

Fig. 36.3. Barrier model. In the barrier model, the capsule wall is thicker and the pores are bigger than in the gatekeeper model. This allows the immune system components to enter the membrane. However, sufficiently smaller pores inside the membrane act as barriers to prevent or delay most of the immune system components from passing all the way through to the inner volume of the capsule, where the islet resides.

through the immunoisolation membrane as a result of "random walk motion." For the sake of simplicity, we shall study the motion in one dimension, and assume the probability to go forward and backward an equal number of steps is the same. After a total of N such steps, the probability of finding the immune system at a position m steps away from the origin is

$$P_N(m) = \frac{N!}{\left(\dfrac{N+m}{2}\right)!\left(\dfrac{N-m}{2}\right)!}\left(\frac{1}{2}\right)^N,$$

(1)

where $m = n_f - n_b$ and $N = n_f + n_b$; n_f represents the step forward, n_b the step backward. This is a Guassian distribution curve centered at zero. The root-mean-square displacement, $\sqrt{\overline{(\delta x)^2}}$, of the immune system after N steps is found to be

$$\sqrt{\overline{(\delta x)^2}} = \sqrt{\overline{(\delta m)^2}}\, R = \sqrt{N}\, R,$$

(2)

where $\delta x = x - \bar{x}$ and $\delta m = m - \bar{m}$. Here R is the average step size and can be approximated to be the average pore size. If we assume the immune system is originally located at the surface of the capsule, and the root-mean-square distance needed to travel is the membrane thickness d. Eq. (2) can be rewritten as

$$\frac{d^2}{R^2} = N.$$

(3)

We can now write the immunisolation time, Γ, defined as the time required for the immune system to pass through the membrane with a thickness d as

$$\Gamma \simeq \frac{d^2}{R^2} f\tau.$$

(4)

Using equipartition theorem, the random-walk motion in three dimensions can be written as

$$\Gamma \simeq 3\frac{d^2}{R^2} f\tau,$$

(5)

where the f is the ratio between surface areas of small pores and large pores, and τ is the time that the immune system can be trapped by the small pores in each collision. The value f can be determined quantitively from the pore surface area distribution, and the value τ for different immune systems can be estimated qualitively from the dextran size exclusion chromatography (SEC) retention time.

Using an antibody complement component as an example, the estimated immunoisolation time for the sodium alginate/poly(lysine)/sodium alginate capsule (110-kDa pore size by dextran standard, 2-μm wall thickness) is on the order of days. This calculated value is in agreement with many of the experimental observations on how long islets encapsulated inside the capsule can survive after transplantation into presensitized animal models, such as NOD mice. More importantly, this model might suggest that the current capsule will have limited success in treating type I diabetic disease in humans.

MASS TRANSPORT

To maintain the viability of islets inside an immunoisolation capsule, the nutrient and oxygen must be allowed to diffuse in, and the insulin out. This is a convective diffusive process. Mass transport across a random network structure is an extremely complicated system. Any attempt to undertake a rigorous calculation of this system is beyond the scope of this chapter. Here, the intent is to examine the order of magnitude power dependence of capsule membrane parameters on the steady-state mass transport rate across a capsule membrane. The diffusion mass transport flux, Q, across the membrane is proportional to the concentration gradient, and membrane porosity.

$$Q \sim \frac{D}{d} \Delta c A_{\text{eff}},\tag{6}$$

where D is the diffusion coefficient, d is the thickness of the membrane, and Δc is the mass concentration difference; here A_{eff} is the effective capsule surface area that participates in mass transport. The diffusion of biomaterial in liquids is under a nonslip condition; D is proportional to the temperature, T, and inversely proportional to the fluid viscosity, η, and the particle radius, r. This is usually called the Stokes velocity. It represents the diffusion of large spherical biomaterial in an open system. To account for the effect of capsule membrane porosity on mass transport, we need to calculate A_{eff}. This problem can be simplified by modeling the polymer membrane as a bundle of interwoven pipes. Those pipes have inner diameters, R, which approximate average pore sizes. The capsule surface area participating in diffusion is NR^2, and N is the number of pipes per unit surface area. The mass flow inside a pipe has a velocity profile of R^2. Thus, the effective diffusive mass flow passing through a cross-section of the pipe per unit of time is proportional to the fourth power of the radius of the pipes. Equation (6) can then be rewritten as

$$Q \sim \frac{R^4 T}{r \eta d} \Delta c N.\tag{7}$$

Equation (7) is phenomenological. However, it might provide some valuable insight on the capsule design optimization.

MECHANICAL STRENGTH

Many immunoisolation devices fail inside animal and/or human hosts because of capsule breakage. This exposes unprotected living cells to the host's immune system. At best, the cells are destroyed; at worst, an immunoreaction can be triggered. Therefore, if the immunoisolation devices are to be used for the long-term cure of hormone deficiency diseases in humans, the mechanical strength of the capsules must be sufficiently great to survive the constant pressure exerted.

Examine the case of a spherical capsule with a radius of R. Under unidirectional pressure, P_x, the capsule distorts the shape to compensate the pressure:

$$P_x = Y d \left(\frac{2}{R_2} - \frac{2}{R_1} \right),\tag{8}$$

where Y is Young's modulus, d is the thickness, and R_1 and R_2 are major and minor axes of the deformed capsule. If the distortion is small, we can approximate $R_1 = R + \delta r$ and $R_2 = R - \delta r$. Equation (8) can be rewritten as

$$\frac{\delta r}{R} = \frac{R}{4 Y d} P_x.\tag{9}$$

Equation (9) shows the advantages that a smaller capsule and a thicker membrane have on resisting deformation, $\delta r/R$, under pressure, P_x.

Equations (5), (7), and (9) clearly show that if the capsule parameters cannot be adjusted independently, it will be very difficult to satisfy all the dichotomy requirements of immunoisolation, mass transport, and mechanical strength.

EXPERIMENT

BIOCOMPATIBILITY

Experiments have demonstrated the feasibility of long-term immunoisolation of islets by artificial poly(acrylonitrile)/poly(vinylchloride) membranes and the long-term biocompatibility of the membrane versus the graft and versus the recipient (Lanza *et al.,* 1991, 1992). Data indicate that islet implants can provide correction of hyperglycemia in dogs and rodents for periods of several months to more than 1 year without the use of immunosuppressive drugs. Diffusion chambers fabricated from permselective acrylic membranes showed little or no evidence of an inflammatory response when implanted intraperitoneally in either spontaneously or streptozotocin-induced diabetic rats (Lanza *et al.,* 1992, 1993). Complications such as abscess formation or intestinal adhesions, which have been observed with other technologies (Andersson, 1979), were not observed with these implants. These studies have been extended to implantation in the peritoneal cavity of a large animal, the dog, with surgically induced diabetes. Histologic examination of the chambers revealed that they were biocompatible. The outer surface showed only scattered foci of macrophages and lymphocytes. Intactness and sterility of the chambers, however, was a crucial factor in the success of the implants. In addition to loss of islet viability, damaged or contaminated membranes were often encapsulated by fibrous tissue that exhibited an interstitial acute and/or chronic inflammatory reaction, and development of granulation tissue was observed. Most of these transplants ultimately failed because of membrane breakage.

BLOOD GLUCOSE CONTROL

The motivation for islet transplantation is to provide physiologic control of blood glucose concentration. *In vitro* and *in vivo* experiments have demonstrated only moderately delayed changes (lag time <10 min) in insulin secretion in response to changes in glucose concentration. Perfusion of encapsulated islets with glucose elicited an approximately fourfold average increase from the basal insulin secretion (Wang *et al.,* 1997; Lanza *et al.,* 1991). There was a delay of only 7 ± 1 min before the insulin concentration in the perfusate began to increase. Although this response is well within a time frame compatible with closed-loop insulin delivery (pharmacokinetic modeling of glucose homeostasis in humans suggests that the lag time of the increase in insulin delivery by an artificial pancreas must be <15 min to avoid the overexcursion of postprandial blood glucose) (Kraegen *et al.,* 1981), a reduction in the volume of the islet cell compartment would further improve the transmission of the glycemic signal from the blood to the islets, and of insulin from the islets to the recipient. Immunoisolated islets lack intimate vascular access, and must be supplied with oxygen and nutrients by diffusion from the nearest blood vessels over distances greater than those normally encountered. In wider-bore membrane chambers, the problem of cell death or dysfunction as a result of oxygen supply limitations, or accumulation of wastes or other cellular products, is likely to be more severe. Chambers with a 4.8-mm inner diameter retrieved from the peritoneal cavity of dogs several months after allotransplantation contained a central core of necrotic islets. Only a rim of islets remained viable within approximately 0.5–1 mm of the inner membrane wall. Similar results were obtained with canine islet implants into rats. These findings are in agreement with the theoretical model of Eq. (7). Clearly, careful attention must be paid to the diffusion distances and transport properties of the membranes. The relatively small volume (and short distances) associated with smaller spherical devices would maximize the transfer of oxygen and nutrients to the islets, and minimize the suppression of function by any islet hormonal or metabolic products that could accumulate in the islet compartment.

HOST SENSITIZATION

Although encapsulation systems serve to block uptake of antibodies and complement, the possibility must be considered that antigens released from the cell compartment could stimulate a host humoral response. Such antibodies could be induced by antigens shed from the cell surface, or by proteins secreted by live cells or liberated after cell death. This could lead to an allergic response and/or immune complex disease. Studies have been performed by Lanza *et al.* (1994a,b) in order to help rule out the possibility of serious immunologic sequelae in recipients. Although these studies were carried out using islets encapsulated inside tubular membrane chambers, the results may be reflective of other cell encapsulation systems as well, including microcapsules, hollow fibers, devices anastomosed to the vascular system as arterioventricular shunts, and other types of implantable diffusion chambers.

In these studies, chambers containing porcine or bovine islets were implanted intraperitoneally into streptozotocin-induced diabetic rats. Sera were collected at various intervals and tested against isolated canine and porcine islets for tissue specificity and interspecies cross-reactivity by fluorescence immunocytochemistry. No immunofluorescence (or only weak background staining) was obtained when islets were exposed to horse sera, or to sera obtained prior to implantation of devices containing xenogeneic islets. Within 2–6 weeks, however, the postimplantation sera showed strong immunoreactivity. The antibodies were found to be reactive to multiple tissues, and to possess little or no interspecies cross-reactivity.

The appearance of these xenoantibodies coincided with the appearance of circulating soluble immune complexes. However, none of the respiratory, cutaneous, or gastrointestinal manifestations that are characteristic of an anaphylactic reaction, or of the diseases of immediate-type hypersensitivity, were observed, even following the intraperitoneal injection of additional naked islet tissue. Renal glomeruli did not stain for IgG or C3 in islet recipients. These results suggest that islet cell antigens crossed the membrane and stimulated antibody formation in the host, although they did not appear to cause renal or immune complex disease during the course of this study.

CAPSULE DEVELOPMENT

ALGINATE/POLY(L-LYSINE) CAPSULES

Capsules fabricated from alginate and poly(L-lysine) (PLL) were originally described by Lim and Sun in 1980, and have been used to reverse diabetes in animals and more recently in a small-scale clinical trial (Soon-Shiong *et al.*, 1993, 1994; Sun *et al.*, 1996). However, the inability to adjust capsule parameters independently (e.g., immunoisolation versus mechanical strength versus permeability) has limited the success of this system. Implanted discordant islet xenografts produced blood glucose control for a shorter period of time. These studies performed in mice usually required adjunctive treatment with immunosuppressive agents (Calafiore *et al.*, 1987; Ricker *et al.*, 1987; Wang *et al.*, 1987). Weber *et al.* (1990) found that alginate/PLL microcapsules containing canine islets functioned for <2 weeks in diabetic NOD mice. With anti-CD4 monoclonal antibody treatment, however, long-term functional survival was observed in many of the recipients. Patrick Soon-Shiong *et al.* (1993) have also reported successful long-term implantation of a microencapsulated allograft in larger animals. They treated spontaneous diabetes in dogs administered low doses of cyclosporin. Using the same technology, Soon-Shiong *et al.* (1994) have also achieved insulin independence in a patient for approximately 1 month. However, this patient was on a regimen of immunosuppressive drugs, and subsequently required exogenous insulin therapy. Calafiore (1992) has also achieved insulin independence in one of three alloxan-induced diabetic dogs and, transiently, in one of two patients without any pharmacologic immunosuppression. However, these microcapsules were coated with PLL and were deposited in artificial prostheses directly anastomosed to blood vessels.

BIODEGRADABLE CAPSULES

Prolongation of discordant xenograft survival in streptozotocin-induced diabetic mice has been achieved without immunosuppression using alginate microspheres without the synthetic poly(L-lysine) membrane (Lanza *et al.*, 1995). Uncoated alginate spheres containing porcine and bovine islets routinely reversed hyperglycemia after intraperitoneal injection into diabetic mice. All but one (13/14) of the grafts functioned for at least 1 month, and 11 animals lived for at least 10 weeks. In comparison, free islet transplants failed to function, or sustained euglycemia for <4 days.

The ability of uncoated microspheres to achieve marked prolongation of discordant xenograft

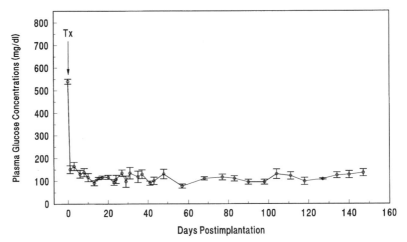

Fig. 36.4. Mean nonfasting plasma glucose levels in four streptozotocin-induced diabetic rats that received intraperitoneal implants of alginate-encapsulated bovine islets. All of the animals received low-dose cyclosporin A (10–20 mg/kg/day, subcutaneous). Immunohistochemical staining of the grafts (>100–150 days) revealed healthy viable islets (80–100%, comparable to the day 0 control specimens), with well-granulated alpha, beta, and delta cells. This method for transplanting discordant islets into rats is simple and inexpensive, and may also be a useful procedure for transplanting other cells and tissues. The spheres can be formed simply by extruding a mixture of cells and sodium alginate with a 16-gauge angiocatheter into a CaCl$_2$ solution. From Lanza et al. (1995b), with permission.

survival is surprising, because the destruction of the grafts might have been expected to occur based on the presence of circulating preformed natural antibodies in the recipients in reaction to cells of the donor (Platt and Bach, 1991). The data suggest that the alginate matrix served as a physical barrier to prevent direct effector cell contact with the donor tissue. However, we cannot rule out the possibility that immunologic effector molecules were excluded from the negatively charged alginate gels based on properties other than simply molecular weight.

Although the results using uncoated alginate microspheres in mice are encouraging, preliminary experiments in diabetic rats suggest that uncoated alginate spheres containing porcine and bovine islets can reverse hyperglycemia for periods of up to more than 175 days (Fig. 36.4) (Lanza et al., 1995). Experiments in spontaneously diabetic dogs have also been performed by Lanza et al. (1995c). Results indicate that long-term survival of canine islet allografts can be achieved by encapsulation inside uncoated alginate spheres (Fig. 36.5). However, these results have been inconsistent and usually have required adjunctive treatment with immunosuppressive agents.

To overcome this limitation, Lanza et al. have designed a hydrogel-based microcapsule based on the uncoated microsphere technology (Lanza et al., 1999). These new composite capsules break

Fig. 36.5. Fasting blood glucose concentrations of a spontaneously diabetic dog that received an intraperitoneal implant of alginate-encapsulated canine islets. From Lanza et al. (1995d), with permission.

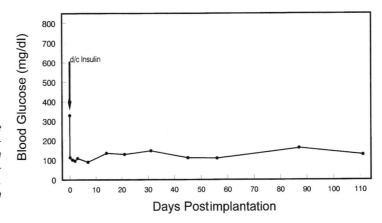

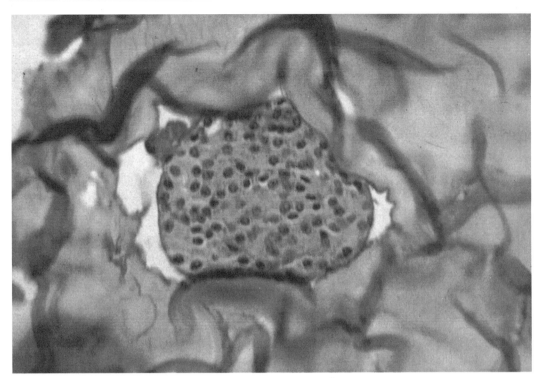

Fig. 36.6. *Biodegradable microreactor. Encapsulated bovine islet retrieved from the peritoneal cavity of a dog 6 weeks after xenotransplantation. These new biodegradable microreactors are injectable and do not appear to require the use of immunosuppressive drugs to prevent rejection.*

down at a rate that can be adjusted to correspond to the functional longevity of the encapsulated cells, and can be engineered to degrade over several weeks to months for short-term drug delivery, or to remain intact and immunoprotective for more extended periods of time. When the supply of cells needs to be replenished, no surgery will be required to localize and remove the old capsules. This system has been tested successfully in rats using discordant porcine and bovine islet tissue. The tissue remained in excellent condition indefinitely (>40 weeks.) The viability was comparable to preimplant control specimens without use of any immunosuppression. Large animal studies are underway (Fig. 36.6).

PMCG CAPSULES

A new multicomponent capsule based on a theoretical model was developed by Wang *et al.* (1997). Over a thousand combinations of polyanions and polycations were studied in search of polymer candidates that would be suitable for entrapment capsules. A combination of sodium alginate (SA), cellulose sulfate (CS), poly(methylene–co-quinidine) (PMCG), calcium chloride (CaCl$_2$), and sodium chloride (NaCl) (Lacik *et al.*, 1998; Wang *et al.*, 1997) was found to be most promising. The polyelectrolyte complexation has distinct advantages. It is very simple and inexpensive and can be processed under extremely mild physiologic conditions. A distinct advantage of this multicomponent system was the ability to adjust the capsule's mechanical properties, permeability (Brissova *et al.*, 1998), and surface properties (Xu *et al.*, 1998) independently, as well as the capsule size for retrieval if needed.

Encapsulated rat islets reversed chemically induced diabetes in mice (strepzotocin) for at least 4–6 months. Similar studies were performed with diabetic NOD female mice and diabetes was reversed for up to 180 days. Wang and Elliott of New Zealand have successfully reversed diabetes (14 weeks) in a NOD mice model with encapsulated porcine islets without the use of any immunosuppression (Fig. 36.7). The capsules were found freely floating in the peritoneal cavity, and showed no significant fibrosis histologically and no tissue reaction. The encapsulated islets continued to maintain long-term function as assessed by their ability to secrete insulin following glucose stimulation. The eventual failure of encapsulated islets transplanted into animals with chem-

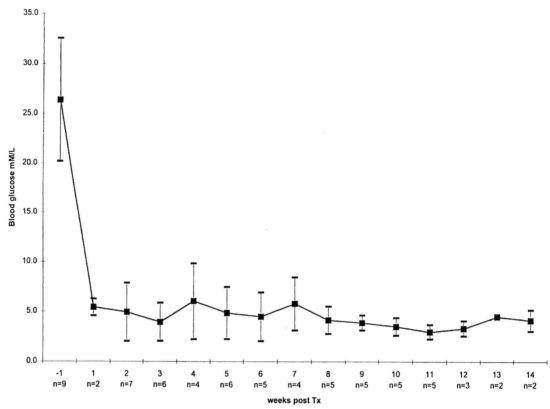

Fig. 36.7. Transplantation of encapsulated porcine islets reverse diabetes. Encapsulated porcine islets were transplanted into the peritoneal cavity of female NOD mice that had developed spontaneous diabetes by the age of 20 weeks. Of the diabetic mice that received transplants, the number of nondiabetic animals at selected time points is shown. This experiment is part of the results of a collaborative effort between T. Wang and R. Elliott of New Zealand.

ically induced diabetes did not result from capsule rupture or from an immune attack. Capsules retrieved from such animals were freely floating and their surface was free of fibrosis. Thus, the failure likely results from gradual islet death within the capsule from either a nutrient deficiency or from islet toxicity because of soluble immune factors (cytokines). In contrast, the failure of encapsulated islets transplanted into NOD mice may have resulted from an immune or autoimmune attack. Capsules retrieved from such animals were clumped and demonstrated marked fibrosis around the capsule surface. Experiments to extend the longevity of the capsules by optimizing the permeability, wall thickness, and surface characteristics are in progress with very promising results. Large animal studies have also been initiated with encouraging preliminary results.

OTHER APPROACHES

Many other ingenious approaches are being tested in other research labs. Macro-Agarose beads (The Rogosin Institute), poly(ethylene gycol) capsules (Neocrin), Stealth microcapsules (Encell), and Islets Sheet (Islets Sheet Medical) all offer interesting possibilities. Unfortunately, there are little or no published data available on these systems at the present time.

DISCUSSION

Organ transplantation is one of the miracles of the twentieth century. However, along with success has come frustration. The supply of human organs is insufficient to satisfy demand. Within the United States alone, thousands of patients await transplants of heart, liver, kidney, and lung, and millions more struggle with diseases such as diabetes. Demand exceeds the supply by two orders of magnitude. So the pressure is mounting to devise ways such as immunoisolation to use animal tissues and cells as possible substitutions for human biologic material, but without the need for immunosuppression and its accompanying side effects. There is considerable excitement among researchers and clinicians who hope to see the introduction of a hybrid pancreas to replace conventional insulin therapy.

Polymer membranes have been the choice to encapsulate living cells for immunoisolation. Data have shown exciting successes in animal models. A new physical picture and a better understanding of the mechanisms of immunoisolation have advanced the systematic studies in preparation for human clinical trials. The next few years should prove to be a great challenge and it is hoped with great excitement that the promise of islet transplantation comes to fruition. It appears likely that in the year 2000 clinical trials utilizing encapsulated islet xenografts to treat diabetic patients will be a reality. This will serve as an important starting point for developing living drug delivery systems for treating a wide range of additional disorders. For patients with diabetes as well as those with other diseases, such as cancer, hemophilia, liver failure, and Parkinson's disease, the coming years hold promise for greatly improved therapy.

ACKNOWLEDGMENTS

The Vanderbilt University studies were supported by grants from the Evans–Gilruth Foundation, Vanderbilt University School of Engineering, and the Vanderbilt Diabetes Research and Training Center (NIH DK20593). T. W. wishes to express gratitude for the valuable contribution provided by Igor Lacik, of the Polymer Institute, Slovak Academy of Sciences; A. V. Anilkumar; Paul LeMaster, of Vanderbilt University; and A. C. Powers, of Vanderbilt University Medical School/VA Hospitals. T. W. also acknowledges the excellent support provided by New Zealand colleagues, in particular, Prof. R. Elliott and Dr. Livia.

REFERENCES

Alejandro, R., Feldman, E. C., Bloom, A. D. *et al.* (1989). Effects of cyclosporin on insulin and C-peptide secretion in healthy beagles. *Diabetes* **38**, 698.

Andersson, A. (1979). Survival of pancreatic islet allografts. *Lancet* **2**, 585.

Bretzel, R. G., Hering, B. J., Schultz, A. O., Geier, C., and Federlin, K. (1996). International Islet Transplant Registry report. *In* "1996/97 Yearbook of Cell and Tissue Transplantation" (R. P. Lanza and W. L. Chick, eds.). Kluwer Academic press, Dordrecht, The Netherlands.

Brissova, M., Lacik, I., Powers, A. C., Anilkumar, A. V., and Wang, T. G. (1998). Control and measurement of permeability for design of microcapsule cell delivery system. *J. Biomed. Mater. Res.* **39**, 61–70.

Cai, Z., Shi, Z., Sherman, M. *et al.* (1989). Development and evaluation of a system of microencapsulation of primary rat hepatocytes. *Hepatology* **10**, 855.

Calafiore, R. (1992). Transplantation of microencapsulated pancreatic human islets for therapy of diabetes mellitus. *ASAIO J.* **38**, 34.

Calafiore, R., Janjic, D., Koh, N. *et al.* (1987). Transplantation of microencapsulated canine islets into NOD mice: Prolongation of survival with superoxide dismutase and catalase. *Clin. Res.* **35**, 499A.

Cole, D. R., Waterfall, M., McIntyre, M., and Baird, J. D. (1992). Microencapsulated islet grafts in the BB/E rat: A possible role for cytokines in graft failure. *Diabetalogia* **35**, 231–237.

Colton, C. K., and Avgoustiniatos, E. S. (1991). Bioengineering in development of the hybrid artificial pancreas. *J. Biomech. Eng.* **113**, 152–170.

Darquy, S., and Reach, G. (1985). Immunoisolation of pancreatic β cells by microencapsulation. *Diabetologia* **28**, 776.

De Vos, P., Wolters, G. H. J., Fritschy, W. M., and van Schilfgaarde, R. (1993). Obstacles in the application of microencapsulation in islet transplantation. *Int. J. Artif. Organs* **16**, 205–212.

Diabetes Control and Complication Trial Research Group (1993). The effect of intensive treatment of diabetes on the development and progression of long-term complications in insulin-dependent diabetes mellitus. *N. Engl. J. Med.* **329**, 977.

Eguchi, S., Chen, S., Rozga, J., and Demetriou, A. A. (1996). Tissue engineering/hybrid tissues: Liver. *In* "1996/97 Yearbook of Cell and Tissue Transplantation" (R. P. Lanza and W. L. Chick, eds.). Kluwer Academic Press, Dordrecht, The Netherlands.

Fan, M. Y., Lum, Z., Levesque, L., Tai, I. T., and Sun, A. M. (1990). Reversal of diabetes in BB rats by transplantation of encapsulated rat islets. *Diabetes* **39**, 519–522.

Freed, W. J. (1996). Neural transplantation: Brain. *In* "1996/97 Yearbook of Cell and Tissue Transplantation" (R. P. Lanza and W. L. Chick, eds.). Kluwer Academic Press, Dordrecht, The Netherlands.

Frim, D. M., Short, M. P., Rosenberg, W. S. *et al.* (1993). Local protective effects of nerve growth factor-secreting fibroblasts against excitotoxic lesions in the rat striatum. *J. Neurosurg.* **78**, 267.

Fu, X. W., and Sun, A. M. (1989). Microencapsulated parathyroid cells as a bioartificial parathyroid: *In vivo* studies. *Transplantation* **47**, 432.

Fung, J. J., Alessiani, M., Abu-Elmagd, K. *et al.* (1992). Adverse effects associated with the use of FK 506. *Transplant. Proc.* **23**, 3105.

Goosen, M. F. A. (1994). *In* "Immunoisolation of Pancreatic Islets" (R. P. Lanza and W. L. Chick, eds.), pp. 21–44. R. G. Landes, Austin, TX.

Gunnarsson, R., Klintmalm, G., Lundgren, G. *et al.* (1983). Deterioration in glucose metabolism in pancreatic transplant recipients given cyclosporin. *Lancet* **2**, 571.

Hefti, F., and Knusel, B. (1988). Chronic administration of nerve growth factor and other neurotropic factors to the brain. *Neurobiol. Aging* **9**, 689.

Hellman, L., Smedsrod, B., Sandberg, H. *et al.* (1989). Secretion of coagulant factor VIII activity and antigen by *in vitro* cultivated rat liver sinusoidal endothelial cells. *Br. J. Hematol.* **73**, 348.

Hering, B. J., Wahoff, D. C., and Sutherland, D. E. R. (1996). Clinical islet transplantation. *In* "1996/97 Yearbook of Cell and Tissue Transplantation" (R. P. Lanza and W. L. Chick, eds.). Kluwer Academic Press, Dordrecht, The Netherlands.

Hori, M. (1988). Will artificial liver therapy ever become a reality? *Artif. Organs* **12**, 293.

Iwata, H., Takagi, T., Kobayashi, K., Oka, T., Tsuki, T., and Ito, F. (1994). Strategy for developing microbeads applicable to islet xenotransplantation into a spontaneous diabetic NOD mouse. *J. Biomed. Mater. Res.* **28**, 1201–1207.

Kahan, B. D. (1989). Cyclosporine. *N. Engl. J. Med.* **321**, 1725.

Kraegen, E. W., Chisholm, D. J., and MacNamara, M. E. (1981). Timing of insulin delivery with meals. *Horm. Metab. Res.* **13**, 365.

Lacik, I., Brissova, M., Anilkumar, A. V., Powers, A. C., and Wang, T. G. (1998). New capsule with tailored properties for the encapsulation of living cells. *J. Biomed. Mater. Res.* **39**, 52–60.

Lanza, R. P., and Chick, W. L., eds. (1994). "Pancreatic Islet Transplantation," Vol. III. R. G. Landes/CRC Press, Austin, TX.

Lanza, R. P., and Chick, W. L. (1995). Encapsulated cell therapy. *Sci. Am. Sci. Med.* **2**, 16.

Lanza, R. P., Butler, D. H., Borland, K. M. *et al.* (1991). Xenotransplantation of canine, bovine, and porcine islets in diabetic rats without immunosuppression. *Proc. Natl. Acad. Sci. U.S.A.* **88**, 11100.

Lanza, R. P., Borland, K. M., Lodge, P. *et al.* (1992). Treatment of severely diabetic, pancreatectomized dogs using a diffusion-based hybrid pancreas. *Diabetes* **41**, 886.

Lanza, R. P., Sullivan, S. J., and Chick, W. L. (1992b). Islet transplantation with immunoisolation. *Diabetes* **41**, 1503.

Lanza, R. P., Butler, D. H., Borland, K. M. *et al.* (1992c). Successful xenotransplantation of a diffusion-based biohybrid artificial pancreas: A study using canine, bovine, and porcine islets. *Transplant. Proc.* **24**, 669.

Lanza, R. P., Borland, K. M., Staruk, J. E. *et al.* (1992d). Transplantation of encapsulated canine islets into spontaneously diabetic BB/Wor rats without immunosuppression. *Endocrinology (Baltimore)* **131**, 637.

Lanza, R. P., Beyer, A. M., Staruk, J. E., and Chick, W. L. (1993). Biohybrid artificial pancreas: Longterm function of discordant islet xenografts in streptozotocin diabetic rats. *Transplantation* **56**, 1067.

Lanza, R. P., Kuhtreiber, W. M., Beyer, A. *et al.* (1994a). Humoral response to encapsulated islets. *Transplant. Proc.* **26**, 3346.

Lanza, R. P., Beyer, A. M., and Chick, W. L. (1994b). Xenogeneic humoral responses to islets transplanted in biohybrid diffusion chambers. *Transplantation* **57**, 1371.

Lanza, R. P., Kuhtreiber, W. M., Ecker, D. *et al.* (1995a). Xenotransplantation of porcine and bovine islets without immunosuppression using uncoated alginate microspheres. *Transplantation* **59**, 1377.

Lanza, R. P., Ecker, D., Kuhtreiber, W. M. *et al.* (1995b). A simple method for transplanting discordant islets into rats using alginate gel spheres. *Transplantation* **59**, 1486.

Lanza, R. P., Lodge, P., Borland, K. M. *et al.* (1995c). Transplantation of islet allografts using a diffusion-based biohybrid artificial pancreas: Long-term studies in diabetic, pancreatectomized dogs. *Transplant. Proc.* **25**, 978.

Lanza, R. P. *et al.* (1995d). *Tissue Eng.* **1**.

Lanza, R. P., Jackson, R., Sullivan, A., Ringeling, J., McGrath, C., Kuhtreiber, W. M., and Chick, W. L. (1999). Xenotransplantation of cells using biodegradable microcapsules. *Transplantation* **67**, 1105–1111.

Lim, F., and Sun, A. M. (1980). Microencapsulated islets as bioartificial endocrine pancreas. *Science* **210**, 908–910.

Lum, Z. P., Tai, P., Krestow, I., Tai, T., and Sun, A. M. (1992). Xenografts of microencapsulated rat islets into diabetic mice. *Transplantation* **53**, 1180–1183.

Makowka, L., Falk, R. E., Rostein, L. E. *et al.* (1980). Cellular transplantation in the treatment of experimental hepatic failure. *Science* **210**, 901.

Mito, M., and Sawa, M. (1996). Hepatocyte transplantation. *In* "1996/97 Yearbook of Cell and Tissue Transplantation" (R. P. Lanza and W. L. Chick, eds.). Kluwer Academic Press, Dordrecht, The Netherlands.

Phelps, C. H., Gage, F. H., Growdon, J. H. *et al.* (1989a). Potential use of nerve growth factor to treat Alzheimer's disease. *Science* **243**, 11.

Phelps, C. H., Gage, F. H., Growdon, J. H. *et al.* (1989b). Potential use of nerve growth factor to treat Alzheimer's disease. *Neurobiol. Aging* **10**, 205.

Platt, J. L., and Bach, F. H. (1991). The barrier to xenotransplantation. *Transplantation* **52**, 937.

Ricker, A., Bhatia, V., Bonner-Weir, S. *et al.* (1987). Microencapsulated xenogeneic islet grafts in NOD mouse: Dexamethasone and inflammatory response. *Cold Spring Harbor Symp.* October, p. 53A.

Roberts, H. R. (1989). The treatment of hemophilia: Past tragedy and future promise. *N. Engl. J. Med.* **321**, 1188.

Sagen, J. (1996). Chromaffin cell transplantation. *In* "1996/97 Yearbook of Cell and Tissue Transplantation" (R. P. Lanza and W. L. Chick, eds.). Kluwer Academic Press, Dordrecht, The Netherlands.

Schlumpf, R., Largiader, F., Uhlschmid, G. K. *et al.* (1986). Is cyclosporine toxic for transplanted pancreatic islets? *Transplant. Proc.* **28**, 1169.

Schumacher, J. M., Short, M. P., Hyman, B. T. *et al.* (1991). Intracerebral implantation of nerve growth factor-producing fibroblasts protects striatum against neurotoxic levels of excitatory amino acids. *Neuroscience* **45**, 561.

Soon-Shiong, P., Lu, Z. N., Lanza, R. P. *et al.* (1990). An *in vitro* method of assessing the immunoprotective properties of microcapsule membranes using pancreatic and tumor cell targets. *Transplant. Proc.* **22**, 754.

Soon-Shiong, P., Feldman, E., Nelson, R., Heintz, R., Yao, Q., Yao, Z. *et al.* (1993). Long-term reversal of diabetes by the injection of immunoprotected islets. *Proc. Natl. Acad. Sci. U.S.A.* **90**, 5843–5847.

Soon-Shiong, P., Heintz, R. E., Merideth, N., Yao, O. X., Yao, Z., Zheng, T. *et al.* (1994). Insulin independence in a Type 1 diabetic patient after encapsulated islet transplantation. *Lancet* **343**, 950–951.

Sun, A. M., Cai, Z., Shi, Z. *et al.* (1986). Microencapsulated hepatocytes as a bioartificial liver. *Trans. Am. Soc. Artif. Intern. Organs* **32**, 39.

Sun, A. M., Cai, Z., Shi, Z. *et al.* (1987). Microencapsulated hepatocytes: An *in vitro* and *in vivo* study. *Biomater. Artif. Cells, Artif. Organs* **15**, 1483.

Sun, Y., Ma, X., Zhou, D., Vacek, I., and Sun, A. M. (1996). Normalization of diabetes in spontaneously diabetic cynomolgus monkeys by xenografts of microencapsulated porcine islets without immunosuppression. *J. Clin. Invest.* **98**, 1417–1422.

Sutherland, D. E. R., Gruessner, R. W. G., and Gores, P. F. (1994). Pancreas and islet transplantation. *Transplant. Rev.* **8**, 185.

Tessler, A. (1996). Neural transplantation: Spinal cord. *In* "1996/97 Yearbook of Cell and Tissue Transplantation" (R. P. Lanza and W. L. Chick, eds.). Kluwer Academic Press, Dordrecht, The Netherlands.

Van Schilfgaarde, R., van der Burg, M. P. M., van Suylichem, H. G. *et al.* (1986). Does cyclosporin influence beta cell function? *Transplant. Proc.* **28**, 1175.

Wang, T. G. (1997). New technologies for artificial cells. *Artif. Organs* **21**, No. 6.

Wang, T. G., Lacik, I., Brissova, M., Anilkumar, A. V., Prokop, A., Hunkeler, D., Green, L., Shahrokhi, K., and Powers, A. C. (1997). An encapsulation system for the immunoisolation of pancreatic islets. *Nat. Biotechnol.* **15**, 358–362.

Wang, Y., Hao, L., Gill, R. *et al.* (1987). Autoimmune diabetes in NOD mouse is L3T4 T-lymphocyte dependent. *Diabetes* **36**, 535.

Weber, C. J., Zabinski, S., Koschitzky, T., Wicker, L., Rajotte, R., D'Agati, V., Peterson, L., Norton, J., and Reemtsma, K. (1990). The role of CD4$^+$ helper T cells in the destruction of microencapsulated islet xenografts in NOD mice. *Transplantation* **49**, 396–104.

Widner, H., Brundin, J., Rehncroma, S. *et al.* (1991). Transplanted allogeneic fetal dopamine neurons survive and improve motor function in idiopathic Parkinson's disease. *Transplant. Proc.* **23**, 793.

Winn, S. R., Hammang, J. P., Emerich, D. F. *et al.* (1994). Polymer-encapsulated cells genetically modified to secrete human growth factor promote the survival of axotomized septal cholinergic neurons. *Proc. Natl. Acad. Sci. U.S.A.* **91**, 23.

Xu, K., Hercules, D., Lacik, I., and Wang, T. G. (1998). Atomic force microscopy used for the surface characterization of microcapsule immunoisolation devices. *J. Biomed. Mater. Res.* **41**, 461–467.

CHAPTER 37

PARATHYROID

C. Hasse, A. Zielke, T. Bohrer, U. Zimmermann, and M. Rothmund

HISTORY OF PARATHYROID ALLOTRANSPLANTATION

Although autotransplantation of parathyroid tissue is a well-established clinical procedure (Wells and Christiansen, 1974), allotransplantation of the parathyroid gland still is at the level of experimental surgery. In 1884 the German physiologist Moritz Schiff was the first to demonstrate that tetany, occurring after total thyroidectomy, could be reversed by replantation of the thyroid (Welbourn, 1990). Six years later, Loeb, who investigated the cramps of isolated muscles submersed in a low-calcium saline solution *in vitro,* demonstrated that addition of calcium to the medium resolved the spasms, thus uncovering the relation between hypocalcemia and tetany (Loeb, 1899–1900). In 1909 MacCallum and Voegtlein discovered that hypocalcemia after thyroidectomy was caused by parathyroidectomy. Hanson in 1924 and Collip in 1925 were the first to try to reverse this effect by intramuscular injections of parathyroid extracts to patients with post-thyroidectomy tetany. For limited periods of time, they were able to restore normocalcemia in these patients.

In 1972, almost 90 years after Schiff's first report of thyroid replantation, Tochilin (1972a,b,c) performed the first allotransplantation of fetal thyroid and parathyroid glands in humans and documented long-term reversal of hypothyroidism, but only short-term reversal of hypoparathyroidism. Tochilin later combined this treatment with simultaneous transplantation of the pancreas, and achieved similar results (Kiprenskii *et al.,* 1990). The first allotransplantation of parathyroid glands with systemic immunosuppression was reported by Groth *et al.* in 1973, in a patient who had previously undergone successful renal transplantation. Because the patient became normocalcemic with no further need for calcium supplementation, the transplant apparently was functioning; however, neither the further course of the patient nor the period of time that normocalcemia was maintained is known. The potential long-term outcome of simultaneous renal and parathyroid allotransplants with systemic immunosuppression was outlined in a case report of 1979. In this particular case, the function of the parathyroid tissue was documented by determination of parathyroid hormone (transplant-bearing versus non-transplant-bearing forearm) as well as by histology from a biopsy of the allotransplanted tissue. Function continued until complete rejection of both organs 30 months postoperatively (Ross *et al.,* 1979). Also in 1979, two children with hypoparathyroidism due to DiGeorge syndrome (see later) received parathyroid allotransplants without immunosuppression. Because of a severe concomitant immune defect, both children died too soon to allow for an assessment of allograft function (Wells *et al.,* 1979).

Kunori *et al.* (1991) also completely ignored the ensuing immunologic reaction following transgenous transplantation. They reported the first successful long-term therapy of postoperative hypoparathyroidism by repetitive allotransplantation of fresh and cryoconserved parathyroid tissue particles in 1991. The patient received initial immunosuppressive medical therapy for the first 2 days postoperatively. For a period of 6 months, the patient reportedly had normal levels of serum calcium and improvement of symptoms with half of the substitutive therapy (Kunori *et al.,* 1991). One year later, Segerberg *et al.* (1992) performed the first successful syngenous transplantation of

Table 37.1 Historic synopsis of significant achievements along the way to parathyroid allotransplantation

Year	Investigator	Achievement
1884	Schiff	Reversal of postthyroidectomy tetany by thyroid replantation
1924/ 1925	Hanson Collip	Treatment of hypoparathyroidism with injections of parathyroid extract
1960	Swan *et al.*	First attempt of parathyroid tissue immunoisolation by coating the tissue with millipore membranes
1967	Fischer *et al.*	Increase of parathyroid graft function by choosing the anterior eye chamber as transplant site
1972	Tochilin	Homologous *en bloc* transplantation of the thyroid–parathyroid complex
1973	Groth *et al..*	First allotransplantation of the parathyroid with systemic immuno-suppression in secondary hypoparathyroidism
1974	Raaf *et al.*	First attempt to reduce parathyroid tissue immunogenicity by tissue culture passage
1977	Starling *et al.*	First use of antidonor serum for allotransplantation of the parathyroids
1979	Ross *et al.*	Survival for 30 months of allotransplanted parathyroid allotransplant with systemic immunosuppressions
1979	Wells *et al.*	First allotransplantation of the parathyroid without systemic immunosuppression in primary hypoparathyroidism
1983	Sollinger *et al.*	Function for 15 months of a human parathyroid following ionizing irradiation and nud mouse *in vivo* passage of the parathyroid tissue
1989	Fu and Sun	First successful immunoisolation of parathyroid cells using micro-encapsulation: function *in vivo* for 12 weeks
1990	Kiprenski *et al.*	First homologous *en bloc* transplantation of parathyroid, thyroid gland, and pancreas
1991	Kunori *et al.*	First successful treatment of postoperative hypoparathyroidism by repeated allotransplantation of parathyroid tissue particles after initial immunosuppression
1992	Segerberg *et al.*	First successful syngenous transplantation of human parathyroids without immunosuppression
1993	Yao *et al.*	Intracerebral parathyroid allotransplant survive for 12 weeks without systemic immunosuppression
1995	Anton *et al.*	First allotransplantation of single parathyroid cells after reduction of MHC I/II antigens: transplants funtion for 3 months with mild immunosuppression
1996	Tolloczko *et al.*	First allotransplantation of single parathyroid cells without immunosuppression: hormonally active for up to 1 year
1997	Hasse *et al.*	First successful clinical allotransplantation of cultured, micro-encapsulated parathyroid tissue without immunosuppression

parathyroid glands in identical twins, one of whom suffered from permanent postoperative hypoparathyroidism. A chronology of parathyroid allotransplantation is given in Table 37.1.

Many subsequent *in vitro* investigations and experimental animal studies have been performed to accomplish parathyroid allotransplantation without the need of postoperative immunosuppression. The earlier assumption, that parathyroid tissue would be less immunogenic than other tissue types, was proved wrong (Wells *et al.,* 1973a,b). In general, two basic lines of research have evolved. Some have tried to transplant parathyroid tissue to regions with apparently little or no immunologic competence. For instance, Fisher *et al.* (1967) used the anterior eye chamber, and a Chinese group reported transplantation to the lateral ventricle of the brain (Dib-Kuri *et al.,* 1975) and to the scrotum. However, all three attempts did not achieve long-term graft function (Fisher *et al.,* 1967; Dib-Kuri *et al.,* 1975; Yao *et al.,* 1993). Others have aimed to reduce the immunogenicity of the tissue to be transplanted. Ultraviolet and ionizing irradiation, various tissue culture methods, coating of the graft with millipore membranes, nude mice *in vivo* passage, as well as treating the recipient with antidonor serum have been employed in an attempt to reduce the immunologic response. However, a reproducible increase of *in vivo* graft survival for more than just several weeks has not yet been accomplished (Swan *et al.*, 1960; Raaf *et al.*, 1974; Starling *et al.*, 1977; Gough and Finnimore, 1980). Only Sollinger and co-workers (1983) were able to obtain functioning parathyroid transplants in two patients without postoperative immunosuppression for a period of 15 months, following pretreatment of parathyroid tissue with ionizing irradiation and subsequent nude mouse *in vivo* passage. The course of both patients and whether the patients further required substitutive therapy are, however, unclear. Although the authors concluded that their procedure was effective and very promising, there have been no further clinical reports on this combination of techniques (Sollinger *et al.*, 1983). Dyna-beads, i.e., magnetic microspheres loaded with specific antibodies, effectively reduce major histocompatability complex (MHC I and MHC II) antigen concentrations of single parathyroid cells. Using this technique, the first successful transplantation of single parathyroid cells was reported. With the patients receiving "mild immunosuppression," transplants were functioning for periods of up to 12 weeks, allowing for a reduction of oral calcium and vitamin D substitution (Anton *et al.*, 1995). A combination of this technique with cell separation and *in vitro* culture passage yielded hormonally active allotransplants without immunosuppression for periods of 3 months and—in one case—for a period of 1 year (Tolloczko *et al.*, 1996). In 1989 Fu and Sun published the first short-term *in vivo* results (12 weeks) of functioning parathyroid allotransplantations without postoperative immunosuppression, by implanting parathyroid cells coated with an alginate, a natural substrate used for immunoisolation of cells. The technique of microencapsulation has since been continuously enhanced. Several long-term animal experiments have documented long-term survival of syn- and transgenous parathyroid transplants without any immunosuppressive therapy (Hasse *et al.*, 1994, 1996, 1997a,b,c, 1998a,b). The first two clinical cases in patients with postoperative hypoparathyroidism receiving successful parathyroid allotransplants without immunosuppression have been reported, utilizing the technique of microencapsulation (Hasse *et al.*, 1997d).

HYPOPARATHYROIDISM

DEFINITION

Hypoparathyroidism is defined as a reduction of the concentration of intact parathyroid hormone and (total) serum calcium to subnormal levels, caused by impaired or absent parathyroid function (Fitzpatrick and Bilezikian, 1990; Aurbach *et al.*, 1992; Habener *et al.*, 1995). Persistent, symptomatic hypoparathyroidism is diagnosed when the parathyroid hormone deficiency syndrome persists for more than 6 months and/or is accompanied by complaints or sequelae commonly associated with hypoparathyroidism.

ETIOLOGY AND CLASSIFICATION

Symptomatic hypoparathyroidism may occur in a primary and a secondary form (Potts, 1995). Cases in which the etiology is unknown or cannot be established with certainty are referred to as idiopathic hypoparathyroidism (Ziegler, 1976). In principle, all of the different forms of symptomatic hypoparathyroidism may be an indication for parathyroid allotransplantation. Therefore, all of the parathyroid hormone deficiency syndromes, including some rare ones, are included in the following discussions (Fig. 37.1).

Primary hypoparathyroidism (congenital hypoparathyroidism)

Congenital forms of persistent hypoparathyroidism comprise three distinct entities: autoimmune hypoparathyroidism, DiGeorge syndrome, and Kenny–Linarelli syndrome. Only few reports of these exist in the world literature, not allowing for estimates of the prevalence of these entities. They are, however, quite rare compared to the acquired parathyroid hormone deficiency syndromes.

Autoimmune hypoparathyroidism

Autoimmune hypoparathyroidism most commonly occurs as a part of the polyglandular autoimmune syndrome, autoimmune polyendocrinopathy/candidiasis/ectodermal dystrophy (APECED). Two types of autoimmune polyendocrinopathy have to be differentiated. Autoim-

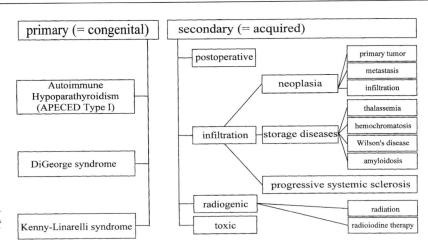

Fig. 37.1. Etiology and classification of diseases of persistent hypoparathyroidism.

mune hypoparathyroidism occurs only in association with type I, in which half of the cases are familial (autosomal recessive) and the other half are sporadic (Windeck and Reinwein, 1992). With variable frequency, type I is associated with pernicious anemia, Hashimoto thyroiditis, early-onset ovarian insufficiency, diabetes mellitus type I, and/or mucocutaneous candidiasis. Other ectodermal problems associated with type I autoimmune polyendocrinopathy include dental or nail dystrophy, alopecia areata, vitiligo, and keratinophathy (Fitzpatrick and Bilezikian, 1990; Aurbach *et al.,* 1992; Windeck and Reinwein, 1992; Habener *et al.,* 1995). It is only in the familial cases that one may also encounter Addison's disease (Windeck, 1996). Persistent hypoparathyroidism is seen only in type I autoimmune polyendocrinopathy, with an incidence of 80%. Patients suffering from type II polyglandular autoimmune syndrome (Schmidt syndrome) have normal parathyroid function (Fitzpatrick and Bilezikian, 1990; Aurbach *et al.,* 1992; Habener *et al.,* 1995).

DiGeorge syndrome

Hypoparathyroidism in DiGeorge syndrome is due to agenesis of the parathyroid and the thymus gland caused by a failure of the development of the third and fourth branchial pouch. It may be associated with severe immunodeficiency, facial dysplasia, and anomalies of the heart (Fitzpatrick and Bilezikian, 1990; Aurbach *et al.,* 1992; Windeck and Reinwein, 1992; Habener *et al.,* 1995; Windeck, 1996).

Kenny–Linarelli syndrome

Kenny–Linarelli syndrome is an autosomal recessively inherited disease characterized by cretinism, delayed occlusion of the fontanels, myopia or hyperopy, cortical hyperostosis of cortical bones, and episodes of hypocalcemic tetanies due to hypoparathyroidism. Exogenous parathyroid hormone has been administered to patients affected by this syndrome, resulting in an increase of cAMP excretion in urine (Fitzpatrick and Bilezikian, 1990; Aurbach *et al.,* 1992; Keck and Kuhlencordt, 1992; Habener *et al.,* 1995).

Secondary hypoparathyroidism (acquired hypoparathyroidism)

Injury to the parathyroid glands during operative procedures in the neck is the main reason for acquired hypoparathyroidism. The parathyroids are either inadvertently removed or their blood supply is impaired by iatrogenic hypo- or devascularization, edema, or hematoma (Ziegler, 1976). The incidence of postoperative hypoparathyroidism is related to the type of surgery performed as well as to the event of prior surgery to the neck. Accordingly, the incidence may vary considerably, from 0 to 33% (Gann and Paone, 1979). Parathyroid hormone deficiency may also very occasionally be caused by neoplastic infiltration of the parathyroid glands. This has been reported for primary tumors of the parathyroid local tumors of various etiologies, and metastases (Ziegler, 1976, 1987; Keck and Kuhlencordt, 1992). Similar to the rare problems of the parathyroids secondary to progressive systemic sclerosis, storing of copper or iron, or amyloidosis, only very few cases have been reported or are anecdotally mentioned in textbooks (Fitzpatrick and Bilezikian,

1990; Aurbach *et al.,* 1992; Keck and Kuhlencordt, 1992; Habener *et al.,* 1995; Windeck, 1996). This is also true for some isolated cases of hypoparathyroidism occurring after irradiation therapy, radioiodine treatment, and chemotherapy (Fitzpatrick and Bilezikian, 1990; Aurbach *et al.,* 1992; Keck and Kuhlencordt, 1992; Habener *et al.,* 1995).

PATHOGENESIS

The concentration of ionized extracellular calcium is regulated by parathyroid hormone and vitamin D (calcitriol) within tight margins of variance. Due to a reduction of parathyroid hormone-dependent osteoclastic bone resorption, chronic parathyroid hormone deficiency invariably results in hypocalcemia. The symptoms of parathyroid hormone deficiency are caused by hypocalcemia (Ziegler, 1976; Fitzpatrick and Bilezikian, 1990; Aurbach *et al.,* 1992; Habener *et al.,* 1995). Serum calcium concentrations may drop to a minimum of 1.5 mmol/liter. Below this threshold range, the gradient between the calcium concentration of the bone and that of the unsaturated extracellular fluid is such that bone-derived calcium will effectively go into solution independent of parathyroid hormone (Ziegler, 1976). In this respect, the often observed hyperphosphatemia, resulting from an increase in tubulorenal reabsorbtion, is of minor importance.

SYMPTOMS

The signs and symptoms of chronic hypoparathyroidism were first described in 1941 by Lachmann. One of the main features of symptomatic hypoparathyroidism is increased neuromuscular excitability resulting from hypocalcemia (Ziegler, 1976; Wagner, 1980; Fitzpatrick and Bilezikian, 1990; Aurbach *et al.,* 1992). Hyperexcitability may cause severe muscular convulsions without loss of consciousness (Keck and Kuhlencordt, 1992; Habener *et al.,* 1995). They are either appreciable as tonoclonic convulsions or isolated muscular spasms, which often occur as symmetric spasms of the extremities (Wagner, 1980; Fitzpatrick and Bilezikian, 1990; Aurbach *et al.,* 1992). The characteristic clinical sign of carpopedal tetany induced by hypocalcemia is referred to as "obstetricians' hand" and talipes in the case of tarsopedal tetany (Ziegler, 1976, 1987; Wagner, 1980; Fitzpatrick and Bilezikian, 1990; Keck and Kuhlencordt, 1992; Habener *et al.,* 1995). Spasms of the perioral muscles cause a typical facies commonly referred to as "fish mouth appearance" (Ziegler, 1987). The bronchi and the respiratory muscles may also be affected, causing laryngeal spasms and severe dyspnoa (Aurbach *et al.,* 1992; Keck and Kuhlencordt, 1992). Occasionally, hypocalcemia may also be responsible for cramping abdominal pain, hence the term visceral tetany, and for stenocardic complaints in some patients (Ziegler, 1987; Fitzpatrick and Bilezikian, 1990; Aurbach *et al.,* 1992; Habener *et al.,* 1995; Windeck, 1996).

All of these spasms may occur without any appreciable precipitating cause, yet more often in situations of increased physical or psychic stress. They may last for only a few minutes as well as for up to several hours. Females are more often affected than males and a periodic increment in frequency as well as severity has been reported, suggesting a relation to the menstruation cycle. Also, these spasms apparently have a tendency to occur more often during spring and autumn (Ziegler, 1976).

The typical clinical appearance of a tetany is easily diagnosed; however, in only one-fourth of patients is it fully appreciable (Ziegler, 1987). This may explain why less obvious cases of "latent" hypoparathyroidism are often overlooked. Muscular hyperexcitability may be discrete and appreciated only as aching, tingling, and numbness of extremities ["tingling paraesthesia" (Ziegler, 1976; Wagner, 1980; Habener *et al.,* 1995)].

A comprehensive synopsis of all the symptoms associated with persistent hypoparathyroidism is given in Table 37.2.

DIAGNOSIS

An in-depth patient history and physical examination are the keystones of making the diagnosis of hypoparathyroidism. Symptoms and sequelae of the disease as described previously are considered; they are either reported by the patient or are found as a result of a meticulous questioning and exam. A number of clinical tests can be employed. For instance, tapping the cheek of the patient just anterior to the ear, may cause reproducible contractions of the facial muscles, a sign commonly referred to as Chvostek's sign (Ziegler, 1987; Keck and Kuhlencordt, 1992; Habener *et al.,* 1995; Windeck, 1996). Such contractions of the facial muscles, conducted by the third

Table 37.2. Symptoms of hypoparathyroidism [a]

Increased neuromuscular excitability
  Convulsions in conscious individuals
    Spasms
      Painful, symmetric, carpopedal tetany: "obstetricians' hand"
      Carpopedal tetany: talipes
      Laryngeal and bronchial spasms, spasms of the respiratiory musculature: dyspnea
      Spasms of perioral muscles: "fish mouth appearance"
      Stenocardia
      Cramping abdominal pain: visceral tetany
  Convulsion
    Tingling paraesthesia (facial, perioral, extremities)
Psychological alterations—endocrine psychosyndrome
  Depressive mood
  Hallucination
  Anxiety, restlessness, confusion, agitation, irritability
  Progressive mental retardation
Impaired hearing, loss of hearing
Pain of joints and bones
Feeling of coldness of the extremities
Diarrhea
Koprostasis, constipation
Muscular pain and weakness
Exhaustion

[a]From Ziegler (1976, 1987), Wagner (1980), Fitzpatrick and Bilezikian (1990), Aurbach et al. (1992), Keck and Kuhlencordt (1992), Habener et al. (1995) and Windeck (1996).

branch of the facial nerve, may also be found in as many of 15% of healthy subjects. They are, however, considered characteristic of tetanic predisposition induced by hypocalcemia, if contractions in the area of distribution of the second and third branch of the facial nerve are elicited (Ziegler, 1976). Trousseau's sign is provoked by inflating a blood pressure cuff placed around the upper arm to pressures above the mean arterial pressure for several minutes. The test is regarded positive if carpopedal spasms are noted within 3 min. The test is positive in less than 1% of normocalcemic subjects (Ziegler, 1987; Keck and Kuhlencordt, 1992; Aurbach et al., 1992; Habener et al., 1995; Windeck, 1996). This sign may also be observed after 3 min of provocative hyperventilation (Ziegler, 1976; Fitzpatrick and Bilezikian, 1990). Both Chvostek's and Trousseau's signs were found in 55–62% of patients with postoperative hypoparathyroidism. Finally, the so-called Lust's sign is represented by reproducible extraplantar dorsiflections of the foot, in response to tapping the fibular nerve at the height of the head of the fibula (Ziegler, 1976).

Laboratory tests will reveal low serum concentrations of total and ionized calcium, parathyroid hormone, and magnesium, as well as an increase of serum phosphate. Urinary excretion of calcium, phosphate, and cAMP is decreased. Normal serum levels of creatinine and urea, as well as normal levels of albumin, exclude renal insufficiency and syndromes of malassimilation as the cause of hypocalcemia, rendering the diagnosis of hypoparathyroidism very likely. Occasionally, electrocardiograms may show prolongation of the QT interval, caused by an increase in ST segments (Keck and Kuhlencordt, 1992; Windeck and Reinwein, 1992). Furthermore, T-wave inversion or peaking may be observed (Keck and Kuhlencordt, 1992).

Radiographic examinations in patients with congenital or long-standing persistent hypoparathyroidism may demonstrate scattered skeletal condensations combined with areas of demineralized bone as well as periarticular osteophytes (Windeck and Reinwein, 1992). Some 20% of patients will have evidence of calcifications of the basal ganglia or gray matter by plain radiograms or by computed tomograms of the skull (Keck and Kuhlencordt, 1992).

The diagnosis should be substantiated by functional tests. The diagnosis of hypoparathy-

Table 37.3. Sequelae of hypoparathyroidism[a]

Psychological alterations—endocrine psychosyndrome
    Psychosis
    Depression
Metastatic calcifications
    Basal ganglia (symmetric; deposits of hydroxyapatite)
        Morbus Parkinson and/or Morbus FAHR:
            Headaches
            Speaking disability
            Dementia
            Pyramidal symptoms (e.g., epileptiformic convulsions)
    Subcutaneous
    Lung
    Lens: cataract (tetanic glaucoma)
    Innenohr: loss of hearing
    Periarticular osteophytes
Ectodermal alterations
    Trophic impairment and infections of the skin and *anhangsorgane*
        Alopecy
        Flaky skin
        Candidiasis
Increase of bone density and calcium content of bones
Cardiomyopathy, unspecific enlargement of the heart
Alterations of dentine, anomalies of the teeth[b]
Deformations of the skeleton together with diminished growth[b]
Osteoporosis
Moon face

[a]From Ziegler (1976, 1987), Wagner (1980), Fitzpatrick and Bilezikian (1990), Aurbach *et al.* (1992), Keck and Kuhlencordt (1992), Windeck and Reinwein (1992), Habener *et al.* (1995), and Windeck (1996).
[b]Mainly in children.

roidism is very likely if administration of intact parathyroid hormone (400 IU i.v.) yields a 5- to 10-fold increase of urinary phosphate excretion. This test is called the Ellsworth–Howard test (Ziegler, 1976; Keck and Kuhlencordt, 1992). If latent hypoparathyroidism is suspected, an ethylendiamintetraacetate test (EDTA test) may be diagnostic. The EDTA chelate binds calcium potently. In healthy subjects, a decrease of ionized calcium following infusion of EDTA stimulates parathyroid (PTH) liberation, resulting in a fairly rapid restoration of normal serum calcium levels. In patients with latent hypoparathyroidism, this autoregulatory loop is deficient. During the test, serum calcium levels are repetitively determined before and after infusion of 50–70 mg sodium-EDTA per kilogram of body weight. In contrast to healthy individuals, a typical patient will not reach basal values of serum calcium within 12 hr after administration of the chelate (Ziegler, 1976). Another functional test can be used to differentiate pseudohypoparathyroidism from hypoparathyroidism. Healthy individuals as well as patients with primary hypoparathyroidism, administered infusions of synthetic human 1-34 PTH (30 µg within 5 min), will show a fourfold increase of 1,25-dihydroxy vitamin D within 24 hr, in contrast to those with pseudohypoparathyroidism, who will have no evidence of increasing levels of vitamin D. Electromyography (EMG) is a useful study to substantiate increased neuromuscular excitability. Doublets and triplets during hyperventilation evidence tetanic predisposition (Ziegler, 1976). If these signs and symptoms of hypoparathyroidism are accompanied by facial dysplasia in a child, the diagnosis of the rare DiGeorge syndrome is made. Persistent hypoparathyroidism combined with chronic mucocutaneous candidiasis leads to the diagnosis of APECED type I (Fitzpatrick and Bilezikian, 1990; Aurbach *et al.*, 1992; Habener *et al.*, 1995). Table 37.3 depicts a complete investigative procedure in patients with suspected hypoparathyroidism.

THERAPY OF HYPOPARATHYROIDISM

Symptomatic therapy

Acutely symptomatic patients experiencing tetanic convulsions readily respond to slowly administered intravenous calcium (calcium gluconate, maximum of 2 ml/min) (Windeck, 1996). Care must be taken not to exceed a maximum dose of 15 mg calcium per kilogram of bodyweight (Ziegler, 1976). Chronic hypocalcemia due to persistent hypoparathyroidism is treated by continuous oral calcium. If relief of symptoms and serum calcium concentrations in the low normal range are not achieved with calcium alone, vitamin D (calcitriol) is given additionally (Windeck, 1996). Until the introduction of vitamin D in the 1930s, hypoparathyroidism was a life-threatening condition. Calcitriol raises serum calcium concentrations by osteolysis and by increasing intestinal absorption of calcium. Regular serum calcium and urinary calcium excretion tests are needed to evaluate the effectiveness of calcitriol therapy. The urinary calcium excretion must not exceed 250 mg/day (Windeck and Reinwein, 1992).

The overall course of persistent hypoparathyroidism mainly depends on how early therapy is initiated, its stringent implementation, and life-long control. Without adequate therapy and control, all of the symptoms and sequelae of the disease are imminent. Long-term therapy of patients in which the disease takes a rather moderate course with only minor symptoms is most challenging. A milder symptomatology is most commonly encountered in patients with postoperative hypoparathyroidism, representing by far the majority of patients with persistent hypoparathyroidism. Unspecific conditions such bone pain, loss of energy, exhaustion, depressive mood, and latent convulsions are either ignored or unrecognized for many years until irreversible consequences arise (Ziegler, 1976, 1987). Permanent care of this large number of patients at risk is of particular importance.

However, even with rigorous, long-term therapy, sequelae of hypoparathyroidism may still occur. Chronic parathyroid hormone deficiency invariably results in a loss of activity of bone metabolism, resulting in osteoporosis and bone pain (Wagner, 1980). On the other hand, calcitriol lacks the full renal calcium-retaining ability of parathyroid hormone. Patients administered vitamin D inadvertently develop hypercalciuria, although their serum calcium levels may be normal or in the low normal range. Accordingly, these patients have an increased risk of nephrolithiasis, nephrocalcinosis, and subsequent impairment of renal function (Potts, 1995). Moreover, these patients have an increased risk of developing calcium depositions in the soft tissues, as well as accelerated arteriosclerosis.

Because the therapeutic range of calcium and vitamin D supplementation is small, patients are at risk of drug overdose at all times, manifesting as hypercalcemia syndrome of variable degree. Also, oral calcium and vitamin D supplements will level out the imbalances of calcium and phosphate metabolism at best. They are, however, unable to replace the many other metabolic actions of parathyroid hormone, which remain permanently untreated. It is therefore very important to provide definitive therapy for parathyroid hormone deficiency.

Causative therapy

Treatment of the parathyroid hormone deficiency syndrome may involve administration of parathyroid hormone or transgenous transplantation of parathyroid tissue. In 1925 Collip described a hormone of the parathyroid glands to be the main factor in calcium homeostasis. However, it was not until 1978 that the hormone was sequenced in its entirety (Keutmann *et al.,* 1978). Meanwhile, genetically engineered human recombinant 1-34 PTH has become available. The first clinical trials indicate that, in principle, permanent hypoparathyroidism can be treated by recombinant 1-34 PTH (Winer *et al.,* 1996). For the time being, the short half-life and the *in vivo* instability of synthetic PTH do not allow for continuous, sufficient treatment. Moreover, resistance to the hormone after repetitive application has been described. Therefore, a major focus of continued research is to develop a tracer-bound hormone preparation, suitable for long-term replacement therapy.

Replacing the parathyroid organ is a much more promising approach if physiologic calcium homeostasis is to be achieved. Currently, parathyroid allotransplantation has been performed only with continuous systemic immunosuppression. This approach is not justified, because hypoparathyroidism is rarely ever a life-threatening condition and because systemic immunosuppression causes significant side effects. The advantages of functional replacement of the organ are undisputed. Therefore, allo- and xenotransplantation of parathyroid glands without a need for systemic immunosuppression are a matter of intense research activity. Several different procedures have progressed from the stage of animal experimentation to that of first clinical trials. Encouraging results have been reported, particularly for those techniques that aim to either reduce the immunogenicity of the tissue to be transplanted or to immunoisolate it. At present, however, only first clinical observations have been published and none of the methods has yet been established as a clinical procedure (Anton *et al.,* 1995; Tolloczko *et al.,* 1996; Hasse *et al.,* 1997). Which of the methods, or combinations thereof, are best suited for long-term treatment of patients with symptomatic postoperative hypoparathyroidism awaits further clinical trials in larger series of patients.

MICROENCAPSULATION

Microencapsulation refers to a process by which gas, fluids, or solid active ingredients are encapsulated by a shell substance, producing microspheres of a size ranging from 1 μm to 7 mm, isolating the incorporated core substance from the surrounding environment (Franjione and Vasishtha, 1996). In principle, the encapsulated substance may be released by three different ways. One is mechanical disruption of the capsule. For instance, microcapsules applied to the back of carbonless copy paper burst and release the incorporated dye when pressure is applied with a pen point. This application of the microencapsulation technology was first introduced by the American chemist Green in the 1930s (Fanger, 1974). Second, the capsule may slowly erode or biodegrade. This mechanism is utilized to encapsulate protein-reactive enzymes in powder detergents, which are released at defined conditions, e.g., a certain temperature. Finally, other applications make use of the fact that the contents of a microcapsule are released slowly over time by diffusion, maintaining the integrity of the capsule. This particular aspect, together with the ability of a microcapsule to isolate the core substance from its environment, have made this technology a major focus of interest in transplant medicine.

Ennis and James in 1950 devised the first apparatus by which they were able to obtain single droplets of a uniform size. They modified the principle of producing small amounts of aqueous and nonaqueous fluid spheres conceived in 1947 by Lane and Levvy and called it "droplet sizer" (Ennis and James, 1950). This apparatus was the progenitor of the microencapsulation technology from which many applications currently used in industry, basic research, and medicine have developed. The first application of this technology in the field of medicine was proposed by in 1964 Chang, who intended to develop artificial cellular systems to be used to replace blood. While investigating different carrier systems to encapsulate hemoglobin and several different enzymes simultaneously by thin nylonlike semipermeable membranes, he developed a method of microencapsulation based on interfacial polymerization. However, the reaction needs to be carried out under severe conditions, resulting in liberation of cytotoxic byproducts. Moreover, the membranes were found to be not entirely biocompatible, and the monomers are chemically active to the extent that necrosis of cells is induced. This prompted extensive efforts for alternative shell substances.

ALGINATES

Alginates are extracted from brown algae (*Macrocystis pyrifera*), which are most commonly found on the shorelines of Scotland and Great Britain. In the seventeenth century it was known that the ashes of brown algae contained high amounts of soda, which was used in the production of soap and glass. In 1811, iodine was discovered in the ashes. After less expensive methods to produce soda had been established, the production of potash and iodine kept the local algae-based industry alive. Today, alginates are used for many different purposes in industry (e.g., food, concrete, fire-resistant synthetic materials) as well as in medicine (wound dressings, prosthetics, orofacial surgery). Commercially extracted alginates are a mixture of uronic acid polymers, composed mainly of mannuroic and guluronic acid. Calcium ions are used as a cross-linking agent. The polymer is a complex three-dimensional network of varying density, depending on the concentration of the two mayor compounds. Grant *et al.* (1973) have referred to the ultrastructure of the cross-linked polymer as an "egg-box model" because of a similar pattern of layering.

In 1977 Kierstan and Bucke were the first to use alginates as a shell substance in microencapsulation. For the first time, microencapsulation of viable cells was achieved. By dropwise addition of cells immersed in sodium alginate into a calcium chloride solution, stable hydrogel beads were formed (Kierstan and Bucke, 1977). A drawback of this technique is the instability of the capsule. Biodegradation occurs as a consequence of dilution of calcium ions reacting with phosphates of the recipient. Therefore, Lim and Sun (1980) proposed to coat the alginate capsules with a second membrane of poly(L-lysine) (PLL membrane). Zimmermann and co-workers (1992) succeeded in integrating more stable ions, e.g., barium, as cross-linking agents. Appropriate incubation conditions provide highly stable, fully biocompatible alginates.

EXPERIMENTAL AND CLINICAL TRANSPLANTATION OF MICROENCAPSULATED PARATHYROID TISSUE

We adopted the well-described technique of microencapsulation with alginates already used in pancreatic islet cell transplantation and modified it extensively for transplantation of parathyroid tissue. Combining refined tissue culture techniques and microencapsulation, iso-, allo-, and xenotransplantation of microencapsulated parathyroid tissue were reproducibly achieved in an animal model of experimental hypoparathyroidism without a need for systemic immunosuppression (Hasse et al., 1994, 1997a).

Prior to the first clinical use, continued analysis of the alginate for the purpose of biocompatibility testing revealed mitogenic properties. Using patented techniques, a purified, amitogenic alginate suitable for clinical use was isolated by the Institute of Biotechnology of Würzburg, Germany (Zimmermann et al., 1992; Klöck et al., 1994). The novel alginate displayed distinctly different biochemical and biophysical properties compared to the mitogenic alginate, which required several adjustments of the purification and the microencapsulation procedures. However, subsequent in vivo studies in syn- and allogenic animal models unequivocally showed that the same results as with the mitogenic alginate were also obtained when the novel amitogenic alginate was used (Hasse et al., 1996, 1997b,c, 1998a,b). These results founded the rationale for a first clinical use of microencapsulated parathyroid allotransplants in patients suffering from symptomatic persistent postoperative hypoparathyroidism. We have published the data on the first two patients treated with this method. Twelve weeks following allotransplantation of cultured and microencapsulated hyperplastic parathyroid tissue, the two patients were normocalcemic and had normal levels of parathyroid hormone without immunosuppression. Both patients reported an impressive improvement of symptoms and sequelae of hypoparathyroidism (Hasse et al., 1997d).

ALTERNATIVE APPROACHES AND PERSPECTIVES

The feasibility of parathyroid allotransplantation to patients receiving systemic immunosuppression has already been outlined. This has repeatedly and successfully been performed during transplantation of the kidney, pancreas, heart, and liver, and, because these patients already require life-long immunosuppression, is an entirely justified approach (Groth et al., 1973; Starling et al., 1977; Wells et al., 1979; Bloom et al., 1987). For the time being, this approach is the only clinically established method that allows for a definitive therapy of persistent symptomatic hypoparathyroidism.

Reducing the immunogenicity of parathyroid tissue has been accomplished utilizing several different methods. However, except for the few already mentioned exceptions, these methods will not be of clinical relevance until long-term success is documented for clinical allografting without prolonged immunosuppression (Anton et al., 1995). It would therefore appear that the only two alternatives to allotransplantation of cultured and microencapsulated parathyroid tissue that may be of relevance for the treatment of hypoparathyroid patients without posttransplant immunosuppression are substitution of parathyroid hormone and implantation of parathyroid cells pretreated by specific tissue culture methods aiming to reduce its antigenicity. All of these approaches are at the stage of first clinical trails.

Earlier attempts to use bovine PTH were hampered by an invariable development of antibodies, resulting in hormone resistance (Melick et al., 1967). Although synthetic human PTH has become available, it appears that development of hormone resistance prevails. In 1990 Stögmann et al. reported their initial experience with two patients treated with subcutaneous injections of synthetic human 1-34 PTH (400 E daily). One patient developed resistance after only 10 weeks and the other was able to maintain normal levels of serum calcium for a period of 7 months. The further history of these patients is unknown (Stögmann et al., 1990). These observations were confirmed in another report of a larger series of patients treated for a period of 10 weeks. During this short-term trial, therapy with 1-34 synthetic PTH was successful. However, the problem of overcoming resistance during long-term treatment remains to be solved (Winer *et al.*, 1996).

In 1991 Tolloczko *et al.* reported several cases in which allografting of cultured parathyroid cells by subfascial injection to the nondominant forearm had been performed. Their series, currently comprising 23 patients, of whom some showed evidence of allograft function for up to 14 months after the transplantation without immunosuppression, has been updated (Tolloczko *et al.*, 1991, 1994, 1996). Unfortunately, the details of how these results were achieved are not entirely clear. At first, hyperplastic parathyroid tissue from patients with renal hyperparathyroidism was assessed for infiltration with passenger lymphocytes and expression of HLA antigens using immunohistochemical methods. Without a detailed description of selection criteria, 10–15 particles of parathyroid tissue that seemed most appropriate were used for *in vitro* culture. Tissue culture was initiated by primary explant technique and maintained for 6 to 12 weeks, apparently without further attempts to select for parathyroid cells. Subsequently, 2×10^6 to $2–3 \times 10^7$ viable cells, assumed to be antigen-depleted parathyroid cells, were allotransplanted (Wozniewicz *et al.*, 1996). However, during the serial subculture an exponential decrease of PTH secretion was noted, which may be explained either by necrosis of parathyroid cells or by fibroblastic overgrowth. Cellular contamination and residual antigenicity were underscored by an increasing posttransplant donor-specific T cell reactivity, with gradual cessation of parathyroid allograft function observed in all recipients (Tolloczko *et al.*, 1994).

At present, there is no validated approach, without immunosuppression, to treat the considerable number of patients with persistent hypoparathyroidism. However, several developments have delivered promising early results. Evidently, a combination of different methods—including those that reduce the immunogenicity and immunoisolate the graft—is needed to achieve long-term *in vivo* graft function without posttransplant immunosuppression. To optimize these different approaches and investigate the clinical utility of their combination is a promising approach that should be further explored.

ACKNOWLEDGMENT

This work was supported by Deutsche Forschungsgemeinschaft (DFG: Ha 2036/03-02).

REFERENCES

Anton, G., Decker, G., Stark, J. H., Botha, J. R., and Margolius, L. P. (1995). Allotransplantation of parathyroid cells. *Lancet* **345**, 124.

Aurbach, G. A., Marx, S. J., and Spiegel, A. M. (1992). Parathyroid hormone, calcitonin and the calciferols. *In* "Williams Textbook of Endocrinology" (J. D. Wilson and D. W. Foster, eds.), Vol 8, pp. 1431–1432. Saunders, Philadelphia.

Bloom, A. D., Economou, S. G., Baker, J. W., Gebel, H. M. (1987). Prolonged survival of rat parathyroid allografts after preoperative treatment with cyclosporine A. *Curr. Surg.* **44**, 205–207.

Chang, T.M.S. (1964). Semipermeable microcapsules. *Science* **1**, 524–525.

Collip, J. B. (1925). The extraction of a parathyroid hormone which will prevent or control parathyroid tetany and which regulates the level of blood calcium. *J. Biol. Chem.* **63**, 293.

Dib-Kuri, A., Revilla, A., and Chavez-Peon, F. (1975). Successful rat parathyroid allografts and xenografts to the testis without immunosuppression. *Transplant. Proc.* **7**, 753–756.

Ennis, W. B., Jr., and James, D. T. (1950). A simple apparatus for producing droplets of uniform size from small volumes of liquids. *Science* **112**, 434–436.

Fanger, G. O. (1974). What good are microcapsules? *Chem.-Tech. (Heidelberg)* **4**, 397–405.

Fisher, B., Fisher, E. R., Feduska, N., and Sakai, A. (1967). Thyroid and parathyroid implantation: An experimental re-evaluation. *Surgery* **62**, 1025.

Fitzpatrick, L. A., and Bilezikian, J. P. (1990). Primary hyperparathyroidism. *In* "Principles and Practice of Endocrinology and Metabolism" (K. L. Becker, ed.), Vol. 1, pp. 431–433. Lippincott, Philadelphia.

Franjione, J., and Vasishtha, N. (1996). "The Art and Science of Microencapsulation Information," p. 1. Southwest Research Institute, San Antonio, TX.

Fu, X. W., and Sun, A. M. (1989). Microencapsulated parathyroid cells as a bioartificial parathyroid. *Transplantation* **47**, 432–435.

Gann, D. S., and Paone, J. F. (1979). Delayed hypocalcemia after thyroidectomy for Graves' disease is prevented by parathyroid autotransplantation. *Ann. Surg.* **190**, 508–513.

Gough, I. R., and Finnimore, M. (1980). Rat parathyroid transplantation. Allograft pretreatment with organ culture and antilymphocyte serum. *Transplantation* **29**, 149–152.

Grant, G. T., Morris, E. R., Rees, D. A., Smith, P.J.C., and Thom, D. (1973). Biological interactions between polysaccharides and divalent cations: The egg-box model. *FEBS Lett.* **32**, 195–198.

Groth, C. G., Popovtzer, M., Hammond, W. S., Cascardo, S., Iwatsuki, S., Halgrimson, C. G., and Starzl, T. E. (1973). Survival of a homologous parathyroid implant in an immunosuppressed patient. *Lancet*, 1082–1085.

Habener, J., Arnold, A., and Potts, J. T. (1995). Hyperparathyroidism. *In* "Endocrinology" (L. J. DeGroot, ed.), Vol. 2, pp. 1048–1055. Saunders, Philadelphia.

Hanson, A. M. (1924). The hydrochloric X sicca: A parathyroid preparation for intramuscular injection. *Mil. Surg.* 54, 218.

Hasse, C., Schrezenmeier, J., Stinner, B., Schark, C., Wagner, P. K., Neumann, K., and Rothmund, M. (1994). Successful allotransplantation of microencapsulated parathyroid in rats. *World J. Surg.* 18, 630–634.

Hasse, C., Klöck, G., Zielke, A., Schlosser, A., Barth, P., Zimmermann, U., and Rothmund, M. (1996). Transplantation of parathyroid tissue in experimental hypoparathyroidism: *In vitro* and *in vivo* function of parathyroid tissue microencapsulated with a novel amitogenic alginate. *Int. J. Artif. Organs* 19, 1–7.

Hasse, C., Zielke, A., Klöck, G., Barth, P., Schlosser, A., Zimmerman, U., and Rothmund, M. (1997a). First successful xenotransplantation of microencapsulated human parathyroid tissue in experimental hypoparathyroidism: Long-term function without immunosuppression. *J. Microencapsul.* 14, 617–626.

Hasse, C., Zielke, A., Klöck, G., Schlosser, A., Zimmermann, U., and Rothmund, M. (1997b). Isotransplantation of microencapsulated parathyroid tissue rats. *Exp. Clin. Endocrinol. Diabetes* 105, 53–56.

Hasse, C., Klöck, G., Zimmermann, U., and Rothmund, M. (1997c). Amitogenes Alginat—Schlüssel zum ersten klinischen Einsatz der Microenkapsulierungstechnologie. *Langenbecks Arch. Chir., Chir. Forum Exp. Klin. Forsch,* pp. 755–759.

Hasse, C., Klöck, G., Schlosser, A., Zimmermann, U., and Rothmund, M. (1997d). Parathyroid allotransplantation without immunosuppression. *Lancet* 350, 1296–1297.

Hasse, C., Zielke, A., Klöck, G., Schlosser, A., Barth, P., Zimmermann, U., Sitter, H., Lorenz, W., and Rothmund, M. (1998a). Amitogenic alginates: Key to first clinical application of microencapsulation technology. *World J. Surg.* 22, 659–665.

Hasse, C., Klöck, G., Barth, P., Stinner, B., Zimmermann, U., and Rothmund, M. (1998b). Heterologic transplantation of human parathyroids after microencapsulation with a novel amitogenic alginate suitable for use in humans: Long term function without immunosuppression in an animal experiment. *Langenbecks Arch. Chir., Chir. Forum Exp. Klin. Forsch.,* pp. 713–718.

Keck, E., and Kuhlencordt, F. (1992). Krankheiten der Nebenschilddrüsen. *In* "Innere Medizin in Praxis und Klinik" (H. Hornbostel, W. Kaufmann, and W. Siegenthaler, eds.), pp. 4.79–4.86. Thieme, Stuttgart.

Keutmann, H. T., Sauer, M. M., Hendy, G. N., O'Riordan, J.L.H., and Potts, J. T. (1978). Complete amino acid sequence of human parathyroid hormone. *Biochemistry* 17, 5724.

Kierstan, M., and Bucke, C. (1977). The immobilization of microbial cells, subcellular organelles and enzymes in calcium alginate gels. *Biotechnol. Bioeng.* 19, 387–397.

Kiprenskii, I. V., Priakhin, I. S., and Podshivalin, A. V. (1990). The transplantation of a segment of pancreas and thyroid–parathyroid complex by using a microsurgical technic. *Vestn. Khir. im. I. I. Grekova* 145, 108–110.

Klöck, G., Frank, H., Houben, R., Zekorn, T., Horcher, A., Siebers, U., Wöhrle, M., Federlin, K., and Zimmermann, U. (1994). Production of purified alginates suitable for use in immunoisolated transplantation. *Appl. Microbiol. Biotechnol.* 40, 638–643.

Kunori, T., Tsuchiya, T., Itoh, J., Watabe, S., Arai, M., Satomi, T., Takakura, K., and Yamaguchi, H. (1991). Improvement of postoperative hypocalcemia by repeated allotransplantation of parathyroid tissue without anti-rejection therapy. *Tohoku J. Exp. Med.* 165, 33–40.

Lachmann, A. (1941). Hypoparathyreoidism in Denmark. *Acta Med. Scand.* 121, 1.

Lane, W. R. (1947). Glass apparatus producing small drops of liquid. *J. Sci. Instrum.* 24, 98.

Lim, F., and Sun, A. M. (1980). Microencapsulated islets as bioartificial endocrine pancreas. *Science* 210, 908.

Loeb, J. (1899–1900). On the different effects of ions upon myogenic and neurogenic rhythmical contractions and upon embryonic and muscular tissue. *Am. J. Physiol.* 3, 383.

MacCallum, W. G., and Voegtlein, C. (1909). On the relation of tetany of the parathyroid glands to calcium metabolism. *J. Exp. Med.* 11, 118.

Melick, R. A., Gill, J. R., Berson, S. A., Yalow, R. S., Bartler, F. C., Potts, J. T., and Aurbach, G. D. (1967). Antibodies and clinical resistance to parathyroid hormone. *N. Engl. J. Med.* 276, 144–147.

Potts, J. (1995). Hypoparathyroidismus. *In* "Harrisons Innere Medizin II" (K. Schmailzl, ed.), pp. 2151–2522. McGraw-Hill, New York.

Raaf, J. H., Farr, H. W., Laird Meyers, W. P., and Good, R. A. (1974). Transplantation of fresh and cultured parathyroid glands in the rat. *Am. J. Surg.* 128, 478–483.

Ross, A. J., III, Dale, J. K., Gunnells, J. C., and Wells, S. A., Jr. (1979). Parathyroid transplantation: Fate of a long-term allograft in man. *Surgery* 85, 382–384.

Segerberg, E. C., Grubb, W. G., and Henderson, A. E. (1992). The first successful parathyroid transplant from an identical twin for the cure of permanent postoperative hypoparathyroidism. *Surgery* 111, 357–358.

Sollinger, H. W., Mack, E., Cook, K., and Belzer, F. O. (1983). Allotransplantation of human parathyroid tissue without immunosuppression. *Transplantation* 36, 599–602.

Starling, J. R., Fidler, R., and Corry, R. J. (1977). Prolongation of survival of rat parathyroid allografts by enhancing serum and tissue culture. *Surgery* 81, 668–675.

Stögmann, W., Bohrn, E., and Woloszczuk, W. (1990). Initial experiences with substitution treatment of hypoparathyroidism with synthetic human parathyroid hormone. *Monatsschr. Kinderheilkd.* 138 141–146.

Swan, H., Hallin, R., and Callaghan, P. (1960). Observation on the function and morphology of the parathyroid grafts in the dog using millipore chambers. *Surg. Forum* 10, 87.

Tochilin, V. I. (1972a). Dynamics of accumulation of radioactive iodine by transplantation and protein-bound iodine of plasma in patients following homotransplantation of the thyroid–parathyroid complex. *Probl. Endokrinol.* **18** 32–35.

Tochilin, V. I. (1972b). Homotransplantation of the thyroid–parathyroid complex in myxedema. *Khirurgiya (Moscow)* **48**, 20–24.

Tochilin, V. I. (1972c). The technic of taking and transplanting a thyroid–parathyroid complex to patient suffering thyroid–parathyroid insufficiency. *Klin. Khir.* **9**, 36–40.

Tolloczko, T., Sawicki, A., and Woniewicz, B. (1991). Clinical results of human cultured parathyroid cell allotransplantation in the treatment of surgical hypoparathyroidism. *Abstr., World Congr. Surg., 34th,* Stockholm, *1991,* p. 304.

Tolloczko, T., Wozniewicz, B., Sawicki, A., Nawrot, I., Migaj, M., Zabitowska, T., and Gorski, A. (1994). Cultured parathyroid cell transplantation without immunosuppression in the treatment of surgical hypoparathyroidism. *Transplant. Proc.* **26**, 1901–1902.

Tolloczko, T., Woniewicz, B., Sawicki, A., Gorski, A., Nawrot, I., Zawitkowska, T., and Migaj, M. (1996). Clinical results of human cultured parathyroid cell allotransplantation in the treatment of surgical hypoparathyroidism. *Transplant. Proc.* **28**, 3545–3546.

Wagner, P. K. (1980). Konservierung und Transplantation der Nebenschilddrüsen. *In* "Hyperparathyreoidismus" (M. Rothmund, ed.), pp. 216–228. Thieme, Stuttgart.

Welbourn, R. B. (1990). The parathyroid glands. *In* "The History of Endocrine Surgery" (R. B. Welbourn, ed.), pp. 217–221. Praeger, New York.

Wells, S. A., Jr., and Christiansen, C. L. (1974). The transplanted parathyroid gland: Evaluation of cryopreservation and other environmental factors which affect its function. *Surgery* **78**, 49.

Wells, S. A., Jr., Burdick, J. F., Christiansen, C. L., Abe, M., Sherwood, L. M., Hattler, B. G., and Davis, R. C. (1973a). Long-term survival of dogs transplanted with parathyroid glands as autografts and as allografts in immunosuppressed hosts. *Transplant. Proc.* **5**, 769–771.

Wells, S. A., Jr., Burdick, J. F., Ketcham, A. S., Christiansen, C., Abe, M., and Sherwood, L. (1973b). Transplantation of the parathyroid glands in dogs. Biochemical, histological, and radioimmunoassay proof of function. *Transplantation* **15**, 179–182.

Wells, S. A., Jr., Ross, A. J., Dale, J. K., and Gray, R. S. (1979). Transplantation of the parathyroid glands: Current status. *Surg. Clin. North Am.* **59**, 167.

Windeck, R. (1996). Hypoparathyreoidismus. *In* "Praktische Endokrinologie," (B. Allolio and H. Schulte, eds.), pp. 289–293. Urban & Schwarzenberg, Munich.

Windeck, R., and Reinwein, D. (1992). Hypoparathyroidismus. *In* "Klinische Endokrinologie und Diabetologie" (D. Reinwein and G. Benker, eds.), pp. 163–166. Schattauer, Stuttgart.

Winer, K. K., Yanovski, J. A., and Cutler, G. B. (1996). Synthetic human parathyroid hormone 1-34 vs calcitriol and calcium in the treatment of hypoparathyroidism. *JAMA, J. Am. Med. Assoc.* **276**, 631–636.

Wozniewicz, B., Migaj, M., Giera, B., Prokurat, A., Tolloczko, T., Sawicki, A., Nawrot, I., Gorski, A., Zabitkowska, T., and Kossakowska, A. E. (1996). Cell culture preparation of human parathyroid cells for allotransplantation without immunosuppression. *Transplant. Proc.* **28**, 3542–3544.

Yao, C. Z., Ishizuka, J., Townsend, C. M., and Thompson, J. C. (1993). Successful intracerebroventricular allotransplantation of parathyroid tissue in rats without immunosuppression. *Transplantation* **55**, 251–253.

Ziegler, R. (1976). Hypoparathyreoidismus. *In* "Praktische Endokrinologie" (A. Jores and H. Nowakowski, eds.), pp. 122–131. Thieme, Stuttgart.

Ziegler, R. (1987). Hypoparathyreoidismus. *In* "Hormon—und stoffwechselbedingte Erkrankungen" (R. Ziegler, ed.), pp. 273–283. VCH, Weinheim.

Zimmermann, U., Klöck, G., Federlin, K., Hannig, K., Kowalski, M., Brezel, R. G., Horcher, A., Entenmann, H., Siebers, U., and Zekorn, T. (1992). Production of mitogen contamination free alginates with variable rations of mannuronic to guluronic acid by free flow electrophoresis. *Electrophoresis* **13**, 269.

PART XIII: GASTROINTESTINAL SYSTEM

CHAPTER 38

ALIMENTARY TRACT

Gregory M. Organ and Joseph P. Vacanti

INTRODUCTION

Tissue engineering is an emerging multidisciplinary effort for therapeutic intervention beyond the current state-of-the-art treatments for a variety of clinical scenarios in which there is insufficient tissue. The rationale and historical background for the current state of tissue engineering neointestine for patients suffering from short bowel syndrome is addressed. Specific emphasis is placed on enterocyte isolation and morphogenesis of tissue-engineered neointestinal mucosa.

The maintenance of life by all organisms is dependent on the acquisition and utilization of food. Nutrients are broken down into simple compounds and provide the energy necessary for the maintenance of bodily functions as well as the materials for growth of the organism and repair of tissues that have been damaged.

Simple forms of life, such as bacteria, absorb their nutrients directly through their outer surface layer. As multicellular animals have evolved, specialized cavities have developed in which food is prepared for absorption by a process of digestion. The alimentary tract in humans is a highly specialized organ that is anatomically and functionally divided into the mouth and pharynx (mastication and swallowing), esophagus (transport), stomach and small intestine (digestion, absorption, endocrine, and immunologic), and colon (absorption of water and evacuation). In addition, the liver and pancreas, glandular organs connected by special ducts to the lumen of the intestine, produce juices that aid in the digestive and absorptive process (Hamilton and McMinn, 1976).

A widely varied pathology can affect the different regions of the alimentary tract, requiring different modes of treatment. The goal of any surgical intervention is to remove diseased or insufficient tissue and to reconstruct the remaining tissues in an anatomically and physiologically meaningful fashion. Surgical ablation can result in significant tissue loss of single or multiple anatomic subdivision(s). In most circumstances successful anatomic reconstruction can be achieved without substantial loss of function. The small intestine is a notable exception due to its unique characteristics for digestion and absorption.

Short bowel syndrome (SBS) refers to a condition of malabsorption and malnutrition following major resection of the small intestine, which may also include some or all of the large intestine. The clinical signs and symptoms of SBS are characterized primarily by intractable diarrhea, steatorrhea, weight loss, dehydration, malnutrition, vitamin and electrolyte deficiencies, as well as macrocytic anemia (Tilson, 1980).

The diseases leading to massive small bowel resection depend on the age group of the patient. In newborns, congenital short small bowel, multiple bowel atresias, extensive aganglionosis (Hirschsprung's disease), gastroschisis, meconium peritonitis, and acquired disease such as midgut volvulous and necrotizing enterocolitis all have the potential for producing insufficient intestinal length. In children and adults the common causes include abdominal trauma, Crohn's disease, radiation enteritis, and mesenteric ischemic vascular disease (Kern *et al.*, 1990). In adults, resection of 70% or more of the small intestine can significantly impair the ability of the patient to maintain normal nutrition and metabolism. These patients require special nutritional care on a long-term or permanent basis, especially with the loss of the terminal ileum and ileocecal valve (Dudrick *et al.,* 1991). Infants with greater than 10–15 cm of jejunoileum with an intact ileocecal valve or 25–40 cm of small bowel without the ileocecal valve are usually given the benefit of long-term parenteral nutrition to allow their residual intestine sufficient time to adapt and permit survival without parenteral nutrition (Williamson, 1978a,b).

Intestinal adaptation occurs due to a variety of luminal, circulating, and cellular influences resulting in increased gut surface area for absorption through mucosal hyperplasia, villous lengthening, deepening of intestinal crypts, and increase in diameter of the remaining intestine (Chaves *et al.,* 1987; Lak and Baylin, 1983; Unger *et al.,* 1968; Weber *et al.,* 1991). These adaptive changes occur slowly and can take up to 2 years before adequate nutrition can be maintained through oral intake alone. Children have greater adaptive potential than adults do because the intestine grows as the child ages (Touloukian and Smith, 1983).

Survival of patients with SBS, particularly infants and children, has improved primarily because of the increased use of total parenteral nutrition (TPN) both during hospitalization and at home (Cooper *et al.,* 1984; Dorney *et al.,* 1985; Grosfeld *et al.,* 1986). However, this therapy is not without consequence. Catheter-relative complications (systemic infection, venous thrombosis, mechanical breakage, and malfunction) are frequent and the vast majority of patients require rehospitalization (Caniano *et al.,* 1989; Thompson, 1990). Progressive bone demineralization is characteristic of patients on chronic TPN (Foldes *et al.,* 1990). Transient liver enzyme abnormalities, cholestasis, and cholelithiasis are common. Progressive cholestatic jaundice and cirrhosis associated with TPN account for more than 50% of deaths among infants and children maintained with chronic TPN and 5% of deaths among adults (Howard *et al.,* 1986). The cost of TPN at home currently exceeds $150,000 per year.

Numerous innovative surgical techniques for increasing intestinal surface area and improving absorption have been developed. Several approaches have been investigated, including tapering (Weber *et al.,* 1982) and lengthening (Pokorny and Fowler, 1991) residual intestine, slowing intestinal transit by creating artificial ileocecal valves (Ricotta *et al.,* 1981) and utilizing reversed intestinal segments (Diego *et al.,* 1982), recirculating loops of bowel and growing intestinal neomucosa (Thompson *et al.,* 1984), and intestinal transplantation (Bueno *et al.,* 1999; Reyes *et al.,* 1998).

None of the currently available surgical alternatives for treatment of SBS is sufficiently safe and effective for routine and widespread use. Transplantation of small intestine seemingly offers the most reasonable alternative in patients with near fatal loss of intestine and therefore no hope for adaptation and in those with intolerance to TPN. Despite recent improvements in immunosuppression and preservation of harvested small intestine, the major clinical obstacle to intestinal transplantation continues to be allograft rejection, graft-versus-host disease, and cytomegalovirus infection (Reyes *et al.,* 1998; Schraut *et al.,* 1986). The development of lymphoproliferative complications is also a major concern. The adverse effects to growth and development of chronic corticosteroid use in children are a significant limitation. Thus, the mortality and morbidity remains high with intestinal transplantation. Furthermore, when these immunologic problems are resolved, this therapy will still be faced with donor scarcity, the other significant problem facing all of the field of whole-organ transplantation.

TISSUE ENGINEERING

Transplanting selected cell populations is a method to create new tissue substitutes for the replacement and repair of lost or damaged tissue (Russell, 1985). The emerging field of tissue engineering involves a multidisciplinary effort, merging the fields of cell and molecular biology, materials science, and surgical reconstruction to engineer new tissue (Langer and Vacanti, 1993).

A variety of strategies have been adopted by numerous investigators to engineer virtually every mammalian tissue. Our laboratories have been utilizing highly porous synthetic polymeric devices to deliver the transplanted cells to a given anatomic location, and to serve as a template for the reorganization of the transplanted cells as they form new tissue (Vacanti *et al.,* 1988). The morphogenesis of the implanted cell–polymer construct involves the ingrowth of fibrovascular elements from the host implantation site into the graft, resorption of the bioerodable scaffold, and expression of a differentiated phenotype by the implanted cells in a permanently engrafted new tissue. Various surgical manipulations have been explored and incorporated into the tissue engineering

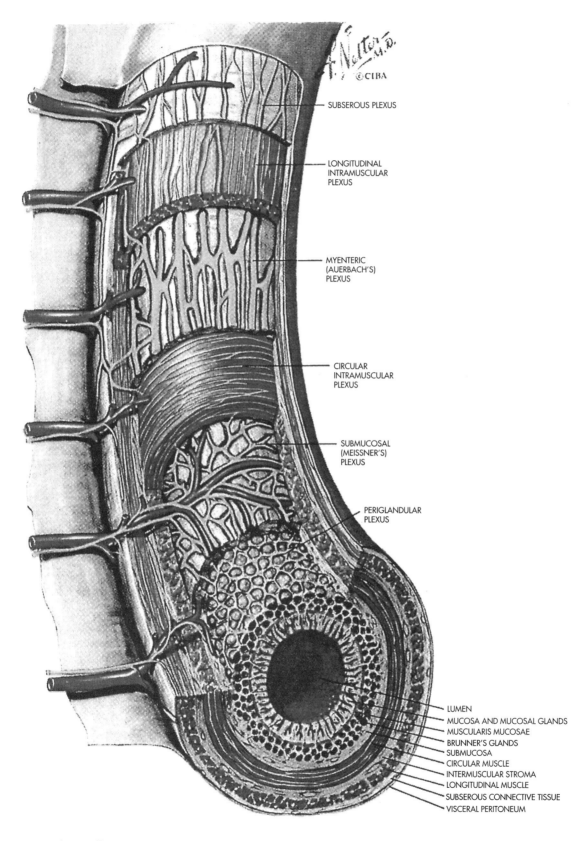

SUBSEROUS PLEXUS

LONGITUDINAL
INTRAMUSCULAR
PLEXUS

MYENTERIC
(AUERBACH'S)
PLEXUS

CIRCULAR
INTRAMUSCULAR
PLEXUS

SUBMUCOSAL
(MEISSNER'S)
PLEXUS

PERIGLANDULAR
PLEXUS

LUMEN
MUCOSA AND MUCOSAL GLANDS
MUSCULARIS MUCOSAE
BRUNNER'S GLANDS
SUBMUCOSA
CIRCULAR MUSCLE
INTERMUSCULAR STROMA
LONGITUDINAL MUSCLE
SUBSEROUS CONNECTIVE TISSUE
VISCERAL PERITONEUM

Fig. 38.1. The small intestine is a laminated, tubular organ composed of an inner mucosal layer for digestion and absorption and an outer neuromuscular layer for peristaltic propulsion of nutrients. From Netter (1962). Copyright 1962. Novartis. Reprinted with permission from The Netter Collection of Medical Illustrations, Vol. 3, Part II, illustrated by Frank H. Netter, M.D., All rights reserved.

strategy to take advantage of adaptive responses. A successfully engrafted and functioning new three-dimensional cell mass would be a true chimera of implanted parenchymal elements from the donor and mesenchymal elements (endothelial cells, fibroblasts, lymphatic and neural cells) from the recipient. The term "chimeric neomorphogenesis" has been coined to describe this approach (Vacanti, 1988).

The small intestine is a laminated, tubular organ composed of an inner mucosal layer for digestion and absorption and an outer neuromuscular layer for peristaltic propulsion of nutrients (Fig. 38.1). The goal of this project is to develop a small animal model for generating new intestinal mucosa, to increase intestinal epithelial surface area for digestion and absorption, as the essential first step to tissue engineer a composite tubular tissue with a neuromuscular coat. The fundamental elements in our approach to engineer new tissue equivalents consist of parenchymal cell isolation, cell–polymer construct formation, cell–polymer construct implantation, and histologic and functional assessment of the engrafted tissue. Perhaps the most critical step in this process is the isolation procedure. A guiding principle of tissue engineering any tissue equivalent is that the harvested parenchymal cells must be viable and maintain their ability to regenerate the tissue of interest *in vivo*.

ENTEROCYTE ISOLATION

The epithelial cells of the small intestine, referred to as intestinal epithelial cells (IECs) or enterocytes, represent a rapidly renewing cell population characterized by continuous growth and differentiation *in vivo* (Carnie *et al.,* 1965a,b; Cheng and Leblond, 1974a,b). Cell replication is limited to the lower two-thirds of the crypts and is balanced by continuous loss of differentiated cells at the villus tip (Fig. 38.2). This pattern of programmed cell death has been termed apoptosis (Potten *et al.,* 1997). IECs have a rapid cell turnover, with a mean cell life-span of 2–3 days in most animals. Approximately 75% of IECs are differentiated, mature absorptive columnar-shaped villus cells and 25% consist of their precursors, the rapidly proliferating undifferentiated cuboidal-shaped crypt cells. Differentiation of these mitotically active crypt cells is accompanied by dramatic changes in enzyme and transport activities along with changes in morphology, including the appearance of a well-organized brush border at the luminal surface and a more columnar cell shape. The four main cell types of the intestinal epithelium—villus columnar, mucous goblet, enteroendocrine, and Paneth cells—originate from pluripotent stem cells located in the crypts of Lieberkuhn (Cheng and Leblond, 1974a,b).

In the first edition of "Principles of Tissue Engineering" we described our initial approach at tissue engineering neointestinal mucosa utilizing suspensions of individual or small clusters of IECs (Organ and Vacanti, 1997). Weiser's chelating method for isolating IECs in a villus-to-crypt cell gradient was modified to harvest IECs on a large scale (Organ *et al.,* 1993a,b; Organ and Vacanti, 1995). A syngeneic Lewis rat model as developed for selective isolation of highly proliferative, crypt cell-enriched fractions of IECs. IECs were seeded onto 10×10 mm^2 sheets of nonwoven fibers of poly(glycolic acid) (PGA) to form constructs. IEC–PGA constructs were rolled around a silastic stent and implanted into various vascular beds. Histologic analysis confirmed that the grafts (1) contained cells that have the cytologic characteristics of IECs and formed a stratified epithelium, (2) had positive cytokeratin immunohistochemical staining features of epithelial cells, (3) showed labeling of implanted IECs with fluorescent vital dyes, confirming the origin of the engrafted cells, and (4) maintained remarkable proliferative cell capacity up to 28 days after implantation. At any of the observed time points out to 60 days, there was not evidence of neomucosa that had remodeled with typical small intestinal villus/crypt morphology. The histologic appearance of the transplanted IECs was reminiscent of fetal intestinal development, arrested at the stratified epithelial stage.

Advances in the field of developmental biology have focused on the importance of the epithelial–mesenchymal interactions, particularly in intestinal organogenesis (Kedinger *et al.,* 1986, 1987). Evans and Potten described an improved method of isolating IECs utilizing neutral protease and collagenase (Evans *et al.,* 1992). This method yields intact crypts with their pericryptal fibroblast, thus maintaining the epithelial–mesenchymal interaction. The resulting crypt cell aggregates (IECs) have been shown to have significantly improved viability, greater *in vitro* proliferation, and expression of differentiated function.

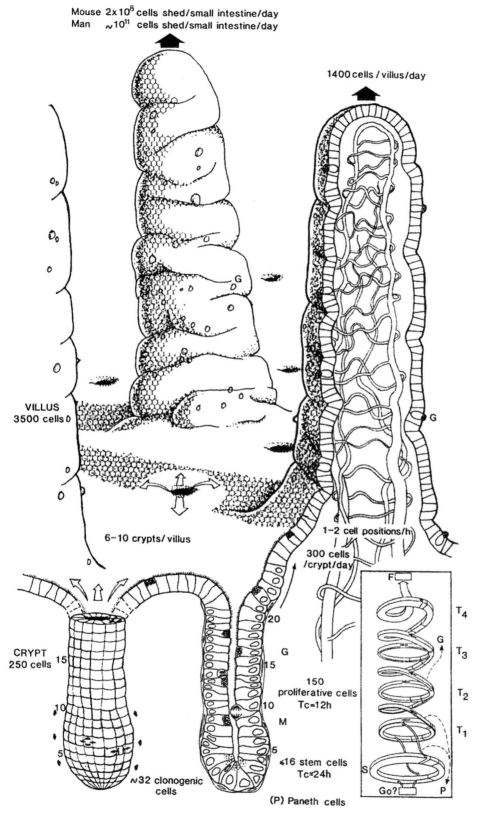

Mouse 2x10⁸ cells shed/small intestine/day
Man   ~10¹¹ cells shed/small intestine/day

1400 cells / villus / day

VILLUS
3500 cells

6–10 crypts/ villus

1–2 cell positions/h

300 cells
/crypt/day

CRYPT
250 cells

~32 clonogenic
cells

20

G

15

10

M

5

150
proliferative cells
Tc=12h

≤16 stem cells
Tc≈24h

(P) Paneth cells

F

T₄

G

T₃

T₂

T₁

S

Go?   P

Fig. 38.2. Anatomy of the villus/crypt axis of the epithelium of the small intestine. From Potten and Loeffler (1990), with permission from the Company of Biologists Ltd.

PROCEDURAL STEPS

IECA ISOLATION

Neonatal (6–26 g) Lewis rats (*n* = 4–8) serve as donors (Organ *et al.*, 1998a,b). Under Enflurane inhalation anesthesia the abdominal viscera are eviscerated through a midline laparotomy. The stomach is intubated, the distal ileum transected, and peroperative lavage performed with 30 ml of 0.9% normal saline at room temperature to remove fecal material and mucous debris. The intestine is lavaged with 10 ml of Dulbecco's Modified Eagle's Media (DMEM) at 0–4°C to institute a state of cellular hypometabolism. The intestine is ligated distal to the ligament of treitz and at the distal ileum. The mesenteric attachments are sharply divided and the bowel is measured and placed in a glass petri dish on ice with Hank's Balance Salt Solution (HBSS). After harvesting all the donor bowel, the small intestine is slit longitudinally and sharply diced into sections 2–3 mm², transferred to a T25-ml flask, and washed eight times in 25-ml changes of HBSS at 0–4°C with vigorous shaking on an orbital shaking platform at 80 cycles/min. The contents are transferred to a sterile petri dish and sharply diced into pieces 1 mm³ and transferred into a T25-ml flask for enzymatic digestion (see Fig. 38.3 for details of isolation).

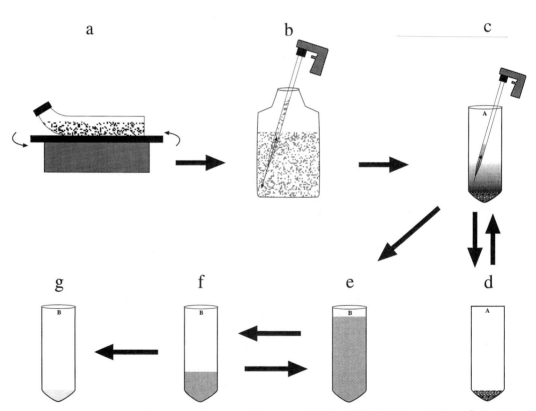

Fig. 38.3. Evans and Potten collagenase/dispase method for IECA isolation. (a) 1 mm³ diced segments of small intestine are placed into a T25 flask with 20 ml of ES on an orbital shaking platform (80 cycles/min × 25 min at 25°C). (b) Pipette suspension vigorously for 30 sec with a 10-ml pipette and transfer to a 60-ml conical tube (tube A). (c) Allow suspension to sediment under gravity for 60 sec. Carefully remove the supernatant and transfer to conical tube B at 0–4°C. (d) Resuspend sediment in tube A with 20 ml of ES and repeat step c twice. (e) Add 10 ml of DMEM-S to conical tube B, mix, and centrifuge at 8 g for 2 min at 0–4°C. Carefully aspirate supernatant and discard. (f) Resuspend pellet in tube B with 20 ml of DMEM-S and repeat step e centrifugation and supernatant aspiration cycles 4 times. (g) Resuspend pellet 100–200 μl) in DMEM 3–5 ml) to form final suspension for construct formation and assays. Media: ES, 0.1 mg/ml dispase (neutal protease type I, Boehringer) and 300 U/ml collagenase (type XI, Sigma) in Hank's Balanced Salt Solution (Life Technologies), pH 7.4; DMEM-S, Dulbecco's Modified Eagle Medium (1 × high glucose with L-glutamine, Life Technologies) + 2.5% fetal calf serum (lot no. 1007757, Life Technologies) + 2% fungizone, 0.25 μg/ml (Biowhittaker). From Evans et al. (1992).

Assays

The final suspension of isolated IECAs is counted in a standard hemocytometer. The dimensions of the IECAs, which are estimated to contain 600–800 cells in a single crypt aggregate, preclude performing hemocytometric counts in the standard fashion (i.e., applying the suspension to the trough of the hemocytometer and allowing the chamber to fill by capillary action). We have modified the procedure by applying 12 µl of the final suspension onto the surface of the hemocytometer, placing the cover slip, and then counting four quadrants in the standard fashion. Hemocytometer counts are performed in triplicate. The final IECA suspension is assayed for cell protein, measured in triplicate, as a separate measure to quantitative cellular mass. Viability is determined by mitochondrial reduction of a tetrazolium sale, 3-(4,5-dimethylthiazol-2-yl)-2,5-diphenyltetrazolium bromide (MTT), to formazan, measured spectrophotometrically (OD 570 nm) in triplicate (Plumb *et al.,* 1989).

To quantify seeding efficiency a cohort of IECA–PGA constructs ($n = 2$–4) are seeded with 500 µl of the final suspension in a separate 24-well plate and incubated for 60 min in a 5% CO_2 incubator at 37°C. Constructs are removed from their original seeding well and placed into a secondary well. MTT activity is measured in the original seeding wells and grafting efficiency is determined by the percentage of MTT activity of the residual IECAs in the original seeding well divided by the MTT activity of the final suspension. IECA cell mass delivered is derived by multiplying the MTT-generated grafting efficiency by the number of IECAs seeded. Donor neonatal intestine yields 779 ± 224* IECA/cm. Employing eight donors, the final suspension typically contains 10,645 ± 2979* IECA/ml with a cell protein concentration of 2.18 ± 0.08* mg/ml. Relative viability of IECA was 25.5 ± 1.5* OD/mg cell protein. Seeding efficiency was 83 ± 1.1%*, resulting in delivery of 4418 ± 1236* IECA/cm² (*, SEM).

IECA–PGA Construct Formation

A volume of 500 µl of the final IECA suspension is gently seeded onto 1-cm² patches ($n =$ 4–8) of nonwoven polymer fibers (diameter, 15 µm; pore size, 150–200 µm; 4-ply at 80–100 µm thickness, Davis & Geck) of poly(glycolic acid) in a 24-well plate (Fig. 38.4). The seeding

Fig. 38.4. Scanning electon micrograph of freshly seeded IECA–PGA construct. The IECA is firmly adherent to the PGA fiber. Courtesy of D. J. Mooney, University of Michigan, Department of Chemical Engineering.

technique for optimal distribution and attachment efficiency is to apply the final IECA suspension a single drop at a time, in a smooth, circular motion starting at the center of the PGA and proceeding toward the edges. An additional 500 μl of DMEM is gently added to each well. IECA–PGA constructs, grouped in cohorts of four constructs per 24 well plate, are placed in a 5% CO_2 incubator at 37°C for 1–2 hr prior to implantation.

IMPLANTATION

Four constructs are implanted as a single cohort. On removal from the incubator, the IECA–PGA constructs are placed on ice. The constructs are gently rolled around a sterile silastic stent (outer diameter, 2.2 mm) and placed on the eviscerated omentum of a male Lewis rat (50–225 g). The omentum is atraumatically folded around the construct and secured with 7-0 Prolene sutures placed in a nonischemic fashion, parallel to the gastroepiploic arcade. The omentum is returned to the peritoneal cavity, 1–2 ml of sterile normal saline is added to the abdominal cavity for postoperative fluid resuscitation, and the incision is closed in two layers.

HISTOLOGIC ASSESSMENT

IECA–PGA grafts are retrieved at serial time points out to 28 days and fixed in 10% buffered formalin. Stents are removed and the implants are cross-sectioned into thirds and embedded in paraffin. Three to four serial 5-μm sections are mounted on glass slides and submitted for histologic evaluation after staining with hematoxylin and eosin. Selected sections were evaluated with periodic acid/Schiff, mucicarmine, and CEA stains.

Histologic evaluations at various time points up to 1 month after implantation demonstrate a distinct pattern of morphogenseis. IECA–PGA constructs stimulate a characteristic pattern of acute inflammation followed by a rich fibrovascular ingrowth from the omental implantation site from day 3 to day 7. During this time period, IECAs form microcystic structures that are identified as early as day 3 and are prominent by day 7. The microcysts appear adherent to the PGA fibers and are surrounded by a fibrovascular stroma (Fig. 38.5). Cysts are found superficial to a layer of fibroblasts situated adjacent to the luminal stent. Evidence of cytologic differentiation is not appreciated during the first week. Depletion of goblet cells is characteristic of the late phase of intestinal epithelial injury and early regeneration (Lewin *et al.,* 1992b).

Day 10 implants are characterized by development of a simple cuboidal/columnar epithelium, with basally oriented nuclei, that appear to be emanating from cryptlike structures (Fig. 38.6a). This neointestinal epithelium develops on a bed of granulation, presumably from the omentum, adjacent to the luminal stent. The thin band of fibroblasts, visualized adjacent to the stent on day 7, seem to have been separated from the underlying granulation tissue by the advancing epithelium. Areas of transition from cuboidal to columnar cells bear a striking resemblance to regenerating intestinal epithelium recovering from a variety of pathologic insults (Lewin *et al.,* 1992a). By day 14 immature villi are present and mitotic figures are seen exclusively in the crypts (Fig. 38.6b). Goblet cells are abundant, and positive staining of the cytoplasmic droplets with mucicarmine and periodic acid/Schiff confirm this level of phenotypic differentiation. The lumen is filled with degenerated cells and debris, suggesting apoptosis is occurring. The PGA fibers are engulfed by a chronic foreign body response.

By day 28 the neointestinal mucosa displays mature villus–crypt axis formation (Fig. 38.7a). Ongoing proliferation is apparent by the presence of multiple mitotic figures isolated to the crypts (Fig. 38.7b). In addition to goblet cells, Paneth cells are identified by their characteristic large, supranuclear, eosinophilic granules in the base of the crypts (Lewin *et al.,* 1992a). Occasional argentaffin cells are also seen. Positive CEA staining was not a feature observed in plants at any of the time points. The grafting efficiency for this study was 54%.

DISCUSSION

The goal of this effort is to tissue engineer a short segment of tissue equivalent to the small intestine as therapy for those afflicted with SBS. In essence, the small intestine is a laminated, tubular organ composed of an inner mucosal layer for digestion and absorption and an outer neuromuscular tube for peristalsis. We have focused on the intestinal epithelium as the fundamental step in this endeavor.

Our initial approach at tissue engineering neointestinal mucosa concentrated on implanting

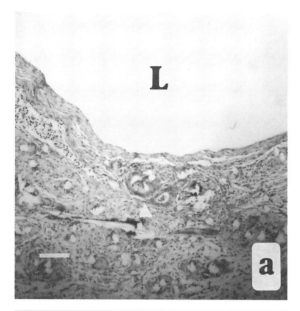

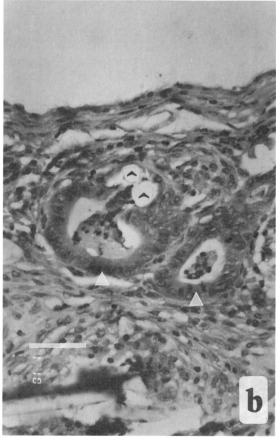

Fig. 38.5. Day 7 IECA–PGA implant. (a) Low-power view demonstrating rich fibrovascular ingrowth from the omentum. The triangle is directed at the IECA forming the microcyst. L, Lumen after the stent is removed. Bar, 100 µm. (b) Triangles identify the implanted IECA forming the adjacent microcystic structures, lined with simple cuboidal/columnar epithelium. Black arrow heads are PGA fibers in cross-section. Bar, 50 µm.

constructs of individual small clusters of IECs from crypt cell-enriched fractions seeded onto matrices of PGA (Organ *et al.,* 1993b). Although the histologic appearance at 1 month resembled normal fetal intestinal development arrested at the stratified epithelial stage, it was not organotypic of mature intestinal mucosa. We adhere to the principle that structure and function are intimately intertwined. To tissue engineer successfully any *functional* organ or tissue, a derived organ or tissue must have a structural form close to, if not identical to, a normal organ or tissue, to be of therapeutic benefit.

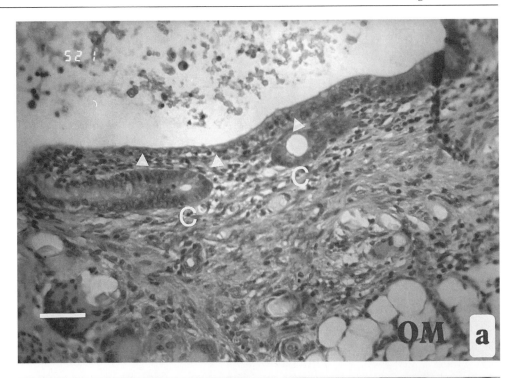

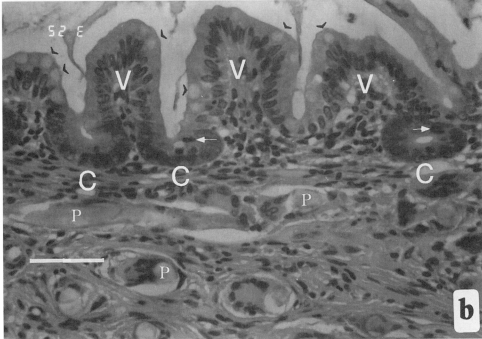

Fig. 38.6. (a) Day 10 implant. Triangles: simple cuboidal/columnar epithelium, with basally oriented nuclei, emanating from cryptlike structures (C) overlying fibrovascular ingrowth from omentum (OM) Bar, 50 μm. (b) Day 14 implant. Immature villi (V) arising from crypts (C) and abundant
goblet cells (arrowheads), indicative of phenotypic differentiation. Arrows: mitotic figures. PGA
fibers (P) are surrounded by granulomatous foreign body reaction. Bar, 50 μm.

The field of developmental biology has illuminated the importance of epithelial–mesenchymal interaction(s) in the morphogenesis of a variety of epithelial organs. Armed with a new method
of IEC isolation that maintains the interaction(s) between endodermal and mesenchymal-derived
cell types of the intestinal epithelium, we hypothesized that changing this sole variable (i.e., IEC
isolation) in our existing model would result in tissue-engineered neointestinal mucosa that is

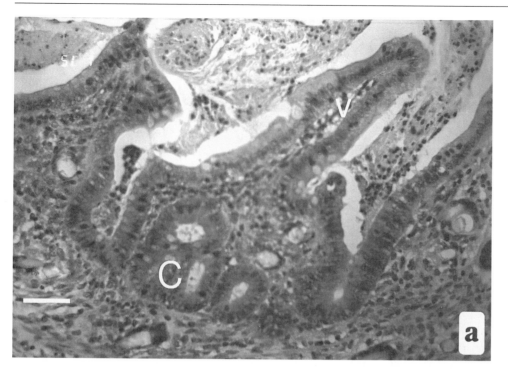

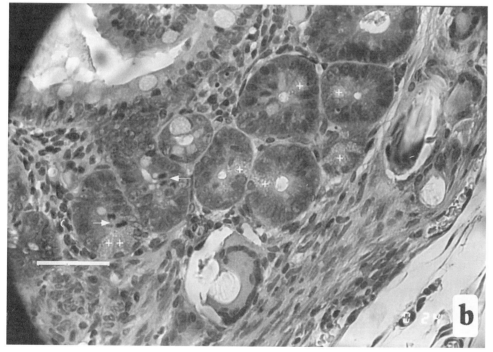

Fig. 38.7. Day 28 implant. (a) Tissue-engineered neointestinal mucosa displays mature villus/crypt axis formation. V, Villi; C, crypts. Bar, 50 μm. (b) Ongoing proliferation is evident from the presence of multiple mitotic figures (arrows). Advanced phenotypic differentiation is evident from the presence of Paneth cells (+). Bar, 50 μm.

organotypic of mature small bowel (Organ *et al.,* 1998b). Prior to testing this hypothesis, 25 experimental IECA isolation procedures were conducted to evaluate, optimize, and demonstrate consistency with this multistep protocol. These results are the culmination of 9 isolation–implantation cohorts to generate satisfactory numbers for meaningful quantitative results and empiric histologic observations. Each cohort requires 4–6 hr of preparation, 8 hr to execute, and an additional 6 hr to conduct assays and tabulate results. We deliberately utilized serum-free media and did not employ growth factors or extracellular matrix proteins with this model, thus allowing determination of base line patterns of vascularization and morphogenesis. In addition, growth factors from other species could be expected to harbor antigenic epitopes that may elicit an incremental inflammatory response.

It is generally accepted that the undifferentiated, pluripotent stem cells of the intestinal epithelium are located in the basal region of the crypts of Lieberkühn. In addition to giving rise to the four main cell types of the intestinal epithelium, these stem cells have a large capacity for epithelial regeneration after tissue injury. Morphometric data from the oncology literature indicate that a surviving stem cell fraction of 1 in 100,000 is capable of regenerating new intestinal epithelium after cytotoxic insults from irradiation or chemotherapy (Moore, 1985; Potten, 1990; Potten *et al.*, 1990). We have utilized these benchmark numbers, which are critical guidelines for determining toxicity in therapeutic oncology, to determine the minimal numbers of IECAs to delivery in this model system.

From detailed modeling of *in vivo* enterocyte kinetics, Potten and Loeffler (1990) have concluded that the small intestinal crypts contain 4–16 actual stem cells in steady state and 30–40 potential stem cells (clonogenic cells) that may take over stem cell properties following perturbations. In this study we delivered $3–5 \times 10^4$ IECA/cm^2, yielding $12–80 \times 10^4$ actual stem cells and $90–200 \times 10^4$ potential stem cells. With the notable exception of cartilage and bone cells, it has been the empiric observation of our laboratories that most transplanted cell types, employing this open-matrix approach to tissue engineering, demonstrate significant decrement in survival in the first 24–48 hr after implantation. This has been attributed to the trauma of the isolation procedure and the ability of the cells to survive on diffusion of oxygen and nutrients from the host implantation site until neovascularization has occurred. If 1 in 10^5 of the stem cell fraction of the implanted IECA survives, this should deliver 0.1–0.8 actual stem cells and 0.9–2.0 potential stem cells. Therefore, we determined that $3–5 \times 10^4$ was the minimum number of IECA/cm^2 to transplant. Our grafting efficiency of 54% for this model suggests this is the minimum acceptable critical mass of IECAs to deliver in this model system.

Histologic evaluation of tissue-engineered neomucosa demonstrates a pattern of morphogenesis identical to intestinal epithelial regeneration from a variety of infectious, inflammatory, ischemic, or dysplastic causes (Lewin *et al.*, 1992b). The IECA–PGA construct elicits an acute inflammatory response evident in the first 3 days after implantation. Days 3–7 are characterized by a rich fibrovascular ingrowth from the omentum. IECAs are identified as microcystic structures with a cuboidal/columnar epithelial lining as early as day 3, and are numerous by day 7. These cryptlike structures begin epithelializing the fibrovascular ingrowth with a simple cuboidal/columnar epithelium by days 10–14. Mature villus/crypt formation with cytologic differentiation of the four cell types of normal intestinal epithelium is present by day 28.

In a complimentary body of work Choi and Vacanti and associates have reported on the successful engraftment and function of tissue-engineered neointestine using a similar model (Choi *et al.*, 1998). Intestinal epithelial organoid units (IEOUs) were isolated utilizing the Evans–Potten technique acquired from sediment of disaggregated smooth muscle, villi, and crypts (see Fig. 39.3, step d). A volume of 30×10^4 to 50×10^4 IEOUs was seeded onto a similar PGA mesh sprayed with 5% poly(lactic acid) to confer rigidity, obviating the need for a luminal stent, and coated with collagen. Serial evaluation out to 14 weeks demonstrated organotypic small intestinal histology in cystic structures with a grafting efficiency of 93%. Immunohistochemistry confirmed sucrase activity at the apical epithelial surface and deposition of laminin in the lamina propria. Positive actin staining demonstrated smooth muscle elements. Neomucosa was mounted in a Ussing chamber and exhibited transepithelial resistive measures comparable to adult rat ileum, thus providing evidence for the functional epithelial integrity of tissue-engineered neomucosa. Important morphometric data of tissue-engineered neomucosa have been documented for base line conditions and in response to 75% small bowel resection, 75% partial hepatectomy, and portacaval shunting at 10 weeks (Kim *et al.*, 1999). Omental implants have been anastamosed in a side-to-side fashion, 3 weeks after implantation, and retrieved at 10 weeks (Kaihara *et al.*, 1999). Patency of 90% or greater of the anastamosis was documented on upper gastrointestinal radiographic examination and direct visualization. Significantly greater villus number, height, and surface length were documented in the anastamosis group, documenting the trophic effects of succus entericus.

The development of tissue-engineered neointestinal mucosa has proceeded in a systematic, step-wise fashion to its current state. Our laboratories have collaborated and employed some of the lessons and observations learned in tissue engineering other tissues. Critical empiric observation, progressively more quantitative assessment of the end point of our work, and application of advances made in the field of epithelial biology have dramatically moved this endeavor closer to clinical application.

The future of tissue engineering neointestinal mucosa must address some important obstacles. Developing methods of IECA isolation for more mature small intestinal epithelium is a rate-limiting obstacle for the whole field of epithelial biology. Because of the significant immunologic issues involved in transplanting intestine, creating methods for isolating IECAs from an SBS patient's own intestine takes on additional importance. Because the intestinal epithelium is the most rapidly proliferating tissue in the human body, this affords a unique opportunity in the field of tissue engineering. Assessing the role of growth factors and extracellular matrix molecules, particularly *in vitro,* will require development of more stringent quantitative measures for these large, multicellular aggregates. Demonstration of the absorptive function of tissue-engineered neomucosa will also play an important role moving this approach toward clinical applicability. Developing a neuromuscular coat for peristalsis will have a profound impact, and is required for a functional, tissue-engineered neointestine.

Although many challenges lie ahead for the entire field of tissue engineering, tremendous strides have been made over the past decade. The foundation for therapeutic implementation of tissue-engineered organs and tissues is being established today. The hope and expectation is that, someday, equipping patients with tissue-engineered organs and tissues may be as routine as coronary bypasses are today (Langer and Vacanti, 1999).

Acknowledgments

The guidance of Marvin McMillan, technical assistance of Harold Robinson, histological evaluation of Betsy Schloo (of the Deborah Heart and Lung Center in Brown Mills, NJ), and the graphics assistance of Jay Alexander are gratefully appreciated. This work was funded by the Department of Surgical Research, Columbia Michael Rees Hospital and Medical Center, Chicago, IL, and a LISS II award from the Living Institute for Surgical Studies, University of Illinois at Chicago College of Medicine, Chicago, IL.

References

Bueno, J., Ohwada, S., Kocoshis, S. *et al.* (1999). Factors impacting the survival of children with intestinal failure referred for intestinal transplantation. *J. Pediatr. Surg.* **34**(1), 27–33.

Caniano, D. A., Stern, J., and Ginn-Pease, M. E. (1989). Extensive short-bowel syndrome in neonates: Outcome in the 1980's. *Surgery* **105**(2, Part 1), 119–124.

Carnie, A. B., Lamerton, L. F., and Steel, G. G. (1965a). Cell proliferation studies in the intestinal epithelium of the rat. I. Determination of kinetic parameters. *Exp. Cell Res.* **39**, 528–538.

Carnie, A. B., Lamerton, L. F., and Steel, G. G. (1965b). Cell proliferation studies in the intestinal epithelium of the rat. II. Theoretical aspects. *Exp. Cell Res.* **39**, 539–553.

Chaves, M., Smith, M. W., and Williamson, R.C.N. (1987). Increased activity of digestive enzymes in ileal enterocytes adapting to proximal small bowel resection. *Gut* **28**, 981–987.

Cheng, H., and Leblond, C. P. (1974a). Origin, differentiation and removal of the four main epithelial cell types in the mouse small intestine. I. Columnar cell. *Am. J. Anat.* **141**, 461–479.

Cheng, H., and Leblond, C. P. (1974b). Origin, differentiation and removal of the four main epithelial cell types in the mouse small intestine. V. Unitarian theory of the origin of the four epithelial cell types. *Am. J. Anat.* **141**, 537–562.

Choi, R. S., Riegler, M., Pothoulakis, C. *et al.* (1998). Studies of brush border enzymes, basement membrane components and electrophysiology of tissue-engineered neointestine. *J. Pediatr. Surg.* **33**(7), 991–997.

Cooper, A., Floyd, T. F., Ross, A., Bishop, H., Templeton, J. M., and Zeigler, M. M. (1984). Morbidity and mortality of short-bowel syndrome acquired in infancy: An update. *J. Pediatr. Surg.* **19**, 711–718.

Diego, M. D., Miguel, E., Lucen, C. M. *et al.* (1982). Short gut syndrome: A new surgical technique and ultrastructural study of the liver and pancreas. *Arch. Surg. (Chicago)* **117**, 789–795.

Dorney, S. F., Ament, M. E., Berquist, W. E., Vargas, J. H., and Hassall, E. (1985). Improved survival in very short small bowel of infancy with use of long-term parenteral nutrition. *J. Pediatr.* **107**, 521–525.

Dudrick, S. J., Latif, R., and Fosnocht, D. E. (1991). Management of short bowel syndrome. *Surg. Clin. North Am.* **71**(3).

Evans, G. S., Flint, N., Somers, A. S., Eyden, B., and Potten, C. S. (1992). The development of a method for the preparation of rat intestinal epithelial cell primary cultures. *J. Cell Sci.* **101**, 219–231.

Foldes, J., Rimon, B., Muggia-Sullam, M. *et al.* (1990). Progressive bone loss during long-term home total parenteral nutrition. *JPEN, J. Parenter. Enteral Nutr.* **14**(2), 139–142.

Grosfeld, J. L., Rescorla, F. J., and West, K. W. (1986). Short bowel syndrome in infancy and childhood: Analysis of survival in 60 patients. *Am. J. Surg.* **151**, 41–46.

Hamilton, W. J., and McMinn, R.M.H. (1976). Digestive system. *In* "Textbook of Human Anatomy" (W. J. Hamilton, ed.), 2nd ed., pp. 329–404. Mosby, St. Louis, MO.

Howard, L., Heaghy, L. L., and Timchalk, M. (1986). A review of the current national status of home parenteral and enteral nutrition from the provider and consumer perspective. *JPEN, J. Parenter. Enteral Nutr.* **10**, 416.

Kaihara, S., Kim, S. S., Benvenuto, M. *et al.* (1999). Successful anastamosis between tissue-engineered intestine and native small bowel. *Transplantation* **67**(2), 241–245.

Kedinger, M., Simmon-Assmann, P., Lacroix, B., Marxer, A., Hauri, H. P., and Haffen, K. (1986). Fetal gut mesenchyme induces differentiation of cultured intestinal endodermal and crypt cells. *Dev. Biol.* **113**, 474–483.

Kedinger, M., Simmon-Assmann, P., Alexandre, E., and Haffen, K. (1987). Importance of a fibroblastic support for *in vitro* differentiation of intestinal endodermal cells and for their response to glucocorticoids. *Cell Differ.* **20**, 171–182.

Kern, I. B., Leece, A., and Bohane, T. (1990). Congenital short gut, malrotation and dysmotility of the small bowel. *J. Pediatr. Gastroenterol. Nutr.* **11**(3), 411–415.

Kim, S. S., Kaihara, S., Benvenuto, M. S. *et al.* (1999). Regenerative signals for intestinal epithelial organoid units transplanted on biodegradable polymer scaffolds for tissue engineering of small intestine. *Transplantation* **67**(2), 227–233.

Lak, G. D., and Baylin, S. B. (1983). Polyamines and intestinal growth—Increased biosynthesis after jejunectomy. *Am. J. Physiol.* **245**, G656–G660.

Langer, R., and Vacanti, J. P. (1993). Tissue engineering. *Science* **260**, 920–926.

Langer, R. S., and Vacanti, J. P. (1999). Tissue engineering: The challenges ahead. *Sci. Am.* **4**, 86–89.

Lewin, K. J., Riddell, R. H., and Weinstein, W. M., eds. (1992a). Small and large bowel structure, developmental and mechanical disorders. *In* "Gastrointestinal Pathology and its Clinical Implications" pp. 703–712. Igaku-Shoin, New York.

Lewin, K. J., Riddell, R. H., and Weinstein, W. M., eds. (1992b). Inflammatory bowel diseases. *In* "Gastrointestinal Pathology and its Clinical Implications" pp. 846–850. Igaku-Shoin, New York.

Moore, J. V. (1985). Clonogenic response of cells of murine intestinal crypts to 12 cytotoxic drugs. *Cancer Chemother. Pharmacol.* **15**, 11–15.

Netter, F. H. (1962). "The CIBA Collection of Medical Illustrations," Vol. 3, Part II, p. 38.

Organ, G. M., and Vacanti, J. P. (1995). Enterocyte transplantation for tissue engineering of neointestinal mucosa. *In* "Methods of Cell Transplantation" (C. Ricordi, ed.), R. G. Landes, Austin, TX.

Organ, G. M., and Vacanti, J. P. (1997). Tissue engineering neointestine. *In* "Principles of Tissue Engineering" (R. P. Lanza, R. Langer, and W. L. Chick, eds.), pp. 441–462. Academic Press, San Diego, CA.

Organ, G. M., Mooney, D. J., Hansen, L. K., Schloo, B., and Vacanti, J. P. (1993a). Enterocyte transplantation using cell–polymer devices causes intestinal epithelial lined tube formation. *Transplant. Proc.* **25**(1), 998–1001.

Organ, G. M., Mooney, D. J., Hansen, L. K., Schloo, B., and Vacanti, J. P. (1993b). Design and transplantation of enterocyte–polymer constructs: A small animal model for neointestinal replacement in short bowel syndrome. *Surg. Forum* **44**, 432–436.

Organ, G. M., Kahn, A. F., Robinson, H. T. *et al.* (1998a). Organotypic differentiation of tissue engineered neointestinal mucosa: The role of mesenchymal–endodermal interaction. *Surg. Forum* **49**, 583–585.

Organ, G. M., Robinson, H. T. *et al.* (1998b). Utilization of crypt cell-enriched aggregates to tissue-engineer intestinal neomucosa. *Tissue Eng.* **4**(4), 482.

Plumb, J. A., Milroy, R., and Kaye, S. B. (1989). Effects of the pH dependence of 3-(4,5-dimethylthiazol-2-yl)-2,5-diphenyl-tetrazolium bromide-formazan absorption on chemosensitivity determined by a novel tetrazolium-based assay. *Cancer Res.* **49**(16), 4435–4440.

Pokorny, W. J., and Fowler, C. L. (1991). Isoperistaltic intestinal lengthening for short bowel syndrome. *Surg. Gynecol. Obstet.* **172**, 39–43.

Potten, C. S. (1990). A comprehensive study of the radiobiological response of the murine (BDF1) small intestine. *Int. J. Radiat. Biol.* **58**, 925–973.

Potten, C. S., and Loeffler, M. (1990). Stem cells: Attributes, cycles, spirals, pitfalls and uncertainties: Lessons for and from the crypt. *Development (Cambridge UK)* **110**, 1001–1020.

Potten, C. S., Owen, G., and Roberts, S. A. (1990). The temporal and spatial changes in cell proliferation within the irradiated crypts of the murine small intestine. *Int. J. Radiat. Biol.* **57** 185–199.

Potten, C. S., Wilson, J. W., and Booth, C. (1997). Regulation and significance of apoptosis in the stem cells of the gastrointestinal epithelium. *Stem Cells* **15**, 82–93.

Reyes, J., Bueno, J., Kocoshis, S. *et al.* (1998). Current status of intestinal transplantation in children. *J. Pediatr. Surg.* **33**(2), 243–254.

Ricotta, J., Zuidema, F. D., Gadacz, R. J., and Sadri, D. (1981). Construction of an ileocecal valve and its role in massive resection of the small intestine. *Surg. Gynecol. Obstet.* **152**, 310–314.

Russell, P. S. (1985). Selective cell transplantation. *Ann Surg.* **201**, 255–262.

Schraut, W. H., Leck, C. W., and Tsujinaku, Y. (1986). Intestinal preservation of small bowel grafts by vascular washout and cold storage. *In* "Small Bowel Transplantation: Experimental and Clinical Fundamentals" (E. Deltz, A. Thiele, and A. Hammelman, eds.), pp. 65–73. Springer-Verlag, Heidelberg.

Thompson, J. (1990). Recent advances in the surgical treatment of the short-bowel syndrome. *Surg. Ann.* **22**, 107–127.

Thompson, J. S., Vanderhoot, J. A., Antonson, D. L. *et al.* (1984). Comparison of techniques for growing small bowel neomucosa. *J. Surg. Res.* **36**(4), 401–406.

Tilson, M. D. (1980). Pathophysiology and treatment of short bowel syndrome. *Surg. Clin. North Am.* **60**(5), 1273–1284.

Touloukian, R. J., and Smith, G.J.W. (1983). Normal intestinal length in pre-term infants. *J. Pediatr. Surg.* **18**, 720–723.

Unger, R. H., Ohneda, A., Valverde, I., Eisentraut, A. M., and Exton, J. (1968). Characterization of the responses of circulating glucagonlike immunoreactivity to intraduodenal and intravenous administration of glucose. *J. Clin. Invest.* **47**, 48–65.

Vacanti, J. P. (1988). Beyond Transplantation, Third annual Jason Mixter lecture. *Arch. Surg. (Chicago)* **23**, 545–549.

Vacanti, J. P., Morse, M. A., Saltzman, W. M., Domb, A. J., Perez-Atayde, A., and Langer, R. (1988). Selective cell transplantation using bioabsorbable artificial polymers as matrices. *J. Pediatr. Surg.* **23**(1), 3–9.

Weber, T. R., Vane, D. W., and Grosfeld, J. L. (1982). Tapering enteroplasty in infants with bowel atresia and short gut. *Arch. Surg. (Chicago)* **117**, 684–688.

Weber, T. R., Tracy, T., and Connors, R. H. (1991). Short-bowel syndrome in children. Quality of life in an era of improved survival. *Arch. Surg. (Chicago)* **126**, 841–846.

Williamson, R.C.N. (1978a). Intestinal adaptation. Part 1. Structural, functional and cytokinetic changes. *N. Engl. J. Med.* **298**(25), 1393–1402.

Williamson, R.C.N. (1978b). Intestinal adaptation. Part II. Mechanisms of control *N. Engl. J. Med.* **298**(26), 1444–1450.

CHAPTER 39

LIVER

Hugo O. Jauregui

INTRODUCTION

Historically, the development of liver support devices has progressed from passive to bioactive systems. Passive systems to remove blood toxins accumulated during liver failure have included hemodialysis, hemoperfusion, hemofiltration, and plasma exchange (Malchesky, 1994). Although some studies have reported transient improvement of some blood parameters or symptoms associated with liver failure (e.g., hepatic encephalopathy), passive methods did not improve patient survival. Simultaneously, in several medical centers, studies of bioreactors or cell-based liver support systems were implemented to replace key functions that prevent hepatic encephalopathy development or assist in its management and thereby improve the outcome of patients with hepatic failure. This chapter presents a critical overview of cell-based liver assist devices, including new prototypes as well as the one that has reached clinical testing. The succinct review of liver anatomy and function, as well as the pathophysiology of hepatic failure, facilitates understanding of this type of cellular therapy.

ANATOMY AND FUNCTION OF THE LIVER

Specialized metabolic, synthetic, and secretory liver functions occur through complex interactions between blood and hepatocytes (parenchymal liver cells). The liver of a 70-kg adult weighs nearly 1.75 kg (~2.5% of total body weight), of which the hepatocyte mass represents approximately 1.2 kg (Arias *et al.*, 1988).

Hepatocytes, in their arrangement as plates of single-cell thickness, receive a dual blood supply derived from the portal vein and the hepatic artery. Combined, these vessels create an extensive vascular space, called the liver sinusoid (Figs. 39.1 and 39.2). Although these sinusoids are lined by endothelial cells, the endothelium is interrupted by pores or fenestrations that allow free passage of most solutes into a pericapillary space (the space of Disse). Hepatocytes project the microvilli of their sinusoidal domain (72% of total cell surface area) into this space, which is rich in basement membrane proteins and proteoglycans, such as collagen type IV, (the major type of collagen of this area), fibronectin, and heparin sulfate. Laminin is not present in this space in adult rodent livers, but is contributory to the embryogenesis of the liver (Martinez-Hernandez *et al.*, 1991). In addition to the sinusoidal domain, hepatocytes present a lateral surface (15% of cell surface) contiguous to other hepatocytes, but interrupted by a canalicular pole (13% of cell surface). Bile canaliculi from the two hemicanaliculi are sealed by tight junctions and present numerous microvilli (Fig. 39.3). This histology constitutes the beginning of the biliary duct system and contributes to the formation of canalicular bile, which is further modified at the bile duct level. This succinct histological description permits appreciation of the structural hepatic organization responsible for a functional organization that allows unidirectional perfusion of polarized hepatocytes arranged in plates. Perfusion studies have shown a hepatocyte distribution that forms a structure called the liver acinus (Rappaport *et al.*, 1954), a microvascular unit in which hepatocytes near the terminal portal venule and hepatic arteriole (periportal cells) are first perfused with blood and

Fig. 39.1. Three-dimensional display of a liver lobule, showing the vascular, biliary, and hepatocyte arrangement. Copyright Eli Lilly and Company. Used with permission.

those near the hepatocyte venule (terminal veins) are perfused secondarily with blood modified previously by upstream hepatocytes. In fact, Rappaport defined a hepatic acinus with three zones (Rappaport *et al.*, 1954): zone one (periportal) receives the highest concentrations of oxygen and

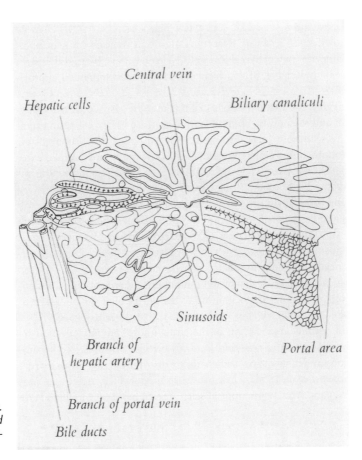

Fig. 39.2. A schematic of Fig. 39.1. Copyright Eli Lilly and Company. Used with permission.

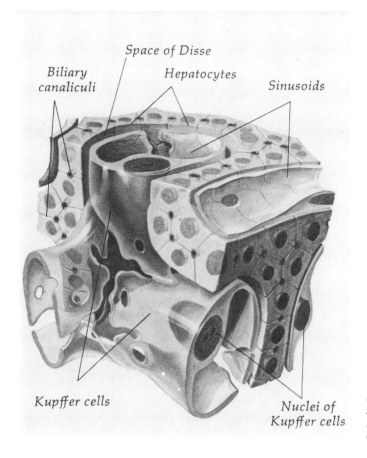

Fig. 39.3. A three-dimensional section of the liver lobule, showing the relationship between the hepatocytes and the Kupffer cells and the sinusoids through the space of Disse. Copyright Eli Lilly and Company. Used with permission.

solutes, zone three (pericentral) meets with the lowest blood concentration of oxygen, and zone two lies between zones one and three.

The concept of zonal hepatocyte distribution may prove important in the design of artificial liver support systems. Liver assist devices (LADs) that incorporate specific zonal hepatocytes with characteristic functions in an arrangement similar to that found *in situ* could be manufactured, provided that (1) hepatocyte functions responsible for the often fatal central nervous system derangement, hepatic coma, could be defined, (2) technology could be developed to isolate zonal hepatocytes in quantities compatible with therapeutic applications, and (3) expression of these functions could be maintained.

Regarding zonal hepatic function, it should be mentioned that whenever toxins, infectious agents, or circulatory distress damage the liver, they typically do so in a zonal pattern of the acinus. Hepatocyte injury commonly generates two processes that also have a zonal distribution, namely, liver regeneration by the remaining hepatocytes and hepatocyte necrosis. It is well known that only three to four hepatocytes situated in proximity to the portal space of the liver acinus have the capacity for mitosis. Although *in vivo* these hepatocytes are responsible for a very slow renewal of liver cells (through the development of proper separation techniques), they may be the source of the cellular component of artificial liver support systems. On the other hand, pericentral hepatocytes are responsible for the detoxification activity of the liver, a process that requires expression of a complex group of enzymes known collectively as the cytochrome oxidase P450 system (Kaplowitz, 1992).

Zonal hepatocyte necrosis frequently stimulates the stromal component of the liver to produce wide areas of fibrous scarring, which ultimately progresses to a general impairment of liver functions. Because the liver receives a dual blood supply, special hemodynamic situations are created by this generalized scarring process (cirrhosis). In this condition, the blood from the portal vein and the hepatic artery is diverted directly into the large hepatic veins (without the usual passage through the sinusoidal circulation) and from there into the inferior vena cava. These shunts conspire against proper blood detoxification because toxins, which should contact liver cells to be metabolized, remain in the systemic circulation and act on other organs, particularly the central nervous system.

Electron microscopic examination of the hepatocyte cytoplasm reveals abundant smooth and rough endoplasmic reticulum cisternae, numerous ribosomes, and mitochondria along with many lysosomal and endocytic vesicles. The major aspects of detoxification occur in the smooth endoplasmic reticulum cisternae.

Total replacement of liver functions by an artificial liver would entail regulation of plasma protein production, metabolism, and storage of vitamins; biotransformation and detoxification of drugs; synthesis of lipoproteins and cholesterol; maintenance of glycogenesis and glycogenolysis; and the conversion of heme to bilirubin and biliverdin (to name just some of the physiological activities of this organ). Although the design of a liver support system probably does not require the provision of all liver functions, the functions essential for liver support are not yet known.

LIVER FAILURE: PATHOPHYSIOLOGICAL CLASSIFICATION

Excessive alcohol consumption is the major cause of morbidity and mortality of liver disease in the United States and most other Western countries. From the 56,773 cases of viral hepatitis reported in 1988 (Zeldis, 1992), 41% were due to hepatitis B, 50% to hepatitis A, and about 5% to hepatitis C. Each virus case presents a similar clinical syndrome, but only hepatitis B and C have the potential to produce chronic hepatitis.

Many drugs are also associated with the production of either chronic or acute liver failure. The metabolism of foreign compounds by the P450 system is invaluable to process xenobiotics, but on occasion this system gives rise to toxic products. It is customary to divide these toxic drug metabolites into electrophiles and drug free radicals. Electrophiles arise from electron sharing of covalent bonds with substances that can donate electrons. Examples of drugs that produce electrophile metabolites, which in turn produce hepatocyte necrosis, are acetaminophen (attacks thiol groups) and halothane (reacts with protein amino groups). Drug free radicals preferentially attack membrane phospholipids, leading to lipid peroxidation.

These various agents produce different types of hepatic failure that are treated either medically or surgically by liver transplantation. The most life-threatening aspect of hepatic failure is the syndrome characterized by a severe impairment of liver function that proceeds to the development of hepatic encephalopathy (HE) (hepatic coma) within 8 weeks of the first symptoms of illness. Bernuau has divided patients with acute liver failure into fulminant hepatic failure (FHF) (appearance of HE within 2 weeks from the onset of jaundice) and subfulminant hepatic failure (HE appearance between 2 weeks and 3 months postjaundice) (Bernuau *et al.*, 1986). The important points to differentiate FHF from chronic hepatic failure (CHF) are that, in FHF, there is no evidence of preexisting liver disease and the mortality rate is about 70–95%. In CHF, mortality is lower, evidence of chronic liver disease exists, and medical management may prove successful, at least in the first stages of HE. Despite widespread disagreement, but because both acute and chronic liver failure may be complicated by HE, there is a tendency to discuss this entity under unifying pathogenic mechanisms (Schenker, 1989).

Probably the most common denominator in FHF and CHF is the shunting of portal–venous

Table 39.1. Clinical observations provide clues to pathogenesis[a]

Neurotoxin origin	Precipitation of encephalopathy
Intestinal (mainly colonic)	Constipation, portal–systemic shunting
Nitrogenous compounds	Protein meal
Bacterial	Infection (nonabsorbable antibiotics improve encephalopathy)
Metabolization by the liver, first-pass extraction	Portal–systemic shunting, portacaval high anastomosis

[a]Based on four postulated mechanisms: ammonia levels, synergistic neurotoxins, false neurotransmitters, and GABA/endogenous benzodiazepines.

blood into the systemic circulation, bypassing the liver. This shunting was discussed above as it pertains to CHF cases resulting from cirrhosis. In FHF, the architecture of the liver is severely affected and the large portal blood flow joins the systemic circulation without the benefit of the detoxification activity of the few remaining hepatocytes. Despite the extensive hepatocyte necrosis, some cases of FHF are known to survive with complete histological recovery of the liver (Jones and Schafer, 1990). Which of the toxic products responsible for hepatic coma should be metabolized by a LAD? Where do they originate? The clinical signs and symptoms of patients with acute or chronic liver failure offer important clues to answer those questions. Table 39.1 lists relationships between encephalopathy and neurotoxin origin (Blei, 1992).

MECHANISMS OF HE IN LIVER FAILURE

AMMONIA

There is no correlation between the grade or stage of hepatic encephalopathy and arterial ammonia levels, but elevated cerebrospinal fluid ammonia does correlate with HE stages (Blei, 1992). In addition, HE may improve if the intestinal NH_3 production common in CHF is reduced. Unfortunately, dialysis treatment of HE patients has failed to awaken these patients in spite of an evident decrease in their levels of blood NH_3. For example, feeding cirrhotic patients methionine (a mercaptain precursor) induced changes in mental state without cocomitant changes in blood NH_3 levels (Blei, 1992).

FALSE NEUROTRANSMITTER HYPOTHESIS

Hepatic coma may occur when brain neurotransmitters (dopamine and noradrenaline) are replaced in cirrhotic patients by weak neurotransmitters (octopamine) generated in the intestine and shunted into the general circulation (Schafer and Jones, 1990). This hypothesis, however, was greatly discredited by studies showing that the administration of octopamine failed to change the mental state of experimental animals and by data showing lower levels of octopamine in the brains of cirrhotic patients than in normal brains (Blei, 1992).

GABA–BENZODIAZEPINE HYPOTHESIS

The decarboxlation of glutamine in the nervous system produces γ-aminobutyric acid (GABA), which causes neuroinhibition through receptor binding. The GABA receptor (subtype A) also binds benzodiazepines and barbiturates. Benzodiazepine action results in the modulation of GABA-ergic neurotransmission. Studies suggest that endogenous benzodiazepines, which are not properly metabolized by the liver, may be associated with the HE that is present in FHF. The benzodiazepine role in HE pathogenesis is supported by the observation that the administration of a benzodiazepine receptor antagonist (flumazenil) improves the status of animals and humans with FHF (Bansky *et al.*, 1989).

An understanding of the possible pathogenic mechanisms of HE is useful to assess current and future development of the LAD in humans and/or experimental animals. Because the pathogenesis of HE is not known, investigators often validate the efficacy of liver-assist devices by testing drug metabolism or hepatic protein synthesis *in vitro*. Two well-established models of acute hepatic failure (AHF) (Blei, 1992) are end-to-side portacaval anastomosis in the rat and the galactosamine-intoxicated rabbit. However, due to the small size of these animals, devices intended for human use are commonly tested in toxin-induced or anhepatic large animal models of liver failure, which may not closely produce the clinical symptoms of HE (Terblanche and Hickman, 1991). In addition, there are no adequate animal models of CHF (Blei, 1992).

MEDICAL NEED FOR LIVER-ASSIST DEVICES

Although several approaches have been used to address liver failure, whole-organ transplantation remains the procedure of choice. However, this approach has several limitations. First, the number of available organs does not meet the demand; second, the procedure carries risks, which increase in certain conditions complicated by AHF; and third, liver transplantation is expensive. Because of the success of liver transplantation, cell-based bioreactors constructed with cells capable of performing liver functions are a potential alternative to treat reversible acute liver failure or a bridge to liver transplantation.

It can be predicted that early use of LADs may result in an improvement of the clinical stages of AHF. Potentially, a total recovery by use of these devices will reduce the typical brain edema found in these critically ill patients. In these cases, LADs should work as a bridge-type therapy by providing the necessary clinical support until a liver becomes available for transplantation or until the native liver regenerates.

BIOLOGICAL COMPONENT OF LIVER-ASSIST DEVICES

Hepatocyte transplantation studies in rats showed that 1.0×10^7 microcarrier-attached hepatocytes (1–2% of liver mass) improved survival of rats with 90% hepatectomy when compared to nontransplanted controls (Demetriou *et al.,* 1988). It was also demonstrated that patients could survive after 90% partial hepatectomy (Monaco *et al.,* 1964). Therefore, an artificial liver should at least provide 1–10% of the liver mass (Demetriou *et al.,* 1988; Monaco *et al.,* 1964; Margulis *et al.,* 1989; Jauregui *et al.,* 1995). An artificial liver support designed for a 70-kg patient and including as much as 10% of the native liver functions would require approximately 120 g of liver tissue or 25×10^9 cells. This figure is controversial, but it should be stated that the liver mass previously employed in artificial liver support systems has ranged from 1 (Margulis *et al.,* 1989) to 15% (Sussman *et al.,* 1994). Due to the limited availability of human livers, the source of the liver cells remains an issue. Alternatives currently being explored include the use of transformed cells, freshly isolated hepatocytes, and cultured hepatocytes.

TOTALLY TRANSFORMED (MALIGNANT) HEPATOCYTES

The primary advantage of totally transformed hepatocytes is the nearly limitless capacity to maintain growth. Under culture conditions, these neoplastic hepatocytes may be subjected to ongoing assessment of purity, function, and pathogenecity. Wolf used a Reuber hepatoma cell line in a hollow-fiber device to treat Gunn rats and demonstrated bilirubin conjugation (Wolf *et al.,* 1979). C3A, a cell line derived from HepG$_2$ (a human hepatoblastoma cell line), has been established and grown in hollow-fiber cartridges (Kelly and Darlington, 1989). Concern arises regarding the potential of totally transformed hepatocytes to exhibit spontaneous changes in chromosome number and in gene expression after a period under culture conditions. Also of concern is the potential of an accidental infusion of tumorigenic cells into the patient's systemic circulation. In addition, the cost associated with cell production (culture supplies, quality control, bioburden monitoring, etc.) may be elevated. For instance, it takes approximately 1 month to grow sufficient C3A cell mass (200 g) for a device (Sussman *et al.,* 1992).

FRESHLY ISOLATED HEPATOCYTES

A second approach is the use of primary hepatocytes, which can be cryopreserved after isolation and thus shipped and stored at the clinical site. Soyer and Eiseman were the first to use cell suspensions or slices to provide hepatic functions (Soyer *et al.,* 1973), but Matsumura, in 1987, was the first to treat successfully a patient in hepatic failure using a suspension of rabbit hepatocytes (Matsumura *et al.,* 1987). LADs containing porcine hepatocytes attached to collagen-coated dextran microcarriers, inoculated into the extracapillary space of cellulose acetate hollow fibers, have been tested in dogs with surgically induced liver ischemia (Rozga *et al.,* 1993a,b). A scaled-up version of this LAD is now in clinical trials. The choice of porcine cells appears justified from previous reports of extracorporeal pig liver perfusion to support patients with liver failure (Abouna *et al.,* 1973; Chari *et al.,* 1994). In addition, there are several attributes in favor of the pig over other animal donors. Pigs have large litters, are inexpensive, rapidly reach maturity (6 months), are relatively disease resistant, and can be raised in controlled herds (e.g., specific pathogen-free herds). Their organs have similarities to human organs with regard to size and physiology, and large numbers of cells (e.g., hepatocytes) can be harvested. Freezing of porcine hepatocytes facilitates their availability in quantities compatible with commercial use. The entire process of isolation and cataloging volumes of porcine hepatocytes demands well-established protocols and reliable methodology that spares any microbiological, virological, and toxicological contamination. A significant effort has been devoted to the cell supply issue through animal sourcing programs. Because the hepatocytes isolated from porcine livers cannot be subsequently sterilized, a comprehensive herd and animal bioburden screening program should be developed to provide adequately "clean" animals, particularly free of potential zoonotic viruses (Gil, 1995). There are certain practical and theoretical objections to the use of porcine hepatocytes. We should mention that there is great concern about the presence of porcine endogenous retroviruses in all pig herds tested to date; the passage of porcine protein into the circulation of the patient and the need to control the source of hepatocytes on a regular basis are additional concerns.

CULTURED HEPATOCYTES

The use of primary cultures of hepatocytes as the biological component of LADs offers some interesting alternatives to the use of freshly isolated hepatocytes, but at the same time presents some complications, particularly the logistics and cost of a massive tissue culture production. One possibility would be to use hollow-fiber devices as tissue culture bioreactors, which would subsequently become LADs by replacing the appropriate tissue culture media with the blood or plasma of the patients. This approach, similar to the one implemented in previous models with tumor cells, is at best not well suited to growing porcine hepatocytes in large quantities, although the system has proved worthwhile to grow rat hepatocytes.

Microcarrier technology to grow hepatocytes in commercial quantities is actively being explored in our laboratory. In this context, it should be stated that short-term monolayer culture of pig hepatocytes produces a cell that metabolizes diazepam, acetaminophen, and ammonia more efficiently compared with freshly isolated cells (Naik, *et al.,* 1996). An important area of improvement to maintain the phenotypic expression of hepatocytes in long-term culture is the identification of new tissue culture (TC) media with defined formulations that will avoid the supplementation with human or animal serum. Following this line, Jauregui *et al.* (1994) have modified a TC medium, Chee's Modified Eagle's Medium (CEM), to cultivate rat (Naik *et al.,* 1992), rabbit (Jauregui and Naik, 1991), and monkey hepatocytes and are actively pursuing the optimal medium for porcine cells.

A controversial subject in hepatocyte culture is the choice of TC substrate. Vitrogen, (Naik *et al.,* 1992), Biomatrix (Rojkind *et al.,* 1980), and Matrigel (Schuetz *et al.,* 1988) (primary laminin) as well as a newly developed polymer (Kobayashi *et al.,* 1986) that binds the so-called asialoglycoprotein receptor, have all been used to maintain long-term viability. Often these substrates are used but not compared with old and well-known substrates that failed to maintain phenotypic expression. Furthermore, there are few if any publications that address the subject in a comprehensive fashion (e.g., by measuring multiple parameters of hepatocyte function).

PARTIALLY TRANSFORMED (IMMORTALIZED) HEPATOCYTES

Implementation of a LAD as a treatment for acute liver failure will be greatly advanced by the availability of porcine hepatocytes that grow indefinitely, with maintenance of functional activity but expressing no tumorogenicity. Likewise, patients with chronic liver failure could be maintained with prolonged treatment with devices harboring partially transformed human hepatocytes. The commercial production of these types of cells has been addressed by the transfer of specific viral oncogenes that can generate cell lines that divide, expressing an intermediate stage of differentiation, a process known as immortalization (Westerman, 1996). As reported by researchers in this field, only few human cells become truly "immortal" following the introduction of viral oncogenes such as the SV40 T antigen (Westerman, 1996). Indeed, although early stages of the immortalization process may expand the life-span of primary human cells in culture, sooner or later these cells senesce and will enter a phase called "crisis," from which only few immortalized cells will evolve as liver cells (Westerman, 1996). Studies with temperature-sensitive forms of oncogenes have shown that oncoprotein inactivation at a nonpermissive temperature allows cells to resume *in vivo* as well as *in vitro* normal differentiation (Renfranz *et al.,* 1991).

The SV40 virus transforms cells of many species. Depending on the genetic properties of the cell in question, the virus infection will be lytic and consequently these cells, called permissive cells, will support full expression of the viral genome and release progeny of viral particles (Chou, 1989). Monkey cells are permissive and are killed after the SV40 infection. On the other hand, mouse, rat, and rabbit cells, which are not permissive and consequently do not support viral replication, express some viral early genes and can become transformed by SV40. Human and porcine hepatocytes are semipermissive, supporting a limited infection with the release of a low number of virions, and thus also become transformed by SV40. To obtain SV40-transformed hepatocyte lines free of infectious virus, we and others use the transforming T antigen, a partial viral particle that is responsible for viral replication but is not related to the viral constituents that are responsible for the viral infectivity (Liu *et al.,* 1999). This viral regulatory protein directs an ordered sequence of events that, starting in early phases of the SV40 infection (prior to the viral DNA replication), proceed into the late phase of infection (Westerman, 1996). This early phase allows the accumulation of T antigens in the cell nucleus, where they alters transcription patterns, stimulating quiescent cells such as hepatocytes in culture to synthesize cellular DNA. How does SV40 T transform cells? This question is still not completely answered but the principal effect is the inactivation of multiple growth suppressors, at least pRb, p53, and an immortalization gene on chromosome 6 (Kim *et al.*, 1998). The HepLiu immortalized porcine hepatocyte line was developed in our laboratory in 1995 and since then has undergone 60 passages and is not tumorigenic when grafted in severe combined immune-deficient (SCID) mice. The HepLiu cell line is the product of several subcloning steps that select for hepatocytes metabolizing diazepam as well as acetaminophen, but not producing albumin (Liu *et al.*, 1999). The latter facilitates the use of these cells in hollow-fiber devices with high-molecular-weight cutoff membranes.

LIVER-ASSIST DEVICE CONFIGURATIONS

Device design focuses on providing an environment for maximum cell viability and function. Criteria for device performance include transport of nutrients and oxygen to maintain cell viability and function, and transport of toxins and metabolites for patient support. Several bioreactors or culture systems have been developed for the cultivation of cells in a three-dimensional architecture: static cultures, matrix cultures, roller bottles, stirred suspension, packed beds, fluidized beds, and hollow-fiber bioreactors (Jessup *et al.*, 1993). Each of these systems has advantages and disadvantages for use as an extracorporeal cell-based liver-assist device. Numerous design variations have been proposed for cell-based LADs (Margulis et al., 1989; Sussman *et al.*, 1992, 1994; Kasai *et al.*, 1994; Dixit, 1994; Rozga *et al.*, 1994; Uchino *et al.*, 1988; Kawamura *et al.*, 1985; Kimura *et al.*, 1980; Yanagi *et al.*, 1989; Li *et al.*, 1993; Wolf and Munkelt, 1975; Gerlach *et al.*, 1994; Nyberg *et al.*, 1993a,b); however, only three main designs (hepatocyte suspension, packed-bed, and hollow-fiber bioreactors) have reached clinical evaluation (Table 39.2).

Hepatocyte Suspension

The bioartificial liver developed by Matsumura was based on the principle of hemodialysis against a suspension of hepatocytes (Matsumura *et al.*, 1987). A modified plate hemodialyzer fitted with a dialysis membrane permeable to middle-weight molecules, but impermeable to albumin, was used. A patient in hepatic failure, secondary to bile duct carcinoma, underwent two treatments of 5 and 4.5 hr each over a period of 3 days. The patient's blood was dialyzed against a flowing suspension of 100 g of previously cryopreserved rabbit hepatocytes. Reduction in serum bilirubin level and improvement in mental states were observed following each treatment, and the patient was discharged from the hospital after the second treatment. However, there has been no further report on the development of this device.

Margulis *et al.* (1989) reported the development of a 20-ml capsule containing a suspension of 40×10^6 porcine hepatocytes and activated charcoal. Each device was used for only 1 hr and a treatment lasted 6 hr; six devices were used per treatment. Mortality in the treated group of 59 patients with acute liver failure was 37% compared to 59% in the control group consisting of 67 patients who received only intensive medical care. However, little information was provided on metabolic support during treatments.

Packed-Bed Bioreactor

Li *et al.* (1993) inoculated hepatocytes in a column filled with glass beads. The packed-bed bioreactor maintained urea and albumin synthesis, and interconnecting three-dimensional structures resembling the cell plate anatomy of the native liver were maintained for up to 15 days in culture. A 10-year-old patient with fulminant hepatic failure was successfully treated with the system. There has been no further report on the development of this device.

Hollow-Fiber Bioreactor

The first hollow-fiber LAD was developed by Wolf and Munkelt (1975). The device contained a rat hepatoma cell line cultured on the exterior surface of semipermeable hollow fibers enclosed within a plastic housing. Sussman *et al.* (1992, 1994) and Sussman and Kelly (1995) developed a similar extracorporeal lever-assist device (ELAD), except that a hepatoma cell line of

Table 39.2. Artificial liver support clinical trials

Investigator	Device configuration	Perfusion medium	Cell selection	Survival rate[a]	
				Treatment	Control
Matusmura *et al.* (1987)	Cell suspension (cellulose membrane 70-kDa cutoff)	Blood (100 g)	Rabbit hepatocytes	100 (1/1)	—
Margulis *et al.* (1989)	Cell suspension $(40 \times 10^6) \times 6$	Blood	Pig hepatocytes	63 (37/59)	41 (27/67)
Li *et al.*[20] (1993)	Packed bed (glass beads)	Plasma	Pig hepatocytes	100 (1/1)	—
Sussman *et al.* (1994)	Hollow-fiber ELAD (cellulose acetate, 70-kDa cutoff)	Blood cell line (200 g)	Human hepatocytes	45 (5/11)	—
Rozga *et al.* (1994)	Hollow-fiber BAL (cellulose acetate, 0.2-μm pore size)	Plasma (5×10^9)	Pig hepatocytes	88 (8/9)	—

[a]Patient survival rates are expressed in percentages; actual numbers are given in parentheses.

human origin (C3A) was used. The ELAD consisted of a cellulose acetate hollow-fiber hemodialyzer (70,000-Da cutoff) containing 200 g of cells grown on the outer surface of the fibers (Sussman *et al.,* 1992, 1994). Blood was perfused through the cartridge. During perfusion, a small amount of plasma was ultrafiltered by the fibers and perfused the cells before returning to the blood path downstream from the ELAD. Of the eleven patients treated with the ELAD, the first two were treated under "Emergency use of unapproved medical devices" and the remaining nine patients were treated according to an investigational device exemption (IDE) (see the Food and Drug Administration web site, http://www.fda.gov/cdrh/manual/idemanul.html). One patient survived, six patients died, and four patients were bridged to orthotopic liver transplantation. No short-term safety problems were associated with the use of the ELAD; however, metabolic support provided by the device was difficult to assess.

In 1985, W.R. Grace & Co. initiated the development of a cell-based liver-assist device in collaboration with Jauregui and co-workers at Rhode Island Hospital. The bioreactor consisted of microporous (0.15-μm pore size) polysulfone hollow fibers assembled in a cylindrical housing. The bioreactor was inoculated with rabbit hepatocytes, placed in an extracorporeal circuit, and perfused with whole blood. The efficacy of the LAD was assessed in two different animal models (Jauregui *et al.,* 1995): (1) normal rabbits injected with diazepam or lidocaine and (2) a galactosamine (Gal)-intoxicated rabbit model of fulminant hepatic failure (FHF).

In the first animal model, the blood levels of both diazepam and lidocaine metabolites clearly increased during treatment, indicating the maintenance of cell activity (P450 function) in the LAD. In the Gal model, a 6-hr treatment significantly increased the survival time and delayed the onset of hepatic encephalopathy when compared to two untreated control groups. Histological evaluations of postmortem livers showed greater hepatocyte regenerative activity in animals treated with the LAD compared with control animals. These findings support the concept that a hollow-fiber LAD containing hepatocytes is able to sustain drug detoxification *in vivo* as well as to modify the course of FHF in a well-characterized animal model. The device was then scaled up for human use and tested for safety using porcine hepatocytes in a sheep model.

Contemporaneous to the research conducted by W.R. Grace & Co., Demetriou and co-workers developed a bioartificial liver (BAL) consisting of hollow fibers and hepatocytes, and was testing the BAL in patients under compassionate use of the device (Rozga *et al.,* 1993b). In 1993, W.R. Grace & Co., and Demetriou and collaborators at Cedars Sinai Medical center joined efforts to further develop the BAL and a liver-assist system (LAS) including the BAL. A comprehensive description of this system is incorporated in this volume.

The hollow-fiber LAD (see Table 39.2) uses membranes of different selectivities (Sussman *et al.,* 1992, 1994). Sussman *et al.* (1994) as well as Matsumura *et al.* (1987) used membranes with low-molecular-weight cutoff (70,000) that reduced passage of serum albumin and excluded proteins of higher molecular weights. Rozga *et al.* (1994) selected microporous (0.2-μm pore size) hollow fibers, which allowed passage of plasma proteins but excluded the passage of cells (i.e., blood cells and hepatocytes). Microporous membranes allow for free exchange of toxins (soluble or protein bound) and large-molecular-weight proteins (e.g., clotting factors) between the blood or plasma and the hepatocytes. On the other hand, the membranes used by Sussman *et al.* (1994) and Matsumura *et al.* (1987) excluded immunoglobulins and provided an immunoprotected environment for the liver cells. However, because BAL treatments (Rozga *et al.*, 1993b) as well as extracorporeal pig liver perfusions have been well tolerated (Chari *et al.*, 1994) and patients have been successfully transplanted posttreatment (Rozga *et al.*, 1993b; Chari *et al.*, 1994), the short-term use (days) of artificial livers that use microporous membranes appears feasible and safe. Consistent with this are the findings of Ye *et al.* (1995), who reported that in primates, allograft transplants following pig heart xenotransplantation as well as pig blood transfusion functioned normally with no histopathologic features of hyperacute humoral or accelerated cellular rejection. This study (Ye *et al.*, 1995) supports the theory that bridging with a concordant or discordant donor xenograft does not prohibit subsequent organ allografting.

Currently, several types of liver-assist systems have been authorized by the Food and Drug Administration to begin human clinical studies to evaluate their safety and biologic activity. Pending the successful outcome of these early studies, scale-up of the manufacturing processes can begin in order to meet the requirements for follow-up multicenter studies and pivotal clinical trials. If progress continues at its current pace, approved liver-assist systems for general distribution will soon be a reality.

REFERENCES

Abouna, G. M., Fisher, L. M., Porter, K. A., and Andres, G. (1973). Experience in the treatment of hepatic failure by intermittent liver hemoperfusions. *Surg. Gynecol. Obstet.* **137**, 741–752.

Arias, I., Jakoby, W., Popper, H., Schachter, D., and Shafritz, D. A. (1988). "The Liver Biology and Pathobiology," 2nd ed., pp. 3–6. Raven Press, New York.

Bansky, G., Meier, P. J., Riederer, E., Walsher, H., Ziegler, W. H., and Schmid, M. (1989). Effect of the benzodiazepine receptor antagonist flumazenil in hepatic encephalopathy in humans. *Gastroenterology* **97**, 744–750.

Bernuau, J., Rueff, B., and Benhamou, J.-P. (1986). Fulminant and subfulminant liver failure: Definitions and causes. *In* "Seminars in Liver Disease" (R. Williams, ed.), p. 97. Thieme, New York.

Blei, A. T. (1992). Hepatic encephalopathy. *In* "Liver and Biliary Diseases" (N. Kaplowitz, ed.), pp. 552–565. Williams & Wilkins, Baltimore, MD.

Chari, R., Collins, B., Magee, J., DiMaio, J. M., Kirk, A. D., Harland, R. C., McCann, R. L., Platt, J. L., and Meyers, W. C. (1994). Treatment of hepatic failure with *ex-vivo* pig–liver perfusion followed by liver transplantation. *N. Engl. J. Med.* **331**, 234–237.

Chou, J. Y. (1989). Differentiated mammalian cell lines immortalized by temperature-sensitive tumor viruses. *Mol. Endo.* **3**(10), 1511–1514.

Demetriou, A. A., Reisner, A., Sanchez, J., Levenson, S. M., Moscioni, A. D., and Chowdhury, J. R. (1988). Transplantation of microcarrier attached hepatocytes into 90% partially hepatectomized rats. *Hepatology* **8**, 1006–1009.

Dixit, V. (1994). Development of a bioartificial liver using isolated hepatocytes. *Artif. Organs* **43**, 645–653.

Gerlach, J., Trost, T., Ryan, C., Meibler, M., Hole, O., Muller, C., and Neuhaus, P. (1994). Hybrid liver support system in a short term application on hepatectomized pigs. *Int. J. Artif. Organs* **17**, 549–553.

Gil, P. (1995). Process considerations for minimizing virus transmissions from primary porcine tissues. *FDA Int. Sci. Conf. Viral Saf. Eval. Viral Clearances Biopharm. Prod.*, Bethesda, MD.

Jauregui, H. O., and Naik, S. (1991). Diazepam metabolic activity in long-term monolayer cultures of rat, rabbit and piglet hepatocytes. *Artif. Organs* **15**, 295.

Jauregui, H. O., Naik, S., Santangini, H., Pan, J., Trenkler, D., and Mullon, C. (1994). Primary cultures of rat hepatocytes in hollow fiber chambers. *In Vitro Cell. Dev. Biol.* **30A**, 23–29.

Jauregui, H. O., Mullon, C.J.P., Trenkler, D., Naik, S., Santangini, H. A., Press, P., Muller, T. E., and Solomon, B. A. (1995). *In vivo* evaluations of a hollow fiber liver assist device. *Hepatology* **21**, 460–469.

Jessup, J., Goodwin, T., and Spaulding, G. (1993). Prospect for use of microgravity-based bioreactors to study three dimensional host–tumor interactions in human neoplasia. *J. Cell. Biochem.* **51**, 290–300.

Jones, E. A., and Schafer, D. R. (1990). Fulminant hepatic failure. *In* "Hepatology: A Textbook of Liver Diseases" (D. Zakim and T. D. Boyer, eds.), pp. 460–492. Saunders, Philadelphia.

Kaplowitz, N. (1992). Drug metabolism and hepatotoxicity. *In* "Liver and Billiary Diseases" (N. Kaplowitz, ed.), pp. 82–97. Williams & Wilkins, Baltimore, MD.

Kasai, S. I., Sawa, M., and Mito, M. (1994). Is the biological artificial liver clinically applicable? A historic review of biological artificial support systems. *Artif. Organs* **18**, 348–354.

Kawamura, A., Takahashi, T., and Kusumoto, K. (1985). The development a new hybrid hepatic support system using frozen liver pieces. *Jpn. J. Artif. Organs* **14**, 253–257.

Kelly, J. H., and Darlington, G. J. (1989). Modulation of the liver specific phenotype in the human hepatoblastoma line HEPG2. *In Vitro Cell. Dev. Biol.* **25**, 217–222.

Kim, S.-H., Banga, S., Jha, K. K., and Ozer, H. L. (1998). SV40-mediated transformation and immortalization of human cells. *In* "Simian Virus (SV40): A Possible Human Polyoma Virus" (F. Brown and A. M. Lewis, eds.), Dev. Biol. Stand. Karger, Basel.

Kimura, K., Gundermann, R., and Lie, T. (1980). Hemoperfusion over small liver pieces for liver support. *Artif. Organs* **4**, 297–301.

Kobayashi, A., Akaike, T., Kobayashi, K., and Sumitomo, H. (1986). Enhanced adhesion and survival efficiency of liver cells in culture dishes coated with a lactose-carrying styrene homopolymer. *Makromol Chem., Rapid Commun.* **7**, 645–650.

Li, A. P., Barker, G., Beck, D., Colburn, S., Monsell, R., and Pellegrin, C. (1993). Culturing of primary hepatocytes as entrapped aggregates in a packed bed bioreactor: A potential bioartificial liver. *In Vitro Cell. Dev. Biol.* **3**, 249–254.

Liu, J., Jing, P., Naik, S., Santangini, H., Thompson, N., Rifai, A., Chowdhury, J. R., and Jauregui, H. O. (1999). Characterization and detoxification functions of a nontumorigenic immortalized hepatocyte cell line (HepLiu). *Cell Transplant.* **18**, 219–232.

Malchesky, P. S. (1994). Nonbiological liver support: Historic review. *Artif. Organs* **18**, 342–347.

Margulis, M. S., Erukhimov, E. A., Andreiman, L. A., and Viksna, L. M. (1989). Temporary organ substitution by hemoperfusion through suspension of active donor hepatocytes in a total complex of intensive therapy in patients with acute hepatic insufficiency. *Resuscitation* **18**, 85–94.

Martinez-Hernandez, A., Delgado, F. M., and Amenta, P. S. (1991). The extracellular matrix in hepatic regeneration. Localization of collagen types I, III, IV, laminin and fibronectin. *Lab. Invest.* **64**, 157–166.

Matsumura, K. N., Guevara, G. R., Huston, H., Hamilton, W. L., Rikimaru, M., Yamasaki, G., and Matsumura, M.S. (1987). Hybrid bioartificial liver in hepatic failure: Preliminary clinical report. *Surgery* **101**, 99–103.

Monaco, A., Hallgrimson, J., and McDermitt, W. (1964). Multiple adenoma of the liver treated by subtotal (90%) resection: Morphological and functional studies of regeneration. *Ann. Surg.* **159**, 513.

Naik, S., Santangini, H., Gann, K., and Jauregui, H. (1992). Influence of different substrates in detoxification activity of adult rat hepatocytes in long-term culture: Implications for transplantation. *Cell Transplant.* **1**, 61–69.

Naik, S., Trenkler, D., Santangini, H., Pan, J., and Jauregui, H. (1996). Isolation and culture of porcine hepatocytes for artificial liver support. *Cell Transplant.* **5**, 107–115.

Nyberg, S., Payne, W., Amiot, B., Shirabe, K., Remmel, R. P., Hu, W.-S., and Cerra, F. B. (1993a). Demonstration of biochemical function by extracorporeal xenohepatocytes in an anhepatic animal mode. *Transplant. Proc.* **25**, 1944–1945.

Nyberg, S. L., Shatford, R. A., Peshvwa, M. V., White, J. G., Hu, W.-S., and Cerra, F. B. (1993b). Evaluation of a hepatocyte entrapment hollow fiber bioreactor: A potential bioartificial liver. *Biotechnol. Bioeng.* **41**, 194–203.

Rappaport, A. M., Borowy, Z. J., Lougheed, W. M., and Lotto, W. N. (1954). Subdivision of hexagonal liver lobules into a structural and functional unit. *Anat. Rec.* **199**, 11–33.

Renfranz, P. J., Cunningham, M. G., and McKay, R.D.G. (1991). Region-specific differentiation of the hippocampal stem cell line HiB5 upon implantation into the developing mammalian brain. *Cell (Cambridge, Mass.)* **66**, 713–729.

Rojkind, M., Gatmaitan, Z., Mackensen, S., Giambrone, M. A., Ponce, P., and Reid, L. M. (1980). Connective tissue biomatrix: Its isolation and utilization for long-term cultures of normal rat hepatocytes. *J. Cell Biol.* **87**, 255–263.

Rozga, J., Holzman, M. D., Ro, M.-S., Griffin, D. W., Neuzil, D. F., Giorgio, T., Moscioni, A. D., and Demetriou, A. A. (1993a). Development of a hybrid bioartificial liver. *Ann. Surg.* **217**, 502–511.

Rozga, J., Williams, F., Ro, M.-S., Neuzil, D. F., Giorgio, T. D., Backfisch, G., Moscioni, A. D., Hakim, R., and Demetriou, A. A. (1993b). Development of a bioartificial liver: Properties and function of a hollow-fiber module inoculated with liver cells. *Hepatology* **17**, 258–265.

Rozga, J., Morsiani, E., LePage, E., Moscioni, A. D., Giorgio, T., and Demetriou, A. A. (1994). Isolated hepatocytes in a bioartificial liver: Single groups view and experience. *Biotechnol. Bioeng.* **43**, 645–653.

Schafer, D. F., and Jones, E. A. (1990). Hepatic encephalopathy. *In* "Hepatology: A Textbook of Liver Diseases" (D. Zakim and T. D. Boyer, eds.), 2nd ed., pp. 447–460. Saunders, Philadelphia.

Schenker, S. (1989). Hepatic encephalopathy: The present and the future. *In* "Hepatic Encephalopathy" (R. F. Butterworth and G. P. Layrargues, eds.), pp. 3–24. Humana Press, Clifton, NJ.

Schuetz, E. G., Li, D., Omiecinski, C. J., Müller-Eberhard, U., Kleinman, H. K., Elswick, B., and Guzelian, P. S. (1988). Regulation of gene expression in adult rat hepatocytes cultured on a basement membrane matrix. *J. Cell. Physiol.* **134**, 309–323.

Soyer, T., Lempinen, M., and Eiseman, B. (1973). *In vitro* extracorporeal liver slices and cell suspensions for temporary hepatic support. *Ann. Surg.* **177**, 393–401.

Sussman, N. L., and Kelly, J. H. (1995). The artificial liver. *Sci. Am.* May/June, pp. 68–77.

Sussman, N. L., Chong, M. G., Koussayer, T., Da-er, H., Shang, T. A., Whisennand, H. H., and Kelly, J. H. (1992). Reversal of fulminant hepatic failure using an extracorporeal liver assist device. *Hepatology* **16**, 60–65.

Sussman, N. L., Gislason, G. T., Conlin, C. A., and Kelly, J. H. (1994). The Hepatix extracorporeal liver assist device: Initial clinical experience. *Artif. Organs.* **18**, 390–396.

Terblanche, J., and Hickman, R. (1991). Animal models of fulminant hepatic failure. *Dig. Dis. Sci.* **36**, 770–774.

Uchino, J., Tsuburaya, T., Kumagai, F., Hase, T., Hamada, T., Komai, T., Funatsu, A., Hashimura, E., Nakamura, K., and Kon, T. (1988). A hybrid bioartificial liver composed of multiplated hepatocyte monolayers. *Trans. Am. Soc. Artif. Intern. Organs* **34**, 972–977.

Westerman, K. A., and LeBoulch, P. (1996). Reversible immortalization of mammalian cells mediated by retroviral transfer and site specific recombination. *Proc. Natl. Acad. Sci. U.S.A.* **93**, 8971–8976.

Wolf, C. F., and Munkelt, B. E. (1975). Bilirubin conjugation by an artificial liver composed of cultured cells and synthetic capillaries. *Trans. Am. Soc. Artif. Intern. Organs* **21**, 16–27.

Wolf, C.F.W., Gans, H., Subramanian, V. A., and McCoy, C. H. (1979). A rat model for study of bilirubin conjugated by a cultured cell/artificial capillary liver assist device. *Int. J. Artif. Organs* **2**, 97–103.

Yanagi, K., Ookawa, K., Mizuno, S., and Oshima, N. (1989). Performance of a new artificial liver support system using hepatocytes entrapped within a hydrogel. *Trans. Am. Soc. Artif. Intern. Organs* **35**, 570–572.

Ye, Y., Luo, Y., Kobayashi, T., Taniguchi, S., Li, S., Niekrasz, M., Kosanke, S., Baker, J., Mieles, L., and Smith, D. (1995). Secondary organ allografting after primary "bridging" xenotransplant. *Transplantation* **60**, 19–22.

Zeldis, J. B. (1992). Molecular biology of viral hepatitis and liver cancer. *In* "Liver and Biliary Diseases" (N. Kaplowitz, ed.), pp. 70–81. Williams & Wilkins, Baltimore, MD.

CHAPTER 40

HEPATASSIST LIVER SUPPORT SYSTEM

Claudy Mullon and Barry A. Solomon

INTRODUCTION

Acute liver failure (ALF) complicated by hepatic encephalopathy is associated with high mortality, approaching 90% (Lee, 1993; Plevris *et al.*, 1998). The most successful treatment of ALF has been orthotopic liver transplantation (OLT), but the overall survival rate for ALF patients continues to be only 50 to 60% due to the shortage of organ donors and the rapid progression of the disease (Plevris *et al.*, 1998; Samuel, 1998). The success of OLT has led to the development of bioartificial liver support systems incorporating hepatocytes (Plevris *et al.*, 1998; McLaughlin *et al.*, 1999; Watanabe *et al.*, 1997a,b; Jauregui *et al.*, 1996). The aim of these systems is to support patients temporarily until recovery occurs or to bridge them to liver transplantation. This chapter introduces a bioartificial liver support system (HepatAssist System) containing porcine hepatocytes for use in the treatment of patients with acute liver failure and presents a summary of the first clinical experience with it.

HEPATASSIST SYSTEM

The HepatAssist System (Fig. 40.1) is an extracorporeal circuit consisting of a hollow-fiber membrane bioreactor containing 7×10^9 viable cryopreserved porcine hepatocytes, two charcoal columns, a membrane oxygenator, and a perfusion pump. The bioreactor contains polysulfone hollow fibers with a nominal 0.15-μm pore size. Prior to treatment, the hepatocytes are thawed and placed into the bioreactor along with collagen-coated dextran beads for cell attachment and spatial distribution. The HepatAssist System is used in combination with a commercially available plasmapheresis machine; plasma is perfused through the system at a flow rate of 400 ml/min. One hepatocyte bioreactor is used per treatment.

CELL PREPARATION

Porcine hepatocytes were isolated from 20- to 45-pound pigs meeting acceptance criteria established in compliance with United States xenotransplantation draft guidelines (U.S. Public Health Service, 1996). The cells were isolated using a modified method based on that reported by Morsiani *et al.* (1995). Typical yields were 145×10^6 hepatocytes per gram of liver with better than 88% cell viability. The cells were cryopreserved after isolation to allow for quality control evaluation and distribution to and storage at medical centers (Cain *et al.*, 1996, 1999). All cell processing steps, including organ procurement, hepatocyte isolation, and cryopreservation, took place under the current good manufacturing practice (CGMP) standards of the Food and Drug Administration (FDA).

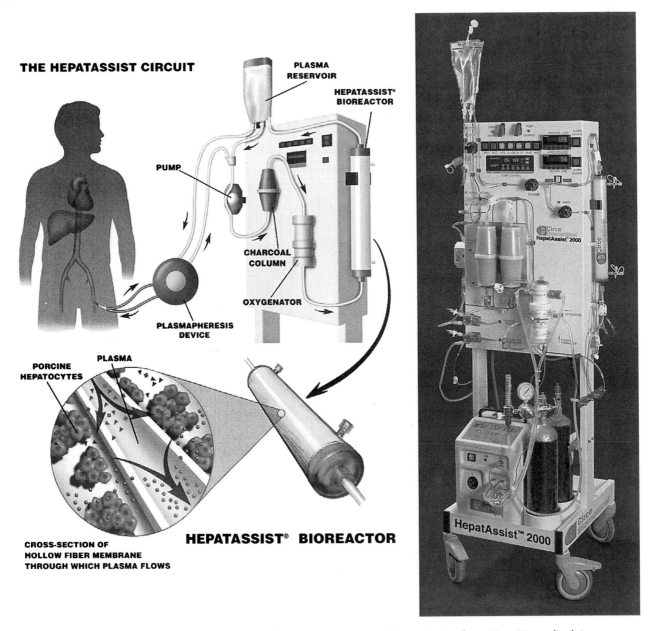

Fig. 40.1. HepatAssist liver support system, with permission from Circe Biomedical, Inc.

Cell quality control (QC) testing was performed at all critical steps of the process (Table 40.1) and the cryopreserved hepatocytes were released for clinical use only when all in-process and final-release testing parameters met established criteria. A minimum of 45 days was required to complete all QC testing and analyses on the animal donor, the final cell product, and the in-process cell and media samples.

CLINICAL STUDY

A phase I/II clinical study to assess the safety and tolerability of treatment with the HepatAssist System was conducted at three centers, two in the United States and one in France, under an Investigational New Drug (IND) protocol from the United States FDA and with approval of the French regulatory authorities. The protocol was an open-label, uncontrolled, multicenter study in patients with acute liver failure and stages III and IV hepatic encephalopathy (Watan-

Table 40.1. Porcine hepatocyte processing quality control

Process steps	Tests and analyses
1. Pigs assessment	Serologic screening
	Blood analysis
	Bacteriology
	Mycoplasma screening
	Virology
	Histopathology
2. Organ procurement,	Bioburden
hepatocyte isolation,	Sterility
hepatocyte cryopreservation	Endotoxin
	Cell viability
	Cell count
3. Final product	Sterility
(cryopreserved hepatocytes)	Endotoxin
	Mycoplasma screening
	Virology
	Cell viability
	Cell count
	Cell function

abe *et al.*, 1997b; Hewitt *et al.*, 1998; Chen *et al.*, 1997). Key patient selection criteria were as follows:

Key inclusion criteria
    1. Primary diagnosis of acute liver failure
    2. Stages III and IV come according to Trey classification
    3. Patients over 12 years of age or 40 kg body weight
Key exclusion criteria
    1. Contraindication for liver transplantation
    2. Malignancy, hypoxemia, irreversible brain damage
    3. Overt bleeding or hematocrit less than 24%
    4. Known hypersensitivity or allergic reaction to pig products
Key treatment withdrawal criteria
    1. Procurement of suitable liver for transplantation
    2. Improvement of hepatic encephalopathy to stage II prior to treatment
    3. Withdrawal of patient from transplant candidate list

Thirty-one patients were enrolled in the study, 17 at Cedars-Sinai Medical Center, Los Angeles, CA; 5 at the University of California in Los Angeles; and 9 at Paul Brousse Hospital, Villejuif, France. Etiologies included 28 patients with acute liver failure and 3 patients with primary non function (PNF) following liver transplantation. An additional 8 patients (7 ALF, 1 PNF) were treated prior to the IND protocol under local institutional review board approval (Watanabe *et al.*, 1997b). Each treatment with the HepatAssist System lasted 6 hr and was administered at the rate of approximately one treatment per day. A total of 80 treatments were administered, or on average 2.1 treatments (range 1 to 5) per patient.

SAFETY

The system was found to be well tolerated (Watanabe *et al.*, 1997b; Hewitt *et al.*, 1998; Chen *et al.*, 1997). The main adverse events were episodes of hypotension in eight patients. In most cases, these events were of mild severity, were readily managed, and did not cause an interruption in the liver support treatment. There was no evidence that treatment with the liver support system resulted in unmanageable adverse events.

Table 40.2 Survival rates of FHF/PNF patients treated with the HepatAssist System

Patients	N^a
Treated with HepatAssist System	39 (100%)
Died waiting for a liver graft	0^b (0%)
Survived without transplantation	6 (15%)
Transplanted	32 (82%)
Died 1 to 30 days post-transplantation	3 (8%)
Survived 30 days and beyond	35 (90%)

[a]Number of patients (percentage survival rate).
[b]One patient died after being taken off the transplant waiting list.

PERFORMANCE

Survival rates for all 39 patients are summarized in Table 40.2. Six patients recovered post-treatment without the need for a graft; all patient candidates for orthotopic liver transplantation were transplanted and the overall patient survival rate at 1 month was 90%. This study did not include a control group; however, the survival of ALF patients, including OLT, is typically 50–60% (Plevris *et al.*, 1998; Samuel, 1998).

Samuel (1998), from Paul Brousse Hospital (one of the three centers in the HepatAssist System clinical study), reported a series of 177 ALF patient candidates for OLT (Table 40.3). Of these, 25 patients (14%) died while waiting for a liver and 2 patients (1%) survived without the need for a graft. Overall survival rate at 1 month was 67 and 58% at 1 year. The principal cause of death before and after liver transplantation was brain death; 95% of the patients were in coma at the time of transplantation and 56% had increased intracranial pressure requiring treatment. In a series of 17 patients treated at Cedars Sinai Medical Center with the HepatAssist System, the intracranial pressure decreased from 17.2 mmHg pretreatment to 9.7 mmHg ($p < 0.01$) posttreatment, and several neurologic parameters improved (Hewitt *et al.*, 1998; Chen *et al.*, 1997). The Glasgow coma score increased from 6.9 to 7.5 ($p = 0.05$), and the comprehensive level of consciousness score increased from 24.8 to 31.2 ($p < 0.01$). In addition, blood ammonia, bilirubin, and liver enzymes improved posttreatment (Hewitt *et al.*, 1998; Chen *et al.*, 1997).

PATIENT SCREENING FOR PORCINE ENDOGENOUS RETROVIRUS

Porcine endogenous retrovirus (PERV) genomic sequences are present in pig cells. Although PERV infectivity has been documented to date only *in vitro* in porcine and human cell line cocultures, the possibility that xenotransplantation with porcine tissue may induce porcine retroviral expression in the recipient is of concern (Bach and Fineberg, 1998; Vogel, 1998; Patience *et al.*, 1997, 1998; Heneine *et al.*, 1998; Chapman *et al.*, 1999). We conducted a retrospective study

Table 40.3. Survival rates of FHF patients admitted at Paul Brousse Hospital[a]

Patients	N
Patients admitted for transplantation	177 (100%)
Died waiting for a liver graft	25 (14%)
Survived without transplantation	2 (1%)
Transplanted	150 (85%)
Died 1 to 30 days posttransplantation	34 (19%)
Survived 30 days and beyond	118 (67%)

[a]Patients were candidates for liver transplantation (Samuel, 1998).

to assess PERV infectivity in 28 patients treated with the HepatAssist System. All patients tested negative for PERV using polymerase chain reaction DNA analysis of peripheral blood mononuclear cells collected 1 to 5 years posttreatment. Of the 28 patients, 25 were transplanted posttreatment and received immunotherapy for up to 5 years; the other 3 patients recovered without the need for a graft. In summary, our results did not support the presence of PERV infectivity in patients treated with this porcine hepatocyte-based bioartificial liver.

In vitro results also showed that our porcine hepatocytes did not produce infectious PERV in cocultures with human kidney cell line 293 (Patience *et al.*, 1997). In addition, *in vitro* experiments using cartridges seeded with the porcine cell line PK-15, which expresses PERV, showed that the membrane in the HepatAssist System decreased PERV transmission by a factor of 10^5 to 10^6.

CONCLUSION

Numerous extracorporeal therapies have been clinically evaluated for their ability to temporarily support patients with acute liver failure (Riordan and Williams, 1997). The success of orthotopic liver transplantation has demonstrated the importance of not only detoxification but also of synthetic and metabolic functions in supporting ALF patients. This led to the advent of cell-based systems or bioartificial livers. There are three potential areas in which a liver support system can benefit ALF patients: (1) increasing patient survival without the need for a graft, (2) bridging patients to transplantation, and (3) decreasing patient mortality posttransplantation.

Initial clinical evaluation of the HepatAssist System demonstrated its safety and tolerability in patients with acute liver failure (Watanabe *et al.*, 1997b; Hewitt *et al.*, 1998; Chen *et al.*, 1997). The study showed a 90% survival rate; a pivotal phase II/III, multicenter, randomized, controlled, parallel group study of the HepatAssist System compared to standard surgical-intensive care in patients with acute hepatic failure has been initiated to further establish system efficacy.

The protocol includes patients with stages III and IV hepatic encephalopathy due to fulminant hepatic failure or primary nonfunction following liver transplantation. The primary end point is 30-day patient survival. The study includes the enrollment of 150 patients with interim analyses. The clinical study was approved in July 1998, and 14 clinical centers in the United States and Europe are currently participating in the study.

The HepatAssist design is based on hollow-fiber membrane bioreactors. Plasma circulates through the lumen of the fibers and the cells are held in the extracapillary space. Although the membrane's porous structure is impermeable to the hepatocytes, it has been designed to maximize mass transfer so that circulating toxins and nutrients reach the cells and metabolites and synthetic products are returned to the patient. However, the membrane also plays a role in decreasing the potential risk of transmission of pathogens. The membrane used in the HepatAssist System decreased by a factor of at least 100,000 the transmission of PERV as measured *in vitro* by the infectivity of human cell line 293. In addition, the membrane is not permeable to porcine cells, therefore significantly reduces the risk of porcine cell infusion into the patient and cell microchimerism in patients treated with the HepatAssist System.

We are currently conducting the first controlled, randomized, multicenter trial in the development of a bioartificial liver. It is our hope that our study, as well as other future studies, will help explore more fully the critical elements of a liver support system (e.g., design, cell sources) and improve our understanding of the pathogenesis of acute liver failure. These systems might also help define an understanding and other potential therapeutic approaches regarding liver regeneration in patients with liver disease.

ACKNOWLEDGMENT

The authors would like to acknowledge and thank Phyllis Ohanian for her help in the preparation and editing of this manuscript.

REFERENCES

Bach, F., and Fineberg, H. V. (1998). Call for moratorium on xenotransplants. *Nature (London)* 391, 326.

Cain, S., Naik, S., Deane, D., Borland, K., Chandler, B., Santangini, H., Trenkler, D., Velazquez, N., Solomon, B., Jauregui, H., and Mullon, C. (1996). Large scale cryopreservation of porcine hepatocytes for the clinical treatment of acute liver failure. *Cell Transplant.* 5(5S-2), 19.

Cain, S. P., Chandler, B. A., Perlman, T. J., Picton, D., Cleversey, G., Deane, D., Pitkin, Z., Solomon, B., and Mullon, C.

(1998). A closed system method for the processing and cryopreservation of porcine hepatocytes for use in a bioartificial liver. *J. Cryobiol.* 37(4), pp. 448–49.

Chapman, L. E., Heneine, W., Switzer, W., Sandstrom, P. and Folks, T. (1999). Xenotransplantation: The potential for xenogeneic infections. *Transplant. Proc.* **31**, 909–910.

Chen, S., Mullon, C., Kahaku, E., Watanabe, F., Hewitt, W., Eguchi, S., Middleton, Y., Arkadopoulos, N., Rozga, J., Solomon, B., and Demetriou, A. (1997). Treatment of severe liver failure with a bioartificial liver. *Ann. N.Y. Acad. Sci.* **831**, 350–360.

Heneine, W., Tibell, A., Switzer, W., Sandstrom, P., Rosales, G., Mathews, A., Korsgren, O., Chapman, L., Folks, T., and Groth, C. (1998). No evidence of infection with porcine endogenous retrovirus on recipients of porcine islet-cell xenografts. *Lancet* **352**, 695–699.

Hewitt, W., Rozga, J., Mullon, C., and Demetriou, A. (1998). Bioartificial liver: Experimental and clinical experience. *In* "Ilots de Langerhans et hépatocytes: Vers une utilisation thérapeutique" (D. Franco, K. Boudjema, and B. Varet, eds.), Vol. 9, pp. 105–118. INSERM, Paris.

Jauregui, H., Mullon, C., and Solomon, B. (1996). Extracorporeal artificial liver support. *In* "Principles of Tissue Engineering" (R. Lanza, W. Chick, and R. Langer, eds.), Vol. 30, pp. 463–79. Springer/R. Lange, Austin, TX.

Lee, W. (1993). Acute liver failure. *N. Engl. J. Med.* **329**, 1862–1872.

McLaughlin, B. E., Tosone, C. M., Custer, L. M., and Mullon, C. (1999). Overview of extracorporeal liver support systems and clinical results. *Ann. N.Y. Acad. Sci.* **875**, 310–325.

Morsiani, E., Rozga, J., Scott, H. C., Lebow, L., Mosaoni, A., Kong, L., McGrath, M., Rosen, H., and Demetriou, A. (1995). Automated liver cell processing facilitates large-scale isolation and purification of porcine hepatocytes. *ASAIO J.* **41**, 155–161.

Patience, C., Takeuchi, Y., and Weiss, R. (1997). Infection of human cells by an endogenous virus of pigs. *Nat. Med.* **3**, 282–286.

Patience, C., Patton, G. S., Takeuchi, Y., Weiss, R., McClure, M., Rydberg, L., and Breiner, M. (1998). No evidence of pig DNA or retroviral infection in patients with short-term extracorporeal connection to pig kidneys. *Lancet* **352**, 699–701.

Plevris, J. N., Schina, M., and Hayes, P. C. (1998). The management of acute liver failure. *Aliment. Pharmacol. Ther.* **12**, 405–418.

Riordan, S. M., and Williams, R. (1997). Experience in design of controlled clinical trials in acute liver failure. *In* Bioartificial Liver Support System: The Critical Issues" (G. Crepaldi, A. A. Demetriou, and M. Muraca, eds.), pp. 104–115. CIC Edizioni International, Rome.

Samuel, D. (1998). La transplantation hépatique orthotopique pour hépatite fulminante et subfulminante. *Lett. Hépato-Gastroentérolo.* **2**, 85–89.

U.S. Public Health Service (1996). Infectious diseases issues in xenotransplantation—Draft guidelines. *Fed. Regist.* **61**(185), 49920.

Vogel, G. (1998). No moratorium on clinical trials. *Science* **279**, 648.

Watanabe, F., Shackleton, C. R., Cohen, S. M., Goldman, D., Arnaout, W., Hewitt, W., Colquhoun, S., Fong, T., Vierling, J., Busuttil, R., and Demetriou, A. (1997a). Treatment of acetaminophen-induced fulminant hepatic failure with a bioartificial liver. *Transplant. Proc.* **29**, 487–488.

Watanabe, F., Mullon, C., Hewitt, W. R., Arkadopoulos, N., Kahaku, E., Eguchi, S., Khalili, T., Arnaout, W., Shackleton, C., Rozga, J., Solomon, B., and Demetriou, A. (1997b). Clinical experience with a bioartificial liver (BAL) in the treatment of severe liver failure: A phase I clinical trial. *Ann. Surg.* **225**(5), 484–494.

CHAPTER 41

LINEAGE BIOLOGY AND LIVER

Arron S. L. Xu,* Thomas L. Luntz,* Jeffrey M. Macdonald,* Hiroshi Kubota,
Edward Hsu, Robert E. London, and Lola M. Reid

INTRODUCTION

Tissue engineering offers a revolution in basic research and in clinical programs in cell and gene therapies, novel forms of plastic surgery, implantable devices, and bioartificial organs. Much of the foundation of this new field is a synthesis of the concepts and technologies in cell and molecular biology, cell and organ culture, extracellular matrix biology and chemistry, polymer sciences, and stem cell and lineage biology. Its greatest impact has been with respect to quiescent tissues, long the bane of investigators, who were forced to study quiescent tissues *in vivo* to achieve any realistic expectations of fully normal and differentiated cells or, if they used them in culture, were forced to offer feeble qualifiers to explain the aberrant behavior *ex vivo* of such differentiated cells.

Despite the extraordinary collective knowledge extant in the fields of cell and molecular biology, investigators have been stymied in trying to maintain tissues *ex vivo* and have struggled to identify conditions under which the cells would express their differentiated functions for more than a few days. In retrospect, past years of confusion are understandable, because the requirements for normalcy in the biology of cells and tissues are multifactorial and complex. Thus, the common approach of utilizing clonal cell populations, desirable for achieving rigorous controls, precluded the essentials of metazoan biology, that is, the biology of communities of cells with dynamic and complicated interactions. It is now appreciated that tissue-specific gene expression cannot be achieved by using a cloned cell population placed onto a foreign, nonbiological surface (glass, plastic) and in a soluble medium (basal medium) with serum that *in vivo* occurs only in a wound. Indeed, cultures on plastic and in serum-supplemented media are now recognized as models of scar tissue formation (cirrhosis, fibrosis) but not of normal physiologic processes.

Tissue engineering comprises a number of new approaches that include using porous, flexible biological substrata (extracellular matrix components) and media defined for specific nutritional and signaling needs; coculturing of multiple cell types with dynamic interactions, using conditions permitting the cells to achieve three-dimensionality; optimizing mass transfer; and recognizing that all tissues are organized as progenitors and maturational lineages. In this chapter we concentrate on these novel approaches, with a special focus on the following requirements:

1. The need for stem cells or progenitors as starting points for cell and gene therapies and for bioartificial organs
2. The importance of recognizing the maturational lineage with its implicit lineage-dependent growth properties and tissue-specific gene expression
3. The development of logical microenvironments using purified and defined components
4. The use of biodegradable scaffoldings to facilitate presentation of that microenvironment in configurations of three-dimensional space
5. An analysis of bioreactor designs to show the failings of most current bioreactors and some possible designs that might work for adherent cell populations

---

Authors contributed equally.

Principles of Tissue Engineering
Second Edition

6. Novel approaches for analyses of bioartificial organs and including nuclear magnetic resonance spectroscopy (NMR) and magnetic resonance imaging (MRI)

To be able to focus on these diverse themes within a document of requisite brevity, we have elected to focus almost exclusively on liver as a model and will use representative literature to support the discussions, referring the reader to more exhaustive reviews for further details. For example, the history of the oval cell field, part of the background for current understandings of liver stem cell biology, has been given in the review in the first edition (Reid, 1997) and in several additional excellent reviews (Grisham and Thorgeirsson, 1997; Sell, 1998), and will not be presented here.

In intact organs and tissues, three interlocking factors—stem cell and maturational lineage mechanisms, soluble signals (hormones, growth factors, nutrients, gas exchange), and insoluble signals (extracellular matrices)—operating independently and interdependently, govern growth and differentiation of all cells. Dynamic and continuous interactions between these determinants are responsible for the complexity, the heterogeneity, and the subtleties of all aspects of tissue phenotype and regulatory processes. In the sections that follow, we will outline approaches we are taking to manage and/or manipulate each of these factors. The first section describes liver biology and the use of stem cells and early progenitors to generate entire tissues. The second section describes use of extracellular matrices with or without biocompatible and/or biodegradable supports that achieve the mass transfer requirements of the tissue, and presents engineering principles of tissue mimicry, optimization, and scale-up. In the third section, we describe the physical and biologic analyses of bioreactor designs that can be used for adherent cell populations such as the liver when trying to establish bioartificial organs. We also discuss methodologies that can be used to analyze cells in three dimensions, including novel NMR and MRI techniques.

LIVER ORGANIZATION AND DEVELOPMENT

Tissue engineering of the liver requires honoring the architecture and cellular paradigms that govern its functions using principles of tissue mimicry. The primary structural and functional unit of the mature liver is the acinus (Fig. 41.1, see color plate), which in cross-section is organized like a wheel around two distinct vascular beds: three to seven sets of portal triads (each with a portal venule, hepatic arteriole, and a bile duct) for the periphery, and with the central vein at the hub. The liver cells are organized as cell plates lined on both sides by fenestrated endothelia, defining a series of sinusoids that are contiguous with the portal and central vasculature. A narrow space, the space of Disse, separates the endothelia from hepatocytes all along the sinusoid. As a result of this organization, hepatocytes have two basal domains, each of which faces a sinusoid, and an apical domain, which is defined by the region of contact between adjacent hepatocytes. The basal domains contact the blood and are involved in the absorption and secretion of plasma components, whereas the apical domain forms bile canaliculi, specialized in the secretion of bile salts, and is associated through an interconnecting network with bile ducts. Blood flows from the portal venules and hepatic arterioles through the sinusoids to the terminal hepatic venules and the central vein.

Based on this microcirculatory pattern, the acinus is divided into three zones: zone 1, the periportal region; zone 2, the midacinar region; and zone 3, the pericentral region (Table 41.1 and Fig. 41.2). Proliferative potential, morphologic criteria, ploidy, and most liver-specific genes are correlated with zonal location (Gebhardt, 1988; Traber et al., 1988; Gumucio, 1989). Gradients in the concentration of blood components (including oxygen) across the acinus, and following the direction of blood flow from the portal triads to the central vein, are responsible for some of this zonation—for example, the reciprocal compartmentation of glycolysis and gluconeogenesis. However, the periportal zonation of the gap junction protein connexin 26 and the pericentral zonation of glutamine synthetase, to name only two, are insensitive to such gradients, are more representative of most tissue-specific genes, and appear to be determined by factors intrinsic to the cells or to variables other than blood flow in the microenvironment.

In addition to hepatocytes, bile duct epithelial cells (cholangiocytes), and endothelial cells, the region between the portal and central tracts contains other cell types, such as Ito cells and Kupffer cells. These play prominent roles in pathogenic conditions of the liver, especially in inflammation and fibrosis, but their direct contribution to the main homeostatic functions of the normal organ is apparently small. Although their biology is of importance, they will not be discussed further here.

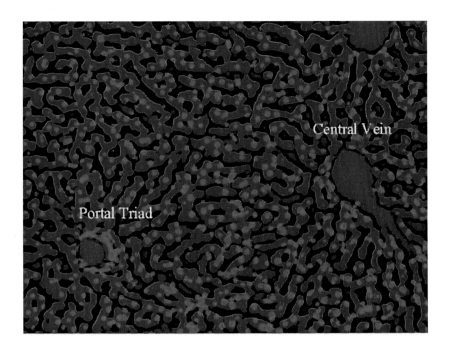

Fig. 41.1. A rat liver section showing a portal triad, central vein, and plates of liver cells extending between them. The section was stained with a DNA dye to reveal nuclei as bright orange. Photograph courtesy of Huifei Liu.

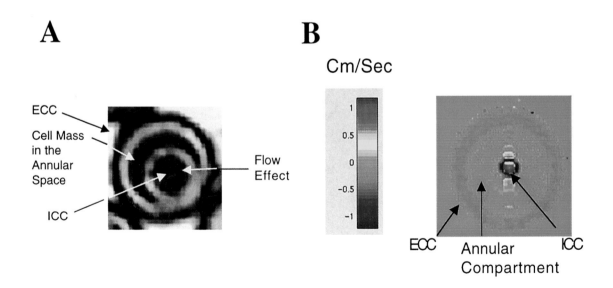

Fig. 41.8. T_2-weighted (A) and velocity-encoded (B) images of a coaxial bioreactor. (A) The effect of flow in the ICC is visible as hypointense, as compared to slower flowing water in the ECC, and the annular space contains rat hepatocytes that appear hypointense. The physical parameters of axial flow (B) in the ICC and ECC are dissected from the T_2-weighted images (A) for empirical input into hydrodynamic models. Cell distribution can be obtained by diffusion-weighted imaging. The T_2-weighted MRI (A) was modified from Macdonald et al., NMR Spectroscopy and MRI Investigation of a Potential Bioartificial Liver. NMR Biomed. **11**, 55–66, © 1998, by permission of John Wiley & Sons, Inc.

Table 41.1 Liver acinus zones

Characteristic	Zone		
	1	**2**	**3**
Ploidy	Diploid	Tetraploid	Mix of tetraploid and octaploid
Average size	7–20 μm	20–30 μm	30–50 μm
Growth	Maximum	Limited (mostly hypertrophic)	Negligible
Extracellular matrix	A gradient in the matrix chemistry located in the space of Disse		
Gene expression	Early	Intermediate	Late

The liver develops as a result of the convergence of a diverticulum formed from the caudal foregut and the septum transversum, part of the splanchnic mesenchyme (reviewed in Zaret, 1998). The formation of the hepatic cells begins after the endodermal epithelium interacts with the cardiogenic mesoderm (Douarin, 1975), probably via fibroblast growth factors (Jung *et al.,* 1999). The specified hepatic cells then proliferate and penetrate into the mesenchyme of the septum transversum with a cordlike fashion, forming the liver anlage. The direct epithelial–mesenchymal interaction is critical in these early developmental stages of the liver and dictates which cells will become hepatocytes or cholangiocytes, and the fenestrated endothelia, respectively. Mutations in the mesenchyme-specific genes *hlx* and *jumonji* block liver development, illustrating the importance of contributions from this tissue (Hentsch *et al.,* 1996; Motoyama *et al.,* 1997). Early in its development, the liver consists of clusters of primitive hepatocytes bounded by a continuous endothelium lacking a basement membrane, and abundant hematopoietic cells. As the endothelium is transformed to become a discontinuous, fenestrated endothelium, the vasculature, especially the portal vasculature, becomes more developed, with the production of basement membranes. The portal interstitium may provide the trigger for the development of bile ducts, and as it surrounds the portal venules, hepatic arterioles, and bile ducts, portal triads are formed. Immature hepatocytes rapidly proliferate and parenchymal plates are formed, probably in response to changes in the amount and distribution of tissue-organizing molecules such as cell–cell adhesion molecule 105 (CCAM 105), Agp110, E-cadherin, and connexins, coincident with the relocation of most, but not all, of the hematopoietic cells to the bone marrow (Stamatoglou and Hughes, 1994). Studies suggest that some hematopoietic progenitors persist in adult quiescent rodent livers (Sigal *et al.,* 1995a), and hematopoietic stem cells have been isolated from both adult human (Crosbie *et al.,* 1999) and murine liver (Taniguchi *et al.,* 1996, 1997). The mature physical organization is achieved within the first weeks after birth in rodents. Metabolic zonation is established according to somewhat different schedules for different enzymes, but becomes evident in the period following birth.

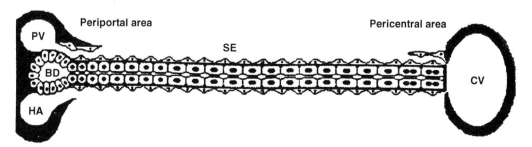

*Fig. 41.2. Schematic of the liver acinus. Information relevant to the pattern of size, ploidy, growth, and gene expression are given in Table 41.1. PV, Portal vein; BD, bile duct; HA, hepatic artery; SE, sinusoidal endothelium over the space of Disse; CV, central vein. (Modified from Brill et al., Proc. Soc. Exp. Biol. Med. (1993). **204,** 261–269, with permission.)*

IDENTIFICATION, ISOLATION, AND SOURCING OF CELLS FOR TISSUE ENGINEERING

STEM CELLS AND COMMITTED PROGENITORS

Tissues are organized as maturational lineages fed, like a spring, by stem cells or early progenitor cell populations (Sigal *et al.,* 1992; Sell, 1993; Potten, 1997). The tissue is defined as going from young, to middle age, to old cells. The maturational process is accompanied by lineage-position-dependent changes in cell size, morphology, antigenic profiles, growth potential, and gene expression. Growth is maximal in the stem cells and early progenitors and wanes with progression through the lineage. Indeed, in many mammalian tissues, the majority of the cells in the adult tissue are polyploid, mostly tetraploid, or octaploid. Even in the liver, one of the most renowned regenerative tissues in the body, less than a third of the cells are diploid. Only the diploid cells have been found capable of undergoing repeated, complete cell division. Data support the concept that the bulk of the regenerative potential in a tissue derives from the diploid cell population and that the older cells contribute to regeneration by increasing cell mass via hypertrophic responses associated with polyploidy (Sigal *et al.,* 1999). Therefore, the best hopes for cell growth, whether in cell or gene therapies or in bioartificial organs, are with the diploid cell population of the tissue.

Defining stem cells is difficult (Loeffler *et al.,* 1997; Morrison *et al.,* 1997). In the older embryological literature, stem cells were defined as primitive cells that self-replicate, that are pluripotent, i.e., produce daughter cells with more than one fate, that grow very, very slowly (are largely in G_0), and that are dependent on signals (undefined) derived from tissue-specific forms of mesenchymal cells. As discussed extensively in an interesting and provocative review (Morrison *et al.,* 1997), this definition does not apply in its entirety to all known stem cells. There are stem cells that cannot be considered the most primitive cells within a tissue, self-replication appears to apply rigorously only to germ cells, pluripotency can be variable depending on the circumstances, and the growth rate of the stem cells is tissue-specific. Despite the increasing complexity in the characterizations of stem cells, there are generic attributes that seem to hold. As Morrison *et al.* (1997) joked, "It is hard to define, but I know it when I see it," a phrase they quoted from United States Supreme Court Justice Byron White's comments with respect to pornography. The definitions and markers for stem cells are likely to emerge in the coming years with the enthusiastic interest now being bestowed on stem cells by many investigators. For now, the terminology and concepts of the stem cell field are as summarized briefly below.

EMBRYONIC STEM CELLS AND OTHER TOTIPOTENT CELLS

Cells with the potential to participate in the development of all cell types, including germ cells, are referred to as "totipotent" and include the zygote and normal embryonic cells up to the 8-cell stage (the morula). Embryonic stem (ES) cells consist of permanent cell populations derived from totipotent normal cells (blastocysts) and were first reported in the early 1980s (Evans and Kaufman, 1981; Martin, 1981). ES cell lines can be cultured *in vitro* with maintenance of totipotency. ES cells are tumorigenic if introduced into immunocompromised hosts in any site other than *in utero,* forming teratocarcinomas. However, when they are injected back into normal blastocysts, they are able to resume embryonic development and participate in the formation of a normal, but chimeric, mouse. Although ES cell lines have been established from many species (mouse, rat, pig, etc.), only the mouse system has been used routinely to generate animals with novel phenotypes (knockouts, transgenics) by merging modified ES cells from culture to blastocysts and then implanting the blastocysts into pseudopregnant hosts. Embryonic germ (EG) cell lines, which show many of the characteristics of ES cells, can be isolated directly *in vitro* from the primordial germ cell population. As with ES cells, the EG cells form teratocarcinomas when injected into immunocompromised mice and contributed to chimeras, including the germ line, when injected into blastocysts (Labosky *et al.,* 1994).

Transformed totipotent cells include embryonal carcinomas and teratocarcinomas. Stevens (1964) identified and described teratocarcinomas in mice and showed that many cell types are present in these tumors. On the basis of cellular characteristics and multipotent developmental activity, the most primitive of teratocarcinoma cells have been designated embryonal carcinoma (EC) cells (Stevens and Pierce, 1975). Ralph Brinster and associates showed that following the transfer of EC cells into blastocysts, the EC cells were able to contribute to donor tissues such as hair (Brinster, 1974). Subsequently, Mintz and Illmensee (1975) demonstrated in a beautiful series of publications that embryonal carcinoma cells, fused with blastocysts and implanted *in utero* in pseudopregnant hosts, are capable of yielding an animal with contributions by the embryonal carcinoma cells to all tissue types. Thus, the malignancy of the cells was thought initially to be due to multipotent cells being in the wrong microenvironment at a critical time (Mintz and Illmensee, 1975). However, even though malignant totipotent cells were able to participate normally in development, findings that support the importance of the microenvironment for normal differentiation, there has always been recognition that an aberrant gene(s) is also a factor, along with microenvironment, in giving rise to malignancies. Consequently, there was a drive toward using normal totipotent cells, efforts that came to fruition in the development of embryonic stem cell lines.

Highly publicized experiments have reported that human ES cell cultures have been established now from human embryos (Thomson *et al.,* 1998). The intent for these human ES cells is to inject them into tissues with the hope that they will be able to reconstitute damaged organs and tissues. Given the findings that ES and EG cells form tumors when injected into sites other than *in utero* (see above), the plan to inoculate human ES cells into patients is unrealistic and carries the grave possibility of creating tumors in the patients. To overcome this impasse, some groups are pursuing the plan of differentiating the ES cells under defined microenvironmental conditions to become determined stem cells that can then be inoculated into patients (Kirby *et al.,* 1996). For example, there is some measure of success in generating hematopoietic progenitors (Kennedy *et al.,* 1997). However, there still remains the concern that residual ES cells in the culture could pose the risk of tumorigenesis, if the cultures are inoculated into a patient. In summary, until research in developmental biology reveals the myriad controls dictating the fates of cells during embryogenesis, the ES cells will remain as an experimental tool with little hope for clinical programs in cell or gene therapies. The only realistic option for clinical programs in cell and gene therapies is to use determined stem cells that have restricted their genetic potential to that for a limited number of cell types. By contrast, the ES cells may hold great promise for bioartificial organs for those tissue types (e.g., hematopoietic cells) that are produced by ES cells under known conditions.

TELOMERASE

Telomeres are nucleoprotein structures that serve a critical function in protecting the ends of linear chromosomes. Telomerase, a ribonucleoprotein reverse transcriptase, plays a key role in the maintenance of telomere length and function. Telomerase is composed of a protein catalytic subunit, telomerase reverse transcriptase (TERT), and an RNA component, telomerase RNA (TR). Only TERT and TR are required for core *in vitro* telomerase activity (Greider, 1998).

Whereas embryonic stem cells have high levels of telomerase and determined stem cells have detectable to moderate levels of telomerase, differentiated somatic cells possess low or undetectable telomerase activity, and telomeres shorten with each cell division *in vivo* and *in vitro.* Accelerated loss of telomeres is associated with cellular senescence. Tumor cells, increasingly recognized to be transformants of stem cells and early progenitors, have high telomerase levels. These observations support the proposition that telomere shortening in the absence of telomerase plays a critical role in cell entry into senescence. In fact, ectopic expression of TERT in primary culture was sufficient to induce telomerase activity and allow unlimited growth (Bodnar *et al.,* 1998). One can expect that stem cell or progenitor populations with high proliferative capacity should have high telomerase activity to maintain the telomere structures. Weissman and colleagues assayed telomerase activity in single hematopoietic cells at different developmental stages, and found that 70% of bone marrow hematopoietic stem cells exhibited detectable telomerase activity (Morrison *et al.,* 1996). However, only a minority of differentiating and transient multipotent progenitors had detectable telomerase activity. These results indicate that telomerase expression is associated with self-renewal potential, rather than proliferative potential or undifferential status of cells. Confirming this, germinal center B cells in the process of clonal expansion were enriched in telomerase-positive cells more than 10-fold relative to splenic B cells (Morrison *et al.,* 1996).

To explore the effect of telomere loss *in vivo,* telomerase-deficient mice (knockouts) were generated by deletion of the mouse TR gene (Blasco *et al.,* 1997). Telomerase deficiency leads to a depletion of male germ cells, diminished hematopoietic colony formation, and impaired mitogen-induced proliferation of primary splenocytes (Lee *et al.,* 1998). These findings indicate an essential role for telomerase in organ systems with high proliferative activity. The effects on splenic lymphocytes demonstrate that the association between self-renewal potential and telomerase expression is not specific to primitive progenitors. The TR knockout mice also show that short telomere length leads to chromosome instability, suggesting that placing normal cells that do not contain telomerase into a proliferative environment subjects them to increased risk of tumorigenic mutations. This is yet further evidence that determined stem cells, not well-differentiated somatic cells or embryonic stem cells, are reasonable candidates for *ex vivo* expansion for tissue engineering or for cell and gene therapies. Moreover, study of determined stem cells should help in defining approaches for directed differentiation of the totipotent cell lines.

Nuclear Transfer Techniques for Cloning of Mammals

Genetic engineering of mammals has long been possible with ES cells, but not with embryonic germ cells or with somatic cells, even determined stem cells. The ability to clone mammals has now been achieved by utilizing nuclear transfer from some but not all somatic cells into enucleated cells, suggesting that cytoplasmic factors are responsible for restricting the genetic potential of the nucleus (Wilmut *et al.*, 1997; Kato *et al.*, 1998; Wakayama *et al.*, 1998; Wells *et al.*, 1999). The groundwork for this realization was pioneered by Gurdon and associates, who developed the techniques of nuclear transplantation (Gurdon, 1962). They hypothesized that nuclei remain totipotent even in a differentiated somatic cell and tested the idea by transferring nuclei from the intestinal endoderm of *Xenopus* tadpoles into activated enucleated eggs. Of the eggs injected, 1.4% of them (10 individuals) developed to the feeding tadpole stage. Serial transplantation, in which the blastula formed as a result of the initial transfer was used as a donor of nuclei for a second transfer into activated eggs, yielded 7% to the feeding tadpole stage, and 7 individuals that metamorphosed into fertile adult frogs (Gurdon, 1962), showing that the original donor nuclei were totipotent. Two major criticisms, that either primordial germ cells, which can migrate through to the gut, were the actual nuclear donors, or that the intestinal epithelia of the feeding-stage tadpole are not truly differentiated, as witnessed by the presence of yolk platelets, were addressed in subsequent experiments. Cultured epithelial cells from adult frog foot webbing, which expressed keratin and were thus considered differentiated, were used as nuclear donors. Serial transplantation yielded numerous tadpoles (Gurdon *et al.*, 1975), demonstrating that the nuclei of some, but not all, somatic cells can be reactivated by transferring the nucleus into an embryonic cytoplasm to yield cells that can mature into neurons, blood cells, muscles, or bones of a swimming tadpole.

Mammalian Cloning

Wilmut and colleagues (1997) have found that viable lambs can be produced by transplanting nuclei from cellular subpopulations of adult mammary glands. By transplanting nuclei from cultured cells from the mammary gland of a 6-year-old ewe in the last trimester of pregnancy at passage numbers 3–6 to an enucleated metaphase II-arrested oocyte, they have created a viable lamb. The success is due largely to the fact that nuclei were taken from cells in the G_0 phase of the cell cycle. Quiescent, diploid donor cells were produced by reducing the concentration of serum in the medium from 10 to 0.5% for 5 days, causing the cells to exit the growth cycle and arrest in G_0 (Wilmut *et al.*, 1997). Moreover, they investigated the telomere length in the cloned sheep. The mean size telomere length decreases in control animals with increasing age, at a mean rate of 0.59 kilobase (kb) per year. The size in the cloned sheep was reduced significantly compared with the age-matched control animals (Shiels *et al.*, 1999). The smaller size is consistent with the age of the progenitor mammary tissue (6 years old) and with the time spent in culture before nuclear transfer. The most likely explanation for the shorter telomere size in the cloned sheep is a reflection of the transferred nucleus, although it is not known whether the actual physiologic age of animals derived by nuclear transfer is reflected accurately by the telomere length (Shiels *et al.*, 1999). It is unknown whether nuclei from any maturational stage of somatic cell can be totipotent, or whether totipotent nuclei are restricted to a discrete subpopulation of maturational stage(s) (e.g., progenitors). However, the important realization at present is that any somatic cell can have a totipotent nucleus.

The cloned sheep has undergone two normal pregnancies and has successfully delivered healthy lambs (Shiels *et al.*, 1999). Mammalian cloning using nuclear transfer techniques with donor nuclei from somatic cells has now been used also to generate mice and calves (Kato *et al.*,

1998; Wakayama *et al.,* 1998; Wells *et al.,* 1999). The results suggest that nuclei of some (all?) somatic cells can be converted to a pluripotent or totipotent state if put through nuclear transfer into an enucleated cell with embryonic cytoplasm. Although the cloning experiments will yield numerous studies on nonhuman species to generate cloned animals of some scientific or commercial significance, their implications for humans are far more limited. Mostly, they will be used to define the sources of controls for differentiation, now shown to occur in the cytoplasm, offering hope by which to dictate the fates of a totipotent cell such as the ES cells. Thus, for now there are no immediate ramifications of this technology for cell or gene therapies or for other tissue engineering goals such as bioartificial organs for humans.

DETERMINED STEM CELLS

Determined stem cells derive from embryonic stem cells but have restricted genetic potential through molecular mechanisms of "cellular commitment" that are understood now to derive from poorly defined changes occurring in the cytoplasm, and occurring during embryogenesis following the 8-cell stage embryos. The one division between the 8-cell and 16-cell stages begins the specialization process toward tissues. One can hypothesize that there may be remnants of the stem cells from the germ layers of the early embryo or determined stem cells producing families of tissues. If germ layer-determined stem cells exist, there would be three types: ectodermal (epidermal and nervous tissues), mesodermal (blood, bones, cartilage, blood vessels), and endodermal (lung, thyroid, intestine, pancreas, liver, etc.). The hypothesis of adult tissues containing determined stem cells equivalent to the germ layer stem cells is untested at present.

Currently, the term "stem cell" persists in being defined as a population of cells that self-replicate, have extensive growth, and are pluripotent. All known determined stem cells have been shown to have extensive growth and to be pluripotent (Potten, 1997). However, there is debate whether any known determined stem cell is capable of self-replication. The debate has been fueled by the studies on telomerase in which determined hematopoietic stem cells from embryos versus adults have distinct telomerase levels. This issue is likely to be the basis of much research in the coming years and so cannot be addressed rigorously at this time.

Although it is implicit that there must be a cascade of stem cells with increasingly narrower phenotypic potential, our knowledge of determined stem cells derives entirely from those thought to be capable of producing the mature cells of a limited number of tissues or even of only one organ or tissue, and which are located in those tissues. Because studies of determined stem cells have focused only on their fates as the mature cells of the designated tissue (epidermal stem cells to yield skin; neuronal stem cells to yield nervous tissue; etc.), their real potential is unknown.

In the past few years, the first real evidence has emerged that identifies determined stem cells with a phenotypic potential intermediate between that of embryonic stem cell and a tissue-specific stem cell. One example will be used as representative: the hemangioblast. The hemangioblast is considered a parental stem cell to determined stem cells for endothelia and hematopoietic cells. The close temporal and spatial relationship of hematopoiesis and vascular development is consistent with the existence of a postulated progenitor. Blood and endothelial progenitors coexpress some molecular markers, such as CD34 and Flk-1 [a receptor for vascular endothelial cell growth factor (VEGF)]. Targeted disruption of Flk-1 results in combined vascular and hematopoietic defects (Shalaby *et al.,* 1995). Flk-1 null embryonic stem cells also fail to contribute to either endothelium or hematopoiesis in chimeric mice (Shalaby *et al.,* 1997). These results show that Flk-1 is essential for both lineages. Furthermore, single Flk-1-positive cells from avian embryos can develop into either hematopoietic or endothelial colonies (Eichmann *et al.,* 1997). These studies are all consistent with the existence of the Flk-1-positive hemangioblast. Such a progenitor may develop under the influence of the stem cell leukemia/T cell acute leukemia-1 (SCL/TAL-1) gene, which was first identified through its translocation in acute T cell leukemia. It encodes a basic helix–loop–helix (bHLH) transcription factor that is expressed specifically in hematopoietic cells, vascular endothelium, and the developing brain. SCL null mice lack embryonic hematopoiesis, and SCL null ES cells also fail to contribute to any hematopoietic lineage in the chimeric mice (Porcher *et al.,* 1996; Robb *et al.,* 1996). Although the role of SCL in endothelial development remains obscure in mice, it was shown that the ectopic expression of the zebrafish SCL homolog in early posterior mesoderm resulted in overproduction of common hematopoietic and endothelial precursors. This gain-of-function experiment in zebrafish indicates that SCL specifies formation of hemangioblasts (Gering *et al.,* 1998). It is worth considering the possible role of bHLH proteins in the control of cell fate in other tissues. For instance, ectopic expression of the myogenic bHLH proteins MyoD, Myf-5, MRF4, or myogenin all induced muscle differentiation (Weintraub, 1993), and the crucial role of the *achaete–scute* complex, bHLH genes in *Drosophila,* in neural development is conserved over species (Guillemot *et al.,* 1993).

The bulk of the literature on determined stem cells is with respect to hematopoietic stem cells (Heimfeld and Weissman, 1991; Marshall and Lord, 1996). However, proof of the existence and the potential for stem cells in other tissues has emerged within the past decade (Potten, 1997). The most well studied of these, independently of the hematopoietic stem cells, are the neuronal stem cells (Snyder and Macklis, 1995; Calof *et al.,* 1998; Kempermann and Gage, 1999), mesenchymal stem cells (Caplan, 1994), skin stem cells (Reynolds and Jahoda, 1994; Cotsarelis *et al.,* 1999), intestinal stem cells (Potten, 1998), and liver stem cells (Sell, 1994; Reid, 1996; Grisham and Thorgeirsson, 1997). Disparate, diverse investigations suggest that stem cells are also in the pancreas (Gittes *et al.,* 1996), lungs (Mason *et al.,* 1997), prostate (De Marzo *et al.,* 1998), and other tissues (Potten, 1997).

Determined Stem Cells in the Liver

In reviews and articles, we and others have summarized the evidence for and against the hypothesis that the liver, even the adult liver, contains determined stem cells that can give rise to cholangiocytes and hepatocytes (Sigal *et al.,* 1992, 1994, 1995b; Brill *et al.,* 1995; Grisham and Thorgeirsson, 1997). The discussion below is much abbreviated, and readers should consult any of the published reviews on hepatic determined stem cells (and see Reid, 1997) to see discussion of the very extensive literature on this subject. In brief, there are currently three competing models with respect to liver biology and regeneration:

Model 1. There are no stem cells. All of the liver cells are coequal and fully capable of clonal growth both *in vitro* and *in vivo* (Farber, 1995; Kennedy *et al.,* 1995; Michalopoulos and DeFrances, 1997). These arguments maintain that the ploidy state of the liver cells is irrelevant, that the tetraploid and octaploid cells should proliferate just as well as diploid cells, and the only variables to consider for proliferative potential are microenvironmental factors.

Model 2. There are hepatic stem cells, but they constitute a small, residual population left over from embryogenesis within the adult liver and are silent except in disease states such as cancer (Sell, 1994; Grisham and Thorgeirsson, 1997; Kay and Fausto, 1997). All liver biology and liver regeneration in the adult occur from adult liver cells that are coequal and capable of clonal growth (largely similar to the opinions expressed in the first model).

Model 3. The stem cells constitute the "cellular spring," located near the portal triads and yielding a maturational lineage of daughter cells that undergo a slow, unidirectional, maturational process ending in terminal differentiation or apoptosis near the central vein (Sigal *et al.,* 1992, 1995a; Reid, 1997). This model is similar to lineage systems in skin, gut, and hematopoietic tissue, but is proposed to occur much more slowly in liver, accounting for the confusion with respect to the role of candidate progenitors in the liver. This model assumes that the diploid cell population have far greater growth potential than do the tetraploid cells, and that the octaploid cells have no or minimal growth potential. Morphology, cell size, growth potential, and the expression of many tissue-specific genes are assumed to be lineage-position dependent.

Wilson and Leduc first postulated the presence of liver stem cells in 1958. However, as in the hematopoietic field, the concept of a liver stem cell gained the most credibility from extensive studies of liver carcinogenesis. A number of experimental protocols have been established that combine a hepatotoxic insult resulting in loss of cells in zones 2 and/or 3 of liver acinus, causing proliferation of small (7–15 μm) cells with scant cytoplasm and ovoid nuclei, termed "oval cells." With chronic insults, these oval cells become altered and can give rise to malignantly transformed cells. Overwhelming experimental evidence has implicated oval cells as partially or completely transformed liver progenitors, and some studies suggest that they are derived from progenitors for liver, gut, and pancreas (reviewed in Grisham and Thorgeirsson, 1997). Although it is experimentally simple to induce the appearance of oval cells, it has proved difficult to identify unequivocally the cells from which they arise. The cells of origin in all likelihood reside in proximity to bile ductules, but whether they are terminal ductule cells or distinct periductular cells remains a matter for debate.

Studies on both normal hepatic progenitors and on their transformed counterparts, oval cells, have been facilitated by the use of monoclonal antibodies (mAbs) raised against antigens characteristic of oval cells (Sell *et al.,* 1987; Hixson *et al.,* 1990). Using mAbs, hepatic progenitors have been found in normal adult livers as well as in fetal and neonatal livers (Sigal *et al.,* 1994, 1995b). Both oval cells and their normal counterparts, hepatic progenitors, are heterogeneous with respect to the binding of the various anti-oval cell antigen mAbs, and can be divided into subpopulations on this basis. Antigenic profiles of hepatic progenitors suggests that three groups exist: bipotential cells capable of differentiation to biliary or hepatocytic cells, precursors committed toward a biliary lineage, and precursors committed toward a hepatocytic lineage (Sigal *et al.,* 1994, 1995a; Brill *et al.,* 1995, 1999; Zvibel *et al.,* 1998). Numerous studies have shown that cells exhibiting combinations of antigenic markers characteristic of hepatocytes and cholangiocytes can be cultured readily from fetal rodent livers (Grisham and Thorgeirsson, 1997). Bipotential cells are present in large numbers in rats through embryonic day 15, when the numbers of the committed biliary and hepatocytic progenitors begin to increase. Subpopulations of hepatic progenitors from rat liver purified by fluorescence-activated cell sorting (FACS) also differ in their fates after introduction into congenic hosts lacking the prevalent liver protein, dipeptidyl peptidase (Sigal *et al.,* 1995b). The data suggest that there exist discrete numbers of antigenically definable, determined hepatic stem cells in fetal, neonatal, and adult liver, despite the view held by some that the liver contains none.

The proposal that there is an active, if slow, lineage system in the liver has been controversial, but increasing evidence has resulted in increasing support for the concept (Sigal *et al.,* 1992, 1995a; Reid, 1996). Parenchymal cells show a gradient of increasing ploidy, granularity, and autofluorescence from the periportal (zone 1) to midacinar (zone 2) to pericentral (zone 3) zones of the liver acinus. These features have been used to define senescing or apoptotic processes in other tissues and are now hypothesized to define an ongoing terminal differentiation process in the liver (Sigal *et al.,* 1995a). Age-related shifts in the parenchymal component of the liver, that is, more cells showing the characteristics of fully mature cells and fewer cells bearing characteristics of young cells, is reminiscent of age-related changes in skin, gut, and blood. Consistent with these observations, zone 1 cells have been observed to progress to a zone 3 phenotype. Periportal hepatocytes transplanted into periportal locations in host livers and translocated to pericentral locations as part of the response to CCl_4-induced liver injury begin to express zone 3-specific cytochrome P450 (Gupta *et al.,* 1999). This demonstrates the biologic potential of periportal hepatocytes to adopt the phenotype of hepatocytes of any zone. The potential of pericentral hepatocytes in this respect is unknown at present, and is predicted by the lineage model to be greatly restricted.

Another prediction of the lineage model is that there should be differences in growth potential correlating with maturational state. Liver repopulation studies have shown that small subpopulations of normal murine parenchymal cells have a proliferative capacity approximating that of hematopoietic stem cells (Overturf *et al.,* 1997). It is highly unlikely that polyploid, multinuclear hepatocytes, which make up a large percentage of the hepatocytes in adult liver, have this capability. In addition, periportal rat hepatocytes respond more strongly in culture to the mitogenic effect of HGF (Rajvanshi *et al.,* 1998), which is in agreement with the proposition that they represent cells of lesser maturity. Therefore, analogous to other lineage systems, it appears that a large reservoir of proliferative capacity resides in a small subpopulation of liver cells, certainly subpopulations within the diploid cells of the liver.

These observations indicate that periportal hepatocytes have properties that make them attractive as cellular feedstocks for artificial livers: they can attain the phenotype of all parenchymal cells in the mature acinus, and they have better proliferative potential than the hepatocyte population in general, which suggests that they may be seeded into devices at a lower density than is possible with more mature cells. The importance of the determined stem cell and lineage model of the liver is that it predicts that there is a population of cells, even in mature liver, whose potential utility is even greater. Hepatic progenitor cells meeting some of the criteria for determined stem cells have already been isolated from rodents, and work is proceeding on accomplishing this from human tissue. The ultimate impact of the harnessing of determined stem cells from liver is that it will overcome difficulties with supply and storage and make it possible to construct and maintain bioartificial liver-assist devices using human cells.

COMMITTED PROGENITORS

Committed progenitors are descendents of the stem cells and have undergone "commitment," a restriction in their genetic potential to limit their fate to that for one cell type. For example, in the liver, there are committed biliary progenitors and hepatocytic progenitors that yield descendents that are bile duct epithelia or hepatocytes, respectively. The committed progenitors of the skin ("transit amplifying cells"), of the intestine, or of the liver are the most rapidly growing cells of the tissue (Reid, 1996; Potten, 1997).

STRATEGIES FOR IDENTIFICATION AND MANIPULATION OF DETERMINED STEM CELLS

From the many studies on determined stem cells, there are some generic approaches that have emerged as well as recognition that markers once thought to be unique to a specific determined stem cell may be generic to stem cells. To find a determined stem cell, one must identify one or more tissue-specific characters to decide their identity and to be used as a guide during purification processes. The markers can be surface markers permitting immunoselection technologies (e.g., CD34) for purification of the cells or can be culture conditions that are selective for the stem cells (e.g., those used for mesenchymal stem cells) (Bruder *et al.*, 1994) or for neuronal stem cells. For example, the hemangioblast is identified by the master gene for the bHLH family of transcription factors, whereas hepatic lineages all have α-fetoprotein and albumin (Zaret, 1998) (see Fig. 41.3). Ideal models for searching for surface markers are transformants of the determined stem cells and consisting of tissue-specific cell lines, whether frankly transformed or only partially transformed (Zvibel *et al.*, 1998). Telomerase activity can be used as an assay for the self-renewal potential of

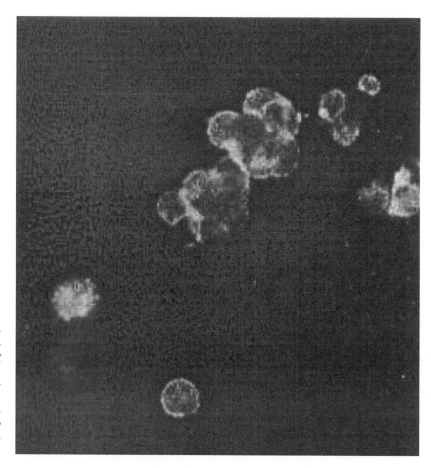

Fig. 41.3. Cytospin of rodent hepatoblast purified by multiparametric flow cytometry (contrast the mature cells shown in Fig. 41.4). The cells are mostly nucleus with a thin rim of cytoplasm, express α-fetoprotein, weakly express albumin, are diploid, are approximately 7–10 μm in diameter, and are agranular.

the candidate stem cells. *Ex vivo* expansion or differentiation conditions should be focused to maintain the tissue-specific identity and the telomerase activity. Markers defining known stem cells include receptors for signaling pathways (e.g., c-kit, CD38), telomerase, developmentally regulated cell adhesion molecules (e.g. CD34), and transcription factors (e.g., bHLH family).

SOURCING OF HUMAN CELLS FOR TISSUE ENGINEERING

A major problem for all basic, clinical, or industrial investigations on human tissue is being able to get it reliably, when needed, and in reasonably good condition. The supply of human tissue is erratic, such differentiated tissues are difficult to cryopreserve and therefore difficult to stockpile, and there is always the concern that a disease in the cells may be transmitted to a gravely ill and/or immunocompromised recipient. These difficulties have resulted in serious consideration of the use of established and well-characterized cell lines (at least for use in bioreactors) or even of the use of nonhuman tissues, such as from pigs (Sussman *et al.,* 1994; Baquerizo *et al.,* 1999).

Cell lines have long been a popular option, given the ease in handling, expansion, and storage. Concern about "metastasis" of tumor-derived cell lines or transmission of uncharacterized infectious agents from tumor cells or animal tissue in bioartificial organs to patients makes it unlikely that devices containing these readily available cells will attain widespread use. Even the use of "normal" cell lines has very limited potential, because their establishment in culture is via processes that inherently result in partially or completely transformed cells, with resulting truncation of the functions to those of their normal counterparts, the progenitors of the tissue from which the cell line derived (Potter, 1978, 1981; Sell, 1994; Zvibel *et al.,* 1998). The cells are blocked in the ability to mature through their lineage to yield cells capable of intermediate or late functions (Potter, 1981; Sell, 1994).

The use of porcine tissues for bioartificial organs and even for some types of cell therapies is another approach that has received very serious attention (Chari *et al.,* 1994; Baquerizo *et al.,* 1997, 1999). The advantages are, of course, that the animals can be maintained under strict, germ-free conditions and can provide a ready supply of healthy tissue. However, concern that transfer of zoonoses that are nonpathogenic in pigs but possibly pathogenic in humans has spawned much controversy and a flurry of intense debates within the Food and Drug Administration (FDA) and among scientists. Clearly, if human tissues were available, the motivation to pursue nonhuman tissues would be eliminated or reduced.

Long-term success after transplantation of whole organs leads one to believe that normal human cells, with appropriate safeguards, present much less risk than either cell lines or porcine cells. By this important criterion, normal human cells are the best choices for cell and gene therapies and for bioartificial organs. Although in the past, procurement of human tissue has proved difficult, the rise of well-established procurement organizations, such as the United Network for Organ Sharing or (UNOS), has made access to human tissue much easier. Some tissues, such as heart valves, corneas, and skin are retrieved postmortem, whereas other types of tissues or organs (kidney, liver, lungs, etc.) are presently thought to be usable only if from abortuses (fetal tissues) or from organ transplantation programs.

Investigators routinely use fetal tissues, because they are easy to get and to handle, are relatively inexpensive, and are readily established in culture; they account for much of the research on human tissues. Yet there is widespread resistance to using fetal tissue in clinical or industrial programs for fear of political, ethical, and/or legal repercussions.

Whole organs for transplantation are collected through organizations such as UNOS (see their web site at http://www.unos.org). In general, organs for transplantation can derive only from donors who are brain dead but who have not undergone cardiac arrest, accounting for approximately 2% of candidate donors. The organs collected for transplantation are evaluated after removal from the donors. Approximately 90% are transplanted into appropriate recipients, leaving 5–10% (this varies considerably from tissue to tissue) that are rejected for reasons such as abnormalities in the vascular anatomy, lesions in the tissue, or presence of a pathogen(s), such as a virus. It is these rejects that are made available for basic, clinical, or industrial investigations. Another variable is the considerable length of time that can pass from the time at which the organ is removed from the donor to the time at which it is received by investigators. On average, it is about 20 hr, a time frame that can result in severe ischemia and autolysis of the tissue. Although postmortem collection of organs has been found unacceptable for whole organ transplantation, it remains an untested option for any organs to be dispersed into cell suspensions and remains as one of the major untapped resources for human cells and tissues.

For a liver-assist device to be effective, a cellular mass representing 10–20% of the liver cell mass of the patient is needed. Because harvesting mature liver cells from donated human livers is relatively inefficient, because the mature cells have limited growth, and because no really successful cryopreservation methods for them yet exist, a limited number of bioartificial livers can be established from each donor organ. Therefore, cellular growth potential will determine the number of bioartificial organs that can be produced and used. This reasserts the issue of defining the cells within the liver with the greatest proliferative potential. A major clue is the fact, discussed in brief above, that the majority of mature liver cells are polyploid (tetraploid or octaploid) (Styles, 1993; Mossin *et al.,* 1994; Seglen, 1997; R. A. Williams *et al.,* 1997; Yin *et al.,* 1998). The extent of polyploidy is developmentally dependent. Fetal and neonatal tissues are entirely diploid. In the liver, there is a gradual appearance of binucleated diploid cells that have undergone nuclear division but with absence of cytokinesis. Subsequently, some of these binucleated cells give rise to daughter cells, and the nuclear membranes of the two nuclei fuse, resulting in tetraploid cells. The majority of the rat and human liver cells are tetraploid (up to 80% in rat livers). A similar process occurs, resulting in binucleated cells with tetraploid nuclei that subsequently convert to octaploid cells. Although polyploidy has been well documented in the liver, it is beginning to be studied in the mature cells of other quiescent tissues (Hamada *et al.,* 1990; Matturri *et al.,* 1991). In general, the polyploid cells undergo limited complete cell division and preferential hypertrophic growth in response to regenerative stimuli. In addition, maturation is associated with increasing autofluorescence, due to changes in amounts of fluorescent cytoplasmic pigments such as lipofuscins, and with increasing cytoplasmic "complexity," as perceived by flow cytometric analyses, due to maturation-dependent increases in mitochondria, ribosomes, and other cytoplasmic organelles. The studies of Gupta and associates have provided suggestive evidence that the periportal diploid cells are highly proliferative (Rajvanshi *et al.,* 1998). Therefore, even if the markers for purification of determined stem cells or early progenitors are not known, it would seem logical to select for diploid cells as candidates for maximal proliferation potential.

In addition to a quantitative benefit (owing to the utilization of feedstock cells with great expansion potential), the developmental potential of progenitor cells—or at the very least diploid cells—should translate into an improved opportunity for cell or gene therapies or for recreating acinar heterogeneity in the bioartificial liver. This in turn may mean that in a patient or in a bioreactor, all isoforms of such clinically important, zone-specific enzymes (e.g., the cytochromes P450) are expressed. The initial cell population should have the developmental capacity to take on the characteristics of any acinar zone, if the bioreactor and extracellular matrix conspire to provide the appropriate environment.

MICROENVIRONMENT

PRINCIPLES

The determinants of a cell's position within the lineage and its phenotype at that position have been shown to be due to a combination of at least three factors:

1. Autonomous intracellular mechanisms that are division number dependent or time dependent, such as chromatin rearrangements or methylation events, controls on nuclear division or cytokinesis, etc. (discussed above).

2. Signal transduction mechanisms activated by (a) gradients of soluble signals (autocrine, paracrine, or endocrine factors), of nutrients, and of gases (O_2/CO_2 levels), in multiple combinations, and/or in synergistic effects with extracellular matrix components, and (b) gradients of extracellular matrix components—a complex, insoluble mixture of proteins and carbohydrate-rich molecules found on the lateral borders (lateral matrix) and on the basal borders (basal matrix) of all cells.

3. Mechanisms affecting mass transfer of nutrients, gases, or signals to the cells. *In vivo* and especially *in vitro,* the ability of signals to access the cells is a critical variable, dictating the gradients of the signals and the biologic responses of the cells. These factors are especially important

when cells are being established under three-dimensional conditions, such as in bioreactors, or even, in simple form, in spheroids.

The issues of mass transfer are more extensively addressed in the discussion on bioreactors (below). Therefore, here will be discussed the other microenvironmental variables regulating the growth and differentiation of cells.

The heterogeneity of gene expression and the growth potential in all tissues is now recognized to be lineage-position dependent. In effect, there are early, intermediate, and late genes activated in a lineage-position-dependent fashion. Identical microenvironmental cues induce distinct genes at different maturational stages. Thus, signals from cell–cell interactions maintain and regulate growth and govern gene expression of a restricted set of genes, those defined by the maturational stage of the cells.

SOLUBLE SIGNALS

Signals from cell–cell interactions consist of a soluble set and an insoluble set. Soluble signals include autocrine factors, produced by a cell and active on the same cell; paracrine factors, produced by one cell and acting on a neighboring cell(s); and endocrine factors, produced by one cell and acting on a target cell at a distance from the source cell and delivered to the target cell(s) through blood or lymph. All of these soluble signals, including the autocrine ones, are highly regulated and under very strict controls. The wealth of knowledge on the myriad hormones and growth factors precludes the possibility of any detailed discussion of individual factors. However, many of them have been reviewed and detailed in various publications (Loughna and Pell, 1996; Ziegler *et al.,* 1997; Dickson and Salomon, 1998; McKay and Brown, 1998). Moreover, the efforts of many investigators have revealed most of the rules with respect to the identification of soluble signals or nutrients required for growth and/or differentiation of specific cell types (Reid, 1990; Reid and Luntz, 1997). Below will be presented simply a summary of the findings and of approaches to generate the microenvironment needed for cells in culture or in bioreactors.

Any cell type can be maintained *ex vivo* if it is provided with an appropriate substratum of extracellular matrix and with a serum-free, hormonally defined medium (HDM) containing purified hormones, growth factors, trace elements, and nutritional factors. The exact composition of any HDM is tissue specific, and very rarely is species specific. Thus, for example, an HDM developed for optimal growth of rat lung tissue will be virtually identical to that for growth of human lung tissue. For the few tissues for which cells at distinct maturational stages have been isolated in bulk (hematopoietic cells, liver), the exact composition of HDM contains subsets that are identical for all maturational stages and others that are distinct. Complementing this is the realization that the exact composition of an HDM is distinct for cells in an optimal growth versus optimal differentiation state.

Classic cell culture conditions, tissue culture plastic and serum-supplemented medium (SSM), produce cell cultures with minimal differentiated functions due to inhibition of transcription of tissue-specific mRNAs and destabilization, i.e., short half-lives, of those mRNAs. Common gene mRNAs (e.g., actin, tubulin) are stabilized by these conditions. Moreover, normal cells, especially epithelial cells, survive for only a few days when cells are plated on tissue culture plastic substrata and in a serum-supplemented medium. Tumor cells and stroma are better able than normal epithelial cells to survive the deleterious effects of serum, a factor in their preferential selection (e.g., stromal cell overgrowth) in classic cell cultures. Serum-free, hormonally defined media and tissue culture plastic result in cell cultures that have far more stable tissue-specific messenger RNAs, hence greatly enhanced differentiated function. However, the cells still cannot synthesize their tissue-specific mRNAs at normal rates. Thus, the cells maintained on tissue culture plastic and in HDM regulate entirely or almost entirely under posttranscriptional regulatory mechanisms. Survival of both normal and tumor cells maintained as primary cultures on tissue culture plastic persists for only a few days, a week at most.

EXTRACELLULAR MATRIX

Insoluble signals include components of the extracellular matrix, an insoluble material produced by all cells and found on the lateral borders between homotypic cells (the lateral matrix chemistry) and on the basal surfaces of cells, separating them from heterotypic cell types. The soluble and insoluble signals are interregulated and also have multiple synergistic effects. That is, soluble signals can induce synthesis of specific matrix components; matrix substrata can dictate output of specific soluble signals or regulate receptors and can also be "hardwired" for soluble signals; and distinct sets of matrix components and soluble signals cooperate in the regulation of specific physiologic processes. Extracellular matrix components turn over slowly, on the order of days to weeks, and serve to stabilize cells in specific configurations of adhesion sites, antigens, hormone receptors, ion channels, etc. Because the extracellular matrix is connected directly to the cytoskeleton via transmembrane molecules, changes in the conformation of one or more matrix components can directly influence cell shape. The soluble signals that have a more rapid turnover, on the order of seconds to minutes, activate signal transduction processes that induce a specific physiologic process such as growth or expression of tissue-specific genes. The effect of a soluble factor is entirely dependent, both qualitatively and quantitatively, on the matrix chemistry associated with the cell.

The purified matrix components, many of them commercially available, produce highly reproducible biological responses in cells, particularly when they are used in combination with HDM (Maher and Bissell, 1993; Knop et al., 1995; Bader et al., 1996; Nishikawa et al., 1996; Gomez-Lechon et al., 1998; Schuppan et al., 1998). The availability of some classes of matrix components, e.g., proteoglycans, in rigorously prepared forms, remains quite limited. No individual, purified matrix component will enable cells to survive and function for more than a week or two. Rather, long-term stability of survival and physiologic responses has been afforded only by extracts enriched in extracellular matrix and, to less extent, mixtures of matrix components. Tissue extracts enriched in extracellular matrices have proved of great importance in popularizing the importance of matrix substrata for normal or near-normal biologic functions in cells. One of the most popular has been Matrigel, a urea-derived extract from murine embryonal carcinoma tumors [Engelbreth–Holm–Swarm (EHS) tumors]. Matrigel contains primarily laminin, type IV collagen, heparan sulfate proteoglycan, and entactin, resembling the composition of basement membranes as found in tumors. Liver cells cultured on Matrigel attach loosely via laminin bridges to the gel and use their lateral matrix chemistry to form spheroidal aggregates that show greatly increased maintenance of differentiated function (Bissell et al., 1987; Moghe et al., 1996; Coger et al., 1997). Other tissue extracts that have also been used extensively have been biomatrices (e.g., Rojkind et al., 1980; Doerr et al., 1989) derived from any tissue, normal or diseased, and amnionitic membrane extracts (Liotta et al., 1980), used often for cultures of endothelia.

A major function of the matrix is to induce three-dimensional states in cells, essential for achieving full normalcy in cellular phenotype and for normal transcription rates of tissue-specific genes (Ingber et al., 1995; Ingber, 1998). The future, therefore, will be in identifying precise mixtures of purified matrix components that confer stable survival and physiologic responses.

CHANGES IN MATRIX CHEMISTRY WITH RESPECT TO LINEAGE POSITION

The lateral matrix chemistry, found between homotypic cells, consists of cell adhesion molecules (CAMs) and proteoglycans, but no collagens, fibronectins, or laminins. The basal matrix chemistry, found between closely associated heterotypic cell types (e.g., epithelia and mesenchymal cells), consists primarily of CAMs, proteoglycans, basal adhesion molecules (fibronectins or laminins), collagens, and other components that can be tissue specific in their distribution either qualitatively or quantitatively (e.g., entactin, tenascin). Collagen scaffoldings in the basal matrix chemistry of two adjacent cell types are cross-linked by an enzyme, lysl hydroxylase, thereby physically connecting the two cell layers.

The chemistry of the basal and lateral matrices changes in predictable ways within a maturational lineage. All known determined stem cells are associated with fetal lateral and basal matrix chemistries consisting of isoforms of CAMs [e.g., CD34, intercellular adhesion molecules (ICAMs)] and proteoglycans (e.g., certain syndecans) and fetal isoforms of laminin, type IV collagen, and often high levels of hyaluronates (Jones and Watt, 1993; Martinez-Hernandez and Amenta, 1993; Watt, 1998). Commitment and subsequent maturation of the cells are associated with a shift, in gradient fashion, in the lateral and basal matrix chemistries. The shift includes a decline or even loss of expression of the basal matrix components, a decline or loss in the expression of hyaluronates, a conversion to fibrillar collagens, and a shift to adult isoforms of basal adhesion proteins, of CAMs (e.g., CCAM 105), and of proteoglycans (e.g., heparin proteoglycans).

Although all tissues studied have similar basal and lateral matrix chemistries within the stem cell compartment, there are three major variations known in basal and/or lateral matrix chemistries occurring with maturation, as discussed below.

Preservation of expression of all major classes of matrix components but with gradients in isoforms within those classes

Some lineages, exemplified in gut and liver, are associated with a preservation in the expression of all classes of matrix components but a shift in which isoforms are expressed. The liver progenitor cells and intestinal stem cells are associated with a classic embryonic basal lamina consisting of laminin, type IV collagen, fetal CAMs, and specific forms of syndecan, a heparan sulfate proteoglycan. Maturation is associated with gradual shift to increasingly more stable forms of fibrillar collagens, fibronectin, and a heparin-like heparan sulfate proteoglycan, located at highest concentrations in association with the most mature cells (for example, around the parenchyma near the central vein in the liver). In the past, the liver was considered an "epithelioid organ" and not a true epithelium, because most of the mature liver cells were associated with a form of extracellular matrix in the space of Disse that is distinct morphologically from basement membranes. Now it is realized that the liver, as well as the gut, are both true epithelia with gradients of matrix chemistry paralleling and coordinate with maturing lineages of cells.

Loss in expression of basal matrix components

The skin and nervous system undergo maturational processes in which there is a loss in the expression of basal matrix components but retention in expression of lateral matrix components. The lateral matrix components undergo the maturational shifts in chemistry known for all lineages.

Loss in expression of collagens and shift to regulated expression of adhesion proteins with cell-binding domains

Hematopoiesis is associated with loss of collagen expression and with loss of expression of exons encoding the cell-binding domains of adhesion proteins in those cells that become free floating. Expression of the isoforms of the adhesion proteins still containing those cell-binding domains is retained by immunocytes but utilized in a highly regulated way for the many complex cellular interactions involved in immunologic responses. Other hematopoietic cell types, especially the monocytic–macrophage lineage, retain the capacity to produce collagens; their matrix maturational patterns are too poorly characterized to permit generalizations.

FEEDER CELLS

Although many of the soluble and matrix signals driving growth and differentiation of cells have been identified and well characterized, new factors are being recognized from the use of cocultures, the mixture of two or more cell types, and assessment of biologic function. The most common cocultures are those in which feeders of stromal cells are used to support various type of epithelia (Kawase *et al.,* 1994; Talbot *et al.,* 1994; Moore *et al.,* 1997). Stromal cell-derived growth factors active on epithelia include the interleukin-6 (IL-6) family and hepatocyte growth factor (HGF) (Rathjen *et al.,* 1990; Montesano *et al.,* 1991a,b). In addition, many of these signals are presented as complexes with glycosaminoglycans or proteoglycans, having carbohydrate chemistry that modulates the biologic responses of the cells to the protein ligands (Zvibel *et al.,* 1991, 1995; Lyon *et al.,* 1994).

Two descriptive studies have provided suggestions of yet other factors that may be relevant to full function of liver cells and involved cocultures of mixtures of endothelial cells, Ito cells, hepatocytes, bile duct cells, and Kupffer cells (Michalopoulos *et al..,* 1999; Mitaka *et al.,* 1999).

EXPANSION VERSUS DIFFERENTIATION OF CELLS

Liver tissue engineering comprises two steps: a growth phase, in which the seeded progenitor cells expand *in vivo* within the tissue of the host or *ex vivo* in culture or a bioreactor, and a differ-

Table 41.2. Summary of common requirements for all lineage stages[a]

Variable	Growth	Differentiation
Substratum	Flexible, porous substratum coated with type IV collagen and laminin	Flexible, porous substratum coated with fibrillar collagens and fibronectins
Calcium	Less than 0.5 mM	Above 0.5 mM
Glycosaminoglycans/ proteoglycans	Ideally, tissue-specific forms of heparan sulfate-PGs or heparan sulfate saccharides	Ideally, tissue-specific forms of heparin-PGs or specific heparin saccharides
Basal medium	Nutrient-rich medium supplemented with nicotinamide and trace elements (tissue specific) that include selenium	

[a]Requirements given are with respect to liver, but are also applicable in most regards to most differentiated epithelia.

entiation phase, in which the cells cease proliferation and assume their fully mature states. These phases occur naturally when cells are inoculated *in vivo.* By contrast, the conditions appropriate for controlling these phases must be carefully designed for *ex vivo* maintenance of liver cells. Detailed descriptions of the conditions for growth versus differentiation have been published (Brill *et al.,* 1994; Reid and Luntz, 1997; Reid, 1997) and are summarized in Tables 41.2 and 41.3. In brief, the particular conditions required to permit the survival and growth of progenitors are (1) serum-free basal medium, (2) attachment to a porous, flexible surface coated with extracellular matrix of a chemistry identical to that in the stem cell compartment *in vivo,* (3) provision of relevant

Table 41.3. Summary of known requirements that are maturational lineage-position dependent

Variable	Stem cells	Committed progenitors	Mature cells
Nutrients	Complex mixture lipids including HDL, phospholipids, and free fatty acids essential for survival		Lineoleic acid suffices (although preferably the complex mixture)
Attachment	Some determined stem cells prefer suspension; anchorage results in differentiation; those that adhere have the requirements as specified in Table 41.2		Adherence to a substratum required for survival
Feeder layers	Embryonic feeders required (can be tissue-specific)	Feeders required but tolerant of feeders from diverse sources	Feeders not required
Strict mitogens	Unidentified signals from feeders (local signaling dominant)		Systemic signaling dominant (for liver: insulin, EGF)
Other mitogens	IGF II, EGF, and transferrin/Fe are common		Insulin, EGF, thyroid hormones, growth hormone, prolactin
Differentiation signals	The repertoire of genes available for regulation is lineage-position dependent; each gene is regulated coordinately by several soluble signals (usually 3–4) in the context of a specific matrix chemistry		
Serum	Usually inhibitory for survival or growth		Inhibitory for tissue-specific gene expression
Gas exchange	Anaerobic conditions tolerated		Critically dependent on aerobic metabolism

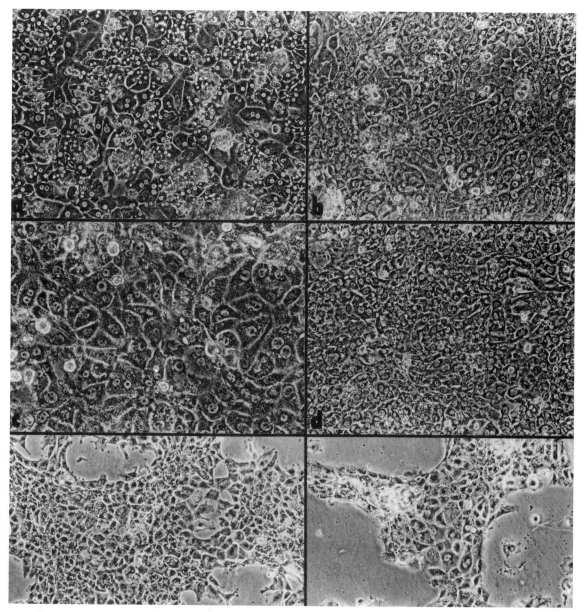

Fig. 41.4. Cultures of mature rat liver in serum-free, hormonally defined media tailored for optimal growth, on purified type I collagen.

nutrients not typically in basal media (e.g., lipids, trace elements), (4) provision of critical soluble signals (hormones, growth factors), and (5) coculture with feeder cells that supply soluble signals that have yet to be identified (e.g., embryonic stromal cells). Experience with rodent cells suggests that differentiation can be achieved by removal of the feeders, or by administering one of several known growth factors [e.g., HGF, FGF, and insulin-like growth factor II (IGF-II)] (Brill *et al.*, 1995; 1999) (see Fig. 41.4).

Maintenance of highly differentiated liver cells, at least short-term, has been achieved with a variety of culture media that often differ in seemingly insignificant ways except that the omission of serum is now recognized as essential (Jefferson *et al.*, 1984). Another stringent factor in culturing is the provision of cell–matrix substrata [discussed above and extensively in reviews by Brill *et al.* (1994), Reid (1997), and Reid and Luntz (1997)]. In addition to the exact chemistry of the matrix substrata, the achievement of three-dimensionality is essential (Wang and Ingber, 1994). This requirement has been shown both by elegant engineering analyses (Ingber *et al.*, 1995; Huang *et al.*, 1998; Ingber, 1998) and by simple culture conditions that elicit three-dimensionality such as spheroid cell cultures (Koide *et al.,* 1989; Tong *et al.,* 1992) and matrix–cell–matrix sandwiches with collagen I gels and Matrigel (LeCluyse *et al.,* 1994; Moghe *et al.,* 1996). Cells grown in such configurations are more cuboidal than those grown on plastic or on matrix proteins coated on plastic, and may have the capability of forming more extensive cell–cell contacts with their neighbors than flattened cells can. The extensive cell–cell contacts achieved by hepatocyte spheroids may similarly account for the high degree of differentiated cell function in this culture system (Hamamoto *et al.,* 1998).

In a bioartificial liver, it is perhaps more important to consider how the medium is brought to all parts of the culture, involving essential issues of mass transfer and thus crucial for achieving high-density cultures. If the developing bioartificial liver is to begin with a small inoculum of cells that proliferate to fill the chamber and proceed to a highly differentiated state in which cell–cell contact is maximized, a combination of biodegradable polymers coated with matrix proteins and sparsely seeded hepatic progenitors should provide conditions that encourage expansion of the cells. As the cells multiply, the biodegradable polymers should disappear to be replaced by a natural extracellular matrix produced by the cells and governed by the cell density and the microenvironment of the bioreactor chamber.

COMPARISON OF NORMAL AND TUMOR CELLS

Tumor cells have *ex vivo* maintenance requirements that overlap extensively with the requirements of normal progenitor cells, their normal counterparts (Zvibel *et al.,* 1998). The fallacy in many of the prior assumptions about tumors has been in the comparison of them with normal, mature cells. For example, the requirements for *ex vivo* maintenance of hepatoblastomas are almost identical to those of hepatoblasts but are quite distinct from those of mature hepatocytes. Similarly, well-differentiated hepatomas or cholangiomas have requirements close to that of committed hepatic or biliary progenitors, respectively (Zvibel *et al.,* 1998). The phenotypic expression of tumors and cultures of them also parallels that in their normal counterparts. Thus, as Potter and Pierce have long argued, tumors are maturationally arrested progenitors, and the expression of fetal genes is simply part of the normal phenotype of the progenitor cells (Potter, 1981; Sell and Pierce, 1994). Proof of the hypothesis of Potter and Pierce has been supported by more additional studies comparing normal hepatic progenitor cells and hepatic tumor cells (Zvibel *et al.,* 1998). For example, matrix/hormonal synergies have been found to regulate fetal (early) genes such as IGF II and TGF, both transcriptionally and posttranscriptionally in tumors (Zvibel *et al.,* 1991) and in their normal counterparts, hepatic progenitors (Zvibel *et al.,* 1998). Identical hormone/matrix synergies regulate adult-onset genes in mature parenchymal cells (Brill *et al.,* 1995). If the tumor cells express any adult-onset genes, they are regulated, invariably, by posttranscriptional mechanisms entirely or with muted transcriptional regulatory mechanisms in combination with the posttranscriptional mechanisms.

BIODEGRADABLE POLYMERS

CRITERIA FOR DESIGN OF BIODEGRADABLE POLYMERS IN TISSUE ENGINEERING OF LIVER

The final realization of tissue engineering of liver depends on the success of research endeavors in (1) liver biology, especially in understanding the extracellular matrix (ECM) biology and chemistry, and liver as a maturational lineage system (Reid, 1997), (2) design and engineering of novel bioreactor systems (Macdonald *et al.,* 1999), and (3) development of degradable extracellular substrata. Biodegradable polymer support in tissue engineering of liver is to provide (1) a surface coated with growth-permissive ECM for attachment/adhesion of anchorage-dependent hepatocytes, (2) a three-dimensional scaffold for localization of cells, which is critical for cell interaction and tissue-specific gene expression (Reid, 1997), (3) a porous structure allowing sufficient nutrient delivery, waste removal and gas exchange, and ingrowth of host tissue, and (4) a vehicle for local delivery of bioactive factors to encourage vascularization and angiogenesis, and thus guiding the growth of hepatic tissue. Polymer supports should have a controllable degradation, thus allowing the role of polymer support to be taken over eventually as a result of tissue growth and ECM formation (Vacanti *et al.,* 1994; Davis and Vacanti, 1996; Kim and Mooney, 1998; Marler *et al.,* 1998). In addition, suitable polymer supports should have sufficient mechanical tensile strength and compressive resistance for transplantation, and desirable physical and chemical characteristics for fabrication into desirable physical formats. Synthetic or modified natural polymers should be biocompatible to minimize or eliminate host inflammatory response. Polymers should be stable both chemically and physically for sterilization.

NATURALLY DERIVED POLYMERS FOR TISSUE ENGINEERING OF LIVER

Naturally derived matrix polymers (e.g., collagens, elastin) may be divided into two classes based on their biological function: fibrous structural proteins, providing important mechanical and organizational support to the tissue, and glycosaminoglycans and derivatives capable of forming hydrated gels through which the fibrous and polysaccharide components of the matrix provide tensile strength and compression resistive support, respectively. Natural ECM components, e.g., collagen and laminin, have been used as cell attachment support by coating on various surfaces such as polystyrene, polyurethane, urethane epoxy on glass, plastic, and cross-linked dextran (Demetriou *et al.,* 1988; Cima *et al.,* 1991; Mooney *et al.,* 1992; Bhatia *et al.,* 1994; Powers and Griffith-Cima, 1996). The advantage of naturally derived polymers is their close compatibility with the ECM environment, thus providing better substrata for *ex vivo* culture of cells. However, they typically lack the necessary mechanical integrity for fabrication into scaffolds or other shaped polymer structures, and suffer from the limitation of sufficient supply and source variation, which limits their use for tissue engineering.

Collagen, a major component of the ECM for both structure and functional support of liver cells, is the most widely used natural polymer for liver cell attachment in *ex vivo* culture. Native and partially digested collagens have been used in various forms, including coating on plastic surfaces to provide attachment support for liver cells (Shimbara *et al.,* 1996); entrapment of hepatocytes and nonparenchymal cells, overlaying to form sandwich systems (Berthiaume *et al.,* 1996; Koike *et al.,* 1996; Kono *et al.,* 1997; Moghe *et al.,* 1997); encapsulation of liver cells to form spheroids (Lazar *et al.,* 1995; Wu *et al.,* 1996; Hu *et al.,* 1997); and entrapment of liver cells onto the surfaces of porous synthetic polymer sponges (Kaufmann *et al.,* 1997; Takeshita *et al.,* 1998). Entrapment and encapsulation of liver cells creates a culture environment allowing direct contact among hepatocytes, which is important to the maintenance of differentiated hepatocyte functions. Encapsulation also function as microfiltration barriers for host immunoglobulins and entrapped hepatocytes while allowing free exchange of metabolites, which is critical for use of allogeneic or exogeneic hepatocytes in extracorporeal liver-support devices (Hu *et al.,* 1997). Despite the importance of collagen for hepatocyte attachment, because of the lack of suitable physical and chemical properties, it is difficult to fabricate collagen into three-dimensional scaffolds to support *ex vivo* culture or transplantation of hepatocytes.

Derived from seaweed, anionic alginate forms gel on cross-linking by Ca^{2+}. It can be readily prepared as microcapsules, threads, or layers of gel. It offers better physical characteristics for entrapment of cells compared to collagen. Alginate encapsulation of liver cells allows the formation of spheroids within the capsules, which is important to hepatocyte differentiated functions such as secretion of albumin and maintenance of P450 activity (Ito and Chang, 1992; Selden *et al.,* 1998). Alginate-encapsulated hepatocytes cultured in bioreactors are used as an extracorporeal liver-assist device (Fremond *et al.,* 1993), and for direct transplantation into rats (Dixit *et al.,* 1990; Gupta *et al.,* 1993; Hirai *et al.,* 1993). Attachment of hepatocytes to alginate also improves the viability and differentiated function of cryopreserved hepatocytes (Guyomard *et al.,* 1996).

Among the naturally derived polymers used in culture of hepatocytes is chitosan, a derivative from partial N-deacetylation of chitin; chitosan contains a binary heteropolysaccharide of [1 → 4]-linked 2-acetamide-2-deoxy-D-glucopyranose and 2-amino-2-deoxy-D-glucopyranose monosaccharide (Draget *et al.,* 1992). It has characteristics similar to glycosaminoglycans, a component of liver ECM, and is biodegradable and nontoxic (Muzzarelli *et al.,* 1988). To improve its biocompatibility and encourage hepatocyte attachment to chitosan-coated surfaces, collagen, albumin, and gelatin have been blended with the polymer (Elcin *et al.,* 1998). Fructose-modified chitosan was found to be better than collagen alone in encouraging hepatocyte attachment and maintenance of differentiated function (Yagi *et al.,* 1997). Chitosan cross-linked by glutaraldehyde was found to improve the mechanical strength of polymers, while maintaining hepatocyte biocompatibility, as demonstrated by cell attachment and maintenance of differentiated function (Kawase *et al.*, 1997).

SYNTHETIC BIODEGRADABLE POLYMERS FOR TISSUE ENGINEERING OF LIVER

Synthetic biocompatible and biodegradable polymer is widely used in surgical sutures and constructs for *ex vivo* cell culture, expansion, and cell transplantation. There is a variety of synthetic biodegradable polymers, including polyesters, polypeptides, hydrogels of large variations in backbone and side chain structures, and copolymer combinations. General discussions of synthetic polymers can be found in Hubbell (1995), Peters and Mooney (1997), and Marler *et al.* (1998). The most widely used synthetic biodegradable polymers, however, are the poly(α-hydroxy acid) family of polymers, including poly(glycolic acid) (PGA), poly(lactic acid) (PLA), poly(D,L-lactide-co-glycolide) (PLGA) copolymer, and their modified derivatives. Because of their wide range of desirable physical properties, synthetic polymers can be precisely constructed in various formats of defined size, shape, and surface and internal morphology (e.g., Peters and Mooney, 1997; Marler *et al.*, 1998) that facilitate cell attachment and localization, mass transfer, gas exchange, and growth of cells. They can be designed and constructed to degrade under *in vitro* and *in vivo* conditions over time by varying the molecular weight, monomer mass distribution in copolymers, chain structure, and surface and internal porous structures (Gilding, 1981). The role of eventually degraded polymer supports may be taken over by the formation of natural ECM and tissue structure with a minimal degree of chronic foreign body responses from the host, which is often caused by nonbiodegradable polymers. However, despite the eventual degradation of biodegradable polymers, their interactions with cells in culture and the initial host response to introduced tissue engineering devices resulted from the immune reaction between synthetic polymers and cellular components present a critical challenge in the design of synthetic polymers (Ignatius and Claes, 1996; Anderson, 1998; Babensee *et al.*, 1998).

Hybrid polymers typically possess both the physical properties of synthetic polymers and the biocompatibility of natural polymers (Hubbell, 1995; Peters and Mooney, 1997; Kim and Mooney, 1998). At present most of the hybrid polymers are based on coupling of cell adhesion peptide sequences or sugar moieties onto synthetic polymers such as poly(ethylene oxide) (PEO), polyurethane, and poly(ethylene glycol) (PEG) (Dai *et al.*, 1994; Griffith and Lopina, 1998). Incorporation of cell-recognizing sequences, e.g., arginine–glycine–aspartic acid (RGD) for the integrin super family of cell adhesion proteins, encourages the attachment of hepatocytes (Hynes, 1992; Dai *et al.*, 1994; Hubbell, 1995; Kim and Mooney, 1998). Carbohydrate-based cell recognition has also been applied to the design of hybrid polymers with improved characteristics of hepatocyte attachment (Weigel *et al.*, 1978; Oka and Weigel, 1986; A. Kobayashi *et al.*, 1992; Cima, 1994; Gutsche *et al.*, 1994; K. Kobayashi *et al.*, 1994; Lopina *et al.*, 1996; Park *et al.*, 1998). The favorable interaction of hepatocytes with galactose-modified surfaces is thought to be attributed to the specific interaction between the polymer ligand (galactose moiety) and a hepatocyte surface receptor, presumably the ASGP-R (Lopina *et al.*, 1996; Griffith and Lopina, 1998). Similarly, lactose and heparin hybrid porous polystyrene foams were found to promote hepatocyte attachment and maintenance of a rounded cellular morphology (Gutsche *et al.*, 1996).

BIOCOMPATIBILITY AND TOXICITY OF SYNTHETIC POLYMERS

To limit chronic foreign body response by host tissue, it is necessary for polymers implanted into host tissue as hepatocyte–polymer conduits and their degraded products to cause low or minimal local and systematic host inflammation. The intensity and duration of the host response determine the biocompatibility of a polymer implant, which is characterized by fibrosis and vascularization of the tissue surrounding the implant device (Anderson, 1998; Babensee *et al.*, 1998). As a result of host response to an implanted device, local reorganization and generation of tissue, as part of the healing response to tissue injury from implantation, lead to formation of fibrovascular tissue and fibrous encapsulation of implant, a key outcome of host response. For hepatocyte transplantation, host tissue ingrowth into the polymer scaffold, vascularization, angiogenesis, and population of cells in the polymer support are important to the survival, growth, and function of transplanted cells. Slow degradation of biodegradable polymer implants allows the formation of tissue vasculature to take over the role of polymer implants, thus alleviating the chronic inflammatory response caused by nondegradable materials.

Natural polymers, e.g., collagens or Matrigel, because of their biologic origin, possess the best biocompatibility. Most synthetic polymer scaffolds for hepatocyte seeding and tissue engineering have largely derived from the poly(α-hydroxy acid) family of biodegradable polymers, including PGA, PLA, and copolymer PLGA. They are, at present, among the safest synthetic polymers commercially available. The degraded products of the poly(α-hydroxy acid) family of biodegradable polymers are lactic and glycolic acids, both metabolites of the Kreb cycle, and thus cleared by respiratory reactions (Gilding, 1981). In principle, they should not induce chronic toxic response by host tissue. There is a concern about the local acidification by lactic and glycolic acids. However, the slow release of lactic and glycolic acids may not be significantly detrimental to the local surrounding tissue, as shown by the grossly normal engineered tissue (Mooney *et al.*, 1997). Metabolic study of the mitochondrial function of a mouse fibroblast cell line also shows no significant cytotoxicity in the presence of low levels of degraded products of PLA and PLGA, a finding supportive to their use as biocompatible materials (Ignatius and Claes, 1996). Although the biocompatibility of the poly(α-hydroxy acid) family of biodegradable polymers as a core material has been investigated, with the rapid advance in the research on synthetic hybrid polymers, the biocompatibility and cytotoxicity of the new biodegradable polymers should be thoroughly evaluated to determine the short- and long-term effects on host tissue and transplanted cells or tissue.

BIODEGRADABLE POLYMER SUPPORTS OF VARIOUS PHYSICAL FORMATS

For culture of metabolically highly active hepatocytes, biodegradable scaffolds ideally should meet the criteria discussed above. Biodegradable fiber meshes or sponges of highly porous surface and internal structure have been developed to meet some of those design criteria (Vacanti *et al.*, 1994; Davis and Vacanti, 1996; Peters and Mooney, 1997; Eiselt *et al.*, 1998; Kim and Mooney, 1998). Encapsulation of hepatocytes in collagen, alginate, or synthetic polymers has also been used to provide an attachment surface and a filtration barrier (Wong and Chang, 1986; Chang, 1992; Nyberg *et al.*, 1993). Porous microspheres may also be used to provide a large surface for hepatocyte attachment. The advantages and shortfalls of the various formats for tissue engineering of liver are discussed below. The techniques for fabrication of biodegradable polymers into various formats have been discussed in Chapter 21 of this volume.

Scaffolds of sponge and fiber mesh

Among various types of biodegradable scaffolds, scaffolds of porous sponge have been most extensively investigated for *ex vivo* culture and transplant of hepatocytes. They can be fabricated into foamlike porous structures (Mikos *et al.*, 1993c, 1994), or templates of complex internal interconnecting networks with spatial resolutions as small as several hundred micrometers. The polymer template can be fabricated to achieve the size and distribution of the channels precisely based on the requirement of the cell seeding distribution, nutrient delivery, and gas exchange of hepatocytes and other cells, and resistance to compressive load (Giordano *et al.*, 1996; Griffith *et al.*, 1997; Kim *et al.*, 1998). Galactose tethered on the surface of the polymer template can encourage hepatocyte attachment (Park *et al.*, 1998). Biodegradable sponges and bonded fiber structure of highly porous surface and internal structure (>83% porosity) have been used to support the growth of hepatocytes *in vitro* and *in vivo* (Mikos *et al.*, 1993a,b,c; Wald *et al.*, 1993; Mooney *et al.*, 1994). To improve seeding of hepatocytes and flow distribution of nutrients throughout the porous sponge, coating PLA or PLGA sponges with surfactant polyvinyl alcohol (PVA) increases the surface hydrophilicity, leading to a greater infiltration of hepatocytes and media flow into the porous sponge structure (Mooney *et al.*, 1995). However, a significant shortfall of porous sponges typically is the insufficient nutrient delivery into the core of the polymer sponge, which can result in necrosis of seeded hepatocytes (Mooney *et al.*, 1997). A number of approaches have been investigated to improve the survival, growth, and maintenance of differentiated function of transplanted hepatocytes. These include increasing hepatic stimulation for vascularization by increasing delivery of portal blood to transplanted cells, prevascularization of polymer sponges, and localized release of epidermal growth factors (Mikos *et al.*, 1993c; Mooney *et al.*, 1996, 1997; Kaufmann *et al.*, 1997). With a highly porous (>90%) structure and a large surface-to-volume ratio (as high as 0.05 μ^{-1}) (Mikos *et al.*, 1993a), biodegradable fiber mesh-based scaffolds represent another class of polymer support for cell culture and transplantation (Gilbert *et al.*, 1993; Vacanti *et al.*, 1994; Takeda *et al.*, 1995). Nonwoven porous polymer tubes fabricated from nonwoven meshes of PGA fiber have been used for seeding with rat hepatocytes and implantation of cells in the small intestinal submucosa (SIS) conduit (Kim and Mooney, 1998).

Hepatocyte encapsulation in polymer supports

The concept of microencapsulation of hepatocytes arose from the need to prevent antibody and antigen interaction in allogeneic or xenogeneic transplantation of hepatocytes. Collagen and alginate form a thin layer, encapsulating the cells in the core region (Wong and Chang, 1986; Chang, 1992; Fremond *et al.*, 1993; Nyberg *et al.*, 1993; Fremond *et al.*, 1996; Hu *et al.*, 1997). Proteins and other macromolecules cannot permeate through the encapsulating membranes, but small metabolites are freely permeable to the barriers and are in free exchange between media and the encapsulated volume. The advantage of microencapsulation over other forms of polymer support is the microfiltration effect of host antibodies, which prevent host rejection of transplanted xenographs. Similar to microcarriers as an attachment support for hepatocytes, encapsulation and entrapment also provide a large area for hepatocyte anchorage in addition to microfiltration of immunoglobulins and other cytotoxic macromolecules. However, the drawback of the microfiltration barrier is the hindrance of mass transfer between the cells and media or blood in perfusion. In addition, encapsulation and entrapment are often highly heterogeneous in the number of cells entrapped. Thus, insufficient nutrient delivery and gas exchange may occur for cells residing in the core region (Peters and Mooney, 1997). However, this may be overcome by decreasing the size of microcapsules or entrapment, or by localizing microentrapment in a narrow nutrient-permeable hollow fiber (Nyberg *et al.*, 1993; Hu *et al.*, 1997; Peters and Mooney, 1997).

Microcarrier polymer support

ECM-coated microcarriers represent another alternative attachment support for hepatocytes. ECM-coated microcarriers offer a large surface for hepatocyte anchorage as a result of their large surface-to-volume ratio. Hepatocytes attached to the microcarriers may be transplanted to recipients by injection or may be cultured *ex vivo* in bioreactors of various physical designs. In addition, protein and other bioactive factors may be incorporated into the polymer or presented on the surface to enhance localized delivery of extracellular signals. However, microcarriers lack the ability to form defined three-dimensional environments in comparison with sponges or fiber-mesh-based scaffolds. Type I collagen-coated cross-linked dextran microspheres have been used for attachment of rat and human hepatocytes for transplantation (Demetriou *et al.*, 1988, 1991; Moscioni *et al.*, 1989) and *ex vivo* culture of hepatocytes in an extracorporeal liver-assist device with limited success in maintaining the hepatocyte differentiated function and encouraging host vascularization (Rozga *et al.*, 1994; Riordan and Williams, 1997; Watanabe *et al.*, 1997). Alternatively, biologically modified poly(2-hydroxyethyl methacrylate) (pHEMA) microcarriers are used to achieve better hepatocyte attachment for culture in a bioreactor (>75%) (Dixit *et al.*, 1990; Dixit, 1994, 1995). Although the exact role of microcarriers in maintaining the survival and function of attached hepatocytes is unclear, it may be attributed to the dependence of hepatocytes on attachment to a collagen matrix for expression and maintenance of differentiated function, and/or the promotion of early vascularization by extracellular matrix-coated microcarriers (Moscioni *et al.*, 1989; Watanabe *et al.*, 1997).

Commercially available nonbiodegradable microcarriers, e.g., heavily cross-linked dextran coated with type I collagen (e.g., Cytodex-3 beads), are used as ECM-coated microcarriers for hepatocyte culture, transplantation, and *ex vivo* culture. However, non-growth-permissive type I collagen-coated nondegradable beads have not been used successfully for long-term *ex vivo* culture or transplantation of hepatocytes. The lack of hepatocyte attachment to a type I collagen-coated microcarrier surface contributes to the difficulty of *ex vivo* culture of rat hepatocytes in hollow-fiber bioreactors with integral oxygen and media supply (A. S. L. Xu, J. M. MacDonald, and L. M. Reid, unpublished data). To provide a nonrigid, growth-permissive ECM-coated biodegradable support for *ex vivo* culture of hepatocytes or hepatic progenitors, porous PLGA beads have been prepared by a polymer emulsion–rapid solvent evaporation method, and a low-temperature-casting/salt-leaching procedure, respectively. Figure 41.5 shows the surface and internal porous structure of

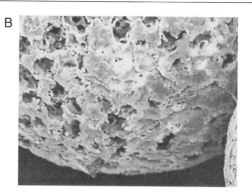

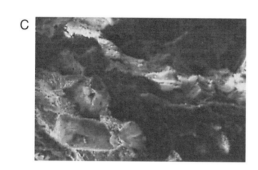

Fig. 41.5. Scanning electron micrograph showing the porous surface morphology of PLGA beads (50/50%) prepared by a polymer emulsion–rapid solvent evaporation method (A and B). (C) The porous internal structure of PLGA beads prepared by a low-temperature casting/salt-leaching procedure. Magnification (A, B, and C, respectively): ×17, ×172, and ×168.

PLGA beads prepared by the two methods. Rat primary hepatocytes were found attached to the type IV collagen-coated surface (Fig. 41.6). Preliminary study of rat primary hepatocytes cultured on PLGA beads in a commercial hollow-fiber bioreactor shows a small but significant increase in survival (A. S. L. Xu, J. M. Macdonald, and L. M. Reid, unpublished data). Thus, biocompatible, biodegradable porous microcarriers coated with growth-permissive ECM may be suitable for *ex vivo* culture and expansion of hepatocytes.

Localized delivery of bioactive factors by polymer support

For both *ex vivo* engineering of functional liver tissue and reconstitution of liver by long-term cell therapy (i.e., hepatocyte transplantation), the diverse function of liver would have to be maintained not only by parenchymal cells, but also nonparenchymal cells, such as sinusoidal endothelia cell, Ito cells, and Kupffer cells (Arias *et al.,* 1994), through their unique *in vivo* tissue structure and cell interaction. A well-developed vasculature for transplanted hepatocytes and nonparenchymal cells, or hepatocytes *ex vivo* cultured in a three-dimensional environment, is needed to achieve sufficient delivery of nutrient, hormones, factors, and oxygen, and thus is critical to cell survival, differentiated function, and growth. Without a developed vasculature, the therapeutic benefit from transplanted hepatocytes and the hepatocytes in bioartificial liver-assist devices may only be short term with limited functional support, because of the critical dependence of hepatocyte growth and expression of differentiated function on formation of tissue vasculature. One of the effective approaches in the design and fabrication of biocompatible and biodegradable polymers is to use polymer support as a vehicle for controlled localized delivery of bioactive factors, to encourage vascularization and angiogenesis. A number of tissue-specific angiogenic factors, such as vascular endothelial growth factor, epidermal growth factor (EGF), and fibroblast growth factor, have been identified as important to angiogenesis and vascularization (Folkman and Klagsburn, 1987; Thomas, 1996; Norrby, 1997; Ajioka *et al.,* 1999; Gerber *et al.,* 1999). However, the cellular origin, molecular mechanism of control, and regulation of angiogenic growth factors under physiologic and pathologic conditions (e.g., increased level in tumor cells) remain to be fully understood. Angiogenic growth factors such as VEGF are present under physiologic conditions at nanomolar to micromolar levels during both initial stages of embryonic development and organ growth in

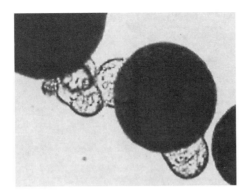
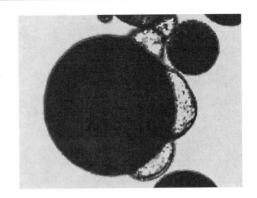
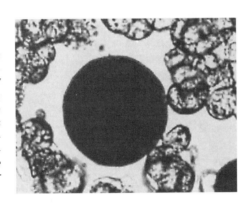
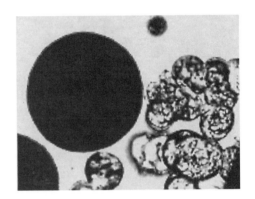

Fig. 41.6. Rat hepatocytes attached to PLGA microcarriers coated with type IV collagen serum-free media containing EGF and insulin (upper panels); hepatocytes incubated with PLGA microcarriers without a coating of ECM in the same serum-free media (lower panel).

lung alveolar cells, kidney glomerular and proximal tubules, hepatic tissue, and brain (Thomas, 1996). Hepatocyte growth factor is another widely recognized multifunctional cytokine that has target receptors among hepatocytes, keratinocytes, kidney tubular epithelia cells, and breast epithelial cells (Michalopoulos, 1995; Zarnegar and Michalopoulos, 1995). In liver, it stimulates hepatocyte DNA synthesis and thus normal hepatic tissue regeneration. Other growth factors, e.g., platelet-derived growth factor, and roles in wound healing, tissue remodeling, and significance to tissue engineering have been discussed in Chapter 12 of this book. Growth factors are tissue specific with a short half-life in the plasma circulation (Davidson *et al.*, 1985; Folkman and Klagsburn, 1987; Thomas, 1996; Norrby, 1997; Baldwin and Saltzman, 1998; Ajioka *et al.*, 1999; Gerber *et al.*, 1999). Under *ex vivo* tissue culture conditions, supplementing media with growth factors is often problematic as a result of factor short half-life and diffusive distribution in plasma and media supply. Under *in vivo* conditions, however, delivery of growth factors to transplanted hepatocytes by microcapillary blood circulation is a formidable task before sufficient tissue vasculature is established among transplanted cells (Assy *et al.*, 1999). This difficulty is compounded further by the very low or inactive state of vascularization under most physiologic conditions (Thomas, 1996; Norrby, 1997). Expression of VEGF in VEGF-transfected hepatocytes was found to induce significant proliferation of endothelial cells, which in turn, led to the increased formation of vascular network in established functional hepatic tissue (Ajioka *et al.*, 1999).

The role of biocompatible and biodegradable polymers in tissue engineering and cell transplantation is not limited to providing a mechanical support for cells. Rather, their usefulness may be attributed to their application to encapsulate or incorporate bioactive factors for controllable localized delivery of factors at the interface between matrix-coated polymer constructs and seeded cells (Hubbell, 1995; Peters and Mooney, 1997). The localized release off factors may promote more effective and locally selective vascularization and angiogenesis, which in turn leads to increased tissue vasculature formation. Delivery of epidermal growth factor by encapsulation with PLGA microspheres was shown to enhance the engraftment of transplanted hepatocytes (Mooney *et al.*, 1996). To increase the survival and transplant engraftment of fetal porcine hepatocytes xenotransplanted into rats, chitosan is used as a scaffold for the hepatocyte attachment, VEGF loading, and delivery (Elcin *et al.,* 1999). Localized delivery of bioactive factors by their physical incorporation or encapsulation in biodegradable polymer constructs depends on the degradation rate of the polymer, a process that is a function of polymer degradation chemistry and the microenvironment in which the polymer resides. Precise delivery of bioactive factors by these approaches may be difficult to control because of an initial burst (or rapid release) of growth factor when encapsulation is used (Ogawa, 1997; Baldwin and Saltzman, 1998), and because of diffusive spread, endocytosis of growth factor, and cell surface receptor complexes. An alternative approach is covalently tethering of a growth factor such as EGF onto the culture polymer surface by coupling EGF with a PEO linker arm, thus providing prolonged and stable cytokine signaling (Kuhl and Griffith-Cima, 1996).

DESIGNING BIOREACTORS FOR TISSUE ENGINEERING

INTRODUCTION

In all tissues, blood flows from the periphery to the central circulation through a capillary bed, and physical, chemical, and biologic gradients exist across the tissue. Figure 41.7A is a two-dimensional schematic representation of blood flow and corresponding gradients in generic tissue. Soluble components such as gases (e.g., oxygen), small molecules (e.g., sugars, amino acids, and fats), and macromolecules (e.g., proteins) have characteristic concentration profiles across the tissue that are described by the appropriate mass transfer equation. Mass transfer depends on the diffusivity, volumetric consumption/production rate, and initial concentration of the soluble component (see Chapter 17, this volume) (Macdonald *et al.,* 1999). Oxygen is typically the biolimiting nutrient in tissues and is especially so in bioreactors, where its relatively high consumption rate and low initial concentration (~0.2 mM at 37°C, 1 atm) combine to make it difficult to provide adequate levels.

A Generalized Flow Dynamics in Tissue

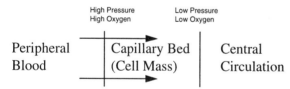

B Tissue Mimicry

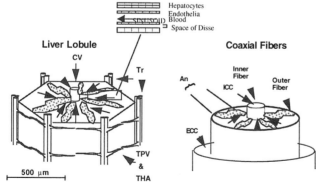

C Scale-up

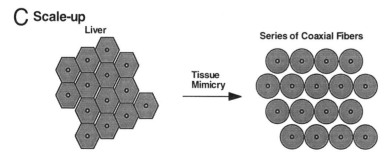

Fig. 41.7. Generalized blood flow in tissue (A), and demonstration of the tissue engineering principles of tissue mimicry (B) and scale-up (C) using coaxial fibers.

Due to the potential for clot formation and the added expense of including hemoglobin additives with perfusion media, bioreactors lack the oxygen-carrying capacity of blood, and in general, oxygen in a bioreactor becomes limiting to cells beyond 200 μm, when mass transfer is dominated by diffusion. This limitation cannot be addressed by significantly increasing the amount of dissolved oxygen, because at high concentrations oxygen is toxic (Macdonald *et al.,* 1999). The presence of an extensive vascular network in natural tissues means that cells are only one or two cell distances from the flowing blood, or nutrient source. This increases the diffusivity term in the mass transfer equation about four orders of magnitude over incoherent diffusion (2×10^{-6} cm²/sec) due to the coherent convective flow of nutrients, which is on the order of 0.01 cm/sec in the sinusoids [red blood cell velocity ranges from 0.25 to 3×10^{-2} cm/sec (McCuskey, 1994)]. The blood velocity increases from normal to cirrhotic tissues and is driven by a pressure drop from 10 to 25 mm Hg in the periportal sinusoid to just a few mm Hg in the pericentral sinusoid (McCuskey, 1994). Because primary cells are adapted to conditions that provide for the optimum mass transfer of nutrients, their use in artificial devices suggests that tissue mimicry should become a primary tissue engineering principle. Tissue mimicry can be divided into two parts: (1) reproducing the smallest organizational unit and (2) scaling up the smallest organizational unit.

TISSUE MIMICRY

Figure 41.7B illustrates the concept of tissue mimicry in tissue engineering of liver (described in detail in Chapter 2 and Part IV). On the left is a depiction of the blood flow in the liver acinus, the smallest physiologic unit of the liver. Peripheral blood from the intestine, spleen, and lungs is passed into the liver acinus by way of portal triads, then through the liver cell mass by way of sinusoids, and back to the central circulation through the central vein. The sinusoids consist of endothelial cells that sandwich a double layer of parenchymal cells, the hepatic plates; endothelia and parenchymal cells are separated by the space of Disse in which the liver's extracellular matrix (i.e., insoluble components) is found. Therefore, endothelial cells experience some level of shear force due to the convective flow of blood perpendicular to its cell surface. By contrast, the parenchymal cells experience very little shear force, and nutrients are delivered primarily by diffusion of interstitial fluid through fenestrated pores (1–2 μm) in the endothelial wall and through the extracellular matrix in the space of Disse. On the right side of Fig. 41.7B is an illustration of the primary functioning unit of a coaxial hollow-fiber bioreactor (Macdonald *et al.,* 1998) that mimics the liver acinus. The arrows depict the flow of media through the bioreactor. If plasma from a patient was being passed through the bioreactor, its flow would be similar to that of the media. As noted, oxygen becomes limiting by 200 μm. However, the radius of the liver acinus is 500 μm and the hepatic plates are less than 200 μm thick (Fig. 41.7B, left side). If radial flow can mimic liver acinar blood flow as depicted in Fig. 41.7B (right side), then the coaxial fibers could be separated by a distance similar to that of the liver acinus radius (i.e., 500 μm). Otherwise, an annular distance similar to that whereby oxygen becomes limiting by diffusion alone (~200 μm) must be used, and thus effectively mimicking liver tissue at the lower tissue level, that of the hepatic plates. Therefore, the *in situ* flow dynamics must be characterized before the optimum coaxial bioreactor dimensions can be determined. The use of noninvasive *in situ* techniques to determine physical parameters in bioartificial organs is discussed below.

SCALE-UP

Typically, one empirically determines or models the optimum mass transfer parameters for the smallest repeatable unit and then scales these units up to a sufficiently sized bioreactor. Some practical problems of scale-up are mass transfer, head pressures, and shear forces. For example, practical aspects of conventional hollow-fiber bioreactor construction can confound the theoretically calculated mass transfer parameters, such as inhomogeneous fiber spacing (Chresand *et al.,* 1988), and large voids within the fiber bundle, and between the fiber bundle and housing. In addition, the large flow rate necessary to alleviate axial oxygen gradients creates large head pressures and shear forces. Cells rest between two fibers in the coaxial bioreactor described in Fig. 41.7B (right side), which prevents hypoxic patches from forming within the fiber bundle and between the housing and fiber bundle.

Figure 41.7C illustrates the principle of scale-up in tissue engineering of liver, and the ease with which the coaxial design can be scaled-up to generate an entire liver necessary for clinical liver support. With present biocompatible materials, this would be an extracorporeal device and a "bioartificial liver" because the bioreactor is composed of an artificial component. Because bioartificial livers are described elsewhere in this book, this will be discussed briefly as they relate to tissue mimicry and scale-up. Biodegradable scaffolds leading to implantable liver tissue were previously discussed in detail.

BIOARTIFICIAL LIVER

Most bioartificial liver (BAL) designs incorporate the principles of tissue mimicry and scale-up and have been extensively reviewed (Nyberg *et al.,* 1992; Kasai *et al.,* 1994; Bader *et al.,* 1995; Demetriou *et al.,* 1995; Anand, 1996; Dixit and Gitnick, 1996; Fremond *et al.,* 1996; Gerlach *et al.,* 1996; Hughes and Williams, 1996a; Sussman and Kelly, 1996; Jaregui *et al.,* 1997; Reid, 1997a; Macdonald *et al.,* 1999). Many other bioartificial liver designs rely on tissue mimicry of the hepatic plates, such as multiplate bioreactors (Yarmush *et al.,* 1992; Bader *et al.,* 1995; Smith *et al.,* 1997), biodegradable scaffolds of a complex interconnecting network (Griffith *et al.,* 1997), spirally wound bioreactors (Flendrig *et al.,* 1997), and hollow-fiber bioreactors (Nyberg *et al.,* 1992). Lower levels of organization, such as spheroids, have been encapsulated (Fremond *et al.,* 1996) and perfused in various bioreactor types, especially for transplantation (Macdonald *et al.,* 1999).

In tissues, hemoglobin acts as an oxygen and carbon dioxide carrier. Confounding the oxygen mass transfer problem in a BAL is the high oxygen requirement of parenchymal cells. A common feature in the most recently developed BALs is integral oxygenation, or oxygenation within the bioreactor proper (Macdonald *et al.,* 1999). Novel multiplate (Smith *et al.,* 1997), spirally wound (Flendrig *et al.,* 1997), and hollow-fiber (Gerlach, 1994; Goffe, 1997) bioreactors all have a compartment for oxygenation within the bioreactor. The more compartments in a BAL design, the more physiologic functions can be performed. The mimicry of arterial–venous flow (Knazek *et al.,* 1980) has introduced convective flow created by the peripheral-to-central circulation (Fig. 41.7A). In conventional two-compartment hollow-fiber bioreactors a convective flow can be created using Starling flow in conjunction with various flow configurations (Macdonald *et al.,* 1999). Starling flow is the convective component of diffusivity responsible for flow of nutrients from blood vessels to capillary beds. However, membranes foul when the bioreactor is completely filled with cells, and convective flow is difficult to maintain, often resulting in burst fibers (Callies *et al.,* 1994). Even in bioreactors with more than two compartments (Gerlach, 1994; Dixit, 1995), membrane fouling occurs, but across the cellular compartment convection can be more easily controlled by creating a pressure drop between two compartments sandwiching the cell compartment. Four functional bioreactor compartments are necessary to have integral oxygenation and regulated peripheral-to-central circulation. The multicompartment hollow-fiber bioreactors such as that described by Gerlach (1994) and others consists of at least four functional compartments (MacDonald *et al.,* 1999).

TECHNICAL ASPECTS OF TISSUE ENGINEERING

CRYOPRESERVATION

Tissue engineering of human tissues and human cell therapies or *ex vivo* gene therapies will be severely constrained until successful cryopreservation methods for stock cells are developed. There is considerable interest in developing these technologies at present. Although bone marrow samples are routinely cryopreserved for clinical programs, the ability to cryopreserve liver cells with retention of function is still at an impasse (Kravchenko *et al.,* 1989; Koebe *et al.,* 1990, 1996; Bischof and Rubinsky, 1993; Ek *et al.,* 1993a,b; Rozga *et al.,* 1993; Bischof *et al.,* 1997; Pazhayannur and Bischof, 1997). Common features to the most successful efforts to date are the use of dimethyl sulfoxide, the addition of Viaspan (University of Wisconsin solution) or the embedding of the cells in collagen gels, careful equilibration of the cells in cryopreservation buffers, and the use of computer-controlled cryopreservation equipment. Although these improvements have resulted in high viabilities (above 75%) of the cells when assessed immediately after thawing and some success with attachment efficiencies onto matrix-coated plates, the cells still do not survive for longer than 24–36 hr after thawing, do not grow, and do not have full functions. Clearly, this is an area in critical need of extensive research efforts.

ANALYSIS OF BIOREACTOR PERFORMANCE

Conventional analyses of physical, chemical, and biologic aspects of bioreactor performance have been somewhat invasive or indirect, or require modification of the bioreactor to accommodate analysis. Recent methods of analysis, such as nuclear magnetic resonance spectroscopy and magnetic resonance imaging permit direct determination of flow dynamics and intracellular metabolites (Fernandez, 1996). The analysis of bioreactors is divided into three areas: determination of physical and biologic factors, and quality assurance.

PHYSICOCHEMICAL MEASURES: FLOW DYNAMICS AND CELL DISTRIBUTION

Flow distribution in hollow-fiber bioreactors has been determined by the use of tracers (Noda *et al.*, 1979; Park and Chang, 1986) and radioopaque dyes in conjunction with X-ray computed tomography (Takesawa *et al.*, 1988), but this is difficult to perform with *in situ* cultures, and to the best of our knowledge, there is no report using these techniques with *in situ* cultures. Diffusion-weighted (Moseley *et al.*, 1990a,b) and velocity-encoded MRI (Lee *et al.*, 1995) have been applied to determine cell distribution (Neeman *et al.*, 1991; Van Zijl *et al.*, 1991; Callies *et al.*, 1994) and flow dynamics (Hammer *et al.*, 1990; Donoghue *et al.*, 1992; Zhang *et al.*, 1995) in hollow-fiber bioreactors. Ideally, empirical measurement of flow dynamics by MRI could be integrated with mass transfer models in hollow-fiber bioreactors (Cima *et al.*, 1990) to accurately describe flow dynamics and mass transfer in bioreactors. In fact, oxygen mass transfer predictions can be evaluated in bioreactors using ^{19}F NMR in conjunction with fluorinated oxygen indicators to obtain oxygen distribution profiles (S. N. O. Williams *et al.*, 1997).

Phase-contrast MRI is a robust, commonly used technique for quantifying flow dynamics (Caprihan *et al.*, 1990). The basic underlying principle of phase-contrast MRI is summarized below. Because the rate of nuclear spin precession is proportional to the magnitude of the local magnetic field, the accumulated phase ϕ of nuclear spins located at location $\mathbf{r}$ (with respect to that of spins located at the spatial origin) in the presence of a linear magnetic field gradient $\mathbf{G}$ is given by

$$\phi = \gamma \int \mathbf{G} \cdot \mathbf{r} \, dt, \tag{1}$$

where γ is the gyromagnetic ratio of the spins. Following a Taylor series expansion of the function $\mathbf{r}$ and assuming zero acceleration and higher orders of motion, the phase term becomes

$$\phi = \gamma \int \simeq \cdot (\mathbf{r}_0 + \mathbf{v}t + \frac{I}{2}\mathbf{a}t^2 + \cdots)dt \cong \gamma \int \mathbf{G} \cdot \mathbf{r}_0 dt + \gamma \int (\mathbf{G} \cdot \mathbf{v})t \, dt. \tag{2}$$

The first term of the right-hand side is the familiar phase shift used for spatial encoding, and the added phase shift is proportional to the magnitude of flow and the first moment of the gradient pulses. Therefore, by encoding velocity in at least two different magnitudes of the gradient pulses, flow velocity can be estimated from the slope of the phase-first moment plot. The encoding gradient pulses are typically implemented as a pair of identical gradient pulses located at either side of the inversion pulse of a spin-echo pulse sequence, and the vector direction of the gradient pulses defines the direction of flow sensitization. Due to the cyclic nature of the phase function, a potential pitfall of the phase-contrast approach is that flow rates higher than the set measurement dynamic range will be "wrapped around" and thus can be erroneously interpreted. Care must be taken to avoid or correct for this complication.

Figure 41.8 shows magnetic resonance images from two coaxial bioreactors using two imaging pulse sequences, T_2 weighted (A) and velocity encoded (B) (see color plate). The T_2-weighted image defines both flow and cell distribution, whereas the velocity-encoded image defines the axial flow dynamics (note the characteristic parabolic laminar flow in the intracellular and extracellular compartments, respectively). Diffusion-weighted imaging can define the cell distribution in bioreactors (Neeman *et al.*, 1991; Van Zijl *et al.*, 1991; Callies *et al.*, 1994) and if this sequence were applied to the coaxial bioreactor depicted in Fig. 41.8 the cell mass would be the most intense portion of the image.

BIOLOGIC MEASURES: GROWTH AND FUNCTIONS

Determination of growth and function in bioreactors using conventional methods has been by analysis of components in media streams. Typically, glucose uptake rate (GUR) [Eq. (3)] and lactate production rate (LPR) [Eq. (4)] in batch mode cultures are determined by the difference in their concentrations between media changes using the following equations:

$$\text{GUR} = F(G_f - G) + V(G_1 - G), \quad G = (G_1 - G_2)/2, \tag{3}$$

$$\text{LPR} = [F(L - L_f) + V(L - L_1)]0.09, \quad L = (L_1 - L_2)/2, \tag{4}$$

where G_f, G_1, and G_2 are the concentrations of glucose in the media initially, at the end of the prior day, and in the current sample, respectively; F is the feed rate; V is the volume of the reservoir; and L_f, L_1, and L_2 are the concentrations of lactate in the media initially, at the end of the prior day, and in the current sample, respectively. Some measure of oxidative metabolism can be determined by the GUR:LPR ratio plotted daily. GUR is not a reliable measure of growth in functional liver cultures because parenchymal cells are glucogenic and one could derive a negative GUR even though there is growth. Synthetic and biotransformatory functions of engineered liver are typically determined by quantification of protein production rates and phase I and II metabolism, respectively (Macdonald *et al.,* 1999). Clinical liver function assays are often used, but their meaning and significance to bioreactor performance are fuzzy. More robust liver function assays are needed (Hughes and Williams, 1996b). Growth and biotransformatory functions can be robustly measured noninvasively by multinuclear NMR spectroscopy (Fernandez, 1996).

Growth and function are generally determined by measuring nutrients, xenobiotics, or proteins in media streams, but NMR spectroscopy can noninvasively measure growth and functions in real-time within bioreactor cultures (Mancuso *et al.,* 1990; Gillies *et al.,* 1991; Macdonald *et al.,* 1998) or by in-line media streams analysis (O'Leary *et al.,* 1987). The design of protocols to test liver functions will be difficult (Hughes and Williams, 1996a), and the optimum and minimum number of tests for liver function and quality control are still being debated. Recently, ^{31}P/^{23}Na, ^{13}C, and ^{19}F nuclear magnetic resonance (NMR) spectroscopies have been applied to measure growth (Mancuso *et al.,* 1990; Gillies *et al.,* 1991), metabolic rates (Mancuso *et al.,* 1998), and metabolic profiles (R. A. Williams *et al.,* 1997), respectively, in hollow-fiber bioreactors.

Multinuclear NMR spectroscopy utilizes stable isotopes and strong magnet fields to noninvasively measure levels of nutrients or wastes inside cells (Fernandez, 1996) or media streams (O'Leary *et al.,* 1987). Growth is typically measured by ^{31}P NMR (Hrovat *et al.,* 1985; Gillies *et al.,* 1991) or ^{23}Na NMR (Mancuso *et al.,* 1990) by monitoring changes in ATP and ^{23}Na levels over the culturing period. ^{19}F NMR can be used in conjunction with fluorinated indicators to measure oxygen profiles (S. N. O. Williams *et al.,* 1997) and intracellular ions (Murphy *et al.,* 1990; London, 1991, 1994), or by direct measure of fluorinated substrates of phase I and II enzymes (Scarfe *et al.,* 1999). ^{13}C NMR in conjunction with isotopomeric analysis can be used to measure primary metabolism in bioreactors (Mancuso *et al.,* 1998), and has been used clinically to determine liver gluconeogenesis and tricarboxylic acid flux (Jones *et al.,* 1998a,b). The reader is directed to several reviews on the use of isotopomeric analysis to determine metabolic fluxes (London, 1988; Malloy *et al.,* 1988) and as input into metabolic control analysis models (Brindle *et al.,* 1997). Indirect detection of ^{15}N NMR has been used to measure ammonia metabolism in rat brain (Kanamori *et al.* 1995), and could be used in a similar fashion for engineered liver.

QUALITY ASSURANCE

Quality assurance is necessary to define the limitations of bioreactor operation. For example, the determination of bubble point and hydraulic permeability in hollow-fiber bioreactors is necessary to assure bioreactor integrity and the maintenance of modeled hydrodynamics. MRI is a powerful tool to determine bioreactor integrity (Callies *et al.,* 1994). A quality assurance protocol should also be developed for step-by-step construction of bioreactors.

FUTURE CONSIDERATIONS

The excitement with respect to tissue engineering has invaded most biologic and clinical fields. The ability to study tissues *ex vivo* with rigorous controls and with maintenance of their normal functions will revolutionize diverse fields. The ramifications are extensive, and the following list includes only some representative future applications:

1. There will likely be much interest in defining and characterizing maturational lineages in many quiescent tissue types.

2. The delineation of expansion conditions for stem cells should greatly facilitate facets of cell and gene therapies.

3. The markers used to define hematopoietic stem cells (e.g., CD34) are proving generic for many types of early progenitors. The implications are that growth regulation and function in early progenitors have common features that cross tissue boundaries.

4. Clinical programs to address degenerative diseases and tissue failure will be revolutionized by forms of cell therapies paralleling bone marrow transplant and dependent on progenitors with high proliferative potential.

5. Gene therapies, about which there has been so much excitement, have not lived up to expectations yet, given the many difficulties in trying to apply them using targeted injectable vectors. It is likely that *ex vivo* gene therapy using modified progenitors will be far more successful.

Bioartificial organs of many tissues will be developed using the combination of new bioreactor designs, highly proliferative progenitor cell populations, and defined hormone and matrix conditions in conjunction with biodegradable polymers. Such bioartificial organs will be useful for scientific and clinical programs in toxicology, drug therapy, vaccine production, and protein production, as well as general cell and molecular biology. In addition, they will be used clinically as assist devices for patients with organ failure and/or who need drastic treatments for diseases such as cancer.

SUMMARY

Tissue engineering requires a merging of biologic and engineering techniques and principles in order to achieve its goal of tissue replacement. We have shown how extensive cell culture data on the effects of numerous soluble and insoluble factors, accumulated by generations of scientists, all attempting to reproduce the natural function of liver cells *ex vivo*, provide a wealth of information to use as a guide in designing three-dimensional culture systems. In addition, knowledge of the development and mature organization of the liver furnishes biologic principles that artificial systems can be made to mimic. Using new concepts in bioreactor design, taking advantage of new biodegradable materials and new methods of fabricating these materials, it is possible to come ever closer to the goal of reproducing natural liver function. Finally, it is clear that, from a practical standpoint, the near-term use of bioartificial livers will require the harvesting and culturing of human hepatic progenitors.

Acknowledgments

Support for the research programs summarized in the review has been provided by grants to L. M. R. from NIH (DK52851), from the North Carolina Biotechnology Center (#1897), by Vertex Pharmaceuticals, and by Renaissance Cell Technologies. Support was also derived from an NIH grant to the Center for Gastrointestinal and Biliary Disease Biology (DK34987, R. Sandler, P. I.) and by grants to J. M. and L. M. R. derived from the Johns Hopkins Center for Alternatives to Animal Testing and an NRSA award (1F32DK09713).

We wish to thank Ms. Cynthia Lodestro for administrative and technical support, Dr. Yi Wei Rong for technical support in liver perfusions, and Mr. Alan Fisher for technical support in flow ctometry. We also thank the Duke University *In Vivo* Magnetic Resonance Microscopy Facility.

Notes Added in Proof

Several recent articles are of special note. An extensive electron micrographic analysis of the Canals of Hering, thought to be the reservoir of hepatic progenitors in the liver, have revealed that they extend in a bottle-brush pattern throughout zone 1 of the liver (Thiese *et al.,* 1999). The ramifications of their findings are to alter potentially the whole issue of "streaming of liver cells" in lever turnover and regenerative processes.

Several investigators have been evaluating heterogeneity in growth potential of subpopulations of adult liver cells. Grompe and his associates have shown that there is variable growth potential of fractionated adult liver cells, although extensive growth potential has been observed with adult hepatocytes with average diameter of about 25μ (Overturf *et al.,* 1999). In addition, Yoshizato and associates (Tateno *et al.,* 2000) have shown that the only adult cells capable of clonal growth *in vitro* are in the size range of so-called "small" hepatocytes and with an average diameter of about 17 μ.

REFERENCES

Ajioka, I., Akaike, T., and Watanabe, Y. (1999). Expression of vascular endothelial growth factor promotes colonization, vascularization, and growth of transplanted hepatic tissues in the mouse. *Hepatology* **29**, 396–402.

Anand, A. C. (1996). Bioartificial livers: The state of the art. *Trop. Gastroenterol.* **17**, 202–211.

Anderson, J. M. (1998). Biocompatibility of tissue-engineered implants. *In* "Frontiers in Tissue Engineering" (C. W. Patrick, Jr., A. G. Mikos, and L. V. McIntire, eds.), pp. 152–165. Pergamon, Oxford and New York.

Arias, I. M., Boyer, J. L., Jakoby, W. B., Fausto, N., Schachter, D., and Shafritz, D. A., eds. (1994). "The Liver: Biology and Pathobiology." Raven Press, New York.

Assy, N., Spira, G., Paizi, M., Shenkar, L., Kraizer, Y., Cohen, T., Neufeld, G., Dabbah, B., Enat, R., and Baruch, Y. (1999). Effect of vascular endothelial growth factor on hepatic regenerative activity following partial hepatectomy in rats. *J. Hepatol.* **30**, 911–915.

Babensee, J. E., Anderson, J. M., McIntire, L. V., and Mikos, A. G. (1998). Host responses to tissue engineered devices. *Adv. Drug Delivery Rev.* **33**, 111–139.

Bader, A., Knop, E., Boker, K., Fruhauf, N., Schuttler, W., Oldhafer, K., Burkhard, R., Pichlmayr, R., and Sewing, K. F. (1995). A novel bioreactor design for *in vitro* reconstruction of *in vivo* liver characteristics. *Artif. Organs* **19**, 368–374.

Bader, A., Knop, E., Kern, A., Boker, K., Fruhauf, N., Crome, O., Esselmann, H., Pape, C., Kempka, G., and Sewing, K. F. (1996). 3-D coculture of hepatic sinusoidal cells with primary hepatocytes-design of an organotypical model. *Exp. Cell Res.* **226**, 223–233.

Baldwin, S. P., and Saltzman, W. M., (1998). Materials for protein delivery in tissue engineering. *Adv. Drug Delivery Rev.* **33**, 71–86.

Baquerizo, A., Mhoyan, A., Shirwan, H., Swensson, J., Busuttil, R. W., Demetriou, A. A., and Cramer, D. V. (1997). Xenoantibody response of patients with severe acute liver failure exposed to porcine antigens following treatment with a bioartificial liver. *Transplant. Proc.* **29**, 964–965.

Baquerizo, A., Mhoyan, A., Kearns-Jonker, M., Arnaout, W. S., Shackleton, C., Busuttil, R. W., Demetriou, A. A., and Cramer, D. V. (1999). Characterization of human xenoreactive antibodies in liver failure patients exposed to pig hepatocytes after bioartificial liver treatment: An *ex vivo* model of pig to human xenotransplantation. *Transplantation* **67**, 5–18.

Berthiaume, F., Moghe, P. V., Toner, M., and Yarmush, M. L. (1996). Effect of extracellular matrix topology on cell structure, function, and physiological responsiveness: Hepatocytes cultured in a sandwich configuration. *FASEB J.* **10**, 1471–1484.

Bhatia, S. N., Toner, M., Tompkins, R. G., and Yarmush, M. L. (1994). Selective adhesion of hepatocytes on patterned surfaces. *Ann. N.Y. Acad. Sci.* **745**, 187–209.

Bischof, J. C., and Rubinsky, B. (1993). Large ice crystals in the nucleus of rapidly frozen liver cells. *Cryobiology* **30**, 597–603.

Bischof, J. C., Ryan, C. M., Tompkins, R. G., Yarmush, M. L., and Toner, M. (1997). Ice formation in isolated human hepatocytes and human liver tissue. *ASAIO J.* **43**, 271–278.

Bissell, D. M., Arenson, D. M., Maher, J. J., and Roll, F. J. (1987). Support of cultured hepatocytes by a laminin-rich gel. Evidence for a functionally significant subendothelial matrix in normal rat liver. *J. Clin. Invest.* **79**, 801–812.

Blasco, M. A., Lee, H. W., Hande, M. P., Samper, E., Lansdorp, P. M., DePinho, R. A., and Greider, C. W. (1997). Telomere shortening and tumor formation by mouse cells lacking telomerase RNA. *Cell (Cambridge, Mass.)* **91**, 25–34.

Bodnar, A. G., Ouellette, M., Frolkis, M., Holt, S. E., Chiu, C. P., Morin, G. B., Harley, C. B., Shay, J. W., Lichtsteiner, S., and Wright, W. E. (1998). Extension of life-span by introduction of telomerase into normal human cells. *Science* **279**, 394–352.

Brill, S., Holst, P. A., Zvibel, I., Fiorino, A., Sigal, S. H., Somasundaran, U., and Reid, L. M. (1994). Extracellular matrix regulation of growth and gene expression in liver cell lineages and hepatomas. *In* "Liver Biology and Pathobiology" (I. M. Arias, J. L. Boyer, N. Fausto, W. B. Jakoby, D. Schachter, and D. A. Shafritz, eds.), pp. 869–897. Raven Press, New York.

Brill, S., Zvibel, I., and Reid, L. M. (1995). Maturation-dependent changes in the regulation of liver-specific gene expression in embryonal versus adult primary liver cultures. *Differentiation (Berlin)* **59**, 95–102, published erratum: **59**(5), 331.

Brill, S., Zvibel, I., and Reid, L. M. (1999). Expansion conditions for early hepatic progenitor cells from embryonal and neonatal rat livers. *Dig. Dis. Sci.* **44**, 364–371.

Brindle, K. M., Fulton, S. M., Gillham, H., and Williams, S. P. (1997). Studies of metabolic control using NMR and molecular genetics. *J. Mol. Recognition* **10**, 182–187.

Brinster, R. L. (1974). The effect of cells transferred into the mouse blastocyst on subsequent development. *J. Exp. Med.* **140**, 1049–1056.

Bruder, S. P., Fink, D. J., and Caplan, A. I. (1994). Mesenchymal stem cells in bone development, bone repair, and skeletal regeneration therapy. *J. Cell. Biochem.* **56**, 283–294.

Callies, R., Jackson, M. E., and Brindle, K. M. (1994). Measurements of the growth and distribution of mammalian cells in a hollow-fiber bioreactor using nuclear magnetic resonance imaging. *Bio/Technology* **12**, 75–78.

Calof, A. L., Mumm, J. S., Rim, P. C., and Shou, J. (1998). The neuronal stem cell of the olfactory epithelium. *J. Neurobiol.* **36**, 190–205.

Caplan, A. I. (1994). The mesengenic process. *Clin. Plast. Surg.* **21**, 429–435.

Caprihan, A., Griffey, R. H., and Fukushima, E. (1990). Velocity imaging of slow coherent flows using stimulated echoes. *Magn. Reson. Med.* **15**, 327–333.

Chang, T. M. (1992). Artificial liver support based on artificial cells with emphasis on encapsulated hepatocytes. *Artif. Organs* **16**, 71–74.

Chari, R. S., Collins, B. H., Magee, J. C., DiMaio, J. M., Kirk, A. D., Harland, R. C., McCann, R. L., Platt, J. L., and Meyers, W. C. (1994). Brief report: Treatment of hepatic failure with *ex vivo* pig-liver perfusion followed by liver transplantation. *N. Engl. J. Med.* **331**, 234–237.

Chresand, T. J., Gillies, R. J., and Dale, B. E., (1988). Optimum fiber spacing in a hollow fiber bioreactor. *Biotech. Bioeng.* **32**, 983–992.

Cima, L. G. (1994). Polymer substrates for controlled biological interactions. *J. Cell. Biochem.* **56**, 155–161.

Cima, L. G., Blanch, H. W., and Wilke, C. R. (1990). A theoretical and experimental evaluation of a novel radial-flow hollow fiber reactor for mammalian cell culture. *Bioprocess. Eng.* **5**, 19–30.

Cima, L. G., Vacanti, J. P., Vacanti, C., Ingber, D., Mooney, D., and Langer, R. (1991). Tissue engineering by cell transplantation using degradable polymer substrates. *J. Biomech. Eng.* **113**, 143–151.

Coger, R., Toner, M., Moghe, P., Ezzell, R. M., and Yarmush, M. L. (1997). Hepatocyte aggregation and reorganization of EHS matrix gel. *Tissue Eng.* **3**, 375–390.

Cotsarelis, G., Kaur, P., Dhouailly, D., Hengge, U., and Bickenbach, J. (1999). Epithelial stem cells in the skin: Definition, markers, localization and functions. *Exp. Dermatol.* **8**, 80–88.

Crosbie, O. M., Reynolds, M., McEntee, G., Traynor, O., Hegarty, J. E., and O'Farrelly, C. (1999). *In vitro* evidence for the presence of hematopoietic stem cells in the adult human liver. *Hepatology* **29**, 1193–1198.

Dai, W., Belt, J., and Saltzman, W. M. (1994). Cell-binding peptides conjugated to poly(ethylene glycol) promote neural cell aggregation. *Bio/Technology* **12**, 797–801.

Davidson, J. M., Klagsbrun, M., Hill, K. E., Buckley, A., Sullivan, R., Brewer, P. S., and Woodward, S. C. (1985). Accelerated wound repair, cell proliferation, and collagen accumulation are produced by a cartilage-derived growth factor. *J. Cell Biol.* **100**, 1219–1227.

Davis, M. W., and Vacanti, J. P. (1996). Toward development of an implantable tissue engineered liver. *Biomaterials* **17**, 365–372.

De Marzo, A. M., Nelson, W. G., Meeker, A. K., and Coffey, D. S. (1998). Stem cell features of benign and malignant prostate epithelial cells. *J. Urol.* **160**, 2381–2392.

Demetriou, A. A., Reisner, A., Sanchez, J., Levenson, S. M., Moscioni, A. D., and Chowdhury, J. R. (1988). Transplantation of microcarrier-attached hepatocytes into 90% partially hepatectomized rats. *Hepatology* **8**, 1006–1009.

Demetriou, A. A., Felcher, A., and Moscioni, A. D. (1991). Hepatocyte transplantation. A potential treatment for liver disease. *Dig. Dis. Sci.* **36**, 1320–1326.

Demetriou, A. A., Rozga, J., Podesta, L., Lepage, E., Morsiani, E., Moscioni, A. D., Hoffman, A., McGrath, M., Kong, L., Rosen, H. *et al.* (1995). Early clinical experience with a hybrid bioartificial liver. *Scand. J. Gastroenterol., Suppl.* **208**, 111–117.

Dickson, R. B., and Salomon, D. S., eds. (1998). "Hormones and Growth Factors in Development and Neoplasia." Wiley-Liss, New York.

Dixit, V. (1994). Development of a bioartificial liver using isolated hepatocytes. *Artif. Organs* **18**, 371–384.

Dixit, V. (1995). Transplantation of isolated hepatocytes and their role in extrahepatic life support systems. *Scand. J. Gastroenterol., Suppl.* **208**, 101–110.

Dixit, V., and Gitnick, G. (1996). Artificial liver support: State of the art. *Scand. J. Gastroenterol., Suppl.* **220**, 101–114.

Dixit, V., Darvasi, R., Arthur, M., Brezina, M., Lewin, K., and Gitnick, G. (1990). Restoration of liver function in Gunn rats without immunosuppression using transplanted microencapsulated hepatocytes. *Hepatology* **12**, 1342–1349.

Doerr, R., Zvibel, I., Chiuten, D., D'Olimpio, J., and Reid, L. M. (1989). Clonal growth of tumors on tissue-specific biomatrices and correlation with organ site specificity of metastases. *Cancer Res.* **49**, 384–392.

Donoghue, C., Brideau, M., Newcomer, P., Pangrle, B., DiBiasio, D., Walsh, E., and Moore, S. (1992). Use of magnetic resonance imaging to analyze the performance of hollow-fiber bioreactors. *Ann. N.Y. Acad. Sci.* **665**, 285–300.

Douarin, N. M. (1975). An experimental analysis of liver development. *Med. Biol.* **53**, 427–455.

Draget, K. I., Varum, K. M., Moen, E., Gynnild, H., and Smidsrod, O. (1992). Chitosan cross-linked with Mo(VI) polyoxyanions: A new gelling system. *Biomaterials* **13**, 635–638.

Eichmann, A., Corbel, C., Nataf, V., Vaigot, P., Breant, C., and Le Douarin, N. M. (1997). Ligand-dependent development of the endothelial and hemopoietic lineages from embryonic mesodermal cells expressing vascular endothelial growth factor receptor 2. *Proc. Natl. Acad. Sci. U.S.A.* **94**, 5141–5146.

Eiselt, P., Kim, B. S., Chacko, B., Isenberg, B., Peters, M. C., Greene, K. G., Roland, W. D., Loebsack, A. B., Burg, K. J., Culberson, C., Halberstadt, C. R., Holder, W. D., and Mooney, D. J. (1998). Development of technologgies aiding large-tissue engineering. *Biotechnol. Prog.* **14**, 134–140.

Ek, S., Ringden, O., Markling, L., and Westgren, M. (1993a). Cryopreservation of fetal stem cells. *Bone Marrow Transplant.* **11**, 123.

Ek, S., Ringden, O., Markling, L., Dahlberg, N., Pschera, H., Seiger, A., Sundstrom, E., and Westgren, M. (1993b). Effects of cryopreservation on subsets of fetal liver cells. *Bone Marrow Transplant.* **11**, 395–398.

Elcin, Y. M., Dixit, V., and Gitnick, G. (1998). Hepatocyte attachment on biodegradable modified chitosan membranes: *In vitro* evaluation for the development of liver organoids. *Artif. Organs* **22**, 837–846.

Elcin, Y. M., Dixit, V., Lewin, K., and Gitnick, G. (1999). Xenotransplantation of fetal porcine hepatocytes in rats using a tissue engineering approach. *Artif. Organs* **23**, 146–152.

Evans, M. J., and Kaufman, M. H. (1981). Establishment in culture of pluripotential cells from mouse embryos. *Nature (London)* **292**, 154–156.

Farber, E. (1995). Cell proliferation as a major risk factor for cancer: A concept of doubtful validity. *Cancer Res.* **55**, 3759–3762.

Fernandez, E. J. (1996). Nuclear magnetic resonance spectroscopy and imaging. *In* "Immobilized Living Cell Systems" (R. G. Willaert, G. V. Baron, and L. DeBacker, eds.), pp. 117–146. Wiley, New York.

Flendrig, L. M., la Soe, J. W., Jorning, G. G., Steenbeek, A., Karlsen, O. T., Bovee, W. M., Ladiges, N. C., te Velde, A. A., and Chamuleau, R. A. (1997). *In vitro* evaluation of a novel bioreactor based on an integral oxygenator and a spirally wound nonwoven polyester matrix for hepatocyte culture as small aggregates. *J. Hepatol.* **26,** 1379–1392.

Folkman, J., and Klagsburn, M. (1987). Angiogenic factors. *Science* **235,** 442–447.

Fremond, B., Malandain, C., Guyomard, C., Chesne, C., Guillouzo, A., and Campion, J. P. (1993). Correction of bilirubin conjugation in the Gunn rat using hepatocytes immobilized in alginate gel beads as an extracorporeal bioartificial liver. *Cell Transplant.* **2,** 453–460.

Fremond, B., Joly, A., Desille, M., Desjardins, J. F., Campion, J. P., and Clement, B. (1996). Cell-based therapy of acute liver failure: The extracorporeal bioartificial liver. *Cell Biol. Toxicol.* **12,** 325–329.

Gebhardt, R. (1988). Different proliferative activity *in vitro* of periportal and perivenous hepatocytes. *Scand. J. Gastroenterol., Suppl.* **151,** 8–18.

Gerber, H. P., Hillan, K. J., Ryan, A. M., Kowalski, J., Keller, G. A., Rangell, L., Wright, B. D., Radtke, F., Aguet, M., and Ferrara, N. (1999). VEGF is required for growth and survival in neonatal mice. *Development (Cambridge, UK)* **126,** 1149–1159.

Gering, M., Rodaway, A. R., Gottgens, B., Patient, R. K., and Green, A. R. (1998). The SCL gene specifies haemangioblast development from early mesoderm. *EMBO J.* **17,** 4029–4045.

Gerlach, J. C. (1994). Use of hepatocyte cultures for liver support bioreactors. *Adv. Exp. Med. Biol.* **368,** 165–171.

Gerlach, J. C., Schnoy, N., Vienken, J., Smith, M., and Neuhaus, P. (1996). Comparison of hollow fibre membranes for hepatocyte immobilisation in bioreactors. *Int. J. Artif. Organs* **19,** 610–616.

Gilbert, J. C., Takada, T., Stein, J. E., Langer, R., and Vacanti, J. P. (1993). Cell transplantation of genetically altered cells on biodegradable polymer scaffolds in syngeneic rats. *Transplantation* **56,** 423–427.

Gilding, D. K. (1981). Biodegradable polymers. *In* "Biocompatibility of Clinical Implant Materials" (D. F. Williams, ed.), Vol. 2, pp. 209–232. CRC Press, Boca Raton, FL.

Gillies, R. J., Scherer, P. G., Raghunand, N., Okerlund, L. S., Martinez-Zaguilan, R., Hesterberg, L., and Dale, B. E. (1991). Iteration of hybridoma growth and productivity in hollow fiber bioreactors using ^{31}P NMR. *Magn. Reson. Med.* **18,** 181–192.

Giordano, R. A., Wu, B. M., Borland, S. W., Cima, L. G., Sachs, E. M., and Cima, M. J. (1996). Mechanical properties of dense polylactic acid structures fabricated by three dimensional printing. *J. Biomater. Sci., Polym. Ed.* **8,** 63–75.

Gittes, G. K., Galante, P. E., Hanahan, D., Rutter, W. J., and Debase, H. T. (1996). Lineage-specific morphogenesis in the developing pancreas: Role of mesenchymal factors. *Development (Cambridge, UK)* **122,** 439–447.

Goffe, R. (1997). High performance cell culture bioreactor and method. United States of America.

Gomez-Lechon, M. J., Jover, R., Donato, T., Ponsoda, X., Rodriguez, C., Stenzel, K. G., Klocke, R., Paul, D., Guillen, I., Bort, R., and Castell, J. V. (1998). Long-term expression of differentiated functions in hepatocytes cultured in three-dimensional collagen matrix. *J. Cell. Physiol.* **177,** 553–562.

Greider, C. W. (1998). Telomeres and senescence: The history, the experiment, the future. *Curr. Biol.* **8,** R178–R181.

Griffith, L. G., and Lopina, S. (1998). Microdistribution of substratum-bound ligands affects cell function: Hepatocyte spreading on PEO-tethered galactose. *Biomaterials* **19,** 979–986.

Griffith, L. G., Wu, B., Cima, M. J., Powers, M. J., Chaignaud, B., and Vacanti, J. P. (1997). *In vitro* organogenesis of liver tissue. *Ann. N.Y. Acad. Sci.* **831,** 382–397.

Grisham, J. W., and Thorgeirsson, S. S. (1997). Liver stem cells. *In* "Stem Cells" (C. S. Potter, ed.), pp. 233–282. Academic Press, London.

Guillemot, F., Lo, L. C., Johnson, J. E., Auerbach, A., Anderson, D. J., and Joyner, A. L. (1993). Mammalian achaete-scute homolog 1 is required for the early development of olfactory and autonomic neurons. *Cell (Cambridge, Mass.)* **75,** 463–476.

Gumucio, J. J. (1989). "Hepatocyte Heterogeneity and Liver Function," Vol. 19. Springer International, Madrid.

Gupta, S., Kim, S. K., Vemuru, R. P., Aragona, E., Yerneni, P. R., Burk, R. D., and Rha, C. K. (1993). Hepatocyte transplantation: An alternative system for evaluating cell survival and immunoisolation. *Int. J. Artif. Organs* **16,** 155–163.

Gupta, S., Rajvanshi, P., Sokhi, R. P., Vaidya, S., Irani, A. N., and Gorla, G. R. (1999). Position-specific gene expression in the liver lobule is directed by the microenvironment and not by the previous cell differentiation state. *J. Biol. Chem.* **274,** 2157–2165.

Gurdon, J. B. (1962). The developmental capacity of nuclei taken from intestinal epithelial cells of feeding tadpoles. *J. Embryol. Exp. Morphol.* **10,** 622–640.

Gurdon, J. B., Laskey, R. A., and Reeves, O. R. (1975). The developmental capacity of nuclei transplanted from keratinized skin cells of adult frogs. *J. Embryol. Exp. Morphol.* **34,** 93–112.

Gutsche, A. T., Parsons-Wingerter, P., Chand, D., Saltzman, W. M., and Leong, K. W. (1994). N-Acetylglucosamine and adenosine derivatized surfaces for cell culture: 3T3 fibroblast and chicken hepatocyte response. *Biotechnol. Bioeng.* **43,** 801–809.

Gutsche, A. T., Lo, H., Zurlo, J., Yager, J., and Leong, K. W. (1996). Engineering of a sugar-derivatized porous network for hepatocyte culture. *Biomaterials* **17,** 387–393.

Guyomard, C., Rialland, L., Fremond, B., Chesne, C., and Guillouzo, A. (1996). Influence of alginate gel entrapment and cryopreservation on survival and xenobiotic metabolism capacity of rat hepatocytes. *Toxicol. Appl. Pharmacol.* **141,** 349–356.

Hamada, S., Namura, K., Fujita, S., Kushima, R., and Hattori, T. (1990). DNA ploidy and proliferative activity of human pulmonary epithelium. *Virchows Arch. B* **58,** 405–410.

Hamamoto, R., Yamada, K., Kamihira, M., and Iijima, S. (1998). Differentiation and proliferation of primary rat hepatocytes cultured as spheroids. *J. Biochem. (Tokyo)* **124,** 972–979.

Hammer, B. E., Heath, C. A., Mirer, S. D., and Belfort, G. (1990). Quantitative flow measurements in bioreactors by nuclear magnetic resonance imaging. *Bio/Technology* **8,** 327–330.

Heimfeld, S., and Weissman, I. L. (1991). Development of mouse hematopoietic lineages. *Curr. Top. Dev. Biol.* **25**, 155–175.

Hentsch, B., Lyons, I., Li, R., Hartley, L., Lints, T. J., Adams, J. M., and Harvey, R. P. (1996). Hlx homeo box gene is essential for an inductive tissue interaction that drives expansion of embryonic liver and gut. *Genes Dev.* **10**, 70–79.

Hirai, S., Kasai, S., and Mito, M. (1993). Encapsulated hepatocyte transplantation for the treatment of D-glactosamine-induced acute hepatic failure in rats. *Eur. Surg. Res.* **25**, 193–202.

Hixson, D. C., Faris, R. A., and Thompson, N. L. (1990). An antigenic portrait of the liver during carcinogenesis. *Pathobiology* **58**, 65–77.

Hrovat, M. I., Wade, C. G., and Hawkes, S. P. (1985). A space-efficient assembly for NMR experiments on anchorage-dependent cell. *J. Magn. Reson.* **61**, 409–417.

Hu, W. S., Friend, J. R., Wu, F. J., Sielaff, T., Peshwa, M. V., Lazar, A., Nyberg, S. L., Remmel, R. P., and Cerra, F. B. (1997). Development of bioartificial liver employing xenogeneic hepatocytes. *Cytotechnology* **23**, 29–38.

Huang, S., Chen, C. S., and Ingber, D. E. (1998). Control of cyclin D1, p27(Kip1), and cell cycle progression in human capillary endothelial cells by cell shape and cytoskeletal tension. *Mol. Biol. Cell* **9**, 3179–3193.

Hubbell, J. A. (1995). Biomaterials in tissue engineering. *Bio/Technology* **13**, 565–576.

Hughes, R. D., and Williams, R. (1996a). Assessment of bioartificial liver support in acute liver failure. *Int. J. Artif. Organs* **19**, 3–6.

Hughes, R. D., and Williams, R. (1996b). Use of bioartificial and artificial liver support devices. *Semin. Liver Dis.* **16**, 435–444.

Hynes, R. O. (1992). Integrins: Versatility, modulation, and signaling in cell adhesion. *Cell (Cambridge, Mass.)* **69**, 11–25.

Ignatius, A. A., and Claes, L. E. (1996). *In vitro* biocompatibility of bioresorbable polymers: poly(L,DL-lactide) and poly(L-lactide-co-glycolide). *Biomaterials* **17**, 831–839.

Ingber, D. (1998). In search of cellular control: Signal transduction in context. *J. Cell. Biochem., Suppl.* **30–31**, 232–237.

Ingber, D. E., Prusty, D., Sun, Z., Betensky, H., and Wang, N. (1995). Cell shape, cytoskeletal mechanics, and cell cycle control in angiogenesis. *J. Biomech.* **28**, 1471–1484.

Ito, Y., and Chang, T. M. (1992). *In vitro* study of multicellular hepatocyte spheroids formed in microcapsules. *Artif. Organs* **16**, 422–427.

Jaregui, H. O., Mullon, C. J.-P., and Soloman, B. A. (1997). Extracorporeal liver support. *In* "Textbook of Tissue Engineering" (R. P. Lanza, R. Langer, and W. L. Chick, eds.), pp. 463–479. R. G. Landes/Academic Press, Austin, TX.

Jefferson, D. M., Clayton, D. F., Darnell, J. E., Jr., and Reid, L. M. (1984). Posttranscriptional modulation of gene expression in cultured rat hepatocytes. *Mol. Cell. Biol.* **4**, 1929–1934.

Jones, J. G., Carvalho, R. A., Franco, B., Sherry, A. D., and Malloy, C. R. (1998a). Measurement of hepatic glucose output, krebs cycle, and gluconeogenic fluxes by NMR analysis of a single plasma glucose sample. *Anal. Biochem.* **263**, 39–45.

Jones, J. G., Solomon, M. A., Sherry, A. D., Jeffrey, F. M., and Malloy, C. R. (1998b). ^{13}C NMR measurements of human gluconeogennic fluxes after ingestion of [U-^{13}C]propionate, phenylacetate, and acetaminophen. *Am. J. Physiol.* **275**, E843–E852.

Jones, P. H., and Watt, F. M. (1993). Separation of human epidermal stem cells from transit amplifying cells on the basis of differences in integrin function and expression. *Cell (Cambridge, Mass.)* **73**, 713–724.

Jung, J., Zheng, M., Goldfarb, M., and Zaret, K. S. (1999). Initiation of mammalian liver development from endoderm by fibroblast growth factors. *Science* **284**, 1998–2003.

Kanamori, K., Ross, B. D., and Kuo, E. (1995). Dependence of *in vivo* glutamine synthetase activity on ammonia concentration in rat brain studies by ^{1}H–^{15}N heteronuclear multiple-quantum coherence transfer NMR. *Biochem. J.* **311**, 681–688.

Kasai, S., Sawa, M., and Mito, M. (1994). Is the biological artificial liver clinically applicable? A historic review of biological artificial liver support systems. *Artif. Organs* **18**, 348–354.

Kato, Y., Tani, T., Sotomaru, Y., Kurokawa, K., Kato, J., Doguchi, H., Yasue, H., and Tsunoda, Y. (1998). Eight calves cloned from somatic cells of a single adult. *Science* **282**, 2095–2098.

Kaufmann, P. M., Heimrath, S., Kim, B. S., and Mooney, D. J. (1997). Highly porous polymer matrices as a three-dimensional culture system for hepatocytes: Initial results. *Transplant. Proc.* **29**, 2032–2034.

Kawase, E., Suemori, H., Takahashi, N., Okazaki, K., Hashimoto, K., and Nakatsuji, N. (1994). Strain difference in establishment of mouse embryonic stem (ES) cell lines. *Int. J. Dev. Biol.* **38**, 385–390.

Kawase, M., Michibayashi, N., Nakashima, Y., Kurikawa, N., Yagi, K., and Mizoguchi, T. (1997). Application of glutaraldehyde-crosslinked chitosan as a scaffold for hepatocyte attachment. *Biol. Pharm. Bull.* **20**, 708–710.

Kay, M. A., and Fausto, N. (1997). Liver regeneration: Prospects for therapy based on new technologies. *Mol. Med. Today* **3**, 108–115.

Kempermann, G., and Gage, F. H. (1999). New nerve cells for the adult brain. *Sci. Am.* **280**, 48–53.

Kennedy, M., Firpo, M., Choi, K., Wall, C., Robertson, S., Kabrun, N., and Keller, G. (1997). A common precursor for primitive erythropoiesis and definitive haematopoiesis. *Nature (London)* **386**, 488–493.

Kennedy, S., Rettinger, S., Flye, M. W., and Ponder, K. P. (1995). Experiments in transgenic mice show that hepatocytes are the source for postnatal liver growth and do not stream. *Hepatology* **22**, 160–168.

Kim, B. S., and Mooney, D. J. (1998). Development of biocompatible synthetic extracellular matrices for tissue engineering. *Trends Biotechnol.* **16**, 224–230.

Kim, S. S., Utsunomiya, H., Koski, J. A., Wu, B. M., Cima, M. J., Sohn, J., Mukai, K., Griffith, L. G., and Vacanti, J. P. (1998). Survival and function of hepatocytes on a novel three-dimensional synthetic biodegradable polymer scaffold with an intrinsic network of channels. *Ann. Surg.* **228**, 8–13.

Kirby, S. L., Cook, D. N., Walton, W., and Smithies, O. (1996). Proliferation of multipotent hematopoietic cells controlled by a truncated erythropoietin receptor transgene. *Proc. Nat. Acad. Sci. U.S.A.* **93**, 9402–9407.

Knazek, R. A., Gullino, P. M., and Frankel, D. S. (1980). Method of simulation of lymphatic drainage utilizing a dual circuit, woven artificial capillary bundle. United States of America.

Knop, E., Bader, A., Boker, K., Pichlmayr, R., and Sewing, K. F. (1995). Ultrastructural and functional differentiation of hepatocytes under long-term culture conditions. *Anat. Rec.* **242**, 337–349.

Kobayashi, A., Kobayashi, K., and Akaike, T. (1992). Control of adhesion and detachment of parenchymal liver cells using lactose-carrying polystyrene as substratum. *J. Biomater. Sci. Polym. Ed.* **3**, 499–508.

Kobayashi, K., Kobayashi, A., and Akaike, T. (1994). Culturing hepatocytes on lactose-carrying polystyrene layer via asialoglycoprotein receptor-mediated interactions. *In* "Methods in Enzymology" (Y. C. Lee and R. T. Lee, eds.), Vol. **247**, pp. 409–418. Academic Press, San Diego, CA.

Koebe, H. G., Dunn, J. C., Toner, M., Sterling, L. M., Hubel, A., Cravalho, E. G., Yarmush, M. L., and Tompkins, R. G. (1990). A new approach to the cryopreservation of hepatocytes in a sandwich culture configuration. *Cryobiology* **27**, 576–584.

Koebe, H. G., Dahnhardt, C., Muller-Hocker, J., Wagner, H., and Schildberg, F. W. (1996). Cryopreservation of porcine hepatocyte cultures. *Cryobiology* **23**, 127–141.

Koide, N., Shinji, T., Tanabe, T., Asano, K., Kawaguchi, M., Sakaguchi, K., Koide, Y., Mori, M., and Tsuji, T. (1989). Continued high albumin production by multicellular spheroids of adult rat hepatocytes formed in the presence of liver-derived proteoglycans. *Biochem. Biophys. Res. Commun.* **161**, 385–391.

Koike, M., Matsushita, M., Taguchi, K., and Uchino, J. (1996). Function of culturing monolayer hepatocytes by collagen gel coating and coculture with nonparenchymal cells. *Artif. Organs* **20**, 186–192.

Kono, Y., Yang, S., and Roberts, E. A. (1997). Extended primary culture of human hepatocytes in a collagen gel sandwich system. *In Vitro Cell. Dev. Biol.: Anim.* **33**, 467–472.

Kravchenko, L. P., Semenchenko, O. A., and Andrienko, A. N. (1989). [The cryopreservation of isolated liver cells]. *Tsitologiya* **31**, 1245–1248.

Kuhl, P. R., and Griffith-Cima, L. G. (1996). Tethered epidermal growth factor as a paradigm for growth factor-induced stimulation from the solid phase. *Nat. Med.* **2**, 1022–1027; published erratum: 3(1), 93 (1997).

Labosky, P. A., Barlow, D. P., and Hogan, L. M. (1994). Mouse embryonic germ (EG) cell lines: Transmission through the germline and differences in the methylation imprint of insulin-like growth factor 2 receptor (*Igf2r*) gene compared with embryonic stem cell lines. *Development (Cambridge, UK)* **120**, 3197–3204.

Lazar, A., Mann, H. J., Remmel, R. P., Shatford, R. A., Cerra, F. B., and Hu, W. S. (1995). Extended liver-specific functions of porcine hepatocyte spheroids entrapped in collagen gel. *In Vitro Cell. Dev. Biol.: Anim.* **31**, 340–346.

LeCluyse, E. L., Audus, K. L., and Hochman, J. H. (1994). Formation of extensive canalicular networks by rat hepatocytes cultured in collagen-sandwich configuration. *Am. J. Physiol.* **266**, C1764–C1774.

Lee, A. T., Pike, G. B., and Pelc, N. J. (1995). Three-point phase–contrast velocity measurements with increased velocity-to-noise ratio. *Magn. Reson. Med.* **33**, 122–126.

Lee, H. W., Blasco, M. A., Gottlieb, G. J., Horner, J. W. N., Greider, C. W., and DePinho, R. A. (1998). Essential role of mouse telomerase in highly proliferative organs. *Nature (London)* **392**, 569–574.

Liotta, L. A., Lee, C. W., and Morakis, D. J. (1980). New method for preparing large surfaces of intact human basement membrane for tumor invasion studies. *Cancer Lett.* **11**, 141–152.

Loeffler, M., Bratke, T., Paulus, U., Li, Y. Q., and Potten, C. S. (1997). Clonality and life cycles of intestinal crypts explained by a state dependent stochastic model of epithelial stem cell organization. *J. Theor. Biol.* **186**, 41–54.

London, R. E. (1988). Carbon-13 labeling in studies of metabolic regulation. *Prog. NMR Spectrosc.* **20**, 337–383.

London, R. E. (1991). Methods for measurement of intracellular magnesium: NMR and fluorescence. *Annu. Rev. Physiol.* **53**, 241–258.

London, R. E. (1994). *In vivo* NMR studies utilizing fluorinated probes. *In* "NMR in Physiology and Biomedicine" (R. Gillies, ed.), pp. 263–277. Academic Press, San Diego, CA.

Lopina, S. T., Wu, G., Merrill, E. W., and Griffith-Cima, L. (1996). Hepatocyte culture on carbohydrate-modified star polyethylene oxide hydrogels. *Biomaterials* **17**, 559–569.

Loughna, P. T., and Pell, J. M., eds. (1996). "Molecular Physiology of Growth." Cambridge University Press, Cambridge, UK and New York.

Lyon, M., Deakin, J. A., Mizuno, K., Nakamura, T., and Gallagher, J. T. (1994). Interaction of hepatocyte growth factor with heparan sulfate. *J. Biol. Chem.* **269**, 11216–11223.

Macdonald, J. M., Grillo, M., Schmidlin, O., Tajiri, D. T., and James, T. L. (1998). NMR spectroscopy and MRI investigation of a potential bioartificial liver. *NMR Biomed.* **11**, 55–66.

Macdonald, J. M., Griffin, J. P., Kubota, H., Griffith, L., Fair, J., and Reid, L. (1999). Bioartificial livers. *In* "Cell Encapsulation Technology and Therapeutics" (W. Kuhtreiber, R. P. Lanza, and W. L. Chick, eds.), pp. 252–286. Birkhaeuser, Boston.

Maher, J. J., and Bissell, D. M. (1993). Cell–matrix interactions in liver. *Semin. Cell Biol.* **4**, 189–201.

Malloy, C. R., Sherry, A. D., and Jeffrey, F. M. (1988). Evaluation of carbon flux and substrate selection through alternate pathways involving the citric acid cycle of the heart by ^{13}C NMR spectroscopy. *J. Biol. Chem.* **263**, 6964–6971.

Mancuso, A., Fernandez, E. J., Blanch, H. W., and Clark, D. S. (1990). A nuclear magnetic resonance technique for determining hybridoma cell concentration in hollow fiber bioreactors. *Bio/Technology* **8**, 1282–1285.

Mancuso, A., Sharfstein, S. T., Fernandez, E. J., Clark, D. S., and Blanch, H. W. (1998). Effect of extracellular glutamine concentration on primary and secondary metabolism of a murine hybridoma: An *in vivo* ^{13}C nuclear magnetic resonance study. *Biotechnol. Bioeng.* **57**, 172–186.

Marler, J. J., Upton, J., Langer, R., and Vacanti, J. P. (1998). Transplantation of cells in matrices for tissue regeneration. *Adv. Drug Delivery Rev.* **33**, 165–182.

Marshall, E., and Lord, B. I. (1996). Feedback inhibitors in normal and tumor tissues. *Int. Rev. Cytol.* **167**, 185–261.

Martin, G. R. (1981). Isolation of a pluripotent cell line from early mouse embryos cultured in medium conditioned by teratocarcinoma stem cells. *Proc. Natl. Acad. Sci. U.S.A.* **78**, 7634–768.

Martinez-Hernandez, A., and Amenta, P. S. (1993). Morphology, localization, and origin of the hepatic extracellular matrix. *In* "Extracellular Matrix: Chemistry, Biology, and Pathobiology with Emphasis on the Liver" (M. Zern and L. M. Reid, eds.), pp. 255–330. Dekker, New York.

Mason, R. J., Williams, M. C., Moses, H. L., Mohla, S., and Berberich, M. A. (1997). Stem cells in lung development, disease, and therapy. *Am. J. Respir. Cell Mol. Biol.* **16**, 355–363.

Matturri, L., Campiglio, G. L., Lavezzi, A. M., Riberti, C., Cavalca, D., and Azzolini, A. (1991). Cell kinetics and DNA content (ploidy) of human skin under expansion. *Eur. J. Basic Appl. Histochem.* **35**, 73–79.

McCuskey, R. A. (1994). The hepatic microvascular system. *In* "Liver Biology and Pathobiology" (I. M. Arias, J. L. Boyer, N. Fausto, W. B. Jakoby, D. Schachter, and D. A. Shafritz, eds.), pp. 1089–1106. Raven Press, New York.

McKay, I. A., and Brown, K. D., eds. (1998). "Growth Factors and Receptors: A Practical Approach." Oxford University Press, Oxford and New York.

Michalopoulos, G. K. (1995). HGF in liver regeneration and tumor promotion. *Prog. Clin. Biol. Res.* **391**, 179–185.

Michalopoulos, G. K., and DeFrances, M. C. (1997). Liver regeneration. *Science* **276**, 60–66.

Michalopoulos, G. K., Bowen, W. C., Zajac, V. F., Beer-Stolz, D., Watkins, S., Kostrubsky, V., and Strom, S. C. (1999). Morphogenetic events in mixed cultures of rat hepatocytes and nonparenchymal cells maintained in biological matrices in the presence of hepatocyte growth factor and epidermal growth factor. *Hepatology* **29**, 90–100.

Mikos, A. G., Bao, Y., Cima, L. G., Ingber, D. E., Vacanti, J. P., and Langer, R. (1993a). Preparation of poly(glycolic acid) bonded fiber structures for cell attachment and transplantation. *J. Biomed. Mater. Res.* **27**, 183–189.

Mikos, A. G., Sarakinos, G., Leite, S. M., Vacanti, J. P., and Langer, R. (1993b). Laminated three-dimensional biodegradable foams for use in tissue engineering. *Biomaterials* **14**, 323–330.

Mikos, A. G., Sarakinos, G., Lyman, M. D., Ingber, D. E., Vacanti, J. P., and Langer, R. (1993c). Prevascularization of porous biodegradable polymers. *Biotechnol. Bioeng.* **42**, 716–723.

Mikos, A. G., Thorsen, A. J., Czerwonka, L. A., Bao, Y., and Langer, R. (1994). Preparation and characterization of poly(L-lactic acid) foam. *Polymer* **35**, 1068–1077.

Mintz, B., and Illmensee, K. (1975). Normal genetically mosaic mice produced from malignant teratocarcinoma cells. *Proc. Natl. Acad. Sci. U.S.A.* **72**, 3585–3589.

Mitaka, T., Sato, F., Mizuguchi, T., Yokono, T., and Mochizuki, Y. (1999). Reconstruction of hepatic organoid by rat small hepatocytes and hepatic nonparenchymal cells. *Hepatology* **29**, 111–125.

Moghe, P. V., Berthiaume, F., Ezzell, R. M., Toner, M., Tompkins, R. G., and Yarmush, M. L. (1996). Culture matrix configuration and composition in the maintenance of hepatocyte polarity and function. *Biomaterials* **17**, 373–385.

Moghe, P. V., Coger, R. N., Toner, M., and Yarmush, M. L. (1997). Cell–cell interactions are essential for maintenance of hepatocyte function in collagen gel but not on Matrigel. *Biotechnol. Bioeng.* **56**, 706–711.

Montesano, R., Schaller, G., and Orci, L. (1991a). Induction of epithelial tubular morphogenesis *in vitro* by fibroblast-derived soluble factors. *Cell (Cambridge, Mass.)* **66**, 697–711.

Montesano, R., Matsumoto, K., Nakamura, T., and Orci, L. (1991b). Identification of a fibroblast-derived epithelial morphogen as hepatocyte growth factor. *Cell (Cambridge, Mass.)* **67**, 901–908.

Mooney, D., Hansen, L., Vacanti, J., Langer, R., Farmer, S., and Ingber, D. (1992). Switching from differentiation to growth in hepatocytes: Control by extracellular matrix. *J. Cell. Physiol.* **151**, 497–505.

Mooney, D. J., Kaufmann, P. M., Sano, K., McNamara, K. M., Vacanti, J. P., and Langer, R. (1994). Transplantation of hepatocytes using porous, biodegradable sponges. *Transplant. Proc.* **26**, 3425–3426.

Mooney, D. J., Park, S., Kaufmann, P. M., Sano, K., McNamara, K., Vacanti, J. P., and Langer, R. (1995). Biodegradable sponges for hepatocyte transplantation. *J. Biomed. Mater. Res.* **29**, 959–965.

Mooney, D. J., Kaufman, P. M., Sano, K., Schwendeman, S. P., Majahokd, K., Schloo, B., Vacanti, J. P., and Langer, R. (1996). Localized delivery of epidermal growth factor improves the survival of transplanted hepatocytes. *Biotechnol. Bioeng.* **50**, 422–429.

Mooney, D. J., Sano, K., Kaufmann, P. M., Majahod, K., Schloo, B., Vacanti, J. P., and Langer, R. (1997). Long-term engraftment of hepatocytes transplanted on biodegradable polymer sponges. *J. Biomed. Mater. Res.* **37**, 413–420.

Moore, K. A., Ema, H., and Lemischka, I. R. (1997). *In vitro* maintenance of highly purified, transplantable hematopoietic stem cells. *Blood* **89**, 4337–4347.

Morrison, S. J., Prowse, K. R., Ho, P., and Weissman, I. L. (1996). Telomerase activity in hematopoietic cells is associated with self-renewal potential. *Immunity* **5**, 207–216.

Morrison, S. J., Shah, N. M., and Anderson, D. J. (1997). Regulatory mechanisms in stem cell biology. *Cell (Cambridge, Mass.)* **88**, 287–298.

Moscioni, A. D., Roy-Chowdhury, J., Barbour, R., Brown, L. L., Roy-Chowdhury, N., Competiello, L. S., Lahiri, P., and Demetriou, A. A. (1989). Human liver cell transplantation. Prolonged function in athymic–Gunn and athymic–analbuminemic hybrid rats. *Gastroenterology* **96**, 1546–1551.

Moseley, M. E., Kucharczyk, J., Mintorovitch, J., Cohen, Y., Kurhanewicz, J., Derugin, N., Asgari, H., and Norman, D. (1990a). Diffusion-weighted MR imaging of acute stroke: Correlation with T2-weighted and magnetic susceptibility-enhanced MR imaging in cats. *AJNR: Am. J. Neuroradiol.* **11**, 423–429.

Moseley, M. E., Mintorovitch, J., Cohen, Y., Asgari, H. S., Derugin, N., Norman, D., and Kucharczyk, J. (1990b). Early detection of ischemic injury: Comparison of spectroscopy, diffusion-, T2-, and magnetic susceptibility-weighted MRI in cats. *Acta Neurochir., Suppl.* **51**, 207–209.

Mossin, L., Blankson, H., Huitfeldt, H., and Seglen, P. O. (1994). Ploidy-dependent growth and binucleation in cultured rat hepatocytes. *Exp. Cell Res.* **214**, 551–560.

Motoyama, J., Kitajima, K., Kojima, M., Kondo, S., and Takeuchi, T. (1997). Organogenesis of the liver, thymus and spleen is affected in jumonji mutant mice. *Mech. Dev.* **66**, 27–37.

Murphy, E., Levy, L., Raju, B., Steenbergen, C., Gerig, J. T., Singh, P., and London, R. E. (1990). Measurement of cytosolic calcium using ^{19}F NMR. *Environ. Health Perspect.* **84**, 95–98.

Muzzarelli, R., Baldassarre, V., Conti, F., Ferrara, P., Biagini, G., Gazzanelli, G., and Vasi, V. (1988). Biological activity of chitosan: Ultrastructural study. *Biomaterials* **9**, 247–252.

Neeman, M., Jarrett, K. A., Sillerud, L. O., and Freyer, J. P. (1991). Self-diffusion of water in multicellular spheroids measured by magnetic resonance microimaging. *Cancer Res.* **51**, 4072–4079.

Nishikawa, Y., Tokusashi, Y., Kadohama, T., Nishimori, H., and Ogawa, K. (1996). Hepatocytic cells form bile duct-like structures within a three-dimensional collagen gel matrix. *Exp. Cell Res.* **223**, 357–371.

Noda, I., Brown-West, D. G., and Gryte, C. C. (1979). Effect of flow maldistribution in hollow fiber dialysis experimental studies. *J. Membr. Sci.* **5**, 209–225.

Norrby, K. (1997). Angiogenesis: New aspects relating to its initiation and control. *APMIS* **105**, 417–437.

Nyberg, S. L., Shatford, R. A., Hu, W. S., Payne, W. D., and Cerra, F. B. (1992). Hepatocyte culture systems for artificial liver support: Implications for critical care medicine (bioartificial liver support). *Crit. Care Med.* **20**, 1157–1168.

Nyberg, S. L., Peshwa, M. V., Payne, W. D., Hu, W. S., and Cerra, F. B. (1993). Evolution of the bioartificial liver: The need for randomized clinical trials. *Am. J. Surg.* **166**, 512–521.

Ogawa, Y. (1997). Injectable microcapsules prepared with biodegradable poly(alpha-hydroxy) acids for prolonged release of drugs. *J. Biomater. Sci., Polym. Ed.* **8**, 391–409.

Oka, J. A., and Weigel, P. H. (1986). Binding and spreading of hepatocytes on synthetic galactose culture surfaces occur as distinct and separable threshold responses. *J. Cell Biol.* **103**, 1055–1060.

O'Leary, D. J., Hawkes, S. P., and Wade, C. G. (1987). Indirect monitoring of carbon-13 metabolism with NMR: Analysis of perfusate with a closed-loop flow system. *Magn. Reson. Med.* **5**, 572–577.

Overturf, K. A.-D. M., Finegold, M., Grompe, M. (1999). The repopulation potential of hepatocyte populations differing in size and prior mitotic expansion. *Am. J. Pathol.* **155**, 2135–2143.

Overturf, K., al-Dhalimy, M., Ou, C. N., Finegold, M., and Grompe, M. (1997). Serial transplantation reveals the stem-cell-like regenerative potential of adult mouse hepatocytes. *Am. J. Pathol.* **151**, 1273–1280.

Park, A., Wu, B., and Griffith, L. G. (1998). Integration of surface modification and 3D fabrication techniques to prepare patterned poly(L-lactide) substrates allowing regionally selective cell adhesion. *J. Biomater. Sci., Polym. Ed.* **9**, 89–110.

Park, J. K., and Chang, H. N. (1986). Flow distribution in the lumen side of a hollow-fiber module. *AIChE J.* **32**, 1937–1947.

Pazhayannur, P. V., and Bischof, J. C. (1997). Measurement and simulation of water transport during freezing in mammalian liver tissue. *J. Biomech. Eng.* **119**, 269–277.

Peters, M. C., and Mooney, D. J. (1997). Synthetic extracellular matrices for cell transplantation. *Mater. Sci. For.* **250**, 43–52.

Porcher, C., Swat, W., Rockwell, K., Fujiwara, Y., Alt, F. W., and Orkin, S. H. (1996). The T cell leukemia oncoprotein SCL/tal-1 is essential for development of all hematopoietic lineages. *Cell (Cambridge, Mass.)* **86**, 47–57.

Potten, C. S., ed. (1997). "Stem Cells." Academic Press, London.

Potten, C. S. (1998). Stem cells in gastrointestinal epithelium: numbers, characteristics and death. *Philos. Trans. R. Soc. London, Ser. B* **353**, 821–830.

Potter, V. R. (1978). Phenotypic diversity in experimental hepatomas: The concept of partially blocked ontogeny. The 10th Walter Hubert Lecture. *Br. J. Cancer* **38**, 1–23.

Potter, V. R. (1981). The present status of the blocked ontogeny hypothesis of neoplasia: The thalassemia connection. *Oncodev. Biol. Med.* **2**, 243–266.

Powers, M. J., and Griffith-Cima, L. (1996). Motility behavior of hepatocytes on extracellular matrix substrata during aggregation. *Biotechnol. Bioeng.* **50**, 392–403.

Rajvanshi, P., Liu, D., Ott, M., Gagandeep, S., Schilsky, M. L., and Gupta, S. (1998). Fractionation of rat hepatocyte subpopulations with varying metabolic potential, proliferative capacity, and retroviral gene transfer efficiency. *Exp. Cell Res.* **244**, 405–419.

Rathjen, P. D., Toth, S., Willis, A., Heath, J. K., and Smith, A. G. (1990). Differentiation inhibiting activity is produced in matrix-associated and diffusible forms that are generated by alternate promoter usage. *Cell (Cambridge, Mass.)* **62**, 1105–1114.

Reid, L. M. (1990). Stem cell biology, hormone/matrix synergies and liver differentiation. *Curr. Opin. Cell Biol.* **2**, 121–130.

Reid, L. M. (1996). Stem cell-fed maturational lineages and gradients in signals: Relevance to differentiation of epithelia. *Mol. Biol. Rep.* **23**, 21–33.

Reid, L. M. (1997). Stem cell/lineage biology and lineage-dependent extracellular matrix chemistry: Keys to tissue engineering of quiescent tissues such as liver. *In* "Principles of Tissue Engineering" (R. P. Lanza, R. Langer, and W. L. Chick, eds.), pp. 481–514. R. G. Landes/Academic Press, Austin, TX.

Reid, L. M., and Luntz, T. L. (1997). *Ex vivo* maintenance of differentiated mammalian cells. *In* "Basic Cell Culture Protocols" (J. W. Pollard and J. M. Walker, eds.), Vol. 75, pp. 31–57. Humana Press, Totowa, NJ.

Reynolds, A. J., and Jahoda, C. A. (1994). Hair follicle stem cells: Characteristics and possible significance. *Skin Pharmacol.* **7**, 16–19.

Riordan, S., and Williams, R. (1997). Bioartificial liver support: Developments in hepatocyte culture and bioreactor design. *Br. Med. Bull.* **53**, 730–744.

Robb, L., Elwood, N. J., Elefanty, A. G., Kontgen, F., Li, R., Barnett, L. D., and Begley, C. G. (1996). The scl gene product is required for the generation of all hematopoietic lineages in the adult mouse. *EMBO J.* **15**, 4123–4129.

Rojkind, M., Gatmaitan, Z., Mackensen, S., Giambrone, M. A., Ponce, P., and Reid, L. M. (1980). Connective tissue biomatrix: Its isolation and utilization for long-term cultures of normal rat hepatocytes. *J. Cell Biol.* **87**, 255–263.

Rozga, J., Williams, F., Ro, M. S., Neuzil, D. F., Giorgio, T. D., Backfisch, G., Moscioni, A. D., Hakim, R., and Demetriou, A. A. (1993). Development of a bioartificial liver: Properties and function of a hollow-fiber module inoculated with liver cells. *Hepatology* **17**, 258–265.

Rozga, J., Podesta, L., LePage, E., Morsiani, E., Moscioni, A. D., Hoffman, A., Sher, L., Villamil, F., Woolf, G., McGrath, M. *et al.* (1994). A bioartificial liver to treat severe acute liver failure. *Ann. Surg.* **219**, 518–544; discussion: pp. 544–546.

Scarfe, G. B., Wright, B., Clayton, E., Taylor, S., Wilson, I. D., Lindon, J. C., and Nicholson, J. K. (1999). Quantitative studies on the urinary metabolic fate of 2-chloro-4-trifluoromethylaniline in the rat using ^{19}F-NMR spectroscopy and directly coupled HPLC-NMR-MS. *Xenobiotica* **29**, 77–91.

Schuppan, D., Schmid, M., Somasundaram, R., Ackermann, R., Ruehl, M., Nakamura, T., and Riecken, E. O. (1998). Collagens in the liver extracellular matrix bind hepatocyte growth factor. *Gastroenterology* **114**, 139–152.

Seglen, P. O. (1997). DNA ploidy and autophagic protein degradation as determinants of hepatocellular growth and survival. *Cell Biol. Toxicol.* **13**, 301–315.

Selden, C., Roberts, E., Stamp, G., Parker, K., Winlove, P., Ryder, T., Platt, H., and Hodgson, H. (1998). Comparison of three solid phase supports for promoting three-dimensional growth and function of human liver cell lines. *Artif. Organs* **22**, 308–319.

Sell, S. (1993). Cellular origin of cancer: Dedifferentiation or stem cell maturation arrest? *Environ. Health Perspect.* **101**, 15–26.

Sell, S. (1994). Liver stem cells. *Mod. Pathol.* **7**, 105–112.

Sell, S. (1998). Comparison of liver progenitor cells in human atypical ductular reactions with those seen in experimental models of liver injury. *Hepatology* **27**, 317–331.

Sell, S., and Pierce, G. B. (1994). Maturation arrest of stem cell differentiation is a common pathway for the cellular origin of teratocarcinomas and epithelial cancers. *Lab. Invest.* **70**, 6–22.

Sell, S., Hunt, J. M., Knoll, B. J., and Dunsford, H. A. (1987). Cellular events during hepatocarcinogenesis in rats and the question of premalignancy. *Adv. Cancer Res.* **48**, 37–111.

Shalaby, F., Rossant, J., Yamaguchi, T. P., Gertsenstein, M., Wu, X. F., Breitman, M. L., and Schuh, A. C. (1995). Failure of blood-island formation and vasculogenesis in Flk-1-deficient mice. *Nature (London)* **376**, 62–66.

Shalaby, F., Ho, J., Stanford, W. L., Fischer, K. D., Schuh, A. C., Schwartz, L., Bernstein, A., and Rossant, J. (1997). A requirement for Flk1 in primitive and definitive hematopoiesis and vasculogenesis. *Cell (Cambridge, mass.)* **89**, 981–990.

Shiels, P. G., Kind, A. J., Campbell, K. H., Waddington, D., Wilmut, I., Colman, A., and Schnieke, A. E. (1999). Analysis of telomere lengths in cloned sheep. *Nature (London)* **399**, 316–317.

Shimbara, N., Atawa, R., Takashina, M., Tanaka, K., and Ichihara, A. (1996). Long-term culture of functional hepatocytes on chemically modified collagen gels. *Cytotechnology* **21**, 31–43.

Sigal, S. H., Brill, S., Fiorino, A. S., and Reid, L. M. (1992). The liver as a stem cell and lineage system. *Am. J. Physiol.* **263**, G139–G148.

Sigal, S. H., Brill, S., Reid, L. M., Zvibel, I., Gupta, S., Hixson, D., Faris, R., and Holst, P. A. (1994). Characterization and enrichment of fetal rat hepatoblasts by immunoadsorption ("panning") and fluorescence-activated cell sorting. *Hepatology* **19**, 999–1006.

Sigal, S. H., Gupta, S., Gebhard, D. F., Jr., Holst, P., Neufeld, D., and Reid, L. M. (1995a). Evidence for a terminal differentiation process in the rat liver. *Differentiation (Berlin)* **59**, 35–42.

Sigal, S. H., Rajvanshi, P., Reid, L. M., and Gupta, S. (1995b). Demonstration of differentiation in hepatocyte progenitor cells using dipeptidyl peptidase IV deficient mutant rats. *Cell. Mol. Biol. Res.* **41**, 39–47.

Sigal, S. H., Rajvanshi, P., Gorla, G. R., Sokhi, R. P., Saxena, R., Gebhard, D. R., Jr., Reid, L. M., and Gupta, S. (1999). Partial hepatectomy-induced polyploidy attenuates hepatocyte replication and activates cell aging events. *Am. J. Physiol.* **276**, G1260–G1272.

Smith, M. D., Airdrie, I., Cousins, R. B., Ekevall, E., Grant, M. H., and Gaylor, J. D. S. (1997). Development and characterization of a hybrid artificial liver bioreactor with integral membrane oxygenation. *In* "Bioartificial Liver Support Systems" (G. Crepaldi, A. A. Demetriou, and M. Muraca, eds.), pp. 27–35. CIC Edizioni Internazionali, Rome.

Snyder, E. Y., and Macklis, J. D. (1995). Multipotent neural progenitor or stem-like cells may be uniquely suited for therapy for some neurodegenerative conditions. *Clin. Neurosci.* **3**, 310–316.

Stamatoglou, S. C., and Hughes, R. C. (1994). Cell adhesion molecules in liver function and pattern formation. *FASEB J.* **8**, 420–427.

Stevens, L. C. (1964). Experimental production of testicularteratomas in mice. *Proc. Natl. Acad. Sci. U.S.A.* **52**, 654–661.

Stevens, L. C., and Pierce, G. B. (1975). Telatomas: Definitions and terminology. *In* "Telatomas and Differentiation" (M. I. Sherman and D. Solter, eds.), pp. 13–14. Academic Press, New York.

Styles, J. A. (1993). Measurement of ploidy and cell proliferation in the rodent liver. *Environ. Health Perspect.* **101**, 67–71.

Sussman, N. L., and Kelly, J. H. (1996). Artificial liver support. *Clin. Invest. Med. - Med. Clin. Exp.* **19**, 393–399.

Sussman, N. L., Gislason, G. T., Conlin, C. A., and Kelly, J. H. (1994). The Hepatix extracorporeal liver assist device: Initial clinical experience. *Artif. Organs* **18**, 390–396.

Takeda, T., Kim, T. H., Lee, S. K., Langer, R., and Vacanti, J. P. (1995). Hepatocyte transplantation in biodegradable polymer scaffolds using the Dalmation dog model of hyperuricosuria. *Transplant. Proc.* **27**, 635–636.

Takesawa, S., Terasawa, M., Sakagami, M., Kobayashi, T., Hidai, H., and Sakai, K. (1988). Nondestructive evaluation by X-ray computed tomography of dialysate flow patterns in capillary dialyzers. *ASAIO Trans.* **34**, 794–799.

Takeshita, K., Bowen, W. C., and Michalopoulos, G. K. (1998). Three-dimensional culture of hepatocytes in a continuously flowing medium. *In Vitro Cell. Dev. Bio.: Anim.* **34**, 482–485.

Talbot, N. C., Pursel, V. G., Rexroad, C. E., Jr., Caperna, T. J., Powell, A. M., and Stone, R. T. (1994). Colony isolation and secondary culture of fetal porcine hepatocytes on STO feeder cells. *In Vitro Cell. Dev. Biol.: Anim.* **30A**, 851–855.

Taniguchi, H., Toyoshima, T., Fukao, K., and Nakauchi, H. (1996). Presence of hematopoietic stem cells in the adult liver. *Nat. Med.* **2**, 198–203.

Taniguchi, H., Sugioka, A., Fukao, K., and Nakauchi, H. (1997). Characterization of hematopoietic stem cells in the adult liver. *Transplant. Proc.* **29**, 1212–1213.

Tateno, C. T.-K. K., Yamasaki, C., Sato, H., and Yoshizato, K. (2000). Heterogeneity of growth potential of adult rat hepatocytes *in vitro. Hepatology,* **31**, 65–74.

Theise, N. (1999). The Canals of Hering and hepatic stem cells in humans. *Hepatology* **30**, 1425–1433.

Thomas, K. A. (1996). Vascular endothelial growth factor, a potent and selective angiogenic agent. *J. Biol. Chem.* **271**, 603–606.

Thomson, J. A., Itskovitz-Eldor, J., Shapiro, S. S., Waknitz, M. A., Swiergiel, J. J., Marshall, V. S., and Jones, J. M. (1998). Embryonic stem cell lines derived from human blastocysts [see comments] *Science* **282**, 1145–1147; published erratum: p. 1827.

Tong, J. Z., De Lagausie, P., Furlan, V., Cresteil, T., Bernard, O., and Alvarez, F. (1992). Long-term culture of adult rat hepatocyte spheroids. *Exp. Cell Res.* **200**, 326–332.

Traber, P. G., Chianale, J., and Gumucio, J. J. (1988). Physiologic significance and regulation of hepatocellular heterogeneity. *Gastroenterology* **95**, 1130–1143.

Vacanti, C. A., Vacanti, J. P., and Langer, R. (1994). Tissue engineering using synthetic biodegradable polymers. *In* "Polymers of Biological and Biomedical Significance" (W. Shalaby, ed.), Vol. 540, pp. 16–34. Am. Chem. Soc., Washington, DC.

Van Zijl, P. C., Moonen, C. T., Faustino, P., Pekar, J., Kaplan, O., and Cohen, J. S. (1991). Complete separation of intracellular and extracellular information in NMR spectra of perfused cells by diffusion-weighted spectroscopy. *Proc. Nat. Acad. Sci. U.S.A.* **88**, 3228–3232.

Wakayama, T., Perry, A. C., Zuccotti, M., Johnson, K. R., and Yanagimachi, R. (1998). Full-term development of mice from enucleated oocytes injected with cumulus cell nuclei. *Nature (London)* **394**, 369–374.

Wald, H. L., Sarakinos, G., Lyman, M. D., Mikos, A. G., Vacanti, J. P., and Langer, R. (1993). Cell seeding in porous transplantation devices. *Biomaterials* **14**, 270–278.

Wang, N., and Ingber, D. E. (1994). Control of cytoskeletal mechanics by extracellular matrix, cell shape, and mechanical tension. *Biophys. J.* **66**, 2181–2189.

Watanabe, F. D., Mullon, C. J., Hewitt, W. R., Arkadopoulos, N., Kahaku, E., Eguchi, S., Khalili, T., Arnaout, W., Shackleton, C. R., Rozga, J., Solomon, B., and Demetriou, A. A. (1997). Clinical experience with a bioartificial liver in the treatment of severe liver failure. A phase I clinical trial. *Ann. Surg.* **225**, 484–491; discussion: pp. 491–494.

Watt, F. M. (1998). Epidermal stem cells: Markers, patterning and the control of stem cell fate. *Philos. Trans. R. Soc. London, Ser. B* **353**, 831–837.

Weigel, P. H., Schmell, E., Lee, Y. C., and Roseman, S. (1978). Specific adhesion of rat hepatocytes to beta-galactosides linked to polyacrylamide gels. *J. Biol. Chem.* **253**, 330–333.

Weintraub, H. (1993). The MyoD family and myogenesis: Redundancy, networks, and thresholds. *Cell (Cambridge, Mass.)* **75**, 1241–1244.

Wells, D. N., Misica, P. M., and Tervit, H. R. (1999). Production of cloned calves following nuclear transfer with cultured adult mural granulosa cells. *Biol. Reprod.* **60**, 996–1005.

Williams, R. A., Baak, J. P., Meijer, G. A., and Charlton, I. G. (1997). Exploring the possibility of DNA ploidy measurements in tissue sections using liver as a model. *Anal. Quant. Cytol. Histol.* **19**, 19–29.

Williams, S. N. O., Callies, R. M., and Brindle, K. M. (1997). Mapping of oxygen tension and cell distribution in a hollow-fiber bioreactor using magnetic resonance imaging. *Biotechnol. Bioeng.* **56**, 56–61.

Wilmut, I., Schnieke, A. E., McWhir, J., Kind, A. J., and Campbell, K. H. (1997). Viable offspring derived from fetal and adult mammalian cells. *Nature (London)* **385**, 810–813; published erratum: **386**, 200.

Wilson, J. W., and Leduc, E. H. (1958). Role of cholangioles in restoration of the liver of the mouse after dietary injury. *J. Pathol. Bacteriol.* **76**, 441–449.

Wong, H., and Chang, T. M. (1986). Bioartificial liver: Implanted artificial cells microencapsulated living hepatocytes increases survival of liver failure rats. *Int. J. Artif. Organs* **9**, 335–336.

Wu, F. J., Friend, J. R., Lazar, A., Mann, H. J., Remmel, R. P., Cerra, F. B., and Hu, W. S. (1996). Hollow fiber bioartificial liver utilizing collagen-entrapped porcine hepatocyte spheroids. *Biotechnol. Bioeng.* **52**, 34–44.

Yagi, K., Michibayashi, N., Kurikawa, N., Nakashima, Y., Mizoguchi, T., Harada, A., Higashiyama, S., Muranaka, H., and Kawase, M. (1997). Effectiveness of fructose-modified chitosan as a scaffold for hepatocyte attachment. *Biol. Pharm. Bull.* **20**, 1290–1294.

Yarmush, M. L., Dunn, J. C., and Tompkins, R. G. (1992). Assessment of artificial liver support technology. *Cell Transplant.* **1**, 323–341.

Yin, L., Ghebranious, N., Chakraborty, S., Sheehan, C. E., Ilic, Z., and Sell, S. (1998). Control of mouse hepatocyte proliferation and ploidy by p53 and p53ser246 mutation *in vivo. Hepatology* **27**, 73–80.

Zaret, K. (1998). Early liver differentiation: Genetic potentiation and multilevel growth control. *Curr. Opin. Genet. Dev.* **8**, 526–531.

Zarnegar, R., and Michalopoulos, G. K. (1995). The many faces of hepatocyte growth factor: From hepatopoiesis to hematopoiesis. *J. Cell biol.* **129**, 1177–1180.

Zhang, J., Parker, D. L., and Leypoldt, J. K. (1995). Flow distributions in hollow fiber hemodialyzers using magnetic resonance Fourier velocity imaging. *ASAIO J.* **41**, M678–M682.

Ziegler, T. R., Pierce, G. F., and Herndon, D. N., eds. (1997). "Growth Factors and Wound Healing: Basic Science and Potential Clinical Applications." Springer, New York.

Zvibel, I., Halay, E., and Reid, L. M. (1991). Heparin and hormonal regulation of mRNA synthesis and abundance of autocrine growth factors: Relevance to clonal growth of tumors. *Mol. Cell. Biol.* **11**, 108–116.

Zvibel, I., Brill, S., and Reid, L. M. (1995). Insulin-like growth factor II regulation of gene expression in rat and human hepatomas. *J. Cell. Physiol.* **162**, 36–43.

Zvibel, I., Fiorino, A. S., Brill, S., and Reid, L. M. (1998). Phenotypic characterization of rat hepatoma cell lines and lineage-specific regulation of gene expression by differentiation agents. *Differentiation (Berlin)* **63**, 215–223.

PART XIV: HEMATOPOIETIC SYSTEM

CHAPTER 42

RED BLOOD CELL SUBSTITUTES

Thomas Ming Swi Chang

INTRODUCTION

Hemoglobin molecules extracted from red blood cells are modified by microencapsulation or cross-linkage to produce red blood cell substitutes. The encapsulation and linkage processes stabilize the hemoglobin molecules and also allow sterilization of the products to remove human immunodeficiency virus (HIV) and other microorganisms. Rapid progress has been made in the past decade toward clinical use of red blood cell substitutes. A number of groups are already well into phase III clinical trials in humans to study safety and efficacy in patients.

In 1956, the first research was initiated on artificial cells, including studies on modified hemoglobin (Chang, 1957, 1964, 1965). Artificial cells (Fig. 42.1) contain biologically active materials such as hemoglobin, enzymes, cells, adsorbent components, and other material (Chang, 1972, 1977a; Chang and Prakash, 1998). The membranes of artificial cells allow permeant molecules such as oxygen and substrates to enter and allow metabolic products, peptides, and other material to leave. In this way the enclosed materials are protected from immunological rejection and from other materials in the external environment.

Until about 1989, most research on artificial cells focused on those containing enzymes, cells, microorganisms, adsorbent components, peptides, and other materials (Chang, 1964, 1965, 1972). Concentrated research and development on modified hemoglobin started after 1987 because of public concerns about HIV-induced autoimmune deficiency syndrome (AIDS) in donor blood recipients. After extraction from red blood cells and before modification, hemoglobin can be sterilized by pasteurization, ultrafiltration, and chemical means. These procedures can remove microorganisms, including those responsible for AIDS, hepatitis, and other blood-borne diseases. There are many situations wherein modified hemoglobin has the potential to substitute for red blood cells (Bowersox and Hess, 1994; Chang, 1997a; Klein, 1994; Winslow, 1994, 1996), including cardiopulmonary bypass surgery, cancer surgery, elective surgery, organ preservation cardioplegia, and in emergency treatment for severe traumatic injuries resulting from traffic accidents and other accidents that cause severe bleeding and hemorrhagic shock. The number of traumatic injuries requiring blood substitutes in civilian use is small when compared to the requirements in major disasters or wars (Bowersox and Hess, 1994). Modified hemoglobin is especially useful in emergency situations such as these. Because modified hemoglobin does not contain red blood cell membrane and therefore no blood group antigens, it can be used without the need for cross-matching or typing. This would save much time and resources and would permit on-the-spot transfusion as required, similar to giving intravenous salt solution. This is further facilitated by the fact that modified hemoglobin can be lyophilized and stored as a stable dried powder that can be reconstituted with the appropriate salt solution just before use. Another application is in treating patients whose religious beliefs do not allow them to use donor blood for transfusion.

WHY DO WE HAVE TO MODIFY HEMOGLOBIN?

Hemoglobin in the red blood cell is responsible for carrying oxygen from the lung and delivering oxygen to tissues. Hemoglobin is a tetramer of four subunits: two α subunits and two β sub-

Fig. 42.1. Artificial cells can be prepared as microcapsules, nanocapsules, or cross-linked proteins. The artificial cell retains biologically active materials, thus protecting the contents from the external immunological system and other destructive elements. At the same time, permeant molecules can enter or leave rapidly, allowing molecules such as oxygen, substrates, metabolites, toxins, and other materials to enter and be acted on by the enclosed materials. Any products derived from the enclosed materials, such as peptides, hormones, metabolites, and oxygen, can readily diffuse out. Reprinted from Chang (1993) by courtesy of Marcel Dekker Inc.

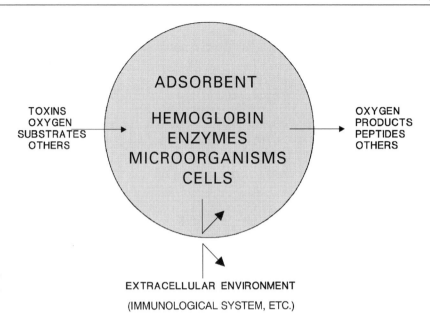

units (Perutz, 1980) (Fig. 42.2). Hemoglobin is in the oxy, related state (R state) when it is carrying oxygen. To release oxygen, the hemoglobin molecule undergoes conformational change with a 15° rotation. The molecule is then in the deoxy, tense state (T state). Red blood cells contain a cofactor 2,3-diphosphoglycerate (2,3-DPG), that facilitates this conformational change. Thus hemoglobin inside red blood cells has a high P_{50}, which allows it to release oxygen readily to tissue at physiological oxygen tensions.

Hemoglobin can be extracted from red blood cells and infused into animals experimentally (Amberson, 1937), but this causes renal toxicity. Removal of the cell membranes to form a stroma-free hemoglobin blood substitute still causes renal toxicity, because of a number of reasons. When infused into the circulation, each four-subunit hemoglobin molecule (tetramer) is rapidly broken down into two subunits (dimers) (Fig. 42.3). The smaller dimers are rapidly excreted by the kidneys, and this also has renal toxicity (Savitsky *et al.*, 1978). Furthermore, without the required 2,3-DPG present inside the red blood cells, hemoglobin cannot readily release oxygen (low P_{50}) until the tissue oxygen tension has reached a lower value.

Fig. 42.2. Inside the red blood cells, each hemoglobin molecule exists as a tetramer of four subunits: two α subunits and two β subunits. Cofactor 2,3-DPG is regained in the red blood cell to facilitate the release of oxygen by hemoglobin, as required by the tissues. Reprinted from Chang (1992) by courtesy of Marcel Dekker Inc.

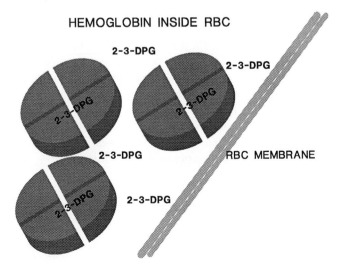

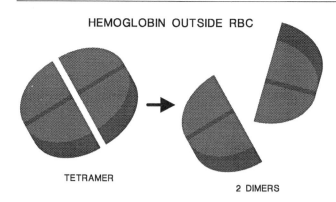

HEMOGLOBIN OUTSIDE RBC

TETRAMER

2 DIMERS

Fig. 42.3. When hemoglobin is outside the red blood cell, the first problem is that 2,3-DPG is no longer available to facilitate the release of oxygen; second, the hemoglobin tetramer breaks down into two dimers, which are rapidly excreted by the kidney. Reprinted from Chang (1992) by courtesy of Marcel Dekker Inc.

MODIFIED HEMOGLOBIN

Biotechnological approaches can be used to either cross-link or encapsulate hemoglobin molecules to prevent the hemoglobin from dissociating into dimers and also to allow hemoglobin to maintain an acceptable P_{50}. Many research groups have contributed to major progress in modifying hemoglobin, especially in the past decade. A number of groups are in phase III clinical trials in humans. The following discussions provide details on the two major subclasses of modified hemoglobin: encapsulated hemoglobin and cross-linked hemoglobin.

ENCAPSULATED HEMOGLOBIN: ARTIFICIAL RED BLOOD CELLS

The first artificially created red blood cells have a P_{50} and an oxygen dissociation curve similar to normal red blood cells, because 2,3-DPG is retainedn inside (Chang, 1957) (Fig. 42.4). Hemoglobin is also retained inside as tetramers. These artificial red blood cells do not have blood group antigens on the membrane and therefore do not aggregate in the presence of blood group antibodies (Chang, 1972). However, the single major problem is the rapid removal of these artificial cells from the circulation. Much of the research since the creation of these cells has been focused on improving their survival in the circulation by decreasing their uptake by the reticuloendothelial system. Because removal of siliac acid from biological red blood cells results in their rapid removal from the circulation (Chang, 1965, 1972), we modified the surface properties of artificial red blood cells using synthetic polymers, cross-linked protein, lipid–protein, lipid–polymer, and other surface modifications (Chang, 1964, 1965, 1972). However, the available technology at the time we conducted these studies allowed production of artificial red blood cells of down to only about 1 μm in diameter.

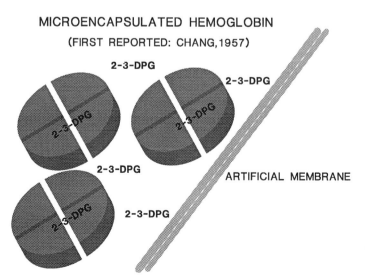

MICROENCAPSULATED HEMOGLOBIN

(FIRST REPORTED: CHANG, 1957)

2-3-DPG

2-3-DPG

2-3-DPG

2-3-DPG

2-3-DPG

2-3-DPG

ARTIFICIAL MEMBRANE

Fig. 42.4. Modification of hemoglobin by microencapsulation inside artificial cells provides an intracellular environment similar to that of biologic red blood cells (Fig. 42.2). Reprinted from Chang (1992) by courtesy of Marcel Dekker Inc.

Later, new technology allowed the preparation of smaller submicron lipid membrane artificial red blood cells with further improvements in circulation time (Djordjevich and Miller, 1980). Circulation time improved by additional surface modifications. Microencapsulated hemoglobin and artificial red blood cells are now being extensively explored by many researchers (Chang, 1997a; Rudolph *et al.,* 1997; Tsuchida, 1995). Developmental advances have increased the average circulation half-time to 20 hor, and modifications using poly(ethylene glycol) have increased the circulation half-time to around 40 hr (Rudolph *et al.,* 1997). It is possible to replace 90% of the red blood cells in rats with artificial red blood cells, and animals with this percentage of exchange transfusion still remain viable. Studies have also reported success in treating hemorrhagic shock. Preliminary studies by a number of groups show that the artificial cells are not toxic; there are no changes in the histology of brain, heart, kidneys, and lungs in rats. Larger scale production is now feasible by using extrusion methods with a microfluidizer. The uptake is mainly by the reticuloendothelial system. To further improve stability and biodegradability, we are now using biodegradable polymer membranes to nanoencapsulate hemoglobin. This results in artificial red blood cells of about 168 nm (0.168 μm) in diameter (Chang and Yu, 1997, 1998; Yu and Chang, 1996).

Microencapsulation of hemoglobin to prepare artificial red blood cells is a rather ambitious approach. Although this attempt to mimic red blood cells has resulted in a complete red blood cell substitute, it is rather complicated. Further research is needed and this approach is now considered to yield a second-generation modified hemoglobin. A simpler cross-linked modified hemoglobin has been developed as a first-generation modified hemoglobin for more immediate clinical applications (Chang, 1997a; Rudolph *et al.,* 1997; Tsuchida, 1995; Winslow, 1994, 1996).

CROSS-LINKED POLYHEMOGLOBIN

Extensive studies of cross-linked hemoglobin have been carried out by many groups, especially in the past decade. Hemoglobin contains many amino groups, most of which are on the surface of the hemoglobin molecule. The use of a bifunctional agent (diacid) to cross-link hemoglobin was first reported by Chang (1964, 1965, 1972) (Fig. 42.5). Diacid was initially used to form cross-linked hemoglobin membranes for artificial red blood cells, but it was found that with decreasing size of artificial cells, all the hemoglobin molecules are cross-linked into polyhemoglobin (Chang, 1964, 1965, 1972). The reaction is as follows:

$$\text{Cl–CO–(CH}_2)_8\text{–CO–Cl} + \text{HB–NH}_2 = \text{HB–NH–CO–(CH}_2)_8\text{–CO–NH–HB}$$

Diacid   Hemoglobin   Cross-linked polyhemoglobin

INTERMOLECULAR(+INTRA) XLINKING POLYHEMOGLOBIN

Fig. 42.5. Cross-linking of hemoglobin using bifunctional agents prevents the breakdown of tetramers into dimers. The addition of pyridoxal phosphates or other 2,3-DPG analogs faciliates oxyhemoglobin in releasing oxygen. Reprinted from Chang (1992) by courtesy of Marcel Dekker Inc.

FIRST CROSSLINKERS USED: SEBACYL CHLORIDE (CHANG,1964) GLUTARALDEHYDE (CHANG,1971)

Another cross-linking reaction, using glutaraldehye as the bifunctional agent to form another type of polyhemoglobin, was studied (Chang, 1971):

$$H-CO-(CH_2)_3-CO-H + HB-NH_2 = HB-NH-CO-(CH_2)_3-CO-NH-HB$$

      dialdehydes   hemoglobin   Cross-linked polyhemoglobin

Cross-linking prevents the breakdown of hemoglobin tetramers into dimers. The addition of a 2,3-DPG analog, pyridoxal phosphate, to cross-linked polyhemoglobin improves the P_{50} (Benesch *et al.*, 1975). This approach has been developed (Dudziak and Bonhard, 1980; Sehgal *et al.*, 1980; DeVenuto and Zegna, 1982). The resulting pyridoxalated glutaraldehyde polyhemoglobin is in phase III clinical trial, and Gould *et al.* have successfully infused 10,000 ml into patients (Gould *et al.*, 1998). Another glutaradehyde cross-linked polyhemoglobin using bovine hemoglobin is also in phase III clinical trial in humans and is already approved by the Food and Drug Administration (FDA) for use in canines (Pearce and Gawryl, 1998). Other cross-linkers are also being developed. Some of these are based on bifunctional dialdehydes derived from oxidizing the ring structures of sugars or nucleotides. These are designed to have the dual function of a bifunctional cross-linker, which also acts as a 2,3-DPG analog. One approach being developed for clinical trial involves the use of a dialdehyde prepared from oxidizing a sugar molecule to form ring-opened raffinose, *o*-raffinose (Hsia, 1991). The *o*-raffinose polymerized hemoglobin has a good P_{50} without the need for an additional 2,3-DPG analog. It is now in phase III clinical trial (Adamson and Moore, 1998).

CROSS-LINKED TETRAMERIC HEMOGLOBIN

The cross-linkers described above can be used for both intermolecular and intramolecular cross-linkage. Studies have been carried out to specifically cross-link hemoglobin molecules intramolecularly to prevent dimer formation after infusion (Bunn and Jandl, 1968). Another example is a bifunctional agent, 2-nor-2-formylpyridoxal 5-phosphate, which is also a 2,3-DPG analog that can intramolecularly cross-link the two β subunits of the hemoglobin molecules (Benesch *et al.*, 1975). Another 2,3-DPG pocket modifier, bis(3,5-dibromosalicyl) fumarate (DBBF), intramolecularly cross-links the two α subunits of the hemoglobin molecule (Walder *et al.*, 1979). This prevents dimer formation and also improves the P_{50}. This product, the so-called α-α-hemoglobin, has been tested up to phase III clinical trial (Estep *et al.*, 1992; Nelson, 1998). Analysis of results has led to cancellation of the clinical trial of this cross-linked tetrameric hemoglobin.

CONJUGATED HEMOGLOBIN

Conjugated hemoglobin is the result of cross-linking hemoglobin to polymers. Initially, hemoglobin was cross-linked to polyamides (Chang, 1964, 1965, 1972). The use of soluble polymers resulted in soluble conjugated hemoglobin with good circulation time (Iwashita *et al.*, 1988; Iwashita, 1992; Nho *et al.*, 1992; Shorr *et al.*, 1996; Wong, 1988). These are being tested in humans in phase II clinical trial.

SOURCES OF HEMOGLOBIN

Where do we obtain all the hemoglobin needed for preparing modified hemoglobin? In addition to hemoglobin from human sources, bovine hemoglobin is being actively investigated for use in glutaradehyde cross-linked bovine polyhemoglobin (Pearce and Gawryl, 1998) and as conjugated hemoglobin (Nho *et al.*, 1992; Shorr *et al.*, 1996). Unlike human hemoglobin, bovine hemoglobin outside of red blood cells has a high P_{50} without requiring 2,3-DPG or its analogs. Bovine hemoglobin has been cross-linked into polyhemoglobin and is being used in phase III clinical trial (Pearce and Gawryl, 1998). Human hemoglobins produced by recombinant technology in microorganisms (Hoffman *et al.*, 1990) and from transgenic animals (O'Donnell *et al.*, 1992) are other exciting potential sources. The recombinant approach using *Escherichia coli* allows the modification of hemoglobin to prevent dimer formation and also affords a good P_{50}. This has been tested up to phase II clinical trial (Freytag and Templeton, 1997), but has been discontinued in order to develop a recombinant hemoglobin that has less affinity for nitric oxide and therefore does not cause vasoconstriction (Doherty *et al.*, 1998). Another potential approach is synthetic heme (Tsuchida, 1995).

CIRCULATING TIME OF MODIFIED HEMOGLOBIN

Removal of polyhemoglobin and conjugated hemoglobin after infusion is mainly accomplished by the reticuloendothelial system. The half-time of polyhemoglobin in the circulation is about 25 to 30 hr. Conjugated hemoglobin stays in circulation slightly longer. Intramolecularly cross-linked tetrameric hemoglobin escapes more rapidly from the circulation and therefore has a shorter circulation time. Survival time of modified hemoglobin in the circulation depends on the dose and the animal species, and is much shorter than that of red blood cells. However, the survival time is adequate for most of the shorter term uses described earlier. For example, all three types of cross-linked hemoglobin are effective in animal studies of hemorrhagic shock and isovolemic exchange transfusions.

SAFETY OF MODIFIED HEMOGLOBIN
BLOOD SUBSTITUTE

How safe is modified hemoglobin when injected into animals and humans? To evaluate this, proper selection of a toxicity animal model is important. This has been described in detail in the FDA publication, "Points to consider in the safety evaluation of hemoglobin-based oxygen carriers" (Frantantoni, 1991). Animal studies using properly prepared cross-linked hemoglobin have not shown any adverse effects on coagulation factors, leukocytes, platelets, or complement activation. Very sensitive immunological studies have been carried out for polyhemoglobin, conjugated hemoglobin, and microencapsulated hemoglobin (Chang *et al.,* 1992; Estep *et al.,* 1992). Results show that modified homologous polyhemoglobins (e.g., rat polyhemoglobin injected into rats) are not immunogenic even with repeated injection. Heterologous polyhemoglobin (e.g., nonrat hemoglobin injected into rats) is not immunogenic initially but is immunogenic after repeated injections with Freud's adjuvant. Conjugation and microencapsulation markedly decreased the antigenicity of heterologous polyhemoglobin. Other studies show that without the use of Freund's adjuvant, repeated subcutaneous and intravenous injections of heterologous cross-linked hemoglobin are much less immunogenic. Because hemoglobin has a high affinity for nitric oxide, extensive studies have been carried out to see the effects of vasoactivity. Tetrameric hemoglobin in the intramolecularly cross-linked form or in the recombinant form causes more vasoactivity than does polyhemoglobin (Chang, 1997c). Those polyhemoglobins with less than 1% tetrameric hemoglobin (Gould *et al.,* 1998) did not show vasoactivity. It is likely that hemoglobin, being less

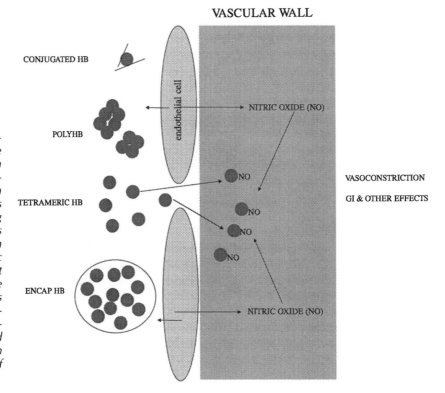

Fig. 42.6. Tetrameric hemoglobin (HB), being small, can move across the intercellular junction of endothelial cells. This hemoglobin binds nitric oxide (NO) in the interstitial space and lowers the NO concentration, resulting in vasoconstriction. Other types of modified hemoglobin contain a varying amount of tetrameric hemoglobin that can also act similarly. Removal of these smaller tetrameric hemoglobins from polyhemoglobin, conjugated hemoglobin, and microencapsulated hemoglobin would prevent this. Reprinted from Chang (1997c) by courtesy of Marcel Dekker Inc.

negatively charged than albumin, can cross the intercellular junction of the endothelial cells of the vasculature (Fig. 42.6). Once there, hemoglobin, with its high affinity for nitric oxide, would remove nitric oxide that is needed for vasodilatation. There has been success in preparing recombinant tetrameric hemoglobins with low affinity for nitric oxide resulting in little or no vasoconstrictor effects (Doherty *et al.,* 1998). Another important area of safety study is the distribution of modified hemoglobin after infusion and also its effect on the reticuloendothelial system.

Cross-linked hemoglobin appears to be safe in animal testing as described above; however, in 1991, the FDA reported unexplained clinical reactions in phase I clinical trials in humans and emphasized the need for further careful safety evaluation (Frantantoni, 1991). One of the problems with safety studies is that a response in animals is not necessarily the same as in humans, especially in the case of immunologic and hypersensitivity types of reactions. We have worked out a simple *in vitro* screening test (Chang and Lister, 1990, 1993a,b, 1994) that consists of adding 0.1 ml of modified hemoglobin to a test tube containing 0.4 ml of human plasma or blood, followed by analysis for complement activation after incubating for 1 hr. This *in vitro* test can be used in research or in industrial scale-up production. It can detect trace contamination of blood group antigens, antibody–antigen complexes, endotoxin, trace fragments of microorganisms, impurities in polymers, and some emulsifiers, for example. This figure has been used to rule out potential problems in industrial scale-up.

EFFICACY OF MODIFIED HEMOGLOBIN BLOOD SUBSTITUTE

Animal studies show that cross-linked hemoglobin is effective in short-term applications, including hemorrhagic shock, hemodilution in surgery, and angioplasty. Animal studies indicate that polyhemoglobin, intramolecularly cross-linked hemoglobin, conjugated hemoglobin, and microencapsulated hemoglobin are all effective in the resuscitation of lethal hemorrhagic shock and in isovolemic exchange transfusion. For obvious reasons this type of severe test of efficacy using lethal animal models cannot be used in human clinical trials. As a result, discussions still continue regarding the types of end points one needs to use for clinical trials of efficacy in humans. The FDA has published a summary on "points to consider" for efficacy designs in efficacy clinical trials in humans (Frantatoni, 1994).

PRESENT STATUS AND FUTURE RESEARCH AND DEVELOPMENT

The different polyhemoglobins now in phase III clinical trials are likely to be the first modified hemoglobins ready for routine clinical use. With these first-generation modified hemoglobin

ISCHEMIA

1. ATP → ADP → AMP → ADENOSINE → INOSINE → HYPOXANTHINE

2. XANTHINE DEHYDROGENASE → XANTHINE OXIDASE

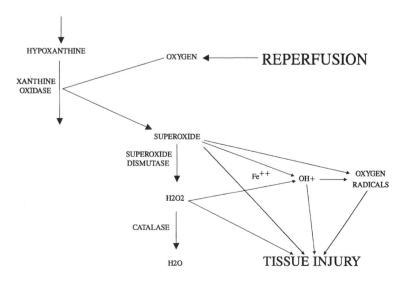

Fig. 42.7. Ischemic reperfusion injuries. Ischemia leads to accumulation of hypoxanthine and activation of xanthine oxidase. Reperfusion brings in oxygen, resulting in superoxide formation. This and other resulting oxidants and oxygen radicals can cause tissue injury. First-generation modified hemoglobin is prepared from ultrapure hemoglobin and therefore does not contain the required deoxidant enzymes. It has been shown that cross-linking polyhemoglobin with superoxide dismutase and catalase can significantly decrease the formation of oxygen radicals (D'Agnillo and Chang, 1998a). Reprinted from Chang (1997c) by courtesy of Marcel Dekker Inc.

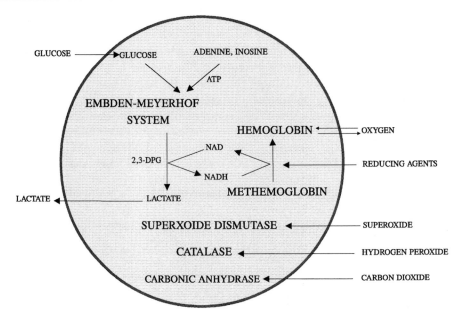

Fig. 42.8. Third-generation modified hemoglobin blood substitute consisting of biodegradable polymeric membrane nanoencapsulated hemoglobin and enzyme systems (Chang, 1997c; Yu and Chang, 1996). Reprinted from Chang (1997c) by courtesy of Marcel Dekker Inc.

blood substitutes being tested in humans, studies have already started on further refinements of cross-linked hemoglobin. One example is the incorporation of sueproxide dismutase and catalase into cross-linked hemoglobin to prevent reperfusion injury due to oxygen radicals (Fig. 42.7) (D'Agnillo and Chang, 1997, 1998a,b; Chang *et al.,* 1998). Artificial red blood cells based on microencapsulation afford a more complete red blood cell substitute and are also being extensively developed. These are more similar to biologic red blood cells because hemoglobin is not exposed to the outside environment. Furthermore, multienzyme systems can also be enclosed. This includes the entrapment of catalase and superoxide dismutase with hemoglobin (Fig. 42.8) (Chang and Yu, 1997). The approach of microencapsulation of multienzyme systems with cofactor recycling is also being used to prevent methemoglobin formation. This brief overview cannot include the numerous ongoing studies and research in this area. Details are available in many published books, journals, and symposia, and on the web site at http://www.artcell.mcgill.ca (Chang, 1997b, 1998; Rudolph *et al.,* 1997; Tsuchida, 1995; Winslow, 1996). There is also a chemical aporoach to red blood cell substitutes based on perfluorochemicals (Keipert, 1998; Reiss, 1998). However, this is not within the scope of this chapter, which describes only the biotechnological approaches. Details on the topic of perfluorochemicals are available in Keipert (1998) and Reiss (1998).

REFERENCES

Adamson, J. G., and Moore, C. (1998). Hemolink TM, an *o*-raffinose crosslinked hemoglobin-based oxygen carrier. *In* "Blood Substitutes: Principles, Methods, Products and Clinical Trials" (T. M. S. Chang, ed.), Vol. 2, pp. 62–79. Karger, Basel.

Amberson, W. R. (1937). Blood substitute. *Biol. Rev. Cambridge Philos. Soc.* **12**, 48.

Benesch, R., Benesch, R. E., Yung, S., and Edalji, R. (1975). Hemoglobin covalently bridged across the polyphosphate binding site. *Biochem. Biophys. Res. Commun.* **63**, 1123.

Bowersox, J. C., and Hess, J. R. (1994). Trauma and military applications of blood substitutes. *Artif. Cells, Blood Substitutes Immob. Biotechnol., Int. J.* **22**, 145–159.

Bunn, H. F., and Jandl, J. H. (1968). The renal handling of hemoglobin. *Trans. Assoc. Am. Physicians* **81**, 147.

Chang, T. M. S. (1957). "Hemoglobin Corpuscles. Report of a Research," pp. 1–25. Medical Library, McIntyre Building, McGill University, Montreal, Quebec; reprinted in *Biomater. Artif. Cells, Artif. Organs* **16**, 1–9 (1988).

Chang, T. M. S. (1964). Semipermeable microcapsules. *Science* **146**, 524–525.

Chang, T. M. S. (1965). Semipermeable aqueous microcapsules. Ph.D. Thesis, McGill University Medical Library, Montreal, Quebec.

Chang, T. M. S. (1971). Stabilization of enzyme by microencapsulation with a concentrated protein solution or by crosslinking with glutaraldehyde. *Biochem. Biophys. Res. Commun.* **44**, 1531–1533.

Chang, T. M. S. (1972). "Artificial Cells," Monogr. Thomas, Springfield, IL.

Chang, T. M. S. (1992). Blood substitutes based on modified hemoglobin prepared by encapsulation or crosslinking. *Biomater., Artif. Cells Immob. Biotechnol., Int. J.* **20**, 154–174.

Chang, T. M. S. (1993). Bioencapsulation in biotechnology. *Biomater., Artif. Cells Immob. Biotechnol., Int. J.* **21**, 291–298.

Chang, T. M. S. (1997a). Artificial cells. *In* "Encyclopedia of Human Biology" (R. Dulbecco, ed.), 2nd ed., pp. 457–463. Academic Press, San Diego, CA.

Chang, T. M. S. (1997b). "Blood Substitutes: Principles, Methods, Products and Clinical Trials," Vol. I, Monogr. Karger, Basel.

Chang, T. M. S. (1997c). Present and future perspectives of modified hemoglobin blood substitutes. *Artif. Cells, Blood Substitutes Immob. Biotechnol., Int. J.* **25**, 1–24.

Chang, T. M. S., ed. (1998). "Blood Substitutes: Principles, Methods, Products and Clinical Trials," Vol. II. Karger, Basel.

Chang, T. M. S., and Lister, C. (1990). A screening test of modified hemoglobin blood substitute before clinical use in patients—Based on complement activation of human plasma. *Biomater. Artif. Cells, Artif. Organs* **18**(5), 693–702.

Chang, T. M. S., and Lister, C. W. (1993a). Screening test for modified hemoglobin blood substitute before use in human. U.S. Pat. 5,200,323.

Chang, T. M. S., and Lister, C. W. (1993b). Use of finger-prick human blood samples as a more convenient way for *in vitro* screening of modified hemoglobin blood substitutes for complement activation: A preliminary report. *Biomater. Artif. Cells Immob. Biotechnol.* **21**, 685–690.

Chang, T. M. S., and Lister, C. (1994). Assessment of blood substitutes: II. *In vitro* complement activation of human plasma and blood for safety studies in research, development, industrial production and preclinical analysis. *Artif. Cells, Blood Substitutes Immob. Biotechnol., Int. J.* **22**, 159–170.

Chang, T. M. S., and Prakash, S. (1998). Microencapsulated genetically engineered cells: Comparison with other strategies and recent progress. *Mol. Med. Today* **4**, 221–227.

Chang, T. M. S., and Yu, W. P. (1997). Biodegradable polymer membrane containing hemoglobin for blood substitutes. U.S. Pat. 5,670,173.

Chang, T. M. S., and Yu, W. P. (1998). Nanoencapsulation of hemoglobin and red blood cell enzymes based on nanotechnology and biodegradable polymer. *In* "Blood Substitutes: Principles, Methods, Products, and Clinical Trials" (T. M. S. Chang, ed.), Vol. 2, pp. 216–231. Karger, Basel.

Chang, T. M. S., Lister, C., Nishiya, T., and Varma, R. (1992). Effects of different methods of administration and effects of modifications by microencapsulation, cross-linkage or PEG conjugation on the immunological effects of homologous and heterologous hemoglobin. *J. Biomater., Artif. Cells Immob. Biotechnol.* **20**, 611–618.

Chang, T. M. S., D'Agnillo, F., and Razack, S. (1998). A second generation hemoglobin based blood substitute with antioxidant activities. *In* "Blood Substitutes: Principles, Methods, Products and Clinical Trials" (T. M. S. Chang, ed.), Vol. 2, pp. 178–186. Karger, Basel/Landes, Austin, TX.

D'Agnillo, F., and Chang, T. M. S. (1997). Modified hemoglobin blood substitute from cross-linked hemoglobin–superoxide dismutase–catalase. U.S. Pat. 5,606,025.

D'Agnillo, F., and Chang, T. M. S. (1998a). Polyhemoglobin–superoxide dismutase: Catalase as a blood substitute with antioxidant properties. *Nat. Biotechnol.* **16**(7), 667–671.

D'Agnillo, F., and Chang, T. M. S. (1998b). Absence of hemoprotein-associated free radical events following oxidant challenge of crosslinked hemoglobin-superoxide dismutase-catalase. *Free Radical Biol. Med.* **24**(6), 906–912.

DeVenuto, F., and Zegna, A. I. (1982). Blood exchange with pyridoxalated–polymerized hemoglobin. *Surg., Gynecol. Obstet.* **155**, 342–346.

Djordjevich, L., and Miller, I. F. (1980). Synthetic erythrocytes from lipid encapsulated hemoglobin. *Exp. Hematol.* **8**, 584.

Doherty, D. H., Doyle, M. P., Curry, S. R., Vali, R. J., Fattor, T. J., Olson, J. S., and Lemon, D. D. (1998). Rate of reaction with nitric oxide determines the hypertensive effect of cell-free hemoglobin. *Nat. Biotechnol.* **16**, 672–676.

Dudziak, R., and Bonhard, K. (1980). The development of hemoglobin preparations for various indications. *Anesthesist* **29**, 181–187.

Estep, T. N., Gonder, J., Bornstein, I., Young, S., and Johnson, R. C. (1992). Immunogenicity of diaspirin crosslinked hemoglobin solutions. *J. Biomater., Artif. Cells Immob. Biotechnol.* **20**, 603–610.

Frantantoni, J. C. (1991). Points to consider in the safety evaluation of hemoglobin based oxygen carriers. *Transfusion (Philadelphia)* **31**(4), 369–371.

Frantantoni, J. C. (1994). Points to consider on efficacy evaluation of hemoglobin and perfluorocarbon based oxygen carriers. *Transfusion (Philadelphia)* **34**, 712–713.

Freytag, J. W., and Templeton, D. (1997). Optro™ (recombinant human hemoglobin): A therapeutic for the delivery of oxygen and the restoration of blood volume in the treatment of acute blood loss in trauma and surgery. *In* "Red Cell Substitutes; Basic Principles and Clinical Application" (A. S. Rudolph, R. Rabinovici, and G. Z. Feuerstein, eds.), pp. 325–334. Dekker, New York.

Gould, S. A., Sehgal, L. R., Sehgal, H. L., DeWoskin, R., and Moss, G. S. (1998). The clinical development of human polymerized hemoglobin. *In* "Blood Substitutes: Principles, Methods, Products and Clinical Trials" (T. M. S. Chang, ed.), Vol. 2, pp. 12–28. Karger, Basel.

Hoffman, S. J., Looker, D. L., and Roehrich, J. M. (1990). Expression of fully functional tetrameric human hemoglobin in *Escherichia coli*. *Proc. Natl. Acad. Sci. U.S.A.* **87**, 8521–8525.

Hsia, J. C. (1991). *o*-Raffinose polymerized hemoglobin as red blood cell substitute. *Biomater. Artif. Cells Immob. Biotechnol.* **19**, 402.

Iwashita, Y. (1992). Relationship between chemical properties and biological properties of pyridoxalated hemoglobin–polyoxyethylene. *J. Biomater., Artif. Cells Immob. Biotechnol.* **20**, 299–308.

Iwashita, Y., Yabuki, A., Yamaji, K., Iwasaki, K., Okami, T., Hirati, C., and Kosaka, K. (1988). A new resuscitation fluid "stabilized hemoglobin," preparation and characteristics. *Biomat. Artif. Cells, Artif. Organs* **16**, 271–280.

Keipert, P. E. (1998). Perfluorochemical emulsions: Future alternatives to transfusion, development, pool. *In* "Blood Substitutes: Principles, Methods, Products and Clinical Trials" (T. M. S. Chang, ed.), Vol. 2, pp. 101–121. Karger/Landes System, Basel.

Klein, H. G. (1994). Oxygen carriers and transfusion medicine. *Artif. Cells, Blood Substitutes Immob. Biotechnol., Int. J.* **22**, 123–135.

Nelson, D. J. (1998). Blood and HemAssistμ (DCLHb): Potentially a complementary therapeutic team. *In* "Blood Substitutes: Principles, Methods, Products and Clinical Trials" (T. M. S. Chang, ed.), Vol. 2, pp. 39–57. Karger, Basel.

Nho, K., Glower, D., Bredehoeft, S., Shankar, H., Shorr, R., and Abuchowski, A. (1992). PEG–bovine hemoglobin: Safety in a canine dehydrated hypovolemic–hemorrhagic shock model. *J. Biomater. Artif. Cells Immob. Biotechnol.* **20**, 511–524.

O'Donnell, J. K., Swanson, M., Pilder, S., Martin, M., Hoover, K., Huntress, V., Karet, C., Pinkert, C., Lago, W., and Logan, J. (1992). Production of human hemoglobin in transgenic swine. *J. Biomater. Artif. Cells Immob. Biotechnol.* **20**(1), 149.

Pearce, L. B., and Gawryl, M. S. (1998). Overview of preclinical and clinical efficacy of Biopure's HBOCs. *In* "Blood Substitutes: Principles, Methods, Products and Clinical Trials" (T. M. S. Chang, ed.), Vol. 2, pp. 82–98. Karger, Basel.

Perutz, M. F. (1980). Stereochemical mechanism of oxygen transport by hemoglobin. *Proc. R. Soc. London, Ser. B* **208**, 135–147.

Reiss, J. G. (1998). Fluorocarbon-based oxygen-delivery: Basic principles and product development pool. *In* "Blood Substitutes: Principles, Methods, Products and Clinical Trials" (T. M. S. Chang, ed.), Vol. 2, pp. 101–121. Karger, Basel.

Rudolph, A. S., Rabinovici, R., and Feuerstein, G. Z., eds. (1997). "Red Blood Cell Substitutes." Dekker, New York.

Savitsky, J. P., Doozi, J., Black, J., and Arnold, J. D. (1978). A clinical safety trial of stroma free hemoglobin. *Clin. Pharmacol. Ther.* **23**, 73.

Sehgal, L. R., Rosen, A. L., Gould, S. A., Sehgal, H. L., Dalton, L., Mayoral, J., and Moss, G. S. (1980). *In vitro* and *in vivo* characteristics of polymerized pyridoxalated hemoglobin solution. *Fed. Proc., Fed. Am. Soc. Exp. Biol.* **39**, 2383.

Shorr, R. G., Viau, A. T., and Abuchowski, A. (1996). Phase 1B safety evaluation of PEG hemoglobin as an adjuvant to radiation therapy in human cancer patients. *Artif. Cells, Blood Substitutes Immob. Biotechnol., Int. J.* **24**, 407 (abstr. issue).

Tsuchida, E., ed. (1995). "Artificial Red Cells." Wiley, New York.

Walder, J. A., Zaugg, R. H., Walder, R. Y., Steele, J. M., and Klotz, I. M. (1979). Diaspirins that crosslink alpha chains of hemoglobin: Bis(3,5-dibromosalicyl) succinate and bis(3,5-dibromosalicyl) fumarate. *Biochemistry* **18**, 4265–4270.

Winslow, R. M. (1994). Symposium volume on "Blood Substitutes: Modified Hemoglobin," *Artif. Cells, Blood Substitutes Immob. Biotechnol., Int. J.* **22**, 360–944.

Winslow, R. M. (1996). "Blood Substitutes in Development," pp. 27–37. Ashley Publications, Boston.

Wong, J. T. (1988). Rightshifted dextran hemoglobin as blood substitute. *Biomater., Artif. Cells Artif. Organs* **16**, 237–245.

Yu, W. P., and Chang, T. M. S. (1996). Submicron polymer membrane hemoglobin nanocapsules as potential blood substitutes: Preparation and characterization. *Artif. Cells, Blood Substitutes Immob. Biotechnol., Int. J.* **24**, 169–184.

CHAPTER 43

LYMPHOID CELLS

Una Chen

INTRODUCTION

For researchers who are interested in cell-based therapy, it is a great challenge to learn how to expand normal lymphoid cells and their precursor cells *ex vivo* under defined conditions. Lymphocytes are defined by cell surface receptors—immunoglobulin (Ig) for B cells, and the T cell receptor (TCR) for T cells. Precursor B cells bear a pre-B receptor, which contains λ5 in mice (Ψ in humans), and precursor T cells bear a pre-TCR-α. No biochemically and genetically defined surface receptors for the progenitors of lymphoid precursors have been reported. It is not entirely clear how committed stem cells differentiate into lymphoid cells, which then mature into effector cells. Many theories have been postulated. Differentiation and maturation involve both antigen-independent and antigen-dependent processes. For the first process, I will explore the sequential commitment model proposed by Brown *et al.* (1985, 1988). For my treatment of the second process, I am indebted to many colleagues who helped me to summarize current views.

Ligand–receptor interactions must play an important role in activating lymphoid cells to proliferate and to differentiate. The term "ligand" means all external signals, including those provided by stroma cells and cytokines during the antigen-independent stage and those provided by antigens and accessory cells during the antigen-dependent stage. The process of turning on and turning off of transcription factors at each stage of differentiation as a consequence of ligand–receptor interaction is the key element controlling both phenotype and function of cells. Thus, hoping that the growth of lymphoid (precursor) cells can be manipulated at will in the future by using genetic tools, I will discuss transcription factors known to be important at various stages of lymphopoiesis. So far, there are only a few successful examples of expansion of normal lymphoid (precursor) cells in culture. All are based on the support of stroma cell lines and cytokines. On removal of these elements, cells either differentiate toward terminal effector cells or die.

I have arbitrarily defined seven stages of lymphopoiesis to discuss in detail. For each one I will discuss the available cellular and molecular markers, our current understanding of cellular phenotype, the potential of these cells to expand *ex vivo,* and what could be done in the future. In general, both somatic and genetic manipulations are possible in animal models. By combining modern cellular technology and available mutant mice, one should be able to grow normal lymphoid (precursor) cells at every stage of lymphopoiesis. Due to ethical and safety considerations, only somatic manipulation is considered for use with human beings.

Despite the efforts in many labs in the past few years, human stroma cell lines that can reproducibly support the growth of different stages of lymphoid (precursor) cells are still not available. To apply inducible regulation of stage-specific and lymphoid-specific transcription factors in controlling the growth of normal lymphoid cells seems to be the key element—a great task for the year 2000 for researchers in this field.

The aim of this chapter is to understand lymphocytes and their precursors. The purpose is to learn how these cells might be manipulated to make them useful in cell-based, somatic gene therapy. I will attempt to address the possibility of growing normal lymphoid cells and their stem cells of mouse and human origin in a controllable manner. That is, cells should proliferate *ex vivo* without becoming malignantly transformed and with little or no differentiation.

We distinguish two main sorts of stem cells: totipotential and multipotential. Totipotential stem cells usually divide symmetrically to give rise to two totipotent daughter stem cells, which are identical in phenotype to their parent. Under appropriate conditions, totipotent stem cells can differentiate into any other form of stem cell. The only known stem cell lines that seem to be close to totipotential are mouse embryonic stem cells (ESCs), and, more recently, human ESCs. Multipotential stem cells, which exist for the lifetime of an organism, undergo primarily asymmetric divisions. One daughter cell is another stem cell like its parent, whereas the other daughter cell is a more highly differentiated cell that performs a tissue-specific function. Due to this unique property, multipotential stem cells are ideal vehicles for cell-based, somatic gene therapy. They will carry the transgene for the lifetime of the organism and they will maintain expression of the transgene in the differentiated cells that they spawn.

Both lymphoid and lymphoid precursor cells are the progenies of hematopoietic stem cells. One of our main tasks here is to review whether lymphoid cells and their progenitors possess stem cell-like properties. Are they self-renewing, and with available culture conditions and technology, can they expand *ex vivo* under controllable growth conditions? Unlimited growth of lymphoid cells and their progenitors is well documented. Lymphoid leukemia cells develop either spontaneously or after infection with viruses. These cell lines are useful for studying lymphoid cells, but not for cell-based therapy because they develop tumors when reintroduced into the organism. Thus, this chapter is devoted to exploring the possibility of growing "normal" lymphoid cells that can be engineered in culture and reimplanted into syngeneic or autologous hosts for therapeutic purposes.

PROPERTIES OF LYMPHOCYTES

Lymphoid precursor cells are bipotential progenitors of pre-T and pre-B cells. The development of lymphoid precursor cells is independent of the presence of antigen (Ag). It is controversial whether lymphoid precursor cells divide asymmetrically with self-renewal. Lymphocyte types are defined by cell surface receptors—immunoglobulin (Ig) for B cells, and the T cell receptor for T cells. Precursor B cells bear a pre-B receptor, which contains $\lambda 5$ in mice, and Ψ in humans, and precursor T cells bear a pre-TCR-α. So far, no biochemically and genetically defined surface receptors for the progenitors of lymphoid precursors have been reported. Lymphocytes and their precursors are similar to cells of other lineages in that they proliferate, differentiate, communicate with other cells, age, and die. However, lymphocytes also possess unique properties that distinguish them from other cells: (1) formation of receptor genes by VDJ recombination, (2) requirement for a thymuslike environment to generate mature CD4$^+$8$^+$ T cells, (3) requirement for somatic education and selection by antigen, (4) somatic hypermutation to generate more diversity in B cells, (5) immunologic memory, (6) Ig heavy chain class switch recombination in B cells, and (7) κ light chain editing somatically to generate new specificities of B cell receptor (Hertz and Nemazee, 1998). Development of techniques to expand *ex vivo,* genetically manipulate, and reimplant into host animals lymphocytes and their precursors requires lymphocyte engineers.

LYMPHOCYTE ENGINEERING: REALITY AND POTENTIAL

Two models describe the sequence of cell commitment during lymphopoiesis: stochastic and inductive. In this section, I will use an inductive model in an attempt to explain lymphocyte commitment during development of the organism and in the adult.

INDUCTIVE MODEL OF SEQUENTIAL CELL COMMITMENT OF HEMATOPOIESIS

This model is divided into two parts. The first part deals with the antigen-independent stage, starting with the differentiation from the null cells (fertilized egg to embryonic stem cells) to mesoderm and then to lymphoid precursor cells. A significant portion of the first part of this model was expanded from Bailey's hypothesis of developmental progress (1986) and modified according to Brown *et al.* (1985, 1988).

Antigen-independent stage of lymphopoiesis—Part 1

As multipotential stem cells are committed to differentiate, it is hypothesized that the differentiation sequence is genetically determined. Cells within the sequence are precommitted in their ability to respond to various inducers of differentiation, and on encountering appropriate factors or a suitable microenvironment, they proliferate and mature along a particular pathway. Brown *et al.* (1985, 1988) suggested that multipotential stem cells that do not receive a signal for differentiation toward mature end cells progress to the next stage in the sequence of development. Alternatively, cells that do not receive a signal die. Only cells that receive proper signals will differentiate further toward the mature end cells. Because commitment is gradual, there is a continuous spectrum of multipotential stem cells. Thus, some of them may be able to respond to inducers of one or another sort. In terms of self-renewal, multipotential stem cell populations continuously occupy each potential for proliferation at any given time and therefore respond to the requirement for each cell type. Some daughter cells will differentiate into the next stage. The continuous development of stem cells may not be diverted entirely toward one cell type, even in the presence of differentiation factors for that type. Because stem cells divide as they respond, some cells are still able to progress to the next stage of commitment.

Antigen-dependent stage of lymphopoiesis—Part 2

The second part of this model deals with the antigen (Ag)-dependent stage, after receptors are expressed on the surface of lymphocytes. There are two main lineages of lymphocytes: T and B cells. T cells can mature into CD8+ cytotoxic T cells and CD4+ helper T cells. Based on their function and the spectrum of cytokines produced, helper T cells can be divided into Th1 and Th2 subpopulations. The main functions of CD4+ T cells are to recognize cell-bound antigen and to communicate with and help cytotoxic T cells and B cells to perform their duties as effector cells. Both lineages generate memory cells. How and when lymphoid cells are committed to differentiate into effector cells or into memory cells remain a mystery.

Diagrams to Explain This Model

Figure 43.1 is a simple diagram to explain this model. Hematopoietic stem cells (HSCs) are multipotential stem cells derived from the mesoderm. Originally they are descended from common stem cells (null, or 0, cells) such as fertilized eggs and from mouse embryonic stem cell lines. The cells diagonally to the right (Fig. 43.1) represent cells committed to hematopoietic lineages, which gradually lose their multipotentiality during embryogenesis. This concept is expressed with numbers (5, 4, 3, and 2), inside the progenitor cells on the diagonal. For example, HSCs (noncycling) enter the proliferating stage to become the hematopoietic progenitor cells (HPCs; cycling). The number 5 cell, for instance, means that the HPCs at this stage have the potential to differentiate into five different types of precursor cells.

The committed, monopotential precursor cells then differentiate toward mature end cells such as megakaryocytes (Me), erythrocytes (E), neutrophils/granulocytes (G), monocytes (M), and lymphocytes (L). Lymphoid precursor cells then differentiate into monopotential precursors for B and T cells. After expression of Ag receptors on the cell surface and on encountering Ag, B cells differentiate into plaque-forming cells (PFCs; i.e., plasma cells) or memory B cells. T cells become CD8+ or CD4+ after thymic education, maturation, and negative and positive selection. Memory T cells are generated later during differentiation. Figures 43.2–43.4 show the markers and events that occur during the development of pre-T cells—the process of maturation, education, and apoptosis of mature CD4+ or CD8+ T cells from CD4+8+ T cells (Kisielow and von Boehmer, 1995; Rodewald and Fehling, 1998). Genes encoding transcription factors and other important genes known to be specifically expressed in lymphopoiesis and which affect function when knocked out include *Brachyury; scl; c-myb; GATA-1, -2,* and *-3; rbtn2; pu.1/spi-1; Ikaros; Tcf-1; E2A, pax 5; Oct-2, Blimp-1; pTα; bcl-x;* and *p53.*

Some Comments on This Model

Practically nothing is known about how totipotent stem cells differentiate into ectoderm, mesoderm, and endoderm, nor how mesoderm differentiates into hematopoietic lineages. These are the central issues of embryogenesis, and this part of the model is intentionally very sketchy. Why and when multipotential, lineage-specific stem cells divide asymmetrically or progress to the next stage of commitment are unknown. The underlying rules may be stochastic (Ogawa, 1993) or inductive as suggested by Brown *et al.* But I think the progression of the stage of commitment might more likely be due to the turning on and off of genes encoding transcription factors.

Brown and co-workers originally formulated their model based on the data from the differ-

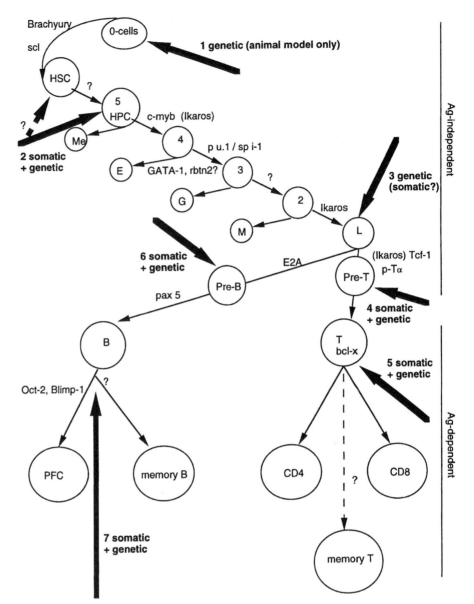

Fig. 43.1. *Proposed sequence of cell commitment during lymphopoiesis and stages of hematopoiesis and lymphopoiesis at which somatic and/or genetic engineering of lymphocytes and their precursor cells might be feasible. Solid block arrows point to the cell stages that might be targeted. The first part of the theory is the sequential commitment model of hematopoiesis of Brown et al., (1985, 1988), modified to include the pattern of expression of transcription factors (H. Singh, M. Koshland, and A. Schimpl, personal communications). The cells depicted diagonally to the right represent hematopoietic stem cells (HSC) committed to multipotential precursor cells or hematopoietic progenitor cells (HPC), which gradually lose pluripotency, as expressed by the numbers 5, 4, 3, and 2, inside the stem cells. Number 5, for example, means that the HPCs at this stage have the potential to differentiate into five different types of precursor cells. The committed monopotential precursor cells can then differentiate toward mature end cells such as megakaryocytes (Me), erythrocytes (E), neutrophils/granulocytes (G), monocytes (M), and finally lymphocytes (L). HSCs are derived from very primitive stem cells (null cells, or 0 cells) such as fertilized eggs, and/or mouse embryonic stem cells (ESC). In the second part of the theory, lymphoid precursor cells differentiate into monopotential precursor B cells (B) and monopotential precursor T cells (T). After the expression of cell surface receptors, and on encountering antigen (Ag), B cells can differentiate to plaque-forming cells (PFC) or memory B cells. T cells can become CD8[+] or CD4[+] cells after education and maturation in the thymus. Memory T cells can be generated at a certain stage of differentiation. The transcription factors are expressed and are known to affect differentiation. The data come mostly from phenotypic defects in knockout mice: Brachyury (Wilson et al., 1993; Kispert and Hermann, 1993, 1994), scl (Robb et al., 1995; Shivdasani et al., 1995), c-myb (Mucenski et al., 1991), GATA-1 (Penvy et al., 1991), rbtn2 (Boehm et al., 1991), pu.1/spi-1 (Scott et al., 1994), Ikaros (Georgopoulos et al., 1994), Tcf-1 (Verbeek et al., 1995), E2A (Bain et al., 1994; Zhuang et al., 1994), pax 5 (Urbanek et al., 1994), Oct-2 (Corcoran et al.,1993; Corcoran and Karvelas, 1994; Chen et al., 1991), and Blimp-1 (Turner et al., 1994) (no knockout mice generated yet). Other examples of gene products that might be critical for engineering are pTα (Fehling et al., 1995), bcl-x (Boise et al., 1993; Ma et al., 1995), and p53 (Bogue et al., 1996; see also Chen, 1996).*

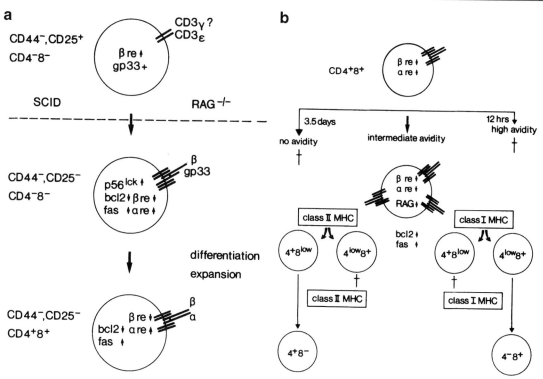

a

CD44⁻,CD25⁺
CD4⁻8⁻

CD3γ ?
CD3ε

β re ↓
gp33+

SCID   RAG⁻/⁻

CD44⁻,CD25⁻
CD4⁻8⁻

p56ˡᶜᵏ ↓
bcl2↓ β re ↓
fas ↓α re ↓

β
gp33

differentiation

expansion

CD44⁻,CD25⁻
CD4⁺8⁺

β re ↑
bcl2 ↓ α re ↑
fas ↑

β
α

b

CD4⁺8⁺

β re ↓
α re ↓

3.5 days   12 hrs ↓
                            high avidity

no avidity   intermediate avidity   †

†

β re ↓
α re ↓
RAG ↓

class II MHC   class I MHC

bcl2 ↓
fas ↓

4⁺8ˡᵒʷ   4ˡᵒʷ8⁺   4⁺8ˡᵒʷ   4ˡᵒʷ8⁺

†   †

class II MHC   class I MHC

4⁺8⁻   4⁻8⁺

Fig. 43.2. (a) Developmental control by the pre-T cell receptor. CD4⁻, CD8⁻, CD44⁻, CD25⁻, thymocytes rearrange the TCR-B locus while expressing the gp33 gene encoding the invariant chain of the pre-T cell receptor. The pre-TCR complex associated with CD3 signal-transducing molecules is inserted in the membrane. Signaling by the pre-TCR, which may or may not involve binding to an intrathymic ligand, results in massive cellular expansion as well as differentiation. Differentiation events may include CD4⁺8⁺ expression, suppression of TCR-β and enhancement of TCR-α rearrangement, down-regulation of the Bcl-2 protein, and up-regulation of the Fas protein. (b) Developmental control by the αβ T cell receptor. CD4⁻ 8⁺ cells with αβ TCRs have a life-span of 3 or 4 days, during which TCR-α rearrangement continues and cells produce new receptors. During this time, if the αβ TCRs do not bind to intrathymic ligands, the cell will die a programmed death (left). If the receptor binds to the specific peptide, cell death will be precipitated and the cell will be eliminated by apoptosis within hours (right). If the receptor binds to a positively selecting ligand different from the MHC-specific peptide, the cell will be rescued from programmed cell death and begin to down-regulate either CD4 or CD8 coreceptors (center). This may be an entirely or partly stochastic event. As a rule, only cells with "fitting" αβ TCR–CD4, CD8 coreceptor combinations, able to bind to the same class I or class II MHC molecule, will fully mature. Cells with a mismatched combination will die. Positive selection will terminate ongoing TCR-α rearrangement and induce higher expression of Bcl-2 protein while down-regulating Fas protein levels. From Kisielow and von Boehmer (1995).

entiation pattern of several myeloid cell lines. The scheme of the development of lymphoid lineages was purely hypothetical in the original model, and the development of mature lymphoid cells (in Part 2) was not included. It was proposed that precursor B cells become committed before precursor T cells. I have modified this step to the precommitment to lymphoid precursor cells. Also, I have changed the commitment of lymphoid precursor cells to occurring after the myeloid precursor cells. However, the committed lymphoid precursor cells should still be viewed as hypothetical, because there are many examples supporting the common lineage development of B cells and myeloid cells, and not precursor T cells. Committed lymphoid precursor cells have never been clearly identified and isolated.

There is ambiguity in the original model. For example, the issue of the timing of each point of decision was not addressed. The original hypothesis implied that stem cells that do not receive the appropriate signals will proceed "spontaneously" to the next stage in the differentiation sequence. Based on much published data and on observations on the abortive development of lym-

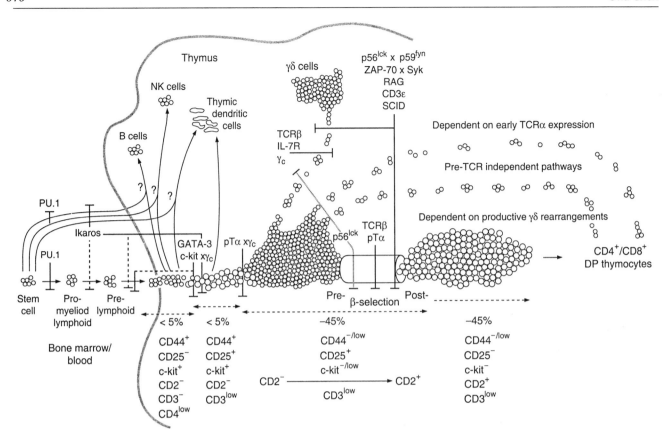

Fig. 43.3. Schematic illustration of early T cell development. Dashed lines with arrows at both ends indicate the four, developmentally successive triple negative (TN) subsets defined by the differential expression of CD25, CD44, and c-kit surface markers. Percentages give the approximate proportion of each respective subset within the TN population. The three prethymic stages of T cell development are defined only functionally, i.e., by their developmental potential. Thin arrows indicate minor pathways within the thymus that give rise to non-T lineage cells. It is not yet clear whether B, NK, and thymic dendritic cells are generated from extra-, and/or intrathymic pathways. The dark bars symbolize points of complete developmental arrest due to null mutations in specific genes mentioned on top of the respective bar. In GATA-3-deficient and c-kit × γ_c double-deficient mice, T cell development may actually be arrested before precursor entry into the thymus. This may also be true for some Ikaros-deficient mice, as indicated by dashed bars. The developmental block in $p56^{lck}$-deficient mice is partial (stippled bar). It has not been investigated to what relative extent the pre-TCR-dependent and the two minor pathways are inhibited by the absence extent the pre-TCR-dependent and the two minor pathways are inhibited by the absence of $p56^{lck}$. Other mutations that result in a partial block of thymopoiesis are not included in the scheme nor are mutations that cause a complete block only during fetal, but not adult, thymopoiesis. γ_c denotes the common cytokine receptor γ chain. The indicated block in thymopoiesis in $pT\alpha × \gamma_c$ double-deficient mice has not yet been published (J. DiSanto, personal communication). From Rodewald and Fehling (1998).

phoid cells from blood islands derived from embryonic stem (ES) embryoid bodies (Chen *et al.,* 1997), and on the essential requirement of cytokines to maintain and to propogate stem cells derived from transgenic fetal liver, I tend not to be in favor of spontaneous progression; appropriate inductive signals for the transcriptional activation of lineage-committed genes are stringently required. Moreover, because programmed cell death plays an important role in the fate of cells during embryogenesis, I have included apoptosis as another important factor. This model has gained great support from the phenotypes of mice in which genes encoding transcription factors have been knocked out. Transcription factors controlling the sequential commitment of HSCs have been summarized (Chen, 1996).

CRITERIA FOR ENGINEERING DEVELOPMENTAL STAGES OF LYMPHOPOIESIS

I will attempt to explore the feasibility of engineering lymphoid cells and their stem cells at several defined stages of differentiation. My criteria for deciding on feasibility are (1) intrinsic self-

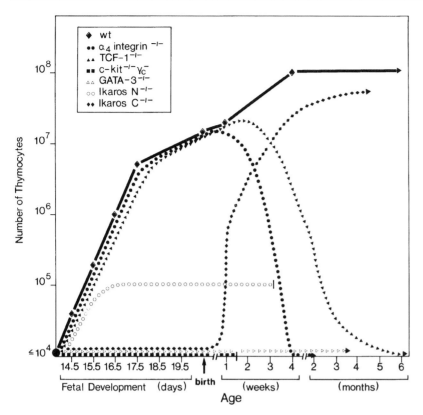

Fig. 43.4. Thymic cellularity as a function of age in wild-type and mutant mice. Total thymic cellularity is shown during both fetal and adult life. Cell numbers are given only as approximations based on data from the following mutant mice: Ikaros N$^{-/-}$ mice lack N-terminal sequences of Ikaros. This mutation appears to act in a dominant negative fashion (Georgopoulos et al., 1994). Ikaros C$^{-/-}$ mice, considered as Ikaros null mutants, carry a deletion of the C terminus of Ikaros (J.-H. Wang et al., 1996). For mutations with an early embryonic lethal phenotype, chimeric mice were generated by RAG complementation. In such chimeric mice, the effects of lack of GATA-3 (Ting et al., 1996) or α4 integrin (Arroyo et al., 1996) on T cell development were determined. Mice lacking c-kit and γ_c (Rodewald et al., 1997) live until ~10 days after birth, precluding analysis of adult thymocyte development. Disruption of TCF-1 in mice (Verbeek et al., 1995) causes a block at the earliest stages in thymopoiesis in late adult, but not fetal or early postnatal, life. From Rodewald and Fehling (1998).

renewal of HSCs, HPCs, and more differentiated cells, (2) the availability of cell surface markers for identification and purification, (3) the supply of recombinant growth factors and appropriate stroma cells, (4) the existence of favorable cell culture conditions that promote growth instead of differentiation, (5) the possibility of educating lymphocytes to become antigen-specific memory cells, and (6) the possibility of introducing, either genetically or somatically, genes of interest along with inducible promoter and regulatory sequences. Technical advances make the engineering of HSCs, HPCs, and lymphoid cells attractive for further manipulation, such as introducing new genes or reprogramming cells.

STAGES OF LYMPHOPOIESIS FOR ENGINEERING

Figure 43.1 shows a schematic diagram of the stages at which lymphoid cells and lymphoid precursor cells might be engineered. The stages will be discussed individually.

Stage 1: 0 cells

Null (0) cells are fertilized eggs or embryonic stem cell lines. Mouse ES cell lines of different strain origin have been available for a long time. Establishment of human ES-like cell lines has been reported in the United States, using expired-date frozen, fertilized eggs. Several lineage-committed cells, such as from bone and cartilage, have been shown to be obtainable through differentiation *in vitro*. However, strictly speaking, the only way to prove that an ES cell line is truly totipotent is to demonstrate germ-line transmission in human beings—a task that should not even be considered. Similarly, I think that both genetic and cellular engineering at this level should be limited to mouse studies. ES cells are stem cell lines that divide symmetrically. Unlike the lineage-committed stem cells, which are rare, ES cells are almost an exception, being readily available. They can proliferate without differentiation if feeder cells and/or cytokines are provided. They can be induced to differentiate by removing the feeder cells and cytokines. Only a few surface molecules are available for characterizing ES cells. Antibodies against stage-specific embryonic antigen-1 (SSEA-1) enable one to distinguish undifferentiated from differentiated ES cells. In recent years, many genes encoding lymphoid-specific functions have been knocked out, and more studies are in progress to develop specific deletions or replacement of genes by various techniques using homologous recombination (Figs. 43.1, 43.3, and 43.4). Vectors with inducible promoters are being used to introduce genes in a controllable, tissue-specific manner. This may be another powerful tool to study the commitment of cells from one lineage to another.

Growth requirement for differentiation of ES cells to lymphoid cells

Under appropriate conditions, ES cells can differentiate *in vitro*. Many different culture systems are available for studying the development of lineage commitment. Because ES cells are totipotential, optimal culture conditions would enable the formation of mature cells and tissues and even organs in well-organized three-dimensional structures *in vitro*. That is, one should be able to generate an artificial fetus in culture without any external cells or other factors. In fact, of course, organogenesis from ES cell culture is but a dream. Nevertheless, with the culture conditions and system that we have developed, mature and embryonic cells in some form of tissuelike organization, though never organs, of certain orderly structures (gutlike, epidermal-like, and thymuslike structures), can be observed. In recent years, many researchers have attempted but failed to study the differentiation of lymphopoietic precursor cells obtained from ES cells. Many observe the development of yolk sac blood islands in culture; these presumably contain HSCs, but only myeloid cells could be produced in the culture systems reported. Only a few systems allow the generation of lymphoid precursor cells, mostly B cells, but also T and B cells, in a quantity sufficient for further manipulation (Chen *et al.*, 1997; Potocnik *et al.*, 1994, 1997a,b).

Reasons for these variations are unclear. It is partially due to the origin and totipotentiality, hence the fragility and variability, of ES cells, but partially due to experimental manipulations. Even in the hands of a single researcher, the potential of lymphopoiesis from the same ES cell lines varies greatly among different experiments, and why this is so remains a mystery. The culture conditions vary, depending on which lineages are chosen for study. For example, methylcellulose culture is preferred for myeloid lineages but is inferior for lymphoid lineages. Several culture conditions that favor the differentiation of ES cells to lymphoid lineages are available.

Through the work of at least five groups, it has become clear that lymphoid precursor cells can be obtained from ESCs in culture. However, the development of mature B plasma cells and mature T cells requires a two-step procedure—a two-step culture procedure for plasma cells, and culture and animal implantation for mature B and T cells. When chimeric bone marrow cells from embryoid body-implanted mice, which contain cells of ESC origin, were injected into the host mice, the bone marrow cells were shown to repopulate the primary and secondary lymphoid organs, i.e., bone marrow, spleen, and lymph nodes. The data indicate that HSCs and HPCs of ESC origin can be transferred. However, further transfer experiments similar to those performed by Harrison *et al.* (1988, 1993) were not done.

Our culture system can also be used to study thymic stroma as well as well-organized, highly differentiated cells of many other lineages, including gut epithelia, skin with hair follicles, bone, muscle, neurons, and glia. The development of endogenic thymic epithelial cells as well as stroma cells inside the ES-embryonic bodies also explains partially the findings of an efficient development of T cells in the absence of exogenous stroma cells. Stroma cell line RP.0.10 is able to support both T and B lymphoid precursor cells derived from ESCs in some labs, but this is not reproducible elsewhere. Stroma cell lines OP9 and S17 support B precursor cells and myeloid but not T precursor cells derived from ESCs. Many other stroma cell lines do not support the differentiation of ESCs to lymphoid precursor cells. Müller-Sieburg and co-workers have systematically compared the S17 stroma cell line, which is widely used in many labs, to many stroma cell lines for the production of cytokines and stroma function, i.e., the stem cell support function (Wineman *et al.*, 1993, 1996) (see also below, Stage 2); there were no correlations in the assays. They postulated that a yet to-be-identified extracellular matrix molecule of S17 cell lineage may play a role in supporting the function of HPCs.

Bauer *et al.* (1998) reported that a novel protein molecule, dlk, an epidermal growth factor-like molecule of 66 kDa, in stroma cell lines influences the requirement for interleukin-7 (IL-7) in supporting mouse pre-B lymphopoiesis. They suggest that dlk may play an important role in the bone marrow HPC microenvironment. Cortex *et al.* (1999) reported the novel high-molecular-weight CD166 (HCA/ALCAM) glycoprotein expressed in human HPCs as well as stroma cells. The data suggest that this molecule is being involved in adhesive interaction between HPCs and stroma cells in the most primary blood-forming organs. Rethinking the seemingly conflicting results and the inconsistent report of differentiation of ES cells to lymphocytes, I wonder if culturing ES-derived lymphoid progenitor cells on stroma cell lines derived from the most primitive stroma environment from the fetus (such as paraaortic cells) may be a better choice to obtain consistent results, rather than using the existing stroma cell lines?

Stage 2: HSCs and HPCs

The terms "hematopoietic" stem cells, "progenitor" cells, and "precursor" cells are loosely used in the literature. Here, I define hematopoietic stem cells and hematopoietic progenitor cells as two populations of stem cells sharing similar surface markers. HSCs are rather quiescent, noncycling stem cells, whereas HPCs are cycling cells. In mice, c-kit$^+$, thy-1^{low}, lin$^-$, sca1$^+$ bone marrow cells are accepted as HSCs or HPCs. Many groups have shown that HSCs and HPCs in bone marrow are heterogeneous in size and in self-renewal capacity. Moreover, they have finite life-spans. In human bone marrow, CD34$^+$, HLA$^-$, lin$^-$, thy-1^{low}, rhodamine 123^{low} stem cells are generally defined as HSCs and HPCs. However, CD34 is also expressed in other lineages, such as vascular endothelial cells. It is well established that mouse HSCs and HPCs can be isolated, cultured, transduced by retroviruses, and used to repopulate animal hosts. In addition to bone marrow, they can be isolated from fetal liver, cord blood, fetal blood, yolk sac, and paraaortic spanchnopleura.

Do HSCs and HPCs self-renew?

The self-renewal of HSCs and HPCs was carefully examined by Harrison and co-workers (1988, 1993). By transferring bone marrow stem cells from one host to the other, they found that the pool of donor bone marrow stem cells was smaller on transfer, and concluded that self-renewal is limited for somatic bone marrow stem cells. However, their results could be due to a dilution effect of the donor stem cells, if a certain absolute number of stem cells were required for the successful implantation of donor stem cells in the bone marrow. Using fluorescence *in situ* hybridization (FISH) and telomerase assays (Allsopp *et al.,* 1992), it was shown that telomere DNA length predicts the age and replication capacity of human fibroblasts. Similarly, Lansdorp and co-workers (Vaziri *et al.,* 1994; Lansdorp, 1995a,b) have correlated telomere length with the replication potential of HSCs and HPCs. In human CD34$^+$, CD71^{low}, CD45RA^{low} bone marrow stem cells, the length of telomeric DNA in HSCs and HPCs is correlated with the age of the donor. The stem cells in adults have shorter telomerases than are found in fetal cells, but both have the same telomerase activity. When the HPCs were expanded with cytokines without the support of stroma cells, there was a loss of telomeric DNA in culture. The loss of telomeric DNA in HSCs and HPCs from older people and cultured stem cells was interpreted to mean that their replication potential is finite. An alternative interpretation would be that the culture conditions that allow stem cells to grow in the presence of cytokines may not favor the maintenance of telomerase activity and hence not the self-renewal of HSCs and HPCs.

Many culture systems for studying the biology of mouse HSCs and HPCs are available. Many mouse stroma cells lines support the growth of HPCs. Studies on human stem cells use mouse stroma cell lines such as S17 (Collins and Dorshkind, 1987) and others (Wineman *et al.,* 1993), or mixed primary stroma populations isolated from human bone marrow or other sources (Dittel and LeBien, 1995), to support purified HSCs and HPCs or unseparated bone marrow cells. There was a report of the establishment of heterogeneous human stromal cell lines with the help of a plasmid containing SV40 T antigen under the control of an inducible metallothionein promoter (Cicuttini *et al.,* 1992), but their ability to support lymphopoiesis could not be demonstrated. Simmons *et al.* (1992) have obtained mixed human stroma cells and cell lines that support human long-term culture initiating cells. However, these human stroma cell lines could not be shown to support the development of human lymphocytes from bone marrow stem cells.

Major problems in this field are the variability of culture conditions, the efficiency of differentiation of the cultured cells to lymphoid lineages, and the poor reproducibility in the hands of different investigators. Also contributing to problems is failure to report specific reagents and ingredients used as well as undescribed procedures, stroma cell conditions, batches of serum, and growth factors required in each system. Of course, the intrinsic multipotentiality of HSCs and HPCs also contributes significantly to the elements of variability. Many have claimed that a cocktail of cytokines alone could promote the differentiation of myeloid lineages from HSCs and HPCs in mouse (Timothy *et al.,* 1995) and human (Henschler *et al.,* 1994) systems. A few have claimed the development of B cells at various stages of maturation, but no mature T cells could be found in such culture conditions. The maintenance of long-term cultures of self-renewing stem cells with potency for lymphopoiesis, especially T lymphopoiesis, requires stroma cells, additional cytokines, and other culture conditions yet to be defined. Several existing culture systems seem to be promising for expansion *ex vivo* and potential differentiation of human (Koller *et al.,* 1993; Dittel and LeBien, 1995) and mouse HSCs. Because large quantities of cultured cells are needed for manipulation and reimplantation, the development of bioreactors and the development of reliable human stromal cell lines for lymphopoiesis are of high priority.

Selected examples of clinical applications of human HSCs and HPCs

Using human HSCs and HPCs as the cell base for gene therapy is a complicated issue. In most clinical protocols, HPCs are transduced *ex vivo* with retroviral vectors and reimplanted into patients. One example currently undergoing phase I clinical trial is the treatment of adenosine deaminase (ADA) deficiency. ADA deficency exists in all cells examined. However, T cells of these patients are selectively missing, causing a server combined immunodeficiency symptom. Despite more than three decades of study, the mechanism by which the ADA defect causes specific deficiency of T cells remains unknown. Four possible mechanisms have been postulated and reviewed by Resta and Thompson (1997). The apoptotic pathway of CD95 (Fas/apo-1)-induced cell death might be the most likely mechanism. Patients suffering from ADA deficiency have been treated using ADA gene transfer therapy (Bordignon, 1993; Blaese *et al.,* 1993). Stem cells were expanded *in vitro* and transduced with retroviral vectors containing ADA genes. Several ADA-deficient children were subsequently followed up using the polymerase chain reaction (PCR) assay to assess the expression of circulating T cells bearing the transduced ADA gene. Peripheral blood carrying the transduced ADA gene was shown to persist for 7 months before disappearing. Bone marrow cells appeared later and became the major source of T cells expressing the transduced ADA gene in the peripheral blood.

Although initial reports of these studies claimed to be successful, careful evaluation of ADA gene therapy protocols shows that there is a major problem with the protocols. During *ex vivo* expansion and manipulation for gene transfer, HPCs lose their capacity to proliferate in favor of differentiation and maturation. Thus there is a great need for repeated treatments of the same patient with retrovirus transduction, and continuous infusion of poly(ethylene glycol)-conjugated ADA enzyme. There is another trial using autologous cord blood CD34⁺ cells in ADA-deficienct babies, and the combined gene and enzyme therapies seem to generate better immune function (Kohn *et al.,* 1998). The different results are due to the fact that very few mature, functional T cells make enough enzyme to detoxify the serum in ADA patients. If stem cells are rare, small differences in technique and uncontrolled variables could make the difference between success and failure. Another, not mutually exclusive, interpretation of the lack of long-term reconstituting stem cells is that the cycling HPCs transduced are not the true stem cells. Human immunodeficiency virus (HIV)-based lentiviral vector has been shown to efficiently transfer foreign genes into postmitotic and noncycling cells, including HSCs. Nevertheless, this vector is still under development, and no package cell line is yet available for lentiviral vectors. Improved culture conditions for growth and maintenance of human stem cells in large quantities are urgently required. The establishment and use of reliable human stroma cell lines that could support the growth of lymphopoietic HPCs *in vitro* might prove to be a breakthrough in this field.

Stage 3: Lymphoid precursor cells

Until recently, no published work described the characterization of lymphoid lineage-committed precursor cells defined in terms of cell surface markers and functional assays of precursors and progenies. One cell type that is described closest to the definition of lymphoid precursor cells is the CD4^low precursor population isolated from adult thymus by Shortman and co-workers (Shortman and Wu, 1996). Because this population of cells is restricted in its potential to differentiate into myeloid cells, it is preferentially committed to differentiate into T and B cells. Amazingly, these cells can also differentiate into thymic dendritic cells. The molecular markers of this population have not been well characterized. It would be interesting to see whether this population of cells can differentiate into myogenic cells, because J. Morgan and T. Partridge (personal communication) did obtain differentiated myogenic cells from adult thymus. Thymus contains lymphoid and presumably mesenchymal precursor cells that can develop into other lineages, such as bone, cartilage, and lung (Pereira *et al.,* 1995).

Nothing is known about the culture conditions for growing this population of cells. It is possible that stroma cell lines such as S17 could be used to expand the CD4^{lo2} population of cells in culture, because Godin *et al.* (1995) was able to use S17 and cytokines to maintain paraaortic spanchnopleura-derived precursor cells possessing the T cell lineage potency in culture for some time. Originally, Ikaros, a lymphoid lineage-specific transcriptional factor (Georgopoulos *et al.,* 1994), was thought to be an important molecular marker of lymphoid precursor cells at this specific stage of development. Ikaros was subsequently shown to have multiple effects on HSCs, lymphoid cells, T precursor cells, and others (Nichogiannopoulou *et al.,* 1998) (see Fig. 43.4).

The human counterpart of the CD4^{low} population is unknown. Dittel and LeBien (1995) reported the expansion of CD34$^+$, CD31$^+$ human bone marrow B progenitor cells and partial differentiation to B precursors in serum-free culture medium in the presence of mixed primary human stroma cells and IL-7 for a duration of 16 days. A few CD34$^+$ precursor cells maintained under their culture condition could differentiate into natural killer (NK) cells and T precursors when subjected to a secondary culture conditions, presumably providing the precursor population with the thymic environment. It is not clear whether T precursor cells are derived from HPCs or are the precommitted precursor cells migrating to bone marrow.

Stage 4: Precursor T cells

Precursor T cells are thy-1$^+$, CD117^{low}, DC3$^-$, pTα^+ (Fehling *et al.,* 1995; Kisielow and von Boehmer, 1995; Fehling and von Boehmer, 1997). They can be derived from fetal and adult blood. Using new markers such as the SCA-2 and CD44 variants, they can be classified into subpopulations (Rodewald and Fehling, 1998) (see Figs. 43.3 and 43.4). Both somatic and genetic manipulations seem to be possible. However, this population of cells is not self-renewing. Large-scale expansion of this population is questionable, because apoptosis occurs easily, and only some cells of this compartment differentiate into double-positive mature T cells. Fetal thymus organ culture systems and suspension cultures with dissociated thymic epithelial cells for studying the differentiation and education of T cells are available (Jenkinson *et al.,* 1982; Jenkinson and Anderson, 1994). However, how to expand precursor T cells *ex vivo* without differentiation and apoptosis will be a challenge to the workers in this field. Changing the O$_2$ concentration in the fetal organ culture system was shown to help the survival of T cells and precursor T cells (Mackinnon and Ceredig, 1986; McLaughlin *et al.,* 1996). Perhaps O$_2$ and other radicals stimulate intracellular molecules that lead to the apoptotic pathway. Reducing the O$_2$ concentration with N$_2$ or a high density of stroma cells favors the survival of (precursor) T cells. Although Godin *et al.* (1995) can transiently expand paraaortic spanchnopleural precursor T cells in their culture conditions, that population of precursor T cells perished after a few generations in culture. Whether the additional cytokines and reducing the O$_2$ concentration could help to keep these cells proliferating longer remain to be tested.

Stage 5: Memory T cells

Using cultures of mature T cells in the presence of antigen-presenting cells and recombinant cytokines, one can study the growth and differentiation of T cells from a variety of sources. Long-term mouse and human T cell clones are also available. For unknown reasons, T cells can be maintained as clones in culture far longer than B cells. However, long-term human T cell clones, even from healthy donors, have not been reimplanted into patients. Human T cells (mainly natural killer cells) grown in the presence of cytokines in a short-term culture have been reimplanted into autologous cancer patients. No study was reported to evaluate their long-term repopulating potentiality in the human severe combined immunodeficient (SCID) model or the self-renewal property of long-term mouse T cell clones *in vivo.* Most CD8$^+$ and CD4$^+$ T cells have a short life-span. Memory T cells have been well studied, but, in reality, they do not exist in abundance and can be demonstrated *in vivo* only by repeated priming with antigen. They are very difficult to define at cellular and molecular levels. There is no isotype 3′-end downstream of the TCR constant region for class switching to occur. Somatic hypermutation of the TCR-β gene has never been claimed, and it is controversial whether the TCR-α gene is hypermutable. Memory T cells are thus defined using criteria such as accelerated cellular responses, distinct pathways of lymphocyte recirculation *in vivo,* distinct DNA motifs of TCR genes, cytokine-producing pattern and diminishing expression of surface markers (Bradley *et al.,* 1993; Dutton *et al.,* 1998), and functional and antigen requirements (Gray and Matzinger, 1991). In mice, memory CD4$^+$ T cells are CD45RO, L-selectin

(MEL-14)[lo2]. There is the equivalent of human CD45Ro, keeping in mind that the CD45R family (A, B, C, and O) may not be the best marker to define naive versus memory T cells. Because foreign antigens do not always quickly elicit unprimed T cells, memory T cells must exist, but the commitment, the mechanism of development, and the maintenance of these cells are unknown. Memory T cells are thought to be generated either when T cells acquire specificity to kill or to help during thymic education, or are generated during the mature stage.

Memory DC8+ T cells

Because cell-bound antigen on antigen-processing cells and more than one signal are required for educating T cells to perform effector functions instead of becoming tolerized, work on the generation of CD8+ memory T cells has been mainly *in vivo,* using viruses. Although the short-term expansion of memory T cells can be demonstrated in culture, it is not clear how long these cells can be maintained in culture. Interesting questions remain to be addressed: Is it possible to prevent cytotoxic T cells from performing their function by releasing granzymes, and, if so, what happens to committed CD8+ T cells? Do they die, or become anergic, or become memory cells? A study addressing the avoidance of granzyme B-induced apoptosis in target cells is interesting in this regard (Greenberg and Litchfield, 1995; Wang *et al.,* 1996). Dephosphorylation of cdc2 was shown to be a critical step in granzyme B-induced apoptosis in the targets of cytotoxic T cells. A nuclear kinase encoded by the *wee1* gene was transiently expressed and shown to induce phosphorylation of the tyrosine residues of cdc2 kinase, and that in turn provoked mitosis and the rescue of target cells. Because cytotoxic T cells are subject to being killed by their colleagues, the apoptosis pathway in these cytotoxic T cells could be similar to that of the target cells. It remains to be demonstrated whether expression of *wee1* or equivalents could be induced to rescue cytotoxic T cells to become memory T cells.

Memory CD4+ T cells

Like memory CD8+ T cells, the generation of memory CD4+ T cells is studied *in vivo.* This population of cells has been well characterized in mice. By using the soluble protein antigen keyhole limpet hemocyanin (KLH), Bradley *et al.* (1993) found effector memory CD4+ cells in mouse spleen to be CD45RO, L-selectin−, CD44+ and to produce elevated levels of IL-4. McHeyzer-Williams and Davis (1995) used a different soluble antigen-priming system to study memory CD4+ T cells. Whether this population of cells can be expanded *ex vivo* has not been reported. Because T cells of IL-2 knockout mice survive much longer than those of conventional mice, they may be useful for studying the development of memory T cells. Moreover, molecules such as Fas and Fas ligands of the apoptotic pathways play critical roles in determining the fate of cells after activation. The Bcl-2/bcl-x family seems to function by helping the survival of cells via action on the intermediate steps of apoptosis (Nagata and Golstein, 1995; Krammer *et al.,* 1994; Korsmeyer, 1995). Genes controlling the cell death pathway may play critical roles in the development of, and the subsequent genetic manipulation of, memory T cells in the future. It is important to improve the conditions for growing CD8+ T cells *in vitro,* because for tumor therapy, the generation of tumor-specific CD8+ T cells is a very critical step.

Although surface markers and life-span of a population of T cells equivalent to those of the mouse system have been documented (Michie *et al.,* 1992), ethical and safety considerations have prevented a systematic study of human memory T cells *in vivo.* Thompson *et al.* (Boise *et al.,* 1995; Noel *et al.,* 1996; Levine *et al.,* 1997) have reported the possibility of producing human CD4+ T cells on a large scale. By stimulating peripheral, resting T cells with cytokines such as IL-2, anti-CD28, and solid-phase anti-CD3, they reported that survival and proliferation of CD4+, but not CD8+, T cells could be greatly promoted. What remains to be shown are the specificity and function of these cells and how they are related to the regulation of the Bcl-2/bcl-x family (Boise *et al.,* 1993; Ma *et al.,* 1995) and CD95 (Fas/apo-1) ligand, and whether they are candidate memory CD4+ T cells or abortive T cells.

Stage 6: Precursor B cells

Mouse and human precursor B cells have been characterized extensively by several groups. Surface markers and molecular events of precursor B cells at intermediate stages of development have been defined (Table 43.1).

Table 43.1. Expression of cellular and molecular markers during stages of B cell development

Human

B Cell	Progenitor B stage		Precursor B stage		Research groups[a]
	I	**II**	**III**	**PreB**	
CD34	+	+	+	−	Cooper, LeBien, Schiff-Fougereau
CD38	nd	+	+	+	
ΨL	−	+	+	+	
CD10	nd	−+	+	+	
CD19	−	−	+	+	
λ-like/Vpre-B	+	+	+	+	
Rag1	+	+	+	+	
TdT	+	+	+	−	
VH/Cμ	−	+	+	+	
Vκ/Cκ	−	+	+	+	
mb1/B29	+/−	+/−	+	+	
cyto-μ	−	−	+/−	+	

Mouse

B Cell	**A**	**B**	**C**	**C'**	**D**	Research groups[a]
B220	+	+	+	+	+	Hardy, Rajewsky
CD43	+	+	+	+	−	
HSA	−	+	+	+	+	
BP1	−	−	+	+	+	
λ5/VpreB	+/−	+	+	+	+	
Rag 1–2	+/−	+	+	+	+	
TdT	+/−	+	+	+/−	−	
mb1	+/−	+	+	+	+	
D-JH		+	+			
VH-D-JH			+ (out)		+(on)	

B Cell	**ProB**	**PreB I**	**PreB II**		Research groups[a]
			Large	**Small**	
CD43	+	+	+ −	−	Rolink, Melchers
c-Kit	+	+	−	−	
CD25	−	−	+	+	
ΨL	+	+	+	+	
λ5/VpreB	+	+	+	+	
Rag 1–2	+	+	−	+	
TdT	+	+	−	−	
cyto-μ	−	−	+	+	
D-JH		+			
VH-D-JH			+	+	
VL-JL				+	

[a]Key references: Dittel and LeBien (1995); Erlich *et al.* (1994); Lassoued *et al.* (1993); Li *et al.* (1993) Loeffert *et al.* (1994); Rolink *et al.* (1994); Melchers *et all.* (1995); Ghia *et al.* (1998).

Mouse pre-B cell lines

In the mouse system, with the help of several stroma cell lines and recombinant growth factors such as IL-7 (Rolink *et al.,* 1991; see Melchers *et al.,* 1995), it becomes feasible to expand mouse pre-B cells without gross differentiation. On release from the stroma cells and IL-7, and in the presence of the bacterial mitogen lipopolysaccharide (LPS), some pre-B cell lines differentiate into plasma cells. Pre-B cell clones have been established from various lymphoid organs of wild-type mice, transgenic mice, and some knockout mice. Thus, both somatic and genetic manipulation of these cell lines becomes possible. Some pre-B cell lines have been used to repopulate SCID mice and RAG-2 knockout mice. Injected cells migrate to bone marrow, lymph nodes, peritoneal cavity, and spleen. Plasma cells, mature B cells, and pre-B cells were detected in the host. The percentage of cells mature into various B compartments seems to vary from experiment to experiment.

However, several questions remain to be answered. Do these pre-B cell lines retain the capacity to expand? Are these cells self-renewing *in vivo,* as stem cells must be? The critical experiment of repeatedly transferring donor pre-B cells, from one host to another, to show that the implanted cells are still precursor B cells, has not yet been done. It is interesting that pre-B cells established from pax $5^{-/-}$ mice (Urbanek *et al.,* 1994) differentiated into T cells, myeloid cells, dendritic cells, and osteoclasts, but not mature B cells, in SCID mice. The data indicate that the microenvironment which plays a role in keeping pre-B cells committed to B cell lineage, may be missing in the pax $5^{-/-}$ mice (A. Rolink, personal communication).

Curiosities of growing human precursor B cells

The question of whether human precursor B cells can expand *ex vivo* is still open. Several reports claim it possible by using either primary mixed human stroma cells (Dittel and LeBien, 1995) or mouse stroma cells (Rawlings *et al.,* 1995). When subjected to mixed stroma cells and cytokines such as IL-7, most human bone marrow cells expand for a limited period, then either perish or differentiate (see Ghia *et al.,* 1998). In other words, the culture conditions established so far are only for short-term expansion of pre-B cells. However, no normal human pre-B cell lines have been established. This distinguishes mouse and human precursor B cells. The study of human pre-B cells cannot be separated completely from the study of human HSCs and HPCs, given the complexity of the cell types involved and the difficulty in establishing human stroma cell lines. The establishment of human stroma cell lines supporting the growth of human pre-B cells is critical. The failure lies in the inability of adherent cells derived from human bone marrow (stroma cells) to proliferate well under normal culture conditions with conventional sera. Methods for immortalizing human cells include transfecting plasmids or retroviruses containing oncogenes such as SV40 T antigen. Cell proliferation is a prerequisite for the stable integration of transgenes into chromosomes and for immortalization. A breakthrough for human HSCs and pre-B cells would be to establish stroma cell lines for B lymphopoiesis, and to optimize the conditions for the growth of stroma cells, as has been done for human mesenchymal stem cells (Yoo *et al.,* 1998).

Stage 7: Memory B cells

B cells with surface Ig isotypes such as IgG, IgA, and IgE, which possess higher affinities for Ag, are generally defined to be memory B cells. In human beings, $CD38^+$, $CD20^+$ germinal center B cells can be distinguished from memory $CD38^-$, $CD20^+$ B cells and $CD38^+$, $CD20^-$ plasma cells. In mice, there are no reliable surface markers to distinguish memory B cells from plasma cells and mature primary B cells. The mechanisms of memory B cell generation are unknown, and the subject of debate for decades. Memory B cells are mature B cells that have encountered antigen, have been activated but not tolerized, and have switched to higher affinity isotypes. IgM-secreting plasma cells are terminal cells, destined to die. It is unknown how memory B cells develop *in vivo. In vitro,* at least two systems may generate memory B cells: the germinal center-like culture system and the suspension culture system, wherein resting B cells are activated by LPS plus anti-μ. These B cells are activated, alive, proliferating, and nonabortive, but they are not plasma cells.

A system stimulating germinal centers in vitro to culture B cells

In a culture environment that mimics the germinal center in lymphoid organs, B cells survive better and live longer. A germinal centerlike environment is a culture system that provides cytokines and supporting cells from either purified follicular dendritic cells (FDCs) from mouse spleen (Kosco *et al.,* 1992), or stroma cells (L cells transfected with CD40 ligand, or fibroblasts) (Flueckiger *et al.,* 1993; Arpin *et al.,* 1995; Wohlleben *et al.,* 1996). Both human and mouse B cells survive for 2 weeks instead of 3–4 days, and absolute cell numbers increase two- to threefold.

The *in vitro* germinal center culture system developed by Kosco and co-workers was designed to study the differentiation of mouse B cells into plasma cells rather than to maintain long-term B cells *in vitro.* Mouse FDCs are nonproliferating, terminally differentiated cells of unknown origin. However, other studies reported that the primary mouse FDCs can be partially replaced by fibroblast cell lines expressing CD40 ligand. The cytokines required to maintain mature B cells in growth phase and differentiation are controversial. For mouse B cells, combined IL-2, IL-4, and IL-5 induce differentiation into mature cells, and for further differentiation into plasma cells, IL-

6 seems to be essential. For human B cells, IL-2 plus IL-10 and combined IL-3, IL-6, and IL-7 have been reported to work. Also, if the CD40 ligand is removed in the secondary culture, human B cells differentiate into plasma cells. These memory-like cells are neither cell lines nor cell clones, rather they are a mixed B cell type with limited life-span (up to a few weeks), and they preferentially switch to certain Ig isotypes, such as IgG and IgA. The data suggest that down-regulation of the J chain may not be essential during the development of memory B cells.

A suspension culture system for stimulating B cells with LPS plus anti-μ

Systems for the short-term culture of primary splenic lymphocytes have long existed (Mishell and Dutton, 1966, 1967; Marbrook, 1967). These systems are invaluable for studying the proliferation and differentiation of B cells, T and B cell interaction, the priming of B cells by antigen, and the mechanism of memory B cell generation.

If B cells could be kept alive and untolerized, but could be prevented from becoming IgM secretory plasma cells, they might become memory B cells. An example is the finding that when stimulated with a bacterial mitogen, lipopolysaccharide, some B cells die and some proliferate and become plasma cells (Andersson *et al.,* 1974). When stimulated with anti-μ, most B cells die right away, some proliferate and exhibit growth arrest at G_1 phase, then die 2 days later, and none become plasma cells. When stimulated with LPS plus anti-μ, most B cells proliferate and none become plasma cells. This nonplasma cell phenomenon has been known as an antidifferentiation effect, and it was postulated to be a way to generate memory B cells (Chen-Bettecken *et al.,* 1985). Through efforts of many, the molecular mechanism of this antidifferentiation phenomenon became clear (Chen-Bettecken *et al.,* 1985; Chen, 1988; Leandersson and Hsu, 1985; Phillips *et al.,* 1996). In the presence of the two stimuli, B cells proliferate maximally, over 90% being in the cell cycle (Seyschab *et al.,* 1989), but IgM secretion is turned off. The block has been shown to be primarily at the level of nuclear RNA processing of the μm-to-μs switch. Inducible nuclear factors binding to the pre-mRNA secretory polyA site have been reported, though the nature of these factors remains unclear because they have not been cloned.

Transcription factors active in the development of plasma cells

On activation of B cells, many transcription factors become engaged in the production and secretion of Ig genes. B cell-specific transcription factors include Oct-2 (Chen *et al.,* 1991; Corcoran *et al.,* 1993, 1994) and Blimp-1 (Turner *et al.,* 1994; Schliephake and Schimpl, 1996; Lin *et al.,* 1997; Messika *et al.,* 1998). Both transcription factors play a role in the decision to switch from μm to μs. Blimp-1 is described as a cofactor of transcription factor pu.1 and was shown to bind to multiple Ig-enhancer motifs and the J chain regulatory element. It is crucial for the transcription of μ, κ, and J chains. Blimp-1 is active in many cell lines and drives the maturation of B cells into Ig-secreting cells in many cell lines. Blimp-1 has been shown to down-regulate c-*myc* in B cell lines.

Transcription factor Oct-2 binds to the μ intron enhancer motif and is essential in transcriptional activation of the μ chain. It was shown that Oct-2 and the J chain are highly expressed in LPS-stimulated B cells and are diminished in LPS + anti-μ-stimulated B cells. It has been shown that Blimp-1 is highly expressed in LPS-stimulated B cells and is diminished in LPS + anti-μ-stimulated B cells. On the other hand, sterile γ chain is highly expressed in the latter system. Transfection of *Blimp-1* into LPS + anti-μ-activated B cells provoked them to become IgM-secreting plasma cells. The data indicated that transcription factors such as Oct-2 and Blimp-1 are tightly regulated in plasma cell development. If one postulates that LPS + anti-μ-stimulated B cells represent some stage in memory B cell development, the down-regulation of Oct-2 and Blimp-1 reflects the specific transcriptional regulation when B cells make the commitment to the memory cell pathway instead of the plasma cell pathway. The identification of such candidate transcription factors that control memory B cell commitment would provide a powerful genetic tool to further manipulate the turning on and off of these lineages at will *in vivo*.

CONCLUDING REMARKS AND PROSPECTS FOR LYMPHOCYTE ENGINEERING

I discussed and summarized the current understanding of the expansion of lymphoid cells and their precursors *ex vivo* at certain stages of lymphopoiesis. I have tried to address the feasibility of expanding lymphoid cells under controlled growth conditions; that is, we need to expand untransformed, nonmalignant cells. In general, in order to maintain the status of cell survival and growth without apoptosis and differentiation, cytokines and cell contact with feeder cells are required. Fundamental questions regarding the engineering of lymphoid cells and their precursors for therapeutic purposes remain, and can be traced to our current understanding of the immune system. Do we ask too much for the survival in culture of cells that are programmed to die? From extensive studies in gene-manipulated mice, it is possible to generate antigen-specific memory T cells; it is still puzzling that there is no good systematic study of human memory cells in culture, although the surface markers are defined.

If there is massive programmed cell death during the development from HSCs to lymphoid precursors and from pre-T to T cell maturation, I wonder whether it is realistic to try to produce enough HSCs, HPCs, and precursor lymphoid cells for therapeutic purposes? In current clinical protocol, expansion of cells *ex vivo* for the purpose of reinfusion into patients is limited to as few passages as possible in order to avoid mutation and contamination *in vitro*. Bioreactors for large-scale production of cells in liquid suspension using cytokines are available. However, they are not designed for coculturing of stem cells with stroma cells. With advances in culture technology and bioreactors, and with increased supply of recombinant cytokines, it becomes possible to obtain a quantity sufficient for reimplantation from 10 ml of bone marrow cells. However, under these condition, very few cells engage in lymphopoiesis. Thus, to grow the HSCs and HPCs consistently and to favor lymphopoiesis, there is a great need for a better way to grow human stem cells using human stroma cell lines.

In view of the massive apoptosis at several stages of lymphopoiesis, I am amazed that mouse precursor B cells can grow normally, become lines and clones, and retain the potential to differentiate *in vitro* and *in vivo*. To grow cells from other stages of lymphopoiesis, it might be advantageous to use cells from the many available mutated mice (such as the $p53^{-/-}$, $Rb^{-/-}$, $IL\text{-}2^{-/-}$, perforin$^{-/-}$, and BCl-2 transgenics). I am still optimistically thinking that the future of cell-based immune system therapy lies in the *ex vivo* expansion of cells. Because the techniques to establish ligand-regulatable vectors are available, the derived cell lines will become available soon and will become valuable resources for many purposes. Other areas remain to be improved, including the search for novel markers of true HSCs and precursor lymphocytes, better sources of HSCs and HPCs such as cord blood and fetal tissues, improved retroviral vectors, and large-scale culture systems for expansion of HSCs and HPCs and precursor lymphoid cells.

With advances in genetic tools and mutated mice technology, it is possible to turn on or off the transcription factors that control the differentiation of cells. Thus, immunology will continue to be a very exciting field in the next decade. One good example is the transcription factor Blimp-1, which is thought to serve as the checkpoint to control the switch from μm to μs, and J chain transcription, a critical juncture for plasma cell versus memory B cell development. The prediction would be that the specific turning on and off of this gene under the control of inducible promoter and regulatory sequences would allow one to switch at will from plasma cell to memory B cell and vice versa.

ACKNOWLEDGMENTS

The author wishes to thank Charley Steinberg, Anneliese Schimpl, and Iris Motta for fruitful discussions as well as critical reading and editing of this chapter. The author's research is supported by grants from the German DFG (No. BE793/8-1) and the European Community (EC-Biotech BIO4-CT95-0284).

REFERENCES

Allsopp, R. C., Vaziri, H., Patterson, C. *et al.* (1992). Telomere length predicts replicative capacity of human fibroblasts. *Proc. Natl. Acad. Sci. U.S.A.* **89**, 10114–10118.

Andersson, J., Bullock, W. W., and Melchers, F. (1974). Inhibition of mitogenic stimulation of mouse lymphocytes by anti-mouse immunoglobulin antibodies. *Eur. J. Immunol.* **4**, 715–722.

Arpin, C., Dechanet, J., Koten, C. V. *et al.* (1995). Generation of memory B cells and plasma cells in vitro. *Science* **268**, 720–722.

Bailey, D. (1986). Genetic programming of development: A model. *Differentiation (Berlin)* **33**, 89–100.

Bain, G., Maandag, E. C. R., Izon, D. J. *et al.* (1994). E2A proteins are required for proper B cell development and initiation of immunoglobulin gene rearrangements. *Cell (Cambridge, Mass.)* **79**, 885–892.

Bauer, S. R., Ruiz-Hidalgo, M. J., Rudikoff, E. K. *et al.* (1998). Modulated expression of the epidermal growth factor-like hemeotic protein dlk influences stromal-preB-cell interactions, stromal cell adipogenesis, and pre-B-cell IL-7 requirements. *Mol. Cell. Biol.* **18**, 5247–5255.

Blaese, R. M., Culver, K. W., Chang, L., Anderson, W. F., Mullen, C., Nienhuis, A., Carter, C., Dunbar, C., Leitman, S., Berger, M., *et al.* (1993). Treatment of scid due to ADA with CD34⁺ selected PBL transduced with ah-ADA. *Hum. Gen. Ther.* **4**, 521–527.

Boehm, T., Foroni, L., Kaneko, Y. *et al.* (1991). The rhombotin family of cysteine-rich LIM-domain oncogenes: Distinct members are involved in T-cell translocations to human chromosomes 11p15 and 11p13. *Proc. Natl. Acad. Sci. U.S.A.* **88**, 4367–4371.

Boise, L. H., Gonzalez-Garcia, M., Postema, C. E. *et al.* (1993). Bcl-x, a bcl-2-related gene that functions as a dominant regulator of apoptotic cell death. *Cell (Cambridge, Mass.)* **74**, 597–608.

Boise, L. H., Minn, A. J., June, C. H. *et al.* (1995). Growth factors can enhance lymphocyte survival without committing the cell to undergo cell division. *Proc. Natl. Acad. Sci. U.S.A.* **92**, 5491–5495.

Bordignon, C. (1993). Progress toward the clinical application of somatic cell gene therapy. *Curr. Opin. Hematol.* **1**, 246–251.

Bradley, L. M., Duncan, D. D., Yoshimoto, K., and Swain, S. L. (1993). Memory effectors: A potent, IL-4-secreting helper T cell population that develops in vivo after restimulation with antigen. *J. Immunol.* **150**, 3119–3130.

Brown, G., Bunce, D. M., and Guy, G. R. (1985). Sequential determination of lineage potentials during haemopoiesis. *Br. J. Cancer* **52**, 681–686.

Brown, G., Bunce, D. M., Lord, J. M., and McConnell, F. M. (1988). The development of cell lineages: A sequential model. *Differentiation (Berlin)* **39**, 83–89.

Chen, U. (1988). Anti-IgM antibodies inhibit IgM expression in lipopolysaccharide-stimulated normal murine B-cells: Study of RNA metabolism and translation. *Gene* **72**, 209–217.

Chen, U. (1996). *In* "Principles of Tissue Engineering," Chapter 33, pp. 527–549. Academic Press, San Diego, CA.

Chen, U., Scheuermann, R., Wirth, T., Gersten, T., Roeder, R. G., Harshman, K., and Berger, C. (1991). Anti-IgM antibodies down modulate μ-enhancer activity and OTF2 level in LPS stimulated mouse splenic B-cells. *Nucleic Acids Res.* **19**, 5981–5989.

Chen, U., Esser, R., Kotlenga, K. *et al.* (1997). Potential application of quasi-totipotent ES cells, a ten-year study of soft-tissue engineering with ES cells. *J. Tissue Eng.* **3**, 321–328.

Chen-Bettecken, U., Wecker, E., Schimpl, A. (1985). IgM RNA switch from membrane to secretory form is prevented by adding anti-receptor antibody to LPS stimulated murine primary B-cell cultures. *Proc. Natl. Acad. Sci. U.S.A.* **82**, 7384–7388.

Cicuttini, F. M., Martin, M., Salvaris, E. *et al.* (1992). Support of humancord blood progenitor cells on human stromal cell lines transformed by SV40 large T antigen under the influence of an inducible (Metallothionein) promoter. *Blood* **80**, 102–112.

Collins, L. S., and Dorshkind, K. (1987). A stromal cell line from myeloid long-term bone marrow cultures can support myelopoiesis and B lymphopoiesis. *J. Immunol.* **4**, 1082–1087.

Corcoran, L. M., and Karvelas, M. (1994). Oct-2 is required early in T cell-independent B cell activation for G1 progression and for proliferation. *Immunity* **1**, 635–645.

Corcoran, L. M., Karvelas, M., Nossal, G. J. V. *et al.* (1993). Oct-2, although not required for early B cell development, is critical for later B cell maturation and for postnatal survival. *Genes Dev.* **7**, 570–582.

Cortes, F., Deschaseaux, F., Uchida, N. *et al.* (1999). HCA, an immunoglobulin-like adhesion molecule present on the earliest human hematopoietic precursor cells, is also expressed by stromal cells in blood-forming tissues. *Blood* **93**, 826–837.

Dittel, B. M., and LeBien, T. W. (1995). The growth response to IL-7 during normal human B cell ontogeny is restricted to B-lineage cells expressing CD34. *J. Immunol.* **154**, 58–67.

Dutton, R., Bradley, L., and Swain, S. (1998). T cell memory. *Annu. Rev. Immunol.* **16**, 201–223.

Ehlich, A., Martin, V. Mueller, W., and Rajewsky, K. (1994). Analysis of the B-cell progenitor compartment at the level of single cells. *Curr. Biol.* **4**, 573–583.

Fehling, H. J., and von Boehmer, H. (1997). Early αβ T cell development in the thymus of normal and genetically atered mice. *Curr. Opin. Immunol.* **9**, 263–275.

Fehling, H. J., Krotkova, A., Saint-Ruf, C., and von Boehmer, H. (1995). Crucial role of the pre-T cell receptor α gene in development of αβ but not γδ T cells. *Nature (London)* **375**, 795–798.

Flueckiger, A. C., Farrone, P., Durand, I. *et al.* (1993). Interleukin 10 (IL 10) upregulates functional high affinity IL-2 receptors on normal and leukemic B lymphocytes. *J. Exp. Med.* **178**, 1473–1481.

Georgopoulos, K., Bigby, M., Wang, J. H., Molnar, A., Wu, P., Winandy, S., and Sharpe, A. (1994). The *Ikoros* gene is required for the development of all lymphoid lineages. *Cell (Cambridge, Mass.)* **79**, 143–156.

Ghia, P., ten Boekel, E., Rolink, A. G. *et al.* (1998). B-cell development, a comparison between mouse and man. *Immunol. Today* **19**, 480–485.

Godin, I., Dieterlen-Lievre, F., and Cumano, A. (1995). Emergence of multipotent hemopoietic cells in the yolk sac and paraaortic splanchnopleura in mouse embryos, beginning at 8.5 dpc. *Proc. Natl. Acad. Sci. U.S.A.* **92**, 773–777.

Gray, D., and Matzinger, P. (1991). T cell memory is short-lived in the absence of antigen. *J. Exp. Med.* **174**, 969–974.

Greenberg, A. H., and Litchfield, D. W. (1995). Granzymes and apoptosis, targeting the cell cycle. *Curr. Top. Microbiol. Immunol.* **198**, 95–119.

Harrison, D. E., Astle, C. M., and Lerner, C. (1988). Number and continuous proliferative pattern of transplanted primitive immunohematopoietic stem cells. *Proc. Natl. Acad. Sci. U.S.A.* **85**, 822–826.

Harrison, D. E., Jordan, C. T., Zhong, R. K., and Astle, C. M. (1993). Primitive hemopoietic stem cells: Direct assay of most productive populations by competitive repopulation with simple binomial, correlation and covariance calcualtions. *Exp. Hematol.* **21**, 206–219.

Henschler, R., Brugger, W., Luft, T. *et al.* (1994). Maintenance of transplantation potential in ex vivo expanded CD34+-selected human peripheral blood progenitor cells. *Blood* **84**, 2898–2903.

Hertz, M., and Nemazee, D. (1998). Receptor editing and commitment in B lymphocytes. *Curr. Opin. Immunol.* **10**, 208–213.

Jenkinson, E. J., and Anderson, G. (1994). Fetal thymic organ cultures. *Curr. Opin. Immunol.* **6**, 293–297.

Jenkinson, E. J., Franchi, L., Kinston, R., and Owen, J. J. T. (1982). Effect of deoxyguanosine on lymphopoiesis in the developing thymus rudiment in vitro: Application in the production of chimeric thymus rudiments. *Eur. J. Immunol.* **12**, 583–587.

Kisielow, P., and von Boehmer, H. (1995). Development and selection of T cells: Facts and puzzles. *Adv. Immunol.* **58**, 87–209.

Kispert, A., and Hermann, B. G. (1993). The *Brachyury* gene encodes a novel DNA binding. *EMBO J.* **12**, 3211–3220.

Kispert, A., and Hermann, B. G. (1994). Immunohistochemical analysis of the Brachyury protein in wild-type and mutant mouse embryos. *Dev. Biol.* **161**, 179–193.

Kohn, D. B., Hershfield, M. S., Carbonaro, D. *et al.* (1998). T lymphocytes with a normal ADA gene accumulate after transplantation of transduced autologous umbilical cord blood CD34+ cells in ADA-deficient SCID neonates. *Nat. Med.* **4**, 775–780.

Koller, M. R., Emerson, S. G., and Palsson, B. O. (1993). Large-scale expansion of human stem and progenitor cells from bone marrow mononuclear cells in continuous perfusion cultures. *Blood* **82**, 378–384.

Korsmeyer, S. J. (1995). Regulators of cell death. *Trends Genet.* **11**, 101–105.

Kosco, M. H., Pflugfelder, E., and Gray, D. (1992). Follicular dendritic cell-dependent adhesion and proliferation of B cells in vitro. *J. Immunol.* **148**, 2331–2339.

Krammer, P. H., Dhein, J. *et al.* (1994). The role of APO-1-mediated apoptosis in the immune system. *Immunol. Rev.* **142**, 175–191.

Lansdorp, P. M. (1995a). Telomere length and proliferation potential of hematopoietic stem cells. *J. Cell Sci.* **108**, 1–6.

Lansdorp, P. M. (1995b). Developmental changes in the function of hematopoietic stem cells. *Exp. Heamtol.* **23**, 187–191.

Lassoued, K., Nunez, C. A., Billips, L. *et al.* (1993). Expression of surrogate light chain receptors is restricted to a late stage in pre-B cell differentiation. *Cell* **73**, 73–86.

Leandersson, T., and Hsu, E. (1985). Anti-IgM treatment influences immunoglobulin heavy and light chain mRNA levels in mitogen-stimulated B lymphocytes. *Eur. J. Immunol.* **15**, 641–643.

Levine, B., Bernstein, W., Connors, M. *et al.* (1997). Effects of CD28 costimulation on long-term proliferation of CD4+ T cells without exogenous feeder cells. *J. Immunol.* **159**, 5921–5930.

Li, Y.-S., Hayakawa, K., and Hardy, R. R. (1993). The regulated expression of B lineage associated genes during B cell differentiation in bone marrow and fetal liver. *J. Exp. Med.* **178**, 951–960.

Lin, Y., Wong, K., and Calame, K. (1997). Repression of c-myc transcription by Blimp-1, an inducer of trminal B cell differentiation. *Science* **276**, 596–599.

Loeffert, D., Schaal, S., Ehlich, A. *et al.* (1994). Eaarly B-cell development in the mouse: Insights from mutations introduced by gene targeting. *Immunol. Rev.* **137**, 135–153.

Ma, A., Pena, J. C., Chang, B. *et al.* (1995). Bcl-x regulates the survival of double-positive thymocytes. *Proc. Natl. Acad. Sci. U.S.A.* **92**, 4763–4767.

Mackinnon, D., and Ceredig, R. (1986). Changes in the expression of potassium channels during mouse T cell development. *J. Immunol.* **164**, 1846–1861.

Marbrook, J. (1967). Primary immune response in cultures of spleen cells. *Lancet* **2**, 1279–1283.

McHeyzer-Williams, M. G., and Davis, M. M. (1995). Antigen-specific development of primary and memory T cells in vivo. *Science* **268**, 106–111.

McLaughlin, K. A., Osborne, B. A., and Goldsby, R. A. (1996). The role of oxygen in thymocyte apoptosis. *Eur. J. Immunol.* **26**, 1170–1174.

Melchers, F., Rolink, A., Grawunder, U., Winkler, T. H., Karasuyama, H., Ghia, P., and Andersson, J. (1995). Positive and nebative selection events during B lymphopoiesis. *Curr. Opin. Immunol.* **7**, 214–227.

Messika, E. J., Lu, P. S., Sung, Y. J. *et al.* (1998). Differential effect of Blimp-1 expression on cell fate during B cell development. *J. Exp. Med.* **188**, 515–525.

Michie, C. A., McLean, A., Alcock, C., and Beverley, P. C. (1992). Lifespan of human lymphocyte subsets defined by CD45 isoforms. *Nature (London)* **360**, 264–265.

Mishell, R. I., and Dutton, R. W. (1966). Immunization of normal mouse spleen cell suspension in vitro. *Science* **153**, 1004–1006.

Mishell, R. I., and Dutton, R. W. (1967). Spleen cell culture from normal mice. *J. Exp. Med.* **126**, 423–428.

Mucenski, M. L., McLain, K., Kier, A. B. *et al.*, (1991). A functional c-myb gene is required for normal murine fetal hepatic hematopoiesis. *Cell (Cambridge, Mass.)* **65**, 677–689.

Nagata, S., and Golstein, P. (1995). The Fas death factor. *Science* **267**, 1449–1456.

Nichogiannopoulou, A., Trevisan, M., Fridrich, C. *et al.* (1998). Ikaros in hemopoietic lineage determination and homeostasis. *Semin. Immunol.* **10**, 119–125.

Ogawa, M. (1993). Differentiation and proliferation of hematopoietic stem cells. *Blood* **81**, 2844–2853.

Penvy, L., Simon, M. C., Robertson, E. *et al.* (1991). Erythroid differentiation in chimaeric mice blocked by a targeted mutation in the gene for transcription factor GATA-1. *Nature (London)* **349**, 257–260.

Pereira, R. F., Halford, K. W., O'Hara, M. D. *et al.* (1995). Cultured adherent cells from marrow can serve as long-lasting precursor cells for bone, cartilage, and lung in irradiated mice. *Proc. Natl. Acad. Sci. U.S.A.* **92**, 4857–4861.

Phillips, C., Schimpl, A., Dietrich-Goetzx, W. *et al.* (1996). Inducible nuclear factors binding the μ-chain pre-mRNA secretory poly(A) site. *Eur. J. Immunol.* **26**, 3144–3152.

Potocnik, A. J., Nielsen, P. J., and Eichmann, K. (1994). In vitro generation of lymphoid precursors from embryonic stem cells. *EMBO J.* **13**, 5274–5283.

Potocnik, A. J., Nerz, G., Kohler, H. *et al.* (1997a). Reconstitution of B cell subsets in Rag−/− mice by transplantation of in vitro differentiated ES cells. *Immunol. Lett.* **57**, 131–137.

Potocnik, A. J., Kohler, H., and Eichmann, K. (1997b). Hemato-lymphoid in vivo reconstitution potential of subpopulations derived from in vitro differentiated embryonic stem cells. *Proc. Natl. Acad. Sci. U.S.A.* **94**, 10295–10300.

Rawlings, D., Quan, S. F., Kato, R. M., and Witte, O. N. (1995). Long-term culture system for selective growth of human B-cell progenitors. *Proc. Natl. Acad. Sci. U.S.A.* **92**, 1570–1574.

Resta, R., and Thompson, L. (1997). SCID, the role of ADA deficiency. *Immunol. Today* **18**, 371–374.

Robb, L., Lyons, I., Li, R. *et al.* (1995). Absence of yolk sac hematopoiesis from mice with a targeted disruption of the *scl* gene. *Proc. Natl. Acad. Sci. U.S.A.* **92**, 7075–7079.

Rodewald, H.-R., and Fehling, H. J. (1998). Molecular and cellular events in early thymocyte development. *Adv. Immunol.* **69**.

Rodewald. H.-R., Ogawa, M., Haller, C., Waskow, W., and DiSanto, J. P. (1997). Prothymocyte expansion by c-kit and the common cytokine receptor γ chain is essential for repertoire formation. *Immunity* **6**, 265–272.

Rolink, A., Kudo, A., Karasuyama, H. *et al.* (1991). Long-term proliferatin early pre-B cll lines and clones with the potential to develop to surface-Ig positive mitogen-reactive B cells 'in vitro' and 'in vivo'. *EMBO J.* **10**, 327–336.

Rolink, A., Karasuyama, H., Haasner, D. *et al.* (1994). Two pathways of B-lymphocyte development in mouse bone marrow and the roles of surrogate L chain in this development. *Immunol. Rev.* **137**, 185–201.

Schliephake, D. E., and Schimpl, A. (1996). Blimp-1 overcomes the block in IgM secretion in LPS/anti-mu F(ab')2-costimulated B lymphocytes. *Eur. J. Immunol.* **26**, 268–271.

Scott, E. W., Simon, M. C., Anastasi, J., and Singh, H. (1994). Requirement of transcription factor PU.1 in the development of multiple haemotopoietic lingeages. *Science* **265**, 1573–1577.

Seychab, H., Friedl, R., Schindler, D., Hoehn, H., Rabinovitch, P. S., and Chen, U. (1989). The effects of bacterial lipopolysaccharide, anti-receptor antibodies and recombinant γ-interferon on mouse B-cell cycle progression using Brdu/Hoechst flow cytometry. *Eur. J. Immunol.* **19**, 1605–1612.

Shivdasani, R. A., Mayer, E. L., and Orkin, S. H. (1995). Absence of blood formation in mice lacking the T-cell leukaemia oncoprotein tal-1/SCL. *Nature (London)* **373**, 432–434.

Shortman, K., and Wu, L. (1996). Early T-progenitors. *Annu. Rev. Immunol.* **14**, 29–47.

Simmons, P. J., Masinovsky, B., Longenecker, B. M. *et al.* (1992). Vascular dell adhesion molecule-1 expressed by bone marrow stromal cells mediates the binding of hematopoietic progenitor cells. *Blood* **80**, 388–395.

Timothy, C., Hirayama, E., and Ogawa, M. (1995). Lymphohematopoietic progenitors of normal mice. *Blood* **85**, 3086–3092.

Ting, C.-N., Olson, M. C., Barton, K. P., and Leiden, J. M. (1996). Transcription factor GATA-3 is required for developmetn of the T-cell lineage. *Nature (London)* **384,** 474–478.

Turner, C. A., Mack, D., and Davis, M. M. (1994). Blimp-1, a noval zinc finger-containing protein that can drive the maturartion of B lymphocytes into immunoglobulin-secreting cells. *Cell (Cambridge, Mass.)* **77**, 297–306.

Urbanek, P., Wang, Z.-Q., Fetka, I., Wagner, E. F., and Busslinger, M. (1994). Complete block of early B cell differentiation and altered patterning of the posterior midbrain in mice lacking Pax5/BSAP. *Cell (Cambridge, Mass.)* **9**, 901–912.

Vaziri, H., Dragowska, W., Allsopp, R. C. *et al.* (1994). Evidence for a mitotic clock in human hematopoietic stem cells: Loss of telomeric DNA with age. *Proc. Natl. Acad. Sci. U.S.A.* **91**, 9857–9860.

Verbeek, S., Izon, D., Hofjuis, F. *et al.* (1995). An HMG-box-containing T cell factor required for thymocyte differentiation. *Nature (London)* **374**, 70–74.

Wang, S., Miura, M., Jung, Y. K. *et al.* (1996a). Identification and characterization of Ich-3, a member of the interleukin-1beta converting enzyme (ICE)/Ced-3 family and an upstream regulator of ICE. *J. Biol. Chem.* **271**, 20580–20587.

Wang, J.-H., Nichogiannopoulou, A., Wu, L., Sun, L., Sharpe, A. H., Bigby, M., and Georgopoulos, K. (1996b). Selective defects in the development of the fetal and adult lymphoid system in mice with an Ikaros null mutation. *Immunity* **5**, 537–549.

Wilson, V., Rashbass, P., and Beddington, R. S. P. (1993). Chimeric analysis of T (Brachyury) gene function. *Development (Cambridge, UK)* **117**, 1321–1331.

Wineman, J. P., Nishikawa, S., and Muller-Sieburg, C. E. (1993). Maintenance of high levels of pluripotent hematopoietic stem cells in vitro: Effect of stromal cells and c-kit. *Blood* **81**, 365–372.

Wineman, J. P., Moore, K., Lemischka, I. *et al.* (1996). Functional heterogeneity of the hematopoietic microenvironment, rare stromal elements maintain long-term repopulating stem cells. *Blood* **87**, 4082–4090.

Wohlleben, G., Gray, D., and Schimpl, A. (1996). In vitro immunization of naive mouse B cells, establishment of IgM secreting hybridomas specific for soluble protein or hapten from B cells cultured on CD40 ligand transfected m-fibroblasts. *Int. Immunol.* **8**, 343–349.

Yoo, J. U., Barthel, T. S., Nishimura, K. *et al.* (1998). The condrogenic potential of hu-BM-derived mesenchymal progenitor cells. *J. Bone J. Surg., Am. Vol.* **80A**, 1745–1757.

Zhuang, Y., Soriano, P., and Weintraub, H. (1994). The helix-loop-helix gene E2A is required for B cell formation. *Cell (Cambridge, Mass.)* **79**, 875–884.

CHAPTER 44

HEMATOPOIETIC STEM CELLS

Anne Kessinger and Graham Sharp

INTRODUCTION

All of the mature functional cells of the hematopoietic and immune systems originate from a very small number of undifferentiated cells called hematopoietic stem cells (HSCs), the most primitive of which constitute about 1 in 10^5 of nucleated hematopoietic cells in bone marrow. HSCs also circulate in blood at a frequency of approximately 1 log less than their counterparts in the marrow. These cells, by a process of sequential clonal amplification and differentiation of their progeny, populate the hematopoietic and immune systems. Manipulation of the total mass of hematopoietic and immune (lymphoid) tissues presents special challenges because these interrelated cellular systems are arranged both as discrete masses of tissue as well as diffuse distributions throughout most other organs of the body. Migration is an essential component of their function (Sharp, 1993).

Hematopoietic precursor cells in blood can establish endothelial cells in culture. In the adult, whether HSC and endothelial precursors are separate and distinct populations or whether they are closely related, perhaps even daughter cells of the same rare primitive stem cell, is not clear. In the embryo, HSCs and vascular stem cells probably are closely related and share a common precursor (Kluppel *et al.,* 1997; Shalaby *et al.,* 1997). Whether distinct precursors to these two cell types are established and their common stem cell is lost or whether a rare common stem cell persists but is dwarfed in frequency by slightly more differentiated lineage-restricted progeny of hematopoietic versus endothelial cells in the adult is as yet unknown. A similar puzzle exists in adult bone marrow that clearly contains HSCs, mesenchymal stem cells (Vogel, 1999), and cells that appear to be able to give rise to muscle (Ferrari *et al.,* 1998). Are these stem cell populations entirely distinct or are they related? If they are related, when during development do they diverge? Is the divergence complete or do rare stem cells persist that are a common precursor? In other words, exactly how much plasticity exists in the most primitive HSC of adults? Neural stem cells can give rise to blood cells (Bjornson *et al.,* 1999). Is the converse true? Some reviews of this topic imply this might be so (Bjorklund and Svendsen, 1999), but for now these important questions remain unanswered. The answers will have profound implications for tissue engineering of the HSC, because HSC collection is not controversial and considerable experience in transplanting them exists. Might HSCs be a noncontroversial alternative source to embryonic stem cells (O'Shea, 1999; Solter and Gearhart, 1999; Vogel, 1999) for a variety of tissues?

In the embryo, hematopoietic stem cells originate in a region associated with the dorsal aorta, the lateral plate mesenchyme, and the yolk sac. These initial HSCs migrate via the embryonic circulation to the liver (which then becomes the primary hematopoietic organ of the fetus), to the spleen, and, before birth, to the developing bone marrow spaces to establish the primary hematopoietic organ of postnatal life (Lansdorp, 1995). The process of HSC migration to the bone marrow involves the expression of the chemokine receptor CXCR4 and interactions with its ligand, stromal-derived factor 1 (SDF1). CXCR4 knockout mice exhibit a defect of bone marrow hematopoiesis (Aiuti *et al.,* 1997; Ma *et al.,* 1998; Nagasawa *et al.,* 1996); SDF1 is also implicated in the successful engraftment of human HSCs in immunodeficient NOD/SCID mice (Peled *et al.,* 1999). The possibility that a small reserve of HSCs remains in the circulation has been postulated but remains unproved. In the adult, the active red marrow retracts largely to the axial skeleton, leaving the long bones populated with yellow (fatty) marrow (Tavassoli, 1989). The progeny of HSCs differentiate, migrate to establish the lympho–hematopoietic component of other organs (thymus, spleen, lymph node, Peyer's patches), and circulate throughout the tissues of the body. They express essential properties of hematopoietic and immune cell effector functions, cytokine secretion, and provision of key enzymes required by tissues for detoxification of harmful materials. Consequently, congenital abnormalities of HSCs can lead to hematopoietic, immune, and systemic diseases. Tissue engineering by way of replacement (transplantation) or manipulation of hematopoietic stem cells offers an approach to the therapy of these diseases.

Because bone marrow and all of the lymphoid components of the body cannot be removed surgically, systemic therapies are required for manipulations of these tissues. These therapies may be either cytotoxic (radiation, chemotherapy, immunotherapy) or regulatory (recombinantly engineered cytokines), and have potential effects and toxicities for uninvolved tissues. Conversely, therapies targeted at nonhematopoietic or nonimmune tissues can damage the hematopoietic and immune systems. Such iatrogenic events can require cellular or tissue engineering to restore hematopoietic and immune function.

The diffuse nature of hematopoietic and immune tissues presents challenges, as well as opportunities. The most primitive HSCs and other bone marrow or lymphoid cells can be harvested from the body, manipulated, and reinfused. The reinfused cells home to appropriate microenvironments and resume their original *in vivo* functions or, if altered in function, begin their new activity. These properties can be exploited to introduce genes into the body to correct congenitally deficient genes, to add entirely new genes, or to alter genes expressed by the infused cells. Although genetic engineering of HSCs has succeeded in principle (Brenner and Rill, 1994), many practical issues remain to be solved before this approach will become an effective therapy. For example, the proportion of primitive HSCs in the cell cycle is a few percent at best (Goodell *et al.,* 1996). Consequently, HSCs are difficult to transduce with vectors that require a proliferating target (Agrawall *et al.,* 1996; Kohn, 1996). The use of cytokines to stimulate proliferation can promote terminal differentiation and cytokine-exposed cells may not engraft well (Kittler *et al.,* 1997). Even if transduction is successful, clonal succession of primitive HSCs can complicate the *in vivo* pattern of gene expression (Lansdorp, 1995). Systemic cytoreductive therapy may or may not be needed to "create space" for the infused manipulated cells. Quesenberry and colleagues have followed up on original observations by Brecher *et al.* (1982) that HSCs (bone marrow) infused into unmanipulated recipient mice competitively establish a significant degree of chimerism (Quesenberry *et al.,* 1994). The key to this process appears to involve daily infusion of donor cells over a few days, rather than a single infusion. If purified primitive HSCs are infused, their competitive repopulating capacity is substantial. Additionally, such cells do divide following transplantation, which might lead to integration of genetic vectors present in the cells (Bradford *et al.,* 1997).

Clearly, the field of tissue engineering using genetically altered HSCs presents many interesting and exciting challenges to overcome before achieving the status of a routine clinical procedure. In this review, an overview of hematopoiesis and its regulation together with the practical aspects of harvesting and processing of HSCs is presented as a basis for the discussion of the tissue engineering of HSCs.

OVERVIEW OF HEMATOPOIESIS: PROPERTIES AND REGULATION OF HEMATOPOIETIC STEM CELLS RELEVANT TO TISSUE ENGINEERING

The hematopoietic stem cells compartment comprises a differentiation hierarchy in which relatively rare and uncommitted peripotential "primitive" stem cells give rise to multiple lineages of mature cells by a linear branching differentiation process (Sharp, 1993). Within each lineage, the cells are increasingly committed stochastically to the type of mature cells they produce (Ogawa, 1994). The existence of several "levels" of differentiation within this compartment potentially leads to confusion over the use of the terms "stem" cell, "progenitor" cell, "precursor" cell, etc. As a simplification, here all of these cell types are considered stem cells.

Hematopoiesis occurs in specific stromal microenvironments (Tavassoli and Takahashi, 1982), which vary in location during development and aging, and in pathologic situations. Stromal cells in these microenvironments have multiple interactions with developing hematopoietic cells (Dexter, 1982; Lambertsen, 1984). These include the provision of extracellular matrix molecules (Anklesaria *et al.,* 1991; Siczkowski *et al.,* 1992; Yoder and Williams, 1995; Zipori *et al.,* 1985), soluble and membrane-bound growth and cell survival factors (Du and Williams, 1994; Whetton and Spooncer, 1998), and adhesion molecules (Conget and Minguell, 1995; Verfaillie *et al.,* 1994), as well as the regulation of the release of cells to the circulation at an appropriate stage of differentiation (Weiss and Geduldig, 1991). The potentials of marrow stromal cells have been reviewed (Owen, 1988). Stromal cell lines have been developed that duplicate many of these functions (Zipori, 1989a). A mesenchymal stem cell from bone marrow has been identified (Pittenger *et al.,* 1999). The use of such cells combined with HSCs might be necessary to solve some of the tissue engineering challenges presented by HSCs. Some of the progeny of primitive stem cells must pass through, or otherwise be influenced by, additional microenvironments, for example, the thymus, for key differentiation events to occur. Collectively, these processes produce the multiple mature cell types needed to sustain the hematologic and immunologic needs of the organism. Attempts to engineer these tissues, therefore, require a duplication of their key components, for example, provision of an appropriate population of stromal cells. The functional status of these stromal cells, i.e., whether they are promoting quiescence versus active replication of their associated HSC, also may be important. Maybe more than one stromal cell type will be necessary to provide a microenvironment "compleat" (Walton, 1653) for HSCs?

ORGANIZATION OF THE STEM CELL COMPARTMENT

Do the most primitive hematopoietic stem cells comprise a "cell renewal" system to provide replacement of any cells lost from this compartment by cell division, or are a fixed number of stem cells available in an organism large enough to provide vast numbers of progeny without becoming significantly depleted in a lifetime (Harrison *et al.,* 1987; Lansdorp, 1995; Micklem *et al.,* 1983)? These two hypotheses of stem cell compartment organization are not necessarily mutually exclusive. Both receive some experimental support, and experiments that test unequivocally which of these alternatives, if either, is correct have been difficult to devise (Van Zant *et al.,* 1997). The answer underlies the likely success or failure of attempts to expand or genetically manipulate most primitive stem cells *ex vivo* for the purposes of tissue engineering (Lansdorp, 1995). The consequences of telomere shortening in this circumstance are a concern (Engelhardt *et al.,* 1997; Lansdorp *et al.,* 1997).

The best available evidence suggests that, in humans, HSCs, as identified by the CD34 antigen, self-renew in fetal liver to establish the physiologically necessary stem cell compartment. In the adult, few, if any, CD34-positive cells self-renew under normal circumstances (Landsdorp *et al.,* 1993). Fetal hematopoietic stem cells may differ significantly in their properties and perhaps in their regulation from stem cells in the adult. Although controversial for clinical use, fetal tissue might offer some advantages in engineering new or replacement tissues (Harrison *et al.,* 1997). Cord blood is an additional and potentially very useful source of stem cells for tissue engineering (Wagner *et al.,* 1995). In the mouse, the most primitive HSCs are CD34-negative cells, and CD34-positive cells are their progeny (Osawa *et al.,* 1996). CD34-negative stem cells likely exist in humans, but their proportion and place in the stem cell hierarchy remain to be defined (Goodell *et al.,* 1997).

STROMAL MICROENVIRONMENTS

Just as hematopoiesis *in vivo* is limited to specific sites during development, hematopoietic cells demand stringent conditions for their maintenance in culture. For many years, the data suggested that mixed stromal cell populations, including large "blanket" cells, were necessary to maintain hematopoietic stem cells in culture (Dexter, 1982; Tavassoli and Takahashi, 1982). Stromal cell lines that support hematopoiesis have now been developed (Itoh *et al.,* 1989; Zipori, 1989a). Although an intimate association of stem cells with stromal cells is probably essential, contact may not be needed for their survival and differentiation (Verfaillie, 1992). Currently, all of the interactive properties of stromal cells with stem cells cannot be completely duplicated with known recombinant survival, growth, and differentiation factors. Consequently, the mechanisms by which cloned stromal cells support hematopoietic stem cells from several species (mouse, swine, and humans) in culture are incompletely defined. Potentially, these stromal cells either produce soluble growth factor(s) or present a membrane-bound growth factor(s), or both, which regulate stem cell survival and production or inhibition of apoptosis and self-renewal versus differentiation (Whetton and Spooncer, 1998; Williams, 1994). Stromal cells or other cell types may also be a source of inhibitors of stem cell differentiation (Waegell *et al.*, 1994; Zipori, 1989b), and the presence and possible adherence of more differentiated hematopoietic cells (Verfaillie *et al.*, 1994) to stromal cells may be a feedback signal necessary to maintain inhibitor levels. If these differentiated cells or the inhibitors are removed, the inhibition of stem cell differentiation ceases and the stem cells may differentiate out of existence. Primitive HSCs likely have several signal transduction pathways that can be triggered by different signals (or combinations of signals). Stromal cells might be involved in regulating asymmetric versus symmetric division, which potentially involves a role for the *notch* gene and its ligands (Morrison *et al.*, 1997; Whetton and Spooncer, 1998). Different signaling pathways may be used for stem cell survival (vs. apoptosis), self-renewal, proliferation, and lineage-specific differentiation. Indeed, regulation of cell production by modification of the rate of apoptosis of hematopoietic cells might be a critical component of the process. Methods to alter (suppress) apoptosis have received little attention in tissue engineering, although this may be one of the mechanisms by which microenvironmental stromal cells act. In myelodysplastic syndromes, increased apoptosis appears to be a critical problem (Mundle *et al.*, 1994).

Other than in very early embryogenesis, the majority of stromal cell precursors are likely distinct from primitive hematopoietic stem cells (Dexter and Allen, 1992; Huang and Terstappen, 1992, 1994). Consequently, in any engineered system, using adult cells might require both hematopoietic stem cells and stromal cells, but in a temporal order, i.e., stromal cells first followed by HSCs might be necessary to establish appropriate stromal niches (Allen, 1981).

ASSAYS OF STEM CELLS

Assays to enumerate stem cells at specific stages of differentiation, especially assays that require the production of mature progeny, must be designed to meet the physiologic demands of the cell types assayed. For example, the most primitive stem cells demand either an *in vivo* environment or a culture system that permits an intimate association with stromal cells. Assays that meet this criterion include the long-term repopulating ability (LTRA) assay using marked cells, e.g., the Y chromosome of male cells transplanted into female recipients (Watt and Visser, 1992), or, secondarily, the assay of long-term culture initiating cells (LTC-ICs) (Sutherland *et al.*, 1989). If the stem cells are grown on stromal lines and not harvested for assay of secondary colonies, but instead are assessed by hematopoietic cobblestone areas in the primary cultures, the assayed cells are designated as long-term culture cobblestone-area-forming cells (LTC-CAFCs) (Ploemacher *et al.*, 1991; Neben *et al.*, 1993). The frequency of CD34+ cells in human bone marrow is at least 100 times greater than the frequency of LTC-ICs (Watt and Visser, 1992). Because the frequency of LTC-ICs correlates most closely with the LTRA in the mouse (the only species in which LTRA frequency has been accurately determined), this implies that in humans only a small fraction of CD34+ cells actually belongs to the most primitive category of stem cells. The frequency of LTC-ICs in the blood of normal individuals is very low, $3/\text{ml}$ or $1/2 \times 10^6$ nucleated cells, or approximately 100 times lower than in bone marrow (Udomsakdi *et al.*, 1992).

More differentiated stem cells will grow in a semisolid medium, e.g., agar or methylcellulose, provided their physiologic regulator(s) of differentiation, i.e., colony-stimulating factors (CSFs) and interleukins (ILs), are provided (Watt and Visser, 1992). Sudden removal of CSFs or ILs can trigger apoptosis of the target cells, indicating the critical importance of these regulators (Williams *et al.*, 1990). Because the differentiation hierarchy from primitive stem cell to the most differentiated progenitor cell is a linear branching system, the mature functional progenies of the more differentiated members of this hierarchy become evident much faster in colony-forming cell (CFC) assays, e.g., mature cells from committed granulocyte–monocyte CFCs (GM-CFC) at 7 days. The mature progenies of intermediately differentiated, high-proliferative potential colony-forming cells (HPP-CFCs) are evident after 14–28 days of culture compared to LTC-ICs, which require 4–8 weeks for assay (Ploemacher *et al.*, 1991). The more differentiated committed progenitors, GM-CFCs, require a single cytokine, GM-CSF, whereas HPP-CFCs need multiple CSFs or ILs and LTC-ICs must be grown on irradiated bone marrow stromal cells or carefully screened cloned stromal cell lines (Watt and Visser, 1992; Sutherland *et al.*, 1989; Ploemacher *et al.*, 1991).

Human HSC Harvesting and Processing

The posterior crests of the pelvic bone provide the only readily accessible site from which marrow HSCs in amounts sufficient to provide a successful hematopoietic graft can be extracted. Marrow cells are collected by aspiration through specially designed needles. Approximately 2×10^8 nucleated cells/kg recipient weight (approximately 11 ml/kg) are sufficient. The aspiration is usually performed while the donor is under the influence of general anesthesia.

Collection of HSCs from the circulation currently requires apheresis, because the numbers of HSCs per milliliter of blood are quite small. The HSC donor can be treated with cytokines or, in the case of autologous donors, with myelosuppressive chemotherapy plus cytokines to increase the number of circulating CD34$^+$ cells (Brugger *et al.*, 1992). When sufficient numbers of these cells are mobilized into the circulation, an adequate graft product can sometimes be generated with a single apheresis procedure (Pettengell *et al.*, 1993). However, autologous collections from patients who have already received substantial amounts of antitumor cytotoxic therapy often require several apheresis procedures to collect a useful graft product. Not all normal donors mobilize successfully. About 10–20% of normal donors are poor mobilizers, which suggests that factors other than prior cytotoxic therapy are implicated in poor mobilization. An undefined circulating inhibitor has been implicated (Kessinger and Sharp, 1998).

Harvested marrow cells or peripheral HSCs are cryopreserved and stored in either a liquid nitrogen freezer or in temperatures at or below $-80°C$, unless they are to be infused within 24–48 hr. If *ex vivo* manipulations of these cells are performed, such procedures generally occur prior to cryopreservation.

At the time of infusion, the cells are thawed by placing their plastic storage bags in a warm water bath. The thawed cells are immediately infused into the circulation of an individual whose marrow function has been ablated with cytotoxic therapy. From there, the cells make their way to the appropriate niches in the marrow stroma to grow and restore hematopoietic function. Approximately 7 days are required after infusion before mature functional cells begin to appear in the circulation. In the meanwhile, patients are supported with red cell and platelet transfusions. Therefore, the first evidence that marrow function is returning after transplantation is the reappearance of circulating white blood cells.

Replacement of congenitally defective HSC

Patients with congenitally defective stem cells present with a myriad of clinical manifestations, depending on the specific defect. The resultant diseases include severe combined immune deficiency (SCID), a disorder of the lymphoid stem cell (Wiscott–Aldrich syndrome), a disorder of the lymphoid and hematopoietic stem cells (thalassemia, or erythroid disorder), Fanconi anemia, and osteopetrosis (a disorder of osteoclasts), among others. Destruction of the defective marrow and replacement with normal marrow from a compatible donor will reverse these otherwise fatal diseases. For patients who lack donors, genetically engineered stem cell grafts are an alternative (Kohn, 1996; Liu *et al.*, 1997). In addition, patients with generalized congenital enzyme deficiencies such as osteopetrosis and Gaucher's disease can benefit from allogeneic hematopoietic stem cell transplant because the transplanted normal marrow cells provide progenies with sufficient amounts of enzyme to control the symptomatology of these disorders. Alternatively, they require engineered autologous grafts (Xu *et al.*, 1994; Dunbar and Kohn, 1996).

Patients with acquired stem cell defects can also be treated with allogeneic stem cell transplants. Disorders such as aplastic anemia, leukemia, myeloma, and dysmyelopoietic syndromes have been cured when the diseased marrow was destroyed and replaced with normal cells. Certain diseases (e.g., aplastic anemia subsequent to radiation or administration of chemotherapeutic agents) might arise from a microenvironmental defect (Testa *et al.*, 1985; Mauch *et al.*, 1995). Such disorders may not be amenable to cure by infusion of stem cells alone but could also require replacement of the microenvironment. Potentially, such defects would be primary targets for therapy with engineered bone marrow (Naughton *et al.*, 1994).

Unfortunately, observations in patients undergoing genetically engineered autologous transplants confirm data from primates and to some extent mice that, using current techniques, the transfection frequency of HSCs with retroviral vectors is low. The quiescent status of the majority of primitive hematopoietic stem cells is believed to be the cause of this problem. Forced cycling of the HSCs by cytokines increases the transfection frequency (Luskey *et al.,* 1992). A concern with this approach, especially in adults, is that if the HSCs that are forced to cycle also differentiate, they may lose some of their "stemness" (Peters *et al.,* 1995). The survival time of such transfected cells on transplantation will be reduced and the loss of transfected gene expression or perhaps even complete loss of the grafted cells might occur. Potentially, the application of an engineered system that more closely duplicates *in vivo* hematopoiesis and in particular provides stromal cells will lead to improved gene transduction successes (Xu *et al.,* 1995; Naughton *et al.,* 1992). This is the attraction of some of the newly developed perfusion systems.

Artificial bone marrow and perfused hematopoietic culture systems

Historical

The evidence suggesting the need for a supporting matrix to maintain hematopoiesis has been reviewed by Fliedner and Calvo (1978). Knospe and colleagues, following the lead of Seki (1973), employed cellulose ester membranes that were coated with various cell types then implanted *in vivo* (Knospe, 1989).

Long-term bone marrow cultures (LTBMCs) (Dexter *et al.,* 1977) have been acknowledged to represent the most physiological *ex vivo* model of hematopoiesis. Hematopoietic colony-forming cells can be detected in human LTBMCs for about 6 months and much longer in cultures from subhuman primates and mice (Dexter *et al.,* 1984). The demonstration that hematopoietic stem cells could be maintained in culture paved the way to the engineering of artificial bone marrow and perfused systems to support and grow stem cells. The application in modern times of tissue engineering to hematopoietic stem cells in bone marrow can be traced to pioneering work with artificial capillary cultures described by Knazek *et al.* (1972; Gullino and Knazek, 1979), who grew mouse fibroblasts and human choriocarcinoma cells, and to studies by Chick *et al.* (1975), who attempted to create an artificial pancreas. Based on this work, starting in the mid-1970s attempts to grow normal hematopoietic cells in artificial capillary cultures were made. The original intent was to grow bone marrow stroma or stromal cell lines in the extracapillary space and circulate hematopoietic cells suspended in medium through the stromal cell-surrounded capillaries. The technical challenges of this approach became quickly apparent. There was a very high cell death rate of the circulating hematopoietic cells. At that time damage by shear forces was thought to be primarily responsible. In retrospect, normal cell death due to physiological apoptosis was also a likely significant contributor. As an alternative, an attempt was made to grow an "artificial marrow" in the extracapillary spaces, which were perfused with medium flowing through the capillaries. Although this was somewhat more successful, the system was far from ideal because the extracapillary volume was relatively large compared to the volume of cells, perhaps because an inadequate matrix was available for normal cell growth. Additionally, the long-term (weeks to months) maintenance of sterility was difficult (Anderson and Sharp, 1981), emphasizing the need for relatively sophisticated engineering solutions to these problems. Consequently, partnerships with companies have become an essential element of tissue engineering of hematopoietic stem cells. In the remainder of this review reference is made to several company products; however, the references are not exhaustive and no endorsement of company products is implied. In order to obtain a better overview of available resources, a search of the Worldwide Web is recommended. The company names mentioned can serve as a measure of the effectiveness of the search engine employed.

Naughton and colleagues have employed nylon mesh templates as a matrix on which to grow rodent hematopoietic cells. Mature cells and late-stage precursors were observed for 39 weeks (Naughton *et al.,* 1987). Subsequently, they described the role of stromal cells in this system using rat, nonhuman primate, and human cells (Naughton and Naughton, 1989). The three-dimensional nylon mesh system supports human bone marrow stem cell differentiation into multiple lineages (Naughton *et al.,* 1990) and can be used to evaluate the effects of drugs on hematopoiesis (Naughton *et al.,* 1991) and for gene transfer studies (Naughton *et al.,* 1992). Naughton *et al.* (1994) evaluated the effects of surgical implantation of bioengineered bone marrow tissue in rats. Active hematopoiesis was observed at sites of implantation for up to 110 days without the addition of exogenous growth factors. A company, Cytomatrix, has also evaluated a porous biocompatible matrix to support hematopoiesis (Rosenzweig, 1999). The problem of shear stress has been addressed by investigators at the National Aeronautics and Space Administration (NASA), who have devised microgravity bioreactors that are distributed by Synthecon (Goodwin *et al.,* 1993). Although this latter approach appears successful, it is unlikely to become a routine laboratory procedure.

Bone marrow organoids

In an attempt to circumvent the high loss of normal cells, established long-term bone marrow culture-derived cells were employed as starting material to establish bone marrow organoids. The majority of differentiated normal hematopoietic cells are lost from long-term cultures during the first few weeks of culture, which permits harvesting of hematopoietic stem cells associated with the adherent layer. The supernatant is discarded from the culture and the remaining cells are scraped from the flask. The cells can be concentrated to a slurry by gently pelleting in a conical centrifuge tube. When cultured for a subsequent period, these cells form bone marrow organoids. A micropipette is used to deposit a volume of this cell slurry onto a polycarbonate filter floating on a "raft" in medium. The hematopoietic reconstituting potential of these organoids can be demonstrated by transplanting them to lethally irradiated syngeneic recipients (Sharp *et al.,* 1985). If these organoids were transplanted under the kidney capsule of intact adult mice, then a fat organ resembling yellow bone marrow was formed. In contrast, in irradiated recipients not only did the organoid show hematopoietic differentiation, but spleen colonies were formed presumably by hematopoietic stem cells that migrated from the organoid (Sharp *et al.,* 1985). Such organoids are also a potential long-term source of hematopoietic regulatory factors (Brockbank and van Peer, 1983). The full scope of the potential of these hematopoietic organoids has yet to be explored. Because there are reports that graft-versus-host disease (GVHD) does not occur with long-term cultured cells (Spooncer and Dexter, 1983), whether organoids or artificial bone marrow alter the chimeric status and extent of GVHD in allogeneic recipients should be examined. Much more information has been gathered following infusion of bone marrow or blood as a cell suspension, which produces regeneration in the context of recipient stroma rather than the organoids, which possess donor stroma.

Perfused systems

Engineering solutions to the problems of shear stress, nutrition, oxygenation, etc. have been applied and this has led to the development of successful perfusion systems in which hematopoietic stem cell numbers can be amplified. Schwartz *et al.* (1991) demonstrated that increased long-term bone marrow perfusion was superior to the traditional approach. Establishment of more optimal perfusion conditions supported the continuous stable generation of human progenitor cells for over 5 months in culture (Schwartz *et al.,* 1991). This information was employed to design and construct a perfusion culture system employing bioreactors, which provides at least 1 log expansion over 14 days of total cells and of various progenitor cell types and almost 1 log expansion of the more primitive LTC-ICs (Koller *et al.,* 1993). The original system required establishment of a stromal layer and represents an automated media exchange system rather than a fully perfused culture system. Even so, an automated version of this system marketed by Aastrom permits from 10- to 15-fold marrow cell expansion under good manufacturing practice (GMP) conditions with minimal operator intervention (Bachier *et al.,* 1999). Stem and progenitor cells expanded using this system are now being evaluated in clinical trials (Koller *et al.,* 1993; Palsson, 1994). Bioreactors for stem cell expansion have also been described by Wang and Wu (1992), providing a severalfold increase in cell output over an 8-week period (Wang *et al.,* 1995). Several other groups are working on perfusion systems (Kompala and Highfill, 1994; Papoutsakis *et al.,* 1994). *Ex vivo* expanded progenitor cells have been shown to contribute to reconstitution of transplanted patients (Brugger *et al.,* 1995).

Newer technologies are approaching the goal of fully perfused culture systems. This approach is being pursued by Cytometrix. Acordis Research has also developed a fully perfused, oxygenated, waste-exchanging, culture perfusion unit (Glockner *et al.,* 1999), which should permit duplication of the environment experienced by blood stem cells. This system also accommodates the establishment of a stromal matrix that might more closely duplicate bone marrow microenvironment. Such systems may provide the environment for controlled proliferation of primitive stem cells without their terminal differentiation, thus duplicating events during the first 2 weeks posttransplantation *in vivo*. Controlled proliferation of primitive stem cells might significantly increase the frequency of gene transduction, which should greatly amplify the potential applications of tissue-engineered hematopoietic stem cells. In addition to the application of gene-marked stem cells (Brenner and Rill, 1994), HSCs engineered for therapy have been employed in the treatment of adenosine deaminase SCID (ADA-SCID) (Kohn, 1996), Gaucher's disease (Dunbar and Kohn, 1996), and Fanconi anemia (Liu *et al.,* 1997), and HSCs engineered to be resistant to the human immunodeficiency virus (HIV) by Systemix are currently in a phase I/II trial (Dutton, 1999). Additionally, work has begun on the development of an artificial thymus (Chen, 1994). After about 25 years in development, tissue-engineered hematopoietic stem cells have finally been administered to humans. The outcome of transplantation of even more primitive stem cells grown and manipulated *ex vivo* is currently awaited (Kessinger, 1995), and the full potential of engineered stem cells has become an exciting topic of both research and contemplation.

Acknowledgments

The authors wish to thank Shirene Seina, who typed this manuscript; Sally Mann, for technical assistance; and the National Institutes of Health, the National Cancer Institute, and the Nebraska Department of Health, for partial support of some of the research described. This overview was updated April 15, 1999.

References

Agrawall, Y. P., Agrawall, R. S., Sinclair, A. M., Young, D., Maruyama, M., Levine, F., and Ho, A. D. (1996). Cell-cycle kinetics and VSV-G pseudotyped retrovirus-mediated gene transfer in blood-derived CD34+ cells. *Exp. Hematol.* **24,** 738–747.

Aiuti, A., Webb, I. J., Bleul, C., Springer, T., and Gutierrez-Ramos, J. C. (1997). The chemokine SDF-1 is a chemoattractant for human CD34+ hematopoietic progenitor cells and provides a new mechanism to explain the mobilization of CD34+ progenitors to periphral blood. *J. Exp. Med.* **185,** 111–120.

Allen, T. D. (1981). Haemopoietic microenvironments *in vitro:* Ultrastructural aspects. *In* "Microenvironments in Haemopoietic and Lymphoid Differentiation," pp. 38–67. Pitman Medical, London.

Anderson, R. W., and Sharp, J. G. (1981). Evaluation of monolayer, liquid, gelfoam sponge and artificial capillary culture systems for the study of hematopoiesis. *Exp. Hematol.* **9**(2), 187–196.

Anklesaria, P., Greenberger, J. S., Fitzgerald, T. J., Sullenbarger, B., Wicha, M., and Campbell, A. (1991). Hemonectin mediates adhesion of engrafted murine progenitors to a clonal bone marrow stromal cell line from S1/S1^d mice. *Blood* **77**(8), 1691–1698.

Bachier, C. R., Gokmen, E., Teale, J., Lanzkron, S., Child, C., Franklin, W., Shpall, E., Douville, J., Weber, S., Muller, T., Armstrong, D., and LeMaistre, C. F. (1999). *Ex-vivo* expansion of bone marrow progenitor cells for hematopoietic reconstitution following high-dose chemotherapy for breast cancer. *Exp. Hematol.* **27,** 615–623.

Bjorklund, A., and Svendsen, C. (1999). Breaking the brain–blood barrier. *Nature (London)* **397,** 569–570.

Bjornson, C. R. R., Rietze, R. L., Reynolds, B. A., Magli, M. C., and Vescovi, A. L. (1999). Turning brain into blood: A hematopoietic fate adopted by adult neural stem cells *in vivo. Science* **283,** 534–537.

Bradford, G. B., Williams, B., Rossi, R., and Bertoncello, I. (1997). Quiescence, cycling, and turnover in the primitive hematopoietic stem cell compartment. *Exp. Hematol.* **25,** 445–453.

Brecher, G., Ansell, J. D., Micklem, H. S., Tjio, J. H., and Cronkite, E. P. (1982). Special proliferative sites are not needed for seeing and proliferation of transfused bone marrow cells in normal syngeneic mice. *Proc. Natl. Acad. Sci. U.S.A.* **79,** 5085.

Brenner, M. K., and Rill, D. R. (1994). Gene marking to improve the outcome of autologous bone marrow transplantation. *J. Hematother.* **3,** 33–36.

Brockbank, K. G. M., and van Peer, C. M. J. (1983). Colony-stimulating activity production by hemopoietic organ fibroblastoid cells *in vitro. Acta Haematol.* **69,** 369–375.

Brugger, W., Bross, K., Frisch, J., Dern, P., Weber, B., Mertelsmann, R., and Kanz, L. (1992). Mobilization of peripheral blood progenitor cells by sequential administration of interleukin-3 and granulocyte–macrophage colony-stimulating factor following polychemotherapy with etoposide, ifosfamide, and cisplatin. *Blood* **79**(5), 1193–1200.

Brugger, W., Heimfeld, S., Berenson, R. J., Mertelsmann, R., and Kanz, L. (1995). Reconstitution of hematopoiesis after high-dose chemotherapy by autologous progenitor cells generated *ex vivo. N. Engl. J. Med.* **333**(5), 283–287.

Chen, U. (1994). Development of artificial thymus, lymphocytes, and other tissues from wild type and transgenic mouse embryonic stem (ES) cells in a combined *in vitro* and *in vivo* differentiation system. *J. Cell. Biochem. Suppl.* **18C,** 274.

Chick, W. L., Like, A. A., and Lauris, V. (1975). Beta cell culture on synthetic capillaries: An artificial endocrine pancreas. *Science* **187,** 847–849.

Conget, P., and Minguell, J. J. (1995). IL-3 increases surface proteoglycan synthesis in haemopoietic progenitors and their adhesiveness to the heparin-binding domain of fibronectin. *Br. J. Haematol.* **89,** 1–7.

Dexter, T. M. (1982). Stromal cell associated haemopoiesis. *J. Cell. Physiol. Suppl.* **1,** 87–94.

Dexter, M., and Allen, T. (1992). Multi-talented stem cells? *Nature (London)* **360,** 709–710.

Dexter, T. M., Allen, T. D., and Lajtha, L. G. (1977). Conditions controlling the proliferation of haemopoietic stem cells *in vitro. J. Cell. Physiol.* **91**, 335–344.

Dexter, T. M., Spooncer, E., Simmons, P. *et al.* (1984). Long-term marrow culture: An overview of techniques and experience. *In* "Long-Term Bone Marrow Culture" (D. G. Wright and J. S. Greenberger, eds.), pp. 57–96. Alan R. Liss, New York.

Du, X. X., and Williams, D. A. (1994). Interleukin-11: A multifunctional growth factor derived from the hematopoietic microenvironment. *Blood* **83**(8), 2023–2030.

Dunbar, C., and Kohn, D. (1996). Retroviral mediated transfer of the cDNA for human glucocerebrosidase into hematopoietic stem cells of patients with Gaucher disease. A phase I study. *Hum. Gene Ther.* **7**, 231–253.

Dutton, G. (1999). Stem cell therapies draw biofirms' interest. *Genet. Eng. News* **19**, 8.

Engelhardt, M., Kumar, R., Albanell, J. Pettengell, R., Han, W., and Moore, M. A. (1997). Telomerase regulation, cell cycle, and telomere stability in primitive hematopoietic cells. *Blood* **90**, 182–193.

Ferrari, G., Cusella-De Angelis, G., Coletta, M., Paolucci, E., Stornaiuolo, A., Cossu, G., and Mavillio, F. (1998). Muscle regeneration by bone marrow-derived myogenic progenitors. *Science* **279**, 1528–1530.

Fliedner, T. M., and Calvo, W. (1978). Hematopoietic stem-cell seeding of a cellular matrix: A principle of initiation and regeneration of hematopoiesis. *In* "Differentiation of Normal and Neoplastic Hematopoietic Cells" (B. Clarkson, P. A. Marks, and J. E. Till, eds.), pp. 757–773. Cold Spring Harbor Lab., Cold Spring Harbor, NY.

Glockner, H., Jonuleit, T., Zimmerer, C., and Lemke, H.-D. (1999). New bioreactor for *in vivo* like cell culture, cell expansion and cell therapy. *Cytotherapy* (abstr.) (in press).

Goodell, M. A., Brose, K., Paradis, G., Conner, A. S., and Mulligan, R. C. (1996). Isolation and functional properties of murine hematopoietic stem cells that are replicating *in vivo. J. Exp. Med.* **184**, 1797–1806.

Goodell, M. A., Rosenzweig, M., Kim, H., Marks, D. F., DeMaria, M., Paradis, G., Grupp, S. A., Sieff, C. A., Mulligan, R. C., and Johnson, R. P. (1997). Dye efflux studies suggest that hematopoietic stem cells expressing low or undetectable levels of CD34 antigen exist in multiple species. *Nat. Med.* **3**, 1337–1345.

Goodwin, T. J., Prewett, T. L., Wolf, D. A., and Spaulding, G. F. (1993). Reduced shear stress: A major component in the ability of mammalian tissues to form three-dimensional assemblies in simulated microgravity. *J. Cell. Biochem.* **51**(3), 301–311.

Gullino, P. M., and Knazek, R. A. (1979). Tissue culture on artificial capillaries. *In* "Methods in Enzymology" (W. B. Jakoby and I. H. Tastan, eds.), Vol. 58, pp. 178–184. Academic Press, New York.

Harrison, D. E., Lerner, C., Hoppe, P. C., Carlson, G. A., and Alling, D. (1987). Large numbers of primitive stem cells are active simultaneously in aggregated embryo chimeric mice. *Blood* **69**, 773–777.

Harrison, D. E., Zhong, R. K., Jordan, C. T., Lemischka, I. R., and Astle, C. M. (1997). Relative to adult marrow, fetal liver repopulates nearly five times more effectively long-term than short-term. *Exp. Hematol.* **25**, 293–297.

Huang, S., and Terstappen, L. W. M. M. (1992). Formation of haematopoietic microenvironment and haematopoietic stem cells from single human bone marrow stem cells. *Nature (London)* **360**, 745–749.

Huang, S., and Terstappen, L. W. M. M. (1994). Correction. *Nature (London)* **368**, 664.

Itoh, K., Tezuka, H., Sakoda, H., Konno, M., Nagata, K., Uchiyama, T., Uchino, H., and Mori, K. J. (1989). Reproducible establishment of hematopoietic supportive stromal cell lines from murine bone marrow. *Exp. Hematol.* **17**, 145–153.

Kessinger, A. (1995). Circulating stem cells—Waxing hematopoietic. *N. Engl. J. Med.* **333**(5), 315–316.

Kessinger, A., and Sharp, J. G. (1998). Mobilization of blood stem cells. *Stem Cells* **16**(Suppl. 1), 139–143.

Kittler, E. L., Peters, S. O., Crittenden, R. B., Debatis, M. E., Ramshaw, H. S., Stewart, F. M., and Quesenberry, P. J. (1997). Cytokine-facilitated transduction leads to low-level engraftment in nonablated hosts. *Blood* **90**, 865–872.

Kluppel, M., Donoviel, D. B., Brunkow, M. E., Motro, B., and Bernstein, A. (1997). Embryonic and adult expression patterns of the *Tec* tyrosine kinase gene suggest a role in megakaryocytopoiesis, blood vessel development, and melanogenesis. *Cell Growth Differ.* **8**, 1249–2156.

Knazek, R. D., Gullino, P. M., Kohler, P. O., and Dedrick, R. L. (1972). Cell culture on artificial capillaries: An approach to tissue growth *in vitro. Science* **178**, 65–67.

Knospe, W. H. (1989). Hematopoiesis in artificial membranes. *In* "Handbook of the Hemopoietic Microenvironment" (M. Tavassoli, ed.), pp. 189–218. Humana Press, Clifton, NJ.

Kohn, D. B. (1996). Gene therapy for hematopoietic and immune disorders. *Bone Marrow Transplant.* **18**(Suppl. 3), S55–S58.

Koller, M. R., Emerson, S. G., and Palsson, B. O. (1993). Large-scale expansion of human stem and progenitor cells from bone marrow mononuclear cells in continuous perfusion cultures. *Blood* **82**, 378–384.

Kompala, D. S., and Highfill, J. G. (1994). Large-scale cultivation of murine bone marrow: The effects of inoculum density and cell type on the productivity of cells grown in an airlift packed bed bioreactor. *J. Cell. Biochem., Suppl.* **18C**, 276.

Lambertsen, R. H. (1984). Interdigitative coupling of presumptive hematopoietic stem cells to macrophage in endocloned marrow colonies. *Blood* **63**(5), 1225–1229.

Lansdorp, P. M. (1995). Developmental changes in the function of hematopoietic stem cells. *Exp. Hematol.* **23**, 187–191.

Lansdorp, P. M., Dragowska, W., and Mayani, H. (1993). Ontogeny-related changes in proliferative potential of human hematopoietic cells. *J. Exp. Med.* **178**(3), 787–791.

Lansdorp, P. M., Poon, S., Chavez, E., Dragowska, V., Zijlmans, M., Bryan, T., Redell, R., Egholm, M., Bacchetti, S., and Martens, U. (1997). Telomeres in the haemopoietic system. *Ciba Found. Symp.* **211**, 209–218.

Liu, J. M., Young, N. S., Walsh, C. E., Cottler-Fox, M., Carter, C., Dunbar, C., Barrett, A. J., and Emmons, R. (1997). Retroviral mediated gene transfer of the Fanconi anemia complementation group C gene to hematopoietic progenitors of group C patients. *Hum. Gene Ther.* **8**, 1715–1730.

Luskey, B. D., Rosenblatt, M., Zsebo, K., and Williams, D. A. (1992). Stem cell factor, interleukin-3, and interleukin-6 promote retroviral-mediated gene transfer into murine hematopoietic stem cells. *Blood* **80**(2), 396–402.

Ma, Q., Jones, D., Borghesani, P. R., Segal, R. A., Nagasawa, T., Kishimoto, T., Bronson, R. T., and Springer, T. A. (1998). Impaired B-lymphopoiesis, myelopoiesis, and derailed cerebellar neuron migration in CXCR4- and SDF-1-deficient mice. *Proc. Natl. Acad. Sci. U.S.A.* **95**, 9448–9453.

Mauch, P., Constine, L., Greenberger, J., Knospe, W., Sullivan, J., Liesveld, J. L., and Deeg, H. J. (1995). Hematopoietic stem cell compartment: Acute and late effects of radiation therapy and chemotherapy. *Int. J. Radiat. Oncol. Biol. Phys.* **31**(5), 1319–1339.

Micklem, H. S., Ansell, J. D., Wayman, J. E., and Forrester, L. (1983). The clonal organization of hematopoiesis in the mouse. *In* "Progress in Immunology V" (Y. Yamamura and T. Tada, eds.), pp. 633–644. Academic Press, Tokyo.

Morrison, S. J., Shah, N. M., and Anderson, D. J. (1997). Regulatory mechanisms in stem cell biology. *Cell (Cambridge, Mass.)* **88**, 287–298.

Mundle, S., Iftikhar, A., Shetty, V., Dameron, S., Wright-Quinones, V., Marcus, B., Loew, J., Gregory, S., and Raza, A. (1994). Novel *in situ* double labeling for simultaneous detection of proliferation and apoptosis. *J. Histochem. Cytochem.* **42**, 1533–1557.

Nagasawa, T., Hirota, S., Tachibana, K., Takakura, N., Nishikawa, S., Kitamura, Y., Yoshida, N., Kikutani, H., and Kishimoto, T. (1996). Defects of B-cell lymphopoiesis and bone-marrow myelopoiesis in mice lacking the CXC chemokine PBSF/SDF-1. *Nature (London)* **382**, 635–638.

Naughton, B. A., and Naughton, G. K. (1989). Hematopoiesis on nylon mesh templates: Comparative long-term bone marrow culture and the influence of stromal support cells. *Ann. N.Y. Acad. Sci.* **554**, 125–140.

Naughton, B. A., Preti, R. A., and Naughton, G. K. (1987). Hematopoiesis on nylon mesh templates. I. Long-term culture of rat bone marrow cells. *J. Med.* **18**, 219–250.

Naughton, B. A., Jacob, L., and Naughton, G. K. (1990). A three-dimensional culture system for the growth of hematopoietic cells. *Prog. Clin. Biol. Res.* **333**, 435–445.

Naughton, B. A., Sibanda, B., Azar, L., and San Roman, J. (1991). Differential effects of drugs upon hematopoiesis can be assessed in long-term bone marrow cultures established on nylon screens. *Proc. Soc. Exp. Biol. Med.* **199**, 481–490.

Naughton, B. A., Dai, Y., Sibanda, B., Scharfmann, R., San Roman, J., Zeigler, F., and Verma, I. M. (1992). Long-term expression of a retrovirally introduced β-galactosidase gene in rodent cells implanted *in vivo* using biodegradable polymer meshes. *Somatic Cell Mol. Genet.* **18**, 451–462.

Naughton, B. A., San Román, J., Sibanda, B., Weintraub, J. P., Morales, D. L., and Kamali, V. (1994). Surgical implantation of bioengineered bone marrow tissue into rats. *In* "Advances in Bone Marrow Purging and Processing: Fourth International Symposium," pp. 711–725. Wiley-Liss, New York.

Neben, S., Anklesaria, P., Greenberger, J., and Mauch, P. (1993). Quantitation of murine hematopoietic stem cells *in vitro* by limiting dilution analysis of cobblestone area formation on a clonal stromal cell line. *Exp. Hematol.* **21**, 438–443.

Ogawa, M. (1994). Hematopoiesis. *J. Allergy Clin. Immunol.* **94**(3), 645–650.

Osawa, M., Hanada, K., Hamada, H., and Nakauchi, H. (1996). Long-term lymphohematopoietic reconstitution by a single CD34-low/negative hematopoietic stem cell. *Science* **273**, 242–245.

O'Shea, K. S. (1999). Embryonic stem cell models of development. *Anat. Rec.* **257**, 32–41.

Owen, M. (1988). Marrow stromal cells. *J. Cell Sci., Suppl.* **10**, 63–76.

Palsson, B. O. (1994). Hematopoietic perfusion bioreactors: Scientific and clinical utility. *J. Cell. Biochem., Suppl.* **18C**, 270.

Papoutsakis, E. T., Koller, M. R., Sandstrom, C. E., Miller, W. M., and Bender, J. G. (1994). *Ex vivo* expansion of primitive hematopoietic cells under perfusion conditions. *J. Cell. Biochem., Suppl.* **18C**, 271.

Peled, A., Petit, I., Kollet, O., Magid, M., Ponomaryov, T., Byk, T., Nagler, A., Ben-Hur, H., Many, A., Shultz, L., Lider, O., Alon, R., Zipori, D., and Lapidot, T. (1999). Dependence of human stem cell engraftment and repopulation of NOD/SCID mice on CXCR4. *Science* **283**, 845–848.

Peters, S. O., Kittler, E. L. W., Ramshaw, H. S., and Quesenberry, P. J. (1995). Murine marrow cells expanded in culture with IL-3, IL-6, IL-11, and SCF acquire an engraftment defect in normal hosts. *Exp. Hematol.* **23**, 461–469.

Pettengell, R., Morgenstern, G. R., Woll, P. J., Chang, J., Rowlands, M., Young, R., Radford, J. A., Scarffe, J. H., Testa, N. G., and Crowther, D. (1993). Peripheral blood progenitor cell transplantation in lymphoma and leukemia using a single apheresis. *Blood* **82**(12), 3770–3777.

Pittenger, M. F., Mackay, A. M., Beck, S. C., Jaiswal, R. K., Douglas, R., Mosca, J. D., Moorman, M. A., Simonetti, D. W., Craig, S., and Marshak, D. R. (1999). Multilineage potential of adult human mesenchymal stem cells. *Science* **284**, 143–147.

Ploemacher, R. E., Van der Sluijs, J. P., van Beurden, C. A. J., Baert, M. R., and Chan, P. L. (1991). Use of limiting-dilution type long-term marrow cultures in frequency analysis of marrow-repopulating and spleen colony-forming hematopoietic stem cells in the mouse. *Blood* **78**, 2527–2533.

Quesenberry, P. J., Ramshaw, H., Crittenden, R. B., Stewart, F. M., Rao, S., Peters, S., Becker, P., Lowry, P., Blomberg, M., Reilly, J. *et al.* (1994). Engraftment of normal murine marrow into nonmyeloablated host mice. *Blood Cells* **20**, 348–350.

Rosenzweig, M. (1999). Three dimensional hematopoietic stem cell engineering. *In* "Advances in Tissue Engineering," IBC Symp.

Schwartz, R. M., Palsson, B. O., and Emerson, S. G. (1991). Rapid medium perfusion rate significantly increases the productivity and longevity of human bone marrow cultures. *Proc. Natl. Acad. Sci. U.S.A.* **88**, 6760–6764.

Seki, M. (1973). Hematopoietic colony formation in a macrophage layer provided by intraperitoneal insertion of a cellulose acetate membrane. *Transplantation* **16**, 544–549.

Shalaby, F., Ho, J., Stanford, A. L., Fischer, K. D., Schuh, A. C., Schwartz, L., Bernstein, A., and Rossant, J. (1997). A requirement for Flk1 in primitive and definitive hematopoiesis and vasculogenesis. *Cell (Cambridge, Mass.)* **89**, 981–990.

Sharp, J. G. (1993). Hematopoiesis and the dynamics of stem cell engraftment. *In* "Peripheral Blood Stem Cells" (D. M. Smith and R. A. Sacher, eds.), pp. 1–18. American Association of Blood Banks, Bethesda, MD.

Sharp, J. G., Udeaja, G., Jackson, J. D., Mann, S. L., Murphy, B. O., and Crouse, D. A. (1985). Transplantation of adherent cells from murine long term marrow cultures. *In* "Leukemia: Recent Advances in Biology and Treatment" (R. P. Gale and D. W. Golde, eds.), pp. 415–426. Alan R. Liss, New York.

Siczkowski, M., Clarke, D., and Gordon, M. Y. (1992). Binding of primitive hematopoietic progenitor cells to marrow stromal cells involves heparan sulfate. *Blood* **80**(4), 912–919.

Solter, D., and Gearhart, J. (1999). Putting stem cells to work. *Science* **283**, 1468–1469.

Spooncer, E., and Dexter, T. M. (1983). Transplantation of long-term cultured bone marrow cells. *Transplantation* **35**(6), 624–626.

Sutherland, H. J., Eaves, C. J., Eaves, A. C., Dragowska, W., and Lansdorp, P. M. (1989). Characterization and partial purification of human marrow cells capable of initiating long-term hematopoiesis *in vitro. Blood* **74**, 1563–1570.

Tavassoli, M. (1989). Fatty involution of marrow and the role of adipose tissue in hemopoiesis. *In* "Handbook of the Hemopoietic Microenvironment" (M. Tavassoli, ed.), pp. 157–188. Humana Press, Clifton, NJ.

Tavassoli, M., and Takahashi, K. (1982). Morphological studies on long-term culture of marrow cells: Characterization of the adherent stromal cells and their interactions in maintaining the proliferation of hematopoietic stem cell. *Am. J. Anat.* **164**(2), 91–111.

Testa, N. G., Hendry, J. H., and Molineux, G. (1985). Long-term bone marrow damage in experimental systems and in patients after radiation or chemotherapy. *Anticancer Res.* **5**, 101–110.

Udomsakdi, C., Lansdorp, P. M., Hogge, D. E., Reid, D. S., Eaves, A. C., and Eaves, C. J. (1992). Characterization of primitive hematopoietic cells in normal human peripheral blood. *Blood* **80**, 2513–2521.

Van Zant, G., de Haan, G., and Rich, I. N. (1997). Alternatives to stem cell renewal from a developmental viewpoint. *Exp. Hematol.* **25**, 187–192.

Verfaillie, C. M. (1992). Direct contact between human primitive hematopoietic progenitors and bone marrow stroma is not required for long-term *in-vitro* hematopoiesis. *Blood* **79**, 2821–2826.

Verfaillie, C. M., Benis, A., Iida, J., McGlave, P. B., and McCarthy, J. B. (1994). Adhesion of committed human hematopoietic progenitors to synthetic peptides from the C-terminal heparin-binding domain of fibronectin: Cooperation between the integrin α4β1 and the CD44 adhesion receptor. *Blood* **84**(6), 1802–1811.

Vogel, G. (1999). Harnessing the power of stem cells. *Science* **283**, 1432–1434.

Waegell, W. O., Higley, H. R., Kincade, P. W., and Dasch, J. R. (1994). Growth acceleration and stem cell expansion in Dexter-type cultures by neutralization of TGF-β. *Exp. Hematol.* **22**, 1051–1057.

Wagner, J. E., Kernan, N. A., Steinbuch, M., Broxmeyer, H. E., and Gluckman, E. (1995). Allogeneic sibling umbilical-cord-blood transplantation in children with malignant and non-malignant disease. *Lancet* **346**, 214–219.

Walton, I. (1653). "The Compleat Angler."

Wang, T.-Y., and Wu, J. H. D. (1992). A continuous perfusion bioreactor for long-term bone marrow culture. *Ann. N.Y. Acad. Sci.* 274–284.

Wang, T.-Y., Brennan, J. K., and Wu, J. H. D. (1995). Multilineal hematopoiesis in a three-dimensional murine long-term bone marrow culture. *Exp. Hematol.* **23**, 26–32.

Watt, S. M., and Visser, J. W. M. (1992). Recent advances in the growth and isolation of primitive human haemopoietic progenitor cells. *Cell Proliferation* **25**, 263–297.

Weiss, L., and Geduldig, U. (1991). Barrier cells: Stromal regulation of hematopoiesis and blood cell release in normal and stressed murine bone marrow. *Blood* **78**, 975–990.

Whetton, A. D., and Spooncer, E. (1998). Role of cytokines and extracellular matrix in the regulation of haematopoietic stem cells. *Curr. Opin. Cell Biol.* **10**, 721–726.

Williams, D. A. (1994). Molecular analysis of the hematopoietic microenvironment. *Pediatr. Res.* **36**(5), 557–560.

Williams, G. T., Smith, C. A., Spooncer, E., Dexter, T. M., and Taylor, D. R. (1990). Haemopoietic colony stimulating factors promote cell survival by suppressing apoptosis. *Nature (London)* **343**, 76–79.

Xu, L., Stahl, S. K., Dave, H. P., Schiffmann, R., Correll, P. H., Kessler, S., and Karlsson, S. (1994). Correction of the enzyme deficiency in hematopoietic cells of Gaucher patients using a clinically acceptable retroviral supernatant transduction protocol. *Exp. Hematol.* **22**, 223–230.

Xu, L. C., Kluepfel-Stahl, S., Blanco, M., Schiffman, R., Dunbar, C., and Karlsson, S. (1995). Growth factors and stromal support generate very efficient retroviral transduction of peripheral blood CD34+ cells from Gaucher patients. *Blood* **86**, 141–146.

Yoder, M. C., and Williams, D. A. (1995). Matrix molecule interactions with hematopoietic stem cells. *Exp. Hematol.* **23**, 961–967.

Zipori, D. (1989a). Cultured stromal cell lines from hemopoietic tissues. *In* "Handbook of the Hemopoietic Microenvironment" (M. Tavassoli, ed.), pp. 287–333. Humana Press, Clifton, NJ.

Zipori, D. (1989b). Stromal cells from the bone marrow: Evidence for a restrictive role in regulation of hemopoiesis. *Eur. J. Haematol.* **42**, 225–232.

Zipori, D., Duksin, D., Tamir, M., Argaman, A., Toledo, J., and Malik, Z. (1985). Cultured mouse marrow stromal cell lines. II. Distinct subtypes differing in morphology, collagen types, myelopoietic factors, and leukemic cell growth modulating activities. *J. Cell. Physiol.* **122**, 81–90.

PART XV: KIDNEY AND GENITOURINARY SYSTEM

CHAPTER 45

RENAL REPLACEMENT DEVICES

H. David Humes

INTRODUCTION

The kidney was the first solid organ whose function was approximated by a machine and a synthetic device. In fact, renal substitution therapy with hemodialysis or chronic ambulatory peritoneal dialysis (CAPD) has been the only successful long-term *ex vivo* organ substitution therapy to date (Iglehart, 1993). The kidney was also the first organ to be successfully transplanted from a donor individual to an autologous recipient patient. However, the lack of widespread availability of suitable transplantable organs has kept kidney transplantation from becoming a practical solution to most cases of chronic renal failure.

Although long-term chronic renal replacement therapy with either hemodialysis or CAPD has dramatically changed the prognosis of renal failure, it is not complete replacement therapy, because it provides only filtration function (usually on an intermittent basis) and does not replace the homeostatic, regulatory, metabolic, and endocrine functions of the kidney. Because of the nonphysiologic manner in which dialysis performs or does not perform the most critical renal functions, patients with end-stage renal disease (ESRD) on dialysis continue to have major medical, social, and economic problems (U.S. Renal Data System, 1998). Accordingly, dialysis should be considered as renal substitution rather than renal replacement therapy.

Tissue engineering of an implantable artificial kidney composed of both biologic and synthetic components will most likely have substantial benefits for the patient by increasing life expectancy, mobility, and quality of life, with less risk of infection and with reduced costs. This approach could also be considered a cure rather than a treatment for patients. A successful tissue engineering approach to the kidney is dependent on a thorough knowledge of the physiologic basis of kidney function.

BASICS OF KIDNEY FUNCTION

Excretory function of the kidney is initiated by filtration of blood at the glomerulus, which is an enlargement of the proximal end of the tubule incorporating a vascular tuft. The structure of the glomerulus is designed to provide efficient ultrafiltration of blood to remove toxic waste from the circulation, yet retain important circulating components, such as albumin. The regulatory function of the kidney, especially with regard to fluid and electrolyte homeostasis, is provided by the tubular segments attached to the glomerulus. The functional unit of the kidney is therefore composed of the filtering unit (the glomerulus) and the regulatory unit (the tubule). Together they form the basic component of the kidney, the nephron. In addition to these excretory and regulatory functions, the kidney is an important metabolic and endocrine organ. Erythropoietin, active forms of vitamin D, renin, angiotensin, prostaglandins, leukotrienes, kallikrein-kinins, various cytokines, and complement components are some of the endocrinologic compounds produced by the kidney.

Because of the efficiency inherent in the kidney as an excretory organ, life can be sustained with only 5–10% of normal renal excretory function. Accordingly, the approach to a tissue engineering construct becomes easier to entertain, especially because only a fraction of normal renal excretory function is required to maintain life (Humes *et al.*, 1997a).

Principles of Tissue Engineering
Second Edition

The process of urine formation begins within the capillary bed of the glomerulus (Brenner and Humes, 1977). The glomerular capillary wall has evolved into a structure with the property to separate as much as one-third of the plasma entering the glomerulus into a solution of a nearly ideal ultrafiltrate. This high rate of ultrafiltration across the glomerular capillary is a result of hydraulic pressure generated by the heart and vascular tone of the preglomerular and postglomerular vessels, as well as the high hydraulic permeability of the glomerular capillary wall. This hydraulic pressure and permeability of the glomerular capillary bed is at least two times and two orders of magnitude higher, respectively, than most other capillary networks within the body (Landis and Pappenheimer, 1965). Despite this high rate of water and solute flux across the glomerular capillary wall, this same structure retards the filtration of important circulating macromolecules, especially albumin, so that all but the lower molecular-weight plasma proteins are restricted in the passage across this filtration barrier (Anderson and Quinn, 1974; Chang *et al.,* 1975; Brenner *et al.,* 1978).

This ultrafiltration process of glomeruli in normal human kidneys forms approximately 100 ml of filtrate every minute. Because daily urinary volume is roughly 2 liters, more than 98% of the glomerular ultrafiltrate must be absorbed by the renal tubule. The bulk of reabsorption, 50–65%, occurs along the proximal tubule. Similar to glomerular filtration, fluid movement across the renal proximal tubule cell is governed by physical forces. Unlike the fluid transfer across the glomerular capillary wall, however, tubular fluid flux is principally driven by osmotic and oncotic pressures rather than hydraulic pressure. Renal proximal tubule fluid absorption is based on active NA^+ transport, requiring the energy-dependent Na^+,K^+-ATPase located along the basolateral membrane of the renal tubule cell to promote a small degree of luminal hypotonicity (Andreoli and Schafer, 1978). This small degree of osmotic difference (2–3 mOsm/kg H_2O) across the renal tubule is sufficient to drive isotonic fluid reabsorption due to the very high diffusive water permeability of the renal tubule cell membrane. Once across the renal proximal tubule cell, the transported fluid is taken up by the peritubular capillary bed due to the favorable oncotic pressure gradient. This high oncotic pressure within the peritubular capillary is the result of the high rate of protein-free filtrate formed in the proximate glomerular capillary bed (Knox *et al.,* 1983). As can be appreciated, an elegant system has evolved in the nephron to filter and reabsorb large amounts of fluid in bulk to attain high rates of metabolic product excretion while maintaining regulatory salt and water balance.

TISSUE ENGINEERING APPROACH TO RENAL FUNCTION REPLACEMENT

In designing an implantable bioartificial kidney for renal function replacement, essential features of kidney tissue must be utilized to direct the design of the tissue engineering project. The critical elements of renal function must be replaced, including the excretory, regulatory transport, and endocrinologic functions. The functioning excretory unit of the kidney, as detailed previously, is composed of the filtering unit, the glomerulus, and the regulatory or transport unit, the tubule. Therefore, a bioartificial kidney requires two main units, the glomerulus and the tubule, to replace excretory function.

BIOARTIFICIAL HEMOFILTER

The potential for a bioartificial glomerulus has been achieved with the use of polysulfone fibers *ex vivo* with maintenance of ultrafiltration in humans for several weeks with a single device (Golper, 1986; Kramer *et al.,* 1977). The availability of hollow fibers with high hydraulic permeability has been an important advancement in biomaterials for replacement function of glomerular ultrafiltration. Conventional hemodialysis for end-stage renal disease has used membranes in which solute removal is driven by a concentration gradient of the solute across the membranes, and is, therefore, predominantly a diffusive process. Another type of solute transfer also occurs across the dialysis membrane via a process of ultrafiltration of water and solutes across the membrane. This convective transport is independent of the concentration gradient and depends predominantly on the hydraulic pressure gradient across the membrane. Both diffusive and convective processes occur during traditional hemodialysis, but diffusion is the main route of solute movement.

The development of synthetic membranes with high hydraulic permeability and solute re-

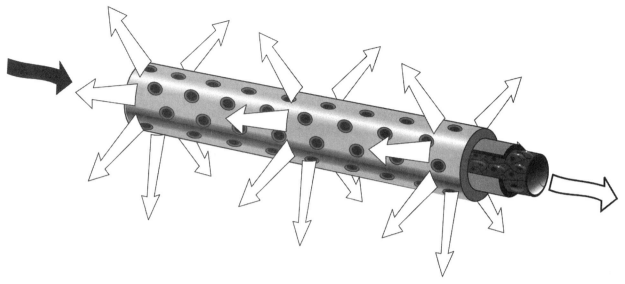

Fig. 45.1. Schematic of a tissue-engineered hemofilter composed of a microporous synthetic bio-compatible hollow fiber, a preadhered extracellular matrix, and a confluent monolayer of autol-ogous endothelial cells lining the luminal surface of the fiber. Arrows refer to vectorial ultrafiltrate formation.

tention properties in convenient hollow fiber form has promoted ESRD therapy based on convective hemofiltration rather than diffusive hemodialysis (Colton *et al.*, 1975; Henderson *et al.*, 1975). Removal of uremic toxins, predominantly by the convective process, has several distinct advantages, because it imitates the glomerular process of toxin removal with increased clearance of higher molecular-weight solutes and removal of all solutes (up to a molecular-weight cutoff) at the same rate. Distinct differences are apparent between diffusive and convective transport across a semipermeable membrane. The clearance of a molecule by diffusion is negatively correlated with the size of the molecule. In contrast, clearance of a substance by convection is dependent on size up to a certain molecular weight. The bulk movement of water carries passable solutes along with it in approximately the same concentration as in the fluid.

Development of an implantable device that mimics glomerular filtration will thus depend on convective transport. This physiologic function has been achieved clinically with the use of polymeric hollow fibers *ex vivo*. Major limitations to the currently available technology for long-term replacement of filtration function include bleeding associated with required anticoagulation, diminution of filtration rate due either to protein deposition in the membrane over time or to thrombotic occlusion, and large amounts of fluid replacement required to replace the ultrafiltrate formed from the filtering unit. The use of autologous endothelial cell-seeded conduits along filtration surfaces may provide improved long-term hemocompatibility and hemofiltration *in vivo* (Kadletz *et al.*, 1992; Schnider *et al.*, 1988; Shepard *et al.*, 1986). A report detailing the capability of isolating angioblasts from peripheral blood makes this approach more readily achievable (Asahara *et al.*, 1997). The initial step to develop a tissue-engineered hemofilter with adequate ultrafiltration rates through an endothelial-lined synthetic hemofiltration membrane has been achieved (Humes *et al.*, 1997b) and is depicted in Fig. 45.1.

A potential rate-limiting step in endothelial cell-lined hollow fibers of small caliber is thrombotic occlusion, which limits the functional patency of this filtration unit. In this regard, gene transfer into seeded endothelial cells for constitutive expression of anticoagulant factors can be envisioned to minimize clot formation in these small-caliber hollow fibers. Because gene transfer for *in vivo* protein production has been clearly achieved with endothelial cells (Zweibel *et al.*, 1989; Wilson *et al.*, 1989), gene transfer into endothelial cells for the production of an anticoagulant protein is clearly conceivable. This strategy has been used with hirudin, a protein from the blood-sucking leech, which is a potent specific inhibitor of thrombin. A replication-defective, amphotropic, recombinant retrovirus containing the hirudin gene has been used to infect endothelial cells, which were then seeded onto a polysulfone membrane (Humes *et al.*, 1997a; Rade *et al.*, 1996). Culture supernatants of the transfected cells had high thrombin inhibitory activity, as determined by enzyme-linked immunosorbent assay.

For differentiated endothelial cell morphology and function, an important role for various components of the extracellular matrix (ECM) has been demonstrated (Carey, 1991; Carley *et al.*, 1988). The ECM has been clearly shown to dictate phenotype and gene expression of endothelial cells, thereby modulating morphogenesis and growth. Various components of the ECM, including collagen type I, collagen type IV, laminin, and fibronectin, have been shown to affect endothelial cell adherence, growth, and differentiation. Of importance, ECM produced by Madin–Darby canine kidney (MDCK) cells, a permanent renal epithelial cell line, promotes endothelial cells to develop fenestrations. Fenestrations are large openings that act as channels or pores for convective transport through the endothelial monolayer and are important in the high hydraulic permeability and sieving characteristics of glomerular capillaries (Carley *et al.*, 1988; Milici *et al.*, 1985). Thus, the ECM component on which the endothelial cells attach and grow may be critical in the functional characteristics of the lining monolayer.

BIOARTIFICIAL TUBULE

As detailed above, the efficiency of reabsorption, even though dependent on natural physical forces governing fluid movement across biologic as well as synthetic membranes, requires specialized epithelial cells to perform vectorial solute transport. Fortunately, a population of cells residing in the adult mammalian kidney have retained the capacity to proliferate and morphogenically differentiate into tubule structures *in vitro* (Humes and Cieslinski, 1992; Humes *et al.*, 1996), and can be used as the key cellular element of a tissue-engineered renal tubule device.

The bioartificial renal tubule can be readily conceived as a combination of living cells supported by polymeric substrata, using epithelial progenitor cells cultured on water- and solute-permeable membranes seeded with various biomatric materials so that expression of differentiated vectorial transport and metabolic and endocrine function is attained (Fig. 45.2). With appropriate membranes and biomatrices, immunoprotection of cultured progenitor cells has been achieved concurrent with long-term functional performance as long as conditions support tubule cell viability. This bioartificial tubule has been shown to transport salt and water effectively along osmotic and oncotic gradients (MacKay *et al.*, 1998).

A bioartificial proximal tubule satisfies a major requirement of reabsorbing a large volume of filtrate to maintain salt and water balance within the body. The need for additional tubule segments to replace other nephronal functions, such as the loop of Henle, to perform more refined homeostatic elements of the kidney, including urine concentration or dilution, may not be necessary. Patients with moderate renal insufficiency lose the ability to regulate salt and water homeostasis finely because they are unable to concentrate or dilute, yet are able to maintain reasonable fluid and electrolyte homeostasis due to redundant physiologic compensation via other mechanisms. Thus, a bioartificial proximal tubule, which reabsorbs isoosmotically the majority of the filtrate, may be sufficient to replace required tubular function to sustain fluid electrolyte balance in a patient with end-stage renal disease.

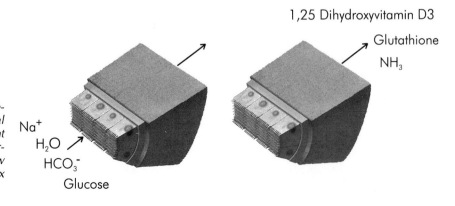

Fig. 45.2. Schematic of a tissue-engineered renal tubule. Renal epithelial cells form a confluent monolayer along the inner surface of a polysulfone hollow fiber with preadhered matrix molecules.

IMPLANTABLE TISSUE-ENGINEERED KIDNEY

The development of a bioartificial filtration device and a bioartificial tubule processing unit would lead to the possibility of an implantable bioartificial kidney, consisting of a filtration device followed in series by the tubule unit. The filtrate formed by this device will flow directly into the tubule unit. This tubule unit should maintain viability, because metabolic substrates and low-molecular-weight growth factors are delivered to the tubule cells from the ultrafiltration unit (Nikolovski *et al.*, 1999). Furthermore, immunoprotection of the cells grown within the hollow fiber is achievable due to the impenetrance of immunologically competent cells through the hollow fiber (O'Neil *et al.*, 1997). Rejection of transplanted cells will, therefore, not occur. This arrangement thereby allows the filtrate to enter the hollow-fiber network internal compartments, which are lined with confluent monolayers of renal tubule cells for regulated transport function.

This device could be used either extracorporeally or implanted within a patient. In this regard, the specific implant site for a bioartificial kidney will depend on the final configuration of both bioartificial filtration and tubule device. As presently conceived, the endothelial-lined bioartificial filtration hollow fibers can be placed into an arteriovenous circuit using the common iliac artery and vein, similar to the surgical connection for a renal transplant. The filtrate is connected in series to a bioartificial proximal tubule, with the reabsorbate transported back into the systematic circulation. The processed filtrate exiting the tubule unit is then connected via tubing to the proximate ureter for drainage and urine excretion via the recipient's own urinary collecting system.

Although the ultimate goal of this approach is to construct a fully implantable bioartificial kidney with a hemofilter and tubular system, the pathway to achieve this goal will most likely occur along a staged developmental strategy, with each intermediate device providing both clinical therapeutic benefit and substantial experience in the use of more elementary tissue-engineered devices (Humes *et al.*, 1997b). In this regard, the initial component of the bioartificial kidney to be developed for clinical evaluation will be an extracorporeal renal tubule-assist device (RAD), to optimize current hemofiltration approaches to treat the clinical disorder of acute renal failure.

DEVELOPMENT OF A RENAL TUBULE-ASSIST DEVICE

Replacement of the multivariate tubular functions of the kidney cannot be achieved with inanimate membrane devices, as has been accomplished with the renal ultrafiltration process, but requires the use of the naturally evolved biologic membranes of the renal tubular epithelium. In this regard, the tissue engineering of a bioartificial renal tubule as a cell therapy device to replace this missing component can be conceived as a combination of living cells supported on synthetic scaffolds (MacKay *et al.*, 1998; Nikolovski *et al.*, 1999). A bioartificial tubule can be constructed utilizing renal tubule progenitor cells (Humes and Cieslinski, 1992; Humes *et al.*, 1996), cultured on semipermeable hollow-fiber membranes on which extracellular matrix has been layered to enhance the attachment and growth of the epithelial cells (Timpl *et al.*, 1979). These hollow-fiber synthetic membranes not only provide the architectural scaffold for these cells but also provide immunoprotection, as has been observed in the long-term implantation of the bioartificial pancreas in a xenogeneic host (O'Neil *et al.*, 1997). The successful tissue engineering of a bioartificial tubule as a confluent monolayer has been achieved in a single hollow-fiber bioreactor system (Fig. 45.3) (MacKay *et al.*, 1998). The successful scale-up from a single-fiber system to a multifiber bioreactor and porcine renal proximal tubule cells has also been successfully engineered (Humes *et al.*, 1999b). Porcine cells were used because the pig is currently considered the best source of organs for both human xenotransplantation and immunoisolation cell therapy due to its anatomic and physiologic similarities to human tissue and the relative ease of breeding pigs in large numbers (Cozzi and White, 1995).

The scale-up from a single hollow fiber to a multifiber bioartificial renal tubule-assist device has proceeded with porcine renal proximal tubule cells grown as confluent monolayers along the inner surface of polysulfone immunoisolating hollow fibers (Humes *et al.*, 1999b). These hollow-fiber cartridges are packaged in bioreactor cartridges with membrane surface areas as large as 0.7 m^2, resulting in a device containing up to 2.5×10^9 cells. *In vitro* studies of these RADs have demonstrated their retention of differentiated active vectorial transport of sodium, bicarbonate, glucose, and organic anions (Humes *et al.*, 1999b). These transport properties were suppressed with specific transport inhibitors. Ammoniagenesis and glutathione metabolism, which are important differentiated metabolic processes of the kidney, were also demonstrated in these devices.

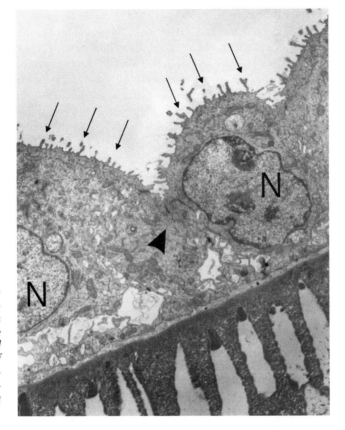

Cells

Extracellular
Matrix

Hollow
Fiber

Fig. 45.3. Electron micrograph of a tissue-engineered bioartificial renal tubule. The nucleus (N) is indicated; arrows delineate apical microvilli (a differentiated morphologic characteristic of proximal tubule cells) and the arrowhead identifies the tight junctional complex of a transporting epithelium.

Synthesis of 1,25-dihydroxyvitamin D3 as a key endocrinologic metabolite was also documented. These metabolic processes were also shown to be regulated by important physiologic parameters and achieved rates comparable to those of a whole kidney.

The critical next step for clinical application of a renal tubule cell therapy device was to ascertain whether the RAD maintained differentiated renal functional performance, similar to that observed *in vitro,* and viability in an extracorporeal hemoperfusion circuit in an acutely uremic dog. The RAD was accordingly placed in a series into a standard hollow-fiber hemofiltration cartridge; the cartridge had ultrafiltrate and postfiltered blood connections, to duplicate the structural anatomy of the nephron and to mimic the functional relationship between the renal globerulus and tubule (see Fig. 45.4). The successful completion of this step has been reported (Humes *et al.,* 1999a). Data show that fluid and electrolyte balances in the animals, as reflected by plasma parameters, were adequately controlled with the bioartificial kidney. In fact, plasma potassium and blood urea nitrogen levels were more easily controlled during RAD treatment compared with sham control conditions. The step-up of oncotic pressure in the postfiltrate blood, which was delivered to the antiluminal space of the bioartificial proximal tubules, allowed the fractional reabsorption of sodium and water to achieve 40–50% of the ultrafiltrate volume, an amount similar to that seen in the nephron *in vivo.* Active transport of potassium, bicarbonate, and glucose by the RAD was demonstrated. To further evaluate the capacity of the RAD to optimize renal function replacement of the bioartificial kidney, three metabolic parameters were evaluated: ammonia excretion, glutathione reclamation, and 1,25-$(OH)_2$D3 production. With increasing time of use in the uremic animal, the RAD displayed incremental ability to excrete ammonia to a level as high as 100/μmol/hr. The RAD demonstrated effective reclamation of filtered glutathione and an improvement in plasma glutathione levels in the uremic animal. Finally, these studies also demonstrated the ability of the RAD to produce 1,25-$(OH)_2$D3, the most active vitamin D metabolite (Rade *et al.,* 1996). This production rate was high enough to return plasma values of this hormone to normal levels from the significantly depressed levels observed in the acutely uremic condition.

It is, in fact, because of the ability of the RAD to replace the numerous metabolic processes

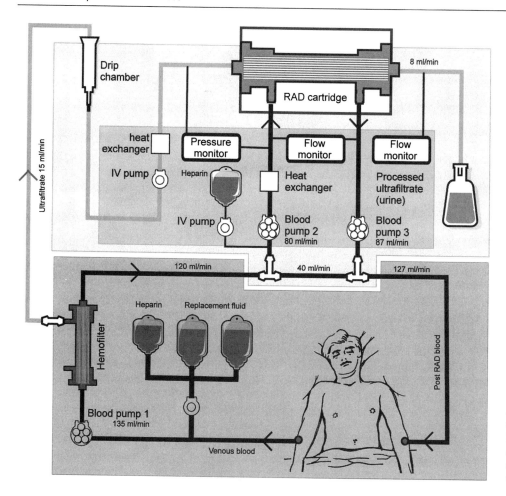

Fig. 45.4. The extracorporeal hemoperfusion circuit of a tissue-engineered kidney for the treatment of a patient with acute renal failure. The synthetic hemofilter is connected in series to the bioartificial renal tubule-assist device (RAD). Heat exchangers, flow and pressure monitors, and multiple pumps are required for optimal functioning of this extracorporeal device.

lost in renal failure that this form of cell therapy may optimize current renal substitution therapy of hemodialysis and hemofiltration. Patients who develop ischemic or nephrotoxic acute renal failure still have mortality rates exceeding 50%, despite hemodialysis or hemofiltration therapy. These patients suffer from high catabolic rates, immunologic dysregulation, and predilection to septic shock, all of which contribute to the high mortality rate (Humes, 1995; Humes *et al.*, 1997a). The ability of tubule cells to metabolize and synthesize critical compounds, such as various cytokines, glutathione, free-radical scavenging enzymes, and 1,25-$(OH)_2$D3, which play important roles in immunologic reactivity, may have the substantial added benefit of altering the natural history of this disease process. Because the pathophysiology of this disorder is due predominantly to proximal tubule cell injury and necrosis, the replacement of proximal tubule cell function with a cell therapy device would appear a logical therapeutic intervention for this condition. The results of this study indicate the technologic feasibility of this form of cell therapy in the acutely uremic state and set the stage to initiate clinical trials with this exciting new therapy in the very near future.

A second intermediate tissue-engineered device that may prove to have clinical utility is an implantable hemofilter with a capacity to produce 2–4 ml/min of ultrafiltrate. Although this rate of filtration is considerably less than the 10–15 ml/min rate required to correct the uremic state in ESRD patients, it is sufficient enough to improve solute clearance to lessen dialysis time and optimize clearance rates of urea and other uremic toxins. In addition, this rate of ultrafiltration from a stand-alone implanted hemofilter is small enough so that, with a urinary bladder capacity of 500–750 ml, a patient can tolerate the frequency of urination required from the filtrate draining directly into the urinary tract system. Both of these intermediate devices will provide substantial experience to improve the durability, efficacy, and efficiency of an implantable hemofilter and a tubule-assist device to achieve the required hemofiltration rate of 15–20 ml/min and selective reabsorption for adequate clearance while avoiding volume depletion and intolerable urinary frequency.

SUMMARY

Similar to gene therapy, cell therapy is an exciting new approach to the treatment of acute and chronic diseases (Gage, 1998). Cell therapy centers on techniques directed toward the expansion of specific cells to perform differentiated tasks and the introduction of these cells or cell products into a patient either within extracorporeal circuits or as implants. This approach leverages the natural abilities of cells to replace critical physiologic functions deranged or lost in clinical disorders. Tissue engineering is a further extension of cell therapy to combine the biologic and engineering disciplines to construct devices to replace tissue or organ functions lost to disease (Langer and Vacanti, 1993). Current approaches have focused on placing cells into hollow-fiber bioreactors or encapsulating membranes. Extracorporeal liver-assist devices and encapsulated islets of Langerhans to treat liver failure and diabetes mellitus are most promising examples (Watanabe *et al.*, 1997; Maki *et al.*, 1996). The progress in this area suggests that the application of these technologies may be successfully applied to renal replacement therapy.

The kidney was the first organ whose function was substituted by an artificial device. The kidney was also the first organ to be transplanted successfully. The ability to replace renal function with these revolutionary technologies in the past was due to the fact that renal excretory function is based on natural physical forces that govern solute and fluid movement from the body compartment to the external environment. The need for coordinated mechanical or electrical activities for renal substitution was not required. Accordingly, the kidney may well be the first organ to be available as a tissue-engineered implantable device that is a fully functional replacement part for the human body.

REFERENCES

Anderson, J. L., and Quinn, J. A. (1974). Restricted transport in small pores. A model for steric exclusion and hindered particle motion. *Biophys. J.* **14**, 130.

Andreoli, T. E., and Schafer, J. A. (1978). Volume absorption in the pars recta: III. Luminal hypotonicity as a driving force for isotonic volume absorption. *Am J. Physiol.* **234**(4), F349.

Asahara, T., Murohara, T., Sullivan, A., Silver, M., van der Zee, R., Li, T., Witzenbichler, B., Schatteman, G., and Isner, J. M. (1997). Isolation of putative progenitor endothelial cells for angiogenesis. *Science* **275**, 964–967.

Brenner, B. M., and Humes, H. D. (1977). Mechanisms of glomerular ultra-filtration. *N. Engl. J. Med.* **297**, 148.

Brenner, B. M., Hostetter, T. H., and Humes, H. D. (1978). Molecular basis of proteinuria of glomerular origin. *N. Engl. J. Med.* **298**, 826.

Carey, D. J. (1991). Control of growth and differentiation of vascular cells by extracellular matrix proteins. *Annu. Rev. Physiol.* **53**, 161.

Carley, W. W., Milica, A. J., and Madri, J. A. (1988). Extracellular matric specificity for the differentiation of capillary endothelial cells. *Exp. Cell Res.* **178**, 426.

Chang, R. L. S., Robertson, C. R., Deen, W. M. *et al.* (1975). Permselectivity of the glomerular capillary wall to macromolecules: I. Theoretical considerations. *Biophys. J.* **15**, 861.

Colton, C. K., Henderson, L. W., Ford, C. A. *et al.* (1975). Kinetics of hemodiafiltration. *In vitro* transport characteristics of a hollow-fiber blood ultrafilter. *J. Lab. Clin. Med.* **85**, 855.

Cozzi, E., and White, D. (1995). The generation of transgenic pigs as potential organ donors for humans. *Nat. Med.* **1**, 965–966.

Gage, F. H. (1998). Cell therapy. *Nature (London)* **392**, 518–524.

Golper, T. A. (1986). Continuous arteriovenous hemofiltration in acute renal failure. *Am. J. Kidney Dis.* **6**, 373.

Henderson, L. W., Colton, C. K., and Ford, C. A. (1975). Kinetics of hemodiafiltration: II. Clinical characterization of a new blood cleansing modality. *J. Lab. Clin. Med.* **85**, 372.

Humes, H. D. (1995). Acute renal failure: Prevailing challenges and prospects for the future. *Kidney Int.* **48**, S-26–S-32.

Humes, H. D., and Cieslinski, D. A. (1992). Interaction between growth factors and retinoic acid in the induction of kidney tubulogenesis in tissue culture. *Exp. Cell Res.* **201**, 8–15.

Humes, H. D., Krauss, J. C., Cieslinski, D. A., and Funke, A. J. (1996). Tubulogenesis from isolated single cells of adult mammalian kidney: Clonal analysis with a recombinant retrovirus. *Am J. Physiol.* **271**(40), F42–F49.

Humes, H. D., MacKay, S. M., Funke, A. J., and Buffington, D. A. (1997a). Acute renal failure: Growth factors, cell therapy, gene therapy. *Proc. Assoc. Am. Physicians* **109**, 547–557.

Humes, H. D., MacKay, S. M., Funke, A. J., and Buffington, D. A. (1997b). The bioartificial renal tubule assist device to enhance CRRT in acute renal failure. *Am. J. Kidney Dis.* **30**, S28–S31.

Humes, H. D., Buffington, D. A., MacKay, S. M., Funke, A. J., and Weitzel, W. F. (1999a). Replacement of renal function in uremic animals with a tissue-engineered kidney. *Nat. Biotechnol.* **17**, 451–455.

Humes, H. D., MacKay, S. M., Funke, A. J., and Buffington, D. A. (1999b). Tissue engineering of a bioartificial renal tubule assist device: *In vitro* transport and metabolic characteristics. *Kidney Int.* **55**, 2502–2514.

Iglehart, J. K. (1993). The American health care system: The End Stage Renal Disease Program. *N. Engl. J. Med.* **328,** 366.

Kadletz, M., Magometshnigg, H., Minar, E. *et al.* (1992). Implantation of *in vitro* endothelialized polyteterafluoroethylene grafts in human beings. *J. Thorac. Cardiovasc. Surg.* **104,** 736.

Knox, F. G., Mertz, J. I., Burnett, J. C. *et al.* (1983). Role of hydrostatic and oncotic pressures in renal sodium reabsorption. *Circ. Res.* **52,** 491.

Kramer, P., Wigger, W., Rieger, J. *et al.* (1977). Arteriovenous hemofiltration: A new and simple method for treatment of overhydrated patients resistant to diuretics. *Klin. Wochenschr.* **55,** 1121.

Landis, E. M., and Pappenheimer, J. R. (1965). Exchange of substances through the capillary walls. *In* "Handbook of Physiology" (W. F. Hamilton and P. Dow, eds.), Sect. 2, Vol. 2, p. 961. Am. Physiol. Soc., Washington, DC.

Langer, R., and Vacanti, J. P. (1993). Tissue engineering. *Science* **260,** 920–926.

MacKay, S. M., Funke, A. J., Buffington, D. A., and Humes, H. D. (1998). Tissue engineering of a bioartificial renal tubule. *ASAIO J.* **44,** 179–183.

Maki, T., Otsu, L., O'Neil, J. J., Dunleavy, K., Mullon, C. J., Solomon, B. A., and Monaco, A. P. (1996). Treatment of diabetes by xenogeneic islets without immunosuppression. Use of a vascularized bioartificial pancreas. *Diabetes* **45**(3), 342–347.

Milici, A. J., Furie, M. B., and Carley, W. W. (1985). The formation of fenestrations and channels by capillary endothelium *in vitro. Proc. Natl. Acad. Sci. U.S.A.* **82,** 6181.

Nikolovski, J., Gulari, E., and Humes, H. D. (1999). Design engineering of a bioartificial renal tubule cell therapy device. *Cell Transplant.* **8**(4), 351–364.

O'Neil, J. J., Stegemann, J. P., Nicholson, D. T., Mullon C. J.-P., Maki, T., Monaco, A. P., and Solomon, B. A. (1997). Immunoprotection provided by the bioartificial pancreas in a xenogeneic host. *Transplant. Proc.* **29,** 2116–2117.

Rade, J. J., Schulick, A. H., and Dichek, D. A. (1996). Local adenoviral-mediated expression of recombinant hirudin reduces neointima formation after arterial injury. *Nat. Med.* **2**(3), 293–298.

Schnider, P. A., Hanson, S. R., Price, T. M. *et al.* (1988). Durability of confluent endothelial cell monolayers of small-caliber vascular prostheses *in vitro. Surgery* **103,** 456.

Shepard, A. D., Eldrup-Jorgensen, J., Keough, E. M. *et al.* (1986). Endothelial cell seeding of small-caliber synthetic grafts in the baboon. *Surgery* **99,** 318.

Timpl, R., Rhode, H., Robey, P. G., Rennard, S. I., Foidart, J. M., and Martin, G. M. (1979). Laminin—A glycoprotein from basement membranes. *J. Biol. Chem.* **254,** 9933–9937.

U.S. Renal Data System (1998). Excerpts from annual data report. *Am. J. Kidney Dis.* **32**(Suppl. 1), 569–580.

Watanabe, F. D., Mullon, C. J., Hewitt, W. R., Arkadopoulos, N., Kahaku, E., Eguchi, S., Khalili, T., Arnaout, W., Shackleton, C. R., Rozga, J., Solomon, B., and Demetriou, A. A. (1997). Clinical experience with a bioartificial liver in the treatment of severe liver failure. A phase I clinical trial. *Ann. Surg.* **225**(5), 484–491.

Wilson, J. M., Birinyi, L. K., Salomon, R. N. *et al.* (1989). Implantation of vascular grafts lined with genetically modified endothelial cells. *Science* **244,** 1344.

Zweibel, J. A., Freeman, S. M., Kantoff, P. W., *et al.* (1989). High-level recombinant gene expression in rabbit endothelial cells transduced by retroviral vectors. *Science* **243,** 220.

CHAPTER 46

GENITOURINARY SYSTEM

Byung-Soo Kim, David J. Mooney, and Anthony Atala

INTRODUCTION

Large numbers of patients suffer from a variety of diseases in the genitourinary system, which is composed of kidneys, ureters, bladder, urethra, and genital organs. Genitourinary diseases include congenital abnormalities, iatrogenic injuries, and disorders such as cancer, trauma, infection, and inflammation. These diseases often involve or result in the loss of tissue structure or function.

Lost or malfunctioning genitourinary tissues have traditionally been reconstructed with native nonurologic tissues or synthetic prostheses. The nonurologic tissues include gastrointestinal segments (Atala and Hendren, 1994; Hendren and Atala, 1994; Leong and Ong, 1972), skin (Draper and Stark, 1956), peritoneum (Hutschenreiter *et al.*, 1978), fascia (Neuhof, 1917), omentum (Goldstein *et al.,* 1967), pericardium (Kambic *et al.,* 1992), and dura (Kelâmi, 1971). Synthetic prostheses have been fabricated from polymers, including silicone (Bogash *et al.,* 1960; Henly *et al.,* 1995), polyvinyl (Kudish, 1957; Ulm and Lo, 1959), and Teflon (Kocvara and Zak, 1962; O'Donnell and Puri, 1984; Politano *et al.,* 1974). Reconstruction with nonurologic native tissues rarely replaces the entire function of the original organ, and typically leads to complications, including metabolic abnormalities, infection, perforation, and malignancy (Atala, 1997; Khoury *et al.,* 1992; Leong and Ong, 1972; McDougal, 1992). Furthermore, the limited amount of autologous donor tissue confines these types of reconstructions. Attempts to use synthetic prostheses have usually failed due to the wide array of complications (e.g., device malfunction, infection, and stone formation) associated with mechanical or biocompatibility problems (Bona and De Gresti, 1966; Henly *et al.,* 1995; Kudish, 1957). Tissue engineering may provide an alternative to current therapies for the genitourinary system. Cell transplantation using three-dimensional, biocompatible scaffolds offers the possibility of creating new functional genitourinary tissues (Atala, 1997, 1999).

TISSUE ENGINEERING STRATEGIES IN THE GENITOURINARY SYSTEM

Tissue engineering integrates cells, scaffolds, and specific signals to create new functional tissues (Kim and Mooney, 1998a). In this approach, cells isolated from a small biopsy and expanded *in vitro* can be seeded onto a suitable scaffold. They are either allowed to develop into new tissue *in vitro* or transplanted into patients to create new functional tissue that is structurally integrated with the body. Various types of cells can be isolated from tissues and greatly expanded *in vitro,* potentially providing an unlimited supply of therapeutic cells. The scaffolds provide an appropriate three-dimensional environment, and guide the development of the desired structure from the cells by providing mechanical support until the newly formed tissues are structurally stabilized. Application of specific signals (e.g., growth factors, cell adhesion molecules, and mechanical strain) during the process of tissue development may induce the appropriate pattern of gene expression in the cells, and may allow the engineered tissues to maintain proper specific functions.

One of the challenges in genitourinary tissue engineering is to expand a small number of genitourinary-associated cells to a clinically useful cell mass. Although cell transplantation has been proposed for the replacement of a variety of tissues, including skin (Bell *et al.,* 1981; Yannas *et al.,* 1982), pancreas (Lim and Sun, 1980), and cartilage (Freed *et al.,* 1994), the concept of transplanting urothelial cells, which line most of the urinary tract, had not been approached in the laboratory setting until the early 1990s, because of the inherent difficulties encountered in expanding urothelial cells in large quantities. Our laboratory has been successful in greatly expanding urothelial cells from small biopsy specimens (Cilento *et al.,* 1994). The specifics of cell culture must be optimized for every cell type. Smooth muscle cells of ureter, bladder, urethra, and corporal cavernosum have also been successfully harvested and expanded in our laboratory (Atala *et al.,* 1993a; Cilento *et al.,* 1995; Kershen *et al.,* 1998; Oberpenning *et al.,* 1999; Park *et al.,* 1999; Yoo *et al.,* 1995, 1998b).

Scaffolds for engineering genitourinary tissues have been fabricated from three classes of biomaterials: naturally derived materials (e.g., collagen and alginate), synthetic polymers [e.g., poly(glycolic acid) (PGA), poly(lactic acid) (PLA), and poly(lactic–co-glycolic acid) (PLGA)], and acellular tissue matrices (e.g., bladder submucosa and small intestinal submucosa). Collagen exhibits excellent biocompatibility and specific cellular interactions, and is amenable to cellular remodeling during the process of tissue formation (Cavallaro *et al.,* 1994). Alginate, a polysaccharide isolated from seaweed, has a great advantage for injectable cell-delivery vehicle applications owing to its hydrogel formation in the presence of calcium ions (Smidsrød and Skjåk-Bræk, 1990). Synthetic polymers can be produced reproducibly on a large scale and can be also processed into a scaffold in which the macrostructure, mechanical properties, and degradation time can be readily controlled and manipulated. Scaffolds fabricated from biodegradable polymers (e.g., PGA, PLA, and PLGA) will eventually erode in the body, avoiding a chronic foreign-body response. Acellular tissue matrices are collagen-rich xenogeneic matrices that are prepared by removing the cellular components of various tissues, such as intestine or bladder. Acellular tissue matrices have been utilized for the regeneration of many types of tissues [e.g., blood vessels (Badylak *et al.,* 1989; Lantz *et al.,* 1990, 1992), urethra (Atala *et al.,* 1999; Chen *et al.,* 1999), and bladder (Piechota *et al.,* 1998; Probst *et al.,* 1997; Yoo *et al.,* 1998b)], with no evidence of immunogenic rejection.

An important role of the scaffold in tissue engineering is to promote the development of new tissues with a predefined configuration and dimension. In these applications it is critical that the scaffold withstands *in vivo* forces exerted by the surrounding tissue and maintains its structure until the engineered tissue has sufficient mechanical integrity to support itself. This can be achieved by an appropriate choice of mechanical and degradative properties of the scaffolds (Kim and Mooney, 1998b). Urinary tissues with tubular structures (e.g., ureters and urethras) may be engineered with tubular scaffolds with central lumens (Atala *et al.,* 1993a). The ability of the tubular scaffolds to resist *in vivo* compressive forces and maintain the predefined structure can be controlled by the polymer type and physical structure (Mooney *et al.,* 1994, 1995, 1996a).

The tissue-specific function of an engineered tissue can be maintained by providing an appropriate combination of specific signals (e.g., growth factors) during the process of tissue development. The expression of genes by cells in engineered tissues may be regulated by interactions with the microenvironment, including interactions with the adhesion surface (Hynes, 1992), with other cells (Parsons-Wingerter and Saltzman, 1993), and with growth factors (Deuel, 1997), and by the mechanical stimuli imposed on the cells (Banes, 1993). These interactions can be controlled by incorporating or integrating a variety of signals, such as cell adhesion peptides (Bouhadir *et al.,* 1999; Hubbell, 1995; Rowley *et al.,* 1999) and growth factors (Mooney *et al.,* 1996b), into the scaffold or subjecting it to mechanical strain (Niklason *et al.,* 1999).

ENGINEERING GENITOURINARY TISSUES

KIDNEYS

The kidneys remove metabolic wastes from the blood, control fluid balance by maintaining homeostasis, and provide important regulatory activities by secreting hormones. End-stage renal failure is a devastating condition that involves multiple organs in affected individuals. Numerous pathologies, such as diabetes, hypertension, glomerulonephritis, and obstructive uropathy, result in end-stage renal disease. Current available therapies include hemodialysis, peritoneal dialysis, and renal transplantation (Amiel and Atala, 1999). Although dialysis can prolong survival for many patients with end-stage renal disease, only renal transplantation can currently restore renal function. However, the morbidity associated with renal transplantation, such as allograft failure, immunosuppression, or operative complications, is not trivial. Furthermore, renal transplantation is severely limited by a donor shortage.

There has been an effort directed toward the development of extracorporeal bioartificial renal units (Cieslinski and Humes, 1994; MacKay *et al.*, 1998). In this approach, the bioartificial kidney consists of two main units, glomeruli and tubules, which replace two critical renal functions, excretion and reabsorption. In the bioartificial glomerulus unit, hollow fibers with high hydraulic permeability facilitate filtration of blood delivered to the lumen of the fibers. The filtrate is then delivered to the lumen of hollow fibers in the bioartificial tubular unit, and epithelial cells on the lumen reabsorb the isoosmotic ultrafiltrate. It has been demonstrated that the combination of a synthetic filtration device and a renal tubule cell therapy device in an extracorporeal perfusion circuit replaces filtration, transport, metabolic, and endocrinologic functions of the kidney in a dog model (Humes *et al.*, 1999). However, the application of these extracorporeal devices may be best reserved for temporary situations rather than a permanent solution.

Augmentation of either isolated or total renal function with kidney cell expansion *in vitro* and subsequent autologous transplantation may be a feasible solution for complete replacement therapy. Our laboratory demonstrated the reconstruction of functional renal units by implanting isolated and cultured renal cells on three-dimensional polymer scaffolds (Atala *et al.*, 1995; Yoo *et al.*, 1996). Renal cells were successfully harvested, expanded *in vitro,* seeded onto polymer scaffolds, and implanted into host animals in which the cells can proliferate and organize into glomeruli and highly organized tubulelike structures (Fig. 46.1). These structures allowed for solute transport by the tubular cells across the membrane, resulting in the excretion of high levels of uric acid and creatinine through a urinelike fluid. Recent successes in harvesting and expanding renal cells *in vitro* and the development of biologically active scaffolds may allow the creation of three-dimensional functioning renal units that can be applied for partial or, eventually, full replacement of kidney function.

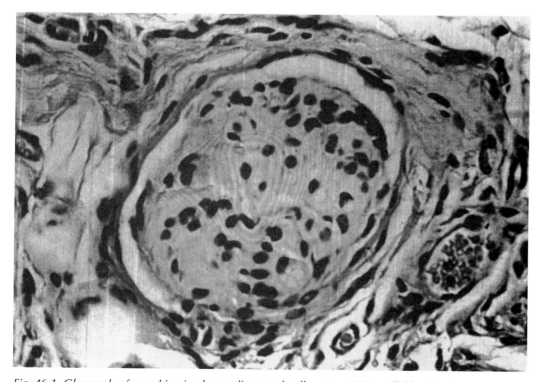

Fig. 46.1. Glomerulus formed in vivo by seeding renal cells onto a PGA scaffold and subsequent implantation (hemotoxylin and eosin; ×400).

Ureters and Urethra

The ureters are tubular conduits that convey urine from the renal pelvis to the bladder by peristaltic contraction. The urethra is a membranous canal conveying urine from the bladder to the exterior of the body. The primary indications for ureteral reconstruction are congenital or acquired obstructions that produce deterioration of renal function, carcinoma, and traumatic injury. Urethral reconstruction is performed often to repair hypospadias, epispadias (congenital defects in which the urethra terminates abnormally in the penis), urethral injuries with stricture formation, and malignant tumors of the urethra.

Various materials have served as a ureteral or urethral substitute. The use of nondegradable, synthetic materials [e.g., vitallium (Lord and Eckel, 1942), tantalum (Lubash, 1947), polyethylene (Tulloch, 1952), polyvinylchloride (Ulm and Lo, 1959), silicone (Blum *et al.,* 1963), Teflon (Kelâmi *et al.,* 1971; Kocvara and Zak, 1962), Dacron (Kocvara and Zak, 1962), and polyethylene vinyl acetate (Ramsay *et al.,* 1986)] has usually failed due to bioincompatibility and infection. Preputial skin is the most commonly used graft for urethral repair, but patients with more severe proximal urethral defects or failed urethral reconstruction may not have sufficient preputial skin for repair. In these instances, several alternatives have been used, including nonpenile skin (Devine and Horton, 1977), bladder mucosa (Koyle and Ehrlich, 1987), buccal mucosa (Dessanti *et al.,* 1992), and composite grafts (Bazeed *et al.,* 1983; Olsen *et al.,* 1992). However, some of these grafts are associated with several complications (e.g., hair growth, sebaceous secretions, stenosis, and salt deposits) and additional procedures for graft retrieval.

Our laboratory demonstrated that xenogeneic acellular collagen matrices (Chen *et al.,* 1999) and PGA mesh (Cilento *et al.,* 1995) are suitable grafts for repairing narrow urethral defects. The acellular collagen matrix was obtained from porcine bladder. Neourethras were reconstructed without any signs of strictures or complications. The animals were able to void through the neourethras. These results were confirmed clinically in a series of patients with severe urethral defects (Fig. 46.2) (Atala *et al.,* 1999). One of the advantages over previous tissue grafts is that the material is "off the shelf." This eliminates the necessity of additional surgical procedures for graft harvesting, which may decrease the operative time, as well as the potential morbidity due to the harvest procedure.

Cultured sheets of urethral epithelium have been used for hypospadias repair in humans (Romagnoli *et al.,* 1990). In this approach, autologous epithelium was cultured *in vitro* from 1- to 2-mm urethral mucosa biopsies and subsequently transplanted. The transplanted cells formed the new urethra with a well-organized epithelium.

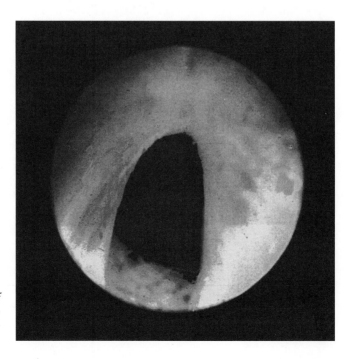

Fig. 46.2. Endoscopic view of urethra reconstructed with a collagen matrix. The urethra maintains patency.

Another approach for urethral and ureteral tissue reconstruction is to transplant urothelial cells using biodegradable polymer scaffolds as delivery vehicles. Our laboratory has succeeded in engineering tubular structures in animal models from cultured urothelial cells and PGA scaffolds (Atala *et al.,* 1992). In one study, urothelial cells isolated from human bladder and expanded *in vitro* were seeded onto PGA scaffolds with tubular configurations and implanted *in vitro*. Tubular structures developed from these implants, and the degradation of the polymer scaffolds resulted in the eventual formation of a new urothelial tissue. Similarly, human urothelial cells and smooth muscle cells were seeded onto PGA tubular scaffolds in another study (Atala *et al.,* 1993a). Following implantation, the urothelial cells proliferated to form a multilayered luminal lining, while smooth muscle cells organized into multilayered structures surrounding the urothelial cells. This study suggested that it was possible to engineer urologic tissues containing multiple cell types. This approach has been expanded to engineer new ureters (Yoo *et al.,* 1995) and urethras (Cilento *et al.,* 1995) by transplanting smooth muscle cells and urothelial cells on tubular polymer scaffolds *in vivo*.

BLADDER

The function of the bladder is to store large amounts of urine at low pressures and to empty efficiently on demand. However, fibrosis, abnormal contraction, and nerve malfunction may result in poor compliance, decreased bladder capacity, and high urinary storage pressure. The causes of bladder malfunction are congenital (posterior urethral valves, bladder extrophy, or epispadias), traumatic (injury, multiple bladder procedures), inflammatory (chronic cystitis, interstitial cystitis, or tuberculosis), radiation induced (cancer of the cervix, rectum, prostate, and bladder), or functional contraction (neurogenic bladder dysfunction or idiopathic bladder instability) (Goldwasser and Webster, 1986). In turn, these problems may lead to recurrent urinary tract infections, incontinence, renal parenchymal damage, and renal impairment and failure (Atala, 1997).

To date, bladder reconstruction has been attempted with both natural materials [gastrointestinal segments (Leong and Ong, 1972), skin (Draper and Stark, 1956), peritoneum (Hutschenreiter *et al.,* 1978), fascia (Neuhof, 1917), dura (Kelâmi, 1971), omentum (Goldstein *et al.,* 1967), and gelatin sponges (Kropp *et al.,* 1996)] and synthetic polymers [silicone (Bogash *et al.,* 1960), polyvinyl (Kudish, 1957; Ulm and Lo, 1959), and Teflon (Bona and De Gresti, 1966)]. However, these attempts have usually failed due to mechanical, structural, or biocompatibility problems. Permanent synthetic materials usually succumbed to mechanical failure and urinary stone formation. The use of natural materials resulted in a reduced bladder capacity or various complications such as excessive mucus production, infection, perforation, stone formation, and tumor development (Atala, 1997; Khoury and Zak, 1962; McDougal, 1992; Leong and Ong, 1972).

A series of studies have shown that small intestinal submucosa (SIS) (Pope *et al.,* 1997; Tsuji *et al.,* 1967) and acellular bladder matrix (Piechota *et al.,* 1998; Probst *et al.,* 1997; Yoo *et al.,* 1998b) can promote morphological and functional regeneration of bladder tissues. SIS is a relatively acellular, collagen-rich biomaterial prepared from small intestine by removing the mucosa from the inner surface and serosa and the tunica muscularis from the outer surface of the intestine. The acellular bladder matrix is prepared by mechanically and chemically removing all cellular components from the tissue. In rat or dog models, partial cystectomy (35–70%) was performed, followed by bladder augmentation with SIS or acellular bladder matrix. All three layers of the bladder (mucosa, smooth muscle, and serosa) were regenerated (Piechota *et al.,* 1998; Pope *et al.,* 1997; Probst *et al,* 1997; Tsuji *et al.,* 1967; Yoo *et al.,* 1998b). The regenerated bladders exhibited compliance similar to that of normal bladders (Yoo *et al.,* 1998b). Although the acellular matrices without seeded cells may be appropriate grafts for narrow defects, there could be limitations for tissue regeneration, especially if the size of the tissue to be regenerated is large. In most of the studies to date, the biomaterials were utilized without cell seeding. In these cases, bladder tissues were regenerated by cell ingrowth from the surrounding tissues. Our laboratory reported that bladder capacity increased significantly (100% increase) when augmented with scaffolds seeded with cells compared to scaffolds without cells (30% increase) (Yoo *et al.,* 1998b). It is hypothesized that this is due to a faster rate of scaffold resorption than that of cell ingrowth and subsequent shrinkage of the scaffold when implanted without cell seeding.

Our laboratory reported on the development of a viable bladder substitute using smooth mus-

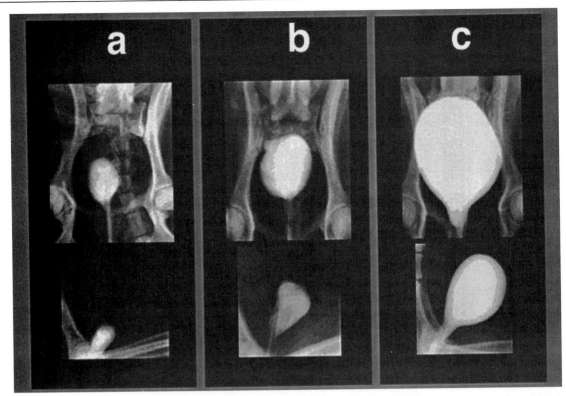

Fig. 46.3. Radiographic cystograms 11 months after subtotal bladder removal. (a) Subtotal bladder removal without a reconstructive procedure; (b) reconstruction with a cell-free bladder-shaped polymer scaffold; (c) reconstruction with a tissue-engineered neobladder. The tissue-engineered neobladder appears normal in size and configuration, but the unaugmented trigone and the polymer-only implant show smaller sized reservoirs. From Oberpenning et al. (1999), with permission.

cle cells and urothelial cells isolated from autologous bladders and cultured on biodegradable polymer scaffolds (Oberpenning *et al.*, 1999). Most of the previous approaches have reconstructed the bladder either by surgical enlargement or by replacement of partial, defective walls using various biomaterials (Piechota *et al.*, 1998; Probst *et al.*, 1997; Pope *et al.*, 1997; Tsuji *et al.*, 1967). In this study, transplantable whole bladders were created using cultured bladder cells and polymer scaffolds. In this approach, urothelial and smooth muscle cells were isolated from 1-cm^2 bladder biopsies, expanded *in vitro*, and seeded on preformed bladder-shaped polymer scaffolds. These neobladders were implanted in dogs that had the majority of their native bladders excised. In functional evaluations for up to 11 months, the bladder neoorgans demonstrated a normal capacity to retain urine, normal compliance, ingrowth of neural structures, and a normal histologic architecture (Figs. 46.3 and 46.4).

Genital Tissues

The management of inadequate genital function (e.g., impotence) and ambiguous genitalia, caused by a rudimentary penis, severe hypospadias, traumatic injury, cancer, infection, or chronic disease (e.g., diabetes and vascular insufficiency), may involve phallic reconstruction (Goodwin and Scott, 1952; Horton and Dean, 1990). Earlier attempts at penile reconstruction involved multiple stages of surgery using autologous rib cartilage as a stiffener. However, the unsatisfactory functional and cosmetic results due to its curvature discouraged its use. Silicone rigid prostheses were popularized in the 1970s. Although silicone penile prostheses are an acceptable treatment in adults, complications, such as erosion and inflammation, remain a problem (Nukui *et al.*, 1997).

Natural penile prostheses created from the patient's own cells may be preferable and may decrease the biocompatibility risks associated with the artificial prostheses. Our laboratory has demonstrated that cartilage rods can be created using chondrocytes seeded on preformed polymer

Fig. 46.4. Histologic analysis of the tissue-engineered neobladder 6 months after reconstructive surgery (hemotoxylin and eosin; ×140). The neobladder tissue consists of a morphologically normal uroepithelial lining over a sheath of submucosa, followed by a layer of multiform smooth muscle bundles.

scaffolds (Yoo *et al.*, 1998a). Biomechanical studies showed that the cartilage rods were readily elastic and could withstand high degrees of pressure. Additional studies demonstrated that autologous chondrocytes seeded on polymer scaffolds are able to form penile prostheses in an animal model (Yoo *et al.*, 1999).

The availability of transplantable autologous corpus carvernosum tissue for use in reconstructive procedures would be beneficial to many patients with congenital and acquired abnormalities of the genital system (Fig. 46.5). It would facilitate enhanced cosmetic results and provide the possibility of reconstructing *de novo* functional erectile tissue. Our laboratory has shown that cultured cavernosum cells seeded on biodegradable polymers formed corporal smooth muscle when implanted *in vivo* (Kershen *et al.*, 1998). Additional studies were performed in order to investigate the possibility of developing corporal tissue *in vivo* by transplanting smooth muscle cells in conjunction with endothelial cells on polymer scaffolds (Park *et al.*, 1999). The cell–polymer constructs formed neocorporal tissues, consisting of corporal smooth muscle and neovasculature formed by both the host and implanted endothelial cells. Similar studies have been performed with clitoral and vaginal tissues (Atala, 1999).

Fetal Tissue Engineering

Engineering of fetal genitourinary tissues would benefit the newborn with prenatally diagnosed birth defects. Treatment of many congenital anomalies at birth is often impeded by lack of normal tissues or organs, either autologous or allologous. The severe donor shortage for tissue or organ transplantation is even more critical during the neonatal period. Autologous grafting is frequently not an option in newborns because of donor site size limitations. Having a ready supply of genitourinary-associated tissue for surgical reconstruction at birth may be advantageous. Fetal tissue can be engineered by a minimally invasive harvest of tissue prenatally. The tissue may be engineered *in vitro* while the pregnancy is allowed to continue so that a one-stage reconstruction can be performed at the time of birth.

Our laboratory utilized this concept to repair bladder defects in a fetal lamb model (Fauza *et al.*, 1998a). Bladder exstrophy (a developmental anomaly marked by absence of a portion of the lower abdominal wall and the anterior urinary bladder wall) was created in 90- to 95-day gestation fetal lambs. A small fetal bladder specimen was harvested via fetoscopy to obtain cells for en-

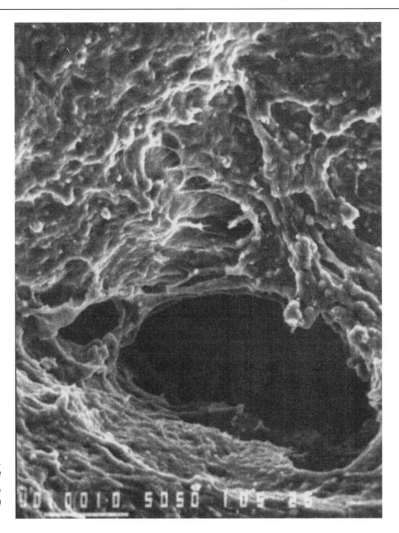

Fig. 46.5. Scanning electron micrograph of penile corporal smooth muscle cells seeded onto an acellular matrix (bar = 100 μm).

gineering fetal bladder tissue (Fig. 46.6). At 7 to 10 days prior to delivery, the expanded cells were seeded onto polymer scaffolds and maintained *in vitro*. After delivery, the lambs had surgical closure of their bladder using the tissue-engineered bladder tissues. At 2 months after the reconstructive surgery, histologic analysis of the engineered tissues showed a normal histologic pattern, indistinguishable from native bladder. In addition, the engineered bladders were more compliant and had a higher capacity than did bladders closed using only the native bladders. This technology was also expanded to engineer other types of fetal tissues (e.g., skin) in an animal model (Fauza *et al.*, 1998b). Replication of this technology in humans may be possible in the future.

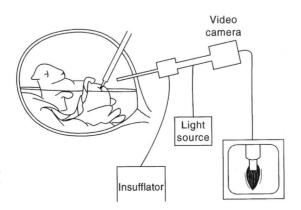

Fig. 46.6. Schematic diagram showing the arrangement for videofetoscopic harvest of fetal tissue.

Injectable Therapies

Endoscopic therapies using injectable bulking agents are attractive procedures to treat urinary incontinence and vesicoureteral reflux. Urinary incontinence results from a deficient or weak musculature in the bladder neck and urethral area. Primary vesicoureteral reflux results from a congenitally deficient longitudinal submucosal muscle of the distal ureter, which leads to an abnormal flow of urine from the bladder to the upper tract. For many years, open surgery [e.g., ureteral reimplantation, pubovaginal sling (McGuire and Lytton, 1978), and the artificial urinary sphincter (Webster *et al.,* 1992)] was the standard of care for the treatment of these diseases owing to intrinsic sphincteric dysfunction. In recent years, minimally invasive techniques for the endoscopic correction of these maladies have been developed and refined, obtaining reasonable success rates with significantly decreased patient morbidity. The endoscopic technique for the correction of both urinary incontinence and vesicoureteral reflux relies on increasing muscle resistance with the injection of bulking agents into the suburethral or subureteral tissues. In general, endoscopic procedures are less time consuming, more cost effective, and more easily performed, often in an outpatient setting.

Searching for an ideal material for injection has been a challenging task for many decades. The materials used for injection therapies include Teflon [poly(tetrafluoroethylene)] microparticles (O'Donnell and Puri, 1984; Politano *et al.,* 1974), poly(vinyl alcohol) (Merguerian *et al.,* 1990), autologous fat (Palma *et al.,* 1997), silicone microparticles (Henly *et al.,* 1995), and collagen (Appell, 1994; Leonard *et al.,* 1990). However, none of these materials is ideal due to migration, granuloma formation, and volume loss (Atala, 1994a,b; Claes *et al.,* 1989; Malizia *et al.,* 1984).

The ideal material for the endoscopic treatment of vesicoureteral reflux and incontinence should be injectable, nonantigenic, nonmigratory, volume stable, and safe for human use (Kershen and Atala, 1999). Our laboratory has investigated the use of chondrocyte–alginate gel suspensions as a possible injectable material for the treatment of vesicoureteral reflux and incontinence (Atala *et al.,* 1993b). Initially, chondrocytes mixed with alginate gel were injected into animals. The injected suspension formed cartilaginous tissues *in vivo* and showed no evidence of cartilage or alginate migration, granuloma formation, or volume loss. Additional studies demonstrated that vesicoureteral reflux could be treated with an autologous chondrocyte–alginate suspension without any evidence of obstruction in an animal model (Fig. 46.7) (Atala *et al.,* 1994). To use this approach in humans, a biopsy could be taken from the symphysis pubis or the cartilaginous part of

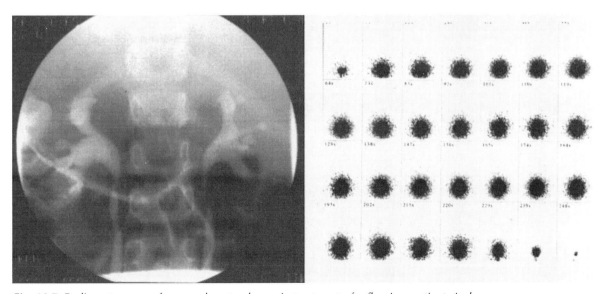

Fig. 46.7. Radiocystograms of pre- and postendoscopic treatment of reflux in a patient. A chondrocyte–alginate suspension was injected endoscopically in the subureteral region. Preoperative fluoroscopic cystogram (left) shows bilateral reflux. Postoperative radionuclide cystogram (right) shows resolution of reflux.

the ear under a local anesthetic. A few weeks later the patient would be treated with an endoscopic injection of the autologous chondrocyte–alginate suspension. The first human application of tissue engineering for the genitourinary system was accomplished with the injection of chondrocytes for the correction of vesicoureteral reflux (Kershen and Atala, 1999). Similar clinical trials with this technology have been initiated for the treatment of urinary incontinence. In addition to its use for the endoscopic treatment of reflux and incontinence, the system of injectable autologous cells may also be applicable for the treatment of other medical conditions, such as rectal incontinence, dysphonia, plastic reconstruction, and wherever an injectable biocompatible material is needed.

FUTURE DIRECTIONS

Studies to identify factors that control proliferation, differentiation, cellular gene expression, and tissue organization may be required to engineer functional genitourinary tissues. This technology may allow for expansion of isolated cells in large quantities with the correct phenotype *in vitro,* while maintaining appropriate cellular gene expression during the process of tissue development. Although studies were performed to identify culture conditions of several types of cells in the genitourinary system, the mechanisms that control gene expression are not well understood. Appropriate bioreactor systems that would facilitate cell seeding onto scaffolds and subsequent culture *in vitro* must also be developed (Kim *et al.,* 1998).

A tremendous effort is being expanded to develop scaffolds that induce functional tissue development. Scaffolds with controlled mechanical and degradative properties (Bouhadir *et al.,* 1999; Kim and Mooney, 1998a) would promote tissue development with predefined configurations. The development of scaffolds with specific cell-recognition sites may induce tissue-specific gene expression of the seeded cells (Rowley *et al.,* 1999) and promote the spatial control of specific cell types in genitourinary tissues with multiple cell types (Hubbell *et al.,* 1991).

Vascularization and innervation of the engineered genitourinary tissues are essential for the successful integration of the construct with the host body. The vascularization of the engineered tissues could be achieved by growth factor-stimulated ingrowth of vascular cells from the surrounding tissue (Folkman and Klagsbrun, 1987) or transplantation of vascular cells (Holder *et al.,* 1997). Innervation may be promoted with graft, guidance channels, and various nerve growth factors (Bellamkonda and Aebischer, 1994).

Genitourinary tissue with genetic defects may be repaired with gene therapy. Genetically modified cells can be transplanted using the scaffold as a gene delivery vehicle. In one study, urothelial cell–polymer constructs were introduced with the DNA plasmid and transplanted (Yoo and Atala, 1997). The transplanted cells formed an organlike structure with functional expression of the transfected genes *in vivo*. This technology may be applied to any type of urological-associated pathology.

Several technologies developed in our laboratory are currently in clinical trials. Vesicoureteral reflux and urinary incontinence are being treated endoscopically with engineered chondrocytes. Urethral tissue replacement surgeries using engineered matrices are being performed in patients with severe disease. Engineered bladder tissue will soon be used in patients who require bladder augmentation. Even though a careful and long-term follow-up will be necessary in order to assess the currently available technologies, it is obvious that tissue engineering may play a vital role in the treatment of several genitourinary conditions.

References

Amiel, G. E., and Atala, A. (1999). Current and future modalities for functional renal replacement. *Urol. Clin. North Am.* **26**, 235–246.

Appell, R. A. (1994). Collagen injection therapy for urinary incontinence. *Urol. Clin. North Am.* **21**, 177–182.

Atala, A. (1994a). Non-autologous substances in VUR and incontinence therapy. *Dial. Pediatr. Urol.* **17**, 11.

Atala, A. (1994b). Non-autologous substance in VUR and incontinence therapy. *Dial. Pediatr. Urol.* **17**, 12.

Atala, A. (1997). Tissue engineering in the genitourinary system. *In* "Synthetic Biodegradable Polymer Scaffolds" (A. Atala and D. J. Mooney, eds.), pp. 149–164. Birkhaeuser, Boston.

Atala, A. (1999). Future perspectives in reconstructive surgery using tissue engineering. *Urol. Clin. North Am.* **26**, 157–165.

Atala, A., and Hendren, W. H. (1994). Reconstruction with bowel segments. *Dial. Pediatr. Urol.* **17**, 7.

Atala, A., Vacanti, J. P., Peters, C. A., Mandell, J., Retik, A. B., and Freeman, M. R. (1992). Formation of urothelial structures *in vivo* from dissociated cells attached to biodegradable polymer scaffolds *in vitro*. *J. Urol.* **148**, 658–662.

Atala, A., Freeman, M. R., Vacanti, J. P., Shepard, J., and Retik, A. B. (1993a). Implantation *in vivo* and retrieval of arti-ficial structures consisting of rabbit and human urothelium and human bladder muscle. *J. Urol.* **150**, 608–612.

Atala, A., Cima, L. G., Kim, W., Paige, K. T., Vacanti, J. P., Retik, A. B., and Vacanti, C. A. (1993b). Injectable alginate seeded with chondrocytes as a potential treatment for vesicoureteral reflux. *J. Urol.* **150**, 745–747.

Atala, A., Kim, W., Paige, K. T., Vacanti, C. A., and Retik, A. B. (1994). Endoscopic treatment of vesicoureteral reflux with a chondrocyte-alginate suspension. *J. Urol.* **152**, 641–643.

Atala, A., Schlussel, R. N., and Retik, A. B. (1995). Renal cell growth *in vivo* after attachment to biodegradable polymer scaffolds. *J. Urol.* **153**, 4.

Atala, A., Guzman, L., and Retik, A. B. (1999). A novel inert collagen matrix for hypospadias repair. *J. Urol.* **162**, 1148–1151.

Badylak, S. F., Lantz, G. C., Coffey, A., and Geddes, L. A. (1989). Small intestinal submucosa as a large diameter vascu-lar graft in the dog. *J. Surg. Res.* **47**, 74–80.

Banes, A. J. (1993). Mechanical strain and the mammalian cell. *In* "Physical Forces and the Mammalian Cell" (J. A. Fran-gos, ed.), pp. 81–123. Academic Press, San Diego, CA.

Bazeed, M. A., Thüroff, J. W., Schmidt, R. A., and Tanagho, E. A. (1983). New treatment for urethral strictures. *Urology* **21**, 53–57.

Bell, E., Ehrlich, H. P., Buttle, D. J., and Nakatsuji, T. (1981). Living tissue formed *in vitro* and accepted as skin-equiva-lent tissue of full thickness. *Science* **211**, 1052–1054.

Bellamkonda, R., and Aebischer, P. (1994). Tissue engineering in the nerve system. *Biotechnol. Bioeng.* **43**, 543–554.

Blum, J. A., Skemp, C., and Reiser, M. (1963). Silicone rubber ureteral prostheses. *J. Urol.* **90**, 276–280.

Bogash, M., Kohler, F. P., Scott, R. H., and Murphy, J. J. (1960). Replacement of the urinary bladder by a plastic reser-voir with mechanical valves. *Urology* **1**, 900–903.

Bona, A. V., and De Gresti, A. (1966). Partial substitution of the bladder wall with Teflon tissue. *Minerva Urol.* **18**, 43.

Bouhadir, K. H., Hausman, D. S., and Mooney, D. J. (1999). Synthesis of cross-linked poly(aldehyde guluronate) hydro-gel. *Polymer* (in press).

Cavallaro, J. F., Kemp, P. D., and Kraus, K. H. (1994). Collagen fabrics as biomaterials. *Biotechnol. Bioeng.* **43**, 781–791.

Chen, F., Yoo, J. J., and Atala, A. (1999). Acellular collagen matrix as a possible "off the shelf" biomaterial for urethral re-pair. *Urology* **54**, 407–410.

Cieslinski, D. A., and Humes, H. D. (1994). Tissue engineering of a bioartificial kidney. *Biotechnol. Bioeng.* **43**, 678–681.

Cilento, B. G., Retik, A. B., and Atala, A. (1995). Urethral reconstruction using a polymer mesh. *J. Urol.* **153** (Suppl.), 371A.

Cilento, B. J., Freeman, M. R., Schneck, F. X., Retik, A. B., and Atala, A. (1994). Phenotypic and cytogenetic character-ization of human bladder urothelia expanded in vitro. *J. Urol.* **152**, 665–670.

Claes, H., Stroobants, D., van Meerbeek, J., *et al.* (1989). Pulmonary migration following periurethral polytetrafluo-roethylene injection for urinary incontinence. *J. Urol.* **142**, 821–822.

Dessanti, A., Rigamonti, W., Merulla, V., *et al.* (1992). Autologous buccal mucosa graft for hypospadias repair: An initial report. *J. Urol.* **147**, 1081–1084.

Deuel, T. F. (1997). Growth factors. *In* "Principles of Tissue Engineering" (R. P. Lanza, R. Langer, and W. L. Chick, eds.), pp. 133–149. Academic Press, San Diego, CA.

Devine, C. J., Jr., and Horton, C. E. (1977). Hypospadias repair. *J. Urol.* **118**, 188–193.

Draper, J. W., and Stark, R. B. (1956). End results in the replacement of mucous membrane of the urinary bladder with thick-split grafts of skin. *Surgery* **39**, 434–440.

Fauza, D. O., Fishman, S. J., Mehegan, K., and Atala, A. (1998a). Videofetoscopically assisted fetal tissue engineering: Bladder augmentation. *J. Pediatr. Surg.* **33**, 7–12.

Fauza, D. O., Fishman, S. J., Mehegan, K., and Atala, A. (1998b). Videofetoscopically assisted fetal tissue engineering: Skin replacement. *J. Pediatr. Surg.* **33**, 357–361.

Folkman, J., and Klagsbrun, M. (1987). Angiogenic factors. *Science* **235**, 442–447.

Freed, L. E., Vunjak-Novakovic, G., Biron, R. J., Eagles, D. B., Lesnoy, D. C., Barlow, S. K., and Langer, R. (1994). Biodegradable polymer scaffolds for tissue engineering. *Bio Technology* **12**, 689–693.

Goldstein, M. B., Dearden, L. C., and Gualtieri, V. (1967). Regeneration of subtotally cystectomized bladder patched with omentum: An experimental study in rabbits. *J. Urol.* **97**, 664–668.

Goldwasser, B., and Webster, G. D. (1986). Augmentation and substitution enterocystoplasty. *J. Urol.* **135**, 215–224.

Goodwin, W. E., and Scott, W. W. (1952). Phalloplasty. *J. Urol.* **68**, 903.

Hendren, W. H., and Atala, A. (1994). Use of bowel for vaginal reconstruction. *J. Urol.* **152**, 752–755.

Henly, D. R., Barrett, D. M., Welland, T. L., *et al.* (1995). Particulate silicone for use in periurethral injections: Local tis-sue effects and search for migration. *J. Urol.* **153**, 2039–2043.

Holder, W. D., Gruber, H. E., Roland, W. D., Moore, A. L., Culberson, C. R., Loebsack, A. B., Burg, K. J. L., and Mooney, D. J. (1997). Increased vascularization and heterogeneity of vascular structures occurring in polyglycolide matrices containing aortic endothelial cells implanted in the rat. *Tissue Eng.* **3**, 149–160.

Horton, C. E., and Dean, J. A. (1990). Reconstruction of traumatically acquired defects of the phallus. *World J. Surg.* **14**, 757–762.

Hubbell, J. A. (1995). Biomaterials in tissue engineering. *Bio Technology* **13**, 565–576.

Hubbell, J. A., Massia, S. P., Desai, N. P., *et al.* (1991). Endothelial cell-selective materials for tissue engineering in the vas-cular graft via a new receptor. *Bio Technology* **9**, 568–572.

Humes, H. D., Buffington, D. A., MacKay, S. M., Funke, A. J., and Weitzel, W. F. (1999). Replacement of renal function in uremic animals with a tissue-engineered kidney. *Nat. Biotechnol.* **17**, 451–455.

Hutschenreiter, G., Rumpelt, H. J., Klippel, K. F., and Hohenfellner, R. (1978). The free peritoneal transplant as a substitute for the urinary bladder wall. *Invest. Urol.* **15**, 375–379.

Hynes, R. O. (1992). Integrins: Versatility, modulation and signaling in cell adhesion. *Cell (Cambridge, Mass.)* **69**, 11–25.

Kambic, H., Kay, R., Chen, J.-F., Matsushita, M., Harasaki, H., and Zilber, S. (1992). Biodegradable pericardial implants for bladder augmentation: A 2.5-year study in dogs. *J. Urol.* **148**, 539–543.

Kelâmi, A. (1971). Lyophilized human dura as a bladder wall substitute: Experimental and clinical results. *J. Urol.* **105**, 518–522.

Kelâmi, A., Korb, G., Ludtke-Handjery, A., Rolle, J., Schnell, J., and Lehnhardt, F. H. (1971). Alloplastic replacement of the partially resected urethra in dogs. *Invest. Urol.* **9**, 55–58.

Kershen, R. T., and Atala, A. (1999). New advances in injectable therapies for the treatment of incontinence and vesicoureteral reflux. *Urol. Clin. North Am.* **26**, 81–94.

Kershen, R. T., Yoo, J. J., Moreland, R. B., Krane, R. J., and Atala, A. (1998). Novel system for the formation of human corpus cavernosum smooth muscle tissue *in vivo. J. Urol.* **159** (Suppl.), 156.

Khoury, J. M., Timmons, S. L., Corbel, L., and Webster, G. D. (1992). Complications of enterocystoplasty. *Urology* **40**, 9–14.

Kim, B. S., and Mooney, D. J. (1998a). Development of biocompatible synthetic extracellular matrices for tissue engineering. *Trends Biotechnol.* **16**, 224–230.

Kim, B. S., and Mooney, D. J. (1998b). Engineering smooth muscle tissue with a predefined structure. *J. Biomed. Mater. Res.* **41**, 322–332.

Kim, B. S., Putnam, A. J., Kulik, T. J., and Mooney, D. J. (1998). Optimizing seeding and culture methods to engineer smooth muscle tissue on biodegradable polymer matrices. *Biotechnol. Bioeng.* **57**, 46–54.

Kocvara, S., and Zak, F. (1962). Ureteral substitution with Dacron and Teflon prosthesis. *J. Urol.* **88**, 365–376.

Koyle, M. A., and Ehrlich, R. M. (1987). The bladder mucosal graft for urethral reconstruction. *J. Urol.* **138**, 1093.

Kropp, B. P., Rippy, M. K., Badylak, S. F., Adams, M. C., Keating, M. A., Rink, R. C., and Thor, K. B. (1996). Regenerative urinary bladder augmentation using small intestine submucosa: Urodynamic and histologic assessment in long-term canine bladder augmentations. *J. Urol.* **155**, 2098–2104.

Kudish, H. G. (1957). The use of polyvinyl sponge for experimental cystoplasty. *J. Urol.* **78**, 232.

Lantz, G. C., Badylak, S. F., Coffey, A., Geddes, L. A., and Blevins, W. E. (1990). Small intestinal submucosa as a small-diameter arterial graft in the dog. *J. Invest. Surg.* **3**, 217–227.

Lantz, G. C., Badylak, S. F., Coffey, A., Geddes, L. A., and Sandusky, G. E. (1992). Small intestinal submucosa as a superior vena cava graft in the dog. *J. Surg. Res.* **53**, 175–181.

Leonard, M. P., Canning, D. A., Epstein, J. I., Gearhart, J. P., and Jeffs, R. D. (1990). Local tissue reaction to the subureteral injection of glutaraldehyde cross-linked bovine collagen in humans. *J. Urol.* **143**, 1209–1212.

Leong, C. H., and Ong, G. B. (1972). Gastrocystoplasty in dogs. *Aust. N. Z. J. Surg.* **41**, 272–279.

Lim, F., and Sun, A. M. (1980). Microencapsulated islets as bioartifical endocrine pancreas. *Science* **210**, 908–910.

Lord, J. W., and Eckel, J. H. (1942). Use of vitallium tubes in the urinary tract of dogs. *J. Urol.* **48**, 412.

Lubash, S. (1947). Experiences with tantalum tubes in the reimplantation of the ureter into the sigmoid in dogs and humans. *J. Urol.* **57**, 1010.

MacKay, S. M., Funke, A. J., Buffington, D. A., and Humes, H. D. (1998). Tissue engineering of a bioartificial renal tubule. *ASAIO J.* **44**, 179–183.

Malizia, A. A., Reiman, H. M., Myers, R. P., Sande, J. R., Barham, S. S., Benson, R. C., Dewanjee, M. K., and Ulz, W. J. (1984). Migration and granulomatous reaction after periurethral injection of polytef (Teflon). *JAMA, J. Am. Med. Assoc.* **251**, 3277–3281.

McDougal, W. S. (1992). Metabolic complications of urinary intestinal diversion. *J. Urol.* **147**, 1199–1208.

McGuire, E. J., and Lytton, B. (1978). Pubovaginal sling procedure for stress incontinence. *J. Urol.* **119**, 82–84.

Merguerian, P. A., McLorie, G. A., Khoury, A. E., Thorner, P., and Churchill, B. M. (1990). Submucosal injection of polyvinyl alcohol foam in rabbit bladder. *J. Urol.* **144**, 531–533.

Mooney, D. J., Organ, G., Vacanti, J. P., and Langer, R. (1994). Design and fabrication of biodegradable polymer devices to engineer tubular tissues. *Cell Transplant.* **3**, 438–446.

Mooney, D. J., Mazzoni, C. L., Breuer, C., McNamara, K., Hern, D., Vacanti, J. P., and Langer, R. (1995). Fabricating tubular tissues with devices of poly(D,L-lactic-coglycolic acid). *Tissue Eng.* **1**, 107–118.

Mooney, D. J., Mazzoni, C. L., Breuer, C., McNamara, K., Hern, D., Vacanti, J. P., and Langer, R. (1996a). Stabilized polyglycolic acid fiber-based devices for tissue engineering. *Biomaterials* **17**, 115–124.

Mooney, D. J., Kaufmann, P. M., Sano, K., Schwendeman, S. P., Majahod, K., Schloo, B., Vacanti, J. P., and Langer, R. (1996b). Localized delivery of epidermal growth factor improves the survival of transplanted hepatocytes. *Biotechnol. Bioeng.* **50**, 422–429.

Neuhof, H. (1917). Fascia transplantation into visceral defects. *Surg., Gynecol. Obstet.* **14**, 383–427.

Niklason, L. E., Gao, J., Abbott, W. M., Hirschi, K. K., Houser, S., Marini, R., and Langer, R. (1999). Functional arteries grown *in vitro. Science* **284**, 489–493.

Nukui, F., Okamoto, S., Nagata, M., Kurokawa, J., and Fukui, J. (1997). Complications and reimplantation of penile implants. *Int. J. Urol.* **4**, 52–54.

Oberpenning, F., Meng, J., Yoo, J. J., and Atala, A. (1999). *De novo* reconstruction of a functional mammalian urinary bladder by tissue engineering. *Nat. Biotechnol.* **17**, 149–155.

O'Donnell, B., and Puri, P. (1984). Treatment of vesicoureteric reflux by endoscopic injection of Teflon. *Br. Med. J.* **289**, 7–9.

Olsen, L., Bowald, S., Busch, C., Carlsten, J., and Eriksson, I. (1992). Urethral reconstruction with a new synthetic absorbable device. *Scand. J. Urol. Nephrol.* **26**, 323–326.

Palma, P. C., Riccetto, C. L., Herrmann, V. *et al.* (1997). Repeated lipoinjections for stress urinary incontinence. *J. Endourol.* **11**, 67–70.

Park, H. J., Yoo, J. J., Kershen, R. T., Moreland, R. B., Krane, R. J., and Atala, A. (1999). Reconstruction of human corpus cavernosum smooth muscle and endothelial cells *in vivo. J. Urol.* **162**, 1106–1109.

Parsons-Wingerter, P. A., and Saltzman, W. M. (1993). Growth versus function in three-dimensional culture of single and aggregated hepatocytes within collagen gels. *Biotechnol. Prog.* **9**, 600–607.

Piechota, H. J., Dahms, S. E., Probst, M., Gleason, C. A., Nunes, L. S., Dahiya, R., Lue, T. F., and Tanagho, E. A. (1998). Functional rat bladder regeneration through xenotransplantation of the bladder acellular matrix graft. *Br. J. Urol.* **81**, 548–559.

Politano, V. A., Small, M. P., Harper, J. M., *et al.* (1974). Periurethral Teflon injection for urinary incontinence. *J. Urol.* **111**, 180–183.

Pope, J. C., Davis, M. M., Smith, E. R., Jr., Walsh, M. J., Ellison, P. K., Rink, R. C., and Thor, K. B. (1997). The ontogeny of canine small intestinal submucosa regenerated bladder. *J. Urol.* **158**, 1105–1110.

Probst, M., Dahiya, R., Carrier, S., and Tanagho, E. A. (1997). Reproduction of functional smooth muscle tissue and partial bladder replacement. *Br. J. Urol.* **79**, 505–515.

Ramsay, J. W. A., Miller, R. A., Crocker, P. R., Ringrose, B. J., Jones, S., Levison, D. A., Whitfield, H. N., and Wickham, J. E. A. (1986). An experimental study of hydrophilic plastics for urological use. *Br. J. Urol.* **58**, 70–74.

Romagnoli, G., Luca, M. D., Faranda, F., Bandelloni, R., Franzi, A. T., Cataliotti, F., and Cancedda, R. (1990). Treatment of posterior hypospadias by the autologous graft of cultured urethral epithelium. *N. Engl. J. Med.* **323**, 527–530.

Rowley, J. A., Madlambayan, G., and Mooney, D. J. (1999). Alginate hydrogels as synthetic extracellular matrix materials. *Biomaterials* **20**, 45–53.

Smidsrød, O., and Skjåk-Bræk, G. (1990). Alginate as an immobilization matrix for cells. *Trends. Biotechnol.* **8**, 71–78.

Tsuji, I., Shiraishi, Y., Kassai, T., Kunishima, K., Orikasa, S., and Abe, N. (1967). Further experimental investigations on bladder reconstruction without using the intestine. *J. Urol.* **97**, 1021–1028.

Tulloch, W. S. (1952). Restoration of continuity of the ureter by means of polyethylene tubing. *Br. J. Urol.* **24**, 42–45.

Ulm, A. H., and Lo, M.-C. (1959). Total bilateral polyvinyl ureteral substitutes in the dogs. *Surgery* **45**, 313.

Webster, G. D., Perez, L. M., Khoury, J. M., *et al.* (1992). Management of type III stress urinary incontinence using the artificial urinary sphincter. *Urology* **39**, 499–503.

Yannas, I. V., Burke, J. F., Orgill, D. P., and Skrabut, E. M. (1982). Wound tissue can utilize a polymeric template to synthesize a functional extension of skin. *Science* **215**, 174–176.

Yoo, J. J., and Atala, A. (1997). A novel gene delivery system using urothelial tissue engineered neo-organs. *J. Urol.* **158**, 1066–1070.

Yoo, J. J., Satar, N., Retik, A. B., and Atala, A. (1995). Ureteral replacement using biodegradable polymer scaffolds seeded with urothelial and smooth muscle cells. *J. Urol.* **153** (Suppl.), 375A.

Yoo, J. J., Ashkar, S., and Atala, A. (1996). Creation of functional kidney structures with excretion of urine-like fluid *in vivo. Pediatrics* **98S**, 605.

Yoo, J. J., Lee, I., and Atala, A. (1998a). Cartilage rods as a potential material for penile reconstruction. *J. Urol.* **160**, 1164–1168.

Yoo, J. J., Meng, J., Oberpenning, F., and Atala, A. (1998b). Bladder augmentation using allogenic bladder submucosa seeded with cells. *Urology* **51**, 221–225.

Yoo, J. J., Park, H. J., Lee, I., and Atala, A. (1999). Autologous cartilage rods for penile reconstruction. *J. Urol.* **162**, 1119–1121.

PART XVI: MUSCULOSKELETAL SYSTEM

CHAPTER 47

STRUCTURAL TISSUE ENGINEERING

Charles A. Vacanti, Lawrence J. Bonassar, and Joseph P. Vacanti

INTRODUCTION

This chapter provides a review of important developments related to the tissue engineering of structural tissues, including bone, cartilage, ligament, and tendon, over the past 10 years, with an emphasis on recent developments. Basic principles of tissue engineering, focusing on cell biology and materials science as used currently in the field, are presented. Early efforts to combine cells with biocompatible materials are described along with more recent endeavors. Applications of the technology are presented with a particular focus on uses in orthopedics, plastic and reconstructive surgery, and maxillofacial surgery. Finally, future challenges are outlined from the perspective of integrating technologies from medicine, biology, and engineering in hopes of translating tissue engineering to clinical applications.

According to legend, the first homotransplantation of an entire limb was performed by two saints, Damian and Cosmas, as depicted by the artist, Fra Angelico. Indeed, the concept was described even earlier: "The Lord God, cast a deep sleep on the man, and while he was asleep, he took out one of his ribs and closed up its place with flesh. The Lord God then built up into a woman the rib that he had taken from the man" (New Jerusalem Bible, 1990).

Although the number of research centers undertaking tissue engineering has greatly increased during the past decade and potential applications have become more widespread, the basic concept that the repair and regeneration of biological tissues can be guided through application and control of cells, materials, and chemoactive proteins remains central. As such, tissue engineering is, at its core, an interdisciplinary field, requiring the interactions of physicians, scientists, and engineers.

Our goal is to utilize cells from either a relatively small specimen of tissue or a universal donor source, expand them *in vitro,* and deliver them associated with a template, in a configuration that will generate a new functional tissue. The ability to achieve this goal requires a keen understanding of the extracellular matrix, the structural framework, and the regulation of cell behavior.

As early as 1908, Lexer described early attempts at tissue engineering of structural tissue in his report of the use of freshly amputated or cadaver allografts for joint reconstruction. Several decades later, work by Huggins resulted in the search for the elusive bone morphogenic proteins. In 1965, Urist demonstrated the generation of one structural tissue, that is, bone, by means of autoinduction. He described "wandering histiocytes, foreign body giant cells, and inflammatory connective cells" being "stimulated by degradation products of dead matrix to grow in and repopulate the area of an implant of decalcified bone." The process, as described by Urist, is followed immediately by autoinduction, in which both the inductor cells and the induced cells are derived from ingrowing cells of the host bed. Differentiation of the osteoprogenitor cell was felt to be elicited by local alterations in cell metabolic cycles.

In more recent times (1970s), Green (1977) described a series of experiments devoted to the generation of new cartilage. In one study, chondrocytes, seeded onto sterile bone spicules, were implanted into nude mice. Although the experiments were not entirely successful, Green correctly postulated that the advent of new biocompatible materials might enable cells to be seeded onto a synthetic scaffolding and implanted into animals to generate new functional tissue.

The past two decades have given rise to a multitude of descriptions involving the replacement of bone and cartilage lost to injury and disease. Living cells have been delivered on alloplastic implants (Ohgushi *et al.,* 1989) to produce bone. Cartilage growth, whether achieved by use of cell suspensions alone (Bentley and Greer, 1971; Chesterman *et al.,* 1968; Grande *et al.,* 1989; Lipman *et al.,* 1983), cell attachment to naturally occurring matrices (Itay *et al.,* 1987; Wakitani *et al.,* 1989) from perichondrium (Upton *et al.,* 1981), or peptide stimulation (Wozney *et al.,* 1988), had been described only in minute quantities until the early 1990s. Until then, cartilage was generated only in association with, and confined to a defect in, underlying bone or cartilage, and was devoid of specific shapes. As such, the purpose of the efforts was to augment the quantity of native tissue to add or restore some function. New hyaline cartilage growth in significant amounts, not associated with endogenous bone or cartilage (Vacanti *et al.,* 1991), has now been achieved using a technology termed "tissue engineering" to generate parenchymal tissue (Vacanti, 1988; Vacanti *et al.,* 1988). During the past decade, the principles of tissue engineering have been applied to virtually every organ system in the body (Fig. 47.1), with a great deal of attention being focused on orthopedic and maxillofacial applications, including engineering of bone, cartilage, tendon, and ligament.

MATERIALS DEVELOPMENT

The development of matrices to serve as templates for cell attachment/suspension and delivery has progressed at a tremendous rate during the past 5–10 years. Initial efforts have focused on the use of collagen as a natural matrix for cell delivery. A large degree of success was reported in the development of skin equivalents, primarily the dermal component. More recent studies have focused on using natural extracellular matrix (ECM) proteins such as collagen as the structural support for cell transplantation. Yannas (1988) first designed an artificial dermis from collagens and glycosaminoglycans. Others quickly followed in attempts at creating new blood vessels (Weinberg and Bell, 1986) and cartilage (Grande *et al.,* 1989; Wakitani *et al.,* 1989). Poor mechanical properties and variable physical properties with different sources of the protein matrices have hampered progress with these approaches. Concerns have arisen regarding immunogenic problems associated with the introduction of foreign collagen. Also, there are inherent biophysical constraints in collagen used as scaffolding. For example, cartilage grown from cells seeded on collagen must be confined in a rigid "well" such as underlying bone or cartilage, because of collagen's nonmoldability.

Following the developmental efforts using naturally occurring polymers as scaffolds, attention turned in the middle 1980s to the use of synthetic naturally occurring polymers as matrices for cell transplantation. It was believed that synthetic scaffoldings had the advantages of being biocompatible as well as biodegradable. It was postulated that the synthetic polymer scaffolds might act as cell anchorage sites and give the transplanted complex intrinsic structure. Employment of synthetic rather than naturally occurring polymers would allow exact engineering of matrix configuration so that the biophysical limitations of mass transfer would be satisfied. Synthetic matrices would also give one the flexibility to alter physical properties and potentially facilitate reproducibility and scale-up. The configuration of the synthetic matrix could also be manipulated to vary the surface area available for cell attachment as well as to optimize the exposure of the attached cells to nutrients. The chemical environment surrounding a synthetic polymer can be affected in a controlled fashion as the polymer is hydrolyzed. The potential exists to deliver continuously nutrients and hormones that can be incorporated into the polymer structure (Folkman and Hochberg, 1973). Much became known about the chemical environment surrounding these compounds as they degraded. The ratio of the surface area to mass can be altered or the porosity of differing configurations can be changed. The size of the pores of polymers of the same porosity can be altered to increase or decrease the intrinsic strength and elasticity of the polymer matrix, as well as compressibility or creep recovery. We can also change the rate of degradation of the polymer matrices and the environment into which the cells are implanted by systematically altering the surface chemistry of the polymers, creating an acidic or basic environment as they degrade. For instance, polyaminocarbonates cause a local basic environment, whereas poly(glycolic acid) and polyanhydrides cause a local acidic environment. Polyanhydrides and poly(ortho esters) show sur-

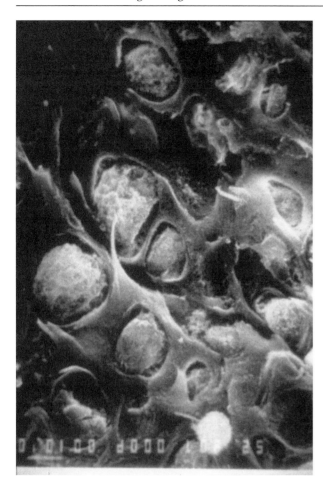

Fig. 47.1. Scanning electron micrograph demonstrating the uniform distribution of cells suspended in the naturally occurring hydrogel calcium alginate.

face erosion; other polymers show bulk erosion. By manipulating design configurations one can increase and decrease the surface area of the polymers. Also, synthetic polymers offer the advantage of consistent reproduction, and thus varying quality is not a problem as it sometimes is with naturally occurring polymers. With synthetic polymers, one has the potential of adding side chains to the polymer structure. Thus one might potentially deliver nutrients and hormones to the cells as the polymer breaks down.

Scaffold materials play a critical role in providing mechanical stability to constructs prior to synthesis of new extracellular matrix by the cells. It is then desirable to match the mechanical properties of the material with that of the tissue. Consequently, scaffolds for bone often contain ceramic hydroxyapatite, which has high stiffness like bone, and scaffolds for cartilage and tendon tend to be made from more compliant polymers. In addition to mechanical stability, scaffold materials often serve to reduce immune response to allogeneic cells. The most notable example of this is encapsulation of pancreatic islet cells for diabetes treatment.

During the past decade, many groups have turned to the use of synthetic biodegradable polymers as templates to which cells adhere and are transplanted. The polymers act as a scaffolding that can be engineered to allow for implantation of transplanted cells within only a few cell layers from the capillaries, and thus allow for nutrition and gas exchange by diffusion until successful engraftment is achieved. In this manner, it is hoped to generate permanent functional new tissues composed of donor cells and recipient interstitium and blood vessels.

A variety of biodegradable materials have been used for tissue scaffolds, including ceramics and polymers. The primary use of ceramics has been in tissue engineering of bone, where porous formulations of hydroxyapatite have been used to carry osteoprogenitor cells derived from periosteum or bone marrow. Typically, ceramic materials have long degradation times *in vivo,* often on the order of years. Synthetic polymers have seen widespread use as scaffold materials, due to their good processing characteristics. These materials have a range of degradation times from very short (days) to long (several months). Typically, polymer scaffolds are in the form of fibrous meshes, porous sponges or foams, or hydrogels. The more common polymers used in fibrous meshes and foams include the linear polyesters, e.g., poly(glycolic acid) (PGA), poly(lactic acid) (PLA), and poly-carpolactone (PCL); poly(ethylene glycol) (PEG); and natural polymers such as collagen and hyaluronic acid (HA). Polymeric hydrogels have the distinct advantage of being injectable, which allows the delivery of the construct to be less invasive and thereby reduces surgical risks. Employment of these types of polymers also ensures delivery of an even distribution of a precise number of cells. They can be configured to provide mechanical support to the cells to maintain their specific phenotype, without inhibiting migration. Common hydrogel substrates include the copolymers of poly(ethylene oxide) and poly(propylene oxide), known as Pluronics, and natural polymers, including alginate and agarose. The delivery of a known concentration of cells is simplified when using a hydrogel, whereby 100% of the cells are encapsulated within the delivery system, as compared to the fibers, for which cell delivery is dependent on cell attachment. The hydrogels also allow the suspended cells to be uniformly distributed throughout the volume of polymer delivered (Fig. 48.1). In contrast, the distribution of cells in the polymer fiber system is not uniform and is difficult to predict.

As the demand for new and more sophisticated scaffolds develops, materials are being designed that have a more active role in guiding tissue development. Instead of merely holding cells in place, these bioactive matrices are designed to encourage cell attachment to the polymer through cell surface adhesion proteins. Toward this end, polymers that have been synthesized have an integrin polypeptide sequence (RGD) in the backbone (Shakesheff *et al.,* 1998; Hern and Hubbel, 1998) or branches (Harrison *et al.,* 1997), or are constructed entirely of polypeptide sequences (Petka *et al.,* 1998). This allows the scaffold effectively to mimic the extracellular matrix and induce attachment of cells directly to the material. This may be particularly important in tissues that bear mechanical loads, because it would allow physical stimuli to be sensed by the cells in the developing tissue in a more physiologic manner.

STRUCTURAL TISSUES

BONE

In initial studies of bone tissue engineering, series of studies were performed to identify substrate features that would allow cells to be anchored and were thus important for maintaining cell function. Identification of these properties allowed us to incorporate them into appropriate polymer scaffolds to be used for transplantation. For example, when some cell types are cultured under conventional conditions, such as on plastic or collagen-coated dishes, gene transcription is depressed and cell-specific mRNA declines, whereas the mRNA of structure-related genes increases many fold. By contrast, when cultured on extracellular matrix, rich in laminin and type IV collagen, some cell types exhibit increased longevity and maintenance of several cell-specific functions. The capability of gene transcription persists (Bucher *et al.,* 1989) when cells are cultured on extracellular gel matrix. We learned that cell shape and function could be controlled by varying the coating density of any of a variety of different naturally occurring ECM substrates on bacteriologic plates (Mooney *et al.,* 1990a). Some cell types exhibit a rounded morphology when cultured on low-density ECM, and an increase in ECM coating density results in more spreading of the cells and epithelial-like morphology. ECM coating density also seems to affect the ability of the cells to enter the synthetic phase (S) and continue through the cell cycle, not exhibiting the S phase in low-density ECM studies but entering the S phase in a high percentage of the high-density ECM studies (Mooney *et al.,* 1990b). Some cell types are able to be switched between programs of growth and differentiation simply by modulating the ECM coating density, thereby altering the substratum's ability to resist cell-generated mechanical load. These and other studies also suggest that cell shape, which might be altered by manipulation of the physicochemical properties of the polymer, may be important in determining cell function. These principles, in combination with our studies, led us to conclude that the matrix to which specific cell types are attached *in vitro* is one of the most important variables in the engineering of new functional tissue.

New tissue-engineered bone is composed of donor cells embedded within recipient interstitium and blood vessels. We elected to work with biodegradable systems for reasons mentioned above. Our experience is based primarily on experience with cells derived from periosteum. We have employed several synthetic scaffolds as matrices. The first was a fiber-based scaffold composed of nonwoven mesh of fibers of poly(glycolic acid), 15 μm in diameter, with average interfiber interspaces of 150–200 μm. This was a polymer construct that had worked very well for the generation of tissue-engineered cartilage. Two other scaffolds for which we have accumulated a large body of data are injectable hydrogel systems, one being calcium alginate and the other being a reverse thermosensitive copolymer, referred to as Pluronics, composed of poly(ethylene oxide) (PEO) and poly(propylene oxide).

We initially chose periosteum as our source of cells to generate tissue-engineered bone. To demonstrate that these cells exhibited an osteoblast-like phenotype, they were nourished in media consisting of Tissue Culture Media 199 (Life Technologies; Baltimore, MD) with 10% fetal calf serum and 50 μg/ml ascorbic acid with L-glutamine (292 μg/ml), penicillin (100 U/ml), and streptomycin (100 μg/ml) and were allowed to incubate *in vitro* at 37°C in the presence of 5% CO_2. Supernatant taken from the culture media was then analyzed for the presence of osteocalcin using a radioimmunoassay kit (Biomedical Tech. Inc.; Stoughton, MA). Our studies revealed that periosteal cells cultured in vitamin D-deficient medium secreted osteocalcin, (1 ± 0.1 fg/cell/day), whereas periosteal cells cultured in medium containing 1,25-dihydroxy vitamin D_3 synthesized and secreted osteocalcin at a higher rate (3 ± 1 fg/cell/day), indicating that these cells indeed exhibited an osteoblast-like phenotype. Based on our experience with chondrocytes, we elected to seed various polymer constructs at a cellular density of 50–100 million cells/ml, at which time the cell–polymer complexes were then either injected/implanted immediately, or maintained *in vitro* for an additional 7–10 days (Vacanti *et al.,* 1993). When PGA fibers were employed, most of the fibers were coated with multiple layers of cells after a week to 10 days, at which time they were implanted subcutaneously. The cell–polymer constructs were excised and evaluated both grossly and histologically after being implanted for increasing periods of time.

As a result of several studies, we learned that cells shed from periosteum multiplied well *in vitro*. Implants seeded with periosteal cells resulted in the generation of what initially appeared to be cartilage, both grossly and histologically, within the first few weeks after implantation. In time, however, all specimens matured to form new bone containing the cellular elements of bone marrow. The rate of the morphogenesis from cartilage to bone correlated with the site into which the cell–polymer construct had been implanted, and seemed to be related to the vascularity of the site. This is in contrast to our experience with polymers seeded *in vitro* with chondrocytes rather than osteocytes. Chondrocyte-seeded implants generated cartilage that did not undergo a morphogenesis to bone, regardless of the site and the duration of time they had been implanted. To support this observation, bone defects surgically created in both weight-bearing and non-weight-bearing bones of male nude (athymic) rats were filled with polymer seeded with chondrocytes, periosteal cells, polymer alone, or nothing at all. Non-weight-bearing defects, 3–4 cm^2, were created in the temporal and parietal bones using a dremel drill. Defects were made through the entire thickness of the bone down to the underlying dura, which was left intact. Weight-bearing defects were approximately 3 cm long and consisted of excision of the middle third of the femoral shaft, which was then bridge plated and filled as described above. Eight weeks after each weight-bearing defect had been created and "repaired," the bone shaft was exposed, the bridge plate removed, and the area in which the defect had been created was examined. Wounds in animals having defects repaired 8 weeks earlier with polymer seeded with either periosteal cells or chondrocytes were then closed and the animals were allowed to recover from anesthesia and ambulate ad lib. We again observed that implants seeded with periosteal cells resulted first in the generation of "cartilage," which ultimately underwent a morphogenesis to become bone. In contrast, polymers that had been seeded *in vitro* with chondrocytes remained hyaline cartilage regardless of the site of implantation, and remained cartilage even when it had been generated within a weight-bearing bone and allowed to ambulate on the tissue-engineered cartilage after the bridge plate had been removed.

These studies have led us to question the currently accepted concepts of endochondral ossification. It is interesting that the wave of advancing ossification in specimens of "cartilage" generated from polymers seeded with periosteal cells and undergoing a metamorphosis to become bone appears histologically to very closely resemble the growth plate, or epiphysis, of growing long bone

Fig. 47.2. Synthetic polymer mold of a rat distal femur seeded with periosteal cells.

in immature animals. In this system, it appears that implanted progenitor cells first develop into chondrocytes, followed by maturation into osteocytes. This provides support for the so-called "transdifferentiation" hypothesis, which was posed as an explanation for the mechanisms for long bone growth (Thesingh *et al.,* 1991).

When using periosteal cells, however, the generation of bone is optimized by the use of calcium alginate. The engineered tissue appears to extract the calcium from the hydrogel as it undergoes morphogenesis to bone. By using calcium alginate as a vehicle for delivery of periosteal cells, we have been able to make molds in specific shapes, such as the distal femur of a rat, which results in the generation of tissue-engineered bone of the same shape (Figs. 47.2 and 47.3). A vascularized pedicle of tissue-engineered bone has been successfully generated by molding a synthetic scaffolding, seeded with periosteal cells, around a vascular bundle. This vascularized pedicle of bone can then be transplanted into a large bone defect. Current efforts in the field are examining the efficacy of this approach in generating bone in large animals using autologous cells, as well as the ability to generate composite structures of bone and cartilage.

CARTILAGE

The focus of cartilage tissue engineering to date has been in two main areas: orthopedic surgery and plastic and reconstructive surgery. In the field of orthopedic surgery, significant interest has been generated in the area of articular cartilage repair, sparked by a report of autologous chondrocyte transplantation in humans (Brittberg *et al.,* 1994). These studies involved harvest of cartilage from those locations of the articular surface that are deemed "low weight bearing," isolation and culture of these cells, and injection of these cells in saline under a periosteal flap sewn over a focal defect. Results of these studies have been encouraging; however, interpretation of the data has been clouded by the inability to assess the possible contributions of cells from the periosteum, which have been used in separate studies to stimulate articular cartilage repair (O'Driscoll *et al.,* 1994).

Numerous other studies have investigated seeding of articular chondrocytes onto various synthetic and natural polymer substrates, including PGA (Freed *et al.,* 1993), PEO (Elisseeff *et al.,* 1999), fibrin glue (Silverman *et al.,* 1999), and collagen gels (Nehrer *et al.,* 1998). In addition, studies have addressed the issue of chondrocyte attachment to native cartilage matrix (Peretti *et al.,* 1998) and the ability of pieces of cartilage to generate mechanically functional repair tissue (Peretti *et al.,* 1999; Reindel *et al.,* 1995). To complement these efforts, enhancement of the car-

A

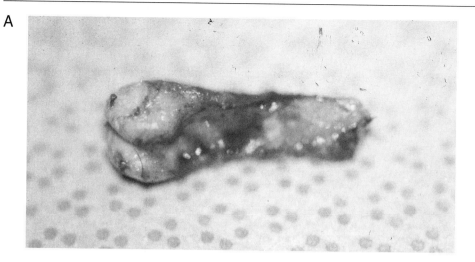

B

C

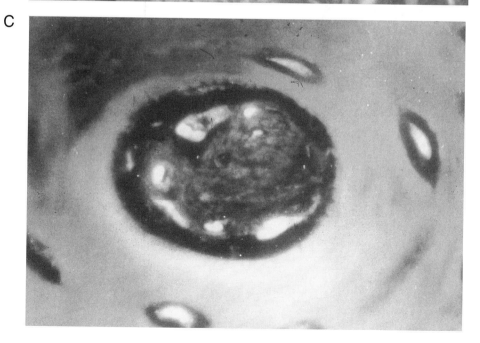

Fig. 47.3. (A) The tissue-engineered femur generated from the seeded mold shown in Fig. 47.2. Note that it has the same dimensions. (B) Histologic appearance, using hematoxylin and eosin stains, of the excised specimen. Note the lamellar structure of the new bone as visualized under polarized light. (C) Note the formation of what appears to be a haversion system in the specimen.

tilage repair process has been attempted using a variety of growth factors, including transforming growth factor β (TGF-β) (Glansbeek *et al.*, 1998) and bone morphogenic protein 2 (BMP 2) (Sellers *et al.*, 1997).

Attempts at tissue engineering cartilage for orthopedic applications have not been restricted only to articular cartilage. Studies documenting tissue engineering of the meniscus (Ibarra *et al.*, 1997) and intervertebral disc (Kusior *et al.*, 1998) demonstrate the potential for regeneration or replacement of other cartilaginous tissues in the musculoskeletal system in addition to articular cartilage.

Efforts in cartilage tissue engineering for plastic and reconstructive surgery have had a slightly different focus than those in orthopedic surgery. Due to the types of tissues being replaced, the mechanical constraints and requirements for cartilage used in non-weight-bearing structures are much less. However, in these applications, new challenges arise around creating constructs in complex geometries that retain their form as they develop into cartilage. Several techniques have evolved specifically for the purpose of creating these forms, including the use of PLA-reinforced PGA scaffolds (Kim *et al.*, 1994) and injection molding of seeded hydrogels (Rowley *et al.*, 1998). Using these approaches, cartilage has been engineered for the replacement or reconstruction of the ear (Cao *et al.*, 1997), temporomandibular disc (Puelacher *et al.*, 1994a), nasal septum (Puelacher *et al.*, 1994b), orbital floor (Kim *et al.*, 1994), and nipple (Cao *et al.*, 1998).

An additional complication for the use of tissue-engineered cartilage for these applications lies in the subcutaneous environment in which implants are placed. Compared to the articular environment, the immune response evoked by implants is much greater in subcutaneous or intramuscular sites. This immune response may result from foreign body reaction to the material scaffold, or may be part of the general wound healing process resulting from implantation surgery. In both of these cases, the cells and the materials of a tissue-engineered construct are exposed to histiocytes and macrophages, as well as a variety of chemical and biochemical factors that may adversely affect the implant. Specifically, these environments are often at lower pH, which will accelerate the degradation of polyesters such as PGA and PLA. Further, inflammatory factors such as tumor necrosis factor α (TNF-α) and interleukins (IL-1α and β) are known to degrade cartilage and compromise the mechanical integrity of the tissue (Bonassar *et al.*, 1997).

For any potential therapeutic application involving tissue engineering of cartilage, expansion of the cell population *in vitro* is an important step in the process of building a construct. In the process, it is important to ensure that the expanded cell population retains its phenotypic function. This is of great concern for chondrocytes, which will dedifferentiate on repeated passaging (Benya and Shaffer, 1982).

The quality of newly engineered tissue is not only a function of optimal matching of the cell type to polymer matrix employed, but also of the animal species studied. In nude recipients for example, the fiber system is an excellent scaffolding for the generation of cartilage, whereas injectable hydrogels seem to be superior when engineering the same type of cartilage in immunocompetent animals using autologous cells. Excellent results are produced when delivering chondrocytes in the Pluronics system. This is in contrast to the use of calcium alginate, whereby the chondrocytes are encapsulated in a vehicle that seems to impair cell replication, and takes a long period of time to degrade. When chondrocytes are delivered in Pluronics, the polymer degrades much more rapidly, allows for cell replication with minimal inflammatory response, and generates excellent cartilage in higher species using autologous cells.

TENDON AND LIGAMENT

Initial efforts in tissue engineering of tendon and ligament began with the use of denatured collagen as a scaffolding for generation of an anterior cruciate ligament repair. This material is implanted, not seeded, with cells, but ultimately is seeded by the host. Synthetic devices to replace or reinforce biologic materials have been used as grafts, with the belief that these materials would act as scaffolds on which fibrous tissue could be induced to proliferate and form ligamentous and tendinous tissues. It has been proposed that, in the presence of a tensile force, immature fibrous tissue so formed will gradually mature, with its fibers oriented functionally (Fujikawa, 1989; Jenkins *et al.*, 1976). However, this is disputed by reports (Andrish and Woods, 1984; Aronczky *et al.*, 1986; McPherson *et al.*, 1985) that the induced tissue is characterized by a chronic foreign body reaction and by disorderly collagen fibers. The use of carbon fiber-derived scaffolds that are eventually replaced by fibrous tissue to ultimately replace the lost function of a tendon or ligament was extensively studied in the 1970s and 1980s (Forster *et al.*, 1978; Jenkins, 1978; Jenkins *et al.*, 1980; Tayton *et al.*, 1982; Rushton *et al.*, 1983). Nevertheless, the fact that carbon particles did not disappear completely, and were eventually found in regional lymph nodes even several years later, discouraged use of such scaffolds (Jenkins *et al.*, 1980). Later, carbon fibers were shown to somehow inhibit tendon fibroblast growth (Ricci *et al.*, 1991). Mendes *et al.* then concluded that collagen fibers did not significantly contribute to the tensile strength of carbon fiber composite ligaments and tendons (Irie *et al.*, 1992). Although some authors continue to report good results with their use (Demmer *et al.*, 1991), no significant differences have been shown in the outcome of patients in whom such devices have been used to reinforce or augment biologic grafts (Strum and Larson, 1985). Other synthetic materials such as Dacron have been used to replace or augment tendinous grafts in the treatment of injuries to the ligaments in the knee. However, histologic and biochemical studies have demonstrated that Dacron-induced fibrous tissue more closely resembles scar or granulomatous tissue than it does neoligament, and it has been shown to elicit a foreign body inflammatory reaction (Irie *et al.*, 1992).

Concerning the use of collagen tendon prostheses (Goldstein *et al.*, 1989; Kato *et al.*, 1991), such prostheses have been shown to have structures and mechanical properties similar to those of autogenous tendon grafts (Kato *et al.*, 1991). These implants are invaded by the host's fibroblasts and are eventually reabsorbed, giving way to the formation of a neotendon, following a biologic pathway similar to those of autografts and allografts. The use of xenografts (Milthorpe, 1994) or animal-derived collagen devices (Tauro *et al.*, 1991) has also been studied. Although they have advantages related to availability, decreased risk of transmission of infectious diseases, and avoiding damage to the patient's healthy tissues, the preparation of the grafts apparently affects significantly their biomechanical properties, and the possibility of rejection can still limit their use. Tendon autografts and allografts are currently the most commonly used substitutes to reconstruct injured ligaments or tendons. Allografts, however, still face the risk of potential transmission of infectious diseases (Conrad *et al.*, 1995; Asselmeier *et al.*, 1993), and some of the techniques used to sterilize them present the possible risk of affecting the biomechanical characteristics of the grafts (Rasmussen *et al.*, 1994). Tendon autografts are perhaps the most commonly transplanted tissue structures. Their biologic and biomechanical characteristics serve their purpose relatively well, considering that usually they rely on creeping substitution by the host's cells of a frequently inert organic scaffold from which most cells have usually disappeared (Aronczky *et al.*, 1982; Rougraff *et al.*, 1993; Bosch and Kasperczyk, 1993; Abe *et al.*, 1993; Lane *et al.*, 1993; Kasperczyk *et al.*, 1993). Autografts, however, present the disadvantage of creating a secondary morbid site during harvesting, and the fact that a functioning structure has to be removed from its original site for use in a nearby or remote site.

BIOMECHANICS

An understanding of the structure of engineered constructs needs to be closely associated with an assessment of tissue function. Indeed, the entire motivation for the field of tissue engineering is to restore function of tissue lost to disease, accident, or malformations. Therefore, it is critical to determine the extent to which the functional properties of generated tissues are similar to those of native tissue. The focus of many tissue engineering applications is on structural tissues, such as bone, cartilage, tendon, skin, and muscle, necessitating the analysis of the biomechanical properties of generated tissues. As with a discussion of tissue biochemistry or molecular biology, an understanding of the biomechanical properties of engineered tissue must start with an understanding of the mechanical properties of the native tissue. And as with the previous discussion, this requires an assessment of the level of sophistication that adequately describes the system. Tissues with nonlinear or time-dependent mechanical responses may not be adequately described by a modulus determined at equilibrium or at a single strain rate. Indeed, soft tissues such as skin, tendon, muscle, and cartilage require viscoelastic or poroelastic parameters in addition to the modulus to describe their mechanical behavior.

Further, it is critical to understand the relationship between the structure of native or engineered tissues and their mechanical properties. Tissues such as tendon and ligament can endure high tensile elongations due to the high content of collagen and elastin in the extracellular matrix. In a similar way, cartilage is composed of a highly charged proteoglycan network and a cross-linked collagen that strengthen the tissue in compression and shear, respectively. The composition of bone includes a variety of extracellular matrix proteins, including collagen, as well as a mineralized component of predominantly hydroxyapatite. This mineral component gives bone not only its extraordinary compressive strength, but also its brittleness in tension or shear.

Adequate understanding of the mechanical properties of structural tissues is not only an important yardstick for evaluating tissue-engineered constructs, but also provides an important set of design criteria for material scaffolds. Even at the initial stages of development, engineered constructs implanted into a defect site must in some way endure the stress, strains, and flows to which the surrounding tissues are exposed. This is, in fact, reflected in the types of materials used in existing efforts to engineer structural tissues: ceramics for bone, polymer fibers and cross-linked proteins for tendons and ligaments, and hydrogels and polymer meshes for cartilage.

CONCLUSION

From these studies we learned that if cell anchorage sites and reasonable structural cues are provided in an appropriate environment, the intrinsic ability of cells to reorganize and generate new tissue is enhanced. The efforts described enable the natural ability of the cells to reconstitute as functional tissue, and intercellular signaling mechanisms and a predetermined expression of genes may be the ultimate driving mechanisms behind the principles of tissue engineering that we have discussed. It is our hope to develop this system to maximize the potential of cells to generate functional replacement tissue as an alternative to conventional organ transplantation and tissue reconstruction. Advances in polymer design are allowing us to utilize this ability fully. We anticipate the onset of human trials with these structural cells within the next few years.

References

Abe, S., Kurosaka, M., Iguchi, T. *et al.* (1993). Light and electron microscopic of remodeling and maturation process in aurogenous graft for anterior cruciate ligament reconstruction. *Arthroscopy* 9(4), 394–405.

Andrish, J. R., and Woods, L. D. (1984). Dacron augmentation in anterior cruciate ligament reconstruction in dogs. *Clin. Orthop. Relat. Res.* 183, 298–302.

Aronczky, S. P., Tarvin, G. B., and Marshall, J. L. (1982). Anterior cruciate ligament replacement using patellar tendon: An evaluation of graft revascularization. *J. Bone Jt. Surg. Am. Vol.* 64-A, 217–224.

Aronczky, S. P., Warren, R. F., and Minei, J. P. (1986). Replacement of the anterior cruciate ligament using a synthetic prothesis: An evaluation of graft biology in the dog. *Am. J. Sports Med.* 14, 1–6.

Asselmeier, M. A., Caspari, R. B., and Bottenfield, S. (1993). A review of allograft processing and sterilization techniques and their role in transmission of the human immunodeficiency virus. *Am. J. Sports Med.* 21, 170–175.

Bentley, G., and Greer, R. G., III. (1971). Homotransplantation of isolated epiphyseal and articular cartilage chondrocytes into joint surfaces of rabbits. *Nature (London)* 230, 385–388.

Benya, P. D., and Shaffer, J. D. (1982). Dedifferentiated chondrocytes reexpress the differentiated collagen phenotype when cultured in agarose gels. *Cell (Cambridge, Mass.)* 30, 215–224.

Bonassar, L. J., Sandy, J. D., Lark, M. W., Plaas, A. H., Frank, E. H., and Grodzinsky, A. J. (1997). Inhibition of cartilage degradation and changes in physical properties induced by IL-1 beta and retinoic acid using matrix metalloproteinase inhibitors. *Arch. Biochem. Biophys.* 344, 404–412.

Bosch, U., and Kasperczyk, W. J. (1993). The healing process after cruciate ligament repair in the sheep model. *Orthopade* 22(6), 366–371.

Brittberg, M., Lindahl, A., Nilsson, A., Ohlsson, C., Isaksson, O., and Peterson, L. (1994). Treatment of deep cartilage defects in the knee with autologous chondrocyte transplantation. *N. Engl. J. Med.* 331, 889–895.

Bucher, N. L. R., Aiken, J., Robinson, G. S., Vacanti, J. P., and Farmer, S. R. (1989). Transcription and translation of liver specific cytoskeletal genes in hepatocytes on collagen and EHS tumor matrices. *Abstr., Am. Assoc. Study Liver Dis.*

Cao, Y., Vacanti, J. P., Paige, K. T., Upton, J., and Vacanti, C. A. (1997). Transplantation of chondrocytes utilizing a polymer–cell construct to produce tissue-engineered cartilage in the shape of a human ear. *Plast. Reconstr. Surg.* 100(2), 297–302.

Cao, Y. L., Lach, E., Kim, T. H., Rodriguez, A., Arevalo, C. A., and Vacanti, C. A. (1998). Tissue-engineered nipple reconstruction. *Plast. Reconstr. Surg.* 102, 2293–2298.

Chesterman, P. J., Reading, and Smith, A. U. (1968). Homotransplantation of articular cartilage and isolated chondrocytes. *J. Bone Jt. Surg., Br. Vol.* 50B, 184–197.

Conrad, E. U., Gretch, D. R., and Obermeyer, K. R. (1995). Transmission of the hepatitis-C virus by tissue transplantation. *J. Bone Jt. Surg., Am. Vol.* 77-A, 214–224.

Demmer, P., Fowler, M., and Marino, A. A. (1991). Use of carbon fibers in the reconstruction of knee ligaments. *Clin. Orthop. Relat. Res.* 271, 225–232.

Elisseeff, J., Anseth, K., Sims, D., McIntosh, W., Randolph, M., and Langer, R. (1999). Transdermal photopolymerization for minimally invasive implantation. *Proc. Natl. Acad. Sci. U.S.A.* 96, 3104–3107.

Folkman, J., and Hochberg, M. M. (1973). Self-regulation of growth in three dimensions. *J. Exp. Med.* 138, 745–753.

Forster, I. W., Rallis, Z. A., McKibbin, B., and Jenkins, D. H. R. (1978). Biological reaction to carbon fiber implants: The formation and structure of a carbon-induced neotendon. *Clin. Orthop. Relat. Res.* 131, 299–307.

Freed, L. E., Marquis, J. C., Nohria, A., Emmanual, I., Mikos, A. G., and Langer, R. (1993). Neocartilage formation *in vitro* and *in vivo* using cells cultured on synthetic biodegradable polymers. *Biomed. Mater. Res.* **27**, 11–23.

Fujikawa, K. (1989). Clinical study of anterior cruciate ligament reconstruction with a scaffold type prosthetic ligament (Leeds–Keio). *J. Jpn. Orthop. Assoc.* **63**, 774–788.

Glansbeek, H. L., van Beuningen, H. M., Vitters, E. L., van der Kraan, P. M., and van den Berg, W. B. (1998). Stimulation of articular cartilage repair in established arthritis by local administration of transforming growth factor-beta into murine knee joints. *Lab. Invest.* **78**, 133–142.

Goldstein, J. D., Tria, A. J., Zawadsky, J. P. *et al.* (1989). Development of a reconstituted collagen tendon prosthesis. *J. Bone Surg.* **8**, 1183–1191.

Grande, D. A., Pitman, M. L., Peterson, L., Menche, D., and Klein, M. (1989). The repair of experimentally produced defects in rabbit articular cartilage by autologous chondrocyte transplantation. *J. Orthop. Res.* **7**(2), 208–218.

Green, W. T., Jr. (1977). Articular cartilage repair: Behavior of rabbit chondrocytes during tissue culture and subsequent allografting. *Clin. Orthop. Relat. Res.* **124**, 237–250.

Harrison, D., Johnson, R., Tucci, M., Puckett, A., Tsao, A., Hughes, J., and Benghuzzi, H. (1997). Interaction of cells with UHM–WPE impregnated with the bioactive peptides RGD, RGE, or poly(L-lysine). *Biomed. Sci. Instrum.* **34**, 41–46.

Hern, D. L., and Hubbell, J. A. (1998). Incorporation of adhesion peptides into nonadhesive hydrogels useful for tissue resurfacing. *J. Biomed. Mater. Res.* **39**, 266–276.

Ibarra, C., Jannetta, C., Vacanti, C. A., Cao, Y., Kim, T. H., Upton, J., and Vacanti, J. P. (1997). Tissue engineered meniscus: A potential new alternative to allogeneic meniscus transplantation. *Transplant. Proc.* **29**, 986–988.

Irie, K., Kurosawa, H., and Oda, H. (1992). Histological and biochemical analysis of the fibrous tissue induced by implantation of synthetic ligament (Dacron): An experimental study in a rat model. *J. Orthop. Res.* **10**(6), 886–894.

Itay, S., Abramovici, A., and Nevo, Z. (1987). Use of cultured embryonal chick epiphyseal chondrocytes as grafts for defects in chick articular cartilage. *Clin. Orthop. Relat. Res.* **220**, 284–303.

Jenkins, D. H. R. (1978). The repair of cruciate ligaments with flexible carbon fibre: A longer term study in the induction of new ligaments and of the fate of the implanted carbon. *J. Bone Jt. Surg., Br. Vol.* **60B**, 520–522.

Jenkins, D. H. R., Forster, I. W., McKibben, B., and Wales, Z. A. R. C. (1976). Induction of tendon and ligament formation by carbon implants. *J. Bone Jt. Surg., Am. Vol.* **58-A**, 1083–1088.

Jenkins, D. H. R. *et al.* (1980). The role of flexible carbon fibre implants as tendon and ligament substitutes in clinical practice: A preliminary report. *J. Bone Jt. Surg., Br. Vol.* **62B**, 497–499.

Kasperczyk, W. J., Bosch, U., Oestern, H. J. *et al.* (1993). Staging of patellar tendon autograft healing after posterior cruciate ligament reconstruction. A biomechanical and histological study in a sheep model. *Clin. Orthop. Relat. Res.* **286**, 271–282.

Kato, Y. P., Dunn, M. G., Zawadsky, J. P. *et al.* (1991). Regeneration of Achilles tendon prosthesis. Results of a one-year implantation study. *J. Bone Jt. Surg.* **73**(4), 561–574.

Kim, W. S., Vacanti, J. P., Cima, L., Mooney, D., Upton, J., Puelacher, W. C., and Vacanti, C. A. (1994). Cartilage engineered in predetermined shapes employing cell transplantation on synthetic biodegradable polymers. *Plast. Reconstr. Surg.* **94**, 233–237.

Kusior, L. J., Vacanti, C. A., Bayley, J. C., and Bonassar, L. J. (1998). Tissue engineering of bovine nucleus pulposus in nude mice. *Tissue Eng.* **4**, 469.

Lane, J. G., McFadden, P., Bowden, K. *et al.* (1993). The ligamentization process: A 4 year case study following ACL reconstruction with a semitendinous graft. *Arthroscopy* **9**(2), 149–153.

Lexer, D. (1908). Die verwendung der freien knochenplastik nebast veruchen uber gelenkversteifung und gelenktransplantation. *Arch. Klin. Chir.* **86**, 939–954.

Lipman, J. M., McDevitt, C. A., and Sokoloff, L. (1983). Xenografts of articular chondrocytes in the nude mouse. *Calcif. Tissue Int.* **35**(6), 767–772.

McPherson, G. K., Mendenhall, H. V., Gibbons, D. F. *et al.* (1985). Experimental mechanical and histological evaluation of the Kennedy ligament augmentation device. *Clin. Orthop. Relat. Res.* **196**, 186–195.

Milthorpe, B. K. (1994). Xenografts for tendon and ligament repair. *Biomaterials* **15**(10), 745–752.

Mooney, D. J., Hansen, L. H., Vacanti, J. P., Langer, R. S., Farmer, S. R., and Ingber, D. E. (1990a). Switching between growth and differentiation in hepatocytes: Control by extracellular matrix. *J. Cell Biol.* **3**(5), 149a.

Mooney, D. J., Langer, R. S., Vacanti, J. P., and Ingber, D. E. (1990b). Control of hepatocyte function through variation of attachment site densities. *Abstr., Am. Inst. Chem. Eng.*

Nehrer, S., Breinan, H. A., Ramappa, A., Hsu, H. P., Minas, T., Shortkroff, S., Sledge, C. B., Yannas, I. V., and Spector, M. (1998). Chondrocyte-seeded collagen matrices implanted in a chondral defect in a canine model. *Biomaterials* **19**, 2313–2328.

New Jerusalem Bible (1990). Genesis 2:21–22. Doubleday, New York.

O'Driscoll, S. W., Recklies, A. D., and Poole, A. R. (1994). Chondrogenesis in periosteal explants. An organ culture model for *in vitro* study. *J. Bone Jt. Surg., Am. Vol.* **76-A**, 1042–1051.

Ohgushi, H., Goldberg, V. M., and Caplan, A. I. (1989). Repair of bone defects with marrow cells and porous ceramic: Experiments in rats. *Acta Orthop. Scand.* **60**(3), 334–339.

Peretti, G. M., Randolph, M. A., Caruso, E. M., Rossetti, F., and Zaleske, D. J. (1998). Bonding of cartilage matrices with cultured chondrocytes: An experimental model. *J. Orthop. Res.* **16**, 89–95.

Peretti, G. M., Bonassar, L. J., Caruso, E. M., Randolph, M. A., Trahan, C. A., and Zaleske, D. J. (1999). Biomechanical analysis of a chondrocyte-based repair model of articular cartilage. *Tissue Eng.* (in press).

Petka, W. A., Harden, J. L., McGrath, K. P., Wirtz, D., and Tirrell, D. A. (1998). Reversible hydrogels from self-assembling artificial proteins. *Science* **281**, 389–391.

Puelacher, W. C., Wisser, J., Vacanti, C. A., Ferraro, N. F., Jaramillo, D., and Vacanti, J. P. (1994a). Temporomandibular joint disc replacement made by tissue-engineered growth of cartilage. *J. Oral Maxillofacial Surg.* **52**, 1172–1177.

Puelacher, W. C., Mooney, D., Langer, R., Upton, J., Vacanti, J. P., and Vacanti, C. A. (1994b). Design of nasoseptal cartilage replacements synthesized from biodegradable polymers and chondrocytes. *Biomaterials* **15**, 774–778.

Rasmussen, T. J., Feder, S. M., Butler, D. L., and Noyes, F. R. (1994). The effects of 4 Mrad of gamma irradiation on the mechanical properties of bone-patellar tendon-bone grafts. *Arthroscopy* **10**(2), 188–197.

Reindel, E. S., Ayroso, A. M., Chen, A. C., Chun, D. M., Schinagl, R. M., and Sah, R. L. (1995). Integrative repair of articular cartilage *in vitro:* Adhesive strength of the interface region. *J. Orthop. Res.* **13**, 751–760.

Ricci, J. L., Gona, A. G., and Alexander, H. (1991). *In vitro* tendon cell growth rates on a synthetic fiber scaffold material and on standard culture plates. *J. Biomed. Mater. Res.* **25**(5), 651–666.

Rougraff, B., Shelbourne, K. D., Gerth, P. K. *et al.* (1993). Arthroscopic and histologic analysis of human patellar tendon autografts used for anterior cruciate ligament reconstruction. *Am. J. Sports Med.* **21**(2), 277–284.

Rowley, J. A., Genes, N., Chang, S., Mooney, D. J., and Bonassar, L. J. (1998). Injection molding of alginate/cartilage constructs. *Tissue Eng.* **4**, 466.

Rushton, N. *et al.* (1983). The clinical, arthroscopic, and histologic findings after replacement of the anterior cruciate ligament with carbon fibre. *J. Bone Jt. Surg., Br. Vol.* **65B**, 308–309.

Sellers, R. S., Peluso, D., and Morris, E. A. (1997). The effect of recombinant human bone morphogenetic protein-2 (rhBMP-2) on the healing of full-thickness defects of articular cartilage. *J. Bone Jt. Surg., Am. Vol.* **79-A**, 1452–1463.

Shakesheff, K., Cannizzaro, S., and Langer, R. (1998). Creating biomimetic micro-environments with synthetic polymer-peptide hybrid molecules. *J. Biomater. Sci. Polym. Ed.* **9**, 507–518.

Silverman, R. P., Passaretti, D., Huang, W., Randolph, M. A., and Yaremchuk, M. J. (1999). Injectable tissue-engineered cartilage using a fibrin glue polymer. *Plast. Reconstr. Surg.* **103**, 1809–1818.

Strum, G. M., and Larson, R. L. (1985). Clinical experience and early results of carbon fiber augmentation of anterior cruciate reconstruction of the knee. *Clin. Orthop. Relat. Res.* **196**, 124–138.

Tauro, J. C., Parsons, J. R., Ricci, J. *et al.* (1991). Comparison of bovine collagen xenografts to autografts in the rabbit. *Clin. Orthop.* **266**, 271–284.

Tayton, K. *et al.* (1982). Long term effects of carbon fibre on soft tissues. *J. Bone Jt. Surg., Br. Vol.* **64B**, 112–114.

Thesingh, C. W., Groot, C. G., and Wassenaar, A. M. (1991). Transdifferentiation of hypertrophic chondrocytes into osteoblasts in murine fetal metatarsal bones, induced by co-cultured cerebrum. *Bone Miner.* **12**, 25–40.

Upton, J., Sohn, S. A., and Glowacki, J. (1981). New cartilage derived from transplanted perichondrium: What is it? *Plast. Reconstr. Surg.* **68**(2), 166–174.

Urist, M. R. (1965). Bone: Formation by autoinduction. *Science* **150**, 893–899.

Vacanti, J. P. (1988). Beyond transplantation. *Arch. Surg. (Chicago)* **123**, 545–549.

Vacanti, J. P., Morse, M. A., and Saltzman, W. M. (1988). Selective cell transplantation using bioabsorbable artificial polymers as matrices. *J. Pediatr. Surg.* **23**, 3–9.

Vacanti, C. A., Langer, R., Schloo, B. *et al.* (1991). Synthetic polymers seeded with chondrocytes provide a template for new cartilage formation. *Plast. Reconstr. Surg.* **87**(5), 753–759.

Vacanti, C. A., Kim, W. S., Upton, J., Vacanti, M. P., Mooney, D., Schloo, B, and Vacanti, J. P. (1993). Tissue engineered growth of bone and cartilage. *Transplant. Proc.* **25**(1), 1019–1021.

Wakitani, S., Kimura, T., Hirooka, A. *et al.* (1989). Repair of rabbit articular surfaces with allograft chondrocytes embedded in collagen gel. *J. Bone Jt. Surg., Br. Vol.* **63B**, 529–538.

Weinberg, C. B., and Bell, E. (1986). A blood vessel model constructed from collagen and cultured vascular cells. *Science* **231**, 397–400.

Wozney, J. M. *et al.* (1988). Novel regulators of bone formation: Molecular clones and activities. *Science* **242**, 1528–1534.

Yannas, I. V. (1988). Regeneration of skin and nerve by use of collagen templates. *In* "Collagen III" (M. E. Nimni, ed.). CRC Press, Boca Raton, FL.

CHAPTER 48

Bone Regeneration through Cellular Engineering

Scott P. Bruder and Arnold I. Caplan

INTRODUCTION

A teenager is found to have a tumor in the midshaft of the femur. To save this young individual's life, a large segment of the bone is excised. Why is it that the patient cannot simply regenerate bone to fill the huge gap? A baby is diagnosed as having osteogenesis imperfecta (OI) (brittle bone disease) caused by a single point mutation in the gene for type I collagen (Byers and Bonadio, 1985). If it is a "mild" penetrance, the parents will see their child grow to be a dwarf with multiple fractures of many of its bones. How can we cure this malady, given that we now know its cause? An elderly woman falls, fractures her hip, and never regains mobility. Osteoporosis has reduced both her total bone mass and her ability to produce enough new bone to heal the fracture. How can we replenish the supply of bone-forming cells in this condition? These three medical problems appear to be quite dissimilar, but potentially, through cell-based therapies, the skeletal tissues of these patients can be engineered to cure themselves by accomplishing what they cannot do normally. The key to any medical intervention is to mobilize and induce the body to accomplish feats that it does not normally have the capacity to do. In the first case, the solution is to grow osteoprogenitor cells in large enough numbers (Bruder and Jaiswal, 1996) to fill the excision gap. For this scenario, it is now possible to grow mesenchymal progenitor cells outside the body and then put them back in such a way that bone will be regenerated in a mass not normally possible. In the second case, the normal type I collagen gene will be molecularly inserted into osteoprogenitor cells, which continually renew themselves, so that when their eventual progeny naturally replace osteoblasts, the bone-forming cells of the body, bone will be fabricated with a normal type I collagen matrix. Alternatively, patients with OI can be exposed to chemotherapeutic drugs to destroy their bone marrow and normal, immunomatched allogeneic marrow containing normal osteoprogenitor cells can be grafted into the OI patient (Horwitz et al., 1999). In this case, these normal osteoprogenitor cells will produce normal progeny that will replace the genetically defective osteoblasts that routinely expire. And in the third case, age-related bone loss and diminished repair potential may be prevented by periodic administration of autologous osteoprogenitor cells.

The key issues that control the successful medical interventions depicted above involve defining the parameters that govern natural regeneration and turnover of bone and engineering new technologies to make use of such insight. The primary assumption is that the principles that govern regeneration and natural turnover are, to some extent, recapitulations of embryonic-formative events and that reparative strategies must be structured to follow these recapitulation parameters.

MARROW: THE SOURCE OF MESENCHYMAL PROGENITOR CELLS

The challenge of mesenchymal tissue regeneration is to bring large numbers of multipotent progenitors to the regeneration site as quickly as possible; the rapid formation of a mesenchymal

683

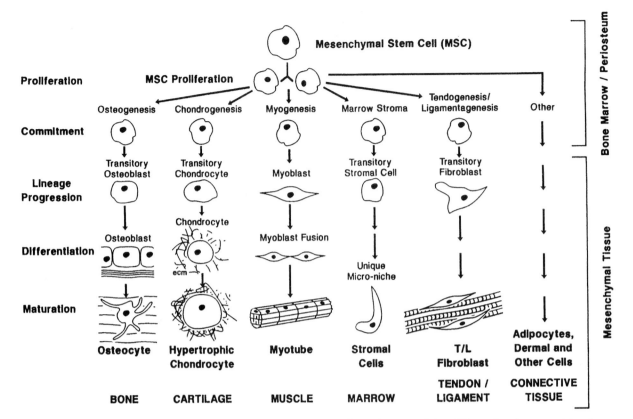

Fig. 48.1. The mesengenic process. MSCs have the potential to differentiate into a variety of mesenchymal tissues, such as bone, cartilage, tendon, muscle, marrow, dermis and fat. Proliferating
MSCs enter a lineage, following their commitment to that particular pathway. The commitment
event involves the action of specific growth factors and/or cytokines, as does the next phase, in
which the lineage-committed cells progress through a number of transitory stages in the lineage
progression process. Terminal differentiation involves the cessation of proliferation and the massive biosynthesis of tissue-specific products. Last, these differentiated cells go through a maturation stage in which they acquire an ability to function in aspects of tissue homeostasis as opposed
to high levels of synthetic activity. All of these end-stage-differentiated cells have fixed half-lives
and can be expected to expire; these cells are replaced by newly differentiated cells arising from
the continuous transition down the lineage pathway. The lineages are arranged from left to right
based on the relative information known about definitive lineage stages.

blastema of critical size is the key step, because the waves of wound site instructional cues are ever
changing and must coordinate with the mesenchymal differentiation sequence. There are several
repositories of such mesenchymal progenitors in adults, with bone marrow and periosteum being
the most accessible (Haynesworth *et al.,* 1992b; Nakahara *et al.,* 1990, 1992). Orthopedic surgeons have long understood that vascular continuity, excess marrow, and intact periosteum function to accelerate bone repair/regeneration (Hamm, 1930; Kojimoto *et al.,* 1988; Brighton, 1984).
Notably, bone marrow has been studied in both human and animal models and has been shown
to house isolatable osteochondral progenitor cells that we operationally refer to as mesenchymal
stem cells (MSCs) (Haynesworth *et al.,* 1992a,b; Friedenstein, 1976; Owen, 1985; Beresford,
1989; Bruder *et al.,* 1990; Owen and Friedenstein, 1988; Caplan, 1991, 1994; Dennis and Caplan, 1993; Caplan *et al.,* 1993). In principle, these MSCs can differentiate into a number of mesenchymal phenotypes by entering discrete lineage pathways, as shown in Fig. 48.1. We have developed the technology to isolate these cells from bone marrow (Fig. 48.2) of patients ranging from
newborns to individuals in their eighth and ninth decade of life (Haynesworth *et al.,* 1992a,b,
1994; Beresford, 1989). Importantly, there appears to be a decrease in the number of MSCs per
total nucleated cells in marrow as a function of age (Oreffo *et al.,* 1998; Muschler *et al.,* 1997;
Kahn *et al.,* 1995; Egrise *et al.,* 1992). This age-related deficit may be the cause of the reduced potential and/or kinetics of skeletal repair in older patients.

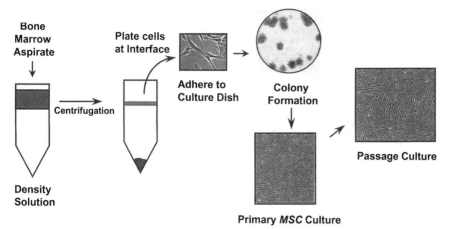

Primary *MSC* Culture

Fig. 48.2. MSC purification. A marrow aspirate of 10–20 ml is obtained with standard clinical biopsy procedures. Large particles of bone chips are removed and the cells applied to Percoll density gradient centrifugation, following the procedures of Haynesworth et al. (1992b). The MSCs are localized in the low-density fraction. The MSCs are recovered at a pooled density of 1.03 g/ml, centrifuged, resuspended in Dulbecco's modified Eagle's medium plus 10% fetal bovine serum from selected lots (Lennon et al., 1996), and seeded into cell culture dishes. The MSCs selectively adhere while most hematopoietic cells remain in suspension. Approximately 1000 to 3000 MSCs are recovered in these primary cultures and they grow as colonies.

Because it is now possible to isolate MSCs from bone marrow, and expand them in culture away from all hematopoietic and vascular cells, one can establish therapeutic paradigms for the treatment of mesenchymal tissue defects. These cells can flourish in culture, and when reintroduced into *in vivo* test sites, MSCs can differentiate into a variety of site-specified phenotypes, including bone-forming cells. That marrow contains uniquely potent mesenchymal progenitors has been clearly documented by isolating clones of such progenitors and documenting that the progenies of these single cells are capable of differentiating into a variety of mesenchymal phenotypes (Dennis and Caplan, 1996; Dennis *et al.*, 1999; Pittenger *et al.*, 1999).

This basic information is the foundation for our vision of regenerative tissue therapy, which capitalizes on the replicative potential of MSCs, their bioactivity following cryopreservation, and their ability to form appropriate tissue types after implantation in a site in need of tissue regeneration (Fig. 48.3). We have previously shown that MSCs, from a small marrow biopsy of an adult, can be passaged for over 30 population doublings *in vitro* and expanded in number by over 1 billion-fold without loss of testable osteogenic potential (Bruder *et al.*, 1997b). Based on experiments indicating that MSCs are present in young adults at a level of 200 per milliliter of whole marrow, this billion-fold expansion is equivalent in MSC number to many thousands of *liters* of harvested

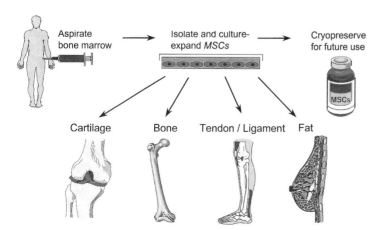

Fig. 48.3. Regenerative tissue therapy. Marrow is harvested and MSCs purified and seeded into cell culture, where they proliferate. These cells can be cryopreserved or used for cartilage resurfacing, bone repair, tendon/ligament repair, or adipocytic implantation. Suspensions of cells may also be injected for the treatment of osteoporosis or as an adjuvant therapy to bone marrow transplantation to improve the quality of marrow stroma.

Table 48.1. Cell surface molecular profile of human mesenchymal stem cells [a]

Molecules present	Molecules absent
Integrins	
α1,α2,α3,α5,α6,aα,αv,β1,β3,β4	Cβ2,α4,αL
Growth factor and cytokine receptors	
bFGFR,PDGFR, IL-1R, IL-3Rα, IL4-R, IL-6R, IL-7R, IFN-γR, TNFIR and TNFIIR, TGF-βIR and TGF-βIIR, bFGFR, PDGFR	EGFR-3, IL-2R, Fas ligand
Cell adhesion molecules	
ICAM-1 and -2, VCAM-1, L-selectin, LFA-3, ALCAM, endoglin, CD44	ICAM-3, Cadherin-5, E-selectin, P-selectin, PECAM-1
Miscellaneous antigens	
Transferrin receptor, CD9, Thy-1, SH-3, SH-4, SB-20, SB-21	CD4,CD14, CD34, CD45, von Willebrand factor

[a]Adapted from Bruder *et al.* (1998a) and Pittenger *et al.* (1999). Abbreviations: bFGFR, basic fibroblast growth factor receptor; PDGFR, platelet-derived growth factor receptor; IL-R, interleukin receptor; IFN-γR, interferon-γ receptor; TNFIR, tumor necrosis factor I receptor; TNFIIR, tumor necrosis factor II receptor; TGF-βIR, transforming growth factor-β I receptor; TGF-βIIR, transforming growth factor-β II receptor; EGFR-3, epidermal growth factor receptor 3; ICAM, intercellular adhesion molecule; VCAM, vascular cell adhesion molecule; LFA-3, lymphocyte function-related antigen-3, ALCAM, leukocyte cell adhesion molecule, PECAM, platelet endothelial cell adhesion molecule; CD, cluster designation.

whole marrow. The entire adult human body contains less than 6 liters of whole blood and obviously much less marrow. Thus, after only two to three subcultivations, enough MSCs could be generated to provide a sufficient number of osteoprogenitor cells ($3–5 \times 10^8$) to fill a massive bony defect in adults; this number of MSCs is more than those present in the entire adult human body. We have extended these observations to provide a detailed analysis of the surface molecules that characterize culture-expanded human MSCs (Table 48.1) (Pittenger *et al.*, 1999; Bruder *et al.*, 1998a). This work stems from our continued effort to document the changes that occur in cell surface architecture as a function of lineage progression (Bruder *et al.*, 1997a, 1998d). The profile of cell adhesion molecules, growth factor and cytokine receptors, and miscellaneous antigens serves to establish the unique phenotype of these cells, and provides a basis for exploring the function of selected molecules during osteogenic and other lineage progression.

In view of the fact that MSCs are believed to be the source of osteoblastic cells during the processes of normal bone growth, remodeling, and fracture repair in humans (Owen, 1985; Bruder *et al.*, 1994; Hollinger *et al.*, 1998), we have used them as a model to study aspects of osteogenic differentiation. When cultivated in the presence of osteogenic supplements (OSs) (dexamethasone, ascorbic acid, and β-glycerophosphate), purified MSCs undergo a developmental cascade defined by the acquisition of cuboidal osteoblastic morphology, transient induction of alkaline phosphatase (APase) activity, and deposition of a hydroxyapatite-mineralized extracellular matrix (Jaiswal *et al.*, 1997) (Fig. 48.4A–C). Gene expression studies illustrate that APase is transiently increased, type I collagen is down-regulated during the late phase of osteogenesis, and osteopontin is up-regulated at the late phase (Bruder *et al.*, 1998a) (Fig. 48.4D). Similarly, bone sialoprotein and osteocalcin (Jaiswal *et al.*, 1997) are up-regulated late in the differentiation cascade, and osteonectin is constitutively expressed. Additional studies detail the growth kinetics and high replicative capacity of these cells, which do not lose their osteogenic potential following a 1 billion-fold expansion or cryopreservation (Bruder *et al.*, 1997b). We have documented that OS-treated MSCs secrete a small-molecular-weight osteoinductive factor into their conditioned medium, which is capable of stimulating osteogenesis in naive cultures (Jaiswal and Bruder, 1997), similar to that reported for rat marrow stromal cells directed into the osteogenic lineage (Hughes and McCulloch, 1991). We have completed a comprehensive series of pulse–chase and transient exposure experiments using dexamethasone to determine which steps of the lineage pathway are dependent on exogenous factors, and which are supported by either (1) paracrine/autocrine fac-

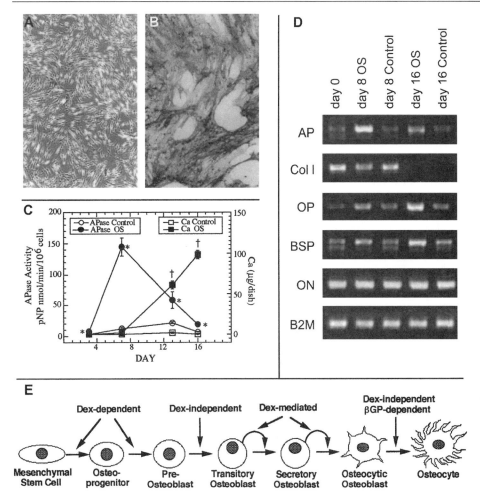

Fig. 48.4. Osteogenic differentiation of human MSCs in vitro. Phase-contrast photomicrographs of (A) human MSC cultures under growth conditions display characteristic spindle-shaped morphology and uniform distribution (unstained; ×18), and (B) human MSC cultures grown in the presence of osteoinductive supplements for 16 days and stained for APase and mineralized matrix. APase staining appears gray in these micrographs (originally red) and mineralized matrix appears dark (APase and von Kossa histochemistry; ×45). (C) Mean APase activity and calcium deposition of MSC cultures grown in control or OS medium and harvested on days 3, 7, 13, and 16 (n = 3). The vertical bars indicate standard deviations; *P < 0.05; †P < 0.005 (compared to control). (D) Expression of characteristic osteoblast mRNAs during in vitro osteogenesis. Reverse transcriptase–polymerase chain reactions using oligonucleotide primers specific for selected bone-related proteins were performed with RNA isolated at the indicated times. (E) Diagrammatic representation of the stages of dexamethasone (Dex)-induced osteogenic differentiation of MSCs in vitro.

tors in culture or (2) sustained lineage progression events following brief exposure to dexamethasone (Bruder and Jaiswal, 1995; Jaiswal and Bruder, 1996). A diagrammatic representation of these results is presented in Fig. 48.4E).

DELIVERY OF OSTEOPROGENITOR CELLS

It is clear that there is a coordinate relationship between bone formation and vasculature (Drushel *et al.,* 1985; Caplan and Pechak, 1987; Collin-Osdoby, 1994). It appears that vasculature provides the orientation for the secretory activity of osteoblasts and, hence, in engineering the delivery of osteoprogenitor cells to a reparative site, adequate space must be provided for the close juxtapositioning of capillaries to sheets of osteoprogenitor cells. Of importance to note here is the observation that if MSCs, at an *in vivo* site, aggregate into a dense mass in the absence of intervening capillaries, they differentiate into cartilage (Dennis and Caplan, 1993). Indeed, no one, to

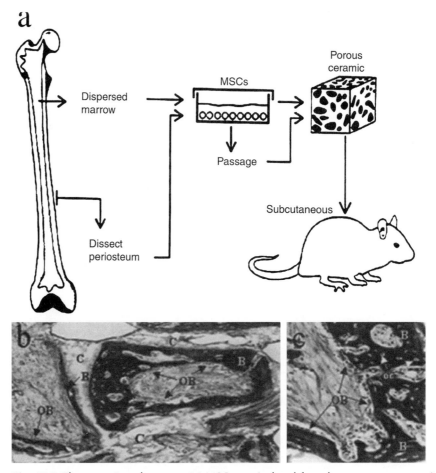

Fig. 48.5. The ceramic cube assay. (a) MSCs are isolated from bone marrow or periosteum and seeded into culture, where they adhere and mitotically expand. As the growing cells reach confluence, they are exposed to trypsin to liberate them from the dish and are either returned to new dishes (where they will continue to divide) or seeded into 3-mm porous calcium phosphate ceramic cubes. The cubes are then implanted into subcutaneous pouches in syngeneic or immuno-compromised rodents. The cubes are harvested 3 to 6 weeks later and scored for the presence of bone and/or cartilage. (b) Histologic appearance of porous calcium phosphate ceramic loaded with MSCs and implanted subcutaneously in a syngeneic rat host. Six weeks after implantation bone is observed lining the surface of a pore. Vasculature is present at the center of the pore, and directs secretory osteoblasts (OB) to secrete bone matrix (B) onto the surface of the ceramic (C), which has been decalcified during histologic processing and thus appears as a grey granular material. Osteocytes are also present in regions of bone formation that are more advanced. (c) Illustration, at higher magnification, of active osteogenesis, with prominent palisades of osteoblasts (OB) laying down new bone matrix (B) with inclusion of osteocytes (oc). Mallory–Heidenhain staining.

our knowledge, has been successful in duplicating or mimicking the natural arrangement of sheets of secretory osteoblasts in a culture system. We suggest that this lack of success is due to our lack of detailed understanding about the key parameters involved in cell–cell, cell–matrix, and/or cell–vasculature relationships involved in osteogenic activity of secretory osteoblasts. Thus, it is important to construct a delivery system that has the potential to place sheets of osteoprogenitor cells in a configuration such that host vasculature can quickly position itself to provide the appropriate cueing. Likewise, the cell attachment face of the delivery vehicle should be a bonelike substance based on the view that such an environment could assist in the orientation of the osteoblasts' secretory products.

The success of such a cell-based approach for tissue engineering of bone is critically dependent on the development of an appropriate matrix scaffold for cell delivery. The ideal matrix formulation should possess several properties: (1) the material should foster uniform cell loading, cell

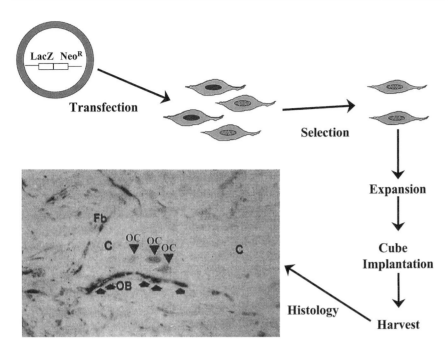

Fig. 48.6. LacZ expression and bone formation in vM5LacZ-transduced human MSCs. Human MSCs were transfected with a lacZ-containing retrovirus, selectively expanded in vitro, combined with ceramic cubes, and implanted in SCID mice. Mice were sacrificed and ceramic–cell composites were harvested 6 weeks later, X-Gal-stained for expression of LacZ, decalcified, embedded, sectioned, and counterstained with Mallory–Heidenhain. Osteoblasts (OB) are seen lining the surface of the ceramic (C), which also contains labeled osteocytes (OC) (×200).

division, and retention; (2) the scaffold should support rapid vascular invasion; (3) the matrix should be designed to orient the formation of new bone in anatomically relevant shapes; (4) the composition of materials should be resorbed and replaced by new bone as it is formed; (5) the material should be radiolucent to allow the new bone to be distinguished radiographically from the original implant; (6) the formulation should encourage osteoconduction with host bone; and (7) it should possess desirable handling properties for the specific clinical indication.

This laboratory long ago explored the use of porous calcium phosphate ceramics together with whole marrow as a delivery vehicle for osteogenic cells at both heterotopic and orthotopic sites (Ohgushi *et al.*, 1989a,b). Based on this experience with whole marrow, we further refined the technology to allow us to efficiently situate culture-expanded MSCs onto the pore surfaces of such ceramic vehicles (Haynesworth *et al.*, 1992b; Dennis and Caplan, 1993; Dennis *et al.*, 1992). Figure 48.5a shows the outline of experiments wherein marrow or periosteal MSCs are obtained, culture expanded, and, with serial subculturing (passages), placed into a 3-mm porous ceramic cube, which is subsequently implanted into subcutaneous pockets in syngeneic or immunocompromised rodent hosts. We have used such ceramic cubes to assay the osteogenic potential of rodent, rabbit, goat, sheep, canine, and human MSCs (Lennon *et al.*, 1996; Kadiyala *et al.*, 1997b). Figure 48.5 shows the histologic appearance of newly forming bone in the pores of such a ceramic. The sheets are oriented, and secretory osteoblasts are clearly visible as new bone. Using cells that have been tagged by retroviral transfection with a reporter gene for β-galactosidase (*lacZ*), labeled osteoblasts and osteocytes derived from these MSCs can be observed in Fig. 48.6. Of importance is the fact that the stained cells were derived from labeled marrow-derived human MSCs, and that the characteristic cytoplasmic processes of mature osteocytes are observed in the newly fabricated bone within the pore of the ceramic vehicle. Notably, the osseous tissue that is formed in these implants is donor derived, and not dependent on the presence of host-derived osteoprogenitors (Allay *et al.*, 1997). This cube assay has now evolved into a semiquantitative assay to judge MSC number and/ or osteochondrogenic potential (Dennis *et al.*, 1998; Cassiede *et al.*, 1996). To observe bone within the pores of the implant ceramic does not require an absolutely pure population of MSCs, because diluting the human MSCs by 50% with human dermal fibroblasts does not seem to affect the amount of bone in the ceramic cubes (Lennon *et al.*, 1999).

BONE REGENERATION BY MSCs

To show clinical applicability of tissue regeneration therapies using MSCs, a series of preclinical studies was performed. These investigations, all aimed at achieving osseous regeneration

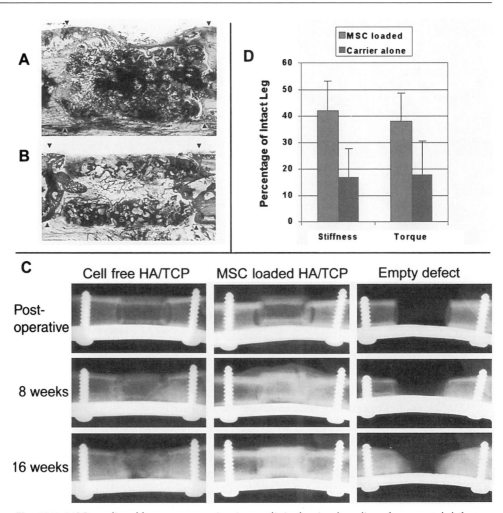

Fig. 48.7. MSC-mediated bone regeneration in preclinical animal studies of segmental defect repair. (A) Rat segmental defects fitted with a HA/TCP carrier loaded with syngeneic rat mesenchymal stem cells form a solid osseous union with the host, and contain substantial new bone throughout the pores by 8 weeks. (B) Defects fitted with a cell-free HA/TCP carrier do not contain bone within the pores of the implant, nor is there significant union at the interfaces, indicated by the arrowheads (stain, toluidine blue-O; ×16). (C) Radiographic appearance of bone healing in a 21-mm canine femoral gap defect 16 weeks after surgery. Animals that did not receive an implant had a fibrous nonunion. Animals that received a mesenchymal stem cell-loaded HA/TCP cylinder had a substantial amount of bone regeneration at the defect site, including a periimplant callus that remodeled to the size of the original bone by 16 weeks. Those animals receiving cell-free implants did not achieve healing of their defects, as is evident by the lack of new bone and the multiple fractures throughout the body of the implant material. (D) Results of biomechanical torsion testing performed on athymic rat femurs 12 weeks after implantation with human mesenchymal stem cell-loaded ceramics. Load to failure and overall stiffness was more than twice that of controls at the same time.

in a critical-sized defect of the femur, were designed to compare the efficacy and effective dosing when the cellular source was escalated from rodents to large animals, and eventually to humans. In the first study, culture-expanded, marrow-derived MSCs were used to repair a segmental defect in the femur of rats (Kadiyala *et al.*, 1997a). Syngeneic MSCs at a density of 7.5×10^6 cells/ml were loaded onto a porous hydroxyapatite/β-tricalcium phosphate cylinder and implanted in an 8-mm-long diaphyseal defect. By 8 weeks, nearly every pore of the implants contained considerable new bone (Fig. 48.7A). In contrast, cell-free implants were well vascularized, but displayed little if any bone formation within the pores (Fig. 48.7B). In the defects treated with MSC-loaded implants, substantial new bone formation occurred at the interface between the host and the implant, leading to a continuous span of bone across the defect. Furthermore, a periosteal callus of bone also was present in samples loaded with MSCs but not with cell-free implants. These studies established the proof of principle for MSC-based tissue regeneration therapy in bone.

The ability of MSC-loaded implants to repair defects in larger animals was examined in a canine femoral gap defect model (Bruder *et al.,* 1998b; Kraus *et al.,* 1999). The healing of a 21-mm osteoperiosteal defect was studied using ceramic implants made from porous hydroxyapatite/β-tricalcium phosphate that had been loaded with autologous culture-expanded MSCs at a density of 7.5×10^6 cells/ml. At 16 weeks, atrophic nonunion occurred in all of the femurs that had untreated defects, and only a small amount of trabecular bone formed at the cut ends of the cortex of the host bone in this group (Fig. 48.7C). In contrast, radiographic union was established rapidly at the interface between the host bone and the implants in samples that had been loaded with MSCs. Numerous fractures, which became more pronounced with time, developed in the implants that had not been loaded with cells. Significantly more bone was found in the pores of the implants loaded with MSCs than in the cell-free implants. In addition, a large collar of bone formed around the implants that had been loaded with cells; this collar became integrated and contiguous with callus that formed in the region of the periosteum of the host bone. The collar of bone was remodeled during the study, ultimately resulting in a size and shape that were comparable with the segment of bone that had been resected. Osseous callus did not develop around the cortex of the host bone or around the implant in any of the specimens in the control groups.

Culture expanded *human* MSCs also have been shown to effect bone repair in a segmental gap defect model (Bruder *et al.* 1998c). Human MSCs were loaded onto ceramic cylinders similar to those described above, and used to repair an identical femoral defect created in athymic rats. Because these rats lack T lymphocytes, transplantation of xenogeneic material does not lead to graft rejection. Like the studies using syngeneic rat MSCs, substantially more bone was found in defects that received cylinders loaded with human MSCs than in those that received cell-free implants. With the MSC-loaded implants, complete union between bone and implant was achieved at 12 weeks. Before that time, there was an increasing amount of bone formed between the 4- and 8-week evaluation points. Biomechanical testing confirmed that femurs implanted with MSC-loaded ceramics were significantly stronger than those that received cell-free ceramics (Fig. 48.7D). Measurement of stiffness and torque showed increases of 245 and 212%, respectively, in the MSC-loaded samples compared with the cell-free carrier samples. These studies show that human and rat MSCs are capable of healing a clinically significant bone defect in a well-established model of bone repair.

The three studies described above show the success of using mesenchymal stem cells to regenerate bone in segmental defects. Using rat, canine, or human mesenchymal stem cells, bone fill ranged from 40 to 47% of available space, a value not likely to be exceeded given the nature of the scaffold and the requirement for vasculature and associated soft tissue (Kadiyala *et al.,* 1997a; Bruder *et al.,* 1998b,c). The cellular doses required to effect bone repair were constant across species, and the process of bone repair remained the same. The ability of the therapeutic approach to scale up in a predictable manner increases one's confidence that MSC-based therapies for bone repair will be effective in the treatment of human disease.

These studies indicate that culture-expanded MSCs can be presented in sufficient numbers in a supportive delivery vehicle to provide bony regeneration of a large, osteoperiosteal femoral defect in both small and large animals. Figure 48.8 depicts how this experience may potentially be used to construct an *autologous* cell–matrix composite for excision repair of femoral gaps and other osseous defects in humans. In the case pictured, we suggest that the use of an external fixator could allow load bearing by the patient shortly after implantation. If the fixator could be constructed to transfer partial load incrementally to the implant, the fixator could be sequentially offloaded so that more and more load can be gradually transferred to the implant. As the MSCs differentiate into osteoblasts and fabricate bone onto the walls of the porous implant, the incremental loading would serve to orient the new bone and stimulate its acquisition of mechanical integrity. Eventually, the implant would carry 100% of the load and the external fixator could be removed so that no metal devices would remain. The delivery vehicle would be naturally replaced as osteoclasts resorb its matrix and osteoblasts replace it with new, host-derived bone as governed by the natural turnover sequence. The selection of appropriate biomaterials and configurations for each formulation will depend on the anatomic site for regeneration, the mechanical loads present at the site, and the desired rate of incorporation.

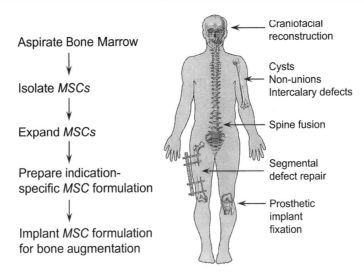

Aspirate Bone Marrow

Isolate *MSCs*

Expand *MSCs*

Prepare indication-
specific *MSC* formulation

Implant *MSC* formulation
for bone augmentation

Craniofacial
reconstruction

Cysts
Non-unions
Intercalary defects

Spine fusion

Segmental
defect repair

Prosthetic
implant
fixation

Fig. 48.8. Bone regeneration paradigm. MSCs are isolated from bone marrow and expanded in number in culture. When sufficient cells are available, they are loaded into an appropriate delivery vehicle and surgically inserted into the site requiring bone augmentation. In the case of segmental defect repair, an external fixator can be applied to initially carry all of the load by a device capable of the stepwise transfer of partial load to the implant. As bone forms in the implant, the load on the implant is sequentially increased to assist in orienting and stimulating bone formation. Eventually, 100% of the load is carried across the regenerated bone, at which time the external fixator is removed. In other clinical settings, fixation may not be necessary, or may take the form of simple casting.

It may also be possible to use allogeneic MSCs for these massive osseous repairs. In this case, the immunoreactive groups on the MSCs would need to be muted or inactive. The availability of allogeneic cell therapy, or universal donor cells, would provide easy access for trauma-related skeletal reconstruction or for gene therapy, as discussed below. In addition, we would propose that such allogeneic cells could be incubated in osteogenic supplement prior to implantation to cause them to all enter the osteogenic pathway. These osteogenic or jump-started cells could be loaded into the delivery vehicle, thereby accelerating the bone formation. In addition, the allo-MSCs would all become either osteocytes or would expire as osteoblasts. This accelerated bone formation would attract host-derived osteogenic cells to the repair site and the bulk of the repair bone would then become of host origin. Eventually, when the repair bone turned over, the allogeneic osteocyte-containing bone would be replaced by host bone, and immunosuppression therapy, if used, could be discontinued.

MSCs AND GENE THERAPY

Figure 48.9 shows the hypothetical treatment of a patient with a genetic defect such as osteogenesis imperfecta. MSCs would be obtained from the patient, and, as they are mitotically expanded in culture, they could be transduced/transfected with a retrovirus containing the gene for normal type I collagen with its appropriate promoters. Such transfections have already been shown to allow the insertion of a selectable gene (*neo*) and a reporter gene (bacterial β-galactosidase) into human MSCs, without loss of the *in vivo* osteogenic potential of these transduced cells, as shown in Fig. 48.5 (Allay *et al.,* 1993, 1997). In order to effect efficient repopulation of the marrow by MSCs, the marrow must be injured. Hematopoietic progenitors could be obtained by current apheresis techniques (Brugger *et al.,* 1995) and the patient could then be given chemotherapy or radiation to compromise the bone marrow, followed by a transplantation of autologous hematopoietic progenitors to which the culture-expanded MSCs have been added. These cells would home to the bone marrow, reestablish themselves (Pereira *et al.,* 1995), and thus provide mesenchymal progenitors with the inserted normal type I collagen gene. As osteoblasts naturally and normally expire, the new osteoblasts would arise from transduced MSCs and would, hence, be "normal" relative to the synthesis of type I collagen and osteoid. This strategy depends on the

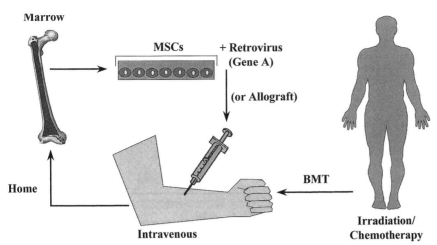

Marrow

MSCs   + Retrovirus
        (Gene A)

        (or Allograft)

        BMT

Home

Intravenous

Irradiation/
Chemotherapy

Fig. 48.9. Autologous gene therapy. MSCs are obtained from bone marrow and, as they proliferate in culture, a retrovirus or other vector is used to insert Gene A into the MSCs. The patient is exposed to radiation or chemotherapy to injure the marrow and to decrease the body pool of MSCs. Both autologous hematopoietic progenitors and MSCs are introduced intravenously, and they home to the bone marrow. Thereafter, as differentiated mesenchymal cells expire, they are replaced by newly differentiated mesenchymal cells, which originate from the MSC pool. These genetically altered MSCs will thus provide genetically altered progenies, which give rise to "normal" differentiated cells that express Gene A. Alternately, allogeneic MSCs with allogeneic hemapoietic progenitors, or allogeneic whole marrow, can be used to introduce genetically normal MSCs, as has been reported for OI patients (Horwitz et al., 1999).

normal local *cueing* and the control of osteogenesis in the recovering and growing young patient. As long as the appropriate genetic control elements for type I collagen are present and the inserted gene does not interfere with other genes, this MSC transduction could, in principle, provide a molecularly engineered treatment for incurable bone diseases such as osteogenesis imperfecta. As previously stated, allogeneic bone marrow transplantation from normal donors could, likewise, accomplish this aim, as has already been reported (Horwitz *et al.,* 1999).

PREVENTION OF AGE-RELATED BONE LOSS

Studies indicate that estrogen deficiency, which is involved in osteoporosis, leads to a diminution in the number of osteoprogenitors by virtue of their decreased proliferative potential (Tabuchi *et al.,* 1986). Similarly, the bone loss that occurs as a function of aging is likely due to the decrease in the number of resident MSCs. In view of this, early replenishment of the osteoprogenitor cell supply could serve to prevent, or even reverse, the bone loss associated with aging and osteoporosis. With this in mind, we have developed techniques and conditions for extensive subcultivation, without lineage progression, of human MSCs (Haynesworth *et al.,* 1992a,b; Lennon *et al.,* 1996). We have also shown that these cells can be cryopreserved and subsequently thawed for later use, with no loss in osteogenic potential (Bruder and Jaiswal, 1996; Bruder *et al.,* 1997b). Although it is advantageous to be able to increase dramatically the number of stemlike cells *ex vivo* for some clinical conditions, it may be desirable in other scenarios to stimulate osteogenic commitment and differentiation *in vitro* prior to reimplantation. This latter treatment could serve to hasten the *in vivo* lineage progression and thereby lead to more rapid and consistent bone formation. In the case of osteoporosis, however, providing culture-expanded autologous MSCs to the patient through an intravenous (Lazarus *et al.,* 1995) or intramedullary route should lead to a renewed source of endogenous MSCs. In this case, *in vitro* induction of osteogenesis would be undesirable, because the commitment of these cells would not allow their self-perpetuation as MSCs following reimplantation or infusion into the host. By repopulating the marrow reservoir with a "youthful" cache of undifferentiated mesenchymal progenitor cells, it should be possible for the body to recruit these cells into and maintain the normal bone remodeling cycle. Preliminary experiments from our laboratories, employing a standardized rat ovariectomy model, indicate that bone loss can be prevented via the systemic administration of syngeneic MSCs in rats (Fig. 48.10). Such a renewed

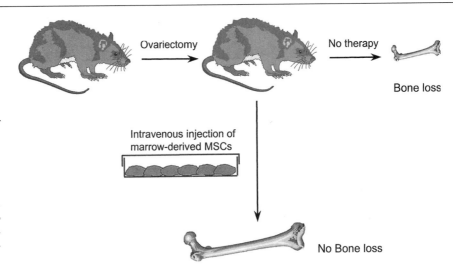

Fig. 48.10. Surgical removal of the ovaries from an adult causes the loss of circulating estrogen and subsequent bone loss. Culture-expanded MSCs can be introduced intravenously within a few weeks following ovariectomy to protect against such bone loss as measured by densitometric scanning or quantitative histomorphometry.

source of endogenous MSCs would also provide the cellular building blocks for the needs of other mesenchymal tissues. The utility of this self-cell therapy is that, because of our ability to manipulate these cells *ex vivo,* patients can indefinitely store their cryopreserved MSCs following a single harvest and then receive multiple bone-maintenance doses at appropriate intervals thereafter. Finally, we hope to engineer these same cell populations molecularly *in vivo* by providing the critical cues that instruct these cells first to divide, to travel to the proper sites, and to then form bone.

CONCLUSION

The key to successful repair/regeneration of bone is to provide the repair site with sufficient osteogenic progenitor cells in a suitable delivery vehicle to ensure osteoblastic differentiation and optimal biosynthetic activity of bone matrix and growth and differentiation factors. Exclusion of interfering tissues, promotion of rapid vascular penetrance, and provision of appropriate mechanical and other instructional cues must be engineered. As we gain more information about the identity and reactivity of all of the cueing factors, it may be possible to orchestrate massive bone regeneration by clever combinations of delivery vehicles and such instructional agents. The management of MSCs, both *in situ* and *ex vivo,* will be crucial to the success of such tissue engineering. It may be that in older patients, whose MSC titers are decreasing with age, it will not be possible to accomplish these feats of local tissue engineering without first increasing the reservoir of available progenitor cells *in situ.* To this end, the *in vitro* expansion and manipulation of human MSCs for self-cell therapy are now being perfected and may become the therapy of choice in the future. The twenty-first century holds the promise of tissue regeneration protocols that we can now only dream about.

ACKNOWLEDGMENTS

We thank our colleagues in the Skeletal Research Center and Osiris Therapeutics, Inc. for contributing to the studies described in this review. We extend special thanks to Victor M. Goldberg, Stephen E. Haynesworth, Neelam Jaiswal, Sudha Kadiyala, James E. Dennis, and Luis Solchaga. Studies reported here were funded in part by the National Institutes of Health and Osiris Therapeutics, Inc. Some of the technologies described herein have been transferred to Osiris Therapeutics, Inc., in which both authors hold an equity interest.

REFERENCES

Allay, J. A., Dennis, J. E., Haynesworth, S. E., Clapp, D. W., Lazarus, H. M., Caplan, A. I., and Gerson, S. L. (1993). Retroviral transduction of marrow-derived mesenchymal precursors. *Blood* **82**(10, Suppl. 1), 4779.

Allay, J. A., Dennis, J. E., Haynesworth, S. E., Majumdar, M., Clapp, D. W., Caplan, A. I., and Gerson, S. L. (1997). LacZ and interleukin-3 expression *in vitro* after retroviral transduction of marrow-derived human osteogenic mesenchymal progenitors. *Hum. Gene Ther.* **8**, 1417–1427.

Beresford, J. (1989). Osteogenic stem cells and the stromal system of bone and marrow. *Clin. Orthop. Relat. Res.* **240**, 270–280.

Brighton, C. T. (1984). The biology of fracture repair. *In* "The American Academy Orthopaedic Surgeons Instructional Course Lectures" (J. A. Murray and C. V. Mosby, eds.), Vol. 33, pp. 60–106. AAOS, St. Louis, MO.

Bruder, S. P., and Jaiswal, N. (1995). Transient exposure of human mesenchymal stem cells to dexamethasone is capable of inducing sustained osteogenic lineage progression. *J. Bone Miner. Res.* **10,** S41.

Bruder, S. P., and Jaiswal, N. (1996). The osteogenic potential of human mesenchymal stem cells is not diminished after one billion-fold expansion *in vitro. Trans. Orthop. Res. Soc.* **21,** 580.

Bruder, S. P., Gazit, D., Passi-Even, L., Bab, I., and Caplan, A. I. (1990). Osteochondral differentiation and the emergence of stage-specific osteogenic cell-surface molecules by bone marrow cells in diffusion chambers. *Bone Miner.* **11,** 141–151.

Bruder, S. P., Fink, D. J., and Caplan, A. I. (1994). Mesenchymal stem cells in bone formation, bone repair, and skeletal regeneration therapy. *J. Cell. Biochem.* **56**(3), 283–294.

Bruder, S. P., Horowitz, M. C., Mosca, J. D., and Haynesworth, S. E. (1997a). Monoclonal antibodies reactive with human osteogenic cell surface antigens. *Bone* **21,** 225–235.

Bruder, S. P., Jaiswal, N., and Haynesworth, S. E. (1997b). Growth kinetics, self-renewal and the osteogenic potential of purified human mesenchymal stem cells during extensive subcultivation and following cryopreservation. *J. Cell. Biochem.* **64**(2), 278–294.

Bruder, S. P., Jaiswal, N., Ricalton, N. S., Kraus, K. H., and Kadiyala, S. (1998a). Mesenchymal stem cells in osteobiology and applied bone regeneration. *Clin. Orthop. Relat. Res.* **355**(Suppl.); S247–S256.

Bruder, S. P., Kraus, K. H., Goldberg, V. M., and Kadiyala, S. (1998b). The effect of implants loaded with autologous mesenchymal stem cells on the healing of canine segmental bone defects. *J. Bone Jt. Surg., Am. Vol.* **80-A,** 985–996.

Bruder, S. P., Kurth, A. A., Shea, M., Hayes, W. C., Jaiswal, N., and Kadiyala, S. (1998c). Bone regeneration by implantation of purified, culture-expanded human mesenchymal stem cells. *J. Orthop. Res.* **16,** 155–162.

Bruder, S. P., Ricalton, N. S., Boynton, R., Connolly, T. J., Jaiswal, N., Zaia, J., and Barry, F. P. (1998d). Mesenchymal stem cell surface antigen SB-10 corresponds to activated leukocyte cell adhesion molecule and is involved in osteogenic differentiation. *J. Bone Miner. Res.* **13**(4), 655–663.

Brugger, W., Heimfeld, S., Berenson, R. J., Mertelsmann, R., and Kanz, L. (1995). Reconstitution of hematopoiesis after high-dose chemotherapy by autologous progenitor cells generated *ex vivo. N. Engl. J. Med.* **333**(5), 283–287.

Byers, P., and Bonadio, J. (1985). The molecular basis of clinical heterogeneity of osteogenesis imperfecta: Mutations in type I collagen genes have different effects on collagen processing. *In* "Genetic and Metabolic Diseases in Pediatrics" (J. Lloyd and C. Scriver, eds.), pp. 56–90. Butterworth, London.

Caplan, A. I. (1991). Mesenchymal stem cells. *J. Orthop. Res.* **9,** 641–650.

Caplan, A. I. (1994). The mesengenic process. *Clin. Plast. Surg.* **21**(3), 429–435.

Caplan, A. I., and Pechak, D. (1987). The cellular and molecular biology of bone formation. *In* "Bone and Mineral Research" (W. A. Peck, ed.), Vol. 5, pp. 117–183. Elsevier, New York.

Caplan, A. I., Fink, D. J., Goto, T., Linton, A. E., Young, R. G., Wakitani, S., Goldberg, V. M., and Haynesworth, S. E. (1993). Mesenchymal stem cells and tissue repair. *In* "The Anterior Cruciate Ligament: Current and Future Concepts" (D. Jackson, S. Arnoczky, S. Woo, and C. Frank, eds.), pp. 405–417. Raven Press, New York.

Cassiede, P., Dennis, J. E., Ma, F., and Caplan, A. I. (1996). Osteochondrogenic potential of marrow mesenchymal progenitor cells exposed to TGF-β or PDGF-BB as assayed *in vivo* and *in vitro. J. Bone Miner. Res.* **11,** 1264–1273.

Collin-Osdoby, P. (1994). Role of vascular endothelial cells in bone biology. *J. Cell. Biochem.* **55,** 304–309.

Dennis, J. E., and Caplan, A. I. (1993). Porous ceramic vehicles for rat-marrow-derived (*Rattus norvegicus*) osteogenic cell delivery: Effects of pre-treatment with fibronectin or laminin. *J. Oral Implant.* **19,** 106–115.

Dennis, J. E., and Caplan, A. I. (1996). Differentiation potential of conditionally immortalized mesenchymal progenitor cells from adult marrow of a H-2K^b–tsA58 transgenic mouse. *J. Cell. Physiol.* **167,** 523–538.

Dennis, J. E., Haynesworth, S. E., Young, R. G., and Caplan, A. I. (1992). Osteogenesis in marrow-derived mesenchymal cell porous ceramic composites transplanted subcutaneously: Effect of fibronectin and laminin on cell retention and rate of osteogenic expression. *Cell Transplant.* **1,** 23–32.

Dennis, J. E., Konstantakos, E. A., Arm, D., and Caplan, A. I. (1998). *In vivo* osteogenic assay: A rapid method for quantitative analysis. *Biomaterials* **19,** 1323–1328.

Dennis, J. E., Meriam, A., Awadallah, A., and Caplan, A. I. (1999). A quadripotential mesenchymal progenitor cell isolated from the marrow of an adult mouse. *J. Bone Miner. Res.* **14,** 1–10.

Drushel, R. F., Pechak, D. G., and Caplan, A. I. (1985). The anatomy, ultrastructure and fluid dynamics of the developing vasculature of the embryonic chick wing bud. *Cell Differ.* **16,** 13–28.

Egrise, D., Martin, D., Vienne, A., Neve, P., and Schoutens, A. (1992). The number of fibroblastic colonies formed from bone marrow is decreased and the *in vitro* proliferation rate of trabecular bone cells increased in aged rats. *Bone* **13,** 355–361.

Friedenstein, A. J. (1976). Precursor cells of mechanocytes. *Int. Rev. Cytol.* **47,** 327–355.

Hamm, A. W. (1930). A histological study of the early phases of bone repair. *J. Bone Jt. Surg.* **12,** 827–844.

Haynesworth, S. E., Baber, M. A., and Caplan, A. I. (1992a). Cell surface antigens on human marrow-derived mesenchymal cells are detected by monoclonal antibodies. *Bone* **13,** 69–80.

Haynesworth, S. E., Goshima, J., Goldberg, V. M., and Caplan, A. I. (1992b). Characterization of cells with osteogenic potential from human marrow. *Bone* **13,** 81–88.

Haynesworth, S. E., Goldberg, V. M., and Caplan, A. I. (1994). Diminution of the number of mesenchymal stem cells as a cause for skeletal aging. *In* "Musculoskeletal Soft-Tissue Aging: Impact on Mobility" (J. A. Buckwalter, V. M. Goldberg, and S. L. Y. Woo, eds.), Sec. 1, Chapter 7, pp. 80–86. AAOS, Rosemont, IL.

Hollinger, J., Buck, D. C., and Bruder, S. P. (1998). Biology of bone healing: Its impact on clinical therapy. *In* "Applications of Tissue Engineering in Dentistry" (S. Lynch, ed.), pp. 17–53. Quintessence Publishing Co., Chicago.

AQ 1537

Horwitz, E. M., Prockop, D. J., Fitzpatrick, L. A., Koo, W. W., Gordon, P. L., Neel, M., Sussman, M., Orchard, P., Marx, J. C., Pyeritz, R.E., and Brenner, M. K. (1999). Transplantability and therapeutic effects of bone marrow-derived mesenchymal cells in children with osteogenesis imperfecta. *Nat. Med.* **5,** 309–313.

Hughes, F., and McCulloch, C. A. G. (1991). Stimulation of the differentiation of osteogenic rat bone marrow stromal cells by osteoblast cultures. *Lab. Invest.* **64,** 617–622.

Jaiswal, N., and Bruder, S. P. (1996). The pleiotropic effects of dexamethasone on osteoblast differentiation depend on the developmental state of the responding cells. *J. Bone Miner. Res.* **11,** S259.

Jaiswal, N., and Bruder, S. P. (1997). Human osteoblastic cells secrete paracrine factors which regulate differentiation of osteogenic precursors in marrow. *Trans. Orthop. Res. Soc.* **22,** 524.

Jaiswal, N., Haynesworth, S. E., Caplan, A. I., and Bruder, S. P. (1997). Osteogenic differentiation of purified, culture-expanded human mesenchymal stem cells *in vitro. J. Cell. Biochem.* **64**(2), 295–312.

Kadiyala, S., Jaiswal, N., and Bruder, S. P. (1997a). Culture-expanded, bone marrow-derived mesenchymal stem cells can regenerate a critical-sized segmental bone defect. *Tissue Eng.* **3,** 173–185.

Kadiyala, S., Young, R. G., Thiede, M. A., and Bruder, S. P. (1997b). Culture-expanded canine mesenchymal stem cells possess osteochondrogenic potential *in vivo* and *in vitro. Cell Transplant.* **6**(2), 125–134.

Kahn, A., Gibbons, R., Perkins, S., and Gazit, D. (1995). Age-related bone loss: A hypothesis and initial assessment in mice. *Clin. Orthop. Relat. Res.* **313,** 69–75.

Kojimoto, H., Yasui, N., Goto, T., Matsuda, S., and Shimomura, Y. (1988). Bone lengthening in rabbits by callus distraction. The role of periosteum and endosteum. *J. Bone Jt. Surg., Br. Vol.* **70B**(4), 543–549.

Kraus, K., Kadiyala, S., Wotton, H. M., Kurth, A., Shea, M., Hannan, M., Hayes, W. C., Kirker-Head, C. A., and Bruder, S. P. (1999). Critically sized osteo-periosteal femoral defects: A dog model. *J. Invest. Surg.* **12,** 115–124.

Lazarus, H. M., Haynesworth, S. E., Gerson, S. L., Rosenthal, N. S., and Caplan, A. I. (1995). *Ex-vivo* expansion and subsequent infusion of human bone marrow-derived stromal progenitor cells (mesenchymal progenitor cells) [MPCs]: Implications for therapeutic use. *Bone Marrow Transplant.* **16,** 557–564.

Lennon, D. P., Haynesworth, S. E., Bruder, S. P., Jaiswal, N., and Caplan, A. I. (1996). Development of a serum screen for mesenchymal progenitor cells from bone marrow. *In Vitro Cell. Dev. Biol.* **32,** 602–611.

Lennon, D. P., Arm, D., Caplan, A. I. (2000). Dilution of human mesenchymal stem cells with dermal fibroblasts and the effects on *in vitro* and *in vivo* osteogenesis. Submitted for publication.

Muschler, G. F., Boehm, C., Easley, K. (1997). Aspiration to obtain osteoblast progenitor cells from human bone marrow: The influence of aspiration volume. *J. Bone Jt. Surg.* **79**(11), 1699–1709.

Nakahara, H., Bruder, S. P., Haynesworth, S. E., Holecek, J., Baber, M., Goldberg, V. M., and Caplan, A. I. (1990). Bone and cartilage formation in diffusion chambers by subcultured cells derived from the periosteum. *Bone* **11,** 181–188.

Nakahara, H., Goldberg, V. M., and Caplan, A. I. (1992). Culture-expanded periosteal-derived cells exhibit osteochondrogenic potential in porous calcium phosphate ceramics *in vivo. Clin. Orthop. Relat. Res.* **276,** 291–298.

Ohgushi, H., Goldberg, V. M., and Caplan, A. I. (1989a). Heterotopic osteogenesis in porous ceramics induced by marrow cells. *J. Orthop. Res.* **7,** 568–578.

Ohgushi, H., Goldberg, V. M., and Caplan, A. I. (1989b). Repair of bone defects with marrow cells and porous ceramic: Experiments in rats. *Acta Orthop. Scand.* **60,** 334–339.

Oreffo, R. O., Bord, S., and Triffitt, J. T. (1998). Skeletal progenitor cells and ageing human populations. *Clin. Sci.* **94**(5), 549–555.

Owen, M. (1985). Lineage of osteogenic cells and their relationship to the stromal system. *In* "Bone and Mineral" (W. A. Peck, ed.), Vol. 3, pp. 1–25. Elsevier, Amsterdam.

Owen, M., and Friedenstein, A. J. (1988). Stromal stem cells: Marrow-derived osteogenic precursors. *In* "Cell and Molecular Biology of Vertebrate Tissues" (D. Evered and S. Harnett, eds.), Ciba Foundation Symposia, 136, pp. 42–60. Wiley, Chichester.

Pereira, R. F., Halford, K. W., O'Hara, M. D., Leeper, D. B., Sokolov, B. P., Pollard, M. D., Bagasva, O., and Prockop, D. J. (1995). Culture adherent cells from marrow can serve as long-standing precursor cells for bone, cartilage, and lung in irradiated mice. *Proc. Natl. Acad. Sci. U.S.A.* **92,** 4857–4861.

Pittenger, M. F., Mackay, A. M., Beck, S. C., Jaiswal, R. K., Douglas, R., Mosca, J. D., Moorman, M. A., Simonetti, D. W., Craig, S., and Marshak, D. R. (1999). Multilineage potential of adult human mesenchymal stem cells. *Science* **284,** 143–147.

Tabuchi, C., Simmons, D. J., Fausto, A., Russell, J., Binderman, I., and Avioli, L. (1986). Bone deficit in ovariectomized rats: Functional contribution of the marrow stromal cell population and the effect of oral dihydrotachysterol treatment. *J. Clin. Invest.* **78,** 637–642.

CHAPTER 49

ARTICULAR CARTILAGE INJURY

John M. McPherson and Ross Tubo

INTRODUCTION

The fact that articular cartilage has little or no capacity for effective repair following traumatic injury has been recognized by clinicians for more than 250 years (Hunter, 1743). Even today the natural history of the progression of joint degeneration following injury is not completely understood, but it is clear that injury to articular cartilage often leads to symptomatic pain and eventually to osteoarthritis (Dandy, 1991; Johnson, 1986b; Johnson-Nurse and Dandy, 1985). It appears that degeneration of articular cartilage following injury may be a consequence of an imbalance in the normally slow remodeling and maintenance processes provided by chondrocytes, which are imbedded in the dense extracellular matrix of the articular surface. The mechanisms responsible for tilting the balance of remodeling in favor of degradation, versus synthesis, are not understood, but may in part reflect local changes in biomechanical loading of chondrocytes following injury that profoundly influence cell metabolism.

It has been postulated that articular cartilage has a limited capacity for repair due to a limited supply of cells in the vicinity of the wound to mediate the repair process (for review, see Gillogly *et al.,* 1998). Unlike skin, for example, in which both vasculature and adjacent tissues provide cells for mediating the wound healing process, articular cartilage contains no vascular supply. In addition, articular chondrocytes, which are normally involved in articular cartilage synthesis and maintenance, are encased in a dense extracellular matrix that limits their mobility and their ability to contribute to the wound healing process. Synovial cells are present in synovial fluid, but apparently their numbers are too few or their biological properties are too limited to mediate adequate repair of any but the most minute cartilage defects (Hunziker and Rosenberg, 1996).

Numerous surgical procedures have been developed during the past 30 years in an effort to deal with the problem of cartilage injury, particularly following the advent of arthroscopic surgery. The most conventional approach to cartilage injury is debridement and lavage (Baumgaertner *et al.,* 1990; Bert and Maschka, 1989; Dandy, 1991; Rand, 1991; Timoney *et al.,* 1990). The primary objective of this procedure is to improve the contour of the cartilage defect by shaving the edges of the lesion and removing the cartilage debris generated either by the original injury or the debridement procedure. In general, clinical reports in the literature suggest that debridement and lavage provide immediate short-term relief of the symptoms in the majority of patients with small cartilage defects. The efficacy provided by the procedure fades with time, particularly in defects of greater than 1 cm², and many patients exhibit evidence of osteoarthritis, as judged by joint space narrowing.

Defects treated with debridement and lavage usually remain devoid of repair tissue unless the subchondral plate is penetrated during the debridement procedure. Based on this observation, surgeons developed several methods to penetrate the subchondral tissue (Fig. 49.1) with the objective of enabling mesenchymal cells from the bone marrow to migrate into the wound site and mediate the repair process. Abrasion arthroplasty involves the use of an arthroscopic burring device to remove tissue in the base of the wound to the level of subchondral bone (Bert and Maschka,

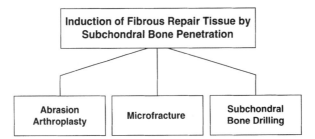

Fig. 49.1. Conventional surgical procedures utilized to treat damaged hyaline articular cartilage.

1989; Friedman *et al.,* 1984; Johnson, 1986a,b; Kim *et al.,* 1991). As with debridement and lavage, short-term results (i.e., less than 3 years) with this procedure appeared to provide satisfactory results in the majority of patients evaluated. Results were generally considered to be disappointing at time points in the time frame of 3 to 5 years, with only approximately 50% of the patients reporting satisfactory results. Alternative strategies for providing bone marrow-derived cells to the defect site have included microfracture and drilling (Buckwalter and Lohmander, 1994; Rodrigo *et al.,* 1994). Microfracture involves use of an awllike device to poke holes through the subchondral plate, and drilling involves arthroscopic drilling of holes through the subchondral plate. Both procedures provide access of blood and marrow-derived cells to the defect site. The clinical results using these procedures have reportedly been similar to those observed with abrasion arthroplasty, i.e., short-term efficacy but limited long-term utility (Gillogly *et al.,* 1998).

It is generally accepted that procedures that result in penetration of the subchondral plate lead to the formation of fibrocartilage. Fibrocartilage is composed primarily of type I collagen and proteoglycans such as versican, which are characteristic of mesenchymal tissues such as dermis. The extracellular matrix composition and organization of fibrocartilage provide significantly lower biomechanical compressive strength as compared to hyaline cartilage. The formation of fibrocartilage in articular defects following subchondral plate penetration is somewhat analogous to the formation of scar in skin following cutaneous injury. The main difference is that although the reduced biomechanical strength of scars in skin rarely leads to wound dehiscence, the reduced biomechanical properties of fibrocartilage in articular defects in most patients ultimately lead to failure of the repair tissue and degeneration of the surrounding tissue in the pathological process of osteoarthritis.

Articular cartilage, on the other hand, is normally composed of hyaline cartilage. Hyaline cartilage is composed of a complex organization of type II collagen and other minor collagens in combination with hyaluronic acid and a cartilage-specific proteoglycan termed aggrecan. Some of the differences in extracellular matrix composition between hyaline cartilage and fibrocartilage are summarized in Table 49.1. The hyaline cartilage-specific extracellular matrix molecules convey the capacity to withstand significant compressive forces without displaying significant deformation.

Table 49.1. Biochemical comparison of hyaline cartilage and fibrocartilage[a]

Hyaline cartilage	Fibrocartilage
Type II collagen	Type I collagen
Type VI collagen	Proteoglycans
Type IX collagen	Hyaluronic acid
Aggrecan	Link protein
Hyaluronic acid	
Link protein	
COMP protein	

[a]Key biochemical components that are useful for distinguishing hyaline articular cartilage from fibrocartilage. The specific biochemistry and organization of extracellular matrix molecules in these two types of cartilage dictate the capacity of the tissues to resist stress.

Hyaline cartilage also provides a very low frictional surface to accommodate joint movement. The physical properties of hyaline cartilage are a direct result of its extracellular matrix composition and organization.

Surgeons have developed alternative strategies to subchondral plate penetration in an effort to deliver a more appropriate "hyaline" cell type to articular cartilage defects, and thus to achieve more durable results. For example, autologous perichondrial grafts from the rib have been used to treat full-thickness articular defects in the knee (Homminga *et al.*, 1990). Perichondrial grafts were cut to fit the defects and were immobilized with fibrin glue in the defect site. By 1 year following transplantation, 18 of 25 patients exhibited excellent functional recovery, and arthroscopic assessment indicated that the defects were completely filled with cartilage-like tissue in 90% of the defects. Unfortunately, long-term follow-up (up to 8 years) with these and other patients revealed that approximately 60% of the patients experienced treatment failure, which in many cases was a consequence of endochondral ossification and graft delamination (Gillogly *et al.*, 1998).

O'Driscoll has utilized autologous periosteal tissue in full-thickness articular defects in an effort to achieve a more durable repair tissue (O'Driscoll *et al.*, 1986). This procedure involved debridement of the defect to a bleeding bed followed by suturing the periosteum to the base of the defect, with the cambium layer facing toward the joint space. Follow-up analysis of 15 patients reportedly treated with this procedure revealed that 9 patients experienced satisfactory results and 6 experienced graft failure within a few years. Other investigators that have utilized periosteal grafts have reported similar results in that their clinical case studies (Angermann, 1994; Hoikka *et al.*, 1990; Rubak *et al.*, 1982). These results have indicated lack of durability of the repair tissue in most patients and the available biopsy data have indicated that the repair tissue in the defect sites was not hyaline-like cartilage.

Another surgical approach to cartilage injury is osteochondral grafting. This approach involves the replacement of the entire bone and cartilage structure of the femoral condyle with size-matched human cadaver tissue (McDermott *et al.*, 1984). The reported clinical success observed using this procedure for repairing osteochondral defects, coupled with the lack of adequate donor tissue, has led to the development of an alternative approach for repair of small cartilage defects involving the use of autologous cylindrical plugs of hyaline cartilage and underlying bone. These plugs are harvested from a minor weight-bearing area of the intercondylar notch or the anterior superior lateral femoral condyle and are press-fitted into matching holes that have been introduced in defects in the femoral condyle (Bobic, 1996; Matsusue *et al.*, 1993). Preliminary results from patients treated with this procedure suggest that it may be useful for the treatment of small articular defects of less than 2 cm^2. However, no long term data are available from patients that have been treated using this procedure, and concern has been expressed regarding the durability of the repair tissue as well as the ultimate fate of the harvest sites.

AUTOLOGOUS CHONDROCYTE TRANSPLANTATION

Given the generally disappointing intermediate to long-term results obtained with the various surgical procedures outlined above for the treatment of full-thickness articular defects, investigators explored alternative strategies to mediate a more durable repair. Because recruitment of cells from bone marrow was observed to induce formation of fibrocartilage, and transplantation of tissues such as perichondrium or periosteum to cartilaginous defects often led to ossification, it was concluded that perhaps a more productive strategy was to isolate and utilize autologous articular chondrocytes to mediate the repair process. This strategy was particularly appealing because articular chondrocytes were the cell type normally involved in the production and maintenance of hyaline cartilage on the articular surface.

Pioneering work by Benya and Shaffer (1982) had demonstrated that it was feasible to isolate rabbit chondrocytes from articular cartilage and propagate them *in vitro*. The work of these investigators had demonstrated that once articular chondrocytes were enzymatically released from hyaline cartilage and cultured on tissue culture plastic, they exhibited a profound change in their phenotype, exhibited by development of a fibroblastic morphology and a switch in production from type II collagen to type I collagen. Importantly, it was demonstrated that once the propagated chondrocytes were released from the tissue culture plastic surface and placed in suspension culture, they went through a process of "redifferentiation" as judged by reexpression of type II collagen and chondroitin sulfate proteoglycans.

Based on these observations it seemed feasible that isolation of chondrocytes from a small biopsy sample of articular cartilage, followed by enzymatic isolation, propagation of the cells, and ultimate reintroduction of the propagated cells into an articular defect, may provide a means to repair the defect with "hyaline-like" cartilage, rather than the fibrocartilage that was observed when cells recruited from bone marrow had been used to mediate the repair process. It seemed likely that if it were possible to produce hyaline-like cartilage in an articular defect, then the quality and durability of the repair tissue would be superior to that provided by fibrocartilage.

It is worth noting that the concept of autologous cell-based therapy for treatment of serious tissue injury was initially developed and reduced to practice by Rheinwald and Green (1975; Compton *et al.*, 1993; Green *et al.*, 1979). These investigators developed methods to isolate keratinocytes from small skin biopsies, propagate the cells to form epidermal grafts, and then use these grafts to provide permanent skin replacement for severe burn victims. This technology was commercialized in 1989 and has been successfully utilized by many surgeons for treatment of catastrophic burn victims.

Armed with this information investigators initiated experiments to test the possibility of using propagated articular chondrocytes for treatment of articular defects in animal models of articular cartilage injury. Grande *et al.* (1989) evaluated the utility of autologous chondrocyte transplantation in a rabbit model of articular cartilage injury. A 3-mm-diameter chondral defect was introduced in the right and left patellas of the animals, chondrocytes were isolated from the cylindrical plugs of cartilage removed from the defect site, and cells were enzymatically released from the harvested tissue and propagated in tissue culture. Between 2 and 3 weeks following the first surgery, a second surgery was performed to introduce the propagated chondrocytes into the defect site of the right knee. The patella of the left knee served as the control in these experiments. A periosteal patch (cambium layer side toward the defect) was sutured in place over the defects in both knees and approximately 1×10^6 cells were introduced underneath the patch of the right knee. Six weeks following implantation, the animals were sacrificed and their joints and defect sites were evaluated grossly and histologically. The results of this study revealed that grafted sites had markedly less synovitis and other degenerative changes as compared to control sites. Histologic evaluation of control sites indicated that there was an average of 18% fill of the control defects with repair tissue as compared to an average of 82% fill of the defects in cell-grafted sites. It was concluded, based on histologic evaluation, that the repair tissue in the cell-grafted defects was very similar to the hyaline cartilage present in surrounding nonwounded articular cartilage. Results of other experiments using propagated chondrocytes that had been radiolabeled with tritiated thymidine provided evidence that the implanted cells were incorporated into the repair tissue and were contributing to the repair process.

During the same period of time that Grande *et al.* were studying autologous chondrocyte implantation in rabbits, Wakitani *et al.* (1989) evaluated the use of allogeneic articular chondrocyte transplantation for treatment of articular cartilage defects in the patellar groove of the right femur of rabbits. These studies were different from those of Grande *et al.* not only in the use of allogeneic versus autologous cells but also by the fact that the allogeneic cells were not propagated and were delivered in a collagen gel versus a cell suspension underneath a periosteal patch. Despite these differences, the results of these studies indicated that defect sites implanted with cells exhibited dramatically improved healing as compared to untreated defects or defects treated with the collagen gel alone. Histologic evaluation of cell implant sites indicated that the repair tissue was composed of hyaline cartilage.

Based on these promising results Brittberg *et al.* initiated clinical evaluation of autologous chondrocyte implantation in human patients using the basic procedure described by Grande *et al.* (Fig. 49.2). The results of the first 16 patients treated with this therapy in femoral condyle defects and followed for 16–66 months are summarized in Table 49.2 (Brittberg *et al.*, 1994). The data indicated that two years after treatment, 14 of 16 patients of these patients experienced good to excellent clinical results. Clinical results for patients with patellar transplants were less impressive, with only 2 of 7 patients experiencing excellent results. Biopsy specimens were obtained for 15 of the 16 patients with femoral condyle transplants. Histologic evaluation of these biopsy specimens indicated that the majority (11 of 15) contained hyaline-like cartilage (Fig. 49.3). Immunohistochemical analysis of 5 biopsy specimens with hyaline-like cartilage provided evidence of type II collagen staining.

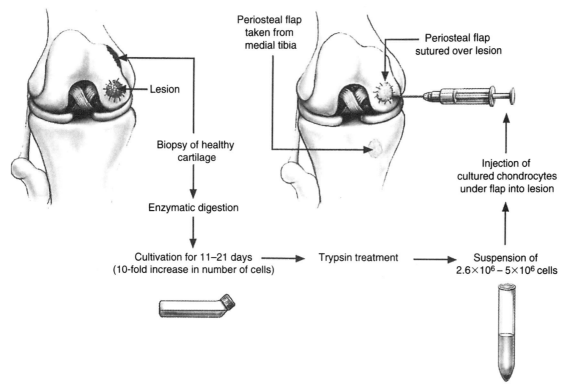

Fig. 49.2. Flow diagram for articular chondrocyte harvest and implantation, as performed in the treatment of deep cartilage defects in the knee with autologous chondrocyte implantation. From Brittberg et al. (1994). Copyright © 1994 Massachusetts Medical Society. All rights reserved.

COMMERCIALIZATION OF AUTOLOGOUS CHONDROCYTE IMPLANTATION

The preclinical and clinical results outlined above for autologous chondrocyte implantation, along with clear evidence of an unmet medical need in the area of clinical management of articular cartilage injury, stimulated interest in providing this treatment on a commercial basis, as had been done previously for autologous keratinocyte transplantation for severe burn victims. Conversion of the methods initially utilized by Brittberg *et al.* to a commercially feasible process presented many technical and quality control challenges. As noted above, autologous chondrocyte implantation relies on the use of cultured cells, initially isolated from a minor weight-bearing surface of the femoral condyle; the cells are expanded *in vitro,* and finally administered to the clinically relevant site for action. Because the biosynthetic profile of cultured animal chondrocytes was shown to be altered during proliferative expansion *in vitro* (Benya and Shaffer, 1982), an evaluation of the ability of propagated human chondrocytes subsequently to reexpress the appropriate cellular phenotype consistent with differentiated articular cartilage, as had been demonstrated for rabbit chondrocytes, was required to ensure the delivery of a functional and reproducible product.

Transcriptional analyses of human chondrocytes propagated on tissue culture plastic revealed that they underwent the reversible "dedifferentiation" process, as had been observed for other mammalian species, with reduction in expression of type II collagen and aggrecan and an up-regulation of type I collagen (Binette *et al.,* 1998). Release of the human chondrocytes from tissue culture plastic and subsequent culture in suspension culture resulted in a reexpression of type II collagen and aggrecan, but importantly showed no evidence of expression of type X collagen, a marker of chondrocyte hypertrophy and ossification (Fig. 49.4). The results of these experiments provided evidence that human chondrocytes, like the chondrocytes of other animal species, possessed the inherent capability to reexpress an articular cartilage phenotype following cellular propagation. This type of assay provided a useful means to assess the effects of various process changes on the capacity of cells to redifferentiate. For example, cultured autologous human articular chon-

Table 49.2. Femoral condylar defects in 16 patients treated with transplanted condrocytes[a]

Patient No.	Age (years)/ sex	Duration of symptoms (years)	Size of defect (cm2)	Microscopic appearance	Histologic appearance	Biopsy (M.A.)[b]	Duration of follow-up (M.A.)[b]	Clinical grade[c]
1	27/M	3	1.6	Not BA	FH	16	16	Poor (2nd operation)
2	24/M	3	2.0	BA, CW	FH	14	48	Good
3	22/M	2	3.0	Not BA, CW	FH	12, 36	36	Poor (2nd operation)
4	48/M	3	2.0	BA, CW	FH	12, 24	48	Good
5	14/F	2	3.0	BA	HL	12, 46	46	Good
6	25/F	1	1.6	BA	HL	12	55	Excellent
7	40/M	3	2.2	BA	HL	22	59	Excellent
8	46/M	2	2.0	BA	HL	16	48	Good
9	22/F	3	4.0	BA	HL	12, 46	46	Excellent
10	26/M	3	2.4	BA	HL	12	36	Excellent
11	27/M	4	2.5	BA	HL	12	54	Good
12	27/F	2	2.0	BA	No biopsy	—	36	Good
13	23/M	6	5.0	BA	HL	17	24	Good
14	18/M	6	4.4	BA	HL	12, 32	32	Good
15	32/F	3	4.5	BA	HL	12	27	Excellent
16	19/M	2	4.0	BA	HL	12, 36	36	Excellent

[a]Patients 3 and 11 had injuries of the lateral femoral condyle, and all other patients had injuries of the medial femoral condyle. The follow-up period for patients 1 and 3 ended at the time of the second operation. Abbreviations: BA, biologically acceptable; FH, fibrous hyaline cartilage; CW, central wear; HL, hyaline-like cartilage.
[b]M.A., Number of months after transplantation.
[c]Note the correlation of clinical grade with the histologic appearance of repair tissue observed in patients treated with autologous chondrocyte implantation.

drocytes for clinical use were first propagated in culture medium composed of Ham's F-12 medium supplemented with 15% (v/v) autologous serum (Brittberg *et al.*, 1994). Cells from each patient were cultured in medium supplemented with their own serum. The use of autologous human serum for propagation of chondrocytes was judged to be technically and economically infeasible. Therefore, the redifferentiation potential of adult articular chondrocytes cultured in donor-matched autologous human serum (Brittberg *et al.*, 1994) or a single lot of commercially

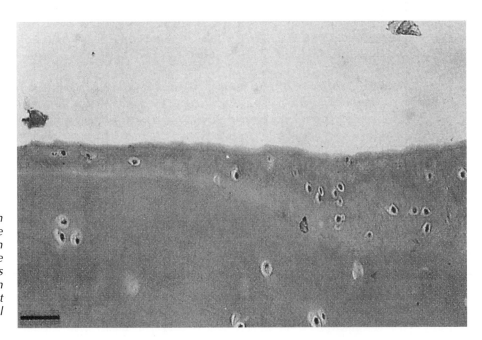

Fig. 49.3. Histologic section from a biopsy of cartilage repair tissue 36 months after treatment of an articular cartilage defect on the femoral condyle with autologous chondrocyte implantation. From Brittberg et al. (1994). Copyright © 1994 Massachusetts Medical Society. All rights reserved.

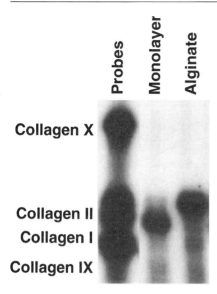

Fig. 49.4. Collagen types I, II, IX, and X gene expression during adult articular chondrocyte redifferentiation in long-term suspension culture. RNA isolated from human articular chondrocytes growing in a proliferative monolayer or in an alginate suspension was subjected to RNase protection assay using antisense RNA probes for types I, II, IX, and X collagen. From Binette et al. (1998).

available bovine serum [Dulbecco's modified eagle's medium (DMEM) supplemented with 10% (v/v) bovine serum] (Aulthouse *et al.*, 1989; Binette *et al.*, 1998; Bonaventure *et al.*, 1994) was compared. The study revealed that redifferentiation potential, or chondrocyte function, was more consistent in medium supplemented with the single lot of fetal bovine serum (FBS) (Fig. 49.5). The inconsistent maintenance of articular chondrocyte redifferentiation potential using matched donor sera was the likely consequence of varying levels of growth and differentiation factors observed in different lots of serum (Freshney, 1994; Yaeger *et al.*, 1997).

Subsequent studies using numerous strains of human chondrocytes propagated with media supplemented with 10% bovine serum revealed that the age of the individual from which the chondrocytes were isolated had a very modest effect on their potential to redifferentiate (Fig. 49.6). Likewise, it was observed that the number of population doublings expended prior to implantation was unaffected by age (Fig. 49.7). Because chondrocytes for implantation are routinely passaged an average of 8 population doublings prior to clinical use, it is worth noting that chondrocytes retain their ability to redifferentiate until just before cell senescence, at about 35–50 population doublings.

Priot to commercialization of autologous chondrocyte implantation, additional animal studies were performed in an effort to better understand and document the process of articular cartilage repair that was mediated by this cell-based therapy. Experiments in dogs revealed that redifferentiation of autologous articular chondrocytes was observed within 6 weeks of implantation into 4-mm cartilage defects created in the femoral condyle or trochlear groove in the stifle joint of the animals (Shortkroff *et al.*, 1996). A chondrocyte nesting phenomenon was observed, where

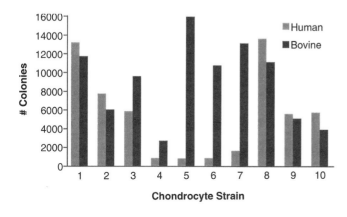

Fig. 49.5. Comparison of chondrocyte culture in autologous human serum vs. bovine serum. Effect of serum source on the ability of human articular chondrocytes to redifferentiate following proliferative expansion in monolayer. Human articular chondrocytes were cultured in culture medium supplemented with fetal bovine serum or donor-matched autologous human serum. Redifferentiation of chondrocytes was assessed by counting colonies in agarose suspension culture, as described by Benya and Shaffer (1982).

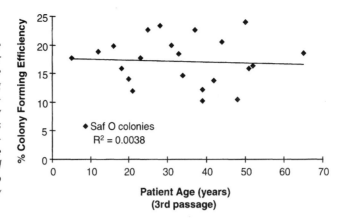

Fig. 49.6. Impact of patient age on the ability of human articular chondrocytes to redifferentiate in suspension culture following proliferative expansion in monolayer. The colony-forming ability of human articular chondrocytes derived from the articular cartilage of 22 donors ranging in age from 5 to 65 years was assessed following monolayer culture into the third passage (approximately 7–10 population doublings).

clusters of implanted chondrocytes in lacunae were located within type II collagen-positive matrix (Fig. 49.8). Chondrocytes having a fibroblastic morphology were observed in type I collagen-positive matrix. A transitional morphology was identified within the cartilage defect, whereby chondrocytic morphology was apparent in the depths of the defect and fibroblastic morphology was present toward the articular surface. Untreated defects without periosteum or cells, and defects treated with periosteum alone, exhibited cellular fill that was consistent with fibrocartilage.

At 6 months in the canine cartilage repair model, the tissue filling the cartilage defects was hyaline-like, with articular chondrocytes in lacunae (Shortkroff *et al.*, 1996). Columnar organization of chondrocytes was observed in some histologic sections (Fig. 49.9). Although a general trend toward more cartilage repair was observed at 6 months in the chondrocyte-implanted defects, the results were complicated by variable levels of spontaneous cartilage repair in the absence of chondrocyte implantation. The positive trend, in terms of improved cartilage repair in cell grafted sites as compared to controls, which was observed at 6 months, was no longer apparent at 12–18 months postimplantation (Breinan *et al.*, 1998). This was a consequence of both spontaneous healing in some of the control sites (no periosteum or cells) and development of degenerative joint disease in both control and treatment sites in many of the animals at these latter time points. Despite these limitations of the animal model, results of the study provided additional evidence for the safety of the procedure and provided insights into the mechanisms of chondrocyte-mediated repair in articular cartilage defects.

Autologous chondrocyte implantation was approved for treatment of articular cartilage defects by the Center for Biologics Evaluation and Research of the Food and Drug Administration in August, 1997, under provisions of accelerated approval guidelines. These regulations provide for approval of biological products for serious or life-threatening illnesses based on surrogate endpoints that are likely to predict clinical benefit. Approval of products under these regulations generally requires additional postapproval clinical studies to confirm that the surrogate end points utilized to provide the basis for approval of the product do indeed correlate with clinical outcomes. Such studies are currently in progress with autologous chondrocyte implantation.

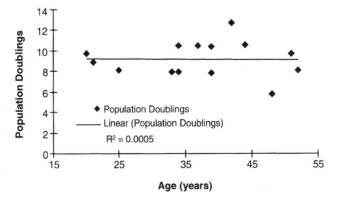

Fig. 49.7. The number of population doublings expended by human articular chondrocytes prior to clinical use. The number of population doublings was calculated as $\log_2$ (number of cells at subculture/number of cells seeded).

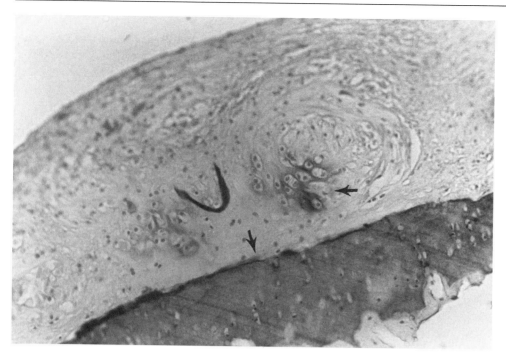

Fig. 49.8. Histologic section of an articular cartilage defect treated with cultured autologous chondrocytes at 6 weeks, stained for type II collagen. Autologous articular chondrocytes were implanted into 4-mm focal defects in the articular cartilage of adult dogs. Animals were sacrificed at 6 weeks and analyzed by histology. Note the type II collagen staining around the nests of chondroid cells (arrow pointing left) and in the calcified cartilage layer (arrow pointing down). Reprinted from Biomaterials, Vol. 17, No. 2. S. Shortkroff, L. Barone, H. P. Hsu, C. Wrenn, T. Gagne, T. Chi, H. Breinan, T. Minas, C. B. Sledge, R. Tubo, and M. Spector, pp. 147–154. Copyright 1996, with permission from Elsevier Science.

In addition to these studies, a patient registry program was voluntarily initiated by Genzyme Tissue Repair, the company responsible for commercializing autologous chondrocyte implantation (ACI). The registry consists of data collected at the cartilage harvest, cell implantation, and periodic follow-up visits to the physician at 6, 12, and 24 months. All patients treated with ACI have been included in the registry. When defects on the femoral condyles have been treated with ACI, significant improvements in joint function and patient symptoms compared to base line have been observed (>2 years of follow-up), with 86% of the patients recorded by themselves and their clinician as improved (Browne *et al.,* 1998; LaPrade and Swiontkowski, 1999).

In addition to these data, the initial positive results reported for femoral condyle defects by

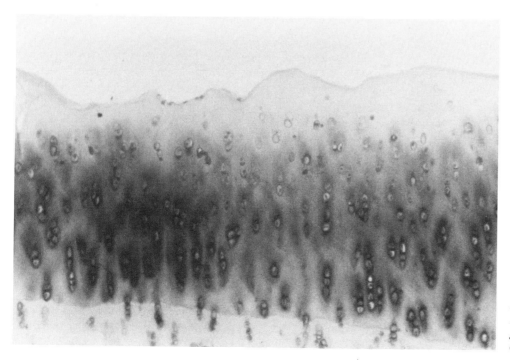

Fig. 49.9. Hyaline articular cartilage observed in a cartilage defect treated with cultured autologous chondrocyte implantation at 6 months. Animals were sacrificed at 6 months and histologic sections were stained with hematoxylin and eosin. Chondrocytes were observed in lacunae structures with columnar organization. This is characteristic of hyaline articular cartilage.

Brittberg *et al.* (1994) were further substantiated by a report on 219 patients at a 2- to 10-year follow-up visit (Peterson, 1998). Functional improvement was observed in 89% of cases with isolated femoral lesions. Moreover, histologic analysis of biopsies removed during second-look arthroscopy revealed that 74% of patients had hyaline-like articular cartilage repair. Good clinical outcome was correlated with hyaline-like repair tissue, supporting the hypothesis that biochemistry equals function in articular cartilage. The longevity of the ACI repair was also demonstrated, in that, 30 of the 31 patients (96%) who initially had good or excellent results at 2 years of follow-up maintained their good or excellent results at an average of 7.4 years postoperatively (Peterson, 1998). Additional reports from orthopedic surgeons have provided independent evidence that the procedure can be useful in the treatment of single large defects of up to 15 cm², or multiple defects of the femoral condyles (Gillogly *et al.,* 1998; Minas, 1998).

ALTERNATIVE STRATEGIES FOR THE DELIVERY OF CELL-BASED THERAPIES FOR CARTILAGE REPAIR

Despite the excellent clinical results with autologous chondrocyte implantation, opportunities for further improvement of this technology clearly exist. Patients who have been treated with autologous chondrocyte implantation do not typically return to full physical activity until at least 1 year postsurgery. The invasive nature of the surgical procedure, an open arthrotomy, and the time interval between chondrocyte implantation and production of a functional repair tissue have been hypothesized to be responsible for the relatively long period of patient rehabilitation. As noted above, the open arthrotomy, commonly known as an open-knee procedure, is required for the surgeon to suture the thin periosteal membrane directly to the margins of the cartilage defect. Therefore, strategies that facilitate the arthroscopic administration of a cartilage repair construct and at the same time accelerate the chondrocyte-mediated repair process are being investigated. The next-generation cartilage repair construct will likely be either a preformed cartilage tissue or will be composed of articular chondrocytes embedded within a biocompatible extracellular matrix supplemented with factors that positively impact chondrocyte growth and differentiation.

Articular chondrocyte viability and redifferentiation has been evaluated with several potential carrier matrices, including agarose (Rahfoth *et al.,* 1998), varying configurations of type I collagen gels and sponges (Ben-Yishay *et al.,* 1992; Frenkel *et al.,* 1997; Grande *et al.,* 1989, 1997; Qi and Scully, 1997; Wakitani *et al.,* 1989), type II collagen sponges (Nehrer *et al.,* 1998), hyaluronic acid derivatives (Solchaga *et al.,* 1999), poly(lactic acid), and poly(glycolic acid) and their derivatives (Freed *et al.,* 1994; Grande *et al.,* 1997; Vacanti *et al.,* 1991), and fibrin (Itay *et al.,* 1987; Sims *et al.,* 1998). The use of these materials to deliver articular chondrocytes to articular cartilage defects has resulted in varying degrees of cartilage repair in small animal models (Ben-Yishay *et al.,* 1992; Grande *et al.,* 1989; Itay *et al.,* 1987; Wakitani *et al.,* 1989). The results obtained in these models were complicated by varying levels of spontaneous repair, presumably due to penetration of the subchondral bone and the subsequent fibrocartilaginous response. However, the importance of cells in the repair of cartilage defects was illustrated by the fact that defects treated with extracellular matrix constructs containing cultured cells typically exhibited enhanced cartilage repair over control matrices without cells.

In terms of accelerating cell-mediated tissue repair, articular chondrocytes respond to a number of growth factors that stimulate proliferation and differentiation, including insulin, insulin-like growth factor I and II (IGF-I and IGF-II), and members of the TGF-β superfamily, including some of the bone morphogenetic proteins (BMP2 and BMP4), basic fibroblast growth factor (bFGF), and platelet-derived growth factor (PDGF) (Martin and Buckwalter, 1996). Moreover, a positive impact on articular cartilage repair *in vivo* has been observed for several of these factors when supplemented into an extracellular matrix delivery vehicle including insulin-like growth factor-I into fibrin matrices (Fortier *et al.,* 1999), or members of the TGF-β superfamily into type I collagen constructs (Hunziker and Rosenberg, 1996).

The paradigm for preformed cartilage tissue was set by osteochondral allografts. Osteochondral allografts have been used successfully for the resurfacing or reconstruction of joints missing large surface areas (Bentley, 1992; Nevo *et al.,* 1991). These bone and cartilage grafts have been shown to integrate very well with the existing bone and provide rapid clinical benefit. However, the supply of such tissue is extremely limited. Methods have been developed for the generation of a predifferentiated cartilage tissue *in vitro* (Kandel *et al.,* 1995; Peel *et al.,* 1998). The pre-differ-

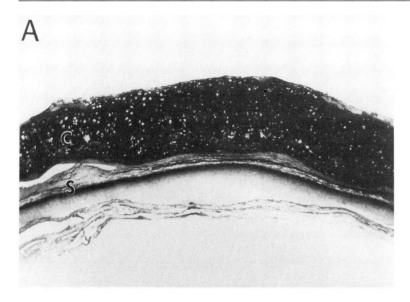

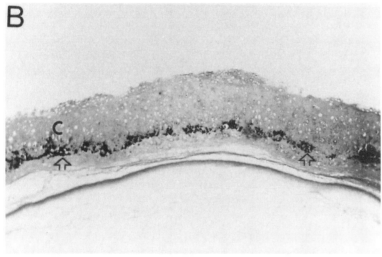

Fig. 49.10. Histologic appearance of a cartilaginous composite graft composed of articular chondrocytes and small intestine submucosa following 8 weeks in culture. (A) Cartilaginous tissue (C) on small intestine submucosa (S) stained with toluidine blue. (B) A zone of mineralization (arrows) is apparent in the lower area of the cartilaginous tissue (C) when cultured under conditions favoring bone formation. This may be helpful for integration with the subchondral bone in vivo. From Peel et al. (1998).

entiated cartilaginous implant can be produced *in vitro* by culturing articular chondrocytes on a type II collagen-coated filter membrane, or on a thin membrane of small intestine submucosa (Kandel *et al.,* 1995, 1997; Peel *et al.,* 1998). After 4–6 weeks in culture a cartilage tissue construct characterized by the histologic hallmarks of hyaline articular cartilage is observed (Fig. 49.10). This hyaline-like cartilage persists when implanted into cartilage defects *in vivo*. Integration of the preformed cartilage implant with the cartilage surrounding of the defect is better than that observed to subchondral bone.

CONCLUSIONS

Taken together, the preclinical and clinical data summarized above provide compelling evidence that chondrocyte implantation can provide superior clinical results in the treatment of articular cartilage defects as compared to alternative cell-based strategies that either generate fibrocartilage or ultimately lead to ossification at the repair site. Scientifically, the use of articular chondrocytes to mediate repair of articular cartilage seems logical, just as it seems logical to use keratinocytes to replace epidermis following massive cutaneous injury. Despite the promising results with autologous chondrocyte implantation, it is clear that significant opportunities for improvement of this procedure exist. As noted above, the current procedure is technically challenging and results in a lengthy rehabilitation process. Thus, it is likely that the next-generation cell-based product to treat articular cartilage damage will be performed using minimally invasive procedures and will be configured in a manner that will significantly reduce the time to generate functionally active tissue in the defect site. It is also clear that the ultimate product to treat articular cartilage injury will be characterized by its ability to be used in large-scale joint resurfacing and will have the added ability to mediate the wound repair process in a manner to reverse the degenerative processes of osteoarthritis. Although such a product seems a distant fantasy, given the scientific progress that has been made in our understanding of cartilage maintenance and repair during the past 20 years, such a dream may be closer than we think.

REFERENCES

Angermann, P. (1994). Osteochondritis dissecans of the femoral condyle treated with periosteal transplantation: A preliminary clinical study of 14 cases. *Orthop. Int.* **2,** 425–428.

Aulthouse, A. L., Beck, M., Griffey, E., Sanford, J., Arden, K., Machado, M. A., and Horton, W. A. (1989). Expression of the human chondrocyte phenotype *in vitro. In Vitro Cell. Dev. Biol.* **25,** 659–668.

Baumgaertner, M. R., Cannon, W. D., Jr., Vittori, J. M., Schmidt, E. S., and Maurer, R. C. (1990). Arthroscopic debridement of the arthritic knee. *Clin. Orthop. Relat. Res.* **253,** 197–202.

Bentley, G. (1992). Articular tissue grafts. *Ann. Rheum. Dis.* **51,** 292–296.

Benya, P. D., and Shaffer, J. D. (1982). Dedifferentiated chondrocytes reexpress the differentiated collagen phenotype when cultured in agarose gels. *Cell (Cambridge, Mass.)* **30,** 215–224.

Ben-Yishay, A., Grande, D. A., Menche, D., and Pitman, M. (1992). Repair of articular cartilage defects using collagen-chondrocyte allografts. *38th Annu. Meet. Orthop. Res. Soc.,* Washington, DC, p. 174.

Bert, J. M., and Maschka, K. (1989). The arthroscopic treatment of unicompartmental gonarthrosis: A five-year follow-up study of abrasion arthroplasty plus arthroscopic debridement and arthroscopic debridement alone. *Arthroscopy* **5,** 25–32.

Binette, F., McQuaid, D. P., Haudenschild, D. R., Yaeger, P. C., McPherson, J. M., and Tubo, R. (1998). Expression of a stable articular cartilage phenotype without evidence of hypertrophy by adult human articular chondrocytes *in vitro. J. Orthop. Res.* **16,** 207–216.

Bobic, V. (1996). Arthroscopic osteochondral autograft transplantation in anterior cruciate ligament reconstruction: A preliminary clinical study. *Knee Surg. Sports Traumatol. Arthroscopy* **3,** 262–264.

Bonaventure, J., Kadhom, N., and Cohen-Solal, L. (1994). Reexpression of cartilage-specific genes by dedifferentiated human articular chondrocytes cultured in alginate beads. *Exp. Cell Res.* **212,** 97–104.

Breinan, H. A., Minas, T., Barone, L., Tubo, R., Hsu, H. P., Shortkroff, S., Nehrer, S., Sledge, C. B., and Spector, M. (1998). Histological evaluation of the course of healing of canine articular cartilage defects treated with cultured autologous chondrocytes. *Tissue Eng.* **4,** 101–114.

Brittberg, M., Lindahl, A., Nilsson, A., Ohlsson, C., Isaksson, O., and Peterson, L. (1994). Treatment of deep cartilage defects in the knee with autologous chondrocyte transplantation. *N. Engl. J. Med.* **331,** 889–895.

Browne, J. E., Fu, F. H., Mandelbaum, B. R., Micheli, L. J., and Moseley, B. (1998). Autogenous chondrocyte implantation for treatment of articular cartilage knee defects 2–4 year multi-center experience. *Transa. European Society of Sports Traumatology, Knee Surgery and Arthroscopy, 8th Congr.,* Nice, France, p. 103.

Buckwalter, J. A., and Lohmander, S. (1994). Operative treatment of osteoarthrosis. Current practice and future development. *J. Bone Jt. Surg., Am. Vol.* **76-A,** 1405–1418.

Compton, C. C., Hickerson, W., Nadire, K., and Press, W. (1993). Acceleration of skin regeneration from cultured epithelial autografts by transplantation to homograft dermis. *J. Burn Care Rehabil.* **14,** 653–662.

Dandy, D. J. (1991). Arthroscopic debridement of the knee for osteoarthritis. *J. Bone Jt. Surg., Br. Vol.* **73B,** 877–888.

Fortier, L. A., Lust, G., and Nixon, A. J. (1999). Insulin-like growth factor-I enhances cell-based articular cartilage resurfacing. *Trans. 45th Annu. Meet. Orthop. Res. Soc.,* p. 58.

Freed, L. E., Grande, D. A., Lingbin, Z., Emmanual, J., Marquis, J. C., and Langer, R. (1994). Joint resurfacing using allograft chondrocytes and synthetic biodegradable polymer scaffolds. *J. Biomed. Mater. Res.* **28,** 891–899.

Frenkel, S. R., Toolan, B., Menche, D., Pitman, M. I., and Pachence, J. M. (1997). Chondrocyte transplantation using a collagen bilayer matrix for cartilage repair. *J. Bone Jt. Surg., Br. Vol.* **79B,** 831–836.

Freshney. (1994). "Serum-free Media. Culture of Animals Cells," pp. 91–99. Wiley, New York.

Friedman, M. J., Berasi, C. C., Fox, J. M., Del Pizzo, W., Snyder, S. J., and Ferkel, R. D. (1984). Preliminary results with abrasion arthroplasty in the osteoarthritic knee. *Clin. Orthop. Relat. Res.* **182,** 200–205.

Gillogly, S. D., Voight, M., and Blackburn, T. (1998). Treatment of articular cartilage defects of the knee with autologous chondrocyte implantation. *J. Orthop. Sports Phys. Ther.* **28,** 241–251.

Grande, D. A., Pitman, M. I., Peterson, L., Menche, D., and Klein, M. (1989). The repair of experimentally produced defects in rabbit articular cartilage by autologous chondrocyte transplantation. *J. Orthop. Res.* **7,** 208–218.

Grande, D. A., Halberstadt, C., Naughton, G., Schwartz, R., and Manji, R. (1997). Evaluation of matrix scaffolds for tissue engineering of articular cartilage grafts. *J. Biomed. Mater. Res.* **34,** 211–220.

Green, H., Kehinde, O., and Thomas, J. (1979). Growth of cultured human epidermal cells into multiple epithelia suitable for grafting. *Proc. Nat. Acad. Sci. U. S. Am.* **76,** 5665–5668.

Hoikka, V. E., Jaroma, H. J., and Ritsila, V. A. (1990). Reconstruction of the patellar articulation with periosteal grafts. 4-year follow-up of 13 cases. *Acta Orthop. Scand.* **61,** 36–39.

Homminga, G. N., Bulstra, S. K., Bouwmeester, P. S., and van der Linden, A. J. (1990). Perichondral grafting for cartilage lesions of the knee. *J. Bone Jt. Surg., Br. Vol.* **72B,** 1003–1007.

Hunter, W. (1743). On structure and diseases of articulating cartilage. *Philos. Trans. R. Soc. London, Ser. B* **24B,** 514–521.

Hunziker, e. B., and Rosenberg, L. C. (1996). Repair of partial-thickness defects in articular cartilage: Cell recruitment from the synovial membrane. *J. Bone Jt. Surg., Am. Vol.* **78-A**, 721–733.

Itay, S., Abramovici, A., and Nevo, Z. (1987). Use of cultured embryonal chick epiphyseal chondrocytes as grafts for defects in chick articular cartilage. *Clin. Orthop. Relat. Res.* **220**, 284–303.

Johnson, L. L. (1986a). Arthroscopic abrasion arthroplasty. *In* "Operative Arthroscopy" (J. B. McGinty, ed.), pp. 341–360. Raven Press, New York.

Johnson, L. L. (1986b). Arthroscopic abrasion arthroplasty historical and pathologic perspective: Present status. *Arthroscopy* **2**, 54–69.

Johnson-Nurse, C., and Dandy, D. J. (1985). Fracture-separation of articular cartilage in the adult knee. *J. Bone Jt. Surg., Br. Vol.* **67B**, 42–43.

Kandel, R. A., Chen, H., Clark, J., and Renlund, R. (1995). Transplantation of cartilagenous tissue generated *in vitro* into articular joint defects. *Artif. Cells Blood Substitutes Immob. Biotechnol.* **23**, 565.

Kandel, R. A., Boyle, J., Gibson, G., Cruz, T., and Speagle, M. (1997). *In vitro* formation of mineralized cartilagenous tissue by articular chondrocytes. *In Vitro Cell. Dev. Biol.* **33**, 174–181.

Kim, H. K., Moran, M. E., and Salter, R. B. (1991). The potential for regeneration of articular cartilage in defects created by chondral shaving and subchondral abrasion. An experimental investigation in rabbits. *J. Bone Jt. Surg., Am. Vol.* **73-A**, 1301–1315.

LaPrade, R. F., and Swiontkowski, M. F. (1999). Update on the clinical experience with autologous chondrocyte implantation. *JAMA, J. Am. Med. Assoc.* **281**, 876–878.

Martin, J. A., and Buckwalter, J. A. (1996). Articular cartilage aging and degeneration. *Sports Med. Arthroscopy Rev.* **4**, 263–275.

Matsusue, Y., Yamamuro, T., and Hama, H. (1993). Arthroscopic multiple osteochondral transplantation to the chondral defect in the knee associated with anterior cruciate ligament disruption. *Arthroscopy* **9**, 318–321.

McDermott, A. G., Langer, F., Pritzker, K. P., and Gross, A. E. (1984). Fresh small-fragment osteochondral allografts. *Clin. Orthop. Relat. Res.* **197**, 96–102.

Minas, T. (1998). Autogenous cultured chondrocyte implantation in the repair of focal chondral lesions of the knee: Clinical indications and operative techniques. *J. Sports Trauma* **20**, 1–13.

Nehrer, S., Breinan, H. H., Ashkar, S., Shortkroff, S., Minas, T., Sledge, C. B., Yannas, I. V., and Spector, M. (1998). Characteristics of articular chondrocytes seeded in collagen matrices *in vitro*. *Tissue Eng.* **4**, 175–183.

Nevo, Z., Robinson, D., and Halperin, N. (1991). The use of grafts composed of cultured cells for the repair and regeneration of cartilage and bone. *In* "Bone" (B. K. Hall, ed.), p. 123. CRC Press, Boca Raton, FL.

O'Driscoll, S. W., Keeley, F. W., and Salter, R. B. (1986). The chondrogenic potential of free autogenous periosteal grafts for biological resurfacing of major full-thickness defects in joint surfaces under the influence of continuous passive motion. An experimental investigation in the rabbit. *J. Bone Jt. Surg., Am. Vol.* **68-A**, 131–140.

Peel, S. A. F., Chen, H., Renlund, R., Badylak, S. F., and Kandel, R. A. (1998). Formation of a SIS-cartilage composite graft *in vitro* and its use in the repair of articular cartilage defects. *Tissue Eng.* **4**, 143–155.

Peterson, L. (1998). Autologous chondrocyte transplantation: 2–10 year follow-up in 219 patients. *6th Annu. Meet., Am. Acad. Orthop. Surg.,* New Orleans, LA.

Qi, W. N., and Scully, S. P. (1997). Extracellular collagen modulates the regulation of chondrocytes by transforming growth factor-beta 1. *J. Orthop. Res.* **15**, 483–490.

Rahfoth, B., Weisser, J., Sternkopf, F., Aigner, T., von der Mark, K., and Brauer, R. (1998). Transplantation of allograft chondrocytes embedded in agarose gel into cartilage defects of rabbits. *Osteoarthritis Cartilage* **6**, 50–65.

Rand, J. A. (1991). Role of arthroscopy in osteoarthritis of the knee. *Arthroscopy* **7**, 358–363.

Rheinwald, J. G., and Green, H. (1975). Serial cultivation of strains of human epidermal keratinocytes: The formation of keratinizing colonies from single cells. *Cell (Cambridge, Mass.)* **6**, 331–343.

Rodrigo, J. J., Steadman, J. R., Silliman, J. F., and Fulstone, H. A. (1949). Improvement of full-thickness chondral defect healing in the human knee after debridement and microfracture using continuous passive motion. *Am. J. Knee Surg.* **7**, 109–116.

Rubak, J. M., Poussa, M., and Ritsila, V. (1982). Effects of joint motion on the repair of articular cartilage with free periosteal grafts. *Acta Orthop. Scand.* **53**, 187–191.

Shortkroff, S., Barone, L., Hsu, H. P., Wrenn, C., Gagne, T., Chi, T., Breinan, H., Minas, T., Sledge, C. B., Tubo, R., and Spector, M. (1996). Healing of chondral and osteochondral defects in a canine model: The role of cultured chondrocytes in regeneration of articular cartilage. *Biomaterials* **17**, 147–154.

Sims, C. D., Butler, P. E., Cao, Y. L., Casanova, R., Randolph, M. A., Black, A., Vacanti, C. A., and Yaremchuk, M. J. (1998). Tissue engineered neocartilage using plasma derived polymer substrates and chondrocytes. *Plast. Reconstr. Surg.* **101**, 1580–1585.

Solchaga, L. A., Lundberg, M., Yoo, J. U., Huibregtse, B. A., Goldberg, V. M., and Caplan, A. I. (1999). Hyaluronic acid-based polymers in the treatment of osteochondral defects. *45th Annu. Meet., Orthop. Res. Soc.,* Anaheim, CA. 56.

Timoney, J. M., Kneisl, J. S., Barrack, R. L., and Alexander, A. H. (1990). Arthroscopy update: 6. Arthroscopy in the osteoarthritic knee. Long-term follow-up. *Orthop. Rev.* **19**, 371–373.

Vacanti, C. A., Langer, R., Schloo, B., and Vacanti, J. P. (1991). Synthetic polymers seeded with chondrocytes provide a template for new cartilage formation. *Plast. Reconstr. Surg.* **88**, 753–759.

Wakitani, S., Kimura, T., Hirooka, A., Ochi, T., Yoneda, M., Owaki, H., Ono, K., and Yasui, N. (1989). Repair of rabbit articular surfaces with allografts of chondrocytes embedded in collagen gels (in Japanese). *Nippon Seikeigeka Gakkai Zasshi* **63**, 529–538.

Yaeger, P. C., Masi, T. L,. de Ortiz, J. L., Binette, F., Tubo, R., and McPherson, J. M. (1997). Synergistic action of transforming growth factor-beta and insulin-like growth factor-I induces expression of type II collagen and aggrecan genes in adult human articular chondrocytes. *Exp. Cell Res.* **237**, 318–325.

CHAPTER 50

TENDONS AND LIGAMENTS

Francine Goulet, Denis Rancourt, Réjean Cloutier, Lucie Germain,
A. Robin Poole, and François A. Auger

INTRODUCTION

The need for human tissues and organs has increased over the past few years, mostly because of population growth. The risks for transmission of contagious diseases have also imposed serious restrictions to the availability of human transplants. Recent advances in tissue and organ engineering have generated much interest among clinicians and patient sin various fields of medicine. This modern approach has already shown successful results with human skin substitutes all over the world.

This chapter is dedicated to ligament and tendon bioengineering. The technical approach developed to produce a human anterior cruciate ligament (ACL) substitute is described. This methodology may eventually be adapted to produce other ligaments or tendons in culture. A bioengineered ACL (bACL) was developed by seeding human ligament fibroblasts in a hydrated collagen matrix. Our bACL is anchored with two bones, because bone-to-bone insertion is reported as the most secure method for ligament fixation. The ACL substitute is a good tool to study connective tissue repair and the environmental and cellular factors that can affect collagen alignment and cross-linking *in vitro*. It may also become a therapeutic alternative for torn ACL replacement.

The effects of cyclic stretching on the histologic features of the bACL are also described. Mechanical stimuli represent one of several concepts to explore in order to optimize bioengineered tissues. Our findings may also be useful to readers involved in the fields of orthopedics and physical therapy. Histologic analyses of bACL cultured under cyclic traction revealed that fibroblasts are surrounded by bundles of collagen fibers organized in a wavy pattern, comparable to histologic features of native ACL. These results suggest that mechanical stimuli have an important impact on the evolution of bioengineered tissue *in vitro*.

THE NEED FOR BIOENGINEERED TENDON AND LIGAMENT SUBSTITUTES

In the United States, at least 120,000 patients per year undergo tendon or ligament repairs (Langer and Vacanti, 1993). So far, the therapeutic options to repair torn ligaments are tissue reconstruction using autograft or allograft, reparation alone or with augmentation, or replacement using a synthetic prosthesis. Unfortunately, none of these surgical alternatives provides a long-term adequate solution. Synthetic material implantation was a very popular surgical technique in the 1980s but it frequently led to implant degeneration and failure (Olson *et al.,* 1988; Woods *et al.,* 1991). Allogeneic grafting of ligaments is still at an experimental stage and may lead to immunologic reactions that prevent good healing of the wound (Jackson and Kurzweil, 1991; Sabiston *et al.,* 1990). For most of the approaches, particular attention has been given to the surgical procedure for ligament fixation to the bones. This step would constitute a critical achievement toward the permanent implantation of bioengineered ACL *in vivo,* as mentioned in the literature (Kurosaka *et al.,* 1987; Olson *et al.,* 1988). Little is known about physiologic and pathologic events that regulate the histologic organization of ACL *in vivo.*

The recent advances in the domain of tendon and ligament bioengineering are summarized below. Tissue bioengineering raises considerable interest in the fields of biology and medicine, notably in orthopedics. Ligaments and tendons are frequently targets of sport and aging trauma. Many clinical studies have strived to develop new therapeutic alternatives such as allografts (Andrews *et al.,* 1994; Arnoczky *et al.,* 1982) and synthetic prostheses (Bessette and Hunter, 1990). However, functional problems frequently arise, despite some progress, because ligaments and tendons support large mechanical stresses *in vivo.*

Over the past few years, attempts have been made to discover new biocompatible materials for ligament and tendon replacement. Collagen alone or with cells isolated from the tissue of origin has been used to produce bioengineered functional organs *in vitro.* The main ideas behind such attempts were to first produce *in vitro* models of ligaments or tendons by tissue engineering, in order to implant them permanently *in vivo,* where their histologic and functional properties could be improved by local and systemic stimuli. Data reported from animal experimentations showed that vascularization, cell migration, tissue remodeling, and extracellular matrix deposition occurred *in vivo* in an acellular composite collagen prosthesis 12 weeks postgrafting (Dunn *et al.,* 1994). Several research groups are presently involved in the development of human bioengineered tissues that would include living cells, to obtain models showing several histologic and functional properties acquired under particular culture conditions prior to transplantation. Such models could also be useful to study various aspects of tendons and ligaments, such as healing and tissue remodeling *in vitro.*

HISTOLOGIC DESCRIPTION OF TENDONS AND LIGAMENTS

Tendons connect bones to muscles, and ligaments join one bone to another. The literature used to describe the histologic structures of fascia, tendons, and ligaments is similar. They are dense connective tissues consisting of fibroblasts surrounded by type I collagen bundles (Amiel *et al.,* 1984). This may explain why patellar tendon and tensor fascia have frequently been used to replace the torn anterior cruciate ligament of the knee joint (Arnoczky *et al.,* 1982). The central quadriceps tendon autograft is reported as one of the best alternatives for ACL reconstruction (Arnoczky *et al.,* 1982; Fulkerson and Langeland, 1995). However, some pain, loss of motion, knee instability, and other problems can be associated with this approach (Tanzer and Lenczner, 1990; Tria *et al.,* 1994; Vergis and Gillquist, 1995), according to each type of trauma and depending on variables that are unique for each individual (Harner *et al.,* 1994; Johnson *et al.,* 1992; Parker *et al.,* 1994). The success of such grafts depends on the revascularization of the transplanted tissues as they are progressively surrounded by a synovial membrane, rich in vessels (Arnoczky *et al.,* 1982). The tendon finally gains some ligament properties and the term "ligamentization" was used by Amiel *et al.* (1986) to describe this physiologic phenomenon studied in *in vivo* postgrafting.

Both ligaments and tendons consist mainly of closely packed and thick collagen fibers (predominantly type I with a small proportion of type II collagen) but also include small quantities of glycosaminoglycans (dermatan sulfate and hyaluronic acid) (Watanabe *et al.,* 1994). The composition of these biochemical constituents is modulated during tissue growth and development, changing also in response to functional requirements (Watanabe *et al.,* 1994).

Tendon collagen fibrillogenesis is initiated in early development by fibroblasts (Trelstad and Hayashi, 1979). Fibrils are embedded in an organized, hydrated, proteoglycan matrix and crosslinked through aldol or Schiff base adducts between aldehydes on one or more of the α chains of the collagen molecules and aldehydes or amino groups on adjacent chains or molecules (Davison, 1989). Such cross-links contribute to the tensile strength of the fibrils (Davison, 1989) and thus to the tensile properties of the whole tendon. The aldehyde-derived cross-links are found in two forms; some are unstable in dilute acids and others are stable (Davison, 1989). The ratio of one form to the other varies in different tissues, generally increasing with aging.

Data reported by Amiel *et al.* (1984) suggested that ligaments are more metabolically active than tendons, having more plump cellular nuclei, higher DNA content, larger amounts of reducible cross-links, and more type III collagen, as compared with tendons. They also contain slightly less total collagen than do tendons and more glycosaminoglycans, particularly close to the joint (Amiel *et al.,* 1984). Extracellular matrix fibers in ligaments are also less regularly arranged than in tendons (Puddu and Ippolito, 1983), although the fibers are oriented parallel to the longitudinal axis of both tissues (Amiel *et al.,* 1984). For example, the anterior cruciate ligament, one of the strongest tissues of the body, is composed of multiple fascicles, the basic unit of which is type I collagen (Bessette and Hunter, 1990). The collagen fibrils are nonparallel, but are themselves arranged into fibers that are oriented roughly in the long axis of the ligament in a wavy, undulating pattern ("crimp"), which slowly straightens out as small loads are applied to the ligament (Bessette and Hunter, 1990). To give an example of the forces applied on such tissues *in vivo,* the ACL usually supports loads of about 169 Newtons (38 lb), and ruptures around 1730 Newtons (about 390 lb) (Bessette and Hunter, 1990). An elongation of only 6% (about 2 mm) of human ACL (about 32 mm long) is reported to be the limit beyond which damage must be expected (Amis and Dawkins, 1991). Thus, the main challenges associated with the production of bioengineered tendons and ligaments are to obtain strong and functional tissues.

PRODUCTION OF BIOENGINEERED TENDONS AND LIGAMENTS

Several approaches are proposed to produce bioengineered tissues *in vitro.* However, the development of such technology always involves the use of biocompatible and, preferably, biodegradable materials that (1) can provide mechanical resistance and (2) can be colonized and reorganized by living cells *in situ* postgrafting. Some tendons and ligaments are at the top of the list in terms of injury frequency and problems to repair, replace, or heal. In contrast with the medial collateral ligament, anterior cruciate ligament regeneration is hampered *in vivo* (Lyon *et al.,* 1991; Ross *et al.,* 1990). Certain anatomic factors may predispose people to ACL injury (Harner *et al.,* 1994). In the young and active population, reconstruction is often the best therapeutic option. However, to overcome the drawbacks associated with ACL repair and healing, several efforts have been put into the production of a bioengineered ACL model over the past decade.

In a composite material, the orientation and density of fibers in the matrix significantly affect its mechanical properties. From a macroscopic point of view, the ACL can be considered as a composite material in which the fibroblasts secrete collagen and elastin, which are organized into fibers. Hence, it is not surprising to see that human ligaments have a very low degree of stiffness in traction, for small strains, because of the cross-linked organization of the fibers (Strocchi *et al.,* 1992).

Collagen remains the basic protein of interest in the field of connective tissue bioengineering. Dunn *et al.* (1992, 1993, 1994), a group actively involved in the production of ligament biocompatible prostheses, have reported successful replacement of the ACL by a cross-linked collagen prosthesis implanted in rabbits. A similar approach was previously reported for the regeneration of the Achilles tendon (Kato *et al.,* 1991). Such a biodegradable prosthesis, prepared by the alignment of 200 cross-linked collagen fibers and fixed by surgically created bone tunnels (2.8 mm diameter) *in situ,* is produced using glutaraldehyde, dehydrothermal-cyanamide (DHTC), or carbodiimide as collagen cross-linking agents (Dunn *et al.,* 1993, 1994; Kato *et al.,* 1991). These collagen fibers have high tensile strengths (30–60 MPa) and small diameters (20–60 mm) (Dunn *et al.,* 1994), similar to fibers of the normal ACL (Danylchuk *et al.,* 1978). Rapid ingrowth (within 8 weeks) of fibrous tissue and bone was observed in the bone tunnels after implantation following degradation of the composite collagen prosthesis (Dunn *et al.,* 1994). Thus, Dunn's group ACL model seems to be one of the promising bioengineering alternatives for human ligament replacement.

A NEW LIVING BIOENGINEERED LIGAMENT MODEL

A different approach for the production of a human ACL model *in vitro* was established in our laboratory through bioengineering. This approach combines skills and concepts associated with tissue culture and biomechanical principles. Such expertise has been successfully developed in our laboratory for other human tissues, such as skin (Auger, 1988; Auger *et al.,* 1993, 1995; Bouvard *et al.,* 1992; López Valle *et al.,* 1992a,b; Rompré *et al.,* 1990; Black *et al.,* 1998), bronchi (Goulet *et al.,* 1996; Paquette *et al.,* 1998), and blood vessels (L'Heureux *et al.,* 1998). Our research programs aim at producing bioengineered tissues that can serve both experimental and clinical interests.

A new bioengineered ACL model was developed. The main differences between our bACL and other ligament models (Bellows *et al.,* 1982; Dunn *et al.,* 1994) consist of the following advantages: (1) Our bACL contains living ACL fibroblasts, which contract, synthesize, and initiate remodeling and organization of the extracellular matrix in which they are initially seeded. (2) Our bACL can be a very useful tool to study connective tissue repair *in vitro* because the cells can maintain their activities for at least 2 months in culture. (3) Our bACL is produced without chemical cross-linking agents and synthetic materials. Such an approach eliminates risks of cytotoxicity and the production of foreign molecules due to mechanical friction, which can occur *in vivo* in the case of its eventual use as ligament prosthesis. (4) Our approach uses bones to anchor the bioengineered tissue; these bones are included with the bACL structure, right from the start of its production, to facilitate its eventual transplantation *in vivo*. Moreover, autologous bone fragments and ligament fibroblasts can be used to produce bACLs for eventual ligament replacement in humans. Such a possibility could greatly reduce the risks of immune reactions and infections, favoring permanent graft integration posttransplantation.

Various research groups have shown that fibroblasts seeded in collagen gels can degrade and reorganize the surrounding extracellular matrix and adopt a specific orientation in a contracted collagen lattice as a function of culture conditions and time (Auger, 1988; Auger *et al.,* 1993, 1995; López Valle *et al.,* 1992b; Rompré *et al.,* 1990; Bell *et al.,* 1979, 1981a,b; Delvoye *et al.,* 1983; Grinnell and Lamke, 1984; Guidry and Grinnell, 1985, 1986, 1987; Huang *et al.,* 1993; Kasugai *et al.,* 1990). Several studies have also shown that mechanical stimuli of biological tissues can produce ultrastructural and histologic content variations. For instance, it is well known that ligament remodeling depends on the mechanical stress to which the ligament is subjected *in vivo* (Amiel *et al.,* 1984). Our experiences also confirmed such behavior with anchored dermal equivalents *in vitro* (Grinnell and Lamke, 1984). The addition of an anchorage to dermal equivalents resulted in cell and collagen alignment in the plane of geometric constraints induced by the anchoring device. In addition, bACLs have collagen and cells aligned in a direction parallel to the stress exerted. Finally, Huang *et al.* (1993) reported that the mechanical properties along with the ultrastructure of ligament equivalents change in response to mechanical strengthening *in vitro*. Similar observations were made following *in vitro* elongation studies on rabbit ACL (Matyas *et al.,* 1994).

Thus, a cyclic stretching machine (Fig. 50.1) was designed and produced to study the effect of cyclic stretching on the mechanical and histologic properties of our bACL *in vitro* (Langelier *et al.,* 1999). The results described in this chapter strongly suggest that it is possible to modulate tissue organization by dynamic mechanical stimuli *in vitro*. Such observations are promising for the production of bioengineered tissues, the ACL in particular.

The first aim of our work was to establish methods for the isolation and culture of ACL fibroblasts to reconstruct *in vitro* a three-dimensional bACL, reproducing the elongated shape of the tissue *in vivo*. Using this biotechnical approach to produce a cultured ligament, we have compared the effects of continuous static and cyclic stretchings on the tissue organization. The technical steps leading to the production and analyses of bACLs are described below. This approach could likely be adapted for the production of bioengineered tendons *in vitro*. Autologous bACLs could eventually be used as ACL replacement prostheses in humans.

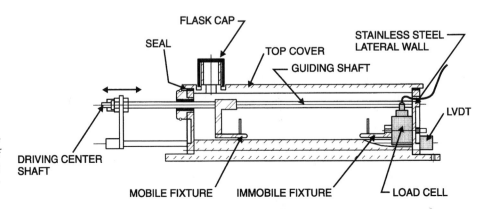

Fig. 50.1. A side schematic view of the sterile culture chamber of the cyclic traction device. From Langelier et al. (1999), with permission.

TECHNICAL STEPS TO PRODUCE AND ANALYZE OUR bACL

STEP 1: HUMAN ACL FIBROBLAST ISOLATION AND CULTURE

Human ACL biopsies are collected from patients (28–65 years old) during total knee surgery. The biopsies were kept at 4°C for no longer than 3 hr before cell isolation. The human ACLs are weighed and cut into small pieces after removal of the periligamentous tissue. The fragments are digested with 0.125% collagenase, containing 2 m*M* $CaCl_2$ (1 ml of enzymatic solution/mg of tissue) for 20 hr under gentle agitation, at 37°C. A 0.1% trypsin solution (1 ml/mg of hydrated tissue) is then added to the cellular suspension for 1 hr. The enzymes are dissolved in Dulbecco's modified Eagle's medium (DMEM), pH 7.4, containing antibiotics. The human ligament fibroblasts (HLFs) isolated from the ACL are cultured in DMEM supplemented with 10% fetal calf serum (FCS), 100 IU/ml penicillin G, and 25 mg/ml gentamicin. When HLF primary cultures reach 85% confluence, the cells are detached from their culture flasks using 0.05% trypsin–0.01% ethylenediaminetetraacetic acid (EDTA) solution (pH 7.8), for about 30 min at 37°C. HLF suspensions are centrifuged twice at 200 *g* for 10 min. The cell pellets are resuspended in complete culture medium and the cells are counted with a Coulter counter and a Multisizer analyzer. The cellular viability is determined using the trypan blue exclusion method.

Several populations of HLFs were isolated and cultured from more than 20 patent biopsies. HLFs appear morphologically slightly bigger and present a less fusiform shape than do dermal fibroblasts in culture. All cells maintain their morphology for more than seven passages in culture. For bACL production, HLFs from passages 3 to 5 are used. Immunofluorescent labeling analysis reveals that HFLs extracted from ACL biopsies express vimentin, fibronectin, types I and III collagens, and elastin (Goulet *et al.,* 1997a).

STEP 2: CULTURE OF bACL

Preparation of the bACL human bone anchors

Partially demineralized human bone pieces are cut into a cylindrical shape (average size, 1 cm in diameter and 2 cm long). A transverse hole (0.125 inch in diameter) is made in each of them. Two anchors are transferred in one sterile 15-ml plastic tube and kept in position by passing a hot metal pin across the tube and through the holes previously made. A bone is fixed at the bottom and another at the top of the tube. Then all the tubes containing the anchors are rinsed with 100% ethanol, dried, and sterilized with ethylene oxide.

Production of the human bACL *in vitro*

A solution of DMEM, 2.7×, containing antibiotics is mixed with a second solution containing heat-inactivated (30 min at 56°C) fetal calf serum (FCS), solubilized bovine type I collagen, and living HLFs (passages 3 to 5). The cells are added at a final concentration of 1×10^6 cells ml. This mixture is quickly poured into a sterile plastic tube containing the two bone anchors. According to our latest protocol, the final concentration of bovine type I collagen is always 1.5 mg/ml in the bACL. All of the bACLs are cultured in DMEM supplemented with 10% FCS, 50 mg/ml ascorbic acid, 100 IU/ml penicillin G, and 25 mg/ml gentamicin. They are maintained in a static vertical position during the first 24 hr of culture.

STEP 3: CULTURE CONDITIONS OF VARIOUS GROUPS OF bACLs

Static tension

Half of the bACLs (total of at least six) are transferred into sterile culture chambers containing two fixed pins, which are passed through the holes made in the osseous anchors. The initial static tension applied in the longitudinal axis of these bACLs is adjusted to the minimal level required to keep each tissue elongated to its full length for several weeks. Figure 50.2A shows a bACL after 10 days. The culture medium is changed every other day.

Cyclic stretching

The other group of bACLs (total of at least six) is cultured in a sterile chamber of the cyclic stretching device (Fig. 50.2B). The device is placed in a standard cell culture incubator, the stretch-

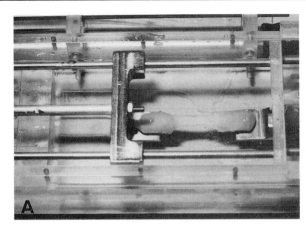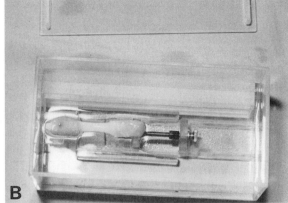

Fig. 50.2. A bACL installed into the sterile culture chamber of (A) the cyclic traction device [from Langelier et al. (1999), with permission] and (B) the static stretching device.

ing chamber allowing sterile exchanges with the surrounding 8% CO_2 atmosphere at 37°C. The bACL is maintained in place in the sterile chamber by inserting the bone anchors in metal pins, one rigidly fixed to a load cell and the other attached to a motion-controlled cursor. By controlling the position of the cursor, the bACL can be subjected to cyclic stretching with amplitudes from 0 to 30 mm at a frequency of up to 1 Hz for any extended period of time. The whole system is controlled via LABview VI software. The experimenter can easily change the experimental conditions and supervise the ongoing tests to make sure that everything is running smoothly. The device was designed to control parameters such as stretching amplitude and frequency. The bACL can be subjected to continuous cyclic stretching for several weeks.

Six experiments have been conducted under similar culture conditions to evaluate the effects of cyclic stretching on the evolution of our bACL model. The cycling frequency was fixed at 1 Hz. During the first 5 days, the bACLs are subjected to a 1-mm stretch per cycle, always returning to their initial length (about 4 cm) to complete each cycle. The amplitude is increased to 2 mm from day 5 to day 10. These conditions were established arbitrarily. We compared two groups of bACLs casted on the same day, seeded with the same ACL cell populations, suspended in the same collagen solution, in order to control as much as possible the cellular parameters involved. This approach allowed us to evaluate the effects of cyclic stretching on the histologic and ultrastructural properties of bACLs.

STEP 4: ACL HISTOLOGIC ANALYSES

Histologic studies were performed after 10 days on bACLs cultured under horizontal static tension compared to bACLs subjected to cyclic stretching. The bACL biopsies were fixed in Bouin's solution, paraffin embedded, and 6-mm-thick tissue sections were stained using the standard Masson's trichrome method, which allows excellent visualization of the matricial collagen fibers. The histologic features of both groups of bACLs were compared.

Culture under cyclic stretching did not induce any detectable change in the macroscopic aspect of the bACL. The alignment of matrix fibers in a direction parallel to the tension applied to a cell-populated collagen lattice has been reported. As expected, Masson's trichrome staining of tissue sections showed that matricial alignment in the control bACL was maintained under static horizontal elongation (Fig. 50.3A). Interestingly, bACLs subjected to cyclic stretching (Fig. 50.3B) contained thick bundles of extracellular matrix fibers, highly organized in a wavy pattern, typical of native ACL histologic features (Fig. 50.3C). Thus, cyclic stretching can induce complex structural organization of the extracellular matrix in bioengineered ligaments *in vitro*. The crimps were not observed in control bACLs. Results were repeatedly similar from one experiment to another.

STEP 5: bACL INDIRECT IMMUNOFLUORESCENCE ANALYSES

To determine the organization of the extracellular matrix, immunohistologic studies were performed on frozen sections of bACLs. The histologic immunostaining was performed on frozen tis-

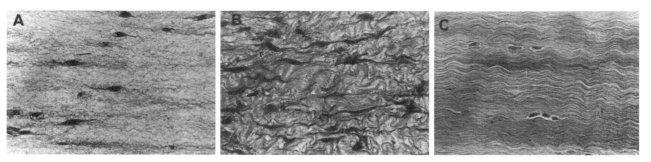

Fig. 50.3. Photomicrographs of bACL histologic sections stained using Masson's trichrome methods. (A) A bACL cultured under static conditions, (B) a bACL submitted to cyclic traction, and (C) native human ACL (×40). Note the wavy pattern of highly organized bundles of collagen fibers induced by cyclic traction.

sue sections (6 mm thick) fixed for 10 min with acetone at −20°C, as described previously (Auger *et al.*, 1995). A monoclonal antichondroitin sulfate antibody, antihuman type I collagen, type III collagen, and elastin antibodies were used to analyze the organization of these extracellular matrix constituents in the cultured bACLs.

The network of matrix fibers is better organized in bACLs subjected to cyclic stretching than in those cultured under static tension. Type I collagen is the major constituent of the bundles of fibers induced by cyclic stretching (Fig. 50.4A), as observed in the native human ACL (Fig. 50.4B). Moreover, the bACLs cultured under cyclic stretching contain more chondroitin sulfate (Fig. 50.4D) than do the bioengineered tissues maintained under static tension (Fig. 50.4F). However, all bACLs contain less chondroitin sulfate than does native ACL (Fig. 50.4E).

Step 6: bACL Ultrastructural Analyses

Biopsies can be processed and mounted for electron microscope analyses, as described previously (Auger *et al.*, 1993, 1995). Electron microscope analyses of bACLs subjected to cyclic traction revealed that the HLFs are surrounded by bundles of collagen fibers (Fig. 50.5A) aligned with the cells in the direction parallel to the long axis of all bACLs (Fig. 50.5B). The typical 67-nm periodicity of the collagen (Fig. 50.5C) is observed in the collagen fibers.

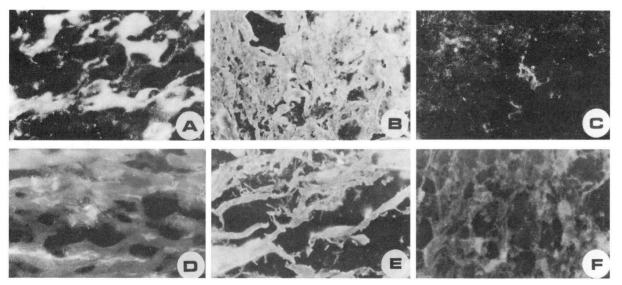

Fig. 50.4. Immunofluorescence labeling of human type I collagen fibers (A, B, and C) and chondroitin sulfate (D, E, and F) in frozen sections of native ACL (B and E) and bACL cultured for 10 days under cyclic (A and D) and static (C and F) elongation (×40). Note the stronger signals of human type I collagen and chondroitin sulfate within the matricial network of bACL submitted to cyclic traction compared to static tension.

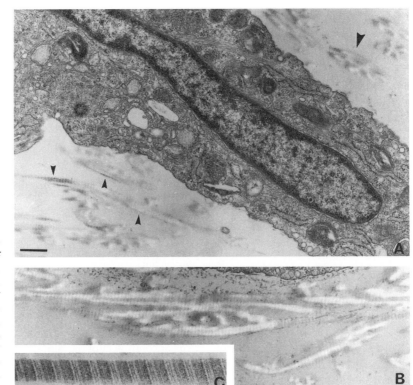

Fig. 50.5. Electron microscopy analyses of ultrathin sections of bACL cultured under cyclic traction. (A) Collagen fibers (arrowheads) surrounding a living HLF (bar: 8.3 mm = 500 nm). (B) Bundles of collagen fibers oriented in a direction parallel to the long axis of the bACL (bar: 9.25 mm = 500 nm). (C) A collagen fiber with the typical 67-nm periodicity.

CONCLUSIONS

Recent advances in bioengineering and cellular biology have led to revolutionary new therapeutic applications in many medical fields. The use of autologous tissue-engineered epidermis and living skin substitutes for the treatment of human burns is an example of such progress (Auger *et al.,* 1993, 1995; Green *et al.,* 1979; Boyce *et al.,* 1991; Contard *et al.,* 1993; Fernyhough *et al.,* 1987; Germain *et al.,* 1993; Hansbrough *et al.,* 1989; López Valle *et al.,* 1992b). The bioengineered ligament model described in this chapter presents features that can be adapted to other culture conditions *in vitro* (our bACL) and very likely to *in vivo* postgrafting in animals (see, for example, Dunn's ligament substitute) (Dunn *et al.,* 1994).

Collagen fibers organized in wavy bundles can be obtained *in vitro* in a human bioengineered living tissue. Our data strongly suggest that living ACL cells seeded into bACL analogs can respond to mechanical stimuli *in vitro.* The stimulation of collagen secretion and the acute matrix remodeling observed in bACLs subjected to cyclic traction are very promising observations. Indeed, such bACLs could be cultured under these conditions for longer periods of time, to promote further development of histologic characteristics. We previously reported that HLFs isolated from human ACL biopsies and seeded into a collagen matrix can secrete and reorganize matricial fibers in response to static tension (Goulet *et al.,* 1997a,b). Our data strongly suggest that cyclic traction can induce the cells to organize thick bundles of collagen fibers into a wavy structural pattern.

Such matricial reorganization might improve the mechanical properties of the bACLs. Matrix strengthening of ligament equivalents (hydrated collagen gels seeded with human dermal fibroblasts) was reported after 12 weeks of mechanical stress *in vitro* (induced by maintaining the slowly contracting equivalent at a fixed length), with the maintenance of cell viability (Huang *et al.,* 1993). These results showed that it is possible to reinforce a bioengineered tissue using static mechanical stimuli. We expect to see comparable or better effects induced by dynamic stimuli, notably cyclic traction, on bACL strength. Cyclic traction may also contribute to maintaining the biomechanical response of the human ACL fibroblasts used to populate our bACL, on culture time.

This work was performed for a given cyclic frequency and stretching amplitude. Extensive studies have to be done to determine the optimal frequency and amplitude that would best reorganize the matrix. These studies raise hope for successful production of functional autologous ACL analogs in culture through new bioengineering approaches. Presently, the strength of our bACL does not compare with the strength of native ACL. However, a bioengineered blood vessel developed in our laboratory, using human endothelial cells, smooth muscle cells, and fibroblasts organized in layers supported by a collagen matrix, can withstand a pressure of at least 2000 mmHg (L'Heureux *et al.*, 1998). The production of this blood vessel equivalent demonstrates that it is possible to obtain strong bioengineered tissues cultured *in vitro*.

A comparable level of strength may be obtained for our bACL in culture and the tissue could be further reinforced *in vivo*, during the animal implantation phase of the project. For instance, only a part of the patellar tendon is grafted when used as an autologous ACL analog. Orthopedic surgeons who practice the bone–patellar ligament–bone autograft for ACL reconstruction have described graft weakening at about 6 weeks posttransplantation. Despite this decrease in functionality, when these grafts are successful the implants are slowly revascularized after about 6 weeks and present most ACL histologic and functional properties after 30 weeks in humans. The graft becomes stronger and is revascularized as the patient slowly returns to normal activities, suggesting that a significant remodeling of our bACL could occur *in vivo*. Such remodeling could bring the bACL mechanical properties to a functional level. We are presently testing this hypothesis using autologous bACLs in goats. Preliminary observations made on the animal's legs only 1 month postimplantation are very encouraging.

Nevertheless, the main advantages of our bACL over other ligament models stem from its histologic organization which is progressively modulated by the living ACL fibroblasts included in its structure. Because fibroblasts seeded in the bACL secrete and reorganize extracellular matrix constituents in culture, it makes it a promising tool to study the role of various cellular elements involved in the maintenance or the induction of ligament organization *in vitro*.

PERSPECTIVES IN TENDON AND LIGAMENT BIOENGINEERING

Combined techniques of cell biology and tissue engineering produce tendons and ligaments in culture. These different bioengineered tissues will indeed have to be optimized prior to human transplantation. However, investigators involved in this area of research have not yet started to exploit all the possibilities offered by this new biotechnological concept. For instance, gene therapy might find some interesting applications by introducing living transfected cells within various types of bioengineered tissues.

Despite the challenges that lie ahead for those hoping to reproduce tissues and organs *in vitro*, there is every reason to feel optimistic, based on the advances already made in various medical fields through tissue bioengineering (Langer and Vacanti, 1993). It is quite probable that in the near future, people will benefit from bioengineered tendon and ligament substitutes. Production may start in a laboratory and reach its final completion in a human body. Currently, a delay of 6 weeks after injury is required before ACL surgery is performed, and studies have shown that it is better to perform surgical intervention of the ACL when the nearly normal range of motion of the knee has returned and inflammation has been mostly eliminated (Johnson *et al.*, 1992). This time frame allows for the production of a bACL *in vitro*. Engineering of an autologous ACL would avoid the need for detrimental harvesting of the patients's tissues. This approach may limit the need for knee surgery, because it may be possible to graft a bACL anchored with bones simply by arthroscopy. In addition, bioengineered tissues cannot be rejected postgrafting when produced with autologous cells. That is the common ultimate goal shared by all researchers and clinicians. Realization of this goal would certainly revolutionize the present therapeutic approach to tendon and ligament implantation.

ACKNOWLEDGMENTS

Dr. Albert Normand died early after the initiation of our work on human ligament bioengineering. Our research team expresses sincere gratitude to this devoted orthopedic surgeon, who was the first to propose cyclic stretching for the development of our bioengineered ACL *in vitro*. The authors are grateful to Jean Lamontagne and Marc Bouchard, for their helpful contribution as residents in orthopedic surgery on the human ACL bioengineering project. The authors thank Louis-Mathieu Stevens and Geneviève Bordeleau, two students who have actively contributed to the latest advances in this work; Julie Tremblay and Julie Bergeron for precious research assistance; Eve Langelier and Christine Lord for the design of the cyclic stretching device; Dominique Robitaille, Gilbert Castilloux, Stéphane Bouchard, François Berthod, and Pierre-Philippe Stevens for computer programming and functional setup for mechanical analyses; Michel Bordeleau and Roland Martineau for designing and producing the static culture device; Réjean Desbiens for helpful discussions as a consultant in mechanical engineering; Aristide Pusteria for sample preparation for electron microscopy analyses; and Claude Marin for photographic assistance. This work was supported by the Medical Research Council of Canada. The Canadian Orthopaedic Association, The Renaud–Lemieux Foundation of CHA, Pavillion Saint-Sacrement, and The Club Richelieu of Limoilou, Quebec.

REFERENCES

Amiel, D., Frank, C., Harwood, F., Fronek, J., and Akeson, W. (1984). Tendons and ligaments: A morphological and biochemical comparison. *J. Orthop. Res.* **1**, 257–265.

Amiel, D., Kleiner, J. B., Roux, R. D., Harwood, F. L., and Akeson, W. (1986). The phenomenon of "ligamentization": anterior cruciate ligament reconstruction with autogenous patellar tendon. *J. Orthop. Res.* **4**, 162–172.

Amis, A. A., and Dawkins, G. P. C. (1991). Functional anatomy of the anterior cruciate ligament: Fibre bundle actions related to ligament replacements and injuries. *J. Bone Jt. Surg., Br. Vol.* **73B**, 260–267.

Andrews, M., Noyes, F. R., and Barber-Westin, S. D. (1994). Anterior cruciate ligament allograft reconstruction in the skeletally immature athlete. *Am. J. Sports Med.* **22**, 48–54.

Arnoczky, S. P., Tarvin, G. B., and Marshall, J. L. (1982). Anterior cruciate ligament replacement using patellar tendon. *J. Bone Jt. Surg., Am. Vol.* **64-A**, 217–224.

Auger, F. A. (1988). The role of cultured autologous human epithelium in large burn wound treatment. *Transplant./Implant. Today,* May, pp. 21–24.

Auger, F. A., Guignard, R., López Valle, C. A., and Germain, L. (1993). Role and inocuity of Tisseel®, a tissue glue, in the grafting process and *in vivo* evolution of human cultured epidermis. *Br. J. Plast. Surg.* **46**, 136–142.

Auger, F. A., López Valle, C. A., Guignard, R., Tremblay, N., Noël, B., Goulet, F., and Germain, L. (1995). Skin equivalents produced using human collagens. *In Vitro Cell. Dev. Biol.* **31**, 432–439.

Bell, E., Ivarsson, B., and Merrill, C. (1979). Production of a tissue-like structure by contraction of collagen lattices by human fibroblasts of different proliferative potential *in vitro. Proc. Natl. Acad. Sci. U.S.A.* **76**, 1274–1278.

Bell, E., Ehrlich, H. P., Buttle, D. J., and Nakatsuji, T. (1981a). Living tissue formed *in vitro* and accepted as skin-equivalent tissue of full thickness. *Science* **211**, 1052–1054.

Bell, E., Ehrlich, H. P., Sher, S., Merrill, C., Sarber, R., Hull, B., Nakatsuji, T., Church, B., and Buttle, D. (1981b). Development and use of a living skin equivalent. *Plast. Reconstr. Surg.* **67**, 386–392.

Bellows, C. G., Melcher, A. H., and Aubin, J. E. (1982). Association between tension and orientation of periodontal ligament fibroblasts and exogenous collagen fibres in collagen gels *in vitro. J. Cell Sci.* **58**, 125–138.

Bessette, G. C., and Hunter, R. E. (1990). The anterior cruciate ligament. *Orthopedics* **13**, 551–562.

Black, A. F., Berthod, F., L'Heureux, N., Germain, L., and Auger, F. A. (1998). *In vitro* reconstruction of a human capillary-like network in a tissue engineered skin equivalent. *FASEB J.* **12**, 1331–1340.

Bouvard, V., Germain, L., Rompré, P., Roy, B., and Auger, F. A. (1992). Influence of dermal equivalent maturation on a cultured skin equivalent. *Biochem. Cell Biol.* **70**, 34–42.

Boyce, S. T., Foreman, T. J., English, K. B., Stayner, N., Cooper, M. L., Sakabu, S., and Hansbrough, J. F. (1991). Skin wound closure in athymic mice with cultured human cells, biopolymers, and growth factors. *Surgery* **110**, 866–876.

Contard, P., Bartel, R. L., Jacobs, L., II, Perlish, J. S., MacDonald, E. D., II, Handler, L., Cone, D., and Fleischmajer, R. (1993). Culturing keratinocytes and fibroblasts in a three-dimensional mesh results in epidermal differentiation and formation of a basal lamina-anchoring zone. *J. Invest. Dermatol.* **100**, 35–39.

Danylchuk, K. D., Finlay, J. B., and Krcek, J. P. (1978). Microstructural organization of human and bovine cruciate ligaments. *Clin. Orthop. Relat. Res.* **131**, 294–298.

Davison, P. F. (1989). The contribution of labile crosslinks to the tensile behavior of tendons. *Connect. Tissue Res.* **18**, 293–305.

Delvoye, P., Nusgens, B., and Lapière, C. M. (1983). The capacity of retracting a collagen matrix is lost by dermatosparactic skin fibroblasts. *J. Invest. Dermatol.* **81**, 267–270.

Dunn, M. G., Tria, A. J., Kato, Y. P., Bechler, J. R., Ochner, R. S., Zawadsky, J. P., and Silver, F. (1992). Anterior cruciate ligament reconstruction using a composite collagenous prosthesis: A biomechanical and histologic study in rabbits. *Am. J. Sports Med.* **20**, 507–515.

Dunn, M. G., Avasarala, P. N., and Zawadsky, J. P. (1993). Optimization of extruded collagen fibers for ACL reconstruction. *J. Biomed. Mater. Res.* **27**, 1545–1552.

Dunn, M. G., Maxian, S. H., and Zawadsky, J. P. (1994). Intraosseous incorporation of composite collagen prostheses designed for ligament reconstruction. *J. Orthop. Res.* **12**, 128–137.

Fernyhough, W., Aukhil, I., and Link, T. (1987). Orientation of gingival fibroblasts in stimulated periodontal spaces *in vitro* containing collagen gel. *J. Periodontol.* **58**, 762–769.

Fulkerson, J. P., and Langeland, R. (1995). An alternative cruciate reconstruction graft: The central quadriceps tendon. *Arthroscopy* **11**, 252–254.

Germain, L., Rouabhia, M., Guignard, R., Carrier, L., Bouvard, V., and Auger, F. A. (1993). Improvement of human keratinocyte isolation and culture using thermolysin. *Burns* **19**, 99–104.

Goulet, F., Boulet, L.-P., Chakir, J., Tremblay, N., Dubé, J., Laviolette, M., and Auger, F. A. (1996). Morphological and functional properties of bronchial cells isolated from normal and asthmatic subjects. *Am. J. Respir. Cell Mol. Biol.* **15**, 312–318.

Goulet, F., Germain, L., Rancourt, D., Caron, C., Normand, A., and Auger, F. A. (1997a). Tendons and ligaments. *In* "Textbook of Tissue Engineering" (R. Lanza, R. Langer, and W. L. Chick, eds.), pp. 633–644. Academic Press, San Diego, CA.

Goulet, F., Germain, L., Caron, C., Rancourt, D., Normand, A., and Auger, F. A. (1997b). Tissue-engineered ligament. *In* "Ligaments and Ligamentoplasties" (L. H. Yahia, ed.), pp. 367–377. Springer-Verlag, Berlin.

Green, H., Kehinde, O., and Thomas, J. (1979). Growth of cultured human epidermal cells into multiple epithelia suitable for grafting. *Proc. Natl. Acad. Sci. U.S.A.* **76,** 5665–5668.

Grinnell, F., and Lamke, C. R. (1984). Reorganization of hydrated collagen lattices by human skin fibroblasts. *J. Cell Sci.* **66,** 51–63.

Guidry, C., and Grinnell, F. (1985). Studies on the mechanisms of hydrated collagen gel reorganization by human skin fibroblasts. *J. Cell Sci.* **79,** 67–81.

Guidry, C., and Grinnell, F. (1986). Contraction of hydrated collagen gels by fibroblasts: Evidence for two mechanisms by which collagen fibrils are stabilized. *Collagen Relat. Res.* **6,** 515–529.

Guidry, C., and Grinnell, F. (1987). Heparin modulates the organization of hydrated collagen gels and inhibits gel contraction by fibroblasts. *J. Cell Biol.* **104,** 1097–1103.

Hansbrough, J. F., Boyce, S. T., Cooper, M. L., and Foreman, T. J. (1989). Burn wound closure with cultured autologous keratinocytes and fibroblasts attached to a collagen-glycosaminoglycan substrate. *JAMA, J. Am. Med. Assoc.* **262,** 2125–2130.

Harner, C. D., Paulos, L. E., Greenwald, A. E., Rosenberg, T. D., and Cooley, V. C. (1994). Detailed analysis of patients with bilateral anterior cruciate ligament injuries. *Am. J. Sports Med.* **22,** 37–43.

Huang, D., Chang, T. R., Aggarwal, A., Lee, R. C., and Ehrlich, H. P. (1993). Mechanisms and dynamics of mechanical strengthening in ligament-equivalent fibroblast-populated collagen matrices. *Ann. Biomed. Eng.* **21,** 289–305.

Jackson, D. W., and Kurzweil, P. R. (1991). Allograft in knee ligament surgery. *In* "Ligament and Extensor Mechanism of the knee: Diagnosis and Treatment" (W. N. Scott, ed.), pp. 349–360. Mosby Year Book, St. Louis, MO.

Johnson, R. J., Beynnon, B. D., Nichols, C. E., and Renstrom, A. F. H. (1992). Current concepts review: The treatment of injuries of the anterior cruciate ligament. *J. Bone Jt. Surg., Am. Vol.* 74-**A,** 140–151.

Kasugai, S., Susuki, S., Shibata, S., Yasui, S., Amano, H., and Ogura, H. (1990). Measurements of the isometric contractile forces generated by dog periodontal ligament fibroblasts *in vitro. Arch. Oral Biol.* **35,** 597–601.

Kato, Y. P., Dunn, M. G., Zawadsky, J. P., Tria, A. J., and Silver, F. H. (1991). Regeneration of achille tendon with a collagen tendon prosthesis. *J. Bone Jt. Surg., Am. Vol.* 73-**A,** 561–574.

Kurosaka, M., Yoshiya, S., and Andrish, J. T. (1987). A biomechanical comparison of different surgical techniques of graft fixation in anterior cruciate ligament reconstruction. *Am. J. Sports Med.* **15,** 225–229.

Langelier, E., Rancourt, D., Bouchard, S., Lord, C., Stevens, P.-P., Germain, L., and Auger, F. A. (1999). Cyclic traction machine for long-term culture of fibroblast-populated collagen gels. *Ann. Biomed. Eng.* **27,** 67–72.

Langer, R., and Vacanti, J. P. (1993). Tissue engineering. *Science* **260,** 920–926.

L'Heureux, N., Pâquet, S., Labbé, R., Germain, L., and Auger, F. A. (1998). A completely biological tissue-engineered human blood vessel. *FASEB J.* **12,** 47–56.

López Valle, C. A., Glaude, P., and Auger, F. A. (1992a). Tie-over dressings: Surgical model for *in vivo* evaluation of cultured epidermal sheets in mice. *Plast. Reconstr. Surg.* **89,** 139–143.

López Valle, C. A., Auger, F. A., Rompré, P., Bouvard, V., and Germain, L. (1992b). Peripheral anchorage of dermal equivalents. *Br. J. Dermatol.* **127,** 365–371.

Lyon, R. M., Akeson, W. H., Amiel, D., Kitabayashi, L. R., and Woo, S. L. (1991). Ultrastructural differences between the cells of the medial collateral and the anterior cruciate ligaments. *Clin. Orthop. Relat. Res.* **272,** 279–286.

Matyas, J., Edwards, P., Miniaci, A., Shrive, N., Wilson, J., Bray, R., and Frank, C. (1994). Ligament tension affects nuclear shape *in situ:* An *in vitro* study. *Connect. Tissue Res.* **31,** 45–53.

Olson, E. J., Kang, J. D., Fu, F. H., Georgescu, H. I., Mason, G. C., and Evans, C. H. (1988). The biomechanical and histological effects of artificial ligament wear particles: *In vitro* and *in vivo* studies. *Am. J. Sports Med.* **16,** 558–570.

Paquette, J. S., Goulet, F., Boulet, L.-P., Tremblay, N., Chakir, J., Germain, L., and Auger, F. A. (1998). Three-dimensional production of bronchi *in vitro. Can. Respir. J.* **5,** 1.

Parker, A. W., Drez, D., and Cooper, J. L. (1994). Anterior cruciate ligament injuries in patients with open physes. *Am. J. Sports Med.* **22,** 44–47.

Puddu, G., and Ippolito, E. (1983). Reconstruction of the anterior cruciate ligament using the semitendinosus tendon. Histologic study of a case. *Am. J. Sports Med.* **11,** 14–16.

Rompreé, P., Auger, F. A., Germain, L., Bouvard, V., López Valle, C. A., Thibault, J., and Le Duy, A. (1990). Influence of initial collagen and cellular concentrations on the final surface area of dermal and skin equivalents: A Box–Behnken analysis. *In Vitro Cell. Dev. Biol.* **26,** 983–990.

Ross, S. M., Joshi, R., and Frank, C. B. (1990). Establishment and comparison of fibroblast cell lines from the medial collateral and anterior cruciate ligaments of the rabbit. *In Vitro Cell. Dev. Biol.* **26,** 579–584.

Sabiston, P., Frank, C. Y., Lam, T., and Shrive, N. (1990). Allograft ligament transplantation. A morphological and biochemical evaluation of a medial collateral ligament complex in a rabbit model. *Am. J. Sports Med.* **18,** 160–168.

Strocchi, R., De Pasquale, V., Gubellini, P., Facchini, A., Marcacci, M., Buda, R., Zaffaghini, S., and Ruggeri, A. (1992). The human anterior cruciate ligament: Histological and ultrastructural observations. *Anat. Embryol.* **180,** 515–551.

Tanzer, M., and Lenczner, E. (1990). The relationship of intercondylar notch size and content to notchplasty requirement in anterior cruciate ligament surgery. *Arthroscopy* **6,** 89–93.

Trelstad, R. L., and Hayashi, K. (1979). Tendon collagen fibrillogenesis: Intracellular subassemblies and cell surface changes associated with fibril growth. *Dev. Biol.* **71,** 228–242.

Tria, A. J., Alicea, J. A., and Cody, R. P. (1994). Patella Baja in anterior cruciate ligament reconstruction of the knee. *Clin. Orthop. Relat. Res.* **299,** 229–234.

Vergis, A., and Gillquist, J. (1995). Graft failure in intra-articular anterior cruciate ligament reconstructions: A review of the literature. *Arthroscopy* **11**, 312–321.

Watanabe, M., Nojima, M., Shibata, T., and Hamada, M. (1994). Maturation-related biochemical changes in swine anterior cruciate ligament and tibialis posterior tendon. *J. Orthop. Res.* **12**, 672–682.

Woods, G. A., Indelicato, P. A., and Prevot, T. J. (1991). The Gore-tex anterior cruciate ligament prosthesis. Two versus three year results. *Am. J. Sports Med.* **19**, 48–55.

CHAPTER 51

MECHANOSENSORY MECHANISMS IN BONE

Stephen C. Cowin and Melvin L. Moss

INTRODUCTION

The mechanosensory mechanisms in bone include (1) the cell system that is stimulated by external mechanical loading applied to the bone, (2) the system that transduces that mechanical loading to a communicable signal, and (3) the systems that transmit that signal to the effector cells for the maintenance of bone homeostasis and for strain adaptation of the bone structure. The effector cells are the osteoblasts and the osteoclasts. These systems and the mechanisms that they employ have not yet been unambiguously identified. We review here candidate systems. In particular, we summarize the current theoretical and experimental evidence that suggests that osteocytes are the principal mechanosensory cells of bone, that they are activated by shear stress from fluid flowing through the osteocyte canaliculi, and that the electrically coupled three-dimensional network of osteocytes and lining cells is a communications system for the control of bone homeostasis and structural strain adaptation.

It has long been known that living adult mammalian bone tissue adapts its material properties, and that whole bones adapt their shape, in response to altered mechanical loading (Wolff, 1870, 1892, 1986; Frost, 1964, 1969; Uhthoff and Jaworski, 1978; Jaworski *et al.,* 1980; Lanyon *et al.,* 1982). Progress is being made in understanding the cellular mechanisms that accomplish the absorption and deposition of bone tissue. The physiologic mechanism by which the mechanical loading applied to bone is sensed by the tissue, and the mechanism by which the sensed signal is transmitted to the cells that accomplish the surface deposition, removal, and maintenance, have not been identified. The purpose of this chapter is to review some of the background research on these mechanosensory mechanisms and to outline candidates for the mechanosensory system [see Burger *et al.* (1998) for an earlier review of similar literature].

All vital cells are "irritable," i.e., capable of perturbation by, and response to, alterations in their external environment (Jacobs, 1998; Junge, 1998). The mechanosensing process(es) of a cell enable it to sense the presence of, and to respond to, extrinsic physical loadings. This property is widespread in uni- and multicellular animals (French, 1992; Kernan *et al.,* 1994; Fraser and Macdonald, 1994; Hamill and McBride, 1995; Hackney and Furness, 1995; Cui *et al.,* 1995), plants (Wildron *et al.,* 1992; Goldsmith, 1994), and bacteria (Olsson and Hanson, 1995). Tissue sensibility is a property of a connected set of cells and it is accomplished by the intracellular processes of mechanoreception and mechanotransduction. "Mechanoreception" is the term used to describe the process that transmits the informational content of an extracellular mechanical stimulus to a receptor cell. "Mechanotransduction" is the term used to describe the process that transforms the content of a mechanical stimulus into an intracellular signal. We employ the term "mechanosensory" to mean both mechanoreception and mechanotransduction. Additional processes of intercellular transmission of transduced signals are required at tissue, organ, and organismal structural levels.

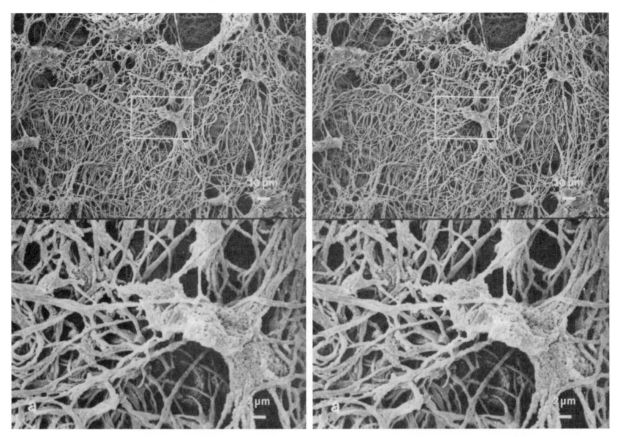

Fig. 51.1. A scanning electron micrograph showing the replicas of lacunae and canaliculi in situ in mandibular bone from a young subject aged 22 years. The inset shows enlarged lacunae identified by a rectangle. This micrograph illustrates the interconnectivity of the CCN. From Atkinson and Hallsworth (1983), with permission.

THE CONNECTED CELLULAR NETWORK

The bone cells that lie on all bony surfaces are osteoblasts, either active or inactive. Inactive osteoblasts are called bone-lining cells; they have the potential of becoming active osteoblasts. The bone cells that are buried in the extracellular bone matrix are the osteocytes. Each osteocyte, enclosed within its mineralized lacuna, has many (perhaps as many as 80) cytoplasmic processes. These processes are approximately 15 μm long and are arrayed three-dimensionally in a manner that permits them to interconnect with similar processes of up to as many as 12 neighboring cells. These processes lie within mineralized bone matrix channels called canaliculi. The small space between the cell process plasma membrane and the canalicular wall is filled with bone fluid and macromolecular complexes of unknown composition. All bone cells except osteoclasts are extensively interconnected by the cell processes of the osteocytes, forming a connected cellular network (CCN) (Cowin *et al.,* 1991; Moss, 1991a,b). The interconnectivity of the CCN is graphically illustrated in Fig. 51.1, which is a scanning electron micrograph showing the replicas of lacunae and canaliculi *in situ* in mandibular bone from a young subject aged 22 years. The inset shows enlarged lacunae (outlined by a rectangle).

The touching cell processes of two neighboring bone cells contain gap junctions (Bennett and Goodenough, 1978; Doty, 1981, 1989; Jones and Bingmann, 1991; Schirrmacher *et al.,* 1992; Jones *et al.,* 1993; Gourdie and Green, 1993; Civitelli, 1995). A gap junction is a channel connecting two cells. The walls of the channel consist of matching rings of proteins piercing the membrane of each cell, and when the rings associated with two cells connect with each other, the cell-to-cell junction is formed. This junction allows ions and compounds of low molecular weight to pass between the two cells without passing into the extracellular space. The proteins making up a gap junction are called connexins; in bone the protein is either connexin 43 or 45, with 43 predominating (the number refers to the size of the proteins calculated in kilodaltons) (Minkoff *et al.,* 1994, 1999; Lecanda *et al.,* 1998). A ring of connexins in one cell membrane is generally called a connexon or hemichannel. Both mechanical strain and fluid shear stress cause increased expression of connexin 43 in vascular tissues (Cowan *et al.,* 1998). In cardiac tissue the turnover rate of connexin 43 is very rapid (Beardslee *et al.,* 1998). The rapid dynamics of gap junction turnover and the plasticity of gap junction expression in response to various stimuli offer the possibility for remodeling of the intercellular circuits, both within and between communication compartments in the cardiovascular system (Spray, 1998). In bone, gap junctions connect superficial osteocytes to periosteal and endosteal osteoblasts. All osteoblasts are similarly interconnected laterally on a bony surface; perpendicular to the bony surface, gap junctions connect periosteal osteoblasts with preosteoblastic cells, and these, in turn, are similarly interconnected. Effectively, each CCN is a true syncytium (Doty, 1989; Schirrmacher *et al.,* 1992; Rodan, 1992; Jones *et al.,* 1993). Gap junctions are found where the plasma membranes of a pair of markedly lapping canalicular processes meet (Rodan, 1992). In compact bone, canaliculi cross the cement lines that form the outer boundary of osteons. Thus extensive communication exists between osteons and interstitial regions (Curtis *et al.,* 1985).

Bone cells are electrically active (Bingmann *et al.,* 1988a,b; Chesnoy-Marchais and Fritsch, 1988; Massass *et al.,* 1990; Rubinacci *et al.,* 1998). In addition to permitting the intercellular transmission of ions and small molecules, gap junctions exhibit both electrical and fluorescent dye transmission (Jeansonne *et al.,* 1979; Schirrmacher *et al.,* 1993; Spray, 1994; Moreno *et al.,* 1994). Gap junctions are electrical synapses, in contradistinction to interneuronal, chemical synapses, and, significantly, they permit bidirectional signal traffic (e.g., biochemical, ionic, electrical). In a physical sense, the CCN represents the hard wiring (Cowin *et al.,* 1991; Moss, 1991a,b; Nowak, 1992) of bone tissue.

MECHANOSENSATION ON THE CCN

Stimuli

The *stimulus* for bone remodeling is defined as that particular aspect of the bone's stress or strain history that is employed by the bone to sense its mechanical load environment and to signal for the deposition, maintenance, or resorption of bone tissue. The bone tissue domain or region over which the stimulus is felt is called the *sensor* domain. When an appropriate stimulus parameter exceeds threshold values, loaded tissues respond by the triad of bone adaptation processes: deposition, resorption, and maintenance. The CCN is the site of intracellular stimulus reception, signal transduction, and intercellular signal transmission. It is thought that stimulus reception occurs in the osteocyte (Cowin *et al.,* 1991), and that the CCN transduces and transmits the signal to the surface lining or osteoblast. The osteoblasts alone directly regulate bone deposition and maintenance, and indirectly regulate osteoclastic resorption (Martin and Ng, 1994). The possible role of the osteoblast as a stimulus receptor has not yet been thoroughly investigated (Owan *et al.,* 1997). Although it is reasonably presumed that initial mechanosensory events occur at the plasma membrane of the osteocytic soma and/or canalicular processes, the initial receptive, and subsequent transductive, processes are not well understood.

It follows that the true biologic stimulus, although much discussed, is not precisely known. A variety of mechanical loading stimuli associated with ambulation (at a frequency of 1 to 2 Hz) have been considered for bone remodeling. The majority have followed Wolff (1892, 1986) in suggesting that some aspect of the mechanical loading of bone is the stimulus. The mechanical stimuli suggested include strain (Cowin and Hegedus, 1976), stress (Wolff, 1892, 1986), strain energy (Fyhrie and Carter, 1986; Huiskes *et al.,* 1987), strain rate (Hert *et al.,* 1969, 1971, 1972; O'Connor *et al.,* 1982; Lanyon, 1984; Goldstein *et al.,* 1991; Fritton *et al.,* 1999), and fatigue microdamage (Carter and Hayes, 1976; Martin and Burr, 1982). In some cases the time-averaged values of these quantities are suggested as the mechanical stimulus, and in others the amplitudes of the oscillatory components and/or peak values of these quantities are the candidates for the mechanical stimulus. Two dozen possible stimuli were compared in a combined experimental and analytic approach (Brown *et al.,* 1990). The data supported strain energy density, longitudinal shear stress, and tensile principal stress or strain as stimuli; no stimulus that could be described as rate dependent was among the two dozen possible stimuli considered in the study [for a consideration of the stimulus in microgravity see Cowin (1998)].

The case for strain rate as a remodeling stimulus has been building over the past quarter century. The animal studies of Hert and co-workers [Hert *et al.,* 1969, 1971, 1972) suggested the importance of strain rate. Experiments (Lanyon, 1984; Goldstein *et al.,* 1991; Fritton *et al.,* 1999) have quantified the importance of strain rate over strain as a remodeling stimulus. The studies (Weinbaum *et al.,* 1991, 1994; Cowin *et al.,* 1995) directed at the understanding of the cellular mechanism for bone remodeling have suggested that the prime mover is the bone strain rate-driven motion of the bone fluid, the signal of which is transduced by osteocytes. In the model of Weinbaum *et al.* (1991, 1994) the shear stress from the bone fluid flow over the osteocytic processes in the canaliculi is a cellular mechanism-based model suggesting strain rate as a stimulus. The study of Gross *et al.* (1997) that showed bone deposition to be related to strain gradients actually, by the model of Weinbaum *et al.* (1991, 1994), also demonstrated a dependence on strain rate.

In experiments with cultured cells it has been shown that osteocytes, but not periosteal fibroblasts, are extremely sensitive to fluid flow, resulting in increased prostaglandin as well as nitric oxide (NO) production (Klein-Nulend *et al.,* 1995a,b). Three different cell populations, namely, osteocytes, osteoblasts, and periosteal fibroblasts, were subjected to two stress regimes, pulsatile fluid flow (PFF) and intermittent hydrostatic compression (IHC) (Klein-Nulend *et al.,* 1995a). IHC was applied at 0.3 Hz with 13-kPa peak pressure. PFF was a fluid flow with a mean shear stress of 0.5 Pa with cyclic variations of 0.02 Pa at 5 Hz. The maximal hydrostatic pressure rate was 130 kPa/sec and the maximal fluid shear stress rate was 12 Pa/sec. Under both stress regimes, osteocytes appeared more sensitive than osteoblasts, and osteoblasts more sensitive than periosteal fibroblasts. However, despite the large difference in peak stress and peak stress rate, PFF was more effective than IHC. Osteocytes, but not the other cell types, responded to 1 hr of PFF treatment with a sustained prostaglandin E_2 up-regulation lasting at least 1 hr after PFF was terminated. By comparison, IHC needed 6 hr of treatment before a response was found. These results suggested that osteocytes are more sensitive to mechanical stress than are osteoblasts, which are again more sensitive than periosteal fibroblasts. Furthermore, osteocytes appeared particularly sensitive to fluid shear stress, more so than to hydrostatic stress. These conclusions are in agreement with the theory that osteocytes are the main mechanosensory cells of bone, and that they detect mechanical loading events by the canalicular flow of interstitial fluid, which results from that loading event. Weinbaum *et al.* (1994) used Biot's porous media theory to relate loads applied to a whole bone to the flow of canalicular interstitial fluid past the osteocytic processes. Their calculations predict fluid-induced shear stresses of 0.8–3 Pa, as a result of peak physiologic loading regimes. The findings that bone cells *in vitro* actually respond to fluid shear stress of 0.2–6 Pa (Reich *et al.,* 1990; Williams *et al.,* 1994; Hung *et al.,* 1995; Klein-Nulend *et al.,* 1995a,b) lend experimental support to the theory.

Osteocytes also rapidly release NO in response to stress (Pitsillides *et al.,* 1995; Klein-Nulend *et al.,* 1995b), and this NO response seems to be required for the stress-related prostaglandin release (Klein-Nulend *et al.,* 1995b). Therefore, the behavior of osteocytes compares to that of endothelial cells, which regulate the flow of blood through the vascular system, and also respond to fluid flow of 0.5 Pa with increased prostaglandin and NO production (Hecker *et al.,* 1993). The response of endothelial cells to shear stress is likely related to their role in mediating an adaptive remodeling of the vasculature, so as to maintain constant endothelial fluid shear stress throughout the arterial site of the circulation (Kamiya *et al.,* 1984).

Skeletal muscle contraction is a typical bone loading event and has been suggested (Moss, 1968, 1969, 1978) as a stimulus. Frequency is one of the critical parameters of the muscle stimulus and it serves to differentiate this stimulus from the direct mechanical loads of ambulation that occur at a frequency of 1 to 2 Hz. The frequency of contracting muscle in tetanus is from 15 Hz to a maximum of 50–60 Hz in mammalian muscle (McMahon, 1984). It has been observed (McLeod and Rubin, 1992; Rodrequez *et al.,* 1993) that these higher order frequencies, significantly related to bone adaptational responses, are (Rubin *et al.,* 1993) "present within the [muscle contraction] strain energy spectra regardless of animal or activity and implicate the dynamics of muscle contraction as the source of this energy band." The close similarity of muscle stimulus frequencies to bone tissue response frequencies is noted below.

Reception and Transduction

The osteocyte has been suggested as the stimulus sensor, the receptor of the stimulus signal (Cowin *et al.,* 1991); histologic and physiologic data are consistent with this suggestion (Aarden *et al.,* 1994; Aarden, 1996). The placement and distribution of osteocytes in the CCN three-dimensional array are architecturally well suited to sense deformation of the mineralized tissue encasing them (Lanyon, 1993). Because only a population of cells, and not an individual receptor (Edin and Trulsson, 1992), can code unambiguously, the osteocytes in the CCN are potential mechanoreceptors by virtue of their network organization.

Osteocytic mechanotransduction may involve a number of different processes or cellular systems. These processes include stretch- and voltage-activated ion channels, cytomatrix sensation–transduction processes, cytosensation by fluid shear stresses, cytosensation by streaming potentials, and exogenous electric field strength. Each of these processes or cellular systems is discussed below.

Stretch- and voltage-activated ion channels

The osteocytic plasma membrane contains stretch-activated ion channels (Guggino *et al.,* 1989; Duncan and Misler, 1989; Keynes, 1994; Hamill and McBride, 1996), which are also found in many other cell types (Morris, 1990; French, 1992; Ghazi *et al.,* 1998). When activated in strained osteocytes, they permit passage of certain ions (Sachs, 1986, 1988; Sackin, 1995), including K^+, Ca^{2+}, Na^+, and Cs^+. Such ionic flow may, in turn, initiate cellular electrical events; for example, bone cell stretch-activated channels may modulate membrane potential as well as Ca^{2+} ion flux (French, 1992; Harter *et al.,* 1995). Rough estimates of osteocytic mechanoreceptor strain sensitivity have been made (Cowin *et al.,* 1991), and the calculated values cover the morphogenetically significant strain range of 0.1 to 0.3% in the literature (Lanyon, 1984; Rubin and Lanyon, 1984, 1987). This appears to be too low a strain to open a stretch-activated ion channel. A model involving bone fluid flow has been suggested for the amplification of the 0.1 to 0.3% strain applied to a whole bone to a value an order of magnitude larger at the osteocytic cytoplasmic process membrane (Cowin and Weinbaum, 1998). Stretch-activated ion channels also occur in osteoblastic cells (Davidson *et al.,* 1990).

As in most cells, the osteocytic plasma membrane contains voltage-activated ion channels, and transmembrane ion flow may be a significant osseous mechanotransductive process (Chesnoy-Marchais and Fritsch, 1988; Ravelsloot *et al.,* 1991; Ferrier *et al.,* 1991; Jan and Jan, 1992). It is also possible that such ionic flow generates osteocytic action potentials, capable of transmission through gap junctions (Schirrmacher *et al.,* 1993).

Cytomatrix sensation–transduction processes

The mineralized matrix of bone tissue is strained when loaded. Macromolecular mechanical connections between the extracellular matrix and the osteocytic cell membrane exist, and these connections may be capable of transmitting information from the strained extracellular matrix to the bone cell nuclear membrane. The basis of this mechanism is the physical continuity of the transmembrane integrin molecule, which is connected extracellularly with the macromolecular collagen of the organic matrix and intracellularly with the cytoskeletal actin. The latter, in turn, is connected to the nuclear membrane (Hughes *et al.,* 1993; Watanabe *et al.,* 1993; Green, 1994; Richardson and Parsons, 1995; Clark and Brugge, 1995; Forgacs, 1995; Shapiro *et al.,* 1995; Dolce *et al.,* 1996; Shyy and Chien, 1997; Salter *et al.,* 1997; Meazzini *et al.,* 1998; Carvalho *et al.,* 1998; Ingber, 1998; Janmey, 1998). It is suggested that such a cytoskeletal lever chain, connecting to the nuclear membrane, can provide a physical stimulus able to activate the osteocytic genome (Jones and Bingmann, 1991), possibly by first stimulating the activity of components such as the c-*fos* genes (Uitto, 1991; Jones and Bingmann, 1991; Dayhoff *et al.,* 1992; Sadoshima *et al.,* 1992; Watanabe *et al.,* 1993; Yanagishita, 1993; Wang *et al.,* 1993; Haskin and Cameron, 1993; Petrov and Usherwood, 1994; Machwate *et al.,* 1995; Salter *et al.,* 1997).

Cytosensation by fluid shear stresses

A hypothesis concerning the mechanism by which the osteocytes housed in the lacunae of mechanically loaded bone sense the load applied to the bone by the detection of dynamic strains was suggested by Weinbaum *et al.* (1991, 1994). It was proposed that the osteocytes are stimulated by relatively small fluid shear stresses acting on the membranes of their osteocytic processes. A hierarchical model of bone tissue structure that related the cyclic mechanical loading applied to the whole bone to the fluid shear stress at the surface of the osteocytic cell process was presented (Weinbaum *et al.,* 1994). In this model the sensitivity of strain detection is a function of frequency; in the physiologic frequency range (1–20 Hz), associated with either locomotion (1–2 Hz) or the maintenance of posture (15–30 Hz), the fluid shear stress is nearly proportional to the product of frequency and strain. Thus if bone cells respond to strains on the order of 0.1% at frequencies of 1 or 2 Hz, they will also respond to strains on the order of 0.01% at frequencies of 20 Hz. The fluid shear stresses would also strain the macromolecular mechanical connections between the cell and the extracellular bone matrix (mentioned in the previous section); thus fluid shear stress is also potentially capable of transmitting information from the strained matrix to the bone cell nuclear membrane, where it can effectively regulate its genomic functions.

Several investigators (Piekarski and Munro, 1977; Johnson *et al.,* 1982; Kufahl and Saha, 1990) have examined other aspects of the lacunar–canalicular porosity using simple circular pore models and have attempted to analyze its possible physiologic importance. These studies have primarily emphasized the importance of the convective flow in the canaliculi between the lacunae as a way of enhancing the supply of nutrients between neighboring osteocytes. Previous studies on the relaxation of the excess pore pressure have been closely tied to the strain-generated potentials (SGPs) associated with bone fluid motion. The SGP studies are briefly reviewed below.

Cytosensation by streaming potentials

The fact that the extracellular bone matrix is negatively charged due to its proteins means that a fluid electrolyte bounded by the extracellular matrix will have a diffuse double layer of positive charges. When the fluid moves, the excess positive charge is convected, thereby developing streaming currents and streaming potentials. The cause of the fluid motion is the deformation of the extracellular matrix due to whole-bone mechanical loading. Pollack and co-workers (Pollack *et al.,* 1984; Salzstein *et al.,* 1987; Salzstein and Pollack, 1987) have laid an important foundation for explaining the origin of strain-generated potentials. However, there is disagreement on the anatomical site in bone tissue that is the source of the experimentally observed SGPs. Salzstein and Pollack (1987) concluded that this site was the collagen–hydroxyapatite porosity of the bone mineral, because small pores of approximately 16 nm radius were consistent with their experimental data if a poroelastic–electrokinetic model with unobstructed and connected circular pores was assumed (Salzstein *et al.,* 1987). Cowin *et al.* (1995), using the model of Weinbaum *et al.* (1994), have shown that the data of Salzstein and Pollack (1987), Scott and Korostoff (1990), and Otter *et al.* (1992) are also consistent with the larger pore space (100 nm) of the lacunar–canalicular porosity being the anatomical source site of the SGPs, if the hydraulic drag and electrokinetic contribution associated with the passage of bone fluid through the surface matrix (glycocalyx) of the osteocytic process are accounted for. The mathematical models of Salzstein *et al.* (1987) and Weinbaum *et al.* (1994) are similar in that they combine poroelastic and electrokinetic theory to describe the phase and magnitude of the SGP. The two theories differ in the description of the interstitial fluid flow and streaming currents at the microstructural level and in the anatomic structures that determine the flow. In Weinbaum *et al.* (1994) this resistance resides in the fluid annulus that surrounds the osteocytic processes, i.e., the cell membrane of the osteocytic process, the walls of the canaliculi, and the glycocalyx (also called the surface matrix or capsule) that exists in this annular region. In Cowin *et al.* (1995) the presence of the glycocalyx increases the SGPs and the hydraulic resistance to the strain-driven flow. The increased SGP matches the phase and amplitude of the measured SGPs. In the Salzstein *et al.* (1987) model this fluid resistance and SGP are achieved by assuming that an open, continuous small pore structure (≈16 nm radius) exists in the mineralized matrix. The poroelastic model of Weinbaum *et al.* (1994) for bone fluid flow has been developed further (Zhang *et al.,* 1997, 1998a,b,c; Wang *et al.,* 1999) and a review of the related poroelastic literature has appeared (Cowin, 1999).

Experimental evidence indicating that the collagen–hydroxyapatite porosity of the bone mineral is unlikely to serve as the primary source of the SGP is obtained from several sources, including the estimates of the pore size in the collagen–hydroxyapatite porosity and permeability studies with different size-labeled tracers in both mineralized and unmineralized bone. Such permeability studies clearly show time-dependent changes in the interstitial pathways as bone matures. At the earliest times, the unmineralized collagen–proteoglycan bone matrix is porous to large solutes. The studies with ferritin (10 nm in diameter) in 2-day-old chick embryos (Dillaman, 1984) show a continuous halo around primary osteons 5 min after the injection of this tracer. The halo passes right through the lacunar–canalicular system, suggesting that, before mineralization, pores of the size predicted in Salzstein and Pollack (1987) (radii ≈ 16 nm) can exist throughout the bone matrix. In contrast, Montgomery *et al.* (1988), using this same tracer in adult dogs, also found a fluorescent halo surrounding the Haversian canals; however, this halo was not continuous but was formed by discrete lines, suggesting that the pathways were limited to discrete pores whose spacing was similar to that observed for canaliculi. This conclusion is supported by the studies of Tanaka and Sakano (1985) in the alveolar bone of 5-day-old rats using the much smaller tracer, microperoxidase (MP) (2 nm). These studies clearly showed that the MP penetrated only the unmineralized matrix surrounding the lacunae and the borders of the canaliculi and was absent from the mineralized matrix. Using more mature rats, the study of Ayasaka *et al.* (1992) confirmed the failure of the small (2 nm) MP tracer to penetrate the mineralized matrix tissue from the bone fluid compartments.

Exogenous electric field strength

Bone responds to exogenous electrical fields (Otter *et al.*, 1998). Although the extrinsic electrical parameters are unclear, field strength may play an important role (Brighton *et al.*, 1992; Otter *et al.*, 1998). A significant parallelism exists between the parameters of exogenous electrical fields and the endogenous fields produced by muscle activity. Bone responds to exogenous fields in an effective range of 1–10 μV/cm, strengths that are on the order of those endogenously produced in bone tissue during normal (muscle) activity (McLeod *et al.*, 1993; 1998).

The uniqueness of osseous mechanosensation

The difference between mechanosensation in bone tissue and mechanosensation in nonosseous processes is revealing and thought to be significant. First, most mechanosensory cells are cytologically specialized (e.g., rods and cones in the retina); bone cells are not. Second, one bone loading stimulus can evoke three adaptational responses (deposition, maintenance, or resorption) whereas nonosseous process stimuli evoke one response (e.g., vision). Third, osseous signal transmission is aneural, whereas all other organismal mechanosensational signals (Moss-Salentijn, 1992) utilize afferent neural pathways (e.g., the visual pathway) (Hackney and Furness, 1995). Fourth, the evoked bone adaptational responses are confined within each "bone" independently (e.g., within a femur); there is no organismal involvement (e.g., touch is a generalized somatic sensation). However, all afferent transductive processes, osseous and nonosseous, share many common mechanisms and processes (Gilbertson, 1998; Wilson and Sullivan, 1998).

SIGNAL TRANSMISSION

From a communications viewpoint the CCN is multiply noded (each osteocyte is a node) and multiply connected. Each osteocytic process is a connection between two osteocytes, and each osteocyte is multiply connected to a number of osteocytes that are near neighbors (see Fig. 51.1). Cell-to-cell communication is considered first below, then we speculate on the ability of the CCN to compute as well as signal. It is useful to note the possibility that bone cells, like neurons, may communicate intercellular information by volume transmission, a process that does not require direct cytological contact, but rather utilizes charges in the environment (Fuxe and Agnati, 1991; Marotti, 1996; Schaul, 1998).

Cell-to-cell communication

In order to transmit a signal over the CCN, one osteocyte must be able to signal a neighboring osteocyte that will then pass the signal on until it reaches an osteoblast on the bone surface. There are a variety of chemical and electrical cell-to-cell communication methods (De Mello, 1987). The passage of chemical signals, such as Ca^{2+}, from cell to cell appears to occur at a rate that would be too slow to respond to the approximately 30-Hz signal associated with muscle firing. We focus here on electrical cell-to-cell communication. Zhang *et al.* (1997, 1998a) have formulated a cable model for cell-to-cell communication in an osteon. The spatial distribution of intracellular electric potential and current from the cement line to the lumen of an osteon was estimated as the frequency of the loading and conductance of the gap junction were altered. In this model the intracellular potential and current are driven by the mechanically induced strain-generated streaming potentials produced by the cyclic mechanical loading of bone. The model differs from earlier studies (Harrigan and Hamilton, 1993) in that it pursues a more physiologic approach in which the microanatomical dimensions of the connexon pores, osteocytic processes, and the distribution of cellular membrane area and capacitance are used to estimate quantitatively the leakage of current through the osteoblast membrane, the time delay in signal transmission along the cable, and the relative resistance of the osteocytic processes and the connexons in their open and closed states. The model predicts that the cable demonstrates a strong resonant response when the cable coupling length approaches the osteonal radius. The theory also predicts that the pore pressure relaxation time for the draining of the bone fluid into the osteonal canal is of the same order as the characteristic diffusion time for the spread of current along the membrane of the osteocytic processes. This coincidence of characteristic times produced a spectral resonance in the cable at 30 Hz. These two resonances lead to a large amplification of the intracellular potential and current in the surface osteoblasts, which could serve as the initiating signal for osteoblasts to conduct remodeling.

Signal processing and integration

When a physical representation of a CCN, such as Fig. 53.1, is viewed by someone familiar with communications, there is often an intuitive response that the CCN may function as a neural network for processing the mechanical loading stimulus signals being felt over the network. That idea is explored here with no justification other than shared intuition. A CCN is operationally analogous to an artificial neural network in which massively parallel, or parallel distributed, signal processing occurs (Edin and Trulsson, 1992; Denning, 1990; Martino *et al.*, 1994). Fortunately, the bases of connectionist theory are sufficiently secure to permit a biologically realistic CCN model (Dayhoff, 1990; Grossberg, 1988; Hinton and Anderson, 1989; McClelland and Rumelhart, 1987; Zorntzer *et al.*, 1990).

A CCN consists of a number of relatively simple, densely interconnected, processing elements (bone cells), with many more interconnections than cells. Operationally these cells are organized into layers: an initial input, a final output, and one or more intermediate or hidden layers. However, such networks need not be numerically complex to be operationally complex (Kupfermann, 1992).

The operational processes are identical, in principle, for all bone cells in all layers. Each cell in any layer may simultaneously receive several weighted (i.e., some quantitative measure) inputs. In the initial layer these are the loading stimuli (mechanoreception). Within each cell independently, "all the weighted inputs are then summed" (Wasserman, 1989). This net sum is then compared, within the cell, against some threshold value. If this liminal value is exceeded, a signal is generated (mechanotransduction in input layer cells) that is then transmitted identically to all the hidden layer cells (adjacent osteocytes) to which each initial layer cell is connected. Similar processes of weighted signal summation, comparison, and transmission occur in these layers until final layer cells (osteoblasts) are reached. The outputs of these surface-situated cells regulate the specific adaptation process of each group of osteoblasts (Parfitt, 1994). All neighboring osteoblasts that carry out an identical bone adaptational process form a communication compartment, a cohort of structurally and operationally similar cells, because all these cells are interconnected by open, functional gap junctions. At the boundary between such compartments that are carrying out different adaptational processes, the intervening gap junctions are closed and are incapable of transmitting information. These boundaries are probably changing continuously, because some of the cells have some down time (Jeansonne *et al.*, 1979; Spray, 1998).

Information is not stored discretely in a CCN, as is the case in conventional computers. Rather it is distributed across all or part of the network, and several types of information may be stored simultaneously. The instantaneous state of a CCN is a property of all of its cells and of their connections. Accordingly, its informational representation is redundant, assuring that the network is fault, or error, tolerant; i.e., one or several inoperative cells cause little or no noticeable loss in network operations (Wasserman, 1989).

CCNs exhibit oscillation; i.e., iterative reciprocal (feedback) signaling between layers enables them to adjustively self-organize. This is related to the fact that CCNs are not preprogrammed, rather they learn by unsupervised training (Fritzke, 1994), a process involving the adaptation of the CCNs to the responses of the cytoskeleton to physical activity (Dayhoff *et al.,* 1992). In this way the CCN adjusts to the customary mechanical loading of the whole bone (Lanyon, 1984). In a CCN, structurally more complex attributes and behavior gradually self-organize and emerge during operation. These are not reducible; they are neither apparent nor predictable from a prior knowledge of the behavior of individual cells.

As noted above, gap junctions as electrical synapses permit bidirectional flow of information. This is the cytologic basis for the oscillatory behavior of a CCN. The presence of sharp discontinuities between groups of phenotypically different osteoblasts is related also to an associated property of gap junctions, i.e., their ability to close and so prevent the flow of information (Kam and Hodgins, 1992; Kodoma and Eguchi, 1994). Significantly, informational networks can also transmit inhibitory signals, a matter beyond our present scope (Marotti *et al.,* 1992).

It is suggested that a CCN displays the following attributes: developmentally it is self-organized and self-adapting and, in the sense that it is epigenetically regulated, it is untrained; operationally it is a stable, dynamic system, whose oscillatory behavior permits feedback—in this regard, it is noted that a CCN operates in a noisy, nonstationary environment, and that it also employs useful and necessary inhibitory inputs.

The CCN permits a triad of histologic responses to a (seemingly) unitary loading event. Although in this chapter, as in almost all the related literature, the organization of bone cells is treated as if it existed only in two dimensions, and as if bone tissue loadings occurred only at certain discrete loci, and that without consideration of loading gradients, the biologic situation is otherwise. Given such a loading event, a three-dimensional bone volume, gradients of deformation must exist, and each osteocyte may sense correspondingly different strain properties. Moreover, it is probable that each osteocyte potentially is able to transmit three different signals in three different directions, some stimulatory and some inhibitory: such states have not yet been modeled (Moss, 1997a,b).

A tentative mechanotransduction synthesis

The molecular lever mechanisms that permit muscle function to regulate directly the genomic activity of strained bone cells, including their phenotypic expression, when combined together with electric field effects and contraction frequency energetics, provide a biophysical basis for an earlier hypothesis of epigenetic regulation of skeletal tissue adaptation (Moss and Young, 1960; Moss, 1962, 1968, 1969; Moss and Salentijn, 1969a,b).

It is probable (Moss, 1997c,d) that electrical and mechanical transductive processes are neither exhaustive nor mutually exclusive. While utilizing differing intermediate membrane mechanisms and/or processes, they share a common final pathway (Bhalla and Iyengar, 1999); i.e., both mechanical and electrical transductions result in transplasma membrane ionic flow(s), creating a signal(s) capable of intercellular transmission to neighboring bone cells via gap junctions (Weinbaum *et al.,* 1994; Zeng *et al.,* 1994; Cowin *et al.,* 1995; Zhang *et al.,* 1997, 1998a). These signals are inputs to a CCN, whose outputs regulate the bone adaptational processes.

The primacy of electrical signals is suggested here, because although bone cell transduction may also produce small biochemical molecules that can pass through gap junctions, the time course of mechanosensory processes is believed to be too rapid for the involvement of secondary messengers (French, 1992; Wildron *et al.,* 1992; Carvalho *et al.,* 1994). As was noted previously, the passage of chemical signals, such as Ca^{2+}, from cell to cell appears to occur at a rate that would be too slow to respond to the approximately 30-Hz signal associated with muscle firing.

QUESTIONS FOR FUTURE RESEARCH; SOCRATIC QUESTIONS

1. Given that several alternative plausible osseous mechanosensory modalities are postulated in the literature, is it reasonable to consider that a possible "common pathway" exists, capable of subsuming them all (cf. Bhalla and Iyengar, 1999; Zuker and Ranganathan, 1999)?

2. Is it useful to consider osseous mechanosensation at the quantum level (cf. Ghazi *et al.,* 1998; Green and Triffit, 1997)?

3. Does NO play a significant role in osseous mechanosensation (cf. Lee *et al.,* 1998; Pitsillides *et al.,* 1995; Klein-Nulend *et al.,* 1995b)?

4. Does the osteocytic syncytium share similar response characteristics with other cellular syncytia; i.e., are they also periodic, chaotic, and functioning at the edge of a state of chaos (cf. Kashimori *et al.,* 1998)?

5. Is bone a "smart material," organized as a "smart structure" (cf. Wang and Kang, 1998; Culshaw, 1996)? Further, in this regard, is bone an "intelligent material" in the sense that this is reflected in the proposed relationship to "strain history"?

6. Is sufficient attention given to the inherent difference between *in vivo* and *in vitro* experimentation (cf. Brown *et al.,* 1998)?

7. Given the rapidly evolving conceptual bases for the complexity sciences, is it appropriate to consider the extent to which bone tissue and cellular responses to mechanical loadings are evidence of a self-adapting and self-organizing biological system (cf. Heylighen, 1997; Wriggers *et al.,* 1998; Cammarota and Onaral, 1998)?

8. Given that a single osteocyte, existing in a synctial cellular network capable of signaling back-propagation, has three-dimensional connectivity, is it possible that in response to a unitary loading event, a unitary osteocyte is capable of simultaneously transmitting three different "signals," one in each direction; i.e., one signal for deposition, one for resorption and one for stasis? If so, how is this accomplished? And if not, why not?

9. What factor(s) regulate the "size" (area) of a "packet" (communication compartment, cohort) of osteoblasts, all engaging simultaneously in the same activity (i.e., deposition, resorption, stasis) on a bone surface, at any one period of time?

10. What is the relationship between the electromagnetic frequencies to which bone responds and those frequencies to which other, nonosseous cells and tissues respond (cf. Shafik, 1998)?

11. Is it possible to develop a comprehensive explication of all aspects of bone tissue response to extrinsic "loadings" that is equally valid at all organizational levels—molecular, microscopic, cellular, tissue, and organismal (cf. Sasaki and Odajima, 1996)?

12. What are the rules of the self-assembly process that governs the formation of structure in bone? How does this process differ in morphogenesis, growth, strain adaptation, fracture healing, and the development of secondary osteons?

REFERENCES

Aarden, E. M. (1996). The third cell—A study of the role of the osteocyte in bone. Thesis, Leiden.

Aarden, E. M., Burger, E. H., and Nijweide, P. J. (1994). Function of osteocytes in bone. *J. Cell. Biochem.* **55**, 287–299.

Atkinson, P. J., and Hallsworth, A. S. (1983). The changing structure of aging human mandibular bone. *Gerodontology* **2**, 57–66.

Ayasaka,, N., Kondo, T., Goto, T., *et al.* (1992). Differences in the transport systems between cementocytes and osterocytes in rats using microperoxidase as a tracer. *Arch. Oral Biol.* **37**, 363–368.

Beardslee, M. A., Laing, J. G., Beyer, E. C. *et al.* (1998). Rapid turnover of connexin 43 in the adult rat heart. *Circ. Res.* **83**, 629–635.

Bennett, M. V. L., and Goodenough, D. A. (1978). Gap junctions. Electronic coupling and intercellular communication. *Neurosci. Res. Program Bull.* **16**, 373–485.

Bhalla, U. S., and Iyengar, R. (1999). Emergent properties of networks of biological signaling pathways. *Science* **283**, 381–383.

Bingmann, D., Tetsch, D., and Fritsch, J. (1988a). Membraneigenschaften von zellen aus knochenexplantaten. *Z. Zahnartzl. Implantol.* **4**, 277–281.

Bingmann, D., Tetsch, D., and Fritsch, J. (1988b). Membrane properties of bone cells derived from calvaria of newborn rats (tissue culture). *Pflueger's Arch.* **S412**, R14–R15.

Brighton, C. T., Okerehe, E., Pollack, S., *et al.* (1992). *In vitro* bone-cell response to a capacitively coupled electrical field. Role of field strength, pulse pattern and duty cycle. *Clin. Orthop. Relat. Res.* **285**, 255–262.

Brown, T. D., Pedersen, D. R., Gray, M. L., *et al.* (1990). Toward an identification of mechanical parameters initiating periosteal remodeling: A combined experimental and analytic approach. *J. Biomech.* **23**, 893–905.

Brown, T. D., Bottlang, M., Pederson, D. R. *et al.* (1998). Loading paradigms—intentional and unintentional—for cell culture mechanostimulus. *Am. J. Med. Sci.* **316**, 162–168.

Burger, E. L., Klein Nulend, J., and Cowin, S. C. (1998). Mechanotransduction in bone. *Adv. Organomet. Biol.* **5a**, 107–118.

Cammarota, J. P., Jr., and Onaral, B. (1998). State transitions in physiologic systems: A complexity model for loss of consciousness. *IEEE Trans. Biomed. Eng.* **45**, 1017–1023.

Carter, D. R., and Hayes, W. C. (1976). Bone compressive strength: The influence of density and strain rate. *Science* **194**, 174–175.

Carvalho, R. S., Scott, J. E., Suga, D. M., *et al.* (1994). Stimulation of signal; transduction pathways in osteoblasts by mechanical strain potentiated by parathyroid hormone. *J. Bone Miner. Res.* **9**, 999–1011.

Carvalho, R. S., Schaffer, J. L., and Gerstenfeld, L. C. (1998). Osteoblasts induce osteopontin expression in response to attachment on fibronectin: Demonstration of a common role for integrin receptors in the signal transduction processes of cell attachment and mechanical stimulation. *J. Cell. Biochem.* **70**, 376–390.

Chesnoy-Marchais, D., and Fritsch, J. (1988). Voltage-gated sodium and calcium currents in rat osteoblasts. *J. Physiol. (London)* **398**, 291–311.

Civitelli, R. (1995). Cell-cell communication in bone. *Calcif. Tissue Int.* **56**(Suppl. 1), S29–S31.

Clark, E. A., and Brugge, J. S. (1995). Integrins and signal transduction pathways: The road taken. *Science* **268**, 233–239.

Cowan, D. B., Lye, S. J., and Langille, B. L. (1998). Regulation of vascular connexin 43 gene expression by mechanical loads. *Circ. Res.* **82**, 786–793.

Cowin, S. C. (1998). On mechanosensation in bone under microgravity. *Bone* **22**, 119S–125S.

Cowin, S. C. (1999). Bone poroelasticity. *J. Biomech.* **32**, 218–238.

Cowin, S. C., and Hegedus, D. M. (1976). Bone remodeling I: A theory of adaptive elasticity. *J. Elasticity* **6**, 313–325.

Cowin, S. C., and Weinbaum, S. (1998). Strain amplification in the bone mechanosensory system. *Am. J. Med. Sci.* **316**, 184–188.

Cowin, S. C., Moss-Salentijn, L., and Moss, M. L. (1991). Candidates for the mechanosensory system in bone. *J. Biomech. Eng.* **113**, 191–197.

Cowin, S. C., Weinbaum, S., and Zeng, Y. (1995). A case for bone canaliculi as the anatomical site of strain generated potentials. *J. Biomech.* **28**, 1281–1296.

Cui, C., Smith, D. O., and Adler, J. (1995). Characterization of mechanosensitive channels in *Eschericia coli* cytoplasmic cell membrane by whole-cell patch clamp recording. *J. Membr. Biol.* **144**, 31–42.

Culshaw, P. (1996). "Smart Structures and Materials." Artech House, Boston.

Curtis, T. A., Ashrafi, S. H., and Weber, D. F. (1985). Canalicular communication in the cortices of human long bones. *Anat. Rec.* **212**, 336–344.

Davidson, R. M., Tatakis, D. W., and Auerbach, A. L. (1990). Multiple forms of mechanosensitive ion channels in osteoblast-like cells. *Eur. J. Physiol.* **416**, 646–651.

Dayhoff, J. (1990). "Neural Network Architecture." Van Nostran-Reinhold, New York.

Dayhoff, J. E., Hameroff, S. R., Lahoz-Beltra, R., *et al.* (1992). Intracellular mechanisms in neuronal learning: Adaptive models. *Int. Jt. Conf. Neural Networks,* pp. 173–178.

De Mello, W. C. (1987). The ways cells communicate. *In* "Cell-to-Cell Communication," pp. 1–20. Plenum, New York.

Denning, P. J. (1990). Neural networks. *Am. Sci.* **80**, 426–429.

Dillaman, R. M. (1984). Movement of ferritin in the 2 day-old chick femur. *Anat. Rec.* **209**, 445–453.

Dolce, C., Kinniburgh, A. J., and Dziak, R. (1996). Immediate early-gene induction in rat osteoblastic cells after mechanical deformation. *Arch. Oral Biol.* **41**, 1101–1108.

Doty, S. B. (1981). Morphological evidence of gap junctions between bone cells. *Calcif. Tissue Int.* **33**, 509–512.

Doty, S. B. (1989). Cell-to-cell communication in bone tissue. *In* "The Biological Mechanism of Tooth Eruption and Root Resorption" (Z. Davidovitch, ed.), pp. 61–69. EBSCO Media, Birmingham, AL.

Duncan, R., and Misler, S. (1989). Voltage-activated and stretch activated Ba^{2+} conducting channels in an osteoblast-like cell line (URM 106). *FEBS Lett.* **251**, 17–21.

Edin, B. B., and Trulsson, M. (1992). Neural network analysis of the information content in population responses from human periodontal receptors. *SPIE, Sci. Neural Networks* **1710**, 257–266.

Ferrier, J., Crygorczyk, C., Grygorczyk, R., *et al.* (1991). Ba^{2+}-induced action potentials in osteoblastic cells. *J. Membr. Biol.* **123**, 255–259.

Forgacs, G. (1995). Biological specificity and measurable physical properties of cell surface receptors and their possible role in signal transduction through the cytoskeleton. *Biochem. Cell Biol.* **73**, 317–326.

Fraser, D. J., and Macdonald, A. G. (1994). Crab hydrostatic pressure sensors. *Nature (London)* **371**, 383–384.

French, A. S. (1992). Mechanotransduction. *Annu. Rev. Physiol.* **54**, 135–152.

Fritton, S. P., McLeod, K. J., and Rubin, C. T. (1999). Quantifying the strain history of bone: Spatial uniformity and self-similarity of low magnitude strains. *J. Biomech.,* in press.

Fritzke, B. (1994). Growing cell structures—A self-organizing network for unsupervised and supervised learning. *Neural Networks* **7**, 1441–1460.

Frost, H. M. (1964). "The Laws of Bone Structure." Thomas, Springfield, IL.

Frost, H. M. (1969). Tetracycline-based histological analysis of bone remodeling. *Calcif. Tissue Res.* **3**, 211–237.

Fuxe, J., and Agnati, L. F., eds. (1991). "Volume Transmission in the Brain." Raven Press, New York.

Fyhrie, D. P., and Carter, D. R. (1986). A unifying principle relating stress to trabecular bone morphology. *J. Orthop. Res.* **4**, 304–317.

Ghazi, A., Berrier, C., *et al.* (1998). Mechanosensitive ion channels and their mode of activation. *Biochemie* **80**, 357–362.

Gilbertson, T. A. (1998). Peripheral mechanisms of taste. *In* "The Scientific Basis of Eating" (R. W. A. Linden, ed.), Vol. 9, pp. 1–28. Karger, Basel.

Goldsmith, P. (1994). Plant stems: A possible model system for the transduction of mechanical information in bone modeling. *Bone* **15**, 249–250.

Goldstein, S. A., Matthews, L. S., Kuhn, J. L., *et al.* (1991). Trabecular bone remodeling: An experimental model. *J. Biomech.* **24**(Suppl. 1), 135–150.

Gourdie, R., Green, C. (1993). The incidence and size of gap junctions between bone cells in rat calvarial. *Anat. Embryol.* **187**, 343–352.

Green, P. B. (1994). Connecting gene and hormone action to form, pattern and organogenesis: Biophysical transductions. *J. Exp. Bot.* **45**(Spec. Issue), 1775–1788.

Green, S. S., and Triffit, T. (1997). The animal brain as a quantal computer. *J. Theor. Biol.* **184**, 385–403.

Gross, T. S., Edwards, J. L., *et al.* (1997). Strain gradients correlate with sites of periosteal bone formation. *J. Bone Miner. Res.* **12**, 982–988.

Grossberg, S. (1988). "Neural Networks and Artificial Intelligence." MIT Press, Cambridge, MA.

Guggino, S. E., LaJeunesse, D., Wagner, J. A., *et al.* (1989). Bone remodeling signaled by a dihydropyridine- and phenylalkylamine-sensitive calcium channel. *Proc. Natl. Acad. Sci. U.S.A.* **8**, 2957–2960.

Hackney, C. M., and Furness, D. N. (1995). Mechanotransduction in vertebrate hair cells: Structure and function of the stereociliary bundle. *Am. J. Physiol.* **268** (Cell Physiol. 37), C1–C13.

Hamill, O. P., and McBride, D. W., Jr. (1995). Mechanoreceptive membrane channels. *Am. Sci.* **83**, 30–37.

Hamill, O. P., and McBride, D. W., Jr. (1996). The pharmacology of mechanogated membrane ion channels. *Pharmacol. Rev.* **48**, 231–248.

Harrigan, T. P., and Hamilton, J. J. (1993). Bone strain sensation via transmembrane potential changes in surface osteoblasts: Loading rate and microstructural implications. *J. Biomech.* **26**, 183–200.

Harter, L. V., Hruska, K. A., and Duncan, R. L. (1995). Human osteoblast-like cells respond to mechanical strain with increased bone matrix protein production independent of hormonal regulation. *Endocrinology (Baltimore)* **136**, 528–535.

Haskin, C., Cameron, I. (1993). Physiological levels of hydrostatic pressure alter morphology and organization of cytoskeletal and adhesion proteins in MG-63 osteosarcoma cells. *Biochem. Cell Biol.* **71**, 27–35.

Hecker, M., Mülsch, A., Bassenge, E., and Busse, R. (1993). Vasoconstriction and increased flow: Two principal mechanisms of shear stress-dependent endothelial autocoid release. *Am. J. Physiol.* **265** (Heart Circ. Physiol. 34), H828–H833.

Hert, J., Liskova, M., and Landgrot, B. (1969). Influence of the long-term continuous bending on the bone. An experimental study on the tibia of the rabbit. *Folia Morphol.* **17**, 389–399.

Hert, J., Liskova, M., and Landa, J. (1971). Reaction of bone to mechanical stimuli. Part I. Continuous and intermittent loading of tibia in rabbit. *Folia Morphol.* **19**, 290–300.

Hert, J., Pribylova, E., and Liskova, M. (1972). Reaction of bone to mechanical stimuli. Part 3. Microstructure of compact bone of rabbit tibia after intermittent loading. *Acta Anat.* **82**, 218–230.

Heylighen, F. (1997). Publications on complex, evolving systems: A citation based survey. *Complexity* **2**, 31–36.

Hinton, G. E., and Anderson, J. A. (1989). "Parallel Models of Associative Memory." Erlbaum, Hillsdale, NJ.

Hughes, D. E., Salter, D. M., Dedhar, S., *et al.* (1993). Integrin expression in human bone. *J. Bone Miner. Res.* **8**, 527–533.

Huiskes, R., Weinans, H. J., Grootenboer, H. J., *et al.* (1987). Adaptive bone remodeling theory applied to prosthetic-design analysis. *J. Biomech.* **20**, 1135–1150.

Hung, C. T., Pollack, S. R., Reilly, T. M., and Brighton, C. T. (1995). Real-time calcium response of cultured bone cells to fluid flow. *Clin. Orthop. Relat. Res.* **313**, 256–269.

Ingber, D. E. (1998). Cellular basis of mechanotransduction. *Biol. Bull. (Woods Hole, Mass.)* **194**, 323–327.

Jacobs, R., ed. (1998). "Osseoperception." Catholic University, Leuven.

Jan, L. Y., and Jan, Y. N. (1992). Tracing the roots of ion channels. *Cell (Cambridge, Mass.)* **69**, 715–718.

Janmey, P. A. (1998). The cytoskeleton and cell signaling: Component localization and mechanical coupling. *Physiol. Rev.* **78**, 763–774.

Jaworski, Z. F. G., Liskova-Kiar, M., and Uhthoff, H. K. (1980). Effect of long term immobilization on the pattern of bone loss in older dogs. *J. Bone Jt. Surg., Br. Vol.* **62B**, 104–110.

Jeansonne, B. G., Feagin, F. F., *et al.* (1979). Cell-to-cell communication of osteoblasts. *J. Dent. Res.* **58**, 1415–1423.

Johnson, M. W., Chakkalakal, D. A., Harper, R. A., *et al.* (1982). Fluid flow in bone. *J. Biomech.* **11**, 881–885.

Jones, D. B., and Bingmann, D. (1991). How do osteoblasts respond to mechanical stimulation? *Cells Mater.* **1**, 329–340.

Jones, S. J., Gray, C., Sakamaki, H., *et al.* (1993). The incidence and size of gap junctions between bone cells in rat calvaria. *Anat. Embryol.* **187**, 343–352.

Junge, D. (1998). "Oral Sensorimotor Function." Medico Dental Media, International, Inc., Pacific, MO.

Kam, E., and Hodgkins, M. B. (1992). Communication compartments in hair follicles and their implication in differentiative control. *Development (Cambridge, UK)* **114**, 389–393.

Kamiya, A., Bukhari, R., and Togawa, T. (1984). Adaptive regulation of wall shear stress optimizing vascular tree function. *Bull. Math. Biol.* **46**, 127–137.

Kashimori, Y., Funakubo, H., and Kambara, T. (1998). Effect of syncytium structure of receptor systems on stochastic resonance by chaotic potential fluctuation. *Biophys. J.* **75**, 1700–1711.

Kernan, M., Cowan, D., and Zuker, C. (1994). Genetic dissection of mechanoreception-defective mutations. *Drosophila Neuron* **12**, 1195–1206.

Keynes, R. D. (1994). The kinetics of voltage-gated ion channels. *Q. Rev. Biophys.* **27**, 339–434.

Klein-Nulend, J., Van der Plas, A., Semeins, C. M., Ajubi, N. E., Frangos, J. A., Nijweide, P. J., and Burger, E. H. (1995a). Sensitivity of osteocytes to biomechanical stress *in vitro.* *FASEB J.* **9**, 441–445.

Klein-Nulend, J., Semeins, C. M., Ajubi, N. E., Nijweide, P. J., and Burger, E. H. (1995b). Pulsating fluid flow increases nitric oxide (NO) synthesis by osteocytes but not periosteal fibroblasts—Correlation with prostaglandin upregulation. *Biochem. Biophys. Res. Commun.* **217**, 640–648.

Kodoma, R., and Eguchi, G. (1994). The loss of gap junctional cell-to-cell communication is coupled with dedifferentiation of retinal pigmented epithelial cells in the course of transdifferentiation into the lens. *Int. J. Dev. Biol.* **38**, 357–364.

Kufahl, R. H., and Saha, S. (1990). A theoretical model for stress-generated fluid flow in the canaliculi-lacunae network in bone tissue. *J. Biomech.* **23**, 171–180.

Kupfermann, I. (1992). Neural networks: They do not have to be complex to be complex. *Behav. Brain Sci.* **15**, 767–768.

Lanyon, L. E. (1984). Functional strain as a determinant for bone remodeling. *Calcif. Tissue Int.* **36**, S56–S61.

Lanyon, L. E. (1993). Osteocytes, strain detection, bone modeling and remodeling. *Calcif. Tissue Int.* **53**(Suppl. 1), S102–S106.

Lanyon, L. E., Goodship, A. E., Pye, C. J., and MacFie, J. H. (1982). Mechanically adaptive bone remodeling. *J. Biomech.* **15**, 141–154.

Lecanda, F., Towler, D. A., *et al.* (1998). Gap junctional communication modulates gene expression in osteoblastic cells. *Mol. Biol. Cell* 9, 2249–2258.

Lee, D. H., Frean, S. P., Lees, P. *et al.* (1998). Dynamic mechanical compression influences nitric oxide by articular chondrocytes. *Biochem. Biophys. Res. Commun.* 251, 580–585.

Machwate, M., Jullienne, A., Moukhtar, M., *et al.* (1995). Temporal variation of c-*fos* protooncogene expression during osteoblasts differentiation and osteogenesis in developing rat bone. *J. Cell. Biochem.* 57, 62–70.

Marotti, G. (1996). The structure of bone tissues and the cellular control of their deposition. *Ital. J. Anat. Embryol.* 101, 25–79.

Marotti, G., Ferretti, M., Muglia, M. A., *et al.* (1992). A quantitative evaluation of osteoblast–osteocyte relationships on growing endosteal surface of rabbit tibiae. *Bone* 13, 363–368.

Martin, R. B., and Burr, D. B. (1982). A hypothetical mechanism for the stimulation of osteonal remodeling by fatigue damage. *J. Biomech.* 15, 137–139.

Martin, T. J., and Ng, K. W. (1994). Mechanisms by which cells of the osteoblastic lineage control osteoclast formation and activity. *J. Cell. Biochem.* 56, 357–366.

Martino, R. L., Johnson, C. A., Suh, E. B., *et al.* (1994). Parallel computing in biomedical research. *Science* 265, 902–907.

Massass, R., Bingmann, D., Korenstein, R., *et al.* (1990). Membrane potential of rat calvaria bone cells: Dependence on temperature. *J. Cell. Physiol.* 144, 1–11.

McClelland, J. L., and Rumelhart, D. E. (1987). "Parallel Distributed Processing. Psychological and Biological Models," Vol. 2. MIT Press, Cambridge, MA.

McLeod, K. J., and Rubin, C. T. (1992). The effect of low-frequency electrical fields on osteogenesis. *J. Bone Jt. Surg., Am. Vol.* 74-A, 920–929.

McLeod, K. J., Donahue, H. J., *et al.* (1993). Electric fields modulate bone cell function in a density-dependent manner. *J. Bone Miner. Res.* 8, 977–984.

McMahon, T. A. (1984). "Muscles, Reflexes, and Locomotion." Princeton University Press, Princeton, NJ.

Meazzini, M. C., Toma, C. D., *et al.* (1998). Osteoblast cytoskeletal modulation in response to mechanical strain *in vitro*. *J. Bone Jt. Surg.* 16, 170–180.

Minkoff, R., Rundus, V. R., Parker, S. B., *et al.* (1994). Gap junction proteins exhibit early and specific expression during intramembranous bone formation in the developing chick mandible. *Anat. Embryol.* 190, 231–241.

Minkoff, R., Bales, E. S., *et al.* (1999). Antisense oligonucleotide blockade of connexin expression during embryonic bone formation: Evidence of functional compensation within a multigene family. *Dev. Genet.* 24, 43–56.

Montgomery, R. J., Sutker, B. D., *et al.* (1988). Interstitial fluid flow in cortical bone. *Microvasc. Res.* 23, 188–200.

Moreno, A. P., Rook, M. B., *et al.* (1994). Gap junction channels: Distinct voltage-sensitive and -insensitive conductance states. *Biophys. J.* 67, 113–119.

Morris, C. E. (1990). Mechanosensitive ion channels. *J. Membr. Biol.* 113, 93–107.

Moss, M. L. (1962). The functional matrix. *In* "Vistas in Orthodontics" (B. Kraus and R. Reidel, eds.), pp. 85–98. Lea & Febiger, Philadelphia.

Moss, M. L. (1968). The primacy of functional matrices in orofacial growth. *Trans. Br. Soc. Study Orthodont. Dent. Pract.* 19, 65–73.

Moss, M. L. (1969). A theoretical analysis of the functional matrix. *Acta Biotheor.* 18, 195–202.

Moss, M. L. (1978). "The Muscle–Bone Interface: An Analysis of a Morphological Boundary." Center for Human Growth and Development, Ann Arbor, MI.

Moss, M. L. (1991a). Bone as a connected cellular network: Modeling and testing. *In* "Topics in Biomedical Engineering" (G. Ross, ed.), pp. 117–119. Pergamon, New York.

Moss, M. L. (1991b). Alternate mechanisms of bone remodeling: Their representation in a connected cellular network model. *Ann. Biomed. Eng.* 19, 636.

Moss, M. L. (1997a). The functional matrix hypothesis revisited. 1. The role of mechanotransduction. *Am. J. Orthod. Dentofacial Orthop.* 112, 8–11.

Moss, M. L. (1997b). The functional matrix hypothesis revisited. 2. The role of an osseous connected cellular network. *Am. J. Orthod. Dentofacial Orthop.* 112, 221–226.

Moss, M. L. (1997c). The functional matrix hypothesis revisited. 3. The genomic thesis. *Am. J. Orthod. Dentofacial Orthop.* 112, 338–342.

Moss, M. L. (1997d). The functional matrix hypothesis revisited. 4. The epigenetic antithesis and the resolving synthesis. *Am. J. Orthod. Dentofacial Orthop.* 112, 410–417.

Moss, M. L., Salentijn, L. (1969a). The primary role of the functional matrices in facial growth. *Am. J. Orthod.* 55, 566–577.

Moss, M. L., and Salentijn, L. (1969b). The capsular matrix. *Am. J. Orthod.* 56, 474–490.

Moss, M. L., and Young, R. (1960). A functional approach to craniology. *Am. J. Phys. Anthropol.* 18, 81–92.

Moss-Salentijn, L. (1992). The human tactile system. *In* "Advanced Tactile Sensing for Robotics" (H. R. Nicholls, ed.), Chapter 4. World Scientific, Singapore.

Nowak, R. (1992). Cells that fire together, wire together. *J. NIH Res.* 4, 60–64.

O'Connor, J. A., Lanyon, L. E., and MacFie, H. (1982). The influence of strain rate on adaptive bone remodeling. *J. Biomech.* 15, 767–781.

Olsson, S., and Hanson, B. S. (1995). Action potential-like activity found in fungal mycelia is sensitive to stimulation. *Naturwissenschaften* 82, 30–31.

Otter, M. W., Palmieri, V. R., Wu, D. D., *et al.* (1992). A comparative analysis of streaming potentials *in vivo* and *in vitro*. *J. Orthop. Res.* 10, 710–719.

Otter, M. W., McLeod, K. J., and Rubin, C. T. (1998). Effects of electromagnetic fields in experimental fracture repair. *Clin. Orthop. Relat. Res.* **355S**, S90–S104.

Owan, I., Burr, D. B., *et al.* (1997&). Mechanotransduction in bone: Osteoblasts are more responsive to fluid forces than mechanical strain. *Am. Physiol. Soc.* **273**, C810–C815.

Parfitt, A. M. (1994). Osteonal and hemi-osteonal remodeling: The spatial and temporal framework for signal traffic in adult human bone. *J. Cell. Biochem.* **55**, 273–286.

Petrov, A. G., and Usherwood, P. N. (1994). Mechanosensitivity of cell membranes. Ion channels, lipid matrix and cytoskeleton. *Eur. Biophys. J.* **23**, 1–19.

Piekarski, K., and Munro, M. (1977). Transport mechanism operating between blood supply and osteocytes in long bones. *Nature (London)* **269**, 80–82.

Pitsillides, A. A., Rawlinson, S. C. F., *et al.* (1995). Mechanical strain-induced NO production by bone cells: A possible role in adaptive bone (re)modeling? *FASEB J.* **9**, 1614–1622.

Pollack, S. R., Petrov, N., *et al.* (1984). An anatomical model for streaming potentials in osteons. *J. Biomech.* **17**, 627–636.

Ravelsloot, J. H., van Houten, R. J., Ypey, D. L., *et al.* (1991). High-conductance anion channels in embryonic chick osteogenic cells. *J. Bone Miner. Res.* **6**, 355–363.

Reich, K. M., Gay, C. V., and Frangos, J. A. (1990). Fluid shear stress as a mediator of osteoblast cyclic adenosine monophosphate production. *J. Cell. Physiol.* **143**, 100–104.

Richardson, A., and Parsons, J. T. (1995). Signal transduction through integrins: A central role for focal adhesion. *BioEssays* **17**, 229–236.

Rodan, G. (1992). Introduction to bone biology. *Bone* **13**, S3–S6.

Rodrequez, A. A., Agre, J. C., *et al.* (1993). Acoustic myography compared to electromyography during isometric fatigue and recovery. *Muscle Nerve* **16**, 188–192.

Rubin, C. T., and Lanyon, L. E. (1984). Regulation of bone formation by applied dynamic loads. *J. Bone Jt. Surg., Am. Vol.* **66-A**, 397–415.

Rubin, C. T., and Lanyon, L. E. (1987). Osteoregulatory nature of mechanical stimuli: Function as a determinant for adaptive bone remodeling. *J. Orthop. Res.* **5**, 300–310.

Rubin, C. T., Donahue, H. J., *et al.* (1993). Optimization of electric field parameters for the control of bone remodeling: Exploitation of an indigenous mechanism for the prevention of osteopenia. *J. Bone Miner. Res.* **8**(Suppl. 2), S573–S581.

Rubinacci, A., Villa, I., *et al.* (1998). Osteocyte-bone lining cell system at the origin of steady ionic current in damaged amphibian bone. *Calcif. Tissue Int.* **63**, 331–339.

Sachs, F. (1986). Biophysics of mechanoreception. *Membr. Biochem.* **6**, 173–195.

Sachs, F. (1988). Mechanical transduction in biological systems. *CRC Rev. Biomed. Eng.* **16**, 141–169.

Sackin, H. (1995). Mechanosensitive channels. *Annu. Rev. Physiol.* **57**, 333–353.

Sadoshima, J., Takahashi, T., *et al.* (1992). Roles of mechano-sensitive ion channels, cytoskeleton and contractile activity in stretch-induced immediate-early gene expression and hypertrophy of cardiac myocytes. *Proc. Natl. Acad. Sci. U.S.A.* **89**, 9905–9909.

Salter, D. M., Robb, J. E., and Wright, M. O. (1997). Electrophysiological responses of human bone cells to mechanical stimulation: Evidence for specific integrin function in mechanotransduction. *J. Bone Miner. Res.* **12**, 1133–1141.

Salzstein, R. A., and Pollack, S. R. (1987). Electromechanical potentials in cortical bone. II. Experimental analysis. *J. Biomech.* **20**, 271–280.

Salzstein, R. A., Pollack, S. R., *et al.* (1987). Electromechanical potentials in cortical bone. I. A continuum approach. *J. Biomech.* **20**, 261–270.

Sasaki, N., and Odajima, S. (1996). Elongation mechanism of collagen fibrils and force-strain relations of tendon at each level of structural hierarchy. *J. Biomech.* **29**, 1131–1136.

Schaul, N. (1998). The fundamental neural mechanisms of electroencephalography. *Electroencephalogr. Clin. Neurophysiol.* **106**, 101–107.

Schirrmacher, K., Schmitz, I., *et al.* (1992). Characterization of gap junctions between osteoblast-like cells in culture. *Calcif. Tissue Int.* **51**, 285–290.

Schirrmacher, K., Brummer, F., *et al.* (1993). Dye and electric coupling between osteoblast-like cells in culture. *Calcif. Tissue Int.* **53**, 53–60.

Scott, G. C., and Korostoff, E. (1990). Oscillatory and step response electromechanical phenomena in human and bovine bone. *J. Biomech.* **23**, 127–143.

Shafik, A. (1998). Electro-oophorogram: A preliminary study of the electrical activity of the ovary. *Gynecol. Obstet. Invest.* **46**, 105–110.

Shapiro, F., Cahill, C., *et al.* (1995). Transmission electron microscopic demonstration of vimentin in rat osteoblast and osteocytic cell bodies and processes using the immunoglod technique. *Anat. Rec.* **241**, 39–48.

Shyy, J. Y., and Chien, S. (1997). Role of integrins in cellular responses to mechanical stress and adhesion. *Curr. Opin. Cell Biol.* **9**, 707–713.

Spray, D. C. (1994). Physiological and pharmacological regulation of gap junction channels. *In* "Molecular Mechanisms of Epithelial Cell Junctions: From Development to Disease" (S. Chi, ed.), pp. 195–215. R. G. Landes, Austin, TX.

Spray, D. C. (1998). Gap junction proteins: Where do they live and how do they die? *Circ. Res.* **83**, 679–681.

Tanaka, T., and Sakano, A. (1985). Differences in permeability of microperoxidase and horseradish peroxidase into alveolar bone of developing rats. *J. Dent. Res.* **64**, 870–876.

Uhthoff, H. K., and Jaworski, Z. F. G. (1978). Bone loss in response to long term immobilization. *J. Bone Jt. Surg., Br. Vol.* **60B**, 420–429.

Uitto, V. J. (1991). Extracellular matrix molecules and their receptors: An overview with special emphasis on periodontal tissues. *Crit. Rev. Oral Biol. Med.* **2**, 323–354.

Wang, L., Fritton, S. P., *et al.* (1999). Fluid pressure relaxation depends upon osteonal microstructure: Modeling of an oscillatory bending experiment. *J. Biomech.* **32**, 663–672.

Wang, N., Butler, J. P., and Ingber, D. E. (1993). Mechanotransduction across the cell surface and through the cytoskeleton. *Science* **260**, 1124–1127.

Wang, Z. L., and Kang, Z. C. (1998). "Functional and Smart Materials." Plenum, New York.

Wasserman, P. D. (1989). "Neural Computation. Theory and Practice." Van Nostrand-Reinhold, New York.

Watanabe, H., Miake, K., and Sasaki, J. (1993). Immunohistochemical study of the cytoskeleton of osteoblasts in the rat calvaria. *Acta Anat.* **147**, 14–23.

Weinbaum, S., Cowin, S. C., and Zeng, Y. (1991). A model for the fluid shear stress excitation of membrane ion channels in osteocytic processes due to bone strain. *In* "Advances in Bioengineering" (R. Vanderby, Jr., ed.), pp. 317–320. Am. Soc. Mech. Eng., New York.

Weinbaum, S., Cowin, S. C., and Zeng, Y. (1994). Excitation of osteocytes by mechanical loading-induced bone fluid shear stresses. *J. Biomech.* **27**, 339–360.

Wildron, D. C., Thain, J. F., *et al.* (1992). Electrical signaling and systematic proteinase inhibitor induction in the wounded plant. *Nature (London)* **360**, 62–65.

Williams, J. L., Iannotti, J. P., Ham, A., Bleuit, J., and Chen, J. H. (1994). Effects of fluid shear stress on bone cells. *Biorheology* **31**, 163–170.

Wilson, D. A., and Sullivan, R. M. (1998). Peripheral mechanisms of smell. *In* "The Scientific Basis of Eating" (R. W. A. Linden, ed.), Vol. 9, pp. 29–39. Karger, Basel.

Wolff, J. (1870). Uber der innere Architektur der Knochen und ihre Bedeutung fur die Frage vom Knochenwachstum. *Arch. Pathol. Anat. Physiol. Klin. Med.* **50**, 389–453.

Wolff, J. (1892). "Das Gesetz der Transformation der Knochen." Hirschwald, Berlin.

Wolff, J. (1986). "The Law of Bone Remodelling." Springer, Berlin.

Wriggers, W., Milligan, R. A., Schulter, K. *et al.* (1998). Self-organized neural networks bridge the biomolecular gap. *J. Mol. Biol.* **284**, 1247–1254.

Yanagishita, M. (1993). Function of proteoglycans in the extracellular matrix. *Acta Pathol. Jpn.* **4**, 283–293.

Zeng, Y., Cowin, S. C., and Weinbaum, S. (1994). A fiber matrix model for fluid flow and streaming potentials in the canaliculi of an osteon. *Ann. Biomed. Eng.* **22**, 280–292.

Zhang, D., Cowin, S. C., and Weinbaum, S. (1997). Electrical signal transmission and gap junction regulation in bone cell network: A cable model for an osteon. *Ann. Biomed. Eng.* **25**, 357–374.

Zhang, D., Cowin, S. C., and Weinbaum, S. (1998a). Electrical signal transmission in a bone cell network: The influence of a discrete gap junction. *Ann. Biomed. Eng.* **26**, 644–659.

Zhang, D., Weinbaum, S., and Cowin, S. C. (1998b). On the calculation of bone pore water pressure due to mechanical loading. *Int. J. Solids Struct.* **35**, 4981–4997.

Zhang, D., Weinbaum, S., and Cowin, S. C. (1998c). Estimates of the peak pressures in the bone pore water. *J. Biomech. Eng.* **120**, 697–703.

Zorntzer, S. F., Davis, J., and Lau, C. (1990). "An Introduction to Neural and Electronic Networks." Academic Press, San Diego, CA.

Zuker, C. S., and Ranganathan, R. (1999). The path to specificity. *Science* **283**, 650–651.

CHAPTER 52

MYOBLAST THERAPY

Joanne C. Cousins, Jennifer E. Morgan, and Terence A. Partridge

INTRODUCTION

Interest in grafting muscle precursor cells (MPCs) initially arose from the desire to discover a treatment for genetic diseases of skeletal muscle, such as Duchenne muscular dystrophy (Partridge *et al.,* 1978). Duchenne muscular dystrophy (DMD), the most common inherited myopathy, is characterized by the lack of the subsarcolemmal protein dystrophin (Hoffman *et al.,* 1987). Muscle fibers that lack dystrophin undergo repeated cycles of necrosis and regeneration, resulting in progressive replacement of muscle fibers by fibrotic tissue and adipose tissue (Dubowitz, 1985). This is reflected in gradual loss of effective muscle function and eventual premature death (Partridge, 1993). Similar but less severe symptoms of muscle weakness and wasting are seen in the rarer Becker muscular dystrophy (BMD), a more phenotypically variable allelic form of muscular dystrophy. A truncated form of the dystrophin protein is present and disease onset is usually later with less severe muscle wasting, allowing some patients to remain ambulant throughout their lives (Dubowitz, 1985). The rationale behind myoblast transplantation therapy (MTT), is that normal myogenic MPCs would participate in the repair and remodeling of the degenerating muscle, making use of the regeneration pathway of normal adult skeletal muscle after injury (Partridge, 1991a; Partridge and Davies, 1995). Normal adult skeletal muscle consists of long, multinucleate postmitotic fibers and quiescent mononuclear MPCs, termed "satellite" cells, situated beneath the basal lamina of each fiber. Following damage to a fiber, satellite cells are activated to proliferate and fuse with the surviving portion of the syncytium to repair the damaged area of the fiber (Bischoff, 1994). MTT would therefore lead to repair of the degenerated region of the muscle fiber; the normal myoblasts and gene products produced by the myonuclei derived from these transplanted cells could compensate for deficient or abnormal host proteins, even if the missing gene or gene product causing the disease were unknown (Partridge and Davies, 1995). Myoblasts are also potentially effective vehicles for transfer of exogenous genes because they are able to divide extensively, migrate, and differentiate. Furthermore, long-term expression of exogenous genes is possible due to the longevity of the postmitotic nuclei derived from the implanted cells. This rationale would also make myoblast transplantation into skeletal muscle an attractive prospect for the synthesis of systemically active peptides and proteins for the treatment of nonmuscle diseases (Barr and Leiden, 1991). Myoblast transplantation has also provided useful insights into how postnatal muscle regenerates and is repaired, which is of vital importance when considering transplantation as a treatment for myopathies.

MYOBLAST-MEDIATED GENE TRANSFER

Myoblast-mediated gene transfer has been investigated as a means to construct a recombinant protein factory within skeletal muscle to synthesize a range of systemically active peptides (Barr and Leiden, 1991). As a target tissue for cell therapy, skeletal muscle has several attractive properties. It is richly vascularized, which would allow efficient systemic delivery of transgene products. Moreover, transplanted myoblasts terminally differentiate on incorporation into the myofiber by

Principles of Tissue Engineering
Second Edition

fusion with each other or with the existing fiber, thus providing a stable tissue for protein production.

C2C12 myoblasts have been used in myoblast transplantation (MT) to express human growth hormone (Barr and Leiden, 1991; Dhawan *et al.,* 1991). Myoblasts engineered in culture to produce recombinant human (Yao and Kurachi, 1992) and canine (Dai *et al.,* 1992) factor IX have given sustained expression of biologically active protein in the serum of transplanted mice. To optimize the procedure, myoblasts were first retrovirally infected with a *lacZ* reporter gene; this also confirmed that the C2C12 cell line used would form muscle on transplantation (Yao and Kurachi, 1992). Persistent factor IX mRNA expression and long-term protein expression were seen following transplantation of engineered primary cells (Yao *et al.,* 1994), and therapeutic levels of human protein were achieved by repeated transplantation (Wang *et al.,* 1997). Stable expression was seen in both studies, and the participation of donor cells in muscle fiber formation was again confirmed by β-galactosidase expression from the *lacZ* reporter gene.

Sufficient and sustained levels of functionally active human erythropoietin have been achieved by transplanting C2C12 myoblasts coexpressing erythropoietin and β-galactosidase; transplanted cells were seen to participate in muscle formation, although tumors were also frequently observed (Hamamori *et al.,* 1995). It should be noted that even though a number of these studies used the C2C12 myogenic cell line (Barr and Leiden, 1991; Dhawan *et al.,* 1991; Yao and Kurachi, 1992; Hamamori *et al.,* 1995), previous work has shown that C2C12 myoblasts generate tumors on transplantation into mouse muscle (Morgan *et al.,* 1992; Wernig *et al.,* 1991). Consequently, the exogenous protein detected could have been secreted by the tumor cells or by mature muscle. This has been addressed in work by Bou-Gharios *et al.* (1999), which demonstrated that mature fibers of donor origin expressed the recombinant protein. Direct intramuscular injection of plasmid DNA was compared to transplantation of *H-2Kᵇ*–tsA58 myoblasts (Jat *et al.,* 1991) engineered *ex vivo* to secrete human α-1-antitrypsin (α1-AT). Immunohistochemistry on the *mdx* nude muscles injected with myoblasts showed that mature donor fibers expressing dystrophin as a marker of donor origin also expressed human α1-AT. Detectable levels of human α1-AT were also observed in serum, although it is unclear whether the α1-AT seen in muscle fibers is directly related to the serum levels of α1-AT.

The problems associated with producing a protein factory within an existing muscle, such as reducing or terminating protein expression, can be overcome by transplanting engineered myoblasts (Irintchev *et al.,* 1998) or bioartificial muscles (organoids); developed in culture from myoblasts (Vandenburg *et al.,* 1996) these products can be surgically introduced into accessible subcutaneous sites. Irintchev *et al.* (1998) showed that as early as 1 week after subcutaneous implantation of myoblasts, muscle of mature morphology and phenotype was formed. These ectopic muscles were vascularized and often connected to underlying skeletal muscle and/or innervated. These are interesting findings in terms of producing recombinant proteins from engineered myoblasts; the subcutaneous space could be exploited for the development of these cells and for delivery of their protein product to the circulation.

THERAPEUTIC MYOBLAST TRANSPLANTATION FOR MYOPATHIES

ANIMAL STUDIES

In vivo MT was first used in the late 1970s to elucidate the basic cell biology of muscle lineages (Lipton and Schultz, 1979; Jones, 1979). To investigate the possibility that donor muscle cells might repair host muscles damaged by myopathy or injury, minced muscle grafts were transplanted between two strains of inbred mice and fusion was observed between donor and host myoblasts (Partridge *et al.,* 1978). Subsequently it has shown that single muscle precursor cells prepared from disaggregated donor muscle can be incorporated into an area of regeneration within the host mouse, forming hybrid fibers from the fusion of donor nuclei with host fibers (Watt *et al.,* 1982). It was postulated that transplantation of myoblasts could provide a possible therapy even without identification of the defective protein or gene responsible for the myopathy (Law *et al.,* 1988; Partridge *et al.,* 1989; Karpati *et al.,* 1989). The first evidence that implanted MPCs could ameliorate a biochemical deficiency was provided in phosphorylase kinase-deficient mice (Morgan *et al.,* 1988). This was followed by the formation of dystrophin-positive fibers in MPC-injected muscular dystrophy (mdx) muscles (Partridge *et al.,* 1989). The mdx mouse (Bulfield *et al.,* 1984) is a biochemical homolog of DMD, lacking the protein dystrophin (Hoffman *et al.,* 1987) due to a point mutation in the dystrophin gene (Sicinski *et al.,* 1989). Characterized by early-onset muscle degeneration and regeneration and mild clinical phenotype, few deleterious effects are present until 20–24 months of age, when many pathological changes similar to DMD are seen (Pastoret and Sebille, 1995a,b). The mdx mice have an almost normal life-span and efficient muscle regeneration, and cannot therefore be considered an ideal disease model (Partridge, 1991b, 1997). The initial experiments showed that dystrophin-positive fibers were produced after transplantation of normal myoblasts into regenerating mdx muscle (Partridge *et al.,* 1989; Karpati *et al.,* 1989), although with current knowledge, some of these fibers (Karpati *et al.,* 1989) were probably due to a reversion from a dystrophin-negative to a dystrophin-positive phenotype (Hoffman *et al.,* 1990) (see later). Blocking the regenerative capacity of the mdx muscle by local high doses of X-irradiation (Wakeford *et al.,* 1991; Quinlan *et al.,* 1995, 1997) and γ-irradiation (Weller *et al.,* 1991) provides a convenient model of the fiber atrophy and connective tissue replacement seen in DMD. Normal myoblasts transplanted into X-irradiated mdx muscles produced large numbers of dystrophin-positive fibers, fiber loss was counteracted, and there was restoration of an almost normal histologic appearance in the muscle (Morgan *et al.,* 1990). MPCs implanted into preirradiated mdx muscles gave rise to long-lived dystrophin-positive fibers and showed better migration when compared with nonirradiated controls (Morgan *et al.,* 1993). Results from a study involving implanting myoblasts overexpressing dystrophin into mdx mice gave an insight into the amount of dystrophin required to repair a fiber (Kinoshita *et al.,* 1998). Although there was a 50-fold overexpression of dystrophin, the membrane area covered by this expression was only increased by 3-fold, and the percentage of normal mRNA by only 4-fold. This can probably be explained by the limited diffusion of dystrophin protein along a fiber from the transgenic nucleus that produced it; however, as there are also mdx nuclei in a mosaic fiber, it does not have to be entirely composed of transplanted normal nuclei.

It has been suggested that lack of dystrophin in the mdx mouse may be compensated by up-regulation of a structurally related protein, utrophin (Deconinck *et al.,* 1997a). Mice lacking both utrophin and dystrophin have a severely myopathic phenotype with ultrastructural, neuromuscular, and myotendinous junction abnormalities (Deconinck *et al.,* 1997a). Mice lacking utrophin alone have subtle neuromuscular defects, and it has been speculated that the relatively mild phenotypes of the dystrophin and utrophin single-mutant mice, compared to the double mutant, reflect the compensation between the two proteins (Grady *et al.,* 1997). In much the same vein it has been suggested that the milder phenotype of the mdx mice compared to DMD patients may be attributable to more successful utrophin compensation, because utrophin is detected around the sarcolemma of regenerating adult mdx skeletal muscle fibers (Deconinck *et al.,* 1997a). Furthermore, expression of a truncated utrophin protein in transgenic mdx mice led to a major functional improvement (Deconinck *et al.,* 1997b) similar to that previously seen following transgenic expression of a truncated dystrophin protein (Dunckley *et al.,* 1993; Deconinck *et al.,* 1996), and expression of a full-length transgenic utrophin protein at the sarcolemma of mdx mice prevented the development of muscular dystrophy (Tinsley *et al.,* 1998). However, because neither the dystrophion nor the utrophin single-mutant mice are phenotypically normal, the two proteins cannot have completely overlapping developmental roles.

Although myoblast transplantation in the mdx mouse has generated large numbers of dystrophin-positive fibers that can persist for long periods of time (Morgan *et al.,* 1993), and restored near-normal muscle histology (Morgan *et al.,* 1990), the conditions required to achieve this are idealized for the animal model and are not directly transferable to human clinical trials. However, numerous studies have determined the feasibility of myoblast transfer therapy in the treatment of primary myopathies, such as Duchenne muscular dystrophy (Partridge, 1991a).

Patient Clinical Trials

Successful myoblast transplantation in mdx mice (Partridge *et al.,* 1989) led to four human MT clinical trials (Law *et al.,* 1990; Gussoni *et al.,* 1992; Huard *et al.,* 1992; Karpati *et al.,* 1993). In the first trial Law *et al.* (1990) injected one patient's extensor digitorum brevis muscle with cloned satellite cells derived from a biopsy taken from the patient's father. The contralateral muscle was not injected with cells and served as a negative control. Western blots and immunocytochemistry detected dystrophin at the sarcolemma in the cell-injected leg muscle, but not in the control muscles. A second trial saw dystrophin mRNA transcripts detected by reverse transcription and polymerase chain reaction (RT–PCR) from three of eight patients injected with cells from a first-degree relative (Gussoni *et al.*, 1992). The success of the transplantation into these three patients with deletions in the dystrophin gene was perhaps a virtue of their relative lack of fibrosis compared to older patients. In a preliminary study, Huard *et al.* (1991) reported dystrophin-positive fibers in a 16-year-old DMD patient following MT. The nine-patient MT clinical trial that followed, between human leukocyte antigen (HLA) class I (A, B, and C) and class II DR-matched donors and hosts, saw dystrophin expression in most patients, but this expression decreased with time (Huard *et al.*, 1992). This may have been due to an immune response because a reaction was observed against muscle proteins, including dystrophin, and antibodies detected in the patient's serum reacted against donor myoblasts and myotubes. Karpati *et al.* (1993), however, saw no significant strength improvement or increase in dystrophin in eight cyclophosphamide-immunosuppressed patients following multiple injections of donor MPCs. There was poor overall therapeutic efficiency of MT, attributed in part to cryptic immunorejection and/or toxic effects of the immunosuppressive agent. Tremblay *et al.* (1993) also saw generation of a humoral immune response following MT in young DMD patients, which suggested that cell (or gene) therapy would require host immunosuppression, because the antigens causing the immune response probably included the dystrophin protein. In a trial in which patients were immunosuppressed with cyclosporine, a moderate strength increase was seen in both cell and placebo-injected muscles, indicating an effect of the cyclosporine on muscle strength (Miller *et al.*, 1997). Previous trials had failed to control adequately for the ameliorating effects of either the cyclosporine immunosuppression (Law *et al.*, 1990) or placebo injection (Law *et al.*, 1993). Furthermore, in a study of Becker MD patients controlled for the effect of the transplantation procedure and cyclosporine immunosuppression, no improvement in strength was seen in myoblast-injected muscles (Neumeyer *et al.*, 1998). After a lull caused by the lack of success of human trials of myoblast therapy for DMD, animal studies have increasingly been used to elucidate the problems seen, and to further our understanding of the principles underlying muscle regeneration following injury or due to myopathic disease.

FACTORS FOR SUCCESSFUL TRANSPLANTATION

A number of problems have been held accountable for the poor outcome of MTT clinical trials to date, ranging from the preparation of the cells prior to implantation, to their poor dispersal and survival once implanted, and to controlling the immune response. The immune reaction seen following MT is being increasingly studied in animal experiments; the availability of immunodeficient animal strains has provided a means of avoiding this immune response for experimental purposes. However, this issue does need to be further evaluated if clinical trials are to succeed. The results of Vilquin *et al.* (1995b) following transplantation of myoblasts transgenically expressing β-galactosidase into mdx mouse muscle under cyclophosphamide immunosuppression mirrored those seen in the Karpati *et al.* (1993) clinical trial. It was concluded that as an immunosuppressant, cyclophosphamide at high concentrations was toxic to injected myoblasts and at low concentrations was not sufficiently immunosuppressive. Human myoblasts transplanted into immunodeficient or immunosuppressed mice were able to fuse and produce dystrophin, although there was evidence of humoral or cellular immune reaction from rejection of some cells in inadequately suppressed mice (Huard *et al.*, 1994b). In a separate study, dystrophin-positive fibers were still present 9 months after MT with histocompatible mouse cells into mdx mice, even though antibodies to dystrophin and/or the muscle membrane were also present in many mice, indicating that dystrophin incompatibility alone does not necessarily induce rejection (Vilquin *et al.*, 1995a). The need for adequate immunosuppression for MT in DMD patients was further highlighted in the failure of MT in normal mice (Guerette *et al.*, 1995). FK-506 was shown to be an effective immunosuppressant in mdx and normal mice (Kinoshita *et al.*, 1994a,b) and was also effective in preventing short-term immune reactions following MT in monkeys (Kinoshita *et al.*, 1996a). The inflammatory reaction can be prevented by treating the host with FK-506 and a monoclonal antibody against a lymphocyte function-associated antigen (LFA-1) (Guerette *et al.*, 1997a), although cell death was not reduced by treatment with FK-506 alone, indicating the mortality was not due to a specific acute reaction (Guerette *et al.,* 1997b). The antiinflammatory property of a transforming growth factor (TGF-β1) on myoblast survival was investigated. TGF-β1 expression from retrovirally infected transplanted myoblasts decreased the inflammatory reaction against the transplanted cells, thus prolonging their survival (Merly *et al.,* 1998). Two approaches, blocking the action of interleukin-1 (IL-1) by engineering transplanted myoblasts to express a cytokine inhibitor, or selection of specific muscle-derived cell populations, have also been suggested to improve transplanted cell survival (Qu *et al.,* 1998).

Although the inflammatory reaction has been proposed as the cause of the cell death seen during the first 3 days posttransplantation, it cannot explain the massive cell death seen immediately posttransplantation in both immunocompetent (Huard *et al.,* 1994a) and nude mice (Beauchamp *et al.,* 1994). Using [^{14}C]thymidine, a semiconserved marker, Beauchamp *et al.* (1997) found that 17% of donor cells survived 48 hr after transplantation, similar to the 20% survival reported by Huard *et al.* (1994a) following transplantation of adenovirally marked cells. In the same study, with an inherited marker (Y chromosome) only 35% of the original number of copies of host genome were present after 48 hr, again indicating a massive loss of donor cells (Beauchamp *et al.,* 1997). In a previous study we estimated the amount of muscle formed from transplantation of a known number of myogenic cells by multiplication of the weight of the muscle by the percentage of dystrophin-positive fibers. Even assuming that all transplanted cells survive, these cells must divide extensively to account for the calculated value of ~10 mg of donor-derived muscle (Morgan *et al.,* 1996). When the actual number of cells surviving transplantation is considered, all muscle of donor origin is formed from a small proportion of the initial injected population, which must, therefore, undergo rapid proliferation (Beauchamp *et al.,* 1999). Whether donor myoblasts were unlabeled or labeled with [^{14}C]thymidine or [^{3}H]thymidine, the outcome was the same, despite [^{3}H]thymidine being toxic to dividing cells. This indicates that the cells that survived transplantation and went on to proliferate and form muscle fibers were not labeled by [^{3}H]thymidine, presumably because they divide slowly in culture and were therefore not dividing during labeling (Beauchamp *et al.,* 1999). This slow division is characteristic of some stem cells (Morrison *et al.,* 1997), which are considered by many to be the ideal cells for MT, giving rise to both fully differentiated muscle and to more precursor cells capable of muscle regeneration following further periods of degeneration. This idea has received a boost by the demonstration that bone marrow-derived cells contribute to the regeneration of mouse skeletal muscle following bone marrow transplantation and also by direct grafting into host muscle (Ferrari *et al.,* 1998) and by the extraction of stemlike cells from muscle (Qu *et al.,* 1998; Gross and Morgan, 1999).

Webster and Blau (1990) showed that neonatal muscle cells have a greater regenerative potential compared with cells from either a teenage or an adult donor. Neonates would therefore be the preferred donor source because 30–40 cell doublings *in vitro* would be required from a muscle biopsy to replace the entire satellite cell pool within an adult (estimated at between 10^{11} and 10^{12} cells), even assuming all of the transplanted cells survived (Partridge and Davies, 1995). Karpati *et al.* (1993) found a loss of myogenicity in normal adult myoblasts after 15-18 doublings *in vitro* prior to implantation; in addition, there was a decrease in proliferative potential with the age of the donor. In parallel with the loss of proliferative capacity, telomere length of satellite cells also decreased during the first two decades of life, whereas telomere length of myonuclei contained within muscle fibers remained constant (Decary *et al.,* 1997). Telomeres are simple repetitive noncoding DNA sequences found at the end of chromosomes, and a decrease in telomeric DNA is associated with somatic cell division. There appears to be an inverse relationship between telomere length and proliferative capacity, and it is therefore supposed that the use of MPCs derived from young donors would be best for MTT due to greater proliferative potential of satellite cells. However, even in young DMD patients the replicative capacity of satellite cells is further reduced by the repeated cycles of degeneration and regeneration that are characteristic of DMD, and this has implications for the idea of *ex vivo* gene therapy in which the patient's own myogenic cells are engineered to produce dystrophin before being expanded *in vitro,* prior to their reimplantation. For experimental MTT in the mouse, the problem of limited proliferative capacity of donor cells in culture has been overcome by utilizing conditionally immortal myoblasts from transgenic animals, such as the *H-2K^b*–tsA58 mouse (Jat *et al.,* 1991; Morgan *et al.,* 1994). The transgene allows these cells to be cultured for long periods of time while still retaining their myogenic properties, and following transplantation they behave as normal MPCs. Long-term culture of these cells enables them to be marked, for example, with a reporter gene, allowing the survival and migration of donor cells to be monitored and facilitating the assessment of success of the MT procedure.

One marker of donor cells following transplantation is the presence of donor-derived normal dystrophin in a dystrophic host. However, it has been reported that dystrophic muscles of both mdx mice and DMD patients contain rare dystrophin-positive fibers, termed "revertant" fibers (Hoffman *et al.*, 1990), such that identification of a dystrophin-positive fiber alone is insufficient to demonstrate successful transplantation. This source of dystrophin-positive fibers was not considered by Law *et al.* (1990), although the trial conducted by Huard *et al.* (1992) had taken it into account. Full-length dystrophin mRNA is a useful marker of donor myoblasts, although in clinical trials Karpati *et al.* (1993) and Morandi *et al.* (1995) failed to identify any full-length dystrophin mRNA, whereas Gussoni *et al.* (1992) identified low levels that had no particular association with dystrophin-positive fibers on biopsy. Mendell *et al.* (1995), using antibodies specific to particular exons, showed up to 10% dystrophin-positive fibers in one DMD patient following MT. A panel of epitope-specific monoclonal antibodies and a polyclonal antibody were used to show the failure of MT in muscle samples from a patient who had received MTT in a clinical trial performed by the Cell Therapy Research Foundation (Memphis, TN) (Partridge *et al.*, 1998). No difference in dystrophin-positive fibers was seen between myoblast- and sham-injected muscles, and moreover, the pattern of dystrophin epitope expression corresponded to the patient's genetic deletion, indicating that the dystrophin-positive fibers were due to a reversion of the patient's fibers from a dystrophin-negative phenotype rather than to expression from normal donor myonuclei. Markers of donor origin have also been used to look at migration of transplanted cells. Fluorescently labeling normal myoblasts prior to implantation into mdx muscle showed that most of the fibers formed from donor cells are near the injection site, implying limited dispersion of the donor cells (Satoh *et al.*, 1993). Moens *et al.* (1996) showed by grafting whole extensor digitorum longus (EDL) muscles between C57 and mdx mice that very few MPCs that are active in these grafted muscles are able to migrate into adjacent muscles, this may have been because the grafted muscles were not injured to further activate satellite cells. This implies that to make MT efficient enough to be utilized as a therapy for myopathies, every affected muscle will need to be given closely spaced injections of myoblasts. It has previously been shown that rat myoblasts in the same muscle may migrate from one fiber to another (Hughes and Blau, 1990), but migration between muscles did not occur unless the connective tissue barriers were damaged (Schultz *et al.*, 1986; Morgan *et al.*, 1987; Watt *et al.*, 1987, 1993, 1994). Migration of myoblasts following grafting into FK-506-immunosuppressed BALB/c and mdx mice has been increased by prior culture in concanavalin A (Con A), a tetravalent plant lectin (Ito *et al.*, 1998). The mechanism is unclear, although up-regulation of collagenase synthesis has been noted in a variety of cell types. A study by Skuk *et al.* (1999) demonstrated that precise delivery of transplanted myoblasts in primates by multiple, closely spaced injections can compensate for their poor migration *in vivo,* and that the presence of a myotoxic phospholipase, such as notexin, also increased the success of the transplantation. (Notexin is a phospholipase known to induce degeneration in muscle by destroying the plasma membrane of mature fibers, but sparing MPCs (Harris and Johnson, 1978; Dixon and Harris, 1996)). These were the best results obtained from myoblast transplantation in primates, which, due to a close phylogenic relationship, provide a better model than the mouse for accurate extrapolation to man. An increased participation in regeneration of donor mouse myoblasts was seen in both normal and mdx mice following irradiation and notexin treatment of host muscle (Kinoshita *et al.,* 1994b). When human myoblasts were transplanted into irradiated and notexin-damaged severe combined immunodeficient (SCID) mice, the majority of fibers formed were of donor origin, although the actual number of fibers present was, at best, one-third of the normal number due to incomplete regeneration (Huard *et al.*, 1994c). Kinoshita *et al.* (1996b) saw a high percentage of fibers of donor origin in mdx mice injected several times with mouse cells cultured in the presence of basic fibroblast growth factor (bFGF), without notexin treatment. The bFGF did not increase the percentage of myogenic cells present within the primary culture because a myogenic clone cultured in the lowest concentration of bFGF showed the same improvement of MT. Gross and Morgan (1999) treated irradiated and cell-injected mdx muscle with notexin to show that some transplanted cells, in addition to forming dystrophin-positive fibers, can persist as muscle precursor cells.

As discussed, studies have shown the importance of the choice of donor cell population (Beauchamp *et al.*, 1999; Gross and Morgan, 1999) as a consideration for the increased efficacy of clinical trials, in addition to methods to increase donor cell survival and migration on transplantation (Moens *et al.*, 1996; Ito *et al.*, 1998; Skuk *et al.*, 1999). The source of donor cells also has implications when myoblasts are used as gene therapy vehicles, and greater success of this technique may be attained with the use of myogenic cells early in the myogenic pathway, such as muscle stem cells (Brown and Miller, 1996). Donor cells with stem cell characteristics have been identified in injected muscles. Yao and Kurachi (1993) were able to extract myogenic cells of donor origin from injected muscles, and Morgan *et al.* (1994) showed that cells isolated from irradiated and injected muscles could be passaged through further rounds of irradiated mouse muscle. There has also been success in differentiating uncommitted cells from nonmuscle tissue into skeletal muscle under appropriate conditions, providing a possible source of material for MT (Young *et al.*, 1995; Gibson *et al.*, 1995). Ferrari *et al.* (1998) showed that transplanted bone marrow-derived cells (possibly originating from multipotent mesenchymal stromal stem cells) migrated into mouse muscle that had been forced to degenerate, and that these cells were able to undergo myogenic differentiation and participate in the regeneration of the damaged fibers.

The existence of populations of cells with the ability to form new muscle, both immediately after transplantation and following future periods of degeneration, may prove crucial to the development of methods to improve the success of human MTT.

REFERENCES

Barr, E., and Leiden, J. M. (1991). Systemic delivery of recombinant proteins by genetically modified myoblasts. *Science* **254**, 1507–1509.

Beauchamp, J. R., Morgan, J. E., Pagel, C. N., and Partridge, T. A. (1994). Quantitative studies of the efficacy of myoblast transplantation. *Muscle Nerve* **17**(Suppl. 1), S261.

Beauchamp, J. R., Pagel, C. N., and Partridge, T. A. (1997). A dual-marker system for quantitative studies of myoblast transplantation in the mouse. *Transplantation* **63**, 1794–1797.

Beauchamp, J. R., Morgan, J. E., Pagel, C. N., and Partridge, T. A. (1999). Dynamics of myoblast transplantation reveal a discrete minority of precursors with stem cell-like properties as the myogenic source. *J. Cell Biol.* **144**, 1113–1121.

Bischoff, R. (1994). The satellite cell and muscle regeneration. *In* "Myology" (A. G. Engel and C. Franzini-Armstrong, eds.), 2nd ed., pp. 97–118. McGraw-Hill, New York.

Bou-Gharios, G., Wells, D. J., Lu, Q. L., Morgan, J. E., and Partridge, T. (1999). Differential expression and secretion of alpha 1 anti-trypsin between direct DNA injection and implantation of transfected myoblast. *Gene Ther.* **6**, 1021–1029.

Brown, R. H., and Miller, J. B. (1996). Progress, problems and prospects for gene therapy in muscle. *Curr. Opin. Rheumatol.* **8**, 539–543.

Bulfield, G., Siller, W. G., Wright, P. A. L., and Moore, K. J. (1984). X-chromosome-linked muscular dystrophy (mdx) in the mouse. *Proc. Natl. Acad. Sci. U.S.A.* **81**, 39–54.

Dai, Y., Roman, M., Naviaux, R. K., and Verma, I. M. (1992). Gene therapy via primary myoblasts: Long term expression of factor IX protein following transplantation *in vivo*. *Proc. Natl. Acad. Sci. U.S.A.* **89**, 10892–10895.

Decary, S., Mouly, V., Hamida, C. B., Saulet, A., Barbet, J. P., and Butler-Browne, G. S. (1997). Replicative potential and telomere length in human skeletal muscle: Implications for satellite cell-mediated gene therapy. *Hum. Gene Ther.* **8**, 1429–1438.

Deconinck, N., Ragot, T., Marechal, G., Perricaudet, M., and Gillis, J. M. (1996). Functional protection of dystrophic mouse (mdx) muscles after adenovirus-mediated transfer of a dystrophin minigene. *Proc. Natl. Acad. Sci. U.S.A.* **83**, 3570–3574.

Deconinck, A. E., Rafael, J. A., Skinner, J. A., Brown, S. C., Potter, A. C., Metzinger, L., Watt, D. J., Dickson, G., Tinsley, J. M., and Davies, K. E. (1997a). Utrophin-dystrophin-deficient mice as a model for Duchenne muscular dystrophy. *Cell (Cambridge, Mass.)* **90**, 717–727.

Deconinck, N., Tinsley, J., De Backer, F., Fisher, R., Kahn, D., Phelps, S., Davies, K. E., and Gillis, J. M. (1997b). Expression of truncated utrophin leads to major functional improvements in dystrophin-deficient muscles of mice. *Nat. Med.* **3**, 1216–1221.

Dhawan, J., Pan, L. C., Pavlath, G. K., Travis, M. A., Lanctot, A. M., and Blau, H. M. (1991). Systemic delivery of human growth hormone by injection of genetically engineered myoblasts. *Science* **254**, 1509–1512.

Dixon, R. W., and Harris, J. B. (1996). Myotoxic activity of the toxic phospholipase, notexin, from the venom of the Australian tiger snake. *J. Neuropathol. Exp. Neurol.* **55**, 1230–1237.

Dubowitz, V. (1985). "Muscle Biopsy; A Practical Approach," 2nd ed. Baillière Tindall, London.

Dunckley, M. G., Wells, D. J., Walsh, F. S., and Dickson, G. (1993). Direct retroviral-mediated transfer of a dystrophin minigene into mdx mouse muscle *in vivo*. *Hum. Mol. Genet.* **2**, 717–723.

Ferrari, G., Cusella-De Angelis, G., Coletta, M., Paolucci, E., Stronaiuolo, A., Cossu, G., and Mavilio, F. (1998). Muscle regeneration by bone marrow-derived myogenic precursors. *Science* **279**, 1528–1530.

Gibson, A. J., Karasinski, J., Relvas, J., Moss, J., Sherratt, T. G., Strong, P. N., and Watt, D. J. (1995). Dermal fibroblasts convert to a myogenic lineage in mdx mouse muscle. *J. Cell Sci.* **108**, 207–214.

Grady, R. M., Teng, H., Nichol, M. C., Cunningham, J. C., Wilkinson, R. S., and Sanes, J. R. (1997). Skeletal and cardiac myopathies in mice lacking utrophin and dystrophin: A model for Duchenne muscular dystrophy. *Cell (Cambridge, Mass.)* **90,** 729–738.

Gross, J. G., and Morgan, J. E. (1999). Muscle precursor cells injected into irradiated mdx mouse muscle persist after serial injury. *Muscle Nerve* **22,** 174–185.

Guerette, B., Asselin, I., Viquin, J. T., Roy, R., and Tremblay, J. P. (1995). Lymphocyte infiltration following allomyoblast and xenomyoblast transplantation in mdx mice. *Muscle Nerve* **18,** 39–51.

Guerette, B., Asselin, I., Skuk, D., Entman, M., and Tremblay, J. P. (1997a). Control of inflammatory damage by anti-LFA-1: Increase success of myoblast transplantation. *Cell Transplant.* **6,** 101–107.

Guerette, B., Skuk, D., Celestin, F., Huard, C., Tardif, F., Asselin, I., Roy, B., Goulet, M., Roy, R., Entman, M., and Tremblay, J. P. (1997b). Prevention by anti-LFA-1 of acute myoblast death following transplantation. *J. Immunol.* **159,** 2522–2531.

Gussoni, E., Pavlath, G. K., Lanctot, A. M., Sharma, K. R., Miller, R. G., Steinman, L., and Blau, H. M. (1992). Normal dystrophin transcripts detected in Duchenne muscular dystrophy patients after myoblast transplantation. *Nature (London)* **356,** 435–438.

Hamamori, Y., Samal, B., Tian, J., and Kedes, L. (1995). Myoblast transfer of human erythropoietin gene in a mouse model of renal failure. *J. Clin. Invest.* **95,** 1808–1813.

Harris, J. B., and Johnson, M. A. (1978). Further observations on the pathological response of rat skeletal muscle to toxins isolated from the venom of the Australian tiger snake, *Notechis scutatus scutatus. Clin. Exp. Pharmacol. Physiol.* **5,** 587–600.

Hoffman, E. P., Brown, R. H., and Kunkel, L. M. (1987). Dystrophin: The protein product of the Duchenne muscular dystrophy locus. *Cell (Cambridge, Mass.)* **51,** 919–928.

Hoffman, E. P., Morgan, J. E., Watkins, S. C., and Partridge, T. A. (1990). Somatic reversion/suppression of the mouse mdx phenotype *in vivo. J. Neurol. Sci.* **99,** 9–25.

Huard, J., Bouchard, J. P., Roy, R., Labrecque, C., Dansereau, G., Lemieux, B., and Tremblay, J. P. (1991). Myoblast transplantation produced dystrophin-positive muscle fibers in a 16-year-old patient with Duchenne muscular dystrophy. *Clin. Sci.* **81,** 287–288.

Huard, J., Roy, R., Bouchard, J. P., Malouin, F., Richards, C. L., and Tremblay, J. P. (1992). Human myoblast transplantation between immunohistocompatible donors and recipients produces immune reactions. *Transplant. Proc.* **24,** 3049–3051.

Huard, J., Acsadi, G., Jani, A., Massie, B., and Karpati, G. (1994a). Gene transfer into skeletal muscles by isogenic myoblasts. *Hum. Gene Ther.* **5,** 949–958.

Huard, J., Roy, R., Guerette, B., Verreault, S., Tremblay, G., and Tremblay, J. P. (1949b). Human myoblast transplantation in immunodeficient and immunosuppressed mice: Evidence of rejection. *Muscle Nerve* **17,** 224–234.

Huard, J., Verreault, S., Roy, R., Tremblay, M., and Tremblay, J. P. (1994c). High efficiency of muscle regeneration after human myoblast clone transplantation in SCID mice. *J. Clin. Invest.* **93,** 586–599.

Hughes, S. M., and Blau, H. M. (1990). Migration of myoblasts across basal lamina during skeletal muscle development. *Nature (London)* **345,** 350–353.

Irintchev, A., Rosenblatt, J. D., Cullen, M. J., Zweyer, M., and Wernig, A. (1998). Ectopic skeletal muscles derived from myoblasts implanted under the skin. *J. Cell Sci.* **111,** 3287–3297.

Ito, H., Hallauer, P. L., Hastings, K. E. M., and Tremblay, J. P. (1998). Prior culture with concanavalin A increases intramuscular migration of transplanted myoblast. *Muscle Nerve* **21,** 291–297.

Jat, P. S., Noble, M. D., Ataliotis, P., Tanaka, Y., Yannoutsos, N., Larsen, L., and Kioussis, D. (1991). Direct derivation of conditionally immortal cell lines from an H-$2K^b$-tsA58 transgenic mouse. *Proc. Natl. Acad. Sci. U.S.A.* **88,** 5096–5100.

Jones, P. H. (1979). Implantation of cultured regenerate muscle cells into adult rat muscle. *Exp. Neurol.* **66,** 602–610.

Karpati, G., Pouilot, Y., Zubrzycka-Gaarn, E., Carpenter, S., Ray, P. N., Worton, R. G., and Holland, P. (1989). Dystrophin is expressed in mdx skeletal muscle fibers after normal myoblast implantation. *Am. J. Pathol.* **135,** 27–32.

Karpati, G., Ajdukovic, D., Arnold, D., Gledhill, R. B., Guttman, R., Holland, P., Koch, P. A., Shoubridge, E., Spence, D., Vanasse, M., Watters, G. V., Abrahamowicz, M., Duff, C., and Worton, R. G. (1993). Myoblast transfer in Duchenne muscular dystrophy. *Ann. Neurol.* **34,** 8–17.

Kinoshita, I., Vilquin, J. T., Guerette, B., Asselin, I., Roy, R., Lille, S., and Tremblay, J. P. (1994a). Immunosuppression with FK-506 insures good success of myoblast transplantation in mdx mice. *Transplant. Proc.* **26,** 3518.

Kinoshita, I., Vilquin, J. T., Guerette, B., Asselin, I., Roy, R., and Tremblay, J. P. (1994b). Very efficient myoblast allotransplantation in mice under FK-506 immunosuppression. *Muscle Nerve* **17,** 1407–1415.

Kinoshita, I., Roy, R., Dugre, F. J., Gravel, C., Roy, B., Goulet, M., Asselin, I., and Tremblay, J. P. (1996a). Myoblast transplantation in monkeys: Control of immune response by FK506. *J. Neuropathol. Exp. Neurol.* **55,** 687–697.

Kinoshita, I., Vilquin, J. T., Roy, R., and Tremblay, J. P. (1996b). Successive injections in mdx mice of myoblasts grown with bFGF. *Neuromuscular Disord.* **6,** 187–193.

Kinoshita, I., Vilquin, J. T., Asselin, I., Chamberlain, J., and Tremblay, J. P. (1998). Transplantation of myoblasts from a transgenic mouse overexpressing dystrophin produced only a relatively small increase of dystrophin-positive membrane. *Muscle Nerve* **21,** 91–103.

Law, P. K., Goodwin, T. G., and Li, H. J. (1988). Histoincompatible myoblast injection improves muscle structure and function of dystrophic mice. *Transplant. Proc.* **20,** 1114–1119.

Law, P. K., Bertorini, T. E., Goodwin, T. G., Chen, M., Fang, Q. W., Li, H. J., Kirby, D. S., Florendo, J. A., Herrod, H. G., and Golden, G. S. (1990). Dystrophin production induced by myoblast transfer therapy in Duchenne muscular dystrophy. *Lancet* **336,** 114–115.

Law, P. K., Goodwin, T. G., Fang, Q., Deering, M. B., Duggirala, V., Larkin, C., Florendo, J. A., Kirby, D. S., Li, H. J., Chen, M., Cornett, J., Li, L. M., Shirzod, A., Quinley, T., Yoo, T. J., and Holcomb, R. (1993). Cell transplantation as an experimental treatment for Duchenne muscular dystrophy. *Cell Transplant.* **2**, 485–505.

Lipton, B. H., and Schultz, E. (1979). Developmental fate of skeletal muscle satellite cells. *Science* **205**, 1292–1294.

Mendell, J. R., Kissel, J. T., Amato, A. A., King, W., Signore, L., Prior, R. N. T. W., Sahenk, Z., Benson, S., McAndrew, P. E., Rice, R., Nagaraja, H., Stephens, R., Lantry, L., Morris, G. E., and Burghes, H. M. (1995). Myoblast transfer in the treatment of Duchenne's muscular dystrophy. *N. Engl. J. Med.* **333**, 832–838.

Merly, F., Huard, C., Asselin, I., Robbins, P. D., and Tremblay, J. P. (1998). Antiinflammatory effect of transforming growth factor-beta 1 in myoblast transplantation. *Transplantation* **65**, 793–799.

Miller, R. G., Sharma, K. R., Pavlath, G. K., Gussoni, E., Mynhier, M., Lanctot, A. M., Greco, C. M., Steinman, L., and Grounds, M. (1997). Myoblast implantation in Duchenne muscular dystrophy: The San Francisco Study. *Muscle Nerve* **20**, 469–478.

Moens, P. D. J., Colson Van-Schoor, M., and Marechal, G. (1996). Lack of myoblasts migration between transplanted and host muscles of mdx and normal mice. *J. Muscle Res. Cell Motil.* **17**, 37–43.

Morandi, L., Bernasconi, P., Gebbia, M., Mora, M., Crosti, F., Mantegazza, R., and Cornelio, F. (1995). Lack of mRNA and dystrophin expression in DMD patients three months after myoblast transfer. *Neuromuscular Disord.* **5**, 291–295.

Morgan, J. E., Coulton, G. R., and Partridge, T. A. (1987). Muscle precursor cells invade and repopulate freeze-killed muscles. *J. Muscle Res. Cell Motil.* **8**, 386–396.

Morgan, J. E., Watt, D. J., Sloper, J. C., and Partridge, T. A. (1988). Partial correction of an inherited biochemical defect of skeletal muscle by grafts of normal muscle precursor cells. *J. Neurol. Sci.* **86**, 137–147.

Morgan, J. E., Hoffman, E. P., and Partridge, T. A. (1990). Normal myogenic cells from newborn mice restore normal histology to degenerating muscles of the mdx mouse. *J. Cell Biol.* **111**, 2437–2449.

Morgan, J. E., Moore, S. E., Walsh, F. S., and Partridge, T. A. (1992). Formation of skeletal muscle *in vivo* from the mouse C2 cell line. *J. Cell Sci.* **102**, 779–787.

Morgan, J. E., Pagel, C. N., Sherratt, T., and Partridge, T. A. (1993). Long-term persistence and migration of myogenic cells injected into pre-irradiated muscles of mdx mice. *J. Neurol. Sci.* **115**, 191–200.

Morgan, J. E., Beauchamp, J. R., Pagel, C. N., Peckham, M., Ataliotis, P., Jat, P. S., Noble, M. D., Farmer, K., and Partridge, T. A. (1994). Myogenic cell lines derived from transgenic mice carrying thermolabile T antigen: A model system for the derivation of tissue-specific and mutation-specific cell lines. *Dev. Biol.* **162**, 486–498.

Morgan, J. E., Fletcher, R. M., and Partridge, T. A. (1996). Yields of muscle from myogenic cells implanted into young and old mdx hosts. *Muscle Nerve* **19**, 132–139.

Morrison, S. J., Shah, N. M., and Anderson, D. J. (1997). Regulatory mechanisms in stem cell biology. *Cell (Cambridge, Mass.)* **88**, 287–298.

Neumeyer, A. M., Cros, D., McKenna-Yasek, D., Zawadzka, A., Hoffman, E. P., Pegaro, E., Hunter, R. G., Munsat, T. L., and Brown, R. H., Jr. (1998). Pilot study of myoblast transfer in the treatment of Becker muscular dystrophy. *Neurology* **51**, 589–592.

Partridge, T. A. (1991a). Myoblast transfer: A possible therapy for inherited myopathies? *Muscle Nerve* **14**, 197–212.

Partridge, T. A. (1991b). Animal models of muscular dystrophy—What can they teach us? *Neuropathol. Appl. Neurobiol.* **17**, 353–363.

Partridge, T. (1993). Pathophysiology of muscular dystrophy. *Br. J. Hosp. Med.* **49**, 26–36.

Partridge, T. A. (1997). Models of dystrophinopathy, pathological mechanisms and assessment of therapies. *In* "Dystrophin Gene Protein and Cell Biology" (S. C. Brown and J. A. Lucy, eds.), pp. 310–331. Cambridge University Press, Cambridge, UK.

Partridge, T. A., and Davies, K. E. (1995). Myoblast-based gene therapies. *Br. Med. Bull.* **51**, 123–137.

Partridge, T. A., Grounds, M., and Sloper, J. C. (1978). Evidence of fusion between host and donor myoblasts in skeletal muscle grafts. *Nature (London)* **273**, 306–308.

Partridge, T. A., Morgan, J. E., Coulton, G. R., Hoffman, E. P., and Kunkel, L. M. (1989). Conversion of mdx myofibres from dystrophin-negative to -positive by injection of normal myoblasts. *Nature (London)* **337**, 176–179.

Partridge, T., Lu, Q. L., Morris, G., and Hoffman, E. (1998). Is myoblast transplantation effective? *Nat. Med.* **4**, 1208–1209.

Pastoret, C., and Sebille, A. (1995a). Age-related differences in regeneration of dystrophic (Mdx) and normal muscle in the mouse. *Muscle Nerve* **14**, 197–212.

Pastoret, C., and Sebille, A. (1995b). Mdx mice show progressive weakness and muscle deterioration with age. *J. Neurol. Sci.* **129**, 97–105.

Qu, Z., Balkir, L., van Deutekom, J. C. T., Robbins, P. D., Pruchnic, R., and Huard, J. (1998). Development of approaches to improve cell survival in myoblast transfer therapy. *J. Cell Biol.* **142**, 1257–1267.

Quinlan, J. G., Lyden, S. P., Cambier, D. M., Johnson, S. R., Michaels, S. E., and Denman, D. L. (1995). Radiation inhibition of mdx mouse muscle regeneration: Dose and age factors. *Muscle Nerve* **18**, 201–206.

Quinlan, J. G., Cambier, D., Lyden, S., Dalvi, A., Upputuri, R. K., Gartside, P., Michaels, S. E., and Denman, D. (1997). Regeneration-blocked mdx muscle: *In vivo* model for testing treatments. *Muscle Nerve* **20**, 1016–1023.

Satoh, A., Huard, J., Labrecque, C., and Tremblay, J. P. (1993). Use of fluorescent latex microspheres (FLMS) to follow the fate of transplanted myoblasts. *J. Histochem. Cytochem.* **41**, 1579–1582.

Schultz, E., Jaryszak, D. L., Gibson, M. C., and Albright, D. J. (1986). Absence of exogenous satellite cell contribution to regeneration of frozen skeletal muscle. *J. Muscle Res. Cell Motil.* **7**, 361–367.

Sicinski, P., Geng, Y., Ryder-Cook, A. S., Barnard, E. A., Darlinson, M. G., and Barnard, P. J. (1989). The molecular basis of muscular dystrophy in the mdx mouse: A point mutation. *Science* **244**, 1578–1580.

Skuk, D., Roy, B., Goulet, M., and Tremblay, J. P. (1999). Successful myoblast transplantation in primates depends on appropriate cell delivery and induction of regeneration in the host muscle. *Exp. Neurol.* **155**, 1–9.

Tinsley, J., Deconinck, N., Fisher, R., Kahn, D., Phelps, S., Gillis, J.-M., and Davies, K. E. (1998). Expression of full-length utrophin prevents muscular dystrophy in mdx mice. *Nat. Med.* **4**, 1441–1444.

Tremblay, J. P., Malouin, F., Roy, R., Huard, J., Bouchard, J. P., Satoh, A., and Richards, C. L. (1993). Results of a triple blind clinical study of myoblast transplantations without immunosuppressive treatment in young boys with Duchenne muscular dystrophy. *Cell Transplant.* **2**, 99–112.

Vandenburgh, H., Del Tatto, M., Shansky, J., Lemaire, J., Chang, A., Payumo, F., Lee, P., Goodyear, A., and Raven, L. (1996). Tissue-engineered skeletal muscle organoids for reversible gene therapy. *Hum. Gen. Ther.* 7, 2195–2200.

Vilquin, J. T., Wagner, E., Kinoshita, I., Roy, R., and Tremblay, J. P. (1995a). Successful histocompatible myoblast transplantation in dystrophin-deficient mdx mouse despite the production of antibodies against dystrophin. *J. Cell Biol.* **131**, 975–988.

Vilquin, J. T., Kinoshita, I., Roy, R., and Tremblay, J. P. (1995b). Cyclophosphamide immunosuppression does not permit successful myoblast transplantation in mouse. *Neuromuscular Disord.* **5**, 511–517.

Wakeford, S., Watt, D. J., and Partridge, T. A. (1991). X-Irradiation improves mdx mouse as a model of myofiber loss in DMD. *Muscle Nerve* **14**, 42–50.

Wang, J.-M., Zheng, H., Blaivas, M., and Kurachi, K. (1997). Persistent systemic production of human factor IX in mice by skeletal myoblast-mediated gene transfer: Feasibility of repeat application to obtain therapeutic levels. *Blood* **90**, 1075–1082.

Watt, D. J., Lambert, K., Morgan, J. E., Partridge, T. A., and Sloper, J. C. (1982). Incorporation of donor muscle precursor cells into an area of muscle regeneration in the host mouse. *J. Neurol. Sci.* **57**, 329–331.

Watt, D. J., Morgan, J. E., Clifford, M. A., and Partridge, T. A. (1987). The movement of muscle precursor cells between adjacent regenerating muscles in the mouse. *Anat. Embryol.* **175**, 527–536.

Watt, D. J., Karasinski, J., and England, M. A. (1993). Migration of lac Z positive cells from the tibialis anterior to the extensor digitorum longus muscle of the X-linked muscular dystrophic (mdx) mouse. *J. Muscle Res. Cell Motil.* **14**, 121–132.

Watt, D. J., Karasinski, J., Moss, J., and England, M. A. (1994). Migration of muscle cells. *Nature (London)* **368**, 406–407.

Webster, C., and Blau, H. M. (1990). Accelerated age-related decline in replicative life-span of Duchenne muscular dystrophy myoblasts: Implications for cell and gene therapy. *Somatic Cell Mol. Genet.* **16**, 557–565.

Weller, B., Karpati, G., Lehnert, S., and Carpenter, S. (1991). Major alteration of the pathological phenotype in gamma irradiated mdx soleus muscle. *J. Neuropathol. Exp. Neurol.* **50**, 419–431.

Wernig, A., Irintchev, A., Hartling, A., Stephan, G., Zimmerman, K., and Starzinski-Powitz, A. (1991). Formation of new muscle fibers and tumours after injection of cultured myogenic cells. *Neurocytology* **20**, 982–997.

Yao, S.-N., and Kurachi, K. (1992). Expression of human factor IX in mice after injection of genetically modified myoblasts. *Proc. Natl. Acad. Sci. U.S.A.* **89**, 3357–3361.

Yao, S.-N., and Kurachi, K. (1993). Implanted myoblasts not only fuse with myofibers, but also survive as muscle precursor cells. *J. Cell Sci.* **105**, 957–963.

Yao, S.-N., Smith, K. J., and Kurachi, K. (1994). Primary myoblast-mediated gene transfer: Persistent expression of human factor IX in mice. *Gene Ther.* **1**, 99–107.

Young, H. E., Mancini, M. L., Wright, R. P., Smith, J. C., Black, A. C., Jr., Reagan, C. R., and Lucas, P. A. (1995). Mesenchymal stem cells reside within the connective tissues of many organs. *Dev. Dyn.* **202**, 137–144.

PART XVII: NERVOUS SYSTEM

CHAPTER 53

PROTECTION AND REPAIR
OF HEARING

Richard A. Altschuler, Yehoash Raphael, Jochen Schacht,
David J. Anderson, and Josef M. Miller

INTRODUCTION

Deafness is the number one disability in the United States. Hearing loss affects more than 35 million individuals in all age groups. It increases with age, affecting one in three persons over the age of 65 and 50% of individuals over the age of 75. With the increased longevity of our population, this disability is also increasing. Approximately one-half of all hearing losses are thought to be of genetic etiology. Acquired deafness affects one in four to five individuals and can occur from a variety of causes, including noise, drugs, disease, and aging. It is obviously important to develop interventions to help to prevent hearing loss or to improve the chances to restore hearing after deafness. The means for inducing regeneration of the sensory hair cells, the loss of which results in profound deafness, currently remains an unrealized goal in mammals. However, recent advances in our understanding of molecular mechanisms as well as new developments in cochlear prostheses raise the promise for new and effective treatments for protection from acquired and genetic deafness and better restoration of hearing. This chapter discusses the basis for the interventions and the methods for achieving the interventions.

BASIS FOR INTERVENTIONS

ACQUIRED DEAFNESS: ENDOGENOUS PROTECTIVE MECHANISMS

Treatments for protection from acquired deafness can be based on enhancement of endogenous protective mechanisms in the inner ear. Three such endogenous mechanisms have been identified: (1) reactive oxygen metabolites, (2) stress shock proteins, and (3) neurotrophic factors.

Regulation of reactive oxygen metabolites

The overproduction of superoxide radicals—and the ensuing chain reaction of formation of other free radicals—now appears to be a common mechanism by which many forms of stress cause damage in the inner ear, including noise and ototoxic drugs (such as aminoglycoside antibiotics). The formation of superoxide radicals occurs early and throughout the traumatizing event. They appear to form in association with intense mitochondrial activity, in response to prolonged tissue hypoxia and a rebound hyperperfusion (e.g., Thorne *et al.*, 1987) and an enhanced release of excitatory amino acids (Puel *et al.*, 1994, 1996). Reactive oxygen metabolites (ROMs) may be generated by stress events hours following the trauma (Yamane *et al.*, 1995). The formation of these radicals, in turn, leads to the initiation of a cascade of intracellular events that lead to cell death. Interventions can be effective at different steps in the sequence of events: before the formation of radicals, by blocking their formation, before they produce downstream effects, by intervention; and during the sequence, so as to reduce or restrict the extent of damage or to enhance repair.

Prevention of aminoglycoside-induced hearing loss

In the case of aminoglycoside antibiotics, Schacht and colleagues have established a molecular mechanism of action that involves free radical formation. Aminoglycoside antibiotics are capable of forming a chelation complex with iron, as demonstrated for gentamicin by nuclear magnetic resonance. The complex, with a 1:1 stoichiometry of gentamicin to iron, is a redox-active compound that will reduce oxygen to the superoxide radical at the expense of an electron donor (Priuska and Schacht, 1995; Priuska *et al.*, 1998; Sha and Schacht, 1999a). This property is apparently shared by all aminoglycoside antibiotics (Sha and Schacht, 1999b) and is responsible for the ototoxic properties of these drugs. The latter can be deduced from the fact that compounds that suppress free radical formation by aminoglycosides *in vitro* are also capable of attenuating the ototoxic affects *in vivo*. Such compounds include well-established antioxidants such as glutathione and iron chelators (Garetz *et al.*, 1994a,b; Song and Schacht, 1996). Indeed, an excellent protection against the ototoxicity of several aminoglycosides was achieved in guinea pigs with the well-established chelating compounds 2,3-dihydroxybenzoate and deferoxamine. Their coadministration attenuated a gentamicin-induced hearing loss of 60–80 db to a negligible loss of around 10 db (Song *et al.*, 1997). Similar protective affects were seen against the cochlear and vestibular toxicity of kanamycin and streptomycin (Song *et al.*, 1998). Independent studies subsequently confirmed antioxidants as protectants against neomycin and amikacin (Conlon and Smith, 1998; Conlon *et al.*, 1999).

This *in vivo* protection both supports the mechanism of free radical involvement and points toward antioxidant treatment as a potential clinical therapy. A clinically feasible prophylactic therapy requires drugs that, alone, are nontoxic and easily administered to the patient. In their search for a practicable application, Sha and Schacht (2000) tested 2-hydroxybenzoate, a precursor of 2,3-dihydroxybenzoate, and found it effective. This compound, commonly known as salicylate, is a weak iron chelator and radical scavenger and may be oxidized under oxidative stress to the more powerful prophylactic 2,3-dihydroxybenzoate. Salicylates have a long medical history as antiinflammatory drugs, mostly in the commonly supplied oral form of aspirin. These animal experiments then imply that aspirin is a suitable medication to protect from aminoglycoside-induced hearing loss. Clinical trials are now underway to determine efficacy in humans.

Prevention of noise-induced hearing loss

The prophylactic approach is currently less feasible for other stresses, for which mechanisms behind ROM formation have yet to be elucidated or when a proactive treatment is not possible. This reservation would apply to clinical cases of noise-induced hearing loss, and studies are therefore currently focusing on interventions at later times. In animal models, however, the potential of antioxidants to limit noise trauma has been demonstrated.

Good evidence supports the view that ROMs are formed in the inner ear following intense sounds; for example, superoxide radicals have been found in the stria vascularis (Yamane *et al.*, 1995). Consistent with such a free radical involvement, hearing loss can be attenuated by scavengers, including superoxide dismutase and allopurinol (Seidman *et al.*, 1993), and lazaroids and lazaroid-related compounds (Quirk *et al.*, 1994), or by the up-regulation of antioxidant enzymes (Hu *et al.*, 1997). These observations were extended by studies showing the involvement of one of the major endogenous antioxidant systems, glutathione, in the limitation of noise trauma. Glutathione (L-γ-glutamyl-L-cysteinylglycine) serves as an important physiologic antioxidant and radical scavenger, because it reacts directly with a number of ROMs, including superoxide anion radicals, hydroxyl radicals, hydrogen peroxides, and other organic peroxides, to form glutathione disulfide. Glutathione attenuates cochlear damage induced by gentamicin (Lautermann *et al.*, 1995a), kanamycin plus ethacrynic acid (Hoffman *et al.*, 1987), amikacin (Nishida and Takumida, 1996), and cisplatin (Lautermann *et al.*, 1995b; Ravi *et al.*, 1995).

Yamasoba *et al.* (1998) manipulated glutathione using L-buthionine-[*S,R*]-sulfoximine (BSO), which depletes glutathione by inhibiting its synthesis. Following exposure to broadband noise (102 dB SPL*) for 3 hr/day for 5 days, glutathione-depleted guinea pigs showed significantly greater threshold shifts than did control subjects, and outer hair cell damage was more pronounced.

SPL is a physical measure of atmospheric pressure equal to 0.0002 dynes/cm².

Similarly, tissue levels of glutathione can be influenced by the dietary availability of one of its precursors for synthesis, the amino acid L-cysteine. Decreased intake decreases cysteine levels in the plasma and glutathione concentration in tissues. By this mechanism, administration of a low-protein diet leads to a reduction of glutathione levels in the cochlea, thereby increasing the susceptibility to drug-induced hearing loss (Lautermann *et al.*, 1995a,b). When guinea pigs were administered a low-protein diet and exposed to noise, they demonstrated significant elevations in auditory brainstem response (ABR) threshold shifts as compared to animals on a normal protein diet (Ohinata *et al.*, 1999). In such glutathione-depleted animals, dietary glutathione ethyl ester can efficiently increase tissue glutathione levels (Puri and Meister, 1983). In fact, oral administration of glutathione ester and vitamin C elevated glutathione levels in the cochlear sensory epithelium in animals on a low-protein diet (Lautermann *et al.*, 1995b). In noise-exposed animals, administration of glutathione ester provided significant ABR protection at all frequencies as compared to their saline-injected controls on a low-protein diet (Ohinata *et al.*, 1999).

Interventions in noise-induced hearing loss may therefore now be based on the accelerated removal of superoxide or derived free radicals, or on the strengthening or supplementation of general detoxification systems.

Heat-shock proteins

Heat-shock proteins (HSPs) provide a natural protective mechanism for cells in numerous systems and species. HSPs achieve their protective role by influencing the stress-related denaturation of proteins (either reducing the denaturation or enhancing renaturation and regaining the correct tertiary structure), as chaperones, or through an influence on cell death cascades. Several families of HSPs (commonly grouped by their molecular weights) include HSP 25/27, HSP 70/72, and HSP 90. Expression of the inducible form of HSP 70 in the cochlea has been shown with heat (Dechesne *et al.*, 1992), with transient ischemia (Myers *et al.*, 1992), and with noise (Lim *et al.*, 1993). This expression is generally found in the sensory hair cells and is transient, peaking 4–6 hr following the stress. HSP 27, on the other hand, is found constitutively (without the need for induction by stress) and has a more widespread distribution, not only in sensory hair cells, but in supporting cells and the lateral wall tissues (Leonova *et al.*, 1998). Stress appears to modulate this expression. A protective role of HSPs in the inner ear has been suggested. In several studies HSP levels were up-regulated by an initial stress, either low-level noise (Mitchell *et al.*, 1997a; Altschuler *et al.*, 1999) or heat (Yoshida and Liberman, 1999), followed by a noise exposure that would normally cause a significant hearing loss. Either preexposure resulted in a significant (30 db) reduction in noise-induced hearing loss and hair cell loss.

Neurotrophic factors

Neurotrophins were so named because during development they induce neuronal differentiation, proliferation, growth, and migration, as well as the growth and migration of neuronal processes (reviewed by Davies, 1994). Different neurotrophic factors act on different classes of neurons (Davies, 1994) and act at different receptors. NGF (nerve growth factor) acts at the tyrosine kinase a (trka) receptor, neurotrophin-4 (NT-4) and brain-derived neurotrophic factor (BDNF) act at the trkb receptor, and neurotrophin-3 (NT-3) acts at the trkc receptor (reviewed by Klein, 1994). Other trophic factors with similar effects, but acting at different receptors, include ciliary neurotrophic factor (CNTF) (Sendtner *et al.*, 1994), fibroblast growth factor (FGF) (Eckenstein, 1994), and glial line-derived neurotrophic factor (GDNF) (Lin *et al.*, 1993). Many neurotrophic factors and/or their receptors have been shown in the normal cochlea, including GDNF (Ylikoski *et al.*, 1998) and NT-3 and BDNF (Ylikoski *et al.*, 1993). Neurotrophic factors may have a protective action through an influence on ROMs or via intervention in the cascade of events induced by ROM formation to block cell death. The factors may reduce oxidative stress-driven increases in intracellular Ca^{2+} (Hegarty *et al.*, 1997; Mattson *et al.*, 1995; Mattson and Furakawa, 1996) through induced expression of calcium-binding proteins (Collazo *et al.*, 1992; Cheng and Mattson, 1992) or antioxidant enzymes (Mattson *et al.*, 1995; Mattson and Furakawa, 1996). The neurotrophic factors may also share a similar biochemical pathway in providing protection and interact with ROM scavengers. Several studies have examined if local infusion of neurotrophins into the inner ear fluids prior to stress will result in protection from hearing loss. When levels of glial line-derived neurotrophic factor or neurotrophin-3 are increased in the inner ear prior to noise overstimulation or prior to the administration of ototoxic drugs there is significant protection, with decreased hearing loss and hair cell loss (Altschuler *et al.,* 1999; Keithley *et al.,* 1998; A. M. Miller *et al.,* 1999; Park *et al.,* 1998; Shoji *et al.,* 2000; Yagi *et al.,* 1999). Thus neurotrophic factors can provide protection. They are less effective if provided following the stress, suggesting a greater role in protection than in repair.

ACQUIRED DEAFNESS: DEAFNESS-RELATED CHANGES AND THEIR MODULATION

Many changes occur following deafness, both in the inner ear and in the central auditory pathways. In the inner ear, following loss of inner hair cells, a series of pathophysiological changes follow, including scar formation, loss of the peripheral processes of the auditory nerve, and, over time, substantial loss of the auditory nerve. The loss of auditory nerve is now believed to be related to a loss of survival factors, including neural activity, causing these neurons to enter into the cell death cycle. Centrally, neuronal death is minimal but changes occur that include decreased neuronal cell size in the auditory brain stem; synaptic changes; reorganization of connections; changes in neurotransmitters, receptors, and ion channels; and changes in glia, cytoskeletal elements, and regulatory mechanisms (Altschuler *et al.,* 1997; Bledsoe *et al.,* 1997; Miller *et al.,* 1991, 1996). Although some of these changes are related to a down-regulation concomitant with decreased activity, other changes may be a reaction within the pathways to compensate for decreased activity. There can be a large impact on these deafness-induced changes when activity is returned to the auditory system, currently through cochlear prostheses and perhaps in the future through hair cell regeneration. Studies have therefore been designed to develop interventions that can modulate the deafness-related changes.

Enhancement of auditory nerve survival following deafness

Cochlear prostheses depend on direct stimulation of the auditory nerve, so auditory nerve survival is of major importance. Neurotrophic factors not only play a role in protection from stress and trauma (see above), but also are very important for cell maintenance, and their absence can lead to cell death (Rich, 1992). Auditory nerve death following inner hair cell loss has been postulated to be the result of an associated loss of critical survival factors, and studies have therefore examined if replacement of survival factors would stop the death of auditory nerve neurons. Indeed, chronic delivery of neurotrophic factors into the fluids of the inner ear has now been shown to enhance auditory nerve survival significantly following noise- or drug-induced deafness (Altschuler *et al.,* 2000; Ernfors *et al.,* 1996; J. M. Miller *et al.,* 1997; A. M. Miller *et al.,* 1999; Staecker *et al.,* 1996; Ylikoski *et al.,* 1998). Effective neurotrophic factors include BDNF and GDNF.

Depolarization has long been recognized as a trophic treatment in cultured cells, increasing cell survival and modulating neural biochemistry to alter neurotransmitter, neuropeptide, and enzyme expression and density of (functional) voltage-gated calcium channels (Franklin and Johnson, 1992). If activity also serves as a survival factor for the auditory nerve, then electrical stimulation should enhance survival. Studies have shown that chronic electrical stimulation of the auditory nerve does significantly enhance its survival following deafness (Lousteau, 1987; Hartshorn *et al.,* 1991; Miller *et al.,* 1991; Miller and Altschuler, 1995; Mitchell *et al.,* 1997b; Leake *et al.,* 1991, 1992, 1995). The electrical stimulation may serve as a survival factor through activation of voltage-gated ion channels and/or through up-regulation of autocrine factors, including neurotrophic factors. *In vitro* results suggest that a combination of treatments may be most effective (Hegarty *et al.,* 1997) and studies to examine this are still ongoing.

Regrowth of auditory nerve processes following deafness

Studies have also examined if neurotrophic factors can induce regrowth of the peripheral processes of the auditory nerve that regress following inner hair cell loss. Some neurotrophic factors that enhance auditory nerve survival following deafness, such as BDNF (see above), will also induce regrowth of peripheral processes (Cho *et al.,* 1998). Other factors such as FGF are not effective in enhancing auditory nerve survival, but if provided along with a factor such as BDNF, which does enhance survival, will induce even greater regrowth than seen with BDNF alone (Cho *et al.,* 1998). Thus neurotrophic factors can serve not only to enhance survival of the auditory nerve but also to restore its peripheral processes, although different factors may be necessary to accomplish the most effective treatment.

GENETIC DEAFNESS

One in 1000 newborns suffers from hearing impairment. In developed countries, more than half of the afflicted babies are deaf due to genetic reasons. Genetic inner ear impairments can be nonsyndromic, affecting only hearing, or syndromic, with multiple defects occurring in several body systems. Many of the genes for deafness have now been identified, as well as their products. Affected proteins of the inner ear include cytoskeletal proteins such as myosin and a tectorial membrane protein (reviewed by Probst and Camper, 1999). Identification of the genes that are mutated in families with genetic inner ear impairments is important not only for diagnostic purposes, but also for the potential for prevention and cure. One exciting study has addressed interventions into genetic deafness, and has identified a mutation in *Myo-15* as the cause for deafness and balance impairments in the shaker 2 deafness mouse (Probst *et al.,* 1999). This mouse is a model for several families in which mutations in the very same gene cause genetic inner ear impairments (Wang *et al.,* 1998). Probst *et al.* (1998) demonstrated that in transgenic mice destined (genetically) to become deaf, insertion of the wild-type (correct) gene sequence into the fertilized egg corrected the genetic deficit, leading to normal inner ear structure and function in the adult mouse. This demonstrated that addition of the correct copy of the gene can rescue the inner ear from genetic deficits. To make such interventions viable for clinical applications, it is necessary to develop the means for delivering genes into the inner ear cells, which are the cells directly affected by the mutations. Gene therapy is one of the most promising technologies for gene delivery.

METHODS OF INTERVENTION

With effective protective measures for treating acquired deafness and measures for treatment following deafness identified, the next challenge is to find methods to provide these interventions.

AUDITORY PROSTHESES

Auditory prostheses can provide direct stimulation of the auditory pathways to promote return of hearing following deafness. Stimulatory activity can also act as a survival factor to enhance auditory nerve survival.

Cochlear prostheses

In severe and profound sensorineural hearing loss the sensory cells of the cochlea are destroyed. Cochlear prostheses bypass the damaged receptor epithelium and directly electrically excite the auditory nerve fibers. The development of the cochlear prosthesis has been the success story of the field of neuroprostheses. It has provided a therapy for the profoundly deaf, where none previously existed. The first Food and Drug Administration-approved inner ear implants, approximately 25 years ago, were single-channel devices that provided crude input and were an aid to lip reading. Now, multichannel devices, combined with many advances in signal-processing strategies, routinely provide open-set speech discrimination and even use of the telephone in the majority of implant recipients. Multichannel prostheses use strategies that attempt to approximate the "normal" encoding. One strategy is to divide the speech frequency range into six bands, with each activating an electrode in an analog fashion. Another strategy divides the speech frequency range into functionally significant regions (formants), with each assigned to clusters of electrodes that are activated differently depending on the distribution of speech energy in the formant region. Electrode arrays can be activated in an analog or digital fashion, with some devices providing simultaneous activation of sites and others only one at a time, cycling among the sites at a high rate. At this time there are no clear data to indicate that one strategy or approach is inherently better than another, and different devices and strategies work better for some subjects than for others.

Central auditory prostheses

Central auditory prostheses can be advantageous when the cochlea is not suitable for implantation or when there is insufficient auditory nerve survival. The dorsal-to-ventral axis of the cochlear nucleus, as with most auditory nuclei, is tonotopically organized and is therefore a candidate for prostheses placed either on its surface or penetrating into its depth (McCreery *et al.,* 1998; Mobley *et al.,* 1995). Such implants are still in the development stage, although success with surface prostheses on the cochlear nucleus (the first central synapse in the auditory pathways) has been reported (e.g., Laszig *et al.,* 1997; Portillo *et al.,* 1993, 1995). Penetrating prostheses, such as the two-dimensional arrays using multichannel, multishank silicon substrate technology of the Michigan type (Anderson *et al.,* 1989) or the Utah Intracortical Electrode Array (Nordhausen *et al.,* 1994, 1996), offer greater access to the spectral organization. Ultimately, three-dimensional arrays constructed from multiple planes of two-dimensional silicon devices being used in animal experiments may offer even greater access to the organization of brain stem structures (Hoogerwerf and Wise, 1994).

Miniosmotic Pumps

The fluid spaces of the cochlea provide for a closed environment well suited to receive local delivery of chemicals. Local delivery provides improved access and avoids the side effects that systemic delivery could entail. A model miniosmotic pump with cannula into the inner ear fluid of scala tympani, or into the middle ear with access through the round window, has been developed in animal models (e.g., Brown *et al.,* 1993; Park *et al.,* 1998), this model has been effectively utilized for local application of chemicals that provide protection from acquired deafness or enhance survival of auditory nerves (see above sections). Devices have also been developed for human applications (e.g., IntraEar, Inc).

Gene Therapy

Gene therapy is used to manipulate levels of specific proteins in cells and tissues. It can be used to inactivate specific proteins or to overexpress them. In most cases, this is accomplished by introducing a foreign gene into the cells. Viral vectors are the most efficient vehicles for gene transfer, and adenovirus, herpes simplex virus, adeno-associated virus, and lentivirus can mediate gene transfer into cells of the inner ear (reviewed by Raphael and Yagi, 1998). When mature mammalian inner ears are inoculated with adenoviral vectors, fibroblasts and other cells of connective tissue origin are transduced with the highest efficiency (Raphael *et al.,* 1996; Yagi *et al.,* 1999). At present this limits the applicability of adenovirus-based gene therapy in the inner ear to genes encoding secreted and diffusible proteins, such as the neurotrophins. Overexpression of the human GDNF transgene with an adenovirus vector has been shown to protect hair cells from noise and ototoxic drug insults in guinea pigs (Yagi *et al.,* 1999). Similar treatment was also efficient in protecting spiral ganglion neurons from degeneration following hair cell loss.

The next major challenge for inner ear gene therapy is to infect hair cells and supporting cells with a vector for delivering genes to these cells. Once this obstacle is overcome, it will be possible to test the function of proteins intrinsic to hair cells, such as Myo15 and Myo7a, and to deliver genes that can cure genetic inner ear impairments.

Combinations

With studies showing that both electrical stimulation and chemical delivery are important interventions, methods that can combine these can be particularly effective. Tissue engineering interventions are therefore being developed using University of Michigan silicon substrate neuroprobe prostheses that will not only stimulate but will provide chemical factors as well.

Prostheses with microchannels

University of Michigan neuroprobes can now be fabricated with microchannels for local delivery of fluids (Chen *et al.,* 1997). This provides the potential for site-specific delivery of neuroactive factors into the brain or inner ear. Because the neuroprobes containing the microchannels are also capable of stimulation and recording, it will be possible to provide both electrical stimulation and drug delivery to highly specific regions and to assess the functional results. This type of device opens new approaches to understanding the role of brain chemistry in information processing in the nervous system. By artificially adjusting levels of neuroactive components, neural circuits will be modified and will therefore change their function. Longer term changes in circuits, such as the population of dendrites, can be prompted by chemical action.

Prostheses with biopolymers

Biopolymers can be synthesized such that they contain both the protein sequence for silk, to provide strength and flexibility, and sequences of functional domains that naturally occur in the extracellular matrix protein, to confer specific biological properties (Cappello, 1990; Cappello *et al.*, 1990). These can be "spun" in microstructured thin films on the surface of prosthetic devices (Anderson *et al.*, 1993). Biopolymers can be produced that contain the sequence of fibronectin or laminin to provide a biologic "glue" (for neurons and epidermal cells, respectively), which can then be coated on implants. This can, in turn, be used to influence and stabilize the neuroprobe tissue interface, or to attach transformed cells for *ex vivo* gene transfer (see below). There is excellent tissue compatibility for these coated probes, with or without attached Schwann cells (O'Shea *et al.*, 1996). The biopolymer can also be used for timed release of specific factors. Thus a prosthesis could be providing electrical stimulation while releasing one or more chemical factors through microchannels (above) and other factors through diffusion out of the biopolymer. Specific placement of biopolymer and microchannels can also provide different factors to different regions.

Prostheses with biopolymers and *ex vivo* gene transfer

With *ex vivo* gene transfer, cultured cells are transformed to produce a specific gene product and are placed into a specific area of the body for region-specific release. One problem has been that the transformed cells can migrate to a different region. With the use of biopolymers with fibronectin or laminin, transformed cells (e.g., fibroblasts and fibronectin; Schwann cells and laminin), can be securely attached by the biopolymer to a neuroprobe or prosthesis, which is then inserted into the region of interest. The cells will then remain in place and their action will be highly localized.

CONCLUSIONS

The past decade has brought major advances in our understanding of the molecular mechanisms underlying deafness and of the factors that influence and modulate its expression and progression. We can expect, as progress in these areas continues, that this knowledge will form the basis for "molecular otology." Novel tissue engineering-based therapeutic interventions may become a major part of the practice of otolaryngology in the twenty-first century.

REFERENCES

Altschuler, R. A., Sato, K., Dupont, J., Bonneau, J. B., and Nakagawa, H. (1997). NMDA and glycine receptors in the auditory brain stem. Diversity and changes with deafness. *In* "Acoustic Signal Processing in the Central Auditory System" (J. Sykka, ed.), pp. 193–202. Plenum, New York.

Altschuler, R. A., Miller, J. M., Raphael, Y., and Schacht, J. (1999). Strategies for protection of the inner ear from noise induced hearing loss. *In* "Cochlear Pharmacology and Noise Trauma" (D. K. Prasher and B. Canlon, eds.). Novartis Reviews, London, UK.

Altschuler, R. A., Cho, Y., Ylikoski, J., Pirvola, U., Magal, E., and Miller, J. M. (1999). Rescue and regrowth of the auditory nerve by neurotrophic factors following deafferentation. *N. Y. Acad. Sci.* **884,** 305–311.

Anderson, D. J., Najafi, K., Tanghe, S. J., Evans, D. A., Levy, K. L., Hetke, J. F., Xue, X. L., Zappia, J. J., and Wise, K. D. (1989). Batch-fabricated thin-film electrodes for stimulation of the central auditory system. *IEEE Trans. Biomed. Eng.* **36,** 693–704.

Anderson, J. P., Nilsson, S. C., Rajachar, R. M., Logan, R., Weissman, N. A., and Martin, D. C. (1993). Bioactive silklike protein polymer films on silicon devices. *Mater. Res. Soc. Symp. Proc.*

Bledsoe, S. C., Nagase, S., Altschuler, R. A., and Miller, J. M. (1997). Changes in the central auditory system with deafness and return of activity via a cochlear prostheses. *In* "Acoustic Signal Processing in the Central Auditory System" (J. Sykka, ed.), pp. 513–528. Plenum, New York.

Brown, J. N., Miller, J. M., Altschuler, R. A., and Nuttall, A. L. (1993). Osmotic pump implants for chronic infusion of drugs into the inner ear. *Hear. Res.* **70,** 167–172.

Cappello, J. (1990). The biological production of protein polymers and their use. *Trends Biotechnol.* **8,** 309–311.

Cappello, J., Crissman, J., Dorman, M., Mikolajczak, M., Textor, G., Marquet, M., and Ferrari, F. (1990). Genetic engineering of structural protein polymers. *Biotechnol. Prog.* **6,** 198–202.

Chen, J., Wise, K. D., Hetke, J. F., and Bledsoe, S. C., Jr. (1997). USA. A multichannel neural probe for selective chemical delivery at the cellular level. *IEEE Trans. Biomed. Eng.* **44,** 760–769.

Cheng, B., and Mattson, M. P. (1992). NT-3 and BDNF protect CNS neurons against metabolic/excitotoxic insults. *Brain Res.* **640,** 56–67.

Cho, Y., Halsey, K., Prieskorn, D. P., Miller, J. M., and Altschuler, R. A. (1998). Regrowth of auditory nerve peripheral processes after hair cell loss with intrascalar neurotrophin infusion: Effects of FGF, BDNF and CNTF. *Soc. Neurosci. Abstr.* **24,** 2003.

Collazo, D., Takahashi, H., and McKay, R. D. (1992). Cellular targets and trophic functions of neurotrophin-3 in the developing rat hippocampus. *Neuron* **9**, 643–656.

Conlon, B. j., and Smith, D. W. (1998). Attenuation of neomycin ototoxicity by iron chelation. *Laryngoscope* **108**, 284–287.

Conlon, B. J., Aran, J.-M., Erre, J.-P., and Smith, D. W. (1999). Attenuation of aminoglycoside-induced cochlear damage with the metabolic antioxidant a-lipoic acid. *Hear. Res.* **128**, 40–44.

Davies, A. M. Z. (1994). Role of neurotrophins in the developing nervous system. *J. Neurobiol.* **24**, 1334–1338.

Dechesne, C. J., Kim, H. N., Nowak, T. S., and Wenthold, R. J. (1992). Expression of heat shock protein, HSP 72 in guinea pig and cat cochlea after hyperthermia: Immunocytochemical and *in situ* hybridization analysis. *Hear. Res.* **59**, 195–204.

Eckenstein, P. (1994). Fibroblast growth factors in the nervous system. *J. Neurobiol.* **25**, 1467–1480.

Ernfors, P., Duan, M. L., El Shamy, W. M., and Canlon, B. (1996). Protection of auditory neurons from aminoglycoside toxicity by NT-3. *Nat. Med.* **2**, 463–467.

Franklin, J. L., and Johnson, E. M. (1992). Suppression of programmed neuronal death by sustained elevation of cytoplasmic calcium. *Trends Neurosci.* **15**, 501–507.

Garetz, S. L., Rhee, D. J., and Schacht, J. (1994a). Sulfhydryl compounds and antioxidants inhibit cytotoxicity to outer hair cells of a gentamicin metabolite *in vitro. Hear. Res.* **77**, 75–80.

Garetz, S. L., Altschuler, R. A., and Schacht, J. (1994b). Attenuation of gentamicin ototoxicity by glutathione in the guinea pig *in vivo. Hear. Res.* **77**, 81–87.

Hartshorn, D. O., Miller, J. M., and Altschuler, R. A. (1991). Protective effect of electrical stimulation on the deafened guinea pig cochlea. *Otolaryngol. Head Neck Surg.* **104**, 311–319.

Hegarty, J. L., Kay, A. R., and Green, S. H. (1997). Trophic support of cultured spiral ganglion neurons by depolarization exceeds and is additive to that by neurotrophins or cAMP and requires elevation of Ca^{++} with a set range. *J. Neurosci.* **17**, 1959–1970.

Hoffman, D. W., Whitworth, C. A., Jones, K. L., and Rybak, L. P. (1987). Nutritional status, glutathione levels, and ototoxicity of loop diuretics and aminoglycoside antibiotics. *Hear. Res.* **31**, 217–222.

Hoogerwerf, A. C., and Wise, K. D. (1994). A three-dimensional microelectrode array for chronic neural recording. *IEEE Trans. Biomed. Eng.* **41**, 1136–1146.

Hu, B. H., Zheng, X. Y., McFadden, S. L., Kopke, R. D., and Henderson, D. (1997). R-Phenylisopropyladenosine attenuates noise-induced hearing loss in the chinchilla. *Hear. Res.* **113**, 198–206.

Keithley, E. M., Ma, C. L., Ryan, A. F., Louis, J. C., and Magal, E. (1998). GDNF protects the cochlea against noise damage. *NeuroReport* **9**, 2183–2187.

Klein, R. (1994). Role of neurotrophins in mouse neuronal development. *FASEB J.* **8**, 738–744.

Laszig, R., Marangos, N., Sollmann, P., Ramsden, R., Fraysse, B., Lenarz, T., Rask-Andersen, H., Bredberg, G., Sterkers, O., Manrique, M., and Nevison, B. (1997). Initial results from the clinical trial of the nucleus 21-channel auditory brain stem implant. *Am. J. Otol.* **18**, S160.

Lautermann, J., McLaren, J., and Schacht, J. (1995a). Glutathione protection against gentamicin ototoxicity depends on nutritional status. *Hear. Res.* **86**, 15–24.

Lautermann, J., Song, B., McLaren, J., and Schacht, J. (1995b). Diet is a risk factor in cisplatin ototoxicity. *Hear. Res.* **88**, 47–53.

Leake, P. A., Hradek, G. T., Rebscher, S. J., and Snyder, R. L. (1991). Chronic intracochlear electrical stimulation induces selective survival of spiral ganglion neurons in neonatally deafened cats. *Hear. Res.* **54**, 251–271.

Leake, P. A., Snyder, R. L., Hradek, G. T., and Rebscher, S. J. (1992). Chronic intracochlear stimulation in neonatally deafened cats: Effects of intensity and stimulating electrode position. *Hear. Res.* **64**, 99–117.

Leake, P. A., Snyder, R. L., Hradek, G. T., and Rebscher, S. J. (1995). Consequences of chronic electrical stimulation in neonatally deafened cats. *Hear. Res.* **82**, 65–80.

Leonova, E., Fairfield, D., Mitchell, A., Lomax, M., and Altschuler, R. A. (1998). Expression of HSP 27 in the rat cochlea. *Abstr. Cell Biol. Meet.*

Lim, H. H., Jenkins, O. H., Myers, M. W., Miller, J. M., and Altschuler, R. A. (1993). Detection of HSP 72 synthesis after acoustic overstimulation in rat cochlea. *Hear. Res.* **69**, 146–150.

Lin, L. F., Doherty, D. H., Lile, J. D., Bektesh, S., and Collins, F. (1993). GDNF: A glial cell line-derived neurotrophic factor for midbrain dopaminergic neurons. *Science* **260**, 1130–1132.

Lousteau, R. J. (1987). Increased spiral ganglion cell survival in electrically stimulated, deafened guinea pig cochleae. *Laryngoscope* **97**, 837–842.

Mattson, M. P., and Furakawa, K. (1996). Programmed cell life; anti-apoptotic signaling and therapeutic strategies for neurodegenerative disorders. *J. Restorative Neurol. Neurosci.* **9**, 191–205.

Mattson, M. P., Lovell, M. A., Furukawa, K., and Markesebery, W. R. (1995). Neurotrophic factors attenuate glutamate-induced accumulation of peroxides, elevation, and neurotoxicity and increase antioxidant enzyme activities in hippocampal neurons. *J. Neurochem.* **65**, 1740–1751.

McCreery, D. B., Shannon, R. V., Moore, J. K., and Chatterjee, M. (1998). Accessing the tonotopic organization of the ventral cochlear nucleus by intranuclear microstimulation. *IEEE Trans. Rehab. Eng.* **6**, 391–399.

Miller, A. M., Yamasoba, T., and Altschuler, R. A. (1999). Hair cell and spiral ganglion neuron preservation and regeneration—Influence of growth factors. *Curr. Opin. Otolaryngol. Head Neck Surg.* **6**, 301–307.

Miller, J. M., and Altschuler, R. A. (1995). Effectiveness of different electrical stimulation conditions in preservation of spiral ganglion cells following deafness. *Ann. Otol., Rhinol., Laryngol.* **166**, 57–60.

Miller, J. M., Altschuler, R. A., Niparko, J. K., Hartshorn, D. O., Helfert, R. H., and Moore, J. K. (1991). Deafness-induced changes in the central auditory system and their reversibility and prevention. *In* "Noise Induced Hearing Loss" (Dancer, D. Henderson, R. j., Salvi, and R. P. Hamernik, eds.). Mosby Year Book, St. Louis, MO.

Miller, J. M., Altschuler, R. A., Dupont, J., Lesperance, M., and Tucci, D. (1996). Consequences of deafness and electrical stimulation on the auditory system. *In* "Auditory Plasticity and Regeneration" (R. J. Salvi, D. Henderson, F. Fiorino, and V. Colletti, eds.), pp. 378–391. Tieman Med Publishers, New York.

Miller, J. M., Chi, d. H., Kruszka, P., Raphael, Y., and Altschuler, R. A. (1997). Neurotrophins can enhance spiral ganglion cell survival after inner hair cell loss. *Int. J. Dev. Neurosci.* **15**, 631–643.

Mitchell, A., Miller, J. M., and Altschuler, R. A. (1997a). Enhanced recovery from noise overstimulation correlates with HSP expression. *Abstr. Assoc. Res. Otolaryngol.* **20**, 283.

Mitchell, A., Miller, J. M., Finger, P., Heller, J., and Altschuler, R. A. (1997b). Effects of chronic high rate electrical stimulation on the cochlea and eighth nerve in the deafened guinea pig. *Hear. Res.* **105**, 30–43.

Mobley, J. P., Huang, J., Moore, J. K., and McCreery, D. B. (1995). Three-dimensional modeling of human brain stem structures for an auditory brain stem implant. *Ann. Otol., Rhinol., Laryngol., Suppl.* **166**, 30–31.

Myers, M. W., Quirk, W. S., Rizk, S. S., Miller, J. M., and Altschuler, R. A. (1992). Expression of the major mammalian stress protein in the cochlea following transient ischemia. *Laryngoscope* **102**, 981–987.

Nishida, I., and Takumida, M. (1996). Attenuation of aminoglycoside ototoxicity by glutathione. *ORL* **58**, 68–73.

Nordhausen, C. T., Rousche, P. J., and Normann, R. A. (1994). Optimizing recording capabilities of the Utah intracortical electrode array. *Brain Res.* **637**, 27–36.

Nordhausen, C. T., Maynard, E. M., and Normann, R. A. (1996). Single unit recording capabilities of a 100 microelectrode array. *Brain Res.* **726**, 129–140.

Ohinata, Y., Yamasoba, T., Schacht, J., and Miller, J. M. (1999). The effect of glutathione monoethyl ester in the protection from noise-induced hearing loss. *Abstr. Assoc. Res. Otolaryngol.* **22**, 601.

O'Shea, K. S., Buchko, C., Shen, Y., Altschuler, R. A., Finger, P., Wiler, J. A., Cappello, J., and Martin, D. C. (1996). Biocompatibility of CNS implants coated with silklike polymers containing elastin, fibronectin or laminin cell binding motifs. *Abstr. Soc. Neurosci.* **22**, 1983.

Park, G. H., Miller, A. L., Tay, H., Yamasoba, T., Prieskorn, D. M., Magal, E., Altschuler, R. A., and Miller, J. M. (1998). Exogenous GDNF delivered onto the round window membrane prevents aminoglycoside-induced sensorineural hearing loss. *Abstr. Assoc. Res. Otolaryngol.* **21**, 241.

Portillo, F., Nelson, R. A., Brackmann, D. E., Hitselberger, W. E., Shannon, R. V., Waring, M. D., and Moore, J. K. (1993). Auditory brain stem implant: Electrical stimulation of the human cochlear nucleus. *Adv. Oto-Rhino-Laryngol.* **48**, 248–252.

Portillo, F., Mobley, P., Moore, J., and McCreery, D. (1995). Feasibility of a central nervous system auditory prosthesis: Penetrating microelectrode insertion force studies. *Ann. Otol., Rhinol., Laryngol., Suppl.* **166**, 31–33.

Priuska, E. M., and Schacht, J. (1995). Formation of free radicals by gentamicin and iron and evidence for an iron/gentamicin complex. *Biochem. Pharmacol.* **50**, 1749–1752.

Priuska, E. M., Clark, K., Pecoraro, V., and Schacht, J. (1998). NMR spectra of iron–gentamicin complexes and the implications for aminoglycoside toxicity. *Inorg. Chim. Acta* **273**, 85–91.

Probst, F. J., and Camper, S. A. (1999). The role of mouse mutants in the identification of human hereditary hearing loss genes. *Hear. Res.* **130**, 1–6.

Probst, F. J., Fridell, R. A., Raphael, Y., Saunders, T. L., Wang, A., Liang, Y., Morell, R. J., Touchman, J. W., Lyons, R. H., Noben-Trauth, K., Friedman, T. B., and Camper, S. A. (1998). Correction of deafness in shaker-2 mice by an unconventional myosin in a BAC transgene. *Science* **280**, 1444–1447.

Probst, F. J., Chen, K. S., Zhao, Q., Wang, A., Friedman, T. B., Lupski, J. R., and Camper, S. A. (1999). A physical map of the mouse shaker-2 region contains many of the genes commonly deleted in Smith–Magenis syndrome (del17p11.2p11.2). *Genomics* **55**, 348–352.

Puel, J. L., Pujol, R., Tribillac, F., Ladrech, S., and Eybalin, M. (1994). Excitatory amino acid antagonists protect cochlear auditory neurons from excitotoxicity. *J. Comp. Neurol.* **341**, 241–256.

Puel, J. L., d'Aldin, G., Saffiedine, S., Eybalin, M., and Pujol, R. (1996). Excitotoxicity and plasticity of IHC-Auditory nerve contributes to both temporary and permanent threshold shift. *In* "Scientific Basis of Noise-Induced Hearing Loss" (A. Axelsson, H. Borchgrevink, R. P. Hamernik, P. A. Hellström, D. Henderson, and R. J. Salvi, eds.), pp. 36–42. Thieme, Stuttgart.

Puri, R. N., and Meister, A. (1983). Transport of glutathione, as gamma-glutamylcysteinylglycyl ester, into liver and kidney. *Proc. Natl. Acad. Sci. U.S.A.* **80**, 5258–5260.

Quirk, W. S., Shivapuja, B. G., Schwimmer, C. L., and Seidman, M. D. (1994). Lipid peroxidation inhibitor attenuates noise-induced temporary threshold shifts. *Hear. Res.* **74**, 217–220.

Raphael, Y., and Yagi, M. (1998). Gene transfer and the inner ear. *Curr. Opin. Otolaryngol., Head Neck Surg.* **6**, 311–315.

Raphael, Y., Frisancho, J. C., and Roessier, B. J. (1996). Adenoviral-mediated gene transfer into guinea pig cochlear cells *in vivo. Neurosci. Lett.* **207**, 137–141.

Ravi, R., Somani, S. M., and Rybak, L. P. (1995). Mechanism of cisplatin ototoxicity: Antioxidant system. *Pharmacol. Toxicol.* **76**, 386–394.

Rich, K. M. (1992). Neuronal death after trophic factor deprivation. *J. Neurotrauma* **9**, 61–69.

Seidman, M. D., Shivapuja, B. G., and Quirk, W. S. (1993). The protective effects of allopurinol and superoxide dismutase on noise-induced cochlear damage. *Otolaryngol. Head Neck Surg.* **109**, 1052–1056.

Sendtner, M., Carrol, P., Holtmann, B., Hughes, R. A., and Thoenen, H. (1994). Ciliary neurotrophic factor. *J. Neurobiol.* **25**, 1436–1453.

Sha, S.-H., and Schacht, J. (1999a). Formation of reactive oxygen species following bioactivation of gentamicin. *Free Radical Biol. Med.* **26**, 341–347.

Sha, S.-H., and Schacht, J. (1999b). Formation of free radicals by aminoglycoside antibiotics. *Hear. Res.* **128**, 112–118.

Sha, S. H., and Schacht, J. (2000). Salicylate attenuates gentamicin-induced ototoxicity. *Lab. Invest.* (in press).

Shoji, F., Yamasoba, T., Magal, E., Dolan, D. F., Altschuler, R. A., and Miller, J. M. (2000). Glial cell line-derived neurotrophic factor (GDNF) protects auditory hair cells in the guinea pig cochlea from noise stress *in vivo. Hear. Res.* (in press).

Song, B.-B., and Schacht, J. (1996). Variable efficacy of radical scavengers and iron chelators to attenuate gentamicin ototoxicity in guinea pig *in vivo. Hear. Res.* **94**, 87–93.

Song, B.-B., Anderson, D. J., and Schacht, J. (1997). Protection from gentamicin ototoxicity by iron chelators in guinea pig *in vivo. J. Pharmacol. Exp. Ther.* **282**, 369–377.

Song, B.-B., Sha, S.-H., and Schacht, J. (1998). Iron chelators protect from aminoglycoside-induced cochleo- and vestibulotoxicity in guinea pig. *Free Radical Biol. Med.* **25**, 189–195.

Staecker, H., Kopke, R., Malgrange, B., Lefebvre, P., and Van De Water, T. R. (1996). NT-3 and BDNF therapy prevents loss of auditory neurons following loss of hair cells. *NeuroReport* **7**, 889–894.

Thorne, P. R., Nuttall, A. L., Scheibe, F., and Miller, J. M. (1987). Sound-induced artifact in cochlear blood flow measurements using the laser Doppler flowmeter. *Hear. Res.* **31**, 229–234.

Wang, A., Liang, Y., Fridell, R. A., Probst, F. J., Wilcox, E. R., Touchman, J. W., Morton, C. C., Morell, R. J., Noben-Trauth, K., Camper, S. A., and Friedman, T. B. (1998). Association of unconventional myosin MYO15 mutations with human nonsyndromic deafness DFNB3. *Science* **280**, 1447–1451.

Yagi, M., Magal, E., Sheng, Z., Ang, K. A., and Raphael, Y. (1999). Hair cells are protected from aminoglycoside ototoxicity by adenoviral-mediated overexpression of GDNF. *Hum. Gene Ther.* **10**, 813–823.

Yamane, H., Nakai, Y., Takayama, M., Konishi, K., Iguchi, H., Nakagawa, T., Shibata, S., Kato, A., Sunami, K., and Kawakatsu, C. (1995). The emergence of free radicals after acoustic trauma and strial blood flow. *Acta Otolaryngol., Suppl.* **519**, 87–92.

Yamasoba, T., Nuttall, A. L., Harris, C., Raphael, Y., and Miller, J. M. (1998). Role of glutathione in protection against noise-induced hearing loss. *Brain Res.* **784**, 82–90.

Ylikoski, J., Pirvola, U., Moshnyakov, M., Palgi, J., Arumae, U., and Saarma, M. (1993). Expression patterns of neurotrophins and their receptor mRNAs in the rat inner ear. *Hear. Res.* **65**, 69–78.

Ylikoski, J., Pirvola, U., Suvanto, P., Liang, X.-Q., Virkkala, J., Magal, E., Altschuler, R. A., Miller, J. M., and Saarma, M. (1998). Guinea pig auditory neurons are protected by GDNF from degeneration after noise trauma. *Hear. Res.* **124**, 17–26.

Yoshida, N., and Liberman, M. C. (1999). Heat-shock stress protects the ear from permanent acoustic injury. *Abstr. Assoc. Res. Otolaryngol.* **22**, 152.

CHAPTER 54

VISION ENHANCEMENT SYSTEMS

Gislin Dagnelie, Mark S. Humayun, and Robert W. Massof

VISUAL SYSTEM: ARCHITECTURE AND (DYS)FUNCTION

Human vision is mediated by one of the most highly developed sensory systems found in nature. The capacity to combine high spatial resolution near the center of fixation with a wide peripheral field of view, accurate depth perception, color discrimination, and light–dark adaptation over 12 orders of magnitude is unparalleled; every stage in the visual system is organized to accomplish this. The photoreceptor layer in the retina provides the high signal amplification of the rods required for night vision, and the dense packing of three cone types provides for detailed central and color vision. The intricate local preprocessing performed by subsequent retinal cell layers augments these functions by performing brightness and color comparisons, and helps to reduce the information stream acquired by over 100 million photoreceptors, allowing transport across a mere 1 million fibers in the optic nerve to the visual centers in the brain, where further parsing and interpretation of the image take place.

From a functional point of view, the visual system can be understood as depicted in Fig. 54.1, with sensors, preamplifiers, preprocessors, transmission lines, and several central processor stages. The two most crucial stages in the visual process—the conversion of light into chemical and electrical signals, and the signal transmission from the eyes toward the brain—are also the most vulnerable ones. Light conversion and signal amplification in the photoreceptors require a highly complex interplay between the molecules inside these cells, undergoing interconnecting cycles of conversion and regeneration, and the surrounding cells—in particular, the retinal pigment epithelium (RPE) cells, which provide nutrients to, and digest cell membrane discarded from, the photoreceptors. This process can easily be disrupted by nutritional deficits, overexposure to short-wavelength light (presumably causing oxidative changes), attacks by pathogens, and especially a genetic miscoding of one or more participating molecules (Heckenlively, 1988). Not only does this directly cause impaired signal transduction, but the additional energy demand, the presence of abnormal molecules, and the excess shed cell membrane may exceed the RPE cells' support capacity, inexorably leading to degeneration of both photoreceptors and RPE cells. These diseases are jointly known as retinal degenerations (Fig. 54.2, A and B). In general, secondary retinal cells (horizontal, bipolar, amacrine, and ganglion cells) are not affected to the same extent, a notable exception to transsynaptic degeneration patterns commonly found in the central nervous system. Morphometric studies of retinal tissue such as that shown in Fig. 54.2B have found average cell survival to be 80 and 30% in the macula (Santos *et al.*, 1997), and 40 and 20% at eccentricities up to 25° (Humayun *et al.*, 1999b) for bipolar and ganglion cells, respectively, in retinas devoid of photoreceptor nuclei.

At the signal transmission stage retinal ganglion cells (RGCs) encode the chemical signals in the form of electrical spike trains that travel along RGC axons toward the thalamic relay nuclei and other brain areas; at this stage the system is vulnerable to insults. Damage can result from injury to the optic nerve (trauma), increased pressure inside the eye, crushing the fragile axon fibers at the optic nerve head (glaucoma), and inflammation (optic neuritis) or impaired blood supply

Structural Component   Functional Equivalent

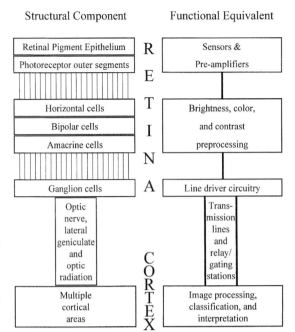

Fig. 54.1. Schematic representation of the visual system. The outer retina (retinal pigment epithelium and photoreceptor layers) forms the sensor array, followed by several inner retinal preprocessing stages, the ganglion cell transmission stages, and further central processing stages in subcortical and cortical brain centers.

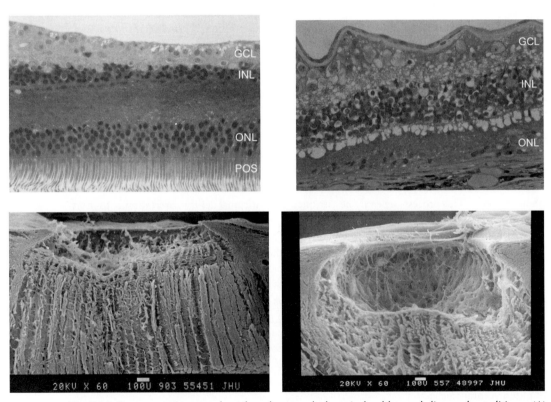

Fig. 54.2. Representative samples of ocular morphology in healthy and diseased conditions. (A) Cross-section through the retina near its center (fovea), showing healthy photoreceptor outer segments (POS), multiple layers of photoreceptor cell nuclei in the outer nuclear layer (ONL), bipolar cell nuclei in the inner nuclear layer (INL), and ganglion cell bodies in the ganglion cell layer (GCL). (B) Retina of a patient with a long history of retinal degeneration, and bare light perception in the last years of life, showing lack of photoreceptor outer segments and cell bodies, in comparison with A. (C) Scanning electron microscope cross-section of the optic nerve head, showing healthy appearance of the support structure, the lamina cribrosa. (D) Optic nerve head from a patient with long history of glaucoma, showing compression of the lamina cribrosa and embedded nerve fibers (retinal ganglion cell axons). Micrographs C and D courtesy of Harry A. Quigley, the Johns Hopkins University (Baltimore, MD).

(ischemic neuropathy) of the optic nerve. Each of these insults can impair or interrupt the signal-carrying capacity of optic nerve fibers (Fig. 54.2, C and D).

In addition to damage occurring at these distinct stages, more generalized damage to the retina can occur. Common mechanisms for this are the leakage of capillaries in diabetic retinopathy and the interruption of the blood supply to the inner retina (retinal vascular occlusive disease). Each of these can lead to widespread cell death in the inner nuclear and ganglion cell layers throughout the affected area; vascular occlusive disease provokes angiogenesis; the new retinal vessels tend to leak and damage the retina's fragile structure. Diabetic retinopathy is a major cause of preventable blindness in the developed world.

POSSIBLE APPROACHES TO VISION RESTORATION

As in all biomedical engineering, approaches to restore function can be based on reversal of the disease process and regeneration of autologous tissue, on tissue grafts, or on hybrid techniques involving tissue as well as synthetic materials and devices. To support the cells affected by disease or injury, neuroprotective substances, growth factors, and genetic modifications of cell function may be used, postponing or preventing further loss of function. In addition, as long as useful function remains, one can strengthen the stimulus and internal response signal, counteracting visual impairment as much as possible.

If little or no sensor function remains, one can seek to restore it through newly grown or transplanted photoreceptors and/or RPE cells, through a prosthetic device that will electrically stimulate remaining secondary retinal cells, or through man-made tissue(s) that mimic photoreceptor function. At the RGC level, substitution for lost signal transmission may be sought through protection of the cells and administration of factors promoting axon regeneration—this may require the use of synthetic tissues and factors enabling reconnection of axons with central structures—or through *in situ* growth of new cells, promoted to differentiate into RGCs and to send new axons through the optic nerve.

In preliminary form, some of these approaches exist in the laboratory, but others are merely ideas. In the two following sections, existing and prospective methods are presented.

CURRENT APPROACHES

OPTOELECTRONIC VISION ENHANCEMENT DEVICES

These devices are based on a combination of optical and electronic image enhancement techniques; their common principle of operation is to improve visibility of the image to the diseased eye, through magnification, contrast, and/or color enhancement, and filtering or feature extraction. Current implementations such as the Low Vision Enhancement System (LVES; Vision systems International; Minneapolis, MN) use optical magnification, zoom, automatic focus, and (in some cases) analog image enhancement (Fig. 54.3) (Massof, 1999). Digital feature enhancement, although possible in the laboratory, is not currently implemented in portable format; however, with the rapidly increasing power of chip-based image-processing circuitry such further improvements are not far off. The same is true for the product of field of view and resolution, i.e., the number of pixels across the screen, which is currently limited to that of standard video and the video graphics adapter (VGA), but which is likely to increase sharply with the introduction of high-resolution video camera and display formats (e.g., high-definition television). Similarly, compactness and color representation will be improved as smaller, lighter, and more luminous flat-panel displays become available.

A useful property of video cameras such as those employed in the LVES is their built-in automatic gain control (AGC), resulting in constant internal image brightness over a wide range of environmental illumination levels. Among other properties, this makes the image acquisition and processing stages of these systems highly suitable as sensors and preprocessors, respectively, for prosthetic and tissue-based image enhancement systems.

VISION PROSTHESES BASED ON ELECTRICAL TISSUE STIMULATION

As was mentioned above, retinas with severe degeneration of the sensor layer retain high numbers of secondary cells. This opens the opportunity to convey pixelized images to the degenerated retina by a prosthetic device stimulating remaining secondary cells with a two-dimensional array

Fig. 54.3. The Low Vision Enhancement System (LVES), developed at the Lions Vision Center (Johns Hopkins University), with support from NASA and the Department of Veterans Affairs, is a characteristic example of optoelectronic vision enhancement systems. This system features binocular orientation cameras, a centrally placed 1.5–10× zoom camera with automatic focus, and a binocular projection path with built-in refractive correction and alignment for the wearer.

of microelectrodes,* not unlike the rastered images provided by a stadium scoreboard (Dagnelie and Massof, 1996). From an engineering point of view this opens a range of possible approaches (to name a few, stimulation electrodes under vs. over the retina; fully integrated photosensor, image processor, and stimulator systems vs. external image capture and processing linked to an intraocular stimulating matrix) and a host of biocompatibility, signal processing, and power management questions.

The most pressing question, however, that of the feasibility to convey visual imagery to patients blind from retinal degeneration, has been answered affirmatively. In a series of experiments started at Duke University in 1992, and continued since 1993 at Johns Hopkins, more than a dozen volunteers with end-stage retinitis pigmentosa (RP)—whose remaining vision was limited to, at best, light perception—have participated in tests whereby, during a surgical procedure under local anesthesia, the inner surface of the retina was electrically stimulated with small and brief biphasic current pulses applied through single or multiple electrodes (Humayun *et al.*, 1996, 1999a). Among the most salient findings are the subjects' ability to see small punctate light flashes (phosphenes), whose perceived location corresponds exactly to that of the stimulation, and the ability to see simple patterns of multiple phosphenes when multiple electrodes are activated simultaneously. Stimulation at rates greater than 40–60 pulses/s is perceived as continuous stimulation, and perceived stimulus intensity increases with pulse duration and amplitude as well as repetition rate. Independent tests in human volunteers and in amphibian retina (Greenberg, 1998) have demonstrated that stimulus pulses 1 ms or longer in duration preferentially stimulate the (deeper) bipolar cells rather than the (more superficial) RGCs.

In a collaborative effort with the group at Johns Hopkins, researchers in the Department of Electrical Engineering at North Carolina State University (NCSU) have developed a modular system for retinal stimulation, with a video image acquisition system and a chipset (radiofrequency encoder/transmitter and decoder/demultiplexer/drivers) to control an intraocular electrode array (Fig. 54.4). These devices are currently being tested in experimental animals with naturally occurring retinal degenerations.

At the present time, several other groups are pursuing similar intraocular prosthetic devices, in two variants: integrated subretinal stimulating "photodiodes," and epiretinal stimulator arrays with external image capture and preprocessing. In the first variant (Chow and Chow, 1997; Zrenner *et al.*, 1997) small units are placed at the level of the missing photoreceptor/RPE layer, and incoming light—which is focused by the optics of the eye at this original photoreceptor plane, thus

---

* *Note the similarity of such a device, in both principle and operation, to the cochlear prosthesis (Clark et al., 1990).*

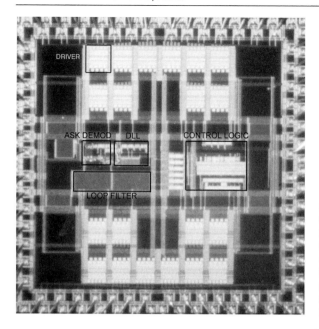

Fig. 54.4. Photomicrograph of the Die3 chip, developed at North Carolina State University, for reception, decoding and driving of 100 intraocular multielectrode signals. A delay-locked loop (DLL) and filter are used to synchronize the amplitude-shift keyed (ASK) digital signal; the control logic extracts the pulse amplitude, sequencing, and duration settings. Each driver unit contains circuitry to drive five electrodes. For design considerations and other information, see Clements et al. (1999) and the web site at http://www.ece.ncsu.edu/erl/erl—eye.html. Photo courtesy of Wentai Liu (NCSU; Raleigh, NC).

providing a sharp image of the outside world—is used to generate electrical impulses that stimulate nearby (bipolar) cells in the overlying retina. The simplicity and elegance of such a system hinge on two important premises: that nutrients and oxygen from the underlying vasculature will still be able to reach the inner retina and that external energy can be provided to produce the necessary pulse generation and signal amplification, especially at low light levels. The developers of such systems are concentrating on finding solutions for these problems, as well as on the design of surgical methods to insert large arrays of such units under the retina.

The second prosthesis type, using an epiretinal stimulator array, is being pursued by two groups, Harvard/Massachusetts Institute of Technology (Wyatt and Rizzo, 1996; Rizzo and Wyatt, 1999) and a consortium of German universities (Eckmiller, 1997), in addition to work at Johns Hopkins/NCSU (Humayun and de Juan, 1998). Each of these groups expects to have working prototypes of a prosthetic device for human implantation within a few years.[†]

In addition to retinal stimulation, two other approaches to electrical stimulation of the visual system are being studied. One of these, direct stimulation of the optic nerve (Veraart *et al.,* 1998), is less invasive surgically than implantation of devices inside the eye, but has the drawback that stimulation of individual optic nerve fibers will require microelectrode arrays much finer than those currently available, and a complex mapping system to establish the correspondence between visual field locations and individual electrodes (see below).

The oldest attempts at vision restoration involved stimulation of the visual cortex through intracranial electrodes, controlled by external image acquisition and signal-conditioning circuitry. Several decades ago, Brindley and Lewin (1968) took the first steps in this direction by implanting a set of electrodes over the visual cortex of a blind volunteer; the phosphenes described by this volunteer were similar to those elicited by intraocular stimulation. More recently, researchers at the National Institute for Neurological Disorders and Stroke (NINDS) have performed tests with over 30 electrodes penetrating the cortical surface (Schmidt *et al.,* 1996); the NINDS group as well as researchers at the University of Utah are preparing for implantation of denser intracortical electrode arrays (Fig. 54.5) (Normann *et al.,* 1996). The Utah researchers have also performed psychophysical studies to establish minimum requirements for prosthetic vision (Cha *et al.,* 1992a,b), and are testing device prototypes in primates.

The cortical prosthesis bypasses both retinal and optic nerve problems, and might therefore

---

[†] *A significant complication in the design of a prosthesis with external image acquisition is the need for eye position monitoring, allowing adjustment of the camera angle as the prosthesis wearer shifts his/her gaze.*

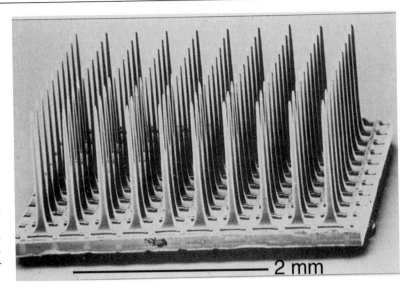

Fig. 54.5. Silicon multielectrode designed by Normann et al. (1996) for penetrating stimulation of the visual cortex or retina. Interelectrode distances between 200 and 400 μm can be manufactured. The 1- to 2-mm-long electrode shafts are insulated, and the tips can be metallized with platinum black or other suitable metals.

be seen as a universal approach to vision restoration. However, cortical stimulation requires complex surgery of an otherwise healthy brain; moreover, the convoluted mapping of the visual cortex complicates mapping of objects and locations in the outside world into a pixelized image that can be understood by the prosthesis wearer. As with the epiretinal prosthesis, eye movement monitoring is required. Algorithms to establish this mapping have been tested through simulations in sighted volunteers (Dagnelie and Vogelstein, 1999), but thus far only for small (32) sets of simulated phosphenes.

CELL TRANSPLANTATION

Cell transplantation in the neurosensory system can, in principle, restore function through two mechanisms: through rescue of threatened cells (e.g., through trophic factors produced by the transplanted cells) or through functional integration into the host tissue. Both mechanisms appear to be operational in retinal cell transplantation, and it is not always clear which mechanism is responsible for functional improvements observed.

Animal experiments performed in the past 15 years provide clear examples of this dilemma. RPE cell transplantation in RCS rats (Li and Turner, 1988; Lopez *et al.*, 1989) can lead to photoreceptor rescue, and morphologic evidence of synapse formation suggests that transplanted photoreceptors in light-damaged rats (del Cerro *et al.*, 1989) and rats with photoreceptor degenerations (Gouras *et al.*, 1991) may be capable of assuming visual function. Yet, behavioral evidence of regained function in light-deprived rats following photoreceptor transplantation (DiLoreto *et al.*, 1996) could be explained by either mechanism, and solid evidence that the observed synapses indeed provide useful vision is still lacking.

Fetal tissue has proved more successful than fully developed retina in forming synaptic connections and retinal morphology resembling that of intact retina (del Cerro *et al.*, 1997a), and has become the tissue of choice in most transplantation attempts. This tissue also carries a lower risk of rejection by the host immune system, at least in photoreceptor grafts; indeed, rejection does not appear to occur in rat or mouse photoreceptor transplantation (del Cerro *et al.*, 1997a). Admittedly, some caution is warranted, because immune reactions in rodents tend to be less severe than in humans. In rat RPE allografts, on the other hand, some inflammation has been observed (del Cerro *et al.*, 1997a), suggesting that immune processes may play a more important role for this cell type.

Preliminary attempts at retinal cell transplantation in blind volunteers, performed primarily to demonstrate safety, have yielded results very similar to those in animals. Allografts of cultured RPE appear to provide protection to functional age-related macular degeneration (AMD) through trophic factors, but in exudative AMD the graft is quickly overwhelmed by an inflammatory reaction (Gouras and Algvere, 1996). Photoreceptor transplants appear to undergo a similar fate (del Cerro *et al.*, 2000), and also appear to convey an improvement in vision in some cases (del Cerro *et al.*, 1997b; Humayun *et al.*, 2000), but whether this is mediated by graft–host synapse formation or by trophic support to the few remaining photoreceptors cannot be distinguished.

Tissue preparation for these early attempts has varied—ranging from cell microaggregates (del Cerro *et al.*, 1997a) and patches of cultured RPE (Gouras and Algvere, 1996) to agar-supported sheets of retinal tissue prepared by vibratome (Silverman and Hughes, 1989) or excimer laser (Liang *et al.*, 1994)—but to date no generally accepted methods have emerged.

Axon Regeneration

Retinal ganglion cells behave as central nervous system neurons in their inability to recover from severe injury: although the axons of peripheral (motor or sensory) neurons can regenerate (and even be reconnected), central nervous system neurons seem to lose that capacity following their original outgrowth during the organism's early development. An important factor preventing later regeneration is provided by Schwann cells or astroglia forming the protective myelin sheath around optic nerve axons. Cut rat optic nerve axons that will not form new neurites in their natural optic nerve environment can be made to regenerate into a peripheral nerve graft (Fukuda *et al.*, 1998; Thanos, 1997). This is a very active field of research, combining the drive to find favorable environments for axon regeneration with studies of nerve growth and neuroprotective factors. The long-term goals are therapies for glaucoma and optic nerve diseases; it undoubtedly offers great potential for tissue engineering approaches, and one may expect tissue engineering approaches to play a major role in achieving these goals.

Drug Delivery

Since 1995 the understanding of photoreceptor and RGC death has changed dramatically. It was previously thought that these cells die because a functional part—the photoreceptor outer segment and the axon, respectively—becomes dysfunctional, but both events are now widely understood to trigger cell death through apoptosis. Better understanding of these apoptotic mechanisms has given rise to the expectation that it may be possible to halt or even reverse this process.

The process of RGC degeneration following transsection or crushing of the optic nerve can be halted, at least in animal models, with the use of neuroprotective agents such as neurotrophin-4/5 (NT-4/5) (Sawai *et al.*, 1996). It turns out, however, that this protection has only a limited duration: Most RGCs die within 2–4 weeks, even with sustained administration of NT-4/5 (Clarke *et al.*, 1998). It seems, therefore, that only axon regeneration may rescue the RGC.

In the case of photoreceptor degenerations, a variety of nerve growth factors and neuroprotective agents have been used, both *in vitro* and in rodent eyes, and several of these have shown promise (LaVail *et al.*, 1998). Clinical trials of some of these agents are planned for the near future, but the optimal route of administration is yet to be decided. Systemic administration is not an option, because these agents do not pass the blood–retina barrier very well, and unwanted side effects might occur elsewhere in the body. Administration as an eye drop is not effective, because the active substance would need to diffuse through the cornea or sclera, and most of it would be washed away in the tear film. Repeated injection into the vitreous is unattractive to the patient, and it is not clear how much of the active substance would reach the outer retina, rather than be washed out of the eye.

The most promising current ideas include slow-release implants and delivery by macromolecules. Slow-release implants could be inserted into the eye through a small incision and attached to its inside wall; this is the most direct approach for protection of the inner retina, i.e., retinal ganglion cells. Macromolecules, e.g., liposomes, to which an active substance is attached, are injected into the bloodstream; the active substance is released in the choroidal space under the retina by irradiation with a low-energy laser beam (Asrani *et al.*, 1997). At the present time this technique is experimentally used for the ablation of new vessel membranes growing into the retina in patients with exudative macular degeneration.

A different form of drug delivery is used in retinal ischemic disease. Just as in cerebral ischemia, if "clotbuster" drugs such as tissue plasminogen activator are used within hours after the ischemic event, much of the damage may be prevented; experimental therapy with these drugs has had promising results (Elman, 1996).

GENETIC INTERVENTIONS

Delivery issues also play a role in the introduction of new genes into the degenerating retina. Most retinal degenerations are genetic in nature, either inherited from one or both parents, or are the consequence of a new mutation. The premise of gene therapy is that intervention at an early stage may restore normal function to the cell, and prevent the degeneration process that might otherwise ensue. Several strategies are needed to combat inherited retinal degenerations:

1. In recessively inherited diseases, both copies of a gene are defective, and introduction of a third copy may be sufficient to achieve adequate function of the molecule encoded by the defective gene. An inactivated (retro)virus is used to introduce a healthy copy of the gene. Small genes have been transfected successfully into photoreceptor and RPE cells using such viral vectors (Dunaief *et al.,* 1995; Sakamoto *et al.,* 1998); however, because the defective genes responsible for retinal degenerations are large, most vectors currently in use are not well suited for this task.

2. In X-linked disease, males carry only one copy of the X chromosome, including the defective gene; here, too, introduction of an additional healthy copy of the gene may suffice to achieve normal function.

3. In dominant disease, a single bad copy of the gene suffices to "poison" the delicate balance of the cellular machinery or, at the very least, prevent its proper function. To curb or prevent this, a therapeutic intervention has to block any step along the transcription pathway from the defective gene through (and including) its product molecule, but not along the path of the good copy of the gene; a gene for such a blocking agent could, in principle, permanently neutralize the defect. This technique has been used *in vitro* (Drenser *et al.,* 1998) and in a transgenic rat model (Lewin *et al.,* 1998), using ribozymes to destroy mRNA produced by the P23H mutant gene defect responsible for one form of autosomal dominant retinitis pigmentosa (RP). In the rat model, a gene for ribozyme production was successfully introduced into photoreceptors.

Gene therapy research is concentrated around RP and related diseases with a clear inheritance pattern. Macular degeneration and optic nerve diseases such as glaucoma and ischemia have no known genetic causes, though genetic predisposition for such diseases exists; other therapeutic approaches are more likely to be effective against these diseases.

STEM CELL RESEARCH

As in other organ systems, use of stem cells, differentiating into various target cell lines depending on trophic factors *in vitro* or *in vivo,* is emerging as an exciting approach to replacement of lost retinal cells, but not much work has been done. Early reports suggest that adult rat hippocampal stem cells survive and differentiate in retina damaged by transient ischemia, but not in healthy control retina (Kurimoto *et al.,* 1999).

APPLICATIONS OF ENGINEERED CELLS AND TISSUES: CHALLENGES AND TENTATIVE SOLUTIONS

Of the current approaches to vision enhancement considered above, only the use of stem cells can strictly be considered a form of tissue engineering. The complexity of ocular structures such as the retina and optic nerve poses daunting challenges to anyone seeking to recreate their function. Although this may explain the lag in progress compared to other organ systems, it should not keep researchers working to restore vision from drawing on the remarkable progress of tissue engineering approaches. We will briefly consider three application areas—corresponding to the aforementioned three sites in the early visual system where severe vision loss may occur.

PHOTOSENSITIVE STRUCTURES

Current efforts in the areas of retinal cell transplantation and intraocular prosthesis design, although promising, are in no way guaranteed to lead to reliable—i.e., long-term stable and of high quality—restoration of useful vision. As was explained previously, each of the methods currently explored has inherent drawbacks.

Epiretinal electrode arrays require external image acquisition and preprocessing, including real-time eye movement compensation, and may well be limited in resolution to ambulatory vision or, at best, visual acuity near legal blindness. The principal cause for this is the distance of at least 200 μm separating the electrodes from the target cells; as a rule of thumb, resolution can be no better than the separation from the target cells, and higher electrode densities are ineffective. However, it may be possible to develop electrodes that make more intimate contact with the target cells, leading to much better resolution; a possible approach would be to use electrodes that penetrate superficial layers of the retina, making close contact with the bipolar cells and potentially improving visual acuity by an order of magnitude.

One can envisage such a penetrating array as the Normann microelectrode shown in Fig. 54.5, but the damage such an array might do to the delicate microphysiology of the retina is a distinct disadvantage. As an alternative one might envisage inserting or growing, *in situ,* an array of parallel "neurites," penetrating the retina until they reach a specific target environment, e.g., the inner nuclear layer; these neurites would act as "tubes," releasing either electrical charge or a neurotransmitter that would activate inner retinal (e.g., bipolar) cells. Such tubes will be much less rigid than silicon, finding their way between retinal cells rather than penetrating them.

Subretinal electrode arrays may overcome the resolution limitation of epiretinal arrays, but this advantage remains to be demonstrated, as does the long-term biocompatibility of these arrays under the primate retina. The most serious limitation, however, is the low yield of semiconductor photoelectric conversion, causing considerable heat production, and demand for high illumination levels or a separate energy source. Consequently, introduction of a high-yield photoelectric process would greatly benefit the development of these devices.

Such high-yield conversions are known to be performed by photosystem I (PSI), a macromolecule present in the membrane of thylakoids, which can be found inside the chloroplasts that give the green color to plant leaves and algae. Experiments with thylakoids (Lee *et al.,* 1996) have shown that it is possible to anchor them onto a metal surface and use them as miniature photovoltaic elements. Lee *et al.* (1997) have also demonstrated that it is possible to chemically modify thylakoid surface membranes to create charge displacement in a specific direction.

The engineering successes of thylakoids and PSI open the opportunity to create "cells" that assume a dipole charge distribution or, with the help of some form of "intracellular electronics," produce a biphasic pulse between the cell's "poles"; in a subsequent stage of development, automatic gain control could be incorporated, as a limited form of light/dark adaptation. If these cells can be made to attach to the outer retinal layers or to migrate into the inner nuclear layer, one could achieve microscopic local current sources with sufficient conversion efficiency to ensure vision at a broad range of (day)light levels: To achieve true night vision, an external device such as a night vision scope could be used.

Note that in this idealized situation the synthetic photoreceptors might be small and sensitive enough to provide good vision without external image preprocessing (e.g., magnification and contrast/edge enhancement). It is to be expected that preliminary forms of such light-sensing units will be neither small nor sensitive enough to provide the dynamic range and resolution required; at this intermediate solution level, external image processing with an advanced LVES-type headset would be a necessary complement to this intraocular light conversion array.

OUTER RETINAL CELL TRANSPLANTATION

Transplantation of RPE and/or photoreceptor cells has not demonstrated full integration of the transplanted cells into the host retina. Also, even if the host immune response appears to be mild, its effects on graft survival—especially of transplanted RPE—cannot be ignored, and this host reaction will have to be effectively controlled, without long-term systemic immune suppression. Both areas can profit from tissue engineering approaches.

Finally, a problem that has not been addressed in any detail is the secondary degeneration of RPE cells in photoreceptor degenerative disease, and of photoreceptors in RPE degeneration. This problem may be addressed adequately only by performing a combined graft of RPE and photoreceptors, presumably prepared in a tissue culture environment and stimulated toward integration with a variety of nerve protection and neurotrophic factors.

Culturing stem cells or differentiated cell lines might allow transplant researchers to create heterogeneous structures such as RPE/photoreceptor double sheets, with multiple photoreceptors over each RPE cell. Culture conditions and growth stage of these sheets should be modulated to create optimal conditions for integration with the host retina. In order to provide sufficient structural support to these fragile sheets, a resorbable polymer layer could be used as a substrate.

CELL MATRICES SUPPORTING AXONAL REGROWTH

As was noted previously, an RGC axon damaged by glaucoma, optic nerve disease, or injury can form new neurites only under the right environmental conditions, essentially mimicking those in a developing organism. At the same time, neuroprotective factors are required to sustain the RGC long enough for the axon to take over this support function. If the diagnosis of axon loss is made early enough, it may be possible to save most of the threatened axons, and thus spare most RGCs and the patient's vision.

The experimental conditions created thus far, using peripheral nerve sheaths to create a substrate for axonal regrowth, are less than ideal, because they do not provide an integrated environment in which regrowing axons combine with intact remaining axons, and in which protection of the threatened RGCs is built into the environment. Tissue engineering approaches may provide both the necessary synthetic materials and a better understanding of the necessary conditions to create such an integrated environment for axonal regrowth.

One approach to such an integrated solution might be that engineered cells could be grown *in situ*, as a loose skeleton of supporting tissue, to follow the course of the optic nerve to the chiasm and optic tract; these cells would be programmed to exude the necessary factors promoting axon growth and RGC protection. Alternatively, it might be possible to influence the normal environment of the optic nerve, prohibiting axonal outgrowth, to become one that (temporarily) allows or even stimulates this growth. Assuming that it would be possible to guide outgrowing axons toward the appropriate cerebral hemisphere, depending on their (hemi)retinal origin, RGCs may actually be capable of restoring their connections to the same midbrain visual centers where normal visual processing occurs.

REPOPULATING ISCHEMIC OR DIABETIC RETINA

New capillaries, formed under the influence of an angiogenic tissue response, tend to be poorly organized and fragile, causing hemorrhaging and thus a great deal of damage to the already stressed retinal tissue. Therefore, the prospects of restoring vision in such retinal areas are, at the present time, poor.

This may change, however, if cell populations can be grown *in vitro* and introduced into the retina under physiologic conditions mimicking those in the embryonal retina. In that case the formation of new blood vessels would follow a much more orderly pattern, and the implanted cells would be in a much better condition to form functional connections. Whether and when it will be possible to recreate embryonal conditions and grow such integrated retinal tissue, from RPE to RGC axons, required to restore vision to the ischemic portion of the retina is difficult to predict; it is a challenge of a magnitude exceeding that of RPE/photoreceptor transplants, functional stimulation of inner retinal cells, and RGC protection/axonal regrowth combined.

TOWARD "2020" VISION

The potential applications of tissue engineering sketched pose enormous challenges, well exceeding the competency of any single group or institution. Concerted research efforts by multidisciplinary groups may allow the implementation of the complex systems required to restore and enhance vision. With improvement of the fundamental understanding of processes such as photoconversion, graft integration, immune regulation, and axon regeneration on the one hand, and the engineering ability to control tissue properties and growth, both *in vitro* and *in situ*, on the other, implementation of partial vision restoration at the RPE/photoreceptor level and at the RGC level is likely to advance to the level of experimental or even clinical therapy. Integration of all these areas to recreate the full range of retinal processing is a much more distant goal.

To accomplish any of these forms of vision restoration, however, funding mechanisms for multidisciplinary research and interest from the corporate sector will have to increase well beyond their current levels. The number of severely visually impaired individuals and the economic impact of vision restoration may not justify that these approaches receive priority over treatment of life-threatening conditions, but the investment required is relatively modest and the improvement in quality of life for (nearly) blind patients can be very significant.

REFERENCES

Asrani, S., Zou, S., D'Anna, S., Lutty, G., Vinores, S. A., Goldberg, M. F., and Zeimer, R. (1997). Feasibility of laser-targeted photoocclusion of the choriocapillary layer in rats. *Invest. Ophthalmol. Visual Sci.* **38**, 2702–2710.

Brindley, G. S., and Lewin, W. S. (1968). The sensations produced by electrical stimulation of the visual cortex. *J. Physiol. (London)* **196**, 479–493.

Cha, K., Horch, K. W., and Normann, R. A. (1992a). Reading speed with a pixelized vision system. *J. Opt. Soc. Am.* **A9**, 673–677.

Cha, K., Horch, K. W., and Normann, R. A. (1992b). Mobility performance with a pixelized vision system. *Vision Res.* **32**, 1367–1372.

Chow, A. Y., and Chow, V. Y. (1997). Subretinal electrical stimulation of the rabbit retina. *Neurosci. Lett.* **225**, 13–16.

Clark, G. M., Tong, Y. C., and Patrick, J. F. (1990). Introduction. *In* "Cochlear Prostheses" (G. M. Clark, Y. C. Tong, and J. F. Patrick, eds.), pp. 1–14. Churchill-Livingstone, Melbourne, Australia.

Clarke, D. B., Bray, G. M., and Aguayo, A. J. (1998). Prolonged administration of NT-4/5 fails to rescue most axotomized retinal ganglion cells in adult rats. *Vision Res.* **38**, 1517–1524.

Clements, M., Vichienchom, K., Liu, W., Hughes, C., McGucken, E., DeMarco, C., Mueller, J., Humayun, M., de Juan, E., Weiland, J., and Greenberg, R. (1999). An implantable neuro-stimulator device for a retinal prosthesis. *IEEE Int. Solid-State Circ. Conf. Proc.*, p. 217.

Dagnelie, G., and Massof, R. W. (1996). Towards an artificial eye. *IEEE Spectrum* **33**(5), 20–29.

Dagnelie, G., and Vogelstein, J. V. (1999). Phosphene mapping procedures for prosthetic vision. *In* "Vision Science and its Applications," OSA Tech. Dig., pp. 294–297. Op. Soc. Am., Washington, DC.

del Cerro, M., Notter, M. F., del Cerro, C., Wiegand, S. J., Grover, D. A., and Lazar, E. (1989). Intraretinal transplantation for rod-cell replacement in light-damaged retinas. *J. Neural Transplant.* **1**, 1–10.

del Cerro, M., Lazar, E. S., and DiLoreto, D., Jr. (1997a). The first decade of continuous progress in retinal transplantation. *Microsc. Res. Tech.* **36**, 130–141.

del Cerro, M., Das, T. P., Lazar, E., del Cerro, C., Sreedharan, A., Sharma, S., and Rao, G. N. (1997b). Neural retinal transplantation into twelve RP patients. *Invest. Ophthalmol. Visual Sci.* **38**, S261 (abstr.).

del Cerro, M., Humayun, M. S., Sadda, S. R., Cao, J. T., Hayashi, N., Green, W. R., del Cerro, C., and de Juan, E., Jr. (2000). Histologic correlation of human neural retinal transplantation. *Invest. Ophthalmol. Visual Sci.* in press.

DiLoreto, D., Jr., del Cerro, M., Reddy, S. V., Janardhan, S., Cox, C., Wyatt, J., and Balkema, G. W. (1996). Water escape performance of adult RCS dystrophic and congenic rats: A functional and histomorphometric study. *Brain Res.* **717**, 165–172.

Drenser, K. A., Timmers, A. M., Hauswirth, W. W., and Lewin, A. S. (1998). Ribozyme-targeted destruction of RNA associated with autosomal-dominant retinitis pigmentosa. *Invest. Ophthalmol. Visual Sci.* **39**, 681–689.

Dunaief, J. L., Kwun, R. C., Bhardwaj, N., Lopez, R., Gouras, P., and Goff, S. P. (1995). Retroviral gene transfer into retinal pigment epithelial cells followed by transplantation into rat retina. *Hum. Gene Ther.* **6**, 1225–1229.

Eckmiller, R. (1997). Learning retina implants with epiretinal contact. *Ophthalmic Res.* **29**, 281–289.

Elman, M. J. (1996). Thrombolytic therapy for central retinal vein occlusion: Results of a pilot study. *Trans. Am. Ophthalmol. Soc.* **94**, 471–504.

Fukuda, Y., Watanabe, M., Sawai, H., and Miyoshi, T. (1998). Functional recovery of vision in regenerated optic nerve fibers. *Vision Res.* **38**, 1545–1553.

Gouras, P., and Algvere, P. (1996). Retinal cell transplantation in the macula: New techniques. *Vision Res.* **36**, 4121–4126.

Gouras, P., Du, J., Gelanze, M., Lopez, R., Kwun, R., Kjeldbye, H., and Krebs, W. (1991). Survival and synapse formation of transplanted rat rods. *J. Neural Transplant. Plast.* **2**, 91–100.

Greenberg, R. J. (1998). Analysis of electrical stimulation of the vertebrate retina—Work towards a retinal prosthesis. Ph.D. Dissertation, Johns Hopkins University, Baltimore, MD.

Heckenlively, J. R., ed. (1988). "Retinitis Pigmentosa." Lippincott, Philadelphia.

Humayun, M. S., and de Juan, E., Jr. (1998). Artificial vision. *Eye* **12**, 605–607.

Humayun, M. S., de Juan, E., Jr., Dagnelie, G., Greenberg, R. J., Propst, R. H., and Phillips, D. H. (1996). Visual perception elicited by electrical stimulation of retina in blind humans. *Arch. Ophthalmol. (Chicago)* **114**, 40–46.

Humayun, M. S., de Juan, E., Jr., Weiland, J. D., Dagnelie, G., Katona, S., Greenberg, R. J., and Suzuki, S. (1999a). Pattern electrical stimulation of the human retina. *Vision Res.* **39**, 2569–2576.

Humayun, M. S., Prince, M., de Juan, E., Jr., Barron, Y., Moskowitz, M. T., Klock, I. B., and Milam, A. H. (1999b). Morphometric analysis of the extramacular retina from post-mortem eyes with retinitis pigmentosa. *Invest. Ophthalmol. Visual Sci.* **40**, 143–148.

Humayun, M. S., de Juan, E., Jr., del Cerro, M., Dagnelie, G., Radner, W., Sadda, S. R., and del Cerro, C. (2000). Human neural retinal transplantation. *Invest. Ophthalmol. Visual Sci.,* in press.

Kurimoto, Y., Shibuki, H., Kaneko, Y., Kurokawa, T., Yoshimura, N., and Takahashi, M. (1999). Transplantation of neural stem cells into the retina injured by transient ischemia. *Invest. Ophthalmol. Visual Sci.* **40**, S727 (abstr.).

LaVail, M. W., Yasumura, D., Matthes, M. T., Lau-Villacorta, C., Unoki, K., Sung, C. H., and Steinberg, R. H. (1998). Protection of mouse photoreceptors by survival factors in retinal degenerations. *Invest. Ophthalmol. Visual Sci.* **39**, 592–602.

Lee, I., Lee, J. W., and Greenbaum, E. (1997). Biomolecular electronics: Vectorial arrays of photosynthetic reaction centers. *Phys. Rev. Lett.* **79**, 3294–3297.

Lee, J. W., Lee, I., and Greenbaum, E. (1996). Platinization: A novel technique to anchor photosystem I reaction centres onto a metal surface at biological temperature and pH. *Biosens. Bioelectron.* **11**, 375–387.

Lewin, A. S., Drenser, K. A., Hauswirth, W. W., Nishikawa, S., Yasumura, D., Flannery, J. G., and LaVail, M. M. (1998). Ribozyme rescue of photoreceptor cells in a transgenic rat model of autosomal dominant retinitis pigmentosa. *Nat. Med.* **4**, 967–971.

Li, L. X., and Turner, J. E. (1988). Inherited retinal dystrophy in the RCS rat: Prevention of photoreceptor degeneration by pigment epithelial cell transplantation. *Exp. Eye Res.* **47**, 911–917.

Liang, F. Q., Bassage, S. D., del Cerro, C., DiLoreto, D., Castillo, B., Lazar, E., Aquavella, J. V., and del Cerro, M. (1994).

A new procedure for obtaining photoreceptor cells for transplantation. *Invest. Ophthalmol. Visual Sci.* **35**, S1523 (abstr.).

Lopez, R., Gouras, P., Kjeldbye, H., Sullivan, B., Reppucci, V., Brittis, M., Wapner, F., and Goluboff, E. (1989). Transplanted retinal pigment epithelium modifies the retinal degeneration in the RCS rat. *Invest. Ophthalmol. Visual Sci.* **30**, 586–588.

Massof, R. W. (1999). Electro-optical head-mounted low vision enhancement. *Pract. Optom.* **9**, 214–220.

Normann, R. A., Maynard, E. M., Guillory, K. S., and Warren, D. J. (1996). Cortical implants for the blind. *IEEE Spectrum* **33**(5), 54–59.

Rizzo, J. F., and Wyatt, J. L., Jr. (1999). Retinal prosthesis. *In* "Age-Related Macular Degeneration" (J. Berger, S. L. Fine, and M. G. Maguire, eds.), pp. 413–432. Mosby, St. Louis, MO.

Sakamoto, T., Ueno, H., Goto, Y., Oshima, Y., Yamanaka, I., Ishibashi, T., and Inomata, H. (1998). Retinal functional change caused by adenoviral vector-mediated transfection of LacZ gene. *Hum. Gene Ther.* **9**, 789–799.

Santos, A., Humayun, M. S., de Juan, E., Jr., Greenberg, R. J., Marsh, M. J., Klock, I. B., and Milam, A. H. (1997). Preservation of the inner retina in retinitis pigmentosa. A morphometric analysis. *Arch. Ophthalmol. (Chicago)* **115**, 511–515.

Sawai, H., Clarke, D. B., Kittlerova, P., Bray, G. M., and Aguayo, A. J. (1996). Brain-derived neurotrophic factor and neurotrophin-4/5 stimulate growth of axonal branches from regenerating retinal ganglion cells. *J. Neurosci.* **16**, 3887–3894.

Schmidt, E. M., Bak, M. J., Hambrecht, F. T., Kufta, C. V., O'Rourke, D. K., and Vallabhanath, P. (1996). Feasibility of a visual prosthesis for the blind based on intracortical microstimulation of the visual cortex. *Brain* **119**, 507–522.

Silverman, M. S., and Hughes, S. E. (1989). Transplantation of photoreceptors to light-damaged retina. *Invest. Ophthalmol. Visual Sci.* **30**, 1684–1690.

Thanos, S. (1997). Neurobiology of the regenerating retina and its functional reconnection with the brain by means of peripheral nerve transplants in adult rats. *Surv. Ophthalmol.* **42**, S5–S26.

Veraart, C., Raftopoulos, C., Mortimer, J. T., Delbeke, J., Pins, D., Michaux, G., Vanlierde, A., Parrini, S., and Wanet-Defalque, M. C. (1998). Visual sensations produced by optic nerve stimulation using an implanted self-sizing spiral cuff electrode. *Brain Res.* **813**, 181–186.

Wyatt, J., and Rizzo, J. (1996). Ocular implants for the blind. *IEEE Spectrum* **33**(5), 47–53.

Zrenner, E., Miliczek, K. D., Gabel, V. P., Graf, H. G., Guenther, E., Hammerle, H., Hoefflinger, B., Kohler, K., Nisch, W., Schubert, M., Stett, A., and Weiss, S. (1997). The development of subretinal microphotodiodes for replacement of degenerated photoreceptors. *Ophthalmic Res.* **29**, 269–280.

CHAPTER 55

Brain Implants

Lars U. Wahlberg

INTRODUCTION

Tissue engineering applied to the central nervous system (CNS) is a relatively young but expanding field. Experimental data strongly support the use of tissue engineering concepts and promise new therapy for patients with poorly treated or untreatable neurological disorders. Parkinson's disease (PD) has been a major target for brain implantation because this disease lacks an optimal therapy and is relatively well studied and understood. The symptoms of PD are caused by the progressive destruction of dopamine-producing neurons, resulting in the hallmark clinical signs of rigidity, tremor, and bradykinesia. Although the detailed toxic mechanisms involved in the destruction of the dopaminergic neurons are not well known, many of the neuroanatomy, neurophysiology, and implant strategies have been elucidated. The discussion of brain implants will therefore focus on this disease.

Despite the success of many drugs for Parkinson's disease, such as L-dihydroxyphenylalanine (L-DOPA) therapy introduced in the late 1960s, most patients on chronic L-DOPA therapy develop progressive symptoms and drug-induced side effects with time. Therefore, surgical treatment strategies developed during the 1950s, such as the ventrolateral pallidotomy for Parkinsonian rigidity, have been rejuvenated. With improvements in imaging and surgical techniques, ablative procedures yield excellent results in select groups of PD patients (Laitinen, 1995; Speelman and Bosch, 1998). Building on the same principles, neural stimulators that inhibit neuronal transmission have been implanted with reports of good therapeutic results (Benabid *et al.,* 1998). Despite the successful resurrection and development of neurosurgical procedures for Parkinson's disease, these procedures are based on the inhibition or destruction of normal neurons within complementary areas of the motor system. They do not address the biology of the underlying disease and, albeit successfully applied in many patients, the destruction of normal tissue is not an optimal treatment for neurodegenerative disorders. There is therefore a need for new treatment strategies that can address the pathology more directly. Fortunately, the accumulated knowledge of the pathological processes, molecular and cell biology, biomaterials, imaging, and surgical procedures makes it now possible to implement tissue engineering concepts to the treatment of PD. This chapter reviews some of the approaches applied, beginning with a discussion of naked cell implants, followed by discussions of encapsulated cells, controlled delivery of biomolecules, axonal guidance, and other disease targets.

NAKED CELL IMPLANTS

PRIMARY TISSUE

Oral L-DOPA therapy is the main treatment for PD. L-DOPA, a precursor to dopamine that passes the blood–brain barrier, is taken up by the dopaminergic neurons that, in turn, convert L-DOPA to dopamine and increase their dopamine production. However, with the progressive loss of dopaminergic neurons, the L-DOPA therapy eventually becomes ineffective and severe fluctuations in the ability to initiate movements occur. Because of the finding that L-DOPA can increase the striatal production of dopamine and alleviate the symptoms of PD, a reasonable therapeutic approach may be to implant dopamine-secreting cells in the striatum. Considering this idea, the first transplantation for Parkinson's disease was done at the Karolinska Hospital in Stockholm, Sweden in 1982 (Backlund *et al.,* 1985). In this experimental procedure, dopamine-secreting adrenal chromaffin cells were harvested from one of the adrenal glands from the patient and transplanted to the putamen (autograft). This type of procedure was adapted very quickly by the neurosurgical community and initial reports indicated good clinical results (Madrazo *et al.,* 1991). However with time, other studies showed poor survival of the cells and minimal positive clinical effects, resulting in the cessation of the procedure in many places (Kordower *et al.,* 1997).

During the late 1970s and early 1980s, a promising cell transplantation strategy for PD was developed by Björklund and co-workers at Lund University in southern Sweden (Björklund and Stenevi, 1979). They collected discarded aborted fetal tissue and dissected out the ventral mesencephalon to create cell suspensions containing developing dopaminergic neurons for transplantation experiments. After several years of extensive validation of the concept in animal models, cells were transplanted to the striatum of two patients in 1987 (Lindvall *et al.,* 1989). The first results were relatively unimpressive but prompted modifications to various parts of the experimental procedure, and a second pair of patients transplanted about 1 year later with the modified techniques fared much batter. These patients showed positive clinical recovery starting about 4 months after the procedure. Positron emission tomography (PET) data indicated that the grafts were able to take up dopamine and survive (Lindvall *et al.,* 1990a,b). Greater than 10 years out from the procedure, one of the patients shows persistent graft viability on PET scanning and is functioning with minimal L-DOPA therapy (Björklund, personal communication). To date, more than 200 patients have been transplanted with fetal ventral mesencephalic tissue at different centers around the world. Although results have varied, there is consistent evidence of survival and efficacy (Freed *et al.,* 1992; Levivier *et al.,* 1997; Olanow *et al.,* 1997; Lindvall, 1998).

On an experimental level, the transplantation of primary cells from fetal dopaminergic tissue has shown good results, indicating that cell transplantation for Parkinson's disease may be an excellent treatment strategy. However, the paucity of suitable tissue, processing problems, and ethical issues prevent the use of primary fetal tissue on a scale needed to make a medical product beneficial to a large patient population. On the other hand, experimental fetal transplantation to the CNS has given us invaluable information for the development of alternative tissue engineering strategies. The fetal transplantation data show that allotransplantation to the CNS is feasible and that transplanted cells can survive and function for many years, even after the withdrawal of immunosuppressive therapy. Although successful, because of the inability to apply the fetal transplant technology to a larger group of patients, many alternatives to human fetal tissue are being explored. One of the simplest replacements is porcine fetal ventral mesencephalon. This tissue can be readily obtained under controlled circumstances but without major ethical concerns (Galpern *et al.,* 1996; Deacon *et al.,* 1997; Edge *et al.,* 1998). However, the recent awareness of the risks of zoonosis and possible immunologic rejection issues make porcine cell transplantation less desirable from both safety and efficacy aspects.

Another interesting primary tissue is the carotid body, which contains dopaminergic glomus cells that could potentially be transplanted in PD (Hansen *et al.,* 1988). In several studies, dopaminergic glomus cells were reported to alleviate Parkinson's disease symptoms in both the 6-hydroxydopamine (6-OHDA) rodent and the 1-methyl-4-phenyl-1,2,3,6-tetrahydropyridine (MPTP) nonhuman primate models of Parkinson's disease (Espejo *et al.,* 1998; Luquin *et al.,* 1999). Behavioral recovery associated with the survival and sprouting of dopaminergic neurons was noted, indicating the possibility to use these cells to treat PD in humans as either auto- or allografts.

PRIMARY CELL LINES

Two of the most important drawbacks to using primary tissue are its limited supply and the difficulty to cryopreserve and bank primary tissues. The fetal transplantation experiments for Parkinson's disease have therefore required fresh tissue from four to eight donors, resulting in procedural difficulties and poor quality control. As mentioned, porcine fetal grafting is not an optimal replacement. Fortunately, new techniques allow for the expansion of populations of human fetal cells. This takes advantage of the human fetal source but removes some of the ethical sensitivity, because one fetal tissue could support the transplantation of thousands or perhaps millions of patients.

The ability to expand and cryopreserve cells in cell banks is paramount to create allogeneic cell replacements for primary fetal tissue. The expansion of normal, genetically unmodified cells that are potentially useful for brain implantation has been achieved using cell isolation and defined culture conditions. These cultures retain a limited number of cell cycles but the cell expansion may still provide enough cells to transplant thousands of patients from a single donor. Cultured cells can be expanded yet retain normal geno- and phenotypes and normal contact inhibition and differentiation behaviors. These cells are therefore relatively safe to use, and the formation of tumors or other abnormal behaviors is unlikely.

One example is mitogen-responsive human neural progenitor cell cultures that can be isolated from first-trimester aborted human tissue and expanded for greater than 1 year *in vitro* (Carpenter *et al.,* 1999). This was accomplished using defined media containing epidermal growth factor (EGF), fibroblast growth factor (FGF-2), and leukemia inhibitory factor (LIF). Similar to previously described EGF-expanded mouse neural progenitor cells (Reynolds *et al.,* 1992), these cells can form the three major phenotypes of the nervous system (neurons, astrocytes, and oligodendrocytes) *in vitro.* These human neural stem cell-containing cultures have been transplanted to various regions in the adult rat and have been shown to survive, integrate, migrate, differentiate, extend neurites, and arborize (Fricker *et al.,* 1999). Unlike mouse and rat stem cell cultures, the human cultures make large numbers of neurons *in vivo,* creating a potential source of cells for implantation in PD and Huntington's disease (HD). Both *in vitro* and *in vivo,* differentiated neurons containing γ-amino butyric acid (GABA) can be observed. Early data indicate that various factors can also induce dopaminergic differentiation of stem cell cultures. This may have great implications for the treatment of PD.

Another interesting source of a dopamine-producing primary cell line is the retinal pigmented epithelial cells found in the retina. Retinal pigmented epithelial (RPE) cells can be harvested from the eyes of organ donors and grown in culture to produce cell lines estimated to be capable of transplanting around 10,000 Parkinson patients per isolation. Attached to gelatin microcarriers to improve survival, these cells have been implanted in the striatum of a rhesus monkey model of MPTP-induced Parkinson's disease. The single injection of the cells resulted in significant improvement in Parkinsonian symptoms that lasted over the course of the 8-month-long study (Subramanian, 1999).

IMMORTAL CELL LINES

Immortal cells are different from cultured primary cells in that they have a genetic change that renders them immortal. Often, tumor cells show aberrant chromosome numbers, lack contact inhibition, and display pleomorphism. However, not all tumor cell lines lose chromosomes and behave erratically. An intriguing tumor cell line has been isolated from a human testicular teratocarcinoma isolated from a metastasis in a patient (Andrews *et al.,* 1982). This immortal cell line is pleuripotent and can be induced to stop dividing and to differentiate into a neuronal phenotype using retinoic acid (Andrews, 1984). This cell line is being investigated in the treatment of ischemic stroke, and animal data suggest that the postinjury transplantation of this cell line into an infarcted area can improve recovery (Saporta *et al.,* 1999). The mechanisms surrounding this effect are unclear. Nevertheless, this cell line is currently being transplanted in the United States into patients with lacunar stroke in a small safety trial (12 patients) at the University of Pittsburg (P. Sanberg, personal communication). The transplantation of a cell line derived from a human cancer has obvious associated risks. However, the approval of this trial demonstrates that cell transplantation for severe neurological disorders is seen as a reasonable strategy by the regulatory agency, as long as safety and efficacy can be demonstrated in animal models.

Other groups are investigating the transplantation of immortal cell lines but are using genetic engineering to immortalize cells. Advances in genetic engineering have made it possible to extend the number of doublings a primary cell line can go through by inserting various oncogenes and cell cycle regulators. This allows for the selection, clonal expansion, and banking of a large number of cells. Besides the genetic modification, these cells retain otherwise normal genotypic characteristics.

One group in England has developed an undifferentiated neuroepithelial stem cell line immortalized with the simian virus (SV40) large T oncogene under temperature-sensitive regulation (Sinden *et al.,* 1997). Using these cells, they have demonstrated positive effects on the sensory neglect and motor asymmetries in a rodent ischemic stroke model. Based on reports in the literature, various human neural cell lines have been created using the same immortalization strategy. Similar to the Pittsburgh study, the initial clinical target is stroke and clinical trials are anticipated to start in late 2000.

Other investigators have immortalized mouse and human neural stem cells using retroviral transduction of the v-*myc* oncogene (Snyder, 1995; Flax *et al.,* 1998). These cells have been transplanted to rodent models and are able to survive, migrate, integrate, and differentiate into neurons, astrocytes, and oligodendrocytes (Snyder *et al.,* 1997; Taylor and Snyder, 1997; Billinghurst *et al.,* 1998).

Engineered immortal cell lines could potentially be made and expanded to treat millions of patients. However, genetically engineered cells may also show a greater propensity to form tumors compared with unmodified cells. To ensure safety, cells may need to be engineered with genes that render the cells mitotically inactive or that commit "suicide" if exposed to particular drugs. Extensive safety studies will be required before these cells can be tested in humans.

GENETICALLY ENGINEERED IMMORTAL CELL LINES

Genetic engineering has also been extended to manipulate cells to secrete specific neurotransmitters or proteins for local delivery in the brain. Again, Parkinson's disease has been one of the disease targets and two basic strategies have been investigated. The first involves replacing the function of lost dopaminergic neurons by implanting engineered L-DOPA or dopamine-producing cells in the striatum. A second strategy involves the protection of transplanted or endogenous dopamine neurons by the delivery of trophic factors, such as glial cell line-derived neurotrophic factor (GDNF), in proximity to the neurons.

It is well known that the rate-limiting enzyme for dopamine synthesis is tyrosine hydroxylase (TH), which converts tyrosine to L-DOPA. An additional enzyme called aromatic amino acid decarboxylase (AADC) is needed to convert L-DOPA to dopamine and is by necessity present in dopaminergic neurons. However, other cells in the striatum may also contain enough AADC to convert L-DOPA to dopamine. Therefore, a relatively simple strategy may be to create and transplant cell lines that secrete L-DOPA, not dopamine. In animal models, the transplantation of various cells secreting L-DOPA and dopamine has been described. Results have shown behavioral improvements with both types of cells (Chen *et al.,* 1991; Fisher *et al.,* 1991; Lundberg *et al.,* 1996; Kaddis *et al.,* 1997). No human trial using this type of approach has been described.

Because Parkinson's disease involves a slow and progressive degeneration of dopaminergic neurons, the ability to arrest or slow the degenerative process could form one therapeutic strategy. Many protein factors have been shown to protect fetal dopaminergic neurons both *in vitro* and *in vivo* and one of the most powerful factors is GDNF (Lin *et al.,* 1993). This factor promotes the survival (neurotrophic effect) and neurite extension (neurotropic effect) of dopaminergic neurons both *in vitro* and *in vivo* (Hoffer *et al.,* 1994; Clarkson *et al.,* 1995; Tomac *et al.,* 1995). Based on positive animal data (Gash *et al.,* 1998), GDNF has been tried in humans; intraventricular bolus injections of GDNF were performed monthly, but the trial was stopped, reportedly because of lack of efficacy (Amgen press-release, April, 1999). This is an unfortunate outcome for patients suffering from PD. However, this is consistent with the idea that potent protein factors such as GDNF may need to be delivered locally and at constant low doses to show efficacy and to avoid toxicity (Rosenblad *et al.,* 1999). The implantation of genetically engineered cells that secrete GDNF may therefore remain applicable to PD. This approach will most likely be tried in the future.

Another interesting application of protein factor-secreting cells is to support the survival and integration of transplanted dopaminergic cells. For example, it has been shown that chromaffin cells show improved survival and convert to a neuronal phenotype when exposed to nerve growth factor (NGF) (Strömberg *et al.,* 1985). Studies using cotransplants of chromaffin cells with engineered cells that secrete NGF (Niijima *et al.* 1995) show enhanced survival. An improvement in the clinical outcome in a PD patient implanted with chromaffin cells perfused with NGF has also been reported (Olson *et al,.* 1991).

Fetal ventral mesencephalic grafts have shown superior survival compared to chromaffin cell implants. However, even in the most optimal grafts, the total fraction of surviving dopaminergic neurons is only about 10–20% (Brundin and Björklund, 1987). This has required a large number of donors (four to eight) to assure enough surviving cells for a clinical effect. Experiments using trophic support of fetal cell transplants have shown excellent results in animal models, with increases in viability, integration, arborization, and functional outcome (Takayama *et al.*, 1995; Clarkson *et al.*, 1998). One relatively simple strategy to improve the outcome and reduce the requirement for primary tissue would thus be to cograft trophic factor-secreting cells with ventral mesencephalic tissue. This has not yet been attempted in the clinic. In this scenario, the application of polymeric or encapsulated delivery devices could also be possible, because the trophic support does not necessarily need to be chronic.

ENCAPSULATED CELL IMPLANTS

The implantation of naked cells has the advantage of allowing for migration, integration, and the formation of neurites and synapses. These may be advantageous features for transplantation in PD, because dopamine needs to be distributed over a relatively large volume in the striatum. However, naked cells cannot be retrieved. Therefore, if a potent protein factor or potentially tumorigenic cells need to be removed, a retrievable encapsulated cell device may be a good option. Encapsulated cell implants are essentially drug delivery devices that provide chronic delivery of biological molecules. One of the most studied devices consists of a tubular polymeric membrane, approximately 1 mm in outer diameter, that surrounds a core of cells suspended in a polymer matrix (Fig. 55.1) (Aebischer *et al.*, 1991b). The polymeric membrane excludes larger molecules and cells but allows for the bidirectional passage of nutrients and secreted products. The encapsulated cells can thus be protected from immune rejection, making xenogeneic transplantation possible and immunosuppressive therapy unnecessary. The host is also protected and the risk of tumor formation or dissemination of cells can be significantly reduced. The macrocapsule can be attached to a polymeric tether making a device that can be easily handled and removed in the case of untoward effects (Fig. 57.1). These devices can be implanted intraparenchymally, intracerebroventricularly (ICV), or intrathecally, depending on the application. Cellular survival and continuous production of neurotransmitters and factors have been demonstrated for at least 6 months in the rodent and guinea pig (Aebischer *et al.*, 1991a) and several months in humans (Aebischer *et al.*, 1996). Encapsulated devices secreting GDNF have been studied in the rodent model (Tseng *et al.*, 1997; Sautter *et al.*, 1998). In these studies, both histologic and behavioral data show neuroprotective effects of encapsulated cell delivery of GDNF on transplanted and intrinsic dopaminergic neurons.

CONTROLLED-RELEASE IMPLANTS

Acellular polymeric brain implants that are able to deliver protein factors to the CNS have also been developed (Tabata *et al.*, 1993; Langer, 1995; Kuo and Saltzman, 1996; Mahoney and Saltzman, 1996; am Ende and Mikos, 1997; Roskos and Maskiewicz, 1997; Yewey *et al.*, 1997). These systems release drugs by degradation- or diffusion-based mechanisms over an extended time (weeks), but cannot sustain release over a long time (months), which is possible with cellular-based systems. The advantage with polymeric release is that cells are not used, the dose can be controlled, and the duration can be set. On the negative side, the formulation and stability of the factor or drug may be problematic. In addition, biodegradable implants often cause a mild inflammatory reaction, which may immunologically sensitize a cellular coimplant.

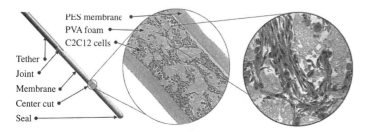

Fig. 55.1. An encapsulated device is illustrated. A growth factor-engineered mouse immortal myoblast line (C2C12) is housed in an extracellular matrix-coated poly(vinylalcohol) (PVA) hydrogel foam contained within an immunoisolatory poly(ethersulfone) (PES) membrane. The foam scaffold promotes diffusion, adherence, differentiation, and survival. The tether allows for handling and removal of the device. Magnifications from left to right are ×1, ×25, and ×100. The sections are stained with hematoxylin and eosin.

Appropriately designed, polymeric controlled-release devices have several possible applications and could, for example, support the survival and integration of transplanted cells. Furthermore, a polymeric system can support the sequential release of growth factors that may be necessary to support fully the stepwise differentiation of immature cells. This concept can be applied to neural stem cells that may lack important embryonic signals in the adult brain (Wahlberg, 1997). Synthetic drugs that cannot be made in cellular-based systems can also be delivered by these implants. The local delivery of steroids (Christenson *et al.,* 1991) and cyclosporin A may be two applications that could be used in combination with cell implants to avoid early immunologic sensitization and rejection.

AXONAL GUIDANCE IMPLANTS

The transplantation of fetal dopaminergic cells to the striatum is called heterotopic transplantation. This means that the dopaminergic cells are transplanted into an area different from their normal location, which is the substantia nigra (SN). The heterotopic implantation of dopaminergic neurons may result in the loss of important normal innervation and feedback loops. Many transplants for PD may thus only function as simple cellular "pumps" that increase the striatal dopamine levels. Although simplicity is desired, an ultimate strategy to treat PD could be to transplant the dopaminergic neurons to their anatomically correct position (homotopic), regenerate the nigrostriatal axonal pathway, and induce sprouting and innervation of the striatal target neurons. This would regenerate the appropriate connections and represent a more physiologic strategy (Fig. 55.2D).

Unfortunately, axonal outgrowth has been limited in the CNS and the ability to regenerate axonal pathways has been difficult. Fortunately, science has made progress in this area; what was thought impossible only a few years ago now appears more feasible. Several molecules are now known to both promote and guide axonal outgrowth. As mentioned, GDNF is a strong promoter of axonal outgrowth of dopaminergic neurons. In addition, extracellular matrix proteins, such as laminin, can guide axonal outgrowth. It has been demonstrated that the creation of laminin bridges promotes the growth of dopaminergic processes from the substantia nigra toward the striatum *in vivo* (Zhou and Azmitia, 1988). Unlike the CNS, the peripheral nervous system is readily able to regenerate axons. Using this fact, the brain implantation of peripheral nerve bridges (Aguayo *et al.,* 1984), and Schwann cell bridges secreting FGF-2 (Brecknell *et al.,* 1996), result in enhanced axonal regeneration. From a tissue engineering point of view, the creation of a degradable regeneration channel releasing survival- and neurite-promoting factors and coated with molecules that facilitate axonal outgrowth may be a future reality (Fig. 57.2D). These types of implants could have great potential use for regeneration in many areas of the CNS, including the spinal cord.

DISEASE TARGETS

As mentioned, Parkinson's disease has been the main target for tissue engineering approaches to treat neurologic disease. However, there are other neurologic indications that should be mentioned.

In Huntington's disease (HD), several neuronal populations slowly degenerate and cause the clinical signs of choreiform movements and progressive dementia. HD is inherited as an autosomal dominant disease and the responsible mutation has been located to chromosome 4 [for a review, see Haque *et al.* (1997)]. Carriers of the disease can therefore be screened for and identified before the onset of symptoms. This makes a protective strategy for HD an attractive possibility, whereby the delivery of neurotrophic factors could prolong the symptom-free interval. One such factor is ciliary neurotrophic factor (CNTF), which protects striatal neurons in both rodent and nonhuman primate models of HD (Emerich *et al.,* 1997a,b). Considering these results, a small clinical trial using an intracerebroventricularly placed encapsulated cell device secreting CNTF is currently ongoing (fig. 57.2A) (M. Peschanski, personal communication).

Based on positive results in animal models [for a review, see Peschanski *et al.,* (1995)], primary fetal striatal tissue transplantation for HD is also being attempted at a handful of centers in the world (Fig. 57.2A). An initial report has described mild improvements in one of the implanted patients (Kopyov *et al.,* 1998). Because the number of implanted patients is few, it is currently difficult to draw any major conclusions regarding the clinical utility of transplantation for HD.

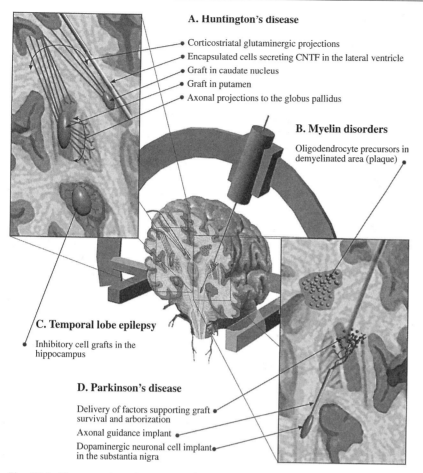

A. Huntington's disease
- Corticostriatal glutaminergic projections
- Encapsulated cells secreting CNTF in the lateral ventricle
- Graft in caudate nucleus
- Graft in putamen
- Axonal projections to the globus pallidus

B. Myelin disorders

Oligodendrocyte precursors in demyelinated area (plaque)

C. Temporal lobe epilepsy

Inhibitory cell grafts in the hippocampus

D. Parkinson's disease

Delivery of factors supporting graft survival and arborization

Axonal guidance implant

Dopaminergic neuronal cell implant in the substantia nigra

Fig. 55.2. Tissue engineering approaches. In the center, an oblique frontal view of a cut section of the human brain shows a stereotactic injection into the putamen, which is the standard method used for fetal transplantation in PD. (A) in HD, both endogenous (corticostriatal) and transplant axonal projections are important and may need extensive regeneration in late stages of the disease. Encapsulated cells delivering CNTF in proximity to the caudate nucleus may protect both endogenous neurons and transplanted cells. Implants may need to be placed in both the putamen and caudate nucleus to achieve the most optimal therapeutic effect. Not all axonal projections need to be extensive, because of the proximity of some of the efferent and afferent nuclei (globus pallidus). (B) Implantation of oligodendrocyte precursors or Schwann cells may remyelinate plaques in myelin disorders such as multiple sclerosis. (C) Temporal lobe epilepsy may be treated with the implantation of cells secreting inhibitory substances such as GABA in the area of the seizure focus. (D) A futuristic treatment strategy for PD is illustrated whereby the dopaminergic precursor cells are homotopically transplanted to the substantia nigra followed by deposit of an axonal guidance implant along the stereotactic tract. Additional local deposits of GDNF-releasing polymers or cells in the striatum may promote survival and arborization.

One possible advantage over PD is that the transplantation for HD involves homotopic implantation, which at least theoretically should allow for the differentiation of the fetal tissue using normal environmental cues. However, in HD, multiple sets of neuronal populations degenerate, including both cortical and striatal neurons. The homotopic transplantation for HD may thus require extensive regeneration of axonal pathways (Fig. 55.2A).

Other diseases that could be amenable to the implantation of cells within the brain are the myelin disorders (Fig. 55.2B). Animal experiments have shown the ability of neural tissue, purified oligodendrocytes, oligodendrocyte precursors, immortalized glial cells, and neural stem cells to remyelinate areas of demyelination [for a review, see Duncan and Milward (1995)]. Although oligodendrocyte precursor cultures can be established from rodent cells, equivalent human cells have not been easy to make.

One substitute may be to transplant growth factor-expanded Schwann cells. This is based on the fact that patients with CNS myelin disorders do not display demyelination of the peripheral nervous system (PNS). In fact, myelination of central axons may spontaneously occur by ingrowth of Schwann cells (the myelinating cell of the PNS) from the periphery (Duncan and Hoffman, 1997). An intriguing therapeutic possibility is therefore to grow Schwann cells from a nerve biopsy and expand these cells in culture. In turn, these cells could be transplanted into demyelinated areas in the same patient. Physiologically, this may not be the best strategy, because Schwann cells myelinate only single axons, whereas oligodendrocytes myelinate multiple axons. However, animal data indicate that, for a limited volume, Schwann cells can remyelinate (Blakemore and Crang, 1985) and restore functions (Felts and Smith, 1992) to central demyelinated areas. Preparation for clinical trials using this strategy are ongoing.

One of the most common neurologic disorders is epilepsy, which affects about 2% of the population. A subgroup of these patients has temporal lobe epilepsy that is generated by a loss of neurons and an imbalance of inhibitory and excitatory neurotransmitters in the hippocampal formation. In intractable cases, this disease is often treated surgically with the removal of the medial hippocampus and the abnormal area. This procedure eliminates or reduces the frequency of seizures in selected patients but involves a major surgical procedure and the ablation of normal tissue. Therefore, a less invasive procedure may be to implant in the seizure focus inhibitory cells that would raise the seizure threshold. This idea is supported by animal experimentation data that indicate that locus coeruleus grafts and the local delivery of inhibitory substances such as GABA can increase the seizure threshold (Bengzon *et al.*, 1990; Kokaia *et al.*, 1994; Lindvall *et al.*, 1994). However, with the lack of suitable cells, such as GABA-producing cells, transplantation for focal epilepsy has not been as extensively studied as for some of the aforementioned disease targets.

Other disease indications that may benefit from brain implant strategies include stroke, brain injury from trauma, Alzheimer's disease, and rare disorders such as cerebellar degeneration and inherited metabolic disorders. These indications will be discussed elsewhere.

SURGICAL CONSIDERATIONS

The surgical implantation of most brain implants involves the use of stereotactic techniques (Fig. 57.2). Stereotaxis in brain surgery was established in the beginning of this century and is now well established in neurosurgical practice (Speelman and Bosch, 1998). It involves attaching a rigid frame (stereotactic frame) to the skull followed by imaging such as magnetic resonance imaging. Attached markers (fiducials) create a three-dimensional coordinate system in which any point in the brain can be defined and related to the frame with high precision. In the operating room, the markers used during imaging are replaced with holders that guide the instruments. It is a relatively simple neurosurgical procedure done under local anesthesia and mild sedation. The procedure is therefore safe and relatively painless. Among the >200 patients transplanted with fetal grafts for Parkinson's disease, rare surgical complications have been reported. The patients are usually discharged from the hospital after an overnight observation.

CONCLUSIONS

In this chapter, various brain implants have been described that may have potential to treat Parkinson's disease and other neurological disorders using tissue engineering strategies. Most of the current literature describes the transplantation of various primary cells, such as fetal tissue and chromaffin cells. Tissue engineering principles have only more recently been introduced. Applications using growth factor support, genetic engineering, scaffolds, extracellular matrices, and encapsulation have all been able to improve the survival and function of the brain implant. The ultimate implants are yet to be developed and may combine genetically modified cells, controlled delivery devices, axonal bridges, scaffolds, and encapsulated cells.

References

Aebischer, P., Schluep, M. *et al.* (1996). Intrathecal delivery of CNTF using encapsulated genetically modified xenogeneic cells in amyotrophic lateral sclerosis patients. *Nat. Med.* 2(6), 696–699. published erratum: 2(9), 1041.

Aebischer, P., Tresco, P. A. *et al.* (1991a). Long-term cross-species brain transplantation of a polymer-encapsulated dopamine-secreting cell line. *Exp. Neurol.* 111(3), 269–275.

Aebischer, P., Wahlberg, L. *et al.* (1991b). Macroencapsulation of dopamine-secreting cells by coextrusion with an organic polymer solution. *Biomaterials* 12(1), 50–56.

Aguayo, A. J., Björklund, A. *et al.* (1984). Fetal mesencephalic neurons survive and extend long axons across peripheral nervous system grafts inserted into the adult rat striatum. *Neurosci. Lett.* **45**(1), 53–58.

am Ende, M. T., and Mikos, A. G. (1997). Diffusion-controlled delivery of proteins from hydrogels and other hydrophilic systems. *Pharm. Biotechnol.* **10**, 139–165.

Andrews, P. W. (1984). Retinoic acid induces neuronal differentiation of a cloned human embryonal carcinoma cell line *in vitro. Dev. Biol.* **103**(2), 285–293.

Andrews, P. W., Goodfellow, P. N. *et al.* (1982). Cell-surface antigens of a clonal human embryonal carcinoma cell line: Morphological and antigenic differentiation in culture. *Int. J. Cancer* **29**(5), 523–531.

Backlund, E. O., Granberg, P. O. *et al.* (1985). Transplantation of adrenal medullary tissue to striatum in parkinsonism. First clinical trials. *J. Neurosurg.* **62**(2), 169–173.

Benabid, A. L., Benazzouz, A. *et al.* (1998). Long-term electrical inhibition of deep brain targets in movement disorders. *Movement Disord.* **13**(Suppl. 3), 119–125.

Bengzon, J., Kokaia, M. *et al.* (1990). Seizure suppression in kindling epilepsy by intrahippocampal locus coeruleus grafts: Evidence for an alpha-2-adrenoreceptor mediated mechanism. *Exp. Brain Res.* **81**(2), 433–437.

Billinghurst, L. L., Taylor, R. M. *et al.* (1998). Remyelination: Cellular and gene therapy. *Semin. Pediat. Neurol.* **5**(3), 211–228.

Björklund, A., and Stenevi, U. (1979). Reconstruction of the nigrostriatal dopamine pathway by intracerebral nigral transplants. *Brain Res.* **177**(3), 555–560.

Blakemore, W. F., and Crang, A. J. (1985). The use of cultured autologous Schwann cells to remyelinate areas of persistent demyelination in the central nervous system. *J. Neurol. Sci.* **70**(2), 207–223.

Brecknell, J. E., Haque, N. S. *et al.* (1996). Functional and anatomical reconstruction of the 6-hydroxydopamine lesioned nigrostriatal system of the adult rat. *Neuroscience* **71**(4), 913–925.

Brundin, P., and Björklund, A. (1987). Survival, growth and function of dopaminergic neurons grafted to the brain. *Prog. Brain Res.* **71**, 293–308.

Carpenter, M. K., Cui, X. *et al.* (1999). *In vitro* expansion of a multipotent population of human neural progenitor cells. *Exp. Neurol.* **158**(2), 265–278.

Chen, L. S., Ray, J. *et al.* (1991). Cellular replacement therapy for neurologic disorders: Potential of genetically engineered cells. *J. Cell. Biochem.* **45**(3), 252–257.

Christenson, L., Wahlberg, L. *et al.* (1991). Mast cells and tissue reaction to intraperitoneally implanted polymer capsules. *J. Biomed. Mater. Res.* **25**(9), 1119–1131.

Clarkson, E. D., Zawada, W. M. *et al.* (1995). GDNF reduces apoptosis in dopaminergic neurons *in vitro. NeuroReport* **7**(1), 145–149.

Clarkson, E. D., Zawada, W. M. *et al.* (1998). Strands of embryonic mesencephalic tissue show greater dopamine neuron survival and better behavioral improvement than cell suspensions after transplantation in parkinsonian rats. *Brain Res.* **806**(1), 60–68.

Deacon, T., Schumacher, J. *et al.* (1997). Histological evidence of fetal pig neural cell survival after transplantation into a patient with Parkinson's disease. *Nat. Med.* **3**(3), 350–353.

Duncan, I. D., and Hoffman, R. L. (1997). Schwann cell invasion of the central nervous system of the myelin mutants. *J. Anat.* **190**(Pt. 1), 35–49; published erratum: **191**(Pt. 2), 318–319.

Duncan, I. D., and Milward, E. A. (1995). Glial cell transplants: Experimental therapies of myelin diseases. *Brain Pathol.* **5**(3), 301–310.

Edge, A. S., Gosse, M. E. *et al.* (1998). Xenogeneic cell therapy: Current progress and future developments in porcine cell transplantation. *Cell Transplant.* **7**(6), 525–539.

Emerich, D. F., Cain, C. K. *et al.* (1997a). Cellular delivery of human CNTF prevents motor and cognitive dysfunction in a rodent model of Huntington's disease. *Cell Transplant.* **6**(3), 249–266.

Emerich, D. F., Winn, S. R. *et al.* (1997b). Protective effect of encapsulated cells producing neurotrophic factor CNTF in a monkey model of Huntington's disease. *Nature (London)* **386**, 395–399.

Espejo, E. F., Montoro, R. J. *et al.* (1998). Cellular and functional recovery of Parkinsonian rats after intrastriatal transplantation of carotid body cell aggregates. *Neuron* **20**(2), 197–206.

Felts, P. A., and Smith, K. J. (1992). Conduction properties of central nerve fibers remyelinated by Schwann cells. *Brain Res.* **574**(1–2), 178–192.

Fisher, L. J., Jinnah, H. A. *et al.* (1991). Survival and function of intrastriatally grafted primary fibroblasts genetically modified to produce L-dopa. *Neuron* **6**(3), 371–380.

Flax, J. D., Aurora, S. *et al.* (1998). Engraftable human neural stem cells respond to developmental cues, replace neurons, and express foreign genes. *Nat. Biotechnol.* **16**(11), 1033–1039.

Freed, C. R., Breeze, R. E. *et al.* (1992). Survival of implanted fetal dopamine cells and neurologic improvement 12 to 46 months after transplantation for Parkinson's disease. *N. Engl. J. Med.* **327**(22), 1549–1555.

Fricker, R. A., Carpenter, M. K. *et al.* (1999). Site-specific migration and neuronal differentiation of human neural progenitor cells after transplantation in the adult rat brain. *J. Neurosci.* **19**, 5990–6005.

Galpern, W. R., Burns, L. H. *et al.* (1996). Xenotransplantation of porcine fetal ventral mesencephalon in a rat model of Parkinson's disease: Functional recovery and graft morphology. *Exp. Neurol.* **140**(1), 1–13.

Gash, D. M., Gerhardt, G. A. *et al.* (1998). Effects of glial cell line-derived neurotrophic factor on the nigrostriatal dopamine system in rodents and nonhuman primates. *Adv. Pharmacol.* **42**, 911–915.

Hansen, J. T., Bing, G. Y. *et al.* (1988). Paraneuronal grafts in unilateral 6-hydroxydopamine-lesioned rats: Morphological aspects of adrenal chromaffin and carotid body glomus cell implants. *Prog. Brain Res.* **78**, 507–511.

Haque, N. S., Borghesani, P. *et al.* (1997). Therapeutic strategies for Huntington's disease based on a molecular understanding of the disorder. *Mol. Med. Today* **3**(4), 175–183.

Hoffer, B. J., Hoffman, A. *et al.* (1994). Glial cell line-derived neurotrphic factor reverses toxin-induced injury to midbrain dopaminergic neurons *in vivo. Neurosci. Lett.* **182**(1), 107–111.

Kaddis, F. G., Clarkson, E. D. *et al.* (1997). Intrastriatal grafting of Cos cells stably expressing human aromatic L-amino acid decarboxylase: Neurochemical effects. *J. Neurochem.* **68**(4), 1520–1526.

Krokaia, M., Aebischer, P. *et al.* (1994). Seizure suppression in kindling epilepsy by intracerebral implants of GABA—but not by noradrenaline-releasing polymer matrices. *Exp. Brain Res.* **100**(3), 385–394.

Kopyov, O. V., Jacques, S. *et al.* (1998). Fetal transplantation for Huntington's disease: Clinical studies. *In* "Cell Transplantation for Neurological Disorders" (T. Freeman and H. Widner, eds.), pp. 95–134. Humana Press, Totowa, NJ.

Kordower, J. H., Goetz, C. G. *et al.* (1997). Dopaminergic transplants in patients with Parkinson's disease: Neuroanatomical correlates of clinical recovery. *Exp. Neurol.* **144**(1), 41–46.

Kuo, P. Y., and Saltzman, W. M. (1996). Novel systems for controlled delivery of macromolecules. *Crit. Rev. Eukaryotic Gene Expression* **6**(1), 59–73.

Laitinen, L. V. (1995). Pallidotomy for Parkinson's disease. *Neurosurg. Clin. North Am.* **6**(1), 105–112.

Langer, R. (1995). 1994 Whitaker Lecture: Polymers for drug delivery and tissue engineering. *Ann. Biomed. Eng.* **23**(2), 101–111.

Levivier, M., Dethy, S. *et al.* (1997). Intracerebral transplantation of fetal ventral mesencephalon for patients with advanced Parkinson's disease. Methodology and 6-month to 1-year follow-up in 3 patients. *Stereotactic Funct. Neurosurg.* **69**(1–4, Pt. 2), 99–111.

Lin, L. F., Doherty, D. H. *et al.* (1993). GDNF: A glial cell line-derived neurotrophic factor for midbrain dopaminergic neurons. *Science* **260**, 1130–1132.

Lindvall, O. (1998). Update on fetal transplantation: The Swedish experience. *Movement Disord.* **13**(Suppl. 1), 83–87.

Lindvall, O., Rehncrona, S. *et al.* (1989). Human fetal dopamine neurons grafted into the striatum in two patients with severe Parkinson's disease. A detailed account of methodology and a 6-month follow-up. *Arch. Neurol. (Chicago)* **46**(6), 615–631.

Lindvall, O., Brundin, P. *et al.* (1990). Grafts of fetal dopamine neurons survive and improve motor function in Parkinson's disease. *Science* **247**, 574–577.

Lindvall, O., Rehncrona, S. *et al.* (1990b). Neural transplantation in Parkinson's disease: The Swedish experience. *Prog. Brain Res.* **82**, 729–734.

Lindvall, O., Bengzon, J. *et al.* (1994). Grafts in models of epilepsy. *In* "Functional Neural Transplantation" (S. B. Dunnett, ed.), pp. 387–413. Raven Press, New York.

Lundberg, C., Horellou, P. *et al.* (1996). Generation of DOPA-producing astrocytes by retroviral transduction of the human tyrosine hydroxylase gene: *In vitro* characterization and *in vivo* effects in the rat Parkinson model. *Exp. Neurol.* **139**(1), 39–53.

Luquin, M. R., Montoro, R. J. *et al.* (1999). Recovery of chronic parkinsonian monkeys by autotransplants of carotid body cell aggregates into putamen. *Neuron* **22**(4), 743–750.

Madrazo, I., Franco-Bourland, R. *et al.* (1991). Autologous adrenal medullary, fetal mesencephalic, and fetal adrenal brain transplantation in Parkinson's disease: A long-term postoperative follow-up. *J. Neural Transplant. Plast.* **2**(3–4), 157–164.

Maloney, M. J., and Saltzman, W. M. (1996). Controlled release of proteins to tissue transplants for the treatment of neurodegenerative disorders. *J. Pharm. Sci.* **85**(12), 1276–1281.

Niijima, K., Chalmers, G. R. *et al.* (1995). Enhanced survival and neuronal differentiation of adrenal chromaffin cells cografted into the striatum with NGF-producing fibroblasts. *J. Neurosci.* **15**(2), 1180–1194.

Olanow, C. W., Freeman, T. B. *et al.* (1997). Neural transplantation as a therapy for Parkinson's disease. *Adv. Neurol.* **74**, 249–269.

Olson, L., Backlund, E. O. *et al.* (1991). Intraputaminal infusion of nerve growth factor to support adrenal medullary autografts in Parkinson's disease. One-year follow-up of first clinical trial. *Arch. Neurol. (Chicago)* **48**(4), 373–381.

Peschanski, M., Cesaro, P. *et al.* (1995). Rationale for intrastriatal grafting of striatal neuroblasts in patients with Huntington's disease. *Neuroscience* **68**(2), 273–285.

Reynolds, B. A., Tetzlaff, W. *et al.* (1992). A multipotent EGF-responsive striatal embryonic progenitor cell produces neurons and astrocytes. *J. Neurosci.* **12**(11), 4565–4574.

Rosenblad, C., Kirik, D. *et al.* (1999). Protection and regeneration of nigral dopaminergic neurons by neurturin or GDNF in a partial lesion model of Parkinson's disease after administration into the striatum or the lateral ventricle. *Eur. J. Neurosci.* **11**(5), 1554–1566.

Roskos, K. V., and Maskiewicz, R. (1997). Degradable controlled release systems useful for protein delivery. *Pharm. Biotechnol.* **10**, 45–92.

Saporta, S., Borlongan, C. V. *et al.* (1999). Neural transplantation of human neuroteratocarcinoma (hNT) neurons into ischemic rats. A quantitative dose–response analysis of cell survival and behavioral recovery. *Neuroscience* **91**(2), 519–525.

Sautter, J., Tseng, J. L. *et al.* (1998). Implants of polymer-encapsulated genetically modified cells releasing glial cell line-derived neurotrophic factor improve survival, growth, and function of fetal dopaminergic grafts. *Exp. Neurol.* **149**(1), 230–236.

Sinden, J. D., Rashid-Doubell, F. *et al.* (1997). Recovery of spatial learning by grafts of a conditionally immortalized hippocampal neuroepithelial cell line into the ischaemia-lesioned hippocampus. *Neuroscience* **81**(3), 599–608.

Snyder, E. Y. (1995). Immortalized neural stem cells: Insights into development; prospects for gene therapy and repair. *Proc. Assoc. Am. Physicians* **107**(2), 195–204.

Snyder, E. Y., Yoon, C. *et al.* (1997&). Multipotent neural precursors can differentiate toward replacement of neurons undergoing targeted apoptotic degeneration in adult mouse neocortex. *Proc. Nat. Acad. Sci. U. S. Am.* **94**(21), 11663–11668.

Speelman, J. D., and Bosch, D. A. (1998). Resurgence of functional neurosurgery for Parkinson's disease: A historical perspective. *Movement Disord.* **13**(3), 582–588.

Strömberg, I., Herrera-Marschitz, M. *et al.* (1985). Chronic implants of chromaffin tissue into the dopamine-denervated striatum. Effects of NGF on graft survival, fiber growth and rotational behavior. *Exp. Brain Res.* **60**(2), 335–349.

Subramanian, T. (1999). *Annu. Meet., Am. Soc. Neural Transplant.* Tampa, FL.

Tabata, Y., Gutta, S. *et al.* (1993). Controlled delivery systems for proteins using polyanhydride microspheres. *Pharm. Res.* **10**(4), 487–496.

Takayama, H., Ray, J. *et al.* (1995). Basic fibroblast growth factor increases dopaminergic graft survival and function in a rat model of Parkinson's disease. *Nat. Med.* **1**(1), 53–58.

Taylor, R. M., and Snyder, E. Y. (1997). Widespread engraftment of neural progenitor and stem-like cells throughout the mouse brain. *Transplant. Proc.* **29**(1–2), 845–847.

Tomac, A., Lindqvist, E. *et al.* (1995). Protection and repair of the nigrostriatal dopaminergic system by GDNF *in vivo.* *Nature (London)* **373**, 335–339.

Tseng, J. L., Baetge, E. E. *et al.* (1997). GDNF reduces drug-induced rotational behavior after medial forebrain bundle transection by a mechanism not involving striatal dopamine. *J. Neurosci.* **17**(1), 325–333.

Wahlberg, L. (1997). "Implantable Bioartificial Hybrids for Targeted Therapy in the Central Nervous System," p. 160. Karolinska Institute; Department of Clinical Neuroscience, Stockholm.

Yewey, G. L., Duysen, E. G. *et al.* (1997). Delivery of proteins from a controlled release injectable implant. *Pharm. Biotechnol.* **10**, 93–117.

Zhou, F. C., and Azmitia, E. C. (1988). Laminin facilitates and guides fiber growth of transplanted neurons in adult brain. *J. Chem. Neuroanat.* **1**(3), 113–146.

CHAPTER 56

NERVE REGENERATION

Eric G. Fine, Robert F. Valentini, and Patrick Aebischer

NERVE REGENERATION AND NEURAL TISSUE ENGINEERING

In human adults, many tissues, including those of the peripheral nervous system (PNS), are capable of healing and regeneration. Neural tissue, due to its complex and numerous long-distance interconnections, must heal by true regeneration, because healing by scar will not reestablish electrical connectivity. In certain situations, underlying disease (such as diabetes or drug-induced neuropathy) or extensive damage (such as large nerve defects or avulsion injuries) leads to negligible recovery. In contrast, central nervous system (CNS) tissues, which include the brain and spinal cord, show limited healing under almost all circumstances. A major impediment to healing is the inability of adult neurons to proliferate *in vivo* and to be cultivated *in vitro*. Neural tissue engineering seeks to provide new strategies to optimize and enhance regeneration through the use of three-dimensional scaffolds, the delivery of growth-promoting molecules, and the use of neuronal support cells or genetically engineered cells. The objectives of this chapter are to provide an overview of the basic biology of neural regeneration (using the PNS as a model), to review biomaterial-based strategies for nerve repair, and to present more recent neural tissue engineering approaches (Table 56.1).

PROBLEMS AND CHALLENGES

Every year, hundreds of thousands of patients are treated for degenerative disease or traumatic PNS and CNS damage, and the attendant social and economic costs are staggering. The adult PNS retains the capacity for regeneration following injury, and many of the molecular/neurotrophic factors are known. Nonetheless, the return of function in the clinical setting is quite variable and motor and sensory deficits invariable persist. Repair of the CNS, which includes the brain and spinal cord, represents an even greater challenge, because almost all injuries lead to an irreversible loss of function and because many victims are young. Refinements in microsurgical techniques and drug therapies have had some beneficial effects on neural repair, but seem to have reached an impasse, suggesting that biologic rather than technical factors are limiting. During World War II, the large caseload of nerve injuries led to the development of nerve repair techniques that involved end-to-end reconnection. Functional outcomes were further improved with advances in microsurgical techniques and instrumentation. In the 1970s and 1980s, advances in polymer science provided synthetic guidance channels for experimental nerve repair, although an ideal clinical material has not been identified. From the 1980s to the present, progress in biotechnology and molecular biology has uncovered some of the molecular and genetic bases of neural diseases, thus opening biological avenues for promoting regeneration. Neural tissue engineering seeks to promote neural repair by combining the potency and specificity of biologically based repair strategies (neurotrophic molecules, extracellular matrix equivalents, transplanted/genetically engineered cells, etc.) with the purity and definition of synthetic technologies (material design/modification, drug delivery systems, microelectronic fabrication, etc.). Neural tissue engineering may provide a means of reducing health care costs and improving the quality of life by accelerating healing, pro-

Table 56.1. Strategies for neural repair

Biomaterial-based approaches
   Surgical suture
   Use of nerve guidance channel
   Fibrin glue
   Poly(ethylene glycol) fusion
Tissue engineering approaches
   Delivery of neurotrophic factors
   Use of functionalized gels
   Seeding with Schwann cells
   Use of genetically engineered cells

viding alternatives to current treatments, and enabling repair where it is not currently feasible. Neural tissue engineering also entails efforts to replace neuroactive molecules in the CNS, to recreate neural circuitry *in vitro,* and to design functional neural interfaces. This chapter concentrates mainly on issues of neural regeneration.

OVERVIEW OF NEURAL REGENERATION

It has long been appreciated that the PNS is capable of regeneration following injury. The central dogma that the adult CNS is not capable of regeneration has been challenged subsequent to reports based on studies suggesting that there is a vigorous capacity for regrowth under optimal conditions. In both nervous systems, regeneration is limited by the fact that adult neurons are not capable of proliferating. In successful regeneration, sprouting axons from proximal nerve tissues traverse the injury site, enter the distal nerve tissue, and make new connections with target structures. For the purposes of this chapter, we will use the PNS as a model for nerve regeneration.

PERIPHERAL NERVOUS SYSTEM

Peripheral nerves are responsible for the innervation of the skin and skeletal muscles and contain electrically conductive fibers, the axons, whose cell bodies reside in or near the spinal cord. Neurons are structurally unique in that axonal processes may extend for considerable distances (e.g., up to 1 m or longer). All axons are ensheathed by Schwann cells, which provide structural and chemical support for all PNS neurons. In fact, neuronal survival is contingent on the continuous, retrograde flow of neurotrophic molecules synthesized by Schwann cells or target tissues. Motor neurons originate in the anterior horn (inner gray matter) of the spinal cord and are responsible for directing voluntary movement. Sensory neurons are localized in dorsal root ganglia found adjacent to the spinal cord and function to sense the local environment through specialized sensors (stretch receptors, Pacinian corpuscles, etc.). Sympathetic neurons are located in sympathetic ganglia found along the length of the cord and are responsible for regulating involuntary functions (sweating, vascular tone, etc.). A single nerve may contain many hundreds of the three neuronal types. Together, nerve fibers, Schwann cells, blood vessels, and surrounding connective tissue constitute a nerve fascicle, which is the basic structural unit of a peripheral nerve. Most peripheral nerves contain one or more fascicles, each containing numerous myelinated and unmyelinated axons. The surgical repair of cut nerves is complicated by the fact that the fascicular tracts branch frequently and follow a tortuous pathway.

PNS REGENERATION

Nerves (including the cell body and axons) react in a characteristic fashion to transection injury (Sunderland, 1991). Injuries occurring close to the spinal cord are more likely to result in neuronal death because a large axonal segment is lost. More peripheral lesions, such as those in the hand or fingers, show a greater degree of successful regeneration. If the neuronal cell body survive the initial injury, it swells and decreases neurotransmitter (i.e., the chemicals that control neuronal signaling) production and increases production of the substrates necessary for reconstituting axonal and membrane components. Immediately after injury, the tip of the proximal nerve stump (i.e., the segment closest to the spinal cord) swells considerably and the severed axons retract several millimeters. After several days the proximal axons begin to sprout vigorously and growth cones emerge. Growth cones are mobile extrusions of the cut axons that flatten and spread when they encounter an adhesive terrain. Growth cones elicit numerous extensions (e.g., fillipodia) that extend outward in all directions until the first sprout reaches an appropriate terrain. The lead sprout is the only one to survive and the others die back in a process of pruning.

In the distal nerve stump, a process termed "Wallerian degeneration" begins within hours after transection. Wallerian degeneration is characterized by the degradation of axons and the proliferation of Schwann cells, which organize into ordered columns, or bands of Bünger. Schwann cell columns provide an optimal environment for growth cone adhesion and axonal outgrowth. In successful regeneration, axons sprouting from the proximal stump bridge the injury gap and encounter an appropriate Schwann cell column (band of Bünger). In humans, axons elongate at an average rate of 1 mm per day. Some of the axons reaching the distal stump traverse the length of the nerve to form functional synapses, but changes in the pattern of muscular and sensory reinnervation invariably occur. The misrouting of growing fibers occurs primarily at the site of injury because once they reach Schwann cell columns their paths do not deviate further. Failures in regeneration may lead to the formation of a neuroma, a dense, irregular tangle of axons, which can cause painful sensations. It should be noted that peripheral nerve transection also leads to marked atrophy of the corresponding targets (e.g., muscle fibers). Accelerated axonal outgrowth and regeneration is an important clinical goal because it results in less muscle wasting and decreased loss of sensory units.

Surgical repair of peripheral nerves

In the absence of reconnection, recovery of function following a nerve injury is negligible. This may be due to one or more factors, including (1) that neurons do not survive due to severe retrograde effects, particularly due to the loss of neurotrophic support, (2) that a gap separating the nerve stumps prevents sprouting axons from finding the distal stump, (3) that scar tissue ingrowth at the injury site acts as a physical barrier to axonal elongation, and (4) that the proximal and distal fascicular patterns differ so much that extending axons cannot find an appropriate distal pathway.

Early attempts to reconnect severed peripheral nerves utilized a crude assortment of materials, and anatomic repair rarely led to an appreciable return of function (Sunderland, 1991). Materials used include biologic materials such as catgut, silk, and cotton, and metal wires. Interventions were often doomed by the high rates of infection resulting from poor sterile techniques and inadequate sterilization of implanted materials. In the 1950s the concept of end-to-end repair, or coaptation, was refined by surgeons who directly reattached individual nerve fascicles or groups of fascicles. Further refinements occurred during the 1960s with the introduction of the surgical microscope and the availability of finer suture materials and instrumentation. In microsurgical nerve repair the ends of severed nerves are apposed and realigned by placing several strands of very fine suture material through the epineurial connective tissue without entering the underlying nerve fascicles. This process is termed epineural repair. In instances in which individual fascicles can be identified, they are often reattached individually in a process called fascicular or perineural repair. Interestingly, perineural repair does not result in significantly better functional recovery when compared to simpler epineural repair. Nerve grafting procedures are performed when nerve retraction or nerve tissue loss prevents direct end-to-end repair. Nerve grafts are also employed when nerve stump reapproximation would create tension along the suture line, a situation that is known to hinder the regeneration process by compromising blood flow. In grafting surgery, an autologous nerve graft from the patient, such as the sural nerve (or other sensory nerves, the removal of which results in acceptable functional deficit), is interposed between the ends of the damaged nerve. Presently, the most commonly used biomaterial in end-to-end repair or nerve grafting is fine monofilament nylon (polyamide) suture hooked to stainless steel needles.

Repair with fibrin glue or fusion with poly(ethylene glycol)

There have also been efforts to develop "gluing" techniques for nerve repair, thus avoiding the disadvantages associated with suturing. Sutures at the injury site can cause tissue reaction and local scar tissue formation, impeding axonal outgrowth. Microsurgical repair techniques usually require expensive instrumentation, take relatively long to perform, and require training and specialized techniques. In addition, some neural tissues, including the CNS, are rather amorphous, making suturing difficult or impossible.

Fibrin glue or "plasma clot" materials have been used for many years in experimental nerve repair (Sunderland, 1991). Fibrin is derived from blood and is a major component of clots and scabs. Fibrin glue is a two-component system composed of concentrated fibrinogen, which is activated and polymerized into fibrin on the addition of thrombin. Some experimental studies show that fibrin glue repair of peripheral nerves is comparable to standard microsurgical repair. Practical advantages include straightforward application, rapid repair time, and easier use in poorly accessible locations. One limitation is in the case of nerve gaps, because the nerve ends cannot be readily approximated.

One unique experimental system uses poly(ethylene glycol) (PEG) to fuse axons of transected nerves in invertebrate animals (earthworms) (Krause and Bittner, 1990). The use of PEG immediately following nerve transection resulted in morphologic fusion of individual axons. The concentration and molecular weight of PEG appear important in determining the extent and quality of fusion. Major limitations of this approach include the need for immediate repair (so that Wallerian degeneration and axonal destruction do not occur) and the requirement that the nerve ends be reapproximated under minimal tension.

Repair with nerve guidance channels

The frequent occurrence of nerve injuries during the World Wars of this century stimulated surgeons and scientists to seek simple, effective means of repairing damaged nerves. A variety of biologic and synthetic materials were shaped into cylinders and used to entubulate cut nerves. Bone, collagen membranes, arteries, and veins were used to repair nerves from the late 1800s through the 1950s. These materials did not enhance the rate of nerve regeneration when compared to standard techniques, and clinical applications were not widespread. Magnesium, rubber, and gelatin tubes were evaluated during World War I and cylinders of parchment paper and tantalum were used during World War II. The poor results achieved with these materials can be attributed to poor biocompatibility, because the channels elicited an intense tissue response that limited the ability of growing axons to reach the distal nerve stump. Following World War II, polymeric materials with more stable mechanical and chemical properties became available. Millipore (cellulose acetate), a microporous filter material, showed early favorable results in experimental models. After long-term implantation (several months) in humans, however, the Millipore material became calcified and eventually fragmented so that its use was discontinued. Silastic (silicone elastomer) tubing was first tested experimentally in the 1960s. Thin-walled Silastic channels were reported to support regeneration over large gaps in several mammalian species. Thick-walled tubing was associated with nerve necrosis and neuroma production. Thin-walled Silastic tubing and bioresorbable poly(glycolic acid) (PGA) tubes are the only synthetic materials currently used, on a very limited basis, in the clinical repair of severed nerves. The subsequent sections discuss results obtained with these and other synthetic materials in experimental nerve repair.

EXPERIMENTAL USE OF GUIDANCE CHANNELS

The availability of a variety of new biomaterials (particularly polymers) led to a resurgence of entubulation studies in the 1980s. In repair procedures using nerve guidance channels, the free or mobilized ends of a severed nerve are positioned within the lumen of a tube and are anchored in place with sutures (Fig. 56.1). Tubulation repair provides several advantages, as listed in Table 56.2. The most important factors include prevention of scar tissue invasion into the regenerating environment, provision of directional guidance for elongating axons, and maintenance of endogenous trophic or growth factors.

Guidance channels are useful in experimental procedures evaluating the process of nerve regeneration because (1) the gap distance between the nerve stumps can be precisely controlled, (2) the fluid and tissue entering the channel can be tested and evaluated, (3) the properties of the channel can be varied, and (4) the channel can be filled with various drugs, gels, etc. Nerves of different sizes and with various dimensions from several mammalian species, including mice, rats, rabbits, hamsters, and nonhuman primates, have been studied. The regenerated tissue in the guidance channel can be evaluated morphologically to quantify the outcome of regeneration. Parameters

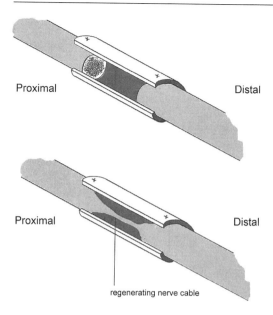

Proximal

Distal

Proximal

Distal

regenerating nerve cable

Fig. 56.1. Nerve guidance channel.

analyzed include the cross-sectional area of the regenerated nerve cable, the number of myelinated and unmyelinated axons, and the relative percentages of cellular constituents (e.g., epineurium, endoneurium, blood vessels). Electrophysiology and functional evaluation can also be performed in studies conducted for long time periods (several weeks or longer).

In the early 1980s the rat sciatic nerve model emerged as a useful system for studying nerve repair and for evaluating guidance channels. The spatial–temporal progress of nerve regeneration across a 10-mm rat sciatic nerve gap repaired with a silicone elastomer tube has been described in detail and provides a point of reference for evaluating other systems (Longo *et al.,* 1984; Williams *et al.,* 1983). During the first hours following repair the tube fills with a clear, protein-containing fluid exuded by the cut blood vessels in the nerve ends. The fluid contains the clot-forming protein, fibrin, as well as neurotrophic factors known to support nerve survival and outgrowth. By the end of the first week the lumen is filled with a longitudinally oriented fibrin matrix that coalesces and undergoes syneresis to form a continuous bridge between the nerve ends. The fibrin matrix is then invaded by cells migrating from the proximal and distal nerve stumps. This includes fibroblasts (which first organize along the periphery of the fibrin matrix), Schwann cells, macrophages, and endothelial cells (which form capillaries and larger vessels). At 2 weeks, axons advancing from the proximal stump are engulfed in the cytoplasm of Schwann cells. After 4 weeks some axons have reached the distal nerve stump and many have become myelinated. The number of axons reaching the distal stump is related to the distance the regenerating nerve has to traverse and the length of original nerve resected. Silicone guidance channels do not support regeneration if the nerve gap is greater than 10 mm and if the distal nerve stump is left out of the guidance channel (Lundborg *et al.,* 1982). In these experimental models, the morphology and structure of the regenerated nerve are far from normal. The size and number of axons and the thickness of myelin sheaths are less than normal. Electromyographic evaluation of nerves regenerated through silicone tubes reveals

Table 56.2. Advantages of nerve guidance channels

Surgical repair is simplified
Tension at the suture line is reduced
Scar tissue invasion is prevented
Outgrowing tissue is guided
Endogenous neuroactive molecules are sequestered
External inhibitory molecules are repelled

Table 56.3. Nerve guidance channel materials

Polymeric materials
  Nonresorbable
    Nonporous
      Ethylene vinyl acetate copolymer (EVA)
      Poly(tetrafluoroethylene) (PTFE)
      Polyethylene (PE)
      Silicone elastomers (SE)
      Poly(vinyl chloride) (PVC)
    Microporous
      Gortex, expanded poly(tetrafluoroethylene) (e-PTFE)
      Millipore (cellulose filter)
    Semipermeable
      Poly(acrylonitrile) (PAN)
      Poly(acrylonitrile) poly(vinyl chloride) (PAN/PVC)
      Polysulfone (PS)
  Bioresorbable
    Poly(glycolic acid) (PGA)
    Poly(L-lactic acid) (PLLA)
    PGA/PLLA blends
    Polycaprolactone (PCL)
    Polyester urethane
Metals
  Stainless steel
  Tantalum
Biologic materials
  Artery
  Collagen
  Hyaluronic acid derivatives
  Mesothelial tubes
  Vein

that axons can make functional synapses with distal targets, although nerve conduction velocities and signal amplitudes are slower than normal, even after many months.

Attempts to improve the success rate and quality of nerve regeneration have led to the use of numerous biomaterials as guidance channels (Table 56.3). Biodurable materials such as acrylic copolymers (Aebischer *et al.,* 1989a; Uzman and Villegas, 1983), and polyethylene (Madison *et al.,* 1988), and bioresorbable materials such as polyglycolides (Molander *et al.,* 1989), polylactides (Nyilas *et al.,* 1983), polyesters (Henry *et al.,* 1985), and collagen (Archibald *et al.,* 1991) have been investigated. There is some concern that biodurable materials may cause long-term complications via compression injury to nerves or soft tissues. Biodegradable materials offer the advantage of disappearing once regeneration is complete, but success thus far has been limited by swelling, increased scar tissue induction, and difficulty in controlling degradation rates. In all cases, these materials have displayed variable degrees of success in bridging transected nerves and the newly formed nerves are morphologically quite different from normal peripheral nerves.

BIOMATERIAL-BASED APPROACHES TO NERVE REPAIR

Studies using nerve guidance systems suggest that neural degeneration is influenced by the polymer system used and that it should not be considered a passive player. For example, manipulating the physical, chemical, and electrical properties of synthetic channels allows control over the regenerating environment and optimization of the regeneration process. The following features of synthetic nerve guidance channels have been studied and will be reviewed: (1) transmural permeability, (2) surface texture/microgeometry, (3) electrical charges, and (4) release of drugs.

Transmural Permeability

The synthetic nerve guidance channel controls the regeneration process by influencing the cellular and metabolic aspects of the regenerating environment. Because transected nerves lose the integrity of their blood–nerve barrier and are in contact with the wound environment, the guidance channel's transmural mass transfer characteristics control solute exchange between the regenerating tissue and the surrounding fluids. Nerves regenerated through semipermeable tubes display more normal morphologic characteristics than do nerves regenerated in impermeable silicone elastomer (SE) and polyethylene (PE) tubes or freely permeable (microporous) expanded poly(tetrafluoroethylene) (e-PTFE) tubes. Nerves found in semipermeable tubes feature more myelinated axons and less connective tissue (Aebischer *et al.,* 1989a; Uzman and Villegas, 1983). The range of permselectivity is important and optimal regeneration is observed with a molecular weight (MW) cutoff range of 50,000–100,000. Permselective poly(acrylonitrile)/poly(vinyl chloride) (PAN/PVC) channels with a MW cutoff of 50,000 support regeneration even in the absence of a distal nerve stump (Aebischer *et al.,* 1988, 1989a).

These observations suggest that controlled solute exchange between the internal regenerative and external wound healing environments is essential in controlling regeneration. The early availability of oxygen and other nutrients in permselective tubes may minimize connective tissue formation in PAN/PVC and polysulfone (PS) tubes. Decreased oxygen levels and waste buildup due to impermeable wall structures may increase connective tissue formation in SE and PE tubes. Regeneration may also be modulated by excitatory and inhibitory factors released by the wound healing environment. Semipermeable channels may sequester locally generated growth factors while preventing the inward flux of molecules inhibitory to regeneration.

Surface Texture/Microgeometry

The microgeometry of the luminal surface of the guidance channel plays an important role in regulating tissue outgrowth. Expanded microfibrillar e-PTFE tubes exhibiting different internodal distances (1, 5, and 10 μm) were compared to smooth-walled, impermeable PTFE tubes. Larger internodal distances result in greater surface irregularity and increased transmural porosity. Highly porous, rough-walled tubes contained isolated fascicles of nerves dispersed within a loose connective tissue stroma. The greater the surface roughness, the greater the spread of fascicles (Valentini and Aebischer, 1993). In contrast, smooth-walled, impermeable PTFE tubes contained a discrete nerve cable delineated by an epineurium and located within the center of the guidance channel. Similar results were observed with semipermeable tubes with either smooth or rough inner walls that were fabricated with the same polymer (PAN/PVC) and exhibited similar mechanical properties and MW cutoffs (50,000) (Aebischer *et al.,* 1990). Nerve regenerated in tubes containing alternating sections of smooth and rough inner walls showed similar morphologies, with an immediate change from single-cable to numerous fascicle morphology at the interface of the smooth and rough segments.

These studies suggest that the microgeometry of the guidance channel lumen also modulates the nerve regeneration process. Wall structure changes may alter the protein and cellular constituents of the regenerating tissue bridge. For example, the orientation of the fibrin matrix is altered in the presence of a rough inner wall. Instead of forming a single, longitudinally oriented bridge connecting the nerve ends, the fibrin molecules remain dispersed throughout the lumen. As a result, cells migrating in from the nerve stumps fill the entire lumen in a loose pattern rather than forming a dense central structure.

Electrical Charge Characteristics

Applied electrical fields and direct current stimulation are known to influence nerve regeneration *in vitro* and *in vivo*. Silicone tubes fitted with electrical leads and used as cuffs to surround lesioned nerves have shown increased regeneration in rat models (Kerns *et al.,* 1991).

Certain dielectric polymers may be used to study the effect of electrical activity on nerve regeneration *in vivo* and *in vitro*. Electret materials are advantageous in that they provide electrical charges without the need for an external power supply, can be localized anatomically, are biocompatible, and can be formed into a variety of shapes including tubes and sheets. Electrets are a class of dielectric polymers that can be fabricated with stable bulk/surface charges because of their unique molecular structure. True electrets, such as PTFE, can be formed to exhibit a static surface charge due to the presence of stable, monopolar charges located in the bulk of the polymer. The sign of the charge is dependent on the poling conditions. Positive, negative, or combined charge patterns may be achieved. Piezoelectric materials such as poly(vinylidene fluoride) (PVDF) display transient surface charges related to dynamic spatial reorientation of molecular dipoles located in the polymer bulk. The amplitude of charge generation is dependent on the degree of physical deformation (i.e., dipole displacement) of the polymer structure. The sign of the charge is dependent on the direction of deformation, and the materials show no net charge at rest.

Negatively and positively poled PVDF and PTFE tubes have been implanted as nerve guidance channels. Poled PVDF and PTFE channels contain significantly more myelinated axons than do unpoled, but otherwise identical, channels (Aebischer *et al.,* 1987; Valentini *et al.,* 1989). In general, positively poled channels contained larger neural cables with more myelinated axons, compared with negatively poled tubes. It is not clear how static or transient charge generation affects the regeneration process. The enhancement of regeneration may be due to electrical influences on ligand binding, ionic fluxes, receptor mobility, growth cone motility, cell migration, etc.

DELIVERY OF NEUROTROPHIC FACTORS

Growth or neuronotrophic factors ensure the survival and general growth of neurons and are produced by support cells (e.g., Schwann cells) or by target organs (e.g., muscle fibers). Some factors support neuronal survival, others support nerve outgrowth, and some do both (Table 56.4) (Raivich and Kreutzberg, 1993). Numerous growth factors have been identified, purified, and synthesized through recombinant technologies (Table 56.4). *In vivo,* growth factors are found in solution in the serum or extracellular fluid or bound to extracellular matrix (ECM) molecules. The release of soluble agents, including growth factors and other substrates from synthetic guidance channels, may improve the degree and specificity of neural outgrowth. Using single or multiple injections of growth factors has disadvantages, including early burst release, poor control over local drug levels, and factor degradation in biologic environments. Guidance channels can be prefilled with drugs or growth factors, but the aforementioned limitations persist. Advantages of using a local, controlled delivery system are that the rate and amount of factor release can be controlled and the delivery can be maintained for long periods of time (several weeks). Channels composed of an ethylene vinyl acetate (EVA) copolymer have been fabricated and are designed to release incorporated macromolecules in a predictable manner (Aebischer *et al.,* 1989b). The amount of drug loaded, size of the drug, and geometry of the drug-releasing structure affect the drug release kinetics (Langer and Folkman, 1976). It is possible to restrict drug release to the luminal side of the guidance channel by coating its outer wall with a film of pure polymer.

Nerve guidance channels fabricated from EVA and designed to slowly release basic fibroblast growth factor (bFGF) support regeneration over a 15-mm gap in a rat model (Aebischer *et al.,*

Table 56.4. Growth factors in neural regeneration

Growth factor	Possible function
Neural factors	
Nerve growth factor (NGF)	Neuronal survival, axon–Schwann cell interaction
Brain-derived neurotrophic factor (BDNF)	Neuronal survival
Ciliary neuronotrophic factor (CNTF)	Neuronal survival
Glia-derived neurotrophic factor (GDNF)	Neuronal survival
Glial growth factor (GGF)	Schwann cell mitogen
Neurotrophin-3 (NT-3)	Neuronal survival
Neurotrophin-4/5 (NT-4/5)	Neuronal survival
General factors	
Insulin-like growth factor 1 (IGF-1)	Axonal growth, Schwann cell migration
Insulin-like growth factor 2 (IGF-2)	Motoneurite, sprouting, muscle reinnervation
Platelet-derived growth factor (PDGF)	Cell proliferation, neuronal survival
Acidic fibroblast growth factor (aFGF)	Neurite regeneration, cell proliferation
Basic fibroblast growth factor (bFGF)	Neurite regeneration, neovascularization

1989b). Control EVA tubes supported regeneration over a maximum gap of only 10 mm, with no regeneration in 15-mm-long gaps. Another study using extruded EVA tubes designed to deliver nerve growth factor (NGF) also supported regeneration over a critical nerve gap (Valentini *et al.,* 1995). The concurrent release of growth factors that preferentially control the survival and outgrowth of motor and sensory neurons may further enhance regeneration, because the majority of peripheral nerves contain both populations. For example, NGF and bFGF control sensory neuronal survival and outgrowth, and brain-derived neurotrophic factor (BDNF), ciliary neuronotrophic factor (CNTF), and glial cell line-derived neurotrophic factor (GDNF) control motor neuronal survival and outgrowth. Other molecules, neurotrophins (NT-3 and NT-4/5), may do both. Factors that promote Schwann cell proliferation [e.g., glial growth factor (GGF)] may be useful in enhancing nerve growth or in culturing transplantable cells. The local release of other pharmacologic agents (e.g., antiinflammatory drugs) may also be useful in enhancing nerve growth. Growth factors released by guidance channels may allow regeneration over large nerve deficits and in sites where regeneration does not normally occur (e.g., brain and spinal cord).

BENCHTOP FABRICATION METHODS OF NERVE GUIDANCE CHANNELS

Trophic factor-releasing tubes

Different experimental methods have been used to tailor the attributes of the nerve guidance channels. These tubes have been made by dipping a mandrel in a polymer solution (Aebischer *et al.,* 1989b; Perego *et al.,* 1994). A volatile solvent allows the first layer of polymer to dry when it is removed from the solution before it is dipped again, forming the next layers. The mandrel is then removed and a hollow cylinder is formed. This technique was improved by coating a horizontally spinning mandrel (Borkenhagen *et al.,* 1998a). The thickness of the tube can easily be controlled and thus its rigidity, strength, toughness, and other mechanical properties are controlled.

If a solvent for the polymer can be chosen such that it does not destroy a protein's activity, different proteins/drugs and carrier molecules can be incorporated into the tube wall for eventual release into the lumen after implantation. Another method incorporates through the length of the tube, a thin polymer rod containing neurotrophic factors, allowing for their long-term sustained release (Bloch *et al.,* 1999).

Thermoplastic polymers can be formed into tubes by extruding them at elevated temperatures and pressures through a die and nozzle (Fine *et al.,* 1991). By mixing relevant proteins into the polymer dope, these tubes can be used as release systems (Valentini *et al.,* 1995).

Semipermeable tubes

There may be some interest in controlling the transfer of moieties from the regenerating environment because optimization of the tube's molecular weight cutoff enhances nerve regeneration (Aebischer *et al.,* 1989a). In addition, it has been shown that a semipermeable channel will allow the regeneration of an axotomized peripheral nerve in the absence of a distal stump, whereas no growth occurs in impermeable Silastic tubes (Aebischer *et al.,* 1988). Semipermeable synthetic polymer fibers can be fabricated by a dry-jet wet-spinning techniques (Cabasso *et al.,* 1976; Michaels, 1971). A polymer dissolved in its solvent is pumped through the outer lumen of a double-bored nozzle and a nonsolvent flows through the inner lumen. As the two liquids come into contact at the exit, a phase inversion occurs. When the solvent and nonsolvent are chosen so that they have a greater affinity for each other than the solvent has for the polymer, the polymer comes rapidly out of solution, forming a semipermeable layer. The nonsolvent continues to move slowly outward past the initial interface, through the polymer solution, forming larger pores (with a trabecular structure). As the forming tube exits the nozzle, it heads towards a nonsolvent bath, which will solidify a dense (tight pored) exterior layer. Altering the parameters of flow, temperature, solvent/nonsolvent pairing, and distance from the nozzle to the nonsolvent bath will modify the resulting hollow fiber formed, and its permeability.

Biodegradable tubes

In trying to produce a more ideal nerve guidance channel that is biodegradable, therefore eliminating the need for a second operation to remove the tubing, many techniques of fabrication are being combined. The desired tube would slowly lose its mass, yet still be strong enough to maintain its shape and protect the regeneration environment during the period of cable reformation before totally resorbing. A balance is sought between degradation rate and the inflammation that degradation products incite. Fine soluble particles such as sugars and salts added to the exterior coatings of mandrel dip-coated tubes form pores when dissolved out, reducing the mass of the bulk polymer (Pennings *et al.,* 1990). Other experimental methods blend in a second filler polymer that is preferentially dissolved away after tube formation. Degradable polyester urethane tubes have been formed with a porous honeycomb structure by casting the polymer solution in a mold and freezing the mold. The frozen solvent crystals are then sublimated under vacuum conditions, leaving a porous structure (Jordan *et al.,* 1999). The pores allow for infiltration of phagocytic cells and create a greater surface area for them to degrade the polymer faster.

TISSUE ENGINEERING APPROACHES TO NERVE REPAIR

Biomaterial-based approaches to nerve repair show promise for certain clinical applications, but may not provide sufficient support to bridge large gaps or to stimulate CNS repair. A higher level of intricacy takes place when using the hollow nerve guidance channel tube as a receptacle for hydrogel scaffolds, in an attempt to enhance regeneration by providing molecular constituents that are present during embryonic development, are found at elevated levels during nerve injury, or are actual components of the nerve tissue to be regenerated. Hydrogels are also used as matrices to seed neuronal support cells or trophic-factor producing genetically engineered cells within the nerve guides. Further advances in understanding the cellular and molecular aspects of neural repair may provide more potent strategies for stimulating regeneration.

Inclusion of Insoluble Gels

As noted earlier, a fibrin cable bridging the nerve stumps serves as a scaffold over which outgrowing cells and axons migrate. Prefilling SE tubes with a longitudinally oriented fibrin gel has been shown to accelerate regeneration across the gap and to increase the distance over which regeneration occurs (Williams *et al.,* 1987). Filling the tube with protein gels also influences the rate and quality of regeneration. Filling guidance channels with fairly viscous gels of laminin and collagen has been shown to improve and accelerate nerve repair in some models. Matrigel (a gel containing collagen type IV), laminin, and glycosaminoglycans (another ECM component), introduced in the guidance channel lumen, support some degree of regeneration over a long gap in adult rats (Madison *et al.,* 1988). SE channels preloaded with templates of collagen and glycosaminoglycans also increase the distance over which regrowth occurs (Yannas *et al.,* 1985). One study using semipermeable tubes filled with a range of collagen and Matrigel showed less regeneration in gel-filled tubes compared with tubes filled only with saline. This suggests that the viscosity and orientation of the gel may be as important as its chemical makeup (Valentini *et al.,* 1987).

Use of Functionalized Gels

The extracellular matrix consists of complex proteins known to control cell shape, position, and function. The availability of artificial ECM analogs will enable control of cell–substrate adhesion, cell–cell communication, and cell guidance. The glycoprotein laminin, an extracellular matrix component present in the basal lamina of Schwann cells, has been reported to promote nerve elongation *in vitro* and *in vivo*. Other ECM products, including the glycoprotein fibronectin and the proteoglycan heparin sulfate, have also been reported to promote nerve extension *in vitro*. Some subtypes of the ubiquitous protein, collagen, also support neural attachment. Inappropriate physicochemical properties of these gels may interfere with the regeneration process. In an alternate approach, the activity of these large, insoluble ECM molecules (up to 10^6 Da) can be mimicked by short peptide sequences of a few amino acids (e.g., GRGDS, YIGSR, SIKVAV) (Table 56.5) (Graf *et al.,* 1987; Iwamoto *et al.,* 1987; Kleinman *et al.,* 1988). The availability of small, soluble, bioactive agents allows more precise control over the chemistry, conformation, and presentation of neuron-specific substances. Additionally, their stability and linear structure facilitate their use compared with labile (and more expensive) proteins that require tertiary structure for activity. *In vitro* two-dimensional work demonstrated that immobilized peptides are indeed capable of promoting cell attachment and differentiation (Ranieri *et al.,* 1993, 1995). Three-di-

Table 56.5. Neuronal attachment and neurite-promoting factors

Factor	Minimal peptide sequence
Collagens	RGD, DGEA
Fibronectin	RGD
Laminin	SIKVAV, YIGSR, RGD
Neural cell adhesion molecule (NCAM)	—
N-Cadherin	—

mensional substrates containing peptide mimics have also been shown to promote neural attachment and regeneration *in vitro* (Bellamkonda *et al.*, 1995a,b) and *in vivo* (Borkenhagen *et al.*, 1998b). This research has centered on the functionalization of naturally occurring gels and polyelectrolytes (e.g., agarose, alginate) with various oligopeptides. Functionalized gels can be placed within guidance channels or implanted along injured neural tracts in the PNS and CNS. Agarose functionalized with laminin-derived peptides has been shown to support neurite elongation *in vitro* and peripheral nerve regeneration *in vivo*. In future studies, it will be important to evaluate the influence of gel viscosity and pore structure as well as the density and gradients of immobilized ligands.

SEEDING NEURONAL SUPPORT CELLS

All axons are ensheathed by support cells, the Schwann cells. Larger axons are surrounded by a multilamellar sheath of myelin, a phospholipid-containing substance that serves as an insulator and enhances nerve conduction. An individual Schwann cell may ensheathe several unmyelinated axons, but only one myelinated axon, within its cytoplasm. Schwann cells are delineated by a fine basal lamina that contains laminin, collagen type IV, and other ECM molecules. The Schwann cell basal lamina is capable of promoting neural outgrowth.

Adding neural support cells to the lumen of guidance channels is another strategy being used to improve regeneration or to make regeneration possible over otherwise irreparable gaps (Table 56.6). Schwann cells can be expanded *in vitro* from small nerve biopsies (Morrisey *et al.*, 1991). For example, Schwann cells cultured in the lumen of semipermeable guidance channels have been shown to improve nerve repair in adult rodents (Guénard *et al.*, 1992; Xu *et al.*, 1997). Cells harvested from inbred rats were first cultured in PAN/PVC tubes using various ECM gels as stabilizers. The Schwann cells and gel formed a cable at the center of the tube after several days in culture. Once implanted, the cells were in direct contact with the nerve stumps. A dose-dependent relationship between the density of seeded cells and the extent of regeneration was noted. The culture and expansion of human Schwann cells have proved difficult and may benefit through the use of GGF and other Schwann cell mitogens. CNS glial cells, such as astrocytes, may also support neural regeneration, although astrocytes obtained from older animals are less supportive than are cells from young animals. The addition of Schwann cells to astrocytes seeded in guidance channels improved the outcome of regeneration (Guénard *et al.*, 1994).

Olfactory ensheathing cells have been harvested and used to help support regeneration of

Table 56.6. Schwann cell support of neural regeneration

Form organized columns following injury (bands of Bünger)
Secrete basal lamina with neurite-promoting properties (laminin, collagen type IV, etc.)
Secrete nerve growth factor and other neurotrophic molecules
Express NGF receptors (sequester NGF at injury site, juxtacrine stimulation of axons)
May be genetically engineered to secrete particular neurotrophins

axons in the central nervous system (Li *et al.,* 1997). These cells were used because the normally occurring support cells of the CNS in adults are inhibitory to regeneration. Contrary to this, olfactory receptor neurons continuously arise from the neuroepithelium and it is believed that they have this capacity, in part, due to the olfactory glia that ensheathe their axons (Ramon-Cueto and Valverde, 1995). For the first time, nerve fibers have crossed a gap in the spinal column and have reentered distal tracks when harvested olfactory ensheathing cells were seeded into the lumen of a nerve guide tube and used as the conduit (Ramon-Cueto *et al.,* 1998).

SEEDING WITH GENETICALLY ENGINEERED CELLS

Another approach toward nerve repair involves the use of cells (Schwann cells, fibroblasts, myoblasts, etc.) that are genetically engineered to secrete growth factors (Gage *et al.,* 1991). The use of support cells that release neuronotrophic and neurite-promoting molecules may enable regeneration over large gaps or in recalcitrant tissues such as the spinal cord. There is significant evidence that PNS elements, especially Schwann cells, are capable of supporting CNS regeneration. For example, Schwann cell-seeded semipermeable channels support regeneration of the optic nerve (R. F. Valentini, and P. Aebischer, unpublished observations) and of the fimbria–fornix pathway (Kromer and Cornbrooks, 1985), which are CNS structures. In addition, fibroblasts engineered to secrete NGF enhance neuronal survival after a lesion of the fimbria–fornix, a CNS structure implicated in Alzheimer's disease (Hoffman *et al.,* 1993). The development of transfection/transduction techniques for the sustained, *in vivo* expression of functional genes (encoding NGF, CNTF, GGF, etc.), is under intense investigation.

SPECIFICITY OF NERVE RECONNECTIONS

Nerve entubulation techniques increase the number of axons and the distance that axons regenerate. Another limitation that must be confronted for nerve repair and regeneration is the mismatch of regenerating nerves to incorrect distal connections, thus limiting functional recovery. In reanastomosis of nerve cables with mixed populations, sensory neurons can haphazardly reinnervate the motor end organs, and the motor axons can reinnervate the sensory organs. Brushart and Seiler (1987) showed in rat experiments that if the proximal motor branch of the femoral nerve was placed in a silicone Y-tube, one-half of the axons made the wrong choice of growing into the end of the tube containing the distal sensory nerve stump, instead of the one containing the distal motor nerve stump, when the gap distance was 2 mm. Interestingly, two thirds of the motor axons grew into the correct branch of the silicone tube when the distance was increased to 5 mm. Despite this increase when lengthening the gap, there is no assurance that the neurons reconnect to their original end organs. Efforts have been initiated to fabricate a microelectronic switchboard that could be inserted into the regenerating environment (Rosen *et al.,* 1990; Zhao *et al.,* 1997). Placed in the nerve guidance channel perpendicular to the regenerating axons, holes in the "nerve chip" would allow neurites to pass through and reconnect to distal targets, and an electronic signal processor could be envisioned that would allow the nerve impulses to be sensed and rerouted to their proper destinations (Dario *et al.,* 1998; Kovacs *et al.,* 1992; Lundborg *et al.,* 1998). This system would have the drawbacks of a chronic long-term implant and the electronics would be subjected to the harsh destructive *in vivo* environment.

SUMMARY

The permeability, textural, and electrical properties of nerve guidance channels can be optimized to impact favorably on regeneration. The release of growth factors, addition of growth substrates, and inclusion of neural support cells and genetically engineered cells also enhances regeneration through guidance channels. Current limitations in PNS repair, especially the problem of repairing long gaps, and in CNS repair, because brain and spinal cord trauma rarely result in appreciable functional return, may benefit from advances in engineering and biology. The ideal guidance system has not been identified, but will most likely be a composite device that contains novel synthetic or bioderived materials and that incorporates genetically engineered cells and new products from biotechnology.

REFERENCES

Aebischer, P., Valentini, R. F., Dario, P., Domenici, C., and Galletti, P. M. (1987). Piezoelectric guidance channels enhance regeneration in the mouse sciatic nerve after axotomy. *Brain Res.* **436,** 165–168.

Aebischer, P., Guénard, V., Winn, S. R., Valentini, R. F., and Galletti, P. M. (1988). Blind-ended semi-permeable guidance channels support peripheral nerve regeneration in the absence of a distal nerve stump. *Brain Res.* **454**, 179–187.

Aebischer, P., Guénard, V., and Brace, S. (1989a). Peripheral nerve regeneration through blind-ended semi-permeable guidance channels: Effect of the molecular weight cutoff. *J. Neurosci.* **9**, 3590–3595.

Aebischer, P., Salessiotis, A. N., and Winn, S. R. (1989b). Basic fibroblast growth factor released from synthetic guidance channels facilitates peripheral nerve regeneration across long nerve gaps. *J. Neurosci. Res.* **23**, 282–289.

Aebischer, P., Guénard, V., and Valentini, R. F. (1990). The morphology of regenerating peripheral nerves is modulated by the surface microgeometry of polymeric guidance channels. *Brain Res.* **531**, 211–218.

Archibald, S. J., Krarup, C., Schefner, J., Li, S. T., and Madison, R. D. (1991). A collagen-based nerve guide conduit for peripheral nerve repair: An electrophysiological study of nerve regeneration in rodents and nonhuman primates. *J. Comp. Neurol.* **306**, 685–696.

Bellamkonda, R., Ranieri, J. P., and Aebischer, P. (1995a). Laminin oligopeptide derivatized agarose gels allow three-dimensional neurite outgrowth *in vitro*. *J. Neurosci. Res.* **41**, 501–509.

Bellamkonda, R., Ranieri, J. P., Bouche, N., and Aebischer, P. (1995b). A hydrogel-based three-dimensional matrix for neural cells. *J. Biomed. Mater. Res.* **29**, 663–671.

Bloch, J., Borkenhagen, M., Tseng, J. L., Zurn, A. D., and Aebischer, P. (1999). NGF and NT-3 releasing guidance channels promote regeneration of the transected dorsal root. *Abstr. Soc. Neurosci.* 402.14.

Borkenhagen, M., Stoll, R., Neuenschwander, P., Suter, U. W., and Aebischer, P. (1998a). *In vivo* performance of a new biodegradable polyester urethane system used as a nerve guidance channel. *Biomaterials* **19**, 2155–2165.

Borkenhagen, M., Clémence, J.-F., Sigrist, H., and Aebischer, P. (1998b). Three-dimensional extracellular matrix engineering in the nervous system. *J. Biomed. Mater. Res.* **40**, 392–400.

Brushart, T. M. E., and Seiler, W. A. (1987). Selective reinnervation of distal motor stumps by peripheral motor axons. *Exp. Neurol.* **97**, 289–300.

Cabasso, I., Klein, E., and Smith, J. K. (1976). Polysulfone hollow fibers. I. Spinning and properties. *J. Appl. Polym. Sci.* **20**, 2377–2394.

Dario, P., Garzella, P., Toro, M., Micera, S., Alavi, M., Meyer, U., Valderrama, E., Sebastiani, L., Ghelarducci, B., Mazzoni, C., and Pastacaldi, P. (1998). Neural interfaces for regenerated nerve stimulation and recording. *IEEE Trans. Rehab. Eng.* **6**, 353–363.

Fine, E. G., Valentini, R. F., Bellamkonda, R., and Aebischer, P. (1991). Improved nerve regeneration through piezoelectric vinylidene trifluoroethylene copolymer guidance channels. *Biomaterials* **12**, 775–780.

Gage, F. H., Kawaja, M. D., and Fisher, L. J. (1991). Genetically modified cells: Applications for intracerebral grafting. *Trends Neurosci.* **14**, 328–333.

Graf, J., Ogle, R. C., Robey, F. A., Sasaki, M., Martin, G. R., Yamada, Y., and Kleinman, H. K. (1987). A pentapeptide from the laminin B1 chain mediates cell adhesion and binds the 67,000 laminin receptor. *Biochemistry* **26**, 6896–6900.

Guénard, V., Kleitman, N., Morrissey, T. K., Bunge, R. P., and Aebischer, P. (1992). Syngeneic Schwann cells derived from adult nerves seeded in semipermeable guidance channels enhance peripheral nerve regeneration. *J. Neurosci.* **12**, 3310–3320.

Guénard, V., Aebischer, P., and Bunge, R. (1994). The astrocyte inhibition of peripheral nerve regeneration is reversed by Schwann cells. *Exp. Neurol.* **126**, 44–60.

Henry, E. W., Chiu, T. H., Nyilas, E., Brushart, T. M. E., Dukkes, P., and Sidman, L. (1985). Nerve regeneration through biodegradable polyester tubes. *Exp. Neurol.* **90**, 652–676.

Hoffman, D., Breakefield, X. O., Short, M. P., and Aebischer, P. (1993). Transplantation of a polymer-encapsulated cell line genetically engineered to release NGF. *Exp. Neurol.* **122**, 100–106.

Iwamoto, Y., Robey, F. A., Graf, J., Sasaki, M., Kleinman, H. K., Yamada, Y., and Martin, G. R. (1987). YIGSR, a synthetic laminin pentapeptide, inhibits experimental metastasis formation. *Science* **238**, 1132–1134.

Jordan, O., Papoloizos, M., Borkenhagen, M., Schmutz, P., Raffoul, W., Neuenschwander, P., Suter, U., and Aebischer, P. (1999). Peripheral nerve regeneration through biodegradable polyesterurethane channels: Influence of 3D characteristics. *Abstr. Soc. Biomater.*, 254.

Kerns, J. M., Fakhouri, A. J., Weinrib, H. P., and Freeman, J. A. (1991). Electrical stimulation of nerve regeneration in the rat: The early effects evaluating by a vibrating probe and electron microscopy. *J. Neurosci.* **40**, 93–107.

Kleinman, H., Ogle, R. C., Cannon, F. B., Little, C. D., Sweeny, T. M., and Luckenbill-Edds, L. (1988). Laminin receptors for neurite formation. *Proc. Natl. Acad. Sci. U.S.A.* **85**, 1282–1286.

Kovacs, G. T. A., Storment, C. W., and Rosen, J. M. (1992). Regeneration microelectrode array for peripheral nerve recording and stimulation. *IEEE Trans. Biomed. Eng.* **39**, 893–902.

Krause, T. L., and Bittner, G. D. (1990). Rapid morphological fusion of severed myelinated axons by polyethylene glycol. *Proc. Natl. Acad. Sci. U.S.A.* **87**, 1471–1475.

Kromer, L. F., and Cornbrooks, C. J. (1985). Transplants of Schwann cell culture cultures promote axonal regeneration in adult mammalian brain. *Proc. Natl. Acad. Sci. U.S.A.* **82**, 6330–6334.

Langer, R., and Folkman, J. (1976). Polymers for sustained release of proteins and other macromolecules. *Nature (London)* **263**, 797–800.

Li, Y., Field, P. M., and Raisman, G. (1997). Repair of adult rat corticospinal tract by transplants of olfactory ensheathing cells. *Science* **277**, 2000–2002.

Longo, L., Hayman, E. G., Davis, G. E., Ruoslahti, E., Engvall, E., Manthorpe, M., and Varon, S. (1984). Neurite-promoting factors and extracellular matrix components accumulating *in vivo* within nerve regeneration chambers. *Brain Res.* **309**, 105–117.

Lundborg, G., Dahlin, L. B., Danielsen, N., Gelberman, R. H., Longo, F. M., Powell, H. C., and Varon, S. (1982). Nerve regeneration in silicone chambers: Influence of gap length and of distal stump components. *Exp. Neurol.* 76, 361–375.

Lundborg, G., Drott, J., Wallman, L., Reimer, M., and Kanje, M. (1998). Regeneration of axons from central neurons into microchips at the level of the spinal cord. *NeuroReport* 9, 861–864.

Madison, R. D., DaSilva, C. F., and Dikkes, P. (1988). Entubulation repair with protein additives increases the maximum nerve gap distance successfully bridged with tubular prostheses. *Brain Res.* 447, 325–334.

Michaels, A. S. (1971). High flow membrane. U.S. Pat. 3,615,024.

Molander, H., Olsson, Y., Engkvist, O., Bowald, S., and Ericksson, I. (1989). Regeneration of peripheral nerve through a polygalactin tube. *Muscle Nerve* 5, 54–57.

Morrissey, T. K., Kleitman, N., and Bunge, R. P. (1991). Isolation and functional characterization of Schwann cells derived from adult nerve. *J. Neurosci.* 11, 2433–2442.

Nyilas, E., Chiu, T. H., Sidman, R. L., Henry, E. W., Brushart, T. M., Kikkes, P., and Madison, R. D. (1983). Peripheral nerve repair with bioresorbable prosthesis. *Trans. Am. Soc. Artif. Intern. Organs* 29, 307–313.

Pennings, A. J., Knol, K. E., Hoopen, H. J., Leenslag, J. W., and van der Lei, B. (1990). A two-ply artificial blood vessel of polyurethane and poly (L-lactide). *Colloid Polym. Sci.* 268, 2–11.

Perego, G., Cella, G. D., Aldini, N. N., Fini, M., and Giordino, R. (1994). Preparation of a new nerve guide from poly(L-lactide-co-caprolactone). *Biomaterials* 15, 189–193.

Raivich, G., and Kreutzberg, G. W. (1993). Peripheral nerve regeneration: role of growth factors and their receptors. *Int. J. Dev. Neurosci.* 11, 311–324.

Ramon-Cueto, A., and Valverde, F. (1995). Olfactory bulb ensheathing glia: A unique cell type with axonal growth-promoting properties. *Glia* 14, 163–173.

Ramon-Cueto, A., Plant, G. W., Avila, J., and Bunge, M. B. (1998). Long-distance axonal regeneration in the transected adult rat spinal cord is promoted by olfactory ensheathing glia transplants. *J. Neurosci.* 18, 3803–3815.

Ranieri, J. P., Bellamkonda, R., Jacob, J., Vargo, T., Gardella, J. A., and Aebischer, P. (1993). Selective neuronal cell attachment to a covalently patterned monoamine on fluorinated ethylene polypropylene films. *J. Biomed. Mater. Res.* 27, 917–927.

Ranieri, J. P., Bellamkonda, R., Bekos, E., Gardella, J. A., and Aebischer, P. (1995). Spatial control of neural cell attachment via patterned laminin oligopeptide chemistries. *J. Biomed. Mater. Res.* 29, 779–785.

Rosen, J. M., Grosser, M., and Hentz, V. R. (1990). Preliminary experiments in nerve regeneration through laser drilled-holes in silicon chips. *J. Restorative Neurol. Neurosci.* 2, 89–102.

Sunderland, S. (1991). "Nerve Injuries and Their Repair." Churchill-Livingstone, London.

Uzman, B. G., and Villegas, G. M. (1983). Mouse sciatic nerve regeneration through semipermeable tubes: A quantitative model. *J. Neurosci. Res.* 9, 325–338.

Valentini, R. F., and Aebischer, P. (1993). The role of materials in designing nerve guidance channels and chronic neural interfaces. *In* "Robotics and Biological Systems: Towards a New Bionics?" (P. dario, G. Sandini, and P. Aebischer, eds.), pp. 625–636. Springer Verlag, New York.

Valentini, R. F., Aebischer, P., Winn, S. R., and Galletti, P. M. (1987). Collagen- and laminin-containing gels impede peripheral nerve regeneration through semipermeable nerve guidance channels. *Exp. Neurol.* 98, 350–356.

Valentini, R. F., Sabatini, A. M., Dario, P., and Aebischer, P. (1989). Polymer electret guidance channels enhance peripheral nerve regeneration in mice. *Brain Res.* 48, 300–304.

Valentini, R. F., DaSilva, M., Akelman, E., Blanks, R., Moore, D., and Shah, K. (1995). Delivery of nerve growth factors by tubular guidance channels allows regeneration across long nerve gaps in adult rats. *Int. Congr. Cell Eng., 2nd.,* 100.

Williams, L. R., Longo, F. M., Powell, H. C., Lundborg, G., and Varon, S. (1983). Spatial-temporal progress of peripheral nerve regeneration within a silicone chamber: Parameters for a bioassay. *J. Comp. Neurol.* 218, 460–470.

Williams, L. R., Danielsen, N., Muller, H., and Varon, S. (1987). Exogenous matrix precursors promote functional nerve regeneration across a 15-mm gap within a silicone chamber in the rat. *J. Comp. Neurol.* 264, 284–290.

Xu, X. M., Chen, A., Guénard, V., Kleitman, N., and Bunge, M. B. (1997). Bridging Schwann cell transplants promote axonal regeneration from both the rostral and caudal stumps of transected adult rat spinal cord. *J. Neurocytol.* 26, 1–16.

Yannas, E. V., Orgill, D. P., Silver, J., Norregaard, T. V., Zervas, N. T., and Schoene, W. C. (1985). Polymeric template facilitates regeneration of sciatic nerve across 15 mm gap. *Trans. Soc. Biomater.* 11, 146.

Zhao, Q., Drott, J., Laurell, T., Wallman, L., Lindstrom, K., Bjursten, L. M., Lundborg, G., Montelius, L., and Danielsen, N. (1997). Rat sciatic nerve regeneration through a micromachined silicon chip. *Biomaterials* 18, 75–80.

CHAPTER 57

TRANSPLANTATION STRATEGIES FOR TREATMENT OF SPINAL CORD DYSFUNCTION AND INJURY

Jacqueline Sagen, Mary Bartlett Bunge, and Naomi Kleitman

INTRODUCTION

The spinal cord occupies a unique position in the central nervous system (CNS) in that it provides our primary conduit for communication and interaction with the world around us. With the exception of the unique sensors in the head, virtually all environmental stimuli are processed by somatosensory systems via the spinal cord. In addition, our responses to and interactions with our environment are under the control of motor systems functioning via final common pathways in the spinal cord. Autonomic sensations and responses also involve the flow of information to and from the spinal cord. Thus the spinal cord plays a critical role in the normal daily functioning of the organism in its environment, and any disruption can have severe consequences in the motor, sensory, and autonomic realms. The use of cellular or tissue transplantation for restoration of function has found potential CNS application in several disease and injury models (Azmitia and Björklund, 1987; Dunnett and Richards, 1990; Freeman and Widner, 1998; Gash and Sladek, 1988). For application to spinal cord dysfunction, several approaches may be envisioned: (1) the use of neural implants as axonal bridges for repair of damaged spinal circuitry, (2) provision of cells for replacement of lost or damaged neuronal or glial elements, and (3) provision of neurotransmitters and/or neurotrophic factors by implanted cellular "minipumps." These approaches are the subject of this chapter. The reader is also referred to other reviews (e.g., Bregman, 1998; Bunge and Kleitman, 1999; Fawcett, 1998; Giménez y Ribotta and Privat, 1998; Plant *et al.,* 2000; Sagen, 1998; Schwab and Bartholdi, 1996; Steeves and Tetzlaff, 1998; Zhang *et al.,* 1999).

SPINAL IMPLANTS AS AXON BRIDGES

The primary goal of implantation of cellular bridges into the spinal cord is to provide a permissive substrate for axonal growth across the injury site to enable repair of cord circuitry and restoration of function following traumatic injury. For this purpose, the most common animal models utilized include compression and contusion (leading to cavitation) injuries, which may closely resemble crush injuries found in the clinic. Hemisection and complete transection models, also much studied, provide information relevant to laceration lesions. The criterion for "success" using various transplantation approaches is a subject of debate. Clearly, the best indication of success would be nearly complete restoration of function. Whereas some indication of functional improvement has been reported using a variety of interventions and a variety of behavioral tests, this appears to be limited, at least for the adult spinal cord, given the presently available technology. Nevertheless, numerous studies have demonstrated that the capacity for regeneration in the adult spinal cord is greater than previously thought, and suggest that restoration of function may ultimately be possible under the right circumstances (e.g., Bregman *et al.*, 1998; Cheng *et al.*, 1996; Feraboli-Lohnherr *et al.*, 1997; Grill *et al.*, 1998; Li *et al.*, 1997; Liu *et al.*, 1999).

Following traumatic spinal cord injury, in addition to obvious cellular destruction at the site of the injury, axons from several sources are damaged or severed. These include segmental axons (e.g., primary axons derived from dorsal root ganglion cells), intersegmental axons (propriospinal or interneuronal), and long axons (e.g., descending axons from the corticospinal tract, monoaminergic axons from the locus coeruleus and raphe nuclei). Approaches toward regeneration of these damaged axons include the use of artificial or biologic matrices and peripheral nerve, Schwann cell, and fetal CNS tissue grafting.

ARTIFICIAL AND BIOLOGIC MATRICES

Materials that have been utilized in bridges in spinal cord injury models include collagen, Matrigel (a basement membrane preparation), Millipore filters, carbon filaments, and fibrin. Polyacrylonitrile/polyvinylchloride and polycarbonate tubes have been utilized to contain bridging cellular transplants [for additional review, see Plant *et al.* (2000)]. Using a cat spinal cord transection model, de la Torre (1982) found that a cell-free collagen matrix could integrate well with host spinal tissue and attract several host cell types, including Schwann cells, fibroblasts, and meningeal cells. Numerous tyrosine hydroxylase-positive fibers were found invading the graft and perhaps reentering the spinal cord, suggesting the possibility of axonal regeneration from brainstem catecholaminergic nuclei. Joosten and colleagues (1995), likewise observing axonal growth along with other cell types into collagen, found that the time of gelation was key for ingrowth of regenerating fibers; corticospinal fibers grew into collagen grafts injected initially as a fluid but not into collagen implanted as a self-assembled gel. Further refinements in this approach include the use of chemical cross-linking of collagen gels (Marchand *et al.*, 1993) and the addition of neurotrophic or growth-promoting substances such as laminin (Goldsmith and de la Torre, 1992). Neurotrophin-3 (NT-3) added to collagen grafts improves corticospinal axon ingrowth (Houweling *et al.*, 1998).

Substrate-bound neurotrophic factors have been added to nitrocellulose implants (Houlé and Johnson, 1989; Houlé and Ziegler, 1994). In spinal transection cavities of adult rats, nerve growth factor (NGF)-treated nitrocellulose strips coimplanted with fetal spinal cord tissue induced extensive ingrowth of axons expressing calcitonin gene-related peptide (CGRP), most probably derived from dorsal root axons. In addition, many fibers appeared to extend into the host spinal cord rostral to the lesion and graft site, suggesting the potential for regrowth of ascending sensory axons using such an approach. Improvement in sensory axon regrowth into the spinal cord has also been obtained using Millipore filters coated with embryonic astroglia (Kliot *et al.*, 1988, 1990).

The future application of artificial and biologic matrices in spinal cord injury is an important and promising area of research. A type of biocompatible and bioresorbable polymer to be further tested is poly(lactic–co-glycolic acid). This compound is well tolerated by Schwann cells and spinal cord tissue (Gautier *et al.*, 1998). Three-dimensional hydrogels are also being investigated in CNS repair. When combined with collagen, they support new tissue growth, thus enabling tissue reconstruction and regeneration (Plant *et al.*, 1995, 1998). Hydrogels can be prepared with added sequences such as the RGD sequence found in extracellular matrix molecules (Plant *et al.*, 1997). Following transplantation into the CNS, such a combination initiates axonal regeneration. Schwann cells may be added to hydrogels as well to improve neurite growth-promoting activity (Woerly *et al.*, 1996).

PERIPHERAL NERVE GRAFTS

It is well established that regeneration of several axonal processes in the CNS is greatly limited in contrast to axonal regeneration in the peripheral nervous system (PNS). Several possible explanations for this have been put forward, including the induction of an astroglial scar barrier (cellular or molecular) in response to CNS injury, the lack of supportive trophic factors and/or matrices, and the presence of growth inhibitory factors produced by CNS oligodendrocytes and the myelin they form. In contrast, peripheral nerves, and particularly Schwann cells, are remarkably supportive for the regrowth of axonal processes, including those of CNS origin. Thus, a rationale for intraspinal implantation of peripheral nerve segments is the provision of a conducive matrix for axon bridging and reconnection in severed CNS pathways following traumatic injury.

Although attempts at using peripheral nerve segments for spinal repair have been reported since the early 1900s, interest in this approach gained momentum in the 1970s and 1980s (for review, see Aguayo, 1985; Vrbová *et al.*, 1994). Peripheral nerve grafts can serve as bridges across transection lesions of the adult spinal cord; fibers enter the PNS grafts from neurons in the host spinal cord rostral and caudal to the lesion (Richardson *et al.*, 1980). Long peripheral nerve bridges circumventing the cord carry CNS axonal processes for remarkable distances, e.g., from the lower medulla to the thoracic cord (as much as 30 mm) (David and Aguayo, 1981). Thus, reconnection of intersegmental and perhaps long axonal pathways may be possible using this approach. Unfortunately, most findings indicate that, once the regenerating axons reach the CNS parenchyma, their further elongation is limited to the immediate vicinity of the graft–host border. One possible impediment to axon regeneration may be the disruption of the astrocytic array that is caused by the transection; this organization may provide an important guidance pathway for axon growth in the stumps (Davies *et al.*, 1994; Li and Raisman, 1994; Reier, 1986). Another, of course, is the neurite growth inhibitory barrier that forms between the peripheral nerve graft and the host cord, such as at the dorsal root entry zone.

Another approach that has been utilized is the implantation of peripheral nerve segments into the lesioned dorsal columns, leaving the rest of the spinal cord intact (Pallini *et al.*, 1988; Wilson, 1984, 1991). Good graft–host junctions were obtained, and some retrograde tracing studies indicated that neurons in the spinal cord sent axonal projections into the graft (Pallini *et al.*, 1988). In addition, there seemed to be some transmission through the grafted peripheral nerve segments, as indicated by electrophysiological recordings, and some improvements in sensation and motor coordination (Waldrope and Wilson, 1986; Wilson, 1991). Neurons in the dorsal root ganglia also can send axonal processes into grafted peripheral nerve fragments, because retrogradely labeled cells were noted in dorsal root ganglia several segments caudal to the graft (Pallini *et al.*, 1988). Other studies have utilized predegenerated peripheral nerve segments as graft material in the dorsal columns (Oudega *et al.*, 1994), the rationale being that their support of regeneration will be improved due to the increased production of neurotrophic factors (Heumann *et al.*, 1987). When combined with lesioning to condition the dorsal root ganglion neurons, more sensory fibers penetrated the peripheral nerve graft (Oudega *et al.*, 1994). When both were combined with neurotrophin infusion beyond the graft, sensory fibers not only extended into grafts but also into the cord beyond (Oudega and Hagg, 1996, 1999). Another study utilizing peripheral nerve implants in spinal dorsal columns demonstrated the capacity of injured dorsal column axons to reinnervate grafted Pacinian corpuscles (Jirmanová *et al.*, 1994).

Peripheral nerve grafts can be utilized to restore segmental motor connectivity by providing axonal bridges from ventral horn motor neurons to skeletal muscle (Bertelli and Mira, 1994; Bertelli *et al.*, 1994; Cullheim *et al.*, 1989; Horvat *et al.*, 1989). In these studies, when peripheral nerve segments were inserted into the spinal cord and denervated skeletal muscle following nerve avulsion injuries, spinal motor neurons were capable of regenerating axonal input to skeletal muscle. As evidence for this, muscle twitch responses could be elicited by electrical stimulation of the implanted nerve segments (Cullheim *et al.*, 1989), functional cholinergic neuromuscular junctions and endplates could be demonstrated (Bertelli and Mira, 1994), and coordinated voluntary and automatic activities could be obtained (Bertelli and Mira, 1994). Carlstedt *et al.* (1995) reported a case study of brachial plexus injury in which avulsed ventral roots were reintroduced directly (C6) and via sural nerve grafts (C7) into the cord just below the surface so as not to interfere with tracts. Voluntary activity in arm muscles was detected electromyographically after 9 months and clinically after 1 year; there was voluntary activity after 3 years. A possible additional benefit of such an approach is the potential survival-promoting effects of peripheral nerve grafts due to their endogenous neurotrophic factors. This is supported by studies demonstrating enhanced motoneuron survival and inhibition of nitric oxide synthase in spinal motor neurons when peripheral nerve segments are implanted at the time of root avulsion; normally root avulsion leads to significant motoneuron loss (W. Wu *et al.*, 1994). In another type of study, taking advantage of survival-promoting factors in muscle, peripheral nerve was used to attach muscle to adult rat cord that received a graft of embryonic spinal cord containing motoneurons (Sieradzan and Vrbová, 1989). (Results of this study are described later, with a discussion of lower motor neurons.)

The above studies indicate the potential for peripheral nerve grafts to aid in the restoration of segmental innervation in the injured spinal cord. Long spinal axon regeneration, e.g., the corticospinal tract, however, is more difficult to achieve using this approach. The work of Cheng *et al.* (1996) is an exception to this. These authors transplanted multiple intercostal nerve grafts with fibrin glue containing a fibroblast growth factor (FGF-1), positioning of the graft endings toward gray matter, and stabilization of the spinal column. The authors reported brainstem and corticospinal growth through and beyond the grafts, and partial recovery of locomotor functions. Corticospinal growth was indicated by both retrograde and anterograde tracing. Combining neurotrophic factors with peripheral nerve transplants improves the regenerative response of subchronically and chronically injured supraspinal nerve cells (Ye and Houle, 1997).

SCHWANN CELL GRAFTS

The Schwann cells of peripheral nerves possess many attributes that predict their capability of fostering CNS axonal regeneration (reviewed in Guénard *et al.*, 1993). There is also the possibility of autotransplanting Schwann cells obtained from a spinal cord-injured person; Schwann cells can be obtained from a piece of their peripheral nerve and expanded in number in culture (see Bunge and Kleitman, 1999). Also, while in culture, they may be genetically modified to generate neurotrophic factors. When cultured Schwann cells are microinjected in small volumes directly into long axon tracts such as the corticospinal tract or dorsal columns, both ascending and descending tracts extend for considerable distances along the tracts and branch into the grafts (Li and Raisman, 1994). Cultured Schwann cell suspensions have also been utilized in bridges in the spinal cord lesioned by various means, including photochemically (Paino and Bunge, 1991; Paino *et al.*, 1994), by inflatable microballoon compression (Martin *et al.*, 1991, 1993), by transection (Wrathall *et al.*, 1984; Xu *et al.*, 1995a; Guest *et al.*, 1997), and by suction (Kuhlengel *et al.*, 1990).

Matrices or guidance channels are often combined with Schwann cells (Paino and Bunge, 1991; Paino *et al.*, 1994; Xu *et al.*, 1995a, 1997). Using semipermeable polymer channels filled with Schwann cells suspended in Matrigel in complete transection lesions, Xu *et al.* (1997) demonstrated in the grafts numerous axons from dorsal root ganglion cells and spinal interneurons as far away as the C3 and S4 levels. The transplanted Schwann cells either myelinated or ensheathed the axons that regenerated into the graft. There was limited regrowth of serotonergic and noradrenergic fibers, whose parent somata are in the brainstem. As in the case of most peripheral nerve transplantation studies, Schwann cell bridges do not engender corticospinal fiber regeneration into the graft and fibers do not leave the graft. The use of a similar type of channel enclosing Schwann cells, but only half the size (i.e., filling a lateral hemisection rather than a complete transection), leads to improved regeneration in that fibers reenter the host cord (Xu *et al.*, 1999). Possibly the restoration of cerebrospinal fluid circulation and more stable cord–graft interfaces (due to a more limited laminectomy) are important factors in the improvement.

Channels containing the Schwann cell grafts that fill complete transection gaps are more effective in a combination strategy. When a neuroprotective compound, methylprednisolone, is administered at the time of Schwann cell transplantation, the Schwann cell bridge contains three times as many myelinated axons, and twice as many spinal neurons respond by extending axons into the graft (Chen *et al.*, 1996). Also, brainstem neurons respond, even though the transplant is in the thoracic region, far from the brainstem nuclei. Moreover, a modest growth of fibers from the graft into the distal cord is observed. When neurotrophins, NT-3, and brain-derived neurotrophic factor (BDNF) are infused into the space around the Schwann cell graft, more myelinated axons are present in the graft and more spinal neurons extend axons into the graft (Xu *et al.*, 1995b). Brainstem neurons also respond to the neurotrophin-containing Schwann cell graft. Genetically modifying the Schwann cells to produce human BDNF leads to an improved regenerative response as well (Menei *et al.*, 1998). Schwann cells genetically modified to secrete high levels of another neurotrophin, nerve growth factor, engender dopaminergic and adrenergic axonal regeneration (Tuszynski *et al.*, 1998). Thus, genetically manipulating Schwann cells to increase their production of neurotrophins improves their ability to support CNS axonal regeneration.

GRAFTS OF OTHER CELL TYPES

Tuszynski *et al.* (1994) have pioneered the use of genetically modified fibroblasts to secrete a variety of neurotrophic factors to improve regeneration in the cord. They have found, for example, that the administration of NT-3 from fibroblasts to a hemisected spinal cord leads to growth of corticospinal tract fibers, not through the fibroblast graft but in the gray matter around the graft in adult rat cord (Grill *et al.,* 1998). An improved locomotor outcome was documented in this work as well. Schnell *et al.* (1994) had initially shown that corticospinal fibers are responsive to NT-3, both during development and after axotomy. Also, transplanted NT-3-producing tissue can rescue axotomized Clarke's nucleus neurons (Tessler *et al.,* 1992). A new paper demonstrates that regeneration of rubrospinal fibers occurs in the wake of transplantation of fibroblasts engineered to secrete BDNF into a cervical hemisection cavity in adult rat cord (Liu *et al.,* 1999). The axons regenerate through and around the transplants, grow for long distances in the white matter below the transplant, and terminate in gray matter regions. These results are accompanied by significant recovery of forelimb usage.

Another cell type that appears to be highly promising in spinal cord regeneration without genetic modification is the ensheathing glial cell. A unique cell in the olfactory system, these cells have been found to support corticospinal tract regeneration in a very localized lesion (Li *et al.,* 1997, 1998) and to enable lengthy axon regeneration and also growth from Schwann cell grafts into the distal cord (Ramón-Cueto *et al.,* 1998). It is not yet known in what ways ensheathing glia affect CNS regeneration: Do they shelter the axon? Do they provide a permissive growth substrate? Or, do they secrete molecules that modify the lesion milieu?

Cells that may make the lesion milieu more favorable for neurite growth are macrophages, which secrete a variety of cytokines and growth factors. Macrophages have been activated *in vitro* by pieces of peripheral nerve and then injected into transected spinal cord (Rapalino *et al.,* 1998). This strategy leads to nerve fiber growth across the transected site, partial recovery of motor function, and evoked muscle responses. Retransection of the cord above the initial lesion site leads to loss of recovery, indicating the involvement of fiber growth across the lesion.

Additional cell types used to bridge lesions, replace lost cells, serve as relay stations, or act as "minipumps" (with added genes) are stem cells. Because they are now known to be present in normal adult CNS tissue, an important area of research will be to determine in what ways they may be influenced to proliferate and populate lesion areas. Or, alternatively, they may be isolated to be used in transplantation. An example showing the potential of neural stem cell transplantation is the intracerebroventricular injection at birth to correct the widespread dysmyelination in the shiverer (*shi*) mouse due to the lack of myelin basic protein (Yandava *et al.,* 1999). The injected cells differentiate into myelin basic protein-expressing oligodendrocytes throughout the *shi* brain, promoting improved widespread myelination with amelioration of behavioral symptoms. An intriguing aspect is that neural stem cells may possess a mechanism whereby their differentiation is directed to replenish deficient cell types. (See further discussion on myelin-forming cells, with discussion of lower motor neurons.)

Fetal CNS Grafts

It has been hypothesized that the fetal CNS, the environment in which CNS neurons grow during development, may be most conducive to encouraging axonal regrowth (Stokes and Reier, 1991). Potential advantages of fetal CNS tissue for grafting in spinal cord include not only the possibility to bridge an injury site with a growth-promoting environment but also the possibility of cellular replacement. In particular, the use of homotypic fetal spinal cord tissue implants may provide a relay via synaptic connections to and from the implant. This approach has been primarily successful when used in neonatal animals (Bregman and Kunkel-Bagden, 1994; Bregman *et al.,* 1993; Kunkel-Badgen and Bregman, 1990; Iwashita *et al.,* 1994; Diener and Bregman, 1998a,b). When spinal hemisections are performed in rats at birth, corticospinal tract axons are spared, because they have not yet grown past the site of injury (Schreyer and Jones, 1983). These developing axons are capable of elongating through fetal spinal cord implants and making appropriate connections with host spinal neurons caudal to the lesion. In addition to corticospinal axons, immunocytochemical and tracing studies revealed that descending serotonergic and noradrenergic axons extend throughout the transplant. CGRP fibers, presumably from dorsal root axons, also grow into and project through the transplant (Bregman and Kunkel-Bagden, 1994; Bregman *et al.,* 1993; Bregman, 1987; Bregman and Bernstein-Goral, 1991; Houle and Reier, 1989). Of particular interest is the apparent restorative capacity of fetal spinal cord transplants in neonates, which improved the maturation of locomotor patterns and shifted the time course of sensorimotor development toward that of normal intact animal (Bregman and Kunkel-Bagden, 1994; Bregman *et al.,* 1993). This includes enhanced recovery evidenced by frequency of contact-placing responses, improved performance on a grid runway, and increased base of support. Iwashita *et al.* (1994) reported long-lasting functional recovery when embryonic spinal cord segments were implanted in the proper orientation in neonatal rats with complete spinal resection. Histologically, in successful cases, the grafts showed remarkable survival and "seamless" integration with the host (14 of 22 cases); the others showed complete degeneration and resorption. Tracing studies indicated bidirectional long axonal projections across successful grafts. In addition, animals with successful replacement exhibited good motor performance with hindlimb/forelimb coordination. Surprisingly, this was not seen when embryonic spinal cord tissue was implanted in the incorrect orientation.

Grafting of embryonic day 14 (ED14) fetal cervical cord tissue after cervical hemisection yielded similar results in neonatal rats (Diener and Bregman, 1998a,b). Improved reflex responses and skilled forelimb activity were associated with growth of raphespinal, corticospinal, brainstem, and dorsal root projections. Raphe projections were seen apposed to propriospinal neurons, suggesting reestablishment of supraspinal input to spinal circuitry below the lesion.

In contrast to neonates, fetal spinal cord implantation in adult injured rat spinal cord has generally yielded less successful outcomes in restoring supraspinal input (Bregman and Kunkel-Bagden, 1994; Bregman *et al.*, 1993; Houle and Reier, 1988, 1989; Bernstein and Goldberg, 1989; Itoh and Tessler, 1990; Jakeman and Reier, 1991; Nothias and Peschanski, 1990; Pallini *et al.*, 1989; Patel and Bernstein, 1983; Reier *et al.*, 1986). Although CGRP-positive dorsal root axons are apparently capable of penetrating the grafted tissue throughout its extent, ingrowth of descending fibers is limited to the immediate host–graft interface. Some functional recovery has been reported, such as improved recovery in stride length and grid runway performance (Bregman and Kunkel-Bagden, 1994; Bregman *et al.*, 1993), although permanent deficits remain and motor behavior does not return to control levels. Slightly better results were reported in adult cats with spinal compression injuries (Wirth *et al.*, 1992), which showed some recovery in walking, gait, and stair climbing. This may be due to later maturation of cat spinal cord, or to the use of a compression, rather than a transection, injury model, which may spare host long axon pathways to some extent. Using a contusion injury model in adult rats, and transplants of fetal spinal tissue in cell suspension rather than solid segments, long-lasting improvements in some motor tasks have been reported, particularly in motivated fine motor behaviors, base of support, stride length, and more normalized inhibitory control of reflex excitability (Stokes and Reier, 1992; Thompson *et al.*, 1993). Suggested mechanisms for these beneficial effects include enhanced anatomic sparing, improved function of remaining systems, or the facilitation of alternative pathways.

Restoration of segmental innervation, particularly from dorsal root axons, is a consistent observation in adult animals with spinal cord injuries (Bernstein and Goldberg, 1989; Bregman *et al.*, 1993; Itoh *et al.*, 1993a,b; Jakeman *et al.*, 1989; Kikuchi *et al.*, 1993; Rhrich-Haddout *et al.*, 1994). Cut central axons of dorsal root ganglion cells readily regenerate into embryonic spinal tissue grafts placed in hemisection cavities, and form synaptic contacts. Numerous fibers immunoreactive for CGRP, which is present in small-diameter primary afferents derived from dorsal root ganglion cells, are found in transplanted tissue for at least 1 year following fetal spinal cord implantation (Itoh *et al.*, 1993b). Further, electrophysiological responses could be evoked in the fetal grafts by regenerated dorsal root axons, indicating the potential for functional segmental recovery using this approach. In addition, embryonic spinal grafts can apparently undergo normal differentiation to some extent, particularly the dorsal horn, because they demonstrate characteristics similar to the adult substantia gelatinosa, including the presence of met- and leuenkephalin, neurotensin, substance P, and somatostatin immunoreactivity (Jakeman *et al.*, 1989). Thus, such grafts may possess the capacity for neuronal replacement if appropriate intraspinal circuitries can be restored.

The absence of regeneration in descending fiber tracts may not be an intrinsic limitation of adult CNS neurons, but an insufficiency of trophic support provided by the fetal CNS grafts. When members of the neurotrophin family were administered concurrently with transplantation, increased serotonergic, noradrenergic and corticospinal ingrowths of axons into fetal spinal cord grafts were observed in adult rats (Bregman *et al.*, 1997). The effect was at least somewhat selective, in that ciliary neuronotrophic factor (CNTF) had no effect on growth of supraspinal axons. This finding is similar to that of Xu *et al.* (1995b), who showed improved supraspinal axonal regeneration into Schwann cell grafts when the neurotrophins BDNF and NT-3 were infused into the graft environment (see above).

Fetal CNS grafts, however, do provide some trophic substances. The observation of some degree of functional recovery, even in the absence of host reinnervation by the graft, has suggested that the recovery may be due to neurotrophic support provided by graft-derived astrocytes or other cellular components (Bernstein and Goldberg, 1989). Support for this comes from studies using heterotopic fetal CNS tissue as graft material. Fetal neocortical transplants in the adult spinal cord can rescue axotomized Clarke's nucleus neurons (Himes *et al.,* 1994), provide graft-derived astrocytes (Bernstein and Goldberg, 1989), and differentiate to produce defined neuronal and glial cell populations (Connor and Bernstein, 1987). Furthermore, studies of fetal spinal transplantation in neonates and adults have suggested a neurotrophic component to observed enhancements in recovery. Following complete midthoracic spinal transection in neonatal rats, Miya *et al.* (1997) observed incomplete integration of the grafts, although some descending serotonergic fibers grew through and beyond the graft. Although anatomic repair was incomplete, some rats were able to demonstrate sufficient coordinated adaptable locomotion to suggest that the effect of the limited supraspinal input was augmented by nonspecific trophic effects of the transplant on the distal cord. Bregman *et al.* (1998) observed that either fetal spinal grafts or neurotrophic factor application reversed atrophy of mature rubrospinal neurons. Both interventions together were more effective than either alone, suggesting the enhancement of graft efficacy may be associated with indirect effects of the trophic factors on the grafts themselves as well as direct effects on the regenerating axons. A report by Broude *et al.* (1999) confirms that both fetal spinal grafts and trophic factors activate neuronal responses, such as c-Jun expression, that may be associated with axonal growth. Thus, it will also be important to consider strategies to activate neuronal expression of regeneration-associated genes (reviewed in Richardson and Lu, 1994; Steeves and Tetzlaff, 1998).

When other fetal CNS tissues, such as cerebellum, neocortex (Himes *et al.,* 1994), and hippocampus (Li and Raisman, 1993), have been grafted, target specificity has been reported in the response of host neurons. Interestingly, "nontarget" CNS grafts (embryonic hippocampus or cortex) did not elicit entry of cut brainstem fibers in a full transection model (Bernstein-Goral *et al.,* 1997). In their hemisected cord model, sprouting of uninjured contralateral axons appeared to account for at least some of the observed ingrowth, leading these authors to suggest that sprouting and regenerating axons may differ in growth responsiveness after injury. The implantation of human fetal spinal cord allografts has undergone initial clinical testing in the United States and Sweden (Falci *et al.,* 1997; Wirth *et al.,* 1998, 1999). This approach has been used in clinical trials in spinal cord-injured patients with recurring syringomyelia (cystic lesions of the cord causing progressive loss of function and increased symptoms such as pain, spasticity, and autonomic dysreflexia). Graft survival and recurrence of cyst expansion are being studied using magnetic resonance imaging and physiologic assessment is underway.

SPINAL IMPLANTS FOR REPLACEMENT OF SPECIFIC CELL POPULATIONS

LOWER MOTOR NEURONS

As discussed above, in addition to providing a supportive growth environment, homotypic grafts of fetal spinal cord could potentially provide for the replacement of lost neuronal populations and restore spinal circuitry via reciprocal host–graft connections. To some extent, the dorsal horn of transplanted embryonic spinal cord appears to differentiate and take on characteristics typical of the adult substantia gelatinosa (Jakeman *et al.,* 1989). However, in general, ventral horn motor neurons do not appear to survive or differentiate well after transplantation. Instead, this region is characterized by small and medium neurons, with a paucity of large neurons, suggesting that the larger motoneurons, which are the earliest born, either selectively die or fail to mature to the adult form. Nevertheless, the conditions necessary to enhance motoneuron replacement and integration in host circuitry could be important in the treatment of neurologic diseases such as amyotrophic lateral sclerosis (ALS) and poliomyelitis. This topic has been the subject of intense investigation by Vrbová and colleagues (1994). This group has identified several factors, both host and graft, that influence the survival of transplanted embryonic motoneurons. First, optimal donor gestational age appears to be after the onset of motoneuron pool proliferation but before axonal extension into the periphery commences. Second, the availability of an appropriate target (i.e., denervated muscle) is essential. Third, the implanted motoneurons must be able to reach the target via a conduit, such as peripheral nerve implanted in the vicinity of the graft. In addition, the authors have found that depletion of the host's own motoneuron pool, similar to the cerebellum of Purkinje cell-deficient mice receiving Purkinje cell transplants (Sotelo and Alvarado-Mallart, 1987), may create necessary "empty niches" for occupation of embryonic motor neurons migrating from the grafts.

Using this approach, Vrbová's group transplanted small fragments of embryonic rat ventral spinal cord (ED11–ED12) into the lumbar segment of hosts whose spinal cord motor neurons had been selectively depleted several weeks earlier by sciatic nerve crush at birth (Sieradzan and Vrbová, 1989). In addition, a host muscle with a length of attached nerve was implanted to create a conduit and target muscle. Donor embryos were prelabeled *in utero* with 5-bromo-2'-deoxyuridine (BrdU) to aid in identification of donor-derived cells following transplantation. Results indicated numerous BrdU-labeled nuclei in the graft as well as host spinal cord, with some labeled cells attaining a larger size than embryonic motoneurons. Retrograde tracing studies indicated that some of these motoneurons, and the expression of choline acetyltransferase (ChAT) and CGRP, which are useful markers for motoneurons, was quite low, again suggesting either selective motoneuron death or the failure to reach full maturity (Sieradzan and Vrbová, 1993). In addition, axons of these grafted neurons were not able to enter vacated endoneurial sheaths of spinal ventral roots to reinnervate muscles deprived of their innervation, and thus recovery of lost motor function was not achieved. In a subsequent study, attempts were made to guide axons from grafted embryonic motoneurons to their target via a reimplanted ventral root (Nógrádi and Vrbová, 1996). Retrograde labeling 3–6 months later revealed axons of graft origin in the ventral ramus and some reversal of the losses in muscle force.

In other studies, fetal spinal neurons have been grafted as a suspension, using cell populations specifically enriched in motoneurons by density-gradient purification or fractionation using the low-affinity NGF receptor as a ligand (Demierre *et al.*, 1990; Peschanski, 1993; Peschanski *et al.*, 1992). These motoneurons survived, received host innervation, and in some cases migrated into host parenchyma. However, as in solid grafts, ChAT expression was generally poor. The neurotoxic lectin, volkensin, has been used to deplete selectively adult motoneurons as a better model of human degenerative motoneuron disease (Nógrádi and Vrbová, 1994). Again, transplanted fetal neurons migrated into host tissue, but few CGRP-positive neurons were found.

Erb *et al.* (1993) implanted dissociated ED14 ventral spinal cord cells into the distal stump of axotomized tibial nerves of adult rats. Large multipolar neurons were observed at the implantation site, with some myelinated axons. Axons were traced to the experimentally denervated gastrocnemius muscle and were seen to form neuromuscular junctions. Completely denervated adult rat gastrocnemius muscles shortened repeatedly, for hours, without electromyographic failure, after reinnervation by ED14–ED15 ventral spinal cord cells and exogenous electrical stimulation (Thomas *et al.*, 1998). Some reinnervated motor units also contract spontaneously, demonstrating some intrinsic excitability of the grafted motoneurons (Sesodia *et al.*, 1999). Using a similar paradigm, Fleetwood *et al.* (1998) reported that stem cells isolated from embryonic mouse spinal cord cultured in FGF-2 and epidermal growth factor (EGF), and prelabeled with 3,3'-dioctadecycloxacarbocyanine perchlorate (DiO), were able to grow axons into denervated sciatic nerve segments. The resultant neuronal cells were immunoreactive for a microtubule-associated protein (MAP-2) and ChAT.

In addition to motoneuron replacement, the provision of appropriate neurotrophic factors by cellular implants may aid in the treatment of neurodegenerative diseases such as ALS by retarding additional motor neuron loss or protecting transplanted embryonic motor neurons. As an example, in animal models of chronic motoneuron degeneration, CNTF can prevent motor neuron loss (Sendtner *et al.*, 1992). Taking advantage of these findings, encapsulated baby hamster kidney (BHK) cells that were genetically engineered to produce CNTF reduced motoneuron death after facial nerve axotomy and prolonged the life-span of mutant mice with progressive mononeuropathy (Sagot *et al.*, 1995). An open-label safety study using encapsulated BHK–hCNTF cells implanted intrathecally in ALS patients revealed detectable levels of CNTF in the cerebrospinal fluid of these patients, with no significant adverse effects noted (Aebischer *et al.*, 1996). However, evaluation of Norris scale slopes before and after implantation indicated that the disease continued to progress. Nevertheless, these findings indicate that it is possible to achieve sustained delivery of a neurotrophic factor intrathecally using cellular implantation.

Myelin-Producing Cells

The efforts of several laboratories have been focused on the potential for grafts of myelin-producing cells to remyelinate axonal pathways, because demyelinating and dysmyelinating disorders such as multiple sclerosis and the leukodystrophies continue to debilitate patients. For many of these studies, mutant animals such as the myelin-deficient (*md*) rat and the shaking pup (*sh*) have been used. Both the *md* rat and *sh* pup have a point mutation in the proteolipid protein gene that results in severe dysmyelination of the CNS and is the molecular basis of dysmyelination in patients suffering from Pelizaeus–Merzbacher disease. Remyelination could theoretically be derived from two distinct cell types: Schwann cells, which myelinate peripheral nerves, and oligodendrocytes, which myelinate CNS axons. Both of these sources have been utilized in attempts to achieve remyelination in the spinal cord. Schwann cells are particularly aggressive in their ability to proliferate and remyelinate axons in the CNS and compete successfully with oligodendrocytes for positions on axonal surfaces, especially in the absence of type 1 astrocytes (Blakemore and Crang, 1989; Blight and Young, 1989). In a sense, this competition for space on axonal surfaces may be detrimental in the long run, because the Schwann cells may displace host oligodendrocytes and astrocytes (Vrbová *et al.,* 1994). Nevertheless, Schwann cell grafts have been shown to be capable of remyelinating spinal cord axons in myelin-deficient mutants (e.g., the quaking mouse, *sh* mouse, and *md* rat), as well as in adult rats experimentally demyelinated using ethidium bromide (Baron-Van Evercooren *et al.,* 1993; Blakemore, 1977, 1980, 1983, 1984; Blakemore *et al.,* 1987; Duncan *et al.,* 1981, 1988). Remyelination by Schwann cells can arise from either peripheral nerve segments placed alongside or inserted into the dorsal columns, or by cultured Schwann cell suspensions. The latter are capable of migrating quite some distance toward the demyelinated regions (Baron-Van Evercooren *et al.,* 1991; Langford and Owens, 1990).

Imaizumi *et al.* (1998) have reported that olfactory ensheathing cells, which have properties resembling both astrocytes and Schwann cells, can remyelinate dorsal column axons following demyelination by X-ray irradiation and ethidium bromide injection. Suspensions of cells from neonatal rat olfactory bulbs were injected. Three weeks later, remyelinated axons showed improved conduction velocity and frequency-response properties associated with PNS-like myelination of axons. Olfactory ensheathing cells are highly migratory, and show evidence of having myelinated areas remote from the injection site. Li *et al.* (1998) reported that suspensions of cells from olfactory bulb injected into small electrolytic lesions of the corticospinal tract at C1–C2 supported regeneration along the tract's original CNS pathway. Regenerating corticospinal axons were associated with myelin sheaths immunopositive for protein-zero (typical of peripheral myelin segments) and exhibiting typical peripheral nervelike morphologies. Peripheral and central myelin segments were observed to be adjacent on some axons at the transition zone between damaged and undamaged host cord (Li *et al.,* 1998).

As mentioned, the invasive nature of Schwann cells, and possibly olfactory ensheathing glia grafts, may limit their usefulness in human application if they cannot be adequately controlled. In contrast, oligodendrocytes may be a better choice, because, together with astrocytes, they are potentially capable of forming a more normal and well-balanced glial environment (Vrbová *et al.,* 1994). As with Schwann cells, oligodendrocyte transplants have been derived either from intact tissue (in this case embryonic spinal cord) or from defined glial cell cultures (Blakemore and Crang, 1988, 1989; Duncan *et al.,* 1988, 1992; Gout *et al.,* 1988; Rosenbluth *et al.,* 1989, 1993; Utzschneider *et al.,* 1994). For example, embryonic spinal cord grafted into the spinal cord of *sh* mice or *md* rats resulted in the formation of normal myelin sheaths with regular periodicity around axons (Rosenbluth *et al.,* 1989). Further, the implantation of spinal glial cell suspensions in *md* rats resulted in a threefold increase in axon conduction velocity and normalized frequency-response properties (Utzschneider *et al.,* 1994). The potential usefulness of oligodendrocytes in spinal transplantation is enhanced by the ability to utilize xenogeneic sources (Rosenbluth *et al.,* 1993; Archer *et al.,* 1994) and cryopreservation techniques (Archer *et al.,* 1994). In addition, transplanted oligodendrocytes can apparently migrate rostrocaudally, particularly toward lesion sites (Duncan *et al.,* 1988; Duncan and Milward, 1995; Gumpel *et al.,* 1989). However, the extent of remyelination has been limited to patches of dysmyelination. A study in canine myelin mutants has indicated that donor tissue derived from fetal sources is most efficacious in myelinating large areas of the CNS (Archer *et al.*, 1997). Using mixed glial cell populations from the brain of an embryonic day (ED) 45 normal canine, injected into the dorsal columns of *sh* pups, extensive remyelination near the implant sites and up to 2 cm rostrally and caudally was observed.

Thus, more recent approaches have utilized progenitor cells as donor sources for remyelination (for review, see Billinghurst *et al.*, 1998; Dubois-Dalcq, 1995; Rogister *et al.*, 1999; Zhang *et al.*, 1999). For example, conditionally immortalized glial precursor cell lines (using the temperature-sensitive SV40 T antigen) were generated from ED14 mouse brain and transplanted in the dorsal columns of rats with ethidium bromide demyelinating lesions (Trotter *et al.*, 1993). Myelin formation with regular periodicity could be observed around demyelinated axons following injections of early passages, although myelin sheaths were thin and showed breaks. The success of remyelination and behavioral recovery after experimental demyelination paradigms is dependent on remaining axonal integrity (Jeffery *et al.*, 1999); however, these models have established the potential for functional remyelination by precursor cell transplants. In another laboratory, a growth factor-dependent cell line of rat oligodendrocyte precursors (CG4) could migrate extensively along axonal tracts from the injection site in *md* rats and myelinate dorsal column axons (Tontsch *et al.*, 1994). Results of grafting CG4 cells into neonatal *md* rats indicated that highly motile, proliferative oligodendrocyte precursors yield the best survival and functional myelination (Espinosa de los Monteros *et al.*, 1997).

A particularly promising approach is the use of epidermal growth factor-responsive neural stem cells, which can be generated from embryonic and adult brain and propagated continuously as spheres (neurospheres) in an undifferentiated state in the presence of EGF (Hammang *et al.*, 1994; McKay, 1997; Reynolds and Weiss, 1992). Under appropriate culture conditions, these cells can be induced to differentiate into oligodendrocytes, astrocytes, or neurons. When transplanted in the undifferentiated stem cell state into the spinal cords of *md* rats, the cells can differentiate into myelinating oligodendrocytes and produce densely packed, normal-appearing myelin (Hammang *et al.*, 1997). Interestingly, although clonal neural stem cells preferentially differentiate into astrocytes *in vitro*, a majority of these multipotential cells may shift toward oligodendroglial lineage in response to environmental cues in myelin-deficient states (Hammang *et al.*, 1997; Yandava *et al.*, 1999). Similar findings have been observed in *sh* pup mutants receiving transplants of EGF-responsive neurospheres derived from the normal canine brain (Milward *et al.*, 1997). Although most studies have utilized progenitor cells derived from embryonic or neonatal animals, multipotent stem cells are also present in the adult brain ependyma and can generate myelin-producing oligodendrocyte progenitors (Zhang *et al.*, 1999). In addition, allogeneic donor sources may be feasible with appropriate immunosuppression (Li and Duncan, 1998), and myelinating oligodendrocyte progenitors may be derived from ectopic sites, such as the optic nerve (Fanarraga *et al.*, 1998).

The first *in vivo* evidence that neurotrophic influences may promote the expansion of oligodendrocyte lineage cells in the spinal cord was attained by McTigue *et al.* (1998). In this study, fibroblasts producing NT-3, BDNF, CNTF, NGF, FGF-2, or β-galactosidase were transplanted subacutely into adult spinal cord injured by contusion. The NT-3 and BDNF groups showed increased numbers of axons and enhanced myelination associated with increased numbers of BrdU-positive oligodendrocytes, suggesting the trophic factors had initiated proliferation and differentiation of endogenous oligodendrocyte precursor cells. Because multipotent neural stem cells could potentially be engineered to produce neurotrophic factors, they may be ideally suited for combination strategies requiring both cell replacement and delivery of therapeutic molecules (Billinghurst *et al.*, 1998).

SPINAL IMPLANTS FOR PROVISION OF NEUROTRANSMITTERS

MONOAMINERGIC CELLS FOR RESTORATION OF MOTOR FUNCTIONS

Several laboratories have utilized serotonergic or noradrenergic cells as graft sources in spinal-injured or -denervated animals. Serotonergic cells have typically been obtained from embryonic mesencephalic or medullary raphe regions (Foster *et al.*, 1985, 1989; Privat *et al.*, 1988, 1989; Reier *et al.*, 1992a,b). In spinal cords of adult rats denervated of serotonergic input by 5,7-dihydroxytryptamine, serotonergic cell suspensions implanted 7 days after the lesion were found to survive for up to 12 months after grafting, with dense fiber outgrowth rostral and caudal to the grafts (Foster et al., 1985, 1989). In general, it appears that better survival of serotonergic neurons is obtained with medullary raphe grafts, possibly because this region is the normal source of serotonergic innervation of the spinal cord. With medullary raphe grafts, many motoneurons could be found surrounded by serotonergic fibers. Serotonin (5-HT) levels in the denervated cord after transplantation were restored to 40% of that in unlesioned animals. In addition, electrical stimulation of the grafts could excite motoneurons, suggesting that functional connections had been formed. Using complete spinal transection, which depletes spinal cord 5-HT, similar survival and integration was reported with embryonic raphe cell suspensions implanted 1 week following the lesion (Privat et al., 1988, 1989). In the early postgrafting period, synaptic afferents to serotonergic neurons were frequently seen. Later, serotonergic innervation was found in normal 5-HT-innervated sites such as the ventral horn near motoneurons, intermediolateral cell column, and superficial dorsal horn. In addition, sexual reflexes, which are under the control of 5-HT and absent in paraplegic rats, were restored in transplanted animals. The window between spinal transection and grafting was opened by Reier et al. (1992a,b), who found that fetal serotonergic grafting delayed from 1 to 3 months posttransection could still project their axons into gray matter regions normally innervated by bulbospinal 5-HT neurons.

Noradrenergic neurons from embryonic brainstem containing the locus coeruleus have also been utilized as intraspinal grafts to restore function in spinal transected or 6-hydroxydopamine (6-OHDA)-denervated animals (Buchanan and Nornes, 1986; Commissiong and Sauvé, 1989; Moorman et al., 1990; Nornes et al., 1988; Nygren et al., 1977; Yakovleff et al., 1989). Bulbospinal noradrenergic systems are involved in a number of functions, including motor activities, autonomic reflexes, and modulation of nociception. Good survival of noradrenergic neurons in locus coeruleus transplants has been reported by inserting grafts directly into spinal cord parenchyma or following placement in subpial cavities. The transplanted cells were found to grow processes and reinnervate the spinal gray matter to nearly normal levels. In particular, immunoreactive noradrenergic processes with terminals were found around motoneurons. In addition, improvements in motor performance tasks involving noradrenergic input were reported, including increased force of hindlimb flexion reflexes (Buchanan and Nornes, 1986; Nornes et al., 1988; Yakovleff et al., 1989), rhythmic locomotion and step frequency (Yakovleff et al., 1989), and motor coordination (Commissiong and Suavé, 1989).

Studies show that embryonic transplants of either raphe or locus coeruleus regions yield dense monoaminergic reinnervation of host lumbar interneurons after a T8–T9 spinal transection, and motoneurons in lumbar segments after grafting at T12–T13 (Yakovleff et al., 1995). Effects on two fictive motor patterns, stepping and hindpaw shaking, could be discriminated 1 to 3 months later: both types of grafts increased the excitability of the spinal stepping generator compared to surgical controls, but neither significantly affected hindpaw shaking. More direct evidence of influence of fetal grafts on spinal stepping generators has been described (Feraboli-Lohnherr et al., 1997). Spinal cord locomotor circuits display plasticity independent of supraspinal input (Giménez y Ribotta and Privat, 1998). This plasticity can be modulated pharmacologically or by transplantation of embryonic monoaminergic neurons. By injecting 6-OHDA to eliminate noradrenergic inputs after transplantation of fetal raphe or coeruleus cells, Feraboli-Lohnherr et al. (1997) showed for the first time that transplanted serotonergic neurons can activate the spinal pattern generator for locomotion independent of adrenergic influence. Electromyographic studies showed that this locomotor-like activity was facilitated by the 5-HT reuptake blocker, zimelidine.

Adrenal chromaffin cells have also been used as intraspinal grafts to restore deficits after spinal catecholamine depletion by 6-OHDA (Pulford et al., 1994). Neonatal rat adrenal medullary cells cultured in the presence of NGF were implanted in the lumbar region of the spinal cord 2 weeks after denervation. These grafts survived for at least 3 months postimplantation. In addition, the long latency component of the hindlimb withdrawal reflex, which is catecholamine modulated, was significantly more forceful than control implanted or unimplanted animals, and equally as effective as fetal noradrenergic implants.

SPINAL CORD GRAFTS FOR PAIN MODULATION

Adrenal medullary chromaffin cells have also been utilized for transplantation in the spinal subarachnoid space to alleviate pain. The choice of chromaffin cells was based on their ability to produce neuroactive substances with known antinociceptive properties, including the catecholamines epinephrine and norepinephrine and opioid peptides metenkephalin and leuen-kephalin. In addition, chromaffin cells produce other agents that could potentially modulate pain, including neurotrophic factors and other neuropeptides. The catecholamines and opioid peptides released from the implanted cells can interact with local host spinal opioid and α-adrenergic receptors. Furthermore, it is thought that decreased pain sensitivity results from the synergistic actions of opioid peptides and catecholamines, because subeffective levels of agents acting at both receptors produce potent analgesia, possibly with reduced tolerance development (Yaksh and Reddy, 1981; Wilcox *et al.*, 1987; Drasner and Fields, 1988; Sherman *et al.*, 1988). Early studies in the Sagen laboratory utilized standard analgesiometric assays—the tail flick, paw pinch, and hot plate tests—to assess whether released neuroactive substances from adrenal medullary allografts could exert behavioral changes (Sagen *et al.*, 1986a). Following nicotinic stimulation to increase release of neuroactive substances from transplanted cells, the level of acute pain tolerance in animals with adrenal medullary transplants was much greater than in control animals. This antinociceptive response was most likely due to the corelease of opioid peptides and catecholamines from transplanted cells, because it could be attenuated by opiate antagonist naloxone or α-adrenergic antagonist phentolamine. Similarly, isolated chromaffin cells obtained from xenogeneic bovine adrenal glands could reduce acute pain sensitivity using these tests (Sagen *et al.*, 1986b). Neurochemical studies of CSF samples collected via push–pull superfusion revealed increased levels of both opioid peptides and catecholamines in the CSF of animals with adrenal medullary implants (Sagen and Kemmler, 1989; Sagen *et al.*, 1991). In addition, good viability of transplanted chromaffin cells, either in solid tissue allografts or isolated suspensions of xenografts, has been consistently demonstrated using immunocytochemistry and electron microscopy.

Behavioral findings using adrenal medullary or chromaffin cell transplants have been extended to chronic pain models, including inflammatory pain (Ortega-Alvaro *et al.*, 1997; Sagen *et al.*, 1990; Siegan and Sagen, 1997; Vaquero *et al.*, 1991; Wang and Sagen, 1995), neuropathic pain (Décosterd *et al.*, 1998; Ginzburg and Seltzer, 1990; Hama and Sagen, 1993, 1994), and central pain models (Brewer and Yezierski, 1998; Hains *et al.*, 1998; Yu *et al.*, 1998a,b). In particular, adrenal medullary transplants have been shown to markedly reduce persistent pain resulting from peripheral or central nervous system injury in rodent models. These pain syndromes, which are difficult to manage clinically using conventional therapies, may be uniquely sensitive to the therapeutic cocktail produced by chromaffin cell transplants, which likely includes neurotrophic factors (Unsicker, 1993) and *N*-methyl-D-aspartate (NMDA) antagonist neuropeptides (Lemaire *et al.*, 1993; Siegan *et al.*, 1997), in addition to catecholamines and opioid peptides. In support for this, recent findings suggest that adrenal medullary transplants may also provide a degree of neuroprotection as a result of either reducing hyperexcitability or secretion of neurotrophic factors. A long-term consequence of persistent noxious activation, such as that resulting from injury to the peripheral or central nervous system, may be a loss of vulnerable inhibitory interneurons, which would otherwise serve to limit pain. For example, peripheral nerve injury results in the appearance of hyperchromatic "dark neurons" indicative of transsynaptic degeneration and a loss of inhibitory GABAergic neurons in the superficial spinal dorsal horn (Sugimoto *et al.*, 1990). Both the appearance of dark neurons and the loss of GABAergic neurons are reduced by adrenal medullary transplants (Hama *et al.*, 1996; Ibuki *et al.*, 1997).

Using three distinct models of spinal cord injury pain, chromaffin cell allografts or xenografts have been shown to reduce hypersensitivity to mechanical and thermal stimuli (Brewer and Yezierski, 1998; Hains *et al.*, 1998; Yu *et al.*, 1998a,b). As an example, an excitotoxic spinal cord injury that results in pathologic changes comparable to ischemic and traumatic clinical spinal cord injury can be produced by intraspinal injections of the mixed $(+/-)$α-amino-3-hydroxy-5-methylisoxazole-4-propionic acid (AMPA)/metabotropic receptor agonist quisqualic acid (Yezierski *et al.*, 1998). This results in a behavioral syndome of mechanical and thermal allodynia and excessive grooming at approximately 10–14 days following quisqualic acid injections. The excessive grooming continued to increase in severity in animals receiving control skeletal muscle transplants (Brewer and Yezierski, 1998). In contrast, in animals with adrenal medullary transplants, the progressive increase in grooming was blocked, and skin areas were completely recovered to normal in a few cases. Both mechanical allodynia and thermal allodynia were also attenuated by adrenal medullary transplants compared to control transplants.

A chronic pain model used in our laboratory is the chronic constriction nerve injury model described by Bennett and Xie (1988). Chronic constriction of the sciatic nerve results in abnormal pain behaviors closely paralleling clinical neuropathic pain syndromes such as causalgia, including allodynia (perception of innocuous stimulus as painful) and hyperalgesia (increased sensitivity to a painful stimulus). Two weeks following induction of peripheral nerve injury, rats were implanted with either adrenal medullary or control tissues. In rats with adrenal medullary allografts, exaggerated responses to noxious thermal and mechanical stimuli induced by nerve constriction can be reversed within 1 week postimplantation. In addition, heightened sensitivity to innocuous cold stimuli and tactile stimuli (von Frey hairs) was reduced in animals with adrenal medullary, but not control, transplants.

Similar reductions in neuropathic pain symptoms were obtained in animals implanted with bovine chromaffin cell xenografts (Hama and Sagen, 1994). These findings suggest that xenogeneic donors are a potential resource for cell transplantation in the therapy of chronic pain. However, the survival of xenogeneic cells in the rodent CNS has required at least short-term immunosuppression with cyclosporin A treatment (Ortega *et al.,* 1992). One approach toward overcoming the problem of immunologic mismatch is the use of encapsulation technology. Polymer capsules made of permselective membranes can allow two-way diffusion of neuroactive substances from encapsulated cells and nutrient/trophic support from the host, while providing a barrier to immunoregulatory cells, antibodies, and complement proteins (Aebischer *et al.,* 1988). Using this approach, isolated bovine chromaffin cells were loaded into capsules and transplanted to the subarachnoid space of the lumbar enlargement (Sagen *et al.,* 1993). Rats were tested using tail flick, paw pinch, and hot plate tests. Results demonstrated that antinociceptive effects of encapsulated bovine chromaffin cells were similar to previous studies with unencapsulated cells, were reproducible throughout the 8-week course of the study, and, like results from either rat adrenal medullary allografts or bovine chromaffin cell xenografts, the antinociception could be attenuated by opioid and α-adrenergic antagonists. Good survival of chromaffin cells in the capsules was confirmed by both neurochemical assays and immunocytochemistry in capsules retrieved from the spinal subarachnoid space at the end of behavioral analysis.

Studies have also suggested that encapsulated bovine chromaffin cells can also alleviate symptoms of chronic spinal cord injury pain in rats (Yu *et al.,* 1998b). A chronic allodynic state was produced using a tunable argon ion laser aimed at the T13 vertebral segment of rats injected with erythrosin B. Animals were implanted with either encapsulated bovine chromaffin cells or control capsules containing alginate matrix only. Using calibrated von Frey hairs to test vocalization thresholds to mechanical pressure and ethyl chloride spray to assess cold allodynia, chromaffin cell-containing capsules were found to totally reverse mechanical allodynia and attenuate cold allodynia, in contrast to control capsules. These effects were stable for at least 2 months postimplantation, suggesting that the encapsulated bovine chromaffin cells survive longer in the rat subarachnoid space than do unencapsulated bovine chromaffin cell suspensions in nonimmunosuppressed animals (Yu *et al.,* 1998a).

Based on the promising results obtained from animal models, clinical studies using adrenal medullary allografts were initiated in limited clinical trials. Initial clinical trials using human adrenal medullary allografts were conducted at the University of Illinois at Chicago in five patients suffering from severe pain secondary to inoperable cancer. Adrenal medullary tissue was obtained from adult adrenal glands via the Regional Organ Bank of Illinois following pathogen screening, and adrenal medullary tissue was dissected, cultured for 5–10 days, and verified for catecholamine synthesis. Tissue from approximately two adrenal glands was implanted in the subarachnoid space of patients via lumbar puncture using a 14-gauge Touhy needle. In follow-up, four of the patients had marked reductions in visual analog scale (VAS) scores and analgesic consumption within 2–8 weeks following adrenal medullary transplantation. Three of these patients remained free of pain throughout the remainder of their lives, two for approximately 1 year [detailed case histories of the patients can be found in Winnie *et al.* (1993)]. Using a similar protocol, Lazorthes *et al.* (1995) performed adrenal medullary allogeneic implants in seven patients suffering from intractable pain due to cancer. A multidisciplinary pain evaluation demonstrated progressively decreased pain scores in six of the patients and opioid analgesic intake was decreased in three of the patients, stabilized in two other patients, and increased in two patients. This study is ongoing and currently includes 15 patients with similar promising results (Y. Lazorthes, personal communication).

Phase I clinical trials have also been conducted at several centers using encapsulated bovine chromaffin cell implants (Aebischer *et al.*, 1994; Buchser *et al.*, 1996; Burgess *et al.*, 1996, 1999). Phase I clinical trials were conducted to assess safety and preliminary efficacy at the University of Lausanne, Switzerland, in seven patients with severe pain, using capsules containing approximately 2 million bovine chromaffin cells. Capsules were implanted under local anesthesia via lumbar puncture, with the active portion of the capsule containing the chromaffin cells in the lumbar cistern, and held in place via a silicone tether sutured to the lumbodorsal fascia. Except for occasional reports of postdural puncture headaches, no adverse effects from the procedure were noted. Of the seven patients, four who were receiving epidural morphine at the time of the implant were reported to have decreased opioid use following the implant, with either a modest improvement or no worsening in pain ratings. Three patients who were not receiving oral or epidural morphine treatment at the time of the implant reported improvement in pain ratings. Histologic examination of the spinal cords, when possible, revealed no remarkable pathology, and chromaffin cell viability was confirmed in retrieved capsules by catecholamine release and immunostaining.

A similar phase I clinical trial has been conducted in the United States (Burgess *et al.*, 1996, 1999). This study was conducted under a commercially sponsored (CytoTherapeutics, Inc.) investigational new drug (IND) application reviewed by the United States Food and Drug Administration and included 19 patients with intractable pain secondary to incurable malignancies and life expectancy less than 5 months. Initially, 15 patients received capsules containing 1×10^6 cells (5.0 cm in length), and an additional 4 patients received capsules containing 3×10^6 cells (7.0 cm in length). Participants selected for the study were advanced cancer patients with chronic pain inadequately relieved by conventional therapies. No serious complications directly attributable to the implant were noted and routine clinical pathology showed no evidence of adverse reactions. The few mild adverse experiences attributable to lumbar puncture, including postlumbar puncture headaches, subcutaneous fluid collections at the implant site, and subarachnoid–cutaneous fistula were easily resolved. Evidence of analgesic efficacy was suggested by reductions in pain scores in 9 patients, and opiate reduction in 8 of the original 15 patients (Burgess *et al.*, 1996). Mean VAS scores showed a gradual decline over time, with an approximate decline of 39% at 4 weeks postimplant. Of the patients exhibiting improved pain control, the majority had pain complaints that were localized to body regions innervated by lumbar and sacral nerves, suggesting that diffusion of pain-reducing neuroactive agents to higher levels may be limited.

Given the initial safety and potential efficacy findings using encapsulated bovine chromaffin cells, a large multicenter, placebo-controlled study in patients with cancer pain was initiated in Switzerland, the Czech Republic, and Poland. The initial evaluation of findings from this trial indicated no difference in pain reduction between encapsulated bovine chromaffin cell implants and placebo control capsules containing matrix only. Although the evaluation has not yet been completed sufficiently to distinguish pain localization (e.g., lumbosacral versus. cervicothoracic) or quality (e.g., neuropathic vs. nociceptive), findings of this study suggest that pain therapies using this approach may be limited by achievable doses provided by limited numbers of encapsulated cells and/or production of natural analgesic agents.

Another possible approach is the utilization of cell lines engineered to produce higher levels of natural or novel analgesic agents. Earlier studies have been reported using cell lines to produce opioid peptides or catecholamines. To provide opioid peptides, the AtT-20 cell line, derived from the mouse anterior pituitary, was implanted intrathecally around rat and mouse spinal cord (H. Wu *et al.*, 1993, 1994b). This cell line releases high levels of opioid peptide, β-endorphin. In addition, a genetically modified version of this cell line, AtT-20/hEnk, which has been transfected with the human proenkephalin gene), was implanted in some animals. This cell line secretes enkephalin in addition to β-endorphin. Neither of these cellular implants altered base-line responses to acute noxious thermal stimuli, but antinociception was rapidly induced by intrathecal injections of the β-adrenergic agonist isoproterenol, presumably via activation of AtT-20 cell surface β receptors. In another study, this group implanted catecholamine-secreting cell lines, derived from the B16 melanoma (H. Wu *et al.*, 1994a). Implantation of the F1C29 clone, which apparently releases higher levels of catecholamines, around the mouse spinal cord enhanced the antinociceptive effect of exogenously administered morphine to a noxious thermal stimulus. This effect could be blocked by α_2-adrenergic antagonists, suggesting mediation via release of catecholamines from implanted cells.

Encapsulated β-endorphin-secreting tumor cell lines implanted in the CSF of rats at the level of the atlanto–occipital junction has also been reported to reduce pain sensitivity as assessed by tail pinch, hot plate, and neuromuscular electrical stimulation (Saitoh *et al.*, 1995). Both the mouse neuroblastoma Neuro2A and the mouse pituitary AtT-20 cell lines transfected with the proopiomelanocortin (POMC) gene were effective, and the antinociception was attenuated by the opiate antagonist naloxone. Encapsulated cells actively secreted peptides and survived for at least 1 month following transplantation.

Another approach, which may address potential limitations of tumor cell lines such as risk of continuous unrestricted cell division, is the use of conditional immortalization strategies. Conditional immortalization is a means to generate an immortalized cell line that can later be disimmortalized to stop cell division. Using the temperature-sensitive mutant of the SV40 large T antigen (tsTag), transfected cells undergo continual cell division at low temperature conditions (e.g., 32–34°C), but differentiate and become postmitotic when the temperature is raised (37–39°C). Conditional immortalization with tsTag has been used to create rodent neuronal cell lines bioengineered with novel genes to deliver potentially antinociceptive molecules following grafting into the subarachnoid space (Eaton *et al.*, 1997, 1999). The transfected genes tested thus far include the synthetic enzymes for the neurotransmitters 5-HT and GABA, the preproenkephalin gene for metenkephalin synthesis, and the preprogalanin gene for the peptide galanin synthesis. Using the chronic constriction injury model of neuropathic pain, 1 million cells expressing bioengineered GABA or 5-HT were transplanted in the rat subarachnoid space at L4–L5 levels. The presence of grafted cells secreting antinociceptive molecules reversed the behavioral hypersensitivity within 1 week of grafting, and these effects lasted at least throughout the 8-week time course of the study. In contrast, the control vector-only cell lines, which do not express the novel genes but survive and differentiate under similar conditions, had no effect on the peripheral neuropathic pain.

In summary, during the past few years, the field of neural transplantation and development of potential strategies for the treatment of spinal cord dysfunction and injury have witnessed rapid growth. Only a short while ago, complications of spinal cord trauma and disease were thought to be irreparable, but recent breakthroughs have demonstrated the regenerative capacity in the adult spinal cord, and suggested that restoration of function may ultimately be attainable under the right circumstances. Numerous promising approaches have been described in this chapter, including the use of artificial and cellular bridges, provision of neutrophic factors, and replacement of specific cell populations. It is likely that the ultimately successful strategy will incorporate a combination of these approaches in the treatment of these debilitating disorders.

REFERENCES

Aebischer, P., Winn, S. R., and Galletti, P. M. (1988). Transplantation of neural tissue in polymer capsules. *Brain Res.* **448**, 364–368.

Aebischer, P., Buchser, E., Joseph, J. M., *et al.* (1994). Transplantation in humans of encapsulated xenogeneic cells without immunosuppression. *Transplantation* **58**, 1275–1277.

Aebischer, P., Schluep, M., Déglon, N., *et al.* (1996). Intrathecal delivery of CNTF using encapsulated genetically modified xenogeneic cells in amyotrophic lateral sclerosis patients. *Nat. Med.* **2**, 696–699.

Aguayo, A. J. (1985). Axonal regeneration from injured neurons in the adult mammalian central nervous system. *In* "Synaptic Plasticity" (C. W. Cotman, ed.), pp. 457–484. Guilford Press, New York.

Archer, D. R., Levén, S., and Duncan, I. D. (1994). Myelination by cryopreserved xenografts and allografts in the myelin-deficient rat. *Exp. Neurol.* **125**, 268–277.

Archer, D. R., Cuddon, P. A., Lipsitz, D., *et al.* (1997). Myelination of canine central nervous system by glial cell transplantation: A model for repair of human myelin disease. *Nat. Med.* **3**, 54–59.

Azmitia, E. C., and Björklund, A. (1987). Cell and tissue transplantation into the adult brain. *Ann. N.Y. Acad. Sci.* **495**, 813.

Baron-Van Evercooren, A., Gansmüller, A., Clerin, E., *et al.* (1991). Hoechst 33342 a suitable fluorescent marker for Schwann cells after transplantation in the mouse spinal cord. *Neurosci. Lett.* **131**, 241–244.

Baron-Van Evercooren, A., Duhamel-Clerin, E., Boutry, J. M., *et al.* (1993). Pathways of migration of transplanted Schwann cells in the demyelinated mouse spinal cord. *J. Neurosci. Res.* **35**, 428–438.

Bennett, G. J., and Xie, Y.-K. (1988). A peripheral mononeuropathy in rats that produces disorders of pain sensation like those seen in human. *Pain* **33**, 87–107.

Bernstein, J. J., and Goldberg, W. J. (1989). Rapid migration of grafted cortical astrocytes from suspension grafts placed in host thoracic spinal cord. *Brain Res.* **491**, 205–211.

Bernstein-Goral, H., Diener, P. S., and Bregman, B. S. (1997). Regenerating and sprouting axons differ in their requirements for growth after injury. *Exp. Neurol.* **148**, 51–72.

Bertelli, J. A., and Mira, J. C. (1994). Brachial plexus repair by peripheral nerve grafts directly into the spinal cord in rats. *J. Neurosurg.* **81**, 107–114.

Bertelli, J. A., Orsal, D., and Mira, J. C. (1994). Median nerve neurotization by peripheral nerve grafts directly implanted into the spinal cord: Anatomical, behavioral and electrophysiological evidences of sensorimotor recovery. *Brain Res.* **644**, 150–159.

Billinghurst, L. L., Taylor, R. M., and Snyder, E. Y. (1998). Remyelination: Cellular and gene therapy. *Semin. Pediatr. Neurol.* **5**, 211–228.

Blakemore, W. F. (1977). Remyelination of CNS axons by Schwann cells transplanted from the sciatic nerve. *Nature (London)* **266**, 68–69.

Blakemore, W. F. (1980). The effect of subdural nerve tissue transplantation on the spinal cord of the rat. *Neuropathol. Appl. Neurobiol.* **6**, 433–447.

Blakemore, W. F. (1983). Remyelination of demyelinated spinal cord axons by Schwann cells. *In* "Spinal Cord Reconstruction" (C. C. Kao, R. P. Bunge, and P. J. Reier, eds.), pp. 281–291. Raven Press, New York.

Blakemore, W. F. (1984). Limited remyelination of CNS axons by Schwann cells transplanted into the subarachnoid space. *J. Neurol. Sci.* **64**, 265–276.

Blakemore, W. F., and Crang, A. J. (1988). Extensive oligodendrocyte remyelination following injection of cultured central nervous system cells into demyelinating lesions in adult central nervous system. *Dev. Neurosci.* **18**, 519–528.

Blakemore, W. F., and Crang, A. J. (1989). The relationship between type-1 astrocytes, Schwann cells and oligodendrocytes following transplantation of glial cell cultures into demyelinating lesions in the adult rat spinal cord. *J. Neurocytol.* **18**, 519–528.

Blakemore, W. F., Crang, A. J., and Patterson, R. C. (1987). Schwann cell remyelination of CNS axons following injection of cultures of CNS cells into areas of persistent demyelination. *Neurosci. Lett.* **77**, 20–24.

Blight, A. R., and Young, W. (1989). Central axons in injured cat spinal cord recover electrophysiological function following remyelination by Schwann cells. *J. Neurol. Sci.* **91**, 15–34.

Bregman, B. S. (1987). Spinal cord transplants permit the growth of serotonergic axons across the site of neonatal spinal cord transection. *Dev. Brain Res.* **34**, 265–279.

Bregman, B. S. (1998). Regeneration in the spinal cord. *Curr. Opin. Neurobiol.* **8**, 800–807.

Bregman, B. S., and Bernstein-Goral, H. (1991). CNS transplants promote anatomical plasticity and recovery of function after spinal cord injury. *J. Restorative Neurol. Neurosci.* **2**, 327.

Bregman, B. S., and Kunkel-Bagden, E. (1994). Potential mechanisms underlying transplant mediated recovery of function after spinal cord injury. *In* "Neural Transplantation, CNS Neuronal Injury, and Regeneration" (J. Marweh, H. Teitelbaum, and K. N. Prasad, eds.), pp. 81–102. CRC Press, Boca Raton, FL.

Bregman, B. S., Kunkel-Bagden, E., Reier, P. J., *et al.* (1993). Recovery from spinal cord injury mediated by antibodies to neurite growth inhibitors. *Nature (London)* **378**, 498–501.

Bregman, B. S., McAtee, M., Dai, H. N., *et al.* (1997). Neurotrophic factors increase axonal growth after spinal cord injury and transplantation in the adult rat. *Exp. Neurol.* **148**, 475–494.

Bregman, B. S., Broude, E., McAtee, M., *et al.* (1998). Transplants and neurotrophic factors prevent atrophy of mature CNS neurons after spinal cord injury. *Exp. Neurol.* **149**, 13–27.

Brewer, K. L., and Yezierski, R. P. (1998). Effects of adrenal medullary transplants on pain-related behaviors following excitotoxic spinal cord injury. *Brain Res.* **798**, 83–92.

Broude, E., McAtee, M., Kelley, M. S., *et al.* (1999). Fetal spinal cord transplants and exogenous neurotrophic support enhance c-Jun expression in mature axotomized neurons after spinal cord injury. *Exp. Neurol.* **155**, 65–78.

Buchanan, J. T., and Nornes, H. O. (1986). Transplants of embryonic brainstem containing the locus coeruleus into spinal cord enhance the hindlimb flexion reflex in adult rats. *Brain Res.* **381**, 225–236.

Buchser, E., Goddard, M., Heyd, B., *et al.* (1996). Immunoisolated xenogeneic chromaffin cell therapy for chronic pain: Initial experience. *Anesthesiology* **85**, 1005–1012.

Bunge, M. B., and Kleitman, N. (1999). Neurotrophins and neuroprotection improve axonal regeneration into Schwann cell transplants placed in transected adult rat spinal cord. *In* "CNS Regeneration: Basic Science and Clinical Advances" (M. H. Tuszynski and J. Kordower, eds.), pp. 631–646. Academic Press, San Diego, CA.

Burgess, F. W., Goddard, M., Savarese, D., *et al.* (1996). Subarachnoid bovine adrenal chromaffin cell implants for cancer pain management. *Am. Pain Soc. Abstr.* **15**, A-33.

Burgess, F. W., Goddard, M., Lewis-Cullinan, C., *et al.* (1999). Bovine adrenal medullary cell implants for cancer pain management. *Pain* (submitted for publication).

Carlstedt, T., Grane, P., Hallin, R. G., *et al.* (1995). Return of function after spinal cord implantation of avulsed spinal nerve roots. *Lancet* **346**, 1323–1325.

Chen, A., Xu, X.-M., Kleitman, N., *et al.* (1996). Methylprednisolone administration improves axonal regeneration into Schwann cell grafts in transected adult rat thoracic spinal cord. *Exp. Neurol.* **138**, 261–276.

Cheng, H., Cao, Y., and Olson, L. (1996). Spinal cord repair in adult paraplegic rats: Partial restoration of hind limb function. *Science* **273**, 510–513.

Commissiong, J. W., and Sauvé, Y. (1989). The physiological basis of transplantation of fetal catecholaminergic neurons in the transected spinal cord of the rat. *Comp. Biochem. Physiol. A* **93A**, 301–307.

Connor, J. R., and Bernstein, J. J. (1987). Expression of peptides and transmitters in neurons and expression of filament proteins in astrocytes in fetal cerebral cortical transplants to adult spinal cord. *Prog. Brain Res.* **71**, 359–371.

Cullheim, S., Carlstedt, T., Linda, H., *et al.* (1989). Motoneurons reinnervate skeletal muscle after ventral root implantation into the spinal cord of the cat. *Neuroscience* **29**, 725–733.

David, S., and Aguayo, A. J. (1981). Axonal elongation into peripheral nervous system "bridges" after central nervous system injury in adult rats. *Science* **214**, 931–933.

Davies, S. J. A., Field, P. M., and Raisman, G. (1994). Long interfascicular axon growth from embryonic neurons transplanted into adult myelinated tracts. *J. Neurosci.* **14**, 1596–1612.

Décosterd, I., Buchser, E., Gilliard, N., *et al.* (1998). Intrathecal implants of bovine chromaffin cells alleviate mechanical allodynia in a rat model of neuropathic pain. *Pain* **76**, 159–166.

de la Torre, J. C. (1982). Catecholamine fiber regeneration across a collagen bioimplant after spinal cord transection. *Brain Res. Bull.* **9**, 545–552.

Demierre, B., Ruiz-Flandes, P., Martinou, J.-C., *et al.* (1990). Grafting of embryonic motoneurones into adult spinal cord and brain. *Prog. Brain Res.* **82**, 233–237.

Diener, P. S., and Bregman, B. S. (1998a). Fetal spinal cord transplants support the development of target reaching and coordinated postural adjustments after neonatal cervical spinal cord injury. *J. Neurosci.* **18**, 763–778.

Diener, P. S., and Bregman, B. S. (1998b). Fetal spinal cord transplants support growth of supraspinal and segmental projections after cervical spinal cord hemisection in the neonatal rat. *J. Neurosci.* **18**, 779–793.

Drasner, K., and Fields, H. F. (1988). Synergy between the antinociceptive effects of intrathecal clonidine and systemic morphine in the rat. *Pain* **32**, 309–312.

Dubois-Dalcq, M. (1995). Regeneration of oligodendrocytes and myelin. *Trends Neurosci.* **18**, 289–291.

Duncan, I. D., and Milward, E. A. (1995). Glial cell transplants: Experimental therapies of myelin diseases. *Brain Pathol.* **5**, 301–310.

Duncan, I. D., Aguayo, A. J., Bunge, R. P., *et al.* (1981). Transplantation of rat Schwann cells grown in tissue culture into the mouse spinal cord. *J. Neurol. Sci.* **49**, 241–252.

Duncan, I. D., Hammang, J. P., Jackson, K. F., *et al.* (1988). Transplantation of oligodendrocytes and Schwann cell into the spinal cord of the myelin-deficient rat. *J. Neurocytol.* **17**, 351–360.

Duncan, I. D., Paino, C., Archer, D. R., *et al.* (1992). Functional capacities of transplanted cell-sorted adult oligodendrocytes. *Dev. Neurosci.* **14**, 114–122.

Dunnett, S. B., and Richards, S. J. (1990). Neural transplantation: From molecular basis to clinical applications. *Prog. Brain Res.* **82**, 743.

Eaton, M. J., Dancausse, H. R., Santiago, D. I., *et al.* (1997). Lumbar transplants of immortalized serotonergic neurons alleviated chronic neuropathic pain. *Pain* **72**, 59–69.

Eaton, M. J., Plunkett, J. A., Martinez, M. A., *et al.* (1999). Transplants of neuronal cells bioengineered to synthesize GABA alleviate chronic neuropathic pain. *Cell Transplant.* **8**, 87–101.

Erb, D. E., Mora, R. J., and Bunge, R. P. (1993). Reinnervation of adult rat gastrocnemius muscle by embryonic motoneurons transplanted into the axotomized tibial nerve. *Exp. Neurol.* **124**, 372–376.

Espinosa de los Monteros, A., Zhao, P., Huang, C., *et al.* (1997). Transplantation of CG4 oligodendrocyte progenitor cells in the myelin-deficient rat brain results in myelination of axons and enhanced oligodendroglial markers. *J. Neurosci. Res.* **50**, 872–887.

Falci, S., Holtz, A., Akesson, E., *et al.* (1997). Obliteration of a posttraumatic spinal cord cyst with solid human embryonic spinal cord grafts: First clinical attempt. *J. Neurotrauma* **14**, 875–884.

Fanarraga, M. L., Griffiths, I. R., Zhao, M., *et al.* (1998). Oligodendrocytes are not inherently programmed to myelinate a specific size of axon. *J. Comp. Neurol.* **399**, 94–100.

Fawcett, J. W. (1998). Spinal cord repair: From experimental models to human application. *Spinal Cord* **36**, 811–817.

Feraboli-Lohnherr, D., Orsal, D., Yakovleff, A., *et al.* (1997). Recovery of locomotor activity in the adult chronic spinal rat after sublesional transplantation of embryonic nervous cells: Specific role of serotonergic neurons. *Exp. Brain Res.* **113**, 443–454.

Fleetwood, I. G., MacDonald, S. C., Sawchuk, M., *et al.* (1998). Survival and differentiation of spinal cord stem cells transplanted into transected sciatic nerves of adult mice. *Abstr. Soc. Neurosci.* **24**, 68.

Foster, G. A., Schultzberg, M., Gage, F. H., *et al.* (1985). Transmitter expression and morphological development of embryonic medullary and mesencephalic raphé neurones after transplantation to the adult rat central nervous system. *Exp. Brain Res.* **60**, 427–444.

Foster, G. A., Roberts, M. H. T., Wilkinson, L. S., *et al.* (1989). Structural and functional analysis of raphe neurone implants into denervated rat spinal cord. *Brain Res. Bull.* **22**, 131–137.

Freeman, T. B., and Widner, H., eds. (1998). "Cell Transplantation for Neurological Disorders: Toward Reconstruction of the Human Central Nervous System." Humana Press, Totowa, NJ.

Gash, D., and Sladek, J. R., Jr. (1988). Transplantation into the mammalian CNS. *Prog. Brain Res.* **78**, 663.

Gautier, S. E., Oudega, M., Fragoso, M., *et al.* (1998). Poly(α-hydroxyacids) for application in the spinal cord: Resorbability and biocompatibility with adult rat Schwann cells and spinal cord. *J. Biomed. Mater. Res.* **42**, 642–654.

Giménez y Ribotta, M., and Privat, A. (1998). Biological interventions for spinal cord injury. *Curr. Opin. Neurol.* **11**, 647–654.

Ginzburg, R., and Seltzer, Z. (1990). Subarachnoid spinal cord transplantation of adrenal medulla suppresses chronic neuropathic pain behavior in rats. *Brain Res.* **523**, 147–150.

Goldsmith, H. S., and de la Torre, J. C. (1992). Axonal regeneration after spinal cord transection and reconstruction. *Brain Res.* **589**, 217–224.

Gout, O., Gansmuller, A., Baumann, N., *et al.* (1988). Remyelination by transplanted oligodendrocytes of a demyelinated lesion in the spinal cord of the adult *shiverer* mouse. *Neurosci. Lett.* **87**, 195–199.

Grill, R., Murai, K., Blesch, A., *et al.* (1998). Cellular delivery of neurotrophin-3 promotes corticospinal axonal growth and partial functional recovery after spinal cord. *J. Neurosci.* **17**, 5560–5572.

Guénard, V., Xu, X.-M., and Bunge, M. B. (1993). The use of Schwann cell transplantation to foster central nervous system repair. *Neurosciences* **5**, 401–411.

Guest, J. D., Rao, A., Olson, L., *et al.* (1997). The ability of human Schwann cell grafts to promote regeneration in the transected nude rat spinal cord. *Exp. Neurol.* **148**, 502–522.

Gumpel, M., Gout, O., Lubetzki, C., *et al.* (1989). Myelination and remyelination in the central nervous system by transplanted oligodendrocytes using the *shiverer* model. *Dev. Neurosci.* **11**, 132–139.

Hains, B. C., Chastain, K. M., Everhart, A. W., *et al.* (1998). Reduction of chronic central pain following spinal cord injury by transplants of adrenal medullary chromaffin cells. *Abstr. Soc. Neurosci.* **24**, 1631.

Hama, A. T., and Sagen, J. (1993). Reduced pain-related behavior by adrenal medullary transplants in rats with experimental pain neuropathy. *Pain* **52**, 223–231.

Hama, A. T., and Sagen, J. (1994). Alleviation of neuropathic pain symptoms by xenogeneic chromaffin cell grafts in the spinal subarachnoid space. *Brain Res.* **651**, 183–193.

Hama, A. T., Pappas, G. D., and Sagen, J. (1996). Adrenal medullary implants reduce transsynaptic degeneration in the spinal cord of rats following chronic constriction nerve injury. *Exp. Neurol.* **137**, 81–93.

Hammang, J. P., Reynolds, B. A., Weiss S., *et al.* (1994). Transplantation of epidermal growth factor-responsive neural stem cell progeny into the murine central nervous system. *In* "Providing Pharmacological Access to the Brain, Methods in Neurosciences" (T. R. Flanagan, D. F. Emerich, and S. R. Winn, eds.), Vol. 21, pp. 281–293. Academic Press, San Diego, CA.

Hammang, J. P., Archer, D. R., and Duncan, I. D. (1997). Myelination following transplantation of EGF-responsive neural stem cells into a myelin-deficient environment. *Exp. Neurol.* **147**, 84–95.

Heumann, R., Korsching, S., Bandtlow, C., and Thoenen, H. (1987). Changes of nerve growth factor synthesis in nonneuronal cells in response to sciatic nerve transection. *J. Cell Biol.* **104**, 1623–1631.

Himes, B. T., Goldberger, M. E., and Tessler, A. (1994). Grafts of fetal central nervous system tissue rescue axotomized Clarke's nucleus neurons in adult and neonatal operates. *J. Comp. Neurol.* **339**, 117–131.

Horvat, J. C., Pecot-Dechavassine, M., Mira, J. C., *et al.* (1989). Formation of functional endplates by spinal axons regenerating through a peripheral nerve graft. A study in the adult rat. *Brain Res. bull.* **22**, 103–114.

Houle, J. D., and Johnson, J. E. (1989). Nerve growth factor (NGF)-treated nitrocellulose enhances and directs the regeneration of adult rat dorsal root axons through intraspinal neural tissue transplants. *Neurosci. Lett.* **103**, 17–23.

Houle, J. D., and Reier, P. J. (1988). Transplantation of fetal spinal cord into the chronically injured adult rat spinal cord. *J. Comp. Neurol.* **269**, 535–547.

Houle, J. D., and Reier, P. J. (1989). Regrowth of calcitonin gene-related peptide (CGRP) immunoreactive axons from the chronically injured rat spinal cord into fetal spinal cord tissue transplants. *Neurosci. Lett.* **103**, 253–258.

Houle, J. D., and Ziegler, M. K. (1994). Bridging a complete transection lesion of adult rat spinal cord with growth factor-treated nitrocellulose implants. *J. Neural Transplant. Plast.* **5**, 115–124.

Houweling, D. A., Lankhorst, A. J., Gispen, W. H., *et al.* (1998). Collagen containing neurotrophin-3 (NT-3) attracts regrowing injured corticospinal axons in the adult rat spinal cord and promotes partial functional recovery. *Exp. Neurol.* **153**, 49–59.

Ibuki, T., Hama, A. T., Wang, X.-T., *et al.* (1997). Loss of GABA immunoreactivity in the spinal dorsal horn of rats with peripheral nerve injury and promotion of recovery by adrenal medullary grafts. *Neuroscience* **76**, 845–858.

Imaizumi, T., Lankford, K. L., Waxman, S. G., *et al.* (1998). Transplanted olfactory ensheathing cells remyelinate and enhance axonal conduction in the demyelinated dorsal columns of the rat spinal cord. *J. Neurosci.* **18**, 6176–6185.

Itoh, Y., and Tessler, A. (1990). Regeneration of adult dorsal root axons into transplants of fetal spinal cord and brain—A comparison of growth and synapse formation in appropriate and inappropriate targets. *J. Comp. Neurol.* **302**, 272–293.

Itoh, Y., Tessler, A., Kowada, M., *et al.* (1993a). Electrophysiological responses in foetal spinal cord transplants evoked by regenerated dorsal root axons. *Acta Neurochir.* **58S**, 24–26.

Itoh, Y., Sugawara, T., Kowada, M., *et al.* (1993b). Time course of dorsal root axon regeneration into transplants of fetal spinal cord: An electron microscopic study. *Exp. Neurol.* **123**, 133–146.

Iwashita, Y., Kawaguchi, S., and Murata, M. (1994). Restoration of function by replacement of spinal cord segments in the rat. *Nature (London)* **367**, 167–170.

Jakeman, L. B., and Reier, P. J. (1991). Axonal projections between fetal spinal cord transplants and the adult rat spinal cord: A neuroanatomical tracing study of local interactions. *J. Comp. Neurol.* **307**, 311–334.

Jakeman, L. B., Reier, R. J., Bregman, B. S., *et al.* (1989). Differentiation of substantia gelatinosa-like regions in intraspinal and intracerebral transplants of embryonic spinal cord tissue in the rat. *Exp. Neurol.* **103**, 17–33.

Jeffery, N. D., Crang, A. J., O'Leary, M. T., *et al.* (1999). Behavioral consequences of oligodendrocyte progenitor cell transplantation into experimental demyelinating lesions in the rat spinal cord. *Eur. J. Neurosci.* **11**, 1508–1514.

Jirmanová, I., Lieberman, A. R., and Zelená, J. (1994). Reinnervation of Pacinian corpuscles by CNS axons after transplantation to the dorsal column: Incidence and ultrastructure. *J. Neurocytol.* **23**, 422–432.

Joosten, E. A., Bar, P. R., and Gispen, W. H. (1995). Collagen implants and cortico-spinal axonal growth after mid-thoracic spinal cord lesion in the adult rat. *J. Neurosci. Res.* **41**, 481–490.

Kikuchi, K., Tessler, A., and Kowada, M. (1993). Roles of embryonic astrocytes and Schwann cells in regeneration of adult rat dorsal root axons: Qualitative observations. *Neurol. Med. Chir. (Tokyo)* **33**, 682–690.

Kliot, M., Smith, G. M., Siegal, J., *et al.* (1988). Engineering the regeneration of sensory fibers into the spinal cord of adult mammals with embryonic astroglia coated millipore implants. *In* "Tissue Engineering" (R. Skalak and C. F. Fox, eds.), pp. 249–256. Alan R. Liss, New York.

Kliot, M., Smith, G. M., Siegal, J. D., *et al.* (1990). Astrocyte-polymer implants promote regeneration of dorsal root fibres into the adult mammalian spinal cord. *Exp. Neurol.* **109**, 57–69.

Kuhlengel, K. R., Bunge, M. B., Bunge, R. P., *et al.* (1990). Implantation of cultured sensory neurons and Schwann cells into lesioned neonatal rat spinal cord. II. Implant characteristics and examination of corticospinal tract growth. *J. Comp. Neurol.* **293**, 74–91.

Kunkel-Bagden, E., and Bregman, B. S. (1990). Spinal cord transplants enhance the recovery of locomotor function after spinal cord injury at birth. iExp. Brain Res. **81**, 25–34.

Langford, L. A., and Owens, G. C. (1990). Resolution of the pathway taken by implanted Schwann cells to a spinal cord lesion by prior infections with a retrovirus encoding β-galactosidase. *Acta Neuropathol.* **80**, 514–244.

Lazorthes, Y., Bès, J. C., Sagen, J., *et al.* (1995). Transplantation of human chromaffin cells for control of intractable cancer pain. *Acta Neurochir.* **64**, 97–100.

Lemaire, S., Shukla, V. J., Rogers, C., *et al.* (1993). Isolation and characterization of histogranin, a natural peptide with NMDA receptor antagonist activity. *Eur. J. Pharmacol.* **245**, 247–256.

Li, D. W., and Duncan, I. D. (1998). The immune status of the myelin deficient rat and its immune responses to transplanted allogeneic glial cells. *J. Neuroimmunol.* **85**, 202–211.

Li, Y., and Raisman, G. (1993). Long axon growth from embryonic neurons transplanted into myelinated tracts of the adult rat spinal cord. *Brain Res.* **629**, 115–127.

Li, Y., and Raisman, G. (1994). Schwann cells induce sprouting in motor and sensory axons in the adult rat spinal cord. *J. Neurosci.* **14**, 4050–4063.

Li, Y., Field, P. M., and Raisman, G. (1997). Repair of adult rat corticospinal tract by transplants of olfactory ensheathing cells. *Science* **277**, 2000–2002.

Li, Y., Field, P. M., and Raisman, G. (1998). Regeneration of adult rat corticospinal axons induced by transplanted olfactory ensheathing cells. *J. Neurosci.* **18**, 10514–10524.

Liu, Y., Kim, D., Himes, B. T., *et al.* (1999). Transplants of fibroblasts genetically modified to express BDNF promote regeneration of adult rat rubrospinal axons and recovery of forelimb function. *J. Neurosci.* **19**, 4370–4387.

Marchand, R., Woerly, S., Bertrand, L., *et al.* (1993). Evaluation of two cross-linked collagen gels implanted in the transected spinal cord. *Brain Res. Bull.* **30**, 415–422.

Martin, D., Schoenen, J., Delree, P., *et al.* (1991). Grafts of syngeneic, adult dorsal root ganglion-derived Schwann cells to the injured spinal cord of adult rats: Preliminary morphological studies. *Neurosci. Lett.* **124**, 44–48.

Martin, D., Schoenen, J., Delree, P., *et al.* (1993). Syngeneic grafting of adult rat DRG-derived Schwann cells to the injured spinal cord. *Brain Res. Bull.* **30**, 507–514.

McKay, R. (1997). Stem cells in the central nervous system. *Science* **276**, 66–71.

McTigue, D. M., Horner, P. J., Stokes, B. T., *et al.* (1998). Neurotrophin-3 and brain-derived neurotrophic factor induce oligodendrocyte proliferation and myelination of regenerating axons in the contused adult rat spinal cord. *J. Neurosci.* **18**, 5354–5365.

Menei, P., Montero-Menei, C., Whittemore, S. R., *et al.* (1998). Schwann cells genetically modified to secrete human BDNF promote enhanced axonal regrowth across transected adult rat spinal cord. *Eur. J. Neurosci.* **10**, 607–621.

Milward, E. A., Lundberg, C. G., Ge, B., *et al.* (1997). Isolation and transplantation of multipotential populations of epidermal growth factor-responsive, neural progenitor cells from the canine brain. *J. Neurosci. Res.* **50**, 862–871.

Miya, D., Giszter, S., Mori, F., *et al.* (1997). Fetal transplants alter the development of function after spinal cord transection in newborn rats. *J. Neurosci.* **17**, 4856–4872.

Moorman, S. J., Whalen, L. R., and Nornes, H. O. (1990). A neurotransmitter specific functional recovery mediated by fetal implants in the lesioned spinal cord of the rat. *Brain Res.* **508**, 194–198.

Nógrádi, A., and Vrbová, G. (1994). The use of embryonic spinal cord grafts to replace identified motoneuron pools depleted by a neurotoxic lectin, volkensin. *Exp. Neurol.* **129**, 130–141.

Nógrádi, A., and Vrobová, G. (1996). Improved motor function of denervated rat hindlimb muscles induced by embryonic spinal cord grafts. *Eur. J. Neurosci.* **8**, 2198–2203.

Nornes, H. O., Buchanan, J., and Björklund, A. (1988). Noradrenaline-containing transplants in the adult spinal cord of mammals. *Prog. Brain Res.* **78**, 181–186.

Nothias, S. F., and Peschanski, M. (1990). Homotypic fetal transplants into an experimental model of spinal cord neurodegeneration. *J. Comp. Neurol.* **301**, 520–534.

Nygren, L.-G., Olson, L., and Seiger, A. (1977). Monoaminergic reinnervation of transected spinal cord by homologous fetal brain grafts. *Brain Res.* **129**, 227–235.

Ortega, J., Sagen, J., and Pappas, G. D. (1992). Short-term immunosuppression enhances long-term survival of bovine chromaffin cell xenografts in rat CNS. *Cell Transplant.* **1**, 33–41.

Ortega-Alvaro, A., Gibert-Rahola, J., Mellado-Fernández, M. L., *et al.* (1997). The effects of different monoaminergic antidepressants on the analgesia induced by spinal cord adrenal medullary transplants in the formalin test in rats. *Anesth. Analg. (Cleveland)* **84**, 816–820.

Oudega, M., and Hagg, T. (1996). Nerve growth factor promotes regeneration of sensory axons into adult rat spinal cord. *Exp. Neurol.* **140**, 218–229.

Oudega, M., and Hagg, T. (1999). Neurotrophins promote regeneration of sensory axons in the adult rat spinal cord. *Brain Res.* **818**, 431–438.

Oudega, M., Varon, S., and Hagg, T. (1994). Regeneration of adult rat sensory axons into intraspinal nerve grafts: Promoting effects of conditioning lesion and graft predegeneration. *Exp. Neurol.* **129**, 194–206.

Paino, C. L., and Bunge, M. B. (1991). Induction of axon growth into Schwann cell implants grafted into lesioned adult spinal cord. *Exp. Neurol.* **114**, 254–257.

Paino, C. L., Fernandez-Valle, C., Bates, M. L., *et al.* (1994). Regrowth of axons in lesioned adult rat spinal cord: Promotion by implants of cultured Schwann cells. *J. Neurocytol.* **23**, 433–452.

Pallini, R., Fernandez, E., Minciacchi, D., *et al.* (1988). Peripheral nerve autografts to the rat spinal cord: A study of the origin of regenerating fibres using fluorescent double labeling. *Acta Neurochir.* **43S**, 210–213.

Pallini, R., Fernandez, E., Gangitano, C., *et al.* (1989). Studies on embryonic transplants to the transected spinal cord of adult rats. *J. Neurosurg.* **70**, 454–462.

Patel, U., and Bernstein, J. J. (1983). Growth, differentiation, and viability of fetal rat cortical and spinal cord implants into adult rat spinal cord. *J. Neurosci. Res.* **9**, 303–310.

Peschanski, M. (1993). Spinal cord transplantation. *J. Neural Transplant. Plast.* **4**, 109–111.

Peschanski, M., Nothias, F., and Cadusseau, J. (1992). Is there a therapeutic potential for intraspinal transplantation of fetal spinal neurons in motoneuronal disease? *J. Restorative Neurol. Neurosci.* **4**, 227.

Plant, G. W., Harvey, A. R., and Chirila, T. V. (1995). Axonal growth within poly(2-hydroxyethyl methacrylate) sponges infiltrated with Schwann cells and implanted into the lesioned rat optic tract. *Brain Res.* **671**, 119–130.

Plant, G. W., Woerly, S., and Harvey, A. R. (1997). Hydrogels containing peptide or aminosugar sequences implanted into the rat brain: Influence on cellular migration and axonal growth. *Exp. Neurol.* **143**, 287–299.

Plant, G. W., Chirila, T. V., and Harvey, A. R. (1998). Implantation of collagen IV/poly(2-hydroxyethyl methacrylate) hydrogels containing Schwann cells into the lesioned rat optic tract. *Cell Transplant.* **7**, 381–391.

Plant, G. W., Ramon-Cueto, A., and Bunge, M. B. (2000). Transplantation of Schwann cells and ensheathing glia to improve regeneration in adult spinal cord. *In* "Nerve Regeneration" (N. A. Ingoglia and M. Murray, eds.). Dekker, New York (in press).

Privat, A., Mansour, H., and Geffard, M. (1988). Transplantation of fetal serotonin neurons into the transected spinal cord of adult rats: Morphological development and functional influence. *Prog. Brain Res.* **78**, 155–166.

Privat, A., Mansour, H., Rajaofertra, N., *et al.* (1989). Intraspinal transplants of serotonergic neurons in the adult rat. *Brain Res. Bull.* **22**, 123–129.

Pulford, B. E., Mihajlov, A. R., Nornes, H. O., *et al.* (1994). Effects of cultured adrenal chromaffin cell implants on hindlimb reflexes of the 6-OHDA lesioned rat. *J. Neural. Transplant. Plast.* **5**, 89–102.

Ramón-Cueto, A., Plant, G. W., Avila, J., *et al.* (1998). Long-distance axonal regeneration in the transected adult rat spinal cord is promoted by olfactory ensheathing glia transplants. *J. Neurosci.* **18**, 3803–3815.

Rapalino, O., Lazarov-Spiegler, O., Agranov, E., *et al.* (1998). Implantation of stimulated homologous macrophages results in partial recovery of paraplegic rats. *Nat. Med.* **4**, 814–821.

Reier, P. J. (1986). Gliosis following CNS injury: The anatomy of astrocytic scars and their influences on axonal elongation. *In* "Astrocytes" (S. Federoff and A. Vernadakis, eds.), Vol. 3, pp. 163–196. Raven Press, New York.

Reier, P. J., Bregman, B. S., and Wujek, J. R. (1986). Intraspinal transplantation of embryonic spinal cord tissue in neonatal and adult rats. *J. Comp. Neurol.* **247**, 275–296.

Reier, P. J., Stokes, B. T., Thompson, F. J., *et al.* (1992a). Fetal cell grafts into resection and contusion/compression injuries of the rat and cat spinal cord. *Exp. Neurol.* **115**, 177–188.

Reier, P. J., Anderson, D. K., Thompson, F. J., *et al.* (1992b). Neural tissue transplantation and CNS trauma: Anatomical and functional repair of the injured spinal cord. *J. Neurotrauma* **9**(1), S223–S248.

Reynolds, B. A., and Weiss, S. (1992). Generation of neurons and astrocytes from isolated cells of the adult mammalian central nervous system. *Science* **255**, 1707–1710.

Rhrich-Haddout, F., Horvat, J.-C., Baillet-Derbin, C., *et al.* (1994). Expression of peripherin in solid transplants of foetal spinal cord and dorsal root ganglia grafted to the injured spinal cord. *Neurosci. Lett.* **170**, 59–62.

Richardson, P. M., and Lu, X. (1994). Inflammation and axonal regeneration. *J. Neurol.* **242**, S57–S60.

Richardson, P. M., McGuinness, U. M., and Aguayo, A. J. (1980). Axons from CNS neurones regenerate into PNS grafts. *Nature (London)* **284**, 264–265.

Rogister, B., Belachew, S., and Moonen, G. (1999). Oligodendrocytes: From development to demyelinated lesion repair. *Acta Neurol. Belg.* **99**, 32–39.

Rosenbluth, J., Hasegawa, M., and Schiff, R. (1989). Myelin formation in myelin-deficient rat spinal cord following transplantation of normal spinal cord. *Neurosci. Lett.* **97**, 35–40.

Rosenbluth, J., Liu, Z., Guo, D., *et al.* (1993). Myelin formation by mouse glia in myelin-deficient rats treated with cyclosporin. *J. Neurocytol.* **22**, 967–977.

Sagen, J. (1998). Transplantation strategies for the treatment of pain. *In* "Cell Transplantation for Neurological Disorders: Toward Reconstruction of the Human Central Nervous System" (T. B. Freeman and H. Widner, eds.), pp. 231–251. Humana Press, Totowa, NJ.

Sagen, J., and Kemmler, J. (1989). Increased levels of met-enkephalin-like immunoreactivity in the spinal cord CSF of rats with adrenal medullary transplants. *Brain Res.* **502**, 1–10.

Sagen, J., Pappas, G. D., and Perlow, M. J. (1986a). Adrenal medullary transplants in the rat spinal cord reduce pain sensitivity. *Brain Res.* **384**, 189–194.

Sagen, J., Pappas, G. D., and Pollard, H. (1986b). Analgesia induced by isolated bovine chromaffin cells implanted in rat spinal cord. *Proc. Natl. Acad. Sci. U.S.A.* **83**, 7552–7526.

Sagen, J., Wang, H., and Pappas, G. D. (1990). Adrenal medullary tissue implants in rat spinal cord reduce nociception in a chronic pain model. *Pain* **42**, 69–79.

Sagen, J., Kemmler, J., and Wang, H. (1991). Adrenal medullary transplants increase spinal cord cerebrospinal fluid catecholamine levels and reduce pain sensitivity. *J. Neurochem.* **56**, 623–627.

Sagen, J., Wang, H., Tresco, P., *et al.* (1993). Transplants of immunologically isolated xenogeneic chromaffin cells provide a long-term source of pain-reducing neuroactive substances. *J. Neurosci.* **13**, 2415–2423.

Sagot, Y., Tan, S. A., Baetge, E., *et al.* (1995). Polymer encapsulated cell lines genetically engineered to release ciliary neurotrophic factor can slow down progressive motor neuronopathy in the mouse. *Nat. Med.* **2**, 696–699.

Saitoh, Y., Taki, T., Arita, N., *et al.* (1995). Analgesia induced by transplantation of encapsulated tumor cells secreting β-endorphin. *J. Neurosurg.* **82**, 630–634.

Schnell, L., Schneider, R., Kolbeck, R., *et al.* (1994). Neurotrophin-3 enhances sprouting of corticospinal tract during development and after adult spinal cord lesion. *Nature (London)* **367**, 170–173.

Schreyer, D. J., and Jones, E. G. (1983). Growing corticospinal axons by-pass lesions of neonatal rat spinal cord. *Neuroscience* **9**, 31–40.

Schwab, M. E., and Bartholdi, D. (1996). Degeneration and regeneration of axons in the lesioned spinal cord. *Physiol. Rev.* **76**, 319–370.

Sendtner, M., Schmalbruch, H., Stockli, K. A., *et al.* (1992). Ciliary neurotrophic factor prevents degeneration of motor neurons in mouse mutant neuronopathy. *Nature (London)* **358**, 502–504.

Sesodia, S., Grumbles, R. M., Erb, D. E., *et al.* (1999). Motor unit and muscle fiber characteristics after reinnervation by E14–15 ventral spinal cord cells. *Abstr. Soc. Neurosci.* **25**, 1149.

Sherman, S. E., Loomis, C. W., Milne, B., *et al.* (1988). Intrathecal oxymetazoline produces analgesia via spinal alpha-adrenoceptors and potentiates spinal morphine. *Eur. J. Pharmacol.* **148**, 371–380.

Siegan, J. B., and Sagen, J. (1997). Attenuation of formalin pain responses in the rat by adrenal medullary transplants in the spinal subarachnoid space. *Pain* **70**, 279–285.

Siegan, J. B., Hama, A. T., and Sagen, J. (1997). Suppression of neuropathic pain by a naturally-derived peptide with NMDA antagonist activity. *Brain Res.* **755**, 331–334.

Sieradzan, K., and Vrbová, G. (1989). Replacement of missing motoneurons by embryonic grafts in the rat spinal cord. *Neuroscience* **31**, 115–130.

Sieradzan, K., and Vrbová, G. (1993). Observations on the survival of grafted embryonic motoneurons in the spinal cord of developing rats. *Exp. Neurol.* **122**, 223–231.

Sotelo, C., and Alvarado-Mallart, R. M. (1987). Reconstruction of the defective cerebellar circuitry in the adult Purkinje cell degeneration mutant mice by Purkinje cell replacement through transplantation of solid embryonic implants. *Neuroscience* **20**, 1–22.

Steeves, J. D., and Tetzlaff, W. (1998). Engines, accelerators, and brakes on functional spinal cord repair. *Ann. N.Y. Acad. Sci.* **860**, 412–424.

Stokes, B. T., and Reier, P. J. (1991). Oxygen transport in intraspinal fetal grafts—Graft–host relations. *Exp. Neurol.* **111**, 312–323.

Stokes, B. T., and Reier, P. J. (1992). Fetal grafts alter chronic behavioral outcome after contusion damage to the adult rat spinal cord. *Exp. Neurol.* **116**, 1–12.

Sugimoto, T., Bennett, G. J., and Kajander, K. C. (1990). Transsynaptic degeneration in the superficial dorsal horn after sciatic nerve injury: Effects of chronic constriction injury, transection, and strychnine. *Pain* **42**, 205–213.

Tessler, A., Himes, B. T., Itoh, Y., *et al.* (1992). Transplant mediated mechanisms of recovery. *J. Restorative Neurol. Neurosci.* **4**, 226.

Thomas, C. K., Erb, D. E., and Bunge, R. P. (1998). Muscle function restored by transplantation of embryonic ventral spinal cord in peripheral nerve. *Exp. Neurol.* **153**, 383.

Thompson, F. J., Reier, P. J., Parmer, R., *et al.* (1993). Inhibitory control of reflex excitability following contusion injury and neural tissue transplantation. *Adv. Neurol.* **59**, 175–184.

Tontsch, U., Archer, D. R., Dubois-Dalcq, M., *et al.* (1994). Transplantation of an oligodendrocyte cell line leading to extensive remyelination. *Proc. Natl. Acad. Sci. U.S.A.* **91**, 11616–11620.

Trotter, J., Crang, A. J., Schachner, M., *et al.* (1993). Lines of glial precursor cells immortalised with a temperature-sensitive oncogene give rise to astrocytes and oligodendrocytes following transplantation into demyelinated lesions in the central nervous system. *Glia* **9**, 25–40.

Tuszynski, M. H., Peterson, D. A., Ray, J., *et al.* (1994). Fibroblasts genetically modified to produce nerve growth factor induce robust neuritic ingrowth after grafting to the spinal cord. *Exp. Neurol.* **126**, 1–14.

Tuszynski, M. H., Weidner, N., McCormack, M., *et al.* (1998). Grafts of genetically modified Schwann cells to the spinal cord: Survival, axon growth, and myelination. *Cell Transplant.* **7**, 187–196.

Unsicker, K. (1993). The trophic cocktail made by adrenal chromaffin cells. *Exp. Neurol.* **123**, 167–173.

Utzschneider, D. A., Archer, D. R., Kocsis, J. D., *et al.* (1994). Transplantation of glial cells enhances action potential conduction of amyelinated spinal cord axons in the myelin-deficient rat. *Proc. Natl. Acad. Sci. U.S.A.* **91**, 53–57.

Vaquero, J., Martinez, R., Oya, S., *et al.* (1988). Transplantation of adrenal medulla into spinal cord for pain relief: Disappointing outcome. *Lancet* **12**(3), 1315.

Vaquero, J., Arias, A., Oya, S., *et al.* (1991). Chromaffin cell allografts into the arachnoid of spinal cord reduce basal pain responses in rats. *NeuroReport* **2**, 149–151.

Vrbová, G., Clowry, G., Nógrádi, A., and Sieradzan, K., eds. (1994). "Transplantation of Neural Tissue into the Spinal Cord," p. 132. R. G. Landes, Austin, TX.

Waldrope, J., and Wilson, D. H. (1986). Peripheral nerve grafting in the spinal cord: A histological and electrophysiological study. *Paraplegia* **24**, 370–378.

Wang, H., and Sagen, J. (1995). Attenuation of pain-related hyperventilation in adjuvant arthritic rats with adrenal medullary transplants in the spinal subarachnoid space. *Pain* **63**, 313–320.

Wilcox, G. L., Carlsson, K.-H., Jochim, A., *et al.* (1987). Mutual potentiation of antinociceptive effects of morphine and clonidine on motor and sensory responses in rat spinal cord. *Brain Res.* **405**, 84–93.

Wilson, D. H. (1984). Peripheral nerve implants in the spinal cord in experimental animals. *Paraplegia* **22**, 230–237.

Wilson, D. H. (1991). Anatomical and physiological assessments of peripheral nerve grafts in the dorsal columns of the spinal cord. *J. Restorative Neurol. Neurosci.* **2**, 251–254.

Winnie, A., Pappas, G. D., Gupta, T. K., *et al.* (1993). Subarachnoid adrenal medullary transplants for terminal cancer pain. *Anesthesiology* **79**, 644–653.

Wirth, E. D., III, Theele, D. P., Mareci, T. H., *et al.* (1992). *In vivo* magnetic resonance imaging of fetal cat neural tissue transplants in the adult cat spinal cord. *J. Neurosurg.* **76**, 261–274.

Wirth, E. D., III, Fessler, R. G., Reier, P. J., *et al.* (1998). Feasibility and safety of neural tissue transplantation in patients with syringomyelia. *Abstr. Soc. Neurosci.* **24**, 70.

Wirth, E. D., III, Fessler, R. G., Reier, P. J., *et al.* (1999). Neural tissue transplantation in patients with syringomyelia: Update on feasibility and safety. *Am. Soc. Neural Transplant. Repair, Abstr.* **5/6**, 22.

Woerly, S., Plant, G. W., and Harvey, A. R. (1996). Cultured rat neuronal and glial cells entrapped within hydrogel polymer matrices: A potential tool for neural tissue replacement. *Neurosci. Lett.* **205**, 197–201.

Wrathall, R., Kapoor, V., and Kao, C. C. (1984). Observation of cultured peripheral nonneuronal cells implanted into the transected spinal cord. *Acta Neuropathol.* **64**, 203–212.

Wu, H., McLoon, S. C., and Wilcox, G. (1993). Antinociception following implantation of mouse pituitary AtT-20 cells and genetically modified AtT-20/hEnk cells in rat spinal cord. *J. Neural Transplant. Plast.* **4**, 15–26.

Wu, H., Lester, B., Sun, Z., *et al.* (1994a). Antinociception following implantation of mouse B16 melanoma cells in mouse and rat spinal cord. *Pain* **56**, 203–210.

Wu, H., Wilcox, G., and McCloon, S. (1994b). Implantation of AtT-20 or genetically modified AtT-20/hENK cells in mouse spinal cord induced antinociception and opioid tolerance. *J. Neurosci.* **14**, 4806–4814.

Wu, W., Han, K., Li, L., *et al.* (1994). Implantation of PNS graft inhibits the induction of neuronal nitric oxide synthase and enhances the survival of spinal motoneurons following root avulsion. *Exp. Neurol.* **129**, 335–339.

Xu, X.-M., Guénard, V., Kleitman, N., *et al.* (1995a). Axonal regeneration into Schwann cell-seeded guidance channels grafted into transected adult rat spinal cord. *J. Comp. Neurol.* **351**, 145–160.

Xu, X.-M., Guénard, V., Kleitman, N., *et al.* (1995b). A combination of BDNF and NT-3 promotes supraspinal axonal regeneration into Schwann cell grafts in adult rat thoracic spinal cord. *Exp. Neurol.* **134**, 261–272.

Xu, X.-M., Chen, A., Guénard, V., *et al.* (1997). Bridging Schwann cell transplants promote axonal regeneration from both the rostral and caudal stumps of transected adult rat spinal cord. *J. Neurocytol.* **26**, 1–16.

Xu, X.-M., Zhang, S.-X., Li, H., *et al.* (1999). Regrowth of axons into the distal spinal cord through a Schwann-cell-seeded mini-channel implanted into hemisected adult rat spinal cord. *Eur. J. Neurosci.* **11**, 1723–1740.

Yakovleff, A., Roby-Brami, A., Guezard, B., *et al.* (1989). Locomotion in rats transplanted with noradrenergic neurons. *Brain Res. bull.* **22**, 112–121.

Yakovleff, A., Cabelguen, J. M., Oral, D., *et al.* (1995). Fictive motor activities in adult chronic spinal rats transplanted with embryonic brainstem neurons. *Exp. Brain Res.* **106**, 69–78.

Yaksh, T., and Reddy, S. V. R. (1981). Studies in the primate on the analgetic effects associated with intrathecal actions of opiates, alpha-adrenergic agonists and baclofen. *Anesthesiology* **54**, 451–467.

Yandava, B. D., Billinghurst, L. L., and Snyder, E. Y. (1999). "Global" cell replacement is feasible via neural stem cell transplantation: Evidence from the dysmyelinated *shiverer* mouse brain. *Proc. Natl. Acad. Sci. U.S.A.* **96**, 7029–7034.

Ye, J. H., and Houle, J. D. (1997). Treatment of chronically injured spinal cord with neurotrophic factors can promote axonal regeneration from supraspinal neurons. *Exp. Neurol.* **143**, 70–81.

Yezierski, R. P., Liu, S., Ruenes, G. L., *et al.* (1998). Behavioral and pathological characteristic of a central pain model following spinal cord injury. *Pain* **75**, 141–155.

Yu, W., Hao, J.-X., Xu, X.-J., *et al.* (1998a). Immunoisolating encapsulation of intrathecally implanted bovine chromaffin cells prolongs their survival and produces anti-allodynic effect in spinally injured rats. *Eur. J. Pain* **2**, 143–151.

Yu, W., Hao, J.-X., Xu, X.-J., *et al.* (1998b). Long-term alleviation of allodynia-like behaviors by intrathecal implantation of bovine chromaffin cells in rats with spinal cord injury. *Pain* **74**, 115–122.

Zhang, S. C., Ge, B., and Duncan, I. D. (1999). Adult brain retains the potential to generate oligodendroglial progenitors with extensive myelination capacity. *Proc. Natl. Acad. Sci. U.S.A.* **96**, 4089–4094.

CHAPTER 58

NEURAL STEM CELLS

Martin P. Vacanti

INTRODUCTION

The general medical doctrine has been that damage to the central nervous system, including the brain and spinal cord, is irreversible. As stated by Valentini and Aebischer (1997), "a major impediment to healing is the inability of adult neurons to proliferate *in vivo* and to be cultivated *in vitro*." The emergence of neural stem cell biology in the adult mammal has broken this barrier. The combination of neural stem cell biology with the applied principles of tissue engineering will revolutionize the medical approach to the treatment of damaged central nervous system tissue, including injuries or diseases of both the brain and the spinal cord. Additionally, current strategies using gene insertion techniques to replace a missing neural product may be enhanced when these corrected cells are delivered using tissue engineering techniques.

BACKGROUND

Neural stem cells qualify as an ideal cell for use in repair strategies of the central nervous system, including the spinal cord. Many of the characteristics that these unique cells possess make them ideal candidates for the creation of functional central nervous system constructs. In the adult mammal these stem cells lay dormant in a quiescent state, but have the capacity to proliferate, producing mature cells with the properties of the damaged or injured tissue (Hines, 1997). As defined by McKay (1997), a neural stem cell must have the capacity to differentiate into the three major cell types of the central nervous system, neurons, astrocytes, and oligodendrocytes. Additionally, the term "progenitor cell," compared with a stem cell, is defined as a cell that has a more restricted potential. A cell isolated from the spinal cord may be restricted to lineage in that it may have the capacity to differentiate only into an oligodendrocyte. In many cases it is impossible to determine if one is dealing with a stem cell or a progenitor cell. To add confusion to the issue, the ultimate fate of a cell may depend on the cell's environment, whether it be *in vivo* or *in vitro*, and the presence or absence of particular cytokines, or combination of cytokines, or even the particular time that the cell is exposed to cytokines. For simplicity's sake, undifferentiated neural cells with the capacity to differentiate into neurons, astrocytes, and oligodendrocytes will be referred to as stem cells. In reality, these cells may actually be a continuum of cell types, from the pure stem cell to the committed progenitor cell.

Reynolds and Weiss (1992) isolated cells from the striatum of the adult mouse. They demonstrated that these cells could be stimulated to proliferate *in vitro* by use of epidermal growth factor (EGF). Before differentiating, these cells expressed nestin, a marker for neuroepithelial stem cells (Lendahl *et al.*, 1990). As the cells divided, they formed clusters of circular cells described as neurospheres. If these cells were allowed to attach to poly(L-ornithine)-coated glass cover slips, they would differentiate into various cells, some with the morphologic characteristics of astrocytes and some with the characteristics of neurons. Immunocytochemistry demonstrated cells with gliofibrillary acidic protein (GFAP) and cells with neuron-specific enolase (NSE), markers of astrocytes and neurons, respectively. If these neurospheres were disassociated and regrown in culture media containing EGF, they would continue to divide and would test positive for the intermediate filament nestin, indicating an undifferentiated state. Additionally, these secondarily derived cells would again differentiate into both astrocytes and neurons if attachment was allowed.

Additional work by Reynolds and Weiss (1996) established stem cell characteristics in cells that were derived from striatum removed from embryonic day 14 mice. Specifically, these cells were EGF responsive and they were able to differentiate into neurons, astrocytes, and oligodendrocytes. When cultured with EGF these cells were passaged 10 times with a billionfold increase in their cell number and still retained their proliferative and multilineage potential. Markers such as β-tubulin, microtubule-associated protein-2 (MAP-2), and neurofilament (NF) indicated neuronal differentiation. Finally, the marker O4 demonstrated the capacity of these stem cells to differentiate into oligodendrocytes.

Gage *et al.* (1995b) isolated fibroblast growth factor-2 (FGF-2)-responsive cells from the hippocampus of the adult rat. In culture these cells were positive for both glia and neuronal cell markers. It was noted on implantation into the adult rat hippocampus that the cells migrated and differentiated both into neurons and into glia. It was also observed that some of these implanted cells were ensheathed by several layers of myelin, indicating maturation. Gage suggested that these hippocampal-derived cells responded to endogenous cues that regulate neurogenesis. This concept is critical in regard to application of tissue engineering practices in the context of implantation of stem cells suspended on a biodegradable scaffolding material: growth, development, and differentiation occur in response to the particular environmental cues at the site of implantation.

Additional work of Weiss *et al.* (1996) demonstrated the presence and responsiveness of adult thoracic spinal cord neural stem cells that were induced to proliferate, self-renew, and expand in the presence of EGF and basic fibroblast growth factor (bFGF). These cells were isolated in all regions of the spinal cord, including the thoracic and lumbosacral areas. They demonstrated stem cell properties in regard to their proliferation and conversion to neurospheres, retaining the capacity to differentiate into all three neural cell lines. Attachment as previously described drove these stem cells toward maturation.

Shihabuddin *et al.* (1997) isolated what appeared to be the same cell type but in the adult rat spinal cord. Essentially, cells were isolated from all regions of the spinal cord and differentiated into neuronal, astroglial, and oligodendrocytic cell lines. Additionally, some large neuronlike cells expressed markers for p75NGFr, possibly implicating them as motor neurons. Neural stem cells seem to be ubiquitous in the central nervous system, including the spinal cord, striatum, and even the cortex.

Davis and Temple (1994) demonstrated neuroectodermal cells in the cortical ventricular zone that had the capacity to generate into the three major cells types of the central nervous system. In the embryonic rat they found that these cells maintained the ventricular zone and were responsible for creation of the subventricular zone that persists into adulthood. They felt that multipotential neural stem cells could be the ancestors of other cortical progenitor cells. Morshead *et al.* (1994) isolated adult mouse neural stem cells that resided in the subependymal area of the lateral ventricle. Undifferentiated, these cells expressed nestin but had the capacity to differentiate into both neurons and glia. The work of Craig *et al.* (1996) demonstrated an endogenous subependymal neural cell population in the adult mouse brain. They demonstrated that EGF stimulated proliferation, and with its withdrawal, these cells were found to differentiate into neurons, astrocytes, and oligodendrocytes. Johansson *et al.* (1999) presented evidence that ependymal cells are neural stem cells, capable of giving rise to a rapidly proliferative cell type that differentiates into neurons and migrates to the olfactory bulb. They also note that after spinal cord injury these same cells proliferate and differentiate into astrocytes. Bjornson *et al.* (1999) demonstrated the versatility of neural stem cells when they are engrafted into irradiated BALB/c mice. With genetic markers these cells differentiated into a variety of blood cell types, including myeloid, lymphoid, and early hematopoietic cells.

Taylor and Snyder (1997) demonstrated successful engraftment into most areas of the brain of what they described as neural progenitor cells, introduced by injection into the lateral ventricle. These cells were derived from the external germinal zone of neonatal mouse cerebellum and were immortalized with the v-*myc* oncogene. They also discuss the potential of those cells for gene product delivery. Shihabuddin *et al.* (1995), using an immortalized cell line, demonstrated differentiation into cells similar to both hippocampus and cerebral cortex cells following transplantation of the cell line into these respective areas in the adult Lewis rat. They discuss the plasticity of the cells in response to local microenvironmental signals. A key concept of this study is that the adult mammalian brain has the capacity to guide differentiation of neuronal cells such that they are similar to neurons endogenous to the area of implantation.

Knowledge derived from studies of neural stem cells is increasing rapidly. Included in such studies are immortalized cell lines that are generally isolated from embryonic central nervous system (CNS) using nontransforming oncogenes such as v-*myc* or the temperature-sensitive T antigen of SV40 (Lundberg *et al.*, 1996). Although much information regarding the basic behavior of neural stem cells will be obtained from these studies, at present we are reluctant to use these cell lines in neural tissue engineering applications because of obvious concerns about conversion to malignancy.

In terms of basic cell biology studies delineating characteristics of mammalian neural stem cells, most of the work has been done on rodent embryos. The study of stem cell biology from cells derived from the adult mammal is an emergent field. This is ideal when considering the applications of tissue engineering, with its focus on the use of autologous cells to generate tissue constructs. Although a generalization, it may be that the behavior of stem cells, whether they are derived from the adult or fetal mammal, is essentially the same, with similar responses to trophic and microenvironmental cues.

Gage *et al.* (1995a) discusses both multipotent cells and lineage-restrictive cells that reside in the mature central nervous system. He emphasizes that they can be maintained, cultured, and stimulated to multiply and differentiate with many of the same factors that stimulate their embryonic counterparts. Realizing that *in vitro* culture systems will not contain all the necessary ingredients to guide the growth and development of the cells, he discusses implantation studies utilizing these precursor cells in specific areas of the developing or adult central nervous system. Both the importance of the microenvironment of the implantation site with respect to cellular differentiation and the adherent plasticity of these cells are emphasized. The therapeutic implications of using these cells, with the understanding of what factors and conditions will result in a desired phenotype, reside in their potential to repair portions of the damaged or diseased brain.

Brustle and McKay (1996) recognized the persistence of neural stem cells in the adult mammalian brain and mentioned the ability of these stem cells to migrate and differentiate into neurons. They feel that the guidance and recruitment of these transplanted cells will be determined by external signals. They see the future of neuronal replacement depending on the development of strategies of using these cells in a reconstructive context.

We feel that combining the use of neural stem cells with the principles of tissue engineering will provide a powerful reconstructive context in regard to neuronal replacement. More precisely, we believe this approach will result in the creation of functioning neural tissue constructs. We foresee applications to effect repair of a damaged or diseased central nervous system. Each particular aspect will have to be developed according to its own unique demands, regarding cell type and specific environmental cues and conditions. For instance, Jankovski *et al.* (1996) found that cerebellum postnatal precursor cells grafted into the cerebellum of adult mice generated only two adult phenotypes, granule cells and molecular layer interneurons. They concluded that these progenitors were strictly specified at the time of grafting. Frantz and McConnell (1996) found that late cortical progenitor cells are not able to generate deep-layer neurons, but are restricted to the upper layers. Alternatively, Snyder *et al.* (1997) suggest that microenvironmental alterations created by using targeted photolytic cell death induced multipotent neural precursors to differentiate into layers 2–3 of neocortical pyramidal neurons. It was noted that this neuronal differentiation usually occurs only during embryonic corticogenesis.

This emphasizes the potential of devising strategies that alter status quo conditions in effecting the development of new neural tissue. Feldman *et al.* (1996) demonstrated that CNS progenitor cells can have their morphologic and electrophysiologic properties modified by culture conditions, including growth factors and attachment substratum. Hammang *et al.* (1997) demonstrated that nestin-positive undifferentiated neural stem cells injected into myelin-deficient rat spinal cord differentiated into oligodendrocytes that produced myelin. These same cells *in vitro* developed into astrocytes. Zigova *et al.* (1998) took subventricular zonal progenitor cells from postnatal rats, and implanted them into the striatum of adult rats. The majority of the surviving cells were found to have a neuronal phenotype. These cells integrated into the striatal area of the rat brain.

Cheng *et al.* (1996) discuss multipotent stem cells serving as progenitors for the development of cortical neurons and glia. They talk of stereotyped division patterns within lineage trees. Both intrinsic cell factors and environmental signals are described as likely to have a key role in a stem cell's behavior. More importantly, they suggest evidence is accumulating that cortical stem cells under the influence of both environmental cues and intrinsic factors change with development. Gaiano and Fishell (1998) described the existence of both pluripotent and unipotent neural progenitors. They also suggest that both intrinsic cell factors and extrinsic cues guide neural cell fate.

Although environmental and intrinsic cell cues appear to guide the fate of these neural stem cells, the role of natural factors also appear to play a major role. Cameron *et al.* (1998) discussed the role of peptide growth factors, neurotransmitters, and neuroactive peptides in regard to the generation of neurons and glia in the developing nervous system. They provide evidence from both *in vivo* and *in vitro* studies. They find that basic fibroblastic growth factor, transforming growth factor α (TGF-α), insulin-like growth factor-1, and the monoamine neurotransmitters stimulate the proliferation of neural precursors. Additionally, they find that glutamate, γ-aminobutyric acid, and opioid peptides play a role in down-regulation. Other factors such as vasoactive intestinal peptide, pituitary adenylate cyclase-activating peptide, platelet-derived growth factor, ciliary neurotrophic factor, and members of the TGF-β family have differential effects on these systems.

The variables involved in the actions of these trophic factors are very complex. Cameron *et al.* (1998) showed that cells grown in 0.1 ng/ml of basic fibroblastic growth factor generated exclusively neurons, whereas a concentration of 10 ng/ml of basic fibroblastic growth factor generated a combination of neurons and glia. Therefore, just varying the concentration of one particular growth factor changes the outcome. A combination of factors may have an effect different in terms of outcome from that when using the factors alone or in sequence. For instance, a population of neural stem cells may be more responsive to the effects of basic fibroblastic growth factor after pretreatment with epidermal growth factor.

For more detailed information concerning the role of growth factors, neural transmitters, and cytokines in the regulation of stem cell development, the reviews by Cameron *et al.* (1998) and Mehler *et al.* (1995) are very informative. If it is possible to make a generalization, it would be that epidermal growth factor stimulates the proliferation of neural stem cells, keeping them in an undifferentiated state. It appears that combining epidermal growth factor and basophilic growth factor has the same effect on mammalian spinal cord stem cells.

To identify the stage of development of neural stem cells, immunocytochemical markers are very useful. The marker nestin, an intermediate filament, is used to identify undifferentiated or neural stem cells. The markers neurofilament, neuron-specific enolase, and microtubule-associated protein identify neurons. O4, a sulfatide, is a good marker of oligodendrocytes. GFAP, an intermediate filament, identifies astrocytes. Gage *et al.* (1995a) provide a complete review of neural markers.

Investigators are recognizing the potential of neural stem cells for gene therapy and neural degenerative conditions. Snyder *et al.* (1995) reports restoring β-glucuronidase along the entire neuroaxis of treated mice with transplanted neural progenitors containing the missing enzyme. They reported widespread correction of this lysosomal storage disease in affected mice by use of this method. Pinkus *et al.* (1998) feel that "progenitors are ideal for genetic manipulation and may be engineered to express exogenous genes for neural transmitters, neurotrophic factors, and metabolic enzymes." They recognize that neural precursors do exist in the adult mammalian brain and that they may be used to develop strategies for central nervous system repair. Tissue engineering may offer an advantage in the quantity and duration of expression of these necessary neural products by creating tissue constructs of genetically corrected neural stem cells.

NEURAL STEM CELL APPLICATIONS IN TISSUE ENGINEERING

Our laboratory is developing functional CNS tissue constructs derived from seeding biodegradable polymer scaffolding materials with neural stem cells suspended in hydrogel, with the goal of implantation into the appropriate site and correction of lost or absent neural function. The biodegradable scaffolding will serve to hold the neural stem cells in the appropriate three-dimensional configuration and also provides an attachment site for differentiation. The local microenvironment should provide the endogenous cues and cytokines needed to orchestrate growth, development, and differentiation of these neural stem cells into functioning neural constructs. The hydrogel suspension allows the infusion of nutrients and oxygen into, and exit of waste products and CO_2 out of, the cells. Conceptually, there are no restrictions placed on these cells in terms of their ability to grow, organize, and differentiate. The ideal scaffolding material should be biologically inert and should disappear over time.

Our model required the isolation of neural stem cells from the adult mammal, with the idea of an autologous application. An autologous source of neural stem cells negates concerns of rejection and immunosuppression. Although others have described the central nervous system as immunoprivileged, CNS autoimmune diseases do exist and the potential for subtle but chronic rejection of allogenic cells cannot be excluded.

Our laboratory has successfully isolated neural stem/progenitor cells from the rodent cortex, striatum, and spinal cord (Fig. 58.1). The stem cells are stimulated to divide and are kept in an undifferentiated state using EGF and bFGF. When these cells are driven to maturation with the addition of serum and/or attachment to the bottom of flask, they take on the morphologic characteristics of neurons, astrocytes, and oligodendrocytes (Fig. 58.2). Immunohistochemistry done on these cells shows differentiation toward neurons, astrocytes, and oligodendrocytes, as indicated by positive marking for neurofilament, GFAP, and O4 respectively. Additionally, we have shown that neural stem cells suspended in the hydrogel Pluronics F-127 at 23% survive and proliferate *in vitro* (Fig. 58.3) (Vacanti *et al.,* 1998). Under these same conditions the phenotypic expression of these stem cells is controlled by growth factors in a fashion similar to stem cells cultured in media alone (Lachyankar *et al.,* 1997).

Our goal was to develop a model in the rodent of neural injury, the correction of which by the use of tissue engineering techniques would be readily demonstrable by gross, microscopic, imaging, and neurologic criteria. Cheng *et al.* (1996) demonstrated partial restoration of hindlimb function in adult rats that sustained complete spinal cord resections that were bridged with multiple intercostal nerve grafts. We adopted this model of injury in the adult rat with the idea of combining spinal cord neural stem cells with tissue engineering techniques to effect repair by the creation of spinal cord tissue constructs.

We first developed a simple technique of isolating spinal cord stem–progenitor cells from the adult Fisher rat, modifying a technique used by Lachyankar *et al.* (1997). *In vitro* these isolated cells started as simple rounded structures from 2 to 4 μm in diameter, with the capacity to differentiate into cells with morphologic characteristics of oligodendrocytes, astrocytes, and mature neurons. Additionally, when these cells were suspended in the hydrogel Pluronics F-127 and were used

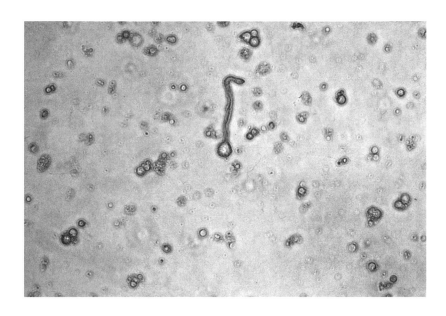

Fig. 58.1. Spinal cord stem cells 3 days after isolation from an adult rat. One appears to be sprouting a process.

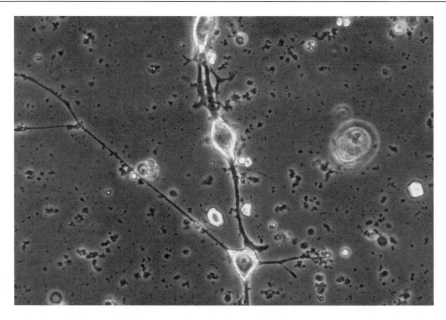

Fig. 58.2. Neurons that have differentiated from stem cells after 7 days in culture.

to saturate small segments of poly(glycolic acid) (PGA), the stem cells were found to survive and attach to the PGA scaffolding material when placed in an incubator at 37°C.

The hydrogel Pluronic F-127 is a copolymer of ethylene and propylene oxide with unique gelation properties. Between 15° and 50°C it is a gel (body temperature is 37°C); it is a liquid above and below this range (Cao *et al.*, 1998). PGA fibers are 15 μm in diameter, with inner spaces averaging 75 to 100 μm (Vacanti *et al.*, 1993). We felt this combination of polymers to be the ideal candidate to serve as our scaffolding material for implantation of the spinal cord neural stem cell constructs. The hydrogel characteristics would allow easy diffusion of oxygen, nutrients, CO_2, and waste products, and the polymer scaffolding material would serve as a site of attachment of the neural stem cells, consequently providing anchorage to drive differentiation, and the large inner spaces would not restrict the growth and development of the neuronal construct.

We hypothesized that spinal cord neural stem cells would grow, develop, and organize, driven by their own intrinsic cell properties and communicating with each other as they orchestrated their development, with guidance by local endogenous cues found in their unique environment,

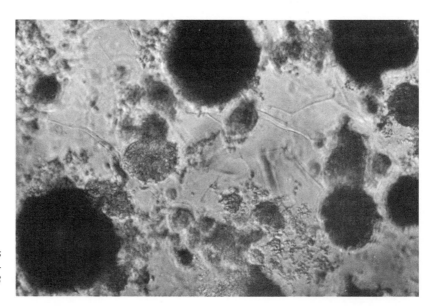

Fig. 58.3. Neural stem cells forming neurospheres in the hydrogel Pluronics F-127 after 3 days of incubation.

i.e., recent surgical injury (a segment of resected spinal cord). Our experimental group consisted of adult Fisher rats, which, under surgical conditions, had 4-mm sections of their spinal cord removed at the T8–T9 segments, creating a clinical condition of paralysis. These animals received implants of adult syngeneic spinal cord progenitor/stem cells suspended in scaffolding as previously described (M. P. Vacanti, J. L. Leonard, B. Dore, L. J. Bonassar, Y. Cao, S. J. Stachelek, J. P. Vacanti, F. O'Connell, C. S. Yu, A. P. Farwell, and C. A. Vacanti, unpublished study).

We realized that the original synaptic connections of the injured spinal cord are extremely complex and extremely difficult, if not impossible, to reduplicate. We felt that any reestablishment of viable neurons and supporting glia would have potential therapeutic benefits because of the inherent plasticity of the central nervous system, which may allow new connections and restoration of function. One unanswered question concerned how these cells would form connections or synapses with what was left of the intact neural tissue after the resection procedure. We felt it was theoretically possible because transected neurons *in vivo* will generate or sprout new dendritic processes. We felt it was logical to assume that these forming processes would synapse with the forming processes of the implanted neural stem cells. If these synapses or connections did occur, we thought functional recovery would result because of the ability of the brain to learn these new connections. Over time this learning process would be reflected in functional neurological recovery.

We used a simple scale to monitor both motor and sensory function of the experimental group as opposed to the control groups. Over an 8-week time period the experimental group had a significant increase in neurological recovery of both sensory and motor function as opposed to the control groups. Gross dissection (Fig. 58.4), magnetic resonance imaging, and histologic evaluation (Fig. 58.5) demonstrated the creation of tissue-engineered spinal cord constructs in the experimental group as opposed to the control group. This model needs much work in its refinement concerning ideal cell numbers, cytokines, and implantation materials. We do feel that it does powerfully demonstrate the potential therapeutic benefit of this approach.

Other future applications may be the creation of models that would demonstrate possible repair of cortical injuries such as stroke or trauma. Additionally, the use of neural stem cells in combination with tissue engineering techniques might be applied to neurodegenerative processes, e.g., Parkinson's disease, or neurogenetic degenerative processes, e.g., Tay Sach's disease, combined with gene insertion technology. Other applications may include repair or replacement of damaged or diseased peripheral nerves, autonomic dysfunction (Hirschsprung's disease), and atonic bladder.

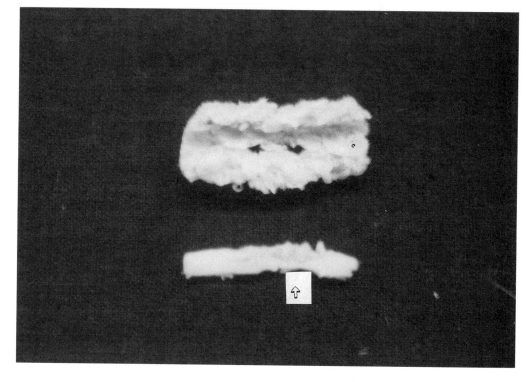

Fig. 58.4. Segment of tissue-engineered spinal cord 3 months after implantation.

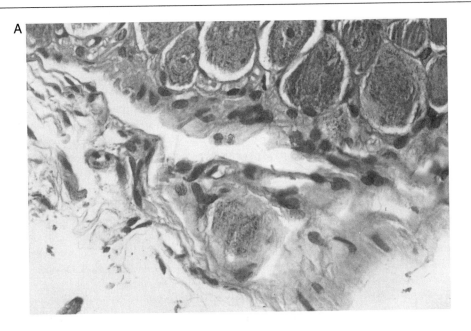

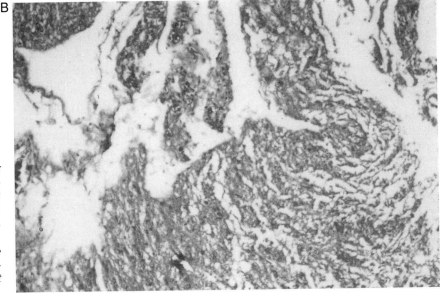

Fig. 58.5. (A) A collection of nerve cell bodies structurally reminiscent of a central canal taken from a tissue construct 3 months postimplantation (hematoxylin and eosin stain; ×100). (B) A transverse section of the same construct highlighting myelin production with luxol fast blue staining.

Neural stem cells may be used to augment the creation of other tissue-engineered constructs, e.g., the innervation of tissue-engineered constructs such as bowel or liver. Finally, these cells may have the ability to guide and orchestrate the growth and development of other cell types used in the creation of tissue-engineered organs.

SUMMARY

Neural stem cell biology and tissue engineering are both emergent fields with great medical potential. We feel that the combination of these disciplines will have a powerful beneficial effect on future treatments of diseased or damaged CNS.

REFERENCES

Bjornson, C. R. R., Rietze, R. L., Reynolds, B. A., Magli, M. C., and Vescovi, A. L. (1999). Turning brain into blood: A hematopoietic fate adopted by adult neural stem cells *in vivo. Science* **283**, 534–537.

Brustle, O., and McKay, R. D. G. (1996). Neuronal progenitors as tools for cell replacement in the nervous system. *Curr. Opin. Neurobiol.* **6**, 688–695.

Cameron, H. A., Hazel, T. G., and McKay, R. D. G. (1998). Regulation of neurogenesis by growth factors and neurotransmitters. *J. Neurobiol.* **36**(2).

Cao, Y. L., Lach, E., and Kim, T. H. (1998). Tissue engineered nipple reconstruction. *Plast. Reconstr. Surg.* **102**(7), 2293–2298.

Cheng, H., Cao, Y., and Olson, L. (1996). Spinal cord repair in adult paraplegic rats: Partial restoration of hind limb function. *Science* **273**, 510–513.

Craig, C. G., Tropepe, V., Morshead, C. M., Reynolds, B. A., Weiss, S., and van DerKooy, D. (1996). *In vivo* growth factor expansion of endogenous subependymal neural precursors cell populations in the adult mouse brain. *J. Neurosc.* **16**(8), 2649–2658.

Davis, A. A., and Temple, S. (1994). A self-renewing multipotential stem cell in embryonic rat cerebral cortex. *Nature (London)* **372**, 263–266.

Feldman, D. H., Thinschmidt, J. S., Peel, A. L., Papke, R. L., and Reier, P. J. (1996). Differentiation of ionic currents in CNS progenitor cells: Dependence upon substraight attachment and epidermal growth factor. *Exp. Neurol.* **140**(2), 206–217.

Frantz, G. D., and McConnell, S. K. (1996). Restriction of late cerebral cortical progenitors to an upper-layer fate. *Neuron* **17**(1), 55–61.

Gage, F. H., Coates, P. W., Palmer, T. D., Kuhn, H. G., Fisher, L. J., Suhonen, J. O., Peterson, D. A., Suhr, S. T., and Ray, J. (1995a). Survival and differentiation of adult neuronal progenitor cells transplanted to the adult brain. *Proc. Nat. Acad. Sci. U.S.A.* **92**, 1179–1183.

Gage, F. H., Ray, J., and Fisher, L. J. (1995b). Isolation, characterization, and use of stem cells from the CNS. *Annu. Rev. Neurosci.* **18**, 159–192.

Gaiano, N., and Fishell, G. (1998). Transplantation as a tool to study progenitors within the vertebrate system. *J. Neurobiol.* **36**(2), 152–161.

Hammang, J. P., Archer, D. R., and Duncan, I. D. (1997). Myelination following transplantation of EGF-responsive neural stem cells into a myelin-deficient environment. *Exp. Neurol.* **147**(1), 84–95.

Hines, P. J. (1997). Frontiers in medicine: Regeneration. *Science* **276**, 59.

Jankovski, A., Rossi, F., and Sotelo, C. (1996). Neuronal precursors in the postnatal mouse cerebellum are fully committed cells: Evidence from heterochronic transplantation. *Eur. J. Neurosci.* **8**(11), 2308–2319.

Johansson, C. B., Momma, S., Clarke, D. L., Risling, M., Lendahl, U., and Frisen, J. (1999). Identification of a neural stem cell in adult mammalian central nervous system. *Cell (Cambridge, mass.)* **96**, 25–34.

Lachyankar, M. B., Condon, P. J., Quesenberry, P. J., Litofski, N. S., Recht, L. D., and Ross, A. H. (1997). Embryonic precursor cells that express TrK receptors: Induction of different cell fates by NGF, BDNF, NT-3, and CNTF. *Exp. Neurol.* **144**, 350–360.

Lacorazza, H. O., Flax, J. D., Snyder, E. Y., and Jendoubi, M. (1996). Expression of human beta-hexosaminidase alpha-subunit gene (the gene defect of Tay–Sachs disease) upon engraftment of transduced progenitor cells. *Nat. Med.* **2**(4), 424–429.

Lendahl, U., Zimmerman, L. B., and McKay, R. D. G. (1990). CNS stem cells express a new class of intermediate filament protein. *Cell (Cambridge, Mass.)* **60**, 585–595.

Lundberg, C., Field, P. M., Ajayi, Y. O., Raisman, G., and Bjorklund, A. (1996). Conditionally immortalized neural progenitor cell lines integrate and differentiate after grafting to the adult rat striatum. A combined autoradiographic and electron microscopic study. *Brain Rev.* **737**(1–2), 295–300.

McKay, R. (1997). Stem cells in the central nervous system. *Science* **276**, 66–71.

Mehler, M. F., Mamur, R., Gross, R., Mabie, P. C., Zang, Z., Papavasiliou, A., and Kessler, J. A. (1995). Cytokines regulate the cellular phenotype of developing neural lineage species. *Int. J. Dev. Neurosci.* **13**(3–4), 213–240.

Morshead, C. M., Reynolds, B. A., Craig, C. G., McBurney, M. W., Staines, W. A., Morassutti, D., Weiss, S., and van Der Kooy, D. (1994). Neural stem cells in the adult mammalian forebrain; a relatively quiescent subpopulation of subependymal cells.

Pinkus, D. W., Goodman, R. R., Fraser, R. A., Neddergaard, M., and Goldman, S. A. (1998). Neural stem and progenitor cells: A strategy for gene therapy and brain repair. *J. Neurosurg.* **42**(4), 858–867.

Reynolds, B. A., and Weiss, S. (1992). Generation of neurons and astrocytes from isolated cells of the adult mammalian central nervous system. *Science* **255**, 1707–1710.

Reynolds, B. A., and Weiss, S. (1996). Clonal and population analyses demonstrate that an EGF-responsive mammalian embryonic CNS precursor is a stem cell. *Dev. Biol.* **175**, 1–13.

Shen, Q., Qian, X., Capela, A., and Temple, S. (1998). Stem cells in the embryonic cerebral cortex: Their role in histogenesis and patterning. *J. Neurobiol.* **36**(2), 162–174.

Shihabuddin, L. S., Hertz, J. A., Holets, V. R., and Whittemore, S. R. (1995). The adult CNS retains the potential to direct region-specific differentiation of a transplanted neuronal precursor cell line. *J. Neurosci.* **15**(10), 6666–6678.

Shihabuddin, L. S., Ray, J., and Gage, F. H. (1997). FGF-2 is sufficient to isolate progenitors found in the adult mammalian spinal cord. *Exp. Neurol.* **148**(2), 577–586.

Snyder, E. Y., and Macklin, J. D. (1995). Multipotent neural progenitor or stem-like cells may be uniquely suited for therapy for some neurodegenerative conditions. *J. Neurosci.* **3**(5), 310–316.

Snyder, E. Y., Taylor, R. M., and Wolfe, J. H. (1995). Neural progenitor cell engraftment corrects lysosomal storage throughout the MPS VII mouse brain. *Nature (London)* **374**, 367–370.

Snyder, E. Y., Yoon, C., Flask, J. D., and Macklis, J. D. (1997). Multipotent neural precursors can differentiate towards replacement of neurons undergoing targeted apoptotic degeneration in adult mouse neocortex. *Proc. Nat. Acad. Sci. U.S.A.* **94**(21), 11663–11668.

Taylor, R. M., and Schnider, E. Y. (1997). Wide-spread engraftment of neural progenitor stem-like cells throughout the mouse brain. *Transplant. Proc.* **29**, 845–847.

Vacanti, C. A., Kim, W., Upton, J., Vacanti, M. P., Mooney, D., Schloo, B., and Vacanti, J. P. (1993). Tissue engineered growth of bone and cartilage. *Transplant. Proc.* **25**(1), 1019–1021.

Vacanti, M., Dore, B., O'Connell, F., Bonassar, L., Engstrom, C., Quesenberry, P., and Vacanti, C. (1998). 3-D culture and phenotypic regulation of neural stem cells grown in pluronic F-127. *Tissue Eng.* **4**, 474.

Valentini, R. F., and Aebischer, P. (1997). Strategies for the engineering of peripheral nervous tissue regeneration. *In* "Principles of Tissue Engineering," Vol. 1, pp. 671–682. Academic Press, San Diego, CA.

Weiss, S., Dunne, C., Hewson, J., Wohl, C., Wheatley, M., Peterson, A. C., and Reynolds, B. A. (1996). Multipotent CNS stem cells are present in the adult mammalian spinal cord and ventricular neural axis. *J. Neurosci.* **16**(23), 7599–7609.

Zigova, T., Pencea, V., Betarbet, R., Wiegand, S. J., Alexander, C., Bakay, R. A., and Luskin, M. B. (1998). Neuronal progenitor cells of the neonatal subventricular zone differentiate and disperse following transplantation into the adult rat striatum. *Cell Transplant.* 7(2), 137–156.

PART XVIII: PERIODONTAL AND DENTAL APPLICATIONS

CHAPTER 59

PERIODONTAL APPLICATIONS

Neal A. Miller, Marie C. Béné, Jacques Penaud, Pascal Ambrosini,
and Gilbert C. Faure

INTRODUCTION

Periodontitis is a disease that results in the destruction of supporting tissues of the teeth, subsequent gingival pocket formation, and ultimately tooth loss. The main etiology is bacterial infection, but occlusal and systemic disorders are considered aggravating factors. Although periodontitis is not life threatening, it is one of the most widespread disorders found among human populations. Most surveys in Europe state that more than 40% of persons in the age group 35–45 years old present periodontal pocketing (Miyazaki *et al.*, 1991). The periodontal status deteriorates as the subjects grow older: in France the percentage of people with periodontitis exceeds 60% in the age group 45–64 years old (Miller *et al.*, 1987). In the United States 36% of the population 19 years and older presents periodontitis; this increases to 52.5% of persons 45–64 years old (Brown *et al.*, 1989). Health costs for this disease have been estimated at $5–6 billion involving 120–133 million hours of treatment (Oliver *et al.*, 1989), even though most patients do not seek treatment.

Initial concern was in arresting the evolution of the disease after restoring lost periodontal material. For several decades, periodontists have been striving to regenerate destroyed periodontium (Hancock, 1989). Among earlier attempts, curettage, open-flap debridment, and diverse bone grafting procedures can be cited. These therapeutic modalities most often result only in partial repair and in residual pockets, primarily because progenitor cells of highly specialized tissues are not easily secured but also because of the particularly difficult wound closing conditions pertaining to periodontal surgery. As in any type of wound, a blood clot is formed very rapidly. In dermal wounds, fibrin bridging is sufficient to close the wound, and epithelial proliferation, which starts within 24 hr, is limited to the epidermal lesion (Clark, 1988). In periodontal lesions, the gingiva can bind only to the vascular surface of the teeth, and very often the adherence is too frail, or the blood clot is resorbed too rapidly. In this case, a long junctional epithelium may develop, between the tooth and the flap, resulting in the formation of a pocket (Polson and Proye, 1983). An additional complication is the accumulation of plaque. Oral microorganisms adhere to the portion of the tooth surface that is next to the marginal gingiva. Because of these wound closure hindrances, complete submersion of the teeth under gingival flaps has been advocated (Klinge *et al.*, 1981) but is seldom feasible in human clinical situations.

New advances in periodontal therapy include the use of membrane barriers to guide the regenerative tissue, and implantation of bone substitutes. These techniques apply to vertical bone lesions and are becoming quite efficacious and predictable. Promoting regeneration at a distance from a potential source of osseous cells remains difficult, but better knowledge of chemoattractants, growth factors, and osteoinduction proteins is likely to help develop clinical applications that enhance healing and facilitate the selection of proper cell populations for bone or gingiva reconstruction. These approaches are detailed in this chapter.

GUIDED TISSUE REGENERATION

Nyman *et al.* (1982), as others, observed that the presence of gingiva is detrimental to the regeneration of the attachment apparatus. However, Nyman and co-workers were the first to find an effective way of excluding gingival epithelial and connective tissue cells from reconstruction sites. Their initial attempt to regenerate periodontium in the human succeeded in forming 5 mm of new attachment with collagen fibers embedded in cementum. They used a Millipore filter, which served as a barrier to keep connective tissue cells out of the resorbed site and avoid downward growth of the gingival epithelium along the root surface. Little bone was produced, perhaps because the filter collapsed against the tooth. Soon after, another material was adopted: expanded poly(tetrafluoroethylene) (e-PTFE), or Teflon. The advantages of e-PTFE are its stiffness, that it allows more room for bone formation, is biocompatible, and the ease with which it can be removed. Numerous studies have proved the ability of e-PTFE membranes to promote repair of intrabony and furcation defects (Becker *et al.*, 1987; Pontoriero *et al.*, 1987). Histologic confirmation of the repair, evidencing neoformation of cementum, periodontal ligament, and alveolar bone, was obtained in both humans (Nyman *et al.*, 1982) and animals (Caffesse *et al.*, 1988b). The material has, however, two major drawbacks: it causes severe gingival recession and, once exposed, accumulates large amounts of bacterial plaque (Becker *et al.*, 1987; Selvig *et al.*, 1990; Grevstad and Leknes, 1993). Vascularization of the marginal gingiva is provided solely by terminal capillaries. The placement of a cell-tight barrier disturbs the blood supply and avoids endothelial anastomosis with blood vessels originating from the underlying tissues. Bacteria actively proliferating in the crevicular part of the tooth will then colonize the e-PTFE membrane. Because of the subsequent inflammation, the Teflon membrane must be removed between 4 and 6 weeks after placement (Becker *et al.*, 1987). To avoid the requirement for second-stage membrane retrieval and to minimize bacterial infection, resorbable membranes (Minabe, 1991) have been developed using collagen or polysaccharide polymers. Collagen membranes used experimentally can be made from rat tail, bovine dermis, or other sources. Industrially produced collagen membranes include collagen from calf endocardium and are in use in human clinics (Penaud *et al.*, 1992). Different types of poly(lactic acid) or poly(galactic acid) have been used to form polymer membranes. A novel product dissolves lactic acid in *N*-methyl-2-pyrolidone. The paste hardens in contact with water. It can be directly applied *in situ* and thus conforms particularly well with the root forms (Bogle *et al.*, 1997; Rosen *et al.*, 1998). When exposed, resorbable membranes accumulate small amounts of bacterial plaque because bacterial and salivary enzymes accelerate their resorption. Thus permeabilized, the membrane is partially protected by the tissue exudate. Many polymer barriers feature pores that allow fluid circulation or even a limited amount of tissue ingrowth (Gottlow *et al.*, 1994). The resorption rate of resorbable barriers can be modulated and thus their effect prolonged if needed (Robert and Frank, 1994). How long a membrane should stay in place has not been determined, but Miller *et al.* (1996) have shown that resorption rates of some products are too fast to assure sufficient cell differentiation. Because of the infections mentioned above, e-PTFE membranes have been removed between 4 and 6 weeks after placement. This is widely considered as being sufficient for specific cells to occupy the defect and avoid ingrowth of the gingival chorion. However, Florès de Jacoby *et al.* (1994) have reported that on removal of e-PTFE membranes, an average 65% of root surface was covered by a neoformed tissue failing to mineralize completely. At 9 to 12 months after placement of the barrier, only 31% of the same root surfaces were still covered. This demonstrates that more than half of the soft red tissue found under an e-PTFE membrane after removal does not mineralize to form either cementum or bone. The neoformed tissue was shown in Taiwan monkeys to be particularly labile when exposed to bacterial plaque (Ling *et al.*, 1994). Advancement of the histologic process is not well known. Amplifying divisions of periodontal ligament cells may have completed by postoperative day 21 (Iglhaut *et al.*, 1988), but these cells must be protected much longer. Sigurdsson *et al.* (1994), using e-PTFE membranes on dogs, demonstrated that no regeneration occurs if the membranes are removed at 3 weeks, whereas at 8 weeks 75% of the bone and 40% of the cementum are repaired. When collagen barriers are used, a period of 10 days is sufficient to avoid epithelial downgrowth (Kodama *et al.*, 1989). When complete resorption occurs before 30 days, neocementum is present, but not new bone (Pitaru *et al.*, 1989). Although membranes exclude gingival tissue they do not enable the selection of periodontal ligament cells or bone cells. It was first postulated that periodontal ligament cells were the slowest to migrate. This was supported by experiments showing that after injury the proliferation of periodontal ligament cells peaked at 2 or 3 days (Aukhil and Iglhaut, 1988). It was suspected that this apparently limited proliferation was caused by a weak angiogenesis and that premature differentiation delayed migration (Aukhil *et al.,* 1986). However, Nyman *et al.* (1987) found no alkylosis on 11 teeth from 10 patients. In their opinion this was proof that periodontal ligament cells reach the root surfaces before bone cells do. Buser *et al.* (1990) placed an implant in contact with a retained root in a dog. Periodontal ligament cells migrated several millimeters along the implant, producing cementum and a periodontal-like ligament. The cells were able to colonize a fairly large surface much faster than the osseous cells crossed the minute space left by the drill. It seems then that these two types of cells do not compete, but primarily restore their intended functions. Whether resorbable or made of e-PTFE, membranes should be positioned under the flap margin. It is recommended to extend their limits at least 2 mm beyond the edges of the osseous defect in order to seal it completely from the gingiva. Gingival covering of the membrane is considered to be of great importance (Tonetti *et al.,* 1993). When second-stage surgery is needed (removal of nonresorbable membrane), care must be taken to cover the neoformed tissue with gingiva, and it is wise to consider the amount and quality of gingiva available before implantation of a membrane (Anderegg *et al.,* 1995). Reported results vary. Pontoriero *et al.* (1987) closed interradicular bony lesions (furcation defects) in 90% of their cases, whereas Demolon *et al.* (1994) closed none. However, Lekovic *et al.* (1989) demonstrated that furcation defects receiving membranes respond much better than those treated by flap surgery and curettage alone. Interdental vertical defects respond quite well, as can be seen in Fig. 59.1. Laurell *et al.* (1994) obtained a mean 4.9-mm gain in pe-

A

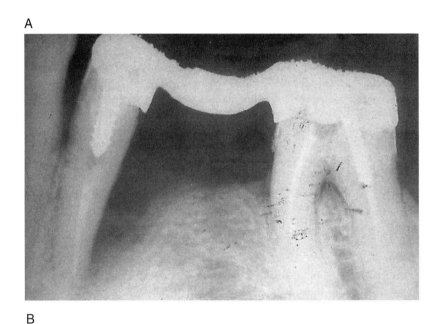

B

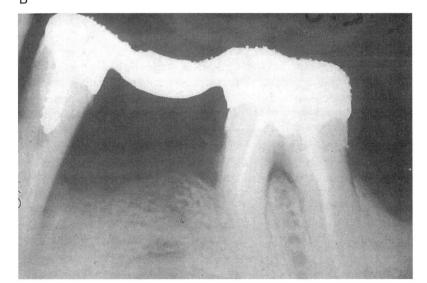

Fig. 59.1. (A) An angular defect has nearly reached the apex of the mandibular bicuspid. (B) Placement of a resorbable membrane promoted bone formation to the alveolar crest.

riodontal attachment using poly(lactic acid) barriers for 47 intrabony defects. In a study limited to mandibular three-walled intraosseous lesions, Becker and Becker (1993) evaluated an average gain in bone filling of 4.3 mm for 24 patients treated with e-PTFE membranes. The appropriate use of barrier membranes promotes predictable osseous repair and histologically verifiable new attachment with neoformation of cementum and periodontal ligament fibers.

BONE SUBSTITUTES

Intraoral autologous bone grafts are, of course, choice material when osseous defects need to be repaired (Becker *et al.,* 1994). However, intraoral autologous bone is not always available in sufficient quantities, particularly when patients present a deficiency of alveolar bone resulting from severe periodontitis. Bone substitutes include synthetic materials such as hydroxyapatite and methyl methacrylate, and modified mineralized biologic substances such as coralline or bovine trabecular bone and lyophilized human bone. These products provide surgical ease and patient comfort, and their various resorption rates are interesting. Opponents of bone grafting techniques affirm that regeneration of the periodontal attachment apparatus is not achievable by, or at least has not been evidenced for, procedures other than membrane implantation (Egelberg, 1987). These assertions overlook the very thorough work of Dragoo (1981), who produced histologic evaluation of human material for every type of regenerative surgery that was known at the time and demonstrated that bone grafts can produce new attachment with neocementum. In an articulate study, Bowers *et al.* (1989a,b) produced histologic evidence of attachment regeneration after implantation of demineralized freeze-dried bone allografts (DFDBAs). The best results were obtained when teeth were submerged under flaps (Bowers *et al.,* 1989a), or when the sites were protected from epithelial downward growth by gingival grafts (Bowers *et al.,* 1989b). Bone substitutes have been suspected to impair the regeneration of connective tissue attachment. Takata *et al.* (1993) have shown that periodontal ligament cells are capable of covering blocks of hydroxyapatite embedded in root surfaces and of subsequently secreting cementum on synthetic material. Production of this newly formed connective attachment with inserted collagen fibers is slightly delayed compared to regeneration on natural dentine surfaces. It is sometimes postulated that new attachment occurs preferentially where new bone is present (Caffesse *et al.,* 1994). All techniques that enhance bone formation would then be beneficial to connective attachment regeneration. Research by Lowenguth *et al.* (1993) exposed a factor that may be important in this view. They implanted surface-demineralized dentine cylinders in rat skin and observed that appropriate approximation of dentine surfaces enhanced the orientation of fibroblasts and collagen fibers perpendicular to the walls of the cylinders. The customary use of bone substitutes has been in infrabony pockets (Yukna, 1994a) and alveolar ridge preservation (Greenstein *et al.,* 1985), but this presents some limitations. If resorbable, the substances can be resorbed by the gingiva. The material can also cause necrosis of the papillar and gingival tissue covering the defects. In such cases the bone substitute is lost and the defect is even harder to treat. Therefore, in spite of the fact that bone substitutes may induce great amounts of bone repair, as can be seen in Fig. 59.2, they are more often associated with resorbable (Penaud *et al.,* 1992) or nonresorbable (Caffesse *et al.,* 1994) membranes. Proponents of the membrane technique often admit that barriers alone do not induce the formation of much alveolar bone, or at least not above the existing crestal level. Only bone grafting techniques with bone or substitute materials promote the formation of new bone, cementum, and periodontal ligament coronal to the ridge (Yukna, 1994b). Another reason to use synthetic materials is as spacers. It is thought that osseous formation under membranes is often limited by the barrier collapsing against tooth structures because of the pressure exerted by the overlaying soft tissues. Bone grafting materials are thus used both as spacers and as bone-conducting substances (Caffesse *et al.,* 1994; Plotzke *et al.,* 1993), sometimes associated with membranes. Dragoo (1981) showed that techniques used in the late 1970s caused epithelial downward growth and residual pockets. The material that blocked apical migration was mostly iliac vital bone, which quickly occupied the root surface and induced ankylosis. Later, blocking epithelial downward growth with a barrier was shown to be a sure and innocuous process. The combined use of grafting material and membranes alone (Lekovic *et al.,* 1989). The choice of a bone substitute will depend on the final goal. Poly(methyl methacrylate) is highly biocompatible, but promotes no bone formation and usually ends up encapsulated in fibrous connective tissue with no increased connective attachment apparatus (Plotzke *et al.,* 1993). The new interest in dental implants now encourages

A

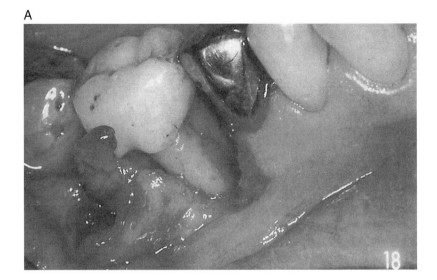

B

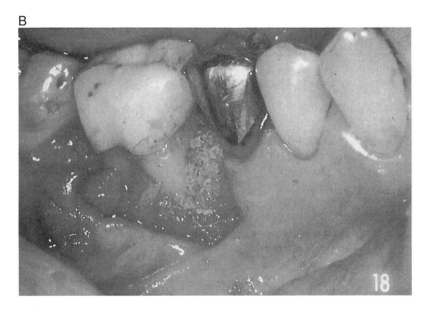

C

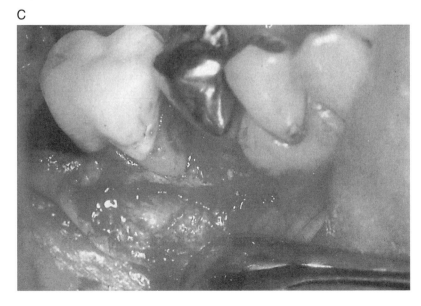

Fig. 59.2. (A) A deep angular bony defect is visible on the mesial aspect of the mandibular molar. (B) The defect has been filled with a bone substitute of bovine origin. (C) Reentry 1 year later: total bone fill has been obtained.

surgeons to use materials that will subsequently allow osteointegration (Seibert and Nyman, 1990; Nevins and Mellonig, 1992), i.e., resorbable bone substitutes ultimately replaceable by new bone. For two- or three-walled defects, rapidly resorbed substances such as demineralized freeze-dried bone allograft can be used. Where rapid osseous repair seems less likely, slowly resorbing materials are preferred, such as freeze-dried bone, use of which appears to result in more bone repair than does DFDBA. In a report by Rummelhart *et al.* (1989), FDBA-treated sites gained 2.4 mm of new bone vs. 1.7 mm in sites treated with DFDBA, even though bone matrix protein was directly available in the demineralized allograft. Slow resorption might also avoid rapid replacement of the material by gingival connective tissue and macrophages. This might be why such materials as coralline calcium carbonate promoted a mean fill of 2.3 mm in the patients treated by Yukna (1994b). The concern that implanted materials of bovine or human origin might contain viruses or prions has implemented the search for new totally synthetic materials. Bioactive glass seems quite promising in its ability to promote bone fill (Lovelace *et al.*, 1998; Price *et al.*, 1997). It also presents antimicrobial properties (Stoor *et al.*, 1998), which could be interesting in treating infected defects in combination with other materials.

DENTAL IMPLANTS

Dental implants replace the dental root, not the periodontium. They are made of inert material, usually titanium, and their placement cannot be considered a tissue engineering technique. Nevertheless, their use often involves an increase of the existing osseous volume. Until 1969, efforts aimed at producing a ligament-like tissue around dental implants, but such fibrous tissue lacked perpendicular collagen fibers embedded in cementum. It rather contained parallel collagen sheets on the surface of the implant, encapsulating it as a foreign substance. In 1969, Bränemark *et al.* introduced the notion of osteointegration. It was then thought that the bone grew in direct contact with the titanium. Later, Albrektsson *et al.* (1985) demonstrated that the interface between implants and bone is a 20- to 40-nm-wide area containing proteoglycans and nonmineralized collagen (Sennerby *et al.*, 1991). Soft tissues around teeth and implants present similarities (Berglundh *et al.*, 1991) but also differences (Buser *et al.*, 1990). A long junctional epithelium connects the implant to the oral epithelium. Apical to the epithelial junction, there is no true connective tissue attachment. The fibers are at first perpendicular to the implant, then take a vertical course as they near its surface, and surround it.

Alveolar bone exists to support the teeth, and this supportive bone begins to resorb if a tooth is extracted. Alveolar atrophy can be severe in edentulous patients, but in many cases, dental implants are difficult to place. Among the most common anatomical obstacles are the sinuses and the mandibular dental nerve. But the height of the crest is not the only problem, because sufficient width is also necessary (Fig. 59.3). Different forms of tissue engineering can be applied in such cases. The most common, guided bone regeneration using a membrane, can be performed before, after, or at the time of implantation, depending on the anatomic situation. The aim of guided bone regeneration in dental implantology is different from guided tissue regeneration in periodontology. When treating infrabony pockets, the periodontist wants to enhance the migration of periodontal ligament cells. When a deficit in osseous volume is the problem, the surgeon merely tries to avoid competition between the bone and the gingiva. Linde *et al.* (1993) and Misch and Dietsch (1993) have shown that gingival fibroblasts cultured in the presence of osseous cells inhibit the production of the latter. Schenk *et al.* (1994) demonstrated that if the membranes were occlusive, they could promote woven bone within 2 months, and distinct cortical and trabecular structures by 4 months. A membrane rigid enough to protect the underlying blood clot can be used alone (Hardwick *et al.*, 1995). However, membranes tend to collapse and must often receive extra support. This can be provided by screws or pins made of titanium or absorbable polymers (Nevins and Mellonig, 1994). These devices are used as tent poles, lifting up the center of the membrane. When made of titanium, they must be removed after ridge augmentation is achieved. Most studies so far have used e-PFTE barriers for guided bone regeneration with less drawbacks than when it is used for periodontal lesions. Flap closure avoids eventual exposure and bacterial colonization, and this nonresorbable material can remain in place as long as necessary, even several years (Dahlin *et al.*, 1991). Barriers can also be used with bone substitutes to achieve ridge augmentation (Nevins and Mellonig, 1992). The materials must at first support the membrane, then be resorbed and replaced by bone. The delay before implantation possibly depends on the nature of the substance applied, and can be quite long. As the material is resorbed the barriers tend to collapse. Reinforced barriers have been designed to avoid this problem. Resorbable membranes disappear long before the material has been transformed into bone. A solution for both concerns is the use of titanium membranes. If bone or bone substitutes are used, the thick titanium mesh allows the passage of blood vessels while maintaining the desired volume (von Arx and Kurt, 1999). When no material is placed, thin nonperforated titanium membranes will provide enough support to keep the gingiva out of the defect and will enable new bone formation. Thus bone is formed much faster than biomaterials can be resorbed and replaced by osseous tissue (Lundgren *et al.*, 1995).

Placement of dental implants in the maxillary premolar and molar regions is often impeded by sinuses. Surgical techniques have been developed to push back the sinus mucous membrane and increase the height of the alveolar bone by autogenous grafting (Springfield, 1992) or by use of other materials such as hydroxyapatite (Misch and Dietsch, 1993).

GROWTH FACTORS AND PLASMA PROTEINS

For many periodontologists, protection of the blood clot that forms after a surgical procedure is the key to periodontium regeneration (Egelberg, 1987), and this can be achieved, among other ways, by membrane barriers. A different type of research is focused on enhancing rapid reorganization of the blood clot. Polson and Proye (1983) have suggested that fibrin linkage between collagen fibers of demineralized dentine matrix and the gingival chorion could enhance and accelerate the formation of a new connective attachment. However, fibrinogen is present in all sites of injury, and studies demonstrating that a supplement of fibrinogen could improve new attachment are lacking. Chemoattraction and cell orientation have also been intensively studied. Fibronectin has been considered a promising promoter of fibroblastic activity. Wikesjo *et al.* (1988) demonstrated that topical application of fibronectin on demineralized roots could be beneficial to healing. Terranova *et al.* (1987) have shown that fibroblast growth factor (FGF) has a potent chemotactic effect on fibroblasts derived from periodontal ligament tissues. FGF is 10,000 times more powerful than fibronectin in inducing fibroblast growth. Platelet-derived growth factor (PDGF) is another chemoattractant with great potential, and no amount of fibronectin can stimulate the proliferation of periodontal ligament fibroblasts to the extent achieved by nanograms of PDGF (Terranova *et al.*, 1987). The usefulness of fibronectin can also be questioned, because it has been shown that it enhances the migration of basal keratinocytes (Woodley *et al.*, 1990). Where epithelial cells colonize the root before connective tissue is formed, there will be no attachment apparatus. Therefore, although fibronectin attracts and orients fibroblasts properly, some of its effects might not be desirable. Plasmatic protein sealants containing large quantities of fibrinogen, but also fibronectin and coagulation factors, have been used with success to avoid suturing, but Caffesse *et al.* (1988a) demonstrated very moderate gains in attachment levels when fibronectin was applied on demineralized root surfaces. Biologic glues have also been proposed as spacers to keep membrane barriers at a distance from the roots (Pini Prato *et al.*, 1988), but the usefulness of fibrinogen and/or fibronectin in periodontal surgery is still inconclusive.

Growth factors nevertheless seem worthy of investigation and clinical trials, for if the membrane technique allows tissue selection and thus favors periodontal attachment, it does not accelerate the healing process. FGF and PDGF both have powerful effects on periodontal ligament fibroblasts. Periodontists have also shown interest in transforming growth factor β (TGF-β), which appears to promote more significant deposition of fibroblast collagen than any of the other growth factors, although it is not a fibroblast chemoattractant (Lynch *et al.*, 1989). The combination of FGF and insulin-like growth factor (IGF-1) seems promising as well, promoting the same collagen deposition as TGF-β and a greater proliferation of ligament fibroblasts than PDGF alone (Lynch, 1994). the PDGF–IGF-1 combination is chemotactic and mitogenic for periodontal fibroblasts; PDGF is chemotactic for osteoblasts, and promotes the synthesis of noncollageneous proteins in bone cultures, and IGF-1 promotes the synthesis of collagen proteins. The PDGF–IGF-1 combination results in greater bone matrix formation than any individual growth factor, and interacts synergistically to yield significant collagen formation and healing in soft tissue wounds (Lynch, 1994). Animal experiments have demonstrated that a gel containing this mixture produces more and faster regeneration in the same procedures performed without these growth factors (Giannobile *et al.*, 1994). The delivery of growth factors through a gel allows the presence

A

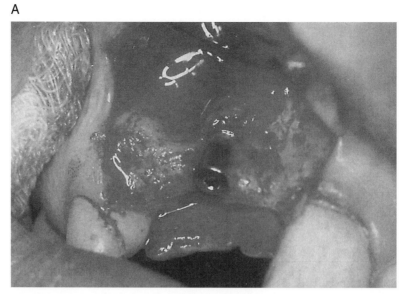

B

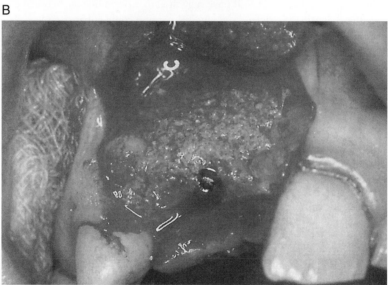

C

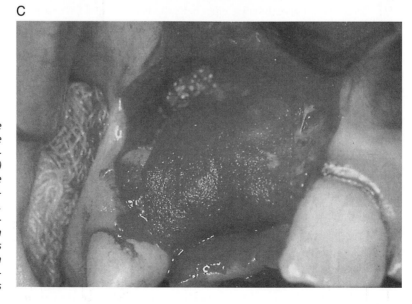

Fig. 59.3. (A) The width of the crestal part of the alveolar ridge is insufficient to house the cervical part of this dental implant. (B) Holes have been drilled into the cortical plate and a bone substitute has been placed as a spacer. (C) A resorbable membrane excludes the gingival chorion from the healing site. (D) Six months later, guided bone regeneration has occurred, completely embedding the implant in osseous tissue.

D

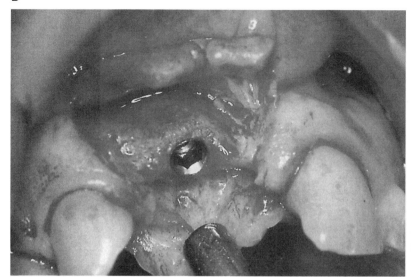

Fig. 59.3. (continued).

of a great amount of PDGF and IGF-1 during a short period of time, and even better results could perhaps be obtained with a slow-release system. Lynch *et al.* 91991) showed that the PDGF–IGF-1 mixture in a methyl cellulose gel had a half-life clearance of 3 to 4.2 hr, with more than 96% of the factors cleared by day 4. Repair requires several weeks and it will be interesting to develop slow-release devices. Other reports have suggested that the type of PDGF used might be important with regard to matrix or cell surface attachment (Kelly *et al.,* 1993), and consequently to the ability of activated fibroblasts to produce new collagen (Cho and Garant, 1981). Additional studies in animal models have supported interest in these compounds and their association to regenerative therapy (Cho *et al.,* 1995; Park *et al.,* 1995). Another group of promising molecules is that of bone morphogenic proteins (BMPs), a mixture of peptides initially identified as a bone growth factor and later shown to comprise at least 13 individual proteins (Wozney, 1995). BMPs can promote new bone formation at the site of implantation, instead of changing the bone growth rate of existing bone. In the early 1970s BMPs were shown to be factors from the extractable protein of bone matrix capable of autoinduction (Urist and Iwata, 1973) and of inducing bone formation in nonosseous environments (Sampath and Reddi, 1981). BMPs induce mesenchymal precursor cells to differentiate in cartilage (Chen *et al.,* 1991) and in bone-forming cells (Yamaguchi *et al.,* 1991). Thus, with recombinant human bone morphogenic protein 2 (rhBMP2), the proximity of residual bone is less important for osseous formation. According to Sigurdsson *et al.* (1995a), rhBMP in a carrier enhances periodontal regeneration. When healing is compared with regeneration following membrane implantation, rhBMP produces more bone, but it is primary woven bone. The membrane promotes a thicker cellular cementum. The use of rhBMP significantly reduced postsurgical resorption. The membrane favors migration and proliferation of already differentiated cells, whereas rhBMP promotes the grouping of mesenchymal stem cells in clusters, later developing in appropriate tissues. Sigurdsson *et al.* (1995b), using rhBMP, managed to promote vertical bone formation in horizontal osseous defects generated in dogs. The abundant bone produced not only covered the denuded half-root but in some cases even grew over portions of the crowns of the teeth. A small 12.5-kDa peptide discovered by Terranova *et al.* (1994), PDL-CTX, also seems to be a potent chemoattractant for periodontal ligament cells, with mitogenic properties.

A totally new approach to reconstructing resorbed periodontal tissues is the use of amelogenin. This growth factor, found in the epithelial cells forming the enamel organ, has been extracted from pigs and results in neoformation of acellular cementum (Hammarström, 1997). *In vitro* experiments show that amelogenin increases proliferation of ligament fibroblasts but not of epithelial cells (Gestrelius *et al.,* 1997). Clinical trials in animals (Zetterström *et al.,* 1997; Araujo and Lindhe, 1998) and humans (Hammarström *et al.,* 1997) have demonstrated significant gains of bone, periodontal ligament, and cementum.

A

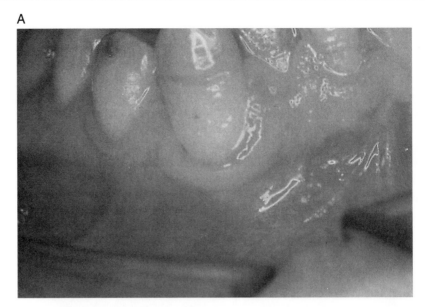

B

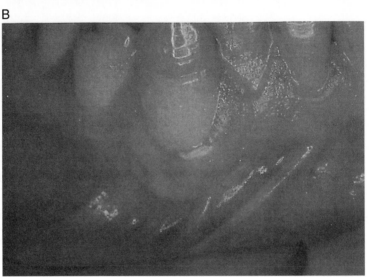

C

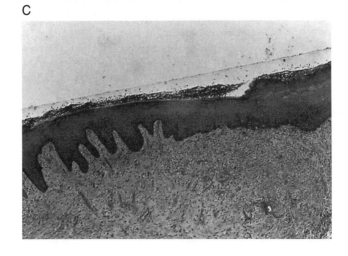

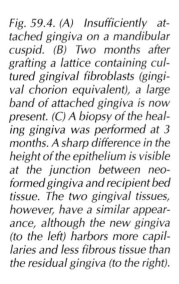

Fig. 59.4. (A) Insufficiently attached gingiva on a mandibular cuspid. (B) Two months after grafting a lattice containing cultured gingival fibroblasts (gingival chorion equivalent), a large band of attached gingiva is now present. (C) A biopsy of the healing gingiva was performed at 3 months. A sharp difference in the height of the epithelium is visible at the junction between neoformed gingiva and recipient bed tissue. The two gingival tissues, however, have a similar appearance, although the new gingiva (to the left) harbors more capillaries and less fibrous tissue than the residual gingiva (to the right).

CELL CULTURE TECHNIQUES

Cell culture techniques have seldom been used to treat periodontal disease. Hanachowicz *et al.* (1989) cultured epithelial cells to treat gingival recession by grafting the cultured epithelial equivalent. The procedure was successful in reinforcing the remaining gingiva, but there was little gain in root coverage. P. Ambrosini and N. A. Miller (unpublished data) applied the method developed by Bell *et al.* (1979) to the culture of human gingiva fibroblasts, and grafted the gingival chorion equivalent they had obtained at the margin of a gingival recession (Fig. 59.4A). Twenty-one days later, a biopsy was performed at the healed wound edge. Histologic examination showed that the gingival chorion equivalent grafted had reconstituted an apparently normal gingival tissue with dense collagen fibers and was covered by keratinized epithelium (Fig. 59.4B). This technique appears to offer interesting advantages. The patient does not experience great discomfort, because an initial biopsy of 8 mm^2 is sufficient to induce the culture. The culture can produce great quantities of chorion, which can be used for the reconstruction of several gingival recession sites. Drawbacks are the time and money needed to sustain such cultures; the use of palate donor sites allows a more rapid reconstruction (Martin *et al.*, 1995). Such techniques, using autologous cultured tissues, could, however, be considered when extensive reconstruction is needed.

Because osseous cells are not as easily harvested as gingival fibroblasts, some researchers have sought to differentiate other cells into osteoblasts. Lecoeur and Ouhayoun (1997) have cultured adipous cells in media containing rhBMP2 and dexamethasone. Thus treated, the adipocytes produced a high output of alkaline phosphatase, suggesting that these cells could undergo an osteogenic differentiation. This has brought hope that such cells could afterward be transplanted into periodontal defects.

CONCLUSION

A certain number of factors impeding spontaneous regeneration of the periodontium have begun to be adequately understood and addressed. For example, the problems of tissue competition between epithelium and connective tissue at the root surface and between gingival chorion and osseous cells in bony defects are rather satisfactorily solved with the use of membranes. Spacing and some degree of osseous conduction can be provided by bone substitutes, and use of these products, associated or alone, is now widespread and partial regeneration of periodontal defects is predictable. The best results are obtained in angular alveolysis. The number of bony walls at the lesion is important: two- or three-walled defects will have better prognosis than one-wall defects. The quality of the bone forming the lesion walls should also be considered: thick walls with cancellous bone will produce more osseous cells compared with cortical bone. The source of bone cells and the distance they must travel are crucial to assess the chances of success of treatment modalities currently in use. Research is opening new horizons. Growth factors seem promising tools to attract the appropriate cells to healing sites and in promoting angiogenesis, an especially interesting feature considering the avascular nature of the dental root. The use of BMPs, amelogenine, or other inductor proteins might even enable induction of stem cells to differentiate into bone or periodontal ligament cells. The biologic effects of cytokines and their carriers still have to be studied. Substances that will not hinder healing but will release biological agents at a suitable rate need to be developed. Animal models demonstrate that horizontal alveolysis can be repaired, and clinical applications will surely emerge with time.

Considering the progress achieved since the early attempts at repairing dental defects with biomaterials, it seems that we are getting closer to the possibility of totally regenerating periodontium destroyed by disease, if not preventing such events.

REFERENCES

Albrektsson, T., Hanson, H. A., and Ivarsson, B. (1985). Interface analysis of titanium and zirconium bone implants. *Biomaterials* **6**, 97–101.

Anderegg, C. A., Metzger, D. G., and Nicoll, B. K. (1995). Gingiva thickness in guided tissue regeneration and associated recession at facial furcation defects. *J. Periodontal.* **66**, 397–402.

Araujo, M. G., and Lindhe, J. (1998). GTR treatment of degree III furcation defects following application of enamel matrix proteins. An experimental study in dogs. *J. Clin. Periodontol.* **25**, 524–530.

Aukhil, I., and Iglhaut, J. (1988). Periodontal ligament cell kinetics following experimental regenerative procedures. *J. Clin. Periodontol.* **15**, 347–382.

Aukhil, I., Pettersson, E., and Suggs, G. (1986). Guided tissue regeneration. An experimental procedure in beagle dogs. *J. Periodontol.* **57**, 727–734.

Becker, W., and Becker, B. E. (1993). Treatment of mandibular 3-wall intra-bony defects by flap debridment and expanded polytetrafluoroethylene barrier membranes. Long term evaluation of 32 treated patients. *J. Periodontol.* **63**, 1138–1144.

Becker, W., Becker, B. E., Prichard, F. *et al.* (1987). Root isolation for new attachment procedures. A surgical and suturing method: Three case reports. *J. Periodontol.* **58**, 819–826.

Becker, W., Becker, B. E., and Caffesse, R. A. (1994). Comparison of demineralized freeze-dried bone to induce bone formation in human extraction sockets. *J. Periodontol.* **65**, 1128–1133.

Bell, E., Ivarsson, B., and Merrill, C. (1979). Production of a tissue-like structure by contraction of collagen lattices by human fibroblasts of different proliferative potential *in vitro*. *Proc. Natl. Acad. Sci. U.S.A.* **76**, 1274–1278.

Berglundh, T., Lindhe, J., Ericson, L. *et al.* (1991). The soft tissue barrier at implant and teeth. *Clin. Oral Implants Res.* **2**, 81–90.

Bogle, G., Garrett, S., Stoller, N. H. *et al.* (1997). Periodontal regeneration in naturally occurring class II furcation defects in Beagle dogs after guided tissue regeneration with bioabsorbable barriers. *J. Periodontol.* **68**, 536–544.

Bowers, G. M., Chadroff, B., Carnavale, R. *et al.* (1989a). Histologic evaluation of new human attachment apparatus formation in humans. Part II: Submerged DFBA did better than nonsubmerged. *J. Periodontol.* **60**, 675–682.

Bowers, G. M., Chadroff, B., Carnavale, R. *et al.* (1989b). Histologic evaluation of new human attachment apparatus formation in humans. Part III: Gingival face grafts limited apical migration of epithelium. *J. Periodontol.* **60**, 683–693.

Branemark, P., Breinen, U., Adell, R. *et al.* (1969). Intraosseous anchorage of dental prostheses: Experimental studies. *Scand. J. Plast. Reconstr. Surg.* **3**, 81–91.

Brown, L. J., Olpuer, R. C., and Loe, H. (1989). Periodontal diseases in the US in 1981: Prevalence, severity extent and role in tooth mortality. *J. Periodontol.* **60**, 363–370.

Buser, D., Warrer, K., and Karring, T. (1990). Formation of a periodontal ligament around titanium implants. *J. Periodontol.* **61**, 597–601.

Caffesse, R. G., and Nasjleti, C. E. (1994). Clinical and histologic results of regenerative procedures. *In* "Periodontal Regeneration. Current Status and Directions" (A. Polson, ed.), pp. 113–135. Quintessence Publishing Co., Chicago.

Caffesse, R. G., Kerry, G. S., Chaves, E. S. *et al.* (1988a). Clinical evaluation of the use of citric acid and autologous fibronectin in periodontal surgery. *J. Periodontol.* **59**, 565–569.

Caffesse, R. G., Smith, B. A., Castelli, W. A. *et al.* (1988b). New attachment achieved by guided tissue regeneration in beagle dogs. *J. Periodontol.* **59**, 589–594.

Caffesse, R. G., Nasjleti, C. E., Plotzke, A. E. *et al.* (1994). Guided tissue regeneration and bone grafts in the treatment of furcation defects. *J. Periodontol.* **64**, 1145–1153.

Chen, P., Carrington, J. L., Hammonds, R. G. *et al.* (1991). Stimulation of chondrogenesis in limb bud mesoderm cells by recombinant human bone morphogenetic protein (BMP-2B) and modulation by transforming growth factor β1 and β2. *Exp. Cell Res.* **195**, 509–515.

Cho, M. I., and Garant, P. R. (1981). An electron microscope radioautographic study of collagen secretion in periodontal ligament fibroblasts of the mouse. *Anat. Rec.* **201**, 577–586.

Cho, M. I., Lin, W. L., and Genco, R. J. (1995). Platelet derived growth factor-modulated guided tissue regenerative therapy. *J. Periodontol.* **66**, 522–530.

Clark, R. A. F. (1988). Overview and general considerations of wound repair. *In* "The Molecular and Cellular Biology of Wound Repair" (R. A. F. Clark and P. M. Henson, eds.), pp. 3–33. Plenum, New York.

Dahlin, C., Lekholm, V., and Linde, A. (1991). Membrane induced bone augmentation and titanium implants: A report on ten fixtures followed from 1–3 years after loading. *Int. J. Periodont. Restorative Dent.* **11**, 273–281.

Demolon, I. A., Persson, G. R., Ammons, W. F. *et al.* (1994). Effects of antibiotic treatment on clinical conditions with guided tissue regeneration: One year results. *J. Periodontol.* **65**, 713–717.

Dragoo, M. R., ed. (1981). "Regeneration of the Periodontal Attachment in Humans." Lea & Febiger, Philadelphia.

Egelberg, J. (1987). Regeneration and repair of periodontal tissues. *J. Periodontol. Res.* **22**, 233–242.

Florès de Jacoby, L., Zimmermann, A., and Tsalikis, L. (1994). Experiences with guided tissue regeneration of advanced periodontal disease. A clinical re-entry study. Part I: Vertical, horizontal and combined vertical horizontal periodontal defects. *J. Clin. Periodontol.* **21**, 113–117.

Gestrelius, S., Andersson, C., Lidstöm, D., *et al.* (1997). *In vitro* studies on periodontal ligament cells and enamel matrix derivative. *J. Clin. Periodontol.* **24**, 685–692.

Giannobile, W. V., Finkelman, R. D., and Lynch, S. E. (1994). Comparison of canine and nonhuman primate animal models for periodontal regenerative therapy: Results following a single administration of PDGF/IGF-1. *J. Periodontol.* **65**, 1158–1168.

Gottlow, J., Laurell, L., Lundgren, D. *et al.* (1994). Periodontal tissue response to GTR therapy using a bioresorbable matrix barrier. A longitudinal study in the monkey. *Int. J. Periodont. Restorative Dent.* **14**, 437–449.

Greenstein, G., Jaffin, R. A., Hilsen, K. L. *et al.* (1985). Repair of anterior gingival defects with durapatite. *J. Periodontol.* **56**, 200–203.

Grevstad, H. J., and Leknes, K. N. (1993). Ultrastructure of plaque associated with polytetrafluoroethylene (PFTE) membranes used for guided tissue regeneration. *J. Clin. Periodontol.* **20**, 193–198.

Hammarström, L. (1997). Enamel matrix, cementum development and regeneration. *J. Clin. Periodontol.* **24**, 658–668.

Hammarström, L., Heijl, L., and Gestrelius, S. (1997). Periodontal regeneration in a buccal dehiscence model in monkeys after application of enamel matrix proteins. *J. Clin. Periodontol.* **24**, 669–677.

Hanachowicz, P. Y., Magloire, H., Joffre, A., *et al.* (1989). Utilisation de cultures épithéliales d'origine humaine en chirurgie muco-gingivale. *J. Parodontol.* **8**, 261–268.

Hancock, E. B. (1989). Regeneration procedures. *In* "Proceedings of the World Workshop on Clinical Periodontitis" (M. Nevius, W. Becker, and K. Kourmian, eds.), VI-1-20. Am. Acad. Periodontol., Chicago.

Hardwick, R., Hayers, B., and Flynn, C. (1995). Devices for dentoalveolar regeneration: An up-to-date literature review. *J. Periodontol.* **66**, 495–505.

Iglhaut, J., Aukhil, I., Simpson, D. M. *et al.* (1988). Progenitor cell kinetics during experimental guided tissue regeneration procedure. *J. Periodontol. Res.* **23**, 107–117.

Kelly, J. L., Sanchez, A., Brown, G. S., *et al.* (1993). Accumulation of PDGF-B and cell binding forms of PDGF-A in the extracellular matrix. *J. Cell biol.* **121**, 1153–1163.

Klinge, B., Nilveus, R., Kiger, R. S. *et al.* (1981). Effect of flap placement and defect size on healing of experimental furcation defects. *J. Periodontol. Res.* **12**, 236–248.

Kodama, T., Minabe, M., Hori, T. *et al.* (1989). The effect of concentrations of collagen barrier of periodontal wound healing. *J. Periodontol.* **60**, 205–210.

Laurell, L., Falk, H., Fornell, J. *et al.* (1994). Clinical use of a bioresorbable matrix barrier guided tissue regeneration therapy. Case series. *J. Periodontol.* **65**, 967–975.

Lecoeur, L., and Ouhayoun, J. P. (1997). *in vitro* induction of osteogenic differenciation from nonosteogenic mesenchymal cells. *Biomaterials* **18**, 989–993.

Lekovic, V., Kenney, E. B., Kovacevic, K. *et al.* (1989). Evaluation of guided tissue regeneration in class II furcation defects. A clinical re-entry study. *J. Periodontol.* **60**, 694–698.

Linde, A., Alberius, P., Dahlin, C. *et al.* (1993). Osteopromotion: A soft tissue exclusion principle using a membrane for bone healing and bone neogenesis. *J. Periodontol.* **64**, 1116–1128.

Ling, L., Lai, Y., Hwang, H. *et al.* (1994). Response of regenerative tissue to plaque: A histological study in monkeys. *J. Periodontol.* **65**, 781–787.

Lovelace, T. B., Mellonig, J. T., Meffert, R. M., *et al.* (1998). Clinical evaluation of bioactive glass in the treatment of periodontal osseous defects in humans. *J. Periodontol.* **69**, 1027–1035.

Lowenguth, R. A., Polson, A., and Capon, J. G. (1993). Oriented cell and fibre attachment systems *in vivo*. *J. Periodontol.* **64**, 330–342.

Lundgren, D., Lundgren, K., Sennerby, L., *et al.* (1995). Augmentation of intramembraneous bone beyond the skeletal envelope using an occlusive titanium barrier. *Clin. Oral Implants Res.* **6**, 67–72.

Lynch, S. E. (1994). The role of growth factors in periodontal repair and regeneration. *In* "Periodontal Regeneration. Current Status and Directions" (A. Polson, ed.), pp. 179–198. Quintessence Publishing Co., Chicago.

Lynch, S. E., Colvine, R. B., and Antoniades, H. N. (1989). Growth factors in wound healing: Single and synergistic effects on partial thickness porcine skin wounds. *J. Clin. Invest.* **84**, 640–646.

Lynch, S. E., De Castilla, G. R., Williams, R. C., *et al.* (1991). The effects of short-term application of a combination of a platelet-derived and insulin-like growth factors on periodontal wound healing. *J. Periodontol.* **62**, 458–467.

Martin, G., Béné, M. C., Molé, N., *et al.* (1995). Synthetic extracellular matrix supports healing of muco-gingival donor sites. *Tissue Eng.* **1**, 279–288.

Miller, N., Penaud, J., Foliguet, B., *et al.* (1996). Comparison of resorption rates of two commercially available resorbable membranes. *J. Clin. Periodontol.* **23**, 1051–1059.

Miller, N. A., Roland, E., Benamghar, *et al.* (1987). An analysis of the CPITN periodontal treatment needs in France. *Commun. Dent. Health* **4**, 415–423.

Minabe, M. (1991). A critical review for the biologic rationale for guided tissue regeneration. *J. Periodontol.* **62**, 171–179.

Misch, C. E., and Dietsch, F. (1993). Bone grafting materials in implant dentistry. *Implant. Dent.* **2**, 158–167.

Miyazaki, H., Pilot, T., and Leclercq, M. H. *et al.*, (1991). Profile of periodontal conditions in adults measured by CPITN. *Int. Dent. J.* **41**, 74–80.

Nevins, M., and Mellonig, J. T. (1992). Enhancement of the damaged edentulous ridge to receive dental implants: A combination of allograft and the Gore-Tex membrane. *Int. J. Periodont. Restorative Dent.* **12**, 97–111.

Nevins, M., and Mellonig, J. T. (1994). The advantages of localized ridge augmentation prior to implant replacement: A stage event. *Int. J. Periodont. Restorative Dent.* **14**, 97–111.

Nyman, S., Lindhe, J., Karring, T. *et al.* (1982). New attachment following surgical treatment of human periodontal disease. *J. Clin. Periodontol.* **9**, 290–296.

Nyman, S., Gottlow, J., Lindhe, J. *et al.* (1987). New attachment formation by guided tissue regeneration. *J. Periodontol. Res.* **22**, 252–254.

Oliver, R. C., Brown, L. J., and Loe, H. (1989). An estimate of periodontal treatment needs in the US based on epidemiologic data. *J. Periodontol.* **60**, 371–380.

Park, J. B., Matsuura, M., Han, K. Y. *et al.* (1995). Periodontal regeneration in class II furcation defects of the beagle dog using guided tissue regenerative therapy with or without platelet-derived growth factor. *J. Periodontol.* **66**, 461–477.

Penaud, J., Martin, G., Membre, H. *et al.* (1992). Influence d'une membrane de collagène expérimentale dans un protocole de régénération de l'attache. Expérimentation chez le chien beagle. *J. Parodontol.* **11**, 55–64.

Pini Prato, G. P., Cortellini, P., Clauser, C., *et al.* (1988). Fibrin and fibronectin sealing system in a guided tissue regeneration procedure. A case report. *J. Periodontol.* **59**, 679–683.

Pitaru, S., Tal, H., Soldinger, M. *et al.* (1989). Collagen membranes prevent apical migration of epithelium and support connective tissue attachment during periodontal wound healing in dogs. *J. Periodontol. Res.* **24**, 247–253.

Plotzke, A. E., Barbosa, C. E., Nasjleti, E. C. *et al.* (1993). Histologic and histometric responses to polymeric composite grafts. *J. Periodontol.* **64**, 393–348.

Polson, A. M., and Proye, M. P. (1983). Fibrin linkage: A precursor of new attachment. *J. Periodontol.* **54**, 141–147.

Pontoriero, R., Nyman, S., Lindhe, J. *et al.* (1987). Guided tissue regeneration in the treatment of furcation defects in man. *J. Clin. Periodontol.* **14**, 618–620.

Price, N., Bendall, S. P., Frondoza, C., *et al.* (1997). Human osteoblast-like cells (MG63) proliferate on a bioactive glass surface. *J. Biomed. Mater. Res.* **37**, 394–400.

Robert, P., and Frank, R. (1994). Periodontal guided regeneration with a new resorbable polylactic acid membrane. *J. Periodontol.* **65**, 414–422.

Rosen, P. S., Reynolds, M. A., and Bowers, G. M. (1998). A technique report on the *in situ* application of Atrisorb as a barrier for combination therapy. *Int. J. Periodont. Restorative Dent.* **18**, 249–255.

Rummelhart, J. M., Mellonig, J. T., Gray, J. L. *et al.* (1989). A comparison of freeze-dried bone allograft and demineralized DFDBA in human periodontal osseous defects. *J. Periodontol.* **60**, 655–663.

Sampath, T. K., and Reddi, A. H. (1981). Dissociative extraction and reconstitution of extracellular matrix components involved in local bone differentiation. *Proc. Natl. Acad. Sci. U.S.A.* **78**, 7599–7603.

Schenk, R. K., Buser, D., Harwick, W. R. *et al.* (1994). Healing pattern of bone regeneration in membrane-protected defects. *Int. J. Oral Maxillofacial Implants* **9**, 13–29.

Seibert, J., and Nyman, S. (1990). Localized ridge augmentation in dogs: A pilot study using membranes and hydroxyapatite. *J. Periodontol.* **61**, 157–165.

Selvig, K., Nilveus, R. E., Fitzmorris, L. *et al.* (1990). Scanning electron microscopic observations of cell population and bacterial contamination of membranes used for guided periodontal tissue regeneration in humans. *J. Periodontol.* **61**, 515–520.

Sennerby, L., Ericson, L. E., Thomsen, P. *et al.* (1991). Structure of the bone-titanium interface in retrieved clinical oral implants. *Clin. Oral Implants Res.* **2**, 103–111.

Sigurdsson, T. J., Marwick, R., Bogle, G. C. *et al.* (1994). Peridontal repair in dogs: Space provision by reinforced ePFTE membranes enhance bone and cementum regeneration in large supra-alveolar defects. *J. Periodontol.* **65**, 350–356.

Sigurdsson, T. J., Lee, M. B., Kubota, K. *et al.* (1995a). Periodontal repair in dogs: Recombinant human bone morphogenetic protein-2 significantly enhances periodontal regeneration. *J. Periodontol.* **66**, 131–138.

Sigurdsson, T. J., Takakis, D. N., Lee, M. *et al.* (1995b). Periodontal regenerative potential of space-providing expanded polytetrafluoroethylene membranes and recombinant bone morphogenetic proteins. *J. Periodontol.* **66**, 511–521.

Springfield, D. S. (1992). Autogenous bone graft: Nonvascular and vascular. *Orthopedics* **15**, 1237–1241.

Stoor, P., Soderling, E., and Salonen, J. I. (1998). Antibacterial effects of a bioactive glass paste on oral microorganisms. *Acta Odontol. Scand.* **56**, 161–165.

Takata, T., Katauchi, K., Miyauchi, M. *et al.* (1993). Periodontal tissue regeneration on the surface of synthetic hydroxyapatite implanted into root surface. *J. Periodontol.* **66**, 125–130.

Terranova, V. P., Hic, S., Franzetti, L., *et al.* (1987). A biochemical approach to periodontal regeneration. AFSCM: Assays for specific cell migration. *J. Periodontol.* **58**, 247–257.

Terranova, V. P., Price, R. M., Nishimura, F. *et al.* (1994). Cellular and molecular biology of periodontal wound healing. *In* "Periodontal Regeneration. Current Status and Directions" (A. Polson, ed.), pp. 167–178. Quintessence Publishing Co., Chicago.

Tonetti, M. S., Pini-Prato, G., and Cortellini, P. (1993). Periodontal regeneration of human intrabony defects. IV: Determinants of healing response. *J. Periodontol.* **64**, 934–940.

Urist, M. R., and Iwata, H. (1973). Preservation and biodegradation of the morphogenetic property of bone matrix. *J. Theor. Biol.* **38**, 155–163.

von Arx, T., and Kurt, B. (1999). Implant placement and simultaneous ridge augmentation using autogenous bone and a microtitanium mesh: A prospective clinical study with 20 implants. *Clin. Oral Implants Res.* **10**, 24–33.

Wikesjo, U. M. E., Claffey, N., Christerson, L. A. *et al.* (1988). Repair of periodontal furcation defects in beagle dogs following reconstructive surgery including root surface demineralization with tetracycline hydrochloride and topical fibronectin application. *J. Clin. Periodontol.* **15**, 73–80.

Woodley, D., Bachmann, P. M., and O'Keefe, E. J. (1990). The role of matrix components in human keratinocyte reepithelialization. *In* "Clinical and Experimental Approaches to Dermal and Epidermal Repair" (A. Barbul *et al.*, eds.), pp. 129–140. Wiley-Liss, New York.

Wozney, J. M. (1995). The potential role of bone morphogenetic proteins in periodontal reconstruction. *J. Periodontol.* **66**, 506–510.

Yamaguchi, A., Katagiri, T., Ikeda, T. *et al.* (1991). Recombinant human bone morphogenetic protein-2 stimulates osteoblastic maturation and inhibits myogenic differentiation *in vitro*. *J. Cell Biol.* **113**, 681–687.

Yukna, R. A. (1994a). Synthetic grafts and regeneration. *In* "Periodontal Regeneration. Current Status and Directions" (A. Polson, ed.), pp. 103–112. Quintessence Publishing Co., Chicago.

Yukna, R. A. (1994b). Clinical evaluation of coralline calcium carbonate as a bone replacement graft material in human periodontal osseous defects. *J. Periodontol.* **65**, 177–185.

Zetterström, O., Andersson, C., Erikson, L., *et al.* (1997). Clinical safety of enamel matrix derivative (Emdogain) in the treatment of periodontal defects. *J. Clin. Periodontol.* **24**, 697–704.

CHAPTER 60

REGENERATION OF DENTIN

R. Bruce Rutherford

INTRODUCTION

The goal of engineering the formation of new dentin is primarily motivated by the clinical need to restore or thicken this mineralized tissue. Dentin surrounds and protects the vital tissues of the tooth that reside in the dental pulp. Dentin is approximately 80% mineralized and provides, in combination with enamel, the major structural mass of the tooth. Like enamel, dentin is a nonvascular tissue; however, unlike enamel, the cells that synthesize dentin (odontoblasts) may remain vital throughout adult life and may synthesize additional dentin. Recent experimental evidence has suggested that, even if the odontoblast cells (which line the dentin pulp interface) are lost, it may be possible to induce the differentiation of odontoblasts from pulp tissue using certain bone morphogenetic proteins and thereby induce the synthesis of new dentin. Successful development of a process to engineer successfully the regeneration of dentin could have commercial applications as an enhanced pulp capping agent, as an alternative to root canal therapy under certain circumstances, and as a potential means of reducing tooth sensitivity often associated with the placement of tooth fillings.

DENTIN

There are several different forms of dentin. Primary and secondary dentins are terms generally applied to dentins that form during the development of teeth. Tertiary dentins are those forming in response to a specific stimulus, such as partial removal of the dentin layer. Tertiary dentin matrices are categorized as reparative or reactionary. Reparative dentin is a tertiary dentin matrix formed by new odontoblast-like cells in response to a specific stimulus after loss of those odontoblasts responsible for synthesis of primary and secondary dentin. Reactionary dentin is formed by original odontoblasts (Smith *et al.*, 1995). We use the terms "reparative" and "reactionary" dentin when referring to tertiary dentin matrices of mammalian teeth. The induction of differentiation of new odontoblasts from undifferentiated tissues may require different strategies than the stimulation of extant odontobasts. It is likely that mature adult pulps contain stromal cells capable of producing reparative dentin if given the appropriate stimulus, in that most data indicate that connective tissue repair is mediated by resident stromal cells near the site of wounding (Prokop, 1997).

BONE MORPHOGENETIC PROTEINS

Bone morphogenetic proteins (BMPs) are a group of signaling molecules that were originally isolated and described for their capacity to induce endochrondral bone formation in subcutaneous, intramuscular, or intrabony sties in mammals (Sampath *et al.*, 1990; Wang *et al.*, 1990). There are now more than 30 BMP-like proteins that have been identified, though only a small subset, at this time, have been demonstrated to be osteogenic (Wozney and Rosen, 1998).

Structurally all BMPs are members of the transforming growth factor β (TGF-β) superfamily and several have been implicated in key stages of embryonic tissue morphogenesis (Hogan, 1996). As such, the early implications for this class of proteins were that their potential role in tissue formation could be possibly broader than a more limited role in cartilage and bone induction. The first nonbone, noncartilage tissue that has been demonstrated to be induced by a bone morphogenetic protein is dentin. The experimental evolution of the use of BMPs to engineer dentin regeneration is described below.

INDUCTION POTENTIAL OF DENTIN

Several investigators have demonstrated that demineralized dentin powder, like demineralized bone powder (Urist, 1965), induces ectopic bone formation (Bang, 1972; Butler *et al.*, 1977; Inoue *et al.*, 1986). However, unlike demineralized bovine bone powder, for which the specific proteins for bone induction have been identified (Sampath *et al.*, 1990), demineralized dentin powders have not been further characterized and the molecular basis for dentin BMP activity has not been established. It is likely that several of the BMPs are involved.

Dentin is also dentinogenic. Mineralized tissue superficially resembling dentin forms in association with dentin fragments introduced into pulp during mechanical exposure (Seltzer and Bender, 1981), and demineralized dentin implanted into dental pulp induces mineralized tissue formation (Tziafas and Kolokuris, 1990; Tziafas *et al.*, 1992). The dentin-associated dentinogenic activity, like the bone-associated osteogenic activity (Sampath *et al.*, 1990), is at least partially soluble in 4 *M* guanidine hydrochloride, implicating molecules similar to the BMPs (Nakashima, 1990). In addition, crude dentin fractions prepared by ethylenediaminetetraacetic acid (EDTA) extraction and collagenase digestion stimulate reparative dentinogenesis in small pulp exposures in ferrets (Smith *et al.*, 1990) or reactive dentinogenesis when applied to freshly cut dentin surfaces (Smith *et al.*, 1994). In addition, dog dental pulps respond to crude allogenic, bone-derived BMP by forming reparative dentin (Nakashima, 1990; Lianjia *et al.*, 1993).

RECOMBINANT HUMAN BMPs INDUCE DENTIN FORMATION

Experiments linking BMPs to reparative dentinogenesis utilized recombinant human BMP2, BMP4, BMP7, TGF-β, and platelet-derived growth factor (PDGF). In a follow-up study to her previous work, Nakashima reported that recombinant human BMP2 and BMP4, but not TGF-β, induce reparative dentinogenesis in dogs when placed on partially amputated dental pulps (Nakashima, 1994). In this study, addition of BMP2 or BMP4 doubled (82 to 42%) the amount of reparative dentin formed in partially amputated dog dental pulps when compared to teeth treated with the collagen-based carrier material after 60 days. The TGF-β was inactive in these experiments. The carrier was allogenic dentin collagen prepared in a manner similar to that described above and mixed with chondroitin 6-sulfate and acid-soluble rat tail tendon collagen.

Qualitatively similar data were obtained from nonhuman primate studies using recombinant human BMP7 (Rutherford *et al.*, 1993, 1994). These studies demonstrated that recombinant human BMP7 complexed with an insoluble type I collagen matrix (CM) delivery vehicle predictably and reliably induced reparative dentinogenesis when placed on freshly cut healthy pulp tissues (Fig. 60.1) in nonhuman primates. The initial experiments in this series tested the hypothesis that recombinant human BMP7-induced tissue formation occurs in the place of and replaces the inductive material. This suggests that the size and shape of the mass of inductive material control the size and shape of the reparative dentin. Placing variable amounts of the BMP7/CM combination in a surgically induced model of pulp exposure resulted in the formation of variable amounts of reparative dentin, demonstrating that the mass of reparative dentin formed is, indeed, proportional to the mass of BMP7/CM placed (Fig. 60.2).

This hypothesis also predicts that the reparative dentin should form largely superficial to, and hence preserve the volume of, the remaining pulp tissue. Initial experiments tested this hypothesis by layering osteogenic protein-1 (OP-1)/CM on the cut surface of the partially amputated pulp and extending it laterally onto the adjacent cut dentin surface. We observed that the reparative dentin formed largely superficial to the existing pulp tissue and extended laterally onto the existing dentin, also superficial to the level of the amputated pulp (Rutherford *et al.*, 1993).

This dentinogenic response appears to be independent of the amount of the pulp tissue removed. We have created variably sized lesions in coronal pulps, ranging from small and superficial (1 mm in diameter and 0.5 mm deep) (R. B. Rutherford, unpublished observations) to large lesions (approximately 1–1.5 × 2–2.5 × 2–3 mm deep), which effectively removed all the dentin covering the pulp chamber (Rutherford *et al.*, 1993), to complete removal of the coronal pulp

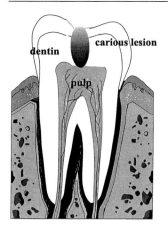

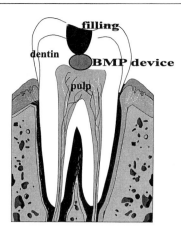

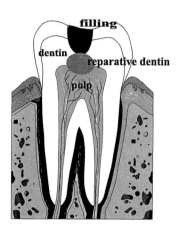

Fig. 60.1. Depiction of the application of the BMP7 device (recombinant human BMP7 with a bovine-derived, type I collagen matrix) to a freshly cut pulpal surface in a surgically induced model of pulp exposure.

(Rutherford *et al.*, 1994). In addition, we have removed approximately 75% of the root pulp from healthy ferret teeth and layered BMP7/CM over the remaining pulp (R. B. Rutherford, unpublished observations). In most cases there was robust formation of reparative dentin. The formation process is described below.

Initially the BMP7/CM is converted to a minimally inflamed, highly vascular granulation tissue that matures to a loose, fibrous connective tissue resembling dental pulp. This new tissue begins an apparently random pattern of mineralization, trapping cells and vascular channels in lacunae. BMP7/CM-induced reparative dentin in monkeys and ferrets is, at first, predominantly

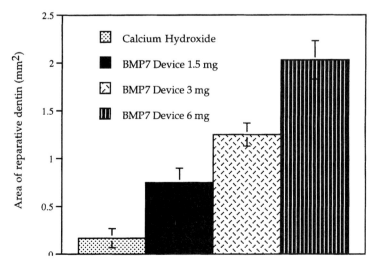

Fig. 60.2. Cross-sectional area of reparative dentin formed in response to varying quantities of the BMP7 device placed in surgically induced primate models of pupal exposure, as determined by histomorphometric analysis (Rutherford et al., 1993).

atubular, with contiguous zones of tubular dentin adjacent to the dental pulp appearing after a few weeks. Odontoblast-like cells are associated with areas of tubular-like dentin with cellular processes apparently extending into the dentinal tubules (Rutherford *et al.,* 1994). Approximately 95% of the original mass of tissue is mineralized by 6 months, yet only about 20% of the surface was composed of tubular dentin and associated columnar odontoblasts, whereas 50% of the pulpal surface is lined with very flat cells that appear metabolically quiescent (Rutherford *et al.,* 1994). The remaining cells appear more cuboidal and are associated with dentin that appears to be atubular.

DELIVERY SYSTEMS

With regard to the choice of delivery systems for application of a dentinogenic substance, the experiments that have been performed to date have utilized naturally sourced, bioresorbable collagen matrices for several important reasons. First and foremost, collagen matrices provide a highly desirable environment for the chemoattraction and retention of local cell types at the site of tissue damage or deficit. Second, collagen matrices provide a temporary scaffold for initiation of the tissue repair/regeneration process. Last and not insignificantly, collagen matrices provide for hemostasis, which is often an important advantage in a site that is likely to exhibit substantial bleeding.

The use of naturally-sourced delivery materials is, of course, not optimal from the standpoint of pharmaceutical development. Inevitably, variations in source material as well as the inherent risk of infectious disease make naturally sourced products or delivery systems less than ideal. However, to date, these naturally sourced delivery materials have provided the most compatible systems for the promotion of tissue regeneration, particularly for the bone- and dentin-promoting activities associated with the BMP subclass of the TGF-β subclass of proteins. Careful selection of source materials and rigorous manufacturing processes can eliminate the potential problems associated with naturally sourced delivery systems. Ultimately, better characterized, biocompatible, bioresorbable, synthetic materials may provide a more desirable alternative.

CLINICAL CONSIDERATIONS

The experiments described above (Nakashima, 1990, 1994; Rutherford *et al.,* 1993, 1994) utilized clinically healthy teeth. No attempt was made to induce inflammation (pulpitis) specifically or to test the capacity of the recombinant proteins to induce reparative dentinogenesis in previously diseased tissues. To determine the effect of a preexisting pulpitis on the response to BMP7/CM, we have developed an animal model of experimental pulpitis by injecting bacterial lipopolysaccharide (LPS; *Salomonella typhimurium,* 5 μg/μl) into small surgical exposures of adult ferret teeth. The teeth were sealed with standard dental materials. After 3 days, most coronal pulps displayed foci of chronic inflammation with variable tissue necrosis in the center of the lesion. All lesions observed ($n = 10$) were limited to the coronal tissues with the subjacent root pulp free of inflammatory cells (R. B. Rutherford, unpublished observations). After 3 days, the teeth were surgically reexposed, the pulp removed to the level of the root canal, and BMP7/CM layered on the freshly amputated pulp tissue. Four concentrations of BMP7 (0, 2.5, 7.5, and 25 μg/mg CM; $n = 20$ teeth per group) were tested. In all cases, no reparative dentin formed in teeth pretreated with LPS, whereas all three BMP preparations induced reparative dentinogenesis in healthy or sham-infected pulps. Most pulps were necrotic after 2 weeks. These data suggest that a single application of recombinant BMP fails to induce reparative dentinogenesis in freshly debrided tissues deep to a focus of inflamed pulp tissue. The reasons for this failure are not clear. It is possible that too few responsive cells are present in the noninflamed pulp deep to the experimental inflammatory lesions. However, this explanation seems unlikely given earlier data suggesting that responsive tissue is present nearly the entire length of the root (see above). More likely, perhaps, is that the effectiveness of BMP7 is attenuated in inflamed tissues to the extent that a single application of up to 25 μg BMP7 is insufficient to induce a reparative response.

It is possible that other clinical conditions exist wherein a single local application of a recombinant protein will prove insufficient to induce tissue regeneration. Therefore, in collaboration with Renny Franceschi, we have developed and are testing a recombinant replication-defective adenovirus that contains a mouse BMP7 gene driven by a cytomegalovirus promoter. Preliminary experiments reveal that the virus transduces dental pulp cells *in vitro* and *in vivo* and induces reparative dentinogenesis in a manner similar to exogenous recombinant protein. Further experiments testing the efficacy of this vector in experimentally inflamed teeth are in progress (R. B. Rutherford, unpublished observations).

ADDITIONAL CONSIDERATIONS

Qualitatively the pattern of reparative dentin formation in response to the crude preparations or to highly purified recombinant proteins described above is similar. The tissue initially appears atubular with cellular and soft tissue inclusions. Subsequently a more tubular form of matrix with associated odontoblast-like cells appears attached to the mass of atubular matrix. This observation supports the idea that some extracellular matrix component is a prerequisite to odontoblast differentiation and tubular dentin formation (Veis, 1985). A detailed study of the structure of the surface of the atubular dentin and the dynamics of the associated odontoblast differentiation may reveal the critical factors involved in this process. Such knowledge may provide the opportunity to regulate odontoblast differentiation and hence the architecture of the tissue. Regulation of dentin architecture may be important. The function of dentinal tubules is not known. The long-term viability of atubular dentin, its caries susceptability, permeability, and the magnitude of the bond strength with existing dentin bonding agents are not known. Optimally engineered reparative dentin may comprise a combination of atubular and tubular dentin. For example, if a layer of tubular-like dentin adjacent the pulp is required for long-term tissue viability, yet atubular dentin is less susceptible to caries, an optimal layer of reparative dentin may be composed of a layer of tubular dentin deep to a layer of atubular dentin.

POTENTIAL APPLICATIONS OF BMP-INDUCED DENTINOGENESIS

Surgical debridement to expose vital pulp and the application of a recombinant BMP (or other dentinogenic molecule) protein or a recombinant viral vector capable of transducing cells to produce a BMP7-based product could lead to several new therapeutic modalities for the preservation of vital and functional teeth. The development of such procedures could provide for improved pulp capping agents with enhanced biological responsiveness as well as potential alternatives to conventional root canal therapy in certain instances.

Another application for the dentinogenic properties of BMPs is the potential reduction of tooth sensitivity via the transdentinal stimulation of reactionary dentin. In this approach BMP7 was applied to freshly cut dentin surfaces in a liquid formulation with the hope that BMP7 might diffuse through the dentinal tubules and stimulate the formation of reactionary dentin within the

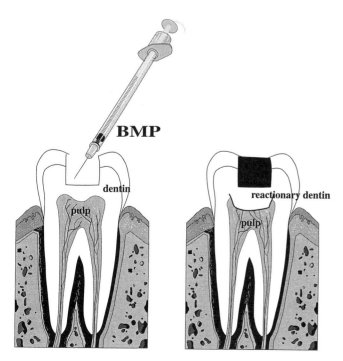

Fig. 60.3. Depiction of the application of recombinant human BMP7 in a liquid formulation to freshly cut dentinal tubules in a deep cavity preparation intended to stimulate the formation of new reactionary dentin.

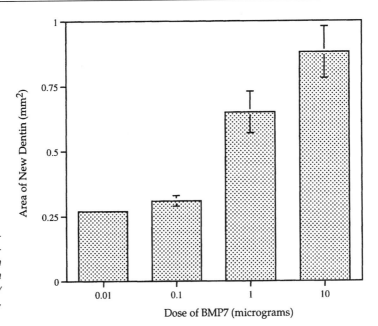

Fig. 60.4. Dose-dependent re-sponse of reactionary dentin for-mation to recombinant human BMP7 under residual dentin in a primate model of deep cavity preparation. (Rutherford et al., 1995).

pulp chamber (Fig. 60.3). Initial experiments appear to indicate that this is possible (Rutherford *et al.,* 1995). Stimulation of reactionary dentin by an ethanol-containing solution of BMP7 revealed a dose-dependent response of the pulpal tissue to the dentinogenic properties of BMP7 (Fig. 60.4). Unfortunately, it is not possible to measure adequately parameters of tooth sensitivity in animal models. As such, the rigorous establishment of a correlation between the capacity of BMP7 to induce a thickening of the residual dentin layer separating a dental filling from the pulp, and thereby, potentially diminish the sensitivity of teeth to newly placed dental fillings, must await clinical investigation in human subjects.

REFERENCES

Bang, G. (1972). Induction of heterotopic bone formation by demineralized dentin in guinea pigs: Antigenicity of the dentin matrix. *J. Oral Pathol.* **1,** 172–185.

Butler, W. T., Mikulski, A., Urist, M. R., Bridges, G., and Uyeno, S. (1977). Noncollagenous proteins of a rat dentin matrix possessing bone morphogenetic activity. *J. Dent. Res.* **56**(3), 228–232.

Hogan, B. L. (1996). Bone morphogenetic proteins in development. *Curr. Opin. Genet. Dev.* **6**(4), 432–438.

Inoue, T., Deporter, D. A., and Melcher, A. H. (1986). Induction of chrondrogenesis in muscle, skin, bone marrow and periodontal ligament by demineralized dentin and bone matrix *in vivo* and *in vitro. J. Dent. Res.* **65,** 12–22.

Lianjia, Y., Yuhao, G., and White, F. (1993). Bovine bone morphogenetic protein induced dentinogenesis. *Clin. Orthop. Relat. Res.* **295,** 305–312.

Nakashima, M. (1990). The induction of reparative dentin in the amputated dental pulp of the dog by bone morphogenetic protein. *Arch. Oral Biol.* **35**(7), 493–497.

Nakashima, M. (1994). Induction of dentin formation on canine amputated pulp by recombinant human bone morphogenetic proteins (BMP)-2 and -4. *J. Dent. Res.* **73**(9), 1515–1522.

Prokop, D. J. (1997). Marrow stromal cells as stem cells for nonhematopoietic tissue. *Science* **276,** 71–14.

Rutherford, R. B., Wahle, J., Tucker, M., Rueger, D., and Charette, M. (1993). Induction of reparative dentin formation in monkeys by recombinant human osteogenic protein-1. *Arch. Oral Biol.* **38,** 571–576.

Rutherford, R. B., Spangberg, L., Tucker, M., Rueger, D., and Charette, M. (1994). Time course of the induction of reparative dentin formation in monkeys by recombinant human osteogenic protein-1. *Arch. Oral Biol.* **39,** 833–838.

Rutherford, R. B., Spangberg, L., Tucker, M., and Charette, M. (1995). Transdentinal stimulation of reparative dentin formation by osteogenic protein-1. *Arch. Oral Biol.* **40,** 681–683.

Sampath, T. K., Coughlin, J. E., Whetstone, R. M., Banach, D., Corbett, C., Ridge, R. J., Ozkaynak, E., Oppermann, H., and Rueger, D. C. (1990). Bovine osteogenic protein is composed of dimers of OP-1 and BMP-2A, two members of the transforming growth factor-beta superfamily. *J. Biol. Chem.* **265**(22), 13198–205.

Seltzer, S., and Bender, I. B. (1981). Pulp capping and pulpotomy. *In* "Biologic Considerations in Dental Procedures" (S. Seltzer and I. B. Bender, eds.), pp. 281–302. Lippincott, Philadelphia.

Smith, A. J., Tobias, R. S., Plant, C. G., Browne, R. M., Lesot, H., and Ruch, J. V. (1990). *In vivo* morphogenetic activity of dentine matrix proteins. *J. Biol. Buccale* **18**(2), 123–129.

Smith, A. J., Tobias, R. S., Cassidy, N., Plant, C. G., Browne, R. M., Begue-Kirn, C., Ruch, J. V., and Lesot, H. (1994). Odontoblast stimulation in ferrets by dentine matrix components. *Arch. Oral Biol.* **39**(1), 13–22.

Smith, A. J., Cassidy, N., Perry, H., Begue-Kirn, C., Ruch, J. V., and Lesot, H. (1995). Reactionary dentinogenesis. *Int. J. Dev. Biol.* **39**(1), 273–280.

Tziafas, D., and Kolokuris, I. (1990). Inductive influences of demineralized dentine and bone matrix on pulp cells: An approach to secondary dentinogenesis. *J. Dent. Res.* **69**, 75–81.

Tziafas, D., Kolokuris, I., Alvanou, A., and Kaidoglou, K. (1992). Short-term dentinogenic response of dog dental pulp tissue after its induction by demineralized or native dentine or predentine. *Arch. Oral Biol.* **37**, 119–128.

Urist, M. R. (1965). Bone: Formation by induction. *Science* **150**, 893–899.

Veis, A. (1985). The role of dental pulp—Thoughts on the session on pulp repair processes. *J. Dent. Res.* **64**, 552–554.

Wang, E. A., Rosen, V., D'Alessandro, J. S., Bauduy, M., Cordes, P., Harada, T., Isreal, D. I., Hewick, R. M., Kerns, K. M., and LaPan, P. (1990). Recombinant human bone morphogenetic protein induces bone formation. *Proc. Natl. Acad. Sci. U.S.A.* **87**, 2220–2224.

Wozney, J. M., and Rosen, V. (1998). Bone morphogenetic protein and bone morphogenetic protein gene family in bone formation and repair. *Clin. Orthop. Relat. Res.* **346**, 26–37.

PART XIX: SKIN

CHAPTER 61

WOUND REPAIR: BASIC BIOLOGY TO TISSUE ENGINEERING

Richard A. F. Clark and Adam J. Singer

INTRODUCTION

The skin is the largest organ in the body and its primary function is to serve as a protective barrier against the environment. Other important functions of the skin include fluid homeostasis, thermoregulation, immune surveillance, sensory detection, and self-healing (Holbrook and Wolff, 1993). Loss of the integrity of large portions of the skin due to injury or illness may result in significant disability or even death. The most common cause of significant skin loss is thermal injury, with an estimated 2.5 million burns each year in the United States alone (Centers for Disease Control, 1982). Other causes of skin loss include chronic ulcerations (secondary to pressure, venous stasis, or diabetes mellitus), trauma, excision of skin tumors, or other dermatology conditions such as pemphigus or toxic epidermal necrolysis. It is estimated that the prevalence of leg ulcers alone is between 0.5 and 1.5%, with an annual cost of nearly $1 billion (Phillips and Dover, 1991). Some experts have estimated that the total number of chronic wounds exceeds 2 million annually in the United States (Medical Data International, 1998). The social and financial tolls of chronic wounds are extremely high.

Principal goals in wound management are to achieve rapid wound closure and a functional and aesthetic scar. Over the past two decades extraordinary advances in cellular and molecular biology have greatly expanded our comprehension of the basic biological processes involved in wound repair and tissue regeneration (Clark, 1996a). Ultimately these strides in basic knowledge will lead to advancements in wound care resulting in accelerated rates of ulcer and normal wound repair, scars of greater strength, and prevention of keloids and fibrosis. Already one growth factor and several skin substitutes have reached the marketplace for second-line therapy of recalcitrant ulcers (Singer and Clark, 1999). Furthermore, because tumor stroma generation is similar to wound healing (Dvorak, 1986), the increased knowledge in wound repair may lead to unexpected advances in tumor therapy. Clearly scientific breakthroughs today in the processes of wound healing and tissue regeneration will lead to future therapeutic successes in wound care and tissue engineering. In this chapter the use of cultured keratinocytes, tissue-engineered skin substitutes, and recombinant growth factors for wound care are discussed in the context of wound healing biology.

Wound repair is not a simple linear process in which growth factors released by phylogistic events activate parenchymal cell proliferation and migration, but is rather an integration of dynamic interactive processes involving soluble mediators, formed blood elements, extracellular matrix, and parenchymal cells. Unencumbered, these wound repair processes follow a specific time sequence and can be temporally categorized into three major groups: inflammation, tissue formation, and tissue remodeling. The three phases of wound repair, however, are not mutually exclusive but rather overlapping in time. In this overview, current ideas about wound repair are presented in a sequence that roughly follows the chronology of wound repair. The reader is referred to "The Molecular and Cellular Biology of Wound Repair" (Clark, 1996a) for a more detailed discussion of the many processes involved in wound healing.

INFLAMMATION

Severe tissue injury causes blood vessel disruption with concomitant extravasation of blood constituents. Blood coagulation and platelet aggregation generate a fibrin-rich clot that plugs severed vessels and fills any discontinuity in the wounded tissue. While the blood clot within vessel lumen reestablishes hemostasis, the clot within the wound space provides a provisional matrix for cell migration.

PLATELETS

Successful hemostasis is dependent on platelet adhesion and aggregation. Platelets first adhere to interstitial connective tissue, then aggregate. In the process of aggregation, platelets release many mediators, including ADP, and express several clotting factors on their membrane surface. Together these platelet products facilitate coagulation and further platelet activation.

When activated platelets discharge their α granules, several adhesive proteins, including fibrinogen, fibronectin, thrombospondin, and von Willebrand factor VIII, are released. The first three act as ligands for platelet aggregation whereas von Willebrand factor VIII mediates platelet adhesion to fibrillar collagens and their subsequent activation (Ruggeri, 1993). Platelet adhesion to all four adhesive proteins is mediated through the platelet glycoprotein (GPIIb/IIIa; integrin αIIbβ3) surface receptor (Ginsberg et al., 1992). Platelet fibrinogen, once converted to fibrin by thrombin, adds to the fibrin clot. In addition, platelets release chemotactic factors for blood leukocytes (Weksler, 1992), and growth factors such as platelet-derived growth factor (PDGF) (Heldin, 1992), transforming growth factor α (TGF-α) (Nanney and King, 1996), and TGF-β (Roberts and Sporn, 1996), which promote new tissue generation.

NEUTROPHILS

The coagulation pathways activate complement pathways and injured parenchymal cells generate numerous vasoactive mediators and chemotactic factors, which together recruit inflammatory leukocytes to the wound site. Infiltrating neutrophils cleanse the wounded area of foreign particles, including bacteria. If excessive microorganisms or indigestible particles have lodged in the wound site, neutrophils will probably cause further tissue damage as they attempt to clear these contaminants through the release of enzymes and toxic oxygen products. When particle clearance has been completed, generation of granulocyte chemoattractants ceases and the remaining neutrophils become effete.

TRANSITION BETWEEN INFLAMMATION AND REPAIR

Whether neutrophil infiltrates resolve or persist, monocyte accumulation continues, stimulated by selective monocyte chemoattractants (Clark, 1996b). Besides promoting phagocytosis and debridement, adherence to extracellular matrix also stimulates monocytes to undergo metamorphosis into inflammatory or reparative macrophages. Wound macrophages, in fact, express TGF-β and PDGF mRNA as well as TGF-α and insulin-like growth factor-1 (IGF-1) mRNA (Rappolee et al., 1988). Because cultured macrophages produce and secrete the peptide growth factors interleukin-1 (IL-1), PDGF, TGF-β, TGF-α, and fibroblast growth factor (FGF), presumably wound macrophages also synthesize these protein products (Clark, 1996b). Such macrophage-derived growth factors may support the initiation and propagation of new tissue formation in wounds. Thus, macrophages appear to play a pivotal role in the transition between inflammation and repair (Riches, 1996).

EPITHELIALIZATION

Reepithelialization of a wound begins within hours after injury. It is clear that rapid reestablishment of any epithelial barrier decreases victim morbidity and mortality. Epithelial cells from residual epithelial structures move quickly to dissect the clot and the damaged stroma from the wound space and to repave the surface of viable tissue.

Migrating wound epidermal cells do not terminally differentiate as do keratinocytes of normal epidermis. These migrating cells do not contain keratin proteins normally found in mature stratified epidermis nor do they contain filaggrin, a matrix protein in which these keratins are embedded. In contrast, cells in all layers of the migrating epidermis contain keratins normally found only in the basal cells of stratified epidermis (Mansbridge and Knapp, 1987). Nevertheless, the phenotype of migrating epidermal cells is not identical to basal cells, in that the migrating cells also contain involucrin, a component of differentiated keratinocyte cell walls, and transglutaminase, an enzyme that cross-links cell wall proteins.

If the basement membrane is destroyed by injury, epidermal cells migrate over a provisional matrix of fibrin(ogen), fibronectin, tenasin, and vitronectin as well as stromal type I collagen (Clark, 1996b). Fibrinogen is interspersed with stromal type I collagen at the wound margin and interwoven with the fibrin clot in the wound space (Clark *et al.*, 1999). Wound keratinocytes express cell surface receptors for fibronectin, tenasin, and vitronectin (Clark, 1990; Juhasz *et al.*, 1993; Larjava *et al.*, 1993; Clark *et al.*, 1996a). These specialized cell membrane receptors are of the integrin superfamily (Table 61.1) (Hynes, 1992; Yamada *et al.*, 1996). In addition, the integrin $\alpha2\beta1$ collagen receptors, which are normally disposed along the lateral sides of basal keratinocytes, redistribute to the basal membrane of wound keratinocytes as they come in contact with type I collagen fibers of the dermis.

The migrating wound epidermis does not simply transit over a wound coated with provisional matrix but rather dissects through the wound, separating desiccated or otherwise nonviable tissue from viable tissue. The path of dissection appears to be determined by the array of integrins that the migrating epidermal cells express on their cell membranes, as described above. In addition, $\alpha v\beta3$, the receptor for fibrinogen/fibrin (Cheresh *et al.*, 1989) and denatured collagen (Davis, 1992), is not expressed on keratinocytes *in vitro* (Adams and Watt, 1991) or *in vivo* (Gailit *et al.*, 1994). Therefore, keratinocytes do not have the capacity to interact with these matrix proteins (Kubo *et al.*, 1999). Furthermore, in the presence of fibrinogen or fibrin, epidermal cells fail to bind fibronectin (Kubo *et al.*, 1999), yet can bind type I collagen. Hence migrating wound epidermis avoids the fibrin-, fibronectin-rich clot while migrating over a bed of fibrinogen, fibronectin, and type I collagen (Fig. 61.1).

Extracellular matrix degradation is clearly required for the dissection of migrating wound epidermis between the collagenous dermis, the fibrin eschar (Bugge *et al.*, 1996), and probably depends on epidermal cell production of both collagenase (Woodley *et al.*, 1986) and plasminogen activator (Grondahl-Hansen *et al.*, 1988) and perhaps stromelysin (Saarialho-Kere *et al.*, 1994). Plasminogen activator activates collagenase as well as plasminogen (Mignatti *et al.*, 1996) and

Table 61.1. Integrin superfamily

Integrins	Ligand	Integrin	Ligand
β1 Integrins		**αv Integrins**	
α1β1	Fibrillar collagen, laminin-1	αvβ1	Fibronectin (RGD), vitronectin
α2β1	Fibrillar collagen, laminin-1	αvβ3	Vitronectin (RGD),
α3β1	Fibronectin (RGD), laminin-5, entactin, denatured collagens		fibronectin, fibrinogen, von Willebrand factor, thrombospondin, denatured collagen
α4β1	Fibronectin (LEDV), VCAM-1	αvβ5	Fibronectin (RGD), vitronectin
α5β1	Fibronectin (RGD)	αvβ6	Fibronectin, tenascin
α6β1	Laminin		
α7β1	Laminin	**β2 Integrins**	
α8β1	Fibronectin, vitronectin	αMβ2	ICAM-1, iC3b, fibronogen, factor X
α9β1	Tenasin	αLβ2	ICAM-1 -2, and -3
Other ECM integrins		αXβ2	iC3b, fibrinogen
αIIbβ3	Same as αvβ3		
α6β4	Laminin		

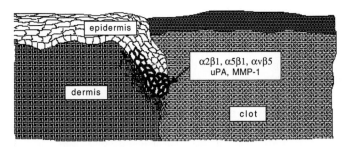

Fig. 61.1. Integrins in early (three-dimensional) reepithelialization. The migrating epidermis in contact with type I dermal collagen expresses a gene set including the collagen receptor α2β1, urokinase plasminogen activator (uPA), and interstitial collagenase (MMP-1). With this armamentarium the epidermis can dissect a path between viable dermis and nonviable clot and denature stroma, ultimately leading to slough of the eschar.

therefore facilitates the degradation of interstitial collagen as well as provisional matrix proteins. Interestingly, keratinocytes in direct contact with collagen greatly increase the amount of interstitial collagenase [matrix metalloproteinase-1 (MMP-1)] they produce compared to that produced when they reside on laminin-rich basement membrane or purified laminin (Petersen *et al.,* 1990). The migrating epidermis of superficial skin ulcers and burn wounds, in fact, expresses high levels of MMP-1 mRNA in areas where it presumably comes in direct contact with dermal collagen (Saarialho-Kere *et al.,* 1992; Stricklin and Nanney, 1994).

At 1 to 2 days after injury, epithelial cells at the wound margin begin to proliferate (Krawczyk, 1971). The stimuli for epithelial proliferation during reepithelialization have not been delineated, but several possibilities exist. Perhaps the absence of neighbor cells at the wound margin signals both epithelial migration and proliferation. The "free edge effect" has been thought to stimulate reendothelialization of large blood vessels after intimal damage (Heimark and Schwartz, 1988) and may relate to dissolution of cadherin junctions on the free edge (Dejana, 1996). Other possibilities, not exclusive of the former, are release of autocrine or paracrine growth factors that induce epidermal migration and proliferation and/or increased expression of growth factor receptors. Leading contenders include the epidermal growth factor (EGF) family (Nanney and King, 1996), especially TGF-α (Barrandon and Green, 1987) and heparin-binding epidermal growth factor (HB-EGF) Higashiyama *et al.,* 1991); the FGF family (Abraham and Klagsbrun, 1996), especially keratinocyte growth factor (KGF) (Werner, 1998); insulin-like growth factor (Galiano *et al.,* 1996); and TGF-β (Zambruno *et al.,* 1995). Although some growth factors, such as IGF, may derive from the circulation and thereby act as a hormone, other growth factors, such as HB-EGF and KGF, derive from macrophages and dermal parenchymal cells, respectively, and act on epidermal cells through a paracrine pathway (Werner, 1998). In contrast, TGF-α and TGF-β, originate from keratinocytes and act directly on the producer cell or adjacent epidermal cells in an autocrine or juxtacrine fashion (Coffey *et al.,* 1987; Brachmann *et al.,* 1989). Many of these growth factors have been shown to stimulate reepithelialization in animal models (Brown *et al.,* 1989; Lynch *et al.,* 1989; Hebda *et al.,* 1990; Staino-Coico *et al.,* 1993) or to lack receptors in models of deficient reepithelialization (Werner *et al.,* 1994), supporting the hypothesis that they are active during normal wound repair.

As reepithelialization ensues, basement membrane proteins reappear in a very ordered sequence from the margin of the wound inward in a zipperlike fashion (Clark, 1996b). Epidermal cells revert to their normal phenotype, once again firmly attaching to reestablished basement membrane through two hemidesmosomal proteins, α6β4 integrin and 180-kDa bullous pemphigoid antigen (Gipson *et al.,* 1993), and to the underlying neodermis through type VII collagen fibrils (Gipson *et al.,* 1988).

Epidermal Autografts and Allografts

Immediate wound coverage, whether permanent or temporary, is one of the cornerstones of wound management. When the epidermis fails to heal normally or when a large surface area has been destroyed, supplemental epidermal coverage can be beneficial. Ideally, wound coverage would be achieved using autologous skin or material having qualities similar to those of autologous skin. Use of skin grafts, however, often requires extensive, invasive harvesting and is of limited quantity.

Autologous cultured keratinocyte grafts were first used in humans by O'Conner *et al.* (1981). Subsequently there has been extensive experience with cultured epidermal grafts for the treatment of burns as well as other acute and chronic wounds (Gallico *et al.,* 1984; Odessey, 1992; Munster, 1996). The major advantage of this technique is the ability to provide autologous grafts capable of covering large areas with reasonable cosmetic results. Another significant advantage of autologous grafts is their ability to serve as permanent wound coverage, because the host does not reject them. Disadvantages include the time interval of 2–3 weeks required before sufficient quantities of keratinocytes are available, the need for an invasive and painful procedure to obtain autologous donor cells, and the large costs, estimated at $13,000 per 1% total body surface area covered (Rue *et al.,* 1993). Furthermore, graft take is widely variable based on wound status, general host status, and operator experience.

Cultured keratinocyte allografts were developed to help overcome the need for a biopsy and a separate cultivation for each patient to produce autologous grafts and to ameliorate the long lag period between epidermal harvest and graft product. Successful keratinocyte allografting was first reported in burns using keratinocytes obtained from cadaver skin (Hefton *et al.,* 1983). Since then, cultured epidermal cells from both cadavers and unrelated adult donors have been used for the treatment of burns (Madden *et al.,* 1986), for skin grafts (Thiovlet *et al.,* 1986), and for treating chronic leg ulcers (Leigh *et al.,* 1987). Importantly, these allografts can result in accelerated and sustained wound healing without any evidence of rejection. This tolerance is probably secondary to the inability of cultured keratinocytes to express major histocompatability complex class II human leukocyte (HLA-DR) antigens (Hefton *et al.,* 1983), and the absence of Langerhan's cells, the major antigen-presenting cells (APCs) of the epidermis (Thiovlet *et al.,* 1986). Nevertheless, these keratinocyte allografts are eventually replaced by recipient cells.

Current allografts utilize neonatal foreskin keratinocytes, which are more responsive than adult cells to mitogens. In addition, they release more growth factors that stimulate adjacent keratinocytes (Gilchrest *et al.,* 1983). Cultured neonatal foreskin allografts promoted accelerated healing and prompt pain relief in a variety of acute and chronic skin ulcers (Phillips *et al.,* 1989). Another major advance was the development of cryopreserved allografts that gave results comparable to those with fresh allografts (Teepe *et al.,* 1990; De Luca *et al.,* 1992). Cryopreserved allografts have been used to cover large wounds with exposed bone and cartilage after Mohs micrographic surgery at a cost of $150 per application (Kolenik and Leffell, 1995). It is hoped that cryopreservation will allow mass allograft production and wide availability.

RECOMBINANT GROWTH FACTORS FOR EPIDERMAL WOUNDS

Epithelial-targeted growth factors, such as EGF and more recently KGF, have been investigated as potential treatment for nonhealing epidermal wounds and/or acute large-defect epidermal wounds. EGF was studied in patients with thermal burns and it was shown to accelerate the healing of donor graft sites by 1.5 days (Brown *et al.,* 1989). Although initial clinical investigations of EGF appeared significant (G. L. Brown *et al.,* 1989; G. Brown *et al.,* 1991), later large clinical trails sponsored by the biotechnology and pharmaceutical industries did not confirm therapeutic benefit. Keratinocyte growth factors are the newest members of the FGF family (Werner, 1998). KGF-1 and -2 are secreted by fibroblasts in response to injury and have both mitogenic and motogenic paracrine effects on keratinocytes, particularly those in hair follicles and sebaceaous and sweat glands (Pierce *et al.,* 1994; Igarashi *et al.,* 1998). Although exogenous administration of KGF-1 has been shown to stimulate keratinocyte proliferation and reepithelialization in partial-thickness porcine burns, it had no effect on reepithelization of full-thickness burns (Danilenko *et al.,* 1995). Likewise, KGF-1 stimulated reepithelialization of partial-thickness, but not full-thickness, excisional porcine wounds; however, the thickness of the neoepidermis in both partial- and full-thickness wounds was significantly increased (Staino-Coico *et al.,* 1993). Animal studies with KGF-2 are promising (Jimenez and Rampy, 1999) and clinical investigations with this factor are now in progress. Clinical investigations with recombinant growth factors that target dermal components of the skin, as well as epidermis, will be discussed at the conclusion of the following section on granulation tissue.

GRANULATION TISSUE

New stroma, often called granulation tissue, begins to form approximately 4 days after injury. The name derives from the granular appearance of newly forming tissue when it is incised and visually examined. Numerous new capillaries endow the neostroma with its granular appearance. Macrophages, fibroblasts, and blood vessels move into the wound space as a unit (Hunt, 1989), which correlates well with the proposed biologic interdependence of these cells during tissue repair. Macrophages provide a continuing source of cytokines necessary to stimulate fibroplasia and angiogenesis, fibroblasts construct new extracellular matrix necessary to support cell ingrowth, and blood vessels carry oxygen and nutrients necessary to sustain cell metabolism. The quantity and quality of granulation tissue depend on biologic modifiers present, the activity level of target cells, and the extracellular matrix environment (Sporn and Roberts, 1986; Juliano and Haskill, 1992). As mentioned previously in the discussion on inflammation, the arrival of peripheral blood monocytes and their activation to macrophages establish conditions for continual synthesis and release of growth factors. In addition, injured and activated parenchymal cells can synthesize and secrete growth factors. The provisional extracellular matrix also promotes granulation tissue formation. Once fibroblasts and endothelial cells express the proper integrin receptors, they invade the fibrin/fibronectin-rich wound space (Fig. 61.2).

FIBROPLASIA

Components of granulation tissue derived from fibroblasts, including the fibroblast cells and the extracellular matrix, are collectively known as fibroplasia. Cytokines, especially PDGF and TGF-β (Heldin and Westermark, 1996; Roberts and Sporn, 1996), in concert with the provisional matrix molecules (Brown *et al.*, 1993; Gray *et al.*, 1993; Xu and Clark, 1996), presumably stimulate fibroblasts of the periwound tissue to proliferate, express appropriate integrin receptors, and

Integrins in 3 day Cutaneous Wound Repair

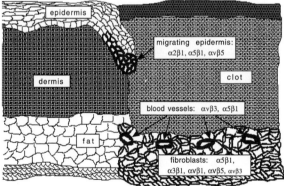

Fig. 61.2. Provisional matrix integrin expression is the rate-limiting step in granulation tissue induction. Once the appropriate integrins are expressed on periwound endothelial cells and fibroblasts on day 3, the cells invade the wound space shortly thereafter (on days 4 and 5). Fibroblasts and endothelial cells express the fibrinogen/fibrin receptor αvβ3 and therefore are able to invade the fibrin clot; the epidermis, however, does not express αvβ3 and therefore dissects under the clot. Ultimately the clot that has not been transformed into granulation tissue by invading fibroblasts and endothelial cells is dissected free of the wound and sloughed as eschar.

Integrins in 5 day Cutaneous Wound Repair

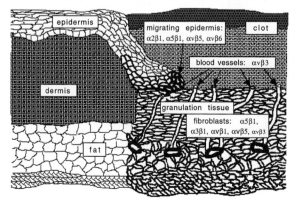

migrate into the wound space. Many of these growth factors are released from platelets and macrophages; however, fibroblasts can produce growth factors, to which they respond in an autocrine fashion (Leof *et al.,* 1986; Raines *et al.,* 1989). Multiple complex, interactive biologic phenomena occur within fibroblasts as they respond to wound cytokines, including the induction of additional cytokines (Leof *et al.,* 1986; Raines *et al.,* 1989) and modulation of cytokine receptor number or affinity) Oppenheimer *et al.,* 1983; Assoian *et al.,* 1984). *In vivo* studies support the hypothesis that growth factors are active in wound repair fibroplasia. Several studies have demonstrated that PDGF, PDGF-like peptides, TGF-β, TGF-α, HB-EGF, and FGF family members are present at sites of tissue repair (Matsuoka and Grotendorst, 1989; Pierce *et al.,* 1992; Werner *et al.,* 1992; Ansel *et al.,* 1993; Levine *et al.,* 1993; Marikovsky *et al.,* 1993; Frank *et al.,* 1996). Furthermore, purified and recombinant-derived growth factors have been shown to stimulate wound granulation tissue in normal and compromised animals (Sporn *et al.,* 1983; Lynch *et al.,* 1989; Greenhalgh *et al.,* 1990; Mustoe *et al.,* 1991; Steed *et al.,* 1992; Shah et al., 1995), and a single growth factor may work both directly and indirectly by inducing the production of other growth factors *in situ* (Mustoe *et al.,* 1991).

Structural molecules of the early extracellular matrix, termed the "provisional matrix" (Clark *et al.,* 1982), contribute to tissue formation by providing a scaffold or conduit for cell migration (fibronectin) (Greiling and Clark, 1997), low impedance for cell mobility (hyaluronic acid) (Toole, 1991), a reservoir for cytokines (Nathan and Sporn, 1991), and direct signals to the cells through integrin receptors (Damsky and Werb, 1992). Fibronectin appearance in the periwound environment as well as the expression of fibronectin receptors appear to be critical rate-limiting steps in granulation tissue formation (Clark *et al.,* 1996b; Gailit *et al.,* 1996; McClain *et al.,* 1996; Xu and Clark, 1996; Greiling and Clark, 1997). In addition, a dynamic reciprocity between fibroblasts and their surrounding extracellular matrix creates further complexity. That is, fibroblasts affect the extracellular matrix through new synthesis, deposition, and remodeling of the extracellular matrix (Kurkinen *et al.,* 1980; Welch *et al.,* 1990), whereas the extracellular matrix affects fibroblasts by regulating their function, including their ability to synthesize, deposit, remodel, and generally interact with the extracellular matrix (Mauch *et al.,* 1988; Grinnell, 1994; Clark *et al.,* 1995; Xu and Clark, 1996). Thus the interactions between extracellular matrix and fibroblasts dynamically evolve during granulation tissue development.

As fibroblasts migrate into the wound space, they initially penetrate the blood clot composed of fibrin and lesser amounts of fibronectin and vitronectin. Fibroblasts presumably require fibronectin *in vivo* for movement from the periwound collagenous matrix into the fibrin/fibronectin-laden wound space, as they do *in vitro* for migration from a three-dimensional collagen gel into a fibrin gel (Greiling and Clark, 1997). Fibroblasts bind to fibronectin through receptors of the integrin superfamily (Table 61.1) (Hynes, 1992; Yamada *et al.,* 1996). The Arg-Gly-Asp-Ser (RGDS) tetrapeptide within the cell-binding domain of these proteins is critical for binding to the integrin receptors α3β1, α5β1, αvβ1, αvβ3, and αvβ5. In addition, the CSIII domain of fibronectin provides a second binding site for human dermal fibroblasts via the α4β1 integrin receptor (Gailit *et al.,* 1993). *In vivo* studies have shown that the RGD-dependent, fibronectin receptors α3β1 and α5β1 are up-regulated on periwound fibroblasts the day prior to granulation tissue formation and on early granulation tissue fibroblasts as they infiltrate the provisional matrix-laden wound (Xu and Clark, 1996). In contrast, the non-RGD-binding α1β1 and α2β1 collagen receptors on these fibroblasts were either suppressed or did not appear to change appreciably (Gailit *et al.,* 1996; Xu and Clark, 1996).

Both PDGF and TGF-β can stimulate fibroblasts to migrate (Seppa *et al.,* 1982; Senior *et al.,* 1985; Postlethwaite *et al.,* 1987) and can up-regulate integrin receptors (Heino *et al.,* 1989; Ahlen and Rubin, 1994; Gailit *et al.,* 1996). Therefore these growth factors may be partially responsible for inducing a migrating-fibroblast phenotype. Interestingly, PDGF increases α3β1 and α5β1 while decreasing α1β1 in cultured human dermal fibroblasts (Gailit *et al.,* 1996). Furthermore, fibronectin- or fibrin-rich environments promote the ability of PDGF to increase α3β1 and α5β1, but not α2β1, mRNA steady-state levels by increasing the stability of these mRNA moieties (Xu and Clark, 1996). This suggests a positive feedback loop between the extracellular matrix and extracellular matrix receptors. *In vitro* fibroblast migration has also been observed in response to a variety of chemoattractants, including fragments of the fifth component of complement (Senior *et al.,* 1988); types I, II, and III collagen-derived peptides (Postlethwaite and Kang, 1976); a fibronectin fragment (Postlethwaite *et al.,* 1981); elastin-derived peptides (Senior *et al.,* 1980); and interleukin-4 (Postlethwaite and Seyer, 1991).

Movement into a cross-linked fibrin blood clot or any tightly woven extracellular matrix may also necessitate an active proteolytic system that can cleave a path for migration. A variety of fibroblast-derived enzymes in conjunction with serum-derived plasmin are potential candidates for this task, including plasminogen activator, interstitial collagenase-1 and -3 (MMP-1 and MMP-13, respectively), the 72-kDa gelatinase A (MMP-2), and stromelysin (MMP-3) (Mignatti *et al.,* 1996; Vaalamo *et al.,* 1997). In fact, high levels of immunoreactive MMP-1 have been localized to fibroblasts at the interface of granulation tissue with eschar in burn wounds (Stricklin and Nanney, 1994) and many stromal cells stain for MMP-1 and MMP-13 in chronic ulcers (Vaalamo *et al.,* 1997). Although TGF-β down-regulates proteinase activity (Laiho *et al.,* 1987; Overall *et al.,* 1989), PDGF stimulates the production and secretion of these proteinases (Circolo *et al.,* 1991).

Once the fibroblasts have migrated into the wound they gradually switch their major function to protein synthesis (Welch *et al.,* 1990). Ultimately the migratory phenotype is completely supplanted by a profibrotic phenotype characterized by decreased α3β1 and α5β1 provisional matrix receptor expression, increased α2β1 collagen receptor expression, abundant rough endoplasmic recticulum, and Golgi apparatus filled with new collagen protein (Welch *et al.,* 1990; Xu and Clark, 1996). The fibronectin-rich provisional matrix is gradually supplanted with a collagenous matrix (Welch *et al.,* 1990; Clark *et al.,* 1995). Under these conditions, PDGF, which is still abundant in these wounds (Pierce *et al.,* 1995), stimulates extremely high levels of α2β1 collagen receptor, but not α3β1 or α5β1 provisional matrix receptors, supporting the contention that the extracellular matrix provides a positive feedback for integrin expression (Xu and Clark, 1996). TGF-β, which is observed in wound fibroblasts at this time (Clark *et al.,* 1995), can induce fibroblasts to produce great quantities of collagen (Ignotz and Massague, 1986; Roberts *et al.,* 1986). IL-4 also can induce a modest increase in types I and III collagen production (Postlethwaite *et al.,* 1992). Because IL-4-producing mast cells are present in healing wounds, as well as fibrotic tissue, they may contribute to collagenous matrix accumulation in these sites.

Once an abundant collagen matrix is deposited in the wound, fibroblasts cease collagen production despite the continuing presence of TGF-β (Clark *et al.,* 1995). The stimuli responsible for fibroblast proliferation and matrix synthesis during wound repair were originally extrapolated from many *in vitro* investigations over the past two decades and then confirmed by *in vivo* manipulation of wounds within the past 10 years (Sprugel *et al.,* 1987; Pierce *et al.,* 1991; Schultz *et al.,* 1991). Less attention had been directed toward elucidating the signals responsible for downregulating fibroblast proliferation and matrix synthesis. Both *in vitro* and *in vivo* studies suggest that γ-interferon may be one such factor (Duncan and Berman, 1985; Granstein *et al.,* 1987). In addition, collagen matrix can suppress both fibroblast proliferation and fibroblast collagen synthesis (Grinnell, 1994; Clark *et al.,* 1995). In contrast, a fibrin or fibronectin matrix has little or no suppressive effect on the mitogenic or synthetic potential of fibroblasts (Clark *et al.,* 1995; Tuan *et al.,* 1996).

Although the attenuated fibroblast activity in collagen gels is not associated with cell death, many fibroblasts in day 10 healing wounds develop pyknotic nuclei (Desmouliere *et al.,* 1995), a cytological marker for apoptosis or programmed cell death (Williams, 1991), as well as other signs of apoptosis. These results in studies of cutaneous wounds and other results in studies of lungs and kidney suggest that apoptosis is the mechanism responsible for the transition from a fibroblast-rich granulation tissue to a relatively acellular scar (Desmouliere *et al.,* 1997). The signal(s) for wound fibroblast apoptosis have not been elucidated. Thus fibroplasia in wound repair is tightly regulated, whereas, in fibrotic diseases such as keloid formation, morphea, and scleroderma, these processes become dysregulated.

NEOVASCULARIZATION

Fibroplasia would halt if neovascularization failed to accompany the newly forming complex of fibroblasts and extracellular matrix. The process of new blood vessel formation is called angiogenesis (Madri *et al.,* 1996), and has been extensively studied in the chick chorioallantoic membrane and the cornea (Folkman and Shing, 1992). The soluble factors that can stimulate angiogenesis in wound repair are gradually being elucidated (Roesel and Nanney, 1995), but the factors that do stimulate wound angiogenesis are less clear. Angiogenic activity can be recovered from activated macrophages as well as various tissues, including the epidermis, soft tissue wounds, and solid tumors. Some years ago acidic or basic fibroblast growth factor (aFGF and bFGF) appeared to be responsible for most of these activities (Folkman and Klagsbrun, 1987), but other molecules have now also been shown to have angiogenic activity. These include vascular endothelial growth factor (VEGF) (Keck *et al.*, 1989), TGF-β (Yang and Moses, 1990), TGF-α (Schreiber *et al.*, 1986), tumor necrosis factor α (TNF-α (Leibovich *et al.*, 1987), platelet factor 4 (PF4) (Montrucchio *et al.*, 1994), angiogenin (Vallee and Riordan, 1997), angiotropin (Hockel *et al.*, 1988), angiopoietin (Suri *et al.*, 1996), interleukin-8 (Koch *et al.*, 1992), PDGF (Battegay *et al.*, 1994), thrombospondin (Nicosia and Tuszynski, 1994), low-molecular-weight substances, including the peptide KGHK (Lane *et al.*, 1994), low oxygen tension, biogenic amines, and lactic acid (Folkman and Shing, 1992), and nitric oxide (NO) (Montrucchio *et al.*, 1997). Some of these factors, however, are intermediaries in a single angiogenesis pathway, for example TNF-α induces PF4, which stimulates angiogenesis through NO (Montrucchio *et al.*, 1997).

Angiogenesis cannot be directly related to proliferation of cultured endothelial cells. In fact, Folkman (Folkman and Shing, 1992) has postulated that endothelial cell migration can induce proliferation. If this is true, endothelial cell chemotactic factors may be critical for angiogenesis. Some factors, of course, may have both mitogenic and chemotactic activities. For example, PDGF (Senior *et al.*, 1985) and EGF (Chen *et al.*, 1994) can be either chemotactic factors or mitogenic factors for dermal fibroblasts.

Besides growth factors and chemotactic factors, an appropriate extracellular matrix is also necessary for angiogenesis. Three-dimensional gels of extracellular matrix proteins have more pronounced effects on cultured endothelial cells than do monolayer protein coats (Feng *et al.*, 1999a), as has been observed with smooth muscle cells (Koyama *et al.*, 1996). Rat epididymal microvascular cells cultured in type I collagen gels with TGF-β produce capillary-like structures within 1 week (Madri *et al.*, 1988). Omission of TGF-β markedly reduces the effect. In contrast, laminin-containing gels in the absence of growth factors induce human umbilical vein and dermal microvascular cells to produce capillary-like structures within 24 hr of plating (Kubota *et al.*, 1988). Matrix-bound thrombospondin also promotes angiogenesis (Nicosia and Tuszynski, 1994), possibly through its ability to activate TGF-β (Schultz-Cherry and Murphy-Ullrich, 1993). We have developed a sprouting angiogenesis assay using neonatal human dermal microvascular endothelial cells (HDMECs) and have shown that fibrin, but not collagen, gels support sprouting angiogenesis (Feng *et al.*, 1999b). These studies support the hypothesis that the extracellular matrix plays an important role in angiogenesis. Consonant with this hypothesis, angiogenesis in the chick chorioallantoic membrane is dependent on the expression of αvβ3, an integrin that recognizes fibrin and fibronectin, as well as vitronectin (Brooks *et al.*, 1994a). Furthermore, in porcine cutaneous wounds αvβ3 is expressed only on capillary sprouts as they invade the fibrin clot (Clark *et al.*, 1996b). *In vitro* studies in fact demonstrate that αvβ3 can promote endothelial cell migration on provisional matrix proteins (Leavesley *et al.*, 1993).

Given the information outlined above, a series of events leading to angiogenesis can be hypothesized. Substantial injury causes tissue cell destruction and hypoxia. Potent angiogenesis factors such as FGF-1 and FGF-2 are released secondary to cell disruption whereas VEGF is induced by hypoxia. Proteolytic enzymes released into the connective tissue degrade extracellular matrix proteins. Specific fragments from collagen, fibronectin, and elastin, as well as many phlogistic agents, recruit peripheral blood monocytes to the injured site, where these cells become activated macrophages that release more angiogenesis factors. Certain angiogenic factors, such as FGF-2, stimulate endothelial cells to release plasminogen activator and procollagenase. Plasminogen activator converts plasminogen to plasmin and procollagenase to active collagenase and in concert these two proteases digest basement membrane constituents.

The fragmentation of the basement membrane allows endothelial cells to migrate into the injured site in response to FGF, fibronectin fragments, heparin released from disrupted mast cells, and other endothelial cell chemoattractants. To migrate into the fibrin/fibronectin-rich wound, endothelial cells express αvβ3 integrin. The newly forming blood vessels first deposit a provisional matrix containing fibronectin and proteoglycans, but ultimately form basement membrane. TGF-β may induce endothelial cells to produce the fibronectin and proteoglycan provisional matrix as well as assume the correct phenotype for capillary tube formation. FGF and other mitogens such as VEGF stimulate endothelial cell proliferation, resulting in a continual supply of endothelial cells for capillary extension. Capillary sprouts eventually branch at their tips and join to form capillary loops through which blood flow begins. New sprouts then extend from these loops to form a capillary plexus.

Within a day or two after removal of angiogenic stimuli, capillaries undergo regression as characterized by mitochondria swelling in the endothelial cells at the distal tips of the capillaries, platelet adherence to degenerating endothelial cells, vascular stasis, endothelial cell necrosis, and ingestion of the effete capillaries by macrophages. Although $\alpha v\beta 3$ has been shown to regulate apoptosis of endothelial cells in culture and in tumors (Brooks *et al.*, 1994b), $\alpha v\beta 3$ is not present on wound endothelial cells as they undergo programmed cell death, indicating another pathway of apoptosis in healing wound blood vessels (X. Feng, R.A.F. Clark, and M. G. Tonnesen, unpublished observations). Thrombospondin appears to be a good candidate for this phenomenon (DiPietro *et al.*, 1996).

RECOMBINANT GROWTH FACTORS FOR WOUND HEALING

With the development of recombinant DNA technology the production of large quantities of growth factors is now possible, leading to an increased interest in their potential therapeutic role in wound healing. The four families of growth factors that have shown the greatest potential for enhancing wound repair include the EGF family, the FGF family, the PDGF family, and the TGF-β family.

The EGF family of growth factors includes EGF, TGF-α, and HB-EGF. As mentioned in the previous discussion of epithelialization, these growth factors are potent keratinocyte mitogens and motogens (promote cellular motility) that are released at the site of injury (Feliciani *et al.*, 1996). Attempts to enhance healing of burns by application of EGF or TGF-α in animals have led to conflicting results. Although some have noted enhancement (Brown *et al.*, 1989), other studies have failed to demonstrate consistent and significant biologic effects when EGF was applied to either partial- or full-thickness burns in pigs (Danilenko *et al.*, 1995). Limited acceleration of healing of chronic venous stasis ulcers has similarly been demonstrated in humans with EGF (Brown *et al.*, 1991; Falanga *et al.*, 1992).

The family of FGF includes multiple polypeptides that stimulate proliferation of both keratinocytes and fibroblasts. In excisional wound animal models, basic FGF (FGF-2) significantly increases both granulation tissue formation and reepithelialization (Pierce *et al.*, 1992). FGF has been shown to enhance the healing of chronic pressure sores in a limited number of paraplegics (Robson *et al.*, 1992a). Larger studies sponsored by the pharmaceutical industry, however, failed to show clinical benefit.

The superfamily of PDGF includes PDGF and VEGF. PDGF occurs in three isoforms PDGF-AA, -AB, and -BB (Heldin, 1992). The latter two isoforms are the most potent and are secreted by platelets (PDGF-AB), macrophages (PDGF-BB), and epidermal cells (PDGF-BB) in response to injury or other perturbations and cause substantial proliferation of fibroblasts (Heldin and Westermark, 1996). Application of PDGF to incisional wounds has resulted in more rapid healing. PDGF has also stimulated reepithelialization and granulation tissue formation in animal excisional wound models and burns (Pierce *et al.*, 1992; Danilenko *et al.*, 1995). Clinical trials using PDGF for the treatment of pressure sores and diabetic ulcers have been encouraging, demonstrating more rapid and complete healing in wounds treated with PDGF (Robson *et al.*, 1992b; Steed, 1995). Recombinant PDGF-BB, in fact, is the first recombinant growth factor to gain approval by the Food and Drug Administration (FDA) with an indication to enhance healing of diabetic foot ulcers. This factor has been placed on the market as Regranex by Ortho-McNeil Pharmaceuticals.

VEGF is secreted by keratinocytes and macrophages and induces vascular permeability as well as angiogenesis (Iruela-Arispe and Dvorak, 1997). It has been shown to enhance healing of gastric and duodenal ulcers (Folkman, 1995) and is currently under investigation for use in the treatment of ischemic tissues (Isner, 1997). Other isoforms of VEGF have been described (Veikkola and Alitalo, 1999), but no animal or clinical data are available at this time.

Within another large family, TGF-β has three isoforms that are chemotactic for macrophages and are potent stimulators of procollagen type I and fibronectin (Roberts and Sporn, 1996). Multiple animal studies have demonstrated the ability of TGF-$\beta 1$ and TGF-$\beta 2$ to enhance wound healing (Davidson *et al.*, 1985; Mustoe *et al.*, 1987; Ksander *et al.*, 1990). Preliminary human studies suggest that TGF-β2 may be of benefit in the treatment of venous stasis ulcers (Robson *et al.,* 1995). Both TGF-β1 and TGF-β2 have been shown to increase scar formation in wound models (Border and Noble, 1994), whereas application of neutralizing antibodies has been shown to reduce scarring (Shah *et al.,* 1994). Interestingly, TGF-β3 has also been shown to reduce scar formation in rat incisions (Shah *et al.,* 1995), but this finding is under debate.

OTHER CYTOKINES

Growth hormone has been studied and found to be of benefit in the treatment of donor sites in both children and adults with extensive burns (Herndon *et al.,* 1990, 1995). Granulocyte–macrophage colony-stimulating factor (GM-CSF) and monocyte colony-stimulating factor (M-CSF) are potent stimulators of monocyte activation and chemotaxis and have been shown to accelerate healing of wounds in rats (Middleton and Aukerman, 1991; Jyung *et al.,* 1994). In contrast, granulocyte colony-stimulating factor (G-CSF) has not had any beneficial effects on wound healing. Interleukin-1β, which is chemotactic for neutrophils and macrophages, has been evaluated in the management of pressure sores (Robson *et al.,* 1994).

Although clinical experience with growth factors and other mediators is limited, overall results have been discouraging. This is not surprising when one considers that wound repair is the result of a complex set of interactions between soluble cytokines, formed blood elements, extracellular matrix, and cells. It is possible that combinations of various mediators given at precisely timed intervals will be more efficacious at promoting wound healing. Indeed, synergistic effects on wound healing have been demonstrated for the following combinations: IGF-1 and PDGF (Lynch *et al.,* 1989; Greenhalgh *et al.,* 1993), TGF-α and PDFG (Brown *et al.,* 1994), and KGF and PDGF (Danilenko *et al.,* 1995).

DERMAL SUBSTITUTES

The use of cultured keratinocytes to enhance wound healing has met with success, but it lacks a dermal component that would help prevent wound contraction and provide greater mechanical stability. Allografts of cadaver skin containing dermis have been used for many years yet provide only temporary coverage due their rejection by the host. However, this skin can be chemically treated to remove the antigenic epidermal cellular elements (Alloderm; Life Cell Corporation; Woodlands, TX) and has been used alone or in combination with cultured autologous keratinocytes for closure of various wounds and burns (Cuono *et al.,* 1986; Kraut *et al.,* 1995).

In 1981 Burke *et al.* developed a composite skin graft made of a collagen-based dermal lattice (containing bovine collagen and chondroitin 6-sulfate) with an outer silicone covering. After placement on the wound, the dermal component is slowly degraded. Several weeks later the silastic sheet is removed and covered with an autograft. This composite graft has been used successfully to treat burns (Heimbach *et al.,* 1988) and has received FDA approval for this indication (Integra; Integra Life Sciences Corporation; Plainsboro, NJ). A similar product, in which the silastic outer covering is replaced with human keratinocytes and the dermal component includes viable fibroblasts, has also been used successfully in burns and chronic wounds (Boyce *et al.,* 1995).

Another rendition of a dermal substitute is Trancyte (Dermagraft-TC) (Advanced Tissue Sciences Inc.; La Jolla, CA). This consists of an inner nylon mesh in which human fibroblasts are embedded together with an outer silastic layer to limit evaporation. The fibroblasts are lysed in the final product by freeze–thawing. Prior to lysis, the fibroblasts had manufactured collagen, matrix proteins, and cytokines, all of which promote wound healing by the host. Trancyte has been used successfully as a temporary wound coverage after excision of burn wounds (Purdue, 1996), and has been approved by the FDA for this indication. Dermagraft is a modification of this product in which a biodegradable polygalactin mesh is used instead of a silastic layer and the fibroblasts remain viable. Use of this dermal substitute has had limited success in the treatment of diabetic foot ulcers (Genzkow *et al.,* 1996).

COMBINED DERMAL AND EPIDERMAL SUBSTITUTES

Bell was the first to describe a bilayered model of skin consisting of a collagen lattice with dermal fibroblasts that was covered with epidermal cells (Bell *et al.,* 1981). Modification of this human skin equivalent (HSE) composite consisting of type I bovine collagen and live allogeneic human skin fibroblasts and keratinocytes has been developed (Apligraf; Organogenesis; Canton,

MA). It has been used successfully in surgical wounds (Eaglstein *et al.*, 1995) and venous ulcers (Sabolinski *et al.*, 1996). In a large multicenter trial this product resulted in accelerated healing of chronic nonhealing venous stasis ulcers when compared to standard compressive therapy (Falanga *et al.*, 1998). Both Apligraf (for venous stasis ulcers) and Dermagraft (for diabetic neurotrophic ulcerations) have been available in Canada since 1997. Apligraf has become available in the United States as well, although Dermagraft is still awaiting approval by the FDA.

Several other composite skin substitutes combining dermal and epidermal elements have been developed. Composite cultured skin (CCS; Ortec International Inc.; New York, NY) is composed of both neonatal keratinocytes and fibroblasts embedded in distinct layers of bovine type I collagen. This product is being evaluated in clinical trials for the treatment of burns and in patients with epidermolysis bullosa. More extensive comparative data of the various biologic dressings have been published (Phillips, 1998).

"Smart" Matrix for Wound Healing

Our laboratory has taken an indirect approach to dermal repair, by creating "smart" (a trademark name) matrix constructs that will provide a reparative dressing or packing material for acute and chronic wounds and in addition facilitate granulation tissue formation (patent issued). The first construct will be a recombinant fibronectin mutant composed of the three cell-binding domains (described below) conjugated to a matrix backbone of hyaluronan. Using a testing system (patent issued) wherein fibroblasts move from a collagen gel onto a surface coated with protein, various combinations of recombinant fibronectin fragments (Barkalow and Schwarzbauer, 1994), as well as the enzymatically derived 120-kDa fibronectin fragment, have been tested for their ability to permit fibroblast movement.

The 120-kDa enzymatically derived fragment, which contains cell-binding peptides limited to the classic RGD domain (Pierschbacher and Ruoslahti, 1984) and the PHSRN synergy site (Obara *et al.*, 1988), does not permit fibroblast movement. Furthermore, recombinant proteins containing the Hep II domain (Heparin II domain; HV0) have little, if any activity. However, combinations of the 120-kDa fragment with the Hep II domain do permit fibroblast motility. A recombinant protein containing both the classic cell-binding domain and the Hep II domain also allows movement, but no better than the combination of the 120-kDa fragment and the Hep II domain. These studies demonstrate that the classic cell-binding domain plus the Hep II domain are necessary and sufficient to permit fibroblast movement, but the movement is submaximal compared to intact fibronectin. Maximum movement, however, is attained when the IIICS domain is present along with the 120-kDa fragment and Hep II domain. A recombinant protein containing the classic cell-binding domain, the Hep II domain, and the IIICS domain also promotes maximal migration. Thus, the classic cell-binding domain plus the Hep II domain plus the IIICS domain are necessary and sufficient to allow maximal fibroblast movement compared to intact fibronectin.

In experiments in which synthetic CS1, CS5, and HI through HV fibronectin peptides were added to the fibrin gel, we found that CS1 and CS5, when added separately to the assay, enhanced PDGF-stimulated fibroblast movement from collagen gels into the fibrin gel, whereas HI, HII, HIV, and HV, but not HIII, inhibited this movement. In contrast, when CS1 and CS5 were added to the *in vitro* assay together, they inhibited fibroblast migration. Thus, HI, HII, HIV, HV, and either CS1 or CS5, together with RGDS and possibly PHSRN, are necessary for mesenchymal cell migration.

Composites will be tested for their ability to permit or promote fibroblast transmigration from a collagen organotypic construct into the interstices of the composites. All composites that have the capacity to permit or promote fibroblast migration *in vitro* will be tested for their capacity to promote cutaneous wound repair *in vivo* (McClain *et al.*, 1996). The matrix will be biocompatible and absorbable over a several-week period.

WOUND CONTRACTION AND EXTRACELLULAR MATRIX REORGANIZATION

During the second and third week of healing, fibroblasts begin to assume a myofibroblast phenotype characterized by large bundles of actin-containing microfilaments disposed along the cytoplasmic face of the plasma membrane and the establishment of cell–cell and cell–matrix linkages (Welch *et al.,* 1990). In some (Darby and Gabbiani, 1990), but not all (Welch *et al.,* 1990), wound situations myofibroblasts express smooth muscle actin. Interestingly, TGF-β can induce cultured human fibroblasts to express smooth muscle actin (Desmouliere *et al.,* 1993) and may also be responsible for its expression *in vivo.*

The appearance of the myofibroblasts corresponds to the commencement of connective tissue compaction and the contraction of the wound. Fibroblasts link to the extracellular fibronectin matrix through α5β1 (Welch *et al.,* 1990), to collagen matrix through α1β1 and α2β1 collagen receptors (Ignatius *et al.,* 1990), and to each other through direct adherens junctions (Welch *et al.,* 1990). Fibroblast α2β1 receptors are markedly up-regulated in 7-day-old wounds (Xu and Clark, 1996), a time when new collagenous matrix is accumulating and fibroblasts are beginning to align with collagenous fibrils through cell–matrix connections (Welch *et al.,* 1990). New collagen bundles in turn have the capacity to join end to end with collagen bundles at the wound edge and ultimately to form covalent cross-links among themselves and with the collagen bundles of the adjacent dermis (Yamauchi *et al.,* 1987; Birk *et al.,* 1989). These cell–cell, cell–matrix, and matrix–matrix links provide a network across the wound whereby the traction of fibroblasts on their pericellular matrix can be transmitted across the wound (Singer *et al.,* 1984).

Cultured fibroblasts dispersed within a hydrated collagen gel provide a functional *in vitro* model of tissue contraction (Bell *et al.,* 1979). When serum is added to the admixture, contraction of the collagen matrix occurs over the course of a few days. When observed with time-lapse microphotography, collagen condensation appears to result from a "collection of collagen bundles" executed by fibroblasts as they extend and retract pseudopodia attached to collagen fibers (Bell *et al.,* 1983). The transmission of these traction forces across the *in vitro* collagen matrix depends on two linkage events: fibroblast attachment to the collagen matrix through the α2β1 integrin receptors (Schiro *et al.,* 1991) and cross-links between the individual collagen bundles (Woodley *et al.,* 1991). This linkage system probably plays a significant role in the *in vivo* situation of wound contraction as well. In addition, cell–cell adhesions and cell–fibronectin linkages appear to provide an additional means by which the traction forces of the myofibroblast may be transmitted across the wound matrix.

F-Actin bundle arrays, cell–cell and cell–matrix linkages, and collagen cross-links are all facets of the biomechanics of extracellular matrix contraction. The contraction process, however, needs a cytokine signal. In fact, cultured fibroblasts mixed in a collagen gel contract the collagen matrix only if serum is added to the medium (Bell *et al.,* 1979). PDGF, the major fibroblast mitogen serum, in fact, stimulates fibroblast contraction of the collagen matrix (Clark *et al.,* 1989). PDGF is abundant in wounds (Ansel *et al.,* 1993), thus it may also provide the signal for wound contraction. TGF-β has also been shown to stimulate fibroblast-driven collagen gel contraction (Reed *et al.,* 1994). Because this factor persists in wound fibroblasts during the time of tissue contraction (Clark *et al.,* 1995), it is an additional candidate for the stimulus of contraction. Perhaps both PDGF and TGF-β signal wound contraction—one more example of the many redundancies observed in the critical processes of wound healing. In summary, wound contraction represents a complex and masterfully orchestrated interaction of cells, extracellular matrix, and cytokines.

Collagen remodeling during the transition from granulation tissue to scar is dependent on continued collagen synthesis and collagen catabolism. The degradation of wound collagen is controlled by a variety of collagenase enzymes from macrophages, epidermal cells, and fibroblasts. These collagenases are specific for particular types of collagens but most cells probably contain two or more different types of these enzymes (Hasty *et al.,* 1986). Three matrix metalloproteinases have been described that have the ability to cleave native collagen: MMP-1, or classic interstitial collagenase, which cleaves types, I, II, III, XIII, and X collagens (Grant *et al.,* 1987); neutrophil collagenase (Hasty *et al.,* 1990); and a novel collagenase produced by breast carcinomas and prominent in chronic wounds (Vaalamo *et al.,* 1997). It is not known whether interstitial collagenases, other than MMP-1, are active in the remodeling stage of human wound repair (Saarialho-Kere *et al.,* 1993; Stricklin and Nanney, 1994).

Other MMPs potentially important in wound repair include two gelatinases, 72-kDa gelatinase A (MMP-2) and 92-kDa gelatinase B (MMP-9), with identical substrate specificity to degrade denatured collagens of all types as well as native types V and XI collagens (Hibbs *et al.,* 1987; Stetler-Stevenson *et al.,* 1989); and stromelysin-1, -2, and -3, which degrade a wide variety of substrates, including types III, IV, V, VII, and IX collagens as well as proteoglycans and glycoproteins

(Saus *et al.,* 1988; Murphy *et al.,* 1991). Although gelatinase A and B are expressed in early wounds (Salo *et al.,* 1994), they may play a lesser role in later tissue remodeling. Gelatinase A is produced constitutively by many cells and is up-regulated only by TGF-β (Overall *et al.,* 1991), whereas gelatinase B is subject it to regulation by a variety of physiologic signals, because its promoter contains two AP-1 regulatory elements (Huhtala *et al.,* 1991). Two other metalloproteinases that do not belong to the above subgroups are macrophage metalloelastase (Shapiro *et al.,* 1993) and a novel transmembrane metalloproteinase that can activate secreted progelatinase A (Sato *et al.,* 1994). MMP enzymatic activities are controlled by various inhibitor counterparts called tissue inhibitors of metalloproteinases (TIMPs), which are finely regulated during wound repair (Madlener *et al.,* 1998).

Cytokines such as TGF-β, PDGF, and IL-1 plus the extracellular matrix clearly play an important role in the modulation of collagenase and TIMP expression *in vivo* (Circolo *et al.,* 1991; Sporn and Roberts, 1992; Huhtala *et al.,* 1995). Interestingly, type I collagen induces MMP-1 expression through the α2β1 collagen receptor while suppressing collagen synthesis through the α1β1 collagen receptor (Langholz *et al.,* 1996). Type I collagen also induces expression of α2β1 receptors (Klein *et al.,* 1991; Xu and Clark, 1996), thus collagen can induce the receptor that signals a collagen degradation–remodeling phenotype. Such dynamic, reciprocal cell–matrix interactions probably occur generally during tissue formation and remodeling processes such as morphogenesis, tumor growth, and wound healing, to name a few.

Wounds gain only about 20% of their final strength by the third week, during which time fibrillar collagen has accumulated relatively rapidly and has been remodeled by myofibroblast contraction of the wound. Thereafter the rate at which wounds gain tensile strength is slow, reflecting a much slower rate of collagen accumulation. In fact, the gradual gain in tensile strength has less to do with new collagen deposition than with further collagen remodeling, with formation of larger collagen bundles and an accumulation of intermolecular cross-links (Bailey *et al.,* 1975). Nevertheless, wounds fail to attain the same breaking strength as uninjured skin. At maximum strength, a scar is only 70% as strong as intact skin (Levenson *et al.,* 1965).

THERAPIES DESIGNED TO REDUCE SCARRING

Fetal healing is characterized by lack of scarring. The relative lack of TGF-β1, a profibrotic cytokine, in fetal skin may help explain why the fetus heals without scarring (Adzick and Lorenz, 1994). Studies using neutralizing antibodies to TGF-β1 and TGF-β2, as well as using TGF-β3, which down-regulates the other TGF-β isoforms, have resulted in reduced scarring, supporting the central role of TGF-β1 in scar formation (Shah *et al.,* 1995). Several strategies are being used to block TGF-β1 and TGF-β2 in wounds in an attempt to attenuate scar formation.

REFERENCES

Abraham, J. A., and Klagsbrun, M. (1996). Modulation of wound repair by members of the fibroblast growth factor family. *In* "The Molecular and Cellular Biology of Wound Repair" (R.A.F. Clark, ed.), pp. 195–248. Plenum, New York.

Adams, J. C., and Watt, F. M. (1991). Expression of β1, β3, β4, and β5 integrins by human epidermal keratinocytes and non-differentiating keratinocytes. *J. Cell Biol.* **115,** 829–841.

Adzick, N. S., and Lorenz, H. P. (1994). Cells, matrix, growth factors, and the surgeon. The biology of scarless fetal wound repair. *Ann. Surg.* **220,** 10–18.

Ahlen, K., and Rubin, K. (1994). Platelet-derived growth factor-BB stimulates synthesis of the integrin α2-subunit in human diploid fibroblasts. *Exp. Cell Res.* **215,** 347–353.

Ansel, J. C., Tiesman, J. P., Olerud, J. E., Krueger, J. G., Krane, J. F., Tara, D. C., Shipley, G. D., Gilbertson, D., Usui, M., and Hart, C. E. (1993). Human keratinocytes are a major source of cutaneous platelet-derived growth factor. *J. Clin. Invest.* **92,** 671–678.

Assoian, R. K., Frolik, C. A., Roberts, A. B., Miller, D. M., and Sporn, M. B. (1984). Transforming growth factor-β controls receptor levels for epidermal growth factor in NRK fibroblasts. *Cell (Cambridge, Mass.)* **36,** 35–41.

Bailey, A. J., Bazin, S., Sims, T. J., LeLeus, M., Nicholetis, C., and Delaunay, A. (1975). Characterization of the collagen of human hypertrophic and normal scars. *Biochim. Biophys. Acta* **405,** 412–421.

Barkalow, F. J., and Schwarzbauer, J. E. (1994). Interactions between fibronectin and chondroitin sulfate are modulated by molecular context. *J. Biol. Chem.* **269,** 3957–3962.

Barrandon, Y., and Green, H. (1987). Cell migration is essential for sustained growth of keratinocytes colonies: The roles of transforming growth factor-α and epidermal growth factor. *Cell (Cambridge, Mass.)* **50,** 1131–1137.

Battegay, E. F., Rupp, J., Iruela-Arispe, L., Sage, E. H., and Pech, M. (1994). PDGF-BB modulates endothelial proliferation and angiogenesis *in vitro* via PDGF-β-receptors. *J. Cell Biol.* **125,** 917–928.

Bell, E., Ivarsson, B., and Merrill, C. (1979). Production of a tissue-like structure by contraction of collagen lattices by human fibroblasts of different proliferative potential *in vitro. Proc. Natl. Acad. Sci. U.S.A.* **76,** 1274–1278.

Bell, E., Ehrlich, H. P., Buttle, D. J., and Nakatsuji, T. (1981). Living tissue formed *in vitro* and accepted as skin-equivalent tissue of full thickness. *Science* 211, 1052–1054.

Bell, E., Sher, S., Hull, B., Merrill, C., Rosen, S., Chamson, A., Asselineau, D., Dubertret, L., Coulomb, B., Lepiere, C., *et al.,* (1983). The reconstitution of living skin. *J. Invest. Dermatol.* 81(Suppl.), 2S–10S.

Birk, D. E., Zycband, E. I., Winkelmann, D. A., and Trelstad, R. L. (1989). Collagen fibrillogenesis *in situ:* Fibril segments are intermediates in assembly. *Proc. Natl. Acad. Sci. U.S.A.* 86, 4549–4553.

Border, W. A., and Noble, N. A. (1994). Transforming growth factor β in tissue fibrosis. *N. Engl. J. Med.* 331, 1286–1292.

Boyce, S. T., Glatter, R., and Kitzmiller, W. J. (1995). Treatment of chronic wounds with cultured skin substitutes: A pilot study. *Wounds* 7, 24–29.

Brachmann, R., Lindquist, P. B., Nagashima, M., Kohr, W., Lipari, T., Napier, M., and Derynck, R. (1989). Transmembrane TGF-α precursors activate EGF/TGF-α receptors. *Cell (Cambridge, Mass.)* 56, 691–700.

Brooks, P. C., Clark, R.A.F., and Cheresh, D. A. (1994a). Requirement of vascular integrin αvβ3 for angiogenesis. *Science* 264, 569–571.

Brooks, P. C., Montgomery, A.M.P., Rosenfeld, M., Reisfeld, R. A., Hu, T., Klier, G., and Cheresh, D. A. (1994b). Integrin αvβ3 antagonists promote tumor regression by inducing apoptosis of angiogenic blood vessels. *Cell* 79, 1157–1164.

Brown, G., Curtsinger, L., Jurkiewicz, M. J., Nahai, F., and Schultz, G. (1991). Stimulation of healing of chronic wounds by epidermal growth factor. *Plast. Reconstr. Surg.* 88, 189–194.

Brown, G. L., Nanney, L. B., Griffen, J., Cramer, A. B., Yancey, J. M., Curtsinger, L. J., Holtzin, L., Schultz, G. S., Jurkiewicz, M. H., and Lynch, J. B. (9189). Enhancement of wound healing by topical treatment with epidermal growth factor. *N. Engl. J. Med.* 321, 76–79.

Brown, L. F., Lanir, N., McDonagh, J., Tognazzi, K., Dvorak, A. M., and Dvorak, H. F. (1993). Fibroblast migration in fibrin gel matrices. *Am. J. Pathol.* 142, 273–283.

Brown, R. L., Breeden, M. P., and Greenhalgh, D. B. (1994). PDGF and TGFα act synergistically to improve wound healing in the genetically diabetic mouse. *J. Surg. Res.* 56, 562–570.

Bugge, T. H., Kombrinck, K. W., Flick, M. J., Daugherty, C. C., Danton, M.J.S., and Degen, J. L. (1996). Loss of fibrinogen rescues mice from the pleiotropic effects of plasminogen deficiency. *Cell (Cambridge, Mass.)* 87, 709–719.

Burke, J. F., Yannas, I. V., Quinby, W. C. *et al.* (1981). Successful use of a physiologically acceptable artificial skin in the treatment of extensive burn injury. *Ann. Surg.* 194, 413–428.

Centers for Disease Control (1982). "Reports of the Epidemiology and Surveillance of Injuries." Centers for Disease Control, Department of Health, Education, and Welfare, Atlanta, GA.

Chen, P., Gupta, K., and Wells, A. (1994). Cell movement elicited by epidermal growth factor receptor requires kinase and autophosphorylation but is separable from mitogenesis. *J. Cell Biol.* 124, 547–555.

Cheresh, D. A., Berliner, S. A., Vicente, V., and Ruggeri, Z. M. (1989). Recognition of distinct adhesive sites on fibrinogen by related integrins on platelets and endothelial cells. *Cell (Cambridge, Mass.)* 58, 945–953.

Circolo, A., Welgus, H. G., Pierce, G. F., Kramer, J., and Strunk, R. C. (1991). Differential regulation of the expression of proteinases/antiproteinases in fibroblasts. Effects of interleukin-1 and platelet-derived growth factor. *J. Biol. Chem.* 266, 12283–12288.

Clark, R.A.F. (1990). Fibronectin matrix deposition and fibronectin receptor expression in healing and normal skin. *J. Invest. Dermatol.* 94(Suppl.), 128S–134S.

Clark, R.A.F. (1996a). "The Molecular and Cellular Biology of Wound Repair." Plenum, New York.

Clark, R.A.F. (1996b). Wound repair: Overview and general considerations. "In "The Molecular and Cellular Biology of Wound Repair" (R.A.F. Clark, ed.), pp. 3–50. Plenum, New York.

Clark, R.A.F., Lanigan, J. M., DellaPelle, P., Manseau, E., Dvorak, H. F., and Colvin, R. B. (1982). Fibronectin and fibrin(ogen) provide a provisional matrix for epidermal cell migration during wound reepithelialization. *J. Invest. Dermatol.* 79, 264–269.

Clark, R.A.F., Folkvord, J. M., Hart, C. E., Murray, M. J., and McPherson, J. M. (1989). Platelet isoforms of platelet-derived growth factor stimulate fibroblasts to contract collagen matrices. *J. Clin. Invest.* 84, 1036–1140.

Clark, R.A.F., Nielsen, L. D., Welch, M. P., and McPherson, J. M. (1995). Collagen matrices attenuate the collagen synthetic response of cultured fibroblasts to TGF-β. *J. Cell Sci.* 108, 1251–1261.

Clark, R.A.F., Ashcroft, G. S., Spencer, M.-J., Larjava, H., and Ferguson, M.W.J. (1996a). Reepithelialization of normal human excisional wounds is associated with a switch from αvβ5 to αvβ6 integrins. *Br. J. Dermatol.* 135, 46–51.

Clark, R.A.F., Tonnesen, M. G., Gailit, J., and Cheresh, D. A. (1996b). Transient functional expression of αvβ3 on vascular cells during wound repair. *Am. J. Pathol.* 148, 1407–1421.

Clark, R.A.F., Spencer, M. J., Herrick, S. A., Ashcroft, G., Bini, A., Kudryk, B., and Ferguson, M. (1999). Fibrinogen clears from epidermal and blood vessel matrix prior to healing of chronic wounds. In preparation.

Coffey, R. J., Derynck, R., Wilcox, J. N., Bringman, T. S., Goustin, A. S., Moses, H. L., and Pittelkow, M. R. (1987). Induction and autoinduction of TGF-α in human keratinocytes. *Nature (London)* 328, 817–820.

Cuono, C., Langdon, R., and McGuire, J. (1986). Use of cultured epidermal autografts and dermal allografts as skin replacement after burn injury. *Lancet* 1, 1123–1124.

Damsky, C. H., and Werb, Z. (1992). Signal transduction by integrin receptors for extracellular matrix: Cooperative processing of extracellular information. *Curr. Opin. Cell Biol.* 4, 772–781.

Danilenko, D. M., Ring, B., Tarpley, J., Morris, B., Van, G. Y., Morawiecki, A., Callahan, W., Goldenberg, M., Hershenson, S., and Pierce, G. F. (1995). Growth factors in porcine full and partial thickness burn repair: Differing targets and effects of keratinocyte growth factor, platelet-derived growth factor-BB, epidermal growth factor, and *neu* differentiation factor. *Am. J. Pathol.* 147, 1261–1277.

Darby, I., and Gabbiani, G. (1990). α-Smooth muscle actin is transiently expressed by myofibroblasts during experimental wound healing. *Lab. Invest.* **63**, 21–29.

Davidson, J. M., Klagsbrun, M., Hill, K. E., Buckley, A., Sullivan, R., Brewer, P. S., and Woodward, S. C. (1985). Accelerated wound repair, cell proliferation, and collagen accumulation are produced by a cartilage-derived growth factor. *J. Cell Biol.* **100**, 1219–1227.

Davis, E. D. (1992). Affinity of integrins for damaged extracellular matrix: αvβ3 binds to denatured collagen type I through RGD sites. *Biochem. Biophys. Res. Commun.* **182**, 1025–1031.

Dejana, E. (1996). Endothelial adherens junctions: Implications in the control of vascular permeability and angiogenesis. *J. Clin. Invest.* **98**, 1949–1953.

De Luca, M., Albanese, E., Cancedda, R., Viacava, A., Faggioni, A., Zambruno, G., and Giannetti, A. (1992). Treatment of leg ulcers with cryopreserved allogeneic cultured epithelium. *Arch. Dermatol.* **128**, 633–638.

Desmouliere, A., Geinoz, A., Gabbiani, F., and Gabbiani, G. (1993). Transforming growth factor beta-1 induces alpha-smooth muscle actin expression in granulation tissue myofibroblasts and in quiescent and growing cultured fibroblasts. *J. Cell Biol.* **122**, 103–111.

Desmouliere, A., Redar, M., Darby, I., and Gabbiani, G. (1995). Apoptosis mediates the decrease in cellularity during the transition between greanulation tissue and scar. *Am. J. Pathol.* **146**, 56–66.

Desmouliere, A., Badid, C., Bochaton-Piallat, M. L., and Gabbiani, G. (1997). Apoptosis during wound healing, fibrocontractive diseases and vascular wall injury. *Int. J. Biochem. Cell Biol.* **29**, 19–30.

DiPietro, L. A., Nissen, N. N., Gamelli, R. L., Koch, A. E., Pyle, J. M., and Polverini, P. J. (1996). Thrombospondin 1 synthesis and function in wound repair. *Am. J. Pathol.* **148**, 1851–1860.

Duncan, M. R., and Berman, B. (1985). Gamma interferon is the lymphokine and beta interferon the monokine responsible for inhibition of fibroblast collagen production and late but not early fibroblast proliferation. *J. Exp. Med.* **162**, 516–527.

Dvorak, H. F. (1986). Tumors: Wounds that do not heal: Similarities between tumor stroma generation and wound healing. *N. Engl. J. Med.* **315**, 1650–1659.

Eaglstein, W. H., Iriondo, M., and Laszlo, K. (1995). A composite skin substitute (graftskin) for surgical wounds: A clinical experience. *Dermatol. Surg.* **21**, 839–843.

Falanga, V., Eaglstein, W. H., Bucalo, B., Katz, M. H., Harris, B., and Carson, P. (1992). Topical use of human recombinant epidermal growth factor (h-EGF) in venous ulcers. *J. Dermatol. Surg. Oncol.* **18**, 604–606.

Falanga, V., Margolis, D., Alvarez, O., Auletta, M., Maggiacomo, F., Altman, M., Jensen, J., Sabolinski, M., and Hardin-Young, J. (1998). Rapid healing of venous ulcers and lack of clinical rejection with an allogeneic cultured human skin equivalent. Human Skin Equivalent Investigators Group. *Arch Dermatol.* **134**, 293–300.

Feliciani, C., Gupta, A. K., and Sauder, D. N. (1996). Keratinocytes and cytokine/growth factors. *Crit. Rev. Oral Biol. Med.* **7**, 300–318.

Feng, X., Clark, R.A.F., Galanakis, D., and Tonnesen, M. G. (1999a). Fibrin and collagen differentially regulate human dermal microvascular endothelial cell integrins: Stabilization of αvβ3 mRNA by fibrin. *J. Invest. Dermatol.* **113**, 913–919.

Feng, X., Clark, R.A.F., Galanakis, D., and Tonnesen, M. G. (1999b). Fibrin, but not collagen, 3-dimensional matrix supports sprout angiogenesis of human dermal microvascular endothelial cells. *Am. J. Pathol.* (submitted for publication).

Folkman, J. (1995). Clinical applications of research on angiogenesis. *N. Engl. J. Med.* **333**, 1757–1763.

Folkman, J., and Klagsbrun, M. (1987). Angiogenic factors. *Science* **235**, 442–448.

Folkman, J., and Shing, T. (1992). Angiogenesis. *J. Biol. Chem.* **267**, 10931–10934.

Frank, S., Madlener, M., and Werner, S. (1996). Transforming growth factors β1, β2, and β3 and their receptors are differentially regulated during normal and impaired wound healing. *J. Biol. Chem.* **271**, 10188–10193.

Gailit, J., Pierschbacher, M., and Clark, R.A.F. (1993). Expression of functional α4 integrin by human dermal fibroblasts. *J. Invest. Dermatol.* **100**, 323–328.

Gailit, J., Welch, M. P., and Clark, R.A.F. (1994). TGF-β1 stimulates expression of keratinocyte integrins during re-epithelialization of cutaneous wounds. *J. Invest. Dermatol.* **103**, 221–227.

Gailit, J., Xu, J., Bueller, H., and Clark, R.A.F. (1996). Platelet-derived growth factor and inflammatory cytokines have differential effects on the expression of integrins α1β1 and α5β1 by human dermal fibroblasts *in vitro*. *J. Cell. Physiol.* **169**, 281–289.

Galiano, R. D., Zhao, L. L., Clemmons, D. R., Roth, S. I., Lin, X., and Mustoe, T. A. (1996). Interaction between the insulin-like growth factor family and the integrin receptor family in tissue repair processes. Evidence in a rabbit ear dermal ulcer model. *J. Clin. Invest.* **98**, 2462–2468.

Gallico, C. G., O'Conner, N. E., Compton, C. C., Kehinde, O. K., and Green, H. (1984). Permanent coverage of large burn wounds with autologous cultured human epithelium. *N. Engl. J. Med.* **311**, 448–451.

Genzkow, G., Iwasaki, S. D., and Hershon, K. S. (1996). Use of dermagraft: A cultured human dermis to treat diabetic foot ulcers. *Diabetes Care* **19**, 350–354.

Gilchrest, B. A., Karassik, R. L., and Wilkins, L. M. (1983). Autocrine and paracrine growth stimulation of cells derived from human skin. *J. Cell. Physiol.* **117**, 235–240.

Ginsberg, M. H., Du, X., and Plow, E. F. (1992). Inside-out integrin signalling. *Curr. Opin. Cell Biol.* **4**, 766–771.

Gipson, I. K., Spurr-Michaud, S. J., and Tisdale, A. S. (1988). Hemidesmosomes and anchoring fibril collagen appear synchronously during development and wound healing. *Dev. Biol.* **126**, 253–262.

Gipson, I. K., Spurr-Michaud, S., Tisdale, A., Elwell, J., and Stepp, M. A. (1993). Redistribution of the hemidesmosomes components α6β4 integrin and bullous pemphigoid antigens during epithelial wound healing. *Exp. Cell Res.* **207**, 86–98.

Granstein, R. D., Murphy, G. F., Margolis, R. J., Byrne, M. H., and Amento, E. P. (1987). Gamma interferon inhibits collagen synthesis *in vivo* in the mouse. *J. Clin. Invest.* **79**, 1254–1258.

Grant, G. A., Eisen, A. Z., Marmer, B. L., Roswit, W. T., and Goldberg, G. I. (1987). The activation of human skin fibroblast procollagenase. Sequence identification of the major conversion products. *J. Biol. Chem.* **262**, 5886–5889.

Gray, A. J., Bishop, J. E., Reeves, J. T., and Laurent, G. J. (1993). Aα and Bβ chains of fibrinogen stimulate proliferation of human fibroblasts. *J. Cell Sci.* **104**, 409–413.

Greenhalgh, D. G., Sprugel, K. H., Murray, M. J., and Ross, R. (1990). PDGF and FGF stimulate wound healing in the genetically diabetic mouse. *Am. J. Pathol.* **136**, 1235–1246.

Greenhalgh, D. G., Hummel, R.P.I., Albertson, S., and Breeden, M. P. (1993). Synbergistic actions of platelet-derived growth factor and the insulin-like growth factors *in vivo*. Enhancement of tissue repair in genetically diabetic mice. *Wound Repair. Regeneration* **1**, 69–81.

Greiling, D., and Clark, R.A.F. (1997). Fibronectin provides a conduit for fibroblast transmigration from a collagen gel into a fibrin gel. *J. Cell Sci.* **110**, 861–870.

Grinnell, F. (1994). Fibroblasts, myofibroblasts, and wound contraction. *J. Cell Biol.* **124**, 401–404.

Grondahl-Hansen, J., Lund, L. R., Ralfkiaer, E., Ottevanger, V., and Danø, K. (1988). Urokinase- and tissue-type plasminogen activators in keratinocytes during wound reepithelilaization *in vivo*. *J. Invest. Dermatol.* **90**, 790–795.

Hasty, K. A., Hibbs, M. S., Seyer, J. M., Mainardi, C. L., and Kang, A. H. (1986). Secreted forms of human neutrophil collagenase. *J. Biol. Chem.* **261**, 5645–5650.

Hasty, K. A., Pourmotabbed, T. F., Goldberg, G. I., Thompson, J. P., Spinella, D. G., Stevens, R. M., and Mainardi, C. L. (1990). Human neutrophil collagenase. A distinct gene product with homology to other matrix metalloproteinases. *J. Biol. Chem.* **265**, 11421–11424.

Hebda, P. A., Klingbeil, C. K., Abraham, J. A., and Fiddes, J. C. (1990). Basic fibroblast growth factor stimulation of epidermal wound healing in pigs. *J. Invest. Dermatol.* **95**, 626–631.

Hefton, J. M., Madden, M. R., and Finkelstein, J. L. (1983). Grafting of burn patients with allografts of cultured epidermal cells. *Lancet* **2**, 428–430.

Heimark, R. L., and Schwartz, S. M. (1988). The role of cell–cell interaction in the regulation of endothelial cell growth. *In* "Molecular and Cellular Biology of Wound Repair" (R.A.F. Clark and P. M. Henson, eds.), pp. 359–371. Plenum, New York.

Heimbach, D., Luterman, J. A., and Burke, J. (1988). Artificial dermis for major burns: A multicenter randomized clinical trial. *Ann. Surg.* **208**, 313–320.

Heino, J., Ignotz, R. A., Hemler, M. E., Crouse, C., and Massague, J. (1989). Regulation of cell adhesion receptors by transforming growth factor-β. Concomitant regulation of integrins that share a common β1 subunit. *J. Biol. Chem.* **264**, 380–388.

Heldin, C.-H. (1992). Structural and functional studies on platelet-derived growth factor. *EMBO J.* **11**, 4251–4259.

Heldin, C.-H., and Westermark, B. (1996). Role of platelet-derived growth factor *in vivo*. *In* "The Molecular and Cellular Biology of Wound Repair" (R.A.F. Clark, ed.), pp. 249–274. Plenum, New York.

Herndon, D. N., Barrow, R. E., Kunkel, K. R., Broemeling, L., and Rutan, R. L. (1990). Effects of recombinant human growth hormone on donor-site healing in severely burned children. *Ann. Surg.* **212**, 424–429; discussion: pp. 430–431.

Herndon, D. N., Hawkins, H. K., Nguyen, T. T., Pierre, E., Cox, R., and Barrow, R. E. (1995). Characterization of growth hormone enhanced donor site healing in patients with large cutaneous burns. *Ann. Surg.* **221**, 649–656; discussion: pp. 656–659.

Hibbs, M. S., Hoidal, J. R., and Kang, A. H. (1987). Expression of a metallo-proteinase that degrades native type V collagen and denatured collagens by cultured human alveolar macorophages. *J. Clin. Invest.* **80**, 1644–1650.

Higashiyama, S., Abraham, J. A., Miller, J., Fiddes, F. C., and Klagsbrun, M. (1991). A heparin-binding growth factor secreted by macrophage-like cells that is related to EGF. *Science* **251**, 936–939.

Hockel, M., Jung, W., Vaupel, P., Rabes, H., Khaledpour, C., and Wissler, J. H. (1988). Purified monocyte-derived angiogenic substance (angiotropin) induces controlled angiogenesis associated with regulated tissue proliferation in rabbit skin. *J. Clin. Invest.* **82**, 1075–1090.

Holbrook, K. A., and Wolff, K. (1993). The structure and development of skin. *In* "Dermatology in General Medicine" (T. B. Fitzpatrick, A. Z. Eisen, K. Wolff, I. M. Freedberg, and K. F. Austen, eds.), pp. 97–145. McGraw-Hill, New York.

Huhtala, P., Tuuttila, A., Chow, L. T., Lohi, J., Keski-Oja, J., and Tryggvason, K. (1991). Complete structure of the human gene for 92-kDa type IV collagenase. *J. Biol. Chem.* **266**, 16485–16490.

Huhtala, P., Humphries, M. J., McCarthy, J. B., Tremble, P. M., Werb, Z., and Damsky, C. H. (1995). Cooperative signaling by α5β1 and α4β1 integrins regulates metalloproteinase gene expression in fibroblasts adhering to fibronectin. *J. Cell Biol.* **129**, 867–879

Hunt, T. K. (1980). "Wound Healing and Wound Infection: Theory and Surgical Practice." Appleton-Century-Crofts, New York.

Hynes, R. O. (1992). Integrins: Versatility, modulation, and signaling in cell adhesion. *Cell (Cambridge, Mass.)* **69**, 11–25.

Igarashi, M., Finch, P. W., and Aaronson, S. A. (1998). Characterization of recombinant human fibroblast growth factor (FGF)-10 reveals functional similarities with keratinocyte growth factor (FGF-7). *J. Biol. Chem.* **273**, 13230–13235.

Ignatius, M. J., Large, T. H., House, M., Tawil, J. W., Barton, A., Esch, F., Carbonetto, S., and Reichardt, L. F. (1990). Molecular cloning of the rat integrin α1-subunit: A receptor for laminin and collagen. *J. Cell Biol.* **111**, 709–720.

Ignotz, R. A., and Massague, J. (1986). Transforming growth factor-β stimulates the expression of fibronectin and collagen and their incorporation into extracellular matrix. *J. Biol. Chem.* **261**, 4337–4340.

Iruela-Arispe, M. L., and Dvorak, H. F. (1997). Angiogenesis: A dynamic balance of stimulators and inhibitors. *Thromb. Haemostasis* 78, 672–677.

Isner, J. M. (1997). Angiogenesis for revascularization of ischaemic tissues. *Eur. J. Cardiol.* 18, 1–2.

Jimenez, P. A., and Rampy, M. A. (1999). Keratinocyte growth factor-2 accelerates wound healing in incisional wounds. *J. Surg. Res.* 81, 238–242.

Juhasz, I., Murphy, G. F., Yan, H.-C., Herlyn, M., and Albelda, S. M. (1993). Regulation of extracellular matrix proteins and integrin cell substratum adhesion receptors on epithelium during cutaneous human wound healing *in vivo. Am. J. Pathol.* 143, 1458–1469.

Juliano, R. L., and Haskill, S. (1992). Signal transduction from the extracellular matrix. *J. Cell Biol.* 120, 577–585.

Jyung, R. W., Wu, L., Pierce, G. F., and Mustoe, T. A. (1994). Granulocyte–macrophage colony-stimulating factor and granulocyte colony-stimulating factor: Differential action on incisional wound healing. *Surgery* 115, 325–334.

Keck, P. J., Hauser, S. D., Krivi, G., Sanzo, K., Warren, T., Feder, J., and Connolly, D. T. (1989). Vascular permeability factor, an endothelial cell mitogen related to PDGF. *Science* 246, 1309–1313.

Klein, C. E., Dressel, D., Steinmayer, T., Mauch, C., Eckes, B., Krieg, T., Bankert, R. B., and Werber, L. (1991). Integrin α2β1 is upregulated in fibroblasts and highly aggressive melanoma cells in three demensional collagen lattices and mediates the reorganization of collagen I fibrils. *J. Cell Biol.* 115, 1427–1436.

Koch, A. E., Polverini, P. J., Kunkel, S. L., Harlow, L. A., DiPietro, L. A., Elner, V. M., Elner, S. G., and Strieter, R. M. (1992). Interleukin-8 as a macrophage-derived mediator of angiogenesis. *Science* 258, 1798–1801.

Kolenik, S.A.I., and Leffell, D. J. (1995). The use of cryopreserved human skin allografts in wound healing following Mohs surgery. *Dermatol. Surg.* 21, 615–620.

Koyama, H., Raines, E. W., Bornfeldt, K. E., Roberts, J. M., and Ross, R. (1996). Fibrillar collagen inhibits arterial smooth muscle proliferation through regulation of Cdk2 inhibitors. *Cell (Cambridge, Mass.)* 87, 1069–1078.

Kraut, J. D., Eckhardt, A. J., and Patton, M. L. (1995). Combined simultaneous application of cultured epithelial autograft and Alloderm. *Wounds* 7, 137–142.

Krawczyk, W. S. (1971). A pattern of epidermal cell migration during wound healing. *J. Cell Biol.* 49, 247–263.

Ksander, G. A., Ogawa, Y., Chu, G. H., McMullin, H., Rosenblatt, J. S., and McPherson, J. M. (1990). Exogenous transforming growth factor-beta 2 enhances connective tissue formation and wound strength in guinea pig dermal wounds healing by secondary intent. *Ann. Surg.* 211, 288–294.

Kubo, M., Van De Water, L., Plantefaber, L. C., Newman, D., Mosesson, M., Simon, M., Tonnesen, M. G., Gailit, J., Taichman, L., and Clark, R.A.F. (1999). Fibrin(ogen) is an anti-adhesive for human keratinocytes: A possible mechanism for fibrin eschar slough during wound repair. *Am. J. Pathol.* (in press).

Kubota, Y., Kleinman, H. K., Martin, G. R., and Lawley, T. J. (1988). Role of laminin and basement membrane in the morphological differentiation of human endothelial cells into capillary-like structures. *J. Cell Biol.* 107, 1589–1598.

Kurkinen, M., Vaheri, A., Roberts, P. J., and Stenman, S. (1980). Sequential appearance of fibronectin and collagen in experimental granulation tissue. *Lab. Invest.* 43, 47–51.

Laiho, M., Saksela, O., and Keski-Oja, J. (1987). Transforming growth factor β induction of type-1 plasminogen activator inhibitor. *J. Biol. Chem.* 262, 17467–17474.

Lane, T. F., Iruela-Arispe, M. L., Johnson, R. S., and Sage, E. H. (1994). SPARC is a source of copper-binding peptides that stimulate angiogenesis. *J. Cell Biol.* 125, 929–943.

Langholz, L., Rockel, D., Mauch, C., Kozlowska, E., Bank, I., Krieg, T., and Eckes, B. (1996). Collagen and collagenase gene expression in three-dimensional collagen lattices are differentially regulated by α1β1 and α2β1 integrins. *J. Cell Biol.* 131, 1903–1915.

Larjava, H., Salo, T., Haapasalmi, K., Kramer, R. H., and Heino, J. (1993). Expression of integrins and basement membrane components by wound keratinocytes. *J. Clin. Invest.* 92, 1425–1435.

Leavesley, D. I., Schwartz, M. A., Rosenfeld, M., and Cheresh, D. A. (1993). Integrin β1- and β3-mediated endothelial cell migration is triggered through distinct signaling mechanisms. *J. Cell Biol.* 121, 163–170.

Leibovich, S. J., Polverini, P. J., Shepard, H. M., Wiseman, D. M., Shively, V., and Nuseir, N. (1987). Macrophage-induced angiogenesis is mediated by tumour necrosis factor-alpha. *Nature (London)* 329, 630–632.

Leigh, I. M., Purkis, P. E., Navasaria, H. A., and Phillips, T. J. (1987). Treatment of chronic venous ulcers with sheets of cultured allogeneic keratinocytes. *Br. J. Dermatol.* 117, 591–597.

Leof, E. B., Proper, J. A., Goustin, A. S., Shipley, G. D., DiCorleto, P. E., and Moses, H. L. (1986). Induction of c-sis RNA and activity similar to platelet-derived growth factor by transforming growth factor-β: A proposed model for indirect mitogenesis involving autocrine activity. *Proc. Natl. Acad. Sci. U.S.A.* 83, 2453–2457.

Levenson, S. M., Geever, E. F., Crowley, L. V., Oates, J.F.R., Berard, C. W., and Rosen, H. (1965). The healing of rat skin wounds. *Ann. Surg.* 161, 293–308.

Levine, J. H., Moses, H. L., Gold, L. I., and Nanney, L. B. (1993). Spatial and temporal patterns of immunoreactive transforming growth factor beta 1, beta 2, and beta 3 during excisional wound repair. *Am. J. Pathol.* 143, 368–380.

Lynch, S. E., Colvin, R. B., and Antoniades, H. N. (1989). Growth factors in wound healing. Single and synergistic effects on partial thickness porcine skin wounds. *J. Clin. Invest.* 84, 640–646.

Madden, M. R., Finkelstein, J. L., and Staiano-Coco, L. (1986). Grafting of cultured allogeneic epidermis on second and third degree burn wounds on 26 patients. *J. Trauma* 26, 955–962.

Madlener, M., Parks, W. C., and Werner, S. (1998). Matrix metalloproteinases (MMPs) and their physiological inhibitors (TIMPs) are differentially expressed during excisional skin wound repair. *Exp. Cell Res.* 242, 201–210.

Madri, J. A., Pratt, B. M., and Tucker, A. M. (1988). Phenotypic modulation of endothelial cells by transforming growth factor-β depends upon the composition and organization of the extracellular matrix. *J. Cell Biol.* 156, 1375–1385.

Madri, J. A., Sankar, S., and Romanic, A. M. (1996). Angiogenesis. *In* "The Molecular and Cellular Biology of Wound Repair" (R.A.F. Clark, ed.), pp. 355–372. Plenum, New York.

Mansbridge, J. N., and Knapp, A. M. (1987). Changes in keratinocyte maturation during wound healing. *J. Invest. Dermatol.* **89**, 253–263.

Marikovsky, M., Breuing, K., Liu, P. Y., Eriksson, E., Higashiyama, S., Farbe, P., Abraham, J., and Klagsbrun, M. (1993). Appearance of heparin-binding EGF-like growth factor in wound fluid as a response to injury. *Proc. Natl. Acad. Sci. U.S.A.* **90**, 3889–3893.

Matsuoka, J., and Grotendorst, G. R. (1989). Two peptides related to platelet-derived growth factor are present in human wound fluid. *Proc. Natl. Acad. Sci. U.S.A.* **86**, 4416–4420.

Mauch, C., Hatamochi, A., Scharffetter, K., and Krieg, T. (1988). Regulation of collagen synthesis in fibroblasts within a three-dimensional collagen gel. *Exp. Cell Res.* **178**, 493–530.

McClain, S. A., Simon, M., Jones, E., Nandi, A., Gailit, J., Newman, D., Tonnesen, M. G., and Clark, R.A.F. (1996). Mesenchymal cell activation is the rate limiting step of granulation tissue induction. *Am. J. Pathol.* **149**, 1257–1270.

Medical Data International (1998). "U.S. Markets for Wound Management Products." Medical Data International, Irvine, CA.

Middleton, S. C., and Aukerman, S. L. (1991). The effect of macrophage colony-stimulating factor (M-CSF) on wound healing in normal and genetically diabetic (*db/db*) mice. *J. Cell. Biochem.* **15F**(Suppl.), 197.

Mignatti, P., Rifkin, D. B., Welgus, H. G., and Parks, W. C. (1996). Proteinases and tissue remodeling. "In "The Molecular and Cellular Biology of Wound Repair." (R.A.F. Clark, ed.), pp. 427–474. Plenum, New York.

Montrucchio, G., Lupia, E., Battaglia, E., Passerini, G., Bussolino, F., Emanuelli, G., and Camussi, G. (1994). Tumor necrosis factor alpha-induced angiogenesis depends on *in situ* platelet-activating factor biosynthesis. *J. Exp. Med.* **180**, 377–382.

Montrucchio, G., Lupia, E., de Martino, A., Battaglia, E., Arese, M., Tizzani, A., Bussolino, F., and Camussi, G. (1997). Nitric oxide mediates angiogenesis induced *in vivo* by platelet-activating factor and tumor necrosis factor-alpha. *Am. J. Pathol.* **151**, 557–563.

Munster, A. M. (1996). Cultured skin for massive burns. A prospective, controlled trial. *Ann. Surg.* **224**, 372–377.

Murphy, G., Cockett, M. I., Ward, R. V., and Docherty, A.J.P. (1991). Matrix metalloproteinase degradation of elastin, type IV collagen and proteoglycan. A quantitative comparison of the activities of 95 kDa and 75 kDa gelatinases, stromelysins-1 and -2 and punctuated metalloproteinase (PUMP). *Biochem. J.* **277**, 277–279.

Mustoe, T. A., Pierce, G. F., Thomason, A., Gramates, P., Sporn, M. B., and Deuel, T. F. (1987). Accelerated healing of incisional wounds in rats induced by transforming growth factor-beta. *Science* **237**, 1333–1336.

Mustoe, T. A., Pierce, G. F., Morishima, C., and Deuel, T. F. (1991). Growth factor-induced acceleration of tissue repair through direct and inductive activities in a rabbit dermal ulcer model. *J. Clin. Invest.* **87**, 694–703.

Nanney, L. B., and King, L. E. (1996). Epidermal growth factor and transforming growth factor-α. *In* "The Molecular and Cellular Biology of Wound Repair" (R.A.F. Clark, ed.), pp. 171–194. Plenum, New York.

Nathan, C., and Sporn, M. (1991). Cytokines in context. *J. Cell Biol.* **113**, 981–986.

Nicosia, R. F., and Tuszynski, G. P. (1994). Matrix-bound thrombospondin promotes angiogenesis *in vitro*. *J. Cell Biol.* **124**, 183–193.

Obara, M., Kang, M. S., and Yamada, K. M. (1988). Site-directed mutagenesis of the cell-binding domain of human fibronectin: Separable, synergistic sites mediate adhesive function. *Cell* (Cambridge, Mass.) **53**, 649–657.

O'Conner, N. E., Mulliken, J. B., Banks-Schlegel, S., Kehinde, O., and Green, H. (1981). Grafting of burns with cultured epithelium prepared from autologous epidermal cells. *Lancet* **1**, 75–78.

Odessey, R. (1992). Addendum: Multicenter experience with cultured epidermal autograft for treatment of burns. *J. Burn Care Rehab.* **13**, 174–180.

Oppenheimer, C. L., Pessin, J. E., Massague, J., Gitomer, W., and Czech, M. P. (1983). Insulin action rapidly modulates the apparent affinity of the insulin-like growth factor II receptor. *J. Biol. Chem.* **258**, 4824–4830.

Overall, C. M., Wrana, J. I., and Sodek, J. (1989). Independent regulation of collagenase, 72kD progelatinase, and metalloendoproteinase inhibitor expression in human fibroblasts by transforming growth factor-β. *J. Biol. Chem.* **264**, 1860–1869.

Overall, C. M., Wrana, J. L., and Sodek, J. (1991). Transcriptional and post-transcriptional regulation of 72 kDa gelatinase/type IV collagenase by transforming growth factor-beta 1 in human fibroblasts. Comparisons with collagenase and tissue inhibitor of matrix metalloproteinase gene expression. *J. Biol. Chem.* **266**, 14064–14071.

Petersen, M. J., Woodley, D. T., Stricklin, G. P., and O'Keefe, E. J. (1990). Enhanced synthesis of collagenase by human keratinocytes cultured on type I or type IV collagen. *J. Invest. Dermatol.* **94**, 341–346.

Phillips, T. (1998). New skin for old: Developments in biological skin substitutes. *Arch. Dermatol.* **134**, 344–349.

Phillips, T. J., and Dover, J. S. (1991). Leg ulcers. *J. Am. Acad. Dermatol.* **25**, 965–987.

Phillips, T. J., Kehinde, O., Green, H., and Gilchrest, B. A. (1989). Treatment of skin ulcers with cultured epidermal allografts. *J. Am. Acad. Dermatol.* **21**, 191–199.

Pierce, G. F., Mustoe, T. A., Altrock, B., Deuel, T. F., and Thomas, A. (1991). Role of platelet-derived growth factor in wound healing. *J. Cell. Biochem.* **45**, 319–326.

Pierce, G. F., Tarpley, J. E., Yanagihara, D., Mustoe, T. A., Fox, G. M., and Thomason, T. A. (1992). Platelet-derived growth factor (BB homodimer), transforming growth factor-β1, and basic fibroblast growth factor in dermal wound healing. *Am. J. Pathol.* **140**, 1375–1388.

Pierce, G. F., Yanagihara, D., Klopchin, K., Danilenko, D. M., Hsu, E., Kenney, W. C., and Morris, C. F. (1994). Stimulation of all epithelial elements during skin regeneration by keratinocyte growth factor. *J. Exp. Med.* **179**, 831–840.

Pierce, G. F., Tarpley, J. E., Tseng, J., Bready, J., Chang, D., Kenney, W. C., Rudolph, R., Robson, M. C., Vande Berg, J., Reid, P., Kaufman, S., and Farrell, C. L. (1995). Detection of platelet-derived growth factor (PDGF)-AA in actively healing human wounds treated with recombinant PDGF-BB and absence of PDGF in chronic nonhealing wounds. *J. Clin. Invest.* **96**, 1336–1350.

Pierschbacher, M. D., and Ruoslahti, E. (1984). Cell attachment activity of fibronectin can be duplicated by small synthetic fragments of the molecule. *Nature (London)* **309** 30–33.

Postlethwaite, A. E., and Kang, A. H. (1976). Collagen and collagen peptide-induced chemotaxis of human blood monocytes. *J. Exp. Med.* **143**, 1299–1307.

Postlethwaite, A. E., and Seyer, J. M. (1991). Fibroblast chemotaxis induction by human recombinant interleukin-4: Identification by synthetic peptide analysis of two chemotactic domains residing in amino acid sequences 70–88 and 89–122. *J. Clin. Invest.* **87**, 2147–2152.

Postlethwaite, A. E., Keski-Oja, J., Balian, G., and Kang, A. (1981). Induction of fibroblast chemotaxis by fibronectin. Location of the chemotactic region to a 140,000 molecular weight nongelatin binding fragment. *J. Exp. Med.* **153**, 494–499.

Postlethwaite, A. E., Keski-Oja, J., Moses, H. L., and Kang, A. H. (1987). Stimulation of the chemotactic migration of human fibroblasts by transforming growth factor-β. *J. Exp. Med.* **165**, 251–256.

Postlethwaite, A. E., Holness, M. A., Katai, H., and Raghow, R. (1992). Human fibroblasts synthesize elevated levels of extracellular matrix proteins in response to interleukin 4. *J. Clin. Invest.* **90**, 1479–1485.

Purdue, G. F. (1996). Dermagraft-TC pivotal safety and efficacy study. *J. Burn Care Rehab.* **18**, S13–S14.

Raines, E. W., Dower, S. K., and Ross, R. (1989). Interleukin-1 mitogenic activity for fibroblasts and smooth muscle cells is due to PDFG-AA. *Science* **243**, 393–396.

Rappolee, D. A., Mark, D., Banda, M. J., and Werb, Z. (1988). Wound macrophages express TGF-α and other growth factors *in vivo:* Analysis by mRNA phenotyping. *Science* **241**, 708–712.

Reed, M. J., Vernon, R. B., Abrass, I. B., and Sage, E. H. (1994). TGF-beta 1 induces the expression of type I collagen and SPARC, and enhances contraction of collagen gels, by fibroblasts from young and aged donors. *J. Cell. Physiol.* **158**, 169–179.

Riches, D.W.H. (1996). Macrophage involvement in wound repair, remodeling and fibrosis. *In* "The Molecular and Cellular Biology of Wound Repair" (R.A.F. Clark, ed.), pp. 95–142. Plenum, New York.

Roberts, A. B., and Sporn, M. B. (1996). Transforming growth factor-β (TGF-β). *In* "The Molecular and Cellular Biology of Wound Repair" (R.A.F. Clark, ed.), pp. 275–310. Plenum, New York.

Roberts, A. B., Sporn, M. B., Assoian, R. K., Smith, J. M., Roche, M. S., Heine, U. F., Liotta, L., Falanga, V., Kehrl, J. H., and Fauci, A. S. (1986). Transforming growth factor beta: Rapid induction of fibrosis and angiogenesis *in vivo* and stimulation of collagen formation. *Proc. Natl. Acad. Sci. U.S.A.* **83**, 4167–4171.

Robson, M. C., Phillips, L. G., Lawrence W. T., Bishop, J. B., Youngerman, J. S., Hayward, P. G., Broemeling, L. D., and Heggers, J. P. (1992a). The safety and effect of topically applied recombinant basic fibroblast growth factor on the healing of chronic pressure sores. *Ann. Surg.* **216**, 401–406.

Robson, M. C., Phillips, l. G., Thomason, A., Robson, L., and Pierce, G. F. (1992b). Platelet-derived growth factor BB for the treatment of chronic pressure ulcers. *Lancet* **339**, 23–25.

Robson, M. C., Abdullah, A., and Burns, B. F. (1994). The safety and effect of topically applied recombinant human interleukin-1β (rhuIL-1β) in the management of pressure sores. *Wound Repair. Regeneration* **2**, 177–181.

Robson, M. C., Phillips, L. G., and Cooper, D. M. (1995). The safety and effect of transforming growth factor-β2 for the treatment of venous stasis ulcers. *Wound Repair Regeneration.*

Roesel, J. F., and Nanney, L. B. (1995). Assessment of differential cytokine effects on angiogenesis using an *in vivo* model of cutaneous wound repair. *J. Surg. Res.* **58**, 449–459.

Rue, L.W.I., Cioffi, W. G., McManus, W. F., and Pruitt, B.A.J. (1993). Wound closure and outcome in extensively burned patients treated with cultured autologous keratinocytes. *J. Trauma* **34**, 662–668.

Ruggeri, Z. M. (1993). von Willebrand factor and fibrinogen. *Curr. Opin. Cell Biol.* **5**, 898–906.

Saarialho-Kere, U. K., Chang, E. S., Welgus, H. G., and Parks, W. C. (1992). Distinct localization of collagenase and tissue inhibitor of metalloproteinases expression in wound healing associated with ulcerative pyogenic granuloma. *J. Clin. Invest.* **90**, 1952–1957.

Saarialho-Kere, U. K., Kovacs, S. O., Pentland, A. P., Olerud, J. E., Welgus, H. G., and Parks, W. C. (1993). Cell–matrix interactions modulate interstitial collagenase expression by human keratinocytes actively involved in wound healing. *J. Clin. Invest.* **92**, 2858–2866.

Saarialho-Kere, U. K., Pentland, A. P., Kirkedal-Hansen, H., Parks, W. C., and Welgus, H. G. (1994). Distinct populations of basal keratinocytes express stromelysin-1 and stromelysin-2 in chronic wounds. *J. Clin. Invest.* **94**, 79–88.

Sabolinski, M. L., Alverez, O., Auletta, M., Mulder, G., and Parenteau, N. L. (1996). Cultured skin as a "smart material" for healing wounds: Experience in venous ulcers. *Biomaterials* **17**, 311–320.

Salo, T., Mäkelä, M., Kylmäniemi, M., Autio-Harmainen, H., and Larjava, H. (1994). Expression of matrix metalloproteinase-2 and -9 during early human wound healing. *Lab. Invest.* **70**, 176–182.

Sato, H., Takino, T., Okada, Y., Cao, J., and Shinagawa, A. (1994). A matrix metalloproteinase expressed on the surface of invasive tumor cells. *Nature (London)* **370**, 61–65.

Saus, J., Quinones, S., Otani, Y., Nagase, H., Harris, E. D., Jr., and Kurkinen, M. (1988). The complete primary structure of human matrix metallo-proteinase-3. Identity with stromelysin. *J. Biol. Chem.* **263**, 6742–6745.

Schiro, J. A., Chan, B.M.C., Roswit, W. R., Kassner, P. D., Pentland, A. P., Hemler, M. E., Eisen, A. Z., and Kupper, T. S. (1991). Integrin α2β1 (VLA-2) mediates reorganization and contraction of collagen matrices by human cells. *Cell (Cambridge, Mass.)* **67**, 403–410.

Schreiber, A. B., Winkler, M. E., and Derynck, R. (1986). Transforming growth factor-alpha: A more potent angiogenic mediator than epidermal growth factor. *Science* **232**, 1250–1253.

Schultz, G., Rotatori, D. S., and Clark, W. (1991). EGF and TGF-α in wound healing and repair. *J. Cell. Biochem.* **45**, 346–352.

Schultz-Cherry, S., and Murphy-Ullrich, J. E. (1993). Thrombospondin causes activation of latent transforming growth factor-beta secreted by endothelial cells by a novel mechanism. *J. Cell Biol.* **122**, 923–932.

Senior, R. M., Griffin, G. L., and Mecham, R. P. (1980). Chemotactic activity of elastin-derived peptides. *J. Clin. Invest.* **66**, 859–862.

Senior, R. M., Huang, J. S., Griffin, G. L., and Deuel, T. F. (1985). Dissociation of the chemotactic and mitogenic activities of platelet-derived growth factor by human neutrophil elastase. *J. Cell Biol.* **100**, 351–356.

Senior, R. M., Griffin, G. L., Perez, H. D., and Webster, R. O. (1988). Human C5a and C5a des arg exhibit chemotactic activity for fibroblasts. *J. Immunol.* **141**, 3570–3574.

Seppa, H.E.J., Grotendorst, G. R., Seppa, S. I., Schiffmann, E., and Martin, G. R. (1982). Platelet-derived growth factor is chemotactic for fibroblasts. *J. Cell Biol.* **92**, 584–588.

Shah, M., Foreman, D. M., and Ferguson, M. W. (1994). Neutralising antibody to TGF-β1,2 reduces cutaneous scarring in adult rodents. *J. Cell Sci.* **107**, 1137–1157.

Shah, M., Foreman, D. M., and Ferguson, M.W.J. (1995) Neutralisation of TGF-β1 and TGF-β2 or exogenous addition of TGF-β3 to cutaneous rate wounds reduces scarring. *J. Cell Sci.* **108**, 985–1002.

Shapiro, S. D., Kobayashi, D. K., and Ley, T. J. (1993). Cloning and characterization of a unique elastolytic metalloproteinase produced by human alveolar macrophages. *J. Biol. Chem.* **268**, 23824–23829.

Singer, A. J., and Clark, R.A.F. (1999). Advances in cutaneous wound healing. *N. Engl. J. Med.* **341**, 738–746.

Singer, I. I., Kawka, D. W., Kazazis, D. M., and Clark, R.A.F. (1984). *In vivo* codistribution of fibronectin and actin fibers in granulation tissue: Immunofluorescence and electron microscope studies of the fibronexus at the myofibroblast surface. *J. Cell Biol.* **98**, 2091–2106.

Sporn, M. B., and Roberts, A. B. (1986). Peptide growth factors and inflammation, tissue repair, and cancer. *J. Clin. Invest.* **78**, 329–332.

Sporn, M. B., and Roberts, A. M. (1992). Transforming growth factor-β: Recent progress and new challenges. *J. Cell Biol.* **119**, 1017–1021.

Sporn, M. B., Roberts, A. B., Shull, J. H., Smith, J. M., Ward, J. M., and Sodek, J. (1983). Polypeptide transforming growth factor isolated from bovine sources and used for wound healing *in vitro*. *Science* **219**, 1329–1331.

Sprugel, K. H., McPherson, J. M., Clowes, A. W., and Ross, R. (1987). Effects of growth factors *in vivo*. *Am. J. Pathol.* **129**, 601–613.

Staino-Coico, L., Krueger, J. G., Rubin, J. S., D'limi, S., Vallat, V. P., Valentino, L., Fahey, T. I., Hawes, A., Kingston, G., Madden, M. R., Mathwich, M., Gottlieb, A. B., and Aaronson, S. A. (1993). Human keratinocyte growth factor effects in a porcine model of epidermal wound healing. *J. Exp. Med.* **178**, 865–878.

Steed, D. L. (1995). Clinical evaluation of recombinant human platelet-derived growth factor for the treatment of lower extremity diabetic ulcers. Diabetic Ulcer Study Group. *J. Vasc. Surg.* **21**, 71–78.

Steed, D. L., Goslen, J. B., Holloway, G. A., Malone, J. M., Bunt, T. J., and Webster, M. W. (1992). Randomized prospective double-blind trial in healing chronic diabetic foot ulcers: CT-102 activated platelet supernatant topical versus placebo. *Diabetes Care* **15**, 1598–1604.

Stetler-Stevenson, W. G., Krutzsch, H. V., Wacher, M. P., Margulies, I.M.K., and Liotta, L. A. (1989). The activation of human type IV collagenase proenzyme. Sequence identification of the major conversion product following organomercurial activation. *J. Biol. Chem.* **264**, 1353–1356.

Stricklin, G. P., and Nanney, L. B. (1994). Immunolocalization of collagenase and TIMP in healing human burn wounds. *J. Invest. Dermatol.* **103**, 488–492.

Suri, C., Jones, P. F., Patan, S., Bartunkova, S., Maisonpierre, P. C., Davis, S., Sato, T. N., and Yancopoulos, G. D. (1996). Requisite role of angiopoietin-1, a ligand for the TIE2 receptor, during embryonic angiogenesis. *Cell (Cambridge, Mass.)* **87**, 1171–1180.

Teepe, R.G.C., Koebrugge, E. J., Ponec, M., and Vermeer, B. J. (1990). Fresh versus cryopreserved cultured allografts for the treatment of chronic skin ulcers. *Br. J. Dermatol.* **122**, 81–89.

Thiovlet, J., Faure, M., and Demiderm, A. (1986). Long term survival and immunological tolerance of human epidermal allografts produced in culture. *Transplantation* **42**, 274–280.

Toole, B. P. (1991). Proteoglycans and hyaluronan in morphogenesis and differentiation. *In* "Cell Biology of the Extracellular Matrix" (E. D. Hay, ed.), pp. 305–341. Plenum, New York.

Tuan, T. L., Song, A., Chang, S., Younai, S., and Nimni, M. E. (1996). *In vitro* fibroplasia: Matrix contraction, cell growth, and collagen production of fibroblasts cultured in fibrin gels. *Exp. Cell Res.* **223**, 127–134.

Vaalamo, M., Mattila, L., Johansson, N., Kariniemi, A., Karjalainen-Lindsberg, M., Kahari, V., and Saarialko-Kere, U. (1997). Distinct populations of stromal cells express collagenase-3 (MMP-13) and collagenase-1 (MMP-1) in chronic ulcers but not in normally healing wounds. *J. Invest. Dermatol.* **109**, 96–101.

Vallee, B. L., and Riordan, J. F. (1997). Organogenesis and angiogenin. *Cell Mol. Life Sci.* **53**, 803–815.

Veikkola, T., and Alitalo, K. (1999). VEGFs, receptors and angiogenesis. *Semin. Cancer Biol.* **9**, 211–220.

Weksler, B. B. (1992). Platelets. *In* "Inflammation: Basic Principle and Clinical Correlates" (J. I. Gallin, I. M. Goldstein, and R. Snyderman, eds.), pp. 727–746. Raven Press, New York.

Welch, M. P., Odland, G. F., and Clark, R.A.F. (1990). Temporal relationships of F-actin bundle formation, collagen and fibronectin matrix assembly, and fibronectin receptor expression to wound contraction. *J. Cell Biol.* **110**, 133–145.

Werner, S. (1998). Keratinocyte growth factor: A unique player in epithelial repair processes. *Cytokine Growth Factor Rev.* **9**, 153–165.

Werner, S., Peters, K. G., Longaker, M. T., Fuller-Pace, F., Banda, M. J., and Williams, L. T. (1992). Large induction of keratinocyte growth factor in the dermis during wound healing. *Proc. Natl. Acad. Sci. U.S.A.* **89**, 6896–6900.

Werner, S., Breeden, M., Hubner, G., Greenhalgh, D. G., and Longaker, M. T. (1994). Induction of keratinocyte growth factor expression is reduced and delayed during wound healing in the genetically diabetic mouse. *J. Invest. Dermatol.* **103**, 469–475.

Williams, G. T. (1991). Programmed cell death: Apoptosis and oncogenesis *Cell (Cambridge, Mass.)* **65**, 1097–1098.

Woodley, D. T., Kalebec, T., Banes, A. J., Link, W., Prunieras, M., and Liotta, L. (1986). Adult human keratinocytes migrating over nonviable dermal collagen produce collagenolytic enzymes that degrade type I and type IV collagen. *J. Invest. Dermatol.* **86**, 418–423.

Woodley, D. T., Yamauchi, M., Wynn, K. C., Mechanic, G., and Briggaman, R. A. (1991). Collagen telopeptides (cross-linking sites) play a role in collagen gel lattice contraction. *J. Invest. Dermatol.* **97**, 580–585.

Xu, J., and Clark, R.A.F. (1996). Extracellular matrix alters PDGF regulation of fibroblast integrins. *J. Cell Biol.* **132**, 239–249.

Yamada, K. M., Gailit, J., and Clark, R.A.F. (1996). Integrins in wound repair. *In* "The Molecular and Cellular Biology of Wound Repair" (R.A.F. Clark, ed.), pp. 311–338. Plenum, New York.

Yamauchi, M., London, R. E., Guenat, C., Hashimoto, F., and Mechanic, G. L. (1987). Structure and formation of a stable histidine-based trifunctional cross-link in skin collagen. *J. Biol Chem.* **262**, 11428–11434.

Yang, E. Y., and Moses, H. L. (1990). Transforming growth factor-β1-induced changes in cell migration, proliferation, and angiogenesis in the chicken chorioallantoic membrane. *J. Cell Biol.* **111**, 731–741.

Zambruno, G., Marchisio, P. C., Marconi, A., Vaschieri, C., Melchiori, A., Giannetti, A., and De Luca, M. (1995). Transforming growth factor-β1 modulates β1 and β5 integrin receptors and induces the *de novo* expression of the αvβ6 heterodimer in normal human keratinocytes: Implications for wound healing. *J. Cell Biol.* **129**, 853–865.

CHAPTER 62

SKIN

Nancy L. Parenteau, Janet Hardin-Young, and Robert N. Ross

INTRODUCTION

The engineering of skin tissue has been at the forefront of tissue engineering for many years and has now yielded some of the first medical products to emerge from this field of work. This is the result of work over the past 25 years in the areas of keratinocyte cell biology, extracellular matrix biology, collagen scaffolds, polymer scaffolds, and tissue equivalents. The repair of skin using tissue engineering has taken on many forms—from simple to complex. In this chapter the various strategies are discussed, with an emphasis on the biologic relevance of each to normal skin structure and function and the healing of wounds.

Skin wounds normally heal by formation of epithelialized scar tissue rather than by regeneration of full-thickness skin. Consequently, strategies for the clinical management of wound healing have depended historically on providing a passive cover to the site of the wound during the period required for the reparative mechanisms of wound healing: reepithelialization, remodeling of granulation tissue, and formation of scar tissue. Therapy could do little more than facilitate these little understood processes. But recent advances in our understanding of fetal wound healing, the concerted action of several growth factors, the role of the extracellular matrix in regulating the healing process, and the demonstrated ability of epidermal sheet grafts to close severe wounds not raise the possibility of intervening therapeutically in tissue repair by providing lost epithelium, stimulating dermal regeneration, and reconstituting full-thickness skin.

Tissue-engineered skin substitutes composed of autologous epidermal cell sheets (Epicel, Genzyme Tissue Repair, Cambridge, MA), dermal substrates (Integra Artificial Skin, Integra Life Sciences, Plainsboro, NJ; Alloderm, Lifecell Corporation, Woodlands, TX; Dermagraft, Advanced Tissue Sciences, La Jolla, CA), or temporary coverings (Transcyte, Advanced Tissue Sciences and Smith and Nephew) and a human skin equivalent (HSE) composed of both living dermal and epidermal components [Apligraf* (Graftskin) HSE, Organogenesis Inc., Canton, MA] are now available, or nearly so, for the treatment of acute and chronic wounds. These products represent the first of their kind and are the result of basic research in the biology of skin and wound healing and clinical experience with skin grafts, cultured keratinocyte grafts, acellular collagen matrices, cellular matrices, and cultured composite grafts (Eaglstein and Falanga, 1997; Mian *et al.,* 1992; Palmieri, 1992; Zacchi, *et al.,* 1998).

The dynamics of a chronic wound may be quite different from those of an acute wound. It is not clear in venous ulcers, for instance, what pathologic mechanisms prevent the normal migration of the epidermis over the wound bed. It is hoped, however, that introducing the appropriate factors and living cells into the engineered skin products will provide needed tissue and/or correct the balance of elements, to promote the healing and repair of deep wounds (e.g., burns), large acute wounds, and long-term chronic wounds (e.g., venous ulcers, diabetic ulcers and pressure ulcers).

* Apligraf is a registered trademark of Novartis

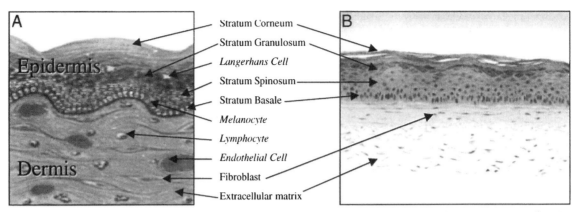

Fig. 62.1. The basic components of skin and an engineered skin equivalent. (A) Diagram showing the major cell types of skin and their organization. Note that stratified keratinocytes make up the epidermis and display distinct morphological phenotypes. (B) A histologic section of Apligraf HSE (hematoxylin and eosin; ×142). Italics indicate cell types present in real skin but not in the engineered skin equivalent.

SKIN STRUCTURE AND FUNCTION

The passive and active functions of skin are carried out by specialized cells and structures located in the two main layers of skin, the epidermis and the dermis (Fig. 62.1A). Complex functional relationships between these two anatomic structures of skin maintain the normal properties of skin. Tissue engineering applications in skin depend on an understanding of the structural components of skin, their spatial organization, and their functional relationships.

THE EPIDERMIS

The skin is a physical barrier between the body and the external environment. The outermost layer of skin, the epidermis, must therefore be tough and impermeable to toxic substances or harmful organisms. It must also control the loss of water from the body to the relatively drier external environment. The epidermis is composed primarily of keratinocytes, which form a stratified squamous epithelium (Fig. 62.1). Proliferating cells in the basal layer of the epidermis anchor the epidermis to the dermis and replenish the terminally differentiated epithelial cells lost through normal sloughing from the surface of the skin. These basal cells stop proliferating and terminally differentiate as they move from the basal layer through the suprabasal layers to the surface of the epidermis. Keratin filaments and desmosomes contribute physical strength in the living layers and maintain the integrity of the epidermis.

The most superficial keratinocytes in the epidermis form the stratum corneum, the dead outermost structure that provides the physical barrier of the skin. In the last stages of differentiation, epithelial cells extrude lipids into the intercellular space to form the permeability barrier. The cells break down their nuclei and other organelles and form a highly cross-linked protein envelope immediately beneath their cell membranes. The physically and chemically resilient protein envelope connects to a dense network of intracellular keratin filaments to provide further physical strength to the epidermis. The cornified envelopes serve as the bricks and the lipids as mortar of the stratum corneum.

Additional cells and structures in the epidermis perform specialized functions (Fig. 62.1A). Skin plays a major role in alerting the immune system to potential environmental dangers. The interacting cells in skin comprise a dynamic network capable of sensing a variety of perturbations (trauma, ultraviolet irradiation, toxic chemicals, and pathogenic organisms) in the cutaneous environment and rapidly sending appropriate signals that alert and recruit other branches of the immune system (Streilein, 1983; Bos and Kapsenberg, 1993). To restore homeostasis in the skin immune system, the multiple proinflammatory signals generated by skin cells must eventually be counterbalanced by mechanisms capable of promoting resolution of a cutaneous inflammatory response. Dendritic cells of the immune system (Langerhans cells) reside in the epidermis and form a network of dendrites through which they interact with adjacent keratinocytes and nerves (Streilein and Bergstresser, 1984). Melanocytes distribute melanin to keratinocytes in the form of melanosomes. Melanin protects the epidermis and underlying dermis from ultraviolet radiation. Sweat glands help to regulate body temperature through evaporation of sweat secreted onto the skin surface. Sebaceous glands associated with hair follicles secrete sebum, an oily substance that lubricates and moisturizes hair and epidermis. Hair keeps the body warm in many mammals, although maintaining body temperature is not an important role for hair in humans. Hair follicles, however, are an important source of proliferating keratinocytes during reepithelialization after severe wounds.

THE DERMIS

The dermis underlies the epidermis. The dermis is divided into two regions: the papillary dermis, which lies immediately beneath the epidermis, and the deeper reticular dermis. The reticular dermis is more acellular and has a denser meshwork of thicker collagen and elastic fibers compared with the papillary dermis. The reticular dermis provides skin with most of its strength, flexibility, and elasticity. Loss of reticular dermis can often lead to excessive scarring and wound contraction.

The dermis provides physical strength and flexibility to skin as well as the connective tissue scaffolding that supports the extensive vasculature, lymphatic system, and nerve bundles. The dermis is relatively acellular, being composed predominantly of an extracellular matrix of interwoven collagen fibrils. Interspersed among the collagen fibrils are elastic fibers, proteoglycans, and glycoproteins.

Fibroblasts, the major cell type of the dermis, produce and maintain most of the extracellular matrix. Endothelial cells line the blood vessels and play a critical role in the skin immune system by controlling the extravasation of leukocytes. Cells of hematopoietic origin in the dermis (e.g., macrophages, lymphocytes) contribute to surveillance function. A network of nerve fibers extends throughout the dermis, which serves a sensory role in the skin (and, to a more limited extent, a motor function). These nerve fibers also secrete neuropeptides that influence immune and inflammatory responses in skin through their effects on endothelial cells, leukocytes, and keratinocytes (Williams and Kupper, 1996).

IMMUNOLOGY AND THE SKIN

The first stage in the induction of a primary immune response in skin is the processing of antigen by dendritic cells (Langerhans cells with dermal dendritic cells), the antigen-presenting cells in skin. These cells process antigen and migrate out of the skin to the draining lymph node, where they can recruit and activate T cells. The interaction of keratinocytes and fibroblasts with T cells has important implications for the use of allogeneic cells in tissue engineering. Under normal conditions, keratinocytes and fibroblasts do not express major histocompatability complex (MHC) class II antigens; they can be induced, however, by interferon-γ to express MHC- class II molecules and thereby acquire the ability to present antigen to T cells. Because the keratinocytes and fibroblasts are deficient in the necessary costimulatory molecules (Nickoloff and Turka, 1994; Phipps *et al.*, 1989), antigen presentation by keratinocytes and fibroblasts does not result in T cell activation. Instead, this antigen presentation can result in T cell nonresponsiveness (Gaspari and Katz, 1988; Bal *et al.*, 1990) or T cell anergy (Gaspari and Katz, 1991). Therefore, the primary mode of skin rejection is likely mediated via an attack on the vasculature present in a normal skin graft (Pober *et al.*, 1986; Moulon *et al.*, 1999).

Autologous skin grafts avoid issues of immunogenicity, of course, but autologous grafts have significant limitations. Growing graft tissues from biopsy takes several weeks, the donor site creates another wound, and in some patients (e.g., severe burn patients) there may be no appropriate donor site. Reproducibly making complex HSE constructs to order from autologous cells would be technically difficult, time consuming, and very costly. Therefore, our ability to effectively use allogeneic human cells is a key element in the success of engineered skin replacements.

THE PROCESS OF WOUND HEALING

The immediate tissue response to wounding is clot formation to stop bleeding. Simultaneously, there is a release of inflammatory cytokines that regulate blood flow to the area, recruit lymphocytes and macrophages to fight infection, and later stimulate angiogenesis and collagen deposition (Williams and Kupper, 1996). These latter processes result in the formation of granulation tissue, a highly vascularized and cellular wound connective tissue. Fibroblasts rich in actin, called myofibroblasts (Desmouliere and Gabbiani, 1996), are recruited through the action of factors such as platelet-derived growth factor (PDGF) and transforming growth factor β (TGF-β). Granulation tissue forms in the wound bed, stimulated by factors such as PDGF. This tissue is gradually replaced by scar tissue through the action of the myofibroblasts and factors such as TGF-β. Keratinocytes are stimulated to proliferate and to migrate into the wound bed to restore epidermal coverage. From our preclinical observations using both full-thickness HSE and dermal matrices, coverage by the epidermis appears to play a key role in the regulation of the underlying inflammatory response. Providing a noninflammatory living connective tissue implant in the wound defect also appears to be beneficial in directing the granulation response.

As this brief description shows, wound healing involves the interaction of many tissue factors and elements. The poor healing response in chronic wounds has been attributed to an imbalance of factors rather than to an insufficiency of any particular factor (Parenteau *et al.*, 1997). However, most, if not all, factor-based approaches have had marginal success. Identification of putative wound healing factors has led to several attempts to speed wound healing by local application of one or more factors that promote cell attachment and migration. Transforming growth factor-β, epidermal growth factor (EGF), vascular endothelial growth factor (VEGF), and platelet-derived growth factor have been candidates for this purpose (McKay and Leigh, 1991; Martin *et al.*, 1992; Abraham and Klagsbrun, 1996; Nanney and King, 1996; Roberts and Sporn, 1996). Of these, only PDGF has shown efficacy in clinical trials and is approved for clinical use (Regranex, Ortho McNeil, Inc., Raritan, NJ). The arginine–glycine–aspartic acid (RGD) matrix peptide sequence has been found to promote the migration of connective tissue cells, and thus stimulate production of a dermal scaffold within the wound bed. This approach has been shown to accelerate healing of sickle cell leg ulcers (Wethers *et al.*, 1994) and diabetic ulcers (Steed *et al.*, 1995), compared with placebo, but not when compared with standard care. In addition, complex cell extracts have been used in hopes of providing the appropriate mixture of elements. These include the use of platelet extract to provide primarily platelet-derived growth factor (Knighton *et al.*, 1989), and the use of keratinocyte extracts to provide a complex mixture of elements of rapidly growing keratinocytes (Duinslaiger *et al.*, 1994); again, effects have been marginal, in part due to the complex nature of the wound healing response (Nathan and Sporn, 1991). In addition, the use of factors is not a sufficient approach, in and of itself, in situations where there is severe or massive loss of skin tissue.

Why should living tissue be different in its response? The epidermal and dermal responses are regulated by inflammatory cytokines and by autocrine and paracrine factors produced by the dermal fibroblasts and epidermal keratinocytes (Ansel *et al.*, 1990; McKay and Leigh, 1991). These factors regulate growth and differentiation of keratinocytes, proinflammatory reaction, angiogenesis, and deposition of extracellular matrix. Living tissue created through tissue engineering can provide complex temporal control of factor delivery and effect and can be used to provide the needed combination of chemical, structural, and, last but not least, normal cellular elements (Marks *et al.*, 1991; Sabolinski *et al.*, 1996).

CELL COMMUNICATION AND REGULATION IN THE SKIN

Regulation of its own function is an essential requirement of skin. Epidermis, for example produces parathyroid hormone-related protein (PTHrP), and plays a role in the regulation of keratinocyte growth and differentiation (Blomme *et al.*, 1998). Keratinocytes produce a large variety of polypeptide growth factors and cytokines; these act as signals between cells, regulate keratinocyte migration and proliferation, and stimulate dermal cells in various ways (e.g., to promote matrix deposition and neovascularization) (McKay and Leigh, 1991; Parenteau *et al.*, 1997). Cytokines produced by keratinocytes are thought to regulate Langerhans cell migration and differentiation (Lappin *et al.*, 1996). Keratinocytes also translate a variety of stimuli into cytokine signals, which are transmitted to the other cells of the skin immune system. Dermal cells produce and respond to cytokines and growth factors to regulate numerous processes critical to skin function (Williams and Kupper, 1996).

Approaches to dermal repair and regeneration center around control of fibroblast repopulation and collagen biosynthesis to limit scar tissue formation. One of the keys to improving dermal repair is control, or redirection, of the wound healing response so that scar tissue does not form. One promising observation is that fetal wounds heal without scarring (Mast *et al.*, 1992). TGF-β1, which is not expressed in the fetus, is a potent stimulator of collagen biosynthesis by fibroblasts in the adult and is thought to be an important inducer of scar formation. The matrix composition of the fetal dermis is also significantly different than that of adult dermis or granulation tissue, therefore providing a better matrix environment for dermal cell repopulation, which may also be key in controlling scar formation.

An engineered skin graft should incorporate as many of these factors as possible: (1) the extracellular matrix, (2) dermal fibroblasts, (3) the epidermis, and (4) a naturally occurring semipermeable membrane, the stratum corneum. These components may act alone, but more importantly they should act synergistically as part of a fully integrated tissue to protect the underlying tissues of a wound bed and to direct healing of the wound (Sabolinski *et al.*, 1996). Dermis containing fibroblasts may be necessary for the maintenance of the epidermal cell population (Lazarus *et al.*, 1994). In turn, the epidermis is necessary for the formation of the so-called neodermis, in the absence of a dermal layer (Compton *et al.*, 1989), and can dramatically influence underlying connective tissue response. The formation of the epidermal barrier also likely influences these processes through control of water loss and its influence on epidermal physiology (Parenteau *et al.*, 1996).

ENGINEERING SKIN TISSUE

Although the epidermis has an enormous capacity to heal, there are situations in which it is necessary to replace large areas of epidermis or in which normal regeneration is deficient. The dermis has very little capacity to regenerate. The scar tissue that forms in the absence of dermis lacks the elasticity, flexibility, and strength of normal dermis. Consequently, scar tissue limits movement, causes pain, and is cosmetically undesirable. Engineered tissues that not only close wounds but also stimulate the regeneration of dermis would provide a significant benefit in human wound healing.

DESIGN CONSIDERATIONS

Tissue engineering has not focused on regenerating certain skin structures, such as hair follicles or sebaceous glands, the loss of which is clinically less significant than the loss of dermis and epidermis, which are needed to cover and protect the underlying tissues. There is some preliminary evidence that sebaceous glands may be possible in the HSE (Wilkins *et al.*, 1994), although the development of functioning adnexal structures is likely to be years away. There has also been little need to stimulate extraneously the regeneration of other dermal components (e.g., blood vessels and cells of the immune system) through tissue engineering methods because these components have the ability to repopulate quickly and to normalize the area of a wound. Langerhans cells, for example, have been shown to migrate and repopulate effectively within months (Desmouliere and Gabbiani, 1996). Control of vascularization is dependent on the makeup of the extracellular matrix and the degree of inflammation present in the wound. Whether modification of vascularization through the use of exogenous factors will be of additional benefit for certain wounds remains to be determined. Although it is technically possible to add melanocytes to HSE for pigmentation (L. Wilkins, personal communication), clinical studies using Apligraf HSE, which essentially lacks melanocytes, have shown repigmentation of the grafted areas through repopulation of the area with host melanocytes, resulting in normal skin color for each individual. Therefore, tissue engineering approaches have primarily focused on providing or imitating structural and biologic characteristics of dermis, epidermis, or both. Some technologies, such as the HSE, have sought to reproduce living, full-thickness tissue for transplantation.

The key features to be replicated in an engineered skin construct are as follows:

1. A dermal or mesenchymal element capable of aiding appropriate dermal repair and epidermal support.
2. An epidermis capable of easily achieving biologic wound closure.
3. An epidermis capable of rapid reestablishment of barrier properties.
4. A permissive milieu for the components of the immune system, nervous system, and vasculature.

5. A tissue capable of achieving normalization of structure and additional function such as reduction of long-term scarring and reestablishment of pigmentation.

EPIDERMAL REGENERATION

Reepithelialization of the wound is a paramount concern. Without epithelial coverage, no defense exists against contamination of the exposed underlying tissue or loss of fluid. The approaches to reestablishing epidermis are numerous, ranging from the use of cell suspensions to full-thickness skin equivalents possessing a differentiated epidermis. Silicone membranes have been used as temporary coverings in conjunction with dermal templates (Heimbach *et al.,* 1988), but living epidermal keratinocytes are necessary to achieve permanent, biologic wound closure.

Green *et al.* (1979; Phillips *et al.,* 1989) developed techniques for growing human epidermal keratinocytes from small patient biopsy samples using coculture methods (Rheinwald and Green, 1975). The mouse 3T3 fibroblast feeder cell system allows substantial expansion of epidermal keratinocytes and can be used to generate enough thin, multilayered epidermal sheets to resurface the body of a severely burned patient (Gallico *et al.,* 1984). Once transplanted, the epidermal sheets quickly form epidermis and reestablish epidermal coverage (Compton, 1993). With time, the cultured epithelial autograft (CEA) stimulates formation of new connective tissue ("neodermis") immediately beneath the epidermis (Compton *et al.,* 1989), but scarring and wound contraction remain significant problems (Sheridan and Tompkins, 1995). Studies have shown that grafting of CEA onto pregrafted cadaver dermis greatly improves graft take (Odessey, 1992). Cultured epithelial autografts (Epicel) have been available since the late 1980s.

DERMAL REPLACEMENT

Human cadaver allograft skin has been used when autologous skin grafts are not possible. Problems associated with human cadaver allografts include the possibility of an immune rejection reaction, potential for infection, and problems of supply and variability in the quality of the material. Decellularized dermal tissue has also been used in an attempt to recapitulate as much of the normal architecture as possible while providing a natural scaffold for reepithelialization (Middelkoop *et al.,* 1995; Langdon *et al.,* 1988; Livesey *et al.,* 1995; Cuono *et al.,* 1986). Cadaver allograft dermis can be processed to make an immunologically inert, acellular dermal matrix with an intact basement membrane to aid the "take" and healing of ultrathin autografts (AlloDerm) (Lattari *et al.,* 1997). Currently, only the upper papillary layer of dermis is used clinically (AlloDerm). One limitation to this approach is that deep dermis and the more superficial papillary layer differ in architecture. The deep reticular dermis is needed to prevent wound contraction. Work is being done to develop a similar type of implant derived from deeper reticular dermis. Providing an appropriate scaffold for deep dermal repair remains a challenge for groups investigating native as well as synthetic matrices.

Tissue engineering has investigated the possibility of redirecting granulation tissue formation through the use of scaffolds and livings cells. In one of the earliest tissue engineering approaches to improving dermal healing, Yannas *et al.* (1982) designed a collagen–glycosaminoglycan sponge to serve as a scaffold or template for dermal extracellular matrix. The goal was to promote fibroblast repopulation in a controlled way that would decrease scarring and wound contraction. A commercial version of this material composed of bovine collagen and chondtroitin sulfate, with a silicone membrane covering (Integra, Integra Life Sciences, Plainsboro, NJ) is currently approved for use in burns (Burke *et al.,* 1981; Heimbach *et al.,* 1988). The dermal layer is slowly resorbed, and the silicone membrane is eventually removed, to be replaced by a thin autograft (Lorenz *et al.,* 1997).

Several variations on the collagen sponge have been studied. Efforts have been made to improve fibroblast infiltration and collagen persistence by collagen cross-linking (Middelkoop *et al.,* 1995; vanLuyn *et al.,* 1995; Cooper *et al.,* 1996), by inclusion of other matrix proteins (Hansbrough *et al.,* 1989; Ansel *et al.,* 1990; de Vries *et al.,* 1993; Murashita *et al.,* 1996), with hyaluronic acid (Cooper *et al.,* 1996), and by modifying porosity of the scaffold (Hansbrough *et al.,* 1989; Yannas *et al.,* 1989).

Although matrix scaffolds have shown some improvement in scar morphology, no acellular matrix has yet been shown to lead to true dermal regeneration. This may be due in part to limits in cell repopulation, the type of fibroblast repopulating the graft (J. Gross, personal communication), and control of the inflammatory and remodeling processes (i.e., the ability of the cells to degrade old matrix while synthesizing new matrix). The inflammatory response must be controlled in dermal repair in order to avoid the formation of scar tissue. Therefore, dermal scaffolds must not be inflammatory and must not stimulate a foreign-body reaction. This has been a problem in the past for some glutaraldehyde cross-linked collagen substrates, for example (de Vries *et al.*, 1993). The ability of the matrix to persist long enough to redirect tissue formation must be balanced with effects of the matrix on inflammatory processes. One way to achieve this is to form a biologic tissue that is recognized as living tissue, not a foreign substance.

There have been advances in the design of artificially grown dermal tissues using human neonatal fibroblasts grown on rectangular sheets of biodegradable mesh (Dermagraft). The fibroblasts propagate among the degrading fibers, producing extracellular matrix in the interstices of the mesh (Hansbrough *et al.*, 1992). Clinical trials of this material are ongoing and the material is commercially available in some European countries and Canada. A related product is a nonviable temporary covering for burns. In this case a nylon mesh, coated with porcine collagen, layered with a nonpermeable silicone membrane (Biobrane, Dow Hickam, Sugarland, TX), serves as a platform for deposition of human matrix proteins and associated factors by the human dermal fibroblast (Transcyte, Dermagraft-TC) (Hansbrough *et al.*, 1997). The material is then frozen to preserve the matrix and factors produced by the fibroblasts. The temporary covering must be removed prior to autografting. It is commercially available for the treatment of second- and third-degree burns (Hansbrough, 1997).

COMPOSITE SKIN GRAFTS

Human skin autograft has been the gold standard for resurfacing the body and closing wounds that are difficult to heal. Cultured epidermal grafts are more likely to take when the dermal bed is relatively intact, probably because dermal factors influence epithelial migration, differentiation, attachment, and growth (Cuono *et al.*, 1987; Phillips *et al.*, 1990; Clark, 1993; Greiling and Clark, 1997). The epidermis and dermis act synergistically to maintain homeostasis (Leary *et al.*, 1992; Parenteau *et al.*, 1997).

Boyce *et al.* have modified the approach first proposed by Yannas *et al.* to form a bilayered composite skin made using a modified collagen–glycosaminoglycan substrate seeded with fibroblasts and overlaid with epidermal keratinocytes (Boyce and Hansbrough, 1988). An autologous form of this composite skin construct has been used to treat severe burns with some success (Hansbrough *et al.*, 1989). An allogeneic form of the construct showed improved healing in a pilot study in chronic wounds (Boyce *et al.*, 1995). A similar technology has been studied in the treatment of patients with genetic blistering diseases (Eisenberg and Llewelyn, 1998).

One of the first attempts to replicate a full-thickness skin graft was by Bell *et al.* (1981), who described a bilayered skin equivalent. The dermal component consisted of a lattice of type I collagen contracted by tractional forces of rat dermal fibroblasts trapped within the gelled collagen. This contracted lattice was then used alone or as a substrate for rat epidermal keratinocytes. Bell's group demonstrated the ability of these primitive skin equivalents to take as a skin graft in rats. This technology has now advanced to enable the production of large amounts of human Apligraf HSE from a single donor (Wilkins *et al.*, 1994). Using methods of organotypic culture, which provides a three-dimensional culture environment that is permissive for proper tissue differentiation, the resulting HSE develops many of the structural, biochemical, and functional properties of human skin (Parenteau *et al.*, 1992; Bilbo *et al.*, 1993; Nolte *et al.*, 1993).

The process for formation of the HSE has been covered in detail many times (Parenteau, 1994; Wilkins *et al.*, 1994) and will not be detailed here. However, there are points to be made about the approach to these procedures. The culture of the HSE proceeds best with minimal intervention. Normal cell populations seem to have an intrinsic ability to reexpress their differentiated program *in vitro* to an extent that is now only beginning to be appreciated (Parenteau, 1999). A medium that supplies adequate amounts of nutrients, lipid precursors, vitamins, and minerals may be all that is required (Wilkins *et al.*, 1994). Another element is the environmental stimulus provided by culture at the air–liquid interface, which promotes differentiation and formation of the epidermal barrier (Nolte *et al.*, 1993).

The immunology of allogeneic tissue-engineered skin grafts is poorly understood. Inconsistencies in the literature are due, in part, to the complexity of biologic and immunologic factors, which determine the ability of a graft to take and to persist over time. Among the properties that determine the immunogenicity of an engineered allogeneic graft are purity of cell populations, antigen-presenting capabilities of graft cells, and the vascularity of the graft.

Purity of cell populations is critical. Differences in cell purity between laboratories may contribute to the conflicting results found in murine and human studies (Hefton *et al.*, 1983; Thivolet *et al.*, 1986; Geilen *et al.*, 1987; Aubock *et al.*, 1988; Cairns *et al.*, 1993). Both culture condition and passage number will affect the purity of cell populations. Very early passages of keratinocyte and fibroblast cultures might be expected to contain contaminating cell populations. The antigen-presenting capabilities of keratinocytes and fibroblasts are also critical to determining the immunogenicity of the HSE grafts. Fibroblasts and keratinocytes are not professional antigen-presenting cells and fail to stimulate the proliferation of allogeneic T cells (Nickoloff *et al.*, 1986; Niederwieser *et al.*, 1988; Gaspari and Katz, 1991; Theobald *et al.*, 1993). Both of these cell populations inherently do not express HLA class II molecules (Nickoloff *et al.*, 1985) or costimulatory molecules such as B7-1 (Nickoloff *et al.*, 1993). The inability of keratinocytes and fibroblasts to induce proliferation of allogeneic T cells is primarily due to the lack of expression of costimulatory molecules, even though aberrant antigen processing and invariant chain expression may also contribute (Nickoloff and Turka, 1994).

The ability to utilize allogeneic cells rather than autologous cells, as in CEA therapy, enables the reproducible manufacture of consistent Apligraf HSE (Wilkins *et al.*, 1994). The inability of epidermal keratinocytes and dermal fibroblasts to stimulate a T cell response, discussed above, permits their use in allogeneic applications. Studies in athymic mice also indicate that the use of a differentiated tissue such as the Apligraf HSE also beneficially affects its ability to engraft successfully (Nolte *et al.*, 1994; Parenteau *et al.*, 1996). Severe combined immunodeficient (SCID) mice lack a functioning immune system and can be successfully transplanted with a functioning human immune system without risk of rejection. SCID mice transplanted with human leukocytes were used as an *in vivo* model to assess the immunogenicity of allogeneic skin grafts (Moulon *et al.*, 1999). In these studies, human skin was rejected but allogeneic tissue-engineered skin graft survival was 100%.

Falanga *et al.* (1998) tested the safety, efficacy, and immunologic impact of Apligraf HSE (Graftskin) in the treatment of venous leg ulcers. There were no signs of rejection or bovine collagen-specific immune responses or response to alloantigens expressed on keratinocytes or fibroblasts. Apligraf HSE (Graftskin) has now been studied clinically in a number of applications, including chronic wounds (Sabolinski *et al.*, 1996), dermatologic excisions, and burns. Its effectiveness in the treatment of venous leg ulcers has been attributed to its ability to interact with the wound in multiple ways (Sabolinski *et al.*, 1996). The integrated tissue architecture of Apligraf HSE provides a robustness needed for clinical manipulation. The pivotal study by Falanga *et al.* (1998) of Apligraf HSE for the treatment of venous leg ulcers showed Apligraf HSE to be safe and effective, healing more ulcers in shorter time when compared to conventional therapy. Data from this study served as the basis for approval by the United States Food and Drug Administration for commercial use. A study of Apligraf HSE (Graftskin) in surgical patients showed take in 12 of 15 patients and also showed no evidence of rejection (Eaglstein *et al.*, 1995). In addition, Apligraf HSE has been studied in burns (Waymack *et al.*, 2000) and diabetic foot ulcers, and most recently, in the treatment of genetic blistering diseases (Falabella *et al.*, 1999).

The living epidermal keratinocytes and the dermal fibroblasts produce cytokines, which serve to regulate themselves, each other, and the cells of the patient. The living dermal extracellular matrix contains collagen and glucosaminoglycans, which serve as noninflammatory living tissue implants and scaffolds for tissue remodeling. The stratum corneum provides physical protection and maintains the proper physical and biologic environment for wound closure. Application of Apligraf HSE (Graftskin) can trigger a cascade of events that lead to normal healing even in the absence of graft take. In this regard, the HSE serves as a "smart material," providing the appropriate factors as needed, responding to the dynamics of the wound healing process, and beneficially changing the condition of the wound, even though the HSE may not always persist as a graft. Our clinical observations and results lead us to believe that all factors are important contributors to the wound healing process (Sabolinski *et al.*, 1996).

CONCLUSION

The routine treatment of skin wounds using engineered skin grafts has become a reality thanks to a firm scientific foundation established over the past 25 years, a readily available cell source (infant foreskin tissue), and the hard work and dedication of many individuals. Perhaps more importantly, our experience with the development of Apligraf HSE demonstrates that making a broadly available, effective medical device through tissue engineering is possible. It also illustrates the important immune tolerance of certain allogeneic cells. In addition, Apligraf HSE demonstrates the flexibility and broad utility of living tissue therapy.

References

Abraham, J. A., and Klagsbrun, M. (1996). Modulation of wound repair by members of the fibroblast growth factor family. *In* "The Molecular and Cellular Biology of Wound Repair" (R.A.F. Clark, ed.), 2nd ed., pp. 195–248. Plenum, New York.

Ansel, J., Perry, P., and Brown, J. (1990). Cytokine modulation of keratinocyte cytokines. *J. Invest. Dermatol.* **94**(Suppl.), 101–107.

Aubock, J., Irschick, E., Romani, N., Kompatscher, P., Hopfl, R., Herold, M., Schuler, G., Bauer, M., Huber, C., and Fritsch, P. (1988). Rejection, after a slightly prolonged survival time, of Langerhans cell-free allogeneic cultured epidermis used for wound coverage in humans. *Transplantation* **45**(4), 730–737.

Bal, V., McIndoe, A., Denton, G., Hudson, D., Lombardi, G., Lamb, J., and Lechler, R. (1990). Antigen presentation by keratinocytes induces tolerance in human T cells. *Eur. J. Immunol.* **20**, 1893–1987.

Bell, E., Ehrlich, P., Buttle, D. J., and Nakatsuji, T. (1981). Living tissue formed in vitro and accepted as skin-equivalent of full-thickness. *Science* **221**, 1052–2054.

Bilbo, P. R., Nolte, C.J.M., Oleson, M. A., Mason, V. S., and Parenteau, N. L. (1993). Skin in complex culture: The transition from 'culture' phenotype to organotypic phenotype. *J. Toxicol.-Cutaneous Ocul. Toxicol.* **12**, 183–196.

Blomme, E. A., Werkmeister, J. R., Zhou, H., Kartsogiannis, V., Capen, C. C., and Rosol, T. J. (1998). Parathyroid hormone-related protein expression and secretion in a skin organotypic culture system. *Endocrine* **8**, 143–151.

Bos, J. D., and Kapsenberg, M. L. (1993). The skin immune system: Progress in cutaneous biology. *Immunol. Today* **14**, 75–78.

Boyce, S. T., and Hansbrough, J. F. (1988). Biologic attachment, growth, and differentiation of cultured human keratinocytes on a graftable collagen and chondroitin-6-sulfate substrate. *Surgery* **103**, 421–431.

Boyce, S. T., Glatter, R., and Kitsmiller, J. (1995). Treatment of chronic wounds with cultured skin substitutes: A pilot study. *Wounds: Compend. Clin. Res. Pract.* **7**, 24–29.

Burke, J. F., Yannas, I. V., Quinby, W. C., Bondoc, C. C., and Jung, W. K. (1981). Successful use of a physiologically acceptable artificial skin in the treatment of extensive burn injury. *Ann. Surg.* **194**, 413–428.

Cairns, B. A., deSerres, S., Kilpatrick, K., Frelinger, J. A., and Meyer, A. A. (1993). Cultured keratinocyte allografts fail to induce sensitization in vivo. *Surgery* **114**(2), 416–422.

Clark, R.A.F. (1993). Basics of cutaneous wound repair. *J. Dermatol. Surg. Oncol.* **19**, 639–706.

Compton, C. C. (1993). Wound healing potential of cultured epithelium. *Wounds: Compend. Clin. Res. Pract.* **5**, 97–111.

Compton, C. C., Gill, J. M., Bradford, D. A., Regauer, S., Gallico, G. G., and O'Connor, N. E. (1989). Skin regenerated from cultured epithelial autografts on full-thickness burn wounds from 6 days to 5 years after grafting. A light, electron microscopic and immunohistochemical study. *Lab. Invest.* **60**(5), 600–612.

Cooper, M. L., Hansbrough, J. F., and Polareck, J. W. (1996). The effect of an arginine-glycine-aspartic acid peptide and hyaluronate synthetic matrix on epithelialization of meshed skin graft interstices. *J. Burn Care Rehab.* **17**, 108–116.

Cuono, C. B., Langdon, R., and McGuire, J. (1986). Use of cultured epidermal autografts and dermal allografts as skin replacement after burn injury. *Lancet* **1**, 1123–1124.

Cuono, C. B., Langdon, R., Birchall, N., Barttelbort, S., and McGuire, J. (1987). Composite autologous allegeneic skin replacement: Development and clinical application. *Plast. Reconstr. Surg.* **80**, 626–637.

Desmouliere, A., and Gabbiani, G. (1996). The role of the myofibroblast in wound healing and fibrocontractive diseases. *In* "The Molecular and Cellular Biology of Wound Repair" (R.A.F. Clark, ed.), 2nd ed., pp. 321–323. Plenum, New York.

deVries, H.J.C., Mekkes, J. R., Middelkoop, E., Hinrichs, W.L.J., Wildevuur, C.H.R., and Westerhof, W. (1993). Dermal substitutes for full-thickness wounds in a one-stage grafting model. *Wound Repair Regeneration* **1**, 244–252.

Duinslaiger, L., Verbeken, G., Reper, P., Delaey, B., Vanhalle, S., and Vanderkelen, A. (1994). Lyophilized keratinocyte cell lysates contain multiple mitogenic activities and stimulate closure of meshed skin autograft-covered burn wounds with efficiency similar to that of fresh allogeneic keratinocyte cultures. *Plast. Reconstr. Surg.* **98**, 110–117.

Eaglstein, W. H., and Falanga, V. (1997). Tissue engineering and the development of Apligraf, a human skin equivalent. *Clin. Ther.* **19**, 894–905.

Eaglstein, W. H., Iriondo, M., and Laszlo, K. (1995). A composite skin substitute (Graftskin) for surgical wounds: A clinical experience. *Dermatol. Surg.* **21**, 839–843.

Eisenberg, M., and Llewelyn, D. (1998). Surgical management of hands in children with recessive dystrophic epidermolysis bullosa: Use of allogeneic composite cultured skin grafts. *Br. J. Plast. Surg.* **51**(8), 608–613.

Falabella, A. F., Schackner, L. A., Valencia, J. C., and Eaglstein, W. H. (1999). The use of tissue-engineered skin (Apligraf) to treat a newborn with epidermolysis bullosa. *Arch. Dermatol.* **135**, 1219–1226.

Falanga, V., Margolis, D., Alvarez, O., Auletta, M., Maggiacomo, F., Altman, M., Jensen, J., Sabolinski, M., and Hardin-

Young, J. (1998). Human Skin Equivalent Investigators Group. Rapid healing of venous ulcers and lack of clinical rejection with an allogeneic cultured human skin equivalent. *Arch. Dermatol.* **134**, 293–300.

Gallico, G. C., III, O'Connor, N. E., Compton, C. C., Kehinde, O., and Green, H. (1984). Permanent coverage of large burn wounds with autologous cultured human epithelium. *N. Eng. J. Med.* **311**, 448–451.

Gaspari, A. A., and Katz, S. I. (1988). Induction and functional characterization of class II MHC (Ia) antigens on murine keratinocytes. *J. Immunol.* **140**, 2956–2963.

Gaspari, A. A., and Katz, S. I. (1991). Induction of in vivo hyporesponsiveness to contact allergens by hapten-modified Ia+ keratinocytes. *J. Immunol.* **147**(12), 4155–4161.

Gielin, V., Faure, M., Mauduit, G., and Thivolet, J. (1987). Progressive replacement of human cultured epithelial allografts by recipient cells as evidenced by HLA class I antigen expression. *Dermatologica* **175**(4), 166–170.

Green, H., Kehinde, O., and Thomas, J. (1979). Growth of cultured human epidermal cells into multiple epithelial suitable for grafting. *Proc. Nat. Acad. Sci. U.S.A.* **76**, 5665–5668.

Greiling, D., and Clark, R. A. (1997). Fibronectin provides a conduit for fibroblast transmigration from collagenous stroma into fibrin clot provisional matrix. *J. Cell Sci.* **110**(Pt. 7), 861–870.

Hansbrough, J. (1997). Dermagraft-TC for partial-thickness burns: a clinical evaluation. *J. Burn Care Rehab.* **18**(1, Pt. 2), S25–S28.

Hansbrough, J. F., Boyce, S. T., Cooper, M. L., and Foreman, T. J. (1989). Burn wound closure with cultured autologous keratinocytes and fibroblasts attached to a collagen-glycosaminoglycan substrate. *JAMA, J. Am. Med. Assoc.* **262**, 2125–2130.

Hansbrough, J. F., Dore, C., and Hansbrough, W. B. (1992). Clinical trails of a living dermal tissue replacement placed beneath meshed, split-thickness skin grafts on excised burn wound. J. Burn Care Rehab. **13**(5), 519–529.

Hansbrough, J. F., Norgan, J., Greenleaf, G., and Underwood, J. (1994). Development of a temporary living skin replacement composed of human neonatal fibroblasts cultured in Biobrane, a synthetic dressing material. *Surgery* **115**, 633–644.

Hansbrough, J. F., Mozingo, D. W., Kealey, G. P., Davis, M., Gidner, A., and Gentzkow, G. D. (1997). Clinical trials of biosynthetic temporary skin replacement, Dermagraft-Transitional covering, compared with cryopreserved human cadaver skin for temporary coverage of excised burn wounds. *J. Burn Care Rehab.* **18**(1, Pt. 1), 43–51.

Hefton, J. M., Madden, M. R., Finkelstein, J. L., and Shires, G. T. (1983). Grafting of burn patients with allografts of cultured epidermal cells. *Lancet* **2** 428–430.

Heimbach, D., Luterman, A., Burke, J. F., Cram, A., Herndon, D., Hunt, J., Jordan, M., McManus, W., Solem, L., and Ward, N. G. (1988). Artificial dermis for major burns: A multi-center randomized clinical trial. *Ann. Surg.* **208**, 313–320.

Knighton, D. R., Fiegel, V. D., Doucette, M. M., Fylling, C. P., and Cerra, F. B. (1989). The use of topically applied platelet growth factors in chronic nonhealing wounds: A review. *Wounds: Compend. Clin. Res. Pract.* **1**, 71–78.

Langdon, R. C., Cuono, C. B., Birchall, N., Madri, J. A., Kuklinska, E., McGuire, J., and Moellmann, G. E. (1988). reconstitution of structure and cell function in human skin grafts derived from cryopreserved allogeneic dermis and autologous cultured keratinocytes. *J. Invest. Dermatol.* **91**, 478–485.

Lappin, M. B., Kimnber, I., and Norval, M. (1996). The role of dendritic cells in cutaneous immunity. *Arch. Dermatol. Res.* **288**, 109–121.

Lattari, V., Jones, L. M., Varcelotti, J. R., Latenser, B. A., Sherman, H. F., and Barrette, R. R. (1997). The use of a permanent dermal allograft in full-thickness burns of the hand and foot: A report of three cases. *J. Burn Care Rehab.* **18**, 147–155.

Lazarus, G. S., Cooper, D. M., Knighton, D. R., Margolis, D. J., Pecorara, R. E., Rodenheaver, G., and Robson, M. C. (1994). Definitions and guidelines for assessment of wound and evaluation of healing. *Wound Repair Regeneration* **130**, 489–493.

Leary, T. Jones, P. L., Appleby, M., Blight, A., Parkinson, K., and Stanley, M. (1992). Epidermal keratinocyte self-renewal is dependent upon dermal integrity. *J. Invest. Dermatol.* **99**, 422–430.

Livesey, S. A., Herndon, D. N., Hollyoak, M. A., Atkinson, Y. H., and Nag, A. (1995). Transplanted acellular allograft dermal matrix. *Transplantation* **60**, 1–9.

Lorenz, C., Petracic, A., Hohl, H. P., Wessel, L., and Waag, K. L. (1997). Early wound closure and early reconstruction. Experience with a dermal substitute in a child with 60 per cent surface area burn. *Burns* **23**, 505–508.

Marks, M. G., Doillon, C., and Silver, F. H. (1991). Effects of fibroblasts and basic fibroblast growth factor on facilitation of dermal wound healing by type I collagen matrices. *J. Biomed. Mater. Res.* **25**, 683–696.

Martin, P., Hopkinson-Woolley, J., and McCluskey, J. (1992). Growth factors and cutaneous wound repair. *Prog. Growth Factor Res.* **4**, 25–44.

Mast, B. A., Nelson, J. M., and Krummel, T. M. (1992). Tissue repair in the mammalian fetus. *In* "Wound Healing: Biochemical and Clinical Aspects" (I. K. Cohen, R. F. Diegelmann, and W. J. Lindblad, eds.), pp. 326–341. Saunders, Philadelphia.

McKay, I. A., and Leigh, I. M. (1991). Epidermal cytokines and their roles in cutaneous wound healing. *Br. J. Dermatol.* **124**, 513–518.

Mian, E., Martini, P., Beconcini, D., and Mian, M. (1992). Healing of open skin surfaces with collagen foils. *Int. J. Tissue React.* **14**, 27–34.

Middelkoop, E., deVries, H.J.C., Ruuls, L., Everts, V., Wildevuur, C.H.R., and Westerhof, W. (1995). Adherence, proliferation, and collagen turnover by human fibroblasts seeded into different types of collagen sponges. *Cell Tissue Res.* **280**, 447–453.

Moulon, K. S., Melder, R. J., Dharmidharka, V., Hardin-Young, J., Jain, R. K., and Briscoe, D. M. (1999). Angiogenesis in the huPBL-SCID model of human transplantation rejection. *Transplantation* **67**, 1626–1631.

Murashita, T., Nakayama, Y., Hirano, T., and Ohashi, S. (1996). Acceleration of granulation tissue in growth by hyaluronic acid in artificial skin. *Br. J. Plast. Surg.* **49**, 58–63.

Nanney, L. B., and King, L. E., Jr. (1996). Epidermal growth factor and transforming growth factor-α. *In* "The Molecular and Cellular Biology of Wound Repair" (R.A.F. Clark, ed.), 2nd ed., pp. 171–194. Plenum, New York.

Nathan, C., and Sporn, M. (1991). Cytokines in context. *J. Cell Biol.* **113**, 981–986.

Nickoloff, B. J., and Turka, L. A. (1994). Immunological functions of non-professional antigen-presenting cells: New insights from studies of T-cell interactions with keratinocytes. *Immunol. Today* **15**(10), 464–469.

Nickoloff, B. J., Basham, T. Y., Merigan, T. C., and Morhenn, V. B. (1985). Keratinocyte class II histocompatibility antigen expression. *Br. J. Dermatol.* **112**(3), 373–374.

Nickoloff, B. J., Basham, T. Y., Merigan, T. C., Torseth, J. W., and Morhenn, V. B. (1986). Human keratinocyte-lymphocyte reactions in vitro. *J. Invest. Dermatol.* **87**(1), 11–18.

Nickoloff, B. J., Mitra, R. S., Lee, K., Turka, L. A., Green, J., Thompson, C., and Shimizu, Y. (1993). Discordant expression of CD29 ligands, BB-1 and B7 on keratinocytes in vitro and psoriatic cells in vivo. *Am. J. Pathol.* **142**(4), 1029–1040.

Niederwieser, D., Aubock, J., Troppmair, J., Herold, M., Schuler, G., Boeck, G., Lotz, J., Fritsch, P., and Huber, C. (1988). IFN-mediated induction of MHC antigen expression on human keratinocytes and its influence on in vitro alloimmune response. *J. Immunol.* **140**(8), 2556–2564.

Nolte, C.J.M., Oleson, M. A., Bilbo, P. R., and Parenteau, N. L. (1993). Development of a stratum corneum and barrier function in an organotypic skin culture. *Arch. Dermatol. Res.* **285**, 466–474.

Nolte, C.J.M., Oleson, M. A., Hansbrough, J. F., Morgan, J., Greenleaf, G., and Wilkins, L. (1994). Ultrastructural features of composite skin cultures grafted onto athymic mice. *J. Anat.* **185**, 325–333.

Odessey, R. (1992). Addendum: Multicenter experience with cultured epidermal autograft for the treatment of burns. *J. Burn Care Rehab.* **13**, 174–180.

Palmieri, B. (1992). Heterologous collagen in wound healing: A clinical study. *Int. J. Tissue React.* **14**(Suppl.), 21–25.

Parenteau, N. L. (1994). Skin equivalents. *In* "Keratinocyte methods" (I. Leigh and F. Watt, eds.), pp. 45–54, Cambridge University Press, Cambridge, UK.

Parenteau, N. L. (1999). Cell differentiation. *In* "Encyclopedia of Animal and Plant Cell Technology" (in press).

Parenteau, N. L., Bilbo, P., Nolte, C.J.M., Mason, V. S., and Rosenberg, M. (1992). The organotypic culture of human skin keratinocytes and fibroblasts to achieve form and function. *Cytotechnology* **9**, 163–171.

Parenteau, N. L., Sabolinski, M., Prosky, S., Nolte, C., Oleson, M., Kriwet, K., and Bilbo, P. (1996). Biological and physical factors influencing the successful engraftment of a cultured human skin substitute. *Biotechnol. Bioeng.* **52**, 3–14.

Parenteau, N. L., Sabolinski, M. L., Mulder, G., and Rovee, D. T. (1997). Wound research. *In* "Chronic Wound Care: A Clinical Source for Healthcare Professionals" (D. Krasner and D. Kane, eds.), 2nd ed., pp. 389–395. Health Management Publications, Wayne, PA.

Phillips, T. J., Kehinde, O., Green, H., and Gilchrest, B. A. (1989). Treatment of skin ulcers with cultured epidermal allografts. *J. Am. Acad. Dermatol.* **21**, 191–199.

Phillips, T. J., Bhawan, J., Leigh, I. M., Baum, H. J., and Gilchrest, B. A. (1990). Cultured epidermal allografts: A study of differentiation and allograft survival. *J. Am. Acad. Dermatol.* **23**, 189–198.

Phipps, R. P., Roper, R. L., and Stein, S. H. (1989). Alternative antigen presentation pathways: Accessory cells which downregulate immune response. *Reg. Immunol.* **2**, 326–339.

Pober, J. S., Collins, T., Bimmbrone, M. A., Jr., Libby, P., and Reiss, C. S. (1986). Inducible expression of class II major histocompatibility complex antigens and the immunogenicity of vascular endothelium. *Transplantation* **41**(2), 141–146.

Rheinwald, J. G., and Green, H. (1975). Serial cultivation of strains of human epidermal keratinocytes: the formation of keratinizing colonies from single cells. *Cell (Cambridge, Mass.)* **6**, 331–344.

Roberts, A. B., and Sporn, M. B. (1996). Transforming growth factor-β. *In* "The Molecular and Cellular Biology of Wound Repair" (R.A.F. Clark, ed.), 2nd ed., pp. 275–308. Plenum, New York.

Sabolinski, M. L., Alvarez, O., Auletta, M., Mulder, G., and Parenteau, N. L. (1996). Cultured skin as a 'smart material' for healing wounds: Experience in venous ulcers. *Biomaterials* **17**, 311–320.

Sheridan, R. L., and Tompkins, R. G. (1995). Cultured autologous epithelium in patients with burns of ninety percent or more of the body surface. *J. Trauma: Injury, Infect. Crit. Care* **38**, 48–50.

Steed, D. L., Ricotta, J. J., Prendergast, J. J., Kaplan, R. J., Webster, M. W., McGill, J. B., and Schwartz, S. L. (1995). Promotion and acceleration of diabetic ulcer healing by arginine-glycine-aspartic acid (RGD) peptide matrix. RGD Study Group. *Diabetes Care* **18**, 39–46.

Streilein, J. W. (1983). Skin-associated lymphoid tissues (SALT): Origins and functions. *J. Invest. Dermatol.* **80**(Suppl.), 12–16.

Streilein, J. W., and Bergstresser, P. R. (1984). Langerhans Cells: Antigen presenting cells of the epidermis. *Immunobiology* **168**, 285–300.

Theobald, V. A., Lauer, J. D., Kaplan, F. A., Baker, K. B., and Rosenberg, M. (1993). "Neutral allografts"—lack of allogeneic stimulation by cultured human cells expressing MHC class I and class II antigens. *Transplantation* **55**(1), 128–133.

Thivolet, J., Faure, M., Demide, A., and Mauduit, G. (1986). Long-term survival and immunological tolerance of human epidermal allografts produced in culture. *Transplantation* **42**(3), 274–280.

van Luyn, M.J.A., Verheul, J., and van Wachem, P. B. (1995). Regeneration of full-thickness wounds using collagen split grafts. *J. Biomed. Mater. Res.* **29**, 1425–1436.

Waymack, P., Duff, R., Sabolinski, M., and the Apligraf Burn Study Group. (2000). The effect of a tissue-engineered bilayered living skin analog, over meshed split-thickness autograff on the healing of excised burn wound. *Burns: J. Intl. Soc. Burn. Inj.,* (in press).

Wethers, D. L., Ramirez, G. M., Koshy, M., Steinberg, M. H., Phillips, G., Jr., Siegel, R. S., Echman, J. R., and Prchal, J. T. (1994). Accelerated healing of chronic sickle-cell leg ulcers treated with RGD peptide matrix. RGD Study Group. *Blood* **84,** 1775–1779.

Wilkins, L. M., Watson, S. R., Prosky, S. J., Meunier, S. F., and Parenteau, N. L. (1994). Development of a bilayered living skin construct for clinical applications. *Biotechnol. Bioeng.* **43,** 747–756.

Williams, I. R., and Kupper, T. S. (1996). Immunity at the surface: Homeostatic mechanisms of the skin immune system. *Life Sci.* **58**(18), 1485–1507.

Yannas, I. V., Burke, J. F., Orgill, D. P., and Skrabut, E. M. (1982). Wound tissue can utilize a polymeric template to synthesize a functional extension of skin. *Science* **215,** 174–176.

Yannas, I. V., Lee, E., Orgill, D. P., and Skrabut, E. M., and Murphy, G. F. (1989). Synthesis and characterization of a model extracellular matrix that induces partial regeneration of adult mammalian skin. *Proc. Nat. Acad. Sci. U.S.A.* **86,** 933–937.

Zacchi, V., Soranzo, C., Cortivo, R., Radice, M., Brun, P., and Abatangelo, G. (1998). In vitro engineering of human skin-like tissue. *J. Biomed. Mater. Res.* **40,** 187–194.

CHAPTER 63

DERMAL EQUIVALENTS

Gail K. Naughton

INTRODUCTION

Tissue engineering, the science of growing living tissues for transplantation, promises to revolutionize a number of aspects of medical care. This science integrates the principles of cell biology, polymer development, biochemistry, bioengineering, and transplantation science to create human physiologic tissues to replace damaged and diseased areas. Tissue engineering offers the ability to create highly tested physiologic human tissues and organs for transplantation. The most advanced area in tissue engineering is the manufacture of skin for the treatment of patients with burns or chronic wounds, with initial products having already received regulatory approval in the United States and other global markets. A number of approaches have been utilized to tissue-engineered skin substitutes. Such approaches have included cellular bovine-based (Yannas and Burke, 1980) human cadaveric materials (Sheridan *et al.,* 1998) and animal collagens containing human cells (Bell *et al.,* 1984; Hansbrough *et al.,* 1989; Eisenberg and Llewelyn, 1998). This chapter describes the development, manufacturing, and clinical development of a human dermal equivalent manufactured by growing newborn allogeneic human dermal fibroblasts on a biodegradable three-dimensional scaffold to produce a human cell and human tissue implant for a variety of wounds (Naughton and Tolbert, 1996).

CELL TESTING AND ESTABLISHMENT
OF MANUFACTURING CELL BANKS

Tissue-engineered products offer advantages over cadaveric materials in that the cells utilized, as well as the final manufactured product, are highly tested for a variety of viruses, contaminants, and pathogens. The human fibroblast cell strains used to produce the human dermal equivalent, Dermagraft (Dg) (Advanced Tissue Sciences, Inc., La Jolla, CA), are established from neonatal foreskin (discarded after surgical circumcision) (Kruse and Patterson, 1973; Jakoby and Pastan, 1979) and cultured by standard methods. Maternal blood samples are tested for exposure to infectious diseases, including human immunodeficiency virus (HIV), human T cell lymphototropic virus (HTLV), herpes simplex virus (HSV), cytomegalovirus (CMV), and hepatitis virus. An initial screen is made of the cultured cells for sterility, mycoplasma, and for the eight human viruses: adeno-associated virus, HSV-1 and -2, CMV, HIV-1 and -2, and HTLV-I and -II. Master cell banks (MCBs) and manufacturer's working cell banks (MWCBs) are created and tested according to applicable sections of the U.S. Food and Drug Administration (FDA) points to consider (1993) and the guidelines from the European Union Committee for Proprietary Medicinal Products (CPMP) (1989). This testing provides a more uniform and highly tested product than is available from cadaveric donors (tissue banks) (Blood *et al.,* 1979; White *et al.,* 1991).

MANUFACTURING A DERMAL REPLACEMENT

Dermagraft is manufactured through the three-dimensional cultivation of human diploid fibroblast cells on a polymer scaffold. The fibroblasts secrete a mixture of growth factors and matrix proteins to create a living dermal structure that, following cryopreservation, remains metabolically active after being implanted into the patient's wound bed (Cooper *et al.,* 1991; Landeen *et al.,* 1992). This dermal product was designed to be shipped and stored at −70°C. Cryopreservation of a tissue-engineered product offers the advantage of extensive sterility and cell and matrix testing, ensuring the release of only sterile, uniform, and reproducible products.

CLOSED MANUFACTURING SYSTEMS

A closed bioreactor system is utilized for the manufacture of Dermagraft. This closed system offers major advantages, including the maintenance of sterility during manufacture, the ability to utilize in-process testing to maintain cell growth and matrix deposition, and the use of automated processes for tissue growth. The rate of deposition of matrix proteins during growth in the bioreactor system is shown in Fig. 63.1, showing that matrix is deposited in both growth-associated (glycosaminoglycans) and non-growth-associated patterns (human collagens).

A series of Dermagraft products is created by seeding the polymer scaffolds at passage 8. This is a population doubling level of approximately 30 (which is about half the life-span for this cell type). Utilization of fibroblasts from neonatal foreskins offers a number of advantages, including the fact that these young cells actively proliferate under routine cell culture conditions and are from a uniform source (same sex, age, and anatomical location), helping reduce variability of starting material as well as final manufactured product. Fibroblasts are seeded into multicavity EVA bags containing a 5 × 7.5 cm piece of knitted lactic acid/glycolic acid copolymer, mesh which is welded into the bag to prevent movement during the growth process or loss of the implant during rinse of the product prior to implantation. Seeded bags are rotated to enhance uniform cell seeding and are cultured in Dulbecco's modified Eagle's medium (DMEM) supplemented with bovine calf serum, glutamine, ascorbate and nonessential amino acids (growth medium), with periodic changes of medium. Tissue is grown under static conditions with medium, changes being made by peristaltic pumping within the closed system. Samples of medium are routinely taken to test for pH and glucose utilization and to predict the precise time for tissue harvest. At harvest, the medium is drained and replaced by a 10% dimethyl sulfoxide-based cryopreservative solution.

Manufacturing, cryopreservation, storage, shipping, and thawing protocols were all established to maintain integrity and metabolic activity of this tissue-engineered implant. The final product is a 2-inch by 3-inch tissue in a laser-welded EVA bioreactor (Fig. 63.2). This bioreactor design allows for use of an EVA container for sterilization, seeding, tissue growth, freezing, shipping, and product storage and prevents any need for product repackaging. The up-scaled bioreactor systems are designed to manufacture over 1000 units of Dermagraft per lot. A cross-section of the dermal implant clearly shows the physiologic characteristics of this product, with human fibroblasts arranged in parallel and surrounded by naturally secreted human matrix proteins.

The bioreactor containing the engineered dermal implant is translucent to allow tracing of the wound and precise trimming of the implant with sterile scissors. Dermagraft is cut to wound size, removed from the ethylvinylacetate (EVA) package, and placed into the wound bed. No su-

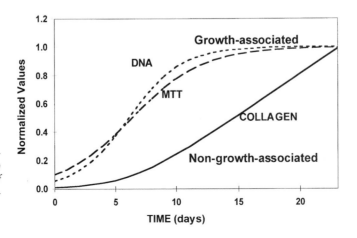

Fig. 63.1. The rate of matrix deposition and cell proliferation during the manufacture of Dermagraft. MTT, 3-(4,5-Dimethylthiazol-2-yl)-2,5-diphenyltetrazolium bromide.

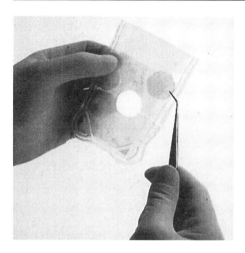

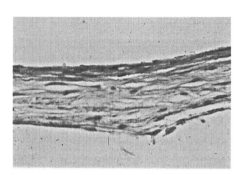

Fig. 63.2. A living bioengineered 2-inch by 3-inch human dermal replacement (Dermagraft) product and a histological cross-section.

tures are required. Dressings are utilized to ensure that the dermal implant remains in place, is kept moist during integration into the patient's wound bed, and functions as a replacement dermis.

PRODUCT CHARACTERIZATION

Dermagraft has been designed to replace the dermal layer of the skin and provide the stimulus for the normal wound healing process. Characterization studies have shown that Dermagraft is potentially a good model to study *in vitro* epithelialization (Contard *et al.,* 1993) and the interactions between the extracellular matrix, fibroblasts, and migrating epithelial cells (Landeen *et al.,* 1992).

Histological characterization of Dermagraft has shown that it has the structure of the papillary dermis of newborn skin. This dermal implant contains normal matrix proteins (Table 63.1), which play an integral role in providing structure as well as enhancing cell growth. A transmission election micrograph of the tissue shows deposition of parallel collagen fibers (Fig. 63.3). These fibers have normal periodicity and normal ratios of collagens type III and type I. In addition, all of the glycosaminoglycans (GAGs) found in young healthy dermis are present in Dermagraft (Table 63.2). These matrix components, vital for cell migration and for the binding of growth factors, are secreted by the newborn fibroblasts during the manufacturing process to yield a dermal implant with an appropriate complement of human GAGs. The fibroblasts, evenly dispersed throughout the tissue, remain metabolically active after implantation and delivery: a variety of growth factors that are key to neovascularization, epithelial migration and differentiation, and integration of the implant into the patient's wound bed. Detection and quantification of specific messenger RNA as well as expression of growth factors have been previously described (Jiang and Harding, 1998; Mansbridge *et al.,* 1998; Naughton *et al.,* 1997). The final product is extensively tested after manufacture for sterility as well as for specific levels of collagens and other matrix proteins. Because Dermagraft was designed to be a living tissue, remaining viable and delivering growth factors and matrix proteins into the wound bed after implantation, testing is performed pre- and postcryopreservation to assess its metabolic activity. A routine 3-(4,5-dimethylthiazol-2-yl)-2,5-diphenyltetrazolium bromide (MTT) assay is utilized for this assessment both at time of manufacture and for testing of product stability over time.

Table 63.1. Matrix proteins in Dermagraft

Matrix proteins	Function
Collagen type I, III	Major structural proteins of dermis
Fibronectin	Cell adhesion, spreading, migration, mitogenesis
Tenascin	Induced wound healing, control of cell adhesion

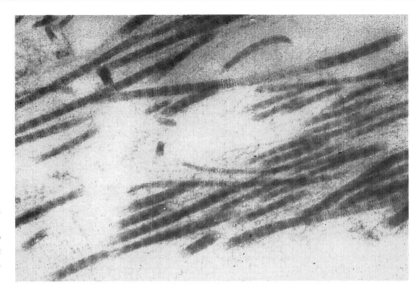

Fig. 63.3. A transmission electron micrograph of Dermagraft. Human collagen fibers are arranged in parallel bundles and have normal banding periodicity.

IMPORTANCE OF GROWTH FACTORS IN A DERMAL REPLACEMENT

Wound healing is a complicated process that involves a number of dynamic cellular and molecular events. Factors regulating wound healing are many and originate from different sources. Key parameters, including cell migration, proliferation, angiogenesis, matrix deposition and degradation, and immune modulation, are essential for this process (Arbiser, 1996; Bennett and Schultz, 1993; Folkman and Klagsbrun, 1987; Marks *et al.*, 1991; Moulin, 1995; Pettet *et al.*, 1996; Raghow, 1994; Schaffer and Nanney, 1996). Improvement of cell motility (e.g., dermal fibroblasts, keratinocytes, melanocytes), angiogenesis (formation of vessels; migration and motility of endothelial cells), and modification of matrix turnover all have impact on wound healing.

A tissue-engineered human dermal replacement can exploit the fact that the skin fibroblast is one of the most important and predominant cell types in the tissue repair process. This cell type appears at an early stage after injury, increasing in number as healing progresses (Arbiser, 1996; Bennett and Schultz, 1993; Folkman and Klagsbrun, 1987; Marks *et al.*, 1991; Moulin, 1995; Pettet *et al.*, 1996; Raghow, 1994). The cell is known to play a central role in the regulation of matrix deposition and degradation in the wound. Fibroblasts, however, also synthesize a rich array of cytokines, regulators of tissue repair in wounds, including interleukins, growth factors, and angiogenic factors.

TISSUE EXPANSION PROPERTIES OF DERMAGRAFT

Jiang and Harding (1998) evaluated the ability of Dermagraft to stimulate tissue growth in an *ex vivo* tissue expansion assay. Extracellular matrix gel was used as a support of tissue in order to test the effect of factors on the tissue expansion from patients with chronic and acute wounds. Briefly, Matrigel was dissolved in culture medium at 500 μg/ml. This was added to a 24-well tissue culture plate. Materials and the complete system were kept ice cold at any given time until the

Table 63.2. Glycosaminoglycans in Dermagraft

GAGs	Function
Versican	Structural, binds hyaluronic acid and collagen
Decorin	Binds growth factors, influences collagen structure
Betaglycan	TGF-β type III receptor
Syndecan	Binds growth factors, enhances activity

next step. Wound tissue was washed vigorously in culture medium and finely minced with a dissecting scalpel and scissors (approximately 0.5 mm in diameter), then further washed and transferred to cool Matrigel solution in a culture well. The temperature for the system was then quickly raised to 37°C to form an irreversible gel, thus allowing the wound tissue to be embedded in the Matrigel. Coculture inserts with Dermagraft were then placed on top of the well for coculturing. The system was monitored over a 4-week period using a Panasonic charge-coupled device (CCD) camera and the distance between the wound edge and the leading front of the expansion was calculated using image analysis software (Optimas; Optimas Ltd., UK).

In Matrigel, wound tissue was able to expand in a three-dimensional manner. Viable tissue obtained from patients with acute and chronic wounds revealed various patterns of cell migration. Briefly, after embedding, polymorphonuclear cells began to migrate out of the wound tissue. However, a large-scale migration of other cells (fibroblasts and endothelial cells) were seen from day 3 onward. The distance between the leading front of the area and the edge of the wound became quantifiable from day 7. Images from these cocultured tissues were then taken with a digital camera (UltraBix; Cambridge Scientific Instrument; Cambridge, UK). The relative distance between this leading edge and wound tissue, termed the "expansion index" in this report, was then quantified from these images with image analysis softward (Optimas). When Dermagraft was included, there was a marked expansion of the wound tissue when compared with tissue without the Dermagraft ($p = 0.0015$). At a higher magnification, the expansion areas represents a mixture of cells and structures. Most cells are fibroblast-like cells and endothelial cells. After a prolonged culture (over 2 weeks), a vessellike structure was clearly seen (Jiang and Harding, 1998).

A number of studies have been performed to correlate growth factor expression and secretion with *in vitro* and *in vivo* wound healing characteristics of Dermagraft (Jiang and Harding, 1998; Mansbridge *et al.*, 1998; Naughton *et al.*, 1997; Newton *et al.*, 1999). The angiogenic-promoting properties of this human dermal replacement have been well characterized and provide a unique tool for wound bed vascularization and promotion of healing.

ANGIOGENIC PROPERTIES OF DERMAGRAFT

Tissue-engineered fibroblast-based tissues are capable of inducing rapid endothelialization and vascularization. Providing such biologically active natural materials has been clinically shown to induce new capillary formation and reduce inflammation in the wound bed of patients with diabetic foot ulcers (Jiang and Harding, 1998; Newton *et al.*, 1999). The tissue-engineered implants secrete a variety of growth factors known to be critical to tissue regeneration and angiogenesis (Table 63.3).

The angiogenic properties of our fibroblast-based engineered tissue have been investigated using a range of techniques, including the chick chorioallantoic membrane assay, the rat aortic ring assay, stimulation of endothelial cell proliferation, chemokinesis, chemotaxis, and the inhibition of apoptosis. These assays cover a wide range of the important individual steps in angiogenesis as well as the overall process.

Analysis of our tissue-engineered human matrix has shown it to contain several components

Table 63.3. Growth factors

Growth factors	Function
PDGF-A	Mitogen for fibroblasts, granulation tissue, chemotactic
IGF	Mitogen for fibroblasts
KGF	Mitogen for keratinocytes
HBEGF	Mitogen for keratinocytes, fibroblasts
TGF-α	Mitogen for keratinocytes, fibroblasts
TGF-β1	Stimulates matrix deposition
TGF-β3	Stimulates matrix deposition, antiscarring
VEGF	Angiogenesis
SPARC	Both anti- and proangiogenic

that may be beneficial in neovascularization. Fibronectin present in the matrix has been shown to stimulate the proliferation of endothelial cells, and the denatured collagen has been proved to be a favorable substrate for human endothelial cell attachment. Bound growth factors in the matrix include transforming growth factor β (TGF-β) and hepatocyte growth factor (HGF), which are important in stimulating new capillary formation and endothelialization. Finally, the matrix contains laminin-1, which can serve to inhibit initial hyperplasia via the YIGSR peptide. The combination of these matrix proteins along with naturally secreted growth factors offers a physiologic solution to promoting the *in vivo* induction of angiogenesis.

Stimulation of angiogenesis in the chick chorioallantoic membrane

Ten-day-old chicken embryos were obtained from McIntyre Farms (Lakeside, CA) and incubated at 37°C. Eggs were candled to locate and mark a target area void of large vessels. Two small

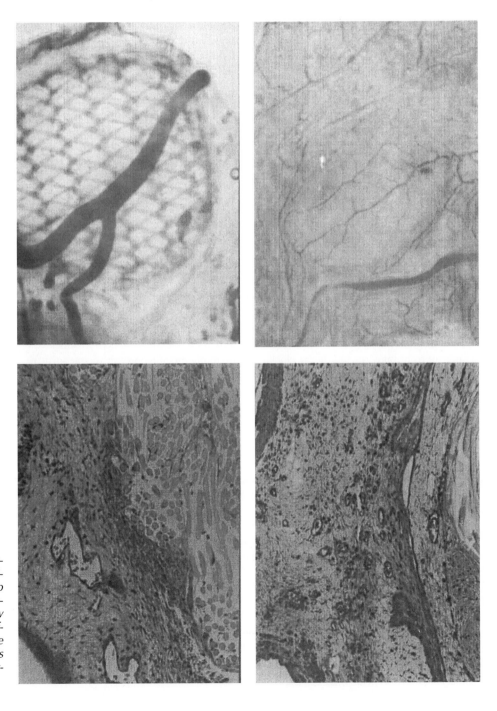

Fig. 63.4. Dermagraft-stimulated angiogenesis in the chorioallantoic membrane assay. Top two panels show the macroscopic view, bottom two panels show histology. Left panels are of scaffold alone (top) or nonviable Dermagraft (bottom); the panels to the right are treated with Dermagraft.

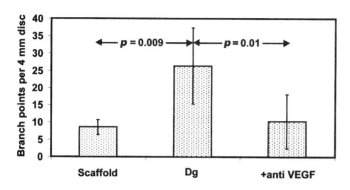

Fig. 63.5. Effect of Dermagraft (Dg) on capillary blood vessel formation in the CAM. Bars represent 95% confidence intervals.

holes were made in the shell with a needle, directly over the air sac and over the target area. Suction was applied to the first hole, causing the chorioallantoic membrane (CAM) to drop away from the marked area. Using a Dremel Moto-Tool, the eggshell was removed from the target area to create a "window." A 4-mm-diameter circular sample (fibroblast-based engineered tissue or control) was then placed on the membrane near, but not on top of, a large blood vessel. The hole was covered with a piece of clear adhesive tape and the eggs were incubated for 72 hr at 37°C to allow blood vessel growth. The treated section of the membrane was then removed, photographed, and fixed in methanol. The number of fine blood vessel branch points in the region of the sample was counted. Biopsy samples were fixed in methanol and sections stained with Masson's Trichrome.

The three-dimensional fibroblast-based tissues induced vessel development in the CAM to a greater extent than control (Fig. 63.4), including both fine capillary development and evidence for increased permeability. The development of capillary blood vessels in CAMs treated with three-dimensional fibroblast cultures was also clearly visible by histology. This type of capillary development is characteristic of vascular endothelial growth factor (VEGF)-induced angiogenesis. It differs from that observed using basic fibroblast growth factor (FGF) stimulation, whereby the vessels show a larger diameter with little or no increase in permeability. When the number of vessels per sample in the CAM was counted, there was a statistically significant difference between the scaffold and the three-dimensional cultures (Fig. 63.5). VEGF stimulated angiogenesis in a dose-dependent manner. The angiogenic activity of the three-dimensional culture was reduced by >90% by preincubation with anti-VEGF's neutralizing antibody prior to placement on the CAM.

Angiogenesis in the rat aortic ring assay

In the aortic ring assay, the ability of the endothelial cells isolated from rat aorta to generate microvessels was used to demonstrate angiogenesis (Jiang and Harding, 1998). Thoracic aortas, removed from 1- to 2-month-old Sprague–Dawley male rats, were transferred to serum-free MCDB131. The periaortic fibroadipose tissue was carefully removed, and the aortas were washed 8 to 10 times and cut into 1-mm lengths. Wells were punched in a 1.5% agarose gel and filled with clotting fibrinogen solution (20 μl of 50 NIH units/ml bovine thrombin in 1 ml fibrinogen). The aortic rings were placed into the centers of the wells. After clotting, the dishes were flooded with serum-free MCDB131. The cultures were incubated at 37°C with 5% CO_2, with medium changes every 3 days. Newly formed microvessels were counted on days 3, 7, and 14. When aortic rings were cocultured with living fibroblast-based engineered tissue, there was a significant increase in the number of microvessels formed (Fig. 63.6).

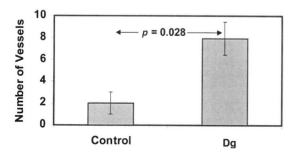

Fig. 63.6. Stimulation of blood vessel formation by Dermagraft (Dg) in the aortic ring assay.

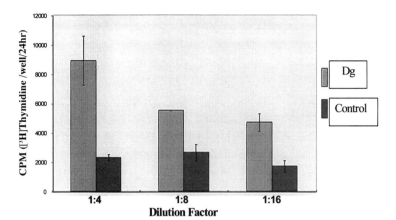

Fig. 63.7. Stimulation of endothelial cell proliferation by Dermagraft (Dg).

Stimulation of endothelial cell proliferation

Endothelial cell proliferation is a critical component of angiogenesis. The ability of fibroblast-based engineered tissue to stimulate this activity was determined by [³H]thymidine incorporation. Various growth factors and concentrated conditioned medium samples were assessed for their influence on the proliferation of human vascular endothelial cells (HUVECs). Confluent cultures were detached and resuspended in HUVEC growth medium to a final concentration of 2.5 × 10⁴ cells/ml. Attachment Factor Solution (Cell Applications, Inc.) was used to pretreat 24-well plates and cells were added, 1 ml of cell suspension per well. Cells were allowed to settle and attach and then were switched to Endothelial Serum Free medium (Cell Applications, Inc.), supplemented with fibroblast culture medium or medium conditioned by monolayer or three-dimensional fibroblast cultures. On day 2, the cells received fresh serum-free medium supplemented as appropriate with 1 μCurie/ml [³H]thymidine. On day 3, medium was removed, cells were washed three times with phosphate-buffered saline (PBS), and 250 μl of 2.3% sodium dodecyl sulfate (SDS) solution was added to solubilize the cells. After 30 min, the SDS extract and 1 ml of a PBS wash were transferred to a scintillation vial. Then 5 ml of ScintiVerse (Fair Lawn, NJ) was added to vials and radioactivity was determined using a Beckman LS6500 Scintillation Counter (Fullerton, CA).

Medium conditioned by incubation with three-dimensional fibroblast cultures stimulated [³H]thymidine incorporation in cultures of endothelial cells (Fig. 63.7). The proliferative effect is dose dependent.

Stimulation of endothelial cell chemokinesis

The ability of our fibroblast-based engineered tissue to stimulate endothelial cell migration was tested in two assays. The first was a chemokinesis assay that determines the stimulation of cell movement without any directional definition. The second measured cell migration toward a stimulation source. The chemokinesis assay was performed and reported by Martin and colleagues (1998). Endothelial cells were grown on Cytodex-2 beads. The assay estimates the dissociation of cells from the beads and reassociation with a culture plate. The cells on the plate were stained and

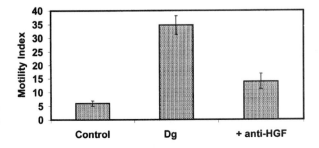

Fig. 63.8. Stimulation of endothelial cell motility by Dermagraft (Dg).

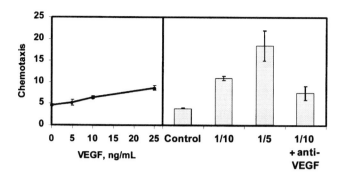

Fig. 63.9. Stimulation of endothelial cell chemotaxis by Dermagraft.

counted. As shown in Fig. 63.8, coculture of the cells with the fibroblast-based engineered tissue gave a marked increase in transfer of cells from bead to plate ($p = 0.0003$). The activity of the three-dimensional fibroblast cultures was inhibited about 60% using anti-HGF neutralizing antibody.

Stimulation of endothelial cell chemotaxis

Cell migration was analyzed with an endothelial cell chemotaxis assay utilizing a Neuro Probe 48-well Boyden chemotaxis chamber (Neuro Probe, Inc.). Polycarbonate membrane filters (Poretics Corporation; 25×80 mm) were soaked in 0.5 M acetic acid overnight, washed three times for 1 hr with water, incubated in a solution of 0.01% calf skin gelatin type III (Sigma; St. Louis, MO) for 12–16 hr, and air dried. HUVECs were detached and resuspended in HUVEC growth medium at a final concentration of 1.0×10^5 cells/ml. The Boyden chamber was assembled as follows: 30 μl/well of sample or standard was added to the bottom wells, the gelatin-coated membrane was placed on top, and 50 μl of cell suspension was added to the upper wells. The chamber was incubated at 37°C for 3 hr. Membranes were then carefully removed from the chamber and the cell side was rinsed in PBS and drawn across a wiper blade to remove nonmigrated cells. The membranes were stained with Wright's Geimsa stain and either the number of cells counted or the density of staining was reported against a standard curve generated with 20, 10, 5.0, and 0 ng/ml purified VEGF.

Medium conditioned by the fibroblast-based engineered tissue greatly stimulated cell migration in a dose-dependent manner (Fig. 63.9). The three-dimensional fibroblast-conditioned medium stimulated HUVEC migration to a greater extent than the positive control using VEGF, even at 50 ng/ml. Anti-VEGF antibody inhibited migration stimulated by three-dimensional fibroblast-culture-conditioned medium by 50%.

Induction of integrin αvβ3

The αvβ3 integrin has been shown to play an important role in angiogenesis, and neutralizing antibodies directed at it are capable of blocking capillary blood vessel formation. Its expression is induced by VEGF and it is thought to play a critical role in endothelial cell migration.

The presence of integrins and cell surface receptors was determined by flow cytometry on a FACStar (Cytometry Research Services; San Diego, CA). Cells were prepared for analysis as follows: HUVECs were trypsinized and the cells resuspended at 1×10^6 cells/ml. Volumes of 250–500 μl of the cell suspensions were washed three times with Hank's Balanced Salt Solution (HBSS; GibcoBRL; Grand Island, NY) and finally resuspended in 10% FBS in HBSS. The cells were incubated for 30 min with primary antibodies diluted to 1 μg/ml in 10% FBS in HBSS, washed three times with HBSS, incubated for 30 min with secondary antibodies diluted to 1 μg/ml in 10% FBS in HBSS, washed three times with HBSS, and fixed in 200 μl of 10% formalin (Baxter; Deerfield, IL) at a density of 10^6 cells/ml.

The presence of αvβ3 integrin on the surface of endothelial cells was analyzed by flow cytometry after treatment with medium conditioned by fibroblast-based engineered tissue. Cultured HUVECs display substantial surface expression of αvβ3 integrin under normal culture conditions. Nonetheless, medium conditioned by the fibroblast-based engineered tissue stimulated a significant increase in expression of this integrin.

Expression of angiogenic growth factors

Several of the angiogenic properties of fibroblast-based engineered tissue described above were demonstrated to be sensitive to neutralizing antibodies against specific growth factors, notably vascular endothelial growth factor and hepatocyte growth factor/scatter factor. Accordingly, experiments were performed to examine the expression of these and other angiogenic factors by the fibroblast-based tissue. Growth factor expression was examined both by estimation of mRNA levels by polymerase chain reaction methods and estimation of the free protein by enzyme-linked immunosorption assay.

Specific messenger RNAs were estimated by quantitative reverse transcriptase and polymerase chain reaction (RT-PCR) using the ABI TaqMan method (Perkin-Elmer; Foster City, CA). Total RNA was extracted from the cells using a Rapid RNA Purification Kit (Amresco; Solon, OH). The RNA was reverse transcribed using Superscript II (Life Technologies; Grand Island, NY) with random hexamer primers (Sigma; St. Louis MO). Amplification of samples of cDNA containing 200 ng of total RNA was detected in real time and compared with the amplification of plasmid-derived standards for specific mRNA sequences using a copy number over a range of five orders of magnitude within 40–4,000,000/reaction. In purification and the efficiency of reverse transcription, mRNA sequences for platelet-derived growth factor (PDGF) B chain, VEGF, or TGF-β_1 were added to RNA isolations, and their yield measured by the TaqMan procedure. The control mRNA sequences were obtained by T7 RNA polymerase transcription of plasmids containing the corresponding sequence. The values were normalized using glyceraldehyde-3-phosphate dehydrogenase as a control.

CLINICAL RESULTS OF WOUND HEALING

CRITICAL PRODUCT SPECIFICATION

Any device designed to act as a replacement tissue must provide critical therapeutic benefits to the patient and function as a physiologic replacement. To conduct such physiological interactions with both the wound bed and the migrating epidermis, it is essential that the dermal implant have a sufficient number of viable, metabolically active fibroblasts remaining after the cryopreservation/storage/thaw process. The pivotal study was performed to examine the use of cryopreserved tissue-engineered living human dermal tissue for the treatment of diabetic foot ulcers. Results from this study elucidated the specific product parameters necessary for a tissue-engineered implant to function physiologically as a dermal replacement (Mansbridge *et al.,* 1998; Naughton *et al.,* 1997; Pollak *et al.,* 1997). A complete analysis of all *in vitro* and clinical data at the conclusion of the study confirmed that the metabolic activity of Dermagraft must lie within an optimal therapeutic range (TR) to ensure that the tissue will be sufficiently active after implantation to affect wound healing by regaining its ability to synthesize and secrete normal dermal proteins, including growth factors. Statistically significant improved healing was observed in patients receiving therapeutically active product at their first two implants and in at least half of their implants (DG-TR group,

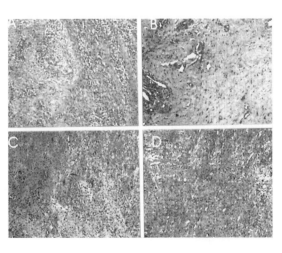

Fig. 63.10. Histologic evaluations of wound site biopsies of patients before and after implantation with Dermagraft show a marked reduction in inflammatory cell infiltration and an increase in blood vessels. Similar results were not evident in the tissue of placebo-treated patients. Magnification, 40×.

Before treatment with Dermagraft

Dermagraft-treated (6 weeks)

Before treatment with placebo

Placebo-treated (6 weeks)

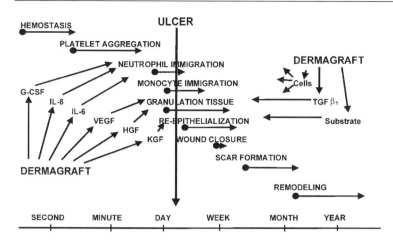

Fig. 63.11. Dermagraft secretes a variety of growth factors and proteins that have been shown to be important in reducing inflammation and enhancing wound bed regeneration and reepithelialization.

$n = 61$). A total of 50.8% of this patient population experienced complete wound closure within 12 weeks ($p = 0.006$). Patients who received every dose of Dermagraft with the correct metabolic activity ($n = 37$) had the highest percentage of complete wound closure (54.1%) by week 12 and were by themselves statistically significantly better than the control group ($p = 0.007$).

In addition to patients being followed through week 12, data were collected for week 32 healing, a full 6 months after the last application of Dermagraft. Because the dosage of Dermagraft was complete at the end of the first 8 weeks, it was expected that the difference between Dermagraft and control healing would narrow considerably at week 32, a full 6 months after the last dose. Despite the long time since the last implantation, however, the Dermagraft-treated patients continued to do much better than the controls. The DG-TR group still had significantly more patients healed than the control group ($p = 0.039$), and the median time to complete healing was twice as fast with Dermagraft (13 weeks for Dermagraft vs. 28 weeks for control). Previous studies have shown that fibroblasts implanted clinically in the Dermagraft product persist for at least 6 months *in vivo* after implantation. The persistence of these cells clearly shows the uniqueness of a tissue-engineered product in wound healing: persistent cells are capable of secreting growth factors and normal matrix proteins long term into the wound bed, thus promoting an on-going physiological healing environment.

SUMMARY

Clinical use of our fibroblast-based tissue-engineered implants has shown rapid vascularization of the wound bed, remodeling of the tissue, and reduction in inflammation (Fig. 63.10). The mechanism of action of this healing has been well studied (Jiang and Harding, 1998; Mansbridge *et al.,* 1998, 1999; Naughton *et al.,* 1999) (Fig. 63.11) and has led clinicians to look for new applications associated with problem wounds whereby this action would be beneficial. This approach offers obvious benefits over a single growth factor approach, such as the local application of exogenous VEGF, because multiple angiogenic factors are naturally secreted by the fibroblast-based tissue-engineered constructs at physiologically relevant ratios. These cells can respond to biological signaling *in vivo* and induce a more physiologic healing process. In addition, the matrix components have been shown to induce endothelial migration and may serve as a source of natural undamaged proteins that can enhance *in vivo* remodeling.

The strong clinical safety profile of this product, along with the physiologic properties that promote rapid healing, make it an excellent candidate for use in the study of revascularization and regeneration of damaged tissue in a variety of chronic and acute wounds.

REFERENCES

Arbiser, J. L. (1996). Angiogenesis and the skin—A primer. *J. Am. Acad. Dermatol.* **34**, 486–497.
Bell, E., Sher, S., and Hull, B. (1984). The living skin-equivalent as a structural and immunological model in skin grafting. *Scanning Electron Microsc.* **4**, 1957–1962.
Bennett, N. T., and Schultz, G. S. (1993). Growth-factors and wound healing—Biochemical-properties of growth-factors and their receptors. *Am. J. Surg.* **165**, 728–737.

Blood, S., Heck, E., and Baxter, C. R. (1979). The importance of the bacterial flora in cadaver homograft burn donor skin. *Proc. 11th Meet., Am. Burn Assoc.,* New Orleans, LA, pp. 79–80.

Cone, D., and Fleischmajer, R. (1993). Culturing keratinocytes and fibroblasts in a three-dimensional mesh results in epidermal differentiation and formation of a basal lamina-anchoring zone. *J. Invest. Dermatol.* **100,** 35–39.

Contard, P., Bartel, R. L., Jacobs, L., II, Perlish, J. S., MacDonald, E. D., II, Handler, L., Cone, D., Fleischmajer, R. (1993). Culturing keratinocytes and fibroblasts in a three-dimensional mesh results in epidermal differentiation and formation of a basal lamina-anchoring zone. *J. Invest. Dermatol.* **100,** 35–39.

Cooper, M. L., Hansbrough, J. F., Spielvogel, R. L., Cohen, R., Bartel, R., and Naughton, G. K. (1991). *In vivo* optimization of a living dermal substitute employing cultured human fibroblasts on a biodegradable polyglycolic acid or polyglactin mesh. *Biomaterials* 12, 243–248.

Eisenberg, M., and Llewelyn, D. (1998). Surgical management of hands in children with recessive dystrophic epidermolysis bullosa: Use of allogeneic composite cultured skin grafts. *Br. J. Plast. Surg.* **51,** 608–613.

European Union Committee for Proprietary Medicinal Products: Ad Hoc Working Party on Biotechnology/Pharmacy (1989). Notes to applicants for marketing authorizations on the production and quality control of monoclonal antibodies of murine origin intended for use in man. *J. Biol. Stand.* **17,** 213.

Folkman, J., and Klagsbrun, M. (1987). Angiogenic factor. *Science* **235,** 442–447.

Hansbrough, J. F., Boyce, S. T., Cooper, M. L., and Foreman, T. J. (1989). Burn wound closure with cultured autologous keratinocytes and fibroblasts attached to a collagen/glycosaminolglycan substrate. *JAMA, J. Am. Med. Assoc.* **262,** 2125–2130.

Jakoby, W. B., and Pastan, I. H., eds. (1979). "Methods in Enzymology," Vol. 58. Academic Press, New York.

Jiang, W. G., and Harding, K. G. (1998). Enhancement of wound tissue expansion and angiogensis by matrix-embedded fibroblast (Dermagraft®), a role of hepatocyte growth factor/scattor factor. *Int. J. Mol. Med.* **2,** 203–210.

Kruse, P. F., Jr., and Patterson, M. K., Jr., eds. (1973). "Tissue Culture Methods and Applications." Academic Press, New York.

Landeen, L. K., Zeigler, F. C., Halberstadt, C., *et al.* (1992). Characterization of a human dermal replacement. *Wounds* 5, 167–175.

Mansbridge, J. N., Liu, K., Patch, R., Symons, K., and Dinny, E. (1998). Three-dimensional fibroblast culture implant for the treatment of diabetic foot ulcers: Metabolic activity and therapeutic range. *Tissue Eng.* **4,** 403–414.

Mansbridge, J. N., Pinney, E. R., Liu, K., and Kern, A. (1999). Comparison of fibroblast properties in scaffold-based and collagen gel three-dimensional culture systems. *J. Invest. Dermatol.* **112**(4), 536.

Marks, M. G., Doillon, C., and Silver, F. H. (1991). Effects of fibroblasts and basic fibroblast growth-factor on facilitation of dermal wound healing by type-I collagen matrices. *J. Biomed. Mater. Res.* **25,** 683–696.

Martin, T. A., Hardin, K. G., and Jiang, W. G. (1998). Regulator of cell motility and angiogenesis: A role for CD44. *Proc. Eur. Conf. Adv. Wound Manage., 8th,* Madrid, Spain.

Moulin, V. (1995). Growth-factors in skin wound healing. *Eur. J. Cell Biol.* **68,** 7211–7216.

Naughton, G. K., and Tolbert, W. R. (1996). Tissue engineering: Skin. *In* "Yearbook of Cell and Tissue Transplantation" (R. P. Lanza and W. L. Chick, eds.), pp. 265–274. Kluwer Academic Publishers, Amsterdam, The Netherlands.

Naughton, G. K., Genztkow, G., and Mansbridge, J. N. (1997). Dermagraft® use in diabetic foot ulcer. *Artif. Organs* **21**(11), 1203–1210.

Naughton, G. K., and Mansbridge, J. N. (1999). Human-based tissue-engineered implants for plastic and reconstructive surgery. *Clinics in Plastic Surgery* **26**(4), 579–586.

Newton, D. J., Khan, F., Leese, G. P., Mitchell, M. R., and Belch, J. J. F. (1999). *Lancet* (in press).

Pettet, G., Chaplain, M., McElwain, D. S., and Byrne, H. M. (1996). On the role of angiogenesis in wound healing. *Proc. R. Soc. London, Ser. B* **263,** 1487–1493.

Pollak, A., Edington, H., Jensen, J., Kroeker, R., and Gentkow, G. (1997). A human dermal replacement for the treatment of diabetic foot ulcers. *Wounds* **9,** 175–183.

Raghow, R. (1994). The role of extracellular-matrix in post-inflammatory wound healing and fibrosis. *FASEB J.* **8,** 823–831.

Schaffer, C. J., and Nanney, L. B. (1996). Cell biology of wound healing. *Int. Rev. Cytol.* **169,** 151–181.

Sheridan, R., Choucair, M., Donelan, M., Lydon, L., Petras, L., and Tompkins, R. (1998). Acellular allodermis in burn surgery: 1-year results of a pilot trial. *J. Burn Rehab.* **19**(6), 528–529.

U.S. Food and Drug Administration (1993). "Points to Consider in the Characterization of Cell Lines used to Produce Biologicals." U.S. Department of Health and Human Services, Bethesda, MD.

White, M. J., Whalen, J. D., Gould, J. A., Brown, G. L., and Polk, H. C. (1991). Procurement and transplantation of colonized cadaver skin. *Am. Surg.* **57,** 402–407.

Yannas, I. V., and Burke, J. F. (1980). Design of an artificial skin. I. Basic design principles. *J. Biomed. Mater. Res.* **14,** 65–81.

PART XX: WOMB

CHAPTER 64

ARTIFICIAL WOMB

Christopher S. Muratore and Jay M. Wilson

A SQUAT grey building of only thirty-four stories. Over the main entrance the words, CEN-TRAL LONDON HATCHERY AND CONDITIONING CENTRE, and, in a shield, the World State's motto, COMMUNITY, IDENTITY, STABILITY. . . .

In the Bottling Room all was harmonious bustle and ordered activity. Flaps of fresh sow's peritoneum ready cut to the proper size came shooting up in little lifts from the Organ Store in the sub-basement. Whizz and then, click! The lift-hatches flew open; the bottle-liner had only to reach out a hand, take the flap, insert, smooth-down, and before the lined bottle had had time to travel out of reach along the endless band, whizz, click! another flap of peritoneum had shot up from the depths, ready to be slipped into yet another bottle, the next of that slow interminable procession on the band.

Next to the Liners stood the Matriculators. The procession advanced; one by one the eggs were transferred from their test-tubes to the larger containers; deftly the peritoneal lining was slit, the morula dropped into place, the saline solution poured in . . . and already the bottle had passed. . . .

Each bottle could be placed on one of fifteen racks, each rack, though you couldn't see it, was a conveyor traveling at the rate of thirty-three and a third centimetres an hour. Two hundred and sixty-seven days at eight metres a day. Two thousand one hundred and thirty-six metres in all. One circuit of the cellar at ground level, one on the first gallery, half on the second, and on the two hundred and sixty-seventh morning, daylight in the Decanting Room. Independent existence—so called.

"But in the interval," Mr. Foster concluded, "we've managed to do a lot to them. Oh, a very great deal." His laugh was knowing and triumphant.

"That's the spirit I like," said the Director once more. "Let's walk around. You tell them everything, Mr. Foster."

Mr. Foster duly told them.

Told them of the growing embryo on its bed of peritoneum. Made them taste the rich blood surrogate on which it fed. Explained why it had to be stimulated with placentin and thyroxin. Told them of the *corpus luteum* extract. Showed them the jets through which at every twelfth metre from zero to 2040 it was automatically injected. Spoke of those gradually increasing doses of pituitary administered during the final ninety-six metres of their course. Described the artificial maternal circulation installed in every bottle at Metre 112; showed them the reservoir of blood-surrogate, the centrifugal pump that kept the liquid moving over the placenta and drove it through the synthetic lung and waste product filter. Referred to the embryo's troublesome tendency to anaemia, to the massive doses of hog's stomach extract and foetal foal's liver with which, in consequence, it had to be supplied. [Aldous L. Huxley, "Brave New World" (932)]

Although Huxley's "Brave New World" describing a futuristic science was written many decades ago, the principles of constructing an artificial womb were remarkably accurate. The womb would require a vessel, biologic tissue for implant (sow's peritoneum), nutrition (blood surrogate), hormones (placentin, thyroxin, pituitary), unknown growth factors (corpus luteum extract, hog stomach extract, fetal foal liver), oxygen, a centrifugal pump, a synthetic lung, a waste product filter, and temperature and light controls. Today, long after Huxley conjured his new world, this list has changed little. Although the "ideal" artificial womb would be a total tissue construct, practically speaking, the CYBORG (cybernetic and organic) approach outlined by Huxley will undoubtedly yield the first practical womb substitute. But, before one begins the task of developing an artificial womb, it is useful to first understand why such a construct might be useful in the first place.

WHAT DOES THE FIELD OF TISSUE ENGINEERING HAVE TO OFFER THE PREMATURE INFANT?

Prematurity is the leading cause of neonatal death and morbidity worldwide, accounting for 60–80% of deaths of infants without congenital abnormalities. The overall neonatal mortality rate in the United States is approximately 7 per 1000, a rate that ranks nineteenth in the world and well below many industrialized and Third World countries. This mortality rate has steadily decreased largely because of improvements in neonatal intensive care, leading to infant survival rates of approximately 50, 80, and 97% for infants born at 24, 28, and 32 weeks of gestation, respectively (normal gestation is 40 weeks (Kipikasa and Bolognese, 1997). However, survival rates alone do not tell the whole story. Many premature survivors have significant morbidity. In one study, children who weighed less than 705 g at birth demonstrated substantially higher rates of mental retardation, cerebral palsy, and visual disabilities as well as neurobehavioral dysfunction and poor school performance. Other complications, including respiratory distress syndrome, intraventricular hemorrhage, necrotizing enterocolitis, bronchopulmonary dysplasia, sepsis, patent ductus arteriosis, and retinopathy of prematurity, increase with decreasing gestational age (Kipikasa and Bolognese, 1997; Dawson *et al.,* 1997).

Currently, the preterm viable infant (>24 weeks of gestation) is cared for in a neonatal intensive care unit (ICU), where it must survive the transition from the protective liquid fetal environs to the harsh invasive realm of the New World. During this transition, rapid and profound physiologic changes enable the infant to make the successful passage from intrauterine to extrauterine life. These changes are developmental tasks that all babies must complete. To accomplish them successfully, the fetus must have relatively mature organ systems, including renal, skin, pulmonary, and cardiovascular. The earlier in gestation the infant is born, the greater the number and magnitude of fetal tasks that are unfinished and the more vulnerable the infant (Rowe, 1998).

Invariably, lung underdevelopment is the most formidable challenge faced by the preterm infant, although significant underdevelopment of the skin, cardiovascular, and immune systems compounds the problem. Although lifesaving, the invasive environment of the modern newborn ICU inflicts an iatrogenic toll that might be avoided if the preterm infant could complete the developmental process while being maintained in a liquid incubation system simulating the protective *in utero* environment.

PREMATURITY—THE CONSEQUENCES

Intraventricular Hemmorhage

The most common form of brain injury in premature infants is intraventricular or periventricular hemorrhage. Risk of intracranial hemorrhage is directly associated with the degree of prematurity. Overall, 46–58% of infants with extremely low birth weight sustain some form of intraventricular hemorrhage; 50% of intraventricular hemorrhages occur on the first day of life and 90% occur by age 1 week. Intraventricular hemorrhage (IVH) is graded on a scale of one to four based on the extent of hemorrhage and the involvement of the surrounding structure. Prognosis is directly related to the severity of the hemorrhage, with mortality rates ranging from 16 to 50% and the incidence of major neurodevelopmental handicaps from 45 to 86%. Extensive hydrocephalus and significant periventricular injuries are associated with more than a 90% risk of major long-term neurodevelopmental sequelae (Rowe, 1998; Dawson *et al.,* 1997). The cause of IVH is thought to be immature development of the brain compounded by the stress of the fetus trying to survive in a postnatal environment for which it is poorly equipped.

HYALINE MEMBRANE DISEASE

Hyaline membrane disease (HMD) contributes significantly to morbidity and mortality among premature infants. The primary biochemical abnormality in HMD is deficiency of pulmonary surfactant, which leads to increased atelectasis, worsening of pulmonary compliance, and progressive respiratory failure. Long-term follow up of survivors of HMD has demonstrated persistent abnormalities in pulmonary function testing, increased susceptibility to pulmonary infections, and development of bronchopulmonary dysplasia, in the most severely affected infants (Dawson *et al.,* 1997).

BRONCHOPULMONARY DYSPLASIA

Bronchopulmonary dysplasia (BPD), or chronic lung disease, in infants is the most common pulmonary sequela of mechanical ventilation of both premature and full-term infants. The incidence of BPD among premature infants is inversely proportional to gestational age and birth weight. Up to 70% of infants with birth weights less than 1000 g will develop BPD. Many factors play a role in the development of BPD, but the most significant are the degree of pulmonary structural and functional immaturity at birth and the degree and length of mechanical ventilation and oxygen support (Dawson *et al.,* 1997). Prevention of HMD and BPD is challenging in the current neonatal ICU setting, but may be achieved utilizing an extrapulmonary source for gas exchange to allow the lungs to mature without injury.

SKIN MATURATION AND TRANSEPITHELIAL AND INSENSIBLE WATER LOSS

Insensible water loss is the invisible loss of water from the lungs and skin. The major component of insensible water loss during the newborn period is transepithelial water loss (TEWL). In the low-birth-weight infant, TEWL can have profound effect on fluid balance and thermoregulation, and it can ultimately effect mortality. TEWL results from water molecules diffusing across the superficial capillaries of the skin. For the extremely low-birth-weight infant with almost no keratinization, water loss can be enormous. As water is lost, so is heat. For every milliliter that evaporates, 0.58 kcal of heat are dissipated. TWEL is inversely related to gestational age and birth weight (Rowe, 1998). Prevention of TEWL is impossible in the current newborn ICU, but might be possible if the preterm infant is cared for in a liquid-based environment.

NECROTIZING ENTERCOLITIS

Since the introduction of neonatal intensive care medicine in the 1960s, and with increasing neonatal survival, necrotizing entercolitis (NEC) has become the most common gastrointestinal tract disease of the newborn (Albanese and Rowe, 1998). The incidence of NEC is approximately 1 to 3 in 1000 live births, but the true incidence is difficult to ascertain. NEC is predominantly a disease of low-birth-weight premature infants. It is believed that the susceptibility of the immunologically immature gut makes it vulnerable to an inciting agent such as endogenous bacteria or hypoxic injury. Enteral feedings also stress the immature gut and provide a substrate for bacterial proliferation, resulting in mucosal injury. The breakdown of the gut barrier initiates an inflammatory cascade, ultimately resulting in intestinal necrosis (Albanese and Rowe, 1998). Prevention may require providing nutrition by a source other than enteral.

LONG-TERM COSTS OF PREMATURITY

Long-term costs of perinatal disabilities secondary to prematurity are dramatic. There are direct cost issues such as hospital bills, physician fees, office visits, laboratory tests, rehabilitation services, and medical expenses, in addition to parental out-of-pocket costs. Indirect costs also reflect foregone opportunities of family members, such as wages lost while seeking medical treatment or while the infant is convalescing (McCormick and Richardson, 1998). The additional costs to society of caring for the complications of prematurity are currently unknown, but are undoubtedly enormous.

THE IDEAL ARTIFICIAL WOMB

Because none of the complications of prematurity occur in a fetus that remains in the womb, one novel method of addressing the problem would be to maintain profoundly premature infants in the fetal environment. No apparatus to achieve this exists, thus we will endeavor to describe the "ideal" artificial womb and then to review the data on preliminary attempts to create it. The ideal artificial womb would have to reproduce many of the functions of the maternal/placental unit, such as gas exchange and nutrition, as well as provide a protective environment for continued development in an environment free of bacterial and viral contamination. The time frame during which such support might be necessary could range from 5 to 15 weeks.

The relatively new field of tissue engineering owes its existence to the understanding of cellular physiology, growth, proliferation, communication, and interaction with biodegradable polymer constructs. On a more global scale, successful maintenance of a fetus mandates that near-perfect fetal physiology is maintained. The goal therefore, for an artificial womb, either completely or partially tissue engineered, would be to mimic as closely as possible the *in utero* environment, maintain fetoplacental circulation, and reproduce the contributions and physiology of the placenta.

RECREATING THE *in Utero* ENVIRONMENT

As depicted in Fig. 64.1, conceptually, the fetus, still in continuity with the umbilical cord and placenta, would be placed in a flexible incubator that permits fetal movement. The incubator is filled with a warm electrolyte solution, closely mimicking amniotic fluid, which is circulated, filtered, and returned to the incubator. Reproduction of a liquid-based protective environment would prevent insensible water loss from the lungs and skin, making fluid balance and thermoregulation easier (Rowe, 1998). Moreover, this liquid-based environment would apply pressure to the outside of the fetal chest, equilibrating and counterbalancing an equal pressure transmitted via an open airway to the lungs (Adzick and Harrison, 1991). This fluid pressure contributes the resultant intrapulmonary distending pressure, which is essential to proper lung development (Adzick and Harrison, 1991). Amniotic fluid is also swallowed and plays an active role in the maintenance and development of the gastrointestinal tract. In order to completely support the fetus, a myriad of growth factors would have to be added to this fluid.

MAINTENANCE OF FETOPLACENTAL CIRCULATION

Normal fetal growth and development are dependent on umbilical blood flow to meet the growing demand of fetoplacental exchange of oxygen, carbon dioxide, nutrients, metabolic wastes, and hormones. Umbilical blood flow increases in direct proportion to the increase in fetal body weight, so that flow remains approximately constant at 110 to 125 ml/min/kg during the third trimester. Furthermore, fetal biventricular cardiac output is near maximal under basal conditions in near-term fetal sheep and the umbilical–placental circulation is near maximally dilated (Adamson *et al.,* 1998). This umbilical blood flow represents about 30% of the fetal biventricular cardiac output. At birth, however, umbilical–placental circulation begins to close. Flow rapidly decreases

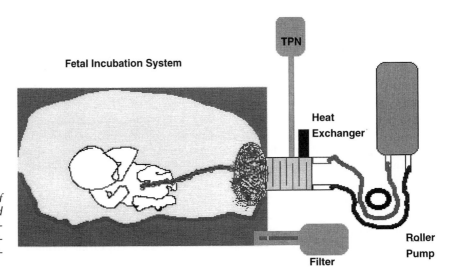

Fig. 64.1. Authors' conception of the ideal artificial womb, based on modifications of current experimental extrauterine fetal support systems. TPN, Total parenteral nutrition.

after delivery to less than 20% of the normal fetal value. This is accompanied by significant decreases in umbilical artery and vein diameters (Adamson *et al.,* 1998). Maintenance of umbilical blood flow and near-maximal placental vascular dilatation will be essential to extrauterine fetal survival. Consequently, understanding the factors that regulate umbilical blood flow is critical to maintaining umbilical–placental vascular patency. The exogenous use of vasoactive factors would be an important adjunct to ensure umbilical–placental blood flow in the extrauterine fetus.

Interestingly, there are no detectable adrenergic or cholinergic nerve fibers in the human placenta or in the human umbilical cord vessels. Thus the absence of the innervation of the extrafetal umbilical cord vessels and the placenta indicates that these vessels are not under neural control (Adamson *et al.,* 1998). Therefore, pharmacologic manipulation directed at adrenergic or cholinergic receptors would not be necessary to maintain umbilical–placental blood flow for the extrauterine fetus. However, there are a number of systemic and placental-derived vasoconstrictors that contribute to umbilical–placental blood flow regulation, including angiotensin II, bradykinin, catecholamines, thromboxane, prostaglandins, endothelin-1, histamine, peptides, and serotonin. In humans, thromboxane and endothelin-1 are potent vasoconstrictors of the placental vessels. Endothelin-1 in particular may increase the sensitivity to other vasoconstrictors and thus modify the vasoactive response of the umbilical–placental circulation to other agents (Adamson *et al.,* 1998). Likewise, a number of hormones have vasodilatory actions on the placenta. Important factors considered to have a vasodilatory role on the umbilical–placental circulation include adenosine, corticotropin-releasing hormone (CRH), endothelial-derived relaxing factor (EDRF), natriuretic peptides, and prostacyclin. In particular, CRH at physiologic concentrations and EDRF appear to play an important role in the maintenance of low vascular resistance in this vascular bed (Adamson *et al.,* 1998).

PLACENTAL PHYSIOLOGY

The importance of the native placenta to the artificial incubator cannot be neglected. Many vasoactive factors are produced and act locally in the placenta; this provides additional justification for including the placenta in the artificial incubation system. The placental membrane surface area is responsible for diffusion of oxygen, carbon dioxide, nutrients, and many systemic and locally produced hormones. Contemporary extracorporeal membrane oxygenator (ECMO) units, whether utilizing a roller pump or centrifugal pump, currently provide adequate gas exchange. The membrane oxygenator can oxygenate and remove CO_2 from either whole blood or a blood substitute such as liquid perfluorocarbon. Nutrition provided in the form of total parenteral nutrition (TPN) can be infused either directly into the placental blood or diffused across the engineered placental–ECMO interface. Likewise, antibiotics and other medications may be given similarly, provided they can cross the interface barrier.

Finally, the development of an autologous placenta–ECMO interface, derived and engineered from human tissue, as a conduit would be far superior to the currently used silastic tubing. To enhance the autologous nature of the ECMO interface membrane, the membrane might be constructed of umbilical vein-derived endothelial cells, seeded on a biodegradable polymer. A tissue-engineered membrane would further obviate the need for anticoagulation or routine circuit changes, both of which are necessary with the conventional ECMO, because the risk of mural thrombi would be less. Moreover, fewer circuit changes translate into a lower risk of infection, and less fetal disturbance or manipulation.

STATE-OF-THE-ART ARTIFICIAL PLACENTA DESIGN

The first efforts to develop long-term extrauterine extracorporeal maintenance of a premature fetus were initiated in the late 1950s and were continued by many groups into the 1960s. The early studies used a pump oxygenator to sustain life or to provide adequate perfusion in previable fetuses. In 1958, Westin *et al.* reported that seven previable human fetuses had been kept alive 5–12 hr by extracorporeal perfusion of the umbilical vein and induction of hypothermia (Callaghan *et al.,* 1956; Westin *et al.,* 1958). In 1961, Lawn and McCance described an incubation system designed to support the isolated fetus and called it an artificial uterus or artificial placenta. The authors used both open and closed tanks filled with isotonic glucose to contain the fetus. In this design the fetal heart pumped the blood through the oxygenator via the umbilical arteries to maintain the circulation. The authors were interested in studying the effects of pressure and flow in the umbilical vessels, and oxygen consumption in the fetus. In 1969, Zapol *et al.* developed an artificial placenta consisting of an extracorporeal circuit and one silicone membrane blood oxygenator. This circuit was connected to the fetus via the umbilical vessels. Arteriovenous perfusion was achieved. Blood from the umbilical artery was collected in a reservoir bag pumped through the membrane lung and returned to the fetal umbilical vein while the fetus was maintained submerged in a warm saline bath. With this apparatus, they were able to maintain exteriorized premature fetuses for up to 55 hr.

In 1987, Kuwabara *et al.* reported the development of an extrauterine fetal incubation system also using an extracorporeal membrane oxygenator. This group exteriorized a preterm fetal goat and cannulated the umbilical vessels with polyvinyl catheters. Following systemic heparinization, the catheters were connected to the ECMO circuit and the fetus was placed into the incubator filled with artificial amniotic fluid consisting of a balanced electrolyte solution, which was warmed to 39.5°C. The fluid was circulated and cleaned by an in-line filter. The ECMO circuit consisted of polyvinyl chloride tubing, a roller pump, and a microporous membrane oxygenator and a heat exchanger. The blood was drained from the umbilical arteries, oxygenated, and returned to the umbilical vein. This system required anticoagulation, which was maintained with a continuous infusion of heparin that was titrated to keep the activated clotting time to approximately 250 sec. Fetal electrocardiogram heart rate, arterial blood pressure, central venous pressure, and electroencephalogram, tracheal pressure, gross movements, and flow rates of perfusate were continuously recorded. Blood samples were sequentially obtained from the circuit for hematologic, biochemical, and blood gas analyses. A catheter was transabdominally implanted in the urinary bladder for the measurement of urine volume, osmotic pressure, and electrolytes.

With this system, fetuses were kept under fairly stable physiologic conditions for periods up to 165 hr. The most common causes of death were circulatory failure and blood clotting. Massive intraperitoneal hemorrhage was noted in some of the cases due to administration of excessive heparin, whereas other cases reported clotting in the oxygenator due to insufficient heparinization. Nevertheless, utilizing this system, reasonable fetal circulation and fetal blood gas profiles were maintained for up to 7 days.

In later studies, Kuwabara *et al.* (1989) reported an improvement in the extrauterine system that allowed for maintenance of the fetus for as long as of 236 hr, with a mean duration of 146.5 hr of extrauterine survival. These improved results were attributed to several improvements in the experimental methods. The flow rate of the extracorporeal circulation was maintained at approximately 100 ml/min/kg. PaO_2 and $PaCO_2$ were kept between 25–35 and 35–45 mmHg, respectively, which was achieved by changing the flow rate of oxygen, nitrogen, and CO_2 gases. Anticoagulation was maintained with continuous infusion of heparin to achieve a lower activated clotting time of approximately 200 sec. All fetuses received a physiologic electrolyte solution intravenously containing 7.5–10% glucose at a rate of 10 ml/hr. An improvement in catheter design, allowing the tips of the catheters to be positioned in the lower part of the abdominal aorta, made it possible to maintain more consistent flow throughout the extracorporeal circuit.

In the previous studies, marked water retention resulting in massive ascites, plural effusions, and subcutaneous edema had appeared as early as the second or third day of the incubation period (Zapol *et al.*, 1969). The etiology for this phenomenon remains unclear but, excess fluid intake by infusion and fetal swallowing and immature renal function together with endocrinologic and or metabolic disturbances while on the ECMO circuit were thought to contribute. To prevent fluid retention on extracorporeal dialysis ultrafiltration method was utilized. This improved the fluid balance, as shown by the fact that no severe water retention was observed at autopsy of any fetus, and the electrolyte balance of the fetus was maintained within the physiologic ranges throughout the study, in contrast to earlier studies. Causes of death, again, in all cases of long-term incubation not associated with mechanical failure, were due to circulatory failure. Circulatory failure was noted to progress slowly throughout the duration of incubation.

In 1990, Unno *et al.* utilized the extrauterine fetal incubation system to study extrauterine fetal physiology compared to *in utero* fetal physiology. Many studies had shown that mammals under physiologic conditions had the ability to maintain a normal oxygen consumption level by increasing the oxygen extraction ratio when oxygen delivery fell to a certain extent. Using extrauterine fetal incubation, Unno *et al.* revealed that the same phenomenon occurred. They determined that the oxygen consumption decreased when the oxygen delivery was less than 10 ml/min/kg and it remained almost constant when the oxygen delivery exceeded 10 ml/min/kg. The relation suggested that this level of oxygen consumption was a basic requirement for the extrauterine goat fetus on ECMO and that the critical oxygen delivery value was approximately 10 ml/min/kg. This study indicated that the oxygen consumption of an isolated goat fetus incubated in an isothermic environment with umbilical arteriovenous extracorporeal membrane oxygenation was almost equal to a fetus *in utero*.

The incubation periods for this experiment were between 58 and 236 hr, a significant improvement from prior attempts. Although these data provided significant information on oxygen delivery and utilization by an extrauterine fetus, the exteriorized animal did not survive for more 10 days. During these studies excessive fetal gross movements and water imbalance due to swallowing of artificial amniotic fluid during the extrauterine fetal incubation were strongly suspected of being the causes of circulatory deterioration and failure during long-term incubation (Unno *et al.*, 1990). To suppress fetal movement and swallowing, pancuronium bromide was administered in subsequent studies (Unno *et al.*, 1993). This was provided either as a continuous infusion or as an intermittent bolus, with the doses adjusted according to fetal activity. The extracorporeal circuit was maintained as previously mentioned. With the administration of muscle relaxant, the externalized goat fetus on ECMO could be supported for approximately 3 weeks. Fetal parameters, including fetal blood gas, O_2 saturation, and PaO_2, $PaCO_2$, and measurements of oxygen consumption all remained stable during long-term extrauterine fetal incubation, demonstrating that the fetus was able to adapt better to extrauterine fetal circulation when fetal activity was suppressed.

Although these physiologic parameters demonstrated that the fetal cardiovascular system was able to adapt to the extracorporeal environment for a lengthy period of time, long-term immobilization is not at all physiologic. When attempts were made to initiate spontaneous lung respiration after 3 weeks of paralysis, spontaneous breathing did not occur. The long-term administration of pancuronium bromide in the preterm state may have caused fetal neuromuscular disturbances that prevented spontaneous breathing, or inadequate muscle development occurred secondary to muscular weakness due to the long-term immobilization.

ORGAN MATURATION DURING MAINTENANCE WITH ARTIFICIAL WOMB

Although the prior studies have shown that long-term paralysis prevented subsequent spontaneous respiration, in 1998, Yasufuku *et al.* demonstrated that the use of an extrauterine support system nevertheless did permit continued lung growth and maturation. Using twin goat fetuses as controls, they compared production of surfactant, maintenance of pulmonary epithelial ion transport, and lung maturation by histologic criteria. To prevent net efflux of lung liquid they ligated the tracheas of their extrauterine fetus. The ductus arteriosus was maintained open with high flow and low oxygen saturation, thereby maintaining fetal circulation. Their results demonstrated normal production of surfactant, maintenance of ion transport of the pulmonary epithelium, increase in lung weight, and increase in mature alveolar type II cells in the ligated animal lungs. This report provided preliminary evidence that the use of an extrauterine fetal incubation system could support the preterm fetus while allowing continued growth and maturation of immature fetal organs such as the fetal lung.

VARIATIONS OF ARTIFICIAL WOMB TECHNOLOGY

All of the above experiments were performed using a fetus without the placenta. An alternative approach, hybrid respiratory support, includes the placenta with extracorporeal membrane placental oxygenation (Akagi *et al.*, 1991). In this conceptual design, a preterm fetus could potentially be connected to a placenta instead of an oxygenator. The placenta would be interfaced with an artificial uterus, namely, the ECMO circuit. The ECMO circuit in this case uses an artificial blood substitute (Perfluorocarbon) to deliver oxygen to the placenta. In *in vitro* experiments, gas exchange took place between the fetal blood and the ECMO circuit across the placental membrane. This hybrid system relied on flow rates of the artificial blood through the placental–ECMO circuit at a fixed rate of 3 liters/min. In these *in vitro* experiments the gas transfer rate increased consistently with the blood flow rates of 60 ml/min per oxygen at 40 ml/min per CO_2 at a blood flow rate of 1 liter/min. These results are comparable to that of clinically available membrane oxygenators for infants today and could represent an alternative method for extrauterine maintenance of a preterm fetus in the future (Akagi *et al.*, 1991).

CONCLUSION

Despite the revolutionary advances that have occurred in modern medicine, premature labor remains an unsolved problem. Consequently, preterm infants are unique patients who require a unique branch of medicine devoted to their care and management. For infants >24 weeks old, the modern newborn ICU is sufficiently equipped to provide care. However, a combination of infant underdevelopment and iatrogenic injury place the premature infant at unique risk. Furthermore, for those infants born <24 weeks of gestation, there is currently no technology that will sustain them. Twenty-four weeks represents a milestone in fetal development because it is approximately the transition period from the pseudoglandular stage of pulmonary development to the terminal saccular stage. The terminal saccular stage represents the earliest histologic and physiologic stages of lung development capable of adequate gas exchange. It is generally accepted that there is no capacity for gas exchange prior to this stage of lung development. As such, attempts to ventilate mechanically an infant <24 weeks of gestation are generally unsuccessful. The development of an artificial womb that could sustain the preterm fetus in a liquid-based environment could potentially salvage the fetus born <24 weeks of gestation.

The invasive environment of the modern newborn ICU inflicts an unintentional iatrogenic toll that might be avoided if the preterm infant could complete the developmental process while being maintained in a protective liquid incubation system. Complications of prematurity, including respiratory distress syndrome, intraventricular hemorrhage, necrotizing enterocolitis, bronchopulmonary dysplasia, sepsis, patent ductus arteriosis, and retinopathy, may be avoided with the development of an artificial womb, thereby making Huxley's vision a reality.

References

Adamson, S. L., Myatt, L., and Byrne, B. (1998). Regulation of umbilical blood flow. *In* "Fetal and Neonatal Physiology" (R. Polin and W. Fox, eds.), 2nd ed., pp. 977–989. Saunders, Philadelphia.

Adzick, N. S., and Harrison, M. (1991). Experimental pulmonary hypoplasia. *In* "The Unborn Patient" (M. Harrison, M. Golbus, and R. Filly, eds.), 2nd ed., pp. 557–564. Saunders, Philadelphia.

Akagi, H., Takano, H., Yoshiyuki, T., *et al.* (1991). Hybrid respiratory support system with extracorporeal placental oxygenation. *ASAIO Trans.* **37**, M409–M410.

Albanese, C., and Rowe, M. (1998). Necrotizing enterocolitis. *In* "Pediatric Surgery" (J. O'Neil, M. Rowe, J. Grosfeld, E. Fonkalsrud, and A. Coran, eds.), 5th ed., pp. 1297–1320. Mosby Year Book, St. Louis, MO.

Callaghan, J., Maynes, E., and Hug, H. (1965). Studies on lambs of the development of an artificial placenta. Review of nine long-term survivors of extracorporeal circulation maintained in a fluid medium. *Can. J. Surg.* **8**, 208–213.

Dawson, J., August, A., and Cole, S. (1997). Complications of premature infants. *In* "Surgery of Infants and Children: Scientific Principles and Practice" (K. Oldham, P. Colombani, and R. Foglia, eds.), pp. 65–72. Lippincott-Raven, Philadelphia.

Huxley, A. L. (1932). *In* "Brave New World." Chatto & Windus, London; Doubleday, Doran & Co., Garden City, NY. [E-text at *http://www.ddc.net/ygg/etext/brave.htm* (accessed Oct. 1, 1999).]

Kipikasa, J., and Bolognese, R. (1997). Obstetrical management of prematurity. *In* "Neonatal–Perinatal Medicine. Disease of the Fetus and Infant" (A. Fanaroff and R. Martin, eds.), 6th ed., pp. 264–284. Mosby Year Book, St. Louis, MO.

Kuwabara, Y., Okai, T., *et al.* (1987). Development of extrauterine fetal incubation system using extracorporeal membrane oxygenator. *Artif. Organs* **11**, 224–227.

Kuwabara, Y., Okai, T., *et al.* (1989). Artificial placenta: Long-term extrauterine incubation of isolated goat fetuses. *Artif. Organs* **13**, 527–531.

Lawn, L., and McCance, R. A. (1961). An artificial placenta. *J. Physiol. (London)* **158**, 2–3.

McCormick, M., and Richardson, D. (1998). Long-term costs of perinatal disabilities. *In* "Avery's Diseases of the Newborn" (H. Taeusch and R. Ballard, eds.), 7th ed., pp. 429–433. Saunders, Philadelphia.

Rowe, M. (1998). The newborn as a surgical patient. *In* "Pediatric Surgery" (J. O'Neil, M. Rowe, J. Grosfeld, E. Fonkalsrud, and A. Coran, eds.), 5th ed., pp. 43–57. Mosby Year Book, St. Louis, MO.

Unno, N., Kuwabara, Y., *et al.* (1990). Development of artificial placenta: Oxygen metabolism of isolated goat fetuses with umbilical arteriovenous extracorporeal membrane oxygenation. *Fetal Diagn. Ther.* **5**, 189–195.

Unno, N., Kuwabara, Y., Okai, T., *et al.* (1993). Development of an artificial placenta: Survival of isolated goat fetuses for three weeks with umbilical arteriovenous extracorporeal membrane oxygenation. *Artif. Organs* **17**, 996–1003.

Westin, B., Nyberg, R., and Enhornig, G. (1958). Technique for perfusion of the previable human fetus. *Acta Paediatr. Scand.* **47**, 339.

Yasufuku, M., Hisano, K., *et al.* (1998). Arterio-venous extracorporeal membrane oxygenation of fetal goat incubated in artificial amniotic fluid (artificial placenta): Influence on lung growth and maturation. *J. Pediatr. Surg.* **33**, 442–448.

Zapol, W. M., Kolobow, T., *et al.* (1969). Artificial placenta: Two days of total extrauterine support of the isolated premature lamb fetus. *Science* **166**, 617.

PART XXI: REGULATORY CONSIDERATIONS

CHAPTER 65

REGULATORY CONSIDERATIONS

Kiki B. Hellman, Ruth R. Solomon, Claudia Gaffey,
Charles N. Durfor, and John G. Bishop III

INTRODUCTION

The United States Food and Drug Administration (FDA) is a science-based regulatory agency in the United States Public Health Service (PHS) whose mission is to promote and protect the public health through regulation of a wide range of products, by assuring the safety of foods, cosmetics, and radiation-emitting electronic products, as well as the safety and effectiveness of human and veterinary pharmaceuticals, biologicals, and medical devices. The FDA recognizes that an important segment of the products that it regulates arises from applications of new technology, such as tissue engineering. Tissue engineering, which applies life sciences and engineering principles to the maintenance, modification, improvement, restoration, or replacement of human tissue or organ function, has led to a broad range of products based on their common source materials, i.e., human tissues or organs (e.g., autologous or allogeneic tissues); animal tissues or organs (e.g., transgenic animals or xenotransplants); and processed, selected, or expanded human or other mammalian cells (e.g., stem/progenitor cells, genetic and somatic cellular therapies), with or without biomaterials and totally synthetic materials of biomimetic design. Representatives of these product classes are in different stages of development. To date, some have been approved for use by the FDA and many are under either preclinical investigation or regulatory evaluation. This chapter provides background on and an overview of the regulatory considerations related to the FDA review and approval of tissue-engineered medical products (TEMPs). It discusses the FDA's legislative authority and product regulatory process together with updates of proposed regulatory approaches, development of standards/guidance, and communications with industry. The review of products is conducted on a case-by-case basis and the product's safety and effectiveness are assessed with respect to its manufacture and clinical performance, as applicable, as well as the manufacturer's claim of intended use. The FDA is divided into six centers, each staffed with the scientific and regulatory expertise appropriate for evaluating the products in its jurisdiction. In addition, each center can apply any of the FDA's statutory authorities to regulate its products. The agency has adopted a cooperative approach across the appropriate FDA centers in developing the regulatory approaches for TEMPs. These approaches have simplified and facilitated the administrative process for evaluating TEMPs, the resolution of product regulatory jurisdiction questions, and the evaluation of product specific safety and effectiveness issues. As tissue engineering technology continues to evolve, leading to new and different products, new and different product-specific issues will also be expected. The FDA will continue to build on current initiatives, apply lessons learned from previous products as applicable, and look to the best scientific minds and methods to achieve innovative, flexible, and appropriate resolutions. A productive dialogue among the FDA, industry, and consumers is the key in this endeavor in order to establish the proper place for TEMPs in clinical medicine.

TISSUE-ENGINEERED MEDICAL PRODUCTS

Range of products

Tissue engineering has been defined as the application of the principles of life sciences and engineering in developing biologic substitutes for the maintenance, modification, improvement, restoration, or replacement of tissue or organ function (Skalak and Fox, 1988; Langer and Vacanti, 1993; Hellman *et al.*, 1998). Products developed through this technology, or tissue-engineered medical products (TEMPs), may be thought of as a broad range of products based on their common source materials—human tissues or organs (e.g., autologous or allogeneic tissues); animal tissues or organs (e.g., transgenic animals or xenotransplants); and processed, selected, or expanded human or other mammalian cells (e.g., stem/progenitor cells, genetic and somatic cellular therapies), in combination with or without biomaterials. Totally synthetic materials of biomimetic design may also be considered tissue engineered.

Product examples

Since its inception, the multidisciplinary technology of tissue engineering, which encompasses advances in cell and molecular biology, materials science, surgical techniques, and engineering, has given rise to a diverse array of potential therapeutic products for many different medical conditions, affecting virtually every organ and system in the body. These products provide either a structural/mechanical function or a metabolic function. Examples, among others published in the scientific literature, include artificial skin constructs, expanded cells for cartilage regeneration, engineered ligament and tendon, bone graft substitutes, products intended for nerve regeneration, cells for spinal cord repair, engineered cornea and lens, products for periodontal tissue repair, engineered products for cardiovascular repair/regeneration, blood substitutes, and encapsulated cells for restoration of tissue or organ function, used either as implants (encapsulated pancreatic islet cells), or *ex vivo* as metabolic support systems (liver-assist devices) (Hellman *et al.*, 1998; Hellman, 1995; Bonassar and Vacanti, 1998). To date, some of these products have been approved for use by the FDA and many are under either preclinical investigation or regulatory evaluation.

LEGISLATIVE AUTHORITY/OVERSIGHT

LAWS

Because TEMPs and their source materials span a broad spectrum of potential clinical applications, the responsibility for overseeing their development and commercialization within the United States federal government has been divided among different regulatory agencies, centers, and programs. The Health Resources Services Administration (HRSA) oversees the National Organ Transplant Program and the National Marrow Donor Program. The remaining tissue-engineered products are regulated by FDA.

The FDA is a science-based regulatory agency in the PHS. The agency's legislative authority for product oversight, premarket approval, and postmarket surveillance and enforcement is derived principally from the Federal Food, Drug, and Cosmetic (FD&C) Act and the Public Health Service Act. Under these authorities, the FDA evaluates and approves medical products for the marketplace, inspects manufacturing facilities sometimes before and routinely during commercial distribution, and takes corrective action to remove products from commerce when they are unsafe, misbranded, or adulterated.

FDA ORGANIZATION

The FDA's mission is to promote and protect the public health through regulation of a wide range of products by assuring the safety of foods, cosmetics, and radiation-emitting electronic products, as well as assuring the safety and effectiveness of human and veterinary pharmaceuticals, biologicals, and medical devices. The FDA's six centers are each staffed with personnel having the scientific and regulatory expertise appropriate to a center's mission. The Center for Biologics Evaluation and Research (CBER) regulates biologics; the Center for Drug Evaluation and Research (CDER) regulates drugs; the Center for Devices and Radiological Health (CDRH) regulates medical devices and radiation-emitting electronic products; the Center for Food Safety and Applied

Nutrition (CFSAN) regulates foods and cosmetics; the Center for Veterinary Medicine regulates animal drugs, devices, and feed; and the National Center for Toxicological Research (NCTR) conducts product-related toxicology research. However, each center can apply any of the statutory authorities to regulate its products. For example, many products reviewed by CBER are regulated under the medical device authority. In addition to the centers, a number of other FDA offices, for example, the Office of Regulatory Affairs (ORA) and the Office of Orphan Products (OOP), provide assistance to the centers on regulatory procedures and facility inspections, as necessary.

PRODUCT EVALUATION AND THE REGULATORY PROCESS

CATEGORIES OF PRODUCTS

Regulations applicable to human medical products depend on whether the product is designated a biologic, drug, medical device, or combination product. Products are reviewed generally by the FDA center with the lead responsibility and jurisdiction for the particular product category. The term "biological product" is defined as a virus, therapeutic serum, toxin, antitoxin, vaccine, blood, blood component or derivative, allergenic product or analogous product, or arsphenamine or its derivatives (or any other trivalent organic arsenic compound), applicable to the prevention, treatment, or cure of diseases or injuries of man [42 U.S.C. 262(a)]. The phrase "analogous product" has been applied to a broad range of products (21 U.S.C. 600.3). A drug is an article intended for use in the diagnosis, cure, mitigation, treatment, or prevention of disease in humans or other animals, and an article (other than food) and other articles intended to affect the structure or any function of the body of humans or other animals [21 U.S.C. 321(g)]. A device is an instrument, apparatus, implement, machine, contrivance, implant, *in vitro* reagent, or other similar or related article, including any component, part, or accessory, which is intended for use in the diagnosis of disease or other conditions, or in the cure, mitigation, treatment, or prevention of disease, in humans or other animals, or intended to affect the structure or any function of the body of humans or other animals, and which does not achieve any of its principal intended purposes through chemical action within or on the body of humans or other animals, and which is not dependent on being metabolized for the achievement of any of its principal intended purposes [21 U.S.C. 201(h)]. When biologics, drugs, and/or devices are combined, such products are designated combination products. Tissue-engineered medical products do not comprise a specific product category under the FDA regulations.

TYPES OF PRODUCT PREMARKET SUBMISSIONS

The Federal FD&C Act requires demonstration of safety and efficacy for new drugs and devices prior to introduction into interstate commerce. The PHS Act requires demonstration of safety, purity, and potency for biological products before introduction into interstate commerce. Consequently, premarket clinical studies with new drugs and devices must be carried out under exemptions from these laws. For drugs and biologics, which are considered drugs under the FD&C Act, the application for the exemption is called an Investigational New Drug (IND) application. For a device, the application is called an Investigational Device Exemption (IDE). The FDA regulations describing the IND application are found in Title 21 of the Code of Federal Regulations (CFR), section 312 (21 CFR 312); the regulations for an IDE are in 21 CFR 812.

The contents of IND and IDE applications are similar. Applications will include a description of (1) the product and manufacturing processes and methods sufficient to allow an evaluation of product safety and (2) preclinical studies that were appropriately designed to assess risks and potential benefits of the product. The IND and IDE applications contain a proposal for a clinical protocol, which describes the indication being treated, proposed patient population, patient inclusion and exclusion criteria, treatment regimen, study end points, patient follow-up methods, and clinical trial stopping rules. Both IND and IDE investigations require Institutional Review Board (IRB) approval before they may commence. Although IND and IDE requirements are somewhat different (e.g., in cost recovery and device risk assessment areas), it is important to note that FDA will apply comparable standards of safety and efficacy for either type of application. When the FDA determines that there is sufficient information to allow clinical investigations to proceed, the IND or IDE exemptions are approved.

The first clinical trials conducted under the IND or IDE applications are often clinical trials involving a small number of individuals (e.g., phase 1/feasibility studies) designed primarily to assess product safety. If these earlier studies indicate reasonable safety, phase 2 studies may be developed to investigate proper and safe dosing and potential efficacy. Phase 3/pivotal studies utilize well-controlled clinical trial designs that support a determination regarding safety and efficacy and lead to an application to FDA for premarket approval of the product.

It should be noted that there can be circumstances in which the first study under an IND or IDE will not be a phase 1/feasibility study. This may occur, for example, when sufficient clinical experience exists to establish the safety of a product after use overseas or in a different patient population. The FDA may review data from clinical studies performed outside the United States in the IND/IDE process and/or in an application for marketing approval. The agency strongly encourages a sponsor to meet with the FDA staff to discuss the clinical protocol, study results, statistical analyses, and applicability of the data to a United States population before submitting the Premarket Approval (PMA) or Biologics License (BLA) applications.

According to the laws and regulations governing commercial distribution of medical products, there are several different types of product premarket submissions. As a general rule, the type of submission for a specific product will depend on the type of product (biologic or device). Tissue-engineered medical products regulated as biologics will require review and approval of a BLA that demonstrates the safety and effectiveness of the product before it may be marketed commercially. If the TEMP is regulated as a device, a PMA demonstrating safety and effectiveness must be approved, or a Premarket Notification [510(k)] must receive clearance. In order to obtain a 510(k) premarket clearance for a product, the sponsor would need to demonstrate substantial equivalence to a legally marketed predicate device. A humanitarian use device (HUD) is a product that may be marketed under an exemption for treatment or diagnosis of a disease or condition that affects fewer than 4000 patients per year in the United States. A Humanitarian Device Exemption (HDE) exempts a HUD from the effectiveness requirements for devices if certain criteria are met as described in Section 529(m)(1) of the FD&C Act, as amended February, 1998.

Postmarket surveillance for TEMPs is an important area of consideration. Manufacturers, user facilities such as hospitals and nursing homes, and health care professionals need to report adverse events through the FDA MedWatch process. Record keeping should be instituted by manufacturers of cellular and tissue therapies to allow tracking, when applicable, of donor tissues or cells and of patients who receive the final product. Postmarketing studies may be necessary when (1) a sponsor seeks a change in labeling, (2) such studies are a condition of the FDA approval, or (3) such studies are necessary to protect the public health or to provide safety or effectiveness data. In addition, postmarket surveillance of a medical device introduced into interstate commerce after January 1, 1991, may be required if it is (1) intended for use in supporting or sustaining human life, (2) presents a potential serious risk to human health, or (3) is a permanent implant, whose failure may cause serious, adverse health consequences or death (see Section 522 of the FD&C Act).

REVIEW OF PRODUCT PREMARKET SUBMISSIONS

The intended use of the product, the claims that the sponsor wants to make for the product, and safety concerns specific to the product will determine the types of data to be developed and submitted for the FDA review. Regulatory evaluation of medical products is conducted on a case-by-case basis. The assessments of safety and effectiveness are the basic elements of premarket review. The manufacturer is responsible for providing evidence of the product's safety and effectiveness.

Product safety and effectiveness are evaluated with respect to product manufacture and the product's clinical performance, as applicable, as well as the manufacturer's claim of intended use (i.e., the patient population to be treated and the role of the product in the diagnosis, prevention, monitoring, treatment, or cure of a disease or condition). For TEMPs, issues of product manufacture can include cell/tissue and biomaterials sourcing, processing and characterization, detection and avoidance of adventitious agents, and product consistency and stability, as well as quality control procedures. Other important considerations include evaluation of the preclinical data, e.g., toxicity testing, biomaterials compatibility, and *in vitro*/animal models for product effectiveness. Collecting data on product performance in humans requires insight into clinical trial design (e.g., patient entry criteria and study end points), study conduct, and subsequent data analyses.

At the request of the sponsor of a new drug, the FDA will facilitate the development and expedite the review of such drug if it is intended for the treatment of a serious or life-threatening condition and it demonstrates the potential to address unmet medical needs for such a condition. The development program for such a drug or biologic may be designated a fast-track development program. The FDA works closely with sponsors of products in fast-track development programs and may apply special procedures such as accelerated approval based on surrogate end points, submission and review of portions of an application, and priority review to facilitate its development and expedite its review, as published in the November 18, 1998, Guidance for Industry: Fast Track Development Programs (see web site at http://www.fda.gov/cber/guidelines.htm).

With regard to the expedited review of medical devices, the FDA published on March 20, 1998, the guidelines entitled "PMA/510(k) Expedited Review Guidance for Industry and CDRH Staff" (http://www.fda.gov.cdrh/modact/expedite.pdf). This guidance document describes the criteria for and procedures under which Premarket Approval Applications (PMAs), PMA Supplements, and Premarket Notification [510(k)] applications undergo expedited review. In general, applications dealing with the treatment or diagnosis of life-threatening or irreversibly debilitating diseases or conditions may be candidates for expedited review if (1) the device represents a clear, clinically meaningful advantage over existing technology, (2) the device is a diagnostic or therapeutic modality for which no approval alternative exists, (3) the device offers a significant advantage over existing approved alternatives, or (4) availability of the device is in the best interest of patients. Granted expedited review status means that the marketing application will receive priority review before other applications. When multiple applications for the same type of medical device (which offer a comparable advantage over existing approved alternatives) have also been granted expedited review, the applications will be reviewed with the priority according to their respective submission due dates.

PRODUCT JURISDICTIONAL DECISIONS

A product jurisdictional decision consists of the determination as to whether a product should be regulated as a biologic, drug, device, or combination product. Jurisdictional decisions for a combination product are based on the product's primary mode of action. In addition, as part of a jurisdictional determination are the decisions as to which center will take the lead for product premarket review and which type of submission is the most appropriate. These determinations may be made on a routine basis when product premarket submissions are filed and reviewed by the individual center's divisions. However, there are times when it is not clear what the product's regulatory path should be. This may often occur with TEMPs because of their novel composition and combination of biologic, drug, and/or device components. In this case, the jurisdictional decision is delegated to the center's management and/or to the FDA Office of Chief Mediator and Ombudsman (OCMO) within the FDA's Office of the Commissioner (e-mail address: ombudsman@oc.fda.gov).

How are combination products regulated?

Section 16 of the Safe Medical Devices Act (SMDA) of 1990 [21 U.S.C. 353(g)] describes how the FDA will determine which center within the agency will have the primary jurisdiction for the premarket review and regulation of any combination product. Designation of the lead FDA center for product review is based on the primary mode of action of the combination product. Designations also frequently will indicate whether a center other than the lead center will be asked to assist the review in a consulting or collaborative capacity.

The regulation on combination products, 21 CFR Part 3, was adopted in November, 1991, and implements Section 16. It defines combination products and discusses the appropriate regulatory approaches. To enhance the efficiency of agency operations, the scope of the regulation was extended to apply to any product for which the jurisdiction is unclear or in dispute. The regulation specifies how a sponsor can obtain an agency determination early in the process before any required filing and which FDA center will have primary jurisdiction for the premarket review and regulation of the product.

FDA InterCenter Agreements

The FDA InterCenter Agreements between CBER, CDER, and CDRH were established in the early 1990s to clarify product jurisdictional issues. The three documents describe the allocation of responsibility for certain products and product categories.

The sponsor seeking a jurisdictional decision may contact the designated center identified in the InterCenter Agreement before submitting an application for product premarket review or to confirm jurisdiction and to discuss the application process. For a combination product not covered by a guidance document or for a product for which the center with primary jurisdiction is unclear or in dispute, the sponsor of an application for premarket review should follow outlined procedures to request a designation of the lead center through a Request for Designation (RFD) to the OCMO, as per CFR Part 3, before submitting applications.

These legislative and regulatory initiatives have simplified and facilitated the resolution of administrative questions such as product regulatory jurisdiction for TEMPs within the FDA. Currently, TEMPs are regulated by both CBER and CDRH. Both centers are involved in the evaluation of these products, with the lead center for product review identified by the product's primary mode of action and the other center acting as a consultant (Hellman *et al.,* 1998; Hellman, 1995). For example, (1) human tissue products (e.g., musculoskeletal tissue, skin, cornea) that are manufactured by methods that do not change tissue function or characteristics are the regulatory responsibility of the Human Tissue Program, Office of Blood Research and Review, CBER; (2) CDRH performs the premarket review of certain cellular wound healing products incorporating a supportive matrix; and (3) xenotransplant, somatic cell, and encapsulated cell and gene therapies are regulated by the Office of Therapeutic Research and Review, CBER. Certain cellular therapies (e.g., *ex vivo* liver-assist devices) are evaluated jointly by CBER and CDRH as combination products, with CBER designated as the lead center.

Tissue Reference Group

The Tissue Reference Group (TRG) was established in 1997 to assist the FDA Ombudsman in determining product jurisdiction for human cellular and tissue-based products (www.fda.gov/cber/tissue/trg.htm). The TRG is an intercenter standing committee composed of six voting members, three representatives from CBER and three from CDRH, and a liaison from OCMO. This committee was established to assist in making jurisdictional decisions and applying consistent policy to the regulation of cellular and tissue-based products. Typically, when an inquiry is directed to the TRG, the sponsor is asked to provide written information about the product in question. At a regularly scheduled meeting, the TRG members discuss this product and render a consensus recommendation, which is then forwarded to the centers for sign off, and relayed to the sponsor in writing. A sponsor may request a meeting with the TRG, either before or after a TRG decision is reached. The sponsor who chooses to address a product jurisdiction inquiry to the TRG has the option of submitting an RFD to the OCMO at any time during this process. In addition to working on specific product jurisdiction, the TRG members discuss general regulatory and scientific issues related to human cellular and tissue-based products and facilitate exchange of information between CBER and CDRH reviewers for these types of products.

ADDITIONAL ONGOING FDA INTERCENTER ACTIVITIES

The FDA, recognizing the complex nature of TEMPs, has taken a cooperative approach across its centers in developing the necessary oversight for these products. These include rulemaking, guidance documents, and long-standing cross-cutting working groups, such as the Wound Healing Clinical Focus Group (WHCFG) and the FDA InterCenter Tissue Engineering Working Group (TEWG). The initiatives of the FDA working groups contribute substantially to this cooperative approach in addressing the scientific issues of TEMPs and providing input/recommendations on regulatory oversight for these products. These groups do not work in isolation, but rather work in a cooperative, complementary way.

Wound Healing Clinical Focus Group

The Wound Healing Clinical Focus Group (WHCFG) was formed in the early 1990s and is composed of approximately 20 members representing CBER, CDER, and CDRH. Its members are, in general, the reviewers of applications to the FDA regarding wound healing products. The mission of the WHCFG is to facilitate the development and assessment of products for wound healing. The group serves to enhance consistency of reviews across centers, share information from all parts of the agency, provide a forum for problem solving, and act as a direct liaison from the FDA to academic and industry organizations. The group has also been charged with preparing a guidance document for industry on wound healing agents. To this end, the WHCFG organized Advisory Committee Meetings to discuss issues associated with clinical trials of patients with chronic (July 15, 1997) and burn (November 14, 1997) wounds.

FDA InterCenter Tissue Engineering Working Group

The FDA InterCenter Tissue Engineering Working Group (TEWG) is composed of scientific research and review staff from five participating FDA centers, CBER, CDER, CDRH, CFSAN, and CVM and certain FDA offices. It was established in July, 1994, to identify and address the emerging scientific and science-based regulatory issues of TEMPs. The group, focusing the agency on TEMPs, facilitates intercenter communication and cooperation among the FDA research, review, and administrative staff in review and regulation of TEMPs, and explores international perspectives on TEMPs (Hellman, 1997; Durfor, 1997). Its goal is to support and strengthen product review, and it works to promote regulatory consistency for TEMPs across the FDA through networking mechanisms on science and review issues.

Communication and education are important functions for the TEWG. Different approaches are being used to disseminate information on TEMPs both within and outside the agency. These include an education/training program in the agency and TEWG-initiated symposia, workshops, short courses, and other projects. The group works in cooperation with other federal agencies such as the National Institutes of Health (NIH), National Science Foundation (NSF), and National Institute of Standards and Technology (NIST), as well as societies, institutes, and academe, to review the scientific and regulatory issues concerning TEMPs (Durfor, 1997).

Recognizing that TEMPs are being developed for a potential worldwide marketplace, the TEWG develops initiatives to gather and assess information on how entities other than the FDA are regulating these products in order to acquire a global perspective for the evaluation of these products. Among others, the TEWG organized a meeting, the Workshop on Tissue Engineering: A Global Regulatory Perspective for Tissue-Engineered Products at the 5th World Congress on Biomaterials, in May, 1996, in order to understand the regulatory approaches for TEMPs by different national regulatory bodies. It was learned that, although there are certain differences in procedures, there are also similarities in requirements for manufacture and performance of TEMPs (Durfor, 1997).

As a result of outreach initiatives, the TEWG and its members are participating in the development of voluntary consensus standards for various aspects of TEMPs through the American Society for Testing and Materials (ASTM), a U.S.-based standard-setting organization, which communicates with the International Standards Organization (ISO). It is believed that standards that meet FDA requirements will be helpful for both the FDA in its regulatory review of TEMPs and for the industry in guiding product development.

FDA Transmissible Spongiform Encephalopathies Working Group

Processing issues related to cell/tissue sourcing and the detection and elimination of adventitious agents are important issues for any medical product, including TEMPs, that incorporates animal or human tissue(s). Following reports of a variant form of Creutzfeldt–Jakob disease (vCJD) in humans in Great Britain in 1996, attention immediately focused on a possible link between human transmissible spoingiform encephalopathy (TSE) disease and bovine spongiform encephalopathy (BSE), of high incidence in the cattle of Great Britain, and on the many products using bovine-derived materials.

The InterCenter FDA TSE Working Group, formed in the late 1980s following the report of BSE in the cattle of Great Britain, had been monitoring the disease and developing initiatives for safeguarding the FDA-regulated products containing or manufactured from bovine-, ovine-, and caprine-derived materials. These initiatives have included recommendations, in the form of letters and guidance to the industry, on the safe sourcing and processing of these animal materials used in the FDA-regulated products. In view of the reports of vCJD, the Group, which is now composed of approximately 30 members for all FDA centers and offices, was charged in March, 1996, with assessing the impact of TSE agents on FDA-regulated products and recommending actions that could be taken to protect animal and human health and to alert industry to appropriate safeguard measures.

Recognizing the importance of the concerns associated with the TSE agents, i.e., resistance to conventional means of inactivation and the lack of a presymptomatic diagnostic test for identification of animal and human cell/tissue donors, the agency sponsored a meeting, the International Workshop on TSE Risks in Relation to Source Materials, Processing, and End-Product Use, in June, 1998. The FDA TSE Working Group has marshaled efforts with international groups, as well as across the FDA, to provide prudent, rational, practical, and timely decisions and actions to safeguard the FDA-regulated products and animal and human health. The Working Group was instrumental in the development of this workshop. The primary objective of the workshop, held under the auspices of the Joint Institute for Food Safety and Applied Nutrition (JIFSAN), a partnership between the University of Maryland and the FDA, was to identify practical guiding principles for evaluating risks posed by animal and human TSE agents in FDA-regulated products, including TEMPs. An outline of critical elements to consider in addressing TSE risk evaluation was developed at the workshop and is available via e-mail (tse@life.umd.edu). This outline is a first step in the development of a generic framework of practical guiding principles for evaluating TSE risks in three interrelated areas: sourcing (selection of raw material), material processing, and the use of final products made from these materials. In parallel, a set of information tools has been developed to facilitate risk evaluation and access to TSE information. This paradigm for decision making may provide a useful model for assessing the impact of other emerging infectious disease agents on medical products.

For sourcing of human cells and tissues, the emergence of vCJD, a human TSE, provides new challenges for the evaluation of risks of possible iatrogenic transmission via human tissues and fluids following tissue transplants or transfusion. Evaluating the risks of vCJD transmission is complicated by several factors: (1) it is not known if vCJD has transmission characteristics similar to CJD, (2) the pathogenesis of vCJD is not yet known, and (3) the number of individuals with subclinical vCJD cannot be determined with certainty. The absence of a definitive diagnostic test for human TSE agents, including the agent of vCJD, means that safe sourcing for human tissues, fluids, and organs must be based on exclusion criteria, which are periodically reviewed and revised (Workshop, 1999).

RECENT DEVELOPMENTS IN PRODUCT EVALUATION

INTERIM RULE AND FINAL RULE: HUMAN TISSUE
INTENDED FOR TRANSPLANTATION

The FDA published an Interim Rule (58 FR 65514, December 14, 1993) that required all facilities engaged in procurement, processing, storage, or distribution of human tissues intended for transplantation to ensure that certain infectious disease testing and screening of the donors of these tissues for human immunodeficiency virus (HIV), hepatitis B, and hepatitis C were performed and that records documenting such testing and screening be available for the FDA inspection. Such tissues are derived from a human body, are intended for administration to another human for the diagnosis, cure, mitigation, treatment, or prevention of any condition or disease, and are recovered, processed, stored, or distributed by methods not intended to change tissue function or characteristics. These tissues included musculoskeletal tissue (e.g., bone, ligaments, tendon, fascia, cartilage), ocular tissue (e.g., cornea), and skin. The rule did not apply to organs, semen, reproductive tissue, human milk, and bone marrow, and did not cover tissue that was currently regulated as a drug, biological product, or medical device. In response to comments received, the FDA clarified and modified many of the provisions in the Interim Rule when it published the Final Rule (62 FR 40429, July 29, 1997). At the same time, the agency issued a Guidance for Screening and Testing of Donors of Human Tissue Intended for Transplantation.

SOMATIC CELL AND GENE THERAPY

In 1993, the FDA published a somatic and gene therapy statement (Application of Current Statutory Authorities to Human Somatic Cell Therapy Products and Gene Therapy Products, Notice 58 FR 53248) outlining how the FDA's statutory authorities governing therapeutic products would apply to somatic cell therapy and gene therapy products. The statement defines somatic cell therapy products as autologous, allogeneic, or xenogeneic cells that have been propagated, expanded, selected, pharmacologically treated, or otherwise altered in biological characteristics *ex vivo* to be administered to humans and applicable to the prevention, treatment, cure, diagnosis, or mitigation of disease or injuries. Gene therapy products are defined as products containing genetic material administered to modify or manipulate the expression of genetic material or to alter the biological properties of living cells. Some gene therapy products (e.g., those containing viral vectors) to be administered to humans fall within the definition of biological products and are subject to the licensing provisions of the PHS Act, as well as to the drug provisions of the FD&C Act. Other gene therapy products, such as chemically synthesized products, meet the definition for a drug and are regulated only under the relevant portions of the FD&C Act. Cellular products intended for use as somatic cell therapy are biological products subject to regulation pursuant to the PHS Act (42 U.S.C. 262) and also fall within the definition of drugs in the FD&C Act [21 U.S.C. 321(g)]. As biologicals, somatic cell therapy products are subject to licensure to ensure product safety, purity, and potency. At the investigational stage, these products must be in compliance with 21 CFR part 312, which stipulates that clinical trials are to be conducted under the code for INDs. The statement also discusses the combination of biological products with drugs or devices.

PROPOSED APPROACH TO REGULATION OF CELLULAR AND TISSUE-BASED PRODUCTS

In the 1990s, the FDA devoted considerable resources to the regulation of human cellular and tissue-based products. In analyzing its regulatory approach to these products, the FDA came to realize that the existing patchwork of regulatory policies did not fully address many issues raised in this rapidly evolving area of product development.

On February 28, 1997, the FDA announced a document entitled, "A Proposed Approach to the Regulation of Cellular and Tissue-Based Products," in conjunction with "Reinventing the Regulation of Human Tissue," the sixth "Reinventing Government" report, produced in conjunction with the Vice President's National Performance Review. A notice of availability was published on March 4, 1997 (62 FR 9721) (see web site, www.fda.gov/cber/tissue/docs.htm). The regulatory framework was designed to provide a comprehensive approach to the regulation of a broad spectrum of human cellular and tissue-based products, both traditional and new. This approach would consist of a tiered, risk-based regulatory program that would protect the public health without either imposing unnecessary government oversight or inhibiting innovation and product development in this rapidly growing field.

Under the proposed approach a human cellular and tissue-based product is defined as a product containing or consisting of human cells or tissues that are intended for implantation, transplantation, infusion, or transfer into a human recipient, e.g., musculoskeletal, ocular, and skin tissue; hematopoietic stem cells; reproductive cells and tissues; human dura mater; human heart valve allografts; somatic cell therapy products; and combination products. The term would not include vascularized human organs for transplantation; whole blood, blood components, or blood derivative products; secreted or excreted human products; minimally manipulated bone marrow; ancillary products used in the manufacture of cellular or tissue-based products, cells, tissues, and organs derived from animals other than humans; and *in vitro* diagnostic products.

The agency identified the principal public health and regulatory concerns for the cellular and tissue-based products, included in the proposed approach, as (1) transmission of communicable disease, (2) processing controls to prevent contamination and preserve product integrity and function, (3) clinical safety and efficacy, (4) promotional claims and labeling, and (5) monitoring and communicating with the industry. The agency also recognized that certain characteristics were major determinants of the degree of risk associated with the use of these products, i.e., (1) source of cells or tissue—autologous or allogeneic, (2) cell and tissue viability, (3) homologous or nonhomologous function, (4) degree of manipulation—minimal or more than minimal, (5) systemic or local effect, (6) storage—in a bank or not stored (unbanked), and (7) combination of cells or tissue with a drug or device.

In analyzing the above concerns and product characteristics, the agency concluded that a cellular or tissue-based product with all of the following characteristics would have the least amount of risk, and thus could be regulated without an actual premarket submission to the FDA—minimal manipulation, homologous function, not combined with a drug or device component, and having a local effect. If a product had any one of the following characteristics, it would present a greater risk and, thus, would be regulated as a biological drug or device requiring a premarket submission to the FDA for evaluation of clinical safety and efficacy—more than minimal manipulation, promoted or labeled for a nonhomologous function, combined with a drug or device component, or having a systemic effect (except in certain specified situations). Certain general requirements would apply to all establishments and products, i.e., registration of the establishment and listing of products, with periodic updates as necessary; determination of donor suitability through donor testing and screening; and current good tissue practices, such as proper handling and processing controls.

As indicated, transmission of communicable disease has been identified as a principal public health and regulatory concern for cellular and tissue-based TEMPs. For a TEMP containing allogeneic cells or tissues, most risks of transmission can be controlled through adequate donor screening and product manufacturing procedures. A lower level of transmission risk would be expected with the use of autologous cells or tissues. However, cell manipulation or processing of autologous tissue may activate latent virus expression, which may present an increased risk when reintroduced to the donor, and inadvertent use of allogeneic tissue may occur if manufacturing and storage procedures do not prevent it. In addition, although certain processing methods may reduce or eliminate the presence of most viruses and, thus, decrease the risk of exposure, no process is currently known to be effective for complete inactivation of all agents without concomitant destruction of the cellular or tissue component.

IMPLEMENTATION OF PROPOSED APPROACH—RULES

As part of the implementation of the proposed approach to the regulation of cellular and tissue-based products, the agency has stated that it will promulgate three proposed rules, which will expand on the existing regulations for human cell and tissue products, both in scope and in content. [See "Human Tissue Intended for Transplantation," Final Rule, 62 FR 40429, July 29, 1997 (www.fda.gov/cber/tissue/docs.htm).]

The first of these proposed rules, "Establishment Registration and Listing for Manufacturers of Human Cellular and Tissue-Based Products," was published on May 14, 1998 (63 FR 26744) (see www.fda.gov/cber/tissue/docs.htm). The FDA proposed a consolidated registration and listing system for establishments engaged in the recovery, screening, testing, processing, storage, labeling, packaging, or distribution of human cellular and tissue-based products. This registration and listing system would establish a database for future communication with this industry. The proposal stated the criteria for determining (1) whether a product would be subject to registration, product listing, and the other proposed regulations, and (2) whether a human cellular or tissue-based product would be subject only to the regulations being promulgated under Section 361 of the PHS Act, or would be regulated as a biological drug or medical device under Section 351 of the PHS Act and/or the FD&C Act. All cell and tissue establishments would use the same registration and listing procedures (as well as the same FDA form) to maintain a unified database. This proposed regulation also specified the details of when, how, and where a manufacturer would register and list, and how a registration number would be assigned.

Two proposed rules are being developed and are expected to be published at a later date: one addressing suitability determination for donors of human cellular and tissue-based products, and the other describing proposed current good tissue practices (GTPs). The agency anticipates that it will also develop Guidance for Industry documents, which would provide information on practices that the FDA would find acceptable in complying with the regulations.

TISSUE ACTION PLAN

In order to develop the regulations and guidance documents needed to implement the proposed approach in a timely manner, the FDA, under CBER leadership, has outlined a tissue action plan that created a core team and several task groups (see www.fda.gov/cber/tissue/tissue.htm). The core team consists of individuals from CBER, CDRH, ORA, and the Office of the Commissioner who meet monthly to oversee progress, make decisions on significant policy issues, and provide final review and sign-off. Although ancillary products were excluded from the proposed approach, the agency has formed a task group to develop guidance for industry in this area.

XENOTRANSPLANTATION ACTION PLAN

The current working PHS definition of xenotransplantation is a procedure that involves the use of live cells, tissue, or organs from a nonhuman animal source (a xenograft), transplanted or implanted into a human or used for *ex vivo* contact with human body fluids, cells, tissues, or organs that are subsequently given to a human recipient. Although many believe that xenotransplantation offers a potentially promising source of therapeutic materials, the use of live animal grafts raises concerns regarding the potential infection of recipients with both known and unknown infectious agents and possible subsequent transmission to their close contacts and into the general human population.

To coordinate activities related to the regulation of xenotransplantation (through the IND and licensing mechanisms), the FDA has established a Xenotransplantation Action Plan and a Xenotransplantation Core Team, which oversees the work of several task groups charged with writing proposed regulations and guidance documents. In addition, the FDA participates with other federal agencies and the Office of Policy within the Office of the Assistant Secretary of Planning and Evaluation in the Department of Health and Human Services (DHHS) in developing policies and guidelines with respect to xenotransplantation. Priorities of this collaborative effort include revision of The Draft Guideline on Infectious Disease Issues in Xenotransplantation (1996) and other initiatives.

INTERACTION BETWEEN FDA HEADQUARTERS AND FIELD OFFICES

TEAM BIOLOGICS

Team Biologics is a plan, implemented in 1997, for reinventing the FDA's ability to optimize compliance of regulated biologics industries, including blood and plasma establishments, plasma fractionators, vaccine manufacturers, licensed *in vitro* diagnostics manufacturers, and the biotechnology industry. The approach involves three elements—a core team of certified ORA investigators, CBER certified inspectors, and specially trained compliance officers, who will actually perform the inspections, with the ORA team member taking the lead role; a steering committee, which will quality control this effort; and an operations group, which will be responsible for actual implementation and daily oversight. Inspectional guidance documents, such as a Compliance Program and an Inspection Guide, are being developed and will be updated periodically.

REGIONAL TRAINING

Because the inspection of tissue establishments is a relatively new area for field investigators, the FDA has made efforts to share the technical expertise at the FDA headquarters with the field. Regional training of field investigators from several districts took place in Baltimore, Maryland in February, 1999, at a 3-day workshop developed by the Central Regional Training Center of the FDA. The workshop included presentations by CBER, CDRH, ORA, the State of Maryland, the American Association of Tissue Banks (AATB), and the Eye Bank Association of America (EBAA).

Future training programs are being planned to enable field investigators to better evaluate the regulatory compliance of tissue establishments in the areas of recovery, screening, testing, processing, storage, and distribution of human tissue. New teaching methodologies, such as interactive video conferences and self-paced learning via electronic mail, are being considered.

DEVELOPMENT OF STANDARDS

HEMATOPOIETIC STEM CELL MODEL

The development of standards intended to ensure the safety and effectiveness of biological products has traditionally occurred in parallel by industry, standard-setting bodies, and the FDA. Such efforts have been labor intensive for all involved. In an attempt to ease the regulatory burden on industry and on the agency, while providing protection to the public, the agency is considering an approach whereby industry would be requested to develop product standards. These standards would be reviewed by the agency and, if found acceptable, would be adopted by the FDA. The agency would develop a process that would include reliance on standards as a basis for licensure.

On January 20, 1998, the FDA published a Request for Proposed Standards for Unrelated Allogeneic Peripheral and Placental/Umbilical Cord Blood Hematopoietic Stem/Progenitor Cell Products; Request for Comments (63 FR 2985) (see www.fda.gov/cber/tissue/docs.htm). This notice requested submission of comments, accompanied by supporting clinical and nonclinical laboratory data and other relevant information, proposing manufacturing standards and product specifications for minimally manipulated hematopoietic stem cells obtained from peripheral blood and cord blood for unrelated allogeneic use (in a recipient unrelated to the donor).

The notice stated that industry and the public would have 2 years to develop and submit data, followed by approximately 1 year for the FDA to review the data. If the FDA determines that the submissions support the development of standards, the FDA intends to issue such standards following the agency's procedures (good guidance practices). If the FDA determines that adequate information for standards development is not available, the agency intends to enforce the traditional IND and license application requirements at the close of the 3-year period.

AMERICAN SOCIETY FOR TESTING AND MATERIALS—FDA PARTICIPATION

As indicated above, the development of standards is envisioned increasingly as an effective vehicle for the FDA to use in its product review process and for the industry to use in its product development. This is consistent with the Food and Drug Administration Modernization Act (FDAMA) of 1997 that contains provisions for standards development for products under the FDA's jurisdiction in order to streamline the product approval process. This legislation allows the FDA to recognize formally appropriate voluntary consensus standards for medical devices developed by nationally and internationally recognized standards development organizations, without making them federal requirements subject to notice, comment, and rulemaking. Under FDAMA, the FDA will continue to select and adopt standards that meet the FDA's regulatory requirements for use in the review process, because it believes that the formal recognition of standards and the manufacturer's ability to conform to such standards will benefit the industry, consumers, and the FDA (Marlowe and Phillips, 1998). Further information on the recognition of consensus standards and declaration of conformity to standards is available in the guidance document entitled "FDA Modernization Act of 1997: Guidance for the Recognition and Use of Consensus Standards; Availability" (63 FR February 25, 1998) (see www.fda.gov/cdrh/modact/k982.html).

Recognizing the importance of standards as well as guidance documents in product development and regulatory review, FDA scientific research and review staffs are active participants/representatives in the recently established American Society for Testing and Materials (ASTM) initiative on TEMPs. For example, the scope of the ASTM effort, in Committee F04, Medical and Surgical Materials and Devices, Division IV: Tissue-Engineered Medical Products, is the development of standards for TEMPs focusing on the components of these products, such as the biological component (cell, tissue, cellular product, and/or biomolecule) and the biomaterial component (biological, biomimetic, and/or synthetic) (see www.astm.org). This division is working with other ASTM committees and other organizations with mutual interests. The division currently includes 10 subcommittees, each working in a different standard area related to TEMPs. Through its voluntary consensus process that encourages input from expert groups and all interested parties, the ASTM procedures aim to ensure that the standards are technically sound and rest on a solid scientific foundation. The standards may include test methods, specifications, practices, guides, classifications, or terminology.

COMMUNICATIONS WITH INDUSTRY

Commensurate with its public health mission in assuring the safety and effectiveness of medical products for human use, the FDA is establishing effective vehicles of communication with the industry on topics ranging from basic research to product development and commercialization. These include FDA-sponsored workshops in emerging research areas such as the annual FDA Science Forums—a session on tissue engineering was a highlight of the December, 1998, FDA Science Forum on Biotechnology—and workshops focused on areas of potential product-related public health concern such as the June, 1998, Workshop on TSE Risks. In addition, the FDA Staff Colleges of the different centers develop courses on generic and specific science and/or product-related areas to provide continuing education of the FDA research, review, and administrative staff. For example, the FDA Staff College course on tissue engineering is currently in its fifth year and has dealt with topics ranging from new technological approaches and product applications to issues important for designing appropriate preclinical and clinical studies for TEMPs. The FDA also develops and participates in a number of national and international meetings focused on TEMP-related scientific and regulatory issues. The FDA considers this participation to be an important vehicle for maintaining an effective dialogue with the industry. In addition, representatives of industry and research and development consortia have been invited to present their research and to discuss both generic and specific product-related regulatory concerns with the FDA staff. Individual meetings with product sponsors are encouraged from the very early stages of product concept through the later stages of product development and approval. Industry representatives are also encouraged to attend the public meetings of the FDA advisory committees on topics of product-related concern, such as the FDA TSE Advisory Committee's deliberations on the safe procurement and processing of animal- and human-derived material used in FDA-regulated products. The industry is also encouraged to comment on proposed regulations and guidance documents that are made publicly available through the Federal Register and the FDA web sites.

THE FDA AND FUTURE PERSPECTIVES FOR TISSUE-ENGINEERED MEDICAL PRODUCTS

As exemplified by some of the initiatives described in this chapter, the FDA will continue to develop science-based rationales with regard to its regulatory oversight of TEMPs that will provide a road map for the FDA reviewers as well as industry. Such an approach will seek to enhance product review and to address questions for manufacturers early on in product development. The initiatives described seek to promote innovation, at the same time ensuring that the proper and appropriate levels of control for safety and effectiveness are maintained during the development of all medical products, including TEMPs. In support of these initiatives, the FDA centers and cross-cutting working groups will continue their programs in research, technology monitoring, guidance/standards, training/education for the FDA staff and industry, and cooperation with public and private groups.

An important part of the FDA assessment for medical products, including TEMPs, will continue to be premarket review, based on product-by-product review. Pivotal in the regulatory review for all products are the manufacturer's claim of intended use, product quality control procedures, and product performance. As tissue engineering technology evolves, leading to new and different products, new and different product-specific issues will also be expected. As a result, the FDA will continue to build on current initiatives in order to resolve such issues through its continuing research, review, and assessment of TEMPs. The lessons learned from previous product categories, such as lessons involving recombinant DNA technology and propagation of cell lines, may be applicable to new products and technologies in some cases. In other cases, the agency will look to the best scientific minds and methods to achieve innovative, flexible, and appropriate resolutions. A productive dialogue among the FDA, industry, and consumers is the key element in this endeavor in order to establish the proper place for TEMPs in the armamentarium of clinical medicine.

Acknowledgments

The authors gratefully acknowledge the expert assistance of Colleen Catron in preparation of this manuscript and the following FDA colleagues for constructive critical review of the manuscript: Jill Warner, Steve Unger, Donald Marlowe, Christine Nelson, Elizabeth Jacobson, Jay Siegel, Susan Alpert, Jay Epstein, Philip Noguchi, and Celia Witten.

References

Bonassar, T. J., and Vacanti, C. A. (1998). Tissue engineering: The first decade and beyond. *J. Cell. Biochem., Suppl.* **30/31,** 297–303.

Durfor, C. N. 91997). Biotechnology biomaterials: A global regulatory perspective for tissue engineered products, summary report and future directions. *Tissue Eng.* **3**(1), 115–120.

Hellman, K. B. (1995). Biomedical applications of tissue engineering technology: Regulatory issues. *Tissue Eng.* **1**(2), 203–216.

Hellman, K. B. (1997). Tissue engineered products and regulatory challenges: An update, who's doing what in tissue engineering: Reality, Regulation and Research Panel. *Ann. Meet., Soc. Biomater.,* New Orleans, LA.

Hellman, K. B., Knight, E., and Durfor, C. N. (1998). Tissue engineering: Product applications and regulatory issues. *In* "Frontiers in Tissue Engineering" (C. W. Patrick, A. G. Mikos, and L. V. McIntire, eds.), pp. 341–366. Elsevier, Amsterdam.

Langer, R., and Vacanti, J. P. (1993). Tissue engineering. *Science* **260**, 920–926.

Marlowe, D. E., and Phillips, P. J. (1998). "FDA Recognition of Consensus Standards in the Premarket Notification Program. Biomedical Instrumentation and Technology," pp. 301–304. Hanley and Belfus, Philadelphia.

Skalak, R., and Fox, C. F., eds. (1988). "Tissue Engineering. Proceedings for a Workshop at Granlibakken, Lake Tahoe, CA. Alan R. Liss, New York.

Workshop (1999). Risks associated with transmissible spongiform encephalopathies. *Emerg. Infect. Dis.* **5**(1).

EPILOGUE

This volume represents a comprehensive compilation of the field of tissue engineering. It captures a moment in time in an area of science and medicine that is continuously growing, gaining new knowledge, and inventing new technologies. Its composition reflects the interdisciplinary nature of the field. There are contributions from clinical surgeons and physicians; biologists expert in cell biology, developmental biology, and molecular biology; and engineers from broad areas of that discipline. The breadth of tissues and organs covered spans the entire mammalian organism. The nature of the science ranges from the chemistry of living systems and biomaterials to the physiology of the whole organism. It includes the understanding of cellular machinery, a hierarchical organization of cells in their local environment, and their association over great distances relative to the entire organism. The organization of the book is intended to provide an in-depth understanding for these diverse areas that is easily accessible. The hope is that it is a compendium for those who are knowledgeable in the field, but can also provide inspiration and instruction to those students who are either interested in or just entering the field.

The amount of new information and the number of experimental systems that are now entering human clinical trials is substantially greater than in the first edition. Because of the exponential rise in interest in the field, this trend is expected to continue. It is not possible to predict the changes that will occur in the next few years, which likely will radically alter the next edition of this text. One of the greatest gifts of Western thought to mankind is that at any given instant in time, humanity has never known so much about the physical world and will never again know so little.

Joseph P. Vacanti

SUBJECT INDEX